WORLD DIRECTORY OF CRYSTALLOGRAPHERS

AND OF OTHER SCIENTISTS

EMPLOYING CRYSTALLOGRAPHIC METHODS

SEVENTH EDITION

1986

GENERAL EDITOR

A. L. BEDNOWITZ

ASSOCIATE EDITOR

A. P. SEGMÜLLER

PUBLISHED FOR THE

INTERNATIONAL UNION OF CRYSTALLOGRAPHY

BY

D. REIDEL PUBLISHING COMPANY

SPRINGER-SCIENCE+BUSINESS MEDIA, B.V.

ISBN 978-90-277-2094-8   ISBN 978-94-017-3701-2 (eBook)
DOI 10.1007/978-94-017-3701-2

The text of this directory was formatted and prepared
on a computer controlled photocomposer printer

through the courtesy of

International Business Machines Corporation
Thomas J. Watson Research Center
Yorktown Heights, NY, U.S.A.

TABLE OF CONTENTS

	Page
Preface	v
How It Was Done	vi

Explanatory Notes

1. Who is included? vii
2. Format and arrangement of the information
    a. Alphabetical order vii
    b. Contents of the biographical entries vii
3. Language ix
4. Abbreviations ix

National lists

Algeria	1	Japan	97
Argentina	1	Korea	110
Australia	2	Libya	111
Austria	7	Malaysia	111
Bangladesh	10	Mexico	112
Belgium	11	Netherlands	113
Bolivia	13	New Zealand	118
Brazil	14	Nigeria	119
Bulgaria	16	Norway	119
Burma	18	Pakistan	121
Canada	18	Peru	122
Chile	23	Philippines	123
China	24	Poland	123
Colombia	30	Portugal	127
Cuba	31	Romania	128
Czechoslovakia	31	Saudi Arabia	129
Denmark	34	Singapore	129
Egypt, Arab Rep.	36	South Africa	130
Finland	37	Spain	132
France	40	Sri Lanka	135
German Dem. Rep.	48	Sudan	135
Germany, Fed. Rep.	55	Sweden	136
Ghana	70	Switzerland	141
Greece	71	Syrian Arab Rep.	144
Hong Kong	73	Taiwan	144
Hungary	73	Tanzania	145
Iceland	75	Thailand	145
India	76	Tunisia	146
Indonesia	84	Turkey	147
Iran	84	USSR	149
Iraq	85	UK	165
Ireland	85	USA	181
Israel	86	Uruguay	216
Italy	88	Venezuela	217
Ivory Coast	96	Yugoslavia	217

Name Index 223

Table of Contents

Appendix

International Union of Crystallography 263
    General Description 263
    Publishing Activities 263
    Administration 264
    Adhering Bodies 264
    Finances 265
    Cooperation with Other International Scientific Organizations 265
Publications of the International Union of Crystallography 266

PREFACE TO THE SEVENTH EDITION

A brief historical account of the background leading to the publication of the first four editions of the *World Directory of Crystallographers* was presented by G. Boom in his preface to the Fourth Edition, published late in 1971. That edition was produced by traditional typesetting methods from compilations of biographical data prepared by national Sub-Editors. The major effort required to produce a directory by manual methods provided the impetus to use computer techniques for the Fifth Edition. The account of the production of the first computer assisted Directory was described by S.C. Abrahams in the preface of the Fifth Edition.

Computer composition, which required a machine readable data base, offered several major advantages. The choice of typeface and range of characters was flexible. Corrections and additions to the data base were rapid and, once established, it was hoped updating for future editions would be simple and inexpensive. The data base was put to other Union uses, such as preparation of mailing labels and formulation of lists of crystallographers with specified common fields of interest.

The Fifth Edition of the *World Directory of Crystallographers* was published in June of 1977, the Sixth in May of 1981. The Subject Indexes for the Fifth and Sixth Editions were printed in 1978 and 1981 respectively, both having a limited distribution.

Beginning in June 1984, Sub-Editors of the Sixth Edition were solicited for their assistance in producing the Seventh Edition. Detailed instructions were provided concerning submission of their entries. Sub-Editors were invited to commence immediately the process of contacting the members of their national crystallographic communities. Very nearly all the original Sub-Editors graciously agreed to undertake the task of updating their entries. In order to provide continuity for future Editions an attempt was made to obtain a Sub-Editor for every country listed.

In the Fourth quarter of 1984 the first updated entries began to arrive. Initially, February 1985 was set as the target date for receipt of updates and additions. This was unrealistic for the larger entries and the date was eventually moved to July, 1985. Efforts were made to ensure that all countries were given the opportunity of crystallographic representation in the Seventh Edition. Four additions and three deletions were made to the list of included countries. In some cases there was a loss of contact due to the movement of previous correspondents out of their country of residence. Where the number of crystallographers in a country was very small it is understandable that a country could lose its representation in the Directory. Individual crystallographers inadvertently omitted from this edition are cordially invited to contact the General Editor for inclusion within the next edition.

The bulk of all biographical data was received by the deadline, although several countries did not provide their data until late in 1985. As soon as most entries for a given country were received, page proofs were produced and sent to the Sub-Editor for correction. This process continued until March 1986 at which time the Name Index compilation was begun and the entries checked for duplications. The production of camera-ready prints commenced in May 1986.

The Name Index permits the country in which a listing appears to be easily identified. In cases where a listing is incomplete the information given reflects all that was available. In order to maintain the accuracy of each entry the General Editor has faithfully tried to reproduce the information furnished to him. Changes were made, however, in the Fields of Interest section, in order to enhance the content of the planned Subject Index.

Errors in this edition are bound to be present. They may be eliminated from the permanent data base now established, with the cooperation of those who note them, if brought to the attention of the General or Associate Editor.

It is a great pleasure to thank the IUCr Executive Secretary Dr. J. N. King, General Secretary Professor V. J. Kurki-Suonio and President Professor Th. Hahn for their continued support, and with appreciation, to recognize all the Sub-Editors, without whose generous and excellent cooperation neither this nor any other edition of the *World Directory of Crystallographers* would be possible.

May 30, 1986

A. L. Bednowitz & A. P. Segmüller
IBM T.J. Watson Research Center
Yorktown Heights, N.Y. 10598
U.S.A.

HOW IT WAS DONE

The text of this directory was prepared by a photocomposer printer at the IBM Thomas J. Watson Research Center in Yorktown Heights, New York. The input for many countries was received in machine readable form; either on standard 80 column punched cards, IBM PC DOS compatible 5 inch flexible disks or IBM compatible 9 Track magnetic tape. The cards, disk and magnetic tape data sets were read into an IBM 3084 operating under the Virtual Machine/Conversational Monitor System (VM/CMS). Special control symbols were used to indicate capitalization, type style and diacritical marks. The Directory entry text was arranged in data sets by country and stored on-line in an IBM 3851 Mass Storage System. Where required, a FORTRAN program was used to convert to full Extended Binary Coded Decimal Interchange Code (EBCDIC) from upper case punch card codes. The Conversational Monitor System (CMS), job control procedure (REX), and text-editing (XEDIT) facilities were used in combination to insert the appropriate (SCRIPT) control symbols for columnar format, headings and diacritical marks.

The addition of diacritical marks required an extensive control symbology. The procedure requested of the Sub-Editors, in order to indicate letters with diacritical marks, uses two-character codes (the first of which is either "/", "?", "¬" or "+"). These codes were changed by a REX program (using the XEDIT MACRO facility) to a control sequence for each mark (Set Symbol) acceptable to the IBM Document Composition Facility (GML - SCRIPT). The largest data set handled was the USA list with 15,630 lines of text.

Final proof readings were done by the General Editor and Sub-Editors before composing the finished product for offset printing. Although many typographical errors were encountered and corrected from the raw data it is expected that many are still embedded in the final text even though it has gone through several stages of proof reading. Even with the powerful assistance of a computerized editing facility there are still classes of error and omission which continue to be unavoidable in this kind of compilation. One serious problem encountered was with the automatic hyphenation of words at the end of a line. In some cases, especially in the handling of non-English words, oddities occurred. With the aid of the Sub-Editors some of these, but not all, were corrected in the final proofs.

EXPLANATORY NOTES

1 WHO IS INCLUDED?

In deciding who should be included in the Seventh Edition of the *World Directory of Crystallographers,* each Sub-Editor was invited to use his own judgement, aided by the general guidelines developed in previous Editions. Table 1 gives the number of entries included for each country in the last three editions. A country is defined in accordance with ICSU statute 8 from which we quote: "A national member adheres to the council either through its principal scientific academy, or its national research council, or any other institution or association of institutions. Such an institution effectively representing the independent scientific activity in a definite territory may be accepted as a national member, provided it can be listed under a name that will avoid any misunderstanding about the territory represented." In accordance with this the words "country" and "territory" are used synonymously in this directory.

Crystallography has been taken in its broadest sense as represented, for example, in the programs of the International Congresses of Crystallography. A crystallographer was recognized as a scientist with an active interest in crystallography, either for its own sake or for the contribution it could make to some other branch of science. It was clearly understood that the Seventh Edition would be of most value if it included all crystallographers. To help ensure none would inadvertently be omitted, it was hoped that Sub-Editors would enroll the aid of all who had previously been listed in their sections. Sub-Editors were provided with a listing of all entries appearing in the Sixth Edition for their section.

Generally, only crystallographers who returned completed Data Entry Forms are included in this edition. An exception was made by including all U.S.A. members of the American Crystallographic Association, as their current addresses were available to their Sub-Editor and their inclusion was regarded as useful. In addition, in those countries where updates were not readily available the entries from the Sixth Edition were included in an attempt to provide continuity.

The total number of entries has increased by 10 percent compared with the previous edition while the total number of represented countries has increased by one. As expected, changes have occurred in the crystallographic population of a number of countries, some increasing considerably with few decreasing significantly.

2. FORMAT AND ARRANGEMENT OF THE INFORMATION

a. *Alphabetical order*

As in previous editions, the sections of this Seventh Edition are arranged in alphabetical order by countries, and by individuals within each country. Prefixes were handled differently depending on whether they are capitalized in spelling. For example, in **De Camp** with a capital "D", the prefix is given preceding the name which is hence placed in the alphabetical "D" group, whereas in **van der Meer** the uncapitalized prefixes follow the name as **Meer, van der** and this is placed in the alphabetical "M" group. Names that contain diacritical marks are generally handled as if there were no such marks, unless a Sub-Editor indicated a different practice for that country.

b. *Contents of the biographical entries*

Complete individual entries contain:

(i) family name, followed by title and given names
(ii) full institutional or correspondence address
(iii) year of birth, in parentheses
(iv) highest degree and field in which the degree was granted
(v) university or institution granting degree,
country if different from that of address in (ii),
and year degree was granted, all in parentheses
(vi) present position
(vii) telephone number, in parentheses
(viii) major fields of scientific interest in the form of Keywords.

The name of each crystallographer listed is contained in the Name Index, together with the country in which the name is listed. Cross references are hence unnecessary, and have not been used.

Explanatory *Notes* precede most country listings, with information provided by the Sub-Editor. Variations in addresses for the same institution may be found: each is as provided by the individual, and is presumed to be acceptable by postal authorities.

The introduction of direct dialing between many countries has led to provision of international telephone country codes in the *Notes* for most countries. In any given country, additional leading digits may be necessary: details for some countries are given in their *Notes*.

Table 1. *Number of Listings by Country*

Country	7th edition 1986	6th edition 1981	5th edition 1977	Country	7th edition 1986	6th edition 1981	5th edition 1977
Algeria	-	2	2	Korea	17	15	11
Argentina	44	28	29	Libya	4	-	-
Australia	197	200	170	Malaysia	30	20	15
Austria	86	90	73	Mexico	42	39	19
Bangladesh	22	25	23	Netherlands	163	166	174
Belgium	86	87	68	New Zealand	42	49	65
Bolivia	11	12	10	Nigeria	9	9	3
Brazil	81	108	109	Norway	68	69	75
Bulgaria	83	74	68	Pakistan	60	31	29
Burma	6	6	4	Panama	-	-	1
Canada	203	165	153	Peru	11	11	4
Chile	44	43	42	Philippines	13	8	4
China	223	10	-	Poland	150	153	136
Colombia	30	28	22	Portugal	20	20	20
Cuba	12	-	-	Romania	31	32	30
Czechoslovakia	103	91	82	Saudi Arabia	14	10	9
Denmark	71	62	62	Singapore	10	6	4
Egypt, Arab. Rep.	71	77	77	South Africa	98	80	72
Ethiopia	-	-	2	Spain	113	113	87
Finland	109	118	119	Sri Lanka	3	5	4
France	312	348	320	Sudan	2	2	1
German Dem. Rep.	374	376	377	Sweden	206	206	207
Germany, Fed. Rep.	589	455	428	Syrian Arab Rep.	2	-	1
Ghana	2	8	4	Switzerland	121	133	132
Greece	88	82	72	Taiwan	30	25	16
Hong Kong	7	6	4	Tanzania	1	-	-
Hungary	74	71	63	Thailand	36	28	20
Iceland	5	5	5	Tunisia	32	18	4
India	370	340	318	Turkey	84	60	27
Indonesia	13	13	4	Uganda	-	-	2
Iran	20	25	23	USSR	697	657	595
Iraq	6	6	2	UK	732	651	598
Ireland	8	5	-	USA	1621	1622	1642
Israel	79	71	56	Uruguay	3	3	4
Italy	344	276	248	Venezuela	13	10	7
Ivory Coast	10	8	6	West Indies	-	-	2
Japan	565	516	446	Yugoslavia	138	128	118
Kenya	-	4	2	Zimbabwe	-	2	5

Totals: 5th edition (1977): 71 countries, 7638 entries.
6th edition (1981): 68 countries, 8174 entries.
7th edition (1986): 69 countries, 8968 entries.

3. LANGUAGE

The language used throughout the bulk of this edition is English. However, addresses, degrees and positions are often given in the national language: in such cases, an explanation may generally be found in the *Notes*.

4. ABBREVIATIONS

Numerous abbreviations have been used in this edition, following the practice of previous editions. Common abbreviations are found in Table 2. Abbreviations peculiar to a given country are explained in their *Notes*. It should be noted that some of the above abbreviations are also valid in other languages, e.g. Ave, Blvd, Dept., Dr, Lab., Labs., Prof., Tel., and U.

Table 2. *English abbreviations*

AB	Bachelor of Arts	Esp.	Especially
Acad. Sci.	Academy of Sciences	Est.	Establishment
AM	Master of Arts	Ext.	Extension
Appl.	Applied	Fac.	Faculty
Apt.	Apartment	Grad.	Graduate
Assoc.	Associate	Inc.	Incorporated
Asst.	Assistant	Inst.	Institute
Ave	Avenue	Instr.	Instructor
BA	Bachelor of Arts	Jr.	Junior
BAgrSc	Bachelor of Agricultural Sciences	Lab.	Laboratory
BASc	Bachelor of Applied Science	Labs.	Laboratories
BEE	Bachelor of Electrical Engineering	Lect.	Lecturer
Bldg.	Building	Ltd.	Limited
Blvd.	Boulevard	MA	Master of Arts
BMet	Bachelor of Metallurgy	Mbr.	Member
BMetE	Bachelor of Metallurgical Engineering	MMet	Master of Metallurgy
BRD	Bundesrepublik Deutschland	MD	Doctor of Medicine
BSc	Bachelor of Science	MSc	Master of Science
BSc(Hons)	Bachelor of Science(Honours)	Nat.	National
C.	College	P.A.	Personal (or private) address
Co.	Company	PhD	Doctor of Philosophy
Coord.	Coordinator	P.O.	Post Office
Corp.	Corporation	Prof.	Professor
CSSR	Czecho-Slovak Socialist Republic	Rep.	Republic
Dept.	Department	Res.	Research or Researcher
Dev.	Development	S., Sch.	School
DDR	Deutsche Demokratische Republik	ScD	Doctor of Science
DDS	Doctor of Dental Surgery	Sci.	Science or Sciences
DEng	Doctor of Engineering	Scient.	Scientific or Scientist
Dev.	Development	ScM	Master of Science
Dir.	Director	Sr.	Senior
Div.	Division	St.	Saint or Street
DPharm	Doctor of Pharmacy	Techn.	Technical or Technology
DPhil	Doctor of Philosophy	Tel.	Telephone
Dr	Doctor	U.	University
DSc	Doctor of Science	UK	United Kingdom
Em.	Emeritus	USA	United States of America
Emb.	Embankment	USSR	Union of Soviet Socialist Republics

ARGENTINA

Sub-Editor: **P.V. Konig de Perazzo**

Notes

1. International telephone country code - 54

2. Degrees confered by Argentine universities are *Doctor* (Dr) (aproximately equivalent to PhD), *Ingeniero* (Ing) (between PhD and MSc), and *Licenciado* (Lic) (aproximately equivalent to MSc).

3. In the list the following abbrevations have been used:
   CINDECA - Centro de investigación y Desarrollo en Procesos Catalíticos,
   CITEFA - Centro de Investigaciones Técnicas de las Fuerzas Armadas
   CONICET - Consejo Nacional de Investigaciones Científicas y Técnicas
   CNEA - Comisión Nacional de Energía Atómica
   LEMIT - Laboratorio de Ensayo de Materiales de Interés Tecnológico.
   PRINSO - Program of Research in Solid State Physics

   UNBA - Universidad Nacional de Buenos Aires
   UN Córdoba - Universidad Nacional de Córdoba
   UN Cuyo - Universidad Nacional de Cuyo
   UNLP - Universidad Nacional de La Plata
   UN del Sur - Universidad Nacional del Sur

Acuña, Dr Rodolfo José Gerencia de Investigaciones y Dessarrollo, Aluar, Aluminio Argentino S.A.I.C., Cangallo 525, Buenos Aires, Argentina. (1941) Dr, physics (UN Córdoba, 1971). Res. scient., crystallography lab. (tel. 01 + 493236). *Metal and alloy structures, X-ray diffraction.*

Alvarez, Dr Alberto Guillermo. Dept. de Física, UNLP, 49 y 115, La Plata, Buenos Aires 1900, Argentina. (1931) Dr, physics (UN Cuyo, 1960). Asst. prof. (tel. 021 + 50831). *X-ray crystallography, X-ray fluorescence.*

Alzari. Dr Pedro María. Dept. de Física, UNLP, 49 y 115, La Plata, Buenos Aires 1900, Argentina. (1956) Dr, physics (UNLP, 1985). Res. scient. (tel. 021 + 39061). *Organic and biological molecules, computing, crystallography.*

Baggio, Dr Ricardo. Dept. de Física, CNEA, Av. del Libertador 8250, Buenos Aires 1429, Argentina. (1946) Dr, physics (UNBA, 1975). Res. scient. (tel. 01 + 707711, ext. 337). *Inorganic and organic crystal structures, X-ray diffraction.*

Baggio, Dr Sergio. Gerencia de Investigaciones y Desarrollo, Aluar, Aluminio Argentino S.A.I.C. Cangallo 525, Buenos Aires, Argentina. (1940) Dr, chemistry (UNBA, 1964). Head, crystallography lab. (tel. 01 + 493236). *Crystal structure, powder methods, X-ray fluorescence.*

Benyacar, de, Lic María Angélica Rodríguez. Dept. de Física, CNEA, Av. del Libertador 8250, Buenos Aires 1429, Argentina. (1928) Lic, chemistry (UNBA, 1952). Head, crystallography div. (tel. 01 + 707711, ext. 337). *Crystal structure, physical properties - structure relationships.*

Bengochea, Dr Amado Leandro. Dept. de Geología, UN del Sur, Avda. Alem 1253, Bahía Blanca, Buenos Aires 8000, Argentina. (1945) Dr, geochemistry (UN del Sur, 1976). Asst. prof., res. scient., Argentine Res. Council. (tel. 25196, ext. 354). *Geochemistry, X-ray powder diffraction, fluid inclusions, Lithogeochemistry.*

Canepa, Dr Horacio Ricardo. PRINSO, CONICET - CITEFA, Zufriategui 4380, Villa Martelli, Buenos Aires 1603, Argentina. (1950) Dr, solid state physics (U. de Rennes, France, 1983). Res. scient. (tel. 01 + 7610031, ext. 212). *Crystal growth, polycrystals, semiconductors IV-VI. defects.*

Casanova, Lic. Jorge Ramón. PRINSO, CONICET - CITEFA, Zufriategui 4380, Villa Martelli, Buenos Aires 1603, Argentina. (1944) Lic., physics (UNBA, 1979). Res. scient. (tel. 01 + 7610331, ext. 240). *Crystal growth, X-ray topography and diffraction, intercalation compounds.*

Chandrasekaran, Dr Muthuswamy. Div. Metales, CNEA, Centro Atomico Bariloche, San Carlos de Bariloche, Rio Negro 8400, Argentina. (1947) PhD, physical metallurgy (PIB, USA, 1974). Res. scient. (tel. 0944 + 22646, ext. 139). *Phase transformations, metal and alloy structures, X-ray diffraction, electron microscopy.*

Cortelezzi, Dr César Rafael. Dept. de Geología, LEMIT, 52 121 y 122, La Plata, Buenos Aires 1900, Argentina. (1926) Dr, geology (UNLP, 1952). Head, res. dept. (tel. 021 + 31141). *Mineralogy, petrography, borate structures.*

Diodati, Dr Francisco Piero. Dept. de Física, Facultad de Ingeniería, UNBA, Paseo Colón 850, Buenos Aires 1063, Argentina. (1940) Dr, physics (UNLP, 1970). Res. scient. (tel. 01 + 346441, ext. 178). *Electron diffraction, gases, solid lasers.*

Dristas, Dr Jorge Anastasio. Dept. de Geología, UN del Sur, Avda. Alem 1253, Bahía Blanca, Buenos Aires 8000, Argentina. (1944) Dr, economic geology - mineralogy (UN del Sur, 1972). Assoc. prof., (tel. 362 + 091285). *X-ray powder diffraction, electron diffraction.*

Fernandez, Mr Juan Carlos. Dept. de Física, Facultad de Ingeniería, UNBA, Paseo Colón 850, Buenos Aires 1063, Argentina. (1951) Ing., electrical engineering (UNBA, 1974). Asst. prof. (tel. 01 + 346441, ext. 156). *Surface science, LEED, electron spectroscopy, Auger spectroscopy, surface crystallography, adsorbates, epitaxy.*

Galloni, Prof. Ernesto. Dept. de Física, Facultad de Ingeniería, UNBA, Paseo Colón 850, Buenos Aires 1063, Argentina. (1906) Ingeniero Civil (UNBA, 1930). Emeritus prof. (tel. 01 + 346441, ext. 178). *Inorganic crystal structures, crystal physics, electron diffraction.*

Gay, Dra Hebe Dina. Dept. de Geología, Facultad de Ciencias Exactas y Naturales, UN Córdoba., Velez Sarfield 299, Córdoba 5000, Argentina. (1927) Dr, geology (UN Córdoba, 1950). Full prof. (tel. 716131). *Crystallography, minerals.*

Guérin, Dr Diego Marcelo Alejandro. Dept. de Física, CINDECA, 47 y 257, La Plata, Buenos Aires 1900, Argentina. (1955) Dr, physics (UNLP, 1985). Res. scient. (tel. 021 + 39061). *Crystallite size.*

Hermida, Lic Jorge Daniel. Dept. de Materiales, CNEA, Av. del Libertador 8250, Buenos Aires 1429, Argentina. (1946) Lic., physics (UNBA, 1971). Res. scient. (tel. 01 + 7550181, ext. 268, telex. 18101 CAC AR). *X-ray diffraction, deformed structures.*

Iñiguez Rodríguez, Dr Adrián Mario. Dept. de Mineralogía y Petrografía, UNLP, Paseo del Bosque, La Plata, Buenos Aires 1900, Argentina. (1937) Dr, geology (UNLP, 1962). Asst. prof. *Mineralogy, clays and mixed layers.*

Ipohorski Lenkiewicz, Dr Miguel. Dept. de Metalurgia, CNEA, Av. del Libertador 8250, Buenos Aires 1429, Argentina. (1939) Dr, physics (UN Cuyo, 1967). Res. scient. (tel. 01 + 7550181, ext. 282). *Electron microscopy, metals, metal physics.*

Konig de Perazzo, Lic Patricia Verónica. Dept. de Física, CNEA, Av. del Libertador 8250, Buenos Aires 1429, Argentina. (1941) Lic, physics (UNBA, 1965). Res. scient. (tel. 01 + 707711, ext. 337). *Inorganic crystal structures.*

Lovey, Dr Francisco Carlos. Div. Metales, CNEA, Centro Atomico Bariloche, San Carlos de Bariloche, Rio Negro 8400, Argentina. (1949) Dr, physics (UN Cuyo, 1981). *Phase transformations, metal and alloy structures, X-ray diffraction, electron microscopy.*

Levi, Dra Laura. Dept. de Física, CNEA, Av. del Libertador 8250, Buenos Aires 1429, Argentina. (1915) Dr, physics (U. Bologna, Italy, 1937). Res. scient. (tel. 01 + 707711, ext. 337). *Solid state, nucleation, cloud physics.*

Mahr von Staszewski, Dr Guillermo. PRINSO, CONICET - CITEFA, Zufriategui 4380, Villa Martelli, Buenos Aires 1603, Argentina. (1949) Dr, physics (UNBA, 1983). Co-Dir., PRINSO. (tel. 01 + 7610212, ext. 212). *Crystal growth.*

Maiza, Dr Pedro José. Dept. de Geología, UN del Sur, Avda. Alem 1253, Bahía Blanca, Buenos Aires 8000, Argentina. (1943) Dr, ore deposits (UN del Sur, 1972). Assoc. prof., res. scient., Argentine Res. Council. (tel. 25196, ext. 254). *Geochemistry, mineralogy, ore geology, X-ray powder diffraction.*

Manghi, Lic Estela Margarita. Dept. de Fisica, CNEA, Avda. del Libertador 8250, Buenos Aires 1429, Argentina. Lic, physics (UNBA, 1964). Res. scient. (tel. 01 + 707711, ext. 337). *Solid state physics, X-ray topography, scanning electron microscopy.*

Marbec, Lic Ema Rosa. Dept. de Química, Instituto Nacional de Tecnología Industrial, CC 157, San Martin, Buenos Aires 1650, Argentina. (1939) Lic, chemistry (UNBA, 1966). Head, X-ray diffraction lab. (tel. 01 + 7556161, ext. 76). *X-ray fluorescence, metals, quantitative analysis by X-ray diffraction, iron oxides, uranyl compounds, organic acids.*

Mas, Dr Graciela Raquel. Dept. de Geología, UN del Sur, Avda. Alem 1253, Bahía Blanca, Buenos Aires 8000, Argentina. (1948) Dr, mineralogy (UN del Sur, 1976). Asst. prof., res. scient., Argentine Res. Council. (tel. 25196, ext. 362). *Zeolite synthesis, X-ray powder diffraction, clay mineralogy.*

Piro, Lic Oscar Enrique. Dept. de Física, Facultad de Ciencias Exactas, UNLP, 49 y 115, La Plata, Buenos Aires 1900, Argentina. (1944) Lic, physics (UNLP, 1970). Lecturer. (tel. 021 + 39061). *Organic and inorganic crystal structure.*

2 Argentina

Pochettino, Dr Alberto Antonio. Dept. de Metalurgia, CNEA, Av. del Libertador 8250, Buenos Aires 1429, Argentina. (1947) Dr, physics (UNLP, 1978). Res. scient. (tel. 01 + 7550181, ext. 282). *Electron microscopy, metals.*

Podjarny, Dr Alberto Daniel. Dept. de Fisica, Facultad de Ciencias Exactas, UNLP, 49 y 115, La Plata, 1900 Buenos Aires, Argentina. (1950) PhD, protein-crystallography, direct methods, (Weizmann Inst. of Sci., Israel, 1976). Sr. scient., Res. Argentine Council. (tel. 021 + 39061). *Protein crystallography, direct methods, biologically interesting macromolecules, protein folding, denaturation.*

Punte, Dr Graciela. Dept. de Fisica, UNLP, 47 y 257, La Plata, Buenos Aires 1900, Argentina. (1944) Dr, physics (UNLP, 1972). Sr. scient., Argentine res. council. (tel. 021 + 39061). *Inorganic and organic crystal structures, X-ray diffraction.*

Rigotti, Dr Graciela. Dept. de Fisica, UNLP, 47 y 257, La Plata, Buenos Aires 1900, Argentina. (1948) Dr, chemistry (UNLP, 1979). Res. scient., scient. res. council, Buenos Aires. (tel. 021 + 39061). *Inorganic and organic crystal structures, X-ray diffraction, solid state kinetics.*

Rivero, Dr Blas Eduardo. Dept. de Fisica, UNLP, 49 y 115, La Plata, Buenos Aires 1900, Argentina. (1942) Dr, physics (UNLP, 1976). Asst. prof. (tel. 021 + 39061). *Organic and inorganic crystal structures.*

Ronco, Dr Alicia Estela. Dept. de Fisica, UNLP, 49 y 115, La Plata, Buenos Aires 1900, Argentina. (1945) PhD, nat. sci. (UNLP, 1974). Res. scient. (tel. 021 + 39061). *Protein crystallography.*

Schmirgeld, Dr Lelia. Dept. de Fisica, CNEA, Av. del Libertador 8250, Buenos Aires 1429, Argentina. (1940) PhD (U. Warwick, UK, 1977). Res. scient. (tel. 01 + 707711, ext. 337). *Electron microscopy, physical properties of crystals.*

Silva, Dr Abelardo Manuel. Dept. de Fisica, UNLP, 49 y 115, La Plata, Buenos Aires 1900, Argentina. (1948) PhD, physics (UNLP, 1978). Res. scient., CONICET. (tel. 021 + 39061). *Protein crystallography, theory and methods in crystallography.*

Spinelli, Lic. Silvia Haydeé Dept. de Fisica, UNLP, 47 y 257, La Plata, Buenos Aires 1900, Argentina. (1954) Lic., physics (UNLP, 1980). Sr. scient., Argentine Res. Council. (tel. 021 + 39061). *Inorganic crystal structures, X-ray diffraction.*

Versaci, Dr Raul Antonio. Dept. de Materiales, CNEA, Av. del Libertador 8250, Buenos Aires 1429, Argentina. (1945) Dr, physics (UNLP, 1979). Res. scient. (tel. 01 + 7550181, ext. 282, telex. 18101 CAC AR). *Metals and alloys, structures, electron microscopy.*

Vidal, Lic Haydée Marta. Dept. de Quimica, Instituto Nacional de Tecnologia Industrial, CC 157, San Martin, Buenos Aires 1650, Argentina. (1938) Lic, chemistry (UNBA, 1970). Res. scient. (tel.01 + 7556161, ext. 76). *Orientation, polymer fibers (natural and synthetic), powder methods.*

Viturro, Dr Pedro Ruben. Dept. de Fisica, Facultad de Ciencias Exactas, UNLP, 49 y 115, La Plata, Buenos Aires 1900, Argentina. (1954) Dr, physics (UNLP, 1985). Res. scient. (tel. 021 + 39061). *Biological molecules, computing, crystallography.*

Walsoe de Reca, Dr Elizabeth Noemi. PRINSO, CONICET - CITEFA, Zufriategui 4380, Villa Martelli, Buenos Aires 1603, Argentina. (1939) Dr, chemistry (UNBA, 1965). Dir., PRINSO. (tel. 01 + 7610081/7610031, ext. 158/145). *TEM, SEM, X-ray diffraction, X-ray topography.*

Zalba, Lic Patricia Eugenia. Div. Geologia, LEMIT, 52 y 121, La Plata, Argentina. (1944) Lic, mineralogy (UNLP, 1970). Head, mineralogy and petrography section. (tel 021 + 31141, ext. 15). *Quantitative analysis by X-ray techniques, minerals.*

Zimmerman, Lic Rosa. Dept. de Fisica del Sólido, CITEFA, Zufrategui y Varela, Villa Martelli, Buenos Aires 1603, Argentina. (1928) Lic, chemistry (UNBA, 1953). Res. scient. (tel. 01 + 7610131, ext. 156). *Thin film structure and properties, microelectronics.*

AUSTRALIA

Sub-Editor: S. L. Mair

Notes

1. International telephone country code - 61.

2. The abbreviation CSIRO is used for the Commonwealth Scientific and Industrial Research Organization.

Anstis, Dr Geoffrey Richard. Dept. of Appl. Physics, New South Wales Inst. of Techn., P.O. Box 123, Broadway, New South Wales 2007, Australia. (1949) PhD, mathematical physics (U. Adelaide, 1975). Lect. (tel. 02 + 218-9552). *Electron diffraction theory, imaging theory.*

Auld, Mr John Hugh. Materials Div., Aeronautical Res. Labs., G.P.O. Box 4331, Melbourne, Victoria 3001, Australia. (1927) MMet, metallurgy (U. Sheffield, UK, 1953). Sr. principal res. scient. (tel. 03 + 647-7513). *Metals, intermediate phase structures.*

Avey, Dr Hugh Philip. School of Appl. Sci., Darling Downs Inst. of Advanced Education, P.O. Darling Heights, Queensland 4350, Australia. (1934) PhD, crystallography (U. London, UK, 1967). Assoc. Dean. (tel. 076 + 30-1300). *Structure and function, biological molecules, methods of crystallography, instrumentation, information science.*

Bagshaw, Dr Anthony Nicholas. Research and Development, ALCOA of Australia Ltd., P.O. Box 161, Kwinana, Western Australia 6167, Australia. (1944) DPhil, chemistry (Oxford U., UK, 1970). Res. supervisor. (tel. 09 + 410-1011, ext. 8473, telex AA 92129). *Automated X-ray powder diffractometry, search-match procedures, oxide mineral systems.*

Bailey, Mr David Eric. Centre for Appl. Sci., Nepean C. of Advanced Education, P.O. Box 10, Kingswood, New South Wales 2750, Australia. (1939) MSc, crystallography (U. Western Australia, 1965). Sr. lect. (tel. 047 + 36-0222, ext. 239). *Organic molecules, accurate cell parameter determinations, X-ray physics.*

Baker, Mr Anthony Thomas. School of Chemistry, U. of New South Wales, P.O. Box 1, Kensington, New South Wales 2033, Australia. (1954) BSc, inorganic chemistry (U. New South Wales, 1975). Grad. student. (tel. 02 + 697-2222, ext. 4567). *Small molecules, inorganic chemistry, disordered structures.*

Bakshi, Dr Edward. St. Vincent's Sch. of Medical Res., 41 Victoria Parade, Fitzroy, Victoria 3065, Australia. (1952) PhD, physics (Monash U., 1983). Sr. res. officer. (tel. 03 + 418-2373). *Diffraction theory, X-ray diffraction, neutron diffraction, magnetism, mathematical modelling, optics.*

Barker, Dr William Wilson. CSIRO Div. of Mineralogy, Private Bag, P.O. Wembley, Western Australia 6014, Australia. (1931) PhD, solid state chemistry (U. Western Australia, 1965). Sr. res. scient. (tel. 09 + 387-4233). *High temperature powder neutron diffraction, new minerals, transmission electron microscopy, microcomputers.*

Barnea, Dr Zwi. Sch. of Physics, U. of Melbourne, Parkville, Victoria 3052, Australia. (1932) PhD, physics (U. Melbourne, 1974). Reader. (tel. 03 + 341-7074). *Crystal physics, diffraction physics, anharmonicity.*

Beale, Dr John Phillip. Computing Services Unit, U. of New South Wales, P.O. Box 1, Kensington, New South Wales 2033, Australia. (1947) PhD, chemical crystallography (U. New South Wales, 1970). Systems analyst. (tel. 02 + 662-3509). *Solid state chemistry, computing, protein crystallography.*

Beretka, Mr Julius. CSIRO Div. of Building Res., Graham Rd., Highett, Victoria 3190, Australia. (1930) MSc, chemistry (U. Adelaide, 1962). Principal res. scient. (tel. 03 + 556-2360). *Materials, reaction between solids, kinetics.*

Bevan, Prof David John Martin. Sch. of Physical Sci., The Flinders U. of South Australia, Sturt Rd., Bedford Park, South Australia 5042, Australia. (1926) PhD, chemistry (Imperial C., London, UK, 1957). Emeritus prof. (tel. 08 + 275-3911, ext. 2357). *Structural inorganic chemistry, solid state chemistry.*

Blake, Mr Ronald George. Materials Div., Australian Atomic Energy Commission Research Establishment, Private Mail Bag, Sutherland, New South Wales 2232, Australia. (1930) BSc, physics (U. New South Wales, 1967). Experimental officer. (tel. 02 + 543-3111). *Materials science, electron microscopy, electron diffraction.*

Boehm, Dr James M. Dept. of Physics, Faculty of Military Studies, U. of New South Wales, Royal Military C., Duntroon, Australian Capital Territory 2600, Australia. (1950) PhD, physics (U. Melbourne, 1978). Professional officer. (tel. 062 + 66-3691). *X-ray diffraction, solid state physics, metallurgy, radiation protection.*

Bowles, Prof. John Stephen. Sch. of Metallurgy, U. of New South Wales, Barker St., Kensington, New South Wales, Australia. (1922) MSc, metallurgy (U. Melbourne, 1942). Res. prof. of physical metallurgy. (tel. 02 + 662-2352, ext. 2352). *Phase transformations, alloys.*

Brown, Dr Roger Norman. Mineralogy-Petrology, Australian Mineral Development Labs., Flemington St., Frewville, South Australia 5063, Australia. (1931) PhD, crystallography (U. Adelaide, 1958). Sr. physicist. (tel. 08 + 79-1662, ext. 279). *General diffraction, clay mineralogy.*

Browne, Mr Ian Bruce. Sietronics Pty. Ltd., P.O. Box 521, Fyshwick, Australian Capital Territory 2609, Australia. (1940) Assoc. Dipl., physics (Royal Melbourne Inst. of Techn.). Director. (tel. 062 + 80-4181, telex AA 62754).

Browne, Dr James David. Dept. of Metallurgy, U. of Newcastle, Newcastle, New South Wales 2308, Australia. (1931) PhD, physics (Monash U., 1968). Sr. lect. *Metal physics, magnetic structures.*

Bursill, Dr Leslie Arthur. Physics Dept., U. of Melbourne, Parkville, Victoria 3052, Australia. (1941) DSc, physics (U. Melbourne, 1981). Reader. (tel. 03 + 341-5431, telex AA 35185, UNIMEL). *High-resolution electron microscopy, structural phase transitions, electron optics, crystal structure, defect structures.*

Cashion, Assoc Prof John Dixon. Dept. of Physics, Monash U., Clayton, Victoria 3168, Australia. (1942) DPhil, physics (Oxford U., UK, 1969). Assoc. Prof. (tel. 03 + 541-0811, ext. 3680). *Coals, minerals, soft lattice modes, divergent beam effects, magnetic ordering.*

Cheary, Dr Robert Winston. Dept. of Physics, New South Wales Inst. of Techn., P.O. Box 123, Broadway, New South Wales 2007, Australia. (1946) PhD, X-ray diffraction (Aston U., Birmingham, UK, 1971). Lect. (tel. 02 + 218-9517). *Line profile analysis for X-rays, diffuse scattering, anharmonicity in crystals.*

Church, Mr William Bret. Dept. of Inorganic Chemistry, U. of Sydney, Sydney, New South Wales 2006, Australia. (1960) BSc, chemistry, applied mathematics (U. New South Wales, 1981). Tutor. (tel. 02 + 692-2830) *Macro-molecular refinement, bio-inorganic chemistry.*

Cockayne, Dr David John Hugh. Electron Microscope Unit, U. of Sydney, Science Rd., Sydney, New South Wales 2006, Australia. (1942) DPhil, materials sci. (Oxford U., UK, 1970). Director. (tel. 02 + 692-2351) *Lattice defects, dynamical scattering.*

Cole, Dr William Frederick. CSIRO Div. of Building Res., Graham Rd., Highett, Victoria 3190, Australia. (1917) DSc, mineralogy (U. Western Australia, 1969). Res. fellow. (tel. 03 + 556-2231) *Mineral structures.*

Colman, Dr Peter Malcolm. CSIRO Div. of Protein Chemistry, 343 Royal Parade, Parkville, Victoria 3052, Australia. (1944) PhD, physics (U. Adelaide, 1969). Chief res. scient. (tel. 03 + 342-4200). *Protein crystallography.*

Colmanet, Mr Silvano. Australian Radiation Lab., Lower Plenty Rd., Yallambie, Victoria 3085, Australia. (1955) MSc, chemistry (U. Melbourne, 1982). Chemist. (tel. 03 + 433-2211, ext. 263). *Structure determination, technetium complexes.*

Colyvas, Mr Kim. Technical Services Labs., BHP Co. Ltd., P.O.Box 196B, Newcastle, New South Wales 2300, Australia. (1950) MSc, chemistry (U. Newcastle, 1982) Development chemist. (tel. 049 + 69-0620) *Powder diffraction, X-ray fluorescence, analytical chemistry, statistics, computers.*

Cook, Prof Allan Cecil. U. of New South Wales, Wollongong U.C., Wollongong, New South Wales 2500, Australia. (1935) PhD, (Cantab, UK). Prof. of geology. (tel. 04 + 229-7311). *Optical properties, coals and cokes.*

Corbett, Dr Madeline. Dept. of Inorganic Chemistry, U. of Melbourne, Parkville, Victoria 3052, Australia. (1941) PhD, crystallography (U. Melbourne, 1970). Res. worker. (tel. 03 + 341-6471). *Inorganic crystal structures, computer programming.*

Cousland, Mr Stuart McKay. Materials Div., Aeronautical Res. Labs., Box 4331, G.P.O., Melbourne, Victoria 3001, Australia. (1940) MSc, physics (U. Melbourne, 1960). Sr. res. scient. (tel. 03 + 647-7517). *Solid-solid phase transformations.*

Cowan, Miss Sandra Wendy. St. Vincent's Sch. of Medical Res., Victoria Parade, Fitzroy, Victoria 3065, Australia. (1962) BSc(Hons), chemistry (U. Melbourne, 1984). Grad. student. (tel. 03 + 418-2373). *Protein X-ray crystallography.*

Coyle, Mr Richard Alan. Materials Div., Aeronautical Res. Labs., 506 Lorimer St., Port Melbourne, Victoria 3207, Australia. (1922) Fellow, physics (Melbourne Techn. C., 1948). Experimental officer class 4. (tel. 03 + 647-7541). *Powder methods, materials science, textures.*

Craig, Mr Donald Chadwick. Sch. of Chemistry, U. of New South Wales, P.O. Box 1, Kensington, New South Wales 2033, Australia. (1936) MSc, geology (U. New South Wales, 1964). Professional officer. (tel. 02 + 697-2222, ext. 4595, telex AA 26054). *Small molecules, diffractometry, computing.*

Creagh, Dr Dudley Cecil. Physics Dept., Faculty of Military Studies, U. of New South Wales, Royal Military C., Duntroon, Australian Capital Territory 2600, Australia. (1935) PhD, physics (U. New South Wales, 1975). Sr. lect. (tel. 062 + 66-3562). *Attenuation coefficients for X-rays, anomalous dispersion corrections, dynamical theory, X-ray and neutron scattering, topography (X-ray and neutron).*

Cuff, Dr Christopher. Electron Microscope Unit and Dept. of Geology, James Cook U. of North Queensland, P.O. Douglas, Townsville, Queensland 4811, Australia. (1944) PhD, mineralogical crystallography (Imperial C., UK, 1971). Director, Electron Microscope Unit. (tel. 077 + 81-4496). *Mineral structures, X-ray and neutron diffraction, silicate site occupancy problems.*

Dance, Dr Ian Gordon. Sch. of Chemistry, U. of New South Wales, Sydney, New South Wales 2013, Australia. (1940) PhD, inorganic chemistry (U. Manchester, UK, 1966). Assoc. prof. (tel. 02 + 662-2581). *Structural inorganic chemistry, metal complexes with sulphur ligands.*

Davis, Dr Ronald Lindsay. AINSE Private Mail Bag, Sutherland, New South Wales 2232, Australia. (1940) PhD, antiferromagnetism in Mn alloys (Monash U., 1970). Res. scient. (tel. 02 + 543-3111) *Magnetic structures, neutron diffraction equipment.*

Dean, Mr Christopher. Dept. of Physical and Inorganic Chemistry, U. of Adelaide, G.P.O. Box 498 (North Terrace), Adelaide, South Australia 5001, Australia. (1960) BSc(Hons), crystallography (U. Adelaide, 1981). Grad. student. (tel. 081 + 228-5712, telex AA 89141, UNIVAD). *Computing, martensitic transformations, computer graphics.*

Delaney, Dr William Timothy. 39 Davies Rd., Claremont, Western Australia 6010, Australia. (1944) PhD, quantum chemistry (U. Western Australia, 1972). Lect. (tel. 09 + 384-2247). *Charge density, structure analysis.*

Donovan, Dr William Francis. Dept. of Appl. Chemistry, Tasman C. of Advanced Education, G.P.O. Box 1415P, Hobart, Tasmania 7001, Australia. (1938) PhD, chemical crystallography (U. Tasmania, 1975). Sr. lect. (tel. 002 + 20-3133). *Structures, transition metal complexes, alkaloids.*

Dowell, Dr Walter Charles Thomas. CSIRO Div. of Chemical Physics, P.O. Box 160, Clayton, Victoria 3168, Australia. (1925) Dr. Rer. Nat., electron microscopy of crystals (Freie U., Berlin, Germany, 1963). Sr. principal res. scient. (tel. 03 + 544-0633, ext. 314). *Transmission electron microscopy, shadow electron microscopy, convergent beam electron diffraction, stereo electron microscopy.*

Eggleton, Dr Richard Anthony. Geology Dept. Australian Nat. U., P.O. Box 4, Canberra, Australian Capital Territory 2601, Australia. (1937) PhD, X-ray crystallography (Wisconsin, USA, 1965). Sr. lect. (tel. 062 + 49-5111, ext. 2060). *Mineralogy, electron microscopy, X-ray diffraction.*

Elcombe, Dr Margaret Marion. Materials Div., Australian Atomic Energy Commission, Private Bag, Sutherland, New South Wales 2232, Australia. (1942) PhD, lattice dynamics (Cambridge U., UK, 1966). Sr. res. scient. (tel. 02 + 543-3111). *Lattice dynamics, neutron diffraction.*

Epstein, Dr Joel. Research Labs., Kodak (Australasia) Pty. Ltd., P.O. Box 90, Coburg, Victoria 3058, Australia. (1950) PhD, chemistry (Carnegie-Mellon U., USA). Res. scient. (tel. 03 + 353-3703) *Molecular charge densities, X-ray scattering theory, diffuse scattering.*

Fallon, Dr Gary David. Dept. of Chemistry, Monash U., Clayton, Victoria 3168, Australia. (1948) PhD, chemical crystallography (Monash U., 1976). Res. asst. (tel. 03 + 541-3614). *Alkali metal oxide systems.*

Field, Dr Donald William. Dept of Physics, Queensland Inst. of Techn., G.P.O. Box 2434, Brisbane, Queensland 4001, Australia. (1947) PhD, physics (U. Adelaide, 1973). Lect. (tel. 07 + 223-2593) *X-ray diffraction.*

Figgis, Prof Brian Norman. Sch. of Chemistry, U. of Western Australia, Nedlands, Western Australia 6009, Australia. (1930) DSc, inorganic chemistry (U. Western Australia, 1966). Prof. of inorganic chemistry. (tel. 09 + 380-3157). *Structure determination, bonding, transition metal complexes.*

Fletcher, Dr Neville Horner. Inst. of Physical Sci., CSIRO, Limestone Ave., Canberra, Australian Capital Territory 2602, Australia. (1930) DSc, solid state physics (U. Sydney, 1973). Director. (tel. 062 + 48-4613, telex AA 62003). *Crystal growth, interfaces, ice physics, acoustics.*

Forwood, Dr Christopher Thomas. CSIRO Div. of Chemical Physics, P.O. Box 160, Clayton, Victoria 3168, Australia. (1940) PhD, physics (U. Bristol, UK, 1966). Principal res. scient. (tel. 03 + 544-0633). *Defect structure of solids.*

Foster, Dr John James. Res. Sch. of Earth Sci., Australian Nat. U., P.O. Box 4, Canberra, Australian Capital Territory 2601, Australia. (1942) PhD, chemistry of chromium compounds (Australian Nat. U., 1973). Microanalyst. (tel. 065 + 49-4370). *Isotope geochemistry, structure, chromium-oxygen networks.*

Fraser, Dr Ronald Bruce. CSIRO Div. of Protein Chemistry, 343 Royal Parade, Parkville, Melbourne, Victoria 3052, Australia. (1924) DSc, biophysics (U. London, UK, 1960). Chief res. scient., asst. chief. (tel. 03 + 342-4200) *Biology, biophysics, keratin structure, collagen structure.*

Freeman, Prof Hans Charles. Dept. of Inorganic Chemistry, U. of Sydney, Sydney, New South Wales 2006, Australia. (1929) PhD, chemistry (U. Sydney, 1957). Prof. of Inorganic Chemistry. (tel. 02 + 692-2222, ext. 2757, telex AA 26169, UNISYD). *Coordination complex structures, metallo-protein structures.*

Gable, Mr Robert William. Dept. of Inorganic Chemistry, U. of Melbourne, Parkville, Victoria 3052, Australia. (1954) BSc(Hons), inorganic chemistry (U. Melbourne, 1979). Grad. res. asst. *Structures, coordination compounds.*

Gardner, Mr Alexander Parker. Dept. of Appl. Chemistry, Swinburne C. of Techn., John St., Hawthorn, Victoria 3122, Australia. (1926) MSc, chemistry (Monash U., 1972). Appl. Sci. Dean. (tel. 819-8179). *Silicates, chemical education, structure determination.*

Garrett, Mr Thomas Peter John. Dept. of Inorganic Chemistry, U. of Sydney, Sydney, New South Wales 2006, Australia. (1959) BSc, chemistry (U. Sydney, 1981). Tutor. (tel. 02 + 692-2222, ext. 2830). *Protein crystallography.*

Gatehouse, Dr Bryan Michael Kenneth. Chemistry Dept., Monash U., Clayton, Victoria 3168, Australia. (1932) DSc, structural inorganic chemistry (U. C., London, UK, 1977). Reader. (tel. 03 + 541-0811, ext. 3561). *Structures of single crystals, mixed metal oxides, inorganic compounds.*

Goodman, Dr Peter. CSIRO Div. of Chemical Physics, P.O.Box 160, Clayton, Victoria 3168, Australia. (1928) DSc, electron diffraction (U. Melbourne, 1978). Chief res. scient. (tel. 03 + 544-0633, ext. 311). *Electron diffraction.*

Graham, Dr James. CSIRO Div. of Mineralogy, P.O. Wembley, Western Australia 6014, Australia. (1929) PhD, physical metallurgy (U. Birmingham, UK, 1956) Principal res. scient. (tel. 092 + 387-0371). *Minerals, sulphides, oxides.*

Grainger, Dr Colin Trevor. Sch. of Physics, U. of New South Wales, Kensington, New South Wales 2033, Australia. (1927). PhD, crystallography (U. New South Wales, 1967). Sr. lect. (tel. 02 + 697-2222, ext. 4567). *Crystal structure methodology, computing.*

Grey, Dr Ian Edward. CSIRO Div. of Mineral Chemistry, P.O. Box 124, Port Melbourne, Victoria 3207, Australia. (1944) PhD, inorganic chemistry (U. Tasmania, 1969). Principal res. scient. (tel. 03 + 647-0211, ext. 268). *Structures, non-stoichiometric oxides, mineral structures, order-disorder, intergrowths.*

Grzinic, Dr Guido. Dept. of Physics, U. of Melbourne, Parkville, Victoria 3052, Australia. (1956) PhD (U. Melbourne, 1984). Visiting res. fellow (tel. 03 + 341-5079 or -5454, telex AA 35185). *Incommensurate superlattices, diffraction theory, high-resolution electron microscopy.*

Guddat, Mr Luke William. St. Vincent's Sch. of Medical Res., Victoria Parade, Fitzroy, Victoria 3065, Australia. (1962) BSc(Hons), chemistry (Monash U., 1983). Grad. student. (tel. 03 + 418-2373). *Protein structures.*

Guss, Dr Jules Mitchell. Dept. of Inorganic Chemistry, U. of Sydney, Sydney, New South Wales 2006, Australia. (1946) PhD, inorganic chemistry (U. Sydney, 1970). Professional officer. (tel. 02 + 692-1122, ext. 2830). *Structures, biological macromolecules, instrumentation.*

Hall, Prof Eric Ogilvie. Dept. of Metallurgy, U. of Newcastle, Newcastle, New South Wales 2308, Australia. (1925) PhD, solid state (Cambridge U., UK, 1952). Prof. (tel. 049 + 68-0401 ext. 425). *Metals, intermediate phase structures.*

Hall, Dr Sydney Reading. Crystallography Centre, U. of Western Australia, Nedlands, Western Australia 6009, Australia. (1939) PhD, crystallography (U. Western Australia, 1964). Assoc. prof. (tel. 09 + 380-2725). *Mineral structures, phasing methodology, computing techniques.*

Hambley, Dr Trevor William. Sch. of Chemistry, U. of Sydney, Sydney, New South Wales 2006, Australia. (1955) PhD, inorganic chemistry (U. Adelaide, 1983). Sr. tutor. (tel. 02 + 692-2830). *Metal complex structures.*

Hamilton, Dr John David Gavin. CSIRO Div. of Mineral Chemistry, 39 Williamstown Rd., Port Melbourne, Victoria 3207, Australia. (1935) DPhil, geology (U. Sydney, 1963). Principal res. scient. (tel. 03 + 647-0211, ext. 295). *Mineralogy, iron ores, iron ore products, clay mineralogy.*

Hartley, Dr Richard H. Optoelectronic Device Physics Group, Weapons Res. Est., P.O. Box 2151, Adelaide, South Australia 5001, Australia. (1939) PhD, thin film physics, electrical engineering (U. Western Australia, 1971). Sr. res. scient. (tel. 08 + 259-5555, ext. 6377). *Solid state physics, thin films.*

Hay, Dr David Gilbert. CSIRO Div. of Materials Sci., Normanby Rd., Clayton, Victoria 3168, Australia. (1948) PhD, chemistry (Latrobe U., 1982). Experimental officer. (tel. 03 + 542-2777). *Structure determination, solid state chemistry.*

Healy, Dr Peter Conrad. Sch. of Sci., Griffith U., Nathan, Queensland 4111, Australia. (1947) PhD, chemistry (U. Western Australia, 1972). Lect. (tel. 07 + 275-7170). *Physical inorganic chemistry.*

Hicks, Dr Trevor John. Physics Dept., Monash U., Wellington Rd., Clayton, Victoria 3168, Australia. (1939) PhD, physics (Monash U., 1966). Reader (tel. 03 + 541-0811, ext. 3681, telex AA 32691, MONASH). *Magnetic structure, neutron diffraction.*

Hill, Dr Roderick Jeffrey. CSIRO Division of Mineral Chemistry, 339 Williamstown Rd., Port Melbourne, Victoria 3207, Australia. (1949) PhD, mineralogy, crystallography (U. Adelaide, Australia, 1976). Sr. res. scient. (tel. 03 + 647-0211). *Crystal chemistry, minerals, bonding in the solid state, powder diffraction.*

Hodge, Mr Leslie Cameron. 118 Preston Point Rd., East Fremantle, Western Australia 6158, Australia. (1928) BSc, geology (U. Western Australia, 1953). *Mineralogy.*

Hogan, Dr Leonard McNamara. Dept. of Mining and Metallurgical Engineering, U. of Queensland, St Lucia, Brisbane, Queensland 4067, Australia. PhD, metallurgy (U. Queensland, 1965). Reader in metallurgy. (tel. 072 + 370-3111). *Physical metallurgy, crystal growth, solidification theory.*

Hons, Mr Alexander. Electronic Systems Div., Plessey Australia Pty. Ltd., Railway Rd., Meadowbank, New South Wales 2114, Australia. (1952) MAppSc, physics (New South Wales Inst. of Techn., 1980). Professional officer. (tel. 02 + 807 0521, telex AA 21471). *Electron microscopy, ceramics, electronic materials.*

Horn, Dr Ernst. Dept. of Physical and Inorganic Chemistry, U. of Adelaide, G.P.O. Box 498, Adelaide, South Australia 5001, Australia. (1952) PhD, chemistry (U. Adelaide, 1983). Res. associate. (tel. 08 + 223-4333, ext. 5712). *X-ray structure determination, inorgano-metallic complexes, organic compounds, metal carbonyl fluoride.*

Hoskins, Dr Bernard Foster. Dept. of Inorganic Chemistry, U. of Melbourne, Parkville, Victoria 3052, Australia. (1935) DSc, chemistry (U. Melbourne, 1976). Reader. (tel. 03 + 341-6471, telex AA 35185, UNIMELB). *Inorganic structures.*

Howard, Dr Christopher John. Appl. Physics Div., Australian Atomic Energy Commission Res. Est., Private Mailbag, Sutherland, New South Wales 2232, Australia. (1942) PhD, nuclear magnetic resonance (U. Nottingham, UK, 1969). Principal res. scient. (tel. 02 + 543-3609). *Neutron diffraction, high resolution neutron powder diffraction, accurate structure determination.*

Hyde, Prof Bruce Godfrey. Res. Sch. of Chemistry, Australian Nat. U., P.O. Box 4, Canberra, Australian Capital Territory 2601, Australia. (1925) DSc, physical chemistry (U. Bristol, UK, 1974). Prof. in inorganic chemistry. (tel. 062 + 49-4401). *Crystal chemistry, electron microscopy.*

Isaacs, Dr Neil William. St. Vincent's Sch. of Medical Res., Victoria Parade, Fitzroy, Victoria 3065, Australia. (1945) PhD, chemistry (U. Queensland, 1970). Sr. res. fellow. (tel. 03 + 418-2211, ext. 2373, telex AA 32229). *Protein crystallography, protein structure refinement.*

Jaeger, Mr Hans. CSIRO Div. of Materials Sci., Normanby Rd., Clayton, Victoria 3168, Australia. (1921) Dipl. Appl. Chem., chemistry (Eng. C., Essen, Germany, 1953). Sr. res. scient. (tel. 03 + 542-2777). *Epitaxy, interface structures.*

James, Dr Veronica Jean. Sch. of Physics, U. of New South Wales, P.O. Box 1, Kensington, New South Wales 2033, Australia. PhD, physics (U. New South Wales, 1970). Sr. lect. (tel. 02 + 663-0351, ext. 3079). *Collagen structure, diseased tissue.*

Johnson, Dr Andrew William Syme. Electron Microscopy Centre, U. of Western Australia, Nedlands, Western Australia 6009, Australia. PhD, physics (U. Western Australia, 1966). Director. (tel. 09 + 380-2764, telex AA 92992). *Electron microscopy, electron diffraction, mineralogy.*

Johnston, Dr Gordon Basil. Electron Microscope Unit, U. of Newcastle, Newcastle, New South Wales 2308, Australia. (1929) PhD, metallurgy (U. Newcastle, 1967). Director. (tel. 049 + 68-0401, ext. 405, telex AA 28194, NEWUN). *Transmission electron microscopy, scanning electron microscopy, electron microprobe X-ray analysis, X-ray and neutron diffraction, magnetic structures.*

Jones, Dr John Brett. Dept. of Geology, U. of Adelaide, Adelaide, South Australia 5001, Australia. (1930) PhD, mineralogy (U. Wisconsin, USA, 1958). Reader. (tel. 08 + 228 5333, ext. 5375, telex AA 8941). *Mineral structures.*

Jostsons, Dr Adam. Materials Div., Australian Atomic Energy Commission Research Establishment, Private Mail Bag, Sutherland, New South Wales 2232, Australia. (1937) PhD, metallurgy (U. New South Wales, 1966). Chief. (tel. 02 + 543-3265, telex AA 24562). *Materials science, electron microscopy, electron diffraction.*

Kelly, Dr Patrick Manning. Deputy Director Research, Australian Atomic Energy Commission, Private Mail Bag, Sutherland, New South Wales 2232, Australia. (1935) ScD, physical metallurgy (Cambridge, UK, 1979). Deputy Director Research. (tel. 02 + 543-3315). *Materials science, electron microscopy, electron diffraction.*

Kennard, Dr Colin Harold Leslie. Dept. of Chemistry, U. of Queensland, St. Lucia, Queensland 4067, Australia. (1935) PhD, crystallography (U. New South Wales, 1961). Reader. (tel. 07 + 377-3296). *Structure analysis, pesticides, metal complexes, teaching methods, neutron diffraction.*

Kennedy, Dr Stanley Wallace. Dept. of Physical and Inorganic Chemistry, U. of Adelaide, Adelaide, South Australia 5001, Australia. PhD, solid state chemistry (U. Belfast, UK, 1954). Reader. (tel. 08 + 228-5943, telex AA 89141). *Structural transformations, martensite crystallography, crystal chemistry, X-ray diffraction, pressure, calorimetry.*

Kennon, Prof Noel Frederick. Dept. of Metallurgy, U. of Wollongong, Northfields Ave., Wollongong, New South Wales 2500, Australia. (1934) PhD, crystallography (U. New South Wales, 1967). Assoc. prof. (tel. 042 + 27-0555, ext. 3012). *Shape memory alloys, crystallography, martensitic transformations, metallography, phase transformation in steels.*

Kucharski, Dr Edward Stanislaw. Sch. of Math. and Physical Sci., Murdoch U., South St., Perth, Western Australia 6150, Australia. (1953) PhD, physical and inorganic chemistry (U. Western Australia, 1984). Res. associate. (tel. 09 + 332-2163). *Charge densities, transition metal complexes, structural studies, magnetically anomalous transition metal complexes, enzyme structures.*

Lambert-Smith, Mr John Ernle Warwick. Collinsville Laboratory, Queensland Electricity Commission, P.O. Box 135, Collinsville, Queensland 4804, Australia. (1931) BSc, inorganic chemistry and X-ray crystallography (U. Sydney, 1957). Station chemist. (tel. 077 + 85-5166, ext. 215, telex AA 47613, QEGCOL). *X-ray crystal structure analysis, crystallographic computer programming, coordination compounds of copper(II).*

Lancucki, Mr Christopher Joseph. CSIRO Div. of Building Res., Graham Rd., Highett, Victoria 3190, Australia. (1935) BSc, physics (U. Western Australia, 1959). Experimental officer. (tel. 03 + 556-2279). *Clay minerals, inorganic crystals, structures.*

Leverett, Dr Peter. Centre for Appl. Sci., Nepean C. of Advanced Education, Second Ave., Kingswood, New South Wales 2750, Australia. (1944) PhD, X-ray crystallography and oxide chemistry (Monash U., 1969). Director. (tel. 047 + 36-0222, ext. 286). *X-ray structure determination, mixed metal oxide systems synthesis, crystal chemistry.*

Lincoln, Dr Francis John. Sch. of Chemistry, U. of Western Australia, Nedlands, Western Australia 6009, Australia. (1935) PhD, chemistry (U. Western Australia, 1967). Sr. lect. (tel. 09 + 380-3142). *Solid state chemistry, electron microscopy, inorganic materials, oxides, minerals, thermodynamic and structural examination.*

Lloyd, Dr Doug. Sci. Education Dept., State C. of Victoria, Osborne St., Bendigo, Victoria 3550, Australia. (1946) PhD, solid state chemistry (Monash U., 1972). Lect. (tel. 054 + 43-9433). *Solid state chemistry, ceramics, salts, X-ray crystallography, structure determination.*

Lucas, Dr Brian William. Dept. of Physics, U. of Queensland, St. Lucia, Brisbane, Queensland 4067, Australia. PhD, physics (U. London, UK, 1965). Sr. lect. (tel. 07 + 377-3421). *Crystal physics, X-ray and neutron diffraction.*

Lukaszewski, Dr George Michael. CSIRO Div. of Mineral Chemistry, P.O. Box 124, Port Melbourne, Victoria 3207, Australia. (1935) PhD, inorganic chemistry

(U. London, UK, 1960). Principal res. scient. (tel. 03 + 647-0211, ext. 210). *Mineral chemistry, sulphide oxidation, thermoanalytical techniques.*

Lynch, Dr Denis Francis. CSIRO Div. of Chemical Physics, P.O. Box 160, Clayton, Victoria 3168, Australia. (1941) PhD, physics (U. New South Wales, 1965). Principal res. scient. (tel. 03 + 544-0633, ext. 221). *Electron diffraction, dynamical scattering.*

Machin, Mr Ken James. St. Vincent's Sch. of Medical Res., Victoria Parade, Fitzroy, Victoria 3065, Australia. (1956) BSc(Hons), biochemistry (Monash U., 1978). Res. officer. (tel. 03 + 418-2375). *Protein crystallography.*

Mackay, Dr Maureen Florence. Dept. of Physical Chemistry, La Trobe U., Bundoora, Victoria 3083, Australia. PhD, crystallography (U. Melbourne, 1968). Reader. (tel. 03 + 479-2520). *Structures, small molecules of biological interest, natural products.*

Mackenzie, Dr James Kenneth. CSIRO Div. of Chemical Physics, P.O.Box 160, Clayton, Victoria 3168, Australia. (1920). PhD, physics (U. Bristol, UK, 1949). Sr. principal res. scient. (tel. 03 + 544-0633). *Crystal physics, crystallography, phase transformations.*

MacRae, Mr Thomas Perry. CSIRO Div. of Protein Chemistry, 343 Royal Parade, Parkville, Victoria 3052, Australia. (1925) MSc, X-ray diffraction of fibrous proteins (U. Manchester, UK, 1954). Sr. principal res. scient. (tel. 03 + 342-4200, ext. 280). *Biology, biophysics, keratin structure, collagen structure.*

Mair, Dr Sylvia Lorraine. CSIRO Div. of Chemical Physics, P.O. Box 160, Clayton, Victoria 3168, Australia. (1948) PhD, physics (U. Melbourne, 1974). Sr. res. scient. (tel. 03 + 544-0633, ext. 322, telex AA 30356). *Thermal vibrations in crystals, structural phase transitions, thermal diffuse scattering.*

Malin, Dr Anthony Samuel. Sch. of Metallurgy, U. of New South Wales, P.O. Box 1, Kensington, New South Wales 2033, Australia. (1932) PhD, metallurgy (U. New South Wales, 1978). Sr. lect. (tel. 02 + 662-2335). *Recrystallization, strain deformation, electron microscopy, X-ray diffraction, metallography.*

Maslen, Dr Edward Norman. Crystallography Centre, U. of Western Australia, Nedlands, Western Australia 6009, Australia. (1935) D.Phil, crystallography (Oxford U., UK, 1960). Director. (tel. 09 + 380-2738). *Electron density distributions.*

Mathieson, Dr Alexander McLeod. CSIRO Div. of Chemical Physics, P.O. Box 160, Clayton, Victoria 3168, Australia. (1920) DSc, chemical crystallography (U. Melbourne, 1956). Chief res. scient. (tel. 03 + 544-0633). *X-ray apparatus, extinction, reflectivity measurement.*

McConnell, Dr Jack Foster. Sch. of Physics, U. of New South Wales, P.O. Box 1, Kensington, New South Wales 2033, Australia. (1918) PhD, crystallography (U. New South Wales, 1958). Retired. (tel. 02 + 663-0351, ext. 3097). *Biological structures.*

McCormick, Ms Robyn Joy. Dept. of Inorganic Chemistry, U. of Melbourne, Parkville, Victoria 3052, Australia. (1954) BSc(Hons) chemistry (U. Melbourne, 1977). Sr. tutor. (tel. 03 + 341-6473). *Structure analysis, Y-O-F system, manganese carbonyl compounds.*

McDougall, Dr Peter George. Sch. of Metallurgy, U. of New South Wales, Anzac Parade, Kensington, New South Wales 2033, Australia. (1930) PhD, metallurgy (U. New South Wales, 1964). Sr. lect. (tel. 02 + 660-0351, ext. 2335). *Phase transformations, metals.*

McKenzie, Dr David Robert. Sch. of Physics, U. of Sydney, Parramatta Rd., Sydney, New South Wales 2006, Australia. (1946) PhD, physics (U. New South Wales, 1972). Sr. lect. (tel. 02 + 692-3180). *Lattice dynamics, electron diffraction, amorphous materials.*

McKenzie, Dr Elwyn Donald. Dept. of Chemistry, U. of Queensland, St. Lucia, Queensland 4067, Australia. (1935) PhD, chemistry (U. New South Wales, 1963). Sr. tutor. (tel. 07 + 377-2306). *Transition metal chemistry, X-ray structure analysis.*

McLaren, Dr Alexander Clark. Dept. of Physics, Monash U., Wellington Rd., Clayton, Victoria 3168, Australia. (1928) ScD, physics (Cambridge U., UK, 1981). Reader. (tel. 03 + 541-0811, ext. 3632 or 3699, telex AA 32691, MONASH). *Defects in solids and minerals, electron microscopy, plastic deformation, minerals, rocks.*

McLaughlin, Mr George Millar. Res. Sch. of Chemistry, Australian Nat. U., P.O.Box 4, Canberra, Australian Capital Territory 2601, Australia. (1947) Grad. R.I.C. (Royal Inst. of Chemistry, UK, 1971). Res. officer. (tel. 062 + 49-3641). *Biologically significant molecules, packing, conformation, computing.*

McLeod, Mr Neil John. Dept. of Geology and Geophysics, U. of Sydney, Sydney, New South Wales 2006, Australia. (1949) BSc(hons), inorganic chemistry, X-ray crystallography (U. Melbourne, 1970). Professional officer. (tel. 02 + 692-2912, ext. 2919). *Inorganic crystal structures.*

Medlin, Dr Edwin Harry. Physics Dept., U. of Adelaide, North Terrace, Adelaide, South Australia 5001, Australia. (1920) PhD, physics (U. Adelaide, 1956). Reader in physics. (tel. 08 + 223-4333, ext. 2310). *Physical optics, crystal physics.*

Millar, Dr John Joseph. Physics Dept, Bendigo C. of Advanced Education, P.O. Box 199, Bendigo, Victoria 3550, Australia. (1946) PhD, physics (U. Melbourne, 1973). Lect. (tel. 054 + 40-3395 or -3405). *Electron microscopy, X-ray diffraction, industrial applications.*

Miller, Ms Sarah Ann. CSIRO Div. of Energy Chemistry, Lucas Heights Res. Labs., Private Mail Bag 7, Sutherland, New South Wales 2232, Australia. (1961) BSc, chemistry (Flinders U. of South Australia, 1982). Experimental officer. (tel. 02 + 543-3111, ext. 3188, telex AA 73341). *Zeolites, industrial minerals, neutron diffraction, X-ray diffraction, powder diffraction.*

Mohyla, Dr Jury. South Australian C. of Advanced Education, Sturt Rd., Bedford Park, Adelaide, South Australia 5042, Australia. (1937) PhD, crystallography (Flinders U. of Sch. Austraia, 1979). Sr. lect. (tel. 08 + 275-5311, ext. 5382). *Solid state physics, education, computing.*

Moodie, Dr Alexander Forbes. CSIRO Div. of Chemical Physics, P.O.Box 160, Clayton, Victoria 3168, Australia. (1923) DSc, physics (Royal Melbourne Inst. of Techn., 1982). Chief res. scient. (tel. 03 + 544-0633 ext. 312). *Electron diffraction, electron microscopy, scattering theory.*

Moon, Dr Anthony Ronald. Dept. of Physics, New South Wales Inst. of Techn., P.O. Box 123, Broadway, New South Wales 2007, Australia. (1945) PhD, physics (U. Melbourne, 1970). Head, Dept. of Physics. (tel. 02 + 2-0930, ext. 9468). *Electron and X-ray diffraction, electron microscopy.*

Moore, Dr Alan James William. CSIRO Div. of Materials Sci., Normanby Rd., Clayton, Victoria 3168, Australia. (1920) PhD, physical chemistry (Cambridge U., UK, 1948). Sr. principal res. scient. (ret.) (tel. 03 + 542-2777). *Surface structures, field ion microscopy.*

Moore, Mr Christopher James. Sch. of Chemistry, Macquarie U., North Ryde, New South Wales 2113, Australia. (1950) BSc(hons), inorganic chemistry (U. Sydney, 1973). Tutor. *Metal complexes.*

Moore, Dr Francis Hugh. Neutron Diffraction Group, Australian Inst. of Nuclear Sci. and Engineering, P.M.B., P.O., Sutherland, New South Wales 2232, Australia. (1933) DPhil, chemical crystallography (Oxford U., UK, 1966). Group leader. (tel. 02 + 543-3607). *Neutron diffraction studies, minerals, biological molecules.*

Morton Dr Allan James. CSIRO Div. of Chemical Physics, P.O. Box 160, Clayton, Victoria 3168, Australia. (1939) PhD, metallurgy (U. New South Wales, 1964). Principal res. scient. (tel. 03 + 544-0633). *Defects in metals, intermediate phase structures, electron microscopy.*

Moss, Dr Barbara Kay. CSIRO Div. of Chemical Physics, P.O. Box 160, Clayton, Victoria 3168, Australia. (1955) PhD, diffraction physics (U. Melbourne, 1983). Postdoctoral fellow. (tel. 03 + 544-0633, ext. 323, telex AA 30356, RIVETT). *Anharmonic thermal motion, electron diffraction, polymers, X-ray diffraction, neutron diffraction.*

Moss, Dr Grant Richard. Physics Dept., U. of Melbourne, Parkville, Victoria 3052, Australia. (1951) PhD, diffraction physics (U. Melbourne, 1978). Res. fellow. (tel. 03 + 341-5451). *Electron density, lattice dynamics.*

Muddle, Dr Barrington Charles. Dept. of Materials Engineering, Monash U., Wellington Rd., Clayton, Victoria 3168, Australia. (1948) PhD, metallurgy (U. New South Wales, 1975). Sr. lect. (tel. 03 + 541-0811, ext. 2051). *Phase transformations, electron microscopy.*

Napier, Mr John Graham. Materials Div., Australian Atomic Energy Commission Research Establishment, Private Mail Bag, Sutherland, New South Wales 2232, Australia. (1933) ASTC, metallurgy (Sydney Techn. C., 1962). Experimental officer. (tel. 02 + 543-3111). *Materials science, electron microscopy, electron diffraction, computation.*

Netherway, Dr David John. Radar Div., Defence Res. Centre Salisbury, Box 2151 G.P.O., Adelaide, South Australia 5001, Australia. (1953) PhD, solid state physics (U. Melbourne, 1982). Res. scient. (tel. 08 + 259-6479). *Modulated structures, electron diffraction, antenna arrays.*

Nimmo, Dr John Kenneth. Physics Dept., U. of Queensland, St. Lucia, Queensland 4067, Australia. (1948) PhD, physics (U. Queensland, 1976). Sr. tutor. (tel. 07 + 377-2364). *Phase transitions, structure determination.*

Norman, Mr Peter David. Dept. of Appl. Physics, Chisholm Inst. of Techn., 900 Dandenong Rd., Caulfield East, Victoria 3145, Australia. (1932) BSc, physics (U. Melbourne, 1953). Lect. (tel. 059 + 75-3040). *Electron diffraction, phase transformations, alloys, Fermiology, Kohn anomalies.*

Nunn, Dr Ernest Keith. Chemistry Dept., Monash U., Clayton, Victoria 3168, Australia. (1935) PhD, chemistry (U. Tasmania, 1963). Lect. (tel. 03 + 541-0811, ext. 3566). *Structures, ionic solids.*

O'Connor, Dr Brian Henry. Dept. of Appl. Physics, Western Australian Inst. of Techn., Kent St., Bentley, Western Australia 6102, Australia. (1941) PhD, crystallography (U. Western Australia, 1964). Principal lect., dept. head. (tel. 09 + 350-7193). *Materials analysis, X-ray powder diffraction, X-ray fluorescence spectrometry.*

Parise, Dr John Baptist. Dept. of Chemistry, New South Wales Inst. of Techn., P.O. Box 123, Broadway, New South Wales 2007, Australia. (1953) PhD, chemistry (James Cook U., 1980). Lect. *Molecular sieves, framework structures, solid state chemistry, crystal structure.*

Parks, Dr Terrence Charles. CSIRO Div. of Mineralogy, Private Bag, P.O. Wembley, Western Australia 6014, Australia. (1940) PhD, physical chemistry (U. Western Australia, 1967). Sr. res. scient. (tel. 09 + 387-4233, ext. 357. telex AA 92178). *Defect structures, crystal chemistry, geochemical processes.*

Phakey, Dr Prem P. Dept. of Physics, Monash U., Wellington Rd., Clayton, Victoria 3168, Australia. (1933) PhD, physics (Monash U., 1968). Sr. lect. (tel. 03 + 541-0811, ext. 3642). *Defects in solids.*

Poppleton, Dr Bruce J. CSIRO Div. of Materials Sci., Normanby Rd., Clayton, Victoria 3168, Australia. (1936) PhD, chemistry (Monash U., 1973). Sr. res. scient. (tel. 03 + 542-2777). *Small organic molecules, computing.*

Prager, Dr Peter Robert. P.O. Box 17, Black Rock, Victoria 3193, Australia. (1944) PhD, physics (U. Melbourne, 1971). (tel. 03 + 589-2550). *Accurate structure analysis, crystal physics.*

Pring, Dr Allan. Dept. of Mineralogy, South Australian Museum, North Terrace, Adelaide, South Australia 5000, Australia. (1956) PhD, high res. electron microscopy (Cambridge U., UK, 1983). Curator of minerals and meteorites. (tel. 08 + 223-8894). *Mineral chemistry, sulphides, electron microscopy, order-disorder, borates, magic angle spinning NMR.*

Rachinger, Prof William Albert. Dept. of Physics, Monash U., Wellington Rd., Clayton, Victoria 3168, Australia. (1927) PhD, metallurgy (U. Melbourne, 1952). Prof. (tel. 03 + 541-0811, ext. 3630). *Low temperature solid state physics, biological composite materials.*

Radoslovich, Dr Edward William. CSIRO Div. of Soils, Private Bag No. 2, Glen Osmond, South Australia 5064, Australia. (1928) DSc, physics (U. Adelaide, 1968). Principal res. scient. (tel. 08 + 274-9309). *Clay-organic complexes, mineral structures.*

Rae, Dr Alan David. Sch. of Chemistry, U. of New South Wales, Kensington, New South Wales 2033, Australia. (1941) PhD, chemistry (U. Auckland, New Zealand, 1964). Assoc. prof. (tel. 02 + 697-2222, ext. 4673, telex AA 26054). *Pseudosymmetric crystal structures, crystal structure refinement techniques, chemical physics.*

Raston, Dr Colin Llewellyn. Dept. of Physical and Inorganic Chemistry, U. of Western Australia, Nedlands, Western Australia 6009, Australia. (1950) PhD, inorganic chemistry (U. Western Australia, 1976). Res. fellow. (tel. 09 + 380-3838). *Structure determination, inorganic, organo-metallic and organic molecules.*

Reid, Dr Allen Forrest. CSIRO Inst. of Energy and Earth Resources, Private Mail Bag 10, Clayton, Victoria 3168, Australia. (1931) DSc, solid state chemistry (Australian Nat. U., 1969). Director of inst. (tel. 03 + 542-2432). *Mineral structures.*

Reynolds, Dr Phillip Andrew. Sch. of Chemistry, U. of Western Australia, Nedlands, Western Australia 6009, Australia. (1947) DPhil, chemistry (Oxford U., UK, 1973). Res. fellow. (tel. 062 + 380-3140). *Molecular crystals, disorder in crystals, transition metal complexes, charge densities, spin densities.*

Ridout, Mr Stephen Charles. Dept. of Physics, U. of Western Australia, Nedlands, Western Australia 6009, Australia. (1952) BSc(Hons), physics (U. Western Australia, 1974). Grad. student. *Electron densities in transition metal complexes.*

Robertson, Dr Glen Bradley. Res. Sch. of Chemistry, Australian Nat. U., G.P.O. Box 4, Canberra, Australian Capital Territory 2601, Australia. (1930) PhD, physics (U. Western Australia, 1966). Sr. fellow. (tel. 062 + 49-4380). *X-ray diffraction, neutron diffraction, organometallic complexes, low temperature analysis, charge density.*

Rossell, Dr Henry John. CSIRO Div. of Materials Sci., Normanby Rd., Clayton, Victoria 3168, Australia. (1936) PhD, physical chemistry (U. Western Australia, 1965). Principal res. scient. (tel. 03 + 542-2777). *Oxides, fluorite-related superstructures, order-disorder.*

Rossouw, Dr Christopher John. CSIRO Div. of Chemical Physics, P.O. Box 160, Clayton, Victoria 3168, Australia. (1951) DPhil, metallurgy and materials sci. (Oxford U., UK, 1977). Sr. res. scient. (tel. 03 + 544-0633, ext. 249). *Electron diffraction, ionization, diffuse scattering, X-ray micro-analysis, electron energy loss spectroscopy.*

Sabine, Prof Terence Murray. Sch. of Physics and Materials, New South Wales Inst. of Techn., P.O. Box 123, Broadway, New South Wales 2007, Australia. (1930) DSc, physics (U. Melbourne, 1971). Head. (tel. 02 + 218-9418). *Neutron crystallography, solar and nuclear energy.*

Sanders, Dr John Veysey. CSIRO Div. of Materials Sci., Normanby Rd., Clayton, Victoria 3168, Australia. (1924) PhD, physics (Cambridge, UK, 1950). Chief res. scient. (tel. 03 + 542-2777). *Surface physics, catalysis.*

Scott, Dr Henry Gordon. CSIRO Div. of Materials Sci., Normanby Rd., Clayton, Victoria 3168, Australia. (1933) PhD, physical chemistry (Cambridge U., UK, 1957). Principal res. scient. (tel. 03 + 542-2777). *Crystal structures, solid state chemistry, refractory oxide systems.*

Segall, Prof Robert Leo. Sch. of Sci., Griffith U., Nathan, Queensland 4111, Australia. (1934) PhD, physics (U. Melbourne, 1954). Prof. of physics. (tel. 07 + 275-7111). *Lattice defects, surface structure.*

Self, Dr Peter Geoffrey. CSIRO Div. of Soils, Private Bag 2, Glen Osmond, South Australia 5064, Australia. (1952) PhD, physics, electron microscopy (U. Melbourne, 1979). Res. scient. (tel. 08 + 274-9311). *Electron microscopy, high resolution electron microscopy, analytical electron microscopy.*

Sellar, Dr Jeffrey Ronald John. Res. Sch. of Chemistry, Australian Nat. U., Canberra, Australian Capital Territory 2601, Australia. (1946) PhD, physics (Arizona State U., 1976). Res. fellow. (tel. 062 + 49-3714). *Electron microscopy, dynamical electron diffraction, solid state chemistry.*

Shields, Dr Kelvin George. CSIRO Div. of Computing Res., 306 Carmody Rd., St. Lucia, Queensland 4067, Australia. (1948) PhD, chemistry (U. Queensland, 1976). Experimental officer. (tel. 07 + 377-0329). *Computing, insecticide structures.*

Siripitayananon, Miss Jintana. Res. Sch. of Chemistry, Australian Nat. U., G.P.O. Box 4, Canberra, Australian Capital Territory 2601, Australia. (1954) MSc, physical chemistry (Mahidol U., Thailand, 1978). Student. (tel. 062 + 49-4239, ext. 4239). *X-ray diffuse scattering.*

Skelton, Dr Brian Warwick. Dept. of Physical and Inorganic Chemistry, U. of Western Australia, Nedlands, Western Australia 6009, Australia. (1948) PhD, chemical crystallography (U. Auckland, New Zealand, 1974). Res. officer. (tel. 092 + 80-3838). *Structures, inorganic and organic compounds.*

Slade, Dr Phillip Garland. CSIRO Div. of Soils, Waite Rd., Glen Osmond, South Australia 5064, Australia. (1941) PhD, chemical mineralogy (U. Adelaide, 1968). Sr. res. scient. (tel. 08 + 79-2749, ext. 302). *Crystal chemistry, structures, clay minerals, clay-organic intercalates.*

Smith, Dr Graham. Dept. of Chemistry, Queensland Inst. of Techn., George St., Brisbane, Queensland 4000, Australia. (1941) PhD, chemical crystallography (U. Queensland, 1978). Lect. (tel. 072 + 221-2411, ext. 293). *Structures, biologically active pesticides, metal complexes.*

Smith, Dr Katherine Leah. Dept. of Physics, New South Wales Inst. of Techn., P.O. Box 123, Broadway, New South Wales 2007, Australia. (1955) PhD, physics (Monash U., 1982). Lect. (tel. 02 + 218-9521). *Mineralogy, physics.*

Smith, Prof Thomas Frederick. Physics Dept., Monash U., Wellington Rd., Clayton, Victoria 3168, Australia. (1939) PhD, physics (U. Sheffield, UK, 1963). Professor, experimental physics. (tel. 03 + 541-3630, telex AA 32691, MONASH). *Lattice dynamics, soft modes, lattice stability.*

Snow, Dr Michael Robert. Dept. of Physical and Inorganic Chemistry, U. of Adelaide, North Terrace, Adelaide, South Australia 5000, Australia. (1940) PhD, inorganic chemistry (U. London, UK, 1966). Reader. (tel. 08 + 228-5559, telex AA 89141, UNIVAD). *Coordination structures, organometallic structures.*

Spackman, Dr Mark Arthur. Dept. of Physics, U. of Western Australia, Nedlands, Western Australia 6009, Australia. (1954) PhD, theoretical chemistry (Western Australia U.). Res. associate. *Electron density studies, electrostatic properties from diffraction data.*

Spink, Mr John Arthur. CSIRO Div. of Materials Sci., Normanby Rd., Clayton, Victoria 3168, Australia. (1925) MSc, physical chemistry (U. Melbourne, 1955). Sr. res. scient. (tel. 03 + 542-2777). *Surface structures, field ion microscopy.*

Steffen, Dr William Lee. CSIRO Div. of Environmental Mechanics, G.P.O. Box 821, Canberra, Australian Capital Territory 2601, Australia. (1947) PhD, inorganic chemistry (U. Florida, USA, 1975). Scientific services officer. (tel. 062 + 46-5648). *Porous media physics, atmospheric physics, science editing and writing.*

Stephens, Dr Frederick Selwyn. Sch. of Chemistry, Macquarie U., North Ryde, New South Wales 2113, Australia. (1938) PhD, chemistry (U. New South Wales, 1963). Sr. lect. (tel. 02 + 888-8000, ext. 9485). *Transition metal complexes.*

Stephenson, Prof Neville Charles. Sch. of Chemical and Earth Sci., New South Wales Inst. of Techn., P.O. Box 123, Broadway, New South Wales 2007, Australia. DSc, chemistry (U. New South Wales, 1970). Head. (tel. 02 + 218-9472). *Solid state chemistry, structural systematics.*

Stevenson, Dr Andrew Wesley. CSIRO Div. of Chemical Physics, P.O. Box 160, Clayton, Victoria 3168, Australia. (1957) PhD, physics (U. Melbourne, 1984). Postdoctoral fellow. (tel. 03 + 544-0633, ext. 322, telex AA 30356, RIVETT). *Anharmonic thermal motion, extinction, thermal diffuse scattering.*

Summerville, Dr Edward. Sch. of Physical Sci., Flinders U. of South Australia, Sturt Rd., Bedford Park, South Australia 5042, Australia. (1942) PhD, chemistry (Flinders U. of South Australia, 1975). Res. associate. (tel. 08 + 275-2497). *Solid state chemistry, X-ray diffraction, electron microscopy, electron diffraction.*

Suzuki, Mr Eikichi. CSIRO Div. of Protein Chemistry, 343 Royal Parade, Parkville, Victoria 3057, Australia. (1932) BEng, chemistry (Yokohama Nat. U., Japan, 1955). Principal res. scient. (tel. 03 + 342-4200, telex PROV AA 33983). *Structure determination, fibrous proteins, image processing.*

Taylor, Dr Donald. Western Australian C. of Advanced Education, P.O. Box 217, Doubleview, Western Australia 6018, Australia. (1947) PhD, physical and inorganic chemistry (Western Australia U.). Educational computing services manager. (tel. 09 + 386-0226). *Chemical crystallography, crystallographic computing.*

Taylor, Dr John Charles. CSIRO Div. of Energy Chemistry, Private Mail Bag 7, Sutherland, New South Wales 2232, Australia. (1935) DSc, chemistry (U. New South Wales, 1982). Sr. principal res. scient. (tel. 02 + 543-3111, ext.3345, telex AA 73341). *Neutron and X-ray crystallography, actinide chemistry, powder methods.*

Taylor, Dr Max Ronald. Sch. of Physical Sci., Flinders U. of South Australia, Sturt Rd., Bedford Park, South Australia 5042, Australia. (1936) PhD, chemistry (U. Sydney, 1964). Sr. lect. in chemistry. (tel. 08 + 275-2467). *Biological metal complexes, computing.*

Thompson, Mr John Gerard. Geology Dept., James Cook U. of North Queensland, Townsville, Queensland 4811, Australia. (1954) BSc(Hons), geochemistry (Victoria U. of Wellington, New Zealand, 1981). Grad. student. (tel. 077 + 81-4836, telex AA 47009). *Solid-state, nuclear magnetic resonance, X-ray, neutron, powder diffraction, Rietveld, clay mineral, intercalation.*

Threadgold, Dr Ian Malcolm. Dept. of Geology and Geophysics, U. of Sydney, Sydney, New South Wales 2006, Australia. (1929) PhD, X-ray crystallography (U. Wisconsin, USA, 1963). *Inorganic crystal structure.*

Tiekink, Mr Edward Richard Thomas. Dept of Inorganic Chemistry, U. of Melbourne, Parkville, Victoria 3052, Australia. (1960) BSc(Hons), chemistry (U. Melbourne, 1982) Grad. student. (tel. 03 + 341-6473). *Metal complexes of sulphur-containing ligands.*

Tietze, Mr Hans Roderick. Dept. of Chemistry, U. of Newcastle, Shortland, New South Wales 2308, Australia. (1921) MSc, chemistry (U. London, UK, 1953). Sr. lect. (tel. 049 + 68-0401). *Inorganic crystal structures.*

Town, Miss Susan Lesley. Physics Dept., Monash U., Wellington Rd., Clayton, Victoria 3168, Australia. (1959) MSc, physics of Heusler alloys (Salford U., UK, 1982). Res. student. (tel. 03 + 541-3652, ext. 3633, telex AA 32691). *Gamma-ray and neutron scattering, Kossel and Kikuchi effects, solid-state crystallography.*

Vagg Dr Robert Sylvester. Sch. of Chemistry, Macquarie U., North Ryde, New South Wales 2113, Australia. (1945) PhD, chemistry (Macquarie U., 1971). Sr. lect. (tel. 02 + 888-8000, ext. 9485). *Transition metal complexes.*

Varghese, Dr Joseph Noozhumurry. CSIRO Div. of Protein Chemistry, 343 Royal Parade, Parkville, Victoria 3057, Australia. (1949) PhD, physics (U. Western Australia, 1974). Sr. res. scient. (tel. 03 + 342-4200). *Protein crystallography.*

Wagenfeld, Dr Heinrich Karsten. Dept. of Appl. Physics, Royal Melbourne Inst. of Techn., 124 Latrobe St., Melbourne, Victoria 3000, Australia. (1928) Dr.Rer.Nat., physics (Freie U. Berlin, BRD, 1958). Dept. head. (tel. 03 + 660-2135, telex AA 36406). *Dynamical scattering, X-ray, neutron, electron scattering, inelastic scattering, ion implantation, secondary electron emission, Auger spectroscopy.*

Watson, Dr Kenneth John. 65 Circe Circle, Dalkeith, Western Australia 6009, Australia. (1935) PhD, physics (U. Western Australia, 1967). Consultant. (tel. 09 + 386-3330). *Chemical bonding, charge density, single crystals, protein structures.*

Watts, Mr John Andrew. CSIRO Div. of Mineral Chemistry, 339 Williamstown Rd., Port Melbourne, Victoria 3207, Australia. (1933) BSc, chemistry (Ballarat Sch. of Mines, 1960). Experimental officer. (tel. 03 + 647-0211, ext. 270). *Structure and chemistry, sulphides and phosphates.*

Welberry, Dr Thomas Richard. Res. Sch. of Chemistry, Australian Nat. U., P.O. Box 4, Canberra, Australian Capital Territory 2601, Australia. (1945) PhD, chemical crystallography (London, UK, 1970). Res. fellow (tel. 062 + 49-4122). *Disorder, statistical models.*

Westphalen, Mr John Arthur. South Australian Inst. of Techn. Levels Campus, Ingle Farm, South Australia 5098, Australia. (1929) MSc, X-ray crystallography (U. Surrey, UK, 1973). Sr. lect. (tel. 085 + 260-2055, ext. 2065). *Structure analysis.*

Whillans. Dr Francis David. Dept. of Appl. Chemistry, Phillip Inst. of Techn., Plenty Rd., Bundoora, Victoria 3083, Australia. (1942) PhD, inorganic crystallography (U. Melbourne, 1971). Lect. (tel. 03 + 468-2200 Ext. 480). *Chemical education.*

White, Dr Allan Henry. Sch. of Chemistry, U. of Western Australia, Nedlands, Western Australia 6009, Australia. (1938) DSc, inorganic chemistry (U. Melbourne, 1981). Assoc. prof. (tel. 092 + 380-3144). *Structure, synthesis, inorganic and coordination chemistry.*

Whitfield, Dr Harold John. CSIRO Div. of Chemical Physics, P.O. Box 160, Clayton, Victoria 3168, Australia. (1931) PhD, chemistry (Victoria U., Wellington, New Zealand, 1967). Principal res. scient. (tel. 03 + 544-0633, ext. 237). *Electron diffraction, high resolution lattice imaging, polytypism, superstructures.*

Wilkins Dr Stephen William. CSIRO, Div. of Chemical Physics, P.O.Box 160, Clayton, Victoria 3168, Australia. (1946) PhD, solid state physics (U. Melbourne, 1972). Principal res. scient. (tel. 03 + 544-0633, ext. 323, telex AA 30356). *Dynamical diffraction, extinction problem, X-ray and neutron optics, Kikuchi effect, information theory, direct methods, thermal vibrations, ordering in alloys, martensitic transformations.*

Williams, Mr Brian Edward. Materials Div., Aeronautical Res. Labs., 506 Lorimer St., Port Melbourne, Victoria 3207, Australia. (1925) Fellowship Dipl, applied physics (Royal Melbourne Inst. of Techn., 1974). Experimental officer. (tel. 03 + 647-7534). *Identification, powder methods.*

Williams, Mr Donald Allan. Central Res. Labs., ICI Australia Operations, Newsom St., Ascot Vale, Victoria 3032, Australia. (1937) MSc, crystallography (La Trobe U., 1974). Sr. res. officer. (tel. 03 + 377-2381). *Analysis, characterisation, computing.*

Williams, Dr Geoffrey Allan. Australian Radiation Lab., Lower Plenty Rd., Yallambie, Victoria 3085, Australia. (1950) PhD, chemistry (U. Melbourne, 1976). Res. scient. (tel. 03 + 433-2211). *Technetium chemistry, X-ray diffraction, neutron diffraction, spin density, charge density, transition metal complexes.*

Williams, Mr Timothy Brendan. Res. Sch. of Chemistry, Australian Nat. U., G.P.O. Box 4, Australian Capital Territory 2601, Australia. (1958) BTech(Hons), materials sci. and techn. (Bradford U., UK, 1982). Grad. student. (tel. 062 + 49-2620). *High resolution TEM, inorganic crystal structures, defects in crystals.*

Wilson, Dr Alan Richard. Materials Div., Aeronautical Res. Labs., Box 4331, G.P.O., Melbourne, Victoria 3001, Australia. (1953) PhD, physics, HREM (U. Melbourne, 1982). Res. Scient. (tel. 03 + 647-7505). *Structure analysis, imaging, energy dispersive X-ray analysis, energy loss spectroscopy, weak-beam electron diffraction.*

Wright, Mr Phillip John. Dept. of Chemistry, Monash U., Wellington Rd., Clayton, Victoria 3168, Australia. (1961) BSc(Hons), chemistry (Monash U., 1984). Res. assistant. (tel. 03 + 541-3641).

Wunderlich Dr Jeffrey Alfred. CSIRO Div. of Mineral Chemistry, P.O.Box 124, Port Melbourne, Victoria 3207, Australia. (1931) Dr. es Sc., chemical crystallography (U. Paris, France, 1958). Principal res. scient. (tel. 03 + 647-0345). *Inorganic and mineral structures.*

Yung, Dr Fook Hong. Western Australia Water Authority, 2 Havelock St., West Perth, Western Australia 6005, Australia. (1943) PhD, electron density (U. Western Australia, 1980). Res. officer. (tel. 09 + 322-0331). *Crystallography, molecular quantum mechanics, electron density, aquifer modelling, non-linear programming, human movement.*

AUSTRIA

Sub-Editor: A. Preisinger

Notes

1. International telephone country code - 43.

2. Austrian universities grant various degrees. Those occurring below are, in ascending order, *Magister* (Mag.), conferred by a university and approximately equivalent to MA; *Diplom-Ingenieur* (Dipl.-Ing.), conferred by a technical university and approximately equivalent to MSc; *Doctor philosophiae* (Dr. phil.), conferred by a university, and *Doctor technicae* (Dr. techn.), conferred by a technical university, both approximately equivalent to PhD; and *Dozent* (Doz.), *venia legendi*, granted by both types of universities.

Aiginger, Prof. Dr Dipl.-Ing. Hannes. Atominst. der Österreichischen Universitäten, Schüttelstrasse 115, A-1020 Vienna, Austria. (1937) Doz., physics of electrons, X-rays and gamma-rays (Techn. U. Vienna, 1970). Full prof. (tel. 0222 + 725136, ext. 61). *Crystal orientation, X-ray metallography, textures, X-ray fluorescence analysis.*

Amthauer, Prof. Dr Georg. Inst. f. Geowissenschaften, Universität, Akademiestr. 26, A-5020 Salzburg, Austria. Prof., mineralogy. (tel. 662 + 44511, ext. 578). *Crystal chemistry, physics of minerals, mineral structures.*

Bauer, Prof. Dr Günther Ernst. Inst. für Physik, Montanu. Leoben, Franz Josefstrasse 18, A-8700 Leoben, Austria. (1942) Doz., physics (Techn. U. Aachen, W. Germany, 1974). Full prof. (tel. 03842 + 42555, ext. 260; telex 033322). *Crystal growth, IV-VI compounds, superlattices, ferroelectric phase transitions, FIR spectroscopy.*

Baumgartner, Dr Dipl.-Ing. Oswald. Inst. für Mineralogie, Kristallographie und Strukturchemie, Techn. U. Wien, Getreidemarkt 9, A-1060 Vienna, Austria. (1948) Dr. techn., chemistry (Techn. U. Vienna, 1977). Asst. (tel. 0222 + 5601, ext. 4743). *Crystal structures, neutron diffraction.*

Becherer, Dr Karl. Inst. für Mineralogie und Kristallographie, U. Wien, Dr. Karl Lueger-Ring 1, A-1010 Vienna, Austria. (1926) Dr. phil., mineralogy-petrography (U. Vienna, 1961). Scient. officer. (tel. 0222 + 4300, ext. 2332). *Analytical chemistry, minerals, crystal optics, general crystallography.*

Beran, Doz. Dr Anton. Inst. für Mineralogie und Kristallographie, U. Wien, Dr. Karl Lueger-Ring 1, A-1010 Vienna, Austria. (1944) Doz., mineralogy (U. Vienna, 1980). Asst. (tel. 0222 + 4300, ext. 2320). *Inorganic crystal structures, ore microscopy, IR spectroscopy.*

Betz, Doz. Dr Gerhard. Inst. für Allgemeine Physik, Techn. U. Wien, Karlsplatz 13, A-1040 Vienna, Austria. (1944) Doz., ionic physics (U. Vienna, 1981). Asst. (tel. 0222 + 5601, ext. 3368). *Surface composition, chemical structure, sputtering, Auger electron spectroscopy.*

Blaschko, Dr Oskar. Inst. für Experimentalphysik, U. Wien, Strudlhofgasse 4, A-1090 Vienna, Austria. (1948) Dr. phil., solid state physics (U. Vienna, 1974). Asst. (tel. 0222 + 342630, ext. 226). *Phonons, phase transformations.*

Boller, Prof. Dr Herbert. Inst. für Physikalische Chemie, U. Wien, Währingerstrasse 42, A-1090 Vienna, Austria. (1937) Doz., physical chemistry

(U. Vienna, 1974). Full prof. (tel. 0222 + 343616, ext. 19). *Solid state chemistry, inorganic compounds, alloys, crystal structures, metallike materials, transition-metal compounds, magnetic properties, magnetic structures.*

Brandstätter Dr Franz. Mineralog.-petrograph. Abteilung, Naturhistorisches Museum, P.O. Box 417, Burgring 7, A-1014 Vienna, Austria. (1953) Dr.phil., mineralogy (U. Vienna, 1979). Res. asst. (tel. 0222 + 934145, ext. 265). *Minerals in meteorites, crystallographic symmetry.*

Derkosch, Prof. Dr Josef. Inst. für Organische Chemie, U. Wien, Währingerstrasse 38, A-1090 Vienna, Austria. (1923) Doz., spectrochemistry (U. Vienna, 1962). Full prof. (tel. 0222 + 344630 ext. 64). *IR spectroscopy, Raman spectroscopy, luminescence spectroscopy.*

Eder, Prof. Dr Dipl.-Ing. Otto Josef. Physikinstitut, Österreichisches Forschungszentrum Seibersdorf, Lenaugasse 10, A-1082 Vienna, Austria. (1936) Doz., experimental physics (U. Vienna, 1979). Head of physics department. (tel. 02254 + 80, ext. 2302). *Solid and liquid state physics, neutron scattering, neutron structure factors, statistical mechanics.*

Effenberger, Dr Herta. Inst. für Mineralogie und Kristallographie, U. Wien, Dr. Karl Lueger-Ring 1, A-1010 Vienna, Austria. (1954) Dr. phil., mineralogy (U. Vienna, 1978). Asst. (tel. 0222 + 4300, ext. 2687). *Crystal synthesis, inorganic crystal structures, crystal chemistry.*

Ernst, Dr Gert. Physikinstitut, Österreichisches Forschungszentrum Seibersdorf, Lenaugasse 10, A-1082 Vienna, Austria. (1944) Dr. phil., physics (U. Vienna, 1971). Res. scient. (tel. 02254 + 80, ext. 2280; telex 014-353). *Solid state physics, phase transformations, lattice dynamics, neutron spectrometry.*

Ettmayer, Prof. Dr Dipl.-Ing. Peter. Inst. für Chemische Technologie Anorganischer Stoffe, Techn. U. Wien, Getreidemarkt 9, A-1060 Vienna, Austria. (1934) Doz., chemical technology (Techn. U. Vienna, 1972). Full prof. (tel. 0222 + 5601, ext. 4799). *Crystal structures, metallic hard materials, powder diffraction.*

Fischer, Dr Richard. Inst. für Mineralogie und Kristallographie, U. Wien, Dr. Karl Lueger-Ring 1, A-1010 Vienna, Austria. (1933) Dr. phil., chemistry (U. Vienna, 1963). Asst. (tel. 0222 + 4300, ext. 2335). *Mineralogy, mineral deposits, inorganic crystal structures, lattice energies.*

Fuith, Dr Armin. Inst. für Experimentalphysik, U. Wien, Strudlhofgasse 4, A-1090 Vienna, Austria. (1949) Dr. phil., metal physics (U. Vienna, 1979). Asst. (tel. 0222 + 342630, ext. 293 or 420; telex 116222). *Phase transitions, crystal growth.*

Glatter, Dr Otto. Inst. für Physikalische Chemie, U. Graz, Heinrichstrasse 28, A-8010 Graz, Austria. (1945) Dr. techn., physics (Techn. U. Graz, 1972). Asst. (tel. 0316 + 380, ext. 5433 or 5439). *Small angle scattering, elastic light scattering, quasi-elastic light scattering, computer programming.*

Götzinger, Dr Michael Alois. Inst. für Mineralogie und Kristallographie, U. Wien, Dr. Karl Lueger-Ring 1, A-1010 Vienna, Austria. (1949) Dr. phil., mineralogy and petrography (U. Vienna, 1976). Asst. (tel. 0222 + 4300, ext. 2688). *Mineralogy, mineral deposits, X-ray diffraction phase analysis, inorganic crystal structures.*

Haditsch, Prof. Dr Johann Georg. Inst. für Geowissenschaften, Montanu. Leoben, Peter Tunnerstrasse 5/I, A-8700 Leoben, Austria. (1934) Doz., mineralogy, economic geology (Montanu. Leoben, 1967). Full prof. (tel. 03842 + 42555, ext. 452). *X-ray diffraction phase analysis, ore microscopy, environmental geology.*

Halwax, Dipl.-Ing. Erich Johann. Inst. für Mineralogie, Kristallographie und Strukturchemie, Techn. U. Wien, Getreidemarkt 9, A-1060 Vienna, Austria. (1951) Dipl.-Ing., chemistry (Techn. U. Vienna, 1976). Asst. (tel. 0222 + 5601, ext. 4742). *Crystal structures, X-ray diffraction phase analysis.*

Hausner, Dipl.-Ing. Robert. Forschungsinstitut, Veitscher Magnesitwerke A.G., Magnesitstrasse 2, A-8707 Leoben, Austria. (1941) Dipl.-Ing., physics (Techn. U. Vienna, 1966). Head of X-ray laboratory (tel. 03842 + 22581, ext.70, telex 33323). *X-ray diffraction, inorganic crystal structures, carbon structures (pitch - coke - graphite), quantitative analysis, refractories, X-ray fluorescence analysis.*

Heritsch, Prof. em. Dr Haymo. Inst. für Mineralogie, Kristallographie und Petrologie, U. Graz, Universitätsplatz 2, A-8010 Graz, Austria. (1911) Doz., mineralogy-petrography (U. Graz, 1939). Prof. em. (tel. 0316 + 380, ext. 5540). *Inorganic crystal structures, crystal chemistry.*

Hiebl, Dr Kurt. Inst. für Physikalische Chemie, U. Wien, Währingerstrasse 42, A-1090 Vienna, Austria. (1948) Dr. phil., physics (U. Vienna, 1973). Asst. (tel. 0222 + 343616, ext. 19). *Magnetism in solids, nuclear magnetic resonance in solids.*

Higatsberger, Prof. Dr Michael Josef. Inst. für Experimentalphysik, U. Wien, Boltzmanngasse 5, A-1090 Vienna, Austria. (1924) Doz., physics (U. Vienna, 1959). Full prof. and head. (tel. 0222 + 345232; telex 116222). *Solid state physics, instrumentation.*

Holub, Dr Fritz. Inst. für Werkstofftechnologie, Österreichisches Forschungszentrum Seibersdorf, Lenaugasse 10, A-1082 Vienna, Austria. (1930) Dr. phil., chemistry (U. Vienna, 1961). Res. scient. (tel. 02254 + 80, ext. 2365). *Inorganic crystal structures, phase analysis, surfaces, applied crystallography.*

Hörl, Prof. Dr Erwin M. Inst. für Werkstofftechnologie, Österreichisches Forschungszentrum Seibersdorf, Lenaugasse 10, A-1082 Vienna, Austria. (1929) Doz., experimental solid state physics (Techn. U. Vienna, 1965). Head. (tel. 02254 + 80, ext. 2367). *Structures, solidified permanent gases, oxygen, stacking faults, computer programming, indexing Debye-Scherrer diagrams, powder patterns of low symmetry.*

Jánosi, Dipl.-Ing. Dr András. Inst. für Physikalische Chemie, U. Graz, Heinrichstrasse 28, A-8010 Graz, Austria. (1929) Dr. phil., chemistry (Katholische U. Löwen, 1959). Scient. (tel. 0316 + 380, ext. 5423). *Small angle X-ray scattering, wide angle X-ray scattering, synthetic and natural polymer systems.*

Kabelka, Dr Heinz I. Inst. für Experimentalphysik, U. Wien, Strudlhofgasse 4, A-1090 Vienna, Austria. (1949) Dr. phil., solid state physics (U. Vienna, 1976). Asst. (tel. 0222 + 342630, ext. 293; telex 116222). *Phase transitions in solids, ultrasonic phase velocity and attenuation, Raman scattering.*

Kahlert, Prof. Dr Hartmut. Inst. für Festkörperphysik, Techn. U. Graz, Petersgasse 16, A-8010 Graz, Austria. (1940) Doz., solid state physics (U. Vienna, 1976). Full prof. and head. (tel. 0316 + 7061, ext. 8460). *Crystal growth, quasi-one-dimensional solids, polymer preparation, polymer crystal structure, polyacethylene, polyparaphenylene, polysulphurnitride.*

Kirchner, Doz. Dr Elisabeth Charlotte. Inst. für Geowissenschaften, U. Salzburg, Akademiestrasse 26, A-5020 Salzburg, Austria. (1935) Doz., mineralogy-petrography-crystallography (U. Salzburg, 1979). Full Prof. (tel. 0662 + 44511, ext. 577). *Powder diffraction, inorganic crystals, structure-symmetry.*

Kohlbeck, Dr Franz. Inst. für Geophysik, Techn. U. Wien, Gusshausstrasse 25, A-1040 Vienna, Austria. (1943) Dr. techn., physics (Techn. U. Vienna, 1974). Scient. officer. (tel. 0222 + 5601, ext. 3803). *Computer progamming, powder diffraction analysis.*

Komarek, Prof. Dr Kurt Ludwig. Inst. für Anorganische Chemie, U. Wien, Währingerstrasse 42, A-1090 Vienna, Austria. (1926) Dr. phil., chemistry (U. Vienna, 1949). Full prof. and head. (tel. 0222 + 345424, ext. 2). *Crystal growth, inorganic crystal structures.*

Kratky, Dr Christoph. Inst. für Physikalische Chemie, U. Graz, Heinrichstrasse 28, A-8010 Graz, Austria. (1946) Dr. techn., chemistry (E.T.H. Zürich, Switzerland, 1976). Asst. (tel. 0316 + 380, ext. 5417). *Organic crystal structure analysis.*

Kratky, Prof. em. Dr Dipl.-Ing. Dr h.c. mult. Otto. Drosselweg 15, A-8010 Graz, Austria. (1902) Dr. techn., chemistry (Techn. U. Vienna, 1929). Prof. em. (tel. 0316 + 42073). *Small angle X-ray scattering.*

Krischner, Prof. Dr Dipl.-Ing. Harald. Inst. für Physikalische und Theoretische Chemie (Strukturforschung), Techn. U. Graz, Rechbauerstrasse 12, A-8010 Graz, Austria. (1930) Doz., physical chemistry (Techn. U. Graz, 1964). Full prof. (tel. 0316 + 7061, ext. 8223). *Inorganic crystal structures, solid state chemistry.*

Kuchar, Doz. Dr Friedemar. Inst für Festkörperphysik, U. Wien, Strudlhofgasse 4, A-1090 Vienna, Austria. (1941) Doz., experimental solid state physics (U. Vienna, 1978). Asst. (tel. 0222 + 342630, ext. 264). *Impurities in semiconductors, IR properties, heterostructures, MOS structures.*

Kunsch, Dr Dipl.-Ing. Barnabas. Physikinstitut, Österreichisches Forschungszentrum Seibersdorf, Lenaugasse 10, A-1082 Vienna, Austria. (1942) Dr. techn., physics (Techn. U. Vienna, 1970). Res. scient. (tel. 02254 + 80, ext. 2305; telex 014-353). *Solid state physics, liquid physics, neutron scattering, neutron structure factors, statistical mechanics.*

Kuzmany, Doz. Dr Hans. Inst. für Festkörperphysik, U. Wien, Strudlhofgasse 4, A-1090 Vienna, Austria. (1940) Doz., experimental physics (U. Vienna, 1976). Asst. (tel. 0222 + 342630, ext. 264). *Solid state physics, Raman scattering, IR spectroscopy, low dimensional systems, piezoelectricity, molecular crystals, phase transitions.*

Laggner, Doz. Dr Peter. Inst. für Röntgenfeinstrukturforschung, Österreichische Akademie der Wissenschaften, Steyrergasse 17, A-8010 Graz, Austria. (1944) Doz., biochemistry-biophysics-physical chemistry (U. Graz, 1978). Director. (tel. 0316 + 71371). *Small angle X-ray scattering, biopolymers in solution, lipoproteins, membranes.*

Lihl, Prof. em. Dr Franz. Inst. für Angewandte Physik, Techn. U. Wien, Karlsplatz 13, A-1040 Vienna, Austria. (1906) Dr. phil., physics (U. Vienna, 1930). Prof. em. (tel. 0222 + 5601, ext. 3237). *Metal structures, magnetic structures, order-disorder and exsolution processes, low temperatures.*

Linke, Dr Walter. Wienerberger Baustoffe, Wienerbergstrasse 11, A-1100 Vienna, Austria. (1944) Dr. phil., mineralogy (U. Vienna, 1970). Scient. (tel. 0222 + 629241, ext. 370). *Crystal growth, crystal structures.*

Ludwiczek, Dr Herbert. Inst. für Mineralogie Kristallographie und Strukturchemie, Techn. U. Wien, Getreidemarkt 9, A-1060 Vienna, Austria. (1940) Dr. phil., physics (U. Vienna, 1973). Scient. (tel. 0222 + 5601, ext. 4749). *Crystal physics, computer programming.*

Mayer, Dr Dipl.-Ing. Helmut. Inst. für Mineralogie, Kristallographie und Strukturchemie, Techn. U. Wien, Getreidemarkt 9, A-1060 Vienna, Austria. (1939) Dr. techn., chemistry (Techn. U. Vienna, 1971). Scient. officer. (tel. 0222 + 5601, ext. 4742). *Inorganic crystal structures and minerals, thermal analysis.*

Mayr, Dr Dipl.-Ing. Michael. Labor für Strukturanalyse (RFP-3), Voest-Alpine, Postbox 02, A-4010 Linz, Austria. (1945) Dr. techn., physics (Techn. U. Vienna, 1975). Head of lab. (tel. 0732 + 585, ext. 6114; telex 2207-461). *X-ray diffraction applications, metallography, non-metallic inclusions, steel, refractories, corrosion, electron microprobe analysis.*

Mereiter, Dr Kurt. Inst. für Mineralogie, Kristallographie und Strukturchemie, Techn. U. Wien, Getreidemarkt 9, A-1060 Vienna, Austria. (1945) Dr. phil., mineralogy and crystallography (U. Vienna,1975). Asst. (tel. 0222 + 5601, ext.

4747). *Mineral structures, inorganic crystal structures, crystal chemistry, morphology, optical crystallography.*
Mikenda, Dr Werner. Inst. für Organische Chemie, U. Wien, Währingerstrasse 38, A-1090 Vienna, Austria. (1946) Dr. phil., chemistry (U. Vienna, 1976). Asst. (tel. 0222 + 344630, ext. 74). *Solid state vibrational spectroscopy, luminescence spectroscopy.*
Mikler, Dr Helga. Inst. für Anorganische Chemie, U. Wien, Währingerstrasse 42, A-1090 Vienna, Austria. (1932) Dr. phil., chemistry (U. Vienna,1958). Asst. (tel. 0222 + 345424, ext. 6). *Inorganic crystal structures.*
Müller, Dr Mag. Karl Werner. Inst. für Röntgenfeinstrukturforschung, Österreichische Akademie der Wissenschaften, Steyrergasse 17, A-8010 Graz, Austria. (1947) Dr. phil., physical chemistry (U. Graz, 1972). Postdoctoral fellow. (tel. 0316 + 74433, ext. 4). *Small angle X-ray scattering, polymers in solution, biopolymers, detergent micelles, gallstones.*
Neckel, Prof. Dr Adolf. Inst. für Physikalische Chemie, U. Wien, Währingerstrasse 42, A-1090 Vienna, Austria. (1926) Doz., physical chemistry (U. Vienna, 1965). Full prof. and head. (tel. 0222 + 343616). *Solid state chemistry, energy band structures and properties.*
Niedermayr, Dr Gerhard. Mineralogisch-Petrographische Abteilung, Naturhistorisches Museum, Burgring 7, A-1014 Vienna, Austria. (1941) Dr. phil., mineralogy and petrology (U. Vienna, 1965). Director of Staatliches Edelsteininstitut. (tel. 0222 + 934541, ext. 274). *Mineral paragenesis, sedimentology, gem materials, general crystallography.*
Pertlik, Doz. Dr Franz. Inst. für Mineralogie und Kristallographie, U. Wien, Dr. Karl Lueger-Ring 1, A-1010 Vienna, Austria. (1943) Doz., mineralogy (U. Vienna, 1979). Asst. (tel. 0222 + 4300, ext. 2329). *Inorganic crystal structures.*
Pilz, Prof. Dr Ingrid Edith. Inst. für Physikalische Chemie, U. Graz, Heinrichstrasse 28, A-8010 Graz, Austria. (1931) Doz., physical chemistry (U. Graz, 1970). Full Prof. (tel. 0316 + 380, ext. 5414) *Small angle scattering, biopolymer structure, immunoglobuline, protein conformational changes, repiratory proteins, hemoglobuline, cyanins.*
Pongratz, Dr Dipl.-Ing. Peter. Inst. für Angewandte Physik, Techn. U. Wien, Karlsplatz 13, A-1040 Vienna, Austria. (1947) Dr. techn., electron microscopy (Techn. U. Vienna, 1980). Asst. (tel. 0222 + 5601, ext. 3245). *Electron diffraction, electron microscopy, contrast theory, defects, semiconductor materials, applied optics, image processing.*
Preisinger, Prof. Dr Anton. Inst. für Mineralogie, Kristallographie und Strukturchemie, Techn. U. Wien, Getreidemarkt 9, A-1060 Vienna, Austria. (1925) Doz., mineralogy and crystal chemistry (U. Vienna, 1956). Full prof. and head. (tel. 0222 + 5601, ext. 4749). *Crystal structures, crystal chemistry, biocrystallography, crystal surfaces, general crystallography.*
Rogl, Doz. Dr Peter Franz. Inst. für Physikalische Chemie, U. Wien, Währingerstrasse 42, A-1090 Vienna, Austria. (1945) Doz., physical chemistry (U. Vienna, 1980). Asst. (tel. 0222 + 343616, ext. 14). *Crystallography, structural chemistry, thermodynamics, phase diagrams, high melting systems, solid state chemistry, alloys, refractories, borides, carbides, crystal structures.*
Sazedj-Khosrawan, Dr Feresteh. Gersthoferstrasse 150/1/8, A-1180 Vienna, Austria. (1954) Dr. phil., mineralogy and petrography (U. Vienna, 1983). Scientist (tel. 0222 + 4759753). *Mineralogy, crystal structures, crystal chemistry.*
Schattschneider, Dr Dipl.-Ing. Mag. Peter. Inst. für Angewandte Physik, Techn. U. Wien, Karlsplatz 13, A-1040 Vienna, Austria. (1950) Dr. techn., physics (Techn. U. Vienna, 1975). Asst. (tel. 0222 + 5601, ext. 3244). *Electron energy loss spectroscopy, plasmons.*
Schroll, Prof. Dr Erich. Geotechnisches Institut, Bundesversuchs- und Forschungsanstalt Arsenal, Franz Grillstrasse 9, Objekt 214; postbox 8, A-1031 Vienna, Austria. (1923) Doz., mineralogy (U. Vienna, 1957). Head. (tel. 0222 + 782531, ext. 475; telex 1-36677). *X-ray fluorescence analysis, X-ray diffraction phase analysis.*
Schuster, Dr Julius Clemens. Inst. für Physikalische Chemie, U. Wien, Währingerstrasse 42, A-1090 Vienna, Austria. (1952) Dr. phil., chemistry (U. Vienna, 1977). Asst. (tel. 0222 + 343616). *Crystal structures, crystal chemistry.*
Schwarz, Prof. Dr Karlheinz. Inst. für Techn. Elektrochemie, Techn. U. Wien, Getreidemarkt 9, A-1060 Vienna, Austria. (1941) Doz., quantum chemistry (Techn. U. Vienna, 1975). Full Prof. (tel. 0222 + 5601, ext. 4754). *Electronic structure of solids, energy band calculations, chemical bonding, electron densities.*
Schwomma, Dr Otto. Entwicklungsabteilung, Österreichische Philips Industrie, Videowerk, Gutheil-Schoder-Gasse 1, A-1230 Vienna, Austria. (1937) Dr. phil., physical chemistry (U. Vienna, 1964). Head. (tel. 0222 + 621361, ext. 397). *Surface physics and chemistry, solid state physics.*
Seeger, Prof. Dr Karlheinz. Ludwig Boltzmanninst. für Festkörperphysik, Kopernikusgasse 15, A-1060 Vienna, Austria. (1927) Dr. rer. nat., physics (U. Heidelberg, W. Germany, 1955). Full prof. and head. (tel. 0222 + 563408, ext. 22). *Semiconductor physics and technology, piezoelectrics, quasi-one-dimensional conductors, intercalated graphite, FIR, Raman and microwave investigations.*
Seidl, Dr Erwin. Inst. für Neutronen und Festkörperphysik, Atominst. der Österreichischen Universitäten, Schüttelstrasse 115, A-1020 Vienna, Austria. (1939) Dr. phil., physics (U. Vienna, 1966). Asst. (tel. 0222 + 725136, ext. 259). *Crystal growth, instrumentation, superconducting single crystal anisotropy, silicon crystals for neutron optics.*
Seifert, Dr Karl Josef. Inst. für Physikalische Chemie, U. Wien, Währingerstrasse 42, A-1090 Vienna, Austria. (1926) Dr. phil., chemistry (U. Vienna, 1962). Scient. officer. (tel. 0222 + 343616, ext. 36). *Inorganic crystal structures, documentation.*
Skalicky, Prof. Dr Peter. Inst. für Angewandte Physik (Elektronenmikroskopie), Techn. U. Wien, Karlsplatz 13, A-1040 Vienna, Austria. (1941) Doz., crystal physics (Techn. U. Vienna, 1973). Full prof. and head. (tel. 0222 + 5601, ext. 3238). *Crystal physics, lattice defects, electron and X-ray diffraction, electron microscopy.*
Sobczak, Doz. Dr Rudolf Josef. Inst. für Physikalische Chemie, Johann Kepler U. Linz, Altenbergerstrasse 69, A-4040 Linz, Austria. (1944) Doz., physical chemistry (U. Linz, 1980). Asst. (tel. 07222 + 231381, ext. 754). *Magnetic measurements, transition element solid compounds, X-ray crystallography, instrumentation, flow birefringence, viscosity.*
Stangler, Prof. Dr Ferdinand Karl Ludwig. Inst. für Festkörperphysik (Tieftemperaturphysik), U. Wien, Strudlhofgasse 4, A-1090 Vienna, Austria. (1928) Doz., experimental physics (U. Vienna, 1962). Full prof. and head. (tel. 0222 + 340673) *Crystal growth, electrical and magnetic properties, metal crystals, plasticity, crystal lattice defects, superconductivity, low temperature instruments.*
Steiner, Doz. Dr Walter. Inst. für Angewandte Physik, Techn. U. Wien, Karlsplatz 13, A-1040 Vienna, Austria. (1942) Doz., low temperature physics (Techn. U. Vienna, 1980). Asst. (tel. 0222 + 5601, ext. 3249). *Inorganic crystal structures, magnetism at low temperatures, Mössbauer spectroscopy.*
Stickler, Prof. Dr Roland. Inst. für Physikalische Chemie - Material Sciences, U. Wien, Währingerstrasse 42, A-1090 Vienna, Austria. (1931) Dr. techn., metallurgy (Techn. U. Vienna, 1958). Full prof. (tel. 0222 + 343616, ext. 26). *Phase stability, phase analysis, defect structures, metals and alloys, semiconductor materials.*
Stumpfl, Prof. Dr Eugen Friedrich. Inst. für Mineralogie, Montanu. Leoben, Franz Josefstrasse 18, A-8700 Leoben, Austria. (1931) Dr. rer. nat., geological science (U. Heidelberg, W. Germany, 1956). Full prof. and head. (tel. 03842 + 42555, ext. 451; telex 33322 mhbleo). *Ore deposits, electron probe analysis, reflected light microscopy.*
Sturm, Prof. Dr Dipl.-Ing. Friedwin. Inst. für Physik (Angewandte Physik), Montanu. Leoben, Franz Josefstrasse 18, A-8700 Leoben, Austria. (1938) Doz., applied physics (Techn. U. Vienna, 1972). Full prof. (tel. 03842 + 42555, ext. 264). *X-ray metallography, computer programming, residual stresses.*
Tuscher, Dr Mag. Engelbert. Inst. für Experimentalphysik, U. Wien, Strudlhofgasse 4, A-1090 Vienna, Austria. (1950) Dr. rer. nat., magnetochemistry (U. Vienna, 1979). Res. scient. (tel. 0222 + 342630, ext. 440). *Hydrogen storage, metals, intermetallic compounds, magnetic properties and structure, hydrides and deuterides.*
Uhl, Dr Eduard. Inst. für Pysikalische Chemie - Magnetochemie, U. Wien, Währingerstrasse 42, A-1090 Vienna, Austria. (1952) Dr. phil., chemistry (U. Vienna, 1980). Asst. (tel. 0222 + 343616) *Magnetic measurements, transition metal compounds, X-ray crystallography.*
Vana, Doz. Dr Dipl.-Ing. Norbert Johannes. Beschleunigerabteilung, Atominst. der Österreichischen Universitäten, Schüttelstrasse 115, A-1020 Vienna, Austria. (1940) Doz., optical and microwave spectroscopy (Techn. U. Vienna, 1975). Sr. scient. (tel. 0222 + 725136, ext. 277). *Lattice defects, radiation damage, color centers, archaeometry.*
Viehböck, Prof. Dr Franz Paul. Inst. für Allgemeine Physik, Techn. U. Wien, Karlsplatz 13, A-1040 Vienna, Austria. (1923) Doz., applied physics (Techn. U. Vienna, 1966). Full prof. and head. (tel. 0222 + 5601, ext. 3370). *Crystal surfaces.*
Völlenkle, Doz. Dr Horst. Inst. für Mineralogie, Kristallographie und Strukturchemie, Techn. U. Wien, Getreidemarkt 9, A-1060 Vienna, Austria. (1938) Doz., structural inorganic chemistry (Techn. U. Vienna, 1980). Asst. (tel. 0222 + 5601, ext. 4742). *Inorganic and organic crystal structures, computer programming.*
Wagendristel, Prof. Dr Alfred Friedrich. Inst. für Angewandte Physik, Techn. U. Wien, Karlsplatz 13, A-1040 Vienna, Austria. (1941) Doz., thin films (Techn. U. Vienna, 1976). Full Prof. (tel. 0222 + 5601, ext. 3242). *Thin film structure, amorphous thin films, diffusion via lattice defects.*
Walitzi, Prof. Dr Eva Maria. Inst. für Mineralogie, Kristallographie und Petrologie (Röntgenabteilung), U. Graz, Universitätsplatz 2, A-8010 Graz, Austria. (1930) Doz., mineralogy-crystallography (U. Graz, 1967). Full prof. (tel. 0316 + 380, ext. 5541). *Inorganic crystal structures, crystal chemistry.*
Warhanek, Prof. Dr Hans. Inst. für Experimentalphysik, U. Wien, Strudlhofgasse 4, A-1090 Vienna, Austria. (1926) Dr. phil., physics (U. Vienna, 1953). Full Prof. (tel. 0222 + 342630, ext. 217; telex 116222). *Phase transitions, crystal growth.*
Weber, Prof. Dr Harald Wolfgang. Atominst. der Österreichischen Universitäten, Schüttelstrasse 115, A-1020 Vienna, Austria. (1944) Doz., low temperature physics (Techn. U. Vienna, 1975). Full Prof. (tel. 0222 + 725136, ext. 240) *Superconductivity, crystallographic properties.*
Wobrauschek, Doz. Dr Dipl.-Ing. Peter. Abteilung für Elektronen- und Röntgenphysik, Atominst. der Österreichischen Universitäten, Schüttelstrasse 115, A-1020 Vienna, Austria. (1939) Doz., X-ray physics (Techn. U. Vienna, 1983). Asst. (tel. 0222 + 725136, ext. 58). *Crystal orientation, X-ray metallography, textures, X-ray fluorescence analysis.*
Zeilinger, Prof. Dr Anton Wolfgang. Abteilung für Neutronen- und Festkörperphysik, Atominst. der Österreichischen Universitäten, Schüttelstrasse 115, A-1020 Vienna, Austria. (1945) Doz., neutron and solid state physics

(Techn. U. Vienna, 1979). Full Prof. (tel. 0222 + 725136, ext. 258). *Dynamical diffraction, perfect crystal neutron optics, neutron interferometry, quantum mechanics foundation.*

Zemann, Prof. Dr Josef. Inst. für Mineralogie und Kristallographie, U. Wien, Dr. Karl Lueger-Ring 1, A-1010 Vienna, Austria. (1923) Doz., mineralogy (U. Vienna, 1951). Full prof. and head. (tel. 0222 + 4300, ext. 2660). *Mineral crystal chemistry, inorganic crystal chemistry, crystal optics, lattice energies.*

Zipper, Dr Peter. Inst. für Physikalische Chemie, U. Graz, Heinrichstrasse 28, A-8010 Graz, Austria. (1941) Dr. phil., chemistry (U. Graz, 1970). Asst. (tel. 0316 + 380, ext. 5415). *Small angle X-ray scattering, wide angle X-ray scattering, biopolymers, synthetic polymers, computer programming.*

Zobetz, Dr Erich. Inst. für Mineralogie, Kristallographie und Strukturchemie, Techn. U. Wien, Getreidemarkt 9, A-1060 Vienna, Austria. (1950) Dr. phil., mineralogy and petrography (U. Vienna, 1983). Asst. (tel. 0222 + 5601, ext. 4742). *Crystal structures, X-ray fluorescence analysis, Dirichlet domains.*

BANGLADESH

Sub-Editor: **Kh. A. I. F. M. Mannan**

Notes

1. International telephone country code - 880.

Ahmad, Mr Raisuddin. Dept. of Geology, U. of Dacca, Ramna, Dacca-2, Bangladesh. (1948) MSc., geology (U. of Dacca, 1970). Lecturer. (tel. 243723). *Structural analysis, minerals and silicates, mineralogy.*

Ahmed, Dr A. H. Moinuddin. Dept. of Biochemistry, Dacca U., Ramna, Dacca-2, Bangladesh. (1938) PhD, inorganic chemistry & crystallography (Aberdeen U., UK, 1969). Assoc. prof. (tel. 245289). *Chemical crystallography, inorganic chemistry, bio-physical chemistry, bio-physics, cement chemistry, electron diffraction, electron microscopy.*

Ahmed, Dr Sultan. Physics Dept., Dacca U., Curzon Hall, Dacca-2, Bangladesh. (1936) PhD, solid state physics, (Southampton U., UK, 1971). Assoc. prof. (tel. Dacca 242935). *Structural studies of crystalline and non-crystalline solids.*

Akhtar, Dr Farida. Dept. of Chemistry, Jahangirnagar U., Savar, Dacca, Bangladesh. (1942) PhD, X-ray crystallography, (London U., UK, 1969). Assoc. prof. (tel. Savar 316071, ext. 19). *Structural investigation of physico-chemical techniques, X-ray crystallography.*

Biswas, Dr Mohommad Alim. Glass and Ceramics Div., BCSIR Labs., Science Lab. Rd., Dacca-2, Bangladesh. (1927) PhD, organic chemistry, (London U., UK, 1958). Head of division, (tel. 315563). *Inorganic crystal structures, clay minerals.*

Chawdhury, Prof. Sadruddin Ahmed. Physics Dept., Rajsahi U., Rajsahi, Bangladesh. (1934) PhD, crystallography (Manchester U., UK,). Prof. *Organic crystal structures, direct methods.*

Chowdhury, Prof. Fazlul Halim. Chemistry Dept., Rajsahi U., Nilkhet, Dacca-2, Bangladesh. (1930) PhD, physical chemistry (U. of Manchester, UK, 1956). Prof. (tel. Dacca 250615 and 250704). *Large molecules, metal structures.*

Haider, Prof. Syed Zahir. Dept. of Chemistry, Dacca U., Ramna, Dacca-2, Bangladesh. (1927) PhD, inorganic chemistry (London U., UK, 1958). Prof. (tel. Dacca 315991). *Metal phosphates, organoboron compounds, catalysts, coordination chemistry, building materials (low cost production).*

Husain, Dr Abul Hasanat Mohammad. Dept. of Physics, U. of Dacca, Curzon Hall, Dacca 2, Bangladesh. (1949) PhD, solid state physics (U. of Exeter, UK, 1977). Assoc. prof. (tel. 242935). *Ultrasonic applications, solar energy research, organic crystals.*

Ibrhim, Dr Muhammad. Physics Dept., Dacca U., Curzon Hall, Dacca-2, Bangladesh. (1945) PhD, surface physics (Southampton U., UK, 1972). Assoc. prof. *Surface structure and electrical properties of solids, thin films, adsorption phenomena.*

Islam, Prof. Aminul. Soil Science Dept., Dacca U., 32-H, Isakhan Rd., Dacca-2, Bangladesh. (1935) PhD, soil science (Michigan State U., USA, 1962). Prof. *Clay mineral structures.*

Islam, Mr Shafiqul. Physics Dept., U. of Rajshahi, Rajshahi, Bangladesh. (1951) MSc, crystallography (U. of Rajshahi, 1972). Lecturer. (tel. Rajshahi 2441, ext. 23). *Structure determination, organic molecules, direct methods.*

Khan, Dr Anwarur Rahman. Dept. of Applied Physics, U. of Dacca, Ramna, Dacca-2, Bangladesh. (1932) PhD, metal physics (U. of London, UK, 1967). Assoc. prof. (tel. 257859). *Electron microscopy, electron diffraction, thin films, structure and properties.*

Mannan, Prof. Dr Kh. A. I. F. Mafizul. Physics Dept., Dacca U., Ramna, Dacca-2, Bangladesh. (1939) DPhil, crystallography (Oxford U., UK, 1965). Prof. (tel. Dacca 242935). *Organic and organo-metallic crystal structures, thin film structure and electrical properties, powder diffraction, amorphous films, jute fibre.*

Manzoor-I-Khuda, Dr Muhammad. T.R.C. Div., Jute Res. Inst., Tejgaon, Dacca-15, Bangladesh. (1933) PhD, organic chemistry (U. of London, UK, 1957). Director. (tel. 310975). *Organic crystal structure, large molecules.*

Quader, Prof. Dr Mohammed Abdul. Physics Dept., Jahangirnagar U., Savar, Dacca, Bangladesh. (1933) DPhil, physics (Calcutta U., 1962). Prof. (tel. Savar 316071). *Powder photography, phase diagrams, phase transformations, defects, faults and strains.*

Rahman, Dr Asadur. Physics Dept., Dacca U., Curzon Hall, Dacca-2, Bangladesh. (1944) PhD, X-ray crystallography (U. of Dundee, UK, 1971). Assoc. prof. (tel. Dacca 242935). *Organic and biological structures, nucleic acid structure and conformation, viruses, macromolecular structures, computer programming for crystallography.*

Rahman, Mr A. F. Md. Maqsudur. Chemistry Dept., Jahangirnagar U., Savar, Dacca, Bangladesh. (1950) MSc, structural inorganic chemistry (Dacca U., 1974). Lecturer. *Structural inorganic chemistry, inorganic coordination complexes, industrial catalysts.*

Rahman, Dr Sheikh Mohammed Mujibur. Physics Dept., U. of Dacca, 59 North Road, Dacca 5, Bangladesh. (1951) PhD, solid state physics (Dacca U., 1976) PhD, solid state physics (Bristol U., UK, 1979). Assoc. prof. (tel. Dacca 242935). *Electronic structure, Binary alloy structures, thermodynamic properties.*

Roy, Dr Ajoy Kumer. Dept. of Physics, U. of Dacca, Ramna, Dacca-2, Bangladesh. (1935) PhD, solid state physics (Leeds U., UK, 1966). Assoc. prof. (tel. Dacca 242935). *Electron paramagnetic resonance spectroscopy, free radicals (oriented), paramagnetic centres.*

Syed, Dr A. Sattar. Industrial Physics Div., C.S.I.R. Sci. Lab. Rd., Dacca-2, Bangladesh. (1935) PhD, solid state physics (U. of British Columbia, Canada, 1964). Sr. res. officer. (tel. Dacca 315563, ext. 05). *Applied crystallography, manganese dioxide, thin films, crystal growth, organometallic complexes.*

Zaman, Dr (Mrs) Nazma, Dept. of Physics, Bangladesh U. of Eng. and Techn., Dacca, Bangladesh. (1946) PhD, X-ray crystallography (U. of Manchester, UK, 1975). Asst. prof. (tel. Dacca 252473). *Organic compounds, crystal structure.*

BELGIUM

Sub-Editor: G.S.D. King

Notes

1. International telephone country code - 32.

2. The academic degrees conferred in Belgium are : *Licencié en sciences* or *licentiaat in de wetenschappen* (Lic) (equivalent to MSc), *Docteur en sciences* or *doctor in de wetenschappen* (DSc) (equivalent to PhD), *Ingénieur civil* or *burgerlijk ingenieur* (Ir), *Agrégé de l'enseignement supérieur* (Agr. Ens. Sup.) or *geaggregeerde voor het hoger onderwijs* (Geaggr. H.O.)

3. The academic positions are : *Professeur ordinaire* or *gewoon hoogleraar*, *Professeur* or *hoogleraar* (both positions are referred to as Prof.), *Chargé de cours* or *docent*, *Professeur associé* or *geassocieerd hoogleraar*, *Chargé de cours associé* or *geassocieerd docent* (both positions are referred to as Assoc. Prof.), *Chef de travaux* or *werkleider*, *Premier assistant* or *eerstaanwezend assistent*, *Assistant* or *assistent* (both positions are referred to as Asst.).

Aernoudt, Prof. Dr Etienne. Dept. Metaalkunde en Toegepaste Materiaalkunde, Katholieke U. Leuven, De Croylaan 2, B-3030 Heverlee, Belgium. (1938) Dr Ir, metallurgy (T. H. Aachen, F. R. Germany, 1966). Prof. (tel. 016 + 220931, ext. 1302). *Physical metallurgy, metal forming, strengthening mechanisms, textures, materials science.*

Amelinckx, Prof. Dr Severin. Studiecentrum voor Kernenergie (SCK), C.E.N. Mol B-2400, Belgium. (1922) DSc, physics (U. Gent, 1948). General Manager. (tel. 014 + 311801, ext. 2100). *Defects, electron microscopy, electron diffraction.*

Baele, Miss Ingrid Albertina Frans Mariette. Fysica van de vaste stof, Rijksuniversitair Centrum Antwerpen, (RUCA), Groenenborgerlaan 171, B-2020 Antwerpen, Belgium. (1961) Lic., physics (U. Antwerp, 1984). Asst. (tel. 03 + 2180495). *Electron microscopy.*

Bauduin, Miss Anne-Marie Ghislaine Gerardine. Facultés Universitaires de Namur, Groupe de Chimie-Physique, Rue De Bruxelles 61, B-5000 Namur, Belgium. (1961) Lic., chemistry (Fac. U. Namur, 1984). Res. student. (tel. 081 + 229061, ext. 2481). *X-ray crystal structure analysis, beta-lactam antibiotics, structure - activity correlations, biophysics.*

Bender, Dr Hugo J.M.R. Centrum voor hoogspanningselektronenmikroskopie, Rijksuniversitair Centrum Antwerpen, (RUCA), Groenenborgerlaan 171, B-2020 Antwerpen, Belgium. (1957) DSc, physics (U. Antwerp, 1984). Asst. (tel. 03 + 2180263). *Electron microscopy, semiconductors.*

Blaton, Dr Norbert Louis. Lab. voor Analytische Chemie en Medicinale Fysicochemie, Inst. voor Farmaceutische Wetenschappen, Katholieke U. Leuven, Van Evenstraat 4, B-3000 Leuven, Belgium. (1945) DSc, chemistry (U. Leuven, 1974). Werkleider. (tel. 016 + 226947). *Structure determination, powder diffraction, medicinal compounds.*

Boesman, Dr Etienne Roland. Lab. voor Kristallografie en Studie van de Vaste Stof, Rijksuniversiteit Gent Krijgslaan 281, B-9000 Gent, Belgium. (1932) DSc, physics (U. Gent, 1962). Docent. (tel. 091 + 225715). *Paramagnetic resonance, colour centres and impurity ions, spin lattice relaxation.*

Bosmans, Prof. Dr Herman Jozef. Lab. voor Analytische en Minerale Scheikunde, Landbouwinstituut, Katholieke U. Leuven, Kardinaal Mercierlaan 92, B-3030 Heverlee, Belgium. (1928) DSc, physical chemistry (U. Leuven, 1966). Prof. (tel. 016 + 220931, ext. 1584). *Oxides and aluminosilicates, analytical chemistry.*

Brasseur, Prof. Dr Henri Alphonse Lambert. Crystallography Dept., U. of Liège, Inst. of Physics B5, Sart Tilman, B-4000 Liège, Belgium. (1905) Agr. Sc., physics (U. Liège, 1934). Retired. (tel. 041 + 525652). *Crystallography.*

Cardon, Prof. Dr Felix. Lab. voor Kristallografie en Studie van de Vaste Stof, Rijksuniversiteit Gent Krijgslaan 281, B-9000 Gent, Belgium. (1935) DSc, physics (U. Gent). Prof. (tel. 091 + 225715). *Solid state physics, photoconductivity, solar cells, semiconductor electrochemistry.*

Clauws, Dr Paul. Laboratorium voor Kristallografie en Studie van de Vaste Stof, Rijksuniversiteit Gent, Krijgslaan 281, B-9000 Gent, Belgium. (1944) DSc, physics (U. Gent, 1973). Werkleider. (tel. 091 + 225715, ext. 2347). *Infrared spectroscopy of crystals.*

Coene, Mr Willem Marie Julia Marcel, Fysica van de vaste stof, Rijksuniversitair Centrum Antwerpen, (RUCA), Groenenborgerlaan 171, B-2020 Antwerpen, Belgium. (1960) Lic., physics (U. Antwerp, 1982). Asst. (tel. 03 + 2180261). *High-resolution electron microscopy, crystal defects.*

Collin, Miss Sonia Bertha Josepha. Facultés Universitaires de Namur, Groupe de Chimie-Physique, Rue De Bruxelles 61, B-5000 Namur, Belgium. (1963) Lic., chemistry (Fac. U. Namur, 1984). Res. student. (tel. 081 + 229061, ext. 2481). *X-ray crystal structure analysis, biological compounds, structure - activity correlations, biophysics, molecular graphics.*

Cornelis, Ir Jozef Frans Elisa. Materials Sci. Dept., S.C.K.-C.E.N., B-2400 Mol, Belgium. (1944) Ir, electro-mechanics (U. Leuven, 1969). Res. Eng. (tel. 014 + 311801, ext. 2728). *Radiation damage, metals and alloys.*

Deblieck, Mr Rudy André Cornelis. Fysica van de vaste stof, Rijksuniversitair Centrum Antwerpen, (RUCA), Groenenborgerlaan 171, B-2020 Antwerpen, Belgium. (1956) Lic., physics (V.U. Brussels, 1979). Asst. (tel. 03 + 2180249). *Electron microscopy, phase transitions, modulated structures.*

Declercq, Prof. Jean Paul. Lab. Chimie-Phys. et Cristallographie, Université Catholique de Louvain, Pl. Pasteur 1, B-1348 Louvain-la-Neuve, Belgium. (1948) DSc, chemistry (U. Louvain, 1972). Chargé de cours. (tel. 010 + 432924). *Direct methods, computer programming.*

De Gryse, Dr Roger Marc. Lab. voor Kristallografie en Studie van de Vaste Stof, Rijksuniversiteit Gent, Krijgslaan 281, B-9000 Gent, Belgium. (1941) Dr, applied sciences (U. Gent, 1973). Werkleider. (tel. 091 + 225715, ext. 2350). *Surface properties of crystals.*

Dekeyser, Prof. Dr Willy Clement. Green Park, Pacificatielaan 63, B-9000 Gent, Belgium. (1910) DSc, physics (U. Gent, 1930). Prof. emeritus U. Gent. *Lattice defects, surface properties.*

Delaey, Prof. Luc J. M. A. E. Dept. Metaalkunde en Toegepaste Materiaalkunde, Katholieke U. Leuven, De Croylaan 2, B-3030 Heverlee, Belgium. (1939) Dr. rer. nat., metallurgy (U. Stuttgart Germany, 1966). Prof. (tel. 016 + 220931, ext. 1272). *Physical metallurgy, phase transformations, electron microscopy.*

Delavignette, Prof. Pierre. Materials Sci. Dept., C.E.N.-S.C.K., B-2400 Mol, Belgium. (1931) DSc, physics (U.L. Brussels, 1962). Physicist, Prof. U.L.B. (tel. 014 + 311801). *Imperfections in crystals, phase transitions, grain boundaries, transmission electron microscopy (TEM).*

Deliens, Dr Michel. Dept. Géologie et Minéralogie, Musée Royal de l'Afrique Centrale, Leuvensesteenweg 13, B-1980 Tervueren, Belgium. (1939) DSc, mineralogy (U. Louvain, 1972). Mineralogist. (tel. 02 + 7675401). *X-ray crystal structure analysis.*

De Ranter, Prof. Dr Camiel Joseph. Lab. voor Analytische Chemie en Medicinale Fysicochemie, Inst. voor Farmaceutische Wetenschappen, Katholieke U. Leuven, Van Evenstraat 4, B-3000 Leuven, Belgium. (1937) DSc, chemistry (U. Leuven, 1964). Prof. (tel. 016 + 226947). *Structure-function studies, small and medium-sized biological molecules.*

Deruyttere, Prof. Dr André. Dept. Metaalkunde en Toegepaste Materiaalkunde, Katholieke U. Leuven, De Croylaan 2, B-3030 Heverlee, Belgium. (1925) PhD, metallurgy (U. Sheffield, UK, 1955). Prof. (tel. 016 + 220931, ext. 1271). *Metallurgy, materials science.*

De Schoenmacker. Ir Dirck Maurice. Fysica van de vaste stof, Rijksuniversitair Centrum Antwerpen, (RUCA), Groenenborgerlaan 171, B-2020 Antwerpen, Belgium. (1961) Ir, applied physics (V.U. Brussels, 1984). Res. asst. (tel. 03 + 2180249). *Electron microscopy, optics.*

De Wolf, Mr Marcus Ludovicus Maria. Chemistry dept., Universitaire Instelling Antwerpen, Universiteitsplein 1, B-2610 Wilrijk, Belgium. (1958) Lic, chemistry (U. Antwerpen 1983). Asst. (tel. 03 + 8282528). *Geochemistry.*

Dideberg, Dr Otto. Crystallography Dept., U. of Liège, Inst. of Physics B5, Sart Tilman, B-4000 Liège, Belgium. (1942) DSc, physics (U. Liège, 1969). Assoc. Prof. (tel. 041 + 561762). *Protein crystallography, biophysics,*

Dupont, Dr Leon. Crystallography Dept., U. of Liège, Inst. of Physics B5, Sart Tilman, B-4000 Liège, Belgium. (1941) DSc, physics (U. Liège, 1969). Chef de travaux. (tel. 041 + 561762). *Crystal structure determination.*

Durant, Prof. François Victor. Facultés Universitaires de Namur, Groupe de Chimie-Physique, Rue De Bruxelles 61, B-5000 Namur, Belgium. (1939) DSc, chemistry (U. Louvain, 1965). Prof. (tel. 081 + 229062). *X-ray crystal structure analysis, biological compounds, structure - activity correlations, biophysics, molecular graphics.*

Elsen, Mr Jan Albrecht. Lab. voor Kristallographie, Katholieke U. Leuven, Celestijnenlaan 200C, B-3030 Leuven, Belgium. (1961) Lic, geology (U. Leuven, 1983). Asst. (tel. 016 + 201015, ext. 1584). *Structure determination, zeolites.*

Evrard, Prof. Guy Henri. Facultés Universitaires de Namur, Groupe de Chimie-Physique, Rue De Bruxelles 61, B-5000 Namur, Belgium. (1943) DSc, chemistry (U. Louvain, 1969). Prof. (tel. 081 + 229061). *X-ray crystal structure analysis, biological organometallic compounds, direct methods, interactive display, data retrieval.*

Feneau-Dupont, Mrs Janine. Lab. Chimie-Phys. et Cristallographie, Université Catholique de Louvain, Pl. Pasteur 1, B-1348 Louvain-la-Neuve, Belgium. (1932) Lic., chemistry (U. Louvain, 1953). Collaboratrice scientifique. (tel. 010 + 432921). *General crystallography.*

Fiermans, Dr Lucien Victor August. Laboratorium voor Kristallografie en Studie van de Vaste Stof, Rijksuniversiteit Gent, Krijgslaan 281, B-9000 Gent, Belgium. (1937) Geaggr. H.O., physics (U. Gent, 1974). Werkleider. (tel. 091 + 225715). *Surface physics, crystallography, defects in semiconductors.*

Fontaine, Dr Frederic Desiré Albert. Exp. Physics Dept., U. of Liège, Inst. of Physics B5, Sart Tilman, B-4000 Liège, Belgium. (1932) DSc, chemistry (U. Liège, 1967). Chef de travaux. (tel. 041 + 561631). *Crystal structures, X-ray small angle scattering.*

Fransolet, Dr André-Mathieu. Inst. for Mineralogy U. of Liège, Place du XX Aout 9-A. 1, B-4000 Liège, Belgium. (1947) DSc, geology and mineralogy (U. Liège, 1975). Res. assoc. FNRS, (tel. 041 + 420080, ext. 429). *Crystallochemistry of minerals.*

Gastuche - Van Oosterwyck, Dr (Mrs) Marie Claire. Musée Royal de l'Afrique Centrale, Steenweg op Leuven, B-1980 Tervuren, Belgium. (1926) Agr. Ens. Sup., mineralogy (1974). Res. sci. (tel. 02 + 7675401, ext. 320). *Mineralogy, X-ray crystallography, clay minerals.*

Geise, Prof. Dr Herman Joseph Victor Heinrich. Chemistry dept., Universitaire Instelling Antwerpen, Universiteitsplein 1, B-2610 Wilrijk, Belgium. (1937) Dr, chemistry (U. Leiden, Netherlands,1964). Prof. (tel. 03 + 8282528, ext. 164) *Conformational analysis, gas phase electron diffraction.*

Germain, Prof. Gabriel. Unité de Chimie physique moléculaire et de Cristallographie, Université Catholique de Louvain, Pl. Pasteur 1, B-1348 Louvain-la-Neuve, Belgium. (1933) DSc, chemistry (U. Louvain, 1958). Chargé de cours. (tel. 010 + 432833). *Direct methods, computer programming.*

Gevers, Prof. Dr Rudolf. Faculty of Sci., Rijksuniversitair Centrum Antwerpen, (RUCA), Groenenborgerlaan 171, B-2020 Antwerpen, Belgium. (1924) DSc, physics (U, Gent, 1953). Prof. (tel. 03 + 2180355). *Electron microscopy, electron diffraction.*

Gibon, Mrs Véronique Julie Jacques Laure. Facultés Universitaires de Namur, Groupe de Chimie-Physique, Rue De Bruxelles 61, B-5000 Namur, Belgium. (1958) Lic., chemistry (Fac. U. Namur, 1980). Asst. (tel. 081 + 229061, ext. 2481). *Polymorphism and phase diagrams, triglycerides.*

Hoogewijs, Dr Robert Richard. Lab. voor Kristallografie en Studie van de Vaste Stof, Rijksuniversiteit Gent, Krijgslaan 281, B-9000 Gent, Belgium. (1950) Geaggr. H.O., Physics (U. Gent, 1980). Werkleider. (tel. 091 + 225715, ext. 2337). *Surface physics.*

Jacobs, Prof. Dr Gilbert. Laboratorium voor Kristallografie en Studie van de Vaste Stof, Rijksuniversiteit Gent, Krijgslaan 281, B-9000 Gent, Belgium. (1925) DSc, chemistry (U. Gent, 1952). Assoc. Prof. (tel. 091 + 225715, ext. 2321). *Defects, ionic crystals.*

Kartheuser, Dr Edward Peter. Theoretical Physics Dept., U. of Liège, Sart Tilman, B-4000 Liège, Belgium. (1938) DSc, physics (U. Liège, 1968). Chef de travaux. (tel. 041 + 561639). *Electron-phonon interaction, ionic crystals, semiconductors, electronic polarisability.*

King, Prof. Geoffrey Stephen Douglas. Laboratorium voor Kristallografie, Katholieke Universiteit Leuven, Celestijnenlaan 200C, B-3030 Leuven, Belgium. (1924) MSc, crystallography (U. London, UK, 1950). Prof. (tel. 016 + 201015, ext. 1582). *Structure determination, computing.*

Lamotte-Brasseur, Dr Josette Marie Louise. Crystallography Dept., U. of Liège, Inst. of Physics B5, Sart Tilman, B-4000 Liège, Belgium. (1943) DSc, physics (U. Liège, 1973). Chef de travaux. (tel. 041 + 561758). *Crystal structure determination, molecular graphics.*

Legrand, Dr Emiel. Materials Sci. Dept., S.C.K.-C.E.N., B-2400 Mol, Belgium. (1931) DSc, physics (U. Leuven, 1966). Res. Physicist. (tel. 014 + 311801). *Neutron diffraction.*

Lenstra, Prof. Dr Albert Teun Hendrik. Chemistry dept., Universitaire Instelling Antwerpen, Universiteitsplein 1, B-2610 Wilrijk, Belgium. (1942) Dr, chemistry (U. Utrecht, Netherlands,1973). Prof. (tel. 03 + 8282528, ext. 218). *Patterson techniques, automation, structure determination.*

Léonard, Dr André Jules Gérard. Groupe de Physico-Chimie Minérale et de Catalyse, Université Catholique de Louvain, Pl. Croix du Sud 1, B-1348 Louvain-la-Neuve, Belgium. (1936) DSc, crystallography (U. Louvain, 1959). Chef de travaux. (tel. 010 + 433588). *Structural analysis, non-crystalline, ill-crystallized materials, crystal physics, clay minerals.*

Maenhout - Van Der Vorst, Dr Mrs Wenefride Marguerite Romain. Lab. voor Kristallografie en Studie van de Vaste Stof, Rijksuniversiteit Gent, Krijgslaan 281, B-9000 Gent, Belgium. (1930) DSc, physics (U. Gent, 1957). Werkleider. (tel. 091 + 225715). *Photography, surface effects, exoelectron emission.*

Mahy, Mr Jan Willem Gaston. Fysica van de vaste stof, Rijksuniversitair Centrum Antwerpen, (RUCA), Groenenborgerlaan 171, B-2020 Antwerpen, Belgium. (1959) Lic., physics (U. Antwerp, 1981). Res. asst. (tel. 03 + 2180248). *Electron microscopy, phase transitions, modulated structures.*

Marcoen, Dr Jean-Marie. Chaire Sc. de la Terre Fac. Agronomie, B-5800 Gembloux, Belgium. (1948) Dr, agronomy (Gembloux, 1977). Maitre de Conférences. (tel. 081 + 612958). *Soil mineralogy, clays, XRD, DTA, IR, EM, solid materials.*

Matthys, Dr Paul Frederik André Edmond. Laboratorium voor Kristallografie en Studie van de Vaste Stof, Rijksuniversiteit Gent, Krijgslaan 281, B-9000 Gent, Belgium. (1949) Dr, physics (U. Gent, 1976). Werkleider. (tel. 091 + 225715, ext. 2365). *EPR, colour centres, alkali halides.*

Meunier - Piret, Dr (Mrs) Jacqueline. Lab. Chimie-Phys. et Cristallographie, U. Catholique de Louvain, Pl. Pasteur 1, B-1348 Louvain-la-Neuve, Belgium. (1934) DSc, chemistry (U. Louvain, 1961). Asst. (tel. 010 + 432922). *Organic and organometallic crystal structures.*

Moreau, Prof. Jules Francois. Lab. de Minéralogie et de Géologie Appliquée, Université Catholique de Louvain, Batiment Mercator, Pl. L. Pasteur 3, B-1348 Louvain-la-Neuve, Belgium. (1931) Mining Eng. (U. Louvain, 1954). Prof. (tel. 010 + 432855). *Mineralogy, economic geology.*

Mortier, Dr Wilfried Jozef. Laboratorium voor Oppervlaktechemie, Katholieke Universiteit Leuven, Kardinaal Mercierlaan 92, B-3030 Heverlee, Belgium. (1946) Geaggr. H.O., chemistry (U. Leuven, 1978). Sr. res. assoc. NFWO. (tel. 016 + 220931, ext. 1593). *Structure, zeolites, intermolecular interactions, catalysis.*

Nagels, Prof. Pieter Jan. Materials Sci. Dept., S.C.K.-C.E.N., B-2400 Mol, Belgium. (1930) DSc, chemistry (U. Gent, 1957). Prof. (tel. 014 + 311801, ext. 2746). *Amorphous materials, Polymer structures.*

Naud, Dr Jean Marcel. Lab. de Minéralogie et de Géologie Appliquée, Université Catholique de Louvain, Batiment Mercator, Pl. L. Pasteur 3, B-1348 Louvain-la-Neuve, Belgium. (1942) DSc, chemistry (U. Louvain, 1968). Chef de travaux. (tel. 010 + 432851). *Powder diffraction, mineralogy, X-ray spectrometry.*

Peeters, Dr Oswald Maurice. Lab. voor Analytische Chemie en Medicinale Fysicochemie, Inst. voor Farmaceutische Wetenschappen, Katholieke U. Leuven, Van Evenstraat 4, B-3000 Leuven, Belgium. (1945) DSc, chemistry (U. Leuven, 1977). Werkleider. (tel. 016 + 226947). *X-ray structure determination, biological and pharmaceutical compounds, structure - activity relations.*

Piret, Prof Paul. Lab. Chimie-Phys. et Cristallographie, Université Catholique de Louvain, Pl. Pasteur 1, B-1348 Louvain-la-Neuve, Belgium. (1932) DSc, chemistry (U. Louvain, 1956). Prof. (tel. 010 + 432769). *Crystal structures, mineralogy.*

Pyckhout. Mr Wim Maurits August. Chemistry dept., Universitaire Instelling Antwerpen, Universiteitsplein 1, B-2610 Wilrijk, Belgium. (1959) Lic, chemistry (U. Antwerpen, 1982). Asst. (tel. 03 + 8282528). *Gas phase electron diffraction.*

Reynaers, Prof. Harry Louis. Dept. Scheikunde, Katholieke Universiteit Leuven, Celestijnenlaan 200F, B-3030 Heverlee, Belgium. (1938) DSc, chemistry (U. Leuven, 1964). Prof. (tel. 016 + 200656, ext. 1496). *Small-angle scattering, polymers (synthetic and biological).*

Rodrique, Dr Luc Willy. Groupe de Physico-Chimie Minérale et de Catalyse (MRAC), Université Catholique de Louvain, Pl. Croix du Sud 1, B-1348 Louvain-la-Neuve, Belgium. (1941) DSc, crystallogrphy (U. Louvain, 1966). Chef de travaux. (tel. 010 + 433665). *Microscopie et diffraction électroniques, microanalyse par sonde électronique.*

Rotti, Mr Marc Maurice. Materials Sci. Dept., S.C.K.-C.E.N., B-2400 Mol, Belgium. (1955) Lic, chemistry (U. Gent, 1978). Asst. (tel. 011 + 251958). *Electrical properties and structural characteristics, polyalkynes.*

Schryvers, Mr Dominique Maurits. Fysica van de vaste stof, Rijksuniversitair Centrum Antwerpen, (RUCA), Groenenborgerlaan 171, B-2020 Antwerpen, Belgium. (1959) Lic., physics (U. Antwerp, 1981). Res. asst. (tel. 03 + 2180495). *Electron microscopy, materials science, alloys, catalysts.*

Sobry, Dr Roger. Exp. Physics Dept., U. of Liège, Inst. of Physics B5, Sart Tilman, B-4000 Liège, Belgium. (1946) DSc, physics (U. Liège, 1972). Chef de travaux. (tel. 041 + 561715). *Diamagnetic susceptibilities, liquid crystals, molecular properties of crystals.*

Spirlet, Dr Marie-Rose. Exp. Physics Dept., U. of Liège, Inst. of Physics B5, Sart Tilman, B-4000 Liège, Belgium. (1946) DSc, chemistry (U. Liège, 1976). Chef de travaux. (tel. 041 + 561758). *Crystal structure analysis.*

Tielemans, Mr Luc. Materials Sci. Dept., S.C.K.-C.E.N., B-2400 Mol, Belgium. (1955) Lic, physics (U. Leuven, 1978). Res. physicist. (tel. 014 + 311801). *Neutron scattering.*

Tinant, Dr Bernard Guy André François. Lab. Chimie-Phys. et Cristallographie, Université Catholique de Louvain, Pl. Pasteur 1, B-1348 Louvain-la-Neuve, Belgium. (1951) DSc, chemistry (U. Louvain, 1978). Chercheur. (tel. 010 + 432924). *Structure determination, molecular dynamics and NMR.*

Tonnard, Dr Victor Edmond. Chaire Sc. de la Terre Fac. Agronomie, B-5800 Gembloux, Belgium. (1929) Dr, agronomy (Gembloux, 1958). Prof. (tel. 081 + 612958). *Clays, mineralogy, X-ray diffraction, DTA, IR, inorganic soil components.*

Toussaint, Prof. Jean. Crystallography Dept., U. of Liège, Inst. of Physics B5, Sart Tilman, B-4000 Liège, Belgium. (1916) DSc, physics (U. Liège, 1945). Retired. (tel. 041 + 561618). *Crystallography.*

Van Alsenoy, Dr Kris. Chemistry dept., Universitaire Instelling Antwerpen, Universiteitsplein 1, B-2610 Wilrijk, Belgium. (1948) DSc, chemistry (V.U. Brussels, 1977). Res. assoc. NFWO. (tel. 03 + 8282528, ext.218). *Quantum chemistry, molecular structure.*

Van Cappellen Mr Eric Edouard Robert. Fysica van de vaste stof, Rijksuniversitair Centrum Antwerpen, (RUCA), Groenenborgerlaan 171, B-2020 Antwerpen, Belgium. (1958) Lic., physics (U. Antwerp, 1981). Asst. (tel. 03 + 2180247). *Electron microscopy.*

Vandenberghe, Dr Robert Emile. Lab. of Magnetism, Rijksuniversiteit Gent, Proeftuinstraat 86, B-9000 Gent, Belgium. (1945) DSc, physics (U. Gent, 1975). Sr. asst. (tel. 091 + 228731, ext. 217). *Metal oxides, spinels, phase transitions, magnetic structure, X-ray and neutron diffractometry, magnetisation, Mössbauer effect.*

Van den Bosch, Dr Adolf. Materials Sci. Dept., S.C.K.-C.E.N., B-2400 Mol, Belgium. (1928) DSc, chemistry (U. Gent, 1963). Res. physicist. (tel. 014 + 311801, ext. 2745). *Radiation damage, magnetism, superconductors.*

Van den Brempt, Dr Guy Ghislain Remy. Crystallography Dept., U. of Liège, Inst. of Physics B5, Sart Tilman, B-4000 Liège, Belgium. (1941) DSc, physics (U. Liège, 1973). Chef de travaux. (tel. 041 + 561763). *Crystallography, diamagnetism.*

Van der Brempt, Miss Christine Marie Paul. Facultés Universitaires de Namur, Groupe de Chimie-Physique, Rue De Bruxelles 61, B-5000 Namur, Belgium. (1960) Lic., chemistry (Fac. U. Namur, 1982). Res. student. (tel. 081 + 229061, ext. 2481). *X-ray crystal structure analysis, electron density, structure - activity correlations, biophysics, molecular graphics.*

Van Dyck, Dr Dirk. Fysica van de vaste stof, Rijksuniversitair Centrum Antwerpen, (RUCA), Groenenborgerlaan 171, B-2020 Antwerpen, Belgium. (1948) DSc., physics (U. Antwerp, 1977). Asst. (tel. 03 + 2180258). *Solid state physics, electron microscopy and diffraction.*

Vanhellemont, Mr Jan Hendrik. Fysica van de vaste stof, Rijksuniversitair Centrum Antwerpen, (RUCA), Groenenborgerlaan 171, B-2020 Antwerpen, Belgium. (1953) Lic., physics (U. Antwerp, 1978). Res. asst. (tel. 03 + 2180248). *Semiconductors, transmission electron microscopy, microelectronics.*

Vanhouteghem, Mr Frankie Marie. Chemistry dept., Universitaire Instelling Antwerpen, Universiteitsplein 1, B-2610 Wilrijk, Belgium. (1961) Lic, chemistry (U. Antwerpen, 1983). Asst. (tel. 03 + 8282528). *Gas phase electron diffraction.*

Van Landuyt, Prof. Joseph Florent. Faculty of Sci., Rijksuniversitair Centrum Antwerpen, (RUCA), Groenenborgerlaan 171, B-2020 Antwerpen, Belgium. (1938) DSc, physics (U. Gent, 1965). Prof. (tel. 03 + 2180259). *Electron microscopy, electron diffraction.*

Van Meerssche, Prof. Maurice. Lab. Chimie-Phys. et Cristallographie, Université Catholique de Louvain, Pl. Pasteur 1, B-1348 Louvain-la-Neuve, Belgium. (1923) DSc, chemistry (U. Louvain, 1948). Prof. (tel. 010 + 432771). *Crystal structure determination, NMR in crystals.*

Van Meervelt, Mr Luc. Lab. voor Kristallographie, Katholieke U. Leuven, Celestijnenlaan 200C, B-3030 Leuven, Belgium. (1958) Lic, chemistry (U. Leuven, 1980). Asst. (tel. 016 + 201015, ext. 1584). *Structure determination, small organic molecules.*

Van Tassel, Prof. Dr René. Prins Albertlei 5ór.4, B-2600 Berchem, Belgium. (1916) DSc, chemistry (U. Leuven, 1939). Retired. (tel. 02 + 6480475). *Minerals.*

Van Tendeloo, Dr Gustaaf. Fysica van de vaste stof, Rijksuniversitair Centrum Antwerpen, (RUCA), Groenenborgerlaan 171, B-2020 Antwerpen, Belgium. (1950) DSc., physics (U. Antwerp, 1974). Asst. (tel. 03 + 2180262). *Electron microscopy, modulated structures, alloy phases.*

Vennik, Prof. Ir Joost. Laboratorium voor Kristallografie en Studie van de Vaste Stof, Rijksuniversiteit Gent, Krijgslaan 281, B-9000 Gent, Belgium. (1927) Electrical Engineer, electronics (U. Gent, 1952). Prof. (tel. 091 + 225715, ext. 2325). *Surface physics, semiconductor physics, materials science.*

Verbist, Prof. Jacques Jozef. Facultés Universitaires de Namur, Département de Chimie, Rue De Bruxelles 61, B-5000 Namur, Belgium. (1943) DSc, chemistry (U. Louvain, 1969). Prof. (tel. 081 + 229061, ext. 2509). *Chemical crystallography, electronic and geometric structures.*

Verlinde, Mr Christophe Louis-Marie Jos. Lab. voor Analytische Chemie en Medicinale Fysicochemie, Inst. voor Farmaceutische Wetenschappen, Katholieke U. Leuven, Van Evenstraat 4, B-3000 Leuven, Belgium. (1958) Pharmacist (U. Leuven, 1981). Asst. (tel. 016 + 226947). *X-ray structure determination, structure - activity relationships, drugs.*

BOLIVIA

Sub-Editor: A. Saavedra M.

Notes

1. International telephone country code - 591.

Alarcón, Mr Hugo. Faculty of Earth Sci., U. Mayor de San Andrés, La Paz, Bolivia. (1941) Lic, geology (U. Federal de Rio de Janeiro, Brazil, 1964). Prof., optical mineralogy (tel. 359581-793392). *Mineralogy, minerography, geochemistry.*

Arduz, Mr Marcelo. Faculty of Earth Sci., U. Mayor de San Andrés, La Paz, Bolivia. (1946) Lic, geological engineering (U. National de la Plata, Argentina). Prof. (tel. 359581-793392). *Optical mineralogy, petrography.*

Arellano, Mr J. Faculty of Earth Sci., U. Mayor de San Andrés, Casilla 5905, La Paz, Bolivia. (1947) Lic, geological engineering (U. Mayor de San Andrés, 1974). (tel. 793392-785262). *Petrography, X-ray crystallography.*

Avila-Salinas, Mr Waldo. Mineralogy and Petrography Div., Dept. of Labs., Geological Service of Bolivia, Fco. Zuazo 1673, Casilla 2729, La Paz, Bolivia. (1941) Lic, geological engineering (U. Mayor de San Andrés, 1965). Chief. (tel. 32692, personal tel.). *Inorganic X-ray crystallography.*

Portugal, Mr Remberto. Dept. of Physics, Faculty of Sci. and Techn., U. Mayor de San Simón, Casilla 2551, Cochabamba, Bolivia. (1947) MSc, physics (U. Estadual de Campinas, Brazil, 1979). Asst. prof. (tel. 042 + 25503, ext 318). *Small angle diffraction.*

Ricaldi, Mr Edgar. Dept. of Physics, U. Mayor de San Andrés, La Paz, Bolivia. (1947) Dipl, geophysics (Freiberg's Academy of Mines, DDR, 1973). Prof. (tel. 799299-792622). *Mineralogy, petrophysics.*

Saavedra, Mr Antonio. Faculty of Sci., U. Mayor de San Andrés, Casilla 604, La Paz, Bolivia. (1939) Lic, geological engineering (U. Mayor de San Andrés, 1964). Dean. (tel. 329701). *Mineralogy, petrography.*

Sanjinés, Mr Orlando. Faculty of Earth Sci., U. Mayor de San Andrés, La Paz, Bolivia. (1944) Lic, geological engineering (U. Mayor de San Andrés, 1968). Prof. (tel. 359581-793392). *Mineralogy.*

Santiváñez, Mr Reynaldo. Faculty of Earth Sci., U. Mayor de San Andrés, Casilla 3698, La Paz, Bolivia. (1944) Lic, geological engineering (U. Mayor de San Andrés, 1978). Prof. (tel. 793392-3598587). *Petrography.*

Villegas, Dr Mario Oscar. Inst. de Investigaciones Físicas, U. Mayor de San Andrés, La Paz, Bolivia. (1943) PhD, metallurgy (U. Nac. del Sur, Argentina, 1975). Res., prof. (tel. 799299-792622). *Phase identification, electron microscopy, X-ray techniques.*

Zelaya, Mr José Miguel. Inst. de Investigaciones Físicas, U. Mayor de San Andrés, La Paz, Bolivia. (1947) MSc, mechanical engineering (U. Estadual de Campinas, Brazil, 1979). Res., prof. (tel. 799299-792622). *Crystal structure determination.*

BRAZIL

Sub-Editor: I.L. Torriani

Notes

1. International telephone country code - 55.

Almeida, Prof. Vasco Nogueira. Dept. de Física, U. Fed. de Goias, Goiania, Goias 74000, Brasil. (1923) MSc, physics (Fac. de Filos. Ciencias, Rio Claro, UNESP, 1966). Prof. adjunto. (tel. 062 + 2613088, ext. 168). *Crystal structures.*

Arguello, Dr Zoraide Primerano. Dept. de Estado Sólido, Inst. de Física, U. Est. de Campinas, Campinas, São Paulo 13100, Brasil. (1938) PhD, physics (U. Est. de Campinas, 1972). Assoc. prof. (tel. 0192 + 391301, ext. 346). *Crystal growth.*

Arruda, Prof. Moacir Rabelo. Inst. de Geociencias, Dept. de Mineralogia e Petrografia, Cidade U., São Paulo, São Paulo 01000, Brasil. (1925) PhD, geology (USP). Instr. (tel. 001 + 2122011). *Instrumentation, crystal optics, inorganic crystal structures.*

Baptista, Mr Augusto. Comissão Nac. de Energia Nuclear, Rio de Janeiro 20000, Brasil. (1930) BSc, chemistry (U. Fed. do Estado da Guanabara). Chemist. (tel. 021 + 2867002). *Organic and inorganic structures.*

Baptista, Mrs Neysa Rocha. Comissão Nac. de Energia Nuclear, Rio de Janeiro, Rio de Janeiro 20000, Brasil. (1930) BSc, chemistry (U. Fed. do Estado da Guanabara). Chemist. (tel. 021 + 2867002). *Inorganic crystal structures.*

Baran, Dr Zbigniew. Inst. de Fisica, U. Fed. da Bahia, Salvador, Bahia 40000, Brasil. (1930) PhD, Physics (U. Warsaw, Poland, 1970). Prof. (tel. 071 + 2472714). *Structural defects in crystalline solids.*

Barelli, Dr Nilso. Dept. de Tecnologia e Química de Aplicaçao, Inst. de Química, UNESP, Araraquara, São Paulo 14800, Brasil. (1944) PhD, mineralogy (Fac. de Filos. Ciencias e Letras de Araraquara, 1974). Asst. prof. (tel. 0162 + 320444, ext. 193). *Mineralogy, crystal growth, morphology, epitaxy.*

Bristoti, Dr Anildo. Dept. de Fisica, U. Fed. do Rio Grande do Sul, Porto Alegre, Rio Grande do Sul 90000, Brasil. (1936) PhD, metallurgical eng. (UCLA, USA, 1970). Assoc. prof. (tel. 0512 + 25-29-22). *X-ray crystallography, powder diffractometry, metallurgy, phase transitions, ferroelectrics.*

Bulhões, Mrs Iseli Angelica M. Dept. de Eng. de Materiais, U. Fed. de São Carlos, São Carlos, São Paulo 13560, Brasil. (1953) MSc, appl. physics (USP, 1979) Grad. student. (tel. 0162 + 718111, ext. 180). *Structure, X-ray diffraction.*

Campelo Farias, Prof. Carlinda. Dept. de Engenharia de Minas, U. Fed. de Pernambuco, Cidade U., Recife, Pernambuco 50000, Brasil. (1939) DSc, mineralogy (U. Fed. de Pernambuco, 1977) Assoc. prof. (tel. 081 + 2271208). *X-ray crystallography, powder diffractometry, minerals.*

Campos, Dr Cicero. Dept. de Estado Sólido, Inst. de Física, U. Est. de Campinas, Campinas, São Paulo 13100, Brasil. (1948) PhD, physics (U. Est. de Campinas, 1983). Asst. prof. (tel. 0192 + 391301, ext. 269). *Multiple X-ray scattering, dynamical theory.*

Carvalho da Silva, Prof. Jair. Dept. de Geologia, Escola Nac. de Minas e Metalurgia, Ouro Preto, Minas Gerais, Brasil. DSc, mineralogy (U. Brasil, 1957). Prof.

Cassedane, Dr Jeannine. Inst. de Geociencias, U. Fed. do Rio de Janeiro, Ilha do Fundão, Rio de Janeiro 20000, Brasil. (1927) PhD, solid state (U. Strasbourg, France, 1969). Prof. (tel. 021 + 2305315). *X-ray crystallography, minerals.*

Castellano, Dr Eduardo Ernesto. Dept. de Fisica e Ciencia dos Materiais, Inst. de Fisica e Quimica de São Carlos USP, C.P. 369, São Carlos, São Paulo 13560, Brasil. (1941) PhD, physics (U. Nac. de La Plata, Argentina, 1968). Assoc. prof. (tel. 0162 + 713365). *Direct methods, macromolecules.*

Caticha Alfonso, Mr Ariel. Dept. do Estado Sólido, Inst. de Física, U. Est. de Campinas, Campinas, São Paulo 13100, Brasil. (1955) MSc, physics (U. Est. de Campinas, 1979). Grad. student. (tel. 0192 + 391301). *Dynamical theory.*

Caticha Ellis, Prof. Stephenson. Dept. do Estado Sólido, Inst. de Física U. Est. de Campinas, Campinas, São Paulo 13100, Brasil. (1927) Eng., engineering (U. de Montivideo, Uruguay, 1954). Prof. (tel. 0192 + 391301, ext. 591). *Crystal physics, imperfections, instrumentation*

Chang, Dr Shih- Lin. Dept. do Estado Sólido, Inst. de Física, U. Est. de Campinas, Cidade U., Campinas, São Paulo 13100, Brasil. (1946) PhD, physics (Polytechnic Inst. of New York, USA, 1975). Assoc. prof. (tel. 0192 + 391301, ext. 269). *Dynamical theory, X-ray optics, Liquid phase epitaxy, thin films, instrumentation, X-ray interferometer, applied crystallography, crystal physics.*

Correia Neves, Prof. José Marques. Inst. de Geociencias, U. Fed. de Minas Gerais, Belo Horizonte, Minas Gerais 30000, Brasil. (1929) DSc, mineralogy, geochemistry (U. Coimbra, Portugal, 1963). Prof. Titular UFMG. *Inorganic crystal structures, mineralogy.*

Costa Gouveia, Prof. Albany H. Dept. de Engenharia de Minas, U. Fed. de Pernambuco, Recife, Pernambuco 50000, Brasil. (1941) DSc, mineralogy (U. Fed. de Pernambuco, 1977). Assoc. prof. (tel. 081 + 2315205). *Optical crystallography, mineralogy.*

Costa Viana, Prof. Carlos Sergio da. Coppe, U. Fed. do Rio de Janeiro, Rio de Janeiro, Rio de Janeiro 20000, Brasil. (1942) PhD, philosophy (U. of Cambridge, UK, 1978). Assoc. prof. (tel. 021 + 2809322, ext. 242). *Texture, formability, anisotropy, mechanical properties.*

Craievich, Dr Aldo Felix. Centro Brasileiro de Pesquisas Físicas, Rua Xavier Sigaud 150, Rio de Janeiro, Rio de Janeiro 22290, Brasil. (1939) PhD, Physics (U. Nac. de Cuyo, Argentina, 1969). Pesquisador titular (tel. 021 + 5410337, ext. 174). *Physical crystallography, small angle X-ray scattering, synchrotron radiation.*

Cusatis, Dr Cesar. Dept. de Física, U. Fed. do Paraná, C.P. 19091, Curitiba, Paraná 80000, Brasil. (1939) PhD, Physics (USP, 1973). Assoc. prof. (tel. 041 + 2669271). *X-ray optics, X-ray interferometry.*

Del Nery, Miss Sheila Maria. Dept. de Estado Sólido, Inst. de Física, U. Est. de Campinas, Campinas, São Paulo 13100, Brasil. (1951) MSc, physics (U. Est. de Campinas, 1979). Grad. student. (tel. 0192 + 391301). *Thin films.*

Dias Rodrigues, Mrs Ana Maria Gonçalves. Dept. de Química e Física Molecular, Inst. de Física e Química de São Carlos, São Paulo 13560, Brasil. (1954) MSc, physical chemistry (USP, 1979). Asst. prof. (tel. 0162 + 721538). *Structure determination.*

Diniz de Carvalho Loyolla, Mr Waldomiro Pelágio. Dept. de Física, Fundação Educacional de Bauru, Av. Luiz E. Coube, Bauru, São Paulo 17100, Brasil. (1955) BSc, physics (U. Est. de Campinas, 1976). Grad. student. *Physical crystallography, crystal defects.*

Ferran, Dr Gustan. Coppe, U. Fed. do Rio de Janeiro, Rio de Janeiro, Rio de Janeiro 20000, Brasil. (1938) PhD, metalurgy (U. Madrid, Spain, 1966). Assoc. prof. (tel. 021 + 2609776). *Preferred orientation, texture, Kossel diffraction.*

Ferreira de Souza, Prof. Milton. Dept. de Fisica e Ciencias dos Materiais, Inst. de Física e Química de São Carlos, Av. Dr Carlos Botelho, 1465 São Carlos, São Paulo, Brasil. (1932) PhD, physics (USP, 1969). Prof. (tel. 0162 + 711016). *Solid state physics, defects in solids, crystal growth.*

Figueiredo Neto, Dr Antonio Martins. Dept. de Física Exp., Inst. de Física da USP, Cidade U., São Paulo 05508, Brasil. (1953) PhD, physics (USP, 1981). Asst. prof. (tel. 0162 + 2114865). *Liquid crystals, small angle X-ray scattering.*

Folgueras Dominguez, Dr Sérvulo. Dept. de Química, U. Fed. de São Carlos, São Carlos, São Paulo 13560, Brasil. (1929) Dr, epitaxy (U. Madrid, Spain, 1979). Assoc. (tel. 0162 + 718111, ext. 164). *Silicate structures, epitaxy.*

Formoso, Dr Milton Luiz. Inst. de Geociencias, U. Fed. do Rio Grande do Sul, Porto Alegre, Rio Grande do Sul 90000, Brasil. (1927) PhD, geology (USP, 1973). Prof. (tel. 0512 + 215422). *Geochemistry, mineralogy, geology.*

Francesconi, Dr Ricardo. Inst. de Geociencias, U. de São Paulo, São Paulo, São Paulo 01000, Brasil. (1941) PhD, mineralogy (USP, 1966). Asst. prof. (tel. 011 + 2122011). *Mineralogy, mineral deposits.*

Francisco, Miss Regina Helena Porto. Dept. de Química e Física Molecular, Inst. de Fisica e Química de São Carlos USP, São Carlos, São Paulo 13560, Brasil. (1952) BSc, quimica (Fac. de Filos. Ciencias e Letras de Rib. Preto, 1973). Instr. (tel. 0162 + 712234, ext. 52). *Structure determination.*

Freire D'aguiar, Mr Manoel Marcos. Inst. Fisica, U. Fed. da Bahia, Salvador, Bahia 40000, Brasil. (1947) MSc, geophysics (U. Fed. da Bahia, 1974). Res. assoc. (tel. 071 + 2472714). *Crystal defects.*

Freire Pimentel, Dr Cecilia A. Inst. de Física, U. de São Paulo, São Paulo, São Paulo 01000, Brasil. (1941) PhD, physics (USP, 1963). Asst. prof. (tel. 011 + 2116955, ext. 228). *Crystal defects, line profiles, Bragg X-ray reflections, small angle X-ray scattering.*

Fujimore, Prof. Kenkichi. Inst. de Astronomia e Geofisica USP, Cidade U., São Paulo, São Paulo 01000, Brasil. (1929) PhD, physics (Tohoku U. Sendai, Japan). Prof. *Inorganic crystal structures.*

Fulfaro, Dr Roberto. Inst. de Pesquisas Energéticas e Nucleares, Cidade U., São Paulo, São Paulo 01000, Brasil. (1938) PhD, neutron physics (U. Est. de Campinas, 1970). Head of center (COURP). (tel. 011 + 2116011, ext. 237). *Neutron scattering, lattice dynamics.*

Gomes, Mr Samuel Irati Novaes. Dept. de Materiais, Escola de Engenharia de São Carlos, São Carlos, São Paulo 13560, Brasil. (1939) BSc, physics (Fac. de Filos. Ciencias, Rio Claro). Instr. (tel. 0162 + 712234, ext. 95). *Metallurgy, defects, metals and alloys.*

Grundig, Prof. Werner. Inst. de Tecnologia, U. Fed. do Rio Grande do Sul, Porto Alegre, Rio Grande do Sul 90000, Brasil. MSc, engineer (U. Fed. do Rio Grande do Sul, 1937). Chief eng.

Herdade, Dr Silvio B. Inst. de Física, U. de São Paulo, C.P. 20516, São Paulo, São Paulo 01000, Brasil. (1926) PhD, physics (U. de Campinas, 1969). Assoc. prof. (tel. 011 + 2116955). *Solid state dielectric track detectors, track formation processes, heavy ions, nuclear physics, geophysics.*

Imakuma, Dr Kengo. Inst. de Pesquisas Energéticas e Nucleares, Cidade U., C.P. 11049, São Paulo, São Paulo 05508, Brasil. (1943) PhD, physics (USP, 1973). Head, physics sect. (tel. 011 + 2116011, ext. 242). *Phase transitions, ceramics, alloys, radiation damage.*

Inglez, Mr Antonio Gabriel. Dept. de Mineralogia e Petrografia, Cidade U., São Paulo, São Paulo 010000, Brasil. (1943) BSc, geology. Instr. *Inorganic crystal structures, optical crystallography, computer programming.*

Kunrath, Mr José Irineu. Inst. de Física, U. Fed. do Rio Grande Do Sul, Porto Alegre, Rio Grande Do Sul 90000, Brasil. (1931) MSc, physics (U. Fed. Do Rio Grande Do Sul, 1960). Asst. prof. (tel. 0512 + 245817). *Solid state.*

Labaki, Ms Lucila Chebel. Dept. de Estado Sólido, Inst. de Física, U. Est. de Campinas, C P. 6165, Campinas, São Paulo 13100, Brasil. (1943) MSc, biophysics (Sophia State U., Bulgaria, 1978) Grad. student. (tel. 0192 + 3913015, ext. 269). *Polymers, small angle X-ray scattering.*

Lechat, Dr Johannes Rudiger. Dept. de Química e Física Molecular, Inst. de Física e Química, USP, São Carlos, São Paulo 13560, Brasil. (1943) PhD, chemistry (USP, 1972). Asst. prof. (tel. 0162 + 712234, ext. 52). *Organic crystal structures.*

Leite, Dr Cirano Rocha. Dept. de Química Tecnológica e de Aplicação, Inst. de Química, UNESP, C.P. 174, Araraquara, São Paulo 14800, Brasil. (1941) PhD, mineralogy (USP, 1969). Prof. (tel.0162 + 320444, ext. 136). *Mineralogy, crystal growth, morphology, epitaxy.*

Madureira Filho, Prof. José Barbosa de. Inst. de Geociencias, Dept. de Mineralogia e Petrografia, U. de São Paulo, C.P. 20899, São Paulo, São Paulo 05508, Brasil. (1940) PhD, mineralogy and petrology (USP, 1983). Asst. Prof. (tel. 011 + 2122011). *Solid solutions*

Mascarenhas, Prof. Yvonne Primerano. Dept. de Física e Ciencia Dos Materiais, Inst. de Física e Química, USP, São Carlos, São Paulo 13560, Brasil. (1931) PhD, physics (USP). Prof. (tel. 0162 + 713365). *Crystal structures.*

Mascarenhas, Prof. Sergio. Dept. de Física e Ciencia dos Materiais, Inst. de Física e Química, Campus de São Carlos USP, C.P. 369, São Carlos, São Paulo 13560, Brasil. (1928) PhD, physics (USP, 1958). Prof. (tel. 0162 + 715381). *Biomolecular structure and function, bioelectrets, biophysics, bound water.*

Mazzaro, Mr Irineu. Dept. de Física, U. Fed. do Paraná, C.P. 19091, Curitiba, Paraná 80000, Brasil. (1953) MSc, physics (U. Est. de Campinas, 1979). Jr. lect. (tel. 041 + 2642855). *X-ray optics, multiple diffraction, crystal defects.*

Medeiros Rodrigues, Dr Maria Mabel. Dept. de Química e Física Molecular, Inst. de Física e Química de São Carlos USP, São Carlos, São Paulo 13560, Brasil. PhD, chemistry (USP, 1968). Asst. prof. (tel. 0162 + 712234, ext. 52). *Crystal structure determinations, organic compounds, biological interaction.*

Murta, Prof. Clecio. Companhia Brasileira de Tecnologia Nuclear, U. Fed. de Minas Gerais, Belo Horizonte, Minas Gerais 30000, Brasil. (1929) MSc, nuclear sci. (U. Fed. Minas Gerais, 1971). Head, instrumental analysis lab. (tel. 42-5422). *Diffractometry, X-ray fluorescence, electron microprobe, thermoanalysis, mineralogy.*

Oliveira Lopes, Prof. Cesar. Inst. de Física, Inst. de Ciencias Exatas, U. Fed. Rural do Rio de Janeiro, Km 47 Ant. Estrada Rio-São Paulo, Rio de Janeiro 23800, Brasil. (1949) MSc, physics (U. Est. de Campinas, 1975). Asst. prof. (tel. 021 + 7821220, ext. 227). *Thin films, ferromagnetic alloys, structure.*

Olivieri, Mr Johnny Rizzieri. Dept. de Física e Ciencia dos Materiais, Inst. de Física e Química de São Carlos USP, São Carlos, São Paulo 13560, Brasil. (1950) BSc, physics (U. de São Paulo, 1976). Grad. student. (tel. 0162 + 712234, ext. 33). *Small angle X-ray scattering, amorphous materials.*

Pavie Cardoso, Dr Lisandro. Dept. do Estado Sólido, Inst. de Física, U. Est. de Campinas, Campinas, São Paulo 13100, Brasil. (1950) PhD, physics (U. Est. de Campinas, 1983). Asst. prof. (tel. 0192 + 391301). *Crystal defects, multiple diffraction.*

Quaranta Cabral, Dr Ubirajara. U. Fed. do Rio de Janeiro, Rio de Janeiro, Rio de Janeiro 20000, Brasil. (1937) DSc, metallurgy (U. Paris, France). Asst. prof. (tel. 021 + 2609776). *Metal physics, metals structures.*

Queiroz do Amaral, Dr Lia. Inst. de Física, USP, São Paulo, São Paulo 01000, Brasil. (1941) PhD, physics (USP, 1972). Asst. prof. (tel. 011 + 2116955). *Liquid crystals, small angle X-ray scattering, phase transitions, membranes.*

Ramos Parente, Dr Carlos Benedicto. Inst. de Pesquisas Energéticas e Nucleares, Cidade U., São Paulo, São Paulo 01000, Brasil. (1937) PhD, neutron diffraction (USP, 1973). Res. (tel. 011 + 2116011, ext. 141). *Neutron diffraction, textures.*

Regueira Teodósio, Prof. Joel. U. Fed. do Rio de Janeiro, Rio de Janeiro 20000, Brasil. (1943) MSc, metallurgy (U. Fed. do Rio de Janeiro, 1973). Asst. prof. (tel. 021 + 2609776). *Crystallography, X-ray diffraction, preferred orientation.*

Ribeiro Franco, Prof. Rui. Inst. de Pesquisas Energéticas e Nucleares, Div. de Física Nuclear, Cidade U., São Paulo, São Paulo 01000, Brasil. (1916) PhD, Petrology (USP, 1944). Head, teaching div. (tel. 011 + 2116011). *Crystal growth, optical crystalllography.*

Riella, Eng. Humberto Gracher. Metallurgy Div., Comissão Nac. de Energia Nuclear, Travessa R 400, São Paulo, São Paulo 05508, Brasil. (1953) Eng. (U. Fed. do Paraná, 1975). Res. (tel. 011 + 2116011, ext. 126). *Diffusion.*

Rodrigues da Silva, Dr Rilson. Dept. de Engenharia de Minas, Centro de Tecnologia, U. Fed. de Pernambuco, Recife, Pernambuco 50000, Brasil. (1932) DSc, mineralogy (U. Strasbourg, France, 1969). Prof., crystallography and mineralogy (tel. 081 + 3612789). *X-ray diffraction, X-ray spectroscopy, mineralogy.*

Rodrigues, Dr Antonio Ricardo Dröher, Dept. de Física e Ciencia dos Materiais, Inst. de Física e Química de São Carlos, C.P. 369, São Carlos, São Paulo 13560, Brasil. (1951) PhD, X-ray optics (King's C., London U., UK, 1979) Lect. (tel. 0162 + 712012). *X-ray optics, instrumentation.*

Rodrigues, Dr Edson. Dept. de Química e Física Molecular, Inst. de Física e Química de São Carlos USP, São Carlos, São Paulo 13560, Brasil. PhD, physics. Prof. (tel. 0162 + 711292). *Crystal physics, magnetic resonance.*

Rolim De Camargo, Prof. William Gerson. Inst. de Geosciences, U. of São Paulo, São Paulo 01000, Brasil. (1920) PhD, mineralogy, crystallography (USP, 1944). Prof., mineralogy. (tel. 011 + 2122011). *Mineralogy, uranium minerals, diamond, crystallography, accurate cell dimensions (methods).*

Santos, Dr Persio de Souza. Chemical Eng. Dept., U. de São Paulo, São Paulo 61348, Brasil. (1928) PhD, physical chemistry (USP, 1960). Head, Ind. chem. dept. (tel. 011 + 8159322). *Minerals (non-metallic), clays, zeolites.*

Santos, Mrs Regina Helena de Almeida. Dept. de Química e Física Molecular, Inst. de Física e Química de São Carlos USP, São Carlos, São Paulo 13560, Brasil. (1947) MSc, physical chemistry (USP, 1974). Asst. prof. (tel. 0162 + 712234, ext. 52). *Structural crystallography.*

Simone, Mr Carlos Alberto de. Dept. de Química, U. Fed. de Alagoas, Campus Universitário, Tabuleiro do Martins, Maceió, Alagoas 57000, Brasil. (1951) MSc, appl. physics (USP, 1983). Asst. prof. (tel. 082 + 2421238). *Crystal structure, natural products, powder data refinement.*

Soledade Jr, Prof. Teomar. Inst. de Física, U. Fed. da Bahia, Salvador, Bahia 40000, Brasil. (1948) MSc, Physics (U. Est. de Campinas, 1976). Res. assoc. (tel. 071 + 2472714). *Crystal defects, divergent beam methods.*

Souza, Prof. Irineu Marques. Inst. de Geociencias USP, Dept. de Geologia Economica, Cidade U., São Paulo, São Paulo 01000, Brasil. (1940) BSc, geology. Instr. (tel. 011 + 2122011). *Crystal optics.*

Souza, de, Mr José Carlos. Dept. de Física, Fundação U. Est. de Maringá, Av. Colombo 3690, Campus Univ., Maringá, Paraná 87100, Brasil. (1948) MSc, physics (U. Est. de Campinas, 1979). Grad. student. *Small angle X-ray diffraction, carbonaceous materials.*

Suzuki, Dr Carlos Kenichi. Solid State and Materials Sci., Inst. de Física U. Est. de Campinas, Campinas, São Paulo 13100, Brasil. (1945) PhD, applied physics engineering (U. Tokyo, Japan, 1980). Asst. prof. (tel. 0192 + 391301, ext. 340). *Dynamical X-ray theory, topography.*

Svisero, Dr Darcy Pedro. Dept. of Mineralogy and Petrology, Inst. of Geosciences U. of São Paulo, São Paulo 01000, Brasil. (1940) PhD, geology (U. of São Paulo, 1971). Asst. prof. (tel. 2122011). *Mineralogy.*

Távora, Prof. Elysiario. Comissão Nac. de Energia Nuclear, Rio de Janeiro, Rio de Janeiro 20000, Brasil. DSc, natural sciences (U. Brasil, 1946). Prof., mineralogy, petrography. (tel. 021 + 2867002). *Theoretical and X-ray crystallography.*

Teixeira Mendes, Prof. Antonio Carlos. Escola Superior de Agricultura Luiz de Queiroz, Piracicaba, São Paulo, Brasil. Geology (Escola Superior de Agricultura Luiz de Queiroz). Prof. *Powder method, clay minerals.*

Tomita, Dr Koichi. Dept. de Fisico-Química, Inst. de Química, Araraquara, São Paulo 14800, Brasil. (1943) PhD, chemistry (Fac. de Filos. Ciencias e Letras de Araraquara, 1967). Asst. prof. (tel. 0162 + 320444, ext. 193). *Crystal structures.*

Torriani, Dr Iris Linares. Dept. de Estado Sólido, Inst. de Física, U. Est. de Campinas, C.P. 6165, Campinas, São Paulo 13100, Brasil. (1934) PhD, physics (U. Nac. de La Plata, Argentina, 1975). Assoc. prof. (tel. 0192 + 391301, ext. 269). *Small angle X-ray diffraction, biological applications, membrane structure.*

Valarelli, Dr José Vicente. Inst. de Geociencias, Dept. de Mineralogia e Petrografia, C.P. 20899, São Paulo, São Paulo 01000, Brasil. (1939) PhD, mineralogy (USP, 1967). Assoc. prof. (tel. 011 + 2122011). *Applied mineralogy, mineral paragenesis, fluid inclusions.*

Varela, Mr José Arana. Dept. de Ciencias Exatas, Fac. de Filos. Ciencias e Letras de Presidente Prudente, Presidente Prudente, São Paulo 19100, Brasil. (1944) MSc, Materials science ceramics (Inst. Tecnológico de Aeronáutica, 1975). Asst. prof. (tel. 0182 + 34116). *Materials science, ceramics, transformations.*

Villarroel, Prof. Hugo Sergio. Dept. Eng. Minas, U. Fed. de Pernambuco, Cidade U. Recife, Pernambuco 50000, Brasil. (1931) MSc, crystallography (U. Austral, Chile, 1970). Prof. (tel. 081 + 2271208). *X-ray diffraction, X-ray spectroscopy, minerals.*

Vinhas, Dr Laercio Antonio. Inst. de Pesquisas Energéticas e Nucleares, Cidade U., São Paulo, São Paulo 01000, Brasil. (1943) PhD, neutron physics (U. Est. de Campinas, 1970). Head res. (tel. 011 + 2116011, ext. 139). *Neutron scattering, neutron diffraction, molecular crystals, lattice dynamics.*

Willig, Prof. Cesar Dorneles. Escola de Geologia, U. Fed. do Rio Grande do Sul, Porto Alegre, Rio Grande do Sul 90000, Brasil. (1940) BSc, geology (U. Fed. do Rio Grande do Sul). Instr. *Inorganic structures, computer programming*

BULGARIA

Sub-Editor: I. Bonev

Notes

1. International telephone country code - 359, area code for Sofia - 02.

2. Degrees conferred in Bulgaria are *graduate* (Grad) (refers to a university diploma in physics, geochemistry, engeneering, etc), *candidate of sciences* (CSc) and *doctor of sciences* (DSc). They are approximately equivalent to MSc, PhD and DSc, respectively. The highest honorary degrees conferred by the Bulgarian Academy of Sciences are *academician* (full member) and *corresponding member* of the Bulgarian Academy of Sciences.

3. The positions at the universities and the equivalent positions at the research institutes are, respectively: *assistant - research associate, docent* (associate professor) *- senior scientist* and *professor - research professor*.

Apostolov, Prof. Andrei. Chair of Solid State Physics, Sofia U., Blvd Anton Ivanov 5, Sofia 1126, Bulgaria. (1935) DSc, physics (Sofia U., 1977). Prof. and head. (tel. 623015). *Solid state physics, magnetism.*

Apostolov, Mr Anton. Central Inst. for Computing Technics, Acad G. Bonchev, Bl. 10, Sofia 1040, Bulgaria. (1951) Grad, physics (Sofia U., 1974). Res. assoc. (tel. 7131, ext. 2708). *Powder diffraction, high temperature X-ray experiments.*

Aslanian, Dr Selma. Geological Inst., Bulgarian Acad. Sci., Sofia 1113, Bulgaria. (1937) Dr.rer.nat., mineralogy (Leipzig U., DDR, 1969). Res assoc. (tel. 7131, ext. 3470). *Crystal growth, crystal chemistry.*

Atanassov, Dr Vassil. Chair of Mineralogy and Petrography, Higher Inst. of Mining and Geology, Sofia 1156, Bulgaria. (1933) CSc, mineralogy (Inst. of Geology and Geophysics, Novosibirsk, USSR, 1973). Docent. (tel. 62581, ext. 385). *Mineral morphology.*

Avramov, Dr Isak. Inst. of Physical Chemistry, Bulgarian Acad. Sci., Sofia 1040, Bulgaria. (1946) CSc, chemistry (Inst. of Physical Chemistry, 1980). Res. assoc. (tel. 7131, ext. 2566). *Crystal growth, glass transitions, thin films.*

Balkanov, Mr Ivan. Central Lab. for Electrochemical Power Sources, Bulgarian Acad. Sci., Sofia 1040, Bulgaria. (1949) Grad, geochemistry (Sofia U., 1974). Res. assoc. (tel. 7131, ext. 2714). *X-ray analysis.*

Bliznakov, Prof. Georgi. Inst. of General and Inorganic Chemistry, Bulgarian Acad. Sci., Sofia 1040, Bulgaria. (1920) Academician, chemistry (Bulgarian Acad. Sci., 1979). Prof., director; vice-pres. (tel. 877783). *Crystal growth, adsorption, catalysis, inorganic synthesis.*

Bonev, Dr Ivan. Geological Inst., Bulgarian Acad. Sci., Sofia 1113, Bulgaria. (1936) CSc, mineralogy (Geological Inst., 1972). Sr. scient. (tel. 7131, ext.2236). *Crystal structures, minerals, crystal growth, epitaxy, whiskers, X-ray crystallography.*

Bostanov, Mr Vesselin. Central Lab. for Electrochemical Power Sources, Bulgarian Acad. Sci., Sofia 1040, Bulgaria. (1933) Grad, metallurgy (Higher Inst. of Chemical Techn., 1956). Res. assoc. (tel. 7131, ext. 2761). *Electrocrystallization.*

Budevski, Prof. Evgeni. Central Lab. for Electrochemical Power Sources, Bulgarian Acad. Sci., Sofia 1040, Bulgaria. (1922) Coresp. member, electrochemistry (Bulgarian Acad. Sci., 1984). Director. (tel. 723454). *Electrocrystallization.*

Budurov, Prof. Stoyan. Chair of Inorganic Chemical Techn., Faculty of Chemistry, U. of Sofia, Blvd Anton Ivanov 1, Sofia 1126, Bulgaria. (1930) DSc, chemistry (Sofia U., 1976). Prof. (tel. 62561,ext. 337). *Crystallization, metals, solid state reactions.*

Delineshev, Dr Svetoslav. Inst. of General and Inorganic Chemistry, Bulgarian Acad. Sci., Sofia 1040, Bulgaria. (1940) CSc, chemistry (Inst. of General and Inorganic Chemistry, 1978). Sr. scient. (tel. 7131, ext. 2542). *Crystal growth, nucleation.*

Dimov, Mr Vergil. Inst. of Appl. Mineralogy, Bulgarian Acad. Sci., Moskowska 6, Sofia 1000, Bulgaria. (1946) Grad, physics (Sofia U., 1973). Res. assoc. (tel. 872450). *Electron microscopy.*

Djarova, Dr Maria. Faculty of Chemistry, Sofia U., Blvd Anton Ivanov 1, Sofia 1126, Bulgaria. (1945) CSc, chemistry (Sofia U., 1979). Res. assoc. (tel. 62561, ext. 479). *Growth and perfection of crystals.*

Dobrev, Dr Dobri. Inst. of Physical Chemistry, Bulgarian Acad. Sci., Sofia 1040, Bulgaria. (1935) CSc, chemistry (Inst. of Physical Chemistry, 1976). Sr. scient. (tel. 719307). *Thin films.*

Draganova, Dr Dragana. Faculty of Chemistry, Sofia U., Blvd Anton Ivanov 1, Sofia 1126, Bulgaria. (1936) CSc, chemistry (Sofia U., 1974). Asst. (tel. 62561, ext. 342). *Growth and perfection of crystals.*

Filizova, Dr Lyudmila. Inst. of Appl. Mineralogy, Bulgarian Acad. Sci., Moskowska 6, Sofia 1000, Bulgaria. (1937) CSc, mineralogy (Inst. of Geology and Geophysics, Novosibirsk, USSR). Res. assoc. (tel. 872450). *Synthesis, minerals, crystal growth.*

Grozdanov, Mr Lyudmil. Geological Inst., Bulgarian Acad. Sci., Sofia 1113, Bulgaria. (1934) Grad, geochemistry (Sofia U., 1957). Res. assoc. (tel. 7131, ext. 2260). *X-ray crystallography, amphiboles.*

Gutzow, Prof. Ivan. Inst. of Physical Chemistry, Bulgarian Acad. Sci., Sofia 1040, Bulgaria. (1933) DSc, chemistry (Inst. of Physical Chemistry, 1972). Res. prof. (tel. 719305). *Nucleation, crystal growth, crystallization in glass-forming systems.*

Iwanov, Mr Dantsho. Inst. of Physical Chemistry, Bulgarian Acad. Sci., Sofia 1040, Bulgaria. (1939) Grad, chemistry (Sofia U., 1967). Res. assoc. (tel. 7131, ext. 3584). *Crystal growth, thin films.*

Kaischew, Prof. Rostislaw. Inst. of Physical Chemistry, Bulgarian Acad. Sci., Sofia 1040, Bulgaria. (1908) Academician, physical chemistry (Bulgarian Acad. Sci., 1961). Prof., director. (tel. 727450). *Physical chemistry, crystal growth, nucleation.*

Kashchiev, Dr Dimcho. Inst. of Physical Chemistry, Bulgarian Acad. Sci., Sofia 1040, Bulgaria. (1942) CSc, chemistry (Inst. of Physical Chemistry, 1975). Sr. scient. (tel. 7131, ext. 2557). *Crystal growth, nucleation.*

Kirkova, Prof. Elena. Faculty of Chemistry, Sofia U., Blvd Anton Ivanov 1, Sofia 1126, Bulgaria. (1923) DSc, chemistry (Sofia U., 1983). Prof. (tel. 62561, ext. 215). *Crystal growth, adsorption.*

Kirov, Mr Georgi Kirilov. Inst. of Appl. Mineralogy, Bulgarian Acad. Sci., Moskowska 6, Sofia 1000, Bulgaria. (1932) Res. assoc. (tel. 885115, ext. 671). *Synthesis, minerals, crystal growth, morphology.*

Kirov, Doc Georgi Nikolov. Chair of Mineralogy and Crystallography, Sofia U., Blvd Russki 15, Sofia 1000, Bulgaria. (1930) Docent. (tel. 8581, ext. 256). *X-ray analysis, minerals.*

Konstantinov, Dr Ivan. Central Lab. of Photographic Processes, Bulgarian Acad. Sci., Sofia 1040, Bulgaria. (1943) CSc, chemistry (Inst. of Physical Chemistry, 1975). Res. assoc. (tel. 7131, ext. 3521). *Crystal growth, crystal physics.*

Kostov, Prof. Ivan. Nat. Natural History Museum, Bulgarian Acad. Sci., Blvd Russki 1, Sofia 1000, Bulgaria. (1913) Academician, mineralogy and crystallography (Bulgarian Acad. Sci., 1966). Prof., director. (tel. 882894). *Crystal growth, morphology, minerals, inorganic crystal structures.*

Kostov, Dr Ruslan. Inst. of Appl. Mineralogy, Bulgarian Acad. Sci., Moskowska 6, Sofia 1000, Bulgaria. (1956) CSc, mineralogy (IGEM, Moscow, USSR, 1984). Res. assoc. (tel. 872450). *Physics of minerals, crystal morphology.*

Kotsev, Dr Iosif. Faculty of Physics, Sofia U., Blvd Anton Ivanov 5, Sofia 1126, Bulgaria. (1942) CSc, physics (Moscow U., USSR, 1976). Sr. scient. (tel. 62561, ext. 458). *Symmetry.*

Kovachev, Dr Peter. Faculty of Chemistry, Sofia U., Blvd Anton Ivanov 1, Sofia 1126, Bulgaria. (1943) CSc, chemistry (Inst. of Physical Chemistry, 1975). Res. assoc. (tel. 62561, ext. 337). *Diffusion in metals.*

Krestev, Mr Venelin. Faculty of Physics, Sofia U., Blvd Anton Ivanov 5, Sofia 1126 Bulgaria. (1940) Grad, physics (Polytechn. Inst. Leningrad, USSR, 1966). Res. assoc. (tel. 62561, ext. 464). *Crystal imperfections, polymer physics.*

Kresteva, Dr Manya. Faculty of Physics, Sofia U., Blvd Anton Ivanov 5, Sofia 1126, Bulgaria. (1943) CSc, physics (Sofia U., 1982). Res. assoc. (tel. 62561, ext. 471). *Polymer physics.*

Maciček, Dr Josef. Inst. of Appl. Mineralogy, Bulgarian Acad. Sci., Moskowska 6, Sofia 1000, Bulgaria. (1953) CSc, chemistry (Moscow U., USSR, 1981). Res. assoc. (tel. 872450). *Crystal structure analysis.*

Maleev, Dr Michael. Inst. of Appl. Mineralogy, Bulgarian Acad. Sci., Moskowska 6, Sofia 1000, Bulgaria. (1941) CSc, mineralogy (Moscow U., USSR, 1968). Docent, director (tel. 872450). *Natural whiskers, electron microscopy.*

Malinowski, Prof. Yordan. Central Lab. of Photographic Processes, Bulgarian Acad. Sci., Sofia 1040, Bulgaria. (1923) Corresp. member, chemistry (Bulgarian Acad. Sci., 1979). Prof., director. (tel. 720073). *Crystal physics, crystal growth, lattice defects.*

Marinov, Dr Miko. Inst. of Physical Chemistry, Bulgarian Acad. Sci., Sofia 1040, Bulgaria. (1939) CSc, chemistry (Inst. of Physical Chemistry, 1981). Res. assoc. (tel. 7131, ext. 2533). *Crystal growth, electron microscopy.*

Markov, Dr Ivan. Inst. of Physical Chemistry, Bulgarian Acad. Sci., Sofia 1040, Bulgaria. (1941) CSc, chemistry (Inst. of Physical Chemistry, 1976). Res. assoc. (tel. 7131, ext. 2557). *Crystal growth and nucleation, epitaxy, thin films.*

Michailov, Mr Evgeni. Inst. of Physical Chemistry, Bulgarian Acad. Sci., Sofia 1040, Bulgaria. (1940) Grad, physics (Leningrad U., USSR, 1966). Res. assoc. (tel. 7131, ext. 2565). *Crystal growth, thin films, adsorption.*

Michailov, Mr Michail. Inst. of Solid State Physics, Bulgarian Acad. Sci., Blvd Lenin 72, Sofia 1184, Bulgaria. (1927) Res. assoc. (tel. 7341, ext. 440). *Electron microscopy, semiconductors, thin films.*

Milchev, Dr Alexander. Inst. of Physical Chemistry, Bulgarian Acad. Sci., Sofia 1040, Bulgaria. (1943) CSc, chemistry (Inst. of Physical Chemistry, 1982). Res. assoc. (tel. 7131, ext. 2558). *Crystal growth, electrocrystallization.*

Milchev, Dr Andrei. Inst. of Physical Chemistry, Bulgarian Acad. Sci., Sofia 1040, Bulgaria. (1946) Dr.rer.nat., physics (Leipzig U., DDR, 1979). Res. assoc. (tel. 7131, ext. 2566). *Phase transition, vitrification, two-dimensional systems.*

Miloshev, Prof. Georgi. Inst. of Geophysics, Bulgarian Acad. Sci., Sofia 1113, Bulgaria. (1933) DSc, physics (Inst. of Geophysics, 1974). Res. prof. (tel. 7131, ext. 3334). *Crystal growth.*

Mincheva-Stefanova, Prof. Yordanka. Mineralogy Section, Geological Inst., Bulgarian Acad. Sci., Sofia 1113, Bulgaria. (1923) Res. prof, section head. (tel. 7131, ext. 2282). *Morphology, minerals, crystal growth, sulphide minerals.*

Moldovanova, Prof. Maria. Chair of Semiconductor Physics, Sofia U., Blvd Anton Ivanov 1, Sofia 1126, Bulgaria. (1919) DSc, physics (Sofia U., 1952). Prof., head. (tel. 62561, ext. 322). *Semiconductor physics and technology.*

Nanev, Dr Christo. Inst. of Physical Chemistry, Bulgarian Acad. Sci., Sofia 1040, Bulgaria. (1938) CSc, chemistry (Inst. of Physical Chemistry, 1973). Sr. scient. (tel. 7131, ext. 2586). *Crystal growth, electrocrystallization, thin films, lattice defects.*

Nenow, Dr Dimiter. Inst. of Physical Chemistry, Bulgarian Acad. Sci., Sofia 1040, Bulgaria. (1931) CSc, chemistry (Inst. of Physical Chemistry, 1970). Sr. scient. (tel. 7131, ext. 2557). *Nucleation, crystal growth, crystal surfaces.*

Nikolaeva, Mrs Rumiana. Faculty of Chemistry, Sofia U., Blvd Anton Ivanov 1, Sofia 1126, Bulgaria. (1936) Grad, chemistry (Sofia U., 1959). Asst. (tel. 62561, ext. 342). *Crystal growth, defects.*

Pangarov, Prof. Nikola. Inst. of Physical Chemistry, Bulgarian Acad. Sci., Sofia 1040, Bulgaria. (1929) DSc, chemistry (Inst. of Physical Chemistry, 1968). Prof. (tel. 7131, ext. 2537). *Crystal growth, thin metal films.*

Pashov, Mr Nikolai. Inst. of Solid State Physics, Bulgarian Acad. Sci., Blvd Lenin 72, Sofia 1184, Bulgaria. (1929) Grad, physics (Sofia U., 1952). Sr. scient. (tel. 7341, ext. 641). *Crystal imperfections, electron microscopy, electron diffraction.*

Paunov, Dr Michael. Inst. of Physical Chemistry, Bulgarian Acad. Sci., Sofia 1040, Bulgaria. (1939) CSc, chemistry (Inst. of Physical Chemistry, 1969). Res. assoc. (tel. 7131, ext. 2586). *Crystal growth, heterogeneous nucleation, surfaces.*

Peneva, Dr Stefka. Faculty of Chemistry, Sofia U., Blvd Atnon Ivanov 1, Sofia 1126, Bulgaria. (1937) PhD, physics (U. of Delhi, India, 1970). Sr. scient. (tel. 62561, ext. 281). *Thin film growth, thin film perfection, thin film properties.*

Peshev, Prof Pavel. Inst. of General and Inorganic Chemistry, Bulgarian Acad. Sci., Sofia 1040, Bulgaria. (1933) DSc, chemistry (Inst. of General and Inorganic Chemistry, 1981). Res. prof. (tel. 7131, ext. 2573). *Inorganic synthesis, crystal growth*

Petrov, Dr Kostadin. Inst. of General and Inorganic Chemistry, Bulgarian Acad. Sci., Sofia 1040, Bulgaria. (1940) CSc, chemistry (Inst. of General and Inorganic Chemistry, 1977). Sr. scient. (tel. 7131, ext. 2587). *X-ray crystallography.*

Petrov, Mr Ognyan. Inst. of Appl. Mineralogy, Bulgarian Acad. Sci., Moskowska 6, Sofia 1000, Bulgaria. (1952) Grad, geochemistry (Sofia U., 1980). Mineralogist. (tel. 872450). *Mineral structures.*

Petrov, Dr Srebri. Chair of Mineralogy and Crystallography, Sofia U., Blvd Russki 15, Sofia 1000, Bulgaria. (1942) CSc, mineralogy (Sofia U., 1984). Res. assoc. (tel. 8581, ext. 408). *X-ray analysis, minerals.*

Petrov, Dr Vasko. Inst. of General and Inorganic Chemistry, Bulgarian Acad. Sci., Sofia 1040, Bulgaria. (1941) CSc, techn. sciences (Moscow Inst. of Energetics, USSR, 1975). Res. assoc. (tel. 7131, ext. 2798). *Crystal growth.*

Philipov, Mr Alexander. Chair of Mineralogy and Crystallography, Sofia U., Blvd Russki 15, Sofia 1000, Bulgaria. (1944) Grad, mineralogy (Sofia U., 1972). Asst. (tel. 8581, ext. 256). *Crystal growth, minerals.*

Platikanova, Dr Vesselina. Central Lab. of Photographic Processes, Bulgarian Acad. Sci., Sofia 1040, Bulgaria. (1939) CSc, chemistry (Inst. of Physical Chemistry, 1971). Sr. scient. (tel. 723713). *Crystal physics, crystal growth.*

Popov, Dr Alexander. Central Lab. for Electrochemical Power Sources, Bulgarian Acad. Sci., Sofia 1040, Bulgaria. (1942) CSc, chemistry (Inst. of Physical Chemistry, 1980). Res. assoc. (tel. 7131, ext. 2758). *Electrocrystallization.*

Poulieff, Mr Christo. Central Lab. for Electrochemical Power Sources, Bulgarian Acad. Sci., Sofia 1040, Bulgaria. (1936) Grad, geochemistry (Sofia U., 1958). Res. assoc. (tel. 7131, ext. 2707). *X-ray powder diffraction, infrared spectra, inorganic materials.*

Rachev, Mr Peter. Central Inst. for Computing Technics, Acad G. Bonchev, Bl. 10, Sofia 1040, Bulgaria. (1957) Grad, physics (Sofia U., 1979). Res. assoc. (tel. 7131, ext. 3991). *Electron microscopy, electron microanalysis.*

Rainov, Dr Nikola. Branch of Lab. Investigations, Geological Survey, G. A. Nassar 16, Sofia 1113, Bulgaria. (1939) CSc, mineralogy (Sofia U., 1975). Head of lab. (tel. 723651, ext. 032). *Feldspars, x-ray powder diffraction, infrared spectra, minerals, instrumentation.*

Rashkova, Dr Diana. Geological Inst., Bulgarian Acad. Sci., Sofia 1113, Bulgaria. (1952) CSc, mineralogy (Geological Inst., 1982). Res. assoc. (tel. 7131, ext. 2236). *Morphology, minerals.*

Russev, Dr Krassimir. Faculty of Chemistry, Sofia U., Blvd Anton Ivanov 1, Sofia 1126, Bulgaria. (1946) CSc, chemistry (Inst. of Physical Chemistry, 1975). Sr. scient. (tel. 62561, ext. 337). *Solid state reactions.*

Simov, Mr Stefan. Inst. of Solid State Physics, Bulgarian Acad, Sci., Blvd Lenin 72, Sofia 1184, Bulgaria. (1934) Grad, physics (Sofia U., 1960). Res. assoc. (tel. 7341, ext. 340). *Crystal growth, whiskers, thin films, electron microscopy.*

Staikov, Dr Georgy. Central Lab. for Electrochemical Power Sources, Bulgarian Acad. Sci., Sofia 1040, Bulgaria. (1943) CSc, chemistry (Inst. of Physical Chemistry, 1981). Res. assoc. (tel, 7131, ext. 2773). *Electrocrystallization.*

Stefanov, Mr Dechko Dimitrov. Geological Inst., Bulgarian Acad. Sci., Sofia 1113, Bulgaria. (1931) Grad, physics (Sofia U., 1954). Sr. scient. (tel. 7131, ext. 2207). *Clay minerals.*

Stefanov, Dr Stefan Rashkov. Inst. of Physical Chemistry, Bulgarian Acad. Sci., Sofia 1040, Bulgaria. (1935) DSc, chemistry (Inst. of Physical Chemistry, 1984). Sr. scient. (tel. 7131, ext. 3537). *Thin metal films, corrosion.*

Stoinova, Dr Margarita. Chair of Mineralogy and Petrography, Higher Inst. of Mining and Geology, Sofia 1156, Bulgaria. (1930) CSc, mineralogy (Sofia U., 1966). Docent. (tel. 62581, ext. 384). *Mineral morphology.*

Stoyanov, Dr Stoyan. Inst. of Physical Chemistry, Bulgarian Acad Sci., Sofia 1040, Bulgaria. (1941) CSc, physics (Inst. of Physical Chemistry, 1977). Sr. scient. (tel. 7131, ext. 2557). *Crystal growth.*

Stoychev, Mr Nikola. Inst. of Metal Sci. and Techn., Bulgarian Acad. Sci., Tchapaev 53, Sofia 1040, Bulgaria. (1939) CSc, physics (Inst. of Metallophysics, Kiev, USSR, 1977). Sr. scient. (tel. 71421, ext. 337). *Crystallization, metals.*

Tchehlarova, Mrs Irina. Branch of Lab. Investigations, Geological Survey, G. A. Nassar 16, Sofia 1113, Bulgaria. (1938) Grad, geochemistry (Sofia U., 1961). Geochemist. (tel. 723651, ext. 033) *X-ray powder diffraction, infrared spectra.*

Tchuneva, Mrs Vassilka. Branch of Lab. Investigations, Geological Survey, G. A. Nasser 16, Sofia 1113, Bulgaria. (1939) Grad, geochemistry (Sofia U., 1962). Geochemist. (tel. 723651, ext. 033). *X-ray powder methods.*

Tomov, Dr Ivan. Inst. of Physical Chemistry, Bulgarian Acad. Sci., Sofia 1040, Bulgaria. (1939) CSc, chemistry (Inst. of Physical Chemistry, 1981). Res. assoc. (tel. 7131, ext. 2533). *Textures, electrolytic layers.*

Topalova-Kalitzova, Mrs Maria. Inst. of Solid State Physics, Bulgarian Acad, Sci., Blvd Lenin 72, Sofia 1184, Bulgaria. (1940) Grad, physics (Sofia U., 1963). Res. assoc. (tel. 7341, ext. 340). *Electron microscopy, implantation, lattice defects.*

Toshev, Dr Alexander. Inst. of General and Inorganic Chemistry, Bulgarian Acad. Sci., Sofia 1040 Bulgaria. (1942). CSc, chemistry (Inst. of General and Inorganic Chemistry, 1978). Res. assoc. (tel. 7131, ext. 2563). *Crystal growth.*

Tsolovski, Dr Ilcho. Inst. of General and Inorganic Chemistry, Bulgarian Acad. Sci., Sofia 1040, Bulgaria. (1937) CSc, physics (Inst. of General and Inorganic Chemistry, 1976). Sr. scient. (tel. 7131, ext 2587). *X-ray crystallography.*

Vassilev, Dr Ivan. Inst. of Solid State Physics, Bulgarian Acad. Sci., Blvd Lenin 72, Sofia 1184, Bulgaria. (1932) CSc, physics (Inst. of Solid State Physics, 1976). Sr. scient. (tel. 7341, ext. 265). *X-ray diffraction, X-ray topography, lattice defects.*

Vesselinov, Mr Iliya. Inst. of Appl. Mineralogy, Bulgarian Acad. Sci., Moskowska 6, Sofia 1000, Bulgaria. (1942) Grad, geochemistry (Sofia U., 1967). Res. assoc. (tel. 885115, ext. 671). *Mineral synthesis, crystal growth, morphology.*

Vitanov, Dr Todor. Central Lab. for Electrochemical Power Sources, Bulgarian Acad. Sci., Sofia 1040, Bulgaria. (1926) CSc, chemistry (Inst. of Physical Chemistry, 1973). Sr. scient. (tel. 7131, ext. 2746). *Electrocrystallization, adsorption.*

Yaneva, Dr Svetlana. Inst. of Metal Sci. and Techn., Bulgarian Acad. Sci., Tchapaev 53, Sofia 1574, Bulgaria. (1942) CSc, chemistry (Sofia U., 1970). Res. assoc. (tel. 71421, ext. 337). *Eutectic crystallization.*

Zidarova, Mrs Bogdana. Inst. of Appl. Mineralogy, Bulgarian Acad. Sci., Moskovska 6, Sofia 1000, Bulgaria. (1943) Grad, geochemistry (Sofia U., 1971). Geochemist. (tel. 885115, ext. 671). *Synthesis, morphology, minerals.*

Zotov, Mr Nikolay, Geological Inst., Bulgarian Acad. Sci., Sofia 1113, Bulgaria. (1957) Grad, physics (Sofia U., 1982). Res. assoc. (tel. 7131, ext. 2232). *X-ray diffraction methods, computing, instrumentation.*

BURMA

Sub-Editor: A.P. In

Notes

1. International telephone country code - 95.

Htoon, Dr Sein. Physics Dept., U. of Rangoon, University P.O., Rangoon, Burma. (1941) FInstP(DSc), physics (Inst. of Physics, UK, 1982). Lecturer. (tel. 31144, ext. Physics). *Theoretical physics, crystallography.*

In, Mr Aung Paik. Physics Dept., U. of Rangoon, University P.O., Rangoon, Burma. (1939) MSc, X-ray crystallography (Rangoon Arts and Science U.). Asst. lecturer. (tel. 31144, ext. Physics). *Theoretical physics, crystallography.*

Kyaw, Dr Htin. Physics Dept., Mandalay U., Mandalay, Burma. (1939) PhD, solid state electronics (U. of Salford, UK, 1971). Lecturer. *Field ion microscopy, crystallography, materials science.*

Mya Mya, Dr Khin. Physics Dept., Inst. of Medicine (I), Rangoon, Burma. (1948) PhD, X-ray crystallography (U. of York, UK, 1976). Asst. lecturer. *Direct methods, X-ray crystallography, Molecular biology (precursory), Astronomy.*

Tun, Mr Saw. Dept. of Analysis, Central Research Organisation, Kabaaye-Pagoda-Kanbe Roads, Rangoon, Yankin P.O., Burma. (1930) BSc, chemistry (U. of Rangoon, 1953). Sr. sci. (tel. 50544, ext. 24). *X-ray spectroscopy, powder diffraction, diffraction methods, analytical chemistry.*

Yin, Dr Soe. Physics Dept., Magwe College, Magwe, Burma. (1941) Dottore, theoretical solid state physics (Trieste U., Italy, 1980). Lecturer. (tel. 31144, ext. Physics). *Solid State Theory, X-ray line broadening.*

CANADA

Sub-Editors: C.P. Huber and M.E. Pippy

Notes

1. International telephone country code - 001.

2. Abbreviations used for the Canadian provinces are as follows: Alta., Alberta; B.C., British Columbia; Man., Manitoba; N.B., New Brunswick; Nfld., Newfoundland; N.S., Nova Scotia; Ont., Ontario; P.E.I., Prince Edward Island; Qué., Québec; Sask., Saskatchewan.

Ahmed, Dr Farid Ramadan. Div. of Biological Sci., Nat. Res. Council of Canada, Ottawa, Ont. K1A 0R6, Canada. (1924) PhD, physical chemistry (U. Leeds, UK, 1953). Principal res. officer. (tel. 613 + 990-0858). *Computing, direct methods, organic and biological structures.*

Allaire, Mr François. Dépt. de Chimie, U. de Montréal, C.P. 6210, Succ. A, Montréal, Qué. H3C 3V1, Canada. (1960) MSc, chemistry (U. de Montréal, 1984). Grad. student. (tel. 514 + 351-1099). *Organometallic chemistry.*

Altermatt, Dr Urs Daniel. Inst. for Materials Res., McMaster U., Hamilton, Ont. L8S 4M1, Canada. (1955) Dr.sc.nat. crystallography (ETH Zürich, Switzerland, 1983). Postdoctoral fellow. (tel. 416 + 525-9140, ext. 4563). *Incommensurate structures, intercalated graphites, inorganic structure systematics, structure prediction.*

Anderson, Dr Wayne Foster. Dept. of Biochemistry, U. of Alberta, Edmonton, Alta. T6G 2H7, Canada. (1948) PhD, molecular biophysics and biochemistry (Yale U., USA, 1975). Assoc. prof. (tel. 403 + 432-2136). *Protein structure and function; nucleic acid binding proteins.*

Araki, Dr Takaharu. Dept. of Geological Sci., McGill U., 3450 University St., Montreal, Que. H3A 2A7, Canada. (1929) DSc, mineralogy (U. Kyoto, Japan, 1961). Res. assoc. (tel. 514 + 392-5840). *Mineral and inorganic structures, computer programming.*

Argo, Dr James L. Canadian Conservation Inst., 1030 Innes Road, Ottawa, Ont. K1A 0M8, Canada. (1937) PhD, solid state chemistry (Cambridge U., UK, 1971). (tel. 613 + 998-3721). *Iron, oxy-hydrides and oxides, corrosion in metals, historic objects, art conservation, bronze, pewter, lead, archeometallurgy.*

Ashmore, Dr John Patrick. 1245 Pebble Road, Ottawa, Ont. K1V 7S1, Canada. (1942) PhD, physics (McMaster U., 1971). (tel. 613 + 998-4751). *Inorganic and organic structure analysis.*

Bagchi, Prof. Subodh Nath. 5550 Bellerive, Brossard, Qué. J4Z 3C8, Canada. (1915) DSc, chemistry (U. Calcutta, India, 1946). Retired. *Amorphous structures, small angle scattering.*

Ball, Dr Richard George. Chemistry Dept., U. of Alberta, Edmonton, Alta. T6G 2G2, Canada. (1950) PhD, chemical crystallography (U. Western Ontario, 1978). Dir., crystallographic lab. *Crystallography, structure.*

Baranyi, Dr Anthony David. Glass & Ceramics Div., Ont. Res. Foundation, Sheridan Park, Mississauga, Ont. L5K 1B3, Canada. (1951) PhD, inorganic chemistry (McGill U., 1976). Assoc. res. sci. (tel. 416 + 822-4111, ext. 323). *Glass, glass - ceramics, ceramics, nuclear waste immobilization, solar thermal energy conversion, toxic materials release.*

Barrington-Leigh, Dr John. Immunology Dept., U. of Alberta, 8-29 Medical Sci. Bldg., Edmonton, Alta. T6G 2H7, Canada. (1943) PhD, natural science, physics (Cambridge U., UK, 1970). Assoc. prof.; Adj. prof. (pharmacy) (tel. 403 + 432-2275). *Synchrotron X-ray Optics, Biological structures, fibre diffraction, energy transduction, structure prediction, immune regulation, cancer diagnostics.*

Bartlett, Dr Michael William. GBG Information Systems, IBM Canada Ltd., 101 Valleybrook Drive, Don Mills, Ont. M3B 3H1, Canada. (1943) PhD, X-ray crystallography (U. Waterloo, 1970). Programmer analyst. (tel. 416 + 443-3100).

Barton, Prof. Richard J. Dept. of Physics, U. of Regina, Regina, Sask. S4S 0A2, Canada. (1928) PhD, chemistry (Iowa State U., USA, 1956). Assoc. prof. (tel. 306 + 584-4653). *Coordination compounds, small molecules.*

Bayliss, Prof. Peter. Dept. of Geology and Geophysics, U. of Calgary, Calgary, Alta. T2N 1N4, Canada. (1936) PhD, mineralogy (U. New South Wales, Australia, 1967). Prof. (tel. 403 + 284-5026). *Mineralogy.*

Beauchamp, Prof. Dr André. Dépt. de Chimie, U. de Montréal, C.P. 6210, Succ. A, Montréal, Qué. H3C 3V1, Canada. (1940) PhD, chemistry (U. de Montréal, 1967). Prof. (tel. 514 + 343-7604). *Structure, bio-inorganic interesting compounds.*

Belanger-Gariepy, Mme Francine. Dépt. de Chimie, U. de Montréal, C.P. 6210, Succ. A, Montréal, Qué. H3C 3V1, Canada. (1955) MSc, chemistry (U. de Montréal, 1981). Res. asst. (tel. 514 + 343-7538). *Syntheses, characterization, crystal structure determination, organometallic and organic compounds.*

Bird, Prof. Peter Hans. Dept. of Chemistry, Concordia U., 1455 de Maisonneuve Blvd. W., Montréal, Qué. H3G 1M8, Canada. (1942) PhD, inorganic chemistry (U. Sheffield, UK, 1966). Assoc. prof. (tel. 514 + 879-4451). *Organometallic complexes of transition metals, small molecules.*

Birnbaum, Dr George I. Div. of Biological Sci., Nat. Res. Council of Canada, Ottawa, Ont. K1A 0R6, Canada. (1931) PhD, chemistry (Columbia U., USA, 1961). Sr. res. officer. (tel. 613 + 990-3245). *Biological and organic structures.*

Birnbaum, Dr Karin Bjåmer. 17 1/2 Cedar Rd., Ottawa, Ont. K1J 6L6, Canada. PhD, crystallography (U. Glasgow, UK, 1968). (tel. 613 + 746-0676). *Biological and organic structures.*

Bluhm, Dr Terry Lee. Xerox Res. Centre of Canada, 2660 Speakman Dr., Mississauga, Ont. L5K 2L1, Canada. (1947) PhD, polymer chemistry (State U. of New York, USA, 1976). Scient. staff member. (tel. 416 + 823-7091). *Polymer crystallography; polysaccharide structures; photoconductors.*

Boldrini, Dr Piero. 44 Robert St., Apt. 202, Hamilton, Ont. L8L 7Z8, Canada.

Boorman, Dr Philip Michael. Chemistry Dept., U. of Calgary, Calgary, Alta. T2N 1N4, Canada. (1939) PhD, inorganic chemistry (U. Nottingham, UK, 1964). Assoc. prof. (tel. 403 + 284 - 5347). *Tungsten complexes with sulfur ligands, coordinated sulfur-donor ligands, structure and reactions.*

Booth, Dr Andrew Donald. c/o Autonetics Res. Assoc. Inc., 5317 Sooke Rd., R.R. 1, Sooke, B.C. VOS 1N0, Canada. (1918) DSc, physics (U. London, UK, 1951). Retired. (tel. 604 + 642-5352). *Computer applications to crystallography, numerical analysis, sound propagation in the ocean.*

Bottomley, Prof. Frank. Dept. of Chemistry, U. of New Brunswick, P.O. Box 4400, Fredericton, N.B. E3B 5A3, Canada. (1941) PhD, inorganic chemistry (U. Toronto, 1968). Prof. (tel. 506 + 453-4777). *Inorganic chemistry, structural chemistry, catalysis.*

Brandon, Dr James Kenneth. Physics Dept., U. of Waterloo, Waterloo, Ont. N2L 3G1, Canada. (1940) PhD, physics (McMaster U., 1967). Assoc. prof. (tel. 519 + 885-1211, ext. 2719). *Inorganic structures, metal and alloy structures.*

Brayer, Dr Gary David. Dept. of Biochemistry, U. of British Columbia, 2146 Health Sciences Mall, Vancouver, B.C. V6T 1W5, Canada. (1953) PhD, biochemistry (U. Alberta, 1979). Asst. prof. (tel. 604 + 228-5216). *Macromolecular crystallography, enzyme structure and function, protein- nucleic acid interactions and complexation phenomena.*

Brierley, Mr Cameron. Dept. of Chemistry, U. of Regina, Regina, Sask. S4S 0A2, Canada. (1953) BSc, chemistry (U. Saskatchewan, 1975). Grad. student. (tel. 306 + 584-4653). *Organic structures.*

Brisse, Dr François. Dépt. de Chimie, U. de Montréal, C.P. 6210 Succ. A, Montréal, Qué. H3C 3V1, Canada. (1935) PhD, chemistry (Dalhousie U., 1967). Prof. (tel. 514 + 343-7604). *Crystal structure determination, model molecules related to polymers, polymer structures (polyesters and polysaccharides), biologically interesting compounds, alkaloids, antibiotics, steroids, electron microscopy, electron diffraction.*

Brisson, Mrs Josée. Dépt. de Chimie, U. de Montréal, C.P. 6210, Succ. A, Montréal, Qué. H3C 3V1, Canada. (1960) MSc, physical chemistry (U. de Montréal, 1982). Grad. student. (tel. 514 + 343-7538). *Polymers, macromolecules, amides.*

Britten, Dr James Francis. Dépt. de Chimie, U. de Montréal, C.P. 6210, Succ. A, Montréal, Qué. H3C 3V1, Canada. (1955) PhD, chemistry (McMaster U., 1984). Postdoctoral fellow. (tel. 514 + 343-7538). *Chemistry, inorganic, bioinorganic crystallography; nuclear magnetic resonance, computer programming.*

Brown, Dr Ian David. Inst. for Materials Res., McMaster U., Hamilton, Ont. L8S 4M1, Canada. (1932) PhD, crystallography (U. London, UK, 1959). Prof. of physics. (tel. 416 + 525-9140, ext. 4710, telex 0618347). *Chemical bonding, inorganic solids, crystallographic data retrieval.*

Bushnell, Prof. Gordon William. Dept. of Chemistry, U. of Victoria, Victoria, B.C. V8W 2Y2, Canada. (1936) PhD, inorganic chemistry (U. West Indies, West Indies, 1966). Assoc. prof. (tel. 604 + 721-7163). *Structural inorganic chemistry, single crystal X-ray diffraction, coordination compounds.*

Buyers, Dr William James Leslie. Neutron & Solid State Physics Div., Atomic Energy of Canada Ltd., Chalk River, Ont. K0J 1J0, Canada. (1937) PhD, physics (U. Aberdeen, UK, 1963). Sr. res. officer. (tel. 613 + 584-3311, ext. 609 or 569). *Neutron scattering, structural phase transitions, phonons, spin waves, disordered materials, liquids, crystalline electric field effects; positron annihilation studies, defects.*

Camerman, Dr Norman. Dept. of Biochemistry, U. of Toronto, Toronto, Ont. M5S 1A8, Canada. (1939) PhD, chemistry (U. British Columbia, 1964). Prof. (tel. 416 + 978-7027). *Biological structures, biological structure - activity relation, molecular design.*

Cameron, Prof. Theodore Stanley. Dept. of Chemistry, Dalhousie U., Halifax, N.S. B3H 4J3, Canada. (1942) DPhil, chemistry (Oxford U., UK, 1969). Prof. (tel. 902 + 424-3305). *Hydrogen bonding, molecular packing, graphical investigation, inorganic and organometallic structures, sulphur structures, silicon chemistry.*

Carty, Prof. Arthur John. Guelph-Waterloo Centre for Graduate Work in Chemistry, Waterloo Campus, U. of Waterloo, Waterloo, Ont. N2L 3G1, Canada. (1940) PhD, inorganic chemistry (U. Nottingham, UK, 1965). Prof. and Dir. (tel. 519 + 885-3544, ext. 3296). *Organometallic chemistry, phosphine and acetylene complexes, phosphinoacetylenes, mercury, cadmium, environmental pollution, structure determination by physical methods.*

Castelliz, Dr Karoline (Lotte) Maria. Dept. of Chemical Engineering, Nova Scotia Techn. C., Halifax, N.S. B3J 2X4, Canada. (1905) PhD, physics & mathematics (U. Vienna, Austria, 1932). Res. prof. (tel. 902 + 429-8300, ext. 266). *Structure - physical properties relationship, ferromagnetic alloys, ferrimagnetic oxides, ferroelectric oxides, semiconducting oxides.*

Černý, Dr Petr. Dept. of Earth Sci., U. of Manitoba, Winnipeg, Man. R3T 2N2, Canada. (1934) PhD, mineralogy (Acad. of Sci. of CSSR, Czechoslovakia, 1966). Prof. (tel. 204 + 474-8252). *Morphology, structure, crystal chemistry, oxide minerals, silicate minerals.*

Chao, Dr George Y. Geology Dept., Carleton U., Ottawa, Ont. K1S 5B6, Canada. (1930) PhD, mineralogy & crystallography (U. Chicago, USA, 1958). Prof. of geology. (tel. 613 + 231-3885). *Structure, minerals.*

Charland, Dr Jean-Pierre. Dépt. de Chimie, U. de Montréal, C.P. 6210, Succ. A, Montréal, Qué. H3C 3V1, Canada. (1957) PhD, chemistry (U. de Montréal, 1984). Student. (tel. 514 + 343-7538). *Chemistry, inorganics, biomolecules, bioinorganic chemistry.*

Cheng, Dr Pei-Tak. Mount Sinai Hospital, Dept. of Labs., 600 University Ave., Toronto, Ont. M5G 1X5, Canada. (1940) PhD, crystallography (U. Toronto, 1972). Staff sci. (tel. 416 + 596-4468). *Physical chemistry, ultrastructure, crystalline deposit diseases, tissue crystallography.*

Chieh, Prof. Chung (Peter). Dept. of Chemistry, U. of Waterloo, Waterloo, Ont. N2L 3G1, Canada. (1939) PhD, chemistry (U. British Columbia, 1969). Assoc. prof. (tel. 519 + 885-1211, ext. 3119). *Coordination, sulfur-containing ligands.*

Clark, Dr Malcolm John Roy. Waste Management Branch, Ministry of Environment, 810 Blanchard St., Victoria, B.C. V8V 1X5, Canada. (1944) PhD, chemistry (U. New Brunswick, 1971). Branch environmental chemist. (tel. 604 + 387-4321, ext. 249). *Environmental chemistry, metals toxicity, computing.*

Cobbledick, Dr Roger Ernest. Bureau of Medical Devices, Environmental Health Centre, Ottawa, Ont. K1A 0Z2 Canada. (1943) PhD, X-ray crystallography (U. Lancaster, UK, 1969). Device evaluator. (tel. 613 + 993-6567). *Organic and organometallic structures, computing, clinical chemistry.*

Codding, Dr Penelope Wixson. Dept. of Chemistry, U. of Calgary, Calgary, Alta. T2N 1N4, Canada. (1946) PhD, chemistry (Michigan State U., USA, 1971). Asst. prof. (tel. 403 + 284-7549). *Structure - activity relationships, drugs, neurotoxins, peptides; DNA and protein crystallography.*

Corbeil, Miss Marie-Claude. Dépt. de Chimie, U. de Montréal, C.P. 6210, Succ. A, Montréal, Qué. H3C 3V1, Canada. (1960) MSc, chemistry (U. de Montréal, 1984). Grad. student. (tel. 514 + 343-7538). *Inorganic chemistry.*

Corlett, Dr Mabel Isobel. Dept. of Geological Sci., Queen's U., Kingston, Ont. K7L 3N6, Canada. (1939) PhD, mineralogy (U. Chicago, USA, 1964). Assoc. prof. (tel. 613 + 547-6218). *Mineral chemistry.*

Cowie, Prof. Martin. Dept. of Chemistry, U. of Alberta, Edmonton, Alta. T6G 2G2, Canada. (1947) PhD, X-ray crystallography (U. Alberta, 1974). Assoc. prof. (tel. 403 + 432-5581). *Small molecule activation and catalysis, transition metal complexes, activation, metal-metal cooperativity, binuclear complexes, polynuclear complexes.*

Curzon, Prof. Albert Edward. Physics Dept., Simon Fraser U., Burnaby, B.C. V5A 1S6, Canada. (1934) PhD, physics (Imperial C., London, UK, 1959). Prof. (tel. 604 + 291-4181). *Electron microscopy, electron diffraction, thin film crystal structures.*

Davis, Mr Alan Ross. Dept. of Chemistry, Simon Fraser U., Burnaby, B.C. V5A 1S6, Canada. (1950) MSc, chemistry (Simon Fraser U., 1975). Grad student. (tel. 604 + 291-4878). *Structural inorganic chemistry.*

Deguire, Mrs Suzanne. Dépt. de Chimie, U. de Montréal, C.P. 6210, Succ. A, Montréal, Qué. H3C 3V1, Canada. (1962) BSc, physical chemistry (U. de Montréal, 1984). Grad. student. (tel. 514 + 343-7538). *Polymers, polyesters.*

Delbaere, Dr Louis Theophil Joseph. Dept. of Biochemistry, U. of Saskatchewan, Saskatoon, Sask. S7N 0W0, Canada. (1943) PhD, chemistry (U. Manitoba, 1970). Prof. (tel. 306 + 966-4366). *Crystal structures, proteins, biologically important molecules; protein structure - function relationship.*

Deslandes, Dr Yves. Xerox Res. Centre of Canada, 2660 Speakman Dr., Mississauga, Ont. L5K 2L1, Canada. (1952) PhD, chemistry (U. de Montréal, 1980). Res. Scient. (tel. 416 + 823-7091). *Electron microscopy, polymer crystallography.*

Dichmann, Dr Klaus. Chemistry Dept. Vanier C., 821 Ste-Croix Blvd, Saint-Laurent, Qué. H4L 3X9, Canada. (1942) PhD, X-ray crystallography (U. Toronto, 1972). Prof. (tel. 514 + 488-2341, ext. 4052). *Crystal structure analysis, structural inorganic chemistry.*

Dion, Mrs Chantal. Dépt. de Chimie, U. de Montréal, C.P. 6210, Succ. A, Montréal, Qué. H3C 3V1, Canada. (1956) MSc, chemistry (U. du Québec à Montréal, 1983). Grad. student. (tel. 514 + 343-7538). *Platinum compounds, antitumoral activity.*

Dolling, Dr Gerald. Physics Div., Atomic Energy of Canada Ltd., Chalk River Nuclear Labs., Chalk River, Ont. K0J 1J0, Canada. (1935) PhD, physics (Cambridge U., UK, 1961). Sr. res. officer. (tel. 613 + 584-3311, ext. 2568, telex. 053-34555). *Neutron and X-ray diffraction; excitations in condensed matter, phonons, magnons; neutron inelastic scattering techniques.*

Donnay, Prof. Gabrielle. Dept. of Geological Sci., McGill U., 3450 University Street, Montréal, Qué. H3A 2A7, Canada. (1920) PhD, crystallography (Massachusetts Inst. of Techn., USA, 1948). Prof. of crystallography. (tel. 514 + 392-5840). *Crystal structure - physical and chemical properties relationships.*

Donnay, Prof. Joseph Désiré Hubert. Dept. of Geological Sci., McGill U., 3450 University Street, Montréal, Qué. H3A 2A7, Canada. (1902) PhD, geology (Stanford U., USA, 1929). Emeritus, The Johns Hopkins U.; res. assoc., McGill U. (tel. 514 + 392-5840). *Crystal morphology, morphology - structure relationships, symmetry, crystalline aggregates.*

Drake, Prof. John E. Chemistry Dept., U. of Windsor, Windsor, Ont. N9B 3P4, Canada. (1936) DSc, inorganic chemistry (U. Southampton, UK, 1978). Prof. & dept. head. (tel. 519 + 253-4232, ext. 2521). *Spectroscopy, structure, organometallics and coordination compounds.*

Duke, Ms Norma Edith. Dept. of Chemistry, U. of Calgary, Calgary, Alta. T2N 1N4, Canada. (1959) BSc, chemistry (U. Calgary, 1982). Grad. student. (tel. 403 + 284-5069).

Edwards, Dr William Donald. Communications Res. Centre, Dept. of Communications, P.O. Box 11490, Station H, Ottawa, Ont. K2H 8S2, Canada. (1926) PhD, physics (U. Birmingham, UK, 1949). Group leader. (tel. 613 + 596-9486). *Electronic crystalline materials, silicon, gallium arsenate, preparation and processing, structural properties, defects, X-ray topography, crystallography.*

Einstein, Prof. Frederick William Boldt. Dept. of Chemistry, Simon Fraser U., Burnaby, B.C. V5A 1S6, Canada. (1940) PhD, inorganic structural chemistry

(U. Canterbury, New Zealand, 1965). Prof. (tel. 604 + 291-3594). *Inorganic structural chemistry, crystallographic computing, low temperature crystallography.*

El-Kabbani, Mr Ossama Ahmed Lofti. Dept. of Biochemistry, U. of Saskatchewan, Saskatoon, Sask. S7N 0W0, Canada. (1959) BSc (Hon.), biochemistry (U. Alexandria, Egypt, 1980). Grad. student. (tel. 306 + 966-4383). *Proteins, biologically important molecules.*

Ercit, Mr Timothy Scott. Dept. of Earth Sci., U. of Manitoba, Winnipeg, Man. R3T 2N2, Canada. (1957) BSc, geology (Laurentian U., 1980). PhD candidate. (tel. 204 + 474-9452). *Pegmatite mineralogy (oxides & phosphates), inorganic structures, paragenetic mineralogy.*

Faerman, Mr Carlos Hugo. Dept. of Chemistry, U. of Toronto, 80 St. George St., Toronto, Ont. M5S 1A1, Canada. (1954) Lic, physics (U. Buenos Aires, Argentina, 1978). Grad. student. (tel. 416 + 978-6275). *Intermolecular forces, molecular crystals.*

Faggiani, Mr Romolo G. Inst. for Materials Res., McMaster U., Hamilton, Ont. L8S 4M1, Canada. (1944) BSc, chemistry (McMaster U., 1968). Technician. (tel. 416 + 525-9140, ext. 2005).

Falk, Dr Michael. Atlantic Res. Lab., Nat. Res. Council of Canada, 1411 Oxford St., Halifax, N.S. B3H 3Z1, Canada. (1931) DSc, physical chemistry (Laval U., 1958). Sr. res. officer. (tel. 902 + 426-8265, ext. 120). *Infrared spectra of crystals, crystalline hydrate structures.*

Faught, Dr John Brian. P.O. Box 35, RR#1 Bell St., Lake Echo, Halifax Co., N.S. B0J 2S0, Canada. (1942) PhD, inorganic chemistry (U. Illinois, USA, 1969). (tel. 902 + 829-3807). *Chemical crystallography, non-metal inorganic compounds.*

Fawcett, Dr John Keith. North Island C., 1413 Island Hwy, Campbell River, B.C. V9W 2E4, Canada. (1940) PhD, chemistry (U. British Columbia, 1965). Instructor. (tel. 604 + 287-2181). *Molecular structure, biological and organic compounds, crystallographic computing.*

Ferguson, Dr George. Chemistry Dept., U. of Guelph, Guelph, Ont. N1G 2W1, Canada. (1936) DSc, crystal structural analysis (U. Glasgow, UK, 1969). Prof. of chemistry. (tel. 519 + 824-4120, ext. 3548 and 3800). *Inorganic and organic structures.*

Ferguson, Prof. Robert Bury. Dept. of Earth Sci., U. of Manitoba, Winnipeg, Man. R3T 2N2, Canada. (1920) PhD, mineralogy (U. Toronto, 1948). Prof. of mineralogy. (tel. 204 + 474-9786). *Crystallography, crystal structures, crystal chemistry, minerals, rock-forming silicates, crystallographic teaching.*

Fitzgerald, Dr Paula Marie Dean. Dept. of Biochemistry, U. of Alberta, Edmonton, Alta. T6G 2H7, Canada. (1949) PhD, biophysics (Johns Hopkins U., USA, 1977). Professional asst. (tel. 403 + 432-4576). *Biological macromolecular structures, molecular replacement.*

Fleet, Dr Michael Edward. Dept. of Geology, U. of Western Ontario, London, Ont. N6A 5B7, Canada. (1938) PhD, geology (U. Manchester, UK, 1963). Prof. (tel. 519 + 679-3135). *Inorganic structures, structure related properties, crystal chemistry.*

Fortier, Dr Suzanne. Dept. of Chemistry, Queen's U., Kingston, Ont. K7L 3N6, Canada. (1949) PhD, crystallography (McGill U., 1976). Asst. prof. (tel. 613 + 547-6231). *Direct methods theory and applications.*

Franklin, Dr Kenneth James. Fuels and Materials Div., AECL, Chalk River, Ont. K0J 1J0, Canada. (1952) PhD, chemistry (McMaster U., 1982). Asst. res. officer. (tel. 613 + 584-3311, ext. 2560). *Structural inorganic chemistry, transition metal complexes, lanthanides, actinides.*

Franklin, Prof. Ursula Martius. Dept. of Metallurgy and Materials Sci., U. of Toronto, Toronto, Ont. M5S 1A4, Canada. (1921) PhD, applied physics (Techn. U., Berlin, BRD, 1948). Prof. (tel. 416 + 978-3012, telex 06-218915). *Characterization, ancient materials, X-ray diffraction techniques, ternary alloy structures.*

Gabe, Dr Eric James. Div. of Chemistry, Nat. Res. Council of Canada, Ottawa, Ont. K1A 0R9, Canada. (1933) PhD, crystallography (U. Wales, UK, 1960). Sr. res. officer. (tel. 613 + 993-2527). *Automation, small computers, structures.*

Gagnon, Miss Carole. Dépt. de Chimie, U. de Montréal, C.P. 6210, Montréal, Qué. H3C 3V1, Canada. (1946) MSc, chemistry (U. de Montréal, 1976). Res. asst. (tel. 514 + 343-6516). *X-ray structure, metal ions complexes (Ag-Cd-Hg), purines and pyrimidines from DNA and RNA.*

Gait, Dr Robert Irwin. Dept. of Mineralogy & Geology, Royal Ontario Museum, 100 Queen's Park, Toronto, Ont. M5S 2C6, Canada. (1938) PhD, mineralogy (U. Manitoba, 1967). Curator of mineralogy. (tel. 416 + 978-3714). *General mineralogy; morphological crystallography; X-ray crystallography; descriptive mineralogy.*

Gaunt, Dr Paul. Dept. of Physics, U. of Manitoba, Winnipeg, Man. R3T 2N2, Canada. (1932) DPhil, metal physics (Oxford U., UK, 1958). Prof. (tel. 204 + 474-9589). *Lattice distortion, order-disorder phenomena, metals and alloys, magnetic properties related to structure, electron microscopy, electron diffraction.*

Gibbons, Dr Cyril Stephen. Dept. of Materials Chemistry, Ontario Res. Foundation, Sheridan Park, Mississauga, Ont. L5K 1B3, Canada. (1945) PhD, structure analysis (U. British Columbia, 1971). Sr. res. sci. (tel. 416 + 822-4111, ext. 244).

Gopal, Dr Ramanathan. Industrial Batteries Div., Gould Manufacturing Canada Ltd., 275 Lewis St., Fort Erie, Ont. L2A 2R3, Canada. (1944) PhD, crystallography & physical chemistry (McMaster U., 1972). Sr. res. sci. (tel. 416 + 871-5600). *Inorganic molecules, small organic molecules, mineral crystallography; electrochemistry, electrocatalysts.*

Graham, Dr Albert Ronald. 515 St. George St. E., Fergus, Ont. N1M 1L1, Canada. (1917) PhD, mineralogy (U. Toronto, 1950). Retired. (tel. 519 + 843-1150). *Economic mineralogy.*

Grattan-Bellew, Dr Patrick Edward. Div. of Building Res., Nat. Res. Council of Canada, Ottawa, Ont. K1A 0R6, Canada. (1934) PhD, experimental mineralogy (Cambridge U., UK, 1969). Res. officer. (tel. 613 + 993-1596). *Cement chemistry, mineralogy in general.*

Grenier, Miss Lucie. Dépt. de Chimie, U. de Montréal, C.P. 6210, Succ. A, Montréal, Qué. H3C 3V1, Canada. (1960) BSc, chemistry (U. de Montréal, 1983). Grad. student. (tel. 514 + 343-7538). *Bioinorganic chemistry.*

Grice, Dr Joel Denison. Mineral Sci. Div., Nat. Museum of Natural Sci., 1926 Merivale Road, Ottawa, Ont. K1A 0M8, Canada. (1946) PhD, mineralogy (U. Manitoba, 1973). Curator of minerals. (tel. 613 + 998-9225). *Crystal systematics, growth of minerals.*

Groat, Mr Lee Andrew. Dept. of Earth Sci., U. of Manitoba, Winnipeg, Man. R3T 2N2, Canada. (1959) BSc, geology (Queen's U., 1982). Grad. student. (tel. 204 + 474-8395). *Mineral structures, optical crystallography, electron microprobe analysis.*

Grundy, Dr Harry Douglas. Dept. of Geology, McMaster U., Hamilton, Ont. L8S 4M1, Canada. (1941) PhD, mineralogy (U. Manchester, UK, 1966). Prof. (tel. 416 + 525-9140, ext. 4516). *Silicate minerals, characterization, inorganic solids.*

Guay, Dr France. Dépt. de Chimie, U. de Québec à Montréal, C.P. 8888, Succ. A, Montréal, Qué. H3C 3P8, Canada. (1957) PhD, inorganic chemistry (U. de Montréal, 1984). Postdoctoral student. (tel. 514 + 282-4314). *Inorganic chemistry, bioinorganic chemistry. heavy metals complexes.*

Hanson, Dr Alfred Wallace. Atlantic Res. Lab., Nat. Res. Council of Canada, 1411 Oxford St., Halifax, N.S. B3H 3Z1, Canada. (1923) PhD, crystallography (U. Manchester, UK, 1953). Sr. res. officer. (tel. 902 + 426-8258). *X-ray crystallography; scanning electron microscopy; microprobe analysis.*

Hassan, Prof. Ishmael. Earth & Planetary Sci., U. of Toronto, Erindale Campus, Mississauga, Ont. L5L 1C6, Canada. (1951) PhD, geochemistry (McMaster U., 1982). Asst. prof. (tel. 416 + 828-5419). *Geologically important materials, structure, chemistry.*

Hawthorne, Dr Frank Christopher. Dept. of Earth Sci., U. of Manitoba, Winnipeg, Man. R3T 2N2, Canada. (1946) PhD, geology (McMaster U., 1973). Assoc. prof. (tel. 204 + 474-9833). *Structural systematics, minerals, graph theory, combinatorial theory, instrumental methods. pyroxenes.*

Hayakawa, Miss Koto. Dept. of Biochemistry, U. of Alberta, Edmonton, Alta. T6G 2H7, Canada. (1940) BSc, science (U. Osaka, Japan, 1966). Technician. (tel. 403 + 432-2422). *Crystallography, biologically interesting compounds.*

Heimann, Prof. Dr Robert Bertram Silvester. Fuel Waste Technology Branch, AECL Whiteshell Nuclear Res. Est., Pinawa, Man. R0E 1L0, Canada. (1938) Dr.rer.nat., mineralogy (Freie U. Berlin, BRD, 1966). Res. geochemist. (tel. 204 + 753-2311, ext. 2439). *Growth and dissolution of crystals; solid state chemistry.*

Hempel, Dr Andrew. Dept. of Biochemistry, U. of Toronto, University Ave., Toronto, Ont. M5S 1A8, Canada. (1944) PhD, organic chemistry (Techn. U. of Gdansk, Poland, 1975). Res. assoc. (tel. 416 + 978-3726). *Biological structures, structure - activity relationships.*

Heyding, Dr Robert Donald. Dept. of Chemistry, Queen's U., Kingston, Ont. K7L 3N6, Canada. (1925) PhD, physical chemistry (McGill U., 1951). Prof. (tel. 613 + 547-2961). *Powder X-ray diffraction, phase transitions, inorganic structures.*

Huber, Dr Carol P. Div. of Biological Sci., Nat. Res. Council of Canada, Ottawa, Ont. K1A 0R6, Canada. (1937) DPhil, chemical crystallography (Oxford U., UK, 1963). Sr. res. officer. (tel. 613 + 990-0856). *Biologically important structures.*

Hubert, Dr Joseph. Chemistry Dept., U. de Montréal, P.O. Box 6210, Station A, Montréal, Qué. H3C 3V1, Canada. (1944) PhD, inorganic chemistry (U. de Montréal, 1974). Prof. (tel. 514 + 343-7056). *Bioinorganic chemistry, inorganic analytical chemistry.*

Hutcheon, Dr Wendy Lou (Brooks). Dept. of Biochemistry, U. of British Columbia, 2146 Health Sciences Mall, Vancouver, B.C. V6T 1W5, Canada. (1946) PhD, crystallography (U. Alberta, 1971). Res. assoc. (tel. 604 + 228-4509). *Protein crystallography, enzyme structure and function, protein-nucleic acid interactions.*

Idler, Miss Kathleen Loralee. Dept. of Chemistry, McMaster U., Hamilton, Ont. L8S 4M1, Canada. (1951) BSc, chemistry (U. British Columbia, 1973). Grad. student. (tel. 416 + 525-9140, ext. 2005). *Inorganic structures.*

James, Prof. Michael N.G. Biochemistry Dept., U. of Alberta, Edmonton, Alta. T6G 2H7, Canada. (1940) DPhil, chemical crystallography (Oxford U., UK, 1966). Prof. (tel. 403 + 432-4550). *Protein structure, enzyme function; biological molecular structures and functions.*

Johari, Dr Gyan Prakash. Dept. of the Environment, Inland Waters Directorate, 562 Booth St., Ottawa, Ont. K1A 0E7, Canada. (1940) PhD, physical chemistry (U. Gorakhpur, India, 1965). Res. sci. (tel. 613 + 994-5098). *Ice physics, amorphous state.*

Jones, Dr Stephen John. AVMRI, Nat. Res. Council of Canada, Ottawa, Ont. K1A 0R6, Canada. (1942) PhD, physics (U. Birmingham, UK, 1967). Sr. res. officer (tel. 613 + 993-2288). *Ice physical properties, sea ice mechanical properties.*

Kaiman, Mr Solomon. 1917 Lauder Drive, Ottawa, Ont. K2A 1A9, Canada. (1920) MA, X-ray crystallography (U. Toronto, 1946). Retired. *X-ray* *diffraction analysis, minerals; uranium mineralogy; mineralogical methods application, ore dressing.*

Kerr, Dr Kathleen Ann. Dept. of Chemistry & Physics, U. of Calgary, Calgary, Alta. T2N 1N4, Canada. (1941) PhD, chemistry (U. Glasgow, UK, 1966). Assoc. prof. (tel. 403 + 284-5395). *Biologically active compounds, diffraction physics.*

Khan, Dr Masood Alam. Atlantic Res. Lab., Nat. Res. Council of Canada, 1411 Oxford St., Halifax, N.S. B3H 3Z1, Canada. (1947) PhD, X-ray crystallography (U. Victoria, 1976). Res. fellow. (tel. 902 + 426-8258). *Structural inorganic, organometallic chemistry.*

King, Prof. Hubert Wylam. Dept. of Engineering Physics, Dalhousie U., Halifax, N.S. B3H 5J3, Canada. (1930) PhD, physical metallurgy (U. Birmingham, UK, 1956). Prof. (tel. 902 + 424-2344). *Low temperature phase transformations, magnetic and superconductive properties.*

Knop, Prof. Osvald. Dept. of Chemistry, Dalhousie U., Halifax, N.S. B3H 4J3, Canada. (1922) DSc, physical chemistry (Laval U., 1957). Prof. (tel. 902 + 424-3317). *Structural inorganic chemistry.*

Kocman, Dr Vladimir. DOMTAR Res. Centre, DOMTAR Inc., Box 300, Senneville, Qué. H9X 3L7, Canada. (1937) Doctorate, inorganic chemistry (U. J.E. Purkyne Brno, Czechoslovakia, 1968). Res. sci. (tel. 514 + 457-6810, ext. 284). *X-ray powder diffraction, X-ray fluorescence, differential thermal analysis, thermogravimetry, instrumental methods.*

Kodama, Dr Hideomi. Agriculture Canada, Chemistry & Biology Res. Inst., CEF, Carling Ave., Ottawa, Ont. K1A 0C6, Canada. (1931) PhD, clay mineralogy (Tokyo U. of Education, Japan, 1961). Sr. res. sci. (tel. 613 + 995-3700, ext. 248). *Structural disorder, layer silicates; amorphous mineral characterization.*

Le Page, Dr Yvon. Chemistry Div., Nat. Res. Council of Canada, Ottawa, Ont. K1A 0R9, Canada. (1943) PhD, physics engineering (U. Montréal, 1974). Asst. res. officer. (tel. 613 + 993-2527). *Accuracy in intensity measurements; symmetry; low temperatures; programming, small computers; crystal structures.*

Lea, Prof. Sydney George. Chemical and Metallurgical Dept., Ryerson Polytechnical Inst., 50 Gould St., Toronto, Ont. M5B 1E8, Canada. (1933) PhD, crystallography (U. London, UK, 1968). Prof. (tel. 416 + 595-5067). *Spectroscopy, inorganic structures, crystal physics.*

Lee, Mrs Florence Lan Fun. Div. of Chemistry, Nat. Res. Council of Canada, Ottawa, Ont. K1A 0R9, Canada. (1938) MEd, education (U. Ottawa, 1966). Res. officer. (tel. 613 + 993-2527). *Structures, programming.*

Lock, Prof. Colin James Lyne. Labs. for Inorganic Medicine, Inst. for Materials Res., McMaster U., Hamilton, Ont. L8S 4M1, Canada. (1933) PhD, inorganic chemistry (Imperial C., London U., UK, 1963). Prof. of chemistry and pathology. (tel. 416 + 525-9140, ext. 4760). *Metal complexes used in medicine, X-ray structural studies.*

Loeb, Dr Stephen Joseph. Dept. of Chemistry, U. of Manitoba, Winnipeg, Man. R3T 2N2, Canada. (1954) PhD, inorganic chemistry (U. Western Ontario, 1982). Asst. prof. (tel. 204 + 786-7811, ext. 548). *Inorganic, organometallic chemistry; catalysis.*

Luo, Mr Yao Guang. Chemistry Dept., U. of Regina, Regina, Sask. S4S 0A2, Canada. (1956) BSc, chemistry (Xiamen U., People's Rep. of China, 1981). Grad. student. (tel. 306 + 584-4923). *Inorganic structures, synthesis.*

Lüth, Dr Hartwig. Res. & Dev. Dept., Fiberglas (Canada) Ltd., Box 3005, Sarnia, Ont. N7T 7M6, Canada. PhD, structural chemistry (U. London, UK, 1968). Res. sci. (tel. 519 + 344-7461, ext. 282). *Powder diffraction, phase - temperature relations in crystals, molecular and crystal structure, inorganic complexes - groups IA and IIA.*

Ma, Dr Lilian Yan Yan. Dept. of Biochemistry, U. of Toronto, Toronto, Ont. M5R 2XA, Canada. (1946) PhD, chemistry (Simon Fraser U., 1971). Res. assoc., professional asst. (tel. 416 + 978-3726). *X-ray crystallography, biological activity - structural basis, molecular design.*

Macek, Dr Josef Jan. Dept. of Energy & Mines, 993 Century St., Winnipeg, Man. R3H 0W4, Canada. (1943) PhD, earth sciences (U. Manitoba, 1979). Geologist. (tel. 204 + 633-9543, ext. 288). *Optical properties, minerals, plagioclase feldspars.*

Mandarino, Dr Joseph Anthony. Dept. of Mineralogy & Geology, Royal Ontario Museum, 100 Queen's Park, Toronto, Ont. M5S 2C6, Canada. (1929) PhD, mineralogy (U. Michigan, USA, 1958). Curator of mineralogy. (tel. 416 + 978-3647). *Crystal optics, descriptive mineralogy, X-ray diffraction, morphological crystallography.*

Marchessault, Dr Robert H. Xerox Res. Centre of Canada, 2660 Speakman Drive, Mississauga, Ont. L5K 2L1, Canada. (1928) PhD, physical chemistry (McGill U., 1954). Vice-president (tel. 416 + 823-7091, ext. 230). *Polymer crystal structure, polymer morphology.*

Martin, Miss Lillian Ruth. Dept. of Chemistry, Simon Fraser U., Burnaby, B.C. V5A 1S6, Canada. (1947) BSc, physical and inorganic chemistry (Simon Fraser U.). Grad. student. (tel. 604 + 291-4408). *Organometallic, coordination and metal-cluster complexes.*

Matthews, Prof. Frederick White. School of Library Service, Dalhousie U., Halifax, N.S. B3H 4H8, Canada. (1915) PhD, physical chemistry (McGill U., 1941). Prof. (tel. 902 + 424-3656). *Powder data; numerical data bases, information systems.*

Maxwell, Prof. George. Mathematics Dept., U. of British Columbia, #121 - 1984 Mathematics Rd., U. Campus, Vancouver, B.C. V6T 1Y4, Canada. (1946) PhD, mathematics (Queen's U., 1970). Assoc. prof. (tel. 604 + 228-6402). *Group theory; geometry; mathematical crystallography.*

Meagher, Dr Edward Patrick. Dept. of Geological Sci., U. of British Columbia, Vancouver, B.C. V6T 2B4, Canada. (1939) PhD, mineralogy (Penn State U., USA, 1967).Prof. (tel. 604 + 228-3508). *Crystal chemistry, bonding theory.*

Médicis, de, Dr Rinaldo M. Faculté de Médecine, U. de Sherbrooke, Chemin de Stooke, Sherbrooke, Qué. J1H 5M4, Canada. (1934) PhD, crystal chemistry (U. de Louvain, Belgium, 1967). Res. asst. (tel. 819 + 565-2144). *Crystal chemistry, X-ray and electron diffraction.*

Michel, Prof. André Gustave. Structural Chemistry Lab., U. de Sherbrooke, 2500 University Blvd., Sherbrooke, Qué. J1K 2R1, Canada. (1944) PhD, physical chemistry (Namur, Belgium, 1976). Prof. (tel. 819 + 565-3668). *X-ray crystal structure analysis, molecular modelling by computer.*

Middlemiss, Dr Nora E. Dept. of Glass & Ceramics, Ontario Res. Foundation, Sheridan Park, Mississauga, Ont. L5K 1B3, Canada. (1947) PhD, physical chemistry (McMaster U., 1978). Assoc. res. sci. (tel. 416 + 822-4111, ext. 247) *Ionic conductors, transition metal oxides, ceramics.*

Mihichuk, Prof. Lynn Michael. Chemistry Dept., U. of Regina, Regina, Sask. S4S 0A2, Canada. (1947) PhD, inorganic chemistry (U. British Columbia, 1975). Asst. prof. (tel. 306 + 584-4793). *Bioinorganic and organometallic structural chemistry.*

Mitchell, Dr Crighton Maurice. 8 Kingsford Cr., Kanata, Ont. K2K 1T3, Canada. (1917) PhD, physics (U. Toronto, 1952). Retired. (tel. 613 + 592-1186). *Deformation, residual stress, preferred orientation, crystal physics.*

Mitchell, Dr Keith A. R. Dept. of Chemistry, U. of British Columbia, 2036 Main Mall, Vancouver, B.C. V6T 1Y6, Canada. (1938) PhD, chemistry (U. London, UK, 1963). Prof. (tel. 604 + 228-5831). *Surface structure, LEED crystallography.*

Montgomery, Prof. (Em.) Henry. 502 - 2910 Cook St., Victoria, B.C. V8T 1B0, Canada. (1916) PhD, inorganic chemistry (U. Washington, USA, 1961). Retired. (tel. 604 + 383-8423). *X-ray structures, inorganics, metal organics.*

Muir, Mr Alastair Kerr. Chemistry Dept., U. of Calgary, Calgary, Alta. T2N 1N4, Canada. (1957) BSc, organic chemistry (U. Calgary, 1978). Grad. student. (tel. 403 + 284-5069). *Benzodiazepine antagonists, drug receptor interactions.*

Natarajan, Dr Mahadevan. 12 Court St.,Antigonish N.S. B2G 1Z6, Canada. (1940) PhD, phase transformations in solids (Indian Inst. of Techn., Kanpur, India, 1970). (tel. 902 + 863-6117). *Phase changes, electrical phenomena, IR, Raman, reflectance spectra. lattice dynamics.*

Ng, Dr Hok-Nam. Inst. for Materials Res., McMaster U., Hamilton, Ont. L8S 4M1, Canada. (1938) PhD, physical chemistry (U. British Columbia, 1971). Postdoctoral fellow. (tel. 416 + 525-9140, ext. 4563). *Phase transitions, electron paramagnetic resonance, oxide crystals, semiconductors.*

Nuffield, Prof. Edward Wilfrid. Dept. of Geology, U. of Toronto, Toronto, Ont. M5S 1A1, Canada. (1914) PhD, mineralogy (U. Toronto, 1944). Prof. (tel. 416 + 978-4931). *Crystal chemistry, ore minerals, X-ray diffraction methods.*

Nyburg, Prof. Stanley C. Dept. of Chemistry, U. of Toronto, Toronto, Ont. M5S 1A1, Canada. (1924) DSc, crystallography & thermodynamics (U. London, UK, 1974). Prof. (tel. 416 + 978-3603 or 978-3625). *Structure and physical properties of crystals, molecular crystals, complex ions, enzymes, intermolecular forces.*

Olivier, M. Marc-J. Dépt. de Chimie, U. de Montréal, C.P. 6210, Succ. A, Montréal, Qué. H3C 3V1, Canada. (1951) MSc, chimie (U. de Montréal, 1980). Agent de recherches. (tel. 514 + 343-6747). *Heavy metal complexes with bases of DNA.*

Owen, Mr Charles Gordon. Div. of Dental Biomaterials Sci., Faculty of Dentistry, Dalhousie U., Halifax, N.S. B3H 3J5, Canada. (1957) MSc, inorganic chemistry (Dalhousie U., 1982). Res assoc. (tel. 902 + 424-7058). *Corrosion products, dental amalgam, microstructure.*

Pagoaga, Dr M. Katherine. Dept. of Chemistry, U. of Calgary, Calgary, Alta. T2N 1N4, Canada. (1952) PhD, geochemistry (U. Maryland, USA, 1983). Postdoctoral fellow. (tel. 403 + 220-5341). *Uranium minerals; crystallographic computing, programming.*

Palmer, Mr Allan D. Dépt. de Chimie, U. de Montréal, C.P. 6210, Succ. A, Montréal, Qué. H3C 3V1, Canada. (1956) MSc, chemistry (U. de Montréal, 1980). Grad. student. (tel. 514 + 343-7538). *Polymers; X-ray, electron diffraction.*

Pandya, Mr Naresh. Chemistry Dept., Brock U., St. Catharines, Ont. L2S 3A1, Canada. (1960) BSc(Hons), chemical physics (U. Sussex, 1982). Grad. asst. (tel. 416 + 688-5550, ext. 3406) *Structure correlation, modelling reaction transition states, crystallography.*

Payne, Dr Nicholas Charles. Chemistry Dept., U. of Western Ontario, London, Ont. N6A 5B7, Canada. (1942) PhD, inorganic chemistry (U. Sheffield, UK, 1967). Prof. (tel. 519 + 679-2392). *Preparation, single crystal X-ray, transition metal complexes, catalytically important complexes, asymmetric synthesis, absolute configuration, Bijvoet absorption edge technique.*

Pazdernik, Prof. LeRoy Joseph. Dépt. de Chimie-Biologie, U. du Québec à Trois-Rivières, C.P. 500, Trois-Rivières, Qué. G9A 5H7, Canada. (1942) PhD, inorganic chemistry (U. Iowa, USA, 1970). Prof. (tel. 819 + 376-5673). *Metal complexes, biological environmental systems, metal coordination compounds, bioinorganic compounds.*

Pearson, Dr William Burton. Dept. of Physics, U. of Waterloo, Waterloo, Ont. N2L 3G1, Canada. (1921) DSc, metal physics (Oxford U., UK, 1968). Prof. (tel. 519 + 885-1211, ext. 3180). *Structures, alloys, phase stability.*

Perrault, Dr Guy. École Polytechnique, C.P. 6079, Montréal, Qué. H3C 3A7, Canada. (1927) PhD, mineralogy (U. Toronto, 1955). Prof. *Crystal structure, minerals, silicates, niobium minerals.*

Peterson, Dr Ronald Charles. Dept. of Geology, Queen's U., Kingston, Ont. K7L 3N6, Canada. (1953) PhD, geology (Virginia Polytechnic Inst., USA, 1981). Asst. prof. (tel. 613 + 597-5816). *Silicate mineralogy, crystal chemistry.*

Pizzey, Mrs Monica Agnes Anastasia. Physics Dept., U. of Regina, Regina, Sask. S4S 0A2, Canada. (1959) Res. asst. (tel. 306 + 584-4653). *Inorganic and organometallic structural chemistry.*

Ploc, Dr Robert Allen. Chemistry and Materials Div., Materials Science Branch, Atomic Energy of Canada Ltd., Chalk River Nuclear Labs., Stn. 82, Chalk River, Ont. K0J 1J0, Canada. (1939) PhD, metallurgy (Cambridge U., UK, 1965). Res. assoc. (tel. 613 + 584-4378, ext. 785). *Electron diffraction (computer analysis), oxide growth on Zr and Zr-based alloys, electron microscopy (transmission and scanning).*

Post, Dr Michael Leonard. Chemistry Div., Nat. Res. Council of Canada, Ottawa, Ont. K1A 0R9, Canada. (1945) PhD, chemistry (U. Surrey, UK, 1971). Assoc. res. officer. (tel. 613 + 993-2506). *Metal alloy hydrides; calorimetry.*

Potworowski, Dr Jean-André. CANMET, Dept. of Energy, Mines & Resources, 555 Booth St., Ottawa, Ont. K1A 0G1, Canada. (1947) PhD, X-ray crystallography (U. Toronto, 1974). Dir., OTT. (tel. 613 + 995-4267). *Administration of Res & Dev., technology transfer.*

Poulin-Dandurand, Mrs Suzie. 5272 Monsabre, Montréal, Qué. H1M 2R2, Canada. (1954) MSc, chemistry (U. du Québec à Montréal, 1978). Res. Assoc. (tel. 514 + 344-4664). *Polymer structures, amorphous silicon devices, electron microscopy.*

Powell, Dr Brian Mathieson. Neutron & Solid State Physics Div., Atomic Energy of Canada Ltd., Chalk River, Ont. K0J 1J0, Canada. (1938) PhD, physics (U. London, UK, 1964). Res. officer. (tel. 613 + 687-5581, ext. 2682). *Neutron diffraction, profile analysis, inelastic neutron scattering.*

Prasad, Dr Lata. Dept. of Chemistry, U. of Toronto, 80 St. George St., Toronto, Ont. M5S 1A1, Canada. (1945) PhD, crystallography (Flinders U. of S. Australia, 1972). Res. assoc. (tel. 416 + 978-6275). *Small molecule structures, instrumentation.*

Pringle, Mr Gordon James. Dept. of Energy, Mines and Resources, Geological Survey of Canada, 601 Booth St., Ottawa, Ont. K1A 0E8, Canada. (1944) MSc, mineralogy (U. New Brunswick, 1972). Electron microprobe analyst. (tel. 613 + 994-5023). *Electron microprobe analysis, feldspar optics, feldspar relations in basaltic rocks, powder diffraction techniques.*

Przybylska, Dr Maria. Div. of Biological Sci., Nat. Res. Council of Canada, Ottawa, Ont. K1A 0R6, Canada. (1923) PhD, X-ray crystallography (U. Glasgow, UK, 1949). Sr. res. officer. (tel. 613 + 990-0857). *Structures, organic compounds, proteins.*

Quail, Dr J. Wilson. Dept. of Chemistry, U. of Saskatchewan, Saskatoon, Sask. S7N 0W0, Canada. (1936) PhD, chemistry (McMaster U., 1963). Prof. (tel. 306 + 966-4663). *Inorganic molybdenum halides; structures, biological activity of nucleosides.*

Rajan, Dr Krishna. Nat. Aeronautical Est., Nat. Res. Council of Canada, Ottawa, Ont. K1A 0R6, Canada. (1952) ScD, materials science (Mass. Inst. of Technology, USA, 1978). Res. officer. (tel. 613 + 993-9280). *High resolution transmission electron microscopy; structure-property relationships in materials; convergent beam electron diffraction, micro-diffraction.*

Ranger, Dr Georges. R & D Div., ERCP Industries, 2 Gibbs Rd., Islington, Ont. M9B 1R1, Canada. (1955) PhD, inorganic chemistry (Wayne State U., USA, 1984). Res. chemist. (tel. 416 + 239-7111, ext. 301). *Inorganic and analytical chemistry, crystal structure determinations, heavy metal complexes.*

Raudsepp, Dr Mati. Dept. of Earth Sci., U. of Manitoba, Winnipeg, Man. R3T 2N2, Canada. (1947) PhD, mineralogy (U. Manitoba, 1984). Postdoctoral fellow. (tel. 204 + 474-9833). *Experimental mineralogy, amphibole crystal chemistry, Rietveld structure analysis, infrared spectroscopy.*

Read, Mr Randy John. Biochemistry Dept., U. of Alberta, Edmonton, Alta. T6G 2H7, Canada. (1957) BSc, biochemistry (U. Alberta, 1970). Grad. student. (tel. 403 + 432-2422). *Protein structure and function.*

Restivo, Dr Roderic John. Dept. of Sci., C.E.G.E.P. de l'Outaouais, P.O. Box 220, Hull, Qué. J8X 3X8, Canada. (1943) PhD, chemistry (U. Waterloo, 1969). Prof., chemistry. (tel. 819 + 778-2270, ext. 22). *Bio-inorganic structural chemistry, organometallic structural chemistry.*

Rettig, Dr Steven John. U. of British Columbia, Chemistry, 2036 Main Mall, U. Campus, Vancouver, B.C. V6T 1Y6, Canada. (1948) PhD, chemistry (U. British Columbia, 1974). Res. assoc. (tel. 604 + 228-4865). *X-ray crystallography, organic and organometallic compounds; organoboron chemistry.*

Richard, Prof. Joseph Albert Pierre. Dépt. de Physique, U. du Québec à Montréal, C.P. 8888, Station A, Montréal, Qué. H3C 3P8, Canada. (1942) PhD, physics engineering (U. de Montréal, 1971). Prof. (tel. 514 + 282-7837). *Crystal structure, organometallics.*

Richardson, Dr John Frederick. Dept. of Chemistry, U. of Calgary, Calgary, Alta. T2N 1N4, Canada. (1954) PhD, chemistry (U. Western Ontario, 1981).Res. assoc. *Structural analysis, small molecules; diffraction, biological macromolecules, viruses.*

Richardson, Dr Mary Frances. Chemistry Dept., Brock U., St. Catharines, Ont. L2S 3A1, Canada. (1941) PhD, inorganic chemistry (U. Kentucky, USA, 1967). Prof. (tel. 416 + 688-5550, ext. 3400) *X-ray structure determination, metal complexes, topochemical reactions, structure correlation approach, chemical reactions.*

Roberts, Mr Andrew Clifford. Dept. of Energy, Mines and Resources, Geological Survey of Canada, 762 - 601 Booth St., Ottawa, Ont. K1A 0E8, Canada. (1950) MSc, mineralogy (Queen's U., 1976). X-ray diffraction mineralogist. (tel. 613 + 992-2802). *Single crystal studies, powder diffraction techniques.*

Robertson, Prof. Beverly Ellis. Dept. of Physics and Astronomy, U. of Regina, Regina, Sask. S4S 0A2, Canada. (1939) PhD, physics (McMaster U., 1967). Prof. (tel. 306 + 584-4264). *X-ray crystallography.*

Rochon, Prof. Fernande D. Dept. de Chimie, U. du Québec à Montréal, C.P. 8888, Station A, Montréal, Qué. H3C 3P8, Canada. PhD, inorganic chemistry (U. de Montréal, 1971). Prof. (tel. 514 + 282-4896 or 282-4119). *Inorganic chemistry, platinum complexes.*

Rose, Dr David Richard. Div. of Biological Sci., Nat. Res. Council of Canada, Ottawa, Ont. K1A 0R6, Canada. (1955) DPhil, molecular biophysics (Oxford U., UK, 1981). Res. assoc. (tel. 613 + 990-0857). *Protein crystallography, antibodies, Fab fragments, molecular immunology.*

Rowland, Mr John Fleming. CANMET, Dept. of Energy, Mines and Resources, 555 Booth St., Ottawa, Ont. K1A 0G1, Canada. (1927) MSc, mineralogy (Queen's U., 1950). Res. sci., crystal structure sect. (tel. 613 + 994-9036). *Mineral structures.*

Rucklidge, Prof. John Christopher. Dept. of Geology, U. of Toronto, Toronto, Ont. M5S 1A1, Canada. (1938) PhD, mineralogy (U. Manchester, UK, 1962). Assoc. prof. (tel. 416 + 978-2061). *Crystal structures, minerals, microanalytical methods, computer applications, mineralogical measurements.*

Ruhl, Miss Barbara Louise. Chemistry Dept., U. of Guelph, Guelph, Ont. N1G 2W1, Canada. (1954) BSc, biology, chemistry (U. Guelph, 1976). Res. asst. (tel. 519 + 824-4120, ext. 3548). *Organic and organometallic compounds.*

Sawyer, Dr Jeffrey Frederick. Lash Miller Chemical Labs., U. of Toronto, 80 St. George St., Toronto, Ont. M5S 1A1, Canada. (1954) PhD, chemistry (U. Warwick, UK, 1977). X-ray systems mgr. (tel. 416 + 978-6275). *Inorganic and organometallic structures, secondary bonding.*

Scott, Dr James Douglas. R & D Div., Kidd Creek Mines Ltd., Suite 5000, Commerce Court W., Toronto, Ont. M5L 1E7, Canada. (1942) PhD, mineralogy (crystallography) (Queen's U., 1970). Sr. staff scient. (tel. 416 + 869-1200, ext. 202). *Sulfide structures, process mineralogy, uranium minerals, atomic substitution in minerals.*

Secco, Prof. Anthony Silvio. Dept. of Chemistry, U. of Manitoba, Winnipeg, Man. R3T 2N2, Canada. (1956) PhD, chemistry (U. British Columbia, 1982). Asst. prof. (tel. 204 + 474-9347). *Organic and inorganic structures; biochemical crystallography.*

Sielecki, Dr Anita R. Biochemistry Dept., U. of Alberta, Edmonton, Alta. T6G 2H7, Canada. (1940) PhD, hydrodynamics (Hebrew U., Israel, 1969). Professional asst. (tel. 403 + 432-2422).

Simard, Mr Michel. Dépt. de Chimie, U. de Montréal, C.P. 6210, Succ. A, Montréal, Qué. H3C 3V1, Canada. (1958) MSc, chemistry (U. de Montréal, 1981). Grad. student. (tel. 514 + 343-7538). *Inorganic chemistry.*

Smith, Prof. Vedene H., Jr. Chemistry Dept., Queen's U., Kingston, Ont. K7L 3N6, Canada. (1935) Fil. dr, quantum chemistry (U. Uppsala, Sweden, 1967). Prof. and dept. head. (tel. 613 + 547-3254). *Electron density distributions (charge - spin - momentum).*

Stanley, Dr Eric. Physics Dept., U. of New Brunswick, P.O. Box 5050, Saint John, N.B. E2L 4L5, Canada. (1924) DSc, physics (U. Saskatchewan, 1972). Prof. (tel. 506 + 657-7310, ext. 242). *Direct methods, refinement, electron density distributions.*

Stephan, Dr D. W. Chemistry Dept., U. of Windsor, Windsor, Ont. N9B 3P4, Canada. (1953) PhD,(U. Western Ontario, 1980). (tel. 519 + 253-4232). *Organometallic and bioinorganic chemistry.*

Sundararajan, Dr Pudupadi R. Xerox Research Centre of Canada, 2660 Speakman Dr., Mississauga, Ont. L5K 2L1, Canada. (1943) DSc, physics (Madras U., India, 1982). Area manager. (tel. 416 + 823-7091, ext. 219). *Polymer crystallography and morphology, polymer conformations in solid state and solution.*

Sunder, Dr Sham. Res. Chemistry Branch, AECL, Whiteshell Nuclear Res. Est., Pinawa, Man. R0E 1L0, Canada. (1942) PhD, physical chemistry (U. Alberta, 1972). Assoc. res. officer. (tel. 204 + 753-2311, ext. 2749). *Surface chemistry, polymorphism, vibrational spectroscopy, ESCA, XPS, phase transitions, solid-liquid interface, membrane structures.*

Sutherland, Dr John Knox. Dept. of Minerals and Materials, Res. and Productivity Council, Box 6000, Fredericton, N.B. E3B 5H1, Canada. (1941) PhD, petrography (U. Manchester, UK, 1965). Mineralogist, chief of analytical services. (tel. 506 + 455-8994, ext. 247). *Mineralogy, petrology, metallurgy, mineral synthesis, analysis methods, rocks, ores, minerals.*

Sygusch, Prof. Jurgen. Dept. of Biochemistry, U. de Sherbrooke, Centre Hospitalier Universitaire, Sherbrooke, Qué. J1H 5N4, Canada. (1945) PhD, chemistry (U. Montréal, 1975). Prof. (tel. 819 + 565-2141). *Protein structure and function, protein crystallography.*

Szymański, Dr Jan Tomasz. CANMET, Dept. of Energy, Mines and Resources, 555 Booth St., Ottawa, Ont. K1A 0G1, Canada. (1938) PhD, inorganic chem-

istry (King's C., London U., UK, 1963). Res. scient., mineralogy section. (tel. 613 + 995-4077). *Mineral structures, sulphides, inorganic complexes.*
Taylor, Dr John Bryan. Div. of Chemistry, Nat. Res. Council of Canada, Ottawa, Ont. K1A 0R9, Canada. (1933) PhD, chemistry (U. Manchester, UK, 1956). Sr. res. officer. (tel. 613 + 993-2506). *Crystal chemistry, rare earth pnictides, intermetallic compounds, high temperature thermodynamics.*
Taylor, Dr Peter. Res. Chemistry Branch, AECL, Whiteshell Nuclear Res. Est., Pinawa, Man. R0E 1L0, Canada. (1949) PhD, inorganic chemistry (U. Birmingham, UK, 1972). Res. chemist. (tel. 204 + 753-2311, ext. 2463). *Structure- solubility- stability relationships, inorganic oxide and salt systems.*
Theophanides, Prof. Theo. Dept. of Chemistry, U. of Montréal, Box 6210, Station A, Montréal, Qué. H3C 3V1, Canada. (1932) PhD, organometallic chemistry (U. Toronto, 1963). Assoc. prof. (tel. 514 + 343-6742). *Platinum coordination complexes, cancer chemotherapy synthesis, structure and bond properties, solubility - structure relationships.*
Traill, Dr Robert James. Dept. of Energy, Mines and Resources, Geological Survey of Canada, 601 Booth St., Ottawa, Ont. K1A 0E8, Canada. (1921) PhD, crystallography (Queen's U., 1956). Asst. dir., central lab. div. (tel. 613 + 994-5023). *Structures, minerals, X-ray diffraction, X-ray fluorescence analysis.*
Trotter, Prof. James. Dept. of Chemistry, U. of British Columbia, Vancouver, B.C. V6T 1Y6, Canada. (1933) DSc, chemistry (U. Glasgow, UK, 1963). Prof. (tel. 604 + 228-4527). *Organic and inorganic crystal structures.*
Tyers, Mr Kenneth George. Dept. of Chemistry, Simon Fraser U., Burnaby, B.C. V5A 1S6, Canada. (1956) MSc, chemistry (U. Western Ontario, 1980). Grad. student. (tel. 604 + 291-4878). *Organometallic coordination compounds, transition metal clusters; synthesis.*
Van Der Heijden, Dr Simon Petrus Nicolaas. Res. & Dev. Center, Saskatchewan Power Corp., 2025 Victoria Ave., Regina, Sask. S4P 0S1, Canada. (1943) PhD, chemistry-crystallography (U. Saskatchewan, 1974). Res. scient. (tel. 306 + 566-2293). *Structural crystallography, crystal chemistry, coal gasification kinetics.*
Van Roode, Dr Johannes Hendricus Gerardus. Dept. of Chemistry, Memorial U. of Newfoundland, Regional C. at Cornerbrook, Cornerbrook, Nfld., Canada. (1944) PhD, chemistry (U. Calgary, 1975). Asst. prof. *Inorganic chemistry, chemical crystallography, environmental chemistry.*
Varughese, Dr Kottayil Iype. Div. of Biological Sci., Nat. Res. Council of Canada, Ottawa, Ont. K1A 0R6, Canada. (1946) PhD, crystallography (Madras U., India, 1974). Res. assoc. (tel. 613 + 990-0856). *Peptides, nucleic acids, natural products.*
Vrielink, Miss Alice. Dept. of Chemistry, U. of Calgary, Calgary, Alta. T2N 1N4, Canada. (1959) BSc, chemistry (U. Calgary, 1982). Grad. student. (tel. 403 + 284-5069). *Pharmacologically related biological molecules, drug structure - activity relationships, protein crystallography.*
Wang, Mr Hong. Chemistry Dept., U. of Regina, Regina, Sask. S4S 0A2, Canada. (1961) BSc, chemistry (Xiamen U., People's Rep. of China, 1981). Grad. student. (tel. 306 + 584-4902). *Inorganic structures, computing methods.*
White, Dr Peter Sutherland. Dept. of Chemistry, U. of New Brunswick, Fredericton, N.B. E3B 5A3, Canada. (1947) PhD, chemistry (Dalhousie U., 1973). Res. assoc. (tel. 506 + 453-4775, telex 014-46202). *Structures, inorganic and organic compounds, mini-computers, direct methods.*
Whitla, Dr William Alexander. Dept. of Chemistry, Mount Allison U., Sackville, N.B. E0A 3C0, Canada. (1938) PhD, inorganic chemistry (McMaster U., 1965). Assoc. prof. (tel. 506 + 536-2040, ext. 471). *X-ray structure determination, inorganic adducts.*
Whitlow, Dr Simon Hugh. Water Quality Branch, Environment Canada, Place Vincent Massey, Ottawa, Ont. K1A 0E7, Canada. (1943) PhD, chemistry (U. British Columbia, 1969). Head, Data and Instrumentation. (tel. 819 + 997-3422). *Inorganic crystal structures, computer programming, data management.*
Wicks, Dr Frederick John. Dept. of Mineralogy & Geology, Royal Ontario Museum, 100 Queen's Park, Toronto, Ont. M5S 2C6, Canada. (1937) D. Phil., mineralogy (Oxford U., UK, 1969). Curator. (tel. 416 + 978-6266). *Serpentine minerals; crystal structures; crystal chemistry; X-ray diffraction; thermal analysis; asbestos.*
Willis, Dr Anthony Creswick. Dept. of Chemistry, Simon Fraser U., Burnaby, B.C. V5A 1S6, Canada. (1951) PhD, physical and inorganic chemistry (U. Western Australia, 1977). Postdoctoral fellow. (tel. 604 + 291-4878). *organometallic, coordination and metal-cluster complexes.*
Wolbaum, Mr Keith Jonathon. Dept. of Physics & Astronomy, U. of Regina, Regina, Sask. S4S 0A2, Canada. (1953) BSc, physics (U. Regina, 1977). Res. asst. (tel. 306 + 584-4653). *Instrumentation.*
Wood, Dr Gordon H. CISTI, Nat. Res. Council of Canada, Ottawa, Ont. K1A 0S2, Canada. (1940) PhD, physics (U. British Columbia, 1969). CAN/SND manager. (tel. 613 + 993-3294). *Crystallographic databases.*

CHILE

Sub-Editor: O. Wittke

Notes

1. International telephone country code - 56.

2. The approximate equivalent of the Chilean degrees *Profesor de Matemáticas y Física* (Prof. de Mat. y Fís.), *Profesor de Física* (Prof. de Fís.), *Profesor de Química* (Prof. de Quím.), *Profesor de Biología y Química* (Prof. de Biol. y Quím.), *Licenciado en Física* (Lic. en Fís.), *Licenciado en Química* (Lic. en Quím.) and *Licenciado en Ciencias Naturales* (Lic. en C. Nat.) is between BSc and MSc in mathematics, physics or chemistry.

Aguilar, Mrs Adela. Lab. de Petrografía, Minera Utah de Chile Inc., Clasificador 11, Correo 10, Santiago, Chile. (1934). Head. (tel. 2313575). *Mineralogy, petrography.*
Almendras, Mrs Eliana. Dept. de Minas, Facultad de Ciencias Físicas y Matemáticas, U. de Chile, Casilla 2777, Santiago, Chile. (1940) Ingeniero de Minas (U. de Chile, 1962). Res. (tel. 6982071, ext. 571). *Mineralogical appraisal.*
Arce, Mrs María Teresa. Sección Metales, Dept. de Ciencias de los Materiales, Facultad de Ciencias Físicas y Matemáticas, U. de Chile, Casilla 1420, Santiago, Chile. (1932) Lic. en Quím. (U. de Chile, 1954). Res. chemist. (tel. 6982071, ext. 418). *Corrosion under stress, metals.*
Barbagelata, Mr Franco. Dept. de Microscopía, Centro de Investigación Minera y Metalúrgica (CIMM), Casilla 170, Santiago, Chile. (1942) Prof. de Quím. (U. de Chile, 1967). Head. (tel. 2289544). *Crystallography, microscopic determination, minerals.*
Barrios, Mr Nelson. Dept. de Análisis Instrumental, Centro de Investigación Minera y Metalúrgica (CIMM), Casilla 170, Santiago, Chile. (1935) MSc, chemistry (U. Nacional Autónoma de México, México, 1971). Head. (tel. 2289544). *X-ray diffraction, emission spectrography.*
Besoain, Dr Eduardo. Lab. de Fisicoquímica y Mineralogía, Inst. de Investigaciones Agropecuarias, Casilla 5427, Santiago, Chile. (1929) Dr. landw., soil mineralogy (U. Bonn, BRD, 1969). Head. (tel. 586061, ext. 13). *Soil mineralogy, X-ray diffraction, X-ray spectrography, infrared spectrophotometry.*
Boys, Dr Daphne. Dept. de Física, Facultad de Ciencias Físicas y Matemáticas, U. de Chile, Casilla 5487, Santiago, Chile. (1941) PhD, X-ray diffraction (Wales U., UK, 1972). Res. scient. (tel. 6960148). *X-ray diffraction, polymers, structures, complexes.*
Cid, Dr Hilda. Dept. de Fisiología, Facultad de Ciencias Biológicas y Recursos Naturales, U. de Conceptción, Casilla 20C, Concepción, Chile. (1933) PhD, crystallography (M.I.T., USA, 1964). Prof., res. scient. (tel. 24985). *Structure and function, biological macromolecules.*
Costamagna, Dr Juan Alberto. Dept. de Química, Facultad de Ciencia, U. de Santiago de Chile, Ecuador 3467, Santiago, Chile. (1940) Dr, chemistry (U. de Buenos Aires, Argentina, 1970). Res. scient. (tel. 95591). *Crystal structure, coordination compounds.*
Donoso, Mr Eduardo. Microscopía Electrónica, Dept. de Ciencias de los Materiales, Facultad de Ciencias Físicas y Matemáticas, U. de Chile, Casilla 1420, Santiago, Chile. (1947) Metalurgista (U. de Chile, 1972). Res. (tel. 6982071, ext. 316). *Electron microscopy (transmission and scanning).*
Escobar, Dr Carmen. Dept. de Física, Facultad de Ciencias Físicas y Matemáticas, U. de Chile, Casilla 5487, Santiago, Chile. (1931) D. 3e cycle, crystallography (U. de Bordeaux, France, 1968). Res. scient. (tel. 6960148). *Structures, complexes, organic crystals.*
Garaycochea-Wittke, Mrs Isabel. Volcán Llaima 7122, Santiago, Chile. (1925) MSc, crystallography (M.I.T., USA, 1966).
Garin, Dr Jorge. Dept. de Metalurgia, Facultad de Ingeniería, U. de Santiago de Chile, Casilla 10233, Santiago, Chile. (1943) PhD, metallurgy and materials science (U. Pennsylvania, USA, 1972). Head. (tel. 90506). *Crystal chemistry, metal physics.*
Garland, Mrs María Teresa. Dept. de Física, Facultad de Ciencias Físicas y Matemáticas, U. de Chile, Casilla 5487, Santiago, Chile. (1944) Lic. en Quím.

(U. de Chile, 1969) Res. scient. (tel. 6960148). *Organic and inorganic crystal structures.*

González, Mr Claudio. Dept. de Física, Facultad de Ciencias Físicas y Matemáticas, U. de Chile, Casilla 5487, Santiago, Chile. (1935) MPhil, physics (London U., UK, 1967). Prof. and Head electron microscopy and diffraction lab. (tel. 6960148). *Thin metallic films.*

González, Mrs Irma. Dept. de Geología, Facultad de Ciencias Físicas y Matemáticas, U. de Chile, Casilla 13518, Santiago, Chile. (1935) Prof. de Biol. y Quím. (U. de Chile, 1963). Res. (tel. 6982071, ext. 447). *Crystallography, minerals.*

González, Mr Yanko. Dept. de Análisis Químico, Centro de Investigación Minera y Metalúrgica (CIMM), Casilla 170, Santiago, Chile. (1954) Lic. en Quí. (U. de Chille, 1980). Res. (tel. 2289544). *X-ray diffraction, X-ray fluorescence.*

Greene, Mr Fernando. Dept. de Microscopía, Centro de Investigación Minera y Metalúrgica (CIMM), Casilla 170, Santiago, Chile. (1943) Prof. de Quím. (U. de Chile, 1968). Assoc. res. (tel. 2289544). *Microscopic determination, minerals, ore dressing products.*

Henríquez, Mr Fernando. Dept. de Minas, Facultad de Ingeniería, U. de Santiago de Chile, Casilla 10233, Santiago, Chile. (1942) MSc, geology (McGill U., Canada, 1972). Res. (tel. 95869). *Ore mineralogy.*

Hervé, Dr Francisco. Dept. de Geología, Facultad de Ciencias Físicas y Matemáticas, U. de Chile, Casilla 13518, Santiago, Chile. (1942) DSc, metamorphic petrology (Hokkaido U., Japan, 1974). Res. scient. (tel. 6982071, ext. 447). *Petrology, structure, metamorphic basement in Chile.*

Infante, Dr Carlos. Dept. de Física, Facultad de Ciencias, U. de Chile, Casilla 653, Santiago, Chile. (1944) DPhil, inorganic chemistry (Oxford U., UK, 1975). Res. scient. (tel. 2712973). *Neutron diffraction.*

Joseph, Dr Günter. Sección Metales, Dept. de Ciencias de los Materiales, Facultad de Ciencias Físicas y Matemáticas, U. de Chile, Casilla 1420, Santiago, Chile. (1929) Dr. rer. nat., physical chemistry of metals (U. Stuttgart, BRD, 1957). Prof., head. (tel. 6982071, ext. 312). *Mechanical properties of materials, corrosion.*

Kittl, Mr Pablo. Microscopía Electrónica, Dept. de Ciencias de los Materiales, Facultad de Ciencias Físicas y Matemáticas, U. de Chile, Casilla 1420, Santiago, Chile. (1934) Lic. en Fis. (U. de Cuyo, Argentina, 1963). Res. (tel. 6982071, ext. 315). *Structure of materials.*

Kremer, Mr Germán. Dept. de Física, Facultad de Ciencias, U. de Chile, Casilla 653, Santiago, Chile. (1941) Prof. de Fís. (U. de Chile, 1966). Res. scient. (tel. 2712973). *Thin metallic films.*

Llanos, Dr Jaime. Dept. de Química, Facultad de Ciencias, U. del Norte, Casilla 1280, Antofagasta, Chile. (1952) Dr. rer. nat. (U. Stuttgart, BRD, 1984). Res. scient. (tel. 222040). *Inorganic crystal structures.*

Manríquez, Dr Víctor. Dept. de Química, Facultad de Ciencias, U. de Chile,, Casilla 653, Santiago, Chile. (1953) Dr. rer. nat. (U. Stuttgart, BRD, 1983). Res. scient. (tel. 2713888). *Inorganic crystal structures.*

Moraga, Mr Luis. Dept. de Física, Facultad de Ciencias, U. de Chile, Casilla 653, Santiago, Chile. (1942) Prof. de Fís. (U. de Chile, 1966). Res. scient. (tel. 2712973). *Solid state physics, electron diffraction.*

Mujica, Dr Carlos. Dept. de Química, Facultad de Ciencias, U. del Norte, Casilla 1280, Antofagasta, Chile. (1954) Dr. rer. nat. (U. Stuttgart, BRD, 1984). Res. scient. (tel. 222040). *Inorganic crystal structures.*

Ossio, Miss Myriam. Sección Metales, Dept. de Ciencias de los Materiales, Facultad de Ciencias Físicas y Matemáticas, U. de Chile, Casilla 1420, Santiago, Chile. (1942) Lic. en Quím. (U. de Chile, 1967). Res. chemist. (tel. 6982071, ext. 314). *Failure analysis, metals and alloys.*

Peña, Miss Luzmila. Dept. de Física, Facultad de Ciencias Físicas y Matemáticas, U. de Chile, Casilla 5487, Santiago, Chile. (1947) Tecnólogo médico (U. de Chile, 1969). Res. assistant. (tel. 6960148). *X-ray crystallography, minerals.*

Pérez, Mrs Carmen. Inst. Investigaciones Tecnológicas (INTEC), Casilla 667, Santiago, Chile. (1945) Ingeniero de Metalurgia Extractiva (U. Técnica del Estado, 1968). Res. (tel. 2289066). *Coal petrography, mineral microscopy.*

Perret, Mr Ramón. Sección Metales, Dept. de Ciencias de los Materiales, Facultad de Ciencias Físicas y Matemáticas, U. de Chile, Casilla 1420, Santiago, Chile. (1932) Lic. en Quím. (U. Católica de Chile, 1957). Res. scient. (tel. 6982071, ext. 312). *Stress corrosion, cracking, metals.*

Rivera, Mr Carlos. Facultad de Física, U. Católica de Chile, Casilla 114-D, Santiago, Chile. (1925) Prof. de Fís. (U. de Chile, 1954). Prof., res. scient. (tel. 2256057). *Crystal optics.*

Schlein, Dr Werner. Centro de Investigación Minera y Metalúrgica (CIMM), Casilla 170, Santiago, Chile. (1936) PhD, chemistry (Brown U., USA, 1971). Director Ejecutivo. (tel. 2289544). *Powder diffractometry.*

Silva, Dr Elisa. Dept. de Física, Facultad de Ciencias Físicas y Matemáticas, U. de Chile, Casilla 5487, Santiago, Chile. (1927) D. 3e cycle, sciences (U. de Paris, France, 1963). Res. scient. (tel. 6982071, ext. 271). *Electron microscopy, bend extinction contours.*

Souza, Mr Carlos. Dept. de Análisis Químico, Centro de Investigación Minera y Metalúrgica (CIMM), Casilla 170, Santiago, Chile. (1954) Lic. en Qui. (U. de Chille, 1980). Res. (tel. 2289544). *X-ray diffraction, X-ray fluorescence.*

Suwalsky, Dr Mario. Dept. de Química, Facultad de Ciencias, U. de Concepción, Casilla 3-C, Concepción, Chile. (1936) PhD, crystallography (Weizmann Inst., Israel, 1969). Prof., res. scient. (tel. 24985, ext. 2171). *Fiber X-ray studies, biologically interesting polymers.*

Varschavsky, Mr Ari. Sección Metales, Dept. de Ciencias de los Materiales, Facultad de Ciencias Físicas y Matemáticas, U. de Chile, Casilla 1420, Santiago, Chile. (1940) Ingeniero Eléctrico (U. de Chile, 1965). Res. scient. (tel. 6982071, ext. 312). *Structure and physical properties, metals, strengthening mechanisms.*

Vera, Mr Rafael. Dept. de Física, Facultad de Ciencias, U. de Concepción, Casilla 20-C, Concepción, Chile. (1927) Ingeniero Químico (U. de Concepción, 1951). Prof. (tel. 24985, ext. 2500). *X-ray spectrography, crystal physics.*

Vogel, Mrs Sonia. Lab. de Rayos X, Servicio Nacional de Geología y Minería, Casilla 10465, Santiago, Chile. (1942) Geólogo (U. de Chile, 1974). Head (tel. 330121). *Powder diffractometry.*

Ward, Mr José. Lab. de Análisis Químico, Centro de Estudios Metalúrgicos, (CESMEC), Casilla 14036, Santiago, Chile. (1934) Prof. de Mat. y Fís. (U. de Chile, 1962). Head. (tel. 746088). *X-ray diffraction, X-ray fluorescence.*

Wittke, Prof. Oscar. Dept. de Física, Facultad de Ciencias Físicas y Matemáticas, U. de Chile, Casilla 5487, Santiago, Chile. (1929) Prof. de Fís. (U. de Chile, 1961). Prof., head, X-ray diffraction lab. (tel. 6960148). *X-ray diffraction.*

Zelada, Mr Gabriel. Lab. de Física, Comisión Chilena de Energía Nuclear, Casilla 188-D, Santiago, Chile. (1955) Lic. en ciencias mención fís. (U. de Chile, 1982). Res. scient. (tel. 2731827, ext. 841). *X-ray diffraction, neutron diffraction.*

Zlosilo, Mr Mario. Dept. de Análisis Instrumental, Centro de Investigación Minera y Metalúrgica (CIMM), Casilla 170, Santiago, Chile. (1943) Prof. de Quím. (U. de Chile, 1967). Assoc. res. (tel. 2289544). *X-ray fluorescence.*

CHINA

Sub-Editor: Xu Xiaojie

Bai, Mr Chun-li. Lab. of Crystal Structure, Inst. of Chemistry, Academia Sinica, Beijing P.O. Box 2709, People's Rep. of China. (1953) MS, structural chemistry (Graduate School, U. of Sci. and Techn. of China, 1981). Res. assoc. *Molecular mechanics, conformation of organic molecules, semiempirical MO calculations, EXAFS, spectroscopy.*

Bi, Dr Ru-chang. Dept. of Protein Crystallogrphy, Inst. of Biophysics, Academia Sinica, Zhong Guan Cun, Beijing 100080, People's Rep. of China. (1940) Dr, molecular biophysics (Leningrad U., USSR, 1965). Res. assoc. (tel. 28-1768). *X-ray structure analysis, protein crystallography, structure and function, biological macromolecules.*

Bi, Mr Yu-run. Dept. of Geology, Peking U., Beijing, People's Rep. of China. (1933). Sr. lect. (tel. 282471). *Crystal structure X-ray analysis.*

Cao, Prof. Ming-zhong. Dept. of Physics, Nankai U., No. 94 Weijin Road, Tianjin, People's Rep. of China. (1935) material sci. Assoc. prof. *Crystal structure analysis, noncrystalline materials.*

Cao, Prof. Zheng-min. Dept. of Geology, Peking U., Beijing, People's Rep. of China. (1933) geological prospecting. Assoc. prof. *Physical properties, mineral crystals, crystal morphology, skarn minerals.*

Chang, Mr Wen-rui. Dept. of Protein crystallography, Inst. of Biophys., Academia Sinica, Zhong Guan Cun, Beijing 100080, Prople's Rep. of China. (1940) chemistry. Res. assoc. (tel. 28-1768). *Protein structure, insulin.*

Chen, Mr Shi-zhi. Dept. of Protein Crystallography, Inst. of Biophys., Academia Sinica, Zhong Guan Cun, P.O. Box No. 349, Beijing 100080, People's Rep. of China. (1935) physical chemistry. Res. assoc. (tel. 28-1768). *Protein crystallography, structural chemistry, X-ray crystallography.*

Chen, Mr Ben-ming. Inst. of Chemistry, Academia Sinica, Zhong Guan Cun, Beijing, People's Rep. of China. (1938). Res. assoc. *Small molecule structures, molecular mechanics.*

Chen, Prof. Dai-zhang. Mineralogy Div., Beijing Graduate School of Wuhan C. of Geology, 29 Chengfu Road, Beijing, People's Rep. of China. (1937) geology. Assoc. prof. *Mineralogy of wall-rock alteration, systematic mineralogy, electron microprobe analysis.*

Chen, Mr Guo-ying. Dept. of Geological Sci., Lanzhou U., Lanzhou, Gansu, People's Rep. of China. (1937) geology. Instructor. *Mineralogy of oxidation zone, powder diffraction.*

Chen, Mr Jing-zhong. Test Center of Rock and Minerals, Wuhan C. of Geology, Yu Jia Shan, Wuchang, Wuhan 430074, Hubei, People's Rep. of China. (1946) MS, mineralogy (Beijing Graduate School of Wuhan C. of Geology, 1981). Lect. (tel. 70481). *Crystal chemistry of minerals, electron microprobe analysis, mineralogy.*

Chen, Prof. Kuang-yuan. Genetic Mineralogy Div., Beijing Graduate School of Wuhan C. of Geology, Chengfu Road 20, Beijing 100083, People's Rep. of China. (1920) Fil. Lic., geology and mineralogy (Uppsala U., Swedan, 1951). Prof. (tel. 277460, ext. 418). *Oxide minerals, silicate minerals, genetic mineralogy, metallogenesis, ore prospecting.*

Chen, Prof. Li-quan. Solid State of Ionics Div., Inst. of Physics, Academia Sinica, Beijing, People's Rep. of China. (1940). Assoc. prof. *Solid state ionics, superionic conductor, crystal growth, crystal physics.*

Chen, Prof. Xian-qiu. Structure Res. Dept., Shanghai Inst. of Ceramics, Academia Sinica, 865 Chang-ning Road, Shanghai 200050, People's Rep. of China. (1925) BS, chemistry (National Zhong Shan U., 1949). Assoc. res. prof. (tel. 522470, telex 33309ASSIC CN). *Crystal growth, optical crystallography, optical mineralogy.*

Chen, Mr Yuan-zhu. Lab. of Structural Chemistry, Fujian Inst. of Res. on the Structure of Matter, Fuzhou, Fujian, People's Rep. of China. (1930) structural chemistry. Lect. *Metal complexes, natural substances, structure analysis.*

Chen, Mr Zhi-xue. Analyzing and Testing Center, Sichuan U., Wang Jia Road, Chengdu 64001, Sichuan, People's Rep. of China. (1935) solid state physics. Instructor. (tel. 316-24401, telex 2345). *Crystal growth, ceramic materials, powder diffraction.*

Chen, Dr Zhong-guo. Lab. for Structure of Matter, Inst. of Physical Chemistry, Peking U., Wei Xiu Yuan, Peking U., Beijing, People's Rep. of China. (1943) PhD, biochemistry (Munich Techn. U., BRD, 1982). Lect. (tel. 282471, ext. 3549). *Protein crystallography, kallikrein, birmann-birk inhibitor, trypsin, pancreatie trypsin inhibitor, superoxide dismutase.*

Cheng, Mr Min-chin. Chemistry Dept., Fudan U., 200 Handan Road, Shanghai 201903, People's Rep. of China. (1939) Lect. (tel. 480906, ext. 280). *Crystal structure analysis, inorganic and metal-organic compounds, drug action, powder diffraction.*

Chiang, Prof. Liang-jun. Geology Div., Central South Inst. of Mining and Metallurgy, Changsha, Hunan, People's Rep. of China. (1913). Prof. (tel. 82811, telex 4349). *Crystallography, spectroscopy, minerals.*

Cia, Mr Jin-hua. Lab. of Structural Chemistry, Fujian Inst. of Res. on the Structure of Matter, Academia Sinica, Fuzhou, Fujian, People's Rep. of China. (1934) Lect. *Transition metal complexes, metal cluster complexes, nitrogen fixation, X-ray crystallography.*

Cui, Prof. Wen-yuan. Dept. of Geplogy, Peking U., Beijing, People's Rep. of China. (1934). Assoc. prof. (tel. 282471, ext. 3227). *Genetic mineralogy.*

Dai, Mr Jin-bi. Dept. of Protein Crystallography, Inst. of Biophys., Academia Sinica, Zhong Guan Cun, Beijing 100080, People's Rep. of China. (1937) mathematics. Res. asst. (tel. 285529). *Direct methods, computing methods, protein crystallography.*

Dong, Mr Ji-he. Mineralogy Div., Qinghai Inst. of Salt Lake, Academia Sinica, No. 8 Xing Ning Street, Xining, Qinghai, People's Rep. of China. (1940). Res. asst. *Cell constants determination, quantitative analysis, powder diffraction.*

Dong, Mr Yi-cheng. Dept. of Protein Crystallography, Inst. of Biophys., Academia Sinica, Zhong Guan Cun, P.O. Box No. 349, Beijing 100080, People's Rep. of China. (1939) physical chemistry. Res. assoc. *Protein crystallography, X-ray crystal structure, structural chemistry.*

Dou, Mr Shi-qi. Dept. of Protein Crystallography, Inst. of Biophys., Academia Sinica, Zhong Guan Cun, P.O. Box No. 349, Beijing 100080, People's Rep. of China. (1940) spectroscopy. Res. assoc. (tel. 28-4483). *Protein crystallography, crystal structure determination, X-ray diffraction.*

Fan, Mr Guang-yu. Chemical Eng. Dept., Beijing Inst. of Techn., Bai Shi Qiao Road, Beijing, P.O. Box 327, People's Rep. of China. (1936) physical chemistry and polymers. Lect. (tel. 89-0321, ext. 605, telex 0055). *Physical chemistry, polymer diffraction, powder diffraction.*

Fan, Prof. Hai-fu. Lab. of X-ray Analysis, Inst. of Physics, Academia Sinica, P.O. Box 603, Zhong Guan Cun, Beijing 100080, People's Rep. of China. (1933) physical chemistry. Prof. *Direct methods, organic natural products, macromolecular structures, electron microscopy.*

Fan, Mr Yu-guo. Inst. of Theoretical Chemistry, Jilin U., 79 Jie Fang Road, Changchun, Jilin, People's Rep. of China. (1940) MS, quantum chemistry (Jilin U., 1981). Lect. *Electron density.*

Fan, Mr Zhao-chang. The Seventeenth Dept., Shanghai Inst. of Organic Chemistry, 345 Lingling Road, Shanghai, People's Rep. of China. (1940) chemical kinetics. Res. assoc. (tel. 313300, ext. 50). *Structural chemistry, crystal structure, organometallic compounds, natural products.*

Fu, Prof. Heng. Lab. of Structural Chemistry, Inst. of Chemistry, Academia Sinica, Zhong Guan Cun, Beijing 100080, People's Rep. of China. (1929) chemistry. Assoc. res. prof. (tel. 284098). *Organic and organometallic structures, bonding in organic compounds.*

Fu, Prof. Ping-qiu. Mineralogy Div., Inst. of Geochemistry, Academia Sinica, Guan Shui Road, Guiyang, Guizhou, People's Rep. of China. (1933). Assoc. prof. (tel. 24757). *Crystal structure, new minerals.*

Fu, Mr Zheng-min. Inst. of Physics, Academia Sinica, P.O. Box 603, Beijing, People's Rep. of China. (1938) physics (Wuhan U.). Res. assoc. *Phase diagrams, phase transitions, crystal structure, DTA, X-ray powder method.*

Fu, Mr Zhu-ji. Lab. of Structural Chemistry, Fujian Inst. of Res. on the Structure of matter, Academia Sinica, Fuzhou, Fujian, People's Rep. of China. (1950) MS, structural chemistry (Fujian Inst. of Res. on the Struture of Matter, 1981). Lect. *X-ray crystallography, metal complexes, protein crystal structure.*

Gao, Mr Yi-guei. Dept. of Protein Crystallography, Inst. of Biophys., Academia Sinica, Zhong Guan Cun, Beijing 100080, People's Rep. of China. (1939) physical chemistry (Nan-Kai U., 1965). Res. assoc. (tel. 28-1768). *Protein crystallography, crystal structure determination.*

Gu, Prof. Xiao-cheng. Dept. of Biology, Peking U., Beijing, People's Rep. of China. (1930) botany. Assoc. prof. (tel. 282471, ext. 3240). *Protein crysatllography.*

Gu, Mr Yuan-xin. Lab. of X-ray Analysis, Inst. of Physics, Academia Sinica, Zhong Guan Cun, Beijing 100080, People's Rep. of China. (1938) physics. Res. assoc. (tel. 281866). *Direct methods, small molecules.*

Guan, Prof. Ya-xian. Crystallography Div., Changchun Inst. of Geology, Changchun, Jilin, People's Rep. of China. (1934) Miner. Dr, mineralogy and crystallography (Crystallography Inst., USSR, 1963). Assoc. prof. (tel. 375-24781). *Mineralogy, X-ray crystallography, geology, mineral crystal structure, crystal chemistry.*

Gui, Ms Lu-lu. Dept. of Crystallography, Inst. of biophys., Academia Sinica, Zhong Guan Cun, Beijing 100080, People's Rep. of China. () analytical chemistry. Res. assoc. *Protein crystallography, growth of protein crystals, structure and function of biological macromolecules.*

Guo, Prof. Dong-yao. Inst. of Theoretical Chemistry, Jilin U., 79 Jiefang Road, Changchun, Jilin, People's Rep. of China. (1935). Assoc. prof. (tel. 23189, ext. 461). *Direct methods, algebraic research, symmetry groups, crystal structure determination.*

Guo, Mrs Fang. Dept. of Crystallography, Inst. of Chemistry, Academia Sinica, Zhong Guan Cun, Beijing, People's Rep. of China. (1939). Res. assoc. *Small molecule structures, computing software, methods of structure determination.*

Han, Mr Fu-son. Lab. of X-ray Analysis, Inst. of Physics, Academia Sinica, Zhong Guan Cun, Beijing 100080, People's Rep. of China. (1946) physics. Res. assoc. (tel. 281866). *Direct methods, small molecules, EXAFS.*

Han, Mr Shao-xu. The Res. Section, Crystal Structure and Crystal Chemistry of Minerals, Beijing Graduate School of Wuhan C. of Geology. Xue Yuan Road 29, Beijing 100083, People's Rep. of China. (1955) MS, crystal structure of minerals (Beijing Graduate School of Wuhan C. of Geology, 1982). Res. asst. (tel. 277461, ext. 560). *Crystal structure, minerals, single crystal diffraction.*

Han, Mr Yu-zhen. Lab. for Structure of Matter, Inst. of Physical Chemistry, Peking U., Beijing, People's Rep. of China. (1940) computational mathematics. Lect. (tel. 282471, ext. 3549). *Crystallographic computing, computational chemistry, computational mathematics.*

He, Prof. Chong-fan. Crystal Lab., Shanghai Inst. of Ceramics, Academia Sinica, 865 Changning Road, Shanghai 200050, People's Rep. of China. (1926) chemistry. Assoc. res. prof. (tel. 522470, telex (761)33309 ASSIC CN). *Single crystal growth, synthetic mica properties, lead molybdate, bismuth germanate, bismuth silicate.*

He, Mr Rei-ling. Geological Sci. Div., Lab. Of Geology and Minerals Bureau of Shanxi Province, He Pin Road, Xi-an, Shanxi, People's Rep. of China. (1936) geology. Engineer (tel. 21024). *X-ray crystallography, powder diffraction, structure analysis.*

Hong, Mr Mao-chun. Lab. of Structural Chemistry, Fujian Inst. of Res. on the Structue of Matter, Academia Sinica, P.O. Box 143, Fuzhou, Fujian, People's Rep. of China. (1953) MS, structural chemistry (Fujian Inst. of Res. on the Structure of Matter, 1981). Res. assoc. *Transition metal complexes, metal cluster compounds, magnetochemistry, X-ray crystallography.*

Hou, Mr Yong-geng. Lab. of Crystal Structure, Inst. of Chemistry, Academia Sinica, Beijing, P.O. Box 2709, People's Rep. of China. (1935) structural chemistry. Res. assoc. *Direct method in crystallography.*

Hu, Mr Heng-liang. Textile Chemistry Dept., Tianjin Textile C. of Techn., Chenglinzhang Road 89, Tianjin 300160, People's Rep. of China. (1942) mathematics. Lect. (tel. 43251). *Crystal structure of macromolecular materials, supermolecular structural parameters of polymers, WAXS and SAXS.*

Hu, Mr Sheng-zhi. Chemistry Dept., Xiamen U., Xiamen, Fuijan, People's Rep. of China. (1932) structure of matter. Sr. lect. *X-ray structure analysis, crystal chemistry.*

Hua, Mr Zi-qian. Dept. of Biology, Peking U., Beijing, People's Rep. of China. (1933) biochemistry. Lect. (tel. 282471, ext. 3240). *Protein crystallography.*

Huang, Mr De-bin. Inst. of Structural Chemistry, Dept. of Chemistry, Fuzhou U., Fuzhou, Fujian, People's Rep. of China. (1951) MS, physical chemistry (Fuzhou U., 1981). Res. asst. *Crystal and molecular structure, transition metal complexes.*

Huang, Prof. Jin-ling. Structural Chemistry Lab., Fujian Inst. of Res. on the Structure of Matter, Academia Sinica, P.O. Box No. 143, Fuzhou, Fujian 350002, People's Rep. of China. (1932) physical chemistry. Res. prof., Inst. deputy director. (tel. 31835). *Crystal and molecular structure, transition metal complexes.*

Huang, Mr Jin-shun. Structural Chemistry Div., Fujian Inst. of Res. on the Structure of Matter, Academia Sinica, Fuzhou, Fujian, People's Rep. of China. (1939). Lect. *Transition metal complexes, metal cluster compounds, X-ray crystallography*

Huang, Mr Liang-ren. Structural Chemistry Div., Fujian Inst. of Res. on the Structure of Matter, Academia Sinica, Fuzhou, Fujian, People's Rep. of China. (1940). Lect. *Transition metal complexes, metal cluster compounds, nitrogen fixation, X-ray crystallography.*

Huang, Mr Tai-shan. Chemistry Dept., Xiamen U., Xiamen, Fujian, People's Rep. of China. (1935) structural chemistry. Lecsturer. *Small crystal crystallography.*

Huang, Mr Zhi-ying. Structural Chemistry Div., Fujian Inst. of Res. on the Structure of Matter, Fuzhou, Fujian, People's Rep. of China. (1941) MS, structural chemistry (Fujian Inst. of Res. on the Structure of Matter, 1981). Lect. *Small organic compound.*

Ji, Prof. Shou-yuan. Dept. of Geology, Nanjing U., Hankou Road, Nanjing 210008, Jinagsu, People's Rep. of China. (1924) BS, geology (Central U., 1947). Assoc. prof. (tel. 34651, ext. 2448). *Crystal chemistry of silicate minerals.*

Jia, Prof. Shou-quan. Crystal Growth Div., Inst. of Physics, Academia Sinica, P.O. Box 603, Beijing 100080, People's Rep. of China. (1930). Assoc. prof., div. head. (tel. 281869). *Crystallography, crystal structure, crystal growth from solutions (including high temperature and high pressure solution), hydrothermal reaction.*

Jiang, Mr An-bei. Laboratory Center of Xiamen U., Si Ming Nan Lu, Xiamen, Fujian, People's Rep. of China. (1939) chemical physics. (tel. 25102-108). *Metal cluster, structure of metal organic and organic compounds.*

Jiang, Mr Han-chen. Physics Dept., Jilin U., Je-fang Road, Chang-chun, Jilin, People's Rep. of China. (1936). Lect. *Electron theory of solids and molecules, structure of metals and alloys.*

Jiang, Mrs Shao-ying. State Bureau of Building Materials, Geological Inst., No. 11 Bei Shun Cheng Street, Xi Zhi Men, Beijing, People's Rep. of China. (1935). Eng., Inst. deputy director. (tel. 551842). *Crystallography, Mineralogy and Physics of minerals.*

Jiang, Mr Xiao-long. Material Physics Div., Shanghai Inst. of Metallurgy, Academia Sinica, 865 Chang Ning Road, Shanghai, People's Rep. of China. (1940) metal physics. Res. assoc. (tel. 520050, ext. 129). *X-ray diffraction, X-ray topography, crystal growth, structure of metal and alloy, phase transition in solids.*

Jiang, Prof. Yan-dao. Crystal Growth Div., Inst. of Physics, Academia Sinica, Beijing, P.O. Box 603, People's Rep. of China. (1936) physics. Assoc. prof., div. head. *Metal growth, imperfection, transport, new material of complex oxide crystals.*

Jin, Dr Wei-qing. Crystal Lab., Shanghai Inst. of Ceramics, Academia Sinica, 865 Chang-ning Road, Shanghai 200050, People's Rep. of China. (1941) Dr, Faculty of Sci. (Tokyo U., Japan, 1984). Res. asst. (tel. 522470, telex (761)33309 ASSIC CN). *Crystal growth mechanism, morphology of crystal, crystal growth from melt.*

Jin, Mr Xiang-lin. Lab. for Structure of Matter, Inst. of Physical Chemistry, Peking U., Beijing, People's Rep. of China. (1940) physical chemistry. Lect. (tel. 282471, ext. 3549). *X-ray single crystal analysis, structural chemistry.*

Ke, Mr Heng-ming. Chemistry Dept., Zhongshan U., Guangzhou, Guangdong, People's Rep. of China. (1948) Polymer Chemistry and Physics. Assoc. lect. (tel. 46300, ext. 218). *Single crystal structures(both of protein and small molecules), powder diffraction, kinetics of crystallization during polymerization.*

Kong, Mr You-hua. Crystal Structure Div., Inst. of Geochemistry, Academia Sinica, Guan Shui Road, Guiyang, Guizhou, People's Rep. of China. (1935). Asst. prof. (tel. 24757). *Crystal structure and crystallochemistry of minerals, diffraction, lattice(or structure) image of minerals, rock-forming minerals, computer programs in above-mentioned fields.*

Kuang, Mrs Bao. Dept. of Biology, Peking U., Beijing, People's Rep. of China. (1943) biophysics. Lect. (tel. 282471, ext. 3240). *Protein Crystallography.*

Kuo, Prof. Ke-hsin. Inst. of Metal Res., Shenyang 110015, Liaoning, People's Rep. of China. (1923) dr tekn h. c., Physical Metallurgy (Royal Inst. of Techn., Sweden, 1980). Prof., Academia Sinica member, foreign member of the Royal Swedish Academy of Eng. Sci. (tel. 483531, ext. 251). *High resolution electron microscopy, surface crystallography.*

Li, Mr Da-ming. R&D Center in Advanced Inorganic Materials, Shanghai Inst. of Ceramics, Academia Sinica, 865 Chang-ning Road, Shanghai 200050, People's Rep. of China. (1933) geology. Head, synthetic diamond res. group. (tel. 522470, telex 33309 ASSIC CN). *Ultra-high pressure and high temperature technology, crystal growth.*

Li, Mr Du. Chemistry Dept., Lanzhou U., Tian Shui Road, Lanzhou, Gansu, People's Rep. of China. (1939) physical chemistry. Lect. (tel. 22991). *Powder diffraction, structure determination, organic molecules, coordination compounds.*

Li, Mr Run-shen. Material Physics Div., Shanghai Inst. of Metallurgy, Academia Sinica, 865 Chang-ning Road, Shanghai, People's Rep. of China. (1943) metal physics. Res. assoc.

Li, Mr Wan-mao. Dept. of Geological Sci., Lanzhou U., 418 Physics Building, Lanzhou, Gansu, People's Rep. of China. (1936) geology. Instructor. *Mineralogy of oxidation zone, powder diffraction.*

Liang, Prof. Dong-cai. Lab. of Protein Crystallography, Inst. of Biophys., Academia Sinica, Beijing 100080, People's Rep. of China. (1932) Cand. chemistry (Acad. of Sci., USSR, 1960). Prof., Inst. director, Academia Sinica member. (tel. 281768). *X-ray crystal structure, protein crystallography, biomolecular structure.*

Liang, Prof. Jing-kui. Crystal Material Div., Fujian Inst. of Res. on the Structure of Matter, Academia Sinica, Fuzhou, Fujian, People's Rep. of China. (1930) material sci. Prof., Inst. director. *Polycrystal structure, inorganic compounds, alloys, phase transition.*

Liang, Ms Li. Dept. of Protein Crystallography, Inst. of Biophys., Academia Sinica, Beijing 100080, People's Rep. of China. (1941) physical chemistry. Res. assoc. (tel. 28-1768). *Structue determination, macromolecules, structure and function, biological molecules, charge density and electrostatic potential.*

Lin, Mr Cheng-yi. Center of Material Analysis, Nanjing U., Hankou Road, Nanjing 210008, Jiangsu, People's Rep. of China. (1938) BS, mineralogy (Leningrad U., U. S. S. R., 1961) Assoc. prof. (tel. 34651, ext. 2523). *Crystal structure analysis, electron microprobe analysis.*

Lin, Prof. Chi-chang. Inst. of Structural Chemistry, Dept. of Chemistry, Fuzhou U., Fuzhou, Fujian, People's Rep. of China. (1933) physical chemistry. Assoc. prof. *Crystal and molecular structure, transition metal complexes.*

Lin, Mr Chuan. Inst. of Physics, Academia Sinica, Beijing, People's Rep. of China. (1940) physics. Inst. deputy director.

Lin, Mr Guang-da. Crystallography Div., Shanghai Inst. of Biochemistry, Academia Sinica, 320 Yue Yang Road, Shanghai 200031, People's Rep. of China. (1937) physics. Res. asst., head, crystallographic group (tel. 374430, telex 3933). *X-ray crystal structure analysis, biological macromolecules.*

Lin, Mr Xian-ti. Structural Chemistry Div., Fujian Inst. of Res. on the Structure of Matter, Academia Sinica, Fuzhou, Fujian, People's Rep. of China. (1938). Lect. *Transition metal complexes, metal cluster complexes, X-ray crystallography.*

Lin, Mrs Yu-juon. Structural Chemistry Div., Fujian Inst. of Res. on the Structure of Matter, Academia Sinica, Fuzhou, Fujian, People's Rep. of China. (1939). Lect. *Transition metal complexes, X-ray crystallography, protein crystal structure.*

Lin, Prof. Zheng-jiong. Dept. of Protein Crystallography, Inst. of Biophys., Academia Sinica, Beijing 100080, People's Rep. of China. (1935) physical chemistry. Assoc. prof. (tel. 28-1768). *Protein crystallography, X-ray diffraction, structure and function, biological macromolecules.*

Liu, Mr Guang-zhao. R&D Center in Advanced Inorganic Materials, Shanghai Inst. of Ceramics, Academia Sinica, 865 Chang-ning Road, Shanghai 200050, People's Rep. of China. (1942) crystal growth. Group Leader, Large Single Diamand Crystal Res. (tel. 522470, telex 33309 ASSIC CN). *Crystal growth, Monte Carlo computer stimulation.*

Liu, Dr Han-qin. Structural Chemistry Div., Fujian Inst. of Res. on the Structure of Matter, Fuzhou, Fujian, People's Rep. of China. (1937) PhD, chemical physics (U. of Chicago, USA, 1967). Assoc. res. prof. *Crystal and molecular structure, magnetic structure, metal cluster complexes, magnetic resonance.*

Liu, Mr Shi-xiong. Inst. of Structural Chemistry, Dept. of Chemistry, Fuzhou U., Fuzhou, Fujian, People's Rep. of China. (1943) physical chemistry. Lect. *Crystal and molecular structure, transition metal complexes.*

Liu, Mr Tian-liang. Inst. of Structural Chemistry, Dept. of Chemistry, Fuzhou U., Fuzhou, Fujian, People's Rep. of China. (1936) physical chemistry. Lect. *Crystal and molecular structure, transition metal complexes.*

Liu, Mr Wan. Geology Div., Geological Inst., The Palace of Geology, Changchun, Jilin, People's Rep. of China. (1935) mineralogy and X-ray crystallography. Instructor. (tel. 375-24781). *Mineralogy, X-ray crystallography, powder diffraction, crystal structure.*

Liu, Mr Xue-lun. Geology and Mineral Resources Div., Lab. of Inner Mongolia Bureau of Geology and Mineral Resources, Geology Building, Hulun South Road No. 4, Huhehaote, Inner Mongolia, People's Rep. of China. (1930) X-ray crystal structure analysis. Engineer (tel. 24512). *X-ray crystallography, optical crystallography, powder diffraction.*

Liu, Mr Zuo-cai. Dept. of Chemical Eng., Beijing Techn. Inst., Bai Shi Qiao Road, Beijing, P.O. Box 327, People's Rep. of China. (1946) MS, structural chemistry (Peking U., 1982). Lect. (tel. 89-0321, ext. 605). *Organic compounds, complexse, solid chemistry.*

Lou, Mrs Mei-zhen. Dept. of Protein Crystallography, Inst. of Biophys., Academia Sinica, Zhong Guan Cun, Beijing 100080, People's Rep. of China. (1935) organic chemistry. Res. assoc. *Protein structure, insulin, crystal growth.*

Lu, Mrs Quang-ying. Dept. of Biology, Peking U., Beijing, People's Rep. of China. (1937) biochemistry. Lect. (tel. 28-2471, ext. 3240). *Protein crystallography.*

Lu, Prof. Jia-xi. Structural Chemistry Lab., Fujian Inst. of Res. on the Structure of Matter, Academia Sinica, P.O. Box No. 143, Fuzhou, Fujian 550002, People's Rep. of China. (1915) PhD, physical chemistry (U. of London, UK, 1939). Res. prof., Academia Sinica member, pres. (tel. 31835). *Transition metal cubanelike clusters, nitrogenase active center model compounds, crystal and molecular structure.*

Lu, Mr Kun-quan. X-ray Analysis Lab., Inst. of Physics, Academia Sinica, Zhong Guan Cun, Beijing, P.O. Box 603, People's Rep. of China. (1939) physics. Res. assoc. (tel. 281866). *EXAFS, X-ray diffraction, crystal growth.*

Lu, Ms Qi. Test Center of Rocks and Minerals, Wuhan C. of Geology, Yu Jia Shan, Wuchang, Wuhan 430074, People's Rep. of China. (1942) MS, mineral crystal chemistry and structure (Beijing Graduate School of Wuhan C. of Geology, 1981). Lect. (tel. 70481). *Crystal chemistry of minerals, X-ray crystallography, clay minerology.*

Lu, Mr Shao-fang. Structural Chemistry Div., Fujian Inst. of Res. on the Structure of Matter, Academia Sinica, Fuzhou, Fujian, People's Rep. of China. (1940).

Lect. *Metal cluster compounds, transition metal complexes, X-ray crystallography, structure-property relations, electron density distribution, transition metal complexes.*

Lu, Prof. Yun-jin. Center of Material Analysis, Nanjing U., Hanlou Road, Nanjing, Jiangsu, People's Rep. of China. (1928) crystallography. Assoc. prof. (tel. 34651, ext. 2523). *Crystal structure, powder diffraction.*

Luo, Prof. Gu-feng. Dept. of Geology, Nanjing U., Hankou Road, Nanjing 210008, Jiangsu, People's Rep. of China. (1933) mineralogy and petrology. Assoc. prof. (tel. 34651, ext. 2448). *X-ray analysis, rock-forming minerals.*

Ma, Prof. Li-dun. Dept. of Chemistry, Fudan U., 200 Handan Road, Shanghai 201903, People's Rep. of China. (1935). Assoc. prof. (tel. 480906, ext. 353). *Structure, inorganic and metal-organic compounds, polycrystalline diffraction, EXAFS.*

Ma, Mr Xing-qi. Dept. of Protein Crystallography, Inst. of Biophys., Academia Sinica, Zhong Guan Cun, Beijing 100080, People's Rep. of China. (1936) physical chemistry. Res. assoc. *X-ray crystal structure, protein crystallography, structural chemistry.*

Ma, Prof. Zhe-sheng. The Res. Section, Crystal Structure and Crystal Chemistry of Minerals, Beijing Graduate School of Wuhan Geological C., Xue Yuan Road 29, Beijing 100083, People's Rep. of China. (1937) crystal structure of minerals. Assoc. prof. (tel. 277461, ext. 560). *Crystal structure, minerals, single crystal diffraction.*

Mai, Prof. Zhen-hong. Crystal Defects Div., Inst. of Physics, Academia Sinica, P.O. Box 603, People's Rep. of China. (1942). Assoc. prof., div. head. (tel. 281869). *Crystallography, crystal defects, X-ray diffraction, topography.*

Meng, Prof. Yi-min. Chemistry Dept., Lanzhou U., Tian Shui Road, Lanzhou, Gansu, People's Rep. of China. (1926) physical chemistry. Assoc. prof. (tel. 22991). *Crystal chemistry, single crystal diffraction.*

Miao, Mr Chun-sheng. Bureau of Geology and Mineral Resources of Liaoning Province, Experiment and Res. Center, Bei Ling Street, Shenyang, Liaoning, People's Rep. of China. (1940). Engineer (tel. 61768, telex 6012). *Powder diffraction, quantitative analysis of X-ray phase, feldspar structure state, crystal lattice measurement.*

Miao, Prof. Fang-ming. Chemistry Dept., Tianjin Normal U., Ba Li Tai, Tianjin, People's Rep. of China. (1935) physical chemistry. Prof., dept. dean. (tel. 23489). *Crystal and molecular structures, small molecules.*

Mu, Mr Xiang-qi. Textile Chemistry Dept., Tianjin Textile C. of Techn., Cheng Lin Zhuang Road 89, Tianjin, People's Rep. of China. (1938) physical chemistry. Lect. (tel. 43251). *Structure of polymers, fiber diffraction methods.*

Ni, Mr Chau-zhou. The Seventeenth Dept., Shanghai Inst. of Organic Chemistry, Abcademia Sinica, 345 Lingling Road, Shanghai, People's Rep. of China. (1937) physical chemistry. Res. assoc. (tel. 313300, ext. 50). *Small molecules, proteins, crystal structure, organometallic compounds, natural products.*

Pan, Prof. Ke-zhen. Structural Chemistry Div., Fujian Inst. of Res. on the Structure of Matter, Academia Sinica, Fuzhou, Fujian, People's Rep. of China. (1933) structural chemistry. Assoc. res. prof. *Metal complexes, X-ray crystallography, proteins, crystal structure.*

Pan, Prof. Zhao-lu. Dept. of Geology, Wuhan C. of Geology, Yu Jia Shan, Wuhan 5378, Hubei, People's Rep. of China. (1925) BS, geology (Qinghua U., 1949). Prof. (tel. 70481). *Crystal chemistry, mineralogy.*

Pan, Mr Zuo-hua. Lab. for Structrue of Matter, Inst. of Physical Chemistry, Peking U., Beijing, People's Rep. of China. (1934) structural chemistry. Lect. (tel. 282471, ext. 3549) *X-ray crystallography, simple crystal structure analysis.*

Peng, Mr Chang-qi. Mineralogy Div., Yichang Inst. of Geology and Mineral Resources. P.O. Box 502, Yichang, Hubei, People's Rep. of China. (1940). Asst. res. fellow. (tel. 21635). *Mineralogy, crystallochemistry, X-ray powder diffraction.*

Peng, Mrs Ming-shen. Geology Div., Central-south Inst. of Mining and Metallurgy, Changsha, Hunan, People's Rep. of China. (1939). Lect. (tel. 82811, telex 4349). *Crystal field theory, spectroscopy, minerals, X-ray crystallography.*

Peng, Prof. Zhi-zhong. Res. Section for Crystal Structure and Crystallochemistry of Minerals, Beijing Graduate School, Wuhan C. of Geology, Xue Yuan Road 29, Beijing, People's Rep. of China. (1932) geology. Head, res. section. (tel. 277461, ext. 560). *X-ray crystallography, mineralogy, geometric crystallography, crystal structure analysis, crystallochemistry of minerals.*

Qi, Dr Zeng-du. R&D Center in Advanced Inorganic Materials, Shanghai Inst. of Ceramics, Academia Sinica, 865 Chang-ning Road, Shanghai 200050, People's Rep. of China. (1937) PhD, physics (Reading U., UK, 1981). Head, high pressure res. dept. (tel. 522470, telex 33309 ASSIC CN). *Crystal growth, high pressure physics.*

Qi, Prof. Zhi-ru. Teaching Res. Group of Petrology and Mineralogy, Xi-an Geological U., 6 Yan Ta Road, Xi-an Shanxi, People's Rep. of China. (1928). Assoc. prof. (tel. 52991). *Powder diffraction, crystal structure.*

Qian, Ms Jin-zi. Lab. of X-ray Analysis, Inst. of Physics, Academia Sinica, Zhong Guan Cun, Beijing 100080, People's Rep. of China. (1940) physics. Res. assoc. (tel. 281489). *Direct methods, small molecules.*

Qian, Mrs Min-xie. Dept. of Crystallography, Inst. of Chemistry, Academia Sinica, Zhong Guan Cun, Beijing, People's Rep. of China. (1949) MS, structural chemistry (Peking U., 1981). Res. assoc. *Structure, small molecule, electron density, molecular mechanics.*

Ren, Prof. Lei-fu. Dept. of Geology, Peking U., Beijing, People's Rep. of China. (1930). Assoc. prof. (tel. 282471, ext. 3354). *Clay mineralogy.*

Shan, Mr Try-seo. Dept. of Chemical Eng., Beijing Inst. of Techn., Beijing, China. (1940) MS, structural chemistry (Peking U., 1981). Engineer. *X-ray crystallography and applications.*

Shang, Mr Mao-yu. Structural Chemistry Div., Fujian Inst. of Res. on the Structure of Matter, Academia Sinica, Fuzhou, Fujian, People's Rep. of China. (1944) MS, structural chemistry (Fujian Inst. of Res. on the Structure of Matter, 1981). Lect. *Transition metal complexes, metal cluster compounds, X-ray crystallography.*

Shao, Prof. Jie-lian. Dept. of Geology, Wuhan C. of Geology, Yu Jia Shan, Wuhan 430074, Hubei, People's Rep. of China. (1929) geology. Assoc. prof. (tel. 70481). *Crystallogeny, ore mineralogy.*

Shao, Prof. Mei-cheng. Lab. for Structure of Matter, Inst. of Physical Chemistry, Peking U., Zhong Guan Cun, Beijing, People's Rep. of China. (1931) structural chemistry. Assoc. prof., head, crystallographic structure res. group (tel. 282471, ext. 3549). *Transition metal complexes, X-ray crystallography.*

Shen, Miss Fu-ling. X-ray Crystal Structure Div., Inst. of Biophys., Academia Sinica, Zhong Guan Cun, Beijing 100080, People's Rep. of China. (1941). *X-ray crystal structure.*

Shen, Mr Cheng. Inst. of Theoretical Chemistry, Jilin U., 79 Jiefang Road, Changchun, Jilin, People's Rep. of China. (1946). Lect. *Crystal structure determination.*

Shen, Prof. Jin-chuan. Test Center of Rocks and Minerals, Wuhan C. of Geology, Yu Jia Shan, Wuchang, Wuhan 430074, Hubei, People's Rep. of China. (1936). Chief eng. (tel. 70481). *Crystal chemistry of minerals, rare earth compounds, X-ray crystallography.*

Shi, Mr Bi-de. Chemistry Dept., Xiamen U., Xiamen, Fujian, People's Rep. of China. (1934) structure of matter. Lect. *X-ray crystallography.*

Shi, Mr Ni-cheng. The Res. Section, Crystal Structure and Crystal Chemistry of Minerals, Beijing Graduate School of Wuhan Geological C., Xue Yuan Road 29, Beijing 100083, People's Rep. of China. (1938) crystal structure of minerals. Lect. (tel. 277461-560). *Crystal structure, minerals, crystallographic computing, single crystal diffraction.*

Shu, Mr Jin-fu. Test Center of Rocks and Minerals, Wuhan C. of Geology, Yu Jia Shan, Wuchang, Wuhan 430074, Hubei, People's Rep. of China. (1942) MS, mineral crystal chemistry and structure (Beijing Graduate School, Wuhan C. of Geology, 1981). Lect. (tel. 70481, telex 5378). *Crystal chemistry of mineral, X-ray crystallography, mineralogy on feldspar.*

Song, Mr Shi-ying. Dept. of Protein Crystallography, Inst. of Biophys., Academia Sinica, Beijing 100080, People's Rep. of China. (1938) physical chemistry. Res. assoc. *Protein crystallography, X-ray diffraction, structure and function, biological macromolecules.*

Sun, Mr Yi-jian. Petrology and Mineralogy Div., Nanjing Inst. of Geology and Mineral Resources, CAGS, 534 East Zhongshan Road, Nanjing, Jinagsu, People's Rep. of China. (1931) X-ray crystallography. Assoc. sr. res. fellow (tel. 5206-43992). *Powder diffraction, rock-forming minerals, feldspars.*

Tan, Prof. Hao-ran. Crystal Lab., Shanghai Inst. of Ceramics, Academia Sinica, 865 Changning Road, Shanghai 200025, People's Rep. of China. (1927) BS, physics (Ling-nan U., 1949). Assoc. prof. (tel. 522740). *Powder diffraction, new crystal materials, acousto-optic properties of crystals.*

Tang, Prof. You-qi. Inst. of Physical Chemistry, Peking U., Beijing 100871, People's Rep. of China. (1920) PhD, chemistry (Caltech, 1950). Prof., Inst. director, Academia Sinica member. (tel. 281471). *X-ray structure analysis, symmetry, structural chemistry.*

Tang, Mr Zhi-kai. Central Lab., Geological Bureau of Sichuan, 101 Renmin Northern Road, Chengdu, Sichuan, People's Rep. of China. (1932) chemistry. Engineer (tel. 31097). *Mineral chemistry, mineral X-ray crystallography, powder diffraction, instrumentation.*

Wang, Prof. Da-cheng. Dept. of Protein Crystallography, Inst. of Biophys., Academia Sinica, Zhong Guan Cun, Beijing 100080, People's Rep. of China. (1940) biophysics. Assoc. prof. *Protein structure, hormones, insulin, enzyme, zymogen.*

Wang, Prof. Gen-yuan. Dept. of Geology, Wuhan C. of Geology, Wuhan 430074, Hubei, People's Rep. of China. (1933) geology. Assoc. prof. (tel. 70481). *History of crystallography- mineralogy and geology in China (especially the ancient period).*

Wang, Mr Guan-xin. X-ray Diffraction Lab., Inst. of Geochemistry, Academia Sinica, Guanshui Road, Guiyang, Guizhou, People's Rep. of China. (1936). Sr. engineer (tel. 24757). *Crystallography and mineralogy, systematic mineralogy, isomorphous substitution of minerals, X-ray diffraction, instrumentation, powder diffraction method.*

Wang, Mr Jia-huai. Dept. of Protein Crystallography, Inst. of Biophys., Acaemia Sinica, Zhong Guan Cun, Beijing 100080, People's Rep. of China. (1941) biophysics. Res. assoc. *Protein crystallography, X-ray crystal structure, structural chemistry.*

Wang, Prof. Kui-ren. Dept. of Earth and Space Sci., U. of Sci. and Techn. of China, 24 Jin Zhai Road, Hefei, Anhui, People's Rep. of China. (1934) X-ray Analysis of Mineral. Assoc. prof., dept. vicehead. *X-ray analysis of minerals, genetical mineralogy, X-ray crystallography.*

Wang, Prof. Pu. Mineralogy Div., Beijing Graduate Inst. of Wuhan C. of Geology, No. 29, Xue Yuan Road, Beijing, People's Rep. of China. (1926) BS, geology

(Tsin Hwa U., 1949). Prof. *Mineralogy of wall-rock alteration, crystal growth, structure, systematic mineralogy.*

Wang, Prof. Shun-jin. Dept. of Geology, Wuhan C. of Geology, Yu Jia Shan, Wuhan 430074, Hubei, People's Rep. of China. (1933) geology. Assoc. prof. (tel. 70481). *Mineralogy of deposits, crystallography of minerals, crystallochemistry of minerals, magnetite, wolframite, scheelite, pyoxenes, amphiboles, plagioclase.*

Wang, Prof. Wen-kui. Dept. of Geology, Wuhan C. of Geology, Yu Jia Shan, Wuhan 430074, Hubei, People's Rep. of China. (1925) BS, geology (Peking U., 1948). Assoc. prof. (tel. 70481). *Morphology, surface microtopography of crystals.*

Wang, Mr Xing-xin. Lab. of Geology, Res. Inst. of Petroleum E&D of Daqing Oil Field, Daqing City, Heilongjiang, People's Rep. of China. (1936). Principle geologist. *Clay minerals, powder diffraction.*

Wang, Mrs Yao-ping. Dept. of Protein Crystallography, Inst. of Biophys., Academia Sinica, Zhong Guan Cun, Beijing 100080, People's Rep. of China. (1939) silicates. Res. assoc. *X-ray crystallography, protein crystallography, structural chemistry.*

Wang, Mr Zhao-zhou. Bureau of Geology and Mineral Resources of Liaoning Province, Exp. and Res. Center, Bei Ling Street, Shenyang, Liaoning, People's Rep. of China. (1937). Engineer (tel. 61768, telex 6012). *Powder diffraction, quantitative X-ray phase analysis, feldspar structure state, crystal lattice measurement.*

Wang, Prof. Zu-tao. Chemistry Dept., Tianjin Advanced Inst. of Sci. and Techn., No. 379 Heping Road, Tianjin, People's Rep. of China. (1924). Assoc. prof. (tel. 3-3467). *Crystal structure, organometallic compounds, drug structure - activity relationship, powder diffraction.*

Wei, Mr Ming-xiu. Dept. of Mineralogy and Petrology, Res. Inst. of Geology for Mineral Resources, San Li Dian, Guilin, Guangxi, People's Rep. of China. (1940) geochemistry. Commiteeman of Mineralogy and Crystallography Society of China. *X-ray diffraction, mineral structure, X-ray fluorescence analysis, mineralogy, petrology, geochemistry.*

Wei, Mr Xin-cheng. Dept. of Biology, Peking U., Beijing, People's Rep. of China. (1937) biophysics. Lect. (tel. 28-2471, ext. 3240). *Protein crystallography.*

Weng, Prof. ling-pao. Mineralogy Dept., Beijing Graduate School of Wuhan C. of Geology, 29 Xue Yuan Road, Beijing, People's Rep. of China. (1932) earth science. Assoc. prof. *Mineralogy of wall-rock alteration, systematic mineralogy, rare element mineralogy.*

Wu, Mr Bo-mu. X-ray Crystal Structure Analysis Div., Inst. of Biophys., Academia Sinica, Zhong Guan Cun, Beijing 100080, People's Rep. of China. (1937) physical chemistry. Res. asst. *X-ray crystal structure analysis, X-ray data collection and treatment.*

Wu, Mr Ding-ming. Structural Chemistry Div., Fujian Inst. of Res. on the Structure of Matter, Academia Sinica, Fuzhou, Fujian, People's Rep. of China. (1943). Lect. *Transition metal complexes, metal cluster compounds, X-ray crystallography.*

Wu, Prof. Qian-zhang. Lab. of Crystallography, Inst. of Physics, Academia Sinica, P.O. Box 603, Beijing, People's Rep. of China. (1910) BS, physics (National Central U., 1933). Assoc. res. prof. (tel. 281869). *High temperature solution crystal growth, structure, phase transition, characterization and application.*

Wu, Mr Shen. Dept. of Protein Crystallography, Inst. of Biophys., Academia Sinica, Zhong Guan Cun, Beijing 100080, People's Rep. of China. (1940) physical chemistry. Res. assoc. *X-ray crystal structure, protein crystallography.*

Wu, Prof. Shou-yu. Dept. of Chemistry, Sichuan U., Wang Jia Road, Chengdu 64001, Sichuan, People's Rep. of China. (1930) chemistry. Assoc. prof. (tel. 24401-266). *Coordination chemistry, chemical bond, crystal structure, powder diffraction.*

Wu, Mr Xin-tao. Structural Chemistry Div., Fujian Inst. of Res. on the Structure of Matter, Academia Sinica, Fuzhou, Fujian, People's Rep. of China. (1939). Lect. *Metal cluster compounds, transition metal complexes, X-ray crystallography, structure-property relations, EXAFS.*

Wu, Mr Zi-wu. Lab. of Structural Chemistry, Fujian Inst. of Res. on the Structure of Matter, Fuzhou, Fujian, People's Rep. of China. (1938) structural chemistry. Lect. *Metal complexes, structural analysis, natural substance, protein crystallography.*

Xia, Mrs Zong-xiang. The Seventeenth Dept., Shanghai Inst. of Organic Chemistry, 345Lingling Road, Shanghai, People's Rep. of China. (1942) physical chemistry. Res. assoc. (tel. 313300, ext. 50). *X-ray crystal structure analysis, proteins, organometallic compounds, natural products.*

Xiao, Prof. Xu-gang. Dept. of Geology, Northeast U. of Techn., Wenhua Raod, Shenyang, Liaoning, People's Rep. of China. (1921) crystallography. Prof. (tel. 483081, telex. 80033 NEIT CN). *Geometry of crystal structure, group theory applied to crystallography, X-ray crystallography, electron microscopy, diffraction physics, defects in mineral and metal crystals, mineralogy.*

Xie, Dr Si-shen. Dept. A-10, Lab. of Phase Transition and Diagram, Inst. of Physics, Academia Sinica, Beijing, P.O. Box 603, People's Rep. of China. (1942) Dr, solid state physics (Inst. of Physics, Academia Sinica, 1983). Res. assoc. (tel. 28-2271). *Phase transformation, phase diagram, crystal structure, physical properties - structure relationship.*

Xie, Prof. Xian-de. Mineralogy Div., Inst. of Geochemistry, Academia Sinica, Guiyang, Guizhou, People's Rep. of China. (1934). Assoc. prof. *X-ray crystallography, lattice distortion of shocked minerals.*

XiMen, Mrs Lu-lu. Geological Sci. Div., Beijing Graduate School of Wuhan Geological C., 29 Xue Yuan Road, Beijing, People's Rep. of China. (1934) geology. Lect. (tel. 27-7461, ext. 560). *Crystal structure, crystal chemistry, minerals, polycrystalline X-ray diffraction, single-crystal analysis.*

Xu, Mr Ji-quan. Physical Chemistry Div., Inst. of Soil Sci., Academia Sinica, 71 Eastern Beijing Road, Nanjing, P.O. Box 821, Jiangsu, People's Rep. of China. (1928). Res. assoc. (tel. 33318). *Mixed layer minerals, layer silicates, powder diffraction.*

Xu, Mr Jing-yang. Material Physics Div., Shanghai Inst. of Metallurgy, Academia Sinica, 865 Chang Ning Road, Shanghai, People's Rep. of China. (1936) metal physics. Res. assoc. (tel. 520050, ext. 129). *X-ray diffraction, X-ray topography, structure, metal and alloys, phase transition in solids.*

Xu, Mr Pei-cang. Mineralogy and Petrology Dept., Xi-an Inst. of Geology and Mineral Resources, Xi-an Inst. of Geol. and Res. Building, West End of You-yi Road, Xi-an, Shan-xi, People's Rep. of China. (1942) physics. Engineer (tel. 5-1266). *Crystal superstructure, minerals, X-ray analysis, spectrum analysis, solid physics, quantum mineralogy, ligand field theory.*

Xu, Prof. Shun-sheng. Materal Physics Div., Shanghai Inst. of Metallurgy, Academia Sinica, 865 Chang Ning Road, Shanghai 200050, People's Rep. of China. (1920) PhD, metal physics (U. of Notre Dame, USA, 1953). Res. prof. (tel. 520050, ext. 129). *X-ray diffraction, X-ray tnpography, crystal growth, structure, metals and alloys, electron microscopy and diffraction, phase transition in solids.*

Xu, Mr Xiao-jie. Lab. for Structure of Matter, Inst. of Physical Chemistry, Peking U., Beijing, People's Rep. of China. (1937). Lect. (tel. 282471, ext. 3549). *Single crystal structure determination, molecular mechanics.*

Xu, Mr Zheng-yi. Optics and Electricity Lab., Inst. of Physics, Academia Sinica, Beijing, P.O. Box 603, People's Rep. of China. (1935) solid state theory. Assoc. prof. (tel. 28-1869). *Ionic conductor, ionic transportation, physical properties of crystals, light scattering in crystals, defect - physical property relations, optical spectrum of crystal.*

Xue, Mrs Ji-yue. Dept. of Geology, Nanjing U., Hankou Road, Nanjing, Jiangsu, People's Rep. of China. (1938) crystallography and mineralogy. Lect. (tel. 34651, ext. 2448). *Crystal structure analysis, X-ray diffraction, minerals.*

Xue, Mr Jun-zhi. Dept. of Geology, Wuhan C. of Geology, Yujiashan, Wuhan 430074, Hubei, People's Rep. of China. (1935) mineralogy. Assoc. prof. (tel. 70481). *Ontogeny and phylogeny, genetic mineralogy, physical-chemistry, natural solid solutions.*

Xue, Prof. Zhi-lin. R&D Center in Advanced Inorganic Materials, Shanghai Inst. of Ceramics, Academia Sinica, 865 Chang-ning Road, Shanghai 200050, People's Rep. of China. (1929) physics. Assoc. res. prof., deputy head, R&D Center (tel. 522470, telex. 33309 ASSIC CN). *Crystal growth, physical properties, characterization, automation, microcomputers.*

Yan, Mr Qi-wei. Neutron Scattering Div., Inst. of Physics, Academia Sinica, P.O. Box 603, Beijing, People's Rep. of China. (1941) MS, neutron scattering (Inst. of Physics, Academia Sinica, 1981). (tel. 285047). *Neutron scattering, neutron inelastic scattering, neutron diffraction.*

Yan, Mr You-wei. Group of X-ray Crystallography, Shanghai Inst. of Biochemistry, Academia Sinica, 320 Yue Yang Road, Shanghai 200031, People's Rep. of China. (1946) MS, crystallography (Shanghai Inst. of Biochemistry, 1983). Res. asst. (tel. 374430, telex 3933). *X-ray protein crystallography.*

Yang, Mr Chuan-zheng. Material Physics Div., Shanghai Inst. of Metallurgy, Academia Sinica, 865 Chang-ning Road, Shanghai, People's Rep. of China. (1939) metal physics. Res. assoc. (tel. 520050, ext. 129). *X-ray diffraction, X-ray topography, crystal growth, structure, metals and alloys, electron microscopy and diffraction, phase transition in solids.*

Yang, Mr Guang-di. Inst. of Theoretical Chemistry, Jilin U., 79 Jiefang Road, Changchun, Jilin, People's Rep. of China. (1942). Lect. *Crystal structure determination.*

Yang, Ms Guang-ming. Test Center of Minerals and Rocks, Wuhan C. of Geology, Yujiashan, Wuchang, Wuhan 430074, Hubei, People's Rep. of China. (1940) MS, crystal chemistry of minerals (Wuhan C. of Geology, 1981). Lect. *Crystal structure, crystallochemistry, minerals, rocks.*

Yang, Mr Hua-guang. Optics and Electricity Lab., Inst. of Physics, Academia Sinica, Beijing, P.O. Box 603, People's Rep. of China. (1936) crystal physics. Res. assoc. (tel. 281869). *Physical properties of crystals, phase transition in solids, light scattering in solids.*

Yang, Mr Hua-hui. Chemistry Dept., Xiamen U., Xiamen, Fujian, People's Rep. of China. (1934) Analytical Chemistry. Lect. *Powder diffraction, EXAFS.*

Yang, Prof. Qi-bin. Inst. of Metal Res., Shenyang 110015, Liaoning, People's Rep. of China. (1938) metal physics. Assoc. prof. (tel. 483531, ext. 251). *Crystal structures, metals and alloys, electron diffraction and microscopy.*

Yang, Ms Qing-chuan. Lab. for Structure of Matter, Inst. of Physical Chemistry, Peking U., Beijing, People's Rep. of China. (1940). Lect. (tel. 282471, ext. 3549). *Structural chemistry, accurate determination of molecular structure.*

Yang, Mr Zuo-sheng. Marine Geology Dept., Shandong C. of Oceanology, 5 Yushan Road, Qindao, P.O. Box 90, Shandong, People's Rep. of China. (1937). Lect. *Powder diffraction, mineralogy.*

Yao, Mr Jia-xing. Lab. of X-ray Analysis, Inst. of Physics, Academia Sinica, Zhong Guan Cun, Beijing 100080, People's Rep. of China. (1941) radio. Res. assoc. (tel. 28-1489). *Direct methods, small molecules.*

Yao, Mr Xin-kan. Test and Computation Center, Nankai U., 94 Weijin Road, Tianjin, People's Rep. of China. (1939). Lect. (tel. 274226). *Small molecules, powder diffraction.*

Ye, Prof. Heng-qiang. Lab. of High-Resolution Electron Microscope, Inst. of Metal Res., Academia Sinica, Wenhua Road, Shenyang 110015, Liaoning, People's Rep. of China. (1940) metal physics. Assoc. prof. (tel. 483531, ext. 242). *Crystal structure, metals and alloys, crystal defects, electron diffraction.*

You, Mr Jun-ming. Dept. of Protein Crystallography, Inst. of Biophys., Academia Sinica, Zhong Guan Cun, Beijing 100080, People's Rep. of China. (1928) chemistry. Res. assoc. (tel. 285529). *Protein structure, insulin, crystal growth.*

You, Prof. Xiao-zeng. Inst. of Coordination Chemistry, Nanjing U., Nanjing, Jiangsu, People's Rep. of China. (1934) physical chemistry. Prof. (tel. 34651, telex 0909). *Crystal structure, coordination chemistry, structural chemistry.*

Yu, Prof. Rui-huang. Physics Dept., Jilin U., Chang-chun, Jilin, People's Rep. of China. (1906) PhD, physics (U. of Manchester, UK, 1938). Prof., Academia Sinica member. *Electron theory of solids and molecules, crystal and molecular structure, X-ray crystal structure analysis, syntheses of X-ray data.*

Yu, Prof. Wei-hai. Dept. of Physics, U. of Sci. and Techn. of China, Hefei, Anhui, People's Rep. of China. (1929). Assoc. prof., dept. vice-chairman. *Solid state physics.*

Yu, Prof. Xiu-fen. Inst. of Structural Chemistry, Dept. of Chemistry, Fuzhou U., Fuzhou, Fujian, People's Rep. of China. (1932) physical chemistry. Assoc. prof. *Crystal and molecular structure, transition metal complexes.*

Zhang Mr Bu-sheng. Applied Physics Div., Geological Inst., Changchun, Jilin, People's Rep. of China. (1940) X-ray crystallography. Instrutor (tel. 375-24781). *Crystallography, crystal physics, crystal construction in minerals, powder diffraction, computing.*

Zhang, Mr Ci-he. Inst. of Biophys., Academia Sinica, Zhong Guan Cun, Beijing 100080, People's Rep. of China. (1942) technical physics. *X-ray diffraction instrumentation, intellectual instruments.*

Zhang, Prof. Dao-biau. Crystal Lab., Shanghai Inst. of Ceramics, Academia Sinica, 865 Chang-ning Road, Shanghai 200050, People's Rep. Of China. (1931) chemical engineering. Assoc. res. prof. (tel. 522470). *Crystal growth from the melt, crystal properties.*

Zhang, Mrs Gen-di. Dept. of Geology, Nanjing U., Hankou Road, Nanjing 210008, Jiansu, People's Rep. of China. (1939) mineralogy and geochemistry. Assoc. Scientist (tel. 34651, ext. 2448). *X-ray diffraction, minerals.*

Zhang, Prof. Guan-ying. Non-metallic Minerals Div., Wuhan Inst. of Building Materials, No. 8 Leshi Road, Wuchang, Wuhan, Hubei, People's Rep. of China. (1933) geology. Assoc. prof. *Non-metallic minerals, X-ray powder diffraction, X-ray fiber photographys, chrysotiles, crystal symmetry.*

Zhang, Mr Guang-rong. State Bureau of Building Materials, Geological Inst., No. 11 Bei Shun Cheng Street, Xi Zhi Men, Beijing, People' s Rep. of China. (1933) . Engineer (tel. 555309). *Crystallography, mineralogy.*

Zhang, Mr Han-hui. Inst. of Structural Chemistry, Dept. of Chemistry, Fuzhou U., Fuzhou, Fujian, People's Rep. of China. (1947) MS, physical chemistry (Fuzhou U., 1981). Res. asst. *Crystal and molecular structure, transition metal complexes.*

Zhang, Prof. Jiang-hong. The Res. Section, Crystal Structure and Crystal Chemistry of Minerals, Beijing Graduate School of Wuhan Geological C., 29 Xue Yuan Road, Beijing 100083, People's Rep. of China. (1936) mineralogy, X-ray crystallography. Assoc. prof. (tel. 277461-560). *X-ray crystallography, minerology, crystal structure, crystal chemistry of minerals, new minerals.*

Zhang, Prof. Le-hui. Lab. of Crystallography, Inst. of Physics, Academia Sinica, P.O. Box 603, Beijing, People's Rep. of China. (1919) BS, physics (Tatung U., 1941). Assoc. res. prof. (tel. 28-1869). *Crystal growth, growth mechanism, high temperature oxide crystals.*

Zhang, Mr Li-xin. Dept. of Structure and Defect of Crystals, Inst. of Metal Research, Academia Sinica, 2-6 Wenhua Road, Shenyang, Liaoning, People's Rep. of China. (1935). Head, structure of surface layer res. group. (tel. 483531, ext. 242). *X-ray metallography, texture, structure of surface layer.*

Zhang, Mrs Rong-ying. Group of X-ray Analysis of Mineral, Hubei Lab. of Geological Sci., Jie Fang Da Dao No. 342, Wuhan, Hubei, People's Rep. of China. (1935) geology. Engineer (tel. 52847). *X-ray powder diffraction, rock and mineral identification.*

Zhang, Mr Rui-lin. Dept. of Physics, Ji-lin U., Je-fang Road, Chang-chun, Ji-lin, People's Rep. of China. (1934). Lect. *Crystal and molecular structure, electron theory of solids and molecules application.*

Zhang, Prof. Shao-hui. Chemistry Dept., Wuhan U., Wuhan, Hubei, People's Rep. of China. (1938). Assoc. prof. *Direct methods in crystallography.*

Zhang, Mr Shi-wei. Lab. for Structure of Matter, Inst. of Physical Chemistry, Peking U., Beijing, People's Rep. of China. (1945) MS, structural chemistry (Peking U., 1981). (tel. 282471, ext. 3549). *Structural chemistry.*

Zhang, Ms Shu-de. Dept. of Molecular Biology, Inst. of Biophys., Academia Sinica, Beijing 100080, People's Rep. of China. (1939). Res. assoc. (tel. 285529). *X-ray crystallography.*

Zhang, Mr Yong-mao. Structural Chemistry Div., Fujian Inst. of Research on the Structure of Matter, Academia Sinica, Fuzhou, Fujian, People's Rep. of China. (1936) structural chemistry. Lect. *X-ray crystallography, metal complexes, protein crystal structure.*

Zhang, Prof. Yuan-long. Crystal Growth Dept., Shanghai Inst. of Ceramics, Academia Sinica, Chang Ning Road 865, Shanghai 200050, People's Rep. of China. (1915) BS, physics (Fu-Jen U., 1937). Res. prof. *Crystal growth theory and practice, phase transitions, structure-growth forms-physical properties correlation.*

Zhang, Mr Yue-ming. X-ray Diffraction Lab., Inst. of Geochemistry, Academia Sinica, Guan Shui Road, Guiyang, Guizhou, People's Rep. of China. (1937) powder diffraction of minerals. Engineer (tel. 24757). *X-ray powder diffraction, isomorphous substitution of minerals, high pressure minerals, programing, powder diffraction method.*

Zhang, Ms Ze-ying. Lab. for Structure of Matter, Inst. of Physical Chemistry, Peking U., Beijing, People's Rep. of China. (1940). Lect. (tel. 282471). *Structural analysis of single crystal.*

Zhang, Prof. Zong. Div. of Mathematics and Physics, Academia Sinica, Beijing, People's Rep. of China. (1929) physics. Prof., div. director. (tel. 868361, ext. 448). *Crystallography, magnetism, neutron diffraction.*

Zhao, Prof. Qi-yuan. Marine Geology Dept., Shandong C. of Oceanology, 5 Yushan Road, Qindao, Shandong, P.O. Box 90, People's Rep. of China. (1929) crystallography and mineralogy. Assoc. prof. *Crystal Synthesis, mineralogy, geochemistry.*

Zheng, Mr Qi-tai. Inst. of Biophys., Academia Sinica, Zhong Guan Cun, Beijing 100080, People's Rep. of China. (1938) physics. Res. assoc. *Determination of crystal structure, computational methods.*

Zheng, Prof. Pei-ju. Chemistry Dept., Fudan U., 200 Handan Road, Shanghai 201903, People's Rep. of China. (1933). Assoc. prof. *Crystal structure analysis, inorganic and metal-organic compounds, X-ray crystallography, powder diffraction, drug action.*

Zheng, Mr Zhe. Dept. of Geology, Peking U., Beijing, People's Rep. of China. (1938). Sr. lect. (tel. 282471). *Symmetry of crystal, microstructures in minerals.*

Zhong, Mr Na-tian. Inst. of Biophys., Academia Sinica, Zhong Guan Cun, Beijing 100080, People's Rep. of China. (1937) physical chemistry. *X-ray diffraction, instrumentation, computing.*

Zhong, Prof. Wei-zhuo. R&D Center in Advanced Inorganic Materials, Shanghai Inst. of Ceramics, Academia Sinica, 865 Chang-ning Road, Shanghai 200050, People's Rep. of China. (1932) crystallography. Assoc. res. prof. (tel. 522470, telex 33309ASSIC CN). *Crystal growth, morphology, defects.*

Zhou, Prof. Gong-du. Lab. for Structure of Matter, Inst. of Physical Chemistry, Peking U., Beijing, People's Rep. of China. (1931) physical chemistry. Assoc. prof. (tel. 282471, ext. 3549). *Structural chemistry.*

Zhou, Mr Gui-en. Central Lab. of Structural Analysis, U. of Sci. and Techn. of China, Hefei, Anhui, People's Rep. of China. (1937). Lect. (tel. 63300). *X-ray crystallography, powder diffraction, polymer structure.*

Zhou, Mr Kang-jing. Lab. of Structural Chemistry, Fujian Inst. of Res. on the Structure of Matter, Academia Sinica, Fuzhou, Fujian, People's Rep. of China. (1939) MS, structural chemistry (Fujian Inst. of Res. on the Structure of Matter, 1981). Lect. *X-ray crystallography, metal complexes, protein crystal structure.*

Zhou, Mr Zhong-yuan. Analysis and Testing Centre of Chengdu Branch, Academia Sinica, Chengdu, Sichuan, People's Rep. of China. (1938). Res. assoc. *Electron density distribution in the bonds, small molecules, EXAFS, powder diffraction.*

Zhou, Mr Zong-hua. Center of Analysis and Test, Sichuan U., Wangjiang Road, Chengdu, Sichuan, People's Rep. of China. (1935). Instructor. (tel. 316-24401). *Powder diffraction, X-ray single crystal structure analysis.*

Zhu, Mr Ji-mu. Dept. of Physics, Sichuan U., Wangjiang Road, Changdu, Sichuan, People's Rep. of China. (1938) solid state physics. Lect. (tel. 24401-260, telex 2345). *Powder diffraction, crystal structure.*

Zhu, Mr Nai-jue. Dept. of Crystallography, Inst. of Chemistry, Academia Sinica, Zhong Guan Cun, Beijing, People's Rep. of China. (1942). Res. assoc. *Structure, small molecules, charge density, neutron diffraction.*

Zhu, Mr Zhong-he. Center of Material Analysis, Nanjing U., Hankou Road, Nanjing Jiangsu, People's Rep. of China. (1939) coordination chemistry. Lect. (tel. 34651, ext. 2523). *Crystal structure.*

Zhuang, Mr Hong-hui. Structural Chemistry Div., Fujian Inst. of Res. on the Structure of Matter, Academia Sinica, Fuzhou, Fujian, People's Rep. of China. (1938). Lect. *Transition metal complexes, metal cluster compounds, X-ray crystallography.*

Zhuang, Mr Jian. Structural Chemistry Div., Fujian Inst. of Res. on the Structure of Matter, Academia Sinica, Fuzhou, Fujian, People's Rep. of China. (1942). Lect. *Metal cluster compounds, magnetochemistry, powder diffraction(X-ray and neutron), crystal growth.*

COLOMBIA

Sub-Editor: E. Posada

Notes

1. International telephone country code - 57.

2. The degrees *ingeniero* (Ing.), *fisico, químico* and *geólogo* are between BSc and MSc. The degrees *licenciado, magister* and *doctor* are equivalent to BSc, MSc and PhD, respectively.

3. For Bogotá D.E. the telephone numbers are preceded by number 2.

Acosta, Prof. Carlos Eduardo. Dept. de Geociencias, U. Nacional de Colombia, Ciudad Universitaria, Bogotá D.E. Colombia. (1919) Géologue, geology (Inst. Catholique de Paris, France, 1967). Prof. (tel. 442810). *Mineralogy.*

Alfonso U., Miss Ana Elena. Asesoría Industrial de la Gerencia, Banco de la República, Calle 16, 5-41, Bogotá D.E. Colombia. (1955) Químico, chemistry (U. Nacional de Colombia, 1977). Res. officer. (tel. 428977). *X-ray diffraction, metallurgy.*

Barriga Villalba, Prof. Antonio María. Museo de Numismática, Calle 11, 4-93, Bogotá D.E. Colombia. (1894) Dr, philosophy and letters, chemistry (Colegio Mayor de Nuestra Señora del Rosario, 1918). Honorary prof. and director. (tel. 437200). *Inorganic crystal structures, metal physics, metals structures.*

Bernal de Ramírez, Prof. Inés. Dept. de Química, U. Nacional de Colombia, Ciudad Universitaria, Bogotá D.E. Colombia. (1940) Químico, chemistry (U. Nacional de Colombia, 1964). Assoc. prof. (tel. 699183). *Clay minerals.*

Brieva, Prof. Jorge Alfonso. Dept. de Geociencias, U. Nacional de Colombia, Ciudad Universitaria, Bogotá D.E. Colombia. (1935) MSc, sedimentary mineralogy, petrology (U. Tulsa, USA, 1963). Assoc. prof. (tel. 442810). *Sedimentary petrography, sedimentary petrology, clay minerals, computer programming.*

Calderón, Prof. Gómez Eduardo. Dept. de Química, U. Nacional de Colombia, Ciudad Universitaria, Bogotá D.E. Colombia. (1923) DSc, pharmacodynamics, pharmacology (U. Paris, France, 1956). Prof. (tel 699183). *Structure, organic compounds.*

Cortés, Dr Abdón. Subdirección Agrológica, Inst. Geográfico Agustín Codazzi, Carrera 30, 48-51, Bogotá D.E. Colombia. (1939) PhD, soil taxonomy (Purdue U., USA, 1971). Sub-director. (tel. 442784). *Soil taxonomy, soil mineralogy, clays.*

Díaz Peraza, Prof. José Milciades. Dept. de Física, U. Nacional de Colombia, Ciudad Universitaria, Bogotá D.E. Colombia. (1940) MSc, physics (U. South Carolina, USA, 1970). Asst. prof. (tel. 442874). *Paramagnetic resonance, X-ray diffraction, crystal growth.*

Erazo Plaza, Mr Antonio David. Colombia Cities Service Petroleoum Corp., Calle 37, 8-43, p11, Bogotá D.E. Colombia. Geólogo, geology (U. Nacional de Colombia, 1963). Geologist. (tel. 451560). *Crystal optics.*

Galvis, Mr Jaime. Integral S.A., Calle 29, 6-94, p6, Bogotá D.E. Colombia. (1941) Geólogo, geology (U. Nacional de Colombia, 1965). Res. officer. (tel. 324513). *Optical crystallography, mineralogy, petrography.*

Hernández, Prof. Luis C. Dept. de Física, U. Nacional de Colombia, Ciudad Universitaria, Bogotá D.E. Colombia. (1942) MSc solid state physics (U. Puerto Rico, USA, 1973). Assoc. prof. (tel. 442874). *X-ray diffraction, paramagnetic resonance.*

Jiménez Crespo, Prof. Augusto. Dept. de Física, U. Nacional de Colombia, Ciudad Universitaria, Bogotá D.E. Colombia. (1947) Físico, physics (U. Nacional de Colombia, 1971). Asst. prof. (tel. 730050). *X-ray diffraction, paramagnetic resonance.*

Lozano, Prof. José A. Dept. de Geociencias, U. Nacional de Colombia, Ciudad Universitaria, Bogotá D.E. Colombia. (1935) PhD, submarine geology (Columbia. U., USA, 1974). Assoc. prof. (tel. 442810). *Sea floor, rocks, sediments, geology.*

Luna, Dr Carlos Alfonso. Lab. de Suelos, Instituto Geográfico Agustín Codazzi, Carrera 30, 48-51, Bogotá D.E. Colombia. Dr, soil science (U. Utrecht, Netherlands, 1969). Chief. (tel. 447719). *Sand, clay minerals, soils.*

Llinás Rivera, Prof. Rubén Darío. Dept. de Geociencias, U. Nacional de Colombia, Ciudad Universitaria, Bogotá D.E. Colombia. (1940) Dipl.Geol., petrography, mineralogy (U. Stuttgart, BRD, 1971). Assoc. prof. (tel. 442810). *Optical crystallography, petrography.*

Macía Sanabria, Prof. Carlos A. Dept. de Geociencias, U. Nacional de Colombia, Ciudad Universitaria, Bogotá D.E. Colombia. (1947) Dr.rer.nat., geochemistry, petrology (U. Stuttgart, BRD, 1978). Asst. prof. (tel. 442810). *Optical crystallography, petrology.*

Malagón Castro, Prof. Dimas. Facultad de Agronomía, U. Nacional de Colombia, Ciudad Universitaria, Bogotá D.E. Colombia. (1938) PhD, pedology (U. Nebraska, USA, 1973). Assoc. prof. (tel. 699111, ext 543). *Clay and sand minerals.*

Mejía Cifuentes, Prof. Leonidas. Centro Interamericano de Fotointerpretación, Carrera 30, 47A-57, Bogotá D.E. Colombia. (1933) MSc, soil taxonomy, mineralogy (U. North Carolina, USA, 1975). Prof. (tel 680300, ext. 64). *Clay mineralogy, sedimentary petrography.*

Merino de Matheus, Prof. Lucía Marina. Dept. de Química, U. Nacional de Colombia, Ciudad Universitaria, Bogotá D.E. Colombia. (1934) Dr, crystallography (U. São Paulo, Brasil, 1975). Asst. prof. (tel. 699183). *Inorganic crystal structures.*

Mora de González, Prof. Nery. Dept. de Química, U. Nacional de Colombia, Ciudad Universitaria, Bogotá D.E. Colombia. (1935) Químico, chemistry (U. Nacional de Colombia, 1956). Prof. (tel. 699183). *Clay minerals, X-ray diffraction.*

Ovalle de Bravo, Prof. Yolanda. Dept. de Química, U. Nacional de Colombia, Ciudad Universitaria, Bogotá D.E. Colombia. (1941) Químico, chemistry (U. Nacional de Colombia, 1965). Asst. prof. (tel. 699111, ext. 787). *Clay minerals.*

Posada G., Prof. Enrique. Instituto de Ciencias Naturales-Museo, U. Nacional de Colombia, Ciudad Universitaria, Bogotá D.E. Colombia. (1909) Dipl.Et.Sup., mineralogy, petrography (Inst. Catholique de Paris, France, 1935). Assoc. prof. (tel. 444403). *Optical crystallography, mineralogy.*

Quevedo, Dr Manuel M. Inst. Investigaciones Científicas y Técnicas, Calle 66A, 17-88 Apartado Aéreo 9403, Bogotá D.E. Colombia. (1915) Dr.rer.nat.habil., mathematics, physics, chemistry (U. Leipzig, DDR, 1969). Director. (tel. 484321). *Inorganic crystal structures, large molecules, crystal physics, crystal optics.*

Rincón Saenz, Mr Luis Felipe. Museo Geológico, Inst. Nacional de Investigaciones Geológicas Mineras, Diagonal 53, 34-35 Colombia. (1922) Chief (tel.443330). *Optical crystallography.*

Rodríguez S., Miss Gloria Inés. Sección de Petrografía y Mineralogía, Inst. Nacional de Investigaciones Geológico Mineras, Diagonal 53, 34-53 Bogotá D.E. Colombia. (1946) Geólogo, geology (U. Nacional de Colombia, 1972). Res. off. (tel. 443330, ext 40) *Mineralogy, optical crystallography.*

Rodríguez Lara, Prof. Jaime. Sección de Estado Sólido, Dept. de Física, U. Nacional de Colombia, Ciudad Universitaria, Bogotá D.E. Colombia. (1933) D. 3e cycle, solid state (U. Lille, France, 1966). Chief. (tel. 442874). *Paramagnetic resonance.*

Rubiano Lamouroux, Prof. Manuel. Dept. de Geociencias, U. Nacional de Colombia, Ciudad Universitaria, Bogotá D.E. Colombia. (1943) D.d'U., crystallography, mineralogy (U. Nancy, France, 1971). Assoc. prof. (tel. 442810). *Crystallography, mineralogy.*

Rubio de Cubides, Prof. Julia. Dept. de Química, U. Nacional de Colombia, Ciudad Universitaria, Bogotá D.E. Colombia. Químico, chemistry (U. Nacional de Colombia, 1956). Assoc. prof. (tel. 699183). *Clay minerals.*

Sanchez V., Mr Alfredo. Asesoría Industrial de la Gerencia, Banco de la República, Calle 16, 5-41, Bogotá D.E. Colombia. (1942) MSc, physical chemistry (U. Moscow, USSR, 1966). Chief. (tel. 428977). *X-ray diffraction, metallurgy.*

Varela Mora, Prof Juan de Dios. Dept. de Física, U. Nacional de Colombia, Ciudad Universitaria, Bogotá D.E. Colombia. (1942) Dr, crystallography, mineralogy (Leningrad State U., USSR, 1981). Assoc. prof. (tel. 440819). *Crystallography, X-ray diffraction, layer silicates.*

CUBA

Sub-Editor: R. Pomés Hernandez

Argüelles, Mr Waldo. Physical Chemistry of Polymers, Inst. of Chemistry and Exp. Biology, 26 th. Ave. Nr. 1605, Nuevo Vedado, Havana, Cuba. (1956) MSc, structure and interaction of lipids (High Inst. of Food, Plovdiv, Bulgaria, 1980). Master in Sci. (tel. 40-8962). *Physical properties and structures, biological structures.*

Blanco, Mrs Magaly. Physical Chemistry of Polymers, Inst. of Chemistry and Exp. Biology, 26 th. Ave. Nr. 1605, Nuevo Vedado, Havana, Cuba. (1949) MSc, electron microscopy (U. Havana, 1972). (tel. 40-8962). *Electron diffraction, electron microscopy, biological structures.*

Callejas, Mr Domingo. Bioorganic Chemistry, Inst. of Chemistry and Exp. Biology, 26 th. Ave. Nr. 1605, Nuevo Vedado, Havana, Cuba. (1949) MSc, organic synthesis (U. Havana, 1971). Master in Sci. (tel. 40-8962). *Organic structures.*

Durruthy, Mr Obel. X-ray Dept., U. of Oriente, Patricio Lumumba s/n, Santiago de Cuba, Cuba. (1945) MSc, X-ray crystallography (U. Oriente, 1977). Asst. prof. (tel. 8056, ext. 14). *Crystal structure determination, precise parameter determination.*

Fajardo, Mr Fabio, X-ray Dept., U. of Oriente, Patricio Lumumba s/n, Santiago de Cuba, Cuba. (1948) MSc, X-ray crystallography (U. Oriente, 1977). Asst. prof. (tel. 8056, ext. 14). *Crystal structure determination, precise parameter determination.*

Garcia, Miss Ileana. Physical Chemistry of Polymers, Inst. of Chemistry and Exp. Biology, 26 th. Ave. Nr. 1605, Nuevo Vedado, Havana, Cuba. (1953) MSc, macromolecules, thermal analysis (U. Havana, 1975). *Biological structures, computation methods, physical properties and structures.*

Mainegra, Mr Virgilio. Faculty of Chemistry and Physics, Central U., Camajuaní Road 10 Km., Santa Clara, Villa Clara, Cuba. (1941) MSc, crystallochemistry, X-rays (State U. "Lomonosov", USSR, 1969). Sr. lector. *X-ray crystallography, mineralogy.*

Oviedo, Miss Danais. Physical Chemistry of Polymers, Inst. of Chemistry and Exp. Biology, 26 th. Ave. Nr. 1605, Nuevo Vedado, Havana, Cuba. (1953) MSc, protein interactions and conformations (U. Havana, 1977). Master in Sci. (tel. 40-8962). *Physical properties and structure, biological structures, computation methods.*

Pomés Hernandez, Prof. Dr Sc. Ramón. X-ray Dept., U. of Oriente, Patricio Lumumba s/n, Santiago de Cuba, Cuba. (1947) DSc, X-ray crystallography (U. Oriente, 1970). Prof. (tel. 8056, ext. 14). *Crystal structure determination.*

Quintana, Dr. Rafael, Inorganic Chemistry Dept., Central U., Camajuaní Road 10 Km., Santa Clara, Villa Clara, Cuba. (1944) PhD, Mineralogy (U. Greifswald, DDR, 1979). Sr. lector. *Mineralogy.*

Ramirez, Mr Edilberto. Physical Chemistry of Polymers, Inst. of Chemistry and Exp. Biology, 26 th. Ave. Nr. 1605, Nuevo Vedado, Havana, Cuba. (1951) MSc, electron microscopy (U. Havana, 1975). Master in sci. (tel. 40-8962). *Electron diffraction, electron microscopy, biological structures.*

Vila, MSc Ileana. Physical Chemistry of Polymers, Inst. of Chemistry and Exp. Biology, 26 th. Ave. Nr. 1605, Nuevo Vedado, Havana, Cuba. (1936) MSc, X-rays (U. Havana, 1974). Master in sci. (tel. 40-8962). *Inorganic crystallography, mineralogy, powder diffraction.*

CZECHOSLOVAKIA

Sub-Editor: M. Čerňanský

Notes

1. International telephone country code - 42.

2. Until 1953 the degree *rerum naturalium doctor* (RNDr., abbreviation Dr) was conferred by the faculties of sciences of the Czechoslovak universities and the degree *doctor of technical sciences* (Dr. techn., abbreviation Dr) by the technical universities. Since 1953 new degrees *candidatus scientiarum* (CSc) and *doctor scientiarum* (DrSc) are conferred by the faculties of sciences of the universities, by the technical universities, and by institutes of the Cz. Acad. Sci. and of the Slovak Acad. Sci. The lowest degree conferred by the faculties of sciences are *graduated physicist, graduated chemist, etc.;* the corresponding lowest degree conferred by technical universities is *inženýr* (Ing). In 1966 the degree RNDr. was re-established. CSc corresponds to PhD and DrSc to DSc in the British system of degrees.

3. At the universities (including the technical universities) persons with an academic training can be appointed in various positions: *profesor, docent* (associated professor; abbreviation doc.), *odb. asistent* (senior lecturer), *asistent* (lecturer).

4. In the list the following abbreviations are used: CSSR - Czechoslovak Socialistic Republic; Cz. Acad. Sci. - Czechoslovak Academy of Sciences; Slovak Acad. Sci. - Slovak Academy of Sciences; em. - (Emeritus) retired.

Baldrian, Dr Josef. Lab. of X-ray Polymer Structure Analysis, Inst. of Macromolecular Chemistry, Cz. Acad. Sci., Heyrovského nám. 2, 162 06 Praha 6, CSSR. (1938) CSc, physical chemistry (Inst. of Macromolecular Chemistry, Cz. Acad. Sci., 1965). Scient. (tel. 360341). *Polymer structure, small angle scattering.*

Barta, Ing Čestmír. Dept. of Material Res., Inst. of Physics, Cz. Acad. Sci., Na Slovance 2, 180 40 Praha 8, CSSR. (1926) CSc, crystal chemistry (Inst. of Chemical Techn., 1954). Head, crystal growth group. (tel. 225589). *Crystal chemistry, material research, cosmical technology.*

Bauer, Doc. Ing Jaroslav. Dept. of Mineralogy, Inst. of Chemical Techn., Suchbátarova 5, 166 28 Praha 6, CSSR. (1920) CSc, mineralogy (Inst. of Chemical Techn., 1962). Doc. of mineralogy. (tel. 332, ext. 4086). *Crystallography, applied mineralogy.*

Bouška, Doc. Dr Vladimír. Dept. of Mineralogy, Geochemistry and Crystallography, Faculty of Sci., Charles U., Albertov 6, 128 43 Praha 2, CSSR. (1933) DrSc, geochemistry (Charles U., 1979). Head of the Department of Geochemistry. (tel. 297541-5, ext. 288). *Crystallochemistry, crystal optics.*

Broul, Ing Miroslav. Dept. of Chemical Engineering, Res. Inst. of Inorganic Chemistry - Chemopetrol, Revoluční 86, 400 60 Ústí nad Labem, CSSR. (1945) CSc, physical chemistry (Inst. of Chemical Techn., 1979). Scient. (tel. 2182510). *Bulk crystallization from solutions.*

Březina, Ing Bohuslav. Dielectric Lab., Inst. of Physics, Cz. Acad. Sci., Na Slovance 2, 180 40 Praha 8, CSSR. (1928) CSc, chemistry (Inst. of Chemical Techn., 1956). Head of the group (tel. 842241, ext. 2641). *Ferroelectric physics and chemistry, monocrystal technology.*

Bubáková, Dr Růžena. Dept. of Structures and Bonding, Inst. of Physics, Cz. Acad. Sci., Na Slovance 2, 180 40 Praha 8, CSSR. (1917) RNDr., physics (Charles U., 1953). Scient. (tel. 354240-9, ext. 87). *Crystal physics, X-ray diffraction on perfect crystals.*

Chalupa, Ing Bohumil. Neutron Physics Dept., Nuclear Physics Inst., Cz. Acad. Sci., 250 68 Rež, CSSR. (1928) Ing., electrotechnics (Techn. U. of Prague, 1956). Head. (tel. 896231, ext. 2383). *Neutron diffraction and spectrometry, slow neutrons, nuclear physics solid state methods.*

Cuchý, Ing Zdeněk. Dept. of Crystal Growth, Monokrystaly, Leninova 175, 511 19 Turnov, CSSR. (1934) CSc, chemistry (Inst. of Chemical Techn., 1957). Head. (tel. 22751). *Optic and spectroscopic properties of single crystals, structure.*

Čapková, Dr Pavla. Dept. of Physics of Semiconductors, Faculty of Mathematics and Physics, Charles U., Ke Karlovu 5, 121 16 Praha 2, CSSR. (1945) CSc, physics (Charles U., 1975). Sr. lect. (tel. 292141, ext. 389). *X-ray structure analysis.*

Čech, Prof. Dr František. Dept. of Mineralogy, Geochemistry and Crystallography, Faculty of Sci., Charles U., Albertov 6, 128 43 Praha 2, CSSR. (1929) CSc, mineralogy (Charles U., 1961). Prof. of mineralogy (tel. 297541-5). *Mineralogy, crystallography.*

Čermák, Dr Jan. Dept. of Metal Physics, Inst. of Physics, Cz. Acad. Sci., Na Slovance 2, 180 40 Praha 8, CSSR. (1926) CSc, physics (Inst. of Solid State Physics, Cz. Acad. Sci., 1963). Scient. (tel. 842241, ext. 2898). *Physics of metals, X-ray diffraction.*

Čerňanský, Ing Marian. Dept. of Metal Physics, Inst. of Physics, Cz. Acad. Sci., Na Slovance 2, 180 40 Praha 8, CSSR. (1946) CSc, physics (Inst. of Physics, Cz. Acad. Sci., 1981) Scient. (tel. 842241, ext. 2898). *Structure of metals, X-ray diffraction, amorphous materials.*

Černohorský, Doc. Dr Martin. Dept. of General Physics, Faculty of Sci., J. E. Purkyně U., Kotlářská 2, 611 37 Brno, CSSR. (1923) CSc, physics (Inst. of Solid State Physics, Cz. Acad. Sci., 1963). Doc. of physics (tel. 51112). *Physics of metals.*

Červeň, Doc. Dr Ivan. Dept. of Physics, Faculty of Electrotechnical Engineering, Slovak Technical U., Mlynská dolina, 812 19 Bratislava, CSSR. (1933) CSc, physics (Slovak Techn. U., 1980). Doc. of physics. *Amorphous structures, symmetry.*

Červinka, Dr Ladislav. Dept. of Structures and Bonding, Inst. of Physics, Cz. Acad. Sci., Na Slovance 2, 180 40 Praha 8, CSSR. (1935) CSc, physics (Inst. of Solid State Physics, Cz. Acad. Sci., 1966). Scient. (tel. 354240-9, ext. 91). *Glasses and amorphous structures, imperfections in crystals.*

Číčel, Ing Blahoslav. Lab. of Crystatl Chemistry and Reactivity of Hydrosilicates, Inst. of Inorganic Chemistry, Centre for Chem. Res., Slovak Acad. Sci., Dúbravská cesta, 842 36 Bratislava, CSSR. (1933) CSc, silicate techn. (Slovak Techn. U., 1963). Scient. (tel. 375488, ext. 2303). *Crystal chemistry, clay minerals, silicate structures.*

Dunaj-Jurčo, Doc. Ing Michal. Dept. of Inorganic Chemistry, Faculty of Chemical Techn., Slovak Technical U., Jánska 1, 880 37 Bratislava, CSSR. (1936) CSc, chemistry (Slovak Techn. U., 1967). Doc. of inorganic chemistry (tel. 56021, ext. 287, 434). *Crystallochemistry and stereochemistry of compounds of transition metals.*

Ďurčanská, Dr Edita. Dept. of Inorganic Chemistry, Faculty of Chemical Techn., Slovak Technical U., Jánska 1, 880 37 Bratislava, CSSR. (1937) CSc, chemistry (Slovak Techn. U., 1970). Sr. lect. (tel. 56021, ext. 290). *Crystallochemistry and stereochemistry of compounds of transition metals.*

Ďurovič, Ing Slavomil. Lab. of Diffraction Methods, Inst. of Inorganic Chemistry, Centre for Chem. Res., Slovak Acad. Sci., Dúbravská cesta, 842 36 Bratislava, CSSR. (1929) CSc, geology and mineralogy (Comenius U., 1962). Scient. (tel. 375488, ext. 2305). *Order disorder structures, polytypes, silicate structures.*

Fiala, Dr. Jaroslav. Dept. of Metallurgy, Central Res. Inst. Škoda, Tylova 46, 316 00 Plzeň, CSSR. (1940) RNDr., solid state physics (Charles U., 1981). Head of the Lab. of X-Ray Diffraction (tel. 211,ext. 2302). *X-ray diffraction phase analysis qualitative and quantitative.*

Fingerland, Dr Antonín. P.A.: Čermákova 1, 120 00 Praha 2, CSSR. (1923) CSc, physics (Inst. of Solid State Physics, Cz. Acad. Sci., 1967). Scient. (tel. 258903). *Theory of diffraction, symmetry, quantum chemistry.*

Ganev, Ing Nikolaj. Dept. of Solid State Physics Engineering, Faculty of Nuclear Science and Physical Engineering, Technical U. of Prague, Břehová 7, 115 19 Praha 1, CSSR. (1953) Ing., solid state physics (Techn. U., of Prague, 1979). Res. asst. (tel. 849951, ext. 413). *Solid state physics, structure analysis, X-ray tensometry, metal physics.*

Garaj, Prof. Dr Ján. Dept. of Analytical Chemistry, Slovak Techn. U., Jánska 1, 880 37 Bratislava, CSSR. (1934) DrSc, inorganic chemistry (Slovak Techn. U., 1976). Prof. of analytical chemistry (tel. 56043). *Analytical chemistry, crystal chemistry, structure analysis of coordination compounds.*

Gosmanová, Dr Galina. Dept. of Solid State Physics Engineering, Faculty of Nuclear Science and Physical Engineering, Technical U. of Prague, Břehová 7, 115 19 Praha 1, CSSR. (1936) RNDr., physics (Charles U., 1976). Sr. lect. *Solid state physics, structure analysis, X-ray tensometry, metal physics.*

Gruber, Dr Boris. Dept. of Mathematical Physics, Faculty of Mathematics and Physics, Charles U., Malostranské nám. 2/25, 118 00 Praha 1, CSSR. (1921) CSc, physics (Charles U., 1965). Scient. (tel. 532132, ext. 82). *Lattice symmetry, programming.*

Gyepes, Doc. Dr Eduard. Dept. of Analytical Chemistry, Faculty of Sci., Comenius U., Mlynská dolina, Pavilón Chémie, 842 15 Bratislava, CSSR. (1935) CSc, chemistry (Comenius U., 1972). Doc. of analytical chemistry. (tel. 320003, ext. 386) *Structure of coordination compounds.*

Gyepesová, Dr Dalma. Lab. of Crystal Chemistry and Reactivity of Hydrosilicates, Inst. of Inorganic Chemistry, Centre for Chem. Res., Slovak Acad. Sci., Dúbravská cesta, 842 36 Bratislava, CSSR. (1936) CSc, chemistry (Comenius U., 1972). Head. (tel. 375 488, ext. 2164). *Silicate structures.*

Handlovič, Ing Milan. Lab. of Diffraction Methods, Inst. of Inorganic Chemistry, Centre for Chem. Res., Slovak Acad. Sci., Dúbravská cesta, 842 36 Bratislava, CSSR. (1932) CSc, chemistry (Inst. of Inorganic Chemistry, Slovak Acad. Sci., 1968). Scient. (tel. 375488, ext. 2310). *Inorganic crystal structures.*

Hanic, Doc. Dr František. Lab. for Phase Analyses, Inst. of Inorganic Chemistry, Centre for Chem. Res., Slovak Acad. Sci., Dúbravská cesta, 842 36 Bratislava, CSSR. (1927) DrSc, inorganic chemistry (Inst. of Chemical Techn., 1966). Scient. (tel. 375 488, ext. 2317). *Crystal chemistry, inorganic compounds, structure and properties.*

Hašek, Dr Jindřich. Lab. of X-Ray Polymer Structure Analysis, Inst. of Macromolecular Chemistry, Cz. Acad. Sci., Heyrovského nám. 2, 162 06 Praha 6, CSSR. (1945) CSc, physics (Charles U., 1972). Scient. (tel. 360341, ext. 387). *Methods of crystal structure determination, molecular structures, quant. chem. calculations, programming.*

Hlavatá, Mrs Drahomíra. Lab. of X-Ray Polymer Structure Analysis, Inst. of Macromolecular Chemistry, Cz. Acad. Sci., Heyrovského nám. 2, 162 06 Praha 6, CSSR. (1942) CSc, physical chemistry (Inst. of Macromolecular Chemistry, Cz. Acad. Sci., 1972). Scient. (tel. 360341). *Organic crystal structures, programming.*

Holý, Dr Václav. Dept. of Solid State Physics, Faculty of Sci., J. E. Purkyně U., Kotlářská 2, 611 37 Brno, CSSR. (1953) CSc, physics (J. E. Purkyně U., 1982). Sr. lect. (tel. 51112, ext. 46). *Dynamical theory, X-ray topography.*

Horváth, Ing Josef. Dept. of Metal Physics, Inst. of Physics, Cz. Acad. Sci., Na Slovance 2, 180 40 Praha 8, CSSR. (1950) CSc, physics (Inst. of Physics, Cz. Acad. Sci., 1982) Scient. (tel. 842241, ext. 2892). *Structure of metals, amorphous materials, theory of scattering.*

Hoschl, Dr Pavel. Inst. of Physics, Charles U., Ke Karlovu 5, 121 16 Praha 2, CSSR. (1938) CSc, physics (Charles U., 1967). Scient. (tel. 292141, ext. 266). *Growth of monocrystals.*

Huml, Dr Karel. Lab. of X-Ray Polymer Structure Analysis, Inst. of Macromolecular Chemistry, Cz. Acad. Sci., Heyrovského nám. 2, 162 06 Praha 6, CSSR. (1934) CSc, physics (Charles U., 1966). Head. (tel. 360341). *Organic crystal structures, programming.*

Hybler, Dr Jiří. Dept. of Structures and Bonding, Inst. of Physics, Cz. Acad. Sci., Na Slovance 2, 180 40 Praha 8, CSSR. (1949) RNDr., crystallography (Charles U., 1973). Res. asst. (tel. 361337,). *Orienting crystals, single crystal and powder studies, minerals, inorganic compounds.*

Ječný, Ing Jiří. Lab. of X-Ray Polymer Structure Analysis, Inst. of Macromolecular Chemistry, Cz. Acad. Sci., Heyrovského nám. 2, 162 06 Praha 6, CSSR. (1938) Ing., instrumentation and automation (Charles U., 1962). Scient. (tel. 360341). *Organic crystal structures, automation.*

Kabešová, Ing Mária. Dept. of Inorganic Chemistry, Faculty of Chemical Techn., Slovak Technical U., Jánska 1, 880 37 Bratislava, CSSR. (1932) CSc, chemistry (Slovak Techn. U., 1971). Sr. lect. (tel. 56021, ext. 290). *Crystallochemistry and stereochemistry of compounds of transition metals.*

Karmazin, Dr Lubomír. Dept. of Phase Transformations, Inst. of Physical Metallurgy, Cz. Acad. Sci., Žižkova 22, 616 62 Brno, CSSR. (1938) CSc, physics (Inst. of Solid State Physics, 1966). Head. (tel. 58111-9, ext. 387). *Structure of metals and alloys, X-ray diffraction, electron probe microanalysis.*

Kellö, Dr Eleonóra. Dept. of Analytical Chemistry, Faculty of Chemical Techn., Slovak Technical U., Jánska 1, 812 37 Bratislava, CSSR. (1949) CSc, chemistry (Slovak Techn. U., 1982). Res. asst. (tel. 56021, ext. 375). *Analytical chemistry, crystal structure, coordination compounds*

Kettmann, Ing Viktor. Dept. of Analytical Chemistry, Faculty of Pharmacy, Comenius U., Odbojárov 10, 832 32 Bratislava, CSSR. (1947) CSc, chemistry (Slovak Techn. U., 1979). Scient. (tel. 60451, ext. 223). *Crystal structure, coordination compounds, conformation of drugs and its relation to the pharmacological activity.*

Kochanovská, Prof. Dr Adéla. Dept. of Solid State Physics Engineering, Faculty of Nuclear Science and Physical Engineering, Technical U. of Prague, Břehová 7, 115 19 Praha 1, CSSR. (1907) DrSc, physics (Inst. of Solid State Physics, 1957); corresponding member of the Cz. Acad. Sci. Em. Prof. of experimental physics (priv. tel. 439 415) *Metal and alloy structures.*

Koman, Ing Marián. Dept. of Inorganic Chemistry, Faculty of Chemical Techn., Slovak Technical U., Jánska 1, 812 37 Bratislava, CSSR. (1953) CSc, chemistry (Slovak Techn. U., 1981). Res. asst. (tel. 56021, ext. 290). *Crystal structure, coordination compounds, inorganic chemistry.*

Komrska, Dr Jiří. Dept. of Electron Optics, Inst. of Instrument Techn., Cz. Acad. Sci., Královopolská 147, 612 64 Brno, CSSR. (1936) CSc, physics (J. E. Purkyně U., 1965). Scient. (tel. 54311, ext. 367). *Electron diffraction, optical diffraction techniques.*

Koreň, Mr Branislav. Dept. of Physics, Faculty of Chemical Techn., Slovak Technical U., Jánska 1, 812 37 Bratislava, CSSR. (1954) graduated physicist, physics (Comenius U., 1977). Sr. lect. (tel. 56021, ext. 343). *Inorganic crystal structures, crystallographic statistics, coordination compounds.*

Kováčová, Dr Katarina. Dept. of Thermic Properties of Metals and Alloys, Inst. of Metal Materials, Slovak Acad. Sci., Februárového víťazstva 75, 801 00 Bratislava, CSSR. (1936) CSc, physical metallurgy (Res. Inst. of Welding, 1971). Scient. (tel. 673351, ext. 340). *Crystallization.*

Kožíšek, Ing Jozef. Dept. of Inorganic Chemistry, Faculty of Chemical Techn., Slovak Technical U., Jánska 1, 812 37 Bratislava, CSSR. (1952) CSc, chemistry (Slovak Techn. U., 1982). Res. asst. (tel. 56021, ext. 290). *Crystal structure, coordination compounds, inorganic chemistry.*

Kožíšková, Ing Zlatica. Dept. of Inorganic Chemistry, Faculty of Chemical Techn., Slovak Technical U., Jánska 1, 812 37 Bratislava, CSSR. (1953) Ing, chemistry (Slovak Techn. U., 1977). Res. asst. (tel. 56021, ext. 290). *Crystal structure, coordination compounds, inorganic chemistry.*

Králik, Dr František. Div. of Thermophysical Properties of Metals, Inst. of Metal Materials, Slovak Acad. Sci., Februárového víťazstva 75, 801 00 Bratislava, CSSR. (1919) RNDr., physics (Masaryk U., 1949). Head of Metal Res. Dept. (tel. 60088). *Physics of metals.*

Králová, Dr Rudolfa. Dept. of Solid State Physics Engineering, Faculty of Nuclear Science and Physical Engineering, Technical U. of Prague, Břehová 7, 115 19 Praha 1, CSSR. (1941) CSc, physics (Techn. U. of Prague, 1984). Sr. lect. (tel. 849911, ext. 414). *Solid state physics, structure analysis, X-ray tensometry, metal physics.*

Kratochvíl, Dr Bohumil. Dept. of Inorganic Chemistry, Faculty of Sci., Charles U., Hlavova 8, 128 40 Praha 2, CSSR. (1949) CSc, inorganic chemistry (Charles U., 1977). Sr. lect. (tel. 292051, ext. 500). *Inorganic crystal structures, crystal chemistry.*

Kraus, Doc. Dr Ivo. Dept. of Solid State Physics Engineering, Faculty of Nuclear Science and Physical Engineering, Technical U. of Prague, Břehová 7, 115 19 Praha 1, CSSR. (1936) CSc, physics (Techn. U. of Prague, 1967). Doc. of experimental physics. *Solid state physics, structure analysis, X-ray tensometry, metal physics.*

Křivý, Dr Ivan. Mathematics Department, College of Education, Dvořákova 7, 701 03 Ostrava 1, CSSR. (1940) CSc, physics (Charles U., 1973). Sr. lect. (tel. 233121, ext. 59). *Programming, statistical methods, crystal lattice dynamics.*

Kuběna, Dr Josef. Dept. of Solid State Physics, Faculty of Sci., J. E. Purkyně U., Kotlářská 2, 611 37 Brno, CSSR. (1935) CSc, physics (J. E. Purkyně U., 1969). Scient. (tel. 51112, ext. 46). *Microdefects in Silicon, X-ray topography.*

Kupka, Dr František. Dept. of Physical Chemistry, Inst. of Mineral Raw Materials, Vítězná 425, 284 03 Kutná Hora, CSSR. (1926) CSc, chemical technology of metals (Inst. of Chemical Techn., 1964). Head. (tel. 2577, ext. 76). *Silicates.*

Kvapil, Ing Jiří. Physical Lab., Monokrystaly, Leninova 175, 511 19 Turnov, CSSR. (1937) CSc, inorganic technology (Inst. of Chemical Techn., 1967). Scient. (tel. 238). *Growth of oxide monocrystals, laser crystals, colour centers.*

Langer, Dr Vratislav. Lab. of X-Ray Polymer Structure Analysis, Inst. of Macromolecular Chemistry, Cz. Acad. Sci., Heyrovského nám. 2, 162 06 Praha 6, CSSR. (1949) CSc, physics (Charles U., 1978). Scient. (tel. 360341, ext. 388). *Organic crystal structures, programming.*

Lokaj, Ing Ján. Dept. of Microanalysis, Faculty of Chemical Techn., Slovak Technical U., Jánska 1, 812 37 Bratislava, CSSR. (1951) CSc, chemistry (Slovak Techn. U., 1982). Scient. (tel. 56021, ext. 206). *X-ray diffraction, crystal structure, coordination compounds, microanalysis.*

Loub, Dr Josef. Dept. of Inorganic Chemistry, Faculty of Sci., Charles U., Hlavova 8/2030, 128 40 Praha 2, CSSR. (1929) CSc, inorganic chemistry (Charles U., 1967). Scient. (tel. 292051-5, ext. 500). *Inorganic chemistry, crystal structures.*

Machajdík, Ing Daniel. Dept. of Superconductivity, Electrotechnical Institute, Slovak Acad. Sci., Dúbravská cesta, 809 32 Bratislava, CSSR. (1944) CSc, technical sciences (Inst. of Inorganic Chemistry, Slovak Acad. Sci., 1978). Scient. (tel. 45741, ext. 2311). *Superconducting materials crystal structure, phase analysis.*

Maďar, Doc. Dr Ján. Dept. of Solid State Physics, Faculty of Mathematics and Physics, Comenius U., Mlynská dolina, Pavilón F2, 842 15 Bratislava, CSSR. (1926) CSc, physics (Comenius U., 1968). Doc. of exp. physics. (tel. 320003). *Structure analysis, instrumentation.*

Majling, Ing Ján. Dept. of Chemical Techn. of Silicates, Faculty of Chemical Techn., Slovak Techn. U., Jánska 1, 880 37 Bratislava, CSSR. (1942) CSc, silicate technology (Inst. of Inorganic Chemistry, Slovak Acad. Sci., 1971). Scient. (tel. 45341, ext. 2575). *Phase analysis, silicates, phosphate systems.*

Malý, Mr Karel. Dept. of Structures and Bonding, Inst. of Physics, Cz. Acad. Sci., Na Slovance 2, 180 40 Praha 8, CSSR. (1952) graduated physicist, physics (Charles U., 1975). Res. asst. (tel. 354240-9, ext. 96). *Structure analysis.*

Mánek, Ing Břetislav. Monokrystaly, Leninova 175, 511 19 Turnov, CSSR. (1933) Ing., inorganic chemistry (Inst. of Chemical Techn., 1956). Technical Director. (tel. 22751). *Crystal growth, oxide monocrystals, laser crystals, colour centers.*

Matherny, Prof. Dr-Ing Mikuláš. Dept. of Chemistry, Metallurgical Faculty, Technical U. in Košice, Švermova 9, 043 85 Košice, CSSR. (1930) DrSc, analytical chemistry (Comenius U., 1980). Prof. of analytical chemistry (tel. 30298). *Analytical chemistry, experimental metallurgy, spectroscopy and spectrochemistry, X-ray diffraction, polycrystals.*

Melka, Dr Karel. X-Ray Lab., Geological Survey, Hradební 9, 110 00 Praha 1, CSSR. (1930) CSc, mineralogy (Charles U., 1964). Head of the X-Ray Lab. (tel. 65446). *Silicates, clay minerals.*

Mikloš, Ing Dušan. Lab. of Diffraction Methods, Inst. of Inorganic Chemistry, Centre for Chem. Res., Slovak Acad. Sci., Dúbravská cesta, 842 36 Bratislava, CSSR. (1944) CSc, inorganic chemistry (Slovak Techn. U., 1976). Head. (tel. 375 488, ext. 2175). *Order disorder structures, structure analysis.*

Moravec, Ing František. Dept. of Solid State Electronics, Inst. of Radio Engineering and Electronics, Cz. Acad. Sci., Lumumbova 1, 182 51 Praha 8, CSSR. (1931) CSc, crystallography (Inst. of Chemical Techn., 1967). Head of the Lab. (tel. 843741, ext. 317). *Crystal growth, single crystals.*

Moravcová, Dr Hana. X-Ray Laboratory, Geological Survey, Hradební 9, 110 15 Praha 1, CSSR. (1950) RNDr., physics (Charles U., 1975). Res. asst. (tel. 65446). *X-ray phase analysis of minerals, crystal structure of silicates.*

Nehasil, Dr Miroslav. P.A.: Novorosijská 14, 100 00 Praha 10, CSSR. (1921) CSc, physics (Techn. U. of Prague, 1981). (tel. 7376203). *X-ray structure analysis, X-ray tensometry.*

Novák, Ing Ctirad. Dept. of Structures and Bonding, Inst. of Physics, Cz. Acad. Sci., Na Slovance 2, 180 40 Praha 8, CSSR. (1921) CSc, mathematics (Inst. of Mathematics, Cz. Acad. Sci., 1961). Scient. (tel. 354240-9, ext. 96). *Crystal geometry, crystallographic computing.*

Nývlt, Ing Jaroslav. Laboratory of Crystallization and Calorimetry, Inst. of Inorganic Chemistry, Cz. Acad. Sci., Majakovského 24, 160 00 Praha 6, CSSR. (1932) DrSc, physical chemistry (Inst. of Chemical Techn., 1967). Head. (tel. 256897). *Bulk crystallization from solutions.*

Pavelčík, Ing František. Dept. of Analytical Chemistry, Faculty of Pharmacy, Comenius U., Odbojárov 10, 832 32 Bratislava, CSSR. (1945) CSc, inorganic chemistry (Slovak Techn. U., 1977). Scient. (tel. 60451, ext. 297). *Crystallographic computing, programming, coordination chemistry, molecular mechanics, biological small molecules.*

Petříček, Dr Václav. Dept. of Structures and Bonding, Inst. of Physics, Cz. Acad. Sci., Na Slovance 2, 180 40 Praha 8, CSSR. (1948) CSc, physics (Inst. of Physics, Cz. Acad. Sci., 1981) Scient. (tel. 354240-9, ext. 96). *Structure analysis.*

Pleštil, Ing Josef. Lab. of X-Ray Polymer Structure Analysis, Inst. of Macromolecular Chemistry, Cz. Acad. Sci., Heyrovského nám. 2, 162 06 Praha 6, CSSR. (1946) CSc, physics (Charles U., 1974). Scient. (tel. 360341). *Polymer structure, small angle scattering.*

Podbrdský, Dr Josef. Dept. of Electron Optics, Inst. of Scientific Instr., Cz. Acad. Sci., Královopolská 147, 612 64 Brno, CSSR. (1945) CSc, physics (J. E. Purkyně U., 1972). Scient. (tel. 54311, ext. 365). *Electron microscopy and diffraction.*

Podlahová, Dr Jana. Dept. of Inorganic Chemistry, Faculty of Sci., Charles U., Albertov 2030, 128 40 Praha 2, CSSR. (1937) CSc, inorganic chemistry (Charles U., 1964). Sr. lect. (tel. 292051-5, ext. 420). *Inorganic chemistry, structure of inorganic compounds.*

Polcarová, Dr Milena. Dept. of Metal Physics, Inst. of Physics, Cz. Acad. Sci., Na Slovance 2, 180 40 Praha 8, CSSR. (1931) CSc, physics (Inst. of Physics, Cz. Acad. Sci., 1961). Scient. (tel. 842241). *X-ray topography, imperfections in crystals.*

Pollert, Ing Emil. Dept. of Magnetism, Inst. of Physics, Cz. Acad. Sci., Na Slovance 2, 180 40 Praha 8, CSSR. (1938) CSc, chemistry (Inst. of Chemical Techn., 1972). Scient. (tel. 354240-9, ext. 67). *Preparation and properties, transition element oxides.*

Rieder, Dr Milan. Inst. of Geological Sciences, Charles U., Albertov 6, 128 43 Praha 2, CSSR. (1940) PhD, geology (Johns Hopkins U., Baltimore, Md., 1968). Scient. (tel. 297541, ext. 295). *Experimental petrology, mineralogy, crystallography.*

Rychlý, Ing Rudolf. State Glass Res. Inst., Škroupova 957, 501 92 Hradec Králové, CSSR. (1941) CSc, physical chemistry (Inst. of Chemical Techn., 1969). Scient. (tel. 32821, ext. 29). *Crystallization in glass, mineralogy.*

Schilder, Ing Jaroslav. Dept. of Electron Microscopy, Electrotechnical Inst., Slovak Acad. Sci., Dúbravská cesta, 809 32 Bratislava, CSSR. (1928) CSc, theoretical electrotechnics (Slovak Techn. U., 1960). Head of the laboratory (tel. 45139, 45651, ext. 2937). *Scanning electron microscopy, thin films.*

Seidl, Ing Vlastimil. Dept. of Mineralogy, Inst. of Chemical Techn., Suchbátarova 5, 166 28 Praha 6, CSSR. (1927) CSc, mineralogy (Inst. of Geochemistry and Mineral Resources, Cz. Acad. Sci., 1963). Scient. (tel. 332, ext. 4087, 4201). *Inorganic crystal structures, programming.*

Sekanina, Prof. Dr Josef. Dept. of Mineralogy and Petrography, Faculty of Sci., J. E. Purkyně U., Kotlářská 2, 602 00 Brno, CSSR. P.A.: Šumavská 30, 602 00 Brno. (1901) DrSc, mineralogy (Cz. Acad. Sci., 1956); corresponding member of the Cz. Acad. Sci. Em. Prof. of mineralogy (tel. 51112). *Mineralogical crystallography, crystal optics.*

Šimerská, Dr Marie. Dept. of Metal Physics, Inst. of Physics, Cz. Acad. Sci., Na Slovance 2, 180 40 Praha 8, CSSR. (1929) CSc, physics (Inst. of Solid State Physics, Cz. Acad. Sci., 1966). Scient. (tel. 842241, ext. 2778). *Metal physics, structure of metals and alloys.*

Sivý, Dr Peter. Dept. of Physics, Faculty of Chemical Techn., Slovak Technical U., Jánska 1, 812 37 Bratislava, CSSR. (1951) RNDr., physics (Comenius U., 1981). Sr. lect. (tel. 56021, ext. 343). *Inorganic crystal structures, crystallographic statistics, crystallochemistry, coordination compounds.*

Soldánová, Ing Jiřina. Dept. of Inorganic Chemistry, Faculty of Chemical Techn., Slovak Technical U., Jánska 1, 812 37 Bratislava, CSSR. (1946) CSc, chemistry (Slovak Techn. U., 1984). Scient. (tel. 56021, ext. 434). *Crystal structure, stereochemistry, coordination compounds.*

Steinhart, Dr Miloš. Lab. of X-ray Polymer Structure Analysis, Inst. of Macromolecular Chemistry, Cz. Acad. Sci., Heyrovského nám. 2, 162 06 Praha 6, CSSR. (1955) RNDr., physics (Charles U., 1979). Scient. (tel. 360341). *Small angle scattering, macromolecular solutions.*

Syneček, Doc. Dr Vladimír. Dept. of Metal Physics, Inst. of Physics, Cz. Acad. Sci., Na Slovance 2, 180 40 Praha 8, CSSR. (1929) CSc, physics (Inst. of Solid State

Physics, Cz. Acad. Sci., 1956). Scient. (tel. 842241, ext. 2778). *Metal physics, structure of metals and alloys.*

Šebo, Dr Pavel. Dept. of Structure Properties of Metals and Alloys, Inst. of Metal Materials, Slovak Acad. Sci., Februárového víťazstva 75, 801 00 Bratislava, CSSR. (1939) CSc, physics (Inst. of Solid State Physics, Cz. Acad. Sci., 1969). Scient. (tel. 673351, ext. 334). *Metal physics.*

Šedivý, Doc. Dr Josef. Dept. of Physics of Semiconductors, Faculty of Mathematics and Physics, Charles U., Ke Karlovu 5, 121 16 Praha 2, CSSR. (1919) RNDr., physics (Charles U., 1952). Doc. of experimental physics (tel. 292141-8, ext. 328). *Structure analysis, lattice dynamics of solids.*

Šíchová, Dr Hana. Dept. of Physics of Semiconductors, Faculty of Mathematics and Physics, Charles U., Ke Karlovu 5, 121 16 Praha 2, CSSR. (1931) CSc, physics (Charles U., 1965). Sr. lect. (tel. 292141-8, ext. 394). *Lattice dynamics of metals by X-ray diffraction.*

Šourek, Dr Zbyněk. Dept. of Structures and Bonding, Inst. of Physics, Cz. Acad. Sci., Na Slovance 2, 180 40 Praha 8, CSSR. (1948) CSc, physics (Inst. of Physics, Cz. Acad. Sci., 1979). Scient. (tel. 354240-9, ext. 92). *Perfect and real crystal X-ray diffraction, X-ray topography.*

Šubrtová, Ing Věra. Dept. of Structures and Bonding, Inst. of Physics, Cz. Acad. Sci., Na Slovance 2, 180 40 Praha 8, CSSR. (1936) Ing., geology (Inst. of Mining and Metallurgy, 1959). Res. asst. (tel. 354240-9, ext. 96). *X-ray structure analysis.*

Ulický, Doc. Ing Ladislav. Dept. of Physical Chemistry, Slovak Techn. U., Jánska 1, 880 37 Bratislava, CSSR. (1931) CSc, physical chemistry (Comenius U., 1961). Doc. of physical chemistry (tel. 56021, ext. 427). *Small angle scattering, X-ray structure analysis.*

Valach, Ing Fedor. Dept. of Physics, Faculty of Chemical Techn., Slovak Technical U., Jánska 1, 812 37 Bratislava, CSSR. (1946) CSc, chemistry (Slovak Techn. U., 1972). Sr. lect. (tel. 56021, ext. 343). *Inorganic crystal structures, crystallographic statistics, crystallochemistry, coordination compounds.*

Valvoda, Dr Václav. Dept. of Physics of Semiconductors, Faculty of Mathematics and Physics, Charles U., Ke Karlovu 5, 121 16 Praha 2, CSSR. (1937) CSc, physics (Charles U., 1968). Sr. lect. (tel. 292141-8, ext. 395). *Lattice dynamics of metals and alloys by X-ray diffraction.*

Vrábel, Ing Viktor. Dept. of Analytical Chemistry, Faculty of Chemical Techn., Slovak Technical U., Jánska 1, 812 37 Bratislava, CSSR. (1947) CSc, chemistry (Slovak Techn. U., 1980). Res. asst. (tel. 56021, ext. 375). *Crystal structure, coordination compounds, analytical chemistry.*

Weiss, Dr Zdeněk. X-Ray Lab., Dept. of Physics and Chemistry, Scientific Coal Res. Inst., Pikartská, 716 07 Ostrava-Radvanice, CSSR. (1942) RNDr., mineralogy and crystallography (Charles U., 1972). Head of X-Ray Lab. (tel. 215444, ext. 432). *Mineralogical crystallography, X-ray diffraction, mathematics.*

Zikmund, Dr Zdeněk. Dept. of Structures and Bonding, Inst. of Physics, Cz. Acad. Sci., Na Slovance 2, 180 40 Praha 8, CSSR. (1939) CSc, physics, (Inst. of Solid State Physics, Cz. Acad. Sci., 1977). Scient. (tel. 354240-9, ext. 96). *Symmetry of crystals and domain structures.*

Žák, Doc. Dr Lubor. Dept. of Mineralogy, Geochemistry and Crystallography, Faculty of Sci., Charles U., Albertov 6, 128 43 Praha 2, CSSR. (1925) CSc, mineralogy (Charles U., 1956). Doc. of mineralogy (tel. 297541, ext. 298). *Mineral physics and chemistry.*

Žák, Dr Zdirad. Dept. of Inorganic Chemistry, Faculty of Sci., J. E. Purkyně U., Kotlářská 2, 611 37 Brno, CSSR. (1941) CSc, inorganic chemistry (J. E. Purkyně U.,1982). Sr. lect. (tel. 51112, ext. 69). *Inorganic crystal structures, inorganic synthesis.*

DENMARK

Sub-Editor: B. Jensen

Notes

1. International telephone country code - 45.

2. Degrees conferred by Danish Universities are *doctor philosophiae, doctor techniches,* etc. (Dr.phil., Dr.techn., etc.) (equivalent to DSc); *magister scientiarum, licentiatus pharmaciae, licentiatus technices,* etc. (Mag.scient., Lic.pharm., Lic.techn., etc.) (equivalent to PhD); a *candidatus magistrii, -pharmaciae, -polytechnices, -scientiarum* (Cand.mag., -pharm., -polyt., -scient.) (equivalent to MSc).

Andersen, Mr Erik Krogh. Dept. of Chemistry, U. of Odense, Campusvej 55, DK-5230 Odense M, Denmark. (1923) Dr.phil., chemistry (Odense U., 1971). Lect. (tel. 09 + 158600, ext. 2538). *Chemistry.*

Andersen, Mrs Inger Grete Krogh. Dept. of Chemistry, U. of Odense, Campusvej 55, DK-5230 Odense M, Denmark. (1929) Cand.pharm. (Roy. Danish Sch. of Pharm., 1954). Res. asst. (tel. 09 + 158600). *Phase identifications.*

Andersen, Dr Niels Hessel. Physics and Metallurgy Dept., Risø National Lab., DK-4000 Roskide, Denmark. (1945) Lic. scient., Physics (Copenhagen U., 1976). Res. assoc. (tel. 02 + 371212, ext. 4700, 5721, telex 43116). *Ionic conductors, solid electrolytes.*

Andersen, Prof. Palle. Dept. of Technology, Royal Dental C., Vennelyst Blvd., DK-8000 Aarhus, Denmark. (1909) Mag.scient., inorganic chemistry (Copenhagen U., 1943). Prof. retired. (tel. 06 + 132533, ext. 255). *Inorganic structures, alloy structures.*

Andersen, Dr Peter. Dept. of Inorg. Chemistry, U. of Copenhagen, Universitetsparken 5, DK-2100 Copenhagen, Denmark. (1938) Cand.polyt., chemistry (Techn. U. of Denmark, 1961). Lect. (tel. 01 + 353133). *Inorganic chemistry, coordination chemistry.*

Andersen, Dr Stig Kjær. Inst. of Electronic Systems, Aalborg U., Strandvejen 19, DK-9000 Aalborg, Denmark. (1947) Lic.scient., experimental physics (Aarhus U., 1977). Sr. res. scient. (tel. 08 + 138788). *X-ray instrumentation, expert systems.*

Bang, Dr Eva Henriette. Dept. of Inorg. Chemistry, U. of Copenhagen, Universitetsparken 5, DK-2100 Copenhagen, Denmark. (1917) Cand.mag., chemistry (Copenhagen U., 1942). Lect. (tel. 01 + 353133). *Inorganic and coordination compound structures.*

Brehm, Dr Lotte. Dept. of Chemistry BC, Royal Danish Sch. of Pharmacy, Universitetsparken 2, DK-2100 Copenhagen, Denmark. (1940) Lic.pharm., organic chemistry (Roy. Danish Pharm. Sch., 1969). Lect. (tel. 01 + 370850, ext. 245). *Organic structures, biological molecules.*

Buchwald, Dr Vagn Fabritius. Inst. of Metallurgy, Techn. U. of Denmark, DTH 204, DK-2800 Lyngby, Denmark. (1929) Dr.scient., metallurgy (Copenhagen U., 1977). Sr. lect. (tel. 02 + 884022, ext. 3237). *Metallurgy, solid state transformations, meteoritic minerals, history (metallurgy), archaeological metals.*

Christensen, Dr Axel Nørlund. Dept. of Chemistry, U. of Aarhus, Langelandsgade 140, DK-8000 Aarhus, Denmark. (1934) Dr.phil., inorganic chemistry (Aarhus U., 1967). Lect. (tel. 06 + 124633). *Crystal growth, inorganic crystal structures.*

Clausen,, Dr Kurt. Physics Dept., Risø National Lab., DK-4000 Roskilde, Denmark. (1952) Lic. techn., Physics (Techn U., 1981). Res. assoc. (tel. 02 + 371212, ext. 4700, telex 43116). *Neutron scattering.*

Cour, la, Dr Troels Frederik Marstrand. Dept. of Chemistry, U. of Aarhus, Langelandsgade 140, DK-8000 Aarhus, Denmark. (1944) Cand.scient., X-ray crystallography (Aarhus U., 1970). Lect. (tel. 06 + 124633, ext. 217). *Biological macromolecular structure and function.*

Danielsen, Dr Jacob. Dept. of Chemistry, U. of Aarhus, Langelandsgade 140, DK-8000 Aarhus, Denmark. (1937) Dr.phil., inorganic chemistry (Aarhus U., 1961). Lect. (tel. 06 + 124633, ext. 230). *Crystal growth, computer programming.*

Drenck, Prof. Kaj. Dept. of Physics, Royal Danish Sch. of Pharmacy, Universitetsparken 2, DK-2100 Copenhagen, Denmark. (1921) Dr.phil., crystallography (Copenhagen U., 1959). Prof. (tel. 01 + 370850, ext. 270). *Apparatus, powders, particle size, ferroelectrics.*

Gajhede, Mr Michael. Chemical Lab. IV, U. of Copenhagen, Universitetsparken 5, DK-2100 Copenhagen, Denmark. (1954) Cand.scient., chemistry (Copenhagen U., 1982). *Electron density determination, crystal structures.*

Gerhard Olsen, Dr Inger Lise. Inst. of Mineral Industry, Techn. U. of Denmark, DTH 204, DK-2800 Lyngby, Denmark. (1938) Lic.techn., physical chemistry (Techn. U. of Denmark, 1967). Lect. (tel. 02 + 882222, ext. 3203). *Optical and infrared mineralogy.*

Gerward, Dr Leif. Lab. of Applied Physics III, Techn. U. of Denmark, DTH 307, DK-2800 Lyngby, Denmark. (1939) Tekn. Dr., physics (Chalmers U. of Techn., 1970). Lect. (tel. 02 + 882488, ext. 2348). *Energy dispersive diffraction, high pressure crystal structures, X-ray interaction with matter.*

Grønlund, Dr Finn. Chemical Lab. IV, U. of Copenhagen, Universitetsparken 5, DK-2100 Copenhagen, Denmark. (1925) DSc, physical chemistry (U. de Paris, 1957). Lect. (tel. 01 + 353133, ext. 242). *Surface structure of crystals, electron diffraction.*

Grundvig, Dr Sidsel. Dept. of Geology, U. of Aarhus, C.F. Møllers Alle, DK-8000 Aarhus, Denmark. (1941) Lic.pharm., crystallography (Roy. Danish Pharm.

Sch., 1967). Lect. (tel. 06 + 128233). *Inorganic crystal structures, X-ray fluorescence analysis, electron microprobe analysis.*

Haagensen, Mr Carl Olaf. Dept. of Chemistry, U. of Aarhus, Langelandsgade 140, DK-8000 Aarhus, Denmark. (1929) Cand.polyt., chemistry (Techn. U. of Denmark, 1955). Lect. (tel. 06 + 124633). *Crystal structures.*

Hazell, Dr Alan Charles. Dept. of Chemistry, U. of Aarhus, Langelandsgade 140, DK-8000 Aarhus, Denmark. (1935) PhD, inorganic and structural chemistry (Leeds U., 1962). Lect. (tel. 06 + 124633, ext. 236). *Crystal structures, platinum metal complexes, sulphur-nitrogen compounds, anomalous scattering.*

Hazell, Mrs Rita Grønbæk. Dept. of Chemistry, U. of Aarhus, Langelandsgade 140, DK-8000 Aarhus, Denmark. (1936) Cand.polyt., chemistry (Techn. U. of Denmark, 1961). Lect. (tel. 06 + 124633, ext. 232). *Crystal structures, methods of determination, refinement methods.*

Honoré, Dr Tage. A/S Ferrosan, Res. Div., Sydmarken 5, DK-2860 Soeborg, Denmark. (1951) Lic.pharm., organic chemistry (Roy. Danish Pharm. Sch., 1978). Head, biochemical. dept. (tel. 01 + 692111, ext. 242). *Neurochemistry.*

Jensen, Mr Aage. Inst. of Mineralogy, U. of Copenhagen, Østervoldgade 10, DK-1350 Copenhagen, Denmark. (1930) Cand.mag., mineralogy (Copenhagen U., 1957). Lect. (tel. 01 + 112232). *Oxides of Fe and Ti, gem minerals.*

Jensen, Dr Birthe. Dept. of Chemistry BC, Royal Danish Sch. of Pharmacy, Universitetsparken 2, DK-2100 Copenhagen, Denmark. (1938) Dr.pharm., organic chemistry (Roy. Danish Pharm. Sch., 1984). Lect. (tel. 01 + 370850, ext. 246). *Organic structures, biological structures, interatomic forces.*

Jensen, Dr Ejnar. Chemistry Dept., Roy. Vetr. and Agricult. U., Thorvaldsensvej 40, DK-1871 Copenhagen, Denmark. (1917) Lic.agro., soil science (Roy. Vetr. and Agricult. U., 1950). Lect. (tel. 01 + 351788). *Clay minerals.*

Jensen, Mr Gunnar Bent. Dept. of Electrophysics, Techn. U. of Denmark, Anker Engelundsvej, DK-2800 Lyngby, Denmark. (1929) Cand.mag., theoretical physics (Copenhagen U., 1955). Lect. (tel. 02 + 881188, ext. 2715). *Crystal physics.*

Jensen, Dr Stig Jorgo. Dept. of Technology, Royal Dental C., Vennelyst Blvd., DK-8000 Aarhus, Denmark. (1932) Dr.phil., chemistry (Aarhus U., 1969). Lect. (tel. 06 + 132533, ext. 256). *Dental materials crystal structures.*

Jerslev Lund, Prof. Bodil. Dept. of Chemistry BC, Royal Danish Sch. of Pharmacy, Universitetsparken 2, DK-2100 Copenhagen, Denmark. (1919) Dr.phil., chemistry (Copenhagen U., 1958). Prof. (tel. 01 + 370850, ext. 240). *Organic structures, organic powder identification.*

Johnsen, Mr Ole. Mineral Collection, Geological Museum, U. of Copenhagen, Østervoldgade 5-7, DK-1350 Copenhagen, Denmark. (1940) Cand.scient., mineralogy (Copenhagen U., 1971). Sr. Lect. (tel. 01 + 135001). *Silicate crystal structures, crystal chemistry, minerals.*

Juul Jensen, Mrs Dorte. Physics and Metallurgy Dept., Risø National Lab., DK-4000 Roskilde, Denmark. (1957) Lic.techn., metallurgy (Techn. U. of Denmark, 1983). Res. assoc. (tel. 02 + 371212, ext. 4756, telex 43116). *Texture, neutron scattering, recrystallization, metals.*

Jørgensen, Mr Ole. Inst. of Mineralogy, U. of Copenhagen, Østervoldgade 10, DK-1350 Copenhagen, Denmark. (1939) Mag.scient., geology (Copenhagen U., 1969). Sr. lect. (tel. 01 + 112232). *Silicate minerals, zeolites, crystal optics, crystal growth, microtopography.*

Kaas, Dr Karen. Chemistry Dept., Roy. Vetr. and Agricult. U., Thorvaldsensvej 40, DK-1871 Copenhagen, Denmark. (1942) Lic.pharm., inorganic chemistry (Roy. Danish Pharm. Sch., 1973). Lect. (tel. 01 + 351788, ext. 2404). *Bridged transition metal complexes.*

Karup-Møller, Dr Sven. Inst. of Mineral Industry, Techn. U. of Denmark, DTH 204, DK-2800 Lyngby, Denmark. (1936) Dr.scient., mineralogy (Copenhagen U.). Lect. (tel. 02 + 882222). *Ore deposit mineralogy.*

Kjeldgaard, Mr Morten. Dept. of Chemistry, U. of Aarhus, Langelandsgade 140, DK-8000 Aarhus, Denmark. (1953) Cand.scient., chemistry (Aarhus U., 1982). Res. fellow. (tel. 06 + 124633, ext. 267). *Protein crystallography, computer graphics.*

Kristensen, Mrs Bente Saustrup. Chemistry Dept. A, Techn. U. of Denmark, DTH 207, DK-2800 Lyngby, Denmark. (1931) Cand.polyt., chemistry (Techn. U. of Denmark, 1956). Lect. (tel. 02 + 883111, ext. 3357). *Inorganic structures.*

Langer, Prof. Ebbe Wang. Dept. for Metallurgy, Techn. U. of Denmark, Lundtoftevej 100, DK-2800 Lyngby, Denmark. (1927) Dr.techn., metallurgy (Techn. U. of Denmark, 1967). Prof. in metallurgy. (tel. 02 + 884022, ext. 3233). *Electron diffraction and electron microscopy, metals.*

Larsen, Mr Finn Krebs. Dept. of Chemistry, U. of Aarhus, Langelandsgade 140, DK-8000 Aarhus, Denmark. (1941) Cand.scient., inorganic chemistry (Aarhus U., 1966). Lect. (tel. 06 + 124633, ext. 241). *Accurate structure analysis, X-ray and neutron diffraction, electron density determination.*

Larsen, Dr Ingrid Kjøller. Dept. of Chemistry BC, Royal Danish Sch. of Pharmacy, Universitetsparken 2, DK-2100 Copenhagen, Denmark. (1935) Lic.pharm., organic chemistry (Roy. Danish Pharm. Sch., 1965). Lect. (tel. 01 + 370850, ext. 244). *Organic structures, biological structures.*

Larsen, Mrs Sine. Chemical Lab. IV, U. of Copenhagen, Universitetsparken 5, DK-2100 Copenhagen, Denmark. (1943) Cand.scient., chemistry (Copenhagen U., 1968). Lect. (tel. 01 + 353133, ext. 252). *Coordination compounds, electron density determinations, intermolecular interactions.*

Lebech, Mrs Bente. Physics Dept., Risø National Lab., DK-4000 Roskilde, Denmark. (1937) Cand.polyt., physics (Techn. U. of Denmark, 1962). Res. assoc. (tel. 02 + 371212, ext. 4705, telex 43116). *Neutron and X-ray scattering, magnetic structures.*

Leffers, Dr Torben. Metallurgy Dept., Risø National Lab., DK-4000 Roskilde, Denmark. (1936) Dr.techn., physical metallurgy (Techn. U. of Denmark, 1975). Sr. scient. (tel. 02 + 371212, ext. 5708). *Structure of metals, deformation of metals, irradiation of metals.*

Leonardsen, Mr Erik Sverre. Inst. of Mineralogy, U. of Copenhagen, Østervoldgade 10, DK-1350 Copenhagen, Denmark. (1934) Cand.real., inorganic chemistry (Oslo U., 1961). Lect. (tel. 01 + 112232). *Mineral structures, mineral identification, computer methods.*

Lindegaard-Andersen, Prof. Asger. Lab. of Applied Physics III, Techn. U. of Denmark, DTH 307, DK-2800 Lyngby, Denmark. (1925) Dr.techn., physics (Techn. U. of Denmark, 1967). Prof. (tel. 02 + 882488, ext. 2345). *Crystal defects, crystal growth, X-ray topography, diffuse X-ray scattering, biomineralization.*

Lindgreen, Dr Holger. Geological Survey of Denmark, Thorasvej 31, DK-2400 Copenhagen NV, Denmark. (1947) Lic.agro., (Roy. Veterinary and Agricultural U., 1974). Sr. scient. (tel. 01 + 106600). *Clay minerals, clay diagnosis.*

Makovicky, Dr Emil. Inst. of Mineralogy, U. of Copenhagen, Østervoldgade 10, DK-1350 Copenhagen, Denmark. (1940) RNDr, mineralogy (Bratislava U., 1967). PhD, geology (Mc Gill U. Montreal, 1970). Lect. (tel. 01 + 112232). *Sulphide and sulphosalt crystal structures, crystal chemistry, mineralogical crystallography, computer programming.*

Micheelsen, Prof. Harry. Inst. of Mineralogy, U. of Copenhagen, Østervoldgade 10, DK-1350 Copenhagen, Denmark. (1931) Dr.phil., mineralogy (Copenhagen U., 1966). Prof. (tel. 01 + 112232). *Mineral submicro structures, crystal optics.*

Møller, Dr Christian Knakkergaard. Chemistry Dept., U. of Odense, Campusvej 55, DK-5230 Odense M, Denmark. (1920) Dr.phil., chemistry (Copenhagen U., 1961). Prof. (tel. 09 + 158600, ext. 2534). *Crystal chemistry, solid state spectroscopy.*

Nellemos Andersen, Mr Jesper. Metallurgy Dept., Risø National Lab., DK-4000 Roskilde, Denmark. (1956) Cand.scient., Physics and math. (Aarhus U.). Scholarship holder. (tel. 02 + 371212, ext. 5728, telex 43116). *Surface structure, texture, neutron scattering, recrystallization, computing.*

Nielsen, Dr Anders. Res. and Dev. Div., Haldor Topsøe A/S, Nymøllevej 55, DK-2800 Lyngby, Denmark. (1919) Dr.techn., chemical engineering (Techn. U. of Denmark, 1950). Manager of res. and develop. div. (tel. 02 + 878100). *Imperfect crystal structures, catalysts.*

Nielsen, Mr Kurt. Chemistry Dept. B, Techn. U. of Denmark, DTH 301, DK-2800 Lyngby, Denmark. (1943) Cand.scient., chemistry (Aarhus U., 1969). Lect. (tel. 02 + 881777, ext. 2112). *Crystal structures, crystallographic methods.*

Noe-Nygaard, Prof. Arne. Geological Museum, U. of Copenhagen, Østervoldgade 10, DK-1350 Copenhagen, Denmark. (1908) Dr.phil., petrology (Copenhagen U., 1937). Prof. (tel. 01 + 112232). *North Atlantic basalt province.*

Norrestam, Prof. Rolf. Chemistry Dept. B, Techn. U. of Denmark, DTH 301, DK-2800 Lyngby, Denmark. (1937) Fil.dr, chemistry (Stockhlom U., 1972). Prof. (tel. 02 + 881777, ext. 2180, telex 37529). *Crystal structures, crystallographic methods, computer graphics.*

Nyborg, Dr Jens. Dept. of Chemistry, U. of Aarhus, Langelandsgade 140, DK-8000 Aarhus, Denmark. (1942) Fil.dr, crystallography (Gothenburg U., 1971). Lect. (tel. 06 + 124633, ext. 217). *Protein and nucleic acid crystallography, direct methods, process control.*

Nørskov-Lauritsen, Dr Leif. Dept. of Chemistry, U. of Aarhus, Langelandsgade 140, DK-8000 Aarhus, Denmark. (1953) Lic.scient., physical organic chemistry (Copenhagen U., 1984). Res. assoc. (tel. 06 + 124633, ext. 267). *Conformational analysis, molecular geometry, molecular graphics.*

Olsen, Dr Janus Staun. Physics Lab. II, H.C. Ørsted Institute, Universitetsparken 5, DK-2100 Copenhagen, Denmark. (1937) Cand.mag. et mag.scient., physics (Copenhagen U., 1963). Lect. (tel. 01 + 353133). *Energy dispersive diffraction, phase transformation, high pressure work.*

Pauly, Prof. Hans. Inst. of Mineral Industry, Techn. U. of Denmark, DTH 204, DK-2800 Lyngby, Denmark. (1921) Dr.phil., mineralogy (Copenhagen U., 1958). Prof. (tel. 02 + 884022, ext. 3216). *Ore-mineralogy, mineralogy, fluoride minerals.*

Petersen, Dr Ole Valdemar. Mineralogical Div., Geological Museum, Østervoldgade 10, DK-1350 Copenhagen, Denmark. (1939) Lic.scient., mineralogy (Copenhagen U., 1970). Sr. lect. (tel. 01 + 112232). *Crystal optics, minerals.*

Plough-Sørensen, Mrs Gudrun. Dept. of Chemistry, U. of Odense, Campusvej 55, DK-5230 Odense M, Denmark. (1938) Cand.pharm. (Roy Danish Pharm. Sch., 1962). Lect. (tel. 09 + 158600). *Crystal structures.*

Rasmussen, Prof. Svend Erik. Dept. of Chemistry, U. of Aarhus, Langelandsgade 140, DK-8000 Aarhus, Denmark. (1925) Dr.phil., chemical chrystallography (Copenhagen U., 1960). Prof. (tel. 06 + 124633, ext. 229). *Coordination compounds, refractory compounds, crystal growth.*

Rindorf, Mrs Grethe. Chemistry Dept. B, Techn. U. of Denmark, DTH 301, DK-2800 Lyngby, Denmark. (1919) Cand.polyt., chemistry (Techn. U. of Denmark, 1945). Lect. (tel. 02 + 881777, ext. 2113). *Inorganic crystal structures.*

Rose-Hansen, Dr John. Inst. of Petrology, U. of Copenhagen, Østervoldgade 10, DK-1350 Copenhagen, Denmark. (1937) Mag.scient., petrology and mineralogy

(Copenhagen U., 1963). Sr. lect. (tel. 01 + 112232). *Petrology, mineralogy, experimental petrology.*
Rønsbo, Mr Jørn. Inst. of Mineralogy, U. of Copenhagen, Østervoldgade 10, DK-1350 Copenhagen, Denmark. (1941) Cand.scient., geology (Copenhagen U., 1971). Sr. lect. (tel. 01 + 112232). *Mineral chemistry, electron microprobe analysis.*
Schmidt-Nielsen, Mr Søren. Sct. Knuds Gymnasium, Lasøegade, DK-5000 Odense, Denmark. (1942) Cand.scient., organic chemistry (Aarhus U., 1971). Lect. (tel. 09 + 901614). *Small angle scattering.*
Simonsen, Mr Ole. Dept. of Chemistry, U. of Odense, Campusvej 55, DK-5230 Odense M, Denmark. (1937) Cand.scient., organic chemistry (Copenhagen U., 1966). Lect. (tel. 09 + 158600). *Organic structures.*
Sørensen, Dr Alex Mehlsen. Dept. of Chemistry BC, Royal Danish Sch. of Pharmacy, Universitetsparken 2, DK-2100 Copenhagen, Denmark. (1935) Lic.pharm., organic chemistry (Roy. Danish Pharm. Sch., 1966). Lect. (tel. 01 + 370850, ext. 226). *Organic structures, powder diffraction.*
Sørensen, Mr Ole. Res. Div., Physic. and Analyt. Dept., Haldor Topsøe A/S, Nymøllevej 55, DK-2800 Lyngby, Denmark. (1935) Cand.polyt., physics (Techn. U. of Denmark, 1960). Res. engineer. (tel. 02 + 878100). *Electron microscopy, electron diffraction, electron microprobe analysis.*
Søtofte, Mrs Inger. Chemistry Dept. B, Techn. U. of Denmark, DTH 301, DK-2800 Lyngby, Denmark. (1940) Cand.mag., chemistry (Copenhagen U., 1965). Lect. (tel. 02 + 881777, ext. 2141). *Crystal structures.*
Thorup, Mr Niels. Chemistry Dept. B, Techn. U. of Denmark, DTH 301, DK-2800 Lyngby, Denmark. (1939) Cand.polyt., chemistry (Techn. U. of Denmark, 1963). Lect. (tel. 02 + 881777, ext. 2140). *Crystal structures, organic conductors, databases.*
Villadsen, Mr Jørgen. Res. and Dev. Div., Haldor Topsøe A/S, Nymøllevej 55, DK-2800 Lyngby, Denmark. (1930) Cand.polyt., chemistry (Techn. U. of Denmark, 1953). Sr. sci., manager of phys. and anal. dept. (tel. 02 + 878100). *Inorganic crystal structures, crystal chemistry, phase analysis.*
Zachau-Christiansen, Stud.lic. Birgit. Fysisk-Kemisk Inst., Techn. U. of Copenhagen, DK-2800 Lyngby, Denmark. (1951) Cand.polyt., chemistry (Techn. U. of Denmark, 1975). Stud.lic. (tel. 02 + 883111, ext. 3405). *Intercalation compounds.*

EGYPT, Arab Rep.

Sub-Editor: M.M. Radwan

Notes

1. International telephone country code - 20.

Abdel Aal, Mr Fawzi Amer. Geological Survey and Mineral Res. Organization, Salah Salem Street, Abbassia, Arab Rep. of Egypt. BSc, geology (Cairo U., 1945). Res. officer. (tel. 830782). *Economic geology.*
Abdel Hady, Prof. Seham. Physics Dept., Faculty of Sci., Helwan U., Helwan, Cairo, Arab Rep. of Egypt. PhD, crystallography (Cairo U., 1965). Dept. head. (tel. 802129). *Organic structures.*
Abdel Kader, Dr Abdel Aziz. X-ray Crystallography Unit, Nat. Res. Centre, Dokki-Cairo, Arab Rep. of Egypt. PhD, geology (London U., UK, 1964). Res. (tel. 802129). *Igneous petrology.*
Abdel Kader, Prof. (Miss) Naima. X-ray Crystallography Unit, Nat. Res. Centre, Dokki-Cairo, Arab Rep. of Egypt. PhD, crystallography (Cairo U., 1965). Res. (tel. 802129). *Organic structures.*
Abdel Kader, Mrs Zeinab Mohamed. Faculty of Sci., U. of Cairo, University Street, Giza-Cairo, Arab Rep. of Egypt. MSc, geology (Cairo U., 1969). Teaching asst. (tel. 841722, 841713). *Economic geology.*
Abdel Mohsen, Dr Hussein. Atomic Energy Commission, Inshas-Cairo, Arab Rep. of Egypt. DSc, geology Nancy U., France, 1959). Prof. (tel. 862013). *Radioactive mineralogy.*
Abdel Rehim, Dr Amin Mohamed. Dept. of Physics, U. of Tanta, Faculty of Sci., Tanta, Arab Rep. of Egypt. PhD, crystallography (London U., UK, 1955). Prof. *Crystal growth.*
Abdu, Prof. Fayez Madi. Faculty of Agriculture, Shoubra el Kheima, U. of Ain Shams, Abbassia-Cairo, Arab Rep. of Egypt. PhD, soils (Ain Shams U., 1957). Prof. (tel. 948716, 942711). *Soil chemistry, soil mineralogy.*
Abou-Saif, Dr Elhamy Aziz. Lab. of Electron Microscopy and Thin Films, Physics Dept., Nat. Res. Cntr., Altahrir Street, Dokki-Cairo, Arab Republic of Egypt. (1939)PhD, solid state physics (Ain-Shams U., 1972). Assoc. prof. *Electron diffraction.*
Ahmed, Dr Mohamed Saleh. Dept. of Physics, U. of Alexandria, Alexandria, Arab Rep. of Egypt. PhD, physics (London U., UK, 1950). Prof. *Diffuse scattering, crystal analysis.*
Akkad, Dr Mohamed Kamal. Dept. of Geology, U. of Tanta, Faculty of Sci., Tanta, Arab Rep. of Egypt. PhD, geology (London U., UK, 1952). Prof. *Petrology.*
Akkad, Mr Salah el Din. Geological Survey and Mineral Res. Organization, Salah Salem Street, Abbassia, Arab Rep. of Egypt. BSc, geology (Cairo U., 1942). Sub-director. (tel. 830782). *Petrology, ore deposits.*
Anwar, Dr Yehia. Faculty of Sci., U. of Alexandria, Moharram Bey, U. of Alexandria, Arab Rep. of Egypt. PhD, crystallography (Durham U., UK, 1950). Prof. *Petrology.*
Arafa, Prof. Salah Arafa Mohamed. Sci. Dept., The American U. in Cairo, 113 Kasr El-Aini St., Cairo, Arab Rep. of Egypt. (1941) PhD, solid state (Cairo U., 1969). Prof. (tel. 22968 169, ext. 5301, telex 92224 AUCAI-UN) *Powder diffraction, crystal growth, optical characterization magnetic characterization, teaching.*
Ashry, Dr Mamdouh. U. of Assiut, Faculty of Sci., Assiut, Arab Rep. of Egypt. PhD, geology (Assiut U., 1967). Lect. *Geochemistry.*
Azer, Dr Nazmi. Faculty of Sci., U. of Cairo, University Street, Giza-Cairo, Arab Rep. of Egypt. Dr.phil, geology (Wien U., Austria, 1956). Asst. prof. (tel. 841722, 841713).

Barakat, Prof. Nayel. Dept. of Physics, Faculty of Sci., U. of Ain Shams, Abbassia-Cairo, Arab Rep. of Egypt. PhD, physics (London U., UK, 1952). Prof. (tel. 821096, 821633). *Spectrography.*
Badr, Dr Yehia Abd-El Hamid. Physics Dept., Faculty of Sci., U. of Cairo, Giza-Cairo, Egypt. (1946) PhD, molecular physics (Leningrad State U., USSR, 1974). Lect. *Phase transitions in ionic crystals.*
Basha, Dr Ahmed Fouad. Physics Dept., Faculty of Sci., U. of Cairo, Cairo, Egypt. (1941) PhD, solid state physics (Moscow State U., USSR, 1974). Lect. *Dielectric properties in molecular solids.*
Bassiouny, Dr Mohamed Khafagi. Faculty of Sci., U. of Ain Shams, Abbassia-Cairo, Arab Rep. of Egypt. Dr.phil., geology (Wien U., Austria, 1969). Lect. (tel. 821096, 821633). *Petrology.*
Elbadri, Dr H. Dept. of Mining, Faculty of Engineering, U. of Cairo, University Street, Giza-Cairo, Arab Rep. of Egypt. PhD, geology (London U., UK, 1951). Prof. (tel. 89 69 26). *Crystal optics, radioactive minerals.*
El Demerdash, Dr Saad. Desert Inst., El Mataria, Cairo, Arab Rep. of Egypt. PhD, soils (Cairo U., 1970). Res. *Soil chemistry, soil mineralogy.*
El Gabi, Dr Sami. U. of Assiut, Faculty of Sci., Assiut, Arab Rep. of Egypt. Dr.rer.nat., geology (U. München, BRD, 1963). Asst. prof. *Igneous petrology.*
El Naggar, Dr Mohamed. U. of Assiut, Faculty of Sci., Assiut, Arab Rep. of Egypt. PhD, geology (London U., UK, 1969). Lect. *Carbonate geochemistry.*
El Ramly, Mr Mohamed Fawzi. Geological Survey and Mineral Res. Organization, Salah Salem Street, Abbassia, Arab Rep. of Egypt. BSc, geology (Cairo U., 1944). Res. officer, (tel. 830782). *Petrology.*
El Sayed, Prof. (Mrs) Karimat. Faculty of Sci., U. of Ain Shams, Abbassia-Cairo, Arab Rep. of Egypt. PhD, physics (London U., UK, 1965). Lect. (tel. 821096, 821633). *X-ray crystallography.*
El Shaabini, Prof. (Mrs) Aida Moustafa. X-ray Crystallography Unit, Nat. Res. Centre, Dokki-Cairo, Arab Rep. of Egypt. MSc, physics (Cairo U., 1969). Res. asst. (tel. 802129). *X-ray metallography.*
El Shanshury, Dr Ismail. Atomic Energy Commission, Inshas-Cairo, Arab Rep. of Egypt. PhD, physics (Rensselaer Polytechnic Inst., USA, 1963). Asst. prof. (tel. 862013). *Metals.*
El Sharkawi, Dr Mohamed Abdel Hamid. Faculty of Sci., U. of Cairo, University Street, Giza-Cairo, Arab Rep. of Egypt. PhD, geology (London U., UK, 1964). Lect. (tel. 841722, 841713). *Ore deposits.*
El Shazli, Dr El Shazli Mohamed. Atomic Energy Commission, Inshas-Cairo, Arab Rep. of Egypt. PhD, geology (London U., UK, 1950). Prof. (tel. 862013). *Ore deposits.*
Elwan, Dr Ahmed Abdel Salam. Desert Inst., El Mataria, Cairo, Arab Rep. of Egypt. PhD, soils (Ain Shams U., 1975). Res. *Soil chemistry, soil mineralogy.*
Fayed, Dr (Mrs) Leila. Faculty of Sci., U. of Cairo, University Street, Giza-Cairo, Arab Rep. of Egypt. PhD, geology (Sheffield U., UK, 1966). Lect. (tel. 841722, 841713). *Rock mechanics.*
Gad, Dr Gamal Mohamed. Refractories & Ceramic Lab., Nat. Res. Centre, Dokki-Cairo, Arab Rep. of Egypt. PhD, geochemistry (London U., UK, 1950). Prof. (tel. 802129). *Clay mineralogy.*

Ghebrial, Dr Mounir Guirgis. Geological Survey and Mineral Res. Organization, Salah Salem Street, Abbassia, Arab Rep. of Egypt. PhD, geology (London U., UK, 1953). Res. officer. (tel. 830782). *Structure petrology.*

Guindi, Dr Amin Riad. Faculty of Sci., Moharram Bey, U. of Alexandria, Alexandria, Arab Rep. of Egypt. PhD, geology (London U., UK, 1950). Prof. *Petrology.*

Gweifel, Dr Ismail. Faculty of Agriculture, Chatby, U. of Alexandria, Arab Rep. of Egypt. PhD, soils (Alexandria U., 1967). Lect. *Soil morphology, soil mineralogy.*

Hamdi, Prof. Hassan Mahmoud. Faculty of Agriculture, Shoubra el Kheima, U. of Ain Shams, Abbassia-Cairo, Arab Rep. of Egypt. DSc, (ETH Zürich, Switzerland, 1942). Prof. (tel. 948716, 942711). *Clay mineral structures.*

Harga, Dr Ahmed Amin. Desert Inst., El Mataria, Cairo, Arab Rep. of Egypt. PhD, soils (Ain Shams U., 1971). Res. *Soil chemistry, soil mineralogy.*

Hassan, Prof. Mohamed Youssef. Faculty of Sci., U. of Ain Shams, Abbassia-Cairo, Arab Rep. of Egypt. PhD, geology (Bristol U., UK, 1951). Prof. (tel. 821096, 821633). *Paleontology, stratigraphy.*

Helmi, Prof. Mohamed Ezzeldin. Faculty of Sci., U. of Ain Shams, Abbassia-Cairo, Arab Rep. of Egypt. PhD, geology (Michigan U., USA, 1951). Prof. (tel. 821096, 821633). *Mineralogy, crystal structure.*

Hinawi, Prof. Essam. Nat. Res. Centre, Dokki-Cairo, Arab Rep. of Egypt. PhD, geology (Cairo U., 1961). Asst. prof., res. (tel. 802129). *Economic geology.*

Kabish, Dr Lotfi. Nat. Res. Centre, Dokki-Cairo, Arab Rep. of Egypt. PhD, geology (Cairo U., 1948). Asst. prof., researcher. (tel. 802129). *Igneous petrology.*

Kamel, Prof. Dr Raafat Wasef. Physics Dept., Faculty of Sci., U. of Cairo, Giza-Cairo, Egypt. (1926) DSc, solid state physics (Cairo U., 1968). Prof. and head. (tel. 894095). *Defect structure in crystalline solids.*

Khadr, Dr Moustafa. Faculty of Agriculture, Chatby, U. of Alexandria, Alexandria, Arab Rep. of Egypt. PhD, soils (Alexandria U., 1956). Prof. *Soil morphology, soil mineralogy.*

Khalifa, Prof. (Mrs) Berlant. Faculty of Sci., U. of Ain Shams, Abbassia-Cairo, Arab Rep. of Egypt. PhD, physics (London U., UK, 1967). Lect. (tel. 821096, 821633). *Electron diffraction.*

Khalifa, Dr (Mrs) B. Abdel Meguid. Atomic Energy Commission, Inshas-Cairo, Arab Rep. of Egypt. PhD, chemistry (London U., UK, 1968). Res. (tel. 862013). *Inorganic structures.*

Khidr, Prof. (Mrs) Fatma Abdel Hakim. X-ray Crystallography Unit, Nat. Res. Centre, Dokki-Cairo, Arab Rep. of Egypt. MSc, physics (Cairo U., 1967). Res. asst. (tel. 802129). *X-ray metallography.*

Kholeif, Dr Mahmoud. Nat. Res. Centre, Dokki-Cairo, Arab Rep. of Egypt. Cand. geol.-min., geology (Moscow U., USSR, 1964). Res. (tel. 802129). *Ore deposits.*

Kishk, Dr Fawzi Mohamed. Faculty of Agriculture, Chatby, U. of Alexandria, Alexandria, Arab Rep. of Egypt. PhD, soils (U. California, Berkeley, USA, 1967). Asst. prof. *Soil chemistry, soil mineralogy.*

Labib, Dr (Mrs) Fawkia. X-ray Crystallography Unit, Nat. Res. Centre, Dokki-Cairo, Arab Rep. of Egypt. PhD, soils (Ain Shams U., 1970). Res. (tel. 802129). *Soil chemistry, soil mineralogy.*

Labib, Dr Tarik. Faculty of Agriculture, U. of Assiut, Assiut, Arab Rep. of Egypt. PhD, soils (U. California, Davis, USA, 1968). Lect. *Soil morphology, soil mineralogy.*

Lashin, Dr A. Mohamed. Faculty of Sci., Moharram Bey, U. of Alexandria, Alexandria, Arab Rep. of Egypt. PhD, crystallography (London U., UK, 1952). Prof. *Crystal structures.*

Lotfy, Dr Mohamed. Faculty of Sci., U. of Cairo, University Street, Giza-Cairo, Arab Rep. of Egypt. PhD, geology (Cairo U., 1957). Asst. prof. (tel. 841722, 841713). *Geology and petrology.*

Mansour, Mr Saber Moustapha. Geological Survey and Mineral Res. Organization, Salah Salem Street, Abbassia, Arab Rep. of Egypt. BSc, geology (Cairo U., 1944). Res. officer. (tel. 830782). *Petrology.*

Naga, Dr Mohamed Abdel Hamid. Faculty of Agriculture, U. of Cairo, Giza-Cairo, Arab Rep. of Egypt. PhD, soils (Cairo U., 1954). Prof. (tel. 89 65 86, 89 67 66). *Soil mineralogy.*

Nakhla, Dr Fakhry. Dept. of Mining, Faculty of Engineering, U. of Cairo, University Street, Giza-Cairo, Arab Rep. of Egypt. PhD, geology (London U., UK, 1951). Prof. (tel. 84 06 55). *Minerals.*

Philipp, Dr George. Faculty of Sci., U. of Cairo, University Street, Giza-Cairo, Arab Rep. of Egypt. PhD, geology (Cairo U., 1956). Asst. prof. (tel. 841722, 841713). *Sedimentary petrology.*

Rabie, Dr (Mrs) Farida Hamed. Faculty of Agriculture, Shoubra el Kheima, U. of Ain Shams, Abbassia-Cairo, Arab Rep. of Egypt. PhD, soils (U. California, Davis, USA, 1967). Asst. prof. (tel. 948716, 942711). *Soil morphology, soil mineralogy.*

Radwan, Dr Mostafa Mohsen Abdel-Razik. Physics Dept., Military Tech. C., Qubri-El-Quba, Cairo, Arab Rep. of Egypt. (1947) PhD, physics (Dundee U., UK, 1982). Lect. (tel. 853983). *Biomolecular structures, virus structures, amorphous structures.*

Ragab, Dr Abdel Ghani. Faculty of Sci., U. of Ain Shams, Abbassia Cairo, Arab Rep. of Egypt. MSc, geology (Ain Shams U., 1968). Teaching asst. (tel. 821096, 821633). *Petrology.*

Sadek, Dr Gamil. Nat. Res. Centre, Dokki-Cairo, Arab Rep. of Egypt. DSc, geology Lyon U., France, 1964). Res. (tel. 802129). *Ore deposits.*

Salem, Dr Safia Mahmoud. Physics Dept., Faculty of Sci., U. of Cairo, Egypt. PhD, X-ray crystallography (Cairo U., 1961). Asst. prof. (tel. 802129). *X-ray crystallography.*

Shoukri, Dr Nasri M. Faculty of Sci., U. of Cairo, University Street, Giza-Cairo, Arab Rep. of Egypt. PhD, geology (London U., UK, 1940). Prof. (tel. 841722, 841713). *Petrology.*

Soliman, Mr F. Abdel Aal. Geological Survey and Mineral Res. Organization, Salah Salem Street, Abbassia, Arab Rep. of Egypt. BSc, geology (Cairo U., 1944). Res. officer. (tel. 830782). *Petrology.*

Soliman, Dr Mohamed Soliman. Faculty of Sci., U. of Ain Shams, Abbassia-Cairo, Arab Rep. of Egypt. PhD, geology (Stanford U., USA, 1958). Asst. prof. (tel. 821096, 821633). *Sedimentation.*

Takla, Mr Maher Azmi. Faculty of Sci., U. of Cairo, University Street, Giza-Cairo, Arab Rep. of Egypt. MSc, geology (Cairo U., 1968). Teaching asst. (tel. 841722, 841713). *Economic geology.*

Thabet, Dr Atef. Geological Survey and Mineral Res. Organization, Salah Salem Street, Abbassia, Arab Rep. of Egypt. PhD, geology (Cairo U., 1963). Res. officer. (tel. 830782). *Petrology.*

Tousson, Dr Salama. Faculty of Sci., Moharram Bey, U. of Alexandria, Alexandria, Arab Rep. of Egypt. DSc, geology Nancy U., France, 1957). Asst. prof. *Mineral structures.*

Youssef, Dr I. Mourad. Faculty of Sci., U. of Ain Shams, Abbassia-Cairo, Arab Rep. of Egypt. PhD, geology (Alexandria U., 1950). Prof. (tel. 821096, 821633). *Structural geology.*

Zaghloul, Dr Mohamed Zaki. U. of Mansoura, Faculty of Sci., Mansoura, Arab Rep. of Egypt. PhD, geology (Bristol U., UK, 1958). Prof. *Crystal structures.*

Zatout, Mr Mohamed Abdel Meguid. General Egyptian Mining Organisation, Geheni Street, Dokki-Cairo, Arab Rep. of Egypt. *Petrology.*

FINLAND

Sub-Editor: **A. Vahvaselkä**

Notes

1. International telephone country code - 358.

Åberg, Prof. Teijo. Lab. of Physics, Helsinki U. of Techn., 02150 Espoo, Finland. (1937) PhD, physics (U. Helsinki, 1969). Assoc. prof. (tel. 0 + 460144). *X-ray spectroscopy.*

Ahlgrén, Dr, Assoc.Prof. Markku Jouko. Dept. of Chemistry, U. of Joensuu, 80130 Joensuu, Finland. (1944) PhD, inorganic chemistry (U. Helsinki, 1979). Assoc. prof. (tel. 73 + 28311, ext. 358). *Metal complexes, clusters.*

Ahtee, Dr Sisko-Maija. Dept. of Physics, U. of Helsinki, 00170 Helsinki, Finland. (1939) PhD, physics (U. Helsinki, 1971). Lect. (tel. 0 + 645021, telex 122229 nuphu sf). *X-ray and neutron crystallography, ferroelectric materials.*

Aikala, Dr Oiva Jaakko Mikael. Dept. of Physical Sciences, U. of Turku, 20500 Turku, Finland. (1940) PhD, theoretical physics (U. Turku, 1977). Asst. (tel. 21 + 645689). *Compton scattering, momentum and charge density theory, LCAO-methods.*

Aksela, Prof. Seppo Olavi. Dept. of Physics, U. of Oulu, Linnanmaa, 90570 Oulu, Finland. (1942) PhD, physics (U. Oulu, 1971). Assoc. prof. (tel. 81 + 345411). *Electron spectrometry.*

Carlson, Dr Sirkka Liisa. Dept. of Geology, U. of Helsinki, 00170 Helsinki, Finland. (1943) PhD, geology and mineralogy (1982). Asst. (tel. 0 + 1914427). *Crystallography, mineralogy.*

Friman, Dr Rauno Kalevi. Dept. of Physical Chemistry, Åbo Akademi, 20500 Åbo, Finland. (1946) PhD, physical chemistry (Åbo Akademi, 1983). Lect.

(tel. 21 + 335133, ext. 255). *Small angle X-ray scattering, soap and detergent micelle structure, biological membranes.*

Haapala, Prof. Ilmari Johannes. Dept. of Geology, U. of Helsinki, 00170 Helsinki, Finland. (1939) PhD, geology and mineralogy (U. Helsinki, 1966). Prof. (tel. 0 + 1914426). *Crystallography, mineralogy.*

Hämäläinen, Dr Reijo Pertti. Dept. of Chemistry, U. of Helsinki, 00100 Helsinki, Finland. (1940) PhD, inorganic chemistry (U. Helsinki, 1972). Docent. (tel. 0 + 650211, ext. 510). *Metal complexes.*

Heleskivi, Dr Jouni Martti. Semiconductor Lab., Techn. Res. Centre of Finland, 02150 Espoo, Finland. (1938) DSc, electron physics (Helsinki U. of Techn., 1972). Director of Lab. (tel. 0 + 4566300). *Semiconductor constants.*

Hiismäki, Dr Pekka Eljas. Reactor Lab., Techn. Res. Centre of Finland, 02150 Espoo, Finland. (1939) DSc, physics (Helsinki U. of Techn., 1970). Deputing director. (tel. 0 + 4566362). *Neutron diffraction, magnetic structures.*

Hiltunen, Mr Lassi Ilmari. Dept. of Chemistry, Helsinki U. of Techn., 02150 Espoo, Finland. (1943) MSc, inorganic chemistry (Helsinki U. of Techn., 1969). Lab. administrator. (tel. 0 + 4512758).

Hocksell, Mr Veli Eerik. Metals Lab.,Ovako Oy Ab, 55100 Imatra, Finland. (1951) MSc, physical metallurgy (Helsinki U. of Techn.,1976).Lab. eng. (tel. 54 + 63688, telex 5711 ovai sf). *Electron microscopy, microanalysis, metallography.*

Hölsä, Dr Jorma Pertti Kalervo. Dept. of Chemistry, Helsinki U. of Techn., 02150 Espoo, Finland. (1952) DTechn., spectroscopy of rare earth elements (Helsinki U. of Techn., 1983). Chief asst. (tel. 0 + 4512758, telex 125161). *Luminescence, structure, thermal stability, rare earth elements, quantum mechanics.*

Hytönen, Dr Kai Kalevi Gabriel. Petrological Dept., Geological Survey of Finland, 02150 Espoo, Finland. (1925) PhD, geology and mineralogy (U. Turku, 1959). State geologist. (tel. 0 + 46931, ext. 263, telex 123185). *Rock-forming minerals of Finland, chemistry, structure, paragenesis.*

Järvinen, Dr Matti Johannes. Dept. of Physics and Mathematics, Lappeenranta U. of Techn., 53850 Lappeenranta, Finland. (1934) PhD, physics (U. Helsinki, 1969). Lab. manager. (tel. 53 + 27570, ext. 2807). *X-ray diffraction, materials science.*

Johanson, Mr Bo Stefan. Geological Survey of Finland, 02150 Espoo, Finland. (1954) MSc, geology and mineralogy (U. Helsinki, 1984). Geophysicist. (tel. 0 + 4693325). *Anorthosites/Gabbros, mineralogy, pyroxenes (PX-termometry).*

Kähkönen, Dr Heikki Antero. Dept. of Physical Sciences, U. of Turku, 20500 Turku, Finland. (1934) PhD, physics (U. Turku, 1968). Lect. (tel. 21 + 645111). *X-ray diffraction.*

Kaihola, Mr Lauri Leo. Dept. of Physical Sciences, U. of Turku, 20500 Turku, Finland. (1944) PhLic., physics (U. Turku, 1975). Asst. (tel. 21 + 645111). *Soft X-ray spectroscopy.*

Kallio, Mr Pekka Yrjö Juhani. Petrological Dept., Geological Survey of Finland, PO Box 237, 70101 Kuopio, Finland. (1937) MSc, geology and mineralogy (U. Helsinki, 1964). Geologist. (tel. 71 + 227677). *Mineralogy, X-ray diffraction.*

Kalliomäki, Dr Martti Salomo. Dept. of Physics, U. of Helsinki, 00170 Helsinki, Finland. (1945) PhD, physics (U. Helsinki, 1982). Asst. (tel. 0 + 650211, telex 122229 nuphu sf). *Phase transformations, high pressure crystal structures.*

Kansikas, Mr Jarno Juhani. Dept. of Chemistry, U. of Helsinki, 00100 Helsinki, Finland. (1947) PhLic., inorganic chemistry (U. Helsinki, 1977). Instructor. (tel. 0 + 650211). *Metal complex crystal structures.*

Karvinen, Mrs Saila Marjatta. Dept. of Chemistry, Helsinki U. of Techn., 02150 Espoo, Finland. (1959) MSc, inorganic chemistry (1984). Asst. (tel. 0 + 4512758, telex 125161 htkk sf). *Crystal structures.*

Ketolainen, Prof. Pertti Pekka Juhani. Dept. of Physics, U. of Joensuu, 80130 Joensuu, Finland. (1937) PhD, physics (U. Turku, 1969). Assoc. prof. (tel. 73 + 28311, ext. 375). *Crystal lattice defects.*

Kettunen, Prof. Pentti Olavi. Inst. of Materials Science, Tampere U. of Techn., 33100 Tampere, Finland. (1932) DTechn., physical metallurgy (Helsinki U. of Techn., 1965). Prof. (tel. 31 + 162280, telex 22313 ttktr sf). *Plasticity, strain hardening, fatigue, electron microscopy, X-ray diffraction, small-angle scattering, microanalyzing.*

Kivekäs, Dr Raikko Terjo Ilari. Dept. of Chemistry, U. of Helsinki, 00100 Helsinki, Finland. (1944) PhD, inorganic chemistry (U. Helsinki, 1977). Instructor. (tel. 0 + 650211). *Inorganic and organoselenium crystal structures.*

Kivilahti, Prof. Jorma Kalevi. Dept. of Mining and Metallurgy, Helsinki U. of Techn., 02150 Espoo, Finland. (1943) DSc, physical metallurgy (Helsinki U. of Techn., 1976). Assoc. prof. (tel. 0 + 4566115, telex 125161). *X-ray metallography, physical metallurgy, titanium and alloys, metal-hydrogen systems.*

Klinga, Mr Martti Evert. Dept. of Chemistry, U. of Helsinki, 00100 Helsinki, Finland. (1942) MA, inorganic and analytical chemistry (U. Helsinki, 1971). Instructor. (tel. 0 + 650211). *Inorganic and organometallic crystal structures.*

Knuuttila, Mrs Hilkka Ritva-Liisa. Dept. of Chemistry, U. of Jyväskylä, 40100 Jyväskylä, Finland. (1946) PhD, inorganic chemistry (U. Jyväskylä, 1983). Asst. (tel. 41 + 291211). *Inorganic and organometallic crystal structures.*

Knuuttila, Mr Pekka Juhani. Dept. of Chemistry, U. of Jyväskylä, 40100 Jyväskylä, Finland. (1944) PhD, inorganic chemistry (U. Jyväskylä, 1982). Asst. (tel. 41 + 291211). *Inorganic and organometallic crystal structures.*

Koikkalainen, Miss Seija Anneli. Dept. of Physics, U. of Helsinki, 00170 Helsinki, Finland. (1957) MSc, physics (U. Helsinki, 1983). Asst. (tel. 0 + 650211, ext. 236 or 562, telex 122229 nuphu sf). *Compton scattering.*

Komu, Mr Markku Eino Sakari. Wihuri Physical Lab., U. of Turku, 20500 Turku, Finland. (1946) MA, physics (U. Turku, 1973). Asst. (tel. 21 + 645944). *Nuclear magnetic resonance, ionic crystals.*

Kontio, Dr Airi Outi. Dept. of Physics, U. of Helsinki, 00170 Helsinki, Finland. (1951) PhD, physics (U. Helsinki, 1984). Asst. (tel. 0 + 650211, ext. 251, telex 122229 nuphu sf). *Charge density.*

Korhonen, Prof. Unto Kalervo. Lab. of Physics, Helsinki U. of Techn., 02150 Espoo, Finland. (1915) PhD, physics (U. Helsinki, 1951). Prof. emeritus. (tel. 0 + 8781845). *Charge density determination.*

Korvenranta, Dr Jorma Artturi. Dept. of Chemistry, U. of Helsinki, 00100 Helsinki, Finland. (1938) PhD, inorganic chemistry (U. Helsinki, 1974). Lect. (tel. 0 + 650211).

Kurittu, Dr Jyrki Veli Einari. Dept. of Physics, U. of Helsinki, 00170 Helsinki, Finland. (1944) PhD, physics (U. Helsinki, 1978). Asst. (tel. 0 + 650211, ext. 418, telex 122229 nuphu sf). *Lattice dynamics, crystallography.*

Kurki-Suonio, Prof. Kaarle Veikko Juhani. Dept. of Physics, U. of Helsinki, 00170 Helsinki, Finland. (1933) PhD, physics (U. Helsinki, 1959). Prof., dept. head. (tel. 0 + 650211, ext. 298, telex 122229 nuphu sf). *Charge density, spin and momentum density, deformation models in analysis of diffraction data.*

Kyröläinen, Mr Antero Johannes. Outokumpu Oy, 95400 Tornio, Finland. (1951) Dipl.eng., physical metallurgy (U. Oulu, 1976). Res. eng. (tel. 80 + 4521, ext. 580). *Electron microscopy, stainless steels.*

Lähdeniemi, Dr Matti Juhani Iisakki. Dept. of Physical Sciences, U. of Turku, 20500 Turku, Finland. (1949) PhD, electronic structure of metals (U. Turku, 1982). Jr. res. assoc. (tel. 21 + 645697). *Electronic structure, metals and alloys, molecular crystals, XPS, UPS, XES.*

Lahti, Dr Seppo Ilmari. Petrological Dept., Geological Survey of Finland, 02150 Espoo, Finland. (1947) PhD, geology and mineralogy (U. Helsinki, 1981). Geologist. (tel. 0 + 4693248). *Mineral crystal structures, single crystal X-ray methods, powder diffraction methods, structure determination of minerals.*

Laiho, Dr Reino Toivo Salomo. Wihuri Physical Lab., U. of Turku, 20500 Turku, Finland. (1941) PhD, physics (U. Turku, 1973). Res. scient. (tel. 21 + 645943). *Optical properties of solids.*

Laine, Dr Ensio Sulo Uolevi. Dept. of Physical Sciences, U. of Turku, 20500 Turku, Finland. (1940) PhD, physics (U. Turku, 1973). Lect. *X-ray crystallography and applications.*

Lehtinen, Dr Martti Kalevi. Dept. of Geology, U. of Helsinki, 00170 Helsinki, Finland. (1941) PhD, geology and mineralogy (U. Helsinki, 1976). Curator of minerals. (tel. 0 + 1914424). *Crystallography, mineralogy.*

Leiro, Dr Jarkko Albert. Dept. of Physical Sciences, U. of Turku, 20520 Turku, Finland. (1949) PhD, physics (U. Turku, 1982). Asst. (tel. 21 + 645698). *Electronic structure, X-ray spectroscopy, photoelectron spectroscopy.*

Leskelä, Dr Markku Antero. Dept. of Chemistry, U. of Oulu, 90570 Oulu, Finland. (1950) DTechn., inorganic chemistry (Helsinki U. of Techn., 1980). Assoc. prof. (tel. 81 + 352207). *Solid state chemistry.*

Levoska, Mr Pentti Juhani. Dept. of Techn. Physics, U. of Oulu, Linnanmaa, 90570 Oulu, Finland. (1942) Techn.Lic., techn. physics (U. Oulu, 1974). Sr. asst. (tel. 81 + 352521). *X-ray diffraction, diffuse scattering.*

Lindqvist, Mr Kristian Vilhelm. Geological Survey of Finland, 02150 Espoo, Finland. (1949) MSc, geology and mineralogy (U. Helsinki, 1981). Geologist. (tel. 0 + 46931). *X-ray diffraction.*

Lindroos, Prof. Veikko Kalervo. Dept. of Mining and Metallurgy, Helsinki U. of Techn., 02150 Espoo, Finland. (1938) DSc, physical metallurgy (Helsinki U. of Techn., 1968). Prof. (tel. 0 + 4512610, telex 125161). *Dislocation theory, electron microscopy, X-ray metallography, phase transformation, silicon technology, semiconductor metallurgy, non- waste technology.*

Lindström, Dr Rauno. Dept. of Math. and Physics, U. of Joensuu, 80100 Joensuu, Finland. (1937) PhD, physics (U. Turku, 1973). Lect. (tel. 73 + 28311, ext. 287). *Crystal defects.*

Lumme, Prof. Paavo Olavi. Dept. of Inorganic Chemistry, U. of Helsinki, 00100 Helsinki, Finland. (1923) PhD, physical chemistry (U. Helsinki, 1956). Prof., dept. head. (tel. 0 + 662134). *Bioinorganics, complexes, magneto-chemistry, organometallics, spectroscopy, structural chemistry, thermal chemistry.*

Mäki, Mr Jouko Kalervo. Lab. of Electron Microscopy, U. of Turku, 20520 Turku, Finland. (1943) MSc, physics (U. Turku, 1970). Lab. eng. (tel. 21 + 513355, ext. 318, telex 62293 tyl sf). *Long period order, binary alloys.*

Manninen, Dr Seppo Olavi. Dept. of Physics, U. of Helsinki, 00170 Helsinki, Finland. (1944) PhD, physics (U. Helsinki, 1972). Lect. (tel. 0 + 650211, ext. 217 or 233, telex 122229 nuphu sf). *Compton scattering, momentum and charge density determination and theory.*

Mansikka, Prof. Kauko Antti. Dept. of Physical Sciences, U. of Turku, 20500 Turku, Finland. (1932) PhD, physics (U. Turku, 1961). Prof. (tel. 21 + 645111, ext. 5680). *Quantum theory of solids, interaction of radiation with matter.*

Martikainen, Mr Hannu Olavi. Dept. of Mining and Metallurgy, Helsinki U. of Techn., 02150 Espoo, Finland. (1952) MSc, physical metallurgy (Helsinki U. of Techn., 1976). Asst. (tel. 0 + 4554122). *Electron microscopy, dislocation theory, grain boundaries.*

Meisalo, Prof. Veijo Pauli Juhani. Dept. of Teacher Education, U. of Helsinki, 00120 Helsinki, Finland. (1938) PhD, physics (U. Helsinki, 1967). Assoc. prof.

(tel. 0 + 645021, ext. 120). *X-ray diffraction, high pressure studies, phase transformations, physics education.*
Merisalo, Dr Matti Juhani. Dept. of Physics, U. of Helsinki, 00170 Helsinki, Finland. (1937) PhD, physics (U. Helsinki, 1967). Lab. manager, docent. (tel. 0 + 650211, telex 122229 nuphu sf). *Charge density, materials science.*
Minni, Dr Erkki Esa Kalervo. Turku Regional Inst. of Occupational Health, 20500 Turku, Finland. (1946) PhD, physics (U. Turku, 1982). Docent. (tel. 21 + 337970). *Materials science, surface science.*
Muhonen, Mr Heikki Juhani. Dept. of Chemistry, U. of Helsinki, 00100 Helsinki, Finland. (1946) PhLic., inorganic chemistry (U. Helsinki, 1978). Instructor. (tel. 0 + 650211, ext. 515). *Crystal structures, transition metal complexes, small organic compounds.*
Mutikainen, Mr Ilpo Pellervo. Dept. of Inorganic Chemistry, U. of Helsinki, 00100 Helsinki, Finland. (1947) MA, inorganic and analytical chemistry (U. Helsinki, 1973). Instructor. (tel. 0 + 650211, ext. 568). *Crystal structures, metal complexes.*
Mutka, Dr Hannu Mika Ilmari. Reactor laboratory, Techn. Research Centre of Finland, 02150 Espoo, Finland. (1952) DSc, physics (U. de Paris-Sud (Orsay), 1982). Sr. Scient. (tel. 0 + 4566354). *Neutron diffraction, electron microscopy, electrical and magnetic properties of solids, defects in solids.*
Näsäkkälä, Dr Matti Eerik. Dept. of Chemistry, U. of Helsinki, 00100 Helsinki, Finland. (1944) PhD, inorganic chemistry (U. Helsinki, 1977). Instructor, docent. (tel. 0 + 650211, ext. 536). *Crystal structures, transition metal complexes.*
Näsänen, Prof. Reino Olavi. Dept. of Chemistry, U. of Helsinki, 00100 Helsinki, Finland. (1908) PhD, chemistry (U. Helsinki, 1939). Prof. emeritus. (tel. 0 + 650211). *Coordination chemistry.*
Nenonen, Mr Pertti Olavi. Lab. of Metallurgy, Techn. Res. Centre of Finland, 02150 Espoo, Finland. (1943) Techn.Lic., physical metallurgy (Helsinki U. of Techn., 1974). Res. eng. (tel. 0 + 4561, ext. 5408). *Electron microscopy.*
Nieminen, Dr Kari Veikko Juhani. Dept. of Chemistry, U. of Helsinki, 00100 Helsinki, Finland. (1946) PhD, inorg. chemistry (U. Helsinki, 1983). Special scient. (tel. 0 + 650211, ext. 567). *Crystal structure, complexes formed by metal ions and some inorganic compounds.*
Niinistö, Dr Lauri. Dept. of Chemistry, Helsinki U. of Techn., 02150 Espoo, Finland. (1941) DSc, inorganic chemistry (Helsinki U. of Techn., 1973). Prof., lab. head. (tel. 0 + 4512750, telex 125161). *Structural inorganic chemistry, solid state chemistry.*
Oksman, Mr Pentti. Wihuri Physical Lab. and Dept. of Chemistry, U. of Turku, 20500 Turku, Finland. (1948) MA, physics (U. Turku, 1973). Lab. eng. (tel. 21 + 513355, ext. 391). *NMR, ionic crystals, mass spectrometry of organic compounds.*
Orama, Dr Olli Antero. Dept. of Chemistry, U. of Helsinki, 00100 Helsinki, Finland. (1944) PhD, inorganic chemistry (U. Helsinki, 1976). Docent. (tel. 0 + 650211, ext. 567). *Crystal structure, transition metal complexes, organometallic compounds.*
Paakkari, Prof. Timo Lauri Päiviö. Dept. of Physics, U. of Helsinki, 00170 Helsinki, Finland. (1937) PhD, physics (U. Helsinki, 1968). Assoc. prof. (tel. 0 + 650211, ext. 245, telex 122229 nuphu sf). *Charge density, momentum and spin density.*
Paalassalo, Mr Pentti Olavi. Wihuri Physical Lab. and Dept. of Physical Sciences, U. of Turku, 20500 Turku, Finland. (1933) PhLic., physics (U. Turku, 1965). Lab. eng. (tel. 21 + 645653). *Electron microscopy, metals and alloys.*
Pajunen, Dr Aarne Veikko. Dept. of Chemistry, U. of Helsinki, 00100 Helsinki, Finland. (1939) PhD, inorganic chemistry (U. Helsinki, 1967). Assoc. prof. (tel. 0 + 650211). *Crystal structures, coordination compounds.*
Pakkanen, Dr Tapani Antti. Dept. of Chemistry, U. of Joensuu, 80120 Joensuu, Finland. (1949) PhD, chemistry (SUNY, Stony Brook, USA, 1977). Assoc. prof. (tel. 73 + 28311, ext. 263). *Cluster chemistry, surface chemistry, catalysis.*
Pakkanen, Dr Tuula Tellervo. Dept. of Chemistry, U. of Joensuu, 80120 Joensuu, Finland. (1949) PhD, chemistry (SUNY, Stony Brook, USA, 1978). Asst. (tel. 73 + 28311, ext. 264). *Cluster chemistry, catalysis, organometallic compounds.*
Papunen, Prof. Heikki Tapani. Inst. of Geology and Mineralogy, U. of Turku, 20500 Turku, Finland. (1936) PhD, geology and mineralogy (U. Turku, 1971). Prof. (tel. 21 + 645480). *Mineralogy.*
Pessa, Prof. Viljo Markus. Dept. of Physics, Techn. U. of Tampere, 33100 Tampere, Finland. (1941) PhD, physics (U. Turku, 1971). Assoc. prof. (tel. 31 + 162111). *Electron and X-ray spectroscopy.*
Piispanen, Dr Risto Anton. Dept. of Geology, U. of Oulu, Linnanmaa, 90570 Oulu, Finland. (1938) PhD, geology (U. Oulu, 1972). Lect. (tel. 81 + 352161). *Statistical methods, geology, mineralogy.*
Pitkänen, Mrs Tuula Esteri. Dept. of Physics, U. of Helsinki, 00170 Helsinki, Finland. (1953) PhLic., physics (U. Helsinki, 1980). Asst. (tel. 0 + 650211, ext. 233, telex 122229 nuphu sf). *Compton scattering.*
Punkkinen, Dr Matti. Wihuri Physical Lab., U. of Turku, 20500 Turku, Finland. (1939) PhD, physics (U. Turku, 1967). Prof. (tel. 21 + 645947). *Nuclear magnetic resonance in solids.*
Pyykkö, Prof. Veli Pekka. Dept. of Chemistry, U. of Helsinki, 00100 Helsinki, Finland. (1941) PhD, physics (U. Turku, 1967). Prof. (tel. 0 + 410566, ext. 448, telex 121199 seism sf). *Relativistic quantum chemistry, heavy element structures, magnetic resonance spectroscopy.*

Ranta, Mr Lasse Kosti. Dept. of Physical Sciences, U. of Turku, 20500 Turku, Finland. (1949) MSc, physics (U. Turku, 1975). Asst. (tel. 21 + 645698). *Crystallography, minerals.*
Rautioaho, Dr Risto Heikki. Inst. of Physical Metallurgy, U. of Oulu, Linnanmaa, 90570 Oulu, Finland. (1945) DTechn., materials science (U. Oulu, 1981). Assoc.prof. (tel. 81 + 353611). *Structure, ceramics.*
Seitsonen, Dr Sulo Iivari. Dept. of Physical Sciences, U. of Turku, 20500 Turku, Finland. (1935) PhD, physics (U. Turku, 1967). Lect. *X-ray diffraction.*
Serimaa, Mrs Ritva Elina. Dept. of Physics, U. of Helsinki, 00170 Helsinki, Finland. (1957) MSc, physics (U. Helsinki, 1982). Asst. (tel. 0 + 650211, ext. 256 and 233, telex 122229 nuphu sf). *Uncompletely crystallized materials, cellulose, computing.*
Siivola, Prof. Jaakko Uolevi. Dept. of Geology and Mineralogy, U. of Helsinki, 00170 Helsinki, Finland. (1938) PhD, geology and mineralogy (U. Helsinki, 1971). Prof. (tel. 0 + 1914457). *Crystallography, mineralogy.*
Sivonen, Mr Seppo Juhani. Inst. of Electron Optics, U. of Oulu, Linnanmaa, 90570 Oulu, Finland. (1942) MSc, techn. physics (U. Oulu, 1967). Lab. eng. (tel. 81 + 352633). *X-ray microanalysis.*
Smolander, Dr Kari Juhani. Dept. of Physics, U. of Helsinki, 00170 Helsinki, Finland. (1947) PhD, physics (U. Helsinki, 1980). Asst. (tel. 0 + 650211, ext. 256, telex 122229 nuphu sf). *Ferroelectricity, Monte Carlo simulations, phase transitions at high pressures.*
Smolander, Dr Kimmo Juhani Nils-Eric. Dept. of Chemistry, U. of Joensuu, 80130 Joensuu, Finland. (1944) PhD, inorganic chemistry (U. Helsinki, 1983). Assoc. prof. (tel. 73 + 28311, ext. 355). *Metal complexes.*
Stubb, Dr Arne Henrik. Semiconductor Lab., Techn. Res. Centre of Finland, 02150 Espoo, Finland. (1946) PhD, physics (U. Helsinki, 1979). Res. officer. (tel. 0 + 4561). *Electronic materials synthesis and characterization.*
Sundius, Dr Tom Robert. Dept. of Physics, U. of Helsinki, 00170 Helsinki, Finland. (1942) PhD, physics (U. Helsinki, 1981). Asst., docent. (tel. 0 + 650211, ext. 400, telex 122229 nuphu sf). *Molecular vibrations and force fields, computer programming.*
Sundström, Dr Lorna Jean. Dept. of Physics, U. of Helsinki, 00170 Helsinki, Finland. (1940) PhD, solid state physics (U. Helsinki, 1968). Asst., docent. (tel. 0 + 650211, ext. 248, telex 122229 nuphu sf). *Solid state theory, energy bands, collective excitations, luminescence.*
Suoninen, Prof. Eero Juhani. Dept. of Physical Sciences, U. of Turku, 20500 Turku, Finland. (1929) PhD, physics (Mass. Inst. of Techn., USA, 1957). Prof. (tel. 21 + 645694). *Electronic structure of crystals, materials research, instrumentation, electron spectroscopy, surface science.*
Suortti, Prof. Pekka. Dept. of Physics, U. of Helsinki, 00170 Helsinki, Finland. (1938) PhD, physics (U. Helsinki, 1967). Assoc. prof. (tel. 0 + 650211, ext. 249, telex 122229 nuphu sf). *Electron distribution in simple solids, synchrotron radiation.*
Tarna, Mr Toivo Mikael. Dept. of Physical Sciences, U. of Turku, 20500 Turku, Finland. (1940) PhLic., physics (U. Turku, 1976). Asst. *Electron density.*
Teuho, Mr Juhani Erkki Tapani. Wihuri Physical Lab. and Dept. of Physical Sciences, U. of Turku, 20500 Turku, Finland. (1948) PhLic., physics (U. Turku, 1973). Asst. (tel. 21 + 645950). *Electron microscopy, metals and alloys.*
Tiitta, Mr Antero Tapani. Reactor Lab., Techn. Res. Centre of Finland, 02150 Espoo, Finland. (1946) Techn.Lic., eng. dept. of techn. physics (Helsinki U. of Techn., 1970). Sr. res. scient. (tel. 0 + 4566350, telex 122972 vttin sf). *Neutron time-of-flight, neutron diffraction, instrument design.*
Tilli, Mr Markku Väinö Kalevi. Dept. of Mining and Metallurgy, Helsinki U. of Techn., 02150 Espoo, Finland. (1950) MSc, physical metallurgy (Helsinki U. of Techn., 1974). Res. scient. (tel. 0 + 4554122, ext. 211, telex 125161). *X-ray topography, semiconductor metallurgy, electron microscopy.*
Toivonen, Mr Jukka Tapio. Dept. of Chemistry, Helsinki U. of Techn., 02150 Espoo, Finland. (1951) MSc, inorganic chemistry (Helsinki U. of Techn., 1976). Asst. (tel. 0 + 4512758). *Inorganic crystal structures.*
Törnroos, Dr Ragnar Fredrik. Geological Survey of Finland, 02150 Esbo, Finland. (1943) PhD, geology and mineralogy (U. Helsinki, 1983). (tel. 0 + 46931, ext. 321). *Analytical mineralogy, descriptive mineralogy.*
Tuomi, Prof. Turkka Olavi. Lab. of Physics, Helsinki U. of Techn., 02150 Espoo, Finland. (1939) DSc, techn. physics (Helsinki U. of Techn., 1968). Assoc. prof. (tel. 0 + 4512145, telex 125161 htkk sf). *Synchrotron X-ray topography, semiconductors, optical properties of solids.*
Turpeinen, Dr Urho Taneli. Dept. of Chemistry, U. of Helsinki, 00100 Helsinki, Finland. (1944) PhD, inorganic chemistry (U. Helsinki, 1977). Instructor. (tel. 0 + 650211). *Crystal structures, metal complexes.*
Turunen, Dr Markus Johannes. Instrument Works, Valmet Corporation, PO Box 237, 33101 Tampere, Finland. (1947) DSc, physical metallurgy (Helsinki U. of Techn., 1975). Docent. (tel. 31 + 650522). *Dislocation theory, plastic deformation modelling, numerical computation, dislocation stress fields.*
Uggla, Dr Rolf Åke Magnus. Dept. of Chemistry, U. of Helsinki, 00100 Helsinki, Finland. (1924) DTechn., physical chemistry (Helsinki U. of Techn., 1960). Docent. (tel. 0 + 650211, ext. 538 or 0 + 460411). *Copper amine complexes, copper hexaoxoiodates, bin(diphenylphosphino)acetylene derivates.*

Finland

Unonius, Mr Lars-Olof. Dept. of Physics, U. of Helsinki, 00170 Helsinki, Finland. (1947) PhLic., physics (U. Helsinki, 1980). Asst. (tel. 0 + 650211, ext. 225, telex 122229 nuphu sf). *X-ray scattering and attenuation.*

Vähäkangas, Mr Jouko Kaarlo. Inst. of Techn. Physics, U. of Oulu, Linnanmaa, 90570 Oulu, Finland. (1948) Techn.Lic., techn. physics (U. Oulu, 1980). Asst. (tel. 81 + 353611). *X-ray diffraction, diffuse scattering, surface structure.*

Vahvaselkä, Dr Aino Margit. Dept. of Physics, U. of Helsinki, 00170 Helsinki, Finland. (1942) PhD, physics (U. Helsinki, 1978). Aman. (tel. 0 + 650211, ext. 240, telex 122229 nuphu sf). *Deformation models, X-ray and neutron diffraction analysis.*

Vahvaselkä, Dr Kaarlo Sakari. Dept. of Physics, U. of Helsinki, 00170 Helsinki, Finland. (1942) PhD, physics (U. Helsinki, 1977). Asst. (tel. 0 + 650211, ext. 250, telex 122229 nuphu sf). *X-ray diffraction, solid and liquid metals.*

Valkonen, Prof. Jussi Uolevi. Dept. of Chemistry, U. of Jyväskylä, 40100 Jyväskylä, Finland. (1947) DTechn., inorganic chemistry (Helsinki U. of Techn., 1979). Prof. (tel. 41 + 291211). *Inorganic chemistry, structural chemistry.*

Visapää, Mr Asko Edvard. Chemical Lab., Techn. Res. Centre of Finland, 02150 Espoo, Finland. (1921) MA, analytical chemistry (U. Helsinki). Res. officer. (tel. 0 + 4561 or 0 + 4565270). *Analytical chemistry, instrumental analysis, spectrometry, cellulose structure, X-ray diffractometry.*

Vorma, Prof., Dr Atso Ilmari. Dept. of Petrology, Geological Survey of Finland, 02150 Espoo, Finland. (1933) PhD, geology and mineralogy (U. Helsinki, 1963). Dept. chief, (tel. 0 + 46931, ext. 266, telex 123185 geolo sf). *Mineralogy, crystal structure determination.*

Ylinen, Dr Eero Elias. Wihuri Physical Lab., U. of Turku, 20500 Turku, Finland. (1944) PhD, physics (U. Turku, 1978). Res. physicist. (tel. 21 + 645944). *Nuclear magnetic resonance, ionic crystals.*

FRANCE

Sub-Editor: **Y. Epelboin**

Notes

1. International telephone country code - 33.

2. Degrees conferred by the French universities are the *doctorat-ès-sciences* (DSc) (approximately equivalent to PhD), the *agrégation* (Agr), the *diplome de docteur-ingénieur* (D.Ing), the *doctorat de 3e cycle* (D.3e cycle) (approximately between MSc and PhD), the *diplome d'études approfondies* (D.ét.ap.), the *diplome d'études supérieures* (D.ét.sup.), the *maîtrise-ès-sciences* (MSc) (approximately equivalent to BSc) and the *licence-ès-sciences* (LSc). The Grandes Ecoles in France confer a *diplome d'ingénieur* (Ing). The Conservatoire National des Arts et Métiers (CNAM) confers a *diplome d'ingénieur* which has a more technological nature. A new system of degrees is now introduced in the french universities.

3. The functions at the Universities are *Professeur* (Professor), *maître-assistant* or *maître de conférences* (associate Professor) and *assistant* (assistant Professor). At the Centre National de la Recherche Scientifique the functions are *directeur-, maître-, chargé-* and *attaché de recherche*. In industrial laboratories the function *ingénieur* (ing.) is generally used; further, *chercheur* (research scientist) is used.

4. The following abbreviations for institutions are used in the French list:

   CEA - Commissariat à l'Energie Atomique
   CEN - Centre d'Etudes Nucléaires
   CENG - Centre d'Etudes Nucléaires Grenoble
   CNET - Centre National d'Etudes des Télécommunications
   CNRS - Centre National de la Recherche Scientifique
   ERA - Equipe de recherche associée (Research group of the CNRS)
   IBMC - Institut de Biologie Moleculaire et Cellulaire

   ILL - Institut Laue-Langevin
   IUT - Institut Universitaire de Technologie
   LETI - Lab. d'électronique et de technologie de l'informatique
   LURE - Lab. pour l'utilisation du rayonnement électromagnétique
   UER - Unité d'enseignement et de recherche
   UA or LA - Unité associée (Lab. of the CNRS and of a university)

5. Additional abbreviations used in this list:

   BP - Boite postale
   Dépt. - Département
   Dir. - Directeur
   Inst. - Institut
   Lab. - Laboratoire
   Miss - Mademoiselle
   Mrs - Madame

   Mr - Monsieur
   Prof. - Professeur
   Rech. - Recherche(s)
   Sup. - Supérieur(e)
   Tel. - Téléphone
   U. - Université

6. The telephone numbers are valid after the October 23, 1985 change.

Aberdam, Dr Daniel Jean. U. Scientifique et Médicale (Grenoble 1), Lab. de Spectrométrie Physique, Domaine Universitaire, BP 53, 38041 Grenoble Cedex, France. (1937) DSc, crystallography (U. de Grenoble, 1971). Maître de rech. CNRS (tel. 76 54 81 52, ext. 571). *Surface physics.*

Aleonard, Dr Suzanne. Lab. de Cristallographie, CNRS Grenoble, 25 avenue des Martyrs, BP 166X, Centre de Tri, 38042 Grenoble Cedex, France. (1926) DSc, physics (U. de Grenoble, 1963). Maître de rech. CNRS (tel. 76 96 98 37). *Crystal chemistry, fluorine compounds.*

Allais, Prof. Gérard. U. de Caen, UER des Sciences, Esplanade de la Paix, 14032 Caen, France. (1935) DSc, physics (U. de Paris, 1967). Prof. (tel. 31 94 59 10, ext. 393). *Inorganic crystal structures, phase transition.*

Alléaume, Dr Marc. Lab. de Cristallographie, U. de Bordeaux 1, Faculté des Sciences, 351 cours de la Libération, 33405 Talence, France. (1934) DSc, physics (U. de Bordeaux, 1967). Maître-asst. (tel. 56 80 69 50, ext. 297). *Biological compounds, structures.*

Andonov, Mrs Paulette. Lab. de Physique des Solides, U. Paris Sud, (Paris 11), Bat. 510, 91405 Orsay, France. (1932) Ing., Chargée de rech. CNRS. (tel. 1 + 69 41 82 50, ext. 31 82). *Amorphous materials, liquids, structures.*

André, Dr Daniel. Lab. de Physicochimie Structurale, U. Paris 12, UER des Sciences, av. du Gal de Gaulle, 94010 Créteil, France. (1946) DSc, physics (U. de Paris (Orsay), 1975). Maître-asst. (tel. 1 + 48 98 91 44, ext. 2445). *Organic structures, plasticity.*

Arnoux, Mrs Bernadette Inst. de Chimie des Substances Naturelles, CNRS, 91190 Gif sur Yvette, France. D.3e cycle, chemistry (U. de Paris 7, 1979). Attaché de rech. CNRS (tel. 1 + 69 07 78 28, ext. 605). *Biologically interesting compounds, structures.*

Aubry, Dr André Roger. Lab. de Minéralogie- Cristallographie, U. de Nancy 1, Faculté des Sciences, Case Officielle 140, 54037 Nancy Cedex, France. (1942) DSc, physical chemistry (U. de Nancy 1, 1976). Chargé de rech. CNRS (tel. 83 28 93 93). *Biological structures.*

Audier, Dr Marc Eugene Raymond. Lab. de Thermodynamique et Physico Chimie Metallurgiques, INPG-ENSEEG, BP 75 Domaine Universitaire, 38402 St. Martin - D'Hères, France. (1947) Chargé de rech. CNRS. (tel. 76 54 41 27, ext. 515). *Crystallization, small particles, phase transformation, Quasicrystals.*

Authier, Prof. André Lab. de Minéralogie- Cristallographie, tour 16, U. Pierre et Marie Curie (Paris 6), 4 place Jussieu, 75230 Paris Cedex 05, France. (1932) DSc, physics (U. de Paris, 1961). Prof. (tel. (1 + 43 36 25 25, ext. 52 24). *X-ray topography, X-ray diffraction, crystal defects.*

Averbuch-Pouchot, Dr Marie-Thérèse. Lab. de Cristallographie, CNRS Grenoble, 25 avenue des Martyrs, BP 166X, Centre de Tri, 38042 Grenoble Cedex, France. (1940) DSc, crystal chemistry (U. de Grenoble, 1974). Ing. CNRS. (tel. 76 96 98 37, ext. 452). *Phosphates, structures.*

Ayroles, Dr René Lab. d'Optique Electronique, CNRS, 29 rue J.Marvig, BP 4347, 31055 Toulouse Cedex, France. (1936) Dsc, physics (U. de Toulouse, 1968).

Maître de Recherches, (tel. 61 52 65 96). *Electron microscopy, structures determination, theory of contrast.*

Badie, Mr Jean-Marie. Lab. Ultra-réfractaires, CNRS, BP 5, 66120 Odeillo Font-Romeu, France. (1943) D.3e cycle, physics (U. de Montpellier, 1968). Attaché de Recherches, (tel. 69 30 10 24). *ceramics.*

Baert, Dr François. Lab. Physique fondamentale, Bldg P5, Université Lille I, 59650 Villeneuve d'Ascq, France. (1935) Dsc, physics (U. de Lille, 1976). Chargé de Recherches, (tel. 20 91 92 22, ext. 2209). *Accurate electronic densities, plastic crystals.*

Baffier, Dr Noel. Lab. de Chimie de la Matière Condensée, Ecole Nationale Supérieure de Chimie, 11 rue P. et M. Curie, 75231 Paris Cedex 05, France. (1935) DSc, (1970). Maître-asst. (tel. 1 + 43 36 25 25, ext. 38 23). *Transition metals, intercalation, lamellar, water.*

Balcou, Dr Yves. U. de Rennes 1, Groupe de Physique Cristalline, av. du Gal Leclerc, 35031 Rennes, France. (1931) DSc, physics (U. de Rennes, 1970). Maître-asst. (tel. 99 36 48 15, ext. 1041). *Molecular crystals.*

Balibar, Prof. Françoise. Lab. de Minéralogie- Cristallographie, tour 16, U. Pierre et Marie Curie (Paris 6), 4 place Jussieu, 75230 Paris Cedex 05, France. (1941) DSc, physics (U. de Paris, 1969). Prof. (tel. 1 + 43 36 25 25, ext. 52 15). *X-ray dynamical theory.*

Bally, Dr Renée. Lab. de Minéralogie- Cristallographie, tour 16, U. Pierre et Marie Curie (Paris 6), 4 place Jussieu, 75230 Paris Cedex 05, France. (1928) DSc, physics (U. de Paris, 1966). Chargée de rech. CNRS. (tel. 1 + 43 36 25 25, ext. 5082). *Biological structures.*

Baltzinger, Dr Christiane. Lab. de Métallurgie Structurale, Ile du Saulcy, Faculté des Sciences de Metz, 57000 Metz, France. (1933) DSc, physics (U. de Metz, 1974). Maître-asst. (tel. 87 30 58 40). *Surfaces, damages.*

Baro, Prof. Raymond. Lab. de Métallurgie Structurale, Ile du Saulcy, Faculté des Sciences de Metz, 57000 Metz, France. (1928) DSc, physics (U. de Strasbourg, 1959). Prof., dir. de lab. et d'IUT. (tel. 87 30 58 40). *Metallurgy, textures.*

Barrans, Dr Yvette. Lab. de Cristallographie et Physique Cristalline, U. de Bordeaux 1, Faculté des Sciences, 351 cours de la Libération, 33405 Talence, France. (1940) DSc, physics (U. de Bordeaux 1, 1971). Chargée de rech. CNRS. (tel. 56 80 84 50, ext. 218). *Biological structures.*

Barraud, Prof. Jean. Lab. de Physique des Structures et des Systèmes Biologiques, U. Paris 7, 12 rue Cuvier, 75005 Paris, France. (1911) DSc, physics (U. de Paris, 1949). Prof. (tel. 1 + 43 36 25 25, ext. 32 10). *Biological compounds, holography.*

Barraud, Mr Jean-Yves. Division Matériaux, Centre de Rech., Compagnie Générale d'Electricité, route de Nozay, 91460 Marcoussis, France. (1941) D.3e cycle, inorganic chemistry (U. de Bordeaux, 1970). Ing. (tel. 1 + 64 49 12 40). *Minerals, organic materials, physicochemistry.*

Baruchel, Dr José ILL, Ave. des Martyrs, BP 156X, F-38042 Grenoble Cedex, France. (1942) DSc, physics (U. de Grenoble, 1980). Attaché de rech. CNRS. (tel. 76 48 70 91). *Magnetism, X-ray topography, neutron topography, neutron diffraction, domains.*

Baucher, Mr Alain. Lab. d'analyse, S.A. Philips, 15 rue Buffon, 75005 Paris, France. (1949) D.3e cycle, materials science (U. de Paris, 1975). Ing. (tel. 1 + 48 30 11 11, ext. 464). *X-ray instrumentation.*

Baudet, Dr Mona. Centre National d'Etudes des Télécommunications, Route de Tregastel, 22301 Lannion, France. (1938) DSc, crystallography (U. de Paris, 1970). Ing. (tel. 96 38 23 43). *X-ray diffraction, thin films, characterization.*

Baudour, Dr Jean-Louis. U. de Rennes 1, Groupe de Physique Cristalline, Campus de Beaulieu, 35042 Rennes, France. (1940) DSc, physical chemistry (U. de Rennes, 1975). Maître-asst. (tel. 99 36 48 15). *Organic crystals, order-disorder transitions.*

Bavoux, Miss Claude. Lab. de Minéralogie- Cristallographie, U. Claude Bernard (Lyon 1), 43 bd du 11 novembre 1918, 69622 Villeurbanne Cedex, France. (1943) D.3e cycle, crystallography (U. de Lyon 1, 1971). Maître de conf. (tel. 78 89 81 24, ext. 32 96). *Organic materials, molecular compounds, polymorphism.*

Becker, Mr Paul. Mesures Physiques, IUT, 57000 Metz, France. (1941) D.3e cycle, physical chemistry (U. de Strasbourg, 1968). Maître-asst. *Metallurgy, crystallography, kinetics, stress.*

Becker, Prof. Pierre. Lab. de Cristallographie, CNRS, BP 166X, 38042 Grenoble, France. (1942) DSc, (U. de Paris, 1973). Prof. (tel. 76 94 41 11). *Electron density, magnetism, incommensurate, diffraction, accurate crystallography.*

Bentley, Dr Graham Arthur. ILL, BP 156X, 38042 Grenoble, France. (1946) PhD, chemistry (U. of Auckland, New Zealand, 1971). Chercheur. (tel. 76 48 72 73). *Molecular biology.*

Bergevin, de, Dr François. Lab. de Cristallographie, CNRS Grenoble, 25 avenue des Martyrs, BP 166X, Centre de Tri, 38042 Grenoble Cedex, France. (1934) DSc, physics (U. de Grenoble, 1968). Maître de rech. (tel. 76 96 98 37, ext.484). *Crystallography, solid state physics.*

Bertaut, Prof. Erwin Félix. Lab. de Cristallographie, CNRS, 25 avenue des Martyrs, BP 166X, Centre de Tri, 38042 Grenoble Cedex, France. (1913) DSc, solid state physics (U. de Grenoble, 1949). Membre de l'Institut, (tel. 76 96 98 37, ext. 484). *Crystallography, magnetism, phase transitions.*

Berthet-Colominas, Dr Carmen. Grenoble Outstation, European Molecular Biology Laboratory, Ave. des Martyrs, BP 156X, F-38042 Grenoble Cedex, France. (1940) Dr, solid state physics (U. de Grenoble, 1967). Responsible, rotating anode X-ray generators. (tel. 76 48 72 76, telex 461613 EMBLD). *X-ray crystallography, adenovirus, virus assembly kinetics.*

Bessiere Dr Michel. CNRS, Centre d'éetudes de chimie métallurgique, 15 rue Georges Urbain, 94400 Vitry sur Seine, France. (1950) Ing. chemistry (ENSCP, 1974). Ing. CNRS. (tel. 1 + 49 41 82 70, ext. 3730). *X-ray scattering, disorder.*

Biais, Mrs Régine. 23 bis route de Saint Brice, 95160 Montmorency, France. *Macromolecules.*

Bideau, Dr Jean-Pierre. Lab. de Cristallographie et Physique Cristalline, U. de Bordeaux 1, Faculté des Sciences, 351 cours de la Libération, 33405 Talence, France. (1937) DSc, physics (U. de Bordeaux, 1971). Maître-asst. (tel. 56 80 76 09). *Biological structures.*

Bienfait, Prof. Michel. Dépt. of Physics, Faculté des Sciences de Luminy, 13288 Marseille, France. (1939) DSc, physics (U. de Nancy, 1965). Prof. (tel. 91 26 91 81). *Surface physics, interfaces.*

Bois, Dr Claudette. Lab. de Chimie Systématique, Université P.M.Curie, 8 rue Cuvier, 75005 Paris, France. (1936) DSc, physics. Maître-asst. *Structural chemistry, statistics.*

Boistelle, Dr Roland. Centre de Recherche sur les Mécanismes de la Croissance Cristalline, campus Luminy, Case 913, 13228 Marseille Cedex 2, France. (1939) DSc, physics (U. de Nancy, 1966). Maître de rech. CNRS. (tel. 91 41 01 52). *Crystal growth.*

Bonnelle, Prof. Christiane. Lab. de Chimie Physique, Université P.M.Curie, 11 rue P. M. Curie, 75231 Paris Cedex 05, France. (1930) DSc, physics (U. de Paris, 1964). Prof. (tel. 1 + 43 36 25 25, ext. 3901). *Spectroscopy, X-Ray excitation, electron distribution, X-Ray emission-absorption.*

Bonnet, Prof. Jean-Jacques. Lab. de Chimie de Coordination, CNRS, 205 route de Narbonne, 31400 Toulouse, France. (1939) DSc, physical chemistry (U. de Toulouse, 1972). Prof., (tel. 61 52 11 66, ext. 328). *Clusters, catalysis.*

Bonnet, Dr Michel. ILL, 156X, 38042 Grenoble Cedex, France. (1944) Dsc Physics, (U. d'Orsay, 1976). Staff member, (tel. 76 48 71 33). *Neutron diffraction, magnetism.*

Bonpunt, Dr Louis. U. de Bordeaux 1, Lab. de physique Cristalline, 351 cours de la Libération, 33405 Talence, France. (1944) Dsc Physics, (U. de Bordeaux 1, 1981). Chargö de Recherches, (tel. 56 80 76 09). *Point defects, scattering*

Bordeaux, Mrs Denise. U. Grenoble 1, Lab. de Spectrométrie Physique, Domaine universitaire, BP 53X, 38402 Grenoble Cedex, France. (1936) Ing., chemistry (ENSEEG, 1960). Ing., (tel. 76 54 81 52). *Synthesis, structures.*

Boucher, Dr Bernard Yves. CEA, CEN Saclay, DPh-G-SRM, BP 2, Orme des Merisiers, 91190 Gif sur Yvette, France. (1931) DSc, physics (U. de Paris, 1968). Chercheur. (tel. 1 + 69 41 88 77, ext. 311-233). *Amorphous materials, metallic alloys.*

Boucherle, Dr Jean-Xavier. CEA Grenoble, 85X, 38041 Grenoble Cedex, France. (1948) DSc, physics (U. de Grenoble, 1977). Chargé de Recherches CNRS. (tel. 76 88 44 00, ext. 311ý33). *Spin density, magnetism.*

Braganza, Dr Lellis Francis. ILL, Ave. des Martyrs, BP 156X, F-38042 Grenoble Cedex, France. (1957) PhD, physics (U. Southamton, UK, 1982). *High pressure, biological membranes, membrane models.*

Brassy, Dr Claude. G.R.C.P.C., U. de Poitiers, Domaine du Déffend, Mignaloux-Beauvoir, 86800 Saint Julien l'Ars, France. (1934) DSc, physics (U. de Poitiers, 1974). Maître-asst. (tel. 49 46 72 17, ext. 36). *Organic, organometallic crystal structures.*

Bravic, Dr Georges. U. de Bordeaux 1, Lab. de Cristallographie, 351 Cours de la Libération, 33405 Talence, France. (1938) DSc, physics (U. de Bordeaux, 1975). Chargé de rech. CNRS, (tel. 56 80 76 09). *ADN, densities.*

Brown, Prof. William Liddle. Lab. de Pétrologie et Géochimie, U. de Nancy 1, Faculté des Sciences, Case Officielle 140, 54037 Nancy cedex, France. (1929) DSc, mineralogy (ETH, Zürich, Suisse, 1958). Prof. (tel. 83 28 93 93, ext. 22 67). *Minerals, textures.*

Brown, Dr Penelope, Jane. ILL, avenue des Martyrs, 166X, 38042 Grenoble Cedex, France. (1932) PhD, physics (Cambridge U., UK, 1958). Chercheur. (tel. 76 48 70 40). *Neutron diffraction, magnetic scattering, magnetic structures, spin density distributions.*

Brunel, Dr Michel. Lab. de Cristallographie, CNRS Grenoble, 25 av. des Martyrs, BP 166X, Centre de Tri, 38042 Grenoble Cedex, France. (1938) DSc, crystallography (U. de Grenoble, 1969). Chargé de rech. CNRS. (tel. 76 96 98 37, ext. 461). *Order-disorder transitions, magnetism, X-ray diffraction.*

Brunie, Dr Simone. CNRS, Centre de Génétique Moléculaire, 91190 Gif sur Yvette, France. (1936) DSc, physics (U. de Lyon, 1970). Chargé de rech. CNRS, (tel. 1 + 69 07 78 28, ext. 820). *Organic crystal structures.*

Burgeat, Dr Jacques. Centre National d'Etudes des Télécommunications, Rech. physiques et Composants, 196 rue de Paris, 92220 Bagneux, France. (1924) DSc, crystallography (U. de Paris, 1969). Ing. (tel. 1 + 46 38 53 47). *Crystal defects, X-ray topography, phase transitions.*

Busetta, Dr Bernard. Lab. de Cristallographie, U. de Bordeaux 1, Faculté des Sciences, 351 cours de la Libération, 33405 Talence, France. (1945) DSc, physics (U. de Bordeaux 1, 1973). Chargé de rech. CNRS. (tel. 56 80 84 50, ext. 271). *Biological structures.*

Cagnon, Dr Maurice. Lab. de Physique des Solides, U. Paris Sud, (Paris 11), Bat. 510, 91405 Orsay, France. (1934) DSc, (1973) Maître-asst. (tel. 1 + 69 07 78 23, ext. 32 71). *Plastic deformation, rheology, liquid crystals.*

Capella, Prof. Lucien. Lab. de Physique Cristalline, U. Aix-Marseille 3, Centre St Jérome, rue Henri Poincaré, 13397 Marseille Cedex 4, France. (1929) DSc, physics (U. d'Alger, Algérie, 1961). Prof. (tel. 91 98 90 10, ext. 312). *Crystal defects, deformation.*

Capelle, Dr Bernard. Lab. de Minéralogie- Cristallographie, tour 16, U. Pierre et Marie Curie (Paris 6), 4 place Jussieu, 75230 Paris Cedex 05, France. (1949) DSc, physics (U. de Paris, 1982). Chargé de rech. (tel. 1 + 43 36 25 25, ext. 52 17). *Phase transitions, X-ray topography, defects, ferroelectricity.*

Capponi, Claude Annie. CNRS, Lab. de Cristallographie, BP 166X, Centre de tri, 38042 Grenoble, France. (1941) DSc, Physics, (U. de Aix-Marseille, 1979). Maître de conf. (tel. 76 96 98 37). *Structures.*

Capponi, Jean-Jacques. CNRS, Lab. de Cristallographie, BP 166X, Centre de tri, 38042 Grenoble, France. (1939) DSc, Physics, (U. de Grenoble, 1973). Chargé de rech. (tel. 76 96 98 37). *Crystal synthesis, high pressures.*

Caranoni, Dr Claude Anny. Lab. de Physique Cristalline, CNRS, rue Henri Poincaré, 13397 Marseille Cedex 4, France. (1941) DSc, physics (U. d'Aix-Marseille 3, 1979). Maître de conf. (tel. 91 98 90 10). *Structures.*

Castaing, Prof. Raymond Bernard René Lab. de Physique des Solides, U. Paris Sud, (Paris 11), Bat. 510, 91405 Orsay, France. (1921) DSc, physics (U. de Paris, 1951). Prof. (tel. 1 + 69 41 82 50, ext. 3345). *Interactions in solids (ions-electrons).*

Cauchois, Prof. Yvette. Lab. de Chimie Physique, U. P. et M. Curie, 11 rue P. M. Curie, 75231 Paris Cedex 05, France. DSc, physical chemistry (U. de Paris, 1933). Prof. (tel. 1 + 43 36 25 25, ext. 39 01). *Structures, Spectroscopy.*

Cazaux, Prof. Jacques. U. de Reims, Dépt. de Physique, Moulin de la Housse, BP 347, 51062 Reims, France. (1934) DSc, physics (U. de Paris, 1970). Prof. (tel. 26 85 23 24, ext. 223). *Microanalysis, spectroscopy, radiography, microfluorescence.*

Cesario, Mrs Michèle. Inst. de Chimie des Substances Naturelles, CNRS, Cristallochimie, 91190 Gif sur Yvette, France. (1942) D. Ing., physics (U. de Paris Sud, 1976). Ing. CNRS. (tel. 1 + 69 07 78 28, ext. 605). *Biological structures.*

Champier, Prof. Georges. Lab. de Physique du Solide, Institut National Polytechnique de Lorraine, Ecole Nationale Supérieure des Mines de Nancy, Parc de Saurupt, 54042 Nancy, France. (1927) DSc, physics (U. de Nancy, 1958). Prof. (tel. 83 51 42 32, ext. 281). *Crystal defects, plasticity, X-ray topography.*

Chanh, Dr Nguyen-Ba. Lab. de Cristallographie et Physique Cristalline, U. de Bordeaux 1, Faculté des Sciences, 351 cours de la Libération, 33405 Talence, France. (1934) DSc, physical chemistry (U. de Bordeaux, 1965). Maître de rech. CNRS. (tel. 56 80 84 50, ext. 296). *Organic crystals, phase transitions, bidimensionnality, thermodynamics.*

Chapelle, Prof. Jean-Pierre. Lab. de Physique Cristalline, U. Paris Sud (Paris 11), 91 Orsay, France. (1914) DSc, physics (U. de Paris, 1949). Prof. *Ionic crystals.*

Charbonneau, Dr Guy Paul. U. de Rennes 1, Groupe de Physique Cristalline, avenue du Gal Leclerc, 35031 Rennes, France. (1937) DSc, solid state physics (U. de Rennes, 1968). Maître-asst. (tel. 99 36 48 15, ext. 22 48). *Cristallography, semiconductors, microelectronics.*

Charbonnier, Dr François. Lab. de Chimie Analytique 2, U. Lyon 1 Claude Bernard, 43 boulevard du 11 Novembre 1918, 69622 Villeurbanne Cedex, France. (1938) DSc, physics (U. de Lyon, 1970). Maître-asst. (tel. 78 89 81 24, ext. 35 27). *Coordination crystal chemistry.*

Charpin, Dr Pierrette. Service de Chimie Physique, CEA, CEN Saclay, BP 2, Orme des Merisiers, 91190 Gif sur Yvette, France. (1927) D. Ing. (Toulouse - Paris, 1960). Dir. du Lab. (tel. 1 + 69 08 32 22). *Structures, diffraction, uranium, actinides.*

Chasseau, Dr Daniel. Lab. de Cristallographie, Université de Bordeaux I, 351 cours de la Libération, 33405 Talence, France. (1946) Dsc, chemistry (U. de Bordeaux, 1979). Maître de conf. (tel. 56 80 84 50, ext. 299). *organic conductors, stress, physical structure of crystals.*

Chenavas, Prof. Jean. CNRS, Lab. de Cristallographie, BP 166X, Centre de Tri, 38042 Grenoble, France. (1938) DSc, physics (U. de Grenoble, 1973). Prof. (tel. 76 96 98 37). *Phase transitions, structures, crystallography.*

Chevalier, Prof. Raymond U. de Clermont-Ferrand II, Lab. de Physique des Matériaux, BP 45, 63170 Aubiere, France. (1941) DSc, physics (U. de Paris, 1973). Prof. (tel. 73 26 41 10, ext. 3305). *Structures, ionic conductors, NMR, glass conductors.*

Chevy, Dr Alain Jean-Pierre. Lab. de Physique des Milieux trés Condensés, U. Pierre et Marie Curie (Paris 6), 4 place Jussieu, 75230 Paris Cedex 05, France. (1945) DSc, physics (U. de Paris, 1982). Maître de conf. (tel. 1 + 43 36 25 25, ext. 44 72). *Crystallogenesis, characterisation.*

Chiaroni, Mrs Angèle. Inst. de Chimie des Substances Naturelles, CNRS, Cristallochimie, 91190 Gif sur Yvette, France. (1942) D.3e cycle, physics (U. de Paris, 1969). Ing. (tel. 1 + 69 07 78 28, ext. 605). *Biological structures.*

Chion, Dr Bernadette. Lab. de Spectronométrie Physique, Université de Grenoble, Domaine Universitaire, 38400 Saint Martin d'Hères, France. (1945) DSc, physics (U. de Grenoble, 1976) Maître-asst. (tel. 76 51 47 59). *X-Ray diffraction, organic materials, chirality, solid solutions.*

Clastre, Prof. José Lab. de Cristallographie et Physique Cristalline, U. de Bordeaux 1, Faculté des Sciences, 351 cours de la Libération, 33405 Talence, France. (1923) DSc, physics (U. de Bordeaux, 1958). Prof. (tel. 56 80 84 50). *Structures.*

Cohen-Addad, Dr Claudine. Lab. de Spectrométrie Physique, U. Scientifique et Médicale (Grenoble 1), Domaine universitaire BP 53, 38041 Grenoble Cedex, France. (1939) DSc, physics (U. de Grenoble, 1969). Chargée de rech. (tel. 76 51 47 60). *Organic crystal structures, charge density.*

Coing-Boyat, Dr Jean Claude. Lab. de Cristallographie, CNRS Grenoble, 25 av. des Martyrs, BP 166X, Centre de Tri, 38042 Grenoble Cedex, France. (1930) DSc, physics (U. de Grenoble, 1966). Chargé de rech. (tel. 76 96 98 37). *Inorganic crystal structures, magnetic structures.*

Colliex, Dr Christian. Lab. de Physique du Solide, Université Paris-Sud, Bldg 510, 91405 Orsay, France. (1944) DSc, physics (U. d'Orsay, 1970). (tel. 1 + 69 41 53 70). Maître de Recherches. *electron microscopy, electron diffraction, scattering, phase transitions, order-disorder.*

Comes, Dr Robert Lab. de Physique du Solide, Bat. 510, U. Paris Sud, 91405 Orsay, France. (1937) DSc, physics (U. de Paris Sud, 1969). Maître de rech., (tel. 1 + 69 41 60 51). *X-ray scattering, neutron scattering, phase transitions, order-disorder.*

Commarond, Dr Marie-Bernard. Unité d'Immunologie Structurale, Institut Pasteur, 28 rue du docteur Roux, 75724 Paris Cedex 15, France. (1955) D.Ing. (U. de Strasbourg, 1981). Attaché de Recherches (tel. 1 + 43 06 19 19, ext. 3136). *Biophysics, immunoglobulines.*

Constant, Prof. Georges. ERA 263, Cristallochimie, Ecole Nationale Supérieure de Chimie de Toulouse, 118 route de Narbonne, 31077 Toulouse Cedex, France. (1937) DSc (U. de Strasbourg, 1963). Prof. (tel. 61 53 14 21). *Thin layers, OMCVD, epitaxy, organominerals.*

Convert, Dr Pierre. ILL, Ave. des Martyrs, BP 156X, F-38042 Grenoble Cedex, France. (1941) physics (U. de Grenoble, 1975). Ing. (tel. 33 76 48 7148). *Position sensitive detectors, diffractometry.*

Coulomb, Dr Pierre. Microscopie et Structure des Matériaux, U. Paul Sabatier, 118 route de Narbonne, 31062 Toulouse Cedex, France. (1929) DSc, physics (U. de Paris, 1960). Prof. (tel. 61 53 11 20). *Alloys, plasticity, phase transitions.*

Courtois, Dr Alain Raymond. Lab. de Minéralogie- Cristallographie, U. de Nancy 1, Faculté des Sciences, Case Officielle 140, 54037 Nancy Cedex, France. (1943) DSc, physical chemistry (U. de Nancy 1, 1976). Maître de conf. (tel. 83 28 93 93, ext. 2263ý248). *Organic, inorganic crystal structures, electron density.*

Curie, Prof. Daniel. Lab. de Luminescence, tour 13, U. Pierre et Marie Curie (Paris 6), 4 place Jussieu, 75230 Paris Cedex 05, France. (1927) DSc, physics (U. de Paris, 1951). Prof. (tel. 1 + 43 36 25 25, ext. 44 57). *Luminescence.*

Curien, Prof. Hubert. Lab. de Minéralogie- Cristallographie, tour 16, U. Pierre et Marie Curie (Paris 6), 4 place Jussieu, 75230 Paris Cedex 05, France. (1924) DSc, physics (U. de Paris, 1952). Prof. (tel. 1 + 43 36 25 25, ext. 5083). *Crystal structures, crystal defects.*

Dahan, Dr Françoise. Lab. de Chimie de Coordination, CNRS, 205 route de Narbonne, 31400 Toulouse, France. (1941) DSc, physics (U. de Paris 6, 1980). Ing. (tel. 61 52 11 66). *Organometallic crystal structures.*

Darces, Mr Jean-François. Faculté des Sciences, Lab. de Cristallographie, 25030 Besancon Cedex, France. (1944) D.3e cycle, physics (U. de Franche Comté, 1970). Asst, (tel. 81 53 81 22). *Crystal growth, X-ray diffraction.*

Dartyge, Dr Elisabeth. Lab. de Physique des Solides, U. Paris Sud, (Paris 11), Bat. 510, 91405 Orsay, France. (1941) DSc, physics (U. de Paris sud, 1979). Maître-asst. (tel. 1 + 69 41 82 70, ext. 33 75). *EXAFS.*

Dartyge, Mr Jean-Marcel. Lab. de Chimie-Physique, I.R.C.H.A., BP 1, 91710 Vert le Petit, France. (1941) D.3e cycle, crystallography (U. de Paris sud, 1967). Ing. (tel. 1 + 64 93 24 75). *Polymers, carbon, composites.*

De Kouchkovsky, Mr Rostislav. CEA, CEN-Saclay, BP 2, Orme des Merisiers, 91190 Gif sur Yvette, France. (1930) LSc, (1959) Ing. (tel. 1 + 69 08 75 16). *X-ray diffraction, phase transitions, amorphous materials structures.*

Delapalme, Dr Alain. Lab. Léon Brillouin, CEA/CEN Saclay BP 2, 91191 Gif sur Yvette Cedex, France. (1930) DSc, solid state physics (U. de Grenoble, 1967). Chercheur. (tel. 1 + 69 08 47 79 , ext. 60 38). *Electron density.*

Delettré, Dr Jean. Lab. de Minéralogie-Cristallographie, tour 16, U. Pierre et Marie Curie (Paris 6), 4 place Jussieu, 75230 Paris Cedex 05, France. (1943) DSc, physics (U. de Paris 6, 1978). Maître-asst. (tel. 1 + 43 36 25 25, ext. 52 36). *Proteins, organic structures.*

Delord, Prof. Pierre. Lab. de Cristallographie, U. des Sciences et Techniques du Languedoc, place Eugène Bataillon, 34000 Montpellier, France. (1937) DSc, crystallography (U. de Montpellier, 1970). Prof. (tel. 67 63 91 44, ext. 510). *Liquid crystals.*

Despujols, Prof. Jacques. Lab. d'Electronique des Rayons X, Université de Reims, Moulin de la Housse, BP 347, 51062 Reims Cedex, France. (1925) DSc, physics (U. de Paris, 1956). Prof. (tel. 26 85 23 24). *Defects, surfaces, thin layers.*

Di-Persio, Dr Jean Anthony. U. des Sciences et Techniques de Lille, UER de Physique Fondamentale, BP 36, 59650 Villeneuve d'Ascq, France. (1940) DSc, physics (U. de Lille, 1980). Maître-asst. (tel. 20 56 92 00, ext. 23 13). *Crystal defects, plasticity, X-ray topography.*

Duchefdelaville, Mr Gérard. Bibliothèque technique, Michelin, 63040 Clermont-Ferrand, France. (1935) D.3e cycle, (1960). Ing. (tel. 73 92 42 21).

Duchemin, Mr Jean-Pierre. Société INEL, 261 rue Louis Blériot, 78530 Buc, France. (1937) Dir. (tel. 1 + 39 56 31 90). *Crystallography, fluorescence, instrumentation.*

Ducros, Prof. Pierre. Lab. de Spectrométrie Physique, Domaine Universitaire, U. Scientifique et Médicale (Grenoble 1), BP 53, 38041 Grenoble Cedex, France. (1931) DSc, physics (U. de Paris, 1960). Prof. (tel. 76 87 85 71, ext. 570). *Surface physics, teaching.*

Ducruix, Dr Arnaud. Inst. de Chimie des Substances Naturelles, CNRS, Cristallochimie, 91190 Gif sur Yvette, France. (1947) DSc, physics (U. de Paris sud, 1976). Attaché de rech. CNRS. (tel. 1 + 69 07 78 28, ext. 605). *Proteins, nucleic acids, crystallography.*

Dugué, Prof. Jérome. Lab. de Chimie Minérale Structurale, Faculté des Sciences Pharmaceutiques et Biologiques, 4 avenue de l'Observatoire, 75270 Paris Cedex 06, France. (1944) DSc, physics (U. de Paris 6, 1978). Prof. (tel. 1 + 43 29 12 08). *Crystal structures, X-Ray diffraction.*

Dumas, Mr Philippe. Lab. de Cristallographie Biologique, IBMC, 15 rue René Descartes, 67084 Strasbourg Cedex, France. (1951) These de 3e cycle, fluid mechanics (U. de Provence, 1979). Attaché de rech. (CNRS). (tel. 88 61 02 02, ext. 387). *Macromolecular crystallography.*

Durif, Dr André Lab. de Cristallographie, CNRS Grenoble, 25 av. des Martyrs, BP 166X, Centre de Tri, 38042 Grenoble cedex, France. (1929) DSc, crystal chemistry (U. de Grenoble, 1958). Dir. scient. (tel. 76 96 98 37, ext. 485). *Phosphates, minerals, structures.*

Eberhart, Prof. Jean-Pierre. Lab. de Minéralogie, U. Louis Pasteur, 1 rue Blessig, 67000 Strasbourg, France. (1930) DSc, crystallography-mineralogy (U. de Strasbourg, 1963). Prof. (tel. 88 35 66 03). *Clay minerals, glasses.*

Elkaim, Mr Erik. Lab. de Minéralogie-Cristallographie, Université Nancy I, 9 rue Charlet, 88000 Epinal, France. (1959) DEA Cristallographie (U. de Nancy, 1982). Student (tel. 29 82 25 23).

Epelboin, Dr Yves. Lab. de Minéralogie- Cristallographie, tour 16, U. Pierre et Marie Curie (Paris 6), 4 place Jussieu, 75230 Paris Cedex 05, France. (1944) DSc, physics (U. de Paris 6, 1974). Maître de conf. (tel. 1 + 43 36 25 25, ext. 52 16). *X-ray topography, defects, 3D images, 3D graphics, computing.*

Esteoule, Prof. Jacques. Lab. de Minéralogie et Géotechnique, Institut National des Sciences Appliquées, 20 av. des Buttes de Coesnes, 35000 Rennes, France. (1932) DSc, (U. de Rennes, 1969). Prof. (tel. 99 36 48 30). *Clay minerals, mineralogy.*

Fauvet, Mr Gérard. Lab de Rayons X, ENRAF-NONIUS FRANCE, 3 rue du Troyon, 75017 Paris, France. (1948) D.3e cycle (1977). Ing. (tel. 1 + 43 80 35 12). *Apparatus.*

Ferey Prof. Gérard. Lab. des Fluorures, UA 449, Faculté des Sciences, route de Laval, 72017 Le Mans Cedex, France. (1941) DSc, physics (U. de Paris, 1977). Prof. (tel. 43 24 72 36, ext. 333). *Crystal growth, structures, magnetism, Mossbauer.*

Fitch, Dr Andrew Nicholas. ILL, Ave. des Martyrs, BP 156X, F-38042 Grenoble Cedex, France. (1956) D.Phill, chemistry (Oxford U., 1981). Contract scient. (tel. 76 48 70 43). *Powder neutron diffraction, ionic conductors, zeolites.*

Fontaine Dr Alain. LURE, Université Paris Sud, Bldg 209, 91405 Orsay, France. (1944) DSc, physics (U. de Paris Sud, 1975). Maître de conf. (tel. 1 + 69 41 82 70, ext. 34 26). *Synchrotron, EXAFS, metallurgy, X-Ray optics.*

Fontaine, Prof. Hubert. U. de Sciences et Techniques de Lille, Lab. de Dynamique des Cristaux Moléculaires, BP 36, Villeneuve d'Ascq, France. (1936) DSc, solid state physics (U. de Paris, 1973). Prof. (tel. 20 91 92 22, ext. 21 74). *Order-disorder, phase transitions.*

Fouret, Prof. René U. des Sciences et Techniques de Lille, UER de Physique Fondamentale, BP 36, 59650 Villeneuve d'Ascq, France. (1925) DSc, solid state physics (U. de Paris, 1963). Prof. (tel. 20 91 92 22, ext. 2173). *Organic crystal, lattice dynamics.*

Fourme, Prof. Roger. LURE, U. Paris Sud, 91405 Orsay, France. (1942) DSc, physics (U. de Paris, 1970). Prof. (tel. 1 + 69 41 82 70, ext. 698). *Protein crystallography, EXAFS, synchrotron radiation, area detectors.*

Frey, Dr Michel. Centre de Recherche sur les Mécanismes de la Croissance Cristalline, CNRS, campus Luminy, case 913, 13288 Marseille Cedex 2, France. (1938) DSc, physics (U. de Caen, 1970). Maitre de rech. CNRS. (tel. 91 41 01 52). *Protein crystallography, crystal growth.*

Friedel, Prof. Jacques. Lab. de Physique des Solides, U. Paris Sud, (Paris 11), Bat. 510, 91405 Orsay, France. (1921) PhD, DSc, solid state physics (Bristol U., UK, 1951; U. de Paris, 1954). Prof. (tel. 1 + 69 41 82 50, ext. 33 66). *Metals, alloys, electronic structures, crystal defects.*

Freund, Dr Andreas Karl. ILL, Ave. des Martyrs, BP 156X, F-38042 Grenoble Cedex, France. (1942) Dr.rer.nat., solid state physics (Techn. U. München, BRD, 1973). Staff scient. (tel. 76 48 71 11, ext. 7083, 7291, telex 320621 F). *Characterization, difraction, X-rays, Gamma-rays, neutron diffraction, instrumentation, synchrotron radiation, monochromators, real crystals, extinction, materials science.*

Gabis, Prof. Victor Michel. Lab. de Minéralogie Appliquée, Ecole Supérieure de l'Energie et des Matériaux, U. d'Orléans, 45046 Orléans, France. (1932) DSc, geochemistry, mineralogy (U. de Paris 6, 1964). Prof. (tel. 38 63 37 03). *Difusion, mineralogy.*

Gadet, Mr Alain. U. Paris 12, UER des Sciences, Lab. de Physicochimie Structurale, avenue du Général de Gaulle, 94010 Créteil, France. (1942) D.3e cycle, crystallography (U. de Paris 6, 1973). Maître de conf. (tel. 1 + 48 98 91 44, ext. 2498). *Proteins.*

Galigné, Dr Jean-Louis. Lab. de Minéralogie- Cristallographie, U. des Sciences et Techniques du Languedoc, place Eugène Bataillon, 34000 Montpellier, France. (1936) DSc, crystallography, solid state physics (U. de Montpellier, 1969). Maître-asst. *Crystal structures.*

Gallois, Mr Bernard. Lab. de Physique Cristalline, U. de Bordeaux 1, Cours de la Libération, 33405 Talence, France. (1953) D.3e cycle, crystallography (U. de Bordeaux, 1980) Attaché de rech. (tel. 56 80 76 09). *Organic conductors, surstructures.*

Gallot, Dr Bernard. Centre de Biophysique, CNRS, av. de la Recherche Scientifique, 45045 Orléans la Source, France. (1935) DSc, physics (U. de Strasbourg, 1965). Maitre de rech. *Polymer structures, biological membranes, zeolites, structure determination.*

Galy, Dr Jean. Lab. de Chimie de Coordination, CNRS, 205 route de Narbonne, 31400 Toulouse cedex, France. (1938) DSc, physical chemistry (U. de Bordeaux 1, 1966). Dir. de rech. CNRS. (tel. 61 52 11 66). *Solid state chemistry, inorganic structures, amorphous.*

Gandais, Dr Madeleine. Lab. de Minéralogie- Cristallographie, tour 16, U. Pierre et Marie Curie (Paris 6), 4 place Jussieu, 75230 Paris Cedex 05, France. (1934) DSc, crystallography (U. de Paris, 1969). Maitre de rech. CNRS. (tel. 1 + 43 36 25 25, ext. 52 12). *Feldspars, deformation, electron microscopy.*

Gasperin, Dr Madeleine. Lab. de Minéralogie- Cristallographie, tour 16, U. Pierre et Marie Curie (Paris 6), 4 place Jussieu, 75230 Paris Cedex 05, France. (1924) DSc, crystallography (U. de Paris, 1960). Maitre de rech. (tel. 1 + 43 36 25 25, ext. 50 82). *Oxides, synthesis, mineral structures.*

Gatineau, Dr Lucien Charles. Centre de Recherche sur les Solides à Organisation Cristalline Imparfaite (CRSOCI), 1bis rue de la Férollerie, 45045 Orléans cedex, France. (1927) DSc, physics (U. de Paris, 1964). Maitre de rech. (tel. 38 63 39 37). *Semicrystalline materials, structures.*

Gauthier, Dr Jean-Pierre. Lab. de Minéralogie- Cristallographie, U. de Lyon, 43 bd du 11 novembre 1918, 69622 Villeurbanne Cedex, France. (1943) DSc, crystallography (U. de Lyon, 1978). Maître-asst. (tel. 78 89 81 24, ext. 32 49). *Polytypism, surfaces, interfaces, electron diffraction, gemms.*

Ghelis, Mrs Marianne. Lab. de Minéralogie- Cristallographie, U. P.M.Curie, 4 place Jussieu, 75230 Paris Cedex 05, France. (1938) D.U., mineralogy (U. de Paris 6, 1980). Chercheur. (tel. 1 + 43 36 25 25, ext. 5076). *Feldspars, mineral structures, synthesis.*

Ghermani, Mr Noureddine. Lab. Minéralogie-Cristallographie, U. de Nancy 1, BP 239, 54506 Vandoeuvre les Nancy Cedex, France. (1957) DEA, crystallography (U. de Nancy 1, 1981). Student. *charges density.*

Gilles, Prof. Jean-Claude. Lab. du Solide Minéral, Ecole Supérieure de Physique et Chimie (ESPCI), 10 rue Vauquelin, 75231 Paris Cedex 05, France. (1937) DSc, physics (U. de Paris, 1965). Prof. (tel. 1 + 43 37 77 00). *Solid state.*

Gillier-Pandraud, Prof. Hélène. U. Paris Nord (Paris 13), UER de Médecine et Biologie Humaine, 74 rue Marcel Cachin, 93012 Bobigny Cedex, France. (1928) DSc, chemistry (U. de Paris, 1955). Prof., Dir. du lab. (tel. 1 + 48 36 55 79, ext. 259). *X-Rays, phosphoranes, glucides, xenobiotics.*

Ginderow, Dr Daria. Lab. de Minéralogie- Cristallographie, tour 16, U. P.M.Curie (Paris 6), 4 place Jussieu, 75230 Paris Cedex 05, France. (1933) DSc, (U. d'Orléans, 1969). Chargée de rech. CNRS. (tel. 1 + 43 36 25 25, ext. 50 59). *Mineral structures, organometallics.*

Gleizes, Dr Alain Nicolas. Lab. de Chimie de Coordination, CNRS, 205 route de Narbonne, 31400 Toulouse, France. (1943) DSc, chemistry (U. de Toulouse, 1974). Maître-asst. (tel. 61 53 05 25). *Complex, transition metals, low dimensionnality structures.*

Godefroy, Prof. Lucien René Lab. Physique des Solides, U. de Dijon, 21104 Dijon Cedex, France. (1927) DSc, physics (U. de Paris, 1963). Prof. (tel. 80 66 64 13). *Ferroelectricity, phase transitions, incommensurables.*

Gondrand, Dr Monique. Lab. de Cristallographie, CNRS, Av. des Martyrs, 166X, 38042 Grenoble Cedex, France. (1931) DSc, physics (U. de Grenoble, 1972). Maître-asst. (tel. 76 96 98 37). *Cristallochemistry, ionic conductivity.*

Graf, Prof. René U. Lab. des Rayons X, Faculté des Sciences de Rouen, BP 67, 76130 Mont-Saint-Aignan, France. (1920) DSc, physics (U. de Paris, 1955). Prof. (tel. 35 98 28 50). *Metallurgy, alloy structures, phase transitions.*

Grand, Dr André Dépt. de Rech. Fondamentales, CEA-CENG, (Centre d'Etudes Nucléaires de Grenoble), 85X, 38041 Grenoble Cedex, France. (1947) DSc, physics (U. de Grenoble, 1980). Asst. (tel. 79 97 41 11, ext. 32 65). *Organometallics, magnetic structures, quantum chemistry.*

Grandjean, Prof. Daniel Lab. de Cristallochimie, U. de Rennes, Campus de Beaulieu, 35042 Rennes Cedex, France. (1939) DSc, physics (U. de Strasbourg, 1966) Professor (tel. 99 36 48 15). *Cristallochemistry, coordination, metal-metal binding.*

Guet, Dr Jean-Michel. Lab. de Cristallographie, U. d'Orléans, 45046 Orléans Cedex, France. (1945) DSc, physics (U. d'Orléans, 1983). Maitre-asst. (tel. 38 63 22 16, ext. 697). *X-Ray diffraction, electron microscopy, mesophases, carbon fibers.*

Guilhem, Dr Jean. Inst. de Chimie des Substances Naturelles, CNRS, Cristallochimie, 91190 Gif sur Yvette, France. (1932) DSc, physics (U. de Paris, 1967). Chargé de rech. CNRS. (tel. 1 + 69 07 78 28, ext. 605). *Natural compounds, biologically interesting structures.*

Guinier, Prof. André Jean. Lab. de Physique des Solides, U. Paris Sud, Bat. 510, 91405 Orsay, France. (1911) DSc, physics (U. de Paris, 1939). Prof. (tel. 1 + 69 41 82 50, ext. 32 73). *X-ray diffuse scattering.*

Guitel, Mr Jean-Claude. Lab. de Cristallographie, CNRS, BP 166X, Centre de Tri, 38042 Grenoble Cedex, France. (1939) Master Physics, (U. de Grenoble, 1961). Maître-asst. (tel. 76 96 98 37, ext. 440). *Computing, inorganic structures, phosphates.*

Habersetzer, Dr Catherine. Centre de Rech. des Mécanismes de Croissance Cristalline, CNRS, Campus de Luminy, Case 913, 13288 Marseille Cedex 9, France. (1940) DSc (1972). Chargé de rech. (tel. 91 41 01 52). *Neurotoxines.*

Haget, Dr Yvette. Lab. de Cristallographie, U. de Bordeaux 1, 315 cours de la Libération, 33405 Talence, France. (1934) DSc, physics (U. de Bordeaux, 1968). Chargé de rech. (tel. 56 80 84 50, ext. 296). *Diffusion, defects, molecular alloys, phase transitions.*

Hardy, Mrs Anne-Marie. Lab. de Cristallochimie minérale, Faculté des Sciences de Poitiers, 40 av. du recteur Pineau, 86022 Poitiers Cedex, France. (1938) D.ét.sup., inorganic chemistry (U. de Rennes, 1961). Collaborateur technique. (tel. 49 42 26 30, ext. 497). *Mineral structures.*

Hardy, Prof. Antoine. Lab. de Cristallochimie Minérale, Faculté des Sciences de Poitiers, 40 av. du recteur Pineau, 86022 Poitiers, France. (1929) DSc, inorganic chemistry (U. de Bordeaux, 1962). Prof. (tel. 49 42 26 30, ext. 637). *Mineral structures, automation, zeolites.*

Haser, Dr Richard Michel. Centre de Recherche sur les Mécanismes de la Croissance Cristalline, CNRS, campus Luminy, case 913, 13288 Marseille Cedex 2, France. (1942) DSc, physics (U. d'Aix-Marseille, 1972). Chargé de rech. CNRS. (tel. 91 41 01 52). *Relation structure- activity, redox proteines, enzymes, X-Ray diffraction.*

Hauw, Dr Christian. Lab. de Cristallographie, Université de Bordeaux 1, 351 Cours de la Libération, 33405 Talence, France. (1934) Dsc, physics (1967). Maître-asst. *Organic structures, conductivity, high pressures.*

Herpin, Prof. Paulette. U. Pierre et Marie Curie (Paris 6), 4 place Jussieu, 75230 Paris Cedex 05, France. (1919) DSc, crystallography, chemistry (U. de Paris, 1956). Dir. du lab. (tel. 1 + 43 36 25 25, ext. 45 49). *X-ray diffraction, neutron diffraction.*

Horn, Prof. Paul. Lab. de Physique Moléculaire, ERA 828, CNRS, U. Nancy 1, Case officielle 140, 54037 Nancy Cedex, France. (1925) DSc, physics (U. de Strasbourg, 1954). Prof. (tel. 83 28 93 93, ext. 27 83). *Biological structures.*

Hospital, Dr Michel. Lab. de Cristallographie, U. de Bordeaux 1, 351 cours de la Libération, 33405 Talence, France. (1935) DSc, physics (U. de Bordeaux 1, 1968). Dir. de rech. CNRS. Dir. du lab. (tel. 56 80 84 50, ext. 218). *Biological structures, macromolecules, peptides.*

Housty, Dr Jacques. Lab. de Cristallographie, U. de Bordeaux 1, 351 cours de la Libération, 33405 Talence, France. (1928) DSc, crystallography (U. de Bordeaux, 1966). Maître-asst. (tel. 56 80 84 50). *Crystal structures, crystal growth.*

Janin, Prof. Joel. Inst. de biochimie, U. Paris Sud, bat. 430, 91405 Orsay Cedex, France. (1943) DSc, physics (U. de Paris Sud, 1969). Prof. (tel. 1 + 69 41 29 73). *Protein structures, molecular graphics, enzymology.*

Jeannin, Prof. Yves. Lab. de Chimie des Métaux de Transition, U. P. et M. Curie (Paris 6), 4 place Jussieu, 75230 Paris Cedex 05, France. (1931) DSc, physics (U. de Paris, 1962). Prof. (tel. 1 + 43 36 25 25, ext. 30 34). *Inorganic, organometallic, coordination compounds, crystal structures.*

Jehanno, Dr Germain Pierre. CEA, CEN-Saclay, DPh./SRM, BP 2, Orme des Merisiers, 91190 Gif sur Yvette, France. (1931) DSc, physics (U. de Paris, 1965). Ing. (tel. 1 + 69 41 81 77, ext. 437). *Phase transformations, modulated structures.*

Joel, Dr Nahum. 46 rue Fabert, 75007 Paris, France. (1924) PhD, physics (Cambridge U., UK, 1959). (tel. 1 + 45 51 97 85). *Optical, elastic properties of crystals, teaching.*

Joubert, Prof. Jean-Claude. Lab. de Cristallographie, CNRS Grenoble, 25 av. des Martyrs, BP 166X, Centre de Tri, 38042 Grenoble Cedex, France. (1939) DSc, physics (U. de Grenoble, 1965). Prof. (tel. 76 96 98 37, ext. 442). *Magnetic materials, pigments, superconductor materials.*

Jouffrey, Dr Bernard. Lab. d'Optique Electronique, 29 rue J.Marvig, 31055 Toulouse Cedex, France. (1936) DSc, physics (U. de Paris-Sud, 1964). Dir. du lab. (tel. 61 52 65 96). *Electron microscopy, crystal defects.*

Jourdan, Dr Claude, René Centre de Recherche des Mécanismes de la Croissance Cristalline, campus de Luminy, 13288 Marseille Cedex 2, France. (1934) DSc, solid state physics (U. de Marseille, 1965). Maître de rech. (tel. 91 41 01 52). *Metals, recrystallization, growth defects, joints boundaries.*

Kahn, Dr Andrée, Solid State Chemistry, ENS de Chimie de Paris, 11 rue P.M.Curie, 75231 Paris Cedex 05, France. (1942) DSc, chemistry (U. de Paris, 1970). Chercheur. (tel. 1 + 43 36 25 25, ext. 3814). *Crystal structures, solid state chemistry.*

Kappenstein, Prof. Charles, Lab. de Chimie Minérale, U. de Poitiers, 40 avenue du Recteur Pineau, 86022 Poitiers Cedex, France. (1946) DSc, physics (U. de Rennes, 1977). Prof. (tel. 49 46 26 30, ext. 660). *Catalysis, crystallography.*

Kern, Prof. Raymond. Centre de Recherche des Mécanismes de la Croissance Cristalline, CNRS, campus Luminy, case 913, 13288 Marseille Cedex 2, France. (1928) DSc, physics (U. de Strasbourg, 1953). Dir. du lab. (tel. 91 41 01 52). *Crystal growth, surfaces, polymers, crystallography.*

Kléman, Dr Maurice. Lab. de Physique des Solides, U. Paris Sud, (Paris 11), Bat. 510, 91405 Orsay, France. (1934) DSc, solid state physics (U. de Paris, 1967). Dir. de rech., CNRS. (tel. 1 + 69 41 82 50, ext. 33 31). *Defects, liquid crystals, magnetic materials, glasses.*

Kuhs, Dr Werner Friedrich. ILL, Ave. des Martyrs, BP 156X, F-38042 Grenoble Cedex, France. (1952) Dr.rer.nat., crystallography (U. Freiburg, BRD, 1978). Scient. (tel. 76 48 7111, ext. 7232, telex 320621 F). *Thermal motion, disorder, ice, crystallography, phase transitions.*

Labbé, Dr Philippe. Lab. de Cristallographie et Chimie du Solide, LA 251, U. de Caen, Esplanade de la Paix, 14032 Caen, France. (1939) DSc, physics (U. de Caen, 1978). Maître-asst. (tel. 31 94 81 40, ext. 3546). *Oxides, crystal structures, channeling, teaching.*

Laberrigue, Prof. André U. de Reims, Dépt. de Physique, Moulin de la Housse, BP 347, 51062 Reims, France. (1924) DSc, physics (U. de Paris, 1958). Prof. (tel. 26 47 82 61, ext. 249). *Electron microscopy, high resolution.*

Lafourcade, Prof. Lucien. Lab. de Physique Structurale, U. Paul Sabatier (Toulouse), 118 route de Narbonne, 31077 Toulouse Cedex, France. (1914) DSc, physics (U. de Toulouse, 1952). Prof. (tel. 61 53 11 20, ext. 534). *Crystal chemistry.*

Lajzerowicz, Prof. Janine. U. Scientifique et Médicale, Lab. de Spectrométrie Physique, Domaine Universitaire, BP 87, 38042 Grenoble Cedex, France. (1932) DSc, physics (U. de Grenoble, 1964). Prof. (tel. 76 51 47 61). *Organic crystal structures, order-disorder.*

Lambert, Prof. Marianne. Lab. Léon Brillouin, CEA, CEN Saclay, 91191 Gif sur Yvette Cedex, France. (1932) DSc, physics (U. de Paris, 1958). Prof. (tel. 1 + 69 08 32 54). *Phase transitions, order-disorder, lattice dynamics.*

Langlet, Dr Gérard André Dépt. de Physico-Chimie, CEA, CEN Saclay, 91191 Gif sur Yvette, France. (1940) DSc, physics (U. de Paris, 1969) Ing. (tel. 1 + 69 41 80 00, ext. 44 95, 47 26). *Crystal geometry, computing.*

Lapasset, Prof. Jacques. Lab. de Cristallographie, U. du Languedoc, place Eugène Bataillon, 34060 Montpellier Cedex, France. (1937) DSc, physics, (USTL, 1972) Prof. (tel. 67 63 43 12). *Molecular structures, transitions, modulated phases.*

Larroque, Prof. Paul. Lab. de Physique Électronique et Ionique, U. Paul Sabatier (Toulouse), 118 route de Narbonne, 31062 Toulouse Cedex, France. (1926) DSc, (1962). Prof. (tel. 61 55 68 88). *Surfaces, epitaxy, clusters, microdiffraction.*

Lartigue, Dr Colette. Chimie Metallurgique des Terres Rares, CNRS, 1 place A. Briand, 92190 Meudon Cedex, France. (1953) Doctorate, physics (U. Paris VI, 1984). (tel. 1 + 45 34 75 50, ext. 2359). *Intermetallic compounds with hydrogen, neutron diffraction, quasi-elastic neutron scattering, hydrogen dynamics in intermetallic hydrides.*

Laruelle, Prof. Pierre Etienne Charles. Lab. de Physique, Faculté des Sciences Pharmaceutiques, 4 avenue de l'Observatoire, 75006 Paris, France. (1923) DSc, physics (U. de Paris, 1956). Prof. (tel. 1 + 43 29 12 08, ext. 340). *Structures, rare earths, chalcogenides.*

Laugier, Mr Jean. CEA-CENG, (Centre d'Etudes Nucléaires de Grenoble), Diffraction Neutronique, av. des Martyrs, BP 85X, 38041 Grenoble Cedex, France. (1932) LSc, (1954). Ing. (tel. 76 97 41 11, ext. 41 46). *X-ray crystallography, crystal structures.*

Laugt, Dr Marguerite. Lab. de Batiments Solaires, CNRS, BP 21, 06562 Valbonne Cedex, France. (1944) DSc, (U. de Grenoble, 1974). Chargé de rech. (tel. 93 74 63 63). *Salt hydrates, surfusion, thermic stocking, crystallization.*

Laurent, Dr Pierre. 37 av. Fourcault de Pavaut, 78000 Versailles, France. (1912) DSc, (U. de Paris, 1944). (tel. 1 + 34 90 12 80). *Solid state physics.*

Laurent, Prof. Yves. Lab. de Chimie Minérale C, Université Rennes 1, Campus de Beaulieu, 35042 Rennes Cedex, France. (1939) DSc, (U. de Rennes, 1968). Prof. (tel. 99 36 48 15, ext. 21 62). *X-Ray diffraction, neutrons, nitrides, glasses, ceramics.*

Le Bars, Mrs Michèle, Lab. de Spectrométrie Physique, LA 08, BP 87, 38042 Saint Martin d'Hères Cedex, France. (1940) Agr., Asst., (tel. 76 51 47 60). *Molecular complexes, phases transitions.*

Le Roux, Mr Guy, Lab. RAG-PMM, CNET, 196 rue de Paris, 92220 Bagneux, France. (1944) D. 3e cycle, physics (U. de Paris Sud, 1970). Ingenieur (tel. 1 + 46 38 52 06). *Materials for telecommunications.*

Leclaire, Dr André Groupe de Cristallographie et de Chimie des Solides, U. de Caen, Esplanade de la Paix, 14032 Caen, France. (1944) Dsc. (U. de Caen, 1976). Chargé de rech. (tel. 31 94 81 40, ext. 3546). *Solvatation, channeling, crystallographic structures, bronzes.*

Lecomte Dr Claude. Lab. de Minéralogie- Cristallographie, Université de Nancy 1, Case officielle 140, 54037 Nancy Cedex, France. (1948) DSc, physics (U. de Nancy, 1979) Maître de conferences. (tel. 8 328 93 93, ext. 2404). *Electron densities, metalloporhyrins, thermal motion, bonding.*

Ledesert, Dr Mariannick. Lab. de Minéralogie- Cristallographie, ISMRA, U. de Caen, Esplanade de la Paix, 14032 Caen, France. (1934) DSc, physics (U. de Caen 1965) Chargé de rech. (tel. 31 94 81 40, ext. 35 09). *Crystal growth, structures.*

Lefebvre, Dr Simone. Chimie Métallurgique, CNRS, 15 rue Georges Urbain, 94400 Vitry sur Seine, France. (1937) DSc, physics (U. P.M. Curie, 1975) Maître-asst. (tel. 1 687 35 93). *Alloys, order- disorder, neutrons.*

Legros, Dr Jean-Pierre. Lab. de chimie de coordination, CNRS, 205 route de Narbonne, 31400 Toulouse, France. (1943) DSc,(1976). Chercheur Enseignant. (tel. 61 52 11 66). *Magnetism, charges densities, conductors.*

Lehmann, Dr Mogens. ILL, avenue des Martyrs, 38042 Grenoble Cedex, France. (1942) licence, (U. de Danemark, 1972) staff scientist. (tel. 76 48 70 89). *Neutrons, biochemistry, water, molecular interactions, instrumentation.*

Leligny, Mr Henri. Lab. de Cristallographie, U. de Caen, Esplanade de la Paix, 14032 Caen, France. (1941) D.3e cycle, crystallography (U. de Paris 6, 1971). Maître-asst. (tel. 31 81 59 10) *Hydrates, order- disorder, modulated structures.*

Lemoine, Mlle Pascale. Faculté de Pharmacie Paris Luxembourg, 4 av. de l'Observatoire, 75006 Paris, France. (1954) D.3e cycle (U. de Paris, 1979) Asst. (tel. 1 + 43 29 12 08, ext. 235). *Rare earths structures, mixed valences, electrical properties.*

Lepicard, Dr Geneviève, Lab. de Minéralogie- Cristallographie, U. de Nancy 1, BP 239, 54056 Vandoeuvre Cedex, France. (1939) DSc, physics (U. de Paris, 1978) Maître-asst. (tel. 83 28 93 93, ext. 22 68). *Structures.*

Leroy, Dr Bernard. Dépt. 1817, IBM France, BP 58, 91102 Corbeil-Essonnes Cedex, France. (1945) DSc, physics (U. de Paris, 1983). Ing. (tel. 1 + 60 88 50 10). *Silicon, defects, diffusion.*

Letort, Mr Marc Yves. 3 av. Paul Doumer, 75016 Paris, France. (1898) Ing. LSc, mineralogy (Ecole Centrale Paris, U. de Paris, 1920). Retired. (tel. 1 + 45 53 29 81). *Aluminum silicates.*

Levalois, Mr Marc. Lab. de Cristallographie, U. de Caen, Esplanade de la Paix, 14032 Caen Cedex, France. (1951) D.3e cycle, crystallography (U. de Caen, 1975). Chercheur Enseignant, (tel. 31 94 81 40, ext. 3509). *Electron densities, cristallography.*

Levelut, Dr Anne-Marie. Lab. de Physique des Solides, U. Paris Sud, (Paris 11), Bat. 510, 91405 Orsay, France. (1936) DSc, physics (U. de Paris, 1968). Chargée de rech. CNRS. (tel. 1 + 69 41 53 94). *Liquid crystal, structures.*

Lewit-Bentley Dr Anita. EMBL, BP 156X, 38042 Grenoble Cedex, France. (1948) RNDr, physical chemistry (Charles U., Prague, Czechoslovakia, 1972). Sci. (tel. 76 48 72 75). *Molecular biology, macromolecular strucutures.*

Lifchitz, Mr Alain. Lab. de Minéralogie- Cristallographie, tour 16, U. Pierre et Marie Curie (Paris 6), 4 place Jussieu, 75230 Paris Cedex 05, France. (1946) D.3e cycle, physics (U. de Paris 6, 1974). Chargé de rech. CNRS (tel. 1 + 43 36 25 25, ext. 45 84). *Biologically interesting compounds, computing.*

Loiseleur, Dr Henri. Lab. de Chimie Analytique, Université de Lyon 1, 43 bd du 11 novembre 1918, 69622 Villeurbanne, France. (1939) DSc, physical chemistry (U. de Lyon, 1965). Maître-asst. (tel. 78 89 81 24, ext. 35 27). *Coordination chemistry.*

Longueville, Mr Willy. UER Physique fondamentale, U. Lille 1, 59655 Villeneuve d'Ascq Cedex, France. (1937) D. 3e cycle, Solid state physics (U. de Lille 1, 1970). Maître-asst. (tel. 20 91 92 22, ext. 23 24). *Plastic crystals, Raman spectroscopy, neutron diffraction, phase transitions.*

Louis, Mr Remy. Inst. de Chimie, U. Louis Pasteur, 1 rue Blaise Pascal, BP 296R8, 67008 Strasbourg Cedex, France. (1943) DSc, physics (U. Louis Pasteur, 1976) Maître-asst. (tel. 88 61 48 02). *Bio-organic structures, X-Ray cristallography.*

Loupias, Dr Geneviève. Lab. de Minéralogie- Cristallographie, U. P.M.Curie, 4 place Jussieu, 75230 Paris Cedex 05, France. DSc, physics (U. de Paris 6, 1977). Maître-asst. (tel. 1 + 43 36 25 25, ext. 45 84). *Momentum density, electron distributions, inelastic scattering.*

Lucas, Prof. Jacques. Lab. de Chimie Minérale C, Université de Rennes Beaulieu, 35042 Rennes Cedex, France. DSc, physics (U. de Rennes, 1964). Prof., (tel. 99 36 48 15). *Glasses, fluorescence, simulation.*

Luzzati, Dr Vittorio. Centre de Génétique Moléculaire, CNRS, 91190 Gif sur Yvette, France. (1923) DSc, physics (U. de Paris, 1950). Dir. de rech., CNRS. (tel. 1 + 69 07 78 28, ext. 825). *Biology, liquid crystals, small angle scattering.*

Malgrange, Prof. Cécile. Lab. de Minéralogie- Cristallographie, tour 16, U. Pierre et Marie Curie (Paris 6), 4 place Jussieu, 75230 Paris Cedex 05, France. (1939) DSc, physics (U. de Paris, 1967). Prof. (tel. 1 + 43 36 25 25, ext. 52 15). *X-ray topography, ferroelectric crystals, neutron topography.*

Marezio, Dr Massimo. Lab. de Cristallographie, CNRS, BP 166X, 38042 Grenoble Cedex, France. Dir. de rech. Dir. du lab. (tel. 76 96 98 37). *Neutrons.*

Marsau, Dr Pierre Michel. Lab. de Cristallographie et Physique Cristalline, U. de Bordeaux 1, 351 cours de la Libération, 33405 Talence, France. (1937) DSc, physics (U. de Bordeaux 1, 1972). Maître-asst. (tel. 56 80 84 50, ext. 219). *Biological structures, molecular graphics.*

Mason, Dr Sax Anton. ILL, BP 156X, 38042 Grenoble. (1946) PhD, chemistry (Melbourne U., Australia, 1971). Staff scient. (tel. 76 48 70 67). *Protein crystallography, neutron diffraction, area detectors, protein dynamics.*

Massaux, Dr Michel Louis. Lab. de Physique des Matériaux, U. de Clermont II, Les Cezeaux, 24 avenue des Landais, 63170 Aubière, France. (1934) DSc, physics (U. de Clermont-Ferrand, 1957). Maître-asst. (tel. 73 26 41 10, ext. 32 22). *Organometallic crystals.*

Meinnel, Prof. Jean. U. de Rennes 1, Groupe de Physique Cristalline, avenue du Gal Leclerc, 35031 Rennes, France. (1926) DSc, physics (U. de Paris, 1958). Prof. (tel. 99 36 48 15, ext. 22 34). *Molecular dynamics, crystal structures, microelectronics.*

Meresse, Dr Alain. Lab. de Cristallographie, U. de Bordeaux 1, 351 Cours de la Libération, 33405 Talence, France. (1943) Dsc physics (U. de Bordeaux, 1981). Maître-asst. (tel. 56 80 76 09, ext. 299). *Organic solids, low dimensionnality, desorder.*

Mérigoux, Prof. Henri. Lab. de Cristallographie, U. de Franche Comté 25030 Besançon, France. (1935) DSc, physics (U. de Paris, 1967). Prof. (tel. 84 53 81 22). *Cristallography, defects, crystal growth.*

Messager, Dr Jean-Claude. U. de Rennes 1, Groupe de Physique Cristalline, avenue du Gal Leclerc, 35031 Rennes, France. (1940) DSc, physics (U. de Rennes, 1976). Maître-asst. (tel. 99 36 48 15). *Organic crystal structures, lattice dynamics.*

Metz, Dr Bernard Jean Claude. Inst. de Chimie, 67008 Strasbourg, France. (1941) DSc, chemical crystallography (U. de Strasbourg, 1972). Maître-asst. *Mineral, organic structures.*

Michel, Prof. Pierre. Lab. de Minéralogie- Cristallographie, U. Claude Bernard (Lyon 1), 43 bd du 11 novembre 1918, 69622 Villeurbanne Cedex, France. (1925) DSc, physics (U. de Strasbourg, 1958). Prof., Dir. du lab. (tel. 78 89 81 24, ext. 32 21). *Crystal structures, LEED, HEED, RHEED, polytypism, surface physics.*

Minari, Prof. Fernand Henri. Lab. de Physique Cristalline, U. Aix-Marseille 3, Centre St Jérome, rue Henri Poincaré, 13397 Marseille Cedex 4, France. (1934) DSc, physics (U. d'Aix-Marseille, 1971). Prof. (tel. 91 98 90 10, ext. 311). *Crystal defects, microdeformation.*

Moineau, Mr Hervé Lab. d'Electronique et Rayons X, U. de Reims, Faculté des Sciences, Moulin de la Housse, BP 347, 51062 Reims, France. (1938) D. 3e cycle, solid state physics (U. de Strasbourg, 1968). Maître-asst. (tel. 26 85 23 24, ext. 254). *X-ray diffraction, minerals, deformation, TSEE.*

Monier, Prof. Jean-Claude. ISMRA, U. de Caen, UER des Sciences, Esplanade de la Paix, 14032 Caen, France. (1925) DSc, crystallography (U. de Paris, 1953). Prof. (tel. 31 94 81 40, ext. 3264). *Crystal growth, inorganic structures, hydrates, oxydes.*

Montmory, Mrs Marie-Claire. Lab. de Cristallographie, CNRS Grenoble, 25 av. des Martyrs, BP 166X, Centre de Tri, 38042 Grenoble Cedex, France. (1934) Agr., (1958). Maître-asst. (tel. 76 87 22 11, ext. 443). *Chemical crystallography, magnetic structures.*

Montmory, Dr Robert. Lab. de Physique des Précipitations, U. Scientifique et Médicale de Grenoble, CNRS, BP 53X, 38041 Grenoble Cedex, France. (1927) DSc, atmospheric physics (U. de Clermond-Ferrand, 1959). Maître de rech. CNRS, (tel. 76 54 81 57, ext. 405). *Ice nucleation, precipitation enhancement, atmospheric pollution.*

Moras, Dr Denis Lab. de Cristallographie Biologique, IBMC, 15 rue Descartes, 67084 Strasbourg Cedex, France. (1944) DSc, chemistry (U. Louis Pasteur, 1971) Maître de rech. (tel. 88 61 02 02, ext. 367). *Biology, structures.*

Moreau, Prof. Jean-Michel. Lab. de Structure de la Matière, U. de Savoie, 9 rue de l'Arc en Ciel, BP 908, 74019 Annecy le Vieux Cedex, France. (1944) Dsc physics (U. de Grenoble, 1976) Prof. (tel. 50 23 29 93). *Metallic alloys, magnetism.*

Morlon, Mr Bernard. Lab. des Dielectriques, Faculté des Sciences de Mirande, BP 138, 21100 Dijon, France. (1936) Agr. (1962). Maître-asst. (tel. 80 66 64 13, ext. 746). *Ferroelectricity, anomal scattering.*

Mornon, Dr Jean-Paul. Lab. de Minéralogie- Cristallographie, tour 16, U. Pierre et Marie Curie (Paris 6), 4 place Jussieu, 75230 Paris Cedex 05, France. (1942) DSc, crystallography (U. de Paris, 1969). Maître de rech. CNRS. (tel. 1 + 43 36 25 25, ext. 52 36). *Proteins, steroids, organic structures, molecular graphics.*

Mosset, Dr Alain. Lab. de Chimie de Coordination, CNRS, 205 route de Narbonne, 31030 Toulouse, France. (1946) D.Sc., physics (U. de Toulouse, 1979). Asst. (tel. 61 52 11 66, ext. 273). *Amorphous materials, complexes, heterometallic chains.*

Naudon, Dr André Lab. de Métallurgie Physique, Faculté des Sciences de Poitiers, 40 av. du recteur Pineau, 86022 Poitiers, France. (1939) DSc, solid state physics (U. de Poitiers, 1971). Maître de rech. CNRS. (tel. 49 46 26 30, ext. 361). *Physical metallurgy, glasses, SAS of X-rays.*

Neuman, Mr Alain. U. Paris Nord (Paris 13), UER de Médecine et Biologie Humaine, 74 rue Marcel Cachin, 93000 Bobigny, France. (1943) D.3e cycle, chemistry (U. de Paris, 1971). Maître-asst. (tel. 1 + 48 33 02 13, ext. 232). *Crystal chemistry, biomolecular crystals.*

Nguyen, Prof Huy Dung. Lab. de Physique, Faculté de Pharmacie, 4 av. de l'Observatoire, 75006 Paris, France. (1939) Dsc (U. de Paris, 1972). Prof. (tel. 1 + 43 29 12 08, ext. 235). *Polytypism, order-disorder, modulated structures.*

Oumous, Mr Hassan. Lab. de Minéralogie Cristallographie, Université de Nancy 1, BP 239, 54506 Vandoeuvre les Nancy Cedex, France. (1956) DEA, physics (U. de Nancy, 1981). Student. *X-Ray diffraction, structures, magnetism.*

Pannetier, Dr Jean. ILL, avenue des Martyrs, BP 166X, 38042 Grenoble Cedex, France. (1947) DSc, physics (U. de Rennes, 1974) Staff scient. (tel. 76 48 70 91, telex 320621 F). *Neutron scattering, solid state chemistry.*

Pascard, Dr Claudine. Inst. de Chimie des Substances Naturelles, CNRS, Cristallochimie, 91190 Gif sur Yvette, France. (1929) DSc, physics (U. de Paris, 1960). Dir. de rech., CNRS. (tel. 1 + 69 07 78 28, ext. 605). *Biologically interesting compounds, structures.*

Pauthenet, Prof. René Service des champs intenses, CNRS, Institut National Polytechnique, BP 166X, 38042 Grenoble Cedex, France. (1925) Dsc physics (1956). Prof. (tel. 76 96 98 37). *Magnetism, neutrons, intense fields.*

Payan, Dr Françoise. Centre de Recherche sur les Mécanismes de la Croissance Cristalline, campus Luminy, case 913, 13288 Marseille Cedex 2, France. (1941) DSc, physics (U. de Provence, 1973). Chargée de rech., CNRS. (tel. 91 41 01 52).

Pèpe, Dr Gérard. Centre de Recherche sur les Mécanismes de la Croissance Cristalline, campus Luminy, case 913, 13288 Marseille Cedex 2, France. (1945) DSc, physics (U. de Provence, 1976). Chargé de rech. CNRS. (tel. 91 41 01 52). *Macromolecules, modelisation, molecular graphics.*

Pérez, Mr Serge. C.E.R.M.A.V., C.N.R.S., 38402 St. Martin D'Héres, France. (1947) DSc, physics (U. de Grenoble, 1978). Res. scient. (tel. 76 54 11 45). *Polymers, oligosaccharides, crystal structures, computer modeling, conformational analysis.*

Perrin, Dr Monique. Lab. de Minéralogie- Cristallographie, U. Claude Bernard (Lyon 1), 43 bd du 11 novembre 1918, 69622 Villeurbanne Cedex, France. (1936) DSc, crystallography (U. de Lyon, 1974). Maître de conf. (tel. 78 89 81 24). *Organic solid state, polymorphism.*

Petiau, Prof. Jacqueline. Lab. de Minéralogie- Cristallographie, tour 16, U. Pierre et Marie Curie (Paris 6), 4 place Jussieu, 75230 Paris Cedex 05, France. (1937) DSc, physics (U. de Paris, 1966). Prof. (tel. 1 + 43 36 25 25, ext. 45 84). *Solid state physics, ferroelectricity, EXAFS.*

Petipas, Prof. Claude. UA 808, Faculté des Sciences, U. de Haute Normandie, 76130 Mont Saint Aignan, France. (1936) DSc, physics (U. de Paris Sud, 1969). Prof. (tel. 35 98 28 50). *Small angle X-ray scattering, colloids, sedimentation, micellar solutions.*

Petroff, Prof. Jean-François. Lab. de Minéralogie- Cristallographie, tour 16, U. Pierre et Marie Curie (Paris 6), 4 place Jussieu, 75230 Paris Cedex 05, France. (1935) DSc, physics (U. de Paris, 1971) Prof. Dir. du lab. (tel. 1 + 43 36 25 25, ext. 52 16). *Crystal physics, synchrotron radiation, X-ray topography, crystal defects.*

Pezerat, Dr Henri. Lab. de Chimie des Solides, U. Pierre et Marie Curie (Paris 6), 4 place Jussieu, 75230 Paris Cedex 05, France. (1928) DSc, physics (U. de Paris, 1967). Chercheur. CNRS. (tel. 1 + 43 36 25 25, ext. 55 60). *Silicates, crystal chemistry.*

Philibert, Prof. Jean. Lab. de Métallurgie Physique, Institut de Sciences des Matériaux, Université Paris Sud, 91405 Orsay Cedex, France. (1928) DSc, physics (U. de Paris, 1955). Prof., (tel. 1 + 69 41 70 20). *Crystal defects, solid state diffusion, materials.*

Pierrot, Dr Marcel. Service de Cristallochimie, Faculté des Sciences de Saint-Jerome, rue Henri Poincaré, 13397 Marseille Cedex 13, France. (1938) DSc, physics (U. d'Aix-Marseille, 1968). Maître de rech. CNRS. (tel. 91 98 32 08). *Diffraction, proteins, crystallochemistry, structures.*

Poljak, Prof. Roberto J. Lab. d'Immunologie Structurale, Inst. Pasteur, 25 rue du Dr Roux, 75724 Paris Cedex 15, France. (1932) DSc, (U. de la Plata, 1956). Prof. (tel. 1 + 43 06 19 19, ext. 3139). *Immunology, biology, crystallography.*

Pouget, Dr. Lab. de Physique des Solides, U. Paris Sud, (Paris 11), Bat. 510, 91405 Orsay, France. (1947) Dr d'Etat, solid state physics (U. Paris Sud, 1974). (tel. 1 + 69 41 60 46, telex FAC ORS 692166F). *X-ray diffuse scattering, neutron scattering, one dimensional conductors, conducting polymers.*

Precigoux, Dr Gilles. Lab. de Cristallographie, U. de Bordeaux 1, 351 Cours de la Libération, 33405 Talence, France. (1946) Dsc physics, (U. de Bordeaux 1, 1978). Chargé de rech. (tel. 56 80 84 50, ext. 271). *X-ray diffraction, biological molecules, peptids.*

Primot, Mr Jacques. CNET, BAG-PMM, 196 rue de Paris, 92220 Bagneux, France. (1931) Ing. (tel. 1 + 46 38 52 06). *Microstrucutures, telecommunications, optics.*

Protas, Prof. Jean. Lab. de Minéralogie- Cristallographie, U. de Nancy 1, BP 239, 54506 Vandoeuvre Cedex, France. (1932) DSc, physics (U. de Paris, 1959). Prof. (tel. 83 28 93 93, ext. 2264, 2266). *Cristallochemistry, organic, inorganic crystals, computing.*

Pulou, Prof. Raymond. Lab. de Minéralogie et Cristallographie, U. Paul Sabatier, 118 route de Narbonne, 31062 Toulouse, France. (1920) DSc, physics (U. de Toulouse, 1949). Prof. *Crystal physics, morphology.*

Quéré, Prof. Yves. Métallurgie, CEN, BP 6, 92 Fontenay aux Roses, France. (1931) DSc, Physical metallurgy (U. d'Orsay, 1954). Maître de conf. Ecole Polytechnique, Chef de section CEA. (tel. 1 + 46 57 13 26, ext. 4149). *Defects, radiation damage, channelling.*

Rambaud, Dr Joëlle. Faculté de Pharmacie, U. de Montpellier 1, Ave. Ch. Fiahault, 34000 Montpellier, France. (1942) DSc, physical chemistry (U. de Montpellier, 1980) Maître de conf. (tel. 67 63 53 60, ext. 446). *X-ray diffraction, polymorphism, drugs, solid state.*

Rees, Dr Bernard. Lab. de Cristallographie Biologique, IBMC, 15 rue Descartes, 67084 Strasbourg Cedex, France. (1940) DSc, chemistry (U. de Strasbourg, 1969). Maître de rech. CNRS. (tel. 88 61 48 30, ext. 369). *Electron density.*

Regourd, Dr Micheline. Microstructures, CERILH, 23 rue de Cronstadt, 75015 Paris, France. (1931) DSc, physics (U. de Paris, 1964). Chef de Dépt. (tel. 1 + 45 32 58 40, ext. 229). *Binders (hydraulic), structures, activity, phase transitions.*

Renaud, Prof. Michel. Lab. de Physicochimie Structurale, Université Paris 12, avenue du Général de Gaulle, 94010 Creteil, France. (1932) DSc, physics (U. de Paris, 1959). Prof. (tel. 1 + 48 98 92 24). *Biologic structures, polymorphism, proteins.*

Renouprez, Dr Albert Jean. Inst. de Catalyse, CNRS, 2 avenue Einstein, 69626 Villeurbanne, France. (1937) Dr (U. de Lyon). Dir. de rech., CNRS. (tel. 79 93 34 71, ext. 265). *Small angle X-ray and neutron scattering, metal grains, EXAFS.*

Rérat, Dr Claude. Lab. de Cristallographie, CNRS, 1 place A. Briand, 92190 Bellevue, France. (1924) DSc, physics (U. de Paris, 1959). Dir. (tel. 1 + 45 34 75 50, ext. 22 28). *Biologically interesting compounds, structures.*

Ricard, Mr Jean Henri. Centre de Rech., Ugine Kuhlmann, 38560 Jarrie, France. (1928) LSc, physics (Lyon, Grenoble, 1961). Chef du service monocristaux. (tel. 76 68 16 11). *Chloride, silicon.*

Riche, Dr Claude. Inst. de Chimie des Substances Naturelles, CNRS, Cristallochimie, 91400 Gif sur Yvette, France. (1939) DSc, physical chemistry (U. de Paris, 1972). Maître de rech. CNRS. (tel. 1 + 69 07 78 28, ext. 605). *Natural products, direct methods.*

Rimsky, Dr Alexandre. Lab. de Minéralogie- Cristallographie, tour 16, U. Pierre et Marie Curie (Paris 6), 4 place Jussieu, 75230 Paris Cedex 05, France. (1922) DSc, (U. de Paris, 1958). Maître de rech. CNRS. (tel. 1 + 43 36 25 25). *Cristallography, biologically interesting compounds, molecular graphics, instrumentation, computing.*

Risler, Dr Jean-Loup. Centre de Génétique Moléculaire, CNRS, 91190 Gif sur Yvette, France. (1943) DSc, biophysics (U. de Paris, 1973). Chargé de rech. (tel. 1 + 69 07 78 28, ext. 820). *Protein crystallography.*

Robert, Mr Marc. Lab. de Physique, Faculté de Pharmacie, 4 avenue de l'Observatoire, 75270 Paris Cedex 06, France. (1953) D. 3e cycle (U. de Paris, 1981). Asst. (tel. 1 + 43 29 12 08, ext. 340). *Structures, scattering, electron microscopy.*

Rolland, Mr Guy. CEA-CENG, Micro Electronique Physique - LETI, av. des Martyrs, BP 85X, 38041 Grenoble Cedex, France. (1947) D. 3e cycle, physics (U. de Grenoble, 1973). Ing. (tel. 76 88 44 00, ext. 31 43). *Solid state physics, characterization, X-Ray topography, X-Ray diffraction.*

Roth, Dr Michel. ILL, avenue des Martyrs, BP 166X, 38042 Grenoble Cedex, France. (1939) DSc, metal physics (U. de Grenoble, 1969) Staff scient. (tel. 76 48 71 80). *Material science, neutron diffraction, protein crystallography.*

Rose, Dr Jean. 3 rue Jules Cousin, 75004 Paris, France. (1915) DSc, physics (U. de Paris, 1947). *Scientific culture promotion.*

Rossat-Mignod, Dr Jean. Dépt. de Rech. Fondamentales, Commissariat à l'Energie Atomique, ave. des Martyrs, 38041 Grenoble Cedex, France. (1944) Dr d'Etat, magnetism and neutron scattering (U. de Grenoble, 1972). Head, Magnetism et Diffraction Neutronique Lab. (tel. 76 88 44 00, ext. 3041, telex 320-323). *Magnetism, rare earth and actinide compounds, neutron scattering, elastic and inelastic neutron studies, superconductivity, low dimensional magnetic systems.*

Roult, Mr Georges. Dépt. de Rech. Fondamentales SPH/S, Centre d'Etudes Nucléaires de Grenoble (CEA-CENG), av. des Martyrs, BP 85X, 38041 Grenoble Cedex, France. (1930) D.ét.sup., crystal growth (U. de Paris, 1954). Head, TOF neutron diffraction group (tel. 76 88 44 00, ext. 39 30). *Time of flight neutron diffraction, high pressure, high temperature, ceramics*

Rousseaux, Dr Françoise. Lab. de Cristallographie, UA 810, Faculté des Sciences, rue de Chartres, 45046 Orléans Cedex, France. (1940) D. 3e cycle, crystallography (U. de Paris, 1968). Chargé de rech. CNRS. (tel. 38 63 22 16, ext. 695). *Lamellar structures, graphite, phase transitions.*

Saludjian, Mr Pedro. Lab. d'Immunologie Structurale, Inst. Pasteur, 25 rue du Dr Roux, 75724 Paris Cedex 15, France. (1931) LSc. (U. de Buenos Aeros, 1958) chercheur. (tel. 1 + 43 06 19 19, ext. 3136). *Macromolecules, physical chemistry, cristallography.*

Saul, Mr Frederic. Lab. d'Immunologie Structurale, Inst. Pasteur, 25 rue du Dr Roux, 75724 Paris Cedex 15, France. (1947) BSc, (U. of Maryland, 1970) chercheur (tel. 1 + 43 06 12 12, ext. 3135). *Macromolecules, refinement.*

Sauvage- Simkin, Mme Michèle. Lab. de Minéralogie- Cristallographie, U. Pierre et Marie Curie (Paris 6), 4 place Jussieu, 75230 Paris Cedex 05, France. (1941) DSc, physics (U. de Paris 6, 1968). Maître de rech. (tel. 1 + 43 36 25 25, ext. 52 15). *X-ray topography, heterojunctions, surfaces, synchrotron.*

Schiller, Dr Claude. Semiconductors, L.E.P., 3 av. Descartes, 91600 Limeil-Brévanne, France. (1940) DSc, physics (U. de Paris, 1967). Ing. (tel. 1 + 69 25 39 10, ext. 360). *Crystal defects, X-ray topography, scanning electron microscopy.*

Schlenker, Prof. Michel. Lab. Louis Néel, CNRS, BP 166, 38042 Grenoble Cedex, France. (1940) DSc, physics (U. de Grenoble, 1970). Prof. (tel. 76 88 10 92). *Magnetic structures, topography, neutrons, electron microscopy.*

Sfez, Mr Gérard. Lab. de Physique, Faculté de Pharmacie, 4 avenue de l'Observatoire, 75006 Paris, France. (1940) D.3e cycle, Ingénieur. (tel. 1 + 43 26 26 80, ext. 235). *Direct methods, proteins, computing.*

Simon, Dr Jean-Paul Henri Maurice. Lab. de Thermodynamique et Physico Chimie Metallurgiques, BP 75 Domaine Universitaire, CNRS - INPG, 38402 St. Martin - D'Héres, France. (1945) DSc, physique (Ecole Nat. Supérieure de Mines de Paris, 1968) Chargé de rech. (tel. 76 54 41 27, ext. 513). *Phase transformations, unmixing, Al alloys, transmission electron microscopy, anomalous small angle X-ray scattering.*

Stansfield, Dr Robert Frank David. ILL, BP 156X, 38042 Grenoble, France. (1954) PhD, crystallography (U. Bristol, 1979). Chercheur. (tel. 76 48 70 90, telex 320621 F). *Crystal structure determination, small molecules, proteins, X-ray and neutron diffraction, position sensitive detectors.*

Stora, Dr Cécile. Lab. de Cristallochimie, tour 44, U. Pierre et Marie Curie (Paris 6), 4 place Jussieu, 75230 Paris Cedex 05, France. (1907) DSc, physics (U. de Paris, 1937). Dir. de rech. (tel. 1 + 43 36 25 25, ext. 55 28). *Stereochemistry, biological molecules.*

Surcouf, Dr Evelyne. Lab. de Minéralogie- Cristallographie, tour 16, U. Pierre et Marie Curie (Paris 6), 4 place Jussieu, 75230 Paris Cedex 05, France. (1949) DSc, physics (U. de Paris 6, 1982). Asst. (tel. 1 + 43 36 25 25). *Biologically interesting compounds, structures.*

Tasset, Dr Francis Joseph Emmanuel. ILL, avenue des Martyrs, BP 166X, 38042 Grenoble Cedex, France. (1944) These d'etat, physics (U. de Grenoble, 1975) Staff scient. (tel. 76 48 70 59). *Polarized neutron diffraction, spin density measurement, low temperature crystallography.*

Taupin, Dr Daniel Gilbert Roger, Lab. de Physique des Solides, Centre Universitaire, Bat. 510, 91400 Orsay, France. (1936) DSc (U. de Paris Sud). Maître de rech., CNRS. (tel. 1 + 69 41 60 79). *Computer science crystallographic applications, probabilities, statistics, protein crystallography.*

Théobald, Dr François Roland. Dépt. de Chimie, U. de Besançon, Faculté des Sciences, La Bouloie, 25030 Besançon, France. (1942) DSc, physics (U. de Besançon, 1975). Maître-asst. (tel. 81 53 81 22, ext. 366). *Structures, heterogeneous catalysis, oxydes, lattice energy.*

Thierry, Dr Jean-Claude. Lab. de Cristallographie Biologique, IBMC, 15 rue Descartes, 67084 Strasbourg Cedex, France. (1942) DSc, Maître de rech. (tel. 88 61 02 02, ext. 367). *Biology, nucleic acids, proteins.*

Thomas-David, Prof. Germaine. U. Claude Bernard (Lyon 1), UER Chimie Biochimie, 43 bd du 11 novembre 1918, 69621 Villeurbanne, France. (1928) DSc, chemistry (U. de Lyon, 1960). (tel. 78 52 07 04, ext. 35 24). *Metallic compound structures.*

Thozet, Mr Alain Maurice. Lab. de Minéralogie- Cristallographie, U. Claude Bernard (Lyon 1), 43 bd du 11 novembre 1918, 69622 Villeurbanne Cedex, France. (1941) D. 3e cycle, physics (U. de Lyon, 1971). Maître-asst. (tel. 78 89 81 24, ext. 32 19). *Organic crystal structures, computing.*

Tomas, Dr Alain. Lab. de Physique, Faculté de Pharmacie de Paris, 4 avenue de l'Observatoire, 75006 Paris, France. (1946) DSc., physics, (U. de Paris 6, 1979). Maître de conf. (tel. 1 + 43 29 12 08 , ext. 340). *Sulfides, electron microscopy, disorder.*

Tougard, Dr Pierre Henri. Dépt. d'immunologie, Inst. Pasteur, 25 rue du Dr Roux, 75015 Paris, France. (1942) DSc, crystal chemistry (U. de Paris 6, 1974). Chargé de rech. CNRS. (tel. 1 + 43 06 19 19). *Crystallochemistry, molecular structures.*

Toupet, Mr Loic. Physique Cristalline (ERA 015), Faculté des Sciences de Rennes, Campus de Beaulieu, 35042 Rennes Cedex, France. (1949) D.3e cycle (U. de Rennes, 1976). Ingénieur (tel. 99 36 48 15, ext. 2247). *Instrumentation, structures, X-ray scattering.*

Tran Huu Dau, Mrs Marie-Elise. Lab. de Chimie des Substances Naturelles, CNRS, 91190 Gif sur Yvette, France. (1938) D.3e cycle, chemistry (U. de Paris Sud, 1973) Ing. (tel. 1 + 69 07 78 28, ext. 605). *Biological structures.*

Tran, Dr Vinh. Lab. de Physicochimie Moléculaire, Institut National de Recherches en Agronomie, Chemin de la Géraudère, 44072 Nantes, France. (1952) DSc., physics (U. de Grenoble, 1983), staff member. (tel. 40 76 23 64). *Amylaces, powder diffraction, molecular graphics.*

Tranqui, Dr Duc. Lab. de Cristallographie, CNRS, BP 166X Centre Tri, 38042 Grenoble Cedex, France. (1938) DSc., physics (U. de Grenoble, 1968). Chercheur CNRS. (tel. 76 96 98 37, ext. 436). *Structures, crystallography, electron density.*

Tronc, Dr Elisabeth. Lab. de Chimie Appliquée État Solide, Ecole Nationale Supérieure de Chimie, 11 rue P. et M. Curie, 75231 Paris Cedex 05, France. (1943) DSc, (1972). Chargé de rech. CNRS. (tel. 1 + 43 36 25 25, ext. 38 28). *Inorganic compounds, crystal growth, physical properties.*

Tsoucaris, Dr Georges. Lab. de Physique, U. Paris Sud (Paris 11), Centre Pharmaceutique, rue J.B. Clément, 92290 Chatenay-Malabry, France. DSc, physics (U. de Paris, 1959). Dir. de rech., CNRS; Prof., Ecole Polytechnique. (tel. 1 + 46 61 33 25, ext. 633). *Direct methods, organic crystal structures, reactivity.*

Turco, Prof. Guy Henri Robert. Lab. de Pétrologie-Minéralogie, Faculté de Nice, av. Valrose, 06034 Nice Cedex, France. (1927) DSc, mineralogy (U. de Paris, 1962). Prof., Dir. (tel. 93 51 91 00, ext. 469). *Mineral inclusions, synthesis, metamorphic rocks.*

Vettier, Dr Christian. ILL, av. des Martyrs, 156X, 38042 Grenoble Cedex, France. (1946) DSc., physics (U. de Grenoble, 1975). Staff member. (tel. 76 48 71 65). *Neutrons scattering, high pressures, magnetism, incommensurate phases.*

Vicat, Dr Jean. Lab. de Cristallographie, CNRS, BP 166X, 38042 Grenoble Cedex, France. (1941) DSc, physics (U. de Grenoble 1977). Maître-asst. (tel. 76 96 98 37, ext. 437). *Diffraction, electrons density, conductivity, ionic order.*

Vidal, Dr Geneviève. Lab. de Dynamique des Phases Condensées, U. des Sciences et Techniques du Languedoc, place Eugène Bataillon, 34060 Montpellier Cedex, France. (1944) DSc, solid state physics (U. de Montpellier, 1975). Maître de conf. (tel. 67 63 91 44, ext. 425). *Multipolar momentum, physical properties.*

Vigneron, Dr Françoise. Lab. Léon Brillouin, C.E.N. Saday, 91191 Gif sur Yvette Cedex, France. *Neutron diffraction, magnetic structures.*

Wade, Dr Richard Harry, Groupe Structures, Service de Physique, DRF/G, Centre D'Etudes Nucleaires de Grenoble, CENG, 38041 Grenoble Cedex, France. (1937) PhD, metal physics (Cambridge U., UK, 1962). Res. scient. (tel. 76 88 44 00, ext. 40.24). *Electron microscopy, structural biology.*

Waintal, Dr Alex. Lab. de Louis Neel, CNRS, 25 avenue des Martyrs, BP 166X, Centre de Tri, 38042 Grenoble Cedex, France. (1934) DSc, physics (U. de Grenoble, 1969). Maître de conf. (tel. 76 96 98 37, ext. 41904). *Diffraction, high pressure, ultrosonics, group theory applications.*

Walter, Dr Hannes Ulrich. Microgravity office/STS, European Space Agency, 8-12 rue Mario Nikis, F-75738 Paris, France. (1938) Ph.D., crystallography (U. Tübingen, 1969). Sr. scient. (tel. 1 + 2737319, telex ESA 202746). *Semiconductors, magnetic materials, superconductors.*

Weigel, Prof. Dominique Jean. Chimie Physique du Solide, Ecole Centrale, Grande Voie des Vignes, 92290 Chatenay-Malabry, France. (1929) DSc, physics (U. de Paris, 1960). Prof. (tel. 1 + 46 61 33 10, ext. 13 23). *Phase transitions, 4D symmetry, incommensurate structures.*

Weiss, Prof. Raymond. Inst. Le Bel, U. Louis Pasteur, 4 rue Blaise Pascal, 67084 Strasbourg, France. (1929) DSc, inorganic chemistry (U. de Nancy, 1959). Prof. (tel. 88 61 48 30, ext. 229). *Transition metals, molecular materials, metalloenzymes, X-Ray crystallography.*

Weulersse, Prof. Philippe. U. de Technologie, BP 233, 60206 Compiègne, France. (1937) DSc, physics (U. de Paris Sud, 1970). Prof. (tel. 4 420 45 02). *X-ray crystallography, materials.*

Wey, Prof. Raymond. Lab. de Chimie Minérale Générale, Ecole Supérieure de Chimie, 3 rue A. Werner, 68093 Mulhouse, France. (1926) DSc, physics (U. de Strasbourg, 1955). Prof. (tel. 89 42 70 20, ext. 57). *Clays, zeolites, silicates.*

Weyl, Dr Colette. Lab. de Physique des Solides, U. Paris sud (Paris 11), Bat. 510, 91405 Orsay, France. (1933) DSc, physical chemistry and solid state physics (U. de Paris, 1973). Maître-asst. (tel. 1 + 69 41 82 50, ext. 30 11). *Metal-insulator transition, organic conductors.*

Whuler, Dr Annick. Lab. de Minéralogie- Cristallographie, U. P.M.Curie, 75230 Paris Cedex 05, France. (1946) DSc, physics (U. P.M.Curie, 1978) Maître-asst. (tel. 1 + 43 36 25 25, ext. 5082) *Structures, polymers.*

Willaime, Prof. Christian. Centre Armoricain d'étude structurale des socles, U. de Rennes, avenue du Général Leclerc, 35042 Rennes Cedex, France. (1940) DSc, mineralogy and crystallography (U. de Paris 6, 1972). Prof. (tel. 99 36 48 15). *Minerals, deformation, defects, electron microscopy.*

Wintenberger, Dr Micheline. Lab. de Chimie Minérale, Faculté de Pharmacie, 4 avenue de l'Observatoire, 75270 Paris Cedex 06, France. (1929) DSc, (1962). Maître de rech. CNRS. (tel. 1 + 43 29 12 08, ext. 174). *Magnetism.*

Witz, Dr Jean. Lab. de Virologie, Inst. de Biologie Moléculaire et Cellulaire, 15 rue Descartes, 67084 Strasbourg, France. (1935) DSc, physics (U. de Strasbourg, 1964). Maître de rech. CNRS. (tel. 88 61 02 02, ext. 318). *Viruses, morphogenesis.*

Wyart, Prof. Jean. Lab. de Minéralogie- Cristallographie, U. Pierre et Marie Curie (Paris 6), 4 place Jussieu, 75230 Paris Cedex 05, France. (1902) DSc, physics (U. de Paris, 1933). Prof. Honoraire. (tel. 1 + 43 54 82 36). *Crystallography, mineralogy, synthesis.*

Zaccai, Dr Giuseppe. ILL, Ave. des Martyrs, BP 156X, F-38042 Grenoble Cedex, France. (1947) PhD, physics (U. Edinburgh, UK, 1972). Dir. de rech., CNRS. (tel. 76 48 70 46). *Proteins structure, nucleic acids, biological membranes, neutrons.*

Zarka, Dr Albert. Lab. de Minéralogie- Cristallographie, U. Pierre et Marie Curie (Paris 6), 4 place Jussieu, 75230 Paris Cedex 05, France. (1942) DSc, crystallography (U. de Paris, 1973). Chargé de rech. CNRS. (tel. 1 + 43 36 25 25, ext. 52 16). *Crystal defects, natural crystals, X-ray topography.*

Zelwer, Dr Charles Marcel. Centre de Génétique Moléculaire, CNRS, 91190 Gif sur Yvette, France. (1940) DSc, chemistry (U. de Paris, 1967). Chargé de rech. (tel. 1 + 69 07 78 28, ext. 820). *Biology, T-ARN synthesis, microdensitometry.*

GERMAN DEMOCRATIC REPUBLIC
DDR

Sub-Editor: **E. Höhne**

Notes

1. International telephone country code - 37.

2. The degrees conferred by the universities are the *Doctor habilitatus* (Dr.habil.) (comparable to DSc), *Doctor rerum naturalium* (Dr.rer.nat.) (equivalent to PhD), *Diplom-Chemiker* (Dipl.Chem.), *Diplom-Metallurge* (Dipl.Met.), *Diplom-Mineraloge* (Dipl.Min.), *Diplom-Physiker* (Dipl.Phys.), *Diplom-Kristallograph* (Dipl. Krist.), *Diplom-Mathematiker* (Dipl.Math.), *Diplom-Geologe* (Dipl.Geol.), (these seven equivalent to MSc). At the *Technische Universität (TU)* and *Technische Hochschulen (TH)* the degrees *Doctor Ingenieur* (Dr.Ing.), *Diplom Ingenieur* (Dipl.Ing.), and *Physik Ingenieur* (Phys.Ing.) can be obtained.

Abelmann, Mr Rolf-Ulrich. Akademie der Wissenschaften der DDR, Inst.für Festkörperphysik und Elektronenmikroskopie, DDR-4010 Halle, Am Weinberg 2, German Dem. Rep. Dipl.-Phys. *Crystallography.*

Adamski, Mrs Hannelore. Vereinigung für Kristallographie (VFK) in der GGW, DDR-1040 Berlin, Invalidenstrasse 43, German Dem. Rep. (1948). Dipl.Krist. *Ceramics, phase analysis.*

Albrecht, Prof. Günter. Friedrich-Schiller-Univ. Jena, Sektion Physik, DDR-69 Jena, Max-Wien-Platz 1, German Dem. Rep. (1930). Dr.habil. *Low temperature physics, solid state physics.*

Alex, Dr Volker. Akademie der Wissenschaften der DDR, Inst. für Physik der Werkstoffbearbeitung, DDR-1211 Falkenhagen, German Dem. Rep. (1939). Dr.rer.nat. *Crystallography.*

Alter, Dr Uwe. Vereinigung für Kristallographie (VFK) in der GGW, DDR-1040 Berlin, Invalidenstrasse 43, German Dem. Rep. (1946). Dr.rer.nat. *Real structures, crystal physics.*

Anders, Mr Rudolf. Vereinigung für Kristallographie (VFK) in der GGW, DDR-1040 Berlin, Invalidenstrasse 43, German Dem. Rep. (1932). Dipl.Phys. *Crystal growth, epitaxy.*

Andree, Mrs Anneliese. Vereinigung für Kristallographie (VFK) in der GGW, DDR-1040 Berlin, Invalidenstrasse 43, German Dem. Rep. (1951). Dipl.Krist. *Mineralogy.*

Andreeft, Prof. Alexander. Bergakademie Freiberg, Sektion Physik, DDR-9200 Freiberg, Gustav-Zeuner-Strasse 5, German Dem. Rep. (1932). Dr.habil. *Radiation damage, implantation.*

Andrehs, Dr Gerhard. Karl-Marx-Univ. Leipzig, Sektion Chemie, Fachrichtung Kristallographie, DDR-7030 Leipzig, Scharnhorststrasse 20, German Dem. Rep. (1933). Dr.rer.nat. *Technical petrography.*

Arnold, Mrs Christine. Akademie der Wissenschaften der DDR, Zentralinstitut für Anorganische Chemie, DDR-1199 Berlin, Rudower Chaussee 5, German Dem. Rep. (1950). Dipl.Krist. *Epitaxy, mineralogy.*

Arnold, Prof. Heinrich. Technische Hochschule Ilmenau, Sektion Physik, DDR-63 Ilmenau, German Dem. Rep. (1933). Dr.habil. *Semiconductors.*

Arnold, Mr Rolf. Technische Hochschule Carl Schorlemmer Merseburg, DDR-4200 Merseburg, Geusärstrasse, German Dem. Rep. (1947). Dipl. Min. *Crystallography.*

Backhaus, Dr Karl-Otto. Akademie der Wissenschaften der DDR, Zentralinstitut für physikalische Chemie, DDR-1199 Berlin, Rudower Chaussee 5, German Dem. Rep. (1936). Dr.rer.nat. *X-ray structure analysis.*

Balarin, Prof. Manfred. Akademie der Wissenschaften der DDR, Zentralinstitut für Kernforschung Rossendorf, DDR-8050 Dresden, PF-19, German Dem. Rep. (1934). Dr.sc. *Solid state physics.*

Barthel, Dr Johannes. Akademie der Wissenschaften der DDR, Zentralinstitut für Festkörperphysik und Werkstofforschung, DDR-8032 Dresden, Helmholtzstrasse 20, German Dem. Rep. (1931). Dr.habil. *Single crystal growth.*

Baumbach, Mr Manfred. Vereinigung für Kristallographie (VFK) in der GGW, DDR-1040 Berlin, Invalidenstrasse 43, German Dem. Rep. (1947). Dipl.Krist. *Real structures.*

Baumgärtel, Mr Rolf. Vereinigung für Kristallographie (VFK) in der GGW, DDR-1040 Berlin, Invalidenstrasse 43, German Dem. Rep. (1934). Dipl.-Min. *Crystal growth.*

Bautsch, Prof. Hans-Joachim. Humboldt-Univ. zu Berlin, Sektion Physik, DDR-1040 Berlin, Invalidenstrasse 43, German Dem. Rep. (1929). Dr.habil. *Crystallography, petrography.*

Becherer, Prof. Gerhard. Wilhelm-Pieck-Univ. Rostock, Sektion Physik, DDR-2520 Rostock, Universitätsplatz 3, German Dem. Rep. (1915). Dr.sc. *X-ray and crystal physics.*

Becker, Dr Claus. Akademie der Wissenschaften der DDR, Zentralinstitut für Werkstofforschung, DDR-8020 Dresden, Wittenbergstrasse 28, German Dem. Rep. (1933). Dr.rer.nat. *Crystallography.*

Becker, Mr Reinhardt. Verlag Junge Welt, Redaktion Wissenschaft und Forschung, DDR-108 Berlin, Kronenstrasse 30-31, German Dem. Rep. (1948). Dipl.Krist. *Crystallography.*

Beckmann, Prof. Günter. Ingenieurhochschule Zittau, DDR-88 Zittau, Strasse der Jungen Pioniere 2, German Dem. Rep. (1927). Dr.habil. *Crystal physics, X-ray structure analysis.*

Beier, Dr Wilfried. Martin-Luther Univ. Halle-Wittenberg, DDR-4020 Halle, Friedemann-Bach-Platz 6, German Dem. Rep. (1936). Dr.rer.nat. *Electron beam microanalysis.*

Berger, Dr Hans. Humboldt-Univ. zu Berlin, Sektion Physik, DDR-1040 Berlin, Invalidenstrasse 43, German Dem. Rep. (1940). Dr.rer.nat. *X-ray diffraction, real structures, crystal growth.*

Bergner, Mr Joachim. Vereinigung für Kristallographie (VFK) in der GGW, DDR-1040 Berlin, Invalidenstrasse 43, German Dem. Rep. (1930). Physik-Ing. *Crystallography.*

Bertram, Mrs Marion. Technische Univ. Dresden, Sektion Physik, DDR-8027 Dresden, Mommsenstrasse 13, German Dem. Rep. (1949). Dipl.Krist. *Real structures.*

Betzl, Dr Manfred. Akademie der Wissenschaften der DDR, Zentralinstitut für Kernforschung, DDR-8051 Dresden, Postfach 19, German Dem. Rep. (1933). Dr.rer.nat. *X-ray physics, structure analysis.*

Binas, Dr Horst. Vereinigung für Kristallographie (VFK) in der GGW, DDR-1040 Berlin, Invalidenstrasse 43, German Dem. Rep. (1934). Dr.rer.nat. *Crystal growth.*

Blankenburg, Dr Hans-Joachim. Bergakademie Freiberg, Sektion Geowissenschaften, DDR-9200 Freiberg, Brennhausgasse 14, German Dem. Rep. (1938). Dr.rer.nat. *Mineralogy, phase analysis.*

Bohm, Dr Joachim. Akademie der Wissenschaften der DDR, Zentralinstitut für Optik und Spektroskopie, DDR-1199 Berlin, Rudower Chaussee 5, German Dem. Rep. (1935). Dr.habil. *Crystal growth, real structures.*

Bornmann, Dr Horst. Vereinigung für Kristallographie (VFK) in der GGW, DDR-1040 Berlin, Invalidenstrasse 43, German Dem. Rep. (1936). Dr.rer.nat. *Crystal physics.*

Bornmann, Dr Peter. Vereinigung für Kristallographie (VFK) in der GGW, DDR-1040 Berlin, Invalidenstrasse 43, German Dem. Rep. (1945). Dr.rer.nat. *X-ray structure analysis.*

Bräutigam, Dr Gunter. Technische Hochschule Carl Schorlemmer Merseburg, DDR-4200 Merseburg, Geusärstrasse, German Dem. Rep. (1941). Dr.rer.nat. *Crystal growth, molten salts.*

Brand, Prof. Paul. Bergakademie Freiberg, Sektion Chemie, DDR-9200 Freiberg, Leipziger Strasse, German Dem. Rep. (1931). Dr.sc. *Crystal chemistry, structures.*

Brauer, Dr Karl-Heinz. Martin-Luther-Univ. Halle-Wittenberg, DDR-4020 Halle, Friedemann-Bach-Platz 6, German Dem. Rep. (1919). Dr.habil. *Crystallography.*

Brauny, Mr Siegfried. Technische Univ. Dresden, Sektion Physik, DDR-8027 Dresden, Salvador-Allende-Platz 3, German Dem. Rep. (1934). Dipl.Phys. *X-ray physics.*

Broosch, Mrs Erika. Vereinigung für Kristallographie (VFK) in der GGW, DDR-1040 Berlin, Invalidenstrasse 43, German Dem. Rep. (1953). Dipl.Krist. *Crystallography.*

Brückner, Dr Winfried. Akademie der Wissenschaften der DDR, Zentralinstitut für Festkörperphysik und Werkstofforschung, DDR-8032 Dresden, Helmholtzstrasse 20, German Dem. Rep. (1939). Dr.rer.nat. *Crystal physics, solid state physics.*

Brühl, Dr Hans-Gerd. Karl-Marx-Univ. Leipzig, DDR-7010 Leipzig, Linnestrasse 5, German Dem. Rep. (1944). Dr.rer.nat. *Crystallography.*

Brümmer, Prof. Otto. Martin-Luther-Univ. Halle-Wittenberg, DDR-4020 Halle, Friedemann-Bach-Platz 6, German Dem. Rep. (1920). Dr.habil. *X-ray physics, crystal physics.*

Bublitz, Mr Günter. Akademie der Wissenschaften der DDR, Inst.für Meereskunde, DDR-253 Warnemünde, Seestrasse 15, German Dem. Rep. (1934). Dipl.Min. *Sedimentation.*

Buchheiser, Dr Klaus. Karl-Marx-Univ. Leipzig, Sektion Chemie, Fachrichtung Kristallographie, DDR-7030 Leipzig, Scharnhorststrasse 20, German Dem. Rep. (1949). Dr.rer.nat. *Crystallography.*

Burkhardt, Mr Wolfgang. Vereinigung für Kristallographie (VFK) in der GGW, DDR-1040 Berlin, Invalidenstrasse 43, German Dem. Rep. (1949). Dipl.Krist. *Thermal analysis, glass chemistry.*

Butter, Prof. Ehrenfried. Karl-Marx-Univ. Leipzig, Sektion Chemie, DDR-701 Leipzig, Liebigstr. 18, German Dem. Rep. (1931). Dr.sc. *Crystal growth, epitaxy.*

Christoph, Mr Arthur. Vereinigung für Kristallographie (VFK) in der GGW, DDR-1040 Berlin, Invalidenstrasse 43, German Dem. Rep. (1921). Dipl.Chem. *Crystal growth.*

Däweritz, Dr Lutz. Akademie der Wissenschaften der DDR, Zentralinstitut für Elektronenphysik, DDR-102 Berlin, Neu Schönhauser Strasse 20, German Dem. Rep. (1943). Dr.rer.nat. *Crystal chemistry.*

Damaschun, Mr Ferdinand. Humboldt-Univ. zu Berlin, Sektion Physik, DDR-1040 Berlin, Invalidenstrasse 43, German Dem. Rep. (1950). Dipl.Krist. *Crystallography.*

Demus, Dr Dietrich. Martin-Luther-Univ. Halle-Wittenberg, DDR-4020 Halle, Mühlpforte 1, German Dem. Rep. (1935). Dr.habil. *Crystallography.*

Dietrich, Dr Burkhard. Friedrich-Schiller-Univ. Jena, Sektion Physik, DDR-69 Jena, Max - Wien - Platz 1, German Dem. Rep. (1939). Dr.rer.nat. *Crystal physics, real structures.*

Dörrfeld, Mr Hans-Georg. Akademie der Wissenschaften der DDR, Zentrum für wissenschaftlichen Gerätebau, DDR-1199 Berlin, Rudower Chaussee 6, German Dem. Rep. (1935). Dipl.Ing. *Crystal growth.*

Dörschel, Dr Jürgen. Akademie der Wissenschaften der DDR, Inst. für Physik der Werkstoffbearbeitung, DDR-1166 Berlin, Seestrasse 82, German Dem. Rep. (1942). Dr.rer.nat. *Real structures.*

Dornics, Mrs Monika. Akademie der Wissenschaften der DDR, Zentralinstitut für Anorganische Chemie, DDR-1199 Berlin, Rudower Chaussee 5, German Dem. Rep. (1949). Dipl.Krist. *Mineralogy, epitaxy, röntgenography.*

Dressler, Dr Ludwig. Friedrich-Schiller-Univ. Jena, Sektion Physik, DDR-69 Jena, Max - Wien - Platz 1, German Dem. Rep. (1944). Dr.rer.nat. *Crystallography, X-ray physics.*

Driesel, Dr Wolfgang. Akademie der Wissenschaften der DDR, Inst.für Festkörperphysik und Elektronenmikroskopie, DDR-4000 Halle, Weinberg 2, German Dem. Rep. (1946). Dr.rer.nat. *Real structures.*

Dünkel, Mr Lothar. Akademie der Wissenschaften der DDR, Zentralinstitut für physikalische Chemie, DDR-1199 Berlin, Rudower Chaussee 5, German Dem. Rep. (1942). Dipl.-Chem. *X-ray structure analysis.*

Eggers, Mr Peter. Vereinigung für Kristallographie (VFK) in der GGW, DDR-1040 Berlin, Invalidenstrasse 43, German Dem. Rep. (1953). Dipl.Krist. *Crystallography.*

Eichhorn, Dr Gerd. Technische Hochschule Ilmenau, Sektion Physik, DDR-63 Ilmenau, German Dem. Rep. (1939). Dr.rer.nat. *Crystal growth, epitaxy.*

Eichler, Dr Klaus. Technische Univ. Dresden, Sektion Physik, DDR-8027 Dresden, Salvador-Allende-Platz 3, German Dem. Rep. (1941). Dr.rer.nat. *Crystal growth.*

Eichler, Dr Wolfgang. Martin-Luther-Univ. Halle-Wittenberg, DDR-4020 Halle, Bachstrasse 6, German Dem. Rep. (1938). Dr.rer.nat. *Real structures.*

Elbinger, Dr German. Akademie der Wissenschaften der DDR, Inst.für Magnetische Werkstoffe, DDR-69 Jena, Helmholtzweg 4, German Dem. Rep. *Crystallography.*

Emons, Prof. Hans-Heinz. Technische Hochschule Carl Schorlemmer Merseburg, DDR-4200 Merseburg, Geusärstrasse, German Dem. Rep. (1930). Dr.sc. *Inorganic and organic techniqüs, chemistry.*

Engel, Dr Aribert. Humboldt-Univ. zu Berlin, Sektion Physik, DDR-1040 Berlin, Invalidenstrasse 43, German Dem. Rep. (1934). Dr.rer.nat. *Crystal growth.*

Engels, Prof. Siegfried. Technische Hochschule Carl Schorlemmer Merseburg, DDR-4200 Merseburg, Geusärstrasse, German Dem. Rep. (1932). Dr.habil. *Solid state chemistry.*

Falkenberg, Dr Wolfgang. Technische Hochschule Carl Schorlemmer Merseburg, DDR-4200 Merseburg, Geusärstrasse, German Dem. Rep. (1945). Dr.rer.nat. *Chemical crystallography, X-ray structure analysis.*

Fanter, Mr Detlef. Akademie der Wissenschaften der DDR, Inst.für Polymerenchemie, DDR-153 Teltow, Kantstrasse 55, German Dem. Rep. (1945). Dipl.Krist. *High polymer structures.*

Faust, Mr Wolfgang. Technische Hochschule Karl-Marx-Stadt, Sektion Physik, DDR-9010 Karl-Marx-Stadt, Reichenhainerstrasse, German Dem. Rep. (1949). Dipl.Krist. *Crystallography.*

Feher, Mr Andreas. Vereinigung für Kristallographie (VFK) in der GGW, DDR-1040 Berlin, Invalidenstrasse 43, German Dem. Rep. (1953). Dipl.Krist. *Crystallography.*

Feher, Mrs Elvira. Vereinigung für Kristallographie (VFK) in der GGW, DDR-1040 Berlin, Invalidenstrasse 43, German Dem. Rep. (1954). *Crystallography.*

Fehling, Mr Wolfgang. Vereinigung für Kristallographie (VFK) in der GGW, DDR-1040 Berlin, Invalidenstrasse 43, German Dem. Rep. (1941). Dipl.Min. *Crystal growth, crystal chemistry, materials science.*

Felbinger, Dr Adolf. Vereinigung für Kristallographie (VFK) in der GGW, DDR-1040 Berlin, Invalidenstrasse 43, German Dem. Rep. (1937). Dr.rer.nat. *Crystallization, crystallography.*

Feltz, Prof. Adalbert. Friedrich-Schiller-Univ. Jena, Sektion Chemie, DDR-69 Jena, August-Bebel-Strasse 2, German Dem. Rep. (1934). Dr.habil. *Solid state chemistry, semiconductor chemistry.*

Fichtner-Schmittler, Dr Helga. Akademie der Wissenschaften der DDR, Zentralinstitut für physikalische Chemie, DDR-1199 Berlin, Rudower Chaussee 5, German Dem. Rep. (1932). Dr.rer.nat. *Crystal structures, powder methods, order-disorder phenomena.*

Fichtner, Dr Konrad. Akademie der Wissenschaften der DDR, Zentralinstitut für physikalische Chemie, DDR-1199 Berlin, Rudower Chaussee 5, German Dem. Rep. (1941). Dr.rer.nat. *Symmetry, computer programming, order-disorder phenomena.*

Fiedler, Dr Gustav. Vereinigung für Kristallographie (VFK) in der GGW, DDR-1040 Berlin, Invalidenstrasse 43, German Dem. Rep. (1932). Dr.rer.nat. *Mineralogy phase-analysis.*

Filscher, Mr Gerold. Friedrich-Schiller-Univ. Jena, Sektion Physik, DDR-69 Jena, Max - Wien - Platz 1, German Dem. Rep. (1933). *Crystallography.*

Fischer, Mr Karl. Vereinigung für Kristallographie (VFK) in der GGW, DDR-1040 Berlin, Invalidenstrasse 43, German Dem. Rep. (1934). Dipl.Min. *Solid state physics, defect structures.*

Fitzl, Mr Günther. Karl-Marx-Univ. Leipzig, Sektion Chemie, DDR-701 Leipzig, Liebigstrasse 18, German Dem. Rep. (1950). Dipl.Krist. *X-ray structure analysis.*

Flögel, Dr Peter. Akademie der Wissenschaften der DDR, Zentralinstitut für Elektronenphysik, DDR-102 Berlin, Neu Schönhauser Strasse 20, German Dem. Rep. (1926). Dr.rer.nat. *Crystal growth.*

Förster, Dr Eckhart. Friedrich-Schiller-Univ. Jena, Sektion Physik, DDR-69 Jena, Max - Wien - Platz 1, German Dem. Rep. (1944). Dr.rer.nat. *Real structures.*

Försterling, Dr Gerd. Technische Univ. Dresden, Sektion Physik, DDR-8027 Dresden, Mommsenstrasse 16, German Dem. Rep. (1940). Dr.rer.nat. *Crystallography.*

Freudenberg, Mr Axel. Akademie der Wissenschaften der DDR, Zentralinstitut für organische Chemie, DDR-1040 Berlin, Invalidenstrasse 44, German Dem. Rep. (1952). Dipl.-Krist. *Crystallography.*

Freydank, Mrs Gisela-Christine. Vereinigung für Kristallographie (VFK) in der GGW, DDR-1040 Berlin, Invalidenstrasse 43, German Dem. Rep. (1944). Dipl.Krist. *Real structures, semiconductors.*

Fröhlich, Dr Fritz. Martin-Luther-Univ. Halle-Wittenberg, DDR-4020 Halle, Friedemann-Bach-Platz 6, German Dem. Rep. (1929). Dr.habil. *Solid state physics.*

Frühauf, Dr Joachim. Technische Hochschule Karl-Marx-Stadt, Sektion Physik, DDR-9010 Karl-Marx-Stadt, PSF 964, German Dem. Rep. (1943). Dr.Ing. *Crystal growth.*

Gast, Mr Roland. Akademie der Wissenschaften der DDR, Inst.für Energetik, DDR-7024 Leipzig, Torgaür Strasse 114, German Dem. Rep. (1947). Dipl.Min. *Corrosion.*

Gause, Dr Hans. Vereinigung für Kristallographie (VFK) in der GGW, DDR-1040 Berlin, Invalidenstrasse 43, German Dem. Rep. (1904). Dr.rer.nat. *Mineralogy, geology.*

Gedicke, Mrs Christine. Humboldt-Univ. zu Berlin, Sektion Physik, DDR-1040 Berlin, Invalidenstrasse 43, German Dem. Rep. (1952). *Crystallography.*

Gernand, Mr Martin. Vereinigung für Kristallographie (VFK) in der GGW, DDR-1040 Berlin, Invalidenstrasse 43, German Dem. Rep. (1938). Dipl.Phys. *Crystallography.*

Gesemann, Prof. Renate. Ingenieurhochschule Mittweida, Sektion Elektronischer Gerätebau, DDR-925 Mittweida, German Dem. Rep. (1936). Dr.rer.nat. *Structure of solid state.*

Geserick, Mrs Sabine. Technische Hochschule Karl-Marx-Stadt, Sektion Physik, DDR-9010 Karl-Marx-Stadt, PSF 964, German Dem. Rep. (1952). *Electron microscopy.*

Göbel, Mr Ralf. Akademie der Wissenschaften der DDR, Zentralinstitut für Festkörperphysik und Werkstofforschung, DDR-8027 Dresden, Helmholzstrasse 20, German Dem. Rep. (1935). Dipl.Phys. *Crystallography.*

Göcke, Dr Wolfhart. Wilhelm-Pieck-Univ. Rostock, DDR-2520 Rostock, Universitätsplatz 3, German Dem. Rep. (1940). Dr.rer.nat. *Crystallography.*

Görnert, Dr Peter. Akademie der Wissenschaften der DDR, Inst.für Magnetische Werkstoffe, DDR-6900 Jena, Helmholzweg 4, German Dem. Rep. (1349). Dr.rer.nat. *Crystal growth, crystallization.*

Götz, Dr Konrad. Friedrich-Schiller-Univ. Jena, Sektion Physik, DDR-69 Jena, Max - Wien - Platz 1, German Dem. Rep. (1938). Dr.rer.nat. *Interference theory.*

Götz, Dr Wolfgang. Friedrich-Schiller-Univ. Jena, Sektion Chemie, DDR-69 Jena, Sellierstrasse 6, German Dem. Rep. (1928). Dr.habil. *Crystallography, crystal chemistry.*

Gottschalch, Dr Volker. Karl-Marx-Univ. Leipzig, Sektion Chemie, DDR-701 Leipzig, Liebigstr. 18, German Dem. Rep. (1945). Dr.rer.nat. *Crystal growth, real structures.*

Grau, Mr Lutz. Akademie der Wissenschaften der DDR, Inst.für Magnetische Werkstoffe, DDR-69 Jena, Helmholtzweg 4, German Dem. Rep. (1947). Dipl.Krist. *Crystallography.*

Griesbach, Mrs Karin. Vereinigung für Kristallographie (VFK) in der GGW, DDR-1040 Berlin, Invalidenstrasse 43, German Dem. Rep. (1940). Dipl.Min. *Crystallography.*

Gülzow, Mr Hansjürgen. Vereinigung für Kristallographie (VFK) in der GGW, DDR-1040 Berlin, Invalidenstrasse 43, German Dem. Rep. Dipl.Min. *Crystal growth.*

Günther, Prof. Fritz. Bergakademie Freiberg, Sektion Metallurgie und Werkstofftechnik, DDR-9200 Freiberg, Gustav-Zeuner-Strasse 5, German Dem. Rep. (1912). Dr.habil. *Real structures.*

Gütt, Mr Rainer. Vereinigung für Kristallographie (VFK) in der GGW, DDR-1040 Berlin, Invalidenstrasse 43, German Dem. Rep. (1943). Dipl.Krist. *Crystallography.*

Hadan, Dr Marianne. Humboldt-Univ. zu Berlin, Sektion Physik, DDR-1040 Berlin, Invalidenstrasse 43, German Dem. Rep. (1939). Dr.rer.nat. *Crystal growth, thin film techniqüs.*

Hähle, Dr Siegfried. Akademie der Wissenschaften der DDR, Inst. für Physik der Werkstoffbearbeitung, DDR-1166 Berlin, Seestrasse 62, German Dem. Rep. (1940). Dr.rer.nat. *Crystal optics, solid state physics.*

Hähnert, Dr Irmela. Humboldt-Univ. zu Berlin, Sektion Physik, DDR-1040 Berlin, Friedrichstrasse 112b, German Dem. Rep. (1941). Dr.rer.nat. *Crystal growth.*

Hähnert, Dr Manfred. Akademie der Wissenschaften der DDR, Zentralinstitut für Anorganische Chemie, DDR-1199 Berlin, Rudower Chaussee 5, German Dem. Rep. (1934). Dr.rer.nat. *Crystal growth, crystal chemistry.*

Hahne, Dr Bodo. Technische Hochschule Carl Schorlemmer Merseburg, DDR-4200 Merseburg, Geusärstrasse, German Dem. Rep. (1941). *Phase analysis.*

Hanold, Mrs Karin. Humboldt-Univ. zu Berlin, Sektion Physik, DDR-1040 Berlin, Invalidenstrasse 43, German Dem. Rep. (1957). *Crystallography.*

Hartmann, Dr Horst. Akademie der Wissenschaften der DDR, Zentralinstitut für Elektronenphysik, DDR-108 Berlin, Mohrenstrasse 40-41, German Dem. Rep. (1934). Dr.rer.nat. *Crystal growth, real structures.*

Hartung, Dr Helmut. Martin-Luther-Univ. Halle-Wittenberg, DDR-4020 Halle, Mühlpforte 1, German Dem. Rep. (1935). Dr.rer.nat. *X-ray structure analysis.*

Hartwig, Dr Jürgen. Friedrich-Schiller-Univ. Jena, Sektion Physik, DDR-6900 Jena, Max - Wien - Platz 1, German Dem. Rep. (1947). Dr.rer.nat. *Real structures.*

Hegenbarth, Prof. Ernst. Technische U. Dresden, Sektion Physik, DDR-8027 Dresden, Mommsenstrasse 16, German Dem. Rep. (1933). Dr.habil. *Solid state physics.*

Heide, Dr Klaus. Friedrich Schiller Univ. Jena, Otto Schott Institut, DDR-69 Jena, Sellierstrasse 6, German Dem. Rep. (1938). Dr.habil. *Crystallography.*

Heim, Dr Joachim. Technische Hochschule Karl-Marx-Stadt, Sektion Physik, DDR-9010 Karl-Marx-Stadt, PSF 964, German Dem. Rep. (1933). Dr.rer.nat. *Crystallography, crystal growth.*

Heinze, Joachim. Vereinigung für Kristallographie (VFK) in der GGW, DDR-1040 Berlin, Invalidenstrasse 43, German Dem. Rep. (1932). Dr.rer.nat. *Crystallography.*

Hellmold, Mr Peter. Technische Hochschule Carl Schorlemmer Merseburg, DDR-4200 Merseburg, Geusärstrasse, German Dem. Rep. (1937). Dr.rer.nat. *Crystallography.*

Hennig, Prof. Klaus. Akademie der Wissenschaften der DDR, Zentralinstitut für Kernforschung Rossendorf, DDR-8050 Dresden, PF 19, German Dem. Rep. (1936). Dr.sc.

Herberger, Dr Jürgen. Technische Hochschule Karl-Marx-Stadt, Sektion Physik, DDR-9010 Karl-Marx-Stadt, Strasse der Nation 62, German Dem. Rep. (1941). Dr.rer.nat. *Solid state physics, crystal growth, X-ray structure analysis.*

Hermoneit, Mr Bernd. Akademie der Wissenschaften der DDR, Zentralinstitut für Optik und Spektroskopie, DDR-1199 Berlin, Rudower Chaussee 5, German Dem. Rep. (1944). Dip.Krist. *Crystallography.*

Herms, Dr Gerhard. Wilhelm-Pieck Universität, Sektion Physik, DDR-2520 Rostock, Universitätsplatz 3, German Dem. Rep. (1932). Dr.sc.rer.nat.

Herrmann, Mrs Christel. Akademie der Wissenschaften der DDR, Zentralinstitut für Werkstoffforschung, DDR-8020 Dresden, Wittenbergstrasse 28, German Dem. Rep. (1953). Dipl. Krist. *Crystallography.*

Herrmann, Dr Frank-Peter. Akademie der Wissenschaften der DDR, Zentralinstitut für Optik und Spektroskopie, DDR-1199 Berlin, Rudower Chaussee 5, German Dem. Rep. (1940). Dr.rer.nat. *Crystallography.*

Herrmann, Prof. Rudolf. Humboldt-Univ. zu Berlin, Sektion Physik, DDR-1040 Berlin, Hessische Strasse 2, German Dem. Rep. (1936).

Heydenreich, Dr Johannes. Akademie der Wissenschaften der DDR, Zentralinstitut für Festkörperphysik und Werkstoffforschung, DDR-4000 Halle, Weinberg 2, German Dem. Rep. (1930). Dr.habil. *Boundary surface physics, solid state physics.*

Heymann, Dr Gunter. Humboldt-Universität-zu Berlin, Sektion Elektronik, DDR-1136 Berlin, Hans-Loch-Strasse, German Dem. Rep. (1940). Dr.sc. *Crystallography.*

Hinz, Dr Dietrich. Akademie der Wissenschaften der DDR, Zentralinstitut für Festkörperphysik und Werkstoffforschung, DDR-8027 Dresden, Helmholzstrasse 20, German Dem. Rep. (1945). Dr.rer.nat. *Crystallography.*

Hinze, Mr Thomas. Ingenieurtechnisches Zentalbüro Böhlen, DDR-7202 Böhlen, German Dem. Rep. (1949). Dipl.Krist. *Crystallography.*

Höbler, Mr Hans-Joachim. Karl-Marx-Univ. Leipzig, Sektion Chemie, Fachrichtung Kristallographie, DDR-7030 Leipzig, Scharnhorststrasse 20, German Dem. Rep. (1951). Dipl.Min. *Crystal growth, mineralogy.*

Höche, Dr Hans-Reiner. Martin-Luther-Univ. Halle-Wittenberg, DDR-4020 Halle, Friedemann-Bach-Platz 6, German Dem. Rep. (1942). Dr.rer.nat. *Crystallography.*

Höche, Dr Hellmut. Akademie der Wissenschaften der DDR, Inst.für Festkörperphysik und Elektronenmikroskopie, DDR-4010 Halle, Am Weinberg 2, German Dem. Rep. (1946). Dr.rer.nat. *Crystallography.*

Höhne, Prof. Ernst. Akademie der Wissenschaften der DDR, Zentralinstitut für Molekularbiologie, DDR-1115 Berlin, Lindenberger Weg 70, German Dem. Rep. (1927). Dr.habil. *Structural chemistry, X-ray structure analysis.*

Hötzsch, Dr Günter. Bergakademie Freiberg, Sektion Metallurgie und Werkstofftechnik, DDR-9200 Freiberg, Gustav-Zeuner-Strasse 5, German Dem. Rep. (1932). Dr.habil. *Real structures, solid state physics.*

Hoff, von, Dr Siegfried. Vereinigung für Kristallographie (VFK) in der GGW, DDR-1040 Berlin, Invalidenstrasse 43, German Dem. Rep. (1940). Dr.rer.nat. *X-ray diffraction, thin film techniqüs.*

Hoffmann, Dr Brigitte. Akademie der Wissenschaften der DDR, Forschungsinstitut für Aufbereitung, DDR-92 Freiberg, Strasse des Friedens 40, German Dem. Rep. (1938). Dr.rer.nat. *Structural change of real crystals.*

Holldorf, Dr Horst. Bergakademie Freiberg, Sektion Chemie, DDR-9200 Freiberg, Leipziger Strasse, German Dem. Rep. (1939). Dr.rer.nat. *Mass crystallization.*

Hoppe, Prof. Günter. Humboldt-Univ. zu Berlin, Sektion Physik, DDR-1040 Berlin, Invalidenstrasse 43, German Dem. Rep. (1919). Dr.sc. *Mineralogy, petrography.*

Hoppe, Prof. Hans. Technische Hochschule Carl Schorlemmer Merseburg, DDR-4200 Merseburg, Geusärstrasse, German Dem. Rep. Dr.rer.nat. *Crystallography.*

Hoyer, Dr Walter. Technische Hochschule Karl-Marx-Stadt, Sektion Physik, DDR-9010 Karl-Marx-Stadt, PSF 964, German Dem. Rep. (1944). Dr.rer.nat. *Crystallography.*

Hübner, Dr Manfred. Hochschule für Bauwesen, DDR-703 Leipzig, R.-Lehmannstrasse 32, German Dem. Rep. (1935). Dr.rer.nat. *Chemistry, boundary surfaces.*

Hultzsch, Mr Rainer. Vereinigung für Kristallographie (VFK) in der GGW, DDR-1040 Berlin, Invalidenstrasse 43, German Dem. Rep. (1940). Dipl. Phys. *Laser physics, crystal testing.*

Ickert, Dr Lars. Humboldt-Univ. zu Berlin, Sektion Physik, DDR-1040 Berlin, Invalidenstrasse 43, German Dem. Rep. (1934). Dr.habil. *Crystal growth.*

Jährling, Mr Thomas. Akademie der Wissenschaften der DDR, Zentralinstitut für Elektronenphysik, DDR-108 Berlin, Mohrenstrasse 41-43, German Dem. Rep. (1952). Dipl. Phys. *Crystallography.*

Jegerlehner, Mr Kurt. Vereinigung für Kristallographie (VFK) in der GGW, DDR-1040 Berlin, Invalidenstrasse 43, German Dem. Rep. (1935). Dipl.Min. *Solid state physics.*

Jenichen, Dr Bernd. Akademie der Wissenschaften der DDR, Zentralinstitut für Elektronenphysik, DDR-1080 Berlin, Mohrenstrasse 41-43, German Dem. Rep. (1953). Dipl.Phys. *Crystallography.*

Jerschkewitz, Dr Hans-Georg. Akademie der Wissenschaften der DDR, Zentralinstitut für physikalische Chemie, DDR-1199 Berlin, Rudower Chaussee 5, German Dem. Rep. Dr.rer.nat. *Crystallography.*

Johansen, Dr Heinrich. Akademie der Wissenschaften der DDR, Inst.für Festkörperphysik und Elektronenmikroskopie, DDR-4010 Halle, Am Weinberg 2, German Dem. Rep. (1939). Dr.rer.nat. *Electron microscopy.*

Jost, Dr Karlheinz. Akademie der Wissenschaften der DDR, Zentralinstitut für Anorganische Chemie, DDR-1199 Berlin, Rudower Chaussee 5, German Dem. Rep. (1927). Dr.habil. *Inorganic structures, solid state chemistry.*

Jurisch, Dr Manfred. Akademie der Wissenschaften der DDR, Zentralinstitut für Festkörperphysik und Werkstoffforschung, DDR-8032 Dresden, Helmholtzstrasse 20, German Dem. Rep. (1940). Dr.-Ing. *Crystallography.*

Kämmel, Dr Thomas. Vereinigung für Kristallographie (VFK) in der GGW, DDR-1040 Berlin, Invalidenstrasse 43, German Dem. Rep. (1931). Dr.habil. *Geological and geochemistry analysis.*

Kaiser, Dr Johannes. Karl-Marx-Univ. Leipzig, Sektion Chemie, DDR-701 Leipzig, Liebigstrasse 18, German Dem. Rep. (1940). Dr.rer.nat. *X-ray structure analysis, crystal chemistry.*

Kaiser, Mrs Ute. Akademie der Wissenschaften der DDR, Zentralinstitut für anorganische Chemie, Bereich Keramik, DDR-1040 Berlin, Invalidenstrasse 44, German Dem. Rep. (1953). Dipl.Krist. *Crystallography.*

Kalinna, Mr Hartmut. Vereinigung für Kristallographie (VFK) in der GGW, DDR-1040 Berlin, Invalidenstrasse 43, German Dem. Rep. (1934). Dip.Min. *Crystal structures, semiconductors.*

Kalweit, Mr Harald. Akademie der Wissenschaften der DDR, Zentralinstitut für Anorganische Chemie, DDR-1199 Berlin, Rudower Chaussee 5, German Dem. Rep. (1945). Dipl. Min. *Mineralogy, petrography.*
Kamprath, Mr Fred-Bodo. Akademie der Wissenschaften der DDR, Zentralinstitut für Anorganische Chemie, DDR-1199 Berlin, Rudower Chaussee 5, German Dem. Rep. (1948). Dipl.Krist. *Crystallography.*
Kanis, Dr Michäl. Vereinigung für Kristallographie (VFK) in der GGW, DDR-1040 Berlin, Invalidenstrasse 43, German Dem. Rep. (1949). Dipl.Krist. *Crystallography.*
Katzschmann, Mr Kurt. Vereinigung für Kristallographie (VFK) in der GGW, DDR-1040 Berlin, Invalidenstrasse 43, German Dem. Rep. (1928). Dipl.Phys. *Crystal optics, thin film techniqüs.*
Kaufmann, Dr Thorsten. Technische Hochschule Ilmenau, Sektion Physik und technisch- elektronische Baülemente, DDR-63 Ilmenau, Weimarerstrasse, German Dem. Rep. (1941). Dr.rer.nat. *Solid state chemistry.*
Keller, Dr Kurt Wolfgang. Akademie der Wissenschaften der DDR, Inst.für Festkörperphysik und Technische Hochschule Carl Schorlemmer Merseburg, DDR-4200 Merseburg, German Dem. Rep. (1936). Dr.rer.nat. *Crystal growth, boundary surface physics.*
Kelling, Mrs Gerhild. Vereinigung für Kristallographie (VFK) in der GGW, DDR-1040 Berlin, Invalidenstrasse 43, German Dem. Rep. (1943). Dipl.Min. *Crystallography.*
Kersten, Mr Friedrich. Ingenieurhochschule Mittweida, DDR-925 Mittweida, Platz der DSF, German Dem. Rep. (1934). Ing. *Crystal growth, real structures.*
Kiedrowski, von, Mr Hartmut. Akademie der Wissenschaften der DDR, Zentralinstitut für Elektronenphysik, DDR-108 Berlin, Mohrenstrasse 41-43, German Dem. Rep. Dipl. Krist. *Crystallography.*
Kieling, Mr Knut. Vereinigung für Kristallographie (VFK) in der GGW, DDR-1040 Berlin, Invalidenstrasse 43, German Dem. Rep. (1951). Dipl.Krist. *Semiconductors, epitaxy.*
Kies, Dr Jörg. Akademie der Wissenschaften der DDR, Zentralinstitut für Molekularbiologie, DDR-1115 Berlin, Lindenberger Weg 70, German Dem. Rep. (1938). Dr.rer.nat. *Structure.*
Kleinert, Dr Peter. Akademie der Wissenschaften der DDR, Inst.für Magnetische Werkstoffe, DDR-69 Jena, Helmholtzweg 4, German Dem. Rep. (1925). Dr.sc. *Crystallography.*
Kleinstück, Prof. Karlheinz. Technische U. Dresden, Sektion Physik, DDR-8027 Dresden, Mommsenstrasse 16, German Dem. Rep. (1929). Dr.habil. *Crystallography.*
Kleint, Dr Christian. Karl-Marx-Univ. Leipzig, DDR-7010 Leipzig, Linnestrasse 5, German Dem. Rep. (1926). Dr.habil. *Crystallography.*
Klimanek, Dr Peter, Bergakademie Freiberg, Sektion Met. und Werkst. Technik, DDR-9200 Freiberg, Gustav Zeunerstrasse 5, German Dem. Rep. (1935). Dr.Ing. *Crystallography.*
Klock, Mr Winfried. Humboldt-Univ. zu Berlin, Sektion Physik, DDR-1040 Berlin, Invalidenstrasse 43, German Dem. Rep. (1948). *Materials science.*
Kneschke, Dr Götz. Akademie der Wissenschaften der DDR, Forschungsinstitut für Aufbereitung, DDR-92 Freiberg, Strasse des Friedens 46, German Dem. Rep. (1937). Dr.rer.nat. *Real structures.*
Köhler, Dr Rolf. Akademie der Wissenschaften der DDR, Zentralinstitut für Elektronenphysik, DDR-1080 Berlin, Mohrenstrasse 41-43, German Dem. Rep. (1942). Dr.rer.nat. *Real structures.*
Köpernik, Mr Horst. Vereinigung für Kristallographie (VFK) in der GGW, DDR-1040 Berlin, Invalidenstrasse 43, German Dem. Rep. (1943). Dipl.Chem. *Zeolites.*
Kötitz, Dr Günther. Vereinigung für Kristallographie (VFK) in der GGW, DDR-1040 Berlin, Invalidenstrasse 43, German Dem. Rep. (1935). Dr.rer.nat. *Crystal growth, solid state physics.*
Kosche, Mrs Ingeborg. Akademie der Wissenschaften der DDR, Zentralinstitut für Anorganische Chemie, DDR-1199 Berlin, Rudower Chaussee 5, German Dem. Rep. (1942). Dipl.Min. *Electron microscopy.*
Krause, Mr Waldefried. Vereinigung für Kristallographie (VFK) in der GGW, DDR-1040 Berlin, Invalidenstrasse 43, German Dem. Rep. (1930). Dipl.Min. *Mineralogy.*
Krause, Mrs Christa. Akademie der Wissenschaften der DDR, Zentrum für Rechentechnik, DDR-1199 Berlin, Rudower Chaussee 5, German Dem. Rep. (1935). Dipl. Math. *Crystallography.*
Krausse, Dr Joachim. Friedrich Schiller Univ. Jena, Sektion Chemie, DDR-69 Jena, Lessingstrasse 10, German Dem. Rep. (1933). Dr.rer.nat. *X-ray structure analysis.*
Kressner, Dr F.Harry. Akademie der Wissenschaften der DDR, Inst. für Physik der Werkstoffbearbeitung, DDR-1166 Berlin, Seestrasse 82, German Dem. Rep. (1936). Dr.rer.nat. *Crystallography.*
Kretschmer, Dr Rolf-Günther. Akademie der Wissenschaften der DDR, Zentralinstitut für Molekularbiologie, DDR-1115 Berlin, Lindenberger Weg 70, German Dem. Rep. (1934). Dr.rer.nat. *Computing methods in crystallography.*
Kuban, Mr Ralf-Jürgen. Akademie der Wissenschaften der DDR, Zentralinstitut für physikalische Chemie, DDR-1199 Berlin, Rudower Chaussee 5, German Dem. Rep. (1953). Dipl. Krist. *Crystallography.*
Kühn, Dr Günther. Karl-Marx-Univ. Leipzig, Sektion Chemie, Fachrichtung Kristallographie, DDR-7030 Leipzig, Scharnhorststrasse 20, German Dem. Rep. (1939). Dr.rer.nat. *Crystal growth, solid state chemistry.*
Kürsten, Dr Hans-Dieter. Akademie der Wissenschaften der DDR, Zentralinstitut für Optik und Spektroskopie, DDR-1199 Berlin, Rudower Chaussee 6, German Dem. Rep. (1945). Dr.rer.nat. *Crystal growth, real structures.*
Kulpe, Dr Siegfried. Akademie der Wissenschaften der DDR, Zentralinstitut für physikalische Chemie, DDR-1199 Berlin, Rudower Chaussee 5, German Dem. Rep. (1927). Dr.sc. *X-ray structure analysis.*
Kutschabsky, Dr Leo. Akademie der Wissenschaften der DDR, Zentralinstitut für Molekularbiologie, DDR-1115 Berlin, Lindenberger Weg 70, German Dem. Rep. (1933). Dr.rer.nat. *X-ray structure analysis, crystal chemistry, crystallography.*
Lammert, Mrs Barbara. Vereinigung für Kristallographie (VFK) in der GGW, DDR-1040 Berlin, Invalidenstrasse 43, German Dem. Rep. (1949). *Astronomy.*
Langer, Dr Hans-Dieter. Vereinigung für Kristallographie (VFK) in der GGW, DDR-1040 Berlin, Invalidenstrasse 43, German Dem. Rep. (1941). Dr.rer.nat. *Solid state physics.*
Lebek, Dr Alexander. Akademie der Wissenschaften der DDR, Zentrum für Wissenschaftlichen Gerätebau, DDR-1199 Berlin, Rudower Chaussee 6, German Dem. Rep. (1926). Dr.rer.nat. *Crystallography.*
Lehmann, Dr Gottfried. Humboldt Univ. zu Berlin, Sektion Exp.Hl.Physik, DDR-1040 Berlin, Hessische Strasse 2, German Dem. Rep. (1924). Dr.rer.nat. *Crystallography.*
Lehmann, Mr Günter. Vereinigung für Kristallographie (VFK) in der GGW, DDR-1040 Berlin, Invalidenstrasse 43, German Dem. Rep. Dipl.Min. *Crystallography.*
Leipert, Mrs Yvonne. Karl-Marx-Univ. Leipzig, Sektion Chemie, Fachrichtung Kristallographie, DDR-7030 Leipzig, Scharnhorststrasse 20, German Dem. Rep. (1953). *Crystallography.*
Lemke, Mr Guntram. Karl-Marx-Univ. Leipzig, DDR-7010 Leipzig, Linnestrasse 5, German Dem. Rep. (1950). Dipl.Krist. *Organic crystals, liquids.*
Leonhardt, Dr Albrecht. Akademie der Wissenschaften der DDR, Zentralinstitut für Werkstofforschung, DDR-8020 Dresden, Winterbergstrasse 20, German Dem. Rep. (1949). Dr.rer.nat. *Crystal growth.*
Leonhardt, Dr Gunter. Karl-Marx-Univ. Leipzig, Sektion Chemie, Fachrichtung Kristallographie, DDR-7030 Leipzig, Scharnhorststrasse 20, German Dem. Rep. (1939). Dr.rer.nat. *Crystallography.*
Leppin, Mrs Christine. Vereinigung für Kristallographie (VFK) in der GGW, DDR-1040 Berlin, Invalidenstrasse 43, German Dem. Rep. (1949). Dipl.Krist. *Crystallography.*
Lilie, Mr Martin. Vereinigung für Kristallographie (VFK) in der GGW, DDR-1040 Berlin, Invalidenstrasse 43, German Dem. Rep. (1949). Dipl.Krist. *Mineralogy.*
Linke, Dr Dietmar. Friedrich-Schiller Univ. Jena, DDR-69 Jena, August Bebel-Strasse 2, German Dem. Rep. (1940). Dr.rer.nat. *Crystal and glass chemistry.*
Lobenstein, Mrs Heidrun. Humboldt-Univ. zu Berlin, Sektion Physik, DDR-1040 Berlin, Invalidenstrasse 43, German Dem. Rep. (1942). Dipl.Chem. *Crystallography.*
Lösche, Prof. Artur. Karl-Marx-Univ. Leipzig, DDR-7010 Leipzig, Linnestrasse 5, German Dem. Rep. (1921). *Crystallography.*
Lutz, Mr Dieter. Vereinigung für Kristallographie (VFK) in der GGW, DDR-1040 Berlin, Invalidenstrasse 43, German Dem. Rep. (1951). Dipl.Krist. *Crystallography.*
Lux, Mr Bernd. Akademie der Wissenschaften der DDR, Zentrum für Wissenschaftlichen Gerätebau, DDR-1199 Berlin, Rudower Chaussee 6, German Dem. Rep. (1949). Dipl.Krist. *Crystal growth.*
Lux, Dr Georg. Martin-Luther-Univ. Halle, DDR-4020 Halle, Weinbergweg, German Dem. Rep. (1929). Dr.rer.nat. *Ideal structures.*
Marx, Prof. Günter. Technische Hochschule Karl Marx Stadt, DDR-9010 Karl-Marx-Stadt, Strasse der Nation 62, PS 964, German Dem. Rep. (1938). Dr.sc. *Crystallography.*
Masche, Mr Wolfgang. Vereinigung für Kristallographie (VFK) in der GGW, DDR-1040 Berlin, Invalidenstrasse 43, German Dem. Rep. (1947). Dipl.Min. *Crystallography.*
May, Prof. Martin. Technische Hochschule Carl Schorlemmer Merseburg, DDR-4200 Merseburg, Geusärstrasse, German Dem. Rep. (1919). Dr.rer.nat. *Crystallography.*
Meisel, Prof. Armin. Karl-Marx-Univ. Leipzig, DDR-7010 Leipzig, Linnestrasse 2, German Dem. Rep. (1926). Dr.sc. *X-ray spectroscopy.*
Metze, Mr Dieter. Vereinigung für Kristallographie (VFK) in der GGW, DDR-1040 Berlin, Invalidenstrasse 43, German Dem. Rep. (1939). Dipl. Phys. *Crystallography.*
Meyer, Prof. Klaus. Friedrich-Schiller-Univ. Jena, Sektion Chemie, DDR-69 Jena, Lessingstrasse 10, German Dem. Rep. (1936). Dr.habil. *Crystallography, chemistry, boundary surfaces.*
Michel, Dr Bernd. Martin-Luther Univ. Halle-Wittenberg, DDR-4020 Halle, Friedemann-Bach-Platz 6, German Dem. Rep. (1949). Dr.rer.nat. *Crystallography.*
Möhling, Dr Werner. Akademie der Wissenschaften der DDR, Zentralinstitut für Elektronenphysik, DDR-1080 Berlin, Mohrenstrasse 41-43, German Dem. Rep. (1934). Dr.rer.nat. *Crystallography, real structures.*

Mohr, Dr Ulrich. Vereinigung für Kristallographie (VFK) in der GGW, DDR-1040 Berlin, Invalidenstrasse 43, German Dem. Rep. (1935). Dr.rer.nat. *Crystal growth, real structures.*
Mothes, Mr Heinrich. Vereinigung für Kristallographie (VFK) in der GGW, DDR-1040 Berlin, Invalidenstrasse 43, German Dem. Rep. (1935). Dipl.Phys. *Crystal growth, electronics.*
Mucha, Mrs Christine. Vereinigung für Kristallographie (VFK) in der GGW, DDR-1040 Berlin, Invalidenstrasse 43, German Dem. Rep. (1948). Dipl.Krist. *Crystal chemistry, structure analysis.*
Mühlberg, Mr Manfred. Humboldt-Univ. zu Berlin, Sektion Physik, DDR-1040 Berlin, Invalidenstrasse 43, German Dem. Rep. (1949). *Crystallography.*
Müller, Dr Bernd. Friedrich-Schiller-Univ. Jena, Sektion Chemie, DDR-69 Jena, Lessingstr. 10, German Dem. Rep. (1943). Dr.rer.nat. *X-ray structure analysis.*
Müller, Dr Brigitte. Akademie der Wissenschaften der DDR, Zentralinstitut für Festkörperphysik und Werkstofforschung, DDR-8032 Dresden, Helmholtzstrasse 20, German Dem. Rep. *Crystallography.*
Müller, Dr Eberhard. Friedrich-Schiller-Univ. Jena, Sektion Chemie, DDR-69 Jena, Lessingstrasse 10, German Dem. Rep. (1942). Dr.rer.nat. *Boundary surface physics and chemistry.*
Müller, Dr Helmut. Wilhelm-Pieck-Universität, Sektion Physik, DDR-2520 Rostock, Universitätsplatz 3, German Dem. Rep. (1928). Dr.rer.nat. *Real structures, glass ceramics.*
Muschner, Dr Wolfgang. Vereinigung für Kristallographie (VFK) in der GGW, DDR-1040 Berlin, Invalidenstr. 43, German Dem. Rep. (1934). *Crystal growth.*
Neels, Prof. Hermann. Karl-Marx-Univ. Leipzig, Sektion Chemie, Fachrichtung Kristallographie, DDR-7030 Leipzig, Scharnhorststrasse 20, German Dem. Rep. (1913). Dr.rer.nat. *Crystal growth, crystallization.*
Neumann, Mr Wolfgang. Akademie der Wissenschaften der DDR, Inst.für Festkörperphysik und Elektronenmikroskopie, DDR-4010 Halle, Am Weinberg 2, German Dem. Rep. (1944). Dipl.Min. *Electron diffraction.*
Nieber, Dr Johannes. Martin-Luther Univ. Halle-Wittenberg, DDR-4020 Halle, Friedemann-Bach-Platz 6, German Dem. Rep. (1948). Dr.rer.nat. *Crystallography.*
Niebsch, Dr Hans-Hermann. Humboldt-Univ. zu Berlin, Sektion Physik, DDR-1040 Berlin, Invalidenstrasse 43, German Dem. Rep. (1933). Dr.rer.nat. *Crystal physics, crystal growth.*
Nitsche, Mr Walter. Vereinigung für Kristallographie (VFK) in der GGW, DDR-1040 Berlin, Invalidenstrasse 43, German Dem. Rep. (1948). Dipl.Krist. *Crystallography.*
Noack, Dr Joachim. Akademie der Wissenschaften der DDR, Zentralinstitut für Elektronenphysik, DDR-108 Berlin, Mohrenstrasse 40-41, German Dem. Rep. (1929). Dr.rer.nat. *Crystal growth.*
Oettel, Dr Heinrich. Bergakademie Freiberg, Sektion Metallurgie, DDR-9200 Freiberg, Zeunerstrasse 5, German Dem. Rep. (1940). Dr.Ing. *Crystallography.*
Oppermann, Mr Dieter. Karl-Marx-Univ. Leipzig, Sektion Chemie, Fachrichtung Kristallographie, DDR-70 Leipzig, Scharnhorststrasse 20, German Dem. Rep. (1938). *Crystallography.*
Oppermann, Dr Heinrich. Akademie der Wissenschaften der DDR, Zentralinstitut für Festkörperphysik und Werkstofforschung, DDR-8027 Dresden, Helmholzstrasse 20, German Dem. Rep. (1934). *Crystal growth, crystallization.*
Osterland, Mrs Martina. Vereinigung für Kristallographie (VFK) in der GGW, DDR-1040 Berlin, Invalidenstrasse 43, German Dem. Rep. (1953). Dipl.Krist. *Petrography, precious stones.*
Pätz, Dr Kurt W. Akademie der Wissenschaften der DDR, Zentralinstitut für Elektronenphysik, DDR-108 Berlin, Mohrenstrasse 41-43, German Dem. Rep. (1905). Dr.phil. *Semiconductor chemistry.*
Pätzke, Mrs Nora. Vereinigung für Kristallographie (VFK) in der GGW, DDR-1040 Berlin, Invalidenstrasse 43, German Dem. Rep. (1950). Dipl.Krist. *Thermal analysis.*
Pätzold, Mrs Rita. Friedrich-Schiller-Univ. Jena, Sektion Chemie, DDR-69 Jena, Lessingstrasse 10, German Dem. Rep. (1953). Dipl.Chem. *Crystal chemistry, spectroscopy.*
Pasemann, Dr Monika. Akademie der Wissenschaften der DDR, Zentralinstitut für Festkörperphysik und Elektronenmikroskopie, DDR-4010 Halle, Am Weinberg 2, German Dem. Rep. (1947). Dr.Ing. *Electron microscopy, real structures.*
Paufler, Prof. Peter. Karl-Marx-Univ. Leipzig, Sektion Chemie, DDR-7030 Leipzig, Scharnhorststrasse 20, German Dem. Rep. (1940). Dr.sc. *Crystallography*
Paul, Mrs Sabine. Vereinigung für Kristallographie (VFK) in der GGW, DDR-1040 Berlin, Invalidenstr. 43, German Dem. Rep. (1945). Dipl.Min. *Crystallography.*
Pechstein, Mrs Gisela. Vereinigung für Kristallographie (VFK) in der GGW, DDR-1040 Berlin, Invalidenstrasse 43, German Dem. Rep. (1949). Dipl.Krist. *Crystallography.*
Peibst, Dr Herbert. Akademie der Wissenschaften der DDR, Zentralinstitut für Elektronenphysik, DDR-108 Berlin, Mohrenstrasse 40-41, German Dem. Rep. (1921). Dr.sc. *Crystallography.*
Penndorf, Mr Jürgen. Karl-Marx-Univ. Leipzig, Sektion Chemie, Fachrichtung Kristallographie, DDR-7030 Leipzig, Scharnhorststrasse 20, German Dem. Rep. (1950). Dipl.Krist. *Crystallography.*

Perthel, Dr Rolf. Akademie der Wissenschaften der DDR, Inst.für Magnetische Werkstoffe, DDR-69 Jena, Helmholtzweg 4, German Dem. Rep. (1929). Dr.sc. *Solid state magnetism.*
Pietsch, Mr Ullrich. Humboldt-Univ. zu Berlin, Sektion Physik, DDR-1040 Berlin, Invalidenstrasse 43, German Dem. Rep. (1952). *Crystallography.*
Pietzsch, Dr Claus. Bergakademie Freiberg, Sektion Physik, DDR-9200 Freiberg, Silbermannstr.1, German Dem. Rep. (1938). Dr.rer.nat. *Mössbaür spectroscopy.*
Poppendieck, Mr Detlef. Vereinigung für Kristallographie (VFK) in der GGW, DDR-1040 Berlin, Invalidenstrasse 43, German Dem. Rep. (1949). Dipl.Krist. *Mineralogy.*
Preiss, Dr Henry, Akademie der Wissenschaften der DDR, Zentralinstitut für physikalische Chemie, DDR-1199 Berlin, Rudower Chaussee 5, German Dem. Rep. (1935). Dr.rer.nat. *Inorganic chemistry, X-ray structure analysis, mass spectroscopy.*
Preuss, Dr Heinz. Pädagogische Hochschule Karl-Liebknecht Potsdam, DDR-15 Potsdam, Am Neuen Palais, German Dem. Rep. (1934). Dr.rer.nat. *Crystallography, real structures.*
Pritzkow, Mr Wolfgang. Humboldt-Univ. zu Berlin, Sektion Physik, DDR-1040 Berlin, Invalidenstrasse 43, German Dem. Rep. (1952). *Crystallography.*
Puff, Mr Manfred. Ingenieurhochschule Zwickau, Abt.Mathematik, DDR-95 Zwickau, Dr.-Friedrich-Ring 2a, German Dem. Rep. (1929). Dipl. Phys. *Real structures, X-ray structure analysis.*
Raidt, Dr Helmut. Vereinigung für Kristallographie (VFK) in der GGW, DDR-1040 Berlin, Invalidenstrasse 43, German Dem. Rep. *Crystallography.*
Rechenberg, Dr Ingrid. Akademie der Wissenschaften der DDR, Zentralinstitut für Optik und Spektroskopie, DDR-1199 Berlin, Rudower Chaussee 6, German Dem. Rep. (1938). Dr.rer.nat. *Crystal growth, crystallization, epitaxy.*
Reck, Dr Günter. Akademie der Wissenschaften der DDR, Zentralinstitut für Molekularbiologie, DDR-1115 Berlin, Lindenberger Weg 70, German Dem. Rep. (1939). Dr.rer.nat. *X-ray structure analysis, crystal chemistry.*
Reiche, Mr Manfred. Akademie der Wissenschaften der DDR, Inst.für Festkörperphysik und Elektronenmikroskopie, DDR-4010 Halle, Am Weinberg 2, German Dem. Rep. (1951). Dipl.Krist. *Crystallography.*
Reinecke, Mrs Kriemhild. Akademie der Wissenschaften der DDR, Zentralinstitut für physikalische Chemie, DDR-1199 Berlin, Rudower Chaussee 5, German Dem. Rep. (1940). Dipl.Krist. *X-ray structure analysis.*
Reinhold, Mrs Ingrid. Vereinigung für Kristallographie (VFK) in der GGW, DDR-1040 Berlin, Invalidenstrasse 43, German Dem. Rep. (1935). Dipl.Min. *Crystal growth.*
Rentsch, Mr Harald. Karl-Marx-Univ. Leipzig, Sektion Chemie, Fachrichtung Kristallographie, DDR-7030 Leipzig, Scharnhorststrasse 20, German Dem. Rep. (1950). Dipl.Krist. *Crystallography.*
Reuter, Mr Dietrich. Vereinigung für Kristallographie (VFK) in der GGW, DDR-1040 Berlin, Invalidenstrasse 43, German Dem. Rep. (1944). Dipl.Chem. *Crystallography.*
Richter, Dr Frank. Akademie der Wissenschaften der DDR, Inst. für Physik der Werkstoffbearbeitung, DDR-1211 Falkenhagen, German Dem. Rep. (1947). *Real structures, epitaxy.*
Richter, Mr Hans. Vereinigung für Kristallographie (VFK) in der GGW, DDR-1040 Berlin, Invalidenstrasse 43, German Dem. Rep. (1939). Dipl.-Met. *Crystallography.*
Richter, Dr Klaus. Vereinigung für Kristallographie (VFK) in der GGW, DDR-1040 Berlin, Invalidenstrasse 43, German Dem. Rep. (1935). Dr.rer.nat. *Röntgenography, catalysis.*
Richter, Dr Rainer. Karl-Marx-Univ. Leipzig, Sektion Chemie, DDR-701 Leipzig, Liebigstrasse 18, German Dem. Rep. (1946). Dr.rer.nat. *Crystallography.*
Richter, Dr Waltraut. Akademie der landwirtschaft der DDR, Inst.für Düngungsforschung, DDR-7022 Leipzig, Gustaw-Kühn-Strasse 8, German Dem. Rep. (1944). Dr.rer.nat. *Crystallography.*
Ringel, Dr Lilli. Vereinigung für Kristallographie (VFK) in der GGW, DDR-1040 Berlin, Invalidenstrasse 43, German Dem. Rep. (1937). Dr.Ing. *Crystallography.*
Ritschel, Mr Manfred. Akademie der Wissenschaften der DDR, Zentralinstitut für Festkörperphysik und Werkstofforschung, DDR-8032 Dresden, Helmholtzstrasse 20, German Dem. Rep. (1947). Dipl. Krist. *Crystal growth.*
Rosin, Dr Horst. Vereinigung für Kristallographie (VFK) in der GGW, DDR-1040 Berlin, Invalidenstr. 43, German Dem. Rep. (1943). Dr.rer.nat. *Real structures.*
Rossner, Mr Johannes. Vereinigung für Kristallographie (VFK) in der GGW, DDR-1040 Berlin, Invalidenstr. 43, German Dem. Rep. (1945). *Crystallography.*
Rost, Mrs Jutta. Technische Hochschule Karl-Marx-Stadt, Sektion Physik, DDR-9010 Karl-Marx-Stadt, PSF 964, German Dem. Rep. (1947). Dipl. Krist. *Interfaces, boundary surface physics, thin film techniqūs.*
Rudolph, Dr Peter. Humboldt-Univ. zu Berlin, Sektion Physik, DDR-1040 Berlin, Invalidenstrasse 43, German Dem. Rep. *Crystallography.*
Ruscher, Prof. Christian. Akademie der Wissenschaften der DDR, Inst.für Polymerenchemie, DDR-153 Teltow-Seehof, Kantstrasse 55, German Dem. Rep. (1928). Dr.habil. *Polymerization, X-ray diffraction.*
Sarodnik, Dr Reinhard. Vereinigung für Kristallographie (VFK) in der GGW, DDR-1040 Berlin, Invalidenstrasse 43, German Dem. Rep. (1942). Dr.rer.nat. *Glasses.*

Schaal, Mr Joachim. Technische Hochschule Carl Schorlemmer Merseburg, DDR-4200 Merseburg, Geusärstrasse, German Dem. Rep. (1952). Dipl.Krist. *Polarization microscopy.*

Schäfer, Mr Peter. Humboldt-Univ. zu Berlin, Sektion Physik, DDR-1040 Berlin, Invalidenstrasse 43, German Dem. Rep. (1953). *Crystallography.*

Scharfenberg, Mr Rudolf. Akademie der Wissenschaften der DDR, Zentralinstitut für Festkörperphysik und Werkstofforschung, DDR-8032 Dresden, Helmholtzstrasse 20, German Dem. Rep. (1935). Dipl.Ing. *Crystal growth.*

Schenk, Dr Manfred. Akademie der Wissenschaften der DDR, Forschungsbereich Werkstoffwissenschaft, DDR-8032 Dresden, Helmholtzstrasse 20, German Dem. Rep. (1934). Dr.habil. *Crystallography.*

Scheschinski, Mrs Karin. Friedrich-Schiller-Univ. Jena, Sektion Chemie, DDR-69 Jena, Lessingstr. 10, German Dem. Rep. (1950). Dipl.Chem. *Crystallography.*

Schilling, Mr Hansjoachim. Akademie der Wissenschaften der DDR, Zentrum für Wissenschaftlichen Gerätebau, DDR-1199 Berlin, Rudower Chaussee 6, German Dem. Rep. (1938). Dipl.Ing. *Crystal growth.*

Schippel, Dr Erhard. Vereinigung für Kristallographie (VFK) in der GGW, DDR-1040 Berlin, Invalidenstrasse 43, German Dem. Rep. (1935). Dr.rer.nat. *Crystallography, structure, germination.*

Schläfer, Dr Dietrich. Akademie der Wissenschaften der DDR, Zentralinstitut für Festkörperphysik und Werkstofforschung, DDR-8032 Dresden, Helmholtzstrasse 20, German Dem. Rep. (1939). Dr.Ing. *Texture, metallography.*

Schläfer, Dr Ursula. Akademie der Wissenschaften der DDR, Zentralinstitut für Festkörperphysik und Werkstofforschung, DDR-8032 Dresden, Helmholtzstrasse 20, German Dem. Rep. (1938). Dr.Ing. *Texture, metallography.*

Schmelzler, Dr Hans-Peter. Akademie der Wissenschaften der DDR, Zentralinstitut für Kernforschung Rossendorf, DDR-8051 Dresden, PF 19, German Dem. Rep. (1947). Dr.rer.nat. *Crystallography.*

Schmidt, Mr Peter. Humboldt-Univ. zu Berlin, Sektion Physik, DDR-1040 Berlin, Invalidenstrasse 43, German Dem. Rep. (1952). Dipl.Krist. *Crystallography.*

Schmidt, Prof. Günter. Martin-Luther-Univ. Halle-Wittenberg, DDR-4020 Halle, Friedemann-Bach-Platz 6, German Dem. Rep. (1921). Dr.sc. *Crystal and solid state physics.*

Schmidt, Mrs Margot. Humboldt-Univ. zu Berlin, Sektion Physik, DDR-1040 Berlin, Invalidenstrasse 43, German Dem. Rep. (1951). Dipl.Krist. *Crystallography.*

Schmidt, Prof. Werner. Pädagogische Hochschule, Sektion Physik, DDR-8060 Dresden, Wigardstrasse 17, German Dem. Rep. (1934). Dr.habil. *Real structures, crystal physics.*

Schmitz, Dr Werner. Karl-Marx-Univ. Leipzig, Sektion Chemie, Fachrichtung Kristallographie, DDR-7030 Leipzig, Scharnhorststrasse 20, German Dem. Rep. (1943). Dr.rer.nat. *Crystallography, crystal growth.*

Schmücker, Mr Jürgen. Vereinigung für Kristallographie (VFK) in der GGW, DDR-1040 Berlin, Invalidenstrasse 43, German Dem. Rep. (1948). Dipl.Krist. *Crystallography.*

Schneider, Prof. Günter. Technische Hochschule Karl-Marx-Stadt, Sektion Physik, DDR-9010 Karl-Marx-Stadt, Strasse der Nationen 62, PSF 964, German Dem. Rep. (1929). Dr.habil. *Epitaxy.*

Schneider, Prof. Herbert. Bergakademie Freiberg, Section Physik, DDR-9200 Freiberg, Silberman Strasse 1, German Dem. Rep. (1926). Dr.habil. *Solid state physics, interfacial physics.*

Schott, Prof. Günter. Wilhelm-Pieck Univ. Rostock, Sektion Chemie, DDR-25 Rostock, Buchbinderstrasse 9, German Dem. Rep. (1921). Dr.habil. *Organic silicon chemistry.*

Schrader, Prof. Richard. Vereinigung für Kristallographie (VFK) in der GGW, DDR-1040 Berlin, Invalidenstrasse 43, German Dem. Rep. (1915). Dr.habil. *Inorganic chemistry, crystal chemistry.*

Schrauber, Mrs Hannelore. Akademie der Wissenschaften der DDR, Zentrum für Rechentechnik, DDR-1199 Berlin, Rudower Chaussee 5, German Dem. Rep. (1934). Dipl.Phys. *Crystallography, computer programming.*

Schreiter, Mr Peter. Karl-Marx-Univ. Leipzig, Sektion Chemie, Fachrichtung Kristallographie, DDR-7030 Leipzig, Scharnhorststrasse 20, German Dem. Rep. (1938). Dr.rer.nat. *Mineralogy, petrography.*

Schröder, Dr Winfried. Akademie der Wissenschaften der DDR, Zentrum für Wissenschaftlichen Gerätebau, DDR-1199 Berlin, Rudower Chaussee 6, German Dem. Rep. (1937). Dr.Ing. *Crystal growth, real structures.*

Schubert, Dr Gernot. Krankenhaus Friedrichshain, Urologische Klinik, DDR-1017 Berlin, Leninallee 171, German Dem. Rep. (1945). Dr.rer.nat. *X-ray structure analysis, order-disorder phenomena.*

Schubert, Mrs Heike-Kristina. Vereinigung für Kristallographie (VFK) in der GGW, DDR-1040 Berlin, Invalidenstrasse 43, German Dem. Rep. (1945). Dipl.Krist. *Microscopy, structure analysis.*

Schulz, Dr Manfred. Vereinigung für Kristallographie (VFK) in der GGW, DDR-1040 Berlin, Invalidenstrasse 43, German Dem. Rep. (1934). Dr.rer.nat. *Crystal physics, semiconductor physics.*

Schulze, Dr Dietrich. Akademie der Wissenschaften der DDR, Zentralinstitut für Festkörperphysik und Werkstofforschung, DDR-8032 Dresden, Helmholtzstrasse 20, German Dem. Rep. (1922). Dr.habil. *Solid state physics, electron microscopy.*

Schulze, Dr Günter. Technische Hochschule Carl Schorlemmer Merseburg, DDR-4200 Merseburg, Geusär Strasse, German Dem. Rep. (1932). Dr.rer.nat. *Crystallography.*

Schumann, Dr Bernd. Karl-Marx-Univ. Leipzig, Sektion Chemie, Fachrichtung Kristallographie, DDR-7030 Leipzig, Scharnhorststrasse 20, German Dem. Rep. (1947). *Real structures.*

Schunk, Mr Wolfgang. Martin-Luther-Univ. Halle, Sektion Chemie, DDR-4020 Halle, Domstrasse 5, German Dem. Rep. (1950). Dipl.Chem. *Crystallography.*

Schwartze, Mrs Gabriele. Karl-Marx-Univ. Leipzig, Sektion Chemie, Fachrichtung Kristallographie, DDR-7030 Leipzig, Scharnhorststrasse 20, German Dem. Rep. (1952). *Crystallography, mineralogy, petrography.*

Sedlacek, Dr Paul. Akademie der Wissenschaften der DDR, Zentralinstitut für physikalische Chemie, DDR-1199 Berlin, Rudower Chaussee 5, German Dem. Rep. (1920). Dr.rer.nat. *X-ray structure analysis.*

Seemann, Dr Hans. Akademie der Wissenschaften der DDR, Zentralinstitut für physikalische Chemie, DDR-1199 Berlin, Rudower Chaussee 5, German Dem. Rep. Dr.rer.nat. *Crystallography.*

Seidowski, Dr Eckart. Ingenieurhochschule Dresden, DDR-8019 Dresden, Hans-Grundig-Strasse 25, German Dem. Rep. (1946). Dr.Ing. *Semiconductor physics and techniqüs.*

Seifert, Dr Wolfgang. Akademie der Wissenschaften der DDR, Zentralinstitut für Physik der Erde, DDR-1500 Potsdam, Telegrafenberg, German Dem. Rep. (1947). Dr.rer.nat. *Mineralogy.*

Seydel, Mrs Renate. Vereinigung für Kristallographie (VFK) in der GGW, DDR-1040 Berlin, Invalidenstrasse 43, German Dem. Rep. (1949). Dipl.Phys. *Crystallography.*

Sieler, Dr Joachim. Karl-Marx-Univ. Leipzig, Sektion Chemie, DDR-701 Leipzig, Liebigstrasse 18, German Dem. Rep. Dr.rer.nat. *Crystallography.*

Sitte, Mrs Jutta. Vereinigung für Kristallographie (VFK) in der GGW, DDR-1040 Berlin, Invalidenstra. 43, German Dem. Rep. (1950). Dipl.Krist. *Crystallography.*

Soa, Dr Ernst-Adolf. Akademie der Wissenschaften der DDR, Zentralinstitut für Optik und Spektroskopie, DDR-69 Jena, Humboldtstrasse 11, German Dem. Rep. (1930). Dr.habil. *Thin film techniqüs.*

Sokoll, Mr Rolf. Friedrich-Schiller-Univ. Jena, Sektion Chemie, DDR-69 Jena, Lessingstrasse 10, German Dem. Rep. (1951). Dipl.Chem. *Crystallography.*

Sommer, Mr Joachim. TU Dresden, Sektion Physik, DDR-8027 Dresden, Mommsenstr. 13, German Dem. Rep. (1934). Dr.rer.nat. *X-ray metallography.*

Sommermann, Mr Günter. Akademie der Wissenschaften der DDR, Zentralinstitut für physikalische Chemie, DDR-1199 Berlin, Rudower Chaussee 5, German Dem. Rep. (1931). Dipl.Phys. *Spectroscopy, electron microscopy.*

Sorge, Dr Georg. Martin-Luther-Univ. Halle-Wittenberg, DDR-4020 Halle, Friedemann-Bach-Platz 6, German Dem. Rep. (1936). Dr.rer.nat. *Solid state physics.*

Spindler, Dr Herbert. Vereinigung für Kristallographie (VFK) in der GGW, DDR-1040 Berlin, Invalidenstrasse 43, German Dem. Rep. (1934). Dr.rer.nat. *X-ray structure analysis.*

Sprenger, Dr Heinz. Akademie der Wissenschaften der DDR, Zentralinstitut für Anorganische Chemie, DDR-1199 Berlin, Rudower Chaussee 5, German Dem. Rep. (1946). Dipl.Krist. *Crystallography.*

Stadermann, Mr Gerd. Akademie der Wissenschaften der DDR, Zentralinstitut für Optik und Spektroskopie, DDR-1199 Berlin, Rudower Chaussee 5, German Dem. Rep. (1947). Dipl.Krist. *Electron beam microanalysis.*

Starke, Dr Rainer. Bergakademie Freiberg, Sektion Geowissenschaften, DDR-9200 Freiberg, Brennhausgasse 14, German Dem. Rep. (1933). Dr.habil. *Mineralogy, crystal chemistry.*

Stecker, Prof. Kurt. Martin-Luther-Univ. Halle-Wittenberg, DDR-4020 Halle, Friedemann-Bach-Platz 6, German Dem. Rep. (1929). Dr.sc. *Semiconductor physics.*

Stegmann, Dr Eleonore. Technische Hochschule Carl Schorlemmer Merseburg, DDR-4200 Merseburg, Geusärstrasse, German Dem. Rep. (1929). Dr.rer.nat. *Chemistry, microscopy.*

Steil, Dr Helmut. Wilhelm-Pieck Univ. Rostock, Sektion Physik, DDR-2520 Rostock, Universitätsplatz 3, German Dem. Rep. (1949). Dr.rer.nat. *Crystallography.*

Steinbruch, Mrs Uta. Vereinigung für Kristallographie (VFK) in der GGW, DDR-1040 Berlin, Invalidenstrasse 43, German Dem. Rep. (1949). Dipl.Chem. *Crystal growth.*

Steinicke, Prof. Ursula. Akademie der Wissenschaften der DDR, Zentralinstitut für physikalische Chemie, DDR-1199 Berlin, Rudower Chaussee 5, German Dem. Rep. (1935). Dr.rer.nat. *Crystal growth, X-ray diffraction.*

Stelzner, Mrs Sabine. Vereinigung für Kristallographie (VFK) in der GGW, DDR-1040 Berlin, Invalidenstrasse 43, German Dem. Rep. (1949). Dipl.Krist. *Solid state physics.*

Stephan, Dr Dieter. TU Dresden, Sektion Physik, DDR-8027 Dresden, Mommsenstr. 13, German Dem. Rep. (1939). Dr.rer.nat. *X-ray diffraction, real structures.*

Stephanik, Dr Heinz. Martin-Luther-Univ. Halle, Sektion Chemie, DDR-4020 Halle, Domstrasse 5, German Dem. Rep. (1925). Dr.rer.nat. *Crystallography.*

Sterneck, Dr Dirk. Bauakademie der DDR, Inst.für Baustoffe, DDR-53 Weimar, Belvederer Allee 17, German Dem. Rep. (1941). Dr.rer.nat. *Silicate materials.*

Steussloff, Mr Peter. Vereinigung für Kristallographie (VFK) in der GGW, DDR-1040 Berlin, Invalidenstrasse 43, German Dem. Rep. (1942). Dipl.Phys. *Semiconductor physics.*

Stewig, Mr Helmut. TH Carl Schorlemmer Merseburg, DDR-4200 Merseburg, Geusärstr., German Dem. Rep. (1947). Dipl.Chem. *Crystallography.*

Storbeck, Prof. Fritz. TU Dresden, Sektion Physik, DDR-8027 Dresden, Mommsenstr. 13, German Dem. Rep. (1937). Dr.sc.nat. *Crystallography.*

Sutter, Dr Dietrich. Karl-Marx-Univ. Leipzig, Sektion Chemie, Fachrichtung Kristallographie, DDR-7030 Leipzig, Scharnhorststrasse 20, German Dem. Rep. *Crystallography.*

Swillens, Mr Eckhard. Vereinigung für Kristallographie (VFK) in der GGW, DDR-1040 Berlin, Invalidenstrasse 43, German Dem. Rep. (1935). Dipl.Min. *Physical chemistry, crystallography.*

Syhre, Dr Hans. Akademie der Wissenschaften der DDR, Zentralinstitut für Werkstoffforschung, DDR-8020 Dresden, Wittenbergstrasse 28, German Dem. Rep. *Crystallography.*

Szulzewsky, Dr Klaus. Akademie der Wissenschaften der DDR, Zentralinstitut für physikalische Chemie, DDR-1199 Berlin, Rudower Chaussee 5, German Dem. Rep. (1940). Dr.rer.nat. *X-ray structure analysis.*

Tänzer, Mr Dietmar. Akademie der Wissenschaften der DDR, Zentralinstitut für Elektronenphysik, DDR-108 Berlin, Mohrenstrasse 40-41, German Dem. Rep. (1940). Dipl.Phys. *Crystallography.*

Tempel, Dr Alfred. Karl-Marx-Univ. Leipzig, Sektion Chemie, Fachrichtung Kristallographie, DDR-7030 Leipzig, Scharnhorststrasse 20, German Dem. Rep. (1941). Dr.rer.nat. *Solid state physics, real physics, electron microscopy.*

Tempelhoff, Dr Klaus. Akademie der Wissenschaften der DDR, Zentralinstitut für Elektronenphysik, DDR-108 Berlin, Mohrenstrasse 41-43, German Dem. Rep. (1938). Dr.rer.nat. *Crystal growth, ideal structures.*

Teresiak, Mrs Angelika. Technische Hochschule Karl-Marx-Stadt, Sektion CWT, DDR-9010 Karl-Marx-Stadt, Scheffelstr. 15, German Dem. Rep. (1949). Dipl.Krist. *Crystallography.*

Tetzner, Dr Gottfried. Bergakademie Freiberg, Sektion Chemie, DDR-9200 Freiberg, Leipziger Strasse, German Dem. Rep. (1926). Dr.habil. *X-ray diffraction, inorganic chemistry.*

Thieme, Dr Wolfgang. Akademie der Wissenschaften der DDR, Zentralinstitut für Optik und Spektroskopie, DDR-1199 Berlin, Rudower Chaussee 6, German Dem. Rep. (1946). Dr.Ing. *Crystallography.*

Trettin, Mr Reinhard. Akademie der Wissenschaften der DDR, Zentralinstitut für Anorganische Chemie, DDR-1199 Berlin, Rudower Chaussee 5, German Dem. Rep. (1951). Dipl.Krist. *Real structures, surface analysis.*

Uecker, Mr Reinhard. Akademie der Wissenschaften der DDR, Zentralinstitut für organische Chemie, DDR-1199 Berlin, Rudower Chaussee 5, German Dem. Rep. (1951). Dipl.Krist. *Crystallography.*

Ulbricht, Prof. Heinz. Wilhelm-Pieck Univ. Rostock, Sektion Physik, DDR-2520 Rostock, Universitätsplatz 3, German Dem. Rep. (1931). Dr.sc.nat. *Crystallography.*

Ullrich, Dr Hans-Jürgen. Technische Univ. Dresden, Sektion Physik, DDR-8027 Dresden, Mommsenstr. 13, German Dem. Rep. (1938). Dr.sc. *Real structures, X-ray structure analysis.*

Unangst, Prof. Dietrich. Friedrich-Schiller-Univ. Jena, Sektion Physik, DDR-69 Jena, Max - Wien - Platz 1, German Dem. Rep. (1931). Dr.habil. *Crystallography, real structures.*

Unger, Prof. Konrad. Karl-Marx-Univ. Leipzig, DDR-7010 Leipzig, Linnestrasse 5, German Dem. Rep. (1934). Dr.sc. *Crystallography.*

Velfe, Mr Hans Dieter. Akademie der Wissenschaften der DDR, Inst.für Festkörperphysik und Elektronenmikroskopie, DDR-4010 Halle, Am Weinberg 2, German Dem. Rep. (1942). Dipl.Phys. *Electron microscopy.*

Vester, Mr Jörg. Akademie der Wissenschaften der DDR, Inst.für Festkörperphysik und Elektronenmikroskopie, DDR-4010 Halle, Am Weinberg 2, German Dem. Rep. (1947). Dipl.Phys. *Solid state physics.*

Voigt, Dr Dieter. Vereinigung für Kristallographie (VFK) in der GGW, DDR-1040 Berlin, Invalidenstr. 43, German Dem. Rep. (1938). Dr.rer.nat. *Crystallography.*

Voigt, Mrs Gabriele. Vereinigung für Kristallographie (VFK) in der GGW, DDR-1040 Berlin, Invalidenstrasse 43, German Dem. Rep. (1952). Dipl.Krist. *Crystallography.*

Voigt, Dr Rita. Akademie der Wissenschaften der DDR, Zentralinstitut für physikalische Chemie, DDR-1199 Berlin, Rudower Chaussee 5, German Dem. Rep. (1941). Dr.rer.nat. *Real structures.*

Vollstädt, Dr Heiner. Akademie der Wissenschaften der DDR, Zentralinstitut für Physik der Erde, DDR-1500 Potsdam, Telegrafenberg, German Dem. Rep. (1939). Dr.rer.nat. *Mineralogy, geology.*

Wadewitz, Dr Heinz. Akademie der Wissenschaften der DDR, Zentralinstitut für Festkörperphysik und Werkstoffforschung, DDR-8032 Dresden, Helmholtzstr. 20, German Dem. Rep. (1921). Dr.rer.nat. *X-ray diffraction, metal physics.*

Wäsch, Dr Elke. Humboldt-Universität zu Berlin, Sektion Physik, DDR-1040 Berlin, Hessischestrasse 4, German Dem. Rep. (1942). *Crystal growth, crystal perfection.*

Wagner, Mr Gerald. Karl-Marx-Univ. Leipzig, Sektion Chemie, Fachrichtung Kristallographie, DDR-7030 Leipzig, Scharnhorststrasse 20, German Dem. Rep. (1950). Dipl.Krist. *Crystal growth, real structures.*

Wagner, Mr Gunther. Vereinigung für Kristallographie (VFK) in der GGW, DDR-1040 Berlin, Invalidenstrasse 43, German Dem. Rep. (1940). Dipl.Min. *Real structures, X-ray structure analysis.*

Wahner, Mrs Christa. Akademie der Wissenschaften der DDR, Zentralinstitut für Anorganische Chemie, Bereich Glas Keramik, DDR-1040 Berlin, Invalidenstrasse 44, German Dem. Rep. (1950). Dipl.Krist. *X-ray structure analysis, X-ray diffraction.*

Walther, Mrs Christa. TH Carl Schorlemmer Merseburg, DDR-4200 Merseburg, Geusärstr., German Dem. Rep. (1934). Dipl.Phys. *X-ray structure analysis.*

Wappler, Dr Gert. Humboldt-Univ. zu Berlin, Sektion Physik, DDR-1040 Berlin, Invalidenstrasse 43, German Dem. Rep. (1935). Dr.rer.nat. *Mineralogy, crystal structures.*

Wawra, Mr Herbert. Vereinigung für Kristallographie (VFK) in der GGW, DDR-1040 Berlin, Invalidenstrasse 43, German Dem. Rep. (1942). Dipl.Krist. *Semiconductor and interface physics.*

Weis, Mr Josef. Vereinigung für Kristallographie (VFK) in der GGW, DDR-1040 Berlin, Invalidenstrasse 43, German Dem. Rep. (1933). Dr.rer.nat. *Crystal growth, X-ray diffraction.*

Weise, Dr Günter. Akademie der Wissenschaften der DDR, Zentralinstitut für Festkörperphysik und Werkstoffforschung, DDR-8032 Dresden, Helmholtzstrasse 20, German Dem. Rep. (1930). Dr.habil. *Crystallography.*

Weiss, Mr Hans-Georg. Akademie der Wissenschaften der DDR, Zentrum für Rechentechnik, DDR-1199 Berlin, Rudower Chaussee 5, German Dem. Rep. (1935). Dipl.Phys. *Computing, computer programming.*

Weiss, Dr Helmut. Vereinigung für Kristallographie (VFK) in der GGW, DDR-1040 Berlin, Invalidenstrasse 43, German Dem. Rep. (1934). Dr.rer.nat. *Crystal growth, crystallography.*

Wendland, Mrs Bettina. Vereinigung für Kristallographie (VFK) in der GGW, DDR-1040 Berlin, Invalidenstrasse 43, German Dem. Rep. (1950). Dipl.Krist. *Metals and material science.*

Werner, Dr Inge. Akademie der Wissenschaften der DDR, Zentralinstitut für Arbeitsmedizin der DDR, DDR-1134 Berlin, Nöldnerstrasse 40-42, German Dem. Rep. (1934). Dr.rer.nat. *X-ray diffraction, phase analysis.*

Werner, Mr Michael. Karl-Marx-Univ. Leipzig, Sektion Chemie, Fachrichtung Kristallographie, DDR-7030 Leipzig, Scharnhorststrasse 20, German Dem. Rep. (1950). *Silicate structures, glasses.*

Wieser, Dr Egbert. Akademie der Wissenschaften der DDR, Zentralinstitut für Kernforschung, DDR-8051 Dresden, Postfach 19, German Dem. Rep. (1938). Dr.rer.nat. *Solid state physics.*

Wilde, Dr Wolfgang. Humboldt-Univ. zu Berlin, Sektion Chemie, DDR-1040 Berlin, Hessischestrasse 1-2, German Dem. Rep. (1935). Dr.rer.nat. *Crystal chemistry, X-ray structure analysis, synthetic chemistry.*

Wildner, Mr Günter. Vereinigung für Kristallographie (VFK) in der GGW, DDR-1040 Berlin, Invalidenstrasse 43, German Dem. Rep. (1932). Dipl.Min. *Petrography, polarization microscopy.*

Windsch, Prof. Wolfgang. Karl-Marx-Univ. Leipzig, DDR-7010 Leipzig, Linnestrasse 5, German Dem. Rep. (1931). *Solid state physics, ferrölectrics.*

Winzer, Dr Achim. TH Carl Schorlemmer Merseburg, DDR-4200 Merseburg, Geusärstr., German Dem. Rep. (1937). Dr.rer.nat. *Crystallography.*

Wolf, Mr Eberhard. Akademie der Wissenschaften der DDR, Zentrum für Wissenschaftlichen Gerätebau, DDR-1199 Berlin, Rudower Chaussee 6, German Dem. Rep. (1932). Dipl.Min. *Crystal growth, X-ray diffraction.*

Worzala, Dr Horst. Akademie der Wissenschaften der DDR, Zentralinstitut für Anorganische Chemie, DDR-1199 Berlin, Rudower Chaussee 5, German Dem. Rep. (1938). Dr.rer.nat. *X-ray structure analysis, structure chemistry.*

Wünsche, Mrs Inez. Humboldt-Univ. zu Berlin, Sektion Physik, DDR-1040 Berlin, Hessischestrasse 2, German Dem. Rep. (1953). Dipl.Krist. *Crystallography.*

Wurl, Mr Bernd. Vereinigung für Kristallographie (VFK) in der GGW, DDR-1040 Berlin, Invalidenstrasse 43, German Dem. Rep. (1947). Dipl.Krist. *Crystal growth.*

Zedler, Dr Achim. Akademie der Wissenschaften der DDR, Zentralinstitut für physikalische Chemie, DDR-1199 Berlin, Rudower Chaussee 5, German Dem. Rep. (1934). Dr.rer.nat. *X-ray structure analysis.*

Zeigan, Dr Dieter. Akademie der Wissenschaften der DDR, Zentralinstitut für physikalische Chemie, DDR-1199 Berlin, Rudower Chaussee 5, German Dem. Rep. (1935). Dr.rer.nat. *X-ray structure analysis.*

Zickert, Mr Kurt. Wilhelm-Pieck Univ. Rostock, Sektion Physik, DDR-2520 Rostock, Universitätsplatz 3, German Dem. Rep. (1936). Dipl.Phys. *Amorphous solid state, X-ray physics.*

Ziemer, Dr Burkhard. Akademie der Wissenschaften der DDR, Zentralinstitut für Anorganische Chemie, DDR-1199 Berlin, Rudower Chaussee 5, German Dem. Rep. *Crystallography.*

Zschach, Dr Siegfried. Vereinigung für Kristallographie (VFK) in der GGW, DDR-1040 Berlin, Invalidenstrasse 43, German Dem. Rep. (1938). Dr.rer.nat. *Instrument manufacture.*

GERMANY, FED. REP.
BRD

Sub-Editors: H. Burzlaff and H.W. Zimmermann

Notes

1. International telephone country code - 49. Exceptions are Austria 060 and Luxemburg 050 instead of 49. These numbers, e.g. 00949 for Denmark or 060 for Austria replace the first digit 0 of the area code.

2. Degrees conferred by the German Universities are the *Doctor honoris causae* (Dr.h.c.) (an honorary degree), *Doctor philosophiae* (Dr.phil.), *Doctor philosophiae naturalis* (Dr.phil.nat.), *Doctor philosophiae rerum naturalium* (Dr.phil.rer.nat.), *Doctor rerum naturalium* (Dr.rer.nat.), and *Doctor scientiae naturalis* (Dr.sc.nat.) (all approximately equivalent to PhD), and the *Diplom-Chemiker* (Dipl.-Chem.), *Diplom-Mineraloge* (Dipl.-Min.) and *Diplom-Physiker* (Dipl.-Phys.) (these three are approximately equivalent to MSc). At the German Technische Hochschulen and Technical Universities the degrees *Doctor-Ingenieur* (Dr.-Ing.), *Doctor rerum technicarum* (Dr.rer.techn.), *Doctor scientiae technicae* (Dr.sc.techn.), and *Diplom-Ingenieur* (Dipl.-Ing.) can be obtained. To get the position of a *Dozent* (equivalent to reader) at a German University, Technische Hochschule, or Technical University the conditions of the *Habilitation* have to be fulfilled corresponding to the degree of *Dr.habil.*

3. At a German University, Technische Hochschule or Techn. University, persons can be appointed in the following positions *Ord. Professor* (equivalent to full professor), *Wiss. Rat und Professor* (equivalent to associate professor), *ausserplanmässiger Professor* (equivalent to assistant professor), *Dozent* (equivalent to reader). Other positions are *Akad. Direktor, Akad. Oberrat, Akad. Rat, Kustos, Konservator, Oberassistent, Assistent (Asst.), Wiss. Angestellter (Ang.), and Wiss. Mitarbeiter (Mit.)* (equivalent to assistant professor, research associate, or research assistant).

4. Further abbreviations used in the German list are
   Abt. - Abteilung (department)
   Ang. - Angestellter (employee)
   Ass. - Assistent (assistant)
   akad. - akademisch (academic)
   angew. - angewandt (applied)
   Apl. Prof. - ausserplanmässiger Prof.
   em. - emeritiert (emeritus)
   FU - Freie Universität at Berlin
   Geb. - Gebäude (building)
   Ges. - Gesellschaft (society)
   GH - Gesamthochschule (University)
   Inst. - Institut
   i.R. - im Ruhestand (retired)
   Mit. - Mitarbeiter (co-worker)
   MPI - Max-Planck-Institut
   Ord. Prof. - ordentlicher Prof.
   Str. - Strasse (street)
   RWTH - Rheinisch-Westfälische Technische Hochschule at Aachen
   TH - Technische Hochschule
   TU - Technische Universität
   U. - Universität
   wiss. - wissenschaftlich (scientific)

5. The postal code consists of D- and a four-digit number, e.g. D-7500 and precedes the name of the town.

6. The telephone numbers included in the code list are normally those of a central switchboard. If the number ends in a '1', it is in general possible to bypass the central switchboard by dialing the extension number instead of the '1'. The first group of digits beginning with '0' is the area code.

7. The alphabetical order of the names in the list is according to German usage. So *von Schnering* is listed under S and vowels with an Umlaut (Ä, Ö, Ü) are placed as if they were written as AE, OE, and UE respectively.

8. The name of the 'present position' is given in English or German, in most cases according to the data presented by the crystallographer.

Abriel, Dr Walter. Inst. f. Anorg. Chemie, Universität, Callinstr. 9, D-3000 Hannover, Germany, Fed. Rep. (1949) Dr.rer.nat., chemistry (U. Regensburg, 1980). Privatdozent. (tel. 0511 + 7621, ext. 3552). *Inorganic solid state chemistry.*

Abs-Wurmbach, Dr Irmgard. Inst. f. Mineralogie, Ruhr- Universität, Universitätsstr. 150, D-4630 Bochum, Germany, Fed. Rep. (1938) Dr.rer.nat., mineralogy (U. Bonn, 1973). Wiss. Ang. (tel. 0234 + 7001, ext. 4378). *Absorption spectroscopy, neutron diffraction, pressure-temperature dependent phase-relations.*

Albers, Miss Ursula. An St. Albertus Magnus 31, D-4300 Essen, Germany, Fed. Rep. (1959)

Alexander, Prof. Dr Helmut. Abt. f. Metallphysik im II. Physik. Inst., Universität, Zülpicher Str. 77, D-5000 Köln 41, Germany, Fed. Rep. (1928) Prof., metal physics Wiss. Abteilungsvorsteher. (tel. 0221 + 4701, ext. 4200). *Electron microscopy, crystal defects, plasticity, electron paramagnetic resonance.*

Allmann, Prof. Dr Rudolf. Fachbereich Geowissenschaften, Universität, Lahnberge, D-3550 Marburg, Germany, Fed. Rep. (1931) Dr.habil., mineralogy and crystallography (U. Marburg, 1968). Prof. (tel. 06421 + 281, ext. 3002). *Double layer structures, heteropoly acids, organic structures, electron density.*

Amberger, Prof. Dr Eberhard. Inst. f. Anorganische Chemie, Universität, Meiserstr. 1, D-8000 München 2, Germany, Fed. Rep. (1928) Dr., inorganic chemistry (U. München, 1956). Prof. (tel. 089 + 5902, ext. 356). *Incommensurate structures, organometallic and organic superconductors*

D' Amour-Sturm, Dr Hedwig. Fachbereich Physik, Universität-GH, Warburgerstr. 100, D-4790 Paderborn, Germany, Fed. Rep. (1948) Dr., crystallography (U. Marburg, 1974). Asst. (tel. 05251 + 602674). *Crystal structures, phase transitions, high pressure techniques.*

Amstutz, Prof. Dr Dr hc Christian Gerhard. Mineralogisch-Petrogr. Inst., Universität, Postfach 104040, D-6900 Heidelberg, Germany, Fed. Rep. (1922) Dr.sc.nat. Dr.h.c., mineralogy, petrology, geology (ETH Zürich, 1952). Prof. and Dir. (tel. 06221 + 561, ext. 2802). *Mineralogy, petrology, ore deposits, igneous and sedimentary petrology, history and philosophy of science, pseudomorphism.*

Angermund, Mr Klaus Peter. MPI f. Kohlenforschung, Lembkestr. 5, D-4330 Mülheim/Ruhr, Germany, Fed. Rep. (1958) Dipl.-Chem. (U. Düsseldorf, 1983). PhD student. (tel. 0208 + 306 ext. 491). *Organometallic compounds, electron deformation density distribution.*

Armbruster, Dr Thomas. Inst. f. Mineralogie, Universität, Universitätsstr. 150, D-4630 Bochum, Germany, Fed. Rep. (1950) Dipl.-Min., mineralogy (U. Mainz, 1975). Wiss. Ang. (tel. 0234 + 7001, ext. 4419). *Inorganic crystal structures, crystal growth.*

Arndt, Prof. Dr Jörg Friedrich. Mineralogisch-Petrogr. Inst., Universität, Wilhelmstr. 56, D-7400 Tübingen, Germany, Fed. Rep. (1938) Habilitation, mineralogy and crystallography (U. Tübingen, 1974). Prof. (tel. 07071 + 291, ext. 6802). *Mineralogy, crystallography, crystal chemistry, high pressure mineralogy, glasses - structure and properties, technical mineralogy, shock metamorphism, lunar and terrestrial materials.*

Arni, Mr Raghuuir Krishnaswamy. Inst. f. Kristallographie, FU Berlin, Takustr. 6, D-1000 Berlin 33, Germany, Fed. Rep. (1953) Diplom, mineralogy/crystallography (TU Berlin, 1984) Wiss. Mit. (tel. 030 + 838-6325). *Crystal physics, optics, X-ray crystallography, high-resolution protein structures.*

Arnold, Prof. Dr Heinrich Günther Alfred. Inst. f. Kristallographie, RWTH, Templergraben 55, D-5100 Aachen, Germany, Fed. Rep. (1930) Dr.rer.nat., crystallography (U. Würzburg, 1964). Dozent. (tel. 0241 + 80, ext. 6901, telex 08/32704 THAC D). *Symmetry, phase transitions, lattice dynamics.*

Attig, Dr Rainer. Chemische Landesuntersuchungsanstalt, Hoffstr. 3, D-7500 Karlsruhe, Germany, Fed. Rep. (1944) Dr.rer.nat., chemistry (TU Braunschweig, 1973). Reg. Chemierat. (tel. 0721 + 1351, ext. 3631). *Inorganic and organic crystal structures, neutron diffraction.*

Axmann, Dr Anton. Bereich C1, Hahn-Meitner-Inst. f. Kernforschung, Glienicker Str. 100, D-1000 Berlin 39, Germany, Fed. Rep. (1937) Dr.rer.nat., physics (RWTH Aachen, 1968). Asst. Group Leader. (tel. 030 + 80091, ext. 2756). *Instrumentation, neutron scattering.*

Baars, Dr-Ing Jan Walter. Abt. Infrarotphysik, Fraunhofer Inst. f. Angew. Festkörperphysik, Eckerstr. 4, D-7800 Freiburg, Germany, Fed. Rep. (1931) Dr.-Ing. solid state physics (TU Berlin, 1967). Branch Head. (tel. 0761 + 2714, ext. 247). *Crystal growth, epitaxy, inorganic crystal structures.*

Babel, Prof. Dr Dietrich. Fachbereich Chemie, Universität, Hans-Meerwein-Str., D-3550 Marburg, Germany, Fed. Rep. (1930) Prof., inorganic chemistry (U. Marburg, 1971). Prof. (tel. 06421 + 28, ext. 5625). *Solid state chemistry, inorganic crystal structures.*

Bade, Mr Dirk. Physik Dept., Techn. Universität, James-Franck-Str., D-8046 Garching, Germany, Fed. Rep. (1950) Dipl.-Phys., physics (TU München, 1974). Wiss. Ang. (tel. 089 + 3209, ext. 2509). *Protein structure analysis, Mössbauer - Rayleigh scattering interference method.*

Bärnighausen, Prof. Dr Hartmut. Inst. f. Anorganische Chemie, Universität, Engesserstr., Geb. Nr. 30.45, Postfach 6380, D-7500 Karlsruhe, Germany, Fed. Rep. (1933) Dr.rer.nat., chemistry (U. Freiburg, 1959). Prof. (tel. 0721 + 608, ext. 3484). *Inorganic crystal structures, crystal chemistry, symmetry.*

Bambauer, Prof. Dr Hans Ulrich. Inst. f. Mineralogie, Universität, Corrensstr. 24, D-4400 Münster, Germany, Fed. Rep. (1929) Dr.rer.nat., mineralogy (U. Mainz, 1957). Full Prof. (tel. 0251 + 83, ext. 3450). *Rock-forming minerals (feldspars, quartz), petrology, environmental geochemistry (fly ashes), electron microscopy.*

Banner, Dr David William. Eur. Mol. Biol. Lab., Meyerhofstr. 1, D-6900 Heidelberg, Germany, Fed. Rep. (1946) D. phil., molecular biophysics (U. Oxford, UK, 1972). Sci. (tel. 06221 + 387, ext. 255). *Protein structure, nucleic acid structure, protein-nucleic acid interactions.*

Baresel, Dr-Ing Detlef Wilhelm Berthold. Forschungszentrum, FCW, Robert Bosch GmbH, Postfach 50, D-7000 Stuttgart 1, Germany, Fed. Rep. (1926) Dr.-Ing., inorganic chemistry (TU Berlin, 1962). Wiss. Referent. (tel. 0711 + 8111, ext. 6552). *Inorganic chemistry, physical chemistry, catalysis, surface chemistry, adsorption, chemisorption, solid state chemistry, surface structure, semiconductors, ionic conductors.*

Bartl, Prof. Dr Hans. Inst. f. Kristallographie, Universität, Senckenberg-Anlage 30, D-6000 Frankfurt/Main, Germany, Fed. Rep. (1933) Dr.habil., crystallography and mineralogy (U. Frankfurt/Main, 1971). Prof. (tel. 069 + 798, ext. 2105, Telex 413932 Unif). *X-ray and neutron diffraction.*

Bartsch, Dr Hans-Hagen. Inst. f. Mineralogie, Corrensstr. 24, D-4400 Münster, Germany, Fed. Rep. (1955) Dr.rer.nat., mineralogy (U. Hamburg, 1984). Wiss. Mit. (tel. 0251 + 83, ext. 3463). *Inorganic and organic crystal structures, crystal disorder.*

Bats, Dr Jan, Willem. Inst. f. Kristallographie, Universität, Senckenberganlage 30, D-6000 Frankfurt/Main, Germany, Fed. Rep. (1949) Dr., X-ray and neutron diffraction (Twente U., Netherlands, 1976). Res. Asst. (tel. 069 + 798, ext. 2293). *High resolution X-ray and neutron diffraction studies, electron density studies.*

Bauer, Prof. Dr Ernst Georg. Physikalisches Inst., TU, Leibnizstr. 4, D-3392 Clausthal-Zellerfeld, Germany, Fed. Rep. (1928) Dr.phil., physics (U. München, 1955). Ord. Prof. (tel. 05323 + 721, ext. 2249). *Surface physics, thin film growth and structure.*

Baumann, Mr Jürgen Rudolf. Inst. f. Kristallographie, TU, Adolf-Römer-Str. 2a, D-3392 Clausthal-Zellerfeld, Germany, Fed. Rep. (1957) Cand.min. (tel. 05323 + 78513). *Single crystals, metals, minerals, neutron diffraction.*

Baumgart, Mr Helmut. MPI f. Festkörperforschung, Heisenbergstr. 1, D-7000 Stuttgart, Germany, Fed. Rep. (1952) M.S., physics (Purdue U., W. Lafayette, Indiana, USA, 1977). Grad. student. (tel. 0711 + 7830, ext. 247). *X-ray topography, semiconductor materials engineering, laser annealing.*

Bayh, Prof. Dr Werner. Mineralogisches Inst., Universität, Wilhelmstr. 56, D-7400 Tübingen, Germany, Fed. Rep. (1928) Dr.rer.nat., physics (U. Tübingen, 1962). Dozent. (tel. 07071 + 29, ext. 2648). *Crystal growth, crystal physics, interface physics.*

Beck, Prof. Dr Horst Philipp. Inst. f. Anorganische Chemie, Universität, Egerlandstr. 1, D-8520 Erlangen, Germany, Fed. Rep. (1941) Dr.habil., inorganic chemistry (U. Karlsruhe, 1980). Prof. (tel. 09131 + 85, ext. 7353). *Crystal chemistry, high pressure.*

Becker, Prof. Dr Gerd. Fak. f. Chemie - Inst. f. Anorg. Chemie, Universität, Pfaffenwaldring 55, D-7000 Stuttgart 80, Germany, Fed. Rep. (1940) Dr.habil., inorganic chemistry (U. Karlsruhe, 1976). Ord. Prof. (tel. 0711 + 685, ext. 4172). *X-ray structure determinations of compounds with elements of the main-groups (esp. P, As, Sb, Bi).*

Behrens, Dr Heinrich. Fachinformationszentrum Energie Physik Mathematik GmbH, Kernforschungszentrum, D-7514 Eggenstein-Leopoldshafen 2, Germany, Fed. Rep. (1937) Dr.rer.nat., physics (U. Karlsruhe, 1966). (tel. 07247 + 821, ext. 4554).

Behrens, Dr Ulrich Hermann. Inst. f. Anorg. u. Angew. Chemie, Universität, Martin-Luther-King-Platz 6, D-2000 Hamburg 13, Germany, Fed. Rep. (1946) Dr.habil., inorganic chemistry (U. Hamburg 1975). Priv.-Doz. (tel. 040 + 4123, ext. 2894). *Organometallic and coordination chemistry, crystal structures.*

Behruzi, Dr Massoud. Inst. f. Kristallographie, RWTH, Templergraben 55, D-5100 Aachen, Germany, Fed. Rep. (1939) Dr., crystallography, mineralogy (RWTH Aachen, 1972). Wiss. Ang. (tel. 0241 + 42, ext. 6906). *Crystal structure, crystal chemistry, silicates and related compounds.*

Belzner, Mr Andreas. Inst. f. Kristallographie, LM-Universität, Theresienstr. 41, D-8000 München 2, Germany, Fed. Rep. (1957) Dipl.Min. (U. München) (tel. 089 + 13941).

Bennett, Dr William, Samuel, Jr. Abteilung Wittmann, MPI f. Molekulare Genetik, Ihnestr. 63-73, D-1000 Berlin, Germany, Fed. Rep. (1949) PhD., molecular biology (Yale U., USA, 1978). Res. scient. (tel. 030 + 8307-1, ext. 344). *Protein structure and function, crystallographic computing.*

Bente, Dr Klaus Alexander. Mineralogisch-Kristallographisches Inst., Universität, Goldschmidtstr. 1, D-3400 Göttingen, Germany, Fed. Rep. (1946) Dr.habil., mineralogy (U. Göttingen). Priv.-Doz. *Physico-chemical crystallography, sulfosalts, crystal growth.*

Berg, Mrs Dr Lieselotte. Gmelin-Inst., Max-Planck-Ges., Varrentrappstr. 40/42, D-6000 Frankfurt/Main, Germany, Fed. Rep. (1933) Dr.rer.nat., crystallography (TU Braunschweig, 1964). Wiss. Mit. (tel. 069 + 7917, ext. 239). *Crystallography, inorganic chemistry, metalorganic chemistry.*

Bergerhoff, Prof. Dr Guenter. Inst. f. Anorg. Chemie, Universität, Gerhard-Domagk-Str. 1, D-5300 Bonn 1, Germany, Fed. Rep. (1926) Dr.rer.nat., inorganic chemistry (U. Bonn, 1954). Prof. (tel. 0228 + 731, ext. 2657). *Inorganic crystal structures, structure documentation.*

Berking, Dr Bernhard. Fachbereich Seefahrt, Fachhochschule Hamburg, Rainvilleterrasse 4, D-2000 Hamburg 52, Germany, Fed. Rep. (1939) Dr.habil., crystallography (U. Hamburg, 1975). Dozent. (tel. 040 + 3802848). *Mineralogy, organic biochemistry.*

Bernotat, Dr Walter Hermann. Inst. f. Mineralogie, Corrensstr. 24, D-4400 Münster, Germany, Fed. Rep. (1939) Dr., mineralogy (U. Zürich, 1969). (tel. 0251 + 831, ext. 3454). *X-ray powder techniques.*

Bernotat-Wulf, Mrs Dr Hannelore. Inst. f. Kristallographie, Universität, Kaiserstr. 12, D-7500 Karlsruhe, Germany, Fed. Rep. (1940) Dr., crystallography (ETH Zürich, 1970). Wiss. Mit. (tel. 0721 + 6083320) *Crystal chemistry, inorganic structures.*

Bertelmann, Mr Dieter Wilhelm. Inst. f. Kristallographie, Universität, Kaiserstr. 12, D-7500 Karlsruhe, Germany, Fed. Rep. (1954) Dipl.Min., mineralogy (U. Karlsruhe). *crystallography.*

Berthold, Prof. Dr Hans Joachim. Inst. f. Anorg. Chemie, Universität, Callinstr. 9, D-3000 Hannover, Germany, Fed. Rep. (1923) Dr.phil., inorganic chemistry (U. Köln, 1952). Prof. C4, Institutsdirektor. (tel. 0511 + 762, ext. 2254). *Inorganic crystal structures, phase transitions.*

Berthold, Mr Thomas. Inst. f. Kristallographie, Universität, Theresienstr. 41, D-8000 München, Germany, Fed. Rep. (1955) (tel. 089 + 23944352). *Anomalous X-ray scattering, crystal growth, crystal defects.*

Betz, Mr Helmut. Arbeitsgruppe f. Chem. Kristallographie, MPI f. Biochemie, Am Klopferspitz, D-8033 Martinsried, Germany, Fed. Rep. (1956) Dipl.Chem., chemistry (TU. München, 1983). (tel. 089 + 85782659). *X-ray structure analysis, organic compounds.*

Betzel, Mr Christian. Inst. f. Kristallographie, FU. Berlin, Takustr. 6, D-1000 Berlin 33, Germany, Fed. Rep. (1956) Dipl.Phys., physics (U. Göttingen, 1982). Res. scient. (tel. 030 + 8386318). *Protein crystallography, neutron diffraction, hydrogen bonding, synchrotron radiation.*

Beyer, Mrs Angelika. Inst. f. Anorganische Chemie, Engesserstr., Geb.-Nr. 30.45, D-7500 Karlsruhe, Germany, Fed. Rep. (1954) Dipl.-Chem., (U. Karlsruhe, 1978). (tel. 0721 + 608, ext. 3485). *Crystal chemistry.*

Biedl, Dr Albrecht. Fachbereich Informatik, TU, Strasse des 17. Juni 135, D-1000 Berlin 12, Germany, Fed. Rep. (1938) Dr.phil., mineralogy (U. Wien, 1963). Akad. Rat. (tel. 030 + 314, ext. 4893/4891). *Computer programming, symmetry.*

Bielen, Prof. Dr Helmut Josef. Neuhausstr. 15, D-6000 Frankfurt/Main 1, Germany, Fed. Rep. (1929) Prof., solid state physics (U.P.R., USA, 1963). Industrial Consultant. (tel. 069 + 593210). *Inorganic crystal structures, magnetic crystal structures.*

Birnstock, Dr Ronald Alfred Harri. Beschleunigerlabor, Universität München, Hochschulgelände, D-8046 Garching, Germany, Fed. Rep. (1934) Dr., crystallography (U. München, 1965). Wiss. Ang. (tel. 089 + 3209, ext. 5273). *Inorganic crystal structures, computer programming.*

Bissert, Mrs Dr Elisabeth Gertrud. Mineralogisches Inst., Universität, Olshausenstr. 40, D-2300 Kiel, Germany, Fed. Rep. (1933) Dr., crystallography (U. Kiel, 1969). Akad. Direktor. (tel. 0431 + 880, ext. 2893). *Inorganic crystal structures, silicate crystal chemistry.*

Blaschke, Prof. Dr Rochus Bruno Albert. Inst. f. Medizinische Physik, Universität, Hüfferstr. 68, D-4400 Münster, Germany, Fed. Rep. (1930) Dr.rer.nat., mineralogy (U. Giessen, 1959). (tel. 0251 + 83, ext. 5113). *Electron microscopy, biomineralogy (urinary calculi), applied mineralogy.*

Bleif, Dr Hans-Jürgen. Kernchemie und Reaktor, Hahn-Meitner-Inst. f. Kernforschung, Glienicker Str. 100, D-1000 Berlin 39, Germany, Fed. Rep. (1945) Dr.rer.nat., solid state physics (U. Tübingen, 1978). Physicist. (tel. 030 + 8009, ext. 2758). *Phase transitions, diffuse scattering.*

Block, Prof. Dr Jochen Hermann. Fritz-Haber-Inst., Max-Planck-Ges., Faradayweg 4-6, D-1000 Berlin 33, Germany, Fed. Rep. (1929) Dr.sc.h.c., physical chemistry. Direktor. (tel. 030 + 8305, ext. 411). *Surface chemistry, field ionization, mass spectrometry, fast chemical reactions.*

Blüthgen, Dipl-Min Waldemar. Fa. Ernst Leitz Wetzlar GmbH, Ernst-Leitz-Str. 30, Pf 2020, D-6330 Wetzlar, Germany, Fed. Rep. (1940) Diplom, crystallography, metal surfaces, epitaxy (U. Bonn, 1968). Patent Engineer. (tel. 06441 + 292466, telex 483849 Leiz D). *Crystal optics, ceramics, optical glasses, microscopy, glass ceramics.*

Bock, Mr Hans. Inst. f. Kristallographie, Universität, Im Stadtwald 15, Fr. 17.3, D-6600 Saarbrücken, Germany, Fed. Rep. (1954) Dipl. Mineraloge. *Ferroelectrics, ceramics, X-ray diffraction.*

Bögge, Dr Hartmut. Fak. f. Chemie, Universität, Universitätsstr., D-4800 Bielefeld, Germany, Fed. Rep. (1953) Dipl.Chem., chemistry (U. Dortmund, 1977). (tel. 0521 + 106, ext. 2906). *Transition metal complexes.*

Boehm, Prof. Dr Hanns-Peter. Inst. f. Anorganische Chemie, Universität, Meiserstr. 1, D-8000 München 2, Germany, Fed. Rep. (1928) Dr.rer.nat., inorganic chemistry (TH Darmstadt, 1953). Ord. Prof. (tel. 089 + 5902355). *Carbon and graphite chemistry, surface chemistry, catalysis.*

Böhm, Dr Horst. Inst. f. Mineralogie, Universität, Corrensstr. 24, D-4400 Münster, Germany, Fed. Rep. (1937) Dr., crystallography (ETH Zürich, 1969). Prof. (tel. 0251 + 83, ext. 3459). *Structure - physical properties relationships, phase transformations, modulated structures.*

Böhme, Mrs Dr Reinhild. Inst. f. Mineralogie, Ruhr-Universität, Universitätsstr. 150, D-4630 Bochum, Germany, Fed. Rep. (1942) Dr.rer.nat.habil., crystallography (U. Erlangen-Nürnberg, 1982). Priv. Doz. (tel. 0234 + 700, ext. 3740). *Structure determination, superstructure effects, crystallographic computing.*

Boese, Dr Roland. Inst. f. Anorg, Chemie, Universität, Universitätsstr. 5-7, D-4300 Essen, Germany, Fed. Rep. (1945) Dr., chemistry (1976). (tel. 0201 + 183, ext. 2416, telex 8579091 Unie D) *Molecular structures, electron density calculations, low temperature techniques, phase transitions.*

Böttcher, Dr Peter. Inst. f. Anorg. Chemie, RWTH, Prof.-Pirlet-Str. 1, D-5100 Aachen, Germany, Fed. Rep. (1939) Dr.rer.nat., Chemie (RWTH Aachen, 1971). Privatdozent. (tel. 0241 + 80, ext. 4664). *Preparation in liquid ammonia, inorganic chalcogenides, crystal structure investigation.*

Bohatý, Dr Ladislav. Inst. f. Kristallographie, Universität, Zülpicher Str. 49, D-5000 Köln 1, Germany, Fed. Rep. (1948) Dr.rer.nat., mineralogy (U. Köln, 1975). Akad. Rat u. Priv. Doz. (tel. 0221 + 4703368). *Crystal physics, crystal growth.*

Bolzenius, Mr Beda Helmut. Mineral. Inst., Universität Bonn, KFA Jülich Postfach 1913, D-5170 Jülich, Germany, Fed. Rep. (1956) Dipl.phys., physics (U. Bonn, 1982). (tel. 02461 + 61, ext. 4054). *X-ray diffraction, profile analysis, structure analysis and refinement.*

Bondza, Mr Harald Werner. Inst. f. Angew. Physik, Lehrstuhl f. Kristallogr., Universität, Bismarckstr. 10, D-8520 Erlangen, Germany, Fed. Rep. (1959) Dipl.phys., physics (U. Erlangen, 1985). Wiss. Mit. (tel. 09131 + 85, ext. 2711). *Dynamical theory, X-ray scattering.*

Bonse, Prof. Dr Ulrich Karl Eberhard. Lehrstuhl f. Experimentelle Physik I, Universität, Postfach 500500, D-4600 Dortmund, Germany, Fed. Rep. (1928) Dr.habil., physics (U. Münster, 1963). Ord. Prof. (tel. 0231 + 755, ext. 3504). *Solid state physics, X-ray and neutron diffraction physics, Compton scattering, synchrotron radiation.*

Borchardt-Ott, Dr Walter. Inst. f. Mineralogie, Universität, Corrensstr. 24, D-4400 Münster, Germany, Fed. Rep. (1933) Dr.rer.nat., mineralogy (U. Münster, 1964). Akad. Direktor. (tel. 0251 + 83, ext. 3453). *Crystal growth.*

Born, Dr Eberhard. Angew. Mineralogie u. Geochemie, Techn. Universität, Lichtenbergstr. 4, D-8046 Garching, Germany, Fed. Rep. (1939) Dr.habil., crystallography (TU München,1974). Prof. (tel. 089 + 3209, ext. 3226). *Crystal defects, X-ray topography, X-ray diffraction methods, textures of metals.*

Born, Dr Liborius. Zentralbereich ZF-D, Bayer AG, D-5090 Leverkusen-Bayerwerk, Germany, Fed. Rep. (1931) Dr., crystallography, physics (U. Marburg, 1961). Industrial Crystallographer. (tel. 0214 + 305852). *Organic crystal structures.*

Born, Mr Reinhard. Inst. f. Kristallographie, Universität, Charlottenstr. 33, D-7400 Tübingen, Germany, Fed. Rep. (1955) Dipl.phys., physics (U. Mainz, 1983). (tel. 030 + 80009, ext. 2708). *Neutron scattering.*

Borrmann, Prof. Dr Ing. Gerhard. Fritz-Haber-Inst., MPI, Faradayweg 4-6, D-1000 Berlin 33, Germany, Fed. Rep. (1908) O. Prof, physics (TH Danzig, 1936). Em.

Bossert, Dipl-chem Werner. Inst. f. Anorganische Chemie, Universität, Engesserstr., Geb. Nr. 30.45, Postfach 6380, D-7500 Karlsruhe, Germany, Fed. Rep. (1948) Dipl.-chem., chemistry (U. Karlsruhe, 1975). Res. Asst. (tel. 0721 + 608, ext. 2974). *Inorganic crystal structures, symmetry.*

Boysen, Dr Hans. Inst. f. Kristallographie und Mineralogie, Universität, Theresienstr. 41, D-8000 München 2, Germany, Fed. Rep. (1941) Dipl.-phys., physics (U. München, 1970). Wiss. Ang. (tel. 089 + 23941). *Neutron diffraction, spectrometry, powder diffraction, phase transitions, lattice dynamics.*

Bradaczek, Prof. Dr Hans Arthur. Inst. f. Kristallographie, Freie U. Berlin, Takustr. 6, D-1000 Berlin 33, Germany, Fed. Rep. (1930) Dr.rer.nat., physics (FU Berlin, 1966). Dir. (tel. 030 + 838, ext. 3461). *Diffraction theory, small angle scattering, single crystal diffractometry, phase transitions, crystal surfaces.*

Brämer, Dr Wulf. Heraeus GmbH, Heraeusstr. 12-14, D-6450 Hanau/Main, Germany, Fed. Rep. (1946) Dipl.-Chem., inorganic chemistry (U. Münster,

Germany, Fed. Rep. 57

1975). Leiter der metallurgischen Entwicklung. (tel. 06181 + 360, ext. 795). *Solid state physics, metals.*

Brandmueller, Prof. Dr Josef Karl August. Sektion Physik, LM Universität, Schellingstr. 4, D-8000 München, Germany, Fed. Rep. (1921) Prof., physics (U. München, 1955). O. Prof. (tel. 089 + 2180, ext. 3211). *Optical solid state spectroscopy, group and representation theory.*

Brandt, Ing Gernot. Abt. Kristallchemie, Fraunhofer Inst. f. Angew. Festkörperphysik, Eckerstr. 4, D-7800 Freiburg, Germany, Fed. Rep. (1936) Dr., physics (U. Freiburg, 1969). (tel. 0761 + 2714, ext. 281). *Crystal growth of II-VI and III-V compounds, chalcopyrites.*

Brauer, Prof. Dr Georg Karl. Chemisches Laboratorium, Universität, Albertstr. 21, D-7800 Freiburg, Germany, Fed. Rep. (1908) Prof., Inorganic chemistry (U. Freiburg, 1932). Prof. Em. (tel. 0761 + 203, ext. 2894). *Inorganic solid state chemistry, crystal structure, metal oxides, fluorides, nitrides, carbides, intermetallic compounds.*

Braun, Dr Dieter Johann. Abteilung Chemie, Universität, Postfach 500500, D-4600 Dortmund, Germany, Fed. Rep. (1951) Dr.rer.nat., chemistry (U. Giessen, 1979). Wiss. Ass. (tel. 0231 + 755, ext. 3749). *Structure and properties, inorganic and intermetallic crystals.*

Braun, Dr Eckart. Inst. f. Mineralogie, FU, Takustr. 6, D-1000 Berlin 33, Germany, Fed. Rep.

Brehler, Prof. Dr Bruno. Inst. f. Mineralogie u. Mineralische Rohstoffe, TU, Adolf-Römer-str. 2a, D-3392 Clausthal-Zellerfeld, Germany, Fed. Rep. (1922) Dr.rer.nat., mineralogy (U. Göttingen, 1951). Ord. Prof. (tel. 05323 + 72, ext. 2207). *Inorganic crystal structures, crystal chemistry, applied mineralogy.*

Breit, Dipl-Min Udo. Inst. f. Mineralogie, Universität, Corrensstr. 24, D-4400 Münster, Germany, Fed. Rep. (1939) Dipl.-Min., mineralogy (U. Münster, 1974). Postgrad. (tel. 0251 + 83, ext. 3405). *Instrumentation, silicate crystal structures.*

Breitinger, Prof. Dr Dietrich Karl. Inst. f. Anorg. Chemie, Universität, Egerlandstr. 1, D-8520 Erlangen, Germany, Fed. Rep. (1935) Dr.rer.nat., inorganic and analytical chemistry (RWTH Aachen, 1964). Section leader, lect. (tel. 09131 + 85, ext. 7352). *Spectroscopy (vibrational - electronic - high pressure), inorganic solids, single crystal spectroscopy.*

Brill, Prof. Dr Rudolf Friedrich. Arzbacher Str. 6, D-8172 Lenggries, Germany, Fed. Rep. (1899) Dr.phil., chemistry (U. Berlin, 1923). Em. scient. member, Max-Planck-Ges. (tel. 08042 + 2618). *Electron distribution in crystals.*

Brill, Mr Wolfgang. Inst. f. Kristallographie, Universität, Im Stadtwald, D-6600 Saarbrücken, Germany, Fed. Rep. (1956) Dipl.phys., physics (U. Saarbrücken, 1982). Wiss. Mit. (tel. 0681 + 3022470). *Solid state physics, computer technology, X-ray diffraction.*

Brodalla, Dipl.-Chem Dieter. Inst. f. Anorg. Chemie und Strukturchemie, Universität, Universitätsstr. 1, D-4000 Düsseldorf, Germany, Fed. Rep. (1947) Dipl.-Chem., chemistry (U. Düsseldorf, 1979). Res. asst. (tel. 0211 + 311, ext. 3068). *Solid state chemistry, crystal growth, inorganic crystal structures, clathrate structures, low temperature crystal structure determination.*

Brodersen, Prof. Dr Klaus. Inst. f. Anorg. Chemie, Universität, Egerlandstr. 1, D-8520 Erlangen, Germany, Fed. Rep. (1926) Dr.phil., inorganic and analytical chemistry (U. Greifswald, 1951). Full Prof. (tel. 09131 + 85, ext. 7350). *Structural chemistry of mercury compounds.*

Brokmeier, Dr Heinz-Günter. Inst. f. Physik, GKSS-Forschungszentrum, Max-Planck-Str., D-2054 Geesthacht, Germany, Fed. Rep. (1952) Dr.rer.nat., crystallography (TU Clausthal-Zellerfeld, 1983). (tel. 04152 + 12257). *X-ray and neutron diffraction, textures, deformation and recrystallisation processes.*

Bronger, Prof. Dr Welf. Inst. f. Anorganische Chemie, RWTH, Prof. Pirlet-Str. 1, D-5100 Aachen, Germany, Fed. Rep. (1932) Dr.rer.nat., chemistry (U. Münster, 1961). Ord. Prof. (tel. 0241 + 80, ext. 4643). *Inorganic solid state chemistry.*

Buck, Prof. Dr Peter. Inst. f. Mineralogie, Universität, Lahnberge, D-3550 Marburg, Germany, Fed. Rep. (1939) Dr.rer.nat., crystallography (U. Freiburg, 1967). Hochschullehrer. (tel. 06421 + 28, ext. 5500). *Crystal growth, defects in crystals, crystal physics.*

Buehner, Dr Manfred. Arbeitsgruppe Röntgenstrukturanalyse, Universität, Am Hubland, D-8700 Würzburg, Germany, Fed. Rep. (1940) Dr.rer.nat., biochemistry (U. Konstanz, 1969). Head, Res. Group. (tel. 0931 + 888-386) *Structure analysis, proteins, enzymology.*

Bülow, Miss Renate. Inst. f. Kristallographie, FU, Takustr. 6, D-1000 Berlin 33, Germany, Fed. Rep. (1956) Dipl.Chem., chemistry (FU Berlin, 1984). (tel. 030 + 8383408). *Single crystal analysis (X-ray and neutron), charge density, empirical methods.*

Bunge, Prof. Dr Dr hc Hans-Joachim. Inst. f. Metallkunde und Metallphysik, Techn. Universität, Grosser Bruch 23, D-3392 Clausthal-Zellerfeld, Germany, Fed. Rep. (1929) Dr.rer.nat.habil, physics (Humboldt U. Berlin, 1964). Prof. (tel. 05323 + 72, ext. 2244). *Material science, X-ray and neutron diffraction, textures.*

Burschka, Dr Christian. Inst. f. Anorg. Chemie, Universität, Am Hubland, D-8700 Würzburg, Germany, Fed. Rep. (1946) Dr.rer.nat., chemistry (RWTH Aachen, 1975). Akad. Rat. (tel. 0931 + 888, ext. 286). *Solid state and inorganic chemistry, computer programming, direct methods, information retrieval.*

Burzlaff, Prof. Dr Hans. Inst. f. Angew. Physik, Lehrstuhl f. Kristallographie, Universität, Bismarckstr. 10, D-8520 Erlangen, Germany, Fed. Rep. (1932)

58 Germany, Fed. Rep.

Dr.habil., crystallography (U. Marburg, 1968). Full Prof. (tel. 09131 + 85, ext. 2700). *Symmetry, structure determination, electron density, crystal physics.*

Buschmann, Dr Juergen Friedrich. Inst. f. Kristallographie, Freie U. Berlin, Takustr. 6, D-1000 Berlin 33, Germany, Fed. Rep. (1939) Dr.rer.nat., crystallography (FU Berlin, 1980). Scient. (tel. 030 + 838-3408). *Single crystal X-ray diffractometry, low temperature, neutron diffraction, organic molecules, deformation density distribution.*

Cammenga, Prof. Dr Heiko Karl. Inst. f. Theoretische Chemie, TU, Hans Sommer-Str. 10, D-3300 Braunschweig, Germany, Fed. Rep. (1938) Dr., physical chemistry (TU Braunschweig, 1967). Prof. (tel. 0531 + 391, ext. 5333). *Phase transitions, solid state reactions, heterogen. kinetics.*

Cemič, Dr Ladislav. Inst. f. Mineralogie und Kristallographie, TU, Hardenbergstr. 42, D-1000 Berlin 12, Germany, Fed. Rep. (1940) Dr.rer.nat., mineralogy (U. Bonn, 1972). Wiss. Mit. (tel. 030 + 314, ext. 2747). *Structure and physical properties, high pressure and temperature.*

Chattopadhyay, Dr Tapan Kumar. MPI f. Festkörperforschung, Heisenbergstr. 1, D-7000 Stuttgart 80, Germany, Fed. Rep. (1942) PhD, physics (Indian Inst. of Techn., Kharagpur, India, 1972). Guest scient. (tel. 0711 + 7830, ext. 4187). *Imperfections in crystals, lattice vibrations, structure, phase stabilities, intermetallic compounds, electron distributions.*

Claus, Mr Karl Heinz. Röntgenlabor, MPI f. Kohlenforschung, Lembkestr. 5, D-4330 Mülheim/Ruhr, Germany, Fed. Rep. (1940) (tel. 0208 + 306, ext. 493). *Crystal data collection, powder diffractometry, instrumentation.*

Cordier, Dr Gerhard. Inst. f. Anorg. Chemie, TH, Hochschulstr. 4, D-6100 Darmstadt, Germany, Fed. Rep. (1949) Dr, chemistry (TH Darmstadt, 1977). (tel. 06151 + 162492). *Structures, inorganic materials.*

Czank, Dr Michael. Mineralogisches Inst., Universität, Olshausenstr. 40-60, D-2300 Kiel, Germany, Fed. Rep. (1941) Dr.rer.nat., crystallography (E.T.H. Zürich, 1973). (tel. 0431 + 880, ext. 2903). *Electron microscopy, realbau of minerals, silicate crystal chemistry.*

Dachs, Prof. Dr Hans. Hahn-Meitner-Inst. f. Kernforschung, Glienicker Str. 100, D-1000 Berlin 39, Germany, Fed. Rep. (1927) Dr.rer.nat., crystallography (U. München, 1956). Prof. (tel. 030 + 8009, ext. 2741). *Neutron diffraction, magnetic structures, phase transitions.*

Dahlkamp, Dr Franz-Joses. Oelbergstr. 10, D-5307 Wachtberg-Liessem, Germany, Fed. Rep. (1931) Dr.phil., Petrogr.-Geol. (U. Graz, Austria, 1958). Independent Consultant. (tel. 0228 + 341904). *Appl. mineralogy, exploration.*

Debaerdemaeker, Dr Tony. Sektion f. Röntgen- u. Elektronenbeugung, Universität, Oberer Eselsberg, D-7900 Ulm, Germany, Fed. Rep. (1945) Dr.phil., physics (U. York, UK, 1971). Res. Asst. (tel. 0731 + 1761). *Direct methods, computer programming, crystal structures.*

Dederer, Dr Bernhard. Abt. f. Strukturforschung I, MPI f. Biochemie, Am Klopferspitz, D-8033 Martinsried, Germany, Fed. Rep. (1949) Dipl.-Chem., chemistry (TU München, 1975). Res. Assoc. (tel. 089 + 8585, ext. 661). *Structure analysis.*

Deiseroth, Dr Hans-Jörg. Fachbereich 8, Anorg. Chemie, Universität-GH, Postfach 101240, D-5900 Siegen, Germany, Fed. Rep. (1945) Dr., inorganic chemistry (U. Giessen, 1972). Prof. (tel. 0271 + 740-4219). *Inorganic chemistry, crystal structures, computer programming, amorphous compounds, EXAFS.*

Depmeier, Dr habil. Wulf Helmut Heinz. Inst. f. Kristallographie, Universität, Kaiserstr. 12, D-7500 Karlsruhe, Germany, Fed. Rep. (1944) Dr.rer.nat., crystallography (U. Hamburg, 1973). Wiss. Mit. (tel. 0721 + 608, ext. 3317). *Inorganic crystal structures, phase transitions, modulated structures, superstructures.*

Deppisch, Dr Bertold. Inst. f. Kristallographie, Universität, Kaiserstr. 12, D-7500 Karlsruhe, Germany, Fed. Rep. (1943) Dr., physics (U. Karlsruhe, 1972). Akad. Rat. (tel. 0721 + 608, ext. 3317). *Crystal structure determination, symmetry.*

Diehl, Dr J. Inst. f. Werkstoffwiss., MPI f. Metallforschung, Seestr. 92, D-7000 Stuttgart 1, Germany, Fed. Rep.

Diehl, Dr Roland. Fraunhofer Inst. f. Angew. Festkörperphysik, Eckerstr. 4, D-7800 Freiburg, Germany, Fed. Rep. (1944) Dr.rer.nat., crystallography (U. Freiburg, 1972). Asst. dir. (tel. 0761 + 2714, ext. 286, telex 07-72510). *Crystal growth, epitaxy of III-V semiconductor compounds, III-V device and process technology.*

Dhlipia, Mr Gursev Singh. Inst. f. Gesteinshüttenkunde, RWTH, Mauerstr. 5, D-5100 Aachen, Germany, Fed. Rep. (1946) Dipl.Min., mineralogy (RWTH Aachen, 1979). Res. asst. (tel. 0241 + 804984). *Electron microscopy, refractories, ceramics.*

Dietrich, Prof. Dr Hans Karl Ernst. Fritz-Haber-Inst., Max-Planck-Ges., Faradayweg 4-6, D-1000 Berlin 33, Germany, Fed. Rep. (1923) Dr.rer.nat., chemistry (U. Heidelberg, 1956). Arbeitsgruppenleiter. (tel. 030 + 83051). *Structure, metallorganic compounds, electron density distribution, chemical bonding.*

Dittmar, Dr Günter. GID, Lyoner Str. 44-48, D-6000 Frankfurt/Main, Germany, Fed. Rep. (1936) Dr.rer.nat., chemistry (TH Darmstadt, 1976). Wiss. Mit. *Inorganic crystal structures, crystal chemistry systematics, computer programming, symmetry.*

Dörr, Dr Friedrich Johannes. Instrumentelle Analytik + Mineralogie, Schott-Glaswerke, Hattenbergstr. 10, D-6500 Mainz, Germany, Fed. Rep. (1924) Dr.rer.nat., mineralogy (U. Mainz, 1955). Leitender Wissenschaftler. (tel. 06131 + 663497, telex 4187920 sm). *X-ray diffractometry and fluorescence analysis, electron-probe microanalysis, structure and properties of silicates, glass systems.*

Dräger, Prof. Dr Martin. Inst. f. Anorg. und Analyt. Chemie, Universität, Johann-Joachim-Becher-Weg 24, D-6500 Mainz, Germany, Fed. Rep. (1940) Dr.rer.nat., chemistry (U. Mainz, 1970). Prof. (tel. 06131 + 39, ext. 5757). *Inorganic and organic crystal structures, conformational analysis, ring systems.*

Durchschlag, Dr Helmut. Inst. f. Biophysik und Phys. Biochemie, Universität, Universitätsstr. 31, D-8400 Regensburg, Germany, Fed. Rep. (1944) Dr.phil., physical chemistry (U. Graz, 1971). Akad.Rat. (tel. 0941 + 943, ext. 3041). *Small angle scattering, biopolymers.*

Eberhard, Prof. Dr Emil. Mineralogisches Inst., Universität, Welfengarten 1, D-3000 Hannover, Germany, Fed. Rep. (1928) Dr., mineralogy (U. Fribourg, France, 1954). Prof. (tel. 0511 + 762, ext. 2443). *Crystal chemistry.*

Eckerlin, Dr Peter. Philips GmbH Forschungslaboratorium, Weisshaus-Str., D-5100 Aachen, Germany, Fed. Rep. (1926) Dr., chemistry (TH Darmstadt, 1955). Res. Chemist. (tel. 0241 + 62071). *Ternary oxides, intermetallic compounds.*

Eckhardt, Prof. Dr Franz-Jörg. Abt. f. Mineralogie und Petrologie, Bundesanstalt f. Geowiss. und Rohstoffe, Postfach 510153, D-3000 Hannover 51, Germany, Fed. Rep. (1929) Dr.phil.nat., mineralogy (U. Frankfurt/Main, 1957). Dir. and Prof. (tel. 0511 + 6468, ext. 559). *Crystallographic methods, mineralogy, petrography, layer silicates, geochemistry, crystal chemistry.*

Eckstein, PhD. Dipl. Ing. Juraj. Kristallogr. Inst., Universität, Hebelstr. 25, D-7800 Freiburg, Germany, Fed. Rep. (1927) PhD, chemistry (U. Prague, ČSSR, 1958). Res. asst. (tel. 0761 + 2034283). *Crystal growth, thermodynamics, analytics.*

Egert, Dr Ernst. Inst. f. Anorg. Chemie, Universität, Tammannstr. 4, D-3400 Göttingen, Germany, Fed. Rep. (1949) Dr.Ing., chemistry (TH Darmstadt, 1979). Hochschulass. (tel. 0551 + 393023). *Patterson search, structure-activity relationships, force-field calculations.*

Ehses, Dr Karl-Heinz. Inst. f. Kristallographie, Dep. f. Physik, Universität, Im Stadtwald, D-6600 Saarbrücken, Germany, Fed. Rep. (1943) Dr., physics (U. d. Saarlandes, 1973). Wiss. Mit. (tel. 0681 + 302, ext. 3460). *Solid state physics, X-ray diffraction.*

Eisenmann, Mrs Dr Brigitte. Abt. II f. Anorganische Chemie, TH, Hochschulstr. 4, D-6100 Darmstadt, Germany, Fed. Rep. (1942) Dr., inorganic chemistry (U. München, 1971). Akad. Oberrätin. (tel. 06151 + 16, ext. 2492). *Inorganic crystal structures.*

Eitel, Dr Manfred. Inst. f. Anorg. Chemie, Universität, Engesserstr., Geb. Nr. 30.45, Postfach 6380, D-7500 Karlsruhe, Germany, Fed. Rep. (1955) Dr.rer.nat., chemistry (U. Karlsruhe, 1985). Ass. (tel. 0721 + 6083485). *Inorg. crystal structures, crystal chemistry.*

Elf, Mr Frank. Mineralogisches Inst., Universität, In KFA Jülich, Postfach 1913, D-5170 Jülich, Germany, Fed. Rep. (1944) Dipl. Phys., physics (U. Bonn, 1980). (tel. 02461 + 61, ext. 6024). *Neutron diffraction, magnetic and crystallographic structure analysis and refinement, profile analysis.*

Ellner, Dr Martin Oliver. Inst. f. Werkstoffwissenschaften, MPI f. Metallforschung, Seestr. 75, D-7000 Stuttgart, Germany, Fed. Rep. (1938) Dr.rer.nat., chemistry (U. Stuttgart, 1971). Sen. res. sci. (tel. 0711 + 2095, ext. 244, telex 723742). *Crystal chemistry, structure, intermetallic compounds, metastable crystalline and amorphous phases, instrumentation.*

Engel, Dr Walter. Fraunhoferinst. f. Treib- u. Explosivstoffe, Hummelberg, D-7507 Berghausen, Germany, Fed. Rep. (1935) PhD, inorg. chemistry (U. Giessen, 1962). (tel. 0721 + 46101) *Phase transitions.*

Englisch, Mr Uwe-Franz. Abt. f. Chemie, MPI f. Experim. Medizin, Hermann-Rein-Str. 3, D-3400 Göttingen, Germany, Fed. Rep. (1954) (tel. 0551 + 303, ext. 356). *Protein crystallography, biochemistry.*

Ensling, Dr Jürgen. Inst. f. Anorg. und Analyt. Chemie, Universität, Jakob-Welder-Weg 11, D-6500 Mainz, Germany, Fed. Rep. (1940) Dr., Mössbauer-spectroscopy (TH Darmstadt, 1970). Res. Asst. (tel. 06131 + 39, ext. 2703). *Mössbauer spectroscopy, solid state chemistry and physics.*

Ermer, Dr Otto. Abt. f. Chemie, Ruhr-Universität, Universitätsstr. 150, Postfach 102148, D-4630 Bochum, Germany, Fed. Rep. (1940) Priv.-Doz., chemistry (U. Bochum, 1981). Akad. Rat. (tel. 0234 + 700, ext. 7702, telex 0825860). *Structure and energy of organic molecules.*

Esselborn, Dr Reiner Ferdinand. Zentrallabor f. Anorganische Chemie, E. Merck, Frankfurter Str. 250, D-6100 Darmstadt, Germany, Fed. Rep. (1927) Dr.rer.nat., chemistry (U. Freiburg, 1958). Head of Res. & Dev. (tel. 06151 + 722296). *Solid state chemistry, thin films, crystal growth methods, properties, high purity materials, solid state technology, fibre optic technology.*

Eulenberger, Dr Günther Richard. Inst. f. Chemie, Universität Hohenheim, Garbenstr. 30, D-7000 Stuttgart, Germany, Fed. Rep. (1936) Ph.D., chemistry (U. Vienna, Austria, 1963). (tel. 0711 + 4501, ext. 2166). *Solid state chemistry, inorganic and organic structures, computer programming.*

Euler, Dr Robert. Kanalstr. 13, D-6452 Hainburg 1, Germany, Fed. Rep. (1925) Dr.phil., mineralogy (U. Marburg, 1958). Lehrbeauftragter. (tel. 06182 + 69231). *Technical mineralogy, metallurgical slags, refractories.*

Eysel, Prof. Dr Walter. Mineralogisch-Petrogr. Inst., Universität, Im Neuenheimer Feld 236, D-6900 Heidelberg, Germany, Fed. Rep. (1935) Dr., crystallography, mineralogy (RWTH Aachen, 1968). Lect. (tel. 06221 + 56, ext. 2807). *Crystal structures, crystal chemistry, oxides, crystal physics, silicates, ionic conductivity.*

Faber, Dr Peter. Anwendungstechnik (Elektrochemie), Rheinisch-Westf. Elektrizitätswerk AG, Kölner Str., D-8757 Karlstein, Germany, Fed. Rep.

(1924) Dr., electrochemistry (U. Strasbourg, France, 1976). Chief, Res. and Dev. Group. (tel. 06188 + 2197). *Solid state electrochemistry, structure, surface reactions.*

Feld, Dr Rainer Hans Helmut. Miat Ges. f. Informationssysteme, Nerobergstr. 1, D-6500 Mainz, Germany, Fed. Rep. (1950) Dr.rer.nat., crystallography (U. Marburg, 1980). Managing Dir. (tel. 06131 + 681038). *Sensor and information technology, solid state physics, image processing.*

Felsche, Prof. Dr Jürgen. Fachbereich Chemie, Universität, Postfach 7733, D-7750 Konstanz, Germany, Fed. Rep. (1939) Dr. Dr.habil., crystallography, crystal chemistry (U. Hamburg, ETH Zürich, 1966, 1971). Prof. (tel. 07531 + 88, ext. 2025). *Solid state chemistry, thermal analysis, high and low temperature X-ray equipment, inorganic crystal structures.*

Fenske, Prof. Dr Dieter. Inst. f. Anorg. Chemie, Universität, Engesserstr. Geb. Nr. 30.45, D-7500 Karlsruhe, Germany, Fed. Rep. (1942) Dr.rer.nat., chemistry (1972) Prof. (tel. 0721 + 6082086). *Complex chemistry (trans. metals), X-ray structure, phosphine ligands.*

Fischer, Dr Carl-Otto. Hahn-Meitner-Inst. f. Kernforschung, Glienicker Str. 100, D-1000 Berlin 39, Germany, Fed. Rep. (1938) Dr.rer.nat., physics (FU Berlin, 1969). Physicist. (tel. 030 + 8009, ext. 2746). *Phase transitions, molecular compounds, liquid structure and dynamics.*

Fischer, Prof. Dr Karl. Inst. f. Kristallographie, Dept. f. Physik, Universität, Im Stadtwald, D-6600 Saarbrücken, Germany, Fed. Rep. (1925) Dr., chemistry (U. Erlangen, 1954). Full Prof. (tel. 0681 + 302, ext. 3410). *Crystal structure determination, X-ray and neutron diffraction, instrumentation, synchrotron radiation.*

Fischer, Mrs Ute Eva-Maria. Inst. f. Kristallographie, Universität, Kaiserstr. 12, D-7500 Karlsruhe, Germany, Fed. Rep. (1957) Dipl.chem., chemistry (U. Karlsruhe, 1983).

Fischer, Prof. Dr Werner. Inst. f. Mineralogie, Universität, Hans-Meerwein-Str., D-3550 Marburg, Germany, Fed. Rep. (1931) Dr.rer.nat., crystallography (U. Kiel, 1959). Prof. (tel. 06421 + 28, ext. 5704). *Mathematical crystallography, crystal chemistry, crystal physics, crystal structure determination, crystallographic computing.*

Flörke, Prof. Dr Otto Wilhelm. Inst. f. Mineralogie, Ruhr-Universität, Universitätsstr. 150, Postfach 102148, D-4630 Bochum, Germany, Fed. Rep. (1926) Dr.phil., crystallography (U. Marburg, 1951). Prof. (tel. 0234 + 700, ext. 3512). *Crystal chemistry, crystal physics, crystal growth, mineralogy, ceramics.*

Follner, Prof. Dr Heinz. Inst. f. Mineralogie u. Mineralogische Rohstoffe, TU, Adolf-Riemer-Str. 2A, D-3392 Clausthal-Zellerfeld, Germany, Fed. Rep. (1938) Dr.rer.nat., crystallography and mineralogy (TU Clausthal, 1971). Prof. (tel. 05323 + 722394). *Crystal growth, structure-physical properties relation.*

Forst, Mr Hans Rainer. Aussenstelle f. Neutronenbeugung, Inst. f. Kristallographie und Mineralogie, Am Coulombwall 1, D-8046 Garching, Germany, Fed. Rep. (1949) Dipl. Min. (U. München, 1975). (tel. 089 + 3209, ext. 5018). *Organic structures, lattice dynamics, phase transitions.*

Frank, Dr Walter. FR 13.1 Anorg. Chemie, Universität, Im Stadtwald, D-6600 Saarbrücken, Germany, Fed. Rep. (1957) Dr.rer.nat., inorg. chemistry (TU Braunschweig, 1985). Wiss. Angest. (tel. 0681 + 302, ext. 2975). *Structural inorganic and organometallic chemistry.*

Freiburg, Dr Johann Christoph. Zentralabt. Chemische Analysen, Kernforschungsanlage Jülich, D-5170 Jülich, Germany, Fed. Rep. (1931) Dr., physical chemistry (U. Bonn, 1962). Scient. employee. (tel. 02461 + 61, ext. 3291). *Powder mixture identification, stoichiometry.*

Frey, Prof. Dr Friedrich. Inst. f. Kristallographie und Mineralogie, Universität, Theresienstr. 41, D-8000 München 2, Germany, Fed. Rep. (1942) Dr.habil., neutron diffraction (U. München, 1980). Prof. (tel. 089 + 2394, ext. 4332). *Neutron diffraction, phase transitions, diffuse scattering.*

Fröhlich, Dr Roland. ENRAF-NONIUS GmbH, Obere Dammstr. 10, D-5650 Solingen, Germany, Fed. Rep. (1952) Dr.rer.nat., inorganic chemistry (U. Köln, 1982). (tel. 02122 + 52062, telex 8514749). *Crystal structures, X-ray equipments.*

Fuess, Prof. Dr Hartmut. Inst. f. Kristallographie, Senckenberg-Anlage 30, D-6000 Frankfurt/Main, Germany, Fed. Rep. (1941) Dr.-Ing., structural chemistry (TH Darmstadt, 1968). Prof. (tel. 069 + 798, ext. 3103, telex 413932 UNIF). *Chemical crystallography, neutron diffraction, crystal physics.*

Gahm, Dr Josef. Mikroskopisches Labor, Fa. Carl Zeiss, D-7082 Oberkochen, Germany, Fed. Rep. (1925) Dr., mineralogy, geology, physics (U. Tübingen, 1953). Wiss. Mit. (tel. 07364 + 20, ext. 3288). *Instrumentation, optics, computer programming, image analysis, microscopy (interference and polarisation).*

Gassmann, Dr Johann. MPI f. Biochemie, Am Klopferspitz, D-8033 Martinsried, Germany, Fed. Rep. (1934) Dr., crystallography (TU München, 1966). Res.asst. (tel. 089 + 8585, ext. 723). *Direct methods, crystallographic computing, electron microscopy, structure refinement, large molecules.*

Gebert, Dr Walter Richard. Inst. f. Mineralogie, Ruhr-Universität, Universitätsstr. 150, D-4630 Bochum, Germany, Fed. Rep. (1938) Dr., mineralogy (U. Bochum, 1970). Wiss. Ang. (tel. 0234 + 700, ext. 4380). *Inorganic crystal structures, crystal growth.*

Gebhardt, Prof. Dr Manfred Adolf Hermann. Mineralogisches Inst., Universität, Poppelsdorfer Schloss, D-5300 Bonn, Germany, Fed. Rep. (1934) Prof., mineralogy and crystallography (U. Bonn, 1972). (tel. 0228 + 73, ext. 3268). *Applied mineralogy, biomineralization, thin films.*

Gehlen, Prof. Kurt von. Inst. f. Geochemie, Universität, Senckenberganlage 28, D-6000 Frankfurt/Main, Germany, Fed. Rep. (1927) Dr.rer.nat., mineralogy (U. Freiburg, 1952). Prof. (tel. 069 + 798, ext. 2102). *Ore deposits, mineralogy, geochemistry, isotope geochemistry, petrology.*

Gerold, Prof. Dr Volkmar. Inst. f. Werkstoffwissenschaften, MPI f. Metallforschung, Seestr. 92, D-7000 Stuttgart, Germany, Fed. Rep. (1922) Dr., physics (U. Stuttgart, 1953). Prof. (tel. 0711 + 2095, ext. 219). *Metal physics.*

Geyer, Mr Andreas. Mineralogisch-Petrogr. Inst., Universität, Im Neuenheimer Feld 236, D-6900 Heidelberg, Germany, Fed. Rep. (1958) Student. *Thermal analysis (TGA,DTA,DSC), crystal growth.*

Gieren, Priv.-Doz. Dr Alfred. Abt. f. Strukturforschung I, MPI f. Biochemie, Am Klopferspitz, D-8033 Martinsried, Germany, Fed. Rep. (1939) Dr.habil., chemistry (TU München, 1974). Head, chem. cryst. group. (tel. 089 + 8585, ext. 662, telex 521740). *X-ray and neutron structure analysis, molecular compounds, natural products, heterocyclic ring systems, S-N multiple bonding systems, stereochemistry, charge densities (X-N).*

Gies, Dr Hermann. Mineralogisches Inst., Universität, Olshausenstr. 40-60, D-2300 Kiel, Germany, Fed. Rep. (1952) Dr., chemistry (U. Kiel, 1982). (tel. 0431 + 8802912). *Silicates, zeolites, inclusion compounds, silicate solutions.*

Goddard, Dr Richard. MPI f. Kohlenforschung, Lembkestr. 5, D-4330 Mülheim/Ruhr, Germany, Fed. Rep. (1952) PhD., structural chemistry (Bristol U., UK, 1977). Res. assoc. (tel. 0208 + 306, ext. 485). *Structural chemistry, theoretical chemistry, electron density determination, organometallic compounds, catalytic reaction products and intermediates.*

Göbel, Dr Herbert Ernst. ZFE FKE 42, SIEMENS AG, Otto-Hahn-Ring 6, D-8000 München 83, Germany, Fed. Rep. (1940) Dr.rer.nat., gamma spectroscopy (TU München, 1969). Laboratory leader. (tel. 089 + 636, ext. 3274, telex 52109-11). *X-ray powder diffraction, materials science, high-temperature diffraction, thin films, position-sensitive detectors.*

Göttlicher, Prof. Dr Siegfried. Fachgebiet Strukturforschung, Inst. f. Phys. Chemie, Petersenstr. 20, D-6100 Darmstadt, Germany, Fed. Rep. (1929) Dr.rer.nat., chemistry (TU Darmstadt, 1961). Prof. (tel. 06151 + 16, ext. 2893). *Electron density determination in crystals; structure determination.*

Gomm, Dr Martin. Inst. f. Angew. Physik, Lehrst. f. Kristallographie, Universität, Bismarckstr. 10, D-8500 Erlangen, Germany, Fed. Rep. (1943) Dr.rer.nat. (U. Erlangen-Nürnberg, 1977). Akad. Rat (tel. 09131 + 85, ext. 2700). *Structure analysis, solid state physics, mycology, honey bees.*

Gonschorek, Dr Walter. Inst. f. Kristallographie, RWTH, Templergraben 55, D-5100 Aachen, Germany, Fed. Rep. (1937) Dr., crystallography (RWTH Aachen, 1971). Priv.-Doz. (tel. 0241 + 80, ext. 6909). *Chemical bond, lattice dynamics, physical properties, minerals, statistics.*

Graetsch, Mr Heribert. Inst. f. Mineralogie, Universität, Universitätsstr. 155, D-4630 Bochum, Germany, Fed. Rep. (1953) Dipl., mineralogy (1981). Wiss. Mit. (tel. 0234 + 7003516). *Ferrites, SiO2: structure and properties.*

Graf, Dr Hans Anton. Kernchemie und Reaktor, Hahn-Meitner-Inst. f. Kernforschung, Glienicker Str. 100, D-1000 Berlin 39, Germany, Fed. Rep. (1945) Dr.rer.nat., chemistry (U. München, 1975). Res. asst., chemist. (tel. 030 + 8009, ext. 2778). *Neutron scattering, magnetic structures.*

Greis, Dr Ortwin. Mineralogisch-Petrogr. Inst., Universität, Im Neuenheimer Feld 236, D-6900 Heidelberg 1, Germany, Fed. Rep. (1941) Dr.rer.nat., inorganic solid state chemistry (U. Freiburg, 1976). Wiss. Ang. (tel. 06221 + 56, ext. 2807). *Inorganic chemistry, solid state chemistry, crystal chemistry, minerals, superstructures, X-ray powder methods, electron microscopy, electron diffraction.*

Grosse, Prof. Dr Peter. I. Physikalisches Inst., RWTH, Templergraben 55, D-5100 Aachen, Germany, Fed. Rep. (1932) Dr.rer.nat.habil., physics (U. Köln, 1969). Ord. Prof. (tel. 0241 + 80, ext. 7155, telex 08/32704). *Solid state spectroscopy IR,FIR, preparation of single crystals of semiconductors and semimetals.*

Grossi, Dr Paolo. Teutonenstr. 50, D-6200 Wiesbaden, Germany, Fed. Rep. (1933) Dr., Geologie und Mineralogie (U. Firenze, Italy, 1966). Wiss. Mit. (tel. 06121 + 812716). *Ecology.*

Grotepass-Deuter, Mrs Margit. Inst. f. Kristallogr., Dept. f. Physik, Universität, Im Stadtwald, D-6600 Saarbrücken, Germany, Fed. Rep. (1958) Dipl., mineralogy (RWTH Aachen, 1984). Grad. stud. (tel. 0681 + 3021). *Structure refinement, synchrotron radiation.*

Grühn, Prof. Dr Reginald. Inst. f. Anorgan. u. Analyt. Chemie, Universität, Heinrich-Buff-Ring 58, D-6300 Giessen, Germany, Fed. Rep. (1929) Dr.rer.nat.habil., inorg. and analyt. chemistry (U. Münster, 1969). Prof. (tel. 0641 + 7025670). *Transmission electron microscopy, inorg. crystal structures, chemical transport.*

Gütlich, Prof. Dr Philipp. Inst. f. Anorg. und Analyt. Chemie, Universität, Jakob-Welder-Weg 11, D-6500 Mainz, Germany, Fed. Rep. (1934) Prof., inorganic and analytical chemistry (U. Mainz, 1975). Ord. Prof. (tel. 06131 + 39, ext. 2373). *Transition metal chemistry, electronic structure, solid state chemistry, theoretical inorganic chemistry (ligand field and MO theories), magnetochemistry, spectroscopy.*

Gupta, Dr Amaresh. Bundesinstitut f. chem.-techn. Untersuchungen (BICT), Grosses Cent, D-5357 Swisttal 1, Germany, Fed. Rep. (1941) Dr.rer.nat., crystallography (U. Bonn, 1975). Wiss. Mit. (tel. 02222 + 6008, ext. 384,telex 8869315 bict d). *Crystal structures, explosives, low and high temperature X-ray diffraction, non-destructive X-ray fluorescence.*

Gussone, Dr Rainer Carl Leonard. Inst. f. Mineralogie und Lagerstättenlehre, RWTH, Wüllnerstr. 2, D-5100 Aachen, Germany, Fed. Rep. (1931) Dr.-Ing., economic geology (RWTH Aachen, 1964). Akad. Oberrat. (tel. 0241 + 80, ext. 5759). *X-ray spectrometry, X-ray diffraction, mineral and rock diagnosis, ore microscopy.*

Guth, Dr Helmut Karl Richard. SFB 127 Universität Marburg, Inst. f. Angew. Kernphysik, KFZ., Postfach 3640, D-7500 Karlsruhe, Germany, Fed. Rep. (1952) Dr.rer.nat. physics (U. Karlsruhe, 1979). Wiss. Mit. (tel. 07247 + 82, ext. 3438). *Neutron and X-ray diffraction, crystal structures, phase transitions.*

Haase, Prof. Dr Wolfgang. Inst. f. Physikalische Chemie, TH, Petersenstr. 20, D-6100 Darmstadt, Germany, Fed. Rep. (1936) Dr.rer.nat., chemistry (U. Jena, 1964). Prof. (tel. 06151 + 16, ext. 3398). *Inorganic crystal structures, transition metal complex structures, spin coupling, liquid crystals.*

Haase-Wessel, Dr Werner. Am Weidenfeld 2C, D-3352 Einbeck, Germany, Fed. Rep. (1941) Dr.rer.nat., physical chemistry (U. Göttingen, 1973). Wiss. Mit. *Inorganic and organic crystal structures.*

Hädicke, Dr Erich Emil Hermann. Anorg. Chemie ZAA/S-M325, BASF AG, D-6700 Ludwigshafen, Germany, Fed. Rep. (1940) Dr.rer.nat., chemistry (TU München, 1969). Lab. head. (tel. 0621 + 60, ext. 4805, telex ZA-TTX 6215934=basf). *Material science, X-ray diffraction, electron diffraction, magnetic materials, textures.*

Hafner, Prof. Dr Stefan S. Fachbereich Geowissenschaften, Universität, Lahnberge, D-3550 Marburg, Germany, Fed. Rep. (1932) Dr.sc.nat., petrography (Fed. Inst. of Techn. Switzerland, 1958). Prof. (tel. 06421 + 28, ext. 5617). *Inorganic crystal structures, spectroscopy, physical properties, minerals.*

Hahn, Prof. Dr Theodor. Inst. f. Kristallographie, RWTH, Templergraben 55, D-5100 Aachen, Germany, Fed. Rep. (1928) Dr.rer.nat., mineralogy (U. Frankfurt/Main, 1952). Prof. (tel. 0241 + 80, ext. 6900). *Inorganic crystal chemistry, crystal physics, symmetry.*

Hanke, Dr Kurt. Mineralogisch-Kristallogr. Inst., Universität, Goldschmidtstr. 1, D-3400 Göttingen, Germany, Fed. Rep. (1935) Dr., mineralogy (U. Göttingen, 1965). Akad. Oberrat. (tel. 0551 + 39, ext. 3937). *Inorganic crystal structures.*

Haq, Mr Anwar-Ul. Inst. f. Angew. Kernphysik, KFZ, D-7500 Karlsruhe, Germany, Fed. Rep. (1947) Diplom, physics (U. Karlsruhe, 1978). Doktorand. (tel. 07247 + 82, ext. 2807). *Polymers, superconducting materials.*

Harbrecht, Dr Bernd. Anorg. Chemie I, Universität, Otto-Hahn-Strasse, Postf. 500500, D-4600 Dortmund 50, Germany, Fed. Rep. (1950) Dr.rer.nat., chemistry (RWTH Aachen, 1981). (tel. 0231 + 7553794, telex 822465). *Solid state chemistry.*

Harms, Dr Klaus. Inst. f. Anorg. Chemie, Universität, Tammannstr. 4, D-3400 Göttingen, Germany, Fed. Rep. (1953) Dr.rer.nat., chemistry (U. Göttingen, 1984). Wiss. Angest. (tel. 0551 + 39, ext. 3073). *Structural organic chemistry, pericyclic reactions.*

Harr, Dr Albrecht Wolfgang Michael. Battelle-Institut, Wiesbadenerstr., D-6000 Frankfurt/Main, Germany, Fed. Rep. (1947) Dr.rer.nat., physics (U. Göttingen, 1977). Princip. res. sci. (tel. 069 + 79082883). *Growth of compound and organic semiconductors.*

Hartl, Prof. Dr Hans. Inst. f. Anorg. u. Analyt. Chemie, Freie Universität, Fabeckstr. 34-36, D-1000 Berlin 33, Germany, Fed. Rep. (1940) Dr.rer.nat., inorganic chemistry (TH München, 1969). Prof. (tel. 030 + 838, ext. 4003). *Inorganic chemistry (halogen-chemistry), structure determination, crystal chemistry.*

Hartmann, Dr Werner Johannes. MPI f. Festkörperforschung, Heisenbergstr. 1, D-7000 Stuttgart, Germany, Fed. Rep. (1946) Dr., physics (U. Stuttgart, 1975). (tel. 0711 + 7830, ext. 247). *X-ray topography in real-time, materials science, semiconductor materials engineering.*

Hauck, Dr Jürgen. Inst. f. Festkörperforschung, Kernforschungsanlage Jülich, D-5170 Jülich, Germany, Fed. Rep. (1941) Dr., inorganic chemistry (U. Frankfurt/Main, 1968). Scient. (tel. 02461 + 61, ext. 4237). *Inorganic crystal structures, metal hydrides, ternary oxides, stoichiometry, phase diagrams, radiochemistry.*

Hausen, Dr Hans-Dieter. Inst. f. Anorg. Chemie, Universität, Pfaffenwaldring 55, D-7000 Stuttgart 80, Germany, Fed. Rep. (1937) Dr.rer.nat., chemistry (U. Stuttgart, 1966). (tel. 0711 + 685, ext. 4220). *Inorganic and organometallic crystal structures.*

Haussühl, Prof. Dr Siegfried Georg. Inst. f. Kristallographie, Universität, Zülpicher Str. 49, D-5000 Köln 1, Germany, Fed. Rep. (1927) Dr.rer.nat., natural sciences (U. Tübingen, 1956). Direktor. (tel. 0221 + 470, ext. 3194). *Crystal growth, thermal properties, mechanical properties, electrical properties, nonlinear optics, nonlinear acoustics, phase transitions, electron diffraction.*

Hecht, Dr Hans-Jürgen. Abt. Röntgenstrukturanalyse, Physiol.-Chem. Inst., Universität, Am Hubland, Zentralbau Chemie, D-8700 Würzburg, Germany, Fed. Rep. (1947) Dr.rer.nat., chemistry (FU Berlin, 1976). Res. assoc. (tel. 0931 + 888, ext. 386). *X-ray crystallography, proteins, small molecules of biological interest.*

Heger, Dr Gernot Wolfgang. Inst. f. Angew. Kernphysik, Kernforschungszentrum, Postfach 3640, D-7500 Karlsruhe, Germany, Fed. Rep. (1943) Dr.rer.nat., physics (U. Tübingen, 1972). Wiss. Mit. (tel. 07247 + 82, ext. 3985). *Neutron diffraction, crystal physics, phase transitions.*

Heide, Dr Helmut. Battelle Inst. e.V., Am Römerhof, D-6000 Frankfurt, Germany, Fed. Rep. (1934) Dr.rer.nat., mineralogy (U. Bonn, 1969). Deputy chief. (tel. 069 + 7908, ext. 2764). *Materials research, biochemical engineering, ceramics, phase relations, sintering kinetics.*

Heim, Dr Harald Josef Robert. Mandelring 67, D-6730 Neustadt/Weinstr. 13, Germany, Fed. Rep. (1951) Dr.rer.nat., chemistry (U. Karlsruhe, 1979). (tel. 06321 + 88103). *Inorganic crystal structures, symmetry, crystallography.*

Heinemann, Dr Udo. Inst. f. Kristallographie, FU, Takustr. 6, D-1000 Berlin 33, Germany, Fed. Rep. (1953) Dr.rer.nat., chemistry (U. Göttingen, 1982). Hochschulass. (tel. 030 + 838, ext. 3463). *Molecular biology, X-ray crystallography of proteins and nucleic acids.*

Hellner, Prof. Dr Erwin E. Fachbereich Geowissenschaften, Universität, Lahnberge, D-3550 Marburg, Germany, Fed. Rep. (1920) Dr.rer.nat., mineralogy (U. Göttingen, 1945). Prof. (tel. 06421 + 28, ext. 2045). *Theoretical crystallography, structure type systematics, high pressure research, charge density.*

Helmreich, Dr Dieter. HELIOTRONIC GmbH, Joh.-Hess-Str. 24, D-8263 Burghausen, Germany, Fed. Rep. (1939) Dr.rer.nat., physics (TU München, 1968). (tel. 08677 + 83, ext. 2736, telex 56923). *Crystallization of silicon, ingot casting, direct sheet growth.*

Henke, Dr Henning. Inst. f. Anorganische Chemie, Universität, Engesserstr., Geb. Nr. 30.45, Postfach 6380, D-7500 Karlsruhe, Germany, Fed. Rep. (1943) Dr.rer.nat., chemistry (U. Karlsruhe, 1971). Wiss. Ang. (tel. 0721 + 608, ext. 2977). *Inorganic crystal structures, hydrates, hydrogen bonding.*

Henkel, Dr Gerald. Anorg.-Chem. Inst., Universität, Corrensstr. 36, D-4400 Münster, Germany, Fed. Rep. (1948) Dr.rer.nat., chemistry (U. Bielefeld, 1976). Priv.-Doz. (tel. 0251 + 83, ext. 3153). *Structural chemistry, inorganic and organic compounds.*

Hentschel, Dr Manfred Paul. Gruppe Parakristallforschung, BAM, Unter den Eichen 44/46, D-1000 Berlin 45, Germany, Fed. Rep. (1943) Dr.rer.nat., physics (FU Berlin, 1981). Wiss. Mit. (tel. 030 + 8104, ext. 0912). *Paracrystals, glassy structures, composites, nondestructive testing, fibers, lipids, protein structure, molecular crystals.*

Herdtweck, Dr Eberhardt. Fachber. Chemie, TU, Lichtenbergstr. 4, D-8046 Garching, Germany, Fed. Rep. (1948) Dr.rer.nat., chemistry (U. Tübingen, 1978). *Fluorides, oxyfluorides and their hydrates.*

Herzberg, Dr Armin. Mineralog.-Petrogr. Inst., TU, Adolf-Roemer-Str. 2A, D-3392 Clausthal-Zellerfeld, Germany, Fed. Rep. (1948) Dr.rer.nat., mineralogy (TU Clausthal, 1977). Wiss. Ass. (tel. 05323 + 72, ext. 2326). *Petrography, granites, autometasomatism, mineralogy, carbonates, chromites, micas, minerals analysis (RFA - AAS - ICP - EMPA).*

Hess, Prof. Dr Heinz. Inst. f. Anorganische Chemie, Universität, Pfaffenwaldring 55, D-7000 Stuttgart 80, Germany, Fed. Rep. (1924) Dr.rer.nat., chemistry (U. Stuttgart, 1958). Prof. (tel. 0711 + 784, ext. 4254). *Inorganic and organometallic crystal structures, mineral structures.*

Hesse, Dr Karl-Friedrich. Mineralogisches Inst., Universität, Olshausenstr. 40-60, D-2300 Kiel, Germany, Fed. Rep. (1939) Dr., mineralogy (U. Kiel, 1973). Wiss. Mit. (tel. 0431 + 880, ext. 2901). *Crystal chemistry, silicates, X-ray crystallography, single crystal X-ray diffractometer.*

Hildebrandt, Prof. Dr Gerhard. Fritz-Haber-Inst., Max-Planck-Ges., Faradayweg 4-6, D-1000 Berlin 33, Germany, Fed. Rep. (1922) Prof., physics (TU Berlin, 1975). Arbeitsgruppenleiter. (tel. 030 + 8305, ext. 280). *X-ray diffraction, topography with X-rays, solid state physics.*

Hildmann, Dr-Ing Bernd Otfried. Inst. f. Kristallographie, RWTH, Templergraben 55, D-5100 Aachen, Germany, Fed. Rep. (1944) Dr.Ing., electrical engineering (RWTH Aachen, 1980). Res. assoc. (tel. 0241 + 80, ext. 6908). *Dielectric crystals, pyro- and piezoelectric single crystals, phase transitions, ferro-electric and -elastic crystals, critical point phenomena, symmetry, relation of properties to crystal structure, crystal growth.*

Hilgenfeld, Dr Rolf. Inst. f. Kristallographie, Freie U. Berlin, Takustr. 6, D-1000 Berlin 33, Germany, Fed. Rep. (1954) Dr.rer.nat., chemistry (U. Göttingen, 1985). Ass. (tel. 030 + 838, ext. 6309). *Protein crystallography, molecular graphics, drug design, membrane protein crystalization, detergents properties.*

Hiller, Dr Wolfgang Paul. Inst. f. Anorg. Chemie, Universität, Auf der Morgenstelle 18, D-7400 Tübingen, Germany, Fed. Rep. (1953) Dr.rer.nat., chemistry (U. Tübingen, 1981). Akad. Rat. (tel. 07071 + 296230). *Crystal structure analysis, Cu-,Ag-,Au-complexes.*

Hinsch, Mr Thorsten Reinhard. Mineralog.-Petrogr. Inst., Universität, Grindelallee 48, D-2000 Hamburg 13, Germany, Fed. Rep. (1959) Dipl., mineralogy (U. Hamburg, 1984). *Inorganic crystal structures, crystal chemistry, crystal growth.*

Hinze, Dr Eckhard. Mineralogisches-Petrologisches Inst. und Museum, Universität, Poppelsdorfer Schloss, D-5300 Bonn, Germany, Fed. Rep. (1934) Privatdozent, mineralogy (U. Bonn, 1978). Priv.-Dozent. (tel. 02221 + 73, ext. 2735). *High pressure X-ray diffraction, crystal chemistry, physical properties, minerals, mantle minerals, phase equilibria, high pressure mineral synthesis.*

Hodenberg, von, Mrs Dr Renate Barbara. Mineralogisches Inst., Universität, Welfengarten 1, D-3000 Hannover, Germany, Fed. Rep. (1926) Dr.rer.nat., mineralogy (U. Bonn, 1956). Akad. Oberrätin. (tel. 0511 + 762, ext. 2326). *Salt minerals, crystal optics, crystal chemistry, inorganic crystal structures.*

Höfer, Dr Hans Hermann. CORNING KERAMIK GmbH & CoKG, Dotzheimerstr. 168, D-6200 Wiesbaden, Germany, Fed. Rep. (1948) Dr., crystallography (RWTH Aachen, 1979). Quality assurance manager. (tel. 06121 + 804-140). *Thermodynamics and kinetics, solid state reactions; solid ion conductors.*

Höfler, Mrs Sabine Ida Gerda. Mineral. Inst., Universität, in KFA Jülich, Postfach 1913, D-5170 Jülich, Germany, Fed. Rep. (1957) Dipl.Min., mineralogy (U. Kiel, 1982). (tel. 02461 + 616024). *Textures, structure, minerals, neutron diffraction.*

Höhling, Prof. Dr Hans Jürgen. Inst. f. Medizinische Physik, Universität, Hüfferstr. 68, D-4400 Münster, Germany, Fed. Rep. (1930) Dr.rer.nat.habil., mineralogy (U. Münster,1964). Prof. (tel. 0251 + 83, ext. 5191). *Crystallography, hard tissues (normal & diseased).*

Hönle, Dr Wolfgang. MPI f. Festkörperforschung, Heisenbergstr. 1, D-7000 Stuttgart, Germany, Fed. Rep. (1947) Dr., inorganic chemistry (U. Münster, 1975). Assistant editor. (tel. 0711 + 6860, ext. 416, telex 7-255555 mpif-d). *Inorganic crystal preparation, crystal structures.*

Hösler, Dr. Inst. f. Kristallographie, Universität, Theresienstr. 41, D-8000 München 2, Germany, Fed. Rep. (1951) Dr., crystallography (tel. 089 + 2394, ext. 4335). *LEED, surface structure, surface topography.*

Hoffmann, Prof. Dr Wolfgang. Inst. f. Mineralogie, Universität, Corrensstr. 24, D-4400 Münster, Germany, Fed. Rep. (1935) Dr.rer.nat., mineralogy, crystallography (U. Hamburg, 1961). Direktor. (tel. 0251 + 83, ext. 3461). *Crystal structure analysis, physical properties, phase transitions.*

Hoffmeister, Dr Wolfgang. Material and Reliability Analysis, IBM Laboratories, Schoenaicher Str. 220, D-7030 Böblingen, Germany, Fed. Rep. (1930) Dr., nuclear chemistry (U. Köln, 1961). Chief chemist. (tel. 07031 + 660, ext. 8307). *Trace analysis, radiotracer application.*

Hofmeister, Dr Wolfgang. Inst. f. Geowissenschaften, Universität, Saarstr. 21, D-6500 Mainz, Germany, Fed. Rep. (1952) Dr.rer.nat., mineralogy (U. Mainz, 1981). Hochschulassistent. (tel. 06131 + 39, ext. 4365). *Structure determination and crystal chemistry of minerals, mineralogy.*

Hoge, Dr Reinhold. Fachrichtung 17.3 Kristallographie, Universität, D-6600 Saarbrücken, Germany, Fed. Rep. (1943) Ph.D., chemistry (British Columbia U., Canada, 1969). Wiss. Mitabeiter. (tel. 0681 + 302, ext. 2470). *Structure determination, methods.*

Hohlwein, Dr Dietmar. Inst. f. Kristallographie, Universität, Charlottenstr. 33, D-7400 Tübingen, Germany, Fed. Rep. (1944) Dr.habil., physics (U. Tübingen, 1981). Dozent. (tel. 07071 + 29, ext. 6058). *Instruments in neutron diffraction, disordered crystalline solids, plastic crystals.*

Holinski, Dr Rüdiger. Res. and Dev. DOW CORNING GmbH, Pelkovenstr. 152, D-8000 München 50, Germany, Fed. Rep. (1939) Dr.rer.nat., chemistry (TU Clausthal, 1968). Dept. Manager. (tel. 089 + 14860, ext. 265, telex 5215654). *Tribology, specialty lubrication, solid lubricants, metallurgy, solid state chemistry, analytical surface investigations.*

Holmes, Prof. Kenneth Charles. Abt. f. Biophysik, MPI f. medizinische Forschung, Jahnstr. 29, D-6900 Heidelberg, Germany, Fed. Rep. (1934) Ph.d., biophysics (U. London, 1959). Dir. of res. (tel. 06221 + 486, ext. 270). *Macromolecular structures, crystallography, muscle physiology.*

Honigmann, Dr Bertold. Hauptlaboratorium, BASF-AG, Carl-Bosch-Str., D-6700 Ludwigshafen, Germany, Fed. Rep. (1921) Dr., physical chemistry (TU Berlin, 1948). Head, res. group. (tel. 0621 + 60, ext. 99481). *Crystal growth, pigments - crystal properties, dispersions - physical and chemical properties.*

Hoppe, Prof. Dr hc Rudolf. Inst. f. Anorg. und Analytische Chemie, Justus-Liebig-Universität, Heinrich-Buff-Ring 58, D-6300 Giessen, Germany, Fed. Rep. (1922) Ord. Prof., inorganic chemistry (U. Giessen, 1965). Dir. (tel. 0641 + 702 ext. 5660). *Crystal chemistry, Madelung factors, metal oxides, synthesis, solid state chemistry, fluorine.*

Hoppe, Dr Walter. Strukturforschung I, MPI f. Biochemie, Am Klopferspitz, D-8033 Martinsried, Germany, Fed. Rep. (1917) Dr.rer.nat., chemistry (Deutsche U. Prag, 1941). Dir., Prof. (tel. 089 + 8585, ext. 656). *Two- and three-dimensional structure determination, electron microscopy, organic crystal structures, instrumentation, large molecules, computer programming.*

Horst, Dr Wolfgang. Tucholskystr. 22, D-6000 Frankfurt/Main, Germany, Fed. Rep. (1948) Dr.phil.nat., physics (U. Frankfurt/Main, 1981). (tel. 069 + 627260). *Modulated structures, rock forming minerals, satellite reflections.*

Horstmann, Prof. Dr Manfred. Universität Osnabrück, Neuer Graben, D-4500 Osnabrück, Germany, Fed. Rep. (1928) Dr.rer.nat., experimental physics (U. Hamburg, 1960). Prof. (tel. 0541 + 608, ext. 4100). *Solid state physics, surface physics.*

Hosemann, Prof. Dr Dr hc Rolf. Gruppe Parakristallforschung, BAM, Unter den Eichen 44-46, D-1000 Berlin 45, Germany, Fed. Rep. (1912) Prof. Dr.phil.nat.habil., physics (U. Freiburg, 1935). Res. Group Leader. (tel. 030 + 8104, ext. 0913). *Equlibrium states, thermodynamics, glass, melts, polymers, catalysts.*

Hoser, Dr Andrzej. Inst. f. Kristallographie, Universität, Charlottenstr. 33, D-7400 Tübingen, Germany, Fed. Rep. (1952) Dr., chemistry (U. Poznań, Poland, 1982). Wiss. Mit. (tel. 07071 + 296058). *X-ray and neutron structure analysis, molecular crystals, phase transitions.*

Hubbert, Mrs Dr Elisabeth. Anorg. Chemie I, Ruhr-Universität, Universitätsstr. 150, D-4630 Bochum, Germany, Fed. Rep. (1934) Dr., inorganic chemistry (U. Münster, 1966). Wiss. Ang. (tel. 0234 + 700, ext. 4153). *Crystal growth, inorganic crystal structures.*

Huber, Prof. Robert. Strukturforschung II, MPI f. Biochemie, Am Klopferspitz, D-8033 Martinsried, Germany, Fed. Rep. (1937) Prof., chemistry (TU München, 1976). Direktor. (tel. 089 + 8585, ext. 677). *Biochemistry, protein crystallography.*

Hubert, Mr Georg. Analytical Systems, E689F, SIEMENS AG, Östliche Rheinbrückenstr. 50, D-7500 Karlsruhe, Germany, Fed. Rep. (1948) Diplom, physics (TH Karlsruhe, 1975). Product Manager X-ray Analysis. (tel. 0721 + 5954265, telex 78255-69). *All field X-ray analysis, instrumentation, software.*

Huch, Dr Volker. Inst. f. Anorg. Chemie, Universität, Im Stadtwald, D-6600 Saarbrücken, Germany, Fed. Rep. (1955) Dr., inorg. chemistry (TU Braunschweig, 1984). Akad. Rat. (tel. 0681 + 3022265). *Inorganic and organometallic crystal structures.*

Hübner, Mr Thomas. Arbeitsgruppe f. chem. Kristallogr., MPI f. Biochemie, Am Klopferspitz, D-8033 Martinsried, Germany, Fed. Rep. (1956) Diplom, chemistry (U. München, 1982). (tel. 089 + 8578, ext. 2659). *X-ray structure analysis, organic compounds, heterocyclic systems, organometallic molecules.*

Hümmer, Prof. Dr Kurt. Inst. f. Angew. Physik, Lehrst. f. Kristallographie, Universität, Bismarckstr. 10, D-8520 Erlangen, Germany, Fed. Rep. (1939) Dr.rer.nat.habil., physics (U. Erlangen, 1978). Prof. (tel. 09131 + 852711). *Experimental crystallography, electron density, multiple scattering and phase problem.*

Hummel, Dr Hans-Ulrich. Inst. f. Anorg. Chemie, Universität, Egerlandstr. 1, D-8520 Erlangen, Germany, Fed. Rep. (1954) Dr.rer.nat., chemistry (1980). Akad. Rat. (tel. 09131 + 857393). *Preparation and structural investigation of inorganic solids (alkaline and earth-alkaline compounds with pseudohalogenides and pseudochalcogenides).*

Hund, Dr Franz Josef. AC-Forschung, Anorg. Wiss. Labor, Bayer AG, Rheinuferstr. 7-9, D-4150 Krefeld, Germany, Fed. Rep. (1916) Dr., inorganic chemistry (U. Heidelberg, 1945). Industrial chemist. (tel. 0215 + 52346). *Inorganic pigments, crystal structure, synthesis.*

Hundt, Dr Rudolf. Inst. f. Anorganische Chemie, Universität, Gerhard-Domagk-Str. 1, D-5300 Bonn 1, Germany, Fed. Rep. (1941) Dr.rer.nat., inorg. chemistry (U. Bonn, 1973). OStR i. HDnst.. (tel. 0228 + 73, ext. 5337). *Crystal structures, crystallographic computing, instrumentation, computer applications in chemistry.*

Ihringer, Dr Jörg, Inst. f. Kristallographie, Universität, Charlottenstr. 33, D-7400 Tübingen, Germany, Fed. Rep. (1943) Dr. (U. München, 1977). (tel. 07071 + 29, ext. 6058). *Phase transitions, X-ray powder diffraction.*

Irngartinger, Dr Hermann. Organisch-Chem. Inst., Universität, Im Neuenheimer Feld 270, D-6900 Heidelberg, Germany, Fed. Rep. (1938) Dr.habil., organic chemistry (U. Heidelberg, 1972). Prof. (tel. 06221 + 56, ext. 2422). *Organic crystal structures, organic solid state chemistry, bonding, electron density distribution.*

Isenberg, Dr Wilhelm. Analytical Systems, E689, SIEMENS AG, Östliche Rheinbrückenstr. 50, D-7500 Karlsruhe, Germany, Fed. Rep. (1953) Diplom, inorg. chemistry (U. Göttingen, 1979). Product manager X-ray diffraction. (tel. 0721 + 5952425, telex 78255-69). *Diffractometry theory, instumentation, software.*

Jacob, Dr Herbert. Res. Div., Wacker-Chemitronic GmbH, Postfach 1140, D-8263 Burghausen, Germany, Fed. Rep. (1928) Dr.rer.nat., chemical thermodynamics (U. Münster, 1958). Forschungsbereichsleiter. (tel. 08677 + 83868). *Industrial crystal growth, epitaxy.*

Jacobi, Dr Hans. Mineralog.-Kristallograph. Inst., TU, Sägemüllerstr. 4, D-3392 Clausthal-Zellerfeld, Germany, Fed. Rep. (1934) Dr.phil., mineralogy (U. Marburg, 1963). Akad. Oberrat. (tel. 05323 + 72, ext. 2392). *Modulated crystal structures, optics.*

Jacobs, Prof. Dr Herbert Ernst Hermann. Anorganische Chemie, Universität, Postfach 500500, Otto-Hahn-Str., D-4600 Dortmund 50, Germany, Fed. Rep. (1936) Dr.rer.nat., chemistry (U. Kiel, 1966). (tel. 0231 + 7553802, telex 822465). *Solid state chemistry.*

Jäger, Dr Hans. Abt. Festkörperphysik, Battelle-Inst. e.V., Am Römerhof 35, D-6000 Frankfurt/Main, Germany, Fed. Rep. (1934) Dr.rer.nat., solid state physics, metal physics (TH Stuttgart, 1963). (tel. 069 + 7908, ext. 2766). *Semiconductor physics, device development, crystal growth, epitaxy, X-ray detection, gamma-ray detection, infrared physics.*

Jäger, Dr Susanne Christine. Gmelin Inst., Joachim-Becher-Str. 2, D-6000 Frankfurt/Main, Germany, Fed. Rep. (1943) Dr.rer.nat., chemistry (RWTH Aachen, 1982). Editorial assoc. (tel. 069 + 564122). *Solid state chemistry.*

Jagodzinski, Prof. Dr Dr h c Heinz Ernst. Inst. f. Kristallographie und Mineralogie, Universität, Theresienstr. 41, D-8000 München 2, Germany, Fed. Rep. (1916) Dr.rer.nat.h.c., physics (1981). Em. Ord. Prof. (tel. 089 + 2394, ext. 4357). *Diffraction methods (X-ray - neutron - slow electrons), crystal physics, order-disorder phenomena, inorganic crystal structures.*

Jahn, Dr Irmin-Rudolf. Inst. f. Kristallographie, Universität, Charlottenstr. 33, D-7400 Tübingen, Germany, Fed. Rep. (1939) Dr.rer.nat., physics (U. Tübingen, 1971). Wiss. Angest. (tel. 07071 + 29, ext. 6394). *Crystal physics.*

Jakobs, Dr Rüdiger-Hasko. VDO Adolf Schindling AG, Sodener Str. 9, D-6231 Schwalbach/Taunus, Germany, Fed. Rep. (1941) Dr., physics (TU Berlin, 1979). (tel. 06196 + 8012710). *General crystallography.*

Jansen, Dr Martin. Inst. f. Anorg.Chemie, Universität, Callinstr.9, D-3000 Hannover 1, Germany, Fed. Rep. (1944) Dr.rer.nat., inorganic and analytical chemistry (U. Giessen, 1973). Prof. (tel. 0511 + 7623696). *Preparative solid state chemistry, inorganic crystal structures, X-ray and neutron diffraction.*

Jarchow, Prof. Dr Otto. Mineralogisch-Petrogr. Inst., Universität, Grindelallee 48, D-2000 Hamburg, Germany, Fed. Rep. (1931) Dr.rer.nat., mineralogy and crystallography (U. Saarbrücken, 1961). Dozent. (tel. 040 + 4123, ext. 2056). *Organic and inorganic crystal structures, crystal chemistry, crystal disorder, crystal growth, symmetry.*

Jauch, Dr Wolfgang. Hahn-Meitner-Inst. f. Kernforschung, Glienicker Str. 100, D-1000 Berlin 39, Germany, Fed. Rep. (1947) Dr.rer.nat., solid state physics (TU Berlin, 1979). Physiker. (tel. 030 + 800912767). *X-ray and neutron structure analysis, structure and physical propetries.*

Jeitschko, Prof. Dr Wolfgang. Anorg.-Chem. Inst., Universität, Corrensstr. 36, D-4400 Münster, Germany, Fed. Rep. (1936) Dr.phil., chemistry (U. Wien, 1964). Dir. (tel. 0251 + 833121). *Structure and properties, inorganic and intermetallic crystals.*

Jex, Dr Hartmut. Inst. f. Kernphysik der Universität, August-Euler-Str. 6, D-6000 Frankfurt, Germany, Fed. Rep. (1940) Habilitation, Physik (U. Frankfurt, 1979). *Lattice dynamics, crystal structure, phase transitions, solitons.*

Joswig, Dr Werner. Inst. f. Kristallographie, Universität, Senckenberg-Anlage 30, D-6000 Frankfurt/Main, Germany, Fed. Rep. (1940) Dr., crystallography (U. Frankfurt/Main, 1972). Wiss. Mit. (tel. 069 + 798, ext. 3502, telex 413932). *Inorganic crystal structures.*

Jung, Dr Detlef. Abt. Forschung SIGRI GmbH, Werner-v.-Siemensstr., D-8901 Meitingen, Germany, Fed. Rep. (1937). (tel. 08271 + 3384).

Jung, Dr Volkhard. Inst. f. Angewandte Kernphysik, KFZ Karlsruhe GmbH, Postfach 3640, D-7500 Karlsruhe 1, Germany, Fed. Rep. (1934) Dr.rer.nat., nuclear physics (U. Heidelberg, 1964). Physicist. (tel. 07247 + 82, ext. 3406). *Phase tranformations, austenitic steels, lattice parameter variation, stressed metals.*

Jung, Dr Walter. Inst. f. Anorg. Chemie, Universität, Greinstr. 6, D-5000 Köln 41, Germany, Fed. Rep. (1940) Dr.rer.nat.habil., inorg. chemistry (U. Köln, 1979). Priv.-Doz. (tel. 0221 + 4703353). *Inorganic solid state chemistry, ternary borides.*

Kaat, te, Prof Dr Erich Heinz. Inst. f. Physik, Universität, Otto-Hahn-Str. D-4600 Dortmund, Germany, Fed. Rep. (1937) Dr., physics (1969). Prof. (tel. 0231 + 755, ext. 3506). *Electron microscopy, radiation damage, ion implantation.*

Kabs, Mr Michael. Abt. PCH-TEV, W.C.Heraeus GmbH, Heraeusstr. 12-14, D-6450 Hanau 1, Germany, Fed. Rep. (1952) Dr.rer.nat., chemistry (TU Berlin, 1982). (tel. 06181 + 35, ext. 5122). *Slags, ceramics, material research.*

Kabsch, Dr Wolfgang. Abt. f. Biophysik, MPI f. medizinische Forschung, Jahnstr. 29, D-6900 Heidelberg, Germany, Fed. Rep. (1941) Dr.rer.nat., physics (U. Heidelberg, 1972). Res. scient. (tel. 06221 + 486, ext. 276, telex 461505). *Protein structure, protein folding.*

Kaerlein, Mr Carsten-Peter. Arbeitsgr. f. chem. Kristallographie, MPI f. Biochemie, Am Klopferspitz, D-8033 Martinsried, Germany, Fed. Rep. (1942) Diplom, physics (U. München, 1981). (tel. 089 + 8578, ext. 2661). *X-ray structure analysis, organic compounds, biologically interesting molecules.*

Kambe, Dr Kyozaburo. Fritz-Haber-Inst., Max-Planck-Ges., Faradayweg 4-6, D-1000 Berlin 33, Germany, Fed. Rep. (1926) Dr.Sci., physics (U. Tokyo, FU Berlin, 1961). Scient. (tel. 030 + 8305, ext. 260). *Electron diffraction, electron microscopy, X-ray topography, diffraction theory, channeling, crystal defects, surface crystallography, electron density.*

Karl, Prof Dr Norbert. Physikalisches Inst., Teil 3, Universität, Pfaffenwaldring 57, D-7000 Stuttgart 80, Germany, Fed. Rep. (1939) Dr.habil., crystal physics (U. Stuttgart, 1975). Apl. Prof. (tel. 0711 + 685, ext. 5195, telex 072-55445). *Crystal growth, organic crystals, high purity, electrical properties, organic laser crystals.*

Kassner, Mr Dethard. Inst. f. Kristallographie, Universität, Senckenberganlage 30, D-6000 Frankfurt/Main 1, Germany, Fed. Rep. (1949) Diplom, physics (U. Frankfurt, 1979). Wiss. Mit. (tel. 069 + 798-2105, telex 413932). *Crystallographic computing.*

Katscher, Dr Hartmut. Gmelin-Inst., Max-Planck-Ges., Varrentrappstr. 40/42, D-6000 Frankfurt/Main 90, Germany, Fed. Rep. (1933) Dr., chemistry (U. Würzburg, 1964). Editor. (tel. 069 + 79171). *Inorganic crystal structures.*

Kaub, Mr Jürgen. FB Chemie, Universität, Erwin-Schrödinger-Str., D-6750 Kaiserslautern, Germany, Fed. Rep. (1958) Diplom, chemistry (1984). Wiss. Mit. (tel. 0631 + 2052962). *Inorganic crystal structures.*

Kaus, Dr Gerhard. IBM, Dept. 0135-7032-47, Tübinger Allee, D-7032 Sindelfingen, Germany, Fed. Rep. (1941) Nat.Sc.D., mineralogy, crystallography (U. Mainz, 1969). Staff assoc. (tel. 07031 + 611, ext. 2269). *Surface analysis methods (SIMS - AES - ESCA), ceramics, semiconductors.*

Keller, Dr Egbert. Kristallogr. Inst., Universität, Hebelstr. 25, D-7800 Freiburg i.Br., Germany, Fed. Rep. (1950) Dr.rer.nat., chemistry (U. Freiburg, 1978). Wiss. Angest. (tel. 0761 + 2034279). *X-ray structure analysis, computer graphics in crystallography.*

Keller, Dr Hans-Lothar. Inst. f. Anorganische Chemie, Universität, Olshausenstr. 40-60, D-2300 Kiel, Germany, Fed. Rep. (1943) Dr.rer.nat., inorganic chemistry (U. Giessen, 1973). Wiss. Ang. (tel. 0431 + 880, ext. 2096). *Inorganic crystal structures.*

Keller, Prof. Dr Heimo Jürgen. Anorganisch-Chemisches Inst., Universität, Im Neuenheimer Feld 270, D-6900 Heidelberg, Germany, Fed. Rep. (1935) Dr.habil., inorganic chemistry (TU München, 1966). Full prof. (tel. 06221 + 56, ext. 2438). *Linear chain transition metal compounds, one-dimensional metals, crystal structure, coordination compounds.*

Keller, Prof Dr Paul. Inst. f. Mineralogie u. Kristallchemie, Universität, Pfaffenwaldring 55, D-7000 Stuttgart 80, Germany, Fed. Rep. (1940) Prof. (U. Stuttgart, 1973). (tel. 0711 + 6854112, telex 7255445). *Crystal chemistry, crystallography, mineralogy.*

Keller, Dr Wolfgang Ludwig. WDHLIZ, SIEMENS AG, Frankfurter Ring 152, D-8000 München 46, Germany, Fed. Rep. (1927) Dr.rer.nat., experimental mineralogy (U. Tübingen, 1954). Lab. leader. (tel. 089 + 3500, ext. 778). *Crystal growth, float zone growth, silicon.*

Kemmler-Sack, Mrs Prof. Dr Sibylle. Inst. f. Anorganische Chemie, Universität, Auf der Morgenstelle 18, D-7400 Tübingen, Germany, Fed. Rep. (1934) Prof., inorganic chemistry (U. Tübingen, 1973). Prof. (tel. 07071 + 29, ext. 2439). *Noble metal compounds, complex oxides, transition metals, rare earths, optical properties.*

Keppler, Dr Ulrich H. IBM, Dept. 0348, 7032-16, Tübinger Allee, D-7032 Sindelfingen, Germany, Fed. Rep. (1933) Dr.rer.nat., solid state physics (U. Hamburg, 1962). Manager. (tel. 07031 + 611, ext. 2444). *Surface analysis, structure analysis.*

Kettler, Mr Peter. Bereich C1, Hahn-Meitner-Inst. f. Kernforschung, Glienicker Str. 100, D-1000 Berlin 39, Germany, Fed. Rep. (1949) Dipl.Phys., solid state physics (TU Berlin, 1976). Physicist. (tel. 030 + 8009, ext. 2771). *Neutron scattering, magnetic structures, phase transitions, solid state physics.*

Kiel, Dr Gertrude Lina. Inst. f. Anorg. u. Analyt. Chemie, Universität, Johann-Joachim-Becher-Weg 24, D-6500 Mainz, Germany, Fed. Rep. (1937) Dr.rer.nat., inorg. chemistry (U. Göttingen, 1967). Akad. Oberrat. (tel. 06131 + 392284). *X-ray structure determination.*

Kirfel, Prof Dr Armin Harald. Mineralog. Inst., Universität, Poppelsdorfer Schloss, D-5300 Bonn, Germany, Fed. Rep. (1943) Dr., mineralogy (U. Bonn, 1972). Prof. (tel. 0228 + 73, ext. 2770). *Electron density, chemical bonding, structure solution, mineral structures.*

Klapper, Dr Helmut. Inst. f. Kristallographie, RWTH, Templergraben 55, D-5100 Aachen, Germany, Fed. Rep. (1937) Dr., crystallography (U. Köln, 1970). Res. asst., Privatdozent. (tel. 0241 + 80, ext. 6902). *Crystal growth, crystal physics, X-ray topography.*

Klaska, Dr Karl-Heinz. Mineralog.-Petrogr. Inst., Universität, Grindelallee 48, D-2000 Hamburg 13. Germany, Fed. Rep. (1943) Dr., mineralogy (U. Hamburg, 1974). Wiss. Mit. (tel. 040 + 41232063). *Organic and inorganic crystal structures, crystal chemistry.*

Klaska, Dr Rolf. Inst. f. Mineralogie, Universität, Universitätsstr. 150, D-4630 Bochum, Germany, Fed. Rep. (1944) Dr., crystallography (U. Hamburg, 1977). (tel. 0234 + 7004375). *Crystal growth, structure and topology, tetrahedral frameworks, oxide crystals, twinning, polytypism, phase intergrowth.*

Klebe, Dr Gerhard, ZHV/D, BASF AG, D-6700 Ludwigshafen, Germany, Fed. Rep. (1954) Dr.rer.nat., chemistry (U. Frankfurt, 1982). (tel. 0621 + 6041966). *Organic crystal structures, molecular modelling, structure prediction, structure and bonding.*

Klee, Prof. Dr Wilfrid Edgar. Inst. f. Kristallographie, Universität, Kaiserstr. 12, D-7500 Karlsruhe, Germany, Fed. Rep. (1935) Dr., physical chemistry (U. Freiburg, 1960). Dozent. (tel. 0721 + 608, ext. 2136). *Vibrational spectroscopy, minerals, graph theory, crystallography.*

Klement, Dr Ulrich. FB Chemie, Universität, Universitätsstr. 31, D-8400 Regensburg, Germany, Fed. Rep. (1930) Dr., crystallography (U. München, 1959). Akad. Dir. *Structure determination.*

Klepp, Dr Kurt Otto. Inst. f. Anorg. Chemie, RWTH, Prof.-Pirlet-Str. 1, D-5100 Aachen, Germay, Fed. Rep. (1949) Ph.D., chemistry (U. Wien, Austria, 1971). Wiss. Angest. (tel. 0241 + 804643). *Structural chemistry of inorganic materials, solid state chemistry, mixed-valent compounds.*

Klessen, Mr Gerhard. Chemistry Division, KKW Gundremmingen Betr. Ges., Postfach 300, D-8871 Gundremmingen, Germany. Fed. Rep. (1949) Diplom, chemistry (RWTH Aachen, 1978). 2nd chemist. (tel. 08224 + 782192, telex 531143). *Intermetallic phases, corrosion products, oxide layers.*

Kniep, Prof Dr Rüdiger. Inst. f. Anorg. Chemie u. Strukturchemie, Universität, Universitätsstr. 1, D-4000 Düsseldorf, Germany, Fed. Rep. (1945) Dr.rer.nat.habil., inorganic chemistry (U. Düsseldorf, 1978). Prof. (tel. 0211 + 311, ext. 3147). *Solid state chemistry, inorganic crystal structures.*

Knöchel, Dr Claus-Dieter. VIC ELO, E.Merck, Frankfurter Str. 250, D-6100 Darmstadt, Germany, Fed. Rep. (1946) Dr.rer.nat., physics (TH Darmstadt, 1979). Tech. manag. monocrystals. (tel. 06151 + 723686, telex 419328-0). *Crystal growth, properties and applications.*

Knoch, Dr Falk A. Anorg.-Chem. Inst., Universität, Gerhard-Domagk-Str. 1, D-5300 Bonn 1, Germany, Fed. Rep. (1953) Dr.rer.nat., inorg. chemistry (U. Bonn, 1984). (tel. 0228 + 73, ext. 3737). *Computers in chemistry, direct methods, crystal structures, phosphorous structural chemistry.*

Knof, Mr Wolfgang Erich. Inst. f. Kristallographie, Dept. f. Physik, Universität, Im Stadtwald, D-6600 Saarbrücken, Germany, Fed. Rep. (1961) Dipl., mathematics (U. Kiel, 1984). Grad. stud. (tel. 0681 + 302, ext. 2470). *X-ray analysis, synchrotron radiation.*

Knorr, Dr Klaus. Inst. f. Kristallographie, Universität, Charlottenstr. 33, D-7400 Tübingen, Germany, Fed. Rep. (1937) Dr.rer.nat., physics (TU München, 1967). Akad. Rat. (tel. 07071 + 29, ext. 6059). *Phase transitions, layer structures, X-ray and neutron diffraction.*

Koch, Mrs Dr Elke. Fachbereich Geowissenschaften, Inst. f. Mineralogie, Universität, Hans-Meerwein-Str., D-3550 Marburg/Lahn, Germany, Fed. Rep.

(1943) Dr.rer.nat., crystallography (U. Marburg, 1972). Wiss. Ang. (tel. 06421 + 28, ext. 5610). *Mathematical crystallography, crystal chemistry, crystal physics, crystallographic computing.*

Koch-Wallraf, Mrs Prof. Dr Maria. Am Prinzenrain 6, D-5300 Bonn, Germany, Fed. Rep. (1920) Dr.rer.nat., physical chemistry (U. Bonn, 1950). Prof. *Inorganic crystal structures, computer programming, crystal growth.*

Kockel, Dr Andreas. Inst. f. Mineralogie, Ruhr-Universität, Universitätsstr. 150, D-4630 Bochum, Germany, Fed. Rep. (1932) Dr., mineralogy (U. Marburg, 1960). Akad. Oberrat. (tel. 0234 + 700, ext. 3514). *Thermal expansion, magnetic oxides formation.*

König, Dr Burkhard. Bundesanstalt f. Geowiss. und Rohstoffe, Stilleweg 2, D-3000 Hannover 51, Germany, Fed. Rep. (1938) Dr.rer.nat., crystallography, mineralogy (TU Clausthal, 1970). (tel. 0511 + 6468, ext. 652). *Inorganic crystal structures.*

Kokkinidis, Dr Michael. EMBL X-Ray Division, Meyerhof-Str. 1, D-6900 Heidelberg, Germany, Fed. Rep. (1952) Dr.rer.nat., crystallography (TU München, 1981). Postdoc. (tel. 06221 + 387308). *Structure-activity relationships, biomolecules.*

Konz, Dr Werner. Inst. f. Kristallographie, Dept. f. Physik, Universität, Im Stadtwald, D-6600 Saarbrücken, Germany, Fed. Rep. (1950) Dr., crystallography (U. Freiburg, 1983). Wiss. Mit. (tel. 0681 + 302-2470). *X-ray crystallography, synchrotron radiation.*

Kopf, Dr Jürgen. Inst. f. Anorganische Chemie, Universität, Martin-Luther-King-Platz 6, D-2000 Hamburg 13, Germany, Fed. Rep. (1942) Dr., crystallography, X-ray structures analysis (U. Hamburg, 1973). Wiss. Oberrat. (tel. 040 + 4123, ext. 2897). *X-ray structure analysis, crystallography, mathematics, computer-controlled single crystal diffractometry, computer science, programming, organometallic chemistry, theoretical chemistry.*

Kosten, Mr Klaus. Inst. f. Kristallographie, RWTH, Templergraben 55, D-5100 Aachen, Germany, Fed. Rep. (1949) Dipl., mineralogy (RWTH Aachen, 1978). *Phase transitions, solid state reactions, powder diffraction, synchrotron radiation.*

Krämer, Prof. Dr Volker. Kristallographisches Inst., Universität, Hebelstr. 25, D-7800 Freiburg, Germany, Fed. Rep. (1940) Dr.rer.nat., (TH München, 1968). Prof. (tel. 0761 + 203, ext. 4277). *Crystal growth, thermal analysis, materials science, structure determination, inorganic crystal structures.*

Kramer, Dr Irmtraud. Batelle-Institut, Am Römerhof 35. D-6000 Frankfurt/Main, Germany, Fed. Rep. (1956) Dr.rer.nat., physics (RWTH Aachen, 1982). Res. scient. (tel. 069 + 79082378). *Crystal growth (organic conductors).*

Krane, Mr Hans-Georg. Inst. f. Kristallographie, Dept. f. Physik, Universität, Im Stadtwald, D-6600 Saarbrücken, Germany, Fed. Rep. (1958) Dipl.-Min., crystallography (U. Saarbrücken, 1985). Grad. stud. (tel. 0681 + 3021). *Mineralogy, synchrotron radiation.*

Krebs, Prof. Dr Bernt. Anorg.-Chem. Inst., Universität, Corrensstr. 36, D-4400 Münster, Germany, Fed. Rep. (1938) Dr.rer.nat., chemistry (U. Göttingen, 1965). Prof. (tel. 0251 + 83, ext. 3131). *Crystal chemistry, inorganic and organic crystal structures.*

Kreutz, Dr Ernst Wolfgang. Inst. f. Angew. Physik, Schlossgartenstr. 7, D-6100 Darmstadt, Germany, Fed. Rep. (1940) Dr.rer.nat., physics (1969). Akad. Oberrat. (tel. 06151 + 161, ext. 2082). *Surface structure, laser lattices.*

Krogmann, Prof. Dr Klaus. Inst. f. Anorganische Chemie, Universität, Engesserstr., Geb. Nr. 30.45, Postfach 6380, D-7500 Karlsruhe, Germany, Fed. Rep. (1925) Dr., inorganic chemistry (U. Stuttgart, 1956). Prof. (tel. 0721 + 608, ext. 2980). *Inorganic crystal structures.*

Kroll, Prof. Dr Herbert. Inst. f. Mineralogie, Universität, Corrensstr. 24, D-4400 Münster, Germany, Fed. Rep. (1940) Dr.rer.nat., mineralogy (U. Münster, 1971). Prof. (tel. 0251 + 83, ext. 3455). *Crystallography in earth sciences.*

Krüger, Prof. Dr Carl. Röntgenlabor, MPI f. Kohlenforschung, Lembkestr. 5, D-4330 Mülheim/Ruhr, Germany, Fed. Rep. (1933) Dr.rer.nat., inorganic chemistry (RWTH Aachen, 1961). Prof., res. dir. (tel. 0208 + 306487). *Structural chemistry, organometallic compounds, catalytic reaction products and intermediates, electron density distribution.*

Krug, Prof. Dr Detlef. Laboratorium f. Anorganische Chemie, Universität, Auf der Morgenstelle 18, D-7400 Tübingen, Germany, Fed. Rep. (1936) Dr., inorganic chemistry (U. Tübingen, 1972). Prof. (tel. 07071 + 29, ext. 6227). *Inorganic solid state reactions, thermal analysis, X-ray investigations, chemical analysis.*

Kuehn, Prof. Dr Robert. Richard-Wagner-Str. 31, D-6901 Wilhelmsfeld, Germany, Fed. Rep. (1911) Dr.phil., mineralogy (U. Kiel, 1938). Prof. (tel. 06220 + 8924). *Geosciences of salts and salt deposits.*

Küppers, Prof. Dr Horst. Mineralogisches Inst., Olshausenstr. 40-60, D-2300 Kiel, Germany, Fed. Rep. (1933) Dr.rer.nat., physics (U. Freiburg). Prof. (tel. 0431 + 880, ext. 2897). *Physical properties of crystals.*

Kupčik, Prof. Dr Vladimir. Inst. f. Mineralogie und Kristallographie, Universität, V.M. Goldschmidtstr. 1, D-3400 Göttingen, Germany, Fed. Rep. (1934) Ph.D., mineralogy and crystallography (U. Bratislava, CSSR, 1965). Prof. (tel. 0551 + 39, ext. 3891). *Crystal structure analysis, crystal chemistry, sulfide compounds, synchrotron radiation, modulated structures.*

Kutoglu, Dr Ali. Fachbereich Geowissenschaften, Universität, Lahnberge, D-3550 Marburg, Germany, Fed. Rep. (1935) Dr.rer.nat., geology (U. Kiel, 1963). Res. scient. (tel. 06421 + 28, ext. 2246). *Inorganic crystal structures, crystal structures, organometallic compounds.*

Labischinski, Dr Harald. Robert-Koch-Institut, Bundesgesundheitsamt, Nordufer 20, D-1000 Berlin, Germany, Fed. Rep. (1948) Dr.rer.nat.habil., biocrystallography (FU Berlin, 1983). (tel. 030 + 4503406). *Biopolymer structure and function, small-angle X-ray scattering.*

Lacmann, Prof. Dr Rolf. Inst. f. Physikalische u. Theoretische Chemie, TU, Hans Sommer-Str. 10, D-3300 Braunschweig, Germany, Fed. Rep. (1927) Prof., physical chemistry (TU Braunschweig, 1974). Ord. Prof. (tel. 0531 + 391, ext. 5326, telex 0952526). *Crystal growth, nucleation, growth kinetic, growth form, thermodynamics of mixtures.*

Lamm, Dr Andreas. Abt. f. Strukturforschung, MPI f. Biochemie, Am Klopferspitz, D-8033 Martinsried, Germany, Fed. Rep. (1950) Dipl.Phys., (TU München, 1977). (tel. 089 + 8585, ext. 660). *Structure analysis.*

Langbein, Prof Dr Werner Dieter. Battelle-Institut, Wiesbadenerstr., D-6000 Frankfurt/Main, Germany, Fed. Rep. (1932) Hon.Prof., physics (U. Frankfurt, 1972). Sen.Res.Sci. (tel. 069 + 79082539). *Convective heat and mass transport theory, stability of multicomponent systems.*

Langer, Prof Dr Klaus. Inst. f. Mineralogie u. Kristallographie, TU, Hardenbergstr. 42, D-1000 Berlin 12, Germany, Fed. Rep. (1936) Prof., inorg. chemistry (U. Kiel, 1965). Prof. (tel. 030 + 3145325). *Crystal chemistry, synthesis, silicates.*

Lauck, Mr Rudolf. Kristall- und Materiallabor, Fak. f. Physik, Universität, Kaiserstr. 12, D-7500 Karlsruhe, Germany, Fed. Rep. (1947) Diplom, physics (U. Saarbrücken, 1974). Wiss. Mit. (tel. 0721 + 608, ext. 3551). *Crystal growth, characterization.*

Lehmann, Prof. Dr Gerhard Rudolf. Inst. f. Physikalische Chemie, Universität, Schlossplatz 4, D-4400 Münster, Germany, Fed. Rep. (1935) Dr.rer.nat., physical chemistry (U. Münster, 1963). Prof. (tel. 0251 + 83, ext. 3427). *Solid state spectroscopy, crystal growth, color centers, defects in solids, ligand field theory.*

Lehmpfuhl, Dr Gunter. Fritz-Haber-Inst., Max-Planck-Ges., Faradayweg 4-6, D-1000 Berlin 33, Germany, Fed. Rep. (1928) Dr., physics (FU Berlin, 1961). Wiss. Oberasst. (tel. 030 + 8305, ext. 261). *Electron diffraction, structure potential measurement, electron microscopy, direct observation of atomic steps.*

Lerf, Dr Anton Eduard. Zentralinstitut f. Tieftemperatur-Forschung, Bayerische Akademie der Wissenschaften, D-8046 Garching, Germany, Fed. Rep. (1948) Dr., chemistry (U. München, 1976). Wiss. Mit. (tel. 089 + 3209, ext. 5218). *Crystal chemistry, superconductors, organic solid state chemistry, intercalation reactions.*

Leusmann, Dr Dietrich Bertold. Inst. f. Mediz. Physik, Universität, Hüfferstr. 68, D-4400 Münster, Germany, Fed. Rep. (1950) Dipl. Min., crystallography (U. Münster, 1975). (tel. 0251 + 83, 5100). *Bio-mineralisation, urinary-tract stones, crystal growth, crystal physics, crystal chemistry.*

Leute, Prof. Dr Volkmar. Inst. f. Physikalische Chemie, Universität, Schlossplatz 4, D-4400 Münster, Germany, Fed. Rep. (1938) Dr.rer.nat., physical chemistry (U. München, 1969). Prof. (tel. 0251 + 83, ext. 3431). *Chalcogenide semiconductors, superstructures, phase diagrams, defects.*

Lewis, Dr James jr. Scientific and Technical Team, Army Material Command-STITEUR, IG-Farben-Hochhaus room 740, Grüneburgplatz, D-6000 Frankfurt/Main 1, Germany, Fed. Rep. (1940) PhD, crystallography (TU Clausthal, 1978). Material scient. (tel. 069 + 151, ext. 8263). *Crystal chemistry, electron density distribution, structure and physical properties, crystal physics, crystal defects.*

Liebau, Prof. Dr Friedrich Karl Franz. Mineralogisches Inst., Universität, Olshausenstr. 40-60, D-2300 Kiel, Germany, Fed. Rep. (1926) Dr.rer.nat., chemistry (Humboldt U. Berlin, 1956). Full prof. (tel. 0431 + 880, ext. 2888). *Inorganic crystal chemistry, silicates, phosphates.*

Liebertz, Prof. Dr Josef. Inst. f. Kristallographie, Universität, Zülpicher Str. 49, D-5000 Köln 1, Germany, Fed. Rep. (1929) Dr rer.nat., mineralogy (U. Bonn, 1961). Prof. (tel. 0221 + 470, ext. 4420). *Crystal growth, crystal chemistry.*

Lindemann, Prof. Dr Willi. Lehrstuhl f. Kristallstrukturlehre, Universität, Am Hubland, D-8700 Würzburg, Germany, Fed. Rep. (1921) Dr.phil.nat., crystallography (U. Erlangen, 1951). Prof. (tel. 0931 + 881, ext. 430). *Mathematical crystallography, structure analysis, computer programming.*

Lipka, Mrs Dr Annegret. Analytisch-Chem. Labor, Sachtleben Bergbau GmbH, Meggener Str., D-5940 Lennestadt, Germany, Fed. Rep. (1948) Dr.rer.nat., chemistry (U. Münster, 1975). Wiss. Ass. (tel. 02721 + 835275). *Crystal chemistry, antimony(III) compounds, stereochemistry.*

Löchner, Dr Ulrich. Postfach 250405, D-4630 Bochum 25, Germany, Fed. Rep. (1948) Dr, inorganic chemistry (U. Karlsruhe, 1980). Res. assoc. *Rare earths, catalysts, multiphase mixtures, textures, solid state, high temperature chemistry.*

Löns, Dr Jürgen. Fachbereich Chemie, Universität, Corrensstr. 24, D-4400 Münster, Germany, Fed. Rep. (1939) Dr., crystallography (U. Hamburg, 1969). Akad. Oberrat. (tel. 0251 + 83, ext. 3456). *Crystallography, structure determination, crystal chemistry.*

Lorenz, Mr Günter. Inst. f. Kristallographie und Mineralogie, Universität, Theresienstr. 41, D-8000 München, Germany, Fed. Rep. (1944) Diplom, mineralogie (U. München, 1978). Wiss. Ang. (tel. 089 + 2394, ext. 4313). *Polytypism, order-disorder structures, lattice dynamics, phase transitions, neutrons, X-rays, gamma-rays.*

Lorenz, Prof. Dr Wolfgang J. Inst. f. Physikalische Chemie und Elektrochemie, Universität, Kaiserstr. 12, D-7500 Karlsruhe, Germany, Fed. Rep. (1933) Prof., physical chemistry (TU, Clausthal-Cellerfeld, 1961). Prof. (tel. 0721 + 6083303). *Physics and chemistry of surfaces and interfaces, electrocrystallization.*

Luger, Prof. Dr Peter. Inst. f. Kristallographie, FU, Takustr. 6, D-1000 Berlin 33, Germany, Fed. Rep. (1943) Dr habil., crystallography (FU Berlin, 1974). Prof. (tel. 030 + 838, ext. 3411). *X-ray and neutron single crystal analysis, conformational analysis, carbohydrates, drug design, computer graphics.*

Mages, Dr Gert Rudolf. FL ALE 3, Siemens AG, Guenther-Scharowsky-Str. 2, D-8520 Erlangen, Germany, Fed. Rep. (1939) Dr.rer.nat., mineralogy, petrology (U. Erlangen, 1970). Scient. (tel. 09131 + 22858). *X-ray analysis, small angle X-ray scattering, inorganic crystal structures.*

Mandelkow, Dr Eckard. Dept. of Biophysics, MPI f. Medical Research, Jahnstr. 29, D-6900 Heidelberg, Germany, Fed. Rep. (1943) Dr, physics (U. Heidelberg, 1973). Res. scient. (tel. 06221 + 486277). *Fibre diffraction, small angle scattering, synchrotron radiation, electron microscopy, image reconstruction.*

Marie, Dr Alain Louis. Abt. f. Strukturforschung I, MPI f. Biochemie, Am Klopferspitz, D-8033 Martinsried, Germany, Fed. Rep. (1946) Ph.D., biochemistry (U. Sherbrooke, Quebec, Canada, 1973). Wiss. Ang. (tel. 089 + 8585, ext. 682). *Enzymology, protein sequencing, protein chemistry, crystallography.*

Mariolacos, Dr Konstantin. Mineralogisch-Kristallogr. Inst., Universität, V.M. Goldschmidtstr. 1, D-3400 Göttingen, Germany, Fed. Rep. (1936) Dr., crystal structures (U. Wien, 1972). Techn. Ang. (tel. 0551 + 39, ext. 3931). *Phase diagrams, crystal synthesis, crystal structures, sulpho-salts, sulphohalogenides.*

Martin, Dr Reinhold. Inst. f. Kristallographie und Mineralogie, Universität, Theresienstr. 41, D-8000 München 2, Germany, Fed. Rep. (1948) Dr.rer.nat., chemistry (U. Heidelberg, 1978). Wiss. Ang. (tel. 089 + 2394, ext. 4313). *Neutron diffraction, phase determination, one-dimensional metals, mineralogy.*

Maslowska, Miss Maria. Inst. f. Kristallographie, FU, Takustr. 6, D-1000 Berlin 33, Germany, Fed. Rep. (1954) Dipl., biology (U. Lodz, Poland, 1979). (tel. 030 + 8383463). *Biology, macromolecules.*

Massa, Dr Werner. Fachbereich Chemie, Universität, Hans-Meerwein-Str., D-3550 Marburg, Germany, Fed. Rep. (1944) Priv. Doz., inorganic chemistry (U. Marburg, 1982). Akad. Rat. (tel. 06421 + 28, ext. 5525). *Structure determination, inorganic solids, fluorometallates, hydrogen bonds, magnetism.*

Mateika, Dr Dieter. Philips GmbH Forschungslaboratorium, Postfach 540840, Vogt-Kölln-Str. 30, D-2000 Hamburg 54, Germany, Fed. Rep. (1935) Dr.rer.nat., mineralogy (U. Bonn, 1969). Scient. (tel. 040 + 5493, ext. 553, telex 21331656). *Crystal growth.*

Mattes, Prof. Dr Rainer. Inst. f. Anorganische Chemie, Universität, Corrensstr. 36, D-4400 Münster, Germany, Fed. Rep. (1937) Dr.habil., inorganic chemistry (U. Münster, 1970). Prof. (tel. 0251 + 83, ext. 3117). *Inorganic crystal structures, structure, oxo and thio complexes.*

Matz, Prof. Dr Günther. Ingenieurabteilung AP, Verfahrenstechnik, Bayer AG, Friedrich Ebert-Str. 217-319, D-5600 Wuppertal-Elberfeld, Germany, Fed. Rep. (1920) Dr., process engineering (U. Frankfurt/Main, 1950). Dept. head. (tel. 0202 + 36, ext. 7618). *Industrial crystallization, crystallization from solution, kinetics of crystallization.*

Matzat, Dr Eckhart. Kristallogr. Inst., Universität, Goldschmidtstr. 1, D-3400 Göttingen, Germany, Fed. Rep. (1938) Dr.rer.nat., mineralogy (U. Göttingen, 1968). (tel. 0551 + 39, ext. 3893). *Structure determination, modulated structures.*

Mayer, Dr Hugo Werner Waldemar. IAK 1, Kernforschungszentrum Karlsruhe, Postfach 3640, D-7500 Karlsruhe, Germany, Fed. Rep. (1943) Dr.rer.nat., chemistry (U. Stuttgart, 1979). Wiss. Ang. (tel. 07247 + 82, ext. 3438). *Crystal structures, phase transitions, elastic neutron scattering.*

Meier, Prof. Dr Hans. Staatl. Forschungsinstitut f. Geochemie, Concordiastr. 28, D-8600 Bamberg, Germany, Fed. Rep. (1927) Dr.rer.nat., physical chemistry (U. Mainz, 1954). Inst. head. (tel. 0951 + 27280). *Organic semiconductors.*

Mende, Prof. Dr Hans Horst. Inst. f. Angewandte Physik, Universität, Roxeller Str 70-72, D-4400 Münster, Germany, Fed. Rep. (1926) Dr.habil., physics (U. Münster). (tel. 0251 + 83, ext. 3512). *Whisker growth and size effects, electrical properties, magnetic properties, mechanical properties, thin metallic wires and whiskers.*

Mertin, Dr Wilhelm. Inst. f. Anorg. u. Analyt. Chemie, Universität, Heinrich-Buff-Ring 58, D-6300 Giessen, Germany, Fed. Rep. (1938) Dr.rer.nat., inorg. chemistry (U. Münster, 1967). Akad. Oberrat. (tel. 0641 + 7025676). *Transmission elctron microscopy, inorganic crystal structures.*

Messner, Dr Dieter. Physik-Labor, Kalle Niederlassung der Hoechst AG, Rheingaustr. 190, D-6200 Wiesbaden, Germany, Fed. Rep. (1928) Dr.rer.nat., physics (TH Stuttgart, 1956). Physicist. (tel. 06121 + 68, ext. 6921). *Physics, high and low molecular weight materials, X-ray structures.*

Metter, Mr Joachim. Inst. f. Kristallographie, Universität, Senckenberganlage 30, D-6000 Frankfurt/Main 1, Germany, Fed. Rep. (1955) Dipl., inorg. chemistry (U. Würzburg, 1983). (tel. 069 + 7982105). *Rare earths, X-ray and neutron diffraction, group V elements, computer aided chemistry.*

Mewis, Dr Albrecht. Inst. f. Anorganische Chemie, Universität, Greinstr. 6, D-5000 Köln 41, Germany, Fed. Rep. (1942) Priv. Doz., inorg. chemistry (U. Köln, 1981). Priv. Doz. (tel. 0221 + 4703267). *Inorganic solid state chemistry, ternary pnictides of transition metals.*

Meyer, Mr Andreas. Inst. f. Anorgan. Chemie, Universität, Kaiserstr. 12, D-7500 Karlsruhe, Germany, Fed. Rep. (1952) *Group theory, crystallography, crystal chemistry.*

Meyer, Dr Gerd Heinrich. Inst. f. Anorg. und Analyt. Chemie, Universität, Heinrich-Buff-Ring 58, D-6300 Giessen, Germany, Fed. Rep. (1949) Dr.rer.nat., inorganic chemistry (U. Giessen, 1976). Priv. Doz. and Akad. Rat. (tel. 0641 + 7025695). *Solid state chemistry, inorganic crystal structures.*

Meyer, Prof. Dr Hans-Jürgen. Mineralog. Inst., Universität, Poppelsdorfer Schloss, D-5300 Bonn, Germany, Fed. Rep. (1927) Dr.habil., mineralogy and crystallography (U. Bonn, 1962). Wiss. Rat und Prof. (tel. 0228 + 73, ext. 2771). *Kinetics of crystal growth, physicochemical crystallography.*

Meyer-Ehmsen, Prof. Dr Gerhard. Fachbereich 4, Universität, Postfach 4469, D-4800 Osnabrück, Germany, Fed. Rep. (1932) Dr.rer.nat., physics (U. Hamburg, 1961). Prof. (tel. 0541 + 608, ext. 2435). *Electron diffraction, surface science.*

Mikhail, Dr Ibrahim Fahmy. Inst. f. Kristallographie, FU. Berlin, Takustr. 6, D-1000 Berlin 33, Germany, Fed. Rep. (1939) Dr.rer.nat., physics (TU Clausthal-Zellerfeld, 1973). Wiss. Ang. (tel. 030 + 838, ext. 3412). *X-ray diffraction, phase transitions, biophysics.*

Milius, Mr Wolfgang. Inst. f. Anorg. Chemie I, Universität, Egerlandstr. 1, D-8520 Erlangen, Germany, Fed. Rep. (1958). (tel. 09131 + 857351).

Millhouse, Dr Arthur Holmes. Inst. f. Kristallographie, Dept. f. Physik, Universität, Im Stadtwald, D-6600 Saarbrücken, Germany, Fed. Rep. (1938) Ph.D., physics (Virg. Polyt. Inst. and State U., USA, 1970). Physicist. (tel. 0681 + 3021). *Synchrotron radiation, magnetic structures.*

Möller, Dr Manfred. Anorg. Chem. Inst., Universität, Corrensstr. 36, D-4400 Münster, Germany, Fed. Rep. (1953) Dr.rer.nat., chemistry (U. Dortmund, 1983). Wiss. Ass. (tel. 0251 + 833116). *Structure and properties, inorganic and intermetallic crystals.*

Moh, Prof. Dr Günter Harald. Mineralogisch-Petrogr. Inst., Universität, Im Neuenheimer Feld 236, Postfach 104040, D-6900 Heidelberg, Germany, Fed. Rep. (1929) Dr.habil., mineralogy and petrology (U. Heidelberg, 1967). Prof. (tel. 06221 + 562810). *Experimental mineralogy, sulfide petrology, crystal chemistry.*

Mootz, Prof. Dr Dietrich. Inst. f. Anorg. Chemie und Strukturchemie, Universität, Universitätsstr. 1, D-4000 Düsseldorf, Germany, Fed. Rep. (1933) Dr.rer.nat., chemistry (TU Berlin, 1959). Prof. (tel. 0211 + 311, ext. 3135). *Inorganic and organic crystal structures, solid state chemistry.*

Moritz, Dr Wolfgang Otto. Inst. f. Kristallographie und Mineralogie, Universität, Theresienstr. 41, D-8000 München 2, Germany, Fed. Rep. (1943) Dr.rer.nat.habil., Kristallographie (U. München, 1976). Akad. Oberrat. (tel. 089 + 2394, ext. 4336). *LEED, surface crystallography.*

Müller, Prof. Dr Gerd. Inst. f. Mineralogie, TH, Schnittspahnstr. 9, D-6100 Darmstadt, Germany, Fed. Rep. (1942) Dr.rer.nat., mineralogy (U. Karlsruhe, 1969). Prof. (tel. 06151 + 165280). *Crystal chemistry, material science, mineralogy.*

Müller, Prof. Horst. Chemisches Laboratorium, Universität, Albertstr. 21, D-7800 Freiburg, Germany, Fed. Rep. (1929) Dr., inorganic chemistry (U. Freiburg, 1958). Prof. (tel. 0761 + 203, ext. 2915). *Solid state chemistry, radio and nuclear chemistry, hydrides.*

Müller, Dr Paul Hubert. Inst. f. Anorg. Chemie, RWTH, Prof.-Pirlet-Str. 1, D-5100 Aachen, Germany, Fed. Rep. (1951) Dr., inorg. chemistry (RWTH Aachen, 1980). (tel. 0241 + 804669). *Inorganic chemistry, neutron diffraction, magnetochemistry.*

Müller, Prof. Dr Ulrich. Fachbereich Chemie, Universität, Hans-Meerwein-Str., D-3550 Marburg, Germany, Fed. Rep. (1940) Dr., chemistry (U. Stuttgart, 1966). Prof. (tel. 06421 + 285686). *Crystal structure systematics, stacking faults in crystals, preparative inorganic chemistry.*

Müller, Prof. Dr Wolfgang Friedrich. Inst. f. Mineralogie, Techn. Universität, Schnittspahnstr. 9, D-6100 Darmstadt, Germany, Fed. Rep. (1939) Dr.rer.nat., mineralogy (U. Tübingen, 1965). Prof. (tel. 06151 + 16, ext. 3380). *Crystal chemistry, petrography, meteorites, crystal defects, electron microscopy.*

Müller-Buschbaum, Prof. Dr Hanskarl. Inst. f. Anorganische Chemie, Universität, Olshausenstr. 40-60, D-2300 Kiel, Germany, Fed. Rep. (1931) Ord. Prof., inorganic chemistry (U. Kiel, 1969). Direktor. (tel. 0431 + 880, ext. 2410). *Solid state reactions, plasma and laser chemistry, high temperatures, solid state chemistry, inorganic crystal structures.*

Müller-Vogt, Dr German. Kristall- und Materiallabor, Fak. f. Physik, Universität, Kaiserstr. 12, D-7500 Karlsruhe, Germany, Fed. Rep. (1943) Dr.rer.nat., physics (U. Karlsruhe, 1971). Leiter des Kristall- und Materiallabors. (tel. 0721 + 608, ext. 3470). *Crystal growth and characterization.*

Müllner, Dr Manfred. Inst. f. Kernphysik, Universität, August-Euler-Str. 6, D-6000 Frankfurt, Germany, Fed. Rep. (1928) Dr., physics (1966). (tel. 0611 + 798, ext. 4238). *Solid state physics, lattice dynamics, structure determination.*

Münninghoff, Dr Günter. XRD, Nicolet Instrument GmbH, Senefelderstr. 162, D-6050 Offenbach, Germany, Fed. Rep. (1953) PhD, crystallography (1978). Product specialist. (tel. 069 + 837001, telex 04185411). *Single crystal diffraction, powder diffraction, X-ray diffraction, neutron diffraction.*

Mullen, Dr Donald Joseph Edgar. FR 17.3 - Kristallographie, Universität, D-6600 Saarbrücken, Germany, Fed. Rep. (1945) Ph.D., crystallography (U. St. Andrews, UK, 1970). Res. scient. (tel. 0681 + 302, ext. 3470). *Electron density in bonds, neutron diffraction, extinction.*

Mundt, Dr Otto. Inst. f. Anorg. Chemie, Universität, Pfaffenwaldring 55, D-7000 Stuttgart, Germany, Fed. Rep. (1950) Dr.rer.nat., chemistry (U. Karlsruhe, 1979). (tel. 0711 + 7854221). *Structural chemistry of org. compounds of main group elements*

Murad, Dr Enver. Lehrstuhl f. Bodenkunde, TU München, D-8050 Freising-Weihenstephan, Germany, Fed. Rep. (1941) Dr.phil.nat., mineralogy (U. Frankfurt, 1970). Wiss. Mit. (tel. 08161 + 71, ext. 735). *Crystal chemistry, Mössbauer spectroscopy.*

Mutter, Mrs Graciela. Electron Microscopy Group, MPI f. Metallforschung, Inst. f. Werkstoffwissenschaften, Seestr. 92, D-7000 Stuttgart 1, Germany, Fed. Rep. (1955) Dipl., mineralogy (U. Heidelberg, 1983). *Ceramics, cordierite, grain boundaries.*

Nägele, Dr Walter. Inst. f. Kristallographie, Universität, Charlottenstr. 33, D-7400 Tübingen, Germany, Fed. Rep. (1949) Dr.rer.nat., physics (U. Tübingen, 1979). Wiss. Mit. (tel. 07247 + 82, ext. 3158). *Crystal structures, magnetic structures, neutron scattering, spin glasses.*

Narita, Dr Hajime. Mineralogisch - Petrogr. Inst. und Museum, Universität, Olshausen str. 40-60, D-2300 Kiel, Germany, Fed. Rep. (1944) DSc, mineralogy (Osaka U., Japan). Wiss. Mit. (tel. 0431 + 880, ext. 2906). *Crystal chemistry, crystallography, mineralogy.*

Neff, Prof. Dr Hans Josef. Abt. E63, Siemens AG, Rheinbrückenstr. 50, D-7500 Karlsruhe, Germany, Fed. Rep. (1920) Dr.rer.nat., physics (U. Karlsruhe, 1951). Manager of application labs. (tel. 0721 + 595, ext. 2656). *Electron microscopy, X-ray diffraction, X-ray fluorescence analysis, gas and HP liquid chromatography.*

Neifeind, Dipl-Ing Axel. IBM, Dept. 0348, 7032-16, Tübinger Allee, D-7032 Sindelfingen, Germany, Fed. Rep. (1941) Dipl.-Ing., metallurgy (U. Stuttgart, 1969). Staff eng. (tel. 07031 + 611, ext. 3114). *Crystal structures, semi conductors, crystal defects, surface analysis, orientation methods, X-ray topography.*

Nelkowski, Prof. Dr Horst. Inst. f. Festkörperphysik, TU, Strasse des 17. Juni 135, D-1000 Berlin 12, Germany, Fed. Rep. (1921) Dr., experimental physics (TU Berlin, 1970). Dir. (tel. 030 + 314, ext. 2247). *Solid state physics, crystal growth (esp. II-VI- and I-III-V I2-compounds), surface physics, semiconductor physics, photoconductivity, luminescence (electroluminescence).*

Nesper, Dr Reinhard Friedrich. MPI f. Festkörperforschung, Heisenbergstr. 1, D-7000 Stuttgart, Germany, Fed. Rep. (1949) Dr.rer.nat., chemistry (tel. 0711 + 6860320). *Solid state chemistry, crystallography, structure and properties, electronic structure of solids.*

Neubüser, Prof. Joachim Franz Friedrich Gerhard. Lehrstuhl D f. Mathematik, RWTH, Templergraben 64, D-5100 Aachen, Germany, Fed. Rep. (1932) Dr.rer.nat., mathematics (U. Kiel, 1957). Prof. (tel. 0241 + 80, ext. 4543). *Crystallographic groups, computational group theory.*

Newesely, Prof. Dr Heinrich. Inst. f. Mineralogie und Kristallographie, TU, Hardenbergstr. 42, D-1000 Berlin, Germany, Fed. Rep. (1933) Dozent, crystal chemistry and micromorphology (TU Berlin, 1964). Prof. (tel. 030 + 314, ext. 2746). *Crystal chemistry, micromorphology (electron microscopy), biocrystallography, fine particle dusts research.*

Nitsche, Prof. Dr Rudolf. Kristallographisches Inst., Universität, Hebelstr. 25, D-7800 Freiburg, Germany, Fed. Rep. (1922) Ph.D., physical chemistry (U. Heidelberg, 1951). Prof. (tel. 0761 + 203, ext. 4280). *Crystal growth, crystal chemistry, phase relations, crystal physics.*

Nitschmann, Dr Günter Max Alfred. Am Entenspiel 1, D-6330 Wetzlar, Germany, Fed. Rep. (1914) Dr., crystallography, mineralogy (U. Breslau, 1940). (tel. 06441 + 23748). *Crystal growth.*

Noll, Prof. Dr Walter Friedrich Heinrich. Mineralogisches Inst., Universität, Zülpicherstr. 47, D-5000 Köln 1, Germany, Fed. Rep. (1907) Dr.phil.nat., mineralogy (U. Jena, 1930). Honorary prof. (tel. 02171 + 31578). *Silicate crystal structures, crystal chemistry, silicate technology, historical aspects of silicates.*

Noltemeyer, Dr Mathias Rolf. Inst. f. Anorg. Chemie, Universität, Tammannstr. 4, D-3400 Göttingen, Germany, Fed. Rep. (1947) Dr., chemistry (U. Göttingen, 1977). Wiss. Angest. (tel. 0551 + 391). *X-ray apparatus.*

Nover, Dr Georg. Inst. f. Werkstoffwissenschaft, MPI f. Metallforschung, Seestr. 75, D-7000 Stuttgart, Germany, Fed. Rep. (1948) Dr.rer.nat., mineralogy (U. Bonn, 1979). Wiss. Ang. (tel. 0711 + 2095, ext. 202). *Crystal structure analysis, inorganic and metallic phases.*

Nowack, Miss Ellen Carla. Inst. f. Kristallographie, RWTH, Templergraben 55, D-5100 Aachen, Germany, Fed. Rep. (1956) Dipl.Min., crystallography (1983). Wiss. Ang. (tel. 0241 + 806918). *Charge density distributions in solids, chemical bonding, crystallographic computing, properties and structural relations.*

Oefner, Mr Christian. X-Ray Division, EMBL, Meyerhofstr. 1, D-6900 Heidelberg, Germany, Fed. Rep. (1956) Dipl., physics (U. Heidelberg, 1982). Grad. student. (tel. 06221 + 387308). *Structure - activity relationships, biomolecules.*

Oehlschlegel, Dr Georg. Inst. f. Mineralogie, Ruhr-Universität, Universitätsstr. 150, D-4630 Bochum, Germany, Fed. Rep. (1938) Dr., mineralogy (U. Hamburg, 1962). Wiss. Ang. (tel. 0234 + 700, ext. 4396). *Glass, ceramics, crystal growth, crystal optics.*

Okrusch, Prof. Dr rer nat Martin. Mineralogisches Inst., Technische Universität, Gauss-Strasse 28-29, D-3300 Braunschweig, Germany, Fed. Rep. (1934) Habilitation, mineralogy (U. Würzburg, 1968). Prof. (tel. 0531 + 391, ext. 2263). *Rock forming minerals.*

Otten, Mr Peter. Gilberstr. 27, D-2000 Hamburg, Germany, Fed. Rep. (1938) Diplom, crystallography (U. Bonn, 1967). Scientific coordinator. (tel. 040 + 3193302). *X-ray diffractometry, phase transitions, equipment development.*

Otto, Dr Hans Hermann. Fakultät f. Physik, Universität, Universitätsstr. 31, Postfach 397, D-8400 Regensburg 2, Germany, Fed. Rep. (1938) Dr.rer.nat., crystallography, mineralogy (TU Berlin, 1970). Res. staff member. (tel. 0941 + 943, ext. 2123). *Inorganic crystal structures, crystal growth, ferroic materials, crystal physics.*

Overkott, Dr Paul Engelbert. Keramisches Laboratorium, Dr. C. Otto und Comp. GmbH, Bochum, Nachtigallenstr. 30, D-5820 Gevelsberg, Germany, Fed. Rep. (1926) Dr.rer.nat., mineralogy (U. Köln, 1959). Head, ceramic lab. (tel. 0234 + 4191, ext. 244). *Industrial ceramics, refractories, X-ray diffraction, REM-EDAX.*

Pachali, Dr Klaus Erich. Analytisches Labor, Deutsche Akzo Coatings GmbH, Magirusstr. 26, D-7000 Stuttgart 30, Germany, Fed. Rep. (1943) Dr.rer.nat., inorganic chemistry (U. Stuttgart, 1970). Wiss. Mit. (tel. 0711 + 895, ext. 379). *Structures, ternary chalkogenides, polyphosphides, chelate complexes, inorganic crystal structures.*

Pähler, Dr Arno. Inst. f. Kristallographie, FU, Takustr. 6, D-1000 Berlin 33, Germany, Fed. Rep. (1949) Dr, physics (U. Göttingen, 1983). Res. assoc. (tel. 030 + 838-6326). *Protein crystallography, synchrotron radiation.*

Pai, Dr Emil Friedrich. Dept. of Biophysics, MPI f. Med. Res., Jahnstr. 29, D-6900 Heidelberg, Germany Fed. Rep. (1950) Dr.rer.nat., chemistry (U. Heidelberg, 1978). Res. scient. (tel. 06202 + 486275, telex. 461505). *Enzyme chemistry, protein structures.*

Pal, Dr Gour Pada, Inst. f. Kristallographie, FU, Takustr. 6, D-1000 Berlin 33, Germany, Fed. Rep. (1953) PhD, protein chemistry (Calcutta U., India, 1980). Wiss. Mit. (tel. 030 + 838-6325). *Proteins, crystal growth, X-ray crystallography.*

Pannhorst, Dr Wolfgang. Res. and Dev., SCHOTT Glaswerke, Hattenbergstr. 10, D-6500 Mainz, Germany, Fed. Rep. (1942) Dr.rer.nat.habil., crystallography (U. Karlsruhe, 1980). General manager. (tel. 06131 + 663676, telex. 4187920). *Glasses, glass ceramics, optical glasses, devitrification, nucleation, crystal growth, fibres, integrated optics, radiation defects, photochromism.*

Parak, Dr Fritz Günther. Physikdep. E 15, TU, James-Franck-Str., D-8046 Garching, Germany, Fed. Rep. (1940) Dr., physics (TU München, 1970). Wiss. Ang. (tel. 089 + 3209, ext. 2509). *Protein structure determination, perfect single crystals, Mössbauer effect, small angle scattering.*

Parge, Dr Hans Erich. Inst. f. Kristallographie, FU, Takustr. 6, D-1000 Berlin 33, Germany, Fed. Rep. (1955) PhD, chemistry (Trinity C. Dublin, Ireland, 1978). Res. assoc. (tel. 030 + 838-6326). *Biomolecules, organometallic structures, organic molecules, single crystal studies.*

Pattison, Dr Philip. Bereich C1, Hahn-Meitner-Inst. f. Kernforschung, Glienicker Str. 100, D-1000 Berlin 39, Germany, Fed. Rep. (1951) PhD, Physics (Warwick U., UK, 1975). Res. asst. (tel. 030 + 8009, ext. 2686). *Gamma-ray scattering (elastic and inelastic), electron charge and momentum density.*

Paulitsch, Prof. Dr Peter. Inst. f. Mineralogie, TH, Schnittspahnstr. 9, D-6100 Darmstadt, Germany, Fed. Rep. (1922) Dr.rer.nat., mineralogy (U. Graz, Austria, 1944). Full prof. (tel. 06151 + 16, ext. 2180). *Experimental petrofabrics.*

Paulus, Dr Erich Friedrich. Angewandte Physik, Hoechst AG, Postfach 800320, D-6230 Frankfurt/Main 80, Germany, Fed. Rep. (1937) Dr., chemistry (U. München, 1965). Scient. (tel. 0611 + 3051, ext. 6360). *Crystal structures, organic and inorganic compounds, powder diffraction, fibres, high polymer structures.*

Pentinghaus, Dr Horst. Dept. of Chemistry, Inst. f. Nuclear Waste Technology, KFZ, Postfach 3640, D-7500 Karlsruhe 1, Germany, Fed. Rep. (1932) Dr.rer.nat., mineralogy (U. Münster, 1970). Head of the department. (tel. 07247 + 824476, telex 7826484). *Experimental crystallography, crystal chemistry.*

Penzkofer, Dr Benno. Rathausstr. 14, D-8025 Unterhaching, Germany, Fed. Rep. (1947) Dr.rer.nat., crystallography (U. München, 1980). Reg. Rat z. A. (tel. 089 + 616527). *Structural crystallography, computing methods, order-disorder phenomena.*

Peterat, Dr Michael. HELIOTRONIC GmbH, Johannes-Hess-Str. 24, D-8263 Burghausen, Germany, Fed. Rep. (1946) Dipl., mineralogy (TU München, 1974). (tel. 08677 + 834026). *Unconventional crystallization techniques from Si-melt, solar-cells, order-disorder phenomena.*

Peters, Dr Karl. MPI f. Festkörperforschung, Heisenbergstr. 1, D-7000 Stuttgart 80, Germany, Fed. Rep. (1940) Dr.rer.nat., inorganic chemistry (U. Münster, 1971). Wiss. Ang. (tel. 0711 + 68601). *Crystal structure determination, organic and inorganic compounds.*

Petzoldt, Dr Jürgen Hugo Hans. Res. & Dev. Dept., Schott Glaswerke, Hattenberg Str. 10, D-6500 Mainz, Germany, Fed. Rep. (1935) Dr.rer.nat., inorganic chemistry (U. Bonn, 1963). Executive Vice President Res. and Dev. (tel. 06131 + 66, ext. 3508, telex 4187920 sm d). *Glass ceramics, glass structure and properties.*

Pfefferkorn, Prof. Dr Gerhard Erich. Inst. f. Medizinische Physik, Universität, Hüfferstr. 68, D-4400 Münster, Germany, Fed. Rep. (1913) Dr.rer.nat., physics (U. Berlin, 1938). Prof. (tel. 0251 + 831, ext. 5101). *Biocrystallography, electron microscopy, X-ray micro analysis.*

Pflugrath, Dr James William. MPI f. Biochemie, Am Klopferspitz, D-8033 Martinsried, Germany, Fed. Rep. (1957) PhD, biochemistry (Rice U., USA,

1984). Res. assoc. (tel. 089 + 8578-2681). *Biological macromolecules, structure and function, computational crystallography.*
Philipsborn, von, Prof. Dr Henning. Abt. f. Kristallographie, Universität, D-8400 Regensburg, Germany, Fed. Rep. (1934) Dr.phil., crystallography (U. Zürich, 1964). Prof. (tel. 0941 + 943, ext. 2481). *Crystal chemistry, crystal growth, electronic materials, X-ray diffraction, symmetry.*
Pickardt, Prof Dr Joachim. Inst. f. Anorg. u. Analyt. Chemie, TU, Str. d. 17.Juni 135, D-1000 Berlin 12, Germany, Fed. Rep. (1939) Dr.Ing., chemistry (TU Berlin, 1971). Apl. Prof. (tel. 030 + 3142469). *inorganic chemistry, structural chemistry, solid state chemistry.*
Pieper, Dr Gerhard. Inst. f. Kristallographie, Universität, Senckenberg-Anlage 30, D-6000 Frankfurt/Main, Germany, Fed. Rep. (1934) Dr.phil.nat., mineralogy and crystallography (U. Frankfurt/Main, 1967). Akad. Oberrat. (tel. 069 + 798, ext. 104). *Inorganic crystal structures.*
Plies, Dr Volker. Inst. f. Anorg. u. Analyt. Chemie, Universität, Heinrich-Buff-Ring 58, D-6300 Giessen, Germany, Fed. Rep. (1945) Dr.rer.nat., inorganic chemistry (U. Giessen, 1975). Akad. Rat. (tel. 0641 + 7025706). *Inorganic crystal structures, TEM.*
Ploog, Dr Klaus. MPI f. Festkörperforschung, Heisenbergstr. 1, D-7000 Stuttgart 80, Germany, Fed. Rep. (1941) Dr., inorganic chemistry (U. München, 1970). Project leader. (tel. 0711 + 6860383). *Double crystal X-ray diffraction, semiconductor superlattices, RHEED - grazing incidence, semiconductor (III-V) surfaces.*
Plust, Dr Heinz-Günther. Deutsche Automobilgesellschaft mbH, Geschäftsführung, Postfach 85, D-7300 Esslingen-Mettingen, Germany, Fed. Rep. (1927) Dr., solid state chemistry (TU Berlin, 1953). Geschäftsführer. (tel. 0711 + 3026877). *Crystal growth, inorganic crystal structures, instrumentation.*
Pohl, Prof. Dr Dieter. Mineralog.-Petrogr. Inst., Universität, Grindelallee 48, D-2000 Hamburg 13, Germany, Fed. Rep. (1940) Dr.rer.nat., physics (U. Hamburg, 1968). (tel. 040 + 41232482). *Crystal physics, inorganic crystal structures.*
Pohl, Prof. Dr Siegfried. Fachbereich Chemie, Universität, CarlvonOssietzky Str., D-2900 Oldenburg, Germany, Fed. Rep. (1943) Dr., inorganic chemistry (U. Bielefeld, 1974). Prof. *Crystal chemistry, main group iodides, iron-sulfur-clusters.*
Polborn, Dr Kurt Volkmar. Inst. f. Anorg. Chemie, Universität, Meiserstr. 1, D-8000 München, Germany, Fed. Rep. (1946) Dr.rer.nat. (U. München, 1975). Akad. Rat. (tel. 089 + 5902250). *Cluster compounds, superstructures.*
Poll, Dr Wolfgang. Inst. f. Anorg. Chemie und Strukturchemie, Universität, Universitätsstr. 1, D-4000 Düsseldorf, Germany, Fed. Rep. (1954) Dr.rer.nat., chemistry (U. Düsseldorf, 1979). Akad. Rat. (tel. 0211 + 3113146). *Solid state chemistry, crystal structures, computer programming, instrumentation.*
Pollmann, Dipl-Min Siegfried. Chemisch-Phys. Labor, Steinmüller GmbH, Postfach 1949-1960, D-5270 Gummersbach, Germany, Fed. Rep. (1931) Dr.rer.nat., mineralogy (U. Münster, 1964). Dept. head. (tel. 02261 + 85, ext. 2761). *Technical mineralogy, crystal chemistry.*
Prandl, Prof. Dr Wolfram. Inst. f. Kristallographie, Universität, Charlottenstr. 33, D-7400 Tübingen, Germany, Fed. Rep. (1935) Dr., crystallography, mineralogy (U. München, 1974). Prof. (tel. 07071 + 29, ext. 6058). *Elastic and inelastic neutron diffraction, magnetic phase transitions, group theoretical methods, phase transitions.*
Preut, Dr Hans. Anorganische Chemie, Universität, Otto-Hahn-Str., D-4600 Dortmund 50, Germany, Fed. Rep. (1940) Dr., inorganic chemistry (U. Dortmund, 1972). Wiss. Ang. (tel. 0231 + 755, ext. 3813). *Inorganic and organic crystal structures, instrumentation, lattice energy calculations, molecular packing analysis.*
Puff, Prof. Dr Heinrich. Anorg.-Chemisches Inst., Universität, Gerhard-Domagk-Str. 1, D-5300 Bonn 1, Germany, Fed. Rep. (1921) Dr.rer.nat., inorganic chemistry (U. Kiel, 1956). Full prof. (tel. 0228 + 73, ext. 2661). *Inorganic crystal structures, computer programming.*
Rabenau, Prof. Dr Albrecht. MPI f. Festkörperforschung, Heisenbergstr. 1, D-7000 Stuttgart 80, Germany, Fed. Rep. (1922) Dr.habil., solid state chemistry (RWTH Aachen, 1963), Direktor. (tel. 0711 + 7830, ext. 720). *Solid state chemistry, materials research.*
Rager, Dr Helmut. FB Geowissenschaften, Inst. f. Mineralogie, Petrologie und Kristallographie, Lahnberge, D-3550 Marburg, Germany, Fed. Rep. (1941) Dr.rer.nat., physical chemistry (U. Münster, 1973). Priv.-Doz. (tel. 06421 + 282232). *NMR, EPR, ligand field spectroscopy.*
Range, Prof. Dr Klaus-Jürgen. Inst. f. Chemie, Universität, Universitätsstr. 31, D-8400 Regensburg, Germany, Fed. Rep. (1938) Dr.rer.nat., inorganic chemistry (U. Heidelberg, 1966). Ord. Prof. (tel. 0941 + 943, ext. 4551, telex 065658 unire d). *Inorganic crystal structures, solid state chemistry, high pressure crystallography, materials research.*
Rath, Prof. Dr Robert. Mineralogisch-Petrogr. Inst., Universität, Grindelallee 48, D-2000 Hamburg, Germany, Fed. Rep. (1924) Prof., mineralogy (U. Hamburg, 1971). Prof. (tel. 040 + 4123, ext. 2053). *Crystal optics, applied mineralogy.*
Recker, Prof. Dr Kurt. Mineralog. Inst., Universität, Poppelsdorfer Schloss, D-5300 Bonn, Germany, Fed. Rep. (1924) Prof., crystallography and mineralogy (U. Bonn, 1970). Wiss. Rat und Prof. (tel. 0228 + 73, ext. 2769). *Crystal growth, crystal chemistry, spectroscopy, color and luminescence of minerals.*
Reimers, Mr Walter. Inst. f. Mineralogie, Universität, Lahnberge, D-3550 Marburg, Germany, Fed. Rep. (1954) Dipl.-Min., mineralogy (U. Marburg, 1977). Wiss. Ang. *Structure determination, magnetic structure determination.*

Reinen, Prof. Dr Dirk. Fachbereich Chemie, Universität, Hans-Meerwein-Str., D-3550 Marburg, Germany, Fed. Rep. (1930) Dr., inorganic chemistry (U. Bonn, 1960). Ord. Prof. (tel. 06421 + 285668, telex 482372). *Structure and bonding, transition metal compounds, spectroscopy, vibronic coupling effects, Jahn-Teller effect.*
Renninger, Prof. Dr Mauritius. Fachbereich Geowissenschaften, Universität, Lahnberge, D-3550 Marburg, Germany, Fed. Rep. (1905) Dr.techn.Sci., physics (TH München, 1930). Retired. (tel. 06421 + 28, ext. 5705). *Real crystals, perfect crystals, diffraction optics, multiple crystal diffractometry, X-ray topography, lattice defects.*
Reuber-Kürbs, Mrs Dr-Ing Ellen. Fritz-Haber-Inst., Max-Planck-Ges., Faradayweg 4-6, D-1000 Berlin 33, Germany, Fed. Rep. (1923) Dr.-Ing., physics (TU Berlin, 1953). Res. scient. (tel. 030 + 8305, ext. 346). *Electron microscopy, structure analysis, images at atomic resolution, image reconstruction methods, biological structures, large molecules, instrumentation.*
Reuter, Prof. Dr-Ing Bertold. Inst. f. Anorg. u. Analyt. Chemie, TU, Strasse des 17. Juni 135, D-1000 Berlin 12, Germany, Fed. Rep. (1916) Dr.-Ing., inorganic chemistry (TH Berlin, 1941). Ord. Prof. (tel. 030 + 314, ext. 2470). *Inorganic crystal chemistry.*
Richter, Mrs Ursula. Morgengraben 1, D-5000 Köln 80, Germany, Fed. Rep. (1953) Diplom, mineralogy (U. Göttingen, 1978). (tel. 0221 + 664241). *Crystal growth, optical and electrical properties of crystals, crystal structure analysis.*
Rickert, Prof. Dr Hans. Physikalische Chemie, Universität, Postfach 500500, D-4600 Dortmund 50, Germany, Fed. Rep. (1928) Prof., physical chemistry (U. Dortmund, 1969). Prof. (tel. 0231 + 755, ext. 3900). *Physical chemistry of solids, chemical thermodynamics, transport reactions, electrochemistry, corrosion.*
Riedel, Prof. Dr Erwin. Inst. f. Anorg. u. Analyt. Chemie, TU, Strasse des 17. Juni 135, D-1000 Berlin 12, Germany, Fed. Rep. (1930) Dr.-Ing., inorganic chemistry (TU Berlin, 1970). Prof. (tel. 030 + 314, ext. 3498). *Solid state chemistry.*
Rösch, Dr Heinrich. Abt. Geoch. u. Mineralogie, Bundesanst. f. Geowiss. u. Rohstoffe, Stilleweg 2, D-3000 Hannover 51, Germany, Fed. Rep. (1935) Dr.phil., mineralogy and crystallography (U. Kiel, 1961). Sr. scient., sr. res. officer. (tel. 0511 + 6468, ext. 562). *Clay minerals, crystallography, silica modifications, quantitative X-ray diffraction analysis.*
Rohmer, Mr Christian. IBM, Dept. 0348, 7032-16, Tübinger Allee, D-7032 Sindelfingen, Germany, Fed. Rep. (1938) Technician. (tel. 07031 + 611, ext. 3963). *X-ray topography, silicon materials, X-ray analysis, single crystals, polycrystalline materials.*
Rossmanith, Mrs. Dr Elisabeth. Mineralogisch-Petrographisches Inst., Universität, Grindelallee 48, D-2000 Hamburg, Germany, Fed. Rep. (1943) Dr.phil.habil., physics, crystallography (U. Hamburg, 1981). Wiss. Ass. (tel. 040 + 4123, ext. 2485). *Lattice dynamics, metals, structure determination, crystal physics.*
Rothbauer, Dr Richard. Weidenstr. 11, D-6234 Hattersheim 3, Germany, Fed. Rep. (1938) Dr.phil.nat., crystallography (U. Frankfurt, 1971). *Methods of structure analysis, single crystal diffractometry, elastic neutron diffraction.*
Rott, Dr Volkwin. Schunk und Ebe GmbH, Rodheimer Str. 59-61, D-6300 Giessen, Germany, Fed. Rep. (1939) Dr., crystallography (TU Clausthal, 1970). Leiter der Patentabteilung. (tel. 0641 + 78081). *Carbon.*
Ruban, Prof. Dr Gerhard. FB 21, WE 5, Inst. f. Kristallographie, FU, Takustr. 6, D-1000 Berlin 33, Germany, Fed. Rep. (1926) Prof., crystallography (FU Berlin, 1971). Prof. (tel. 030 + 838, ext. 3462). *X-ray structure analysis, crystal chemistry.*
Rudert, Mr Rainer. Inst. f. Kristallographie, FU, Takustr. 6, D-1000 Berlin 61, Germany, Fed. Rep. (1956) Diplom, physics (FU Berlin, 1983). Physicist. (tel. 030 + 8383460). *Mössbauer spectroscopy, electron density.*
Ruiz Perez, Mrs Catalina. Arbeitsgruppe f. chem. Kristallographie, MPI f. Biochemie, Am Klopferspitz, D-8033 Martinsried, Germany, Fed. Rep. (1957) Diplom, physics (U. Valencia, Spain, 1983). (tel. 089 + 85782661). *X-ray structure analysis, organic compounds.*
Ruppersberg, Prof. Dr Henner. Fachbereich Angewandte Physik, Universität, D-6600 Saarbrücken, Germany, Fed. Rep. (1933) Dr.rer.nat.,Prof. (tel. 0681 + 302, ext. 3448). *Amorphous phases, chemical short-range order, X-ray stress analysis.*
Saalfeld, Prof. Dr Horst. Mineralogisch-Petrogr. Inst., Universität, Grindelallee 48, D-2000 Hamburg, Germany, Fed. Rep. (1920) Prof., mineralogy (U. Saarbrücken, 1960). Direktor. (tel. 040 + 4123, ext. 2050). *Crystal chemistry, clay mineralogy.*
Sabrowsky, Prof. Dr Horst. Anorganische Chemie I, Ruhr-Universität, D-4630 Bochum, Germany, Fed. Rep. (1934) Prof. (tel. 0234 + 700, ext. 4151). *Solid state chemistry, structures, magneto chemistry, high-pressure thermogravimetry.*
Saenger, Prof. Dr Wolfram H. E. Inst. f. Kristallographie, FU, Takustr. 6, D-1000 Berlin 33, Germany, Fed. Rep. (1939) Prof., chemistry (Darmstadt, 1965). Prof. (tel. 030 + 838-3412, telex 184019). *Protein structures, Nucleic acids, oligosaccharides, hydrogen bonding, inclusion complexes.*
Salje, Prof. Dr Ekhard. Mineralogisches Inst., Universität, Welfengarten 1, D-3000 Hannover, Germany, Fed. Rep. (1946) Prof. crystallography (U. Hannover, 1977). Prof. (tel. 0511 + 762, ext. 3583 or 2222). *Thermodynamics, crystal physics, phase transitions, spectroscopy, crystal chemistry.*

Saur, Prof. Dr-Ing Eugen. Inst. f. Angew. Physik, Universität, Leihgesterner Weg 106, D-6300 Giessen, Germany, Fed. Rep. (1910) Dr.-Ing., physics (TU Stuttgart, 1936). Full prof. (tel. 0641 + 702, ext. 2791). *Superconductivity.*

Schäfer, Prof. Dr Herbert Leo. Abt. II f. Anorganische Chemie, TH, Hochschulstr. 4, D-6100 Darmstadt, Germany, Fed. Rep. (1933) Dr., inorganic chemistry (TH Darmstadt, 1971). Prof. (tel. 06151 + 162292). *Inorganic crystal structures.*

Schäfer, Dr Wolfgang. Mineralog. Inst., Universität Bonn, KFA, Postfach 1913, D-5170 Jülich, Germany, Fed. Rep. (1942) Dr., physics (U. Bonn, 1971). (tel. 02461 + 616024). *Structure analysis and refinement, neutron diffraction, magnetic ordering.*

Schanda, Dr Friedrich. Abt. f. Strukturforschung I, MPI f. Biochemie, Am Klopferspitz, D-8033 Martinsried, Germany, Fed. Rep. (1947) Diplom, physics (TU München, 1975). Res. assoc. (tel. 089 + 8585, ext. 660). *X-ray and neutron structure analysis, bonding electron density studies.*

Scharfenberger, Miss Ulrike. Mineralog.-Petrogr. Inst., Universität, Im Neuenheimer Feld 236, D-6900 Heidelberg, Germany, Fed. Rep. (1961) Student.

Scheringer, Prof Dr Christian Andreas. Inst. f. Mineralogie, Universität, Lahnberge, D-3550 Marburg, Germany, Fed. Rep. (1930) Dr.habil., crystallography (RWTH Aachen, 1971). Honorarprof. (tel. 06421 + 28, ext. 3009). *Crystal physics, lattice dynamics and thermal motions, electron density distributions.*

Schildkamp, Dr Wilfried. Fachrichtung Kristallographie, Universität, D-6600 Saarbrücken, Germany, Fed. Rep. (1949) Dr.rer.nat., crystallography (U. Saarbrücken, 1979). Res. assoc. (tel 0681 + 3021, ext. 2470). *Structural phase transitions, instrumentation, X-ray and neutron scattering methods.*

Schimanski, Dr Uwe Lothar. Materials Lab. I, Dept. 4627,65-05, IBM GmbH, Hechtsheimerstr. 2, D-6500 Mainz 1, Germany, Fed. Rep. (1953) Dr.rer.nat., chemistry (U. Bielefeld, 1983). Specialist f. surface analysis. (tel. 06131 + 842038). *SEM, STEM, surface analysis, powder diffractometry, crystal structure determination.*

Schirmer, Dr Ulrich. VGB, Techn. Vereinigung der Gross-kraft-werks-betrei-ber, Klinkestr. 27-31, D-4300 Essen, Germany, Fed. Rep. (1949) Dr.rer.nat., technical mineralogy (U. Münster, 1979). (tel. 0201 + 198, ext. 281). *Instrumentation, polymorphic transformations, ceramics, refractories, waste combustion.*

Schliephake, Dr Rolf-Werner. Abt. Mineralogie und Petrographie, Bergbau-Forschung GmbH, Frillendorfer Str. 351, D-4300 Essen 13, Germany, Fed. Rep. (1925) Dr.rer.nat., chemistry (TH München, 1958). Leiter des Röntgenlabors. (tel. 0201 + 105, ext. 9381). *X-ray powder diffractometry, phase analysis, instrumentation.*

Schloemer, Prof. Dr Hermann J. Technische Mineralogie, Universität, Im Stadtwald, D-6600 Saarbrücken, Germany, Fed. Rep. (1923) Prof., mineralogy (U. Saarbrücken, 1968). Prof. (tel. 0681 + 302, ext. 2912). *Crystal growth, high pressure and temperature research, cement chemistry, mineral mobility, coal gasification.*

Schmahl, Mr Wolfgang Wilhelm. Mineralog.- Petrogr. Inst., Universität, Olshausenstr. 40-60, D-2300 Kiel, Germany, Fed. Rep. (1958) Diplom, mineralogy (U. Mainz, 1981). (tel. 0431 + 8802906). *Phase transitions, crystal chemistry, theoretical crystallography.*

Schmalle, Dr Helmut Willi. Mineralogisch-Petrographisches Inst., Universität, Grindelallee 48, D-2000 Hamburg 13, Germany, Fed. Rep. (1941) Dr.rer.nat., mineralogy (U. Hamburg, 1977). Wiss. Mit. (tel. 040 + 4123, ext. 2068). *Structure determination, organic molecules, naturally occurring allergens, antibiotics, structure-activity relationships.*

Schmetzer, Dr Karl. Mineralogisch-Petrogr. Inst., Universität, Im Neuenheimer Feld 236, D-6900 Heidelberg, Germany, Fed. Rep. (1951) Dr.rer.nat., mineralogy (U. Heidelberg, 1978). (tel. 06221 + 56, ext. 2805). *Inorganic crystal structures, crystal chemistry, spectroscopy, colour of inorganic materials.*

Schmidt, Mr Bertram Felix Paul. Kristall-und Materiallabor, Fak. f. Physik, Universität (TH), Kaiserstr. 12, D-7500 Karlsruhe, Germany, Fed. Rep. (1953) Diplom, physics (U. Karlsruhe, 1979). Wiss. Mit. (tel. 0721 + 608, ext. 3471). *Crystal growth and characterization, X-ray topography.*

Schneider, Dr Hartmut. Forschungsinst. d. Feuerfest-Industr., An der Elisabethkirche 27, D-5300 Bonn 1, Germany, Fed. Rep. (1941) Dr.rer.nat. (tel. 0228 + 211051, telex 886533). *Technical ceramics, high temperature chemistry.*

Schneider, Dr Jochen Richard. Hahn-Meitner-Inst. f. Kernforschung, Glienicker Str. 100, D-1000 Berlin 39, Germany, Fed. Rep. (1941) Dr.rer.nat., physics (U. Hamburg, 1973). Physicist. (tel. 030 + 8009, ext. 2768). *Electron charge, momentum density, metal hydrides, ionic conductors, semiconductors, imperfect single crystal diffraction, structural phase transitions.*

Schneider, Dr Julius. Inst. f. Kristallographie und Mineralogie, Universität, Theresienstr. 41, D-8000 München 2, Germany, Fed. Rep. (1942) Dr.rer.nat., solid state physics, neutron scattering (T.U. München, 1975). (tel. 089 + 23944354). *Crystallography, solid state physics.*

Schneider, Dr Walter. Inst. f. Material- u. Festkörperforschung, KFZ, Postfach 3640, D-7500 Karlsruhe, Germany, Fed. Rep. (1936) Dr.rer.nat., mineralogy (U. Göttingen, 1960). Gruppenleiter. (tel. 07247 + 824903). *Electron microscopy, electron diffraction, metals, radiation defects, precipitation in steels.*

Schnering, von, Prof. Dr Dr hc Hans Georg. MPI f. Festkörperforschung, Heisenbergstr. 1, D-7000 Stuttgart 80, Germany, Fed. Rep. (1931) Dr.rer.nat., habil., inorganic chemistry (U. Münster, 1960). Direktor. (tel. 0711 + 6860560). *Solid state chemistry, general structural chemistry, cluster compounds, structure and bonding in solids.*

Schöllhorn, Prof. Dr Robert. Anorganisch-Chemisches Inst., Universität, Gievenbecker Weg 9, D-4400 Münster, Germany, Fed. Rep. (1935) Ph.D., inorganic chemistry (U. Heidelberg, 1963). Prof. (tel. 0251 + 83, ext. 3115). *Crystal structure, intercalation compounds, topotactic processes, surface structures, low-temperature phase transitions, crystal growth from ionic melts.*

Scholz, Dr Heinz Werner. Philips GmbH Forschungslaboratorium, Weisshausstr., D-5100 Aachen, Germany, Fed. Rep. (1930) Dr., chemistry (U. Münster, 1959). Res. chemist. (tel. 0241 + 621). *Crystal growth.*

Schomburg, Dr Dietmar. Ges. f. Biotechn. Forschung, Mascheroder Weg 1, D-3300 Braunschweig, Germany, Fed. Rep. (1950) Dr.rer.nat., chemistry (TU Braunschweig, 1976). Priv.-Doz. (tel. 0531 + 618371). *Structural chemistry, hypervalent main group compounds, biologically interesting compounds.*

Schramm, Dr Volker. FR 17.3 Kristallographie, Universität, D-6600 Saarbrücken 11, Germany, Fed. Rep. (1939) Dr., mineralogy (U. Saarbrücken, 1972). (tel. 0681 + 302, ext. 3470). *Crystal structures, organometallic complexes, computer programming, teaching.*

Schröcke, Prof Dr Helmut. Inst. f. Kristallographie und Mineralogie, Universität, Theresienstr. 41, D-8000 München 2, Germany, Fed. Rep. (1922) Prof., physico-chemical mineralogy (U. München, 1951). Abteilungsvorstand. (tel. 089 + 2394, ext. 4331). *Physicochemical mineralogy, mineral deposits, mineral systematics.*

Schröder, Dr Friedrich Anton. Gmelin-Inst., Max-Planck-Ges., Varrentrappstr. 40/42, D-6000 Frankfurt/Main 90, Germany, Fed. Rep. (tel. 069 + 79171).

Schröpfer, Dr Lothar Maximilian. Inst. f. Kristallographie, Universität, Senckenberg-Anlage 30, D-6000 Frankfurt/Main, Germany, Fed. Rep. (1941) Dr.phil.nat., mineralogy (U. Frankfurt/Main, 1971). Akad. Oberrat. (tel. 069 + 7982103). *Crystal chemistry, inorganic crystal structures, electron microscopy.*

Schubert, Mr Helmut. Powder Metallurgy Lab., MPI f. Metallforschung, Heisenbergstr. 5, D-7000 Stuttgart 80, Germany, Fed. Rep. (1951) Dipl.Min., materials science (RWTH Aachen, 1981). (tel. 0711 + 2095628). *Zirconia, powder preparation and characterisation, martensitic transformations, microstructural design.*

Schubert, Prof. Dr Konrad. Inst. f. Werkstoffwissenschaften, MPI f. Metallforschung, Seestr. 75, D-7000 Stuttgart, Germany, Fed. Rep. (1915) Prof., crystallography (U. Stuttgart, 1960). Scient. coworker. (tel. 0711 + 2095210). *Crystal structures, metallic phases, bonding, inorganic phases.*

Schülke, Prof Dr Winfried. Inst. f. Physik, Universität, Otto-Hahn-Str., D-4600 Dortmund 50, Germany, Fed. Rep. (1935) Dr.rer.nat.habil. (U. Halle, 1971). Prof. (tel. 0231 + 7553507, telex 822465). *X-ray diffraction, inelastic X-ray scattering, gamma-Compton-scattering, dynamical X-ray diffraction theory, synchrotron radiation applications.*

Schüller, Prof. Dr Karl-Heinz. Fachbereich Werkstofftechnik, Ohm-FH, Kesslerplatz 12, D-8500 Nürnberg, Germany, Fed. Rep. (1928) Prof., technical mineralogy (U. Erlangen-Nürnberg, 1972). Prof. (tel. 0911 + 5880216). *Technical mineralogy, ceramic technology, ceramic microstructures, properties of ceramic raw materials.*

Schuermann, Dr Kay Uwe. Fachbereich Geowissenschaften, Universität, Lahnberge, D-3550 Marburg, Germany, Fed. Rep. (1939) Dr.rer.nat., mineralogy (U. Marburg, 1966). Res. scient. (tel. 06421 + 282228). *Silicate structures, Mg-Fe-distribution (inter- and intracrystalline).*

Schultze-Rhonhof, Dr Ernst. KHD Humboldt-Wedag AG, Drachenfelsweg 19, D-5300 Bonn 3, Germany, Fed. Rep. (1934) Dr., chemistry (U. Bonn, 1964). Chemiker. (tel. 0228 + 465641). *Inorganic crystal structures, computer programming.*

Schulz, Dr Georg Eberhardt Bruno. Abt. f. Biophysik, MPI f. medizinische Forschung, Jahnstr. 29, D-6900 Heidelberg, Germany, Fed. Rep. (1939) Prof., biophysics (U. Heidelberg, 1973). Wiss. Ang. (tel. 06221 + 486274). *Protein structure, biochemistry.*

Schulz, Prof. Dr Heinz Hermann. Inst. f. Kristallographie u. Mineralogie, Universität, Theresienstr. 41, D-8000 München, Germany, Fed. Rep. (1935) Dr.rer.nat., physics (U. Saarbrücken, 1964). Prof. (tel. 089 + 23944311, telex univm 529815). *Surface crystallography, high-pressure crystallography, fast ionic conductors, thermal motion.*

Schur, Dr Karl. Inst. f. Medizinische Physik, Universität, Hüfferstr. 68, D-4400 Münster, Germany, Fed. Rep. (1929) Dr.rer.nat., mineralogy (U. Giessen, 1971). Dr. (tel. 0251 + 83, ext. 5100). *Applied electron microscopy, crystal growth.*

Schuster, Prof. Dr Hans-Uwe. Inst. f. Anorganische Chemie, Universität, Greinstr. 6, D-5000 Köln 41, Germany, Fed. Rep. (1930) Dr., inorganic solid state chemistry (U. Kiel, 1967). Ord. Prof. (tel. 0221 + 470, ext. 3262). *Intermetallic compounds, ternary compounds (metallic or semiconducting), crystal structures, chemical bonding, magnetic properties.*

Schwarz, Dr Wolfgang. Inst. f. Anorg. Chemie, Universität, Pfaffenwaldring 55, D-7000 Stuttgart 80, Germany, Fed. Rep. (1939) Dr.rer.nat., chemistry (U. Stuttgart, 1973). (tel. 0711 + 6854238). *Inorganic and organometallic crystal structures.*

Schwarzmann, Mrs Dr Sigrid. Mineralogische Sammlung, Staatl. Naturw. Sammlung, Theresienstr. 41, D-8000 München 2, Germany, Fed. Rep. (1928)

Dr.rer.nat., mineralogy, crystallography (U. Göttingen, 1956). Oberkonservatorin. (tel. 089 + 2394, ext. 4308). *Inorganic crystal structures, crystal optics, crystal growth.*

Schweinsberg, Dr Heinz Friedrich. Zentrale Technik, Thyssen Stahl AG, Postfach 110067, D-4100 Duisburg 11, Germany, Fed. Rep. (1936) Dr.rer.nat., crystallography (U. Kiel, 1971). Res. manager. (tel. 0203 + 5224733). *Crystal chemistry, refractories.*

Seidel, Dr Peter. Inst. f. Mineralogie, Universität, Corrensstr. 24, D-4400 Münster, Germany, Fed. Rep. (1938) Dr., crystallography (U. Münster, 1976). Wiss. Ang. (tel. 0251 + 833458). *Phase transitions, physical properties, instrumentation.*

Seifert, Prof. Dr Hans-Joachim. Inst. f. Anorg. Chemie, Gesamthochschule, Heinrich-Plett-Str. 40, D-3500 Kassel, Germany, Fed. Rep. (1930) Prof., inorganic chemistry (U. Giessen, 1969). Prof. (tel. 0561 + 8044760). *Inorganic crystal structures, thermochemistry.*

Seifert, Prof. Dr Karl-Friedrich. Mineralog. Inst., Universität, Poppelsdorfer Schloss, D-5300 Bonn, Germany, Fed. Rep. (1927) Dr.rer.nat., mineralogy (U. Münster, 1957). Akad. Oberrat. (tel. 0228 + 732768). *High pressure, crystallography, mineralogy, physical properties, biomineralization.*

Serafin, Dr Michael. Inst. f. Anorg. u. Analyt. Chemie, Universität, Heinrich-Buff-Ring 58, D-6300 Giessen, Germany, Fed. Rep. (1949) Dr.rer.nat., inorg. chemistry (U. Giessen, 1980). Akad. Rat. (tel. 0641 + 7025666). *Solid state chemistry, inorg. crystal structures.*

Sheldrick, Prof. Dr George Michael. Inst. f. Anorg. Chemie, Universität, Tammannstr. 4, D-3400 Göttingen, Germany, Fed. Rep. (1942) PhD, chemistry (U. Cambridge, UK, 1966). Prof. (tel. 0551 + 393021). *Structural chemistry, crystallographic computing.*

Sheldrick, Prof Dr William Stephen. FB Chemie, Universität, Erwin-Schrödinger-Str., D-6750 Kaiserslautern, Germany, Fed. Rep. (1945) PhD, inorganic chemistry. (U. Cambridge, UK, 1969). Prof. (tel. 0631 + 2052986). *Inorganic structural chemistry, bio-inorganic chemistry.*

Siebels, Mr Hansjörg. Abt. f. Strukturforschung I, MPI f. Biochemie, Am Klopferspitz, D-8033 Martinsried, Germany, Fed. Rep. (1950) Dipl.-Phys., physics (TU München, 1975). Res. assoc. (tel. 089 + 8585, ext. 660). *X-ray and neutron structure analysis.*

Sieber, Mr Norbert Hermann Wilhelm. Mineralogisches Inst., Universität, Am Hubland, D-8700 Würzburg, Germany, Fed. Rep. (1960) Dipl., mineralogy (U. Mainz, 1985). (tel. 0931 + 888431). *Crystal chemistry, crystal structure analysis, X-ray crystallography.*

Sievers, Dr Rolf. Inst. f. Anorganische Chemie, Universität, Gerhard-Domagk-Str. 1, D-5300 Bonn, Germany, Fed. Rep. (1938) Dr., inorganic chemistry (U. Kiel, 1968). Akad. Oberrat. (tel. 0228 + 73, ext. 5336). *Crystal structures, crystallographic computing, instrumentation, computer applications in chemistry.*

Simon, Prof. Dr Arndt. MPI f. Festkörperforschung, Heisenbergstr. 1, D-7000 Stuttgart, Germany, Fed. Rep. (1940) Dr., chemistry (U. Münster, 1966). Direktor. (tel. 0711 + 6860640, ext. 2640). *Inorganic crystal structures, preparation, crystal growth, structure and physical properties, instrumentation.*

Sirtl, Prof. Dr Erhard. R & D Heliotronic GmbH, Postfach 1129, D-8263 Burghausen, Germany, Fed. Rep. (1928) Apl. Prof., inorganic and physical chemistry (LMU München, 1968). R and D manager. (tel. 08677 + 83, ext. 2580). *Crystal growth, crystal defects.*

Sondermann, Dr Ulrich. Inst. f. Mineralogie, Lahnberge, D-3550 Marburg, Germany, Fed. Rep. (1936) Dr.rer.nat., physics (U. Marburg, 1970). (tel. 06421 + 28, ext. 2226). *Magnetism, crystal physics.*

Sowa, Dr Heidrun. Inst. f. Kristallographie u. Mineralogie, Universität, Theresienstr. 41, D-8000 München, Germany, Fed. Rep. (1954) Dr.rer.nat., mineralogy (U. Marburg, 1983). Akad. Rätin. (tel. 089 + 23944313, telex 529815). *High-pressure crystallography.*

Stadler, Mr Maximilian Wolfgang. Inst. f. Anorg. Chemie, Universität, Universitätsstr. 31, D-8400 Regensburg, Germany, Fed. Rep. 1956).

Steeb, Prof. Dr Siegfried. Inst. f. Werkstoffwissenschaften, MPI f. Metallforschung, Seestr. 92, D-7000 Stuttgart, Germany, Fed. Rep. (1931) Prof. metal-physics (U. Stuttgart, 1969). Head, sci. dept. (tel. 0711 + 20951). *X-ray diffraction, electron and neutron diffraction, melts and amorphous solids.*

Steffen, Dipl. Chem Michael Georg. Inst. f. Anorg. Chemie u. Strukturchemie, Universität, Universitätsstr. 1, D-4000 Düsseldorf, Germany, Fed. Rep. (1950) Dipl.-Chem., chemistry (U. Düsseldorf, 1977). Res. asst. (tel. 0211 + 3113857). *Low temperature crystal growth, inorganic crystal structures.*

Stegemann, Dr-Ing Jürgen. Nicolet Instrument GmbH, Senefelderstr. 162, D-6050 Offenbach/Main, Ger. Fed. Rep. (1947) Dr.-Ing., X-ray structure analysis (TU Darmstadt, 1980). Application engineer. (tel. 0611 + 837001). *X-ray structure analysis, organic compounds, semi-empirical and force-field calculations, intra- and inter-molecular structures.*

Steiner, Dr Michael. Hahn-Meitner-Inst. f. Kernforschung, Glienicker Str. 100, D-1000 Berlin 39, Germany, Fed. Rep. (1943) Dr.rer.nat.habil., experimental physics (TU Berlin, 1978). Privatdozent. (tel. 030 + 80091, ext. 2757). *Magnetic structures, phase transitions, magnetic systems, static and dynamic properties, low-dimensional magnets, scattering from polarized nuclei.*

Steuhl, Dr Hans Hermann. Inst. f. Mineralogie, Corrensstr. 24, D-4400 Münster, Germany, Fed. Rep. (1922) Dr.rer.nat., physics (T.U. München, 1958). Prof. (tel. 02461 + 615774). *Neutron scattering, molecular solids, liquid crystals.*

Stöckelmann, Dr Diedrich. Inst. f. Mineralogie, Universität, Corrensstr. 24, D-4400 Münster, Germany, Fed. Rep. (1938) Dr., mathematics, mineralogy (U. Münster, 1983). Wiss. Ass. (tel. 0251 + 833493, telex 892529 unims d). *Crystallographic computing.*

Strähle, Prof. Dr Joachim. Inst. f. Anorganische Chemie, Universität, Auf der Morgenstelle 18, D-7400 Tübingen, Germany, Fed. Rep. (1937) Ord. Prof., inorg. chemistry (U. Tübingen, 1976) Ord. Prof., chair of inorganic chemistry. (tel. 07071 + 296102). *Inorganic chemistry, inorganic and organic crystal structures.*

Strell, Mrs Dr Irmtraud. MPI f. Biochemie, Am Klopferspitz, D-8033 Martinsried, Germany, Fed. Rep. (1922) Dr.rer.nat., physical chemistry (TU München, 1949). (tel. 089 + 8585697). *Organic molecules, crystal growth, biochemical derivatives.*

Strocka, Dr Bernhard. Philips GmbH Forschungslaboratorium, Vogt-Kölln-Str. 30, D-2000 Hamburg 54, Germany, Fed. Rep. (1943) Dr.-Ing., solid state chemistry (TU Berlin, 1971). Wiss. Mit. (tel. 040 + 5493, ext. 550ě68). *Precision measurement of lattice constants, X-ray topography, lattice defects, single-crystal characterization.*

Strübel, Prof. Dr Günter. Mineralogisch-Petrologisches Inst., Justus Liebig-Universität, Senckenbergstr. 3, D-6300 Giessen, Germany, Fed. Rep. (1932) Prof. Dr. habil., natural sciences (Justus Liebig U. Giessen, 1962). Prof. (tel. 0641 + 7028372). *Technical mineralogy, applied mineralogy, hydrothermal mineralization, mineral raw materials.*

Strumpel, Dr Marianna Katona. Inst. f. Kristallographie, FU, Takustr. 6, D-1000 Berlin 33, Germany, Fed. Rep. (1939) Dr.rer.nat., crystallography (FU Berlin, 1983) Wiss. Mit. (tel. 030 + 838-3457). *Small organic molecules, conformational calculations, programming, vector search techniques.*

Strunz, Prof. Dr Hugo. Inst. f. Mineralogie, TU, Hardenbergstr. 42, D-1000 Berlin 12, Germany, Fed. Rep. (1910) Dr.phil.habil., Dr.sc.techn., mineralogy (U., TU München, 1933, 1935). Ord. Prof. (tel. 030 + 3142659). *Mineral classification, pegmatite minerals, crystal chemistry, minerals.*

Stuhrmann, Prof. Dr Heinrich B. HASYLAB, DESY, Notkestr. 85, D-2000 Hamburg 52, Germany, Fed. Rep. (1938) Prof., physical chemistry (U. Mainz, 1970). Prof. (tel. 040 + 89983008). *Resonant X-ray scattering, synchrotron radiation, polarized neutron scattering, dynamically polarized nuclei.*

Suck, Dr Dietrich. Eur. Mol. Biol. Lab., Postfach 102209, Meyerhofstr. 1, D-6900 Heidelberg, Germany, Fed. Rep. (1944) Dr, structure analysis (Braunschweig U., 1971). Group leader. (tel. 06221 + 387307). *Protein structure and function, protein-nucleic acid interactions.*

Süsse, Prof. Dr Peter. Mineralogisch-Kristallogr. Inst., Universität, Goldschmidtstr. 1, D-3400 Göttingen, Germany, Fed. Rep. (1939) Prof., mineralogy (U. Göttingen, 1978). Prof. (tel. 0551 + 393936). *Crystal structures, minerals, organic dyes.*

Suhre, Miss Ursula. Anorg. Chem. Inst., Universität, Greinstr. 6, D-5000 Köln 41, Germany, Fed. Rep. (1956) Diplom, chemistry (1983). (tel. 0221 + 4702913). *Structure, coordination compounds, CO and PF3-complexes.*

Sussieck-Fornefeld, Mrs Cornelia. Mineralog. Inst., Universität, Im Neuenheimer Feld 236, D-6900 Heidelberg, Germany, Fed. Rep. (1951) Dipl. Min., petrology (U. Bochum, 1975). (tel. 06221 + 562805). *Synthesis of silver-baring oxides of germanium and silica, thermodynamics in mineralogy.*

Tapfer, Dr Leander. MPI f. Festkörperforschung, Heisenbergstr. 1, D-7000 Stuttgart 80, Germany, Fed. Rep. (1956) Dr.rer.nat., physics (U. Stuttgart, 1984). MTS. (tel. 0711 + 68601, telex 7255555 mpif d). *Double crystal X-ray diffractometry, X-ray topography, structure and properties, semiconductors, molecular beam epitaxy.*

Taxer, Dr Karlheinz Jürgen. Hochschulrechenzentrum, Universität, Gräfstr. 38, D-6000 Frankfurt/Main, Germany, Fed. Rep. (1939) Dr. (TU Berlin, 1970). (tel. 069 + 7988106). *Crystal structure analysis, crystal chemistry, order-disorder substructures, derivative structures.*

Tebbe, Prof. Dr Karl-Friedrich. Inst. f. Anorg. Chemie, Universität, Greinstr. 6, D-5000 Köln, Germany, Fed. Rep. (1941) Dr.rer.nat., inorganic chemistry (U. Münster, 1970). Prof. (C3). (tel. 0221 + 407, ext. 3285). *Inorganic, crystal and solid state chemistry, structure determination, computer programming.*

Tennyson, Mrs Prof. Dr Christel. Inst. f. Mineralogie u. Kristallographie, TU, Hardenbergstr. 42, D-1000 Berlin 12, Germany, Fed. Rep. (1925) Dr.rer.nat., mineralogy (TU Berlin, 1953). Prof. (tel. 030 + 314, ext. 2192). *Crystal chemistry, inorganic crystal structures, mineral classification and chemistry.*

Teske, Dr Christoph Ludwig. Inst. f. Anorg. Chemie, Olshausenstr. 40-60, D-2300 Kiel, Germany, Fed. Rep. (1942) Dr.rer.nat., (U. Giessen, 1970). Akad. Oberrat. (tel. 0431 + 880, ext. 2408). *Solid state reactions, sulfides, crystal structures, crystal chemistry.*

Thewalt, Prof. Dr Ulf. Sektion Röntgenbeugung, Universität, Oberer Eselsberg, D-7900 Ulm, Germany, Fed. Rep. (1939) Dr.rer.nat., chemistry (U. Heidelberg, 1965). Sektionsleiter. (tel. 0731 + 176, ext. 2554). *Coordination chemistry, organometallic chemistry, symmetry.*

Thiele, Prof. Dr Gerhard. Inst. f. Anorg. Chemie, Universität, Albertstr. 21, D-7800 Freiburg, Germany, Fed. Rep. (1935) Dr.rer.nat., inorganic chemistry

(RWTH Aachen, 1964). Prof. (tel. 0761 + 2032894). *Structural inorganic chemistry.*

Thurn, Dr Herbert. Inst. f. Anorganische Chemie, Universität, Pfaffenwaldring 55, D-7000 Stuttgart 80, Germany, Fed. Rep. (1937) Dipl.-Chem., chemistry (U. Stuttgart, 1962). Oberasst. (tel. 0711 + 784, ext. 4237). *Inorganic crystal structures, symmetry.*

Tillmann, Dr Bruno. Mineralog.-Kristallogr. Inst., Universität, Goldschmidtstr. 1, D-3400 Göttingen, Germany, Fed. Rep. (1936) Dr.rer.nat., physics (TU Hannover, 1967). Wiss. Ang. (tel. 0551 + 393892). *Inorganic and organic crystal structures, crystallographic computing.*

Tillmanns, Prof. Dr Ekkehart. Mineralog. Inst., Universität, Am Hubland, D-8700 Würzburg, Germany, Fed. Rep. (1941) Dr.rer.nat., mineralogy (U. Bochum, 1968). Prof. (tel. 0931 + 888431). *Crystal structure analysis, X-ray crystallography, crystal chemistry.*

Töpel-Schadt, Dr Jutta. Mineralog.-Petrogr. Inst., Universität, Zülpicherstr. 49, D-5000 Köln 1, Germany, Fed. Rep. (1952) Dr.phil.nat., mineralogy (U. Frankfurt, 1980). Wiss. Mit. (tel. 0221 + 4703190). *Electron microscopy, real structures in crystals, phase transitions.*

Tolksdorf, Prof. Dr Wolfgang. Philips GmbH Forschungslaboratorium, Vogt-Kölln-Str. 30, Postfach 540840, D-2000 Hamburg 54, Germany, Fed. Rep. (1932) Prof., inorganic chemistry, applied mineralogy (U. Frankfurt, 1978). Group leader in res. (tel. 040 + 5493548, telex 21331656). *Ferrimagnetic materials, semiconductors, crystal growth, dielectrics, ceramics.*

Treimer, Dr Wolfgang. Hahn-Meitner-Institut, Glienickerstr. 100, D-1000 Berlin 39, Germany, Fed. Rep. (1950) Dr. phil., physics (U. Vienna, Austria, 1975). Asst. (tel. 030 + 8009221). *Neutron interferometry, dynamical diffraction, fundamental physics, magnetic domains and Bloch walls.*

Trömel, Prof. Dr Martin Gerhard. Inst. f. Anorganische Chemie, Universität, Niederurseler Hang, D-6000 Frankfurt/Main 50, Germany, Fed. Rep. (1934) Dr., inorganic chemistry (U. Frankfurt/Main, 1963). Prof. (tel. 069 + 58009159). *Inorganic crystal structures, crystal chemistry.*

Trost, Dr Friedrich Karl. Res.& Develop., Joh.Schaefer Kalkwerke, Brandenburgerstr. 38, D-6252 Diez/Lahn, Germany, Fed. Rep. (1931) Dr.rer.nat. (U. Tübingen, 1964). (tel. 06432 + 2327). *Physics, chemistry, crystallography, calcium (carbonates - sulphates - silicates).*

Trumm, Dr Alfons. Inst. f. Kristallographie und Mineralogie, Universität, Theresienstr. 41, D-8000 München 2, Germany, Fed. Rep. (1940) Dr., physics (U. München, 1971). Wiss. Ass. (tel. 089 + 23944330). *Physicochemical mineralogy, electrochemistry, high pressure.*

Tsay, Dr Yi-Hung. Röntgenlabor, MPI f. Kohlenforschung, Lembkestr. 5, D-4330 Mülheim/Ruhr, Germany, Fed. Rep. (1939) Ph.D., chemistry (Georgetown U., USA, 1970). (tel. 0208 + 306484). *Molecular structure determination, organic and organometallic compounds.*

Tsernoglou, Prof. Demetrius. Eur. Mol. Biol. Lab., Meyerhofstr. 1, D-6900 Heidelberg, Germany, Fed. Rep. (1935) PhD, biophysics (Yale U., 1966) Sr. scient. (tel. 06221 + 387270, telex embld 461613) *Proteins.*

Tschulena, Dr Guido. Sensorik, Battelle-Inst. e.V., Am Römerhof 35, D-6000 Frankfurt/Main, Germany, Fed. Rep. (1945) Dr. phil., solid state physics (U. Vienna, Austria, 1971). *Thin solid films, solid state sensors, electronic materials, process control.*

Urban, Dr Heinz. Nichtmetallische Werkstoffe, TU, Zehntner Str. 2A, D-3392 Clausthal-Zellerfeld, Germany, Fed. Rep. (1928) Dr.habil., applied mineralogy (TU Clausthal, 1957). Prof. (tel. 05323 + 1091). *Applied mineralogy, structural ceramics.*

Vahrenkamp, Prof. Dr Heinrich. Inst. f. Anorganische Chemie, Universität, Albertstr. 21, D-7800 Freiburg, Germany, Fed. Rep. (1940) Dr.rer.nat., inorganic chemistry (U. München, 1967). Full prof. (tel. 0761 + 203, ext. 2913). *Inorganic crystal structures.*

Valeton, Mrs Prof. Dr Ida Walburga Jakobine. Geologisch- Paläontologisches Inst., Universität, Bundesstr. 55, D-2000 Hamburg, Germany, Fed. Rep. (1922) Prof., (U. Hamburg, 1944). Head, sedimentology dept. (tel. 040 + 41235042). *Crystallography, clay minerals, glauconites, bauxit-minerals (gibbsite - boehmite - diaspore).*

Veith, Prof. Dr Michael. Inst. f. Anorg. Chemie, Universität, Im Stadtwald, D-6600 Saarbrücken, Germany, Fed. Rep. (1944) Dr.rer.nat., chemistry (U. München, 1971) Prof. (tel. 0681 + 3023415). *Structural inorganic and metalorganic chemistry, crystal structure analysis, low temperature crystal structures.*

Vielhaber, Dr Edmund Antonius. Inst. f. Anorg. und Analyt. Chemie, Universität, Heinrich-Buff-Ring 58, D-6300 Giessen, Germany, Fed. Rep. (1931) Dr.rer.nat., inorganic chemistry (U. Münster, 1964). Akad. Oberrat. (tel. 0641 + 70255693). *Crystal growth, inorganic crystal structures.*

Viswanathan, Prof. Dr Krishnamoorthy. Mineralog. Inst., TU, Gauss-Str. 29, D-3300 Braunschweig, Germany, Fed. Rep. (1936), Dr.sci.nat., mineralogy (ETH Zürich, Switzerland, 1967). Prof. (tel. 0531 + 3912263). *Inorganic crystal structures, minerals, crystal physics.*

Vorbach, Dr Angelika Irene. Siemensstr. 1A, D-7550 Rastatt, Germany, Fed. Rep. (1953) Dr.rer.nat., mineralogy (U. Karlsruhe, 1983). (tel. 07222 + 25161). *Gem stones, experimental mineralogy, petrography, geochemistry.*

Vorderwisch, Dr Peter. Hahn-Meitner-Inst. f. Kernforschung, Glienicker Str. 100, D-1000 Berlin 39, Germany, Fed. Rep. (1940) Dr.rer.nat., physics (FU Berlin, 1974). Physicist. (tel. 030 + 80092747). *Solid state physics, neutron scattering, lattice dynamics, molecular dynamics, crystal-field theory, spin dynamics.*

Wacker, Dr Friedel Klaus. Kristallogr. Inst., Universität, Hebelstr. 25, D-7800 Freiburg i. Br., Germany, Fed. Rep. (1954) Dr., crystallography (U. Marburg, 1983). (tel. 0761 + 2034287). *Physical properties of crystals.*

Wagner, Dr Ernst-Heinz. Fritz-Haber-Inst., Max-Planck-Ges., Faradayweg 4-6, D-1000 Berlin 33, Germany, Fed. Rep. (1925) Dr. (1952) Dr.rer.nat., physics (TH Stuttgart, 1954). Wiss. Mit. (tel. 030 + 8305271). *Crystal physics.*

Walcher, Dr Herbert. Inst. f. Angew. Festkörperphysik, Fraunhofer Ges., Eckerstr. 4, D-7800 Freiburg i. Br., Germany, Fed. Rep. (1949) Dr., crystallography (U. Freiburg, 1982). Staff crystallographer. (tel. 0761 + 2714288, telex 07-72510). *Crystal growth, II-VI compounds, low-temperature methods, microgravity influence on crystal growth.*

Walker, Dr Nigel P. C. ZHV/D - A30, BASF AG, D-6700 Ludwigshafen, Germany, Fed. Rep. (1955) PhD, protein crystallography (U. Bristol, UK, 1983). Chemiker (tel. 0621 + 6042638). *Protein crystallography, molecular modelling, small molecule crystallography, computer programming.*

Wallis, Dr Julian Mark. Anorg.-Chem. Inst., TU, Lichtenbergstr. 4, D-8046 Garching, Germany, Fed. Rep. (1959) D. phil., organomet. chemistry (U. Oxford, UK, 1984). Res. Assoc. (tel. 089 + 3130, telex 17898174). *Metal vapour synthesis, organometallic compounds, structures.*

Wallrafen, Dr Franz. Mineralog. Inst., Universität, Poppelsdorfer Schloss, D-5300 Bonn, Germany, Fed. Rep. (1938) Dr.rer.nat., mineralogy and crystallography (U. Bonn, 1972). Wiss. Ang. (tel. 0228 + 732961). *Crystal growth.*

Wang, Dr Naiding. Mineralogisch-Petrogr. Inst., Universität, Im Neuenheimer Feld 236, D-6900 Heidelberg, Germany, Fed. Rep. (1935) Dr., crystallography (U. Heidelberg, 1968). Res. asst. (tel. 06221 + 562804). *Crystallography, sulfides, synthesis, phase relations.*

Wartchow, Dr Rudolf. Inst. f. Anorg. Chemie, Universität, Callinstr. 9, D-3000 Hannover, Germany, Fed. Rep. (1940) Dr.rer.nat., inorganic chemistry (TU Hannover, 1975). Akad. Rat. (tel. 0511 + 7622216). *Inorganic crystal structures, computer programming, symmetry.*

Weber, Dr Hans-Jürgen. Inst. f. Physik, Universität, Otto-Hahn-Str., D-4600 Dortmund, Germany, Fed. Rep. (1943) Dr.rer.nat.habil., crystallography (U. Köln, 1983). Priv.-Doz. (tel. 0231 + 7553526, telex 822445). *Crystal physics, crystal growth, crystal optics.*

Weber, Prof. Dr Kurt. Inst. f. Mineralogie u. Kristallographie, TU, Hardenbergstr. 42, D-1000 Berlin 12, Germany, Fed. Rep. (1926) Dr.rer.nat.habil., crystallography (TU Berlin, 1967). Prof. (tel. 030 + 314, ext. 3919). *Crystal growth, physical properties, inorganic crystal structures.*

Weckert, Mr Edgar. Inst. f. Angew. Physik, Lehrst. f. Kristallographie, Universität, Bismarckstr. 10, D-8520 Erlangen, Germany, Fed. Rep. (1960) Dipl.-Ing., material sciences (U. Erlangen, 1984). Wiss. Mit. (tel. 09131 + 852705). *Multiple X-ray scattering.*

Wegener, Dr Joachim Rolf. Chemie-Labor, Philips GmbH, Valvo RHW, Stresemannallee 101, D-2000 Hamburg 54, Germany, Fed. Rep. (1942) Dr.rer.nat., inorganic chemistry (U. Göttingen, 1970). Chief Chemist. (tel. 040 + 5613297). *Silicon, SiO2, Si3N4, thin metals.*

Weiner, Dr Karl Ludwig. Inst. f. Kristallographie und Mineralogie, Theresienstr. 41, D-8000 München 2, Germany, Fed. Rep. (1922) Dr.rer.nat., crystallography, mineralogy (U. Bonn, 1954). Akad. Direktor. (tel. 089 + 23944355). *Instrumentation, X-ray diffraction, thin films, high and low temperature, powder diffraction, applied crystallography, applied mineralogy, archaeometry.*

Weiss, Prof. Dr Alarich. Inst. f. Physikalische Chemie, TH, Petersenstr. 20, D-6100 Darmstadt, Germany, Fed. Rep. (1925) Dr.rer.nat., physics (TH Darmstadt, 1955). Prof. (tel. 06151 + 162607). *Solid state physical chemistry.*

Weiss, Prof. Dr Erwin Ludwig. Inst. f. Anorg. und Angew. Chemie, Universität, Martin-Luther-King-Platz 6, D-2000 Hamburg, Germany, Fed. Rep. (1926) Prof., inorganic chemistry (TH München, 1965). Prof. (tel. 040 + 41233103). *Organometallic and coordination chemistry, crystal structures.*

Weiss-Nowak, Mr Christian. Inst. f. Kristallographie, FU, Takustr. 6, D-1000 Berlin, Germany, Fed. Rep. (1957). (tel. 030 + 8383455). *X-ray structure analysis, biomolecules.*

Weitzel, Dr Hans. Inst. f. Physikal. Chemie, Strukturforschung, TH, Petersenstr. 20, D-6100 Darmstadt, Germany, Fed. Rep. (1941) Dr.rer.nat., physics (U. Tübingen, 1969). Akad. Oberrat. (tel. 06151 + 162298). *Neutron diffraction, crystal structures, magnetic structures, phase transitions.*

Wendl, Dr Wolfgang. Kristall- und Materiallabor, Fak. f. Physik, Universität, Kaiserstr. 12, D-7500 Karlsruhe, Germany, Fed. Rep. (1943) Dr.rer.nat., chemistry (U. Heidelberg, 1970). Wiss. Mit. (tel. 0721 + 6083558). *Crystal growth, chemical characterization of crystals.*

Wenig, Prof Dr Werner. Lab. f. Angew. Physik, Universität, Lotharplatz, D-4100 Duisburg, Germany, Fed. Rep. (1943) Dr.rer.nat., physics (U. Ulm, 1974). Prof. (tel. 0203 + 3792386). *X-ray analysis, polymers, amorphous substructures.*

Wilhelm, Dr Eberhard. Balthasar-Schönfelderstr. 2, D-8550 Forchheim, Germany, Fed. Rep. (1948) Dr.rer.nat., chemistry (U. Saarbrücken, 1975). *Organic crystal structures, stereochemistry, NMR spectroscopy.*

Wilke, Prof. Dr Wolfgang. Abt. f. Exper. Physik, Universität, Oberer Eselsberg, D-7900 Ulm, Germany, Fed. Rep. (1934) Apl. Prof., physics (U. Ulm, 1974). Prof. (tel. 0731 + 1762510). *Crystal physics, polymer structure, defects in crystals.*

Wilken, Dr Gerdt. Exploration and Producing, Mobil Oil, Burggrafstr.1, D-3100 Celle, Germany, Fed. Rep. (1940) Dr.rer.nat., physics (U. Saarbrücken, 1970). (tel. 5141 + 15228). *Crystal growth, X-ray analysis, equilibrium studies, precipitation from aqueous solutions.*

Will, Prof. Dr Georg. Mineralog. Inst., Universität, Poppelsdorfer Schloss, D-5300 Bonn, Germany, Fed. Rep. (1930) Dr.rer.nat., crystallography (TU München, 1958). Prof. (tel. 0228 + 732761). *Crystal structure analysis, X-rays and neutrons, high pressure structures, low-temperature structures, magnetic structures, chemical bonding.*

Winter, Prof Dr Werner. Res. Center, Grünenthal GmbH, Zieglerstr. 6, D-5100 Aachen, Germany, Fed. Rep. (1943) Dr. habil., chemistry (U. Tübingen, 1978). (tel. 0241 + 51022326). *Medicinal chemistry, molecular modelling.*

Witte, Prof. Dr Helmut Hermann Wolfgang. Inst. f. Physikalische Chemie, TH, Petersenstr. 20, D-6100 Darmstadt, Germany, Fed. Rep. (1909) Dr.phil., physics (U. Göttingen, 1933). Em. ord. Prof. (tel. 06151 + 165107). *Inorganic crystal structures, physical chemistry, metals and alloys.*

Wögerbauer, Dr Rupert. RME, Lab. f. Röntgenographie, Mineralogie u. Edelsteinforschung, Am Schlehenbusch 15, D-8711 Mainstockheim, Germany, Fed. Rep. (1941) Dr.rer.nat., crystallography (U. Würzburg, 1974). (tel. 09321 + 5628). *X-ray diffraction, crystal growth, crystal physics.*

Wölfel, Prof. Dr Erich Richard. Fachgebiet Strukturforschung, TH, Petersenstr. 15, D-6100 Darmstadt, Germany, Fed. Rep. (1922) Dr.rer.nat., chemistry (TH Darmstadt, 1952). Prof. (tel. 06151 + 162298). *Instrumentation, valency electron distribution.*

Wolf, Dr Dieter. Inst. f. Kristallographie und Mineralogie, Universität, Theresienstr. 41, D-8000 München 2, Germany, Fed. Rep. (1939) Dr.rer.nat.habil., physics (U. München, 1972). Priv.-Doz. (tel. 089 + 23944333, telex 529815). *LEED, surface crystallography, surface disorder.*

Wondratschek, Prof. Dr Hans. Inst. f. Kristallographie, Universität, Kaiserstr. 12, D-7500 Karlsruhe, Germany, Fed. Rep. (1925) Dr.rer.nat., mineralogy (U. Bonn, 1953). Prof. (tel. 0721 + 6083320). *Symmetry, crystal chemistry.*

Wunderlich, Dr Hartmut. Inst. f. Anorg. Chemie u. Strukturchemie, Universität, Universitätsstr. 1, D-4000 Düsseldorf, Germany, Fed. Rep. (1938) Dr.rer.nat., physics (TU Braunschweig, 1969). Akad. Oberrat. (tel. 0211 + 3113144). *Inorganic and organic crystal structures, computer programming, instrumentation, teaching.*

Zaki, Mr Chakib. Inst. f. Kristallographie, FU, Takustr. 6, D-1000 Berlin 33, Germany, Fed. Rep. (1955) Dipl., physics (TU Berlin, 1983). (tel. 030 + 8383408). *Single crystal analysis, charge density, computer graphics. laser pulse modulation and compression.*

Zeyfang, Dr Rolf Robert. Forschungsinstitut, AEG-Telefunken, Goldsteinstr. 235, D-6000 Frankfurt/Main, Germany, Fed. Rep. (1938) Dr.rer.nat., physics (U. Stuttgart, 1967). Group leader. (tel. 0611 + 6679221). *Inorganic crystal structures, crystal growth, ceramics, ferroelectrics, epitaxy.*

Ziegler, Prof. Dr Manfred Ludwig. Anorg.-Chem. Inst., Universität, Im Neuenheimer Feld 270, D-6900 Heidelberg, Germany, Fed. Rep. (1936). (tel. 06221 + 562468). *Metalorganic chemistry, X-ray structure analysis, molecular compounds.*

Zigan, Prof. Dr Franz Martinus. Inst. f. Kristallographie, Universität, Senckenberg-Anlage 30, D-6000 Frankfurt/Main, Germany, Fed. Rep. (1928) Dr.phil., physics (U. Marburg, 1958). Prof. (tel. 069 + 7982100). *Crystal physics, neutron diffraction, inorganic crystal structures.*

Zimmer, Mr Alfons. Inst. f. Kristallographie und Mineralogie, Universität, Theresienstr. 41, D-8000 München 2, Germany, Fed. Rep. (1947) Dipl.-Phys., physics (U. München, 1973). Wiss. Mit. (tel. 089 + 23944335). *Surface structure, LEED.*

Zimmermann, Dr Helmuth Walter. Inst. f. Angew. Physik, Lehrstuhl f. Kristallographie, Universität, Bismarckstr. 10, D-8520 Erlangen, Germany, Fed. Rep. (1945) Dr.rer.nat.habil., crystallography (U. Erlangen-Nürnberg, 1982). Priv.-Doz. (tel. 09131 + 852700). *Mathematical crystallography, X-ray crystallography, direct methods, crystallographic computing, crystal structures.*

Zobel, Dr Dieter. Inst. f. Kristallographie, FU, Takustr. 6, D-1000 Berlin 31, Germany, Fed. Rep. (1941) Dipl.-Phys., physics (1968). Wiss. Ass. (tel. 030 + 8383453). *Crystal structures, physical properties, organic and polymer semiconductors.*

Zorn, Dr Gerhard. Res. Laboratories, SIEMENS AG, Otto-Hahn-Ring 6, D-8000 München 83, Germany, Fed. Rep. (1956) Dr., crystallography (tel. 089 + 6363274). *X-ray diffraction, diffraction software.*

Zulehner, Dr Werner. EK, Wacker-Chemitronic GmbH, Postfach 1140, D-8236 Burghausen, Germany, Fed. Rep. (1941) Dr.rer.nat., physics (U. Saarbrücken, 1973). Dev. manager. (tel. 8677 + 832547). *Solid state physics, semiconductor physics, crystal growth, crystallography.*

GHANA

Sub-Editor: F. L. Phillips

Notes

1. International telephone country code - 233.

Baeta, Dr Robert Domingo. Dept. of Physics, U. of Ghana, P.O. Box 63, Legon, Accra, Ghana. (1939) PhD, physics (U. Bristol, UK, 1969). Sr. lect. (tel. Accra + 75381, ext. 8427). *Electron microscopy, defects, X-ray topography.*

Phillips, Dr Frederick Lloyd. Dept. of Chemistry, U. of Ghana, Legon, Accra, Ghana. (1949) PhD, chemical crystallography (U. London, UK, 1975). Lect. (tel. Accra + 75381, ext. 8329). *Coordination complexes.*

GREECE

Sub-Editor: P. J. Rentzeperis

Notes

1. International telephone country code - 0030.

2. Greek Universities confer the following degrees: a) The basic degree of the corresponding field, at the end of the undergraduate studies (4 - 6 years), equivalent to BA or BSc. b) The *Doctor's* degree, equivalent to DSc or PhD. c) At some Universities a degree equivalent to MA or MSc, at the end of a postgraduate course in a special field, e.g. Electronics.

3. The regular teaching and research positions at Greek Universities are equivalent to *Professor, Associate Professor, Assistant Professor, Lecturer, Chief Assistant, Scientific Assistant* and *Teaching Assistant*. A specialist, holding a *Doctor's* degree, may be appointed temporarily to the position of a *Special Scientist* for teaching purposes.

Alexandropoulos, Prof. Nikos. Physics Dept., U. of Ioannina, Ioannina, Greece. (1934) PhD, Physics (Athens U., 1964). Prof. (tel. 0651 + 91-396). *X-ray physics, electron distribution, dynamical diffraction, instrumentation.*

Alexandropoulos, Mrs Tina. Physics Dept., U. of Ioannina, Ioannina, Greece. (1934) BSc, Physics (Athens U., 1961). Instructor. (tel. 0651 + 91-950). *X-ray physics.*

Alexopoulos, Prof. (Emeritus) Kessar. Spefsipou 7, Athens 139, Greece. (1909) DSc, physics (E.T.H. Zurich, Switzerland, 1935). Retired from Athens U. (tel. 01 + 738-442). *Crystal physics, structure defects.*

Antonopoulos, Prof. John. Physics Dept., U. of Thessaloniki, Thessaloniki, Greece. (1939) DSc, electron microscopy (Thessaloniki U., 1972). Assoc. prof. (tel. 031 + 99-2806). *Crystal structures, defects, elastic properties, fatigue.*

Basilakis, Mr Michael. Physics Lab., U. of Patras, Patras, Greece. (1947) BSc, physics (Patras U., 1973). Asst. (tel. 061 + 429-713). *Liquid crystals.*

Boyiatzis, Mr Ioannis. Physics Lab., U. of Patras, Patras, Greece. (1951) BSc, physics (Patras U., 1972). Asst. (tel. 061 + 429-713). *Diffuse X-ray scattering, inorganic crystal structures.*

Bozopoulos, Mr Anastasios Panayiotis. Appl. Physics Lab., U. of Thessaloniki, Thessaloniki, Greece. (1944) BSc, physics (Thessaloniki U., 1970). Asst. (tel. 031 + 99-2803). *Inorganic and organic crystal structures.*

Calamiotou, Dr Maria. Physics Dept., U. of Athens, Athens, Greece. (1949) DSc, physics (Athens U., 1978). Lect. (tel. 01 + 363-3413). *Crystal physics, dynamical diffraction theory, archaeometry.*

Christidis, Prof. Panayiotis Chrysostomos. Appl. Physics Lab., U. of Thessaloniki, Thessaloniki, Greece. (1942) DSc, crystallography (Thessaloniki U., 1975). Asst. prof. (tel. 031 + 99-2803). *Inorganic and organometallic crystal structures, crystal physics.*

Economou, Prof. Nicolaos Alkiviadis. Physics Lab., U. of Thessaloniki, Thessaloniki, Greece. (1926) PhD, physics (Wayne State U., USA, 1959). Prof. (tel. 031 + 99-1439). *Structure and defects, semiconductors.*

Euthymiou, Prof. Paraskevi. Physics Lab., U. of Athens, Athens, Greece. (1923) DSc, physics (Athens U., 1952). Asst. prof. (tel. 01 + 3611-927). *Semiconductor physics, transport phenomena in crystals.*

Evangelidou, Miss Christina. Electrical Eng. Dept., U. of Thrace, Xanthi, Greece. (1950) BSc, Physics (Thessaloniki U., 1973). Asst. (tel 0541 + 26-475). *Crystal structures, magnetic structures, rare earths.*

Fanariotis, Mr Iakovos. Appl. Physics Lab., U. of Thessaloniki, Thessaloniki, Greece. (1951) BSc, chemistry (Thessaloniki U., 1974). Asst. (tel. 031 + 991-2803). *Inorganic crystal structures, crystal growth.*

Filippakis, Dr Sophokles. Physics Lab., N. R. C. Demokritos, Athens, Greece. (1933) PhD, crystallography (Weizmann Inst. of Sci., Israel, 1966). Sr. Scient. (tel. 01 + 6513-111). *Crystal structures, crystal physics, structure of metals.*

Ftikos, Dr Christos. Chemistry Lab., Technical U. of Athens, Athens, Greece. (1949) DSc, inorg. chemistry (Techn. U. of Athens, Athens, 1972). Asst. prof. (tel. 01 + 369-1385). *X-ray diffraction, electron microscopy, cement chemistry and technology.*

Galinos, Prof. Andreas. Physics Lab., U. of Patras, Patras, Greece. (1925) DSc, inorganic chemistry (Athens U., 1955). Prof. *Inorganic crystal structures.*

Gangas, Prof. Nicolas - Hercule. Physics Lab., U. of Ioannina, Ioannina, Greece. (1937) DPhil, nuclear physics (Wien U., Austria, 1960). Prof. (tel. 0651 + 25-928). *Crystal physics.*

Gountsidou, Mrs Vasiliki. Solid State Physics Dept., U. of Thessaloniki, Thessaloniki, Greece. (1955) MSc, Electronics, (Thessaloniki U., 1978). Res. Asst. (tel. 031 + 99-1434). *Measurements, semiconductors.*

Grigoriades, Prof. Panayotis. Physics Lab., U. of Thessaloniki, Thessaloniki, Greece. (1943) DSc, physics (Thessaloniki U., 1980). Asst. prof. (tel. 031 + 99-1400). *Crystal lattice defects.*

Hamodrakas, Dr Stavros. Physics Lab., N. R. C. Demokritos, Athens, Greece. (1947) PhD, biophysics (Leeds U., 1974). Res. asst. (tel. 01 + 6513-111). *Molecular biology, pharmacology, proteins.*

Hountas, Dr Athanasios. Physics Lab., N. R. C. Demokritos, Athens, Greece. (1945) DSc, X-ray diffraction (Thessaloniki U., 1978). Res. asst. (tel. 01 + 6513-111). *Crystal physics, dynamical diffraction theory.*

Kagarakis, Prof. Constantine. Electrical Eng. Dept., Techn. U. of Athens, Athens, Greece. (1933) DSc, chemical eng. (Techn. U. of Athens, 1955). Prof. (tel. 01 + 362-3770). *Semiconductors.*

Kambas, Prof. Kostas. Solid State Physics Dept., U. of Thessaloniki, Thessaloniki, Greece. (1945) PhD, solid state physics (Thessaloniki U., 1981). Asst. prof. (tel. 031 + 99-2850). *Lattice dynamics, Raman spectroscopy, superionic conductors.*

Kanellis, Prof. George. Physics Lab., U. of Thessaloniki, Thessaloniki, Greece. (1942) DSc, physics (Thessaloniki U., 1977). Asst. prof. (tel. 031 + 99-1421). *Lattice dynamics, Raman spectroscopy, superionic conductors, phase transitions.*

Katagas, Dr Christos. Geology Lab., U. of Patras, Patras, Greece. (1944) PhD, geology (U. of Manchester, U.K., 1975). Lect. (tel. 061 + 991-972). *Structure, minerals.*

Katsanos, Mr Demetrios Evangelos. Solid State Physics Dept., U. of Ioannina, Ioannina, Greece. (1959) BSc, Physics. Asst. (tel. 0651 + 91-234). *Incoherence X-ray spectroscopy.*

Kavounis, Mr Konstantinos. Appl. Physics Lab., U. of Thessaloniki, Thessaloniki, Greece. (1945) BSc, physics (Thessaloniki U., 1971). Asst. (tel. 031 + 99-2688). *Organic crystal structures.*

Kokkou, Prof. Socrates Constantinos. Appl. Physics Lab., U. of Thessaloniki, Thessaloniki, Greece. (1940) DSc, crystallography (Thessaloniki U., 1975). Asst. prof. (tel. 031 + 99-2688). *Organic crystal structures, direct methods.*

Konguetsof, Dr Helen. Electrical Eng. Dept., U. of Thrace, Xanthi, Greece. (1938) PhD, Physics (Sussex U., England, 1.1.). Lect. (tel. 0541 + 26475). *Crystal structures, mgnetic structures, rare earths.*

Kosmopoulos, Dr John. Physics Lab., U. of Patras, Patras, Greece. (1945) DSc, physics (U. of Patras,1975). Chief Asst. (tel. 061 + 429-713). *Liquid crystals.*

Kotsanidis, Mr Panayotis. Electrical Eng. Dept., U. of Thrace, Xanthi, Greece. (1946) BSc, Physics (Thessaloniki U., 1970). Asst. (tel. 0541 + 26475). *Crystal structures, mgnetic structures, rare earths.*

Kotsis, Mr Konstantinos. Physics Dept., U. of Ioannina, Ioannina, Greece. (1959) BSc, Phycics (Thessaloniku U., 1980). Instructor. (tel. 0651 + 91-950). *Dynamical diffraction theory.*

Koumelis, Dr Christos. Physics Lab., U. of Athens, Athens, Greece. (1931) DSc, physics (Athens U., 1963). Lect. (tel. 01 + 3633-412). *Inorganic crystal structures, solid state physics.*

Kountouris, Mr Costas. Physics Lab., U. of Patras, Patras, Greece. (1950) BSc, physics (Patras U., 1972). Asst. (tel. 061 + 429-713). *Diffuse X-ray scattering, inorganic crystal structures.*

Kyriakos, Prof. Demetrius. Physics Lab., U. of Thessaloniki, Thessaloniki, Greece. (1941) DSc, solid state physics (Thessaloniki U., 1978). Asst. prof. (tel. 031 + 99-2807). *Transport phenomena in solids, crystal growth.*

Leventouri, Dr Dora. Physics Lab., U. of Athens, Athens, Greece. (1943) DSc, physics (Athens U., 1972). Teaching asst. (tel. 01 + 363-3413). *Solid state physics, X-ray spectroscopy.*

Loizos, Mr Zafiris. Chem. Eng. Dept., Techn. U. of Athens, Athens, Greece. (1950) BSc, chem. eng. (Techn. U. of Athens, 1973). Asst. (tel. 01 + 369-1244). *Inorganic crystal structures.*

Londos, Dr Charalampos. Physics Lab., U. of Athens, Athens, Greece. (1947) DSc, physics (Athens U., 1979). Asst. (tel. 01 + 363-3413). *Solid state physics, X-ray inelastic scattering.*

Manolikas, Prof. Konstantinos. Solid State Physics Dept., U. of Thessaloniki, Thessaloniki, Greece. (1944) PhD, Electron microscopy (Thessaloniki U., 1976). Assoc. prof. (tel. 031 + 99-1421). *Transmission electron microscopy.*

Mavridis, Prof. Aristides. Chemistry Dept., U. of Athens, Athens, Greece. (1944) PhD, crystallography (Michigan State U., USA, 1975). Asst. prof. (tel. 01 + 360-6529). *Structural chemistry, theoretical chemistry, small molecules.*

Miliotis, Prof. Demitrios Menelaos. Appl. Physics Lab., U. of Ioannina, Ioannina, Greece. (1931) DSc, physics (Athens U., 1965). Prof. (tel. 0651 + 22-956). *Inelastic X-ray scattering.*

Mourikis, Dr Stamatios. Physics Lab., U. of Athens, Athens, Greece. (1928) DSc, physics (Athens U., 1963). Chief asst. (tel. 01 + 363-3412). *Metal physics.*

Moustakali Mavridis, Dr Irene. Physics Lab., N. R. C. Demokritos, Athens, Greece. (1946) PhD, physical chemistry (U. Michigan, USA, 1975). Res. asst. (tel. 01 + 6513-111). *Molecular structure, natural products, micromolecular structures, electron distributions.*

Panagos, Prof. Athanasios. Geology Lab., U. of Patras, Patras, Greece. (1926) PhD, crystallography (E.T.H. Zurich, Switzerland, 1960). Prof. (tel. 061 + 429-714). *Inorganic structures.*

Papadakis, Prof. Alexander. Mineralogy Lab., U. of Thessaloniki, Thessaloniki, Greece. (1928) DSc, mineralogy and petrology (Thessaloniki U., 1965). Asst. prof. (tel. 031 + 99-2818). *Crystal optics.*

Papadimitraki Chlichlia, Prof. Helena. Physics Lab., U. of Thessaloniki, Thessaloniki, Greece. (1930) DSc, solid state physics (Thessaloniki U., 1960). Prof. (tel. 031 + 99-2405). *Electrical and thermal conductivity, hall effect, structure of materials.*

Papadopoulos, Mr Demetrius. Physics Lab., U. of Thessaloniki, Thessaloniki, Greece. (1945) BSc, physics (Thessaloniki U., 1969). Asst. (tel. 031 + 99-2851). *Electron microscopy, crystalline materials.*

Papathanassopoulos, Dr Constantinos. Physics Lab., N. R. C. Demokritos, Athens, Greece. (1934) DSc, physics (T. Hochschule, Braunschweig, 1964). Res. asst. (tel. 01 + 6513-111). *Radiation damage, crystal growth.*

Papazoglou, Mr Aristides. Appl. Physics Lab., U. of Thessaloniki, Thessaloniki, Greece. (1954) BSc, physics (Thessaloniki U., 1977). Res. asst. (tel. 031 + 99-2803). *Inorganic crystal structures, crystal growth.*

Parissakis, Prof. George. Chemistry Lab., Technical U. of Athens, Athens, Greece. (1929) PhD, chemistry (E.T.H. Zurich, Switzerland, 1955). Prof. *Inorganic crystal structures, structure, metals.*

Perdikatsis, Dr Basilios. Mineralogy Lab., Inst. of Geol. and Min. Res., Athens, Greece. (1942) PhD, crystallography (U. Erlangen, BRD, 1972). Head, X-ray dept. (tel. 01 + 779-8412). *Inorganic and organic crystal structures, powder diffraction, microanalysis.*

Polychroniades, Prof. Efstathios. Physics Lab., U. of Thessaloniki, Thessaloniki, Greece. (1946) DSc, solid state physics (Thessaloniki U., 1972). Asst. prof. (tel. 031 + 99-1427). *Crystal lattice defects.*

Priftis, Prof. George. Physics Lab., U. of Patras, Patras, Greece. (1937) DSc, physics (Athens U., 1969). Asst. prof. *Diffuse X-ray scattering, inorganic crystal structures.*

Profi, Mrs Stella. Physics Lab., N. R. C. Demokritos, Athens, Greece. (1949) BSc, natural sciences (Athens U., 1972). Res. asst. (tel. 01 + 6513-111). *Inorganic and mineral structure, mineral chemistry.*

Rentzeperis, Prof. Panayiotis Ioannis. Appl. Physics Lab., U. of Thessaloniki, Thessaloniki, Greece. (1928) DSc, crystallography, mineralogy and petrography (Thessaloniki U., 1956). Prof. and head. (tel. 031 + 99-2444). *Inorganic and organic crystal structures, crystal physics.*

Rigopoulos, Prof. Rigas. Physics Lab., U. of Patras, Patras, Greece. (1929) PhD, physics (Birmingham U., UK, 1963). Prof. (tel. 061 + 429-713). *Liquid crystals.*

Rocophyllou Agathonikou, Dr Elsa - Helena. Physics Lab., N. R. C. Demokritos, Athens, Greece. (1936) DSc, physical chemistry (U. Paris, France, 1964). Res. officer. (tel. 01 + 6513-111). *Radiation damage, metal physics, structure, metals.*

Roilos, Prof. Minas. Physics Lab., U. of Patras, Patras, Greece. (1930) DSc, solid state physics (Athens U., 1970). Prof. (tel. 061 + 99-1764). *Inorganic crystal structures, crystal growth, amorphous semiconductors.*

Sahalos, Prof. John. Physics Dept., U. of Thessaloniki, Thessaloniki, Greece. (1943) PhD, Appl. electromagnetics (Thessaloniki U., 1974). Prof. (tel. 031 + 99-1400). *Magnetic materials, ferrites, microwave applications of ferrites.*

Sakellaridis, Prof. Paul. Chemistry Lab., Techn. U. of Athens, Athens, Greece. (1920) DSc, chemistry (U. Paris, France, 1953). Prof. *Inorganic crystal structures.*

Sakkopoulos, Dr Sotirios. Physics Lab., U. of Patras, Patras, Greece. (1945) DSc, solid state physics (Patras U., 1974). Chief asst. (tel. 061 + 429-713). *Electric properties, metals, semiconductors.*

Sandalaki, Dr Zefi. Physics Lab., U. of Athens, Athens, Greece. DSc, physics (Athens U., 1953). Chief asst. (tel. 01 + 363-3412). *Crystal physics.*

Semitelou, Mrs Julia. Electrical Eng. Dept., U. of Thrace, Xanthi, Greece. (1956) BSc, Physics (Thessaloniki U., 1980). Asst. (tel. 0541 + 26475). *Crystal structures, mgnetic structures, rare earths.*

Sianou, Miss Anna. Physics Lab., U. of Thessaloniki, Thessaloniki, Greece. (1945) MA, history of science (Cornell U., USA, 1970). Asst. (tel. 031 + 99-1456). *Electrical and thermal conductivity in crystals, structure materials.*

Siapkas, Prof. Demetrios John. Physics Lab., U. of Thessaloniki, Thessaloniki, Greece. (1934) DSc, physics (Thessaloniki U., 1970). Asst. prof. (tel. 031 + 99-1427). *F.I.R. and Raman spectroscopy, phase transition, ferroelectricity.*

Soldatos, Prof. Constantinos. Mineralogy Lab., U. of Thessaloniki, Thessaloniki, Greece. (1924) DSc, mineralogy and petrology (Thessaloniki U., 1955), DPhil, (E.T.H. Zurich, Switzerland, 1960). Prof. (tel. 031 + 99-1445). *Feldspars, crystal optics.*

Spyrellis, Dr Nicolaos. Chemistry Lab., Technical U. of Athens, Athens, Greece. (1943) DSc, appl. chemistry (U. P. et M. Curie, Paris, France, 1974). Lect. (tel. 01 + 369-1244). *Crystal growth, electrocrystallization.*

Spyridelis, Prof. John. Physics Lab., U. of Thessaloniki, Thessaloniki, Greece. (1929) DSc, solid state physics (Thessaloniki U., 1964). Prof. (tel. 031 + 99-1437). *Crystal physics, electron microscopy.*

Stergiou, Dr Anagnostis Charalambos. Appl. Physics Lab., U. of Thessaloniki, Thessaloniki, Greece. (1941) DSc, crystallography (Thessaloniki U., 1984). Lect. (tel. 031 + 99-2643). *Inorganic and organic crystal structures.*

Stergioudis, Dr Georgios Asterios. Appl. Physics Lab., U. of Thessaloniki, Thessaloniki, Greece. (1946) DSc, crystallography (Thessaloniki U., 1984). Lect. (tel. 031 + 99-2803). *Inorganic crystal structures.*

Stoimenos, Prof. John Nikolaos. Physics Lab., U. of Thessaloniki, Thessaloniki, Greece. (1934) DSc, physics (Thessaloniki U., 1969). Prof. (tel. 031 + 99-1427). *Crystal lattice defects.*

Terzis, Dr Aristides. Physics Dept., N. R. C. Democritos, Athens, Greece. (1941) PhD, Inorganic chemistry, (Princeton U., USA, 1970). Res. assoc. (tel. 01 + 651-3111, ext. 123, telex 21-6199). *Transition metal complexes, unidirectional compounds, low temperatures.*

Theodoridou, Dr Irini. Physics Dept., U. of Ioannina, Ioannina, Greece. (1948) PhD, physics (Ioannina U., 1983). Lect. (tel. 0651 + 91-951). *X-ray physics, electron distribution.*

Theodoropoulos, Prof. Dimitrios. Physics Lab., Techn. U. of Athens, Athens, Greece. (1926) DSc, chemistry (Athens U., 1951). Prof. (tel. 01 + 363-3412). *Organic crystal structures.*

Theodossiou, Prof. Alexandros. Physics Lab., U. of Patras, Patras, Greece. (1918) DSc, solid state physics (Athens U., 1950). Prof. *Crystal physics.*

Tsatis, Mr Demetrius. Physics Lab., U. of Patras, Patras, Greece. (1939) MSc, nuclear physics (Carleton U., Canada, 1968) Asst. (tel. 061 + 272-945). *Electric and thermal properties of crystals.*

Tsimberis, Mr Nikolaos. Physics Lab., U. of Patras, Patras, Greece. (1946) BSc, physics (Patras U., 1973). Asst. (tel. 061 + 429-713). *Liquid crystals.*

Tsoli Kataga, Dr (mrs) Panayota. Geology Lab., U. of Patras, Patras, Greece. (1946) DSc, geology (Patras U., 1979). Lect. (tel. 061 + 991-972). *Structure of clay minerals.*

Tsoukalas, Prof. John. Physics Lab., U. of Thessaloniki, Thessaloniki, Greece. (1941) DSc, solid state physics (Thessaloniki U., 1975). Assoc. prof. (tel. 031 + 99-1456). *Hall effect, metals and alloys.*

Valassiades, Prof. Odysseus. Physics Lab., U. of Thessaloniki, Thessaloniki, Greece. (1944) DSc, physics (Thessaloniki U., 1976). Asst. prof. (tel. 031 + 99-1400). *Defects, semiconductors, transport phenomena.*

Venetopoulos, Dr Cleanthis. Appl. Physics Lab., U. of Thessaloniki, Thessaloniki, Greece. (1929) DSc, crystallography (Thessaloniki U., 1975). Res. assoc. (tel. 031 + 99-2803) *Inorganic crystal structures, computer programming.*

Vgenopoulos, Prof. Andreas. Mineralogy Dept., Techn. U. of Athens, Athens, Greece. (1940) PhD, mineralogy (Techn. U. Athens). Assoc. prof. (tel. 01 + 369-1307). *Powder diffraction.*

Voliotis, Prof. Stavros. Chemistry Lab. U. of Patras, Patras, Greece. (1934) DSc, chemistry (U. Paris VII, France, 1975). Prof. (tel. 061 + 99-1973). *Structural crystallography, crystal chemistry.*

Voutsas, Dr George Panayiotis. Appl. Physics Lab., U. of Thessaloniki, Thessaloniki, Greece. (1945) DSc, crystallography (Thessaloniki U., 1983). Lect. (tel. 031 + 99-2803). *Inorganic and organometallic crystal structures.*

Vradis, Mr Alexandros. Physics Lab., U. of Patras, Patras, Greece. (1950) BSc, physics (Patras U., 1974). Asst. (tel. 061 + 429-713). *Diffuse X-ray scattering, inorganic crystal structures.*

Yakinthos, Prof. John. Physics Lab., U. of Thrace, Xanthi, Greece. (1937) DSc, solid state physics (Grenoble U., France, 1971). Prof. (tel. 0541 + 26-944). *Crystal structure, magnetic structure, rare earths.*

Zardas, Mr George. Physics Lab., U. of Athens, Athens, Greece. (1944) BSc, physics (Athens U., 1970). Teaching asst. (tel. 01 + 363-3413). *Solid state physics, elastic properties of crystals.*

Zenginoglou, Mr Charalambos. Physics Lab., U. of Patras, Patras, Greece. (1946) BSc, physics (Athens U., 1970). Asst. (tel. 061 + 429-713). *Liquid crystals.*

HONG KONG

Sub-Editor: T. F. Lai

Notes

1. International telephone country code - 852.

2. Degrees conferred by the Hong Kong universities are the *Master of Science* (MSc), *Master of Philosophy* (MPhil) and *Doctor of Philosophy* (PhD).

3. The university appointments are similar to the UK equivalents.

Cheng, Mr Graham Cheng-hsun. Taching Petroleum Co. Ltd., 2B Boundary Crest, 177-177A Boundary St., Kowloon, Hong Kong. (1936) MSc, geology (U. of Manchester, UK, 1962). Director. (tel. 3 + 386143). *Inorganic crystal structures, crystal physics, phase transitions, high temperature properties.*

Cheung, Dr Kung Kai. Chemistry Dept., U. of Hong Kong, Hong Kong. (1935) PhD, crystallography (U. of Glasgow, UK, 1965). Lect. (tel. 5 + 8592161). *Single-crystal and powder diffractometry.*

Hon, Dr Ping-Kay. Chemistry Dept., Chinese U. of Hong Kong, Shatin N.T., Hong Kong. (1936) PhD, chemistry (U. of Illinois, USA, 1964). Lect. (tel. 12 + 633111, ext. 329). *Organometallic compounds, crystal structures, analytical chemistry, instrumentation.*

Lai, Dr Ting Fong. Chemistry Dept., U. of Hong Kong, Hong Kong. (1930) DPhil, crystallography (U. of Oxford, UK, 1960). Sr. lect. (tel. 5 + 8592159). *Organic and organometallic structures.*

Leung, Mr Wilhelm Kei Hong. Analytical & Advisory Services Division, Air Chemistry Section, Government Lab., Oil Street, North Point, Hong Kong. (1950) MSc, pharmaceutical analysis (U. of Manchester, UK, 1981). Chemist. (tel. 5 + 719203). *Powder diffractometry - X-ray.*

Mak, Prof. Thomas Chung-wai. Chemistry Dept., The Chinese U. of Hong Kong, Shatin, New Territories, Hong Kong. (1936) PhD, chemistry (U. of British Columbia, Canada, 1963). Prof. & chairman (tel. 12 + 633111, ext. 282). *Crystal structure analysis, clathrates, hydrates, hydrogen bonded molecular adducts, metal complexes.*

Wong, Dr Yau-Shing. Forensic Div., Government Lab. of Hong Kong, 17th. Floor, May House, Arsenal Street, Wan Chai, Hong Kong. (1946) PhD, chemistry (U. of Waterloo, Canada, 1975). Chemist. (tel. 5 + 284284, ext. 749). *Amino acid - heavy metal complexes, Platinum group - thiocyanate complexes.*

HUNGARY

Sub-Editor: T. Ungár

Notes

1. International telephone country code - 36.

2. The university degrees *Dr. phil., Dr. techn.* and *Dr. rer. nat.* are approximately equivalent to or lower than PhD. The degrees *okl. fizikus, okl. mérnök, okl. tanár* and *okl. vegyész* refer to a university diploma in physics, engineering, teaching or chemistry, respectively, and they are approximately equivalent to MSc. The degrees conferred by the Hungarian Academy of Sciences are *a MTA rendes tagja* (akadémikus) (member of the Hungarian Academy of Sciences), *a MTA levelezö tagja* (lev. tag) (corresponding member of the Hungarian Academy of Sciences), *a ... tudományok doktora* (Dokt.) (doctor in the ... sciences), and *a ... tudományok kandidátusa* (Kand.) (candidate in the ... sciences). The latter two degrees are approximately equivalent to DSc and PhD, respectively. However, the abbreviations Dokt. and Kand. are not in general use in Hungary.

3. Special abbreviations used for institutions which confer degrees: BME - Budapesti Müszaki Egyetem (Technical University of Budapest), ELTE - Eötvös Loránt Tudományegyetem (L. Eötvös University, Budapest), MTA - Magyar Tudományos Akadémia (Hungarian Academy of Sciences).

Arató, Dr Péter. Dept. Metal Physics, Res. Inst. Techn. Physics, Hung. Acad. Sci., H-1325 Budapest P.O.B. 76, Hungary. (1941) Kand., physics (MTA, 1974). Res. scient. (tel. 1 + 692-100). *X-ray diffraction, real structures.*

Argay, Mr Gyula. Dept. X-ray Diffraction, Central Res. Inst. Chemistry, Hung. Acad. Sci., H-1525 Budapest P.O.B. 17, Hungary. (1939) Okl. fizikus, physics (ELTE, 1962). Res. scient. (tel. 1 + 353-735, ext. 327). *Organic crystal structures, computer programming.*

Banizs, Dr Károly. Div. Physical Metallurgy, Res. Eng. Prime Contr. Centre, Hung. Aluminium Corp., ALUTERV-FKI, H-1509 Budapest P.O.B. 5, Hungary. (1940) Kand., physics (MTA, 1979). Res. scient. (tel. 1 + 669-311) *Electron microscopy, diffraction, Al-alloys.*

Barna, Dr Péter. Dept. Thin Film Physics, Res. Inst. Techn. Physics, Hung. Acad. Sci., H-1325 Budapest P.O.B. 76, Hungary. (1928) Kand., physics (MTA, 1966). Head. (tel. 1 + 693-541). *Solid state physics, thin film physics, thin film growth, structure research.*

Bodor, Prof. Géza. Polymer Micromorphology Dept., Polymer Res. Inst., H-1950 Budapest, Hungary. (1930) Dokt., chemistry (MTA, 1969). Head. (tel. 1 + 634-200, ext. 146). *Polymer structures.*

Bognár, Dr László. Dept. Mineralogy, ELTE, Múzeum krt 4/A, H-1088 Budapest, Hungary. (1936) Dr.rer.nat. mineralogy (ELTE, 1969). Asst. lect. (tel. 1 + 338-795). *Structure determination, polycrystalline mineral-systems.*

Bottyán, Dr László. Dept. Material Sci., Res. Eng. Prime Contr. Centre, Hung. Aluminium Corp., ALUTERV-FKI, H-1509 Budapest P.O.B. 5, Hungary. (1951) Dr.rer.nat., physics (ELTE, 1978). Res. scient. (tel. 1 + 669-311, ext. 268). *Powder diffraction methods, inorganic structures, computer science.*

Csanády-Bokody, Mrs Ágnes. Dept. Material Sci., Electronoptical Lab., Res. Eng. Prime Contr. Centre, Hung. Aluminium Corp., ALUTERV-FKI, H-1509 Budapest P.O.B. 5, Hungary. (1935) Okl. vegyész, chemistry (ELTE, 1958). Lab. head. (tel. 1 + 669-028). *Electron beam methods, oxidation processes, metallurgy, aluminium and alloys, metallic and nonmetallic materials, physico-chemical investigations.*

Csordás, Dr László. Dept. Solid State Physics, ELTE, Múzeum krt. 6-8. H-1088 Budapest, Hungary. (1929) Kand., physics (MTA, 1968). Assoc. prof. (tel. 1 + 189-833). *Organic structures.*

Csordás-Tóth, Dr Anna. Dept. Material Sci., Res. Eng. Prime Contr. Centre, Hung. Aluminium Corp., ALUTERV-FKI, H-1509 Budapest P.O.B. 5, Hungary. (1947) Dr. rer. nat., physics (JATE, 1977). Res. scient. (tel. 1 + 669-311, ext. 279). *Real structures, non-metallic materials, scanning electron diffraction, microanalytical investigations, bauxites.*

Cziráki, Dr (Miss) Ágnes. Dept. Solid State Physics, Eötvös U., Múzeum krt 6-8, H-1088 Budapest, Hungary. (1944) Dr, physics (ELTE, 1972). Sr. asst. (tel. 1 + 189-833). *Electron microscopy, amorphous and crystalline materials.*

Czugler, Dr Mátyás. Dept. X-ray Diffraction, Central Res. Inst. Chemistry, Hung. Acad. Sci., H-1525 Budapest P.O.B. 17, Hungary. (1948) Dr. techn., chemistry (BME, 1977). Res. scient. (tel. 1 + 353-735, ext. 327). *Organic crystal structures.*

Dankházi, Mr Zoltán. Dept. Solid State Physics, Eötvös U., Múzeum krt 6-8, H-1088 Budapest, Hungary. (1960) Okl. fizikus, physics (ELTE, 1979). Res. asst. (tel. 1 + 189-833). *X-ray fluorescent spectroscopy, X-ray diffraction.*

Farkas-Jahnke, Dr Mária. X-ray Dept., Res. Inst. Techn. Physics, Hung. Acad. Sci., H-1325 Budapest P.O.B. 76, Hungary. (1932) Kand., physics (MTA,

1974). Head. (tel. 1 + 692-100). *Real structure of crystals, phase transformations, polytypes.*
Frigyik, Mr Gábor. Dept. Mech. Techn., Mikolc U., H-3515 Miskolc, Hungary. (1948) Okl. mérnök, metallurgy (Miskolc U., 1967). Res. scient. (tel. 46 + 65111). *Fracture, X-ray diffraction.*
Fuchs, Dr Erik. Metallurgical Dept., Res. Inst. Ferrous Metallurgy, H-1509 Budapest P.O.B. 14, Hungary. (1930) Dokt., physics (MTA, 1973). Head. (tel. 1 + 250-020, ext. 103). *Metallurgical investigation technology, properties of metals.*
Gaál, Dr István. Dept. Light Sources, Res. Inst. Techn. Physics, Hung. Acad. Sci., H-1325 Budapest P.O.B. 76, Hungary. (1936) Kand., physics (MTA, 1976). Res. scient. (tel. 1 + 692-100). *Physical metallurgy, X-ray scattering, lattice defects.*
Gadó, Dr Pál. Dept. Material Sci., Res. Eng. Prime Contr. Centre, Hung. Aluminium Corp., ALUTERV-FKI, H-1509 Budapest P.O.B. 5, Hungary. (1933) Kand., physics (MTA, 1971). Sci. adviser. (tel. 1 + 251-837). *Real structure, minerals, inorganic materials, powder diffractometry, electron microscopy (structural), phase analysis, automation.*
Geleji-Neubauer, Mrs Irén. Dept. Material Sci., Res. Eng. Prime Contr. Centre, Hung. Aluminium Corp., ALUTERV-FKI, H-1509 Budapest P.O.B. 5, Hungary. (1938) Okl. fizikus, physics (ELTE, 1962). Res. scient. (tel. 1 + 250-020, ext. 227). *Structure, inorganic crystals and metals, powder methods, inorganic phase transformations, crystal physics.*
Gergely, Dr Márton. Metallurgical Dept., Res. Inst. Ferrous Metallurgy, H-1509 Budapest P.O.B. 14, Hungary. (1943) Dr. rer. nat., metallurgy (Techn. U., Miskolc, 1972). Head of lab. (tel. 1 + 250-020, ext. 103). *Diffraction methods, phase transformations.*
Griger, Dr Ágnes. Dept. Material Sci., Res. Eng. Prime Contr. Centre, Hung. Aluminium Corp., ALUTERV-FKI, H-1509 Budapest P.O.B. 5, Hungary. (1946) Dr. rer. nat., physics (ELTE, 1976). Res. scient. (tel. 1 + 669-028, ext. 260). *Crystallographic computing, X-ray measurements, quantitative phase analysis, real structures.*
Grosz, Mr Tamás. X-ray Dept., Res. Inst. Techn. Physics, Hung. Acad. Sci., H-1325 Budapest P.O.B. 76, Hungary. (1949) Okl. fizikus, physics (ELTE, 1973). Res. scient. (tel. 1 + 692-100). *Crystal orientation, phase transformations, solid state reactions.*
Hajdu, Dr Ferenc. Dept. X-ray Diffraction, Central Res. Inst. Chemistry, Hung. Acad. Sci., H-1525 Budapest P.O.B. 17, Hungary. (1926) Kand., chemistry (MTA, 1976). Sr. res. scient. (tel. 1 + 353-735). *Diffraction methods, liquids, solutions, amorphous solids.*
Hajdu, Mr János. Dept. Solid State Physics, ELTE, Múzeum krt 4/A, H-1088 Budapest, Hungary. (1932) Okl. fizikus, physics (ELTE, 1958). Asst. lect. (tel. 1 + 336-316). *Metals and alloys.*
Hargittai, Prof István. Dept. Str. Studies, Res. Lab. Inorganic Chemistry, Hung. Acad. Sci., H-1431 Budapest P.O.B. 117, Hungary. (1941) Dokt., chemistry (MTA, 1976). Head. (tel. 1 + 334-929). *Structural chemistry (exp. and theoretical), electron diffraction.*
Hargittai, Dr Magdolna. Dept. Str. Studies, Res. Lab. Inorganic Chemistry, Hung. Acad. Sci., H-1431 Budapest P.O.B. 117, Hungary. (1945) Kand., chemistry (MTA, 1978). Sr. res. scient. (tel. 1 + 334-929). *Inorganic molecular structures, electron diffraction.*
Harsányi, Dr László. Dept. Str. Studies, Res. Lab. Inorganic Chemistry, Hung. Acad. Sci., H-1431 Budapest P.O.B. 117, Hungary. (1952) Dr. rer. nat., chemistry (ELTE, 1981). Res. scient. (tel. 1 + 334-929). *Molecular structures, quantum chemistry, electron diffraction.*
Hartmann, Dr Ervin. Res. Lab. Crystal Physics, Hung. Acad. Sci., Budaörsi út 45, H-1112 Budapest, Hungary. (1935) Kand., physics (Moscow Inst. of Steels and Alloys, USSR, 1968). Sr. res. scient. (tel. 1 + 850-777, ext. 361). *Crystal growth, physical properties of crystals.*
Imre-Baán, Mrs Irén. Dept. Material Sci., Res. Eng. Prime Contr. Centre, Hung. Aluminium Corp., ALUTERV-FKI, H-1509 Budapest P.O.B. 5, Hungary. (1938) Okl. vegyész, chemistry (BME, 1962). Res. scient. (tel. 1 + 669-311, ext. 279). *Microanylitical investigations, fused materials.*
Kádár, Dr György. Inst. Microelectronics, Central Res. Inst. Physics, Hung. Acad. Sci., H-1525 Budapest P.O.B. 49, Hungary. (1942) Kand., physics (MTA, 1978). Res. scient. (tel. 1 + 698-566) *Magnetic structures.*
Kálmán, Prof. Alajos. Dept. X-ray Diffraction, Central Res. Inst. Chemistry, Hung. Acad. Sci., H-1525 Budapest P.O.B. 17, Hungary. (1935) Dokt., chemistry (MTA, 1975). Head. (tel. 1 + 353-735, ext. 160). *Organic structures, clay minerals, X-ray diffraction.*
Kálmán, Dr Erika. Dept. X-ray Diffraction, Central Res. Inst. Chemistry, Hung. Acad. Sci., H-1525 Budapest P.O.B. 17, Hungary. (1942) Kand., chemistry (Techn. U. of Dresden, DDR, 1970) Sr. res. scient, electron optical group. (tel. 1 + 334-329, ext. 91). *Liquid structure, electron diffraction.*
Kardos, Mrs Jutta. Dept. Semiconductor Development, TUNGSRAM Co., H-1340 Budapest, Hungary. (1939) MSc, crystallography (Humboldt U., DDR, 1963). Head of res. group. (tel. 1 + 680-628). *Semiconductor single crystal topography and orientation, diamond structures.*
Kertész, Dr László. Dept. Solid State Physics, ELTE, Múzeum krt 6-8, H-1088 Budapest, Hungary. (1925) Kand., physics (MTA, 1966). Assoc. prof., head. (tel. 1 + 336-316). *Metals and alloys.*

Klug, Mrs Annamária. Dept. Material Sci., Res. Eng. Prime Contr. Centre, Hung. Aluminium Corp., ALUTERV-FKI, H-1509 Budapest P.O.B. 5, Hungary. (1947) PhD, chemistry (Stockholm U., Sweden, 1977). Res. scient. (tel. 1 + 669-311, ext. 227). *Real structures, inorganic materials, minerals, powder diffraction methods, X-ray spectrometry.*
Koritsánszky, Mr Tibor. Dept. X-ray Diffraction, Central Res. Inst. Chemistry, Hung. Acad. Sci., H-1525 Budapest P.O.B. 17, Hungary. (1954) Okl. vegyész, chemistry (ELTE, 1979). Res. scient. (tel. 1 + 353-735). *Organic crystal structures, theoretical crystallography.*
Kormány, Dr Teréz. Electronic Components Div., Res. Inst. Telecommunication, H-1525 Budapest P.O.B. 15, Hungary. (1936) Kand., technical sciences (MTA, 1969). Head res. group. (tel. 1 + 353-900). *Thin films, semiconductors, crystal defects.*
Krén, Dr Emil. Physics Dept. I., Central Res. Inst. Physics, Hung. Acad. Sci., H-1525 Budapest P.O.B. 49, Hungary. (1935) Kand., physics (MTA, 1974). Director. (tel. 1 + 698-566). *Magnetic structures, phase transitions.*
Lányi, Dr Péter., W, Mo Lab., Tungsram RT, H-1340 Budapest Váci ut 77, Hungary. (1943) Kand., physics (Acad. Sci., DDR, 1970). Head. (tel. 1 + 692-800). *Powder diffractometry, spectrometry.*
Lovas, Dr György Antal. Dept. Mineralogy, ELTE, Múzeum krt 4/A, H-1088 Budapest, Hungary. (1947) Dr. rer. nat., chemistry (ELTE, 1985). Sr. asst. (tel. 1 + 189-833). *Crystal and molecular structure determination.*
Malicskó, Dr László. Res. Lab. Crystal Physics, Hung. Acad. Sci., Budaörsi út 45, H-1112 Budapest, Hungary. (1934) Kand., physics (MTA, 1968). Sr. res. scient. (tel. 1 + 850-777, ext. 363). *Crystal growth, real structure of crystals.*
Medgyaszay, Dr Márton. Res. Lab. Crystallography, EGyT Pharmacochemical Works, Keresztúri ut 30-38, H-1106 Budapest, Hungary. (1936) Dr, pharm. chemist (BOTE, 1959). Head. (tel. 1 + 835-315). *Organic structures, powder diffractometry, morphology, technological properties, pharmaceutical crystallized products.*
Menczel, Dr György. Dept. Solid State Physics, ELTE, Múzeum krt. 6-8. H-1088 Budapest, Hungary. (1921) Kand., physics (MTA, 1973). Assoc. prof. (tel. 1 + 189-833). *Organic structures.*
Nemetz, Prof. Ernö. Dept. Mineralogy, Techn. U. Chemical Engineering, Veszprém, H-8201 Veszpréh P.O.B. 28, Hungary. (1920) Dr. phil., chemistry (Pázmány Péter U. Sci., 1944). Head of dept. and rector of U. (tel. 80 + 12-550, ext. 130). *X-ray crystallography, clay and zeolite minerals.*
Oszko, Mr Albert Zoltán. Dept. Solid State Physics, Eötvös U., Múzeum krt 6-8, H-1088 Budapest, Hungary. (1959) Okl. fizikus, physics (ELTE, 1978). Res. asst. (tel. 1 + 189-833). *Electron microscopy, amorphous and crystalline materials.*
Pál, Mrs Edith. X-ray Dept., Res. Inst. Techn. Physics, Hung. Acad. Sci., H-1325 Budapest P.O.B. 76, Hungary. (1948) Okl. fizikus, physics (ELTE, 1971). Res. scient. (tel. 1 + 692-100). *X-ray topography.*
Pálinkás, Mr Gábor. Dept. X-ray Diffraction, Central Res. Inst. Chemistry, Hung. Acad. Sci., H-1525 Budapest P.O.B. 17, Hungary. (1941) Okl. fizikus, physics (ELTE, 1968). Res. scient. (tel. 1 + 353-735). *Liquid structure, electro diffraction, X-ray diffraction.*
Párkányi, Dr László. Dept. X-ray Diffraction, Central Res. Inst. Chemistry, Hung. Acad. Sci., H-1525 Budapest P.O.B. 17, Hungary. (1940) Kand., chemistry (MTA, 1982). Head, res. group. (tel. 1 + 353-735, ext. 220). *Organic and organometallic crystal structures.*
Radnai, Dr Tamás. Dept. X-ray Diffraction, Central Res. Inst. Chemistry, Hung. Acad. Sci., H-1525 Budapest P.O.B. 17, Hungary. (1948) Dr. rer. nat., physics (ELTE, 1977). Res. scient. (tel. 1 + 353-735). *Liquid structure, small angle X-ray scattering, X-ray diffraction.*
Rischák, Mr Géza. Mineralogical and Petrographical Dept., Hungarian State Geological Survey, H-1442 Budapest P.O.B. 106, Hungary. (1931) Okl. vegyész, chemistry (József Attila U., Szeged, 1955). Head. (tel. 1 + 837-940, ext. 106). *Powder diffraction methods, fluorescent spectroscopy, clay minerals.*
Rozsondai, Dr Béla. Dept. Str. Studies, Res. Lab. Inorganic Chemistry, Hung. Acad. Sci., H-1431 Budapest P.O.B. 117, Hungary. (1934) Kand., chemistry (MTA, 1977). Sr. res. scient. (tel. 1 + 334-929). *Molecular structures, gas electron diffraction, computational methods.*
Sasvári, Dr Judit. Dept. Material Sci., Res. Eng. Prime Contr. Centre, Hung. Aluminium Corp., ALUTERV-FKI, H-1509 Budapest P.O.B. 5, Hungary. (1945) Dr. rer. nat., physics (ELTE, 1972). Res. scient. (tel. 1 + 669-028). *Phase analysis, X-ray topography, computer science.*
Sasvári, Dr Kálmán. Dept. X-ray Diffraction, Central Res. Inst. Chemistry, Hung. Acad. Sci., H-1525 Budapest P.O.B. 17, Hungary. (1912) Dr phyl., physics (Pázmány Péter U., 1938). Retired. (tel. 1 + 661-881) *Organic crystal structures, computer programming.*
Schultz, Dr György. Dept. Str. Studies, Res. Lab. Inorganic Chemistry, Hung. Acad. Sci., H-1431 Budapest P.O.B. 117, Hungary. (1938) Kand., chemistry (MTA, 1982). Sr. res. scient. (tel. 1 + 334-929). *Molecular structures, electron diffraction.*
Schuszter, Dr Ferenc. Dept. Solid State Physics, Eötvös U., Múzeum krt 6-8, H-1088 Budapest, Hungary. (1937) Dr. rer. nat., physics (ELTE, 1971). Sr. asst. (tel. 1 + 189-833). *Electron microscopy, amorphous and crystalline materials.*
Simon, Dr Kálmán. Physical Chemistry Dept., Chinoin Pharmaceutical and Chemical, H-1325 Budapest P.O.B. 110, Hungary. (1946) Kand., chemistry (MTA,

1981). Head, res. group. (tel. 1 + 691-900, ext. 643). *Crystal structures, biological applications, direct methods.*
Stefániay, Mr Vilmos. Central Measuring Dept., Res. Eng. Prime Contr. Centre, Hung. Aluminium Corp., ALUTERV-FKI, H-1509 Budapest P.O.B. 5, Hungary. (1939) Okl. mérnök, electrical eng. (BME, 1967). Res. scient. (tel. 1 + 250-020, ext. 279). *Lattice defects in semiconductors, scanning electron beam methods, X-ray topography.*
Szemethy, Miss Andrea. Dept. X-ray Diffraction, Hungarian Geological Inst., H-1442 Budapest P.O.B. 106, Hungary. (1947) MSc, geology (ELTE, 1972). Res. scient. (tel. 1 + 837-940, ext. 106). *X-ray phase analysis, zeolite structures, carbonates.*
Sztrókay, Prof. Kálmán. Dept. Mineralogy, ELTE, Múzeum krt 4/A, H-1088 Budapest, Hungary. (1907) Dokt., geology (MTA, 1958). Prof. (tel. 1 + 141-874). *Crystallography, structure and physical properties, mineralogy, phase composition.*
Tardy, Dr Pál. Metallurgical Dept., Res. Inst. Ferrous Metallurgy, H-1509 Budapest P.O.B. 14, Hungary. (1940) Kand., technical sci. (MTA, 1974). Head of lab. (tel. 1 + 250-020, ext. 103). *Electron microscopy, ferrous metallurgy.*
Tarján, Prof. Imre. Dept. Biophysics, Semmelweiss Medical U., Budapest H-1088 Budapest Puskin utca 9, Hungary. (1912) Akadémikus, physics (MTA, 1976). Head. (tel. 1 + 339-599). *Crystal physics, crystal growth, biological macromolecules.*
Tóth, Mr Lajos. Dept. Thin Films Physics, Res. Inst. Techn. Physics, Hung. Acad. Sci., H-1325 Budapest P.O.B. 76, Hungary. (1952) Okl. fizikus, physics (ELTE, 1975). Res. scient. (tel. 1 + 692-100). *Thin films, structural research.*
Tremmel, Mr János. Dept. Str. Studies, Res. Lab. Inorganic Chemistry, Hung. Acad. Sci., H-1431 Budapest P.O.B. 117, Hungary. (1927) Okl. mérnök, electrical eng. (BME, 1959). Res. scient. (tel. 1 + 334-929). *Electro optical techniques, high temperature, gas electron diffraction, instrumentation.*
Turmezey, Dr Tibor. Dept. Material Sci., Res. Eng. Prime Contr. Centre, Hung. Aluminium Corp., ALUTERV-FKI, H-1509 Budapest P.O.B. 5, Hungary. (1948) Dr. rer. nat., physics (ELTE, 1974). Head. (tel. 1 + 669-028). *Material science, electron diffraction, metal physics.*
Ungár, Dr Tamás. Inst. for General Physics, ELTE, Múzeum krt. 6-8., H-1088 Budapest, Hungary. (1943) Kand., physics (MTA, 1980). Assoc. prof. (tel. 1 + 189-833). *Metals and alloys, small angle X-ray scattering, X-ray diffraction.*
Vajda, Dr Erzsébet. Dept. Str. Studies, Res. Lab. Inorganic Chemistry, Hung. Acad. Sci., H-1431 Budapest P.O.B. 117, Hungary. (1938) Dr. rer. nat., chemistry (ELTE, 1971). Res. scient. (tel. 1 + 334-929). *Molecular structures, electron diffraction.*
Varga, Dr László. Dept. Light Sources, Res. Inst. Techn. Physics, Hung. Acad. Sci., H-1325 Budapest P.O.B. 76, Hungary. (1935) Dr. techn., metallurgy (BME, 1972). Res. scient. (tel. 1 + 692-100). *Metals and alloy structures.*
Varga, Mr László. Dept. Mech. Techn. Mat. Sci., BME, Goldmann Gy. tér 3. H-1111 Budapest, Hungary. (1944) Okl. mérnök, mechanical eng. (U. Heavy Industry, Miskolc, 1967). Res. scient. (tel. 1 + 664-011, ext. 28-83). *Crystal defects, internal stress analysis, metallurgy.*
Verö, Dr Balázs. Metallurgical Dept., Res. Inst. Ferrous Metallurgy, H-1509 Budapest P.O.B. 14, Hungary. (1944) Dr. rer. nat., metallurgy (Techn. U., Miskolc, 1972). Head of lab. (tel. 1 + 250-020, ext. 103). *Ferrous metallurgy.*
Viczián, Dr István. Dept. X-ray Diffraction, Hungarian Geological Inst., H-1442 Budapest P.O.B. 106, Hungary. (1940) Kand., geology (MTA, 1976). Res. scient. (tel. 1 + 637-600). *X-ray phase analysis, clay mineralogy and petrology.*
Vizi, Dr Béla. Dept. Gen. Inorg. Chemistry, Veszprém U. of Chemical Engineering, H-8201 Veszprém P.O.B. 28, Hungary. (1936) Dr. techn., chemistry (Veszprém U. of Chemical Engineering, 1966). Sr. lect. (tel. 80 + 12-550, ext. 229). *Normal coordinate analysis, vibrational amplitudes.*
Zábráczki, Mr Jósef. Dept. Material Sci., Res. Eng. Prime Contr. Centre, Hung. Aluminium Corp., ALUTERV-FKI, H-1509 Budapest P.O.B. 5, Hungary. (1948) Okl. fizikus, physics (ELTE, 1971). Res. scient. (tel. 1 + 669-311, ext. 227). *Powder diffraction methods, computer science. X-ray spectrometry.*
Zsoldos, Mrs Éva. Inst. Microelectronics, Central Res. Inst. Physics, Hung. Acad. Sci., H-1525 Budapest P.O.B. 49, Hungary. (1934) Okl. fizikus, physics (ELTE, 1957). Head, lab. (tel. 1 + 698-566). *X-ray topography, magnetic crystals.*
Zsoldos, Dr Lehel. Dept. of Structure Res., Res. Inst. Techn. Physics, Hung. Acad. Sci., H-1325 Budapest P.O.B. 76, Hungary. (1931) Kand., physics (MTA, 1965). Head. (tel. 1 + 690-315). *Defects, phase transformations, instrumentation.*

ICELAND

Sub-Editor: **H. Kristmannsdóttir**

Notes

1. International telephone country code - 354.

Eiríksson, Dr Vésteinn Runi. Hamrahlid Col., Reykjavík, Iceland. (1944) PhD, physics, crystallography (U. Edinburgh, UK, 1974). Lecturer. (tel. 91 + 44710). *Phase transitions and structure, ferroelectrics.*
Kristmannsdóttir, Cand.Real. Hrefna. National Energy Authority, Dept. of Natural Heat, Grensásvegur 9, Reykjavík, Iceland. (1944) Cand.Real., mineralogy, petrology (U. Oslo, Norway, 1970). Section leader. (tel. 91 + 83600). *Clay minerals, feldspars, zeolites.*
Sigvaldason, Dr Gudmundur. The Nordic Volcanologic Institute, Reykjavik, Iceland. (1932) Dr.Rer.Nat., geochemistry, mineralogy (U. Göttingen, BRD, 1959). Director. (tel. 91 + 25088). *Crystal optics, silicate crystal structures, clay minerals.*
Steinthórsson, Dr Sigurdur. Dept. of Geology, U. of Iceland, Reykjavik, Iceland. (1940) PhD, geochemistry, petrology (Princeton U., USA, 1974). Assoc. prof. (tel. 91 + 21340). *Crystal optics, silicate crystal structures, ice crystal growth and chemistry.*
Tómasson, Cand.Real. Jens. National Energy Authority, Dept. of Natural Heat, Grensásvegur 9, Reykjavík, Iceland. (1925) Cand.Real., mineralogy, petrology (U. Oslo, Norway, 1962). Section leader. (tel. 91 + 83600). *Low metamorphic minerals.*

INDIA

Sub-Editor: G. B. Mitra

Notes

1. International telephone country code - 91.

2. The Indian Universities and Institutes confer the degrees of DPhil, PhD, and DSc. The first two degrees are equivalent to the PhD degree of the UK Universities and the last one is a higher degree.

3. Abbreviations in this section include:
   ATIRA - Ahmedabad Textile Industries Research Association
   C.G.C. - Crystal Growth and Characterization
   C.S.S. - Central Scientific Services
   I.A.C.S. - Indian Association for the Cultivation of Science
   I.I.S. - Indian Institute of Science, Bangalore
   I.I.T. - Indian Institute of Technology

Agarwal, Dr Bhagwatiprasad. Physics Dept., U. Sch. of Sci., Ahmedabad 380009, India. (1943) PhD, physics (Sardar Patel U., 1972). Lect. (tel. 40929) *Growth of single crystals, defect properties of crystals.*

Agarwal, Dr Ramesh Chandra. Cntr. for Appl. Res. in Electronics, Indian Inst. of Techn., New Delhi 110029, India. (1947) PhD, electrical engineering (Rice U., USA, 1974). Res. Principal scient. officer. (tel. 665674) *Least-squares refinement, large structures, fast fourier transform applications, fast convolution techniques, digital signal processing applications.*

Aggarwal, Dr Prem Sarup. Central Glass & Ceramic Res. Inst., Calcutta 700032, India. (1934) PhD, chemistry (Poona U., 1958). Scient. (tel. 463496) *X-ray and electron diffraction, ceramics and raw materials, thermal analysis.*

Agrawal, Mr Jawahar Lal. Physics Dept., Ranchi U., Ranchi 834008, India. (1950) PhD, physics (Ranchi U., 1976). Res. scholar. *X-ray crystallography, organometallic compounds.*

Ali, Mr Mohsin. Dept. of Physics, Handique Girls C., Guwahati 781001, India. (1950) MSc, physics (Gauhati U., 1971). Lect. *Crystal structure analysis.*

Ali, Dr Sultana Zulfiqar. X-ray Group, Nat. Physical Lab., Hill Side Road, New Delhi 110012, India. (1922) PhD, X-ray crystallography (Leeds U., UK, 1949). Em. scient. *Semiconductor chacogenides, defects, X-ray fluorescence.*

Amirthalingam, Dr V. Chemistry Div., BARC, Trombay, Bombay. 400085, India. (1929) PhD, physics (Wales U., UK, 1962). Scient. officer. *Biocrystallography, phase transition.*

Anantha Murthy, Dr Rayasa V. Lang Camera Group, Nat. Physical Lab., Hill Side Road, New Delhi 110012, India. (1948) PhD, X-ray crystallography (I.I.T., Madras, 1976). Scient. 'B'. (tel. 58 + 1733). *Crystal structure analysis, organic and metal complexes; crystal growth, characterization.*

Anantharaman, Prof. Tanjore Ramachandra. Inst. of Techn., Banaras Hindu U., Varanasi 221005, India. (1927) DPhil, metallurgy (Oxford U., 1954). Prof. & dir. *Structural changes, structural imperfections, metallic structures.*

Aravindakshan, Cheethambadi. Physical Res. Wing, Projects & Dev. India Ltd., P.O. Sindri, Dist. Dhanbad 828122, India. (1934) PhD, X-ray crystallography (Madras U., 1974). Deputy superintendent. *Metal physics, metals and alloys, corrosion, crystal structure, fertilizer materials.*

Arjunan, P. Dept. of Organic Chemistry, Indian Inst. of Sci., Bangalore 560012, India. (1956) MSc, chemistry (Madurai Kamraj U., 1979). Sr. res. student. *X-ray structure analysis, photoreactive crystals.*

Arora, Mr Narinder Kumar. Materials Div., Nat. Physical Lab., Hillside Road, New Delhi 110012, India. (1952) MSc, physics (Panjab U., 1973). Sr. tech. asst. (tel. 581455). *Semiconductors, ferroelectric materials.*

Awasthi, Dr Santosh Kumar. Radiochemistry Div., BARC, Trombay, Bombay 400085, India. (1944) PhD, chemistry (Bombay U., 1974). Scient. officer. (tel. 523321, ext. 456) *Inorganic and organometallic compounds, oxides, X-ray spectrometry.*

Balasingh, Mr C. Materials Sci. Div., Nat. Aeronautical Lab., Bangalore 560017, India. (1941) MSc, physics (Annamalai U., 1963). Scient. 'B'. *Powder diffractometry, stress - strain analysis, instrumentation.*

Balasubramanian, Dr R. Dept. of Crystallography and Biophysics, U. of Madras, A. C. C. Campus, Madras 600025, India. (1943) PhD, molecular biophysics (Madras U., 1973). Lect. (tel. 432248, ext. 16). *Theoretical studies on crystal structures.*

Bandopadhyay, Mrs Tapati. Crystallography & Molecular Biology Div., Saha Inst. of Nuclear Physics, 92 Acharya P. C. Rd., Calcutta 700009, India. (1950) MSc, physics (Calcutta U., 1973). Res. fellow. (tel. 354281). *Crystal structure, biologically important compounds.*

Banerjee, Dr Asok. Dept. of Biophysics, Bose Inst., P-1/12 C.I.T. Scheme, Calcutta, India. (1942) PhD, physics (Calcutta U., 1974). Reader. *Molecular biophysics, macromolecular structure, function dynamics.*

Banerjee, Dr Krishna. Physics Dept., Ranchi U., Ranchi, Bihar 834008, India. (1950) PhD, X-ray crystallography (Ranchi U., 1982). Res. scholar. *Crystal structure determination, X-ray diffraction techniques.*

Banerjee, Dr Srikumar. Metallurgy Div., Bhabha Atomic Res. Cntr., Trombay, Bombay 400085, India. (1946) PhD, metallurgy (I.I.T., Kharagpur, 1975). Scient. officer. (tel. 523321, ext. 242). *Phase transformation, electron microscopy, radiation damage.*

Bansigir, Prof. K. Goswami. Sch. of Studies in Physics, Jiwaji U., Gwalior, M. P., India. (1923) PhD, solid state physics (Osmania U., 1959). Prof. *Dislocations, X-ray topography, dissolution studies, colour centres, lattice vibrations, photo-elastic effect.*

Basak, Prof. Bejoysanker. Physics Dept., Presidency C., College St. Calcutta 700073, India. (1921) PhD, physics (Calcutta U., 1950). Principal(retd), Presidency C. *Crystal structure, organic compounds, bond lengths.*

Basak, Dr Madan Gopal. Physics Dept., Presidency C., Calcutta 700073, India. (1923) DPhil, X-ray crystallography (Calcutta U., 1960). Asst. prof. (tel. 341121). *Organic crystal structures.*

Betal, Mr Badal Kumar. Physics Dept., Presidency C., Calcutta 700073, India. (1943) MSc, physics (Calcutta U., 1963). Lect. (tel. 341121). *Sterol structures.*

Bhadbhade, Mr Mohan Madhav. Organic Chemistry Dept., Indian Inst. of Sci., Bangalore, Karnataka 560012, India. (1956) MSc, physics (I.I.T., Kharagpur, 1977). Res. scholar. (tel. 34411, ext. 403). *Drug molecules, steroids, organic molecules (moderately strained), X-ray studies, molecular mechanics calculations.*

Bhaduri, Dr Debabrata. Crystallography & Molecular Biology Div., Saha Inst. of Nuclear Physics, 92 A. P. C. Road, Calcutta 700009, India. (1946) PhD, physics (Calcutta U., 1976). Res. fellow. (tel. 354281, ext. 51). *Biomolecules, bioenergetics.*

Bhagvantam, Prof. Suri. SriNiket Tarnaka, Hyderabad 500017, India. DSc(Hon.), physics (Andhra U., 1941; Osmania U., 1969). *Crystal physics*

Bhagwat, Dr Vasant. Dept. of Chemistry, Vikram U., Ujjain, Madhya Pradesh 456001, India. (1947) PhD, Chemistry (Vikram U., 1970). Lect. *X-ray crystallographic studies, alkali and akaline earth complexes of macrocyclic compounds.*

Bhakay-Tamhane, Mrs Sandhya Nitin. Neutron Physics Div., Bhabha Atomic Res. Cntr., Trombay, Bombay 400085, India. (1954) MSc, solid state physics (Bombay U., 1975). Scient. officer. (tel. 521441). *Charge density studies, X-ray and neutron diffraction, LCAO-MO calculations.*

Bhaktapriya, Dr S. R. Y. Physics Dept., Ranchi U., Ranchi 834008, India. (1935) PhD, X-ray crystallography (Ranchi U., 1981) Prof. *Organo-metallic compounds, structures.*

Bhargava, Dr L. R. Dept. of Atomic Energy, X-ray Diffraction Lab., Amd Complex, Begumpet, Hyderabad, India. (1943) PhD, Geology (Saugor U., 1973). Scient. officer. *Inorganic crystal structure.*

Bhat, Dr Laxminarayana H. Physics Dept., Indian Inst. of Sci., Bangalore 560012, India. (1943) PhD, (Sardar Patel U., 1973). Asst. prof. (tel. 5 - 34411, ext. 315). *Crystal growth, crystal defects, ferroelectricity, phase transition.*

Bhat, Mr T. Narayana. Physics Dept., Indian Inst. of Sci., Bangalore 560012, India. (1948) MSc, physics (Sardar Patel U., 1972). Sr. res. fellow. *Biologically important compounds, complexes, medically important compounds.*

Bhatia, Mr Subhash Chandra. Chemistry Dept., Kurukshetra U., Kurukshetra 132119, India. (1950) MSc, physical chemistry (Kurukshetra U., 1972). Sr. Res. fellow. *X-ray structure determination.*

Bhatt, Dr Vaikunthray Promodray. Physics Dept., Faculty of Sci., M.S.U., Baroda 390002, India. (1929) PhD, solid state physics (M.S.U., Baroda, 1963). Reader. *Growth, dissolution and mechanical properties, metal and alloy single crystals.*

Bhattacharjee, Mrs Lilabati. Mineral Physics Div., Geological Survey of India, 29 Chowringhee Road, Calcutta 700016, India. MSc, physics (Calcutta U., 1951). Sr. mineralogist. (tel. 238321, ext. 8). *Structural crystallography, optical transform methods, computer programming, phase transformations, crystal growth, topography, instrumentation.*

Bhattacharya, Archana. Dept. of Physics, Jadavpur U., Calcutta 7000329, India. (1951) MSc, physics (Jadavpur U., 1974). Res. scholar. *X-ray crystallography.*

Bhattacharya, Dr Ramendranarayan. Dept. of Magnetism, Indian Assoc. for Cultivation of Sci., Calcutta 700032, India. (1934) PhD, physics (Calcutta U. 1975). Lect. *Crystal physics.*

Bhattacharyya, Dr Subodh Chandra. Crystallography & Molecular Biology Div., Saha Inst. of Nuclear Physics, Sector-1 Block-AF, Bidhannagar, Calcutta 70064, India. (1933) PhD, physics (Calcutta U., 1975). Lect. *Structure and function, drug molecules.*

Bhattacherjee, Mr Santi Brata. Physics Dept., Serampore C., College Street, Serampore, W. Bengal, India. (1925) MSc, physics (Dacca U., Bangladesh, 1948). Prof. *Crystal structure, phase transformation, X-ray spectroscopy.*

Bhattacherjee, Dr Satyananda. Physics Dept., Indian Inst. of Techn. Kharagpur 721302, India. (1935) PhD, X-ray & structure of matter (I.I.T., Kharagpur, 1970). Lect. *Clay minerals, curved lattices, elastic properties, thermal properties.*

Bhawalkar, Dr Ramkrishna Haribhau. X-ray Group, Nat. Physical Lab., Hillside Road, New Delhi 110012, India. (1933) PhD, luminescence (Sagar U., 1963), Scient. (tel. 584179). *Powder diffractometry, X-ray fluorescence analysis, crystallography.*

Bindal, Mr Arvind. Materials Div., Nat. Physical Lab., Hillside Road, New Delhi 110012, India. (1953) MSc, physics (Kurukshetra U., 1974). Res. scholar. (tel.581455). *Crystal growth.*

Bindlish, Mr Jag Mohan. Chemistry Dept., Kurukshetra U., Kurukshetra 132119, India. (1949) PhD, X-ray crystallography (Kurukshetra U.). Sr. res. fellow. *X-ray crystallography.*

Bist, Dr B. M. S. Physics Dept., Benaras Hindu U., Varanasi 221005, India. (1950) PhD, physics (Banaras Hindu U., 1974). Postdoctoral fellow. *Thin films, electron microscopy.*

Biswas, Goutam. Dept. of Biophysics, Bose Inst., P-1/12, C.I.T. Scheme Kankurgachi, Calcutta, India. (1960) MSc, physics (Calcutta U., 1982). Res. Scholar. *Organic crystal structure, biological macromolecules.*

Biswas, Mr Subhash Chandra. C.S.S. Dept., Indian Assoc. for the Cultivation of Sci., Calcutta 700032, India. (1939) MSc, physics (Dacca U., Bangladesh, 1962). Lect. *Organic structures, disorder, computer programming.*

Biswas, Dr Sundar Gopal. Physics Dept. Visva Bharati U., Santiniketan 731235, India. (1931) PhD, crystallography (Calcutta U., 1961). Lect. *Organic compounds of biological importance.*

Bose, Mr Shyamal Kumar. X-ray Crystallography Group, Nat. Metallurgical Lab., Jamshedpur 831007, India. (1936) MSc, physics (Allahabad U., 1957). Scient. 'C'. *Metals and alloys, structure and transformation.*

Chacko, Dr K. K. Dept. of Crystallography and Biophysics, U. of Madras, Guindy Campus, Madras 600025, India. (1942) PhD, physics (Madras U., 1971). Reader. (tel. 432248). *biologically interesting compounds, medically interesting compounds, anomalous dispersion.*

Chadha, Dr Gopal Krishan. Dept. of Physics and Astrophysics, Delhi U., Delhi 110007, India. (1942) PhD, physics (Delhi U., 1968). Reader. *Crystal growth, X-ray diffration, defects in crystals, polytypism.*

Chakrabarty, Dr Mrs Chandana. Crystallography & Molecular Biology Div., Saha Inst. of Nuclear Physics, Sector-1 Block-AF, Bidhannagar, Calcutta 700064, India. (1951) PhD, physics (Calcutta U. 1984). Res. assoc. *Biologically important compounds, proteins.*

Chakrabarty, Dipak Kumar. Dept. of Physics, X-ray Lab., Jadabpur U., Calcutta 700032, India. (1953) MSc, physics (Calcutta U., 1976). Lect. *Crystal structure, lattice dynamics.*

Chakrabarty (Chatterjee), Mrs Ela. Dept. of General Physics and X-ray, I.A.C.S., Calcutta 700032, India. (1957) MSc, physics (I.I.T., Bombay, 1979). Res. scholar. *Crystallography, microstructure, thin films.*

Chakrabarty, Subhasis. Dept. of Biophysics, Bose Inst. P-1/12, C.I.T. Scheme, Kankurgach, Calcutta 700054, India. (1959) MSc, physics (Calcutta U., 1982). Res. scholar. *Molecular biophysics.*

Chakrabarty, Mr Sugoto. Physics Div., Bhabha Atomic Res. Cntr., Bombay 400085, India. (1957) MSc, physics (Delhi U., 1979). Scient. officer. (tel. 551-3848). *Protein crystallography.*

Chakraborty, Dr Suchit Chandra. Physics Dept., U. of Burdwan, Burdwan, W. Bengal, India. (1928) PhD, physics (Allahabad U., 1958). Prof. *Crystal structure, thermal diffuse scattering, lattice vibrations, phase transitions.*

Chanda, Dr Gopal Krishan. Dept. of Physics and Astrophysics, Delhi U. Delhi 110007, India. (1942) PhD, polytypism in vapour grown crystals (Delhi U., 1968). Lect. (tel. 228993). *Crystal growth, structure, defects in crystals.*

Chandra, Dr Suresh. Physics Dept., Allahabad U., Allahabad 211002, India. (1949) DPhil, diffuse X-ray studies (Allahabad U., 1973). Res. fellow. *Dynamical study of solids, X-ray diffraction.*

Chandrasekaran, Prof. Katuputhur Sarma. Physics Dept., Madurai U., Madurai 625021, India. (1925) PhD, X-ray crystallography (Madras U., 1956). Prof. and head. *Crystalline solid solutions, lattice defects, electron density, texture & perfection of crystals.*

Chandrasekaran, Dr R. Molecular Biophysics Unit., Indian Inst. of Sci., Bangalore 560012, India. (1939) PhD, X-ray crystallography (Madras U., 1968). Asst. prof. (tel. 34411, ext.459). *Biopolymers, small molecules.*

Chandrasekhar, Prof. Sivaramakrishna. Liquid Crystals Lab., Raman Res. Inst., Bangalore 560006, India. (1930) PhD, crystallography (Cambridge U., UK, 1957). Prof. (tel. 30124). *X-ray diffraction, crystal optics, liquid crystals.*

Chandrasekharaiah, Dr M. N. Dept. of Metallurgical Eng., Inst. of Techn., Benaras Hindu U., Varanasi 221005, India. (1945) PhD physics (Benaras Hindu U., 1972). Res. assoc. *Crystal growth, geometrical crystallography, defects (planar).*

Chandy, Dr K. C. Mineral Physics, Geochronology and Isotope Geology Div., Geological Survey of India, 29 Chowringhee Road, Calcutta 700016, India. (1928) PhD, physics (Saugar U., 1957). Dir. (tel. 238321, ext. 6). *Crystal structure, minerals.*

Chatterjee, Dr Amitava. Dept. of General Physics & X-ray, I.A.C.S., Calcutta 700032, India. (1948) PhD, physics (Calcutta U., 1978). Pool officer, CSI. *Macromolecular structure, conformational analysis.*

Chatterjee, Dr Sanat Kumar. Physics Dept., Regional Eng. C., Durgapur 731209, India. (1945) PhD, physics (Calcutta U., 1976). Lect. *Lattice imperfections, liquid metals, mechanical properties.*

Chaudhuri, Dr Ahindra Kumar. Solid State Lab., Central Glass & Ceramic Res. Inst., Calcutta 700032, India. (1941) PhD, physics (I.I.T., Kharagpur, 1974). Scient. *Nucleation and crystallisation of glass, order-disorder, thin films.*

Chaudhuri, Prof. Bhumidhar. Physics Dept., Gauhati U., Gauhati 781014, India. (1922) PhD, X-ray crystallography (Manchester U., UK, 1956). Prof. (tel. 88531). *Crystal structure analysis, optical transforms.*

Chaudhuri, Mr Siddhartha. Crystallography & Molecular Biology Div., Saha Inst. of Nuclear Physics, 92 A. P. C. Road, Calcutta 700009, India. (1945) MSc, physics (Calcutta U., 1969). Res. fellow. (tel. 354281). *Biologically important compounds, computing methods.*

Chawla, Dr Krishan Lal. Radiochemistry Div., Bhabha Atomic Res. Cntr., Bombay 400085, India. (1939) PhD, inorganic chemistry (Rajasthan U., 1968). Scient. officer. (tel. 523321, ext. 456). *Crystal chemistry, actinides.*

Chetal, Prof. Amritlal R. Dept. of Applied Sci., Indian Sch. of Mines, Dhanbad 826004, India. (1938) PhD, physics, chemistry (Poona U., 1966). Prof. *Crystal structure.*

Chidambaram, Dr Rajagopala. Physics Group, Bhabha Atomic Res. Cntr., Trombay Bombay 400085, India. (1936) PhD, physics (I.I.S., Bangalore, 1962). Dir., physics group. *Neutron diffraction, hydrogen bonding, high pressure physics.*

Chopra, Prof. Kasturilal. Physics Dept., Indian Inst. of Techn., New Delhi 110029, India. (1933) PhD, solid state physics (British Columbia U., Canada, 1957). Sr. prof. and head. (tel. 79163, ext. 257,268). *Thin films, amorphous semiconductors, electron diffraction, electron microscopy, vacuum technology.*

Dadel, Mrs Snehlata. Physics Dept., Ranchi U., Ranchi 834008, India. (1939) MSc, physics (Ranchi U., 1962). Lect. *Crystallography, organo-metallic compounds.*

Das, Mr Birendra Nath. Physics Dept., Vivekananda C., D. H. Road, Calcutta 700063, India. (1941) MSc, physics (Calcutta U., 1962). Lect. (tel. 453657). *Organic compounds of biological interest.*

Das, Dr Indu Mohan. Physics Dept., Gauhati U., Gauhati 781014, India. (1935) PhD, physics (Gauhati U., 1967). Reader. (tel. 88531). *X-ray crystallography, X-ray fluorescence spectroscopy.*

Das, Pratap Kumar. Dept. of Physics, Surendranath C., Calcutta 700009, India. (1947) MSc, physics (Calcutta U., 1969). Lect. *Organic and organometallic structure, disorder, phase transition.*

Das, Dr Sabita. Physics Dept., Victoria Inst. (college branch), Calcutta 700009, India. (1934) PhD, physics (Calcutta U., 1969). Lect. *Theoretical crystallography, molecular biophysics.*

Das Gupta, Mr Prabal. C.S.S. Dept., Indian Assoc. for the Cultivation of Sci., Jadvpur, Calcutta 700032, India. (1952) MSc, chemistry (Jadavpur U., 1977). Techn. Superintendent. *Structure - properties relationships.*

Datta, Mr Amal Kumar. Physics Dept., Indian Inst. of Techn., Kharagpur, West Bengal 721302, India. (1953) D.I.I.T. Postgraduate diploma, X-ray crystallography (I.I.T., Kharagpur, 1977). Sr. res. asst. *X-ray microscopy, small angle X-ray scattering.*

Dattagupta, Dr Jiban Kanti. Crystallography & Molecular Biology Div., Saha Inst. of Nuclear Physics, Sector-1 Block-AF, Bidhannagar, Calcutta 700064, India. (1945) PhD, physics (Calcutta U., 1972). Reader. *Structure & conformation, medicinal compounds, protein crystallography.*

Dave, Prof. Jatashanker Sadashiv. Chemistry Dept., Faculty of Sci., M.S.U., Baroda 390002. India. (1913) PhD, physical chemistry (London U., UK, 1954). U. G. C. retired teachers scheme awardee. *Liquid crystals.*

Dayal, Dr Radha Raman. Special Materials Div., Defence Sci. Lab., Matcalfe House, Delhi 110054, India. (1941) PhD, phase equilibria (Aberdeen U., UK, 1965). Sr. scient. officer. (tel. 221521, ext. 12). *Phase equilibria, crystal growth, crystal chemistry.*

De, Mr Adhip Kanti. Physics Dept., Indian Inst. of Techn., Kharagpur 721302, India. (1951) MSc, solid state physics (Jabalpur U., 1975). Sr. res. asst. *Clay minerals.*

De, Dr Madhusudan. Dept. of General Physics & X-rays, Indian Assoc. for the Cultivation of Sci., Calcutta 700032, India. (1941) PhD, physics (Calcutta U., 1970). *Metal physics, lattice imperfections, thermal expansion.*

Deopura, Dr B. L. Textile Techn. Dept., Indian Inst. of Techn., New Delhi 110016, India. (1946) PhD, physics (I.I.T., Kanpur, 1972). Asst. Prof. *Small angle X-ray scattering, polymers, fibres, X-ray diffraction.*

Dhanaraj, Mr V. Molecular Biophysics Unit, Indian Inst. of Sci., Bangalore 560012, India. (1957) MSc, chemistry (Madras U., 1979). Res. scholar. *Molecular biophysics.*

Dhaneshwar, Dr Narayandatta Nagesh. Physical Chemistry Div., Nat. Chemical Lab., Pashan Road, Poona 411008, India. (1936) PhD, X-ray crystallography (Poona U., 1973). Sr. scient. officer. (tel. 56451). *Low temperature crystallography, heterocyclic compounds, small organic molecules, natural products.*

Dhar, Rakesh. C.G.C. Sect., Nat. Physical Lab., Hillside road, New Delhi 110012, India. (1960) MSc, (Jammu U., 1982). Res. scholar. *X-ray diffraction, diffuse X-ray scattering studies.*

Dhawan, Mrs Urmil. Specialized Techniques (X-rays) Div., Nat. Physical Lab., Hillside Road, New Delhi 110012, India. (1939) MSc, physics (Panjab U., 1964). Scient. (tel. 584179). *Inorganic structures, defects, diffraction theory, experimental methods & apparatus.*

Duraipandianadar, Mr P. P. Sch. of Physics, Madurai Kamraj U., Madurai 625021, India. (1956) MSc, physics (Annamalai U., 1980). Res. scholar. *Crystal growth, solid solutions*

Durairaj, Kanagapushpam. Organic Chemistry Dept., Indian Inst. of Sci., Bangalore 560012, India. (1961) MSc, chemistry (Madurai Kamraj U., 1983). Res. scholar. *Organic crystal structure, structure - reactivity relationships.*

Dutta, Dr Bishnu Pada. Physics Dept., Science C., Patna 800005, India. (1942) PhD, X-rays & structure of matter (Ranchi U., 1975). Lect. *X-ray crystallography.*

Dutta, Dr Sachindra Nath. Dept. of Education of Sci. & Math., Nat. Inst. of Education, Sri Arbindo Marg, New Delhi 110016, India. (1928) PhD, X-ray crystallography (Manchester U., UK, 1959). Reader. (tel. 79546, ext. 225). *Crystal structure, solid state physics, liquid physics.*

Dweltz, Dr Neville Edwin. Physics Div., ATIRA, Ahmedabad 380015, India. (1933) PhD, crystallography (Madras U., 1962). Deputy dir. *Fibre structure, macromolecular materials research, small angle scattering.*

Dwivedi, Dr Ganpat Lal. Mineral Physics Div., Geological Survey of India, 29 Chowringhee Road, Calcutta 700016, India. (1942) PhD, X-ray crystallography (I.I.T., Kanpur, 1971). (tel. 238321, ext. 8). *Organic structures, inorganic and mineral crystal structures.*

Eswara Prasad, Mr Gummuluri. Metallurgy Div., Bhabha Atomic Res. Cntr., Trombay, Bombay 400085, India. (1940) MSc, chemistry (Bombay U., 1970). Scient. officer. (tel. 523321, ext. 242). *Structure and properties, oxide ceramics.*

Fernandes, Mr Jacob Richard. Liquid Crystals Lab., Raman Res. Inst., Bangalore 560006, India. (1951) MSc, physics (I.I.T., Kharagpur, 1973). Res. fellow. (tel. 30124). *Liquid crystals.*

Ganesan, V. Dept. of Crystallography and Biophysics, Madras U., Madras 600025, India. (1959) MSc, physics (Madras U.). *Lattice dynamics.*

Ghosh, Dr Mrs Minakshi. Dept. of Physics, Presidency C., Calcutta 700073, India. (1948) PhD, physics (Calcutta U., 1981). Lect. *Organic and Inorganic structures, phase transitions, disorder.*

Ghosh, Dr Sujit Kumar. Physical Res. Wing, Projects & Dev. India Ltd., P.O. Sindri, Dist. Dhanbad 828122, India. (1932) PhD, X-ray crystallography (Indiana U., USA, 1960). Superintendent. *Crystal structures, organic molecules, crystal chemistry, catalysts, line profile analysis, computer programming, mineral characterization and beneficiation, fertilizer materials.*

Ghosh, Miss Sutapa. Crystallography & Molecular Biology Div., Saha Inst of Nuclear Physics, Sector-1 Block-AF, Bidhannagar, Calcutta 700064, India. (1959) MSc, physics (Calcutta U., 1981). Res. scholar. *Crystal structure, biomolecules.*

Ghosh, Mr Timir Baran. Physics Dept., Pachhunga U. C., Aizawl 796001, India. (1949) PhD, physics (I.I.T., Kharagpur, 1982). Lect. *Crystal lattice defects, liqui-sol-quenching, amorphous materials.*

Ghouse, Dr Khaja Mohd. X-ray Div., Regional Res. Lab., Hyderabad 500009, India. (1931) PhD. physics (Osmania U., 1967). Scient. (tel. 71351). *X-ray crystallography, organic and inorganic compounds.*

Giri, Mr Anit K. C.S.S. Dept., Indian Assoc. for the Cultivatuon of Sci., Jadavpur, Calcutta 700032, India. (1960) MSc, physics (Calcutta U., 1982). Res. scholar. *Structure - properties relationships, amorphous materials.*

Giri, Mr Siba Narayan. Dept. of Physics, Bidhan Chandra Krishi Viswavidyalaya Mohanpur, Nadia, India. (1936) MSc, physics (Allahbad U., 1960). Reader. *Crystal structure, organic compounds.*

Girirajan, Dr K. S. Dept. of Biophysics and Crytallography, U. of Madras, Guindy Campus, Madras 600025, India. (1942) PhD, physics (I.I.T., Madras, 1974). Lect. *Thermal vibrations, crystallography.*

Gnanaguru, Mr K. Dept. of Organic Chemistry, Indian Inst. of Sci., Bangalore 560012, India. (1948) MSc, chemistry (I.I.S., 1972). Res. scholar. *Structure - photoreactivity relationships.*

Godavarthi, Bhagavannarayana. C.G.C. Sect., Nat. Physical Lab., New Delhi 110012, India. (1955) MSc, physics (Andhra U., Waltair, 1979). Scient. *High resolution X-ray diffraction.*

Gomes, Albert Cardinal. Dept. of Biophysics, Bose Inst., P-1/12, C.I.T. Scheme 7 M, Kankurgachi, Calcutta 700054, India. (1947) MSc, molecular Biology (Calcutta U., 1972). Res. scholar. *Organic crystal structure, biological macromolecules.*

Goswami, Dr Anilprasanna. Physical Chemistry Dept., Nat. Chemical Lab., Poona 411008, India. (1919) DSc, science (London U., UK, 1970). Asst. dir. (tel. 56451, ext. 10). *Thin film physics, crystal growth, electron diffraction, electron microscopy.*

Goswami, Dr Kaidar Nath. Physics Dept., Jammu U., Jammu 180001, India. (1934) PhD, solid state physics (Sardar Patel U., 1963). Reader. (tel. 6340). *X-ray crystallography, medicinal compounds, phase transitions.*

Goswami, Dr (Mrs) S. N. N. Materials Characterization Div., Nat. Physical Lab. New Delhi, 110012, India. (1949) PhD, physics (Jiwaji U., Gwalior, 1975). Res assoc. *Single crystal characterization, perfection, X-ray diffraction.*

Gupta, Dr Kinkar Prosad. Metallurgical Eng. Dept., Indian Inst. of Techn., Kanpur 208016, India. (1933) PhD, metallurgical engineering (Illinois U., USA, 1962) Prof. *Complex structures, phase transformations, theory & properties, metals and alloys, magnetic materials.*

Gupta, Mr Manoj Kumar. Physics Dept., Lucknow U., Lucknow 226007, India. (1955) MSc, physics (Kanpur U., 1973). Res. scholar. (tel. 25140). *Structure and conformation, polypeptides and biomolecules.*

Gupta, Prof. Manoranjan Prasad. Physics Dept., Ranchi U., Ranchi 834008, India. (1926) PhD, crystallography (London U., UK, 1953). Prof. and Head. (tel. 22336). *Crystal structure, crystal physics.*

Gupta, Mr Satish Chandra. Neutron Physics Div., Bhabha Atomic Res. Cntr., Trombay, Bombay 400085, India. (1951) BSc(Hons.), physics (Delhi U., 1971). Scient. officer. (tel. 523848). *High pressure crystallography, shock wave phenomena in materials.*

Gupta, Miss Sunita. Physics Dept. Lucknow U., Lucknow 226007, India. (1952) MSc, physics (Lucknow U., 1972). Res. scholar. (tel. 25140). *Lattice dynamics, intermolecular forces.*

Gupta, Mr Vijai Prakash. Physics Dept., Lucknow U., Lucknow 226007, India. (1950) MSc, physics (Banaras Hindu U., 1972). Sr. res. fellow. (tel. 25140). *Conformation, polypeptide and biomolecular conformational transitions.*

Gupta, Prof. Vishwambhar Dayal. Physics Dept., Lucknow U., Lucknow 226007, India. (1934) DPhil, X-ray structure of cellulose (Allahabad U., 1958). Prof. (tel. 25140). *Biomolecules, small angle X-ray scattering, phase transitions, lattice dynamics.*

Gururow, Dr Tayur N. Physical Chem Div., Nat. Chem. Lab., Pune 411008, India. (1951) PhD, X-ray crystallography (I.I.S., 1976). Scient. *Crystal structures, organic and inorganic compounds, electron density, computer programming.*

Halder, Dr Sujit Kumar. Crystal Growth and Characterization Sect., Nat. Physical Lab., New Delhi 110012, India. (1948) PhD, Physics (Calcutta U., 1976). Scient. *X-ray diffraction, microstructure, alloys, thin films, crystal growth, characterization.*

Hariharan, Meena. Dept. of Crystallography and Biophysics, U. of Madras, Madras 60000, India. (1957) MSc, (U. Madras). Res. scholar. *Biocrystallography.*

Hemkar, Dr Mangla Prasad. Physics Dept., Allahabad U., Allahabad, UP, India. (1932) DPhil, lattice dynamics (Allahabad U., 1962). Lect. *X-ray crystallography, fibre structures, diffuse X-ray scattering, lattice dynamics.*

Iyenger, Dr Leela. Physics Dept., C. of Sci., Osmania U., Hyderabad 500007, India. (1943) PhD, X-ray crystallography (Osmania U., 1970). Lect. *Thermal expansion, crystal structures.*

Jagannadham, Dr Adibhatla Vankata. Physics Dept., Government C., Ajmer 305001, India. (1924) PhD, crystalline field parameters by EPR (I.I.T., Kanpur, 1971). Postgrad. head. *X-ray crystallography, electron paramagnetic resonance, lasers.*

Jain, Dr Prem Chand. Chemistry Dept., B. N. C. U., Kurukshetra 132119, India. (1931) PhD, studies of colloids (Saugar U., 1960). Reader. *X-ray crystallography.*

Jakkal, Vasant Shankar. Water Chemistry Div, Bhabha Atomic Res Cntr., Trombay Bombay 400085, India. (1936) MSc, physics (Bombay U., 1961) Scient. officer. *Organic crystal structure.*

Jayadevan, Dr Naduviledath Chennuvittil. Radiochemistry Div., Bhabha Atomic Res. Cntr., Bombay 400085, India. (1936) PhD, inorgnic chemistry (McMaster U., Canada, 1968). Scient. officer. *Transition metal complexes, actinide complex, mixed oxides, mixed carbides.*

Jayanty, Dr Ashok Physics Dept., Ranchi U., Ranchi 834008, India. (1951) PhD, X-ray crystallography (Ranchi U., 1980) Lect. *Crystal srtucture determination.*

Jayashree, Ms A.N. Molecular Biophysics unit, Indian Inst. of Sci., Bangalore 560012, India. (1963) MSc, Physics (Karnatak U., 1984). Res. Student. *Biological crystallography.*

Joshi, Prof. Ramesh Vinayak. Applied Physics Dept., M.S.U., Baroda 390001, India. (1925) PhD, physics (Leeds U., UK, 1958). Prof. *Crystal defects.*

Joshi, Dr Shri Krishna. Physics Dept., Roorkee U., Roorkee 247667, India. (1935) DPhil, crystal physics (Allahabad U., 1962). Prof. & head. (tel. 743). *Solid state theory, electrons & phonons in disordered systems, magnetism.*

Kakati, Dr Kandarpa Kumar. Physics Dept., Gauhati U., Gauhati 781001, India. (1937) PhD, physics (London U., UK, 1969). Reader. (tel. 88531). *Thin films, X-ray crystallography.*

Kar(Roy), Dr (Mrs) Tanusree Dept. of General Physics & X-ray, Indian Association for the Cultivation of Sci., Calcutta 700032, India. (1954) PhD, physics (Calcutta U., 1984). Res. scholar. *Structures, organic and biological moleculess, Molecular conformation, crystal physics.*

Kalyanaraman, Dr A. R. Physics Dept., SITRA, Avinashi Road, Coimbatore 641014, India. (1937) PhD, X-ray crystallography (Madras U., 1967). Res. assoc. (tel. 28367, ext. 36). *Crystallography, fibre diffraction, computer programming, texture.*

Kannan, Dr Kazhiur Kothandapani. Neutron Physics Div., Bhabha Atomic Res. Cntr., Bombay 400085, India. Scient. (tel. 523848).

Kashyap, Dr Ram Prasad. Chemistry Dept., Guru Nanak Dev U., Amritsar 143005, India. (1947) PhD, crystallography (Kurukshetra U., 1974). Lect. *X-ray crystallography.*

Kaillathe, Padmanabhan. Dept. of Organic Chemistry, Indian Inst. of Sci., 560012, India. (1958) MSc, chemistry (Calicut U., 1981). Res. scholar. *Chemical crystallography, organic crystal structures, inclusion complexes, organic solid state reactions, telarion reactions, structure and chemical reactivity.*

Kasturi, Prof Tirumali R. Organic Chemistry Dept., Indian Inst. of Sc., Bangalore 560012, India. (1931) PhD, organic chemistry (Bombay U., 1957). Prof.; Chairman, div. of chemical & biological sci. *Crystal structure, organic molecules.*

Kini, Mr Ullal Devappa. Liquid Crystals Lab., Raman Res. Inst., Bangalore 560006, India. (tel. 30124). *Liquid crystals (hydrodynamics).*

Kodandapani, Mr R. Molecular Biophysics Unit, Indian Inst of Sci., Bangalore 560012, India. (1961) MSc, physics (Karnataka U., 1983). Res. scholar. *Biological crystallography.*

Kohli, Dr Vijay Kumar. Crystal Growth & Characterization Sect., Nat. Physical Lab., New Delhi 110012, India. (1960) MSc, chemistry (Guru Nanak Dev U., 1981) Scient. officer. *Crystal growth, topography, instrumentation, X-ray diffraction techniques.*

Krishna, Prof. Padmanabhan. Physics Dept., Banaras Hindu U., Varanasi 221005, India. (1938) PhD, physics (Banaras Hindu U., 1962). Prof. (tel. 54291, ext. 371). *Crystal growth and perfection, phase transformation, polymorphism and polytypism, crystallography.*

Krishna Rao, Prof. K. V. Physics Dept., C. of Sci., Osmania U., Hyderabad 500007, India. (1920) PhD, X-ray crystallography (London U., UK, 1958). Prof. *X-ray crystallography, photoelastic effect, solid state physics.*

Krishnaswamy, Mr S. Dept. of Crystallography and Biophysics, U. of Madras, Madras 60025, India. (1959) MSc, physics (Madras U., 1981). Res. scholar. *Molecular biophysics, crystal structure, organic molecules.*

Krishnaiah, Mr Musali. Physics Dept., Sri Venkateswara U., Tirupati, Andra Pradesh 517502, India. (1950) MSc, physics (Sri Venkateswara U., 1974). *X-ray crystallographic investigations, Inorganic and organic compounds, phosphozenes and phosphorus compounds.*

Krishnan, Dr Rangachari. Metallurgical Div., Bhabha Atomic Res. Cntr., Bombay 400085, India. (1935) PhD, physical metallurgy (Bombay U., 1966). Group leader. (tel. 523321, ext. 336). *Phase transformations, electron microscopy.*

Kumar, Dr Rajendra. Nat. Metallurgical Lab., Jamshedpur 831007, India. (1929) PhD, metallurgy (Sheffield U., 1956). Scient. dir. *Crystal structure, metallic phases.*

Kumar, Mr Vinay Neutron Physics Div., Bhabha Atomic Res. Cntr., Bombay 400085, India. (1960) MSc, chemistry (Guru Nanak Dev U., 1981). Scient. officer. *Macromolecular crystallography.*

Kundra, Mr Krishan Dev. Specialized Techniques (X-rays) Div., Nat. Physical lab., Hillside Road, New Delhi 110012, India. (1934) MSc, physics (Panjab U., 1957). Scient. (tel. 584179). *Diffraction theory, inorganic structures, defects, experimental methods & apparatus.*

Kuppuswamy, Dr Nagarajan. Res. & Dev., Searle (India) Ltd., Thane-Bhelapur Road, Thane 400601, India. (1930) PhD, (Madras U., 1954). Dir. *Organic crystal structure.*

Lahari, Dr Barendra Nath. Physics Dept., Burdwan U., Burdasan, West Bengal, India. (1938) PhD, X-ray crystallography (Calcutta U., 1973). Lect. (tel. 341121). *X-ray crystallography, silicate minerals, textile fibres.*

Lal, Dr Krishan. Nat. Physical Lab., Hillside Road, New Delhi 110012, India. (1941) PhD, (Delhi U., 1969) Deputy dir. *Crystal growth, lattice imperfections, high resolution X-ray diffraction techniques, topography, multi-crystal X-ray diffractometry, diffuse X-ray scattering.*

Lele, Prof. Shrikant Dept. of Metallurgical Eng., Banaras Hindu U., Varanasi 221005, India. (1943) PhD, physical metallurgy (Banaras Hindu U., 1967). Prof., physical metallurgy. (tel. 64491, ext. 250). *Phase transitions.*

Lohar, Dr Jayanarayan Mangaliprasad. Appl. Chemistry Dept., M.S.U., Baroda 390001, India. (1928) PhD, liquid crystals (M.S.U., Baroda, 1961). Reader. *Liquid crystals.*

Lokanatha, Mr S. Physics Dept., Indian Inst. of Techn., Kharagpur 721302, India. (1958) MSc, (Bangalore U., 1981). Res. scholar. *Structure - properties relationship, minerals.*

Mahanta, Dr Bhubaneswar. Physics Dept., Ranchi U., Ranchi 834008, India. (1928) PhD, X-ray crystallography (Ranchi U., 1979) Res. scholar. *X-ray crystallography, organo-metallic compounds.*

Madhusudana, Dr N. V. Liquid Crystals Lab., Raman Res. Inst., Bangalore 560006, India. (1944) PhD, liquid crystals (Mysore U., 1971). Scient. (tel. 30124). *Liquid crystals.*

Mahata, Dr Akhil Physics Dept., Ranchi U., Ranchi 834008, India. (1939) PhD, X-ray crystallography (Ranchi U., 1978) Res. scholar. *X-ray crystallography, organo-metallic compounds.*

Maiti, Dr Gobinda Chandra. Physical Res. Wing, Projects & Dev. India Ltd., P.O. Sindri, Dist. Dhanbad 828122, India. (1944) PhD, chemistry (Calcutta U., 1977). Technologist. *Catalysis, crystal chemistry, heterogeneous catalysts, solid solution, structural changes.*

Majumdar, Mr Kanti Lal. Crystallography & Molecular Biology Div., Saha Inst. of Nuclear Physics, 92 A. P. C. Road, Calcutta 700009, India. (1936) MSc, crystallography (London U., UK, 1971). Res. fellow. (tel. 354281). *Biologically important compounds, amino acids.*

Majumdar, Dr Sunil Kumar. Crystallography & Molecular Biology Div., Saha Inst. of Nuclear Physics, 92 A. P. C. Road, Calcutta 700009, India. (1933) PhD, X-ray crystallography (Madras U., 1964). Lect. (tel. 354281). *X-ray crystallography, proteins, nucleic acid, biological molecules.*

Malik, Alpana. Crystal Growth and Characterization Sect., Nat. Physical Lab., New Delhi 110012, India. (1960) MPhil, physics (Delhi U., 1982). Res. scholar. *X-ray diffraction, diffuse X-ray scattering, crystal perfection.*

Mande, Prof. Chintamani. Physics Dept., Nagpur U., Nagpur 440010, India. (1925) DSc, X-ray spectroscopy (Paris U., France, 1958). Prof. and head. (tel. 31946). *X-ray spectroscopy, intermetallic compounds, chemical bonding, spinel structures.*

Mande, Mr Sekhar Chintamani. Molecular Biophysics Unit, Indian Inst. of Sci., Bangalore 560012, India. (1962) MSc, physics (Nagpur U., 1984). Res. scholar. *Molecular biophysics.*

Mani, Mr A. Materials Sci. Div., Nat. Aeronautical Lab., Bangalore 560017, India. (1950) BSc, physics (Madurai U., 1971). Sr. lab. asst. *Organic structures, crystallographic methods and apparatus.*

Manickkavachgam, Ramanathan. Sch. of Physics, Madurai Kamraj U., Madurai 625021, India. (1955) MSc, physics (Madras U., 1979). Res. scholar. *Crystal structure, organic and inorganic compounds, phase transition.*

Manohar, Dr Hattikudur. Dept. of Inorganic & Physical Chemistry, Indian Inst. of Sci., Bangalore 560012, India. (1929) PhD, physics (I.I.S., Bangalore, 1963). Prof. (tel. 34411, ext. 382). *Crystallography, bio-coordinated compounds, Cyclophosphazenes, organometallics; solid state reactions.*

Mathur, Dr Balbir Kumar. Physics Dept., Indian Inst. of Techn., Kharagpur 721302, India. (1948) PhD, physics (I.I.T., Kharagpur, 1978). Sr. scient. asst. *Lattice defects, optical transforms, computer programming.*

Mathur, Dr Rajendra Kumar. Material Sci. & Techn. Wing, Nat. Metallurgical Lab., Jamshedpur 831007, India. (1929) DMet, metallurgical engineering (Sheffield U., UK, 1974). Deputy dir. (tel. 4284). *Structure of rapidly solidified metallic phases, liquid metals, solidification, aluminium conductors, high temperature materials.*

Misra, Dr Nirmal Kumar. Dept. of Physics & Meteorology, Indian Inst. of Techn., Kharagpur 721302, India. (1939) PhD, physics (I.I.T., Kharagpur, 1967). Asst. prof. *Crystal growth, defect crystal structure, thin films, metallic glass.*

Misra, Prof. Somnath. Principal, Regional Eng. C., Rourkela 769008, India. (1936) ScD, metallurgy (MIT, USA, 1963). Principal. (tel. 5050). *Intermetallic compounds, order-disorder transformations.*

Misra, Dr Tripurari. Dept. of Physics, Regional Eng. C., Rourkela 769008, India. (1933) PhD, physics (Sambalpur U., 1973). Asst. Prof. *Molecular biophysics.*

Mitra, Prof. Girija Bhushan. C.S.S. Dept., Indian Assoc. for the Cultivation of Sci., Calcutta 700032, India. (1923) DSc, X-ray studies of lattice defects (Calcutta U., 1967). Em. scient. *Lattice defects, lattice vibrations, dynamical diffraction theory, material characterization, structure, instrumentation, small angle X-ray scattering, structure - properties relationships.*

Mohan Rao, Mr Vattipalli. Inorganic and Physical Chemistry Dept., Indian Inst. of Sci., Bangalore, Karnataka 560012, India. (1955) MSc, chemistry (Berhampur U., 1976). Res. scholar. (tel. 34411, ext. 382). *Magnetic and crystallographic studies, transition metal complexes, structural studies, solid state reactions.*

Mohanlal, Dr Sembu Krishnaiyer. Sch. of Physics, Madura Kamraj U., Madurai625021, India. (1940) PhD, X-ray crystallography (Madurai U., 1971). Reader. *X-ray diffraction, solid solutions, lattice defects, electron density distribution, instrumentation.*

Mukherjee, Dr Alok Kumar. Dept. of Physics, Jadavpur U., Calcutta 700032, India. (1950) PhD, physics (Visva-Bharati U., 1978). Lect. *Organic and organometallic structure, disorder.*

Mukherjee, Dr Amal Bikash Dept. of Geology & Geophysics, Indian Inst. of Tech., Kharagpur 721302, India. (1935) PhD, geology (I.I.T., Kharagpur, 1961). Prof. *Crystal growth, structure, minerals.*

Mukherjee, Dr Biswanath. X-ray Crystallography Dept., Central Glass & Ceramic Res. Inst., Calcutta 700032, India. (1936) PhD, physics (Calcutta U., 1965). Asst. dir. *Material science, crystal stability, structure.*

Mukherjee(mondal), Dr (Mrs) Monika. Dept. of Magnetism, Indian Assoc. for the Cultivation of Sci., Calcutta 700032, India. (1947) PhD, physics (Calcutta U., 1977). Res Asst. *Organic and organometallic structure, disorder, phase transition.*

Mukherjee, Dr Partha Sarathi. Materials Div., Regional Res. Lab., Trivandrum 695019, India. (1952) PhD, physics (I.I.T., Kharagpur, 1982). Scient. *Characterization, semicrystalline materials, polymers, minerals.*

Mukhopadhyay, Anuradha. Dept. of Magnetism, Indian Assoc. for the Cultivation of Sci., Calcutta 700032, India. (1955) MSc, physics (Calcutta U., 1977). Res. scholar. *Coordination complexes, organic compounds, disorder, phase transition.*

Mukhopadhyay, Mr Bishnu Prasad. Crystallography & Molecular Biology Div., Saha Inst. of Nuclear Physics, Sector-1 Block-AF, Bidhannagar, Calcutta 700064, India. (1954) MSc, chemistry (Burdwan U., 1977). Res. scholar. *Crystal structure, Biomolecules.*

Mukhopadhyay, Mr Pradip. Metallurgy Div., Bhabha Atomic Res. Cntr., Bombay 400085, India. (1943) PhD, physics (Bombay U., 1970). Scient. officer. (tel. 523321, ext. 242). *Phase transformation, defect structure - properties correlation, electron microscopy, electron diffraction.*

Munirathinam, Nethoji. Dept. of Crystallography & Biophysics, Madras U., Madras 60025, India. (1958) MSc (Madras U., 1981). Res. scholar. *Crystal structure, biomolecules.*

Munshi, Mr Sanjeeu Kumar. Molecular Biophysics Unit, Indian Inst. of Sci., Bangalore 560012, India. (1962) MSc, biochemistry (Kashmir U., 1984). Res. scholar. *Molecular biophysics.*

Muralidharan, Mr K. V. Chemistry Div., Bhabha Atomic Res. Cntr., Bombay 400085, India. (1937) MSc, physics (Bombay U., 1971). Scient. officer. (tel. 523321, ext. 288). *Biologically important structures.*

Murali, Mr R. Dept. of Crystallography & Biophysics, Madras U., Madras 60025, India. (1958) MSc, physics (Madras U., 1980). Res. scholar. *Molecular biophysics.*

Murthy, Mr G. S. Dept. of Organic Chemistry, Indian Inst. of Sci., Bangalore 560012, India. (1959) MSc, chemistry (Central U., Hyderabad, 1981). Res. scholar. *Organic chemical crystallography, reactions in solid state, molecular mechanics, functional groups - interaction in solid state.*

Murthy, Dr Mathur R. N. Molecular Biophysics Unit, Indian Inst. of Sci., Bangalore 560012, India. (1950) PhD, X-ray crystallography (I.I.S., 1977). Asst. prof. *Molecular biophysics.*

Nag, Dr Dilip Kumar. Mineral Physics Div., Geological Survey of India, 29 Chowringhee Road, Calcutta 700016, India. (1945) PhD, X-ray crystallography (Calcutta U., 1974). Mineralogist. (tel. 238321, ext. 8) *Organic and inorganic structures, mineral crystal structures, instrumentation, crystal physics.*

Nag, Miss Jhumjhumi. Dept. of General Physics & X-rays, Indian Assoc. for the Cultivation of Sci., Calcutta 700032, India. (1955) MSc physics (Calcutta U., 1976). Res. scholar. *Structural characterization, amorphous materials, thin films crystallography, X-ray diffraction.*

Nagabhushana Rao, Mr Chemboli. Metallurgy Div., Bhabha Atomic Res. Cntr., Bombay 400085, India. (1941) MTech, metallurgy (I.I.T., Kanpur, 1973). Scient. officer. *X-ray metallography, texture studies.*

Nagapal, Mr Kailash Chander. X-ray Sect., Nat. Physical Lab., Hillside Road, New Delhi 110012, India. (1932) MSc, physics (Delhi U., 1954). Scient. (tel. 584179). *X-ray crystallography, diffractometry, X-ray fluorescence analysis, experimental methods & apparatus, inorganic structures.*

Nagendra Nath, Prof. N. S. Parimala, Site No. 2363, II Stage, Rajaninagar, Bangalore 5610010, India. (1912) PhD, physics (Cambridge U., UK, 1939). Retired Prof. *Crystallography.*

Naik, Mrs Uma Murlidhar. Metallurgy Div., Bhabha Atomic Res. Cntr., Bombay 400085, India. (1940) MSc, X-ray metallography (Banaras Hindu U., 1961). Scient. officer. *X-ray metallography, electron microscopy studies.*

Naik, Mr Vaman Madhusudanrao. Inorganic & Physical Chemistry Dept., Indian Inst. of Sci., Bangalore 560012, India. (1950) MSc, physics (Karnatak U., 1973). Res. scholar. *Crystal structure, inorganic compounds.*

Nandi, Asok Kumar. X-ray Sect., Central Glass & Ceramic Res. Inst., Calcutta 700032, India. (1944) MSc, (Calcutta U., 1965). Scient. *Crystal structure, organic compounds, amorphous materials.*

Nandi, Dr Ranjan Kumar. Flat Products Group, Steel Authority of India Ltd., Ranchi, Bihar, India. (1950) PhD, physics (Calcutta U., 1978). Res. eng. *Crystal imperfections, thin film structure.*

Narasimha Murthy, Mr Mattur R. Organic Chemistry Dept., Indian Inst. of Sci., Bangalore 560012, India. (1950) MSc, physics (I.I.T., Madras, 1972). Res. fellow. (tel. 30441, ext. 403). *Structural crystallography, methods of crystallography, molecular forces.*

Narasimhan, Dr P. Physics Dept., A. M. Jain C., Madras 600061, India. (1944) PhD, physics (Madras U., 1984). Prof. *Crystal structure, organic compounds, dipeptides, small molecules.*

Narayan, Mr Ramesh. Materials Sci. Div., Nat. Aeronautical Lab., Bangalore 560017, India. (1950) MSc, physics (Bangalore U., 1973). Res. fellow. *Line profile analysis, scattering by liquids, amorphous solids, high pressure crystallography.*

Narayanan, Prof. Palamadi Sundaram. Physics Dept., Indian Inst. of Sci., Bangalore, Karanataka 560012, India. (1926) PhD, crystal physics (Madras U., 1951). Prof., chairman, div. of physics and math. sciences. (tel. 34411, ext., 469). *Ferroelectricity, structural phase transitions, crystal growth.*

Natarajan, Dr S. Physics Dept. Anna U., Madras 600025, India. (1941) PhD, physics (I.I.S., 1969). Prof. and Head. *Crystal structure, high pressure X-ray diffraction.*

Natarajan, Dr Subramanian. Sch. of Physics, Madurai Kamraj U., Madurai 625021, India. (1949) PhD, physics (Madurai Kamraj U., 1979). Lect. *Crystal structure, amine complexes, crystal growth.*

Natesan, Mr Elango. Dept. of Crystallography and Biophysics, Madras U., Madras 60025, India. (1957) MSc (Madras U., 1980). Res. scholar. *Theoretical X-ray crystallograpy, structures.*

Nath, Mr Kashi. Physics Dept., Lucknow U., Lucknow 226007, India. (1951) MSc, physics (Lucknow U., 1973). Res. fellow. (tel. 25140). *Conformation and conformational transitions, polypeptides and biomolecules.*

Nigli, Selina. Dept. of Physics & Astrophysics, U. of Delhi, Delhi 110007, India. (1949) MPhil (Delhi U., 1982). Res. scholar. *Crystal growth, X-ray diffraction, defects, physical properties.*

Noor, Sahina Begum. Inorganic and Physical Chemistry Dept. Indian Inst. of Sci., Bangalore 560012, India. (1960) MSc (Bangalore U., 1982). Res. scholar. *X-ray crystallography, metal nucleotide complexes.*

Padmanabhan, Dr V. M. Neutron Physics Div., Bhabha Atomic Res. Cntr., Trombay, Bombay 400085, India. PhD, crystallography (Bombay U.). *Crystallography, light scattering.*

Pahwa, Mr Des Raj. Lang Camera Group, Nat. Physical Lab., Hillside Road, New Delhi 110012, India. (1938) MSc, physics (Punjab U., 1964). Scient. 'B'. (tel. 581442). *X-ray topography, crystal growth, infrared spectroscopy.*

Pan, Dr Nitya Ranjan. Physics Dept., Calcutta U., 92 A. P. C. Road, Calcutta 700009, India. (1935) PhD, physics (Calcutta U., 1975). Lect. (tel. 359186). *Electret orientation & related properties.*

Pandey, Dr Dhananjai. Sch. of Material Sci. & Techn., Banaras Hindu U., Varanasi 221005, India. (1952) PhD, physics (Banaras Hindu U., 1976). Reader. *Structural imperfections, solid state transformations, ferroelectrics.*

Pandya, Prof. Janardhan Rameshchandra. Dept. of Physics, M.S.U., Baroda 390002, India. (1934) PhD, physics (M.S.U., 1961). Prof. *Crystal growth, dissolution (etch phenomenon), hardness, electrical conductivity.*

Pant, Dr Arun Kumar. Physics Dept. Gorakhpur U., Gorakhpur 273001, India. (1940) PhD, X-ray crystallography (Calcutta U., 1964). Lect. *Organic and inorganic structures, lattice dynamics.*

Pant, Dr Lalit Mohan. Physical Chemistry Div., Nat. Chemical Lab., Pashan Road, Poona 411008, India. (1928) PhD, crystallography (London U., UK, 1958). Scient. *Structure analysis, disorder in crystals.*

Papavinasam, Mr E. Dept. of Physics, Yadava C., Madurai 625014, India. (1943) MPhil, physics (Madurai Kamraj U., 1977). Prof. *Crystal growth, structure analysis.*

Parasnis, Prof. Arawind Shripad. Physics Dept., Indian Inst. of Techn., Kanpur 208016, India. (1928) PhD, crystal physics (Bristol U., UK, 1960). Prof. (tel. 40066). *Intercrystalline interfaces, surface free energies of crystals, dislocations, metal-insulator transition, electron microscopy.*

Parthasarathi, Dr V. Dept. of Crystallography & Biophysics, Madras U., A.C.C. Campus, Madras 600025, India. (1946) PhD, X-ray crystallography (Madras U., 1975). Postdoctoral res. fellow. (tel. 432248, ext. 16). *Theoretical crystallography, crystal structure analysis.*

Parthasarathy, Dr Soundarajan. Dept. of Crystallography & Biophysics, Madras U., A. C. C. Campus, Madras 600025, India. (1940) PhD, crystallography (Madras U., 1967). Prof. (tel. 432248, ext. 16). *Theoretical crystallography, crystal structure analysis.*

Patel, Prof. Ambalal Ranchhodbhai. Dept. of Physics, Sardar Patel U., Vallabh Vidyanager, Gujrat, India. (1917) PhD, physics (London U., UK, 1958). Retired prof. *Mineral crystals.*

Patel, Dr Prabhudas Revandas. Silical Lab., Sandhporepardi, Valsad 396001, India. (1938) PhD, liquid crystals (M.S.U., Baroda, 1967). Partner. (tel. 2281, 2154). *Structural changes, organic molecules, liquid crystals.*

Patel, Dr Tankadhar. Dept. of Physics, Regional Eng. C., Rourkela 769008, India. (1943) PhD, physics (I.I.T., Bombay, 1982). Lect. *Crystal structure, organic and inorganic substances.*

Pattanayek, Mrs Rekha Rani Crystallography & Molecular Biology Div., Saha Inst. of Nuclear Physics, Sector-18 Block-AF, Bidhannagar, Calcutta 700064, India. (1953) MSc, physics (I.I.S.). *Crystal structure, biomolecules.*

Pathak, Prof. Pushkrrai Dalpatram. Physics Dept., U. Sch. of Sci., Ahmedabad 380009, India. (1916) MSc, physics (Bombay U., 1946). Prof. (tel. 40929). *X-ray diffraction, defects, thermal properties.*

Pattabhi, Dr (Mrs) Vasantha. Crystallography & Biophysics, U. of Madras, A.C.C. Campus, Madras 600025, India. (1944) PhD, X-ray crystallography (Madras U., 1972). Lect. (tel. 432248, ext. 16). *Crystallography, bio-molecules, conformational analysis.*

Podder, Dr (Mrs) Aloka. Crystallography & Molecular Biology Div., Saha Inst. of Nuclear Physics, Sector-1 Block-AF, Bidhannagar, Calcutta 700064, India. (1938) PhD, physics (Calcutta U., 1977) Lect. *Biomolecular structure and conformation.*

Poojary, Dr M. Damodara Inorganic and Physical Chemistry Dept., Indian Inst. of Sci., Bangalore 560012, India. (1955) PhD, physics (I.I.S., 1984). Scient. Asst. *X-ray crystallography, metal-nucleotide complexes.*

Pradhan, Dukhabandhu. Dept. of Physics, Indian Inst. of Techn., Kharagpur, 721302, India. (1948) MSc, (Sambalpur U., 1971). Res. scholar. *Crystal structure, organic compounds, intensity statistics.*

Prasad, Dr Narayan, Physics Dept., Ranchi U., Ranchi 834008, India. (1944) PhD, X-ray crystallography (Ranchi U., 1978) Lect. *X-ray crystallography, organometallic compounds, hydrogen bonds.*

Prasad, Dr Ravindra. Geochronology & Isotope Geology Div., Geological Survey of India, 29 Jawaharlal Nehru Road, Calcutta 700016, India. (1943) PhD, defects (Banaras Hindu U., 1971). Mineralogist. (tel. 238321, ext. 30). *Defects in* *polytypic crystals, crystal growth, structure, phase transformations, stacking fault energies.*

Prasad, Dr Satya Murti. Physics Dept., Ranchi U., Ranchi 834008, India. (1943) PhD, X-ray crystallography (Ranchi U., 1970). Lect. *X-ray crystallography, organic and organo-metallic compounds.*

Prasad, Dr Y. R. Ananth. Materials Div., Nat. Physical Lab., Hillside Road, New Delhi 110012, India. (1942) PhD, semiconductors (I.I.T., Delhi, 1970). Scient. 'C'. (tel. 581455) *Growth of single crystals, ferroelectric crystals, diffusion technology, ion-implantation and microscopy.*

Prayaga, Dr Chandra Sekhar. Dept. of Physics, Indian Inst. of Sci., Bangalore 560012, India. (1948) PhD, physics (I.I.S., 1975). Asst. Prof. *Crystal growth, structure.*

Puranik, Dr (Mrs) Vedavati Gururaj. Physical Chemistry Div., Nat. Chemical Lab., Poona 411008, India. (1953) PhD, physics (Bangalore U., 1983). Res. Assoc. *Crystal structure, organic and inorganic molecules, organometallic molecules, biologically important molecules.*

Raghavacharyulu, Dr Iyyunni Venkata Veera. Nuclear Physics Div., Bhabha Atomic Res. Cntr., Bombay 400085, India. (1934) DSc, group theory & representation of space groups (Andhra U., 1958). Scientific Officer (E). (tel. 523321, ext. 352). *Mathematical crystallography, solid state physics.*

Raghunatha, Chary. Dept. of Physics, Indian Inst. of Sci., Bangalore 560012, India. (1955) MPhil, physics (Hyderabad U., 1979). Res. student. *Crystal growth, crystal structure.*

Raghurama, Mr G. Physics Dept., Indian Inst. of Sci., Bangalore 560012, India. (1958) MSc, physics (I.I.T., Madras, 1980). Res. scholar. *Crystal structure, crystal binding.*

Rajagopal, Mr Hariharasubramonia Iyer. Neutron Physics Div., Cirus Bhabha Atomic Res. Cntr., Bombay 400085, India. (1941) BSc, physics (Kerala U., 1962). Scient. officer. *Neutron diffraction, computer programming, software development.*

Rajan, Mr R. D. Dept. of Physics, Anna U., Madras 600025, India. (1945) MPhil, physics (Anna U., 1984). Lect. *Crystal structure, organic compounds.*

Rajan, Dr S. S. Crystallography & Biophysics Dept., Madras U., Madras 60025, India. (1949) PhD, physics (Madras U., 1977). Lect. *Crystal structure analysis, organic compounds, proteins.*

Rajaram, Dr Ramasamy Karunandam. Sch. of Physics, Madurai Kamraj U., Madurai 625021, India. (1951) PhD, (Madras U., 1978). Reader. *X-ray crystallography, hydrogen bonding, phase transitions.*

Raju, Dr K. S. Dept. of Crystallography & Biophysics, Guindy Campus, Madras U., Madras 600025, India. (1937) PhD, (Sardar Patel U., 1969). Lect. *Crystal growth.*

Raju, Dr I. V. K. Bhagaban. Dept. of Physics, Kakatiya U., Warangal 506009, India. (1946) PhD, physics (Osmania U., 1973). Reader. *Crystal growth, lattice defects plastic flow, materials.*

Ram, Dr Purushottam. Physics Dept., Ranchi U., Ranchi 834008, India. (1948) PhD, physics (Ranchi U., 1979). Lect. *X-ray crystallography, organo-metallic compounds.*

Rama Rao, Dr B. X-ray Crystallography Sect., Regional Res. Lab., Hyderabad 500009, India. (1930) Dr.rer.nat., mineralogy & crystallography (Göttingen, BRD, 1961). Asst. dir. *Organic and inorganic crystal structures, X-ray powder analysis.*

Ramachandran, Prof. Gopalasamudram N. Mathematical Philosophy Group, Indian Inst. of Sci., Bangalore 560012, India. (1922) DSc, crystallography (Madras U., 1949). INSA Albert Einstein Prof. *Theoretical crystallography.*

Ramakrishnan, Dr Chandrasekhara. Molecular Biophysics Unit, Indian Inst. of Sci., Bangalore 560012, India. (1939) PhD, biophysics (Madras U., 1966). Asst. prof. (tel. 34411, ext. 460). *X-ray crystallography, bio-molecules, computer programming.*

Ramanadham, Dr Muthyala. Neutron Physics Div., Bhabha Atomic Res. Cntr., Trombay, Bombay 400085, India. (1945) PhD, physics (Bombay U., 1975). Scient. 'E'. (tel. 5513848). *Biological crystallography, proteins, crystallographic computing, neutron diffraction, hydrogen bonding, X-ray diffraction.*

Ramaswami, Dr Krishnamachari. Physics Dept., Annamalai U., Annamalai Nagar 608101, India. (1935) PhD, molecular spectroscopy (Annamalai U., 1961). Head. *Bio-molecular structures, Raman spectroscopy, infrared spectroscopy, nuclear magnetic resonance, X-ray techniques.*

Ramesh, Mr Chellaswamy. Physics Dept., Madurai U., Madurai 625021, India. (1954) MSc, physics (Madurai U., 1975). Res. scholar. *X-ray diffraction, solid solutions.*

Ramkumar, Mr Sambatur. Organic Chemistry Dept., Indian Inst. of Sci., Bangalore 560012, India. (1949) MSc, physics (I.I.T., Madras, 1971). Res. scholar. *Crystallography, biomolecules, drug molecules, highly strained molecules.*

Ranganath, Dr G. S. Liquid Crystals Lab., Raman Res. Inst., Bangalore 560006, India. (1944) PhD, crystal optics (Bangalore U., 1974). Scient. (tel. 30124). *Crystal optics, liquid crystals.*

Ranganathan, Prof. Srinivasa. Dept. of Metallurgy, Indian Inst. of Sci., Bangalore 560012, India. (1941) PhD, metallurgy (Cambridge U., UK, 1965). Prof. *Crystal defects.*

Rao, Dr J. Krishna Mohana. Physics Dept., Madurai U., Madurai 625021, India. (1943) PhD, X-ray crystallography (I.I.S., Bangalore, 1971). Lect. *X-ray crystallography, phase transitions, hydrogen bonding, coordination chemistry, organic and inorganic materials, biological materials.*

Rao, Dr Keshavamurthy Narayana Swamy. Physics Dept. Indian Inst. of Techn., Kanpur 208016, India. (1934) PhD, crystal physics (Kanpur U., 1974). Sr. res. asst. (tel. 40066, ext. 30). *Crystal growth & characterization, crystal defects, intercrystalline boundaries.*

Rao, Dr (Miss) Leea Madhava. Physics Dept., Indian Inst. of Techn., Bombay 400076, India. (1929) PhD, Crystallography (London U., UK, 1958). Asst. prof. (tel. 581421). *Biological interactions, protein metal interactions, solid state physics.*

Rao, Mr N. Nagaraja. Atomic Minerals Div., Dept. of Atomic Energy, 1-11-252Å Begumpet, Hyderabad 500016, India. (1923) MSc, crystallography & mineralogy (Mysore U., 1949). Scient. officer-in-charge, X-ray Lab., (tel. 76152, ext. 1). *Mineral identification, inorganic crystal structure.*

Rao, Prof. P. Rama. Defense Metallurgical Res. Lab., Hyderabad, India. (1942) PhD, metallurgical eng. (Banaras Hindu U., 1968). Dir. *Structure, intermetallic phases.*

Rao, Mr Sudhakar S. P. Inorganic and Physical Chemistry Dept., Indian Inst. of Sci., Bangalore 560012, India. (1958) MSc, chemistry (Mysore U., 1980). Res. scholar. *Bioinorganic chemistry, X-ray crystallography.*

Rao, Usha K. Inorganic and Physical Chemistry Dept., Indian Inst. of Sci., Bangalore 560012, India. (1962) MSc, chemistry (Mangalore U., 1984). Res. scholar. *Biomolecular crystallography.*

Ratho, Prof. Trinath. Physics Dept., Regional Eng. C., Rourkela 769008, India. (1923) DPhil, physics (Graz U., Austria, 1959). Prof. and head. (tel. 5050). *X-ray crystallography, macromolecular structure, small angle X-ray scattering.*

Ratna, Miss B. R. Liquid Crystals Lab., Raman Res. Inst., Bangalore 560006, India. (1949) MTech, physical engineering (I.I.S., 1972). Res. fellow. (tel. 30124). *Liquid crystals.*

Ravi Kumar, Mr K. Dept. of Crystallography & Biophysics, U. of Madras, Madras 60025, India. (1958) MSc, physics (Madras U., 1981). Res. scholar. *Bio crystallography, crystal structure, organic molecules.*

Ravichandran, Mrs V. Dept. of Crystallography & Biophysics, U. of Madras, Madras 60025, India. (1959) MSc, physics (Madras U., 1981). Res. scholar. *Molecular biophysics.*

Ray, Dr (Mrs) Gouri. Radon House (P) Ltd., 7 Sirdar Sankar Road, Calcutta 700026, India. (1935) PhD, physics (Calcutta U., 1966). Managing dir. (tel. 461773). *Crystal physics, instrumentation.*

Ray, Dr Pankaj Narayan. Crystallography & Molecular Biology Div., Saha Inst. of Nuclear Physics, 92 A. P. C. Road, Calcutta 700009, India. (1934) PhD, physics (Calcutta U., 1974). Res. asst. (tel. 354281). *Amino acids, bio-compounds.*

Ray, Dr Pradip Kumar. Physics Div., Jute Techn. Res. Lab., Indian Council of Agricultural Res., 12, Regent Park, Calcutta 700040, India. (1938) PhD, fibre structure (Calcutta U., 1968). Scient. (tel. 468764). *Cellulose fibre structure.*

Ray, Dr Siddhartha. Dept. of Magnetism, Indian Assoc. for the Cultivation of Sci., Calcutta 700032, India. (1932) DSc, physics (Calcutta U., 1974). Reader. *Inorganic structures, phase transitions, disorder.*

Roychowdhury, Dr Priyobroto. Physics Dept., Calcuta U., 92 A.P,C. Road, Calcutta 700009, India. (1940) PhD, X-ray crystallography (Calcutta U., 1974). Lect. *Molecular structure.*

Ruban, G. Albert. Dept. of Crystallography & Biophysics, U. of Madras, Madras 60025, India. (1960) MSc, biophysics (Madras U., 1984). Res. scholar. *Crystal structure analysis.*

Saha, Mr Ajay Prakash. Physics Dept., Ranchi U., Ranchi 834008, India. (1951) MSc, physics (Ranchi U., 1972). Lect. *X-ray crystallography, organic compounds.*

Saha, Mr Bishwa Nath. Physics Dept., Ranchi U., Ranchi 834008, India. (1936) MSc, physics (Bihar U., 1960). Res, Scholar. *X-ray crystallography, organic and organometallic compounds.*

Saha, Prof. Narendra Nath. Crystallography & Molecular Biophysics Div., Saha Inst. of Nuclear Physics, Sector-3 Block-AF, Bidhannagar, Calcutta 700064, India. (1922) PhD, biophysics (Leeds U., UK, 1952) Retd. Prof. *Molecular biophysics, structure & function, biomolecules, drug molecules.*

Sahani, Dr Jiwan Lal. Surgery Dept., Inst. of Medical Sciences, Banaras Hindu U., Varanasi 221005, India. (1950) MS, surgery (Banaras Hindu U., 1976). Sr. resident. *Crystallographic studies, gallstones.*

Sahu, Dr Bholanath. Physics Dept., Ranchi U., Ranchi 834008, India. (1943) PhD, X-ray crystallography (Ranchi U., 1970). Lect. *X-ray crystallography, organic structures.*

Sahu, Dr Mahendra. Physics Dept., Ranchi U., Ranchi 834008, India. (1928) PhD, X-ray crystallography (Ranchi U., 1970). Lect. *X-ray crystallography, organic compounds.*

Sahu, Dr N. C. Dept. of Physics, Post Graduate Regional Eng. C., Rourkela 769008, India. (1937) PhD, physics (Sambalpur U., 1972). Dept. head. *Crystal structure, fibres, molecular biophysics.*

Sahu, Dr Ram Gopal. Dept. of Physics, G.A.L.C., Daltonganj 822102, India. (1934) PhD, physics (Ranchi U., 1968). Reader. *Crystal growth, crystal structure, organic crystals.*

Sahu, Dr Ramdhani. Physics Dept., Ranchi U., Ranchi 834008, India. (1927) PhD, X-ray crystallography (Ranchi U.1982) Lect. *Crystal structure determination.*

Sahaymary, Mrs J. James Dept. of Biophysics & Crystallography, A.C. C., Madras U., Madras 60025, India. (1958) MSc, physics (Madras U., 1981). Res. scholar. *Molecular biophysics.*

Salunke, Dr Dinakar M. Molecular Biophysics Unit, Indian Inst. of Sci., Bangalore 560012, India. (1955) PhD, molecular biophysics (I.I.S., 1983). Res. assoc. *Molecular biophysics.*

Samanta, Chitra. Dept. of Physics, Jadavpur U., Calcutta 700032, India. (1949) MSc, Applied mathematics (Jadavpur U., 1971). Res. scholar. *X-ray crystallography, molecular lattice dynamics, computer simulation.*

Samantaray, Dr Biswas Kumar. Dept. of Physics, Indian Inst. of Techn., Kharagpur 721302, India. (1947) PhD, physics (I.I.T., Kharagpur, 1977). Res asst. *Applied crystallography, thin films, mineralogy.*

Sankar, Mr B. N. Dept. of Physics, C. of Eng., Anna U., Madras 60025, India. (1951) MPhil physics (Anna U., 1983).Lect. *Crystal structure, organic compounds.*

Sarkar, Chitrita. Dept. of Physics, Jadavpur U., Calcutta 700032, India. (1951) MSc, physics (Jadavpur U., 1975). Res. scholar. *X-ray crystallography.*

Sarkar, Dr Satyabrata. Dept. of physics Jadavpur U., Calcutta 700032, India. (1944) PhD, physics (Jadavpur U., 1983). Techn. Superintendent. *X-ray crystallography, diffuse scattering, lattice dynamics, Structure solution, computer simulation.*

Sasisekharan, Prof. Visvanathan. Molecular Biophysics Unit, Indian Inst. of Sci., Bangalore 560012, India. (1933) PhD, crystallography & biophysics (Madras U., 1959). Prof. and chairman. (tel. 34411, ext. 458). *Crystal structure, conformation theory, quantum chemistry.*

Sastry, Dr Bommakanti Sri Rama. Industrial Ceramics Div., Regional Res. Lab., Hyderabad 500009, India. (1923) PhD, ceramic techn. (Pennsylvania State U., USA, 1958). Asst. dir. (tel. 71351). *High temperature technology, industrial ceramics, material science, phase equilibria.*

Sastry, Dr G. V. S. Dept. of Metallurgical Eng., Banaras Hindu U., Varanasi, 221005, India. (1952) PhD, metallurgical eng. (Banaras Hindu U., 1982) Reader. *Crystal structure, alloy metastable phase, electron difraction*

Sastry, Dr Medury Dattatreya. Radiochemistry Div., Bhabha Atomic Res. Cntr., Bombay 400085, India. (1942) PhD, physics (I.I.T., Kanpur, 1967). Scientific Officer. (tel. 523321, ext. 456). *Crystallographic studies, magnetic resonance, Mössbauer spectroscopy, X-ray diffraction.*

Satyanarayana Murthy, Dr Keta. Physics Dept., Nizam C., Hyderabad 500001, India. (1940) PhD, X-ray crystallography (Osmania U., 1971). Lect. (tel. 34231). *X-ray crystallography, solid state physics, biophysics.*

Savithramma, Miss K. L. Liquid Crystals Lab., Raman Res. Inst., Bangalore 560006, India. (1952) MSc, theoretical physics (Mysore U., 1974). Res. fellow. (tel. 30124). *Liquid crystals.*

Seal, Alpana. Magnetism Dept., Indian Assoc. for the Cultivation of Sci., Calcutta 700032, India. (1955) MSc, physics (Calcutta U., 1976). Res. scholar. *Coorination complexes, organic compounds, disorder, phase transition.*

Seal, Prof. Arun Kumar. Bengal Engineering C., Howrah 711103, India. (1928) PhD, physics (Sheffild U., 1956). Principal. *Crystal structure, metals.*

Seetharaman, Mr Venkataramakrishanan. Metallurgy Div., Bhabha Atomic Res. Cntr., Bombay 400085, India. (1950) BTech, metallurgy (I.I.T., Madras, 1971). Scient. officer. (tel. 523321, ext. 242). *Phase transformations, electron microscopy, microstrain, radiation damage.*

Sekar, Mr K. Dept. of Crystallography & Biophysics, U. of Madras, Madras 60025, India. (1961) MSc, biophysics (Madras U., 1984). Res. scholar. *X-ray crystallography.*

Sen, Mrs Nirupa. Dept. of Organic Chemistry, Indian Inst. of Sci., Bangalore 560012, India. (1957) MSc, chemistry (Gorakhpur U., 1980). Res. scholar. *Chemical crystallography, organic crystal structures, polarized twisted ethylenes, organic solid state reactions, chemical reactivity.*

Sen, Dr Deb Kumar. Mineral Physics Div., Geological Survey of India, 29 Jawaharlal Nehru Road, Calcutta 700016, India. (1942) PhD, X-ray crystallography (Calcutta U., 1972). Mineralogist (Jr.). (tel. 238321, ext. 8). *X-ray crystallography, organic compounds, minerals, electron microprobe analysis.*

Sen, Miss Mina. Physics Dept., Presidency C., Calcutta 700073, India. (1947) MSc, physics (Calcutta U., 1970). Lecturer in Vidyasagar C. (tel. 341121). *X-ray crystallography, organic and bio- molecules.*

Sen, Dr Ranjit Kumar Dept. of General Physics & X-rays, Indian Assoc. for the Cultivation of Sci., Calcutta 700032, India. (1919) DSc, physics (Allahabad U., 1956). Em. scient. *Organic and organometallic structures, lattice vibrations, Inelastic X-ray - phonon scattering.*

Sen, Mrs Suchitra. General Physics & X-rays Dept., Indian Assoc. for the Cultivation of Sci., Calcutta 700032, India. (1950) MSc, physics (I.I.T., Kharagpur, 1971). Sr. res. scholar. (tel. 469371, ext. GX). *Lattice imperfections, thin films.*

Sen Gupta, Prof. Siba Prasad. Dept. of General Physics and X-rays, Indian Assoc. for the Cultivation of Sci., Calcutta 700032, India. (1941) DPhil, physics (Calcutta U., 1968). Prof. and Head. *Organic and biological structures, lattice defects, thin films, crystal growth, characterization, topography, electron microscopy, amorphous materials.*

Senadhi, Mr Vijay Kumar. Physics Dept., Indian Inst. of Techn., Powai, Bombay 400076, India. (1952) MSc, X-ray crystallography (I.I.T. Bombay, 1974). Res. scient. (tel. 584141, ext. 287).

Sequeira, Dr Anisbert Stanislaus. Neutron Physics Sect., Bhabha Atomic Res. Cntr., Bombay 400085, India. (1938) PhD, crystallography (Bombay U., 1970). Scient. 'E'. (tel. 521441, ext. 221). *Neutron diffraction, biological crystallography, automation.*

Sharma, Mr Braj Bhushan. Solid State Physics Lab., Ministry of Defence, Lucknow Road, Delhi 110007, India. (1941) MSc, physics (Agra U., 1961). Sr. scient. officer. (tel. 226919). *X-ray studies of semiconductors, X-ray topography.*

Sharma, Dr Surinder Dutt. Crystal Growth Characterization Sect., Nat. Physical Lab., New Delhi 110012, India. (1947) PhD, physics (Allahabad U., 1975). Scient. *Crystal growth, X-ray topography.*

Shashidhar, Dr R. Liquid Crystals Lab., Raman Res. Inst., Bangalore 560006, India. (1946) PhD, liquid crystals (Mysore U., 1972). Scient. (tel. 30124). *Liquid crystals.*

Shivaprakash, Dr N. C. Instrumentation & Service Unit, Indian Inst. of Sci., Bangalore 560012, India. (1955) PhD, physics (Mysore U., 1982). Scient. *Liquid crystals, structure.*

Shrivastava, Prof. Hari Narayan. Chemistry Dept., Indian Inst. of Techn., Bombay 400076, India. (1927) PhD (Glasgow U., UK, 1960). Prof. (tel. 581421, ext. 416). *X-ray crystallography, natural products, acid salts, electron spin resonance, infrared spectroscopy.*

Sikka, Dr Satinder Kumar. Neutron Physics Div., Bhabha Atomic Res. Cntr., Bombay 400085, India. (1942) PhD, physics (Bombay U., 1970). Scient. 'E'. (tel. 523848, ext. 211). *High pressure crystallography, phase problem, biostructures, computations.*

Singh, Dr Anil Kumar. Materials Sci. Div., National Aeronautical Laboratory, Bangalore 560017, India. (1939) PhD, Physics (I.I.T., Madras, 1966). Scient. *Instrumentation, high pressure crystallography.*

Singh, Dr Bachchan. Laser Div., Defence Sci. Lab., Matcalfe House, Delhi 110006, India. (1930) PhD, physics (Pennsylvania State U., USA, 1962). Principal Scient. officer. (tel. 221521, ext. 28). *Crystal growth & evaluation.*

Singh, Dr Bhanu Pratap. High Pressure Techn. Div., Nat. Physical Lab., New Delhi 110012, India. (1947) PhD, physics (Delhi U., 1980). Scient. *Defect characterization, diffuse X-ray scattering, high pressure phase transformation.*

Singh, Dr Govind. Physics Dept., Banaras Hindu U., Vanarasi 221005, India. (1940) PhD, physics (Banaras Hindu U., 1967). Reader. *X-ray crystallography, crystal growth & imperfections, electron microscopy.*

Singh, Mr K. D. P. Atomic Minerals Div., Dept. of Atomic Energy, X-ray Diffraction Lab., AMD complex, Begumpet, Hyderabad 500016, India. (1944) MSc, (Banaras Hindu U., 1967). Scient. officer. *Inorganic crystal structure.*

Singh, Mr Surendra Prakash. Physics Dept., Ranchi U., Ranchi 834008, India. (1954) MSc, physics (Ranchi U., 1976). Res. scholar. *X-ray crystallography, organo-metallic compounds.*

Singru, Prof. Ramesh Madhao. Physics Dept., Indian Inst. of Techn., Kanpur 208016, India. (1935) PhD, physics (Purdue U., USA, 1963). Prof. (tel. 40066). *Electron momentum density.*

Sinha, Dr Umesh Chandra. Physics Dept., Indian Inst. of Techn., Bombay 400076, India. (1936) DPhil, X-ray crystallography (Allahabad U., 1961). Asst. prof. *X-ray diffraction, crystallography, instrumentation.*

Sirdeshmukh, Dr Dinker. Physics Dept., Kakatiya U., Warangal 506009, India. (1935) PhD, X-ray crystallography (Osmania U., 1964). Prof. (tel. 7403). *X-ray crystallography, crystal growth, defects, thermal & mechanical properties.*

Sivakumar, Mr K. Dept. of Physics, C. of Eng., Anna U., Madras 60025, India. (1960) MSc, physics (Anna U., 1982). Res. scholar. *Crystal structure determination.*

Soman, Dr K. V. Molecular Biophysics Unit, Indian Inst. of Sci., Bangalore 560012, India. (1954) PhD, biophysics (I.I.S., 1984). Res. scholar. *Molecular biophysics.*

Srinivasa, Dr Vishwanathapuram Kalasa. Physical Res. Wing, Projects & Dev. India Ltd., P.O Sindri, Dhanbad 828122, India. (1934) PhD, physics (Indian Sch. of Mines, Dhanbad, 1978). Deputy superintendent. (tel. 2613, telex: 0629-215-216 FPDIL) *High & Low temperature X-ray diffraction, powder diffraction techniques, X-ray cameras and instrumentation, fertilizer materials, nitrogen fixation, solid solutions, double salts, adducts, stress analysis, phase transformation.*

Sreekantan, Dr Arunachalam. Inorganic and Physical Chemistry Dept., Indian Inst. of Sci., Malleswaram, Bangalore, Karnataka 560012, India. (1953) PhD, chemistry (I.I.T., 1980). Post doctoral fellow. (tel. 34411, ext. 382). *Spectroscopy, crystallography, nucleo base nucleoside and nucleotide coordination compounds.*

Srinivasan, Dr Annankoil Renganatha. Dept. of Crystallography & Biophysics, U. of Madras, A.C.C. Campus, Madras 600025, India. (1946) PhD, biomolecular physics (Madras U., 1974). Lect. (tel. 432248, ext. 16). *Conformation of biomolecules, crystallography.*

Srinivasan, Prof. R. Dept. of Crystallography & Biophysics, U. of Madras, A.C.C. Campus Madras 600025, India. (1933) PhD, physics (Madras U., 1958). Sr. prof. and head. (tel. 432248, ext. 16). *X-ray diffraction, statistical application, protein crystallography, nuclear magnetic resonance (wide-line).*

Srinivasan, Dr Sampat. Radiochemistry Div., Bhabha Atomic Res. Cntr., Bombay 400085, India. (1943) PhD, inorganic chemistry (I.I.T., Madras, 1971). Scient. officer. (tel. 523321, ext. 456). *Crystal chemistry, actinides, oxides and carbides.*

Sriramadas, Prof. Aluru. Geology Dept., Andhra U., Visakhapatnam 530003, India. (1925) PhD, geology (Harvard U., USA, 1955). Sr. prof. and head. (tel. 4627). *Crystal chemistry, crystal optics, silicates, pyroxenes.*

Srivastava, Dr Ramesh Chandra. Physics Dept., Indian Inst. of Techn., Kanpur 208016, India. (1936) DPhil, physics (Allahabad U., 1960). Asst. prof. *Organic & inorganic structures, metal oxides, diffuse X-ray scattering.*

Srivastava, Dr Surendra Nath. Physics Dept., Allahabad U., Allahabad 211002, India. (1932) DPhil, X-ray crystallography (Allahabad U., 1963). Lect. *Liquid state.*

Subhadra, Dr K. G. Physics Dept., Kakatiya U., Warangal 506009, India. (1947) PhD, physics (Osmania U., 1976). Reader. *Crystal growth, inorganic crystal structure, chemical crystallography.*

Subramanian, Dr K. Physics Dept., C. of Eng., Anna U., Madras 60025, India. (1947) PhD, physics (Madras U., 1983). Lect. *Crystal structure determination.*

Sudhakar, Mr Varanasi. Molecular Biophysics Unit, Indian Inst. of Sci., Bangalore 560012, India. (1952) MSc, physics (I.I.T., Kharagpur, 1974). Res. scholar. (tel. 34411, ext. 337). *X-ray crystallography, amino acids, co-enzymes, molecular evolution, proteins.*

Sundaramoorthy, Mr M. Molecular Biophysics Unit, Indian Inst. of Sci., Bangalore 560012, India. (1960) MSc, material science (Anna U., 1983). Res. scholar. *Molecular biophysics*

Suresh, Mr C. G. Molecular Biophysics Unit, Indian Inst. of Sci., Bangalore 560012, India. (1955) post MSc, Dip. biophysics (I.I.S., 1980). Res. scholar. *Molecular biophysics.*

Suresh, Mr K. A. Liquid Crystals Lab., Raman Res. Inst., Bangalore 560006, India. (1948) MSc, solid state physics (Mysore U., 1969). Res. fellow. (tel. 30124). *Liquid crystals.*

Suri, D. K. X-ray Sect., Material Characterization Div., Nat. Physical Lab., New Delhi 110012, India. (1953) MSc, physics (Meerut U., 1972). Scient. *Crystal structure, phase transformation, inorganic materials.*

Suryanarayana, Dr Challapalli. Dept. of Metallurgical Eng., Banaras Hindu U., Varanasi 221005, India. (1945) PhD, physical metallurgy (Banaras Hindu U., 1970). Reader. (tel. 54290, ext. 250). *Crystallography, electron microscopy, defects.*

Suryanarayana, Dr Shambhuni V. Physics Dept., C. of Sci., Osmania U., Hyderabad 500007, India. (1943) PhD, X-ray crystallography (Osmania U., 1971). Lect. (tel. 71251, ext. 242). *X-ray crystallography, thermal expansion, Debye temperatures, alloy phases.*

Suta, Elizabeth Dept. of Physics, Indian Inst. of Sci., Bangalore 560012, India. (1959) MSc, physics (Kerala U., 1983) Scient. asst. *Crystal growth, characterization.*

Talapatra, Dr S. K. Dept. of Physics, Jadavpur U., Calcutta 700032, India. (1930) PhD, physics (Calcutta U., 1968). Reader. *X-ray crystallography, molecular lattice dynamics, computer simulations.*

Talukdar, Dr Amarendra Nath. Dept. of Physics, Gauhati U., Gauhati 781014. (1934) PhD, physics (Gauhati U., 1974). Reader. *Crystal structure analysis.*

Tavale, Dr Sudam Shankar. Physical Chemistry Div., Nat. Chemical Lab., Pushan Road, Poona 411008, India. (1934) PhD, X-ray crystallography (Poona U., 1964). Scient. 'C'. *X-ray crystallography, bio-compounds.*

Tewari, Dr Raghavendra. Computer Cntr., Aligarh Muslim U., Aligarh 202001, India. (1945) PhD, X-ray crystallography (I.I.T., Kanpur, 1973). Lect. (tel. 4809). *X-ray crystallography.*

Trigunayat, Prof. Govind Chandra. Dept. of Physics, U. of Delhi, Delhi 110007, India. (1936) PhD, physics (Delhi U., 1960). Prof. *Crystal growth, defects, polytypism.*

Vadakkanthara, Gopalakrishnan Thailambal. Dept. of Crystallography & Biophysics, Madras U., Madras 60025, India. (1959) MSc, (I.I.T., Madras, 1981). Res. scholar. *Crystal structure, biomolecules.*

Vani, Miss G. V. Liquid Crystal Lab., Raman Res. Inst., Bangalore 560006, India. (1949) MSc, solid state physics (Mysore U., 1970). Res. fellow. (tel. 30122, ext. 25). *Liquid crystals.*

Vasanth, Dr K. L. Chemistry Dept., PSG C. of Techn., Coimbatore 641004, India. (1940) PhD, chemistry (M.S.U., Baroda, 1968). Prof. and head. *Liquid crystals, metal & alloy electrodeposit structures.*

Vedavati, Miss B. M. Materials Sci. Div., Nat. Aeronautical Lab., Bangalore 560017, India. (1953) MSc, physics (Bangalore U., 1974). Res. fellow. *Crystallographic methods, programming, organic & bio-structures.*

Veerapandian, Mr B. Molecular Biophysics Unit, Indian Inst. of Sci., Bangalore 560012, India. (1951) MSc, physics (Annamalai U., 1975). Res. scholar. *Molecular biophysics*

Velmurugan, Dr D. Dept. of Crystallography & Biophysics, Madras U., Madras 60025, India. (1954) PhD, physics (Madras U., 1985). Lect. *Theoretical crystallography, structure, X-ray crystallography.*

Venkatesan, Prof. Kailash. Organic Chemistry Dept., Indian Inst. of Sci., Bangalore 560012, India. (1932) PhD, X-ray crystallography (Madras U., 1959). Prof. (tel. 34411, ext. 403). *X-ray crystallography, bio-molecules, strained ring organic molecules, crystallography methods.*

Venkatarama, Mr R. Dept. of Crystallography & Biophysics, Madras U., Madras 60025, India. (1954) MSc, physics (Madras U., 1975). Res. asst. *Molecular biophysics.*

Venkatesan, S. Dept. of Crystallography & Biophysics, Madras U., Madras 60025, India. (1959) MSc, physics (Madras U., 1981). Res. scholar. *Molecular biophysics.*

Venkatasubramanian, Dr K. Diffraction and Computer Facilities, Central Salt & Marine Chemicals Res. Inst., Bhavnagar 364002, India. (1938) PhD, (Calcutta U., 1984). Scient. *Crystal structure, organic and organometallic coordination compounds, theoretical methods, molecular biophysics.*

Venkobarao, Mr H. N. Instrumentation Div., Central Electrochemical Res. Inst., Karaikudi 623006, India. (1925) MSc, physics (Mysore U., 1949). Scient. in charge. *X-ray crystallography, instrumentation, solid state physics.*

Venugopalan, Dr S. Liquid Crystals Lab., Raman Res. Inst., Bangalore 560006, India. (1944) PhD, solid state spectroscopy (Purdue U., USA, 1973). Scient. (tel. 30124). *Liquid crystals.*

Verma, Dr Ajit Ram Nat. Physical Lab., Hillside Road, New Delhi 110012, India. (1921) DSc, physics (London U., 1969). Em. scient. *Crystal growth, lattice imperfection.*

Vijayakar, Mr Suresh Jaywant. Metallurgy Div., Bahbha Atomic Res. Cntr., Bombay 400085, India. (1931) MSc, chemistry (Bombay U., 1968). Scient. officer. (tel. 523321, ext. 242). *Phase transformations, electron microscopy.*

Vijayalaksmi, Mrs J. Dept. of Crystallography & Biophysics, Madras U., Madras 60025, India. (1950) MSc, physics (Madras U., 1971). Res. scholar. *Molecular biophysics.*

Vijayan, Dr (Mrs) Kalyani. Materials Sci. Div., Nat. Aeronautical Lab., Bangalore 560017, India. (1942) PhD, X-ray crystallography (I.I.S., 1969). Scient. 'B'. *Crystallographic methods, organic & bio-structures, liquid crystals.*

Vijayan, Prof. Mamannamana. Molecular Biophysics Unit, Indian Inst. of Sci., Bangalore 560012, India. (1941) PhD, physics (I.I.S., 1967). Prof. (tel. 34411, ext. 337). *Biological crystallography, molecular biophysics.*

Viswamitra, Prof. Mysore Ananthamurthy. Physics Dept., Indian Inst. of Sci., Bangalore 560012, India. (1932) PhD, X-ray crystallography (I.I.S., 1960). Prof. *Biological crystallography, high temperature crystallography.*

Vora, Dr Rasiklal Amulakhbhai. Appl. Chemistry Dept., Faculty of Techn. & Eng., M.S.U., Baroda 390001, India. (1938) PhD, liquid crystals (M.S.U., Baroda, 1975). Lect. *Liquid crystals.*

Vyas, Mr K. Inorganic & Physical Chem. Dept., Indian Inst. of Sci., Bangalore 560012, India. (1958) MSc, Ed. chemistry (Mysore U., 1980). Res. scholar. *Crystallography, small molecules, topochemistry, solid state reactions.*

Wadhawan, Dr Vinod Kumar. Neutron Physics Sect., Bhabha Atomic Res. Cntr., Bombay 400085, India. (1944) PhD, physics (Bombay U., 1976). Scient. officer. *Organic & inorganic crystallography, computations, phase transitions, high pressure crystallography.*

Yadav, Mr Asheshwar. Physics Dept., Ranchi U., Morabadi Campus, Ranchi, Bihar 834008, India. (1945) MSc, X-ray crystallography (Ranchi U., 1966). Lect. *Structure determination, X-ray diffraction methods.*

Yadav, Dr Vijay Singh. Neutron Physics Div., Bhabha Atomic Res. Cntr., Bombay 400085, India. (1942) PhD, crystallography (Bombay U., 1974). Scient. officer. (tel. 523321, ext. 265). *Organic & bio-molecular crystallography, X-ray & neutron diffraction.*

Yadav, Mr Tapaswi. Physics Dept., Ranchi U., Morabadi Campus, Ranchi, Bihar 834008, India. (1939) MSc, X-ray crystallography (Ranchi U., 1967). Lect. *Crystal structure determination, X-ray diffraction techniques.*

Yadava, Dr Bishwanath Physics Dept., Ranchi U., Ranchi 834008, India. (1935) PhD, X-ray crystallography (Ranchi U., 1981) Res. scholar.

INDONESIA

Sub-Editor: W. Loeksmanto

Notes

1. International telephone country code - 62.

Amilius, Mr Zuharli. Physics Lab., Bandung Reactor Centre, Jln. Tamansari 71, Bandung, West Java, Indonesia. (1936) MSc, experimental physics (Bandung Inst. of Techn., 1964). Head. (tel. 82799, ext. 24). *Molecular spectroscopy with neutrons, magnetic neutron diffraction.*

Baraba, Miss Widad. Ceramics Res. Centre, Jln. Jenderal A. Yani 392, Bandung, West Java, Indonesia. (1946) MSc, chemical engineering (Gajah Mada U., 1977). Res worker. *Ceramic raw materials, X-ray diffraction, X-ray fluorescence.*

Djonoputro, Mr Bernard Darmawan. Dept. of Physics, Bandung Inst. of Techn., Jalan Ganesya 10, Bandung, West Java, Indonesia. (1925) MSc, solid state physics (Northwestern U., USA, 1963). Lect. *X-ray diffraction.*

Gandadisastra, Mr Samsa. Lab. and Experimental Div., Mineral Techn. Dev. Centre, Jln. Jenderal Sudirman 623, Bandung, West Java, Indonesia. (1932) MSc, mineral processing (Hokkaido U., Japan, 1964). Head. (tel. 613 + 483-86 ext. 139). *Mineral studies, X-ray diffraction, X-ray spectrometry.*

Kartiwa, Mr Sumadi. Lab. and Experimental Div., Mineral Techn. Dev. Centre, Jln. Jenderal Sudirman 623, Bandung, West Java, Indonesia. (1945) MSc, analytical chemistry (New South Wales U., Australia, 1977). *Mineral analysis, X-ray spectrometry.*

Loeksmanto, Dr Waloejo. Dept. of Physics, Bandung Inst. of Techn., Jalan Ganesya 10, P.O.Box 273, Bandung, Indonesia. (1933) D.3e cycle, structural chemistry (U. Sciences et Techn., Languedoc(USTL), France, 1979). Assoc. prof. *Structure - electrical conductivity relationship, one dimensional conductors, solid electrolytes - LISICON.*

Marsongkohadi, Mr. Dept. of Physics, Bandung Inst. of Techn., Jalan Ganesya 10, Bandung, West Java, Indonesia. (1932) MSc, experimental physics (Bandung Inst. of Techn., 1958). Sr. lect. (tel. 82943). *Structural studies, neutron diffraction, neutron scattering, condensed matter.*

Muliawati, Mrs G. S. Physics Dept., Faculty of Science, U. of Indonesia, Jln. Salemba 4, Jakarta, Indonesia. (1944) MSc, physics and engineering science (1979). Lect. *Structural studies, X-ray diffraction.*

Parangtopo, Dr. Physics Dept., Faculty of Science, U. of Indonesia, Jln. Salemba 4, Jakarta, Indonesia. (1935) PhD, physics and math. (Lomonosov U., USSR, 1971). Sr. lect. *X-ray diffuse scattering, phonon dispersion, alloys, phase transitions.*

Sartono, Mr. Dept. of Physics, Bandung Inst. of Techn., Jalan Ganesya 10, Bandung, West Java, Indonesia. (1955) MSc, physics (1979). Jr. lect. *Physics of materials.*

Soepono, Mr R. Physics Dept., Faculty of Science, Gadjah Mada U., Bulaksumr, Yogyakarta, Central Java, Indonesia. (1928) MSc, physics (Nebraska U., USA, 1979). Sr. lect. *Solid state physics, X-ray diffraction.*

Wirjosoedirdjo, Mr Hardjatmo. Lab. and Experimental Div., Mineral Techn. Dev. Centre, Jln. Jenderal Sudirman 623, Bandung, West Java, Indonesia. (1942) BE, mining exploration (Acad. of Mining and Geology, 1974). *Mineral analysis, X-ray diffractometry.*

Wirjosumarto, Dr Harsono. Dept. of Mechanical Eng., Bandung Inst. of Techn., Jalan Ganesya 10, Bandung, West Java, Indonesia. (1935) PhD, metallurgy (Kentucky U., USA, 1973). Sr. lect. *Physical and mechanical metallurgy.*

IRAN

Sub-Editor: E. Arzi

Notes

1. International telephone country code - 98.

2. At the Iranian universities persons with an academic training can be appointed to the following academic positions: *Professor* (Prof.), *Associate Professor* (Assoc. prof.), *Assistant Professor* (Asst. prof.) and *Instructor*.

3. Degrees conferred by Iranian Universities in science and technology are BSc and MSc. Degrees conferred by other countries such as the UK, Germany, etc. are quoted in their original form. The descriptions of such other degrees can be found in the *Notes* of the appropriate country.

Alavi, Dr Mehdi. Chemistry Dept., Faculty of Sci., U. of Isfahan, University Street, Isfahan, Iran. (1936) Dr rer.nat., mineralogy and crystallography (Techn. U. München, BRD, 1969). Asst. prof. (tel. 031 + 40060-69, ext. 267). *X-ray analysis, minerals.*

Amighian, Dr Jamshid. Physics Dept., Faculty of Sci., U. of Isfahan, University Street, Isfahan, Iran. (1945) PhD, solid state physics (U. Durham, UK, 1975). Asst. prof. (tel. 031 + 40060-69 ext. 270). *Magnetic and electrical properties, single crystals.*

Arzi, Dr Ezatollah. Physics Dept., U. of Tehran, P.O. Box 11365-7693, Tehran, Iran. (1944) PhD, solid state physics (Queen Mary C., U. of London, UK, 1975). Asst. prof. (tel. 021 + 635776, ext. 19) *Neutron diffraction, X-ray diffraction, low temperature techniques, crystallographic computing.*

Asadi, Dr Parviz. Physics Dept., Sharif U. of Techn., P.O.Box 3406, Tehran, Iran. (1921) DSc, solid state physics (Sorbonne, France, 1967). Assoc. prof. (tel. 021 + 972001-9 ext. 164). *Electrical conductivity, crystalline structure.*

Azarnia, Dr Nezhat. Dept. of Basic Sci., Tehran Polytechnic, Hafez Avenue, Tehran, Iran. (1945) PhD, crystallography (U. Pittsburgh, USA, 1971). Asst. prof., chemistry. *Carbohydrates, biological molecules, vitamins, heavy atom molecules.*

Banaii, Dr Nasser. Atomic Energy Organization of Iran, Nuclear Research Center, P.O. Box 11365-8486, Tehran, Iran. (1941) PhD, Solid State Physics (Ghent State U., Belgium, 1976) Asst. prof., spectroscopy div. head. (tel. 021 + 638025-28, ext. 284-279, telex 212165). *Color centers, crystal growth, X-ray diffraction, topography, rocking curve.*

Beglari, Dr Ing. Parviz. Geology Dept., Faculty of Sci., U. of Azarabadegan, Tabriz, Iran. (1934) Dr.Ing., petrography (T.H. Aachen, BRD, 1975). Asst. prof. *Petrography.*

Etemadi-Abdolabadi, Dr Bijan. Geology Dept., C. of Arts and Sci., Shiraz U., Shiraz, Iran. (1951) PhD, crystal structure (Birkbeck C., U. London, UK, 1981). Asst. prof. *Structure analysis, small molecules, large molecules.*

Forughi, Dr Ali-Assghar. Geology Dept., Faculty of Sci., U. of Azarabadegan, Tabriz, Iran. (1935) Dr rer.nat, geology (U. Mainz, BRD, 1974). Asst. prof. SiO_2-Al_2O_3-P_2O_5 *system.*

Gourang, Dr Mansour. Geology Dept., Faculty of Sci., U. of Isfahan, Hezardjarib, Isfahan, Iran. (1938) PhD, mineralogy and petrography (U. Vienna, Austria, 1972). Asst. prof. (tel. 031 + 40061-9 ext. 271). *X-ray crystallography, differential thermal analysis.*

Khandar-Shahabad, Dr Ali-Akbar. Chemistry Dept., Faculty of Sci., Tabriz U., Daneshgah Street, Tabriz, Iran. (1951) PhD, X-ray crystallography (U. East Anglia, UK, 1980). Asst. prof. (tel. 041 + 30081, ext. 227). *Crystal structure analysis, phase problem.*

Manutchehr-Danai, Dr Mohsen. Geology Dept., Faculty of Sci., U. of Meshed, Asrar, Meshed, Iran. (1939) Dr, mineralogy (U. Mainz, BRD, 1968). Prof. (tel. 051 + 35468). *Inorganic crystal structures.*

Modjtahedi, Dr Mansour. Geology Dept., Faculty of Sci., U. of Azarabadegan, Tabriz, Iran. (1946) PhD, geology (U. Vienna, Austria, 1973). Asst. prof. *Optical crystallography.*

Noorbehesht, Dr Iradj. Geology Dept., Faculty of Sci., U. of Isfahan, Hezardjerib, Isfahan, Iran. (1943) Dr rer.nat., mineralogy, crystallography (Darmstadt, BRD, 1975). Asst. prof. (tel. 031 + 40061-9 ext. 271). *X-ray crystallography.*

Rostami, Dr Farzaneh. Geology Dept., Faculty of Sci., U. of Azarabadegan, Khomeini Street, Tabriz, Iran. (1935) PhD, geology (U. Vienna, Austria, 1972). Asst. prof. (tel. ext. 22397). *Mineralogy, crystallography.*

Rosukhi, Dr Siamak. Geology Dept., Jundi Shapur U., Ahwaz, Iran. (1938) PhD, mineralogy and petrography (U. Vienna, Austria, 1970). Assoc. prof. *Structural crystallography.*

Saadi-Nam, Dr Abolfazle. Chemistry Dept., Faculty of Sci., U. of Birjand, Khorasan, Iran. (1933) D.3e, crystallography (U. Bordeaux, France, 1974). Asst. prof. (tel. 0561 + 4804, 7044, ext. 35). *Crystal structure analysis.*

Safa, Dr Mehdi. Dept. of Physics, Isfahan U. of Techn., Isfahan, Iran. (1946) PhD, physics (U. Durham, UK, 1978). Lecturer. *Solid state physics, crystal growth, perfection and domains of crystals, X-ray topography.*

Schams-Ashtiani, Dr Ahmad. Petrology Div., Ministry of Roads, Mechanical & Soil Lab., North Amirabad, Tehran, Iran. (1939) PhD, geochemistry and petrography (U. Mainz, BRD, 1973). Geochemist and petrographer. (tel. 630031 ext. 37). *Crystal chemistry, petrography.*

Tajabor, Dr Nasser. Physics Dept., Faculty of Sci., U. of Mashhad, Mashhad, Iran. (1947) PhD, X-ray topography (U. Essex, UK, 1976). Asst. prof. (tel. 051 + 30131, ext. 64). *X-ray topography, X-ray crystallography, crystal growth.*

IRAQ

Sub-Editor: N.N. Rammo

Notes

1. International telephone country code - 964.

Alabdulla, Dr Ihsan Abdul Ghani. Chemistry Dept., C. of Sci., U. of Mosul, Mosul, Iraq. (1944) PhD, inorganic structural chemistry (Leeds U., UK, 1974). Lect. *Structure determination, transition metal complexes, X-ray diffraction methods.*

Al-Karaghouli, Dr Abdulrazzak Hamoudi. Chemistry Dept., Nuclear Res. Inst., Tuwaitha, Baghdad, Iraq. (1941) PhD, inorganic chemistry (Southampton U., UK, 1970). Res. *Structural chemistry, X-ray diffraction, neutron diffraction, metal ion complexes, biologically interesting compounds.*

Ibrahim, Dr T. K. Atomic Energy Commission, P.O. Box 765, Tuwaitha-Baghdad, Iraq.

Mahmoud, Dr Mouyed Mohamed. Physics Dept., C. of Sci., Al-Mustansiryha U., Baghdad, Iraq. (1947) PhD, physics (Nottingham U., UK, 1976). Lect. *Crystallography, crystal structure analysis.*

Obaid, Dr Yassin Najim. Physics Dept., C. of Sci., Al-Mustansiryha U., Baghdad, Iraq. (1937) D.3e cycle, physics (U. de Bordeaux, France, 1976). Asst. lect. *Crystallography.*

Rammo, Dr Nabil Naim. Physics Dept., C. of Sci., U. of Baghdad, Jadryia - Baghdad, Iraq. (1946) PhD, polymer crystallography (Manchester U., UK, 1977). Lect. (tel. 01 + 20001, ext. 210). *Polymer crystallography, molecular biology.*

IRELAND

Sub-Editor: B. J. Hathaway

Notes

1. International telephone country code - 353.

Byrne, Mr Peter G. Analytical Chemistry Dept., Inst. for Industrial Res. and Standards, Ballymum Road, Dublin 9, Ireland. (1946) MSc, analytical chemistry (London U., UK, 1972). Sr. scient. officer. (tel. 370101, ext. 621, telex 25449). *X-ray Analysis.*

Cardin, Dr Christine Janet. Dept. of Chemistry, Trinity C., Dublin 2, Ireland. (1947) DPhil., organometallic synthesis and structures (Sussex U., 1973). Lect. (tel. 353 + 1-772941, ext. 1357) *Organometallic and coordination chemistry, bioinorganic chemistry, single crystal X-ray diffraction studies.*

Cunningham, Dr Patrick Desmond. Dept. of Chemistry, Inorganic Div., U.C. Galway, Galway, Ireland. (1942) PhD, inorganic/physical chemistry (Nat. U., U. C. Dublin, 1966). Lect. (tel. 091 + 24411, ext. 483). *Tin chemistry (structural and synthetic), Mössbauer spectroscopy.*

Hathaway, Prof. Brian John. Dept. of Chemistry, U.C. Cork, Cork, Ireland. (1929) PhD, inorganic chemistry (Nottingham U., UK, 1954). Prof. (tel. 26871, ext. 2270). *Electronic properties, stereochemistry, copper (II) complexes, zinc (II) complexes, X-ray crystallography.*

Kelly, Dr Thomas C. Inorganic Materials Dept., IIRS, Dublin 9, Ireland. (1954) PhD, inorganic/physical chemistry (Nat. U., U.C. Galway, 1980) Sr. scient.officer. (tel. 01-370101, ext. 615). *Inorganic materials, ceramic materials, X-ray identification, crystallite size and strain measurements.*

Kiely, Dr Patrick Vincent. The Agricultural Inst., Johnstown Castle, Wexford, Ireland. (1934) PhD, soil science (Wisconsin U., USA, 1964). Res. officer. (tel. 053 + 22888) *Clay minerals, crystal optics.*

McArdle, Dr Patrick. Dept. of Chemistry, Inorganic Div., U.C. Galway, Galway, Ireland. (1945) PhD, inorganic/physical chemistry (Nat. U., U. C. Dublin, 1968). Lect. (tel. 091 + 24411, ext. 487). *Organometallic iron chemistry (structural and synthetic).*

Strogen, Mr Peter. Dept. of Geology, U. C. Dublin, Nat. U. of Ireland, Belfield, Stillorgan Road, Dublin 4, Ireland. (1937) BSc, geology (London U., UK, 1958). Lect. (tel. 693244, ext. 244). *X-ray analysis, natural mineral composition.*

ISRAEL

Sub-Editors: M. Shoham and M. Harel

Notes

1. International telephone country code - 972.

Adan, Mr Nachum. Intel Electronics, P.O.B. 3173, Jerusalem 91031, Israel. (1947) BSc, geology (Hebrew U., 1971). Process control group leader. (tel. 02 + 887211). *X-ray diffraction, electron microscopy, silicon, microelectronics fabrication.*

Agmon, Mrs Ilana. Chemistry Dept., Technion, Israel Inst. of Techn., Haifa 32000, Israel. (1949) MSc, crystallography (Technion, 1980). Res. assoc. (tel. 04 + 293716). *Crystal structure, phase transformation.*

Ashkenazi, Dr Joseph. Physics Dept., Technion, Israel Inst. of Techn., Haifa 32000, Israel. (1944) PhD, physics (Hebrew U., 1975). Sr. res. assoc. (tel. 04 + 293521). *Electronic structure of solids.*

Atzmony, Prof. Uzi. Physics Dept., Nuclear Res. Center-Negev, P.O.B. 9001, Beer-Sheva, Israel. (1937) PhD, physics (Hebrew U., 1967). Assoc. prof. and group leader. (tel. 057 + 967706). *Mossbauer effect, magnetic and electric crystal fields, interactions in solids.*

Azmon, Prof. Emanuel. Dept. of Geology and Mineralogy, Ben-Gurion U., P.O.B. 653, Beer-Sheva 84105, Israel. (1929) PhD, geology (U. Southern California, USA, 1960). Assoc. prof. (tel. 057 + 61338). *Sedimentology, clay mineralogy, clay ceramics.*

Bar, Mrs Ilana. Chemistry Dept., Ben-Gurion U., P.O.B. 653, Beer-Sheva 84105, Israel. (1950) MSc, chemical crystallography (Ben-Gurion U., 1976). PhD student. (tel. 057 + 64379). *Organic crystal structures, conformational polymorphism.*

Bart, Prof. Dr Jan J. C. Structural Chemistry Dept., Weizmann Inst., Rehovot 76100, Israel. (1941) PhD, structural chemistry (U. Amsterdam, Netherlands, 1967). Meyerhoff visiting prof. (tel. 08 + 842384, telex 361900 WIX IL). *Inorganic crystal structures, structure - property relations, X-ray absprption spectroscopy, heterogeneous catalysis, solid state chemistry.*

Batt, Mr Ahron. Jerusalem C. of Techn., Jerusalem, Israel. (1943) MSc, crystallography (Polytechnic Inst. of Brooklyn, USA, 1961). Director. (tel. 02 + 423131).

Ben Yair, Dr Moshe Pinhas. The Standards Inst. of Israel, Ramat Aviv, Israel. (1922) PhD, geochemistry (Hebrew U., 1957). Res. coordinator. (tel. 03 + 422811). *Crystal growth, clay minerals.*

Benghiat, Dr Victor. Biological Services, Weizmann Inst., Rehovot 76100, Israel. (1932) PhD, crystallography (Weizmann Inst., 1971). Res. coordinator. (tel. 08 + 483223). *Ultrastructures, electron microscopy.*

Berkovitch-Yellin, Dr Ziva. Structural Chemistry Dept., Weizmann Inst., Rehovot 76100, Israel. (1946) PhD, structural chemistry (Weizmann Inst., 1976). Sr. sci. (tel. 08 + 482631). *Theoretical and experimental studies, molecular structures, electron spin densities.*

Bernstein, Prof. Joel. Chemistry Dept., Ben-Gurion U., P.O.B. 653, Beer-Sheva 84105, Israel. (1941) PhD, physical chemistry (Yale U., USA, 1967). Assoc. prof. (tel. 057 + 664379). *Chemical crystallography, organic structures, solid state chemistry, polymorphism, crystal forces and molecular conformation.*

Bino, Dr Avi. Dept. of Inorganic Chemistry, Hebrew U., Jerusalem 91904, Israel. (1949) PhD, inorganic chemistry (Hebrew U., 1978). Lect. (tel. 02 + 585482). *Inorganic crystal structures.*

Blech, Prof. Ilan Asriel. Materials Engineering Dept., Technion, Israel Inst. of Techn., Haifa, Israel. (1936) DSc, metallurgy (M.I.T., USA, 1964). Prof. (tel. 04 + 230159). *Dynamical diffraction, X-ray topography.*

Brown, Mr Jerry Howard. Structural Chemistry Dept., Weizmann Inst., Rehovot 76100, Israel. (1962) BSc, chemistry (Brandeis U., USA, 1983). Student. (tel. 08 + 482541). *Protein crystallography.*

Cohen, Dr Shmuel. Dept. of Inorganic Chemistry, Hebrew U., Jerusalem 91904, Israel. (1944) PhD, chemistry (Hebrew U., 1976). Crystallographer. (tel. 02 + 585482). *Solid state chemistry.*

Dariel, Prof. Moshe Pierre. Materials Engineering Dept., Ben-Gurion U., Beer-Sheva 84501, Israel. (1935) PhD, crystallography (Weizmann Inst., 1968). Prof. (tel. 057 + 968736). *Physical metallurgy, intermetallic compounds, hydrogen absorbers.*

Deutsch, Dr Moshe. Physics Dept., Bar-Ilan U., Ramat-Gan 52100, Israel. (1946) PhD, physics (Bar-Ilan U., 1979). Lect. (tel. 03 + 718433). *Mathematical methods in crystallography, X-ray physics, X-ray optics, liquid crystals, surface physics.*

Eisenberg, Prof. Henryk. Polymer Dept., Weizmann Inst., Rehovot 76100, Israel. (1921) PhD, chemistry (Hebrew U., 1951). Prof. (tel. 08 + 483362). *Nucleic acids, enzymes, halophilic proteins, nucleoproteins, chromatin, small angle X-ray scattering, neutron scattering, light scattering.*

Eisenstein, Dr Miriam. Structural Chemistry Dept., Weizmann Inst., Rehovot 76100, Israel. (1951) PhD, structural chemistry (Weizmann Inst., 1981). Fellow. (tel. 08 + 482672). *Electron density.*

Erez, Dr Gidon. Physics Dept., Nuclear Res. Center-Negev, Beer-Sheva, P.O.B. 9001, Israel. DSc, physics (Technion, 1960). (tel. 057 + 57111). *Metal physics, metal structure.*

Felsteiner, Prof. Joshua. Physics Dept., Technion, Israel Inst. of Techn., Haifa 32000, Israel. (1938) PhD, physics (Toronto U., Canada, 1967). Assoc. prof. (tel. 04 + 293869). *Magnetic structure of solids, charge and momentum densities.*

Fraenkel, Prof. Benjamin. Racah Inst. of Physics, Hebrew U., Jerusalem, Israel. (1923) PhD, physics (Hebrew U., 1955). Chairman, Plasma Section. (tel. 02 + 584335). *Highly ionized atoms, X-ray spectroscopy, far ultra violet.*

Frolow, Dr Felix. Structural Chemistry Dept., Weizmann Inst., Rehovot 76100, Israel. (1947) PhD, crystallography (Weizmann Inst., 1980). (tel. 08 + 482631). *X-ray and neutron diffraction, crystal physics.*

Goldberg, Dr Israel. Chemistry Dept., Tel-Aviv U., Ramat Aviv, Israel. (1945) PhD, physical chemistry (Tel-Aviv U., 1974). Sr. lect. (tel. 03 + 420258). *Organic crystal structures, inclusion compounds, clathrates, structure - properties relationships.*

Goldstein(Choshen), Mr Ehud. Chemistry Dept., Ben-Gurion U., P.O.B. 653, Beer-Sheva 84105, Israel. (1953) MSc, chemical crystallography (Ben-Gurion U.). PhD student. (tel. 057 + 64379). *Crystal structures, direct methods.*

Goshen, Dr Shmuel Yehuda. Physics Dept., Nuclear Res. Center-Negev, P.O.B. 9001, Beer-Sheva 84105, Israel. (1933) PhD, nuclear physics (Weizmann Inst., 1966). *Phase transitions, plasma.*

Grünbaum, Prof. Enrique. School of Engineering, Tel-Aviv U., Ramat Aviv, Israel. (1926) MSc, physics (London U., UK, 1957). Assoc. prof. (tel. 03 + 422692). *Growth and structure, thin metal and semiconductor films, electron diffraction and microscopy.*

Gurewitz, Dr Eitan. Physics Dept., Nuclear Res. Center-Negev, P.O.B. 9001, Beer-Sheva 84190, Israel. (1938) PhD, physics (Weizmann Inst., 1976). Res. assoc. (tel. 057 + 967357). *Magnetic structures of solids, magnetic and crystallographic symmetry, neutron scattering, phase transitions, symmetry.*

Harel, Dr Michal. Structural Chemistry Dept., Weizmann Inst., Rehovot 76100, Israel. (1940) PhD, crystallography (Weizmann Inst., 1975). (tel. 08 + 482631). *Protein crystallography.*

Haran, Ms Tali. Structural Chemistry Dept., Weizmann Inst., Rehovot 76100, Israel. MSc, chemistry (Weizmann Inst., 1980). PhD student. (tel. 08 + 482647). *Biological macromolecules, DNA.*

Heller-Kallai, Prof. Lisa. Geology Dept., Hebrew U., Jerusalem, Israel. (1926) PhD, crystallography (London U., UK, 1951). Prof. (tel. 02 + 584883). *Clay minerology.*

Herbstein, Prof. Frank Herzl. Chemistry Dept., Technion, Israel Inst. of Techn., Haifa, Israel. (1926) DSc, physical and chemical crystallography (Capetown U., S. Africa, 1967). Prof. (tel. 04 + 22511 ext. 415). *Crystal structures, phase transformations, solid state reactions.*

Hirshfeld, Prof. Fred. Structural Chemistry Dept., Weizmann Inst., Rehovot 76100, Israel. (1927) PhD, crystallography (Hebrew U., 1956). Prof. (tel. 08 + 483320). *Organic crystal structures, electron distribution, computer programming.*

Kaftory, Dr Menahem. Chemistry Dept., Technion, Israel Inst. of Techn. Haifa 32000, Israel. (1943) DSc, chemistry (Technion, 1973). Sr. lect. (tel. 04 + 293761). *Chemical crystallography, solid state chemistry, reaction paths.*

Kalb(Gilboa), Dr Aaron Joseph. Biophysics Dept., Weizmann Inst., Rehovot 76100, Israel. (1937) PhD, chemistry (U. California, USA, 1963). Sr. sci. (tel. 08 + 483609). *Protein structures.*

Kalman, Prof. Zwi Heinrich. Racah Inst. of Physics, Hebrew U., Jerusalem 91904, Israel. (1924) PhD, physics (Hebrew U.,1965). Assoc. prof. (tel. 02 + 584646). *X-ray diffraction methods, materials science, solid state physics.*

Kapon, Dr Moshe. Chemistry Dept., Technion, Israel Inst. of Techn., Haifa 32000, Israel. (1941) DSc, chemistry (Technion, 1974). Dept. crystallographer. (tel. 04 + 293716). *Organic polyiodide structures, thermal decomposition of materials.*

Katz, Dr Gerald. 21 Ra'anan St., Haifa 34384, Israel. PhD, solid state science (Pennsylvania State U., USA, 1965). *Crystal growth, topotaxy, epitaxy, optical spectroscopy, rare earth doped glasses.*

Kazinets, Dr Maria. SEM Lab., Ben-Gurion U., P.O.B. 1025, Beer-Sheva 84110, Israel. (1935) PhD, solid state physics (Inst. of Physics, Azerb. USSR, 1969). Engineer. (tel. 057 + 664308). *Crystal growth, phase transitions, thin films, electron microscopy, diffraction.*

Kimmel, Dr Giora. Metallurgical Lab., Nuclear Res. Center-Negev, Beer-Sheva 84190, P.O.B. 9001, Israel. (1939) DSc, materials science (Technion, 1973). Head, physical metallurgy group. (tel. 057 + 967017). *Alloy phases, powder crystallography, applied crystallography.*

Kisch, Prof. Hanan Joseph. Dept. of Geology and Mineralogy, Ben-Gurion U., P.O.B. 653, Beer-Sheva 84105, Israel. (1935) PhD, geology and mineralogy (Amsterdam U., Netherlands, 1962). Prof. (tel. 057 + 661291). *Metamorphism, coalification, fluid inclusions, clay minerals, amphiboles, X-ray diffraction, mineralogy, structural geology, diagenesis, metamorphic assemblages.*

Landau, Mr Alex. Metallurgy Dept., Nuclear Res. Center-Negev, P.O.B. 9001, Beer-Sheva 84190, Israel. (1954) BSc, materials science (Ben-Gurion U., 1981). Physical metallurgist. (tel. 057 + 967017). *Alloy phases, powder crystallography, applied crystallography.*

Leiserowitz, Prof. Leslie. Structural Chemistry Dept., Weizmann Inst., Rehovot 76100, Israel. (1934) PhD, crystallography (Hebrew U., 1965). Prof. (tel. 08 + 482631). *Molecular packing modes, electron density distribution, solid state and surface chemistry.*

Leser, Dr Julio. Radiation Protection Dept., Ministry of Health, 27 Prof. Shor St., Tel-Aviv, Israel. (1940) PhD, crystallography (Weizmann Inst., 1975). Head physicist. (tel. 03 + 428818). *Neutron diffraction, organic crystal structure, radiation protection.*

Low, Prof. William. Div. of Microwave Physics, Hebrew U., Jerusalem, Israel. (1922) PhD, physics (Columbia U., USA, 1965). Prof. (tel. 02 + 584240). *Inorganic crystal structure, crystal growth, crystal physics.*

Maltz, Dr Abraham. Chemistry Dept., Ben-Gurion U., P.O.B. 653, Beer-Sheva 84105, Israel. (1947) DSc, chemical crystallography (Glasgow U., Scotland, 1979). Post doctoral fellow. (tel. 057 + 64379). *Crystal structures, direct methods.*

Mayer, Prof. Itzchak. Chemistry Dept., Hebrew U., Jerusalem, Israel. (1927) PhD, chemistry (Hebrew U., 1960). Assoc. prof. (tel. 02 + 585214). *Inorganic crystal structures, apatite structures, powder diffraction.*

Mayorzik, Mrs Hemda. Chemistry Dept., Tel-Aviv U., Ramat Aviv, Israel. (1955) BSc, chemistry (Tel-Aviv U., 1978). Res. asst. (tel. 03 + 420258). *Crystalline charge transfer compounds, structure.*

Melamud, Dr Mordechai. Physics Dept., Nuclear Res. Center-Negev, P.O.B. 9001, Beer-Sheva 84190, Israel. (1943) PhD, physics (Weizmann Inst., 1976). Researcher. (tel. 057 + 967468). *Neutron diffraction.*

Minkoff, Prof. Isaac. Materials Engineering Dept., Technion, Israel Inst. of Techn., Haifa 32000, Israel. DSc, metallurgy (M.I.T., USA, 1957). Prof. (tel. 04 + 292167). *Crystal growth.*

Nadiv, Prof. Shmuel. Materials Engineering Dept., Technion, Israel Inst. of Techn., Haifa, Israel. (1929) DSc, physical metallurgy (1964). Assoc. prof. (tel. 04 + 230165). *Metals and alloys structure, phase transitions, solid state reactions.*

Nathan, Dr Yaakov. Geochemistry Dept., Geological Survey of Israel, 30 Malchei Israel St., Jerusalem, Israel. (1931) PhD, geology (Hebrew U., 1969). Sr. sci. (tel. 02 + 286121). *Inorganic crystal structure, clay minerals, phosphates.*

Pinto, Mr Haim. Physics Dept., Nuclear Res. Center-Negev, P.O.B. 9001, Beer-Sheva 84190, Israel. (1937) BSc, physics (Ben-Gurion U., 1976). Researcher. (tel. 057 + 967468). *Neutron diffraction.*

Rabin, Mr Baruch. Dept. 1603, Israel Aircraft Industries, Lod 70100, Israel. (1929) MSc, physics of metals (Ural Polytechnical Inst., USSR, 1969). Manager, materials and process engineering. (tel. 03 + 973940). *Metal structure, phase analysis.*

Rabinovich, Prof. Dov. Structural Chemistry Dept., Weizmann Inst., Rehovot 76100, Israel. (1929) PhD, crystallography (Hebrew U., 1964). Assoc. prof. (tel. 08 + 482672). *Organic crystal structures, solid state chemistry, computer programming, flash X-ray diffraction, DNA structures.*

Reisner, Dr George. Chemistry Dept., Technion, Israel Inst. of Techn., Haifa 32000, Israel. (1943) PhD, chemistry (Technion, 1966). Sr. res. fellow. (tel. 04 + 293716).

Regev, Dr Haya. Chemistry Dept., Ben-Gurion U., P.O.B. 653, Beer-Sheva 84105, Israel. (1941) DSc, chemistry (Technion, 1978). Information sci. (tel. 057 + 64432). *Molecular complexes, organic crystal structures, information science.*

Rudman, Prof. Peter S. Physics Dept., Technion, Israel Inst. of Techn., Haifa, Israel. (1926) DSc, metallurgy (M.I.T., USA, 1955). Prof. (tel. 04 + 22511 ext. 366). *Alloys, order-disorder, liquids.*

Saper, Dr Mark A. Structural Chemistry Dept., Weizmann Inst., Rehovot 76100, Israel. (1954) PhD, biochemistry (Rice U., USA, 1983). Post doctoral fellow. (tel. 08 + 482541). *Protein and DNA structure, computer graphics, synchrotron radiation.*

Sariel, Mr Joseph. Metallurgy Dept., Nuclear Res. Center-Negev, P.O.B. 9001, Beer-Sheva 84190, Israel. (1947) MSc, materials science (Ben-Gurion U., 1981). Metallurgist. (tel. 057 + 968786). *Alloy phases, powder crystallography, crystal structure.*

Schieber, Prof. Michael. Dept. of Experimental Physics, Hebrew U., Jerusalem, Israel. PhD, material sci. (Hebrew U., 1962). Prof. (tel. 02 + 584693). *Inorganic crystal structures, crystal growth, crystal physics, metal physics.*

Shaanan, Dr Boaz. Structural Chemistry Dept., Weizmann Inst., Rehovot 76100, Israel. (1946) PhD, physical chemistry (Tel-Aviv U., 1979). Sr. sci. (tel. 08 + 482647). *Macromolecular structure - function - dynamics, molecular structure, lattice dynamics.*

Shaked, Prof. Hagai. Physics Dept., Ben-Gurion U., P.O.B. 653, Beer-Sheva 84105, Israel. (1931) PhD, engineering sci. (U. California, Berkeley, USA, 1963). Prof. (tel. 057 + 664291). *Magnetism, symmetry, neutron diffraction, phase transition, group theory.*

Shakked, Prof. Zippora. Structural Chemistry Dept., Weizmann Inst., Rehovot 76100, Israel. (1943) PhD, crystallography (Weizmann Inst., 1975). Assoc. prof. (tel. 08 + 482672). *Organic structures, DNA structures, protein-DNA interactions.*

Shamir, Dr Noah. Physics Dept., Nuclear Res. Center-Negev, P.O.B. 9001, Beer-Sheva 84190, Israel. (1947) PhD, physics (Weizmann. Inst., 1977). Res. assoc. (tel. 057 + 967795). *Magnetic structure of crystals, magnetic interactions in solids, electric crystal fields in solids.*

Shmueli, Prof. Uri. Chemistry Dept., Tel-Aviv U., Ramat Aviv, Israel. (1928) PhD, crystallography (Weizmann Inst., 1966). Assoc. prof. (tel. 03 + 420258). *Crystalline molecular compounds, computing methods, motion in crystals.*

Shoham, Dr Gil. Dept. of Inorganic Chemistry, Hebrew U., Jerusalem 91904, Israel. (1952) PhD, crystallography (Harvard U., USA, 1984). Lect. (tel. 02 + 585610). *Biological compounds, macromolecules.*

Shoham, Dr Menachem. Structural Chemistry Dept., Weizmann Inst., Rehovot 76100, Israel. (1944) PhD, chemistry (Weizmann Inst., 1979). Sr. sci. (tel. 08 + 482647). *Protein and nucleic acid structure.*

Shpigler, Mr Bilu. Metalurgical Lab., Israel Inst. of Metals, Technion, Haifa 32000, Israel. (1932) MSc, metallurgy (Technion, 1963). Lab. head. (tel. 04 + 235104). *Metals micro-structure, instrumentation, electron optics, fracture mechanics, color metallography, failure analysis, composite materials, hot isostatic pressing.*

Steinberg, Prof. Itzhak. Physics Dept., Hebrew U., Jerusalem, Israel. PhD, physics (Hebrew U., 1957). Prof. (tel. 02 + 584648). *Inorganic crystal structures, crystal growth, crystal physics.*

Sussman, Prof. Joel Leonard. Structural Chemistry Dept., Weizmann Inst., Rehovot 76100, Israel. (1943) PhD, biophysics (M.I.T., USA, 1972). Assoc. prof. (tel. 08 + 482638). *Proteins, nucleic acids, computer graphics, model building, structure determination, refinement.*

Traub, Prof. Wolfie. Structural Chemistry Dept., Weizmann Inst., Rehovot 76100, Israel. (1927) PhD, physics (London U., UK, 1955). Prof. (tel. 08 + 482668). *Biological macromolecules, mineralized tissues.*

Yahalom, Prof. Joseph. Materials Engineering Dept., Technion, Israel Inst. of Techn., Haifa 32000, Israel. (1933) PhD, metallurgy (Cambridge U., UK, 1963). Prof., corrosion lab. head. (tel. 04 + 292177). *Oxides, corrosion, electrodeposition.*

Yonath, Prof. Ada. Structural Chemistry Dept., Weizmann Inst., Rehovot 76100, Israel. PhD, crystallography (Weizmann Inst., 1969). Assoc. prof. (tel. 08 + 482541). *Biological macromolecules, ribosomes.*

Zeldes, Mr Nathan. Intel Electronics, P.O.B. 3173, Jerusalem 91031, Israel. MSc, physics (Hebrew U., 1977). Quality assurance group leader. (tel. 02 + 887211). *Electron microscopy, silicon, microelectronics fabrication.*

Zellingher, Mr Naphtaly. Mechanical Engineering Dept., Ben-Gurion U., P.O.B. 653, Beer-Sheva 84105, Israel. (1927) MSc, physics (C.I.Parhon U., Romania, 1958). Sr. lect. (tel. 057 + 61321). *Whisker growth.*

Zevin, Prof. Lev. Inst. for Applied Res., Ben-Gurion U., P.O.B. 1025, Beer-Sheva 84110, Israel. (1930) PhD, applied crystallography (Moscow Inst. for Crystallography, USSR, 1963). Assoc. prof. (tel. 057 + 664308). *Intermetallic crystal structures, powder diffraction applications, diffraction apparatus.*

Zilber, Dr Raphael. Solid State Physics Dept., Soreq Nuclear Res. Center, Yavne, Israel. (1941) PhD, physics (Grenoble U., France, 1969). (tel. 03 + 952075 ext. 343). *Crystallography, order-disorder, phase transition.*

ITALY

Sub-Editors: M. Mammi and G. Zanotti

Notes

1. International telephone country code - 39.

2. The first degree awarded by Italian Universities for scientific subjects is *Dottore* (Dr). It is obtained after a 4 year course, with the exceptions of the degrees in chemistry (5 years), engineering (5 years), and medicine (6 years). The previous training is given by a 5 + 3 + 5 year course at the elementary, secondary and higher schools, respectively. At some Universities various diplomas for special training may be awarded to Dr after courses of instruction. The doctorate *(Dottore di Ricerca)* is obtained after regular courses and research lasting three-four years beyond the Dr, and is awarded by the Ministry of Education.

3. Positions and functions of teachers and researchers at an Italian University are at present *approximately* equivalent to: *Professore ordinario* (Prof.) - Full Professor; *Professore associato* (Prof. assoc.) - Associate Professor; *Ricercatore* - Researcher; *Professore Incaricato* (Prof. inc.) - Lecturer; *Assistente* (Asst.) - Research Assistant; *Tecnico Laureato* (Tecn. laur.) - Technical Research Assistant; *Borsista* - Postdoctoral Research Fellow. New appointments cannot be made to the position of *Libero Docente* (Lib. doc.) - (Reader) since 1970.

4. At the Consiglio Nazionale delle Ricerche (CNR - National Research Council) and other public institutions, persons can be appointed: *Collaboratore Tecnico-Professionale* (Coll. tecn. pr.) - Research officer, and *Borsista* - Postdoctoral Research Fellow.

5. Other functions in the above and other institutions and companies are *Direttore* - Director (of Institute, Div., Centre, etc.); *Capo-reparto* - Head (of Dept., Group, Lab., etc.); *Ricercatore* - Research Scientist.

6. Abbreviations used in this section include: Dip. for Dipartimento, Tecn. for Tecnologia, and Ist. for Istituto.

Abbona, Prof. Francesco. Dip. di Scienze della Terra, U. di Torino, Via S. Massimo 24, 10123 Torino, Italy. (1937) Dr, chemistry (U. di Torino, 1961). Prof. assoc. (tel. 011 + 832193). *Crystal growth from solutions.*

Accorsi, Prof. Carla Alberta. Dip. di Chimica, U. di Ferrara, Via Luigi Borsari 46, 44100 Ferrara, Italy. (1941) Dr, chemistry (U. di Ferrara, 1966). Prof. assoc. (tel. 0532 + 36374). *Crystal growth, habit modifications.*

Ajó, Dr David. Ist. di Chimica e Tecn. dei Radioelementi, C.N.R., Corso Stati Uniti, 35100 Padova, Italy. (1946) Dr, chemistry (U. di Roma, 1970). Coll. tecn. pr. (tel. 049 + 845373, telex 430302 CNRPDI). *Biologically important substances, structure and function, conformation, polymers, polymorphism.*

Albano, Prof. Vincenzo Giulio. Ist. Chimico 'Ciamician', U. di Bologna, Via F. Selmi 2, 40126 Bologna, Italy. (1937) Dr, chemistry (U. di Bari, 1960). Prof., inorganic chemistry. (tel. 051 + 235292). *Coordination compounds.*

Alberti, Prof. Alberto. Ist. di Mineralogia e Petrologia, U. di Modena, Via S. Eufemia 19, 41100 Modena, Italy. (1938) Dr, physics (U. di Modena, 1962). Prof. assoc. (tel. 059 + 218062, ext. 16). *Structures, inorganic compounds, minerals.*

Albinati, Prof. Alberto. Ist. di Chimica Farmaceutica, U. di Milano, Viale Abruzzi 42, 20131 Milano, Italy. (1945) Dr, chemistry (U. di Milano, 1970) Prof. assoc. (tel. 02 + 209197, telex 320484 UNIMII). *Organometallic compounds, small organic molecules, neutron diffraction, synchrotron radiation, powder diffraction.*

Alietti, Prof. Andrea. Ist. di Mineralogia e Petrologia, U. di Modena, Via S. Eufemia 19, 41100 Modena, Italy. (1923) Dr, chemistry (U. di Pavia, 1947). Prof., mineralogy. (tel. 059 + 218062). *Clay minerals.*

Allegra, Prof. Giuseppe. Ist. di Chimica, Politecnico di Milano, Piazza L. Da Vinci 32, 20133 Milano, Italy. (1933) Dr, chemical engineering (Politecnico di Milano, 1958). Prof., chemistry. (tel. 02 + 230845-6-7-8-9, telex 333467 POLIMI I). *Direct methods, organic molecular structures, conformational analysis, statistical mechanics, polymers.*

Amicarelli, Prof. Vincenzo. Ist. di Chimica Applicata, Facolta' di Ingegneria, U. di Bari, Via Re David 200, 70125 Bari, Italy. (1930) Dr, Industrial chemistry (U. di Napoli, 1956). Prof., appl. chemistry. (tel. 081 + 228014). *Engineering materials.*

Andreetti, Prof. Giovanni Dario. Ist. di Strutturistica Chimica, U. di Parma, Via M. D'Azeglio 85, 43100 Parma, Italy. (1939) Dr, chemistry (U. di Parma, 1963). Prof., organic chemical crystallography. (tel. 0521 + 22432). *Crystal structure analysis, organic and organometallic compounds, electron density distributions, computer programming.*

Antolini, Prof. Luciano. Dip. di Chimica, U. di Modena, Via Campi 183, 41100 Modena, Italy. (1942) Dr, chemistry (U. di Modena, 1967). Prof. assoc. (tel. 059 + 362243). *Crystal structure determination, inorganic compounds, coordination compounds.*

Antonini, Prof. Marcello. Dip. di Fisica, U. di Modena, Via Campi 213, 41100 Modena, Italy. (1939) Dr, physics (U. di Rome, 1962). Prof. assoc. (tel. 059 + 361142 ext.83). *Atomic defects, radiation damage, metals, insulators, microstructure determination, glasses.*

Antonione, Prof. Carlo. Ist. di Chimica Generale ed Inorganica, Fac. di Farmacia, U. di Torino, Via P. Giuria 9, 10125 Torino, Italy. (1931) Dr, chemistry (U. di Torino, 1958). Prof. assoc. (tel. 011 + 657000). *Solid state chemistry, structure, crystalline metals, amorphous metals.*

Aquilano, Prof. Dino. Dip. di Scienze della Terra, U. di Torino, Via S. Massimo 24, 10123 Torino, Italy. (1940) Dr, physics (U. di Torino, 1963). Prof. assoc. (tel. 011 + 832193). *Crystal growth, crystal defects.*

Armigliato, Dr Aldo. Ist. LAMEL, C.N.R., Via de' Castagnoli 1, 40126 Bologna, Italy. (1940) Dr, physics (U. di Padova, 1965). Coll. tecn. pr. (tel. 051 + 519593, ext. 216). *Electron microscopy, materials, electronics devices, X-ray microanalysis.*

Artioli, Dr Gilberto. Ist. di Mineralogia e Petrologia, U. di Modena, Via S. Eufemia 19, 41100 Modena, Italy. (1957) Dr, Geological sciences (U. di Modena, 1980). Ricercatore (tel. 059 + 218062). *Zeolite crystal structures, neutron diffraction.*

Bachechi, Dr Fiorella. Ist. di Strutturistica Chimica 'Giordano Giacomello', C.N.R., C.P. 10 - 00016 Monterotondo Stazione, Roma, Italy. (1939) PhD, crystallography (Toronto U., Canada, 1970). Coll. tecn. pr. (tel. 06 + 9005641, ext 616). *Organometallic and coordination compounds, structure - chemical properties relationship.*

Balzarotti, Prof. Adalberto. Dip. di Fisica, U. di Roma II 'Tor Vergata', Via O. Raimondo, 00173 Roma, Italy. (1939) Dr, physics (U. di Pavia, 1961). Prof., physics. (tel. 06 + 79791 ext. 2305, telex 611462 UNIVRM I). *Optical properties of solids, synchrotron radiation, X-ray spectroscopy.*

Bandoli, Prof.Giuliano. Dip. di Scienze Farmaceutiche, U. di Padova, Via Marzolo 5, 35131 Padova, Italy. (1941) Dr, chemistry, pharmacy (U. di Padova, 1966, 1975). Prof. assoc. (tel. 049 + 831634). *Inorganic crystal structures.*

Barbieri, Prof. Renato. Ist. di Chimica Generale, U. di Palermo, 28 Via Archirafi, 90123 Palermo, Italy. (1930) Dr, chemistry (U. di Padova, 1956). Prof. (tel. 091 + 230474). *Mössbauer spectroscopy, inorganic and organometallic tin compounds, infrared spectroscopy, X-ray diffractometry, inorganic and organometallic compounds.*

Bardi, Prof. Renato. Dip. di Chimica Organica, U. di Padova, Via Marzolo 1, 35100 Padova, Italy. (1924) Dr, chemistry (U. di Padova, 1957). Prof. assoc. (tel. 049 + 831111 or 831311). *Organic compounds, biologically important substances.*

Barret, Mr Nicholas. Divisione di Fisica, CCR EURATOM, Ispra, 21020 Varese, Italy. (1962) BSc Hons 1st Class (Strathclyde U., UK, 1983). Res. student. (tel. 0332 + 789111, ext. 5878). *EXAFS, monochromator, synchrotron, fluorescence, glasses, radiation damage.*

Basso, Prof. Riccardo. Ist. di Mineralogia, U. degli Studi, Palazzo delle Scienze, Corso Europa, 16132 Genova, Italy. (1947) Dr, mathematics (U. di Genova, 1971). Prof., appl. mineralogy. (tel. 010 + 518184). *Inorganic crystal structures.*

Battaglia, Prof. Luigi Pietro. Ist. di Chimica Generale ed Inorganica, U. di Parma, Via M. D'Azeglio 85, 43100 Parma, Italy. (1943) Dr, chemistry (U. di Parma, 1970). Prof. assoc. (tel. 0521 + 34400). *Crystal structures, inorganic, organometallic and bioinorganic compounds.*

Battezzati, Prof. Livio. Ist. di Chimica Generale ed Inorganica, U. di Torino, Via P. Giuria 9, 10125 Torino, Italy. (1950) Dr, chemistry (U. di Torino, 1974). Prof. assoc. (tel. 011 + 657000). *Rapid solidification, metallic glasses, metallurgy.*

Bedarida, Prof. Federico. Ist. di Mineralogia, U. di Genova, Corso Europa (Palazzo delle Scienze), 16100 Genova, Italy. (1924) Dr, physics and mathematics (U.

di Genova, 1958). Prof. (tel. 010 + 518184 ext. 1). *Crystal growth, coherant light crystallographic applications.*

Bellon, Prof. Pier Luigi. Ist. di Chimica Generale ed Inorganica, U. di Milano, 21, Via Veneziani, 20133 Milano, Italy. (1931) Dr, industrial chemistry (U. di Padova, 1955). Prof., general chemistry. (tel. 02 + 2361410). *Crystallography of semiconductor materials, low resolution protein structures, laboratory automation.*

Benedetti, Prof. Ettore. Dip. di Chimica, U. di Napoli, Via Mezzocannone 4, 80134 Napoli, Italy. (1940) Dr, chemistry (U. di Napoli, 1965). Prof. (tel. 081 + 206800). *Synthesis, structure, conformation, peptides - linear and cyclic; solution and solid state characterization, natural and synthetic peptides; model compounds; macromolecules.*

Benetollo, Dr Franco. Ist. di Chimica e Tecn. dei Radioelementi, C.N.R., Corso Stati Uniti, 35100 Padova, Italy. (1947) Dr, pharmacy (U. di Padova, 1981) Coll. tecn. pr. (tel. 049 + 845111 or 845368). *Inorganic crystal structures.*

Benna, Dr Piera. Dip. di Scienze della Terra, U. di Torino, Via S. Massimo 24, 10123 Torino, Italy. (1954) Dr, geological sciences (U. di Torino, 1978) Ricercatore. (tel. 011 + 832193). *Crystal chemistry, experimental mineralogy.*

Bertolasi, Prof. Valerio. Dip. di Chimica, U. di Ferrara, Via L. Borsari 46, 44100 Ferrara, Italy. (1949) Dr, chemistry (U. di Ferrara, 1973). Prof. assoc. (tel. 0532 + 33522). *Determination of crystal structures, molecular systematics, crystal packing, molecular pharmacology.*

Biagini Cingi, Prof. Marina. Ist. di Chimica Generale ed Inorganica, U. di Parma, Via M. D'Azeglio 85, 43100 Parma, Italy. (1925) Dr, chemistry (U. di Parma, 1949). Prof., general chemistry. (tel. 0521 + 34400). *Crystal structures, inorganic compounds, bioinorganic compounds, coordination compounds.*

Bianchi, Dr Riccardo. Centro di Studio per la Relazione tra Struttura e Reattivita' Chimica, C.N.R., Via Golgi 19, 20133 Milano, Italy. (1942) Dr, mathematics (U. di Milano, 1972). *Charge density distribution in solids, chemical bonding, low temperature crystallography, crystallographic computing.*

Bianchi Orlandini, Dr Annabella. Ist. Stereochimica, C.N.R., Via F. D. Guerrazzi 27, 50132 Firenze, Italy. (1944) Dr, chemistry (U. di Firenze, 1968). Coll. tecn. pr. (tel. 055 + 243990). *Structures, metal complexes.*

Bigi Dr Adriana. Ist. di Chimica 'G. Ciamician', U. di Bologna, Via Selmi 2, 40126 Bologna, Italy. (1951), Dr, chemistry (U. di Bologna, 1975). Ricercatore. (tel. 051 + 235292). *Crystal structures, biological structures.*

Bigoli, Prof. Francesco. Ist. di Chimica Generale ed Inorganica, U. di Parma, Via M. D'Azeglio 85, 43100 Parma, Italy. (1936) Dr, chemistry (U. di Parma, 1964). Prof. assoc. (tel. 0521 + 34400). *Crystal structures, inorganic - organic and organometallic compounds.*

Bisi Castellani, Prof. Carla. Dip. di Chimica Generale, U. di Pavia, Via Taramelli 14, 27100 Pavia, Italy. (1931), Dr, chemistry (1954). Prof. assoc. (tel 0382 + 24714). *Coordination and Bioinorganic chemistry.*

Blasi, Prof. Achille. Dip. di Scienze della Terra, U. degli Studi, Via Botticelli 23, 20133 Milano, Italy. (1940) Dr, geological sciences (U. di Milano, 1968). Prof. assoc. (tel. 02 + 293994). *Rock-forming minerals, especially feldspars, crystal optics, X-ray crystallography, computer programming.*

Bocelli, Dr Gabriele. Centro di Studio per la Strutturistica Diffrattometrica, C.N.R., Via M. D'Azeglio 85, 43100 Parma, Italy. (1942) Dr, chemistry (U. di Parma, 1970). Coll. tecn. pr. (tel. 0521 + 22432). *Crystal structure analysis, organic and organometallic compounds, correlation between NMR spectroscopic data and crystal structure results.*

Bolis, Dr Vera Maria. Ist. di Chimica Generale ed Inorganica, Fac. di Farmacia, U. di Torino, via P. Giuria 9, 10125 Torino, Italy. (1950) Dr, chemistry (U. di Torino, 1974) Ricercatore. (tel. 011 + 657000). *Materials science, surfaces, adsorption, catalysis.*

Bolognesi, Prof. Martino. Sezione di Cristallografia, Dip. di Genetica e Microbiologia, U. di Pavia, Via Taramelli 16, 27100 Pavia, Italy. (1951) Scuola perfezionamento, chemistry-biochemistry (U. di Pavia, 1974-78). Prof. assoc. (tel. 0382 + 422394). *Proteins - Proteolytic enzymes, metallo proteins, carrier proteins.*

Bombieri, Prof. Gabriella. Ist. di Chimica Farmaceutica, U. di Milano, viale Abruzzi 42, I-20131 Italy. (1936) Dr, chemistry (U. di Pavia, 1962). Prof. (tel. 02 + 222041 ext 602, telex 320484UNIMII). *Inorganic pharmaceutical structural chemistry, coordination compounds, actinide and lanthanide derivatives stereochemistry.*

Bonamartini-Corradi, Prof. Anna. Ist. di Chimica Generale ed Inorganica, U. di Parma, Via M. D'Azeglio 85, 43100 Parma, Italy. (1942) Dr, chemistry (U. di Parma, 1967). Prof. assoc. (tel. 0521 + 34400). *Crystal structures, inorganic compounds, organic and organometallic compounds.*

Bonamico, Dr Mario. Ist. Teoria e Struttura Elettronica e Comportamento Spettrochimico dei Composti di Coordinazione, C.N.R., Area della Ricerca di Roma, 00016 Monterotondo Stazione, Roma, Italy. (1930) Dr, chemistry (U. di Roma, 1956). Coll. tecn. pr. (tel. 06 + 90020341). *X-ray crystal structure.*

Bovio, Prof. Bruna. Dip. di Chimica Generale, U. di Pavia, Viale Taramelli 12, 27100 Pavia, Italy. (1928) Dr, chemistry (U. di Pavia, 1960). Prof. inc. stab. of general chemistry. (tel. 0382 + 29270). *Inorganic and organic structures.*

Braga, Dr Dario. Ist. Chimico 'G. Ciamician', via Selmi 2, 40126 Bologna, Italy. (1953) Dr, chemistry (U. di Bologna, 1977). Ricercatore. (tel. 051 + 235292 ext. 17). *Organometallic compounds, metal clusters.*

Braibanti, Prof. Antonio. Ist. di Chimica Farmaceutica, Sezione Chimica Fisica, U. di Parma, Via La Spezia 73, 43100 Parma, Italy. (1927) Dr, chemistry (U. di Parma, 1951). Prof., physical chemistry. (tel. 0521 + 592207). *Structures, metal complexes, drugs, equilibria in solutions, thermodynamics.*

Bresciani Pahor, Prof. Nevenka. Dip. di Scienze Chimiche, U. di Trieste, Piazzale Europa 1, 34127 Trieste, Italy. (1949) Dr, chemistry (U. di Trieste, 1973). Prof. assoc. (tel. 040 + 51172). *Structural inorganic chemistry.*

Brigatti, Dr Maria Franca. Ist. di Mineralogia dell' Universita', Largo S. Eufemia 19, 41100 Modena, Italy. (1946) Dr, geological Science (U. di Modena, 1970). Ricercatore. (tel. 059 + 218062). *Clay minerals.*

Brückner, Prof. Sergio. Dip. di Chimica, Politecnico di Milano, Piazza L. da Vinci 32, 20133 Milano, Italy. (1943) Dr, chemistry (U. di Trieste, 1967) Prof. assoc. (tel. 02 + 230845, ext. 256). *Crystallography, conformational analysis.*

Bruno, Prof. Emiliano. Dip. di Scienze della Terra, U. di Torino, Via S. Massimo 24, 10123 Torino, Italy. (1938) Dr, geological sciences (U. di Genova, 1962). Prof. (tel. 011 + 832193). *Crystal chemistry, experimental mineralogy.*

Bruzzone, Prof. Giacomo. Ist. di Chimica Fisica, U. di Genova, Corso Europa (Palazzo delle Scienze), 16132 Genova, Italy. (1931) Dr, chemistry (U. di Genova, 1959). Prof., physical chemistry. (tel. 010 + 515076). *Structure and physical properties, intermetallic compounds.*

Burla, Dr Maria Cristina. Dip. di Scienze della Terra, Sezione di Cristallografia, U. di Perugia, Piazza Università, 06100 Perugia, Italy. (1953) Dr, mathematics (U. di Perugia, 1976). Ricercatore (tel. 075 + 23822). *Crystallographic computing, direct methods.*

Busetti, Prof. Vilma. Dip. di Chimica Organica, U. di Padova, Via Marzolo 1, 35100 Padova, Italy. (1934) Dr, chemistry (U. di Padova, 1960). Prof. assoc. (tel. 049 + 831273). *Organic and inorganic compounds.*

Buttinelli, Prof. Dante. Dip. di Ingegneria Chimica, dei Materiali, delle Materie prime e Metallurgia, U. di Roma 'La Sapienza', Via Eudossiana 18, 00184 Roma, Italy. (1935) Dr, chemistry (U. di Roma, 1958). Prof. inc. stab. (tel. 06 + 464286). *Coherent and non coherent precipitates, martensitic phase, stainless steels.*

Caglioti, Prof. Dr Giuseppe. Laboratorio Materiali, CESNEF - Ist. Ingegneria Nucleare del Politecnico, Via Ponzio 34/3, 20133 Milano, Italy. (1931) Dr, physics (U. di Roma, 1953). Prof., solid state physics. (tel. 02 + 2360386, ext. 10). *Crystal physics, materials science, lattice dynamics, mechanical properties of materials.*

Calestani, Dr Gianluca. Ist. di Strutturistica Chimica, U. di Parma, Via Massimo D'azeglio 85, 43100 Parma, Italy. (1952) PhD (U. Saarlandes, BRD, 1980). Ricercatore. (tel. 0521 + 22432). *Crystal structure analysis, inorganic and organic compounds, crystal growth, EPR spectroscopy.*

Calleri, Prof. Mariano Bernardino. Dip. di Scienze della Terra, U. di Torino, Via San Massimo 24, 10123 Torino, Italy. (1934) Dr, chemistry (U. di Torino, 1958). Prof. (tel. 011 + 832193). *Structure and conformation, germacranes, crystal growth and polymorphism.*

Calligaris, Prof. Mario. Dip. di Scienze Chimiche, U. di Trieste, Piazzale Europa 1, 34127 Trieste, Italy. (1939) Dr, chemistry (U. di Trieste, 1964). Prof. assoc. (tel. 040 + 51172). *Crystallography - Inorganic chemistry.*

Camalli, Dr Mercedes. Ist. di Strutturistica Chimica 'G. Giacomello', C.N.R., C.P. 10, Monterotondo Stazione, 00016 Roma, Italy. (1947) Dr, chemistry (Buenos Aires U., Argentina, 1974, U. di Napoli, 1978). Coll. tecn. pr. (tel. 06 + 9005142). *Direct methods, computer programming, coordination complexes.*

Cameroni, Prof. Riccardo. Dip. di Scienze Farmaceutiche, U. di Modena, Via S. Eufemia 19, 41100 Modena, Italy. (1926) Dr, chemistry and pharmacy (U. di Modena, 1950 and 1954). Prof., appl. farmaceutical chemistry. (tel. 059 + 219093). *Polymorphic forms, drugs, solvation.*

Campanelli, Dr Anna Rita. Dip. di Chimica, U. di Roma 'La Sapienza', Piazzale A. Moro 5, 00185 Roma, Italy. (1953) Dr, chemistry (U. di Roma, 1978) Ricercatore. (tel. 06 + 4991, ext. 322). *Inclusion compounds, micellar aggregates.*

Candeloro De Sanctis, Prof. Sofia. Dip. di Chimica, U. di Roma 'La Sapienza', Piazzale A. Moro, 00185 Roma, Italy. (1942) DPhil, crystallography (U. di Oxford, UK, 1970). Prof. assoc. (tel. 06 + 4991, ext. 322). *Molecules of biological interest, inclusion compounds, micellar aggregates, packing determination, potential energy calculations.*

Cannas, Prof. Mario. Dip. di Scienze Chimiche, U. di Cagliari, Via Ospedale 72, 09100 Cagliari, Italy. (1926) Dr, chemistry (U. di Cagliari, 1951). Prof., crystallography. (tel. 070 + 668047). *Materials sciences, small organic molecules.*

Cannillo, Dr per Elio. Centro di Studio per la Cristallografia Strutturale, C.N.R., Via A. Bassi 4, 27100 Pavia, Italy. (1938) Dr, chemistry (U. di Pavia, 1962). Coll. tecn. pr. (tel. 0382 + 36689). *Crystal- chemistry of rock-forming minerals; computer programming.*

Capasso, Prof. Sante. Dip. Chimico, U. di Napoli, Via Mezzocannone 4, 80134 Napoli, Italy. (1944) Dr, chemistry (U. di Napoli, 1969). Prof. assoc. (tel. 081 + 205730). *Biological structures, proteins.*

Capotorto, Dr Concetta. Dip. di Ingegneria Chimica, dei Materiali, delle Materie prime e Metallurgia, U. di Roma 'La Sapienza', Via Eudossiana 18, 00184 Roma, Italy. (1936) Dr, chemistry (U. di Roma, 1967). Asst. (tel. 06 + 464286). *Coherent and non-coherent precipitates, martensitic phase, stainless steels.*

Carbonin, Dr Susanna. Ist. di Mineralogia e Petrologia, U. di Padova, Corso Garibaldi 37, 35100 Padova, Italy. (1948) Dr, Natural sciences (U. di Padova, 1972). Ricercatore. (tel. 049 + 663122) *Inorganic crystal structures.*

Carteni-Farina, Prof. Maria. Ist. di Biochimica delle Macromolecole, U. di Napoli, Via Costantinopoli 16, 80138 Napoli, Italy. (1945) Dr, chemistry (U. di Napoli,

1969). Prof. assoc. (tel. 081 + 217052-213668). *Chemistry, sulfonium compounds, polyamine biosynthesis, structure and function, enzymes.*

Caruso, Dr Francesco. Ist. di Strutturistica Chimica 'G. Giacomello', C.N.R., C.P. 10, Monterotondo Stazione, 00016 Roma, Italy. (1947), Dr, chemisty (U. di Buenos Aires, 1974, U. of Napoli, 1977). Coll. tecn. pr. (tel. 06 + 9005142). *Single crystal analysis, coordination compounds, NMR - crystal structure correlation.*

Casalone, Dr Gianluigi. Centro di Studio per le Relazioni tra Struttura e Reattività Chimica, C.N.R., Via Golgi 19, 20133 Milano, Italy. (1939) Dr, industrial chemistry (U. di Milano, 1965). Coll. tecn. pr. (tel. 02 + 292900). *Surface crystallography, low energy electron diffraction (LEED), organic crystal structures.*

Cascarano, Dr Giovanni Luca. Dip. Geomineralogico, U. di Bari, Via Amendola, 70121 Bari, Italy. (1953) Dr, computer science (U. di Bari, 1979). (tel. 080 + 369100 or 214761) *Direct methods, computer science.*

Casellato, Dr Umberto. Ist. di Chimica e Tecn. dei Radioelementi, C.N.R., Corso Stati Uniti, 35100 Padova, Italy. (1942) Dr, chemistry (U. di Padova, 1970). Coll. Tecn. Pr. (tel. 049 + 845111) *Inorganic structural chemistry, coordination compounds, radio-element stereochemistry.*

Catti, Prof. Michele. Dip. di Chimica Fisica ed Elettrochimica, U. di Milano, Via Golgi 19, 20133 Milano, Italy. (1945) Dr, chemistry (U. di Torino, 1969). Prof., crystal chemistry. (tel. 02 + 295229). *Structures, minerals and inorganic materials, lattice energy calculations, elastic properties, phase transitions.*

Cavalca, Prof. Luigi. Ist. di Strutturistica Chimica, U. di Parma, Via M. D'Azeglio 85, 43100 Parma, Italy. (1911) Dr, chemistry (U. di Parma, 1945) Prof., structural chemistry. (tel. 0521 + 22432). *Structure determination methods, inorganic and organic compounds.*

Cellai, Dr Luciano. Ist. di Strutturistica Chimica 'Giordano Giacomello', C.N.R., C.P. 10 - 00016 Monterotondo Stazione (Roma), Italy. (1944) Dr, chemistry (U. di Roma, 1969). Coll. tecn. pr. (tel. 06 + 9005142). *Biologically important compounds.*

Celotti, Dr Giancarlo. Ist. di Chimica e Tecn. dei Materiali e Componenti per l'Elettronica (LAMEL), C.N.R., Via de'Castagnoli 1, 40126 Bologna, Italy. (1943) Dr, physics (U. di Bologna, 1966). Coll. tecn. pr. (tel. 051 + 519593, ext. 218). *Defects, semiconductors, structure determination, inorganic and organic compounds, precise lattice parameter determination, crystallography, perturbed structures.*

Cerrini, Dr Silvio. Ist. di Strutturistica Chimica 'Giordano Giacomello', C.N.R., C.P. 10 - 00016 Monterotondo Stazione (Roma), Italy. (1937) Dr, physics (U. di Roma, 1963). Coll. tecn. pr. (tel. 06 + 9005142). *Biologically important substances, diffraction physics, computer programming.*

Cesari, Prof. lib. doc. Marco. ENIRICERCHE, Via F. Maritano 26, 20097 San Donato Milanese, Milano, Italy. (1930) Dr, chemistry (U. di Parma, 1954). Capo-reparto, lib. doc. at U. of Parma. (tel. 02 + 5205897). *Structures, organometallic and inorganic compounds, structure and morphology, polymers.*

Chiari, Prof. Giacomo. Dip. di Scienze della Terra, U. Via San Massimo 24, 10123 Torino, Italy. (1943) Dr, chemistry (U. di Torino, 1967). Prof. assoc. (tel. 011 + 832193). *Crystal structures, minerals and organic compounds, computer programming.*

Chiesi-Villa, Prof. Angiola. Ist. di Strutturistica Chimica, U. di Parma, Via M. D'Azeglio 85, 43100 Parma, Italy. (1940) Dr, chemistry (U. di Parma, 1964). Prof. assoc. (tel. 0521 + 22432). *Crystal structures, coordination compounds.*

Ciani, Prof. Gianfranco. Ist. di Chimica Generale ed Inorganica, U. degli Studi di Milano, Via G. Venezian 21, 20133 Milano, Italy. (1944) Dr, chemistry (U. di Milano, 1968). Prof. assoc. (tel. 02 + 235120). *Transition metal complexes, metal cluster compounds.*

Cimino, Prof. Alessandro. General Inorganic Chemistry, U. of Rome, Città Universitaria, 00100 Roma, Italy. (1926) Dr, chemistry (U. di Roma, 1948). Prof. (tel. 06 + 4991, ext. 1251). *Solid state chemistry.*

Cini, Prof. Renzo. Dip. di Chimica, U. di Siena, Pian dei Mantellini 44, 53100 Siena, Italy. (1949) Dr, chemistry (U. di Pisa, 1974). Prof. assoc. (tel. 0577 + 47054). *Structures, organometallic compounds, bioinorganic chemistry.*

Cipriani, Prof. Curzio. Ist. di Mineralogia, U. di Firenze, Via La Pira 4, 50121 Firenze, Italy. (1927) Dr, chemistry (U. di Firenze, 1950). Prof. (tel. 055 + 265190). *Crystal chemistry, borates, sulfides, native elements.*

Cirafici, Prof. Salvino S. Ist. di Chimica Fisica, U. di Genova, Corso Europa (Palazzo delle Scienze), 16132 Genova, Italy. (1944) Dr, chemistry (U. di Genova, 1971). Prof. assoc. (tel. 010 + 515076). *Structure and physical properties, intermetallic compounds.*

Clemente, Prof. Dore Augusto. Ist. di Chimica Generale ed Inorganica, Via Loredan 4, 35100 Padova, Italy. (1940) Dr, chemistry (U. di Padova, 1966). Prof. assoc. (tel. 049 + 831279). *Inorganic crystal structures, theoretical molecular geometry.*

Coda, Prof. Alessandro. Dip. di Genetica e Microbiologia, Sez. di Cristallografia, U. di Pavia, Via Taramelli 16, 27100 Pavia, Italy. (1936) Dr, chemistry (U. di Pavia, 1959). Prof., crystallography. (tel. 0382 + 422394). *Organic crystal structures, bio- or pharmacologically interesting structures.*

Coiro, Dr Vincenza Maria. Ist. di Strutturistica Chimica 'Giordano Giacomello', C.N.R., C.P. 10 - 00016 Monterotondo Stazione (Roma), Italy. (1937) Dr, (U. di Napoli, 1961). Coll. tecn. pr. (tel. 06 + 9005641). *Biologically interesting molecules, phase problem solution, semi-empirical potential function calculations, inclusion compounds.*

Cojazzi, Dr Gianna. Centro di Studio per la Fisica delle Macromolecole, C.N.R., Via Selmi 2, 40126 Bologna, Italy. (1935) Dr, chemistry (U. di Padova, 1961). Coll. tecn. pr. (tel. 051 + 235292-3-4-5-6). *Macromolecular physics, crystal structure, polymers.*

Cola, Prof. Mario Luigi. Dip. di Chimica Generale, U. di Pavia, Viale Taramelli 12, 27100 Pavia, Italy. (1922) Dr, chemistry (U. di Pavia, 1950). Prof. assoc. (tel. 0382 + 24714). *Inorganic crystallography.*

Colapietro, Prof. Marcello. Ist. di Strutturistica Chimica 'Giordano Giacomello', C.N.R., C.P. 10 - 00016 Monterotondo Stazione (Roma), Italy. (1937) Dr, physics (U. di Roma, 1962). Coll. tecn. pr. (tel. 06 + 9005641). *Synchrotron radiation, methodology, crystallography of small molecules.*

Colombo, Dr Arturo. Ist. di Chimica delle Macromolecole, C.N.R., Via Bassini 15/A, 20133 Milano, Italy. (1935) Dr, chemistry (U. di Pavia, 1960). Sr. scient. (tel. 02 + 296071). *Structures, organic and natural compounds, computer programming.*

Corchia, Dr Massimo. Divisione Scienza dei Materiali, E.N.E.A. - C.R.E. Casaccia, S. P. Anguillarese 301, Casella Postale N. 2400 00100 Roma, Italy. (1944) Dr, physics (U. di Pavia, 1967). Ricercatore. (tel. 06 + 69484355, telex ENEACAI 613296). *X-ray metallography, solid state physics, inorganic crystal structures, metallic materials.*

Corradini, Prof. Paolo. Dip. di Chimica, U. di Napoli, Via mezzocannone 4, 80134 Napoli, Italy. (1930) Dr, chemistry (U. di Roma, 1951). Prof. (tel. 081 + 205730, ext. 121). *Crystal structure, macromolecules, structure - properties relationship, statistical thermodynamics, polymeric materials.*

Costa-Bizzarri, Prof. Paolo. Ist. di Chimica degli Intermedi, U. di Bologna, Viale Risorgimento 4, 40136 Bologna, Italy. (1941) Dr, industrial chemistry (U. di Bologna, 1966). Prof. assoc. (tel. 051 + 422848). *Electrical conductivity, polymers, doping, charge transfer complexes.*

Cremona, Ing. Luigi. Assing S.p.A., Via A. de Pretis 70, 00184 Roma, Italy. (1945) Dr, engineering (U. di Rome, 1969). Techn. manager. (tel. 06 + 4750341). *High technology equipment, X-ray diffractometry.*

Croatto, Prof. Ugo. Chemistry Dept., U. di Padova, 4 Via Loredan, 35100 Padova, Italy. (1914) Dr, chemistry (U. di Padova, 1936). Prof., inorganic chemistry. (tel. 049 + 650133). *Coordination chemistry, metallorganic chemistry, nuclear chemistry.*

Cubiotti, Prof. Gaetano. Ist. di Fisica Teorica, U. di Messina, Via dei Verdi, 98100 Messina, Italy. (1939) Dr, physics (U. di Messina, 1962). Prof., physics. (tel. 090 + 714492). *Structural properties of disordered systems.*

Dal Negro, Prof. Alberto. Ist. di Mineralogia e Petrologia, U. di Padova, Corso Garibaldi 37, 35100 Padova, Italy. (1941) Dr, chemistry (U. di Pavia, 1964). Prof., crystallography. (tel. 049 + 663122). *Inorganic crystal structures.*

Dapporto, Prof. Paolo. Dip. di Energetica, U. di Firenze, Via Santa Marta 3, 50139 Firenze, Italy. (1942) Dr, chemistry (U. di Firenze, 1966). Prof. (tel. 055 + 4796209). *Structures, metal complexes.*

Davoli, Dr Paolo. Ist. di Mineralogia e Petrologia, U. di Modena, Via S. Eufemia 19, 41100 Modena, Italy. (1958) Dr, Physics (U. di Modena, 1982). (tel. 059 + 218069). *Inorganic crystal structures.*

De Angelis, Prof. Giuseppe. Dip. di Scienze della Terra, U. di Roma 'la Sapienza', Piazzale A. Moro 5, 00185 Roma, Italy. (1944) Dr, geological sciences (U. di Roma, 1970). Asst. (tel. 06 + 4991, ext. 308). *Single crystal diffraction, powder diffraction, Mössbauer methods, crystallographic computer applications.*

Della Casa, Prof. Carlo. Ist. di Chimica degli Intermedi, U. di Bologna, Viale Risorgimento 4, 40136 Bologna, Italy. (1941), Dr, Industrial chemistry (U. di Bologna, 1964). Prof. assoc. (tel. 051 + 422848). *Electrical conductivities of solids, polymers, doping, charge transfer complexes.*

Della Giusta, Prof. Antonio. Ist. di Mineralogia e Petrologia, U. di Padova, Corso Garibaldi 37, 35100 Padova, Italy. (1941) Dr, geology (U. di Genova, 1964). Prof. assoc. (tel. 010 + 505218). *Inorganic crystal structures.*

Dellea Dr Rita. SGEL - Geochimica, AGIP S.p.A., via Bonarelli, San Donato Milanese, 20097 Milano, Italy. (1957) Dr, chemistry (U. di Milano, 1982). (tel. 02 + 520 ext. 38072). *Clay minerals, X-ray fluorescence.*

Del Monte, Prof. Marco Emiliano. Ist. di Geologia, Via Zamboni 67, 40126 Bologna, Italy. (1939) Dr, Geological sciences (U. di Bologna, 1964). Prof., mineralogy and geology (tel. 051 + 228810). *Building stones weathering, anthigene minerals, X-ray powder diffraction, atmospheric aerosol.*

Del Piero, Dr Gastone. ENIRICERCHE, via F. Maritano 76, 20097 San Donato Milanese, Milano, Italy. (1943) Dr, chemistry (U. di Padova, 1968). Ricercatore. (tel. 02 + 52024904) *Structures, inorganic and organometallic compounds, computer programming.*

Del Pra, Prof. Antonio. Ist. di Chimica Farmaceutica, U. di Milano, Viale Abruzzi 42, I-20131 Italy. (1932) Dr, chemistry (U. di Padova, 1962). Prof. *Pharmaceutical structural chemistry.*

Demartin, Dr Francesco. Dip. di Chimica Inorganica e Metallorganica, U. di Milano, Via G. Venezian 21, 20133 Milano, Italy. (1953), Dr, chemistry (U. di Milano, 1978). Ricercatore (tel. 02 + 235120). *Transition metal complexes, metal cluster compounds.*

Demontis, Dr Pierfranco. Ist. di Chimica Fisica, U. di Sassari, Via Vienna 2, 07100 Sassari, Italy. (1954) Dr, Chemistry (U. di Sassari, 1979). Ricercatore. (tel. 079 + 218415) *Molecular dynamics, disordered and amorphous solids, silicates.*

De Pol Blasi, Prof. Carla. Dip. di Scienze della Terra, U. degli Studi, Via Botticelli 23, 20133 Milano, Italy. (1936) Dr, geological sciences (U. di Milano, 1960). Prof. assoc. (tel. 02 + 293994). *Rock-forming minerals, especially feldspars, X-ray crystallography.*

De Santis, Prof. Pasquale. Dip. di Chimica, U. di Roma, Piazzale A. Moro 5, 00100 Roma, Italy. (1935) Dr, chemistry (U. di Bari, 1959). Prof., physical chemistry. (tel. 06 + 4952993). *Conformational analysis, structure and properties, biological macromolecules and model compounds.*

Dessy, Dr Giulia. Ist. Teoria e Struttura Elettronica e Comportamento Spettrochimico dei Composti di Coordinazione, C.N.R., Area della Ricerca di Roma, 00016 Monterotondo Stazione, Roma, Italy. (1936) Dr, chemistry (U. di Milano, 1964). Coll. tecn. pr. (tel. 06 + 90020341). *X-ray crystal structure, coordination compounds.*

Destro, Prof. Riccardo. Dip. di Chimica Fisica ed Elettrochimica, U. degli Studi di Milano, Via Golgi 19, 20133 Milano, Italy. (1940) Dr, chemistry (U. di Milano, 1964). Prof. assoc. (tel. 02 + 292900 or 293805). *Molecular structure, chemical reactivity, charge density distributions, low temperature crystallography.*

Di Blasio, Prof. Benedetto. Dip. di Chimica, U. di Napoli, Via Mezzocannone 4, 80134 Napoli, Italy. (1945) Dr, physics (U. di Napoli, 1971). Prof. assoc. (tel. 081 + 206800). *Conformational analysis, peptides.*

D'Ilario, Prof. Lucio. Laboratori Ricerche di Base, Assoreni, Via Ramarini, 00015 Monterotondo (Roma), Italy. (1946) Dr, chemistry (U. di Rome, 1969). Scient. *Polymer morphology, polymer structure, conformational analysis.*

Di Vaira, Prof. Massimo. Dip. di Chimica, U. di Firenze, Via Maragliano 77, 50144 Firenze, Italy. (1940) Dr, chemistry (U. di Firenze, 1963). Prof. (tel. 055 + 2476956). *Structures, organometallic compounds, metal complexes.*

Domeneghetti, Dr Maria Chiara. Centro di Studio per la Cristallografia Strutturale, C.N.R., Via Bassi 4, 27100 Pavia, Italy. (1954) Dr, natural sciences (U. di Pavia, 1978). Coll. tecn. pr.(tel. 0382 + 36689). *Mineral crystal-chemistry.*

Domenicano, Prof. Aldo. Dip. di Chimica, U. di Roma 'La Sapienza', Città Universitaria, 00185 Roma, Italy. (1938) Dr, chemistry (U. di Roma, 1965). Prof. assoc. (tel. 06 + 6374803). *Crystallography of small molecules, gas electron diffraction, substituent effects on molecular structure.*

Domenici, Dr Marcello. R & D, Dynamit Nobel Silicon DNS S.p.A., Viale Gherzi 31, 28100 Novara, Italy. (1932) Dr, physics (U. di Pisa, 1958). R & D director. (tel. 321 + 442, ext. 376, telex 200486). *Neutron diffraction, crystal growth and characterization.*

Domiano, Prof. Paolo. Ist. di Strutturistica, U. di Parma, Via M. D'Azeglio 85, 43100 Parma, Italy. (1936) Dr, chemistry (U. di Parma, 1960). Prof. assoc. (tel. 0521 + 22432). *Computer programming, crystal structures.*

Donati, Dr Donato. Ist. di Chimica Organica, U. di Siena, Pian dei Mantellini 44, 53100 Siena, Italy. (1946) Dr, chemistry (U. di Firenze, 1970). Prof. assoc. (tel. 0577 + 47054). *Structures, organic compounds, photochemistry.*

Dovesi, Prof. Roberto. Ist. di Chimica Teorica, U. di Torino, Via P. Giuria 5, 10125 Torino, Italy. (1947) Dr, chemistry (U. di Torino, 1971). Prof. assoc. (tel. 011 + 657274). *Solid state theoretical chemistry.*

Emiliani, Prof. Francesco. Ist. di Mineralogia, U. di Parma, Via Gramsci 9, 43100 Parma, Italy. (1923) Dr, chemistry (U. di Bologna, 1949). Prof., mineralogy. (tel. 0521 + 208174, ext. 15). *Mineralogy, geochemistry.*

Fagherazzi, Prof. Giuliano. Dip. di Chimica Fisica, U. Ca' Foscari di Venezia, D.D. 2137, 30123 Venezia, Italy. (1938) Dr, physics (U. di Padova, 1961). Prof., solid state physical chemistry. (tel. 041 + 27554). *Submicrostructural studies by small and wide angle X-ray scattering on inorganic polycristalline and amorphous materials.*

Fagnani, Prof. Gustavo. Dip. di Scienze della Terra, U. degli Studi, Via Botticelli 23, 20133 Milano, Italy. (1917) Dr, natural sciences (U. di Milano, 1942). Prof. assoc. (tel. 02 + 293994). *Alpine mineral morphology.*

Fanfani, Prof. Luca. Dip. di Scienze della Terra, U. di Cagliari, Via Trentino 51, 09100 Cagliari, Italy. (1941) Dr, chemistry (U. di Firenze, 1964). Prof., mineralogy. (tel. 070 + 290505). *Inorganic compounds.*

Fares, Dr Vincenzo. Ist. Teoria e Struttura Elettronica e Comportamento Spettrochimico dei Composti di Coordinazione, C.N.R., Area della Ricerca di Roma, 00016 Monterotondo Stazione, Roma, Italy. (1942) Dr, chemistry (U. di Roma, 1966). Coll. tecn. pr. (tel. 06 + 90020285). *X-ray crystal structure, coordination compounds.*

Favretto, Prof. Luciano. Ist. di Merceologia, U. di Trieste, Piazzale Europa, 34100 Trieste, Italy. (1932) Dr, chemistry (U. di Trieste, 1956). Prof. (tel. 040 + 54889). *Environmental chemistry, clay mineralogy.*

Fedeli, Prof. Walter. Dip. di Chimica, Ingegneria Chimica e Materiali, U. dell'Aquila, Via Assergi 4, 67100 L'Aquila, Italy. (1933) Dr, chemistry (U. di Roma, 1961). Prof., general chemistry. (tel. 0862 + 25387). *Biologically important substances.*

Ferracini, Prof. Elena. Ist. di Chimica, U. di Bologna, Via Selmi 2, 40126 Bologna, Italy. (1925) Dr, industrial chemistry (U. di Bologna, 1962). Prof. assoc. (tel. 051 + 235292-3-4-5-6). *SAS, WAXS, natural and synthetic polymers.*

Ferrari-Belicchi, Prof. Marisa. Ist. di Strutturistica Chimica, U. di Parma, Via Massimo D'Azeglio 85, 43100 Parma, Italy. (1942) Dr, chemistry (U. di Parma, 1966). Prof. assoc. (tel. 0521 + 22432). *Crystal structures.*

Ferraris, Prof. Giovanni. Dip. di Scienze della Terra, U. di Torino, Via S. Massimo 24, 10123 Torino, Italy. (1937) Dr, physics (U. di Torino, 1960). Prof., crystallography. (tel. 011 + 832193). *Silicate minerals, inorganic crystal chemistry, hydrogen bonding.*

Ferrero Rognoni, Prof. adele. Ist. Chimico 'G. Ciamician', U. di Bologna, via F. Selmi 2 40126 Bologna, Italy. (1925), Dr, chemistry (U. di Bologna, 1950). Prof. assoc. (tel. 03951 + 235292). *SAS, WAXS, natural and synthetic polymers.*

Ferretti, Dr Valeria. Dip. di Chimica, U. di Ferrara, Via L. Borsari 46, 44100 Ferrara, Italy. (1958) Dr, chemistry (U. di Ferrara, 1981). Ricercatore. (tel. 0532 + 33522, ext. 15). *Determination of crystal structures, molecular systematics, crystal packing, molecular pharmacology.*

Fichera, Dr Anna Maria. Centro di Studio per la Fisica delle Macromolecole, C.N.R., Via Selmi 2, 40126 Bologna, Italy. (1939) Dr, chemistry (U. di Bologna, 1965). Coll. tecn. pr. (tel. 051 + 235292-3-4-5-6). *Macromolecular physics, crystal structure, polymers.*

Filippini, Dr Giuseppe. Centro di Studio per le Relazioni tra Struttura e Reattività Chimica, U. di Milano, Via Golgi 19, 20133 Milano, Italy. (1940) Dr, industrial chemistry (U. di Milano, 1968). Coll. tecn. pr. (tel. 02 + 292900). *Lattice dynamics, molecular crystals, intermolecular forces, molecular structure, chemical reactivity.*

Foresti, Prof. Elisabetta. Ist. di Tecn. Chimiche Speciali, Facolta' di Chimica Industriale, Viale Risorgimento 4, 40136 Bologna, Italy. (1943) Dr, chemistry (U. di Bologna, 1968). Prof. assoc. (tel. 051 + 412951). *Crystal structures, biological apatites.*

Fornasini, Prof. Maria Luisa. Ist. di Chimica Fisica, U. di Genova, Corso Europa (Palazzo delle Scienze), 16132 Genova, Italy. (1941) Dr, chemistry (U. di Genova, 1965). Prof. assoc. (tel. 010 + 515076). *Structure and physical properties, intermetallic compounds.*

Forni, Prof. Flavio. Dip. di Scienze Farmaceutiche, U. di Modena, Via S. Eufemia 19, 41100 Modena, Italy. (1951) Dr, Pharmacy and Biological sciences (U. di Modena, 1973, 1978). Prof. assoc. (tel. 059 + 219093). *Polymorphic forms, drugs, structure.*

Forsellini, Dr Eleonora. Ist. di Chimica e Tecn. dei Radioelementi, C.N.R., Corso Stati Uniti, 35100 Padova, Italy. (1934) Dr, chemistry (U, di Padova, 1965). Coll. tecn. pr. (tel. 049 + 760933). *Inorganic structural chemistry, coordination compounds, radioelement stereochemistry.*

Forti, Prof. Paolo, Ist. di Geologia, U. di Bologna, Via Zamboni 67, 40127 Bologna, Italy. (1945) Dr, chemistry (U. di Bologna, 1969). Prof. assoc. (tel. 051 + 228810). *Mineral diagenesis in natural caves.*

Franceschi, Prof. Enrico. Ist. di Chimica Fisica, U. di Genova, Corso Europa (Palazzo delle Scienze), 16132 Genova, Italy. (1942) Dr, chemistry (U. di Genova, 1965). Prof. assoc. (tel. 010 + 515076). *Structure and physical properties, intermetallic compounds.*

Franchini, Prof. Marinella. Dip. di Scienze della Terra, U. di Torino, Via San Massimo 24, 10123 Torino, Italy. (1936) Dr, natural sciences (U. di Torino, 1959). Prof. assoc. (tel. 011 + 832193). *Crystal growth, clay mineralogy.*

Franzini, Prof. Marco. Dip. di Scienze della Terra, U. di Pisa, Via S. Maria 53, 56100 Pisa, Italy. (1938) Dr, geological sciences (U. di Pisa, 1960). Prof., mineralogy. (tel. 050 + 501457). *Optical crystallography, low energy radiation interaction with condensed matter.*

Franzosi, Dr Paolo. MASPEC Ist. Materiali Speciali per Elettronica e Magnetismo, C.N.R., Via Chiavari 18/A, 43100 Parma, Italy. (1948) Dr, physics (U. di Parma, 1972). Coll. tecn. pr. (tel. 0521 + 96841). *X-ray topography, electron microscopy, electronics materials, X-ray microanalysis.*

Frigeri, Dr Cesare. Ist. MASPEC, C.N.R., Via Chiavari 18/A, 43100 Parma, Italy. (1950) Dr, Physics (U. di Bologna). Coll. tecn. pr. (tel. 0521 + 96841). *X-ray diffraction analysis, electron microscopy, materials science.*

Fumi, Prof. Fausto Gherardo. Dip. di Fisica, U. di Genova, Viale Dodecaneso 33, 16146 Genova, Italy. (1924) Dr, physics (U. di Genova, 1948). Prof., solid state physics. (tel. 010 + 59931, ext. 222, telex 211154 INFNGE I). *Crystal physics, ionic solids, lattice imperfections.*

Galli, Prof. Ermanno. Ist. di Mineralogia e Petrologia, U. di Modena, Via S. Eufemia 19, 41100 Modena, Italy. (1937) Dr, geological sciences (U. di Modena, 1963). Prof., mineralogy (tel. 059 + 218062). *Zeolites, silicates, Inorganic crystal structures.*

Ganazzoli, Dr Fabio. Dip. di Chimica, Politecnico di Milano, Piazza L. da Vinci 32, 20133 Milano, Italy. (1955) Dr, chemistry (U. di Milano, 1978). Ricercatore. (tel. 02 + 230845). *X-ray crystal structures, semiempirical quantum studies, polymers.*

Garbassi, Dr Fabio. Ist. di Ricerche ,G. Donegani, Montedison S.p.A., Via Fauser 4, 28100 Novara, Italy. (1942) Dr, chemistry (U. di Trieste, 1965). Ricercatore. (tel. 0321 + 24701). *Structure, catalytically interesting materials, clean surfaces, adsorbate layers, low energy electron diffraction.*

Gasparri-Fava, Prof. Giovanna. Ist. di Chimica Generale, U. di Parma, Via Massimo D'Azeglio 85, 43100 Parma, Italy. (1930) Dr, chemistry (U. di Parma, 1955). Prof., general chemistry. (tel. 0521 + 34400). *Crystal structures.*

Gastaldi, Dr Leonardo. Ist. Teoria e Struttura Elettronica e Comportamento Spettrochimico dei Composti di Coordinazione, C.N.R., Area della Ricerca di Roma, 00016 Monterotondo Stazione, Roma, Italy. (1944) Dr, chemistry (U. di Roma, 1969). Coll. tecn. pr. (tel. 06 + 90020343). *Coordination compounds.*

Gatti, Dr Giuseppina. Dip. di Genetica e Microbiologia, Sez. di Cristallografia, U. di Pavia, via Taramelli 16, 27100 Pavia, Italy. (1954) Dr, chemistry (U. di Pavia, 1978). Tecn. laur. (tel. 0382 + 422394). *Proteins, organic compounds.*

Gauzzi, Prof. Franco. Dip. di Ingegneria Meccanica, U. di Roma Tor Vergata, Via O. Raimondo, 00173 Roma, Italy. (1931) Dr, chemistry (U. di Roma, 1955). Prof. inc. stab. (tel. 06 + 464286). *Martensitic transformations, metallic and non metallic materials, composite materials, quantitative measurements, defects in crystals, recrystallization.*

Gavezzotti, Prof. Angelo. Dip. di Chimica Fisica ed Elettrochimica, U. di Milano, Via Golgi 19, 20133 Milano, Italy. (1944) Dr, chemistry (U. di Milano, 1968). Prof. assoc. (tel. 02 + 292900). *Non-bonded interactions, organic solid state reactivity.*

Gavuzzo, Dr Enrico. Ist. di Strutturistica Chimica 'Giordano Giacomello', C.N.R., C.P. 10 - 00016 Monterotondo Stazione (Roma), Italy. (1943) Dr, chemistry (U. di Roma, 1972). Coll. tecn. pr. (tel. 06 + 9005142). *Biologically important substances.*

Gazzano, Dr Massimo. Istituto di Chimica 'G. Ciamician', U. di Bologna, Via Selmi 2, 40126 Bologna, Italy. (1958) Dr, chemistry (U. di Bologna, 1982) (tel. 051 + 235292). *Crystal structures and biological structures.*

Gazzoni, Dr Giuseppe. Dip. di Scienze della Terra, U. di Torino, Via San Massimo 22, 10123 Torino, Italy. (1937) Dr, physics (U. di Torino, 1963). Coll. tecn. pr. (tel. 011 + 832193). *Crystal chemistry, crystal structure, framework silicates and germanates.*

Gervasio, Prof. Giuliana. Ist. di Chimica Generale ed Inorganica, U. di Torino, Corso M. D'Azeglio 48, 10125 Torino, Italy. (1943) Dr, chemistry (U. di Torino, 1967). Prof. assoc. (tel. 011 + 655831-3). *Crystal staructures, coordination and organometallic compounds.*

Ghezzi, Prof. Carlo. Dip. di Fisica, U. di Parma, Via M. D'Azeglio 85, 43100 Parma, Italy. (1940) Dr, physics (U. di Milano, 1963). Prof. (tel. 0521 + 96841). *Diffuse X-ray scattering, lattice defects, semiconductors, semiconductor physics.*

Ghilardi, Dr Carlo Alfredo. Ist. Stereochimica, C.N.R., Via F. D. Guerrazzi 27, 50132 Firenze, Italy. (1943) Dr, chemistry (U. di Firenze, 1968). Coll. tecn. pr. (tel. 055 + 243990). *Structures, metal complexes.*

Giacovazzo, Prof. Carmelo. Dip. Geomineralogico, U. di Bari, Via Amendola, 70121 Bari, Italy. (1940) Dr, chemistry (U. di Bari, 1965). Prof., mineralogy. (tel. 080 + 369100, ext. 353). *Direct methods for phase solution.*

Giglio, Prof. Edoardo. Dip. di Chimica, U. di Roma, Piazzale delle Scienze, 00185 Roma, Italy. (1931) Dr, chemistry (U. di Bari, 1955). Prof., practical physical chemistry. (tel. 06 + 4952993). *Crystal packing determination, potential energy calculations, synthetic and biological polymers, inclusion compounds, conformational analysis, micellar aggregates.*

Gilli, Prof. Gastone. Dip. di Chimica, U. di Ferrara, Via L. Borsari 46, 44100 Ferrara, Italy. (1937) Dr, chemistry (U. di Ferrara, 1962). Prof., general chemistry. (tel. 0532 + 33522, ext. 20). *Teaching, crystal structures determination, crystal packing, molecular systematics, molecular pharmacology, powder diffraction.*

Giordani, Prof. Marino. Ist. di Chimica, Facoltà di Ingegneria, U. di Genova, Fiera del Mare, Padiglione D, Piazzale J.F. Kennedy, 16129 Genova, Italy. (1944) Dr, chemistry (U. di Genova, 1968). Prof. assoc. (tel. 010 + 566427). *Material science.*

Giordano, Prof. Federico. Dip. di Chimica, U. di Napoli, Via Mezzocannone 4, 80134 Napoli, Italy. (1939) Dr, chemistry (U. di Napoli, 1964). Prof. assoc. (tel. 081 + 206800). *Organic and biological crystal structures, proteins.*

Giordano Orsini, Prof. Paolo. Dip. di Ingegneria dei Materiali e della Produzione, U. di Napoli, Piazzale Tecchio, 80125 Napoli, Italy. (1926) Dr, industrial chemistry (U. di Napoli, 1949). Prof., chemistry. (tel. 081 + 610437, telex INGENA I 722392). *Science of materials, microstructre; morphology of solids, solid state transformations.*

Giunchi, Dr Giovanni. Ist. G. Donegani S.p.A., Via Fauser 4, 28100 Novara, Italy. (1946), Dr, physics (U. di Bologna, 1970). Head department. (tel 0321 + 24701). *Crystallography. Single crystals, polymer structures, theory diffraction by disordered structures, conformational analysis of macromolecules.*

Giuseppetti, Prof. Giuseppe. Dip. di Scienze della Terra, U. di Pavia, Via A. Bassi 4, 27100 Pavia, Italy. (1923) Dr, chemistry and pharmacy (U. di Camerino, 1947 and 1949). Prof., mineralogy. (tel. 0382 + 21135). *Inorganic and organic crystal structures.*

Gottardi, Prof. Glauco. Ist. di Mineralogia e Petrologia, U. di Modena, Via Santa Eufemia 19, 41100 Modena, Italy. (1928) Dr, chemistry (U. di Pisa, 1951). Prof., mineralogy. (tel. 059 + 218062). *Zeolites, feldspatoids, apparatus.*

Gramaccioli, Prof. Carlo Maria. Dip. di Scienze della Terra, U. di Milano, Via Botticelli 23, 20133 Milano, Italy. (1935) Dr, industrial chemistry (U. di Milano, 1959). Prof., physical chemistry (tel. 02 + 293994). *Thermodynamicsal properties and spectra of minerals.*

Graziani, Prof. Rodolfo. Ist. di Chimica Generale, U., Via Loredan 4, 35100 Padova, Italy. (1937) Dr, chemistry (U. di Padova, 1964). Prof. assoc. (tel. 049 + 831279). *Structure and bonding, metal complexes.*

Guastini, Prof. Carlo. Ist. di Strutturistica Chimica, U. di Parma, Via M. D'Azeglio 85, 43100 Parma, Italy. (1931) Dr, chemistry (U. di Parma, 1962). Prof. assoc. (tel. 0521 + 22432). *Crystal structures, coordination compounds.*

Guerreschi, Dr Luigi Giuseppe. Dip. di Chimica, Città Universitaria, 00185 Roma, Italy. (1921) Dr, chemistry (U. di Roma, 1949). Coll. tecn. pr., Prof. inc. stab. (tel. 06 + 4991, ext. 297).

Guzman, Dr Luis Alberto. Ist. per la Ricerca Scientifica e Tecn. (I.R.S.T.), 38050 Povo, Trento, Italy. (1947) Dr, Physics (U. Geneva, Switzerland, 1972). Sr. scient. (tel. 0461 + 810105). *X-ray crystallography, metals and alloys, surface phases, ion implantation, ion-beam-mixing.*

Iandelli, Prof. Aldo. Ist. di Chimica Fisica, U. di Genova, Corso Europa (Palazzo delle Scienze), 16132 Genova, Italy. (1912) Dr, chemistry (U. di Firenze, 1933). Prof., physical chemistry, inst. dir. (tel. 010 + 515076). *Structure and physical properties, intermetallic compounds.*

Ianelli, Dr Sandra. Ist. di Chimica Generale, U. di Parma, Via M. D'Azeglio 85, 43100 Parma, Italy. (1951) Dr, Chemistry (U. di Parma, 1976). Ricercatore. (tel. 0521 + 34400). *Coordination and organometallic chemistry.*

Immirzi, Prof. Attilio. Ist. di Ingegneria, Fac. di Scienze, Via Baronissi, 84100 Salerno, Italy. (1938) Dr, industrial chemistry (U. di Napoli, 1961). Prof. *Transition metal compounds, structures, natural organic substances, crystallographic computing.*

Imperatori, Dr Patrizia. Ist. Teoria e Struttura Elettronica e Comportamento Spettrochimico dei Composti di Coordinazione, C.N.R., Area della Ricerca di Roma, 00016 Monterotondo Stazione, (Roma), Italy. (1957) Dr, chemistry (U. di Roma, 1981). Coll. tecn. pr. (tel. 06 + 90020341). *X-ray crystal structure, coordination compounds.*

Isetti, Prof. Giovanni. Ist. di Mineralogia, U. di Genova, Corso Europa, Palazzo delle Scienze, 16132 Genova, Italy. (1925) Dr, chemistry (U. di Genova, 1949). Prof., mineralogy. (tel. 010 + 518184). *Crystal physics, lattice defects.*

Ivaldi, Dr Gabriella. Dip. di Scienze della Terra, U. di Torino, Via S. Massimo 24, 10123 Torino, Italy. (1950) Dr, natural sciences (U. di Torino, 1974). Ricercatore. (tel. 011 + 832193). *Structures, inorganic compounds and minerals.*

Kovács, Prof. Alessandro L. Dip. di Medicina Sperimentale, U. di Roma, Città Universitaria, 00100 Roma, Italy. (1936) Dr, physics (U. di Bologna, 1961). Prof., physics. (tel. 06 + 493971). *Molecular dynamics, mathematical modelling, information theory.*

Krajewski, Dr Adriano. Dip. di Ceramiche Classiche, Ist. di Ricerche Tecnologiche per la Ceramica, C.N.R., Via Granarolo 64, 48018 Faenza (Ravenna), Italy. (1947) Dr, chemistry (U. di Bologna, 1971). (tel. 0546 + 46147, ext. 87). *Structure - activity correlation in solid state, numerical analysis, crystallography, crystal growth, sintering, glazes, bioceramics, refractories, electronic ceramics, ancient ceramics.*

Lagomarsino, Dr Stefano. Ist. di Elettronica dello Stato Solido, C.N.R., Via Cineto Romano 42, 00156 Roma, Italy. (1948) Dr, physics (U. di Roma). Coll. tecn. pr. (tel. 06 + 4124345). *X-ray diffraction, materials characterization.*

Lamba, Dr Doriano. Ist. di Strutturistica Chimica 'G. Giacomello', C.N.R., C. P. 10, 00016 Monterotondo Stazione (Roma), Italy. (1955) Dr, chemistry (U. di Roma, 1980). Coll. tecn. pr. (tel 06 + 9005142). *Direct methods, organic crystal structures, bio or pharmacologically interesting structures.*

Lanfranchi, Dr Maurizio. Ist. di Chimica Generale ed Inorganica, U. di Parma, Via M. D'Azeglio 85, 43100 Parma, Italy. (1942) Dr, chemistry (U. di Parma, 1985). Tecn. laur. (tel. 0521 + 34400). *Crystal structures, inorganic, organic and coordination compounds.*

Leoni, Prof. Leonardo. Dip. di Scienze della Terra, U. di Pisa, Via S. Maria 53, 56100 Pisa, Italy. (1944) Dr, geological sciences (U. di Pisa, 1967). Prof. assoc. (tel. 050 + 501417). *Optical crystallography, low energy radiation interaction with condensed matter.*

Leporati, Prof. Enrico. Ist. di Chimica Generale ed Inorganica, U. di Parma, Via M. D'Azeglio 85, 43100 Parma, Italy. (1934) Dr, chemistry (U. di Bologna, 1964). Prof. assoc. (tel. 0521 + 34400). *Crystal structures, inorganic - organic and organometallic compounds.*

Licci, Dr Francesca. MASPEC - C.N.R., Via Chiavari 18/A 43100 Parma, Italy. (1945) Dr, chemistry (U. di Bologna, 1969). Coll. tecn. pr. (tel. 0521 + 96841) *Crystal growth, magnetic material, chemical analysis, DTA analysis.*

Licheri, Prof. Giovanni. Dip. di Scienze della Terra, U. di Cagliari, Via Ospedale 72, 09100 Cagliari, Italy. (1939) Dr, physics (U. di Cagliari, 1962). Prof. (tel. 070 + 668047). *EXAFS. Amorphous materials.*

Liquori, Prof. Alfonso Maria. Ist. di Chimica Fisica, U. di Roma II 'Tor Vergata', Via O. Raimondo, 00173 Roma, Italy. (1926) Dr, chemistry (U. di Roma, 1948). Prof., physical chemistry. (tel. 06 + 79791, ext. 2420). *Molecular conformation, macromolecules, conformational transition, tertiary structure, proteins, crystal packing.*

Locchi, Prof. Stelio Giovanni. Dip. di Chimica Generale, U. di Pavia, Viale Taramelli 12, 27100 Pavia, Italy. (1929) Dr, chemistry (U. di Pavia, 1952). Prof. assoc. (tel. 0382 + 29270). *Inorganic and organic crystal structures.*

Loreto, Prof. Lucio. Ist. di Mineralogia e Petrografia, U. di Roma, Piazzale delle Scienze, Città Universitaria, 00185 Roma, Italy. (1937) Dr, geological science (U. di Roma, 1965). Prof. assoc. (tel. 06 + 4991, ext. 308). *Symmetry, crystal growth, morphology, crystallographic computer applications, metamictic state, radiation damage in crystals.*

Magini, Dr Mauro. ENEA - Divisione Chimica, Via Anguillarese, 00060 CRE-Casaccia, Roma, Italy. (1941) Dr, chemistry (U. di Roma, 1965). Res. (tel. 06 + 69483391). *Solutions, liquids, amorphous structures.*

Malpezzi, Dr Luciana. Dip. di Chimica, Politecnico di Milano, Piazza L. da Vinci 32, 20133 Milano, Italy. (1946) Dr, Chemistry (U. di Bologna, 1970). Ricercatore. (tel. 02 + 230845, ext. 256). *Crystallography, conformational analysis.*

Malta, Dr Viscardo. Centro di Studio per la Fisica delle Macromolecole, C.N.R., Via Selmi 2, 40126 Bologna, Italy. (1938) Dr, chemistry (U. di Bologna, 1968). Coll. tecn. pr. (tel. 051 + 235292-3-4-5-6). *Macromolecular physics, crystal structure, polymers.*

Mammi, Prof. Mario. Dip. di Chimica Organica, U. di Padova, Via Marzolo 1, 35100 Padova, Italy. (1932) Dr, chemistry (U. di Padova, 1956). Prof., structural chemistry, dir., Biopolymer Res. Centre of C.N.R. (tel. 049 + 663736 and 831-234 or 232 or 111, telex 430176 UNIPADU I). *Biologically important molecules, organic compounds, instrumentation.*

Manassero, Prof. Mario. Ist. di Chimica Generale ed Inorganica, U. degli Studi di Milano, Via G. Venezian 21, 20133 Milano, Italy. (1944) Dr, chemistry (U. di Milano, 1968). Prof. assoc. (tel. 02 + 235120). *Transition metal complexes, metal cluster compounds, crystallographic computing.*

Mancini, Dr Annamaria. Dip. di Fisica, U. di bari, Via Amendola 173, 70126 Bari, Italy. (1944) Dr, chemistry (U. di Bari, 1970). Prof. assoc. (tel. 080 + 331044). *Crystal growth - Structural and physical characterization.*

Mangani, Dr Stefano. Dip. di Chimica, U. di Siena, Pian dei Mantellini 44, 53100 Siena, Italy. (1951) Dr, chemistry (U. di Firenze, 1977). Ricercatore. (tel. 0577 + 47054). *X-ray diffraction, EXAFS, bioinorganic chemistry.*

Mangia, Prof. Alessandro. Ist. di Chimica Generale ed Inorganica, U. di Parma, Via D'Azeglio 85, 43100 Parma, Italy. (1941) Dr, chemistry (U. di Parma, 1965). Prof. assoc. (tel. 0521 + 34400). *Organometallic and coordination compounds.*

Manotti-Lanfredi, Prof. Anna Maria. Ist. di Chimica Generale ed Inorganica, U. di Parma, Via M. D'Azeglio 85, 43100 Parma, Italy. (1933) Dr, chemistry (U. di Parma, 1957). Prof. assoc. (tel. 0521 + 34400). *Crystal structures, inorganic, bioinorganic and organometallic compounds.*

Mantovani, Prof. Giorgio. Ist. Chimico, U. di Ferrara, Via Luigi Borsari 46, 44100 Ferrara, Italy. (1925) Dr, chemistry (U. di Ferrara, 1948). Prof., industrial chemistry. (tel. 0532 + 36374). *Crystal growth, habit modifications.*

Marchetti, Dr Fabio. Dip. di Chimica e Chimica Industriale, U. di Pisa, Via Risorgimento 35, 56100 Pisa, Italy. (1950) Dr, chemistry (U. di Pisa, 1974). Ricercatore. (tel. 050 + 28238). *Coordination chemistry, crystal chemistry.*

Marigo, Prof. Antonio. Dip. di Chimica Inorganica, Organometallica ed Analitica, U. di Padova, Via Loredan 4, 35100 Padova, Italy. (1950) Dr, chemistry (U. di Padova, 1974) Prof. assoc. (tel. 049 + 831287). *Macromolecular physics, crystal structure, macromolecules, structural investigations, polyolefins and plyammides.*

Marongiu, Prof. Giaime. Dip. di Scienze Chimiche, U. di Cagliari, Via Ospedale 72, 09100 Cagliari, Italy. (1939) Dr, chemistry (U. di Cagliari, 1963). Prof. assoc. (tel. 070 + 668047). *Materials sciences, small organic molecules.*

Martinelli, Prof. Giuliano. Ist. di Fisica, U. di Ferrara, Via Paradiso 12, 44100 Ferrara, Italy. (1938) Dr, physics (U. di Ferrara, 1969). Prof. assoc. (tel. 0532 + 39292). *Electron diffraction studies, lattice defects, single crystal X-ray diffraction, precise lattice parameter determination.*

Martorana, Dr Antonino. Dip. di Chimica Inorganica, Organometallica ed Analitica, U. di Padova, Via Loredan 4, 35100 Padova, Italy. (1952) Dr, physics (U. di Padova, 1979). Ricercatore. (tel. 049 + 831287). *Macromolecular physics, crystal structure, structural disorder, macromolecules, structural investigations, polyolefins and polyammides.*

Massarotti, Prof. Vincenzo. Dip. di Chimica Fisica, U. di Pavia, Viale Taramelli 16, 27100 Pavia, Italy. (1944) Dr, chemistry (U. di Pavia, 1968). Prof. assoc. (tel. 0382 + 27082). *Solid state kinetics, Defects in ionic crystals, structural properties.*

Mattia, Prof. Carlo. Dip. di Chimica, U. di Napoli, Via Mezzocannone 4, 80134 Napoli, Italy. (1946) Dr, chemistry (U. di Napoli, 1970). Prof. assoc. (tel. 081 + 206800). *Biological structures, proteins.*

Mattias, Prof. Pierpaolo. Ist. di Geologia appl. e Giacimenti Minerari, Fac. di Ingegneria, Via Eudossiana 18, 00184 Roma, Italy. (1936) Dr, geological sciences (U. di Roma, 1960). Prof. assoc. (tel. 06 + 461810 or 465913). *Clays and clay minerals.*

Mazza, Dr Fernando. Ist. di Strutturistica Chimica 'Giordano Giacomello', C.N.R., C.P. 10 - 00016 Monterotondo Stazione (Roma), Italy. (1938) Dr, chemistry (U. di Roma, 1964). Coll. tecn. pr., prof. inc. (tel. 06 + 9005641). *Organic structures, biological structures, inclusion compounds, Potential energy calculations.*

Mazzarella, Prof. Lelio. Dip. di Chimica, U. di Napoli, Via Mezzocannone 4, 80134 Napoli, Italy. (1938) Dr, chemistry (U. di Napoli, 1961). Prof. (tel. 081 + 206800). *Biological structures, proteins.*

Mazzi, Prof. Fiorenzo. Dip. di Scienze della Terra, U. di Pavia, Via A. Bassi 4, 27100 Pavia, Italy. (1924) Dr, chemistry (U. di Firenze, 1947). Prof., mineralogy. (tel. 0382 + 21135). *Inorganic crystal structures.*

Mealli, Prof. Carlo. Ist. per lo Studio della Stereochimica ed Energetica dei Composti di Coordinazione, ISSECC - C.N.R., Via Guerrazzi 27, 50132 Firenze, Italy. (1946) Dr, chemistry (U. di Firenze, 1969). Coll. tecn. pr. (tel. 055 + 243990). *Structure determination, transition metal complexes, organometallic compounds.*

Mellini, Dr Marcello. Dip. di Scienze della Terra, C.N.R., Via S. Maria 53, 56100 Pisa, Italy. (1949) Dr, chemistry (U. di Pisa, 1974). Coll. tecn. pr. (tel. 050 + 501457). *Structural mineralogy, electron microscopy.*

Menchetti, Prof. Silvio. Dip. di Scienze della Terra, U. di Firenze, Via La Pira 4, 50121 Firenze, Italy. (1937) Dr, geological sciences (U. di Firenze, 1961). Prof. of mineralogy (tel. 055 + 287140). *Inorganic crystal structures, crystal chemistry.*

Menzinger, Prof. Filippo. Dip. di Fisica, U. di Roma II, via O. Raimondo, 00173 Roma, Italy. (1937) Dr, physics (U. di Roma, 1961). Prof. (tel. 06 + 479791, ext. 2321). *Spin density in metals, lattice dynamics, static and dynamical structure, liquids and amorphous materials, neutron diffraction, spectroscopy.*

Meriani, Dr Sergio. Ist. di Chimica Applicata, U. di Trieste, Via Valerio 2, 34127 Trieste, Italy. (1941) Dr, inorganic chemistry (U. di Trieste, 1966). Prof. assoc. *Applied chemistry, solid state reactions, ceramics.*

Merlini, Prof. Alfonso Enrico. Physics Div., C.C.R. Euratom (Joint Res. Center of the European Community), 21020 Ispra, Varese, Italy. (1926) PhD, metallurgical engineering (Illinois U., USA, 1954). Head, physics div. (tel. 0332 + 789809, telex 380042,380058 EUR I). *Dynamical diffraction theory, Fine structure of X-ray absorption (EXAFS), synchrotron light applications.*

Merlino, Prof. Stefano. Dip. di Scienze della Terra, U. di Pisa, Via S. Maria 53, 56100 Pisa, Italy. (1938) Dr, chemistry (U. di Pisa, 1962). Prof., crystallography. (tel. 050 + 501457). *Crystal chemistry, structural mineralogy.*

Merlo, Prof. Franco. Ist. di Chimica Fisica, U. di Genova, Corso Europa (Palazzo delle Scienze), 16132 Genova, Italy. (1940) Dr, chemistry (U. di Genova, 1963). Prof. assoc. (tel. 010 + 515076). *Structure and physical properties, intermetallic compounds.*

Molin, Prof. Gianmario. Ist. di Mineralogia e Petrologia, Corso Garibaldi 37, 35100 Padova, Italy. (1948) Dr, geological sciences (U. di Padova, 1972). Prof. assoc. (tel. 049 + 663122) *Inorganic crystal structures.*

Monaco, Dr Hugo Luis. Centro di Studio sui Biopolimeri, Ist. di Chimica Organica, U. di Padova, Via Marzolo 1, 35100 Padova, Italy. (1947) Ph.D., chemistry (Harvard U., USA, 1978). Coll. tecn. pr. (tel. 049 + 831327). *Protein crystallography.*

Monari, Dr Magda. Istituto di Chimica 'G. Ciamician', U. di Bologna, Via Selmi 2, 40126 Bologna, Italy. (1955) Dr, industrial chemistry (U. di Bologna, 1982). (tel. 051 + 235292). *Coordination compounds.*

Mongiorgi, Prof. Romano. Ist. di Mineralogia, U. di Bologna, Piazza di Porta S. Donato 1, 40127 Bologna, Italy. (1940) Dr, geological sciences (U. di Bologna, 1969). Prof. assoc. (tel. 051 + 231962, ext. 92). *Structure - activity relationship, intermolecular forces, crystal growth, morphology, kidney stones, biological apatites.*

Montenero, Prof. Angelo. Ist. di Strutturistica Chimica, U. di Parma, Via M. D'Azeglio 85, 43100 Parma, Italy. (1944) Dr, chemistry (U. di Parma, 1968). Prof. assoc. (tel. 0521 + 22432). *Powder X-ray diffraction, glass structure.*

Morandi, Prof. Noris. Ist. di Mineralogia e Petrografia, U. di Bologna, Piazza San Donato 1, 40127 Bologna, Italy. (1938) Dr, geology (U. di Bologna, 1961). Prof. assoc. (tel. 051 + 231961, ext. 56). *Mineralogy, crystal chemistry, clay minerals, X-ray diffractometry, I.R. spectroscopy.*

Morelli, Prof. Gianluca. Via Bordolano 3/B, 20097 San Donato Milanese, Milano, Italy. (1925) Dr, chemistry (U. di Roma, 1951). Consultant. (tel. 02 + 5273085). *Clay minerals.*

Motta, Dr Nunzio. Dip. di Fisica, U. di Roma, Via O. Raimondo, 00173 Roma, Italy. (1957), Dr, physics (U. di Roma, 1981). Ricercatore. (tel. 06 + 7979, ext. 2305). *Semiconductors, metals, crystal structure, EXAFS.*

Mugnoli, Prof. Angelo. Ist. di Chimica Fisica, U. di Genova, Palazzo delle Scienze, Corso Europa, 16132 Genova, Italy. (1933) Dr, industrial chemistry (U. di Milano, 1958). Prof., structural chemistry (tel. 010 + 516016). *Molecular structure, chemical reactivity, organic compounds.*

Mura, Dr Pasquale. Ist. di Strutturistica Chimica 'Giordano Giacomello', C.N.R., C.P. 10 - 00016 Monterotondo Stazione (Roma), Italy. (1944) Dr, chemistry (U. di Roma, 1975). Coll. tecn. pr. (tel. 06 + 9005142). *Organometallic and coordination compounds, synthesis, structure - properties relationship.*

Musatti, Prof. Amos. Ist. di Strutturistica Chimica, U. di Parma, Via M. D'Azeglio 85, 43100 Parma, Italy. (1935) Dr, chemistry (U. di Parma, 1962). Prof. assoc. (tel. 0521 + 22432). *Crystal structures, computer programming.*

Napolitano, Dr Roberto. Dip. di Chimica, U. di Napoli, Via Mezzocannone 4, 80134 Napoli, Italy. (1949) Dr, chemistry (U. di Napoli, 1974). Ricercatore. (tel. 081 + 206450). *Conformational analysis, crystal structure, polymers.*

Nardelli, Prof. Mario. Ist. di Chimica Generale ed Inorganica, U. di Parma, Via M. D'Azeglio 85, 43100 Parma, Italy. (1922) Dr, chemistry (U. di Parma, 1946). Prof., general chemistry (tel. 0521 + 34400 or 206258, telex 530327 UNIV PR I). *Structure determination methods, inorganic and organic structures, accurate electron density, precise atomic parameters, crystallographic computing.*

Nardin, Prof. Giorgio. Dip. di Scienze Chimiche, U. di Trieste, Piazzale Europa 1, 34127 Trieste, Italy. (1940) Dr, chemistry (U. di Trieste, 1964). Prof. assoc. (tel. 040 + 51172). *Crystallography - Inorganic chemistry.*

Navarra, Dr Gabriele. Dip. di Scienze Chimiche, U. di Cagliari, Via Ospedale 72, 09100 Cagliari, Italy. (1954) Dr, chemistry (U. di Cagliari, 1980). Ricercatore. (tel. 070 + 668047). *Material sciences, X-ray crystallography, EXAFS spectroscopy.*

Nunzi, Prof. Antonio. Dip. di Scienze della Terra, Sez. di Cristallografia, U. di Perugia, Piazza Università, 06100 Perugia, Italy. (1943) Dr, chemistry (U. di Perugia, 1966). Prof. assoc. (tel. 075 + 23822). *Crystallographic computing, direct methods.*

Oberti, Dr Roberta. Centro di Studio per la Cristallografia Strutturale, C.N.R., Via Bassi 4, 27100 Pavia, Italy. (1952) Dr, chemistry (U. di Pavia, 1976). Coll. tecn. pr. (tel. 0382 + 36689). *Crystal-chemistry of rock-forming minerals, organic crystal structures, computer programming.*

Olcese, Prof. Giorgio L. Ist. di Chimica Fisica, U. di Genova, Corso Europa (Palazzo delle Scienze), 16132 Genova, Italy. (1933) Dr, chemistry (U. di Genova, 1957). Prof., physical chemistry. (tel. 010 + 515076). *Structure and physical properties, intermetallic compounds.*

Orioli, Prof. Pierluigi. Dip. di Chimica, U. di Firenze, via Maragliano 27, 50100 Firenze, Italy. (1933) Dr, chemistry (U. di Firenze, 1956). Prof., structural chemistry. (tel. 055 + 2476949). *X-ray diffraction and EXAFS, bioinorganic chemistry, coordination compounds.*

Palenzona, Prof. Andrea. Ist. di Chimica Fisica, U. di Genova, Corso Europa (Palazzo delle Scienze), 16132 Genova, Italy. (1935) Dr, chemistry (U. di Genova, 1961). Prof., inorganic chemistry. (tel. 010 + 515076). *Structure and physical properties, intermetallic compounds.*

Paorici, Prof. Carlo. Dip. di Fisica, U. di Parma, Via D'Azeglio 85, 43100 Parma, Italy. (1936) Dr, chemistry (U. di Roma, 1962). Prof. assoc. (tel. 0521 + 96841, telex 531639 MASPEC I). *Semiconductors, crystal growth, crystal perfection assessment, physical characterization.*

Pasero, Dr Marco. Dip. di Scienze della Terra, Via S. Maria 53, 56100 Pisa, Italy. (1958) Dr, geological sciences (U. di Pisa, 1982). (tel. 050 + 501457). *Electron microscopy, inorganic crystal structures.*

Passaglia, Prof. Elio. Ist. di Mineralogia, U. di Ferrara, Corso Ercole I D'Este 32, 44100 Ferrara, Italy. (1941) Dr, geological sciences (U. di Modena, 1965). Prof. of mineralogy (tel. 0532 + 32987). *Crystal chemistry of sylicates, minerals, inorganic crystal structures.*

Pasti, Mr Fabio. Ital Structures, Area Industriale, 38066 Riva del Garda, Trento, Italy. (1937) Diploma, electrotechnique (Higher Sch., 1957). Maker of scientific instruments. (tel. 0464 + 513426). *Scientific instruments, crystal growth.*

Pavel, Dr Nicolae Viorel. Dip. di Chimica, U. di Roma 'La Sapienza', Piazzale A. Moro 5, 00185 Roma, Italy. (1949) Dr, chemistry (U. fo Roma, 1973). Ricercatore. (tel. 06 + 4991, ext. 322). *Inclusion compounds, conformational analysis, synthetic and biological polymers, micellar aggregates.*

Pedone, Prof. Carlo. Ist. Chimico, U. di Napoli, Via Mezzocannone 4, 80134 Napoli, Italy. (1938) Dr, chemistry (U. di Napoli, 1961). Prof. (tel. 081 + 206800). *Crystal structure, organic and metallo-organic molecules.*

Pelizzi, Prof. Corrado. Ist. di Chimica Generale ed Inorganica, U. di Parma, Via Massimo D'Azeglio 85, 43100 Parma, Italy. (1942) Dr, chemistry (U. di Parma, 1966). Prof. inc. stab. (tel. 0521 + 34400 or 66258). *Organometallic and coordination compounds.*

Pellinghelli, Prof. Maria Angela. Ist. di Chimica Generale ed Inorganica, U. di Parma, Via M. D'Azeglio 85, 43100 Parma, Italy. (1943) Dr, chemistry (U. di Parma, 1966). Prof. assoc. (tel. 0521 + 34400). *Crystal structures, inorganic - organic and organometallic compounds.*

Penco, Prof. Anna Maria. Ist. di Mineralogia, U. di Genova, Corso Europa, 16132 Genova, Italy. (1927) Dr, natural sciences (U. di Genova, 1953). Prof., mineralogy. (tel. 010 + 518184). *Crystallography, optical crystallography, morphology, X-ray crystallography.*

Perego, Dr Giovanni. ENIRICERCHE, Via F. Maritano 26, 20097 San Donato Milanese, Milano, Italy. (1938) Doctorat d'Université, sciences (U. de Strasbourg, France, 1971). Ricercatore. (tel. 02 + 5207543). *Structure, organometallic compounds, inorganic compounds, morphology, polymers.*

Petraccone, Prof. Vittorio. Dip. di Chimica, U. di Napoli, Via Mezzocannone 4, 80134 Napoli, Italy. (1943) Dr, chemistry (U. di Napoli, 1966). Prof. assoc. (tel. 081 + 205730 or 206450). *Conformational analysis, crystal structure, polymers.*

Peyronel, Prof. Giorgio. Dip. di Chimica, Sez. Inorganica, U. di Modena, Via G. Campi 183, 41100 Modena, Italy. (1913) Dr, chemistry (U. di Milano, 1936). Prof., general and inorganic chemistry. (tel. 059 + 362243). *Crystal structure determination, inorganic compounds, coordination compounds.*

Piazzesi, Prof. AnnaMaria. Dip. di Chimica Organica, U. di Padova, Via Marzolo 1, 35100 Padova, Italy. (1927) Dr, chemistry (U. di Padova, 1963). Prof. assoc. (tel. 049 + 831311). *Organic compounds, biologically important substances.*

Pifferi, Dr Augusto. Ist. di Strutturistica Chimica 'G. Giacomello', C.N.R., 00016 Monterotondo Stazione, Roma, Italy. (1955) Dr, physics (U. di Roma, 1979). Coll. tecn. pr. (tel. 06 + 9005142). *Synchrotron radiation, methodology, powder diffraction.*

Pignedoli, Prof. Anna. Dip. Chimica Generale, sez. Inorganica, U. di Modena, Via G. Campi 183, 41100 Modena, Italy. (1924) Dr chemistry (U. di Modena, 1947). Prof. assoc. (tel. 059 + 362243). *Crystal structure determination, inorganic compounds, coordination compounds.*

Pilati, Dr Tullio. Centro di Studio per le relazioni tra Struttura e reattivita' Chimica, C.N.R., via Golgi 19, 20133 Milano, Italy. (1946) Dr, chemistry (U. di Milano, 1971). Coll. tecn. pr. (tel. 02 + 292900). *Organic crystal structures, minerals, crystallographic computing.*

Pirozzi, Prof. Beniamino. Dip. di Chimica, U. di Napoli, Via Mezzocannone 4, 80134 Napoli, Italy. (1947) Dr, chemistry (U. di Napoli, 1970). Prof. assoc. (tel. 081 + 206450). *Conformational analysis, crystal structure, polymers.*

Pisani, Prof. Cesare. Ist. di Chimica Teorica, U. di Torino, Via P. Giuria 5, 10125 Torino, Italy. (1938) Dr, physics (U. di Milano, 1963). Prof. (tel. 011 + 657274). *Solid state theoretical chemistry.*

Polidori, Dr Giampiero. Dip. di Scienze della Terra, Sez. Cristallografia, U. di Perugia, Piazza Università, 06100 Perugia, Italy. (1950) Dr, mathematics (U. di Perugia, 1973). Ricercatore. (tel. 075 + 23822). *Crystallographic computing, direct methods.*

Polo, Dr Adriano. Ist. di Chimica e Tecn. dei radioelementi, C.N.R., Corso Stati Uniti 4, 35100 Padova, Italy. (1956), Dr, chemistry (U. di Padova, 1982). Coll. tecn. pr. (tel. 049 + 845111 or 845368). *Inorganic crystal structures.*

Pompa, Dr Francesco. Divisione Scienza dei Materiali, ENEA - C.R.E. Casaccia, S. P. Anguillarese - C. P. 2400, 00100 Roma, Italy. (1931) Dr, chemistry (U. di Bari, 1958). Ricercatore (tel. 06 + 6948, ext. 3117, telex ENEACA I 613296). *Instrumentation, computer programming, inorganic crystal structure, metallic materials.*

Poppi, Prof. Luciano. Ist. di Mineralogia e Petrografia, U. di Bologna, P.zza S. Donato 1, 40127 Bologna, Italy. (1940) Dr, geology (U. di Modena, 1968). Prof. assoc. (tel. 051 + 231961, ext. 57). *Crystal chemistry, clay minerals.*

Porta, Prof. Piero. Dip. di Chimica, U. di Roma 'La Sapienza', Piazzale A. Moro 5, 00185 Roma, Italy. (1933) Dr, chemistry (U. di Bari, 1958). Prof. (tel. 06 + 493202, ext. 52). *Solid state chemistry, inorganic structures, heterogeneous catalysis.*

Portalone, Dr Gustavo. Dip. di Chimica, U. di Roma 'La Sapienza', Città Universitaria, 00100 Roma, Italy. (1952) Dr, pharmaceutical chemistry and technology (U. di Roma, 1977). Ricercatore. (tel. 06 + 8128323). *Crystallography of small molecules, gas electron diffraction substituent effects on molecular structure.*

Porzio, Dr William. Ist. di Chimica delle Macromolecole, C.N.R., Via A. Corti 12, 20133 Milano, Italy. (1951) Dr, industrial chemistry (U. di Milano, 1976). Scient. (tel. 02 + 296071, ext. 17). *Organometallic chemistry, synthetic polymers.*

Puliti, Dr Raffaella. Ist. per la Chimica di Molecole di Interesse Biologico, C.N.R., Via Toiano 2, 80072 Arco Felice, Napoli, Italy. (1936) Dr, chemistry (U. di Bari, 1961). Coll. tecn. pr. (tel. 081 + 8601444 or 205730). *Biological compounds.*

Quagliata, Prof. Claudio. Dip. di Chimica, U. di Roma, Piazzale A. Moro 5, 00185 Roma, Italy. (1945) Dr, chemistry (U. di Roma, 1970). Prof. assoc. (tel. 06 + 4952993). *Potential energy calculations, inclusion compounds, conformational analysis, micellar aggregates.*

Quartieri, Dr Simona. Ist. di Mineralogia e Petrologia, U. di Modena, Via S. Eufemia 19, 41100 Modena, Italy. (1955) Dr, chemistry (U. di Modena, 1979). (tel. 059 + 218062, ext. 16). *Structures, inorganic compounds and minerals, molecular dynamic.*

Randaccio, Prof. Lucio. Dip. di Scienze Chimiche, U. di Trieste, Piazzale Europa 1, 34127 Trieste, Italy. (1940) Dr, chemistry (U. di Napoli, 1963). Prof. (tel. 040 + 51172). *Structural inorganic chemistry.*

Ravaglioli, Dr Antonio. Dip. di Ceramiche Classiche, Ist. di Ricerche Tecnologiche per la Ceramica, C.N.R., Via Granarolo 64, 48018 Faenza (Ravenna), Italy. (1938) Dr, chemistry (U. di Bologna, 1967). Coll. tecn. pr. (tel. 0546 + 46147). *Crystal growth, sintering, bioceramics, glazes, refractories, ceramic raw materials, electronic ceramics, ancient ceramics.*

Riganti, Prof. Vincenzo. Dip. di Chimica Generale, U. di Pavia, Viale Taramelli 12, 27100 Pavia, Italy. (1932) Dr, chemistry (U. di Pavia, 1955). Prof. assoc. (tel. 0382 + 24714, ext. 12). *Structural chemistry, merceology, ecological problems.*

Rigault de la Longrais, Prof. Germain. Dip. di Scienze della Terra, U. di Torino, Via San Massimo 24, 10123 Torino, Italy. (1930) Dr, chemistry (U. di Torino, 1953). Prof., mineralogy. (tel. 011 + 832193). *Symmetry, crystal optics, morphology.*

Rinaldi, Prof. Romano. Ist. di Mineralogia e Petrologia, U. di Modena, Via S. Eufemia 19, 41100 Modena, Italy. (1944) Dr, geological sciences (U. di Modena, 1969). Prof. assoc. (tel. 059 + 218062). *Structures, X-ray diffraction, electron microscopy, X-ray microanalysis, minerals, materials.*

Rinaudo, Dr Caterina. Dip. di Scienze della Terra, U. di Torino, Via S. Massimo 24, 10123 Torino, Italy. (1951) Dr, natural sciences (U. di Torino, 1974). Ricercatore. (tel. 011 + 832193). *Crystal growth from solution.*

Ripamonti, Prof. Alberto. Ist. di Chimica 'G. Ciamician', U. di Bologna, Via Selmi 2, 40126 Bologna, Italy. (1930) Dr, chemistry (U. di Roma, 1953). Prof., general chemistry. (tel. 051 + 235292). *Crystal structures, biological structures.*

Ripamonti, Dr Carlo. Dip. di Fisica, U. di Genova, Via Dodecaneso 33, 16146 Genova, Italy. (1950) Dr, physics (U. di Genova, 1975). Ricercatore. *Crystal physics.*

Riva di Sanseverino, Prof. Lodovico. Ist. di Mineralogia, U. di Bologna, Piazza di Porta S. Donato 1, 40127 Bologna, Italy. (1939) Dr, chemistry (U. di Firenze, 1962). Prof., mineralogy. (tel. 051 + 231962, ext. 50). *Structure - activity correlation, intermolecular forces.*

Rosa, Dr Rodolfo. IST. LAMEL, C.N.R., Via de' Castagnoli 1, 40126 Bologna, Italy. (1944) Dr, physics and philosophy (U. di Bologna, 1968, 1977). Coll. tecn. pr. (tel. 051 + 519593). *Computer simulation, electron interaction with matter, X-ray microanalysis.*

Rossi, Dr Giuseppe. Centro di Studio per la Cristallografia Strutturale, C.N.R., Via A. Bassi 4, 27100 Pavia, Italy. (1938) Dr, geological sciences (U. di Pavia,

1961). Coll. tecn. pr. (tel. 0382 + 36689). *Inorganic and organic crystal structures.*

Roveri, Prof. Norberto. Ist. di Chimica 'G. Ciamician', U. di Bologna, Via Selmi 2, 40126 Bologna, Italy. (1947) Dr, chemistry (U. di Bologna, 1972). Prof. assoc. (tel. 051 + 235292). *Crystal structures, biological structures.*

Rubbo, Dr Marco. Dip. di Scienze della Terra, U. di Torino, Via San Massimo 24, 10123 Torino, Italy. (1946) Dr, chemistry (U. di Torino, 1971). Ricercatore. (tel. 011 + 832193). *Crystal growth.*

Sabat, Dr Michal. Ist. per lo Studio della Stereochimica ed Energetica dei Composti di Coordinazione, C.N.R., Via Guerrazzi 27, 50132 Firenze, Italy. (1947) PhD, Chemistry (U. Wroclaw, Poland, 1976). Coll. tecn. pr. (tel. 055 + 243990). *Structure determination, molecular orbital methods, transition metal compounds, organometallic and bioinorganic compounds.*

Sabatino, Dr Piera. Ist. di Scienze Chimiche, Fac. di Farmacia, U. di Bologna, Via San Donato 15, 40127 Bologna, Italy. (1950) Dr, pharmacy (U. di Bologna, 1974). Ricercatore. (tel. 051 + 223748). *Organic crystal structures, biologically and pharmacologically important compounds.*

Sabbioni, Dr Cristina. FISBAT, C.N.R., Via Castagnoli 1, 40126 Bologna, Italy. (1954) Dr, physics (U. di Bologna, 1978). Coll. tecn. pr. (tel. 051 + 519593). *Atmospheric aerosol, building stone weathering, electron microscopy.*

Sabelli, Dr Cesare. Centro di Studio per la Mineralogia e la Geochimica dei Sedimenti, C.N.R., Via La Pira 4, 50121 Firenze, Italy. (1934) Dr, geological sciences (U. di Firenze, 1959). Coll. tecn. pr. (tel. 055 + 287140). *Inorganic crystal structures, crystal chemistry.*

Sacconi, Prof. Luigi. Ist. di Chimica Generale e Inorganica, U. di Firenze, Via J. Nardi 39, 50132 Firenze, Italy. (1911) Dr, chemistry (U. di Parma, 1942). Prof., general and inorganic chemistry. (tel. 055 + 572291 or 572292). *Coordination chemistry, inorganic chemistry.*

Sacerdoti, Prof. Michele. Ist. di Mineralogia, U. di Ferrara, Corso Ercole I d'Este 32, 44100 Ferrara, Italy. (1935) Dr, geological sciences (U. di Padova, 1959). Prof. assoc. (tel. 0532 + 32987). *Crystal structure determination, refinements, inorganic compounds, plastic deformation, minerals.*

Salviati, Dr Giancarlo. Ist. MASPEC, C.N.R., Via Chiavari 18/A, 43100 Parma, Italy. (1950) Dr, physics (U. di Parma). Ricercatore. (tel. 0521 + 96841). *Electron microscopy, crystal characterization.*

Sansoni, Prof. Mirella. Ist. di Chimica Generale ed Inorganica, U. degli Studi di Milano, Via G. Venezian 21, 20133 Milano, Italy. (1939) Dr, physics (U. di Milano, 1964). Prof. assoc. (tel. 02 + 235120). *Crystallographic computing, metal cluster compounds, transition metal complexes.*

Sartori, Prof. Franco. Dip. di Scienze della Terra, U. di Pisa, Via S. Maria 53, 56100 Pisa, Italy. (1938) Dr, geological sciences, (U. di Pavia, 1962). Prof. assoc. (tel. 050 + 501663). *Phyllosilicates, clay minerals.*

Scandale, Prof. Eugenio. Dip. Geomineralogico, U. di Bari, Via Amendola, 70121 Bari, Italy. (1943) Dr, physics (U. di Bari, 1971). Prof. assoc. (tel. 080 + 369100, ext. 402). *X-ray topography, natural single crystals.*

Scaramuzza, Dr Lucio. Ist. Teoria e Struttura Elettronica e Comportamento Spettrochimico dei Composti di Coordinazione, C.N.R., Area della Ricerca di Roma, 00016 Monterotondo Stazione, Roma, Italy. (1935) Dr, chemistry (U. di Roma, 1965). Coll. tecn. pr. (tel. 06 + 4991, ext. 322). *Coordination compounds, instrumentation.*

Scatturin, Prof. Vladimiro. Ist. di Chimica Generale e Inorganica, U. degli Studi di Milano, Via G. Venezian 21, 20133 Milano, Italy. (1922) Dr, chemistry (U. di Padova, 1946). Prof., general chemistry. (tel. 02 + 235288). *Structural chemistry, teaching.*

Schiavinato, Prof. Giuseppe. Dip. di Scienze della Terra, U. degli Studi, Via Botticelli 23, 20133 Milano, Italy. (1915) Dr, geological sciences (U. di Padova, 1939). Prof., mineralogy. (tel. 02 + 293994). *Inorganic crystal structures, instrumentation, clay minerals, crystal optics.*

Scordari, Prof. Fernando. Dip. Geomineralogico, U. di Bari, Campus Universitario, 70100 Bari, Italy. (1944) Dr, geological sciences (U. di Bari, 1968). Prof. assoc. (tel. 080 + 369100). *Inorganic crystal structures.*

Secco, Dr Luciano. Ist. di Mineralogia e Petrologia, U. di Padova, Corso Garibaldi 37, 35100 Padova, Italy. (1955) Dr, geological sciences (U. di Padova, 1980). Ricercatore. (tel 049 + 663122). *Inorganic crystal structures.*

Sersale, Prof. Riccardo. Dip. di Ingegneria dei Materiali e della Produzione, U. di Napoli, P.le Tecchio, 80125 Napoli, Italy. (1927) Dr, chemistry (U. di Napoli, 1943). Prof., appl. chemistry. (tel. 081 + 624238, telex INGENA I 722392). *Chemistry of cement, chemistry of industrial silicates.*

Servidori, Dr Marco. Ist. LAMEL, CNR, Via de' Castagnoli 1, 40126 Bologna, Italy. (1943) Dr, industrial chemistry (U. di Bologna, 1967). Coll. tecn. pr. (tel. 051 + 519593, ext. 216). *X-ray topography, electron microscopy, electronics materials and devices, multiple crystal X-ray diffraction.*

Sgarabotto, Prof. Paolo. Ist. di Strutturistica Chimica, U. degli Studi di Parma, Via M. D'Azeglio 85, 43100 Parma, Italy. (1940) Dr, chemistry (U. di Parma, 1967). Prof. assoc. (tel. 0521 + 22432). *Crystal structure analysis, organic and organometallic compounds, X-ray diffractometry, instrumentation.*

Sgarlata, Prof. Francesco. Ist. di Mineralogia e Petrografia, U. di Roma, Piazzale A. Moro, Città Universitaria, 00185 Roma, Italy. (1926) Dr, physics (U. di Roma, 1949). Prof., crystallography. *Solid state physics, X-ray diffraction, gamma resonance spectrometry, phases in natural heterogeneous systems.*

Sgualdino, Dr Giulio. Ist. Chimico, U. di Ferrara, Via L. Borsari 46, 44100 Ferrara, Italy. (1947) Dr, chemistry (U. di Ferrara, 1974). Ricercatore. (tel. 0532 + 36374). *Crystal growth, habit modifications.*

Simonetta, Prof. Massimo. Dip. di Chimica Fisica ed Elettrochimica, Via Golgi 19, 20133 Milano, Italy. (1920) Dr, industrial chemistry (U. di Milan, 1943). Prof., physical chemistry. (tel. 02 + 292900). *Quantum chemistry, physical organic chemistry, crystallography.*

Sironi, Prof. Angelo. Ist. Chimica Generale ed Inorganica, U. degli Studi di Milano, Via G. Venezian 21, 20133 Milano, Italy. (1948) Dr, chemistry (U. di Milano, 1972). Prof. assoc. (tel. 02 + 235120). *Transition metal complexes, metal cluster compounds, crystallographic computing.*

Spadon, Prof. Paola. Dip. di Chimica Organica, U. di Padova, Via Marzolo 1, 35100 Padova, Italy. (1947) Dr, chemistry (U. di Padova, 1971). Prof. assoc. (tel. 049 + 831327 or 831236). *Biologically important molecules, protein crystallography.*

Spagna, Dr Riccardo. Ist. di Strutturistica Chimica 'Giordano Giacomello', C.N.R., C.P. 10 - 00016 Monterotondo Stazione (Roma), Italy. (1941) Dr, chemistry (U. di Roma, 1967). Coll. tecn. pr. (tel. 06 + 9005142). *Crystal structures, computer programming.*

Stasi, Dr Francesca. Dip. Geomineralogico, U. di Bari, Via Amendola, 70121 Bari, Italy. (1939), Dr, physics (U. di Bari, 1971). Tecn. laur. (tel. 080 + 369100). *X-ray topography, electron microscopy.*

Suffritti, Prof. Giuseppe Baldovino. Ist. di Chimica Fisica, U. di Sassari, Via Vienna 2, 07100 Sassari, Italy. (1947) Dr, physics (U. di Milano, 1972). Prof. assoc. (tel 079 + 218415). *Molecular dynamics, disordered and amorphous solids, silicates.*

Tadini, Prof. Carla. Dip. di Scienze della Terra, U. di Pavia, Via A. Bassi 4, 27100 Pavia, Italy. (1924) Dr, chemistry (U. di Pavia, 1956). Prof. assoc. (tel. 0382 + 21135). *Inorganic and organic crystal structures.*

Tazzoli, Prof. Vittorio. Dip. di Scienze della Terra, U. di Pavia, Via A. Bassi 4, 27100 Pavia, Italy. (1938) Dr, chemistry (U. di Pavia, 1963). Prof. assoc. (tel. 0382 + 21135). *Mineral crystal chemistry.*

Tieghi, Prof. Giuseppe. Dip. di Chimica Industriale ed Ingegneria Chimica, Politecnico di Milano, Piazza L. da Vinci 32, 20133 Milano, Italy. (1943) Dr, chemical engineering. (U. di Milano, 1968). Prof. assoc. (tel. 02 + 292105, ext. 18). *Macromolecular material science.*

Tiripicchio, Prof. Antonio. Ist. di Chimica Generale ed Inorganica, U. di Parma, Via M. D'Azeglio 85, 43100 Parma, Italy. (1936) Dr, chemistry (U. di Parma, 1959). Prof., general chemistry. (tel. 0521 + 34400). *Crystal structures, inorganic, bioinorganic and organometallic compounds.*

Tiripicchio-Camellini, Prof. Marisa. Ist. di Chimica Generale ed Inorganica, U. di Parma, Via M. D'Azeglio 85, 43100 Parma, Italy. (1938) Dr, natural sciences (U. di Parma, 1962). Prof. assoc. (tel. 0521 + 34400). *Crystal structures, inorganic, bioinorganic and organometallic compounds.*

Tomassini, Dr Marco. Dip. di Scienze della Terra, Sez. di Cristallografia, U. di Perugia, Piazza U., 06100 Perugia, Italy. (1949) Dr, chemistry (U. di Perugia, 1974). Prof. inc. (tel. 075 + 23822). *Potential energy calculations in crystals, computer programming, computer applications.*

Tosi, Prof. Giorgio. Dip. di Scienze dei Materiali e della Terra, Fac. di Ingegneria, Via della Montagnola 30, 60100 Ancona, Italy. (1942) Dr, Industrial chemistry (U. di Bologna, 1966). Prof. assoc. (tel. 071 + 5893469). *Molecular associations structure, chemical and physical relationship, clusters.*

Trosti-Ferroni, Prof. Renza. Dip. di Scienze della Terra, U. di Firenze, Via La Pira 4, 50121 Firenze, Italy. (1947) Dr, natural sciences (U. di Firenze, 1969). Asst. (tel. 055 + 287140). *Inorganic crystal structures, crystal chemistry, mineralogy.*

Ugliengo, Dr Piero. Ist. di Chimica Fisica, U. di Torino, Corso M. D'Azeglio 48, 10125 Torino, Italy. (1957) Dr, chemistry (U. di Torino, 1981). Ricercatore. (tel. 011 + 6505102, ext. 15). *Computational crystallography, direct methods.*

Ugozzoli, Dr Franco. Ist. di Strutturistica Chimica, Via M. D'Azeglio 85, 43100 Parma, Italy. (1947) Dr, physics (U. di Parma, 1983). Tecn. laur. (tel. 0521 + 22432). *Crystal structure analysis, organic and organometallic compounds, X-ray spectroscopy, Mossbauer spectroscopy, amorphous systems, Computer programming.*

Ungaretti, Prof. Luciano. Dip. di Scienze della Terra, U. di Pavia, Via A. Bassi 4, 27100 Pavia, Italy. (1942) Dr, chemistry (U. di Pavia, 1965). Prof., mineralogy. (tel. 0382 + 21135). *Mineral crystal-chemistry.*

Vaccari, Dr Giuseppe. Ist. Chimico, U. di Ferrara, Via Luigi Borsari 46, 44100 Ferrara, Italy. (1948) Dr, chemistry (U. di Ferrara, 1972). Prof. inc. (tel. 0532 + 36374). *Crystal growth, habit modifications.*

Vaciago, Prof. Alessandro. Ist. di Fisica, U. di Roma, Città Universitaria, 00100 Roma, Italy. (1931) Dr, physics (U. di Roma, 1953). Prof., structural chemistry. (tel. 06 + 4976353, telex 613255 INFNRO I). *Organic and biological crystal structures, quantum chemistry, instrumentation, synchrotron radiation.*

Valdrè, Prof. Ugo. Dip. di Fisica, U. di Bologna, Via Irnerio 46, 40126 Bologna, Italy. (1926) Dr, physics (U. di Bologna, 1951). Professor of Physics. (tel. 051 + 260991 or 272063, ext. 108, telex 211664 INFN BO). *Electron microscopy, instrumentation, semiconductors, superconductors, low temperatures, organic materials, nanoanalysis.*

Valigi, Prof. Mario. Dip. di Chimica, U. di Roma, Piazzale A. Moro 5, 00100 Roma, Italy. (1936) Dr, chemistry (U. di Roma, 1961). Prof. assoc. (tel. 06 + 4990, ext. 863). *Solid state chemistry, inorganic and structural chemistry, surface chemistry, heterogeneous catalysis.*

Valle, Dr Giovanni. Centro di Studio sui Biopolimeri, C.N.R., Dip. di Chimica Organica, Via Marzolo 1, 35100 Padova, Italy. (1930) Dr, industrial chemistry (U. di Padova, 1960). Coll. tecn. pr. (tel. 049 + 831229). *Organic compounds, biological molecules.*

Venturello, Porf. Giovanni. Ist. di Chimica Generale, Fac. di Farmacia, U. di Torino, Via P. Giuria 9, 10125 Torino, Italy. (1912) Dr, chemistry (U. di Torino, 1935). Prof. (tel 011 + 657000). *Material science and amorphous catalysis.*

Verdini, Dr Brunella. Dip. di Ingegneria Chimica, dei Materiali, delle Materie Prime e Metallurgia, 00184 Roma, Italy. (1931) Dr, physics (U. di Roma, 1960). Coll. tecn. pr. (tel. 06 + 464286). *Martensitic transformations, metallic and non metallic materials, composite materials, quantitative measurements, defects in crystals, recrystallization.*

Vezzalini, Dr Maria Giovanna. Ist. di Mineralogia e Petrologia, U. di Modena, Via S. Eufemia 19, 41100 Modena, Italy. (1951) Dr, geological sciences (U. di Modena, 1974). Ricercatore. (tel. 059 + 218062, ext. 16). *Structures, inorganic compounds and minerals.*

Vitali, Dr Francesca. Ist. di Chimica Generale, U. di Parma, Via M. D'Azeglio 85, 43100 Parma, Italy. (1957) Dr, Pharmaceutical chemistry (U. di Parma, 1983). Ricercatore. (tel. 0521 + 34400). *Structural coordination chemistry, pharmacologically interesting molecules.*

Vitali, Dr Gianfranco. Dip. di Energetica, U. di Roma 'La Sapienza', Via A. Scarpa 14, 00161 Roma, Italy. (1934) Dr, physics. (tel. 06 + 425787). *Electron microscopy. Laser annealing of semiconductors.*

Viterbo, Prof. Davide Lazzaro Marco. Ist. di Chimica-Fisica, U. di Torino, Corso Massimo D'Azeglio 48, 10125 Torino, Italy. (1939) Dr, chemistry (U. di Torino, 1962). Prof. assoc. (tel. 011 + 6505102, ext. 15). *Direct methods for structure determination, small organic structures, heterocycles.*

Zagari, Prof. Adriana. Dip. di Chimica, U. di Napoli, Via Mezzocannone 4, 80134 Napoli, Italy. (1946) Dr, chemistry (U. di Napoli, 1969). Prof. assoc. (tel. 081 + 205730). *Biological structures, proteins.*

Zanazzi, Prof. Pier Francesco. Dip. di Scienze della Terra, Sez. di Cristallografia, U. di Perugia, Piazza U., 06100 Perugia, Italy. (1939) Dr, chemistry (U. di Firenze, 1962). Prof., crystallography. (tel. 075 + 23822). *Inorganic and organic crystal structures.*

Zangrando, Dr Ennio. Dip. di Scienze Chimiche, U. di Trieste, P.le Europa 1, 34127 Trieste, Italy. (1950) Dr, chemistry (U. di Trieste, 1974). Ricercatore. (tel 040 + 51172). *Structures. Inorganic chemistry.*

Zannetti, Prof. Roberto. Dip. di Chimica Inorganica, Metallorganica ed Analitica, U. di Padova, Via Loredan 4, 35100 Padova, Italy. (1929) Dr, chemistry (U. di Padova, 1953). Prof., general chemistry. (tel. 049 + 662972). *Macromolecular physics, crystal structure, macromolecules, structural investigations, polyolefins and polyamides.*

Zanotti, Prof. Giuseppe. Dip. di Chimica Organica, U. di Padova, Via Marzolo 1, 35100 Padova, Italy. (1950) Dr, chemistry (U. di Padova, 1974). Prof. assoc. (tel. 049 + 831229 or 831293). *Protein crystallography.*

Zanotti, Dr Lucio. Ist. MASPEC, C.N.R., Via Chiavari 18/A, 43100 Parma, Italy. (1944) Dr, chemistry (U. di Bologna, 1969). Coll tecn. pr. (tel. 0521 + 96841). *Crystal growth. Crystal defect analysis. Chemical analysis.*

Zappia, Prof. Vincenzo. Ist. di Biochimica delle Macromolecole, U. di Napoli, Via Costantinopoli 16, 80138 Napoli, Italy. (1939) Dr, medicine (U. di Napoli, 1963). Prof., biological chemistry. (tel. 081 + 217052-213668). *Chemistry, sulfonium compounds, polyamine biosynthesis, structure and function, enzymes.*

Zefiro, Dr Livio. Ist. di Mineralogia, U. degli Studi, Palazzo delle Scienze, Corso Europa, 16132 Genova, Italy. (1948) Dr. physics (U. di Genova, 1972). Ricercatore. (tel. 010 + 518184). *Crystal growth, coherent optics in crystallography, holographic interferometry, hydrodinamics of solution.*

Zerbi, Prof. Giuseppe. Dip. di Chimica Industriale, Politecnico di Milano, Piazza L. da Vinci 32, 20133 Milano, Italy. (1933) Dr, chemistry (U. di Pavia, 1956). Prof., appl. chemistry (tel. 02 + 292105 or 292125). *Lattice dynamics, organic crystals, vibrational spectroscopy.*

Zocchi, Prof. Marcello. Dip. di Chimica Inorganica e Metallorganica, U. di Milano, Via G. Venezian 21, 20133 Milano, Italy. (1929) Dr, chemistry (U. di Roma, 1956). Prof. assoc. (tel. 02 + 292125, ext. 40). *Inorganic stereochemistry, materials science, accurate intensity measurements.*

Zosi, Prof. Gianfranco Luigi. Ist. di Fisica Generale, U. di Torino, Corso M. D'Azeglio 46, 10125 Torino, Italy. (1940) Dr, physics (U. di Torino, 1962). Prof. inc. stab. (tel. 011 + 655162). *Fundamental constants, X-ray interferometry.*

IVORY COAST

Sub-Editor: **R. Viani**

Notes

1. International telephone country code - 225.

Bonnaud, Dr Bernard Henri. Dépt. de Physique, Faculté des Sciences, U. d'Abidjan, 04 BP 322 Abidjan 04, Côte d'Ivoire. (1946) D.3e cycle, structural science (U. de Marseille - Provence, France, 1985). Teacher. (tel. 225 + 43 90 00, ext. 3125, telex RECTUNIV 3469). *Organic structures, liquid crystals.*

Briard, Mrs Pierrette. Dépt. de Physique, Faculté des Sciences, U. d'Abidjan, 04 BP 322 Abidjan 04, Côte d'Ivoire. (1940) D.3e cycle, metallurgy (U. de Toulouse, France, 1973). Maitre-asst. (tel. 225 + 43 90 00, ext. 3125, telex RECTUNIV 3469). *Organic structures, X-ray diffraction.*

Charbonnier, Mrs Sylvie. Dépt. de Physique, Faculté des Sciences, U. d'Abidjan, 04 BP 322 Abidjan 04, Côte d'Ivoire. (1954) Doctorat Spec., material science (U. de Nance, France). Asst. (tel. 225 + 43 90 00, ext. 3121, telex RECTUNIV 3469). *Structure determination, organic structures, molecular mechanics.*

Cossu, Dr Michéle Josette. Dépt. de Physique, Faculté des Sciences, U. d'Abidjan, 04 BP 322 Abidjan 04, Côte d'Ivoire. (1943) DSc, physics (U. de Provence, France, 1983). Maître de conf. (tel. 225 + 43 90 00, ext. 3129, telex RECTUNIV 3469). *Organic structures, strained molecules.*

Ebby, Dr N' Dédé. Dépt. de Physique, Faculté des Sciences, U. d'Abidjan, 04 BP 322 Abidjan 04, Côte d'Ivoire. (1941) DSc, physics (U. d'Abidjan, 1980). Maître de conf. (tel. 225 + 43 90 00, ext. 3141, telex RECTUNIV 3469). *Organic structures, X-ray diffraction.*

Mansilla - Koblavi, Mrs Frédérica Gbédégbé Marie Sonia. Dépt. de Physique, Faculté des Sciences, U. d'Abidjan, 04 BP 322 Abidjan 04, Côte d'Ivoire. (1958) Doctorat Spec., molecular chemistry (U. de Toulouse Paul Sabtier, France). Asst. (tel. 225 + 43 90 00, ext. 3125, telex RECTUNIV 3469). *Organometallic structures, X-ray diffraction.*

Ouattara, Mr Drissa. Dépt. de Physique, Faculté des Sciences, U. d'Abidjan, 04 BP 322 Abidjan 04, Côte d'Ivoire. (1956) DEA, structural science (U. d'Abidjan, 1985). Teacher. (tel. 225 + 43 90 00, ext. 3128, telex RECTUNIV 3469). *Organic structures, strained molecules.*

Tenon, Mr Abodou Jules. Dépt. de Physique, Faculte des Sciences, U. d'Abidjan, 04 BP 322 Abidjan 04, Côte d'Ivoire. (1955) DEA, structural science (U. d'Abidjan, 1984). Student. (tel. 225 + 43 90 00, ext. 3125, telex RECTUNIV 3469). *Organic structures, X-ray diffraction.*

Toure, Dr Siaka. Dépt. de Physique, Faculte des Sciences, U. d'Abidjan, 04 BP 322 Abidjan 04, Côte d'Ivoire. (1947) DSc, physics (U. d'Abidjan, 1979). Maître de conf. (tel. 225 + 43 90 00, ext. 3122, telex RECTUNIV 3469). *Organic compounds, structure determination, X-ray diffraction, natural products.*

Viani, Dr Robert Antoine. Dépt. de Physique, Faculté des Sciences, U. d'Abidjan, 04 BP 322 Abidjan 04, Côte d'Ivoire. (1942) DSc, physics (U. de Provence, France, 1978). Maître de conf. (tel. 225 + 43 90 00, ext. 3129, telex RECTUNIV 3469). *Organic structures, strained molecules.*

JAPAN

Sub-Editor: Y. Iitaka

Notes

1. International telephone country code - 81.

2. At the Universities, persons with an academic training can be appointed in various functions ranging from *professor, assistant professor, lecturer* to *research associate*. At some Universities the use of the English translations *associate professor* and *research assistant* instead of assistant professor and research associate, respectively, is preferred although in Japanese these differences do not occur.

Abe, Prof. Hideo. Dept. of Metallurgy and Materials Sci., Faculty of Eng., U. of Tokyo, 7-3-1 Hongo Bunkyo-ku, Tokyo 113, Japan. (1924) DEng, metallurgy (Tokyo U., 1960). Prof. (tel. 03 + 812-2111, ext. 7159). *Texture, recrystallization in metals.*

Abe, Prof. Ryuji. Dept. of Applied Physics, Faculty of Eng., Nagoya U., Furo-cho Chikusa-ku, Nagoya 464, Japan. (1922) DSc, physics (Nagoya U., 1959). Prof. (tel. 052 + 781-5111, ext. 4465). *Ferroelectrics.*

Abe, Mr Shuich. Second Forensic Sci. Div. National Res. Inst. of Police Sci., 6 Sanban-cho Chiyoda-ku, Tokyo 102, Japan. (1927) BSc, physics (Tohoku U., 1951). Techn. official. (tel. 03 + 261-9986, ext. 2474). *X-ray crystallography, solid state physics.*

Adachi, Prof. Kengo. Faculty of Eng., Nagoya U., Furo-cho Chikusa-ku, Nagoya 464, Japan. (1926) DSc, physics (Tohoku U., 1958). Prof. of physics. (tel. 052 + 781-5111, ext. 3567). *Magnetism, order-disorder phenomena, alloys.*

Aihara, Prof. Ariyuki. Dept. of Materials Sci., U. of Electro-Communications, 5-1 Chofugaoka 1-chome Chofu, Tokyo 182, Japan. (1920) DSc, chemistry (Tokyo U., 1954). Prof. (tel. 0424 + 83-2161). *Dielectric spectroscopy, molecular behavior in crystals, NMR.*

Aikawa, Dr Nobuyuki. Dept. of Geosciences, Faculty of Sci., Osaka City U., 459 Sugimoto-cho Sumiyoshi-ku, Osaka 558, Japan. (1944) DSc, mineralogy (Tokyo U., 1975). Res. assoc. (tel. 06 + 692-1231). *Crystal structure, minerals, diffuse X-ray reflections.*

Aizaki, Mr Nao-aki. Central Res. Labs., Nippon Electric Co. Ltd., 4-1-1 Miyazaki Takatsu-ku, Kawasaki Kanagawa 213, Japan. (1945) MEng, applied physics (Tokyo U., 1971). Res. member. (tel. 044 + 855-1111). *Semiconductors, X-ray topography.*

Akamatu, Prof. Hideo. Inst. for Molecular Sci., 38 Nishigonaka Myodaiji Okazaki, Aichi 444, Japan. (1910) DSc, chemistry (Tokyo U., 1942). Director general; Em. prof. (Tokyo U.). (tel. 0564 + 52-9771). *Electronic structure and behavior, molecular crystals.*

Akao, Dr Masaru. Inst. for Med. Dent. Eng., Tokyo Med. Dent. U., 2-3-10 Kanda Surugadai Chiyoda-ku, Tokyo 101, Japan. (1945) DEng, chemical eng. (Tokyo Inst. Techn., 1975). (tel. 03 + 291-3721). *Mineralogy, bioceramics.*

Akimitsu, Dr Jun. C. of Sci. and Eng., Aoyama Gakuin U., 16-1 Chitosedai 6-chome Setagaya-ku, Tokyo 157, Japan. (1940) DSc, physics (Tokyo U., 1972). Asst. prof. (tel. 03 + 307-2888). *Neutron diffraction, low energy electron diffraction.*

Akizuki, Dr Mizuhiko. Inst. of Mineralogy, Petrology and Economic Geology, Tohoku U., Aramaki Aza-Aoba, Sendai 980, Japan. (1937) DSc, mineralogy (Tohoku U., 1968). Asst. prof. (tel. 0222 + 22-1800, ext. 4227). *Texture, minerals, twins.*

Amemiya, Dr Yoshiyuki. Photon Factory, Nat. Lab. for High Energy Physics. Oho-machi, Tsukuba, Ibaraki 305, Japan. (1952) DEng, applied physics (Tokyo U., 1979) Res. assoc. (tel. 0298 + 64-1171, ext. 5025). *Synchrotron radiation, instrumentation, small angle scattering, time-resolved X-ray measurement, position sensitive X-ray detector.*

Ando, Dr Masami. Photon Factory, Nat. Lab. for High Energy Physics. Oho-machi, Tsukuba, Ibaraki 305, Japan. (1942) DEng., applied physics (Tokyo U., 1974). Assoc. prof. (tel. 0298 + 64-1171, ext. 5040). *Synchrotron radiation, instrumentation.*

Ando, Dr Yoshinori. Dept. of Physics, Meijo U., Tempaku-cho Tempaku-ku, Nagoya 468, Japan. (1942) DEng, applied physics (Nogoya U., 1970). Asst. prof. (tel. 052 + 832-1151, ext. 571). *Crystal growth, X-ray and electron diffraction, electron microscopy, ceramics.*

Annaka, Prof. Shoichi. Tokyo U. of Mercantile Marine, 1-6 Etchujima 2-chome Koto-ku, Tokyo 135, Japan. (1924) DSc, physics (Tokyo U. of Education, 1958). Prof. (tel. 03 + 641-1171). *Crystal physics, TDS.*

Aoe, Mr Hiroyuki. Technical Operations, Res. Center, Sanyo Electric Co. Ltd., 18-13 Hashiridani 1-chome Hirakata, Osaka 573, Japan. (1943) MSc, physics (Nagoya U., 1967). (tel. 0720 + 41-1161).

Aoki, Dr Katsuyuki. Inst. of Physical and Chemical Res., 1 Hirosawa 2-chome Wako, Saitama 351, Japan. (1945) DPharm, chemistry (Tokyo U., 1978). Res. (tel. 0484 + 62-1111). *Inorganic structural biochemistry, organometallic structural chemistry.*

Aoki, Dr Yoshikazu. Dept. of Geology, Faculty of Sci., Kyushu U., 10-1 Hakozaki 6-chome, Higashi-ku, Fukuoka 812, Japan. (1939) DSc, mineralogy (Kyushu U., 1977). Assoc. prof. (tel. 092 + 641-1101, ext. 4312). *Mineralogy, crystal growth.*

Asada, Prof. Eiichi. Dept. of Materials Sci., Toyohashi U. of Techn., 1-1 Hibarigaoka Tempaku-cho Toyohashi 440 Japan. (1924) DEng, applied chemistry (Tokyo U., 1979). Prof. (tel. 0532 + 47-0111, ext. 442). *X-ray diffraction, X-ray spectroscopy, crystal chemistry, analytical chemistry.*

Asai, Dr Takeshi. Inst. of Scient. and Industrial Res., Osaka U., Yamadakami Suita, Osaka 565, Japan. (1942) DSc, chemistry (Osaka U., 1971). Res. assoc. (tel. 06 + 877-5111, ext. 3551). *Crystal chemistry, inorganic solid materials, nuclear magnetic resonance, magnetism.*

Ashida, Dr Sakichi. Yokohama Works, Hitachi Ltd. 292 Yoshida-cho Totsuka-ku, Yokohama Kanagawa 244, Japan. (1935) DSc, mineralogy (Tohoku U., 1965). Sr. eng. (tel. 045 + 881-1241, ext. 2056). *Crystal growth, piezoelectric materials, solid state physics.*

Ashida, Prof. Tamaichi. Dept. of Applied Chemistry, Faculty of Eng., Nagoya U., Furo-cho Chikusa-ku, Nagoya 464, Japan. (1933) DSc, chemistry (Osaka U., 1964). Prof. (tel. 052 + 781-5111, ext. 3339). *Organic crystal structures, protein structure, computer programming.*

Azuma, Mr Nagao. Faculty of General Education, Ehime U., 3 Bunkyo-cho Matsuyama, Ehime 790, Japan. (1943) MSc, chemistry (Kyoto U., 1972). Lect. in chemistry. (tel. 0899 + 24-7111, ext. 3862). *Organic structural chemistry, magnetic behavior, organic radicals.*

Bunno, Mr Michiaki. The University Museum, U. of Tokyo, 3-1 Hongo 7-chome Bunkyo-ku, Tokyo 113, Japan. (1942) MEng, applied mineralogy (Tokyo U., 1970). Curator. (tel. 03 + 812-2111, ext. 7845). *Descriptive mineralogy.*

Chihaya, Prof. Tadashi. Ibaraki Techn. C., 866 Fukayatsu Nakane Katsuta, Ibaraki 312, Japan. (1910) President; Em. prof. (Ibaraki U.). (tel. 0292 + 72-5201). *Metallurgy, physics.*

Chikaura, Dr Yoshinori. Dept. of Physics, Kyushu Inst. of Techn., 1 Sensui-cho Tobata-ku, Kitakyushu Fukuoka 804, Japan. (1946) DSc, physical metallurgy (Tokyo Inst. of Techn., 1973). Asst. prof. (tel. 093 + 871-1931, ext. 486). *X-ray topography, magnetic domains, crystal growth, crystal imperfection, phase transition.*

Chikawa, Dr Jun-ichi. Photon Factory, Nat. Lab. for High Energy Physics, Oho-machi, Tsukuba, Ibaraki 305, Japan. (1930) DSc, physics (Kyoto U., 1961). Prof. (tel. 0298 + 64-1171). *X-ray topography, magnetic domains, crystal growth, crystal imperfection, phase transition.*

Dohi, Prof. Shoso. Dept. of Applied Physics, National Defense Academy, 10-20 Hashirimizu 1-chome Yokosuka, Kanagawa 239, Japan. (1927) DSc, physics (Hiroshima U., 1961). Prof. (tel. 0468 + 42-3810, ext. 540). *Physical properties, less-common metals.*

Doi, Dr Kenji. Physics Div., Japan Atomic Energy Res. Inst., 2-4 Shirakata-Shirane Tokaimura Nakagun, Ibaraki 319-11, Japan. (1929) DSc, mineralogy, solid state physics, (Tokyo U., 1959). Sr. physicist. (tel. 02928 + 2-5731). *X-ray and neutron diffraction, neutron diffraction topography, electron microscopy.*

Ejiri, Dr Koichi. Techn. Div., Ricoh Company, 3-6 Naka-magome 1-chome Ota-ku, Tokyo 143, Japan. (1945) DSc, mineralogy (Tokyo U., 1973). Res. (tel. 03 + 772-8111). *Crystal growth, semiconductor physics, ceramics.*

Fujii, Dr Satoshi. Faculty of Pharmaceutical Sci., Osaka U., 1-6 Yamada-oki Suita, Osaka 565, Japan. (1946) DPharm, pharmacy (Osaka U., 1978). Res. assoc. (tel. 06 + 877-5111, ext. 6213). *Biophysical sciences, organic crystal structure.*

Fujii, Mr Tetsuo. Res. Lab. of Eng. Materials, Tokyo Inst. of Techn. Nagatsuta-machi 4259 Midori-ku, Yokohama 227, Japan. (1957) MSc, materials science (Tokyo Inst. of Techn., 1982). Postgraduate student (tel. 045 + 922-1111, ext. 2311). *Inorganic crystal structures.*

Fujii, Assoc. Prof. Yasuhiko. Faculty of Eng. Sci., Osaka U., Machikaneyama 1-1 Toyonaka, Osaka 560, Japan. (1943) DSc, physics (Osaka U., 1973). Assoc.

prof. (tel. 06 + 844-1151, ext. 4686). *X-ray diffraction, neutron scattering, phase transitions, high pressure physics, nonequilibrium physics.*

Fujiki, Prof. Yoshibumi. Inst. of Physics, Kyoto Prefectural U. of Medicine, Daishogun Nishitakatsukasa-cho Kita-ku, Kyoto 603, Japan. (1914) DSc, physics (Kyoto U., 1960). Prof. (tel. 075 + 462-6932). *Metal physics, crystal growth, metal films.*

Fujime, Dr Satoru. Biophysics Div., Mitsubishi-Kasei Inst. of Life Sciences, 11 Minamiooya Machida, Tokyo 194, Japan. (1936) DSc, physics (Tohoku U., 1967). Div. chief. (tel. 0427 + 26-1211, ext. 288). *Dynamical properties, biological macromolecules.*

Fujimoto, Prof. Fuminori. Inst. of Physics, C. of General Education, U. of Tokyo, Komaba, Meguro-ku, Tokyo 153, Japan. (1928) DSc, physics (Tokyo U., 1959). Prof. (tel. 03 + 467-1171, ext.268). *Electron diffraction, electron channeling, channeling radiation, high energy electron microscopy, ion beam physics, nuclear and high energy physics - crystal applications.*

Fujimoto, Prof. Hirofumi. Miyagi U. of Education, Aramaki-Aoba, Sendai 980, Japan. (1923) DSc, physics (Tohoku U., 1961). Prof. of physics. (tel. 0222 + 22-1021). *Brems-radiation, materials structure.*

Fujino, Dr Kiyoshi. Dept. of Earth Sci., Faculty of Sci., Ehime U. Bunkyo-cho 2-5 Matsuyama, Ehime 790, Japan. (1945) DSc, mineralogy (Tokyo U., 1974). Assoc. prof. (tel. 0899 + 24-7111, ext. 3597). *X-ray crystallography, electron microscopy, rock forming minerals.*

Fujino, Dr Nobukatsu. Fundamental Res. Section, Central Res. Labs., Sumitomo Metal Industries Ltd., 3 Nishinagasu Hondori 1-chome Amagasaki, Hyogo 660, Japan. (1935) DSc, chemistry (Osaka U., 1967). Sr. res. engineer. (tel. 06 + 401-6201, ext. 311). *X-ray spectroscopy, electron spectroscopy, X-ray diffraction, metallurgical applications.*

Fujita, Prof. Francisco Eiichi. Dept. of Material Sci., Fac. of Eng. Sci., Osaka U., Machikaneyama 1-1 Toyonaka, Osaka 560, Japan. (1925) DSc, physics (Hokkaido U., 1959). Prof. of metal physics. (tel. 06 + 844-1151, ext. 4660). *Metal physics, lattice defect, Mössbauer spectroscopy, electron microscopy.*

Fujita, Prof. Hiroshi. Res. Center, Ultra- High Voltage Electron Microscopy, Osaka U., Yamada-kami Suita, Osaka 565, Japan. (1926) DEng, metal physics (Osaka U., 1958). Prof. (tel. 06 + 877-5111, ext. 4131,4133). *Metal physics, lattice imperfection behavior, electron microscopy applications, materials science.*

Fujiwara, Prof. Kunio. Inst. of Phys., C. of Arts & Sci., U. of Tokyo Komaba 3-8-1 Meguro-ku, Tokyo 153, Japan. (1932) DSc, Phys. (Tohoku U., 1962). Prof. (tel. 03 + 467-1171, ext. 556). *Positrons in crystals.*

Fujiwara, Prof. Hiroshi. Faculty of Sci., Hiroshima U., 1-89 Higashisendamachi 1-chome, Hiroshima 730, Japan. (1928) DSc, physics (Tokyo U., 1959). Prof. (tel. 0822 + 41-1221, ext. 438). *Materials under high pressure, high pressure X-ray diffraction, thin film crystallography, magnetic properties, metals and alloys.*

Fujiwara, Dr Takaji. Faculty of Pharmaceutical Sci., Osaka U., 1-6 Yamada-oka Suita, Osaka 565, Japan. (1936) DSc, chemistry (Osaka U., 1971). Asst. prof. (tel. 06 + 877-5111, ext. 6212). *Organic crystal structures, large molecules.*

Fujiyoshi, Dr Yoshinori. Crystal and Powder Chemistry, Inst. for Chemical Res., Kyoto U., Gokasho Uji, Kyoto-Fu 611, Japan. (1948) Dr, chemistry (Kyoto U., 1982). Techn. assoc. (tel. 0774 + 32-3111, ext. 2263). *Electron diffraction, X-ray diffraction analysis, high resolution electron microscopy.*

Fukamachi, Dr Tomoe. Dept. of Electronic Eng., Saitama Inst. of Techn., 1690 Fusaiji Okabe, Saitama 369-02, Japan. (1943) DSc, physics (Tokyo U., 1976). Asst. prof. (tel. 0485 + 85-2521). *Energy-dispersive diffractometry, anomalous scattering, Compton scattering.*

Fukano, Prof. Yasushige. C. of General Education, Nagoya U., Furo-cho Chikusa-ku, Nagoya 464, Japan. (1927) DSc, physics (Nagoya U., 1961). Prof. (tel. 052 + 781-5111, ext. 4846). *Physics, fine particles, metals, alloys, semiconductors.*

Fukuda, Dr Tsuguo. Electron Devices Lab., Toshiba Res. and Dev. Center, Tokyo Shibaura Electric Co. Ltd., 1 Komukai Toshiba-cho Saiwai-ku, Kawasaki Kanagawa 210, Japan. (1939) DSc, crystallography (Tokyo U., 1971). Res. staff. (tel. 044 + 511-2111, ext. 2389). *Crystal growth, ferroelectrics, semiconductors, optics, crystallography, mineralogy.*

Fukuhara, Dr Akira. Central Res. Lab., Hitachi Ltd., 280 Higashi-Koigakubo 1-chome, Kokubunji Tokyo 185, Japan. (1933) DSc, physics (Tokyo U., 1961). (tel. 0423 + 23-1111). *Electron and X-ray diffraction.*

Fukuma, Dr Keiichi. Faculty of Eng., Tottori U., 101 Minami 4-chome Koyama-cho, Tottori 680, Japan. (1949) DSc, chemistry (Osaka U., 1979). Res. assoc. (tel. 0857 + 28-0321, ext. 4222). *Protein crystallography.*

Fukuyama, Dr Tsutomu. Atmospheric Environment Division, Nat. Inst. for Environmental Studies, 16-2 Onogawa, Yatabe-cho, Tsukuba-gun Ibaraki 305, Japan. (1942) DSc, chemistry (Tokyo U., 1971) Sect. chief. (tel. 0298 + 51-6111, ext. 698). *Aerosol chemistry, nucleation.*

Furukawa, Dr Kozo. Dept. of Metal Sci. and Techn., Faculty of Eng., Kyoto U., Yoshida Honmachi Sakyo-ku, Kyoto 606, Japan. (1925) DEng, metallurgy (Kyoto U., 1967). Asst. prof. (tel. 075 + 751-2111, ext. 5467). *Lattice defects, metals, X-ray diffraction.*

Furusaki, Dr Akio. Dept. of Chemistry, Faculty of Sci., Hokkaido U., Kita 10-jo Nishi 8-chome Kita-ku, Sapporo 060, Japan. (1939) DSc, chemistry (Osaka U., 1968). Lect. (tel. 011 + 711-2111, ext. 2703).

Goto, Prof. Masaru. Faculty of Eng., Oita U., 700 Dannoharu, Oita 870-11, Japan. (1932) DSc, physics (Hiroshima U., 1965). Prof. of physics. (tel. 0975 + 69-3311, ext. 426, 425). *Mechanical properties (elastic and non-elastic), metals, defects, dislocations, internal friction.*

Goto, Mr Yoshiaki. Dept. of Chemistry, Faculty of Techn., Gunma U., 5-1 Tenjincho 1-chome Kiryu, Gunma 376, Japan. (1940) MSc, chemistry (Gunma U., 1967). Res. assoc. (tel. 0277 + 22-3181).

Gyobu, Mr Atsuo. Dept. of Mechanical Eng., Niihama Techn. C., 7-1 Yakumo-cho Niihama, Ehime 792, Japan. (1945) BEng, Res. assoc. (tel. 0897 + 37-1240). *Mineralogy.*

Haga, Dr Nobuhiko. Mineralogical Inst., Faculty of Sci., U. of Tokyo, 3-1 Hongo 7-chome Bunkyo-ku, Tokyo 113, Japan. (1945) DSc, mineralogy (Tokyo U., 1973). Res. assoc. (tel. 03 + 812-2111, ext. 4548). *Crystal structure, minerals.*

Haisa, Prof. Masao. Dept. of Chemistry, Faculty of Sci., Okayama U., Tsushima Okayama 700, Japan. (1923) DSc, chemistry (Osaka U., 1961). Prof. (tel. 0862 + 52-1111, ext. 441). *Organic crystal structures, large molecules, crystal physics, polymer chemistry.*

Hamamura, Assoc. prof. Kenji. Dept. of Physics, Inst. of Voc. Training, 1960 Aihara Sagamihara 229, Japan. (1941) DEng, metallurgy (Tokyo Inst. of Techn., 1975). Assoc. prof. (tel. 0427 + 61-2111, ext. 346). *X-ray crystallography, crystal growth.*

Hamauzu, Dr Yoshihiro. Dept. of Industrial Chemistry, Numazu C. of Techn., 3600 Ooka Numazu, Shizuoka 410, Japan. (1944) DSc, physics (Hokkaido U., 1974) Assoc. prof. (tel. 0559 + 21-2700, ext. 455). *Atom diffraction, surface crystallography.*

Harada, Prof Jimpei. Dept. of Applied Physics, Faculty of Eng., Nagoya U., Chikusa-ku, Nagoya 4648, Japan. (1931) DSc, physics (Tokyo Inst. of Techn., 1964). Prof. (tel. 052 + 781-5111, ext. 4464). *X-ray and neutron diffraction, solid state physics.*

Harada, Mr Shigeharu. Inst. for Protein Res., Osaka U., Yamada Kami Suita 565, Japan. (1954) MSc, chemistry (Osaka U., 1979). Grad. student. (tel. 06 + 877-5111, ext. 3837). *Protein crystallography.*

Harada, Mr Yoshiyasu. Electron Optics Div., JEOL Ltd., 1418 Nakagami Akishima, Tokyo 196, Japan. (1943) BSc, physics (Nihon U., 1966). Res. (tel. 0425 + 43-1111, ext. 321). *Electron microscopy.*

Harada, Dr Zyunpei. 4426 Ikuta Tama-ku, Kawasaki 214, Japan. (1898) DSc, mineralogy (Tokyo U., 1939). Em. prof. (tel. 044 + 911-8177). *Mineralogy, physical crystallography.*

Hariya, Dr Yu. Dept. of Geology and Mineralogy, Hokkaido U., Kita 10-jo Nishi 8-chome Kita-ku, Sapporo 060, Japan. (1929) DSc, mineralogy (Hokkaido U., 1961). Asst. prof. (tel. 011 + 711-2111, ext. 2728). *Geochemistry of manganese, phase equilibrium, minerals at high pressures.*

Hasebe, Mr Tooru. Dept. of Chemistry, Faculty of Education, Fukushima U., 12-23 Hamada-cho, Fukushima 960, Japan. (1947) MSc, physical chemistry (Kanazawa U., 1972). Lect. (tel. 0245 + 48-5151). *Molecular motions, molecular interactions in solid state.*

Hashimoto, Prof. Hatsujiro. Dept. of Applied Physics, Osaka U., 133-1 Yamadakami Suita, Osaka 565, Japan. (1921) DSc, physics (Kyoto U., 1953). Prof. (tel. 06 + 877-5111, ext. 4657). *Dynamical theory of electron diffraction, electron microscopy, crystal growth, oxides, sulphides.*

Hashimoto, Mr Hideki. Res., Toray Res. Cntr. Inc., Sonoyama, Otsu, Shiga 520, Japan. (1958) MSc, X-ray topography (Nagoya U., 1983). (tel. 0775 + 37-0700, ext. 4133). *EXAFS.*

Hashimoto, Mr Katsunobu. Dept. of Physics, Osaka Kyoiku U., 43 Minami-Kawahori-cho Tennoji-ku, Osaka 543, Japan. (1950) MSc, physics (Osaka Kyoiku U., 1976). Grad. student. (tel. 06 + 771-8131).

Hashizume, Assoc. Prof. Hiroo. Res. Lab. of Eng. Materials, Tokyo Inst. of Techn., Nagatsuta, Midori-ku, Yokohama 227, Japan. (1940) DSc(eng.), applied physics (Tokyo U., 1970). Assoc. prof. (tel. 045 + 922-1111, ext. 2333). *Material sciences, dynamical X-ray diffraction, single crystal characterization, X-ray optics, instrumentation.*

Hasiguti, Prof. Ryukiti. Faculty of Eng., The Science U. of Tokyo, Kagurazaka Shinjuku-ku, Tokyo 162, Japan. (1914) DEng, metallurgy (Tokyo U., 1953). Prof. of materials science. (tel. 03 + 260-4271, ext. 439). *Crystal lattice defects, radiation damage, crystal growth.*

Hata, Dr Yasuo. Inst. for Protein Res., Osaka U., 5311 Yamada-kami Suita, Osaka 565, Japan. (1951) DSc, polymer sci. (Osaka U., 1979). Asst. (tel. 06 + 877-5111, ext. 3837). *X-ray crystallography, protein crystallography.*

Hayakawa, Dr Kazunobu. Central Res. Lab., Hitachi Ltd., 280 Higashi-Koigakubo 1-chome, Kokubunji Tokyo 185, Japan. (1936) DSc, physics (Tokyo U., 1970). Sr. res. (tel. 0423 + 23-1111, ext. 378). *LEED, surface physics.*

Hayashi, Dr Kooya. Dept. of Chemistry, Faculty of Sci., Okayama U. of Sci., 1 Ridai-cho 1-chome, Okayama 700, Japan. (1947) DEng, solid state chemistry (Tokyo Inst. of Techn.,1975). Lect. (tel. 0862 + 52-3161, ext. 315). *Inorganic synthetic chemistry, structure analysis.*

Hayashi, Prof. Mitsuhiko. Dept. of Physics, Toyama Medical and Pharmaceutical U., 2-11 Shibazono-cho 3-chome, Toyama 930, Japan. (1930) DSc, physics (Tokyo Inst. of Techn., 1971). Prof. (tel. 0764 + 42-3300, ext. 52). *Crystal growth, dielectric properties of matter, magnetic resonance.*

Hayashi, Dr Yasunori. Dept. of Iron and Steel Metallurgy, Kyushu U., 3576 Hakozaki, Higashi-ku, Fukuoka 812, Japan. (1939) DEng, applied physics

(Tokyo U., 1971). Asst. prof. (tel. 092 + 641-1101, ext. 3413). *Surface physics and chemistry, metal surfaces.*

Hidaka, Mr Tsuneo. Res. and Dev. Div., Asahi Optical Co. Ltd., 36-9 Maeno-cho 2-chome, Itabashi-ku, Tokyo 174, Japan. (1940) MSc, ceramics (MIT, USA, 1974). Chief eng. (tel. 03 + 960-5151, ext. 371). *Optics, materials science.*

Higashi, Prof. Akira. Dept. of Applied Physics, Faculty of Eng., Hokkaido U., Kita 13-jo Nishi 8-chome Kita-ku, Sapporo 060, Japan. (1922) DSc, physics (Hokkaido U., 1951). Prof. (tel. 011 + 711-2111, ext. 6635). *Crystal growth, crystal defects, ice crystals.*

Higuchi, Assoc. Prof. Taiichi. Dept. of Chemistry, Faculty of Sci., Osaka City U., 3-3-138 Sugimoto-cho, Sumiyoshi-ku, Osaka 558, Japan. (1929) DSc, chemistry (Osaka City U., 1967). Assoc. prof. (tel. 06 + 692-1232, ext. 3244). *Organic and organometallic structures, large molecules, solid state chemistry.*

Hikichi, Dr Kunio. Dept. of Polymer Sci., Faculty of Sci., Hokkaido U., Kita 10-jo Nishi 8-chome Kita-ku, Sapporo 060, Japan. (1936) DSc, physics (Hokkaido U., 1964). Asst. prof. (tel. 011 + 711-2111, ext. 2770). *Physics, biophysics, polymer physics.*

Hirabayashi, Prof. Makoto. Res. Inst. for Iron, Steel and Other Metals, Tohoku U., 1-1 Katahira 2-chome, Sendai 980, Japan. (1925) DSc, metal physics (Tohoku U., 1959). Prof. (tel. 0222 + 27-6200, ext. 2943). *Metal physics, metals structure.*

Hirahara, Dr Eiji. Sendai Radio Techn. C., Miyagi-cho Miyagi-gun, Miyagi 989-31, Japan. (1912) DSc, physics (Hiroshima U., 1949). President. (tel. 022386 + 4761). *Solid state physics, magnetism, semiconductors, phase transition.*

Hirakawa, Prof. Kinshiro. Neutron Diffraction Div., Inst. for Solid State Physics, U. of Tokyo, 22-1 Roppongi 7-chome Minato-ku, Tokyo 106, Japan. (1926) DSc, physics (Osaka U., 1959). Prof. (tel. 03 + 402-6231, ext. 687, 695). *Neutron diffraction.*

Hirano, Prof. Shin-ichi. Dept. of Applied Chemistry., Faculty of Eng., Nagoya U., Furocho Chikusaku, Nagoya 464, Japan. (1942) D. Eng, applied chemistry (Nagoya U., 1970). Prof. (tel. 052 + 781-5111, ext. 3343). *Hydrothermal crystal growth, crystallization, hydrolysis, organometallic compounds, ceramic processing.*

Hirayama, Mr Noriaki. Tokyo Res. Lab., Kyowa Hakko Kogyo Co. Ltd., 3-6-6 Asahimachi Machida, Tokyo 194, Japan. (1948) DSc, chemistry (Tokyo Inst. of Techn., 1981). Scient. (tel. 0427 + 25-2555, ext. 256). *Drug - macromolecule interaction, protein crystallograpy, polymorphism.*

Hirokawa, Prof. Sakutaro. Dept. of Chemistry, National Defense Academy, 10-20 Hashirimizu 1-chome Yokosuka, Kanagawa 239, Japan. (1919) DSc, chemistry (Osaka U., 1955). Prof. (tel. 0468 + 41-3810, ext. 525). *Crystal structures, organic compounds.*

Hirokawa, Prof. Tomoo. Physics Lab., C. of Eng., Nihon U., Tamura-machi Koriyama, Fukushima 963, Japan. (1914) DSc, physics (Hiroshima U., 1961). Prof. (tel. 0249 + 44-1300, ext. 313). *Crystal defects, dislocation mobility, X-ray crystallography.*

Hirotsu, Dr Ken. Dept. of Chemistry, Faculty of Sci., Osaka City U. Sugimoto, Sumiyoshi-ku, Osaka 558, Japan. (1942) DSc, chemistry (Osaka City U., 1974). Lect. (tel. 06 + 692-1231, ext. 3244). *Organic crystal structures, large molecules.*

Hirotsu, Dr Yoshihiko. Technological U. of Nagoya, 1603-1 Nagamine Kamitomioka-cho, Nagaoka 949-54, Japan. (1945) DEng (Tokyo Inst. of Techn., 1974). Asst. prof. (tel. 0258 + 46-6000). *Structure, metals and alloys, electron diffraction, electron microscopy.*

Homma, Mr Shigeru. Nat. Inst. for Res. in Inorganic Materials, Sakura-mura Niihari-gun, Ibaraki 305, Japan. (1936) BSc, mineralogy (Hokkaido U., 1962). Head res. (tel. 0298 + 51-3351, ext. 280). *X-ray diffraction topography, crystal growth.*

Homma, Dr Teiichi. Dept. of Applied Physics and Mechanics, Inst. of Industrial Sci., U. of Tokyo, 22-1 Roppongi 7-chome Minato-ku, Tokyo 106, Japan. (1931) DEng, metallurgy (Tokyo U., 1966). Asst. prof. (tel. 03 + 402-6231, ext. 272). *Metals, surface science, high temperature oxidation.*

Honjin, Prof. Ryohei. Dept. of Anatomy, School of Medicine, Kanazawa U., 13-1 Takara-machi, Kanazawa 920, Japan. (1921) MD, biophysics (Kanazawa U., 1950). Prof. (tel. 0762 + 62-8151, ext. 230). *Yolk platelets crystalline lattice structure, myelin membrane molecular organization, nerve fibers.*

Honjo, Prof. Goro. Physics Dept., Tokyo Inst. of Techn., Oh-okayama Meguro-ku, Tokyo 152, Japan. (1918) DSc, physics (Nagoya U., 1954). Em. prof. (tel. 03 + 726-1111). *Crystal growth, lattice defects, surfaces.*

Horiuchi, Dr Hiroyuki. Inst. of Scient. and Industrial Res., Osaka U., Yamadakami Suita, Osaka 565, Japan. (1940) DSc, mineralogy (Tokyo U., 1969). Res. assoc. (tel. 06 + 877-5111, ext. 3546). *Crystal structures, inorganic materials, minerals.*

Horiuchi, Dr Shigeo. Nat. Inst. for Res. in Inorganic Materials, Sakura-mura Niihari-gun, Ibaraki 305, Japan. (1939) DEng, metallurgy (Tokyo U., 1967). Chief res. (tel. 0298 + 51-3351, ext. 283). *Structure analysis, high resolution electron microscopy.*

Hosaka, Dr Masahiro. Gemmology, Gemmology & Jewelry Arts of Yamanashi. Tokojicho 1955-1, Kofu Yamanashi 400, Japan. (1944) PhD, crystal growth, inorganic chemistry (Yamanashi U., 1967). Asst. prof. (tel. 0552 + 32-6672). *Hydrothermal growth, Inorganic chemistry.*

Hoshino, Prof. Sadao. Inst. for Solid State Physics, U. of Tokyo, 22-1 Roppongi 7-chome Minato-ku, Tokyo 106, Japan. (1926) DSc, physics (Osaka U., 1958). Prof. (tel. 03 + 402-6231, ext. 661). *Phase transition, neutron diffraction, ferroelectricity.*

Hosoya, Mr Masahiko. Dept. of Physics, U. of the Ryukyus, Senbaru-1 Nisihara, Okinawa 903-01, Japan. (1943) MSc, Physics (Hokkaido U., 1967). Assoc. prof. (tel. 09889 + 5-2221, ext. 2635). *solid state physics.*

Hosoya, Prof. Em. Sukeaki. Bunka Women's U., 22-1 Yoyogi 3 chome, Shibuya-ku, Tokyo 151, Japan. (1924) PhD, physics (Wales U., UK, 1958) DSc, physics (Tokyo U., 1961). Prof.; Em. prof. (Tokyo U.). (tel. 03 + 370-9828). *Teaching crystallography, anomalous dispersion, Compton scattering, molecular biology.*

Ibata, Dr Koichi. Central Res. Labs., Kuraray Co. Ltd., 2045-1 Aoeyama Sakazu Kurashiki, Okayama 710 Japan. (1947) DSc, chemistry (Tokyo Inst. of Techn., 1975). Res. worker. (tel. 0864 + 23-2271, ext. 333). *Chemistry, astronomy.*

Ichikawa, Dr Mizuhiko. Dept. of Physics, Faculty of Sci., Hokkaido U., Kita 10-jo Nishi 8-chome, Kita-ku, Sapporo 060, Japan. (1940) DSc, chemistry (Osaka U., 1979). Res. assoc. (tel. 011 + 716-2111, ext. 5427, Telex 932510HOKUSC J). *X-ray and neutron analysis, structure determination, structural phase transition, ferroelectric and ferroelastic transitions, hydrogen bonding in crystals.*

Ichimiya, Dr Ayahiko. Dept. of Applied Physics, Nagoya U., Furo-cho Chikusa-ku, Nagoya 464, Japan. (1940) DSc, physics (Nagoya U., 1966). Lect. (tel. 052 + 781-5111, ext. 4459). *Electron diffraction, electron microscopy, crystal growth, surface physics.*

Ichimura, Mr Takeo. Glass Manufacturing Div., Nippon Kogaku K.K., 1773 Asamizodai Sagamihara, Kanagawa 228, Japan. (1929) BSc, applied chemistry (Tokyo U., 1954). (tel. 0427 + 45-3311).

Iida, Dr Atsuo. Dept. of Industrial Chemistry, Faculty of Eng., U. of Tokyo, 7-3-1 Hongo Bunkyo-ku, Tokyo 113, Japan. (1948) DEng, crystal characterization (Tokyo U., 1977). Res. asst. (tel. 03 + 812-2111, ext. 7234). *X-ray crystallography, X-ray spectrometry, X-ray optics.*

Ichinokawa, Prof. Takeo. Dept. of Applied Physics, Waseda U., 3-4-1 Ohkubo Shinjuku-ku, Tokyo 160, Japan. (1928) DSc, physics (Waseda U., 1958). Prof. (tel. 03 + 209-3211, ext. 3559). *Electron microscopy, surface science.*

Iijima, DSc Kinya. Dept. of Chemistry, Shizuoka U., 836 Oya, Shizuoka 422, Japan. (1941) DSc, chemistry (Hokkaido U., 1976). Res. assoc. (tel. 0542 + 37-1111, ext. 552). *Structure of molecules.*

Iijima, Prof. Takao. Faculty of Sci., Gakushuin U., 1-5-1 Mejiro, Toshima-ku, Tokyo 171 Japan. (1934) DSc, chemistry (Tokyo U., 1962). Prof. (tel. 03 + 986-0221, ext. 424).

Iishi, Dr Kazuaki. Dept. of Mineralogical Sci. and Geology, Faculty of Sci., Yamaguchi U., 1677-1 Yoshida, Yamaguchi 753, Japan. (1942) DSc, mineralogy (Hiroshima U., 1973). Asst. prof. (tel. 0839 + 22-6111, ext. 385). *Lattice dynamics of solids, crystal growth, electron microscopy.*

Iitaka, Prof. Yoichi. Faculty of Pharmaceutical Sci. U. of Tokyo, 3-1 Hongo 7-chome Bunkyo-ku, Tokyo 113, Japan. (1927) DSc, crystallography (Tokyo U., 1959). Prof. (tel. 03 + 812-2111, ext. 4840). *Organic crystal structures, instrumentation, large molecules, biological substances, computer programming.*

Iizuka, Dr Masakatsu. Faculty of Education, Bunkyo U., 3337 Minamiogishima Koshigaya Saitama 343, Japan. (1933) DSc, structural crystallography (Tokyo U. of Education, 1972). Lect. (tel. 0489 + 74-8811). *Crystal structure, crystal chemistry.*

Iizumi, Dr Masashi. Dept. of Physics, Japan Atomic Energy Res. Inst., Tokai, Ibaraki 319-11, Japan. (1935) DSc, Physics (Tokyo U., 1974). Deputy dir. (tel. 0292 + 82-5476). *Neutron scattering, phase transition.*

Ikemoto, Dr Isao. Dept. of Chemistry, Faculty of Sci., U. of Tokyo, 3-1 Hongo 7-chome Bunkyo-ku, Tokyo 113, Japan. (1940) DSc, chemistry (Tokyo U., 1968). Lect. (tel. 03 + 812-2111, ext. 3644). *Crystal structure, molecular complexes, X-ray photoelectron spectroscopy.*

Imado, Dr Sataro. First Assay Dept., Products Control Lab., Tanabe Seiyaku Co. Ltd., 16-89 Kashima 3-chome Yodogawa-ku, Osaka 532, Japan. (1923) DPharm, pharmaceutical sciences (Kumamoto Pharm. U., 1961). Manager. (tel. 06 + 301-1221). *Metal chelate compounds, pharmaceuticals.*

Imanishi, Dr Yasuhiro. Res. and Dev. Div., Toshiba Ceramics Co. Ltd., 30 Soya Hatano, Kanagawa, Japan. (1947) DSc, mineralogy (Tokyo U., 1977). Res. (tel. 0463 + 81-8407). *Phase transformation, minerals, crystal growth, semiconductor materials, physical properties.*

Imura, Prof. Toru. Dept. of Metallurgy, Faculty of Eng., Nagoya U., Furo-cho Chikusa-ku, Nagoya 464, Japan. (1924) DSc, physics (Osaka U., 1957). Prof. (tel. 052 + 781-5111, ext. 3350). *Metal physics, electron and X-ray diffraction, electron microscopy, crystal growth, crystal characterization, deformation and strength, amorphous metal structure and stability.*

Ino, Dr Shozo. Res. Inst. for Iron Steel and Other Metals, Tohoku U., 2-1-1 Katahira-cho, Sendai 980, Japan. (1936) DSc, physics (Tohoku U., 1970). Asst. prof. (tel. 0222 + 27-6200, ext. 2917). *Electron diffraction, surface physics, thin film physics.*

Ino, DSc Tadashi. Dept. of Physics, Faculty of Sci., Osaka City U., Sugimoto-cho Sumiyoshi-ku, Osaka 558, Japan. (1923) DSc, physics (Nagoya U., 1961). Prof. (tel. 06 + 692-1231, ext. 3232). *Gas molecular structures, amorphous substances, microcrystals.*

Inoue, Dr Morio. Res. Lab., Matsushita Electronics Corporation, 1-1 Saiwai-cho Takatsuki, Osaka 569, Japan. (1937) DSc, chemistry (Kyoto U., 1972). Chief engineer. (tel. 0726 + 82-5521, ext. 276). *Crystal growth, semiconductor materials, solid state chemistry.*

Inoue, Dr Zenzaburo. Third Res. Group, Nat. Inst. for Res. in Inorganic Materials, Namiki 1-1, Sakura-mura Niihari-gun, Ibaraki 305, Japan. (1940) DSc, mineralogy (Tokyo U., 1974). Sr. res. officer. (tel. 0298 + 51-3351, ext. 373). *Polytypism, phase transition in crystals, high temperature X-ray crystallography, ceramic crystal structures, crystal growth.*

Inui, Prof. Teturo. 1-26 Wakabacho 1-chome Chofu, Tokyo 182, Japan. (1905) DSc, physics (Tokyo U., 1941). Em. prof. (tel. 03 + 300-4596). *Solid state physics, group-theoretic research, crystallography, phase transition, mathematical physics.*

Ishibashi, Dr Yoshihiro. Synthetic Crystal Res. Lab., Faculty of Eng., Nagoya U., Furo-cho Chikusa-ku, Nagoya 464, Japan. (1935) DSc, physics (Tokyo U., 1963). Asst. prof. (tel. 052 + 781-5111, ext. 3597). *Ferroelectricity, phase transition, crystal growth.*

Ishida, Prof. Kohtaro. Dept. of Physics, Faculty of Sci. and Techn., Science U. of Tokyo, Noda-shi, Chiba 278, Japan. (1940) DSc, physics (Kyoto U., 1970). Assoc. prof. (tel. 0471 + 24-1501, ext. 287). *X-ray and electron diffraction, crystallography.*

Ishida, Dr Toshimasa. Physical Chemistry, Osaka C. of Pharmacy, 2-10-65 Kawai Matsubara, Osaka 580, Japan. (1946) Dr, pharmacy (Osaka U., 1977). Assoc prof. (tel. 0723 + 32-1015, ext. 298). *Structure - function relationships, bioactive substances, drugs.*

Ishiguro, Mr Takashi. Dept. of Inorganic Materials, Faculty of Eng., Tokyo Inst. of Techn., O-okayama Meguro-ku, Tokyo 152, Japan. (1954) MEng, inorganic materials (Tokyo Inst. of Techn., 1979). Student of doctor course. (tel. 03 + 726-1111, ext. 2518). *Crystal growth, semiconductor physics, ceramics.*

Ishihara, Mr Nobukazu. Dept. of Physics, C. of Humanities and Sci. Nihon U., 25-40 Sakurajosui 3-chome Setagaya-ku, Tokyo 156, Japan. (1933) BSc, chemistry (1956). Asst. prof. (tel. 03 + 302-8131, ext. 283). *Crystal growth, polymer crystal, electron microscopy.*

Ishikawa, Prof. Yoshikazu. Physics Dept., Tohoku U., Aramaki Aza-Aoba, Sendai 980, Japan. (1930) DSc, physics (Tokyo U., 1958). Prof. (tel. 0222 + 29-1800, ext. 5353). *Neutron scattering, magnetism.*

Ishizuka, Mr Kazuo. Lab. of Crystal and Powder Chemistry, Inst. for Chemical Res., Kyoto U., Gokasho Uji 611, Japan. (1947) MPharm, chemistry (Kyoto U., 1972). Postdoctoral fellow. (tel. 0774 + 32-3111, ext. 351). *X-ray and electron diffraction analysis, high-resolution electron microscopy.*

Isobe, Dr Mitsumasa. Nat. Inst. for Res. in Inorganic Materials, 1-1 Namiki Sakura-mura Niihari-gun, Ibaraki 305, Japan. (1944) DEng, chemistry (Tokyo Inst. of Techn., 1973). Res. (tel. 0298 + 51-3351, ext. 254). *Crystal chemistry.*

Itai, Dr Akiko. Faculty of Pharmaceutical Sci. U. of Tokyo, 7-3-1 Hongo Bunkyo-ku, Tokyo 113, Japan. (1941) Dr, chemistry (Tokyo U., 1969). Res. assoc. (tel. 03 + 812-2111, ext. 4842). *X-ray crystallography, energy analysis.*

Ito, Mr Kazuomi. Ashikaga Inst. of Techn., 1-268 Oomae-cho Ashikaga, Tochigi 326, Japan. (1943) MSc, electronics (Tokyo Electric Eng. C., 1974). Res. assoc. (tel. 0284 + 62-0605). *Crystal growth, semiconductors, plasmas.*

Ito, Mr Kohei. Magnetic & Electronic Materials Res. Lab., Hitachi Metals Ltd., 5200 Mikajiri Kumagaya, Saitama 360, Japan. (1945) BSc, physics (Tokyo U., 1969). Res. (tel. 0485 + 32-2211). *Crystal growth.*

Ito, Prof. Masatoki. Chemistry Dept. Faculty of Sci. and Eng., Keio U. 3-14-1 Hiyoshi, Kohoku-ku, Yokohama 223, Japan. (1942) DSc, chemistry (Tokyo U., 1970) Assoc. prof. (tel. 044 + 63-1141, ext. 3911). *Surface physics, catalysis.*

Ito, Dr Tetsuzo. Crystal Physics Lab., Inst. of Physical and Chemical Res., 1 Hirosawa 2-chome Wako, Saitama 351, Japan. (1936) DSc, chemistry (Tokyo U., 1965). Sr. scient. (tel. 0484 + 62-1111). *Molecular crystals, electron density.*

Ito, Dr Yuji. Inst. for Solid State Physics, U. of Tokyo, Roppongi Minato-ku, Tokyo 106, Japan. (1936) PhD, physics (MIT, USA, 1967). Assoc. prof. (tel. 03 + 402-6231, ext. 688). *Neutron scattering, solid state physics, biophysics.*

Iwanaga, Mr Hiroshi. Faculty of Liberal Arts, Nagasaki U., 1-14 Bunkyo-machi, Nagasaki 852, Japan. (1938) BEdu, education (Nagasaki U., 1961). Asst. prof. (tel. 0958 + 47-1111). *Growth mechanism, imperfections, ZnO crystals grown by chemical reaction.*

Iwasaki, Dr Fujiko. Dept. of Materials Sci., U. of Electro-Communications, 5-1 Chofugaoka 1-chome Chofu, Tokyo 182, Japan. (1937) DSc, chemistry (Tokyo U., 1966). Asst. prof. (tel. 0424 + 83-2161, ext. 443). *Organic crystal structures, crystal chemistry.*

Iwasaki, Dr Hiroshi. Res. Div., Nippon Telegraph and Telephone Corp., 9-11 Midori-cho 3-chome Musashino, Tokyo 180, Japan. (1933) DSc, chemistry (Osaka U., 1963). (tel. 0422 + 59-2616). *Ferroelectric crystals, crystal growth.*

Iwasaki, Prof. Hiroshi. Res. Inst. for Iron, Steel and Other Metals, Tohoku U., 1-1 Katahira 2-chome, Sendai 980, Japan. (1933) DSc, physics (Tohoku U., 1966). Prof. (tel. 0222 + 27-6200, ext. 2916). *Structure, metals and alloys, phase transformation, high pressure X-ray diffraction.*

Iwasaki, Dr Hitoshi. Crystal Physics Lab., Inst. of Physical and Chemical Res., 1 Hirosawa 2-chome Wako, Saitama 351, Japan. (1935) DSc, chemistry (Tokyo U., 1965). Chief scient. (tel. 0484 + 62-1111, ext. 3341). *Inorganic crystal structures, liquids and solutions, anomalous scattering.*

Iwata, Mrs Miyuki. Inst. for Solid State Physics, U. of Tokyo, 22-1 Roppongi 7-chome Minato-ku, Tokyo 106, Japan. (1940) MSc, chemistry (Tokyo U., 1964). (tel. 03 + 402-6231, ext. 663). *Accurate electron density determination, diffraction in general, molecular science.*

Iwata, Dr Yutaka. Res. Reactor Inst., Kyoto U., Kumatori-cho Sennan-gun, Osaka 590-04, Japan. (1937) DSc, physics (Kyushu U., 1977). Res. assoc. (tel. 07245 + 2-0901, ext. 2348). *Ferroelectric structures.*

Iwayanagi, Prof. Shigeo. Faculty of Techn., Gunma U., 5-1 Tenjin-cho 1-chome Kiryu, Gunma 376, Japan. (1916) DSc, physics (Tokyo U. of Education, 1955). Prof. (tel. 0277 + 22-3181, ext. 333). *Polymer physics, biophysics.*

Izui, Dr Kazuhiko. Chemistry Div., Japan Atomic Energy Res. Inst., 2-4 Shirakata-Shirane Tokaimura Nakagun, Ibaraki 319-11, Japan. (1929) DSc, physics (Hiroshima U., 1965). Principal scient. (tel. 02928 + 2-5523). *Crystal structure, lattice defects, radiation damage, electron microscopy.*

Izumi, Prof. Takatoshi. Dept. of Eng. Physics, Faculty of Eng., U. of Chubu, 1200 Matsumoto-cho, Kasugai Aichi 487, Japan. (1938) DEng, BSc, physics (Osaka Gakugei U., 1966). Prof. (tel. 0568 + 51-1111, ext. 446). *X-ray crystallography, ferroelectrics.*

Izumi, Dr Yoshinobu. Dept. of Polymer Sci., Faculty of Sci., Hokkaido U., Kita 10-jo Nishi 8-chome Kita-ku, Sapporo 060, Japan. (1946) DSc, physics (Hokkaido U., 1972). Res. assoc. (tel. 011 + 711-2111). *Polymers, biophysics.*

Kai, Dr Yasushi. Dept. of Applied Chemistry, Osaka U., Yamadakami Suita, Osaka 565, Japan. (1943) DEng, chemistry (Osaka U., 1973). Res. assoc. (tel. 06 + 877-5111, ext. 4323). *Organic crystal structures.*

Kainuma, Prof. Yoshiro. Physics Lab., Dept. of General Education, Nagoya U., Furo-cho Chikusa-ku, Nagoya 464, Japan. (1922) DSc, physics (Nagoya U., 1955). Prof. (tel. 052 + 781-5111, ext. 3517). *Electron and X-ray diffraction, crystal physics.*

Kaito, Dr Chihiro. Dept. of Physics, Kyoto Inst. of Techn. Matsugasaki, Sakyo-ku, Kyoto 606, Japan. (1943) BSc, Physics (Ritsumeikan U., 1965). Asst. prof. (tel. 075 + 791-3211, ext. 445). *Crystal growth, crystal structure, smoke particle coalescence, electron microscopy.*

Kaji, Dr Keisuke. Dept. of Polymer Sci., Faculty of Eng., Kyoto U., Yosida Honmachi Sakyo-ku, Kyoto 606, Japan. (1939) DEng, polymer chemistry (Kyoto U., 1970). Res. assoc. (tel. 075 + 751-2111, ext. 5628).

Kakehi, Dr Masahiro. Dept. of Electrical Eng., Mie U., Kamihama-cho Tsu Mie 514, Japan. (1942) DEng, physics (Nagoya U., 1971). Asst. prof. (tel. 0592 + 32-1211, ext. 3816). *Crystal growth, semiconductors physics.*

Kakinoki, Prof. Jiro. Dept. of Physics, Faculty of Sci., Osaka City U., 459 Sugimoto-cho Sumiyoshi-ku, Osaka 558, Japan. (1912) DSc, physics (Osaka U., 1944). Prof. (tel. 06 + 692-1231, ext. 3132). *Disordered structure, polytypes, high polymers, amorphous structure, gases.*

Kakitani, Prof. Satoru. Inst. of Geology and Mineralogy, Faculty of Sci., Hiroshima U., 1-89 Higashisendamachi 1-chome, Naka-ku, Hiroshima 730, Japan. (1924) DSc, mineralogy (Hiroshima U., 1961). Prof. (tel. 0822 + 41-1221, ext. 339). *Inorganic crystal structures, crystal growth, clay minerals, optical absorption spectra.*

Kakudo, Prof. Masao. Inst. for Protein Res., Osaka U., Yamadakami Suita, Osaka 565, Japan. (1918) DSc, chemistry (Osaka U., 1953). Prof. (tel. 06 + 877-5111, ext. 3836). *Proteins, biological substances, structure.*

Kakumoto, Dr Kenichi. Koyo Res. Center, Koyo Seiko Co. Ltd., 1-24 Kokubu Higanjo-cho Kashiwara, Osaka 582, Japan. (1946) DSc, physics Res. worker. (tel. 0729 + 77-1111, ext. 316). *Surface physics, lubrication in vacuum, ion plating.*

Kamijo, Dr Nagao. Government Industrial Res. Inst. Osaka, 1-8-31 Midorigaoka Ikeda, Osaka 563, Japan. (1936) DSc, chemistry, solid state physics (Kwansei Gakuin U., 1978). Sr. res. (tel. 0727 + 51-8351). *X-ray crystallography, solid state physics, phase transition.*

Kamiya, Dr Kazuhide. Chemical Res. Labs., Central Res. Div., Takeda Chemical Industries Ltd., Jusohonmachi Yodogawa-ku, Osaka 532, Japan. (1939) DSc, crystallography (Tokyo U., 1980). Res. (tel. 06 + 301-1231, ext. 2401). *X-ray crystallography.*

Kamiya, Dr Nobuo. Photon Factory, Nat. Lab. for High Energy Physics. Ohomachi, Tsukuba, Ibaraki 305, Japan. (1953) DSc, chemistry (Nagoya U., 1984) (tel. 0298 + 64-1171, ext. 5027). *Protein X-ray crystallography, synchrotron radiation, instrumentation.*

Kamiya, Dr Yoshihiro. Dept. of Physics, Nagoya U., Furo-cho Chikusa-ku, Nagoya 464, Japan. (1932) DSc, physics (Nagoya U., 1960). Asst. prof. (tel. 052 + 781-5111, ext. 2448). *Electron diffraction, electron microscopy.*

Kasai, Prof. Nobutami. Dept. of Applied Chemistry, Osaka U., Yamadakami Suita, Osaka 565, Japan. (1929)DEng, chemistry (Osaka U., 1962). Prof. (tel. 06 + 877-5111, ext. 4321). *Organic crystal structures, organometallic compounds, large molecules, instrumentation.*

Kashino, Dr Setsuo. Dept. of Chemistry, Faculty of Sci., Okayama U., Tsushima Okayama 700, Japan. (1937) DSc, chemistry (Osaka U., 1973). Asst. prof. (tel. 0862 + 52-1111, ext. 391). *Organic crystal structures, solid phase organic reaction.*

Kashiwase, Dr Yasuji. Physics Lab., Dept. of General Education, Nagoya U., Furo-cho Chikusa-ku, Nagoya 464, Japan. (1932) DSc, physics (Nagoya U., 1965). Assoc. prof. (tel. 052 + 781-5111, ext. 4845). *Lattice vibration, X-ray crystallography, solid state physics.*

Kasuga, Prof. Dr Masanobu. Dept. of Electronics, Faculty of Eng., Yamanashi U., 3-11 Takeda 4-chome, Kofu 400, Japan. (1941) DEng, electronics (Nagoya U., 1971). Prof. (tel. 0552 + 52-1111, ext. 5245). *Crystal growth, CVD, semiconductors (II-VI).*

Katada, Prof. Kinya. Dept. of Physics, U. of Osaka Prefecture, Mozu Umemachi 4-chome Sakai, Osaka 591, Japan. (1925) DSc, chemistry (Osaka U., 1961). Prof. (tel. 0722 + 52-1161). *Structure, phase stability, metals and alloys.*

Katagawa, Dr Takeshi. Dept. of electronics, Faculty of Eng., Nagoya U., Furo-cho Chikusa-ku, Nagoya 464, Japan. (1941) DEng, applied physics (Nagoya U., 1977). Res. asst. (tel. 052 + 781-5111, ext. 5889). *X-ray crystallography.*

Katayama, Mr Chuji. Dept. of Chemistry, Faculty of Sci., Nagoya U., Furo-cho Chikusa-ku, Nagoya 464, Japan. (1944) MSc, chemistry (Kwansei Gakuin U., 1968). Res. assoc. (tel. 052 + 781-5111, ext. 3550). *Structure analysis, molecular crystals.*

Katayama, Prof. Mikio. Dept. of Pure and Applied Sci. C. of General Education, U. of Tokyo, 8-1 Komaba 3-chome Meguro-ku, Tokyo 153, Japan. (1926) DSc, chemistry (Tokyo U., 1957). Prof. (tel. 03 + 467-1171, ext. 323). *Laser spectroscopy.*

Kato, Dr Akira. Dept. of Geology, National Science Museum, 23-1 Hyakunin-cho 3-chome Shinjuku, Tokyo 160, Japan. (1931) DSc, geology (Tokyo U., 1959). Res. officer. (tel. 03 + 364-2311, ext. 331). *Descriptive mineralogy, mineral classification.*

Kato, Mr Ichiro. Consumer Goods Testing Lab., Tokyo Shibaura Electric Co., 14-8 Omori-nishi 1-chome Ota-ku, Tokyo 143, Japan. (1931) MSc, mineralogy (Tokyo U. 1955). Manager. (tel. 03 + 762-6844). *Applied physics, crystallography, perfect crystal.*

Kato, Dr Katsuo. Nat. Inst. for Res. in Inorganic Materials, 1-1 Namiki Sakura-mura Niihari-gun, Ibaraki 305, Japan. (1938) Dr.rer.nat. habil., mineralogy and crystallography (U. Hamburg, BRD, 1972). Group leader. (tel. 0298 + 51-3351, ext. 375). *Inorganic crystal structures.*

Kato, Prof. Masanori. Dept. of Inorganic Materials, Faculty of Eng., Tokyo Inst. of Techn., 12-1 Oh-okayama 2-chome Meguro-ku, Tokyo 152, Japan. (1928) DEng, chemistry (Tokyo Inst. of Techn., 1966). Prof. (tel. 03 + 726-1111, ext. 2518). *Inorganic materials.*

Kato, Prof. Norio. Dept. of Applied Physics, Faculty of Eng., Nagoya U., Furo-cho Chikusa-ku, Nagoya 464, Japan. (1923) DSc, physics (Nagoya U., 1954). Prof. (tel. 052 + 781-5111). *Diffraction theory, crystal perfection, crystal growth.*

Kato, Prof. Toshio. Inst. of Earth Sci., Yamaguchi U., 1677-1 Yoshida, Yamaguchi 753, Japan. (1931) DSc, mineralogy (Tokyo U., 1958). Prof. (tel. 0839 + 22-6111, ext. 518). *Crystal structures, layer silicates.*

Kato, Dr Yoshihiro. Dept. of Physics, Osaka Kyoiku U., 43 Minami-Kawahori-cho Tennoji-ku, Osaka 543, Japan. (1934) BEd, physics (Osaka Kyoiku U., 1958). Asst. prof. (tel. 06 + 771-8131, ext. 295). *Structural disorder, phase transition, organic crystals.*

Katsube, Prof. Yukiteru. Inst. for Protein Res., Osaka U. Yamada-oka 3-2 Suita, Osaka 565, Japan. (1930) DSc, chemistry (Osaka U., 1963). Prof. (tel. 06 + 877-5111, ext. 3836). *Organic and protein crystal structure.*

Kawada, Dr Isao. Group 2, Nat. Inst. for Res. in Inorganic Materials, 1-1 Namiki Sakura-mura Niihari-gun, Ibaraki 305, Japan. (1935) DSc, mineralogy (Tokyo U., 1964). Group leader. (tel. 0298 + 51-3351, ext. 319). *Inorganic crystal structures.*

Kawado, Dr Seiji. Res. Ctr., Sony Corp., 174 Fujitsuka-cho, Hodogaya-ku, Yokohama 240, Japan. (1940) DEng, applied physics (Tokyo U., 1982). Sr. res. scient. (tel. 045 + 351-1271, ext. 461). *X-ray topography, crystal imperfections, semiconductors.*

Kawahara, Dr Akira. Dept. of Earth Sci. Faculty of Sci., Okayama U., Tsushima Okayama 700, Japan. (1932) DSc, mineralogy (Tokyo U., 1962). Asst. prof. (tel. 0862 + 52-1111, ext. 439). *Crystal structure, minerals.*

Kawai, Mr Toshiaki. R & D Group, Eng. Div., Hamamatsu TV Co. Ltd., 1126-1 Ichino-cho, Hamamatsu 435, Japan. (1942) MEng, crystal growth (Shizuoka U., 1974). (tel. 0534 + 34-3311). *Crystal growth, semiconductor physics, infrared physics, X-ray diffraction.*

Kawai, Dr Yoriyoshi. Dept. of Physics, Faculty of Sci., Gakushuin U. 1-5-1 Mejiro Toshima-ku, Tokyo 171, Japan. (1937) DSc, Physics (Gakushuin U., 1981). Res. (tel. 03 + 986-0221, ext. 454). *Physical acoustics in solids, oxide magnetism.*

Kawaminami, Dr Masaru. Dept. of Physics, C. of Liberal Arts, Kagoshima U., 21-30 Korimoto 1-chome, Kagoshima 890, Japan. (1941) DSc, physics (Kyushu U., 1970). Asst. prof. (tel. 0992 + 54-7141, ext. 5793). *Phase transition, precision lattice constants.*

Kawamori, Prof. Asako. School of Sci., Kwansei Gakuin U., Uegahara Nishinomiya, Hyogo 662, Japan. (1935) DSc (Osaka U.). Prof. (tel. 0798 + 51-2407). *Magnetic resonance, solid state physics, biophysics.*

Kawamura, Dr Tsutomu. Central Res. Labs., Nippon Electric Co. Ltd., 1-1 Miyazaki 4-chome Takatsu-ku, Kawasaki Kanagawa 213, Japan. (1931) DSc, mineralogy (Tokyo U., 1966). Res. manager. (tel. 044 + 855-2111). *Crystal growth, crystal defects, inorganic crystal structures.*

Kawano, Assoc. Prof. Shigeaki. C. of General Education, Kyushu U., 4-2-1 Ropponmatsu Chuo-ku, Fukuoka 810, Japan. (1931) BSc, physics (Kyushu U., 1954). Asst. prof. (tel. 092 + 771-4161, ext. 359). *X-ray crystallography, crystallographic database system, universal program system, solid state physics.*

Kawata, Dr Hiroshi. Photon Factory, Nat. Lab. For High Energy Physics, Oho-machi, Tsukuba, Ibaraki 305, Japan. (1955) DSc, physics (Tokyo Inst. of Techn., 1982). Asst. prof. (tel. 0298 + 64-1171, ext. 5030). *X-ray crystallography, ferroelectrics.*

Kaya, Prof. Seiji. 34-16 Jingumae 5 Shibuya-ku, Tokyo 150, Japan. (1898) DSc, physics (Tohoku U., 1929). Em. prof. (U. of Tokyo); Member, Japan Academy. (tel. 03 + 407-1434).

Kihara, Dr Kuniaki. Dept. of Earth Sci., Kanazawa U., 1-1 Marunouchi, Kanazawa 920, Japan. (1943) DSc, mineralogy (Tokyo U. of Education, 1972). Res. assoc. (tel. 0762 + 62-4281, ext. 576). *X-ray crystallography.*

Kikuchi, Dr Seishi. Applied Metallurgical Eng., Tokyo Inst. of Techn., 12-1 Oh-okayama 2-chome Meguro-ku, Tokyo 152, Japan. (1935) DEng, metallurgy (Tokyo Inst. of Techn., 1966). Asst. prof. (tel. 03 + 726-1111, ext. 3138). *Structures, metals and alloys.*

Kikuta, Prof. Seishi. Applied Physics Dept., Faculty of Eng., U. of Tokyo, 3-1 Hongo 7-chome Bunkyo-ku, Tokyo 113, Japan. (1938) DSc, physics (Tokyo U., 1970). Asst. prof. (tel. 03 + 812-2111, ext. 6826). *Dynamical diffraction phenomena of X-rays and neutrons, X-ray optics, X-ray holography, surface physics, low energy electron diffraction, ion scattering, photo-electron emission.*

Kimura, Prof. Masao. Dept. of Chemistry, Faculty of Sci., Hokkaido U., Kita 10-jo Nishi 8-chome Kita-ku, Sapporo 060, Japan. (1921) DSc, chemistry (Nagoya U., 1949). Prof. (tel. 011 + 711-2111, ext. 3501). *Gas electron diffraction.*

Kiriyama, Prof. Hideko. Dept. Chemistry, Faculty of Sci., Kobe U. Rokkodai-cho Naka-ku, Kobe 657, Japan. (1923) DSc, chemistry (Osaka U., 1960). Prof. (tel. 078 + 881-1212, ext. 4402). *Crystal chemistry, inorganic compounds.*

Kiriyama, Prof. Ryoiti. Inst. of Scient. and Industrial Res., Osaka U., Yamadakami Suita, Osaka 565, Japan. (1913) DSc, chemistry (Osaka U., 1949). Prof. (tel. 06 + 877-5111, ext. 3550). *Inorganic crystal chemistry, solid state reaction.*

Kiryu, Prof. Setsuo. Faculty Pharm. & Pharmaceutical Sci., Fukuyama U. 985 Higashimura-cho Fukuyama, Hiroshima 729-02, Japan. (1933) DPharm, chemistry (Kyushu U., 1971). Prof. (tel. 08485 + 8-2111). *Physical pharmacy.*

Kishi, Dr Kiyoshi. Dept. of Applied Physics, Science U. of Tokyo, 1-3 Kagurazaka, Tokyo 162, Japan. (1933) DSc, physics (Science U. of Tokyo, 1974). Asst. prof. (tel. 03 + 260-4271). *Crystal growth, surface physics, electronic instrumentation.*

Kishino, Dr Seigo. Central Res. Lab., Hitachi Ltd., 280 Higashi-Koigakubo 1-chome, Kokubunji Tokyo 185, Japan. (1938) DSc, applied physics (Tokyo U., 1972). Res. (tel. 0423 + 23-1111, ext. 576). *X-ray diffraction, lattice defects, crystal growth.*

Kitahama, Dr Katsuki. Inst. of Scient. and Industrial Res., Osaka U., Yamadakami Suita, Osaka 565, Japan. (1941) DSc, chemistry (Osaka U., 1972). Res. assoc. (tel. 06 + 877-5111, ext. 3551). *X-ray analysis.*

Kitano, Mr Yukishige. Pioneering Res. and Dev. Labs., Toray Industries Inc., 2-1 Sonoyama 3-chome Otsu, Shiga 510, Japan. (1941) BSc, chemistry (Osaka U., 1965). (tel. 0775 + 37-0600). *Organic crystal structures, large molecules.*

Kobayashi, Dr Akiko. Dept. of Chemistry, Faculty of Sci., U. of Tokyo, 3-1 Hongo 7-chome Bunkyo-ku, Tokyo 113, Japan. (1943) DSc, chemistry (Tokyo U., 1972). Res. assoc. (tel. 03 + 812-2111). *Inorganic chemistry, solid state chemistry.*

Kobayashi, Dr Hayao. Dept. of Chemistry, Toho U., 542 Miyama-cho Funabashi, Chiba 274, Japan. (1942) DSc, chemistry (Tokyo U., 1970). Asst. prof. (tel. 0474 + 72-1141). *Solid state chemistry.*

Kobayashi, Prof. Jinzo. Dept. of Applied Physics, Waseda U., 3-4-1 Okubo, Shinjuku-ku, Tokyo 160, Japan. (1925) DEng, applied physics (Waseda U., 1960). Prof. (tel. 03 + 209-3211, ext. 3564). *Ferroelectrics, crystal optics.*

Kobayashi, Prof. Nobuyuki. Applied Physics, Toyama U., Gofuku 3190, Toyama 930, Japan. (1942) DEng, crystal growth (Nagoya U., 1972). Prof. (tel. 0764 + 41-1271, ext. 837). *Crystal growth, fluid mechanics, heat transfer, mass transfer, numerical analysis, computer simulation, material science, solid state physics, thermal stress.*

Kobayashi, Dr Tadashi. Dept. of Pharmacy, Hokuriku U., 3 Ho Kanagawa-machi, Kanazawa 920-11, Japan. (1944) DSc, physics (Kyushu U., 1972). Asst. prof. (tel. 0762 + 29-1161, ext. 229). *Magnetic resonance, X-ray diffraction, ferroelectricity, ferroelasticity, quantum chemistry.*

Kobayashi, Dr Takaaki. Res. Inst. for Iron, Steel and Other Metals, Tohoku U., 1-1 Katahira 2-chome, Sendai 980, Japan. (1944) DSc, chemistry (Tohoku U., 1973). Res. assoc. (tel. 0222 + 27-6200, ext. 2335). *Crystal growth, metals and compounds, inorganic chemistry, rare earth elements.*

Kobayashi, Dr Takashi. Inst. for Chemical Res., Kyoto U., Gokasho Uji 611, Japan. (1938) DSc, chemistry (Kyoto U., 1970). Res. assoc. (tel. 0774 + 32-3111, ext. 414). *Electron optics, crystal growth, semiconductors, structure analysis, electron microscopy, electron microscopy, surface physics.*

Koda, Mr Shigetaka. Res. Labs., Fujisawa Pharmaceutical Co. Ltd., 1-6 Kashima 2-chome Yodogawa-ku, Osaka 532, Japan. (1946) MSc, chemistry (Osaka City U., 1971). Res. staff. (tel. 06 + 301-1271, ext. 532).

Kohra, Prof. Kazutake. Photon Factory, National Lab. for High Energy Physics, Oho-cho Tsukuba-gun, Ibaraki-ken 305, Japan. (1921) DSc, physics (Kyushu U., 1954). Director. (tel. 0298 + 64-1171). *Dynamical diffraction, X-ray optics.*

Kohyama, Mr Masaki. Analytical Lab., Res. Center, Mitsui Petrochemical Industries Ltd., Waki-cho Kuga-gun, Yamaguchi 740, Japan. (1950) MSc, polymer science (Osaka U., 1975). Res. (tel. 0827 + 22-4111, ext. 485). *Polymer crystallography, polymer physics.*

Koide, Dr Tsutomu. Dept. of Chemistry, Osaka Kyoiku U., 43 Minami-Kawahori-cho Tennoji-ku, Osaka 543, Japan. (1930) DSc, chemistry (Osaka U., 1955).

Asst. prof. (tel. 06 + 771-8131). *Phase transition, organic compounds, phase transition effects of crystal particle size.*

Koizumi, Dr Hideo. Res. Div., Musashino Electrical Communication Lab., Nippon Telegraph and Telephone Corp., 9-11 Midori-cho 3-chome Musashino, Tokyo 180, Japan. (1928) DSc, physics (Osaka U., 1962). Sr. scient. (tel. 0422 + 59-3120). *Crystal structure analysis, crystal physics, structure, laser materials, ferroelectric materials.*

Koizumi, Prof. Mitsue. Inst. Scient. Industrial Res., Osaka U. Mihogaoka 8-1 Ibaraki, Osaka 567, Japan. (1923) DSc, High pressure synthesis (Tokyo U., 1958). Prof. of mineral science (tel. 06 + 877-5111, ext. 3535, Telex 5286213ISIROU-J). *High pressure, ceramics, zeolites.*

Komatsu, Prof. Hiroshi. Res. Inst. for Iron, Steel and Other Metals, Tohoku U., 1-1 Katahira 2-chome, Sendai 980, Japan. (1935) DSc, mineralogy (Tokyo U. of Education, 1964). Prof. (tel. 0222 + 27-6200, ext. 2908). *Crystal growth, surface microtopography, interferometry, optical microscopy.*

Komori, Dr Tetsuya. Dept. of Pharmaceutical Techn., Kyushu U., 1-1 Maedashi 3-chome Higashi-ku, Fukuoka 812, Japan. (1929) DPharm, chemistry (Tokyo U., 1960). Asst. prof. (tel. 092 + 641-1151, ext. 4152). *Natural product chemistry, biochemistry, crystallography.*

Komoto, Dr Tadashi. Dept. of Polymer Techn., Tokyo Inst. of Techn., 12-1 Ohokayama 2-chome Meguro-ku, Tokyo 152, Japan. (1944) DEng, polymer science (Tokyo Inst. of Techn., 1972). Res. assoc. (tel. 03 + 726-1111, ext. 2133).

Komura, Prof. Shigehiro. Faculty of Integrated Arts & Sci. Hiroshima U., 1-89 Higashisendamachi 1-chome, Hiroshima 730, Japan. (1933) DSc, physics (Tokyo U., 1967). Prof. (tel. 0822 + 41-1221, ext. 565, 317). *Neutron diffraction, solid state physics, magnetism, metals and alloys, biophysics.*

Komura, Prof. Yukitomo. Dept. of Materials Sci., Faculty of Sci., Hiroshima U., 1-89 Higashisendamachi 1-chome, Hiroshima 730, Japan. (1924) DSc, physics (Osaka U., 1961). Prof. (tel. 0822 + 41-1221, ext. 568). *Structure, metals and alloys, intermetallic compounds, stacking faults, short-range ordering.*

Konaka, Dr Shigehiro. Dept. of Chemistry, Faculty of Sci., Hokkaido U., Kita 10-jo Nishi 8-chome, Kita-ku, Sapporo 060, Japan. (1939) DSc, chemistry (Hokkaido U., 1969). Asst. prof. (tel. 011 + 711-2111, ext. 3533). *Electron diffraction, molecular structure, charge density.*

Konno, Dr Michiko. Inst. for Solid State Physics, U. of Tokyo, 22-1 Roppongi 7-chome Minato-ku, Tokyo 106, Japan. (1946) DSc, chemistry (Tokyo U., 1974). Techn. assoc. (tel. 03 + 402-6231, ext. 662). *X-ray structure analysis.*

Kotani, Prof. Masao. 2nd Div. Japan Academy, Ueno Park, Daito-ku, Tokyo 110, Japan. (1906) DSc, (U. Tokyo, 1942). Member, Japan Acad. (tel. 03 + 822-2101). *Protein structure.*

Koto, Dr Kichiro. Inst. of Scient. and Industrial Res., Osaka U., Mihoga-oka, Ibaraki, Osaka 567, Japan. (1936) DSc, mineralogy (Tokyo U., 1969). Asst. prof. (tel. 06 + 877-5111, ext. 3547, Telex 5286213ISIROUJ). *Crystal structure, superstructure, EXAFS, inorganic compounds, minerals, superionic conductors.*

Koyama, Dr Hirozo. Dept. of Physical Chemistry, Shionogi Res. Lab., 12-4 Sagisu 5-chome Fukushima-ku, Osaka 553, Japan. (1924) DSc, chemistry (Osaka U., 1962). Res. officer. (tel. 06 + 458-5861, ext. 311). *Structure analysis methods, organic structures.*

Koyama, Dr Yasumasa. Dept. of Metallurgy, Faculty of Eng., Tokyo Inst. of Techn., 2-12-1 Ookayama, Meguro-ku, Tokyo 152, Japan. (1952) DEng, metallurgy (Tokyo Inst. of Techn., 1981). Res. assoc. (tel. 03 + 726-1111, ext. 3145). *Phase transformation, electron microscopy, X-ray diffraction.*

Koyano, Mr Kazuo. Central Res. Inst., Teijin Limited, 3-2 Asahigaoka 4-chome Hino, Tokyo 191, Japan. (1932) MSc, chemistry. Res. assoc. (tel. 0425 + 81-4321). *Structure, biopolymers.*

Kozaki, Prof. Shigeru. The Inst. of Vocational Training, 1960 Aihara Sagamihara, Kanagawa 229, Japan. (1934) DEng, applied physics (Tokyo U., 1974). Prof. (tel. 0427 + 61-2111).

Kubo, Prof. Ikumaro. Faculty of Education, Nagasaki U., 1-14 Bunkyo-machi, Nagasaki 852, Japan. (1912) DSc, physics (Hiroshima U., 1963). Prof. (tel. 0958 + 47-1111). *Crystal growth, educational technology.*

Kubo, Prof. Teruichiro. Dept. of Chemistry, Musashi Inst. of Techn., 28-1 Tamazutsumi 1-chome Setagaya-ku, Tokyo 158, Japan. (1907) DSc, chemistry (Tokyo Inst. of Techn., 1940). Prof.; Em. prof. (Tokyo Inst. of Techn.). (tel. 03 + 703-3111, ext. 490). *Mechanochemistry, powder technology, solid state chemistry.*

Kuchitsu, Prof. Kozo. Chemistry Dept., Faculty of Sci., U. of Tokyo, 7-3-1 Hongo Bunkyo-ku, Tokyo 113, Japan. (1927) DSc, chemistry (Tokyo U., 1958). Prof. (tel. 03 + 812-2111, ext. 4334, Telex UTYOSCIJ33659). *Gas electron diffraction, vibration - rotation spectroscopy, chemical processes by electronic and atomic impact.*

Kudoh, Dr Yasuhiro. Mineralogical Inst., Faculty of Sci., U. of Tokyo, 3-1 Hongo 7-chome Bunkyo-ku, Tokyo 113, Japan. (1947) DSc, mineralogy (Tokyo U., 1975). Res. assoc. (tel. 03 + 812-2111, ext. 4548). *Inorganic crystal structures, high pressure X-ray diffraction.*

Kumagawa, Dr Masashi. Res. Inst. of Electronics, Shizuoka U., 5-1 Johoku 3-chome, Hamamatsu 432, Japan. (1938) DEng, material science (Tohoku U., 1967). Asst. prof. (tel. 0534 + 71-1171, ext. 431). *Crystal growth, semiconductors, mechanisms of crystal growth, device fabrication.*

Kumao, Mr Akihiro. Physics Lab., Kyoto Inst. of Techn., Matsugasaki, Sakyo-ku, Kyoto 606, Japan. (1941) DSc, physics (Hiroshima U., 1983). Assoc. prof. (tel. 075 + 791-3211, ext. 446). *Electron microscopy, crystal growth.*

Kurahashi, Dr Masayasu. Corrosion Div., National Res. Inst. for Metals, 3-12 Nakameguro 2-chome Meguro-ku, Tokyo 153, Japan. (1943) DSc, chemistry (Osaka City U., 1975). Sr. res. officer. (tel. 03 + 719-2271, ext. 262). *Inorganic and organic crystal structures, LEED,*

Kuribayashi, Mr Shunsuke. Government Industrial Res. Inst., 1-8-31 Midorigaoka Ikeda, Osaka 563, Japan. (1923) BSc, chemistry (Osaka U., 1949). Res. (tel. 0727 + 51-8351, ext. 279). *Structure, polymers and related compounds.*

Kuroda, Prof. Haruo. Dept. of Chemistry, Faculty of Sci., U. of Tokyo, 3-1 Hongo 7-chome Bunkyo-ku, Tokyo 113, Japan. (1931) DSc, chemistry (Tokyo U., 1958). Prof. (tel. 03 + 812-2111, ext. 2447). *Physical properties, crystal structure, molecular complexes, radical salts, X-ray photoelectron spectroscopy.*

Kurosawa, Dr Kou. Dept. of Electronics, U. of Osaka Prefecture, Mozu Umemachi 4-chome Sakai, Osaka 591, Japan. (1946) DEng, crystal growth (Osaka U. Prefecture, 1976). Res. assoc. (tel. 0722 + 52-1161, ext. 2282). *Crystal growth, X-ray topography, lattice defect.*

Kuroya, Prof. Hisao. Dept. of Chemistry, Okayama U. of Sci., 1-1 Ridai-cho Okayama, Okayama 700, Japan. (1916) DSc, chemistry (Osaka U., 1949). Prof. (tel. 0862 + 52-3161, ext. 407). *Crystal and molecular structure, metal complexes.*

Kushi, Dr Yoshihiko. Dept. of Chemistry, Faculty of Sci., Hiroshima U., 1-89 Higashisendamachi 1-chome, Hiroshima 730, Japan. (1937) DSc, chemistry (Osaka City U., 1968). Asst. prof. (tel. 0822 + 41-1221, ext. 801). *Crystal structure, coordination compounds.*

Kusunoki, Dr Masami. Crystallographic Res. Cntr., Inst. for Protein Res., Osaka U. Yamadaoka 3-2 Suita, Osaka 565, Japan. (1953) DSc, chemistry (Osaka U., 1980). Asst. prof. (tel. 06 + 877-5111, ext. 3912). *X-ray crystallography, proteins, nucleic acids.*

Kuwabara, Prof. Shigeya. Electronics Dept., Faculty of Sci. and Eng., Saga U., 1 Honjomachi, Saga 840, Japan. (1923) DSc, physics (Hiroshima U., 1958). Prof. (tel. 09522 + 4-5191, ext. 2655). *Electron diffraction, electron microscopy, electron optics, instrumentation.*

Kyotani, Mr Mutsumasa. Dept. of Polymer Physics, Res. Inst. for Polymers and Textiles, Sawatari 4, Kanagawa-ku, Yokohama 221, Japan. (1938) BEng, chemistry. (tel. 045 + 311-5901). *Crystallization, polymers.*

Mannami, Dr Michi-hiko. Dept. of Physics, Faculty of Sci., Kyoto U., Oiwake-cho Kitashirakawa Sakyo-ku, Kyoto 606, Japan. (1935) DSc, physics (Kyoto U.). Asst. prof. (tel. 075 + 751-2111, ext. 3754). *Radiation damage in solids, electron microscopy, electron microscopy, ion channelling.*

Maruha, Prof. Juro. C. of Liberal Arts, Kanazawa U., 1-1 Marunouchi, Kanazawa 920, Japan. (1921) DSc, chemistry (Tohoku U., 1960). Prof. (tel. 0762 + 62-4281 ext. 648). *Physical chemistry.*

Marukawa, Dr Kenzaburo. Dept. of Applied Physics, Faculty of Eng., Hokkaido U., Kita 13-jo Nishi 8-chome Kita-ku, Sapporo 060, Japan. (1937) DSc, physics (Kyoto U., 1967). Asst. prof. (tel. 011 + 711-2111, ext. 6644). *Lattice defects, metal crystals.*

Marumo, Prof. Fumiyuki. Res. Lab. of Eng. Materials, Tokyo Inst. of Techn., Nagatsuta-machi 4259 Midori-ku, Yokohama 227, Japan. (1931) DSc, mineralogy (Tokyo U., 1960). Prof. (tel. 045 + 922-1111, ext. 2312). *Inorganic crystal structures.*

Maruno, Dr Shigeo. Materials Res. Lab., Nagoya Inst. of Techn., Gokiso-cho Showa-ku, Nagoya 466, Japan. (1934) DSc, physics (Hokkaido U., 1969). Asst. prof. (tel. 052 + 732-2111, ext. 584). *Amorphous materials, X-ray spectroscopy.*

Maruse, Prof. Susumu. Dept. of Electronics, Nagoya U., Furo-cho Chikusa-ku, Nagoya 464, Japan. (1926) DEng, electrical engineering (Nagoya U., 1962). Prof. (tel. 052 + 781-5111, ext. 4436). *Electron optics.*

Maruyama, Prof. Saiyu. Div. of Physics, Dept. of Natural Sci., Osaka Women's U., Daisen-cho Sakai, Osaka 590, Japan. (1923) DSc, physics (Kyoto U., 1961). Prof. (tel. 0722 + 22-4811, ext. 335). *Long period structure, thin films, electron microscopy.*

Masaki, Dr Norio. Faculty of Pharmaceutical Sci. Kyoto U., Yoshida Shimoadachi-cho Sakyo-ku, Kyoto 606, Japan. (1931) DSc, physics (Osaka U., 1961). Asst. prof. (tel. 075 + 751-2111, ext. 4533). *Structure, biological membranes and related molecules.*

Masakuni, Dr Mayumi. First Dept. of Biochemistry, National Defense Medical C., 2-3 Namiki, Tokorozawa, Saitama 359, Japan. (1934) MD, biochemistry (Nihon U., 1983). Asst. (tel. 0429 + 95-1211, ext. 2293). *Protein structure and function, enzymes.*

Matsuda, Dr Ikuo. Materials Sci. Lab., Pioneering Res. and Dev. Labs., Toray Industries Inc., 2-1 Sonoyama 3-chome Otsu, Shiga 520, Japan. (1931) DSc, chemistry (Kyushu U., 1962). Manager. (tel. 0775 + 37-0600, ext. 430). *Molecular spectroscopy, structure and properties, polymers, analytical chemistry.*

Matsubara, Prof. Takeo. Dept. of Physics, Faculty of Sci., Kyoto U., Oiwake-cho Kitashirakawa Sakyo-ku, Kyoto 606, Japan. (1921) DSc, physics (Osaka U., 1951). Prof. (tel. 075 + 751-2111, ext. 3741). *Lattice dynamics, phase transition, random systems, surface physics.*

Matsuda, Prof. Hidehiko. Dept. of Applied Physics, Kyushu Inst. of Techn., 1 Sensuicho Tobata-ku Kitakyushu, Fukuoka 804, Japan. (1931) DEng, metallurgy (Kyushu U., 1972). Prof. (tel. 093 + 871-1931, ext. 467). *Physics, metals and alloys.*

Matsui, Dr Masanori. Chemical Lab., Kanazawa Medical U., Uchinada, Kahokugun, Ishikawa 920-02, Japan. (1949) DSc (Kwansei Gakuin U., 1982). Lect. (tel. 0762 + 86-2211, ext. 7108). *Modeling crystal structures, high temperature, high pressure.*

Matsui, Mr Toshiro. Electron Devices Lab., Toshiba Res. and Dev. Center, Tokyo Shibaura Electric Co. Ltd., 1 Komukai Toshiba-cho Saiwai-ku, Kawasaki Kanagawa 210, Japan. (1943) BSc, mineralogy (Tokyo U., 1967). Res. (tel. 044 + 511-2111).

Matsumoto, Dr Takeo. Dept. of Earth Sci. Faculty of Sci., Kanazawa U., 1-1 Marunouchi, Kanazawa Ishikawa 920, Japan. (1932) DSc, mineralogy (U.of Tokyo, 1962). Asst. prof. (tel. 0762 + 62-4281, ext. 568). *Inorganic crystal structures, mathematical crystallography, symmetry.*

Matsuo, Dr Munetsugu. Fundamental Res. Labs., Nippon Steel Corporation, 1618 Ida Nakahara-ku, Kawasaki Kanagawa 211, Japan. (1936) DEng, metallurgy (Tokyo U., 1967). Sr. res. (tel. 044 + 777-4111). *Metal physics, texture.*

Matsusaki, Dr Hideo. Fine Chemicals Lab., Dye Div., San-Ei Chemicals Co. Ltd., 1-11 Sanwa-cho 1-chome Toyonaka, Osaka 561, Japan. (1942) DEng, physical chemistry (Kyoto U., 1971). Res. assoc. (tel. 06 + 333-0521, ext. 16). *Liquid crystals, lasers, LSI, crystal growth.*

Matsushita, Dr Tadashi. Photon Factory, Nat. Lab. for High Energy Physics. Oho-machi, Tsukuba, Ibaraki 305, Japan. (1945) DEng., applied physics (Tokyo U., 1972). Assoc. prof. (tel. 0298 + 64-1171, ext. 5039). *X-ray optics, instrumentation, synchrotron radiation.*

Matsuura, Dr Yoshiki. Inst. for Protein Res., Osaka U.,. Yamada-oka Suita, Osaka 565, Japan. (1943) DSc, chemistry (Osaka U. 1976). Res. assoc. (tel. 06 + 877-5111, ext. 3912). *Proteins, immunoglobulins, crystallography.*

Matsuzaki, Mr Takao. Central Res. Labs., Mitsubishi Chemical Industries Ltd., 1000 Kamoshida, Midori-ku, Yokohama 227, Japan. (1945) MPharm. chemistry (Tokyo U., 1970). Sr. res. scient. (tel. 045 + 962-1211, ext. 3312). *Biological molecules, computer programming.*

Mihama, Prof. Kazuhiro. Dept. of Applied Physics, Faculty of Eng., Nagoya U., Furo-cho Chikusa-ku, Nagoya 464, Japan. (1927) Dr, science (Universite de Paris, France, 1960). Prof. (tel. 052 + 781-5111, ext. 4457). *High resolution electron microscopy, crystal growth, thin films.*

Miida, Mr Rokuro. Dept. of Physics, Tohoku U., Aramaki Aza-Aoba, Sendai 980, Japan. (1938) BSc, physics (Science U. of Tokyo, 1961). Res. assoc. (tel. 0222 + 22-1800, ext. 5345). *Structure, alloys.*

Miki, Dr Kunio. Dept. of Applied Chemistry, Faculty of Eng., Osaka U., Yamadakami Suita, Osaka 565, Japan. (1952) DEng, chemistry (Osaka U., 1981). Res. assoc. (tel. 06 + 877-5111, ext. 4322). *Protein crystallography, protein structure and function, organometallic crystal structures.*

Min, Miss Eungi. Mineralogical Inst., U. of Tokyo, 3-1 Hongo 7-chome Bunkyo-ku, Tokyo 113, Japan. (1949) MSc, mineralogy (Tokyo U., 1976). Grad. student. (tel. 03 + 812-2111, ext. 4483).

Minagawa, Dr Teruaki. Dept. of Physics, Osaka Kyoiku U., 43 Minami-Kawahori-cho Tennoji-ku, Osaka 543, Japan. (1942) DSc, physics (Osaka City U., 1971). Res. assoc. (tel. 06 + 771-8131, ext. 295). *Polytypes, phase transformations.*

Minato, Prof. Hideo. Inst. of Earth Sci. and Astronomy, C. of General Education, U. of Tokyo, 8-1 Komaba 3-chome Meguro-ku, Tokyo 153, Japan. (1921) DSc, mineralogy (Tokyo U., 1952). Prof. (tel. 03 + 467-1171, ext. 310). *Mineralogy, mineral chemistry, clay mineralogy, descriptive mineralogy, analytical chemistry, minerals, occurrence mode of minerals, clay minerals, zeolites.*

Minomura, Prof. Shigeru. Inst. for Solid State Physics, U. of Tokyo, Roppongi Minato-ku, Tokyo 106, Japan. (1923) DSc, chemistry (Kyoto U.). Prof. of chemistry. (tel. 03 + 402-6231, ext. 677). *High pressure physics and chemistry.*

Mitsuda, Dr Hiromichi. Wireless Res. Lab., Matsushita Electric Industrial Co. Ltd., 1006 Kadoma Osaka 571, Japan. (1935) DSc, physics (Osaka U., 1967). Chief engineer. (tel. 06 + 908-1291, ext. 313). *X-ray structure analysis, X-ray spectroscopy.*

Mitsuda, Dr Takeshi. Materials Res. Lab., Nagoya Inst. of Techn., Gokiso-cho Showa-ku, Nagoya 466, Japan. (1931) DSc, mineralogy (Hokkaido U., 1961). Asst. prof. (tel. 052 + 732-2111). *Chemistry, cements and allied materials.*

Mitsui, Prof. Toshio. Dept. of Biophysical Eng., Faculty of Eng. Sci., Osaka U., 1 Machikaneyama-cho 1-chome Toyonaka, Osaka 560, Japan. (1926) DSc, physics (Hokkaido U., 1960). Prof. (tel. 06 + 856-1151, ext. 3202). *Small angle X-ray scattering, biomembranes and other biological systems.*

Mitsui, Dr Yukio. Faculty of Pharmaceutical Sci. U. of Tokyo, 3-1 Hongo 7-chome Bunkyo-ku, Tokyo 113, Japan. (1938) DPharm, chemistry (Tokyo U., 1966). Res. assoc. (tel. 03 + 812-2111, ext. 7940). *Protein crystallography, fiber crystallography, molecular biology, molecular pharmacology, biophysics, instrumentation, computer technology.*

Mitsuishi, Prof. Tomokuni. Jobu U., 634-1 Toyazuka-cho Isezaki, Gunma 372, Japan. (1917) DSc, physics (Tokyo U., 1967). Prof. (tel. 0270 + 32-1011). *Semiconductors.*

Miura, Mr Naoki. Dept. of Mineral Industries, School of Sci. and Eng., Waseda U., 170 Nishiokubo 4-chome Shinjuku-ku, Tokyo 160, Japan. (1951) BEng, mineralogy (Waseda U., 1975). Grad. student. (tel. 03 + 209-3211, ext. 372).

Miura, Dr Yasuhiro. Dept. of Metallurgy, Faculty of Eng., Kyushu U., 10-1 Hakozaki 6-chome Higashi-ku, Fukuoka 812, Japan. (1941) PhD, physical metallurgy (U. California, Berkeley, USA,1970). Asst. prof.(tel. 092 + 641-1101, ext. 3238). *Plasticity and mechanical properties, metals and alloys.*

Miura, Dr Yasunori. Dept. of Mineralogical Sci. and Geology, Faculty of Sci., Yamaguchi U., 1677-1 Yoshida, Yamaguchi 753, Japan. (1946) DSc, mineralogy (Tohoku U., 1976). Res. assoc. (tel. 0839 + 22-6111, ext. 382). *Ion and electron microprobe analyses, inorganic crystal structure, X-ray crystallography, electron microscopy, crystal optics, computer programming.*

Miyaji, Mr Hirofumi. Materials Group, National Res. Inst. for Metals, 3-12 Nakameguro 2-chome Meguro-ku, Tokyo 153, Japan. (1939) BSc, physics (Science U. of Tokyo, 1963). Principal res. officer. (tel. 03 + 719-2271). *Phase transformation, recrystallization.*

Miyake, Prof. Shizuo. School of Sci. and Eng., Science U. of Tokyo, 2641 Higashikameyama Yamazaki, Noda 278, Japan. (1911) DSc, physics (Tokyo U., 1942). Prof. (tel. 0471 + 24-1501). *Crystal physics, dynamical diffraction (HEED - LEED - X-rays).*

Miyake, Dr Yasuhiro. Dept. of Polymer Sci., Faculty of Sci., Hokkaido U., Kita 10-jo Nishi 8-chome Kita-ku, Sapporo 060, Japan. (1925) DSc, physics (Hokkaido U., 1963). Asst. prof. (tel. 011 + 711-2111). *Polymers, biophysics.*

Miyamae, Dr Hiroshi. Faculty of Sci., Josai U., Keyakidai 1-1, Sakado, Saitama 350-02, Japan. (1950) DSc, chemistry (Tokyo U., 1978). Lect. (tel. 0492 + 86-2233, ext. 525). *X-ray crystallography, transition metal complexes, intercalation.*

Miyamoto, Dr Masamichi. Pure and Applied Sci., C. of Arts and Sci., U. of Tokyo, 3-8-1, Komaba, Meguro-ku, Tokyo 153, Japan. (1949) Res. assoc. (tel. 03 + 467-1171, ext. 402). *Computer simulation, Mineralogy, Crystallography.*

Miyata, Dr Takeshi. Gemmology, Inst. of Gemmology & Jewelry Arts of Yamanashi, 1955-1, Tokojicho, Kofu Yamanashi 400, Japan. (1949) PhD, mineralogy (Tohoku U. 1980). Lect. (tel. 0552 + 32-6671, ext. 36). *Crystal growth, morphology, gemmology.*

Miyazawa, Dr Shintaro. Functional Device Development Div., Atsugi Electrical Communication Lab., N.T.T. 1839 Ono, Atsugi-shi, Kanagawa 243-01, Japan. (1942) DEng, electronics (Tohoku U., 1978). Head, device materials section (tel. 0462 + 40-2720). *Crystal growth of III-V, characterization, new materials.*

Miyoshi, Dr Tadahiko. Materials Res. Dept., Hitachi Res. Lab., Hitachi Ltd., 4026 Kuji-cho Hitachi-shi, Ibaraki 319-12, Japan. (1943) DSc, chemistry (Tokyo U., 1971). Res. (tel. 0294 + 52-5111, ext. 284). *Metal oxide semiconductors, ceramics.*

Mizota, Mr Tadato. Dept. of Mining and Mineral Eng., Faculty of Techn., Yamaguchi U., Tokiwadai Ube, Yamaguchi 755, Japan. (1941) MSc, mineralogy (Tohoku U., 1966). Lect. (tel. 0836 + 31-5100, ext. 236). *Mineralogy, crystallography, mineral processing.*

Mizuno, Dr Hiroshi. Dept. of Molecular Biology, National Inst. of Agrobiological Resources, Kan-nondai 1-2, Yatabe Tsukuba Science City, Ibaraki 305, Japan. (1943) DPharm, chemistry (Osaka U., 1972). Sr. res. (tel. 02975 + 6-7014). *Macromolecular crystallography, molecular biology.*

Mizuno, Prof. Joji. Dept. of Electronics, Tohoku Inst. of Techn., 19 Koeji, Nagamachi Sendai 982, Japan. (1918) DSc, crystallography (Tohoku U., 1962). Prof. (tel. 0222 + 29-1151, ext. 253). *Inorganic crystal structures.*

Mori, Dr Saburo. Wireless Res. Lab., Matsushita Electric Industrial Co. Ltd., 1006 Kadoma Osaka 571, Japan. (1927) DSc, physics (Kyoto U., 1970). Chief of analytical section. (tel. 06 + 908-1291, ext. 311). *Solid reaction, crystal phase analysis.*

Morikawa, Dr Hideki. Res. Lab. of Eng. Materials, Tokyo Inst. of Techn. Nagatsuta 4259, Midori-ku, Yokohama 227, Japan. (1942) DEng, ceramics (Tokyo Inst. of Techn. 1973). Assoc. prof. (tel. 045 + 922-1111, ext. 2311). *Amorphous structure.*

Morikawa, Dr Hiroshi. Dept. of Eng. Sci. Nagoya Inst. of Techn., Gokiso-cho Showa-ku, Nagoya 466, Japan. (1942) DSc, chemistry (Kyoto U., 1972). Res. assoc. (tel. 052 + 732-2111, ext. 634). *Solid surfaces, crystal growth, corrosion, electron microscopy, field ion microscopy.*

Morimoto, Dr Jun. Dept. of Applied Physics, National Defense Academy. 1-10-20 Hashirimizu Yokosuka, Kanagawa 239, Japan. (1950) DEng, information processing (Tokyo Inst. of Techn, 1983). Lect. (tel. 0468 + 41-3810, ext. 2459). *Crystal growth, semiconductor physics.*

Morimoto, Prof. Nobuo. Dept. of Geology and Mineralogy, U. of Kyoto, Sakyo-ku, Kyoto 606, Japan. (1925) DSc, mineralogy (Tokyo U., 1954). Prof. (tel. 075 + 751-2111, ext. 4150). *Minerals, inorganic materials, nonstoichiometric compounds, superstructures in minerals.*

Morino, Prof. Yonezo. Sagami Chemical Res. Center, 4-1 Nishi-Ohnuma 4-chome Sagamihara, Kanagawa 229, Japan. (1908) DSc, chemistry (Tokyo U., 1937). President; Em. prof. (Tokyo U.). (tel. 0427 + 42-4791). *Molecular structure.*

Morita, Prof. Takeo. Dept. of Physics, Fukuoka U., 11 Nanakuma, Fukuoka 814, Japan. (1907) DSc, physics (Hiroshima U., 1952). Prof. (tel. 09403 + 6-6369). *Crystal growth.*

Motegi, Dr Hiroshi. Dept. of Materials Sci., Faculty of Sci., Hiroshima U., 1-89 Higashisendamachi 1-chome, Hiroshima 730, Japan. (1939) DSc, physics (Tokyo U., 1973). Lect. (tel. 0822 + 41-1221, ext. 802). *Ferroelectricity.*

Mukai, Prof. Tadasuke. Lab. of Physics, Fukuoka U. of Education, Akama Munakata-machi, Fukuoka 811-14, Japan. (1917) DSc, physics (Hiroshima U., 1959). Prof. (tel. 09403 + 2-2381, ext. 360). *Crystal growth, plastic deformation, metals.*

Murakami, Dr Takashi. Dept. of Environmental Safety Res., Japan Atomic Energy Res. Inst., Shirakata Tokai, Ibaraki 319-11, Japan. (1951) DSc, Mineralogy (Tokyo U., 1980) Res. Scient. (tel. 0292 + 82-5872). *High level waste management.*

Muraoka, Dr Hisashi. Toshiba Res. and Dev. Center, 72 Horikawa-cho Saiwai-ku, Kawasaki 210, Japan. (1924) DSc, mineralogy (Tokyo U., 1956). Corporate fellow. (tel. 044 + 522-2111, ext. 830). *Semiconductors, crystal growth.*

Murata, Prof. Yoshitada. Inst. for Solid State Physics, U. of Tokyo, 7-22-1 Roppongi Minato-ku, Tokyo 106, Japan. (1935) DSc, chemistry (Tokyo U., 1964). Prof. (tel. 03 + 478-6811, ext. 5301, Telex ISSP UT J 32469). *Surface science, low energy electron diffraction.*

Nagai, Prof. Ryutaro. Dept. of Physics, Tokyo Gakugei U., 1-1 Nukuikita-machi 4-chome Koganei, Tokyo 184, Japan. (1916) DSc, physics (Hiroshima U., 1961). Prof. (tel. 0423 + 21-1741, ext. 343). *Crystal physics, metal physics.*

Nagakura, Dr Ichiro. Dept. of Physics, Faculty of Education, Gunma U., 1375 Aramaki-cho, Maebashi 371, Japan. (1936) DSc, physics (Tohoku U., 1964). Asst. prof. (tel. 0272 + 32-1611). *Solid state physics.*

Nagakura, Prof. Saburo. Inst. for Solid State Physics, U. of Tokyo, 22-1 Roppongi 7-chome Minato-ku, Tokyo 106, Japan. (1920) DSc, chemistry (Tokyo U., 1953). Prof. (tel. 03 + 402-6231, ext. 649). *Structures, optical properties, molecular crystals.*

Nagakura, Prof. Sigemaro. Dept. of Metallurgy, Tokyo Inst. of Techn., 12-1 Ohokayama 2-chome Meguro-ku, Tokyo 152, Japan. (1926) DSc, physics (Kyoto U., 1959). Prof. (tel. 03 + 726-1111, ext. 3144). *Structure, metals and alloys, crystal growth, electron diffraction, electron microscopy, X-ray diffraction, topography.*

Nagano, Dr Kozo. U. of Tokyo, 7-3-1 Hongo Bunkyo-ku, Tokyo 113, Japan. (1933) DPharm (Tokyo U., 1962). Assoc. prof. (tel. 03 + 812-2111, ext. 4841). *Tertiary structure prediction, biological macromolecules.*

Nagashima, Dr Seiichi. Dept. of Physics, C. of Eng., Nihon U. 1 Nakagawara Tokusada, Tamura-cho Koriyama, Fukushima 963, Japan. (1949) DEng, electrical engineering (Nihon U., 1977). Lect. (tel. 0249 + 44-1300, ext. 312). *Epitaxial growth, thin films, LEED.*

Nagata, Dr Fumio. 4th Dept. Central Res. Lab., Hitachi Ltd., 280 Higashi-Koigakubo 1-chome, Kokubunji Tokyo 185, Japan. (1940) DEng, applied physics (Nagoya U., 1971). (tel. 0423 + 23-1111). *Electron microscopy.*

Naiki, Prof. Toshio. Dept. of Physics, Kyoto Techn. U., Matsugasaki Sakyo-ku, Kyoto 606, Japan. (1925) DSc, physics (Kyoto U., 1962). Prof. (tel. 075 + 791-3211, ext. 441). *Crystal growth, electron microscopy, thin films.*

Nakahigashi, Dr Kiyotaka. Faculty of General Education, U. of Osaka Prefecture, Mozu Umemaci 4-chome Sakai, Osaka 591, Japan. (1941) BSc, physics (Osaka City U., 1964). Lect. (tel. 0722 + 52-1161). *Structure, metals and alloys.*

Nakai, Dr Hisayoshi. Dept. of Chemistry, Hyogo C. of Medicine, 1 Mukogawa 1-chome Nishinomiya, Hyogo 663, Japan. (1942) DSc, chemistry (Osaka City U., 1971). Asst. prof. (tel. 0798 + 45-6442). *Coordination chemistry, inorganic biochemistry.*

Nakajima, Dr Tetuo. Photon Factory, Nat. Lab. For High Energy Physics, Ohomachi, Tsukuba, Ibaraki 305, Japan. (1935) DSc, metal physics (1964) Assoc. prof. (tel. 0298 + 64-1171, ext. 5027). *Magnetic substances, low temperature, neutron scattering, synchrotron radiation.*

Nakajima, Dr Yoshiharu. Inst. of Scient. and Industrial Res., Osaka U., Yamadakami Suita, Osaka 565, Japan. (1946) DSc, inorganic and physical chemistry (Osaka U., 1975). Res. assoc. (tel. 06 + 877-5111, ext. 3546). *Inorganic crystal structure, X-ray crystallography, high resolution microscopy.*

Nakamura, Dr Kazuo. Faculty of Pharmaceutical Sci. U. of Tokyo, 3-1 Hongo 7-chome Bunkyo-ku, Tokyo 113, Japan. (1945) DPharm, pharmaceutical sciences (Tokyo U., 1974). Res. assoc. (tel. 03 + 812-2111, ext. 4842). *Protein crystallography, computer programming, microcomputers.*

Nakamura, Dr Minoru. 73 Div., Hitachi Lab., 4026 Omika Kuji-cho Hitachi, Ibaraki 319-12, Japan. (1941) DSc, chemistry (Osaka U., 1973). Res. (tel. 0294 + 52-5111). *Molecular physics, spectroscopy.*

Nakamura, Mr Naotake. Dept. of Chemistry, Ritsumeikan U., 27 Tojiin-Kitamachi Kita-ku, Kyoto 603, Japan. (1943) BEng, chemistry (Ritsumeikan U., 1966). Res. assoc. (tel. 075 + 463-1131, ext. 335). *Structures and physical properties, normal long chain compounds.*

Nakamura, Mr Osamu. 1st Div., Government Industrial Res. Inst., 8-31 Midorigaoka 1-chome Ikeda, Osaka 563, Japan. (1946) (tel. 0727 + 51-8351, ext. 408).

Nakamura, Prof. Terutaro. The Inst. for Solid State Physics, U. of Tokyo, 22-1 Roppongi 7 Minato-ku, Tokyo 106, Japan. (1923) DSc, physics (Tokyo U., 1961). Prof. (tel. 03 + 402-6231, ext. 624). *Dielectric physics, ferroelectric phase transition, light scattering in solids, amorphous state.*

Nakamura, Dr Toshio. Dept. of Anatomy, School of Medicine, Kanazawa U., 13-1 Takara-machi, Kanazawa 920, Japan. (1929) MD, anatomy Asst. prof. (tel. 0762 + 62-8151, ext. 231). *X-ray crystallographical studies, biological membranes.*

Nakanishi, Prof. Norihiko. Dept. of Chemistry, Faculty of Sci., Konan U., 9-1 Okamoto 8-chome, Kobe 658, Japan. (1928) DSc, chemistry. Prof. (tel. 078 + 431-4341, ext. 229, 282). *Solid state chemistry, physical metallurgy, powder metallurgy.*

Nakano, Dr Shigeru. Dept. of Physics, Faculty of Sci., Chiba U., 33 Yayoicho 1-chome, Chiba 280, Japan. (1929) DSc, physics (Tokyo U., 1965). Asst. prof. (tel. 0472 + 51-1111, ext. 2610). *X-ray diffraction, superconductivity.*

Nakata, Mr Kazuaki. Dept. of Physics, Osaka Kyoiku U., 43 Minami-Kawahori-cho Tennoji-ku, Osaka 543, Japan. (1946) MEd, physics (Osaka Kyoiku U., 1971). Res. assoc. (tel. 06 + 771-8131, ext. 216). *Disorder, phase transition, organic crystals.*

Nakatani, Prof. Kazumi. Dept. of Chemistry, Faculty of Sci., Kwansei Gakuin U., Uegahara Nishinomiya, Hyogo 662, Japan. (1928) DSc, chemistry (Osaka U., 1961). Prof. (tel. 0798 + 51-3301). *Structural chemistry, organic and organometallic compounds; photographic science.*

Nakatsu, Dr Hiromoto. Nat. Inst. for Res. in Inorganic Materials, 1 Namiki Sakura-mura Niihari-gun, Ibaraki 305, Japan. (1940) DSc, inorg. chemistry (Osaka U.). Sr. Res. officer. (tel. 0298 + 51-3351, ext. 321). *X-ray crystallography, mineralogy.*

Nakazumi, Dr Yoshihide. Nakazumi Crystals Corporation, 1 Sakae-machi 2-chome Ikeda, Osaka 563, Japan. (1919) DEng, chemistry (Osaka U., 1961). President of Nakazumi Crystals. (tel. 0727 + 51-0118). *Crystal growth, fine ceramics, geology.*

Namba, Dr Yoshiyuki. Dept. of Chemistry, Osaka Kyoiku U., 43 Minami-Kawahori-cho Tennoji-ku, Osaka 543, Japan. (1924) DSc, physics (Osaka U., 1962). Asst. prof. (tel. 06 + 771-8131). *Organic crystal structure, computer application.*

Namikawa, Mr Kazumichi. Crystallography I Div., Inst. for Solid State Physics, U. of Tokyo, 22-1 Roppongi 7-chome Minato-ku, Tokyo 106, Japan. (1944) BSc, physics (Tokyo U. of Education, 1968). Techn. assoc. (tel. 03 + 402-6231, ext. 658). *LEED, phase transition, surface physics.*

Nawata, Dr Yoshiharu. Res. Labs., Chugai Pharmaceutical Co. Ltd., 41-8 Takada 3-chome, Toshima-ku, Tokyo 171, Japan. (1934) DPharm, pharamcy (Tokyo U., 1984). Sr. scient. (tel. 03 + 987-7111, ext. 219). *X-ray crystallography, molecular structure analysis, natural organic products.*

Naya, Prof. Shigeo. Dept. of Physics, Faculty of Sci., Kwansei Gakuin U., 1-155 Uegahara 1-bancho Nishinomiya, Hyogo 662, Japan. (1927) DSc, physics (Osaka U., 1959). Prof. (tel. 0798 + 51-3301). *Phase transition, crystal physics.*

Niimura, Dr Nobuo. Lab. of Nuclear Sci., Faculty of Sci., Tohoku U., Mikamine Sendai 982, Japan. (1942) DSc, physics (Tokyo U., 1970). Res. assoc. (tel. 0222 + 45-2151, ext. 22). *Neutron scattering, instrumentation.*

Nishi, Dr Fumito. Mineralogical Inst., Faculty of Sci., U. of Tokyo, 7-3-1 Hongo Bunkyo-ku, Tokyo 113, Japan. (1949) DSc, mineralogy (Tokyo U., 1978). (tel. 03 + 812-2111, ext. 4545). *Inorganic crystal structures.*

Nishida, Dr Isao. Physics Lab., Dept. of General Education, Nagoya U., Furo-cho Chikusa-ku, Nagoya 464, Japan. (1933) DSc, physics (Tokyo Inst. of Techn., 1971). Assoc. prof. (tel. 052 + 781-5111, ext. 4846). *Metal fine particles, electron microscopy, crystal morphology.*

Nishida, Dr Takashi. C. of Arts and Sci. Chiba U., 1-33 Yayoi-cho Chiba-city, Chiba 260, Japan. (1938) DSc, mineralogy (Tokyo U., 1970). Asst. prof. (tel. 0472 + 51-1111, ext. 2282). *Inorganic crystal structures, polytypism, twinning, crystal growth.*

Nishiguchi, Mr Munehiro. Inst. of Scient. and Industrial Res., Osaka U., Yamadakami Suita, Osaka 565, Japan. (1953) BSc, chemistry (Osaka U., 1975). Grad. student. (tel. 06 + 877-5111, ext. 3546). *Inorganic chemistry.*

Nishikawa, Dr Masana. Eng. & Dev. Div., Mitsubishi Atomic Power Ind. Inc., 297 Kitabukuro 1-chome Omiya Saitama 330, Japan. (1942) DSc, geophysics (Tokyo U., 1971). Chief. (tel. 0486 + 41-5111, ext. 292, 370). *Nuclear fusion experiment facility design, high energy particle - solid interactions.*

Nishinaga, Prof. Tatau. Dept. of Electronics, Tokyo U., 7-3-1 Hongo Bunkyo-ku, Tokyo 113, Japan. (1939) DEng, electronics (Nagoya U., 1967). Prof. (tel. 03 + 812-2111, ext. 6673). *Crystal growth - theory and experiment.*

Nishino, Mr Yoichi. Third Dept., Central Res. Lab., Hitachi Ltd., 1-280 Higashi-Koigakubo, Kokubunji, Tokyo 185, Japan. (1955) DEng, materials science (Nagoya U., 1983). Res. (tel. 0423 + 23-1111, ext. 3307). *X-ray crystallography, dislocations, semiconductors.*

Nishiyama, Dr Tsutomu. Natural Sci. Lab., Toyo U., 28-20 Hakusan 5-chome Bunkyo-ku, Tokyo 112, Japan. (1939) DSc, mineralogy (Tokyo U. of Education, 1971). Asst. prof. (tel. 03 + 945-7392). *Clay minerals.*

Nishiyama, Prof. Zenji. 391-19 Shimoda-cho Kohoku-ku, Yokohama 223, Japan. (1901) DSc, physics (Tohoku U., 1932). Em. prof. (Osaka U.). (tel. 044 + 61-7774). *Metal physics.*

Nittono, Dr Osamu. Dept. of Metallurgy, Tokyo Inst. of Techn., 12-1 Oh-okayama 2-chome Meguro-ku, Tokyo 152, Japan. (1941) DEng, metallurgy (Tokyo Inst. of Techn., 1970). Asst. prof. (tel. 03 + 726-1111, ext. 3145). *X-ray and electron diffraction, X-ray topography, crystal growth and characterization.*

Noda, Prof. Tokiti. 15-3 Shikannonmichi Nishi Tashiro-cho Chikusa-ku, Nagoya 464, Japan. (1903) DEng, applied chemistry (Tokyo U., 1940). Em. prof. (Nagoya U. and Mie U.). (tel. 052 + 711-2959). *Crystal growth, inorganic materials, carbon and graphite.*

Noda, Dr Yasutoshi. Dept. of Materials Sci., Faculty of Eng., Tohoku U., Aoba Aramaki, Sendai 980, Japan. (1942) DEng, materials science (Tohoku U., 1970). Assoc. prof. (tel. 0222 + 22-1800, ext. 4464). *Solid state chemistry.*

Noda, Mr Yoshitoshi. Faculty of Education, Oita U., 700 Dannoharu, Oita 870-11, Japan. (1939) MEng, electrical engineering (1964). Asst. prof. (tel. 0975 + 69-4652). *Lasers.*

Noda, Mr Yukio. C. of General Education, Osaka U., 1 Machikaneyama-cho 1-chome Toyonaka, Osaka 560, Japan. (1948) MSc, physics (Osaka U., 1973). Postgraduate student. (tel. 06 + 856-1151, ext. 2745). *X-ray and neutron diffraction, phase transition.*

Nonaka, Dr Kohzo. Central Res. Lab., Kyoto Ceramic Co. Ltd., 11-17 Kogahonmachi Fushimi-ku, Kyoto 612, Japan. (1924) DSc, physics (Osaka U., 1962). General manager. (tel. 075 + 933-5121). *Electronic ceramics, thin films, alloys and compounds.*

Nukui, Dr Akihiko. Group 11, Nat. Inst. for Res. in Inorganic Materials, Namiki-1 Sakura-mura Niihari-gun, Ibaraki 305, Japan. (1944) DEng, ceramics (Tokyo Inst. of Techn., 1973). Sr. res. (tel. 0298 + 51-3351, ext. 254). *Phase transition, X-ray diffraction, EXAFS.*

Oda, Mr Isao. Res. and Dev. Lab., NGK Insulators Ltd., 2-56 Suda Mizuho-ku, Nagoya 467, Japan. (1939) BSc, mineralogy (Tohoku U., 1962). Sr. res. engineer. (tel. 052 + 882-7731). *Crystal growth.*

Oda, Prof. Tsutomu. Dept. of Chemistry, Osaka Kyoiku U., 43 Minami-Kawahoricho Tennoji-ku, Osaka 543, Japan. (1916) DSc, physical chemistry (Osaka U., 1945). Prof. (tel. 06 + 771-8131). *Crystal chemistry.*

Odajima, Prof. Akira. Dept. of Applied Physics, Faculty of Eng., Hokkaido U., Kita 13-jo Nishi 8-chome Kita-ku, Sapporo 060, Japan. (1922) DSc, physics. Prof. (tel. 011 + 711-2111, ext. 6620). *Amorphous structure, lattice dynamics, polymer structure.*

Ogawa, Mr Katsumi. Dept. of Mineral Industries, School of Sci. and Eng., Waseda U., 170 Nishiokubo 4-chome Shinjuku-ku, Tokyo 160, Japan. (1950) MEng, mineralogy (Waseda U., 1975). Grad. student. (tel. 03 + 209-3211, ext. 374).

Ogawa, Prof. Kazuhide. Inst. of Chemistry, C. of General Education, Osaka U., 1 Machikaneyama-cho 1-chome Toyonaka, Osaka 560, Japan. (1922) DSc, physics (Osaka U., 1962). Prof. (tel. 06 + 844-1151, ext. 5285). *Inorganic and organic crystal structures.*

Ogawa, Prof. Shiro. Dept. of Metallurgy, Shibaura Inst. of Techn., 9-14 Shibaura 3-chome Minato-ku, Tokyo 108, Japan. (1912) DSc, physics (Tohoku U., 1924). Prof. (tel. 03 + 452-3201, ext. 331) *Structure, metals and alloys, films, surfaces, metal physics.*

Ogawa, Prof. Tomoya. Dept. of Physics, Faculty of Sci., Gakushuin U., 5-1 Mejiro 1-chome Toshima-ku, Tokyo 171, Japan. (1930) DEng, electronics (Tokyo U., 1966). Prof. (tel. 03 + 986-0221, ext. 459). *Crystal growth, texture.*

Ogura, Prof. Iwao. C. of Eng., Nihon U., Tamura-machi Koriyama, Fukushima 963, Japan. (1922) DSc, physics (Hiroshima U., 1961). Prof. (tel. 0249 + 44-1300, ext. 314). *Thin films, surface science, material science and engineering, electron microscopy.*

Ogura, Prof. Kiyosi. Dept. of Physics, Faculty of Sci. and Techn., Kinki U., 4-1 Ko-Wakae 3-chome Higasi-Osaka, Osaka 577, Japan. (1924) Prof. (tel. 06 + 721-1332). *Theory of solids, hydrogen bond, water.*

Ohachi, Dr Tadashi. Dept. of Electronics, Doshisha U., Karasuma-Imadegawa, Kyoto 602, Japan. (1941) DEng, solid state electro chemistry and electronics (Doshisha U., 1975). Asst. prof. (tel. 075 + 211-2311). *Crystal growth, ionic conduction in solid.*

Ohama, Dr Nobuhiko. Dept. of Physics, Kyushu U., 10-1 Hakozaki 6-chome Higashi-ku, Fukuoka 812, Japan. (1940) DSc, physics (Kyushu U., 1971). Res. assoc. (tel. 092 + 641-1101, ext. 4180). *Crystal physics, instrumentation, structural phase transition.*

Ohashi, Dr Yuji. Lab. of Chemistry for Natural Products, Tokyo Inst. of Techn., 12-1 Oh-okayama 2-chome Meguro-ku, Tokyo 152, Japan. (1941) DSc, chemistry (Tokyo U.). Res. assoc. (tel. 03 + 726-1111, ext. 3076). *Organic chemistry, reaction mechanisms, biochemistry.*

Ohba, Dr Shigeru. Dept. of Chemistry, Faculty of Sci. & Techn., Keio U. Hiyoshi 3, Kohoku-ku, Yokohama 223, Japan. (1953) DSc, chemistry (Tokyo U., 1981). Asst. (tel. 044 + 63-1141, ext. 3912). *Electron density distribution, transition metal complexes.*

Ohgaki, Mr Masataka. Res. Lab. of Eng. Materials, Tokyo Inst. of Techn. 4259 Nagatsuta-cho, Midori-ku, Yokohama Kanagawa 227, Japan. (1959) MSc, material science (Yokohama Nat. U., 1983). Postgraduate student (tel. 045 + 922-1111, ext. 2311). *Crystal structure analysis, electron density determination.*

Ohkawa, Mr Tokio. Dept. of Applied Physics, Faculty of Eng., Osaka U., Yamadakami Suita, Osaka 565, Japan. (1934) BSc, Grad. student. (tel. 06 + 877-5111). *Physics.*

Ohmasa, Dr Masaaki. Inst. of Materials Sci., U. of Tsukuba, Sakura-mura Niihari-gun, Ibaraki 305, Japan. (1935) DSc, mineralogy (Tokyo U., 1964). Asst. prof. (tel. 0298 + 53-5012). *Crystal structures, phase transition.*

Ohno, Dr Tamotsu. Dept. of Physics, Aichi Medical C., Yazako Nagakute-cho, Aichi 480-11, Japan. (1937) DEng, physics (Nagoya U., 1975). Asst. prof. (tel. 05616 + 2-3311, ext. 2055). *Biophysics, radiation biology, electron optics.*

Ohshima, Dr Ken-ichi. Dept. of Applied Physics, Faculty of Eng., Nagoya U., Furo-cho Chikusa-ku, Nagoya 464, Japan. (1946) DSc, physics (Tohoku U., 1975). Res. assoc. (tel. 052 + 781-5111, ext. 4453). *Diffraction physics.*

Ohsumi, Dr Kazumasa. Mineralogical Inst., Faculty of Sci., U. of Tokyo, 3-1 Hongo 7-chome Bunkyo-ku, Tokyo 113, Japan. (1943) DSc, mineralogy (Tokyo U., 1971). Res. assoc. (tel. 03 + 812-2111, ext. 4547). *Symmetry, inorganic crystal structures, phase relation.*

Ohta, Dr Takao. Toshiba Res. and Dev. Center, Tokyo Shibaura Electric Co. Ltd., 1 Komukai Toshiba-cho Saiwai-ku, Kawasaki Kanagawa 210, Japan. (1941) DSc, mineralogy (Tokyo U., 1970). Res. (tel. 044 + 511-2111, ext. 2291). *Inorganic crystal structure, crystal growth.*

Ohta, Mr Tsutomu. Mineralogical Inst., Faculty of Sci., U. of Tokyo, 3-1 Hongo 7-chome Bunkyo-ku, Tokyo 113, Japan. (1950) MSc, mineralogy (Tokyo U., 1975). Grad. student. (tel. 03 + 812-2111, ext. 3728). *Crystal chemistry.*

Ohtsuka, Mr Yasukuni. Dept. of Physics, Tohoku Inst. of Techn., 19 Koeji Nagamachi Sendai 982, Japan. (1935) MSc, physics (Tohoku U., 1961). Asst. prof. (tel. 0222 + 29-1151). *Thin films, epitaxy.*

Okabe, Dr Toshio. Dept. of Physics, Faculty of Literature and Sci., Toyama U., 3190 Gofuku, Toyama 930, Japan. (1942) DSc, physics (Kyoto U., 1974). Asst. prof. (tel. 0764 + 41-1271, ext. 314). *Crystal growth, structure, amorphous materials, diffuse scattering of electrons and X-rays.*

Okada, Mr Kenji. System Dev. Div., Ricoh Co. Ltd., 3-6 1-chome, Naka-magome, Ohta-ku, Tokyo 143 Japan. (1942) BSc, applied chemistry (Sci. U. of Tokyo, 1965). Acting manager. (tel. 03 + 777-8111, telex 246-6201).

Okada, Dr Kiyoshi. Dept. of Inorganic Materials, Faculty of Eng., Tokyo Inst. of Techn., 12-1 Oh-okayama 2-chome, Meguro-ku, Tokyo 152, Japan. (1948) DEng, inorganic chemistry (Tokyo Inst. of Techn., 1976). Res. assoc. (tel. 03 + 726-1111, ext. 2524). *Crystal chemistry, clay mineralogy, amorphous substances.*

Okada, Prof. Masakazu. Dept. of Applied Physics, Faculty of Applied Biology Sci., Hiroshima U., Midorimachi, Fukuyama-city 720, Japan. (1928) DSc, physics (Tokyo Sci. u., 1956). Prof. (tel. 0849 + 24-6211, ext. 320). *Crystal physics, electron crystallography.*

Okada, Mr Yasumasa. Fundamental Sci. Div., Electrotechnical Lab., Umezono Sakura-mura Niihari-gun, Ibaraki 305, Japan. (1940) BSc, solid state physics (Sci. U. of Tokyo, 1965). Sr. res. (tel. 0298 + 54-5136). *X-ray crystallography, solid state physics, crystal defects.*

Okazaki, Prof. Atsushi. Dept. of Physics, Kyushu U., 10-1 Hakozaki 6-chome Higashi-ku, Fukuoka 812, Japan. (1931) DSc, physics (Kyushu U., 1961). Prof. (tel. 092 + 641-1101, ext. 4177). *Crystal physics, structural phase transition, instrumentation.*

Okamura, Dr Fujio Peter. 15th Res. Group, Nat. Inst. for Res. in Inorganic Materials Namiki 1-1 Sakura-mura Niihari-gun, Ibaraki 305, Japan. (1939) DSc, mineralogy (Tokyo U., 1969). Sr. res. officer (tel. 0298 + 51-3351, ext. 289). *Crystal chemistry, phase prediction, CAD of materials.*

Okunuki, Mr Masahiko. Semiconductor Equipment Dev. Div., Canon Ltd. Morinosato Wakamiya 5 Atsugi, Kanagawa 243-01, Japan. (1946) MSc, physics (Gakushuin U., 1971). (tel. 0462 + 47-2111). *X-ray diffraction, X-ray lithography, focused ion beam, micro lithography.*

Okuyama, Dr Kenji. Polymer Eng., Tokyo U. of Agriculture and Techn. Nakamachi 2-24-16 Koganei, Tokyo 184, Japan. (1946) DSc, polymer chemistry (Osaka U., 1977). Assoc. prof. (tel. 0423 + 81-4221, ext. 236). *Biological substances, structure.*

Onuma, Dr Shigeki. Dept. of Chemistry, Shizuoka U., 836 Oya, Shizuoka 422, Japan. (1937) DSc, chemistry (Tohoku U., 1966). Asst. prof. (tel. 0542 + 37-1111, ext. 554).

Ookawa, Prof. Akiya. Dept. of Physics, Faculty of Sci., Gakushuin U., 1-5 Mejiro Toshima-ku, Tokyo 171, Japan. (1918) DSc, physics (Tokyo U., 1958). Prof. (tel. 03 + 986-0221, ext. 485). *Crystal growth, statistical thermodynamics.*

Osaka, Dr Toshiaki. Nat. Inst. for Res. in Inorganic Materials, Kurakake Sakura-mura Niihari-gun, Ibaraki 300-31, Japan. (1941) DEng, metallurgy (Waseda U., 1972). Res. (tel. 0298 + 57-3351). *Thin films.*

Osakabe, Mr Nobuyuki. Central Lab. Hitachi, 1-280 Higashikoigakubo Kokubunji-shi, Tokyo 185, Japan. (1955) MSc, physics (Tokyo Inst. of Techn., 1980). Res. (tel. 0423 + 23-1111). *Surface physics, electron microscopy.*

Osaki, Dr Kenji. Tsukahara 6-11-23 Takatsuki, Osaka 569, Japan. (1920) DSc, physics (Osaka U., 1958). Em. prof. (Kyoto U.). (tel. 0726 + 95-3116). *Crystal structures, molecular interactions, crystallographic information system.*

Otsuka, Prof. Ryohei. Dept. of Mineral Industry, Waseda U., 170 Nishiokubo 4-chome Shinjuku-ku, Tokyo 160, Japan. (1922) DEng, mineralogy (Waseda U., 1957). Prof. (tel. 03 + 209-3211, ext. 372). *Clay minerals.*

Oyanagi, Dr Hiroyuki. Fundamental Sci. Div., Electrotechn. Lab., 1-1-4 Umezono Sakuramura Niiharigun, Ibaraki 305, Japan. (1952) PhD, physical chemistry (Tokyo U. 1976). Chief res. (tel. 0298 + 54-5112). *Synchrotron radiation, X-ray spectroscopy, EXAFS, organic conductors, thin films.*

Ozawa, Dr Tohru. Mineralogical Inst., Faculty of Sci., U. of Tokyo, 7-3-1 Hongo Bunkyo-ku, Tokyo 113, Japan. (1940) DSc, mineralogy (Tokyo U., 1968). Lect. (tel. 03 + 812-2111, ext. 4546). *Crystal chemistry, minerals, electron microscopy, inorganic materials.*

Sadanaga, Prof. Ryoichi. 3-18-16 Chuo, Nakano-ku, Tokyo 164, Japan. (1920) DSc, mineralogy (Tokyo U., 1953). Em. prof. (Tokyo U.); Member, Japan Academy. (tel. 03 + 369-4768). *Mathematical crystallography, inorganic and organic crystal structures.*

Sagawa, Prof. Takasi. Dept. of Physics, Faculty of Sci., Tohuku U., Aoba Aramaki, Sendai 980, Japan. (1926) DSc, physics (Tohoku U., 1961). Prof. (tel. 0222 +

22-1800, ext. 3238). *Soft X-ray vacuum UV spectroscopy, photoelectron spectroscopy.*

Saito, Prof. Norio. Chemical Lab., Meiji-gakuin U., 2-37 Shiroganedai 1-chome Minato-ku, Tokyo 108, Japan. (1935) DSc, biochemistry (Tokyo U. of Education, 1964). Prof. (tel. 03 + 443-8231, ext. 254). *Plant pigments, biosynthesis.*

Saito, Prof. Yoshihiko. Dept. of Chemistry, Faculty of Sci. & Techn., Keio U., 14-1 Hiyoshi 3-chome, Kohoku-ku, Yokohama 223, Japan. (1920) DSc, Chemistry (Osaka U., 1952). Prof. (tel. 044 + 63-1141, ext. 3910). *Inorganic and organic crystal structures, absolute configuration, accurate electron density distribution. teaching solid state physics,*

Saito, Mr Yoshio. Dept. of Physics, Kyoto Inst. of Techn., Matsugasaki, Sakyo-ku, Kyoto 606, Japan. (1944) BEE, electronics (Osaka Inst. of Techn., 1968). Asst. (tel. 075 + 791-3211, ext. 448). *Crystal growth, crystal structure, electron microscopy.*

Saito, Mr Yoshiyuki. Res. Lab., Kawasaki Steel Corporation, 8-4 Ritsurincho 1-chome, Takamatsu 760, Japan. (1948) MSc, physics (1974). (tel. 0878 + 62-2853).

Saka, Dr Takashi. Dept. of Crystalline Materials Sci., Faculty of Eng., Nagoya U., Furo-cho Chikusa-ku, Nagoya 464, Japan. (1946) DEng, applied physics (Nagoya U., 1974). Res. asst. (tel. 052 + 781-5111, ext. 5889). *X-ray crystallography.*

Sakabe, Dr Noriyoshi. Dept. of Chemistry, Nagoya U., Furo-cho Chikusa-ku, Nagoya 464, Japan. (1934) DSc, chemistry (Nagoya U., 1966). Asst. prof. (tel. 052 + 781-5111, ext. 3551). *Protein crystallography, molecular biophysics.*

Sakaki, Prof. Em. Yoneichiro. 105 Yayoigaoka, Tempaku-ku, Nagoya 468, Japan. (1913) DEng, electron microscopy (Nagoya U., 1951). Prof. em. - Nagoya U. (tel. 052 + 831-1787). *Electron optics.*

Sakamoto, Dr Yosio. SAKAMOTO's Niggli-&-Born, Inst. for Studying Crystal Structure Types. 15-5 Higashikasumi-machi, Minami-ku Hiroshima, Hiroshima 734, Japan. (1919) DSc, chemistry (Hiroshima U. of Lit. & Sci., 1957). Res. crystallographer; former Prof. Hiroshima U. (tel. 082 + 282-6513). *Codification, crystal structure types, lattice & point complex method; crystal lattice potential.*

Sakata, Dr Makoto. Dept. of Applied Physics, Faculty of Eng., Nagoya U., Furocho, Chikusa-ku, Nagoya Aichi 464, Japan. (1944) PhD, chemistry (Tokyo U of Education, 1974). Assoc. prof. (tel. 052 + 781-5111, ext. 4453). *Solid state physics, X-ray crystallography, neutron diffraction.*

Sakuma, Dr Takashi. Dept. of Physics, Faculty of Sci., Ibaraki U., Mito 310, Japan. (1951) DSc, physics, (Tokyo U., 1978). Res. assoc. (tel. 0292 + 26-1621, ext. 478). *X-ray and neutron diffraction.*

Sakurai, Dr Junji. Dept. of Materials Sci., Faculty of Sci., Hiroshima U., 1-89 Higashisendamachi 1-chome, Hiroshima 730, Japan. (1936) DSc, chemistry (Kyoto U., 1964). Asst. prof. (tel. 0822 + 41-1221, ext. 654). *Magnetism.*

Sakurai, Dr Kiichi. Japan. (1916) DSc, physics (Osaka U., 1959). (tel. 078 + 593-6211, ext. 247). *Structure analysis, structural disorder.*

Sakurai, Dr Tosio. Crystal Physics Lab., Inst. of Physical and Chemical Res., 1 Hirosawa 2-chome Wako, Saitama 351, Japan. (1926) DSc, physics (Tokyo U., 1962). Sr. scient. (tel. 0484 + 62-1111, ext. 3343). *Crystallographic computing, structure, biological molecules.*

Sasada, Prof. Yoshio. Lab. of Chemistry for Natural Products, Tokyo Inst. of Techn., Nagatsuta Midori-ku, Yokohama 227, Japan. (1926) DSc, chemistry (Osaka U., 1958). Prof. (tel. 045 + 984-1111, ext. 2386). *Structural chemistry, organic compounds, solid state reaction, biological molecular structures.*

Sasaki, Prof. Dr Akio. Dept. of Electrical Eng., Kyoto U., Kyoto 606, Japan. (1932) PhD, (U. California, Berkley, USA, 1966) DEng, electrical engineering (Kyoto U., 1976). Prof. (tel. 075 + 751-2111, ext. 5296). *Solid state electronics.*

Sasaki, Dr Kyoyu. C. of Medical Techn., 1-1-20 Daiko-minami, Higashi-ku, Nagoya 461, Japan. (1940) DSc, chemistry (Nagoya U., 1973). Assoc. Prof. (tel. 052 + 723-1111, ext. 241). *Protein crystallography, computing techniques.*

Sasaki, Dr Satoshi. Photon Factory, Nat. Lab. For High Energy Physics, Ohomachi, Tsukuba, Ibaraki 305, Japan. (1951) DSc, mineralogy (Tokyo U., 1979). Res. assoc. (tel. 0298 + 64-1171, ext. 5023, Telex 3652-534). *X-ray diffraction, synchrotron radiation research, crystallography, mineralogy.*

Sasaki, Prof. Yukiyoshi. Dept. of Chemistry, Faculty of Sci., U. of Tokyo, 7-3-1 Hongo Bunkyo-ku, Tokyo 113, Japan. (1928) DSc, chemistry (Tokyo U., 1960). Prof. (tel 03 + 812-2111, ext. 4359). *Polyanion, inorganic structural chemistry.*

Sato, Mr Hiroki. Musashi Works, Hitachi Ltd., 1450 Josui-honmachi Kodaira, Tokyo 187, Japan. (1944) MSc, mineralogy (Tohoku U., 1970). Engineer. (tel. 0423 + 43-8623). *Surface or interface science.*

Sato, Mr Kazuo. Res. & Dev. Div., Japan Information Processing Service, 1-3 Kudan Kita 4-chome Chiyoda-ku, Tokyo 102, Japan. (1946) MSc, physical chemistry (Kwansei Gakuin U.). (tel. 03 + 265-1361, ext. 231). *Information retrieval.*

Sato, Assoc. Prof. Mitsuo. Dept. of Chemistry, Faculty of Techn., Gunma U., Tenkincho 1-5-1, Kiryu, Gunma 376, Japan. (1932) DSc, mineralogy (Tokyo Kyoiku U., 1962). Assoc. prof. (tel. 0277 + 22-3181, ext. 429). *Structural chemistry, inorganic materials.*

Sato, Prof. Shin'ichi. Dept. of Applied Physics, Faculty of Eng., Hokkaido U., Kita 13-jo Nishi 8-chome, Kita-ku, Sapporo 060, Japan. (1925) DSc, physics (Osaka U., 1962). Prof. (tel. 011 + 716-2111, ext. 6643). *Diffraction crystallography, martensitic phase transformation.*

Sato, Dr Shoichi. Inst. for Solid State Physics, U. of Tokyo, 7-22-1 Roppongi Minato-ku, Tokyo 106, Japan. (1930) DSc, chemisrty (Tokyo U., 1980). Res. assoc. (tel. 03 + 402-6231, ext. 662). *X-ray and neutron diffraction, transition metal complexs, electron distribution in crystals, ferroelectrics.*

Sato, Mr Tomohiro. Shionogi Res. Labs., Shionogi and Co. Ltd., Fukushima-ku, Osaka 553, Japan. (1939) MPharm (Tokyo U., 1965). Res. (tel 06 + 458-5861). *X-ray crystallography.*

Satow, Dr Yoshinori. Photon Factory, Nat. Lab. for High Energy Physics, Ohomachi, Tsukuba, Ibaraki 305, Japan. (1949) DPharm, pharmaceutical sciences (Tokyo U., 1977). Asst. (tel. 0298 + 64-1171, ext. 5025). *X-ray crystallography, protein crystallography, synchrotron radiation.*

Sawa, Dr Isao. Dept. of Electronics, Faculty of Eng., Kansai U., 3 Yamate Suita, Osaka 564, Japan. (1938) DEng, electronics (Kansai U.). Lect. (tel. 06 + 388-1121).

Sawada, Dr Akikatsu. Synthetic Crystal Res. Lab., Nagoya U., Furo-cho Chikusa-ku, Nagoya 464, Japan. (1941) DEng, physics (Nagoya U., 1968). Res. assoc. (tel. 052 + 781-5111, ext. 3597). *Ferroelectrics, phase transition.*

Sawada, Prof. Masao. Dept. of Solid State Electronics, Osaka Electro-Communication U., 18-8 Hatsu-cho Neyagowa, Osaka 572, Japan. (1903) DSc, physics (Kyoto U., 1933). Prof.; Em. prof. (Osaka U.). (tel. 0723 + 65-0074). *Crystal physics, X-ray spectroscopy, electronic structure of matter.*

Sawada, Dr Toshiyuki. Mineralogical Inst., Faculty of Sci., U. of Tokyo, 3-1 Hongo 7-chome Bunkyo-ku, Tokyo 113, Japan. (1949) MSc, mineralogy (Tokyo U., 1975). Grad. student. (tel. 03 + 812-2111, ext. 3287). *Crystal structures.*

Sawada, Mr Yasuaki. Dept. of Information Sci., Kanazawa Inst. of Techn., 7-1 Ogigaoka Nonoichi-machi, Ishikawa 921, Japan. (1941) MSc, mathematics (Kanazawa U., 1967). Asst. prof. (tel. 0762 + 48-1100). *Symmetry.*

Sawaguchi, Prof. Etsuro. Dept. of Physics, Faculty of Sci., Hokkaido U., Kita 10-jo Nishi 8-chome Kita-ku, Sapporo 060, Japan. (1925) DSc, physics (Tokyo U., 1960). Prof. (tel. 011 + 711-2111, ext. 2680). *Crystal structure analysis, ferroelectric phenomena.*

Seiyama, Prof. Tetsuro. Dept. of Materials Sci. & Techn., Faculty of Eng., Kyushu U., 10-1 Hakozaki 6-chome Higashi-ku, Fukuoka 812, Japan. (1920) DEng, applied chemistry (Kyushu U., 1954). Prof. (tel. 092 + 641-1101, ext. 3198). *Crystal growth, metal oxides, sulphides, electrocrystallization of metals, relation between structure and properties of solids.*

Seki, Prof. Syuzo. Dept. of Chemistry, Kwansei Gakuin U., 1-1-155 Uegahara, Nishinomiya 662, Japan. (1915) DSc, chemistry (Osaka U., 1945). Prof. (tel. 0798 + 51-2407). *Physical chemistry, thermodynamical studies, phase transition in solids.*

Sekizaki, Dr Masao. C. of Liberal Arts, Kanazawa U., 1-1 Marunouchi, Kanazawa 920, Japan. (1941) DSc, chemistry (Nagoya U., 1972). Asst. prof. (tel. 0762 + 62-4281, ext. 650). *Structures, transition metal complexes.*

Sera, Mr Masaaki. Analytical Centre, Nippon Light Metal Res. Lab. Ltd., 4540 Kambara-cho Ihara-gun, Shizuoka 421-32, Japan. (1936) BSc, applied physics (Osaka U., 1960). Chief of Instrument Analysis Lab. (tel. 05438 + 5-2121, ext. 232). *Instrumental analysis, X-ray diffraction, X-ray fluorescence, EPMA, SEM, electron microscope, ESCA and IMMA, physical metallurgy.*

Shibata, Prof. Noboru. Faculty of Liberal Arts, Nagasaki U., 1-14 Bunkyo-machi, Nagasaki 852, Japan. (1924) BSc, physics (Osaka U., 1946). Prof. (tel. 0958 + 47-1111). *Growth mechanisms, imperfections, ZnO crystals grown by chemical reaction.*

Shibata, Prof. Shuzo. Dept. of Chemistry, Shizuoka U., 836 Oya, Shizuoka 422, Japan. (1924) DSc, chemistry (Nagoya U., 1958). Prof. (tel. 0542 + 37-1111, ext. 553). *Gas electron diffraction, inorganic crystal structure.*

Shibuya, Prof. Iwao. Div. of Slow Neutron Physics, Res. Reactor Inst., Kyoto U., 1052 Noda Kumatori-cho, Osaka 590-04, Japan. (1930) DSc, solid state physics (Kyushu U., 1961). Asst. prof. (tel. 07245 + 2-0901, ext. 2322, 2282). *Neutron diffraction studies, ferroelectrics.*

Shichiri, Dr Takaki. Dept. of Physics, Faculty of Sci., Osaka City U., Sugimoto-cho Sumiyoshi-ku, Osaka 558, Japan. (1933) DTechn., applied physics (Nagoya U., 1972). Asst. prof. (tel. 06 + 692-1231, ext. 3232). *Crystal growth.*

Shigenari, Dr Takeshi. Dept. of Eng. Physics, U. of Electro-Communications, 5-1 Chofugaoka 1-chome Chofu, Tokyo 182, Japan. (1939) DSc, physics (Tokyo U., 1970). Asst. prof. (tel. 0424 + 83-2161, ext. 502). *Laser light scattering spectroscopy, phase transition.*

Shimanouchi, Dr Hirotaka. Faculty of Sci., Tokyo Inst. of Techn., 12-1 Ohokayama 2-chome Meguro-ku, Tokyo 152, Japan. (1939) DSc, chemistry (Tohoku U., 1967). Asst. prof. (tel. 03 + 726-1111, ext. 3076). *Crystal and molecular structure, X-ray crystallography.*

Shimaoka, Prof. Kohji. Dept. of Mathematics and Physics, Ritsumeikan U., 28-1 Tojiin-Kitamachi, Kita-ku, Kyoto 603, Japan. (1928) DSc, physics (Tokyo Inst. of Techn., 1958). Prof. (tel. 075 + 463-1131, ext. 326). *Crystal structures, neutron diffraction, phase transition.*

Shimazaki, Dr Yoshihiko. Mineral Deposits Dept., Geological Survey of Japan, 8 Kawada-cho Shinjuku-ku, Tokyo 162, Japan. (1928) PhD, mineralogy (Stanford U., USA, 1957); DSc, mineralogy (Tokyo U., 1962). Chief, Mineralogy Sec. (tel. 03 + 341-7131). *Mineralogy.*

Shimazu, Dr Masaji. Group 10, Nat. Inst. for Res. in Inorganic Materials, 1-1 Namiki Sakura-mura Niihari-gun, Ibaraki 305, Japan. (1930) DSc, mineralogy

(Tokyo U. of Education, 1965). Leader in 10th res. group. (tel. 0298 + 51-3351). *X-ray crystallography, ferroelectrics.*

Shimizu, Prof. Kenichi. Inst. of Scient. and Industrial Res., Osaka U., Yamadakami Suita, Osaka 565, Japan. (1928) DSc, physics (Nagoya U., 1962). Prof. (tel. 06 + 877-5111, ext. 3555). *Physical metallurgy, crystallography, metals.*

Shimura, Mr Fumio. Material Res. Lab., Central Res. Labs., Nippon Electric Co. Ltd., 1753 Shimonumabe Nakahara-ku, Kawasaki Kanagawa 211, Japan. (1948) MEng, crystallography (Nagoya Inst. of Techn., 1974). Res. member. (tel. 044 + 433-1111, ext. 2839). *Growth theory of crystals, crystal growth, crystals grown for electronics.*

Shinnaka, Prof. Yasuhiro. Dept. of Physics, Faculty of Literature and Sci., Yamaguchi U., 1677-1 Yoshida, Yamaguchi 753, Japan. (1926) DSc, physics (Kyoto U., 1960). Prof. (tel. 08392 + 2-6111, ext. 371). *Phase transition, X-ray diffraction.*

Shinohara, Dr Kenichi. 5-16-14 Mejiro Toshima-ku, Tokyo 171, Japan. (1905) DSc, physics (Kyoto U., 1939). *Radiation physics and chemistry.*

Shintani, Dr Ryuichi. Dept. of Physics, Faculty of Sci., Kwansei Gakuin U., 1-155 Uegahara 1-bancho Nishinomiya, Hyogo 662, Japan. (1928) DSc, chemistry (Osaka U., 1960). Prof. (tel. 0797 + 22-9674). *Organic semiconductors.*

Shiojiri, Prof. Makoto. Dept. of Physics, Kyoto Techn. U., Matsugasaki Sakyo-ku, Kyoto 606, Japan. (1936) DSc, chemistry (Kyoto U., 1967). Prof. (tel. 075 + 791-3211, ext. 444). *Crystal growth, crystal physics.*

Shirai, Prof. Shunji. 99-30 Judayu Nagareyama, Chiba 270-01, Japan. (1902) DSc, physics (Tokyo U., 1942). Em. prof. (Tohoku U.). (tel. 0471 + 53-2627). *Growth and structure, single-crystal films.*

Shirai, Mr Yoshihiro. Lab. of Crystal and Powder Chemistry, Inst. for Chemical Res., Kyoto U., Gokasho Uji 611, Japan. (1950) MSc, chemistry (Kyoto U., 1976). Grad. student. (tel. 0774 + 32-3111, ext. 352). *Epitaxy, interaction between elementary particles and materials.*

Shiro, Dr Motoo. Dept. of Physical Chemistry, Shionogi Res. Lab., 12-4 Sagisu 5-chome Fukushima-ku, Osaka 553, Japan. (1931) DSc, chemistry (Osaka City U., 1979). Sr. res. chemist. (tel. 06 + 458-5861, ext. 262). *Crystal structure analysis, organic compounds from physicochemical and biochemical viewpoint.*

Shirozu, Prof. Haruo. Dept. of Geology and Mineralogy, Kyushu U., 10-1 Hakozaki 6-chome Higashi-ku, Fukuoka 812, Japan. (1925) DSc, mineralogy (Kyushu U., 1960). Prof. (tel. 092 + 641-1101, ext. 4142). *Clay mineralogy.*

Shoda, Prof. Tokugoro. Senshu U., 4764 Ikuta Tama-ku, Kawasaki Kanagawa 214, Japan. (1913) DSc, mineralogy (Tokyo U., 1957). Prof. (tel. 044 + 911-7131). *Optical crystallography.*

Somiya, Dr Shigeyuki. Lab. for Hydrothermal Syntheses, Res. Lab. of Eng. Materials, Tokyo Inst. of Techn., 4259 Nagatsuta Midori, Yokohama 227, Japan. (1928) Prof.; director (Hydrothermal Syn. Lab.). (tel. 045 + 984-1111, ext. 2309). *Hydrothermal synthesis, phase equilibria.*

Sonoda, Mr Minoru. X-Ray Film Dept., Fuji Photo Film Co. Ltd., 200 Onakazato Fujinomiya, Shizuoka 418, Japan. (1932) BSc, chemistry (Kyoto U., 1954). Manager. (tel. 05442 + 7-1211, ext. 400). *Photographic science, physical chemistry.*

Sonoike, Prof. Sanemi. Dept. of Physics, Chuo U., Kasuga Bunkyo-ku, Tokyo 112, Japan. (1922) DEng, applied physics (Tokyo U., 1958). Prof. (tel. 03 + 813-4175, ext. 363). *Defects in crystalline solids, physical properties of silver halide.*

Sudo, Prof. Toshio. Geological and Mineralogical Inst., Fac. of Sci., Tokyo U. of Education, 29-1 Otsuka 3-chome Bunkyo-ku, Tokyo 112, Japan. (1911) DSc, mineralogy (Tokyo U., 1944). Em. prof. (tel. 03 + 946-2151). *Crystal chemistry, clay minerals.*

Suemune, Dr Yasutaka. Res. and Dev. Bureau, Nippon Telegraph and Telephone Corp., 9-11 Midori-cho 3-chome Musashino, Tokyo 180, Japan. (1936) DSc, physics (Kyushu U., 1967). Sr. staff engineer. (tel. 0422 + 59-2208). *Solid state physics.*

Sueno, Dr Shigeho. Inst. of Geoscience, U. of Tsukuba, Sakura-mura Niihari-gun, Ibaraki 300-31, Japan. (1937) DSc, physics (Tokyo U., 1966). Asst. prof. (tel. 0298 + 53-4466). *Inorganic crystal chemistry.*

Sugaike, Dr Suezo. Toshiba Ceramics Co. Ltd., Chemical Eng. Div., P.O. box 3012, Shinjuku Nomura Bldg. Shinjuku, Tokyo 160, Japan. (1921) DSc, mineralogy (Tokyo U., 1957). Techn. consultant. (tel. 0463 + 81-8407) *Ceramics, solid electrolyte.*

Sugawara, Dr Yoko. Crystal Physics Lab., RIKEN (The Inst. of Physical and Chemical Res.) Wako Saitama 351-01, Japan. (1952) DPharm, physical chemistry (Tokyo U., 1980). Res. (tel. 0484 + 62-1111, ext. 3342). *Biophysical chemistry.*

Sugihara, Dr Akio. Dept. of Biochemistry, Osaka Municipal Techn. Res. Inst., 38 Kita-ogimachi Kita-ku, Osaka 530, Japan. (1943) DSc, chemistry (1970). Res. worker. (tel. 06 + 312-6551). *Protein structure and function.*

Sugiura, Prof. Seiji. Dept. of Earth Sci., Kanazawa U., 1-1 Marunouchi, Kanazawa 920, Japan. (1920) DSc, mineralogy (Tokyo U., 1961). Prof. (tel. 0762 + 62-4281, ext. 562). *Clay mineralogy, sulphate mineralogy.*

Suita, Prof. Tokuo. Atomic Energy Commission, 2-1 Kasumigaseki 2-chome Chiyoda-ku, Tokyo 100, Japan. (1911) DSc, physics, DEng, engineering (Osaka U., 1947, 1949). Commisioner; Em. prof. (Osaka U.). (tel. 03 + 581-2585).

Suito, Dr Eiji. 30 Kamiikeda-cho Kitashirakawa Sakyo-ku, Kyoto 606, Japan. (1912) DSc, chemistry (Kyoto U., 1942). President of Maizuru Techn. C. (tel. 0773 + 62-5600, ext. 001). *Electron microscopy, microcrystals (colloid or powder).*

Sunagawa, Prof. Ichiro. Inst. of Mineralogy, Petrology and Economic Geology, Tohoku U., Aramaki Aza-Aoba, Sendai 980, Japan. (1924) DSc, mineralogy (Hokkaido U., 1957). Prof. (tel. 0222 + 22-1800, ext. 3433). *Crystal growth mechanism, crystal morphology, characterization, natural minerals.*

Suzuki, Prof. Hideji. Dept. of Physics, Faculty of Sci., U. of Tokyo, 3-1 Hongo 7-chome Bunkyo-ku, Tokyo 113, Japan. (1924) DSc, physics (Tohoku U., 1955). Prof. (tel. 03 + 812-2111, ext. 4521). *Lattice defects, mechanical properties of crystals, phase transition in solids, liquid physics.*

Suzuki, Dr Ikuo. Dept. of Eng. Sci., Nagoya Inst. of Techn., Gokiso-cho Showa-ku, Nagoya 466, Japan. (1941) DEng, physics (Nagoya U., 1968). Asst. prof. (tel. 052 + 732-2111, ext. 570). *Ferroelectricity, magnetic resonance.*

Suzuki, Prof. Michio. C. of Arts and Sci. Tohoku U., Kawauchi Sendai 980, Japan. (1924) DSc, physics (Tohoku U., 1961). Prof. (tel. 0222 + 23-1181, ext. 4161). *X-ray scattering, X-ray lasers.*

Suzuki, Dr Shigeo. Dept. of Physics, Tokyo Inst. of Techn., 12-1 Oh-okayama 2-chome Meguro-ku, Tokyo 152, Japan. (1939) DSc, physics (Tokyo Inst. of Techn., 1969). Res. assoc. (tel. 03 + 726-1111, ext. 2463). *Crystal physics, X-ray topography, ferroelectrics, phase transition.*

Suzuki, Prof. Tadasu. Dept. of Physics, Sophia U., Kioi-cho Chiyoda-ku, Tokyo 102, Japan. (1925) DSc, physics (Tokyo U., 1960). Prof. (tel. 03 + 265-9211, ext. 296). *Electrons with spin and momentum density in solids.*

Suzuki, Dr Teruo. Dept. of Mining and Eng. Minerallogy, Tohoku U., Aramaki Aza-Aoba, Sendai 980, Japan. (1925) DSc, minerallogy (Tohoku U.). Asst. prof. (tel. 0222 + 22-1800, ext. 3504). *Semiconducting properties, ore minerals, habit change and occurrence, pyrite.*

Suzuki, Dr Toshimasa. Dept. of System Eng., Faculty of Eng., Nippon Inst. of Techn. 4-1 Gakuendai Miyashiro, Minami-Saitama, Saitama 345, Japan. (1948) DEng, applied physics (Tokyo Inst. of Techn., 1979). Lect. (tel. 0480 + 34-4111, ext. 559). *Crystal growth, semiconductors.*

Suzuki, Prof. Yoshio. Inst. of Geoscience, U. of Tsukuba, Sakura-mura Niihari-gun, Ibaraki 300-31, Japan. (1927) DSc, petrology (Hokkaido U., 1958). Prof. (tel. 0298 + 53-2111). *Petrology, crystal growth, material science.*

Tabata, Dr Hideyo. Ceramic Eng. Div., Gov. Ind. Res. Inst. of Nagoya, 1-1 Hirate-machi, Kita-ku, Nagoya Aichi 462, Japan. (1941) DSc, chemistry (Osaka U., 1976). Res. manager. (tel. 052 + 911-2111, ext. 477). *Crystal growth, ceramic sintering.*

Tadokoro, Prof. Hiroyuki. Dept. of Polymer Sci., Faculty of Sci., Osaka U., 1 Machikaneyama-cho 1-chome Toyonaka, Osaka 560, Japan. (1920) DSc, chemistry (Osaka U., 1959). Prof. (tel. 06 + 856-1151, ext. 2517). *Structure, crystalline polymers, X-ray diffraction, infrared spectroscopy, Raman spectroscopy, energy calculations.*

Tagai, Dr Tokuhei. Mineralogical Inst., Faculty of Sci., U. of Tokyo, 7-3-1 Hongo Bunkyo-ku, Tokyo 113, Japan. (1943) DSc, mineralogy (Tokyo U., 1972). Asst. prof. (tel. 03 + 812-2111, ext. 4544). *real structure of crystals, extraterrestrial minerals.*

Takagi, Dr Mieko. Tokyo Kasei Gakuin C., 2600 Aihara Machida, Tokyo 194-02, Japan. (1919) DSc, physics (Osaka U., 1956). Prof. (tel. 0427 + 82-9811, ext. 541). *X-ray topography, ferroelectrics.*

Takagi, Prof. Satio. School of Sci. and Eng., Science U. of Tokyo, 2641 Higashikameyama Yamazaki, Noda 278, Japan. (1916) DSc, physics (Tokyo U., 1958). Prof. (tel. 0471 + 24-1501). *Diffraction theory, X-ray topography, electron diffraction, electron microscopy.*

Takagi, Prof. Yutaka. Daido Inst. of Techn., 2-21 Daido-cho Minami-ku, Nagoya 457, Japan. (1914) DSc, physics (Tokyo U., 1944). Prof. (tel. 052 + 611-0511, ext. 416). *Phase transition, ferroelectricity, ferroelasticity, mechanical properties of solids.*

Takahashi, Mr Shoichi. Res. Lab., Toshiba Ceramics Co. Ltd., 30 Soya Hatano, Kanagawa 257, Japan. (1938) BSc, mineralogy (Tokyo U., 1962). Chief res. (tel. 03 + 334-7818). *Mineralogy, crystallography, ceramics, crystal growth.*

Takahashi, Dr Toshio. Dept. Applied Physics, Faculty of Eng., U. of Tokyo, 7-3-1 Hongo Bunkyo-ku, Tokyo 113, Japan. (1950) DEng, applied physics (Tokyo U., 1979). Res. asst. (tel. 03 + 812-2111, ext. 6828) *Dynamical diffraction, X-rays and neutrons, photoelectron spectroscopy.*

Takahashi, Prof. Yasuhiro. Dept. of Macromolecular Sci., Faculty of Sci., Osaka U. Machikaneyama Toyonaka, Osaka 560, Japan. (1941) DSc, macromolecular science (Osaka U., 1973). Asst. prof. (tel. 06 + 844-1151, ext. 4252). *Macromolecules, disorder, phase transformation, molecular motion, X-ray crystallography.*

Takai, Dr Mitsuo. Dept. of Applied Chemistry, Faculty of Eng., Kita 13-jo Nishi 8-chome Kita-ku, Sapporo Hokkaido 060, Japan. (1941) DEng, physical chemistry (Hokkaido U., 1969). Res. assoc. (tel. 011 + 711-2111, ext. 6570). *Fine structure, biosynthesis, cellulose.*

Takaki, Dr Yoshito. Dept. of Physics, Osaka Kyoiku U., 43 Minami-Kawahori-cho Tennoji-ku, Osaka 543, Japan. (1931) DSc, physics (Osaka U., 1963). Prof. (tel. 06 + 771-8131, ext. 228). *Structural disorder, phase transformation in crystals.*

Takamura, Prof. Jin-ichi. Dept. of Metal Sci. and Techn., Faculty of Eng., Kyoto U., Yoshida Honmachi Sakyo-ku, Kyoto 606, Japan. (1921) DEng, metallurgy (Kyoto U., 1953). Prof. (tel. 075 + 751-2111, ext. 5461). *Lattice defects, strength of metals, metal physics.*

Takano, Dr Tsunehiro. Faculty of Pharmaceutical Sci., Setsunan U. 45-1 Nagaotoge cho, Hirakata, Osaka 573-01, Japan. (1936) DSc, physical chemistry (Osaka U., 1965). Prof. (tel. 0720 + 68-7000, ext. 428). *Proteins, nucleic acids.*

Takano, Prof. Yasumasa. Dept. of Electro-science, Fac. of Sci., Okayama U. of Sci., 1 Ridai-cho 1-chome, Okayama 700, Japan. (1921) DSc, physics (Hiroshima U., 1961). Prof. (tel. 0862 + 52-3161, ext. 335). *Radiation detection, crystal physics, X-ray chemical analysis.*

Takano, Dr Yukio. Dept. of Pure and Applied Sci. C. of General Education, U. of Tokyo, 8-1 Komaba 3-chome Meguro-ku, Tokyo 153, Japan. (1924) DSc, mineralogy (Tokyo U., 1959). Asst. prof. (tel. 03 + 467-1171, ext. 402). *Crystal morphology, crystal growth, inorganic crystal structures.*

Takasu, Dr Shin-ichiro. Toshiba Res. and Dev. Center, Tokyo Shibaura Electric Co. Ltd., 1 Komukai Toshiba-cho Saiwai-ku, Kawasaki Kanagawa 210, Japan. (1928) DSc, mineralogy (Tokyo U., 1958). Sr. res. member. (tel. 044 + 511-2111, ext. 2267). *Crystal growth, morphology, X-ray crystallography, optical crystallography, Silicon, semiconducting materials, inorganic materials.*

Takayanagi, Prof. Kunio. Physics Dept., Tokyo Inst. of Techn., Oh-okayama 1-12-1, Meguro-ku, Tokyo 152, Japan. (1947) DSc, surface physics, electron microscopy (Tokyo Inst. of Techn., 1969). Assoc. prof. (tel. 03 + 726-1111, ext. 2079). *Surface structure, surface phase transition, crystal growth, thin films, electron diffraction, electron microscopy, X-ray crystallography.*

Takeda, Dr Hirofumi. Coating Resin Dept., Synthetic Resin Techn. Res. Lab., Dainippon Ink and Chemicals Inc., 3-1 Takasago Takaishi-shi, Osaka 592, Japan. (1950) PhD, chemistry (Osaka U., 1978). (tel. 0722 + 68-3111, ext. 486, Telex 5374-795).

Takeda, Prof. Hiroshi. Mineralogical Inst., Faculty of Sci., U. of Tokyo, 3-1 Hongo 7-chome Bunkyo-ku, Tokyo 113, Japan. (1934) DSc, mineralogy (Tokyo U., 1962). Prof. (tel. 03 + 812-2111, ext. 4543). *Polymorphism, polytypism, twinning, crystal chemistry, meteoritic and lunar minerals.*

Takenaka, Dr Akio. Faculty of Sci., Tokyo Inst. of Techn., Nagatsuta, Midori-ku, Yokohama 227, Japan. (1942) DSc, chemistry (Kwansei Gakuin U., 1971). Res. Assoc. (tel. 045 + 922-1111). *X-ray crystallography, organic and inorganic crystal structures, biological molecular structures.*

Takeoka, Mr Yoshikatsu. Toshiba Res. and Dev. Center, Tokyo Shibaura Electric Co. Ltd., 1 Komukai Toshiba-cho Saiwai-ku, Kawasaki Kanagawa 210, Japan. (1945). (tel. 044 + 511-2111, ext. 2329).

Takeuchi, Prof. Yoshio. Mineralogical Inst., Faculty of Sci., U. of Tokyo, 3-1 Hongo 7-chome Bunkyo-ku, Tokyo 113, Japan. (1924) DSc, mineralogy (Tokyo U., 1953). Prof. (tel. 03 + 812-2111, ext. 4542). *Crystal chemistry, minerals, inorganic materials.*

Takeyama, Prof. Yoshio. Dept. for Liberal Arts, Shizuoka U., 836 Oya, Shizuoka 422, Japan. (1923) BSc, physics (Kyushu U., 1948). Prof. (tel. 0542 + 37-1111). *X-ray topography.*

Tamada, Mr Osamu. Inst. of Earth Sci., C. of General Arts, Kyoto U., Yoshida Nihonmatsu-cho Sakyo-ku, Kyoto 606, Japan. (1944) MSc, mineralogy and geology (Kyoto U., 1973). Res. assoc. (tel. 075 + 751-2111, ext. 6865). *Mineralogy, binding energy in minerals.*

Tamaki, Dr Shozo. Electronics Div., Osaka Prefectural Industrial Res. Inst., Enoko-jima Kamino-cho Nishi-ku, Osaka 550, Japan. (1942) DSc, physics (Osaka U., 1971). Res. worker. (tel. 06 + 443-1121, ext. 269). *Solid surfaces, mass spectrometry.*

Tamura, Dr Chihiro. Analytical & Metabolic Res. Lab., 2-58 Hiromachi 1-chome, Shinagawa-ku, Tokyo 140, Japan. (1930) DSc, chemistry (Tokyo Metropolitan U., 1963). Res. officer. (tel. 03 + 492-3131, ext. 599). *Organic crystal structures, biological molecular structures.*

Tanabe, Dr Kazuya. Semiconductor Eng. Div., SONY Corporation, 14-1 Asahi-cho 4-chome Atsugi, Kanagawa 243, Japan. (1941) DSc, mineralogy (Tohoku U., 1970). (tel. 0462 + 20-5533). *Crystal growth.*

Tanaka, Mr Isao. Dept. of Applied Chemistry, Faculty of Eng., Nagoya U., Furo-cho Chikusa-ku, Nagoya 464, Japan. (1948) MSc, chemistry (Osaka U., 1973). Res. assoc. (tel. 052 + 781-5111, ext. 6715). *Biochemistry.*

Tanaka, Prof. Jiro. Dept. of Chemistry, Faculty of Sci., Nagoya U., Furo-cho Chikusa-ku, Nagoya 464, Japan. (1929) DSc, chemistry (Tokyo U., 1957). Prof. (tel. 052 + 781-5111, ext. 2481). *Intermolecular interaction, electronic spectra, molecular crystals, absolute configuration, chiral molecules.*

Tanaka, Dr Kiyoaki. Res. Lab. of Eng. Materials, Tokyo Inst. of Techn., 4259 Nagatsuta, Midori-ku, Yokohama 227, Japan. (1946) DSc, chemistry (Tokyo U., 1975). Res. assoc. (tel. 045 + 922-1111, ext. 2312). *Crystal structure analysis, accurate electron density determination.*

Tanaka, Dr Michiyoshi. Dept. of Physics, Faculty of Sci., Tohoku U., Aramaki Aza-Aoba, Sendai 980, Japan. (1938) DSc, physics (Tokyo Inst. of Techn., 1965). Assoc. prof. (tel. 0222 + 22-1800, ext. 3296). *Electron diffraction, electron microscopy, condensed matter.*

Tanaka, Dr Nobuo. Inst. of Protein Res., Osaka U., Yamada-kami Suita-shi, Osaka 565, Japan. (1941) DSc, chemistry (Osaka U., 1971). Asst. prof. (tel. 06 + 877-5111, ext. 3837). *Protein crystallography.*

Tanaka, Prof. Tetsuro. Dept. of Electronics, Fac. of Eng., Kyoto U., Yoshida Honmachi Sakyo-ku, Kyoto 606, Japan. (1916) DEng, physics (Kyoto U., 1951). Prof. (tel. 075 + 751-2111, ext. 5278). *Semiconductors, ferroelectrics, ferromagnetics.*

Tanemura, Dr Sakae. 5th Div. Solar Res. Lab., Government Industrial Res. Inst., Nagoya, 1-1 Hirate-machi Kita-ku, Nagoya 462, Japan. (1943) DEng, applied physics (Nagoya U., 1971). Sr. res. assoc. (tel. 052 + 911-2111, ext. 432). *Solar energy collecting materials, spectroscopic properties, bulk crystal properties, thin filmed stacks, X-ray diffraction.*

Tani, Dr Katsuhiko. Mineralogical Inst., Faculty of Sci., U. of Tokyo, 3-1 Hongo 7-chome Bunkyo-ku, Tokyo 113, Japan. (1944) DSc, mineralogy (Tokyo U., 1976). (tel. 03 + 812-2111, ext. 2418). *Symmetry.*

Taniguchi, Mr Tomohiko. Dept. of Physics, Osaka Kyoiku U., 43 Minami-Kawahori-cho Tennoji-ku, Osaka 543, Japan. (1936) BEd, physics (Osaka Kyoiku U., 1962). Lect. (tel. 06 + 771-8131, ext. 295). *Phase transformation, molecular crystals.*

Tanisaki, Prof. Sigetosi. Faculty of Literature and Sci., Yamaguchi U., 1677-1 Yoshida, Yamaguchi 753, Japan. (1922) DSc, physics (Kyoto U., 1960). Prof. (tel. 08392 + 2-6111, ext. 370). *Ferroelectricity, X-ray diffraction.*

Tanishiro, Mr Yasumasa. Physics Dept., Tokyo Inst. of Techn., Oh-okayama Meguro-ku, Tokyo 152, Japan. (1955) MSc, physics (Tokyo Inst. of Techn., 1980). Student in DSc course. (tel. 03 + 726-1111, ext. 2079). *Surface physics, crystal growth, electron microscopy, electron spectroscopy.*

Taoka, Prof. Tadami. Dept. Fundamental Eng., Inst. Vocational Training, 1960 Aihara Sagamihara, Kanagawa 229, Japan. (1916) DSc, physics (Tokyo U., 1957). Prof. (tel. 0427 + 61-2111, ext. 313). *Metal physics.*

Tate, Prof. Isao. Industrial Chemistry, Fac. of Eng., Shinshu U., 500 Wakasato, Nagano 380, Japan. (1923) DEng, industrial chemistry (Nagoya U., 1971). Prof. (tel. 0262 + 26-4101, ext. 394). *Applied mineral chemistry, crystal growth.*

Tatekawa, Prof. Masahisa. Faculty of Education, Inst. of Earth Sci., Shiga U., Ohtsu, Shiga 520, Japan. (1919) DSc, mineralogy (Kyoto U., 1960). Prof. (tel. 0775 + 37-0081, ext. 242). *Mineralogy, petrology, geochemistry.*

Tatsuka, Dr Kiyoaki. Analytical Section, Import Div., Osaka Custom House, 10-3 Chikko 4-chome Minato-ku, Osaka 552, Japan. (1929) DSc, mineralogy (Tohoku U., 1974). Supervisory analyst. (tel. 06 + 572-5321). *Experimental mineralogy.*

Tatsuzaki, Prof. Itaru. Ferroelectrics section, Res. Inst. of Applied Electricity, Hokkaido U., Kita 12-jo Nishi 6-chome Kita-ku, Sapporo 060, Japan. (1925) DSc, physics (Hokkaido U., 1960). Prof. (tel. 011 + 711-2111, ext. 2882, 3664). *Phase transition, ferroelectrics, dielectrics.*

Terauchi, Dr Hikaru. Faculty of Sci., Kwansei Gakuin U., 1-155 Uegahara 1-bancho Nishinomiya, Hyogo 662, Japan. (1942) DSc, physics (Kwansei Gakuin U., 1972). Asst. prof. (tel. 0798 + 51-2407). *X-ray diffuse scattering, ferroelectrics, liquid crystals, metals.*

Togawa, Dr Sen-ichi. Tokyo U. of Mercantile Marine, 1-6 Etchujima 2-chome Koto-ku, Tokyo 135, Japan. (1926) DSc, physics (Tokyo U., 1965). Prof. (tel. 03 + 641-1171). *Solid state physics.*

Tokonami, Prof. Masayasu. Mineralogical Inst., Faculty of Sci., U. of Tokyo, 7-3-1 Hongo Bunkyo-ku, Tokyo 113, Japan. (1933) DSc, mineralogy (Tokyo U., 1966). Prof. (tel. 03 + 812-2111, ext. 4541). *Inorganic crystal structures, crystal physics.*

Tokushita, Mr Motoyuki. Explosive Plant, Asahi Chemical Industry, 6-1 Mizushiri-cho Nobeoka, Miyazaki 882, Japan. (1936) BSc, physics (Hokkaido U., 1960). Asst. manager. (tel. 09823 + 33-6141). *Analytical chemistry, metallurgy.*

Tomeoka, Mr Kazushige. Mineralogical Inst., Faculty of Sci., U. of Tokyo, 3-1 Hongo 7-chome Bunkyo-ku, Tokyo 113, Japan. (1952) BSc, mineralogy (Tokyo U., 1975). Grad. student. (tel. 03 + 812-2111, ext. 2418). *Sulfide mineralogy, crystal structure.*

Tomimitsu, Mr Hiroshi. Solid State Physics - 1, Japan Atomic Energy Res. Inst., 2-4 Shirakata-Shirane Tokaimura, Nakagun Ibaraki 319-11, Japan. (1942) MSc, physics (Tokyo U., 1969). Res. staff. (tel. 02928 + 2-5466). *X-ray crystallography, neutron diffraction, electron diffraction.*

Tomisaka, Prof. Takeshi. Dept. of Mineralogical Sci. & Geology, Faculty of Sci., Yamaguchi U., 1677-1 Yoshida, Yamaguchi 753, Japan. (1921) DSc, mineralogy (Hokkaido U., 1957). Prof. (tel. 0839 + 22-6111, ext. 383). *Crystal physics, crystal growth.*

Tomita, Dr Katsutoshi. Dept. of Geology & Mineralogy, Faculty of Sci., Kyoto U., Oiwake-cho Kitashirakawa Sakyo-ku, Kyoto 606, Japan. (1934) DSc, mineralogy (Kyoto U., 1965). Res. assoc. (tel. 075 + 751-2111, ext. 4162). *Crystal structure, minerals, micro-texture in crystals, rock forming minerals.*

Tomita, Prof. Ken-ichi. Faculty of Pharmaceutical Sci. Osaka U., 1-6 Yamadakaoka Suita, Osaka 565, Japan. (1928) DSc, physical chemistry (Osaka U., 1959). Prof. (tel. 06 + 877-5111, ext. 6211). *Organic crystal structures, large molecules, biological substances.*

Tomita, Dr Takanori. Dept. of Physics, Faculty of Sci., Saitama U., 255 Shimo-ohkubo, Urawa 338, Japan. (1923) DSc, physics (Osaka U., 1961). Asst. prof. (tel. 0488 + 52-2111, ext. 463, 456). *Crystal growth (esp. anomalous structures), crystal defects, EPR sensitive defects, X-ray diffraction micrography, gamma ray and neutron irradiation effects.*

Tonomura, Dr Akira. Central Res. Lab., Hitachi Ltd., 280 Higashi-koigakubo 1-chrome, Kokubunji Tokyo 185, Japan. (1942) DEng, applied physics. (Nagoya U., 1973). Sr. res. (tel. 0423 + 23-1111, ext. 228) *Electron microscopy, electron diffraction.*

Toraya, Dr Hideo. Ceramic Eng. Res. Lab., Nagoya Inst. of Techn. 10-6-29, Asahigaoka Tajimi, Gifu 507, Japan. (1949) DSc, material science (Tokyo Inst. of Techn., 1980). Lect. (tel. 0572 + 27-6811). *X-ray diffraction analysis, crystal chemisty, inorganic materials.*

Toriumi, Mr Koshiro. Inst. for Solid State Physics, U. of Tokyo, 22-1 Roppongi 7-chome Minato-ku, Tokyo 106, Japan. (1949) MSc, chemistry (Tokyo U., 1974). Grad. student. (tel. 03 + 402-6231, ext. 663). *Accurate electron density determination, metal complexes.*

Toyoda, Dr Koichi. Res. Inst. of Electronics, Shizuoka U., 5-1 Johoku 3-chome, Hamamatsu 432, Japan. (1933) BSc, physics (Konan U., 1956). Asst. prof. (tel. 0534 + 71-1171, ext. 435). *Ferroelectric crystals, information storage and retrieval on solid state data.*

Tsuda, Mr Noritoshi. Dept. of Mathematics and Physics, Ritsumeikan U., 28-1 Tojiin-Kitamachi Kita-ku, Kyoto 603, Japan. (1940) MSc, physics (Ritsumeikan U., 1966). Res. assoc. (tel. 075 + 463-1131, ext. 326). *Inorganic crystal structures, phase transition.*

Tsuji, Mr Kazuhiko. Central Res. Lab., Matsushita Electric Industrial Co. Ltd., 3-15 Yakumo-Nakamachi Moriguchi, Osaka 570, Japan. (1947) MEng, physics (Nagoya U.,). Engineer. (tel. 06 + 909-1121, ext. 538). *Semiconductor devices and technology, crystal growth.*

Tsuji, Mr Koji. Res. Labs., Ashigara, Fuji Photo Film Co. Ltd., 210 Nakanuma Minamiashigara, Kanagawa 250-01, Japan. (1944) MSc, chemistry (Kwansei Gakuin U., 1969). Sr. res. (tel. 0465 + 74-1111, ext. 347).

Tsukihara, Dr Tomitake. Faculty of Eng., Tottori U., 1-1 Koyama, Tottori 680, Japan. (1944) DSc, biochemistry (Osaka U., 1974). Lect. (tel. 0857 + 28-0321, ext. 468). *Protein crystallography.*

Tsunekawa, Dr Shin. Inst. for Molecular Sci., Myodaiji Okazaki, Aichi 444, Japan. (1943) DSc, chemistry (Nagoya U., 1979). Res. asst. (tel. 0564 + 54-1111, ext. 436). *Ferroelastics, superplasticity, superconductivity, toughening materials, ultra fine particles.*

Uechi, Mr Tetsuo. General Education Dept., Kyushu U., 2-1 Ropponmatsu 4-chome Chuo-ku, Fukuoka 810, Japan. (1936) Res. assoc. (tel. 092 + 771-4161). *Crystal structure analysis.*

Ueda, Prof. Ikuhiko. C. of General Education, Kyushu U., 2-1 Ropponmatsu 4-chome Chuo-ku, Fukuoka 810, Japan. (1921) DSc, physics (Kyushu U., 1963). Prof. (tel. 092 + 771-4161, ext. 277). *Structural analysis of complex molecules, heterocyclic and chelate compounds.*

Uefuji, Mr Tateki. Electronics Dept., Faculty of Sci. and Eng., Saga U., 1 Honjomachi, Saga 840, Japan. (1945) MSc, physics (1970). Res. assoc. (tel. 09522 + 4-5191). *Electron diffraction, lattice defects, metals.*

Ueki, Dr Tatzuo. Dept. of Biophysical Eng., Faculty of Eng. Sci., Osaka U., 1 Machikaneyama-cho 1-chome Toyonaka, Osaka 560, Japan. (1940) DSc, physical chemistry (Osaka U., 1968). Asst. prof. (tel. 06 + 844-1151, ext. 4766). *Structures, biological membranes (supra-molecular systems), protein molecules.*

Ueno, Mr Tsunehisa. Semiconductor Div., Tokyo Shibaura Electric Co. Ltd., 1 Komukai Toshiba-cho Saiwai-ku, Kawasaki Kanagawa 210, Japan. (1948) MEng, mineralogy (Tohoku U., 1973). (tel. 044 + 511-3111, ext. 691, 473). *Crystallography, applied physics, mineralogy, spectroscopy, electronics, photographic engineering.*

Uesu, Prof. Yoshiaki. Dept. of Physics, Waseda U., 170 Nishiokubo 4-chome Shinjuku-ku, Tokyo 160, Japan. (1942) DSc, physics (Waseda U., 1971). Asst. prof. (tel. 03 + 209-3211, ext. 352). *Solid state physics, ferroelectrics, biophysics.*

Ukaji, Prof. Takeshi. Dept. of Process Chemical Eng., Ikutoku Techn. U., 1030 Shimo-ogino Atsugi, Kanagawa 243-02, Japan. (1922) DSc, chemistry (Tokyo U. of Education, 1958). Prof. (tel. 0462 + 41-1211, ext. 254). *Molecular structure determination, gas electron diffraction.*

Umegaki, Prof. Yoshiharu. Faculty of Sci., Okayama C. of Sci., 1 Ridai-cho 1-chome, Okayama 700, Japan. (1909) DSc, mineralogy (Kyoto U., 1952). Prof.; Em. prof. (Hiroshima U.). (tel. 0862 + 52-3161, ext. 345). *Force field in crystal structure.*

Umeno, Dr Masataka. Dept. of Precision Eng., Faculty of Eng., Osaka U., 133-1 Yamadakami Suita, Osaka 565, Japan. (1939) DEng, applied physics (Osaka U., 1967). Asst. prof. (tel. 06 + 877-5111, ext. 4602). *X-ray diffraction, crystal defects.*

Uno, Prof. Ryosei. Dept. of Physics, C. of Humanities and Sci. Nihon U., 25-40 Sakurajosui 3-chome Setagaya-ku, Tokyo 156, Japan. (1924) DSc, physics (Kyushu U., 1947). Prof. (tel. 03 + 329-1151, ext. 283). *X-ray diffraction, semiconductors, solid state physics.*

Uragami, Dr Takuyuki. General Education, Okayama U. of Sci. 1-1 Ridai-cho, Okayama 700, Japan. (1938) DSc, physics (Tokyo U., 1971). Assoc. prof. (tel. 0862 + 52-3161, ext. 344). *Structure analysis, dynamical theory.*

Uyeda, Prof. Natsu. Inst. for Chemical Res., Kyoto U., Gokasho Uji 611, Japan. (1924) DSc, chemistry (Kyoto U., 1958). Prof. (tel. 0774 + 32-3111, ext. 350). *Epitaxial growth, thin crystalline films, organic semiconductors, high resolution electron microscopy in atomic order.*

Uyeda, Prof. Ryozi. Dept. of Physics, Meijo U., Tempaku, Nagoya 468, Japan. (1911) DSc, physics (Tokyo U., 1944). Prof. (tel. 052 + 832-1151, ext. 570). *Electron diffraction, electron microscopy.*

Wada, Prof. Eiichi. Sci. and Techn., Nihon U., 1-8-14 Surugadai Chiyoda-ku, Tokyo 101, Japan. (1914) DEng, physics (Tokyo U., 1955). Prof. (tel. 03 + 293-3201, ext. 842). *Polymer physical chemistry, small angle X-ray scattering.*

Wada, Dr Takeo. Res. Labs., Chemical Products Div., Takeda Chemical Industries Ltd., 17-85 Juso-honmachi 2-chome, Yodogawa-ku, Osaka 532, Japan. (1934) DSc, inorganic and physical chemistry (Osaka U., 1963). Chief res. (tel. 06 + 871-4432). *Clay minerals.*

Wakabayashi, Dr Katsuzo. Dept. of Biophysical Eng., Faculty of Eng. Sci., Osaka U., 1 Machikaneyama-cho 1-chome Toyonaka, Osaka 560, Japan. (1943) DSc, biophysics (Hokkaido U., 1971). Res. assoc. (tel. 06 + 856-1151, ext. 3215). *X-ray structure analysis, X-ray and neutron small angle scattering, biological materials.*

Watanabe, Dr Akiteru. Fourth Group, Nat. Inst. for Res. in Inorg. Materials, 1-1 Namiki, Sakura-mura Niihari-gun, Ibaraki 305, Japan. (1945) DEng, crystal chemistry (Tokyo Inst. Techn., 1983). Sr. res. (tel. 0298 + 51-3351, ext. 231). *X-ray crystallography, solid state chemistry, mixed bismuth oxide.*

Watanabe, Prof. Denjiro. Dept. of Physics, Faculty of Sci., Tohoku U., Aramaki Aza-Aoba, Sendai 980, Japan. (1926) DSc, physics (Tohoku U., 1960). Prof. (tel. 0222 + 22-1800, ext. 3295). *Metal and alloy structures, electron diffraction, electron microscopy.*

Watanabe, Prof. Hiroshi. Res. Inst. for Iron, Steel and Other Metals, Tohoku U., 1-1 Katahira 2-chome, Sendai 980, Japan. (1916) DSc, physics (Tokyo U.), 1960). Prof. (tel. 0222 + 27-6200, ext. 2901). *Magnetism, neutron diffraction.*

Watanabe, Dr Masaru. Electron Optics Div., JEOL Ltd., 1418 Nakagami Akishima, Tokyo 196, Japan. (1922) DEng, physics (Tokyo Inst. of Techn., 1962). Director, General Manager of Electron Optics Div. (tel. 0425 + 43-1111, ext. 417). *Instrumentation, electron optics, electron diffraction.*

Watanabe, Dr Takashi. Dept. of Geology and Mineralogy, Geoscience, Joetsu U. of Education, Yamayashiki Joetsu, Niigata 943, Japan. (1940) DSc, mineralogy (Tokyo U. of Education, 1970). Assoc. prof. (tel. 0255 + 22-2411, ext. 442). *Clay mineralogy.*

Watanabe, Prof. Takeo. Geological Inst., Faculty of Sci., U. of Tokyo, 3-1 Hongo 7-chome Bunkyo-ku, Tokyo 113, Japan. (1907) DSc, geology (Tokyo U., 1943). Em. prof. (Tokyo U.); Member, Japan Academy. (tel. 03 + 812-2111, ext. 2427). *Paragenesis of minerals, mineral deposits, boron and manganese minerals.*

Watanabe, Prof. Tokunosuke. P. A., 2-22-7 Shinsenri-kitamachi, Toyonaka 565, Japan. (1904) DSc, chemistry (Osaka U., 1939). Em. prof. (Osaka U.). (tel. 06 + 872-3790). *Phase transition.*

Watanabe, Mr Yasunari. Crystal Physics Lab., Inst. of Physical and Chemical Res., 1 Hirosawa 2-chome Wako, Saitama 351, Japan. (1936) BSc, physics (Science U. of Tokyo, 1963). Scientist. (tel. 0484 + 62-1111, ext. 3342). *X-ray crystallography.*

Watanabe, Prof. Yasuyoshi. Dept. of Physics, Faculty of Sci., Chiba U., 33 Yayoicho 1-chome, Chiba 280, Japan. (1920) DSc, physics (Tokyo U. of Education, 1956). Prof. (tel. 0472 + 51-1111, ext. 2632). *Electron diffraction, crystal growth, LEED.*

Watase, Dr Hideo. R & D Dept., Teijin Chemicals Ltd., 6-21 Nishi-Shinbashi 1-chome Minato-ku, Tokyo 105, Japan. (1928) DSc, chemistry (Osaka U., 1959). Manager. (tel. 03 + 506-4721). *High polymers.*

Yagi, Prof. Katsumichi. Physics Dept., Tokyo Inst. of Techn., Oh-okayama Meguro-ku, Tokyo 152, Japan. (1939) DSc, physics (Tokyo Inst. of Techn., 1967). Prof. (tel. 03 + 726-1111, ext. 2078). *Diffraction crystallography, surface physics, thin film physics, electron microscopy, electron microscopy, phase transition.*

Yakushi, Mr Kyuya. Dept. of Chemistry, Faculty of Sci., U. of Tokyo, 3-1 Hongo 7-chome Bunkyo-ku, Tokyo 113, Japan. (1945) MSc, chemistry (Tokyo U., 1970). Res. assoc. (tel. 03 + 812-2111). *Structure and physical properties, molecular crystals.*

Yamaguti, Prof. Tasaburo. B-208 353-1 Eda Midori-ku, Yokohama Kanagawa 227, Japan. (1904) DSc, physics (Tokyo U., 1938). Em. prof. (Chiba U.). (tel. 045 + 911-8596). *Crystal growth, diffusion, surfaces, films, deposition.*

Yamamoto, Mr Akiji. Nat. Inst. for Res. in Inorganic Materials, Namiki Sakura-mura Niihari-gun, Ibaraki 305, Japan. (1945) MSc (Tokyo Inst. of Techn., 1971). Sr. res. (tel. 0298 + 51-3351, ext. 371). *Modulated-structure analysis.*

Yamamoto, Dr Naoki. Dept. of Physics, Tokyo Inst. of Techn., Oh-okayama Meguro-ku, Tokyo 152, Japan. (1950) DSc, physics (Tokyo Inst. of Techn., 1979). Res. asst. (tel. 03 + 726-1111, ext. 2079). *Electron microscopy, ferroelectricity, phase transition.*

Yamanaka, Dr Takamitsu. Mineralogical Inst., Faculty of Sci., U. of Tokyo, 3-1 Hongo 7-chome Bunkyo-ku, Tokyo 113, Japan. (1942) DSc, mineralogy (Tokyo U., 1972). Res. assoc. (tel. 03 + 812-2111, ext. 4547). *Structural crystal chemistry.*

Yamane, Dr Takashi. Dept. of Applied Chemistry, Faculty of Eng., Nagoya U., Furo-cho Chikusa-ku, Nagoya 464, Japan. (1946) DSc, chemistry (Osaka U., 1975). Res. assoc. (tel. 052 + 781-5111, ext. 3342). *Structure, biological macromolecules.*

Yamashita, Mr Shuji. Physics Div., Tokyo Metropolitan Isotope Res. Centre, 11-1 Fukazawa 2-chome Setagaya-ku, Tokyo 158, Japan. (tel. 03 + 702-3111). *Crystal physics.*

Yamashita, Prof. Tadayoshi. Dept. of Physics, National Defense Academy, 10-20 Hashirimizu 1-chome, Yokosuka 239, Japan. (1919) DSc, physics (Hiroshima U., 1955). Prof. (tel. 0468 + 41-3810, ext. 352). *Metal physics.*

Yashiro, Prof. Yuzo. Dept. of Eng. Sci. Nagoya Inst. of Techn., Gokiso-cho Showa-ku, Nagoya 466, Japan. (1915) DSc, chemistry (Kyoto U., 1959). Prof. (tel. 052 + 732-2111, ext. 561). *Field ion microscopy, field emission microscopy.*

Yasuoka, Dr Noritake. Inst. for Protein Res., Osaka U., Yamadakami Suita, Osaka 565, Japan. (1936) DSc, chemistry (Osaka U., 1968). Asst. prof. (tel. 06 + 877-5111, ext. 3837). *Organic crystal structures, proteins, biological substances.*

Yoda, Dr Osamu. Takasaki Res. Est., Japan Atomic Energy Res. Inst., 1233 Watanuki-machi Takasaki, Gunma 370-12, Japan. (1944) DEng, chemistry (Hokkaido U., 1979). Res. scient. (tel. 0273 + 46-1211, ext. 382). *X-ray and neutron diffraction, polymers.*

Yokomori, Dr Yoshinobu. Dept. of Chemistry, Nat. Defense Academy, 1-10-20 Hashirimizu, Yokosuka, Kanagawa 239, Japan. (1950) DEng, chemistry (Tokyo U., 1978). Res. assoc. (tel. 0468 + 41-3810, ext. 2413). *X-ray crystallography, oligopeptides.*

Yonei, Dr Moto'o. Central Res. Lab., Chugai Pharmaceutical Co. Takada 3-chome, Toshima-ku, Tokyo 171, Japan. (1950) DPharm, chemistry (Tokyo U., 1979). Res. (tel. 03 + 987-7111, ext. 564). *Organic crystal structures, biological substances, pharmaceutical substance*

Yoshida, Dr Kentaro. Faculty of Eng., Kobe U., 1 Rokkodai-machi Nada-ku, Kobe 657, Japan. (1935) DSc, physics (Kyoto U., 1973). Asst. prof. (tel. 078 + 881-1212). *Electron microscopy, electron diffraction, metastable structures, thin alloy films, microstructure in rocks and minerals.*

Yoshida, Prof. Sanae. Dept. of Chemistry, Faculty of Sci., Gakushuin U., 5-1 Mejiro 1-chome Toshima-ku, Tokyo 171, Japan. (1906) DSc, chemistry (Tokyo U., 1941). Prof. (tel. 03 + 986-0221, ext. 478). *Physical chemistry.*

Yoshimatsu, Prof. Mitsuru. Applied Physics, Faculty of Sci., Fukuoka U. 8-19-1 Nanakuma, Johnan-ku, Fukuoka 814-01, Japan. (1926) DSc, physics (Kyushu U., 1962). Prof. (tel. 092 + 871-6631, ext. 6159). *X-ray crystallography, crystal growth.*

Yoshimura, Dr Junichi. Inst. of Inorganic Synthesis, Yamanashi U., 3-11 Takeda 4-chome Kofu, Yamanashi 400, Japan. (1943) DEng, applied physics (Tokyo U., 1975). Lect. (tel. 0552 + 52-1111, ext. 571). *Crystal physics, crystal chemistry, crystal growth, crystal defects, diffraction theory.*

Yoshimura, Mr Yukio. Dept. of Mathematics and Physics, Ritsumeikan U., 28-1 Tojiin-kitamachi Kita-ku, Kyoto 603, Japan. (1941) BSc, physics (Ritsumeikan U., 1964). Res. assoc. (tel. 075 + 463-1131, ext. 326). *Crystal physics, inorganic crystal structures, phase transition.*

Yoshioka, Prof. Hide. Dept. of Physics, Nagoya U., Furo-cho Chikusa-ku, Nagoya 464, Japan. (1922) DSc, physics (Nagoya U., 1958). Prof. (tel. 052 + 781-5111). *Low temperature physics.*

KOREA

Sub-Editor: W. Shin

Notes

1. International telephone country code - 82.

Ahn, Prof. Dr Choong Tai. Chemistry Dept., Hankuk U. of Foreign Studies, Yongin-gun, Kyunggi-do 170-41, Korea. (1931) PhD, crystallography (Seoul Nat. U., 1979). Prof. (tel. Seoul 389-8632). *X-ray diffraction, organic compounds.*

Cho, Prof. Sung-Il. Chemistry Dept., City C. of Seoul, Jeonnong-dong, Dongdaemun-gu, Seoul 131, Korea. (1947) PhD, physical chemistry (Seoul Nat. U., 1981). Asst. prof. (tel. 245-8111). *X-ray diffraction, biological compounds (small and medium size).*

Chung, Prof. Dr Su Jin. Inorganic Materials Eng. Dept., Seoul Nat. U., Sinrim-dong, Gwanag-gu, Seoul 151, Korea. (1938) Dr.rer.nat., crystallography (Tech. Hochschule Aachen, BRD, 1972). Assoc. prof. (tel. 877-0101, Ext. 3518). *Inorganic crystal chemistry, X-ray diffraction, electron diffraction, symmetry.*

Kim, Dr Hoon Sup. P.O. Box 35, Daejun, Chungnam, Korea. (1935) PhD, crystallography (Pittsburgh U., USA, 1968). *X-ray diffraction, biological compounds (small and medium size).*

Kim, Dr Key Soo. Electro-Optics Lab., Korea Atomic Energy Res. Inst., 170-2 Gongneung-dong, Dobong-gu, Seoul, Korea. (1929) DSc, solid state physics (Seoul Nat. U., 1972). Head. (tel. 96 5083). *X-ray crystallography, electron microscopy, crystal growth, radiation damage, nuclear materials.*

Kim, Prof. Moon-Jib. Physics Dept., C. of Sci., Soon Chun Hyang U., Onyang, Chungnam 331, Korea. (1954) MSc, physics (Chungnam Nat. U., 1981). Prof. (tel. 0418 + 2-4751). *Solid state physics.*

Kim, Prof. Dr Soo Jin. Geological Sciences Dept., Seoul Nat. U., Sinrim-dong, Gwanag-gu, Seoul 151, Korea. (1939) PhD, mineralogy (Seoul Nat. U., 1971), Dr.rer.nat. (U. Heidelberg, BRD, 1979). Assoc. prof. (tel. 877-0101, Ext. 2604). *X-ray diffraction, crystal chemistry, minerals, crystal optics.*

Kim, Prof. Dr Yang. Chemistry Dept., Pusan Nat. U., Dongrae-gu, Pusan 607, Korea. (1940) PhD, chemistry (Hawaii U., USA, 1979). Assoc. prof. (tel. 051 + 56-0171). *Crystal structure analysis, zeolites.*

Kim, Prof. Dr Yang Bae. Manufacturing Pharmacy Dept., Seoul Nat. U., Sinrim-dong, Gwanag-gu, Seoul 151, Korea. (1940) PhD, pharmacy (Seoul Nat. U., 1974). Assoc. prof. (tel. 877-0101, Ext. 2884). *Crystal structure analysis, biological compounds.*

Koo, Prof. Dr Chung Hoe. Chemistry Dept., Seoul Nat. U., Sinrim-dong, Gwanag-gu, Seoul 151, Korea. (1922) DSc, crystallography (Seoul Nat. U., 1960). Prof. (tel. 877-0101, Ext. 2595). *Crystal structure analysis, organic compounds, X-ray diffraction.*

Namgung, Prof. Dr Hae. Chemistry Dept., Kukmin U., Sungbuk-gu, Seoul 132, Korea. (1942) Dr.rer.nat. (U. Bonn, BRD, 1978) Prof. *Crystal structure, crystal chemistry, crystal physics.*

Park, Prof. Young Ja. Chemistry Dept., Sook-Myung Women's U., Yongsan-ku, Seoul 140, Korea. (1942) PhD, crystallography (Pittsburgh U., USA, 1970). Prof. (tel. 713 9391, ext. 301). *X-ray diffraction, molecular mechanics, biological compounds (small and medium size).*

Shin, Prof. Hyun Sum. Chemistry Dept., Dong-Guk U., Pil-dong, Chung-gu, Seoul 110, Korea. (1937) PhD, crystallography (Seoul Nat. U., 1979). Assoc. prof. (tel. 261 8131). *X-ray diffraction, organic compounds.*

Shin, Prof. Dr Whanchul. Chemistry Dept., Seoul Nat. U., Sinrim-dong, Gwanag-gu, Seoul 151, Korea. (1950) PhD, crystallography (Pittsburgh U., USA, 1978). Assoc. prof. (tel. 877-0101, Ext. 3318) *X-ray crystallography, biologically active small molecules, proteins, crystallographic computing.*

Suh, Prof. Dr Ill-Hwan. Physics Dept., Chungham Nat. U., Daejun, Chungnam 300-01, Korea. (1936) PhD, crystallography (Korea U., 1976). Prof. (tel. 042 + 45-0101). *Crystal structure determination, crystal physics.*

Suh, Prof. Jung Sun. Chemistry Dept., Myong-Ji U., Namgajwa-dong, Seodaemun-gu, Seoul, Korea. (1942) PhD, physical chemistry (Seoul Nat. U., 1979). Assoc. prof. (tel. 602-1536). *X-ray diffraction, pharmacological compounds.*

Suh, Prof. Se Won. Chemistry Dept., Seoul Nat. U., Sinrim-dong, Gwanag-gu, Seoul 151, Korea. (1951) PhD, chemistry (UCLA, USA, 1980). Asst. prof. (tel. 877-0101, ext. 3249). *Protein structure and function.*

LIBYA

Sub-Editor: **M.M. Almajdub**

Notes

1. International telephone country code - 218.

Ajaal, Mr Tawfik Taher. Material Sci. Dept., Tajura Nuclear Res. Center, P.O. Box 30878, Tajura - Tripoli, Libya. (1958) BSc, material science (Alfateh U., 1983). Eng. (telex 13615) *Neutron diffractometry, solid state research, corrosion.*

Almajdub, Mr Musbah Meftah. Solid State Physics Lab., Tajura Res. Center, T.N.R.C, P.O. Box 397, Tripoli, Libya. (1954) BSc, physics (Tennessee Techn. U., USA, 1980) Res. (telex 20792 TAJ RC LY) *Neutron diffractometry, crystal structure, crystal dynamics, inelastic neutron scattering, phonon resonance, alloys.*

Elzawi, Mr Rajab Abdulla. Dept. of Physics, Solid State Lab., Tajura Res. Center, T.N.R.C, P.O. Box 30878, Tajura-Tripoli, Libya. (1954) BSc, physics (U. Oklahoma, USA, 1980) Res. asst. (tel. 607052, telex 20792 TAJ RC LY) *Neutron diffraction, magnetic structure analysis, nuclear structure, Mössbauer spectroscopy.*

Shihub, Mr Salahedin I. Solid State Physics Lab., Tajura Nuclear Res. Center, P.O. Box 397, Tripoli, Libya. (1954) BSc, physics (U. Washington, USA, 1980) Res. asst. (telex 20792 TAJ RC LY) *Neutron spectrometry, crystal dynamics.*

MALAYSIA

Sub-Editor: **Chatar Singh**

Notes

1. International telephone country code - 60

2. The science degrees conferred by Malaysian universities are the PhD, MSc, and BSc and correspond to British degrees.

3. The following abbreviations have been used in this section:
   U.M. - *Universiti Malaya* (University of Malaya).
   U.S.M. - *Universiti Sains Malaysia* (University of Science of Malaysia).
   U.K.M. - *Universiti Kebangsaan Malaysia* (National University of Malaysia).
   U.P.M. - *Universiti Pertanian Malaysia* (University of Agriculture of Malaysia).
   U.T.M. - *Universiti Teknologi Malaysia* (University of Technology of Malaysia).
   M.I.T. - Mara Institute of Technology.
   M.A.R.D.I. - Malaysian Agricultural Research and Development Institute.
   S.I.R.I.M. - Standards and Industrial Research Institute of Malaysia.
   P.O.R.I.M. - Palm Oil Research Institute of Malaysia.

Abdul Aziz, Mr Abdul Halim bin. Sch. of Physics, U.S.M., Penang, Malaysia. (1958) MSc, X-ray crystallography (U. London, UK, 1983). Lect. (tel. 04 + 883822, ext. 654). *X-ray crystal structure determination.*

Abu Bakar, Mr Ismail. Central Res. Labs. Div., M.A.R.D.I., P.O.Box 12301, Kuala Lumpur 01-02, Malaysia. (1952) MSc, mineralogy (Ghent U, Belgium, 1981). Res. officer. (tel. 03 + 356601, ext. 478). *Malaysian soils, mineralogy, soil survey, micromorphology.*

Almashoor, Dr Syed Sheikh. Dept. of Geology, U.K.M., Bangi, Selangor, Malaysia. (1944) PhD, igneous petrology (Penn. State U., U.S.A., 1983). Lect. (tel. 03 + 350011, ext. 2390). *Geochemistry, mineralogy, petrography.*

Ang, Dr Ha Ming. Chemical Eng. Dept., U.M., Pantai Valley, Kuala Lumpur 22-11, Malaysia. (1946) PhD, bulk crystallization (U. London, UK, 1973). Assoc. prof. (tel. 03 + 553466, ext. 293). *Crystal characteristics, bulk crystallization.*

Baba, Mrs Jasmin. Mechanical Eng. Faculty, U.T.M., Gurney Road, Kuala Lumpur 15-01, Malaysia. (1954) MSc, material science (Cranfield Inst. of Techn., UK, 1981). Lect. (tel. 03 + 929033, ext. 533). *Electron microscopy, X-ray diffraction techniques.*

Cheang Dr Kok Keong. Mineral Resources Eng., Sch. of Eng. Sci. and Industrial Techn., U.S.M., Penang, Malaysia. (1949) PhD, geology, geochemistry (U. Georgia, USA, 1982). Lect. (tel. 04 + 883822, ext. 614). *X-ray crystallography, mineralogy, igneous petrology, exploration geochemistry.*

Chen, Dr Wei. Chemistry Dept., U.M., Pantai Valley, Kuala Lumpur 22-11, Malaysia. (1948) PhD, X-ray crystallography (U. New South Wales, Australia, 1976). Lect. (tel. 03 + 555466). *Single crystal diffractometry, crystal structure, inorganic and organic compounds.*

Faqir, Dr Gul. Sch. of Eng., M.I.T., Shah Alam, Selangor, Malaysia. (1951) PhD, metallurgical eng. (U. Liverpool, UK, 1982). Lect. (tel. 03 + 362311). *Mechanical metallurgy, materials.*

Fun, Dr Hoong Kun. Sch. of Physics, U.S.M., Penang, Malaysia. (1946) PhD, solid state physics (Purdue U., USA, 1974). Lect. (tel. 04 + 883822, ext. 652). *Electron spin resonance, ENDOR, radiation damage, semiconductors, insulators, X-ray crystallography.*

Gan, Mr Ah Sai. Mineralogy and Petrology Div., Geological Survey of Malaysia, Scrivenor Road, Ipoh, Perak, Malaysia. (1945) BSc(Hons), applied geology (U.M., 1970). Sr. geologist. (tel. 05 + 557644, ext. 116). *Mineralogy, ore microscopy, powder diffraction.*

Hassan, Dr Wan Fuad bin Wan. Geology Dept., U.K.M., Bangi, Selangor, Malaysia. (1948) PhD, mineralogy and geochemistry (U. Leeds, UK, 1982). Lect. (tel. 03 + 350001, ext. 2657). *Geochemistry, mineralogy, ore minerals, granites.*

Hutchison, Prof. Dr Charles Strachan. Geology Dept., U.M., Pantai Valley, Kuala Lumpur 22-11, Malaysia. (1933) PhD, petrology and mineralogy (U.M., 1966). Prof. (tel. 03 + 555466, ext. 203). *Mineral chemistry, petrology.*

Lee, Dr Chnoong Kheng, Chemistry Dept., U.P.M., Serdang, Selangor, Malaysia. (1948) PhD, inorganic chemistry (U. Aberdeen, UK, 1972). Lect. (tel. 03 + 355425, ext. 445). *Mineral chemistry, X-ray crystallography.*

Mohamad, Dr Hamzah. Geology Dept., U.K.M., Bangi, Selangor, Malaysia. (1951) PhD, geochemistry, petrology (U. Strathclyde, Glasgow, UK, 1980). Lect. (tel. 03 + 350001, ext. 2664). *Mineral chemistry, petrology.*

Ng, Dr Wee Lam. Chemistry Dept., U.M., Pantai Valley, Kuala Lumpur 22-11, Malaysia. (1943) PhD, solid state chemistry (U. Western Ontario, Canada, 1971). Assoc. prof. (tel. 03 + 555466, ext. 249). *Electrical conduction, thermal decomposition, crystallization, structural defects, crystal structures.*

Oh, Mr Chuan Ho, Flingoh. Chemistry and Techn. Div., P.O.R.I.M., 6 Persiaran Institusi, Bandar Baru Bangi, Selangor, Malaysia. (1947) MSc, crystallization (U.M., 1980). Sr. res. officer. (tel. 03 + 335775, ext. 1086, 1084). *Crystallization of oil and fat, powder diffraction.*

Othman, Dr Abdul Hamid bin. Chemistry Dept., U.K.M., Bangi, Selangor, Malaysia. (1948) PhD, X-ray crystallography (U. Reading, UK, 1977). Head. (tel. 03 + 350001, ext. 2420). *Coordination compounds, structures.*

Othman, Dr Radzali. Materials Eng. Div., Sch. of Eng. Sci. and Industrial Techn., U.S.M., Penang, Malaysia. (1954) PhD, ceramics (U. Sheffield, UK, 1982). Lect. (tel. 04 + 883822, ext. 613). *Structural ceramics, clay mineralogy.*

Ong, Mr Yeoh Han. Minerals Clearance Project, Geological Survey Dept., Scrivenor Road, Ipoh, Perak, Malaysia. (1948) BSc, geology (U.M., 1972). Geologist. (tel. 05 + 557644, ext.7). *Mineralogy, gemology, X-ray methods.*

Rao, Mr Nutakki Nageswara. Materials Eng. Div., Sch. of Eng. Sci. and Industrial Techn., U.S.M., Penang, Malaysia. (1949) MTechn., metallurgy (I.I.T., Madras, India, 1973). Lect. (tel. 04 + 883822, ext. 613). *Powder metallurgy, fracture analysis, welding, corrosion.*

Malaysia

Salleh, Dr Mansor bin Haji. S.I.R.I.M., P.O.Box 35, Shah Alam, Selangor, Malaysia. (1944) PhD, corrosion (U. Manchester, UK, 1978). Controller. (tel. 03 + 592635). *Metallurgy, materials, corrosion.*

Salleh, Dr Mohammad Nawi. Sch. of Eng., M.I.T., Shah Alam, Selangor, Malaysia. (1946) PhD, metallurgical eng. (Colorado Sch. of Mines, USA, 1979). Sr. lect. (tel. 03 + 362464). *Mechanical metallurgy, metal forming.*

Silong, Dr Sidik bin. Chemistry Dept., U.P.M., Serdang, Selangor, Malaysia. (1953) PhD, inorganic chemistry (U. Reading, UK, 1982). Lect. (tel. 03 + 356101, ext. 546). *Structural determination, macrocyclic compounds.*

Singh, Dato' Prof. Dr Chatar. Sch. of Physics, U.S.M., Penang, Malaysia. (1929) PhD, X-ray crystallography and physics (Cambridge U., UK, 1961). Prof. (tel. 04 + 883822, ext. 659). *Organic crystal structure analysis, crystallography.*

Teh, Dr Guan Hoe. Geology Dept., U.M., Pantai Valley, Kuala Lumpur 22-11, Malaysia. (1946) Dr.rer.nat., mineralogy, geomicrobiology, experimental petrology (U. Heidelberg, BRD, 1979). Lect. (tel. 03 + 560022). *Economic geology, mineralogy, experimental petrology, analytical methods.*

Teh, Dr Ser Kok. Mechanical Eng. Dept., U.M., Pantai Valley, Kuala Lumpur 22-11, Malaysia. (1947) PhD, materials science (U. London, UK, 1976). Lect. (tel. 03 + 553466, ext. 265). *Crystallography, Aluminum oxide, creep and fracture, metallurgy.*

Teoh, Mr Lay Hock. Geological Survey of Malaysia, Scrivenor Road, Ipoh, Perak, Malaysia. (1950) BSc(Hons), geology and geochemistry (U. Victoria, New Zealand, 1972). Geologist. (tel. 05 + 557644). *Mineral identification, petrology.*

Tuan Sarif, Mr Tuan Besar. Mineral Resources Eng. Div., Sch. of Eng. Sci. and Industrial Techn., U.S.M., Penang, Malaysia. (1957) MSc, geochemistry, (Iowa State U., U.S.A., 1983). Lect. (tel. 04 + 883822, ext. 614). *Mineralogy, crystallography, geochemistry, mining, geotechnical engineering.*

Yahaya, Dr Muhamad. Physics Dept., U.K.M., Bangi, Selangor, Malaysia. (1947) PhD, electronic properties (Monash U., Australia, 1979). Head. (tel. 03 + 350001, ext. 2900). *Transition metals electronic structure and properties.*

Yong, Mr Swee Kee. Technical Support Services, Geological Survey Dept., Scrivenor Road, Ipoh, Perak, Malaysia. (1942) BSc, geology (U. Adelaide, Australia, 1964). Geologist. (tel. 05 + 557644, ext. 7). *Mineralogy, mineragraphy, petrography, economic geology.*

MEXICO

Sub-Editor: L.E. Rendón Diaz Mirón

Notes

1. International telephone country code - 52.

2. Degrees conferred by the Mexican Universities are the *Doctor en Ciencias* (Dr.C.)(equivalent to PhD. *Maestro en Ciencias* (M.C.)(equivalent to MSc), and *Ingeniero* (Ing.) or *Licenciado en Ciencias* (Lic.), both equivalent to BSc.

3. At a Mexican University, persons can be appointed in the pollowing positions not belonging to the regular formation of personnel: *Profesor* (Titular, Asociado, Ayudante), *(investigador)* (Titular, Asociado, Ayudante), and *Tecnico Academico* (Titular, Asociado, Ayudante).

Aguilera Herrera, Prof. Nicolás. Depto. de Biologia, Facultad de Ciencias, U. Nac. Aut. de México, Circuito Exterior, Ciudad Universitaria, Mexico 04510 D.F. (1920) MSc, soil science (U. Wisconsin, USA, 1953). Prof. (tel.550-5913). *Clay, soil, mineralogy, edaphology.*

Bosch Giral, Dr Pedro. Instituto Mexicano del Petróleo, Ap. Post. 14-805, Ave. 100 Metros 152, Mexico 14 D.F. (1948) DSc, crystal chemistry and catalysis (U. Claude Bernard, France, 1946). Investigador cientifico. (tel. 5-67-66-00, ext. 2377) *Catalysis, X-ray diffraction.*

Cabrera Bravo, Prof. Enrique. Depto. Estado Solido, Inst. de Fisica, U. Nac. Aut. de México, Ap. Post. 20-364, Ciudad Universitaria, Mexico 04510 D.F. (1946) MSc, theoretical physics (Bieloruss. St. U., USSR, 1970). Res. worker. (tel. 5-48-81-92). *Electron microscopy, crystallography with electron microscopes, diffusion in solids.*

Castellanos Guzman, Dr A. Guillermo. Centro de Investigación en Ciencias Básicas, U. de Colima, Ap. Post. 2-1694, Colima 28000, Mexico. (1939) PhD, physics (London U., UK, 1981). Scient. (tel. 331 + 2-58-18, telex. UCOLMEX 62248) *Crystal growth, dialectric and optical properties, polar crystals, X-ray structure determination.*

Castellanos Román, Mrs Maria Asunción. Div. Estudios de Postgrado, Fac. de Quimica, U. Nac. Aut. de México, Ciudad Universitaria, Mexico 04510 D.F. (1943) MSc, (Aberdeen U., UK, 1979). Jefe lab. Rayos X (tel. 5-48-82-10). *Crystal chemistry, oxide complexes.*

Chapela Castañares, Dr Víctor Manuel. Inst. de Investigaciones de Materiales, U. Nac. Aut. de México, Ap. Post. 70-360, Ciudad Universitaria, Mexico 04510 D.F. (1944) PhD, (Imperial C. London U., UK, 1979). Jefe depto. bajas temperaturas. (tel. 550-5215, ext. 4735). *Intercalation of aminoacids in dichalcogenides.*

Cano Corona, Dr Octavio. Inst. de Fisica, U. Nac. Aut. de México, Ap. Post. 70-364, Ciudad Universitaria, Mexico 04510 D.F. (1921) PhD, physics (Pennsylvania State U., USA, 1954). Acad. techn. (tel. 550-5215, ext. 5940). *20-ray powder diffraction, crystal studies.*

Cordero, Dr Adolfo. Inst. de Fisica, U. Nac. Aut. de México, Ap. Post. 70-364, Ciudad Universitaria, Mexico 04510 D.F. (1950) Dr (U. Nac. Aut. de México) Res. (tel. 550-5215, ext. 5940). *Crystal growth, crystal properties.*

Cota Araiza, Mr Leonel Susano. Depto. Estado Solido, Inst. de Fisica, U. Nac. Aut. de México, Ap. Post. 20-364, Ciudad Universitaria, Mexico 04510 D.F. (1944) MPh, surface physics (Warwick U., UK, 1974). Res. assoc. (tel. 5-48-81-92, ext. 332). *Low energy electron diffraction, Auger electron spectroscopy, photoemission.*

De Pablo Galan, Dr Linerto. Inst. de Geologia, U. Nac. Aut. de México, Ap. Post. 70-296, Ciudad Universitaria, Mexico 04510 D.F. (1934) PhD, (Ohio State U., USA). Jefe del Depto. de Geoquimica. (tel. 550-5215, ext. 4268). *Mineralogy, crystallography.*

Dominguez Esquivel, Dr José Manuel. Instituto Mexicano del Petróleo, Ap. Post. 14-805, Ave. 100 Metros 152, Mexico 14 D.F. (1948) DSc, physics (U. Claude Bernard, France, 1977). Investigador cientifico. (tel. 5-67-66-00, ext. 2377) *Catalysis, surface science.*

Echavarri Hernandez, Dr Ariel. Direccion de Mineria, Geologia Y Energeticos, Gobierno del Estado de Sonora, Paseo de la Arboleda 30, Hermosillo Sonora 83000, Mexico. (1939) Dr, petrology (U. de Paris, France, 1967). Dir. (tel. 621 + 31-968). *Mineralogy, petrology.*

Fabregat Guinchard, Dr Francisco José. Inst. de Geologia, U. Nac. Aut. de México, Ap. Post. 70-296, Ciudad Universitaria, Mexico 04510 D.F. (1909) Dr, ciencias naturales (U. Central Madrid, Spain, 1948). Invest. tit., prof. (tel. 550-5215, ext. 4270). *inorganic crystal structures, computer programming, mineralogy, morphology.*

Fernández González, Dr Alonso. Rectory, U. Aut. Metropolitana, Unidad Iztapalapa, Ap. Post. 55-532, Mexico 13 D.F. (1927) PhD, solid state physics (Manchester U., UK, 1958). Prof. (tel. 5-81-52-06). *Crystal growth, electrical breakdown, conductivity in crystals, semiconductors.*

Gómezdaza Almendaro, Mr Mariano. Inst. de Investigaciones de Materiales, U. Nac. Aut. de México, Ap. Post. 70-360, Ciudad Universitaria, Mexico 04510 D.F. (1944) BSc, chemical engineering (Fac. Quimica, UNAM, 1969). Investigador. (tel. 550-5215, ext. 4747). *Semiconductors, sulfides.*

Gomez Ramirez, Dr Ricardo. Depto. Estado Solido, Inst. de Fisica, U. Nac. Aut. de México, Ap. Post. 20-364, Ciudad Universitaria, Mexico 04510 D.F. (1944) PhD, materials science (Stanford U., USA, 1971). Investigador titular. (tel. 5-48-81-92). *Plastic deformation, creep, nucleation in solid-solid deformation.*

Huanosta Tera, Mr Alfonso. Inst. de Investigaciones de Materiales, U. Nac. Aut. de México, Ap. Post. 70-360, Ciudad Universitaria, Mexico 04510 D.F. (1944) MSc, (Fac. de Ciencias, UNAM, 1978) Investigador. (tel. 550-5215, ext. 4746). *Electron microscopy, alloys of Cu-Al.*

José Yacaman, Dr Miguel. Inst. de Fisica, U. Nac. Aut. de México, Ap. Post. 20-364, Ciudad Universitaria, Mexico 04510 D.F. (1946) DrSc, (Fac. de Ciencias, UNAM, 1972). Head Director. (tel. 550-5215, ext. 5940). *Crystallography, microcrystals.*

Lara Magaña, Mrs Maria Eugenia. Ciencia e Ingenieria de Materiales al Servico de la Industria, Ojito no.34 Coyoacan Mexico 04000 D.F. (1950) BSc, pharmacy (U. Texas, USA, 1975). Investigadora. (tel. 554-5945) *Crystal chemistry, organic materials obtained from plants.*

Lee Moreno, Dr José Luis. Consejo de Recursos Minerales., Ninos Heroes 139, Mexico 7 D.F. (1939) PhD, geological engineering (U. Arizona, USA, 1972). Manager of special studies. (tel. 5-78-59-42). *Geochemistry, minerals exploration, fluid oclusions, computer applications.*

Muñoz Picone, Mr Eduardo. Depto. Estado Solido, Inst. de Fisica, U. Nac. Aut. de México, Ap. Post. 20-364, Ciudad Universitaria, Mexico 04510 D.F. (1937)

MSc, physics (U. Nac. Aut. de México, 1968). Investigador, prof. (tel. 5-48-81-92). *Crystal growth.*

Piña de Noyola, Prof. Maria Cristina. Depto. Estado Solido, Inst. de Fisica, U. Nac. Aut. de México, Ap. Post. 20-364, Ciudad Universitaria, Mexico 04510 D.F. (1946) MSc, Physics (U. Nac. Aut. de México, 1976). Res. worker. (tel.5-48-81-92). *Thermodynamics of solids, diffusion, metallurgy.*

Quintana Owen, Mrs Patricia. Division Estudios de Postgrado. Fac. de Quimica, U. Nac. Aut. de México, Ciudad Universitaria, Mexico 04510 D.F. (1951) MSc, chemistry (UNAM, 1977). Investigadora. (tel. 5-48-52-10). *Phase diagrams, oxide system Li-Zr-Si.*

Ramos Bernal, Dr Sergio. Centro de Estudios Nucleares, U. Nac. Aut. Mexico, Circuito exterior, Ciudad Universitaria, Mexico 04510 D.F. (1945) PhD, (U. Manchester, UK, 1947). Secretario academico. (tel. 5-48-45-69). *Radiation damage, magnetic materials.*

Rendón Diaz Mirón, Dr Luis Emilio. Inst. de Investigaciones de Materiales, U. Nac. Aut. de México, Ap. Post. 70-360, Ciudad Universitaria, Mexico 04510 D.F. (1946) PhD, materials science and engineering (U. Texas, USA, 1977). Jefe proyecto materiales para electronica.(tel. 550-5215, ext. 4747). *Crystal chemistry, transition metal sulfides and selenides.*

Rios Jara, Dr David. Inst. de Investigaciones de Materiales, U. Nac. Aut. de México, Ap. Post. 70-360, Ciudad Universitaria, Mexico 04510 D.F. (1950) Dr, (Inst. Nac. des Ciences Appl. de Lyon, France). Res. (tel. 550-5215, ext. 4746). *Phase transformation, electron microscopy, alloys.*

Rivera Moras, Mr Vicente. Inst. de Investigaciones de Materiales, U. Nac. Aut. de México, Ap. Post. 70-360, Ciudad Universitaria, Mexico 04510 D.F. (1948) MSc, (Fac. de Ciencias, UNAM, 1979). Investigador. (tel. 550-5215, ext. 4746). *Electron microscopy (lorentz), magnetic materials.*

Riveros Rotgé, Dr Héctor Gerardo. Depto. Estado Solido, Inst. de Fisica, U. Nac. Aut. de México, Ap. Post. 20-364, Ciudad Universitaria, Mexico 04510 D.F. (1940) Dr, solid state (U. Nac. Aut. de México, 1973). Investigador titular. (tel. 5-48-81-92). *Crystal growth.*

Romero, Dr Miguel. Romero S. Hnos. S. A., Calle 7 Norte 356, Tehuacan, Puebla, Mexico. (1925) Dr, organic chemistry (U. Nac. Aut. Mexico, 1964). General director. (tel. 238ý-15-80). *Chemistry, mineralogy, crystallography, geology, nutrition.*

Rouffignac, Dr Eric de. Instituto Mexicano del Petróleo, Ap. Post. 14-805, Ave. 100 Metros 152, Mexico 14 D.F. (1945) PhD (U. Texas, USA, 1978). Jefe depto. de adsorcion, IBP. (tel. 5-67-81-67) *Thermodynamic properties of crystals, adsorption, chemisorption.*

Ruiz Mejia, Dr Carlos. Depto. Estado Solido, Inst. de Fisica, U. Nac. Aut. de México, Ap. Post. 20-364, Ciudad Universitaria, Mexico 04510 D.F. (1939) DrSc, crystallography (U. Nac. Aut. de México, 1964). Investigador, Prof. (tel. 5-48-81-92). *Crystal growth, crystal optics.*

Reyes Chumacero, Mr Antonio. Div. Estudios de Postgrado, Fac. de Quimica, U. Nac. Aut. de México, Ciudad Universitaria, Mexico 04510 D.F. (1940) Ing, (Fac. de Quimica, UNAM, 1964). Jefe Depto. fisico-quimica. (tel. 5-48-02-49). *Thermodynamics of condensed phases.*

Romero Romo, Dr Mario. Depto. Ciencia de Materiales, U. Autonoma. Metropolitana, Av. San Pablo 180, Mexico 02200 D.F. (1947) PhD, (U. Liverpool). Res. (tel. 382-5000, ext. 235). *Phase transformation, properties, oxides, alloys.*

Salas, Dr Guillermo Armando. Depto. de Geologia, U. de Sonora, Hermosillo, Sonora, Mexico. (1942) PhD, geology, (Stanford U. USA, 1971). Chairman. (tel. 621-4390, ext. 149). *Geology, mineralogy, geochemistry, ore deposits.*

Solorio Munguia, Prof. José Gregorio. Inst. de Geologia, U. Nac. Aut. de México, Ap. Post. 70-296, Ciudad Universitaria, Mexico 04510 D.F. (1922) QuimMet, metallurgy (U. Nac. Aut. de México, 1958). Invest. Prof. (tel. 550-5215, ext. 4270). *Mineralogy, minerals separation, crystallography, geochemistry.*

Soriano Garcia, Dr Manuel. Inst. de Quimica, U. Nac. Aut. de México, Circuito Exterior, Delagación Coyoacán, Mexico City 04510, Mexico. (1947) PhD, biophysics (Suny at Buffalo, USA, 1976) Res. prof. (tel. 5-50-5215, ext. 2456). *Crystal structure determination, biologically interesting compounds, protein crystallography, crystal chemistry.*

Téllez Ortiz, Mrs Minerva Estela. Div. Estudios de Postgrado, Fac. de Quimica, U. Nac. Aut. de México, Ciudad Universitaria, Mexico 04510 D.F. (1943) MSc, (Fac. de Quimica, UNAM, 1977). Tecnico academico. (tel 5-48-82-10) *X-ray diffraction, X-ray spectroscopy.*

Torres Villaseñor, Dr Gabriel. Inst. de Investigaciones de Materiales, U. Nac. Aut. de México, Ap. Post. 70-360, Ciudad Universitaria, Mexico 04510 D.F. (1944) PhD, (Case Western U., USA, 1972). Project chief. (tel. 550-5215, ext. 4746). *Alloys of Cu, domains in Cu gamma phases.*

Toscano, Mr Ruben Alfredo. Inst. de Quimica, U. Nac. Aut. de México, Circuito Exterior, Ciudad Universitaria, Mexico 04510 D.F. (1958) BSc (U. Nac. Aut. de México) Acad. Techn. (tel. 548-5448). *X-ray structure determination, infrared spectroscopy.*

Valenzuela Monjarás, Dr Raúl Alejandro. Inst. de Investigaciones de Materiales, U. Nac. Aut. de México, Ap. Post. 70-360, Ciudad Universitaria, Mexico 04510 D.F. (1946) DSc, (Fac. des Sciences, Paris, 1974). Jefe depto. ceramica y metalurgia. (tel. 550-5215, ext. 4746). *Magnetic properties, ceramic materials.*

Vera Calderón, Mrs Gloria. Div. Estudios de Postgrado, Fac. de Quimica, U. Nac. Aut. de México, Ciudad Universitaria, Mexico 04510 D.F. (1938) Qumica Industrial (U. de Guanajuato, 1962). Tecnico academico. (tel. 5-8210). *X-ray diffraction, organics.*

Villafuerte Castrejón, Mrs María Elena. Inst. de Investigaciones de Materiales, U. Nac. Aut. de México, Ap. Post. 70-360, Ciudad Universitaria, Mexico 04510 D.F. (1948) MSc, inorganic chemistry (Fac. de Quimica, UNAM, 1979). Investigadora. (tel. 550-5215, ext. 4747). *Crystal chemistry, ceramic materials.*

NETHERLANDS

Sub-Editor: R. Olthof-Hazekamp

Notes

1. International telephone country code - 31.

2. Degrees conferred by the Netherlands universities are *doctor* (Dr) (approximately equivalent to PhD at British universities), *doctorandus* (Drs) and *ingenieur* (Ir) (these latter two between MSc and PhD).

3. At universities, persons with an academic training can be appointed in various positions ranging from *hoogleraar* (professor) via *wetenschappelijk hoofdmedewerker, wetenschappelijk medewerker* and *hoofdassistent,* to *assistent* (approximately equivalent to assistant-lecturer).

Admiraal, Dr Gerrit. Vakgroep ASKA, Gorlaeus Lab., U. of Leiden, P. O. Box 9502, 2300 RA Leiden, The Netherlands. (1953), Dr, chemistry (U. Groningen, 1981). Res. scient. (tel. 071 + 148333, ext. 4415). *Peptide structures, DNA.*

Aerts, Dr Jozef. Corporate Res. Dept., AKZO, Velperweg 76, 6800 AB Arnhem, The Netherlands. (1957) PhD, chemistry (U. Leuven, Belgium, 1983). Res. scient. (tel. 085 + 662744). *Structures and physical properties of polymers.*

Altona, Prof. Cornelis. Dept. of Organic Chemistry, Gorlaeus Lab., P. O. Box 9502, 2300 RA Leiden, The Netherlands. (1931) Dr, organic chemistry (U. Leiden, 1964). Prof. (tel. 071 + 148333, ext. 3812). *Conformational analysis, biomolecules (nucleotides and steroids).*

Baak, Ing. Leonardus Cornelis. Lab. PXR, ENRAF-NONIUS, Röntgenweg 1, P. O. Box 483, 2600 AL Delft, The Netherlands. (1944) Ing., electrical engineering. Development. (tel. 015 + 569230, ext. 310) *Computer programming.*

Bartels, Drs Willem Jan. Philips Res. Lab., Ned. Philips Bedrijven BV, P. O. box 80000, 5600 JA Eindhoven, The Netherlands. (1948) Drs, chemistry (U. Groningen, 1971). Res. scient. (tel. 040 + 742208). *X-ray topography, double crystal diffractometry.*

Bastin, Dr Ir Guillaume F. Dept. of Physical Chemistry, Eindhoven U. of Techn., P. O. Box 513, 5600 MD Eindhoven, The Netherlands. (1944) Dr, chemistry (Eindhoven U. of Techn., 1972). Wetensch. hoofdmedew. (tel. 040 + 473049). *Powder diffraction, texture, metals.*

Behm, Dr Helmut Johannes Julius. Lab. voor Kristallografie, U. of Nijmegen, Toernooiveld, 6525 ED Nijmegen, The Netherlands. (1947) Dr, chemistry (U. Freiburg, BRD, 1976). Res. scient. (tel. 080 + 558833, ext. 2596, telex 48228). *Borates, crystal chemistry, Patterson methods.*

Bennema, Prof. Dr Pieter. Science Dept., U. of Nijmegen, Toernooiveld, 6525 ED Nijmegen, The Netherlands. (1932) Dr, crystal growth (Delft U. of Techn., 1965). Prof. (tel. 080 + 558833). *Crystal growth, morphology.*

Berg, van den, Ir Adrianus Johannes. Afd. Technische Natuurkunde, Delft U. of Techn., Lorentzweg 1, 2628 CJ Delft, The Netherlands. (1940) Ir, physical chemistry (Delft U. of Techn, 1966). Docent. (tel. 015 + 782481). *Crystal physics, powder diffraction.*

Berger, Dr Rolf Anders. Lab. voor Anorganische Chemie, U. of Groningen, Nijenborgh 16, 9747 AG Groningen, The Netherlands. (1946) Dr, chemistry

(Uppsala U., Sweden, 1978). Wetensch. ambt. (tel. 050 + 117140). *Transmission metal compounds.*
Bergsma, Dr Jitze. Physics Dept., Netherlands Energy Res. Foundation, E. C. N., P. O. Box 1, 1755 ZG Petten NH, The Netherlands. (1932) Dr, physics (U. Leiden, 1970). Head physics dept. (tel. 02246 + 4949). *Neutron diffraction, neutron inelastic scattering, crystal dynamics.*
Beurskens, Dr Gezina. Lab. voor Kristallografie, U. of Nijmegen, Toernooiveld, 6525 ED Nijmegen, The Netherlands. (1936) Dr, chemistry (U. Utrecht, 1961). (tel. 080 + 558833, ext. 2188). *Crystal structure determination, direct methods.*
Beurskens, Prof. Dr Paul T. Lab. voor Kristallografie, U. of Nijmegen, Toernooiveld, 6525 ED Nijmegen, The Netherlands. (1934) Dr, chemistry (U. Utrecht, 1965). Prof. (tel. 080 + 558833, ext. 2188 or 2875) (telex 48228). *Crystal structure determination, direct methods, Patterson methods, computer programming, automation.*
Beyer, Dr Ir Jenö. Dept. of Mechanical Engineering, Twente U. of Techn., P. O. Box 217, 7500 AE Enschede, The Netherlands. (1942) Dr, mat. sci. (Twente U. of Techn., 1982). Wetensch. medew. (tel. 053 + 894232). *Kinetics, crystallography, martensitic transformations.*
Birker, Dr Paul J. M. W. L. Physical & Analytic Sci. Div., Unilever Res. Lab., Olivier van Noortlaan 120, 3133 AT Vlaardingen, The Netherlands. (1947) Dr, chemistry (U. Nijmegen, 1974). Head, X-ray diffraction and calorimetry dept. (tel. 010 + 605513, telex 23261). *Powder diffraction, small angle scattering, structure determination.*
Bleeker, Drs Ernö Johan. Physics Dept., Netherlands Energy Res. Foundation, E. C. N., P. O. Box 1, 1755 ZG Petten NH, The Netherlands. (1927) Drs, physics (U. Amsterdam, 1962). Res. scient. (tel. 02246 + 4514). *Neutron diffraction.*
Bloem, Prof. Dr Jan. Lab. Vaste Stof Chemie, U. of Nijmegen, Toernooiveld, 6525 ED Nijmegen, The Netherlands. (1924) Dr, chemistry (U. Utrecht, 1956). Prof. (tel. 080 + 558833, ext. 2584). *Crystal growth, chemical vapour deposition, segregation, physical properties, defects in solids.*
Boer, de, Dr Jan Louwert. Lab. voor Anorganische Chemie, U. of Groningen, Nijenborgh 16, 9747 AG Groningen, The Netherlands. (1936) Dr, chemistry (U. Groningen, 1970). Wetensch. hoofdmedew. (tel. 050 + 117120). *Inorganic and organic crystal structures, diffuse scattering, instrumentation.*
Bolhuis, van, Mr Fré. Afd. Participatie onderzoek, U. of Groningen, Nijenborgh 16, 9747 AG Groningen, The Netherlands. Sr. res. asst. (tel. 050 + 117045). *X-ray diffraction, instrumentation.*
Boom, Dr Geert. Vakgroep Techn. Fysica, U. of Groningen, Nijenborgh 18, 9747 AG Groningen, The Netherlands. (1933) Dr, crystallography (U. Groningen, 1966). Wetensch. hoofdmedew. (tel. 050 + 117059). *Scanning electron microscopy, transmission electron microscopy, microprobe elemental analysis.*
Bosman, Drs Wilhelmus P. J. H. Lab. voor Kristallografie, U. of Nijmegen, Toernooiveld, 6525 ED Nijmegen, The Netherlands. (1937) Drs, chemistry (U. Nijmegen, 1969). Wetensch. medew. (tel. 080 + 558833, ext. 2591) (telex 48228). *Computer programming, inorganic crystal structures, direct methods.*
Bouwmeester, Drs Hennie J. M. Lab. voor Anorganische Chemie, U. of Groningen, Nijenborgh 16, 9747 AG Groningen, the Netherlands. (1954) Drs, chemistry (U. Groningen, 1982). Wetensch. asst. (tel. 050 + 117059). *Inorganic structures, intercalation compounds, ionic conductors, anharmonic thermal motion.*
Braam, Dr Adrianus Wilhelmus Maria. Dept. of Physical Chemistry, CRO-DSM, P. O. Box 16, 6160 MD Geleen, The Netherlands. (1952) Dr, chemistry and physics (U. Groningen, 1981). Head of X-ray dept. (tel. 04494 + 65307 or 66782). *X-ray crystallography, inter- and intra-molecular interactions, morphology, polymers, small angle X-ray scattering.*
Braun, Dr Poul Bernard. Bremdreef 6, 5571 AD Bergeyk, The Netherlands. (1917) Dr, crystallography (U. Amsterdam, 1956). Sr. sci. (tel. 04975 + 1814). *Organic and inorganic crystal structures, Patterson methods.*
Bronsema, Drs Klaas Derk. Lab. voor Anorganische Chemie, U. of Groningen, Nijenborgh 16, 9747 AG Groningen, The Netherlands. (1957) Drs, chemistry (U. Groningen, 1981). Adj. wetensch. ambt. (tel. 050 + 117140). *X-ray diffraction, modulated structures (commensurate and incommensurate), neutron diffraction.*
Bruggen, van, Dr Christiaan Frans. Lab. voor Anorganische Chemie, Materialen Studie Centrum, U. of Groningen, Nijenborgh 16, 9747 AG Groningen, The Netherlands. (1934) Dr, chemistry (U. Groningen, 1969). Wetensch. hoofdmedew. (tel. 050 + 117138). *Transition element solid compounds, structure - physical properties relation.*
Bruins Slot, Drs Hilbert Jan. Lab. voor Kristallografie, U. of Nijmegen, Toernooiveld, 6525 ED Nijmegen, The Netherlands. (1956) Drs, chemistry, (U. Utrecht, 1982). Wetensch. medew. (tel. 080 + 558833, ext. 2596) (telex 48228). *Crystal structure determination, direct methods, Patterson methods, computer programming, automation.*
Buschow, Dr Kurt Heinz Jürgen. Philips Res. Lab., Ned. Philips Bedrijven BV, P. O. Box 80000, 5600 JA Eindhoven, The Netherlands. (1934) Dr, physical chemistry (Free U. of Amsterdam, 1963). Sr. res. (tel. 040 + 742052). *Intermetallic compounds, crystal structure, magnetic properties, intermetallics, ternary hydrides, amorphous alloys.*
Bijen, Dr Jan. Intron BV, Het Rondeel 18, 6219 PG Maastricht, The Netherlands. (1948) Dr, chemistry (U. Utrecht, 1974). Director. (tel. 043 + 54577, ext. 11). *Gas electron diffraction, microwave spectroscopy, crystallographic structures, inorganic bonding materials.*

Cras, Dr Joannes Antonius. Lab. voor Anorganische Chemie, U. of Nijmegen, Toernooiveld, 6525 ED Nijmegen, The Netherlands. (1930) Dr, chemistry (U. Leiden, 1961). Docent, crystal chemistry. (tel. 080 + 558833, ext. 2186). *Crystal chemistry, structures, coordination compounds, organometallic compounds, inorganic solid state compounds.*
Delhez, Dr Ir Eric Jan. Tussenafd. der Metaalkunde, Delft U. of Techn., Rotterdamseweg 137, 2628 AL Delft, The Netherlands. (1940) Dr, chemistry (Delft U. of Techn., 1978). Wetensch. hoofdmedew. (tel. 015 + 782261). *X-ray diffraction, line profile analysis, diffusion zone diffraction effects, rapidly quenched alloys.*
Doesburg, Dr Hendrikus M. Lab. Automation, Duphar BV, C. J. van Houtenlaan 36, P. O. Box 2, 1380 AA Weesp, The Netherlands. (1953) Dr, chemistry (U. Nijmegen, 1984). Automation expert. (tel. 02940 + 79651). *Computer programming, crystal structure determination.*
Drenth, Prof. Dr Jan. Lab. voor Chemische Fysica, U. of Groningen, Nijenborgh 16, 9747 AG Groningen, The Netherlands. (1925) Dr, chemistry (U. Groningen, 1957). Prof. (tel. 050 + 117106). *Structure and action, biological macromolecular systems.*
Driessen, Drs René A. J. Lab. voor Kristallografie, U. of Amsterdam, Nieuwe Achtergracht 166, 1018 WV Amsterdam, The Netherlands. (1959) Drs, chemistry (U. Leiden, 1982). Wetensch. medew. (tel. 020 + 5224040). *Direct methods.*
Duisenberg, Drs Albert Jozef Maria. Lab. voor Kristal- en Structuurchemie, U. of Utrecht, Padualaan 8, P. O. Box 80050, 3505 TB Utrecht, The Netherlands. (1935) Drs, chemistry (U. Utrecht, 1964). Wetensch. hoofdmedew. (tel. 030 + 533127). *Instrumentation, computer programming, organic crystal structures.*
Dijk, van, Dr Cornelis. Physics Dept., Netherlands Energy Res. Foundation, E. C. N., P. O. Box 1, 1755 ZG Petten NH, The Netherlands. (1934) Dr, physics (U. Leiden, 1970). Group leader. (tel. 02246 + 4576). *Static and dynamic structure determination, neutron diffraction.*
Dijkstra, Drs Bauke Wiepke. Lab. voor Chemische Fysica, U. of Groningen, Nijenborgh 16, 9747 AG Groningen, The Netherlands. (1948) Drs, chemistry (U. Groningen, 1976). Wetensch. medew. (tel. 050 + 117109). *Biopolymers.*
Enckevort, van, Dr Wilhelmus Johannus Petrus. Lab. voor Vaste Stof Chemie, U. of Nijmegen, Toernooiveld, 6525 ED Nijmegen; Res. Div., Drukker Internationaal, Beversestraat 20, 5431 SH Cuyk, The Nethelands. (1952) Dr, chemistry (U. Nijmegen, 1976). Wetensch. medew. (tel. 080 + 558833). *Crystallography, X-ray diffraction topography, surface microtopography, solid state physics, research on diamonds.*
Feil, Prof. Dr Dirk. Chemical Physics Lab., Twente U. of Techn., P. O. Box 217, 7500 AE Enschede, The Netherlands. (1933) Dr, chemistry (U. Utrecht, 1961). Prof. (tel. 053 + 892661). *Electron density, molecular interaction, quantum chemistry.*
Felius, Dr Robert Onno. Inst. voor Aardwetenschappen, U. of Utrecht, Budapestlaan 4, P. O. Box 80021, 3508 TA Utrecht, The Netherlands. (1938) Dr, geology (U. Leiden, 1976). Wetensch. hoofdmedew. (tel. 030 + 535097). *Physical properties and crystal structure of minerals, applied mineralogy.*
Fleischmann, Dr Klaus Dietrich. Scientific Instruments PXR, ENRAF-NONIUS, Röntgenweg 1, P. O. Box 483, 2600 AL Delft, The Netherlands. (1937) Dr, physical chemistry (Munich U. of Techn., FRD, 1970). Vice president. (tel. 015 + 569230, ext. 118). *Instrumentation, organic structures.*
Frikkee, Dr Evert. Physics Dept., Netherlands Energy Res. Foundation, E. C. N., P. O. Box 1, 1755 ZG Petten NH, The Netherlands. (1934) Dr, physics (U. Leiden, 1973). Res. scient. (tel. 02246 + 4527). *Spin waves, magnetic structures.*
Geelen, van, Dr Bernard. Werkgemeenschap voor Kristal- en Structuuronderzoek, Koningin Sophiestraat 124, 2595 TM 's Gravenhage, The Netherlands. (1924) Dr, chemistry (U. Utrecht, 1958). Director SON, executive secretary WKSO. (tel. 070 + 824381).
Geerestein, van, Drs Vincent Johan. Lab. voor Kristal- en Structuurchemie, U. of Utrecht, Padualaan 8, P. O. Box 80050, 3508 TB Utrecht, The Netherlands. (1959) Drs, chemistry (U. Utrecht, 1983). Wetensch. asst. (tel. 030 + 532865). *Bio-molecular structures, structure - activity relation.*
Gellings, Prof. Dr Paul Johann. Dept. of Inorganic Chemistry and Materials Science, Twente U. of Techn., P. O. Box 217, 7500 AE Enschede, The Netherlands. (1927) Dr, chemistry (U. Amsterdam, 1963). Prof. (tel. 053 + 892482). *Structure and bonding, coordination compounds, transition metals, catalysts, oxidation products.*
Goedkoop, Prof. Dr Jacob A. Netherlands Energy Res. Foundation, E. C. N., P. O. Box 1, 1755 ZG Petten NH, The Netherlands. (1921) Dr, chemistry (U. Amsterdam, 1952). (tel. 02246 + 4949). *Neutron diffraction.*
Gomes de Mesquita, Dr Albert Hijman. Information Systems and Automation (ISA), Ned. Philips Bedrijven BV, P. O. Box 80000, 5600 JA Eindhoven, The Netherlands. (1930) Dr, chemistry (U. Amsterdam, 1962). Leader of automation group. (tel. 040 + 783155). *Automation.*
Gorter, Drs Ing. Sybout. Vakgroep ASKA, Gorlaeus Lab., U. of Leiden, P. O. Box 9502, 2300 RA Leiden, The Netherlands. (1940) Drs, chemistry (U. Leiden, 1974). Wetensch. ambt. 1, (tel. 071 + 148333, ext. 4415 or 4420). *Programming, measurement of intensities, inorganic and organic structures, development of apparatus.*

Goubitz, Drs Kees. Lab. voor Kristallografie, U. of Amsterdam, Nieuwe Achtergracht 166, 1018 WV Amsterdam, The Netherlands. (1953) Drs, chemistry (U. Amsterdam, 1981). Wetensch. medew. (tel. 020 + 5224038). *Crystal structure determination.*

Graaff, de, Dr Rudolf Adriaan Gerard. Vakgroep ASKA, Gorlaeus Lab., U. of Leiden, P. O. Box 9502, 2300 RA Leiden, The Netherlands. (1941) Dr, X-ray crystallography (U. Leiden, 1974). Wetensch. medew. (tel. 071 + 148333, ext. 4414). *Direct methods, accurate structure factor determination, computer programming, organic crystal structures.*

Grampel, van de, Dr Johan Christoph. Lab. voor Anorganische Chemie, U. of Groningen, Nijenborgh 16, 9747 AG Groningen, The Netherlands. (1934) Dr, chemistry (U. Groningen, 1967). Wetensch. hoofdmedew. (tel. 050 + 117143). *Structure, bonding, main-group elements.*

Harkema, Dr Sybolt. Chemical Physics Lab., Twente U. of Techn., P. O. Box 217, 7500 AE Enschede, The Netherlands. (1940) Dr, chemistry (Twente U. of Techn., 1971). Wetensch. hoofdmedew. (tel. 053 + 892671). *Electron density, structure, crown-ethers.*

Hartman, Prof. Dr Piet. Inst. voor Aardwetenschappen, U. of Utrecht, Budapestlaan 4, P. O. Box 80021, 3508 TA utrecht, The Netherlands. (1922) Dr, crystallography (U. Groningen, 1953). Prof. (tel. 030 + 535065). *Crystal growth, X-ray phase analysis, mineral structures.*

Havinga, Dr Edsko Enno. Philips Res. Lab., Ned. Philips Bedrijven BV, P. O. Box 80000, 5600 JA Eindhoven, The Netherlands. (1932) Dr, physical chemistry (U. Groningen, 1957). Sr. sci. (tel. 040 + 742547). *Structure - physical properties relation, organic molecules, organic crystals.*

Heinerman, Dr Jacobus Johannes Leonardus. Koninklijke - Shell Lab., Badhuisweg 3, 1031 CM Amsterdam, The Netherlands. (1951) Dr, chemistry (U. Utrecht, 1977). Res. scient. (tel. 020 + 302988). *Crystal structure determination, direct methods, catalysis, catalytic cracking.*

Helmholdt, Dr Robert Barteld. Physics Dept., Netherlands Energy Res. Foundation, E. C. N., P. O. Box 1, 1755 ZG Petten NH, The Netherlands. (1943) Dr, chemistry (U. Groningen, 1975). Res. scient. (tel. 02246 + 4529). *Electron density distribution, thermal motion, computer programming, neutron diffraction.*

Heijnen, Drs Wilhelmus Marinus Maria. Inst. voor Aardwetenschappen, U. of Utrecht, Budapestlaan 4, P. O. Box 80021, 3508 TA Utrecht, The Netherlands. (1956) Drs, geology (U. Leiden, 1980). Wetensch. asst. (tel. 030 + 535062). *Crystal growth, morphology, calcium oxalates and carbonates, X-ray crystallography.*

Hol, Dr Wim. Lab. voor Chemische Fysica,, U. of Groningen, Nijenborgh 16, 9747 AG Groningen, The Netherlands. (1945) Dr, protein crystallography (U. Groningen, 1971). Wetensch. medew. (tel. 050 + 117107). *Structure and action, biological macromolecular systems.*

Hornstra, Drs Jan. Philips Res. Lab., Ned. Philips Bedrijven BV, P. O. Box 80000, 5600 JA Eindhoven, The Netherlands. (1927) Drs, chemistry (U. Groningen, 1952). Sr. sci. (tel. 040 + 742021). *Crystal structure determination, diffractometry, lattice defects.*

Huiszoon, Dr Cornelis. Chemical Physics Lab., Twente U. of Techn., P. O. Box 217, 7500 AE Enschede, The Netherlands. (1933) Dr, microwave spectroscopy (U. Nijmegen, 1966). Wetensch. medew. (tel. 053 + 892673). *Intermolecular forces, quantum mechanics.*

Hummel, van, Ing. Gerrit Jan. Chemical Physics Lab., Twente U. of Techn., P. O. Box 217, 7500 AE Enschede, The Netherlands. (1945). Sr. res. asst. (tel. 053 + 892674). *Electron density, organic structures, computer programming, instrumentation.*

Janner, Prof. Dr Aloysio. Inst. voor Theoretische Fysica, U. of Nijmegen, Toernooiveld 1, 6525 ED Nijmegen, The Netherlands. (1928) PhD, philosophy (U. Zürich, Switzerland, 1962). Prof. (tel. 080 + 558833, ext. 2981, telex 48228 WINAT). *Solid state physics, group theory, incommensurate structures.*

Jellinek, Prof. Dr Franz. Lab. voor Anorganische Chemie, U. of Groningen, Nijenborgh 16, 9747 AG Groningen, The Netherlands. (1925) Dr, chemistry (U. Utrecht, 1957). Prof. (tel. 050 + 117141). *Structure - physical properties - composition relation, inorganic and organometallic structures.*

Kalk, Mr Kornelis Harm. Lab. voor Chemische Fysica, U. of Groningen, Nijenborgh 16, 9747 AG Groningen, The Netherlands. (1944). Sr. res. asst. (tel. 050 + 117116). *Biologically important molecules, instrumentation.*

Kamphuis, Dr Irenus Gerhardus. Techn. Wetenschappelijk Rekencentrum, Netherlands Energy Res. Foundation, E. N. R., 1755 ZG Petten (NH), The Netherlands. (1952) Dr, protein crystallography (U. Groningen, 1983). Software eng. (tel. 02246 + 4099). *Protein crystallography.*

Kanters, Dr Jan. Lab. voor Kristal- en Structuurchemie, U. of Utrecht, Padualaan 8, P. O. Box 80050, 3508 TB Utrecht, The Netherlands. (1928) Dr, chemistry (U. Utrecht, 1958). Wetensch. hoofdmedew. (tel. 030 + 533410). *Hydrogenbond pattern, conformation, saccharides, organic acids.*

Keulen, Dr Evert. Applic. Lab. X-ray Diffraction, Philips Science and Industry, Lelyweg 1, 7602 EA Almelo, The Netherlands. (1932) Dr, chemistry (U. Groningen, 1969). Res. scient. (tel. 05490 + 18291, ext. 446) (telex 36591 NLUALSB). *Instrumentation, precision structure analysis, powder diffraction.*

Keijser, de, Dr Ir Thomas Henri. Tussenafd. der Metaalkunde, Delft U. of Techn., Rotterdamseweg 137, 2628 AL Delft, The Netherlands. (1937) Dr, chemistry (Delft U. of Techn., 1977). Wetensch. hoofdmedew. (tel. 015 + 784105). *Crystallography, diffraction, phase transformations, rapidly quenched alloys.*

Kiers, Dr Conradus Theodorus. Res. & Dev. Lab., ENRAF-NONIUS, Röntgenweg 1, P.O. Box 483, 2600 AL Delft, The Netherlands. (1947) Dr, chemistry (U. Groningen, 1976). Software product specialist. (tel. 015 + 569230, ext. 432, telex 38083). *Crystallgraphic computing.*

Kinneging, Drs Albertus Jacobus. Vakgroep ASKA, Gorlaeus Lab., U. of Leiden, P. O. Box 9502, 2300 RA Leiden, The Netherlands. (1959) Drs, chemistry (U. Leiden, 1982). Wetensch. asst. (tel. 071 + 448833, ext. 4414). *Direct methods, computer programming.*

Klop, Drs Enno Anton. Lab. voor Kristal- en Structuurchemie, U. of Utrecht, Padualaan 8, P. O. Box 80050, 3508 TB Utrecht, The Netherlands. (1960) Drs, chemistry (U. Utrecht, 1984). Wetensch. asst. (tel. 030 + 532533). *Anomalous scattering, X-ray crystal structure determination.*

Knippenberg, Dr Wilhelmus Franciscus. Philips Res. Lab., Ned. Philips Bedrijven BV, P. O. Box 80000, 5600 JA Eindhoven, The Netherlands. (1930) Dr, mathematics and physics (U. Leiden, 1963). Res. group leader. (tel. 040 + 742621). *Crystal growth, crystal structures, material characterization.*

Kok, de, Dr Anthonie Johannes. Vakgroep ASKA, Gorlaeus Lab., U. of Leiden, P. O. Box 9502, 2300 RA Leiden, The Netherlands. (1942) Dr, chemistry (U. Leiden, 1976). Res. asst. (tel. 071 + 148333, ext. 4415). *Inorganic and organic structures.*

Kolster, Prof. Dr Ir Benjamin Harry. Dept. of Fundamental Res., Metal Inst. TNO, Laan van Westenenk 501, P. O. Box 541, 7300 AM Apeldoorn, The Netherlands. (1938) Dr, crystallography (Delft. U. of Techn., 1968). Res. coordinator, Prof., material science, Twente U. of Techn. (tel. 055 + 773344). *Corrosion, structure - mechanical properties relation, powder metallurgy, non-waste technology.*

Koningsveld, van, Dr Hendrikus. Afd. der Technische Natuurkunde, Delft U. of Techn., Lorentzweg 1, 2628 CJ Delft, The Netherlands. (1942) Dr, physical chemistry (U. Utrecht, 1970). Wetensch. hoofdmedew. (tel. 015 + 782605). *Crystal structures.*

Koopmans, Prof. Dr Kasper. Afd. der Mijnbouwkunde, Delft U. of Techn., Mijnbouwstraat 120, 2628 RX Delft, The Netherlands. (1927) Dr, chemistry (Eindhoven U. of Techn., 1971). Prof. (tel. 015 + 785001). *Powder diffraction, ores, minerals, rocks, line profile analysis.*

Koch, Dr Beatrix. Lab. voor Kristallografie, Werkgroep Röntgenfluorescentie Spectrometrie en Poederdiffractie, U. of Amsterdam, Nieuwe Achtergracht 166, 1018 WV Amsterdam, The Netherlands. Dr, crystallography (U. Amsterdam, 1975). Sr. staff member. (tel. 020 + 5223574). *Mineralogical crystallography, powder diffraction, X-ray spectroscopy, carotenoid structures.*

Krabbendam, Drs Hendrik. Lab. voor Kristal- en Structuurchemie, U. of Utrecht, Padualaan 8, P. O. Box 80050, 3508 TB Utrecht, The Netherlands. (1934) Drs, chemistry (U. Utrecht, 1959). Wetensch. hoofdmedew. (tel. 030 + 533414). *Structure analysis methods, organic crystal structures, protein crystal structures.*

Krever, Drs Maarten. Lab. voor Kristallografie, U. of Amsterdam, Nieuwe Achtergracht 166, 1018 WV Amsterdam, The Netherlands. (1955) Drs, chemistry (U. Leiden, 1983). Wetensch. medew. (tel. 020 + 5224039). *Direct methods, crystal structure determination.*

Kroon, Prof. Dr Jan. Lab. voor Kristal- en Structuurchemie, U. of Utrecht, Padualaan 8, P. O. Box 80050, 3508 TB Utrecht, The Netherlands. (1937) Dr, chemistry (U. Utrecht, 1964). Prof. (tel. 030 + 532383 or 533209, telex 70353 BITRA NL). *Crystal and molecular structure, structure determination methods, conformational analysis, hydrogen bonding, biological structure-activity relation.*

Kroon-Batenburg, Dr Louise Maria Johanna. Lab. voor Kristal- en Structuurchemie, U. of Utrecht, Padualaan 8, P. O. Box 80050, 3508 TB Utrecht, The Netherlands. (1956) Drs, chemistry (U. Utrecht, 1985). Wetensch. asst. (tel. 030 + 533130). *Crystal and molecular structure, carbohydrates, hydrogen bonds, molecular mechanics.*

Kummer, Drs Ernst Albertus. Fysisch Geografisch en Bodemkundig Lab., U. of Amsterdam, Dapperstraat 115, 1093 BS Amsterdam, The Netherlands. (1932) Drs, physical geography (U. Amsterdam, 1962). Clay mineralogist. (tel. 020 + 923030, ext. 25). *X-ray diffraction, clay mineralogy, mineralogy, weathering and soilforming processes.*

Loopstra, Prof. Dr Bert Onno. Lab. voor Kristallografie, U. of Amsterdam, Nieuwe Achtergracht 166, 1018 WV Amsterdam, The Netherlands. (1928) Dr, crystallography (U. Amsterdam, 1958). Prof. (tel. 020 + 5224029). *Neutron diffraction, crystal structure determination.*

Lugt, van der, Dr Willem. Solid State Physics Dept., U. of Groningen, Melkweg 1, 9718 EP Groningen, The Netherlands. (1929) Dr, physics (U. Leiden, 1961). Prof. (tel. 050 + 115427). *Liquid metals, structural research, electronic transport properties, nuclear magnetic resonance.*

Maaskant, Prof. Dr Willem Johannes Albert. Lab. voor Anorganische Chemie, Gorlaeus Lab., U. of Leiden, P. O. Box 9502, 2300 RA Leiden, The Netherlands. (1932) Dr, chemistry (U. Leiden, 1963). Prof. (tel. 071 + 148333, ext. 4214). *Theoretical physics, solid state physics, solid state chemistry, physical and theoretical chemistry, organic and inorganic chemistry, crystallography, applied mathematics.*

Macgillavry, Prof. Dr Carolina H. Mensinge 63, 1083 HE Amsterdam, The Netherlands. (1904) Dr, natural sciences (U. Amsterdam, 1937). Em., U. Amsterdam. (tel. 020 + 443999). *General crystallographic interest.*

Meetsma, Drs Auke. Afd. Participatie onderzoek, U. of Groningen, Nijenbourgh 16, 9747 AG Groningen, The Netherlands. (1946) Drs, physical chemistry (U. Groningen, 1980). Res. asst. (tel. 050 + 117675). *Crystal structure determination.*

Meurs, van, Dr Ir Frank. Scientific Instruments PXR, ENRAF-NONIUS, Röntgenweg 1, P. O. Box 483, 2600 AL Delft, The Netherlands. (1946) Dr, chemistry (Delft U. of Techn., 1978). Applications. (tel. 015 + 569230, ext. 151). *Computer programming, structures, instrumentation, organometallics, conformational analysis.*

Mittemeijer, Dr Eric Jan. Tussenafd. der Metaalkunde, Delft U. of Techn., Rotterdamseweg 137, 2628 AL Delft, The Netherlands. (1950) Dr, chemistry (Delft U. of Techn., 1978). Project leader. (tel. 015 + 782207). *X-ray diffraction, line profile analysis, diffraction from diffusion zones, electron diffraction, diffusion, thin films, rapidly quenched alloys, surface coatings.*

Mijlhoff, Dr Frans Cornelis. Vakgroep ASKA, Gorlaeus Lab., U. of Leiden, P. O. Box 9502, 2300 RA Leiden, The Netherlands. (1932) Dr, chemistry (U. Amsterdam, 1964). Wetensch. hoofdmedew. (tel. 071 + 148333, ext. 4211 or 4412). *Gas phase molecular structure, electron microscopy, EXAFS.*

Noordik, Dr Jan Hendrik. Lab. voor Kristallografie, U. of Nijmegen, Toernooiveld, 6525 ED Nijmegen, The Netherlands. (1944) Dr, chemistry (U. Nijmegen, 1971). Wetensch. hoofdmedew. (tel. 080 + 558833, ext. 2875) (telex 48228). *Crystal structure determination, computer programming, automation in chemistry.*

Northolt, Dr Ir Maurits Gerhard. Corporate Res. Dept., AKZO, Velperweg 76, 6824 BM Arnhem, The Netherlands. (1939) Dr, crystallography (U. Amsterdam, 1968). Res. scient. (tel. 085 + 664056). *Polymer crystal structures, diffraction, structure - mechanical properties relation.*

Nota, Dr Dirk Johannes Gregorius. Dept. of Soil Science and Geology, Wageningen U. of Agriculture, Duivendaal 10, P. O. Box 37, 6700 AA Wageningen, The Netherlands. (1926) Dr, geology (U. Utrecht, 1958). Wetensch. medew. (tel. 08370 + 82418). *Hydrogeology, sedimentology, sedimentary petrology.*

Nijveldt, Ir Dick. Lab. voor Anorganische Chemie, U. of Groningen, Nijenbourgh 16, 9747 AG Groningen, The Netherlands. (1951) Ir, techn. physics (Twente U. of Techn., 1979). Wetensch. ambt. (tel. 050 + 117102). *Accurate electron density, structure determination.*

Oen, Prof. Dr Ing Soen. Vakgroep Ertskunde-Petrologie-Mineralogie, U. of Amsterdam, Nieuwe Prinsengracht 130, 1018 VZ Amsterdam, The Netherlands. (1928) Dr, geology and mineralogy (U. Amsterdam, 1958). Prof. (tel. 020 + 5222839). *Mineralogy.*

Olthof, Dr Gerrit Jan. Afd. Epidemologie en Informatica, Ministerie van WVC, Dr Reijerstraat 12, P. O. Box 439, 2260 AK Leidsendam, The Netherlands. (1950) Dr, chemistry (U. Amsterdam, 1981). Informatica adv. *Direct methods.*

Olthof-Hazekamp, Drs Roeli. Lab. voor Kristal- en Structuurchemie, U. of Utrecht, Padualaan 8, P. O. Box 80050, 3508 TB Utrecht, The Netherlands. (1937) Drs, chemistry (U. Groningen, 1963). Wetensch. medew. (tel. 030 + 532385 or 532869). *Computer programming.*

Overweel, Dr Cornelis Johannes. Inst. voor Prehistorie, U. of Leiden, Reuvensplaats 4, P. O. Box 9515, 2300 RA Leiden, The Netherlands. (1921) Dr, agricultural science (U. Wageningen, 1977). Wetensch. hoofdmedew. (tel. 071 + 148333, ext. 2429). *Prehistoric ceramics, X-ray powder diffraction, optical mineralogy.*

Peerdeman, Prof. Dr Antonius Franciscus. Lab. voor Kristal- en Structuurchemie, U. of Utrecht, Padualaan 8, P. O. Box 80050, 3508 TB Utrecht, The Netherlands. (1921) Dr, X-ray crystallography (U. Utrecht, 1955). Prof. (tel. 030 + 533124). *X-ray crystallography, apparatus, direct methods, anomalous scattering, molecular conformations, intermolecular interactions, thermodynamics.*

Perdok, Prof. Dr Wiepko Gerhardus. Dental School, Lab. of Material Technica, Ant. Deusinglaan 1, 9713 AV Groningen, The Netherlands. (1914) Dr, chemistry, crystallography (U. Groningen, 1942). Em., U. Groningen. (tel. 050 + 117869). *Crystal optics, crystal growth, X-ray diffraction applications.*

Peterse, Ir Wilhelmus J. A. M. Afd. der Technische Natuurkunde, Delft U. of Techn., Lorentzweg 1, 2628 CJ Delft, The Netherlands. (1934) Ir, physics (Delft U. of Techn., 1963). Wetensch. hoofdmedew. (tel. 015 + 782405 or 784276). *Inorganic crystal structures, crystal physics, computer programming, instrumentation.*

Plas, van der, Prof. Dr Leendert. Dept. of Soil Science and Geology, Wageningen U. of Agriculture, Duivendaal 10, P. O. Box 37, 6700 AA Wageningen, The Netherlands. (1928) Dr, petrography and mineralogy (U. Leiden, 1959). Prof. (tel. 08370 + 84415). *Clay mineralogy, feldspar properties, clay geochemistry, zeolites, salt minerals.*

Peschar, Drs René. Lab. voor Kristallografie, U. of Amsterdam, Nieuwe Achtergracht 166, 1018 WV Amsterdam, The Netherlands. (1956) Drs, chemistry (U. Amsterdam, 1980). Wetensch. medew. (tel. 020 + 5224040). *Direct methods.*

Pontenagel, Dr Willibrordus Maria Gertrud Franciscus. Analytical Dept. CRO, DSM, P. O. Box 18, 6160 MD Geleen, The Netherlands. (1954) Dr, chemistry (U. Utrecht, 1983). Res. scient. (tel. 04494 + 66394). *Direct methods, X-ray crystal structure determination, atomic spectroscopy.*

Poot, Mr Simon. Scientific Instruments, ENRAF-NONIUS, Röntgenweg 1, P. O. Box 483, 2600 AL Delft, The Netherlands. (1927) Product specialist. (tel. 015 + 569230, ext. 117). *Instrumentation, low temperatures.*

Popma, Prof. Dr Theo Johan August. Afd. Technische Natuurkunde, Twente U. of Techn., P. O. Box 217, 7500 AE Enschede, The Netherlands. (1941) Dr, solid state chemistry (U. Groningen, 1970). Prof. (tel. 053 + 893444). *Materials science.*

Postma, Drs Johannes Petrus Maria. Lab. voor Chemische Fysica, U. of Groningen, Nijenborgh 16, 9747 AG Groningen, The Netherlands. (1951) Drs, chemistry (U. Groningen, 1979). Wetensch. asst. (tel. 050 + 117042). *Computer graphics, molecular dynamics.*

Prick, Dr Petrus Antonius Johannes. Hiddemaheerd 16, 9737 JN Groingen, The Netherlands. (1952) Dr, chemistry (U. Nijmegen, 1979). *Direct methods, refinement, protein structures.*

Prins, Prof. Dr Jan Albert. Gen. Foulkesweg 297, Wageningen, The Netherlands. (1899) Dr, mathematics (U. Groningen, 1927). Em., Delft U. of Techn. (tel. 08370 + 13151). *Physics and chemistry.*

Putten, van der, Dr Nicolaas H. J. J. R & D Lab., ENRAF-NONIUS, Röntgenweg 1, P. O. Box 483, 2600 AL Delft, The Netherlands. (1950) Dr, crystallography (U. Amsterdam, 1975). Software product specialist. (tel. 015 + 569230, ext. 432) (telex 38083). *Computer programming, direct methods, proteins.*

Renetseder, Mr Roland. Lab. voor Chemische Fysica, U. of Groningen, Nijenborgh 16, 9747 AG Groningen, The Netherlands. (1956) Dipl. Natw. (E. T. H. Zürich, Switzerland, 1980). Wetensch. asst. (tel. 050 + 117108). *Biomacromolecular structure.*

Reiss, Drs Céleste A. Lab. voor Kristallografie, U. of Amsterdam, Nieuwe Achtergracht 166, 1018 WV Amsterdam, The Netherlands. (1952) Drs, chemistry (U. Amsterdam, 1979). Wetensch. medew. (tel. 020 + 5224038). *Direct methods, crystal structure determination.*

Rieck, Prof. Dr Gerard Daniel. Mecklenburglaan 5, 2243 HN Waalre, The Netherlands. (1911) Dr, chemistry (U. Utrecht, 1945). Em., Eindhoven U. of Techn. (tel. 04904 + 3768). *Reactions, recrystallization, grain growth, textures, metals, oxidec solids.*

Rietveld, Dr Hugo Marie. Dienst Technische en Wetenschappelijke Informatie, Netherlands Energy Res. Foundation, E. C. N., P.O. Box 1, 1755 ZG Petten, The Netherlands. (1932) Dr, physics (U. Western Aust., Australia, 1964). Head. (tel. 02246 + 4365). *Data base technology, information retrieval.*

Romers, Prof. Dr Cornelis. Nachtegaallaan 17, 2172 JP, Sassenheim, The Netherlands. (1919) Dr, chemistry (U. Amsterdam, 1948). Em., U. Leiden.

Rutten-Keulemans, Drs Elisabeth Wilhelmina Maria. Vakgroep ASKA, Gorlaeus Lab., U. of Leiden, P. O. Box 9502, 2300 RA Leiden, The Netherlands. (1932) Drs, chemistry (U. Leiden, 1959). Wetensch. hoofdmedew. (tel. 071 + 148333, ext. 4220). *Computer programming.*

Seal, Dr Michael. D. Drukker & ZN. N.V., 12 Sarphatikade, 1017 WV Amsterdam, The Netherlands. (1930) PhD, physics (Cambridge U., UK, 1957). Res. dir. (tel. 020 + 267321, telex 14143). *Diamonds, thermal conductivity, crystal growth, infrared spectroscopy, microscopy, X-ray crystal orientation methods, natural crystal isotope distribution, laser cutting, SEM.*

Schapink, Dr Frederik Willem. Tussenafd. der Metaalkunde, Delft U. of Techn., Rotterdamseweg 137, 2628 AL Delft, The Netherlands. (1931) Dr, physics (Delft U. of Techn., 1969). Wetensch. hoofdmedew. (tel. 015 + 782272). *Electron diffraction, electron microscopy, X-ray diffraction, physical metallurgy.*

Schierbeek, Mr Abraham Johan. Lab. voor Chemische Fysica, U. of Groningen, Nijenborgh 16, 9747 AG Groningen, The Netherlands. (1955) *Protein crystallography.*

Schenk, Prof. Dr Hendrik. Lab. voor Kristallografie, U. of Amsterdam, Nieuwe Achtergracht 166, 1018 WV Amsterdam, The Netherlands. (1939) Dr, chemistry (U. Amsterdam, 1969). Prof. (tel. 020 + 5224035, telex 16460 FACWN NL). *Direct methods, crystal structure determination.*

Schoone, Prof. Dr Jean C. Comeniuslaan 117, 3706 XE Zeist, The Netherlands. (1919) Dr, chemistry (U. Utrecht, 1950). Em., U. Utrecht. (tel. 03403 + 57702). *Computer programming.*

Schooneveld, van, Ing. Marinus. Scientific Instruments PXR, ENRAF-NONIUS, Röntgenweg 1, P. O. Box 483, 2600 AL Delft, The Netherlands. (1943) Ing., electrical engineering. Product specialist. (tel. 015 + 569230, ext. 119). *Instrumentation.*

Schreuder, Drs Herman Antony. Lab. voor Chemische Fysica, U. of Groningen, Nijenborgh 16, 9747 AG Groningen, The Netherlands. (1958) Drs, biochemistry (U. Groningen, 1983). Wetensch. asst. (tel. 050 + 117041) *Protein crystallography.*

Schreurs, Drs Antonius Mathias Maria. Lab. voor Kristal- en Structuurchemie, U. of Utrecht, Padualaan 8, P. O. Box 80050, 3508 TB Utrecht, The Netherlands. (1955) Drs, chemistry (U. Utrecht, 1981). Wetensch. medew. (tel. 030 + 533902). *Computer programming, computer graphics, crystal structure statistics.*

Schuyff, Prof. Dr Abraham. Lab. voor Kristal- en Structuurchemie, U. of Utrecht, Padualaan 8, P. O. Box 80050, 3508 TB Utrecht, The Netherlands. (1927) Dr, chemistry (U. Utrecht, 1962). Prof. (tel. 030 + 533125). *Phase transitions.*

Smaalen, van, Drs Sander. Lab. voor Anorganische Chemie, U. of Groningen, Nijenborgh 16, 9747 AG Groningen, The Netherlands. (1958) Drs, chemistry (U. Groningen, 1981). Wetensch. asst. (tel. 050 + 117140). *Modulated structures, physical properties of solids.*

Smit, Dr Paul. Science and Industry, Philips' Export BV, Lelyweg 1, 7602 HV Almelo, The Netherlands. (1949) Dr, physical chemistry (U. Utrecht, 1978).

Product manager. (tel. 05490 + 18291). *Powder diffraction, stress analysis, crystallography.*

Smits, Mr Johannes Martinus Maria. Lab. voor Kristallografie, U. of Nijmegen, Toernooiveld, 6525 ED Nijmegen, The Netherlands. (1948). Techn. asst. (tel. 080 + 558833, ext. 2030). *Inorganic and organic crystal structures, computer programming.*

Soest, van, Dr Teunis Cornelis. Unilever Res. Lab., Olivier van Noortlaan 120, 3133 AT Vlaardingen, The Netherlands. (1931) Dr, physical chemistry (U. Utrecht, 1969). Head, computer dept. (tel. 010 + 606933). *Computer programming.*

Sonneveld, Mr Eduard Jan. X-ray Dept., Technisch Physische Dienst TNO-TH, Stieltjesweg 1, P. O. Box 155, 2600 AD Delft, The Netherlands. (1945). Res. asst. (tel. 015 + 787005). *Powder diffraction.*

Spek, Dr Anthony Louis. Lab. voor Kristal- en Structuurchemie, U. of Utrecht, Padualaan 8, P. O. Box 80050, 3508 TB Utrecht, The Netherlands. (1944) Dr, chemistry (U. Utrecht, 1975). Wetensch. hoofdmedew. (tel. 030 + 532538). *Direct methods, automation, computer programming, computer graphics, organic and organometallic structures.*

Stam, Dr Casper Hendrik. Lab. voor Kristallografie, U. of Amsterdam, Nieuwe Achtergracht 166, 1018 WV Amsterdam, The Netherlands. (1925) Dr, chemistry (U. Amsterdam, 1963). Wetensch. hoofdmedew. (tel. 020 + 5224033). *Crystal structure determination, crystal optics.*

Stevels, Dr Albert Leendert Nicolaas. Dept. Glas, Ned. Philips Bedrijven BV, P. O. Box 80000, 5600 JA Eindhoven, The Netherlands. (1944) Dr, chemistry (U. Groningen, 1969). Res. scient. (tel. 040 + 788014). *Crystallography, inorganic compounds, structure - lumenescence phenomena relation.*

Stouten, Drs Pieter F. W. Lab. voor Kristal- en Structuurchemie, U. of Utrecht, Padualaan 8, P. O. Box 80050, 3508 TB Utrecht, The Netherlands. (1959) Drs, chemistry (U. Utrecht, 1984). Wetensch. asst. (tel. 030 + 532866). *Conformational analysis, crystal structure statistics, molecular dynamics, molecular mechanics.*

Straver, Drs Leonardus Hendrikus. Scient. Instruments PXR, ENRAF- NONIUS, Röntgenweg 1, P.O. Box 483, 2600 AL Delft, The Netherlands. (1954) Drs, chemistry (U. Utrecht, 1980). Manager, applications. (tel. 015 + 569230, ext. 310). *Computer programming, instrumentation, organometallics, conformational analysis.*

Struikmans, Drs Rink. Afd. der Technische Natuurkunde, Delft U. of Techn., Lorentzweg 1, 2628 CJ Delft, The Netherlands. (1941) Drs, physics (Free U. of Amsterdam, 1970). Wetensch. hoofdmedew. (tel. 015 + 784098). *Phase transitions, optical crystallography, crystal physics, instrumentation.*

Struijs, van der, Drs Johannes P. Lab. voor Kristallografie, U. of Amsterdam, Nieuwe Achtergracht 166, 1018 WV Amsterdam, The Netherlands. (1941) Drs, chemistry (U. Amsterdam, 1967). Wetensch. medew. (tel. 020 + 5224032). *Biofibres crystallography.*

Stuut, Ir Harmannus Aaldrik. AKZO Res. Lab. Arnhem, Velperweg 76, 6824 BM Arnhem, The Netherlands. (1946) Ir, metallurgy (Delft U. of Techn., 1975). Res. scient. (tel. 085 + 662289). *Polymers, textures, structure - mechanical properties relation, electron diffraction.*

Thijsse, Dr Barend Jan. Tussenafd. der Metaalkunde, Delft U. of Techn., Rotterdamseweg 137, 2628 AL Delft, The Netherlands. (1950) Dr, physics (U. Leiden, 1978). Wetensch. medew. (tel. 015 + 782221). *X-ray diffraction, neutron diffraction, non-crystalline solids, metallic glasses, structural relaxation.*

Tuinstra, Prof. Dr Ir Fokke. Afd. der Technische Natuurkunde, Delft U. of Techn., Lorentzweg 1, 2628 CJ Delft, The Netherlands. (1934) Dr, physics (Delft U. of Techn., 1967). Prof. (tel. 015 + 786112 or 784276). *Structure - properties relation, crystal physics.*

Veen, Prof. Dr Arthur Willem Lourens. Vakgroep Fysische Geografie en Bodemkunde, U. of Groningen, Melkweg 1, 9718 EP Groningen, The Netherlands. (1942) Dr, geology and mineralogy (U. Amsterdam, 1970). Prof. (tel. 050 + 115415, ext. 5452). *Clay mineralogy.*

Veen, van der, Dr Adriaan Hendrik. Billiton Research B. V., Westervoortsedijk 67 D, P. O. Box 40, 6800 AA Arnhem, The Netherlands. (1925) Dr, mineralogy, petrology (Delft U. of Techn., 1963). Staff consultant petrological-mineralogical res. (tel. 085 - 654316, telex 75026 BIRES NL). *Quantitative X-ray diffraction (compositions), high temperature X-ray diffraction, differential thermal analysis, thermal gravimetric analysis, quantitative microscopy (stereology), petrology, mineralogy, ore-microscopy, geochemistry.*

Veld, In 't, Ir Gerard Adriaan. LAB PHI, ENRAF-NONIUS, P. O. Box 483, 2600 AL Delft, The Netherlands. (1956) Ir, electronics (Delft U. of Techn., 1982). Head product specific group. (tel. 015 + 569230, ext. 432) (telex 38083). *Hardware development, computers, software.*

Verschoor, Dr Gerrit Christiaan. Vakgroep ASKA, Gorlaeus Laboratories, U. of Leiden, P. O. Box 9502, 2300 RA Leiden, The Netherlands. (1929) Dr, chemistry (U. Groningen, 1967). Wetensch. hoofdmedew. (tel. 071 + 148333, ext. 4411). *Inorganic and organic structures, disorder.*

Versteeve, Dr Abraham Jan. Refractories Dept., Billiton Res. BV, Westervoortsedijk 67, P.O. Box 40, 6800 AA Arnhem, The Netherlands. (1942) Dr, petrology (U. Utrecht, 1974). Geologist. (tel. 085 + 654911). *Natural and industrial mineralogy, petrology, ceramics, gemmology.*

Visser, Dr Rudolph Joseph Jacobus. Lab. voor Anorganische Chemie, U. of Groningen, Nijenborgh 16, 9747 AG Groningen, The Netherlands. (1953) Dr, chemistry (U. Groningen, 1984). Wetensch. medew. (tel. 050 + 117102). *Diffuse X-ray scattering, phase transitions, one-dimensional structures.*

Visser, Drs Jan Willem. X-ray Dept., Technisch Physische Dienst TNO-TH, Stieltjesweg 1, P. O. Box 155, 2600 AD Delft, The Netherlands. (1925) Drs, chemistry (U. Amsterdam, 1955). Head X-ray Dept. (tel. 015 + 787004 or 787130) (telex 38091). *Powder diffraction, automation.*

Vonk, Dr Christ Gysbertus. Beatrixlaan 12, 6165 CX Geleen, The Netherlands. (1925) Dr, physical chemistry (U. Groningen, 1957). (tel. 04494 + 42153). *Polymers, small angle X-ray scattering.*

Vos-Looyenga, Dr Aafje. Nieuwe Plantage 68, 2611 XL Delft, The Netherlands. (1928) Dr, structural chemistry (U. Groningen, 1955). (tel. 015 + 135303). *Accurate structure determinations, pseudo one-dimensional crystals, oligopeptides.*

Vries, de, Drs Johan Louis. Humperdincklaan 51, 5654 PB Eindhoven, The Netherlands. (1920) Drs, chemistry (U. Amsterdam, 1950). (tel. 040 + 522102). *Instrumentation, powder diffraction in industry.*

Vucht, van, Dr Johannes Hendrikus Nicolaas. Isodorusweg 23, 5624 KD Eindhoven, The Netherlands. (1924) Dr, technical sciences (Eindhoven U. of Techn., 1963). Sr. res. (tel. 040 + 446196). *Intermetallic compounds, inorganic compounds, structure - physical properties relation.*

Waal, van de, Ir Benjamin Willem. Chemical Physics Lab., Twente U. of Techn., P. O. Box 217, 7500 AE Enschede, The Netherlands. (1936) Ir, physics (Delft U. of Techn., 1966). Wetensch. medew. (tel. 053 + 892664). *Molecular packing, intermolecular forces.*

Wagner, Dr Anton Johan. Vakgroep Chemische Fysica, U. of Groningen, Nijenborgh 16, 9747 AG groningen, The Netherlands. (1933) Dr, chemistry (U. Groningen, 1966). Wetensch. hoofdmedew. (tel. 050 + 117104). *X-ray diffraction, molecular structure, quantum chemistry.*

Wal, van der, Dr Robert Jan. SARA, Ondersteuning, U. of Amsterdam, Kruislaan 415, 1098 SJ Amsterdam, The Netherlands. (1955) Dr, chemistry (U. Groningen, 1982). Sr. system progr. (tel. 020 + 5923000). *X-ray crystallography, accurate electron density determination, electron density distribution interpretation, supercomputers.*

Wevers, Drs Joice. Inst. voor Aardwetenschappen, U. of Utrecht, Budapestlaan 4, P. O. Box 80021, 3508 TA Utrecht, The Netherlands. (1945) Drs, geology (U. Utrecht, 1969). Wetensch. hoofdmedew. (tel. 030 + 535098). *Spectroscopy (visible and infra-red) of minerals, X-ray diffraction.*

Wiebenga, Prof. Dr Eelco Herman. 5 Allée des Cimes (la Pinède), 83420 La Croix-Valmer, France. (1913) Dr, chemistry (U. Utrecht, 1940). Em., U. Groningen. (tel. 94 + 796853). *Quantum chemistry.*

Wierenga, Dr Rikkert Klaas. Lab. voor Chemische Fysica, U. of Groningen, Nijenborgh 16, 9747 AG Groningen, The Netherlands. (1949) Dr, chemistry (U. Groningen, 1978). Wetensch. medew. (tel. 050 + 117100). *Protein crystallography, drug design, enzyme design.*

With, de, Dr Gijsbertus. Philips Res. Lab., Ned. Philips Bedrijven BV, P. O. Box 80000, 5600 JA Eindoven, The Netherlands. (1950) Dr, chemical physics (Twente U. of Techn., 1978). Res. scient. (tel. 040 + 742132). *Oxides, nitrides, mechanical and chemical properties.*

Woensdregt, Drs Cornelis Franciscus. Inst. voor Aardwetenschappen, U. of Utrecht, Budapestlaan 4, P. O. Box 80021, 3508 TA Utrecht, The Netherlands. (1937) Drs, geology (U. Leiden, 1963). Wetensch. hoofdmedew. (tel. 030 + 535070). *Crystal growth, crystal morphology, electron microscopy, X-ray diffraction, computer programming.*

Wiegers, Dr Gerrit Adriaan. Lab. voor Anorganische Chemie, U. of Groningen, Nijenborgh 16, 9747 AG Groningen, The Netherlands. (1930) Dr, chemistry (U. Groningen, 1963). Wetensch. hoofdmedew. (tel. 050 + 117033). *Inorganic crystal structures, transition element chalcogenides, phase transitions, powder diffraction.*

Wolff, de, Dr Pieter Maarten. Meermanstraat 126, 2614 AM Delft, The Netherlands. (1919) Dr, physics (Delft U. of Techn., 1951). Em., Delft U. of Techn. (tel. 015 + 120396). *Symmetry, phase transitions.*

Zeedijk, Ir Hendrik Bastiaan. Metaalinst. TNO, Laan van Westenenk 501, 7334 DT Apeldoorn, The Netherlands. (1936) Ir, physical chemistry (Delft U. of Techn., 1960). Sr. sci. (tel. 055 + 773344). *Electron microscopy, metallography, micro-analysis.*

Zwaan, Dr Pieter Cornelis. National Museum of Geology and Mineralogy, Hooglandse Kerkgracht 17, 2312 HS Leiden, The Netherlands. (1928) Dr, mineralogy (U. Leiden, 1955). Wetensch. hoofdmedew, keeper of minerals. (tel. 071 + 124741). *X-ray diffraction, instrumentation, gem minerals.*

NEW ZEALAND

Sub-Editor: C.E.F. Rickard

Notes

1. International telephone country code - 64

2. Degrees conferred by New Zealand universities are generally similar to British Degrees.

3. Department of Scientific and Industrial Research is abbreviated to D.S.I.R.

Aldridge, Dr Laurence Philip. Chemistry Div., D.S.I.R., Private Bag, Petone, New Zealand. (1945) PhD, chemistry (Otago U., 1971). Res. chemist. (tel. 4 + 666-919, ext. 470). *Zeolites, cement.*

Baker, Dr Edward Neill. Chemistry Dept., Massey U., Palmerston North, New Zealand. (1942) PhD, chemistry (Auckland U., 1968). Sr. lect. (tel. 63 + 69-089, ext. 773). *Protein structures.*

Bates, Prof. Richard Heaton Tunstall. Dept. of Electrical Eng., U. of Canterbury, Private Bag, Christchurch, New Zealand. (1929) DSc, engineering (London U., UK, 1972). Prof. (tel. 3 + 71-649, ext. 336). *Applied electromagnetics, computationally-orientated diffraction theory, image processing, computer modelling, man-machine interaction, applied Fourier theory.*

Beckingsale, Mr Peter Gerard. Chemistry Dept., U. of Auckland, Private Bag, Auckland, New Zealand. (1952) MSc, chemistry (Auckland U., 1976). Technician. (tel. 9 + 792-300, ext. 9274). *Computing, organic and organometallic compounds.*

Brown, Dr Kevin Laurie. Chemistry Div., D.S.I.R., Private Bag, Taupo, New Zealand. (1946) PhD, chemistry (Auckland U., 1972). Res. chemist. (tel. 074 + 48211, ext. 846). *Inorganic and organic crystal structures.*

Couldwell, Dr Margaret Claire. Computer Centre, Massey U., Palmerston North, New Zealand. (1950) PhD, chemistry (Canterbury U., 1974). Computer Scient. (tel. 63 + 69099, ext. 8564). *Structural chemistry, organometallic compounds, molecular complexes.*

Childs, Dr Cyril Walter. Soil Bureau, D.S.I.R., Private Bag, Lower Hutt, New Zealand. (1941) PhD, chemistry (Otago U., 1967). Sci. (tel. 4 + 673-119, ext. 875). *Soil mineralogy, Mössbauer spectroscopy.*

Churchman, Dr Gordon John. Soil Bureau, D.S.I.R., Private Bag, Lower Hutt, New Zealand. (1944) PhD, chemistry (Otago U., 1970). Res. chemist. (tel. 4 + 673-119, ext. 876). *Clay mineralogy, surface chemistry.*

Claridge, Dr Graeme Geoffrey. Soil Bureau, D.S.I.R., Private Bag, Lower Hutt, New Zealand. (1931) PhD, chemistry (Auckland U., 1955). Res. chemist. (tel. 4 + 673-119). *Clay mineral formation in soils.*

Clark, Dr George Raymond. Chemistry Dept., U. of Auckland, Private Bag, Auckland, New Zealand. (1942) PhD, chemistry (Auckland U., 1968). Assoc. prof. (tel. 9 + 737999, ext. 8294). *Organometallic structures, bioinorganic structures.*

Coombs, Prof. Douglas Saxon. Geology Dept., U. of Otago, P.O. Box 56, Dunedin, New Zealand. (1924) Hon. DSc, (Geneva U., Switzerland, 1974). Prof. (tel. 24 + 771-640, ext. 844). *Mineralogy, metamorphic and volcanic rocks, zeolites.*

Cooper, Dr Alan Frederick. Geology Dept., U. of Otago, P.O. Box 56, Dunedin, New Zealand. (1945) PhD, petrology (Otago U., 1970). Lect. (tel. 24 + 771-640, ext. 601). *Mineralogy, metamorphic petrology, carbonatite mineralogy and petrology.*

Cutfield, Dr John Franklin. Biochemistry Dept., U. of Otago, P.O. Box 56, Dunedin, New Zealand. (1945) PhD, chemistry (Auckland U., 1970). Sr. lect. (tel. 24 + 771-640, ext. 618). *Protein structure-function studies.*

Evans, Mr David Lindsay. 3a Snowdon Road, Christchurch, New Zealand. New Zealand. (1949) MSc, chemistry (Canterbury U., 1981). Medical scientist. (tel. 3 + 516-134). *Charge transfer complexes, structure and physical properties, computer management techniques, data analysis.*

Freeman, Dr Alan George. Chemistry Dept., Victoria U. of Wellington, Private Bag, Wellington, New Zealand. (1935) PhD, chemistry (Aberdeen U., UK, 1962). Sr. lect. (tel. 4 + 721-000, ext. 772). *Solid state reactions, layer structures, defect structures.*

Gainsford, Dr Graeme John. Chemistry Div., D.S.I.R., Private Bag, Petone, New Zealand. (1945) PhD, chemistry (Canterbury U., 1969). Scientist. (tel. 4 + 666-919, ext. 682). *Computing techniques, crystal structure, powder diffraction.*

Hall, Dr David. U. Grants Committee, P.O. Box 12-348, Wellington, New Zealand. (1928) DSc, chemistry (Auckland U., 1969). Chairman. (tel. 4 + 728-600). *Molecular packing, conformation.*

Jones, Dr Tony Cristofer. Chemistry Dept., U. of Auckland, Auckland, New Zealand. (1955) PhD, chemistry (York U., UK, 1980). Techn. officer. (tel. 9 + 737-999, ext. 8274). *Computing, pulsed NMR, metal hydrides, hydrogen energy.*

Kawachi, Dr Yosuke. Geology Dept., U. of Otago, P.O. Box 56, Dunedin, New Zealand. (1932) PhD, geology (Otago U., 1970). Sr. res. officer. (tel. 24 + 771-640, ext. 722). *Petrology, metamorphism, mineralogy.*

Kirkman, Dr John Henry. Soil Sci. Dept., Massey U., Palmerston North, New Zealand. (1938) PhD, mineralogy (Aberdeen U., UK, 1965). Sr. lect. (tel. 63 + 69-099). *Clay mineralogy, soils and volcanic ash.*

Lyons, Dr Karen. Food Techn. Dept., Massey U., Private Bag, Palmerston North, New Zealand. (1955) PhD, chemistry (Auckland U., 1981). Res chemist (tel. 63 + 69-099, ext. 2448). *Organometallic structures.*

Lyons, Mr Paul John. Computer Sci. Dept., Massey U., Palmerston North, New Zealand. (1953) MSc, chemistry (Auckland U., 1977). Lect. (tel. 63 + 69-099). *Molecular packing analysis, artificial intelligence applied to chemistry, protein structure, programming languages.*

March, Dr Frank Conroy. Physics & Eng. Lab., D.S.I.R., Private Bag, Petone, New Zealand. (1944) PhD, crystallography (Canterbury U., 1970). Scientist. (tel. 4 + 666-919). *Organometallic molecular structures.*

Maslen, Dr Hugh Stafford. Chemistry Dept., U. of Auckland, Private Bag, Auckland, New Zealand. (1924) PhD, chemistry (Auckland U., 1974). Sr. lect. (tel. 9 + 737999, ext. 8291). *Structural chemistry, molecular conformation.*

Oliver, Dr Peter John. N.Z. Geological Survey, P.O. Box 30368, Lower Hutt, New Zealand. (1948) PhD, geochemistry (Canterbury U., 1977). Geologist. (tel. 4 + 699-059). *Petrology and geochemistry, volcanic rocks, hydrothermal geochemistry and crystal chemistry, paleomagnetism.*

Page, Dr Campbell Thomas. Textile Chemistry Dept., Wool Research Organisation of N.Z., Private Bag, Christchurch, New Zealand. (1951) PhD, chemistry (Canterbury U., 1979). Res. scientist. (tel. 3 + 252-421). *Structure, transition metal complexes containing sulphur donor ligands.*

Parry, Dr David Anthony Dougall. Biophysics Dept., Massey U., Palmerston North, New Zealand. (1942) PhD, biophysics (London U., UK, 1966). Reader. (tel. 63 + 69-099, ext. 473). *Structural and functional studies, proteins, muscle, collagen, keratin.*

Penfold, Prof. Bruce Russell. Chemistry Dept., U. of Canterbury, Private Bag, Christchurch, New Zealand. (1927) PhD, chemistry (Cambridge U., UK, 1952). Prof. (tel. 3 + 482-009). *Organometallic and inorganic structures, computer aided teaching, information retrieval.*

Percival, Dr Henry Joseph. NZ Soil Bureau, DSIR, Eastern Hutt Rd., Taita, Private Bag, Lower Hutt, New Zealand. (1943) PhD, chemistry (Victoria U. of Wellington, 1970). Scientist. (tel. 4 + 673-119, ext. 857). *Soils - physical chemistry, clay mineralogy.*

Rickard, Dr Clifton Edward Frank. Chemistry Dept., U. of Auckland, Private Bag, Auckland, New Zealand. (1941) PhD, chemistry (Auckland U., 1967). Sr. lect. (tel. 9 + 737999, ext. 8289). *Structural chemistry, coordination compounds of the actinide elements.*

Robinson, Dr Ward Thomas. Chemistry Dept., U. of Canterbury, Private Bag, Christchurch, New Zealand. (1937) PhD, chemistry (Canterbury U., 1964). Reader. (tel. 3 + 482-009, ext. 294). *Structure analysis, symmetry, model building, computing.*

Rodgers, Dr Kerry Anthony. Geology Dept., U. of Auckland, Private Bag, Auckland, New Zealand. (1942 PhD, geology (Auckland U., 1972). Assoc. prof. (tel. 9 + 737-999 ext 7414). *Crystal chemistry, minerals.*

Rumball, Dr Sylvia Vine. Chemistry Dept., Massey U., Palmerston North, New Zealand. (1939) PhD, chemistry (Auckland U., 1966). Sr. lect. (tel. 63 + 69-089, ext. 7958). *Protein three-dimensional structure.*

Shelley, Dr David. Geology Dept., U. of Canterbury, Private Bag , Christchurch, New Zealand. (1940) PhD, geology (Bristol U., UK, 1964). Sr. lect. (tel. 3 + 482-009, ext. 693). *Crystal growth, silicate mineralogy.*

Smale, Mr David. N.Z. Geological Survey, U. of Canterbury, Private Bag, Christchurch, New Zealand. (1939) MSc, geology (Auckland U., 1962). Sedimentary petrologist. (tel. 3 + 482009, ext. 593). *Detrital mineralogy, sedimentary petrology.*

Smalley, Dr Ian James. Soil Bureau, D.S.I.R., Private Bag, Lower Hutt, New Zealand. (1936) PhD, materials science (City U., London, UK, 1966). Sci. (tel. 4 + 673-119, ext. 883). *Clay minerals, quartz particles in sediments.*

Stewart, Dr Robert Bruce. Soil Sci. Dept., Massey U., Palmerston North, New Zealand. (1951) PhD, soil science (Massey U, 1983). Lect. (tel. 63 + 69-089, ext. 2454). *Mineralogy, weathering processes, isotope distribution.*

Waters, Assoc. Prof. Joyce Mary. Chemistry Dept., U. of Auckland, Private Bag, Auckland, New Zealand. (1931) PhD, chemistry (New Zealand U., 1960). Assoc. prof. (tel. 9 + 792-300, ext. 9273). *Inorganic and organic structures, large molecules.*

Waters, Dr Thomas Neil Morris. Massey U., Palmerston North, New Zealand. (1931) DSc, (Auckland U., 1969). Vice-Chancellor. (tel. 63 + 69-099, Ext. 733). *Inorganic and organic structures, large molecules.*

Watters, Dr William Asher. N.Z. Geological Survey, P.O. Box 30368, Lower Hutt, New Zealand. (1926) PhD, petrology (Cambridge U., UK, 1956). Chief petrologist. (tel 4 + 699-059). *Mineralogy of rock-forming minerals.*

Weaver, Dr Stephen Donald. Geology Dept., U. of Canterbury, Christchurch, New Zealand. (1947) PhD, geology (London U., UK, 1973). Lect. (tel. 3 + 482-009, ext. 569). *Silicate minerals.*

Whimp, Dr Peter Olaf. Physics and Eng. Lab., D.S.I.R., Private Bag, Lower Hutt, New Zealand. (1942) PhD, chemistry (Victoria U. of Wellington, 1967). Sci. (tel. 4 + 666-919, ext. 547). *Coordination chemistry, organometallics, computing.*

NIGERIA

Sub-Editor: O.O. Adewoye

Notes

1. International telephone country code - 234.

Adetunji, Dr Jacob. Physics Dept., Ahmadu Bello U., Main Campus, Zaria, Kaduna State, Nigeria. (1944) PhD, physics (Essex U., UK, 1976). Lect. *Electron microscopy, radiation damage in crystalline solids.*

Adewoye, Dr Olusegun Oyeleke. Metallurgical and Materials Eng. Dept., U. of Ife, Faculty of Techn., Ile-Ife, Oyo State, Nigeria. (1947) PhD, materials science (Cambridge U., UK, 1976). Sr. lect. (tel. Ife 2290-2299, telex IFEVASITY IFE). *Structure determination, crystallography, ceramics, deformation modes, coal diffraction.*

Aladekomo, Prof. Johnson Bandele. Physics Dept., U. of Ife, Ife Campus, Ile-Ife, Oyo State, Nigeria. (1938) PhD, physics (Manchester U., UK, 1965). Prof. *Exciton diffusion, molecular crystals, Luminescence effects, crystal impurities.*

Dubey, Dr Ram Janam. Chemistry Dept., U. of Maiduguri, P.M.B. 1069, Maiduguri, Nigeria. (1941) DSc, crystallography (Laval U., Canada, 1973). Asst. prof. *Synthesis, structure, bonding, organometallics; protein structure, impurities.*

Koshy, Dr Jacob. Physics Dept., U. of Ibadan, 1 Parry Road, Ibadan, Nigeria. (1942) PhD, solid state physics (Sardar Patel U., 1970). Sr. lect. (tel. 462550, ext. 1042). *Crystal imperfections, crystal growth, electron microscopy, epitaxial growth, thin films.*

Onyeagocha, Dr Anthony Chukwuma. Geology Dept., U. of Nigeria, Nsukka Campus, Nsukka, Anambra State, Nigeria. (1942) PhD, geology (U. Washington, USA, 1973). Lect. *Petrology, crystal chemistry, phase equilibria, geochemistry, mineralogy.*

Sanni, Dr Bamidele. Chemistry Dept., U. of Benin, Benin City, Nigeria. (1943) PhD, chemistry (Ibadan U., 1974). Lect. (tel. Benin City 343). *Large molecules, organic structures.*

Sharma, Dr (Mrs) Aysel. Physics Dept., U. of Benin, B28 Ugbowo Campus, Benin City, Bendel State, Nigeria. (1941) PhD, Physics (St. Andrews U., UK, 1973). Lect. *Crystal structure analysis, organic compounds, X-ray and neutron diffraction.*

Sharma, Dr Vinod Chander. Physics Dept., U. of Benin, B28 Ugbowo Campus, Benin City, Bendel State, Nigeria. (1940) PhD, Physics (St. Andrews U., UK, 1973). Sr. lect. *Kinematic and dynamic X-ray diffraction, topography.*

NORWAY

Sub-Editor: P. Groth

Notes

1. International telephone country code - 47.

2. Degrees conferred by the Norwegian universities are the *Doctor philosophiae* (Dr philos.) and *Doctor technicae* (Dr techn.)(both approximately equivalent to the English DSc), *Doctor scientiarium* (Dr scient.) and *Doctor ingenieur* (Dr ing.)(both equivalent to PhD), *Candidatus realium* (Cand. real.) and *Magister scientiarium* (Mag. scient.)(both range between PhD and MSc), *Candidatus scientiarum* (Cand.scient.) and *Sivil ingenieur* (siv.ing.)(approximately equivalent to MSc).

3. The position *førstelektor* and *førsteamanuensis* both correspond to senior lecturer, whereas *amanuensis* corresponds to lecturer.

Almenningen, Mr Arne. Dept. of Chemistry, U. of Oslo, 0315 Oslo 3, Norway. (1921) Cand. real., physics (Oslo U.), 1952). Førsteamanuensis. (tel. 02 + 455408). *Electron diffraction, instrumentation.*

Almlöf, Prof. Jan. Dept. of Chemistry, U. of Oslo, 0315 Oslo 3, Norway. (1945) Fil.dr, chemistry (Uppsala U., Sweden, 1970). Prof. (tel. 02 + 455430) *Organic crystal structures, hydrogen bonds, theoretical chemistry.*

Alver, Mr Eyvind. Dept. of Chemistry, U. of Bergen, N-5000 Bergen, Norway. (1922) Cand. real., chemistry (Oslo U.), 1954). Sr. lect. (tel. 05 + 213544). *Structural chemistry, philosophy of scientific method.*

Andersen, Dr Per. Dept. of Chemistry, U. of Oslo, 0315 Oslo 3, Norway. (1919) Dr philos., chemistry (Oslo U., 1968). Førsteamanuensis. (tel. 02 + 455458). *Electron diffraction, free radicals.*

Andresen, Dr Arne Fridtjof. Dept. of Neutron Physics, Inst. for Energy Techn., P.O. Box 40, 2007 Kjeller, Norway. (1926) Dr philos., physics (Oslo U., 1972). Asst. div. head. (tel. 02 + 712560 ext. 294). *Neutron diffraction, inorganic crystal structures, magnetic properties.*

Åse, Mr Kjell. Dept. of Chemistry, U. of Bergen, 5014 Bergen, Norway. (1939) Cand. real., chemistry (Bergen U., 1966). Lect. (tel. 05 + 213561) *Inorganic crystal structures.*

Bastiansen, Prof. Otto Christian Astrup. Dept. of Chemistry, U. of Oslo, 0315 Oslo 3, Norway. (1918) Dr philos., physical chemistry (Oslo U., 1949). Prof. of physical chemistry. (tel. 02 + 455401). *Electron diffraction, molecular structure.*

Björnevåg, Mr Sturle Vegard. Dept. of Chemistry, U. of Bergen, 5014 Bergen, Norway. (1952) Cand. real., inorganic chemistry (Bergen U., 1980). Res. asst. (tel. 05 + 213547). *Selenium complexes.*

Bremer, Dr Johannes. Dept. of Physics and Mathematics, U. of Trondheim-NTH, N-7034 Trondheim-NTH, Norway. (1949) Dr ing., (U. Trondheim-NTH, 1979). Førsteamanuensis. (tel. 07 + 593582). *Spectroscopy, scattering effects, surface physics.*

Bye, Dr Erik. Inst. of Occupational Health, P.O. Box 8149, 0033 Oslo 1, Norway. (1945) Dr philos., chemistry (Oslo U., 1976). Scient. (tel. 02 + 466850 ext. 784). *Biologically active molecules, powder diffraction, mineral dust, inorganic dust.*

Dahl, Dr Tor. Dept. of Chemistry, Inst. of Mathematical and Physical Sci., U. of Tromsö, P.O. Box 953, N-9001 Tromsö, Norway. (1938) Dr philos., chemistry

(Tromsö U., 1976). Sr. lect. (tel. 083 + 81688, ext. 332). *Crystal structures, organic charge transfer compounds.*
Fernholt, Mrs Liv. Dept. of Chemistry, U. of Oslo, 0315 Oslo 3, Norway. (1915) Siv. ing, chemistry (Norges Tekniske Högskole, 1939). Res. asst. (tel. 02 + 455406). *Electron diffraction.*
Fjaer, Dr Erling. Dept. of Physics and Mathematics, U. of Trondheim-NTH, N-7034 Trondheim-NTH, Norway. (1951) Dr ing., physics (U. Trondheim-NTH, 1983). Res. asst. (tel. 07 + 593583). *X-ray diffraction, modulated crystal structures.*
Fjeldberg, Mr Torgny. Dept. of Chemistry, U. of Trondheim, N-7055 Dragvoll, Norway. (1953) Cand. real., chemistry (Oslo U., 1978). Res. assoc. (tel. 07 + 596222). *Sterically hindered metal amides and -alkyls, gas-phase electron diffraction.*
Foss, Prof. Olav. Dept. of Chemistry, U. of Bergen, 5014 Bergen, Norway. (1918) Dr techn., chemistry (Norges Tekniske Högskole, 1947). Prof. (tel. 05 + 213549). *Inorganic crystal structures.*
Furuseth, Mrs Sigrid. Dept. of Chemistry, U. of Oslo, 0315 Oslo 3, Norway. (1939) Cand. real., chemistry (Oslo U., 1964). Försteamanuensis. (tel. 02 + 455561). *Alloy structures.*
Gjönnes, Prof. Jon Kjell. Dept. of Physics, U. of Oslo, 0316 Oslo 3, Norway. (1931) Dr philos., physics (Oslo U., 1967). Prof. (tel. 02 + 456490). *Electron diffraction, microscopy, inorganic materials.*
Grjotheim, Prof. Kai. Dept. of Chemistry, U. of Oslo, 0315 Oslo 3, Norway. (1919) Dr techn., chemistry (Norges Tekniske Högskole, 1956). Prof. (tel. 02 + 455039). *Inorganic crystal structures.*
Groth, Mr Per Arne. Dept. of Chemistry, U. of Oslo, 0315 Oslo 3, Norway. (1934) Cand. real., chemistry (Oslo U., 1960). Res. fellow. (tel. 02 + 455692). *Organic crystal structures, programming.*
Grönvold, Prof. Fredrik. Dept. of Chemistry, U. of Oslo, 0315 Oslo 3, Norway. (1924) MSc, metallurgy (Michigan U., USA, 1951). Prof. (tel. 02 + 455599). *Inorganic structures and transitions.*
Gundersen, Miss Grete. Dept. of Chemistry, U. of Oslo, 0315 Oslo 3, Norway. (1940) Cand. real., chemistry (Oslo U., 1967). Amanuensis. (tel. 02 + 455695). *Electron diffraction, molecular structure.*
Haaland, Dr Arne. Dept. of Chemistry, U. of Oslo, 0315 Oslo 3, Norway. (1936) Dr philos., chemistry (Oslo U., 1969). Sr. lect. (tel. 02 + 455407). *Molecular structure, organometallic compounds.*
Hadler, Mrs Eva. Dept. of Pharmacy, U. of Oslo, 0316 Oslo 3, Norway. (1921) Cand. real., chemistry (Oslo U., 1950). Försteamanuensis. (tel. 02 + 456580). *Inorganic and organic crystal structures, large molecules and molecular biology.*
Hagen, Dr Kolbjörn. Dept. of Chemistry, U. of Trondheim, N-7055 Dragvoll, Norway. (1943) Dr philos., physical chemistry (Trondheim U., 1979). Sr. lect. (tel. 07 + 596223). *Determination of structure and conformation of gas-phase molecules, electron diffraction.*
Hansen, Mr Lars Kristian. Dept. of Chemistry, Inst. of Mathematical and Physical Sci., U. of Tromsö, P.O. Box 953, N-9001 Tromsö, Norway. (1944) Cand. real., (Bergen U., 1971). Res. fellow. (tel. 083 + 81688, ext.338). *Organic crystal structures, protein crystallogrphy.*
Hauback, Mr Björn Christian. Dept. of Physics and Mathematics, U. of Trondheim-NTH, N-7034 Trondheim-NTH, Norway. (1957) Siv. ing., physics (U. Trondheim-NTH, 1981). Res. asst. (tel. 07 + 593584). *X-ray diffraction, spectroscopy, electron density distribution.*
Hauge, Dr Sverre. Dept. of Chemistry, U. of Bergen, 5014 Bergen, Norway. (1932) Dr philos., chemistry (Bergen U., 1978). Sr. lect. (tel. 05 + 213545). *Inorganic complexes.*
Hordvik, Prof. Asbjörn. Dept. of Chemistry, Inst. of Mathematical and Physical Sci., U. of Tromsö, P.O. Box 790, N-9001 Tromsö, Norway. (1928) Dr philos., (Bergen U., 1968). Prof. (tel. 083 + 81688 + ext. 327). *Inorganic and organic structures, protein crystallography.*
Hough, Dr Edward. Dept. of Chemistry, Inst. of Mathematical and Physical Sci., U. of Tromsö, P.O. Box 953, N-9001 Tromsö, Norway. (1941) PhD, X-ray crystallography (London U., UK, 1975). Sr. lect. (tel. 083 + 81688, ext. 262). *Biologically important organic molecules, biological metal complexes, protein crystallography.*
Husebye, Prof. Steinar. Dept. of Chemistry, U. of Bergen, 5014 Bergen, Norway. (1933) PhD, chemistry (Tulane U., USA, 1963). Prof. (tel. 05 + 213551). *Complexes with central Se and Te, sterical influence of lone pairs in ions.*
Hvoslef, Dr Jan. Dept. of Chemistry, U. of Oslo, 0315 Oslo 3, Norway. (1926) Dr philos., chemistry (Oslo U., 1972). Sr. lect. (tel. 02 + 455413). *Structures, biologically important molecules, ascorbic acid and derivatives.*
Höier, Prof. Ragnvald. Dept. of Physics and Mathematics, U. of Trondheim-NTH, N-7034 Trondheim-NTH, Norway. (1938) Dr philos., physics (Oslo U., 1973). Prof. (tel. 07 + 593588). *X-ray diffraction, electron diffraction, electron microscopy.*
Jynge, Mr Knut. Dept. of Chemistry, Inst. of Mathematical and Physical Sci., U. of Tromsö, P.O. Box 953, N-9001 Tromsö, Norway. (1933) Cand. real., chemistry (Tromsö U., 1976). Res. asst. (tel. 083 + 81688, ext. 328). *Physical chemistry, biologically important structures, protein crystallography.*
Kjekshus, Prof. Arne. Dept. of Chemistry, U. of Oslo, 0315 Oslo 3, Norway. (1932) Dr philos., (Oslo U., 1971). Prof. (tel. 02 + 455560). *Inorganic crystal structures, metals structures, magnetic structures.*

Klewe, Mr Bernt. Dept. of Chemistry, U. of Oslo, 0315 Oslo 3, Norway. (1933) Cand. real., chemistry (Oslo U., 1961). Sr. lect. (tel. 02 + 455463). *Organic crystal structures.*
Maartman-Moe, Mr Knut. Dept. of Chemistry, U. of Bergen, 5014 Bergen, Norway. (1928) Cand. real., inorganic chemistry (Bergen U., 1961). Sr. lect. (tel. 05 + 213446). *X-ray diffractometry, computing, organic crystal structures.*
Markali, Mr Joar. Dept. of Materials Res., Central Inst. for Industrial Res., Forskningsveien 1, 0371 Oslo 3, Norway. (1921) Cand. real., chemistry (Oslo U., 1951). Head, div. of characterization of materials. (tel. 02 + 452010). *Solide state chemistry, physical chemistry, high temperature oxidation, electron microscopy, electron diffraction.*
Martinsen, Mr Knut. Dept. of Physics and Mathematics, U. of Trondheim-NTH, N-7034 Trondheim-NTH, Norway. (1956) Siv. ing., physics (U. Trondheim-NTH, 1981). Univ. scholar. (tel. 07 + 593589). *Electron diffraction, x-ray diffraction, solid state physics.*
Maröy, Dr Kjartan. Dept. of Chemistry, U. of Bergen, 5014 Bergen, Norway. (1930) Dr philos., inorganic chemistry (Bergen U., 1976). Sr. lect. (tel. 05 + 213546). *Thelurium complexes, ploythionates.*
Mladeck, Mr Micael Hiorth. Dept. of Geology, U. of Oslo, 0315 Oslo 3, Norway. (1927) Dr phil. nat., mineralogy (Bern U., Switz., 1962) Amanuensis. (tel. 02 + 456678). *Minerals, crystal chemistry, crystal structures.*
Mo, Prof. Frode. Dept. of Physics and Mathematics, U. of Trondheim-NTH, N-7034 Trondheim-NTH, Norway. (1937) Dr techn., crystallograpy (U. Trondheim-NTH, 1980). Prof. (tel. 07 + 593585). *Electron density in solids, accurate structure studies, x-ray diffraction, multiple scattering effects.*
Mostad, Mr Arvid. Dept. of Chemistry, U. of Oslo, 0315 Oslo 3, Norway. (1929) Cand. real., chemistry (Oslo U., 1959). Sr. lect. (tel. 02 + 455415). *Molecular structure - biological activity relationships.*
Nicholson, Dr David Graham. Dept. of Chemistry, U. of Trondheim, N-7055 Dragvoll, Norway. (1944) PhD, inorganic chemistry (London U., UK, 1969). Sr. lect. (tel. 07 + 596204). *Main-group chemistry, lower oxidation states.*
Nordenson, Mr Svein. Dept. of Chemistry, U. of Oslo, 0315 Oslo 3, Norway. (1949) Cand. real., chemistry (Oslo U., 1976). Res. fellow. (tel. 02 + 455452). *Organic crystal structures.*
Norman, Prof. Nico. Dept. of Physics, U. of Oslo, 0316 Oslo 3, Norway. (1919) Dr philos., (Oslo U., 1956). Prof. (tel. 02 + 456434). *Crystal structures, imperfections.*
Olsen, Dr Arne. Dept. of Physics, U. of Oslo, 0316 Oslo 3, Norway. (1944) Dr philos., physichs (Oslo U., 1978). Försteamanuensis. (tel. 02 + 456432). *Electron diffraction, microscopy, inorganic materials.*
Pedersen, Dr Berit Fjærtoft. Dept. of Pharmacy, U. of Oslo, 0316 Oslo 3, Norway. (1933) Dr philos., (Oslo U., 1969). Sr. lect. (tel. 02 + 455694). *Organic and inorganic crystal structures.*
Pedersen, Prof. Björn. Dept. of Chemistry, U. of Oslo, 0315 Oslo 3, Norway. (1933) Dr philos., chemical physics (Oslo U., 1964). Prof. (tel. 02 + 455690). *Crystal structures, atomic motion, nuclear magnetic resonance.*
Raade, Mr Gunnar. Mineralogical-Geological Museum, U. of Oslo, Sarsgate 1, 0562 Oslo 5, Norway. (1944) Cand. real., mineralogy (Oslo U., 1973). Curator of minerals. (tel. 02 + 686960 ext. 147). *Minerals.*
Riste, Prof. Tormod. Dept. of Physics, Inst. for Energy Techn., P.O. Box 40, 2007 Kjeller, Norway. (1925) Dr philos., physics (Oslo U., 1961). Prof. (tel. 02 + 712560). *Neutron diffraction, solid state physics, statistical physics.*
Rosenqvist, Prof. Ivan Thoralf. Dept. of Geology, U. of Oslo, 0316 Oslo 3, Norway. (1916) Dr philos., mineralogy (Oslo U., 1945). Prof. (tel. 02 + 456652). *Clay mineralogy, rock forming minerals.*
Römming, Prof. Christian. Dept. of Chemistry, U. of Oslo, 0315 Oslo 3, Norway. (1928) Dr philos., chemistry (Oslo U., 1968). Prof. (tel. 02 + 455403). *Molecular structure - properties relationships.*
Röst, Mr Erling. Dept. of Chemistry, U. of Oslo, 0315 Oslo 3, Norway. (1924) Cand. real., chemistry (Oslo U., 1954). Sr. lect. (tel. 02 + 455613). *Alloy structures, phase equilibria.*
Samdal, Mr Svein. Oslo Engineering School, Cort Adelersgate 30, 0254 Oslo 2, Norway. (1945) Cand. real., chemistry (Oslo U., 1973). Lect. (tel. 02 + 566680). *Electron diffraction, conformational analysis.*
Samuelsen, Prof. Emil J. Dept. of Physics and Mathematics, U. of Trondheim-NTH, N-7034 Trondheim-NTH, Norway. (1937) Dr philos., physics (Oslo U., 1971). Prof. (tel. 07 + 593412). *Partly disordered solids, x-ray and neutron scattering, Raman spectroscopy.*
Schei, Dr Svanhild Helene. Dept. of Chemistry, U. of Trondheim, N-7055 Dragvoll, Norway. (1945) Dr philos., physical chemistry (Trondheim U., 1984). Res. assoc. (tel. 07 + 596219). *Molecular structure and conformation, gas-phase electron diffraction.*
Semmingsen, Dr Dag. Rogaland Distriktshögskole, P.O. Box 2540, Stavanger, N-4001, Norway. (1940) Dr philos., chemistry (Oslo U., 1976). Sr. lect. (tel. 04 + 874248). *Structure determinations by X-rays and neutrons, organic and inorganic compounds, phase transitions and cooperative phenomenons.*
Skjerpe, Mr Per Martin. Dept. of Physics, U. of Oslo, 0316 Oslo 3, Norway. (1953) Cand. real., physichs (Oslo U., 1980). Res. asst. (tel. 02 + 455049). *Transmission Electron Microscopy in Metallurgy.*

Sletten, Dr Einar. Dept. of Chemistry, U. of Bergen, 5014 Bergen, Norway. (1939) Dr philos., chemistry (Bergen U., 1979). Sr. lect. (tel. 05 + 213352). *Structures, metal complexes, nucleic acid compounds.*

Sletten, Dr Jorunn. Dept. of Chemistry, U. of Bergen, 5014 Bergen, Norway. (1941) Dr philos., chemistry (Bergen U., 1976). Sr. lect. (tel. 05 + 213562). *Organic crystal structures.*

Strand, Dr Tor Gogstad. Dept. of Chemistry, U. of Oslo, 0315 Oslo 3, Norway. (1934) Dr philos., (Oslo U., 1968). Sr. lect. (tel. 02 + 455411). *Electron diffraction.*

Strömme, Dr Knut Olaf. Dept. of Chemistry, U. of Oslo, 0315 Oslo 3, Norway. (1928) Dr philos., chemistry (Oslo U., 1974). Sr. lect. (tel. 02 + 455454). *Order-disorder in solids.*

Stölevik, Dr Reidar. Dept. of Chemistry, U. of Trondheim, N-7055 Dragvoll, Norway. (1938) Dr philos., physical chemistry (Oslo U., 1975). Sr. lect. (tel. 07 + 596224). *Structure and conformation, electron diffraction, molecular mechanics calculations.*

Svinning, Dr Torgeir. Div. of Materials and Processes, SINTEF, N-7034 Trondheim-NTH, Norway. (1948) Dr ing., crystallography (U. Trondheim-NTH, 1978). Res. scient. (tel. 07 + 592928). *Structure and mechanical properties, metal working.*

Sæthre, Mr Leif Jarle. Dept. of Chemistry, Inst. of Mathematical and Physical Sci., U. of Tromsö, P.O. Box 953, N-9001 Tromsö, Norway. (1945) Cand. real., (Bergen U., 1971). Sr. lect. (tel. 083 + 81688). *Molecular structures, ESCA, Auger spectroscopy.*

Thorkildsen, Dr Gunnar. Rogaland distriktshöyskole, P.O. Box 2540, Ullandshaug, N-4001 Stavanger, Norway. (1953) Dr ing., crystallography (U. Trondheim-NTH, 1983). Försteamanuensis. (tel. 04 + 874257). *X-ray diffraction and spectroscopy, charge density in solids, solid state physics.*

Tibballs, Dr John Earl. Central Inst. for Industrial Res., Forskningsveien 1, 0371 Oslo 3, Norway. (1947) PhD, physics (Melburn U., Australia, 1974). Res. scient. (tel. 02 + 452978). *Phase transitions, electron microscopy, welds, diffuse scattering, intermetallic phases.*

Trætteberg, Prof. Marit. Dept. of Chemistry, U. of Trondheim, N-7055 Dragvoll, Norway. (1930) Dr philos., physical chemistry (Trondheim U., 1970). Prof. (tel. 07 + 596225). *Physical organic chemistry, molecular structure, unsaturated compounds, conformational analysis, gas electron diffraction.*

PAKISTAN

Sub-Editors: S.S.H. Rizvi & A. Saghir

Notes

1. International telephone country code - 92.

2. Abbreviations used for the Pakistan entries include:
    PCSIR - Pakistan Council of Scientific and Industrial Research
    NPSL - National Physical and Standards Laboratory
    GSP - Geological Survey of Pakistan

Ahmad, Mr Anwaruddin M. GSP, 22- Ali Block, New Garden Town, Lahore- 16, Pakistan. (1934) MSc, geophysics (Punjab U., 1955). Superintending geophysicist. (tel. 85+ 5232). *Geophysics, mineral explorations, seismology.*

Ahmad, Mr Dabir. NPSL Centre, PCSIR Labs., Off University Road, Karachi, Pakistan. (1938) MSc, physics (Karachi U., 1961). Sr. res. officer. (tel. 460101, ext. 21). *Organic crystal structure, direct methods, instrumentation.*

Ahmad, Dr Zulfiqar. Centre of Excellence in Mineralogy, U. of Baluchistan, Sariab Road, Quetta, Pakistan. (1945) PhD, geology (London U., UK, 1982). Dir. *Mineralogy, crystallography, petrology, economic geology, geological mapping*

Akhtar, Mr Mohammad. GSP, 22- Ali Block, New Garden Town, Lahore- 16, Pakistan. (1944) MSc, geology (Sind U., 1965). Dpty. dir. (tel. 042 + 855922). *Stratigraphy, sedimentology.*

Akhtar, Mr Javed. GSP, 22- Ali Block, New Garden Town, Lahore- 16, Pakistan. (1955) MSc, geophysics (Punjab U., 1978). Asst. geophysicist. (tel. 042 + 855816). *Exploration geophysics.*

Ali, Dr Syed Wajahat. NPSL Centre, PCSIR Labs., off University Road, Karachi, Pakistan. (1937) PhD, physical chemistry (Karachi U., 1981). Sr. res. officer. (tel. 461371). *Solutions, liquid structures.*

Anwar, Mr Muhammad. GSP, 22- Ali Block, New Garden Town, Lahore- 16, Pakistan. (1953) MSc, geology (Punjab U., 1976). Asst. dir. (tel. 042 + 855922) *Micropaleontology.*

Baqri, Dr Syed Rafiqul-Hassan. Earth Sci. Div., Pakistan Museum of Natural History, Al-Markaz, F-7/2, Islamabad, Pakistan. (1945) PhD, X-ray diffraction (Southhampton U., UK, 1977). Dir. (tel. 82439). *X-ray diffraction, geochemical analysis.*

Butt, Dr Khursheed Alam. Atomic Energy Mineral Centre, Ferozepur Road, Lahore, Pakistan. (1947) PhD, petrology (New Brunswick U., Canada, 1976). Head. (tel. 042 + 870237, ext. 07) *Minerology, X-ray diffraction and spectroscopy, ore microscopy, texture, petrology.*

Butt, Mr Muhammad Hafeez. GSP, 22- Ali Block, New Garden Town, Lahore- 16, Pakistan. (1945) MSc. geology (Punjab U., 1968). Geophysicist. (tel. 042 + 855232). *Seismology, exploration geophysics.*

Chaudhary, Dr Abdul Majid. Nuclear Materials Div., Pakistan Inst. of Nuclear Sci. and Techn., Nilore, Rawalpindi, Pakistan. (1945) PhD, amorphous solids (New England U., Australia, 1978). Sr. scient. (tel. 42812). *X-ray diffraction, electrical properties, amorphous solids.*

Chaudhary, Dr Mr G. Sarwar Alam. GSP, (Punjab Div.), 14- Canal Park, Gulberg, Lahore- 11, Pakistan. (1944) MSc, mineralogy (Punjab U., 1967). Dpty. dir. (tel. 042+ 881192). *Igneous and metamorphic rocks, metallic and nonmetallic minerals.*

Chaudhry, Mr Mohammad Anwar. GSP, 14- Canal Park, Gulberg, Lahore- 11, Pakistan. (1944) MSc, geology (Peshawar U., 1972). Asst. dir. (tel. 042 + 881192). *Mineral exploration, sedimentary and igneous rocks.*

Chauhan, Mr Ehsanul Haq. Chem. Div., GSP, Sariab Road, Quetta, Pakistan. (1931) MS, geology (Idaho U., USA, 1968). Chief chemist. (tel. 081 + 72617). *Crystallography, petrology, mineralogy, exploration geochemistry.*

Chowdhry, Mrs Khursheed. CTS Div., PCSIR Labs., Off University Road, Karachi, Pakistan. (1941) MSc. physics (Peshawar U., 1968). Sr. res. officer. (tel. 460101), *X-ray spectroscopy.*

Elahi, Mr Manzoor. Metallurgy Group, Defence Sci. and Techn. Organization, Chaklala, Rawalpindi1, Pakistan. (1937) M.Tech., metallurgical quality control (Brunel U., UK, 1980). Res. officer. (tel. 64746, ext. 58). *Metallic materials, alloys, mathematical techniques in structure determination, crystal structure.*

Faruqi, Dr Fazal Ahmad. Glass and ceramics Div.,PCSIR Labs., Roomi Road Lahore- 16, Pakistan. (1927) PhD, glass and ceramics (London U., UK, 1960). Director (tel. 880763). *Mineralogy, non-metallic minerals, glass, ceramics, refractories, building materials.*

Fatmi, Prof. Ali Nasir. GSP, 22- Ali Block, New Garden Town, Lahore- 16, Pakistan. (1930) PhD, stratigraphy and paleontology (Wales U., 1968). Dpty. dir. general. (tel. 042+ 852826). *Mesozoic stratigraphy, paleontology.*

Fazal, Dr Muhammad. Investment Promotion Bureau, Kandawala Building, M.A. Jinah Road, Karachi, Pakistan. (1932) PhD, X-ray diffraction (Strathclyde U., UK, 1966). Dpty. dir. general. (tel. 718681). *Crystal structure, alloys, X-ray diffraction.*

Gillani, Mr Jamshed Ali. GSP, 22- Ali Block, New Garden Town, Lahore- 16, Pakistan. (1956) MSc, geochemistry (Karachi U., 1977). Asst. dir. (tel. 042 + 855816). *Geological mapping, sedimentary rocks.*

Habib, Mr Syed Abbas. GSP, 22- Ali Block, New Garden Town, Lahore- 16, Pakistan. (1936) MSc, sedimentology (Victoria U., New Zealand, 1971). Dpty. dir. (tel. 042 + 855923). *Sedimentology.*

Hasan, Dr Faizul. Metallurgical Eng. Dept., U. of Eng. and Techn., G.T. Road, Lahore, Pakistan. (1948) PhD, metallurgy (Manchester U., 1984). Assoc. prof. (tel. 042 + 339207) *X-ray diffraction, electron diffraction, crystallography, precipitation in alloys.*

Hussain, Dr Khadim. Centre For Solid State Physics, Punjab U. New Campus, Lahore- 20, Pakistan. (1947) PhD, ice nucleation (Victoria U., Manchester, UK, 1981). Asst. prof. (tel 854113). *Crystal structure analysis, X-ray diffraction.*

Ikram, Dr Nazma. Centre For Solid State Physics, Punjab U., Lahore, Pakistan. (1949) PhD, theoretical solid state physics (Cambridge U., U.K, 1976). Asst. prof. (tel. 854113). *Crystal structure analysis, electron diffraction, X-ray diffraction.*

Iqbal, Mr Mir Waseluddin Ahmad. GSP, 22- Ali Block, New Garden Town, Lahore- 16, Pakistan. (1926) MSc, geology (U. California, USA, 1964). Dir. (tel. 852547). *Tertiary bivalve and gastropod fauna, stratigraphy, regional correlations.*

Pakistan

Jafry Mr Syed Qamar Abbas. GSP, 22- Ali Block, New Garden Town, Lahore- 16, Pakistan. (1953) MSc, mineralogy (Punjab U., 1976). Asst. dir. (tel. 042 + 855922). *Geological mapping, sedimentary rocks.*

Khalid, Mr Mohammad. NPSL Centre, PCSIR Labs., Off University Road, Karachi, Pakistan. (1940) MSc, physics (Karachi U., 1963). Sr. res. officer. (tel. 460101, ext. 21). *Crystal structure determination methods, X-ray spectroscopy.*

Khan, Dr Ainul Hassan. R & D Div. PCSIR, Shahrahe-Kamal Ataturk, Karachi Pakistan. (1932) PhD, geochemistry (Manchester U., U.K, 1965). Dir. (tel. 213454). *X-ray diffraction.*

Khawaja, Mr Mahmood-ul-Hassan. GSP, 14- Canal Park, Gulberg, Lahore- 11, Pakistan. (1944) MSc, geology (Punjab U., 1966). Dpty. dir. (tel. 042 + 881192). *Geological mapping, mineral evaluation, mineralogical studies.*

Khwaja, Dr Farid Akhtar. Physics Dept., Quaid-i-Azam U., Islamabad, Pakistan. (1949) PhD, solid state physics (Moscow State U., USSR, 1976). Asst. prof. (tel. 29472). *Order- disorder transition, X-ray and neutron diffraction, electronic and physical properties of materials.*

Mahmood, Mr Khursheed. Dept. of Mech. Eng., NED U. of Eng. and Techn., University Road, Karachi, Pakistan. (1952) MSc, metallurgical eng. (Cranfield Inst. of Tech., UK, 1978). Asst. prof. (tel. 461866). *Structural analysis, high temp. corrosion, mechanical properties of materials.*

Mian, Dr Mohammad Ashraf. Inst. of Applied Geology, Azad Jammu and Kashmir U., Muzaffarabad, Azad Kashmir, Pakistan. (1938) PhD, Geochemistry (Punjab U., 1976). Teacher, appl. geology. (tel. 058 + 2706). *Economic geology, mineralogy, geochemical exploration.*

Mian, Mr Muhammad Asghar. GSP, 22- Ali Block, New Garden Town, Lahore- 16, Pakistan. (1951) MSc, geology (Punjab U., 1976). Geophysicist. (tel. 042 + 855816). *Geophysics.*

Mir, Mr Jan Mohammad. Dept. of Geology, U. of Karachi, University Road, Karachi, Pakistan. (1937) MSc, sedimentology, mineralogy (Manitoba U., Canada, 1972). Asst. prof. (tel. 460211, ext. 95). *Crystal morphology, crystal structure, minerals, carbonate petrography.*

Munir, Mr Mohammad. Centre of Excellence in Mineralogy, U. of Baluchistan, Sariab Road, Quetta, Pakistan. (1950) MSc, geology (Baluchistan U., 1974). Lect. *Crystallography, mineralogy, igneous and metamorphic rock, petrology.*

Naqvi, Dr Syed Ali Anwar. Dept. of Urology, Dow Medical College, Karachi, Pakistan. (1947) MS, urology (Karachi U., 1984). Sr. registrar. (tel. 219551, ext. 247). *X-ray diffraction, renal stones.*

Nasreen, Miss Shagufta. Mineral Res. Div., PCSIR Labs., Jamrud Road, Peshawar, Pakistan. (1955) MSc, physical chemistry (Peshawar U., 1979). Res. officer. (tel. 8817). *X-ray diffraction.*

Pathan, Dr Muhammad Taqee. Geology Dept. U. of Sind, Jamshoro, Pakistan. (1938) PhD, geology, mineralogy (Moscow State U., USSR, 1972). Prof. (tel. 71291, ext. 16). *Optical crystallography, hard rocks, petrology, mineralogy.*

Pervaiz, Mr Rashed. GSP, 22- Ali Block, New Garden Town, Lahore- 16, Pakistan. (1952) MSc, geophysics (Quad-i-Azam U. Islamabad, 1975). Geophysicist. (tel. 042 + 855816) *Exploration geophysics.*

Qaiser, Mr Mohammad Ali. Mineral Res. Div., PCSIR Labs., Jamrud Road, Peshawar, Pakistan. (1940) MSc, physics (Bihar U., India, 1962). Sr. res. officer. (tel. 41191, ext. 19). *Crystal structure, minerals, mathematical techniques in structure determination.*

Qurashi, Dr Mazhar Mahmood. NPSL, PCSIR, No. 16, Sector H- 9, Islamabad, Pakistan. (1925) DSc, physics, X-ray crystallography (Manchester U., UK, 1962). Director general. (tel. 96 + 843680). *Crystal structure, alloys, minerals, liquid structure, scientometrics, mathematical techniques, structure determination, history and philosophy of science.*

Qureshi, Mr Khalid Mahmood. Mineralogy Div., Atomic Energy Mineral Centre, Ferozpur Road, Lahore, Pakistan. (1953) MSc, physics (Punjab U., 1980). Sci. officer. (tel. 99 + 870276, ext. 04). *X-ray diffractometery.*

Qureshi, Mr Mohammed Kaleem Akhtar. GSP, 22- Ali Block, New Garden Town, Lahore- 16, Pakistan. (1945) MSc, geology (Punjab U., 1968). Dpty. dir. (tel. 042 + 855816). *Sedimentology, geotectonics, structural geology.*

Rahman, Mr Mohammad Abdul. Geology Div., Pakistan Atomic Energy Commission, D.G.Khan, Pakistan. (1941) MSc, mineralogy (Aberdeen U., UK, 1979). Principal geologist. *Urainium geology, ore mineralogy.*

Rana, Mr Riaz Ahmad. GSP, 14- Canal Park, Gulberg, Lahore- 11, Pakistan. (1950) MSc, geology (Punjab U., 1977). Asst. dir. (tel. 042 + 881192). *Tectonics, sedimentary environments.*

Rizvi, Prof Adibul Hasan. Dept. of Urology, Dow Medical College, Karachi, Pakistan. (1938) FRCS, surgery(urology) (Royal C. of Surgeons, UK, 1967). Prof. (tel. 219551, ext. 247). *X-ray diffraction, renal stones analysis.*

Rizvi, Dr Syed Sadrul Hassan. NPSL Centre, PCSIR Labs., Off University Road, Karachi, Pakistan. (1933) PhD, physics (Manshester U., UK, 1962). Chief. scient. officer. (tel. 460101, ext. 21). *Crystal structure determination methods, X-ray spectrometry, instrumentation.*

Russell, Mr Nazirullah. GSP, 22- Ali Block, New Garden Town, Lahore- 16, Pakistan. (1948) MSc, geophysics (Punjab U., 1970). Geophysicist. (tel. 042 + 855816). *Exploration geophysics.*

Saeed, Mr Syed Mohammad. NPSL Centre, PCSIR Labs., Off University Road, Karachi, Pakistan. (1943) MSc, physics (Karachi U., 1971). Res. officer. (tel. 460101, ext. 21). *Instrumentation, X-ray diffraction.*

Saghir, Mr Ahmad. NPSL Centre, PCSIR Labs., Off University Road, Karachi, Pakistan. (1947) MSc, physics (Karachi U., 1968). Sr. res. officer. (tel. 460101, ext. 21). *Organic crystal structure, powder diffraction.*

Shah, Prof Muzaffar Ali. Centre for Solid State Physics, Punjab U. New Campus, Lahore 20, Pakistan. PhD, physics (U. C., London U., UK, 1965). Prof. (tel. 854113). *Ceramics, solid state research, phase diagrams.*

Shahi, Mr Ghulam Nabi. Geology Dept.,U. of Baluchistan, Sariab Road, Quetta, Pakistan. (1953) MSc, geology (Baluchistan U., 1976). Asst. prof. (tel. 081 + 73484). *Mineralogy, petrology.*

Shaikh, Mr Mohammad Iqbal. GSP, 22- Ali Block, New Garden Town, Lahore- 16, Pakistan. (1951) MSc, geology (Punjab U., 1971). Asst. dir. (tel. 042 + 855922). *Micropaleontology.*

Shaikh, Mr Qameruddin. NPSL Centre, PCSIR Labs., Off University Road, Karachi, Pakistan. (1947) MSc, physics (Karachi U., 1974). Exp. officer. (tel. 460101, ext. 21). *X-ray diffraction, instrumentation.*

Shaikh, Mr Mohammad Sualehin. NPSL Centre PCSIR Labs., Off University Road, Karachi, Pakistan. (1939) MSc, physics (Manchester U., UK, 1966). Sr. res. officer. (tel. 460101, ext. 21). *Mathematical physics, crystal structure determination, thin film optics.*

Shuja, Mr Tauqir Ahmad. GSP, 22- Ali Block New Garden Town, Lahore- 16, Pakistan. (1943) MSc, geology (Punjab U., 1967). Dpty. dir. (tel. 042 + 855922). *Geothermal energy, groundwater, engineering geology.*

Siddiqui, Mr Jawed Ahmad. Centre of Excellence in Mineralogy, U. of Baluchistan, Sariab Road, Quetta, Pakistan. (1950) MSc, geology (Karachi U., 1977). Lect. *Carbonate rocks, fluvial environment, petroleum geology.*

Siddiqui, Dr Rafiq Ahmad. Solar Energy Div., PCSIR Labs., Roomi Road, Lahore, Pakistan. (1930) PhD, physical chemistry (British Columbia U., Canada, 1961). Principal scient. officer. (tel. 871349). *Solar energy storage systems, crystal growth.*

Yousufzai, Mr Inayatullah Khan. X-ray and Microscopy Div., Pakistan Inst. of Cotton Res. and Techn., Moulvi Tameezuddin Khan Road, Karachi, Pakistan. (1938) MSc, fibre science (Strathcylde U., UK, 1966). Sr. res. officer. (tel. 552007). *Textile technology, fibre structure, fibre microscopy.*

Yusaf, Dr Mohammad. Glass and Ceramics Div., PCSIR Labs., Roomi Road, Lahore-16, Pakistan. (1940) PhD, analytical chemistry (Charles U., Czechoslovakia, 1973). Principal scient. officer. (tel. 870324). *Minerals, X-ray diffraction.*

PERU

Sub-Editor: R. Salazar Orrego

Notes

1. International telephone country code - 51.

Asmat, Dr Humberto. Physics Dept., U. Nac. de Ingenieria, Ave. Tupac Amaru s/n Rimac, Lima 25, Peru. (1944) D. 3e cycle, solid state physics (U. Scient. et Medicale de Grenoble, France, 1977). Prof. (tel. 811070, ext. 136). *Solid state physics, magnetism.*

Avalos, Dr Jaime. Physics Dept., U. Nac. de Ingenieria, Ave. Tupac Amaru s/n Rimac, Lima 25, Peru. (1947) D. 3e cycle, solid state physics (U. Scient. et Medicale de Grenoble, France, 1977). Prof. (tel. 811070, ext. 137). *Magnetic resonance.*

Cavero Ghersi, Dr César Augusto. Lab. de Quimica Inorganica, U. Nac. de Ingenieria, Ave. Tupac Amaru s/n Rimac, Lima 25, Peru. (1942) Dr, structural chemistry (U. Scient. et Medicale de Grenoble, France, 1975). Assoc. prof. (tel. 811070, ext. 216). *Structural crystallography, crystal chemistry.*

Cisneros Ramos, Prof. Luis. Dept. de Fisica, U. Nac. de Ingenieria, Casilla 1301, Lima, Peru. (1945) Dr, physics (U. Scient. et Medicale de Grenoble, France, 1975). *Solid state physics, mineralogy, ceramics.*

Espinoza, Prof. Odon. Chemistry Dept., U. Nac. Mayor de San Marcos, Ave. Venezuela s/n Lima, Lima, Peru. (1920) chemistry - chemical engineering (U. Nac. de San Marcos, 1958). Prof. (tel. 525635). *Mineralogy.*

Horn, Prof. Manfred Josef. Dept. de Fisica, U. Nac. de Ingenieria, Casilla 1301, Lima, Peru. (1938) PhD, physics (British Columbia U., Canada, 1971). Assoc. prof. (tel. 811070, ext. 281). *Solid state physics, phase transitions, mineralogy, solar energy conversion (selective surfaces), X-ray diffraction.*

Linares, Dr Jorge. Dept. of Sci., Pontifica U. Catolica, Ave. Bolivar s/n Pueblo Libre, Lima, Peru. (1949) D. 3e cycle, solid state physics (U. Scient. et Medicale de Grenoble, France, 1978). Prof. (tel. 622540, ext. 239). *Magnetism, crystallography, Mössbauer spectroscopy, X-rays.*

Salazar Orrego, Prof. Ramon. Dept. de Fisica, U. Nac. de Ingenieria, Casilla 1301, Lima, Peru. (1932) MSc, physics (U. Nac. de Ingenieria, 1977). Principal prof. (tel. 811070, ext. 137). *Solid state physics, electron paramagnetic resonance.*

Valera, Dr Anibal Abel. Physics Dept., U. Nac. de Ingenieria, Ave. Tupac Amaru s/n Rimac, Lima, Peru. (1950) Dr.Rer.Nat., solid state physics (U. Stuttgart, BRD, 1979). Prof. (tel. 811070, ext. 136). *Solar energy, photovoltaic cells.*

Vega, Prof. Juan. Physics Dept., U. Nac. de Ingenieria, Ave. Tupac Amaru s/n Rimac, Lima, Peru. (1944) MSc, solid state physics (U. Nac. de Ingenieria, 1978). Prof. (tel. 811070, ext. 137). *Magnetic resonance.*

Velasquez, Prof. Jaime. Physics Dept., U. Nac. de Ingenieria, Ave. Tupac Amaru s/n Rimac, Lima, Peru. (1943) Lic(BS), solid state physics (U. Nac. de Ingenieria, 1974). Prof. (tel. 811070, ext. 136). *Solar energy, photovoltaic, cells, selective absorbtion coatings.*

PHILIPPINES

Sub-Editor: **B. S. Austria**

Notes

1. International telephone country code - 63.

Austria, Dr Benjamin Suarez. Nat. Inst. of Geological Sci., U. of The Philippines, Diliman, Quezon City 3004, Philippines. (1946) PhD, geology (Harvard U., USA, 1975). Prof. (tel. 97-60-61, ext. 455). *Geochemistry, ore deposits, ore mineralogy.*

Cejalvo, Prof. Flor. Dept. of Mathematics, U. of The Philippines, Diliman, Quezon City 3004, Philippines. (1933) MS, mathematics (Stanford U., USA, 1964). Prof. (tel. 95-14-71). *Crystallographic groups.*

Felix, Dr Rene P.. Dept. of Mathematics, U. of The Philippines, Diliman, Quezon City 3004, Philippines. (1950) PhD, mathematics (U. Philippines, 1981). Prof. (tel. 95-14-71). *Crystallographic groups.*

Fernandez, Prof. Aurora Reyes. Dept. of Mathematics, U. of The Philippines, Diliman, Quezon City 3004, Philippines. (1938) MS, mathematics (U. Detroit, USA, 1963). Prof. (tel. 95-14-71). *Crystallographic groups.*

Llaguno, Dr Elma C. Dept. of Chemistry, U. of The Philippines, Diliman, Quezon City 3004, Philippines. (1947) PhD, physical chemistry (Illinois U., USA, 1973). Prof. (tel. 50-46-11, ext. 275). *Organic crystal structures, molecular orbital calculations, water chlorination.*

Mallari-Kaballo, Mrs Paz P. Dept. of Mathematics, U. of The Philippines, Diliman, Quezon City 3004, Philippines. (1950) MS, statistics (U. Philippines, 1976). Asst. prof. (tel. 95-14-71). *Crystallographic groups.*

Natera, Dr Manolito Garcia. Material Sci. Res. Inst., Gen. Santos Ave. Bicutan, Taguig, Metro Manila, Philippines. (1945) PhD, solid state physics (Bombay U., India, 1970). Nuclear res. supervisor. *Neutron diffraction, X-ray diffraction, neutron inelastic scattering, liquid crystals, magnetism, magnetic materials, semiconductors, crystal growth, cryogenics.*

Nochefranca, Miss Luz R. Dept. of Mathematics, U. of The Philippines, Diliman, Quezon City 3004, Philippines. (1953) MS, mathematics (Ateneo de Manila U., 1981). Instructor. (tel. 95-14-71). *Crystallographic groups.*

Quibilan, Mr Edelmiro I., Solar Energy Sect., Philippine Nat. Oil Co., Energy Res. and Dev. Center, Don M. Marcos Ave., Diliman, Quezon City, Philippines. (1946) MS, physics (Pennsylvania State U., USA, 1973). Analyst. *Solid state physics, nuclear magnetic resonance, photovoltaics.*

Soriano-Calix, Mrs Virginia B. Physics Res. Div., Philippine Atomic Energy Commission, Don M. Marcos Ave., Diliman, Quezon City, Philippines. (1944) MS, physics (U. Kansas, USA, 1970). Res. assoc. III. *Solid state physics, electron paramagnetic resonance.*

Trance, Dr Aurora Serrana. Dept. of Mathematics, U. of The Philippines, Diliman, Quezon City 3004, Philippines. (1944) PhD, mathematics (Ateneo de Manila U., 1980). Prof. (tel. 95-14-71). *Crystallographic groups.*

Valencia, Dr Iluminado G. Nuclear Training Dept., Philippine Atomic Energy Commission, Don M. Marcos Ave., Diliman, Quezon City, Philippines. (1926) PhD, mineralogy (U. Wisconsin, USA, 1962). Dept. chief. (tel. 97-60-11 to 15). *Clay mineralogy, fission product adsorption, clay analysis and identification.*

Victorio-Gervasio, Mrs Visitacion. Nat. Inst. of Geological Sci., U. of The Philippines, Diliman, Quezon City 3004, Philippines. (1929) BS, chemistry (U. Philippines, 1951). Instructor. *Mineralogy, petrography, petrology.*

POLAND

Sub-Editor: **A. Pietraszko**

Notes

1. International telephone country code - 48.

2. The Polish equivalent of MSc is *magister* (Mgr), which is also used as a title. *Adiunkt* is equivalent to reader and *docent* to assistant professor. The degree of *doktor habilitowany* (Dr hab.) is next higher than PhD.

Adamiak, Dr Dorota Anna. Crystallography Dept., A. Mickiewicz U., Grunwaldzka 6, Poznań 60780, Poland. (1948) PhD, chemistry (A. Mickiewicz U., 1975). Adiunkt. (tel. 699-181, ext. 489). *Organic crystal structures, biological structures.*

Anulewicz, Mrs Romana. Crystallography Lab., Fundamental Problems in Chem. Inst., Warszawa U., Pasteura 1, Warszawa 02093, Poland. (1937) MSc, crystallochemistry (Warszawa U.). (tel. 222-892). *Organic crystal structures.*

Auleytner, Prof. Julian Jan. Central Lab., X-ray and Electron Microscopy, Inst. of Physics, Pol. Acad. of Sci., Al. Lotników 32/46, Warszawa 02668, Poland. (1922) Prof. dr hab., physics (Inst. of Physics, 1962). Ord. prof. and head. (tel. 436-034). *X-ray physics, electron microscopy, real structure of crystalline materials.*

Bąk, Dr Jadwiga. Central Lab., X-ray and Electon Microscopy, Inst. of Physics, Pol. Acad. of Sci., Al. Lotników 32/46, Warszawa 02668, Poland. (1942) PhD, physics (Pol. Acad. of Sci., 1974). Scient. worker. (tel. 436-034). *Real crystal structure, X-ray studies.*

Barcik, Dr Jan. Inst. of Physics and Chemistry of Metals, Silesian U., Bankowa 12, Katowice 40007, Poland. (1933) PhD, solid state physics (Silesian U., 1968). Adiunkt. (tel. 40-007). *X-ray metallography, electron microscopy (TEM & SEM), metal physics, phase transformation.*

Barszcz, Doc Edward. Physics Metallurgy, Inst. of Ferrous Metallurgy, K. Miarki 12, Gliwice 44-100, Poland. (1936) PhD, physics (Silesian U., 1969). Adiunkt. (tel. 914-051, ext. 327). *Phase transformations, metals.*

Bartczak, Dr Tadeusz Jan. Dept. of Structure Res. and Crystal Chem., Inst. of General Chemistry, Techn. U. of Łódź Żwirki 36, Łódź, Poland. (1935) Dr, chemistry (Techn. U. of Łódź, 1965). docent. (tel. 65522) *X-ray crystallography, organic and complex compounds.*

Bednarski, Dr Stanisław. Solid State Physics, Inst. of Nuclear Res., Świerk Res. Est., Otwock 05400, Poland. (1930) MSc, physics (Wrocław U., 1955). Adiunkt. (tel. 798, ext. 684). *Crystal growth.*

Bedyńska, Dr Teresa Ewa. Central Lab., X-ray and Electron Microscopy, Inst. of Physics, Pol. Acad. of Sci., Al. Lotników 32/46, Warszawa 02668, Poland. (1932) PhD, physics (Inst. of Physics, 1963). Scient. worker. (tel. 436-034). *Dynamical X-ray diffraction theory.*

Blinowski, Dr Konrad. Solid State Physics, Inst. of Nuclear Res., Świerk Res. Est., Otwock 05400, Poland. (1928) PhD, physics (Świerk Inst. of Nuclear Res., 1965). Adiunkt. (tel. 798, ext.805). *Neutron diffraction, magnetic structures.*

Bogucka-Ledóchowska, Dr Maria. Dept. of Pharm. Techn. and Biochem., Techn. U., Majakowskiego 11/12, Gdańsk 80-952, Poland. (1927) PhD, organic chemistry (Techn. U. Gdańsk, 1965). Res. worker. (tel. 539-652). *X-ray structure analysis, biologically active compounds, biological activity - chemical structure relation.*

Bojarski, Prof. Zbigniew. Inst. of Physics and Chemistry of Metals, Silesian U., Bankowa 12, Katowice 40007, Poland. (1921) PhD, solid state chemistry (Silesian Techn. U., 1957). Head. (tel. 596-929). *X-ray methods, phase transformations, structure defects, crystal growth.*

Bołd, Dr Tadeusz. Physical Metallurgy, Inst. of Ferrous Metallurgy, K. Miarki 12, Gliwice 44100, Poland. (1934) PhD, physical metallurgy (Inst. of Ferrous Metallurgy, 1972). Head of X-ray Lab. (tel. 914-051, ext. 327). *Phase transformations, metals and alloys.*

Borowiak, Dr hab. Teresa. Dept. of Crystallography, A. Mickiewicz U., Grunwaldzka 6, Poznań 60780, Poland. (1939) Dr hab., structural chemistry (A. Mickiewicz U., 1968). Asst. prof. (tel. 699-181, ext. 489). *Organic crystal structures, crystal chemistry.*

Bronowska, Dr Wiesława Maria. Inst. of Physics, Techn. U. of Wrocław, Wybrzeże Wyspiańskiego 27, Wrocław 50370, Poland. (1944) Dr, physics (Inst. for Low Temp. and Structure Res., 1978). (tel. 203278). *ferroelectrics phase transitions, crystal structure analysis.*

Brzozowski, Andrzej Marek. Dept. of Crystallography, Inst. of Chemistry, U. of Łódź, Nowotki 18, Łódź 91-416, Poland. (1953) PhD, chemistry (Łódź U., 1976). Adiunt. (tel. 332365). *Protein structure, crystallography, biologically active compounds.*

Bukowska-Strzyzewska, Dr hab Maria. Dept. of Structure Res. and Crystal Chem., Inst. of General Chemistry, Techn. U. of Łódź Zwirki 36, Łódź 90924, Poland. (1929) dr hab, chemistry (Techn. U. of Łódź, 1976). docent. (tel. 65522) *X-ray crystallography, organic and complex compounds.*

Chełkowski, Prof. August Jan. Physics Dept., Silesian U., Uniwersytecka 4, Katowice 40007, Poland. (1927) Dr hab., solid state physics (A. Mickiewicz U., 1959). Prof. (tel. 598-764). *Metal physics, molecular physics.*

Ciechanowicz-Rutkowska, Dr Maria. Regional Lab., Physicochem. Analysis and Structural Res., Jagiellonian U., Krupnicza 41, Kraków 30060, Poland. (1941) PhD, chemistry (Imperial C. London, UK, 1971). Res. asst. (tel. 36-377, ext. 453). *Crystal structure, polymorphic compounds, organic compounds.*

Ciunik, Dr Zbigniew. Inst. of Chemistry, U. of Wrocław, Joliet-Curie 14, Wrocław 50383, Poland. (1949) Dr, crystallography (Wrocław U., 1979) Asst. *X-ray crystallography, structural chemistry, amino acids, peptides.*

Cygler, Dr Mirosław. Dept. of Crystallography, Inst. of Chemistry, U. of Łódź, Nowotki 18, Łódź 91-416, Poland. (1947) PhD, crystallography (Łódź U., 1976). Adiunkt. (tel. 36-891). *Organic crystal structures, computer programming, structure determination methods.*

Czachor, Dr hab Andrzej. Dept. of solid state physics, Inst. of Nuclear Res., Świerk Res. Est., Otwock 05400, Poland. (1934) Dr hab. physics (Świerk Nuclear Res. Inst.,1975). Asst. prof. (tel. 798649). *Crystal lattice dynamics theory.*

Damm, Prof. Józef Zbigniew. Dept. of Crystal Defects, Inst. for Low Temp. and Struc- ture Res., Pl. Katedralny 1, Wrocław 50950, Poland. (1924) Dr hab., chemistry (Inst. of Physical Chemistry PAS, 1969). Head. (tel.221-071, ext. 29). *Structural defects in ionic crystals.*

Derewenda, Zygmunt Stanisław. Dept. of Crystallography, Inst. of Chemistry, U. of Łódź, Nowotki 18, Łódź 91-416, Poland. (1953) PhD, chemistry (Łódź U., 1976). Adiunt. (tel. 332365). *Crystallography, biologically active compounds.*

Dauter, Dr Zbigniew. Dept. of Pharm. Techn. and Biochem., Techn. U., Majakowskiego 11/12, Gdańsk 80952, Poland. (1948) PhD, organic chemistry (Techn. U. Gdańsk, 1975). Adiunkt. (tel. 471-618). *Stereochemistry, biological activity - molecular structure relations.*

Dobrowolska, Dr Wanda. Dept. of Structure Res., Inst. of General Chemistry, Techn. U. of Łódź, Zwirki 36, Łódź 90 924, Poland. (1945) PhD, chemistry (Inst. of General Chemistry, 1976). Adiunkt. (tel. 65-522). *Organic crystal structures.*

Dobrzyński, Dr hab. Ludwik. Solid State Physics, Inst. of Nuclear Res., Świerk Res. Est., Otwock 05400, Poland. (1941) Dr hab., physics (Świerk Inst. of Nuclear Res., 1975). Adiunkt. (tel. 798, ext. 805). *Neutron diffraction, magnetic structures.*

Drzymala, Dr Janusz. Physics Dept., Silesian U., Uniwersytecka 4, Katowice 40007, Poland. (1946) Dr, chemistry (A. Mickiewicz U.). Adiunkt (tel. 588216). *Solid state physics, crystal structure analysis.*

Durski, Dr Zygmunt. Dept. of General Chem. & Inorg. Techn., Techn. U. of Warszawa, Noakowskiego 3 Warszawa 00662, Poland. (1931) PhD, chemistry (Techn. U. of Warszawa, 1970). Adiunkt. *Inorganic and organic crystal structures, X-ray crystallography.*

Dynowska, Mrs Elżbieta Grażyna. Central Lab., X-ray and Electron Microscopy, Inst. of Physics PAS, Al. Lotnikow 32/46, Warszawa 02668, Poland. (1944) MSc, physics of solid state (Warszawa U., 1969). Asst. (tel. 436-034). *X-ray diffraction methods, crystal studies.*

Figielski, Dr hab Tadeusz. Inst. of Physics PAS, Aleja Lotników 32/46, Warszawa 00-681, Poland. Dr hab., physics (Inst. of Physics). Asst. prof. *Crystal physics.*

Gałązka, Dr hab. Robert. Inst. of Physics PAS, Aleja Lotników 32/46, Warszawa 00-681, Poland. Dr hab., physics (Inst. of Physics). Asst. prof. (tel. 436-034). *Crystal growth.*

Gałdecki, Dr hab Zdzisław. Dept. of Structure Res. and Crystal Chem., Inst. of General and Inorganic Chemistry, Techn. U., Zwirki 36, Łódź 90539, Poland. (1924) PhD, chemistry (Techn. U. of Łódź, 1960). Head. (tel. 65-522, ext. 595). *Crystal structure analysis, computer programming.*

Gawron, Dr Marian Ryszard. Dept. of Crystallography, A. Mickiewicz U., Grunwaldzka 6, Poznań 60780, Poland. (1950) MSc, chemistry (A. Mickiewicz U., 1973). Asst. (tel. 699-181, ext. 489). *Organic crystal structures.*

Gdaniec, Dr Maria. Dept. of Crystallography, A. Mickiewicz U., Grunwaldzka 6, Poznań 60780, Poland. (1951) MSc, chemistry (A. Mickiewicz U., 1974). Postgraduate student. (tel. 699-181, ext. 489). *Organic crystal structures.*

Giebułtowicz, Dr Tomasz Mieczysław. Nuclear Methods, Solid State Physics Dept., Inst. of Exp. Physics, Warszawa U., Hoża 69, Warszawa 00-681, Poland. (1945) PhD, physics (Warszawa U., 1975). Adiunkt. (tel. 283-031, ext. 166). *Neutron studies of solids.*

Główka, Dr Marek. Dept. of Structure Res., Inst. of General Chemistry, Techn. U., Zwirki 36, Łódź 90539, Poland. PhD, chemistry (Techn. U. of Łódź, 1971). Adiunkt. (tel. 65-522). *Inorganic crystal structures.*

Głowiak, Doc Tadeusz. Inst. of Chemistry, U. of Wrocław, Joliot-Curie 14, Wrocław 50383, Poland. (1935) PhD, X-ray crystallography (Wrocław U., 1969). Dept. head. *X-ray crystallography, crystal chemistry, coordination compounds, structural chemistry.*

Godwod, MSc Krzysztof Jan. Central Lab., X-ray and Electron Microscopy, Inst. of Physics PAS, Al. Lotników 32/46, Warszawa 02668, Poland. (1938) MSc, exp. physics (Warszawa U., 1962). Specialist in X-ray spectroscopy. (tel. 436-034). *X-ray diffraction dynamical effects, dynamical theory, X-ray spectrometry, X-ray diffractometry.*

Goliński, Dr Bohdan. Dept. of Structure Res., Inst. of General Chemistry, Techn. U., Zwirki 36, Łódź 90539, Poland. (1931) PhD, chemistry (Techn. U. of Łódź, 1965). Adiunkt. (tel.65-522). *Crystal structures, computer programming.*

Górkiewicz, Mgr Zbigniew. Dept. of Structure Res., Inst. of General Chemistry, Techn. U., Zwirki 36, Łódź 90 924, Poland. (1936) MSc, textile technology (Techn. U. of Łódź, 1968). Asst. (tel. 65-522).

Górski, Dr Ludwik. Plasma Physics and Techn., Inst. of Nuclear Res., Świerk Res. Est., Otwock 05400, Poland. (1936) MSc, physical chemistry (Warszawa U., 1960). Scient. worker. *phase transitions, thermal and pressure treatment, nonstoichiometric and unstable phases, plasma spray and similar processes, organic structures with biological activity, X-ray and neutron diffraction, hydrogen bonds.*

Grabowski, Dr hab. Mieczysław Jerzy. Dept. of Crystallography, Inst. of Chemistry, U. of Łódź, Nowotki 18, Łódź 91-416, Poland. (1928) PhD, crystallography (Łódź U., 1969). Asst. prof. (tel. 36-891). *Crystal structure and properties, X-ray analysis.*

Grochowski, Dr Jacek Maciej. Inst. of Chemistry, Jagiellonian U., Krupnicza 41, Kraków 30060, Poland. (1943) PhD, chemistry (Jagiellonian U., 1975). Res. asst. (tel. 36-377, ext. 457,465).

Grochulski, Dr Pawel. Inst. of Physics, Techn. U. of Łódź, Zwirki 36, Łódź 90924, Poland. (1955) Mgr, physics (Techn. U. of Łódź, 1979). Sr. asst. (tel. 65522). *crystal structure analysis, phase transitions*

Gronkowski, Dr Jerzy. Dept. of Structure Res., Inst. of Experimental Physics, U. of Warsaw, Hoża 69, Warszawa 00-681, Poland. (1949) PhD, physics (Warsaw U., 1979) *Dynamical x-ray diffraction, real crystal structure, X-ray topography.*

Grylicki, Dr Mirosław. Dept. of Pharm. Techn. and Biochem., Techn. U., Majakowskiego 11/12, Gdańsk 80952, Poland. (1926) PhD, solid state chemistry (Acad. of Mining and Metallurgy, Kraków, 1963). Asst. prof. (tel. 471-750). *Crystallography, silicates and related compounds, solid state physical chemistry.*

Grzymek, Prof. Jerzy. Akademia Górniczo-Hutnicza, Krzemionki 11, Kraków 30525, Poland. (1908) Dr hab., crystallography (Akademia Górniczo-Hutnicza, 1957). Prof. (tel. 58-740). *Polymorphic transitions, inorganic crystals.*

Habla, Dr Halina. Inst. of Physics and Chemistry of Metals, Silesian U. Bankowa 12, Katowice 40007, Poland. (1939) Dr, solid state physics (Silesian U., 1970). Adiunkt (tel. 588211, ext. 471) *Structure of solid state, X-ray microanalysis.*

Hodorowicz, Dr Stanisław. Inst. of Chemistry, Jagiellonian U., Krupnicza 41, Kraków 30060, Poland. (1941) PhD, chemistry (Jagiellonian U., 1973). Res. asst. (tel. 36-377, ext. 468). *Crystal chemistry, isopolymolybdates, crystal structures, biologically active compounds.*

Holas, Dr hab Andrzej. Dept. of solid state physics, Inst. of Nuclear Res., Świerk Res. Est., Otwock 05400, Poland. (1940) Dr hab, physics (Inst. of Nuclear Res., 1980). Adiunkt. (tel. 798649). *Crystal lattice dynamics theory.*

Horn, Dr Jerzy. Inst. of Inorganic Chemistry, Techn. U. of Wrocław, Wybrzeże Wyspianskiego 27, Wrocław 50370, Poland. (1935) PhD, chemistry (Inst. for Low Temp. and Structure Res., 1972). Adiunkt. *Inorganic crystal structures.*

Horyń, Dr Roman. Inst. for Low Temp. and Structure Res., Pol. Acad. of Sci., Pl. Katedralny 1, Wrocław 50950, Poland. (1936) Dr, inorganic chemistry (Techn. U., 1967). Adiunkt. (tel. 221-071, ext.280). *Phase equilibria, crystal structure, crystal growth, intermetallics, inorganic materials.*

Isakow, Mrs Zofia Stanisława. Silesian U., Inst. of Physics and Chemistry of Metals, Bankowa 12, Katowice 40007, Poland. (1947) MSc, physics (Silesian U., 1970). Asst. (tel. 587-231, ext. 845). *X-ray methods, computerization, X-ray phase analysis.*

Jackowski, Dr Józef Wojciech. Zakład Metaloznawstwa, Instytut Metalurgii AGH, Mickiewicza 30, Kraków 30059, Poland. (1942) PhD, metal science (Acad. of Mining and Metallurgy, 1971). Adiunkt. (tel. 38-100, ext. 2628). *Mechanisms of deformation and recrystallzation, metals and alloys, texture formation.*

Janik, Prof. Jerzy. Inst. of Nuclear Physics, Radzikowskiego 152, Kraków 31342, Poland. (1927) Dr hab., physics (U. Jagiellonian, 1950). Dept. head. *Molecular crystal dynamics.*

Janko, Dr Andrzej. Dept. Catalysis on Metals, Inst. of Physical Chemistry PAS, Kasprzaka 44/52, Warszawa 01224, Poland. (1925) PhD, physics (PAS, 1965). Adiunkt. (tel. 323-221, ext. 284). *Metals, structures, phase processes, metal catalysts, instrumentation.*

Jaskólski, Dr Mariusz. Dept. of Crystallography, A. Mickiewicz U., Grunwaldzka 6, Poznań 60780, Poland. (1952) MSc, chemistry (A. Mickiewicz U., 1976). Postgrad. student. (tel. 699-181, ext. 489). *Organic crystal structures, programming.*

Kajzar, Dr Franciszek, Dept. of Condensed Phase Physics, Inst. of Physics and Nuclear Techniques, Acad. of Mining and Metallurgy, Al. Mickiewicza 30, Kraków 30059, Poland. (1942) PhD, theoretical physics (Jagiellonian U. Kraków, 1970). Asst. prof. (tel. 39-100, ext. 2955). *Magnetic interaction theory.*

Kalicińska-Karut, Mgr Jarosława. Dept. of Crystallography, Inst. for Low Temp. and Structure Res., Plac Katedralny 1, Wrocław 50950, Poland. (1941) Mgr, chemistry (Techn. U. of Wrocław, 1967). Chemist. (tel. 221-071, ext. 54). *Ferroelectric crystal structures, phase transitions.*

Kałuski, Dr hab. Zygmunt. Chemistry Dept., U. Poznań, Grunwaldzka 6, Poznań 60780, Poland. (1920) Dr hab., chemistry (A. Mickiewicz U., Poznań, 1967). Asst. prof. (tel. 699-181, ext. 443). *X-ray crystallography, organic compounds.*

Karniewicz, Dr hab. Jan. Inst. of Physics, Techn. U., Zwirki 36, Łódź 90 924, Poland. Dept. head. *Crystal growth.*

Karolak-Wojciechowska, Dr Janina. Dept. of Structure Res., Inst. of General Chemistry, Techn. U., Zwirki 36, Łódź 90 924, Poland. (1942) PhD, chemistry (Techn. U. of Łódź, 1972). Adiunkt. (tel. 65-522). *Organic crystal structures.*

Karp, Prof. Dr Jan. Instytut Metalografii, Akademia Górniczo-Hutnicza, Al. Mickiewicza 30, Kraków 30059, Poland. (1922) Prof., physical metallurgy (Akademia Górniczo-Hutnicza, 1961). Prof. (tel. 33-823). *Texture, metals, relative orientations in metals.*

Keller, Dr hab Wlodzimierz Aleksander. Inst. of Material Science, Warsaw Techn. U. Narbutta 85, Warszawa, Poland. (1929) Dr hab, crystallography (Warsaw Techn. U., 1972). Docent. (tel. 499929) *Thermal motion in crystals, dynamical diffraction theory, precise parameter measurement.*

Kociński, Prof. Dr Jerzy. Inst. of Physics, Politechnika Warszawska, Koszykowa 75, Warszawa 00628, Poland. *Phase transitions.*

Kołakowski, Dr Andrzej. Magnetic Div., Inst. of Physics, Pol. Acad. of Sci., Al. Lotników 32/46, Warszawa 02668, Poland. (1941) MSc, crystal physics (Łódź U., 1975). Physicist. *Organic crystal structure, computer programming.*

Kołakowski, Dr Bogdan Józef. Magnetic Div., Inst. of Physics, Pol. Acad. of Sci., Al. Lotników 32/46, Warszawa 02668, Poland. (1933) PhD, crystal physics (Inst. of Physics, Pol. Acad. Sci., 1968) Adiunkt. *Symmetry theory, phase transitions, organic crystal structure.*

Konitz, Mr Antoni. Dept. of Pharm. Techn. and Biochem., Techn. U., Majakowskiego 11/12, Gdańsk 80952, Poland. (1948) MSc, organic chemistry (Techn. U. of Gdańsk, 1972). Res. worker. (tel. 471-618). *Organic chemistry, biochemistry, crystallography.*

Konopka, Dr Danuta Cecylia. Physics Dept., Silesian U., Uniwersytecka 4, Katowice 40007, Poland. (1935) PhD, physics (Silesian U., 1960). Adiunkt. (tel. 598-764). *X-ray structures, solid state physics.*

Kosturkiewicz, Dr hab. Zofia. Dept. of Crystallography, A. Mickiewicz U., Grunwaldzka 6, Poznań 60780, Poland. (1928) Dr hab., crystal chemistry (A. Mickiewicz U., 1969). Head. (tel. 699-181, ext. 488). *Organic crystal structures.*

Koziol, Dr Anna. Dept. of Crystallpgraphy, M. C. Skçodowska U., Pl. Marii Curie-Skçodowskiej 3, Lublin 20031, Poland. (1951) Dr, chemistry (A. Mickiewicz U.). Adiunkt. (tel. 33261) *organic crystal structure.*

Kozłowska, Mrs Krystyna. Dept. of Structure Res., Inst. of General Chemistry, Techn. U., Zwirki 36, Łódź 90539, Poland. (1948) Mgr, mathematics (Łódź U., 1972). Sr. asst. (tel. 65-522). *Computer programming, structure analysis.*

Krajewski, Dr hab. Janusz. Zakład Syntezy Organicznej, Instytut Chemii Organicznej PAN, Kasprzaka 44/52, Warszawa 01224, Poland. (1929) Dr hab., physical chemistry (Instytut Chemii Organicznej PAN, 1969). Asst. prof. *Organic crystal structures.*

Krukowski, Dr Marek. Dept. of Physico-chemistry of Solid State, Inst. of Physical Chemistry, Pol. Acad. Sci., Kasprzaka 44/52, Warszawa 01224, Poland. (1946) MSc, light organic technology (Warszawa Techn. U., 1969). Scient. co-worker.

(tel. 323-221, ext. 262). *Structural and physico-chemical properties in solid state, metals and alloys, high pressures, high temperature, low temperature.*

Krygowski, Prof. Tadeusz, Marek. Crystallochemistry Lab., Dept. of Chemistry, U. of Warsaw, Pasteura 1, Warszawa 02-093, Poland. (1946) Dr hab., chemistry (Warszawa U., 1973). Lab. head. (tel. 222892). *Organic crystal chemistry, physical organic chemistry..*

Kubiak, Doc Ryszard. Dept. of Crystallography, Inst. for Low Temp. and Structure Res., Plac Katedral. ofy 1, Wrocław 50950, Poland. (1944) PhD, chemistry (Inst. for Low Temp. and Structure Res. PAS, 1974). Adiunkt. (tel. 221-071, ext. 50). *Low temperature X-ray analysis, metal compounds.*

Kubiak, Dr Maria. Inst. of Chemistry, U. of Wrocław, Joliet-Curie 14, Wrocław 50383, Poland. (1945) MSc, crystallography (Wrocław U., 1968). *X-ray crystallography, crystal chemistry, coordination compounds, structural chemistry.*

Kucab, Dr Marian. Dept. of Condensed Phase Physics, Inst. of Physics and Nuclear Techniques, Acad. of Mining and Metallurgy, Al. Mickiewicza 30, Kraków 30059, Poland. (1946) PhD, theoretical physics (Jagiellonian U. Kraków, 1973). Asst. prof. (tel. 39-100, ext. 2955). *Group theory applications, magnetic structures.*

Kucharczyk, Dr Damian. Dept. Crystallography, Inst. for Low Temp. and Structure Res., Plac Katedralny 1, Wrocław 50950, Poland. (1951) Dr, physics (Inst. for low Temperature and Structure Res., 1978). Adiunkt *Phase transitions, ferroelectric materials.*

Lągiewka, Dr Eugeniusz Antoni. Silesian U., Inst. of Physics and Chemistry of Metals, Bankowa 12, Katowice 40007, Poland. (1939) PhD, physics of metals (Silesian U. of Katowice, 1974). Adiunkt. (tel. 596-929). *X-ray methods, electrocrystallization, metals and alloys, epitaxy, structure defects.*

Łappa, Dr Ryszard Włodzimierz. Dept. of Chemistry, U. Teachers C., 3 Maja 54, Siedlce 08110, Poland. (1920) PhD, solid state physics (Inst. of Physics PAS, 1969). Adiunkt. (tel. 431-645). *Physics of crystal growth, nucleation phenomena, experimental & theoretical studies, molecular beam technology.*

Leciejewicz, Prof. Janusz. Solid State Physics, Inst. of Nuclear Res., Świerk Res. Est., Otwock 05400, Poland. (1928) Dr hab., physics (Świerk Inst. of Nuclear Res., 1975). Head. (tel. 798, ext. 642). *Neutron diffraction, magnetic structures.*

Lefeld-Sosnowska Dr hab. Maria, Stefania. Dept. of Structure Res., Inst. of Experimental Physics, U. of Warsaw, Hoża 69, Warszawa 00-681, Poland. (1934) Dr hab., physics (Warsaw U., 1979) *Dynamical x-ray diffraction, real crystal structure, X-ray topography.*

Ligenza, Dr Sylwester. Solid State Physics, Inst. of Nuclear Res., Świerk Res. Est., Otwock 05400, Poland. (1935) PhD, physics (Świerk Inst. of Nuclear Res., 1971). Adiunkt. (tel. 798, ext. 324). *Neutron diffraction, magnetic structures.*

Lipkowski, Dr Janusz. Dept. of Physico-chemical Methods of Analysis, Inst. of Physical Chemistry, Pol. Acad. of Sci., Kasprzaka 44/52, Warszawa 01224, Poland. (1943) PhD, inclusion chemistry (Inst. of Physical Chem. PAS, 1972). Dept. head. (tel. 323-221, ext. 213). *Chemical crystallography, inclusion compounds, host - guest complexes, coordination complexes, organometallic compounds.*

Lis, Dr Tadeusz. Inst. of Chemistry, U. of Wrocław, J. Curie 14, Wrocław 50383, Poland. (1947) PhD, chemistry (Wrocław U., 1973). Adiunkt. *Rhenium and manganese chemistry, X-ray crystallography.*

Łukaszewicz, Prof. Dr Kazimierz. Dept. of Crystallography, Inst. for Low Temp. and Structure Res., Plac Katedralny 1, Wrocław 50950, Poland. (1927) Dr hab., physics (Inst. of Physical Chemistry PAS, 1968). Dept. head. (tel. 221-071, ext. 79). *Crystal structure analysis, phase transitions, precision lattice parameters.*

Maciosowski, Dr Andrzej. Physical Metallurgy, Inst. of Ferrous Metallurgy, K. Miarki 12, Gliwice 44101, Poland. (1944) PhD, physical melallurgy (Acad. of Mining and Metallurgy Kraków, 1973). Adiunkt. (tel. 914-051, ext. 288). *Crystallography, phase transformations in solid state, structure - properties relation, steel, texture.*

Majchrzak, Dr Stanisław. Dept. of Catalysis on Metals, Inst. of Physical Chemistry PAS, Kasprzaka 44/52, Warszawa 01224, Poland. (1925) PhD, chemistry (Inst. of Physical Chem. PAS, 1969). Adiunkt. (tel. 323-221, ext. 284). *High pressure X-ray diffraction, hydrides, clathrates, phase transformations, instrumentation.*

Malinowski, Dr Mariusz. Dept. of Crystallography, Inst. for Low Temp. and Structure Res., Plac Katedralny 1, Wrocław 50950, Poland. (1950) Dr, physics (Inst. for low Temperature and Structure Res., 1980). Adiunkt *Phase transitions, ferroelectric materials.*

Maliszewski, Dr Edward. Solid State Physics, Inst. of Nuclear Res., Świerk Res. Est., Otwock 05400, Poland. (1930) PhD, physics (Świerk Inst. of Nuclear Res., 1967). Adiunkt. (tel. 798, ext. 303). *Crystal lattice dynamics, neutron scattering.*

Maluszynska, Dr Hanna. Chemistry Dept., A. Mickiewicz U. Poznań, Grunwaldzka 6, Poznań 60780, Poland. (1947) MSc, physics (A. Mickiewicz U., 1971). Asst. (tel. 699-181, ext. 443). *X-ray crystallography, organic compounds.*

Matyja, Dr Przemysław. Inst. of Physics and Chemistry of Metals, Silesian U., Bankowa 12, Katowice 40007, Poland. (1939) PhD, physics of metals (Silesian U. of Katowice, 1976). Adiunkt. (tel. 587-231, ext. 442). *X-ray metallography.*

Mizera, Dr Elżbieta. Central Lab., X-ray and Electron Microscopy, Inst. of Physics, Pol. Acad. of Sci., Al. Lotników 32/46, Warszawa 02668, Poland. (1938) PhD, techn. sci. (Polytechnic of Warszawa, 1972). Scient. worker. (tel. 437-001, ext. 143). *Crystal defects, electron microscopy.*

Modrzejewski, Dr hab. Antoni. Dept. XIV, Inst. of Nuclear Res., Świerk Res. Est., Otwock 05400, Poland. (1931) Dr hab., physics. Asst. prof. *Crystal growth.*

Murasik, Dr Andrzej. Solid State Physics, Inst. of Nuclear Res., Świerk Res. Est., Otwock 05400, Poland. (1931) PhD, physics (Świerk Inst. of Nuclear Res., 1970). Adiunkt. (tel. 798, ext. 324). *Neutron diffraction, magnetic structures.*

Nizioł, Dr Stanisław Marian. Dept. of Condensed Phase Physics, Inst. of Physics and Nuclear Techniques, Acad. of Mining and Metallurgy, A. Mickiewicza 30, Kraków 30059, Poland. (1941) PhD, exp. physics (Acad. of Mining and Metallurgy, Kraków, 1972). Asst. prof. (tel. 39-100, ext. 2960). *Magnetic structure determination, neutron diffraction method.*

Olejnik, Dr Stanisław Julian. Dept. of Crystallography, Inst. for Low Temp. and Structure Res., Plac Katedralny 1, Wrocław 50950, Poland. (1947) PhD, chemistry (Inst. for Low Temp. and Structure Res. PAS, 1975). Adiunkt. (tel. 221-071, ext. 54). *Solid state, ferroelectrics, crystal structure analysis.*

Oleksyn, Dr Barbara Jadwiga. Inst. of Chemistry, Jagiellonian U., Krupnicza 41/43, Kraków 30060, Poland. (1940) PhD, chemistry (Jagiellonian U., 1972). Res. asst. (tel. 36-377, ext. 457). *Crystal structures, organo-metallic complexes, biologically active compounds, alkaloids.*

Oles, Prof. Andrzej Władysław. Dept. of Condensed Phase Physics, Inst. of Physics and Nuclear Techniques, Acad. of Mining and Metallurgy, Al. Mickiewicza 30, Kraków 30059, Poland. (1923) Prof., exp. physics (Jagiellonian U. Kraków, 1965). Head. (tel. 39-100, ext. 2955). *Magnetic structures.*

Paciorek, Mgr inz. Wlodzimierz. Dept. of Crystallography, Inst. for Low Temp. and Structure Res., Plac Katedralny 1, Wrocław 50950, Poland. (1953) Mgr inz, physics (Techn. U. of Wrocław, 1977) Postgraduate Student. (tel. 221071 ext. 44) *Phase transition theory, ferroelectric materials.*

Pajak, Mgr Lucjan. Inst. of Physics and Chemistry of Metals, Silesian U., Bankowa 12, Katowice 40007, Poland. (1947) Mgr, chemistry (Jagielonian U., 1970). Asst. (tel. 588211, ext. 442). *Small angle X-ray scattering.*

Pawlak, Dr Stanisław Jerzy. Physical Metallurgy, Inst. of Ferrous Metallurgy, K. Miarki 12, Gliwice 44101, Poland. (1944) PhD, physics of metals and physical metallurgy (Acad. of Mining and Metallurgy, Kraków, 1973). Adiunkt. (tel. 914-051, ext. 752). *Crystallography, martensitic transformation, metals, texture, anisotropic properties, structure - properties relations, high strength steels.*

Pielaszek, Dr Jerzy. Dept. of Catalysis on Metals, Inst. of Physical Chemistry PAS, Kasprzaka 44/52, Warszawa 01224, Poland. (1941) PhD, physics (PAS, 1972). Adiunkt. (tel. 323-221, ext. 205). *Real structure of metals, crystal lattice defects, instrumentation.*

Pietraszko, Dr Adam. Dept. of Crystallography, Inst. for Low Temp. and Structure Res., Plac Katedralny 1, Wrocław 50950, Poland. (1943) PhD, physics (Inst. for Low Temp. and Structure Res. PAS, 1974). Adiunkt. (tel. 221-071, ext. 50). *X-ray crystal structure analysis, phase transitions, ferroelectrics, subconductors.*

Pietraszko, Mrs Donata. Instytut Materiałoznawtwa, Politechnika Wrocławska, Smoluchowskiego 25, Wrocław 50370, Poland. (1941) Mgr, physics (Wrocław U., 1966). Sr. asst. *Inorganic crystal structures.*

Ratuszek, Dr Wiktoria Maria. Inst. of Physical Metallurgy, Acad. of Mining and Metallurgy, Al. Mickiewicza 30, Kraków 30059, Poland. (1936) PhD, metal science (Acad. of Mining and Metallurgy, 1971). Adiunkt. (tel. 38-100, ext. 2628). *Stacking fault energy, deformation and recrystallization mechanisms, metals and alloys, texture formation.*

Ratuszna, Dr Alicja Maria. Physics Dept., Silesian U., Uniwersytecka 4, Katowice 40007, Poland. (1947). Dr, physics (Silesian U., 1978). Adiunkt. (tel. 588-211). *Solid state physics, X-ray studies.*

Rychlewska, Dr Urszula. Dept. of Crystallography, A. Mickiewicz U., Grunwaldzka 6, Poznań 60780, Poland. (1948) PhD, chemistry (A. Mickiewicz U., 1976). Adiunkt. (tel. 699-181, ext. 489). *Organic crystal structures, crystal chemistry.*

Sawka-Dobrowolska, Dr Wanda. Inst. of Chemistry, U. of Wrocław, Joliot-Curie 14, Wrocław 50383, Poland. (1945) Dr, chemistry (Wrocław U., 1979). Adiunkt. *X-ray crystallography, crystal chemistry, coordination compounds, structural chemistry.*

Sikora, Dr Wiesława Antonina. Dept. of Condensed Phase Physics, Inst. of Physics and Nuclear Techniques, Acad. of Mining and Metallurgy, A. Mickiewicza 30, Kraków 30059, Poland. (1945) PhD, theoretical physics (Jagiellonian U. Kraków, 1974). (tel. 39-100, ext. 2955). *Magnetic group theory, symmetry of magnetic structures.*

Skarżyński Dr Tadeusz. Dept. of Crystallography, Inst. of Chemistry, U. of Łódź, Nowotki 18, Łódź 91-416, Poland. (1953) PhD, chemistry (Łódź U., 1981). Res. worker. (tel. 36-891). *Organic crystal structures, computer programming, macromolecular crystallography.*

Skoweranda, Dr Jolanta. Dept. of Structure Res., Inst. of General Chemistry, Tech. U., Zwirki 36, Łódź 90539, Poland. (1945) PhD, chemistry (Techn. U. of Łódź, 1976). Adiunkt. (tel. 65-522). *Organic crystal structures.*

Skrzat, Dr Zofia. Dept. of Mineralogy and Crystallography, U. of Toruń, Gagarina 7, Toruń 87100, Poland. PhD, natural science (Toruń U., 1960). Adiunkt. (tel. 27-051). *Inorganic crystal structures.*

Slebarski, Dr Andrzej. Silesian U., Inst. of Physics, Uniwersytecka 4, Katowice 40007, Poland. (1950) MSc, physics (Silesian U., 1973). Asst. (tel. 588-211). *X-ray physics, solid state physics.*

Sosnowska, Dr hab. Izabela. Nuclear Methods, Solids State Physics Dept., Inst. of Exp. Physics, Warszawa U., Hoża 69, Warszawa 00-681, Poland. (1939) Dr hab., physics (Warszawa U., 1973). Asst. prof. and head. (tel. 287-252). *Structure and dynamics of crystal lattices, neutron scattering.*

Sosnowski, Dr hab. Jerzy. Solid State Physics, Inst. of Nuclear Res., Świerk Res. Est., Otwock 05400, Poland. (1936) Dr hab., physics (Świerk Inst. of Nuclear Res., 1975). Adiunkt. (tel. 798, ext. 303). *Crystal lattice dynamics, neutron scattering.*

Stadnicka, Dr Katarzyna. Inst. of Chemistry, Jagiellonian U., Krupnicza 41, Kraków 30060, Poland. (1943) PhD, chemistry (Jagiellonian U., 1973). Res. asst. (tel. 36-377, ext. 457). *Crystal structures, pharmaceutical agents.*

Staliński, Prof. Bohdan. Inst. for Low Temp. Structure Res., PAS, Pl. Katedralny 1, Wrocław 50950, Poland. (1924) Dr hab., chemistry (Techn. U. of Wrocław, 1956). Head. (tel. 221-071, ext. 32). *Crystal structures and physical properties, magnetic compounds.*

Stepień-Damm, Dr Julia. Dept. of Crystallography, Inst. for Low Temp. and Structure Res., Plac Katedralny 1, Wrocław 50950, Poland. (1945) Dr, physics (Inst. for low Temperature and Structure Res., 1980). Adiunkt. (tel. 221-071, ext. 50). *Crystal lattice defects.*

Stepień, Dr Andrzej. Dept. of Crystallography, Inst. of Chemistry, U. of Łódź, Nowotki 18, Łódź 91-416, Poland. (1945) PhD, crystallography (Łódź U., 1973). Adiunkt. (tel. 36-891). *Organic crystal structures, X-ray analysis.*

Stroz, Mgr Danuta. Inst. of Physics and Chemistry of Metals, Silesian U., Bankowa 12, Katowice 40007, Poland. (1951) Mgr, physics of metals (Silesian U., 1974). Asst. (tel. 588211 ext. 617) *electron microscopy, precipitation processes in alloys.*

Surowiec, Dr Marian Ryszard. Inst. of Physics and Chemistry of Metals, Silesian U. Bankowa 12, Katowice 40007, Poland. (1948) Dr, physics (Silesian U., 1977). Adiunkt (tel. 587231 ext. 855) *X-ray diffraction methods, lattice defects in single crystals, X-ray diffraction topography, crystal growth.*

Szarras, Dr Stanisław. Solid State Physics, Inst. of Nuclear Res., Świerk Res. Est., Otwock 05400, Poland. (1924) PhD, chemistry (Techn. U. Warszawa, 1966). Adiunkt. (tel. 798, ext.648). *Crystal perfection.*

Szmid, Dr Zofia. Solid State Physics, Inst. of Nuclear Res., Świerk Res. Est., Otwock 05400, Poland. (1921) PhD, physics (Świerk Inst. of Nuclear Res., 1967). Adiunkt. (tel. 798, ext.648). *Crystal perfection.*

Szummer, Dr hab. Andrzej. Inst. of Materials Ingineering, Techn. U. of Warszawa, Nowowiejska 24, Warszawa 00665, Poland. *Phase transformations in metals.*

Tomaszewski, Dr Pawel Edward. Dept. of Crystallography, Inst. for Low Temp. and Structure Res., Plac Katedralny 1, Wrocław 50950, Poland. (1952) Mgr, physics (Wrocław U., 1976). Sr. asst. (tel. 221071, ext. 44). *phase transitions, ferroelectric materials, crystal structure analysis.*

Tosik, Dr Anita. Dept. of Structure Res., Inst. of General Chemistry, Techn. U., Zwirki 36, Łódź 90539, Poland. (1947) MSc, chemistry (Łódź U., 1970). Sr. asst. (tel. 65-522). *Inorganic crystal structures.*

Tykarska, Dr Ewa. Crystallography Dept., A. Mickiewicz U., Grunwaldzka 6, Poznań 60780, Poland. (1957) PhD, chemistry (A. Mickiewicz U.). (tel. 699-181). *Organic crystal structures.*

Urbanczyk, Prof. Grzegorz. Inst. of Fiber Physics and Textile Finishing, Techn. U. of Łódź, Gdańska 155, Łódź, Poland. (1928) dr hab, fiber physics (Techn. U. of Łódź, 1967). Inst. dir. (tel. 62762) *Fiber textures, fiber crystallinity.*

Urbanowicz, Dr Ewa Renata. Dept. of Crystallography, Inst. for Low Temp. and Structure Res., Plac Katedralny 1, Wrocław 50950, Poland. (1947) Mgr, electronic (Politechnika Wrocławska, 1970). Sr. asst. (tel. 221-071, ext. 44). *Computer programming, mathematical and statistical methods, crystallographic data interpretation, optimization of experiments.*

Wajsman, Dr Elżbieta Henryka. Dept. of Crystallography, Inst. of Chemistry U. of Łódź, Nowotki 18, Łódź 91-416, Poland. (1938) PhD, crystallography (Łódź U., 1975). Adiunkt. (tel. 36-891). *Crystal structures, X-ray analysis.*

Warczewski, Doc Jerzy Zdzisław. Dept. of Condensed Phase Physics, Inst. of Physics and Nuclear Techniques, Acad. of Mining and Metallurgy, Al. Mickiewicza 30, Kraków 30059, Poland. (1939) PhD, exp. physics (Acad. of Mining and Metallurgy, 1969). Asst. prof. (tel. 39-100, ext. 2955). *Crystal structure determination, phase transitions in crystals, displacive modulation in crystals, powder diffraction data error analysis.*

Wardzyński, Prof. Wiesław. Inst. for Experimental Physics, U. of Warszawa, Hoża 69, Warszawa 00-681, Poland. Dept. head. *Crystal growth.*

Warmiński, Dr Tadeusz Piotr. Central Lab., X-ray and Electron Microscopy, Inst. of Physics, Pol. Acad. of Sci., Al. Lotników 32/46, Warszawa 02668, Poland. (1940) Dr hab., physics (Pol. Acad. of Sci., 1974). Scient. worker. (tel. 436-034). *Electron microscopy (mirror - transmission - scanning), material science, X-ray and electron microprobe analysis, real structure of crystalline materials.*

Waskowska, Dr Alicja. Dept. of Crystallography, Inst. for Low Temp. and Structure Res., Plac Katedralny 1, Wrocław 50950, Poland. (1942) PhD, natural sciences (Inst. of Immunology and Exp. Therapy PAS, 1975). Adiunkt. (tel. 221-071, ext. 54). *Solid state, ferroelectrics, crystal structure analysis.*

Wawrzak, Dr Zdzisław. Inst. of Physics, Techn. U. of Łódź, Zwirki 36, Łódź 90924, Poland. (1955) Mgr, physics (Techn. U. of Łódź, 1979). Sr. asst. (tel. 65522, ext. 894) *crystal structure analysis, phase transitions.*

Węglowski, Dr Stanisław. Inst. of Inorganic Chemistry, Techn. U. of Wrocław, Wybrzeze Wyspianskiego 27, Wrocław 50370, Poland. (1929) PhD, chemistry (Techn. U. of Wrocław, 1963). Adiunkt. *Inorganic crystal chemistry.*

Wieteska, Dr Krzysztof. Dept. of Solid State Physics, Inst. of Nuclear Res., Świerk Res. Est., Otwock 05400, Poland. (1946) Dr, physics (Świerk Inst. of Nuclear Res., 1980). Adiunkt. *Crystal defects, X-ray scattering.*

Więckowski, Mr Tadeusz. Crystallography Lab., Fundamental Problems in Chem. Inst., Warszawa U., Pasteura 1, Warszawa 02093, Poland. (1946) MSc, crystallochemistry (Warszawa U., 1975). Asst. (tel. 222-892). *Layer silicate structures, interlayer bonding (layer silicates), crystal chemistry.*

Wieczorek, Dr Michał. Dept. of Structure Res., Inst. of General Chemistry, Techn. U. of Łódź, Żwirki 36, Łódź 90539, Poland. (1937) PhD, chemistry (Inst. of General Chemistry, 1970). Adiunkt. (tel. 65-522). *Organic crystal structures.*

Wiewióra, Dr hab. Andrzej. Crystallography Lab., Fundamental Problems in Chem. Inst., Warszawa U., Pasteura 1, Warszawa 02093, Poland. (1933) Dr hab., mineralogy (Warszawa U., 1972). Asst. prof. (tel. 222-892). *Layer silicate structures, crystallochemistry, mixed-layer minerals.*

Wokulska, Dr Krystyna Barbara. Inst. of Physics and Chemistry of Metals, Silesian U., Bankowa 12, Katowice 40007, Poland. (1945) Dr, physics of solid state (Silesian U., 1978). Adiunkt. (tel. 587231, ext. 521). *crystal growth and dissolution, imperfections in crystals.*

Wokulski, Dr Zygmunt. Inst. of Physics and Chemistry of Metals, Silesian U. Bankowa 12, Katowice 40007, Poland. (1940) Dr, physics of solid state (Silesian U., 1978). Adiunkt (tel. 587231, ext. 521). *crystal growth, crystal perfection characterization, X-ray diffraction techniques.*

Wolcyrz, Dr Marek. Dept. of Crystallography, Inst. for Low Temp. and Structure Res., Plac Katedralny 1, Wrocław 50950, Poland. (1952) Mgr inz, electronics (Techn. U. of Wrocław, 1976). Sr. asst. (tel. 221071, ext. 54). *Precise lattice parameter determination, crystal structure analysis.*

Zielińska-Rohozińska, Dr hab. Elżbieta. Dept. of Structure Res., Inst. of Experimental Physics, U. of Warsaw, Hoza 69, Warszawa 00-681, Poland. (1938) Dr hab., physics (Warsaw U., 1984) *Dynamical x-ray diffraction, real crystal structure, X-ray topography.*

Ziółowska, Dr Blanka. Inst. of General Chem. & Inorg. Techn., Techn. U. of Warszawa, Noakowskiego 3, Warszawa 00664, Poland. (1928) PhD, chemistry (Techn. U. of Warszawa, 1964). Adiunkt. *Organic crystal structures.*

Ziołowski, Dr Zbigniew. Dept. of Structure Res., Inst. of Iron Metallurgy, Miarki 12/14, Gliwice 44100, Poland. (1919) PhD, techn. sci. (Techn. U. of Gliwice, 1960). Asst. prof. *Metals and metallurgical materials - structure.*

Żmija, Prof. Józef. Techn. Acad. of Military, Lazurowa, Warszawa 00908, Poland. Dept. head. (tel.366-661, ext. 33-31). *Crystal growth.*

PORTUGAL

Sub-Editor: **J. Lima-de-Faria**

Notes

1. International telephone country code - 351.

2. Degrees conferred by the Portuguese Universities are *Doutor* (Dr), approximately equivalent to PhD at British Universities, and *Licenciado* (Lic. or Eng. for engineering), approximately equivalent to MSc at British universities.

3. University positions include (with approximate British equivalents added in parentheses), *Professor catedrático* (professor), *Professor associado* (reader), *Professor auxiliar* (lecturer), *Assistente* (demonstrator). Research positions mentioned here are *Investigador* (research officer), and *Naturalista* (curator).

Aires-Barros, Prof. Luis. Secção de Mineralogia, Instituto Superior Técnico, Av. Rovisco Pais, 1000 Lisboa, Portugal. (1931) Dr, mineralogy (Instituto Superior Técnico, Lisboa, 1964). Prof. catedrático. (tel. 19 + 800111). *Mineralogical applications, X-ray crystallography, clay minerals, crystal optics.*

Almeida, de, Prof. Maria José Marques. Depto. de Física, Fac. de Ciencias e Tecnologia, 3000 Coimbra, Portugal. (1946) PhD, physics (Cambridge U., UK, 1975). Prof. associado. (tel. 39 + 23675, ext. 254). *Alloy structures, electron and spin densities.*

Alte da Veiga, Prof. Luis. Depto. de Física, Fac. de Ciencias e Tecnologia, 3000 Coimbra, Portugal. (1932) Dr, crystallography (Cambridge U., UK, 1964). Prof. catedrático. (tel. 39 + 23675, ext. 251). *Alloy structures, electron densities.*

Andrade, Ms Lourdes Rodrigues. Depto. de Física, Fac. de Ciencias e Tecnologia, 3000 Coimbra, Portugal. (1954) Lic., physics (Coimbra U., 1976). Asst. (tel. 39 + 23675, ext. 244). *Inorganic crystal structures.*

Borges, Prof. Frederico. Depto. de Mineralogia e Geologia, Fac. de Ciencias, 4000, Porto, Portugal. (1942) Dr, geology (London U., UK, 1978). Prof. associado. (tel. 29 + 310290). *Recrystallization, intracrystalline deformation, electron microscopy.*

Bravo, Prof. Manuel. Depto. de Geociencias, Fac. de Ciencias e Tecnologia, Universidade Nova, Olivais, 1899 Lisboa Codex, Portugal. (1933) PhD, experimental petrology (Edinburgh U., UK, 1973). Prof. associado. (tel. 19 + 2512660). *Mineral crystal structures.*

Carrondo, Prof. Maria Arménia A. F. C. Teixeira. Centro de Química Estrutural, Complexo Interdisciplinar, Instituto Superior Tecnico, Av. Rovisco Pais, 1000 Lisboa, Portugal. (1948) PhD, chemical crystallography (London U., UK, 1978). Prof. associado. (tel. 19 + 572076). *X-ray structural characterization, organometallic and organic compounds, instrumental methods of general analysis.*

Costa, Prof. M. Margarida Ramalho. Depto. de Física, Fac. de Ciencias e Tecnologia, 3000 Coimbra, Portugal. (1945) PhD, physics (Cambridge U., UK, 1974). Prof. associado. (tel. 39 + 23675, ext. 256). *Alloy structures, electron and spin densities.*

Figueiredo, Dr Maria Ondina. Centro de Cristalografia e Mineralogia, Inst. de Investigação Cientifica Tropical, Alameda D. Afonso Henriques 41-4Esq., 1000 Lisboa, Portugal. (1938) Dr, geology (Techn. U. of Lisbon, 1980). Investigador. (tel. 19 + 534596). *Inorganic structure systematics, mathematical crystallography, phase transformations.*

Fortes, Prof. Manuel Amaral. Depto. de Metalurgia, Instituto Superior Técnico, Av. Rovisco Pais, 1000 Lisboa, Portugal. (1938) PhD, physics (Cambridge U., UK, 1969). Prof. catedrático. (tel. 19 + 802045). *Lattice defects, surface structure and phenomena, plastic deformation, metals.*

Gama Carvalho, Dr Frederico. Lab. Nacional de Engenharia e Tecnologia Industriais, Estrada Nacional N-10, 2685 Sacavem, Portugal. (1936) Dr, physics (U. Karlsruhe, BRD, 1967). Investigador. (tel. 19 + 2510021). *Crystal structure, neutron diffraction, solid state physics, statistical physics.*

Lima-de-Faria, Dr José. Centro de Cristalografia e Mineralogia, Inst. de Investigação Científica Tropical, Alameda D. Afonso Henriques 41-4Esq., 1000 Lisboa, Portugal. (1925) PhD, crystallography (Cambridge U., UK, 1962). Investigador. (tel. 19 - 534596). *Inorganic structure systematics, condensed models, phase transformations, minerals.*

Lopes-Vieira, Prof. António. Centro de Química Estrutural, Complexo Interdisciplinar, Inst. Superior Técnico, Av. Rovisco Pais, 1000 Lisboa, Portugal. (1929) PhD, mineralogy (Oxford U., UK, 1967). Prof. associado. (tel 19 - 534596). *Inorganic crystal structures, mineralogy.*

Matos Beja, Ms Ana Maria. Depto. de Física, Fac. de Ciencias e Tecnologia, 3000 Coimbra, Portugal. (1949) Lic., physics (Coimbra U., 1972). Investigadora. (tel. 39 + 23675). *Alloy structures.*

Montenegro de Andrade, Prof. Miguel. Depto. de Mineralogia e Geologia, Fac. de Ciencias, 4000 Porto, Portugal. (1918) Dr, petrology (Coimbra U., 1955). Prof. catedrático. (tel. 29 + 21208). *Crystal optics.*

Pureza, Mr Fausto. Depto. de Mineralogia e Geologia, Fac. de Ciencias e Tecnologia, 3000 Coimbra, Portugal. (1925) Lic. geology (Coimbra U., 1956). Investigador. (tel. 39 + 23022). *Clay minerals.*

Quadrado, Prof. Ricardo. Depto. de Mineralogia e Geologia, Fac. de Ciencias, Rua da Escola Politécnica, 1200 Lisboa, Portugal. (1920) Dr, crystallography (Madrid U., Spain, 1967). Prof. catedrático. (tel. 19 + 605850). *Crystal physics, teaching crystallography.*

Salgado, Dr José. Lab. Nacional de Engenharia e Tecnologia Industriais, Estrada Nacional N-10, 685 Sacavem, Portugal. (1940) Dr, physics (U. Karlsruhe, BRD, 1974). Investigador. (tel. 19 + 2510021). *Neutron diffraction, solid state physics, structures, dynamics of solids.*

Salvado Canelhas, Mrs Maria da Graça. Museu e Lab. Mineralógico e Geológico, Fac. de Ciencias, Rua da Escola Politécnica, 1200 Lisboa, Portugal. (1940) Lic., geology (Lisboa U., 1964). Naturalista. (tel. 19 + 605850). *Mineral identification.*

Torre de Assunção, Prof. Carlos. Estrada de Mem Martins 222, 2725 Mem Martins, Portugal. (1901) Dr, petrology (Lisboa U., 1938). Prof. catedrático (retired). (tel. 19 + 2910570). *Crystal optics, mineralogical applications, X-ray crystallography.*

ROMANIA

Sub-Editor: J. Ionescu

Notes

1. International telephone country code - 40.

2. Degrees conferred by the Romanian universities are the *doctor docent în ştiinţe* (Dr.doc.st.), *doctor în ştiinţe* (Dr.st.), and *licentiat în ştiinţe* (Lic.st.), equivalent approximately to DSc, MSc and BSc respectively.

3. Other abbreviations used are: Acad. - academician, member of the Academia R.S.R.; Acad. R.S.R. - Academia Republucii Socialistă Românja; and Conf. - conferenţiar universitar (assistant professor).

Anton, Mr Liviu. Ministry of Mines & Geology, Geological & Geophysical Inst. of Romania, 1 Caransebes Street, Bucarest 8, Romania. (1942) Scient. res. (tel. 657530). *Crystallography, mineralogy, geochemistry, experimental works.*

Apostolescu, Eugenia Rodica. Crystallography and Mineralogy Lab., Inst. Politehnic 'Gh. Gheorghiu-Dej', Str. Polizu 1, Bucureşti, Romania. (1930) Ing., geology (Mining and Geology Inst., Bucureşti, 1953). Asst. (tel. 139440). *Crystallography, minerals.*

Balan, Mr Mihai. Ministry of Mines & Geology, Geological & Geophysical Inst. of Romania, 1 Caransebes Street, Bucarest 8, 7000 Romania. (1940) Dr, geological sciences (U. Bucarest, 1975). Scient. res. (tel. 657530). *Crystallography, mineralogy, geochemistry, experimental works.*

Bally, Conf. Dorel. Comitetul de Stat pentru Energia Nucleară- Bucureşti, Inst. of Physics, Bucureşti, Romania. (1923) Dr.st., physics (Moscow State U., USSR, 1953). Dr.doc.st. (U. de Bucureşti, 1967). Div. chief, Neutron Physics Lab., Căsuta Poştală 35, Bucureşti. (tel. 23 68 60). *Neutron diffraction, metal physics.*

Baltă, Conf. Petru. Glass Chemistry and Techn. Lab., Institutul Politehnic 'Gh. Gheorghiu-Dej', Calea Grivitei 132, Bucureşti, Romania. (1930) Dr.Ing., silicate chemistry (Polytechnic Inst. Bucureşti, 1956). Conf. glass techn. (tel. 139440). *Glass, silicates.*

Cioflica, Prof. Graţian. Dept. of Geology & Geography, Blvd. N. Bălcescu 1, Universitatea Bucureşti, Bucureşti, Romania. (1927) Dr.st., mineralogy and petrography (U. de Bucureşti, 1958). Prof. of ore deposits. *Crystal optics.*

Constantinescu, Mr Radu. Ministry of Mines and Geology, Geological & Geophysical Inst. of Romania, 1 Caransebes Street, Bucarest 8, 7000 Romania. (1945) Scient. res. (tel. 657530). *Crystallography, mineralogy, geochemistry, experimental works.*

Cruceanu, Mr Eugen. Comitetul de Stat pentru Energia Nucleară, Inst. of Physics, Bucureşti, Romania. (1931) Dr.st., semiconductor crystals and crystallography (Inst. of Metals 'M. Kalinin', Moscow, USSR, 1960). Res. scient. (tel. 16 66 50). *Crystal growth, crystal structures.*

Dinescu, Prof. Radu. Ceramics and Refractory Materials, Faculty of Chemical Industry, Institutul Politehnic 'Gh. Gheorghiu-Dej', Calea Grivitei 132, Bucureşti, Romania. (1917) Dr.Ing., ceramics (Polytechnic Inst., Bucureşti, 1943). Prof. (tel. 139440). *Minerals, ceramics and refractory materials.*

Draghici, Mr Iosif. Mineralogy Dept., U. Bucharest, 1 N. Balcescu, Bucureşti, Romania. (1930) BSc, Geology and Geography (U. Bucharest). Reader. *Hydrothermal deposits.*

Dumitrescu, Aurelia. Crystallography & Mineralogy Lab., Inst. Politehnic 'Gh. Gheorghiu-Dej', Str. Polizu 1, Bucureşti, Romania. (1928) Mining engineer (Mining Inst., Bucureşti, 1953). Asst. (tel. 139440). *Thermal analysis.*

Giuşcă, Prof. Dan. Dept. of Geology & Geography, Blvd. N. Bălcescu 1, Universitatea Bucureşti, Bucureşti, Romania. (1904) Dr.doc.st., crystal chemistry (U. Cluj, 1927). Acad.; prof., petrology. *Crystal chemistry.*

Ianovici, Prof. Virgil. Ministerul Minelor, Petrolului şi Geologei, Institutul Geologic, Str. Mendeleev 36, Bucureşti 1, Romania. (1900) Dr.doc.st., mineralogy and petrography (U. Iaşi, 1929). Acad.; prof., crystallography, mineralogy. (tel. 333187). *Crystallography, mineralogy.*

Imreh, Mr Iosif. Mineralogy Lab., Universitatea Babeş-Bolyai, Str. Rogălniceanu, 1 Cluj, Romania. (1924) Dr.st., mineralogy and crystallography (U. Iaşi, 1957). Reader. (tel. 3001). *Morphology - structure relationship.*

Ionescu, Mrs Jeana. Mineralogical Lab., Ministerul Minelor, Petrolului şi Geologei, Institutul Geologic, Str. Caransebes 1, Bucureşti, Romania. (1924) Dr.doc.st., crystallography and mineralogy (Acad.R.S.R., 1960). Chief. (tel. 657530). *Inorganic crystal structures, crystal optics, clay minerals.*

Jude, Dr (Mrs) Lidia. Dept. of Geology and Geography, Universitatea Bucureşti, Blvd. N. Balcescu 1, Bucureşti 78344, Romania. (1930) Dr.doc.st., mineralogy and geology (U. Bucharest, 1962). (tel. 90 + 468673). *Crystal optics, crystal physics, crystal growth.*

Kissling, Mr Alexandru. Inst. of Oil and Nat. Gas, Ploieşti, Romania. (1921) Dr.st., mineralogy (U. de Bucureşti, 1964). Conf. *Radiocrystallography, crystal chemistry.*

Lazar, Mr Constantin. Inst. of Geology and Geophysics, Str. Caransebes 1, Bucureşti 78344, Romania. (1935) Trainer for Dr, mineralogy (U. Bucharest). Sr. res. scient. (tel. 90 + 657530). *Mineralogy, ore minerals, crystallography, geochemistry.*

Mănăilă, Mrs Rodica. X-ray Lab., Comitetul de Stat pentru Energia Nucleară, Inst. of Physics, Bucureşti, Romania. (1935) Lic.st., physics (U. de Bucureşti, 1957). Res. scient. (tel. 166550). *Semiconductor crystal structure, amorphous semiconductor structures.*

Mastacan, Prof. Gheorghe. Dept. of Geology and Geography, Universitatea Bucureşti, Blvd N. Bălcescu 1, Bucureşti, Romania. (1907) Dr.st., mineralogy and petrography (U. Iaşi, 1948). Reader. (tel. 127796 and 230754). *Crystal physics, structures, clay minerals, crystal optics.*

Mirzu-Ghergariu, Mrs Lucretia. Mineralogy Lab., Universitatea Babeş-Bolyai, Str. Kogălniceanu, 1 Cluj, Romania. (1932) Lic.st., chemistry, crystallography and mineralogy (U. Cluj, 1954). Asst. (tel. 3001). *Crystal optics, clay minerals.*

Petreus, Mr Ion. Mineralogy Lab., U. 'Al. I. Cuza', Calea 23 August 20A, Iaşi, Romania. (1939) Lic.st., geology (U. de Bucureşti). Asst. *Crystal morphology, clay minerals.*

Popescu, Conf. Ion. Dept. of Geology and Geography, Universitatea Bucureşti, Blvd. N. Bălcescu 1, Bucureşti, Romania. (1906) Dr.st., chemistry (U. de Bucureşti, 1943). Reader. (tel. 151798). *Inorganic crystal structures, clay minerals, crystal optics.*

Radulescu, Prof. Dan. Sedimentary Petrography Lab., Dept. of Geology and Geography, Universitatea Bucureşti, Blvd. N. Bălcescu 1, Bucureşti, Romania. (1928) Dr.doc.st., mineralogy (U. Bucureşti, 1957). Prof. *Crystal optics.*

Rosca, Mr Liviu. Crystallography and Mineralogy Lab., Inst. Politehnic 'Gh. Gheorghiu-Dej', Str. Polizu 1, Bucureşti, Romania. (1914) Lic.st., (U. Cernăuti, 1941). Reader. (tel. 139440). *Crystallography, mineralogy.*

Segal, Mr Eugen. Physical Chemistry Lab., Universitatea Bucureşti, Blvd. Republicii 13, Bucureşti, Romania. (1933) Dr.st., chemistry (U. de Bucureşti, 1964). Asst. prof. (tel. 157980).

Stiopol, Prof. Victoria. Crystallography and Geochemistry Lab., Universitatea Bucureşti, Bucureşti, Romania. (1928) Dr.st., mineralogy (U. de Bucureşti, 1960). Prof. (tel. 156713). *Crystallography, mineralogy.*

Stoicovici, Prof. Eugen. Mineralogy and Geochemistry Lab., Universitatea Babeş-Bolyai, Str. Kogălniceanu, 1 Cluj, Romania. (1906) Dr.doc.st., mineralogy and crystallography (U. Cluj, 1954). Prof. (tel. 3001). *Inorganic crystal structures, crystal optics, crystal physics, metals structure, clay minerals.*

Udubasa, Dr Gheorghe. Institutul de Geologie şi Geofizica, Str. Caransebes nr. 1, Bucureşti 7000, Romania. (1938) Dr, mineralogy of ore deposits (U. Heidelberg, Germany, 1972). Scient. res. (tel. 657530). *Mineralogy, geochemistry and structure of the ore minerals.*

Vanghelie, Mr Iulian. Inst. de Geologie şi Geofizica, Str. Caransebes nr. 1, Bucureşti 7000, Romania. (1946) Lic.st., chemistry (U. de Bucureşti, 1969). Scient. res. (tel. 90 + 657530). *X-ray crystallography, inorganic and organic crystal structures, X-ray diffraction.*

Vlad, Dr Serban-Nicolae. Inst. de Geologie şi Geofizica, Str. Caransebes nr. 1, Bucureşti 78344, Romania. (1941) Dr, mineralogy (U. de Bucureşti, 1971). Sr. res. scient. (tel. 90 + 657530). *Skarn silicates, ore minerals, mineralogy, crystallography.*

SAUDI ARABIA

Sub-Editor: M.S. Hussain

Notes

1. International telephone country code - 966.

Al-Shanti, Prof. Ahmed Mahmoud. Economic Geology Dept., Faculty of Earth Sci., King Abdulaziz U., P.O. Box 1744, Jeddah 21441, Saudi Arabia. (1932) PhD, mining geology (Imp. C., U. London, UK, 1973). Chairman. (tel. Jeddah - 6653735, Telex 401141 KAUNI). *Crystal optics (incident light on opaque minerals).*

Avci, Dr Recep. Physics Dept., U. of Petroleum & Minerals, UPM Box 2018, Dhahran 31261, Saudi Arabia. (1950) PhD, solid state physics (U. Illinois, Urbana, USA, 1978). Asst. prof. (tel. 3 + 860-2292, telex 601060 UPMSI). *Photoelectron spectroscopy, LEED, Auger spectroscopy, EELS, Photoemmission (inverse) spectroscopy, interfaces, materials research.*

Bakr, Dr Abdel Razak. Geology Dept., Faculty of Sci., King Abdulaziz U., P.O. Box 1540, Jeddah, Saudi Arabia. PhD, petrology (U. Leeds, UK, 1973). Head. *Crystal optics in transmitted light.*

El-Mahdi, Dr Omar. Inst. of Applied Geology, King Abdulaziz U., P.O. Box 1744, Jeddah, Saudi Arabia. (1938) PhD, economic geology (U. Utah, USA, 1966). Assoc. Prof. (tel. Jeddah - 24263). *Crystal optics (incident light on opaque materials).*

Haque, Prof. Mazhar-ul. Chemistry Dept., U. of Petroleum & Minerals, UPM Box 1830, Dhahran 31261, Saudi Arabia. (1936) PhD, chemistry (Imperial C., U. London, UK, 1964). Prof. (tel. 03 + 860-2378). *Inorganic and organic crystal structures, natural products, phosphetans.*

Horne, Mr William. U. of Petroleum & Minerals, Dhahran Airport Box 144, Dhahran, Saudi Arabia. (1945) BSc(Hons), chemistry (U. Leeds, UK, 1966). Lect. (tel. 3 + 860-3827). *Computer programming, crystal structures, phosphetans, phosphorinanes.*

Hussain, Dr M. Sakhawat. Chemistry Dept., U. of Petroleum & Minerals, UPM Box 1830, Dhahran 31261, Saudi Arabia. (1939) PhD, chemistry, chemical crystallography (U. California, Davis, USA, 1968). Assoc. prof. (tel. 3 + 860-3828). *Organometallic compounds, crystal structures, complexes with short hydrogen bonds, biologically significant compounds.*

Hussain, Dr Zahid. Physics Dept., U. of Petroleum & Minerals, UPM Box 580, Dhahran 31261, Saudi Arabia. (1949) PhD, chemical physics (U. Hawaii, USA, 1979). Asst. prof. (tel. 3 + 860-2292). *Electron spectroscopy, XPS, UPS, LEED, Auger spectroscopy, EELS, syncrotron radiation instrumentation, EXAFS, surface crystallography, photoelectron diffraction.*

Kenaan, Dr Feisal. Directorate General of Mineral Resources, P.O. Box 345, Jeddah, Saudi Arabia. PhD, petrology (Colorado School of Mines, USA, 1976). *Crystal optics in transmitted light.*

Khattak, Dr Guldad Khattak. Physics Dept., U. of Petroleum & Minerals, UPM Box 1854, Dhahran 31261, Saudi Arabia. (1948) PhD, solid state physics (Purdue U., USA, 1978). Asst. prof. (tel. 3 + 860-2260). *Specific heat, magnetic susceptibility, resistivity, low temperature investigation, semimagnetic semiconductors, oxides, inorganic complexes of Ni & Cu.*

Khawaja, Dr Ehsan Ellahi. Physics Dept., U. of Petroleum & Minerals, UPM Box 1987, Dhahran 31261, Saudi Arabia. (1945) PhD, solid state physics (U. Adelaide, Australia). Asst. prof. (tel. 3 + 860-2267). *Semiconductor physics, thin film structure, X-ray diffraction, electron diffraction.*

Koyama, Dr Kazutoshi. Inst. of Applied Geology, King Abdulaziz U., P.O. Box 1744, Jeddah, Saudi Arabia. (1946) PhD, X-ray crystallography (U. Tokyo, Japan, 1976). Assoc. UNESCO Expert. (tel. Jeddah - 24263). *Crystal structures.*

Naseif, Dr Abdulah. King Abdulaziz U., P.O. Box 1540, Jeddah, Saudi Arabia. PhD, petrology (U. Leeds, UK, 1972). Deputy President. *Crystal optics in transmitted light, minerals.*

Tahoun, Prof. Salah. Inst. of Applied Geology, King Abdulaziz U., P.O. Box 1744, Jeddah, Saudi Arabia. (1937) PhD, clay mineralogy (Michigan State U., USA, 1965). Prof. of soil mineralogy. (tel. Jeddah - 24263). *Structure and identification, clay minerals.*

SINGAPORE

Sub-Editor: Chatar Singh

Notes

1. International telephone country code - 65

2. The science degrees conferred by the National University of Singapore are the PhD, MSc and BSc and correspond to British degrees.

Chowdari, Dr B.V.R. Physics Dept., Nat. U. of Singapore, Kent Ridge, Singapore 0511, Singapore. (1943) PhD, physics (I.I.T., Kanpur, India, 1968). Sr. lect. (tel. 7756666, ext. 2607). *Transport properties of solids, Optical properties of solids, magnetic properties of solids.*

Chung, Dr Mui Fatt. Physics Dept., Nat. U. of Singapore, Kent Ridge, Singapore 0511, Singapore. (1936) PhD, physics (U. New South Wales, Australia, 1966). Sr. lect. (tel. 7756666, ext. 2615). *Surface structure of solids.*

Fong, Dr Hock Sun. Mechanical Eng., Nat. U. of Singapore, Kent Ridge, Singapore 0511, Singapore. (1941) PhD, physical metallurgy, materials science (U. Birmingham, UK, 1969). Assoc. prof. (tel. 7756666, ext. 2211). *Crystallography, phase transformations.*

Hosea, Dr Thomas Jeffrey Cockburn. Physics Dept., Nat. U. of Singapore, Kent Ridge, Singapore 0511, Singapore. (1952) PhD, solid state physics (U. Edinburgh, Scotland, 1978). Lect. (tel. 7756666, ext. 2629). *Raman and brillouin spectroscopy, phase transitions.*

Koh, Dr Lip Lin. Faculty of Sci., Nat. U. of Singapore, Kent Ridge, Singapore 0511, Singapore. (1935) PhD, physical chemistry (Boston U., USA, 1964). Dean, Fac. of Sci. (tel. 7756666, ext. 2774). *Crystallography.*

Kuok, Dr Meng Hau. Physics Dept., Nat. U. of Singapore, Kent Ridge, Singapore 0511, Singapore. (1951) PhD, solid state physics (U. Canterbury, New Zealand,1978). Lect. (tel. 7756666, ext. 2609). *Raman spectroscopy, infrared spectroscopy.*

Ng, Dr Ser Choon. Physics Dept., Nat. U. of Singapore, Kent Ridge, Singapore 0511, Singapore. (1937) PhD, solid state physics (McMaster U., Canada, 1968). Sr. lect. (tel. 7756666, ext. 2610). *X-ray diffraction, neutron diffraction.*

Rajaratnam, Prof. Arthur. Physics Dept., Nat. U. of Singapore, Kent Ridge, Singapore 0511, Singapore. (1927) PhD, spectroscopy (U. London, UK, 1958). Prof. (tel. 7756666, ext. 2628). *Spectroscopy, structure of matter.*

Tan, Dr Hock Siew. Physics Dept., Nat. U. of Singapore, Kent Ridge, Singapore 0511, Singapore. (1950) PhD, solid state physics (U. Rochester, USA, 1980). Lect. (tel. 7756666, ext. 2614). *Structure of solids.*

Teh, Dr Hung Chuan. Information Systems and Computer Sci. Dept., Nat. U. of Singapore, Kent Ridge, Singapore 0511, Singapore. (1941) PhD, solid state physics (McMaster U., Canada, 1972). Sr. lect. (tel. 7756666, ext. 2782). *X-ray diffraction, neutron diffraction, photon correlation spectroscopy, rayleigh-brillouin spectroscopy, data acquisition, instrumentation, computer interfacing.*

SOUTH AFRICA

Sub-Editor: G.J. Kruger

Notes

1. International telephone country code - 27.

Adendorff, Mr Keith Trevor. Nat. Inst. for Materials Res., CSIR, P.O.Box 395, Pretoria 0001, South Africa. (1957) BSc(Hons), geology (Rhodes U., 1983). Res. (tel. 012 + 869211). *Mineralogy, inorganic structures, powder diffraction.*

Adrian, Dr Herbert Wilhelm Werner. Safety Div., Nuclear Dev. Corp., Private Bag X256, Pretoria 0001, South Africa. (1940) DSc, physics (U. Pretoria, 1978). Chief sci. (tel. 012 + 213311, ext. 719). *Neutron diffraction, hydrogen positions, metal compounds.*

Alberts, Prof. Hermanus Lambertus. Physics Dept., Rand Afrikaans U., P.O. Box 524, Johannesburg 2000, South Africa. (1941) PhD, physics (Rand Afrikaans U., 1970). Prof. (tel. 011 + 7265000, ext. 330). *Magnetic and elastic properties.*

Ashworth, Dr Terence Vincent. N.C.R.L., CSIR, P.O. Box 395, Pretoria 0001, South Africa. (1950) PhD, inorganic chemistry (U. South Africa, 1977). Chief res. (tel. 012 + 869211, ext. 2652). *Organometallic chemistry, synthesis, structure, catalysis.*

Archer, Mr Steven James. Physical Chemistry Dept., U. of Cape Town, Private Bag, Rondebosch 7700, Cape Town, South Africa. (1958) BSc(Hons), chemistry (U. Cape Town, 1980). Res. officer. (tel. 021 + 698531, ext. 542). *Small molecule structures.*

Auf Der Heyde, Mr Thomas Paul Edwin. Chemistry Dept., U. of the Western Cape, Private Bag X17, Belville 7530, South Africa. (1958) MSc, physical-inorganic chemistry (U. Cape Town, 1984). Lect. (tel. 021 + 976161, ext. 263). *Structural correlation.*

Ball, Prof. Anthony. Dept. of Materials Engineering, U. of Cape Town, Private Bag, Rondebosch 7700, Cape Town, South Africa. (1939) PhD, physical metallurgy (U. Birmingham, UK, 1964). Head. (tel. 021 + 698531, ext. 320). *Materials deformation and structure.*

Basson, Dr Stephen Smuts. Chemistry Dept., U. of the Orange Free State, P.O. Box 339, Bloemfontein 9300, South Africa. (1942) DSc, chemistry (U. Orange Free State, 1969). Sr. lect. (tel. 051 + 70711, ext. 348). *Complex cyanides, rhodium and iridium chemistry.*

Beukes, Prof Dr Gerhardus Johannes. Geology Dept., U. of the Orange Free State, P.O. Box 339, Bloemfontein 9300, South Africa. (1943) DsC, geology (U. Orange Free State, 1973). Assoc. prof. (tel. 051 + 70711, ext. 393). *X-ray diffractometry, X-ray fluorescence spectrometry, energy dispersive X-ray spectroscopy, applied mineralogy.*

Boeyens, Prof. Jan Christoffel Antonie. Chemistry Dept., U. of the Witwatersrand, 1 Jan Smuts Ave., Johannesburg 2001, South Africa. (1934) DSc, physical and theoretical chemistry (U. Pretoria, 1964). Prof. (tel. 011 + 7162076). *Structural theory, molecular mechanics, disorder, phase transformations.*

Boonstra, Prof. Eelco Gerrit. U. of the Orange Free State, P.O. Box 339, Bloemfontein 9300, South Africa. (1935) PhD, physics (U. Natal, 1966). Vice Rector. (tel. 051 + 70711, ext. 661). *Crystal structures, automation, computing, surface structures.*

Bond, Mrs Dianne Ruth. Physical Chemistry Dept., U. of Cape Town, Private Bag., Rondebosch 7700, Cape Town, South Africa. (1958) MSc, x-ray crystallography (U. Cape Town, 1982). Asst. Lect. (tel. 021 + 698531, ext. 542). *Phosphorus compounds.*

Brown, Prof Michael Ewart. Chemistry Dept., Rhodes U., P.O. Box 94, Grahamstown 6140, South Africa. (1938) PhD, physical chemistry (Rhodes U., 1966). Assoc. prof. (tel. 0461 + 3223). *Solid phase reactions, kinetics and mechanisms.*

Caira, Dr Mino Rodolfo. Physical Chemistry Dept., U. of Port Elizabeth, P.O. Box 1600, Port Elizabeth 6000, South Africa. (1949) PhD, chemistry (U. Cape Town, 1975). Sr. lect. (tel. 041 + 5311147). *Organic and organometallic structures, drug structure-reactivity relationships.*

Caveney, Dr Robert John. De Beers Industrial Diamond Div., P.O. Box 916, Johannesburg 2000, South Africa. (1941) PhD, physics (U. Witwatersrand, 1970). Deputy director of res. (tel. 011 + 8353232, ext. 334). *Crystal growth, defects, diamond-structure, composite materials.*

Clark, Dr James Brian. Nat. Inst. for Materials Res., CSIR, P.O. Box 395, Pretoria 0001, South Africa. (1949) DSc, physics (U. Pretoria, 1973). Chief director. (tel. 012 + 869211, ext. 2862). *Structure, inorganic materials, phase transitions.*

Coetzer, Dr Johan. Zebra Power Systems, P.O.Box 14003, Verwoerdburg 0140, South Africa. (1941) PhD chemistry (U. Indiana, USA, 1968). Managing director. (tel. 012 + 647348). *Superionic conductors, solid state batteries, structural chemistry.*

Comins, Dr Neville Raymond. Metals Div., Nat. Inst. for Materials Res., CSIR, P.O. Box 395, Pretoria 0001, South Africa. (1945) PhD, physics (Cambridge U., UK, 1971). Deputy Director. (tel. 012 + 869211, ext. 3386). *Electron microscopy, metals, weak-beam em investigations, dislocation structures and interactions.*

Copperthwaite, Dr Richard George. Chemistry Dept., U. of the Witwatersrand, 1 Jan Smuts Ave., Johannesburg 2000, South Africa. (1945) PhD, chemistry (U.of London, UK, 1971). Sr. lect. (tel. 011 + 7162262). *Solid state and surface chemistry, heterogeneous catalysis.*

Coville, Dr Neil John. Chemistry Dept, U. of the Witwatersrand, 1 Jan Smuts Ave., Johannesburg 2000, South Africa. (1945) PhD, chemistry (McGill U., Canada, 1973). Sr. lect. (tel. 011 + 7162371). *Organometallic chemistry.*

Crawford, Mr John Lawrence. Physics Dept., U. of the Witwatersrand, 1 Jan Smuts Ave., Johannesburg 2001, South Africa. (1937) BSc Hons., physics (U. Witwatersrand, 1959). Lect. (tel. 011 + 394011, ext. 405). *Electron microscopy, defects in solids.*

Davies, Dr Geoffrey John. De Beers Diamond Res. Lab., P.O. Box 916, Johannesburg 2000, South Africa. (1948) PhD, physics (U. Reading, UK, 1972). Principal scient. officer. (tel. 011 + 8392470, ext. 121). *High pressure and diamond physics.*

Davies, Dr Gladstone. Nat. Building Res. Inst., CSIR, P.O.Box 395, Pretoria 0001, South Africa. (1952) PhD, geology (U. Witwatersrand, 1983). Sr. res. (tel. 012 + 869211, ext. 4438). *Materials research, powder diffraction.*

Denner, Mr Louis. Chemistry Dept, U. of the Witwatersrand, 1 Jan Smuts Ave., Johannesburg 2000, South Africa. (1959) MSc, chemistry (U. Witwatersrand, 1983). Res. officer. (tel. 011 + 7162165). *Molecular conformation, hydrogen bonding,organic structures.*

De Villiers, Prof. Johan Pieter Roos. MINTEK/RAU Applied Mineralogy Res. Group, Dept of Geology, Rand Afrikaans U., P.O. Box 524, Johannesburg 2000, South Africa. (1942) PhD, mineralogy (U. Illinois, USA, 1969). Chief sci. (tel. 011 + 7265000, ext. 304). *Mineral structures, inorganic phase chemistry.*

De Wet, Prof. Julius Ferdinand. Chemistry Dept., U. of Port Elizabeth, P.O. Box 1600, Port Elizabeth 6000, South Africa. (1926) DSc, physical chemistry (U. Pretoria, 1960). Prof. (tel. 041 + 5311928). *Crystal structure determination, crystallographic programming.*

Dillen, Dr Jan Louis Maria. Organic Chem. Div., Nat. Chemical Res. Lab., CSIR, P.O.Box 395, Pretoria 0001, South Africa. (1955) PhD, physical chemistry (U. Antwerp, Belgium,1981). Sr. Res. (tel. 012 + 869211, ext 2628). *Conformational analysis, molecular dynamics, inter- and intramolecular forces.*

Dobson, Mrs Susan Mary. Chemistry Dept, U. of the Witwatersrand, 1 Jan Smuts Ave., Johannesburg 2000, South Africa. (1959) BSc(Hons), chemistry (U. Witwatersrand, 1981). Res. officer. (tel. 011 + 7162101). *Coordination compounds, macrocyclic ligands, conformational analysis.*

Dunlevey, Mr John Norman. Nat. Building Res. Inst., CSIR, P.O.Box 395, Pretoria 0001, South Africa. (1949) MSc, geology (Stellenbosch U., 1978). Chief Res. (tel. 012 + 869211, ext. 2506). *Petrology, geochemistry, materials science.*

Du Plessis, Dr Michael Peter. Res. Dept., AECI Explosives and Chemicals (ltd), P.O. North Rand 1645, South Africa. (1954) PhD, physical chemistry (U. Cape Town, 1982). Sr. res. officer. (tel. 011 + 6081201, ext. 450). *Solid state chemistry, energetic materials, fast reactions.*

Du Plessis, Prof. Paul de Villiers. Physics Dept., Rand Afrikaans U., P.O. Box 524, Johannesburg 2000, South Africa. (1940) DSc physics (U. Orange Free State, 1966). Prof. (tel. 011 + 7265000, ext. 328). *Magnetic and elastic properties, rare earths, actinides, neutron scattering (elastic and inelastic).*

Eales, Prof. Hugh Victor. Geology Dept., Rhodes U., P.O. Box 94, Grahamstown 6140, South Africa. (1929) PhD, geology (Rhodes U., 1961). Prof., dept. head. (tel. 0461 + 2023, ext. 11). *Mineralogy of spinel group minerals, petrology and geochemistry of basic rocks, XRF spectrometry and microprobe techniques, reflected light microscopy.*

Engel, Prof. Dennis Walter. Physics Dept., U. of Durban-Westville, Private Bag X54001, Durban 4000, South Africa. (1939) Dr. rer. nat., physics (Tech. U. München, BRD, 1971). Prof. (tel. 031 + 821211, ext. 226). *Anomalous scattering, powder diffraction.*

English, Dr Robert Bertram. Chemistry Dept., Rhodes U., P.O.Box 94, Grahamstown 6140, South Africa. (1948) PhD, organometallic chemistry (U.

Cape Town, 1977). Sr. lect. (tel. 0461 + 3223). *Metal cluster chemistry, platinum group metal nitrosyl chemistry, alkoxides and carboxylates of Zr,Ti,Al,Pb.*

Esterhuyse, Miss Suzette. Res. Div., SASTECH, P.O.Box 1, Sasolburg 9570, South Africa. (1962) BSc(Hons), chemistry (Stellenbosch U., 1983). Res. Scient. (tel. 016 + 6889111,ext. 2940). *Alkane structures, waxes.*

Field, Dr John Stainer. Chemistry Dept., U. of Natal, P.O. Box 375, Pietermaritzburg 3200, South Africa. (1946) PhD, X-ray crystallography (Cambridge U., UK, 1973). Lect. (tel. 0331 + 63320). *Organometallic preparative chemistry, X-ray and neutron diffraction.*

Förtsch, Prof. Dr Erich Bernhard. Geology Dept., U. of Pretoria, Pretoria 0002, South Africa. (1934) Dr. rer. nat., mineralogy (U. Freiburg, BRD, 1964). Assoc. prof. (tel. 012 + 4202238). *Mineral structures, morphology.*

Fourie, Dr Jacobus Theodor. Nat. Inst. for Materials Res., CSIR, P.O. Box 395, Pretoria 0001, South Africa. (1930) DSc, physics (U. Pretoria, 1956). Chief specialist sci. (tel. 012 + 869211, ext. 3386). *Transmission electron microscopy, plastic deformation of metals.*

Gafner, Dr Geoffrey. Internat. Gold Corp., P.O. Box 61809, Marshalltown 2107, South Africa. (1930) DSc, physics (U. Pretoria, 1960). Industrial R & D Manager. (tel. 011 + 8388211, ext. 532). *Gold.*

Glasser, Prof. Leslie. Chemistry Dept., U. of the Witwatersrand, 1 Jan Smuts Ave., Johannesburg 2001, South Africa. (1935) PhD D.I.C., chemical engineering (Imperial Coll., U. of London, 1960). Prof. of phys. chem., dept. head. (tel. 011 + 7162219). *Electrical properties, materials, hydrogen bonding, chemometrics.*

Hart, Dr Stewart. Ceramics, Glass and Phase Studies Div., Nat. Inst. for Materials Res., CSIR, P.O. Box 395, Pretoria 0001, South Africa. (1937) PhD, physics (U. Aberdeen, UK, 1967). Deputy director. (tel. 012 + 869211, ext. 2783). *Inorganic crystal structure and orientation, high pressure and temperature phase diagrams.*

Heckroodt, Prof. Renier Oelof. Dept. of materials engineering, U. of Cape Town, Private Bag, Rondebosch 7700, Cape Town, South Africa. (1935) DSc, geology (U. Pretoria, 1968). Prof. (tel. 021 + 698531). *Clay mineralogy.*

Heyns, Prof. Anton Michal. Chemistry Dept., U. of Pretoria, Hatfield, Pretoria 0002, South Africa. (1939) PhD, chemistry (U. South Africa, 1968). Prof. (tel. 012 + 4202516). *IR spectroscopy, Raman spectroscopy, X-ray powder diffraction, ionic solids.*

Horsfield, Mr Edgar Charles. Physics Dept., U. of Durban-Westville, Private Bag X54001, Durban 4000, South Africa. (1942) MSc, crystallography (U. Natal, 1969). Lect. (tel. 031 + 821211, ext. 163). *Anomalous scattering, powder diffraction.*

Hutchings, Dr Ronald. Dept of materials engineering, U. of Cape Town, Private Bag, Rondebosch 7700, South Africa. (1951) PhD, physical metallurgy (U. Birmingham, UK, 1976). Sr. lect. (tel. 021 + 698531, ext. 740). *Structure - property relationship, electron diffraction, electron microscopy.*

Irving, Dr Anne. Physical Chemistry Dept., U. of Cape Town, Private Bag., Rondebosch 7700, Cape Town, South Africa. (1940) PhD, X-ray crystallography (U. Leeds, UK, 1969). Lect. (tel. 021 + 698531, ext. 643). *Small molecule structures.*

Kingon, Mr Angus Ian. Nat. Inst. for Materials Res., CSIR, P.O. Box 395, Pretoria 0001, South Africa. (1954) PhD, physical chemistry (U. South Africa, 1981). Chief res. (tel. 012 + 869211, ext. 3395). *Phase studies, inorganic synthesis, ceramics, ferroelectric materials, piezoelectric materials, powder diffraction.*

Kritzinger, Prof. Serfontein. Physics Dept., U. of Stellenbosch, Stellenbosch 7600, South Africa. (1940) PhD physical metallurgy (U. Birmingham, UK, 1967). Prof. of solid state physics. (tel. 02231 + 6037, ext. 15). *Crystallography, defects, metals and alloys.*

Kruger, Dr Gert Jacobus. Chemistry dept., Rand Afrikaans U., P.O.Box 524, Johannesburg 2000, South Africa. (1943) DSc, chemistry (Potchefstroom U., 1970). Prof. (tel. 011 + 7265000, ext. 368). *Crystallographic computing, direct methods, powder diffraction.*

Laing, Dr Mary Elizabeth. 61 Baines Road, Durban 4001, South Africa. (1935) PhD, chemistry (U. California, Los Angeles, USA, 1964). Part-time lect. (tel. 031 + 251951). *Organic and inorganic structures.*

Laing, Prof. Michael John. Chemistry Dept., U. of Natal, King George V Ave, Durban 4001, South Africa. (1937) PhD, inorganic chemistry (U. California, Los Angeles, USA, 1965). Assoc. prof. (tel. 031 + 8163103). *Coordination compounds, strained and aromatic organic compounds, polymorphism in organic crystals.*

Leipoldt, Prof. Johann Gotlieb. Chemistry Dept., U. of the Orange Free State, P.O. Box 339, Bloemfontein 9300, South Africa. (1940) DSc, inorganic chemistry (U. Orange Free State, 1969). Assoc. prof. (tel. 051 + 70711, ext. 497). *Crystal structure, transition metal complexes.*

Le Roux, Dr Johannes Hendrik. Res. Div., SASOL, P.O. Box 1, Sasolburg 9570, South Africa. (1934) PhD, physical chemistry (U. South Africa, 1968). Res. specialist. (tel. 016 + 62431, ext. 7150). *Structure, waxes, coal derivatives, molecular spectroscopy.*

Le Roux, Dr Stephanus David. Physical Metallurgy Div., Nuclear Dev. Corp., Private Bag X256, Pretoria 0001, South Africa. (1947) PhD, solid state physics (Purdue U., USA, 1975). Subdivision head. (tel. 012 + 747811, ext. 533). *Crystallography, uranium compounds, X-ray fluorescence, neutron diffraction, texture analysis, computing methods.*

Levendis, Mr Demetrius Christos. Chemistry Dept., U. of the Witwatersrand, 1 Jan Smuts Ave., Johannesburg 2000, South Africa. (1957) MSc, crystallography (U.

Witwatersrand, 1982). Res. officer. (tel. 011 + 7164133). *Disordered structures, lattice energy and dynamics calculations.*

Liles, Mr David Charles. Inorganic Chem. Div., Nat. Chemical Res. Lab., CSIR, P.O.Box 395, Pretoria 0001, South Africa. (1950) BSc(Hons), chemistry (Loughborough U., 1973). Sr. res. (tel. 012 + 869211, ext. 2628). *Structure, transition metal complexes.*

Lombard, Mr Anthonie van Altena. Cemistry dept., Rand Afrikaans U., P.O.Box 524, Johannesburg 2000, South Africa. (1958) MSc, chemistry (Rand Afrikaans U., 1981). Res. asst. (tel. 011 + 7265000, ext. 368). *Organometallic chemistry.*

Lotz, Dr Simon. Chemistry Dept., U. of Pretoria, Hatfield, Pretoria 0002, South Africa. (1948) PhD, organometallic chemistry (Rand Afrikaans U., 1979). Prof. (tel. 012 + 4209111, ext. 2626). *Organometallics.*

Markwell, Dr Anthony James. Chemistry Dept., U. of the Witwatersrand, 1 Jan Smuts Ave., Johannesburg 2000, South Africa. (1945) PhD, crystallography (U. London, UK, 1974). Lect. (tel. 011 + 7162327). *Structures and properties, inorganic complexes.*

Maske, Prof. Siegfried. Geology Dept., U. of the Witwatersrand, 1 Jan Smuts Ave., Johannesburg 2001, South Africa. (1928) DSc, geology (Stellenbosch U., 1964). Prof. of mining geology. (tel. 011 + 394011, ext. 526). *Ore genesis, mineralogy, geochemistry, crystal structure, sulphide minerals.*

McDougall, Dr Gloria Jeanne. ACIX Div., Nat. Chemical Products, Power street, Industries East, Germiston 1400, South Africa. (1949) PhD, inorganic chemistry (U. Witwatersrand, 1975). Div. manager. (tel. 011 + 8253330, ext. 2307). *Coordination chemistry, ion-exchange, solvent extraction, activated carbon.*

Moore, Dr Alan Charles. Geology Dept., U. of Cape Town, Private Bag, Rondebosch 7700, Cape Town, South Africa. (1941) PhD, petrology (U. Adelaide, Australia, 1970). Sr. lect. (tel. 021 + 698531, ext. 525). *Exsolution structure, spinels.*

Moore, Miss Madeleine. Physical Chemistry Dept., U. of Cape Town, Private Bag, Rondebosch 7700, Cape Town, South Africa. (1960) BSc(Hons), chemistry (U. Natal, 1983). Jr. res. fellow. (tel. 021 + 698531, ext. 542). *Inorganic complexes.*

Nabarro, Prof. Frank Reginald Nunes. U. of the Witwatersrand, 1 Jan Smuts Ave., Johannesburg 2001, South Africa. (1916) DSc, FRS, metallurgy(dislocation theory) (U. Birmingham, UK, 1953). Prof. em. (tel. 011 + 7162175). *Crystal defects.*

Nassimbeni, Prof. Luigi Renzo. Physical Chemistry Dept., U. of Cape Town, Private Bag, Rondebosch 7700, Cape Town, South Africa. (1939) PhD, physical chemistry (U. Cape Town, 1969). Prof. (tel. 021 + 698531, ext. 538). *Organic and inorganic crystal structures.*

Netterberg, Dr Frank. N.I.T.R.R., CSIR, P.O. Box 395, Pretoria 0001, South Africa. (1938) PhD, geology (engineering) (U. Witwatersrand, 1970). Head, Soil Eng. Grp. (tel. 012 + 869211, ext. 2918). *Engineering geology, clay mineralogy, pedogenic materials.*

Niven, Dr Margaret Lilian. Physical Chemistry Dept., U. of Cape Town, Private Bag, Rondebosch 7700, Cape Town, South Africa. (1954) PhD, chemistry (U. Cape Town, 1980). Res. officer. (tel. 021 + 698531, ext. 540). *Small molecules, disorder.*

Nolte, Dr Margaretha Johanna. Zebra Power Systems, P.O.Box 14003, Verwoerdburg 0140, South Africa. (1943) PhD, chemistry (U. Natal, 1976). Project leader. (tel. 012 + 647348). *Structural chemistry, applied electrochemistry.*

Oberholster, Dr Rupert Egbert. N.B.R.I., CSIR, P.O. Box 395, Pretoria 0001, South Africa. (1934) DSc, soil science U. of Pretoria, 1966). Sr. chief res. officer. (tel. 012 + 869211, ext. 2504). *Materials research.*

Paige-Green, Mr Philip. N.I.T.R.R., CSIR, P.O. Box 395, Pretoria 0001, South Africa. (1952) MSc, geology (U. Natal, 1975). Engineering geologist. (tel. 012 + 869211, ext. 2924). *X-ray diffraction analysis, minerals, clays, road materials.*

Pipkin, Dr Noel John. Physics Dept., De Beers Diamond Res. Lab., P.O. Box 916, Johannesburg 2000, South Africa. (1942) PhD, metallurgy (U. Newcastle-upon-Tyne, UK, 1967). Principal scient. officer. (tel. 011 + 8353232, ext. 240). *Graphite, diamonds, cubic boron nitride, characterisation.*

Pretorius, Dr Jan Andries. Analytical Group, Res. Dept., AECI Explosives and Chemicals (ltd), P.O.North Rand 1645, South Africa. (1949) PhD, chemistry (U. South Africa, 1978). Chief res. officer. (tel. 011 + 6081201, ext. 203). *Powder diffraction, X-ray fluorescence, automation, crystallographic computing.*

Retief, Dr Johannes Jacobus. Res. Div., Sasol Technology, P.O. Box 1, Sasolburg 9570, South Africa. (1941) PhD, physics (U. Orange Free State, 1978). Principal res. sci. (tel. 016 + 6882940). *Structure, catalysts, waxes, carbons, clay mineral identification.*

Reynhardt, Prof. Eduard Christiaan. Physics Dept., U. of South Africa, P.O. Box 392, Pretoria 0001, South Africa. (1944) PhD, physics U. of South Africa, 1971). Prof. (tel. 012 + 4401527). *Molecular reorientation in solids, phase transitions in solids.*

Richter, Dr Paul Wilhelm. Nat. Inst. for Materials Res., CSIR, P.O. Box 395, Pretoria 0001, South Africa. (1946) PhD, physical chemistry (U. South Africa, 1971). Chief res. (tel. 012 + 869211, ext. 2434). *Materials research, high pressure, thermal analysis, ceramics, crystal growth.*

Rodgers, Dr Allen Lawrence. Physical Chemistry Dept., U. of Cape Town, Private Bag, Rondebosch 7700, Cape Town, South Africa. (1946) PhD, chemistry (U. Cape Town, 1974). Sr. lect. (tel. 021 + 698531, ext. 558). *X-ray powder diffraction, scanning electron microscopy, calculi.*

South Africa

Roux, Dr Jacobus Paul. Physics Dept., U. of the Orange Free State, P.O. Box 339, Bloemfontein 9300, South Africa. (1935) PhD, physics (U. South Africa, 1970). Prof. (tel. 051 + 70711, ext. 192). *Surface structure.*

Saggerson, Prof. Edward Phillips. Geology Dept., U. of Natal, King George V Ave., Durban 4001, South Africa. (1924) DSc, petrology (U. Newcastle-upon-Tyne, UK, 1965). Assoc. prof. (tel. 031 + 8169111, ext. 2517). *Igneous and metamorphic rock, petrology.*

Schmidt, Mr Elias Rudolph. Consulting clay mineralogist, 7 Twelfth Street, Menlo Park, Pretoria 0081, South Africa. (1923) MSc, petrology (U. Pretoria, 1947). Consultant. (tel. 012 + 462138). *Clay mineralogy, ceramics.*

Schoch, Dr Aylva Ernest. Geology Dept., U. of Stellenbosch, Stellenbosch 7600, South Africa. (1933) DSc, igneous petrology (Stellenbosch U., 1972). Sr. lect. (tel. 02231 + 70028). *Order-disorder relations, rock-forming minerals, petrogenesis, granite rocks, natural processes.*

Schoening, Prof. Friedrich Richard Ludwig. Physics Dept., U. of the Witwatersrand, 1 Jan Smuts Ave., Johannesburg 2001, South Africa. (1923) PhD, physics (U. Witwatersrand, 1959). Reader in crystallography. (tel. 011 + 7161111,ext. 2132). *Diffraction physics, crystal defects, non-crystalline materials.*

Schutte, Prof. Casper Jan Hendrik. Chemistry Dept., U. of South Africa, P.O. Box 392, Pretoria 0001, South Africa. (1934) Dr, physical chemistry (U. Amsterdam, Netherlands, 1960). Prof. (tel. 012 + 4402355). *IR spectroscopy, Raman spectroscopy, molecular vibrations of solids.*

Sewell, Mr Bryan Trevor. Biochemistry Dept., U. of Cape Town, Rondebosch 7700, Cape Town, South Africa. (1953) MSc, biophysics (U. Witwatersrand, 1976). Lect. (tel. 021 + 698531, ext. 782). *Protein crystallography, computer graphics, structure, chromatin.*

Smuts, Dr Jacques. Materials Res. Div., ISCOR, P.O. Box 450, Pretoria 0001, South Africa. (1936) DSc, physics (Stellenbosch U., 1962). Principal res. officer. (tel. 012 + 414111, ext. 2881). *Metallography, mineralogy, TEM, SEM, X-ray diffraction, X-ray fluorescence.*

Sommerville, Mrs Polly Baker Melville. Chemistry Dept., U. of Natal, King George V Ave., Durban 4001, South Africa. (1924) MSc, chemistry (U. Natal, 1970). Res. asst. (tel. 031 + 253411). *Organic structures.*

Spalding, Dr Dennis Raymond. Physics Dept., U. of Natal, King George V Ave., Durban 4001, South Africa. (1942) PhD, physics (Cambridge U., UK, 1969). Sr. lect. (tel. 031 + 253411). *Radiation damage, metals, dislocation-point defect interactions.*

Strydom, Dr Ockert Andries Wilhelm. Material Science Dept., NUCOR, Private Bag X256, Pretoria 0001, South Africa. (1933) PhD, physics (Rand Afrikaans U., 1969). Div. head. (tel. 012 + 747811, ext. 524). *Materials science.*

Subramony, Mr Loganathan. Dept of Physics, U. of Durban-Westville, Private Bag X54001, Durban 4000, South Africa. (1946) MSc, physics (U. Durban-Westville, 1980). Jr. lect. (tel. 031 + 821211, ext. 163). *Organometallic compounds.*

Taylor, Mr Michael William. Physical Chemistry Dept., U. of Cape Town, Private Bag., Rondebosch 7700, Cape Town, South Africa. (1959) BSc(Hons), chemistry (U. Cape Town, 1984). Sr. demonstrator. (tel. 021 + 698531, ext. 542). *Werner clathrates.*

Thackeray, Dr Michael Makepeace. Nat. Inst. for Materials Res., CSIR, P.O. Box 395, Pretoria 0001, South Africa. (1949) PhD, chemistry (U. Cape Town, 1977). Sr. chief res. officer. (tel. 012 + 869211, ext. 3304). *Solid electrolytes, solid solution electrodes, non-stoichiometric compounds.*

Thom, Ms Vivienne Joyce. Chemistry Dept, U. of the Witwatersrand, 1 Jan Smuts Ave., Johannesburg 2000, South Africa. (1960) BSc(Hons), chemistry (U. Witwatersrand, 1979). Res. officer. (tel. 011 + 7162278). *Coordination chemistry, macrocycles, conformational analysis.*

Vale, Dr Roger John. Synthesis Div., De Beers Industrial Diamond Res. Labs., P.O. Box 916, Johannesburg 2000, South Africa. (1948) PhD, physical metallurgy (U. Birmingham, UK, 1976). Sr. res. sci. (tel. 011 + 8392470, ext. 121). *X-ray topography, defects, cubic boron nitride.*

Van Dyk, Miss Martha Sophia. Chemistry Dept., Rand Afrikaans U., P.O.Box 524, Johannesburg 2000, South Africa. (1959) MSc, chemistry (Rand Afrikaans U., 1982). Res. asst. (tel. 011 + 7265000, ext. 361). *Organic structures, regioselectivity, stereoselectivity, NMR analysis.*

Van Rooyen, Dr Petrus Hendrik. N.C.R.L., CSIR, P.O. Box 395, Pretoria 0001, South Africa. (1949) PhD, chemistry (Rand Afrikaans U., 1979). Chief res. (tel. 012 + 869211, ext. 3959). *Organic and inorganic structures, conformational analysis.*

Van Schalkwyk, Prof. Theunis Gabriel Dirkse. Physics Dept., U. of the Western Cape, Private Bag X17, Bellville 7530, South Africa. (1920) MSc, physics (Stellenbosch U., 1943). Prof. (tel. 021 + 976161, ext. 323). *X-ray crystallography, dislocations, teaching - physics.*

Waters, Dr David John. Geology Dept., U. of Cape Town, Private Bag, Rondebosch 7700, Cape Town, South Africa. (1950) DPhil, geology (Oxford U., UK, 1976). Lect. (tel. 021 + 698531, ext. 653). *Mineralogical applications, X-ray crystallography, texture analysis, teaching - undergraduate crystallography.*

Zagt, Mr Simon. Geology Dept., U. of Pretoria, Pretoria 0002, South Africa. (1917) MSc, geology (U. Pretoria, 1941). Sr. lect. (tel. 012 + 746051, ext. 456). *Crystallography, physical geology, groundwater.*

Zemke, Mr Klaus Jurgen. Physical Chemistry Dept., U. of Cape Town, Private Bag., Rondebosch 7700, Cape Town, South Africa. (1962) BSc(Hons), chemistry (U. Cape Town, 1983). Asst. lect. (tel. 021 + 698531, ext. 542). *Metal complexes, disorder.*

SPAIN

Sub-Editor: M. Font-Altaba

Notes

1. International telephone country code - 34.

2. The degrees conferred by Spanish Universities are *doctor* (Dr) (approximately equivalent to PhD, in some cases to DSc), *graduado* (Grad.) (approximately equivalent to MSc), and *licenciado* (Lic) (approximately equivalent to BSc).

3. In C.S.I.C., research officers are classified into one of the following ranks (in order of increasing seniority): *Colaborador cientifico, Investigador cientifico and Profesor de investigacion.*

Alonso Lopez, Dr Jose. Res. and Dev. Dept., S. E. A. Tudor Co., Azuqueca de Henares, Guadalajara, Apartado 2, Spain. (1924) Dr, physical chemistry (U. Comp. Madrid, 1972). Dept. head. (tel. 34-11 + 260390). *X-ray diffraction, polymorphism, inorganic compounds.*

Alvarez Peres, Dr Aurelio. Dept. de Cristalografia, U. Autonoma de Barcelona, Torre de Geologia, Bellaterra, Barcelona, Spain. (1932) PhD, geology (Barcelona U., 1974). Asst. prof. (tel. 34-3 + 2920200, ext. 1257). *Mineralogy, mineral structures.*

Amigo, Prof. Jose Ma. Dept. Cristalografia y Mineralogia, U. de Valencia, Valencia, Spain. (1940) PhD, geology (Barcelona U., 1966). Prof. (tel. 346 + 3630011). *Solid solution crystallography, quantitative analysis, X-ray powder diffraction, experimental mineralogy.*

Amoros, Prof. Jose Luis. Dept. Cristalografia y Mineralogia, Fac. Geologicas, U. Complutense, 28003 Madrid, Spain. (1920) PhD, Ciencias (Madrid U., 1945). (tel. 34-91 + 2433468). *Crystal growth, crystal dynamics, instrumentation.*

Aragon de la Cruz, Prof. Francisco. Inst. de Quimica Inorganica, C.S.I.C, Fac. de Ciencia, Ciudad U., 28003 Madrid, Spain. (1933) Dr, Chemistry (Madrid U., 1960). Prof. de Investigacion *Clay minerals, interlamellar compounds, clathrates.*

Arana, Prof. Rafael. Dept. Cristalografia y Mineralogia, Fac. Ciencias, U. Murcia, Spain. (1942) PhD, crystallography and mineralogy (U. Granada, 1972). Prof. *Ore deposits, ore microscopy.*

Arbunies Andreu, Dr Manuel. Escuela de Gemologia, U. de Barcelona, Gran Via 585, 08007 Barcelona, Spain. (1928) PhD, gemmology (Barcelona U.). Asst. prof. (tel. 34-3 + 3186666). *Gemmology, mineral crystallography.*

Arribas, Prof. Dr Antonio. Dept. de Cristalografia, Fac. de Ciencias, U. Salamanca, Salamanca, Spain. (1923) PhD, mineralogy (Madrid U., 1961). Prof. (tel. 34-23 + 213619). *Mineral crystalline structures.*

Artus, Dr Luis. Dept. de Cristalografia y Mineralogia, U. de Barcelona, Gran Via 585, 08007 Barcelona, Spain. (1951) PhD, physics (U. de Barcelona, 1974). Asst. prof. (tel. 34-3 + 3186666). *Differential thermal analysis, calorimetry, organic compounds.*

Ausio Casas, Mr. Dept. Quimica Macromolecular, Escuela Tècnica Superior de Ingenieros Industriales, U. Politècnica de Barcelona, Diagonal 647, 08028 Barcelona, Spain. (Politecnic U.). Civil eng. *Macromolecular structures.*

Ayora Ibañez, Dr Carlos. Dept. Cristalografia y Mineralogia, Barcelona U., Gran Via 585, 08007 Barcelona, Spain. (1953) PhD, geology (Barcelona U., 1976). Asst. prof. (tel. 34-3 + 318666). *Sulphide petrology.*

Ausio Casas, Mr. Dept. Quimica Macromolecular, Escuela Técnica Superior de Ingenieros Industriales, U. Politécnica de Barcelona, Diagonal 647, 08028 Barcelona, Spain. (Politécnic U.). Civil eng. *Macromolecular structures.*

Ayora Ibañez, Dr Carlos. Dept. Cristalografia y Mineralogia, Barcelona U., Gran Via 585, 08007 Barcelona, Spain. (1953) PhD, geology (Barcelona U., 1976). Asst. prof. (tel. 34-3 + 318666). *Sulphide petrology.*

Azorin Marin, Mr F. Dept. Quimica Macromolecular, Escuela Técnica Superior de Ingenieros Industriales, U. Politécnica de Barcelona, Diagonal 647, 08028 Barcelona, Spain. (Politécnic U.). Civil eng. *Macromolecular structures.*

Balcazar, Dr Jose Luis. Dept. Geologia y Geoquimica, Fac. de Ciencias, Valladolid U., Prado de la Magdalena, Valladolid, Spain. (1929) Dr, geoquimica (U. de Valladolid, 1960). Assoc. prof. (tel. 34-83 + 257296). *Crystal chemistry.*

Balta Calleja, Prof. Francisco Jose. Polymer Group, Inst. de Estructura de la Materia, Serrano 119, 28006 Madrid, Spain. (1936) PhD, physics (Bristol U., UK, 1963). Head, polymer group. (tel. 34-91 + 2619400). *X-ray diffraction, microstructure and physical properties of polymers.*

Batlle Sales, Mr Jorge. Dept. de Geologia, Fac. de Ciencias, U. Autonoma de Madrid, Cantoblanco, 28034 Madrid, Spain.

Bernalte, Prof. Antonio. Dept. de Fisica del Estado Solido, Uned U., Madrid, Spain. (1927) PhD, physics (U. California, Berkeley, USA, 1968). Prof., (tel. 34-44 + 4695200, ext. 314). *Solid state physics, mathematical methods in physics.*

Besteiro Rafales, Dra Josefina. Dept. de Cristalografia y Mineralogia, U. de Zaragoza, Zaragoza, Spain. (1945) PhD, geology (Barcelona U., 1970). Asst. prof. (tel. 34-76 + 415126). *Structural mineralogy, crystal optics.*

Bosch Figueroa, Dr Jose Ma. Dept. de Cristalografia, Inst. Jaime Almera, C.S.I.C., Gran Via 585, 08007 Barcelona, Spain. (1916) DSc, Pharmacy (Barcelona U., 1963) Retired. (tel. 34-3 + 3186666) *Optics (non-linear), absorbent crystals, gem crystallography.*

Brianso, Prof. Jose Luis. Dept. Cristalografia, U. Autonoma de Barcelona, Torre de Geologia, Bellaterra, Barcelona, Spain. (1944) DSc, geology (Barcelona U., 1972). Prof. (tel. 34-3 + 2920200, Ext., 1257). *Organic crystal structures.*

Bru, Prof. Luis. Fac. de Ciencias Fisicas, U. Complutense de Madrid, Ciudad U., 28003 Madrid, Spain. (1909) DSc, physics. Retired prof. solid state physics. (tel. 34-1 + 2442858). *Electron microscopy, defects in materials.*

Caballero Lopez-Lendinez, Prof. Manuel Antonio. Dept. de Cristalografia y Mineralogia, Fac. de Ciencia, U. de Cadiz, Cadiz, Spain. (1942) Dr, geology (U. Complutense de Madrid, 1972) Prof. *Imperfections in crystals.*

Cabezuelo Huertas, Miss Maria. Dept. de Rayos X, Inst. de Quimica-Fisica Rocasolano, Serrano 119, 28006 Madrid, Spain. (1952) MSc, physics (Valencia U., 1976). (tel. 34-1 + 2619400, ext. 318). *Crystal structure determination.*

Calvet Pallas, Mrs Maria Teresa. Dept. de Cristalografia y Mineralogia, U. de Barcelona, Gran Via 585, 08007 Barcelona, Spain. (1960) Grad. (U. de Barcelona, 1985). Jr. res. (tel. 34-1 + 3186666). *Differential thermal analysis, calorimetry, organic compounds.*

Calvo Calvo, Prof. Felipe Angel. Dept. de Metalurgia, Fac. de Ciencias Quimicas, U. Complutense de Madrid, Ciudad U., 28006 Madrid, Spain. (1919) Dr, chemistry and philosophy (Madrid U., 1956, Cambridge U., UK, 1957). Prof. (tel. 34-1 + 2442867). *Metallurgy in all its aspects, geochemistry.*

Campa, Prof. Juan Antonio. Dept. de Cristalografia y Mineralogia, Fac. de Geologia, U. Complutense de Madrid, 28006 Madrid, Spain. (1942) Dr, mineralogy (Barcelona U., 1972). Prof. (tel. 34-1 + 2433468). *Mineralogy, crystallographic methods.*

Canut, Prof. Marisa. Dept. de Cristalografia y Mineralogia, Fac. de Ciencias Geologicas, U. Complutense de Madrid, 28006 Madrid, Spain. (1927) PhD, physics (Barcelona U., 1952). Sr. res. (tel. 34-1 + 2433468). *Computing, optical transforms, crystal thermodynamics.*

Cardellach Lopez, Dr Esteban. Dept. de Cristalografia, U. Autonoma de Barcelona, Torre de Geologia, Bellaterra, Barcelona, Spain. (1948) PhD, geology (Barcelona U., 1977). Asst. prof. (tel. 34-3 + 2920200, ext. 1257). *Mineralogy, ores.*

Carrasco Cantos, Mr Francisco. Dept. de Geologia, U. de Malaga, Malaga, Spain.

Casas Sainz de Aja, Dr José. Dept. de Geologia, Fac. de Ciencias, U. Autonoma de Madrid, Cantoblanco, 28034 Madrid, Spain. PhD, geology (Madrid U., 1976). Jr. res. *Clay minerals.*

Clavaguera Plaja, Dr Narciso. Dept. de Fisica del Estado Solido, Fac. de Fisica, U. de Barcelona, Diagonal 647, 08028 Barcelona, Spain. PhD, physics (Barcelona U.). Asst. prof. (tel. 34-3 + 3307311). *Crystal physics.*

Conde, Prof. Alejandro. Dept. de Optica, Fac. de Ciencias, Palos de la Frontera s/n, Sevilla, Spain. (1947) Dr, physics (Sevilla U., 1972). Prof. (tel. 34-54 + 211923). *Organic structures, thin films properties.*

Coy Yll, Prof. Ramon. Dept. de Cristalografia y Mineralogia, Fac. de Geologia, U. Complutense, 28006 Madrid, Spain. (1940) DSc, geology, crystallography (Barcelona U., 1964; Montreal U., Canada, 1970). Prof. (tel. 34-1 + 2437195). *Crystal physics, mineral crystallography.*

Cuevas Diarte, Dr Miguel Angel. Dept. Cristalografia y Mineralogia, Fac. de Geologia, U. de Barcelona, Gran Via 585, 08007 Barcelona, Spain. (1948) PhD, geology (Barcelona U., 1976). Asst. prof. (tel. 34-3 + 3186666). *Differential thermal analysis and calorimetry, solid solutions, organic compounds.*

Cumbrera Hernandez, Dr Francisco. Dept. de Optica, U. de Sevilla, Palos de Moguer s/n, Sevilla, Spain. PhD, physics (Sevilla U.). Asst. prof. (tel. 34-54 + 211923). *Crystal physics.*

Estop, Dra Eugenia. Dept. de Cristalografia y Mineralogia, U. de Barcelona, Gran Via 585, 08007 Barcelona, Spain. (1950) PhD, geology (Barcelona U., 1978). Asst. prof. (tel. 34-3 + 3186666). *X-ray dynamical theory, X-ray topography, crystalline perfection.*

Evole Martil, Dr Nieves. Dept. de Quimica Inorganica, Inst. Quimica Inorganica Elhuyar, Fac. de Ciencias, Ciudad U., 28030 Madrid, Spain. (1945) Grad., chemistry (U. Complutense de Madrid, 1970). Asst. prof. *Silicates and interlaminar compounds.*

Faraco Muños, Mr Nicolas. Equipo Medico Hospitalario, Sandoz, S. A. E., Ayala 70, 28030 Madrid, Spain. (1939) Lic (U. Complutense de Madrid, 1966). (tel. 34-1 + 4336138). *Instrumentation.*

Fayos, Dr Jose. Dept. de Rayos X, Inst. de Quimica Fisica Rocasolano, C.S.I.C., Serrano 119, 28006 Madrid, Spain. (1940) PhD, physics (Madrid U., 1963). Res. assoc. (tel. 34-1 + 2619400) *Crystal structure determination, natural products.*

Fenoll Hach-Ali, Prof. Purificacion. Dept. de Cristalografia y Mineralogia, U. de Granada, Fac. de Ciencias, Granada, Spain. (1935) Prof., crystallography, mineralogy, (Ministerio de Educacion y Ciencia, 1970). Prof. (tel. 34-58 + 2728859). *Crystallography, X-ray diffraction, mineralogy, ores, clays.*

Fernandez Nieto, Dra Constanza. Dept. de Cristalografia y Mineralogia, Fac. de Ciencias, U. de Zaragoza, Zaragoza, Spain. PhD, geology (Zaragoza U.). Asst. prof. (tel. 34-76 + 415126). *Mineralogy.*

Florencio, Dr Feliciana. Dept. Rayos X, Inst. de Quimica-Fisica Rocasolano, C.S.I.C., Serrano 119, 28006 Madrid Spain. (1924) ScD, chemistry (Madrid U., 1968). Res. fellow. (tel. 34-1 + 2619400). *Crystal structure determination.*

Foces Foces, Dr Concepcion. Dept. Rayos X, Inst. de Quimica-Fisica Rocasolano, C.S.I.C., Serrano 119, 28006 Madrid, Spain. (1946) PhD, physics (Madrid U., 1974) Colaborador. (tel. 34-1 + 2619400). *Crystal structure determination.*

Font-Altaba, Prof. Manuel. Dept. de Cristalografia y Mineralogia, Fac. de Ciencias Geologicas, U. de Barcelona, Gran Via 585, 08007 Barcelona, Spain. (1923) PhD, chemistry, pharmacy (Barcelona U., 1954, Madrid U., 1950). Prof. of crystallography and mineralogy. (tel. 34-3 + 3186666). *General crystallography, crystal growth, structural mineralogy.*

Fuente Cullell, Dr Carlos de la. Dept. de Cristalografia y Mineralogia, Fac. de Geologia, U. de Barcelona, Gran Via 585, 08007 Barcelona, Spain. (1941) PhD, geology (Barcelona U., 1972). Asst. prof. (tel. 34-3 + 3186666). *Ceramic structures and properties, clay materials.*

Fuentes Perez, Mr Manuel. Centro de Investigaciones Técnicas de Guipuzcoa, Barrio de Ibaeta s/n, San Sebastian, Spain.

Galan Huertas, Prof. Emilio. Dept. de Cristalografia y Mineralogia, Fac. de Ciencias, U. de Sevilla, Sevilla, Spain. (1940) PhD, geology (Madrid U., 1974). Prof. (tel. 34-5 + 4215266). *Clay minerals, crystallography.*

Gali, Dr Salvador. Dept. de Cristalografia y Mineralogia, Fac. de Geologia, U. de Barcelona, Gran Via 585, 08007 Barcelona, Spain. (1949) DSc, geology (Barcelona U., 1976). Asst. prof. (tel. 34-3 + 3186666). *Crystal structures.*

Garcia Blanco, Prof. Severino. Dept. de Rayos X, Inst. de Quimica-Fisica Rocasolano, Serrano 119, 28006 Madrid, Spain. (1922) DSc, chemistry (Madrid U., 1952). Res. prof. (tel. 34-1 + 2619400) *Crystal structure determination, crystal chemistry, phases analysis.*

Garcia Guinea, Mr Javier. Dept. de Cristalografia y Mineralogia, Fac. de Ciencias, U. de Extremadura, Badajoz, Spain.

Garcia Martinez, Dr Oscar. Inst. de Quimica Inorganica, C.S.I.C., Fac. Ciencias Quimicas, Ciudad U., 28003 Madrid, Spain. (1929) Dr, inorganic chemistry (U. Complutense Madrid, 1965). Investigador Cientifico. (tel. 34-1 + 4491850). *Basic salts.*

Garcia Degano, Mrs Ma. Josefa. Dept. de Rayos X, Inst. de Quimica Inorganica Elhuyar, C.S.I.C., Fac. Quimicas, Ciudad U., 28003 Madrid, Spain. (1945) Grad., chemistry (U. Madrid, 1969). Post grad. student (tel. 34-1 + 4491850, ext. 7). *Crystal structure determination.*

Garcia Vicente, Prof. Dr Jose. Inst. de Edafologia y Biologia Vegetal, Dept. de Fisico-Quimica, Serrano 115, 28006 Madrid, Spain. (1917) Dr, physical chemistry (U. de Zaragoza, 1949). Retired prof. (tel. 34-1 + 2625020, ext. 19). *Clay minerals.*

Garrido, Prof. Dr Julio. Dept. Geologia, U. Autonoma de Madrid, 28034 Madrid, Spain. (1911) DSc, (U. de Madrid, 1933). Retired prof. (tel. 34-1 + 2615595). *Geometric crystallography, crystal structure.*

Gomez Ruimonte, Dr Florentino. Dept. de Rayos X, Inst. de Quimica-Fisica Rocasolano, C.S.I.C., Serrano 119, 28006 Madrid, Spain. DSc, chemistry (U. de Madrid, 1954). (tel. 34-1 + 2619400, ext. 115). *Chemical analysis, ceramics.*

Gonzalez Garcia, Mrs Victoria. Inst. of Inorganic Chemistry, C.S.I.C., Fac. de Ciencias, 28003 Madrid, Spain. (1945) Lic, quimica (U. de Granada, 1970). Becario. (tel. 34-1 + 4491850). *Uranium compounds (structure).*

Gonzalez Lopez, Dr Eng. Jose. Seccion de Minas, Delegacion de Industria, Tomas de Aquino 1, Cordoba, Spain. (1927) Dr (1967). Ingeniero Subalterno. (tel. 34-57 + 239100, ext. 21). *Radioisotope applications in industry.*

Gonzalez Martinez, Dr José. Dept. de Cristalografia y Mineralogia, Fac. de Ciencias, U. de Zaragoza, Zaragoza, Spain. PhD, geology (Zaragoza U.). Asst. prof. (tel. 34-76 + 415126). *Crystallography.*

Guerrero Laverat, Mr Alejandro. Inst. de Quimica Inorganica, C.S.I.C., Fac. de Quimicas, Ciudad U., 28003 Madrid, Spain. (1927) Lic., quimica inorganica

(Fac. de Ciencias, U. Complutense, 1958). Colaborador cientifico. (tel. 34-1 + 4491850). *Inorganic compounds (structure).*
Gutierrez Puebla, Prof. Enrique. Dept. de Rayos X, Inst. de Quimica Inorganica Elhuyar, Fac. de Quimicas, U. Complutense de Madrid, 28003 Madrid, Spain. (1952) Grad., chemistry (U. Complutense Madrid, 1975). Prof. (tel. 34-1 + 4491850). *Crystal structure determination.*
Gutierrez Rios, Prof. Dr Enrique. Dept. de Quimica Inorganica, U. Complutense, Fac. Cienc. Quimicas, Ciudad U., 28003 Madrid, Spain. (1915) Dr en Ciencias, quimica inorganica (U. Complutense, Madrid, 1943). Prof. (tel. 34-1 + 4491850). *Inorganic chemistry.*
Hernandez Cano, Dr Felix. Dept. de Rayos X, Inst. Rocasolano, C.S.I.C., Serrano 119, 28003 Madrid, Spain. (1941) PhD, crystal structure analysis (U. Complutense, Madrid, 1969). Res. (tel. 34-1 + 2619400). *Methods of resolution, analysis of results.*
Hoyos de Castro, Prof. Angel. Cat. de Edafologia, Fac. Farmacia, U. Complutense Madrid, Ciudad U., 28003 Madrid, Spain. (1913) Dr, ciencias quimicas (U. Madrid, 1940). Retired prof. (tel. 34-1 + 2434863). *Crystallography, mineralogy, clays, soils.*
Hoyos Guerrero, Prof. Miguel Angel. Dept. de Geologia, Fac. de Ciencias, U. Autonoma de Madrid, Cantoblanco, 28034 Madrid, Spain. (1945) PhD, chemistry (Madrid U., 1974). Prof. (tel. 34-1 + 7242450). *Crystal chemistry.* **Iglesias,** Dr Juan Eugenio. Inst. de Fisico-Quimica Mineral, C.S.I.C., Serrano 115 bis, 28006 Madrid, Spain. (1942) PhD, chemical engineering, materials science (U. Texas at Austin, USA, 1971). Prof. de investigacion. (tel. 91 + 262-4526). *Crystal chemistry, diffraction symmetry, IR spectroscopy.*
Jimenez Garay, Dr Rafael. Dept. de Optica, Fac. Ciencias, U. de Sevilla, Palos de la Frontera s/n, Sevilla, Spain. (1946) PhD, ciencias fisicas (U. de Sevilla, 1969). Asst. prof. (tel. 34-54 + 211923). *Organic structures, thin films.*
Labrador Carrasco, Mr Manuel. Dept. de Cristalografia y Mineralogia, U. de Barcelona, Gran Via 585, 08007 Barcelona, Spain. (1952) Grad. (U. de Barcelona, 1984). Jr. res. (tel. 34-1 + 3186666). *Organic alloys, solid solutions, syncrystallization.*
Leguey Gimenez, Prof. Santiago. Dept. de Geologia, Fac. de Ciencias, U. Autonoma de Madrid, Cantoblanco, 28034 Madrid, Spain. (1934) PhD, (Madrid U., 1966). Prof. (tel. 34-1 + 7242450). *Mineral crystallography.*
Lopez Aguayo, Prof. Francisco. Dept. Cristalografia y Mineralogia, Fac. Cienc. Geologicas, U. de Zaragoza, Zaragoza, Spain. (1945) Dr, ciencias geologicas (U. Complutense de Madrid, 1973). Prof. *Crystal chemistry, clays, polymorphism and polytypism.*
Lopez Castro, Prof. Amparo. Dept. Investigaciones Fisicas y Quimicas, U. Sevilla, C.S.I.C., Palos de la Frontera s/n, Sevilla, Spain. (1928) PhD, chemistry (U. Sevilla, 1954). Res. prof. (tel. 34-54 + 211923). *X-ray diffraction, crystal structure, c-nucleosides.*
Lopez Gonzalez, Prof. Juan de Dios. Dept. Quimica Inorganica, U. Granada, Poligono Universitario, Granada, Spain. (1924) PhD, chemistry (U. Madrid, 1949). Prof., Director. (tel. 34-58 + 272878). *Structural inorganic chemistry, surface chemistry, carbon, clay minerals, mixed oxides, coordination chemistry.*
Lopez de Lerma, Dr Julian. Dept. de Rayos X, Inst. de Quimica-Fisica Rocasolano, C.S.I.C., Serrano 113, 28006 Madrid, Spain. (1928) ScD, chemistry (U. Madrid, 1963). Res. assoc. (tel. 34-1 + 2660107). *Crystal structure determination.*
Lopez Soler, Dr Angel. Dept. Cristalografia, Inst. Jaime Almera, C.S.I.C., Alcarria s/n, 08028 Barcelona, Spain. (1940) PhD, geology (U. Barcelona, 1968). Res. (tel. 34-3 + 3302713). *Optical properties of solids.*
Marquez, Prof. Rafael. Dept. Optica, Dep. Investigacion Fisica y Quimica, U. Sevilla, Fac. Ciencias, Palos de la Frontera s/n, Sevilla, Spain. (1929) PhD, chemistry (U. Madrid, 1957). Prof. of physics. (tel. 34-54 + 211923). *X-ray diffraction, electron diffraction, electron microscopy.*
Marti, Dr Jaime. Inst. de Catalisis y Petroleoquimica, Serrano 119, 28006 Madrid, Spain. (1945) ScD, chemistry (U. Complutense Madrid, 1976). (tel. 34-1 + 2619400). *Powder methods, surface analysis, catalysis.*
Martin Pozas, Prof. José Ma. Dept. de Cristalografia y Mineralogia, Fac. de Ciencias, U. de Salamanca, Salamanca, Spain. (1932) PhD, chemistry (Granada U.). Prof. *Mineral crystallography.*
Martinez Carrera, Prof. Sagrario. Dept. de Rayos X, Inst. de Quimica-Fisica Rocasolano, C.S.I.C., Serrano 119, 28006 Madrid, Spain. ScD, chemistry (U. Madrid, 1955). Res. prof. (tel. 34-1 + 2619400, ext. 105). *Crystal structure determination, computer programming.*
Martinez Garcia, Mrs Ma. Luisa. Dept. Rayos X, Inst. de Quimica-Fisica Rocasolano, Serrano 119, 28006 Madrid, Spain. (1946) Grad., chemistry (U. Complutense Madrid, 1969). Grad. student. (tel. 34-1 + 2619400 ext. 113). *Crystal structure determination.*
Martinez Ripoll, Dr Martin. Dept. Rayos X, Inst. de Quimica-Fisica Rocasolano, C.S.I.C., Serrano 119, 28006 Madrid, Spain. (1946) ScD, chemistry (U. Valencia, 1970). Res. (tel. 34-1 + 2619400) *Crystal structure determination, computing programs.*
Maurer, Prof. Enrique. Cristalofisica C.I.F., L. Torres Quevedo, C.S.I.C., Serrano 144, Madrid 6, Spain. (1939) Dr, physics (U. Complutense Madrid, 1972). Res. prof. (tel. 34-1 + 2618806 ext. 52). *Crystal physics, ferroelectricity, ceramics.*
Mendiola Diaz, Dr Jesus. Cristalofisica C.I.F., L. Torres Quevedo, C.S.I.C., Serrano 144, 28006 Madrid, Spain. (1936) Dr, physics (U. Complutense de Madrid, 1966). Sr. res. (tel. 34-1 + 2618806, ext.54). *Crystal physics, ceramic science, ferroelectricity.*
Miguel Alonso, Prof. Santiago. Quimica Inorganica, Inst. de Quimica Inorganica Elhuyar, Fac. Ciencias, Ciudad U., 28003 Madrid, Spain. (1948) Lic. en ciencias quimicas (U. Complutense Madrid, 1972). Prof. *Clathrates, Silicates and interlaminar compounds.*
Miravitlles, Prof. Carlos. Inst. Jaime Almera, C.S.I.C., Alcarria s/n, 08028 Barcelona, Spain. (1942) DSc, pharmacy (U. Barcelona, 1972). Res. prof. (tel. 34-3 + 3302713). *Crystal structures.*
Monge Bravo, Mrs Ma. Angeles. Dept. de Rayos X, Inst. de Quimica Inorganica Elhuyar, U. Complutense de Madrid, Fac. Quimicas, 28003 Madrid, Spain. (1951) Grad., chemistry. (U. Complutense de Madrid, 1976) Grad. student. (tel. 34-1 + 4491850). *Crystal structure determination.*
Montoriol Pous, Prof. Joaquin. Cristalografia y Mineralogia, Fac. de Geologia, U. Barcelona, Gran Via 585, 08007 Barcelona, Spain. (1924) PhD, geology (U. Barcelona, 1964). Prof. (tel. 34-3 + 3186666). *Mineral physics.*
Moreiras Blanco, Dr Damaso. Dept. Cristalografia y Mineralogia, Fac. de Ciencias, U. de Oviedo, Oviedo, Spain. (1953) PhD, geology (Oviedo U., 1980). Asst. prof. (tel. 34-85 + 233200). *Structural crystallography.*
Moreno Echevarria, Dra Esperanza. Dept. de Optica, Fac. de Ciencias, U. de Sevilla, Palos de Moguer s/n, Sevilla, Spain. (1925) PhD, chemistry (Sevilla U.). Prof. (tel. 34-54 + 211923). *Structural crystallography.*
Nogues Carulla, Dr Joaquin Ma. Dept. Cristalografia y Mineralogia, U. de Barcelona, Gran Via 585, 08007 Barcelona, Spain. (1946) Dr, geology (U. Barcelona, 1971). Asst. prof. (tel. 34-3 + 3175982). *Optical properties of solids, optical crystallography.*
Perales Alcon, Dr Aurea. Dept. Rayos X, Inst. de Quimica-Fisica Rocasolano, C.S.I.C., Serrano 119, 28006 Madrid, Spain. (1931) ScD, chemistry (U. Valencia, 1968). Res assoc. (tel. 34-1 + 2619400) *Crystal structure determination, biological crystal structure.*
Perez Alonso, Dr Julio. Laboratorio Central, Centro de Investigacion y Desarrollo Asland, Villaluenga de la Sagra, Toledo, Spain. (1926) Dr, ciencias (U. Madrid, 1951). Dir. (tel. 34-25 + 131137138) *Crystallography.*
Perez Garrido, Mr Simeon. Dept. de Optica, Fac. de Ciencias, Palos de la Frontera s/n, Sevilla, Spain. (1943) Grad., physics (Sevilla U., 1971). Assoc. prof. (tel. 34-91 + 211923). *Organic structures.*
Perez Salazar, Mrs Adela. Dept. de Rayos X, Inst. de Quimica-Fisica Rocasolano, Serrano 119, 28006 Madrid, Spain. (1950) Grad., chemistry (U. Complutense de Madrid, 1973). Grad. student. (tel. 34-1 + 2619400, ext. 113). *Crystal structure determination.*
Perez del Villar, Lic. Luis. Geologia y Mineria, Junta de Energia Nuclear (J.E.N.), P. de la Alameda s/n, Molina de Aragon, Guadalajara, Spain. (1949) Lic. en ciencias geologicas (Granada U., 1972). Geologo Contratado. (tel. 34-11 + 674). *Quantitative analysis, phyllosilicates, X-ray diffraction.*
Plana Llevat, Dr Feliciano. U.E.7 X-ray Inst. Jaime Almera, C.S.I.C., Aliarria, 08028 Barcelona, Spain. (1946) PhD, crystallography (Barcelona U., 1974). Res. fellow, (tel. 34-3 + 3302713). *Crystal structures, powder diffraction.*
Puigjaner, Prof. Luis C. Dept. de Quimica Macromolecular, E.T.S.I.I.B, Diagonal 647, 08028 Barcelona, Spain. (1935) PhD, engineering (Polytechnic U. of Madrid, 1968). Prof. (tel. 34-3 + 2496400, ext. 336). *Chromatin structures, DNA-protein complexes, synthetic polymers.*
Ramos Fernandez, Prof. Felicisimo. Dept. of Physics, U. of Extremadura, Av. Elvas s/n, Badajoz, Spain. (1925) Dr, chemistry (U. de Madrid, 1956, Imperial C., London U., UK, 1962). Full prof. in thermodynamics and physics. (tel. 34-24 + 220545, ext. 25). *Crystal growth, special materials; solar energy absorbtion; materials structure, diffraction techniques.*
Rodriguez Clemente, Dr Rafael. Inst. de Geologia, C.S.I.C., Paseo de la Castellana 84, 28006 Madrid, Spain. (1948) PhD, geology (U. Barcelona, 1974). Res. (tel. 34-1 + 2612513). *Crystal growth, crystal perfection, crystal morphology.*
Rodriguez Gallego, Prof. Manuel. Dept. de Cristalografia y mineralogia, Fac. Ciencias, U. Granada, Fuente Nueva s/n, Granada, Spain. (1935) PhD, pharmacy (U. Granada, 1960). Prof. (tel. 34-58 + 272885). *Silicates.*
Rojas Lopez, Dr Rosa Ma. Dept. Inorganic Chemistry, C.S.I.C., Ciudad U., 28003 Madrid, Spain. (1944) Dr, inorganic chemistry (U. Madrid, 1974). Colaborador cientifico (tel. 34-1 + ,4491850). *Uranium compounds (structure).*
Rueda Bravo, Dr Daniel-Reyes. Lab. de Polimeros Cristalinos, Inst. de Estructura de la Materia, Serrano 119, 28006 Madrid, Spain. (1948) Dr, chemistry (U. Madrid, 1975). Colaborador contratado. (tel. 34-1 + 2619400). *Synthetic polymers, biopolymer structures.*
Ruiz Amil, Dr Antonio. Inst. de Quimica Inorganica Elhuyar, C.S.I.C., Fac. Ciencias Quimicas, Ciudad U., 28003 Madrid, Spain. (1927) DSc, physics (U. Complutense Madrid, 1965) Investigador cientifico. (tel. 34-1 + 4491850). *Clay minerals, inorganic compounds, structure.*
Sainz Amor, Dra Emma. Dept. de Cristalografia y Mineralogia, Fac. de Geologia, U. de Barcelona, Gran Via 585, 08007 Barcelona, Spain. (1926) DSc, (Barcelona U., 1956). Sr. res. (tel. 34-3 + 3186666). *Sediments crystallography.*
Serratosa, Prof. Dr Jose Ma. Inst. de Edafologia y Biologia Vegetal, C.S.I.C., Serrano 115, 28006 Madrid, Spain. (1924) DSc, chemistry (U. Madrid, 1953). Res. prof. (tel. 34-1 + 2625020, est. 27). *Physical chemistry of solids, minerals, clay mineralogy, surface chemistry.*

Smith Verdier, Dr Pilar. Dept. Rayos X, Inst. de Quimica-Fisica Rocasolano, C.S.I.C., Serrano 119, 28006 Madrid, Spain. (1916) ScD, chemistry (U. Madrid, 1963). Res. (tel. 34-1 + 2619400, ext. 105). *Crystal structure determination, computer programming.*

Solans Huget, Prof. Joaquin Ma. Dept. de Cristalografia y mineralogia, U. de Barcelona, 08007 Barcelona, Spain. (1940) Dr, geology (U. Barcelona, 1966). Prof. (tel. 34-3 + 3186666). *Physical properties of inorganic crystals, crystal growth.*

Solans, Prof. Xavier. Cristalografia y Mineralogia, Fac. Geologia, U. Barcelona, Gran Via 585, Barcelona, Spain. (1949) DSc, physics (U, Barcelona. 1976). Prof. (tel. 34-3 + 3186666). *Crystal structure determination.*

Subirana Torrent, Prof. José Antonio. Dept. de Quimica Macromolecular, Escuela Superior Técnica de Ingenieros Industriales, U. Politécnica de Barcelona, Diagonal 647, 08028 Barcelona, Spain. (1932) DSc, chemistry (Barcelona U.). Prof., dept. head. *Macromolecular structures.*

Tauler Ferre, Dr Esperança Dept. de Cristalografia y Mineralogia, U. de Barcelona, Gran Via 585, 08007 Barcelona, Spain. (1953) PhD, crystallography (U. de Barcelona, 1983). Asst. prof. (tel. 34-1 + 3186666). *X-ray diffraction, organic alloys, solid solutions.*

Terol, Dr Salvador. Inst. de Optica, C.S.I.C., Serrano 121, 28006 Madrid, Spain. (1916) DSc, solid state chemistry (U. Central Madrid, 1948). Head, luminescence section. *Solid state chemistry and physics, luminescence, phosphors, pigments, sensors, catalysts.*

Traveria Cros, Dr Adolfo. Dept. de Cristalografia, C.S.I.C., Alearria s/n, 08028 Barcelona, Spain. (1928) PhD, geology (Barcelona U., 1964). Res. (tel. 34-3 + 3302713). *Crystallography, X-ray spectroscopy.*

Valin Alberdi, Mrs Maria Luz. Dept. de Cristalografia y Mineralogia, U. de Oviedo, 33080 Oviedo, Spain. (1956) Lic (U. de Oviedo, 1983). Asst. prof. (tel. 34-85 + 233200). *X-ray diffraction, crystal structure determination.*

Vega, Mrs Rosario. Dept. de Optica, Fac. de Ciencias, Palos de Moguer s/n, Sevilla, Spain. (1921) Lic., physics (Sevilla U., 1952). Assoc. prof. (tel. 34-1 + 211923). *Organic structures.*

Vegas Molina, Dr Angel. Dept. de Rayos X, Inst. de Quimica Inorganica Elhuyar, Fac. de Quimicas, U. Complutense de Madrid, 28003 Madrid, Spain. (1947) ScD, chemistry (U. Complutense de Madrid, 1975). Postdoctoral fellow. (tel. 34-1 + 4491850). *Crystal structure determination.*

Vendrell Saz, Dr Mario. Dept. de Cristalografia y Mineralogia, U. de Barcelona, Gran Via 585, 08007 Barcelona, Spain. (1949) PhD, geology (Barcelona U., 1976), Asst. prof. (tel. 34-3 + 3175982). *Optical properties of solids, optical crystallography.*

Viton Barbolla, Dr Carmen. Dept. de Quimica Inorganica, Inst. de Quimica Inorganica Elhuyar, C.S.I.C., Fac. Ciencias Quimicas, 28003 Madrid, Spain. (1945) Dr, ciencias quimicas (U. Complutense, 1974). Prof. *Silicates.*

SRI LANKA

Sub-Editor: H. W. Dias

Notes

1. International telephone country code - 94.

Amarasena, Mr Kandage Don. Industrial Metallurgy Section, C.I.S.I.R., 363 Bauddhaloka Mawatha, P.O. Box 787, Colombo 7, Sri Lanka. (1946) BSc, chemistry (Sri Jayawardenapura U., 1968). Res. officer (tel. 01 + 93807-9). *Metallurgy.*

Dias, Dr Hanwellage Wijayapala. Chemistry Dept., U. of Peradeniya, Peradeniya, Sri Lanka. (1936) PhD, inorganic chemistry (Leeds U., UK, 1964). Prof. (tel. 08 + 88018). *X-ray crystallography, mineral chemistry.*

Gunawardane, Dr Richard Pemasiri. Chemistry Dept., U. of Peradeniya, Peradeniya, Sri Lanka. (1945) PhD, inorganic chemistry (Aberdeen U., UK, 1974). Assoc. prof. (tel. 08 + 88018). *Silicate chemistry.*

SUDAN

Sub-Editor: S. el D. Hamad

Ali, Dr E. M. Dept. of Physics, Faculty of Education, U. of Khartoum, P.O. Box 406, Omdurman, Sudan. (1936) PhD, physics (U. of Cambridge, UK, 1970). Assoc. prof. (tel. 72271, ext. 298-299). *Crystallography.*

Hamad, Dr Sa'ad El Din. Dept. of Geology, U. of Khartoum, P.O. Box 321, Khartoum, Sudan. (1936) PhD, mineralogy (U. of Cambridge, UK, 1970). Assoc. prof. (tel. 72271, ext. 298-299). *Experimental mineralogy; crystallography; low temperature hydration-dehydration, minerals.*

SWEDEN

Sub-Editor: **L. Jahnberg**

Notes

1. International telephone country code - 46. Omit the zero in the area code after the country code.

2. Degrees conferred by Swedish Universities are *filosofie doktor* (fil.dr.)(approximately equivalent to PhD) and *högskoleexamen* (högsk.ex.)(approximately equivalent to BSc) at Faculties of Science, *medicine doktor* (med.dr.) and *medicine kandidat* (med.kand.) at Faculties of Medicine and at the Institutes of Technology *teknologie doktor* (tekn.dr.) and *civilingenjör* (civ.ing.) or *bergsingenjör* (bergsing.)(at the School of Mines). The older degrees *filosofie licentiat* (fil.lic.), *filosofie magister* (fil.mag.) and *filosofie kandidat* (fil.kand.) are no longer given but are approximately equivalent to PhD, MSc and BSc respectively.

3. *Docent* is either a title given by a faculty to a person with a scientific competence well above the doctor's level or a position for a person performing independent academic research. *Research associate* and *research assistant* are positions for research at lower levels and are often combined with teaching elementary courses.

4. The use of full names is not frequent in Sweden.

Åberg, Dr Märtha M. Dept. of Inorganic Chemistry, Royal Inst. of Techn., S-10044 Stockholm, Sweden. (1942) Tekn.dr., chemistry (Royal Inst. of Techn., 1971). Res. assoc. (tel. 08-7878150). *Inorganic crystal structures, liquid solution structures.*

Adelsköld, Dr Volrath. Dept. of Structural Chemistry, Arrhenius Lab., U. of Stockholm, S-10691 Stockholm, Sweden. (1911) Fil.lic., chemistry (Stockholm U., 1939). Sr. sci. (tel. 08-162393). *Inorganic crystal structures.*

Agrell, Dr Ingela. National Swedish Board for Technical Dev., Box 43200, S-10072 Stockholm, Sweden. (1938) Fil.dr., inorg. crystal structures (Göteborg U., 1971). (tel. 08-7445100). *Inorganic crystal structures.*

Albertsson, Dr Jörgen. Div. of Inorganic Chemistry 2, Chemical Center, U. of Lund, Box 124, S-22100 Lund, Sweden. (1939) Fil.dr., chemistry (Lund U., 1972). Docent. (tel. 046-108223). *Solid state chemistry, instrumentation.*

Aldén, Dr Karl Inge. Dept. of Chemistry, Swedish U. of Agricultural Sciences, S-75007 Uppsala, Sweden. (1937) Fil.lic., chemistry (Uppsala U., 1965). Lect. (tel. 018-171000, ext. 2219). *Biological crystal structures.*

Aleby, Dr Stig E. Dept. of Inorganic Chemistry, Chalmers U. of Techn. and U. of Göteborg, S-41296 Göteborg, Sweden. (1932) Fil.dr., structural chemistry (Göteborg U., 1969). Docent. (tel. 031-810100, ext. 2009). *Crystal structures, metal complexes, organic compounds, crystallographic teaching, instrumentation.*

Andersson, Dr Inger A. Dept. of Molecular Biology, Swedish U. of Agricultural Sciences, Uppsala Biomedical Center, Box 590, S-75124 Uppsala, Sweden. (1949) Dr.rer.nat., biochemistry (U. des Saarlandes, Germany, Fed. Rep., 1980). Res. assoc. (tel. 018-174523). *Biological macromolecules, enzyme catalysis.*

Andersson, Prof. Sten. Div. of Inorganic Chemistry 2, Chemical Center, U. of Lund, Box 124, S-22100 Lund, Sweden. (1931) Fil.dr., chemistry (Stockholm U., 1967). Prof. (tel. 046-108227). *Inorganic chemistry.*

Andersson, Dr Yvonne. Inst. of Chemistry, U. of Uppsala, Box 531, S-75121 Uppsala, Sweden. (1947) Fil.dr., inorg. chemistry (Uppsala U., 1983). Res. assoc. (tel. 018-183730). *Metallic and semiconducting phases, structure, crystal growth.*

Annehed, Mr Håkan. Div. of Inorganic Chemistry 2, Chemical Center, U. of Lund, Box 124, S-22100 Lund, Sweden. (1949) Civ.ing., chemistry (Lund U., 1975). Res. asst. (tel. 046-108233). *Silicate chemistry, crystal structures.*

Annersten, Prof. Hans. Dept. of Geology, Div. of Mineralogy and Petrology, U. of Uppsala, Box 555, S-75122 Uppsala, Sweden. (1940) Fil.dr., mineralchemistry (Uppsala U., 1973). Prof. (tel. 018-120360, ext. 51). *Mineral chemistry, crystal chemistry.*

Antti, Dr Britt-Marie. Dept. of Minerals and Prospecting Techn., Section of Mineral Dressing, U. of Luleå, S-95187 Luleå, Sweden. (1945) Fil.dr., chemistry (Umeå U., 1976). Sr. res. engineer. (tel. 0920-91000, ext. 312). *Inorganic and organic crystal structures.*

Arnberg, Dr Lars. Dept. of Casting and Powder Metallurgy, Swedish Inst. for Metals Res., Drottning Kristinas väg 48, S-11428 Stockholm, Sweden. (1947) Fil.dr., inorg. chemistry (Stockholm U., 1979). Docent. (tel. 08-243330). *Metal and alloy structures.*

Aronsson, Prof. Bertil. Res. & Dev., AB Sandvik Hard Materials, Box 42056, S-12612 Stockholm, Sweden. (1929) Fil.dr., chemistry (Uppsala U., 1960). Vice president. (tel. 08-452620). *Metal physics and structure.*

Åsbrink, Dr Gudrun. Dept. of Structural Chemistry, Arrhenius Lab., U. of Stockholm, S-10691 Stockholm, Sweden. (1930) Fil.lic., chemistry (Stockholm U., 1971). Res. asst. (tel. 08-162387). *Small angle scattering.*

Åsbrink, Dr Stig. Dept. of Inorganic Chemistry, Arrhenius Lab., U. of Stockholm, S-10691 Stockholm, Sweden. (1929) Fil.dr., chemistry (Stockholm U., 1973). Docent. (tel. 08-162387). *Phase transitions, synchrotron radiation, crystal physics, high pressure on single crystals.*

Asplund, Miss Milja. Dept. of Inorganic Chemistry, Chalmers U. of Techn. and U. of Göteborg, S-41296 Göteborg, Sweden. (1954) Fil.kand., chemistry (Göteborg U., 1980). Res. asst. (tel. 031-810100, ext. 1698). *Inorganic crystal structures.*

Åström, Prof. Hans U. Dept. of Solid State Physics, Royal Inst. of Techn., S-10044 Stockholm, Sweden. (1926) Tekn.dr., metal physics (Royal Inst. of Techn., 1958). Prof. (tel. 08-7877300). *Low temperature physics, magnetism, lattice defects.*

Aurivillius, Prof. Bengt. Div. of Inorganic Chemistry 2, Chemical Center, U. of Lund, Box 124, S-22100 Lund, Sweden. (1918) Fil.dr., inorg. chemistry (Stockholm U., 1951). Prof. em. (tel. 046-108230). *Inorganic crystal structures, heavy metals, block structures.*

Bäckerud, Dr Lennart S. Dept. of Structural Chemistry, Arrhenius Lab., U. of Stockholm, S-10691 Stockholm, Sweden. (1932) Fil.dr., inorg. chemistry (Uppsala U., 1968). Adjunct prof. (tel. 08-162383). *Nucleation and growth of crystals.*

Berggren, Dr Jan. Dept. of Inorganic Chemistry, Chalmers U. of Techn. and U. of Göteborg, S-41296 Göteborg, Sweden. (1936) Tekn.dr., chemistry (Chalmers U. of Techn., 1971). Lect. (tel. 031-810100, ext. 1557). *Inorganic crystal structures.*

Björnberg, Dr Arne A. Res. & Dev., Boliden Metall AB, S-93200 Skellefthamn, Sweden. (1951) Fil.dr., inorg. chemistry (Umeå U., 1980). Vice president R&D. (tel. 0910-31500, ext. 3742). *Inorganic crystal structures.*

Boiwe, Mr Torne R. Dept. of Chemistry, Swedish U. of Agricultural Sciences, S-75007 Uppsala, Sweden. (1943) Fil.kand., chemistry (Uppsala U., 1968). Res. asst. (tel. 018-171000, ext. 1555). *Biological macromolecules, photosynthesis.*

Boström, Mr N. Dan. Dept. of Inorganic Chemistry, U. of Umeå, S-90187 Umeå, Sweden. (1954) Fil.kand., chemistry (Umeå U., 1981). Res. asst. (tel. 090-165445). *Inorganic crystal structures.*

Bovin, Dr Jan-Olov. Div. of Inorganic Chemistry 2, Chemical Center, U. of Lund, Box 124, S-22100 Lund, Sweden. (1943) Fil.dr., inorg. chemistry (Lund U., 1975). Docent. (tel. 046-108231). *Solid state chemistry, high resolution electron microscopy.*

Brandberg, Dr Ola. Dept. of Inorganic Chemistry, Arrhenius Lab., U. of Stockholm, S-10691 Stockholm, Sweden. (1929) Fil.lic., chemistry (Stockholm U., 1969). Lect. (tel. 08-163705). *Inorganic crystal structures.*

Brändén, Prof. Carl-Ivar. Dept. of Molecular Biology, Swedish U. of Agricultural Sciences, Uppsala Biomedical Center, Box 590, S-75124 Uppsala, Sweden. (1934) Fil.dr., chemistry (Uppsala U., 1964). Prof. (tel. 018-174478 or 136459). *Biological macromolecules, enzyme catalysis, computer programming.*

Brusewitz, Dr (Mrs) Ann Marie. Geochemistry div., Geological Survey of Sweden, Box 670, S-75128 Uppsala, Sweden. (1918) Fil.lic., chemistry (Stockholm U., 1950). (tel. 018-179000). *Clay minerals.*

Calais, Dr Jean-Louis. Quantum Chemistry Group, U. of Uppsala, Box 518, S-75120 Uppsala, Sweden. (1932) Fil.dr., quantum chemistry (Uppsala U., 1965). Res. position at the Swedish Natural Sciences Res. Council. (tel. 018-183264). *Solid state theory, electronic structure, ionic crystals, transition metal compounds.*

Carlsson, Dr Roger. Swedish Inst. for Silicate Res., Box 5403, S-40229 Göteborg, Sweden. (1943) Tekn.lic., chemistry (Chalmers U. of Techn., 1971). Director. (tel. 031-162318). *Inorganic crystal structures, clay minerals.*

Carlström, Prof. Diego G. Dept. of Medical Biophysics, Karolinska Inst., S-10401 Stockholm, Sweden. (1922) Med.dr., medical biophysics (Karolinska Inst., 1955). Prof. (tel. 08-340560, ext. 1511). *Organic crystal structures.*

Cassel, Dr Anders Ö. Surface Chemistry Div., Berol Kemi AB, Box 851, S-44401 Stenungsund, Sweden. (1946) Fil.dr., inorg. chemistry (Lund U., 1979). R&D group manager. (tel. 0303-85554). *Metal halide phosphines, solid state structures, sulfide mineral flotation.*

Cedergren-Zeppezauer, Dr Eila S. Dept. of Molecular Biology, Swedish U. of Agricultural Sciences, Uppsala Biomedical Center, Box 590, S-75124 Uppsala, Sweden. (1937) Fil.lic., biochemistry (Uppsala U., 1972). Res. assoc. (tel. 018-174000). *Protein crystallography.*

Collini, Prof. Bengt H. E. Inst. of Geology, U. of Uppsala, Box 555, S-75122 Uppsala, Sweden. (1917) Fil.lic., mineralogy and petrology (Uppsala U., 1943), fil.dr.h.c. (Uppsala U., 1983). Prof. em. (tel. 018-182557). *Sedimentary petrology.*

Csöregh, Dr Ingeborg. Dept. of Structural Chemistry, Arrhenius Lab., U. of Stockholm, S-10691 Stockholm, Sweden. (1942) Fil.dr., structural chemistry (Stockholm U., 1983). Res. assoc. (tel. 08-162381). *Organic crystal structures.*

Dagerhamn, Dr Tore. Dept. of Inorganic Chemistry, Arrhenius Lab., U. of Stockholm, S-10691 Stockholm, Sweden. (1930) Fil.lic., chemistry (Stockholm U., 1965). Lect. (tel. 08-162352). *Inorganic crystal structures, metals structure.*

Dahlén, Dr Birgitta. Res. & Dev., KabiVitrum AB, S-11287 Stockholm, Sweden. (1943) Fil.dr., chemistry (Göteborg U., 1972). Docent. (tel. 08-138420). *Biological crystal structures.*

Delaplane, Dr Robert G. Inst. of Chemistry, U. of Uppsala, Box 531, S-75121 Uppsala, Sweden. (1942) PhD, phys. chemistry (Northwestern U., USA, 1969). Res. assoc. (tel. 018-183773). *Neutron diffraction of liquids, electron density studies.*

Edström, Mrs Kristina. Inst. of Chemistry, U. of Uppsala, Box 531, S-75121 Uppsala, Sweden. (1958) Högsk.ex., chemistry (U. of Uppsala). Res.asst. (tel. 018-183775). *Crystal structure - physical properties relationships.*

Eklund, Dr Hans. Dept. of Molecular Biology, Swedish U. of Agricultural Sciences, Uppsala Biomedical Center, Box 590, S-75124 Uppsala, Sweden. (1940) Fil.dr., chemistry (Swedish U. of Agricultural Sciences, 1976). Res. assoc. (tel. 018-174000). *Biological macromolecules.*

Ekström, Dr Thommy. Ceramic Materials Project, AB Sandvik Hard Materials, Box 42056, S-12612 Stockholm, Sweden. (1942) Fil.dr., inorg. chemistry (Stockholm U., 1975). R&D manager. (tel. 08-452620). *Materials research, structure and properties of inorganic or ceramic materials.*

Elgenmark, Miss Ingegerd. Dept. of Inorganic Chemistry, Arrhenius Lab., U. of Stockholm, S-10691 Stockholm, Sweden. (1959) Högsk.ex., chemistry (Stockholm U., 1982). Res. asst. (tel. 08-162365). *Inorganic crystal structures, electron microscopy.*

Enflo, Mrs Anita. Inst. of Theoretical Physics, U. of Stockholm, Vanadisvägen 9, S-11346 Stockholm, Sweden. (1943) Fil.lic., phys. chemistry (Helsinki U., Finland, 1970). Res. asst. (tel. 08-228160, ext. 211). *Large molecules, metalloorganic compounds, biological structures.*

Engström, Dr Ingvar O. J. Inst. of Chemistry, U. of Uppsala, Box 531, S-75121 Uppsala, Sweden. (1934) Fil.dr., chemistry (Uppsala U., 1970). Docent. (tel. 018-183740). *Metallic phase structures, high pressure X-ray diffraction.*

Ericsson, Prof. S. Torsten. Dept. of Mechanical Engineering, Linköping U., Fack, S-58183 Linköping, Sweden. (1938) Tekn.dr., phys. metallurgy (Royal Inst. of Techn., 1970). Prof. (tel. 013-111700, ext. 1168). *X-ray residual stress measurements, fatigue, high temperature coatings.*

Eriksson, Mr Anders. Inst. of Chemistry, U. of Uppsala, Box 531, S-75121 Uppsala, Sweden. (1945) Fil.dr., chemistry (Uppsala U., 1981). Res. assoc. (tel. 018-183766). *Vibrational spectroscopy.*

Eriksson, Dr Birgitta. Dept. of Structural Chemistry, Arrhenius Lab., U. of Stockholm, S-10691 Stockholm, Sweden. (1945) Fil.dr., structural chemistry (Stockholm U., 1982). Res. assoc. (tel. 08-163730). *Inorganic crystal structures.*

Eriksson, Mr Lars. Dept. of Structural Chemistry, Arrhenius Lab., U. of Stockholm, S-10691 Stockholm, Sweden. (1960) Res. assoc. (tel. 08-162393). *Computers (systems, programming), crystal structures.*

Eriksson, Mr Sven. Analyskonsult AB, Bredablicks väg 7, S-18142 Lidingö, Sweden. (1925) Ing. Manager X-ray Analysis. (tel. 08-7679170). *Instrumentation.*

Ersson, Mr Nils Olov. Inst. of Chemistry, U. of Uppsala, Box 531, S-75121 Uppsala, Sweden. (1942) Fil.mag., chemistry (Uppsala U., 1966). Res. asst. (tel. 018-183728). *Instrumentation, computer programming, inorganic structures.*

Fälth, Dr Lars. Div. of Inorganic Chemistry 2, Chemical Center, U. of Lund, Box 124, S-22100 Lund, Sweden. (1942) Fil.dr., chemistry (Lund U., 1976). Docent. (tel. 046-108232). *Silicate chemistry, crystal structures.*

Fischer-Hjalmars, Prof. Inga M. Inst. of Theoretical Physics, U. of Stockholm, Vanadisvägen 9, S-11346 Stockholm, Sweden. (1918) Fil.dr., molecular quantum mechanics (Stockholm U., 1952). Prof. em., theoretical physics. (tel. 08-228160, ext. 177). *Electronic structure of molecules.*

Flodmark, Dr Stig. Inst. of Theoretical Physics, U. of Stockholm, Vanadisvägen 9, S-11346 Stockholm, Sweden. (1926) Fil.dr., solid state physics (Stockholm U., 1959). U. lect. (docent). (tel. 08-228160, ext. 219). *Solid state theory, group theory, quantum theory.*

Forslund, Dr S. Bertil. Dept. of Inorganic Chemistry, Arrhenius Lab., U. of Stockholm, S-10691 Stockholm, Sweden. (1943) Fil.dr., inorg. chemistry (Stockholm U., 1984). Res. assoc. (tel. 08-162353). *Inorganic materials, crystal growth.*

Giesecke, Dr Johan. Dept. of Medical Biophysics, Karolinska Inst., S-10401 Stockholm, Sweden. (1949) Med.dr., medical biophysics (Karolinska Inst., 1979). Res. assoc. (tel. 08-340560, ext. 1511). *Organic crystal structures.*

Glaser, Dr Julius. Dept. of Inorganic Chemistry, Royal Inst. of Techn., S-10044 Stockholm, Sweden. (1948) Fil.dr., inorg. chemistry (Royal Inst. of Techn., 1981). Res. assoc. (tel. 08-7878151). *Complex ions, structure and dynamics in solution, X-ray diffraction, NMR.*

Glehn, von, Dr Marianne. Swedish Natural Science Res. Council, Box 6711, S-11385 Stockholm, Sweden. (1941) Fil.dr., chemistry (Stockholm U., 1971). Docent. (tel. 08-151580, ext. 174). *Bioorganic crystal structures.*

Grenthe, Prof. Ingmar. Dept. of Inorganic Chemistry, Royal Inst. of Techn., S-10044 Stockholm, Sweden. (1933) Fil.dr., inorg. and phys. chemistry (Lund U., 1964). Prof. (tel. 08-7878144). *Structure and bonding, coordination compounds, reactivity in solids.*

Grimvall, Dr Siv H. KomVux, Box 220, S-18323 Täby, Sweden. (1942) Fil.dr., chemistry (Göteborg U., 1979). Lect. (tel. 08-7680385). *Inorganic crystal structures.*

Grins, Dr Jekabs. Dept. of Inorganic Chemistry, Arrhenius Lab., U. of Stockholm, S-10691 Stockholm, Sweden. (1952) Fil.dr., chemistry (Stockholm U., 1980). Res. assoc. (tel. 08-162365). *Solid state chemistry, properties of materials.*

Gullman, Mr Jan O. Res. & Dev. Dept., Swedish Corrosion Inst., Box 5607, S-11486 Stockholm, Sweden. (1943) Fil.kand., chemistry (Uppsala U., 1970). Res. sci. *Corrosion, inorganic crystal structures.*

Gustafsson, Mr Torbjörn. Inst. of Chemistry, U. of Uppsala, Box 531, S-75121 Uppsala, Sweden. (1949) Fil.kand., chemistry (Uppsala U., 1973). Res. asst. (tel. 018-183767). *Inorganic crystal structures, liquid structures, hydrogen bonding, instrumentation.*

Hägg, Prof. Gunnar. Inst. of Chemistry, U. of Uppsala, Box 531, S-75121 Uppsala, Sweden. (1903) Fil.dr., chemistry (Stockholm U., 1929). Prof. em., inorg. chemistry. (tel. 018-183772). *Inorganic crystal structures, metallic phases, instrumentation.*

Hansen, Mr Ernst F. R. Dept. of Inorganic Chemistry, Royal Inst. of Techn., S-10044 Stockholm, Sweden. (1941) Res. engineer. (tel. 08-7878151). *Instrumentation.*

Hansen, Mr Staffan S. Div. of Inorganic Chemistry 2, Chemical Center, U. of Lund, Box 124, S-22100 Lund, Sweden. (1954) Civ.ing., chemistry (The Lund Inst. of Techn., 1978). Res. asst. (tel. 046-108233). *Inorganic crystal structures, silicates.*

Hansson, Dr Arne E. Inst. of Chemistry, U. of Uppsala, Box 531, S-75121 Uppsala, Sweden. (1926) Fil.lic., chemistry (Uppsala U., 1959). Director of studies in chemistry. (tel. 018-183713). *Inorganic and organic crystal structures.*

Hansson, Dr Eva. Div. of Physical Chemistry 1, Chemical Center, U. of Lund, Box 124, S-22100 Lund, Sweden. (1940) Fil.dr., chemistry (Lund U., 1973). Docent. (tel. 046-108156). *Metallo-organic crystal structures, solution X-ray work.*

Haraldson, Dr Stig H. W. Dept. of Solid State Physics, Inst. of Techn., Box 534, S-75121 Uppsala, Sweden. (1923) Fil.dr., physics (Uppsala U., 1973). Docent. (tel. 018-183130). *Ferromagnetic resonance, electron paramagnetic resonance.*

Hårsta, Mr Anders. Inst. of Chemistry, U. of Uppsala, Box 531, S-75121 Uppsala, Sweden. (1952) Fil.kand., chemistry (Uppsala U., 1974). Res. asst. (tel. 018-183729). *Alloys - structure.*

Hassler, Dr Eivind. Inst. of Chemistry, U. of Uppsala, Box 531, S-75121 Uppsala, Sweden. (1939) Fil.lic., chemistry (Uppsala U., 1970). Res. assoc. (tel. 018-183729). *Inorganic crystal structures.*

Hebert, Dr Hans. Dept. of Medical Biophysics, Karolinska Inst., S-10401 Stockholm, Sweden. (1951) Tekn.dr., medical biophysics (Karolinska Inst., 1979). Res. assoc. (tel. 08-340560, ext. 1561). *Biological macromolecules, electron microscopy, image processing.*

Herbertsson, Dr B. Harald. Materials Science, Technical U. of Luleå, S-95187 Luleå, Sweden. (1940) Fil.dr., inorg. chemistry (Royal Inst. of Techn., 1976). Res. assoc. (tel. 0920-91000, ext. 233). *Dicarboxylic acids and their alkali hydrogen salts, silicon nitride sintering.*

Hermansson, Dr Kersti. Inst. of Chemistry, U. of Uppsala, Box 531, S-75121 Uppsala, Sweden. (1951) Fil.dr., inorganic chemistry (Uppsala U., 1984). (tel. 018-182500). Res. assoc. *Molecular dynamics, electron distribution, hydrates.*

Hermansson, Dr Leif Å. G. ASEA CERAMA AB, S-91500 Robertsfors, Sweden. (1947) Tekn.dr., chemistry (Chalmers U. of Techn., 1977). Res. manager (docent). (tel. 0934-10845). *Inorganic crystal structures, high performance ceramics.*

Hermodsson, Dr Yngve. Inst. of Chemistry, U. of Uppsala, Box 531, S-75121 Uppsala, Sweden. (1929) Fil.dr., chemistry (Uppsala U., 1969). Docent. (tel. 018-183701). *Inorganic crystal structures.*

Hesse, Dr Rolf S. Inst. of Chemistry, U. of Uppsala, Box 531, S-75121 Uppsala, Sweden. (1923) Fil.dr., chemistry (Uppsala U., 1963). Docent. (tel. 018-183725). *Coordination compounds, crystal structures.*

Hjertén, Dr Inger. Dept. of Structural Chemistry, Arrhenius Lab., U. of Stockholm, S-10691 Stockholm, Sweden. (1937) Fil.lic., chemistry (Stockholm U., 1972). Res. asst. (tel. 08-162381). *Inorganic crystal structures.*

Holmberg, Dr Bo. National Defence Res. Inst., Section 234, Dept. 2, Box 27322, S-10254 Stockholm, Sweden. (1931) Fil.lic., chemistry (Stockholm U., 1961). Res. assoc. *Inorganic crystal structures.*

Hong, Dr Sam-Hyo. IC Div., Res. & Dev. Dept., Rifa AB, Isafjordsgatan 10-16, S-16381 Stockholm, Sweden. (1940) Fil.dr., inorg. chemistry (Stockholm U., 1982). Res. assoc. (tel. 08-7522690, ext. 4690). *Inorganic crystal structures, phase transitions.*

Horjales, Mr Eduardo. Dept. of Molecular Biology, Swedish U. of Agricultural Sciences, Uppsala Biomedical Center, Box 590, S-75124 Uppsala, Sweden. (1947) MSc, physics (U. de la República, Uruguay, 1977). Res. asst. (tel. 018-174554). *Computer programming, protein-ligand interactions, protein crystallography.*

Hovmöller, Dr Sven. Dept. of Structural Chemistry, Arrhenius Lab., U. of Stockholm, S-10691 Stockholm, Sweden. (1947) Fil.dr., chemistry (Stockholm U., 1980). Docent. (tel. 08-162380). *Electron microscopy and image processing, membrane proteins, inorganic crystals, direct methods.*

Humble, Dr Sten G. Dept. of Solid-State Physics, Royal Inst. of Techn., S-10044 Stockholm, Sweden. (1925) Tekn.dr., solid-state physics (Royal Inst. of Techn., 1971). Lect. (tel. 08-7877000, ext. 7303). *Metal physics, metals structure.*

Hutchinson, Dr William Bevis. Swedish Inst. for Metals Res., Drottning Kristinas väg 48, S-11428 Stockholm, Sweden. (1944) DSc, metallurgy (Manchester U., UK, 1965). Res. group leader. (tel. 08-243330, ext. 201). *Preferred crystallographic orientation, texture, materials processing and properties.*

Ingri, Prof. Nils. Dept. of Inorganic Chemistry, U. of Umeå, S-90187 Umeå, Sweden. (1929) Fil.dr., chemistry (Stockholm U., 1963). Prof. (tel. 090-165260). *Inorganic crystal structures, computer programming.*

Ivarsson, Dr Gun J. M. Dept. of Inorganic Chemistry, U. of Umeå, S-90187 Umeå, Sweden. (1943) Fil.dr., chemistry (Umeå U., 1983). Lect. (tel. 090-165410). *Inorganic and organic crystal structures.*

Jagner, Dr Susan. Dept. of Inorganic Chemistry, Chalmers U. of Techn. and U. of Göteborg, S-41296 Göteborg, Sweden. (1940) Fil.dr., chemistry (Göteborg U., 1970). Docent. (tel. 031-810100, ext. 1616). *Inorganic crystal structures, order-disorder structures.*

Jahnberg, Dr Lena. Dept. of Inorganic Chemistry, Arrhenius Lab., U. of Stockholm, S-10691 Stockholm, Sweden. (1937) Fil.dr., chemistry (Stockholm U., 1972). Res. assoc. (tel. 08-162368). *Inorganic crystal structures, solid state chemistry.*

Jansson, Mr Kjell. Dept. of Inorganic Chemistry, Arrhenius Lab., U. of Stockholm, S-10691 Stockholm, Sweden. (1959) Högsk.ex., chemistry (Stockholm U., 1982). Res. asst. (tel. 08-162372). *Structure of amorphous metals, electron microscopy.*

Jennische, Dr Per. National Swedish Lab. for Agricultural Chemistry, Box 7004, S-75007 Uppsala, Sweden. (1943) Fil.dr., inorg. chemistry (Uppsala U., 1976). Sr. chemist. (tel. 018-171000). *Coordination chemistry, analytical chemistry.*

Joelson, Mr Thorleif. Dept. of Molecular Biology, Swedish U. of Agricultural Sciences, Uppsala Biomedical Center, Box 590, S-75124 Uppsala, Sweden. (1950) Fil.kand. (U. of Uppsala). Res. asst. (tel. 018-174550). *Biological macromolecules.*

Johansson, Dr Georg. Dept. of Inorganic Chemistry, Royal Inst. of Techn., S-10044 Stockholm, Sweden. (1925) Tekn.dr., inorg. chemistry (Royal Inst. of Techn., 1963). Docent. (tel. 08-7878156). *Complexes, structure in solution, crystal structures.*

Johansson, Mr Karl-Erik. Dept. of Structural Chemistry, Arrhenius Lab., U. of Stockholm, S-10691 Stockholm, Sweden. Res. engineer. (tel. 08-162389). *Instrumentation.*

Johansson, Dr Lars-Gunnar. Dept. of Inorganic Chemistry, Chalmers U. of Techn. and U. of Göteborg, S-41296 Göteborg, Sweden. (1952) Fil.dr., inorg. chemistry (Göteborg U., 1982). Res. assoc. (tel. 031-810100, ext. 1320). *Synthesis and structure of inorganic sulfur compounds.*

Jones, Dr T. Alwyn. Dept. of Molecular Biology, Biomedical Center, U. of Uppsala, Box 590, S-75124 Uppsala, Sweden. (1947) PhD, biophysics (London U., UK, 1973). Docent. (tel. 018-174000, ext. 4566). *Biological macromolecules, crystallographic computing, computer graphics.*

Jonsson, Dr Arne. Dept. of Inorganic Chemistry, Arrhenius Lab., U. of Stockholm, S-10691 Stockholm, Sweden. (1921) Fil.lic., chemistry (Stockholm U., 1970). Lect. (tel. 08-163744). *Inorganic crystal structures, metals structure.*

Jönsson, Dr Per-Gunnar. Dept. 4, National Defence Res. Inst., Cementvägen 20, S-90182 Umeå, Sweden. (1937) Fil.dr., chemistry (Uppsala U., 1973). Sr. res. sci. (tel. 090-189230). *Structure and properties of materials.*

Karlsson, Dr Bengt E. Swedish Chemical Society, Upplandsgatan 6A, S-11123 Stockholm, Sweden. (1948) Fil.dr., chemistry (Stockholm U., 1978). Docent. (tel. 08-115260). *Organic crystal structures, X-ray analysis, biological specimen structure, electron microscopy and image reconstruction.*

Kierkegaard, Prof. Peder. Dept. of Structural Chemistry, Arrhenius Lab., U. of Stockholm, S-10691 Stockholm, Sweden. (1928) Fil.dr., chemistry (Stockholm U., 1962). Prof. (tel. 08-162385). *Inorganic and organic structures, protein and vitreous structures, instrumentation.*

Kihlborg, Prof. Lars. Dept. of Inorganic Chemistry, Arrhenius Lab., U. of Stockholm, S-10691 Stockholm, Sweden. (1930) Fil.dr., chemistry (Uppsala U., 1964). Prof. (tel. 08-162370). *Inorganic structures, structural defects, electron microscopy, solid state chemistry.*

Knight, Mr Stefan. Dept. of Molecular Biology, Swedish U. of Agricultural Sciences, Uppsala Biomedical Center, Box 590, S-75124 Uppsala, Sweden. (1957) Högsk.ex., chemistry (Uppsala U., 1984). Res. asst. (tel. 018-174524). *Biological macromolecules, protein engineering.*

Larsson, Dr Sven. Dept. of Physical Chemistry, Chalmers U. of Techn. and U. of Göteborg, S-41296 Göteborg, Sweden. (1941) Fil.dr., quantum chemistry (Uppsala U., 1972). Docent. (tel. 031-810100). *Inorganic structures, metalloproteins, materials research.*

Leijonmarck, Dr Marie. Dept. of Molecular Biology, Biomedical Center, U. of Uppsala, Box 590, S-75124 Uppsala, Sweden. (1946) Fil.dr., chemistry (Stockholm U., 1977). Res. assoc. (tel. 018-174000, ext. 4544). *Biological macromolecules.*

Lenner, Dr Magnus. Dept. of Inorganic Chemistry, Chalmers U. of Techn. and U. of Göteborg, S-41296 Göteborg, Sweden. (1944) Fil.dr., chemistry (Göteborg U., 1980). Res. assoc. (tel. 031-810100, ext. 1515). *Actinide complexes - structure.*

Liem, Dr D. Hay. Dept. of Inorganic Chemistry, Royal Inst. of Techn., S-10044 Stockholm, Sweden. (1932) Tekn.dr., chemistry (Royal Inst. of Techn., 1971). Docent. (tel. 08-7878330). *Metallo-organic compounds, complexes.*

Liljas, Dr Anders. Dept. of Molecular Biology, Biomedical Center, U. of Uppsala, Box 590, S-75124 Uppsala, Sweden. (1939) Fil.dr., chemistry (Uppsala U., 1971). Docent. (tel. 018-174000, ext. 4544). *Biological macromolecules.*

Liljas, Dr Lars. Dept. of Molecular Biology, Biomedical Center, U. of Uppsala, Box 590, S-75124 Uppsala, Sweden. (1947) Fil.dr., biochemistry (Uppsala U., 1977). Docent. (tel. 018-174000). *Biological macromolecules.*

Liminga, Prof. Rune. Inst. of Chemistry, U. of Uppsala, Box 530, S-75121 Uppsala, Sweden. (1932) Fil.dr., chemistry (Uppsala U., 1968). Prof. (tel. 018-183770). *Crystal structure - physical properties relationships, materials.*

Lindahl, Dr Tommie. Ceaverken AB, Box 174, S-15201 Strängnäs, Sweden. (1937) Fil.lic., chemistry (Stockholm U., 1969). *X-ray diffraction film.*

Lindner, Dr Peter W. Dept. of Quantum Chemistry, U. of Uppsala, Box 518, S-75120 Uppsala, Sweden. (1937) Fil.dr., quantum chemistry (Uppsala U., 1970). Docent. (tel. 018-155400). *X-ray scattering theory, interaction between X-rays and molecules.*

Lindqvist, Dr. Bengt. Dept. of Mineralogy, Swedish Museum of Natural History, Box 50007, S-10405 Stockholm, Sweden. (1927) Fil.dr., mineralogy and petrology (Uppsala U., 1966). Curator (docent). (tel. 08-150240, ext. 291). *Mineralogy.*

Lindqvist, Prof. Ingvar. Dept. of Chemistry 1, Swedish U. of Agricultural Sciences, S-75007 Uppsala, Sweden. (1921) Fil.dr., inorg. chemistry (Uppsala U., 1951). Prof. (tel. 018-301563). *Structural biochemistry.*

Lindqvist, Prof. Oliver. Dept. of Inorganic Chemistry, Chalmers U. of Techn. and U. of Göteborg, S-41296 Göteborg, Sweden. (1943) Fil.dr., chemistry (Göteborg U., 1973). Prof. (tel. 031-810100, ext. 1517). *Inorganic crystal structures, glass and liquid structures.*

Lindqvist, Dr Ylva Ch. Dept. of Molecular Biology, Swedish U. of Agricultural Sciences, Uppsala Biomedical Center, Box 590, S-75124 Uppsala, Sweden. (1947) Fil.dr, chemistry (Swedish U. of Agricultural Sciences, 1981). Res. assoc. (tel. 018-174523). *Protein crystallography, photosynthesis.*

Ljungström, Dr Evert B. Dept. of Inorganic Chemistry, Chalmers U. of Techn. and U. of Göteborg, S-41296 Göteborg, Sweden. (1949) Fil.dr., chemistry (Göteborg U., 1979). Docent. (tel. 031-810100, ext. 1614). *Instrumentation, inorganic and organometallic crystal structures.*

Löfgren, Dr Percy. Dept. of Inorganic Chemistry, Arrhenius Lab., U. of Stockholm, S-10691 Stockholm, Sweden. (1927) Fil.dr., chemistry (Stockholm U., 1974). Res. assoc. (tel. 08-162353). *Inorganic crystal structures.*

Löfgren, Dr Tor H. Inst. of Chemistry, U. of Uppsala, Box 531, S-75121 Uppsala, Sweden. (1926) Fil.dr., chemistry (Uppsala U., 1974). Lect. (docent). (tel. 018-183731). *Inorganic crystal structures.*

Lundberg, Dr Bruno. Dept. of Inorganic Chemistry, U. of Umeå, S-90187 Umeå, Sweden. (1939) Fil.dr., chemistry (Umeå U., 1972). Docent. (tel. 090-165262). *Inorganic and organic crystal structures, education.*

Lundberg, Dr Monica. Dept. of Inorganic Chemistry, Arrhenius Lab., U. of Stockholm, S-10691 Stockholm, Sweden. (1938) Fil.dr., chemistry (Stockholm U., 1971). Docent. (tel. 08-162368). *Inorganic crystal structures, image lattice technique, solid state chemistry.*

Lundgren, Dr Jan-Olof. Inst. of Chemistry, U. of Uppsala, Box 531, S-75121 Uppsala, Sweden. (1940) Fil.dr., chemistry (Uppsala U., 1974). Docent. (tel. 018-183771). *Accurate structure analysis, X-ray diffraction, neutron diffraction, computer programming.*

Lundgren, Mr Lennart. National Board of Occupational Safety and Health, Res. Dept., Aerosol section, S-17184 Solna, Sweden. (1950) Fil.kand., chemistry (Uppsala U., 1973). Res. asst. (tel. 08-7309000, ext. 9424). *Powder diffraction, aerosol chemistry.*

Lundström, Dr Torsten. Inst. of Chemistry, U. of Uppsala, Box 531, S-75121 Uppsala, Sweden. (1929) Fil.dr., inorg. chemistry (Uppsala U., 1969). Docent.

(tel. 018-183722). *Instrumentation, crystal growth, structure and properties of materials.*
Lyxell, Mr Dan-Göran. Dept. of Inorganic Chemistry, U. of Umeå, S-90187 Umeå, Sweden. (1945) Fil.mag., physics (Umeå U., 1969). Res. asst. (tel. 090-165445). *Large angle X-ray scattering, liquids, inorganic crystal structures.*
Magnéli, Prof. Arne. Dept. of Inorganic Chemistry, Arrhenius Lab., U. of Stockholm, S-10691 Stockholm, Sweden. (1914) Fil.dr., chemistry (Uppsala U., 1950). Prof. em. (tel. 08-162417, also 018-118650). *Inorganic crystal structures, solid state chemistry.*
Magnusson, Dr Bo. Res. & Dev., ESAB AB, Box 8004, S-40277 Göteborg, Sweden. (1943) Fil.lic., chemistry (Göteborg U., 1971). Res. asst. (tel. 031-509000). *Organic crystal structures.*
Marinder, Dr Bengt-Olov. Dept. of Inorganic Chemistry, Arrhenius Lab., U. of Stockholm, S-10691 Stockholm, Sweden. (1927) Fil.lic., inorg. chemistry (Stockholm U., 1962). Res. asst. (tel. 08-162417). *Inorganic crystal structures.*
Nenner, Miss Ann-Marie. Dept. of Inorganic Chemistry, U. of Umeå, S-90187 Umeå, Sweden. (1953) Fil.kand., chemistry (Umeå U., 1975). Res. asst. (tel. 090-166327). *Inorganic crystal structures.*
Nilsson, Mrs Karin I. Div. of Inorganic Chemistry 1, Chemical Center, U. of Lund, Box 124, S-22100 Lund, Sweden. (1954) Civ.ing., chemistry (Lund U., 1979). Res. asst. (tel. 046-108103). *Metallo-organic crystal structures.*
Nilsson, Dr Rolf O. Div. Kemiteknik, Boliden Kemi AB, Box 902, S-25109 Helsingborg, Sweden. (1927) Tekn.dr., inorg. chemistry (Chalmers U. of Techn., 1958). Chief chemist (docent). (tel. 042-139100). *Liquid structures.*
Nord, Dr Anders G. Dept. of Mineralogy, Swedish Museum of Natural History, Box 50007, S-10405 Stockholm, Sweden. (1942) Fil.dr., chemistry (Stockholm U., 1974). Scientist (docent). (tel. 08-150240, ext. 208). *Inorganic and mineral crystal structures, solid solutions, crystal growth, computer programming.*
Nordlund, Mr Pär L. Dept. of Molecular Biology, Swedish U. of Agricultural Sciences, Uppsala Biomedical Center, Box 590, S-75124 Uppsala, Sweden. (1958) Civ.ing., techn. physics (1984). Res. asst. (tel. 018-174550). *Protein crystallography.*
Norén, Dr Bertil. Div. of Inorganic Chemistry 1, Chemical Center, U. of Lund, Box 124, S-22100 Lund, Sweden. (1931) Fil.dr., chemistry (1970). Docent. (tel. 046-107000, ext. 8109). *Metallo-organic crystal structures, coordination compounds.*
Noréus, Dr Dag. Dept. of Structural Chemistry, Arrhenius Lab., U. of Stockholm, S-10691 Stockholm, Sweden. (1951) Tekn.dr., reactor physics (Royal Inst. of Techn., 1982). Res. assoc. (tel. 08-162391). *Neutron scattering, X-ray diffraction, metal hydrides.*
Norin, Dr Rolf. Dept. of Chemistry, Linköping U., S-58183 Linköping, Sweden. (1930) Fil.dr., inorg. chemistry (Göteborg U., 1970). Lect. (tel. 013-281380). *Inorganic crystal structures.*
Norrby, Dr Lars-Johan. Dept. of Inorganic Chemistry, Arrhenius Lab., U. of Stockholm, S-10691 Stockholm, Sweden. (1938) Fil.dr., inorg. and structural chemistry (Stockholm U., 1970). U. lect. (tel. 08-162417). *Raman-laser spectroscopy and crystal structures of complex salts, chemical education.*
Nygren, Dr Mats. Dept. of Inorganic Chemistry, Arrhenius Lab., U. of Stockholm, S-10691 Stockholm, Sweden. (1938) Fil.dr., chemistry (Stockholm U., 1972). Docent. (tel. 08-162366). *Solid state chemistry.*
Öfverstedt, Dr Lars-Göran W. Dept. of Molecular Biology, Biomedical Center, U. of Uppsala, Box 590, S-75124 Uppsala, Sweden. (1953) Fil.dr., biochemistry (Uppsala U., 1983). Res. assoc. (tel. 018-174000). *Biological macromolecules.*
Olovsson, Mr Gunnar. Inst. of Chemistry, U. of Uppsala, Box 531, S-75121 Uppsala, Sweden. (1953) Fil.kand., chemistry (Uppsala U., 1984). Res. asst. (tel. 018-183775). *Accurate structure analysis, X-ray and neutron diffraction, hydrogen bonding in solids, cation radical salts ("organic metals").*
Olovsson, Prof. Ivar. Inst. of Chemistry, U. of Uppsala, Box 531, S-75121 Uppsala, Sweden. (1928) Fil.dr., chemistry (Uppsala U., 1960). Prof. (tel. 018-183721). *Accurate structure analysis, X-ray diffraction, neutron diffraction, hydrogen bonding in solids, electron density.*
Olson, Mrs Solveig. Dept. of Inorganic Chemistry, Chalmers U. of Techn. and U. of Göteborg, S-41296 Göteborg, Sweden. (1944) Res. asst. (tel. 031-810100, ext. 1816). *Inorganic crystal structures.*
Olsson, Miss Carin. Dept. of Inorganic Chemistry, Chalmers U. of Techn. and U. of Göteborg, S-41296 Göteborg, Sweden. (1961) Fil.kand., chemistry (Göteborg U., 1984). Res. asst. (tel. 031-810100). *Biological macromolecules.*
Olsson, Mr Per-Olof. Dept. of Inorganic Chemistry, Arrhenius Lab., U. of Stockholm, S-10691 Stockholm, Sweden. (1957) Högsk.ex., chemistry (Stockholm U., 1983). Res. asst. (tel. 08-162372). *Inorganic crystal structures, electron microscopy.*
Oskarsson, Dr Åke. Dept. of Chemistry, U. College of Sundsvall, Box 860, S-85124 Sundsvall, Sweden. (1942) Fil.dr., chemistry (Lund U., 1974). Lect. (tel. 060-154260, ext. 505). *Coordination compounds.*
Österberg, Prof. Ragnar. Dept. of Medical Biochemistry, U. of Göteborg, Box 33031, S-40033 Göteborg, Sweden. (1932) Med.dr., biochemistry (Göteborg U., 1966). Res. docent in the Swedish Natural Science Res. Council. (tel. 031-822587). *Supramacromolecular biostructures in solution, small angle X-ray and neutron scattering.*
Pap, Mrs Sarolta. Dept. of Medical Biochemistry, , U. of Göteborg, Box 33031, S-40033 Göteborg, Sweden. (1944) Fil.kand., chemistry (Göteborg U., 1977).

Res. asst. (tel. 031-853000, ext. 3457). *Small angle X-ray and neutron scattering, biomolecular structure in solution.*
Pascher, Dr Irmin. Dept. of Structural Chemistry, Faculty of Medicine, U. of Göteborg, Box 33031, S-40033 Göteborg, Sweden. (1935) Phil.Dr., chemistry (Graz U., Austria, 1963). Docent. (tel. 031-223758). *Membrane lipids, structure and function.*
Persdotter, Miss Ingeborg M. Dept. of Inorganic Chemistry, Chalmers U. of Techn. and U. of Göteborg, S-41296 Göteborg, Sweden. (1956) Fil.kand., chemistry (Göteborg U., 1978). Res. asst. (tel. 031-810100, ext. 1519). *Inorganic crystal structures.*
Pilotti, Dr Anne-Marie. Dept. of Structural Chemistry, Arrhenius Lab., U. of Stockholm, S-10691 Stockholm, Sweden. (1942) Fil.dr., chemistry (Stockholm U., 1971).Docent. (tel. 08-162284). *Organic crystal structures.*
Ribbing, Dr Carl-Gustaf. Dept. of Solid State Physics, Inst. of Techn., Box 534, S-75121 Uppsala, Sweden. (1942) Fil.dr., solid state physics (Uppsala U., 1973). Lect. (docent). (tel. 018-183133). *Optical coatings, solar optical selectivity.*
Rundgren, Mr Kent A. Dept. of Inorganic Chemistry, Chalmers U. of Techn. and U. of Göteborg, S-41296 Göteborg, Sweden. (1958) Högsk.ex., chemistry (Göteborg U., 1981). Res. asst. (tel. 031-810100, ext. 1515). *Large angle X-ray scattering, powder diffraction.*
Rundqvist, Prof. Stig O. Inst. of Chemistry, U. of Uppsala, Box 531, S-75121 Uppsala, Sweden. (1929) Fil.dr., chemistry (Uppsala U., 1963). Prof. (tel. 018-183718). *Metallic and semiconducting phases - structure, crystal growth.*
Sahle, Dr Wubeshet. Dept. of Inorganic Chemistry, Arrhenius Lab., U. of Stockholm, S-10691 Stockholm, Sweden. (1949) Fil.dr., inorg. chemistry (Stockholm U., 1983). Res. assoc. (tel. 08-162368). *Inorganic structural defects, electron microscopy.*
Sandström, Dr Magnus K. E. Dept. of Inorganic Chemistry, Royal Inst. of Techn., S-10044 Stockholm, Sweden. (1945) Tekn.dr., inorg. chemistry (Royal Inst. of Techn., 1978). Docent. (tel. 08-7878156). *Complex ions and coordination compounds, structures in solution, crystal structures, liquid structures, X-ray diffraction, neutron diffraction.*
Schneider, Dr Gunter. Dept. of Molecular Biology, Swedish U. of Agricultural Sciences, Uppsala Biomedical Center, Box 590, S-75124 Uppsala, Sweden. (1953) Dr.rer.nat., chemistry (Saarbrücken U., Germany, Fed. Rep., 1983). Res. assoc. (tel. 018-174524). *Biological macromolecules, protein crystallography.*
Sedzik, Dr Jan. Dept. of Molecular Biology, Biomedical Center, U. of Uppsala, Box 590, S-75124 Uppsala, Sweden. (1946) PhD, natural sciences (Poznań Medical Academy, Poland, 1978). Res. assoc. (tel. 018-174000). *Biological macromolecules, biomembranes, computing.*
Sharma, Mrs Renu. Dept. of Inorganic Chemistry, Arrhenius Lab., U. of Stockholm, S-10691 Stockholm, Sweden. (1952) MSc, inorg. chemistry (Stockholm U., 1984). Res. asst. (tel. 08-162368). *Inorganic crystal structures, electron microscopy (HREM).*
Sjöberg, Prof. Bo. Dept. of Medical Biochemistry, U. of Göteborg, Box 33031, S-40033 Göteborg, Sweden. (1941) Fil.dr., chemistry (Göteborg U., 1974). Prof. (tel. 031-853458). *Biomolecular structure, hydration and dynamics in solution; small-angle X-ray and neutron scattering.*
Sjögren, Miss Agneta. Dept. of Structural Chemistry, Arrhenius Lab., U. of Stockholm, S-10691 Stockholm, Sweden. (1956) Fil.kand., chemistry (Stockholm U., 1980). Res. asst. (tel. 08-162384). *Electron microscopy and image processing, crystalline bacteria cell-walls, membrane proteins.*
Sjölin, Dr H. Lennart G. Dept. of Inorganic Chemistry, Chalmers U. of Techn. and U. of Göteborg, S-41296 Göteborg, Sweden. (1949) Fil.dr., chemistry (Göteborg U., 1979). Docent. (tel. 031-810100, ext. 1514). *Protein crystallography, neutron diffraction.*
Skoglund, Dr B. Ulf. Dept. of Medical Cell Genetics, Karolinska Inst., Box 60400, S-10401 Stockholm, Sweden. (1950) Fil.dr., protein crystallography (Stockholm U., 1979). Docent. (tel. 08-309669). *Protein structures, nucleic acid structures, three dimensional reconstruction.*
Söderberg, Dr Bengt-Olof. Dept. of Molecular Biology, Swedish U. of Agricultural Sciences, Uppsala Biomedical Center, Box 590, S-75124 Uppsala, Sweden. (1940) Fil.dr., chemistry (Swedish U. of Agricultural Sciences, 1975). Res. assoc. (tel. 018-174535). *Biological macromolecules.*
Söderholm, Dr Anne Charlotte. Grafisk Färg AB, Box 502, S-16215 Vällingby, Sweden. (1950) Fil.dr., chemistry (Stockholm U., 1978). Lab. manager. (tel. 08-362600). *Organic structural chemistry.*
Söderholm, Dr Margareta. Dept. of Medical Biophysics, Karolinska Inst., S-10401 Stockholm, Sweden. (1953) Fil.dr., medical biophysics (Karolinska Inst., 1983). Res. assoc. (tel. 08-340560, ext. 1561). *Organic crystal structures.*
Söderlund, Mr Gustaf S. Dept. of Molecular Biology, Swedish U. of Agricultural Sciences, Uppsala Biomedical Center, Box 590, S-75124 Uppsala, Sweden. (1937) Fil.kand., biochemistry (Uppsala U., 1968). Res. asst. (tel. 018-171554). *Biological macromolecules, protein sequencing.*
Sonnerstam, Dr Ulf C. Dept. of Chemistry and Molecular Biology, Swedish U. of Agricultural Sciences, Box 7015, S-75007 Uppsala, Sweden. (1941) Fil.dr., molecular biology (Swedish U. of Agricultural Sciences, 1984). Res. assoc. (tel. 018-171000). *Protein crystallography, structure and function, computer programming.*

Ståhl, Dr Kenny. Div. of Inorganic Chemistry 2, Chemical Center, U. of Lund, Box 124, S-22100 Lund, Sweden. (1953) Tekn.dr., inorg. chemistry (Lund U., 1983). Res. assoc. (tel. 046-108117). *Metallo-organic crystal structures, synchrotron crystallographic studies.*

Stålhandske, Dr Claes. Div. of Inorganic Chemistry 2, Chemical Center, U. of Lund, Box 124, S-22100 Lund, Sweden. (1941) Tekn.dr., inorg. chemistry (Lund U., 1980). Docent. (tel. 046-108234). *Inorganic crystal structures.*

Stefanidis, Mr Theodoros. Dept. of Structural Chemistry, Arrhenius Lab., U. of Stockholm, S-10691 Stockholm, Sweden. (1955) Fil.kand., chemistry (Stockholm U., 1978). Res. asst. (tel. 08-162382). *Phosphates - crystal studies.*

Stenberg, Dr Lars. Div. of Inorganic Chemistry 2, Chemical Center, U. of Lund, Box 124, S-22100 Lund, Sweden. (1949) Tekn.dr., inorg. chemistry (Lund U., 1984). Res. assoc. (tel. 046-108233). *Semiconductors, syntheses, electron microscopy.*

Stensland, Dr Birgitta. Dept. of Structural Chemistry, Arrhenius Lab., U. of Stockholm, S-10691 Stockholm, Sweden. (1938) Fil.lic., chemistry (Stockholm U., 1970). Res. asst. (tel. 08-162381). *Peptide structures, biologically interesting molecules, electron micrographic 3-D reconstruction.*

Stomberg, Prof. Rolf. Dept. of Inorganic Chemistry, Chalmers U. of Techn. and U. of Göteborg, S-41296 Göteborg, Sweden. (1933) Tekn.dr., inorg. chemistry (Chalmers U. of Techn., 1965). U. lect. (docent). (tel. 031-810100, ext. 1816). *Inorganic crystal structures.*

Strandberg, Prof. Bror E. Dept. of Molecular Biology, Biomedical Center, U. of Uppsala, Box 590, S-75124 Uppsala, Sweden. (1930) Fil.dr., chemistry (Uppsala U., 1967). Prof. (tel. 018-113453). *Structure and function, proteins, nucleic acids, viruses, protein - nucleic acid interactions.*

Strandberg, Dr Rolf. Dept. of Inorganic Chemistry, U. of Umeå, S-90187 Umeå, Sweden. (1938) Fil.dr., chemistry (Umeå U., 1974). Res. assoc. (tel. 090-165467). *Inorganic crystal structures.*

Strid, Dr Karl-Gustav. The Inst. for Applied Biotechnology, Box 33053, S-40033 Göteborg, Sweden. (1940) Tekn.dr., physics (Chalmers U. of Techn., 1976). Docent. (tel. 031-415455). *Diffraction by perfect crystals, image science, instrumentation, materials science, crystallographic applications.*

Sundberg, Dr Margareta. Dept. of Inorganic Chemistry, Arrhenius Lab., U. of Stockholm, S-10691 Stockholm, Sweden. (1944) Fil.dr., inorg. chemistry (Stockholm U., 1981). Res.assoc. (tel. 08-162368). *Inorganic crystal structures, electron microscopy.*

Sundell, Dr Lars Staffan. Dept. of Structural Chemistry, Faculty of Medicine, U. of Göteborg, Box 33031, S-40033 Göteborg, Sweden. (1944) Fil.dr., chemistry (Göteborg U., 1983). Res. assoc. (tel. 031-853456). *Biological crystal structures.*

Svensson, Mr Anders. Dept. of Inorganic Chemistry, Chalmers U. of Techn. and U. of Göteborg, S-41296 Göteborg, Sweden. (1959) Högsk.ex., chemistry (Göteborg U., 1983). Res. asst.(tel. 031-810100, ext. 1514). *Protein crystallography.*

Svensson, Dr Christer. Div. of Inorganic Chemistry 2, Chemical Center, U. of Lund, Box 124, S-22100 Lund, Sweden. (1945) Tekn.dr., inorg. chemistry (Lund U., 1978). Docent. (tel. 046-108117). *Crystal structures, properties, diffraction methods, computer programming.*

Svensson, Mr Göran. Div. of Inorganic Chemistry 2, Chemical Center, U. of Lund, Box 124, S-22100 Lund, Sweden. (1955) Civ.ing., chemistry (The Lund Inst. of Techn., 1983). Res. asst. (tel. 046-108112). *Crystal structures, coordination compounds.*

Svensson, Dr Ing-Britt A. Dept. of Chemical Engineering, Chalmers U. of Techn., S-41296 Göteborg, Sweden. (1942) Fil.lic., chemistry (Göteborg U., 1971). Director of studies. (tel. 031-810100, ext. 1303). *Inorganic crystal structures.*

Szentivanyi-Hansson, Dr Helga. Dept. of Inorganic Chemistry, Chalmers U. of Techn. and U. of Göteborg, S-41296 Göteborg, Sweden. (1947) Fil.dr., inorg. chemistry (Göteborg U., 1984). Res. assoc. (tel. 031-810100). *Inorganic crystal structures.*

Tegman, Dr Ragnar. Swedish Inst. of Production Engineering Res., Regnbågsallén, S-95187 Luleå, Sweden. (1943) Fil.dr., chemistry (Umeå U., 1974). Res. and dev. engineer. (tel. 0920-91000, ext. 770). *Inorganic crystal structures, nitrides, non-oxide ceramics, materials research, hot isostatic pressing (HIP).*

Tellgren, Dr I. G. Roland. Inst. of Chemistry, U. of Uppsala, Box 531, S-75121 Uppsala, Sweden. (1930) Fil.dr., chemistry (Uppsala U., 1975). Docent. (tel. 018-183776). *Neutron diffraction, metal hydrides, electron density studies, ferroelectrics.*

Thirup, Mr Søren. Dept. of Molecular Biology, Biomedical Center, U. of Uppsala, Box 590, S-75124 Uppsala, Sweden. (1956) Cand.scient., chemistry (Aarhus U., Denmark, 1984). Res. asst. (tel. 018-174000, ext. 4544). *Biological macromolecules.*

Thomas, Dr John O. Inst. of Chemistry, U. of Uppsala, Box 531, S-75121 Uppsala, Sweden. (1944) PhD, crystallography (London U., UK, 1969). Docent. (tel. 018-182500, ext. 3763). *Electron density, solid electrolytes, metal hydrides, powder technique development.*

Thomasson, Mr Ronnie. Div. of Inorganic Chemistry 2, Chemical Center, U. of Lund, Box 124, S-22100 Lund, Sweden. (1959) Högsk.ex., chemistry (Lund U., 1984). Res. asst. (tel. 046-108229). *Silicate chemistry, crystal structures.*

Törnroos, Mr Karl W. Dept. of Structural Chemistry, Arrhenius Lab., U. of Stockholm, S-10691 Stockholm, Sweden. (1956) Högsk.ex., structural chemistry (Stockholm U., 1983). Res. asst. (tel. 08-162379). *Small proteins, pharmacological molecules.*

Trysberg, Dr Lennart A. G. Dept. of Inorganic Chemistry, Chalmers U. of Techn. and U. of Göteborg, S-41296 Göteborg, Sweden. (1928) Fil.lic., chemistry (Göteborg U., 1972). Res. asst. (tel. 031-810100, ext. 1557). *Crystal structures, crystal growth.*

Unge, Dr K. Torsten. Dept. of Molecular Biology, Biomedical Center, U. of Uppsala, Box 590, S-75124 Uppsala, Sweden. (1945) Fil.dr., biochemistry (Uppsala U., 1979). Res. assoc. (tel. 018-174000). *Biological macromolecules.*

Vannerberg, Prof. Nils-Gösta. Eka AB, S-44501 Surte, Sweden. Also Dept. of Inorganic Chemistry, Chalmers U. of Techn. and U. of Göteborg, S-41296 Göteborg, Sweden. (1930) Tekn.dr., inorg. chemistry (Chalmers U. of Techn., 1959). Prof., Director of R&D. (tel. 0303-98000, ext. 209). *Inorganic crystal structures.*

Vingsbo, Prof. Olof. Materials Science, Inst. of Techn., Villavägen 4, Box 534, S-75121 Uppsala, Sweden. (1931) Fil.dr., solid state physics (Uppsala U., 1967). Prof. (tel. 018-100470, ext. 42). *Metal physics, electron diffraction.*

Wadsten, Dr Tommy. Dept. of Inorganic Chemistry, Arrhenius Lab., U. of Stockholm, S-10691 Stockholm, Sweden. (1934) Fil.lic., chemistry (Stockholm U., 1969). Res. asst. (tel. 08-162367). *Inorganic crystal structures, crystal growth, electron diffraction, biologically active molecules.*

Wägner, Dr Anna. Dept. of Structural Chemistry, Arrhenius Lab., U. of Stockholm, S-10691 Stockholm, Sweden. (1951) Fil.dr., chemistry (Stockholm U., 1980). Res. assoc. (tel. 08-162381). *Biological and pharmacological crystal structures.*

Wahlberg, Mr Anders. Inst. of Chemistry, U. of Uppsala, Box 531, S-75121 Uppsala, Sweden. (1945) Fil.mag., chemistry (Uppsala U., 1970). Res. asst. (tel. 018-183723). *Net-work formations, crystal structures, 1:1 salts.*

Wahlberg, Dr Olof. Dept. of Structural Chemistry, Arrhenius Lab., U. of Stockholm, S-10691 Stockholm, Sweden. (1936) Fil.dr., inorg. solution chemistry (Stockholm U., 1971). (tel. 08-162390). *Bio-inorganic chemistry, geochemistry.*

Wahlström, Dr Ebba. Dept. of Structural Chemistry, Arrhenius Lab., U. of Stockholm, S-10691 Stockholm, Sweden. (1938) Fil.dr., chemistry (Stockholm U., 1970). Res. assoc. (tel. 08-162417). *Inorganic structures.*

Wallenberg, Mr Reine. Div. of Inorganic Chemistry 2, Chemical Center, U. of Lund, Box 124, S-22100 Lund, Sweden. (1957) Civ.ing., chemistry (Lund U., 1983). Res. asst. (tel. 046-108231). *Inorganic crystal structures, electron microscopy.*

Waller, Prof. Ivar. Dept. of Theoretical Physics, U. of Uppsala, Thunbergsvägen 3b, S-75238 Uppsala, Sweden. (1898) Fil.dr., theoretical physics (Uppsala U., 1925). Prof. em. (tel. 018-115159). *X-ray scattering, neutron scattering, crystal physics, liquid physics.*

Waltersson, Dr Kjell. National Defence Res. Inst., Section 234, Dept. 2, Box 27322, S-10254 Stockholm, Sweden. (1947) Fil.dr., chemistry (Stockholm U., 1976). Docent. (tel. 08-631500, ext. 1021). *Inorganic crystal structures.*

Werner, Dr Per-Erik. Dept. of Structural Chemistry, Arrhenius Lab., U. of Stockholm, S-10691 Stockholm, Sweden. (1931) Fil.dr., chemistry (Stockholm U., 1971). Docent. (tel. 08-162393). *Powder diffraction, crystal structures, computer programming.*

Westdahl, Miss Marianne. Dept. of Structural Chemistry, Arrhenius Lab., U. of Stockholm, S-10691 Stockholm, Sweden. (1951) Res. asst. (tel. 08-162393). *Powder diffraction, crystal structures, computer programming.*

Westin, Dr Leif. Res. & Dev., Kloster Speedsteel AB, S-81060 Söderfors, Sweden. (1939) Fil.dr., inorg. chemistry (Stockholm U., 1972). Res. assoc. (tel. 0293-30500). *Powder metallurgy, sintering, metallic phases.*

Westman, Ms Ingrid. Inst. of Chemistry, U. of Uppsala, Box 531, S-75121 Uppsala, Sweden. (1955) Fil.kand., chemistry (Uppsala U., 1980). Res. asst. (tel. 018-183724). *Structure and properties, alloys, crystal growth.*

Westman, Dr Sven. Dept. of Structural Chemistry, Arrhenius Lab., U. of Stockholm, S-10691 Stockholm, Sweden. (1933) Fil.dr., inorg. chemistry (Stockholm U., 1972). Docent. (tel. 08-162390). *Chemical education, inorganic structures.*

Wickman, Prof. Frans Erik. Dept. of Geology, U. of Stockholm, S-10691 Stockholm, Sweden. (1915) Fil.dr., mineralogy (U. of Stockholm). Prof. em., mineralogy, petrology and geochemistry. (tel. 08-340860, ext. 264). *Crystal chemistry, minerals.*

Wilhelmi, Dr Karl-Axel. Dept. of Inorganic Chemistry, Arrhenius Lab., U. of Stockholm, S-10691 Stockholm, Sweden. (1923) Fil.dr., inorg. chemistry (Stockholm U., 1966). Docent. (tel. 08-163728). *Inorganic crystal structures.*

Ymén, Dr B. Ingvar. Technical Res. & Dev. Dept., SUPRA AB, Box 516, S-26124 Landskrona, Sweden. (1954) Fil.dr., inorg. chemistry (Lund U., 1983). Res. assoc. (tel. 0418-76100). *Inorganic compounds, fertilizers.*

SWITZERLAND

Sub-Editor: M. Dobler

Notes

1. International telephone country code - 0041.

2. The degrees *Dr.sc.nat.*, *Dr.sc.techn.*, *Dr.sc.math.*, *Dr.phil.* and *Dr.phil.nat.* are approximately equivalent to a PhD. *Dipl.Ing.Chem.*, *Dipl.Chem.*, *Dipl.Phys.* and *Dipl.Nat.* are first degrees requiring at least 4 years of course work including some research, approximately equivalent to MSc. *Dipl.Ing.HTL* (french *Ing.ETS*) are degrees from Technical Schools, approximately equivalent to Polytechnical Schools in Britain.

3. Appointments at universities other than professors are *Oberassistent* (french *Maître assistent*) (oberasst.), about equivalent to lecturer. *Chargé de cours* or *Chargé de recherche* is equivalent to assistant. The title *PD* (Privatdozent) is about equivalent to reader.

4. Abbreviations:
   Inst. - Institut (department)
   Lab. - Laboratorium (as Institut)
   ETH (french EPF) - Eidgenössische Technische Hochschule
   HTL (french ETS) - Höhere Technische Lehranstalt

Arend, Prof. Hanns. Lab. für Festkörperphysik, ETH-Hönggerberg, CH-8093 Zürich, Switzerland. (1922) Dr, physical chemistry (U. Prague, Czechoslovakia, 1952). Prof. (tel. 01 + 3772329). *Crystal growth, dielectrics, ferroelectrics, phase transitions.*

Ascher, Dr Edgar. Dép. de physique théorethique., 32 boulevard d Yvoy, CH-1211 Genève, Switzerland. (1921) Dr, exp. physics (U. de Lausanne, 1955). Sen. Res. Fellow (tel. 022 + 430726). *Symmetry, phase transitions.*

Bärlocher, Dr Christian. Inst. für Kristallographie und Petrographie, ETH-Zentrum, CH-8092 Zürich, Switzerland. (1944) PhD, physical chemistry (London U., UK, 1973). wiss. Adjunkt (tel. 01 + 2563749). *X-ray Rietveld refinement, zeolites, computer programming.*

Bayer, Prof. Gerhard. Inst. für Kristallographie und Petrographie, ETH-Zentrum, CH-8092 Zürich, Switzerland. (1923) Dr.sc.nat., crystallography (ETH Zürich, 1961). Prof. (tel. 01 + 2563735). *Crystal chemistry, oxides, silicates, crystallization, glasses, ceramics, industrial minerals.*

Bensch, Dr Wolfgang Johannes Georg. Anorganisch-chemisches Inst., U. Zürich, Winterthurerstr. 190, CH-8057 Zürich, Switzerland. (1953) Dr, inorganic chemistry (U. München, 1983). Res. asst. (tel. 01 + 2574650). *Inorganic crystal structures, catalysis.*

Bernardinelli, Dr Gérald. Lab. de Cristallographie aux Rayons X, U. de Genève, 24 quai Ernest-Ansermet, CH-1211 Genève, Switzerland. (1945) PhD, crystallography (U. de Genève, 1977). Chef de travaux. (tel. 022 + 219355, ext. 2372). *Crystallography, organic compounds.*

Berset, Mr Guy Alexis. Chimie minérale analytique et appliquée, U. de Genève, 30 quai Ernest-Ansermet, CH-1211 Genève, Switzerland. (1956) Ing., chimie (U. de Genève, 1983). PhD student. (tel. 022 + 219355, ext. 2036). *Inorganic structures.*

Bezinge, Mr Alex. Lab. de Cristallographie aux Rayons X, U. de Genève, 24 quai Ernest-Ansermet, CH-1211 Genève, Switzerland. (1957) Dipl.phys., physics (ETH Zürich, 1982). Asst. (tel. 022 + 219355, ext. 2372). *Borides, magnetic properties.*

Bochud, Mr Jean-Daniel. Comadur Ltd., CH-1784 Courtepin, Switzerland. (1955) Ing. ETS, mechanics (ETS-Fribourg, 1977). Techn. asst. (tel. 037 + 341545, ext. 16). *Crystallography, Verneuil process.*

Bohac, Dr Petr. Ind. Res. Unit, ETH-Hönggerberg, CH-8093 Zürich, Switzerland. (1942) Dr.sc.techn., inorganic chemistry (ETH Zürich, 1979). Res. scient. (tel. 01 + 3772175). *Crystal growth, fused salt electrolysis, materials research.*

Brinkmann, Prof. Detlef. Physik Inst., U. Zürich, Schönberggasse 9, CH-8001 Zürich, Switzerland. (1931) Dr, exp. physics (U. Zürich, 1961). Prof. (tel. 01 + 2572930). *NMR applications, crystallography, mineralogy, solid state physics.*

Bürgi, Prof. Hans-Beat. Lab. für chemische und mineralogische Kristallographie, U. Bern, Freiestr. 3, CH-3012 Bern, Switzerland. (1942) Dr, chemistry (ETH Zürich, 1969). Prof. (tel. 031 + 654282). *Structure analysis, organic and inorganic compounds, reaction paths, structural correlations, molecular mechanics.*

Burkhard, Dr Andreas. Physics Dept. Klybeck, CIBA-GEIGY AG, CH-4002 Basel, Switzerland. (1947) Dr Phil., mineralogy (U. Basel, 1977). Res. scient. (tel. 061 + 364014). *X-ray powder diffraction, microscopy, alpine mineralogy.*

Bürki, Dr Hans. Lab. für chemische und mineralogische Kristallographie, U. Bern, Baltzerstr. 1, CH-3012 Bern, Switzerland. (1921) Dr.phil.nat., chemistry (U. Bern, 1950). Oberasst. (tel. 031 + 658495). *Crystal structures, biological structures.*

Busch, Prof. Georg Adolf. Lab. für Festkörperphysik, ETH-Hönggerberg, CH-8093 Zürich, Switzerland. (1908) Dr., solid state physics (ETH-Zürich, 1938). Prof. em. (tel. 01 + 3772240). *Condensed matter physics, electrochemistry.*

Cenzual, Mrs Karin Margareta. Lab. de Cristallographie aux Rayons X, U. de Genève, 24 quai Ernest-Ansermet, CH-1211 Genève, Switzerland. (1954) Dipl., chemistry (U. de Genève, 1979). Asst. (tel. 022 + 219355, ext. 2236). *Intermetallic compounds.*

Chabot, Dr Bernard André. Lab. de Cristallographie aux Rayons X, U. de Genève, 24 quai Ernest-Ansermet, CH-1211 Genève, Switzerland. (1948) Dr, crystallography (U. Aix-Marseille III, 1974). Maître-asst. (tel. 022 + 219355, ext. 2236). *X-ray structure analysis, intermetallic compounds, minerals.*

Chapuis, Prof. Gervais Constant. Inst. de Cristallographie, U. de Lausanne, B.S.P. Dorigny, CH-1015 Lausanne, Switzerland. (1944) Dr, crystallography (ETH Zürich, 1971). Prof. (tel. 021 + 462350). *Symmetry, structure determination, programming, phase transitions.*

Chollet, Dr Lucien-Francois. Materials Science and Micromechanics, Centre Suisse d'electronique et de microtechnique, CSEM, rue Breguet 2, CH-2000 Neuchatel, Switzerland. (1931) Dr, physics (U. de Neuchatel, 1960). Scient. (tel. 038 + 245566). *Structure, metals and alloys.*

Curtis, Dr Bernard. RCA Laboratories, Badenerstr. 569, CH-8048 Zürich, Switzerland. (1933) PhD, inorganic chemistry (Imperial C. London, UK, 1958). Member techn. staff. (tel. 01 + 4926350). *Crystal growth from vapour phase, plasma chemistry and plasma etching.*

Daly, Dr John Joseph. ZFE, F. Hoffmann-LaRoche AG, CH-4002 Basel, Switzerland. (1931) PhD, chemistry (Leeds U., UK, 1959). Scient. (tel. 061 + 276046). *Chemistry and structure.*

Delaloye, Prof. Michel. U. de Genève, 13, rue des Maraichers, CH-1211 Genève, Switzerland. (1936) Dr, mineralogy (U. de Genève, 1966). Prof. (tel. 022 + 219355). *Geochemistry, geochronometry.*

Dobler, PD Max. Lab. für organische Chemie, ETH-Zentrum, CH-8092 Zürich, Switzerland. (1937) Dr.sc.techn., chemical crystallography (ETH Zürich, 1963). PD. (tel. 01 + 2564509). *Crystallography, biological molecules.*

Dreiding, Prof. André. Organisch-chemisches Inst., U. Zürich, Winterthurstr. 190, CH-8057 Zürich, Switzerland. (1919) PhD, organic chemistry (Michigan U., USA, 1947). Prof. (tel. 01 + 2574231). *Structural organic chemistry.*

Dubler, PD Erich. Anorganisch-chemisches Inst., U. Zürich, Winterthurerstr. 190, CH-8057 Zürich, Switzerland. (1939) Dr.phil., chemistry (U. Zürich, 1970). PD. (tel. 01 + 326241). *Inorganic crystal structures, biocrystallography, thermal analysis.*

Dunitz, Prof. Jack David. Lab. für organische Chemie, ETH-Zentrum, CH-8092 Zürich, Switzerland. (1923) PhD, chemistry (Glasgow U., UK, 1947). Prof. (tel. 01 + 2562892). *Crystal and molecular structure.*

Egli, Mr Martin. Lab. für organische Chemie, ETH-Zentrum, CH-8092 Zürich, Switzerland. (1961) Dipl.chem., chemistry (ETH-Zürich, 1984). PhD student. (tel. 01 + 2564510). *Crystal structures.*

Emmenegger, Prof. Franzpeter. Inorganic Chemistry Inst., U. de Fribourg, Pérolles, CH-1700 Fribourg, Switzerland. (1935) Dr, chemistry (ETH Zürich, 1963). Prof. (tel. 037 + 826422). *Crystal growth, chemical transport.*

Engel, Ms Nora. Lab. de Cristallographie aux Rayons X, U. de Genève, 24 quai Ernest-Ansermet, CH-1211 Genève, Switzerland. (1953) Dipl.nat., geology (ETH Zürich, 1976). Asst. (tel. 022 + 219355, ext. 2372). *Mineral structures.*

Engel, PD Peter. Lab. für chemische und mineralogische Kristallographie, U. Bern, Freiestr. 3, CH-3012 Bern, Switzerland. (1942) Dr, crystallography (ETH Zürich, 1968). PD. (tel. 031 + 654273). *Crystal structure determination, mathematical crystallography.*

Epprecht, Prof. Willfried Th. Ottenbergstr. 45, CH-8049 Zürich, Switzerland. (1918) Dr.sc.nat. (ETH Zürich). Prof. em. (tel. 01 + 422986). *Physical metallurgy, nuclear reactor materials, creep, fatigue.*

Erdoes, Mr Ernst Gyula. Res. Dept., Sulzer Bros. Ltd., CH-8401 Winterthur, Switzerland. (1919) Ing.Chem., chemistry (ETH Zürich, 1944). Ret. res. eng. (tel. 052 + 815141). *Corrosion, physical metallurgy.*

Fehlmann, PD Melchior. Inst. für Kristallographie und Petrographie, ETH-Zentrum, CH-8092 Zürich, Switzerland. (1940) Dr.phil., crystallography (U. Zürich, 1965). Oberasst. (tel. 01 + 2563747). *Accurate electron density determination, dynamical X-ray diffraction, X-ray topography.*

Fischer, Dr Peter. Lab. für Neutronenstreuung, ETHZ, CH-5303 Würenlingen, Switzerland. (1937) Dr, neutron diffraction (ETH Zürich, 1966). Physicist. (tel. 056 + 981741). *Neutron scattering, solid state physics, structure research, magnetism.*

Flack, Dr Howard David. Lab. de Cristallographie aux Rayons X, U. de Genève, 24 Quai Ernest-Ansermet, CH-1211 Genève, Switzerland. (1943) PhD, crystallography (U. London, UK, 1968). Chargé de recherche. (tel. 022 + 219355, ext. 2249). *Crystallography.*

Forster, Dr Martin. Res. dep., Cerberus AG, alte Landstr. 411, CH-8708 Männedorf, Switzerland. (1945) Dr.sc.nat., natural sciences (ETH-Zürich, 1978). Project leader. (tel. 01 + 9226476). *Single crystals, metal oxides.*

Galetti, Dr Giulio. Inst. de Minéralogie, U. de Fribourg, Pérolles, CH-1700 Fribourg, Switzerland. (1937) Dr, geochemistry (U. di Padova, Italy, 1971). Lecturer. (tel. 037 + 826268). *X-ray analysis, inorganic structures.*

Garavito, PD R. Michael. Abt. Strukturbiologie, Biozentrum der Universität Basel, Klingelbergstr. 70, CH-4056 Basel, Switzerland. (1952) PhD, biology, biophysics (Purdue U., 1978). PD. (tel. 061 + 253880, ext. 276). *Membrane protein structure, protein crystallography.*

Gelato-Volders, Dr (Mrs) Louise Marie. Lab. de Cristallographie aux Rayons X, U. de Genève, 24 quai Ernest-Ansermet, CH-1211 Genève, Switzerland. (1935) Dr, mathematics (U. di Bologna, Italy, 1965). Asst. (tel. 022 + 219355, ext. 2236). *Computer programming.*

Gerdil, Prof. Raymond. U. de Genève, 30 quai Ernest Ansermet, CH-1211 Genève 4, Switzerland. (1929) Dr.sc.techn., organic physical chemistry (ETH Zürich, 1957). Prof. (tel. 022 + 219355). *Clathrates, electron distribution, organic molecules.*

Gilson, Mr Pierre. 55 A, rue du Centre, CH-1025 Saint-Sulpice, Switzerland. (1914) Ing. ICAM. (tel. 021 + 342159, telex 25443 CH Turgi). *Emerald, opal, turquoise, lapislazuli, ruby, saphir, BGO, silicon for semiconductors.*

Giovanoli, Prof. Rudolf. Inorganic Chemistry Inst., U. Bern, Freiestr. 3, CH-3000 Bern 9, Switzerland. (1936) Dr, chemistry (U. Bern, 1965). Prof. (tel. 031 + 654317). *Finely divided solids, metal oxides and oxidehydroxides, interconversion reactions, topotaxy, structure-texture relationships.*

Girgis, Dr Kamal. Inst. für Kristallographie und Petrographie, ETH-Zentrum, CH-8092 Zürich, Switzerland. (1936) Dr.sc.nat., crystallography (ETH Zürich, 1969). Head materials res. group. (tel. 01 + 2563770). *Crystal chemistry, crystal physics, structure, intermetallic compounds, crystal growth, superconductivity, magnetic structures.*

Gotthardt, Dr Rolf. Physics Dept., EPF Lausanne, Ecublens, CH-1015 Lausanne, Switzerland. (1941) Dr.rer.nat., physics (U. Stuttgart, BRD, 1977). Physicist. (tel. 021 + 473392). *Dislocation mobility, martensitic transformation, electron microscopy.*

Gramlich, Dr Rahel. Inst. für Kristallographie und Petrographie, ETH-Zentrum, CH-8092 Zürich, Switzerland. (1953) Dr.sc.nat., crystallography (ETH Zürich, 1981). *Zeolites.*

Gramlich, Dr Volker. Inst. für Kristallographie und Petrographie, ETH-Zentrum, CH-8092 Zürich, Switzerland. (1941) Dr, crystallography. Oberasst. (tel. 01 + 2563756). *Crystallography.*

Gränicher, Prof. Walter Hans Heini. Lab. für Festkörperphysik, ETH-Hönggerberg, CH-8093 Zürich, Switzerland. (1924) Dr.sc.nat., solid state physics (ETH Zürich, 1959). Prof., head Swiss Federal Reactor Techn. Inst. (tel. 01 + 3772330 or 056 + 992111). *Phase transitions in crystals, ferroelectrics, hydrogen bonded crystals.*

Guenter, Prof. John Ralph. Anorganisch-chemisches Inst., U. Zürich, Winterthurerstr. 190, CH-8057 Zürich, Switzerland. (1943) Dr, chemistry (U. Zürich, 1970). Asst. Prof. (tel. 01 + 2574646). *Inorganic crystal structures, crystal chemistry, topotactic reactions, electron microscopy.*

Hauger, Mr Rudi. Lab. für Festkörperphysik, ETH-Hönggerberg, CH-8093 Zürich, Switzerland. (1930) Dipl.Chem., physical chemistry (U. Saarbrücken, BRD, 1962). Oberasst. (tel. 01 + 3772249). *Synthesis, solar cell materials, vapour phase crystal growth, chemical transport reactions.*

Hepp, Dr Alfred. Trutztobel, CH-7074 Malix, Switzerland. (1939) Dr.phil., crystallography (U. Zürich, 1981). *Powder X-ray crystallography, crystallographic computing, silicate structures.*

Hibma, Dr Tjipke. Brown Boveri Res. Centre, CH-5405 Baden-Dättwil, Switzerland. (1943) Dr, physical chemistry (Rijksuniv. Groningen, Netherlands, 1974). Scient. staff member. (tel. 056 + 848411, ext. 8132). *Superionic conductors, order-disorder phenomena.*

Hintermann, Dr Hans-Erich. Materials Science and Micromechanics, Centre Suisse d'electronique et de microtechnique, CSEM, rue A.L. Breguet 2, CH-2000 Neuchatel, Switzerland. (1929) Dr.sc.nat., physical chemistry (ETH Zürich, 1957). Director. (tel. 038 + 245566). *Composite materials, metallurgy, surfaces, tribology, corrosion, catalysis, high temperature materials.*

Hovestreydt, Mr Eric Robert. Lab. de Cristallographie aux Rayons X, U. de Genève, 24 quai Ernest-Ansermet, CH-1211 Genève, Switzerland. (1957) Dipl.chem., chemistry (U. Leiden, Netherlands, 1980). Asst. (tel. 022 + 219355, ext. 2372). *Inorganic and metallic structures, ternary phases.*

Jansonius, Prof. Johan Nomdo. Biozentrum der Universität Basel, Klingelbergstr. 70, CH-4056 Basel, Switzerland. (1932) Dr, chemistry (U. Groningen, Netherlands, 1967). Prof. (tel. 061 + 253880, ext. 265). *Protein structures.*

Jenkins, Dr John Anthony. Abt. Strukturbiologie, Biozentrum der Universität Basel, Klingelbergstr. 70, CH-4056 Basel, Switzerland. (1949) DPhil, biochemistry (Sussex U., 1980). Postdoctoral fellow. (tel. 061 + 253880). *Protein crystallography.*

Kaldis, PD Emanuel. Lab. für Festkörperphysik, ETH-Hönggerberg, CH-8093 Zürich, Switzerland. (1931) Dr.rer.nat., physical chemistry (U. München, 1962). Head materials res. group. (tel. 01 + 3772251). *Solid state chemistry and physics, crystal growth, high temperature chemistry, ultra pure materials.*

Karlsson, Dr Rolf. Abt. Strukturbiologie, Biozentrum der Universität Basel, Klingelbergstr. 70, CH-4056 Basel, Switzerland. (1942) PhD, chemistry (U. Stockholm, 1977). Asst. (tel. 061 + 253880, ext. 274). *Protein crystallography.*

Karrer, Mr Andreas. Lab. für organische Chemie, ETH-Zentrum, CH-8092 Zürich, Switzerland. (1957) Dipl.Chem., chemistry (ETH Zürich, 1981). PhD student. (tel. 01 + 2564510). *Chemical crystallography, computing.*

Kellenberger, Prof. Eduard. Abt. Mikrobiologie, Biozentrum der Universität Basel, Klingelbergstr. 70, CH-4056 Basel, Switzerland. (1920) Dr, biophysics (U. de Genève, 1953). Prof. (tel. 061 + 253880, ext. 284). *Supramolecular structures, electron microscopy methods.*

Keller, Ms Eva Barbara. Inst. für Kristallographie und Petrographie, ETH-Zentrum, CH-8092 Zürich, Switzerland. (1956) Dipl.nat., crystallography (ETH Zürich, 1981). Asst. (tel. 01 + 2563775). *Zeolites, zeolite-like materials.*

Kostorz, Prof. Gernot. Inst. für angewandte Physik, ETH-Hönggerberg, CH-8093 Zürich, Switzerland. (1941) Dr.rer.nat., metal physics (U. Göttingen, FRG, 1968). Prof. (tel. 01 + 3773399). *Alloys, defects, plasticity, short-range order, phase transformations, surfaces, diffuse, small angle, neutron scattering.*

Kubel, Dr Frank. Lab. de Cristallographie aux Rayons X, U. de Genève, 24 quai Ernest-Ansermet, CH-1211 Genève, Switzerland. (1953) Dr, transition metal complexes (U. Tübingen, FRG, 1983). Asst. (tel. 022 + 219355, ext. 2372). *Transition metal complexes, superconductors, crystallographic theory.*

Lévy, Dr Francis. Inst. de physique appliquée, EPF-Lausanne, Ecublens, CH-1015 Lausanne, Switzerland. (1940) Dr.sc.nat., physics (ETH Zürich, 1969). Adj. scient., lecturer. (tel. 021 + 471111). *Synthesis, crystal growth, thin films, semiconductors, physical characterization, optical, electrical properties.*

Ludi, Prof. Andreas. Inorganic Chemistry Inst., U. Bern, Freiestr. 3, CH-3012 Bern, Switzerland. (1936) Dr.phil., inorganic chemistry (U. Bern, 1962). Prof. (tel. 031 + 654244). *Structure, polynuclear cyanides, complexes, mixed valence compounds.*

Meier, Prof. Walter M. Inst. für Kristallographie und Petrographie, ETH-Zentrum, CH-8092 Zürich, Switzerland. (1926) PhD, physical chemistry (London U., UK, 1957). Prof. (tel. 01 + 2563730). *Crystal chemistry, structural chemistry, zeolites, computer simulated crystal structures, model construction.*

Mez, Dr Hans-Christian. Messtechnik und Automation, CIBA-GEIGY AG, Schwarzwaldallee 215, CH-4002 Basel, Switzerland. (1935) Dr.sc.techn., chemical crystallography (ETH Zürich, 1961). Group leader. (tel. 061 + 374950). *Methods.*

Moeckli, Dr Pedro. Inst. de Cristallographie, U. de Lausanne, B.S.P. Dorigny, CH-1015 Lausanne, Switzerland. (1940) Dr.sc.nat., crystallography (ETH Zürich, 1983). Asst. (tel. 021 + 462354). *Structure determination, electron density, instrumentation.*

Moor, Dr Robert. Swiss federal propellant plant, CH-3752 Wimmis, Switzerland. (1948) Dr.sc.nat., crystallography (ETH Zürich, 1983). Consulting eng. (tel. 033 + 552296). *Theoretical crystallography.*

Mooser, Prof. Emanuel. Inst. de physique appliquée, EPF-Lausanne, Ecublens, CH-1015 Lausanne, Switzerland. (1925) Dr.sc.nat., physics (ETH Zürich, 1953). Director. (tel. 021 + 474471). *Semiconductor physics, optoelectronics, biomedical engineering.*

Müller, Dr Rudolf O. CIBA-GEIGY AG, Physik Klybeck, CH-4000 Basel, Switzerland. (1929) Dr.phil.nat., mineralogy and petrology (U. Bern, 1958). Scient. expert. (tel. 061 + 364235). *Polymorphism, identification.*

Ngo, Dr Thong. Georges-Favre 4, CH-2400 Le Locle, Switzerland. (1954) PhD, crystallography (U. de Lausanne, 1981). Group leader. *Crystal structures, electronic ceramics, computer programming.*

Nickel, Prof. Erwin Julius Konstantin. Inst. de Minéralogie, U. de Fribourg, Pérolles, CH-1700 Fribourg, Switzerland. (1921) Dr, mineralogy and petrography (U. de Fribourg). Head. (tel. 037 + 826261). *Crystal growth.*

Niggli, Prof. Alfred. Inst. für Kristallographie und Petrographie, ETH-Zentrum, CH-8092 Zürich, Switzerland. (1922) Dr.phil., physical chemistry (U. Zürich, 1952). Prof. (tel. 01 + 2563731). *Mathematical crystallography, crystal structures.*

Niggli, Prof. Ernst. U. Bern, Baltzerstr. 1, CH-3012 Bern, Switzerland. (1917) Dr.phil.nat., mineralogy and petrography (U. Zürich, 1944). (tel. 031 + 658782). *Clay minerals, crystal optics, ore minerals, minerals in alpine fissures.*

Nissen, Dr Hans-Ude. Lab. für Festkörperphysik, ETH-Hönggerberg, CH-8093 Zürich, Switzerland. (1932) Dr.rer.nat., geology (U. Münster, BRD, 1960). (tel.

01 + 3772262). *Electron microscopy, minerals, inorganic materials, mineralogy, petrology.*

Nowacki, Prof. Werner. Mineral. Inst., U. Bern, Baltzerstr. 1, CH-3012 Bern, Switzerland. (1909) Dr.sc.math., mathematical crystallography (ETH Zürich, 1935). Prof. em. (tel. 031 + 235880). *Crystallography.*

Oberhänsli, Dr Willi E. Central Res. Units, F. Hoffmann-LaRoche AG, Grenzacherstr. 124, CH-4002 Basel, Switzerland. (1930) PhD, chemistry (U. Wisconsin, USA, 1964). (tel. 061 + 273564). *Direct methods, natural products.*

Parthé, Prof. Erwin. Lab. de Cristallographie aux Rayons X, U. de Genève, 24 quai Ernest-Ansermet, CH-1211 Genève, Switzerland. (1928) Dr.phil., chemistry (U. Wien, Austria, 1954). Prof. (tel. 022 + 219355, ext. 2208). *Inorganic and metallic crystal structures, crystal chemistry.*

Pauptit, Dr Richard A. Abt. Strukturbiologie, Biozentrum der Universität Basel, Klingelbergstr. 70, CH-4056 Basel, Switzerland. (1954) PhD, chemical crystallography (British Columbia U., Canada, 1981). Res. asst. (tel. 061 + 253880, ext. 273). *Protein crystallography.*

Peng, Dr Kuoching. Hirzenbachstr. 9, CH-8051 Zürich, Switzerland. Dr.phil., crystallography (U. Zürich, 1955). *X-ray metallography.*

Petcher, Dr Trevor James. Preclinical Res. Dept., Sandoz AG, CH-4002 Basel, Switzerland. (1943) PhD, chemical crystallography (Sheffield U., UK, 1967). Mol. pharmacologist, part time crystallographer. (tel. 061 + 241111, ext. 4851). *Biological structure-activity relationships, drugs, small hormones, drug design, conformational analysis, molecular pharmacology, receptor binding studies.*

Petter, Prof. Walter. Inst. für Kristallographie und Petrographie, ETH-Zentrum, CH-8092 Zürich, Switzerland. (1926) Dr.sc.nat., physics (ETH Zürich, 1969). Prof. (tel. 01 + 2563752). *Crystal chemistry, inorganic crystal structures, instrumentation.*

Priestle, Dr John P. Abt. Strukturbiologie, Biozentrum der Universität Basel, Klingelbergstr. 70, CH-4056 Basel, Switzerland. (1954) PhD, biochemistry (U. Texas, Austin, USA, 1982). Postdoctoral fellow. (tel. 061 + 253880, ext. 256). *Biological macromolecules, crystallographic computing, structure refinement.*

Restori, Mr Renzo. Inst. de Cristallographie, U. de Lausanne, B.S.P. Dorigny, CH-1015 Lausanne, Switzerland. (1957) Asst. (tel. 021 + 462360). *Electron density.*

Rieger, Dr Hans Wolfhart. Forschung und Entwicklung, Schweiz. Aluminium AG, CH-8212 Neuhausen, Metoxit AG, CH-8240 Thayngen, Switzerland. (1939) Dr.phil., chemistry (U. Wien, Austria, 1965). Dept. head. (tel. 053 + 20221, ext. 352). *Crystal chemistry, intermetallic compounds, oxides, ceramics.*

Rihs, Mrs Greti. Central Res. Function, CIBA-GEIGY AG, Postfach, CH-4000 Basel, Switzerland. (1943) Dipl.Nat., organic chemistry (ETH Zürich, 1966). (tel. 061 + 364004). *Organic crystal structures, organic conductors.*

Rinderer, Prof. Leo. Inst. de physique expérimentale, U. de Lausanne, CH-1015 Lausanne, Switzerland. (1927) Dr.sc.nat., physics (ETH-Zürich, 1952). Prof. (tel. 021 + 462320, 461111 ext. 2320). *Crystal growth.*

Rossi, Mr Franco Antonio. Lab. für Biochemie I, ETH-Zentrum, CH-8092 Zürich, Switzerland. (1955) Dipl.bot., plant physiology (U. Zürich, 1982). Asst. (tel. 01 + 2563141). *Macromolecular crystallography, protein structure and function.*

Rüegg, Dr Andreas. Dept. EKS, Brown Boveri Cie. AG, Fabrikstr. 3, CH-5600 Lenzburg, Switzerland. (1946) Dr.sc.nat., crystallography (ETH Zürich, 1977). Dev. head. (tel. 064 + 504399). *Inorganic structures, phase transitions, ceramics.*

Sakellariou, Mr Evangelos. Lab. für chemische und mineralogische Kristallographie, U. Bern, Freiestr. 3, CH-3012 Bern, Switzerland. (1953) Dipl.chem., chemistry (U. Basel, 1976). (tel. 031 + 654274). *Hydrogen bonding, anion complexes, structures, chemical applications.*

Schenk, Dr Kurt Johann. Inst. de Cristallographie, U. de Lausanne, B.S.P. Dorigny, CH-1015 Lausanne, Switzerland. (1951) Dr, crystallography (U. de Lausanne, 1984). Res. asst. (tel. 021 + 462352). *Phase transitions, disordered and modulated structures, structure and physical properties, crystallographic computing.*

Schicker, Mr Peter. Inst. für Kristallographie und Petrographie, ETH-Zentrum, CH-8092 Zürich, Switzerland. (1952) Dipl.nat., chemistry (ETH Zürich, 1982). Asst. (tel. 01 + 2563721). *Zeolites, powder crystallography.*

Schmelczer, Dr Robert. Inst. für angewandte Physik, ETH-Hönggerberg, CH-8093 Zürich, Switzerland. (1946) Dr, crystallography (U. de Lausanne, 1984). Res. asst. (tel. 01 + 3772136). *Metal physics, small angle scattering, structure and physical properties.*

Schmid, Prof. Hans. Dép. de chimie minérale, analytique et appliquée, U. de Genève, 30, quai Ernest-Ansermet, CH-1211 Genève, Switzerland. (1931) Prof., inorganic chemistry (U. de Genève, 1977). Applied chemistry lab. head. (tel. 022 + 219355, ext. 2405). *Magnetic ordering, ferroelectrics, ferroics.*

Schobinger-Papamantellos, Dr (Mrs) Penelope. Inst. für Kristallographie und Petrographie, ETH-Zentrum, CH-8092 Zürich, Switzerland. (1937) Dr.phil.chem., physical chemistry and crystallography (U. Wien, Austria, 1962). Res. asst. (tel. 01 + 2563773). *Magnetic structures, neutron diffraction, X-ray structure analysis.*

Schönholzer, Mr Peter. ZFE, F. Hoffmann-LaRoche AG, Grenzacherstr. 124, CH-4002 Basel, Switzerland. (1937) Dipl.Phys., physics (U. Bern, 1968). (tel. 061 + 272902). *Organic molecules, single crystals, powder diffraction.*

Schwarzenbach, Prof. Dieter. Inst. de Cristallographie, U. de Lausanne, B.S.P. Dorigny, CH-1015 Lausanne, Switzerland. (1936) Dr, crystallography (ETH Zürich, 1965). Prof. (tel. 021 + 462349). *Crystal structures, electron density, computer programming, crystal chemistry.*

Schweizer, Dr Wolfhard Bernd. Lab. für organische Chemie, ETH-Zentrum, CH-8092 Zürich, Switzerland. (1947) Dr.sc.nat., chemical crystallography (ETH Zürich, 1977). Res. Assoc. (tel. 01 + 2564507). *Reaction paths from crystallographic data.*

Seiler, Mr Paul. Lab. für organische Chemie, ETH-Zentrum, CH-8092 Zürich, Switzerland. (1945) Res. Assoc. (tel. 01 + 2564508). *Crystallography.*

Smit, Dr Jan Derk Geert. Lab. für Biochemie, ETH-Zentrum, CH-8092 Zürich, Switzerland. (1945) Dr, chemistry (U. Groningen, Netherlands, 1973). Oberasst. (tel. 01 + 2563141). *Protein crystallography, structure and function, molecular evolution.*

Staehlin, Dr Walter. Forschung und Entwicklung, Schweiz. Aluminium AG, CH-8212 Neuhausen, Switzerland. (1942) Dr, chemistry (U. Zürich, 1970). Dept. head. (tel. 053 + 20221). *Mass crystallization, alumina, crystal growth.*

Stoeckli-Evans, Dr (Mrs) Helen Margaret. Inst. de Chimie, U. de Neuchatel, av. de Bellevaux 51, CH-2000 Neuchatel, Switzerland. (1944) PhD, spectroscopy (Salford U., UK, 1969). Chargée de cours. (tel. 038 + 252815, ext. 46). *Pyrrolizidine alkaloids and active metabolites, small structures, inorganic and organic compounds.*

Strickler, Dr Peter. Abt. Chemie, Kantonsschule Zürcher Oberland, CH-8620 Wetzikon, Switzerland. (1936) Dr, chemical crystallography (ETH Zürich, 1966). Instr. (tel. 01 + 9321933). *Inorganic and organic structural chemistry.*

Trojer, Dr Felix John. Industrial Techn. Center., Battelle Inst., 7 route de Drize, CH-1227 Carouge, Switzerland. (1939) PhD, crystallography (M.I.T., USA, 1969). Manager. (tel. 022 + 439831, ext. 2373). *Glass ceramics, bioceramics, ceramics for electronic applications, glass fibers for cement reinforcement.*

Tschery, Dr Viktor. Lonza AG, Münchensteinerstr. 38, CH-4002 Basel, Switzerland. (1937) Dr, crystallography (ETH Zürich, 1971).

Vedani, Dr Angelo. Lab. für organische Chemie, ETH-Zentrum, CH-8092 Zürich, Switzerland. (1952) PhD, inorg. crystallography (U. Zürich, 1981). Postdoctoral fellow. (tel. 01 + 2564510). *Molecular modelling, drug design, structure-function relations, protein - small molecule complexes.*

Veprek, PD Stanislav. Anorganisch-chemisches Inst., U. Zürich, Winterthurerstr. 190, CH-8057 Zürich, Switzerland. (1939) Dr.phil., physics and chemistry (U. Zürich, 1972). PD. (tel. 01 + 326241, ext. 4651). *Crystal growth, thin films, plasma chemistry, mass spectrometry, photoelectron spectrometry, surface chemistry, surface processes.*

Vincent, Dr Michael Grange. Abt. Strukturbiologie, Biozentrum der Universität Basel, Klingelbergstr. 70, CH-4056 Basel, Switzerland. (1946) PhD, crystallography (London U., UK, 1975). Res. Assoc. (tel. 061 + 253880, ext. 275). *Crystallography.*

Vuagnat, Prof. Marc. Dép. de Minéralogie, U. de Genève, 13, rue des Maraichers, CH-1211 Genève, Switzerland. (1922) Dr, earth sciences (U. de Genève, 1944). Head. (tel. 022 + 219355). *Mineralogy.*

Waldmann, Dr Hans. Rheinstr. 2, CH-4127 Birsfelden, Switzerland. (1906) Dr.sc.nat., chemistry and physics (ETH Zürich, 1935). *Chemical microscopy, optical identification methods, chemical crystallography, instrumental optics.*

Walkinshaw, Dr Malcolm Douglas. Preclinical Res., Pharmaceutical Div., Sandoz AG, CH-4002 Basel, Switzerland. (1950) PhD, physical chemistry (Edinburg U., UK, 1975). Res. scient. (tel. 061 + 241111). *Molecular recognition; protein - drug interactions, single crystal structure, fibre protein structure.*

Wallis, Dr John Douglas. Lab. für organische Chemie, ETH-Zentrum, CH-8092 Zürich, Switzerland. (1954) D.Phil, organic chemistry, crystallography (Oxford U., UK, 1979). Postdoctoral co-worker. (tel. 01 + 2562909). *Organic intermediates, reaction coordinates, medicinal chemistry.*

Weber, Dr Hans Peter. Preclinical Res., Sandoz AG, CH-4002 Basel, Switzerland. (1936) Dr.sc.nat., X-ray analysis (ETH Zürich, 1964). Head. (tel. 061 + 244343). *X-ray analysis, biological molecules, molecular modelling, drug design, molecular mechanics, quantum chemistry.*

Werk, Dr Margit L. Chimie appliquée, U. de Genève, 30 quai Ernest-Ansermet, CH-1211 Genève, Switzerland. (1956) Dr, mineralogy (U. Hamburg, FRG, 1984). Postdoctoral fellow. (tel. 022 + 219355, ext. 2036). *Inorganic crystal structures, crystal chemistry, phase transitions.*

Winkler, Dr Fritz Karl. ZFE, F. Hoffmann-LaRoche, Grenzacherstr., CH-4002 Basel, Switzerland. (1944) Dr.sc.techn., structural organic chemistry (ETH-Zürich, 1973). Scient. *Structures, macromolecules and assemblies.*

Wüest, Mr Hermann. Lab. für Festkörperphysik, ETH-Hönggerberg, CH-8093 Zürich, Switzerland. (1952) Technician. (tel. 01 + 3772336). *Crystal growth.*

Wurtz, Dr Michel. Biozentrum der Universität Basel, Klingelbergstr. 70, CH-4056 Basel, Switzerland. (1940) Dr, biology (U. de Strasbourg, France, 1969). Asst. (tel. 061 + 253880). *Electron microscopy, biocrystals, image processing.*

Yvon, Prof. Klaus. Lab. de Cristallographie aux Rayons X, U. de Genève, 24 quai Ernest-Ansermet, CH-1211 Genève, Switzerland. (1943) PhD, structural chemistry (U. Wien, Austria, 1967). Prof. (tel. 022 + 219355, ext. 2231). *Structural chemistry, condensed matter physics.*

Zehnder, PD (Mrs) Margareta. Inst. für anorganische Chemie, U. Basel, Spitalstr. 51, CH-4056 Basel, Switzerland. (1942) PhD, (1973). Akad. Adjunkt. (tel. 061 + 571557, ext. 821). *Structure, metal complexes.*

Zolliker, Mr Peter. Lab. de Cristallographie aux Rayons X, U. de Genève, 24 quai Ernest-Ansermet, CH-1211 Genève, Switzerland. (1958) Dipl.phys., physics

(ETH Zürich, 1982). Asst. (tel. 022 + 219355, ext. 2372). *Hydrides, powders, neutron diffraction, X-ray scattering.*
Zschokke-Gränacher, Prof. (Mrs) Iris. Inst. für Physik, U. Basel, Klingelbergstr. 82, CH-4056 Basel, Switzerland. (1933) Dr, physics (U. Basel, 1960). Prof. (tel. 061 + 442040). *Organic semiconductors, molecular crystals, molecular glasses, solid solutions.*

SYRIAN ARAB REP.

Sub-Editor: S.E. Ali

Notes

1. International telephone country code - 90.

Abou Ghaloun, Dr Omar Farouk. Chemistry Dept., Faculty of Sci., Aleppo U., Aleppo, Syrian Arab Republic. (1950) PhD, physical chemistry (Nantes U., France, 1983). Lect. *Ionic conductors, semiconductors, crystal structure.*

Ali, Prof. Shams El Din. Physics Dept., Faculty of Sci., Aleppo U., Aleppo, Syrian Arab Republic. (1939) PhD, X-ray crystallography (Hull U., UK, 1969). Prof. *Solid state physics, crystallography.*

TAIWAN

Sub-Editor: S. B. Lin

Notes

1. International telephone country code - 886.

Chang, Mr Tien-show. Mining and Metallurgical Eng. Dept., Provincial Taipei Inst. of Techn., 3 Sec.1 South Shih-sheng Rd., Taipei, Taiwan 106, China. (1941) MSc, mineralogy and petrology (Pennsylvania State U., USA, 1974). Assoc. prof. (tel. 02 - 7712123). *Mineralogy, X-ray crystallography.*

Chen, Prof. Pei-yuan. Earth Sci. Dept., Nat. Taiwan Normal U., Taiwan 107, China. (1920) PhD, mineralogy and geology (U. Texas, USA, 1968). Prof. (tel. 02 - 9317511, ext. 256). *Clay mineralogy, sedimentology, powder diffraction.*

Chen, Dr Ruey-Hong. Physics Dept., Nat. Taiwan Normal U., 88 Roosevelt Road section 5, Taipei, Taiwan 117, China. (1947) PhD, crystallography (U. Pittsburgh, USA, 1977). Prof. (tel. 02 + 931-7511, ext. 52). *Solid state physics, diffraction physics, valence charge density.*

Chen, Mrs Yueh-Hua. Power Res. Lab. Taiwan Power Co., 198 Sec. 4, Roosevelt Rd., Taipei, Taiwan 107, China. (1948) BSc, chemistry (Nat. Cheng Kung U., 1970). Chemist (tel. 02 + 3216959 ext. 48). *X-ray powder diffraction, fluorescence analysis.*

Chung, Dr Beingtau. Physics Dept., Chung Yuan C. of Sci. and Eng., Chung-Li, Taiwan, China. (1944) PhD, applied science (Southern Methodist U., USA, 1974). Assoc. prof. (tel. 034 - 427171, ext. 259). *Electronic properties of materials, crystal structure.*

Houng, Prof. Kun-Huang. Agricultural Chemistry Dept., Nat. Taiwan U., Taipei, Taiwan 107, China. (1932) PhD, soil science (U. Hawaii, USA, 1964). Prof. (tel. 02 - 7818467). *Clay mineralogy, soil science.*

Hsu, Dr Shu-En. Mechanical Eng. Dept., Nat. Taiwan U., Taipei, Taiwan 107, China. (1929) PhD, materials science (Stanford U., USA, 1972). Prof. (tel. 02 + 3510231, ext. 2409). *X-ray crystallography.*

Huang, Mrs Chi-yung. Power Res. Lab. Taiwan Power Co., 198 Sec. 4, Roosevelt Rd., Taipei, Taiwan 107, China. (1948) BEng., metallurgy (Nat. Cheng Kung U., 1971). Metallurgist (tel. 02 + 3216959 ext. 50). *X-ray powder diffraction, fluorescence analysis, metallography.*

Huang, Prof. Chun-Kiang. Geology Dept., Nat. Taiwan U., 245 Choushan Rd., Taipei, Taiwan 107, China. (1916) DSc, mineralogy and mineral deposits (Tohoku U., Japan, 1959). Prof. (tel. 02 - 3510231, ext. 343). *Morphological and X-ray crystallography, optical mineralogy, gemology, mineral deposits.*

Huang, Mr Tung-woo. Material Testing Analysis Section, Materials Center, Chung-Shan Inst. of Sci. and Techn., P.O.Box 1-26-5, Lung-Tan, Taiwan 325, China. (1948) MSc, mineralogy (Chinese Culture U., 1975). Res. Analyst. (tel. 02 - 3814014, ext. 2610). *X-ray powder diffraction, ceramics, semiconductors.*

Jan, Mr Gwo-jen. Physics Dept., Nat. Taiwan U., Taipei, Taiwan 107, China. (1946) MSc, Physics. (Nat. Taiwan U., 1970). Lect. (tel. 02 - 3510231). *Solid state physics, X-ray instrumentation.*

Juang, Dr Tzo-chuan. Plant Nutrition Dept., Taiwan Sugar Res. Inst., 54 Sheng Chan Road, Tainan, Taiwan 700, China. (1931) PhD, soil science (U. Hawaii, USA, 1970). Sr. soil chemist. (tel. 062 - 26121). *Clay mineralogy, soil chemistry, plant nutrition.*

Lee, Prof. Wang Chihming. Geology Dept., Nat. Taiwan U., Taipei, Taiwan 107, China. (1932) PhD, mineralogy (U. Bochum, BRD, 1968). Prof. (tel. 02 - 3510231, ext. 387). *Mineralogy, crystal chemistry, petrology.*

Lin, Dr Hsi-che. Fine Ceramics Div., Materials Res. Lab., Inst. of Industrial Techn., 1021 Kuang-fu Rd., Hsinchu, Taiwan 300, China. (1936) PhD, mineralogy (Ohio State U., USA, 1967). Res. scient. (tel. 035 - 713141). *Phase equilibria, X-ray crystallography, fine ceramics, mineral synthesis.*

Lin, Prof. Szu Bin. Geology Dept., Nat. Taiwan U., 245 Choushan Rd., Taipei, Taiwan 107, China. (1938) PhD, crystallography and mineralogy (McMaster U., Canada, 1971). Prof. (tel. 02 - 3510231, ext. 2341). *X-ray diffraction, fluorescence, single-crystal structure analysis, mineralogical sciences.*

Liu, Dr Ling-Kang. Inst. of Chemistry, Academia Sinica, 128 Sec. 2 Yen-Chiu-Yuan Rd., Taipei, Taiwan 115, China. (1950) PhD, physical chemistry (U. Texas, Austin, USA, 1978). Res. fellow. (tel. 02 + 7619194). *Structure analysis, organic and organometallic compounds, phase problem.*

Lu, Dr Tian-Huey. Physics Dept., Nat. Tsing Hua U., 855 Kuang-Fu Rd., Hsinchu, Taipei, Taiwan 300, China. (1939) MSc, Physics (Nat. Tsing Hua U., 1965). Prof. (tel. 035 + 715131, ext. 430). *X-ray structure determination, protein crystallography, coordination chemistry.*

Ma, Dr Che-bao. Nuclear Energy Res. Inst., A.E.C., P.O. Box 3, Lung-tan, Taoyuan, Taiwan 325, China. (1941) PhD, mineralogy and petrology (Harvard U., USA, 1973). Res. fellow. (tel. 02 - 3814014, ext. 353). *Mineralogy, petrology (experimental), crystallography, neutron diffraction.*

Peng, Dr Shie-ming. Chemistry Dept., Nat. Taiwan U., Taipei, Taiwan 107, China. (1949) PhD, inorganic chemistry (U. Chicago, USA, 1975). Prof. (tel. 02 - 5215039). *Inorganic synthesis, crystal structure.*

Shen, Dr Pooyan. Inst. of Materials Sci. and Eng., Nat. Sun Yat-Sen U., Kaosiung, Taiwan 800, China. (1952) PhD, geophysics (Cornell U., USA, 1982). Assoc. prof. (tel. 07 - 5316171, ext. 246). *Mineral physics, intermetallic compounds, electron diffraction, electron microscopy.*

Tang, Dr Chia-Pin. Chemistry Dept., Chung-Shan Inst. of Sci. and Techn., P.O.Box 1-4-11, Lung-Tan, Taiwan 325, China. (1946) PhD, structural chemistry (Weizmann Inst. of Sci., Israel, 1979). Assoc. sci. (tel. 02 + 3931621). *Molecular packing, solid state chemistry.*

Tseng, Prof. Poh-kun. Physics Dept., Nat. Taiwan U., Taipei, Taiwan 107, China. (1930) PhD, nuclear eng. (U. Michigan, USA, 1968). Prof. (tel. 02 - 3510231). *Solid state physics, X-ray instrumentaion.*

Wan, Mr Hsien-ming. X-ray Lab., Mining Res. and Service Organization, Industrial Techn. Res. Inst., 1 Tun-Hwa S. Rd., Taipei, Taiwan 105, China. (1933) MSc, mineralogy and petrology (Pennsylvania State U., USA, 1975). Res. fellow. (tel. 7518341, ext. 68). *Clay mineralogy, X-ray diffraction, fluorescence.*

Wang, Miss Chiung-jane. Chemistry Dept., Chung-shan Inst. of Sci. and Techn., P.O. Box 1-4, Lung-Tan, Tao-Yuan, Taiwan 325, China. (1947) BSc, chemistry (Chung-Yan C. of Sci. and Eng., 1969). Eng. asst. (tel. 02 - 3814014, ext. 522). *Organic crystal structure, powder diffraction, programming.*

Wang, Prof. Yu. Chemistry Dept., Nat. Taiwan U., Taipei, Taiwan 107, China. (1943) PhD, chemistry (U. Illinois, USA, 1973). Prof. (tel. 02 + 3510231, ext. 3314). *Computational aspects of crystallography, metal alloys, Transition metal complexes, computerized data file base.*

Wu, Prof. Nan-chung. Material Sci. Dept., Nat. Cheng Kung U., Taiwan 700, China. (1936) MSc, physics (Tohoku U., Japan, 1972). Assoc. prof. (tel 064 + 2361111) *X-ray diffraction, electronic ceramics.*

Yang, Prof. Houng-Yi. Earth Sci. Dept., Nat. Cheng Kung U., Ta-Hsue Rd., Tainan, Taiwan 700, China. (1938) PhD, mineralogy (Ohio State U., USA, 1970). Prof. (tel. 062 - 24141, ext. 408). *Mineralogy, petrology.*

Yang, Prof. Tse Chun. Soil Sci. Dept., Nat. Chung Hsing U., 250 Kuo-Kuang Rd., Taichung, Taiwan, China. (1932) PhD, soil sci. (Michigan State U., USA, 1970). Prof. (tel. 042 - 223037, ext. 340). *Soil physics, physical chemistry, mineralogy.*

Yang, Dr Yui-whei. Chemistry Dept., Chung-yuan C. of Sci. and Eng., Chung-Li, Taiwan, China. (1948) PhD, chemistry (SUNY-Buffalo, USA, 1976). Asst. prof. (tel 034 + 427171, ext. 258). *X-ray crystallography, electron density.*

Yu, Prof. Shu-cheng. Earth Sci. Dept. Nat. Cheng Kung U., Ta-Hsue Rd., Tainan, Taiwan 700, China. (1942) PhD, mineralogy (Pennsylvania State U., USA, 1976). Prof. (tel 06 + 2361111, ext. 409). *Minerals and materials, structure, high pressure mineralogy, microstructure analysis.*

TANZANIA

Sub-Editor: M.E. Kamwaya

Notes

1. International telephone country code - 255

Kamwaya, Dr Mombo E. Dept. of Physics, U. of Dar es Salaam, Dar es Salaam 00255, Tanzania. (1941) PhD, crystallography (Free U. Berlin, BRD, 1980). Lect. *X-ray crystal analysis, powder diffraction techniques, optical crystallography, solid state physics, biophysics, fluorescence spectroscopy, computing.*

THAILAND

Sub-Editor: S. Pramatus

Notes

1. International telephone country code - 66.

2. The numbers following Bangkok should be used in mail addresses for quicker delivery.

3. At a Thai University, persons can be appointed to the following academic positions: *lecturer, assistant professor, associate professor* and *professor*. The functions are approximately equivalent to the US system, but with *lecturer* equivalent to *instructor*.

Anantachai, Mrs Suda Yasarawana. Physics Dept., Chiangmai U., Huey Keow Road, Chiangmai 50002, Thailand. (1940) MSc, crystallography (London U., UK, 1973). Lect. (tel. 053 + 22-1934, ext. 51). *Damage in semiconductors.*

Anugul, Mrs Surang. Chemistry Dept., Chulalongkorn U., Phya Thai Road, Bangkok 10500, Thailand. (1935) MS, inorganic chemistry (Oregon State U., USA, 1961). Assoc. prof. (tel. 02 + 252-7019). *Inorganic structures, alloys.*

Busaracome, Mr Suwin. Geology Dept., Khon Kaen U., Khon Kaen 40002, Thailand. (1948) MSc, geochemistry (Victoria U. of Wellington, New Zealand, 1978). Lect. (tel. 043 + 23-6199, ext. 1320). *Crystal growth, mineralogy, petrology, gemmology, X-ray crystallography.*

Chaichit, Dr Narongsak. Physics Dept., Silpakorn U., Nakorn Pathom 73000, Thailand. (1947) PhD, inorganic chemistry - X-ray crystallography (Monash U., Australia, 1982). Lect. (tel. 034 + 24-2072, ext. 15). *Natural-products, Organic and organometallic structures.*

Chaikum, Dr Nitirampai Latavalya. Chemistry Dept., Mahidol U., Rama 6 Road, Bangkok 10400, Thailand. (1947) PhD, X-ray crystallography (The Flinders U. of South Australia, Australia, 1976). Asst. prof. (tel. 02 + 281-5800, ext. 156). *Crystal structure analysis.*

Chaikum, Dr Nopadol. Chemistry Dept., Mahidol U., Rama 6 Road, Bangkok 10400, Thailand. (1949) PhD, geochemistry (Otago U., New Zealand, 1976). Assoc. prof. (tel. 02 + 281-5800, ext. 124). *Clays and clay minerals.*

Choosang, Mrs Pilai. Chemistry Dept., Chulalongkorn U., Phya Thai Road, Bangkok 10500, Thailand. (1937) BSc, chemistry (Chulalongkorn U., 1959). Asst. prof. (tel. 02 + 252-7019). *Alloy structures.*

Jinawath, Dr Supatra. Materials Sci. Dept., Chulalongkorn U., Phya Thai Road, Bangkok 10500, Thailand. (1945) PhD, mineral science (Leeds U., UK, 1974). Asst. prof. (tel. 02 + 251-1954). *Mineralogy, ceramics, cement chemistry.*

Jirajesda, Mr Jate. Geological Survey Div., Dept. of Mineral Resources, Rama 6 Road, Bangkok 10400, Thailand. (1955) BSc, chemistry (Khon Kaen U., 1980). Scient. (tel. 02 + 282-1164). *Powder diffraction, fluorescence.*

Kamolchote, Mr Poonsak. Chemistry Dept., Silpakorn U., Nakorn Pathom 73000, Thailand. (1953) MSc, physical chemistry (Chiangmai U., 1977). Lect. (tel. 034 + 24-2072). *Polymer structures.*

Keankeo, Miss Watcharaporn. Geology Dept., Khon Kaen U., Khon Kaen 40002, Thailand. (1958) BSc, geology (Chiangmai U., 1980). Lect. (tel. 043 + 23-6199, ext. 1320). *Mineral crystallography.*

Keow-kam-nerd, Dr Kanchana. Chemistry Dept., Chiangmai U., Huey Keow Road, Chiangmai 50002, Thailand. (1937) PhD, chemical technology - inorganics (U. de Besancon, France, 1970). Assoc. prof. (tel. 053 + 22-1934, ext. 21). *Ceramics, silicate technology.*

Khantaprab, Dr Chaiyudh. Geology Dept., Chulalongkorn U., Phya Thai Road, Bangkok 10500, Thailand. (1942) PhD, sedimentology (geology) (Imperial C., London U., UK, 1972). Asst. prof. (tel. 02 + 252-5931). *Sedimentology, sediment crystallography, environmental geology.*

Kritayakirana, Mrs Rungsri. Physics Dept., Chulalongkorn U., Phya Thai Road, Bangkok 10500, Thailand. (1940) MS, physics (Northeastern U., USA, 1969). Asst. prof. (tel. 02 + 252-9987). *Powder diffraction.*

Mitrprachachon, Dr Pachanee. Chemistry Dept., Khon Kaen U., Khon Kaen 40002, Thailand. (1950) PhD, inorganic chemistry - X-ray crystallography (Bristol U., UK, 1980). Lect. (tel. 043 + 23-7606, ext. 1269). *Crystal structure, crystallographic computing.*

Nimgirawath, Mrs Kloy. Physics Dept., Silpakorn U., Nakorn Pathom 73000, Thailand. (1944) MSc, X-ray crystallography (U. New South Wales, Australia, 1975). Asst. prof.(tel. 034 + 24-2072). *Organic structures.*

Padmasuta, Mrs Soontari. Geological Survey Division, Dept. of Mineral Resources, Rama 6 Road, Bangkok 10400, Thailand. (1939) BSc, physics (Chulalongkorn U., 1963). Scient. (tel. 02 + 282-1164). *Mineral identification, powder diffraction, fluorescence.*

Pakawatchai, Dr Chaveng. Chemistry Dept., Prince of Songkla U., Haad Yai, Songkla 90112, Thailand. (1951) PhD, crystallography (U. Western Australia, Australia, 1984). Lect. (tel. 074 + 24-4877, ext. 204). *Inorganic structures.*

Phaovibul, Dr Orapin. Chemistry Dept., Mahidol U., Rama 6 Road, Bangkok 10400, Thailand. (1941) Dr.rer.nat., physics (The Free U., Berlin, BRD, 1971). Assoc. prof. (tel. 02 + 281-5040, ext. 147). *Physical properties, liquid crystals, polymers, structure - properties relations.*

Phavanantha, Dr Phathana. Physics Dept., Chulalongkorn U., Phya Thai Road, Bangkok 10500, Thailand. (1942) PhD, crystallography (Imperial C., London U., UK, 1970). Assoc. prof. (tel. 02 + 252-9987). *Inorganic structures, natural-products, applied crystallography.*

Pisutha-Arnond, Dr Visut. Geology Dept., Chulalongkorn U., Phya Thai Road, Bangkok 10500, Thailand. (1951) PhD, inorganic geochemistry (Penn

State U., USA, 1982). Asst. prof. (tel. 02 + 252-5931). *Ore mineralogy, ore deposit research.*

Pongsapich, Dr Wasant. Geology Dept., Chulalongkorn U., Phya Thai Road, Bangkok 10500, Thailand. (1942) PhD, geology (U. Washington, USA, 1974). Asst. prof. (tel. 02 + 252-7989). *Chemical analyses, mineralogical determination.*

Pontchour, Miss Cha-on. Chemistry Dept., Chulalongkorn U., Phya Thai Road, Bangkok 10500, Thailand. (1937) BSc, chemistry (Chulalongkorn U., 1959). Asst. prof. (tel. 02 + 252-7019). *Inorganic structures, alloys.*

Pramatus, Miss Supanich. Physics Dept., Chulalongkorn U., Phya Thai Road, Bangkok 10500, Thailand. (1933) MSc, crystallography (U.C., London U., UK, 1968). Assoc. prof. (tel. 02 + 252-9987). *Inorganic structures, powder diffraction.*

Ratanasthien, Dr Benjavun. Geological Sci. Dept., Chiangmai U., Huey Keow Road, Chiangmai 50002, Thailand. (1946) PhD, geochemistry (Aston U., Birmingham, UK, 1975). Assoc. prof. (tel. 053 + 22-1699, ext. 129). *Clays, clay minerals, X-ray and electron diffraction, crystallography.*

Satittada, Miss Gannaga. Physics Dept., King Mongkut's Inst. of Techn. Thonburi, Suksawad Road, Bangkok 10140, Thailand. (1953) MSc, physics (Chulalongkorn U., 1977). Lect. (tel. 02 + 462-5719). *X-ray crystallography, powder diffraction.*

Siriratwatanakul, Mr Narin. Physics Dept., Chiangmai U., Huey Keow Road, Chiangmai 50002, Thailand. (1952) MSc, physics (Chiangmai U., 1981). Asst. prof. (tel. 053 + 22-1934, ext. 51). *Damage in semiconductors.*

Sriratanaprasithi, Mr Khin. Physics Dept., King Mongkut's Inst. of Techn. Thonburi, Suksawad Road, Bangkok 10140, Thailand. (1955) MSc, physics (Chulalongkorn U., 1982). Lect. (tel. 02 + 462-5719). *X-ray crystallography.*

Sukapaddhanadhi, Mr Narong. Eng. Techn. Service, Thai Oil Refinery Co. Ltd., Km 124 1/2 Sukhumvit Road, Au Udom, Sriracha, Chonburi 20210, Thailand.

(1941) MSc, X-ray crystallography (London U., UK, 1970). Sr. Metallurgist. (tel. 038 + 31-1070, ext. 1511). *Metallography, Electron diffraction, dislocations, polycrystalline texture.*

Thanomkul, Dr Srinuan Chaiwasie. Physics Dept. Chulalongkorn U., Phya Thai Road, Bangkok 10500, Thailand. (1936) Dr ing., X-ray crystallography (U. Trondheim, Norway, 1974). Assoc. prof. (tel. 02 + 252-9987). *Organic and inorganic structures.*

Thinapong, Dr Pongchan Chananont. Chemistry Dept., Mahidol U., Rama 6 Road, Bangkok 10400, Thailand. (1948) PhD, chemistry (Birmingham U., UK, 1981). Lect. (tel. 02 + 281-5800, ext. 154). *Structure and activity, biologically active compounds.*

Tontrakoon, Mr Jeerapong. Physics Dept., Chiangmai U., Huey Keow Road, Chiangmai 50002, Thailand. (1950) MSc, physics (Chiangmai U., 1978). Asst. prof. (tel. 053 + 22-1934, ext. 51). *Damage in semiconductors.*

Tooptakong, Mrs Uncharee Methong. Chemistry Dept., Silpakorn U., Nakorn Pathom 73000, Thailand. (1954) MSc, inorganic chemistry (Chulalongkorn U., 1976). Lect. (tel. 034 + 24-2072). *Organic and inorganic structures.*

Tunkasiri, Dr Tawee. Physics Dept., Chiangmai U., Huey Keow Road, Chiangmai 50002, Thailand. (1943) PhD, physics (U. Surrey, UK, 1975). Assoc. prof. (tel. 053 + 22-1934, ext. 51). *Damage in semiconductors.*

Uttamasil, Dr Lek. Materials Sci. Dept., Chulalongkorn U., Phya Thai Road, Bangkok 10500, Thailand. (1943) PhD, ceramic engineering (Ohio State U., USA, 1971). Asst. prof., head. (tel. 02 + 251-1954). *Clay minerals, high temperature materials.*

Wongshaiboon, Dr Sajee. Physics Dept., Chulalongkorn U., Phya Thai Road, Bangkok 10500, Thailand. (1946) PhD, chemistry (Uppsala U., Sweden, 1981). Asst. prof. (tel. 02 + 252-9987). *General crystallography, crystal structure, physical properties.*

TUNISIA

Sub-Editor: M. Ghedira

Notes

1. International telephone country code - 216.

2. The degrees granted and functional positions at Tunisian universities are patterned upon the French academic designations.

Amara, Dr Mongi. Faculté de Pharmacie, Dépt de Chimie, 5000 Monastir, Tunisia. Dr d'Etat (U. Bordeaux, France) Maître de conf. *Structures, phase transitions.*

Ariguib, Dr Najia. Directrice I.N.R.S.T. Borj Cédria, B.P. 95, Hammam-Lif, Tunisia. (1937) DSc, chemistry (U. Paris, France). Prof. *Structures, phase transitions.*

Belhadj, Dr Ali. Dépt. de Physique, Ecole Normale Supérieure, 7029 Bizerte, Tunisia. (1938) DSc, physics (U. Paris, France, 1974). Prof. *Clays, structure.*

Bel Hassen, Miss Dalila. Dépt. de Chimie, Ecole Normale Supérieure, 7029 Bizerte, Tunisia. (1955) Dipl., chemistry (Ecole Normale Supérieure de Tunis). Asst. *Structures, phase transitions.*

Belkheria, Mr Salah. Dépt. de Chimie, Faculté des Sci. et Techniques, 5000 Monastir, Tunisia. MSc (U. Tunis). Asst. *Structures.*

Ben Amor, Dr. Dépt. de Chimie, Ecole Nat. d'Ingénieurs, 6029 Gabès, Tunisia. Dr 3e cycle. Maître-asst.

Ben Brahim, Dr Jemaiel. Dépt. de Physique, Ecole Normale Supérieure, 7029 Bizerte, Tunisia. (1954) Dr 3e cycle, physique (Faculté des Sci. de Tunis). Maître-asst. *Clays, structure.*

Ben Ghozlane, Dr Hédi. Dépt. de Physique, Ecole Nat. d'Ingénieurs de Sfax, Route de la Soukra, B.P.W. 3038 Sfax, Tunisia. (1951) DSc, physique (Faculté des Sci. de Tunis). Maître de conf. *Complexes, structure.*

Ben Romdhane, Dr. Dépt. de Chimie, Ecole Nat. d'Ingénieurs, 6029 Gabès, Tunisia. Dr d'Etat (France). Maître-asst.

Ben Salah, Dr Abdelhamid. Dépt. de Chimie, Ecole Nat. d'Ingénieurs de Sfax, Route de la Soukra, B.P.W. 3038 Sfax, Tunisia. (1950) DSc, physique. Maître de conf. (tel. 216 - 42088, ext. 04). *Structure determination, phase transitions.*

Billiet, Prof. Yves. Dépt. de Physique, Ecole Nat. d'Ingénieurs, Route de la Soukra, B.P.W. 3038 Sfax, Tunisia. (1936) DSc, sciences physiques (U. Paris Sud, France, 1969). Prof. (tel. 216 - 443473). *Symmetry, group theory, theoretical crystallography, phase transition.*

Bizid, Dr Abdelmalek dit Youssef. Dépt. de Physique, Faculté des Sci. de Tunis, Tunis, Tunisia. (1940) DSc, physics (U. Paris VI, France). Prof. *Structures.*

Bouraoui, Prof. Ahmed. Dépt. de Physique, Ecole Normale Supérieure, 7029 Bizerte, Tunisia. (1931) DSc, physiques (U. Paris, France, 1965). *Radiocrystallography, solid state physics.*

Cheikhrouhou, Dr Abdelwaheb. Dépt. de Physique, Ecole Nat. D'Ingénieurs, Route de la Soukra, B.P.W. 3038 Sfax, Tunisia. (1948) Dr d'Etat es Sci., physique (Faculté de Sci. de Tunis). Maître de conf. *Complexes, structure.*

Dabbabi, Dr Mongi. Dépt. de Chimie, Faculté des Sci. et Techniques, 5000 Monastir, Tunisia. Dr d'Etat (U. Paris, France). Prof. *Structure determination.*

Damak, Dr Mabrouk. Dépt. de Physique, Ecole Nat. d'Ingénieurs de Sfax, Route de la Soukra, B.P.W. 3038 Sfax, Tunisia. (1943) Dr Ing (U. Dijon, France). Maître asst. *Structures.*

Daoud, Prof. Abdelaziz. Dépt. de Chimie, Ecole Nat. d'Ingénieurs, Route de la Soukra, B.P.W. 3038 Sfax, Tunisia. (1939) Dr d'Etat, crystallography (U. Dijon, France, 1970). Prof. (tel. 216 - 42088, ext. 04). *Structures, phase transitions.*

Driss, Dr Ahmed. Dépt. de Chimie, Fac. des Sci. de Tunis, Tunis, Tunisia. (1950) Dr 3e cycle, chimie (Fac. des Sci. de Tunis). Maître-asst. *Structure determination.*

Fakhar, Miss Noura. Dépt. de Chimie, Fac. des Sci. de Tunis, Tunis, Tunisia. (1955) MSc, chemistry (Fac. des Sci. de Tunis). Asst. *Structure determination.*

Ghedira, Dr Mounir. Dépt. de Physique, Faculté des Sci. et Techniques, 5000 Monastir, Tunisia. (1951) Dr d'Etat es Sci., physique (U. Sci. et Medicale, Grenoble, France) Maître de conf. (tel. 216 + 361766). *Structure determination, phase transitions, charge localization.*

Halouani, Dr Foued. Dépt. de Chimie, Ecole Nat. d'Ingénieurs de Sfax, Route de la Soukra, B.P.W. 3038 Sfax, Tunisia. Dr d'Etat (U. Caen, France). Maître de conf. *Structures.*

Jouini, Dr Amor. Dépt. de Chimie, Faculté des Sci. et Techniques, 5000 Monastir, Tunisia. Dr 3e cycle (U. Dijon, France). Maître-asst. *Structure determination.*

Jouini, Dr M. Dépt. de Chimie, Ecole Nat. d'Ingénieurs, 6029 Gabès, Tunisia. Dr d'Etat, Maître de conf.

Jouini, Dr Tahar. Dépt. de Chimie, Faculté des Sci. de Tunis, Tunis, Tunisia. (1939) DSc, chemistry (France). Prof. *Structure determination.*

Kallal, Dr Ahmed. Dépt. de Physique, Faculté des Sci. de Tunis, Tunis, Tunisia. (1937) DSc, physics (U. Grenoble, France). Prof. *Structures.*

Maaref, Dr. Saida. Dépt. de Chimie, Ecole Nat. d'Ingénieurs, 6029 Gabès, Tunisia. Dr 3e Cycle. Maître-asst. *Radio-crystallography.*

Mhiri, Dr Tahar. Dépt. de Chimie, Ecole Nat. d'Ingénieurs de Sfax, Route de la Soukra, B.P.W. 3038 Sfax, Tunisia. Dr 3e cycle (U. Paris, France). Maître asst. *Structures.*

Mlik, Dr Youssef. Dépt. de Physique, Ecole Nat. d'Ingénieurs de Sfax, Route de la Soukra, B.P.W. 3038 Sfax, Tunisia. (1950) Dr d'Etat es Sci., physique (Faculé des Sci. de Tunis). Maître de conf. *Complexes, structure.*

Omezzine, Dr Belgacem. Dépt. de Chimie, Faculé des Sci. de Tunis, 1060 Tunis, Tunisia. (1943) Dr 3e cycle, chimie (Ecole Normale Supérieure). Asst. *Structures, phase transitions.*
Omrani, Dr Hédi. Dépt. de Chimie, Faculté des Sci. et Techniques, 5000 Monastir, Tunisia. Dr 3e cycle (U. Besacon, France) Maître-asst. *Structure determination.*
Rzaigui, Dr Mohamed. Dépt. de Chimie, Ecole Normale Supérieure, 7029 Bizerte, Tunisia. (1948) Dr d'Etat, chemistry (France). Maître-asst. *Structures, phase transitions.*
Trabelsi, Dr Malika. Dépt. de Chimie, Ecole Normale Supérieure, 7029 Bizerte, Tunisia. (1947) DSc, chemistry (France). Prof. *Structures, phase transitions.*

TURKEY

Sub-Editor: N. Armağan

Notes

1. International telephone country code - 90.

2. O.D.T.Ü. (Orta Doğu Teknik Üniversitesi) is the Turkish equivalent of M.E.T.U. (Middle East Technical University).

Adıgüzel, Dr Osman. Fizik Böl., Fırat Ü. Fen-Ed. Fak., Elazığ, Turkey. (1952) PhD, physics (Diyarbakır U., 1980). Asst. prof. (tel. 811-1109, ext. 259). *Phase transformations, metals, crystal structure, powder diffraction.*
Aka, Mr Yavuz. ASELSAN, Ankara, Turkey. (1952) PhD, physics (Ankara U., 1977). Res. scient. (tel. 41-157506, ext. 45). *Thick films.*
Akgün, Mr İrfan. Fizik Böl., Gazi Ü. Fen-Ed. Fak., Ankara, Turkey. (1955) PhD, physics (Fırat U., 1981). Asst. prof. (tel. 41-135538). *Phase transformations, metals.*
Aksoy, Mr İlhan. Fizik Böl., İnönü Ü., Fen-Ed. Fak., Malatya, Turkey. (1955) MSc, physics (İnönü U., 1980). Asst. (tel. 821-21871, ext. 258). *Clay minerals, X-ray diffraction.*
Akyüz, Mr Tanıl. MTA., Ankara, Turkey. (1941) MSc, physics (Ankara U., 1984). Res. scient. (tel. 41-234255). *X-ray diffraction, X-ray Fluorescence, Clay minerals.*
Alkan, Mr Atilla. T. Çim. Müs. Bir., P. K. 2, Bakanlıklar, Ankara, Turkey. (1953) MSc, physics (Ankara U., 1977). Asst. prof. (tel. 41-236515, ext. 66). *X-ray microanalysis, SEM, polarization microscopy, cement industry.*
Alp, Mr Esen Ercan. Metalurji Müh. Böl., O.D.T.Ü., Ankara, Turkey. (1954) MSc, metallurgy (M.E.T.U., 1978). Asst. prof. (tel. 41-237100, ext. 2522). *Magnetic susceptibility.*
Alpaut, Prof. Dr Okyay. Fizikokimya Böl., Hacettepe Ü. Müh. Fak., Ankara, Turkey. (1928) Prof., physical metallurgy (Hacettepe U., 1971). Dean. *Physical chemistry, metals.*
Ankara, Prof. Dr Alpay. Metalurji Müh. Böl., O.D.T.Ü., Ankara, Turkey. (1938) PhD, materials science (London U., UK, 1964). Lect. (tel. 41-237100, ext. 2559). *Materials science.*
Arda, Dr Oğuz. Jeoloji Müh. Böl., O.D.T.Ü., İnönü Bulvarı, Ankara, Turkey. (1941) PhD, mineralogy-petrography (Sheffield U., UK, 1975). Asst. prof. (tel. 41-237100, ext. 2679). *Gemstones, zeolites, radioactive minerals, carbonates, minerals, igneous rocks.*
Armağan, Prof. Dr Nizamettin. Fizik Böl., İnönü Ü. Fen-Ed. Fak., Malatya, Turkey. (1942) PhD, physics (Wales U., UK, 1970). Prof. (tel. 821-24088). *Crystal structure, thermal vibrations, powder diffraction, crystallographic programming, clay minerals.*
Artüz, Prof. Dr Samime. Jeoloji Müh. Böl., İstanbul Ü. Yerbilimleri Fak., Vezneciler, İstanbul, Turkey. (1925) Prof., petrography (İstanbul U., 1972). Prof. (tel. 11-207510, ext. 316). *Petrography.*
Atasoy, Prof. Ertuğrul. Metalurji Müh. Böl., Gazi Ü., Ankara, Turkey. (1941) PhD, physical metallurgy (Sheffield U., UK, 1971). Prof. (tel. 237100, ext. 2559). *Nitriding Fe-V alloys.*
Aydın, Dr Nihal. Mineraloji Kürsüsü, Ankara Ü. Fen Fak., Beşevler, Ankara, Turkey. (1950) PhD, petrography (Ankara U., 1980). Lect. (tel. 236550, ext. 0145). *Petrography.*
Aydınol, Dr Mahmut. Fizik Böl., Dicle Ü. Fen-Ed. Fak., Diyarbakır, Ankara, Turkey. (1948) PhD, physics (Dicle U., 1980). Asst. prof. (tel. 831-18740). *X-ray spectroscopic methods.*
Aydınuraz, Doç. Dr Arsin. Fizik Böl., Ankara Ü. Fen Fak., Beşevler, Ankara, Turkey. (1941) Doç., physics (Ankara U., 1976). Lect. (tel. 236550, ext. 0168). *X-ray diffraction analysis.*
Aypar, Doç. Dr Abidin. Fizik Böl., Ankara Ü. Fen Fak., Beşevler, Ankara, Turkey. (1943) Doç., physics (Ankara U., 1980). Lect. (tel. 236550). *X-ray diffraction, electron spin resonance, molecular physics, crystal growth, thermoluminesence.*
Aytaş, Doç. Dr S. Işık. Fizik Böl., Marmara Araş Enst., Gebze, İzmit, Turkey. (1942) PhD, physics (Surrey U., UK, 1971). Res. scient. *Lattice defects.*
Baysal, Prof. Dr Orhan. Yerbilimleri Uygulama ve Araştırma Merkezi, Hacettepe Ü., Ankara, Turkey. (1940) Prof., mineralogy (Hacettepe U., 1978). Lect. (tel. 41-139456). *Mineralogy, clay minerals, X-ray diffraction, mineral deposits.*
Bayvas, Dr Orhan. ANAM, Beşevler, Ankara, Turkey. (1935) PhD, physics (İstanbul U., 1981). Lect. (tel. 41-139456). *Crystallography, nuclear physics, X-ray diffraction, neutron diffraction.*

Buket, Mr Ersen. Yerbilimleri Böl., Hacettepe Ü., Beytepe, Ankara, Turkey. (1944) MSc, geology (M.E.T.U., 1969). Asst. (tel. 41-235168). *Mineralogy, petrography.*
Büyükgüngör, Dr Orhan. Fizik Böl., Ondokuzmayis Ü. Fen-Ed. Fak., Samsun, Turkey. (1954) PhD, physics (Hacettepe U., 1983). Asst. prof. (tel. 361-19680, ext. 656). *Crystallography.*
Cebe, Prof. Dr Mustafa. Kimya Böl., Uludağ Ü. Fen-Ed. Fak., Bursa, Turkey. (1949) Prof., physical chemistry (Uludağ U., 1984). Lect. (tel. 241-14080). *Molecular structure analysis, conformation analysis, hydrogen bonding, computer programs.*
Ceylan, Dr Mehmet. Fizik Böl., Fırat Ü. Fen-Ed. Fak., Elazığ, Turkey. (1950) PhD, physics (Diyarbakır U., 1980). Asst. prof. (tel. 811-11904). *Crystal structure, austenite - martensite phase transformations, twinning, dislocations, metals and alloys, electron microscopy.*
Ceylan, Mr Kazim. Fizik Böl., Gazi Ü. Fen-Ed. Fak., Beşevler, Ankara, Turkey. (1952) PhD, physics (A.D.M.M.A., 1981). Lect. *Phase transformations, metals.*
Çalışkan. Mrs Nezihe. Fizik Böl., İnönü Ü. Fen-Ed. Fak., Malatya, Turkey. (19562) MSc, physics (İnönü U., 1980). Asst. (tel. 821-21871, ext. 274). *Clay minerals, X-ray diffraction.*
Çapan, Dr Z. Ussal. Yerbilimleri Böl., Hacettepe Ü., Müh. Fak., Ankara, Turkey. (1943) PhD, geology (Hacettepe U. 1981). Lect. (tel. 41-235168). *Mineralogy, petrography.*
Çolakoğlu, Doç. Dr Kemal. Fizik Böl., Gazi Ü. Fen-Ed. Fak., Ankara, Turkey. (1947) PhD, physics (Diyarbakır, 1978). Lect.(tel. 41-135538). *Crystallography.*
Danacıözbey, Mrs Süheyla. Fizik Böl., Hacettepe Ü. Müh. Fak., Beytepe, Ankara, Turkey. (1957) MSc, physics (Hacettepe U., 1982). Asst. (tel. 41-230391). *Crystal structure, biological small molecules, powder diffraction.*
Derici, Dr Rifat. Toprak Bilimi Böl., Çukurova Ü. Ziraat Fak., Adana, Turkey. (1945) PhD, chemistry (U. Michigan, USA, 1975). Asst. prof. (tel. 711-12010, ext. 68). *X-ray diffraction instrumentation.*
Dikici, Dr Mustafa. Fizik Böl., İnönü Ü. Fen-Ed. Fak., Malatya, Turkey. (1946) PhD, physics (Fırat U., 1982). Lect. (tel. 4821-21871, ext. 280). *Phase transformations, martensitic transformations, powder diffraction, crystallographic computer programs, clay minerals.*
Dinçer, Dr Muharrem. Fizik Böl., Ondokuzmayis Ü. Fen-Ed. Fak., Samsun, Turkey. (1955) PhD, physics (Fırat U., 1981). Asst. prof. (tel. 361-19680). *Martensitic Phase transformations, alloys, crystal structure, powder diffraction.*
Doğan, Dr Ali. Fizik Böl., Fırat Ü. Fen-Ed. Fak., Elazığ, Turkey. (1952) PhD, physics (Fırat U., 1984). Lect. (tel. 811-11904, ext. 278). *Phase transformations, nucleation and growth, crystal defects, kinetics of transformation, thermodynamics.*
Durlu, Doç. Dr Tahsin Nuri. Fizik Böl., Ankara Ü. Fen Fak., Beşevler, Ankara, Turkey. (1945) D. Phil, physical metallurgy (Oxford U., UK, 1974). Lect. (tel. 41-236550, ext. 0147). *Phase transformations, metals.*
Ekmekçi, Dr Servet. Fizik Böl., İnönü Ü. Fen-Ed. Fak., Malatya, Turkey. (1952) PhD, physics (Catholic U., USA, 1981). Asst. prof. (tel. 821-21871, ext. 289). *Phase transformations, metamagnets, clay minerals.*
Elerman, Dr Yalçin. Fizik Böl., Ankara Ü. Fen Fak., Beşevler, Ankara, Turkey. (1951) PhD, physics (Ankara U., 1978). Asst. prof. (tel. 41-232105, ext. 17). *Crystallography.*
Erdoğmuş, Dr Muktim. Fizik Böl., Fırat Ü. Fen-Ed. Fak., Elazığ, Turkey. (1939) PhD, physics (Ankara U., 1963). Lect. (tel. 811-11904, ext. 383). *Nuclear physics, thermoluminescence method, radiation dosimetry, crystal vibrations, crystal structures.*
Erdönmez, Dr Ahmet. Fizik Böl., Ondokuzmayis Ü. Fen-Ed. Fak., Samsun, Turkey. (1950) PhD, physics (Hacettepe U., 1980). Lect. 361-19747). *Crystal structure, powder diffraction, crystallographic computer programs.*
Ergin, Dr Ömer. Fizik Böl., Atatürk Ü. Fen-Ed. Fak., Erzurum, Turkey. (1949) PhD, physics (Atatürk U., 1982). Asst. prof. (tel. 011-14120, ext. 855). *Structure determination.*

Gedikoğlu, Doç. Dr Adil. Fizik Böl., Ankara Ü. Fen Fak., Beşevler, Ankara, Turkey. (1945) PhD, physics (U. of Hamburg, BRD, 1973). Lect. (tel. 236550, ext. 0154). *Mössbauer effect.*

Gündoğdu, Dr Niyazi. Jeoloji Böl., Hacettepe Ü., Müh. Fak., Ankara, Turkey. (1951) PhD, geology (Hacettepe U., 1982). Lect. (tel. 41-236730, ext. 1586). *Mineralogy, petrography.*

Güzel, Doç. Dr Nuri. Toprak Bilimi Böl., Çukurova Ü., Ziraat Fak., Adana, Turkey. (1936) Doç., clay mineralogy (Çukurova U., 1975). Lect. (tel. 711-12010, ext. 68). *X-ray diffraction.*

Göğüş, Mrs Gülderen. T. Çim. Müs. Bir., P. K. 2, Bakanliklar, Ankara, Turkey. (1954) MSc, physics (Hacettepe U., 1979). Res. scient. (tel. 41-236515, ext. 68). *Differential thermal analysis, cement, crystal structure.*

Hökelek, Mr Tuncer, Fizik Böl., Hacettepe Ü. Müh. Fak., Beytepe, Ankara, Turkey. (1957) MSc, physics (Hacettepe U., 1980). Asst. (tel. 41-230391). *Crystal structure, organometallic compounds, powder diffraction, thin films.*

İşçi, Doç. Dr Coşkun. Fizik Böl., Ege Ü. Fen Fak., İzmir, Turkey. (1950) Doç., physics (Ege U., 1983). Lect. (tel. 180110, ext. 2383). *Crystal structure analysis, phase transitions.*

Kapur, Dr Selim. Toprak Bilimi Böl., Çukurova Ü., Ziraat Fak., Adana, Turkey. (1946) PhD, mineralogy (Aberdeen U., UK, 1976). Asst. prof. (tel. 711-12010). *Soil mineralogy.*

Kaynak, Dr Uğur. Jeofizik Böl., Yildiz Ü., Müh. Fak., İzmit, Turkey. (1939) PhD, physics (İstanbul U., 1977). Lect. *Induced polarization, mössbauer spectroscopy, phase transformations, crystal structures.*

Kendi, Dr Engin. Fizik Böl., Hacettepe Ü., Müh. Fak., Ankara, Turkey. (1945) PhD, physics (Hacettepe U., 1974). Assoc. prof. (tel. 41-230391). *Crystal structure, biological small molecules, powder diffraction.*

Kizilyalli Doç. Dr Meral. Kimya Böl., O.D.T.Ü. Fen-Ed. Fak. Ankara, Turkey. (1935) PhD, chemistry (London U., UK, 1973). Lect. (tel. 41-237100, ext. 3208). *Solid state reactions, materials preparation, powder diffraction, crystal structure determination.*

Konocak, Doç. Zeki. Fizik Böl., Firat Ü. Fen-Ed. Fak., Elaziğ, Turkey. (1937) Doç., physics (E.D.M.M.A., 1976). Lect. (tel. 811-11904) *Crystal structure, crystal growth.*

Kökçe, Dr Ali. Fizik Böl., Firat Ü. Fen-Ed. Fak., Elaziğ, Turkey. (1953) PhD, physics (Firat U., 1982). Lect. (tel. 811-11904, ext. 278). *Phase transformations, nucleation and growth, crystal defects*

Kumbasar, Prof. Işik. Mineraloji ve Maden Yataklari Kürsüsü, İstanbul Tek. Ü., Maçka, İstanbul, Turkey. (1934) Prof., mineralogy (İstanbul Tek. U., 1977). Lect. (tel. 11-433100, ext. 673). *Mineralogy.*

Munsuz, Prof Nuri. Toprak Bilimleri Böl., Ankara Ü., Ziraat Fak., Ankara, Turkey. (1933) Prof, mineralogy (Ankara U., 1972). Lect. (tel. 41-162109, ext. 58). *Clay mineralogy.*

Oğurtani, Prof. Dr Tarik Ömer. Metalurji Müh. Böl., O.D.T.Ü., Ankara, Turkey. (1934) PhD, materials science (Stanford U., USA, 1964). Prof. (tel. 41-237100, ext. 2558). *Crystal defects.*

Oktik, Dr Şener. Fizik, Böl., Selçuk Ü. Fen-Ed. Fak., Konya, Turkey. (1955) PhD, physics (Durham U., UK, 1982). Lect. (tel. 331-20461). *Crystal growth, X-ray diffraction, electron microscopy, structure determination.*

Özenbaş, Mr A. Macit. Metalurji Müh. Böl., O.D.T.Ü., Ankara, Turkey. (1951) MSc, metallurgy (M.E.T.U., 1975). Inst. (tel. 41-237100, ext. 2523). *Materials science, internal friction.*

Özkan, Doç. Dr Hüsnü. Fizik Böl., O.D.T.Ü. Fen-Ed. Fak., Ankara, Turkey. (1944) Doç., physics (M.E.T.U., 1981). Lect. (tel. 41-237100, ext. 3279). *Phase transitions, elastic properties, radiation damage, high pressure diffraction.*

Özkaplan, Dr Habip. Fizik Böl., Ondokuzmayis Ü. Fen-Ed. Fak., Samsun, Turkey. (1945) PhD, physics (U. Louisville, USA, 1978). Lect. (tel. 361-19680, ext. 657). *Semiconductors, crystal structure, thin films.*

Öztunali, Prof. Dr Önder. Jeoloji Müh. Böl., İstanbul Ü. Yerbilimleri Fak., Vezneciler, İstanbul, Turkey. (1935) Prof., mineralogy (İstanbul U., 1977). Prof. (tel. 11-281639). *Crystallography.*

Polat, Mr Hamza. Fizik Böl., İnönü Ü. Fen-Ed. Fak., Malatya, Turkey. (1955) MSc, physics (Hacettepe U., 1982). Asst. (tel. 821-21871, ext. 290). *Crystal structure, X-ray diffraction.*

Salanci, Prof. Dr Berkin. Maden Böl., Hacettepe Ü. Müh. Fak., Ankara, Turkey. (1939) Prof., mineralogy (Hacettepe U., 1980). Lect. (tel. 41-236730, ext. 1570). *Mineral deposits, mineralogy.*

Sayin, Doç. Dr Mahmut. Toprak Böl., Çukurova Ü. Ziraat Fak., Adana, Turkey. (1946) Doç., mineralogy (Çukurova U., 1983). Lect. (tel. 711-10810, ext. 2214). *Clay minerals, powder diffraction, infrared spectroscopy, mineralogy.*

Sencer, Mr Osman. Fizyopatoloji Kürsüsü, Elektron Mikroskopi Merkezi, Ankara Ü. Tip Fak., Sihhiye, Ankara, Turkey. (1948) MSc, physics (Ankara U., 1973). Res. scient. (tel. 41-244120). *X-ray microanalysis, electron microscopy.*

Sonaer, Mr Kenan. Laboratuvarlar Şubesi, M.T.A., Ankara, Turkey. (1944) MSc, physics (Ankara U., 1971). Res. scient. (tel. 41-234255, ext. 731). *Clay analysis, X-ray diffraction.*

Soylu, Doç. Dr Hüseyin. Fizik Enst., Hacettepe Ü. Müh. Fak., Ankara, Turkey. (1933) Doç., physics (Hacettepe U., 1979). Lect. (tel. 41-230391). *Crystal structure analysis.*

Şahin, Mr Mehmet. T. Çim. Müs. Bir., P. K. 2, Bakanliklar, Ankara, Turkey. (1954) MSc, physics (Hacettepe U., 1980). Res. scient. (tel. 41-236515). *Crystal structure, scanning electron microscopy, clinker minerals, polarization microscopy, cement, X-ray diffraction.*

Taner, Dr Akin. Mineraloji Kürsüsü, Ankara Ü. Fen Fak., Beşevler, Ankara, Turkey. (1937) PhD, mineralogy (Ankara U., 1968). Asst. prof. (tel. 41-236550, ext. 0140). *Optical mineralogy.*

Tarimci, Dr Çelik. ANAM. Beşevler, Ankara, Turkey. (1945) PhD, crystallography (U. Pittsburgh, USA, 1975). Res. scient. (tel. 41-233208). *Molecular structure analysis.*

Tokay, Dr Nesrin. Fizikokimya Böl., Hacettepe Ü. Kimya Fak., Ankara, Turkey. (1951) PhD, physics (Hacettepe U., 1980). Asst. prof. *Physical metallurgy.*

Tonak, Miss Tulin. T. Çim. Müs. Bir., P. K. 2, Bakanliklar, Ankara, Turkey. (1951) MSc, chemistry (Ankara U., 1983). Res. scient. (tel. 41-236515, ext. 64). *X-ray diffraction, differential thermal analysis, cement.*

Tunç, Dr Cemil. Fizik Böl., Karadeniz Ü., Fen-Ed. Fak., Trabzon, Turkey. (1940) PhD, physics (Karadeniz U., 1982). Lect. (tel. 31-16920, ext. 2554). *crystal growth, crystal defects, crystallography.*

Turan, Mrs Canan. T. Çim. Müs. Bir., P. K. 2, Bakanliklar, Ankara, Turkey. (1958) BSc, geology (M.E.T.U., 1981). Res. scient. (tel. 41-236515, ext. 66). *Polarization microscopy, scanning electron microscopy, X-ray diffractometry, cement.*

Unan, Doç. Dr Coşkun. Jeoloji Müh. Böl., O.D.T.Ü., İönü Bulvari, Ankara, Turkey. (1936) Doç., geology (M.E.T.U., 1976). Lect. (tel. 41-237100, ext. 2678). *Mineralogy, ore microscopy, statistical petrology, geochemistry, instrumentation.*

Usanmaz, Doç. Dr Ali. Kimya Böl., O.D.T.Ü. Fen-Ed. Fak., Ankara, Turkey. (1945) PhD, chemistry (PINY, USA, 1974). Assoc. prof. (tel. 41-237100, ext. 3225). *Structure determination.*

Uygur, Prof. Dr M. Eti. Metalurji Müh. Böl., Gazi Ü., Ankara, Turkey. (1941) PhD, metallurgy (M.E.T.U., 1971). Prof. (tel. 41-135538). *Powder metallurgy, physical metallurgy, X-ray diffraction.*

Ülkü, Prof. Dr Dinçer. Fizik Böl., Hacettepe Ü., Müh. Fak., Ankara, Turkey. (1940) Dr. rer. nat., crystallography (U. München, BRD, 1965). Prof. (tel. 41-230391). *Crystal structure analysis.*

Ünak, Doç. Dr Turan. Fizik Böl., Ege Ü. Fen Fak., İzmir, Turkey. (1945) PhD, chemistry (Louis Pasteur U., France, 1971). Lect. (tel. 51-180110, ext. 2920). *X-ray radiation, Auger effect, photoelectric excitation.*

Vardar, Mr Bülent. T. Çim. Müs. Bir., P. K. 2, Bakanliklar, Ankara, Turkey. (1949) MSc, chemistry (Hacettepe U., 1971). Res. scient. (tel. 41-236515, ext. 65). *X-ray quantometry, quantitative analysis, cement.*

Yağbasan, Dr Rahmi. Fizik Böl., İnönü Ü. Fen-Ed. Fak., Malatya, Turkey. (1949) PhD, physics (Hacettepe U., 1980). Lect. (tel. 821-21871, ext. 276). *Crystal structure, powder diffraction, clay minerals.*

Yağci, Dr Osman. Biofizik Bilim Dali, Akdeniz U. Tip Fak., Antalya, Turkey. (1942) PhD, physics (Akdeniz U., 1984). Lect. (tel. 3111-15995). *Soft X-ray absorption, emission spectroscopy.*

Yeşilsoy, Prof. Dr M. Sefik. Toprak Bilimi Böl., Çukurova Ü., Ziraat Fak., Adana, Turkey. (1932) Prof., mineralogy (Çukurova U., 1977). Prof. (tel. 711-12010, ext. 68). *X-ray diffraction.*

Yilmaz, Doç. Dr Osman. Jeoloji Böl., Hacettepe Ü. Müh. Fak., Ankara, Turkey. (1943) Doç. petrology (Hacettepe U., 1980). Lect. (tel. 41-235168, ext. 1586). *Petrography, mineralogy.*

Yücel, Dr Atila. ASELSAN., Ankara, Turkey. (1949) PhD, physics (Ankara U., 1978). Res. scient. (tel. 41-157506). *Thick films.*

Yüksel, Mr Hikmet. M.T.A., Ankara, Turkey. (1950) MSc, physics (Ankara U., 1973). Res. scient. (tel. 41-234255, ext. 731). *Clay mineral analysis.*

UNION OF SOVIET SOCIALIST REPUBLICS

Sub-Editor: E. N. Belova

Notes

1. International telephone country code - 7.

2. The degree *doctor* (Dr) and *candidate* (Cand.) are approximately equivalent to DSc and PhD, respectively.

3. Degrees can be conferred by the Universities, by Institutes belonging to the Academy of Sciences of the USSR (abbreviated as Acad. Sci. USSR) or belonging to Academies of Sciences of the individual Soviet Socialist Republics, and by other Scientific Institutes. The following abbreviations are used to indicate where the degrees were obtained: GORNY - Leningrad Mining Institute; INCRYS - Institute of Crystallography, Academy of Sciences of the USSR; INEOS - Institute of Elemento-Organic Compounds, Academy of Sciences of the USSR; IONCH - Institute of General and Inorganic Chemistry, Academy of Sciences of the USSR; IMGRE - Institute of Mineralogy, Geochemistry, Crystal Chemistry of Rare Elements, Academy of Sciences of the USSR; IGEM - Institute of Geology, Mineralogy and Petrography, Academy of Sciences of the USSR; KARPOV - Karpov Physical Chemistry Institute; LGU - Leningrad State University; MGU - Moscow State University.

4. It should be remarked that more than one transliteration of names, from the non-Latin characters used in the USSR, is often possible. For example, names which in this Edition (as well as in the previous editions) have 'y' at the end may alternatively terminate in 'ii' as in the English translation of the journal *Kristallografiya*.

Abdullaev, Dr Abdulkhamid Aliyevich. Inst. of Crystallography, Acad. of Sci. of the USSR, Leninsky pr. 59, Moscow 117333, USSR. (1936) Cand, physics and mathematics (INCRYS, 1971). Sr. scient. *Crystal growth, optical properties of crystals*

Abdullaev, Dr Gusi Kara ogly. Inst. of Inorganic and Physical Chemistry, Acad. of Sci. of the Azerbaidzhan SSR, Narimanov Pr. 29, Baku 370143, USSR. (1929) Cand, geology and mineralogy (Azizbekov Industrial Inst., 1957). Sr. scient. *X-ray structure analysis, crystal chemistry, borates.*

Afanasyev, Dr Igor' Ivanovich. Leningrad Mining Inst., 21st Liniya 2, Leningrad 199026, USSR. (1935) Cand, geology and mineralogy (GORNY, 1966). Sr. scient. *Crystal growth, crystallography, crystal physics.*

Afonina, Dr Nataliya Nikolayevna. Moscow State U., Dept. of Chemistry, Leninskiye Gory, Moscow 117234, USSR. (1945) Cand, chemistry (MGU, 1973). Scient. *Crystal chemistry.*

Agre, Dr Valeriya Moiseyevna. Inst. of Chemical Reagents and Pure Substances, Bogorodsky Val 3, Moscow 107258, USSR. (1931) Cand, chemistry (IONCH, 1969). Lab. head. *Crystal chemistry, X-ray structure analysis, complex and inorganic compounds.*

Akchurin, Dr Marat Shikhapovich. Inst. of Crystallography, Acad. of Sci. of the USSR, Leninsky pr. 59, Moscow 117333, USSR. (1947) Cand, physics and mathematics (INCRYS, 1983). Scient. *Real structure, mechanical properties.*

Akhmetov, Dr Spartak Fatykhovich. Res. Inst. for Synthesis of Mineral Raw Materials, Institutskaya St. 1, Alexandrov 601600, Vladimirskaya Oblast', USSR. (1938) Cand, geology and mineralogy (Kazakh Polytechnic Inst., 1965). Sr. scient. *Crystal growth, crystal optics, X-ray structure analysis.*

Akhmetova, Dr Galina Leonidovna. Res. Inst. for Synthesis of Mineral Raw Materials, Institutskaya St. 1, Alexandrov 601600, Vladimirskaya Oblast', USSR. (1937) Cand, technics (Inst. of Metallurgy, Acad. Sci. Kazakh SSR, 1966). Sr. scient. *Synthesis of single crystals, crystall-optical methods, X-ray diffraction methods, physical chemistry, melts.*

Akselrud, Dr Lev Grigoryevich. Dept. of Chemistry, Lvov State U., University St. 1, Lvov 290602, USSR. (1948) Cand, chemistry (Lvov U., 1980). Eng. *X-ray structure analysis, crystal chemistry, intermetallic compounds.*

Al'shits, Dr Vladimir Iosifovich. Inst. of Crystallography, Acad. of Sci. of the USSR, Leninsky pr. 59, Moscow 117333, USSR. (1941) Dr, physics and mathematics (INCRYS, 1977). Sr. scient. *Dislocation theory.*

Alaverdova, Dr Olga Georgiyevna. Kharkov Polytechnic Inst., Frunze St. 21, Kharkov 310002, USSR. (1938) Cand, technics (Kharkov Polytechnic Inst., 1975). Docent. *Metals and alloys, structure, crystal structure defects, methods for defect measurement.*

Aldoshin, Dr Sergey Mikhailovich. Branch Inst. of Chemical Physics, Acad. Sci. of the USSR, Chernogolovka 142432, Noginsky Rayon, Moskovskaya Oblast', USSR. (1953) Cand, chemistry (IONCH, 1977). Sci. *Structural photochemistry, superconductors, structure.*

Aleshko-Ozhevsky, Dr Oleg Pavlovich. Inst. of Crystallography, Acad. of Sci. of the USSR, Leninsky pr. 59, Moscow 117333, USSR. (1934) Cand, physics and mathematics (INCRYS, 1969). Sr. scient. *Real structure of crystals.*

Alexandrov, Prof. Alexander Danilovich. Mathematical Inst., Acad. of Sci. of the USSR, Siberian Dept., Novosibirsk 630090, USSR. (1912) Full member of the Acad. Sci. USSR; Dr, physics and mathematics (LGU, 1936). *Geometrical crystallography.*

Alexandrov, Prof. Kirill Sergeyevich. Inst. of Physics, Acad. of Sci. of the USSR, Siberian Dept., Akademgorodok, Krasnoyarsk 660036, USSR. (1931) Full member of the Acad. Sci. USSR; Dr, physics and mathematics (INCRYS, 1967). Dir. *Crystal physics, phase transitions in crystals, ferroelectricity.*

Alexandrov, Prof. Leonid Naumovich. Inst. of Semiconductor Physics, Acad. of Sci. of the USSR, Siberian Dept., Prospekt Nauki 13, Novosibirsk 630090, USSR. (1923) Dr, physics and mathematics (1965) Lab. head. *Crystal growth, nucleation mechanisms, thin film growth, epitaxy.*

Alexandrov, Dr Vladimir Borisovich. Inst. of Mineralogy, Geochemistry and Crystal Chemistry of Rare Elements, Sadovnicheskaya Emb. 71, Moscow 113127, USSR. (1931), Cand, geology and mineralogy (MGU, 1964). Sr. scient. *X-ray structure analysis, crystal chemistry, minerals.*

Alexandrova, Dr Inga Petrovna. Inst. of Physics, Acad. of Sci. of the USSR, Siberian Dept., Akademgorodok, Krasnoyarsk 660036, USSR. (1934) Cand, physics and mathematics (Inst. of Physics, Acad. Sci. USSR, Siberian Dept., 1972). Sr. scient. *Ferroelectricity, phase transitions.*

Aliev, Dr Fazil Isa ogly. Inst. of Physics, Acad. of Sci. of the Azerbaidzhan SSR, Narimanov Pr. 33, Baku 370143, USSR. (1937) Cand, physics and mathematics (Inst. of Physics, Acad. Sci. Azerbaidzhan SSR, 1969). Sr. scient. *Amorphous films, crystallization, structure.*

Aliev, Dr Zainutdin Gasanovich. Branch Inst. of Chemical Physics, Acad. of Sci. of the USSR, Chernogolovka 142432, Noginsky Rayon, Moskovskaya Oblast', USSR. (1939) Cand, physics and mathematics (Inst. of Chemical Physics, Acad. Sci. USSR, 1974). Scient. *Transition metal complex compounds, structure.*

Alikhanov, Dr Ruben Abramovich. Inst. of High Pressure Physics, Acad. of Sci. of the USSR, Akademgorodok, Podol'sky Rayon, Moskovskaya Oblast' 142092, USSR. Cand, physics and mathematics (Inst. of Physical Problems, Acad. Sci. USSR, 1959). Sr. scient. *High pressure, low temperature, structure and dynamics, magnetic crystals, molecular crystals.*

Alyavdin, Dr Vladimir Fedorovich. Leningrad Mining Inst., 21st Liniya 2, Leningrad 199026, USSR. (1913) Cand, geology and mineralogy (Inst. of Geology, Acad. Sci. USSR, 1947). Docent. *Crystal morphology, goniometry.*

Amirov, Dr Savalan Teimur ogly. Inst. of Inorganic and Physical Chemistry, Acad. of Sci. of the Azerbaidzhan SSR, Narimanov Prospekt 29, Baku 370143, USSR. (1939) Cand, chemistry (INCRYS, 1968). Sr. scient. *X-ray structure analysis, crystal chemistry, silicates.*

Andreeva, Dr Nataliya Sergeyevna. Inst. of Molecular Biology, Acad. of Sci. of the USSR, Vavilov St. 32, Moscow 117312, USSR. Dr, physics and mathematics (INCRYS, 1970). Lab. head. *Protein crystallography, fibrillous structures.*

Andrianov, Dr Valery Ivanovich. Inst. of Crystallography, Acad. of Sci. of the USSR, Leninsky pr. 59, Moscow 117333, USSR. (1938) Cand, physics and mathematics (INCRYS, 1969). Sr. scient. *Computing methods in structure analysis.*

Anikin, Dr Igor' Nikolayevich. Moscow State U., Dept. of Geology, Leninskiye Gory, Moscow 117234, USSR. (1929) Cand, geology and mineralogy (MGU, 1955). Sr. scient. *Growth and formation of crystals, instrumentation, physical chemistry, melts.*

Anistratov, Dr Anatoly Tikhonovich. Inst. of Physics, Acad. of Sci. of the USSR, Siberian Dept., Akademgorodok, Krasnoyarsk 660036, USSR. (1935) Cand, physics and mathematics (Krasnoyarsk State Pedagogic Inst., 1967). Sr. scient. *Crystal optics, phase transitions.*

Antipin, Dr Mikhail Yuvenaliyevich. Inst. of Elemento-Organic Compounds, Acad. of Sci. of the USSR, Vavilov St. 28, Moscow 117813, USSR. (1951) Cand, chemistry (INEOS, 1980). Scient. *Organic and elemento-organic compounds (crystal structure), low temperature X-ray structure analysis.*

Antonov, Dr Petr Iosifovich. Physico-Techn. Inst., Acad. of Sci. of the USSR, Politekhnicheskaya St. 26, Leningrad 194021, USSR. (1935) Cand, physics and mathematics (Physico-Techn. Inst., Acad. Sci. USSR, 1968). Sr. scient. *Crystal growth, growth from melts by Stepanov method, crystal structure and properties.*

Antsishkina, Dr Alla Sergeyevna. Inst. of General and Inorganic Chemistry, Acad. of Sci. of the USSR, Leninsky pr. 31, Moscow 117071, USSR. (1926) cand, chemistry (IONCH, 1959). Sr. scient. *Crystal chemistry, stereochemistry, coordination compounds.*

Apinitis, Dr Smuidris Karlovich. Riga Polytechnic Inst., Boulevard Kronvalda 4, Riga 226828, USSR. (1933) Cand, chemistry (Riga Polytechic Inst., 1970). Docent. *X-ray structure analysis.*

Arkhipenko, Dr Diana Konstantinovna. Inst. of Geology and Geophysics, Acad. of Sci. of the USSR, Siberian Dept., Prospekt Nauki 3, Novosibirsk 630090, USSR. (1928) Dr, physics and mathematics (INCRYS, 1963). Lab. head. *Mineral structures, X-ray diffraction methods, infrared spectroscopy.*

Arzumanyan, Dr Gennadiy Ashotovich. Inst. of Crystallography, Acad. of Sci. of the USSR, Leninsky pr. 59, Moscow 117333, USSR. (1953) Cand, chemistry (Moscow Chem. Techn. Inst., 1980). Scient. *Physical chemistry, high temperature oxide crystallization processes, mass spectrometry.*

Asadchikov, Dr Viktor Evgen'yevich. Inst. of Crystallography, Acad. of Sci. of the USSR, Leninsky pr. 59, Moscow 117333, USSR. (1948) Cand, physics and mathematics (INCRYS, 1982). Scient. *X-ray small angle scattering, apparatus, experimental methods.*

Asadov, Dr Yusif Gazanfar ogly. Inst. of Physics, Acad. of Sci. of the Azerbaidzhan SSR, Narimanov Prospekt 33, Baku 370143, USSR. (1934) Cand, physics and mathematics (Azerbaidzhan State U., 1964). Lab. head. *Crystal growth, phase transformations.*

Ashirov, Dr Aman. Physical Technical Inst., Acad. of Sci. of Turkmen SSR, Golgol St. 15, Ashkhabad, USSR. (1935) Cand, physics and mathematics (INCRYS, 1963). Lab. head. *X-ray structure analysis.*

Askhabov, Dr Askhab Magomedovich. Inst. of Geology, Acad. of Sci. of the USSR, Komi Department, Kommunisticheskaya 28, Syktyvkar 167000, USSR. (1948) Cand, geology and mineralogy (GORNY, 1977). Lab. head. *Crystallography.*

Aslanov, Prof. Leonid Alexandrovich. Moscow State U., Dept. of Chemistry, Leninskiye Gory, Moscow 117234, USSR. (1938) Dr, chemistry (MGU, 1973). Prof. *X-ray structure analysis, crystal chemistry.*

Atabayeva, Dr Eleonora Yakubovna. Inst. of High Pressure Physics, Acad. of Sci. of the USSR, Akademgorodok, Podol'sky Rayon, Moskovskaya Oblast' 142092, USSR. Cand, physics and mathematics (Inst. of Earth Physics, 1975). Scient. *High pressure effect on crystal structure.*

Atovmyan, Prof. Lev Oganovich. Branch Inst. of Chemical Physics, Acad. of Sci. of the USSR, Chernogolovka 142432, Noginsky Rayon, Moskovskaya Oblast', USSR. (1928) Dr, chemistry (IONCH, 1971). Lab. head. *Crystal chemistry, coordination compounds.*

Avdiyenko, Dr Klavdiya Ilyinishna. Inst. of Semiconductor Physics, Acad. of Sci. of the USSR, Siberian Dept., Prospekt Nauki 13, Novosibirsk 630090, USSR. Cand, physics and mathematics (Inst. of Semiconductor Physics, Acad. Sci. USSR, Siberian Dept., 1970). Sr. scient. *Crystal growth, structure - physical properties relationship.*

Avilov, Dr Anatoly Sergeyevich. Inst. of Crystallography, Acad. of Sci. of the USSR, Leninsky pr. 59, Moscow 117333, USSR. (1943) Cand, physics and mathematics (INCRYS, 1973). Sr. scient. *Electron diffraction, structure analysis, crystal chemistry.*

Babareko, Dr Alesya Adamovna. Inst. of Metallurgy, Acad. of Sci. of the USSR, Leninsky Prospekt 49, Moscow 117334, USSR. (1928) Cand, techn. (Inst. of Metallurgy, 1978). Sr. scient. *Plastic deformation, crystal growth.*

Bagdasarov, Dr Khachik Saakovich. Inst. of Crystallography, Acad. of Sci. of the USSR, Leninsky pr. 59, Moscow 117333, USSR. (1929) Dr, physics and mathematics (INCRYS, 1972). Lab. head. *Crystal growth at high temperatures.*

Bakakin, Dr Vladimir Vasilyevich. Inst. of Inorganic Chemistry, Acad. of Sci. of the USSR, Siberian Dept., Prospekt Nauki 3, Novosibirsk 630090, USSR. (1933) Dr, geology and mineralogy (MGU, 1963). Lab. head. *Crystal chemistry, inorganic compounds, minerals, teaching crystallography.*

Balagurov, Mr Anatoly Mikhailovich. Joint Inst. for Nuclear Res., Dubna 141980, Moskovskaya Oblast', USSR. (1945). Scient. *Neutron physics.*

Balakirev, Dr Vladimir Georgiyevich. Res. Inst. for Synthesis of Mineral Raw Materials, Institutskaya St. 1, Alexandrov 601600, Vladimirskaya Oblast', USSR. Cand, geology and mineralogy (IGEM, 1977). Sr. scient. *Solid state physics, defects in crystals.*

Balitsky, Dr Vladimir Sergeyevich. Res. Inst. for Synthesis of Mineral Raw Materials, Institutskaya St. 1, Alexandrov 601600, Vladimirskaya Oblast', USSR. (1932) Dr, geology and mineralogy (IGEM, 1971). Dept. head. *Crystal growth in hydrothermal systems.*

Barabanov, Prof. Vladimir Fedorovich. Leningrad State U., University Emb. 7/9, Leningrad 199164, USSR. (1918) Dr, geology and mineralogy (LGU, 1961). Head of Chair. *Crystallography, morphology, typomorphism, solid state physics.*

Baranov, Dr Anatoliy Ivanovich. Inst. of Crystallography, Acad. of Sci. of the USSR, Leninsky pr. 59, Moscow 117333, USSR. (1947) Cand, physics and mathematics (INCRYS, 1973). Sr. scient. *Structural phase transitions, ferroelectricity, superionic conductivity.*

Baranova, Dr Raisa Vladimirovna. Inst. of Crystallography, Acad. of Sci. of the USSR, Leninsky pr. 59, Moscow 117333, USSR. (1936) Cand, physics and mathematics (INCRYS, 1969). Scient. *Electron diffraction, crystal chemistry, structure analysis.*

Barsukova, Dr Marina Leonidovna. Inst. of Crystallography, Acad. of Sci. of the USSR, Leninsky Prospekt 59, Moscow 117333, USSR. (1943) Cand, chemistry (INCRYS, 1980). Scient. *Crystal growth.*

Bartoshinsky, Dr Zbignev Vladislavovich. Lvov State U., Dept. of Mineralogy, Shcherbakova St. 4, Lvov 290005, USSR. (1929) Cand, geology and mineralogy (Lvov U., 1962). Docent. *Crystal morphology, minerals, diamond crystallography and mineralogy.*

Bataliyeva, Dr Nataliya Glebovna. Inst. of Mineralogy, Geochemistry and Crystal Chemistry of Rare Elements, Sadovnicheskaya Emb. 71, Moscow 113127, USSR. (1931) Cand, geology and mineralogy (MGU, 1971). Scient. *Crystal chemistry, minerals, synthetic compounds, rare earth compounds.*

Batsanov, Dr Andrey Stepanovich. Inst. of Elemento-Organic Compounds, Acad. of Sci. of the USSR, Vavilova 28, Moscow 117813, USSR. (1955) Cand, chemistry (INEOS, 1983). Scient. *X-ray structure analysis, structural chemistry, organometallic compounds, coordination compounds, bioactive compounds.*

Bekrenev, Dr Anatoly Nikolayevich. Kuibyshev Polytechnic Inst., Pervomaiskaya St. 18, Kuibyshev 443002, USSR. (1944) Dr, physics and mathematics (Kharkov U., 1971). Prof. *Crystal structure, deformed crystals.*

Bel'sky, Dr Vitaly Konstantinovich. Karpov Physical Chemistry Inst., Obukha St. 10, Moscow 107120, USSR. (1943) Cand, chemistry (MGU, 1969) Sr. scient. *Crystal chemistry, organic compounds, symmetry, X-ray structure analysis.*

Belikova, Dr Galina Sergeyevna. Inst. of Crystallography, Acad. of Sci. of the USSR, Leninsky pr. 59, Moscow 117333, USSR. (1928) Cand, chemistry (INCRYS, 1968). Sr. scient. *Organic crystals, growth.*

Belokoneva, Dr Elena Leonidovna. Moscow State U., Dept. of Geology, Leninskiye Gory, Moscow 117234, USSR. Cand, geology and mineralogy (MGU, 1975). Sci. *Crystal chemistry, inorganic compounds, X-ray structure analysis.*

Belova, Dr Elizaveta Nikolayevna. Inst. of Crystallography, Acad. of Sci. of the USSR, Leninsky pr. 59, Moscow 117333, USSR. Cand, physics and mathematics (INCRYS, 1949). Sr. scient. *Structure analysis.*

Belugina, Dr Nataliya Vasilyevna. Inst. of Crystallography, Acad. of Sci. of the USSR, Leninsky pr. 59, Moscow 117333, USSR. (1941) Cand, physics and mathematics (INCRYS, 1978). Scient. *Real structure of crystals, crystal growth - real structure relations.*

Belyaeva, Mrs Klara Fedorovna. Inst. of Applied Physics, Acad. of Sci. of the Moldavian SSR, Akademicheskaya 5, Kishinev 277028, USSR. (1933). Scient. *X-ray structure analysis methods, coordination compounds.*

Belyustin, Dr Aleksey Vsevolodovich. Gorky State U., Gagarin Pr. 23, Gorky 603022, USSR. (1913) Cand, physics and mathematics (Gorky U., 1945). Docent. *Crystal growth from solutions.*

Bendeliani, Dr Nikolay Alexandrovich. Inst. of High Pressure Physics, Acad. of Sci. of the USSR, Akademgorodok, Podol'sky Rayon, Moskovskaya Oblast' 142092, USSR. Cand, geology and mineralogy (MGU, 1967). Sr. scient. *High pressure effect on crystal structure.*

Beresnev, Dr Leonid Alekseyevich. Inst. of Crystallography, Acad. of Sci. of the USSR, Leninsky pr. 59, Moscow 117333, USSR. (1947) Cand, physics and mathematics (Inst. of Solid State Physics, Acad. Sci. USSR, 1979). Sr. scient. *Structure and properties, ferroelectric liquid crystal systems.*

Berezhkova, Dr Galina Vasilyevna. Inst. of Crystallography, Acad. of Sci. of the USSR, Leninsky pr. 59, Moscow 117333, USSR. (1933) Cand, physics and mathematics (INCRYS, 1964). Sr. scient. *Defects in crystals, mechanical properties of crystals.*

Bershov, Dr Leonid Viktorovich. Inst. of Geology, Mineralogy and Petrography (IGEM), Acad. of Sci. of the USSR, Staromonetny 35, Moscow 109017, USSR. (1935) Dr, geology and mineralogy (IGEM, 1973). Lab. head. *Minerals (physics), crystal field theory, electron paramagnetic resonance.*

Bersuker, Prof. Isaak Borukhovich. Inst. of Chemistry, Acad. of Sci. of the Moldavian SSR, Akademicheskaya 3, Kishinev 277028, USSR. (1928) Dr, physics and mathematics (LGU, 1964). Dept. head. *Crystal chemistry, ferroelectricity, structural phase transitions.*

Betsofen, Dr Sergey Yakovlevich. Inst. of Metallurgy, Acad. of Sci. of the USSR, Leninsky pr. 49, Moscow 117334, USSR. (1946) Cand, technics (Inst. of Metallurgy, Acad. Sci. USSR, 1978) Sr. scient. *Applied crystallography, structure, amorphous materials, diffraction apparatus.*

Beznosikov, Dr Boris Valeriyanivich. Inst. of Physics, Acad. of Sci. of the USSR, Siberian Dept., Akademgorodak, Krasnoyarsk 660036, USSR. (1930) Cand, physics and mathematics (Krasnoyarsk Inst. of Physics, Acad. Sci. USSR, 1978). Scient. *Crystal chemistry, crystal growth.*

Bichurin, Dr Rinnat Chingizkhanovich. Inst. of Crystallography, Acad. of Sci. of the USSR, Leninsky pr. 59, Moscow 117333, USSR. (1952) Cand, chemistry (Inst. of Chemistry, Acad. Sci. Tadzhik SSR). Scient. *Crystal growth, properties, ferroelectrics.*

Biyushkin, Dr Victor Nikolayevich. Inst. of Applied Physics Moldavian SSR Akademicheskaya 5, Kishinev 277028, USSR. (1935) Cand, physics and mathematics (INCRYS, 1969). Sr. scient. *X-ray structure analysis methods, organic compounds.*

Bleidelis, Dr Yanis Yazepovich. Inst. for Organic Synthesis, Acad. of Sci. of the Latvian SSR, Aizkraukles St. 21, Riga 226006, USSR. (1926) Cand, chemistry (IONCH, 1957). Sr. scient. *Crystal chemistry, organic and elemento-organic compounds.*

Blinov, Dr Lev Mikhailovich. Inst. of Crystallography, Acad. of Sci. of the USSR, Leninsky pr. 59, Moscow 117333, USSR. (1939) Dr, physics and mathematics (INCRYS, 1977). Lab. head. *Liquid crystals, structure and properties.*

Blokhin, Prof. Mikhail Arnol'dovich. Rostov State U., Dept. of Physics, Stachki pr. 192, Rostov-on-Don 344061, USSR. (1908) Dr, physics and mathematics (Kiev U., 1955). Head of Chair. *Electronic energy-producing structure of crystals.*

Bodak, Dr Oksana Ivanovna. Lvov State U., Dept. of Chemistry, University St. 1, Lvov 290602, USSR. (1942) Dr, chemistry (Lvov U., 1981). Docent. *X-ray structure analysis, crystal chemistry, intermetallic compounds.*

Boiko, Prof. Boris Timofeyevich. Kharkov Polytechnik Inst., Frunze St. 21, Kharkov 310002, USSR. (1930) Dr, physics and mathematics (Kharkov U., 1971). Prof. *Crystal structure, metals and alloys, defects in crystal structure.*

Boikova, Dr Alexandra Ivanovna. Inst. of Silicate Chemistry, Acad. of Sci. of the USSR, Makarov Emb. 2, Leningrad 199164, USSR. (1926) Cand, chemistry (Inst. of Silicate Chemistry, Acad. Sci. USSR, 1955). Sr. scient. *Crystal chemistry, natural and synthetic minerals.*

Bokiy, Prof. Georgy Borisovich. Inst. of Geology, Mineralogy and Petrography (IGEM), Acad. of Sci. of the USSR, Staromonetny 35, Moscow 109017, USSR. (1909) Corresp. member of the Acad. Sci. USSR, Dr, chemistry (IONCH, 1942). Lab. head. *Crystal chemistry, inorganic compounds, minerals.*

Bondar', Dr Iraida Adamovna. Inst. of Silicate Chemistry, Acad. of Sci. of the USSR, Makarov Emb. 2, Leningrad 199164, USSR. Dr, chemistry (Inst. of Silicate Chemistry, Acad. Sci. USSR, 1967). Lab. head. *Crystal chemistry, crystal growth.*

Bondars, Dr Bruno Yanovich. Inst. of Inorganic Chemistry, Acad. of Sci. of the Latvian SSR, Miyera St. 34, Riga 229021, USSR. (1951) Cand, chemistry (Inst. of Inorganic Chemistry, Acad. Sci. Latvian SSR, 1981). Sr. scient. *Powder X-ray diffraction.*

Borisanova, Dr Lidiya Mikhailovna. Moscow State U., Dept. of Chemistry, Leninskiye Gory, Moscow 117234, USSR. (1940) Cand, chemistry (MGU, 1971). Asst. *Crystal chemistry.*

Borisov, Dr Stanislav Vasilyevich. Inst. of Inorganic Chemistry, Acad. of Sci. of the USSR, Siberian Dept., Prospekt Nauki 3, Novosibirsk 630090, USSR. (1930) Dr, physics and mathematics (INCRYS, 1974). Sr. scient. *Structure determination, methods, inorganic compounds, superstructures.*

Borisov, Dr Vsevolod Vasilyevich. Inst. of Crystallography, Acad. of Sci. of the USSR, Leninsky Prospekt 59, Moscow 117333, USSR. (1937) Cand, physics and mathematics (INCRYS, 1975). Sr. scient. *Protein crystallography.*

Botoshansky, Dr Mark Meyerovich. Inst. of Applied Physics, Acad. of Sci. Moldavian SSR, Akademicheskaya 5, Kishinev 277028, USSR. (1947) Cand, physics and mathematics (Inst. of Applied Physics, Acad. Sci. Moldavian SSR, 1977). Scient. *Crystal chemistry, coordination compounds, organic and bio-organic compounds.*

Boyarskaya, Prof. Yuliya Stanislavovna. Inst. of Applied Physics, Acad. of Sci. of the Moldavian SSR, Akademicheskaya 5, Kishinev 277028, USSR. (1928) Dr, physics and mathematics (INCRYS, 1974). Sr. scient. *Mechanical properties of crystals, crystal lattice defects.*

Brainin, Dr Boris Matveyevich. Petrozavodsk State University, Lenin pr. 33, Petrozavodsk 185018, USSR. (1937) Cand, physics and mathematics (Petrozavodsk State University, 1967). Docent. *Structure of real crystals.*

Brovkin, Dr Anatoly Afanasyevich. Inst. of Mineral Raw Materials, Staromonetny 29, Moscow 109017, USSR. (1937) Cand, geology and mineralogy (Yakutsk Branch of Siberian Dept., Acad. Sci. USSR, 1966). Sr. scient. *Crystal chemistry, borates, X-ray diffraction, quantitative phase analysis.*

Bud'ko, Dr Ivetta Alexandrovna. Inst. 'Mekhanobr', 21st Liniya 8a, Leningrad 199026, USSR. (1932) Cand, geology and mineralogy (LGU, 1968). Sr. scient. *X-ray analysis, crystal chemistry.*

Bukin, Dr Alexander Sergeyevich. Inst. of Geology, Acad. of Sci. of the USSR, Pyzhevsky per. 7, Moscow 109017, USSR. (1947) Cand, physics and mathematics (MGU, 1975). Scient. *Polytypism, layer silicates, order-disorder, isomorphically substituted silicates.*

Bukvetsky, Dr Boris Vladimirovich. Inst. of Chemistry, Far East Scient. Center, Acad. of Sci. of the USSR, Pr. of the 100th Aniv. of Vladivostok 159, Vladivostok 690022, USSR. (1944) Cand, physics and mathematics (INCRYS, 1977). Sr. scient. *Crystal chemistry, inorganic compounds, hydrogen bonds, phase transitions.*

Bulakh, Dr Andrey Glebovich. Leningrad State U., Dept. of Geology, University Emb. 7/9, Leningrad 199164, USSR. (1933) Cand, geology and mineralogy (LGU, 1962). Docent. *Crystal morphology, goniometry.*

Burdina, Dr Valentina Ivanovna. Inst. of Crystallography, Acad. of Sci. of the USSR, Leninsky pr. 59, Moscow 117333, USSR. (1927) Cand, physics and mathematics (Inst. of Mathematics, Acad. Sci. USSR, 1953). Scient. *Computing methods in crystallography.*

Burnasheva, Dr Veniana Venediktovna. Inst. of New Chemical Problems, Acad. of Sci. of the USSR, Chernogolovka 142432, Moskovskaya Oblast', USSR. (1940) Cand, chemistry (Lvov State U., 1970). *Crystal structure, intermetallic compounds, hydride phases.*

Burshtein, Dr Izya Fridelevich. Inst. of Applied Physics, Acad. of Sci. of the Moldavian SSR, Akademicheskaya 5, Kishinev 277028, USSR. (1942) Cand, physics and mathematics (Inst. of Applied Physics, Moldavian Acad. Sci., 1977). Sr. scient. *Computing methods, crystal chemistry, coordination compounds, bioinorganic compounds.*

Butman, Dr Lev Abramovich. Inst. of General and Inorganic Chemistry, Acad. of Sci. of the USSR, Leninsky pr. 31, Moscow 117071, USSR. (1930) Cand, physics and mathematics (INCRYS, 1971). Sr. scient. *Crystal chemistry, stereochemistry, coordination compounds, X-ray diffractometry.*

Chaban, Dr Nadezhda Fedorovna. Lvov State U., Dept. of Chemistry, University St. 1, Lvov 290602, USSR. (1942) Cand, chemistry (Lvov U., 1973). Scient. *X-ray structure analysis, crystal chemistry, intermetallic compounds.*

Chashchinov, Dr Yury Mikhailovich. Leningrad Mining Inst., 21st Liniya 2, Leningrad 199026, USSR. (1939) Cand, geology and mineralogy (GORNY, 1972). Sr. scient. *Crystal growth, defects.*

Chentsova, Dr Leonila Gavrilovna. Inst. of Crystallography, Acad. of Sci. of the USSR, Leninsky pr. 59, Moscow 117333, USSR. (1902) Cand, physics and mathematics (INCRYS, 1950). *Crystal optics, spectroscopy of crystals.*

Cheremskoy, Dr Petr Grigoryevich. Kharkov Polytechnic Inst., Frunze St. 21, Kharkov 310002, USSR. (1942) Cand, technics (Kharkov Polytechnic Inst., 1973). Sr. scient. *Metals and alloys, structure, crystal structure defects.*

Cherepanova, Dr Tamara Alekseyevna. Latvian State University, Boulevard Rainisa 19, Riga 226050, USSR. (1944) Dr, physics and mathematics(Inst. of Physics, Akad. Sci. Latvian SSR, 1985). Lab. head. *Crystal growth, crystallography theory.*

Chernov, Prof. Alexander Alexandrodich. Inst. of Crystallography, Acad. of Sci. of the USSR, Leninsky pr. 59, Moscow 117333, USSR. (1931) Dr, physics and mathematics (INCRYS, 1970). Lab. head. *Crystal growth, surface phenomena, solid state theory.*

Chernysheva, Dr Marina Alexandrovna. Inst. of Crystallography, Acad. of Sci. of the USSR, Leninsky pr. 59, Moscow 117333, USSR. (1911) Cand, physics and mathematics (INCRYS, 1955). Sr. scient. *Real structure, mechanical properties of crystals.*

Chernysheva, Mrs Valentina Fedorovna. Leningrad State U., Dept. of Geology, University Emb. 7/9, Leningrad 199164, USSR. (1933) Asst. *Crystallography, crystal optics, crystal chemistry, goniometry.*

Chetkina, Dr Larisa Arkadyevna. Karpov Physical Chemistry Inst., Obukha St. 10, Moscow 107120, USSR. (1932) Cand, physics and mathematics (INCRYS, 1966). Sr. scient. *X-ray structure analysis, crystal physics, crystal chemistry, organic compounds*

Chiragov, Dr Mamed Isa ogly. Azerbaidzhan State U., Patrisa Lumumba St. 23, Baku 370073, USSR. (1937) Cand, geology and mineralogy (MGU, 1969). Docent. *X-ray structure analysis, crystal chemistry, silicates.*

Chirgadze, Dr Yury Nikolayevich. Inst. of Protein, Acad. Sci. of the USSR, Pushchino, Serpukhovsky Rayon, Moskovskaya Oblast' 142292, USSR. (1935) Dr, physics and mathematics (Inst. of Chemical Physics, 1983). Lab. head. *X-ray structure analysis, globular proteins.*

Chudinova, Dr Svetlana Alekseyevna. Petrozavodsk State U., Lenin St. 33, Petrozavodsk 185018, USSR. (1939) Cand, physics and mathematics, (Ural U.). Docent. *metal oxides, structure, order-disorder transformations, solid solutions.*

Chukhovsky, Dr Felix Nikolayevich. Inst. of Crystallograrhy, Acad. of Sci. of the USSR, Leninsky pr. 59, Moscow 117333, USSR. (1940) Cand, physics and mathematics (INCRYS, 1968). Sr. scient. *X-ray optics of crystals.*

Chuprunov, Dr Evgeniy Vladimirovich. Gorky State University, Sverdlova 37, Gorky 603000, USSR. (1951) Cand, physics and mathematics (Inst. of Applied Physics, Acad. Sci. Moldavian SSR, 1979). Sr. lect. *X-ray structure analysis, symmetry theory.*

D'yachenko, Dr Oleg Anatolyevich. Branch Inst. of Chemical Physics, Acad. of Sci. of the USSR, Chernogolovka 142432, Noginsky Rayon, Moskovskaya Oblast', USSR. (1939) Cand, physics and mathematics (Bransh Inst. of Chem. Physics, 1970). Sr. scient. *Crystal chemistry, organic compounds.*

D'yakon, Dr Ivan Andreyevich. Inst. of Applied Physics, Acad. of Sci. of the Moldavian SSR, Akademicheskaya 5, Kishinev 277028, USSR. (1934) Cand, physics and mathematics (INCRYS, 1970). Sr. scient. *X-ray structure analysis, instrumentation, inorganic compounds, structure.*

Datt, Dr Igor Daudovich. Moscow Chemico-Tech. Inst., Miusskaya sq. 9, Moscow 125820, USSR. (1941) Cand, physics and mathematics (KARPOV, 1973) Docent. *Structure analysis, neutron diffraction.*

Davydchenko, Dr Anatoliy Georgiyevich. Res. Inst. for Synthesis of Mineral Raw Materials, Institutskaya St. 1, Alexandrov 601600, Vladimirskaya Oblast', USSR. (1934) Cand, geology and mineralogy (IGEM, 1966). Dept. head. *Crystal growth, natural and artificial crystal formation, physical chemistry.*

Dedegkayev, Dr Tazaret Temurkanovich. Physical Techn. Inst., Acad. of Sci. of the USSR, Zapovednaya St. 51, Leningrad 194037, USSR. (1934) Cand, physics and mathematics (Sukhumi Physical Techn. Inst., 1969). Sr. scient. *Thermoelectric materials, complex semiconductor compounds.*

Dem'yanets, Dr Lyudmila Nikolayevna. Inst. of Crystallography, Acad. of Sci. of the USSR, Leninsky pr. 59, Moscow 117333, USSR. (1939) Cand, chemistry (INCRYS, 1966). Sr. scient. *Crystal chemistry, crystal growth.*

Dembo, Dr Alexander Teodorovich. Inst. of Crystallography, Acad. of Sci. of the USSR, Leninsky Pr. 59, Moscow 117333, USSR. (1939) Cand, biology (Inst. of Mol. Biology and Genetics, Acad. Sci. Ukrainian SSR, 1977). Scient. *Structure, biopolymers, viruses, small-angle scattering, X-ray and neutron scattering.*

Denisenko, Dr Georgy Alexandrovich. Inst. of Crystallography, Acad. of Sci. of the USSR, Leninsky pr. 59, Moscow 117333, USSR. (1945) Cand, physics and mathematics (Kazan' U., 1975). Scient. secretary. *Crystal spectroscopy.*

Dimitrova, Dr Ol'ga Vladimirovna. Moscow State U., Dept. of Geology, Leninskiye Gory, Moscow 119899, USSR. (1948) Cand, geology and mineralogy (MGU, 1977). Scient. *Crystal chemistry, rare earth compounds.*

Distler, Dr Grigory Isaakovich. Inst. of Crystallography, Acad. of Sci. of the USSR, Leninsky pr. 59, Moscow 117333, USSR. (1920) Dr, chemistry (INCRYS, 1970). Lab. head. *Crystal nucleation, thin films, epitaxy, electron microscopy*

Dmitrieva, Dr Tatyana Vladimirovna. Inst. of Crystallography, Acad. of Sci. of the USSR, Leninsky pr. 59, Moscow 117333, USSR. (1933) Cand, physics and mathematics (INCRYS, 1975). Scient. *Magnetic properties of crystals, Mössbauer effect.*

Dmitriyeva, Dr Margarita Timofeyevna. Inst. of Geology, Mineralogy and Petrography, Acad. of Sci. of the USSR, Staromonetny 35, Moscow 109017, USSR. (1932) Cand, geology and mineralogy (IGEM, 1977). Scient. *X-ray structure analysis, minerals.*

Dodokin, Dr Anatoly Petrovich. Inst. of Crystallography, Acad. of Sci. of the USSR, Leninsky pr. 59, Moscow 117333, USSR. (1943) Cand, physics and mathematics (Physico-Technical Inst., Acad. Sci. USSR, 1972). Sr. scient. *Magnetic properties of crystals, Mössbauer effect.*

Dolivo-Dobrovol'skaya, Dr Galina Ilyinishna. Leningrad Mining Inst., 21st Liniya 2, Leningrad 199026, USSR. (1935) Cand, geology and mineralogy (GORNY, 1964). Sr. scient. *Crystal morphology, defects.*

Dolivo-Dobrovol'skaya, Mrs Elena Maximovna. Leningrad State U., Dept. of Geology, University Emb. 7/9, Leningrad 199164, USSR. (1927). Scient. *Crystal chemistry, X-ray structure analysis, symmetry.*

Dorfman, Dr Moisey Davydovich. Fersman Mineralogical Museum, Leninsky pr. 14/16, Moscow 117071, USSR. (1908) Dr, geology and mineralogy (IGEM, 1962). Sr. scient. *Crystal chemistry, minerals.*

Dorokhova, Dr Galina Igorevna. Moscow State U., Dept. of Geology, Leninskiye Gory, Moscow 119899, USSR. (1952) Cand, geology and mineralogy (MGU, 1983). Asst. *X-ray structure analysis, crystal chemistry, inorganic compounds, morphology, minerals.*

Doroshinsky, Dr Alexander Leibovich. Inst. of New Chemical Problems, Acad. of Sci. of the USSR, Chernogolovka 142232, Moskovskaya Oblast', USSR. (1933) Cand, chemistry (KARPOV, 1973). Scient. *Crystal chemistry, coordination compounds.*

Drits, Dr Victor Anatolyevich. Inst. of Geology, Acad. of Sci. of the USSR, Pyzhevsky per. 7, Moscow 109017, USSR. (1932) Dr, geology and mineralogy (IGEM, 1905). Lab. head. *Layer minerals, structure, diffraction methods.*

Drozdov, Dr Yury Nikolayevich. Gorky State U., Gagarin Prospekt 23, Gorky 603022, USSR. (1947) Cand, physics and mathematics (Gorky U., 1974). Sr. scient. *X-ray structure analysis.*

Dubov, Dr Petr Lvovich. Leningrad State U., University Emb. 7/9, Leningrad 199164, USSR. (1943) Cand, geology and mineralogy (LGU, 1971). Sr. lect. *Symmetry theory and history, geometrical crystallography.*

Dudarev, Dr Vasily Yakovlevich. Karpov Physical Chemistry Inst., Obukha St. 10, Moscow 107120, USSR. (MGU, 1963). Lab. head. *Structure analysis (X-ray - neutrons - electrons).*

Duderov, Dr Nikolay Grigoryevich. Inst. of Crystallography, Acad. of Sci. of the USSR, Leninsky pr. 59, Moscow 117333, USSR. (1945) Cand, chemistry (INCRYS, 1975). Scient. *Crystal growth, crystal chemistry.*

Dudkevich, Dr Vladimir Petrovich. Rostov State U., Dept. of Physics, Prospekt Stachki 192, Rostov-on-Don 344061, USSR. (1935) Cand, physics and mathematics (Rostov U., 1968). Docent. *Thin films structure and physical properties, phase transitions (ferroelectric).*

Dukova, Dr Elena Dmitriyevna. Inst. of Crystallography, Acad. of Sci. of the USSR, Leninsky pr. 59, Moscow 117333, USSR. (1925) Cand, physics and mathematics (INCRYS, 1956). Sr. scient. *Crystal morphology, crystal growth.*

Dvorkin, Dr Alexander Arkadyevich. Inst. of Applied Physics, Acad. of Sci. of the Moldavian SSR, Akademicheskaya 5, Kishinev 277028, USSR. (1947) Cand, physics and mathematics (Inst. of Applied Physics, Acad. Sci. Moldavian SSR, 1975). Scient. *X-ray structure analysis, crystal chemistry.*

Efremov, Dr Valery Alexandrovich. Inst. of Chemical Reagents and pure Substances, Acad. of Sci. of the USSR, Bogorodsky Val 3, Moscow 107258, USSR. (1950) Cand, chemistry (MGU, 1976). Sr. scient. *Crystal chemistry, inorganic compounds with EO_4 anions.*

Egorov-Tismenko, Dr Yury Klavdiyevich. Moscow State U., Dept. of Geology, Leninskiye Gory, Moscow 117234, USSR. (1938) Cand, geology and mineralogy (MGU, 1973). Scient. *Crystal chemistry, inorganic compounds, X-ray structure analysis.*

Egorov, Dr Vladimir Mikhailovich. Inst. of Crystallography, Acad. of Sci. of the USSR, Leninsky pr. 59, Moscow 117333, USSR. (1941) Cand, chemistry (INCRYS, 1975). Scient. *Crystal growth, crystal chemistry.*

Eliseev, Dr Erik Nikolayevich. Inst. of Geochronology of Pre-Cambrian, Makarov Emb. 2, Leningrad 199164, USSR. (1926) Cand, geology and mineralogy (GORNY, 1954). Sr. scient. *Crystal chemistry.*

Esipova, Dr Nataliya Georgiyevna. Inst. of Molecular Biology, Acad. of Sci. of the USSR, Vavilov St. 32, Moscow 117312, USSR. Cand, physics and mathematics (Inst. of Biophysics, Acad. Sci. USSR). Scient. *Fibrillous structures.*

Fedorenko, Prof. Anatoly Ivanovich. Kharkov Polytechnic Inst., Frunze St. 21, Kharkov 310002, USSR. (1937) Dr, technics (Kharkov Polytechnic Inst., 1978). Docent. *Thin film structure, defects, methods for defect study.*

Fedorov, Dr Boris Alexandrovich. Inst. of Protein, Acad. of Sci. of the USSR, Pushchino 142292, Serpukhovsky Rayon, Moskovskaya Oblast', USSR. (1939) Cand, physics and mathematics (Inst. of High Molecular Compounds, Acad. Sci. USSR, 1966). Sr. scient. *X-ray structure analysis, proteins, diffuse X-ray scattering, macromolecules in solution.*

Fedorov, Dr Pavel Pavlovich. Inst. of Crystallography, Acad. of Sci. of the USSR, Leninsky pr. 59, Moscow 117333, USSR. (1950) Cand, chemistry (Moscow Inst. of Fine Chem. Techn., 1977). Sr. scient. *Crystal growth, phase diagrams.*

Fedotov, Dr Alexander Fedorovich. Inst. of Geology - Mineralogy and Petrography, Acad. of Sci. of the USSR, Staromonetny 35, Moscow 109017, USSR. (1924) Cand, geology and mineralogy (IGEM, 1974). Sr. scient. *Structural mineralogy, electron diffraction.*

Feigin, Dr Lev Abramovich. Inst. of Crystallography, Acad. of Sci. of the USSR, Leninsky pr. 59, Moscow 117333, USSR. (1927) Dr, physics and mathematics (INCRYS, 1976). Sr. scient. *Biopolymer structures, X-ray diffraction.*

Fesenko, Prof. Evgeny Grigoryevich. Rostov State U., Dept. of Physics, Prospekt Stachki 192, Rostov-on-Don 344061, USSR. (1918) Dr, physics and mathematics (Rostov U., 1973). Dir., Physics Inst. *Crystal chemistry, complex oxides, structure and physical properties of crystals.*

Fesenko, Dr Oleg Evgenyevich. Rostov State U., Prospekt Stachki 192, Rostov-on-Don 344090, USSR. (1950) Cand, physics and mathematics (Rostov U., 1978). Lab. head. *Ferroelectric crystal physics.*

Filatov, Dr Stanislav Konstantinovich. Leningrad State U., Dept. of Geology, University Emb. 7/9, Leningrad 199164, USSR. (1940) Cand, geology and mineralogy (LGU, 1969). Asst. *Crystallography, crystal chemistry, X-ray structure analysis.*

Filip'yev, Dr Victor Semenovich. Rostov State U., Dept. of Physics, Prospekt Stachki 192, Rostov-on-Don 344061, USSR. (1937) Cand, physics and mathematics (Rostov U., 1966). Docent. *Crystal chemistry, complex oxides, structure and physical properties of crystals.*

Filipenko, Dr Olga Savelyevna. Branch Inst. of Chemical Physics, Acad. of Sci. of the USSR, Chernogolovka 142432, Noginsky Rayon, Moskovskaya Oblast', USSR. (1940) Cand, chemistry (MGU, 1971). Scient. *Crystal chemistry, organic compounds.*

Finkel'shtein, Dr Aleksey Vital'yevich. Inst. of Protein, Acad. of Sci. of the USSR, Pushchino 142292, Serpukhovsky Rayon, Moskovskaya Oblast', USSR. (1947) Cand, physics and mathematics (Moscow Physico-Techn. Inst., 1976). Sr. scient. *Structure, proteins, nucleic acids.*

Flerov, Dr Igor' Nikolayevich. Inst. of Physics, Acad. of Sci. of the USSR, Siberian Dept., Akademgorodak, Krasnoyarsk 660036, USSR. (1942) Cand, physics and mathematics (Krasnoyarsk Inst. of Physics, Acad. Sci. USSR, 1978). Scient. *Thermal properties of crystals, phase transitions (structural).*

Fotchenkov, Dr Anatoly Andreyevich. Res. Inst. for Synthesis of Mineral Raw Materials, Institutskaya St. 1, Alexandrov 601600, Vladimirskaya Oblast', USSR. (1925) Cand, physics and mathematics (INCRYS, 1960). Dept. head. *Physical properties of crystals, piezoelectric and dialectric properties, acoustic and elastic properties, optical properties.*

Frank-Kamenetskaya, Dr Olga Victorovna. Leningrad State U., Dept. of Geology, University Emb. 7/9, Leningrad 199164, USSR. (1945) Cand, geology and mineralogy (LGU, 1973). Scient. *Structure analysis, crystal chemistry.*

Frank-Kamenetsky, Prof. Victor Al'bertovich. Leningrad State U., University Emb. 7/9, Leningrad 199164, USSR. (1915) Dr, geology and mineralogy (GORNY, 1962) Head of Chair. *Crystallography, crystal chemistry, X-ray structure analysis, layer silicates.*

Franke, Dr Valeriya Dmitriyevna. Leningrad State U., University Emb. 7/9, Leningrad 199164, USSR. (1945) Cand, geology and mineralogy (LGU, 1982). Sr. scient. *Crystal growth, morphology.*

Fridkin, Prof. Vladimir Mikhailovich. Inst. of Crystallography, Acad. of Sci. of the USSR, Leninsky pr. 59, Moscow 117333, USSR. (1929) Dr, physics and mathematics (INCRYS, 1963). Sr. scient. *Physical properties of crystals, phase transitions.*

Fundamensky, Mr Vladimir Semenovich. NPO Burevestnik, Stakhanovtsev St. 1, Leningrad 195112, USSR. (1946) Lab. head. *Methods of X-ray structural analysis, crystal chemistry, organic and complex compounds.*

Furmanova (Bokiy), Dr Nina Georgievna. Inst. of Crystallography, Acad. of Sci. of the USSR, Leninsky pr. 59, Moscow 117333, USSR. (1939) Cand, chemistry (INEOS, 1968). Sr. scient. *Crystal chemistry, organic and organometallic compounds.*

Fursenko, Dr Boris Alexandrovich. Inst. of Geology and Geophysics, Acad. of Sci. of the USSR, Siberian Dept., University Prospekt 3, Novosibirsk 630090, USSR. (1946) Cand, geology and mineralogy (Inst. of Geology and Geophysics, Acad.

Sci. USSR, Siberian Dept., 1973). Sr. scient. *Properties of crystals, X-ray structure analysis, high pressures.*

Fykin, Dr Leonid Efimovich. Karpov Physical Chemistry Inst., Obukha St. 10, Moscow 107120, USSR. (1936) Cand, physics and mathematics (Moscow Inst. of Eng. and Physics, 1972). Sr. scient. *Neutron diffraction, instrumentation, methods, structure analysis.*

Gabuda, Prof. Svyatoslav Petrovich. Inst. of Inorganic Chemistry, Acad. of Sci. of the USSR, Siberian Dept., Lavrent'yev Prospekt 3, Novosibirsk 630090, USSR. (1936) Cand, physics and mathematics (Acad. Sci. USSR, Siberian Dept., 1970). Lab. head. *Non-diffraction structure methods, NMR, crystal chemistry of hydrogen.*

Galitsky, Dr Nikolay Mikhailovich. Inst. of Bio-organic Chemistry, Acad. of Sci. of the Belorussian SSR, Leninsky pr. 68, Minsk 220600, USSR. (1950) Cand, chemistry (Inst. of Bio-organic Chemistry Belorussian Acad. Sci., 1978). Group head. *X-ray structure analysis, peptide-protein compounds.*

Galiulin, Dr Ravil Vagizovich. Inst. of Crystallography, Acad. of Sci. of the USSR, Leninsky pr. 59, Moscow 117333, USSR. (1940) Dr, physics and mathematics (INCRYS, 1978). Scient. *Fundamentals of crystallography, mathematical crystallography.*

Galstyan, Dr Viktor Gaikovich. Inst. of Crystallography, Acad. of Sci. of the USSR, Leninsky pr. 59, Moscow 117333, USSR. (1940) Dr, physics and mathematics (MGU, 1971). Sr. scient. *Surfaces, real structure of crystals.*

Garashina, Dr Lyudmila Solomonovna. Inst. of Crystallography, Acad. of Sci. of the USSR, Leninsky pr. 59, Moscow 117333, USSR. (1939) Cand, chemistry (IONCH, 1969). Scient. *Isomorphism.*

Gavrilova, Dr Irina Vladimirovna. Inst. of Crystallography, Acad. of Sci. of the USSR, Leninsky pr. 59, Moscow 117333, USSR. (1924) Cand, physics and mathematics (INCRYS, 1973). Scient. *Crystal growth.*

Geguzina, Dr Galina Alexandrovna. Rostov State U., Dept. of Physics, Prospekt Stachki 192, Rostov-on-Don 344061, USSR. (1945) Cand, physics and mathematics (Rostov U., 1975). Sr. scient. *Crystal chemistry, complex oxides, structure and physical properties.*

Gel'man, Dr Yury Alexandrovich. Inst. of Crystallography, Acad. of Sci. of the USSR, Leninsky pr. 59, Moscow 117333, USSR. (1938) Cand, physics and mathematics (INCRYS, 1978). Scient. *crystal growth from gas phase, sublimation, surface studies.*

Genin, Dr Yakov Vladimirovich. Inst. of Elemento-Organic Compounds, Acad. of Sci. of the USSR, Vavilov St. 28, Moscow 117813, USSR. (1941) Cand, physics and mathematics (INCRYS, 1976). Scient. *Polymer X-ray diffraction, polymer physics.*

Gilinskaya, Dr Emma Abramovna. Inst. of Scient. and Techn. Information, Acad. of Sci. of the USSR, Baltiyskaya St. 14, Moscow 125219, USSR. (1922) Cand, chemistry (IONCH, 1953). Sr. scient. *Crystal chemistry, coordination compounds.*

Givargizov, Dr Evgeny Inviyevich. Inst. of Crystallography, Acad. of Sci. of the USSR, Leninsky pr. 59, Moscow 117333, USSR. (1934) Dr, physics and mathematics (INCRYS, 1976). Sr. scient. *Crystal growth, whisker-crystal growth from vapor phase.*

Gladkikh, Dr Liliya Ivanovna. Kharkov Polytechnic Inst., Frunze St. 21, Kharkov 310002, USSR. (1934) Cand, technics (Kharkov Polytechnic Inst., 1966). Docent. *Metal and alloy structures, defects and methods of investigation.*

Gladky, Dr Vsevolod Vladimirovich. Inst. of Crystallography, Acad. of Sci. of the USSR, Leninsky pr. 59, Moscow 117333, USSR. (1934) Dr, physics and mathematics (INCRYS, 1985). Sr. scient. *Ferroelectrics (physics), phase transitions (ferroelectric).*

Gladyshevsky, Prof. Evgeny Ivanovich. Lvov State U., University St. 1, Lvov 290602, USSR. (1924) Dr, chemistry (MGU, 1967). Head of Chair. *X-ray structure analysis, crystal chemistry, intermetallic compounds.*

Glazov, Dr Aleksey Ivanovich. Leningrad Mining Inst., 21st Liniya 2, Leningrad 199026, USSR. (1942) Cand, geology and mineralogy (GORNY, 1976). Sr. scient. *Crystal morphology, X-ray diffraction.*

Gliki, Dr Nataliya Vladimirovna. Inst. of Crystallography, Acad. of Sci. of the USSR, Leninsky pr. 59, Moscow 117333, USSR. (1927) Cand, physics and mathematics (INCRYS, 1953). *Crystal growth, crystal morphology, defects in crystals.*

Glikin, Mr Arkady Eduardovich. Leningrad State U., Dept. of Geology, University Emb. 7/9, Leningrad 199164, USSR. (1943) Cand, geology and mineralogy (LGU, 1978). Scient. *Crystal growth, crystal morphology.*

Godovikov, Prof. Alexander Alexandrovich. Fersman Mineralogical Museum, Leninsky pr. 18, Moscow 117071, USSR. (1927) Dr, geology and mineralogy (MGU, 1970). Dir. *Theoretical and experimental mineralogy, mineral synthesis.*

Goilo, Dr Eduard Al'bertovich. Leningrad State U., Dept. of Geology, University Emb. 7/9, Leningrad 199164, USSR. (1941) Cand, geology and mineralogy (LGU, 1970). Scient. *X-ray and electron diffraction, layer silicates, crystal chemistry.*

Golovachev, Dr Vladimir Pavlovich. Gorky State U., Gagarin Prospekt 23, Gorky 603022, USSR. (1930) Cand, physics and mathematics (INCRYS, 1972). Docent. *X-ray structure analysis.*

Golovastikov, Dr Nikolay Ivanovich. Inst. of Crystallography, Acad. of Sci. of the USSR, Leninsky pr. 59, Moscow 117333, USSR. (1915) Cand, physics and mathematics (INCRYS, 1953). Sr. scient. *Structure analysis.*

Golovina, Dr Nina Ivanovna. Branch Inst. of Chemical Physics, Acad. of Sci. of the USSR, Chernogolovka, Noginsky Rayon, Moskovskaya Oblast' 142432, USSR. (1934) Cand, chemistry (Branch Inst. of Chemical Physics, 1969). Scient. *Metallo-organic compounds, structure.*

Golubev, Dr Alexander Mikhailovich. Inst. of Crystallography, Acad. of Sci. of the USSR, Leninsky pr. 59, Moscow 117333, USSR. (1948) Cand, chemistry (MGU, 1975). Scient. *Precision structures, inorganic compounds.*

Golubkov, Dr Alexander Vasilyevich. Physico-Techn. Inst., Acad. of Sci. of the USSR, Zapovednaya 51, Leningrad 194047, USSR. Cand, chemistry (Inst. of Silicate Chemistry, Acad. Sci. USSR, 1969). Sci. *Rare earth element compounds.*

Goncharov, Dr Georgy Nikolayevich. Leningrad State U., Dept. of Geology, University Emb. 7/9, Leningrad 199164, USSR. (1941) Cand, geology and mineralogy (LGU, 1968). Sr. scient. *Physics and chemistry, minerals, Mössbauer spectroscopy.*

Gorbunova, Dr Yuliya Efimovna. Inst. of General and Inorganic Chemistry, Acad. of Sci. of the USSR, Leninsky pr. 31, Moscow 117071, USSR. (1932) Cand, chemistry (IONCH, 1971). Scient. *Crystal chemistry, inorganic compounds.*

Gordiyenko, Dr Vladimir Vasilyevich. Leningrad State U., University Emb. 7/9, Leningrad 199164, USSR. (1934) Cand, geology and mineralogy (LGU, 1966). Sr. scient. *Mineralogical crystallography, geochemistry, X-ray studies, minerals.*

Gorogotskaya, Dr Lydumila Ivanova. Inst. of Geochemistry and Physics of Minerals, Acad. of Sci. Ukrainian SSR, Palladin Prospekt 34, Kiev 252068, USSR. (1935) Cand, geology and mineralogy (IGEM, 1967). Sr. scient. *Crystal chemistry, crystal structure, minerals.*

Gorina, Dr Iza Ivanovna. Inst. of Crystallography, Acad. of Sci. of the USSR, Leninsky pr. 59, Moscow 117333, USSR. (1936) Cand, chemistry (Inst. of Oil-Chem. Synthesis, Acad. Sci. USSR, 1966). Sr. scient. *Liquid crystals, physical chemistry.*

Gorshkov, Dr Anatoly Ivanovich. Inst. of Geology, Mineralogy and Petrography (IGEM), Staromonetny 35, Moscow 109017, USSR. (1929) Dr, geology and mineralogy (IGEM, 1971). Lab. head. *Mineral structure, morphology, phase transformation.*

Goryunov, Dr Alexander Ivanovich. Inst. of Molecular Biology, Acad. of Sci. of the USSR, Vavilov St. 32, Moscow 117312, USSR. (1938) Cand, physics and mathematics (Inst. of Biophysics, Acad. Sci. USSR, 1974). Scient. *Protein crystallography.*

Grebenshchikov, Prof. Roman Georgiyevich. Inst. of Silicate Chemistry, Acad. of Sci. of the USSR, Makarov Emb. 2, Leningrad 199164, USSR. (1929) Dr, chemistry (Inst. of Silicate Chemistry, Acad. Sci. USSR, 1967). Lab. head. *Crystal chemistry, X-ray crystallography, inorganic compounds, silicates.*

Grechushnikov, Dr Boris Nikolayevich. Inst. of Crystallography, Acad. of Sci. of the USSR, Leninsky pr. 59, Moscow 117333, USSR. (1925) Cand, physics and mathematics (INCRYS, 1953). Lab. head. *Crystal optics, optical spectroscopy, radio spectroscopy of crystals.*

Grigorov, Dr Sergey Nikolayevich. Kharkov Polytechnic Inst., Frunze St. 21, Kharkov 310002, USSR. (1944) Cand, technics (Kharkov Polytechnic Inst., 1971). Docent. *Crystal structure defects, thin film growth, electron microscopy.*

Grigoryev, Prof. Dimitry Pavlovich. Leningrad Mining Inst., 21st Liniya 2, Leningrad 199026, USSR. (1909) Dr, geology and mineralogy (IGEM, 1943). Lab. head. *Crystal growth.*

Grigoryeva, Dr Tamara Nikolayevna. Inst. of Geology and Geophysics, Acad. of Sci. of the USSR, Siberian Dept., Prospekt Nauki 3, Novosibirsk 630090, USSR. (1933) Cand, physics and mathematics (Irkutsk U., 1971). Scient. *Mineral structures, X-ray structure analysis.*

Grin', Dr Yury Nikolayevich. Dept. of Chemistry, Lvov State U., University St. 1, Lvov 742388, USSR. (1955) Cand, chemistry (Lvov U., 1980). Asst. *X-ray structure analysis, crystal chemistry, intermetallic compounds.*

Grinberg, Dr Svetlana Arnol'dovna, Inst. of Crystallography, Acad. of Sci. of the USSR, Leninsky pr. 59, Moscow 117333, USSR. (1944) Cand, physics and mathematics (INCRYS, 1979). Scient. *Crystal growth.*

Gurin, Dr Vladimir Nikolayevich. Physico-Techn. Inst., Acad. of Sci. of the USSR, Politekhnicheskaya St. 26, Leningrad 194021, USSR. (1936) Cand, chemistry (Physico-Techn. Inst., Acad. Sci. USSR, 1968). Sr. scient. *Crystal growth.*

Gurskaya, Dr Galina Victorovna. Inst. of Molecular Biology, Acad. of Sci. of the USSR, Vavilov St. 32, Moscow 117312, USSR. Cand, physics and mathematics (INCRYS, 1964). Sr. scient. *Biological materials, structure.*

Guseinov, Dr Gakhraman Gusein ogly. Inst. of Physics, Acad. of Sci. of the Azerbaidzhan SSR, Narimanov Prospekt 33, Baku 370143, USSR. (1937) Cand, chemistry (Azerbaidzhan Inst. of Inorg. and Physical Chemistry, 1968). Sr. scient. *Crystal structure, phase transformations, semiconductor compounds.*

Guseinova, Dr Maya Kara kyzy. Inst. for Petrochemical Processes, Acad. of Sci. of the Azerbaidzhan SSR, Telnov St. 30, Baku 370025, USSR. *X-ray structure analysis, crystal chemistry, complex and organic compounds.*

Harutunyan, Dr Emil' Haikovich. Inst. of Crystallography, Acad. of Sci. of the USSR, Leninsky pr. 59, Moscow 117333, USSR. (1935) Dr, chemistry (INCRYS, 1984). Sr. scient. *Protein structure, X-ray analysis.*

Ikornikova, Prof. Nina Yuryevna. Inst. of Crystallography, Acad. of Sci. of the USSR, Leninsky pr. 59, Moscow 117333, USSR. (1913) Dr, geology and mineralogy (INCRYS, 1970). *Hydrothermal synthesis of crystals, solutions - physical and chemical studies.*

Ilyinsky, Dr Alexander Lvovich. Moscow State U., Dept. of Chemistry, Leninskiye Gory, Moscow 117234, USSR. (1937) Cand, chemistry (MGU, 1975). Scient. *X-ray structure analysis, instrumentation, crystal chemistry.*

Ilyushin, Dr Alexander Sergeyevich. Moscow State U., Dept. of Physics, Leninskiye Gory, Moscow 119899, USSR. (1943) Cand, physics and mathematics (MGU, 1971). Sr. scient. *Phase transitions, low-temperature X-ray diffraction.*

Imamov, Dr Rafik Mamedovich. Inst. of Crystallography, Acad. of Sci. of the USSR, Leninsky pr. 59, Moscow 117333, USSR. (1938) Dr, physics and mathematics (INCRYS, 1978). Lab. head. *Electron diffraction, structure analysis, crystal chemistry.*

Indenbom, Prof. Vladimir Lvovich. Inst. of Crystallography, Acad. of Sci. of the USSR, Leninsky pr. 59, Moscow 117333, USSR. (1924) Dr, physics and mathematics (Leningrad Physical Techn. Inst., 1964). Lab. head. *Real crystals (physics), phase transition theory, X-ray optics, internal strain theory, dislocations in crystals.*

Ionov, Dr Pavel Victorovich. Inst. of Crystallography, Acad. of Sci. of the USSR, Leninsky pr. 59, Moscow 117333, USSR. (1941) Cand, physics and mathematics (INCRYS, 1975). Scient. *Photo-stimulated processes in solids.*

Ionov, Dr Vladislav Mikhailovich. Moscow State U., Dept. of Chemistry, Leninskiye Gory, Moscow 117234, USSR. (1934) Cand, chemistry (MGU, 1973). Sr. scient. *X-ray structure analysis, crystal chemistry.*

Iskhakova, Dr Lyudmila Dmitriyevna. Inst. of Chemical Reagents and Pure Substances, Bogorodsky Val 3, Moscow 107258, USSR. (1942) Cand, chemistry (Moscow Inst. of Fine Chem. Eng., 1970). Sr. scient. *Crystal chemistry, inorganic compounds, tetrahedral anions, rare earth salts.*

Ismailzade, Prof. Ibragim Gasan ogly. Inst. of Physics, Acad. of Sci. Azerbaidzhan SSR, Narimanov Prospekt 33, Baku 370143, USSR. (1912) Dr, physics and mathematics (Azerbaidzhan State U., 1968). Lab. head. *Crystal chemistry, ferroelectric crystals.*

Ivanov, Dr Nikolay Rafailovich. Inst. of Crystallography, Acad. of Sci. of the USSR, Leninsky pr. 59, Moscow 117333, USSR. (1938) Cand, physics and mathematics (INCRYS, 1967). Sr. scient. *Optical properties, phase transitions, ferroelectricity, ferroelasticity.*

Ivanova, Dr Irina Vadimovna. Inst. of Physics, Acad. of Sci. of the Azerbaidzhan SSR, Narimanov Prospekt 33, Baku 370143, USSR. (1936) Cand, physics and mathematics (Inst. of Physics, Acad. Sci. Azerbaidzhan SSR, 1981). Sr. scient. *Semiconductor films, thin film formation, structure and physical properties.*

Kabalkina, Prof. Sara Samsonovna. Inst. of High Pressure Physics, Acad. of Sci. of the USSR, Akademgorodok, Podol'sky Rayon, Moskovskaya Oblast' 142092, USSR. Dr, physics and mathematics (INCRYS, 1975). Lab. head. *High pressure effects, crystal structure.*

Kachalov, Dr Oleg Viktorovich. Inst. of Crystallography, Acad. of Sci. of the USSR, Leninsky Pr. 59, Moscow 117333, USSR. (1942) Cand, physics and mathematics (INCRYS, 1973). Scient. *Lattice dynamics, vibrational spectroscopy.*

Kachinsky, Dr Vitol'd Nikolayevch. Inst. of Crystallography, Acad. of Sci. of the USSR, Leninsky pr. 59, Moscow 117333, USSR. (1931) Cand, physics and mathematics (INCRYS, 1964). Scient. *Electronic structure, phase transitions, high pressures.*

Kaganer, Dr Vladimir Mikhailovich. Inst. of Crystallography, Acad. of Sci. of the USSR, Leninsky pr. 59, Moscow 117333, USSR. (1956) Cand, physics and mathematics (INCRYS, 1984). Scient. *Dynamic diffraction, X-ray and electron diffraction, neutron diffraction.*

Kaidalova, Dr Taisiya Alexandrovna. Inst. of Chemistry, Far East Scient. Center, Acad. of Sci. of the USSR, Pr. of the 100th Aniv. of Vladivostok 159, Vladivostok 690022, USSR. (1940) Cand, chemistry (IONCH, 1974). Sr. scient. *Crystal chemistry, inorganic compounds, physical properties of crystals.*

Kalinin, Dr Vladimir Ivanovich. Inst. of Crystallography, Acad. of Sci. of the USSR, Leninsky pr. 59, Moscow 117333, USSR. (1948) Cand, physics and mathematics (MGU, 1977). Scient. *Electrical conductivity in crystals, physical properties - crystal growth relationship.*

Kalinkina, Dr Irina Nikolayevna. Inst. of Crystallography, Acad. of Sci. of the USSR, Leninsky pr. 59, Moscow 117333, USSR. (1930) Cand, physics and mathematics (Inst. of Physical Problems, Acad. Sci. USSR, 1963). Scient. *Spectroscopy of crystals.*

Kalychak, Dr Yaroslav Mikhailovich. Lvov State U., Universitetskaya St. 1, Lvov 290602, USSR. (1947) Cand, chemistry (Lvov State U., 1977). Docent. *Crystal chemistry, intermetallic compounds.*

Kamentsev, Dr Igor' Evgenyevich. Leningrad State U., Dept. of Geology, University Emb. 7/9, Leningrad 199164, USSR. (1933) Cand, geology and mineralogy (LGU, 1964). Docent. *Crystal chemistry, X-ray structure analysis.*

Kaminsky, Dr Alexander Alexandrovich. Inst. of Crystallography, Acad. of Sci. of the USSR, Leninsky pr. 59, Moscow 117333, USSR. (1934) Dr, physics and mathematics (INCRYS, 1974). Sr. scient. *Crystal physics, laser crystals.*

Kaminsky, Dr Vladimir Fedorovich. Branch Inst. of Chemical Physics, Acad. of Sci. of the USSR, Chernogolovka 142432, Noginsky Rayon, Moskovskaya Oblast', USSR. (1941) Cand, physics and mathematics (Inst. of Applied Physics, Acad. Sci. Moldavian SSR, 1974). Scient. *Organic compounds, direct methods, structure determination.*

Kantor, Dr. Matvey Matveyevich. Inst. of Metallurgy, Acad. of Sci. of the USSR, Leninsky pr. 49, Moscow 117234, USSR. (1936) Cand, technics (Inst. of Metallurgy, Acad. Sci. USSR, 1970). Sr. scient. *Electron microscopy, electron diffraction.*

Kaplunnik, Dr Lidiya Nikolayevna. Dept. of Geology, Moscow State U., Leninskiye Gory, Moscow 117234, USSR. (1947) Cand, geology and mineralogy (MGU, 1978). Asst. *Crystal chemistry, inorganic compounds, X-ray structure analysis.*

Kardashev, Dr Boris Konstantinovich. Physico-Techn. Inst., Acad. of Sci. of the USSR, Politekhnicheskaya St. 26, Leningrad 194021, USSR. (1941) Cand, physics and mathematics (Physico-Techn. Inst., Acad. Sci. USSR, 1974). Scient. *Crystal structure defects, elasticity and plasticity, crystal physics.*

Karpinsky, Dr Oleg Georgiyevich. Inst. of Metallurgy, Acad. of Sci. of the USSR, Leninsky pr. 49, Moscow 117334, USSR. (1923) Cand, physics and mathematics (Moscow Inst. of Eng. and Physics, 1958). Sr. scient. *Inorganic compounds, structure.*

Karyakina, Dr Tatyana Alexandrovna. Leningrad Mining Inst., 21st Liniya 2, Leningrad 199026, USSR. (1939) Cand, geology and mineralogy (GORNY, 1969). Sr. scient. *Mineralogical crystallography, crystal growth, defects, quartz.*

Kashaev, Dr Anvar Akhyarovich. State Pedagogical Inst., Nizhnyaya Naberezhnaya 6, Irkutsk 664011, USSR. (1932) Cand, physics and mathematics (Irkutsk State U., 1968). Sr. scient. *Crystal chemistry, X-ray structure analysis, inorganic compounds.*

Kats, Dr Moisey Sukherovich. Inst. of Applied Physics, Acad. of Sci. of the Moldavian SSR, Grosula St. 5, Kishinev 277028, USSR. (1941) Cand, physics and mathematics (Physico-Techn. Inst. Acad. Sci. USSR, 1976). Sr. scient. *Mechanical properties of crystals, defects in crystals.*

Katsnel'son, Prof. Al'bert Anatolyevich. Moscow State U., Dept. of Physics, Leninskiye Gory, Moscow 117234, USSR. (1930) Dr, physics and mathematics (MGU, 1968). Prof. *Solid state physics, X-ray crystallography, X-ray scattering theory.*

Kayushina, Dr Renata Lvovna. Inst. of Crystallography, Acad. of Sci. of the USSR, Leninsky pr. 59, Moscow 117333, USSR. Cand, physics and mathematics (INCRYS, 1965). Sr. scient. *Biopolymer structural studies, X-ray diffraction method.*

Kemme, Dr Andrey Andreyevich. Inst. of Organic Synthesis, Acad. of Sci. of the Latvian SSR, Aizkraukles St. 21, Riga 226006, USSR. (1941) Cand, chemistry (Inst. of Organic Synthesis, Acad. Sci. Latvian SSR, 1977). Scient. *Organic crystal chemistry.*

Kessenikh, Dr Galina Georgiyevna. Inst. of Crystallography, Acad. of Sci. of the USSR, Leninsky pr. 59, Moscow 117333, USSR. (1938) Cand, physics and mathematics (INCRYS, 1972). Sr. scient. *Phase transitions, elastic properties.*

Khachaturyan, Dr Armen Gurgenovich. Inst. of Crystallography, Acad. of Sci. of the USSR, Leninsky pr. 59, Moscow 117333, USSR. (1935) Dr, physics and mathematics (Inst. of Metal Physics, Acad. Sci. Ukrainian SSR, 1970). Sr. scient. *X-ray structure analysis, protein structures, liquid crystals, phase transformation theory, order-disorder transitions.*

Khadzhi, Dr Valentin Evstafyevich. Res. Inst. for Synthesis of Mineral Raw Materials, Institutskaya St. 1, Alexandrov 601600, Vladimirskaya Oblast', USSR. (1932) Cand, geology and mineralogy (INCRYS, 1968). Dept. head. *Crystal growth, hydrothermal solutions.*

Khaikin, Dr Leonid Solomonovich. Moscow State U., Dept. of Chemistry, Leninskiye Gory, Moscow 117234, USSR. (1937) Cand, chemistry (MGU, 1969). Sr. scient. *Elemento-organic compounds, structure, electron diffraction by gases.*

Kharchenko, Dr Lyudmila Yulianovna. Inst. of Inorganic Chemistry, Acad. of Sci. of the USSR, Siberian Dept., Prospekt Nauki 3, Novosibirsk 630090, USSR. (1937) Cand, chemistry (Inst. of Inorganic Chemistry, Acad. Sci. USSR, Siberian Dept., 1968). Sr. scient. *Crystal growth, crystal chemistry, inorganic compounds.*

Kharchenko, Dr Olga Ivanovna. Dept. of Chemistry, Lvov State U., University St. 1, Lvov 742388, USSR. (1946) Cand, chemistry (Lvov U., 1978). Sr. scient. *X-ray structure analysis, crystal chemistry, intermetallic compounds.*

Kharitonov, Dr Yury Alexandrovich. Inst. of Crystallography, Acad. of Sci. of the USSR, Leninsky pr. 59, Moscow 117333, USSR. (1940) Cand, geology and mineralogy (MGU, 1971). Scient. *Crystal structure, crystal growth, structure - properties relationships.*

Khatanova, Dr Nina Abdulovna. Moscow State U., Dept. of Physics, Leninskiye Gory, Moscow 117234, USSR. Cand, physics and mathematics (MGU, 1968). Scient. *X-ray crystallography, phase transformations.*

Kheiker, Dr Daniel' Moiseyevich. Inst. of Crystallography, Acad. of Sci. of the USSR, Leninsky pr. 59, Moscow 117333, USSR. (1930) Dr, physics and mathematics (INCRYS, 1972). Lab. head.. *X-ray structure analysis, instrumentation.*

Kheirov, Dr Mamed Bekovch. Azerbaidzhan Res. Inst., Oil Petrolium Industry, Aganeimatulla St. 39, Baku 370033, USSR. (1925) Dr, geology and mineralogy (Azerbaidzhan Inst. of Oil Chemical Industry, 1975). Sr. scient. *Crystal chemistry, clay minerals, structure, transformation.*

Khetchikov, Dr Lev Nikolayevich. Central Res. Geological Inst. of Non-Ferrous and Noble Metals, Varshavskoye Chaussee 58, Moscow 115430, USSR. (1946) Dr, geology and mineralogy (IMGRE, 1975). Lab. head. *Defect structure in crystals.*

Khisina, Dr Nataliya Rafailovna. Inst. of Geochemistry and Analytical Chemistry, Acad. of Sci. of the USSR, Kosygin pr. 19, Moscow 117334, USSR. (1945)

Cand, geology and mineralogy (Inst. of Geochemistry and Analytical Chemistry, Acad. Sci. USSR, 1978). Scient. *Mineral solid solution decay, structure, cation ordering.*

Khitrova, Dr Valentina Ivanovna. Inst. of Crystallography, Acad. of Sci. of the USSR, Leninsky pr. 59, Moscow 117333, USSR. (1928) Cand, physics and mathematics (INCRYS, 1963). Sr. scient. *Electron diffraction, structure analysis, crystal chemistry.*

Khodashova, Dr Tatyana Semenovna. Inst. of General and Inorganic Chemistry, Acad. of Sci. of the USSR, Leninsky pr. 31, Moscow 117071, USSR. (1928) Cand, chemistry (IONCH, 1963). Sr. scient. *Crystal chemistry, stereochemistry, coordination compounds.*

Khotsyanova, Dr Tatyana Lvovna. Inst. of Elemento-Organic Cmpounds, Acad. of Sci. of the USSR, Vavilov St. 28, Moscow 117312, USSR. (1924) Cand, chemistry (INCRYS, 1952). Sr. scient. *Nuclear quadrupole resonance, structure determination, X-ray structure analysis.*

Khurshudyan, Dr Era Khristoforovna. Inst. of Geological Sciences, Acad. of Sci. Armenian SSR, Barekamutyan St. 24a, Erevan 375200, USSR. (1934) Cand, geology and mineralogy (Erevan State U., 1972). Lab. head. *Crystal chemistry, X-ray structure analysis.*

Kidyarov, Dr Boris Ivanovich. Inst. of Semiconductor Physics, Acad. of Sci. of the USSR, Siberian Dept., Prospekt Nauki 13, Novosibirsk 630090, USSR. (1938) Cand, physics and mathematics (Siberian Dept., Acad. Sci. USSR, 1974). Scient. *Crystal growth, nucleation, defects, physical properties.*

Kiosse, Dr Georgy Alexandrovich. Inst. of Applied Physics, Acad. of Sciences Moldavian SSR, Akademicheskaya 5, Kishinev 277028, USSR. (1932) Cand, physics and mathematics (Inst. of Applied Physics, Acad. Sci. Moldavian SSR, 1975). Sr. scient. *X-ray structure analysis, crystal chemistry, inorganic compounds, phase transitions.*

Kirichenko, Dr Valentina Vasilyevna. Inst. of Crystallography, Acad. of Sci. of the USSR, Leninsky pr. 59. Moscow 117333, USSR. (1932) Cand, physics and mathematics (INCRYS, 1975). Scient. *Crystal lattice defects.*

Kirikov, Dr Vladimir Arkadyevich. Inst. of Crystallography, Acad. of Sci. of the USSR, Leninsky pr. 59, Moscow 117333, USSR. (1941) Cand, physics and mathematics (INCRYS, 1976). Scient. *Ferroelectrics (physics), phase transitions (ferroelectric).*

Kirkinsky, Dr Vitaly Alekseyevich. Inst. of Geology and Geophysics, Acad. of Sci. of the USSR, Siberian Dept., Novosibirsk 630090, USSR. (1937) Dr, geology and mineralogy (Inst. of Geochemistry and Analytical Chemistry, Acad. Sci. USSR, 1984). Lab. head. *Isomorphism, polymorphism, high-pressure crystal chemistry.*

Kirpichnikova, Dr Lyubov' Fedorovna. Inst. of Crystallography, Acad. of Sci. of the USSR, Leninsky pr. 59. Moscow 117333, USSR. (1944) Cand, physics and mathematics (INCRYS, 1972). Scient. *Structural phase transitions, ferroelectrics, ferroelastics.*

Kiselev, Dr Nikolay Andreyevich. Inst. of Crystallography, Acad. of Sci. of the USSR, Leninsky pr. 59, Moscow 117333, USSR. (1928) Coresp. member of the Acad. Sci. USSR; Dr, biology (Inst. of Biochemistry, Acad. Sci. USSR, 1964). Lab. head. *Protein structures, nucleic acids, viruses.*

Kislovsky, Dr Lev Dmitriyevich. Inst. of Crystallography, Acad. of Sci. of the USSR, Leninsky pr. 59, Moscow 117333, USSR. (1924) Cand, physics and mathematics (State Optical Inst., 1960). Sr. scient. *Infrared spectroscopy, quantum crystal chemistry, inorganic compounds.*

Klassen-Neklyudova, Prof. Marina Victorovna. Inst. of Crystallography, Acad. of Sci. of the USSR, Leninsky pr. 59, Moscow 117333, USSR. (1904) Dr, physics and mathematics (Leningrad Physico-Techn. Inst., 1936). Consulting scient. *Mechanical properties of solids.*

Klechkovskaya, Dr Vera Vsevolodovna. Inst. of Crystallography, Acad. of Sci. of the USSR, Leninsky pr. 59, Moscow 117333, USSR. (1938) Cand, physics and mathematics (INCRYS, 1974). Scient. *Structure analysis, electron diffraction, crystal chemistry.*

Kleshchinsky, Dr Leonid Innokentyevich. Irkutsk Inst. of Railway Engineers, Kurchatov St. 10, Irkutsk 664028, USSR. (1934) Cand, physics and mathematics (Leningrad Pedagogical Inst., 1968). Head of Chair. *Electron density and momentum density in solids, real crystals (X-ray diffraction), powder diffractometry.*

Klevtsov, Dr Petr Vasilyevich. Inst. of Inorganic Chemistry, Acad. of Sci. of the USSR, Siberian Dept., Prospekt Nauki 3, Novosibirsk 630090, USSR. (1930) Cand, physics and mathematics (INCRYS, 1955). Sr. scient. *Crystal growth, crystal chemistry, inorganic compounds, polymorphism.*

Klevtsova, Dr Rimma Fedorovna. Inst. of Inorganic Chemistry, Acad. of Sci. of the USSR, Siberian Deprtment, Prospekt Nauki 3, Novosibirsk 630090, USSR. (1928) Cand, physics and mathematics (INCRYS, 1954). Sr. scient. *Crystal chemistry, inorganic compounds, X-ray structure analysis, polymorphism.*

Klimova, Dr Anna Yuryevna. Inst. of Crystallography, Acad. of Sci. of the USSR, Leninsky pr. 59, Moscow 117333, USSR. Cand, chemistry (INCRYS, 1976). Scient. *Optical properties, optical activity of crystals.*

Kliya, Dr Maya Ottovna. Inst. of Crystallography, Acad. of Sci. of the USSR, Leninsky pr. 59, Moscow 117333, USSR. (1927) Cand, geology and mineralogy (INCRYS, 1952). Sr. scient. *Crystal growth, crystal morphology.*

Klyavin, Dr Oleg Vladimirovich. Physico-Techn. Inst., Acad. of Sci. of the USSR, Politekhnicheskaya St. 26, Leningrad 194021, USSR. (1931) Cand, physics and mathematics (Leningrad Polytechnic Inst., 1962). Sr. scient. *Defects in crystals, plasticity physics, strength of crystals.*

Knab, Dr Galina Grigoryevna. Inst. of Crystallography, Acad. of Sci. of the USSR, Leninsky pr. 59, Moscow 117333, USSR. (1934) Cand, physics and mathematics (INCRYS, 1974). Scient. *Defects, optical properties, mechanical properties.*

Kobzareva, Dr Svetlana Alekseyevna. Inst. of Crystallography, Acad. of Sci. of the USSR, Leninsky pr. 59, Moscow 117333, USSR. Cand, chemistry (INCRYS, 1966). Scient. *Crystal nucleation, epitaxy, electron microscopy.*

Kocharov, Dr Alexander Georgiyevich. Inst. of Crystallography, Acad. of Sci. of the USSR, Leninsky pr. 59, Moscow 117333, USSR. (1945) Cand, physics and mathematics (INCRYS, 1972). Scient. *Neutron diffraction, crystal structures, magnetic structures.*

Kolesova, Dr Rimma Vladimirovna. Rostov State U., Dept. of Physics, Prospekt Stachki 192, Rostov-on-Don 344061, USSR. Cand, physics and mathematics (Rostov U., 1967). Docent. *Crystal structure.*

Kolobyanina, Dr Tatyana Nikolayevna. Inst. of High Pressure Physics, Acad. of Sci. of the USSR, Akademgorodok, Podol'sky Rayon, Moskovskaya Oblast' 142092, USSR. Cand, physics and mathematics (MGU, 1974). Scient. *High pressure effect on crystal structure.*

Kolodiyeva, Dr Svetlana Vasilyevna. Res. Inst. for Synthesis of Mineral Raw Materials, Institutskaya St. 1, Alexandrov 601600, Vladimirskaya Oblast', USSR. (1936) Cand, physics and mathematics (MGU, 1979). Sr. scient. *Defects - physical properties interrelation, real crystals.*

Kolontsova, Dr Ekaterina Vasilyevna. Moscow State U., Dept. of Physics, Leninskiye Gory, Moscow 117234, USSR. Dr, physics and mathematics (MGU, 1970). Sr. scient. *Diffuse X-ray scattering, radiation effects, phase transformations in single crystals.*

Kolpakov, Dr Andrey Vasilyevich. Moscow State U., Dept. of Physics, Leninskiye Gory, Moscow 117234, USSR. (1941) Cand, physics and mathematics (MGU, 1968). Asst. *X-ray diffraction theory, group theory.*

Komov, Dr Igor' Leontyevich. Res. Inst. for Synthesis of Mineral Raw Materials, Institutskaya St. 1, Alexandrov 601600, Vladimirskaya Oblast', USSR. (1934) Cand, geology and mineralogy (Rostov U., 1965). Sr. scient. *Morphology, quartz crystals.*

Kon, Dr Aviv Yuliseyevich. Kishinev Polytechnic Inst., Lenin pr. 168, Kishinev 277004, USSR. (1929) Cand, physics and mathematics (Gorky U., 1967). Docent. *X-ray structure analysis methods, crystal chemistry, coordination compounds.*

Kondrashev, Dr Yury Dmitriyevich. State Inst. of Applied Chemistry, Vatny Ostrov 2, Leningrad, USSR. (1916) Cand, chemistry (State Inst. of Applied Chemistry, 1949). Sr. scient. *Organic crystal structure, structure determination methods.*

Kondratyeva, Dr Victoria Victorovna. Leningrad State U., Dept. of Geology, University Emb. 7/9, Leningrad 199164, USSR. (1932) Cand, geology and mineralogy (LGU, 1966). Scient. *Crystallography, crystal chemistry, X-ray structure analysis, borates, borosilicates.*

Konstantinova, Dr Alisa Fedorovna. Inst. of Crystallography, Acad. of Sci. of the USSR, Leninsky pr. 59, Moscow 117333, USSR. (1936) Cand, physics and mathematics (INCRYS, 1969). Scient. *Crystal optics, optical activity in crystals.*

Koptsik, Prof. Vladimir Alexandrovich. Moscow State U., Dept. of Physics, Leninskiye Gory, Moscow 117234, USSR. (1924) Dr, physics and mathematics (MGU, 1963). Prof. *Symmetry, crystal physics.*

Koreshkov, Dr Boris Dmitriyevich. Kolomna Pedagogical Inst., Zelenaya 30, Kolomna 140410, Moskovskaya Oblast', USSR. (1940) Cand, physics and mathematics (Moscow State Pedagogical Inst., 1968). Head of Chair. *Organic molecular crystals, thermodynamic properties.*

Kornev, Dr Aleksey Nikolayevich. Inst. of Biological Physics, Acad. of Sci. of the USSR, Pushchino 142292, Moskovskaya Oblast', USSR. (1944) Cand, physics and mathematics (INCRYS, 1973). Sr. scient. *Structure analysis, biologically active compounds.*

Korsukova, Dr Mariya Mikhailovna. Physico-Techn. Inst., Acad. of Sci. of the USSR, Politekhnicheskaya St. 26, Leningrad 194021, USSR. (1945) Cand, chemistry (Lvov State U. 1976). Scient. *Crystal growth.*

Koryagin, Mr. Vyacheslav Filippovich. Inst. of Crystallography, Acad. of Sci. of the USSR, Leninsky pr. 59, Moscow 117333, USSR. (1921). Scient. *Spectroscopy of crystals, magnetic resonance.*

Kosevich, Prof. Vadim Markovich. Kharkov Polytechnic Inst., Frunze St. 21, Kharkov 310002, USSR. (1931) Dr, physics and mathematics (Physico-Techn. Inst. of Low Temperatures, Acad. Sci. Ukrainian SSR, 1969). Head of Chair. *Real structure of crystals, imperfections in crystal structure, thin film growth, electron microscopy.*

Koshel', Dr Olga Stepanova. Dept. of Chemistry, Lvov State U., University St. 1, Lvov 742388, USSR. (1942) Cand, chemistry (Lvov U., 1977). Sr. scient. *X-ray structure analysis, crystal chemistry, intermetallic compounds.*

Kosmachev, Dr Sergey Mikhailovich. Kharkov Polytechnic Inst., Frunze St. 21, Kharkov 310002, USSR. (1946) Cand, technics (Kharkov Polytechnic Inst., 1974). Docent. *Crystal structure defects, thin film growth, electron microscopy.*

Kosova, Dr Tatyana Borisovna. Moscow State U., Dept. of Geology, Leninskiye Gory, Moscow 117234, USSR. (1940) Cand, geology and mineralogy (MGU, 1973). Scient. *Non-diffraction methods of studies, crystal growth.*

Kosterin, Dr Evgeny Andreyevich. Dept. of Physics, Ivanovo State Medical Inst., Engels St. 8, Ivanovo 153462, USSR. (1935) Cand, physics and mathematics (Ivanovo State Medical Inst.). Head of chair. *Liquid crystals, structure, properties.*

Kostyukova, Dr Evgeniya Prokofyevna. Machinery Inst., Acad. of Sci. of the USSR, Griboyedov St. 4, Moscow 101000, USSR. (1926) Cand, physics and mathematics (Petrozavodsk U., 1963). Sr. scient. *X-ray topography, lattice defects.*

Kotel'nikova, Dr Elena Nikolayevna. Leningrad State U., Dept. of Geology, University Emb. 7/9, Leningrad 199164, USSR. (1945) Cand, geology and mineralogy (LGU, 1982). Scient. *Crystal chemistry, laminated minerals.*

Kotov, Dr Nikolay Vladimirovich. Leningrad State U., Dept. of Geology, University Emb. 7/9, Leningrad 199164, USSR. (1935) Dr, geology and mineralogy (LGU, 1974). Sr. scient. *Crystal chemistry, high pressure synthesis, mineral structural transformation under high pressure.*

Kotur, Dr Bogdan Yaroslavovich. Dept. of Chemistry, Lvov State U., University St. 1, Lvov 742388, USSR. (1952) Cand, chemistry (Lvov U., 1978). Asst. *X-ray structure analysis, crystal chemistry, intermetallic compounds.*

Koval'chuk, Dr Mikhail Valentinovich. Inst. of Crystallography, Acad. of Sci. of the USSR, Leninsky pr. 59, Moscow 117333, USSR. (1946) Cand, physics and mathematics (INCRYS, 1978). Scient. *Dynamic X-ray scattering, real structure of crystals.*

Kovda, Prof. Leonid Mikhailovich. Dept. of Chemistry, Moscow State U., Leninskiye Gory, Moscow 117234, USSR. (1932) Cand, chemistry (MGU, 1971). Prof. *Crystal chemistry, inorganic compounds.*

Kovyev, Dr Ernest Konstantinovich. Inst. of Crystallography, Acad. of Sci. of the USSR, Leninsky pr. 59, Moscow 117333, USSR. (1941) Cand, physics and mathematics (MGU, 1974). Sr. scient. *X-ray diffraction, real structure of crystals.*

Koz'ma, Dr Alexander Alekseyevich. Kharkov Polytechnic Inst., Frunze St. 21, Kharkov 310002, USSR. (1939) Cand, technics (Kharkov Polytechnic Inst., 1968). Sr. scient. *Metals and alloys, structure, defects, methods for defect study.*

Koz'min, Dr Petr Alekseyevich. Inst. of General and Inorganic Chemistry, Acad. of Sci. of the USSR, Leninsky pr. 31, Moscow 117071, USSR. (1929) Cand, chemistry (IONCH, 1965). Sr. scient. *Crystal chemistry, inorganic compounds.*

Kozlenkov, Dr Alexander Ivanovich. Inst. of Metallurgy, Acad. of Sci. of the USSR, Leninsky pr. 49, Moscow 117334, USSR. (1932) Cand, physics and mathematics (Rostov U., 1965). Sr. scient. *Electronic states in crystals, lattice symmetry - electronic states relation, spectroscopy of crystals.*

Kozlova, Dr Olga Gerasimovna. Moscow State U., Dept. of Geology, Leninskiye Gory, Moscow 117234, USSR. (1927) Cand, geology and mineralogy (MGU, 1957). Docent. *Crystal growth, crystal morphology.*

Krasochka, Dr Oleg Nikolayevich. Branch Inst. of Chemical Physics, Acad. of Sci. of the USSR, Chernogolovka, Noginsky Rayon, Moskovskaya Oblast' 142432, USSR. (1943) Cand, chemistry (Branch Inst. of Chemical Physics, Acad. Sci. USSR, 1973). Scient. *Crystal chemistry, transition metal complex compounds.*

Krivandina, Dr Elena Alekseyevna. Inst. of Crystallography, Acad. of Sci. of the USSR, Leninsky pr. 59, Moscow 117333, USSR. (1938) Cand, physics and mathematics (INCRYS, 1980). Scient. *Crystal growth from melt, fluoride - single crystal growth.*

Krivoglaz, Prof. Mikhail Alexandrovich. Inst. of Metal Physics, Acad. of Sciences Ukrainian SSR, Vernadsky St. 36, Kiev 252142, USSR. (1929) Dr, physics and mathematics (Kharkov U., 1962). Dept. head. *X-ray scattering, thermal neutron scattering, phase transformations, defects in crystals.*

Krivokoneva, Dr Galina Kirillovna. Inst. of Mineral Raw Materials(VIMS), Staromonetny 29, Moscow 109017, USSR. (1938) Cand, geology and mineralogy (VIMS, 1971). Sr. scient. *X-ray structure analysis, polycrystals, crystal chemistry, minerals.*

Kruglik, Dr Anatoliy Ivanovich. Inst. of Physics, Acad. of Sci. of the USSR, Siberian Dept., Akademgorodok, Krasnoyarsk 660036, USSR. (1947) Cand, physics and mathematics (Inst. of Physics, Acad. Sci. USSR, Siberian Dept., 1981). Sr. scient. *Structure, ferroelectrics, ferroelastics, X-ray and neutron diffraction.*

Krutova, Dr Glafira Ivanova. Dept. of Geology, Moscow State U., Leninskiye Gory, Moscow 117234, USSR. (1931) Cand, geology and mineralogy (MGU, 1971). Scient. *Crystal chemistry, inorganic compounds, X-ray diffraction.*

Krymov, Dr Vladimir Mikhailovich. Physical Techn. Inst., Acad. of Sci. of the USSR, Politekhnicheskaya St. 26, Leningrad 194021, USSR. (1948) Cand, physics and mathematics (Physical Techn. Inst., Acad. Sci. USSR, 1979). Scient. *Semiconductors, dielectrics, crystal physics, crystallization from melt.*

Kudryavtseva, Dr Galina Petrova. Dept. of Geology, Moscow State U., Leninskiye Gory, Moscow 117234, USSR. (1947) Cand, geology and mineralogy (MGU, 1973). Sr. scient. *Crystal chemistry, real crystals.*

Kukharenko, Prof. Alexander Alexandrovich. Leningrad State U., Dept. of Geology, University Emb. 7/9, Leningrad 199164, USSR. (1914) Dr, geology and mineralogy (LGU, 1954). Head of Chair. *Crystal chemistry.*

Kukina, Dr Galina Alexandrovna. Inst. of General and Inorganic Chemistry, Acad. of Sci. of the USSR, Leninsky pr. 31, Moscow 117071, USSR. (1927) Cand, chemistry (IONCH, 1963). Scient. *Crystal chemistry, stereochemistry, coordination compounds, X-ray structure analysis techniques.*

Kukovsky, Prof. Evgeny Georgiyevich. Inst. of Geochemistry and Physics of Minerals, Acad. of Sci. Ukrainian SSR, Palladina pr. 34, Kiev 252068, USSR. (1925) Dr, geology and mineralogy (Inst. of Geology, Acad. Sci. Ukrainian SSR, 1965). Dept. head. *Crystal structures, submicroscopic structures, silicate microcrystals, phase transformations.*

Kukuy, Dr Anatoly Lvovich. Leningrad Mining Inst., 21st Liniya 2, Leningrad 199026, USSR. (1939) Cand, geology and mineralogy (GORNY, 1970). Scient. *Morphology, crystal growth.*

Kuntsevich, Dr Tamara Serafimovna. Gorky State U., Gagarin Prospekt 23, Gorky 603022, USSR. (1932) Cand, physics and mathematics (Gorky U., 1970). Sr. scient. *X-ray structure analysis, symmetry groups.*

Kupriyanov, Dr Mikhail Fedotovich. Rostov State U., Dept. of Physics, Prospekt Stachki 192, Rostov-on-Don 344061, USSR. (1937) Cand, physics and mathematics (Rostov U., 1968). Docent. *Phase transitions, complex oxides, crystal chemistry.*

Kuranova, Dr Inna Petrovna. Inst. of Crystallography, Acad. of Sci. of the USSR, Leninsky pr. 59, Moscow 117333, USSR. (1933) Cand, chemistry (MGU, 1963). Sr. scient. *Protein structures.*

Kurazhkovskaya, Dr Victoriya Semenovna. Moscow State U., Dept. of Geology, Leninskiye Gory, Moscow 117234, USSR. (1944) Cand, geology and mineralogy (MGU, 1971). Scient. *Crystal chemistry, intermetals and inorganic compounds.*

Kurbanov, Dr Khakim Mamadaliyevich. Physico-Techn. Inst., Acad. of Sci. of the Tadzhik SSR, Akademgorodok, Dushanbe, USSR. (1935) Cand, physics and mathematics (INCRYS, 1964). Lab. head. *Crystal structure, crystal growth.*

Kurdyumov, Prof. Georgy Vyacheslavovich. Inst. of Metal Physics, 2nd Bauman St. 9/23, Moscow 107005, USSR. (1902) Full member of the Acad. Sci. USSR, Dr, physics and mathematics (1937). Dir. *Metal physics, real structure of crystals.*

Kurkutova, Prof. Evdokiya Nikitichna. Vladimir State Pedagogical Inst., Prospekt Stroiteley 11, Vladimir 600024, USSR. (1930) Dr, physics and mathematics (INCRYS, 1978). Head of Chair. *X-ray structure analysis.*

Kuz'ma, Prof. Yury Bogdanovich. Lvov State U., Dept. of Chemistry, University St. 1, Lvov 290602, USSR. (1934) Dr, chemistry (Lvov U., 1974). Head of Chair. *X-ray structure analysis, crystal chemistry, intermetallic compounds.*

Kuz'min, Prof. Eduard Alekseyevich. Gorky State U., Gagarin Prospekt 23, Gorky 603022, USSR. (1939) Dr, physics and mathematics (Rostov U., 1974). *X-ray structure analysis.*

Kuz'min, Prof. Runar Nikolayevich. Moscow State U., Dept. of Physics, Leninskiye Gory, Moscow 117234, USSR. (1932) Dr, physics and mathematics (MGU, 1970). Prof. *Crystallography, solid state.*

Kuz'mina, Dr Irina Pavlovna. Inst. of Crystallography, Acad. of Sci. of the USSR, Leninsky pr. 59, Moscow 117333, USSR. (1932) Cand, chemistry (INCRYS, 1968). Sr. scient. *Crystal growth, crystal chemistry.*

Kuznetsov, Dr Alexander Victorovich. Petrozavodsk State U., Lenin pr. 33, Petrozavodsk 185018, USSR. (1932) Cand, physics and mathematics (MGU, 1963). Docent. *Dynamical theory of X-ray scattering.*

Kuznetsov, Prof. Fedor Andreyevich. Inst. of Inorganic Chemistry, Acad. of Sci. of the USSR, Siberian Dept., Prospekt Nauki 3, Novosibirsk 630090, USSR. (1932) Dr, chemistry (Scient. Council for Chemical Sci., Acad. Sci. USSR, Siberian Dept., 1972). Assoc. dir. *Crystal growth, Thin film growth.*

Kuznetsov, Prof. Vasily Grigoryevich. Inst. of General and Inorganic Chemistry, Acad. of Sci. of the USSR, Leninsky pr. 31, Moscow 117071, USSR. (1906) Dr, chemistry (IONCH, 1969). *Crystal chemistry, coordination compounds, semiconductors.*

Kuznetsov, Dr Victor Andreyevich. Inst. of Crystallography, Acad. of Sci. of the USSR, Leninsky pr. 59, Moscow 117333, USSR. (1938) Cand, geology and mineralogy (INCRYS, 1967). Sr. scient. *Crystal growth, crystal chemistry.*

Kvasnitsa, Dr Victor Nikolayevich. Inst. of Geochemistry and Physics of Minerals, Acad. of Sci. of the Ukrainian SSR, Palladina Pr. 34, Kiev 252068, USSR. (1942) Cand, geology and mineralogy (Inst. of Geochemistry and Physics of Minerals, Acad. Sci. Ukrainian SSR, 1974). Scient. *Mineralogical crystallography.*

Kyutt, Dr Reginal'd Nikolayevich. Physical Techn. Inst., Acad. of Sci. of the USSR, Zapovednaya St. 51, Leningrad 194037, USSR. (1944) Cand, physica and mathematics (Physical Techn. Inst., Acad. Sci. USSR, 1979). Scient. *Defects, diffractometry - single crystal.*

Lazarenkov, Prof. Vadim Grigor'yevich. Leningrad Mining Inst., 21st Liniya 2, Leningrad 199026, USSR. (1933) Dr, geology and mineralogy (GORNY, 1980). Head of chair. *Mineralogical crystallography.*

Lazarev, Dr Eduard Mikhailovich. Inst. of Metallurgy, Acad. of Sci. of the USSR, Leninsky pr. 49, Moscow 117334, USSR. (1937) Cand, technics (Inst. of Metallurgy, Acad. Sci. USSR, 1967). Scient. *Inorganic compounds, structure.*

Lebedev, Prof. Vasily Ilyich. Leningrad State U., Dept. of Geology, University Emb. 7/9, Leningrad 199164, USSR. (1911) Dr, geology and mineralogy (LGU, 1955). Lab. head. *Structural crystallography, crystal chemistry, mineralogy, geochemistry.*

Lebedeva, Dr Marina Vladimirovna. Kharkov Polytechnic Inst., Frunze St. 21, Kharkov 310002, USSR. (1941) Cand, technics (Kharkov Polytechnic Inst., 1973). Sr. lect. *Metals and alloys, structure, defects, methods for defect study.*

Leonyuk, Dr Lidiya Ivanovna. Moscow State U., Dept. of Geology, Leninskiye Gory, Moscow 117234, USSR. (1950) Cand, geology and mineralogy (MGU, 1978). Scient. *Crystal growth, morphology.*

Leonyuk, Dr Nikolay Ivanovich. Moscow State U., Dept. of Geology, Leninskiye Gory, Moscow 117234, USSR. (1941) Cand, geology and mineralogy (MGU, 1972). Sr. scient. *Crystal growth, crystal chemistry, borates.*

Levanyuk, Prof. Arkady Petrovich. Inst. of Crystallography, Acad. of Sci. of the USSR, Leninsky pr. 59, Moscow 117333, USSR. (1933) Dr, physics and mathematics (INCRYS, 1977). Sr. scient. *Polymorphic transformations in crystals, ferroelectrics.*

Lim, Dr Valery Irovich. Inst. of Protein, Acad. of Sci. of the USSR. Pushchino, Serpukhovsky Rayon, Moskovskaya Oblast' 142292, USSR. (1943) Cand, physics and mathematics (INCRYS, 1973). Sr. scient. *Biopolymer crystallography.*

Lindin', Dr Lauma Felixovna. Riga Polytechnical Inst., Kronvald Boulevard 4, Riga 226828, USSR. (1935) Cand, technics (Riga Polytechn. Inst., 1972). Sr. scient. *Crystal chemistry, crystal phase formation, inorganic oxide systems.*

Liopo, Dr Valery Alexandrovich. Pedagogic Inst., Sovetskaya 8, Brest 664005, USSR. (1939) Cand, physics and mathematics (Irkutsk State U., 1968). Prorector. *Crystallographic group theory, teaching, structure and properties of crystals.*

Lisoivan, Dr Vladimir Ivanovich. Inst. of Inorganic Chemistry, Acad. of Sci. of the USSR, Siberian Dept., Prospekt Nauki 3, Novosibirsk 630090, USSR. (1936) Cand, physics and mathematics (Inst. of Semiconductor Physics, Acad. Sci., USSR, Siberian Dept., 1971). Sr. scient. *Crystal lattice defects, X-ray structure analysis.*

Litvin, Dr Alexander Lukich. Inst. of Geochemistry and Physics of Minerals, Acad. of Sci. Ukrainian SSR, Palladina pr. 34, Kiev 252068, USSR. (1927) Dr, geology and mineralogy (Kiev U., 1978). Sr. scient. *Structural typomorphism, rock-forming minerals.*

Litvin, Dr Boris Nikolayevich. Moscow State U., Dept. of Geology, Leninskiye Gory, Moscow 117234, USSR. (1934) Dr, chemistry (IONCH, 1978). Sr. scient. *Crystal growth, crystal chemistry.*

Litvinskaya, Mrs Galina Petrovna. Moscow State U., Dept. of Geology, Leninskiye Gory, Moscow 117234, USSR. (1920) Sr. lect. *Geometrical crystallography, crystal chemistry, inorganic compounds, structure.*

Lityagina, Dr Lyudmila Mitrofanovna. Inst. of High Pressure Physics, Acad. of Sci. of the USSR, Akademgorodok, Podol'sky Rayon, Moskovskaya Oblast' 142091, USSR. Cand, geology and mineralogy (MGU, 1976). Scient. *High pressure effects, crystal structure.*

Lobkovsky, Dr Emil' Borisovich. Inst. of New Chemical Problems, Acad. of Sci. of the USSR, Chernogolovka 142232, Moskovskaya Oblast', USSR. (1941) Cand, chemistry (MGU, 1974). Scient. *Crystal chemistry, transition metal hydrides and tetrahydroborates.*

Loshmanov, Dr Arkady Andreyevich. Inst. of Crystallography, Acad. of Sci. of the USSR, Leninsky pr. 59, Moscow 117333, USSR. (1929) Cand, physics and mathematics (Inst. of Metallurgy of Ferrous Metals, 1967). Sr. scient. *Neutron diffraction, non-crystalline and crystalline state.*

Lukashev, Dr Alexander Nikolayevich. Res. Inst. for Synthesis of Mineral Raw Materials, Institutskaya St. 1, Alexandrov 601600, Vladimirskaya Oblast', USSR. (1926) Cand, geology and mineralogy (GORNY, 1965). Sr. scient. *Morphology and composition - relationship to geological conditions of formation.*

Lvov, Dr Yury Mikhailovich. Inst. of Crystallography, Acad. of Sci. of the USSR, Leninsky pr. 59, Moscow 117334, USSR. (1952) Cand, physics and mathematics (MGU, 1978). Scient. *Small angle X-ray scattering.*

Lyakhovitskaya, Dr Vera Aronovna. Inst. of Crystallography, Acad. of Sci. of the USSR, Leninsky pr. 59, Moscow 117333, USSR. (1929) Cand, chemistry (INCRYS, 1967). Sr. scient. *Crystal growth from melt and vapour phase.*

Lyubalin, Dr Mark Dmitriyevich. Leningrad Mining Inst., 21st Liniya 2, Leningrad 199026, USSR. (1937) Cand, technics (GORNY, 1970). Sr. scient. *Morphology and crystal growth.*

Lyubimov, Dr Vasily Nikolayevich. Karpov Physical Chemistry Inst., Obukha St. 10, Moscow 107120, USSR. (1936) Cand, chemistry (KARPOV, 1965). Sr. scient. *Theoretical crystal physics.*

Lyubitov, Dr Yury Naumovich. Inst. of Crystallography, Acad. of Sci. of the USSR, Leninsky pr. 59, Moscow 117333, USSR. (1932) Cand, physics and mathematics (INCRYS, 1978). *Crystal growth.*

Lyubutin, Dr Igor' Savelyevich. Inst. of Crystallography, Acad. of Sci. of the USSR, Leninsky pr. 59, Moscow 117333, USSR. (1938) Dr, physics and mathematics (INCRYS, 1975). Sr. scient. *Magnetism, Mössbauer spectroscopy.*

Lyutin, Dr Vladimir Ivanovich. Res. Inst. for Synthesis of Mineral Raw Materials, Institutskaya St. 1, Alexandrov 601600, Vladimirskaya Oblast', USSR. (1948) Cand, physics and mathematics (INCRYS, 1974). Sr. scient. *Crystal growth, X-ray structure analysis, crystal chemistry.*

Lyutzau, Dr Vsevolod Grigoryevich. Machinery Inst., Acad. of Sci. of the USSR. Griboyedov St. 4, Moscow 101000, USSR. (1922) Dr, technics (Inst. of Steel and Alloys, 1972). Lab. head. *Real crystals (substructure), lattice defects.*

Maiyer, Prof. Alexander Artemyevich. Moscow Chemical Techn. Inst., Miusskaya Square 9, Moscow 125820, USSR. (1927) Dr, chemistry (Moscow Chemical Techn. Inst., 1970). *Crystallography, crystal chemistry, selenites, chromites, molybdates, tungstates.*

Makarenko, Dr Igor' Nikolayevich. Inst. of Crystallography, Acad. of Sci. of the USSR, Leninsky pr. 59, Moscow 117333, USSR. (1938) Cand, physics and mathematics (INCRYS, 1971). Sr. scient. *High pressure phase transitions.*

Makarov, Prof. Evgeny Sergeyevich. Inst. of Geochemistry and Analytical Chemistry, Acad. of Sci. of the USSR, Kosygin pr. 19, Moscow 117334, USSR. (1911) Dr, chemistry (IONCH, 1954). Consulting prof. *X-ray diffraction, crystal chemistry, inorganic compounds.*

Malakhova, Dr Lyudmila Fedorovna. Inst. of Crystallography, Acad. of Sci. of the USSR, Leninsky pr. 59, Moscow 117333, USSR. (1941) Cand, physics and mathematics (INCRYS, 1976). Sr. scient. *X-ray diffractometry, apparatus and methods.*

Malinenko, Dr Inna Avramovna. Petrozavodsk State U., Lenin Prospekt 33, Petrozavodsk 185018, USSR. (1941) Cand, physics and mathematics (Moscow Inst. of Steel and Alloys, 1971). Docent. *Point defects, lattice dynamics in crystals.*

Malinovsky, Dr Stanislav Tadeushevich. Inst. of Applied Physics, Acad. of Sci. of the Moldavian SSR, Akademicheskaya 5, Kishinev 277028, USSR. (1949) Cand, physics and mathematics (INCRYS, 1978). Scient. *Structural chemistry, coordination compounds, organic and bio-organic compounds.*

Malinovsky, Dr Yury Alexandrovich. Inst. of Crystallography, Acad. of Sci. of the USSR, Leninsky pr. 59, Moscow 117333, USSR. (1947) Cand, geology and mineralogy (MGU, 1976). Sr. scient. *Crystal structure.*

Malinovsky, Prof. Tadeush Iosifovich. Inst. of Applied Physics, Acad. of Sciences Moldavian SSR, Akademicheskaya 5, Kishinev 277028, USSR. (1921) Full member of the Acad. Sci. Moldavian SSR, Dr, physics and mathematics (INCRYS, 1967). Lab. head. *Crystallography, crystal chemistry, X-ray crystal structure analysis.*

Malyushitskaya, Dr Zinaida Vladimirovna. Inst. of High Pressure Physics, Acad. of Sci. of the USSR, Akademgorodok, Podol'sky Rayon, Moskovskaya Oblast' 142092, USSR. Cand, chemistry (MGU, 1974). Scient. *High pressure effect on crystal structure.*

Mamedov, Prof. Kerim Panakh ogly. Inst. of Physics, Acad. of Sci. of the Azerbaidzhan SSR, Narimanov Prospekt 33, Baku 370143, USSR. (1916) Dr, physics and mathematics (Azerbaidzhan State U., 1968). Lab. head. *Phase transitions in solids.*

Mamedov, Prof. Khudu Surkhay ogly. Inst. of Inorganic and Physical Chemistry, Acad. of Sci. of the Azerbaidzhan SSR, Narimanov pr. 29, Baku 370143, USSR. (1927) Corresp. member of the Acad. Sci. Azerbaidzhan SSR, Dr, chemistry (Inst. of Geology, Acad. Sci. Azerbaidzhan SSR, 1970). Lab. head. *Structure analysis and crystal chemistry, complexes, biological compounds, silicates.*

Man, Dr Lucia Ivanovna. Inst. of Crystallography, Acad. of Sci. of the USSR, Leninsky pr. 59, Moscow 117333, USSR. Cand, physics and mathematics (INCRYS, 1970). Scient. *Electron diffraction, semiconductors, crystal chemistry, structure analysis.*

Marfunin, Dr Arnol'd Sergeyevich. Inst. of Geology, Mineralogy and Petrography, Acad. of Sci. of the USSR, Staromonetny 35, Moscow 109017, USSR. (1926) Dr, geology and mineralogy (IGEM, 1962). Lab. head. *Minerals (physics), crystal field theory, electron paramagnetic resonance, electron-hole centers.*

Martyshev, Dr Yury Nikolayevich. Inst. of Crystallography, Acad. of Sci. of the USSR, Leninsky pr. 59, Moscow 117333, USSR. (1931) Cand, physics and mathematics (INCRYS, 1971). Scient. *Optical properties of crystals.*

Mastryukov, Dr Vladimir Saidovich. Moscow State U., Dept. of Chemistry, Leninskiye Gory, Moscow 117234, USSR. (1935) Cand, chemistry (MGU, 1966). Sr. scient. *Structure, elemento-organic compounds, electron diffraction by gases.*

Matkovsky, Dr Orest Ilyarovich. Lvov State U., Shcherbakov St. 4, Lvov 290005, USSR. (1929) Cand, geology and mineralogy (Lvov U., 1957). Head of Chair. *Mineralogical crystallography.*

Matveeva, Mrs Rimma Georgiyevna. Inst. of Crystallography, Acad. of Sci. of the USSR, Leninsky pr. 59, Moscow 117333, USSR. Scient. *X-ray structure analysis, computing.*

Mavlonov, Dr Sharaf. Physical Techn. Inst., Acad. of Sci. of the Tadzhik SSR, Akademgorodok, Dushanbe 734630, USSR. (1935) Cand, physics and mathematics (Azerbaidzhan U., 1962). Lab. head. *Semiconductors, crystal structure.*

Maximov, Dr Boris Alekseyevich. Inst. of Crystallography, Acad. of Sci. of the USSR, Leninsky pr. 59, Moscow 117333, USSR. (1941) Cand, physics an mathematics (INCRYS, 1969). Sr. scient. *X-ray structure analysis.*

Mazus, Dr Mark Davidovich. Inst. of Applied Physics, Acad. of Sciences Moldavian SSR, Akademicheskaya 5, Kishinev 277028, USSR. (1937) Cand, physics and mathematics (Inst. of Applied Physics, Acad. Sci. Moldavian SSR, 1974). Sr. scient. *X-ray structure analysis, coordination and organic compounds.*

Mel'nikov, Dr Oleg Konstantinovich. Inst. of Crystallography, Acad. of Sci. of the USSR, Leninsky pr. 59, Moscow 117333, USSR. (1940) Cand, geology and mineralogy (INCRYS, 1968). Sr. scient. *Crystal growth, crystal chemistry.*

Mel'nikov, Dr Vitaly Alexandrovich. Inst. of Crystallography, Acad. of Sci. of the USSR, Leninsky pr. 59, Moscow 117333, USSR. (1937). Cand, physics and mathematics (INCRYS, 1982). Computing center head. *Automation of crystallographic investigations.*

Mel'nikova, Dr Alina Mikhailovna. Inst. of Crystallography, Acad. of Sci. of the USSR, Leninsky pr. 59, Moscow 117333, USSR. (1940). Cand, physics and mathematics (INCRYS, 1982). Scient. *Crystal growth, growth theory, transfer processes.*

Melekh, Dr Bernard Abu-Talibovich. Physical Techn. Inst., Acad. of Sci. of the USSR, Zapovednaya St. 51, Leningrad 194037, USSR. (1937) Cand, chemistry (Moscow Inst. of Steel and Alloys, 1967). Sr. scient. *Thermoelectric materials, oxide compounds.*

Meleshina, Dr Valentina Alexandrovna. Inst. of Crystallography, Acad. of Sci. of the USSR, Leninsky pr. 59, Moscow 117333, USSR. Cand, physics and mathematics (INCRYS, 1967). Scient. *Real structure of crystals, optical microscopy, electron microscopy.*

Melik-Adamyan, Dr Vil'yam Rafailovich. Inst. of Crystallography, Acad. of Sci. of the USSR, Leninsky pr. 59, Moscow 117333, USSR. (1937) Cand, physics and mathematics (INCRYS, 1969). Sr. scient. *Protein structures.*

Merinev, Dr Boris Vladimirevich. Inst. of Crystallography, Acad. of Sci. of the USSR, Leninsky pr. 59, Moscow 117333, USSR. (1954) Cand, physics and mathematics (INCRYS, 1981). Scient. *X-ray structure analysis, solid electrolytes.*

Metsik, Prof. Mikhail Stepanovich. Irkutsk State U., Dept. of Physics, Karl Marx St. 1, Irkutsk 664003, USSR. (1918) Dr, physics and mathematics (Physical Chemistry Inst., 1965). Head of chair. *Layer silicates, structure, thin water films.*

Mikhailov, Dr Al'bert Mikhailovich. Inst. of Crystallography, Acad. of Sci. of the USSR, Leninsky pr. 59, Moscow 117333, USSR. (1939) Cand, physics and mathematics (INCRYS, 1971). Scient. *Protein structures, virus structures, X-ray structure analysis, electron microscopy.*

Mikhailov, Dr Igor' Fedorovich. Kharkov Polytechnic Inst., Frunze St. 21, Kharkov 310002, USSR. (1949) Cand, technics (Kharkov Polytechnic Inst., 1975). Scient. *Metals, structure, semiconductors, defects, methods for defect study.*

Mikhailov, Dr Vladimir Ivanovich. Inst. of Crystallography, Acad. of Sci. of the USSR, Leninsky pr. 59, Moscow 117333, USSR. (1944) Cand, physics and mathematics (INCRYS, 1978). Scient. *Crystal growth from gas phase, sublimation, surface studies.*

Mikhailov, Dr Yuriy Nikolayevich. Inst. of General and Inorganic Chemistry, Acad. of Sci. of the USSR, Leninsky pr. 31, Moscow 117071, USSR. (1932) Cand, chemistry (IONCH, 1969). Sr. scient. *Inorganic structures, coordination compounds (chemistry).*

Mikhalenko, Dr Svetlana Ivanovna. Lvov State U., Dept. of Chemistry, University St. 1, Lvov 290602, USSR. (1946) Cand, chemistry (Lvov U., 1976). Scient. *X-ray structure analysis, crystal chemistry, intermetallic compounds.*

Mikheeva, Dr Irina Victorovna. Inst. 'Mekhanobr', 21st Liniya 8a, Leningrad 199026, USSR. (1921)Cand, geology and mineralogy (GORNY, 1952). Sr. scient. *X-ray studies, crystal chemistry.*

Millionova, Dr Margarita Ivanovna. Inst. of Molecular Biology, Acad. of Sci. of the USSR, Vavilov St. 32, Moscow 117312, USSR. Cand, physics and mathematics (Inst. of Biophysics, Acad. Sci. USSR, 1964). Scient. *Fibrillous structures.*

Minacheva, Dr Lidiya Khabibovna. Inst. of General and Inorganic Chemistry, Acad. of Sci. of the USSR, Leninsky pr. 31, Moscow 117071, USSR. (1938) Cand, chemistry (IONCH, 1971). Scient. *Crystal chemistry, stereochemistry, coordination compounds*

Mineeva, Dr Rimma Mikhailovna. Inst. of Geology, Mineralogy and Petrography, Acad. of Sci. of the USSR, Staromonetny 35, Moscow 109017, USSR. (1938) Cand, physics and mathematics (Kazan U., 1967). Scient. *Minerals (physics), crystal field theory, electron paramagnetic resonance.*

Mints, Prof. Rafail Isaakovich. Ural Polytechnic Inst., Sverdlovsk 620002, USSR. (1931) Dr, technics (Ural Polytechnic Inst., 1965). Dept. head. *Structure phase transformations, metastable states, ordered condensed systems.*

Miuskov, Dr Vasily Fedorovich. Inst. of Crystallography, Acad. of Sci. of the USSR, Leninsky pr. 59, Moscow 117333, USSR. (1909) Cand, physics and mathematics (Leningrad Polytechnic Inst., 1941). Sr. scient. *X-ray topography, X-ray moire, crystal growth, defects in crystals.*

Model', Dr Mariya Samuilovna. Inst. of Metallurgy, Acad. of Sci. of the USSR, Leninsky pr.49, Moscow 117334, USSR. (1920) Cand, chemistry (IONCH, 1951). Sr. scient. *Inorganic compounds, structure.*

Mokeeva, Dr Valentina Ivanovna. Inst. of Geochemistry and Analytical Chemistry, Acad. of Sci. of the USSR, Kosygin pr. 19, Moscow 117334, USSR. (1923) Cand, physics and mathematics (INCRYS, 1952). Scient. *Inorganic compounds, structure, isomorphism.*

Molchanov, Dr Vladimir Nikolayevich. Inst. of Crystallography, Acad. of Sci. of the USSR, Leninsky pr. 59, Moscow 117333, USSR. (1952) Cand, chemistry (INCRYS, 1982). Scient. *X-ray structure analysis, high pressure.*

Moroz, Dr Ella Mikhailovna. Inst. of Catalysis, Acad. of Sci. of the USSR, Siberian Dept., Prospekk Nauki 5, Novosibirsk 630090, USSR. (1939) Cand, physics and mathematics (Rostov U., 1971). Scient. *Inorganic compounds, structure, phase transformation.*

Moskvin, Dr Valentin Vasilyevich. Inst. of Crystallography, Acad. of Sci. of the USSR, Leninsky pr. 59, Moscow 117333, USSR. (1941) Cand, chemistry (INCRYS, 1979). Scient. *Nucleation, thin films, epitaxy, electron microscopy.*

Mukhtarova, Dr Nina Nikolayevna. Inst. of Nuclear Physics, Acad. of Sci. of the Uzbek SSR, Tashkent 702132, USSR. Cand, physics and mathematics (INCRYS, 1981). Scient. *Structural crystallography, crystal chemistry, inorganic compounds.*

Muradyan, Dr Lyudmila Andranikovna. Inst. of Crystallography, Acad. of Sci. of the USSR, Leninsky pr. 59, Moscow 117333, USSR. (1936) Cand, (INCRYS). Sr. scient. *Crystallographic computing methods.*

Musayev, Dr Faig Nasib ogly. Inst. of Inorganic and Physical Chemistry, Acad. of Sci. of the Azerbaidzhan SSR, Narimanov Pr. 29, Baku 370143, USSR. (1952) Cand, chemistry (Tbilisi State U., 1979). Sr. scient. *Structure, coordination compounds.*

Mustafayev, Dr Nariman Mustafa ogly. Inst. of Inorganic and Physical Chemistry, Acad. of Sci. of the Azerbaidzhan SSR, Narimanov Pr. 29, Baku 370143, USSR. (1929) Cand, geology and mineralogy (INCRYS, 1966). Sr. scient. *Hydrothermal synthesis, crystal chemistry, silicates.*

Myasnikova, Dr Rimma Mikhailovna. Inst. of Elemento-Organic Compounds, Acad. of Sci. of the USSR, Vavilov St. 28, Moscow 117312, USSR. (1932) Dr, physics and mathematics (INCRYS, 1981). Sr. scient. *Crystal structure and properties, binary organic systems.*

Myl'nikova, Dr Irina Evgenyevna. Physico-Techn. Inst., Acad. of Sci. of the USSR, Zapovednaya St. 51, Leningrad 194037, USSR. (1922) Cand, technics (Inst. of Silicate Chemistry, Acad. Sci. USSR, 1956). Sr. scient. *Crystal growth, single crystal growth methods, properties of single crystals.*

Mys'kiv, Dr Mar'yan Grigoryevich. Lvov State U., Dept. of Chemistry, University St. 1, Lvov 290602, USSR. (1947) Cand, chemistry (Lvov U., 1973). Asst. *X-ray structure analysis, crystal chemistry, intermetallic compounds.*

Myshlyayev, Prof. Mikhail Mikhailovich. Inst. of Solid State Physics, Acad. of Sci. of the USSR, Chernogolovka, Noginsky Rayon, Moskovskaya Oblast' 142432, USSR. (1934) Dr, physics and mathematics (INCRYS, 1982). Sr. scient. *Electron microscopy (contrasting defects), defects, plastic deformation mechanisms.*

Naboka, Dr Marat Nikolayevich. Kharkov Polytechnic Inst., Frunze St. 21, Kharkov 310002, USSR. (1933) Cand, technics (Kharkov Polytechnic Inst., 1969). Sr. scient. *Metals and alloy structures, defects and methods of measurement.*

Nadezhina, Dr Tamara Nikolayevna. Moscow State U., Dept. of Geology, Leninskiye Gory, Moscow 117234, USSR. (1946) Cand, geology and mineralogy (MGU, 1975). Scient. *X-ray structure analysis, minerals and synthetic compounds, geocrystal chemistry.*

Nagaitsev, Dr Yury Valeryevich. Leningrad State U., Dept. of Geology, University Emb. 7/9, Leningrad 199164, USSR. (1937) Cand, geology and mineralogy (LGU, 1965). Sr. scient. *Crystal chemistry, metamorphic minerals.*

Nefedova, Dr Elena Vasilyevna. Leningrad Mining Inst., 21st Liniya 2, Leningrad 199026, USSR. (1934) Cand, geology and mineralogy (GORNY, 1971). Sr. scient. *Crystal growth, morphology.*

Neronova, Dr Nina Nikolayevna. All-Union Correspondence Inst. of Eng. and Construction, Lesnaya St. 5, Kostroma 156021, USSR. Cand, physics and mathematics (INCRYS, 1965). Docent. *Symmetry, teaching.*

Nesterova, Dr Yaroslava Mikhailovna. Moscow State U., Dept. of Chemistry, Leninskiye Gory, Moscow 177234, USSR. (1937) Cand, chemistry (MGU, 1973). Scient. *Crystal chemistry, coordination compounds, X-ray structure analysis.*

Nikanorov, Dr Stanislav Prokhorovich. Physico-Techn. Inst., Acad. of Sci. of the USSR, Politekhnicheskaya St. 26, Leningrad 194021, USSR. (1928) Dr, physics and mathematics (Phsico-Techn. Inst., Acad. Sci. USSR, 1978). Assoc. dir. *Crystal growth, structure imperfections, elasticity and plasticity, crystal physics.*

Nikiforov, Dr Igor' Yakovlevich. Rostov State U., Prospekt Stachki 192, Rostov-on-don 344061, USSR. (1930) Cand, physics and mathematics (Rostov U., 1966). Docent. *Perfection studies of single crystals, three-crystal X-ray spectrometers-diffractometers.*

Nikishova, Dr Lidiya Vasilyevna. Inst. of Geology, Acad. of Sci. of the USSR, Yakutsk Branch of Siberian Dept., Lenin Pr. 39, Yakutsk 677007, USSR. (1938) Cand, physics and mathematics (Irkutsk U., 1976). Scient. *Isomorphism, polymorphism, polytypism, electron diffraction, microdiffraction.*

Nikitenko, Prof. Valerian Ivanovich. Inst. of Solid State Physics, Acad. of Sci. of the USSR, Chernogolovka 142432, Noginsky Rayon, Moskovskaya Oblast', USSR. (1937) Dr, physics and mathematics (Inst. of Solid State Physics, 1972). Lab. head. *Defects, crystal structure, physical properties - defects relationship.*

Novozhilov, Dr Alexander Ivanovich. Res. Inst. for Synthesis of Mineral Raw Materials, Institutskaya St. 1, Alexandrov 601600, Vladimirskaya Oblast', USSR. (1939) Cand, physics and mathematics (INCRYS, 1974). Sr. scient. *Optical spectroscopy of crystals, electron paramagnetic resonance.*

Nozik, Dr Yury Zinovyevich. Inst. of Geochemistry and Analytical Chemistry, Acad. of Sci. of the USSR, Kosygin pr. 19, Moscow 117334, USSR. (1933) Dr, physics and mathematics (INCRYS, 1979). Sr. scient. *Neutron diffraction, hydrogen bonds.*

Nuriyev, Dr Idayat Ragim ogly. Inst. of Physics, Acad. of Sci. of the Azerbaidzhan SSR, Narimanov Pr. 33, Baku 370143, USSR. (1941) Cand, physics and mathematics (Inst. of Physics, Acad. Sci. Azerbaidzhan SSR, 1971). Lab. head *Thin film growth, structure and physical properties, epitaxial films, semiconductors.*

Onishchina, Dr Ninel' Mitrofanovna. Leningrad Mining Inst., 21st Liniya 2, Leningrad 199026, USSR. (1935) Cand, geology and mineralogy (GORNY, 1975). Sr. teacher. *Crystal morphology, crystal growth.*

Opekunov, Dr Victor Nikolayevich. Inst. of Crystallography, Acad. of Sci. of the USSR, Leninsky pr. 59, Moscow 117333, USSR. (1950) Cand, physics and mathematics (MGU, 1982). Scient. *Real structure, defects, mechanical properties of crystals.*

Orekhova, Dr Valentina Petrovna. Inst. of Crystallography, Acad. of Sci. of the USSR, Leninsky pr. 59, Moscow 117333, USSR. (1940) Cand, physics and mathematics (INCRYS, 1975). Scient. *Optical spectroscopy, doped crystals.*

Organova, Dr Nataliya Ivanovna. Inst. of Geology, Mineralogy and Petrography, Acad. of Sci. of the USSR, Staromonetny 35, Moscow 109017, USSR. (1929) Cand, geology and mineralogy (IGEM, 1972). Sr. scient. *Minerals, structure, atomic ordering, hydrogen bonding.*

Osip'yan, Prof. Yury Andreyevich. Inst. of Solid State Physics, Acad. of Sci. of the USSR, Chernogolovka, Noginsky Rayon, Moskovskaya Oblast' 142432, USSR.

(1931) Full member of the Acad. Sci. USSR; Dr, physics and mathematics; Director. *Crystal structure, semiconductors, dielectrics.*

Ostanevich, Dr Yury Mechislavovich. Joint Inst. for Nuclear Res., Dubna 141980, Moskovskaya Oblast', USSR. (1936) Dr, physics and mathematics (Joint Inst. for Nuclear Res., 1972). Dept. head. *Neutron physics.*

Otroshchenko, Dr Lyudmila Petrovna. Inst. of Crystallography, Acad. of Sci. of the USSR, Leninsky pr. 59, Moscow 117333, USSR. (1940) Cand, geology and mineralogy (MGU, 1981). Scient. *X-ray structure analysis.*

Ovchinnikov, Dr Yury Kuz'mich. Karpov Physical Chemistry Inst., Obukha St. 10, Moscow 107120, USSR. (1940) Cand, chemistry (KARPOV, 1969). Scient. *Polymers (crystalline and amorphous), structure, liquid crystals.*

Ozerin, Dr Alexander Nikiforovich. Karpov Physical Chemistry Inst., Obukha St. 10, Moscow 107120, USSR. (1952) Cand, physics and mathematics (Moscow Physico-Techn Inst., 1977). Sr. scient. *X-ray structure analysis, polymers.*

Ozerov, Prof. Ruslan Pavlovich. Moscow Chemico-Technological Inst., Miusskaya sq. 9, Moscow 125820, USSR. (1926) Dr, physics and mathematics (INCRYS, 1969). Head of Chair. *X-ray diffraction, neutron diffraction, solid state physics.*

Ozola, Dr Astrida Davovna. Inst. of Inorganic Chemistry, Acad. of Sci. of the Latvian SSR, Meistaru St. 10, Riga 226934, USSR. Cand, chemistry (Inst. of Inorganic Chemistry Acad. Sci. Latvian SSR, 1977). Sr. scient. *Crystal chemistry.*

Ozolin'sh, Dr Gerkhard Vladimirovich. Inst. of Inorganic Chemistry, Acad. of Sci. of the Latvian SSR, Meistaru 10, Riga 226934, USSR. (1934) Cand, physics and mathematics (Latvian State U., 1969). Lab. head. *Instrumentation and techniques for X-ray crystallography, lattice parameter precision determination.*

Ozols, Dr Yan Karlovich. Inst. of Inorganic Chemistry, Acad. of Sci. of the Latvian SSR, Meistaru 10, Riga 226934, USSR. (1915) Cand, chemistry (Latvian State U., 1949). Sr. scient. *Crystal chemistry, borates, chelate compounds.*

Pakhomov, Dr Vladimir Ivanovich. Inst. of General and Inorganic Chemistry, Acad. of Sci. of the USSR, Leninsky pr.31, Moscow 117071, USSR. (1932) Dr, chemistry (IONCH, 1973). Sr. scient. *Crystal chemistry, inorganic compounds.*

Palatnik, Prof. Lev Samoilovich. Kharkov Polytechnik Inst., Frunze St. 21, Kharkov 310002, USSR. (1909) Dr, physics and mathematics (Kharkov U., 1952). Head of Chair. *Metals, structure, sub-structure, semiconductors, dielectrics, films (single- & poly-crystal), amorphous films.*

Palistrant, Dr Alexander Filippovich. Kishinev U., Sadovaya 60, Kishinev 277003, USSR. (1933) Dr, physics and mathematics (INCRYS, 1984). Docent. *Symmetry theory (generalization and applications).*

Palkina, Dr Kapitolina Kapitonovna. Inst. of General and Inorganic Chemistry, Acad. of Sci. of the USSR, Leninsky pr. 31, Moscow 117071, USSR. (1932) Cand, chemistry (IONCH, 1963). Sci. *Crystal chemistry, coordination compounds, semiconductors.*

Panchekha, Dr Petr Alekseyevich. Kharkov Polytechnic Inst., Frunze St. 21, Kharkov 310002, USSR. (1938) Cand, technics (Kharkov Polytechnic Inst., 1968). Sr. scient. *Metals and alloy structures, defects, defect investigation methods.*

Pavlishin, Dr Vladimir Ivanovich. Inst. of Geochemistry and Physics of Minerals, Acad. of Sci. Ukrainian SSR, Palladina Prospekt 34, Kiev 252068, USSR. (1940) Cand, geology and mineralogy (Lvov U., 1966). Sr. scient. *Crystallography of minerals, silicates.*

Pavlovsky, Dr Alexander Grigoryevich. Inst. of Crystallography, Acad. of Sci. of the USSR, Leninsky pr. 59, Moscow 117333, USSR. (1947) Cand, chemistry (Inst. of Molecular Biology, Acad. Sci. USSR, 1979). Scient. *Protein crystallography.*

Pech, Dr Lucia Yanovna. Inst. of Inorganic Chemistry, Acad. of Sci. of the Latvian SSR, Meistaru St. 10, Riga 226934, USSR. Cand, chemistry (Inst. of Inorganic Chemistry Acad. Sci. Latvian SSR, 1977). Sr. scient. *Crystal chemistry.*

Perekalina, Dr Tatyana Mikhailovna. Inst. of Crystallography, Acad. of Sci. of the USSR, Leninsky pr. 59, Moscow 117333, USSR. (1922) Dr, physics and mathematics (INCRYS, 1973). Sr. scient. *Magnetic properties, magnetic phase transitions.*

Perekalina, Dr Zoya Borisovna. Inst. of Crystallography, Acad. of Sci. of the USSR, Leninsky pr. 59, Moscow 117333, USSR. (1929) Cand, physics and mathematics (INCRYS, 1969). Sr. scient. *Optical properties, optical activity.*

Pershin, Dr Vitaly Konstantinovich. Ural Polytechnic Inst., Sverdlovsk 620002, USSR. Cand, physics and mathematics (Ural Polytechnic Inst., 1976). Asst. *Structural phase transformations, metastable states, organic systems.*

Pertsin, Dr Alexander Iosifovich. Inst. of Elemento-Organic Compounds, Acad. of Sci. of the USSR, Vavilov St. 28, Moscow 117813, USSR. (1948) Cand, physics and mathematics (Inst. of Applied Physics, Acad. Sci. Moldavian SSR, 1977). Scient. *Phase transitions, organic crystal thermodynamics.*

Peskin, Dr Vladimir Fedorovich. Inst. of Crystallography, Acad. of Sci. of the USSR, Leninsky pr. 59, Moscow 117333, USSR. (1937) Cand, geology and mineralogy (MGU, 1981). Scient. *Crystal growth.*

Petropavlov, Dr Nikolay Nikolayevich. Inst. of Biological Physics, Acad. of Sci. of the USSR, Pushchino 142292, Moskovskaya Oblast', USSR. (1938) Cand, physics and mathematics (Moscow State Pedagogical Inst., 1971). Sr. scient. *Crystal structure and properties, biologically important substances, phase transitions.*

Petrov, Dr Thomas Georgiyevich. Leningrad State U., Dept. of Geology, University Emb. 7/9, Leningrad 199164, USSR. (1931) Cand, geology and mineralogy (GORNY, 1962). Sr. scient. *Crystallography, crystal growth, defects in crystals.*

Petrova, Dr Irina Vladimirovna. Dept. of Geology, Moscow State U., Leninskiye Gory, Moscow 117234, USSR. (1949) Cand, geology and mineralogy (MGU, 1980). Scient. *Crystal chemistry, inorganic compounds, X-ray diffraction.*

Petrunina, Dr Alla Anastas'yevna. Inst. of Geochemistry and Physics of Minerals, Acad. of Sci. Ukrainian SSR, Palladina pr. 34, Kiev 252068, USSR. (1930) Cand, physics and mathematics (INCRYS, 1972). Sr. scient. *Structural typomorphism, rock-forming minerals.*

Petukhov, Dr Boris Vladimirovich. Inst. of Crystallography, Acad. of Sci. of the USSR, Leninsky pr. 59, Moscow 117333, USSR. (1941) Cand, physics and mathematics (Inst. of Theoretical Physics, Acad. Sci, USSR, 1972). Scient. *Crystal lattice defects, strength and plasticity of crystals.*

Pidzhyan, Prof. Grigory Oganesovich. Inst. of Geology, Acad. of Sciences Armenian SSR, Barekamutyan St. 24a, Erevan 375200, USSR. (1919) Dr, geology and mineralogy (GORNY, 1969). Dept. head. *Crystal chemistry, sulphides and sulphosalts.*

Pikin, Dr Sergey Alekseyevich. Inst. of Crystallography, Acad. of Sci. of the USSR, Leninsky pr. 59, Moscow 117333, USSR. (1941) Dr, physics and mathematics (Moscow Inst. of Engineering and Physics, 1978). Sr. scient. *Theoretical physics, liquid crystal theory, phase transition theory.*

Pinsker, Dr. Garry Zinov'yevich. Inst. of General and Inorganic Chemistry, Acad. of Sci. of the USSR, Leninsky pr. 31, Moscow 117071, USSR. (1929) Dr, physics and mathematics (Inst. of Physics, Acad. Sci. Latvian SSR, 1984). Scient. *Short range order, symmetry, diffraction, structure, amorphous bodies.*

Pinsker, Prof. Zinovy Grigoryevich. Inst. of Crystallography, Acad. of Sci. of the USSR, Leninsky pr. 59, Moscow 117333, USSR. (1904) Dr, chemistry (IONCH, 1943). Consulting scient. *Structure analysis, atomic and real structure, dynamic scattering, X-ray and electron scattering.*

Pisarevsky, Dr Yury Vladimirovich. Inst. of Crystallography, Acad. of Sci. of the USSR, Leninsky pr. 59, Moscow 117333, USSR. (1940) Cand, physics and mathematics (INCRYS, 1974). Sr. scient. *Acoustic and elastic properties, piezoelectric properties.*

Plakhov, Dr Gennady Fedorovich. Moscow State U., Dept. of Geology, Leninskiye Gory, Moscow 117234, USSR. (1938) Cand, geology and mineralogy (MGU, 1976). Scient. *Crystal chemistry, inorganic compounds.*

Pletnev, Dr Vladimir Zakharovich. Inst. of Bioorganic Chemistry, Acad. of Sci. of the USSR, Vavilov St. 32, Moscow 117312, USSR. (1944) Cand, chemistry (Inst. of Bioorganic Chem., 1970). Sr. scient. *Peptide-protein related structures.*

Plyasova, Dr Lyudmila Mikhailovna. Inst. of Catalysis, Acad. of Sci. of the USSR, Siberian Dept., Prospekt Nauki 5, Novosibirsk 630090, USSR. Cand, physics and mathematics (Gorky U., 1967). Sr. scient. *Crystal chemistry, inorganic compounds.*

Plyusnina, Dr Inga Ivanovna. Moscow State U., Dept. of Geology, Leninskiye Gory, Moscow 117234, USSR. (1931) Cand, geology and mineralogy (MGU, 1959). Sr. scient. *Infrared spectroscopy, minerals, inorganic compounds.*

Pobedimskaya, Dr Elena Alexandrovna. Moscow State U., Dept. of Geology, Leninskiye Gory, Moscow 117234, USSR. (1925) Cand, geology and mineralogy (MGU, 1962). Docent. *X-ray structure analysis, minerals, synthetic compounds, geocrystal chemistry of elements.*

Podberezskaya, Dr Nina Vasilyevna. Inst. of Inorganic Chemistry, Acad. of Sci. of the USSR, Siberian Dept., Prospekt Nauki 3, Novosibirsk 630090, USSR. (1938) Cand, physics and mathematics (Gorky U., 1971). Sr. scient. *Structure determination methods, crystal chemistry, inorganic compounds, structure.*

Pokrovsky, Dr Nikolay Leonidovich. Moscow State U., Dept. of Physics, Leninskiye Gory, Moscow 117234, USSR. (1909) Dr, chemistry (MGU, 1968). Sr. scient. *Surface phenomena (physics and chemistry).*

Polchovskaya, Dr Tatyana Mikhailovna. Inst. of Crystallography, Acad. of Sci. of the USSR, Leninsky pr. 59, Moscow 117333, USSR. (1944) Cand, physics and mathematics (INCRYS, 1975). Scient. *Crystal growth from melt.*

Polyanskaya, Dr Tamara Mikhailovna. Inst. of Inorganic Chemistry, Acad. of Sci. of the USSR, Siberian Dept., Prospekt Nauki 3, Novosibirsk 630090, USSR. (1942) Cand, physics and mathematics (Gorky U., 1971). Scient. *Structure determination methods, crystal chemistry, inorganic compounds.*

Polyansky, Dr Evgeny Vasilyevich. Res. Inst. for Synthesis of Mineral Raw Materials, Institutskaya St. 1, Alexandrov 601600, Vladimirskaya Oblast', USSR. (1944) Cand, geology and mineralogy (MGU, 1973). Sr. scient. *Crystal growth, crystal optics.*

Polynova, Dr Tamara Nikitichna. Moscow State U., Dept. of Chemistry, Leninskiye Gory, Moscow 117234, USSR. (1930) Cand, chemistry (MGU, 1963). Docent. *Crystal chemistry, coordination compounds, X-ray structure analysis.*

Ponomarev, Dr Vasily Ivanovich. Branch Inst. of Chemical Physics, Acad. of Sci. of the USSR, Chernogolovka 142432, Noginsky Rayon Moskovskaya Oblast', USSR. (1940) Cand, chemistry (INCRYS, 1971). Sr. scient. *Crystallography, crystal physics.*

Ponyatovsky, Prof. Evgeny Genrikhovich. Inst. of Solid State Physics, Acad. of Sci. of the USSR, Chernogolovka, Noginsky Rayon, Moskovskaya Oblast' 142432, USSR. (1930) Dr, physics and mathematics. Lab. head. *Phase transformations at high pressures, high pressure synthesis, metal hydrides.*

Popolitov, Dr Vladislav Ivanovich. Inst. of Crystallography, Acad. of Sci. of the USSR, Leninsky pr. 59, Moscow 117333, USSR. (1938) Cand, chemistry (INCRYS, 1969). Sr. scient. *Crystal growth.*

Popova, Dr Anastasiya Arsentyevna. Inst. of Crystallography, Acad. of Sci. of the USSR, Leninsky pr. 59, Moscow 117333, USSR. (1916) Cand, chemistry (INCRYS, 1963). *Crystal growth from melt.*

Popova, Dr Svetlana Vladimirovna. Inst. of High Pressure Physics, Acad. of Sci. of the USSR, Akademgorodok, Podol'sky Rayon, Moskovskaya Oblast' 142092, USSR. Dr, physics and mathematics (INCRYS, 1983). Lab. head. *High pressure effects, crystal structure.*

Porai-Koshits, Prof. Mikhail Alexandrovich. Inst. of General and Inorganic Chemistry, Acad. of Sci. of the USSR, Leninsky pr. 31, Moscow 117071, USSR. (1918) Corresp. member of the Acad. Sci. USSR, Dr, physics and mathematics (INCRYS, 1960). Lab. head. *Crystal chemistry, stereochemistry, coordination compounds, methods of structure analysis.*

Porai-Koshits, Prof. Evgeny Alexandrovich. Inst. of Silicate Chemistry, Acad. of Sci. of the USSR, Makarov Emb. 2, Leningrad 199164, USSR. (1907) Dr, physics and mathematics (LGU, 1953). Lab. head. *Disordered systems, glass structure, nature of the glass-like state.*

Portnov, Dr Vadim Nikolayevich. Gorky State U., Gagarin Prospekt 23, Gorky 603022, USSR. (1934) Cand, physics and mathematics (Gorky U., 1966). Docent. *Crystal growth.*

Povarennykh, Prof. Alexander Sergeyevich. Inst. of Geochemistry and Physics of Minerals, Acad. of Sci. Ukrainian SSR, Palladina Pr. 34, Kiev 252068, USSR. (1915) Full member of the Acad. of Sci. of the Ukrainian SSR, Dr, geology and mineralogy (IGEM, 1958). Dept. head. *Crystal chemistry in general, mineral structure, physical properties.*

Prevarsky, Dr Anatoly Petrovich. Lvov State U., Dept. of Chemistry, University St. 1, Lvov 290602, USSR. (1924) Cand, chemistry (Lvov U., 1973). Sr. scient. *X-ray structure analysis, crystal chemistry, intermetallic compounds.*

Prikhod'ko, Dr Leonid Vasilyevich. Inst. of Crystallography, Acad. of Sci. of the USSR, Leninsky pr. 59, Moscow 117333, USSR. (1938) Cand, physics and mathematics (INCRYS, (1971). Scient. *High temperature crystallization.*

Provotorov, Dr Mikhail Viktorovich. Moscow Chemico-Techn. Inst., Miusskaya sq. 9, Moscow 125047, USSR. (1946) Cand, chemistry (Moscow Chemico-Techn. Inst., 1976). Docent. *Synthesis of new crystalline phases, crystal growth from melts.*

Ptitsyn, Prof. Oleg Borisovich. Inst. of Protein, Acad. of Sci. of the USSR, Pushchino, Serpukhovskoy Rayon, Moskovskaya Oblast' 142292, USSR. (1929) Dr, physics and mathematics (Inst. of High Molecular Compounds, Acad. Sci. USSR, 1962). Assoc. dir. *X-ray structure analysis, proteins, diffuse X-ray scattering, biological macromolecules.*

Pudovkina, Dr Zoya Vasilyevna. Inst. of Mineralogy, Geochemistry and Crystal Chemistry of Rare Elements, Sadovnicheskaya Emb. 71, Moscow 113127, USSR. (1933) Cand, geology and mineralogy (MGU, 1968). Scient. *Crystal chemistry, X-ray structure analysis, inorganic compounds, minerals.*

Pugachev, Dr Anatoly Tarasovich. Kharkov Polytechnic Inst., Frunze St. 21, Kharkov 310002, USSR. (1940) Cand, technics (Kharkov Polytechnic Inst., 1967). Docent. *Metals and alloy structures, defects, methods for defect study.*

Punin, Dr Yury Olegovich. Leningrad State U., University Emb. 7/9, Leningrad 199164, USSR. (1941) Cand, geology and mineralogy (LGU, 1970). Scient. *Crystal growth, defect formation.*

Pushcharovsky, Dr Dmitry Yuryevich. Moscow State U., Dept. of Geology, Leninskiye Gory, Moscow 117234, USSR. (1944) Dr, geology and mineralogy (IGEM, 1984). Sr. scient. *Crystal chemistry, silicates and analogs, physical properties, crystal structure.*

Pyatenko, Dr Yury Andreyevich. Inst. of Mineralogy, Geochemistry and Crystal Chemistry of Rare Elements, Acad. of Sci. of the USSR, Sadovnicheskaya Emb. 71, Moscow 113127, USSR. (1928) Dr, geology and mineralogy (IGEM, 1969). Lab. head. *Crystal chemistry, minerals.*

Radautsan, Prof. Sergey Ivanovich. Inst. of Applied Physics, Acad. of Sci. of the Moldavian SSR, Akademicheskaya 5, Kishinev 277028, USSR. (1926) Full member of the Acad. Sci Moldavian SSR, Dr, technics (Leningrad Polytechnic Inst., 1966). Lab. head. *Crystal growth, semiconductor materials, glasses, amorphous semiconductors.*

Rakova, Dr Elena Vasilyevna. Inst. of Crystallography, Acad. of Sci. of the USSR, Leninsky pr. 59, Moscow 117333, USSR. (1941) Cand, chemistry (INCRYS, 1978). Scient. *Thin film growth and structure.*

Rannev, Dr Nikolay Vasilyevich. Physical Chemistry Inst., Obukha St. 10, Moscow 107120, USSR. (1931) Cand, chemistry (KARPOV, 1967). Sr. scient. *X-ray structure analysis, crystal chemistry, inorganic compounds, neutron diffraction.*

Rastsvetaeva, Dr Ramiza Kerarovna. Inst. of Crystallography, Acad. of Sci. of the USSR, Leninsky pr. 59, Moscow 117333, USSR. (1936) Cand, geology and mineralogy (MGU, 1971). Scient. *X-ray structure analysis, crystal chemistry, inorganic crystals.*

Rau, Dr Tamara Fedorovna. Vladimir Pedagogical Inst., Prospekt Stroiteley 11, Vladimir 600024, USSR. (1940) Cand, physics and mathematics (Gorky U., 1972). Lect. *X-ray crystal structure analysis.*

Rau, Dr Valery Georgiyevich. Vladimir Pedagogical Inst., Prospekt Stroiteley 11, Vladimir 600024, USSR. (1940) Dr, physics and mathematics (INCRYS, 1985). Docent. *X-ray crystal structure analysis.*

Regel', Dr Vadim Robertovich. Inst. of Crystallography, Acad. of Sci. of the USSR, Leninsky pr. 59, Moscow 117333, USSR. (1917) Cand, physics and mathematics (INCRYS, 1965). Lab. head. *Mechanical properties of crystals, surface physics, strength of solids.*

Rogacheva, Dr Evelina Danilovna. Gorky Agricultural Inst., Gagarin Prospekt 97, Gorky 603078, USSR. (1928) Cand, physics and mathematics (Gorky U., 1966). Head of Chair. *Crystal formation from solutions.*

Roginskaya, Dr Yuliana Eremeyevna. Physical Chemistry Inst., Obukha St. 10, Moscow 107120, USSR. (1937) Cand, chemistry (KARPOV, 1965). Sr. scient. *Crystal chemistry, complex metal oxides.*

Romanov, Dr Gennady Vasilyevich. Moscow State U., Dept. of Chemistry, Leninskiye Gory, Moscow 117234, USSR. (1937) Cand, chemistry (MGU, 1967). Sr. scient. *Inorganic compounds, structure, electron diffraction by gases.*

Rozhansky, Dr Vladimir Nikolayevich. Inst. of Crystallography, Acad. of Sci. of the USSR, Leninsky pr. 59, Moscow 117333, USSR. (1923) Dr, physics and mathematics (Inst. of Metal Physics, 1969). Lab. head. *Real structure of crystals.*

Rozhdestvenskaya, Dr Ira Vasilyevna. NPO Burevestnik, Malookhtensky Prospekt 78, Leningrad 195112, USSR. (1938) Cand, geology and mineralogy (LGU,1975). Sr. scient. *X-ray crystal structure analysis, crystal chemistry, silicates.*

Rumanova, Dr Iskra Mikhailovna. Fersman Mineralogical Museum, Acad. of Sci. of the USSR, Leninsky pr. 18, Moscow 117071, USSR. Dr, physics and mathematics (INCRYS, 1971). Sr. scient. *Inorganic crystal structures, methodology, structure - properties relationships.*

Ryaboshapka, Dr Karl Petrovich. Inst. of Metallophysics, Acad. of Sci. of the Ukrainian SSR, Vernadsky Prospekt 36, Kiev 252142, USSR. (1931) Cand, physics and mathematics (Joint Scient. Council Acad. Sci. Ukranian SSR, 1964). Sr. scient. *X-ray scattering, dislocation distorted crystals.*

Ryzhenkov, Dr Alexander Pavlovich. Kolomna Pedagogical Inst., Zelenaya St. 30, Kolomna 140410, Moskovskaya Oblast', USSR. (1935) Cand, physics and mathematics (V.I. Lenin Moscow State Pedagogical Inst., 1969). Docent. *Organic crystals, structure, physical properties.*

Sadikov, Dr Georgiy Georgievich. Inst. of General and Inorganic Chemistry, Acad. of Sci. of the USSR, Leninsky pr. 31, Moscow 117071, USSR. (1932) Cand, chemistry (IONCH, 1969). Scient. *X-ray crystal structure analysis, fluorides.*

Sadova, Dr Nina Ivanovna. Moscow State U., Dept. of Chemistry, Leninskiye Gory, Moscow 117234, USSR. (1937) Cand, chemistry (MGU, 1968). Scient. *Elemento-organic compounds, structure, electron diffraction by gases.*

Sakharov, Dr Boris Alexandrovich. Inst. of Geology, Acad. of Sci. of the USSR, Pyzhevsky per. 7, Moscow 109017, USSR. (1944) Cand, geology and mineralogy (Inst. of Geology Acad. Sci. USSR, 1974). Scient. *Diffraction applications, defects, minerals.*

Sal'dau, Dr El'ga Petrovna. Leningrad Mining Inst., 21st Liniya 2, Leningrad 199026, USSR. (1930) Cand, geology and mineralogy (GORNY, 1958). Docent. *Crystal chemistry, X-ray structure analysis.*

Samarskaya, Dr Valentina Dmitriyevna. Kolomna Pedagogical Inst., Zelenaya 30, Kolomna 140410, Moskovskaya Oblast', USSR. (1936) Cand, physics and mathematics (Moscow Pedagogical Inst., 1971). Docent. *Crystallization processes, physical properties and phase state, organic molecular crystals.*

Samoilovich, Dr Lidiya Alexandrovna. Res. Inst. for Synthesis of Mineral Raw Materials, Institutskaya St. 1, Alexandrov 601600, Vladimirskaya Oblast', USSR. (1934) Cand, chemistry (MGU, 1968). Sr. scient. *Crystal growth, real structure of crystals, crystal chemistry.*

Samoilovich, Prof. Mikhail Isaakovich. Res. Inst. for Synthesis of Mineral Raw Materials, Institutskaya St. 1, Alexandrov 601600, Vladimirskaya Oblast', USSR. (1937) Dr, physics and mathematics (Kazan State U., 1973). Lab. head. *Crystal growth, real structure of crystals, crystal chemistry.*

Samus', Dr Ivan Dmitriyevich. Kishinev Polytechnic Inst., Lenin Pr. 168, Kishinev 277004, USSR. (1926) Cand, physics and mathematics (INCRYS, 1967). Head of Chair. *X-ray structure analysis methods, crystal chemistry, coordination compounds.*

Sanadze, Prof. Vladimir Vladimirovich. Polytechnic Inst., Lenin St. 77, Tbilisi 360015, USSR. (1920) Dr, physics and mathematics (INCRYS, 1962). Head of Chair. *Phase transformations, structure analysis, metals and alloys.*

Sandomirsky, Dr Pavel Alexandrovich. Inst. of Geochemistry and Analytical Chemistry, Acad. of Sci. of the USSR, Kosygin pr. 19, Moscow 117334, USSR. (1951) Cand, geology and mineralogy (MGU, 1978). Scient. *Crystal chemistry, inorganic compounds, polymorphism, order-disorder structures.*

Sannikov, Dr Daniil Grigoryevich. Inst. of Crystallography, Acad. of Sci. of the USSR, Leninsky pr. 59, Moscow 117333, USSR. (1931) Cand, physics and mathematics (INCRYS, 1964). Sr. scient. *Ferroelectrics (physics), phase transitions (ferroelectric).*

Sarkisov, Dr Stepan Ervandovich. Inst. of Crystallography, Acad. of Sci. of the USSR, Leninsky pr. 59, Moscow 117333, USSR. (1948) Cand, physics and mathematics (INCRYS, 1979). Scient. *Crystal spectroscopy, crystal growth from melt.*

Sarin, Dr Victor Anatol'yevich. Karpov Physical Chemistry Inst., Obukha St. 10, Moscow 107120, USSR. (1947) Cand, physics and mathematics (Inst. of Appl. Physics, Moldavian SSR, 1978). Lab. head *Neutron diffraction, structures.*

Sedmalis, Prof. Uldis Yanovich. Riga Polytechnic Inst., Kronvalda Boulevard 4, Riga 226828, USSR. (1933) Dr, technics (Byelorussian Polytechnic Inst., 1970). Prof. *Geometrical crystallography, crystallo-optical and X-ray phase analyses.*

Semenova, Dr Tat'yana Fedorovna, Leningrad State U., Dept. of Geology, University Emb. 7/9, Leningrad 19916, USSR. (1951) Cand, geology and mineralogy (LGU, 1978). Sr. scient. *Crystal chemistry, structure analysis, laminated silicates.*

Semenovskaya, Dr Svetlana Victorovna. Inst. of Elemento-Organic Compounds, Acad. of Sci. of the USSR, Vavilov St. 28, Moscow 117312, USSR. (1934) Dr, physics and mathematics (Inst. of Metal Physics, Acad. Sci. Ukrainian SSR, 1976). Scient. *Diffuse X-ray scattering, solid solutions, solid state physics.*

Semiletov, Prof. Stepan Alekseyevich. Inst. of Crystallography, Acad. of Sci. of the USSR, Leninsky pr. 59, Moscow 117333, USSR. (1925) Dr, physics and mathematics (INCRYS, 1970). Lab. head. *Electron diffraction, thin film structure and properties.*

Serdyuk, Dr Igor' Nikolayevich. Inst. of Protein, Acad. of Sci. of the USSR, Pushchino, Serpukhovsky Rayon, Moskovskaya Oblast' 142292, USSR. (1939) Cand, physics and mathematics (Inst. of High Molecular Compounds, Acad. Sci. USSR, 1968). Sr. scient. *Diffuse X-ray scattering, biological macromolecules.*

Serebryakov, Dr Anatoly Valeryevich. Inst. of Solid State Physics, Acad. of Sci. of the USSR, Chernogolovka, Noginsky Rayon, Moskovskaya Oblast' 142432, USSR. (1935) Cand, technics. Lab. head. *Phase transformations.*

Serebryanaya, Dr Nadezhda Ruvimovna. Inst. of High Pressure Physics, Acad. of Sci. of the USSR, Akademgorodok, Podol'sky Rayon, Moskovskaya Oblast' 142092, USSR. Cand, chemistry (INCRYS, 1970). Scient. *High pressure effects, crystal structure.*

Sergeeva, Dr Valeriya Mikhailovna. Physico-Techn. Inst., Acad. of Sci. of the USSR, Zapovednaya 51, Leningrad 194037, USSR. (1932) Cand, chemistry (LGU, 1973). Scient. *Variable composition compounds, nonstoichiometric compounds, chalcogenides of rare earth elements.*

Sergienko, Dr Vladimir Semenovich. Inst. of General and Inorganic Chemistry, Acad. of Sci. of the USSR, Leninsky pr. 31, Moscow 117071, USSR. (1941) Cand, chemistry (IONCH, 1973). Sci. *Crystal chemistry, stereochemistry, coordination compounds.*

Sevast'yanov, Dr Boris Konstantinovich. Inst. of Crystallography, Acad. of Sci. of the USSR, Leninsky pr. 59, Moscow 117333, USSR. (1930) Cand, physics and mathematics (Moscow Physical Techn. Inst., 1962). Assoc. dir. *Optical spectroscopy, magnetic properties of doped ionic crystals.*

Shafizade, Dr Rafik Bekhbud ogly. Inst. of Physics, Acad. of Sci. of the Azerbaidzhan SSR, Narimanov Prospekt 33, Baku 370143, USSR. (1934) Cand, physics and mathematics (Inst. of Physics, Acad. Sci. Azerbaidzhan SSR, 1965). Lab. head. *Semiconductor thin film structure, thin films (phase formation and transformation).*

Shafranovsky, Prof. Ilarion Ilarionovich. Leningrad Mining Inst., 21st Liniya 2, Leningrad 199026, USSR. (1910) Dr, geology and mineralogy (LGU, 1942). Head of Chair. *Morphology, geometrical crystallography.*

Shaldin, Dr Yury Vitalyevich. Inst. of Crystallography, Acad. of Sci. of the USSR, Leninsky pr. 59, Moscow 117333, USSR. (1935) Cand, physics and mathematics (Inst. of Steel and Alloys, 1967). Sr. scient. *Non-linear crystal properties.*

Shamburov, Dr Vladimir Alekseyevich. Inst. of Crystallography, Acad. of Sci. of the USSR, Leninsky pr. 59, Moscow 117333, USSR. (1920) Cand, physics and mathematics (Machinery Inst., Acad. Sci. USSR, 1950). Sr. scient. *Ferroelectrics (physics), crystal optics.*

Shamray, Dr Vladimir Fedorovich. Inst. of Metallurgy, Acad. of Sci. of the USSR, Leninsky pr. 49, Moscow 117334, USSR. (1937) Cand, technics (Inst. of Metallurgy, Acad. Sci. USSR, 1970). Sr. scient. *X-ray structure analysis, superconducting compounds.*

Shashkin, Dr Dmitry Petrovich. Inst. of Chemical Physics, Acad. of Sci. of the USSR, Vorobyevskoye Chaussee 2-b, Moscow 117334, USSR. (1936) Cand, geology and mineralogy (MGU, 1970). Sr. scient. *Crystal structure analysis.*

Shchedrin, Dr Boris Mikhailovich. Moscow State U., Computing Center, Leninskiye Gory, Moscow 117234, USSR. (1934) Cand, physics and mathematics (INCRYS, 1967). Sr. scient. *Computing methods, structure analysis.*

Shcherbakova, Dr Mira Yakovlevna. Inst. of Geology and Geophysics, Acad. of Sci. of the USSR, Siberian Dept., Novosibirsk 630090, USSR. (1926) Cand, physics and mathematics (Tomsk Polytechnic Inst., 1961). Lab. head. *Electron paramagnetic resonance, defects and impurities, minerals, synthetic materials.*

Sheftal', Prof. Nikolay Naumovich. Inst. of Crystallography, Acad. of Sci. of the USSR, Leninsky pr. 59, Moscow 117333, USSR. (1902) Dr, geology and mineralogy (IGEM, 1953). Sr. scient. *Crystal growth, symmetry.*

Sheka, Prof. Elena Fedorovna. Inst. of Solid State Physics, Acad. of Sci. of the USSR, Chernogolovka 142432, Noginsky Rayon, Moskovskaya Oblast', USSR. (1934) Dr, physics and mathematics (Physical Inst., 1972). Lab. head. *Structure, molecular crystals, lattice dynamics, neutron spectroscopy.*

Shekhtman, Dr Veniamin Sholomovich. Inst. of Solid State Physics, Acad. of Sci. of the USSR, Chernogolovka, Noginsky Rayon, Moskovskaya Oblast' 142432, USSR. (1929) Cand, technics (Moscow Inst. of Steel and Alloys, 1962). Lab. head. *X-ray structure analysis, phase and structural changes.*

Shepelev, Dr Yury Fedorovich. Inst. of Silicate Chemistry, Acad. of Sci. of the USSR, Makarov Emb. 2, Leningrad 199164, USSR. (1939) Cand, physics and mathematics (LGU, 1971). Sr. scient. *Crystal chemistry, inorganic compounds.*

Shibaeva, Dr Rimma Pavlovna. Branch Inst. of Chemical Physics, Acad. of Sci. of the USSR, Chernogolovka 142432, Noginsky Rayon, Moskovskaya Oblast', USSR. Dr, physics and mathematics (INCRYS, 1977). Sr. scient. *Organic and metallo-organic compounds, structure analysis, electrical and magnetic properties, direct methods.*

Shishova, Dr Tatyana Gennadiyevna. Gorky Agricultural Inst., Gagarin Prospekt 97, Gorky 603078, USSR. (1949) Cand, physics and mathematics (INCRYS, 1977). Asst. *X-ray structure analysis, organic compounds, crystal growth.*

Shivrin, Dr Oleg Mikolayevich. Petrozavodsk State U., Lenin Prospekt 33, Petrozavodsk 185018, USSR. (1923) Cand, physics and mathematics (MGU, 1961). Docent. *Order-disorder transformations, radiation defects, crystal lattice dynamics.*

Shklover, Dr Valery Efimovich. Inst. of Elemento-Organic Compounds, Acad. of Sci. of the USSR, Vavilov St. 28, Moscow 117312, USSR. (1946) Cand, chemistry (INEOS, 1974). Sci. *Structural chemistry, metallo-organic compounds, organic and bio-organic compounds, X-ray structure analysis.*

Shkol'nikova, Dr Larisa Mikhailovna. Inst. of Chemical Reagents and Pure Substances, Bogorodsky Val 3, Moscow 107258, USSR. (1930) Cand, chemistry (KARPOV, 1960). Sr. scient. *X-ray structure analysis, complex and organic compounds.*

Shnulin, Dr Anatoly Nikolayevich. Inst. of Inorganic and Physical Chemistry, Acad. of Sci. of the Azerbaidzhan SSR, Narimanov Pr. 29, Baku 370143, USSR. (1936) Cand, physics and mathematics (Azerbaidzhan State U., 1960). Sr. scient. *Structure analysis, complexes, biological compounds.*

Shmyt'ko, Dr Ivan Mikhailovich. Inst. of Solid State Physics, Acad. of Sci. of the USSR, Chernogolovka 142432, Noginsky Rayon, Moskovskaya Oblast', USSR. (1946) Cand, physics and mathematics (Moscow Physico-Techn. Inst., 1976). Sr. scient. *Crystallography, phase transitions, X-ray diffraction optics, real crystals.*

Shternberg, Dr Aleksey Alexandrovich. Inst. of Crystallography, Acad. of Sci. of the USSR, Leninsky pr. 59, Moscow 117333, USSR. (1911) Dr, physics and mathematics (INCRYS, 1969). Consultant. *Crystal growth.*

Shul'pina, Dr Iren Leonidovna. Physico-Techn. Inst., Acad. of Sci. of the USSR, Zapovednaya St. 51, Leningrad 194037, USSR. (1936) Cand, physics and mathematics (Inst. of Semiconductors, Acad. Sci. USSR, 1968). Sr. scient. *Crystal lattice defects, X-ray methods, dynamic scattering of X-rays.*

Shumyatskaya, Dr Ninel' Grigoryevna. Inst. of Mineralogy, Geochemistry and Crystal Chemistry of Rare Elements, Sadovnicheskaya Emb. 71, Moscow 113127, USSR. (1930) Cand, geology and mineralogy (MGU, 1974). Scient. *Crystal chemistry, minerals, synthetic compounds.*

Shustov, Dr Alexander Vsevolodovich. Leningrad Mining Inst., 21st Liniya 2, Leningrad 199026, USSR. (1937) Cand, geology and mineralogy (GORNY, 1967). Sr. scient. *Crystal morphology, optical crystallography.*

Shutskever, Dr Nataliya Efimovna. Inst. of Molecular Biology, Acad. of Sci. of the USSR, Vavilov St. 32, Moscow 117312, USSR. Cand, physics and mathematics (MGU, 1972). Scient. *Protein crystallography.*

Shuvalov, Prof. Lev Alexandrovich. Inst. of Crystallography, Acad. of Sci. of the USSR, Leninsky pr. 59, Moscow 117333, USSR. (1923) Dr, physics and mathematics (INCRYS, 1972). Lab. head. *Physical properties of crystals, structural phase transitions, ferroelectricity and ferroelasticity.*

Shvelashvili, Prof. Arsen Eristovich. Inst. of Physical and Inorganic Chemistry, Acad. of Sci. of the Georgian SSR, Dzhikiya St. 5, Tbilisi 380086, USSR. (1935) Dr, chemistry (Tbilisi State U., 1974). Lab. head. *Stereochemistry, coordination compounds.*

Shulakov, Dr Evgeniy Vladimirovich. Inst. of Solid State Physics, Acad. of Sci. of the USSR, Chernogolovka 142432, Noginsky Rayon, Moskovskaya Oblast', USSR. (1949) Cand, physics and mathematics (Inst. of Solid State Physics, Acad. Sci USSR, 1978). Sr. scient. *X-ray diffraction optics, perfect crystals, real crystals; computer modelling of diffraction patterns.*

Sidorenko, Dr Galina Alexandrovna. Inst. of Mineral Raw Materials (VIMS), Staromonetny 29, Moscow 109017, USSR. (1926) Dr, geology and mineralogy (VIMS, 1976). Lab. head. *Crystal chemistry, minerals, X-ray crystallography, polycrystals.*

Sigayev, Dr Vladimir Nikolayevich. Inst. of Crystallography, Acad. of Sci. of the USSR, Leninsky pr. 59, Moscow 117333, USSR. (1945) Cand, physics and mathematics (INCRYS, 1975). Scient. *Neutron diffraction, liquids and glasses.*

Silin', Dr Elga Yanovna. Inst. of Inorganic Chemistry, Acad. of Sci. of the Latvian SSR, Miyera St. 34, Salaspils 229021, Riga District, USSR. (1941) Cand, chemistry (Inst. of Inorganic Chemistry, Acad. Sci. Latvian SSR, 1978). Sr. scient. *Crystal chemistry.*

Sil'vestrova, Dr Iraida Mikhailovna. Inst. of Crystallography, Acad. of Sci. of the USSR, Leninsky pr. 59, Moscow 117333, USSR. (1924) Cand, physics and mathematics (INCRYS, 1963). Sr. scient. *Physical properties, piezoelectricity, elasticity, acoustic properties.*

Simonov, Dr Mikhail Alexandrovich. Moscow State U., Dept. of Geology, Leninskiye Gory, Moscow 117234, USSR. (1940) Cand, physics and mathematics (INCRYS, 1969). Docent. *Crystal chemistry, inorganic compounds, X-ray structure analysis.*

Simonov, Dr Valentin Ivanovich. Inst. of Crystallography, Acad. of Sci. of the USSR, Leninsky pr. 59, Moscow 117333, USSR. (1930) Dr, physics and mathematics (INCRYS, 1972). Assoc. dir. *Structure analysis methods, computing.*

Simonov, Dr Yury Alexandrovich. Inst. of Applied Physics, Acad. of Sciences Moldavian SSR, Akademicheskaya 5, Kishinev 277028, USSR. (1937) Cand, physics and mathematics (Gorky U., 1967). Sr. scient. *X-ray structure analysis, methods, inorganic compounds.*

Sirota, Dr Mikhail Isaakovich. Inst. of Crystallography, Acad. of Sci. of the USSR, Leninsky pr. 59, Moscow 117333, USSR. (1945) Cand, physics and mathematics (INCRYS, 1975). Scient. *Computer programming, structure analysis.*

Sizova, Dr Nataliya Leonidovna. Inst. of Crystallography, Acad. of Sci. of the USSR, Leninsky pr. 59, Moscow 117333, USSR. (1937) Cand, physics and mathematics (INCRYS, 1974). Scient. *Crystal lattice defects, mechanical properties of crystals.*

Skolozdra, Dr Roman Vladimirovich. Lvov State U., Dept. of Chemistry, University St. 1, Lvov 290602, USSR. (1941) Cand, chemistry (Lvov U., 1967). Docent. *X-ray structure analysis, crystal chemistry, intermetallic compounds.*

Smetannikova, Dr Olga Gennadiyevna. Leningrad State U., Dept. of Geology, University Emb. 7/9, Leningrad 199164, USSR. (1947) Cand, geology and mineralogy (LGU, 1974). Scient. *Crystal chemistry, X-ray structure analysis.*

Smirnov, Dr Aleksey Evgen'yevich. Inst. of Crystallography, Acad. of Sci. of the USSR, Leninsky pr. 59, Moscow 117333, USSR. (1946) Cand, chemistry (INCRYS, 1982). Scient. *Real structure of crystals, mechanical properties of crystals.*

Smirnov, Dr Yury Mstislavovich. Kalinin State U., Zhelyabova St. 33, Kalinin 170013, USSR. (1932) Cand, technics (GORNY, 1969). Lab. head. *Crystal growth, morphology.*

Smirnova, Dr Nina Lvovna. Moscow State U., Dept. of Geology, Leninskiye Gory, Moscow 117234, USSR. (1926) Cand, chemistry (INCRYS, 1961). Sr. scient. *Crystal chemistry.*

Smolin, Dr Yury Ivanovich. Inst. of Silicate Chemistry, Acad. of Sci. of the USSR, Makarov Emb. 2, Leningrad 199164, USSR. (1930) Dr, physics and mathematics (INCRYS, 1974). Sr. scient. *Crystal chemistry, inorganic compounds.*

Smotrakov, Dr Valery Georgiyevich. Rostov State U., Prospekt Stachki 192, Rostov-on-Don 344090, USSR. (1944) Cand, chemistry (Rostov U., 1971). Scient. *Crystal growth.*

Sobolev, Dr Boris Pavlovich. Inst. of Crystallography, Acad. of Sci. of the USSR, Leninsky pr. 59, Moscow 117333, USSR. (1936) Dr, chemistry (INCRYS, 1978). Lab. Head. *Crystal growth, crystal chemistry, inorganic compounds.*

Sobolev, Dr Chingis Sergeyevich. Leningrad Mining Inst., 21st Liniya 2, Leningrad 199026, USSR. (1931) Cand, geology and mineralogy (GORNY, 1965). Scient. *Crystallography, crystal morphology.*

Soboleva, Dr Lidiya Victorovna. Inst. of Crystallography, Acad. of Sci. of the USSR, Leninsky pr. 59, Moscow 117333, USSR. (1927) Cand, chemistry (IONCH, 1954). Scient. *Inorganic crystal growth.*

Soboleva, Dr Svetlana Vsevolodovna. Inst. of Geology, Mineralogy and Petrography, Acad. of Sci. of the USSR. Staromonetny 35, Moscow 109017, USSR. (1937) Cand, geology and mineralogy (IGEM, 1967). Sr. scient. *Structural mineralogy, polytypism, electron diffraction.*

Sokol, Dr Anatoly Afanasyevich. Kharkov Polytechnic Inst., Frunze St. 21, Kharkov 310002, USSR. (1937) Cand, technics (Kharkov Polytechnic Inst., 1970). Docent. *Structure, defects, thin film growth, electron microscopy.*

Sokol, Dr Valentina Ivanovna. Inst. of General and Inorganic Chemistry, Acad. of Sci. of the USSR. Leninsky pr. 31, Moscow 117071, USSR. (1927) Cand, chemistry (IONCH, 1965). Scient. *Crystal chemistry, stereochemistry, coordination compounds.*

Sokolov, Dr Yury Alexandrovich. Inst. of Crystallography, Acad. of Sci. of the USSR, Leninsky pr. 59, Moscow 117333, USSR. (1940) Cand, physics and mathematics (INCRYS, 1979). Scient. *Crystal spectroscopy, nuclear magnetic resonance, vibrational spectroscopy.*

Sokolova, Dr Nataliya Gavrilovna. Leningrad Mining Inst., 21st Liniya 2, Leningrad 199026, USSR. (1939) Cand, geology and mineralogy (GORNY, 1969). Sr. scient. *X-ray crystallography, morphology.*

Soldatov, Dr Evgeniy Alexandrovich. Gorky State University, Sverdlova 37, Gorky 603000, USSR. (1949) Cand, physics and mathematics (INCRYS, 1979). Docent. *Mathematical methods, structure analysis, symmetry theory.*

Solo'vyev, Dr Sergey Petrovich. Karpov Physical Chemistry Inst., Obukha St. 10, Moscow 107120, USSR. (1932) Dr, physics and mathematics (Physico-Energetic Inst., 1976). Lab. head. *X-ray and neutron structure analysis, crystal lattice dynamics.*

Solovyeva, Dr Lidiya Pavlovna. Inst. of Catalysis, Siberian Dept. Acad. of Sci. of the USSR, Lavrent'yeva pr. 5, Novosibirsk 630090, USSR. (1935) Cand, geology and mineralogy (MGU, 1965). Sr. scient. *X-ray diffraction, methods, programming, structure, zeolites, layered compounds.*

Sonin, Prof. Anatoliy Stepanovich. Res. Inst. of Semi-Products and Dye-Stuffs, Bol'shaya Sadovaya 1, Moscow 103787, USSR. (1931) Dr, physics and mathematics (Dnepropetrovsk State U., 1972). Sr. scient. *Symmetry, crystallo-physics, liquid crystals.*

Sorokin, Dr Lev Mikhailovich. Physico-Techn. Inst., Acad. of Sci. of the USSR, Zapovednaya St. 51, Leningrad 194037, USSR. (1937) Cand, physics and mathematics (Inst. of Semiconductors, Acad. Sci. USSR, 1968). Sr. scient. *Defect structure, defect dynamics, methods for defect study, dynamic electron and X-ray scattering.*

Sosfenov, Dr Nikita Ilyich. Inst. of Crystallography, Acad. of Sci. of the USSR, Leninsky pr. 59, Moscow 117333, USSR. (1932) Cand, physics and mathematics (INCRYS, 1972). Sr. scient. *Protein crystallography, instrumentation for X-ray structure analysis.*

Spiridonov, Prof. Victor Pavlovich. Moscow State U., Dept. of Chemistry, Leninskiye Gory, Moscow 117234, USSR. (1931) Dr, chemistry (MGU, 1969). Lab. head. *Inorganic compounds, structure, electron diffraction by gases.*

Spitsyna, Dr Valentina Danilovna. Inst. of Crystallography, Acad. of Sci. of the USSR, Leninsky pr. 59, Moscow 117333, USSR. (1941) Cand, chemistry (INCRYS, 1975). Scient. *Crystallization processes in multicomponent systems.*

Starikova, Dr Zoya Alexandrovna. Inst. of Chemical Reagents and Pure Substances, Bogorodsky Val 3, Moscow 107258, USSR. (1934) Cand, chemistry (IONCH, 1968). Sr. scient. *Crystal chemistry, complexes, organic compounds.*

Starostina, Dr Lyudmila Sergeyevna. Inst. of Crystallography, Acad. of Sci. of the USSR, Leninsky pr. 59, Moscow 117333, USSR. (1933) Cand, physics and mathematics (INCRYS, 1964). Scient. *Spectroscopy, crystal growth.*

Stepanova, Dr Alla Nikolayevna. Inst. of Crystallography, Acad. of Sci. of the USSR, Leninsky pr. 59, Moscow 117333, USSR. (1934) Cand, physics and mathematics (INCRYS, 1974). Scient. *Crystal growth, thin film growth, whisker growth.*

Stishov, Dr Sergey Mikhailovich. Inst. of Crystallography, Acad. of Sci. of the USSR, Leninsky pr. 59, Moscow 117333, USSR. Dr, physics and mathematics (INCRYS, 1974). Lab. head. *High pressure crystallography.*

Struchkov, Dr Yury Timofeyevich. Inst. of Elemento-Organic Compounds, Acad. of Sci. of the USSR, Vavilov St. 28, Moscow 117312, USSR. (1926) Dr, chemistry (INEOS, 1978). Lab. head. *Structural chemistry, organometallic compounds, organic and bio-organic compounds, X-ray analysis.*

Strukov, Prof. Boris Anatolyevich. Moscow State U., Dept. of Physics, Leninskiye Gory, Moscow 117234, USSR. (1935) Dr, physics and mathematics (MGU, 1975). Docent. *Ferroelectricity, structural phase transitions, Raman scattering.*

Suvorov, Dr Ernest Vitalyevich. Inst. of Solid State Physics, Acad. of Sci. of the USSR, Chernogolovka, Noginsky Rayon, Moskovskaya oblast 142432, USSR. (1937) Cand, physics and mathematics. Sr. scient. *Dynamic X-ray scattering, real crystals.*

Svergun, Dr Dmitry Ivanovich. Inst. of Crystallography, Acad. of Sci. of the USSR, Leninsky pr. 59, Moscow 117333, USSR. (1954) Cand, physics and mathematics (INCRYS, 1982). Scient. *X-ray and neutron small angle scattering, diffraction theory, methods.*

Sviridov, Prof. Dmitry Timofeyevich. Inst. of Crystallography, Acad. of Sci. of the USSR, Leninsky pr. 59, Moscow 117333, USSR. (1931) Dr, physics and mathematics (INCRYS, 1973). Sr. scient. *Crystal optics, spectroscopy of crystals, crystal structures.*

Tarkhova, Dr Tatyana Nikolayevna. Gorky State U., Gagarin Prospekt 23, Gorky 603022. USSR. Cand, physics and mathematics (INCRYS, 1949). Docent. *X-ray structure analysis, crystal chemistry.*

Tarnopol'sky, Dr Boris Lvovich. Branch Inst. of Chemical Physics, Acad. of Sci. of the USSR, Chernogolovka 142432, Noginsky Rayon, Moskovskaya Oblast', USSR. (1924) Cand, physics and mathematics (INCRYS, 1965). Sr. scient. *Computing methods, structure analysis.*

Tatarchenko, Dr Vitaly Antonovich. Inst. of Solid State Physics, Acad. of Sci. of the USSR, Chernogolovka, Noginsky Rayon, Moskovskaya Oblast' 142432, USSR. (1938) Cand, physics and mathematics. Dept. head. *Crystallization from melt.*

Tatarinova, Dr Lyudmila Ivanovna. Inst. of Crystallography, Acad. of Sci. of the USSR, Leninsky pr. 59, Moscow 117333, USSR. (1903) Cand, physics and mathematics (INCRYS, 1953). Scient. *Synthetic polypeptide structures, amorphous substances, electron diffraction, X-ray diffraction.*

Tatarsky, Prof. Vitaly Borisovich. Leningrad State U., Dept. of Geology, University Emb. 7/9, Leningrad 199164, USSR. (1907) Dr, geology and mineralogy (LGU, 1953). Prof. *Crystal optics, microscopic phase-analysis, goniometry.*

Telegina, Dr Inna Vasilyevna. Moscow State U., Dept. of Physics, Leninskiye Gory, Moscow 117234, USSR. Cand, physics and mathematics (MGU, 1968). Scient. *Diffuse X-ray scattering, small angle X-ray scattering, radiation effects in crystals.*

Teslenko, Dr Valery Fedorovich. Kolomna Pedagogical Inst., Zelenaya 30, Kolomna 140410, Moskovskaya Oblast', USSR. (1937) Cand, physics and mathematics (Moscow Pedagogical Inst., 1967). Docent. *Crystallization processes, physical properties and phase state, organic molecular crystals.*

Tikhomirova, Dr Nataliya Alexandrovna. Inst. of Crystallography, Acad. of sciences of the USSR, Leninsky pr. 59, Moscow 117333, USSR. (1932) Cand, physics and mathematics (INCRYS, 1966). Sr. scient. *Phase transitions in solids, liquid crystals, high pressure properties.*

Tikhonova, Dr Anna Andreyevna. Inst. of Crystallography, Acad. of Sci. of the USSR, Leninsky pr. 59, Moscow 117333, USSR. (1934) Cand, physics and mathematics (INCRYS, 1973). Scient. *Thin film growth and structure.*

Timchenko, Dr Tamara Iosifovna. Moscow State U., Dept. of Geology, Leninskiye Gory, Moscow 117234, USSR. (1930) Cand, chemistry (IGEM, 1962). Sr. scient. *Crystal growth, synthesis.*

Timofeeva, Dr Valentina Alexandrovna. Inst. of Crystallography, Acad. of Sci. of the USSR, Leninsky pr. 59, Moscow 117333, USSR. (1923) Cand, chemistry (Inst. of Chemistry, Acad. Sci. Kazakh SSR, 1949). Sr. scient. *Crystal growth from fluxed melts.*

Timofeyeva, Dr Tat'yana Vladimirovna. Inst. of Elemento-Organic Compounds, Acad. of Sci. of the USSR, Vavilov St. 28, Moscow 117813, USSR. (1947)

Cand, chemistry (INEOS, 1982). Scient. *Molecular packing energy calculations, conformational analysis.*

Tishchenko, Dr Galina Nikolayevna. Inst. of Crystallography, Acad. of Sci. of the USSR, Leninsky pr. 59, Moscow 117333, USSR. Dr, chemistry (INCRYS, 1984). Sr. scient. *Structure analysis, proteins and cyclic peptides.*

Tkachev, Dr Valery Vladimirovich. Branch Inst. of Chemical Physics, Acad. of Sci. of the USSR, Chernogolovka 142432, Moskovskaya Oblast', USSR. (1943) Cand, physics and mathematics (Inst. of Chemical Physics, Acad. Sci. USSR). Scient. *Crystal chemistry, ion-conducting compounds, coordination compounds, crystal structure, phospho-organic compounds.*

Tobelko, Mr Konstantin Ivanovich. Inst. of Geochemistry and Analytical Chemistry, Acad. of Sci. of the USSR, Kosygin pr. 19, Moscow 117334, USSR. (1923). Scient. *Isomorphism, minerals, inorganic compounds, X-ray analysis.*

Tomashpol'sky, Dr Yury Yakovlevich. Karpov Physical Chemistry Inst., Obukha St. 10, Moscow 107120, USSR. (1937) Cand, chemistry (KARPOV, 1965). Lab. head. *Complex oxides, structure, surfaces, thin layers.*

Topor, Dr Nikolay Dmitriyevich. Moscow State U., Dept. of Geology, Leninskiye Gory, Moscow 117234, USSR. (1915) Cand, geology and mineralogy (MGU, 1946). Sr. scient. *Geocrystal chemistry, thermal properties, minerals.*

Tovbis, Dr Alexander Borisovich. Inst. of Crystallography, Acad. of Sci. of the USSR, Leninsky pr. 59, Moscow 117333, USSR. (1940) Cand, physics and mathematics (INCRYS, 1971). Sr. scient. *Computing methods, structure analysis.*

Treivus, Dr Evgeny Borisovich. Leningrad State U., Dept. of Geology, University Emb. 7/9, Leningrad 199164, USSR. (1934) Cand, geology and mineralogy (LGU, 1965). Scient. *Crystal growth, crystal morphology.*

Treushnikov, Dr Evgeniy Nikolayevich. Dept. of Chemistry, Moscow State U., Leninskiye Gory, Moscow 117234, USSR. Cand, physics and mathematics (INCRYS, 1970). Sr. scient. *Electron density distribution (diffraction method).*

Triodina, Dr Nina Sergeyevna. Inst. of Crystallography, Acad. of Sci. of the USSR, Leninsky pr. 59, Moscow 117333, USSR. (1941) Cand, chemistry (INCRYS, 1979). Scient. *Crystal growth.*

Trunov, Dr Vadim Konstantinovich. Inst. of Chemical Reagents and Pure Substances, Bogorodsky Val 3, Moscow 107258, USSR. (1936) Dr, chemistry (MGU, 1972). Dept. head. *Crystallography, oxide compounds, double salts.*

Tseitlin, Dr Mikhail Nevakhovich. Physico-Technical Inst., Acad. of Sci. of the Tadzhik SSR, Akademgorodok, Dushanbe 773630, USSR. (1945) Cand, chemistry (INCRYS, 1974). Sr. scient. *Crystal growth, crystal chemistry.*

Tsikhotsky, Dr Evgeny Stanislavovich. Rostov State U., Dept. of Physics, Prospekt Stachki 192, Rostov-on-Don 344061, USSR. (1946) Cand, physics and mathematics (Rostov U., 1975). Sr. scient. *Perfection studies in crystals.*

Tsinober, Dr Leonid Iosifovich. Res. Inst. for Synthesis of Mineral Raw Materials, Institutskaya St. 1, Alexandrov 601600, Vladimirskaya Oblast', USSR. (1924) Cand, geology and mineralogy (INCRYS, 1962). Lab. head. *Real structure of crystals, structural mineralogy, X-ray crystallography, electron microscopy of crystals.*

Tsintsadze, Prof. Givi Vasilyevich. Polytechnic Inst., Lenin St. 77, Tbilisi 360015, USSR. (1933) Dr, chemistry (Tbilisi U., 1971). Head of Chair. *Crystal chemistry, coordination compounds.*

Tsirel'son, Dr Valadimir Grigoryevich. Moscow Chemical Techn. Inst., Miusskaya Square 9, Moscow 125820, USSR. (1948) Cand, physics and mathematics (Inst. of Applied Physics, Acad. Sci. Moldavian SSR). Scient. *Precision diffraction measurement, electron density distribution.*

Tsuprun, Dr Vladimir Lvovich. Inst. of Crystallography, Acad. of Sci. of the USSR, Leninsky pr. 59, Moscow 117333, USSR. (1948) Cand, physics and mathematics (Inst. of Proteins, Acad. Sci. USSR, 1979). Scient. *Protein crystallography.*

Tsvankin, Dr Daniel' Yakovlevich. Inst. of Elemento-Organic Compounds, Acad. of Sci. of the USSR, Vavilov St. 28, Moscow 117312, USSR. (1929) Dr, physics and mathematics (Inst. of Macromolecular Compounds, Acad. Sci. USSR, 1971). Sr. scient. *Polymer physics, small angle X-ray scattering.*

Tsyganov, Dr Evgeny Matveyevich. Res. Inst. for Synthesis of Mineral Raw Materials, Institutskaya St. 1, Alexandrov 601600, Vladimirskaya Oblast', USSR. (1919) Cand, geology and mineralogy (Lvov State U., 1951). Lab. head. *Synthesis of crystals.*

Tumanyan, Dr Vladimir Gayevich. Inst. of Molecular Biology, Acad. of Sci. of the USSR, Vavilov St. 32, Moscow 117312, USSR. (1938) Cand, physics and mathematics (Physico-Techn. Inst., Acad. Sci. USSR, 1965). Sr. scient. *Fibrillous structures, conformational calculations.*

Tyapunina, Dr Nataliya Alexandrovna. Dept. of Chemistry, Moscow State U., Leninskiye Gory, Moscow 117234, USSR. (1922) Dr, physics and mathematics (MGU, 1972). Sr. scient. *Defects, physical properties of defect crystals.*

Tyvanchuk, Dr Anna Teodorovna. Lvov State U., Dept. of Chemistry, University St. 1, Lvov 290602, USSR. (1947) Cand, chemistry (Lvov State U., 1980). Scient. *Crystal chemistry, intermetallic compounds.*

Udalova, Dr Valentina Vasilyevna. Inst. of Crystallography, Acad. of Sci. of the USSR, Leninsky pr. 59, Moscow 117333, USSR. (1932) Cand, physics and mathematics (INCRYS, 1974). Sr. scient. *Electron diffraction, structure analysis.*

Umansky, Prof. Mark Moiseyevich. Moscow State U., Dept. of Physics, Leninskiye Gory, Moscow 117234, USSR. (1906) Dr, physics and mathematics (INCRYS, 1957). Sr. scient. *X-ray structure analysis, instrumentation.*

Umansky, Prof. Yakov Semenovich. Moscow Inst. of Steel and Alloys, Leninsky pr. 6, Moscow 117049, USSR. (1905) Dr, technics (Mining Inst., Acad. Sci. Kazakh SSR, 1943). *Metal physics, metals and alloys, structure.*

Urusov, Dr Vadim Sergeyevich. Moscow State U., Dept. of Geology, Leninskiye Gory, Moscow 119899, USSR. (1936) Dr, chemistry (Inst. of Geochemistry and Analytical Chemistry, Acad. Sci. USSR, 1975). Head of chair. *Crystal chemistry theory, energetic crystal chemistry, isomorphism, polymorphism.*

Urusovskaya, Dr Aida Alexandrovna. Inst. of Crystallography, Acad. of Sci. of the USSR, Leninsky pr. 59, Moscow 117333, USSR. (1929) Cand, physics and mathematics (INCRYS, 1955). Sr. scient. *Defects, mechanical properties of crystals.*

Usov, Dr Oleg Alekseyevich. Physical Techn. Inst., Acad. of Sci. of the USSR, Zapovednaya St. 51, Leningrad 194037, USSR. (1936) Cand, physics and mathematics (Inst. of Semiconductors, Acad. Sci. USSR, 1967). Sr. scient. *Structural and dynamical properties, crystallographic computing methods.*

Uyukin, Dr Evgeny Mikhailovich. Inst. of Crystallography, Acad. of Sci. of the USSR, Leninsky pr. 59, Moscow 117333, USSR. (1946) Cand, physics and mathematics (INCRYS, 1980). Scient. *Optical and electrical properties of crystals.*

Vainshtein, Prof. Boris Konstantinovich. Inst. of Crystallography, Acad. of Sci. of the USSR, Leninsky pr. 59, Moscow 117333, USSR. (1921) Full member of the Acad. Sci. USSR, Dr, physics and mathematics (INCRYS, 1955). Dir. *X-ray crystallography, electron microscopy, biological macromolecules, diffraction theory, crystal structure analysis theory.*

Val'kovskaya, Dr Margarita Ivanovna. Inst. of Applied Physics, Acad. of Sci. of the Moldavian SSR, Akademicheskaya 5, Kishinev 277028, USSR. (1938) Cand, physics and mathematics (Kishinev U., 1966). Sr. scient. *Physical crystallography, mechanical properties of crystals.*

Vasil'yev, Dr Alexander Borisovich. Inst. of Crystallography, Acad. of Sci. of the USSR, Leninsky pr. 59, Moscow 117333, USSR. (1951) Cand, physics and mathematics (INCRYS, 1979). Scient. *Optics of solids.*

Vasilyev, Dr Evgeny Konstantinovich. Inst. of Earth Crust, Acad. of Sci. of the USSR, Siberian Dept., Lermontov St. 128, Irkutsk 664033, USSR. (1922) Cand, physics and mathematics (Irkutsk State U., 1966). Sr. scient. *Crystal chemistry, X-ray structure analysis, inorganic compounds, isomorphism, instrumentation.*

Velikodnyi, Dr Yury Andreyevich. Inst. of Chemical Reagents and Pure Substances, Bogorodsky Val 3, Moscow 107258, USSR. (1941) Cand, chemistry (MGU, 1975). Sr. scient. *Crystal chemistry, transition metals, double salts.*

Venevtsev, Prof. Yury Nikolayevich. Karpov Physical Chemistry Inst., Obukha St. 10, Moscow 107120, USSR. (1926) Dr, physics and mathematics (Inst. of Physics, Acad. Sci. USSR, 1970). Lab. head. *Crystallography, crystal chemistry, ferroelectrics.*

Veremeichik, Dr Tamara Fedorovna. Inst. of Crystallography, Acad. of Sci. of the USSR, Leninsky pr. 59, Moscow 117333, USSR. (1945) Cand, physics and mathematics (INCRYS, 1977). Scient. *Crystalline field theory, spectroscopy of crystals with impurities.*

Verkhovskaya, Dr Kira Alexandrovna. Inst. of Crystallography, Acad. of Sci. of the USSR, Leninsky pr. 59, Moscow 117333, USSR. (1940) Cand, physics and mathematics (INCRYS, 1968). Scient. *Ferroelectrics, optical properties.*

Vilkov, Dr Lev Vasilyevich. Moscow State U., Dept. of Chemistry, Leninskiye Gory, Moscow 117234, USSR. (1931) Dr, chemistry (MGU, 1969). Sr. scient. *Elemento-organic compounds, structure, electron diffraction by gases.*

Vinokurov, Prof. Vladimir Mikhailovich. Kazan' State U., Lenina 18, Kazan' 420008, USSR. (1921) Dr, geology and mineralogy (IGEM, 1966). Head of Chair. *Radiospectroscopy, physical properties.*

Vistin', Dr Leonard Kazimirovich. Inst. of Crystallography, Acad. of Sci. of the USSR, Leninsky pr. 59, Moscow 117333, USSR. (1933) Cand, physics and mathematics (INCRYS, 1972). Scient. *Crystal physics, liquid crystals.*

Vlasov, Dr Vasily Platonovich. Inst. of Crystallography, Acad. of Sci. of the USSR, Leninsky pr. 59, Moscow 117333, USSR. (1941) Cand, physics and mathematics (INCRYS, 1979). Scient. *Surface physicochemical properties, nucleation, epitaxy, Electron and ionic spectroscopy.*

Voitsekhovsky, Dr Vladimir Nikolayevich. Leningrad Mining Inst., 21st Liniya 2, Leningrad 199026, USSR. (1931) Cand, geology and mineralogy (GORNY, 1966). Sr. scient. *Crystal growth, crystal morphology.*

Vol'kenshtein, Prof. Mikhail Vladimirovich. Inst. of Molecular Biology, Acad. of Sci. of the USSR, Vavilov St. 32, Moscow 117312, USSR. (1912) Corresp. member of the Acad. Sci. USSR; Dr, physics and mathematics (Tomsk U., 1942). Lab. head. *Polymers, macromolecular compounds.*

Volk, Dr Tat'yans Rafailovna. Inst. of Crystallography, Acad. of Sci. of the USSR, Leninsky pr. 59, Moscow 117333, USSR. (1942) Cand, physics and mathematics (INCRYS, 1972). Scient. *Phase transitions, ferroelectrics, radiation effects, crystal properties.*

Volodina, Dr Alexander Petrovich. Inst. of Crystallography, Acad. of Sci. of the USSR, Leninsky pr. 59, Moscow 117333, USSR. (1950) Cand, physics and mathematics (Inst. of Physical Problems, Acad. Sci. USSR, 1978). Scient. *EPR spectroscopy, activated single crystals, low temperatures.*

Volodina, Dr Galina Fedorovna. Inst. of Applied Physics, Acad. of Sci. Moldavian SSR, Akademicheskaya 5, Kishinev 277028, USSR. (1935) Cand, physics and mathematics (INCRYS, 1964). Sr. scient. *X-ray structure analysis methods, coordination and inorganic compounds.*

Voronova, Dr Alexandra Alekseyevna. Inst. of Crystallography, Acad. of Sci. of the USSR, Leninsky pr. 59, Moscow 117333, USSR. Cand, physics and mathematics (INCRYS, 1971). Scient. *X-ray structure analysis, proteins.*

Voskresenskaya, Dr Inna Evgenyevna. Inst. of Crystallography, Acad. of Sci. of the USSR, Leninsky pr. 59, Moscow 117333, USSR. (1937) Cand, geology and mineralogy (MGU, 1968). Sr. scient. *Crystal growth.*

Voznyak, Dr Dmitry Konstantinovich. Inst. of Geochemistry and Physics of Minerals, Acad. of Sci. of the Ukrainian SSR, Palladina Prospekt 34, Kiev 252068, USSR. (1938) Cand, geology and mineralogy (Inst. of Geology, Acad. Sci. Ukrainian SSR, 1971). Sr. scient. *Mineralogical crystallography.*

Vozzhennikov, Dr Valery Mikhailovich. Karpov Physical Chemistry Inst., Obukha St. 10, Moscow 107120, USSR. (1936) Cand, chemistry (KARPOV, 1970). Sci. *Crystal chemistry, organic compounds.*

Vrublevskaya, Dr Zoya Vasilyevna. Inst. of Geology, Mineralogy and Petrography, Acad. of Sci. of the USSR, Staromonetny 35, Moscow 109017, USSR. (1940) Cand, geology and mineralogy (IGEM, 1974). Scient. *Structural mineralogy, polytypism, electron diffraction.*

Yakhontova, Prof. Liya Konstantinovna. Moscow State U., Dept. of Geology, Leninskiye Gory, Moscow 117234, USSR. (1925) Dr, geology and mineralogy (MGU, 1973). Docent. *Structural mineralogy, crystal chemistry, minerals.*

Yakovenko Dr Sergey Sergeyevich. Inst. of Crystallography, Acad. of Sci. of the USSR, Leninsky pr. 59, Moscow 117333, USSR. (1945) Cand, physics and mathematics (INCRYS, 1979). Scient. *Optical properties, inhomogeneous media.*

Yakovlev, Prof. Ivan Alekseyevich. Moscow State U., Dept. of Physics, Leninskiye Gory, Moscow 117234, USSR. (1912) Dr, physics and mathematics (MGU, 1968). Head of Chair. *Phase transitions in crystals, molecular optics of solids.*

Yakubovich, Dr. Ol'ga Vselodovna. Moscow State U., Dept. of Geology, Leninskiye Gory, Moscow 119899, USSR. (1950) Cand, geology and mineralogy (MGU, 1978). Scient. *X-ray structure analysis, crystal chemistry, inorganic compounds.*

Yamnova, Dr Nataliya Arkadyevna. Moscow State U., Dept. of Geology, Leninskiye Gory, Moscow 117234, USSR. (1950) Cand, geology and mineralogy (MGU, 1976). Scient. *Crystal chemistry, inorganic compounds.*

Yanson, Dr Tamara Ivanovna. Lvov State U., Dept. of Chemistry, University St. 1, Lvov 290602, USSR. (1947) Cand, chemistry (Lvov U., 1975). Scient. *X-ray structure analysis, crystal chemistry, intermetallic compounds.*

Yanulov, Dr Kirill Paskalyevich. Inst. of Geology, Acad. of Sci. of the USSR, Komi Dept., Kommunisticheskaya St. 28, Siktivkar 167007, USSR. (1920) Cand, geology and mineralogy (LGU, 1950). Lab. head. *Isomorphism, epitaxial growth.*

Yarmolyuk, Dr Yaroslav Petrovich. Lvov State U., Dept. of Chemistry, University St. 1, Lvov 290602, USSR. (1942) Cand, chemistry (Lvov U., 1972). Docent. *X-ray structure analysis, crystal chemistry, intermetallic compounds.*

Yasinskaya, Dr Angelina Andreyevna. Lvov State U., Dept. of Geology, Shcherbakov St. 4, Lvov 290005, USSR. (1922) Cand, geology and mineralogy (Lvov State U., 1951). Docent. *Mineralogical crystallography.*

Yurin, Dr Vladimir Alexandrovich. Inst. of Crystallography, Acad. of Sci. of the USSR, Leninsky pr. 59, Moscow 117333, USSR. (1927) Cand, physics and mathematics (INCRYS, 1964). Sr. scient. *Ferroelectrics (physics), phase transitions (ferroelectric).*

Yushin, Dr Yury Yakovlevich. Inst. of Crystallography, Acad. of Sci. of the USSR, Leninsky pr. 59, Moscow 117333, USSR. (1937) Cand, physics and mathematics (Inst. of High Energy Physics, 1968). Scient. *Electromagnetic radiation interaction with crystals.*

Yushkin, Prof. Nikolay Pavlovich. Inst. of Geology, Acad. of Sci. of the USSR, Komi Department, Kommunisticheskaya 28, Syktyvkar 167000, USSR. (1936) Dr, geology and mineralogy (GORNY, 1968). Dept. head. *Mineralogical crystallography, crystallogeny, Earth's crust.*

Zadorozhnaya, Dr Lyudmila Alexandrovna. Inst. of Crystallography, Acad. of Sci. of the USSR, Leninsky pr. 59, Moscow 117333, USSR. (1944) Cand, geology and mineralogy (MGU, 1977). Scient. *Crystal growth.*

Zagal'skaya, Dr Yudif' Gertsevna. Moscow State U., Dept. of Geology, Leninskiye Gory, Moscow 117234, USSR. (1921) Cand, geology and mineralogy (MGU, 1966). Docent. *Geometrical crystallography, crystal chemistry, inorganic compounds.*

Zaitsev, Dr Sergey Mikhailovich. Rostov State U., Prospekt Stachki 192, Rostov-on-Don 344090, USSR. (1951) Cand, physics and mathematics (Rostov U., 1979). Scient. *X-ray structure analysis.*

Zaitseva, Dr Mariya Panteleimonovna. Inst. of Physics, Acad. of Sci. of the USSR, Siberian Dept., Akademgorodok, Krasnoyarsk 660036, USSR. (1930) Cand, physics and mathematics (Inst. of Physics, Acad. Sci. USSR, Siberian Dept., 1968). Sr. scient. *Crystal physics, phase transitions (ferroelectric), ferroelectrics (nonlinear electromechanical properties).*

Zakharchenko, Dr Irina Nikolayevna. Rostov State U., Inst. of Physics, Stachki pr. 194, Rostov-on-Don 344090, USSR. (1946) Cand, physics and mathematics (Rostov State U., 1978). Sr. scient. *X-ray structure analysis, defects, phase transitions, two-dimensional crystals.*

Zakharov, Dr Nikolay Dmitriyevich. Inst. of Crystallography, Acad. of Sci. of the USSR, Leninsky pr. 59, Moscow 117333, USSR. (1944) Cand, physics and mathematics (INCRYS, 1976). Scient. *Real structure of crystals, electron microscopy.*

Zakharova, Prof. Mariya Ivanovna. Moscow State U., Dept. of Physics, Leninskiye Gory, Moscow 117234, USSR. (1904) Dr, physics and mathematics (MGU, 1949). Prof. *X-ray crystallography, phase transformations.*

Zalessky, Dr Andrey Vladimirovich. Inst. of Crystallography, Acad. of Sci. of the USSR, Leninsky pr. 59, Moscow 117333, USSR. (1930) Dr, physics and mathematics (INCRYS, 1985). Sr. scient. *Magnetic properties of crystals, nuclear magnetic resonance.*

Zalutsky, Dr Ivan Ilyich. Lvov State U., Dept. of Chemistry, University St. 1, Lvov 290602, USSR. (1935) Cand, chemistry (Lvov U., 1968). Docent. *X-ray structure analysis, crystal chemistry, intermetallic compounds.*

Zamorzayev, Dr Alexander Mikhailovich. Kishinev U., Sadovaya 60, Kishinev 277003, USSR. (1927) Dr, physics and mathematics (INCRYS, 1971). Prof. *Symmetry theory (generalization and applications).*

Zarechnyuk, Dr Oleg Safonovich. Lvov State U., Dept. of Chemistry, University St. 1, Lvov 290602, USSR. (1923) Cand, chemistry (Lvov U., 1968). Docent. *X-ray structure analysis, crystal chemistry, intermetallic compounds.*

Zasorin, Dr Evgeny Zotikovich. Moscow State U., Dept. of Chemistry, Leninskiye Gory, Moscow 117234, USSR. (1934) Cand, chemistry (MGU, 1966). Sr. scient. *Structure, inorganic compounds, electron diffraction by gases.*

Zav'yalova, Anna Alexeyevna. Inst. of Crystallography, Acad. of Sci. of the USSR, Leninsky pr. 59, Moscow 117333, USSR. (1937) Cand, geology and mineralogy (INCRYS, 1970). Scient. *Electron diffraction, thin film structures, crystal chemistry.*

Zayakina, Dr Nadezhda Viktorovna. Inst. of Geology, Acad. of Sci. of the USSR, Siberian Dept., Lenin Pr. 39, Yakutsk 677982, USSR. (1943) Cand, geology and mineralogy (MGU, 1976). Sr. scient. *Crystal structure determination, crystal chemistry, silicates, typomorphism, minerals.*

Zhdanov, Prof. German Stepanovich. Moscow State U., Dept. of Physics, Leninskiye Gory, Moscow 117234, USSR. (1906) Dr, physics and mathematics MGU, 1941). Head of Chair. *Crystal structure, physical properties of crystals.*

Zheludev, Prof. Ivan Stepanovich. Inst. of Crystallography, Acad. of Sci. of the USSR, Leninsky pr. 59, Moscow 117333, USSR. (1921) Dr, physics and mathematics (MGU, 1961). Lab. head. *Ferroelectrics (physics), phase transitions (ferroelectric), symmetry in physics.*

Zheludeva, Dr Svetlana Ivanovna. Inst. of Crystallography, Acad. of Sci. of the USSR, Leninsky pr. 59, Moscow 117333, USSR. (1948) Cand, physics and mathematics (MGU, 1976). Scient. *Thin film growth, thin film properties.*

Zhidkov, Dr Nikolay Petrovich. Moscow State U., Computing Center, Leninskiye Gory, Moscow 117234, USSR. (1918) Cand, physics and mathematics (MGU, 1949). Docent. *Computing methods in structure analysis.*

Zhmurova, Dr Zinaida Ivanovna. Inst. of Crystallography, Acad. of Sci. of the USSR, Leninsky pr. 59, Moscow 117333, USSR. (1930) Cand, physics and mathematics (INCRYS, 1970). Scient. *Crystal growth from melt.*

Zhukhlistov, Dr Anatoliy Pavlovich. Inst. of Geology, Mineralogy and Petrography, Acad. of Sci. of the USSR, Staromonetny 35, Moscow 109017, USSR. (1938) Cand, geology and mineralogy (IGEM, 1977). Sr. scient. *Structural crystallography, mineralogy, polytipy, electron diffraction, electron microscopy.*

Zhuze, Prof. Vladimir Panteleimonovich. Physico-Techn. Inst., Acad. of Sci. of the USSR, Zapovednaya St. 51, Leningrad 194037, USSR. (1904) Dr, physics and mathematics (Inst. of Semiconductors, Acad. Sci. USSR, 1955). Sr. scient. *Phase transitions, non-stoichiometric compounds.*

Zinenko, Dr Victor Ivanovich. Inst. of Physics, Acad. of Sci. of the USSR, Siberian Dept., Akademgorodok, Krasnoyarsk 660036, USSR. (1942) Dr, physics and mathematics (Inst. of Physics, Acad. Sci. USSR, Siberian Dept.). Scient. *Phase transitions in crystals.*

Zorky, Prof. Petr Markovich. Moscow State U., Dept. of Chemistry, Leninskiye Gory, Moscow 117234, USSR. (1933) Dr, chemistry (MGU, 1973). Lab. head. *Symmetry, crystal chemistry.*

Zubenko, Dr Vasily Vasilyevich. Moscow State U., Dept. of Physics, Leninskiye Gory, Moscow 117234, USSR. (1930) Cand, physics and mathematics (MGU, 1968). Sr. scient. *X-ray crystallography, instrumentation and methods.*

Zubov, Dr Yuriy Alexandrovich. Karpov Physical Chemistry Inst., Obukha St. 10, Moscow 107120, USSR. (1932) Dr, chemistry (KARPOV, 1976). Sr. scient. *X-ray structure analysis, polymers.*

Zvezdinskaya, Dr Larisa Vsevolodovna. Inst. of Geology, Mineralogy and Petrography, Acad. of Sci. of the USSR, Staromonetny 35, Moscow 109017, USSR. (1948) Cand, geology and mineralogy (MGU, 1978). Scient. *Crystal chemistry, classification, inorganic compounds, minerals.*

Zviedre, Dr Irena Ilyinichna. Inst. of Inorganic Chemistry, Acad. of Sci. of the Latvian SSR, Meistaru St. 10, Riga 226934, USSR. (1938) Cand, chemistry (Dept. of Biological and Chemical Sciences, Acad. Sci. Latvian SSR). Sr. scient. *Crystal chemistry, borates.*

Zvinchuk, Dr Rostislav Alekseyevich. Leningrad State U., Dept. of Chemistry, 14th Liniya 29, Leningrad 199178, USSR. (1929) Cand, chemistry (LGU, 1964). Docent. *Diffuse phase transitions.*

Zvyagin, Dr Boris Borisovich. Inst. of Geology, Mineralogy and Petrography, Acad. of Sci. of the USSR, Staromonetny 35, Moscow 109017, USSR. (1921) Dr, physics and mathematics (INCRYS, 1963). Lab. head. *Structural crystallography, structural mineralogy, polytypism, electron diffraction.*

UNITED KINGDOM

Sub-Editor: D.W. Penfold

Notes

1. International telephone country code - 44

2. Unless otherwise stated, the exchange name for the telephone number is the place name in the address. Most exchanges now have direct dialling (STD) codes (separated from the rest of the number by '+'). The STD code may not apply to relatively short distance calls (e.g. London to Welwyn Garden); consult the local booklet of dialling codes.

3. In the biographic data, the position held (e.g. Head, Prof.) relates to the Div. or Dept. in the address, unless further details are given.

4. In general the bachelor's degree (BA, BSc) is the first degree awarded by universities in the U.K.; the holder of an Oxford or Cambridge BA may proceed to the MA after the passage of time and payment of money. Qualifications of first degree standard awarded by other bodies include Dip. Tech. (Diploma in Technology), HND (Higher National Diploma) and graduateship, licentiateship or membership of various professional institutions. The Higher National Certificate (HNC) is similar to HND but with a narrower range of subjects. Higher degrees include master's degrees (MSc, MPhil) and various diplomas awarded after courses of instruction or research lasting one or two years. The doctorate (PhD, DPhil) is obtained after research lasting (normally) three years. The senior doctorate (DSc, ScD) is awarded on the basis of published contributions to knowledge.

5. Colleges of the University of London include:

   Bedford
   Birkbeck
   Chelsea
   Imperial (Imp.)

   King's (KQC)
   Queen Mary (QMC)
   University (UC)

6. Abbreviations used for counties or regions are:

   Beds. - Bedfordshire
   Berks. - Berkshire
   Bucks. - Buckinghamshire
   Hants. - Hampshire
   Herts. - Hertfordshire
   Lancs. - Lancashire
   Leics. - Leicestershire

   Middx. - Middlesex
   Northants. - Northamptonshire
   Oxon - Oxfordshire
   Staffs.- Staffordshire
   Wilts. - Wiltshire
   Worcs. - Worcestershire
   Yorks. - Yorkshire

7. The following are grades in the Scientific Civil Service and in some universities:

   CSO - Chief Scientific Officer
   DCSO - Deputy Chief Scientific Officer
   SPSO - Senior Principal Scientific Officer
   PSO - Principal Scientific Officer

   SSO - Senior Scientific Officer
   HSO - Higher Scientific Officer
   EO - Experimental Officer
   SEO - Senior Experimental Officer

8. Other abbreviations used include:

   AERE - Atomic Energy Research Establishment
   AWRE - Atomic Weapons Research Establishment
   BR - British Rail
   CEGB - Central Electricity Generating Board
   CERL - Central Electricity Research Laboratories
   CNAA - Council for National Academic Awards
   GEC - General Electric Company
   ICI - Imperial Chemical Industries
   Lect. - Lecturer
   MRC - Medical Research Council

   Off. - Officer
   plc - public liability company
   RAE - Royal Aircraft Establishment
   RMCS - Royal Military College of Science
   RSRE - Royal Signals and Radar Establishment
   Roy. Soc. - Royal Society
   Sect. - Section
   SERC - Science and Engineering Research Council
   UKAEA - United Kingdon Atomic Energy Authority
   UMIST - University of Manchester Inst. of Science and Technology

Abell, Dr John Stuart. Department of Metallurgy and Materials, U. of Birmingham, PO Box 363, Birmingham B15 2TT, England. (1944) PhD, solid state physics (U. Surrey, 1969). Res. fellow. (tel. 021 + 472-1301, ext. 3446). *Single crystal growth, rare-earth-based intermetallic compounds, opto-electronic device material characterization (oxide and fluoride), structure and physical properties, Nb-H dilute alloys, metal deformation and transformation behaviour.*

Acharaya, Dr Ravindra. Lab. of Molecular Biophysics, U. of Oxford, S. Parks Rd., Oxford OX1 3QU, England. (1955) PhD, crystallography (U. Bangalore, India, 1983). Res. asst. (tel. 0865 + 56733, ext. 415). *Experimental crystallography, computer programming, computer modelling.*

Adam, Dr C.D. Unilever Res., Port Sunlight Lab., Bebington, Wirral LG3 3JW, England.

Adam, Dr Jerzy. Timber Top, Brightwell-cum-Sotwell, Wallingford, Oxon OX10 0RG, England. (1918) PhD, physics (U. St. Andrews, 1949). Retired. (tel. 0491 + 36045). *X-ray analysis and crystallography, structures, nuclear reactor materials, computing.*

Adams, Dr Margaret Joan. Lab. of Molecular Biophysics, Zoology Dept., Oxford U., Rex Richards Building, South Parks Rd., Oxford OX1 3QU, England. (1939) DPhil, protein crystallography (U. Oxford, 1968). Fellow, tutor in chemistry, Somerville C. (tel. 0865 + 56733, ext. 426). *X-ray methods, protein structures, adenine nucleotide dependent enzymes, carboxylation enzymes.*

Adams, Dr M.J. Lab. of Molecular Biophysics, S. Parks Rd., Oxford OX1 3QU, England.

Ahlers, Mr N.H.E. 294 Histon Rd., Cambridge CB4 3HS, England.

Ainsworth, Mr Leonard Ralph. Springfields Nuclear Power Dev. Lab., UKAEA (N. Div.), Preston PR4 0RR, England. (1932) CChem, FRSC, Chemistry (Royal Society of Chemistry, 1960). SSO (tel. 0772 + 728262, ext. 31284). *X-ray powder diffraction, computer methods, vitreous state.*

Alcock, Dr Nathaniel Warren. Dept. of Chemistry, U. of Warwick, Coventry CV4 7AL, England. (1939) PhD, chemistry (U. Cambridge, 1963). Reader. (tel. 0203 + 24011, ext. 2228). *Inorganic crystal structures, absorption corrections.*

Al-Farhan, Mr Khalid A. M. Dept. of Chemistry, UMIST, Manchester M60 1QD, England. (1961) BSc, chemistry (King Saud U., Saudi-Arabia, 1982). Res. Student. (tel. 061 + 434-8262).

Allen, Dr A.J. Materials Res. Div., Building 5212, AERE Harwell, Didcot, Oxon OX11 0RA, England.

Allen, Dr Frank Harmsworth. University Chemistry Lab., U. of Cambridge, Lensfield Rd., Cambridge CB2 1EW, England. (1944) PhD, chemistry (U. London, 1968). Sr. asst. in res. (tel. 0223 + 66499, ext. 376). *Organic crystal structures, crystallographic databases, computer programming.*

Allen, Mr Keith William. Chemistry Dept., The City University, Northampton Square, London EC1V 0HB, England. (1926) MSc, chemistry (U. London,

1958). Lect. (tel. 01 + 253-4399, ext. 3506, telex 263896). *Inorganic chemistry, adhesion science.*

Allman, Miss J.M. Prospect House, 101 Kingsdown Parade, Cotham, Bristol BS6 5UJ, England.

Andrews, Dr S.J. 82 Blackshots Lane, Grays, Essex RM16 2JX, England.

Arndt, Dr Ulrich Wolfgang. Structural Studies Div., MRC Lab. of Molecular Biology, Hills Road, Cambridge CB2 2QH, England. (1924) PhD, physics (U. Cambridge, 1949). Scient. staff, MRC. (tel. 0223 + 248011, telex 81532). *Instrumentation, biological structures, synchrotron radiation.*

Artymiuk, Dr Peter Joseph. Lab. of Molecular Biophysics, Oxford U., Richards Building, South Parks Road, Oxford OX1 3QU, England. (1952) DPhil, Chemistry and molecular biophysics (U. Oxford, 1979). (0865 + 56733, ext. 404). *Protein crystallography, refinement, solvent structure, hydrogen bonding.*

Ashwell, Dr Geoffrey Joseph. Dept. of Chemistry, Sheffield City Poly., Pond Street, Sheffield S1 1WB, England. (1947) PhD, organic semiconductors (U. Nottingham, 1972). Lect. (tel. 0742 + 20911, ext. 380). *One-dimensional metals, organic semiconductors, molecular rectifiers.*

Avery, Dr A.J. Physics Branch, RMCS, Shrivenham, Swindon, Wilts, England.

Bacon, Prof. George Edward. Windrush Way, Guiting Power, Cheltenham, GL54 5US, England. (1917) ScD, physics and crystallography (U. Cambridge, 1964). Em. prof., physics, U. Sheffield. (tel. 04515 + 631). *Neutron diffraction.*

Badawi, Dr H.M. 3 Rainsford Rd., Stanstead, Essex CM24 8DU, England.

Badger, Mr John. Dept. of Physics, U. of York, Heslington, York YO1 5DD, England. (1962) BSc, physics (U. Durham, 1983). Res Student. (tel. 0904 + 59861, ext. 5503). *Ribosomal proteins, refinement, structure.*

Bagley, Mr Arthur George. Director, Hiltonbrooks Ltd., Yew Tree Cottage, Knutsford Rd., Cranage, Holmes Chapel, Cheshire CW4 8EP, England. (1947). Dir. (tel. 0477 + 32687). *Analytical X-ray equipment.*

Bailey, Prof. David Kenneth. Dept. of Geology, U. of Reading, Whiteknights, Reading RG6 2AB, England. (1931) PhD, geology (U. London, 1959). Prof. (tel. 0734 + 875123, ext. 7871). *Petrology, mineralogy, geochemistry.*

Bailey, Dr Neil Anthony. Dept. of Chemistry, U. of Sheffield, Brook Hill, Sheffield S3 7HF, England. (1940) PhD, chemistry (Imp. C., U. of London, 1964). Lect. (tel. 0742 + 78555, ext. 4464). *X-ray crystal structure analysis, binucleating and compartmental acyclic and macrocyclic ligands, metal complexes, organometallic molybdenum, tungsten structural chemistry.*

Bailey, Ms S. Dept. of Crystallography, Birkbeck C., Malet St., London WC1, England.

Bain, Dr Derek Charles. Dept. of Mineral Soils, Macaulay Inst. for Soil Res., Craigiebuckler, Aberdeen AB9 2QJ, Scotland. (1944) PhD, geology (U. Aberdeen, 1974). PSO. (tel. 0224 + 38611). *Soil mineralogy and weathering, X-ray fluorescence spectroscopy.*

Bainbridge, Mr John Evelyn. Materials Physics and Metallurgy Div., AERE Harwell, Didcot, Oxon OX11 0RA, England. (1933) AIM, metallurgy (Inst. of Metallurgists, 1965). SSO. (tel. 0235 + 24141, ext. 4134). *Scanning electron microscopy with analysis, transmission electron microscopy.*

Baker, Mr Patrick Julian. Biochemistry Dep., U. of Sheffield, Western Bank, Sheffield S10 2TN, England. (1961) BSc, Chemistry (U. York, 1982). Res. Student. (tel. 0742 + 78555, ext. 4242). *Protein crystallography.*

Baker, Mr R.W. U. of London Computer Centre, 20 Guildford St., London WC1 1DZ, England.

Baker, Mr Thomas Wilfred. Tirrold Scientific and Technical Services Ltd., Berry Croft, Spring Lane, Aston Upthorpe, Didcot, Oxfordshire OX11 9EH, England. (1923) BSc, physics (U. Manchester, 1949). Dir. & consultant, X-ray diffraction. (tel. 0235 + 850264). *X-ray diffraction applications, materials, powder diffraction, line broadening, preferred orientation, high and low temperatures, high precision, automatic control, computer data processing, X-ray diffraction instruments.*

Balchin, Dr Anthony Arthur. Dept. of Physical Sci., Brighton Polytechnic, Lewes Rd., Moulsecoomb, Brighton BN2 4GJ, England. (1932) PhD, crystallography (U. London, 1968). Sr. lect. (tel. 0273 + 693655, ext. 2498). *Chelation of cations by keto-gluconic acids, layer compounds (structure and characterisation), cadmium iodide types, molybdenum di-sulphide types, optical - X-ray diffraction analogies.*

Balyuzi, Dr Hushang H.M. Dept. of Physics, King's C. (KQC), Strand, London WC2R 2LS, England. (1942) PhD, X-ray diffraction (U. London, 1970). Lect. (tel. 01 + 836-5454, ext. 2145). *Amorphous systems - structure investigation, X-ray diffraction.*

Banbery, Mrs H.J. 102 Warstones Rd., Penn, Wolverhampton WV4 4LP, England.

Bandy, Dr Judith Ann. Chemical Crystallography Lab., U. of Oxford, 9 Parks Road, Oxford OX1 3PD, England. (1957) PhD, Crystallography and Coordination Chemistry (CNAA, 1981). Post-Doctoral res. asst. (tel. 0865 + 53424, ext. 293). *Structural chemistry, X-ray diffraction, organic molecules, metal complexes.*

Barnes, Dr John Conquest. Dept. of Chemistry, U. of Dundee, Dundee DD1 4HN, Scotland. (1935) PhD, chemistry (U. Wales, 1960). Sr. lect. (tel. 0382 + 23181, ext. 284). *Solid state chemistry, crystallography, extended systems, lanthanides.*

Barnes, Dr Paul. Dept. of Crystallography, Birkbeck C., Malet St., London WC1E 7HX, England. (1942) PhD, physics (U. Cambridge, 1968). Sr. lect. (tel. 01 + 580-6622, ext. 417). *Powder X-ray and neutron diffraction, electron microscopy, computer simulation, aqueous systems, alumino-silicates, cements, ceramics.*

Barrett, Mr S.D. Materials Sci. Centre, Birmingham U, Birmingham B15 2TT, England.

Barron, Dr Hugh Wilson Taylor. Dept. of Natural Philosophy, U. of Aberdeen, Aberdeen AB9 2UE, Scotland. (1943) PhD, physics (U. Aberdeen, 1971). Lect. (tel. 0224 + 40241, ext. 284). *Phonon - X-ray scattering, high field transport, semi-conductors.*

Barrow, Dr Michael John. Dept. of Appl. Chemical Sci., Napier C., Edinburgh EH10 5DT, Scotland. (1946) PhD, chemistry (U. Manchester, 1972). Lect. (tel. 031 + 447-7070, ext. 623). *Chemical crystallography, computing, structural chemistry, low-temperature crystallography.*

Bassett, Prof. David Clifford. Physics Dept., U. of Reading, Whiteknights, PO Box 220, Reading, Berks, RG6 2AF, England. (1937) ScD, polymer physics (U. Cambridge, 1981). Prof., physics. (tel. 0734 + 875123, ext. 369, telex 847813). *Polymer crystallization, crystal growth, high pressures.*

Bassett, Mr G. Alan. Dept. of Physics, U. of Warwick, Coventry CV4 7AL, England. (1926) MSc, physics (U. London, 1950). Sr. tutor., physics. (tel. 0203 + 24011, ext. 2375). *Crystal surfaces, electron microscopy, decoration, mechanical properties, crack propagation.*

Batchelder, Dr David Neville. Dept. of Physics, Queen Mary C., Mile End Rd., London E1 4NS, England. (1938) PhD, physics (U. Illinois, USA, 1965). Lect. (tel. 01 + 980-4811, ext. 4004). *Structure and properties, polymer single crystals, conjugated polymers.*

Bates, Dr David Ronald. Analytical Div., British Gas Corp., London Res. Station, Michael Road, London SW6 2AD, England. (1955) PhD, physical chemistry (U. Nottingham, 1982). Res. Scient. (tel. 01 + 736-3344, ext. 4059). *X-ray diffraction, industrial applications; automatic powder diffractometry; crystallite size studies, catalysis.*

Bates, Dr Peter Arthur. Div. of Physics and Astronomy, Lancashire Polytechnic, Corporation St., Preston, Lancs. PR1 2TQ, England. (1945) PhD, physics (U. of Wales). Sr. lect. in physics. (tel. 0772 + 22141, ext. 2182). *X-ray diffraction, instrumentation, magnetic fluids, fine particle magnetic systems.*

Battey, Dr Maurice Hugh. Dept. of Geology, U. of Newcastle upon Tyne, Newcastle upon Tyne NE1 7RU, England. (1922) PhD, petrology (U. Cambridge, 1953). Sr. lect. *Mineralogy and petrology.*

Baxter, Mr Colin. Materials Res. & Dev. Lab., Rolls-Royce Ltd., PO Box 31, Derby DE2 8BJ, England. (1949) HND, metallurgy, (Derby and District C. of Techn., 1971). Principal Eng. (tel. 0332 + 42424, ext. 520). *Aerospace industry applications, qualitative analysis, preferred orientation measurement, single-crystal orientation measurement, residual stress measurement.*

Beagley, Dr Brian. Dept. of Chemistry, UMIST, PO Box 88, Manchester M60 1QD, England. (1936) DSc, structural chemistry (U. Birmingham, 1981). Reader. (tel. 061 + 236-3311, ext. 2567). *Molecular structure determination, electron diffraction (gas-phase), spectroscopic techniques, theoretical interpretation of results, X-ray crystallographic studies, structural chemistry, EXAFS.*

Beamson, Dr G. 34 Westbourne Grove, Goole, North Humberside DN14 6NB, England.

Beddell, Dr Christopher Raymond. Physical Chemistry Dept., Wellcome Res. Labs., Langley Court, Beckenham, Kent BR3 3BS, England. (1944) DPhil, molecular biophysics (U. Oxford, 1971). Researcher. (tel. 01 + 658-2211, ext. 406). *Macromolecule structure and function, macromolecule - ligand interactions, pharmacophore definition.*

Beddoes, Mr Roy L. Dept. of Chemistry, U. of Manchester, Brunswick St., Manchester M13 9PL, England. (1938) MA, physics (U. Cambridge, 1965). Sr. exp. off. (tel. 061 + 273-7121, ext. 5286). *Structure determination.*

Bednar, Mr E. Manchester U. Sch. of Education, De La Salle C., Manchester, England.

Been, Dr J.M. 6 Hayling Gdns., High Salvington, Worthing, W. Sussex BN13 3AJ, England.

Beevers, Dr Cecil Arnold. Dept. of Chemistry, U. of Edinburgh, West Mains Rd., Edinburgh EH9 3JJ, Scotland. (1908) DSc, X-ray crystallography (U. Liverpool, 1938). Em. Reader. (tel. 031 + 667-1081, ext. 3405). *X-ray crystallography, model making.*

Begley, Dr Michael John. Dept. of Chemistry, U. of Nottingham, University Park, Nottingham NG7 2RD, England. (1944) PhD, crystallography (U. Surrey, 1970). Lect. (tel. 0602 + 506101, ext. 2392). *Organic crystal structures, solid state reactions.*

Bell, Mr Anthony Martin Thomas. Microstructural Studies Sect., UKAEA, Springfields, Preston PR4 0RR, Lancs, England. (1963) BSc, chemistry (U. Sheffield, 1984). SO. (tel. 0772 + 728262, ext. 31450, telex 67545). *X-ray powder diffraction, inorganic chemistry.*

Bellamy, Mr Brian Arthur. Materials Dev. Div., AERE Harwell, Didcot, Oxon OX11 0RA, England. (1936) MInstP (Inst. of Physics,1936). SSO. (tel. 0235 + 24141, ext. 4524). *X-ray diffraction applications, materials.*

Bellard, Dr Sharon Ann. Dept. of Organic and Inorganic Chemistry, U. of Cambridge, Lensfield Rd., Cambridge CB2 1EW, England. (1950) PhD, chemistry (U. Cambridge, 1979). Res. asst. (tel. 0223 + 66499, ext. 314). *Information storage and retrieval, data base management, organic and organometallic crystal structures, low temperature crystallography.*

Bennett, Miss S. 484 Archway Rd, London N6, England.

Bevis, Prof. M.J. Dept of Materials Tech., Brunel U., Uxbridge, Middx. UB8 3PH, England.

Bhuva, Dr V.J. 43 St Thomas' Square, Cambridge CB1 3TG, England.

Bishop, Dr Arthur Clive. Dept. of Mineralogy, British Museum (Natural History), Cromwell Rd., London SW7 5BD, England. (1930) PhD, petrology (U. London, 1954). Deputy dir., Keeper, mineralogy. (tel. 01 + 589-6323, ext. 226). *Mineralogy, petrology.*

Black, Prof. Paul Joseph. Chelsea C., Centre for Sci. Education, Bridges Place, London SW6 4HR, England. (1930) PhD, crystallography (U. Cambridge, 1954). Dir. (tel. 01 + 736-3401). *Crystalline intermetallic compounds, Mössbauer and X-ray resonance scattering, physics education.*

Blake, Dr Alexander John. Dept. of Chemistry, U. of Edinburgh, West Mains Rd., Edinburgh EH9 3JJ, Scotland. (1954) PhD, structural chemistry of phosphenes (U. Aberdeen, 1980). Res. Fellow. (tel. 031 + 667-1081, ext. 3405). *Crystal structures, low-melting compounds (especially phosphorus and silicon), powder data.*

Blake, Dr Antony Brian. Dept. of Chemistry, U. of Hull, Hull HU6 7RX, England. (1933) PhD, chemistry (U. London, 1960). Lect. (tel. 0482 + 46311, ext. 7433). *Transition-metal chemistry, magnetic properties, polynuclear complexes, exchange interactions.*

Bland, Dr J.A. Coach House, Matson Drive, Remenham, Henley-on-Thames, Oxford, England.

Bloor, Dr David. Dept. of Physics, Queen Mary C., Mile End Rd., London, E1 4NS, England. (1937) PhD, physics (U. London, 1961). Reader. (tel. 01 + 980-4811, ext. 334). *Structure - properties and preparation, diacetylene monomer, polymer single crystals.*

Blow, Prof David Mervyn. Blackett Lab., Imperial C. of Sci. and Techn., London SW7 2BZ, England. (1931) PhD, physics (U. Cambridge, 1957). Prof., biophysics. (tel. 01 + 589-5111, ext. 6721, telex 261503). *Enzyme structure and activity, protein engineering, protein crystallography techniques.*

Bloxam, Dr Thomas Wallace. Dept. of Geology, University C. Swansea, Swansea SA2 8PP, Wales. (1926) DSc, geology (U. Glasgow & U. Edinburgh, 1981). U. Wales Reader, geology. (tel. 0792 + 295146, ext. 5146). *Mineralogy, petrology and geochemistry.*

Blundell, Dr David James. Petrochemicals and Plastics Div., ICI plc, PO Box 90, Wilton, Middlesborough, Cleveland, TS6 8JE, England. (1940) PhD, physics (U. Bristol, 1967). Sr. res. physicist. (tel. 0642 + 455522, ext. 2007, telex 587461). *Polymer morphology, small-angle X-ray scattering, wide-angle X-ray scattering, thermal analysis.*

Blundell, Prof. Thomas Leon. Dept. of Crystallography, Birkbeck C., Malet St., London WC1E 7HX, England. (1942) DPhil, crystallography (U. Oxford, 1967). Prof. (tel. 01 + 580-6622, ext. 284). *Biological molecular structures, hormones, enzymes, protein crystallography techniques.*

Boles, Dr Michael Owen. Dept. of Environmental Sci., Plymouth Polytechnic, Drake Circus, Plymouth, Devon PL4 8AA, England. (1941) PhD, crystallography (U. London, 1967). Sr. lect. (tel. 0752 + 21312, ext. 5565). *Structure and properties, antibiotic materials, stress analysis.*

Borkakoti, Dr Nivedita Neera. Dept. of Crystallography, Birkbeck C., Malet St., London WC1E 7HX, England. (1949) PhD, crystallography (U. London, 1978). Res. fellow. (tel. 01 + 580-6622, ext. 330). *Organic and biological molecules, computing.*

Bowen, Dr A.W. Materials and structures Dept, RAE, Farnborough, Hants GU14 6TD, England.

Bowen, Dr David Keith. Dept. of Engineering, U. of Warwick, Coventry CV4 7AL, England. (1940) DPhil, metallurgy (U. Oxford, 1962). Chairman, Dept. of Engineering. (tel. 0203 + 24011, ext. 2133, telex 31406). *X-ray topography, synchrotron radiation, materials science, X-ray metrology, electronic materials.*

Bowen-Jones, Dr J. Lower Bawdon Farm, Charley, Loughborough, Leics LE12 9XL, England.

Bown, Dr Michael George. Dept. of Earth Sci., U. of Cambridge, Downing St., Cambridge CB2 3EQ, England. (1928) PhD, crystallography (U. Cambridge, 1955). Lect. (tel. 0223 + 355463, ext. 280). *Crystal structures, minerals, defect structures, crystal physics.*

Boyle, Dr Lewis Laurence. University Chemical Lab., Canterbury, Kent CT2 7NH, England. (1942) DPhil, chemistry (U. Oxford, 1966). Sr. lect. (tel. 0227 + 66822, ext. 584, telex 965449). *Theoretical chemistry, phase transitions, plastic crystals, internal rotation in the solid state, group theory.*

Bradley, Ms S.M. Hope Cottage, Eggington, Nr Leighton Buzzard, Beds LU7 9PF, England.

Briant, Dr Clive Edward. 131 Drayton Rd., Sutton Courtenay, Oxon OX14 4HA, England. (1954) PhD, chemistry (U. Bradford, 1983). *Stuctural chemistry, organic and organometallic compounds, polycyclics, nucleosides, metal clusters.*

Brice, Dr John Chadwick. Solid State Electronics Div., Philips Res. Labs., Redhill, Surrey RH1 5HA, England. (1934) PhD, materials science (U. Cambridge, 1956). Principal physicist. (tel. 02934 + 5544, telex 877261). *Crystal growth, perfection.*

Bright, Dr Alan Aubrey Samuel. Chemical Synthesis Group, FBC Ltd., Chesterford Pk. Res. Stn., Nr. Saffron Walden, Essex CB10 1XL, England. (1948) PhD, chemistry (U. Dundee, 1974). Team Leader (Spectroscopy/Physical Chemistry). (tel. 0799 + 30123). *Molecular structure, biologically important chemicals, pesticides, agrochemicals.*

Brown, Miss Betty Rosina. Crystallography Group, Materials Characterisation Div., Materials Sci. Lab., GEC Res. Labs., Hirst Res. Centre, East Lane, Wembley HA0 7PP, Middx, England. (1926) BSc, physics and mathematics (U. Reading, 1945). Sci. res. staff. (tel. 01 + 904-1262, ext. 281). *Crystal orientation, multiple diffraction, epitaxial layers.*

Brown, Dr Cedric John. Dept. of Metallurgy and Materials, City of London Polytechnic, Central House, Whitechapel High St., London E1 7PF, England. (1915) DSc, crystallography (U. Birmingham, 1955). Res. fellow. (tel. 01 + 283-1030, ext. 479). *Organic structures, diffuse scattering.*

Brown, Mr D. Stoe & Cie GmbH, 21 Dorset Ave, Southall,Middx. UB2 4HF, England.

Brown, Dr David Summers. Dept. of Chemistry, Loughborough U. of Techn., Loughborough, Leics. LE11 3TU, England. (1937) PhD, crystallography (U. Nottingham, 1962). Sr. lect. (tel. 0509 + 263171, ext. 369, telex 34319 UNITEC). *Inorganic and organic structures, small angle X-ray scattering of polymer systems.*

Bruce, Dr P.G. Dept of Chemistry, Heriot-Watt U., Riccarton, Edinburgh, Scotland.

Bryant, Mr P.K. 43 Fraser St, Bilston, W Midlands WV14 7PD, England.

Buckley, Dr Christopher Paul. Polymer Engineering Div., Dept. of Mechanical Engineering, UMIST, PO Box 88, Manchester M60 1QD, England. (1946) DPhil, engineering science (U. Oxford, 1968). Lect. (tel. 061 + 236-3311, ext. 2716). *Polymers, structure - properties and processing.*

Bullen, Dr G.J. Tanglewood, Church Lane, Lexden, Colchester CO3 4DX, England.

Bullen, Mr Henry Eric. Res. and Dev. Lab. (U.K.), Gillette, 454 Basingstoke Rd., Reading RG2 0QE, England. (1930) MSc, crystallography (U. London, 1957). Res. physicist. (tel. 0734 + 875222, ext. 326). *Structures, steels, polymers.*

Bulpett, Mrs S.E. 8 Grenville Ave, Wendover, Aylesbury, Bucks HP22 6AG, England.

Bunn, Dr Charles William. 6 Pentley Park, Welwyn Garden City, Herts. AL8 7RU, England. (1905) DSc, crystallography (U. Oxford, 1953). Retired res. fellow, Royal Inst. (tel. 07073 + 23581). *Crystal growth, X-ray diffraction, structures, organic crystals, macromolecules, physical properties - structure relationship.*

Burge, Prof. Ronald Edgar. Dept. of Physics, King's C. London, Strand, London WC2R 2LS, England. (1932) DSc, physics and biophysics (U. London, 1975). Head. (tel. 01 + 836-5454, ext. 2514). *Structure and properties, natural and synthetic polymers, polymer associations; theory of scattering; image analysis; electron microscopy instrumentation; soft X-ray microscopy.*

Burns, Mr K. Chemical Crystallography Lab., 9 Parks Rd, Oxford OX1 3PD, England.

Bush, Dr Michael Anthony. Wilkinson Sword Ltd., Totteridge Rd., High Wycombe, Bucks. HP13 6EJ, England. (1943) PhD, chemistry (U. Bristol, 1967). Marketing manager. (tel. 0494 + 33300). *Structural inorganic chemistry.*

Bushnell-Wye, Dr Graham. Dept. of Crystallography, Birkbeck C., Malet St., London WC1E 7HX, England. (1950) PhD, crystallography (U. London, 1983). Res. Officer. (tel. 01 + 584-6622, ext. 285). *X-ray scattering, computer modelling, organisation of dense matter, crystal growth, interfaces, random packing.*

Butler, Dr Barry Conrad Milne. Dept. of Earth Sci., U. of Oxford, Parks Rd., Oxford OX1 3PR, England. (1932) PhD, petrology (U. Cambridge, 1960). Lect. (tel. 0865 + 54511). *Mineralogy, industrial materials.*

Butler, Dr Stephen Andrew. Swinden Labs., British Steel Corp., Moorgate, Rotherham, South Yorks. S60 3AR, England. (1946) PhD, mineralogy and crystallography (U. Cambridge, 1973). Sr. Investigator. (tel. 0709 + 60166, ext. 3283). *Qualitative and quantitative phase analysis, X-ray diffractometry, electron probe microanalysis, scanning electron microscopy.*

Buxton, Dr B.F. GEC Hirst Res. Centre, East Lane, Wembley, Middx. HA9 7PP, England.

Cahn, Prof. R.W. 6 Storeys Way, Cambridge CB3 0DT, England.

Cain, Mr Peter Maurice. Chemistry General Div., Metropolitan Police Forensic Science Lab., 109 Lambeth Rd., London SE1 7LP, England. (1944) BSc, chemistry (U. Reading, 1966). HSO. (tel. 01 + 230-6243). *Forensic applications, powder diffraction methods, instrumentation, computing, radiography.*

Cameron, Dr Allan Forbes. Dept. of Chemistry, U. of Glasgow, Glasgow G12 8QQ, Scotland. (1943) PhD, crystallography (U. Glasgow, 1968). Sr. lect. (tel. 041 + 339-8855, ext. 7133). *Structure, organic compounds, ylides, biologically active molecules, structure and properties (chemical - physical - biological).*

Campbell, Dr John Wilson. Dept. of Computer Systems and Electronics, SERC Daresbury Lab., Keckwick Lane, Daresbury, Warrington, Cheshire WA4 4AD, England. (1944) PhD, chemistry (U. Edinburgh, 1969). Principal res. assoc. (tel. 0925 + 65000, ext. 528). *Protein crystallography, computation in crystallography.*

Cardwell, Mr D.A. 2 Smith Drive, Langley Mill, Notts NG16 4GF, England.

Carlisle, Prof. Charles Harold. 12 Marney Rd., London SW11 5EP, England. (1911) DPhil, crystallography (U. Oxford, 1943). Retired Head, Dept. of Crystallography, Birkbeck C., London. *Organic structures, biologically important molecules.*

Carter, Mr Trevor John. Analytical Services Dept., BICC Res. and Eng. Ltd., 38 Wood Lane, London W12 7DX, England. (1938) BSc, physics (U. Leicester, 1959). Technical off. (tel. 01 + 743-1212, ext. 385). *Structure, metals, minerals, polymers.*

Cartwright, Dr Michael. Dept. of Chemistry, U. of Bath, Claverton Down, Bath, Avon, England. (1940) PhD, chemistry (U. London, 1974). EO. (tel. 0225 +

61244, ext. 508). *Inorganic structural chemistry, defect structures, radiation damage.*

Cebula, Dr D. 18 Blakes Way, Bushmead Fields, Eaton Socon, Cambs, England.

Cernik, Dr Robert Joseph. Process Analysis Dept., Ferranti Electronics Ltd., Fields New Rd., Chadderton, Oldham OL9 8NP, England. (1954) PhD, physics (U. Wales, 1984). Res. physicist. (tel. 061 + 682-6844, ext. 226). *Crystal structure analysis, lattice defects in semiconductor crystals.*

Champion, Dr John Anthony. Div. of Materials Applications, Nat. Physical Lab., Teddington, Middx. TW11 0LW, England. (1930) PhD, physics (U. London, 1961). SPSO. (tel. 01 + 977-3222, ext. 4284, telex 262344). *Crystal physics, surface effects, crystal growth.*

Champness, Dr John Norman. Physical Chemistry Dept., Wellcome Res. Labs., Langley Court, Beckenham, Kent BR3 3BS, England. (1943) PhD, biophysics (U. London, 1968). Sr. scient. staff. (tel. 01 + 658-2211). *Biologically significant molecules (large and small), structure and function, computer simulation, molecular models, film methods, automatic microdensitometers, fibre diffraction.*

Champness, Dr Pamela Eileen. Dept. of Geology, U. of Manchester, Oxford Rd., Manchester M13 9PL, England. (1942) PhD, mineralogy (U. Cambridge, 1968). Sr. lect., crystallography. (tel. 061 + 273-7121, ext. 5588). *Transmission electron microscopy, X-ray diffraction, phase transformations, minerals.*

Cheetham, Dr Anthony Kevin. Chemical Crystallography Lab., U. of Oxford, 9 Parks Rd., Oxford OX1 3PD, England. (1946) DPhil, chemistry (U. Oxford, 1971). Lect. (tel. 0865 + 57387, ext. 66). *Solid-state chemistry, X-ray and neutron diffraction, electron microscopy.*

Cherns, Dr David. H.H. Wills Physics Lab., Bristol U., Tyndall Ave., Bristol BS8 1TL, England. (1948) PhD, Physics (U. Cambridge, 1974). Lect. (tel. 0272 + 24161, ext. 762). *Crystal physics, semiconductors, metal - semiconductor contacts, TEM microanalysis, lattice defects, diffraction theory.*

Chisholm, Dr James Edwin. Dept. of Mineralogy, British Museum (Natural History), Cromwell Rd., London SW7 5BD, England. (1945) PhD, mineralogy (U. Manchester, 1973). SSO. (tel. 01 + 589-6323, ext. 516). *Mineral structures, structural defects, electron microscopy.*

Christian, Prof. John Wyrill. Dept. of Metallurgy and Science of Materials, U. of Oxford, Parks Rd., Oxford OX1 3PH, England. (1926) DPhil, metallurgy (U. Oxford, 1949). Prof. of physical metallurgy. (tel. 0865 + 59981, ext. 227). *Phase transformations, metals and alloys, lattice defects.*

Claringbull, Sir Gordon Frank. Langley House, Main Street, Ash, Martock, Somerset TA12 6PB, England. (1911) PhD, petrology (U. London, 1934). Retired museum dir. (tel. 0935 + 822983). *Crystal structures, minerals.*

Clayden, Miss D.A. Dept of Biochemistry, Medical Sch., University Walk, Bristol, Avon, England.

Clegg, Dr William. Dept. of Inorganic Chemistry, The University, Newcastle upon Tyne NE1 7RU, England. (1949) PhD, chemistry (U. Cambridge, 1973). Lect. (tel. 0632 + 328511, ext. 2649). *Crystal structure determination, coordination compounds, diffractometer control, crystallographic computing.*

Clewer, Mr Peter John. Jules Thorn Lighting Lab., Chemical Res. Dept., Thorn EMI Lighting Ltd., Gt. Cambridge Rd., Enfield, Middx. EN1 1UL, England. (1930) MSc, physics (U. London, 1959). Res. officer. (tel. 01 + 363-5353, ext. 2418). *Phase identification, inorganic systems, X-ray diffraction techniques.*

Cochran, Prof. William. Dept. of Physics, U. of Edinburgh, Mayfield Rd., Edinburgh EH9 3JZ, Scotland. (1922) PhD, chemistry (U. Edinburgh, 1946). Prof. (tel. 031 + 667-1081, ext. 2771). *Solid state physics.*

Cohen, Dr L. 9 Limewood Close, London W13 8HL, England.

Colby, Miss J. 157 Rawnsley Rd, Rawnsley, Cannock, Staffs WS12 5JQ, England.

Collyer, Mr S. 11 Dovedale Close, Prenton, Birkenhead, Merseyside, England.

Cook, Dr David Stanley. Dept. of Metallurgy and Materials Engineering, City of London Polytechnic, Central House, Whitechapel High St., London E1 7PF, England. (1938) PhD, crystallography (U. London, 1968). Sr. lect. (tel. 01 + 283-1030, ext. 470). *X-ray crystal structure analysis, computing, teaching.*

Cooper, Dr Malcolm John. Dept. of Physics, U. of Warwick, Coventry CV4 7AL, England. (1944) PhD, physics (U. Cambridge, 1964). Sr. lect. (tel. 0203 + 24011, ext. 2379). *Compton scattering, X-rays, gamma-rays, synchrotron radiation.*

Cooper, Dr Martyn John. Materials Physics and Metallurgy Div., Building 521.2, AERE Harwell, Didcot, Oxon OX11 0RA, England. (1935) PhD, physics (U. Cambridge, 1962). PSO. (tel. 0235 + 24141, ext. 5184, telex 83135). *Composite properties, high-accuracy diffraction techniques, reliability of techniques.*

Corney, Mr David John. Environmental Health and Science Dept., Tottenham C. of Techn., High Rd., London N15, England. (1946) MSc, crystallography (U. London, 1971). Lect. in Chemistry. (tel. 01 + 802-3111, ext. 38). *Crystal and molecular structure, alkaloids, biologically significant organic compounds.*

Cornfield, Ms L. Dept of Chemistry, U. of York, Heslington, York YO1 5DD, England.

Costello, Nr B.A.D. 3 Traemore Court, 81 Knollys Rd, Streatham, London SW16 2JW, England.

Cousins, Mr Christopher Stanley George. Dept. of Physics, U. of Exeter, Stocker Rd., Exeter EX4 4QL, England. (1934) MA, physics (U. Oxford, 1958). Sr. lect. (tel. 0392 + 77911, ext. 455). *Structure, stress or electric field effects, electron density distribution, synchrotron radiation applications.*

Cowlam, Dr Neil. Dept. of Physics, The Hicks Building, U. of Sheffield, Sheffield S3 7RH, England. (1941) PhD, magnetism (U. Sheffield, 1968). Lect. (tel. 0742 + 78555, ext. 4295, telex Unilib Sheff. 54348). *Magnetic crystallography, neutron diffraction, X-ray diffraction, metals and alloys, metallic glasses.*

Cox, Dr Philip John. Sch. of Pharmacy, Robert Gordon's Inst. of Techn., Schoolhill, Aberdeen AB9 1FR, Scotland. (1947) PhD, crystallography (U. Glasgow, 1972). Lect., physical pharmaceutical chemistry. (tel. 0224 + 633611, ext. 495). *Natural products, sesquiterpenoids, drug molecules, steroids, peptides.*

Craig, Mr G.R. 44 Litherland Park, Litherland, Merseyside L21 9HR, England.

Crennell, Mrs K.M. Atlas Centre, Rutherford Appleton Lab, Chilton, Didcot, Oxon OX11 0QX, England.

Cressey, Dr Barbara Anne. 19 Crendon Court, Caversham, Reading RG4 8BE, England. (1953) PhD, electron microscopy (U. Manchester, 1977). (tel. 0734 + 483335). *Hydrous silicate structures.*

Cressey, Dr Gordon. Dept. of Mineralogy, British Museum (Natural History), Cromwell Rd., London SW7 5BD, England. (1952) PhD, mineralogy (U. Manchester, 1979). Sr. res. fellow. (tel. 01 + 589-6323, ext. 385). *Silicate garnets, epidotes.*

Cringean, Mr J.K. 11 Sidegate, Durham, England.

Critchell, Mr J.W. 1 Tudor Ave, Maidstone, Kent ME14 5HH, England.

Crocker, Prof. Alan Godfrey. Dept. of Physics, U. of Surrey, Guildford, Surrey GU2 5XH, England. (1935) DSc, physics (U. London, 1971). Reader. (tel. 0483 + 71281, ext. 553). *Theory, crystal defects, metals, crystallography, deformation twinning, phase transformations, computer simulation, porosity.*

Cruickshank, Prof. Durward William John. Dept. of Chemistry, UMIST, PO Box 88, Manchester M60 1QD, England. (1924) ScD, crystallography (U. Cambridge, 1961). Em. prof. (tel. 061 + 236-3311, ext. 2647, telex 666094). *Crystal structures, electron diffraction by gases, molecular wave functions.*

Cruickshank, Mr M.C. Posnet, Fintray, Aberdeen AB2 0JE, Scotland.

Cullen, Mr F.L. Building 393, MDD, AERE Harwell, Didcot, Oxon OX11 0RA, England.

Dacombe, Mr Michael H. International Union of Crystallography, 5 Abbey Square, Chester CH1 2HU, England. (1950) BSc, chemistry & earth sciences (U. Leeds, 1972). Tech. Editor. (tel. 0244 + 42878, telex 669755 OFFICE G, Attn. UNICRYSTAL).

Darlington, Dr Charles Nicholas Wright. Dept. of Physics, U. of Birmingham, Birmingham B15 2TT, England. (1945) PhD, physics (U. Cambridge, 1971). Res. fellow. (tel. 021 + 472-1301, ext. 2545 or 3470). *Phase transitions in solids, scattering from solids, Mössbauer gamma rays in X-ray crystallography, critical phenomena.*

Dave, Miss S. 23 Oakwood Ave, Southall, Middx, England.

Davies, Dr John Edward. University Chemical Lab., Lensfield Rd., Cambridge CB2 1EW, England. (1947) PhD, crystallography (Monash U., Australia, 1974). Res. asst. (tel. 0223 + 66499, ext. 313). *Crystal structure determination, crystallographic data storage and retrieval.*

Davison, Mr G. 34 Beechfields, Doctors Lane, Eccleston, Chorley, Lancs, England.

Dekker, Mr Henri. Enraf-Nonius Ltd., High View House, 165-7 Station Rd., Edgware, Middx. HA8 7JU, England. (1945). (Tel. 01 + 952-1643). *Instrumentation.*

Delf, Dr Brian William. Dept. of Physics, University C. Cardiff, PO Box 78, Cardiff CF1 1XL, Wales. (1935) PhD, physics (U. Wales, 1962). Sr. lect. (tel. 0222 + 44211, ext. 2379). *X-ray powder diffractometry, thin films.*

Denham, Mr A.W. BR Technical Centre, 203-2B London Rd, Derby DE2 8UP, England.

Dent Glasser, Dr Lesley Scott. Dept. of Chemistry, U. of Aberdeen, Old Aberdeen AB9 2UE, Scotland. (1932) DSc, crystallography (U. Aberdeen, 1972). Reader. (tel. 0224 + 40241, ext. 5658, telex 73458 UNIABN G). *Inorganic structures, silicates and aluminates, teaching crystallography.*

Derwent, Mr Frank William. The Midhurst Medical Res. Centre, Midhurst, W. Sussex GU29 0BL, England. (1912) MSc, physics (U. London, 1948). Extramural res. assoc. (tel. 0428 + 722784, home). *Biological interactions at inorganic & organic interfaces.*

Diamond, Dr Robert. MRC Lab. of Molecular Biology, Hills Rd., Cambridge CB2 2QH, England. (1929) PhD, physics (U. Cambridge, 1956). Sr. scient. staff. (tel. 0223 + 248011, ext. 210, telex 81532). *Protein crystallography, mathematical methods.*

Dineen, Mr C. GEC Res. Labs., Hirst Res. Centre, East Lane, Wembley HA9 7PP, England.

Dingley, Dr David Joseph. Dept. of Physics, U. of Bristol, Tyndall Ave., Bristol BS8 1TL, England. (1939) PhD, physical metallurgy (U. London, 1962). Lect. (tel. 0272 + 24161, ext. 765). *Crystal symmetry determination, electron microscopy, image processing.*

Dobson, Dr Peter James. Solid State Electronics, Philips Res. Lab., Cross Oak Lane, Redhill, Surrey RH1 5HA, England. (1942) PhD, physics (U. Southampton, 1968). Sr. Principal Scient. (tel. 02934 + 5544). *Surface structure, thin film structure, X-ray diffraction - novel aspects, LEED, RHEED, Raman scattering, photoluminescence.*

Dodson, Mrs Eleanor. Dept. of Chemistry, U. of York, Heslington, York YO1 5DD, England. (1936) BA, mathematics (U. Melbourne, Australia, 1957). Res. fellow. (tel. 0904 + 59861, ext. 343). *Protein theoretical crystallography, insulin structure, refinement procedures.*

Dodson, Dr Guy George. Dept. of Chemistry, U. of York, Heslington, York YO1 5DD, England. (1937) PhD, chemistry, crystallography (U. New Zealand, 1962). Reader. (tel. 0904 + 59861, ext. 335). *Protein crystallography, protein structure and function, hormone structure, insulin, haemoglobin, ribonuclease.*

Doig, Dr P. CEGB, SE Region, Scientific Services Dept., Canal Rd, Gravesend, Kent DA12 2RS, England.

Donaldson, Prof John Dallas. Dept. of Chemistry, The City U., Northampton Square, London EC1V 0HB, England. (1935) DSc, chemistry (U. London, 1970). Prof. (tel. 01 + 253-4399). *Structures, main group elements (compounds in lower oxidation states).*

Dougill, Dr Maryon W. 5 Fraser Ave., Horsforth, Leeds LS18 5EA, England. PhD, crystallography (U. Leeds, 1953). Retired. (tel. 0532 + 582472). *Structures, inorganic compounds.*

Dover, Dr Stanley David. Dept. of Biophysics, King's C., 26-29 Drury Lane, London WC2B 5RL, England. (1943) PhD, biophysics (U. London, 1968). Lect. (tel. 01 + 836-8851). *Electron microscopy, three-dimensional reconstruction, image processing, electron microscope tomography, holography.*

Downie, Mr George. Dept. of Geology and Mineralogy, U. of Aberdeen, Marischal C., Broad St., Aberdeen AB2 2HJ, Scotland. (1931) BSc, geology and mineralogy (U. Aberdeen, 1956). Lect. (tel. 0224 + 40241, ext. 380). *Polymorphism, sheet silicates, metamorphic rocks, reaction-melting effects between xenoliths and magma.*

Drew, Dr Michael George Brindley. Dept. of Chemistry, U. of Reading, Whiteknights, Reading, Berks. RG6 2AD, England. (1941) PhD, chemistry (U. London, 1966). Lect. (tel. 0734 + 85123, ext. 7952). *X-ray crystallography.*

Duckett, Mr G.R. 21 Glen Ave, Springboig, Glasgow G32 0DL, Scotland.

Duke, Mr J.R.C. Heatherley, Gypsy Lane, Great Amwell, Ware, Herts, England.

Dunham, Prof A.C. Dept of Geology, U. of Hull, Cottingham Rd, Hull HU6 7RX, England.

Dunning, Mr Anthony John. Directorate-General XIII (Information Market and Innovation), Commission of European Communities, Bâtiment Jean Monnet, Plateau du Kirchberg, BP 1907, Luxembourg. (1939) BSc, physics (U. London, 1961). Principal Administrator. (tel. + 352 + 4301-2964, telex 2752 EURDOC LU). *Data banks, information retrieval, data base management systems, crystal structures, hydrocarbon chain compounds, refinement, crystal structures.*

Dyson, Mr David John. Electron Metallography Div., British Steel Corp., Moorgate, Rotherham, S. Yorks. S60 3AR, England. (1939) BSc, physics (U. Durham, 1960). Div. head. (tel. 0709 + 60166, ext. 3290, telex 547279). *X-ray and electron diffraction methods, transmission, scanning and high voltage electron microscopy, iron and steel technology, quantitive phase analysis, dust analysis, texture studies, automatic collection and reduction of data.*

Edmondson, Mr M. 22 Passmonds Crescent, Rochdale, Lancs. OL11 5AW, England.

Edwards, Dr Anthony John. Dept. of Chemistry, U. of Birmingham, P.O. Box 363, Birmingham B15 2TT, England. (1936) DSc, chemistry (U. Birmingham, 1973). Sr. lect. (tel. 021 + 472-1301, ext. 2370). *Synthesis, structure determination, inorganic compounds, fluorides.*

Edwards, Dr Ian Arthur Samuel. Northern Carbon Res. Labs., Dept. of Physical Chemistry, U. of Newcastle upon Tyne, Newcastle upon Tyne NE1 7RU, England. (1941) PhD, crystallography (U. Newcastle upon Tyne, 1968). Sr. experimental asst. (tel. 0632 + 28511, ext. 3130; from spring 1986, 091 + 2328511). *Structure determination, X-ray methods, surface studies, metals, carbon deposits, LEED, AES; teaching crystallography; carbon science; structure and properties, coals, cokes, graphites and carbons.*

Eeles, Mr Wilfred Trefor. Electricity Council Res. Centre, Capenhurst, Chester CH1 6ES, England. (1930) MSc, physics (U. Wales, 1953). Head, materials sci. sect. (tel. 051 + 339-4181, ext. 309). *Powder diffraction identification, defect structure diffraction, graphite and its intercalates.*

Elder, Dr D.P. 37 Bartongate Drive, Barnton, Edinburgh EH4 8BE, Scotland.

Elder, Dr Michael. SERC, Daresbury Lab., Daresbury, Warrington, Cheshire WA4 4AD, England. (1943) PhD, chemistry (U. Canterbury, New Zealand, 1967). SPSO. (tel. 0925 + 65000, ext. 542, telex 62609). *Computational aspects of crystallography, microdensitometry, information retrieval.*

Elias, Mr E.E. Biophysics Dept., U. of Leeds, Leeds LS2 9JT, England.

Eliopoulos, Dr Elias Edward. Dept. of Biophysics, U. of Leeds, Woodhouse Lane, Leeds LS2 9JT, England. (1958) PhD, physics and biophysics (U. Leeds, 1985). Protein crystallographer. (tel. 0532 + 431751, ext. 7207). *Protein crystallography, molecular graphics, protein folding, small molecule conformation analysis.*

Elliott, Dr Gerald Frank. Biophysics Group, Open U. Res. Unit, Foxcombe Hall, Boars Hill, Oxford, England. (1931) PhD, biophysics (U. London, 1960). Prof. of physics. (tel. 0865 + 730031, ext. 285). *Low angle X-ray and neutron diffraction, biological polyelectrolytes - muscle and cornea.*

Elliott, Dr James Cornelis. Dept. of Biochemistry, The London Hospital Medical C., Turner St., London E1 2AD, England. (1937) PhD, crystallography (U. London, 1964). Sr. lect. in biophysics. (tel. 01 + 377-8800, ext. 32). *Biological minerals, X-ray absorption, instrumentation.*

Elliott, Dr Robert Brian. Dept. of Geology, U. of Nottingham, University Park, Nottingham NG7 2RD, England. (1921) PhD, petrology (U. Nottingham, 1952). Reader(retired). (tel. 0602 + 56101, ext. 3159). *Minerals, volcanic rocks and metabasites, X-ray diffractometry.*

Embrey, Mr Peter Godwin. Dept. of Mineralogy, British Museum (Natural History), Cromwell Rd., London SW7 5BD, England. (1929) MA, chemistry, mineralogy (U. Oxford, 1954). Curator of minerals (PSO). (tel. 01 + 589-6323, ext. 567). *Descriptive mineralogy, optical and morphological crystallography, history of science.*

Evans, Dr Anthony Meredith. Dept. of Geology, University, Leicester LE18 7RH, England. (1929) PhD, petrology and mining geology (Queen's U., Ontario, 1962). Sr. lect. (tel. 0533 + 554455, ext. 148). *Ore microscopy, industrial mineralogy, mining geology.*

Evans, Mr D.M. 17 Beckman Rd, Stourbridge, W Midlands DY9 0TZ, England.

Evans, Dr E.M.H. 3 Waverley Close, Llandough, Penarth, S Glamorgan, Wales.

Evans, Dr John Hedley. Materials Development Div., AERE Harwell, Didcot, Oxon OX11 0RA, England. (1941) PhD, physics (U. Wales, 1966). PSO. (tel. 0235 + 24141, ext. 4166). *Electron microscopy, radiation damage, metals.*

Evans, Dr P.A. 72 Chester Rd, Audley, Stoke-on-Trent, Staffs ST7 8JF, England.

Evans, Dr Robert Crispin. 55 Boxworth Rd., Elsworth, Cambridge CB3 8JQ, England. (1909) PhD, physics (U. Cambridge, 1934). Retired. (tel. 095 47 + 282). *Crystal chemistry*

Evans, Mr R.R. Rhos Dawel, Trerhyngyll, Nr Cowbridge, S Glamorgan, Wales.

Fairclough, Dr D.P. Nicolet Instruments Ltd, Budbrooke Rd, Warwick CV34 5XH, England.

Falshaw, Dr C.P. Chemistry Dept, U. of Sheffield, Sheffield S3 7HF, England.

Farmer, Dr Victor Colin. Dept. of Spectrochemistry, Macauley Inst. for Soil Res., Craigiebuckler, Aberdeen AB9 2QJ, Scotland. (1920) PhD, spectrochemistry (U. Aberdeen, 1947). SPSO. (tel. 0224 + 38611). *Infrared spectroscopy, minerals, clay and oxide surfaces.*

Farrants, Dr George William. Dept. of Biochemistry, U. of Sheffield, Sheffield, S. Yorks S10 2TN, England. (1957) PhD, biochemistry (U. Cambridge, 1981). Post-Doc. Res. asst. (tel. 0742 + 78555, ext. 4241, telex 54348 ULSHEF G). *Proteins, macromolecules, X-ray diffraction, electron microscopy, crystallographic computing, image processing.*

Farrar, Prof. Roy Alfred. Dept. of Mechanical Engineering, U. of Southampton, University Rd., Southampton SO9 5NH, England. (1939) PhD, metallurgy (U. London, 1967). Prof. (tel. 0703 + 559122, ext. 2891). *Non-stoichiometry in II-VI compounds, phase transformations, metallic systems, intermetallic phases in weld metals.*

Farrugia, Dr L.J. Chemistry Dept, U. of Glasgow, Glasgow G12 8QQ, Scotland.

Faruqi, Dr A.R. MRC Molecular Biology Lab., Hills Rd, Cambridge CB2 2QH, England.

Fawcett, Dr John. Chemistry Dept., Leicester U., University Road, Leicester LE1 7RH, England. (1947) PhD, Inorg. chemistry (U. Leicester, 1980). EO. (tel. 0533 + 554455, ext. 20). *Chemical crystallography.*

Fejer, Miss Eleonora Eva. Dept. of Mineralogy, British Museum (Natural History), Cromwell Rd., London SW7 5BD, England. (1927) SSO, head, X-ray sect. (tel. 01 + 589-6323, ext. 447/274). *Crystallography, mineralogy.*

Fenn, Dr Ruth Helen. Dept. of Physics, Portsmouth Polytechnic, King Henry I St., Portsmouth PO1 2DZ, England. (1938) PhD, crystallography (U. London, 1964). Sr. lect. (tel. 0705 + 827681, ext. 102). *Applied crystallography.*

Ferguson, Dr Ian Forster. Springfields Nuclear Power Dev. Lab., UKAEA, (N. Div.), Preston PR4 0RR, England. (1931) PhD, chemistry (U. London, 1961). PSO i/c microstructural studies sect. (tel. 0772 + 728262, ext. 31219, telex 67545). *X-ray powder diffraction - all aspects, computer methods, corrosion, ceramics, surface analysis, materials characterization, automation.*

Fewster, Dr Paul Frederick. Solid State Electronics Div., Philips Res. Labs., Cross Oak Lane, Redhill, Surrey RH1 5HA, England. (1950) PhD, crystallography (U. London, 1977). Principal Scient. (tel. 02934 + 5544, ext. 312, telex 877261). *Semiconductor single-crystal studies, bulk and multilayer structures, Laue diffractometry, multiple crystal diffractometry, X-ray diffractometry, topography, computer simulation, instrument automation.*

Fielding, Mr W.D. 5 Openfield Croft, Water Orton, Warwicks B46 1RE, England.

Finney, Dr John Leslie. Dept. of Crystallography, Birkbeck C., Malet St., London WC1E 7HX, England. (1943) PhD, crystallography (U. London, 1968). Reader. (tel. 01 + 580-6622, ext. 420). *Liquids and amorphous solids, complex biological systems, water and aqueous (biological) systems.*

Fisher, Mr Graham Richard. Aplied Physics Div., GEC Avionics Ltd., Elstree Way, Borehamwood, Herts WD6 1RX, England. (1951) BSc, physics (U. Salford, 1973). Chief Engineer. (tel. 01 + 953-2030, ext. 6739). *X-ray topography, X-ray physics, material science.*

Fisher, Mr Leslie Ernest, Enraf-Nonius Ltd, Highview House, 165-7 Station Rd., Edgware, Middx HA8 7JU, England. (1928). Techn. Manager. (Tel. 01 + 952-1643). *Instrumentation.*

Fitzgerald, Dr Alexander Grant. Dept. of Physics, U. of Dundee, Dundee DD1 4HN, Scotland. (1939) PhD, physics (U. Cambridge, 1964). Lect. (tel. 0382 + 23181, ext. 297). *Electron spectroscopy.*

Fletcher, Dr R.O.W. 29 St Mary's Green, E Finchley, London N2 0UZ, England.

Fletcher, Dr Steven Reginald. Technical Dept., ICI plc, Mond Div., P.O. Box 8, The Heath, Runcorn, Cheshire WA7 4QD, England. (1946) PhD, chemical crystallography (U. London, 1972). Sr. res. scient. (tel. 0928 + 513445). *Powder diffraction, line profile analysis, powder structure analysis.*

Flewitt, Dr Peter Edwin John. Scient. Services Dept., CEGB South Eastern Region, Canal Rd., Gravesend, Kent DA12 2RS, England. DSc, metallurgy (U. London, 1980). Res. officer. (tel. 0474 + 51122, ext. 413). *High temperature deformation, phase transformations, stress corrosion.*

Flower, Mr S.C. 625A Wellsway, Bath BA2 2TY, England.

Fones, Mr M.D. Building 393, MDD, AERE Harwell, Didcot, Oxon OX11 0RA, England.

Ford, Dr Geoffrey Charles. Dept. of Biochemistry, U. of Sheffield, Western Bank, Sheffield, S. Yorks S10 2TN, England. (1942) DPhil, mathematics and crystallography (U. Oxford, 1969). Wellcome Trust sen. lect. (tel. 0742 + 78555, ext. 4241; telex 54348 ULSHEF G). *Proteins, macromolecules, crystallographic computing.*

Forsyth, Dr John Bruce. Neutron Div., SERC Rutherford Lab., Chilton, Didcot, Oxon OX11 0QX, England. (1932) PhD, physics (U. Cambridge, 1959). SPSO. (tel. 0835 + 21900, ext. 6116). *Neutron scattering, magnetism, magnetic moment density distributions, computer control, data acquisition and assessment, position-sensitive neutron detectors.*

Forsyth, Mr V.T. Physics Dept, Keele U., Keele, Staffs, England.

Fox, Mr B.E. Analytical Dept, Raychem Ltd, Faraday Rd, Dorcan, Swindon SN3 5HH, England.

Francis, Mr John Godfrey. Dept. of Mineralogy, British Museum (Natural History), Cromwell Rd., London SW7 5BD, England. (1941) BSc, geology (U. London, 1966). SSO. (tel. 01 + 589-6323, ext. 274). *Minerals, identification, X-ray powder diffraction, structures.*

Frank, Prof. Sir Frederick Charles. H.H. Wills Physics Lab., U. of Bristol, Royal Fort, Bristol BS8 1TL, England. (1911) DPhil, chemistry (U. Oxford, 1937). Em. prof. (tel. 0272 + 24161). *Crystal growth, defects, polymers, geophysics.*

Franks, Dr Albert. Mechanical and Optical Metrology Div., Nat. Physical Lab., Teddington, Middx. TW11 0LW, England. DSc, physics (U. London, 1976). DCSO. (tel. 01 + 977-3222, ext. 3515). *X-ray optics, X-ray microscopy, instrumentation.*

Franks, Dr Joseph. Ion Tech Ltd., 2 Park St., Teddington, Middx. TW11 0LT, England. (1924) PhD, physics (U. London, 1952). Managing dir. (tel. 01 + 977-9306). *Ion equipment, X-ray optics.*

Fraser, Dr G.V. Nicolet Instruments Ltd, Budbrooke Rd, Warwick CV34 5XH, England.

Freeman, Mr Walter Gerard. Hirst Res. Centre, GEC, East Lane, Wembley, Middx, HA9 7PP, England. (1956) MSc, crystallography (U. London, 1980). Res. Scient. (tel. 01 + 904-1262, ext. 281, telex 923429). *Powder diffraction, RHEED.*

Freundlich, Dr A. Biophysics Sect., Physics Dept, Imperial C., London SW7 2BZ, England.

Fuller, Prof. W. Physics Dept, Keele U, Staffs STY5 5BG, England.

Fulton, Dr William Stephen. H.H. Wills Physics Lab., U. of Bristol, Tyndall Ave, Bristol BS8 1TL, England. (1953) PhD, Biophysics (U. Bristol, 1980). Res. assoc. (Tel. 0272 + 24161, ext. 105) *Biopolymers, polymers, crystallography, molecular modelling.*

Gale, Dr Brian. Div. of Mechanical and Optical Metrology, Nat. Physical Lab., Teddington, Middx. TW11 0LW, England. (1928) PhD, physics (U. Bristol, 1952). PSO. (tel. 01 + 977-3222, ext. 4295). *Electromagnetic scattering theory, optical instrument design theory, crystal surface theory, interface structure theory.*

Gallagher, Dr Kevin Joseph. Dept. of Chemistry, University C. of Swansea, Singleton Park, Swansea SA2 8PP, Wales. (1928) PhD, chemistry (Queen's U. Belfast, 1950). Sr. lect. (tel. 0792 + 205678, ext. 5272). *Inorganic solid-state chemistry, neutron diffraction, proton transfer in solids.*

Galloy, Dr Jean. University Chemical Lab., Lensfield Rd., Cambridge CB2 1EW, England. (1949) PhD, chemistry (U. de Louvain, Belgium, 1975). Res. asst. (tel. 0223 + 66499, ext. 314). *Computer programming, chemical graphics.*

Gard, Dr John Alan. Dept. of Chemistry, U. of Aberdeen, Old Aberdeen AB9 2UE, Scotland. (1919) DSc, chemistry (U. Aberdeen, 1973). Res. lect. (tel. 0224 + 40241, ext. 5661). *Electron microscopy, electron (and X-ray) diffraction, minerals and inorganic phases, zeolites and calcium silicates.*

Gare, Mr Terence. Primary Operations Dept, British Steel Corp., Teesside Labs., PO Box 11, Grangetown, Middlesbrough, Cleveland TS6 6UB, England. (1949) MSc, crystallography (London, 1974). Res. officer. (tel. 0642 + 467144, ext. 411). *Iron compounds, iron ores, iron oxides, iron, reduction, gases.*

Garner, Prof. Christopher David. Dept. of Chemistry, U. of Manchester, Brunswick St., Manchester M13 9PL, England. (1941) PhD, chemistry (U. Nottingham, 1966). Sr. lect. (tel. 061 + 273-7121, ext. 5293). *Inorganic biochemistry, coordination chemistry, electronic structure, metal-metal bonds.*

Garratt, Mr R.C. Dept of Crystallography, Birkbeck C., Malet St, London WC1E 7HX, England.

Geddes, Dr Alexander John. Astbury Dept. of Biophysics, U. of Leeds, Leeds LS2 9JT, England. (1941) PhD, biophysics (U. Leeds, 1965). Sr. lect. (tel. 0532 + 431751, ext. 7088). *Protein structure and function, structure-activity relationships, biologically active molecules, drug design.*

Gilmartin, Mr M.G.M. Flat 2D, Lennox Court, 14 Sutherland Ave, Mosshead, Bearsden, Glasgow, Scotland.

Gilmore, Dr Christopher John. Dept. of Chemistry, U. of Glasgow, Glasgow G12 8QQ, Scotland. (1946) PhD, chemical crystallography (U. Bristol, 1971). Lect. (tel. 041 + 339-8855, ext. 506). *Direct methods of solving crystal structures, organic molecules.*

Glasser, Prof. Fred Paul. Dept. of Chemistry, U. of Aberdeen, Old Aberdeen AB9 2UE, Scotland. (1929) DSc, chemistry (U. Aberdeen, 1969). Prof. (tel. 0224 + 40241, ext. 5640). *Phase equilibria, chemical crystallography, oxide chemistry.*

Glasson, Dr Douglas Royston. Dept. of Environmental Sci., Plymouth Polytechnic, Drake's Circus, Plymouth PL4 8AA, England. (1926) PhD, chemistry (U. London, 1949). Res. adviser (surface and solid-state science). (tel. 0752 + 21312, ext. 5387). *Calcareous materials (road - building - soil - etc.), non-oxide ceramics and refractories; nitrides, borides, carbides, silicides, mineral ores, industrial processing.*

Glazer, Dr Anthony Michael. Clarendon Lab., U. of Oxford, Parks Rd., Oxford OX1 3PU, England. (1943) PhD, crystallography (U. London, 1968). Lect., physics fellow of Jesus C. (tel. 0865 + 59291). *Phase transitions, structure - property relationships, disorder, symmetry, instrumentation, synchrotron radiation and new techniques, neutron diffraction, Raman scattering, critical phenomena theory.*

Glen, Dr John Wallington. Dept. of Physics, U. of Birmingham, Birmingham B15 2TT, England. (1927) DSc, physics (U. Birmingham, 1981). Reader, ice physics. (tel. 021 + 472-1301, ext. 3471). *Ice physics, glaciology, editing.*

Glidewell, Dr Christopher. Dept. of Chemistry, U. of St. Andrews, North Haugh, St. Andrews, Fife KY16 9ST, Scotland. (1944) PhD (U. Cambridge, 1970). Reader. (tel. 0334 + 76161, ext. 8399). *Oxo compounds, organometallics, heterocycles, Jahn-Teller effects, clusters.*

Glover, Dr I.D. Crystallography Dept, Birkbeck C., Malet St, London WC1E 7HX, England.

Goaman, Dr Llawenydd Constance Gwynne. Dept. of Physics, Portsmouth Polytechnic, King Henry I St., Portsmouth PO1 2DZ, England. PhD, physics (U. Wales, 1962). Sr. lect. (tel. 0705 + 827681, ext. 102). *Biologically interesting organic molecules.*

Goodfellow, Dr Julia Mary. Dept. of Crystallography, Birkbeck C., U. of London, Malet St., London WC1, England. (1951) PhD, biophysics (Open U., 1975). Lect. (tel. 01 + 580-6622, ext. 421). *Hydration, proteins, nucleic acids.*

Goodman, Prof. C.H.L. 5 Hollies End, Mill Hill Village, London NW7 2RY, England.

Gould, Dr Robert Ozburn. Dept. of Chemistry, U. of Edinburgh, West Mains Rd., Edinburgh EH9 3JJ, Scotland. (1938) PhD, chemistry (U. St. Andrews, 1963). Sr. lect. (tel. 031 + 667-1081, ext. 3649). *Direct methods, structures, carbohydrates, coordination compounds, silicates.*

Gould, Dr Sheila Elizabeth Buchan. Beevers Miniature Models Unit, Chemistry Dept, U. of Edinburgh, Edinburgh EH9 3JJ, Scotland. (1940) PhD, chemistry (U. Edinburgh, 1965). Administrator. (tel. 031 + 667-1081, ext. 3405; telex 727442 UNIVED G). *Biological compounds, model making.*

Grant, Dr Douglas Frank. Warlawbank Cottage, Reston, Berwickshire TD14 5LW, Scotland. (1923) PhD, physics (U. Durham, 1950). Hon. Sr. lect. in physics, St Andrews U. (tel. 03904 + 385). *Crystallographic computing.*

Grant, Dr William Kenneth. 41 Brondesbury Villas, Kilburn, London NW6, England. (1934) PhD, chemistry (U. Glasgow, 1960). *X-ray structure analysis.*

Green, Mr R.S. GTP Engineering Ltd, Station Industrial Estate, Sheppard St, Swindon, Wilts, England.

Griffiths, Mr P.J.F. 10 Crystal Ave, Heath, Cardiff CF2 5QJ, Wales.

Grimes, Dr N.W. 246 May Lane, Kings Heath, Birmingham 14, England.

Groves, Mr P. Brasenose C., Oxford OX1 4AJ, England.

Grubb, Mr K.A. Cookson Group Res. Labs, 7 Wadsworth Rd, Perivale, Greenford, Middx, England.

Gunning, Mr G.R. 75 Greys Rd, Henley, Oxon RG9 1TD, England.

Gutteridge, Mr Walter Alfred. Materials Dept., Res. and Dev. Div., Cement and Concrete Association, Wexham Springs, Slough SL3 6PL, England. (1931) MSc, crystallography (Birkbeck C., U. of London, 1967). Principal Scient. (tel. 028 16 (Fulmer) + 2727, ext. 444). *Cement minerals and silicates, quantitative powder diffractometry.*

Habash, Mr Jarjis. Physics Dept., U. of Keele, Keele ST5 5BG, England. (1943) PhD, chemistry (U. Sheffield, 1981). Res. fellow. (tel. 0925 + 65000, ext. 388). *Crystal chemistry, actinide compounds, protein crystallography.*

Hails, Dr J.E. 16 Redewater Rd, Fenham, Newcastle-upon-Tyne NE4 9UD, England.

Halfpenny, Dr Joan Christine. Dept. of Appl. Chemical Sci., Napier C., Colinton Road, Edinburgh EH10 5DT, Scotland. (1954) PhD, crystallography (U. Lancaster, 1978). Lect. (tel. 031 + 447-7070, ext. 503). *Crystal structures, co-ordination in mercury compounds, catalysts.*

Hall, Dr Ivan Harold. Dept. of Pure and Appl. Physics, UMIST, PO Box 88, Sackville St., Manchester M60 1QD, England. (1928) PhD, physics (U. London, 1965). Sr. lect. (tel. 061 + 236-3311, ext. 2969). *Crystalline polymer structures, interaction between structure and bulk properties.*

Hall, Mr Norman Michael. Res. and Dev. Lab. (U.K.), Gillette, 454 Basingstoke Rd., Reading RG2 0QE, England. (1939) MA, science of metals (U. Oxford, 1960). Principal scientist. (tel. 0734 + 875222, ext. 355). *Precipitation phenomena, steels, thin film structures, electron diffraction.*

Halliwell, Mrs Mary Ann Griffiths. Materials Div., Group R3.1.2, British Telecom Res. Labs, Martlesham Heath, Ipswich IP5 7RE, England. (1942) MSc, solid state physics (Chelsea C., U. of London, 1967). Head of group. (tel. 0473 +

643640). *Heteroepitaxial layers, semi-conductor device processing, double crystal diffractometry, X-ray topography, multiple diffraction.*

Hammond, Dr Christopher. Dept. of Metallurgy, U. of Leeds, Leeds LS2 9JT, England. (1942) PhD, metallurgy (U. Leeds, 1968). Sr. lect. (tel. 0532 + 431751, ext. 445). *Electron diffraction, shear transformations.*

Hammonds, Mr T.G. Green Lane, Churt, Farnham, Surrey GU10 2LT, England.

Hamor, Dr Thomas Andrew. Dept. of Chemistry, U. of Birmingham, PO Box 363, Birmingham B15 2TT, England. (1930) DSc, chemistry (U. Birmingham, 1974). Lect. (tel. 021 + 472-1301, ext. 3283). *Organic and inorganic crystal structures, fluorinated organic molecules, biologically active molecules.*

Harding, Mr J.W. BR 131B Railway Technical Centre, London Rd, Derby DE2 8UP, England.

Harding, Dr Marjorie Mary. Dept. of Inorganic, Physical & Industrial Chemistry, U. of Liverpool, Donnan Labs., Grove St., Liverpool L69 3BX, England. (1934) DPhil, crystallography (U. Oxford, 1961). Lect. (tel. 051 + 709-6022, ext. 2567). *Organic structures, peptides, polysaccharides, proteins.*

Harding, Dr Roger Robertson. Geological Museum, Exhibition Rd., London SW7 2DE, England. (1938) DPhil, geology (U. Oxford, 1962). PSO. (tel. 01 + 589-3444). *Gemmology, mineralogy.*

Hardy, Dr Andrew David. Chemistry Dept., Glasgow U., Glasgow G12 8QG, Scotland. (1946) DPhil, crystallography (U. Sussex, 1971). Tutor-Instructor. (tel. 041 + 339-8855, ext. 427). *X-ray powder diffraction, X-ray single-crystal diffraction, small angle neutron scattering.*

Hargreaves, Dr A. 26 Ley Hey Rd, Marple, Cheshire, England.

Harper, Mr W.H. 4 Katherine Drive, Woodview, Toton, Nottingham, England.

Harries, Mr J.E. 9 New Hey Rd, Cheadle, Cheshire SK8 2AQ, England.

Harris, Mr A.J. 325 Milton Rd, Cambridge, England.

Harris, Miss B.L. 2 Hamilton Rd, Cockfosters, Herts EN4 9EU, England.

Harris, Ms G.W. Dept of Crystallography, Birkbeck C., Malet St, London WC1E 7HX, England.

Harrison, Prof. Pauline May. Dept. of Biochemistry, U. of Sheffield, Western Bank, Sheffield S10 2TN, England. (1926) DPhil, crystallography (U. Oxford, 1953). Prof. (tel. 0742 + 78555, ext. 4242). *Protein structure, metalloproteins, iron metabolism.*

Hart, Mr D.G. Plessey Res. (Caswell) Ltd, Towcester, Nothants, England.

Hart, Prof. Michael. Dept. of Physics, Manchester U., Manchester M13 9PL, England. (1938) DSc, physics (U. Bristol, 1970). Prof. (tel. 061 + 273-7121, ext. 14; telex 668932). Joint appointment as Res. Program Coordinator, SERC Daresbury Lab., Warrington WA4 4AD, England. (tel. 0925 + 65000; telex 629609). *Bragg reflection X-ray and neutron optics, X-ray interferometry, polarimetry, synchrotron radiation.*

Harvey, Dr T.A. 57 Guildford Park Ave, Guildford, Surrey GU2 5NN, England.

Hatt, Mr B.A. 23 Tancred Rd, High Wycombe, Bucks, England.

Hatton, Dr Peter David. Synchrotron Radiation Source, SERC Daresbury Lab., Warrington WA4 4AD, England. (1957) PhD, High Pressure Spectroscopy (Leicester U., 1983). Exp. Sci. (tel. 0925 + 65000, ext. 238; telex 629609). *High-pressure crystallography, phase transitions, synchrotron radiation, powder diffraction.*

Hausermann, Mr D. Flat 20, Bristol House, Southampton Row, London WC1, England.

Hawley, Dr D.M. Bankhead of Keir, Keir, Thornhill, Dumfriesshire DG3 5EB, Scotland.

Haworth, Dr Colin William. Dept. of Metallurgy, U. of Sheffield, St. George's Square, Sheffield S1 3JD, England. (1932) DPhil, metallurgy (U. Oxford, 1958). Sr. lect. (tel. 0742 + 78555, ext. 5509). *Intermetallic phases, precipitation phenomena in alloys, diffusion.*

Hay, Dr James Neilson. Chemistry Dept., U. of Birmingham, Birmingham B15 2TT, England. (1935) DSc, chemistry (U. Birmingham, 1955). Sr. lect. (tel. 021 + 472-1301, ext. 2719). *Polymer crystal structure, melting and crystallization rate studies, amorphous materials.*

Hearmon, Mr R.F.S. Cherry Tree, New Rd, Princes Risborough, Bucks, England.

Heavens, Prof. Oliver Samuel. Dept. of Physics, U. of York, Heslington, York YO1 5DD, England. (1922) DSc, physics (U. London, 1964). Prof. (tel. 0904 + 59861, ext. 5522). *Surface physics - surface structure and chemistry of crystals, lasers, LEED.*

Helliwell, Dr John Richard. SERC Daresbury Lab., Warrington WA4 4AD, England. (1953) DPhil, protein crystallography (U. Oxford, 1978). SSO (tel. 0925 + 65000, ext. 237; telex 629609). *Protein structure, X-ray crystallography, X-ray scattering, synchrotron X-radiation and applications.*

Henderson, Dr Christopher Michael Bradford. Dept. of Geology, U. of Manchester, Manchester M13 9PL, England. (1938) PhD, geochemistry (U. London, 1964). Sr. lect. (tel. 061 + 273-7121, ext. 5594). *Crystal chemistry, thermal expansion, framework silicates; geochemistry, alkaline igneous rocks.*

Herring, Mr R.N. 96 St Fabian's Drive, Chelmsford, Essex CM1 2PR, England.

Highcock. Dr Rose Margaret. Analytical Support & Res. Div., BP Res. Centre, Chertsey Road, Sunbury-on-Thames, Middx. TW16 7LN, England. (1958) PhD, Chemistry (Bristol U., 1982). Chemist. (tel. 09327 + 62080 or 09327 + 81234, ext. 2080; telex 296041 BPSUNA G). *Structure determination (mainly single crystal), powder pattern decomposition, Reitveld refinements, zeolites.*

Hill, Mr Christopher Peter. Chemistry Dept., U. of York, Heslington, York YO1 5DD, England. (1958) BA, Chemistry (U. York, 1980). Res. asst. (tel. 0904 + 59861, ext. 337). *Protein structure, enzyme catalysis.*

Hilleard, Dr Ronald James. Dept. of Physical Sci., The Polytechnic, Wulfruna St., Wolverhampton WV1 1LY, England. (1936) PhD, physics (U. Aston, 1973). Sr. lect. (tel. 0902 + 27371, ext. 126). *Crystallography, magnetic properties, ferrites.*

Hillman, Dr Harold. Unity Lab., U. of Surrey, Guildford, Surrey GU2 5XH, England. (1930) PhD, physiology (U. London, 1958); biochemistry (U. London, 1963). Lab. dir., Reader, physiology. (tel. 0483 + 571281, ext. 573; telex 859331). *Membrane - ionic reactions, tissue - ion affinity.*

Hinde, Dr R.M. 102 Hempstead Rd, Watford, Herts, England.

Hine, Dr Raymond. Dept. of Physics, University C. Cardiff, PO Box 78, Cardiff CF1 1XL, Wales. (1934) PhD, crystallography (U. Wales, 1958). Sr. lect. (tel. 0222 + 44211, ext. 2125). *Phase changes, computer programming.*

Hirsch, Prof. Sir Peter. Metallurgy Dept., U. of Oxford, Parks Road, Oxford, England.

Hitchcock, Dr P.B. Sch. of Molecular Sci., Sussex U., Falmer, Brighton BN1 9QJ, England.

Hockly, Dr M. Group R3 1 3 British Telecom Res. Lab., Martlesham Heath, Ipswich, Suffolk IP5 7RE, England.

Hodgkin, Prof. Dorothy Mary Crowfoot. Chemical Crystallography Lab., U. of Oxford, 9 Parks Rd., Oxford OX1 3PD, England. (1910) PhD, X-ray crystallography (U. Cambridge, 1936). Em. prof. (tel. 0865 + 53387, ext. 293). *Large molecules, biologically interesting molecules.*

Hodgkinson, Mr R.A. Queen Mary C., U. of London, Mile End Rd, London E1 4NS, England.

Hogg, Mr C. I. & A.P., Building 347-2, AERE Harwell, Didcot, Oxon OX11 0RA, England.

Hogg, Dr Joshua Herbert Christopher. Dept. of Physics, U. of Hull, Cottingham Rd., Hull HU6 7RX, England. (1940) PhD, crystallography (U. Hull, 1966). Lect. (tel. 0482 + 46311, ext. 7389 or 7819). *Inorganic and organic crystal structures, liquid crystals, X-ray topographic studies of crystals.*

Holben, J. Neutron Div., Marconi Avionics Ltd, Elstree Way, Borehamwood, Herts WD6 1RX, England.

Holland, Miss S. Lab. of Molecular Biophysics, Richards Building, Oxford OX1 3QU, England.

Holland, Mr S.K. Belgrave New Rd, Youlegreave Bakewe, Derbyshire DE4 1WP, England.

Holmes, Mr R.J. 60 Maze Green Rd, Bishop's Stortford, Herts CM23 2PL, England.

Holt, Mrs J. GEC Res. Labs., Hirst Res. Centre, East Lane, Wembley, Middx, England.

Holt, Dr Ronald Stanley. Dept. of Physics, U. of Warwick, Coventry CV4 7AL, England. (1953) PhD, physics (U. Warwick, 1978). Sr. postdoctoral fellow. (tel. 0203 + 24011, ext. 2377). *Compton X-ray & gamma-ray scattering, Compton scatter imaging & densitometry applications.*

Hornby, Mr M.R. Tioxide UK Ltd, Central Labs, Portrack Lane, Stockton on Tees TS18 2NQ, England.

Hornung, Dr George. Dept. of Earth Sci., U. of Leeds, Leeds LS2 9JT, England. (1934) PhD, mineralogy (U. Leeds, 1961). Sr. lect. (tel. 0532 + 431751, ext. 6472). *Optical mineralogy, X-ray diffraction, X-ray spectrometry, electron microprobe, mineral deposits of economic value, volcanic rocks.*

Howard, Dr Judith Ann Kathleen. Dept. of Inorganic Chemistry, U. of Bristol, Cantocks Close, Bristol BS8 1TS, England. (1945) DPhil, chemistry (U. Oxford, 1971). Res. Fellow. (tel. 0272 + 24161). *Organometallic structures, deformation density studies, polyhydride complexes, neutron diffraction.*

Howard, Mrs S. 15 Farndale Rd, Grove Hill, Middlesborough, Cleveland TS4 2PN, England.

Howes, Mr A.J. 8 Laburnham Close, Pelsall, Walsall, W Midlands WS3 4LD, England.

Howie, Dr R. Alan. Dept. of Chemistry, U. of Aberdeen, Meston Walk, Aberdeen AB9 2UE, Scotland. (1940) PhD, chemistry (U. Aberdeen, 1972). Res. officer. (tel. 0224 + 40241, ext. 5630, 5640; telex 73458 UNIABN G). *Inorganic crystal structures, Mössbauer spectroscopy.*

Howie, Prof. Robert Andrew. Dept. of Geology, King's C., Strand, London WC2R 2LS, England. (1923) ScD, mineralogy (U. Cambridge, 1974). Prof. of mineralogy. (tel. 01 + 836-5454, ext. 2521). *Mineralogy, composition-cell parameter relationships, rock-forming minerals.*

Howlin, Dr B. 3 Bettons Park, Stratford, London E15 3JN, England.

Hudd, Mr R.C. ? Looks Court, Porthcawl, Mid Glamorgan CF36 3JJ, Wales.

Hughes, Dr Antony Elwyn. Materials Physics and Metallurgy Div., AERE Harwell, Didcot, Oxon OX11 0RA, England. (1941) DPhil, physics (U. Oxford, 1966). Division Head. (tel. 0235 + 24141, ext. 5273). *Defect structures in solids, radiation damage, superionic oxides, diffusion, positron annihilation.*

Hughes, Mr David John. Analytical Dept., Pye Unicam Ltd, York Street, Cambridge CB1 2PX, England. (1941) BSc, chemistry (U. London, 1964). Technical manager - X-ray diffraction. (tel. 0223 + 358866, ext. 323; telex 817331). *X-ray diffraction techniques in general, powder diffractometry.*

Hughes, Dr David Lewis. AFRC Unit of Nitrogen Fixation, U. of Sussex, Brighton BN1 9RQ, England. (1941) PhD, chemical crystallography (U. British Columbia, Canada, 1971). PSO. (tel. 0273 + 606755, ext. 244). *Single crystal X-ray structure analysis, complexes of dinitrogen and related ligands with transition metals.*

Hughes, Mr Thomas Ernest. Chem. Services Dept., British Nuclear Fuels plc, Springfields Works, Salwick, Preston PR4 0XJ, England. (1929) HNC, chemistry (Union of Lancs. and Cheshire Insts., 1953). HSO. (tel. 0772 + 728262, ext. 31158). *X-ray powder diffraction - all aspects, powder cameras, automation, electron microprobe analysis.*

Hukin, Dr David Ainsworth. Clarendon Lab., U. of Oxford, Parks Rd., Oxford OX1 3PU, England. (1936) DPhil, crystal growth (U. Oxford, 1966). Sr. res. off. (tel. 0865 + 59291, ext. 271). *Crystal growth, rare earth metals, alloys, intermetallic compounds, high temperature crystal growth techniques, high temperature equipment design and development.*

Hukins, Dr David William Laurence. Dept. of Medical Biophysics, U. of Manchester, Oxford Rd., Manchester M13 9PT, England. (1947) PhD, biophysics (King's C., U. of London, 1972). Sr. lect. (tel. 061 + 273-8241, ext. 120). *Diffraction methods, biology and medicine.*

Hull, Dr Stephen Edward. Fraser Williams (Scientific Systems) Ltd., London House, London Road South, Poynton, Cheshire SK12 1YP, England. (1948) PhD, chemistry (U. Sheffield, 1972). Computer Systems Consultant. (tel. 0625 + 871126, ext. 8). *Chemical and crystallographic data bases.*

Hulme, Mr Ralph. Dept. of Chemistry, U. of St. Andrews, North Haugh, St. Andrews, Fife KY16 9ST, Scotland. (1924) MA, chemistry (U. Oxford, 1948) Sr. lect.(semi-retired). (tel. 0334 + 76161, ext. 8210). *Organometallic structures, antimony and iodine chemistry.*

Humphreys, Prof. Colin John. Dept. of Metallurgy and Materials Science, U. of Liverpool, PO Box 147, Liverpool L69 3BX, England. (1941) PhD, physics (U. Cambridge, 1967). Prof. of Materials Eng., Head of Dept. (tel. 051 + 709-6022, ext. 2030; telex 627095). *Electron diffraction and microscopy, semiconductor materials, electron-beam lithography.*

Hurley, Mr Patrick Walter. Analytical Dept., Pye Unicam, York St., Cambridge CB1 2PX, England. (1937) HNC, chemistry (Birmingham C. of Advanced Techn., 1959). Sales Manager. (tel. 0223 + 358866, ext. 335). *Analytical chemistry, X-ray diffraction, X-ray spectrometry, optical emission spectrometry.*

Hursthouse, Dr Michael Barry. Dept. of Chemistry, Queen Mary C., Mile End Rd., London E1 4NS, England. (1941) PhD, chemistry (U. London, 1965). Reader, structural chemistry. (tel. 01 + 980-4811, ext. 3717; telex 893750). *X-ray structure analysis, inorganic complexes, complexes with bulky ligands, natural and synthetic organic compounds, inorganic molecular mechanics.*

Husain, Dr Jasmine Tickle. Dept. of Crystallography, Birkbeck C., Malet St., London WC1E 7HX, England. (1951) PhD, crystallography (U. London, 1981) MRC Res. asst. (tel. 01 + 580-6622, ext. 448). *Molecular biology, molecular dynamics, mechanism of receptor binding, peptides, drugs.*

Hutchings, Dr M.T. Materials Physics Div., Building 521.2, AERE Harwell, Didcot, Oxon OX11 0RA, England.

Hutchison, Dr John Laird. Metallurgy and Science of Materials, Oxford U., Parks Road, Oxford OX1 3PH, England. (1945) PhD, chemistry (U. Glasgow, 1971). Postdoctoral fellow. (tel. 0865 + 59981, ext. 247). *Oxides, silicates, semiconductors, electron microscopy, lattice images.*

Hutton, Dr Alan Thomas. Dept. of Chemistry, University C. of Swansea, Singleton Park, Swansea SA2 8PP, Wales. (1953) PhD, chemistry (U. Capetown, South Africa, 1980) Lect. (tel. 0792 + 205678, ext. 5185). *Chemical crystallography, organometallic and coordination chemistry.*

Huxley, Dr Hugh Esmor. Dept. of Structural Studies, MRC Lab. of Molecular Biology, Hills Rd., Cambridge CB2 2QH, England. (1924) ScD, biology (U. Cambridge, 1964). Scient. staff, MRC. (tel. 0223 + 48011). *Molecular biology, physiology.*

Iball, Prof. John. Dept. of Chemistry, U. of Dundee, Dundee DD1 4HN, Scotland. (1907) DSc, X-ray crystallography (U. Wales, 1939). Hon. res. fellow. (tel. 0382 + 23181, ext. 284). *Crystallography, organic compounds.*

Ingram, Mrs Lorna. 'Kenmore', 40 Manor Place, Cults, Aberdeen AB1 9QN, Scotland. (1943) MSc, crystallography (U. Aberdeen, 1968). Part-time consultant. (tel. 0224 + 861183). *Crystal structure, transition metal oxides, computer programs for crystal structure determination.*

Isaac, Dr D.H. Dept of Metallurgy, U. C. of Swansea, Swansea SA2 8PP, Wales.

Isherwood, Dr Brian James. Materials Science Lab., GEC Hirst Res. Centre, East Lane, Wembley, Middx. HA9 7PP, England. (1941) PhD, physics (Brunel U., 1970). Lab. Manager. (tel. 01 + 904-1262, ext. 291). *X-ray diffraction, poly- and single crystal characterization, growth and perfection of single crystals, epitaxial layers.*

Isherwood, Mrs S.A. 19 Ilmington Rd, Kenton Harrow, Middx, England.

Jack, Prof. Kenneth Henderson. Wolfson Lab., Dept. of Metallurgy and Engineering Materials, U of Newcastle upon Tyne, Newcastle upon Tyne NE1 7RU, England. (1918) ScD, Materials Science (U. Cambridge, 1978). Em. prof., appl. crystal chemistry. (tel. 0632 + 328511, ext. 3201). *Nitrogen ceramics, nitrogen steels, interstitial alloys, inorganic and metallurgical structures, glasses.*

Jakubovics, Dr John Paul. Dept. of Metallurgy and Science of Materials, U. of Oxford, Parks Rd, Oxford OX1 3PH, England. (1938) PhD, physics (U. Cambridge, 1965) U. lect., metallurgy. (tel. 0865 + 59981). *Electron microscopy, magnetism, magnetic materials, micromagnetism, field ion microscopy, atom probe microanalysis.*

Jaswon, Prof. Maurice Arthur. Dept. of Mathematics, The City University, St. John St., London EC1, England. (1922) PhD, crystal physics (U. Birmingham, 1949). Head. (tel. 01 + 253-4399). *Elasticity theory, crystal physics, mathematical crystallography.*

Jayaweera, Dr Shanath Amarasiri Arumabadu. Chemistry Dept., Teeside Polytechnic, Borough Road, Middlesbrough, Cleveland TS1 3BA, England. (1938) PhD, chemistry (U. London, 1969). Principal lect., inorganic chemistry div. head. (tel. 0642 + 218121). *Inorganic complexes, biologically important molecules, industrial processes, solid reactions, thermal analysis.*

Jeffreys, Dr John Alexander David. Dept. of Chemistry, U. of Strathclyde, Glasgow G1 1XL, Scotland. (1927) DPhil, chemistry (U. Oxford, 1952). Lect. (tel. 041 + 552-4400, ext. 2259). *Molecular structure determination, X-ray diffraction, alkaloids of grasses, perennial rye grass alkaloids.*

Johnson, Dr David Julian. Dept. of Textile Industries, U. of Leeds, Leeds LS2 9JT, England. (1936) PhD, textile physics (U. Leeds, 1965). Reader, textile physics. (tel. 0532 + 431751, ext. 6026). *Fibre structures, high modulus organic and inorganic fibres, profile analysis in diffraction patterns, high resolution electron microscopy.*

Johnson, Dr Louise N. Lab. of Molecular Biophysics, Dept. of Zoology, U. of Oxford, South Parks Rd., Oxford OX11 3PS, England. PhD, biophysics (U. London, 1965). Lect. *Protein crystallography, electron microscopy.*

Johnson, Dr Michael William. Neutron Div., SERC Rutherford Lab., Chilton, Didcot, Oxon. OX11 0QX, England. (1944) PhD, crystallography (U. London, 1971). Res. scientist. (tel. 0235 + 21900, ext. 5418). *Molecular solids, phase transitions, liquid structure-dynamics.*

Johnson, Dr N.P. Chemistry Dept, Portsmouth Polytechnic, White Swan Rd, Portsmouth, Hants, England.

Johnson, Mr Owen. Dept. of Inorganic Chemistry, U. of Bristol, Cantocks Close, Bristol BS8 1TS, England. (1958) MA, chemistry (U. Oxford, 1980). Res. asst. (tel. 0272 + 24161, ext. 539). *Organic and organometallic structures, X-ray diffraction.*

Johnson, Dr Peter Anthony Victor. Fuel and Materials Div., UKAEA Springfields Nuclear Power Dev. Labs., Salwick, Nr Preston, Lancs. PR4 0RR, England. (1953) PhD, physics (U. Reading, 1981). Res. chemist. (tel. 0772 + 72862, ext. 31012). *Oxide glasses, structure; neutron diffraction.*

Jones, (Milne) Dr Angela Alice. Dept. of Soil Science, U. of Reading, London Rd., Reading, Berks. RG1 5AQ, England. (1927) PhD, mineralogy-soil science (U. Aberdeen, 1952). SEO. (tel. 0734 + 875234, ext. 279). *Clay minerals, soils, plant - soil mineralogy.*

Jones, Mr D. Nicolet Instruments Ltd, Budbrooke Rd, Warwick CB34 5HX, England.

Jones, Prof. Derry Wynn. Sch. of Chemistry, U. of Bradford, Bradford BD7 1DP, England. (1928) DSc, chemistry (U. Bradford, 1980). Prof., appl. structural chemistry. (tel. 0274 + 733466, ext. 317 or 480; telex 51309 University Brad). *X-ray and neutron diffraction, NMR, fuel constituents, carcinogenic polycyclics, inorganic hydrates, nucleosides, carbides, polymers.*

Jones, Dr G.R. E707 RSRE, St Andrews Rd, Great Malvern, Worcs WR14 3PS, England.

Jones, Mr P.L. 37 Station Rd, Bynea, Llanelli, Dyfed SA14 9PS, Wales.

Jones, Dr Terry. 6 Tyn-y-Cymmer Close, Porth, Rhondda, Mid-Morgan, Wales. (1952) PhD, Chemistry and crystallography (Sheffield City Polytechnic, 1979). Postdoctoral fellow. *Transition metal complexes, organometallics, porphyrins.*

Jones, Dr W. Physical Chemistry Dept, Lensfield Rd, Cambridge CB2 1EP, England.

Kamminga, Dr H. 178 Sturton St, Cambridge CB1 2QF, England.

Kay, Dr Herbert Frederick. Dept. of Physics, U. of Bristol, Tyndall Ave., Bristol BS8 1TL, England. (1923) PhD, physics (U. Manchester, 1947). Sr. lect. (tel. 0272 + 24161, ext. 103). *Electrical and magnetic materials, liquids, physics of sailing, electronics.*

Kelly, Dr Anthony. U. of Surrey, Guildford, Surrey GU2 5XH, England. (1929) ScD, physics (U. Cambridge, 1967). Vice chancellor. (tel. 0483 + 571281, ext. 696). *Ceramics, metals.*

Kelly, Mr Eric. Electricity Council Res. Centre, Capenhurst, Chester CH1 6ES, England. (1939) BA, Earth Science (The Open U., 1976). Technologist. (tel. 051 + 339-4181, ext. 266). *Industrial applications of crystallography, corrosion.*

Kempster, Dr Charles John Edgar. Dept. of Pure and Appl. Physics, UMIST, PO Box 88, Manchester M60 1QD, England. (1932) PhD, physics (U. Cambridge, 1958). Lect. (tel. 061 + 236-3311, ext. 2050). *Polymer crystallography, polymer blends.*

Kennard, Dr Olga. University Chemical Lab., Lensfield Rd., Cambridge CB2 1EW, England. (1924) ScD, crystallography (U. Cambridge, 1973). External staff, MRC; hon. dir., Crystallographic Data Centre. (tel. 0223 + 66499, ext. 331). *Biological structure databases and their utilisation.*

Kennedy, Miss D.A. 57 Roaman Gardens, Glen Rd, Belfast BT11 8LN, N. Ireland.

Kerr, Dr Ian Segrave. Dept. of Chemistry, Imperial C., London SW7 2AY, England. (1929) PhD, electron diffraction (U. London, 1955). Lect. (tel. 01 + 589-5111, ext. 4508). *Structure, alumino-silicates, organometallic complexes and clathrates.*

Killean, Dr Reginald Cameron Gordon. Sch. of Physical Sci.s, U. of St. Andrews, North Haugh, St. Andrews, Fife KY16 9SS, Scotland. (1934) PhD, physics (U. St. Andrews, 1962). Sr. lect. (tel. 0334 + 76161, ext. 8402). *Solid state physics.*

King, Dr James Newington. International Union of Crystallography, 5 Abbey Square, Chester CH1 2HU, England. (1937) PhD, physics (Imp. C., London, 1963). Executive secretary. (tel. 0244 + 42878, telex 669755 OFFICE G, attn. UNICRYSTAL). *International co-operation in crystallography, publication of crystallographic works.*

Kinsella, Mr A. Siemens Ltd, Siemens House, Varey Rd, Congleton, Cheshire CW12 1PH, England.

Kipling, Miss Susan Jane. Catalyst Group Res. Dept., ICI Agricultural Div., Billingham, Cleveland, England. MA, mineralogy (U. Cambridge, 1970). Tech. off. (tel. 0642 + 553601, ext. 5419). *Catalysts, line profile analysis.*

Klug, Dr Aaron. Div. of Structural Studies, MRC Lab. of Molecular Biology, Hills Rd., Cambridge CB2 2QH, England. (1926) PhD, physics (U. Cambridge, 1952). Scient. staff, MRC. (tel. 0223 + 48011). *Biological macromolecular structures, viruses, nucleic acids and chromatin; image reconstruction techniques in electron microscopy.*

Knight, Mr K.S. GEC Hirst Res. Centre, East Lane, Wembley, Middx. HA9 7PP, England.

Knight, Mr Robert. Dept. of Physics, U. of Hull, Cottingham Rd., Kingston on Hull, N. Humberside HU6 7RX, England. (1956) BSc, chemistry (U. Hull, 1978). Res. technician. (tel. 0482 + 497389). *Organic and inorganic crystal structures, liquid crystals, polmer liquid crystals, topography.*

Knott, Mr P.R. 68 Dorset Ave, Great Baddow, Chelmsford, Essex CM2 9UA, England.

Krohn, Dr A. 36 Park View Gardens, London NW4, England.

Kuroda, Dr Reiko. Dept. of Biophysics, King's C., 26-29 Drury Lane, London WC2B 5RL, England. (1947) PhD, chemistry (U. Tokyo, 1975). Hon. lect., res. fellow. (tel. 01 + 836-8851). *Crystal and molecular structures, chiral compounds, oligonucleotides, biologically active compounds, chiroptical spectroscopy, chiral discrimination, drug-nuclei acid interactions.*

Ladd, Dr Marcus Frederick Charles. Sub-Dept. of Chemical Physics (Chemistry), U. of Surrey, Stag Hill, Guildford, Surrey GU2 5XH, England. (1926) DSc, solid state chemistry (U. London, 1979). Reader (head of sub-dept.). (tel. 0487 + 71281, ext. 427, telex 859331). *Solid state chemistry, crystallographic computing, structure and bonding, ionic crystals, crystal chemistry.*

Lake, Mr P.G. 79 Carlisle Rd, Dartford, Kent DA1 1XJ, England.

Laker, Mr Thomas James. Analytical Div., British Gas Corp., London Res. Station, Michael Rd., London SW6 2AD, England. (1945) HNC, chemistry (Borough Poly., 1967). Sr. scientist. (tel. 01 + 736-3344, ext.4059). *Industrial applications of x-ray diffraction, automatic powder diffractometry, crystallite size studies.*

Lalies, Mr A.A. Ground Floor Flat, 68 Laleham Rd, Catford, London SE6, England.

Lang, Prof. Andrew Richard. H.H. Wills Physics Lab, U. of Bristol, Tyndall Ave., Bristol BS8 1TL, England. (1924) PhD, physics (U. Cambridge, 1953). Prof. (tel. 0272 + 24161, ext. 761). *Crystal physics, diffraction theory, lattice imperfections.*

Langford, Dr John Ian. Dept. of Physics, U. of Birmingham, PO Box 363, Birmingham B15 2TT, England. (1935) PhD, crystallography (U. Wales, 1965). Res. fellow. (tel. 021 + 472-1301, ext. 3498). *Powder diffractometry, crystal imperfections, profile fitting, total pattern size-strain analysis, industrial applications.*

Langford, Miss M. 3 Rowland Ave, Mapperley, Nottingham NG3 6BZ, England.

Last, Mr Paul Edward. Analytical Chemistry Dept., Central Res., Pfizer Ltd, Ramsgate Road, Sandwich, Kent CT13 9NJ, England. (1954) GRSC (Royal Society of Chemistry, 1979). Sr. res. scient. (tel. 0304 + 616621). *Pharmaceutical applications of crystallography, polymorphism, phase changes.*

Lawson, Mr Robert John. Petrographical Dept., British Geological Survey, Murchison House, West Mains Rd., Edinburgh EH9 3LA, Scotland. (1925) BSc, geology (U. Edinburgh, 1952). Principal geologist. (tel. 031 + 667-1000). *X-ray and electron probe, rocks, minerals.*

Leadbetter, Prof. Alan James. Rutherford-Appleton Lab., Chilton, Didcot, Oxon OX11 0QX, England. (1934) DSc, chemistry (U. Bristol, 1971). Assoc. dir. (tel. 0235 + 445124). *Neutron scattering, molecular crystals, liquid crystals.*

Leake, Dr John Anthony. Dept. of Metallurgy and Materials Science, U. of Cambridge, Pembroke St., Cambridge CB2 3QZ, England. (1939) PhD, crystallography (U. Cambridge, 1965). Lect. (tel. 0223 + 65151, ext. 343). *Alloys, intermetallic compounds, ordering, rapid-quenching, amorphous materials, ceramics - technical.*

Leal-González, Dr Javier. Clarendon Lab., U. of Oxford, Parks Rd., Oxford OX1 3PU, England. (1947) PhD, chemistry (U. Sheffield, 1979). Res. asst. (tel. 0865 + 59291, ext. 334). *Crystal growth, inorganic crystal structures, optical properties, phase transitions.*

Lee, Dr John David. Dept. of Chemistry, Loughborough U. of Techn., Loughborough, Leics. LE11 3TU, England. (1931) PhD, chemistry, crystallography (U. Nottingham, 1959). Sr. lect. (tel. 0509 + 263171, ext. 362). *Crystal structure determination, computer programming.*

Lees, Mr M.J. Industrial Chemicals Res., CIBA-Geigy plc, Tenax Rd, Manchester M31 1UQ, England.

Leslie, Dr Andrew Greig William. Biophysics Sect., Dept. of Physics, Imperial C., Prince Consort Rd., London SW7 2BZ, England. (1949) PhD, crystal structure (U. Manchester, 1974). Res. fellow. (tel. 01 + 589-5111, ext. 6732). *X-ray structural studies, enzymes.*

Lewis, Dr D. 49 Ashenden Rd, Guildford, Surrey GU2 5XE, England.

Lewis, Dr Eric Leslie Vallance. Dept. of Physics, U. of Leeds, Leeds LS2 9JT, England. PhD, physics. (U. Birmingham, 1967). Res. fellow. (tel. 0532 + 31751). *Wide and small angle diffraction from polymers, polymer structure and mechanical properties, phase transitions in organic crystals.*

Lewis, Prof. Sir J. University Chemical Labs, Lensfield Rd, Cambridge CB2 1EW, England.

Lewis, Dr Michael Harold. Dept. of Physics, U. of Warwick, Coventry CV4 7AL, England. (1938) DPhil, materials science (U. Oxford, 1963). Reader, physics. (tel. 0203 + 24011, ext. 2392). *Ceramics, electron microscopy, microanalysis, electron diffraction, MAS NMR spectroscopy.*

Liddington, Mr Robert Colin. Chemistry Dept., U. of York, Heslington, York YO1 5DD, England. (1959) BA, chemistry (U. Oxford, 1981). Res. fellow. (tel. 0904 + 59861, ext. 337). *Haemoglobin, allosteric mechanism, synchrotron radiation.*

Lindley, Dr M.W. Audlem Lodge, 2 Ridgeway, Broadstone, Poole, Dorset BH18 8EA, England.

Lindley, Dr Peter Frank. Dept. of Crystallography, Birkbeck C., Malet St., London WC1E 7HX, England. (1942) PhD, chemistry (U. Bristol, 1966). Sr. lect. (tel. 01 + 580-6622, ext. 422). *Protein crystallography, data acquisition and processing, plasma and eye-lens proteins, heterocyclic structures.*

Lipson, Prof. Henry, FRS. Dept. of Physics, UMIST, PO Box 88, Manchester M60 1QD, England. (1910) DSc, physics (U. Liverpool, 1930). Em. prof. (tel. 061 + 236-3311, ext. 2743). *Crystal structure determination, X-ray optics.*

Logan, Dr N. Chemistry Dept, Nottingham U., University Park, Nottingham, England.

Lorimer, Dr Gordon Winston. Dept. of Metallurgy, U. of Manchester, Grosvenor St., Manchester M1 7HS, England. (1941) PhD, metallurgy (U. Cambridge, 1968). Sr. lect. (tel. 061 + 236-3311, ext. 8127). *Transmission (high voltage) analytical electron microscopy; phase transformations, metals, minerals.*

Lovell, Mrs S.E. 131B Railway Technical Centre, London Rd, Derby DE2 8UP, England.

Low, Dr John Nicolson. Depts. of Physics and Chemistry, U. of Dundee, Nethergate, Dundee DD1 4HN, Scotland. (1940) PhD, crystallography (U. Dundee, 1982). SSO. (tel. 0382 + 23181, ext. 4562). *Plant virus structures, nucleoside structures, nucleotide structures.*

Lowde, Dr Raymond Douglas. Materials Physics Div., AERE Harwell, Didcot, Oxon. OX11 0RA, England. (1923) DSc, solid state physics (U. London, 1968). SPSO. (tel. 0235 + 24141, ext. 5101). *Solid state physics, metal physics, magnetism, phase transitions, neutron scattering.*

Lowe, Dr Philip Richard. Dept. of Pharmaceutical Sciences, U. of Aston in Birmingham, Goster Green, Birmingham B4 7ET, England. (1948) PhD, crystallography (U. Aston in Birmingham, 1984). Dept. Superintendent. (tel. 021 + 359-3611, ext. 4190). *Small molecule structure determinations.*

Lowe, Miss Susan Elizabeth. International Union of Crystallography, 5 Abbey Square, Chester CH1 2HU, England. (1950) BSc, biophysics (U. Leeds, 1974). Asst. Tech. Editor. (tel. 0244 + 42878, telex 669755 OFFICE G attn. UNICRYSTAL). *Editing.*

Lydon, Dr John Ennis. Astbury Dept. of Biophysics, U. of Leeds, Leeds LS2 9JT, England. (1940) PhD, chemistry (U. Leeds, 1966). Lect. (tel. 0532 + 431751, ext. 6443). *Structural investigations, liquid crystal systems, mesogenic compounds.*

Lyons, Dr Michael Hamilton. R3.14.2, British Telecom Res. Labs., Martlesham Heath, Ipswich, Suffolk IP5 7RE, England. (1953) PhD, chemistry (U. London, 1982). Executive Engineer. (tel. 0473 + 642619). *Crystal growth, semiconductors, dynamical theory.*

Macdonald, Mr George Leslie. Central Materials Lab., Mullard Mitcham, New Rd., Mitcham, Surrey CR4 4XY, England. (1924) MA, physics (U. Cambridge, 1949). Deputy chief analyst. (tel. 01 + 648-3471, ext. 301). *Physical methods of analysis.*

Machin, Ms Penelope Anne. SERC Daresbury Lab., Daresbury, Warrington, Cheshire WA4 4AD, England. (1946) BSc, physics (U. Bristol, 1967). SSO. (tel. 0925 + 65000, ext. 528, telex 62609). *Computational aspects of crystallography, protein crystallography, information retrieval.*

Mackay, Dr Alan Lindsay. Dept. of Crystallography, Birkbeck C., Malet St., London WC1E 7HX, England. (1926) PhD, physics-crystallography (U. London, 1951). Reader. (tel. 01 + 580-6622, ext. 370). *Geometry, symmetry and systematics of structure; electron microscopy, iron oxides and hydroxides, large component crystallography.*

MacKenzie, Prof. William Scott. Dept. of Geology, U. of Manchester, Oxford Rd., Manchester M13 9PL, England. (1920) PhD, mineralogy and petrology (U. Cambridge, 1953). Prof. (tel. 061 + 273-7121, ext. 5590). *Mineralogy, petrology, geochemistry.*

Mahendrasingham, Mr A. Dept of Physics, U. of Keele, Keele, Staffs, England.

Main, Dr Peter. Dept. of Physics, U. of York, Heslington, York YO1 5DD, England. (1939) PhD, physics (U. Manchester, 1963). Lect. (tel. 0904 + 59861, ext. 5540). *Direct methods, biophysics.*

Mallinson, Dr Paul Raymond. Dept. of Chemistry, U. of Glasgow, Glasgow G12 8QQ, Scotland. (1943) PhD, chemistry (U. Essex, 1971). Computer programmer. (tel. 041 + 339-8855, ext. 409). *Crystal structures, cyclic and inclusion compounds, computer programming.*

Malone, Dr John Francis. Dept. of Chemistry, Queen's U., Stranmillis Rd., Belfast BT9 5AG, N. Ireland. (1944) PhD, chemical crystallography (U. Leeds, 1969). Lect. (tel. 0232 + 661111, ext. 4423). *Crystal structure analysis, organic and organometallic compounds, conformational analysis, physiologically active compounds, computer graphics, molecular modelling.*

Manojlović-Muir, Dr Ljubica. Dept. of Chemistry, U. of Glasgow, Glasgow G12 8QQ, Scotland. (1931) PhD, physical chemistry (U. Belgrade, Yugoslavia, 1963). Lect. (tel. 041 + 339-8855, ext. 506). *Structure and bonding, inorganic and organometallic compounds; Hydrogen bonding in solids, crystal structure analysis, X-ray and neutron diffraction.*

Martin, Dr John Wilson. Dept. of Metallurgy and Science of Materials, U. of Oxford, Parks Rd., Oxford OX1 3PH, England. (1926) PhD, metallurgy (U. Cambridge, 1978). Lect. (tel. 0865 + 59981, ext. 226). *Multi-phase alloys; optical and electron microscopy; electron and X-ray diffraction.*

Mason, Dr Kenneth George. Dept. of Chemistry, Loughborough U. of Techn., Loughborough, Leics. LE11 3TU, England. (1928) PhD, organic chemistry (U. Nottingham, 1955). Lect. (tel. 0509 + 263171, ext. 603). *Organic structures.*

Mason, Prof. Sir Ronald, FRS. Sch. of Molecular Sciences, U. of Sussex, Brighton, Sussex BN1 9QJ, England. (1930) PhD, crystallography (U. London, 1953). Prof. of chemistry. (tel. 0273 + 606755). *Structural chemistry, surface chemistry.*

Mason, Prof. Stephen Finney. Dept. of Chemistry, King's C., Strand, London WC2R 2LS, England. (1923) DSc, chemistry (U. Oxford, 1945). Prof. (tel. 01 + 836-5454, ext. 2259). *Crystal and molecular structure, chiral compounds, chiroptical spectroscopy, chiral discrimination.*

Matthew, Dr James Andrew Davidson. Dept. of Physics, U. of York, Heslington, York YO1 5DD, England. (1938) PhD, physics (U. Aberdeen, 1964). Reader. (tel. 0904 + 59861, ext. 5563). *Surface physics, low energy electron diffraction, Auger electron and electron loss spectroscopy, atomic scattering factors.*

Maung, Mr N. 76 Snowdon Drive, Ty Gwyn, Wrexham, Clwyd LL11 2YA, Wales.

Mazey, Dr David John. B393 Materials Development Div., AERE Harwell, Didcot, Oxon OX11 0RA, England. (1929) DPhil, physical metallurgy (U. Salford, 1975). PSO. (tel. 0235 + 24141, ext. 4592). *Electron microscopy (TEM & STEM); microanalysis and microdiffraction in SEM; defects, precipitates, radiation damage, metals.*

McAdam, Dr A. 55 Ennismore Ave, Greenford, Middx, England.

McAllister, Mr Patrick Brian. Analytical Services, BICC Res. and Engineering Ltd., 38 Wood Lane, London W12 7DX, England. (1934) chartered engineer (M. Inst. Metal., 1979). Sect. leader. (tel. 01 + 743-1212, ext. 264). *Structural studies of materials, metals, polymers.*

McCall, Dr Maxine June. University Chemical Lab., Lensfield Road, Cambridge CB2 1EW, England. (1950) PhD, chemistry (Flinders U., Australia, 1980). Post-doc. fellow. (tel. 0223 + 66499, ext. 486). *Sequence-dependent structure, DNA, Drug complexes with DNA.*

McCarthy, Miss A.E. 108 Dower Rd, Four Oaks, Sutton Coldfield, W Midlands, England.

McDonald, Dr Walter Stanley. Dept. of Inorganic and Structural Chemistry, U. of Leeds. Leeds LS2 9JT, England. (1933) PhD, crystallography (U. Glasgow, 1965). Sr. lect. (tel. 0532 + 31751, ext. 6062). *Structural chemistry.*

McEwen, Mr A.B. Birkbeck C., U. of London, Malet St, London WC1E 7HX, England.

McHardy, Dr William James. Dept. of Mineral Soils, Macaulay Inst. for Soil Res., Craigiebuckler, Aberdeen AB9 2QJ, Scotland. (1936) PhD, chemistry (U. Aberdeen, 1962). PSO. (tel. 0224 + 38611, ext. 255). *Electron microscopy (SEM & TEM), electron probe microanalysis, soil minerals.*

McKie, Dr Christine Hilary. Dept. of Earth Sci., U. of Cambridge, Downing St., Cambridge CB2 3EQ, England. (1931) PhD, crystallography (U. Cambridge, 1958). Lect. (tel. 0223 + 335463, ext. 279). *Mineral structures, solid-state transformations.*

McKie, Dr Duncan. Dept. of Earth Sci., U. of Cambridge, Downing St., Cambridge CB2 3EQ, England. (1930) PhD, mineralogy (U. Cambridge, 1962). Lect. (tel. 0223 + 335463, ext. 294). *Mineral structures, polymorphism, solid state reactions.*

McKinnon, Miss F.J. 62 Mercia House, Lower Precinct, Coventry, W Midlands, England.

McPartlin, Dr Mary. Dept. of Chemistry, The Polytechnic of North London, Holloway, London N7 8DB, England. PhD, crystallography (U. New South Wales, Australia, 1966). Sr. lect. (tel. 01 + 607-2789, ext. 2142). *Chemical crystallography.*

Meader, Dr D. 45 Beaulands Close, De Frevelle Ave, Cambridge CB4 1JA, England.

Megaw, Dr Helen Dick. 22 Dunamallaght Rd., Ballycastle, Co. Antrim BT54 6PB, N. Ireland. (1907) ScD, crystallography (U. Cambridge, 1967). Retired. *Perovskites, feldspars, structural physics.*

Mellon, Mr T.P.A. 17 Ormesby Way, Queens Park, Bedford MK40 4LJ, England.

Mendelssohn, Dr Monica Jutta. Crystallography Unit, Geology Dept., U. C. London, Gower St., London WC1E 6BT, England. (1943) PhD, physics (U. London, 1971). Res. fellow. (tel. 01 + 387-7050, ext. 445). *Divergent beam measurements, diamonds, computing.*

Mercer, Dr William Duncan. Dept. of Biochemistry, Medical Biology Centre, Queen's U. of Belfast, Belfast BT9 7BL, N. Ireland. (1947) PhD, biochemistry (U. Bristol, 1972). Lect. (tel. 0232 + 29241, ext. 2795). *Enzymes, structure determination, structure display, protein structure prediction and analysis.*

Merriman, Mr Richard James. Mineralogy and Petrology Res. Group, British Geological Survey, Keyworth, Nottingham NG12 5GG, England. (1943) BSc, geology (U. London, 1970). PSO. (tel. 06077 + 6111, ext. 3130). *Mineralogy, petrology.*

Metcalfe, Dr Edward. Materials Div., Central Electricity Res. Labs., Kelvin Ave., Leatherhead KT22 7SE, England. (1947) PhD, physics, metallurgy (U. Cambridge, 1973). Res. officer. (tel. 037 23 + 74488, ext. 167). *High temperature oxidation, surface analytical techniques, order-disorder phenomena.*

Michell, Mr Ernest William John. Cookson Group plc, 7 Wadsworth Road, Perivale, Greenford, Middx. UB6 7JQ, England. (1935) MSc, crystallography (U. London, 1965). Head of crystallography. (tel. 01 + 997-5635, ext. 15). *X-ray powder diffraction, crystalline phase composition analysis.*

Middleton, Dr Andrew Philip. Res. Labs., British Museum, Gt. Russell St., London WC1B 3DG, England. (1949) DPhil, mineralogy (U. Oxford, 1974). Scient. (tel. 01 + 636-1555, ext. 282). *X-ray diffraction, scanning microscopy, optical microscopy, archeological materials.*

Milburn, Dr George Henry William. Appl. Chemical Sci., Napier C., Colinton Rd., Edinburgh EH10 5DT, Scotland. PhD, crystallography (U. Leeds, 1966). Head. (tel. 031 + 447-7070, ext. 501). *Metal complexes, novel monomers and polymers (crystalline).*

Miles, Mr J.A.C. 28 Filsham Rd, St Leonards-on-Sea, Sussex, England.

Miles, Miss J.M. Flixton, Manchester M31 1EP, England.

Miles, Dr M.J. Food Res. Inst., Colney Lane, Norwich NR4 7UA, England.

Milledge, Dr H. Judith. Crystallography Unit, Dept. of Geology, U. C. London, Gower St., London WC1E 6BT, England. (1927) DSc, crystallography (U. London, 1963). Reader, crystallography. (tel. 01 + 387-7050, ext. 431). *Automation of structure analysis, oscillation photographs, unstable crystals, solid state reactions, high pressure diffraction, high temperature diffraction, diamonds.*

Miller, Prof. Andrew. Dept. of Biochemistry, U. of Edinburgh, Hugh Robson Building, George Square, Edinburgh EH8 9XD, Scotland. (1936) PhD, X-ray crystallography (U. Edinburgh, 1962). Prof. (tel. 031 + 667-1011, ext. 2336; telex 727442 UNIVED G). *X-ray and neutron scattering, biological fibres, virus solutions, molecular graphics.*

Mills, Mr Owen S. Dept. of Chemistry, U. of Manchester, Manchester M13 9PL, England. BSc, chemistry (U. Liverpool, 1945). Reader. (tel. 061 + 273-7121, ext. 5285). *Chemical crystallography, data retrieval, computer graphics.*

Millward, Dr G.R. Physical Chemistry Dept., Cambridge U., Lensfield Rd, Cambridge CB2 1EW, England.

Mole, Mrs J. 10 Kirkdale Gdns, Eaton, Nottingham, England.

Moore, Miss Alice Elizabeth. Tinkers Coppice, Cliff Bridge, Shanklin, Isle of Wight PO37 6QL, England. (1921) MSc, crystallography (Birkbeck C., U. of London, 1954). Retired from Materials Dept., Cement and Concrete Res. Assoc. (tel. 0983 + 864270). *Cement mineralolgy, anhydrous and hydrated cement.*

Moore, Dr Anthony Moreton. Dept. of Physics, Royal Holloway C., U. of London, Egham, Surrey TW20 0EX, England. (1943) PhD, physics (U. Bristol, 1973). Lect. (tel. 0784 (Egham) + 35351, ext. 36; telex 935504). *X-ray topography, crystal defects, disorder, diamond, growth and dissolution, symmetry.*

Moore, Dr John Carlton. Dept. of Sci. and Techn., Slough C. of Higher Education, Wellington St., Slough, Berks SL1 1YG, England. (1934) PhD, crystallography (U. London, 1985). Sr. lect., appl. physics. (tel. 0753 + 34585, ext. 34). *Solid-state phase changes.*

Moore, Mr Peter Leonard. Techn. Centre, British Steel Corp. Tubes Div., Corby Works, Corby, Northants. NN17 1UA, England. (1941) HNC, applied physics (Northern Polytechnic, London, 1964). Superintendent, electron microscopy, X-ray diffraction. (tel. 053 66 + 2121, ext. 4583). *Powder diffraction techniques, quantitative and qualitative analysis, occupational hygiene, electron microscopy.*

Morffew, Dr Andrew James. Molecular Graphics Project, IBM UK Scient. Centre, Athelstan House, St Clement St., Winchester, Hants SO23 9UT, England. (1950) PhD, protein crystallography (U. London, 1981). Sr. Assoc. scient. (tel. 0962 + 68191, ext. 234). *Molecular graphics, refinement, restrained least squares, protein conformation analysis, peptides (small) structures.*

Morgan, Dr Colin Harris. Dept. of Computer Studies, U. of Hull, Cottingham Rd., Hull HU6 7RX, England. (1937) PhD, crystallography (U. St. Andrews, 1964). Lect. (tel. 0482 + 46311, ext. 7295). *Computing problems in crystallography.*

Morris, Dr Donald Frank Charles. Nuclear Science, Chemistry Dept., Brunel U., Uxbridge, Middx. UB8 3PH, England. (1928) DPhil, chemistry (U. Oxford, 1953). Reader. (tel. 0895 + 37188, ext. 537). *Structure and thermodynamics, inorganic crystals, radiochemistry.*

Morrison, Mr N.S. 146 Pennyland Drive, Thurso, Caithness KW14 7PN, Scotland.

Moseley, Dr Patrick. Materials Dev. Div., AERE Harwell, Didcot, Oxon OX11 0RA, England. (1943) PhD, chemical crystallography (U. Durham, 1968). Res. staff. (tel. 0235 + 24141, ext. 4262). *Structural and solid state inorganic chemistry.*

Moss, Dr David Stanley. Dept. of Crystallography, Birkbeck C., Malet St., London WC1E 7HX, England. (1941) PhD, chemical crystallography (U. London, 1967). Sr. lect. (tel. 01 + 580-6622, ext. 368). *Protein structure and dynamics, computing, statistics.*

Motherwell, Dr William David Samuel. Automation Office, Cambridge U. Library, West Rd., Cambridge CB3 9DR, England. (1941) PhD, chemistry (U. St. Andrews, 1967). Automation Officer. (tel. 0223 + 61441, ext. 234). *Computer programming, information retrieval, crystal structures, molecular packing, packing energy calculation.*

Mudd, Mr K.R. The Cottage, Bildeston Rd, Little Finborough, Stowmarket, Suffolk, England.

Muir, Dr Kenneth Walter. Dept. of Chemistry, U. of Glasgow, Glasgow G12 8QQ, Scotland. (1941) PhD, crystallography (U. Glasgow, 1967). Sr. lect. (tel. 041 + 339-8855, ext. 7345). *Transitional metal structural chemistry, crystallographic computing, teaching.*

Muirhead, Dr Hilary. Dept. of Biochemistry, U. of Bristol, Bristol BS8 1TD, England. (1937) PhD, protein crystallography (U. Cambridge, 1964). Reader. (tel. 0223 + 24161, ext. 1127). *Protein crystallography, enzyme structure and function.*

Murray-Rust, Dr Judith. Dept. of Crystallography, Birkbeck C., Malet St., London WC1E 7HX, England. (1946) PhD, chemistry (U. Stirling, 1971). Technician. *Organic and biological molecules.*

Murray-Rust, Dr Peter. Dept. of Chemistry, Glaxo Group Res., Greenford Rd., Middx. UB6 0HE, England. (1941) DPhil, chemistry (U. Oxford, 1967). Head, molecular graphics sect. (tel. 01 + 422-3434). *Molecular geometry analysis, organic molecules, crystallographic data file use, structural correlations, reaction pathways, drug design.*

Narduzzo, Mr V.G. Enraf Nonius Ltd, Highview House, 165-7 Station Rd, Edgware, Middx. HA8 7TU, England.

Nave, Dr Colin. SERC Daresbury Lab., Warrington WA4 4AD, Cheshire, England. (1949) PhD, crystallography (U. London, 1974). SSO (tel. 0925 + 65000, ext. 237; telex 629609). *X-ray fibre diffraction, solution scattering, biological structures, synchrotron radiation, X-ray detectors.*

Nawaz, Dr Rab. Geology Dept., Ulster Museum, Botanic Gardens, Belfast BT9 5AB, N. Ireland. (1940) PhD, mineralogy and petrology (Queen's U., Belfast, 1975). Curator. (tel. 0232 + 668251, ext. 272). *X-ray diffraction, optical mineralogy, gemmology, igneous and metamorphic petrology, zeolites, thermal metamorphism, ore petrology.*

Neidle, Dr Stephen. CRC Biomolecular Structure Unit, Inst. of Cancer Res., Block F, Clifton Ave., Belmont, Sutton, Surrey SM2 5PX, England. (1946) PhD, chemical crystallography (Imp. C., London, 1970). Dir. (tel. 01 + 643-8901). *Biological molecules, nucleic acids, drug design.*

Neisser, Mr Jerzy Zbigniew. Regional Computer Centre, U. of Manchester, Oxford Road, Manchester M13 9PL, England. (1955) PhD, polymer physics (U. Manchester, 1980). Programmer. (tel. 061 + 273-7121). *Polymer crystallography, computational crystallography.*

Nelmes, Dr Richard J. Dept. of Physics, U. of Edinburgh, Mayfield Rd., Edinburgh EH9 3JZ, Scotland. (1943) DSc, physics (U. Cambridge, 1982). Sr. lect. (tel. 031 + 667-1081, ext. 2743). *High-resolution X-ray and neutron crystallography, methods and instrumentation, structural phase transitions.*

Nelson, Dr James Bowman. McCrone Res. Assocs. Ltd., 2 McCrone Mews, Belsize Lane, London NW3 5BG, England. (1918) PhD, physics (U. Cambridge, 1952). Managing dir. (tel. 01 + 435-2282). *Ultramicrostructural applications to industrial problems, x-ray diffraction techniques, reflected light microscopy, colour measurement, gemstones, ore minerals, paint flakes.*

Nicholas, Mr David Michael. Div. of Materials Sci., Thames Polytechnic, Wellington St., London SE18 6PF, England. (1936) MSc, crystallography (U. London, 1965). Sr. lect. (tel. 01 + 854-2030, ext. 369). *Amorphous structures, crystallite size, powder diffraction.*

Nicholls, Dr Reginald A. Analytical Dept., Pye-Unicam, Philips, York St., Cambridge CB1 2PX, England. (1952) PhD, clay mineral geochemistry (U. Bristol, 1979). Sr. appl. XRD. (tel. 0223 + 358866, ext. 322; telex 817331). *Clay mineral diffraction, computation in diffraction.*

Nicol, Dr Alastair William. Ande Scientific, 12 Greenhill Rd., Moseley, Birmingham B13 9SR, England. (1936) PhD, chemistry (U. Aberdeen, 1962). Sci. consultant, Industry and Education. (tel. 021 + 449-1418). *General analysis, materials characterization, X-ray diffraction analysis.*

Nieduszynski, Dr Ian Alexander. Dept. of Biological Sciences (Biochemistry), U. of Lancaster, Bailrigg, Lancaster LA1 4YQ, England. (1944) PhD, biophysics (U. Leeds, 1969). Lect. (tel. 0524 + 65201, ext. 4662). *X-ray fibre diffraction, physico-chemical studies, polysaccharides, glycosaminoglycans.*

Nix, Mr E.L. 24 Keats Way, W Drayton, Middx. UB7 9DS, England.

North, Prof. Anthony Charles Thomas. Astbury Dept. of Biophysics, U. of Leeds, Leeds LS2 9JT, England. (1931) PhD, biophysics (U. London, 1955). Prof. (tel. 0532 + 431751, ext. 6130). *Enzyme structure and function, drug design, protein engineering, protein crystallography methods, interactive molecular graphics.*

Norval, Dr Stephen Vynne. Res. and Techn. Dept., ICI Petrochemicals and Plastics Div., PO.Box 90, Wilton, Middlesborough, Cleveland TS6 8JE, England. (1949) PhD, chemistry (Glasgow U., 1976). Res. scientist. (tel. 0642 + 455522, ext. 2005; telex 587461). *X-ray diffraction, electron microscopy, heterogeneous catalysts.*

Nowell, Dr Ian William. Dept. of Chemistry, Sheffield City Polytechnic, Pond St., Sheffield S1 1WB, England. (1944) PhD, chemistry (U. Leicester, 1969). Principal lect. (tel. 0742 + 20911, ext. 2458). *Coordination complexes, macrocyclic complexes, organometallic complexes.*

Nye, Prof. John Frederick. H.H. Wills Physics Lab., U. of Bristol, Tyndall Ave., Bristol BS8 1TL, England. (1923) PhD, physics (U of Cambridge, 1948). Prof. (tel. 0272 + 24161, ext. 775). *Glaciology, ice physics, plastic deformation, wave physics.*

O'Connor, Dr Denis Arthur. Dept. of Physics, U. of Birmingham, PO Box 363, Birmingham B15 2TT, England. (1927) DSc, physics (U. Birmingham, 1968). Reader, crystal physics. (tel. 021 + 472-1301, ext. 2545). *Crystal phase transitions, dynamical diffraction theory, Mössbauer effect.*

Oliva, Mr G. Dept of Crystallography, Birkbeck C., Malet St, London WC1E 7HX, England.

Orpen, Dr Anthony Guy. Dept. of Inorganic Chemistry, Bristol U., Cantock's Close, Bristol UB8 3PH, England. (1955) PhD, crystallography (U. Cambridge, 1979). Lect. (tel. 0272 + 24161, ext. 539). *Molecular structure, diffraction techniques, organometallic chemistry, neutron diffraction.*

Osborne, Mr J.F. GEC Avionics Ltd, Appl. Physics Div, Borehamwood, Herts WD6 1RX, England.

Owen, Dr John David. 31 Batford Road, Harpenden, Herts. AL5 5AT, England. (1946) PhD, crystallography (U. Newcastle, 1971).

Owston, Dr Philip George. Sch. of Chemistry, The Polytechnic of North London, Holloway, London N7 8DB, England. (1921) DSc, crystallography, structural chemistry (U. London, 1975). Head. (tel. 01 + 607-2789, ext. 2140). *Coordination compounds, structure - activity relationship.*

Page, Dr James Ernest. 127 Northumberland Rd., Harrow, Middx. HA2 7RB, England. (1915) DSc, chemistry (U. London, 1957). Retired (tel. 01 + 866-8871). *Physical methods of analysis, organic compounds, organic structures.*

Page, Dr Trevor Francis. Dept. of Metallurgy and Materials Science, U. of Cambridge, Pembroke St., Cambridge CB2 3QZ, England. (1946) PhD, metallurgy-materials science (U. Cambridge, 1971). Lect. (tel. 0223 + 65151, ext. 388 or 375). *Electron microscopy (TEM, HREM, SEM); field ion microscopy; optical microscopy; grain boundary and interface structures; crystallography, microstructure, defect structures, phase transformations; ceramic engineering materials; hardness; wear.*

Palmer, Dr Rex Alfred. Dept. of Crystallography, Birkbeck C., Malet St., London WC1E 7HX, England. (1936) PhD, crystallography (U. London, 1962). Sr. lect. (tel. 01 + 580-6622, ext. 330). *Biological molecules, methods of structure determination, teaching crystallography, drug design, quantum calculations, pharmacological activity.*

Pamplin, Dr Brian Randall. Sch. of Physics, U. of Bath, Claverton Down, Bath BA2 7AY, England. (1933) PhD, physics (U. Cambridge, 1960). Sr. lect., editor in chief of 'Progress in Crystal Growth and Characterization'. (tel. 0225 + 61244, ext. 445). *Semiconducting compounds, crystal growth, characterization.*

Papiz, Dr Miroslav Zenko. (Physics Dept., U. of Keele) SERC Daresbury Lab., Warrington WA4 4AD, England. (1955) PhD, protein crystallography (CNAA - Napier C., Edinburgh, 1982). Res. fellow. (tel. 0925 + 65000, ext. 388). *Synchrotron radiation, protein crystallography, anomalous dispersion, electronic area detectors.*

Parker, Dr Andrew. Dept. of Geology, U. of Reading, Whiteknights, Reading RG6 2AB, England. (1941) PhD, geochemistry (U. Reading, 1969). Principal Res. fellow. (tel. 0734 + 875123, ext. 7805). *Clay mineralogy, geochemistry.*

Parker, Mr Michael William. Lab. of Molecular Biophysics, Dept. of Zoology, U. of Oxford, Rex Richards Building, S. Parks Rd., Oxford OX1 3QU, England. (1959) BSc, chemistry (Australian National U., 1981). Res. asst. (tel. 0865 + 56733). *Biochemical and medical applications of crystallography.*

Parpia, Dr Dawood Yusuf. Dept. of Physics, U. of Durham, South Road, Durham DH1 3LE, England. (1945) PhD, physics (Cambridge U., 1974). Sr. res. asst. (tel. 0385 + 64971, ext. 244). *X-ray topography, crystal growth, crystal defects, ferroelectric domains.*

Parry, Mr G. 7 Sch. Walk, Letchworth, Herts SG6 1QD, England.

Parry, Dr George Sparling. Dept. of Chemical Eng. and Chemical Techn., Imperial C., London SW7 2BY, England. (1924) PhD, crystallography (U. London, 1950). Sr. lect. (part-time) in crystallography. (tel. 01 + 589-5111, ext. 4310). *Intercalation phenomena, periodic lattice deformations, phase transformations.*

Parsons, Prof. Ian. Dept. of Geology and Mineralogy, U. of Aberdeen, Marischal C., Aberdeen AB9 1AS, Scotland. (1939) PhD, petrology (U. Durham, 1963). Prof. (tel. 0224 + 40241, ext. 277). *Polymorphism, unmixing, alkali feldspars (synthesis), X-ray diffraction, transmission electron microscopy, feldspars, igneous rocks, igneous petrology, alkaline rocks.*

Parsons, Dr Paul Donald. Springfields Nuclear Labs., UKAEA, Springfields Works, Preston PR4 0RR, England. (1937) PhD, metallurgy (U. Leeds, 1965). Sect. leader. (tel. 0772 + 728262, ext. 31218). *Corrosion, oxidation, materials science, mechanical properties, metals, zircaloy, zirconia, creep.*

Pashley, Prof. Donald William. Dept. of Metallurgy and Materials Sci., Imperial C., Prince Consort Rd., London SW7 2AZ, England. (1927) PhD, physics (Imp. C., London, 1950). Prof. of materials, Dept. head. (tel. 01 + 589-5111, ext. 5901, telex 261503). *Electron microscopy, electron diffraction, thin film growth, epitaxy.*

Paton, Mr John Dennis. Dept. of Chemistry, U. of Dundee, Dundee DD1 4HN, Scotland. (1946) MSc, crystallography (U. Dundee, 1979). SO. (tel. 0382 + 23181, ext. 4334). *Powder diffraction, structure determination, diffractometers.*

Pawley, Prof. G. Stuart. Dept. of Physics, U. of Edinburgh, Mayfield Rd., Edinburgh EH9 3JZ, Scotland. (1937) PhD, physics (U. Cambridge, 1962). Prof. (tel. 031 + 667-2122, ext. 2742). *Computer simulations, molecular lattice dynamics, neutron scattering, constrained refinements.*

Paxton, Mr A.T. Woolfson C., Oxford, England.

Peacock, Dr Norman. Dept. of Fibre Science, U. of Strathclyde, George St., Glasgow G1 1XW, Scotland. (1925) PhD, textile physics (U. Leeds, 1958). Sr. lect. (tel. 041 + 552-4400, ext. 2134). *Molecular and fine structure, textile fibres.*

Peacock, Prof. R.D. Chemistry Dept, U. of Leicester, University Rd, Leicester, England.

Pearce, Mr I.R. CERL, Kelvin Ave, Leatherhead, Surrey GU1 1HX, England.

Penfold, Dr David William. Text Management Services, 105 Fitzwilliam Street, Huddersfield HD1 5PS, England. (1944) PhD, materials science (Imp. C., London, 1971). Managing dir. (tel. 0484 + 510178) *Metals, low temperatures, information science.*

Pennington, Mr Mark. Chemistry Dept., Warwick U., Coventry CV4 7AL, England. (1961) MSc(res), metal-sulphur-chelated complexes (Warwick U., 1985). Res. student. (tel. 0203 + 24011, ext. 2204). *Actinide structural chemistry, sulphur-chelate complexes of post-transition metals.*

Perrins, Dr D.H.G. 58 Widmore Rd, Bromley, Kent BR1 3BD, England.

Perutz, Dr Max Ferdinand. MRC Lab. of Molecular Biology, Hills Rd., Cambridge CB2 2QH, England. (1914) PhD, crystallography (U. Cambridge, 1940). (tel. 0223 + 48011, ext. 216). *Protein crystallography, molecular biology.*

Phillips, Prof. Sir David Chilton. Lab. of Molecular Biophysics, Dept. of Zoology, U. of Oxford, The Rex Richards Building, South Parks Rd., Oxford OX1 3QU, England. (1924) PhD, physics, crystallography (U. Wales, 1951). Prof. (tel. 0865 + 56733, ext. 401). *Crystal structure analysis, biological macromolecules, apparatus developments.*

Phillips, Mr G.I. I. & A.P., Building 347-2, AERE Harwell, Didcot, Oxon OX11 0RA, England.

Phillips, Mr R.J. Chemistry Dept, UMIST, Sackville St, PO Box 88, Manchester M60 1QD, England.

Phillips, Dr Simon Edward Victor. Dept. of Biophysics, U. of Leeds, Leeds LS2 9JT, England. (1950) PhD, chemistry (U. London, 1974). Lect. (tel. 0532 + 431751, ext. 7581). *Protein crystallography.*

Pierce-Butler, Dr Melanie Anne. Crystallography Sect., Royal Armaments Res. and Dev. Est., Powdermill Lane, Waltham Abbey, Essex EN9 1BP, England. (1949) PhD, chemistry (U. Warwick, 1975). Sect. head. (tel. 0992 (Lea Valley) + 713030, ext. 305). *Crystallography, explosives, structural chemistry, crystallisation.*

Pigram, Dr W.J. Physics Dept, U. of Keele, Keele, Staffs ST5 5BG, England.

Pirie, Dr John Douglas. Dept. of Natural Philosophy, U. of Aberdeen, Old Aberdeen AB9 2UE, Scotland. (1939) PhD, physics (U. Aberdeen, 1965). Lect. (tel. 0224 + 40241, ext. 5433; telex 73458 UNIABN G). *Thermal X-ray scattering.*

Pitrola, Dr R. 37 Corporation Ave, Hounslow, Middx. TW4 6AX, England.

Pitts, Dr James Edwin. Dept. of Crystallography, Birkbeck C., Malet St., London WC1E 7HX, England. (1954) DPhil, protein crystallography (U. Sussex, 1980). Roy. Soc. Res. fellow. (tel. 01 + 580-6622, ext. 284). *Protein crystallography, protein engineering.*

Plant, Dr John Stewart. Computing Centre, Keele U., Keele, Staffs ST5 5BG, England. (1945) PhD, neutron diffraction (U. Sheffield, 1970). Res. systems analyst. (tel. 0782 + 621111). *Antiferromagnetic structure (detailed), electron density distributions, computer modelling.*

Plesken, Dr Wilhelm. Sch. of Mathematics, Queen Mary C., U. of London, Mile End Road, London E1 4NS, England. (1950) Dr.rer.nat., mathematics (RWTH Aachen, 1974). Lect. (tel. 01 + 9804811, ext. 3889). *Crystallographic groups, algebra, group theory, computational algebra.*

Pointer, Dr David John. Dept. of Chemistry, Teesside Polytechnic, Borough Rd., Middlesborough TS1 3BA, England. (1937) PhD, chemistry (U. Wales, 1965). Principal lect. (tel. 0642 + 218121, ext. 4179). *Organic molecules, biological activity.*

Pollard, Dr David Ronald. Dept. of Education and Science, Elizabeth House, York Rd., London SE1, England. (1942) PhD, crystallography (U. Glasgow, 1968). Principal. (tel. 01 + 934-9000, ext. 9400). *Computing methods, single crystal analysis, large organic molecules, molecular complexes.*

Pond, Dr Robert Charles. Dept. of Metallurgy and Materials Sci., U. of Liverpool, PO Box 147, Liverpool L69 3BX, England. (1946) PhD, materials science (U. Bristol, 1973). Sr. lect. (tel. 051 + 709-6022, ext. 2028, telex 627095). *Electron microscopy, structure and properties, interfaces, semiconductors.*

Potter, Dr Reginald. Dept. of Physics, University C. Cardiff, PO Box 78, Cardiff CF1 1XL, Wales. (1931) PhD, physics, crystallography (U. Wales, 1962). Sr. lect. (tel. 0222 + 44211, ext. 2342). *Instrumentation, accurate intensities.*

Potterton, Dr Elizabeth Anne. Dept. of Physics, U. of York, York YO1 5DD, England. (1956) PhD, Biophysics (U. Leeds, 1984). Res. asst. (tel. 0904 + 59861, ext. 5547). *Protein crystallography, computer graphics, protein - drug interactions.*

Povey, Dr David Christopher. Dept. of Chemistry, U. of Surrey, Stag Hill, Guildford, Surrey GU2 5XH, England. (1946) PhD, chemical physics (U. Surrey, 1974). Lect. (tel. 0483 + 571281, ext. 338). *X-ray structure analysis, inorganic crystal chemistry, microelectronics applications, crystallographic instruments, microcomputers, crystallographic teaching, biological structures.*

Powell, Prof. Herbert Marcus. Hertford C., Catte St., Oxford, England. (1906) MA, chemistry (U. Oxford, 1927). Retired. *Chemistry, crystallography.*

Prakash-Varma, Dr S. 95 Montbelle Rd, New Eltham, London SE9 3NY, England.

Price, Dr Geoffrey David. Dept. of Geology, U. C. London, Gower St., London WC1E 6BT, England. (1956) PhD, mineralogy (U. Cambridge, 1980). Res. fellow. (tel. 01 + 387-7050, ext. 433). *Silicates, phase transformations, deformation, high pressures.*

Pritchard, Dr Robin Gavin. Dept. of Chemistry, UMIST, PO Box 88, Manchester M60 1QD, England. (1952) PhD, chemistry (U. Wales, 1978). EO. (tel. 061 + 236-3311, ext. 2646, telex UMIST 666094). *Peroxy compounds, hydrogen bonding, solid state interaction, gas phase molecular conformations.*

Prout, Dr Charles Keith. Chemical Crystallography Lab., U. of Oxford, 9 Parks Rd., Oxford OX1 3PD, England. (1934) DPhil, chemistry (U. Oxford, 1959). Lect., fellow of Oriel C. (tel. 0865 + 57542). *Structural chemistry.*

Putnis, Dr Andrew. Dept. of Earth Sci., U. of Cambridge, Downing Place, Cambridge CB2 3EW, England. (1947) PhD, mineralogy and petrology (U. Cambridge, 1976). Lect. (tel. 0223 + 355463, ext. 272). *Transformation behaviour, minerals.*

Puttick, Prof. Keith Ernest. Dept. of Physics, U. of Surrey, Guildford, Surrey GU2 5XH, England. (1926) PhD, physics (U. Bristol, 1957). Prof. (tel. 0483 + 571281, ext. 741, telex 859331). *Mechanical properties (scaling laws, hardness, fracture), ceramics, particulate crystals, polymers, composites, metals.*

Puxley, Dr David Charles. Analytical Div., British Gas Corp., London Res. Station, Michael Rd., London SW6 2AD, England. (1946) PhD, inorganic chemistry (U. London, 1972). Principal scientist. (tel. 01 + 736-3344, ext. 4051). *Industrial X-ray diffraction, automatic powder diffractometry, crystallite size studies, catalysis, position-sensitive detectors.*

Rabinowich, Prof. D. University Chemical Lab, Lensfield Rd, Cambridge CB2 1EW, England.

Rae, Dr Alan William James Melville. Technical Ceramics Div., Anzon Ltd., Cookson House, Willington Quay, Wallsend NE28 6UQ, England. (1951) PhD, metallurgy (U. Newcastle, 1976). General Manager. (tel. 0632 + 622211, telex 537357). *Technical ceramics, inorganic refractory compounds.*

Rae, Dr Alastair Ian Maxwell. Dept. of Physics, U. of Birmingham, PO Box 363, Birmingham B15 2TT, England. (1938) PhD, physics (U. W. Australia, 1963). Lect. (tel. 021 + 472-1301, ext. 3460). *Crystallography, intermolecular forces, phase changes.*

Raftery, Dr James. Dept of Biochemistry, George Square, Edinburgh EH8 9XD, Scotland. (1952) PhD, crystallography (U. London, 1981). Res. fellow. (tel. 031 + 667-1011, ext.2410). *Parallel computers, proteins, data bases.*

Raithby, Dr Paul Robert. Dept. of Inorganic Chemistry, U. of Cambridge, Lensfield Rd., Cambridge CB2 1EW, England. (1951) PhD, chemistry (U. London, 1976). Res. off. (tel. 0223 + 66499, ext. 220). *X-ray crystallography, cluster compounds, osmium chemistry.*

Ralph, Prof. Brian. Dept. of Metallurgy and Materials Sci., University C. Cardiff, Newport Road, Cardiff CF2 1TA, Wales. (1939) ScD, materials science (U. Cambridge, 1980). Dept. head. (tel. 0222 + 44211, ext. 7003; telex 498635). *Grain boundaries, interfaces, electron diffraction, electron microscopy.*

Ramdas, Dr S. Analytical Div., BP Res. Centre, Chertsey Rd, Sunbury-on Thames, Middx. TW16 7LN, England.

Ransom, Mr H. London Valuation Centre, St Dunstan's House, Carey Lane, London EC2V 8AB, England.

Raper, Mr Eric Salvin. Dept. of Chemical and Life Sciences, Newcastle upon Tyne Polytechnic, Ellison Pl., Newcastle upon Tyne NE1 8ST, England. (1933) MSc, crystallography (U. Durham, 1968). Sr. lect. (tel. 0632 + 326002, ext. 3516). *Heterocyclic molecules, coordination compounds.*

Redhouse, Dr Alan David. Dept. of Chemistry and Appl. Chemistry, U. of Salford, Salford M5 4WT, England. (1940) PhD, inorganic chemistry (U. Bristol, 1964). Lect. (tel. 061 + 736-5843, ext. 643). *Crystal structure analysis.*

Reid, Dr John Sinclair. Dept. of Natural Philosophy, U. of Aberdeen, Old Aberdeen AB9 2UE, Scotland. (1942) PhD, physics (U. Aberdeen, 1970). Lect. (tel. 0224 + 40241, ext. 5438). *Diffuse X-ray scattering, lattice dynamics, teaching, meteorological optics, history of scientific instruments.*

Rendle, Dr David Forbes. Chemistry (General) Div., Metropolitan Police Forensic Science Lab., 109 Lambeth Rd., London SE1 7LP, England. (1946) PhD, chemistry (U. Guelph, Canada, 1972). SSO. (tel. 01 + 230-6243). *Forensic applications of powder diffraction methods, instrumentation.*

Reynolds, Mr C.C. 15 Gibson Drive, Paignton, S Devon TQ4 7AL, England.

Reynolds, Dr C.D. Biophysics Lab, Dept of Physics, Liverpool Poly., Liverpool L3 3AF, England.

Rhodes, Prof. Rene George. Dept. of Engineering, U. of Warwick, Coventry CV4 7AL, England. (1916) PhD, physics (U. Cambridge, 1950). Prof. (tel. 0203 + 24011, ext. 2128). *Semiconductors, superconductors, magnetic levitation.*

Rice, Dr David William. Dept. of Biochemistry, U. of Sheffield, Western Bank, Sheffield S10 2TN, England. (1952) DPhil, X-ray crystallography of proteins (U. Oxford, 1979). Lect. (tel. 0742 + 78555, ext. 4242). *Protein crystallography, crystallization of macromolecules, structural homology.*

Richards, Dr Brian Peter. Materials Science Div., GEC Hirst Res. Centre, East Lane, Wembley, Middx. HA9 7PP, England. (1939) PhD, crystallography (U. London, 1974). Manager Materials Characterization Div. (tel. 01 + 904-1262, ext. 280). *Characterisation of crystalline and amorphous materials, carbons and graphites, defects in semiconductor materials.*

Richards, Dr John Philip Gerald. Dept. of Physics, University C. Cardiff, PO Box 78, Cardiff CF1 1XL, Wales. (1932) PhD, physics (U. Wales, 1960). Sr. lect. (tel. 0222 + 44211, ext. 2360). *Molecular motions, phase transitions, thin films, surfaces.*

Richardson, Dr C.H. SERC Daresbury Lab., Daresbury, Warrington WA4 4AD, England.

Richardson, Dr R.M. Sch. of Chemistry, Cantock's Close, Bristol, Avon BS8 1TS, England.

Rickards, Mr A.L. 123 Bury & Rochdale Old Rd, Birtle, Bury, Lancs, England.

Roberts, Dr Kevin John. Dept. of Pure and Appl. Chemistry, U. of Strathclyde, 295 Cathedral St., Glasgow G1 1XL, Scotland. (1950) PhD, material science, crystallography (Portsmouth Poly, 1979). Lect. (tel. 041 + 552-4400, ext. 2265). *Crystal growth (theoretical & experimental studies), molecular crystals, surface crystallography, X-ray diffraction, novel uses of synchrotron radiation.*

Robertson, Dr John Harry. Sch. of Chemistry, U. of Leeds, Leeds LS2 9JT, England. (1923) PhD, crystallography (U. Edinburgh, 1949). Sr. lect. (tel. 0532 + 431751, ext. 7083). *Inorganic and organic structures, physical properties of the crystalline state, philosophy of scientific method.*

Robertson, Prof. John Monteath, FRS. 11A Eriskay Rd., Inverness IV2 3LX, Scotland. DSc, chemistry (U. Glasgow, 1933). Em. prof. (tel. 0463 + 225561). Gregorie Aminoff Gold Medal, Royal Swedish Academy, June 1982. *Chemistry.*

Robertson, Mr Robert Hugh Stannus. Director, Resource Use Inst. Ltd., Dunmore, 25 Bonnethill Rd., Pitlochry, Perthshire PH16 5ED, Scotland. (1911) MA, mineralogy, geology and chemistry (U. Cambridge, 1932). Dir. (tel. 0796 + 2569). *Properties of crystals & affect on applications.*

Roberts-Sengier, Dr Lieve. 31 Church St., Wye, Ashford, Kent TN25 5BN, England. (1955) Doctor, crystallography (K.U. Leuven, Belgium, 1984). *Structure determination, small biological molecules.*

Robinson, Dr K. Physics Dept, U. of Reading, Reading RG6 2AF, England.

Robinson, Mr P.T. 15 Allerton Park, Leeds LS7 4ND, England.

Roebuck, Dr Peter Hamish Athey. Technical Dept., Kanthal Ltd., Inveralmond Industrial Estate, Perth PH1 3EE, Scotland. (1952) PhD, materials science - ceramics (U. Newcastle, 1978). Technical Manager. (tel. 0738 + 20931, ext. 125; telex 76460). *Silicon carbide, high temperature materials (electrical).*

Rogers, Prof. Donald. 11 Salvington Crescent, Bexhill on Sea, E Sussex, TN39 3NP, England. (1921) PhD, physics (U. London, 1944). Em. prof. (tel. 0424 + 222157). *Absolute configuration assignment, automated setting of single crystals.*

Rogers, Mr K.D. Epidemology Dept, WNSM, Heath Park, Cardiff CF4 4XN, Wales.

Rollett, Dr John Sydney. Computing Lab., U of Oxford, 8-11 Keble Rd., Oxford OX1 3QD, England. (1927) PhD, chemistry (U. Leeds, 1952). U. lect. in numerical maths. (tel. 0865 + 54141, ext. 317). *Crystallographic computing, numerical analysis (optimization, linear algebra).*

Ross, Dr Donald Keith. Dept. of Physics, U. of Birmingham, Birmingham B15 2TT, England. (1939) PhD, physics (U. Birmingham, 1972). Lect. (tel. 021 + 472-1307, ext. 3467 or 2078). *Hydrogen in metals, neutron diffraction, quasielastic scattering, clay-water systems, intermetallic systems.*

Rowley, Dr Colin Raymond. Dept. of Geology, Portsmouth Polytechnic, Burnaby Rd., Portsmouth PO1 3QL, England. (1938) PhD, geology (U. Durham, 1965). Principal lect. (tel. 0705 + 827681, ext. 254). *Crystal structures and chemistry, clay minerals, rock-forming carbonates.*

Roys, Mr W.B. 9 Whitburn Rd, Toton Beeston, Nottingham NG9 6HP, England.

Rule, Mr Stephen A. SERC Daresbury Lab., Warrington, Cheshire WA4 4AD, England. (1955) BSc, physics (U. Keele, 1981). Post-doc. Res. asst. (tel. 0925 + 65000, ext. 388). *Synchrotron radiation, protein crystallography.*

Russell, Dr David Robin. Dept. of Chemistry, U. of Leicester, Leicester LE1 7RH, England. (1939) PhD, inorganic chemistry (U. Glasgow, 1963). Sr. lect. (tel. 0533 + 554455). *Chemical crystallography.*

Russell, Dr Paul Robert. MIT Sect., Unilever Res. Lab., Colworth House, Sharnbrook, Beds, England. (1948) PhD, chemistry (Loughborough U. of Tech., 1976).

Rutland, Ms J. 24 Woodrush Way, Chadwell Heath, Romford, Essex RM6 5BL, England.

Saldin, Dr D.K. Solid State Theory Group, Imperial C., Prince Consort Rd, London SW7 2BZ, England.

Salem, Mr M. Chemistry Dept, Queen's U., Belfast BT9 5AG, N Ireland.

Salt, Mr P.D. ECLP & Co. Ltd., John Keay House, St Austell, Cornwall PL25 4DJ, England.

Sampson, Mr Christopher. Materials Physics Div., AERE Harwell, Didcot, Oxon OX11 0RA, England. (1935) Nat. Cert., applied physics (1960). HSO. (tel. 0235 + 24141, ext. 5280). *Lattice parameter determination, powder cameras, powder diffractometry, low temperature distortions, single crystals and powders.*

Sándor, Dr Endre Elek. 1 Fairlawn Drive, Woodford Green, Essex IG8 9AW, England.

Sawyer, Dr Lindsay. Dept. of Biochemistry, U. of Edinburgh, Hugh Robson Building, George Square, Edinburgh EH8 9XD, Scotland. (1944) PhD, protein crystallography (U. Edinburgh, 1971). Lect. (tel. 031 + 667-1011, ext. 2363; telex 727442 UNIVED G). *Molecules of biological interest, proteins.*

Schwalbe, Dr Carl Hellmuth Walter. Dept. of Pharmaceutical Sciences, U. of Aston, Gosta Green, Birmingham B4 7ET, England. (1942) PhD, chemistry (U. Harvard, 1970). Sr. lect. (tel. 021 + 359-3611, ext. 4201, telex 336997). *Structures, drug molecules, single-crystal X-ray diffraction, molecular orbital calculations.*

Schwarzenberger, Mrs D.R. Engineering Dept, Warwick U., Coventry CV4 7AL, England.

Scouloudi, Dr Helen. Lab. of Molecular Biophysics, Dept. of Zoology, U. of Oxford, The Rex Richards Building, South Parks Rd., Oxford OX1 3QU, England. PhD, physics (U. London, 1951). Res. officer. (tel. 0865 + 56733, ext. 404). *Protein structure, biologically interesting molecules.*

Scrimgeour, Dr Sheelagh Nicoll. Dept. of Chemistry, U. of Dundee, Nethergate, Dundee DD1 4HN, Scotland. (1947) PhD, chemistry and crystallography (U. Dundee, 1973). Res. fellow. (tel. 0382 + 23181, ext. 287). *Carcinogens, anticancer agents.*

Scutcher, Dr W. SCM Chemicals Ltd, PO Box 26, Grimsby, S Humberside DN37 8EY, England.

Seddon, Dr John Michael. Dept. of Chemistry, The University, Southampton SO9 5NH, England. (1953) PhD, Biophysics (U. London, 1980). (tel. 0703 + 559122, ext. 2156). *Diffraction, lyotropic liquid crystals, lipid membranes.*

Shackleton, Miss Judith Mary. X-ray Methods, Materials Evaluation, Lucas Res. Centre, Shirley, Solihull, West Midlands B90 4JJ, England. (1958) MSc, physical methods of analysis (U. Aston in Birmingham, 1984). Res. Officer. (tel. 021 + 744-8522, ext. 202). *Powder diffraction methods, phase identification, phase changes, failure analysis, line profile analysis.*

Shah, Dr Jitendra Shantilal. H.H. Wills Physics Lab., U. of Bristol, Royal Fort, Bristol BS8 1TL, England. (1939) PhD, physics (U. Bath, 1967). (tel. 0272 + 24161, ext. 139; telex 444174). *Fibre diffraction, biological deposits, biological crystallization, powder diffraction.*

Shah, Miss V.K. 39'4 Cloveston Drive, Edinburgh EH14 3BE, Scotland.

Sharp, Prof. David William Arthur. Chemistry Dept., U. of Glasgow, Glasgow G12 8QQ, Scotland. (1931) PhD, chemistry (U. Cambridge, 1957). Prof. (tel. 041 + 339-8855, ext. 418, telex 778421). *Inorganic complexes.*

Shaw (née Gözen), Dr Leylâ Süheylâ. Queen Mary C., Mile End Road, London E1 4NS, England. (1951) PhD, chemical crystallography (CNAA, 1982). Hon. res. fellow. (tel. 01 + 980-4811, ext. 3711). *Structure - property (NMR, NQR, basicity) relationship, small molecules, medium-sized molecules, phosphorus-nitrogen compounds, organic nitrogen heterocycles.*

Sheldrick, Dr Bernard. Astbury Dept. of Biophysics, U. of Leeds, Leeds LS2 9JT, England. (1929) PhD, chemical crystallography (U. Leeds, 1964). Sr. lect. (tel. 0532 + 31751, ext. 6104). *Biological molecular structures, computer methods and applications.*

Shepperd, Dr Christine Mary. Analytical Div., British Gas Corp. London Res. Station, Michael Rd., London SW6 2AD, England. (1947) DPhil, structural studies of glasses (U. Oxford, 1980). Sr. res. scient. (tel. 01 + 736-3344, ext. 4054). *Industrial X-ray diffraction, quantitative analysis, amorphous materials.*

Sherwood, Prof. John Neil. Dept. of Pure and Appl. Chemistry, U. of Strathclyde, 295 Cathedral St., Glasgow G1 1XL, Scotland. (1933) DSc, chemistry (U. Durham, 1976). Prof. (tel. 041 + 552-4400, ext. 2288). *Lattice defects in solids, crystal growth (defect influence), chemical and physical properties of solids.*

Sikorski, Dr J. 37 Moor Park Villas, Leeds 6, England.

Silver, Dr Jack. Dept. of Chemistry, U. of Essex, Wivenhoe Park, Colchester, Essex, England. (1948) PhD, inorganic chemistry (U. London, 1973). Lect. *Inorganic chemistry, bioinorganic chemistry, iron.*

Silvester, Mr Philip. Central Res., CIBA Geigy plc, Tenax Rd., Trafford Park, Manchester M17 1WT, England. (1950) MSc, solid state chemistry (U. Bradford, 1975). Sr. res. physicist. (tel. 061 + 872-2323, ext. 3314). *Microstructural evaluation of materials, powder X-ray diffraction.*

Sim, Prof. George Andrew. Dept. of Chemistry, U. of Glasgow, Glasgow G12 8QQ, Scotland. (1929) PhD, chemistry (U. Glasgow, 1955). Prof. (tel. 041 + 339-8855, ext. 419). *Conformation, organic molecules, molecular mechanics.*

Skapski, Dr Andrzej Czeslaw. Chemical Crystallography Lab., Dept. of Chemistry, Imperial C., London SW7 2AY, England. (1938) PhD, chemistry (U. London, 1963). Lect. (tel. 01 + 589-5111, ext. 4609). *Inorganic crystal structures, metal binding sites, nucleic acid components.*

Skarnulis, Dr Anthony Jerome. Dept. of Chemical Crystallography, U. of Oxford, 9 Parks Rd., Oxford OX1 3PD, England. (1948) PhD, chemistry (Arizona State U., 1975). Res. assoc. (tel. 0865 + 53424, ext. 295). *Crystallography, electron microscopy, electron diffraction, X-ray diffraction, computer graphics.*

Skellett, Mr C.A. Appl. Physics Div., GEC Avionics Ltd, Borehamwood, Herts WD2 1RX, England.

Skelly, Miss J.V. 30 Carmelite Rd, Harrow, Middx. HA3 5LR, England.

Skinner, Dr J.M. 18 Highfield Ave, Eaglescliffe, Stockton-on-Tees, Teeside TS16 0DN, England.

Small, Dr Ronald W.H. Dept. of Chemistry, U. of Lancaster, Bailrigg, Lancaster LA1 4YA, England. (1921) DSc, crystallography (U. Birmingham, 1982). Reader. (tel. 0524 + 65201, ext. 4033). *Hydrogen bonding in crystals, intermolecular compounds, phase studies, organomercury compounds.*

Smart, Mr D.W. 64 Essex Drive, Glasgow G14 9LU, Scotland.

Smart, Dr Lesley Elizabeth. Dept. of Chemistry, Open U., Walton Hall, Milton Keynes, MK7 6AA, England. (1947) PhD, chemistry (U. Southampton, 1972). Lect. (tel. 0908 + 653933). *Organometallic structures, halides, fluorides and oxide-fluorides of transition and non-transition elements.*

Smith, Dr Arnold John. Dept. of Chemistry, U. of Sheffield, Sheffield S3 7HF, England. (1931) PhD, inorganic chemistry (U. London, 1957). Lect. (tel. 0742 + 78555, ext. 4476, telex 54348 ULSHEF G). *Inorganic crystal structures, actinide compounds, oxide phases.*

Smith, Dr Bryan Edward. Dept. of Mechanical Engineering, Brunel U., Uxbridge, Middx. UB8 3PH, England. (1936) PhD, physical metallurgy (Brunel U., 1974). Lect. (tel. 0895 + 37188, ext. 417). *Lattice defects, line broadening X-ray techniques, solar collectors, surfaces, coatings, high temperature materials. microstructure - mechanical properties relationships.*

Smith, Mr Gallienus William. Dept. of Chemistry, U. of Surrey, Guildford, Surrey, England. (1924) MSc, crystallography (U. London, 1953). Visiting scientist. (tel. 0483 + 571281). *Structure determination, chromium-vanadium compounds, inorganic compounds, organic compounds, crystallographic teaching.*

Smith, Mr John Michael Andrew. Dept. of Biochemistry, U. of Sheffield, Sheffield S10 2TN, England. (1956) PhD, crystallography (U. Sheffield, 1981). Res. contract staff. (tel. 0742 + 78555, ext. 4242; telex 54348 ULSHEF G). *Single crystal X-ray diffraction studies, biological macromolecules, protein structure - function relationships, inorganic crystal structures, organic crystal structures, powder diffraction, instrumentation, synchrotron radiation.*

Smith, Dr Robert Carr. Director, Kingston Polytechnic, Penrhyn Rd., Kingston upon Thames, Surrey KT1 2EE, England. (1935) PhD, physics (U. London, 1961). Dir. (tel. 01 + 549-1366, ext. 200). *Nonlinear optics, ternary semiconducting compounds.*

Smith, Dr Sidney Herbert. Clarendon Lab., U. of Oxford, Parks Rd., Oxford OX1 3PU, England. (1938) PhD, physics (U. London, 1970). Res. support II. (tel. 0865 + 59291, ext. 275). *Crystal growth - flux.*

Somerville, Mr R.G. 23 Park Crescent, Hornchurch, Essex RM11 1BJ, England.

Sowerby, Mr R.G. Chemistry Dept, Nottingham U., Nottingham NG7 2RD, England.

Spooner, Mr Francis John. Appl. Physics Group, Royal Military C. of Sci., Shrivenham, Swindon, Wilts. SN6 8LA, England. (1932) MSc, physics (U. London, 1964). Sr. res. off. (tel. 0793 + 782551, ext. 2238). *Defect studies, radiation effects in materials, X-ray techniques, ion beam analysis, ion implantation.*

Sprackling, Dr Michael Thomas. Dept. of Physics, King's C. London, Strand, London WC2R 2LS, England. (1934) PhD, physics (U. Bristol, 1959). Sr. lect. (tel. 01 + 836-5454, ext. 2119). *Dislocations, ionic crystals, photography.*

Spraget, Mr H. 18 Mill Farm Close, Dunchurch, Rugby, Warwicks, England.

Spratt, Mr S.B.D. Johnson Matthey Res. Centre, Blounts Court, Sonning Common, Reading, Berks RG4 9NH, England.

Spreadborough, Dr John. John Spreadborough and Co. Ltd., 30 Clarence Rd., Windsor, Berks SL4 5AQ, England. (1933) DPhil, physical metallurgy (U. Oxford, 1958). Managing dir. (tel. 07535 + 61552). *Physical metallurgy, X-ray diffraction, crystallography, crystal structure.*

Spriggs, Dr Paul Humphrey. Dept. of Metallurgy, U. of Manchester, Grosvenor St., Manchester M1 7HS, England. (1931) PhD, metallurgy (U. Manchester, 1967). Fellow. (tel. 061 + 236-3311, ext. 2234). *Binary and ternary compounds, transition metals, transition metal alloy theory.*

Squire, Dr John Michael. Biophysics Sect., Blackett Lab., Imperial C., London SW7 2BZ, England. (1945) PhD, biophysics (U. London, 1969). Reader. (tel. 01 + 589-5111, ext. 6741). *X-ray diffraction, fibrous proteins, muscle, collagen, image analysis, computing.*

Stadler, Dr Hans Peter. Crystallography Lab., U. of Newcastle upon Tyne, Newcastle upon Tyne NE1 7RU, England. (1921) PhD, crystallography (U. Leeds, 1948). Sr. lect. (tel. 0632 + 328511, ext. 3203). *Crystal structure determination, transform methods, radiation protection.*

Stammers, Dr David Kingsley. Biochemistry Dept., Wellcome Res. Labs., Langley Court, Beckenham, Kent BR3 3RS, England. (1949) PhD, crystallography (U. Bristol, 1974). Sr. Scient. (tel. 01 + 658-2211, ext. 263). *Protein structure and function.*

Stanford, Mr Michael John. 92 Peregrine Rd., Sunbury-on-Thames, Middx. TW16 6JP, England. (1949) MSc, crystallography (U. Cape Town, S. Africa, 1979). Continuing student. (tel. 01 + 768-3473). *Symmetry, phase problem, crystallographic computing, molecular biology.*

Steeds, Prof. John Wickham. Dept. of Physics, U. of Bristol, Tyndall Av., Bristol BS8 1TL, England. (1940) PhD, electron microscopy (U. Cambridge, 1965). Prof. (tel. 0272 + 24161). *Anisotropic elasticity, convergent beam electron diffraction, crystal structure determination, low temperature transformations, incommensurate phases.*

Steigmann, Dr Gottfried Albert. Dept. of Physics, U. of Hull, Cottingham Rd., Hull HU6 7RX, England. (1938) PhD, crystallography (U. Hull, 1963). Lect. (tel. 0482 + 46311, ext. 7545 or 7389). *Crystal structures, inorganic compounds, computation.*

Steward, Prof. Edward George. Dept. of Physics, The City U., Northampton Sq., London EC1V 0HB, England. (1923) DSc, crystallography (U. London, 1975). Prof. (tel. 01 + 253-4399, ext. 4403). *Structure - property relationships, molecular medicine.*

Stonebridge, Dr Brian Richard. Dept. of Computer Science, Sch. of Maths., U. of Bristol, University Walk, Bristol BS8 1TW, England. (1940) DPhil, crystallography (U. Oxford, 1968). Lect. (tel. 0272 + 24161, ext. 533). *Mathematical programming, logic programming.*

Stothart, Dr Philip Hamilton. Food Res. Inst., Shinfield, Reading, Berks. RG2 9AT, England. (1946) PhD, physics (U. Reading, 1978). Res. physicist. (tel. 0734 + 883103). *Small-angle X-ray scattering, small-angle neutron scattering, protein structure.*

Strickland, Mr Peter R. International Union of Crystallography, 5 Abbey Square, Chester CH1 2HU, England. (1956) BSc, chemistry (U. Sheffield, 1977). Asst. Tech. Editor. (tel. 0244 + 42878, telex 669755 OFFICE G attn. UNICRYSTAL). *Editing.*

Stryjak, Dr Andrew Jules. UKRDL Gillette UK Ltd., 454 Basingstoke Rd., Reading, Berks RG2 0QE, England. (1950) PhD, glass-ceramics, (U. Warwick, 1977). Res. scient. (tel. 0734 + 85222, ext. 322). *Metal and metallic alloy structures, electron diffraction, X-ray diffraction, glass ceramics.*

Stubbs, Mr Milton. Dept. of Biochemistry, U. of Edinburgh Medical Sch., Hugh Robson Building, George Square, Edinburgh EH8 9XD, Scotland. (1962) BSc, physics (U. Durham, 1983). Res. Student (U. of Oxford). (tel. 031 + 667-1011, ext. 2410). *Small-angle X-ray scattering, small-angle neutron scattering, viruses.*

Sullivan, Dr Richard Arthur. Dept. of Physics, U. of Bath, Claverton Down, Bath BA2 7AY, England. (1936) PhD, crystallography (U. Manchester, 1960). Lect. (tel. 0225 + 61244). *Geophysics, x-ray diffraction.*

Sundaresan, Dr Thiagarajan. Dept. of Science, Maths and Computing, Leigh C., Marshall St., Leigh WN7 4HX, England. (1939) PhD, crystallography (U. Nottingham, 1972). Lect. (tel. 0942 + 608811, ext. 222). *Fibre diffraction, synthetic fibres.*

Sutherland, Dr Hector Howieson. Dept. of Physics, U. of Hull, Cottingham Rd., Hull HU6 7RX, England. (1935) PhD, crystallography and mineralogy (U. St. Andrews, 1962). Sr. lect. (tel. 0482 + 46311, ext. 7389 or 7820). *Inorganic and organic crystal structures, liquid crystals, topography.*

Sutor, Dr Dorothy June. Dept. of Crystallography, Birkbeck C., Malet St., London WC1, England. (1929) PhD, chemistry (Auckland U.C., New Zealand, 1954) PhD, crystallography (U. Cambridge, 1958). Res. fellow., Hon. Sr. lect., Inst. of Urology. (tel. 01 + 580-6622, ext. 453). *Pathological crystals (structure - composition - formation).*

Sutton, Dr A.L. Appl. Physics Dept, The City U., St John St, London EC1V 0HB, England.

Sutton, Dr A.P. 5 Sidney St, Oxford CX4 3AG, England.

Sutton, Dr Brian John. Lab. of Molecular Biophysics, Dept. of Zoology, U. of Oxford, The Rex Richards Building, S. Parks Rd., Oxford OX1 3QU, England. (1954) DPhil, protein crystallography (U. Oxford, 1980). Roy. Soc. Res. fellow. (tel. 0865 + 56733, ext. 413). *Protein crystallography, antibody structure, immunology, enzyme structure and mechanism, antibiotic resistance, glycoprotein and carbohydrate structure.*

Sutton, Mr J.D. Oxford Instruments Ltd, Osney Mead, Oxford OX2 0DX, England.

Swallow, Dr Arnold Graham. Slough C. of Further Education, Wellington St., Slough SL1 1YG, England. (1934) PhD, chemistry (U. Leeds, 1961).

Swindells, Dr David Campbell Neil. Clarendon Lab., U. of Oxford, Parks Rd., Oxford OX1 3PU, England. (1952) PhD, chemistry (U. New Brunswick, Canada, 1981). Res. asst. (tel. 0865 + 59291, ext. 334). *Structure - property relationships, crystal growth, zeolites, synchrotron radiation.*

Tait, Dr John Mervyn. Dept. of Mineral Soils, Macaulay Inst. for Soil Res., Craigiebuckler, Aberdeen AB9 2QJ, Scotland. (1947) PhD, chemistry (U. Aberdeen, 1973). SSO. (tel. 0224 + 38611, ext. 255). *Transmission electron microscopy, electron diffraction, soil minerals.*

Tanner, Dr Brian Keith. Dept. of Physics, U. of Durham, South Rd., Durham DH1 3LE, England. (1947) DPhil, metallurgy (U. Oxford, 1972). Sr. lect. (tel. 0385 + 64971, ext. 372). *X-ray diffractometry (precision), instrumentation, magnetic materials, semiconductor materials.*

Tanner, Dr H.M. 5 Carling Gate, Timperley, Altrincham, Cheshire WA15 7SL, England.

Tarling, Mr S.E. Dept. of Crystallography, Birkbeck C., Malet St, London WC1E 7HX, England.

Tasker, Mr Michael Peter. XM2 Div., RARDE Fort Halstead, Sevenoaks, Kent TN14 7BP, England. (1939) MSc, crystallographic and spectroscopic techniques (CNAA, 1976). HSO. (tel. 0959 + 32222, ext. 3381). *Materials science.*

Tasker, Dr Peter Anthony. Sch. of Chemistry, The Polytechnic of North London, Holloway, London N7 8DB, England. (1944) DPhil, chemistry (U. York, 1968). Sr. lect. (tel. 01 + 607-2789, ext. 2153). *Coordination chemistry, metal ions (unusual coordination numbers and geometries), biological inorganic chemistry, catalysis, selective complexation, metals extraction.*

Tate, Dr Cecil. Physics Dept., U. of York, Heslington, York YO1 5DD, England. (1934) PhD, theoretical nuclear physics (U. London, 1960). Res. fellow. (tel. 0904 + 59861, ext. 5555). *Mathematical and computational methods, structure determination.*

Taylor, Prof. Charles Alfred. 9 Hill Deverill, Warminster, Wilts BA12 7EF, England. (1922) DSc, physics (U. Manchester, 1959). Em. prof. (tel. 0985 + 40574). *Teaching crystallography, optical analogues.*

Taylor, Dr Derek. Doulton Industrial Products, Filleybrooks, Stone, Staffs ST15 0PU, England. (1939) PhD, mineralogy (U. Manchester, 1966). Tech. Dir. (tel. 0785 + 813241, telex 36277). *Thermal expansion, framework structures, ceramic science.*

Taylor, Dr Garry Lindsay. Lab. of Molecular Biophysics, U. of Oxford, S. Parks Rd., Oxford OX1 3QU, England. (1954) PhD, biophysics (U. London, 1978). Res. asst. (tel. 0865 + 56733, ext. 424). *Protein crystallography.*

Taylor, Prof. Harry Francis West. Dept. of Chemistry, U. of Aberdeen, Old Aberdeen AB9 2UE, Scotland. (1923) DSc, chemistry (U. London, 1957). Prof. (tel. 0224 + 40241, ext. 481). *Silicates, chemistry of cements, mineralogical chemistry.*

Taylor, Dr R. Sartoria, Bruisyard Rd, Rendham, Saxmundham, Suffolk IP17 2AH, England.

Tempest, Dr Paul Anthony. Res. Div. CEGB, Berkeley Nuclear Labs., Berkeley, Gloucestershire GL11 5PH, England. (1947) PhD, science of materials (U. London, 1974). Res. off. (tel. 0453 + 810451, ext. 146). *X-ray absorption, preferred orientation, stainless steel oxidation products, hyperstoichiometric uranium dioxide (structure), spinel structures.*

Thatcher, Mr J.S. GTP Engineering Ltd, Station Ind. Est., Sheppard St, Swindon, Wilts, England.

Theocharis, Dr C.R. Chemistry Dept, Brunal U., Uxbridge, Middx. UB8 3PH, England.

Thomas, Prof John Meurig, FRS. Dept. of Physical Chemistry, U. of Cambridge, Lensfield Rd., Cambridge CB2 1EP, England. (1932) DSc, chemistry (U. C. Swansea, 1954), Hon LlD. Head. (tel. 0223 + 66499, ext. 381). *Solid-state chemistry, catalysis (heterogeneous), imperfections in solids, surface characterization techniques, solid state NMR.*

Thomas, Mr Kenneth. Div. of Materials Applications, Nat. Physical Lab., Teddington, Middx. TW11 0LW, England. (1930) BSc, physics (U. Wales, 1952). PSO. (tel. 01 + 977-3222, ext. 3841). *Structure - properties, processing, materials, plastics.*

Thomas, Mr M.A. Cookson Group plc, 7 Wadsworth Rd, Perivale, Greenford, Middx. CB6 7JQ, England.

Thomas, Dr W.A. 28 Milton Ave, Eaton Ford, St Neots, Cambridgeshire, England.

Thompson, Dr Derek Parr. Crystallography Lab., U. of Newcastle upon Tyne, Newcastle upon Tyne NE1 7RU, England. (1945) PhD, mineralogy, crystallography (U. Cambridge, 1972). Lect. (tel. 0632 + 28511, ext. 3202). *Crystal chemistry, inorganic compounds, high temperature materials (containing Si - Al - O - N), powder X-ray techniques, phase investigations.*

Thornley, Dr Frank Richard. Dept. of Appl. Physics, U. of Strathclyde, John Anderson Bldg., 107 Rottenrow, Glasgow G4 0NG, Scotland. (1945) PhD, physics and crystallography (U. Cambridge, 1972). Lect. (tel. 041 + 552-4400, ext. 3359). *Extinction, EXAFS, glasses.*

Thornton-Pett, Dr M.A. 4 Westfield Court, Westfield Rd, Leeds, England.

Tickle, Dr Ian James. Dept. of Crystallography, Birkbeck C., Malet St., London WC1E 7HX, England. (1947) DPhil, crystallography (U. Oxford, 1972). Res. assoc. (tel. 01 + 580-6622). *Crystallography, peptides, proteins.*

Tobin, Mr A.N. 54 Cedars Rd, St Leonards, Exeter, Devon, England.

Tofield, Dr Bruce C. Materials Dev. Div., AERE Harwell, Didcot, Oxon OX11 0RA, England. (1943) DPhil, chemistry (U. Oxford, 1965). Group Leader - Materials and surface chemistry group. (tel. 0235 + 24141, ext. 4453; telex 83135). *Solid state chemistry, gas detectors, lithium batteries, oxidation, surface analysis.*

Tollin, Dr Patrick. Carnegie Lab. of Physics, U. of Dundee, Dundee DD1 4HN, Scotland. (1938) PhD, crystallography (U. Cambridge, 1963). Reader. (tel. 0382 + 23181, ext. 4561). *Structure determination methods, virus structures, biological molecule structures.*

Tomkeieff, Mr Michael Vamime. 3 Osgathorpe Drive, Pitsmoor, Sheffield, S. Yorks S4 7AP, England. (1934) MA, mineralogy and crystallography (U. Cambridge, 1962). (tel. 0742 + 388560). *Crystallography, metallurgy, mineralogy, petrology, statistics, electron beam instruments.*

Townsend, Dr Stephen Phillip. Dept. of Computing Science, U. of Aberdeen, Dunbar St., Old Aberdeen AB9 2TY, Scotland. (1948) DPhil, numerical analysis (U. Oxford, 1977). Lect. (tel. 0224 + 40241, ext. 6417). *Numerical analysis, theoretical crystallography.*

Toy, Dr Mark. Dept. of Physics and Mathematics, Manchester Polytechnic, Chester St., Manchester M1 5GD, England. (1932) PhD, physics (UMIST, 1981). Sr. lect. (tel. 061 + 228-6171, ext. 2321). *Textural changes in polymers with stress.*

Treharne, Miss Annemarie Cerise. Dept. of Crystallography, Birkbeck C., U. of London, Malet St., London WC1E 7HX, England. (1960) BSc, biophysics (U. Leeds, 1983). Res. Student. (tel. 01 + 580-6622, ext. 455). *Peptide hormone analogs, molecular mechanics, refinement.*

Truter, Prof. Mary Rosaleen. Dept. of Chemistry, U. C. London, 20 Gordon St., London WC1 0AJ, England. (1925) DSc, chemistry (U. London, 1965). Visiting prof. (tel. 01 + 387-7050, ext. 450). *Molecular structure, coordination chemistry.*

Tun, Mr Zin. Dept. of Physics, U. of Edinburgh, The King's Buildings, Mayfield Rd., Edinburgh EH9 3JZ, Scotland. (1957) BSc, physics (Arts and Science U., Rangoon, Burma, 1978). Res. Assoc. (tel. 031 + 667-1081, ext. 2743). *Incommensurate materials, disordered materials, order-disorder transitions, high-quality structure determination, anharmonic effects.*

Varma, Dr Satya Prakash. Res. Div., Blue Circle Ind. plc., 305 London Rd., Greenhithe, Kent DA9 9JQ, England. (1936) PhD, physics (Banaras Hindu U., India, 1963). Res. scient. i/c X-ray lab. (tel. 0322 + 842244, ext. 309). *Cement and related products.*

Vickers, Miss Mary Elizabeth. X-ray Diffraction Group, BP International, Chertsey Rd., Sunbury-on-Thames, Middx. TW16 7LN, England. (1953) BSc, chemical physics (U. Bristol, 1975). Physicist. (tel. 09327 + 62079). *Polymers, orientation studies, catalysts.*

Vincent, Dr Roger. H. H. Wills Physics Lab.,U. of Bristol, Tyndall Ave., Bristol BS8 1TL, England. (1944) PhD, physics (U. Cambridge, 1969). Res. assoc. (tel. 0272 + 24161, ext, 8504). *Crystal physics, electron diffraction, structure images.*

Wade, Prof. K. Chemistry Dept, Durham U., South Rd, Durham City DH1 3LE, England.

Walker, Dr Peter Jonathan. Clarendon Lab., U. of Oxford, Parks Rd., Oxford OX1 3PU, England. (1947) PhD, chemistry (U. Liverpool, 1971). Res. chemist. (tel. 0865 + 59291, ext. 330). *Purification, preparation, halide single crystals, Bridgman-Stockbarger technique, Czochralski technique, oxygen-17 enriched materials.*

Wallwork, Dr Stephen Collier. Dept. of Chemistry, U. of Nottingham, University Park, Nottingham NG7 2RD, England. (1925) DPhil, chemical crystallography (U. Oxford, 1950). Assoc. Reader. (tel. 0602 + 506101, ext. 2348). *Organic complexes, radical ion salts and complexes, anhydrous metal nitrates, organonitogenmetal complexes.*

Walton, Miss A.R. Dept of Crystallography, Birkbeck C., Malet St, London WC1E 7HX, England.

Ward, Mr Roger Charles Chavannes. Marconi Infrared Devices Lab., GEC Hirst Res. Centre, East Lane, Wembley, Middx. HA9 7PP, England. (1948) DPhil, physics (U. Oxford, 1981). Res. assoc. (tel. 01 + 904-1262, ext. 499). *X-ray diffraction techniques, thin film growth and analysis, IR materials, crystal growth.*

Waring, Dr J.R.S. 3 Fop St, Uley, Nr Dursley, Glos, England.

Watkin, Dr David John. Chemical Crystallography Lab., U. of Oxford, 9 Parks Rd., Oxford OX1 3PD, England. (1942) PhD, crystallography (U. Birmingham, 1967). Res. asst. (tel. 0865 + 53424, ext. 264). *Chemical crystallography.*

Watson, Dr David Gilfillan. University Chemical Lab., U. of Cambridge, Lensfield Rd., Cambridge CB2 1EW, England. (1934) PhD, chemical crystallography (U. Glasgow, 1960). Asst. dir. of res. (tel. 0223 + 66499, ext. 317). *Organic crystal structures, crystallographic data storage and retrieval.*

Watson, Dr Herman Charles. Dept. of Biochemistry, U. of Bristol, University Walk, Bristol BS8 1TD, England. (1933) DSc, physics (U. Manchester, 1972). Reader. (tel. 0272 + 24161, ext. 305). *Protein crystallography.*

Weakley, Dr Timothy John Ruffer. Dept. of Chemistry, U. of Dundee, Dundee DD1 4HN, Scotland. (1933) DPhil, chemistry (U. Oxford, 1959). Lect. (tel. 0382 + 23181, ext. 284). *Polyoxoanions of V and Mo sub-groups, inorganic crystal chemistry.*

Webster, Dr Michael. Dept. of Chemistry, U. of Southampton, Southampton SO9 5NH, England. (1938) PhD, chemistry (U. London, 1962). Lect. (tel. 0703 + 559122). *Chemical crystallography.*

Welch, Dr A.J. Chemistry Dept, Edinburgh U, Edinburgh EH9 3JJ, Scotland.

Welch, Mrs D.A. Chemistry Dept, Edinburgh U, Edinburgh EH9 3JJ, Scotland.

West, Dr Anthony Roy. Dept. of Chemistry, U. of Aberdeen, Old Aberdeen AB9 2UE, Scotland. (1947) DSc, solid state chemistry (U. Aberdeen, 1984). Sr. lect. (tel. 0224 + 40241, ext. 6176). *Silicates, high temperature oxides, phase equilibria, crystal chemistry, phase transitions, solid electrolytes.*

West, Mr M.P. Nicolet Instruments Ltd, Budbrooke Rd, Warwick CV34 5XH, England.

West, Dr N.G. Health and Safety Executive, 403 Edgware Rd, London NW2 6LN, England.

Wheatley, Dr Peter Jaffrey. Dept. of Physical Chemistry, U. of Cambridge, Lensfield Rd., Cambridge CB2 1EP, England. (1921) DPhil, physical chemistry (U. Oxford, 1950). Lect., fellow of Queens' C. (tel. 0223 + 66499, ext. 428). *Chemical crystallography.*

Whelan, Dr Michael John. Dept. of Metallurgy and Science of Materials, U. of Oxford, Parks Rd., Oxford OX1 3PH, England. (1931) PhD, physics (U. Cambridge, 1958). Reader. (tel. 0865 + 59981, ext. 204). *Electron microscopy, electron diffraction, energy loss spectroscopy, Auger electron spectroscopy, metal physics.*

Whiston, Dr Clive David. Sch. of Appl. Sci., The Polytechnic, Wulfruna St., Wolverhampton WV1 1LY, England. (1937) PhD, crystallography (U. Sheffield, 1963). Sr. lect. in inorganic chemistry. (tel. 0902 + 27371, ext. 129). *Crystal structures, anti-cancer drugs.*

Whitaker, Dr Alan. Dept. of Physics, Brunel U., Kingston Lane, Uxbridge, Middx. UB8 3PH, England. (1932) PhD, crystallography (Birkbeck C., London, 1965). Lect. (tel. 0985 (Uxbridge) + 37188, ext. 406). *Organic crystal structures, pigments.*

White, Dr David Nathaniel James. Dept. of Chemistry, U. of Glasgow, Glasgow G12 8QQ, Scotland. (1946) DPhil, chemical crystallography (U. Sussex, 1970). Reader. (tel. 041 + 339-8855, ext. 7168). *Molecular conformation and mechanics, computer graphics, polypeptides, proteins, drug design.*

White, Dr Janice Larraine. Dept. of Biochemistry, U. of Sheffield, Western Bank, Sheffield, S. Yorks S10 2TN, England. (1948) PhD, biology - protein crystallography (Purdue U.,USA, 1976). Post-doc. res. asst. (tel. 0742 + 78555, ext. 4241, telex 54348 ULSHEF G). *Protein structure, macromolecular structure and function.*

Whitney, Mr R.A. 51 Lansdowne Rd, Stanmore, Middx. MA7 2RZ, England.

Whittaker, Dr Eric James William. Dept. of Earth Sci., U. of Oxford, Parks Rd., Oxford OX1 3PR, England. (1921) PhD, crystallography (U. London, 1956). Retired. (tel. 0865 + 54511). *Silicate structures, disordered structures, four-dimensional crystallography.*

Whitworth, Dr Robert William. Dept. of Physics, U. of Birmingham, Birmingham B15 2TT, England. (1932) PhD, physics (U. Cambridge, 1958). Sr. lect. (tel. 021 + 472-1301). *Dislocations, point defects, ionic crystals, ice.*

Wild, Mr G.A. 9 Kirkham St, Plumstead, London SE18, England.

Wilford, Dr John Bernard. Dept. of Computer Science, Teesside Polytechnic, Borough Rd., Middlesborough TS1 3BA, England. (1940) PhD, chemistry (U. Bristol, 1966). Sr. lect. (tel. 0642 + 218121). *Crystal structures, industrially interesting compounds, computer programming (scientific and educational).*

Wilkinson, Dr Clive. Dept. of Physics, King's C., Strand, London WC2R 2LS, England. (1941) PhD, crystallography (U. Cambridge, 1966). Lect. (tel. 01 + 836-5454, ext. 2586). *Neutron diffraction, X-ray diffraction, magnetic materials.*

Wilkinson, Dr D. Quantum Tech. Comunications Lab., 447 Chester Rd., Manchester M16 1EW, England.

Williams, Dr R.A. 14 St Cybi Ave, Llangybi, Usk, Gwent NP5 1TT, Wales.

Willis, Dr Bertram Terence Martin. Chemical Crystallography Lab., U. of Oxford, 9 Parks Rd., Oxford OX1 3PD, England. (1927) DSc, physics (U. London, 1968). Sr. res. fellow. (tel. 0865 + 53424). *Powder diffraction, neutron diffraction.*

Wilson, Prof. Arthur James Cochran. Crystallographic Data Centre, University Chemical Lab., Cambridge CB2 1EW, England. (1914) PhD, physics (MIT, 1938; U. of Cambridge, 1942). Em. fellow. *Crystallographic statistics, International Tables for Crystallography, data, information.*

Wilson, Prof. Herbert Rees. Physics Dept., U. of Stirling, Stirling FK9 4LA, Scotland. (1929) PhD, physics (U. Wales, 1952). Prof. and Head of Dept. (tel. 0786 + 73171, ext. 2008). *Biomolecules, nucleic acids, virus structure.*

Wilson, Mr J.D. 49 Murray Rd, High Howdon, Wallsend on Tyne, Tyne and Wear NE28 0LY, England.

Wilson, Dr Keith Sanderson. Dept. of Physics, U. of York, Heslington, York YO1 5DD, England. (1949) DPhil, Chemistry (U. Oxford, 1971). Lect. (tel. 0904 + 59861, ext. 5507). *Protein crystallography, macromolecular structure, biophysics.*

Wilson, Dr Michael Jeffrey. Dept. of Mineral Soils, Macaulay Inst. for Soil Res., Craigiebuckler, Aberdeen AB9 2QJ, Scotland. (1937) DSc, geology and soil science (U. Wales, 1984). Head of Dept. (tel. 0224 + 38611, ext. 241). *Mineralogy, rocks and soils, clay mineralogy.*

Wilson, Dr S.J. 9 St Peters St, Duxford, Cambridge, England.

Windle, Dr Alan Hardwick. Dept. of Metallurgy and Materials Science, U. of Cambridge, Pembroke St., Cambridge CB2 3QZ, England. (1942) PhD, metallurgy (U. Cambridge, 1966). Lect. (tel. 0223 + 65151, ext. 333). *Polymer physics, structured liquids, metal-polymer adhesion, composites.*

Windsor, Dr Colin George. Materials Physics and Metallurgy Div., B418, AERE Harwell, Oxon OX11 0RA, England. (1938) DPhil, physics (U. of Oxford. 1960). Group leader, neutron physics. (tel. 0235 + 24141, ext. 4025; telex 83135). *Neutron scattering.*

Wonacott, Dr Alan John. Biophysics Sect., Blackett Lab., Imperial C., Prince Consort Rd., London SW7 2BZ, England. (1941) PhD, biophysics (U. London, 1966). Res. fellow. (tel. 01 + 589-5111, ext. 6725) *Protein crystallography.*

Wood, Mr Dermott. X-ray Lab. Analytical Branch, British Petroleum Co. Ltd., Chertsey Rd., Sunbury-on-Thames, Middx. TW16 7LN, England. (1940) BSc, general (National U. of Ireland, 1961). Technologist. (tel. 093 27 + 85533, ext. 8033). *Catalysts, poorly crystalline materials, powder diffraction, line profile analysis.*

Wood, Dr Ian George. Soils and Plant Nutrition Dept., Rothamsted Experimental Station, Harpenden, Herts AL5 2JQ, England. (1952) PhD, crystallography (U. London, 1977). HSO. (tel. 05827 + 63133, ext. 311). *Mineralogy, phase transitions, disorder, powder diffraction.*

Wood, Dr Raymond Maurice. Dept. of Appl. Physics, Sheffield City Polytechnic, Pond St., Sheffield S1 1WB, England. (1927) PhD, metallurgy (U. Sheffield, 1973). Sr. lect. (tel. 0742 + 20911, ext. 228). *Diffusionless transformations, structures, liquids, liquid crystals.*

Woods, Dr Geoffrey Steward. CSO Valuations Ltd, 17 Charterhouse St., London EC1N 6RA, England. (1939) PhD, physics (U. Witwatersrand, South Africa, 1971). Res. physicist. (tel. 01 + 404-4444, ext. 3181). *Diamonds, electron microscopy, infrared spectroscopy, optical spectroscopy, defects in solids, diffraction methods, radiation damage.*

Woolfson, Prof. Michael Mark. Dept. of Physics, U. of York, Heslington, York YO1 5DD, England. (1927) DSc, physics (U. Manchester, 1961). Prof. (tel. 0904 + 59861, ext. 5550). *Direct methods, small biological molecules.*

Wooster, Mr Antony Martin. Christie and Wooster Res. Ltd., 91 North St., Burwell, Cambridge CB5 0BB, England. (1935) Dir. (tel. 0638 + 741315). *Globular proteins, structure, automatic diffractometers.*

Wright, Mr C.P. 67 Gorsy Bank Rd, Hockley, Tamworth, Staffs B77 5HU, England.

Wright, Dr Helen. Dept. of Computing Science, U. of York, Heslington, York YO1 5DD, England. (1957) DPhil, crystallography (U. York, 1983). Appl. programmer. (tel. 0904 + 59861, ext. 493). *Direct methods, computing methods.*

Wright, Dr John Albert. Dept. of Civil Engineering and Construction, U. of Aston in Birmingham, Gosta Green, Birmingham B4 7ET, England. (1936) PhD, metallurgy (U. Sheffield, 1961). Lect. (tel. 021 + 359-3611, ext. 5195). *Environmental cracking, hydrogen embrittlement, safe-life predictions of structures, damage-tolerant design.*

Wright, Dr John Dalton. University Chemical Lab., U. of Kent, Canterbury, Kent CT2 7NH, England. (1941) DPhil, chemistry (U. Oxford, 1965). Lect. (tel. 0227 + 66822, ext. 519). *Crystal structure, electrical properties and spectra, molecular complexes, molecular crystals.*

Wylie, Mr Thomas Smith. 59 South Beach, Troon, Ayrshire KA10 6EG, Scotland. (1916) BSc, naval architecture (U. Strathclyde, 1946). Retired lect. (tel. 0292 + 312133). *Crystal orientation, biological tissue.*

Young, Mr Brian Raymond. British Geological Survey, Nicker Hill, Keyworth, Nottingham NG12 5GG, England. (1927) MSc, crystallography (U. London, 1959). Head of X-ray unit. (tel. 06077 + 6111). *Mineralogy, clay mineralogy.*

Zussman, Prof. Jack. Dept. of Geology, U. of Manchester, Manchester M13 9PL, England. (1924) PhD, crystallography (U. Cambridge, 1952). Prof. (tel. 061 + 273-7121, ext. 5585). *Mineralogy.*

UNITED STATES OF AMERICA

Sub-Editor: **R.C. Taylor**
Asst. Sub-Editor: **M.P. Wintermeyer**

Notes

1. International telephone country code - 101. In the ten-digit telephone numbers given below, the first three digits are the regional area code; local calls within a regional area require only the last seven digits. Extension (ext.) numbers are used within the institution or company.

2. In the references to universities at which degrees were conferred, the following special abbreviations are used:

CIT - California Institute of Technology
MIT - Massachusetts Institute of Technology
PINY/PIB - Polytechnic Institute of New York/Brooklyn

PSU - Pennsylvania State University
UCLA - University of California at Los Angeles
SUNY - State University of New York

3. In the addresses, the following two-letter abbreviations are used for states and territories:

AL Alabama	IA Iowa	MT Montana	RI Rhode Island
AK Alaska	ID Idaho	NC North Carolina	SC South Carolina
AR Arkansas	IL Illinois	ND North Dakota	SD South Dakota
AZ Arizona	IN Indiana	NE Nebraska	TN Tennessee
CA California	KS Kansas	NH New Hampshire	TX Texas
CO Colorado	KY Kentucky	NJ New Jersey	UT Utah
CT Connecticut	LA Louisiana	NM New Mexico	VA Virginia
CZ Canal Zone	MA Massachusetts	NV Nevada	VI Virgin Islands
DC District of Columbia	MD Maryland	NY New York	VT Vermont
DE Delaware	ME Maine	OH Ohio	WA Washington
FL Florida	MI Michigan	OK Oklahoma	WI Wisconsin
GA Georgia	MN Minnesota	OR Oregon	WV West Virginia
GU Guam	MO Missouri	PA Pennsylvania	WY Wyoming
HI Hawaii	MS Mississippi	PR Puerto Rico	

4. The three degrees nearly always awarded by U.S.A. colleges and universities to graduates in scientific subjects are BS (four year program), MS (additional one or two years of courses which may include research), and PhD (three to five years beyond the BS, including research and dissertation; in some universities the MS is an intermediate requirement). Occasionally the BA degree is awarded to bachelor's graduates in science. The DSc or ScD is generally an honorary degree not indicating professional training in science.

Abad-Zapatero, Dr Celerino. Dept. of Biological Sci., Purdue U., Lilly Hall of Life Sci., West Lafayette, IN 47907, USA. (1947) PhD, biological sciences (U. Texas, Austin, 1978). Asst. res. scient. (tel. 317 + 494-4910). *Viral structure and assembly, proteins, structure and function, energy transducing proteins.*

Abboud, Mr Khalil A. Louisiana State U., P. O. Box 23961, Baton Rouge, LA 70893, USA.

Abdel-Meguid, Sherin S. 4 Falls River Circle, Ivoryton, CT 06442, USA.

Abel, Mr James E. 265 W. Shore Trail, Sparta, NJ 07871, USA. (1915) MS, chemistry (Stevens Inst. of Techn., 1954). Retired.

Abola, Dr Enrique E. Chem. Dept., Brookhaven Nat. Lab., Bldg. 555, Upton, NY 11973, USA. (1947) PhD, crystallography (U. Pittsburgh, 1973). Asst. chemist. (tel. 516 + 345-4382). *Macromolecular structure, crystallographic computing.*

Abola, Dr Jaime Esteva. Crystallography Dept., U. of Pittsburgh, Fifth Ave., Pittsburgh, PA 15260, USA. (1947) PhD, crystallography (U. Pittsburgh, 1973). Res. asst. prof. (tel. 412 + 624-4366). *Protein crystallography, transferrin structure.*

Abraham, Prof. Donald James. Medicinal Chemistry Dept., U. of Pittsburgh, 725 Salk Hall, Pittsburgh, PA 15261, USA. (1936) PhD, organic chemistry (Purdue U., 1963). Prof., chairman. (tel. 412 + 624-3261). *Medicinal chemistry, sickle cell anemia, Alzheimer's desease, drug design, drug protein interactions, neurochemistry, X-ray crystallography.*

Abrahams, Dr Sidney Cyril. AT&T Bell Labs., Murray Hill, NJ 07974, USA. (1924) DSc, crystallography (U. Glasgow, UK, 1957). Distinguished techn. staff member. (tel. 201 + 582-4730, telex 219348 BELL UR). *Physics of condensed matter, atomic arrangement and displacement, crystallographic accuracy.*

Achari, Dr Aniruddha. Dept. of Biochemistry, U. of Chicago, 920 East 58th St., Chicago, IL 60637, USA.

Adams, Prof. Richard Darwin. Chemistry Dept., U. of South Carolina, Columbia, SC 29208, USA. (1947) PhD, inorganic chemistry (MIT, 1973). Prof. (tel. 803 + 777-5104). *Inorganic chemistry; clusters and catalysts.*

Adams, Dr Walter Wade. Polymer Branch, Air Force Materials Lab., AFWAL/MLBP, AF Wright Aeronautical Labs., Wright-Patterson Air Force Base, OH 45433, USA. (1946) PhD, polymer science and engineering (U. Massachusetts, 1984). Material res. eng. (tel. 513 + 255-2340). *Polymer morphology, structure determination.*

Adler, Mr George. Shoreham, NY 11786, USA. (1920) MA, physical chemistry (Brooklyn C., 1952). Retired. *Organic solid state chemistry, polymers.*

Adman, Dr Elinor Thomson. Dept. of Biological Structure, S. of Medicine, U. of Washington - SM20, Seattle, WA 98115, USA. (1941) PhD, physical chemistry (Brandeis U., 1967). Res. assoc. prof. (tel. 206 + 543-6589). *Macromolecular structures, electron transfer proteins.*

Agard, Dr David Andrew. Dept. of Biochemistry, U. of California at San Francisco, Parnassus Ave., San Francisco, CA 94143, USA. (1953) PhD, biological chemistry (CIT, 1980). Asst. prof. (tel. 415 + 666-2521). *X-ray crystallography, 3-D image reconstruction, chromosome structure, protein structure.*

Agron, Mr Paul A. 102 Wilderness Ln., Oak Ridge, TN 37830, USA.

Akers, Dr Charles Kenton. 73 Oakgrove Dr., Williamsville, NY 14221, (Environmental Sci. Dept., Calspan Corp., P. O. Box 400, Buffalo, NY 14221), USA. (1942) PhD, biophysics (SUNY Buffalo, 1972). Sr. chemist. (tel. 716 + 632-7500, ext. 769). *Biomedical science, surface science, small angle X-ray scattering.*

Alber, Dr Tom. Inst. of Molecular Biology, U. of Oregon, Eugene, OR 97403, USA. (1954) PhD, (MIT, 1981). Res. assoc. (tel. 503 + 686-5176). *Protein stability and folding, enzyme activity, protein crystallography.*

Albert, Mr Charles W. Glidden Co., 3901 Hawkins Point Rd., Baltimore, MD 21226, USA.

Alden, Dr Richard Allen. Dept. of Chemistry, B-017, U. of California, San Diego, La Jolla, CA 92093, USA. (1935) PhD, chemistry (U. Washington, 1962). Specialist. (tel. 714 + 452-4229). *Structure and biological function, enzymes, computational methods.*

Alessandrini, Miss Eileen I. Res. Div., IBM T. J. Watson Res. Center, P. O. Box 218, Yorktown Hgts., NY 10598, USA. (1921) AB,BS, physics (Barnard C., Columbia U., 1943). Res. staff member. (tel. 914 + 945-1206). *Thin film physics, electron diffraction and microscopy, film structures.*

Alexander, Prof. Em. Leroy Elbert. 68401 Hill St., Route 2, Sturgis, MI 49091, USA. (1910) PhD, physical chemistry (U. Minnesota, 1943). Retired. (tel. 616 + 651-2850). *Organic and polymer structures, X-ray diffraction.*

Alkire, Randy W. MS H805 Los Alamos Nat. Lab., Los Alamos, NM 87545, USA.

Allcock, Prof. Harry Rex. Chemistry Dept., Pennsylvania State U., University Park, PA 16802, USA. (1932) PhD, chemistry (U. London, UK, 1956). Prof. (tel. 814 + 865-3527). *Inorganic chemistry, polymer chemistry.*

Allen, Joseph H. 3629 Swallow Lane, Irving, TX 75062, USA.

Allersma, Mr Ties. Glass Res. Center, PPG Industries, Box 11472, Pittsburgh, PA 15238, USA. (1936) Ir, physics (Techn. U. of Delft, Netherlands, 1965). Res.

assoc. physicist. (tel. 412 + 665-8500). *Crystallization, glass-ceramics, optics as related to glass.*

Amin, Dr Ahmed A. Hussein. Advanced Dev. Lab., Texas Instruments Inc., 34 Forest, Attleboro, MA 02703, USA. (1945) PhD, solid state science (PSU, 1979). Member tech. staff. (tel. 617 + 699-1094). *Ferroelectrics, structure - properties relationship, grain boundary phenomena, device materials applications, instrumentation, automatic data acquisition.*

Amma, Prof. Elmer Louis. Dept. of Chemistry, U. of South Carolina, Columbia, SC 29208, USA. (1929) PhD, physical chemistry (Case Inst. of Tech., 1952). Prof. (tel. 803 + 777-2542). *Inorganic structural chemistry, solid state metal NMR, protein crystallography.*

Ammon, Prof. Herman L. Chemistry Dept., U. of Maryland, College Park, MD 20742, USA. (1936) PhD, chemistry (U. Washington, 1963). Prof. (tel. 301 + 454-2634). *Small molecules, protein crystallography.*

Amrein, Mr Robert Eugene. Analytical Services, Physical Properties Lab., Fine Particle Res., Cabot Corp., Billerica Techn. Center, Billerica, MA 01821, USA. (1931) BS, liberal arts, chemistry and biology (Kent State U., 1955). Lead scient. (tel. 617 + 272-3500, ext. 255). *Electron microscopy, optical microscopy, spectroscopy, crystallography, fine particle materials, carbon black, silicas.*

Amy, Dr Joseph André. Jet Propulsion Lab., CIT, 4800 Oak Grove Dr., Pasadena, CA 91107, USA. (1932) PhD, physical chemistry (U. Michigan, 1963). Member tech. staff. (tel. 818 + 354-3759). *Physical metallurgy.*

Amzel, Dr Leon Mario. Biophysics Dept., Johns Hopkins S. of Medicine, 725 N. Wolfe St., Baltimore, MD 21205, USA. (1942) Doctor, physical chemistry (U. Buenos Aires, Argentina, 1968). Asst. prof. (tel. 301 + 955-3955). *Proteins, pharmacological compounds.*

Anantha Narayanan, Prof. V. Dept. of Physics, Savannah State C., P.O. Box 20473, Savannah, GA 31404, USA. (1936) PhD, physics (Indian Inst. of Sci., 1962). Prof. (tel. 912 + 356-2317). *Vibrational spectra of crystals, hydrogen bond vibrations in crystals, crystal field calculations, structure parameters from evaluation of spectra of molecules and crystals.*

Anderegg, Prof. John William. Biophysics Lab., U. of Wisconsin, 1525 Linden Dr., Madison, WI 53706, USA. (1923) PhD, physics (U. Wisconsin, 1952). Prof. of physics and biophysics. (tel. 608 + 262-4536). *Biological macromolecules, small angle X-ray scattering.*

Anderson, Ms Christine Alexis Francis. Dept. of Chemistry, Hofstra U., Hempstead, NY 11550, USA. (1955) MS, mineralogy (PSU, 1980). Manager. Chem. Labs. (tel. 516 + 560-5541). *Uranium mineralogy, crystal growth, hydrothermal alterations.*

Anderson, Dr Gary Don. Chemistry Dept. Marshall U., Huntington, WV 25701, USA. (1943) PhD, organic chemistry (Florida State U., 1972). Chairman, assoc. prof. *Organic synthesis, natural products chemistry, computers in chemistry, X-ray crystallography.*

Anderson, Prof. Oren Paul. Dept. of Chemistry, Colorado State U., Fort Collins, CO 80523, USA. (1942) PhD, chemistry (Northwestern U., 1968). Assoc. prof. (tel. 303 + 491-6339). *Coordination chemistry, polydentate chelates, bio-inorganic chemistry, mixed-valence compounds.*

Anex, Prof. Basil G. Dept. of Chemistry, U. of New Orleans, Lakefront, New Orleans, LA 70122, USA. (1931) PhD, chemistry (U. Washington, 1959). Prof. (tel. 504 + 286-6848). *Electronic spectroscopy of single crystals.*

Angilello, Mr Joseph. Res. Div., IBM T. J. Watson Res. Center, P.O. Box 218, Yorktown Hgts., NY 10598, USA. (1929) Sr. res. eng. (tel. 914 + 945-1509). *X-ray diffraction, topography, instrumentation.*

Ansell, Dr Gerald Brian. Analytical Div., Exxon Res. and Eng. Co., Route 22, Clinton Township, NJ 08801, USA. (1936) PhD, crystallography (U. Essex, UK, 1966). Res. assoc. (tel. 201 + 730-2106). *Single crystal, powder diffraction, catalysts, inorganics, organometallics, organic small molecules.*

Antal, Dr John Joseph. Materials Characterization Div., Army Materials & Mechanics Res. Cntr., Arsenal St. AMXMR-OM, Watertown, MA 02172, USA. (1926) PhD, physics (Saint Louis U., 1952). Sup. res. physicist. (tel. 617 + 923-5454). *Neutron scattering, material characterization by neutrons, instrumentation.*

Anthony, Dr John W. Dept. Geosciences, U. of Arizona, Tucson, AZ 85721, USA. (1920) PhD, geology (Harvard U., 1965). Prof. (tel. 602 + 626-2973). *Mineralogy, crystal structures, epitaxy, oxidation zone minerals.*

Appleman, Dr Daniel E. Dept. of Mineral Sci., Smithsonian Institution, NHB 119, Room E-408A, Washington, DC 20560, USA. (1931) PhD, geology, crystallography (Johns Hopkins U., 1956). Crystallographer. (tel. 202 + 381-5916). *Crystal chemistry, silicates, crystal structures, complex polytypic minerals.*

Arai, Dr Gerda Johanna. Eng. Dept., Zenith Electronics Corp., 1000 Milwaukee Ave., Glenview, IL 60025, USA. PhD, crystallography (U. Leiden, Netherlands, 1960). Section manager, analytical chem. group. (tel. 312 + 391-8564). *Inorganic structures, quantitative analysis, X-ray fluorescence.*

Archer, Dr Ronald D. Chemistry Dept., U. of Massachusetts, Amherst, MA 01003, USA. PhD, chemistry (U. Illinois, Urbana, 1959). Prof., dept. head. (tel. 413 + 545-2291). *Structures, coordination compounds.*

Arem, Dr Joel Edward. Multifacet Inc., P. O. Box 5056, Laytonsville, MD 20760, USA. (1943) PhD, mineralogy (Harvard U., 1970). Pres. (tel. 301 + 977-0335). *Mineralogy, crystal growth, synthetic gemstones.*

Arents, Ms Gina. Biophysics Dept., Johns Hopkins U., Charles and 34th Sts., Baltimore, MD 21218, USA. (1944) BA, physics (Goucher C., 1975). Grad. student. (tel. 301 + 338-7912). *Protein crystallography, hemoglobins, myoglobins.*

Argos, Dr Patrick. Biological Sci., Purdue U., Lilly Hall of Life Sci., West Lafayette, IN 47907, USA. (1942) PhD, physics (Saint Louis U., 1968). Asst. prof. (tel. 317 + 494-8333). *Proteins, viruses, X-ray crystallography, molecular evolution, protein structure prediction, protein folding analysis.*

Arif, Dr Atta Mahmood. Dept. of Chemistry, U. of Texas at Austin, Austin, TX 78712, USA. (1953) PhD, (U. London, Queen Mary C., UK, 1983) Res. asst. (tel. 512 + 471-7710). *Structure, organometallic compounds, X-ray analysis.*

Armendarez, Mr Peter X. Physics Dept., Brescia Coll., Owensboro, KY 42301, USA.

Armstrong, Prof. Ronald William. Dept. of Mechanical Eng., U. of Maryland, College Park, MD 20742, USA. (1934) PhD, metallurgical engineering (Carnegie Mellon U., 1958). Prof. (tel. 301 + 454-8881). *Dislocations, X-ray topography, grain boundaries, strength properties. X-ray diffraction microscopy, Berg-Barrett method, Lang topography, Borrmann technique.*

Arnold, Dr Edward Van Dyke. Dept. of Biological Sci., Purdue U., Lilly Hall, W. Lafayette, IN 47907, USA. (1957) PhD, organic chemistry (Cornell U., 1982). Res. assoc. (tel. 317 + 494-6766). *Chemical and biological structure, structure - function relationships, natural products.*

Arnone, Prof. Arthur. Dept. of Biochemistry, U. of Iowa, Iowa City, IA 52242, USA. (1942) PhD, physical chemistry (MIT, 1970). Prof. (tel. 319 + 353-6072). *Macromolecular crystallography.*

Arnott, Prof. Struther. Office of the Vice Pres. for Res., Hovde Hall, Purdue U., West Lafayette, IN 47907, USA. (1934) PhD, chemistry (Glasgow U., UK, 1960). V.P., Res. and Dev., Grad. S., biology prof. (tel. 317 + 494-2604). *Fibrous structures, nucleic acids, polysaccharides.*

Arora, Prof. Satish Kumar. Drug Dynamics Inst., U. of Texas, Austin, TX 78712, USA. (1942) PhD, chemistry (U. Poona, India, 1970). Sr. scient. (tel. 512 + 471-9267). *Biological structures.*

Arrington, Mr Wendell. 175 E. Kenilworth, Newton Square, PA, 19073, USA.

Artioli, Gilberto. Dept. of Physical Sciences, U. of Chicago, 5734 S. Ellis Ave., Chicago, IL 60637, USA.

Aruffo, Mr Alejandro Antonio. Dept. of Molecular Biology, Massachusttes General Hospital, Fruit St., Boston, MA 02114, USA. (1959), chemistry (U. Washington). *Inorganic and coordination compounds, structure.*

Atkinson, Dr David. Biophysics Inst., Boston U., S. of Medicine, 80 East Concord St., Boston, MA 02118, USA. (1944) PhD, biophysics (Council For Nat. Academic Awards, London, UK, 1975). Assoc. prof., medicine and biochemistry. (tel. 617 + 247-6217). *X-ray and neutron scattering, biological macromolecular diffraction, biophysics, lipids, proteins, lipoproteins, membranes.*

Atoji, Dr Masao. 702 86th Place, Downers Grove, IL, 60516, USA. (1925) DSc, physical chemistry and crystallography (Osaka U., Japan, 1956). Dir., chemical lab., Motorola Inc., communication sect. (tel. 312 + 576-0654). *Semiconductors, electronics materials, neutron diffraction, magnetic structures, metals and alloys, crystal growth.*

Attard, Dr Alfred E. 5434 Phelps Luck Dr., Columbia, MD 21045, USA. (1926) PhD, physics (Illinois Inst. of Techn., 1962). *Solid state physics, solid state chemistry, optics, electro-optics.*

Atwood, Prof. Jerry Lee. Chemistry Dept., U. of Alabama, University, AL 35486, USA. (1942) PhD, inorganic chemistry (U. Illinois, 1968). Assoc. prof. (tel. 205 + 348-5979). *Organometallic chemistry, natural products, liquid structure.*

Au, Dr Andrew Yu-Chung. Geophysical Lab., Carnegie Inst. of Washington, 2801 Upton St., NW, Washington, DC 20008, USA. (1952) PhD, geophysics (SUNY at Stony Brook, 1984). Res. assoc. (tel. 202 + 966-0334). *Mineral physics, elasticity.*

Augustin, Mr Rolf M., Jr. Polaroid Corporation, 575 Technology Square, Cambridge, MA 02139, USA.

Austerman, Mr Stanley Boone. Austerman Associates, 18112 Stratford Circle, Villa Park, CA 92667, USA. (1924) BA, physics (Purdue U., 1949). Consultant (tel. 714 + 639-2742). *X-ray crystallography, topography, crystal growth, crystal characterization (thermal- electrical - physical).*

Averbach, Prof. B. L. Dept. of Materials Sci. and Eng., Massachusetts Inst. of Techn., 77 Massachusetts Ave., Cambridge, MA 02139, USA. (1919) ScD, metallurgy (MIT, 1947). Prof. (tel. 617 + 253-3320). *Amorphous materials structure, small angle X-ray scattering, magnetic neutron scattering.*

Aykan, Mr Kamran. 45 Ocean Ave., Apt. 9-D, Monmouth Beach, NJ 07750, USA. (1930) MSc, chemistry (U. Istanbul, Turkey, 1954). *Heterogeneous catalysis, applied crystallography, solid state properties, X-ray diffraction, precious metals (chemistry - alloys - industrial applications).*

Azaroff, Prof. Leonid V. Inst. of Materials Sci., U. of Connecticut, Storrs, CT 06268, USA. (1926) PhD, crystallography (MIT, 1954). Dir., prof. of physics. (tel. 203 + 486-4623,4). *Electronic structure, alloys, X-ray spectroscopy, diffraction studies of solids.*

Babich, Prof. Michael Wayne. Dept. of Chemistry, Florida Inst. of Techn., 150 W. University Blvd., Melbourne, FL 32901-6988, USA. (1945) PhD, chemistry (U. Nevada, 1974). Assoc. prof. (Tel. 305 + 768-8046, Ext. 7376). *Solid state chemistry, reaction kinetics in the solid phase, coordination chemistry, thermal analysis, structural chemistry.*

Baenziger, Prof. Norman C. Dept. of Chemistry, U. of Iowa, Iowa City, IA 52240, USA. (1922) PhD, physical chemistry (Iowa State U., 1948). Prof. (tel. 319 + 353-4688). *Crystal structures, X-ray diffraction.*

Bailey, Marcia F. Dept. of Chemistry, Central Michigan U., Mt. Pleasant, MI 48859, USA.

Bailey, Prof. Sturges Williams. Dept. of Geology and Geophysics, U. of Wisconsin, Weeks Hall, 1215 W. Dayton St., Madison, WI 53706, USA. (1919) PhD, crystallography (U. Cambridge, UK, 1955). Prof. (tel. 608 + 262-1806). *Layer silicate structures, feldspars.*

Baird, Prof. Herbert Wallace. Dept. of Chemistry, Wake Forest U., Reynolds Station, Winston-Salem, NC 27109, USA. (1936) PhD, physical chemistry (U. Wisconsin, Madison, 1963). Prof. (tel. 919 + 761-5325). *Crystal and molecular structures.*

Baker, Mr Kenneth Neil. Dept. of Chemistry, U. of Mississippi, University, MS 38677, USA. (1957) BA chemistry (U. Mississippi, 1979). Grad. student. (tel. 601 + 232-7301). *Transition metal fluorides, magnetic transition state.*

Baldwin, Mr Kenneth John. Dept. of Earth and Space Sci., SUNY at Stony Brook, Stony Brook, NY 11794-2100, USA. (1948) MS, geochemistry (SUNY, Stony Brook, 1973). Res. assoc. (tel. 516 + 246-8381). *Automation, instrumentation, computer systems, programming, crystallography.*

Bale, Prof. Harold D. Physics Dept., U. of North Dakota, Grand Forks, ND 58202, USA. (1927) PhD, physics (U. Missouri, 1959). Prof. (tel. 701 + 772-9293). *Small angle X-ray scattering, liquids, non-crystalline solids.*

Bales, Mr Howard E. Res. Services, Wright State U., Col. Glenn Hwy., Dayton, OH 45435, USA. (1912) BSc, education (Wilmington C., 1934). Retired. (tel. 513 + 372-3688). *Education, chemistry, instrumental analysis, research funding.*

Banaszak, Prof. Leonard J. Dept. of Biological Chemistry, Washington U. Medical S., 660 South Euclid Ave, St. Louis, MO 63110, USA. (1933) PhD, biochemistry (Loyola U., 1961). Prof. (tel. 314 + 362-3341). *Protein crystallography, lipid-protein systems.*

Banerjee, Dr Bani Ranjan. Res. Dept., Ingersoll Rand Co., 1 Sycamore Lane, Skillman, NJ 08558, USA. (1925) PhD, materials engineering (Yale U., 1950). Asst. dir. for res. (tel. 609 + 921-9103). *Materials science and engineering, crystallography, diffraction.*

Banks, Prof. Ephraim. Dept. of Chemistry, Polytechnic Inst. of New York, 333 Jay St., Brooklyn, NY 11201, USA. (1918) PhD, inorganic chemistry (PIB, 1949). Prof. (tel. 718 + 643-4757). *Crystal chemistry, crystal growth, spectra of solids, magnetic and electrical properties, luminescent properties.*

Barber, Prof. Patrick George. Dept. of Natural Sci., Longwood C., Farmville, VA 23901, USA. (1942) PhD, Physical chemistry (Cornell U., 1969). Prof., chemistry. (tel. 804 + 392-9352, ext. 32). *Liquid crystals, crystal growth.*

Bardhan, Pronob. Ceramic Res. Dept., Corning Glass Works, Corning, NY 14830, USA.

Barkigia, Dr Kathleen M. Dept. of Applied Sci., Bldg. 815, Brookhaven Nat. Lab., Upton, NY 11993, USA. (1951) PhD, Chemistry (Georgetown U., 1978) Assoc. scient. (tel. 516 + 282-4382). *X-ray diffraction, photosynthetic pigments, modeling.*

Barnes, Dr Charles Leslie. Chemistry Dept., U. of Puerto Rico, Rio Piedras, PR 00931, USA. (1949) PhD, biochemistry (U. Tennessee, 1980). Asst. prof. (tel. 809 + 764-0000, ext. 2375). *Crystallography, natural products, peptide conformation, structure - function relationships.*

Barnett, Dr Bobby L. Miami Valley Res. Lab., Procter and Gamble, P. O. Box 39175, Cincinnati, OH 45247, USA. (1939) PhD, physical chemistry (U. Texas, Austin, 1970). Group leader, microscopy and X-ray. (tel. 513 + 972-2321). *Coordination complexes, radiopharmaceuticals, drug design microanalysis.*

Barney, Ms. Elsa Pauline. 312 Burton St., Bath, NY 14810, USA. (1922) MS, geology (MIT, 1947). Tax map tech. (tel. 607 + 776-7457). *Mineral structures.*

Barnhart, Dr David Merle. Dept. of Physical Sci., Eastern Montana C., Billings, MT 59101, USA. (1933) PhD, physical chemistry (Oregon State U., 1964). Prof. (tel. 406 + 657-2341). *Organo-metallic complexes.*

Barrett, Prof. Charles Sanborn. Metallurgy and Materials Sci., Denver Res. Inst., U. of Denver, Denver, CO 80208, USA. (1902) PhD, physics (U. Chicago, 1928). Sr. res. scient., adjunct prof. (tel. 303 + 871-3529). *Stress analysis, metals, polymerics, composites, diffraction, instrumentation, transformations, imperfections, ordering, phase identification.*

Barrick, Dr James Clinton. 564 S. Selby Blvd., Worthington, OH 43085, USA. (1940) PhD, inorganic chemistry (McMaster U., Canada, 1972). Information scient. (tel. 614 + 436-7724). *Single crystal analysis, powder X-ray crystallography, materials research, alloys, organic compounds, cobalt complexes, computer program applications.*

Bartell, Prof. Lawrence Sims. Chemistry Dept., U. of Michigan, Ann Arbor, MI 48109, USA. (1923) PhD, chemistry (U. Michigan, 1951). Prof. (tel. 313 + 764-7375). *Electron diffraction, electron holography, electron distribution, molecular structure, molecular vibrations and force fields, isotope effects, quantum chemistry.*

Bartlett, Prof. Neil. Chemistry Dept., U. of California, Berkeley, CA 94720, USA. (1932) DSc, inorganic chemistry (U. Newcastle Upon Tyne, UK, 1958). Prof. (tel. 415 + 642-7259). *Solid-state chemistry, fluorine chemistry, noble-gas chemistry, graphite intercalation.*

Barton, Dr C. J. Dorr-Oliver Inc., 77 Havemeyer Lane, P. O. Box 9312 Stamford, CT 06904, USA. (1936) PhD, metallurgy (Rensselaer Polytechnic Inst., 1966). Pres. (tel. 203 + 358-3200). *Structure - properties relationship.*

Barton, Dr Randolph Jr. Textile Fibers Dept., E. I. du Pont de Nemours and Co., Experimental Station, Wilmington, DE 19898, USA. (1941) PhD, physical chemistry (Johns Hopkins U., 1968). Sr. res. chemist. (tel. 302 + 772-2578). *Fiber structure, morphology.*

Basu, Dr Sankar Prasad. Natural Science Dept., Coppin State C., 2500 W. North Ave., Baltimore, MD 21216, USA. (1941) PhD, biophysics (U. Oklahoma, 1977). Assoc. prof. (tel. 301 + 997-4336). *Structure, X-ray crystallography, biological molecules, proteins, virus.*

Bateman, Dr Linda Ratner. 1608 Turkey Run Rd., Wilmington, DE 19803, USA. (1942) PhD, physical chemistry (U. Wisconsin, 1969). *Organometallic chemistry, transition metal chemistry, elastomers.*

Bates, Prof. Robert Brown. Chemistry Dept., U. of Arizona, Tucson, AZ 85721, USA. (1933) PhD, chemistry (U. Wisconsin, 1957). Prof. (tel. 602 + 884-1662). *Organic chemistry, natural products, carbanions.*

Batterman, Prof. Boris William. S. of Appl. and Eng. Physics, Chess-Synchrotron Radiation Lab., Cornell U., Ithaca, NY 14853, USA. (1930) PhD physics (MIT., 1956). Dir. (tel. 607 + 256-5161). *Synchrotron radiation, X-ray and neutron diffraction, solid state physics, dynamical X-ray diffraction, anharmonic vibrations in crystals.*

Bau, Prof. Robert. Dept. of Chemistry, U. of Southern California, Exposition Blvd., Los Angeles, CA 90007, USA. (1944) PhD, inorganic chemistry (UCLA, 1968). Prof. (tel. 213 + 743-8800, telex 674-803). *Neutron diffraction, structure determination, transition metal hydride complexes, metal-nucleotide complexes.*

Baughman, Mr Richard Joseph. Division 5154, Sandia Labs., P. O. Box 5800, Albuquerque, NM 87185, USA. (1927) BS, chemistry-biology (Mount Union C., 1950). Techn. staff member. (tel. 505 + 264-6337). *Crystal growth, materials preparation.*

Baur, Prof. Werner Heinz. Dept. Geological Sci., U. of Illinois at Chicago, Box 4348, Chicago, IL 60680, USA. (1931) Dr.rer.nat., crystallography (U. Göttingen, BRD, 1956). Prof. (tel. 312 + 996-3154 or 6088). *Powder diffraction, bond length prediction, computer simulation, solid state chemistry, crystal chemistry, zeolite studies, mineral and inorganic crystal structures, hydrogen bonding.*

Bear, Prof. Richard Scott. 614 Morgan Creek Rd., Chapel Hill, NC 27514, USA. (1908) PhD, chemistry (U. California, Berkeley, 1933). Prof. em. (tel. 919 + 929-8337). *Natural fibers, membranes, structure, optics, X-ray diffraction.*

Beard, Mr Donald W. Measuring Systems Marketing Div., Siemens-Allis, 1 Computer Dr., Cherry Hill, NJ 08034, USA. (1928) BS, chemistry (Allegheny C., 1950). Mgr., appl. lab. (tel. 609 + 424-9210, ext. 261). *X-ray diffraction, x-ray spectroscopy, instrumentation.*

Beasley, Prof. Wayne Machon. Materials Sci. Div., Mechanical Eng. Dept., U. of New Hampshire, Durham, NH 03824, USA. (1922) SM, ceramics (MIT, 1965). Prof. em. (tel. 603 + 332-6375). *Inorganic crystal structures, quantitative sterology.*

Becker, Prof. Joseph Whitney. Developmental and Molecular Biology Dept., The Rockefeller U., 1230 York Ave., New York, NY 10021-6399, USA. (1943) PhD, chemistry (Stanford U., 1970). Assoc. prof. (tel. 212 + 570-8183). *Protein crystallography, image reconstruction.*

Bedarkar, Dr Sudhir. Div. of Biological and Medical Res., Argonne Nat. Lab., 9700 S. Cass Ave., Argonne, IL 60439, USA. (1951) PhD, protein crystallography (U. London, UK, 1982). Postdoctoral Fellow (tel. 312 + 972-3825). *Structure & function, proteins, nucleic acids, protein crystallography, biochemistry.*

Bednowitz, Dr Allan Lloyd. Res. Div., IBM T. J. Watson Res. Center, P.O. Box 218, Yorktown Hgts., NY 10598, USA. (1939) PhD, chemical physics (PIB, 1966). Res. staff member. (tel. 914 + 945-1529, telex 137456). *Computer graphics, computer programming, automation, direct determination.*

Bell, Mr Jeffrey A. Cornell U., 263 Clark Hall, Ithaca, NY 14853, USA.

Belt, Dr Roger F. Airtron, Litton Industries Inc., 200 E. Hanover Ave., Morris Plains, NJ 07950, USA. (1929) PhD, physical chemistry (State U. of Iowa, 1956). Res. dir. (tel. 201 + 539-5500, ext. 309). *Crystal growth, crystal structure, perfection of crystals, X-ray techniques, optical and magnetic materials.*

Ben-Hussein, Mr Ahmed Othan. 1201 S. Courthouse Rd., Apt. 119 Arlington, VA 22204, USA.

Benci, Mr Pierluigi. The Chester Engineers, 845 Fourth Ave., Coraopolis, PA 15108, USA.

Bennett, Mr Dennis W. Dept. of Chemistry, U. of Wisconsin, Milwaukee, WI 53201, USA.

Bennett, Dr John Michael. Central Scient. Labs., Union Carbide Corp., Tarrytown Techn. Center, Tarrytown, NY 10591, USA. (1939) PhD, chemistry (U. Aberdeen, UK, 1966). Res. scient. (tel. 914 + 789-3604). *Molecular sieve materials, X-ray diffraction, electron diffraction, neutron diffraction, powder diffraction*

Beno, Dr Mark A. Chemistry Div., Argonne Nat. Lab., 9700 S. Cass Ave., Argonne, IL 60439, USA. (1951) PhD, physical chemistry (The Ohio State U., 1979). Chemist. (tel. 312 + 972-3507). *X-ray and neutron diffraction, instrumentation, low temperature diffractometry.*

Benson, Mr James Edward. Chemistry Dept., Ames Lab.-DOE, Iowa State U., 43 Spedding Hall, Ames, IA 50011, USA. (1933) MS, physical chemistry (Iowa State U., 1963). Assoc chemist. (tel. 515 + 294-8444). *X-ray diffraction, coal and coal related minerals; single crystals, automation.*

Beres, Mr John J. 6014 Echodell NW, North Canton, OH 44720, USA.

Bergmann, Mrs Margot Eisenhardt. Dept. of Physics, Polytechnic Inst. of New York, 640 Riverside Dr., New York, NY 10031, USA. (1913) MS, physical chemistry (Rutgers U., 1940). Res. consultant. (tel. 212 + 643-8997, or 212 + 926-2745). *X-ray diffraction, electron diffraction, electron microscopy, microbeam X-ray camera development, polymers, biological materials.*

Berkebile, Prof. C. Alan. Geology Dept., Corpus Christi State U., Corpus Christi, TX 78412, USA. (1938) PhD, mineralogy (Boston U.). Prof. *Crystal growth, mineralogy, mineral structures, marine geology.*

Berkey, Mr W. W. 9501 Bonnie Dale Rd., Richmond, VA 23229, USA.

Berman, Dr Helen M. Inst. for Cancer Res., 7701 Burholme Ave., Philadelphia, PA 19111, USA. (1943) PhD, crystallography (U. Pittsburgh, 1967). Member. (tel. 215 + 728-2548). *Crystallography, biological molecules, nucleic acid interactions and conformations.*

Bernal, Prof. Ivan. Dept. of Chemistry, U. of Houston, Cullen Blvd., Houston, TX 77004, USA. (1931) PhD, chemical physics (Columbia U., 1963). Prof. (tel. 713 + 749-2618 or 749-2108). *Absolute configuration, organometallics and coordination compounds, mechanisms, chiral resolutions, spontaneous resolution, conglomerate crystallizations, energetics.*

Bernard, Prof. William H. Physics Dept., Louisiana Tech. U., Ruston, LA 71272, USA. (1932) PhD, physics (Tulane U.) Prof. (tel. 318 + 257-4627).

Bernheim, Prof. Marguerite May Yevitz. Dept. of Computer Sci., Pennsylvania State U., 333 Whitmore Lab., University Park, PA 16802, USA. (1946) PhD, inorganic chemistry, crystallography (PSU, 1976). Res. assoc. (tel. 814 + 865-1554 or 865-4041). *Bridged metallic compounds, organometallic systems, structure determination methods.*

Bernstein, Ms Frances C. Protein Data Bank, Dept. of Chemistry, Brookhaven Nat. Lab., Upton, NY 11973, USA. (1942) MS, mathematics (New York U., 1965). Programmer analyst. (tel. 516 + 282-4382). *Macromolecular data bases, macromolecular structures, computer programming.*

Bernstein, Dr Herbert Jacob. Courant Inst. of Mathematical Sciences, New York U., 251 Mercer St., New York, NY 10012, USA. (1944) PhD, mathematics (New York U., 1968). Sr. res. scient. (tel. 212 + 460-7269). *Scientific computing, theoretical crystallography, data acquisition.*

Bernstein, Mr Joel L. Chemistry Dept., Cornell U., Ithaca, NY 14850, USA. (1940) MS, physics (New York U., 1968).

Berry, Dr Chester Ridlon. 37 Heritage Dr., S. Orleans, MA 02662, USA. (1919) PhD, physics (Cornell U., 1946). Consultant. (tel. 617 + 255-6206). *Mechanisms, microcrystalline nucleation and growth, crystal imperfection methods, small particle optical behavior, photographic sensitivity theories.*

Bertrand, Prof. Joseph Aaron. S. of Chemistry, Georgia Inst. of Techn., Atlanta, GA 30332, USA. (1933) PhD, inorganic chemistry (Tulane U., 1961). Prof. (tel. 404 + 894-4003). *Transition metal complexes, X-ray diffraction.*

Bethge, Dr Paul Herman. Dept. of Physiology and Biophysics, Washington U., Medical S., 660 South Euclid Ave., St. Louis, MO 63110, USA. (1945) PhD, chemistry (Harvard U., 1973). Res. assoc. (tel. 314 + 454-3142). *Biological macromolecular crystallography.*

Betts, Dr Foster. Res. Lab., The Hospital for Special Surgery, Cornell U. Medical C., 535 East 70th St., New York, NY 10021, USA. (1932) PhD, electrical eng. (Stanford U., 1972). Assoc. scient. (tel. 212 + 606-1436). *Amorphous materials.*

Bhandary, Dr Krishna K. Dept. of Biophysics, Roswell Park Memorial Inst., 666 Elm St., Buffalo, NY 14263, USA. (1946) PhD, chemistry (Indian Inst of Sci., India, 1974). Cancer res. scient. (tel. 716 + 845-2362). *Cyclic peptides, cardiotonic agents, ionophores.*

Bhardwaj, Mr Jayant. 412 Halifax Ct., Martinez, GA 30907, USA.

Bhat, Dr Narayana Talapady. Lab. of Molecular Biology, Nat. Inst.s of Health, Bldg. 2, Rm. 312, 9000 Rockville Pike, Bethesda, MD 20205, USA. PhD, physics (Indian Inst. of Sci., India, 1977). Visiting assoc. (tel. 301 + 496-4205). *Phase problem, macromolecular structures.*

Bhattacharjee, Mr Sovan K. Chemistry Dept., U. of Maryland, College park, MD 20742, USA.

Bienenstock, Prof. Arthur Irwin. Stanford Synchrotron Radiation Lab., Stanford U., P. O. Box 4349, Bin 69, Stanford, CA 94305, USA. (1935) PhD, applied physics (Harvard U., 1962). Prof., dir. (tel. 415 + 854-3300, ext. 3153). *Physical properties, atomic arrangement, non-crystalline solids, poorly crystallized solids, synchrotron radiation.*

Bigelow, Prof. Wilbur Charles. Dept. of Materials and Metallurgical Eng., U. of Michigan, Dow Bldg., Ann Arbor, MI 48109, USA. (1923) PhD, physical chemistry (U. Michigan, 1951). Prof. (tel. 313 + 764-3321). *Electron diffraction, electron microscopy.*

Bilderback, Dr Donald Heywood. S. of Appl. and Eng. Physics, Cornell U., Ithaca, NY 14853, USA. (1947) PhD, solid state physics (Purdue U., 1975). Operations Manager for CHESS. (tel. 607 + 256-7163). *Diffraction physics, multiwire X-ray detectors, synchrotron radiation.*

Binnie, Dr William Polson. The Carborundum Co., Bldg. 1-2, P.O. Box 1054, Niagara Falls, NY 14302, USA. (1924) PhD, chemistry (U. Glasgow, UK, 1948). Scient. *Mineral structure, mineral analysis.*

Birks, Mr L. S. Code 6680, Naval Res. Lab., Washington, DC 20375, USA. (1919) MS, physics (U. Maryland, 1951). Retired, part-time consultant. *Spectrometer crystals, defects in crystals, crystal diffraction efficiency.*

Bish, Dr David Lee. Earth and Space Sci. Div., Los Alamos Nat. Lab., Mail Stop J978, Los Alamos, NM 87545, USA. (1952) PhD, mineralogy, (PSU, 1977). Staff mineralogist. (tel. 505 + 667-4337). *Mineralogy, clay mineralogy, X-ray powder diffraction.*

Bjorkman, Ms Pamela J. Dept. of Biochemistry, Harvard U., F Divinity Ave., Cambridge, MA 02138, USA.

Blake, Mr J. W. 3214 Montavesta Rd., Lexington, KY 40502, USA.

Blanton, Mr Thomas Nelson. Res. Lab., Eastman Kodak Co., Bldg. 82, Room C-206 Res. Labs., Rochester, NY 14650, USA. (1959) MS, analytical chemistry (Emory U., 1981). Res. chemist. (tel. 716 + 477-6701). *X-ray diffraction, solid state, thin films.*

Blessing, Dr Robert Harry. Molecular Biophysics Dept., Medical Foundation of Buffalo, 73 High St., Buffalo, NY 14203, USA. (1941) PhD, chemistry (Ohio U., 1971). Res. scient. (tel. 716 + 856-9600, ext. 460). *Small biological molecules, hydrogen bonding, accurate structure analysis, electron density distributions.*

Block, Dr Stanley. Crystallography Section, Nat. Bureau of Standards, Washington, DC 20234, USA. (1926) PhD, chemistry (Johns Hopkins U., 1955). Chief. (tel. 301 + 921-2837). *High pressures, inorganic compounds, powder identification, single crystal identification, structure.*

Blount, Dr John Franklin. Hoffman-La Roche Inc., Nutley, NJ 07110, USA. (1937) PhD, chemistry (U. Wisconsin, 1965). Res. group chief. (tel. 201 + 235-3580). *Organic and organometallic crystal structures, automated diffractometer software, computer programming, crystal structure analysis.*

Boehme, Mr Richard Frederick. Physical Sci. Dept., Res. Div., IBM T.J. Watson Res. Center, P. O. Box 218, Yorktown Heights, NY 10598, USA. (1950) MA, chemistry (Boston U., 1976). Sr. assoc. eng. (tel. 914 + 945-1820). *Electron density determination and modeling, low temperature crystallography, EXAFS, atomic structure.*

Boggs, Dr Rita Rose. American Res. and Testing Inc., 144 1/1 W. Gardena Blvd., Gardena, CA 90248, USA. (1938) PhD, physical chemistry (U. Pennsylvania, 1973). Pres. (tel. 213 + 538-9709). *Crystal structure, small molecules, direct methods.*

Bolin, Mr Jeffrey T. Dept. of Biological Sci., Lily Hall of Sci., Purdue U., West Lafayette, IN 47907, USA.

Bonham, Prof. Russell Aubrey. Chemistry Dept., Indiana U., Bloomington, IN 47401, USA. (1931) PhD, physical chemistry (Iowa State C., 1958). Prof. (tel. 812 + 337-4843). *Charge and momentum determination, electron diffraction, high energy electron impact spectroscopy, electron beam time of flight spectroscopy, secondary electron spectroscopy, negative ion resonance spectroscopy.*

Boo, Mr W. O. J. Chemistry Dept., U. of Mississippi, University, MS 38677, USA.

Bordner, Prof. Jon. Chemistry Dept., North Carolina State U., Raleigh, NC 27607, USA. (1940) PhD, organic chemistry (U. California, Berkeley, 1966). Prof. (tel. 919 + 737-2942). *Biologically significant molecular structures, organic synthesis, organic separations.*

Borie, Dr Bernard Simon. Material Science Dept., U. of Tennessee, Knoxville, TN 37996, USA. (1924) PhD, physics (MIT, 1956). Prof. (tel. 615 + 483-6816). *Diffraction crystallography.*

Boskey, Dr Adele Ludin. Res. Div., Hospital for Special Surgery, Cornell U. Medical C., 535 East 70th Street, New York, NY 10021, USA. (1943) PhD, chemistry (Boston U., 1970). Dir., Lab. for Ultrastructural Biochemistry. (tel. 212 + 606-1453). *Calcification mechanisms, biologic calcification.*

Boss, Dr James William. 134 North 54th St., Philadelphia, PA 19139, USA. (1932) PhD, physical chemistry (Ohio State U., 1966). (tel. 215 + 472-1137). *X-ray diffraction, electron spin resonance, disordered structures, glasses.*

Boudreau, Prof. Sharon Martin. Chemistry Dept., Wheaton C., Norton, MA 02766, USA. PhD, inorganic chemistry (U. New Hampshire, 1979). Asst. prof. (tel. 617 + 285-7722, ext. 447). *Crystal growth of metal anthranilates, interaction compounds.*

Bourne, Mr Philip Eric. Cancer Center Computing Facility, Columbia U., 630 W. 168th St., New York, NY 10032 USA.

Bowman, Mr Allen L. 10 Encino, Los Alamos, NM 87544, USA.

Box, Mr Harold C. Roswell Park, MML Inst. 666 Elm St., Buffalo, NY 14263, USA.

Boyko, Prof. Edward Raymond. Chemistry Dept., Providence C., Providence, RI 02918, USA. (1930) PhD, physical chemistry (Rutgers U., 1956). Prof. (tel. 401 + 865-2108). *Coordination compounds, organic crystal structures.*

Boyle, Mr Paul. Dept. of Chemistry, U. of Minnesota, 207 Pleasant St., SE, Minneapolis, MN 55455, USA.

Braden, Dr Bradford Carl. Dept. of Biophysics, The John Hopkins U., 34th and Charles St., Baltimore, MD 21218, USA. (1951) PhD, biophysics (Indiana U., 1978). Res. assoc. (tel. 301 + 338-7250). *Protein structures, peptide ionophores.*

Brader, Mr James J. 911 Wildwood Dr. W., Prospect Heights, IL 60070, USA.

Brady, Mr George W. Div. Lab and Res., NY State Dept. Health, Albany, NY 12201, USA.

Brady, Mr James Henry. Materials Eng. Dept., David W. Taylor Naval Ship R & D Cntr., Annapolis, MD 21402, USA. (1931) BS, chemistry (U. Southwestern Louisiana, 1952). Project eng. (tel. 301 + 267-3754). *Metal physics, powder diffraction.*

Bragg, Prof. Robert Henry. Dept. of Materials Sci. and Mineral Eng., U. of California, Hearst Memorial Mining 210, Berkeley, CA 94720, USA. (1919)

PhD, physics (Illinois Inst. of Techn., 1960). Prof. (tel. 415 + 642-7393). *Structure, electrical properties, carbon materials, graphite intercalation compounds, diffraction physics, small-angle X-ray and neutron scattering.*

Brathovde, Prof. James Robert. Dept. of Chemistry, Northern Arizona U., Flagstaff, AZ 86011, USA. (1926) PhD, physical chemistry, X-ray crystallography (U. Washington, 1956). Prof. em. (tel. 602 + 567-9493). *Geothermal.*

Brech, Mr Frederick. P. O. Box 145, Dover, MA 02030, USA.

Brennan, Dr Richard Gerald. Inst. of Molecular Biology, U. of Oregon, Eugene, OR 97403-1229, USA. (1955) PhD, biochemistry (U. Wisconsin, Madison, 1984). Res. assoc. (tel. 503 + 686-5176). *Protein-nucleic acid complexes, nucleic acids (modified).*

Brennan, Prof. Thomas Francis. Gynecologic Endocrine Lab., Boston U. Medical S., 720 Harrison Ave., Suite 900, Boston, MA 02118, USA. (1943) PhD inorganic chemistry, crystallography (SUNY, Stony Brook, 1970). Asst. dir. (tel. 617 + 247-6295). *Neuropeptide structure and function, catechol estrogens, DNA-drug interactions, models for hormone-receptor binding.*

Brenner, Mr Stephen A. Naval Res. Lab., Code 6030, Washington, DC 20375, USA. (1937) MA, physical chemistry (Boston U., 1962). Res. chemist. (tel. 202 + 767-2735). *Mathematics, chemistry, physics, computer science.*

Brickenkamp, Dr Carroll Shelton. 12405 Beall Spring Rd., Potomac, MD 20854, USA. (1945) PhD, crystallography (U. Pittsburgh, 1970). *Moisture measurement in materials, compliance sampling, prepackaged commodities, borophosphate structure and chemistry.*

Bright, Dr William M. 2500 Wisconsin Ave., N.W., Apt. 960, Washington, DC 20007, USA. PhD, chemistry (Georgetown U., 1974).

Briguglio, Mr James. 177 W. 18th. St., Bayonne, NJ 07002, USA. (1954) MS, inorganic, analytical chemistry (Fairleigh Dickinson U., 1980). Asst. (tel. 201 + 437-9073). *Analytical chemistry, crystallography.*

Britton, Prof. Doyle. Chemistry Dept., U. of Minnesota, Minneapolis, MN 55455, USA. (1930) PhD, chemistry (CIT, 1955). Prof. (tel. 612 + 373-2382). *Intermolecular interactions.*

Broach, Dr Robert William. Dept. of Physical Chemistry and Surface Sci., Signal UOP Res. Ctr., 50 UOP Plaza. Des Plaines, IL 60016, USA. (1949) PhD, chemistry (U. Wisconsin, Madison, 1977). Group leader. (tel. 312 + 391-3313). *X-ray and neutron diffraction, EXAFS, catalysts and adsorbents (heterogeneous - petrochemical - exhaust gas conversion).*

Brock, Prof. Carolyn Pratt. Chemistry Dept., U. of Kentucky, Lexington, KY 40506, USA. (1946) PhD, chemistry (Northwestern U., 1972). Assoc. prof. (tel. 606 + 257-1959). *Molecular packing in crystals, thermal motion.*

Brown, Prof. Bruce Elliot. Geological and Geophysical Sci., U. of Wisconsin, Milwaukee, WI 53201, USA. (1930) PhD, geology (U. Wisconsin, 1960). Assoc. prof. (tel. 414 + 963-4972). *Mineralogy, geochemistry, limnology.*

Brown, Prof. Bruce Willard. Chemistry Dept., Portland State U., P. O. Box 751, Portland, OR 97207, USA. (1927) PhD, chemistry (U. Washington, 1961). Prof. (tel. 503 + 229-3811). *Coordination compounds, computer programming.*

Brown, Dr George Marshall. Chemistry Div., Oak Ridge Nat. Lab., P. O. Box X, Oak Ridge, TN 37830, USA. (1921) PhD, physical chemistry (Princeton U., 1949). Res. staff member. (tel. 615 + 574-4989). *X-ray and neutron crystal structure analysis.*

Brown, Prof. Glenn H. Liquid Crystal Inst., Kent State U., Kent, OH 44242, USA. (1915) PhD, chemistry (Iowa State U., 1951). Dir. and Regents prof. of chemistry. (tel. 216 + 672-2654). *Liquid crystals, structure - properties relationship.*

Brown, Prof. Gordon Edgar, Jr. Geology Dept., Stanford U., Stanford, CA 94305, USA. (1943) PhD, mineralogy and geology (Virginia Polytechnic Inst., 1970). Asst. prof. of mineralogy. (tel. 415 + 497-3518). *Crystal chemistry, mineralogy, rock forming silicate minerals, X-ray and neutron crystallography, high & low temperatures, silicate glass structures, bonding in minerals.*

Brown, Dr Joe Ned, Jr. Physics Dept., Celanese Res. Co., 86 Morris Ct., Summit, NJ 07901, USA. (1947) PhD, physical chemistry (Louisiana State U., New Orleans, 1972). Sr. res. chemist. (tel. 201 + 522-7789). *Peptide conformation, polymer morphology, oxidation catalysis.*

Brown, Dr Leo Dale. Synthetic Fuels Div., Exxon Res. and Eng., P. O. Box 4255, Baytown, TX 77520, USA. (1948) PhD, inorganic chemistry (U. California, Berkeley, 1974). Sr. res. chemist. (tel. 713 + 425-5290). *Structural inorganic chemistry, silicate minerals, crystal chemistry, X-ray powder diffraction, high and low temperature.*

Brumberger, Prof. Harry. Dept. of Chemistry, U. of Syracuse, Syracuse, NY 13210, USA. (1926) PhD, chemistry (PIB, 1955). Prof. (tel. 315 + 423-2359). *Small angle scattering, catalysts.*

Bryan, Prof. Robert Finlay. Chemistry Dept., U. of Virginia, McCormick Rd., Charlottesville, VA 22901, USA. (1933) PhD, chemistry (U. Glasgow, UK, 1957). Prof. (tel. 804 + 924-3619). *Liquid crystals and their precursors, structure determination, crystallographic computing.*

Bryden, Prof. John Heilner. Chemistry Dept., California State U., Fullerton, 800 N. State College Blvd., Fullerton, CA 92634, USA. (1920) PhD, physical chemistry (UCLA, 1951). Prof. (tel. 714 + 773-2184). *Organic and inorganic crystal structures, computer programming, crystallography.*

Buchanan, Prof. David R. Dept. of Textile Eng. and Sci., North Carolina State U., P. O. Box 8301, Raleigh, NC 27695-8301, USA. (1934) PhD, physical chemistry (Ohio State U., 1962). Prof. (tel. 919 + 737-3481). *Polymers, fibers, structure - properties relationship, small angle X-ray scattering.*

Buerger, Prof. Martin Julian. Earth and Planetary Sci., Room 24-412 MIT, Cambridge, MA 02139, USA. (1903) PhD, mineralogy (MIT, 1929). Inst. prof. em. (tel. 617 + 259-8204). *Crystallography, mineralogy.*

Buerger, Dr Newton Weber. Del Mesa Carmel, #244, Carmel, CA 93921, USA. (1907) PhD, crystallography and metallurgy (MIT, 1939). Prof. em., consultant metallurgist Viking Metallurgical Corp. (tel. 714 + 831-9008). *Aerospace forging metallurgy, exotic metals, titanium alloys, vacuum melt process (CVAR), manufacture of specification Ti-alloys from Ti scrap.*

Bugg, Prof. Charles E. Dept. of Biochemistry, U. of Alabama, University Station, Birmingham, AL 35233, USA. (1941) PhD, chemistry (Rice U., 1965). Prof. (tel. 205 + 934-5329). *Biological crystallography.*

Bunick, Mr Gerard J. 147 S. Columbia Dr., Oak Ridge, TN 37830, USA.

Burbank, Mr Robinson D. 45 Woodland Ave., Summit, NJ 07901, USA.

Burnett, Prof. Roger MacDonald. Dept. of Biochemistry and Molecular Biophysics, C. of Physicians and Surgeons, 630 West 168th St., New York, NY 10032, USA. (1941) PhD, protein crystallography (Purdue U., 1970). Assoc. prof. (tel. 212 + 694-3882). *Macromolecular structures & interactions, virus assembly, structure solution methods.*

Burnham, Prof. Charles Wilson. Hoffman Lab., Dept. of Geological Sci., Harvard U., 20 Oxford St., Cambridge, MA 02138, USA. (1933) PhD, mineralogy and cystallography (MIT, 1961). Prof., mineralogy. (tel. 617 + 495-2484). *Minerals, crystal structure analysis, crystal chemistry, crystal physics, high temperature and pressure, physical properties - structure relations.*

Burns, Dr John Howard. Chemistry Div., Oak Ridge Nat. Lab., P.O.Box X, Oak Ridge, TN 37831, USA. (1930) PhD, physical chemistry (Rice Inst., 1955). Sr. res. staff member. (tel. 615 + 574-5018). *Inorganic crystal structures, X-ray and neutron diffraction, liquid structure.*

Burton, Dr Benjamin Paul. Metallurgy, Div. 450, B150/223, Nat. Bureau of Standards, Gaithersburg, MD 20899, USA. (1949) PhD, earth science (SUNY, Stony Brook, 1982). Res. assoc. (tel. 301 + 921-2917).

Busing, Dr William Richard. Chemistry Div., Oak Ridge Nat. Lab., P.O. Box X, Oak Ridge, TN 37831, USA. (1923) PhD, physical chemistry (Princeton U., 1949). Sr. res. staff member. (tel. 615 + 574-4976). *Neutron diffraction, crystallographic computing, molecular modeling, crystal modeling.*

Butcher, Prof. Raymond John. Chemistry Dept., Howard U., College St., Washington, DC 20059, USA. (1945) PhD, chemistry (U. Canterbury, New Zealand, 1974). Asst. prof. (tel. 202 + 636-6829). *Structure and magnetism, polynuclear complexes; bio-inorganic chemistry, molybdenum; enzymes containing copper (model complexes); iron (III) complexes, magnetic properties.*

Butler, Dr William M. Chemistry Dept., U. of Michigan, Ann Arbor, MI 48109, USA. (1943) PhD, inorganic chemistry (U. Arizona, 1972). Dept. crystallographer. (tel. 313 + 763-2009). *Small molecules, X-ray crystallography, computer aided instruction, computer graphics.*

Butler, Mr William O. Res. Center, SDS Biotech Corp., P. O. Box 348, Painesville, OH 44077, USA. (1939) BS, physics (Miami U., 1961). Res. chemist. (tel. 216 + 352-9311, ext. 280). *Materials, surface analysis techniques, electron microscopy (scanning and transmission), optical microscopy, X-ray diffraction, X-ray spectroscopy, ESCA, Auger electron spectroscopy.*

Byram, Mrs Susan Katherine. X-ray Group, Nicolet Analytical Instruments, 5225-5 Verona Rd., Madison, WI 53711, USA. (1945) MSc, crystallography (U. Toronto, Canada, 1970). Product manager. (tel. 608 + 271-3333, ext. 2013, telex 910-286-2528). *Crystallographic programming, real time instrument control.*

Byrn, Prof. Stephen Robert. Medicinal Chem. & Pharmacognosy Dept., S. of Pharmacy and Pharmacological Sci., Purdue U., West Lafayette, IN 47906, USA. (1944) PhD, chemistry (U. Illinois, 1971). Prof. (tel. 317 + 494-1460). *Mechanisms, solid state reactions, conformation comparison (solid state - solution), biologically active compounds, structure.*

Cady, Dr Howard Hamilton. M-1, Mail Stop C920, Los Alamos Scient. Lab., P. O. Box 1663, Los Alamos, NM 87545, USA. (1931) PhD, chemistry (U. California, Berkeley, 1957). Project leader. (tel. 505 + 667-4992). *Optical and X-ray crystallography, explosives, solid state phase studies, solid state physical chemistry.*

Cagle, Prof. Fredric William, Jr. Dept. of Chemistry, U. of Utah, Salt Lake City, UT 84112, USA. (1924) PhD, chemistry (U. Illinois, 1946). Prof. (tel. 801 + 581-7749). *Inorganic and organic crystal structures.*

Calabrese, Dr Joseph C. Central Res. Div., E. I. DuPont de Nemours and Co., Experimental Station, E356/247, Wilmington, DE 19898, USA. (1943) PhD, chemistry (U. Wisconsin, 1971). Res. scient. (tel. 302 + 772-3952). *Structure/function, structure determination of inorganic, organometallic, organic and macromolecular compounds, methodology, computing, data bases, instrumentation.*

Calandra, Mr Peter M. Innovative Tech. Inc., 205 Willow St., South Hamilton, MA 01982, USA.

Callahan, Dr Kenneth Paul. Occidental Res. Corp., P. O. Box 19601, Irvine, CA 92713, USA. (1943) PhD, inorganic chemistry (U. California, Riverside,1969). Sr. res. chemist. (tel. 714 + 957-7263). *Synthesis, structure, inorganic molecules.*

Camerman, Prof. Arthur. Dept. of Medicine (Neurology), RG-27, U. of Washington, Seattle, WA 98195, USA. (1939) PhD, chemistry (U. British

Columbia, Canada, 1964). Prof. (tel. 206 + 543-2340). *Mechanisms, drug action, drug design, structure - activity relationships, biological molecules.*

Campana, Dr Charles F. X-ray Instrument Div., Nicolet Instrument Corp., 5225-5 Verona Rd., Madison, WI 53711, USA. (1947) PhD, inorganic chemistry (U. Wisconsin, Madison, 1976). Sr. scient. (tel. 608 + 271-3333, ext. 2658). *X-ray crystallography, inorganic chemistry, organometallic chemistry.*

Cantrell, Prof. Joseph Sires. Chemistry Dept., Miami U., Oxford, OH 45056, USA. (1932) PhD, physical chemistry (Kansas State U., 1961). Assoc. prof. (tel. 513 + 529-3013). *X-ray crystallography, organic and biological structures, solar energy for chemical applications, electron diffraction, photo electrochemistry.*

Capano, Mr Michael A. Room 13-4077, MIT, 77 Massachusetts Ave., Cambridge, MA 02139, USA.

Cargill, Dr George Slade, III. Res. Div., IBM T. J. Watson Res. Center, P. O. Box 218, Yorktown Heights, NY 10598, USA. (1943) PhD, applied physics (Harvard U., 1969). Res. staff member. (tel. 914 + 945-1958). *Atomic scale structure, magnetic properties, amorphous solids, small angle scattering.*

Carlson, Dr Ernest Howard. Geology Dept., Kent State U., Kent, OH 44242, USA. (1933) PhD, geology (McGill U., Canada, 1966). Assoc. prof. (tel. 216 + 672-3778). *X-ray crystallography, mineralogy, exploration geochemistry.*

Carnahan, Dr Gary Ellis. Div. of Lab. Medicine, Box 8118, Washington U., Sch. of Medicine, 660 S. Euclid Ave., St. Louis, MO 63110, USA. (1950) MD, PhD, biochemistry (Vanderbilt U., 1982). Resident physician. (tel. 314 + 362-3340). *Biochemically interesting structures, proteins.*

Caron, Dr Aimery Pierre. Office of Community Services, C. of the Virgin Islands, St. Thomas, VI 00801, USA. (1930) PhD, chemistry (U. Southern California, 1962). Dir. (tel. 809 + 774-1252, ext. 291). *Molecular structures, X-ray diffraction.*

Carpenter, Mr Dewey K. Chemistry Dept., Louisiana State U., Baton Rouge, LA 70803, USA.

Carpenter, Dr Donald Allmand. Union Carbide Nuclear Div., Y-12 Plant, Building 9203, Mail Stop 001, Oak Ridge, TN 37830, USA. (1941) PhD, inorganic chemistry (Georgia Inst. of Techn., 1968). Dev. chemist. (tel. 615 + 574-0931). *Materials science, ceramics, metallurgy.*

Carpenter, Prof. Gene B. Dept. of Chemistry, Brown U., Providence, RI 02912, USA. (1922) PhD, physical chemistry (Harvard U., 1947). Prof. (tel. 401 + 863-3389). *Crystal and molecular structure, X-ray diffraction techniques.*

Carperos, Mr William E. 2933 Paces Lake Dr. NW, Atlanta, GA 30339, USA.

Carrell, Dr Horace L. The Inst. for Cancer Res., Fox Chase Cancer Center, 7701 Burholme Ave., Philadelphia, PA 19111, USA. (1940) PhD, chemistry (U. Southern California, 1966). Sr. res. assoc. (tel. 215 + 728-2220). *Biologically interesting molecules, biological activity - molecular structure relationship, X-radiation effects on organic molecules.*

Carrithers, Mr Charles H. Enraf-Nonius Service Corp., 390 Central Ave., Bohemia, NY 11716, USA. (tel. 516 + 589-2885).

Carter, Prof. Charles Williams, Jr. Biochemistry Dept., U. of North Carolina, Faculty Lab. and Office Bldg., 231H Chapel Hill, NC 27514, USA. (1945) PhD, biology (U. California-San Diego, 1972). Assoc. prof. (tel. 919 + 966-3263). *Macromolecular structure (function and evolution), genetic molecular apparatus.*

Carter, Dr Forrest Lee Chem. Div., Code 6170, Naval Res. Lab., Washington, DC 20375-5000, USA. (1930) PhD, chemistry (CIT, 1957). Res. chemist. (tel. 202 + 767-2100). *Molecular electronics, valence bonding, high coordination compounds, bonding surfaces, atomic volumes in crystals.*

Carter, Dr William S. 303 Clinton St., Fayetteville, NY 13066, USA. (1941) PhD, chemistry (Syracuse U., 1969). Dir. of res., King Labs. (tel. 315 + 471-8123). *Intermetallic compounds, liquid crystalline compounds.*

Cartz, Prof. Louis. C. of Eng., Marquette U., 1515 W. Wisconsin Ave., Milwaukee, WI 53233, USA. (1926) PhD, crystallography (U. London, UK, 1954). Prof. (tel. 414 + 224-3517). *X-ray crystallography, radiation damage, minerals, ceramics, graphitic materials.*

Case, Mr J. A. M. 2404 N. 41st St., Milwaukee, WI 53210, USA. (tel. 414 + 444-5735).

Caslavsky, Mr Jaroslav L. 244 East St., Lexington, MA 02173, USA.

Caspar, Prof. Donald L. D. Rosenstiel Basic Medical Sci. Center, Brandeis U., 415 South St., Waltham, MA 02254, USA. (1927) PhD, biophysics (Yale U., 1955). Prof. (tel. 617 + 647-2465). *Structural biology of viruses and membranes, X-ray diffraction and electron microscopy, macromolecular assemblies.*

Caughlan, Prof. Charles N. Dept. of Chemistry, Montana State U., Bozeman, MT 59715, USA. (1915) PhD, physical chemistry (U. Washington, 1941). Prof. (tel. 406 + 994-2571). *X-ray diffraction structures, organo-phosphorus compounds, organic molecules, structure - chemical reactivity relation, solution structure - crystal structure relation.*

Cavin, Mr Odis Burl. Metals and Ceramics, Oak Ridge Nat. Lab., P.O. Box X, Oak Ridge, TN 37830, USA. (1929) MS, metallurgy (U. Tennessee, 1959). Res. assoc. (tel. 615 + 483-8611, ext. 3-1460). *X-ray diffraction studies, powders and single crystals.*

Chakoumakos, Dr Bryan Charles. Dept. of Geology, U. of New Mexico, Northrop Hall, Albuquerque, NM 87131, USA. (1955) PhD, mineralogy (Virginia Inst. Tech., 1984). Post-doctoral fellow. (tel. 505 + 277-9447). *Mineralogy, crystal chemistry, materials science.*

Chamberland, Mr B. L. Dept. of Chemistry, U-60, U. of Connecticut, Storrs, CT 06268, USA.

Chambers, Mr John L. ARACOR, 1223 E. Arques Ave., Sunnyvale, CA 94086, USA.

Champion, Mr William C. Chemistry Dept., Colorado C., Colorado Springs, CO 80903, USA.

Chandrasekhar, Mr K. 8306 14th Ave., Apt. 202, Hyattsville, MD 20783, USA.

Chandross, Prof. Ronald Jay. Dept. of Physics, U. of Virginia, Physics Bldg., McCormick Rd., Charlottesville, VA 22901, USA. (1935) PhD, physical chemistry (MIT, 1961). Sr. res. assoc. (tel. 804 + 924-6803). *Proteins, small angle scattering, instrumentation.*

Chaney, Dr Michael Owen. Physical Chemistry Dept., MC 525, Eli Lilly Res. Lab., 307 E. McCarty St., Indianapolis, IN 46206, USA. (1943) PhD, chemistry (Indiana U., 1969). Res. scient. (tel. 317 + 261-4135). *X-ray crystallographic studies, biologically important molecules, structure - activity relationships, molecular modeling.*

Chang, Dr Chong-Hwan. Biological and Medical Res. Div., Argonne Nat. Lab., 9700 S. Cass Ave., Argonne, IL 60439, USA. (1950) PhD, crystallography (U. Pittsburgh, 1982). (tel. 312 + 972-3887). *Macromolecular structure determination, immunoglobulin, photo-reaction centers, interactive computer graphics.*

Chang, Prof. Shih-Chi. Physics Dept., Duquesne U., Pittsburgh, PA 15219, USA. (1933) PhD, physics (Kansas State U., 1963). Assoc. prof. (tel. 412 + 434-6353). *Structure analysis, X-ray diffraction.*

Chasen, Prof. Edith. Physics Dept., St. John's U., Grand Central and Utopia Parkways, Jamaica, NY 11439, USA. (1947) MA, geology crystallography (Brown U., 1970). Instructor. (tel. 718 + 990-6161, Ext. 6289). *Mineralogy, crystallography.*

Chastain, Dr Roger Vernon, Jr. Automation Management Office, 97th General Hospital, Frankfurt Army Regional Medical Center, APO NY 09757, Germany. (1938) PhD, chemistry (U. Washington, 1965). Major. *Data communications, data acquisition, data analysis, process control.*

Chaudhuri, Mr Jharna. 42 Heather Dr., Somerset, NJ 08873, USA.

Chayka, Mr Paul V. 2696 Van Buren St., Weedsport, NY 13166, USA.

Cheer, Prof. Clair James. Chemistry Dept., U. of Rhode Island, Kingston, RI 02881, USA. (1937) PhD, organic chemistry (Wayne State U., 1964). Assoc. prof. (tel. 401 + 792-2103). *Organic crystallography, structurally novel substances, biologically interesting small molecules.*

Chen, Dr Cheng-San. Storage System, Advanced Dev., Digital Equipment Corp., 333 South St., SHR-3/E29, Shrewsbury, MA 01545, USA. (1943) PhD, biophysics (SUNY, Buffalo, 1977). Sr. eng. (tel. 617 + 841-3254). *Small molecules, macromolecules, crystal structures, biological activities, magnetic recording, molecular modeling, molecular simulation.*

Chen, Mr Haydn H. Dept. of Metallurgy and Mining, U. of Illinois, 1304 W. Green St., Urbana, IL 61801, USA.

Chiang, Dr Michael Yen-Nan. Dept. of Chemistry, Columbia U., 119th St. and Broadway, New York, NY 10027, USA. (1954) PhD, chemistry (U. Southern California, 1984). Manager, X-ray lab. (tel. 212 + 280-8402). *Neutron diffraction, X-ray diffraction, small molecules.*

Chidester, Ms Connie. Physical and Analytical Res., Upjohn Co., 301 Henrietta St., Kalamazoo, MI 49001, USA. (1937) MA, mathematics (Western Michigan U., 1968). Res. scient. (tel. 616 + 385-7624). *Organic and biological crystal structures, computer programming.*

Childs, Dr Jerry D. Halliburton Services, Halliburton Co., Chemical Res. and Dev., Duncan, OK 73536, USA. (1943) PhD, physical chemistry (U. Oklahoma, 1972). Chemist. (tel. 405 + 251-3907). *Cement, spectroscopy, thermodynamics.*

Chiong, Ms Pauline 1700 Maxwell Dr., Yorktown Heights, NY 10598, USA.

Chipman, Dr David Randolph. Materials Characterization Div., Army Materials and Mechanics Res. Ctr., Watertown, MA 02171, USA. (1928) ScD, metallurgy (MIT, 1955). Res. physicist. (tel. 617 + 923-5392). *Charge and momentum density, crystal imperfections, amorphous metal alloys.*

Chiu, Ms Celia C. 10137 Pasture Gate Lane, Columbia, MD 21044, USA.

Chiu, Prof. Wah. Dept. of Biochemistry, U. of Arizona, Biosciences West Bldg., Room 308, Tucson, AZ 85721, USA. (1947) PhD, biophysics (U. California, Berkeley, 1975). Assoc. prof. (tel. 602 + 621-7524). *Structural biophysics, electron crystallography, macromolecules.*

Choi, Dr Chang Sun. Reactor Div., Nat. Bureau of Standards, Bldg. 235, Gaithersburg, MD 20899, USA. (1926) PhD, physics (Kyung Pook U., Korea, 1968). Res. physicist. (tel. 301 + 921-3634). *Neutron diffraction, X-ray diffraction, applied crystallography, crystal structures.*

Christofferson, Dr Glen D. Analytical Res. and Services Div., Chevron Res. Co., 576 Standard Ave., Richmond, CA 94802, USA. (1931) PhD, physical chemistry (UCLA, 1958). Sr. res. assoc. (tel. 415 + 620-2837). *Crystal chemistry, powder diffraction.*

Christoph, Dr Gary Gordon. Neutron Scattering Group (P-8), Los Alamos Nat. Lab., P. O. Box 1663 Mailcode H805, Los Alamos, NM 87545, USA. (1945) PhD, chemical physics (U. Chicago, 1971). Sr. scient. (tel. 505 + 667-6069). *Neutron diffraction, catalysis, powder diffraction methods (Reitveld analysis).*

Chu, Prof. Shirley Shan-Chi. Electrical Eng. Dept., S. of Eng. and Appl. Sci., Southern Methodist U., Dallas, TX 75275, USA. (1929) PhD, physical chemistry, crystallography (U. Pittsburgh, 1961). Assoc. prof. (tel. 214 + 692-3024). *X-ray crystal structure determination, organic compounds, photovoltaic solar energy conversion, crystal growth, electronic materials structure characterization.*

Churchill, Prof. Melvyn Rowen. Dept. of Chemistry, State U. of New York at Buffalo, Buffalo, NY 14214, USA. (1940) PhD, inorganic chemistry (Imperial C., London, UK, 1964). Prof. (tel. 716 + 831-3906). *Organo-transition metal structural chemistry.*

Clancy, Ms. Laura Lee. Dept. of Crystallography, U. of Pittsburgh, Pittsburgh, PA 15260, USA. (1950) BS, medical technology (Michigan State U., 1977) Grad. student. (tel. 412 + 624-4366). *Protein structure, bioenergetics, metal salt hydrates, immunobiology.*

Clardy, Prof. Jon Christel. Chemistry Dept., Baker Lab., Cornell U., Ithaca, NY 14853, USA. (1943) PhD, chemistry (Harvard U., 1969). Prof. (tel. 607 + 256-7583). *Natural products, organic chemistry.*

Clark, Ms Connie M. 2391 W. Rapallo Way, Tucson, AZ 85741, USA. (1945) MS, geochemistry and computer science (PSU, 1972 and 1974). Computer programmer. *Computer programming, geochemistry, mineralogy, crystallography.*

Clark, Prof. Edward Shannon. Polymer Eng., U of Tennessee, Knoxville, TN 37916, USA. (1930) PhD, chemistry (U. California, Berkeley, 1956). Prof. (tel. 615 + 974-5340). *Crystalline polymers, structure - properties relationship.*

Clark, Mrs Joan Robinson. 56 Citation Dr., Los Altos, CA 94022-7136, USA. (1920) PhD, crystallography (Johns Hopkins U., 1958). Retired (tel. 415 + 960-0628). *Crystallography, mineralogy.*

Clark, Prof. John Robert. Dept. of Physics, P. O. Box 293, Naval Postgraduate S., Pebble Beach, CA 93953, USA. (1911) ScD, metallurgy (MIT, 1942). Consultant. (tel. 408 + 624-7872). *Physical properties, inorganic materials.*

Clarke, Prof. Roy. Physics Dept., U. of Michigan, Ann Arbor, MI 48109, USA. (1947) PhD, physics (U. London, UK, 1973). Assoc. prof. (tel. 313 + 764-4466). *Graphite intercalation compounds, phase transitions, ferroelectrics, X-ray scattering, heterostructures.*

Claus, Mr Albert C. 105 N. Lancaster St., Mount Prospect, IL 60056, USA.

Clayton, Mr William Rex. Techn. Div., International Fertilizer Dev. Center, P. O. Box 2040, Muscle Shoals, AL 35660, USA. (1938) PhD, physical chemistry (Texas A and M U., 1971). Res. chemist. (tel. 205 + 381-6600, ext. 294). *Inorganic structures.*

Clearfield, Prof. Abraham. Chemistry Dept., Texas A and M U., College Station, TX 77843, USA. (1927) PhD, physical chemistry, crystallography (Rutgers U., 1954). Prof. *X-ray and neutron powder diffraction, inorganic ion exchangers, solid electrolytes, catalysis.*

Clifton, Prof. Donald Frederic. Dept. of Mining Eng. and Metallurgy, U. of Idaho, C. of Mines, Moscow, ID 83843, USA. (1917) PhD, metallurgy (U. Utah, 1957). Prof. (tel. 208 + 885-6376). *Transformations, structure - properties relationship.*

Clinger, Mr Kent. Dept. of Chemistry, U. of Texas, Austin, TX 78712, USA.

Cochran, Prof. Todd G. S. of Pharmacy, U. of Montana, Missoula, MT 59812, USA. (1943) PhD, pharmaceutical chemistry (U. Washington, 1970). Asst. prof. (tel. 406 + 243-6495). *Biological small molecules.*

Cody, Dr Vivian. Molecular Biophysics Dept., Medical Foundation of Buffalo, 73 High St., Buffalo, NY 14203-1196, USA. (1943) PhD, chemistry (U. Cincinnati, 1969). Assoc. res. scient. (tel. 716 + 856-9600, ext. 434). *Molecular endocrinology, structure - function relationships, thyroid - hormones, antifolates, drug design, biological molecules, crystal structures.*

Cohen, Prof. Carolyn. Rosenstiel Basic Medical Sci. Res. Center, Brandeis U., 415 South St., Waltham, MA 02154, USA. (1929) PhD, biophysics (MIT, 1954). Prof., chairman, grad. biophysics program. (tel. 617 + 647-2466). *Structural biology, protein assemblies in the cell.*

Cohen, Mr Charles I. Materials Analysis Lab., Owens Corning Fiberglas Corp., Techn. Center, Granville, OH 43023, USA. (1933) MS, geology (Columbia U., 1961). Sr. res. assoc. (tel. 614 + 582-0610, ext. 228). *Non-routine X-ray diffraction analyses.*

Cohen, Dr Gerson H. Nat. Insts. of Health, 9000 Wisconsin Ave., Bethesda, MD 20014, USA. (1939) PhD, chemistry (Cornell U., 1965). Res. chemist. (tel. 301 + 496-4295). *Protein crystallography, protein structure refinement, computer graphics, structure display and refinement, crystallographic computing, minicomputers, automated diffractometry and densitometry.*

Cohen, Ms Janet Paula (Lentz). Fish and Neave, 875 Third Ave., New York, NY 10708, USA. (1951) J.D. (Juris Doctor) (Pace U., 1982). Lawyer. (tel. 212 + 715-0600). *Patent law.*

Cohen, Prof. Jerome Bernard. Materials Sci. and Eng., Northwestern U., 2145 Sheridan Road, Evanston, IL 60201, USA. (1932) ScD, metallurgy (MIT, 1957). Prof. (tel. 312 + 491-7817). *X-ray diffraction, neutron diffraction, deformation, phase transformations, residual stresses, catalysis, oxides, alloys, local order, clustering.*

Cole, Dr Henderson. IBM Instruments Inc., Orchard Park, P.O.Box 332, Danbury, CT 06810, USA. (1924) PhD, physics (MIT, 1952), Sr. techn. specialist. (tel. 203 + 796-2400). *Perfect crystal diffraction, laboratory automation.*

Colella, Prof. Roberto. Physics Dept., Purdue U., West Lafayette, IN 47907, USA. (1935) PhD, physics (U. Milan, Italy, 1958). Prof. (tel. 317 + 494-3029, telex 272396). *Diffraction physics, perfect crystals, phonons, diffuse scattering, charge densities, charge density waves.*

Collins, Prof. Douglas MacPherson. Dept. of Chemistry, Texas A. and M. U., College Station, TX 77843, USA. (1939) PhD, chemistry (Rutgers U., 1966). Asst. prof. (tel. 409 + 845-2011). *Structural chemistry, large biological molecules, electron density representations, chemical bonding, crystal structure determination theory.*

Collins, Dr Richard Christopher. Res. and Dev., Occidental Chemical Co., Box 300, White Springs, FL 32096, USA. (1946) PhD, analytical chemistry (U. Texas, Austin, 1977). Analytical chemist. (tel. 904 + 397-8186). *Solid state chemistry, X-ray methods of analysis.*

Comey, Mr Paul Van A. Consultant, 151 Apt., 340 Eastern Promenade, Portland, ME 04101, USA. (1898) BChE, (NYU, 1954). Retired.

Conant, Dr John W. Los Alamos Scient. Lab., U. of California, P. O. Box 1663,Los Alamos, NM 87545, USA. (1924) PhD. chemistry (U. Iowa, 1955). Staff member.

Condren, Mr S. M. Box 20, Christian Brothers C., 650 Parkway South, Memphis, TN 38104, USA.

Cook, Dr William Joseph Dept. of Pathology, U. of Alabama at Birmingham, University Station, Birmingham, AL 35294, USA. (1949) MD (U. Alababama, Birmingham, 1974, 1976). Assoc. prof. (tel. 205 + 934-4880). *Biological crystallography.*

Cook, Dr William R., Jr. P. O. Box 17157, Cleveland, OH 44110, USA. (1927) PhD, geology (Case Western Reserve U., 1971). Corporate secretary. (tel. 216 + 486-6100). *Solid state chemistry, piezoelectricity and ferroelectricity, mineralogy.*

Cooper, Mrs Ann S. Solid State Res. Dept., AT&T Bell Labs., Mountain Ave., Murray Hill, NJ 07974, USA. (1929) BS, chemistry (St. Lawrence U., 1950). Member techn. staff. (tel. 201 + 582-6921). *Powder X-ray diffraction, crystal growing.*

Cooper, Ms Jeanette. General Electric Lamp Div., 131 Nela Park, Cleveland, OH 44112, USA.

Copeland, Prof. Richard Franklin. Div. of Sci. and Math., Bethune-Cookman C., Daytona Beach, FL 32015, USA. (1938) PhD, chemistry (Texas A. and M. U., 1965). Prof. (tel. 904 + 255-1401, ext. 328). *Organometallic complexes, scientific and technical information, crystallographic computing.*

Coppens, Prof. Philip. Chemistry Dept., State U. of N.Y. at Buffalo, Acheson Hall, Buffalo, NY 14221, USA. (1930) PhD, physical chemistry (U. Amsterdam, Netherlands, 1960). Prof. (tel. 716 + 831-3911). *Charge density distributions in solids, chemical bonding, low temperature crystallography, accurate measurements, crystallographic computing, low dimensional conductors, structure - physical properties relationships.*

Cordes, Prof. A. Wallace. Chemistry Dept., U. of Arkansas, Fayetteville, AR 72701, USA. (1934) PhD, inorganic chemistry (U. Illinois, 1959). Prof. (tel. 501 + 575-4601). *Inorganic ring and cage molecules, chemistry, arsenic - sulphur - selenium, heterocyclic structures.*

Corey, Prof. Eugene Ray. Dept. of Chemistry, U. of Missouri, St. Louis, 8001 Natural Bridge Road, St. Louis, MO 63121, USA. (1935) PhD, inorganic chemistry (U. Wisconsin, 1963). Prof. (tel. 314 + 553-5717). *Structural chemistry, organometallic compounds.*

Corfield, Prof. Peter William Reginald. Chemistry Dept., The Kings C., Briarcliff Manor, NY 10510, USA. (1937) PhD, chemistry (Durham U., UK, 1963). Prof. (tel. 914 + 941-7200, ext. 203). *Crystal structures, coordination compounds; computer software.*

Corliss, Dr Lester Myron. Chemistry Dept., Brookhaven Nat. Lab., Upton, NY 11973, USA. (1919) PhD, chemical physics (Harvard U., 1949). Sr. chemist. (tel. 516 + 345-4376). *Neutron scattering, magnetism, phase transformations, critical phenomena, diffuse motion in solids.*

Cotton, Prof. Frank Albert. Dept. of Chemistry, Texas A. and M. U., College Station, TX 77843, USA. (1930) PhD, chemistry (Harvard U., 1955). Robert A. Welch distinguished prof. (tel. 713 + 845-4432). *Inorganic chemistry, protein structures.*

Coulman, Ms Betty Ann. Philips Res. Lab.s, M/S 2165, Signetics Corp., 811 E. Arques Ave., Sunnyvale, CA 94088, USA.

Coulter, Dr Charles L. Biotechnology Resources Program, Nat. Inst. of Health, Bethesda, MD 20205, USA. (1933) PhD, chemistry (UCLA,1960). Head, biological structure section. (tel. 301 + 496-5411). *Structural biochemistry, instrumentation.*

Cowley, Prof. John M. Dept. of Physics, Arizona State U., Tempe, AZ 85287, USA.

Cox, Dr David Ernest. Physics Dept., Brookhaven Nat. Lab., Upton, NY 11973, USA. (1934) PhD, inorganic chemistry (Royal C. of Sci., London, UK, 1959). Physicist. (tel. 516 + 282-3818, telex 6852443). *Chemical crystallography, solid-state chemistry, complex oxides, Rietveld analysis, neutron and X-ray powder diffraction, synchrotron X-ray powder diffraction, zeolites, magnetic structures.*

Cox, Dr John Wesley, Jr. Wendt-Sonis Div., TRW, 205 N. 13th St., Rogers, AR 72756, USA. (1940) PhD, structural inorganic chemistry (U. Cincinnati, 1969). Dir., res. and dev. (tel. 501 + 636-1515, ext. 237). *Structure - properties relationship, refractory materials; crystallographic applications, ceramics, carbide metallurgy.*

Craston, Mr Dennis F. 10834 N. 44th St., Phoenix, AZ 85028, USA. (1925) MSc analytical chemistry (Fordham U., 1975). Sr. physicist. (tel. 212 + 340-0127, ext. 120 - 123). *Forensic toxicology, X-ray diffraction, forensic spectroscopy, drug identification.*

Craven, Prof. Bryan Maxwell. Dept. of Crystallography, U. of Pittsburgh, Pittsburgh, PA 15260, USA. (1932) PhD, chemistry (U. New Zealand, 1957). Prof. (tel. 412 + 624-4366). *Biological structure.*

Criasia, Mr Ronald T. Chemistry-Physics Dept., Kean C., Union, NJ 07083, USA.

Crist, Prof. Buckley, Jr. Dept. of Materials Sci. and Eng., Northwestern U., Tech. Inst., Evanston, IL 60201, USA. (1941) PhD, chemistry (Duke U., 1966) Prof. (tel. 312 + 491-3279). *Polymers, morphology, mechanical properties, X-ray diffraction, small-angle X-ray and neutron scattering, optical properties.*

Croft, Dr William J. Materials Characterization Div., Army Materials & Mechanics Res. Cntr., Arsenal St., Watertown, MA 02172, USA. (1926) PhD, crystallography (Columbia U., 1954). Chemist. (tel. 617 + 923-5358). *Electron diffraction, X-ray diffraction, ceramic materials.*

Cromer, Dr Don Tiffany. INC-4 MS C346, Los Alamos Scient. Lab., P. O. Box 1663, Los Alamos, NM 87545, USA. (1923) PhD, chemistry (U. Wisconsin, 1953). Staff member. (tel. 505 + 667-6045). *Intermetallic structures, charge density, computer programming, scattering factors, anomalous dispersion.*

Cucka, Dr Paul. Group Techn. Center, BOC, 100 Mountain Ave., Murray Hill, NJ 07974, USA. (1928) PhD, X-ray crystallography (U. Birmingham, UK, 1952). Dir., Materials R & D. (tel. 201 + 464-8100, ext. 4404). *Materials.*

Cullen, Prof. David Lawrence. Chemistry Dept., Connecticut C., New London, CT 06320, USA. (1940) PhD, chemistry (U. Washington, 1969). Asst. prof. (tel. 203 + 442-5391, ext. 280). *Porphyrin structures, bile pigments, biologically interesting molecules; coordination chemistry.*

Cummings, Dr John Patrick. Administrative Div., Owens-Illinois, P. O. Box 1035, Toledo, OH 43666, USA. (1933) PhD, chemistry (U. Texas, 1968) JD, law (U. of Toledo, 1972). Manager, Environmental Affairs. (tel. 419 + 247-2609). *Environmental, inorganic crystals, trace material studies.*

Curtin, Prof. David Yarrow. Dept. of Chemistry, U. of Illinois, 1209 W. California, Urbana, IL 61801, USA. (1920) PhD, chemistry (U. Illinois, 1945). Prof. (tel. 217 + 333-0797). *Chemistry, organic compounds in the solid state.*

Cuthill, Dr John R. 12700 River Rd., Potomac, MD 20854, USA. (1918) PhD, metallurgy (Purdue U., 1952).

Czerwinski, Prof. Edmund William. Div. of Biochemistry, U. of Texas Medical Branch, Galveston, TX 77550, USA. (1940) PhD, biochemistry (Indiana U., 1971). Asst. prof. (tel. 409 + 761-3287). *Protein crystallography, biologically important small molecules.*

D'Addario, Dr Anthony Paul. U.S. Navy, Navy Drug Screening Lab., San Diego, CA 92134, USA. (1941) PhD, chemistry (Case Inst. of Techn., 1971). Dir., techn. services. (tel. 619 + 233-2349).

Dahl, Prof. Lawrence F. Chemistry Dept., U. of Wisconsin, Madison, WI 53706, USA. (1929) PhD, physical chemistry (Iowa State U., 1956) Prof. (tel. 608 + 262-3183). *Synthesis, structure, bonding, transition metal compounds; organometallics; metal cluster systems.*

Dalley, Prof. Nelson Kent. Chemistry Dept., Brigham Young U., Provo, UT 84602, USA. (1935) PhD, chemistry (U. Texas, Austin, 1968). Prof. (tel. 801 + 378-3434). *Crystal structures, cyclic polyethers (derivatives and cation complexes).*

Danko, Mr Andrew William. 371 Southcroft Rd., Springfield, PA 19064, USA. (1917) BS, chemical engineering (U. Pittsburgh, 1940). Retired. (tel. 215 + 544-7515). *Metallurgy, chemistry.*

Dann, Mr Jeffrey Neil. Chemical and Metallurgical Div., GTE Products Corp., Towanda, PA 18848, USA. (1946) BS, physics (The Cooper Union, 1967). Sr. eng., X-ray diffraction lab. (tel. 717 + 265-2121, ext. 425). *X-ray powder diffraction, computer applications.*

Dantonio, Mr Peter. Code 6030, U.S. Naval Res. Lab., 4555 Overlook Ave., S.W. Washington, DC 20375, USA.

Darling, Mr Stephen D. Dept. of Chemistry, U. of Akron, Akron, OH 44325, USA.

Das, Dr Badri Narayan. Condensed Matter & Radiation Sci., Code 6632, U.S. Naval Res. Lab., 4555 Overlook Ave. S.W., Washington, DC 20375, USA. (1927) PhD, metallurgy (Illinois Inst. of Techn., 1964). Metallurgist. (tel. 202 + 767-3614). *Materials processing; crystal growth; structure, physical properties, magnetic properties, superconducting properties.*

Davies, Dr David R. Lab. of Molecular Biology, Nat. Insts. of Health, Bldg.2, Room 316, Bethesda, MD 20205, USA. (1927) PhD, chemical crystallography (Oxford U., UK, 1952). Chief, molecular structure sect. (tel. 301 + 496-4295). *Proteins, nucleic acids, fibers.*

Davis, Dr Briant LeRoy. Inst. of Atmospheric Sci., South Dakota S. of Mines and Techn., Rapid City, SD 57701, USA. (1936) PhD, geology (UCLA, 1964). Res. prof., geophysics. (tel. 605 + 394-2291). *Cloud physics, nucleation processes, air pollution chemistry and physics.*

Davis, Prof. Phillip Howard. Chemistry Dept., U. of Tennessee at Martin, Martin, TN 38238, USA. (1946) PhD, chemistry (U. Illinois, 1972). Asst. prof. (tel. 901 + 587-7456). *Magnetic resonance, X-ray crystallography, molecular and electronic structure, inorganic materials.*

Davis, Prof. Raymond Edward. Dept. of Chemistry, U. of Texas, Austin, TX 78712, USA. (1938) PhD, chemistry (Yale U., 1965). Prof. (tel. 512 + 471-3097). *Molecular structure determination, organometallic compounds, organic structures.*

Day, Dr Cynthia Ann Secauer. Crystalytics Co., P.O. Box 82286, Lincoln, NE 68501, USA. (1952) PhD. chemistry (U. Nebraska). Pres. (tel. 402 + 489-6393). *Structure, small molecules; organoactinides (novel), inorganic coordination complexes, biological model systems.*

Day, Prof. Roberta Ogilvie. Chemistry Dept., U. of Massachusetts, Amherst, MA 01003, USA. (1941) PhD, physical chemistry (MIT, 1971). Asst. prof. (tel. 413 + 545-2422).

Day, Prof. Victor Warren. Chemistry Dept., U. of Nebraska, Lincoln, NE 68588, USA. (1943) PhD, physical chemistry (Cornell U., 1969). Assoc. prof. (tel. 402 + 472-3531). *Structure and bonding, organic - inorganic and organometallics, structure - function relationships, biological macromolecules.*

De la Camp, Prof. Ulrich Otto. Chemistry Dept., California State C., Dominguez Hills, Carson, CA 90747, USA. (1929) PhD, chemistry (U. California, Davis, 1966). Prof. (tel. 213 + 515-3376). *Structure and conformation, small organic molecules, anomalous scattering of X-rays.*

De Boer, Dr Barry Goodwin. Lighting Products Div., GTE Products Corp. (Sylvania), 60 Boston St., Salem, MA 01970, USA. (1942) PhD, chemistry (U. California, Berkeley, 1968). Eng. specialist. (tel. 617 + 777-1900, ext. 3568). *Methods (experimental and computational), X-ray diffraction, high temperature studies, solid state chemistry, phosphors, luminescent materials.*

De Camp, Dr Wilson H. II. Food and Drug Administration, HFN-180 200 C St., S. W., Washington, D. C. 20204, USA. (1936) PhD, physical chemistry (U. Maryland, 1970). Chemist. (tel. 202 + 245-2750). *Organic crystal structures, absolute configuration, natural products, crystallographic computing, conformational analysis, powder diffraction, drug polymorphism.*

De Fontaine, Prof. Robert Didier. Materials Sci. Dept., Hearst Mining Bldg., U. of California at Berkeley, Berkeley, CA 94720, USA. (1931) PhD, materials science (Northwestern U., 1967). Prof. (tel. 415 + 642-8177). *Phase transformations in alloys, thermodynamics of solids.*

De Haven, Dr Patrick William. IBM Corp., Zip 41C, Route 52, Hopewell Junction, NY 12533, USA. (1949) PhD, physical chemistry (Iowa State U., 1976). Staff eng. (tel. 914 + 894-6859). *X-ray powder diffraction (high temperature).*

De Jarnette, Miss F. Elaine. Molecular Biophysics, The Medical Foundation of Buffalo, 73 High St., Buffalo, NY 14203, USA. (1931) BS, education (SUNY, Buffalo, 1970). Asst. res. scient. (tel. 716 + 856-9600, ext. 53). *Crystallography.*

De Lucia, Dr Mary Lou. 61D-300/2 Kimberly-Clark Corp., 1400 Holcomb Bridge Rd., Roswell, GA 30076, USA. (1947) PhD, physical chemistry (SUNY Buffalo, 1978). Res. scient. (tel. 413 + 243-1000, ext. 233). *Paper structure and performance, biological structure and function.*

De Maggio, Mr Gregory B. 22521 Euclid, St. Clair Shores, MI 48082, USA.

De Meester, Dr Patrice. S. of Eng. and Appl. Sci., SMU, 11464 Pagemill Rd., Dallas, TX 75243, USA. (1943) PhD, crystallography (U. London, UK, 1973). Res. fellow. (tel. 214 + 340-9140). *Chemical crystallography.*

De Rosier, Prof. David John. Dept. of Biology, Brandeis U., 215 South St., Waltham, MA 02254, USA. (1939) PhD, biophysics (U. Chicago, 1965). (tel. 617 + 647-2126). *Actin structure, bacterial flagella, image analysis.*

De Titta, Dr George Thomas. Molecular Biophysics Dept., Medical Foundation of Buffalo, 73 High St., Buffalo, NY 14203, USA. (1947) PhD, biochemistry-crystallography (U. Pittsburgh, 1973). Assoc. res. scient. (tel. 716 + 856-9600, ext. 472). *Biocrystallography, prostaglandins, biotin vitamins, lipid related molecules.*

De Vries, Dr Adriaan. Liquid Crystal Inst., Kent State U., Kent, OH 44242, USA. (1931) PhD, physical chemistry, X-ray crystallography (State U. of Utrecht, Netherlands, 1963). Sr. res. fellow, adjunct assoc. prof. of physics. (tel. 216 + 672-2654). *Organic crystal structures, liquid crystals, liquids (complex molecules).*

Deadwyler, Mr Daniel A. Dept. of Chemistry, U. of North Carolina, Charlotte, NC 28223, USA.

Dean, Mr Johnny Clyde. Res. Dept., Denka Chemical Corp., 8701 Park Place Blvd., Houston, TX 77017, USA. (1928) BS, chemistry (U. Southwestern Louisiana, 1952). Staff chemist. (tel. 713 + 477-8821, ext. 553). *X-ray diffraction, polycrystalline materials.*

Delaney, Prof. Matthew Sylvester. Office of the Academic Dean, Mount St. Mary's C., 12001 Chalon Rd., Los Angeles, CA 90049, USA. (1927) PhD, mathematics (The Ohio State U., 1971). Academic dean. (tel. 213 + 476-2237, ext. 3504). *Mathematical crystallography.*

Delgado, Mr Jose Miguel. MIT, Room 13-4069, 77 Massachusetts Ave., Cambridge, MA 02139, USA.

Delord, Mr Terry. Dept. of Chemistry, Texas A & M U., College Station, TX 77843, USA.

Depalma, Mr Vincent M. Zip 81K, IBM Corp., East Fishkill, Route 52, Hopewell Junction, NY 12533, USA.

Deroski, Ms Betty Rolfs. Suffolk County Community C., 533 College Rd., Selden, NY 11784, USA.

Desmeules, Mr Peter J. 700 Parnassus Ave., Apt. 3, San Francisco, CA 94122, USA.

Desper, Dr C. Richard. Army Materials and Mechanics Res. Center, AMXMR-OX, Watertown, MA 02172, USA. (1937) PhD, chemistry (U. Massachusetts, Amherst, 1966). Res. chemist. (tel.617 + 923-5391). *Synthetic polymers.*

Dewan, Dr John C. Dept. of Chemistry, MIT, 77 Massachusetts Ave., Cambridge, MA 02139, USA. (1949) PhD, inorganic chemistry (U. Western Australia, 1974). Dept. crystallographer. (tel. 617 + 253-1884). *Macromolecular crystallography, small-molecule crystallography, bio-inorganic chemistry.*

Dexter, David D. Computer Center, Alma C., Alma, MI 48801, USA.

Dickerson, Prof. Richard Earl. Molecular Biology Inst., UCLA, Los Angeles, CA 90024, USA. (1931) PhD, physical chemistry (U. Minnesota, 1957). Prof. and Dir. (tel. 213 + 825-5864). *DNA structure, DNA-drug binding, protein-DNA recognition, molecular evolution.*

Dickinson, Dr Charles. Detonation Physics Branch, Naval Surface Weapons Center, Silver Spring, MD 20903-5000, USA. (1937) PhD, chemistry (U. Maryland, 1972). Branch head. (tel. 301 + 394-1180). *Computing systems, small molecules, structure - properties relationship.*

Dickman, Mr Michael H. 1017 Lockman Ave., Cincinnati, OH 45238, USA.

Dieterich, Dr David Allan. Res. Labs., Eastman Kodak Co., 1669 Lake Ave., Rochester, NY 14615, USA. (1946) PhD, chemistry (U. Illinois 1973). Res. lab. head. (tel. 315 + 458-1000, ext. 7-6277). *Photographic science, organic and physical chemistry.*

Dobrott, Dr Robert D. International Assembly, Mostek, 1429 Lamp Post Lane, Richardson, TX 75080, USA. (1932) PhD, physical chemistry (Harvard U., 1964). Materials eng. management. (tel. 214 + 466-7751). *Solid state electronic materials.*

Dodd, Dr Charles Gardner. Connecticut Techn. Consultants Inc., P. O. Box 524, Stratford, CT 06497, USA. (1915) PhD, physical chemistry (U. Michigan, 1948). Pres. (tel. 203 + 375-5015). *Surface chemistry and physics, surface microanalysis, X-ray microanalysis, soft X-ray spectrometry, chemical bonding, ion implantation.*

Dodge, Prof. Richard Patrick. Chemistry Dept., U. of the Pacific, 3601 Pacific Ave., Stockton, CA 95211, USA. (1932) PhD, physical chemistry (U. California, Berkeley, 1958). Prof. (tel. 209 + 946-2272). *Molecular structure, computer applications, quantum chemistry.*

Doedens, Prof. Robert John. Chemistry Dept., U. of California, Irvine, CA 92717, USA. (1937) PhD, chemistry (U. Wisconsin, 1965). Prof. (tel. 714 + 856-6605). *Structural chemistry, inorganic and organometallics, coordination chemistry.*

Doherty, Dr Ruth Marie. Code R11, Bldg. 30, Room 110, Naval Surface Weapons Center, 10901 New Hampshire Ave., Silver Spring, MD 20903-5000, USA. (1948) PhD, chemistry (U. Maryland, 1980). Res. chemist. (tel. 202 + 394-2745). *Small molecule structures, energetic materials.*

Dollase, Prof. Wayne A. Geology Dept., U. of California, Los Angeles, CA 90024, USA. (1938) PhD, crystallography (MIT, 1966). Prof. (tel. 213 + 825-3823). *Crystal chemistry, mineralogy, solid state chemistry.*

Doman, Dr Robert Charles. Res. and Dev., Materials Res., Corning Glass Works, Sullivan Science Park FR-5-1, Corning, NY 14831, USA. (1930) PhD, geology (U. Wisconsin, 1961). Manager. (tel. 607 + 974-3279). *Ceramics, glassceramics, mineralogy, feldspar cordieribe, micas, beta-spodumene, beta-quartz.*

Donohue, Prof. Jerry. Chemistry Dept., U. of Pennsylvania, Philadelphia, PA 19104, USA. (1920) PhD, chemistry (CIT, 1947). Prof. (tel. 215 + 898-7994). *Crystal structures.*

Donohue, Dr Terence. Laser Physics Branch, Naval Res. Lab., Washington, DC 20375, USA. (1946) PhD, physical chemistry (Cornell U., 1973). Res. chemist. (tel. 202 + 767-2175). *Photochemistry, energy transfer, solar energy conversion.*

Doris, Mr Edward. MGR-XRD, Philips Electron Instruments Inc., 85-91 McKee Dr., Mahwah, NJ 07430, USA.

Dorset, Dr Douglas Lewis. Electron Diffraction Dept., Medical Foundation of Buffalo Inc., 73 High Street, Buffalo, NY 14203, USA. (1942) PhD, biophysics (U. Maryland, 1971). Principal res. scient., dept. head (tel. 716 + 856-9600, ext. 475). *Electron diffraction, crystal structure analysis, biomembrane structure, phase transitions, polymer crystal growth.*

Downs, Dr James Winston. Dept. of Geology and Mineralogy, The Ohio State U., 291 Watts Hall, 104 W. 19th St., Columbus, OH 43210, USA. (1952) Phd, mineralogy, (Virginia Poly. Tech., 1983). Asst. prof. (tel. 614 + 422-6290). *Mineralogy, charge density distributions in minerals, crystal chemistry, minerals.*

Dowty, Mr Eric. Dept. of Min. Sci., American Museum of Natural History, Central Park W. & 79th St., New York, NY 10024, USA.

Doyle, Prof. John Robert. Chemistry Dept., U. of Iowa, Iowa City, IA 52242, USA. (1924) PhD, chemistry (Tulane U., 1955). Prof. (tel. 319 + 353-3585). *Inorganic and organometallic chemistry, structures, organometallic compounds.*

Doyne, Prof. Thomas H. Dept. of Chemistry, Villanova U., Villanova, PA 19085, USA. (1927) PhD, biochemistry (PSU, 1957). Prof. of biochemistry, chairman. (tel. 215 + 527-2215, ext. 480). *Peptides, biologically interesting small molecules.*

Dragsdorf, Prof. Russell Dean. Physics Dept., Kansas State U., Cardwell Hall, Manhattan, KS 66506, USA. (1922) PhD, physics (MIT, 1948). Prof. (tel. 913 + 532-6809). *Defect structure, ion implantation, X-ray and electron diffraction.*

Drendel, Mr William B. 511 W. Main St., Apt. 103, Madison, WI 53703, USA.

Drickman, Dr Myra Vivian. 1114 Princeton St., Apt. 5, Santa Monica, CA 90403, USA. (1942) MD, medicine (New York U., 1976). *Medical imaging systems.*

Druyan, Prof. Mary Ellen. Dept. of Biochemistry, Loyola U., S. of Dentistry, 2160 South First Ave., Maywood, IL 60153, USA. (1938) PhD, biochemistry (U. Chicago, 1972). Asst. prof. (tel. 312 + 531-3578). *Single crystal diffraction (large and small molecules), low angle neutron scattering, biologically interesting systems.*

Duax, Dr William Leo. Molecular Biophysics Dept., Medical Foundation of Buffalo, 73 High Street, Buffalo, NY 14203, USA. (1939) PhD, physical chemistry (U. Iowa, 1967). Head. (tel. 716 + 856-9600, ext. 61). *Conformation, hormones, drugs, antibiotics; biological function - structure relationship; direct methods, data storage, data retrieval and analysis; computer graphics.*

Dubin, Mr Robert R. 4669 Persimmon Pl., San Jose, CA 95129, USA.

Duchamp, Dr David James. Pharmaceutical Res. and Dev., The Upjohn Co., 301 Henrietta St., Kalamazoo, MI 49001, USA. (1939) PhD, chemistry (CIT, 1965). Sr. scient. (tel. 616 + 385-7766). *Organic and biological crystal structures, crystallographic computing, molecular mechanics, potential energy calculations.*

Duesler, Ms Eileen N. Chemistry Dept., U. of New Mexico, Albuquerque, NM 87131, USA. Staff scient. (tel. 505 + 277-0505). *Small molecule structures.*

Dumke, Prof. Warren Lloyd. Dept. of Physics and Physical Sci., Marshall U., Huntington, WV 25703, USA. (1928) PhD, physical chemistry (U. Nebraska, 1965). Assoc. prof. (tel. 304 + 696-6764). *X-ray diffraction, neutron diffraction, quantum mechanical calculations.*

Dunn, Mr Harris William. Analytical Chemistry, Oak Ridge Nat. Lab., P. O. Box X, Bldg. 4500N, Room E-10, Oak Ridge, TN 37763, USA. (1918) BS, chemistry and physics (Western Kentucky U., 1942). Res. assoc. (tel. 615 + 574-4877). *X-ray diffraction, X-ray fluorescence, transmission electron microscopy, optical microscopy, scanning electron microscopy.*

Dunn, Mr Karl L., Jr. 1044 Joe Quick Rd., Hazel Green, AL 35750, USA. (tel. 205 + 828-0136).

Dunsieth, Ms Dana G. R. R. 2, Antioch Rd., Leesburgh, OH 45135, USA.

Durbetaki, Mr Antony J. FMC Corp., US 1 PO Box 8, Princeton, NJ 08540, USA.

Dwiggins, Dr Claudius William, Jr. 1211 S. Keeler St., Bartlesville, OK 74003, USA. (1933) PhD, physical chemistry (U. Arkansas, 1958). Res. chemist. (tel. 918 + 336-8546). *Small angle X-ray scattering, scattering theory, non-crystalline materials, ultracentrifuge, X-ray fluorescence.*

Dwight, Mr Austin Elbert. Physics Dept., Northern Illinois U., Faraday Hall, DeKalb, IL 60115, USA. (1919) MS, metallurgical engineering (U. Michigan, 1950). Res. scient. (tel. 815 + 753-0285). *Intermetallic structures, alloy crystal structure, chemistry, alloys and intermetallic compounds.*

Dyke, Mr Maurice. P. O. Box 32077, Aurora, CO 80041, USA.

Dytrych, Mr William J. 5107 S. Blackstone Ave., Apt. 802 Chicago, IL 60615, USA.

Ealick, Dr Steven Edward. Inst. of Dental Res., U. of Alabama in Birmingham, University Station SDB-13, Birmingham, AL 35294, USA. (1951) PhD, physical chemistry (U. Oklahoma, 1976). Res. fellow. (tel. 205 + 934-4259). *Structure and conformation of peptides, protein crystallography.*

Eanes, Dr Edward David. Lab. of Biological Structure, Nat. Inst. of Dental Res., Nat. Insts. of Health, Bethesda, MD 20014, USA. (1934) PhD, physical chemistry-crystallography (The Johns Hopkins U., 1961). Chief, molecular structure section. (tel. 301 + 496-2023). *Crystal chemistry, calcium phosphates, biological mineralization.*

Eddy, Prof. Lowell Perry. Dept. of Chemistry, Western Washington U., Bellingham, WA 98225, USA. (1920) PhD, chemistry (Purdue U., 1952). Assoc. prof. (tel. 206 + 676-3070). *Inorganic coordination chemistry, coordination compounds, structure.*

Edmonds, Dr James William. Phillips Electronic Instruments, 85 McKee Dr., Mahwah, NJ 07430, USA. (1943) PhD, chemistry (Rice U., 1968). Manager, market dev. (tel. 201 + 529-3800, Ext. 453). *X-ray powder diffraction, computer automated diffractometry.*

Edmundson, Mr Allen B. Dept. of Biochemistry and Biology, U. of Utah, 410 Chipeta Way, Salt Lake City, UT 84108, USA.

Edwards, Dr Brian Francis Peregrine. Dept. of Biochemistry, Wayne State U., 540 E. Canfield, Detroit, MI 48201, USA. (1947) PhD, chemistry (Harvard U., 1975). Asst. prof. (tel. 313 + 577-5107). *Protein structure, blood coagulation proteins, proteins from thermophiles.*

Egan, Mr Robert W. Dept. of Biochemistry, U. of Miami, S. of Medicine, P. O. Box 016129, Miami, FL 33101, USA.

Eggleston, Dr Drake Stephen. Analytical- Physical and Structural Chemistry, Smith Kline and French Labs., 1500 Spring Garden St., Philadelphia, PA 19101, USA. (1954) PhD, chemistry (U. North Carolina, Chapel Hill, 1983). Assoc. (tel. 215 + 751-7351). *Peptides, structure, X-ray crystallography, NMR, inorganic chemistry.*

Eichhorn, Dr Edgar Leo. 5522 E. Bavarian Pass, Fridley, MN 55432, USA. (1923) DSc, math and physics (Gemeentelijke U. van Amsterdam, Netherlands). Retired. (tel. 818 + 988-5547). *General crystallography, computational software.*

Eick, Prof. Harry A. Dept. of Chemistry, Michigan State U., Chemistry Building, East Lansing, MI 48824-1322, USA. (1929) PhD, inorganic chemistry (U. Iowa, 1956). Prof., assoc. dean (tel. 517 + 353-4511). *Single crystal X-ray diffraction; solid state inorganic chemistry; lower oxidation states in lanthanide phases; high temperature preparatory chemistry, lanthanides, chalcogens, halogens and oxygen.*

Eigenbrot Dr Charles Weaver, Jr. Dept. of Chemistry, U. of Notre Dame, Notre Dame, IN 46556, USA. (1954) PhD, chemistry (U. California, Berkeley, 1981) Departmental crystallographer. (tel. 219 + 239-6220). *Small molecules, single crystal X-ray diffraction.*

Eilerman, Dr Donna Paige. 2020 Van Roo Ave., Merrick, NY 11566, USA. (1951) PhD, chemistry-crystallography (Adelphi U., 1979). Sr. chemist. *Orientationally disordered plastic crystals, gel-sol transformations, gelatin mixtures.*

Einck, Dr James J. Analytical Lab., Carter-Glogau Labs. Inc., 5160 W. Bethany Home Rd., Glendale, AZ 85301, USA. (1945) PhD, organic chemistry (Arizona State U., 1976). Chief analytical chemist. (tel.602 + 939-7565, Ext. 66). *Small molecule, organic, natural products, software developments.*

Einspahr, Dr Howard Martin. Physical and Analytical Chemistry The Upjohn Co., Kalamazoo, MI 49001, USA. (1943) PhD, chemistry (U. Pennsylvania, 1970). Sr. res. scient. (tel. 616 + 385-5492). *Macromolecular crystallography.*

Eisenberg, Prof. David. Molecular Biology Inst., U. of California, 405 Hilgard Ave., Los Angeles, CA 90024, USA. (1939) D. Phil. chemistry (Oxford U., 1964). Prof. (tel. 213 + 825-3754). *Protein structure and function, water and aqueous solutions, protein folding.*

Eisenberg, Prof. Richard. Dept. of Chemistry, U. of Rochester, River Campus, Rochester, NY 14627, USA. (1943) PhD, chemistry (Columbia U., 1967). Prof. (tel. 716 + 275-5573). *Inorganic chemistry, homogeneous catalysis, structure - reactivity relationships, organometallic systems, catalytically active complexes.*

Elder, Prof. Richard C. Chemistry Dept., U. of Cincinnati, Mail Location 172, Cincinnati, OH 45221, USA. (1939) PhD, chemistry (MIT, 1964). Prof. (tel. 513 + 475-3070). *Transition metal complexes, structural trans effect, arthritis drugs, structure - redox properties relationship.*

El Saffar, Prof. Zuhair M. Physics Dept., De Paul U., 2219 N. Kenmore Ave., Chicago IL 60614, USA. (1934) PhD, nuclear resonance, chemical physics (U. Wales, UK, 1960). Prof. (tel. 312 + 321-8178). *Molecular motion, ferroelectrics, hydrates, nuclear magnetic resonance.*

Eller, Dr P. Gary. Isotopes and Nuclear Chemistry Div., Los Alamos Scient. Lab., Mail Stop 346, NM 87544, USA. (1947) PhD, chemistry (Ohio State U., 1972). Assoc. group leader. *Actinides, fluorine compounds, structure.*

Emerson, Prof. Kenneth. Chemistry Dept., Montana State U., Bozeman, MT 59715, USA. (1931) PhD, physical chemistry (U. Minnesota, 1961). Prof. (tel. 406 + 994-4801). *Synthesis, complex halide salts of transition metals, structures, transition metal halides, low dimensional compounds.*

Emerson, Mr Merle T. P.O. Box 1245, U. of Alabama, Huntsville, AL 35807, USA.

Emge, Dr Thomas James. Divs. of Chemistry & Material Sci. & Techn., Argonne Nat. Lab., 9700 S. Cass Ave., Argonne, IL 60439, USA. (1955) PhD, physical chemistry (Johns Hopkins U., 1981). Postdoctoral res. assoc. (tel. 312 + 972-3509). *X-ray and neutron diffraction, synthetic organic materials.*

Enemark, Prof. John Henry. Chemistry Dept., U. of Arizona, Tucson, AZ 85721, USA. (1940) PhD, chemistry (Harvard U., 1966). Prof. (tel. 602 + 621-2245). *Inorganic and bio-inorganic chemistry, molybdenum compounds, heteronuclear NMR.*

Eng-Wilmot, Dr David Lawrence. Dept. of Chemistry, Rollins C., P. O. Box 2695, Winter Park, FL 32789, USA. (1947) PhD, inorganic chemistry (U. South Florida, 1978). Assoc. prof. (tel. 305 + 646-2520). *Coordination compounds, iron transport, siderophores, ionophores, molecular structure, small molecules, bio-inorganics, conformational analysis, stereochemistry.*

Enwall, Dr Eric Lee. Chemistry Dept., U. of Oklahoma, 620 Parrington Oval, Norman, OK 73019, USA. (1940) PhD, chemistry (Montana State U., 1969). Dir., analytical services center (tel. 405 + 325-2843). *Structural inorganic chemistry, chemical instrumentation.*

Eppelsheimer, Dr Daniel Snell, Jr. Energy Lab., MIT, Room 8-115, 77 Massachusetts Ave., Cambridge, MA 02139, USA. (1941) Dr.rer.nat., mineralogy (U. Heidelberg, BRD, 1981). Postdoctoral assoc. (tel. 617 + 253-3252 and 5069). *Crystal physics.*

Eppelsheimer, Prof. Daniel Snell. Metallurgical and Nuclear Eng., U. of Missouri-Rolla, Box 299, Rolla, MO 65401, USA. (1909) DSc, physical metallurgy (Harvard U., 1935). Prof. (tel. 314 + 341-4711). *Single crystals, pole figures.*

Epperson, Dr John Ernest. Materials Sci. Div., 212, Argonne Nat. Lab., 9700 S. Cass Ave., Argonne, IL 60439, USA. (1933) PhD, metallurgical engineering (U. Tennessee, 1968). Metallurgist. (tel. 312 + 972-4971). *Diffuse scattering studies, short range order, small angle scattering, alloy decomposition.*

Eriks, Prof. Klaas. Dept. of Chemistry, Boston U., 685 Commonwealth Ave., Boston, MA 02215, USA. (1922) PhD, chemistry (U. Amsterdam, Netherlands, 1952). Prof. (tel. 617 + 353-2497). *Amino acid and polypeptide complexes, inorganic and organic phosphates, pyrimidines and purines.*

Ernst, Dr Stephen Richard. Dept. of Chemistry, U. of Texas, Austin, TX 78712, USA. (1939) PhD, physical chemistry (U. Utah, 1972). Res. scient. (tel. 512 + 471-1105). *Structure - activity relations, crystallographic computing.*

Estes, Dr Eva Dixon. Environmental Chemistry Dept., Res. Triangle Inst., P. O. Box 12194, Bldg. 6, Res. Triangle Park, NC 27709, USA. (1949) PhD, inorganic chemistry (U. North Carolina, Chapel Hill, 1975). Res. chemist. (tel. 919 + 541-5926). *Ion chromatography, environmental source assessment, air monitoring.*

Etter, Dr Margaret Eleanor Cairns. 119 Smith Hall, Dept. of Chemistry, U. of Minnesota, 207 Pleasant St., S.E., Minneapolis, MN 55455, USA. (1943) PhD, organic chemistry (U. Minnesota, 1974). Asst. prof. (tel. 612 + 373-5575). *Solid-state organic chemistry, crystallography, solid-state NMR, hydrogen-bond interactions, dye chemistry.*

Evans, Dr Doris Louise. Ceramic Res., Corning Glass Works, Sullivan Park, Corning, NY 14832, USA. (1923) PhD, crystallography (U. London, UK, 1969). Sr. res. assoc. (tel. 607 + 974-3261). *Structure - properties relationship, phase transformations, silicates, glasses.*

Evans, Ms Eloise Humez. Crystallography Section, Nat. Bureau of Standards, Matls. Bldg. Rm. A221, Washington, DC 20234, USA. (1921) BSc, mathematics (MIT, 1942). Res. assoc., Joint Committee on Powder Diffraction Standards, N.B.S. fellow. (tel. 301 + 921-2921). *Standard X-ray diffraction powder patterns, powder patterns (experimental and calculated).*

Evans, Dr Howard Tasker, Jr. U.S. Geological Survey, National Center Stop 959, Reston, VA 22092, USA. (1919) PhD, inorganic chemistry (MIT, 1948). Physicist. (tel. 703 + 860-6666). *Inorganic and mineral crystal structures.*

Extine, Dr Michael Wayne. Molecular Structure Corp., 3304 Longmire Dr., College Station, TX 77840, USA. (1950) PhD, chemistry (Princeton U. 1976). Crystallographer. (tel. 713 + 696-9729). *Organometallic chemistry, X-ray structure analysis.*

Faber, Dr John, Jr. Materials Sci. Div., Argonne Nat. Lab., 9700 S. Cass Ave., Argonne, IL 60439, USA. (1941) PhD, materials science (Marquette U., 1963). Staff scient. (312 + 972-4969). *Structure - properties relationship, high temperature materials, nonstoichiometry effects, low temperature magnetic properties, phase transitions.*

Fackler, Prof. John Paul, Jr. Chemistry Dept., Texas A & M U., College Station, TX 77843, USA. (1934) PhD, chemistry (MIT, 1960). Dean, C. of sci., prof. (tel. 409 + 845-7361). *Transition metal organometallics, coordination compounds, transition metal sulfur compounds.*

Fahey, Prof. James A. Chem. Dept., Bronx Community C., 181 St., W. University Ave., Bronx, New York, NY 10453, USA. (1941) PhD, chemistry (U. Tennessee, Knoxville, 1971). Prof. (tel. 212 + 220-6218, ext. 6903). *Radial-distribution studies, nuclear waste glasses, powder structural analysis, lanthanide and actinide compounds.*

Fair, Dr Carolyn Kay. B. A. Frenz and Assocs., 900 E. Harvey Rd., Suite 16, College Station, TX 77840, USA. (1945) PhD, inorganic chemistry (U. Arkansas, 1973). Computer consultant. (tel. 409+ 764-3999). *Neutron diffraction, software development.*

Fairhurst, Prof. Carl Wayne. Dental Physical Sci., Medical C. of Georgia, S. of Dentistry, 1459 Gwinnett St., Augusta, GA 30902, USA. (1926) PhD, materials science (Northwestern U., 1966). Regents prof., coordinator. (tel. 404 + 828-3354). *Dental materials (physics - chemistry - metallurgy).*

Falvello, Mr Lawrence R. Dept. of Chemistry, Texas A & M U., College Station, TX 77843, USA.

Fang, Prof. Jen Ho. Geology Dept., Southern Illinois U., Carbondale, IL 62901, USA. (1929) PhD, geochemistry (PSU, 1961). Prof. (tel. 618 + 453-3351, ext. 49). *Powder diffraction.*

Fasiska, Dr Edward J. Mellon Inst., 4400 5th Ave., Pittsburgh, PA 15213, USA. (1937) PhD, crystallography (U. Pittsburgh, 1972). *Materials science.*

Faulk, Mr John Warren. Davison Chemical Div., W. R. Grace and Co., P. O. Box 3247, Lake Charles, LA 70601, USA. (1940) BS, physics (McNeese State, 1965). Plant superintendent. (tel. 318 + 527-5228, ext. 325). *X-ray diffraction.*

Faust, Dr George Tobias. 80 S. Maple Ave., P. O. Box 411, Basking Ridge, NJ 07920, USA. (1908) PhD, mineralogy (U. Michigan, 1934). Retired. (tel. 201 + 766-5397). *Geology, petrology, Watchung basalt flows of New Jersey.*

Fawcett, Mr Timothy G. Analytical Labs., 1602 Bldg., Dow Chemical Co., Midland, MI 48640, USA.

Fay, Prof. Robert Clinton. Chemistry Dept., Cornell U., Ithaca, NY 14853, USA. (1936) PhD, inorganic chemistry (U. Illinois, 1962). Prof. (tel. 607 + 256-3636). *Stereochemistry, configurational rearrangements, metal chelate compounds.*

Feldman, Dr Robert Edward. New York Inst. of Tech., Old Westbury, NY 11568, USA. (1939) PhD, physics (PIB, 1969). Asst. prof. (tel. 516 + 686-7535). *X-ray physics, crystal structures.*

Feldmann, Mr Richard Joseph. Div. of Computer Res. and Techn., Nat. Insts. of Health, Bethesda, MD 20014, USA. (1939) MS, electrical engineering (PIB, 1962). Computer specialist. (tel. 301 + 496-1100). *Protein crystallography, molecular structure data retrieval, macromolecular surface representation.*

Fenna, Dr Roger Edward. Biochemistry Dept., U. of Miami Medical S., Box 016129, Miami, FL 33101, USA. (1947) D.Phil., protein crystallography (Corpus Christi C., Oxford, UK, 1973). Asst. prof. (tel. 305 + 547-6564). *Structure and function, biological macromolecules; photosynthesis.*

Fernando, Prof. Quintus. Chemistry Dept., U. of Arizona, Tucson, AZ 85721, USA. (1926) PhD, analytical chemistry (U. Louisville, 1953). Prof. (tel. 602 + 626-2105). *Metal complexes, trace element analysis.*

Fillers, Dr James Paul. Dept. of Chemistry, Michigan State Univ., East Lansing, MI 44824, USA. (1951) PhD, biochemistry (U. Tennessee, 1978). *Structure - activity relationships, drugs, macromolecules.*

Finger, Dr Larry W. Geophysical Lab., 2801 Upton St. N.W., Washington, DC 20008, USA. (1940) PhD, geology (U. Minnesota, 1967). Crystallographer. (tel. 202 + 966-0334). *Mineral structures, high temperature and pressure, crystallographic computing.*

Fink, Prof. Robert. Dept. of Chemistry, Oklahoma City U., 2501 N. Blackwelder, Oklahoma City, OK 73106, USA. (1941) PhD, physical chemistry (U. Oklahoma, 1973). Assoc. prof. (tel. 405 + 525-5411, ext. 2501). *Single crystal X-ray diffraction.*

Fink, Dr William LaVilla. Alcoa Res. Labs., New Kensington, PA, USA. (1896) PhD, chemistry, metallurgy (U. Michigan, 1926). Retired. (tel. 412 + 781-4216). *Chemistry, metallurgy, X-ray diffraction, crystallography.*

Finlayson, Mr Kier. Video Display Div., RCA Corp., New Holland Ave., Lancaster, PA 17604, USA. (1928) BSc, physics (MIT, 1950). Techn. staff. (tel. 717 + 295-6213). *Materials characterization, X-ray diffraction, electron probe microanalysis.*

Fischer, Mr Gerhard Richard. Techn. Staffs Div., Corning Glass Works, Corning, NY 14830, USA. (1926) BSc, physics (U. Jena, DDR, 1951). Physicist. (tel.

607 + 974-3342). *High temperature X-ray diffraction, structure - properties relationship, automated powder diffraction.*

Fischer, Dr Reinhard X. Dept. of Geological Sciences, U. of Illinois at Chicago, P. O. Box 4348, Chicago, IL 60680, USA. (1954) Dr.rer.nat., mineralogy (??, 1983) Res. assoc. (tel. 312 + 996-6218). *Neutron and X-ray structure analysis, zeolites, crystallographic computing optimization.*

Fisher, Dr Richard G. Analytical - Physical and Structural Chemistry, Smith Kline and French/F32, 1500 Spring Garden St., Philadelphia, PA 19101, USA. (1952) PhD, molecular biology (UCLA, 1980). Assoc. sr. investigator. (tel. 215 + 751-7113). *Computer graphics, molecular modeling, structural prediction, protein-DNA macromolecular structures, macromolecular techniques, computer methods.*

Fisher, Dr Robert M. Center for Advanced Materials, Lawrence Berkeley Lab., 506 Cyclotron Rd., Berkeley, CA 94720, USA. (1927) PhD, metallurgy (U. Cambridge, UK, 1962). Acting prof. (tel. 415 + 486-4760). *High voltage electron diffraction.*

Fita, Mr Iguacio. c/o Michael G. Rossman, Lilly Hall of Life Sci., Purdue U., West Lafayette, IN 47907, USA.

Fitzgerald, Dr Alvin. Dept. of Chemistry, Saint Martin's C., Lacey, WA 98503, USA. (1940) PhD, chemistry (Montana State U., 1973). Assoc. prof. (tel. 206 + 491-4700). *Small molecules of biological interest, accurate bond angles and distances, electron density distributions.*

Fitzwilliam, Dr James William. Project Planning Organization, Bell Telephone Labs., 600 Mountain Ave., Murray Hill, NJ 07974, USA. (1918) PhD, physics (MIT, 1947). Executive dir. (tel. 201 + 582-5960). *Order-disorder, clathrates, telephone design.*

Fletterick, Dr Robert J. Biochemistry and Biophysics Dept., U. of California, San Francisco, CA 94143, USA. (1943) PhD, physical chemistry (Cornell U., 1970). Assoc prof. (tel. 415 + 666-5080 or 666-5051). *Protein structure and function.*

Flippen-Anderson, Ms Judith Lee. Lab. for the Structure of Matter, Naval Res. Lab., Code 6030, Washington, DC 20375, USA. (1941) MS, chemistry (Arizona State U., 1966). X-ray crystallographer. (tel. 202 + 767-2624).

Florio, Dr John Victor. AT&T Techn. Inc., P. O. Box 241, Reading, PA 19603, USA. (1925) PhD, physical chemistry (Iowa State U., 1952). Sr. eng. (tel. 215 + 939-6058). *Surface structure, surface chemistry, liquid phase epitaxy.*

Folting-Streib, Mrs Kirsten. Chemistry Dept., Molecular Structure Center, Indiana U., Bloomington, IN 47405, USA. (1932) Lic. pharm., pharmacy (Royal Danish S. of Pharm., Denmark, 1964). Staff crystallographer. (tel. 812 + 335-6604). *Organic crystal structures.*

Foord, Mr Eugene E. U.S. Geological Survey, Mail Stop 905, Denver Federal Center, Lakewood, CO 80225, USA.

Foreman, Mr Dennis W., Jr. 351 Highgate Ave., Worthington, OH 43085, USA.

Foris, Mr C. M. Central Res. and Dev. Dept., E. I. du Pont de Nemours and Co., Exp. Station Bldg. 356, Wilmington, DE 19898, USA. (tel. 302 + 772-3687). *Inorganic crystal chemistry, powder diffraction techniques.*

Fornoff, Mr Mario M. Rigaku/USA Inc., 1 Pebble Brook Rd., Atkinson, NH 03811, USA. (1932) ChE, chemical engineering (U. Cincinnati, 1957). Marketing manager. *Powder diffraction, XRF, electron optics, microanalysis.*

Foss, Ms Linda I. Chemistry Dept., Sci. 2048, U. of New Orleans, Lakefront Campus, New Orleans, LA 70148, USA.

Foster, Prof. Alfred F. Dept. of Chemistry, U. of Toledo, Toledo, OH 43606, USA. (1915) PhD, physical chemistry (Ohio State U., 1950). Prof. *X-ray crystal structures.*

Foster, Ms Beryl Ann. 1634 Monument Ave., Apt. 1, Richmond, VA 23220, USA.

Fox, Dr Robert O., Jr. Dept. of Cellular Biology, Stanford U. Medical S., Fairchild Bldg., Stanford, CA 94305, USA. (1954) PhD, molecular biophysics and biochemistry (Yale U., 1981). Asst. prof. (tel. 415 + 497-6589). *Protein folding and engineering, protein crystallography.*

Foxman, Prof. Bruce Mayer. Dept. of Chemistry, Brandeis U., South St., Waltham, MA 02254, USA. (1942) PhD, inorganic chemistry (MIT, 1968). Asst. prof. (tel. 617 + 647-2441). *Solid state reactions, structural chemistry, sterically crowded molecules.*

Franzen, Prof. Hugo Friedrich. Chemistry Dept., Iowa State U., Ames, IA 50011, USA. (1934) PhD, physical chemistry (U. Kansas, 1962). Prof. (tel. 515 + 294-5773). *Phase transitions, vacancy ordering, heterogeneous equilibria, transition metal chalcogenide structures.*

Fratini, Prof. Albert V. Dept. of Chemistry, U. of Dayton, Dayton, OH 45469, USA. (1939) PhD, chemistry (Yale U., 1965). Assoc. prof. (tel. 513 + 229-2849). *Small organic molecules, structure, mechanisms, solid state reactions, crystalline polymers.*

Frazer, Dr Benjamin Chalmers. Physics Dept., Brookhaven Nat. Lab., Mailing address: 9529 Ash Hollow Place Gaithersburg, MD 20879, USA. (1922) PhD, physics (PSU, 1952). Sr. physicist. *X-ray and neutron scattering.*

Fredericks, Dr Robert J. Res. and Dev., Johnson and Johnson Dental Products Co., 20 Lake Dr., East Windsor, NJ 08520, USA. (1934) PhD, chemistry (Lehigh U., 1965). Vice-pres. (tel. 609 + 443-3300, ext. 247). *Polymer crystallography and morphology.*

Fredrich, Mr Michael F. General Electric NMR, 255 Fourier Ave., Fremont, CA 94539, USA.

Freed, Prof. Robert Lowell. Geology Dept., Trinity U., 715 Stadium Drive, San Antonio, TX 78284, USA. (1938) PhD, mineralogy (U. Michigan, 1966). Prof, chairman. (tel. 512 + 736-7607). *Inorganic structures, clay minerals.*

Freeman, Dr Gerald Richard. Dept. of Pediatrics, Kings Daughter's Clinic, 2205 South - Loop 363, Temple, TX 76502, USA. (1943) PhD, MD, physical chemistry-pediatrics (North Texas State U. 1970; U. of Alabama 1974). Pediatrician. (tel. 817 + 778-8426). *Crystallography, pharmacology.*

Freer, Dr Stephan T. Eng. Dept., MIA-Com Linkabit Inc., 3033 Science Park Rd., San Diego, CA 92121, USA. (1933) PhD, biochemistry (U. Washington, Seattle, 1964). Sr. eng. (tel. 619 + 457-2340). *Macromolecules.*

French, Dr Alfred Dexter. Southern Regional Res. Center, USDA, P.O. Box 19687, New Orleans, LA 70179, USA. (1943) PhD, physical chemistry (Arizona State U., 1971). Res. chemist. (tel. 504 + 589-7597, FTS 682-7597). *Polysaccharide crystal structures, computer modeling.*

French, Mr Robert D. 40 Reed St., Lexington, MA 02173, USA.

Frenz, Dr Bertram Anton. B. A. Frenz and Associates Inc., 900 E. Harvey Rd., Suite 16, College Station, TX 77840, USA. (1945) PhD, chemistry (Northwestern U., 1971). Pres. (tel. 409 + 764-3999). *Computer programming.*

Frevel, Dr Ludo Karl. 1205 W. Park Dr., Midland, MI 48640, USA. (1910) PhD, physical chemistry (Johns Hopkins U., 1934). Fellow by courtesy J.H.U. (tel. 517 + 832-8983). *X-ray crystallography, catalysis, applied mathematics.*

Friedlander, Dr Peter H. Gibbs Hill Inc., 393 7th. Ave., New York, NY 10001, USA.

Friedman, Dr Lawrence Boyd. Analytical Labs., Polaroid Corp., 100 Duchaine Blvd., New Bedford, MA 02745, USA. (1939) PhD, chemistry (Harvard U., 1966). General supervisor. (tel. 617 + 995-9523, ext. 480). *Structural - properties relationship, structural analysis.*

Fritchie, Prof. Charles Julius, Jr. Chemistry Dept., Tulane U., 6823 St. Charles Ave., New Orleans, LA 70118, USA. (1936) PhD, chemistry (CIT, 1962). Prof. (tel. 504 + 865-4713). *Charge-transfer complexes, structure and spectra; metalloorganic crystallography.*

Fronczek, Dr Frank R. Dept. of Chemistry, Louisiana State U., Baton Rouge, LA 70803, USA. (1948) PhD, chemistry (CIT, 1975). Res. assoc. (tel. 504 + 388-8270). *Crystal structure, macrocycles, transition metal complexes, natural products.*

Frueh, Prof. Alfred Joseph. Dept. of Geology and Geophysics, U. of Connecticut, U-45, Storrs, CT 06268, USA. (1919) PhD, mineralogy (MIT, 1949). Prof. (tel. 203 + 486-4433). *Mineral and inorganic structures, order-disorder imperfections.*

Fullam, Mr Ernest F. Ernest F. Fullam Inc., 900 Albany-Shaker Rd., Latham, NY 12110, USA. (1910) AB chemistry (Cornell U., 1937). Retired. (tel. 518 + 785-5533). *Chemical microscopy, micro-probe analysis, X-ray and electron diffraction.*

Fullenwider, Mr Malcolm A. P. O. Box 2, Whitehall, PA 18052, USA.

Fuoss, Dr Paul Henry. AT&T Bell Labs., Crawford Corner Rd., Holmdel, NJ 07733, USA. (1953) PhD, materials science (Stanford U., 1980). Member tech. staff. (tel. 201 + 949-3581). *Surface crystallography.*

Furdanowicz, Mr Waldemar. Dept. of Materials Sci., MIT, Room 13-4069, Cambridge, MA 02139, USA.

Furey, Dr William F., Jr. Biocrystallography Lab., P.O.Box 12055, V.A. Medical Center University Drive C, Pittsburgh, PA 15240, USA. (1952) PhD, crystallography - physical chemistry (Rutgers, 1977). Res. assoc. (tel. 412 + 683-3000, ext. 517). *Crystallographic computing, macromolecular crystallography, molecular dynamics, conformational energy analysis, direct methods, computer graphics.*

Furnas, Dr Thomas Coleman, Jr. Molecular Data Corp., 2869 Scarborough Rd., Cleveland Heights, OH 44118, USA. (1922) PhD, physics (MIT, 1952). Pres. (tel. 216 + 932-4718). *X-ray instrumentation applications (analytical - industrial - medical), single crystal X-ray diffraction, powder diffraction, small angle scattering, monochromators; X-ray diffraction systems, X-ray spectrographic systems, energy and wave length dispersive systems.*

Fuzek, Dr John Frank. Res. Labs., Eastman Chemicals Div., Eastman Kodak Co., P. O. Box 1972, Kingsport, TN 37664, USA. (1921) PhD, physical chemistry (U. Tennessee, 1947). Res. assoc. (tel. 615 + 229-4183). *Fiber structure, polymer structure.*

Gaier, Dr James Richard. Electro-Physics Section, NASA Lewis Res. Center, 21000 Brookpark Rd., Cleveland, OH 44135, USA. (1952) PhD, chemistry (Michigan State U., 1983). Physicist. (tel. 216 + 433-4000 ext. 755). *Intercalated graphite, protein-protein interactions.*

Gaines, Mr James Matthew. Dept. of Electrical Eng., U. of Colorado, Boulder, CO 80309, USA. (1958) MS, electrical engineering (Virginia Polytechnic Inst., 1983). Grad. student. (tel. 303 + 492-7682). *Structure - properties relationship.*

Gall, Prof. William Einar. The Rockefeller U., 1230 York Ave., New York, NY 10021, USA. (1942) PhD, biochemistry (The Rockefeller U., 1969). Assoc. prof. (tel. 212 + 360-1175). *Protein structure, electron microscopy, three-dimensional reconstruction techniques.*

Gallucci, Dr Judith Chlastawa. Dept. of Chemistry, The Ohio State U., 140 W. 18th Ave., Columbus, OH 43210, USA. (1953) PhD, inorganic chemistry (U. Massachusetts, Amherst, 1979). Dept. crystallographer. (tel. 614 + 422-4039). *Chemical crystallography, organic and organometallic structures, molecular mechanics modeling.*

Gambone, Mr Michael J. 25205 Cypress St., Apt. 13, Lomita, CA 90717, USA.

Gangulee, Dr Amitava. Res. Div., IBM T. J. Watson Res. Center, P.O. Box 218, Yorktown Heights, NY 10598, USA. (1941) ScD, metallurgy and materials science (MIT,1967). Res. staff member. (tel. 914 + 945-2042). *Structure - properties relationship, amorphous & crystalline thin films; thin film phase transformation, low temperature diffusion.*

Gantzel, Dr Peter Kellogg. GA Technologies, Box 85608, San Diego, CA 92138, USA. (1934) PhD, physical chemistry (UCLA, 1962). Staff sci. (tel. 619 + 455-2949). *Powder X-ray diffraction, computer programs, inorganic compounds, interstitial carbides, ceramic constituents.*

Garafalo, Mr Alfred R. Massachusetts C. of Pharmacy, 179 Longwood Ave., Boston, MA 02115, USA.

Garbauskas, Dr Mary E. Corporate Res. and Dev., General Electric, P. O. Box 8, Bldg. K-1, Schenectady, NY 12345, USA. (1953) PhD, chemistry (UCLA, 1979) Staff scient. (tel. 518 + 385-8041). *X-ray powder diffraction, X-ray spectrometry, inorganic and organic structure determination.*

Gardner, Dr Kenn Corwin H. Central Res. and Dev. Dept., E. I. DuPont de Nemours and Co. Inc., Exp. Station Bldg. 356, Wilmington, DE 19898, USA. (1947) PhD, macromolecular science (Case Western Reserve U., 1974). Res. scient. (tel. 302 + 772-2408). *Fibrous polymers diffraction, structure - properties - morphology relationship, polymers.*

Garvey, Prof. Roy George. Chemistry Dept., No. Dakota State U., Fargo, ND 58105, USA. (1941) PhD, chemistry (U. Utah, 1966). Assoc prof. (tel. 701 + 237-8697). *Inorganic chemistry, coordination chemistry.*

Gash, Dr Alfred G. South Carolina Marine Resources Res. Inst., P. O. Box 12559, Charleston, SC 29412, USA. (1947) PhD, chemistry (U. South Carolina, 1975). Dir., computer service. (tel. 803 + 795-6350, ext. 319). *Crystallography, small molecules, scientific computing.*

Gaynor, Mr Gary R. 88 Orlin Ave., SE, Minneapolis, MN 55414, USA.

Geckle, Mr Raymond J. AMP Inc., 425 Prince St., Harrisburg, PA 17105, USA.

Gehringer, Mr Ronald. Chemistry Dept., North Dakota State U., Fargo, ND 58105, USA.

Geib, Prof. Irving George. 1809 N. Salisbury St., West Lafayette, IN 47906, USA. (1907) PhD, physics (Purdue U., 1948). Prof. em., physics, Purdue U. (tel. 317 + 463-5658).

Geil, Prof. Phillip Herbert. Polymer Group, Dept. of Metallurgy, Room 211, Metals and Mineralogy Bldg., Urbana, IL 61801, USA. (1930) PhD, physics (U. Wisconsin, 1956). Prof. (tel. 217 + 333-0149). *Structure - properties - morphology relationship, synthetic macromolecules, biological macromolecules.*

Geisinger, Ms Karen L. Chemistry Dept., Arizona State U., Tempe, AZ 85287, USA.

Geiss, Dr Roy Howard. Res. Div., IBM, K32/281, 5600 Cottle Rd., San Jose, CA 95193, USA. (1937) PhD, applied physics (Cornell U., 1967). Res. staff member. (tel. 408 + 256-7045). *Electron microscopy (scanning and transmission), diffraction, inorganic materials, thin films, metals and alloys, corrosion, magnetic properties, defects.*

Geller, Prof. Seymour. Dept. of Electrical Eng., U. of Colorado, Boulder, CO 80309, USA. (1921) PhD, physical chemistry (Cornell U., 1949). Prof. (tel. 303 + 492-7157). *Structure - properties relationship, magnetic materials, solid electrolytes, superconducting materials, pressure induced phases, phase transformations, magnetic structures.*

George, Mr Clifford. Naval Res. Labs., 4555 Overlook Ave., SW, Washington, DC 20375, USA.

Gerdes, Dr Reiner Josef. Continental Labs., Continental Telecom Inc., 270 Scientific Dr., Techn. Park - Atlanta, Norcross, GA 30092, USA. (1935) Dr.rer.nat., physical chemistry (Techn. U., Hannover, BRD, 1963). Dir. (tel. 404 + 448-2206). *Structure - properties relationship; surface structure, catalysis.*

Gerhard, Mr F. Bruce, Jr. 38 Spring Ln., Canton, MA 02021, USA.

Gerst, Dr Kenneth Mark. Electronics Res. Center, Rockwell International, 3370 Miraloma Ave., Anaheim, CA 92803, USA. (1944) PhD, chemistry (U. Michigan, 1976). Res. staff. (tel. 714 + 632-5773). *Biological molecules, electronic materials.*

Getzoff, Dr Elizabeth Dickinson. Molecular Biology Div., MB3, Res. Inst. of Scripps Clinic, 10666 North Torrey Pines Rd., La Jolla, CA 92037, USA. (1954) PhD, biochemistry (Duke U., 1982). Asst. member. (tel. 619 + 455-9100, ext. 2526). *Macromolecular interaction and assembly, protein structure determination, analysis and prediction, computer graphics.*

Gheith, Prof. Mohamed A. Geology Dept., Boston U., 725 Commonwealth Ave., Boston, MA 02215, USA. (1925) PhD, geochemistry and mineralogy, (U. Minnesota, 1952). Dir. of special external programs for the middle east (tel. 617 + 353-2616) *Geological education, economic geology, geochemistry, ore deposits, sulfide mineralogy.*

Ghose, Prof. Subrata. Dept. of Geological Sci., U. of Washington, Seattle, WA 98195, USA. (1932) PhD, mineralogy, crystallography (U. Chicago, 1959). Prof. (tel. 206 + 543-7378). *Oxides and silicates, high pressure and temperature structures; crystal chemistry, rock forming silicates; structure determination, complex silicates and borates; chemical bonding, physical properties, minerals.*

Ghosh, Dr Debashis. Molecular Biophysics, Medical Foundation of Buffalo Inc., 73 High St., Buffalo, NY 14203, USA. (1952) PhD, crystallography (U. Pittsburgh, 1981). Res. scient. (tel. 716 + 856-9600, ext. 477). *Macromolecular crystallography.*

Gibbs, Mr Gerald V. Dept. of Geological Sci., Virginia Polytechnic Inst. and State U., Blacksburg, VA 24060, USA. (1929) PhD, mineralogy and geochemistry (PSU, 1962). Prof., mineralogy. (tel. 703 + 961-6330). *Mineralogy.*

Giese, Mr Rossman F., Jr. 144 Walton Dr., Snyder, NY 14226, USA.

Giessen, Prof. Bill Cormann. Dept. of Chemistry, Northeastern U., 360 Huntington Ave., Boston, MA 02115, USA. (1932) Dr.sc.nat., metallurgy (U. Göttingen, BRD, 1958). Prof., assoc. dir., inst. of chemical analysis, appl. & forensic sci. (tel. 617 + 437-2827). *Intermetallic phases (structure), alloy chemistry, metastable alloys, amorphous metals, high-speed X-ray diffraction methods, analytical applications, X-ray diffraction, forensic chemistry.*

Gilardi, Dr Richard Dean. Lab. for the Structure of Matter, Naval Res. Lab., Code 6030, Washington, DC 20375, USA. (1940) PhD, physical chemistry (U. Maryland, 1966). Res. chemist. (tel. 202 + 767-2624). *Diffraction analysis methods, conformation energy calculations.*

Gilfrich, Mr John Valentine. Code 6680G, Condensed Matter Physics Branch, Naval Res. Lab., Washington, DC 20375, USA. (1927) BA, chemistry (American International C., 1949). Retired, part-time consultant. (tel. 202 + 767-2154). *X-ray spectroscopy, crystal diffraction.*

Gillies, Dr Donald Chalmers. Microelectronics Center, McDonnell Douglas Corp., 8905 Airport Rd., Berkeley, MO 63134, USA. (1939) PhD, crystallography (U. London, UK, 1969). Lead eng. (tel. 314 + 234-8072). *Crystal growth, electronic materials, space processing.*

Gilliland, Mr Gary L. Bldg. 76, Room 1516, Hoffmann- LaRoche Inc., 340 Kingsland St., Nutley, NJ 07110, USA.

Ginell, Prof. Robert. Chemistry Dept., Brooklyn C. of the City U. of New York, Ave. H and Bedford Ave., Brooklyn, NY 11210, USA. (1912) PhD, chemistry (PIB, 1943). Prof. em. (tel. 718 + 780-5753). *Liquids, solids, equations of state, association theory, nucleation.*

Ginell, Dr Stephan Lawrence. Molecular Structure Group, Inst. for Cancer Res., 7701 Burholme Ave., Philadelphia, PA 19111, USA. (1949) PhD, biophysics (SUNY, Buffalo, Roswell Park Div., 1980). Res. assoc. (tel. 215 + 728-2548). *Structure and function, proteins and nucleic acids, hydration in biological systems, radiation damage.*

Giordano, Mr Joseph. 43 Deerfield Rd., Parlin, NJ 08859, USA.

Giorgi, Dr Angelo Louis. CMB-3, Los Alamos Scient. Lab., Los Alamos, NM 87545, USA. (1917) PhD, physical chemistry (U. New Mexico, 1957). Staff member. (tel. 505 + 667-5815). *High temperature chemistry, superconductivity.*

Glaeser, Prof. Robert Martin. Dept. of Biophysics, U. of California, Donner Lab., Berkeley, CA 94720, USA. (1937) PhD, biophysics (U. California, Berkeley, 1964). Prof. (tel. 415 + 642-4131). *Electron diffraction, electron microscopy, membrane structure.*

Glass, Dr Howard L. Science Center, 031-BA01, Rockwell International, P. O. Box 3105, Anaheim, CA 92803, USA. (1942) PhD, physics (Rutgers U., 1969). Techn. staff member. (tel. 714 + 632-3691, telex 678437). *Crystal growth, magnetism, epitaxy, X-ray diffraction.*

Gleason, Dr William Bourke. Central Res. Lab., 3M Co., 201-1W, St. Paul, MN 55144, USA. (1945) PhD, organic chemistry (U. Minnesota, 1974). Sr. res. chemist. (tel. 612 + 733-0720). *Organic and organometallic crystal structures.*

Glen, Dr Gerald Leonard. Materials Res. Section, Owens-Illinois Inc., 1700 N. Westwood Ave., Toledo, OH 43666, USA. (1935) PhD, physical chemistry (Cornell U., 1962). Chief. (tel. 419 + 242-6543, ext. 33-124). *Materials science, structural chemistry, electron spectroscopy, X-ray diffraction and scattering.*

Glick, Prof. Milton Don. C. of Arts and Sci., U. of Missouri-Columbia, 210 Jesse Hall, Columbia, MO 65210, USA. (1937) PhD, chemistry (U. Wisconsin, Madison, 1965). Dean. (tel. 314 + 882-4421). *Structural chemistry, X-ray crystallography, computing.*

Glinka, Mr Charles Joseph. Reactor Div., Nat. Bureau of Standards, Gaithersburg, MD 20899, USA.

Glusker, Dr Jenny Pickworth. Dept. of Molecular Structure, The Inst. for Cancer Res., 7701 Burholme Ave., Philadelphia, PA 19111, USA. (1931) DPhil, chemistry (Oxford U., UK, 1957). Sr. member. (tel. 215 + 728-2220). *Enzyme mechanisms, chemical carcinogenesis and mutagenesis, metal chelation.*

Go, Mr Kuan Tee. 228 Woodcrest Dr., Amherst, NY 14226, USA.

Godycki, Mr L. Edward. 1060 Granada Ave., San Marino, CA 91108, USA.

Goehner, Mr Raymond Philip. Analytical Systems, Siemens-Allis, 1 Computer Dr., Cherry Hill, NJ 08034, USA. (1945) MSc, physics (Rensselaer Polytechnic Inst., 1971). Product manager. (tel. 607 + 424-9210). *X-ray powder diffraction, X-ray fluorescence.*

Goldberg, Prof. Stephen Z. Chemistry Dept., Adelphi U., Garden City, NY 11530, USA. (1947) PhD, chemistry (U. California, Berkeley, 1973). Assoc. prof. (tel. 516 + 294-8700, ext. 7519). *Structural inorganic chemistry.*

Goldberg, Mr Martin Jeffry. 679 Coney Island Ave., Brooklyn, NY 11218, USA.

Goldish, Dr Elihu. Res. Center, Union Oil Co. of California, P. O. Box 76, Brea, CA 92621, USA. (1928) PhD, chemistry (CIT, 1956). Res. chemist. (tel. 714 + 528-7201, ext. 412). *X-ray diffraction, powder diffractometry, X-ray spectrometry (analytical applications).*

Goldsmith, Dr Elizabeth Jane. Dept. of Biochemistry and Biophysics, 964 Science Bldg., S. of Medicine, U. of California, San Francisco, CA 94143, USA. (1945) PhD, physical chemistry (UCLA, 1971). Asst. molecular biologist III. (tel. 415

+ 666-5051). *Protein crystallography, allostery, enzyme mechanism, enzyme structure.*
Goldsmith, Prof. Julian Royce. Dept. of Geophysical Sci., U. of Chicago, 5734 Ellis Ave., Chicago, IL 60637, USA. (1918) PhD, geochemistry (U. Chicago, 1947). Charles E. Merriam distinguished service prof. (tel. 312 + 962-8155). *Geochemistry, phase equilibria, crystal chemistry, silicates and carbonates.*
Goldstein, Dr Barry Michael. Dept. of Pharmacology, S. of Medicine and Dentistry, U. of Rochester, 601 Elmwood Ave., Rochester, NY 14642, USA. (1952) MD, PhD (U. Rochester, 1981, 1982). James P. Wilmot fellow in cancer res. (tel. 716 + 275-5305). *Structure - function relationships, pharmacologically active agents, active agent structures, X-ray diffraction, NMR (solution structure).* (tel. 716 + 275-3841).
Golikeri, Mr Ganesh D. Res. Ctr., Lever Bros. Co., 45 River Rd., Edgewater, NJ 07020, USA.
Gong, Dr Ping-Po. Detector and Eng. Dept., 4115, EG&G/EM Inc., P. O. Box 1912, Las Vegas, NV 89109, USA. (1950) PhD, X-ray physics (PINY, 1983). Scient. specialist. (tel. 702 + 295-2556). *X-ray measurements, X-ray optics, fluorence, diffraction, X-ray systems.*
Gordon, Ms Janice T. 1250 Upper Gulph Rd., Radnor, PA 19087, USA.
Gougoutas, Dr Jack Zanos. Sci. Information (Molecular Modeling), E. R. Squibb and Sons Inc., P. O. Box 4000, Princeton, NJ 08540, USA. (1939) PhD, chemistry (Harvard U., 1963). Sen res. fellow. (tel. 201 + 921-4562). *Solid state reactions, topotaxy, organic compounds, structure - properties relationship.*
Graeber, Dr Edward J. Dept. 5822, Sandia Labs., Albuquerque, NM 87185, USA. (1934) PhD, geology (U. New Mexico, 1970). Staff member. (tel. 505 + 264-5671). *Mineral structures, inorganic structures.*
Graham, Mr Robert Albert. Shock Wave Explosive Physics - 1131, Sandia Nat. Lab., Albuquerque, NM 87185, USA. (1931) MS, engineering mechanics (U. Texas, Austin, 1958). Distinguished member of the techn. staff. (tel. 505 + 844-1931). *Crystal physics, crystal chemistry, shock compression.*
Grasso, Mr Michael. Rm 6D-202, AT&T Bell Telephone Labs., Mountain Ave., Murray Hill, NJ 07974, USA. BS, chemistry (Bucknell U., 1946) (tel. 201 + 582-4492). *Inorganic crystal growth, optical fibers, glass.*
Graves, Mr Bradford J. Dept. of Chemistry, U. of California, Berkeley, CA 94720, USA.
Greenberg, Dr Berton Laurence. Materials Characterization Res., Philips Labs., 345 Scarborough Rd., Briarcliff Manor, NY 10510, USA. (1940) PhD, materials science. (Stevens Inst. of Techn., 1979). Project manager. (tel. 914 + 945-6074). *X-ray powder diffraction, single crystal structure determination, texture analysis, X-ray spectrometry.*
Greenblatt, Prof. Martha. Chemistry Dept., Rutgers U., P. O. Box 939, Piscataway, NJ 08854, USA. (1941) PhD, inorganic chemistry (PIB, 1967). Assoc. prof. (tel. 201 + 932-3277). *Solid state chemistry, crystal growth, structure - properties relationship, bronzes, solid electrolytes, insertion compounds.*
Greenhouse, Mr Harold M. Eng. Dept., Bendix Communications, Dept. 480, East Joppa Rd., Baltimore, MD 21204, USA. (1924) MS, physical chemistry (Ohio State U., 1951). Sr. staff eng. (tel. 301 + 823-2200, ext. 394). *Material science, microelectronics.*
Greer, Dr Jonathan. Physical Biochemistry Lab., Pharmaceutical Products Div., Abbott Labs., D-47E Abbott Park, IL 60064, USA. (1943) PhD, molecular biology (U. Cambridge, UK, 1970). Sr. project leader. (tel. 312 + 937-6933). *Structure and function, proteins, biological macromolecules, protein - ligand interactions, comparative molecular modeling.*
Gregg, Mr R. Q. 3207 Henrietta, Bartlesville, OK 74003, USA.
Gremillion, Prof. Alcuin Florian. Dept. of Chemistry, U. of Arkansas at Little Rock, 33rd. & University Ave., Little Rock, AR 72204, USA. (1925) PhD, inorganic chemistry (Tulane U., 1958) Assoc. prof. (tel. 501 + 569-3152, ext. 22). *Inorganic structures.*
Gress, Dr Mary Edith. Chemical Sci. Div., ER-141 GTN, Office of Basic Energy Sci., Dept. of Energy, Washington, DC 20545, USA. (1946) PhD, physical chemistry (Iowa State U., 1973). Asst. techn. manager. (tel. 301 + 353-5820). *Inorganic and organic crystal structures.*
Grev, Prof. Dennis Merle. Chemistry Dept., Columbia C., 10th and Rogers, Columbia, MO 65216, USA. (1935) MS, chemistry (U. Missouri, Columbia, 1963). Prof. (tel. 314 + 449-0531, ext. 332). *Crystal structures, environmental analytical chemistry, methodology.*
Grieger, Mr Gene R. Res. Div., Signal Res. Center, 50 UOP Plaza, Des Plaines, IL 60016, USA. (1932) BS, physics-math (U. Michigan, 1956). Sr. res. chemist. (tel. 312 + 391-3424). *Electron diffraction.*
Griffin, Dr Jane Flanigen. Molecular Biophysics, Medical Foundation of Buffalo, 73 High St., Buffalo, NY 14203, USA. (1933) PhD, chemistry (SUNY Buffalo, 1974), Res. scient. (tel. 716 + 856-9600). *Structure - activity relationship, cardiac glycosides and opiates, steroid structure, crystallographic information dissemination.*
Griffith, Dr Elizabeth Ann Hall. Chemistry Dept., U. of South Carolina, Columbia, SC 29205, USA. (1935) PhD, physical chemistry (U. South Carolina, 1970). Res. assoc. (tel. 803 + 777-2542). *X-ray crystallography, physical methods, structural chemistry, biologically significant compounds.*
Grill, Mr Charles M. UOP Inc., 10 UOP Plaza, Corp. Res. Center, Des Plaines, IL 60016, USA.
Gromek, Dr Jack Michael. 1032 Callahan Ave., Yeadon, PA 19050, USA. (1953) PhD, chemistry (U. Pennsylvania, 1983). Postdoctoral fellow. (tel. 519 + 824-4120). *Neutron diffraction, small molecular studies, non-crystalline diffraction, thermal motion analysis.*
Grossie, Mr David Alan. Chemistry Dept., Baylor U., Waco, TX 76706, USA.
Gschneidner, Prof. Karl A., Jr. Ames Lab., Dept. of Materials Sci. and Eng., Iowa State U., 255 Spedding Hall, Ames, IA 50011, USA. (1930) PhD, physical chemistry (Iowa State U., 1957). Distinguished Prof. in Sci. and Humanities. (tel. 515 + 294-2272). *Alloy theory, metallic systems, metallurgy, metal physics, rare earth metals and alloys, high purity metals preparation, crystal growth, metal crystals, intermetallic compounds, low temperature heat capacity, magnetic susceptibility, electrical resistivity.*
Gude, Mr Arthur James, III. U.S. Geological Survey, MS-917, Bldg. 25, Federal Center, Lakewood, CO 80225, USA. (1917) MSc, geology (mineralogy) (Colorado S. of Mines, 1949). Res. geologist-mineralogist. (tel. 303 + 234-2991). *Geology, mineralogy, authigenic zeolites, low temperature and low pressure minerals, zeolites, silicate minerals.*
Guentert, Dr Otto Johann. 131 Spring St., Lexington, MA 02173, USA. (1924) PhD, physics (MIT, 1956). (tel. 617 + 860-3021). *Diffraction physics, SEM microscopy, surface analysis.*
Guggenheim, Dr Lloyd Joseph. Polymer Products Dept., E. I. DuPont De Nemours and Co., Experimental Station, Wilmington, DE 19898, USA. (1939) PhD, chemistry (Iowa State U., 1965). Sr. supervisor. (tel. 302 + 772-2664). *Polymers, structure - properties relationship, organometallic chemistry.*
Guggenheim, Prof. Stephen. Dept. of Geological Sci., U. of Illinois at Chicago Circle, Box 4348, Chicago, IL 60680, USA. (1948) PhD, geology (U. Wisconsin-Madison, 1976). Assoc. prof. (tel. 312 + 996-3154). *X-ray and electron diffraction methods, geologic problems.*
Guttmann, Mr Geoffrey D. 122 Wilson St., Bldg. #58, Albany, NY 94710, USA.
Guven, Prof. Necip. Geosciences, Texas Tech U., P. O. Box 4109, Lubbock, TX 79409, USA. (1936) Dr.rer.nat., mineralogy (U. Göttingen, BRD, 1962). Prof., geology. (tel. 806 + 742-3278). *X-ray diffraction, electron diffraction, clay minerals, micas.*
Guy, Mr Joseph Thomas, Jr. 3260A N. Newhall, Milwaukee, WI 53211, USA.
Haas, Dr David Jean. Temtec Inc., Box 59, Ramsey, NJ 07446, USA. (1939) PhD, biophysics (SUNY Buffalo, 1965). Pres. (tel. 914 + 357-3447). *X-ray systems (analytical - industrial - security), NDT, security screening systems.*
Hackert, Prof. Marvin LeRoy. Chemistry Dept., U. of Texas, Austin, TX 78712, USA. (1944) PhD, chemistry (Iowa State U., 1970). Assoc. prof. (tel. 512 + 471-1105). *Structure and function of proteins, protein crystallography, instrumentation, supramolecular assemblies - structure.*
Haeffner, Mr Dean R. Dept. of Materials Sci., Northwestern U., 2145 Sheridan Ave., Evanston, IL 60201, USA.
Haendler, Prof. Helmut M. Chemistry Dept., U. of New Hampshire, Parsons Hall, Durham, NH 03824, USA. (1913) PhD, inorganic chemistry (U. Washington, 1940). Prof. em. (tel. 603 + 862-1550). *Inorganic chemical structural applications.*
Hagler, Dr Arnold T. Biophysics Dept., Agouron Inst., 505 Coast Blvd., South, La Jolla, CA 92117, USA. (1942) PhD, biophysics (Cornell U., 1970). Chairman, Biophysics. (tel. 619 + 456-1623). *Peptide hormones, protein structure, molecular dynamics, protein and peptide design, crystal structure and dynamics.*
Hale, Dr Danforth Rawson. Box 23, Aurora, OH 44202, USA. (1901) PhD, physical chemistry (Cornell U., 1928). Consultant. (tel. 216 + 562-6275). *Crystal growth, quartz; electronic properties, optical properties.*
Hall, Prof. Lowell Headley II. Chemistry Dept., Eastern Nazarene C., 23 E. Elm Ave., Quincy, MA 02170, USA. (1937) PhD, chemistry. (Johns Hopkins U.) Head. (tel. 617 + 773-6350, etx. 280). *Physical chemistry, molecular structure, structure - activity relationships.*
Haller, Dr Kenneth James. Dept. of Chemistry, U. of Wisconsin, Madison, WI 53706, USA. (1951) PhD,, chemistry (U. Arizona, 1978). Dir., X-ray crystallography. (tel. 608 + 262-1486). *Transition metal chemistry.*
Halpern, Dr B. David. RTD, Polysciences/NL, 400 Valley Rd., Warrington, PA 18976, USA. (1921) PhD, organic chemistry (Notre Dame U., 1949). Pres. (tel. 215 + 343-6484, telex 510-665-8542). *Photographic emulsions for crystallography.*
Haltiwanger, Mr Ralph Curtis. Chemistry Dept., Campus Box 215, U. of Colorado, Boulder, CO 80309, USA. (1947) MS, chemistry (U. Virginia, 1971). Res. chemist. (tel. 303 + 492-7239). *X-ray crystallography, organic and inorganic compounds, computer programming.*
Hamill, Dr Gregory Prince. Materials Evaluation Group, GTE Labs. Inc., 40 Sylvan Rd., Waltham, MA 02154, USA. (1949) PhD, applied physics-material science (California Inst. of Techn., 1978). Techn. staff member. (tel. 617 + 890-8460, ext. 748). *X-ray powder diffraction, phase transitions, structure - properties relationship, automation, computer programming, defect analysis, X-ray topography.*
Hamilton, Prof. Robert David. Geology Dept., Colorado S. of Mines, Golden, CO 80401, USA. (1942) PhD, geology (Colorado S. of Mines, 1978). Asst. prof. (tel. 303 + 279-0300, ext. 2817). *Inorganic structures, structure - properties relationship.*
Hamlin, Dr Ronald Craig. Dept. of Physics, Mail Code B-019, U. of California at San Diego, La Jolla, CA 92093, USA. (1946) PhD, physics, biophysics (U.

California, San Diego, 1975). Res. physicist. (tel. 714 + 452-2565). *Position sensitive X-ray detectors, macromolecular structures.*

Hanawalt, Prof. Joseph Donald. Materials and Metallurgical Eng. Dept., U. of Michigan, Ann Arbor, MI 48109, USA. (1902) PhD, physics (U. Wisconsin, 1929). Prof. em. (tel. 313 + 763-5469). *X-ray powder diffraction techniques.*

Hanson, Dr Jonathan C. Chemistry Dept., Brookhaven Nat. Lab., Upton, NY 11973, USA. (1941) PhD, chemistry (U. Michigan, 1969). Sr. computer analyst. (tel. 516 + 282-4378). *Molecular graphics, computer control.*

Hanson, Dr Louise Karle. Energy and Environment Dept., Brookhaven Nat. Lab., Upton, NY 11973, USA. (1946) PhD, chemistry (U. Washington, 1973). Asst. res. scient. (tel. 516 + 345-7709). *Molecular spectroscopy of single crystals.*

Hanson, Dr Marvin Wayne. Bayou State Oil Co., p. O. Box 158, Hosston, LA 71043, USA. (1928) PhD, physical-organic chemistry (U. Houston, 1964). *Molecular structure, kinetics, thermodynamics.*

Hardcastle, Prof. Kenneth Irvin. Chemistry Dept., California State U., 18111 Nordhoff, Northridge, CA 91330, USA. (1931) PhD, inorganic and physical chemistry (U. Southern California, 1961). Prof. (tel. 213 + 885-3381 or 885-3371). *Metal hydrides, structures, organometallics, metal cluster compounds, small biological molecules.*

Hardgrove, Prof. George Lind, Jr. Dept. of Chemistry, St. Olaf C., Northfield, MN 55057, USA. (1933) PhD, chemistry (U. California-Berkeley, 1959). Prof. (tel. 507 + 663-3404). *Organic structures, diffraction methods, magnetic resonance methods.*

Harker, Dr David. Dept. of Molecular Biophysics, Medical Foundation of Buffalo Inc., 73 High St., Buffalo, NY 14203, USA. (1906) PhD, chemistry (CIT, 1936). Res. scient. em. (tel. 716 + 856-9600). *Crystal structure, molecular structure, color symmetry, structural chemistry.*

Harlow, Dr Richard Leslie. Central Res. and Dev., E. I. Dupont de Nemours, E356/317A, Wilmington, DE 19711, USA. (1942) PhD, chemistry (Syracuse U., 1971). Supervisor. (tel. 302 + 772-2097). *Structures; organic compounds, organometallic compounds, inorganic compounds, X-ray powder diffraction, crystallographic computing, minicomputers.*

Haromy, Mr Tuli Patrick. 1716 Kendall Ave., Madison, WI 53705, USA.

Harper, Prof. Richard A. Physics Dept., Rensselaer Polytechnic Inst., Eighth St., Troy, NY 12181, USA. (1936) PhD, physics (New York U., 1970). Assoc. prof and dir., lab. for crystallographic biophysics. (tel. 518 + 270-6434). *Programming applications, linear and quadratic techniques, direct determination, crystal structure refinement, powder diffraction data analysis.*

Harris, Prof. David R. Computer Sci. Dept., CSUC, Chico, CA 95926, USA. (1932) PhD, physical chemistry (U. Colorado, 1963). Prof. (tel. 916 + 895-5884). *Organic and biological structures, computer applications, artificial intelligence.*

Harrison, Dr Robert Wilson. Chemical Physics, Nat. Bureau of Standards, Gaithersburg, MD 20899, USA. (1957) PhD, molecules, biophysics (Yale U., 1985). (tel. 301 + 921-2785). *Protein crystallography, computational methods, enzyme structure, molecular calculations.*

Harrison, Prof. Stephen Coplan. Gibbs Lab. 101, Harvard U., 12 Oxford St., Cambridge, MA 02138, USA. (1943) PhD, biophysics (Harvard U., 1968). Prof., biochemistry. (tel. 617 + 495-4090). *Macromolecular structure and assembly, viruses, protein - nucleic acid interactions, methods development, noncrystallographic symmetry, very large unit cells.*

Hart, Dr Haskell Vincent. Analytical Dept., Shell Dev. Co., P. O. Box 481, Houston, TX 77001, USA. (1943) PhD, chemistry (Harvard U., 1973). Sr. res. chemist. (tel. 713 + 663-2159). *Mineralogy, powder diffraction, electron diffraction, crystallographic database.*

Hartsuck, Dr Jean Ann. Lab. of Protein Studies, Okla. Med. Res. Fndn., 825 N.E. 13th St., Okla. City, OK 73104, USA. (1939) PhD, chemistry (Harvard U., 1964). Assoc. member. (tel. 405 + 271-7293). *Protein crystallography, enzyme structure and function.*

Hastings, Dr Jerome Biller. Physics Dept., Brookhaven Nat. Lab., Upton, NY 11973, USA. (1948) PhD, applied physics (Cornell U., 1975). Assoc. physicist. (tel. 516 + 345-3930). *X-ray physics, synchrotron radiation diffraction applications, phase transitions.*

Hastings, Dr Julius Mitchell. Chemistry Dept., Brookhaven Nat. Lab., Upton, NY 11772, USA. (1920) PhD, physical chemistry (Cornell U., 1945). Sr. scient. (tel. 516 + 345-4377). *Neutron scattering, phase transitions.*

Hau, Dr Herbert H. K. Restorative Dentistry, Tufts S. of Dental Medicine, 1 Kneeland St., Boston, MA 02111, USA. (1941) PhD and DMD, chemistry and dentistry (Boston U., 1970 and Harvard U., 1970). Asst. clinical prof. (tel. 617 + 423-5655). *Crystal structure, inorganic molecules.*

Hauptman, Prof. Herbert Aaron. Medical Foundation of Buffalo, 73 High St., Buffalo, NY 14203, USA. (1917) PhD, mathematics (U. Maryland, 1955). Res. dir. (tel. 716 + 856-9600, ext. 447). *Direct methods, organic crystal structures.*

Hayden, Dr Thomas Day. Construction Products Div., W. R. Grace and Co., 62 Whittemore Ave., Cambridge, MA 02140, USA. (1944) PhD, chemistry (Boston U., 1976). Res. assoc. (tel. 617 + 876-1400, ext. 133). *Structural chemistry, silicates, aluminates, portland cement structure and performance.*

Hazen, Dr Robert Miller. Geophysical Lab., Carnegie Inst. of Washington, 2801 Upton St., NW Washington, DC 20008, USA. (1948) PhD, mineralogy-crystallography (Harvard U., 1975). Exp. mineralogist. (tel. 202 + 966-0334). *Crystal structure, variation with pressure - temp. - composition; physical properties.*

He, Dr Xiao-Min. Dept. of Biology, U. of Utah, Salt Lake City, UT 84112, USA. (1944) PhD, crystallography (U. Pittsburgh, 1984). Res. assoc. (tel. 801 + 581-5445). *Molecular thermal motion, charge density, molecular structure determination.*

Heath, Mr James R. Dept. of Chemistry, Rice U., P. O. Box 1892, Houston, TX 77251, USA.

Heckman, Mr Francis A. Cabot Corp., Concord Rd., Billerica, MA 01821, USA.

Hedberg, Prof. Kenneth Wayne. Chemistry Dept., Oregon State U., Corvallis, OR 97331, USA. (1920) PhD, physical chemistry (CIT, 1948). Prof. (tel. 503 + 754-2371). *Gas phase electron diffraction, free molecules, structure, molecular dynamics and force fields.*

Hedman, Dr Gun-Britt Margareta. Stanford Synchrotron Radiation Lab., Stanford U., SLAC Bin 69, P. O. Box 4349, Stanford, CA 94305, USA. (1949) PhD, chemistry (UMEA U., 1978). Sr. res. assoc. (tel. 415 + 854-3300, ext. SSRL 2874). *Bio-inorganic chemistry, synchrotron radiation, X-ray absorption spectroscopy, protein crystallography using anomalous dispersion.*

Helis, Mr Howard Morrell. Cole Layer Trumble Co., 5757 Woodway, Houston, TX 77057, USA. (1952) MA chemistry (U. North Carolina, Chapel Hill, 1977). Project supervisor. (tel. 713 + 977-7010, ext. 23). *Structure, transition metal complexes, statistical calculations, computer programming.*

Hemily, Dr Philip W. P. O. Box 57160, Washington, DC 20037, USA. (1922) Doctorat, physical chemistry (U. de Paris, France, 1953). Consultant. (tel. 202 + 296-0749).

Hendricks, Dr Robert Wayne. Technology for Energy Corp., One Energy Center, Pellissippi Pkwy., Knoxville, TN 37922, USA. (1937) PhD, materials science (Cornell U., 1964). Manager, product planning and dev. (tel. 615 + 966-5856, telex 810-570-1770). *Position-sensitive detectors, X-ray stress analysis, small-angle scattering, X-ray physics, diffuse scattering.*

Hendrickson, Prof. Wayne A. Dept. of Biochemistry and Molecular Biophysics, Columbia U., New York, NY 10032, USA. (1941) PhD, biophysics (Johns Hopkins U., 1968). Prof. (tel. 212 + 305-3456). *Biological macromolecules, diffraction methods.*

Henslee, Dr Walter Warren. Inorganic Res., Dow Chemical Co., B-1402, Freeport, TX 77541, USA. (1946) PhD, chemistry (U. Texas Austin, 1974). Group leader. (tel. 409 + 238-1364). *X-ray powder diffraction, ceramics, catalysts, metals, mixed metal oxides and hydroxides.*

Herman, Prof. Herbert. Materials Sci. Dept., SUNY, Stony Brook, NY 11794, USA. (1934) PhD, materials science (Northwestern U., 1961). Prof. (tel. 516 + 246-5984). *Small angle neutron scattering, powder diffractometry, stress analysis.*

Hermans, Prof. Jan. Dept. of Biochemistry, U. of North Carolina, Chapel Hill, NC 27514, USA. (1933) PhD, chemistry (U. Leiden, 1958). Prof. (tel. 919 + 966-4644). *Proteins, modeling, thrombosis and hemostasis.*

Herriott, Prof. Jon Roger. Dept. of Biochemistry, U. of Washington, Seattle, WA 98195, USA. (1937) PhD, biophysics (The Johns Hopkins U., 1967). Assoc. prof. (tel. 206 + 543-9484). *Protein structure and function, photosynthesis, hemoglobin.*

Hewston, Ms Terrell Ann. U-60 Chemistry, U. of Connecticut, Storrs, CT 06268, USA.

Heyn, Prof. Anton Nicolaas Johannes. Biology Dept., Louisiana State U., New Orleans, LA 70122, USA. (1906) PhD, (Utrecht U., Netherlands, 1931). Prof. em. (tel. 504 + 288-8098). *Biopolymer structure and conformation, cellulose fibers, small angle X-ray scattering, polymers, molecular biology, molecular biophysics, cell biology, electron microscopy.*

Hibbard, Dr Lyndon Stanley. Dept. of Anesthesia, Hersey Medical Center, PSU, Hershey, PA 17033, USA. (1947) PhD, physical chemistry (Michigan State U., 1977). Res. asst. prof. (tel. 717 + 534-8433). *Biological macromolecule structure and function.*

Higgins, Prof. John Britt. Central Res. Lab., Mobil Res. and Dev. Corp., P. O. Box 1025, Princeton, NJ 08540, USA. (1947) PhD, geology (Virginia Polytech. Inst., 1978). Sr. res. chemist. (tel. 609 + 737-4215). *Mineralogy, crystallography, rock forming minerals, silicate crystal chemistry, phase transitions, microstructures.*

Hinch, Mr Ralph J., Jr. Walter C. McCrone Associates Inc., 2820 S. Michigan Ave., Chicago, IL 60616, USA. (1926) BS, chemistry (Elmhurst C., 1950). Res. scient. (tel. 312 + 842-7100, ext. 21). *Optical and X-ray crystallography, X-ray diffraction analysis, optical microscopy analysis.*

Hingerty, Dr Brian Edward. Health and Safety Res. Div., Oak Ridge Nat. Lab., P.O. Box Y, Oak Ridge, TN 37831, USA. (1948) PhD, physics, biophysics (Princeton U., 1974). Res. staff. (tel. 615 + 574-1253). *Biologically important structures, nucleic acids, polysaccharides, supercomputers.*

Hirschman, Prof. Albert. Anatomy Dept., Downstate Medical Center, SUNY, 450 Clarkson Ave., Brooklyn, NY 11203, USA. (1921) PhD, chemistry (PIB, 1952). Assoc. prof. (tel. 718 + 270-1021). *X-ray diffraction, bone, invertebrate calcifying systems; enzymes, structural proteins, calcifying tissues.*

Hirshfield, Mr Jordan M. P.O. Box 2000, Merck and Co. Inc., Biophysics Dept., Rahway, NJ 07065, USA.

Hirth, Prof. John Price. Met. Engr. Dept., Ohio State U., 116 W. 19th Ave., Columbus, OH 43210, USA. (1930) PhD, metallurgical engineering (Carnegie-Mellon U., 1957). Prof. (tel. 614 + 422-0176). *Metal physics, dislocation theory, surfaces*

Hite, Prof. Gilbert James. Medicinal Chemistry, U. of Connecticut S. of Pharmacy, Storrs, CT 06268, USA. (1931) PhD, medicinal chemistry (U. Wisconsin, 1959). Prof. (tel. 203 + 486-3350). *Organic crystal structures, structure - reaction mechanism - stereochemistry relationship, stereochemical modes of drug action, optical rotatory dispersion.*

Hoard, Prof. James Lynn. Dept. of Chemistry, Cornell U., Ithaca, NY 14853, USA. (1905) PhD, chemistry (CIT, 1932). Prof. em. (tel. 607 + 256-3646). *Structural chemistry.*

Hoard, Dr Laurence Graham. International Paper Co., P. O. Box 797, Tuxedo Park, NY 10987, USA. (1940) PhD, chemistry (U. Michigan, 1977). Sr. res. assoc. (tel. 914 + 351-2101).

Hodgson, Prof. Derek John. Dept. of Chemistry, U. of North Carolina, Chapel Hill, NC 27514, USA. (1942) PhD, chemistry (Northwestern U., 1969). Prof. (tel. 919 + 966-1566). *Magnetically condensed systems, nucleic acid constituents and their analogs, metal complexes, purines, pyrimidines, nucleotides, peptides.*

Hodgson, Prof. Keith Owen. Dept. of Chemistry, Stanford U., Stanford, CA 94305, USA. (1947) PhD, inorganic chemistry (U. California, Berkeley, 1972). Prof. (tel. 415 + 497-1328). *Bio-inorganic chemistry, anomalous dispersion, synchrotron radiation, X-ray absorption, spectroscopy.*

Hodsdon, Dr John Marshall. Longridge Farm, Meredith, NH 03253, USA. (1938) PhD, biochemistry (U. California, Berkeley, 1970). (tel. 603 + 279-6126). *Proteins, structure, small molecule - protein interactions, refinement methods, accuracy, molecular graphics.*

Hoffman, David W. Rt 6, Box 685, Hillsborough, NC 27278, USA.

Hoggins, Dr James Thomas. Diamond Techn. Center, Norton Christensen Inc., 2532 South 3270 West, Salt Lake City, UT 84119, USA. (1942) PhD, materials science (U. Texas, Austin, 1975). Sr. scient. (tel. 801 + 972-3140). *Diamonds, X-ray crystallography, ceramics.*

Holbrook, Dr Stephen Roy. Lab. of Chemical Biodynamics, (Bldg. 3), Lawrence Berkeley Lab., U. of California, Berkeley, CA 94720, USA. (1948) PhD, physical chemistry (U. Oklahoma, 1974). Staff scient. (tel. 415 + 486-4304). *Nucleic acid structure - dynamics, crystallographic refinement, molecular modeling.*

Holden, Ms Hazel M. Inst. of Molecular Biology, U. of Oregon, Eugene, OR 97403, USA.

Holden, Dr James Richard. Energetic Materials, Naval Surface Weapons Center, 10901 New Hampshire Ave., Silver Spring, MD 20903-5000, USA. (1928) PhD, physical chemistry (State U. of Iowa, 1955). Res. chemist. (tel. 202 + 394-2745). *Organic crystal structures, crystallographic computing.*

Holland, Dr Hans J. Techn. Staff Div., Corning Glass Works, Sullivan Park FR-18, Corning, NY 14830, USA. (1929) PhD, chemistry (U. Utah, 1963). Sr. res. chemist. (tel. 607 + 974-3266). *Glass ceramic structures, low temperature X-ray diffraction, automation (XRD).*

Hollander, Dr Frederick J. Dept. of Chemistry, U. of California, Berkeley, CA 94720, USA. (1946) PhD, physical chemistry (U. California, Berkeley, 1972). Postgrad. res. chemist. (tel. 415 + 642-5589). *Small molecules, crystallography, coordination compounds.*

Holmes, Mr F. E. Drawer D, Accokeek, MD 20607, USA.

Holser, Prof. William Thomas. Geology Dept., U. of Oregon, Eugene, OR 97403, USA. (1920) PhD, geology (Columbia U., 1950). Prof. (tel. 503 + 686-4575). *Crystal physics, minerals, symmetry theory, geochemistry.*

Holt, Dr Elizabeth Manners. Chemistry Dept., Oklahoma State U., Stillwater, OK 74078, USA. (1939) PhD, chemistry (Brown U., 1966). Assoc. prof. (tel. 405 + 624-5949). *Fluorescent Cu(I) systems, calcium-allergen interactions, metal cluster systems.*

Holtzberg, Dr Frederic. Res. Div., IBM T. J. Watson Res. Center, P.O. Box 218, Yorktown Hgts., NY 10598, USA. (1922) PhD, physical chemistry (PIB, 1952). Res. staff member. (tel. 914 + 945-1045). *Solid state physics, magnetism, phase transitions, rare earth compounds, materials research.*

Hom, Dr Tommy. Eng. Dept., Philips Electronic Instruments Inc., 85 McKee Dr., Mahwah, NJ 07430, USA. (1949) PhD, physics (PINY, 1979). Scient. (tel. 201 + 529-3800, ext. 261). *Diffraction physics, X-ray optics, automated instrumentation, computer programming.*

Honkonen, Mr Robert S. Chemistry Dept., U. of South Carolina, Columbia, SC 29208, USA.

Honzatko, Prof. Richard E. Dept. of Biochemistry and Biophysics, Iowa State U., Gilman Hall, Ames, IA 50011, USA. (1954) PhD, physical chemistry (Harvard U., 1982). Asst. prof. (tel. 515 + 294-7103). *Macromolecular structure determination, catalysis mechanisms, allostery, anomalous scattering.*

Hoogsteen, Dr Karst. Dept. of Biophysics and Pharmacology, Merck Inst. for Therapeutic Res., Rahway, NJ 07065, USA. (1923) PhD, crystallography (U. Groningen, Netherlands, 1957). Dept. Dir. (tel. 201 + 574-6765, 6766). *Organic and inorganic crystal structure.*

Hope, Prof. Håkon. Dept. of Chemistry, U. of California, Davis, CA 95616, USA. (1930) Cand. real., chemistry (U. Oslo, Norway, 1958). Prof. (tel. 916 + 752-0957,0953). *X-ray diffraction methods, accurate structure analysis, valence electron distribution, absolute configuration; instrumentation, computer programming, natural products.*

Horrigan, Ms Jane Akerlund (Bruce). RADC - ESM, Hanscom AFB, Bedford, MA 01731, USA. (1937) MS, chemistry (Northeastern U., 1965). Res. chemist. (tel. 617 + 861-2215). *Defect structures, X-ray topography, materials and device characterization.*

Horsey, Dr Richard Stephen. Res. and Dev., Keystone Carbon Co., 1935 State St., St. Marys, PA 15857, USA. (1950) PhD, ceramic science (PSU, 1981). Supervisor, thermistor div. (tel. 814 + 781-1591). *Spinels, thick films, thermistors, titanates, ionic conductors.*

Hossain, Mr Mohammed B. Chemistry Dept., Oklahoma U., Norman, OK 73019, USA.

Houska, Prof. Charles Robert. Materials Eng., Virginia Polytechnic Inst., Holden Hall, Blacksburg, VA 24061, USA. (1927) ScD, metallurgy (MIT, 1957). Prof. (tel. 703 + 951-5652). *X-ray diffraction, atomic diffusion, physical metallurgy.*

Howard, Dr Scott A. Dept. of Ceramics, U. of Missouri at Rolla, Rolla, MO 65401, USA. (1958) PhD, ceramic science (New York State C. of Ceramics, 1984). Asst. prof. (tel. 314 + 341-4403). *X-ray and neutron powder diffraction, crystallographic computing.*

Howatson, Prof. John. Chemistry Dept., U. of Wyoming, Laramie, WY 82071, USA. (1920) PhD, chemistry (U. Wisconsin, 1950). Prof. (tel. 307 + 766-4363). *Inorganic and mineral structures.*

Howe, Ms Donna-Beth. 10104 Gardiner Ave., Silver Spring, MD 20902, USA.

Howell, Dr Peter Adam. I.S.D., 3M Co., Bldg. 230-15-07, 3M Center, St. Paul, MN, 55144, USA. (1928) PhD, physical chemistry (U. Minnesota, 1955). Res. specialist. (tel. 612 + 733-9007). *Alumino silicates, zeolites, crystal growth, polymers, structures, glass.*

Hsu, Prof. I-Nan. Chemistry Dept., California State U., 18111 Nordhoff St., Northridge, CA 91330, USA. (1939) PhD, physical chemistry (U. Oklahoma, 1971). Assoc. prof. (tel. 818 + 885-3366). *Protein structure and function, molecular and metal complexes.*

Hsu, Dr Leh-Yeh Ruth. 2634 Cedar Lake Dr., Dublin, OH 43017, USA. (1948) PhD, chemistry (U. Louisville, 1980). *Structure determination, intermolecular forces.*

Huang, Dr Ting Chun. Res. Div., IBM, 5600 Cottle Rd., San Jose, CA 95193, USA. (1942) PhD, physics (PIB, 1972). Res. staff member. (tel. 408 + 256-3993). *X-ray powder diffraction, laboratory automation, computer techniques, X-ray fluorescence, electron microprobe, X-ray thin film analysis, dynamical theory, multiple diffraction, double-crystal diffraction, applied crystallography.*

Hubbard, Dr Camden Richards. Center for Materials Sci., Nat. Bureau of Standards, A221 MATL, Washington, DC 20899, USA. (1944) PhD, physical chemistry (Iowa State U., 1971). Res. chemist. (tel: 309 + 921-2921). *Powder diffraction, materials characterization, data evaluation.*

Hubbell, Mr John Howard. Radiation Physics Div., Nat. Bureau of Standards, Washington, DC 20234, USA. (1925) MSc, engineering physics (U. Michigan, 1950). Dir., X-ray & ionizing radiation data center. (tel. 301 + 921-2685). *X-ray attenuation coefficients, X-ray interactions with atoms, Rayleigh coherent scattering, Compton X-ray scattering, atomic form factors, incoherent scattering functions.*

Huber-Buser, Dr Effi H. Artemis Systems Inc. 125 Berry Corner Lane, Carlisle, MA 01741, USA. (1934) Dr.sc. nat. (ETH Zurich, Switzerland, 1961). Pres. (tel. 617 + 369-8282).

Huddle, Prof. Benjamin Paul. Dept. of Chemistry, Roanoke C., Salem, VA 24153, USA. (1941) PhD, physical chemistry (U. North Carolina, 1968). Assoc. prof. (tel. 703 + 389-2351). *Small molecule structures, inorganic complexes, computer methods.*

Hudgens, Dr Bruce A. USG Corp., 700 N. Highway 45, Libertyville, IL 60048, USA. (1945) PhD, chemistry (U. South Carolina, 1974). Res. assoc. (tel. 312 + 362-9797, ext. 293). *Analytical applications, powder diffraction, X-ray fluorescence, computer applications.*

Hudgens, Mr Claude R. Mound Lab., Miamisburg, OH 45342, USA. (tel. 513 + 435-2251).

Huffman, Dr John Curtis. Molecular Structure Center, Chemistry Dept., Indiana U., Bloomington, IN 47405, USA. (1941) PhD, chemistry (Indiana U., 1974). Sr. scient. (tel. 812 + 335-6742). *Instrumentation, computer graphics, low temperature crystallography.*

Huggins, Dr Maurice Loyal. 135 Northridge Lane, Woodside, CA 94062, USA. (1897) PhD, chemistry (U. California, 1922). (tel. 415 + 368-5386). *Structure - properties relationship.*

Hughes, Dr Edward Wesley. Chemistry and Chemical Eng., California Inst. of Techn., Pasadena, CA 91125, USA. (1904) PhD, chemistry (Cornell U., 1935). Sr. res. assoc. em. (tel. 818 + 356-6527). *X-ray diffraction.*

Hughes, Prof. Robert Edward. Chemistry Dept., Cornell U., Ithaca, NY 14853, USA. (1924) PhD, physical chemistry (Cornell U., 1953). Prof. (tel. 607 + 256-4129). *Inorganic and biochemical structures, macromolecular structures.*

Hughes, Prof. William Eugene. Physics and Astronomy Dept., U. of Southern Mississippi, Hattiesburg, MS 39401, USA. (1932) PhD, physics (U. Alabama, 1963). Chairman. (tel. 601 + 266-7206). *Magnetic resonance, astronomy, crystallography.*

Hunt, Prof. Gary Webb. Dept. of Chemistry and Physics, Shorter C., Shorter Hill, Rome, GA 30161, USA. (1942) PhD, inorganic chemistry (U. Arkansas, 1971). Asst. prof. (tel. 404 + 232-2463, ext. 63). *Inorganic chemistry, environmental chemistry.*

Hurst, Prof. Vernon James. Dept. of Geology, U. of Georgia, GGS Building, Athens, GA 30601, USA. (1923) PhD, geology (Johns Hopkins U., 1954). Res.

prof. (tel. 404 + 542-2652, ext. 23). *X-ray crystallography, petrology (experimental), crystal growth.*

Hutchings, Mr Alan E. Barbeau-Hutchings, 10 S. Franklin Turnpike, Ramsey, NJ 07446, USA. (1933) AB, geology (U. Arizona, 1958). Secretary-treasurer, (tel. 201 + 327-6611). *Analytical instruments.*

Hutchinson, Mr John P. Chemistry Dept., U. of Utah, Box 31, Salt Lake City, UT 84112, USA.

Hybl, Prof. Albert. Dept. of Biophysics, U. of Maryland S. of Medicine, 660 W. Redwood St., Baltimore, MD 21201, USA. (1932) PhD, chemistry and mathematics (CIT, 1961). Assoc. prof. (tel. 301 + 528-7940). *Biological structures, membrane structure and function, computer applications in the biosciences.*

Hyde, Dr C. Craig. Lab. of Molecular Biology, NIADDK, Nat. Inst. of Health, Bldg. 2, Room 316, Bethesda, MD 20205, USA. (1956) PhD, biochemistry (U. Iowa, 1985). Staff fellow. (tel. 301 + 496-4295). *X-ray crystallography, biological macromolecules, protein crystallography.*

Ibers, Prof. James A. Chemistry Dept., Northwestern U., Evanston, IL 60201, USA. (1930) PhD, chemistry (CIT, 1954). Prof. (tel. 312 + 491-5449). *Coordination compounds, organometallic compounds, ternary chalcogenides, metalloporphyrins.*

Ice, Dr Gene Emery. Metals and Ceramics, Oak Ridge Nat. Lab., P.O. Box X, Oak Ridge, TN 37830, USA. (1950) PhD, physics (U. Oregon, 1977). Staff scient. (tel. 615 + 574-4640). *X-ray scattering, synchrotron radiation, anomalous scattering, inner shell cross sections.*

Inniss, Mr Daryl. 3212 Sawtelle Blvd., Apt. 2, Los Angeles, CA 90066, USA.

Inouye, Mr Hideyo. Children's Hospital, 300 Longwood Ave., Box 84, Boston, MA 02115, USA.

Intrater, Mr Josef. 125 Demarest Ave., Englewood, NJ 07632, USA.

Jackobs, Dr John Joseph. Computer Center, Coe C., 1220 1st Ave. NE, Cedar Rapids, IA 52402, USA. (1939) PhD, physical chemistry (Arizona State U., 1967). Registrar, computer center dir., assoc. prof. (tel. 319 + 398-8526). *Computer programming, instrumentation, biological structures.*

Jacobs, Prof. Gerald Daniel. Chemistry Dept., Northern Michigan U., Marquette, MI 49855, USA. (1935) PhD, physical chemistry (Michigan State U., 1961). Prof., head. (tel. 906 + 227-2912). *Crystallography, thermodynamics, kinetics.*

Jacobson, Prof. Robert Andrew. Chemistry Dept., Ames Lab., Iowa State U., 42 Spedding, Ames, IA 50011, USA. (1932) PhD, chemistry (U. Minnesota, 1959). Prof., sr. chemist. (tel. 515 + 294-1144). *Crystallographic computing, automation, metal complexes, insecticides, Patterson methods.*

Jaidong, Mr Ko. Dept. of ESS, SUNY at Stony Brook, Stony Brook, New York, NY 11794, USA.

Jain, Mr Sanjeev. Biochemistry Dept., U. of Wisconsin, 420 Henry Mall, Madison, WI 53706, USA.

Jain, Dr Shri C. Dept. Radiation Biology and Biophysics, U. of Rochester, S. of Medicine and Dentistry, Rochester, NY 14642, USA. (1940) PhD, physical chemistry, crystallography (Poona U., India, 1967). Sr. res. assoc. (tel. 716 + 275-4240). *Crystal structure determination, biological and organic molecules, drug-nucleic acid interactions, nucleic acids structure and function, computer programming.*

James, Prof. William J. Grad. Center for Materials Res., U. of Missouri, Rolla, MO 65401, USA. (1922) PhD, chemistry (Iowa State U., 1953). Prof., sr. investigator. (tel. 314 + 341-4324). *Neutron diffraction, magnetic structures, rare earth alloys.*

Jameson, Prof. Geoffrey Brind. Dept. of Chemistry, Georgetown U., 37 and O St., N.W., Washington, D.C. 20057, USA. (1952) PhD, chemistry (Canterbury, New Zealand, 1977). Asst. prof. (tel. 202 + 625-4319). *Bioinorganic chemistry, twinning, topotaxy.*

Jandacek, Dr Ronald James. Procter and Gamble Co., Miami Valley Labs., PO Box 39175, Cincinnati, OH 45218, USA. (1942) PhD, chemistry (U. Texas, 1968). Chemist. (tel. 513 + 245-2767). *Lipid chemistry.*

Janiak, Dr Martin J. Rigaku/USA Inc., 3 Electronics Ave., Danvers, MA 01923, USA. (1947) PhD, biophysics (Boston U., 1976) Manager, Appl. and Dev. (tel. 617 + 777-2446). *Structure, biological systems, X-ray analysis applications.*

Jasinski, Prof. Jerry P. Dept. of Chemistry, Keene State C., 229 Main St., Keene, NH 03773, USA. (1940) PhD, (U. Wyoming, 1974) Assoc. prof. (tel. 603 + 352-1909, ext. 495). *Inorganic chemistry, biological molecules, polymers.*

Jeffrey, Prof. George Alan. Dept. of Crystallography, U. of Pittsburgh, 304 Thaw Hall, Pittsburgh, PA 15260, USA. (1915) DSc, chemistry (U. Birmingham, UK, 1953). Prof. (tel. 412 + 624-4366). *Structure, small molecules, carbohydrates, hydrates, hydrogen bonding.*

Jejjala, Dr Krishna Mohana Rao. Biological Sciences Dept., Purdue U., W. Lafayette, IN 47907, USA. (1943) PhD, X-ray crystallography (Indian Inst. of Sci., Bangalore, 1971). Res. assoc. (tel. 317 + 494-4911). *Virus crystallography, secondary structure predictions, small molecules, crystallography teaching.*

Jendrek, Dr Eugene F. Monsanto Res. Corp., Mound Facility, Miamisburg, OH 45342, USA. (1949) PhD, physical chemistry (U. Maryland, 1979). Sr. res. chemist. (tel 513 + 865-4205). *Crystallographic computing, inorganic crystal chemistry.*

Jenkins, Dr Ron. JCPDS-ICDD, 1601 Park Lane, Swarthmore, PA 19081, USA. (1932) PhD, chemical physics (PINY, 1981). Principal scient. (tel. 215 + 328-9403). *X-ray diffraction, powder diffractometry.*

Jenks, Ms Janice M. Texas Air Control Board, Lab. Div., 6330 Highway 290E, Austin, TX 78723, USA.

Jennings, Dr Laurence Duane. Materials Characterization Div., Army Materials and Mechanics Res. Center, Watertown, MA 02172-0001 USA. (1929) PhD, physics (MIT, 1955). Physicist. (tel. 617 + 923-5375). *Diffraction techniques, characterization.*

Jensen, Prof. Lyle Howard. Dept. of Biological Structure, U. of Washington, Seattle, WA 98195, USA. (1915) PhD, chemistry (U. Washington, 1943). Prof. (tel. 206 + 543-1983). *Structure and function, biological molecules, accurate bond lengths and angles.*

Jensen, Prof. William Phelps. Chemistry Dept., South Dakota State U., Brookings, SD 57006, USA. (1937) PhD, inorganic chemistry (U. Iowa, 1964). Prof. (tel. 605 + 688-5151). *Coordination compounds, small organic molecules.*

Jesser, Prof. William Augustus. Dept. of Materials Sci., U. of Virginia, Thornton Hall, Charlottesville, VA 22901, USA. (1939) PhD, physics (U. Virginia, 1966). Prof. (tel. 804 + 924-6349). *Thin films, electron microscopy, epitaxy, radiation damage.*

Jircitano, Mr Alan J. 81 Smedley St., North East, PA 16428, USA.

Johnson, Dr Carroll K. Chemistry Div., Oak Ridge Nat. Lab., Bldg. 4500N, M.S. C-18, P.O. Box X, Oak Ridge, TN 37830, USA. (1929) PhD, biophysics (MIT, 1959). Res. crystallogrpher. (tel. 615 + 574-4975). *Crystallographic computing, computer graphics, thermal motion, modulated structures, neutron diffraction, artificial intelligence.*

Johnson, Dr Frank Bacchus. Center for Advanced Medical Education, (ADE), Armed Forces Inst. of Pathology, Washington DC 20306, USA. (1919) MD, medicine (Howard U., 1944). Assoc. dir. for education. (tel. 202 + 576-2934,9). *Inorganic crystal structures, microstructure.*

Johnson, Prof. Gerald Glenn, Jr. Computer Sci. Dept., Pennsylvania State U., 164 MRL, University Park, PA 16801, USA. (1939) PhD, solid state science (PSU, 1965). Assoc. prof. (tel. 814 + 865-1637). *Powder diffraction, search and match techniques, indexing techniques; laboratory automation, crystallographic result presentation, computer graphics, teaching, pattern recognition.*

Johnson, Dr Harold Arthur. Lab. Res. Associates, 2635 W. Cedar Crest Rd., Minnetonka, MN 55343, USA. (1920) chemistry (U. Minnesota, 1951). Res. chemist. (tel. 612 + 545-6601). *Medical research, biochemistry, clinical chemistry, structure determination, metabolites.*

Johnson, Dr John Emil. Biological Sci., Purdue U., Life Sci. Bldg., W. Lafayette, IN 47907, USA. (1945) PhD, physical chemistry (Iowa State U., 1972). Assoc. prof. (tel. 317 + 494-5911). *Virus structure, macromolecular assembly, protein crystallography, biophysical chemistry.*

Johnson, Dr Paul Lorentz. Techn. Information Services, Nat. Energy Software Center, Argonne Nat. Lab., Argonne, IL 60439, USA. (1941) PhD, physical chemistry (Washington State U., 1968). Computer scient. (tel. 312 + 972-4043). *X-ray and neutron diffraction.*

Johnson, Dr Quintin C. Chemistry and Materials Science Dept., Lawrence Livermore Lab., L-370, Livermore, CA 94550, USA. (1935) PhD, chemistry (U. California, Berkeley, 1961). Section leader. (tel. 415 + 422 - 6346). *Powder pattern analysis, materials characterization, micro-organized materials.*

Johnson, Mrs Suzanne Marie. 849 Cathedral Dr., Sunnyvale, CA 94087, USA. (1944) BS, chemistry (U. Arizona, 1966). Manager, techn. support. *Computer applications, semiconductor design.*

Jones, Prof. Daniel Silas. Dept. of Chemistry, U. of North Carolina, UNCC Station, Charlotte, NC 28223, USA. (1943) PhD, physical chemistry (Harvard U., 1971). Assoc. prof. (tel. 704 + 597-4438). *Transition metal complexes.*

Jones, Dr Morton Edward. Physical Sci. Res. Lab., Texas Instruments Inc., P. O. Box 225936 MS 145, Dallas, TX 75265, USA. (1928) PhD, chemistry (CIT, 1953). Dir. (tel. 214 + 238-2468 or 2470). *Crystal growth and characterization, electronic materials, growth of crystals from compounds.*

Jones, Dr Noel Duane. Physical Chemistry Res. Div., Eli Lilly and Co., 307 East McCarty, Indianapolis, IN 46206, USA. (1937) PhD, chemistry (CIT, 1964). Res. scient. (tel. 317 + 261-4668). *Crystal structures, organic molecules.*

Jordan, Prof. Truman H. Chemistry Dept., Cornell C., Mount Vernon, IA 52314, USA. (1937) PhD, chemistry (Harvard U., 1964). Prof. (tel. 319 + 895-8811). *Tin (II) chemistry (phosphates), titanium phosphates.*

Jorgensen, Mr James D. Materials Science and Techn. Div., Argonne Nat. Lab., Bldg. 223, Room D221, 9700 Cass Ave., Argonne, IL 60439, USA.

Julian, Prof. Maureen O'Donnell. Dept. of Geological Sciences, Virginia Polytechnic Inst. and State U., Blacksburg, VA 24061, USA. PhD, physical chemistry (Cornell U., 1966). Adjunct prof. (tel. 703 + 961-6521). *Silicon nitrides, anthracene dimerization, ab initio calculations, deformation.*

Jurnak, Prof. Frances Anne. Biochemistry Dept., U. of California, Riverside, CA 92521, USA. (1946) PhD, chemistry (U. California, Berkeley, 1973). Asst. prof. (tel. 714 + 787-4245). *Macromolecular crystallography, protein elongation factors, complexes, antibiotics, proteins.*

Kadlec, Mr Robert J. Harvard Rd., - Aux 2, Lancaster, MA 01523, USA.

Kaduk, Dr James Albert. Res. and Devel., Standard Oil Co., P. O. Box 400, Naperville, IL 60566, USA. (1952) PhD, inorganic chemistry (Northwestern U., 1977). Staff res. chemist. (tel. 312 + 420-4547). *Catalysts, minerals, zeolites;.*

Kamb, Prof. W. Barclay. Div. of Geological and Planetary Sci., California Inst. of Techn., Pasadena, CA 91125, USA. (1931) PhD, geology (CIT, 1956). Prof. and chairman. (tel. 213 + 795-6811, ext. 2108). *Structural crystallography, minerals, polymorphs, ice, X-ray and neutron diffraction.*

Kanatzidis, Dr Mercouri G. Dept. of Chemistry, U. of Michigan, Ann Arbor, MI 48109, USA. (1957) PhD, (U. Iowa, 1984). Res. fellow. (tel 313 + 764-7361). *Bio-inorganic chemistry, Fe-S proteins, sulfur chemistry, low dimensional materials.*

Kapecki, Dr Jon Alfred. Res. Labs., Eastman Kodak Co., Rochester, NY 14650, USA. (1942) PhD, physical organic chemistry, X-ray crystallography (U. Illinois, 1969). Sr. res. chemist, U. Rochester lect. (tel. 716 + 458-1000, ext. 72056). *Cycloaddition processes, molecular orbital theory, structure - reactivity relationships, sterically crowded molecules, computer modeling, reaction mechanisms, computer applications, chemistry.*

Karcher, Dr Barbara Ann. Chemistry Dept., Michigan State U., East Lansing, MI 48824-1322, USA. (1953) PhD, physical chemistry (Iowa State U., 1981). Res. assoc. (tel. 517 + 353-4505). *Crystallographic computing.*

Karipides, Prof. Anastas. Chemistry Dept., Miami U., Oxford, OH 45056, USA. (1937) PhD, chemistry (U. Illinois, 1964). Prof. (tel. 513 + 529-2813). *Structural inorganic chemistry, solid state chemistry, coordination chemistry, crystal growth, non-bonded interactions.*

Karle, Dr Isabella L. Code 6030, Naval Res. Lab., Washington, DC 20375, USA. (1921) PhD, physical chemistry (U. Michigan, 1944). Head, X-ray crystallography. (tel. 202 + 767-2624). *Structure analysis methods, biologically interesting molecules, polypeptides, photo rearrangement products.*

Karle, Prof. Jerome. Code 6030, Naval Res. Lab., Washington, DC 20375, USA. (1918) PhD, physical chemistry (U. Michigan, 1944). Chief scient., Lab. for the Structure of Matter. (tel. 202 + 767-2665). *Structure analysis methods, diffraction applications (X-ray - electron - neutron).*

Karlsson, Mr Haraldur R. Dept. of Geophysical Sci., U. of Chicago, S. Ellis St., Chicago, IL 60637, USA.

Kartha, Dr Gopinath. Crystallography Center, Roswell Park Memorial Inst., 666 Elm St., Buffalo, NY 14203, USA. (1927) PhD, physics (Madras U., India, 1953). Principal cancer res. scient. (tel. 716 + 845-2362). *X-ray diffraction techniques, biological molecular structures, crystallographic computing, phase problem, macromolecular structures.*

Kasper, Dr John S. Physical Sci. Branch, General Electric Co. Corporate R. and D., Room 4A49-K-1, Schenectady, NY 12345, USA. (1915) PhD, chemistry (Johns Hopkins U., 1941). Physical chemist. (tel. 518 + 385-8432). *Crystal structures, inorganic compounds, metals and alloys, high pressure phases, magnetic structures.*

Kastner, Ms Margaret E. Chemistry Dept., Bucknell U., Lewisburg, PA 17837, USA.

Katz, Dr Bradley Alan. Pharmaceutical Chemistry Dept., U. of California, San Francisco, CA 94117, USA. Contact address: Genentech Inc., 460 Point San Bruno Blvd., South San Francisco, CA 94080, USA. (1953) PhD, inorganic chemistry (UCLA, 1979). Visiting prof. (tel. 415 + 952-1000, ext. 6263). *Biotechnology, neutron crystallography, X-ray crystallography.*

Katz, Mr Henry. Dept. of Molecular Structure, Inst. for Cancer Res., 7701 Burholme Ave., Philadelphia, PA 19111, USA. (1927) MS, chemistry (U. Pennsylvania, 1955). Res. specialist. (tel. 215 + 728-2220). *Small molecule structures, instrumentation.*

Katz, Prof. J. Lawrence. Dept. of Biomedical Eng., Rensselaer Polytechnic Inst., Troy, NY 12181, USA. (1927) PhD, physics (PIB, 1957). Prof., dir. (tel. 518 + 270-6547). *Biomechanical properties and structure, calcified and connective tissues, X-ray diffraction, ultrasonic studies, bone and teeth, scanning electron microscopy, biomedical materials, strain in biological inorganic crystals.*

Katz, Prof. Lewis. Chemistry Dept., U. of Connecticut, Storrs, CT 06268, USA. (1923) PhD, chemistry (U. Minnesota, 1951). Prof. (tel. 203 + 486-3219 or 2012). *Inorganic crystal structures, complex metal oxides, molecular crystals.*

Katz, Dr Louis. Dept. of Biolocigal Sciences, Columbia U., 630 W 168 St., New York, NY 10032, USA. (1932) PhD, physics (U. Wisconsin, 1959). *Nucleic acid and protein structures, computer graphics, computer art, data base management.*

Kaufman, Prof. Hershall William. Dept. of Oral Biology, S. of Dental Medicine, SUNY, Stony Brook, NY 11733, USA. (1940) PhD, oral biology (U. Manitoba, Canada, 1967). Assoc. prof. (tel. 516 + 444-2870). *calcium phosphate crystallography, bone and teeth crystal structure.*

Kay, Dr Mortimer I. 70 Oak Shade Rd., Gaithersburg, MD 20878, USA. (1930) PhD, physical chemistry (U. Connecticut, 1958). *Ferroelectrics, time resolved effects, materials, energy conversion, ocean thermal energy.*

Keder, Dr Nancy Lynn. Molecular Biology Inst., UCLA, Los Angeles, CA 90024, USA. (1955) PhD, chemistry (UCLA, 1984). Post-doctoral assoc. (tel. 213 + 825-8901). *Crystallographic computing.*

Keefe, Prof. William Edward. Biostatistics Dept., Medical C. of Virginia, P.O. Box 678, Medical C. Station, Richmond, VA 23220, USA. (1923) PhD, biophysics (Medical C. of Virginia, 1967). Assoc. prof. (tel. 804 + 786-9824). *Structure, biological molecules, image analysis techniques.*

Keeling, Prof. Rolland O., Jr. Dept. of Physics, Michigan Techn. U., Houghton, MI 49931, USA. (1925) PhD, physics (PSU, 1958). Prof.

Keem, Mr John Edward. Energy Conversion Devices, 1675 W. Maple, Troy, MI 48084, USA.

Kehl, Mr William Louis. 725 Bellaire, Louisville, OH 44641, USA. (1915) MS, physics (U. Iowa, 1941). Retired. (tel. 216 + 875-1335). *Catalysis, structure - properties relationship, inorganic solids.*

Kehres, Mr Paul William. Res. and Dev., A. O. Smith Corp., P. O. Box 584, Milwaukee, WI 53201, USA. (1922) MS, chemistry (Marquette U., 1948). Project coordinator-plastics res. (tel. 414 + 447-4597). *Materials analysis, plastics.*

Keller, Mr Ludwig. Carnet Res. Inc., 318-12th St., Santa Monica, CA 90402, USA.

Kellerman, Dr Martin. Chemistry Dept., California Polytechnic State U., San Luis Obispo, CA 93422, USA. (1932) PhD, chemistry (U. Washington, 1966). Assoc. prof. (tel. 805 + 546-2796). *Crystal structures, metal chelate compounds, small molecules, organic 1-dimensional 'metals'.*

Kelly, Mrs Carol J. (Korsmo). Central Lab. Services, Ford Motor Co., 15000 Century Dr., Dearborn, MI 48020, USA. MS, management (U. Michigan, 1982). Manager, Metallurgy Dept. (tel. 313 + 322-1613). *X-ray diffraction.*

Kelly, Dr Judith Ann. Biochemistry and Biophysics Section, Biological Sciences Group, U. of Connecticut, Storrs, CT 06268, USA. (1944) PhD, biophysics (U. Connecticut, 1977). Asst. prof. (tel. 203 + 486-4622 or 486-4353, telex 994484/UCONNCOOP STOR). *Enzyme structure and function, drug/protein interactions, interactive computer graphics.*

Kestigian, Dr Michael. Sperry Res. Center, North Road, Sudbury, MA 01776, USA. (1928) PhD, inorganic chemistry (U. Connecticut, 1956). Dept. manager. (tel. 617 + 369-4000, ext. 279). *Inorganic chemical reactions and mechanisms, epitaxial film deposition, growth of single crystal electronic materials, structure and characterization, thin films and bulk crystals, penetration phosphor syntheses, wear evaluation, corrosion, lubrication.*

Kim, Ms Eunice E. Chemistry Dept., Boston U., Boston, MA 02215, USA.

Kim, Mr Jung Ja P. Biochemistry Dept., Medical Coll. Wisconsin, 561 N. 15th St., Milwaukee, WI 53233, USA.

Kim, Dr Nancy Ellen Kime. Div. of Labs. and Res., New York State Dept. of Health, ESP-Tower Bldg., Albany, NY 12201, USA. (1942) PhD, inorganic chemistry (Northwestern U., 1969). Res. scient. (tel. 518 + 473-1494). *Bioinorganic crystallography, structure - activity correlations, toxicology.*

Kim, Mr Sangsoo. Iowa State U., 38 Spedding, Ames, IA 50011, USA.

Kim, Mr Sukyoung. NYS College of Ceramics, Alfred U., Alfred, NY 14802, USA.

Kim, Prof. Sung-Hou. Dept. of Chemistry, U. of California, Berkeley, CA 94720, USA. (1937) PhD, chemistry (U. Pittsburgh, 1966). Prof. (tel. 415 + 642-8270). *Structure - function relationships, biological molecules.*

Kimball, Dr Martha R. Dept. of Mineral Sciences, American Museum of Natural History, Central Park West & 79th St., New York, NY 10024, USA. (1945) PhD, crystallography (U. Cambridge, UK, 1974). *Crystallography, biological macromolecules.*

King, Dr Hubert Ellis. Exxon Res. and Eng. Co., Route 22 East, Annandale, NJ 08801, USA. (1949) PhD, earth and space sciences (SUNY, Stony Brook, 1979). Res. physicist. (tel. 201 + 730-2888). *Pressure and temperature effects upon materials, lattice dynamics, mineralogy.*

King, Dr Murray Vernon. Ultrastructure Analysis Section, Center for Labs. and Res., New York State Dept. of Health, Albany, NY 12201, USA. (1922) PhD, physical chemistry (U. Minnesota, 1949). Sr. res. scient. (tel. 518 + 474-7048). *Structural molecular biology, contractile systems; electron microscopy, electron and X-ray diffraction, design of materials and devices.*

Kingman, Ms Priscilla Ward. USABRL, APG, 1115 High Country Rd., Towson, MD 21204, USA. (1934) mechanics, metallurgy (Johns Hopkins U., 1963). Physical metallurgist. (tel. 301 + 278-4269). *Imperfections, twinning, phase transformations, diffraction physics.*

Kirchhoff, Ms Pamela Moore. DP&C, Dow Chemical Co., 1710 Bldg., Midland, MI 48640, USA.

Kirchner, Prof. Richard Martin. Chemistry Dept., Manhattan C., 4513 Manhattan College Parkway, Bronx, NY 10471, USA. (1941) PhD, chemistry (U. Washington, 1971). Assoc. prof. (tel. 212 + 920-0206). *Crystallography, inorganic chemistry, synthesis and characterization, coordination compounds.*

Kirn, Mr J. F. 4324 Southampton Rd., Richmond, VA 23235, USA.

Kirz, Prof. Janos. Physics Dept., SUNY, Stony Brook, NY 11794, USA. (1937) PhD, physics (U. California-Berkeley, 1963). Prof. (tel. 516 + 246-8293). *X-ray optics, X-ray microscopy, high energy physics.*

Kiss, Dr Klara. Analytical Div., Stauffer Chem. Co., Livingstone Ave., Dobbs Ferry, NY 10522, USA. (1930) PhD, analytical chemistry (U. Budapest for Sci and Tech., 1982). Sr. res. assoc. (tel. 914 + 693-1200, ext. 2096). *Powder X-ray diffraction, microbeam analysis, electron diffraction, materials science, laser chemistry, instrumental analysis.*

Kissinger, Charles R. Dept. of Biological Structures, SM-20, U. of Washington, Seattle, WA 98195, USA.

Kissinger, Mr Homer Everett. Materials Dept., Battelle Memorial Inst., Pacific Northwest Labs., P. O. Box 999, Richland, WA 99352, USA. (1923) MS, physics (Kansas State U., 1950). Sr. res. scient. (tel. 509 + 376-3484). *Radiation damage (crystallographic and microstructural aspects), metals.*

Kistenmacher, Dr Thomas John. Applied Physics Lab., The Johns Hopkins U., Johns Hopkins Rd., Laurel, MD 20707, USA. (1943) PhD, chemistry (U. Illinois, 1970). Sen staff chemist. (tel. 301 + 953-6215). *Synthetic organic* *metals, amorphous materials, phase transition phenomena, structure - properties relationship.*

Klanderman, Prof. Kent Arlen. Chemistry Dept., State U. of New York C., Cortland, NY 13045, USA. (1936) PhD, physical chemistry (U. Wisconsin, 1965). Assoc. prof. (tel. 607 + 753-4323). *Bonding and structure, transition metal complexes, organometallic compounds.*

Klein, Prof. Cheryl Lynn. Dept. of Chemistry, Xavier U., 7325 Palmetto St., New Orleans, LA 70125, USA. (1956) PhD, physical chemistry (U. New Orleans, 1982). Asst. prof. (tel. 504 + 486-7411, ext. 377). *Crystallography, drug molecules, neuroleptics, analgesics, charge density, magnetism.*

Klock, Prof. Peter Allan. Dept. of Chemistry, Manchester Community C., 60 Bidwell St, Mail Station 6, Manchester, CT 06040, USA. (1943) PhD, JD, biophysics (Johns Hopkins U., 1973). (tel. 203 + 742-5419). *Structure and function, proteins, immunology.*

Klug, Prof. Harold Philip. 1703-27th Ave. South, Moorhead, MN 56560, USA. (1902) PhD, physical chemistry (Ohio State U., 1928). Retired. (tel. 218 + 233-8373). *Inorganic and organic X-ray crystallography.*

Knobler, Dr Carolyn B. Chemistry Dept., U. of California, 405 Hilgard Ave., Los Angeles, CA 90024, USA. (1934) PhD, chemistry (PSU, 1959). Assoc. res. chemist. (tel. 213 + 825-4330 or 206-6626). *Crystal structures.*

Knox, Prof. James Russell. Biochemistry and Biophysics Section, U. of Connecticut, U-125, Storrs, CT 06268, USA. (1941) PhD, physical chemistry (Boston U., 1967). Prof. (tel. 203 + 486-3133 or 4622). *Protein crystallography, penicillin-binding enzymes, radiation damage.*

Knox, Prof. Kerro. Chemistry Dept., Cleveland State U., 24th St. and Euclid Ave., Cleveland, OH 44115, USA. (1924) PhD, physical chemistry (Yale U., 1950; Cambridge U., 1952). Prof. (tel. 216 + 687-2454). *Eclectic.*

Koch, Dr H. William. Dir.'s Office, American Inst. of Physics, 335 East 45 St., New York, NY 10009, USA. (1920) PhD, physics (U. Illinois, 1944). Dir. (tel. 212 + 685-1940, ext. 284). *Nuclear physics; publication of scientific journals, computerized information systems development.*

Koehler, Dr Wallace Conrad. Solid State Div., Oak Ridge Nat. Lab., Oak Ridge, TN 37830, USA. (1920) PhD, physics (U. Tennessee, 1953). Dir., Nat. Center for Small Angle Scattering Res. (tel. 615 + 574-5232). *Neutron scattering, magnetism.*

Koenig, Dr Donald Frederick. Biology Dept., Bldg. 463, Brookhaven Nat. Lab., Upton, NY 11973, USA. (1927) PhD, biophysics, crystallography (Johns Hopkins U., 1962). Scient. (tel. 516 + 282-3422). *Biophysics, phase determination, crystallographic theory and computations.*

Koeppe, Dr Roger E., II. Dept. of Chemistry, U. of Arkansas, Fayetteville, AR 72701, USA. (1949) PhD, chemistry and biochemistry (CIT, 1976). Assoc. prof. (tel. 501 + 575-4601). *Protein structure and function.*

Koetzle, Dr Thomas Frederick. Chemistry Dept., Brookhaven Nat. Lab., Building 555, Upton, NY 11973, USA. (1943) PhD, chemistry (Harvard U., 1970). Chemist. (tel. 516 + 345-4384). *Neutron diffraction, organometallic compounds, metal hydrides, synchrotron radiation studies.*

Kohn, Prof. Jack Arnold. 65 Wigwam Rd., Locust, NJ 07760, USA. (1925) PhD, mineralogy (U. Michigan, 1950). Adjunct prof., electronic materials. (tel. 201 + 872-2295). *Crystallography of electronic materials, twinning polytypism, polymorphism.*

Koknat, Prof. Friedrich Wilhelm. Dept. of Chemistry, Youngstown State U., 410 Wick Ave., Youngstown, OH 44555, USA. (1938) Dr.rer.nat., chemistry (U. Giessen, BRD, 1965). Assoc. prof. (tel. 216 + 742-3668). *Inorganic crystal structures.*

Kokotailo, Prof. George Thomas. Dept. of Physics, Drexel U.; Mailing address: 98 N. American St., Woodbury, NJ 08096, USA. (1919) PhD, physics (Temple U., 1955). Prof., consultant. (tel. 609 + 845-6508). *Crystal structures, catalyst characterization, crystal growth, zeolite synthesis.*

Kolks, Dr Gary. Dept. of Chemistry, Manhattan C., Riverdale, NY 10471, USA.

Kong, Dr Eric Siu Wai. Manufacturing Res. Center, Hewlett-Packard Labs., 3500 Page Mill Rd., Bldg. 26U, Palo Alto, CA 94304-1209, USA. (1953) PhD, chemistry (Rensselaer Polytechnic Inst., 1978). Member tech. staff. (tel. 415 + 857-8529). *Polymers.*

Konnert, Dr John H. Lab. for the Structure of Matter, Naval Res. Lab., 4555 Overlook Ave., Washington, DC 20375, USA. (1941) PhD, physical chemistry (U. Minnesota, 1967). Res. chemist. (tel. 202 + 767-2735). *Glassy materials, macromolecules.*

Konnert, Mrs Judith A. U.S. Geological Survey, National Center, Stop 959, Sunrise Valley Dr., Reston, VA 22092, USA. (1941) BA, chemistry (Wooster C., 1963). Chemist. (tel. 703 + 860-6666).

Kopka, Mrs Mary Lou. Molecular Biology Inst., UCLA, 405 Hilgard, Los Angeles, CA 91024, USA. (1938) BS, chemistry. Res. assoc. (tel. 213 + 206-8278). *DNA, protein, cancer drugs.*

Korp, Prof. James Douglas. Chemistry Dept., U. of Houston, University Park, Houston, TX 77004, USA. (1950) PhD, analytical chemistry (U. Texas, Austin, 1975). Asst. prof. and departmental crystallographer. (tel. 713 + 749-2108). *Optical activity, computer programming, graphics.*

Kosel, Mr George Eugene. American Gas and Chemical Co. Ltd., 220 Pegasus Ave., Northvale, NJ 07656, USA. (1923) MS, pharmacology (U. Rochester, 1951). Chief chemist. (tel. 201 + 767-7300). *Electrophotography, inorganic crystallography.*

Kosiur, Dr David Richard. Chevron Oil Field Res. Co., P. O. Box 446, La Habra, CA 90631, USA. (1950) PhD, geochemistry (UCLA, 1978) Sr. res. geochemist. (tel. 213 + 694-7361). *Clay mineralogy, small-angle scattering (X-rays and neutrons).*

Kostiner, Prof. Edward Stephen. Inst. of Materials Sci., U. of Connecticut, Storrs, CT 06268, USA. (1940) PhD, chemistry (PIB, 1960). Prof. (tel. 203 + 486-4615). *Solid state inorganic chemistry, crystal chemistry, crystal growth.*

Koszelak, Mr Stanley N. Lab. of Protein Studies, Oklahoma Medical Res. Foundation, 825 Northeast 13th St., Oklahoma City, OK 73104, USA. (1953) BS, microbiology (U. Oklahoma, 1976). Grad. student. (tel. 405 + 235-8331). *Macromolecular structure and function.*

Kountz, Dr Dennis James. Textile Fibers Div., E. I. du Pont de Nemours and Co., Dacron Yarn Res. and Dev., Kinston, NC 28501, USA. (1956) PhD, physical chemistry (Ohio State U., 1984) Res. chemist. (tel. 919 + 522-6795). *Fibre structure, natural and synthetic polymers, ultraweak solids, aggregates, organometallic small molecules.*

Kraatz, Dr Paul. One Res. Park, Northrop Res. and Techn. Center, Palos Verdes Peninsula, CA 90274, USA. (1940) PhD, geology, mineralogy (U. Minnesota, 1972). Res. techn. staff member. (tel. 213 + 377-4811). *Physical properties, crystal growth, thin film, structure - properties - growth relationship.*

Krasner, Prof. Saul. Physical and Ocean Sci., U.S. Coast Guard Academy, New London, CT 06320, USA. (1929) PhD, physics (PIB, 1970). Assoc. prof. (tel. 203 + 443-8463, ext. 372). *Solid state physics, education.*

Kraut, Prof. Joseph. Chemistry Dept., U. of California, San Diego, La Jolla, CA 92093, USA. (1926) PhD, chemistry (CIT, 1954). Prof. (tel. 714 + 452-3366). *Biological macromolecules (structure - function - evolution).*

Krawitz, Prof. Aaron David. Dept. of Mechanical and Aerospace Eng., U. of Missouri, 1006 Engineering Bldg., Columbia, MO 65211, USA. (1943) PhD, materials science (Northwestern U., 1972). Assoc. prof. (tel. 314 + 882-7671). *Neutron diffraction, X-ray diffraction, phase relations, mechanical behavior, alloys, composites, ceramics.*

Kretsinger, Prof. Robert Harvey. Biology Dept., U. of Virginia, Charlottesville, VA 22901, USA. (1937) PhD, biophysics (MIT, 1964). Prof. (tel. 804 + 924-7039). *Protein structure, protein evolution, data measurement by multiwire proportional counters.*

Krieger, Dr Monty. Biology Dept., Whitaker C., MIT, 45 Carlton St., Room E25-236, Cambridge, MA 02139, USA. (1950) PhD, chemistry (CIT, 1976). Asst. prof. (tel. 617 + 253-6793). *Receptor-mediated endocytosis, cell biology, somatic cell genetics, human genetic diseases, proteases.*

Krimm, Prof. Samuel. Dept. of Physics and Biophysics Res. Div., U. of Michigan, 1261 IST, Ann Arbor, MI 48109, USA. (1925) PhD, physical chemistry (Princeton U., 1950). Prof., Dir., biophysics res. div. (tel. 313 + 764-5257). *X-ray diffraction, membrane structure, vibrational spectroscopy, macromolecules.*

Kullnig, Dr Rudolph Karl. 2168 McClellan Rd., R. D. 1 Nassau, NY 12123, USA. (1918) PhD, chemistry (U. Ottawa, Canada, 1958). (tel. 518 + 766-3827). *Structure determinations, X-ray diffraction, spectroscopy, theoretical chemistry.*

Kumosinski, Thomas F. 2400 Chestnut St., Apt. 3207, Philadelphia, PA 19103, USA.

Kuriyama, Dr Masao. Center for Materials Sci., Nat. Bureau of Standards, Gaithersburg, MD 20899, USA. (1931) DSc, physics (U. Tokyo, Japan, 1958). Physicist. (tel. 301 + 921-2986). *Dynamical diffraction theory, imperfect crystals, X-ray dynamical diffraction topography, crystal growth (metals and alloys), X-ray inelastic scattering, phase transformations.*

Kurtz, Dr Stewart Kendall. Clairol Appliances, Res. and Dev. Div., 1 Blachley Rd., Stamford, CT 06902, USA. (1931) PhD, physics (Ohio State U., 1960). Vice-pres. (tel. 203 + 357-5209). *Optical properties, non-centrosymmetric crystals, ferroelectrics, phase transitions, crystal growth, new materials.*

Kvick, Dr Ake H. Chemistry Dept., Brookhaven Nat. Lab., Bldg. 555, Upton, NY 11973, USA. (1942) PhD, chemistry (U. Uppsala, Sweden, 1974). Chemist. (tel. 516 + 282-4381). *X-ray diffraction, synchrotron radiation, neutron diffraction, molecular sieves, dielectric structures, hydrogen bonding.*

La Placa, Mr Sam Joseph. Res. Div., IBM T. J. Watson Res. Center, P.O. Box 218, Yorktown Hgts., NY 10598, USA. (1937) BSc, chemistry (PIB, 1957). Res. staff member. (tel. 914 + 945-2048). *X-ray diffraction, crystal structures, high pressure, low dimensional electronic and ionic conductors.*

La Prade, Dr Marie D.. Chemistry Dept., Rutgers U., Wright Labs., New Brunswick, NJ 08903, USA. (1942) PhD, chemistry (MIT, 1969). Visiting asst. prof. (tel. 201 + 932-3762). *Organometallic structures, crystallographic computing.*

La Rocca, Mr Edward W. 115 San Luis Way, Placentia, CA 92670, USA.

Ladell, Dr Joshua. Materials Characterization Res. Group, Philips Labs., 345 Scarborough Rd., Briarcliff Manor, NY 10510, USA. (1923) PhD, physics (PIB, 1954). Sr. res. scient. (tel. 914 + 945-6332, telex 646326 PHILAB bfrf). *Computer control, X-ray instrumentation, experimental phase determination, crystal diffractometry.*

Laderman. Dr Stephen Stromberg. Hewlett-Packard Labs., Hewlett Packard, 1501 Page Mill Rd., Palo Alto, CA 94304-1181, USA. (1955) PhD, materials science (Stanford U., 1983). Member techn. staff. (tel. 415 + 857-1501, ext. 3202). *Crystalline semiconductors, amorphous materials, diffraction methods, X-ray topography, synchrotron radiation.*

Lake, Prof. James A. Molecular Biology Inst., U. of California, Los Angeles, CA 90024, USA. (1941) PhD, physics, molecular biology (U. Wisconsin, 1967). Prof. (tel. 213 + 825-2546). *Biological structures.*

Lalancette, Prof. Roger A. Dept. of Chemistry, Rutgers U., Olson Labs., 73 Warren St., Newark, NJ 07102, USA. (1939) PhD, analytical chemistry (Fordham U., 1967). Assoc. prof. (tel. 201 + 648-5329). *Macromolecules, small peptides, proteins, hormones, small molecules.*

Lando, Prof. Jerome B. Dept. of Macromolecular Sci., Case Western Reserve U., University Circle, Cleveland, OH 44106, USA. (1932) PhD, physical chemistry (PIB, 1963). Prof. (tel. 216 + 368-4284). *Solid state reactions, polymer crystal structures, pyroelectric and piezoelectric polymers, conformation transitions, polymers in solution.*

Landy, Dr Richard Allen. Res. Dept., North American Refractories Co., Res. Center, Curwensville, PA 16833, USA. (1931) PhD, mineralogy and petrology (PSU, 1961). Dir. (tel. 814 + 236-3890, ext. 112). *Refractory materials.*

Lange, Dr Bruce Ainsworth. W. R. Grace & Co., 62 Whittemore Ave., Cambridge, MA 02140, USA. (1948) PhD, chemistry (U. New Hampshire, 1974). Group leader. *X-ray powder diffraction studies, single crystal X-ray structure analyses, radiopharmaceuticals, surface chemistry, cement chemistry, computer modeling.*

Langridge, Prof. Robert. Computer Graphics Lab., U. of California, 926 Medical Sci., San Francisco, CA 94143, USA. (1933) PhD, crystallography (U. London). Prof. (tel. 415 + 666-2630). *Molecular structure, computer graphics, drug design.*

Langs, Dr David Alan. Dept. of Molecular Biophysics, Medical Foundation of Buffalo, 73 High St., Buffalo, NY 14203, USA. (1941) PhD, inorganic chemistry (SUNY, Buffalo, 1968). Assoc. res. scient. (tel. 716 + 856-9600, ext. 463). *Drug/hormone receptor binding, calcium channel drugs, prostaglandins, direct methods, molecular replacement methods.*

Larson, Dr Allen C. Physics Div., P-8, MS-H805, Los Alamos Nat. Lab., Los Alamos, NM 87545, USA. (1928) PhD, chemistry (Washington U., 1956). Self employed crystallographic consultant. (tel. 505 + 988-5210). *Structural chemistry, computational crystallography, materials science, computer applications, chemistry and physics.*

Larson, Dr Bennett Charles. Solid State Div., Oak Ridge Nat. Lab., P. O. Box X, Oak Ridge, TN 37830, USA. (1941) PhD, physics (U. Missouri, 1970). Physicist. (tel. 615 + 483-8611, ext. 3-1189). *Radiation damage in metals, X-ray diffuse scattering, X-ray topography.*

Larson, Mr Steven B. 2100 Blue Meadow Dr., Austin, TX 78744, USA.

Lashewycz-Rubycz, Dr Romana Alexandra. Chemistry Dept., Hobart and William Smith C., Geneva, NY 14456, USA. (1952) PhD, inorganic chemistry (SUNY, Buffalo, 1979). Asst. prof. (tel. 315 + 789-5500, ext. 555). *Organo-transition metal chemistry, structural and kinetic studies, X-ray diffraction, spectrophotometric and stopped-flow techniques.*

Lattman, Prof. Eaton Edward. Dept. of Biophysics, Johns Hopkins U., S. of Medicine, Baltimore, MD 21205, USA. (1940) PhD, biophysics (Johns Hopkins U., 1969). Assoc. prof. (tel. 301 + 955-8388). *X-ray diffraction analysis, large biological molecules, protein structure and function.*

Laudise, Dr Robert Alfred. Bell Telephone Labs., Murray Hill, NJ 07974, USA. (1930) PhD, chemistry (MIT, 1956). Dir. of res., physical and inorganic chemistry. (tel. 201 + 582-6220). *Crystal growth, hydrothermal chemistry, materials science, quartz.*

Lawless, Prof. Kenneth Robert. Dept. of Materials Sci., U. of Virginia, Thornton Hall, Charlottesville, VA 22901, USA. (1922) PhD, physical chemistry (U. Virginia, 1951). Prof., chairman. (tel. 804 + 924-3264). *Surface structure and properties, electron microscopy, electron and X-ray diffraction, oxidation, epitaxy, magnetic properties.*

Lawson, Dr Charles Alden. M. S. 959, Branch of IGP, U.S. Geological Survey, 12201 Sunrise Valley Dr., Reston, VA, 22092, USA. (1951) PhD, geology (Princeton U., 1981). Geologist. (tel. 703 + 860-6595). *Geology, mineralogy, petrology, rock magnetism.*

Lawton, Mr Stephen Latham. Res. Dept., Mobile Res. and Dev. Corp., Paulsboro, NJ 08066, USA. (1939) MS, chemistry (Iowa State U., 1966). Assoc. (tel. 609 + 423-1040, ext. 2755). *Crystallography, organometallic and inorganic crystal structures, computer science.*

Lebioda, Dr Lukasz. Dept. of Chemistry, U. of South Carolina, Columbia, SC 29208, USA. (1943) PhD, physics (Jagiellonian U., Poland, 1972). Res. asst. prof. (tel. 803 + 777-2140). *Protein crystallography, glycolytic enzymes.*

Ledbetter, Dr Hassel M. Fracture and Deformation Div. (430), Nat. Bureau of Standards, Boulder, CO 80303, USA. (1937) PhD, metallurgy (U. Illinois at Champaign-Urbana, 1969). Res. metallurgist. (tel. 303 + 499-3443). *Phase transformations, elastic properties, physical properties, stacking faults.*

Lee, Prof. Byungkook. DCRT/PSL, Nat. Inst. of Health, Room 2007, Bldg. 12A, Bethesda, MD 20205, USA. (1941) PhD, physical chemistry (Cornell U., 1967). (tel. 301 + 496-1135). *Protein structure, computer graphics, role of solvent.*

Lee, Mr Han Sik. 3140 Sawtelle, Apt 1, Los Angeles, CA 90066, USA.

LeGeros, Prof. Racquel Z. Dental Materials Sci., New York U., 345 East 24th St., New York, NY 10010, USA. (1935) PhD, biochemistry (NYU, 1067). Prof., project dir. (tel. 212 + 481-5831). *Calcium phosphates (synthetic and biological).*

Lenhert, Prof. P. Galen. Physics Dept., Vanderbilt U., Box 1807, Station B, Nashville, TN 37235, USA. (1933) PhD, biophysics (Johns Hopkins U., 1960).

Prof. (tel. 615 + 322-2830). *Crystal structure, inorganic and biological, computer programming for data collection.*

Lentz, Prof. Paul J. Jr. Biology Dept., King's C., 133 N. River St., Wilkes Barre, PA 18702, USA. (1944) PhD, molecular biology (Purdue U., 1971). Asst. prof. (tel. 717 + 824-9931, ext. 236). *Protein structure, virus structure.*

Leonowicz, Dr Michael Edward. Analytical Div., Exxon Res. and Eng. Co., Route 22 East, Annandale, NJ 08801, USA. (1949) PhD, physical chemistry (Cornell U., 1976). Staff chemist. (tel. 201 + 730-2105). *Inorganic and zeolite structures, electron diffraction and microscopy, powder diffraction.*

Lesk, Prof. Arthur Mallay. Dept. of Chemistry, Fairleigh Dickinson U., 1000 River Rd., Teaneck, NJ 07666, USA. (1941) PhD, physics and physical chemistry (Princeton U., 1966). Prof. (tel. 201 + 836-6300, ext. 375). *Biophysics, theoretical chemistry, computer graphics.*

Lessinger, Prof. Leslie. Chemistry Dept., Barnard C., New York, NY 10027, USA. (1943) PhD, chemistry (Harvard U., 1972). Asst. prof. (tel. 212 + 280-5480). *Direct methods, natural products, biologically active molecules, solid state reactions, polyvalent iodine chemistry.*

Lessor, Dr Arthur Eugene, Jr. Res. Div., IBM Corp., East Fishkill Facility, Route 52, Hopewell Junction, NY 12533, USA. (1925) PhD, chemistry (Indiana U., 1955). Manager, semiconductor products. (tel. 914 + 897-8145). *Josephson tunneling, thin films.*

Leung, Peter C. W. Chemistry Div., Argonne Nat. Lab., 9700 South Cass Ave., Argonne, IL 60439, USA.

Levan, Mr Keith R. Dept. of Chemistry, UCLA, Los Angeles, CA 90024, USA.

Levien, Dr Louise. Exxon Production Res. Co., P. O. Box 2189, Houston, TX 77252-2189, USA. (1952) PhD, earth and space sciences (SUNY at Stony Brook, 1979). Res. specialist. (tel. 713 + 966-6041). *Mineralogy, crystallography, mineral elasticity.*

Levy, Dr Henri A. 116 Meadow Rd., Oak Ridge, TN 37830, USA. (1913) PhD, chemistry (CIT, 1935). Consultant. *Neutron diffraction, small molecules, structural chemistry.*

Leyerle, Mr Richard W. UOP Inc., 10 UOP Plaza, Des Plaines, IL 60016, USA.

Li, Dr Chi-Tang. Advanced Ceramic Program, Dow Corning Corp., 3901 S. Saginaw Rd., P. O. Box 1592, Midland, MI 48640, USA. (1934) PhD, physical chemistry (Montana State U., 1964). Res. specialist. (tel. 517 + 496-6058). *Ceramic fiber, glass-ceramics, fuel cell, material science.*

Liang, Mr Keng-San. 226 Windmill Ct. Bridgewater, NJ 08807, USA.

Licklider, Mr Robert A. 4020 5th Ave., NE, Seattle, WA 98105, USA.

Liebman, Dr Michael N. Dept. of Physiology-Biophysics, Mount Sinai Medical Center, 101st and Fifth Ave., New York, NY 10029, USA. (1947) PhD, physical chemistry (Michigan State U., 1977). Asst. prof. (tel. 212 + 650-6353). *Macromolecular modeling, structure - function analysis, design and structure prediction, artifical intelligence, computer graphics.*

Lii, Mrs Sue-Lein Wang. Dept. of Chemistry, Ames Lab., 44 Spedding Hall, Iowa State U., Ames, IA 50011, USA. (1953) PhD candidate, . Grad. asst. (tel. 515 + 294-8444). *X-ray phase problems, Patterson-superposition techniques, RDF (XRD and EXAFS).*

Lind, Dr Maurice David. Sci. Center, Rockwell International, 1049 Camino Dos Rios, Thousand Oaks, CA 91360, USA. (1934) PhD, chemistry (Cornell U., 1962). Techn. staff member. (tel. 805 + 498-4545, ext. 190). *X-ray crystallography, crystal growth.*

Lindenmeyer, Dr Paul Henry. 165 Lee St., Seattle, WA 98109, USA. (1921) PhD, physical chemistry (Ohio State U., 1951). Consultant. (tel. 206 + 284-1283). *Materials research.*

Lingafelter, Prof. Edward Clay. Dept. of Chemistry, U. of Washington, BG-10, Seattle, WA 98195, USA. (1914) PhD, chemistry (U. California, Berkeley, 1939). Prof. em. (tel. 206 + 543-1686). *Inorganic and coordination compound structures.*

Lippard, Prof. Stephen James. Dept. of Chemistry, 18-207, Massachusetts Inst. of Techn., 77 Massachusetts Ave., Cambridge, MA 02139, USA. (1940) PhD, chemistry (MIT, 1965). Prof. (tel. 617 + 253-1892). *Inorganic chemistry, synthesis, structure determination, transition metal bimetallic centers, chemistry and biology, antitumor platinum drugs, heavy atom labeling, biopolymers, coupling reactions, organometallic chemistry, higher coordinate metal complexes.*

Lippert, Dr Ernest L., Jr. Corporate Techn. Div., Owens-Illinois Inc., P. O. Box 1035, Toledo, OH 43666, USA. (1931) PhD, inorganic and structural chemistry (U. Leeds, UK, 1965). Chief, analytical chemistry. (tel. 419 + 242-6543, ext. 33-288). *Scattering theory, scientific computing, X-ray diffraction and fluorescence, analytical instrumentation.*

Lippman, Dr Robert. Chemistry Dept., Adelphi U., Garden City, NY 11530, USA. (1943) PhD, chemistry (Adelphi U., 1977). *Low temperature X-ray diffraction instrumentation, crystal structure analysis.*

Lipscomb, Prof. William Nunn. Chemistry Dept., Harvard U., 12 Oxford St., Cambridge, MA 02138, USA. (1919) PhD, chemistry (CIT, 1946). Abbott and James Lawrence prof.(tel. 617 + 495-4098). *Enzymes, proteins, inorganic and organic compounds, low temperatures.*

Little, Prof. Robert Greenwood. Chemistry Dept., U. of Maryland-Baltimore County, 5401 Wilkins Ave., Baltimore, MD 21228, USA. (1942) PhD, inorganic chemistry (SUNY Buffalo, 1969). Asst. prof. (tel. 301 + 455-2527). *Bioinorganic chemistry, porphyrins, nicotinamides, flavins.*

Litvin, Prof. Daniel Bernard. Dept. of Physics, PSU, Berks Campus, P.O. Box 2150, Reading, PA 19608, USA. (1940) PhD, physics, (Technion - Israel Inst. of Techn., 1971). Assoc. prof. (tel. 215 + 375-4211, ext. 56). *Solid state physics, mathematical crystallography, phase transitions.*

Liu, Mr Hung-Yu. 27 Kiowa Ct., Coram, NY 11727, USA.

Lockhart, Steven H. Chemistry Dept., U. of Virginia, McCormick Rd., Charlottesville, VA 22901, USA.

Loeb, Dr Arthur L. Dept. of Visual and Env. Studies, Harvard U., Carpenter Center, Cambridge, MA 02138, USA. (1923) PhD, chemical physics (Harvard U., 1949). Sr. lect., curator and master of Dudley House. (tel. 617 + 495-3251). *Mathematical crystallography, design science.*

Loehlin, Prof. James Herbert. Dept. of Chemistry, Wellesley C., Wellesley, MA 02181, USA. (1934) PhD, physical chemistry (MIT, 1960). Prof. (tel. 617 + 235-0320, ext. 3043). *Molecular crystals, bonding and intermolecular forces.*

Loghry, Dr Ray Allen. Chemical Res. Dept., Halliburton Services Co., Box 1431, Duncan, OK 73533, USA. (1949) PhD, analytical chemistry -- crystallography (U. Texas, 1976). Sr. chemist - powder diffraction lab. (tel. 405 + 251-3363). *Powder diffraction studies, characterization, industrial scales, iron phosphides.*

Lok, Mr Charles. Dept. of Chemistry, Baylor U., Waco, TX 76703, USA.

Lokken, Prof. Donald Arthur. Chemistry Dept., U. of Alaska, Fairbanks, AK 99701, USA. (1937) PhD, inorganic chemistry (Iowa State U., 1970). Assoc. prof. (tel. 907 + 479-7525). *Growth, mineral crystals, metals in unusual oxidation states, X-ray crystallography.*

Loll, Mr Patrick J. Dept. of Biophysics, Johns Hopkins S. of Medicine, 725 N. Wolfe St., Baltimore, MD 21205, USA. (1958) BChE, chemical engineering (Catholic U. of America, 1981). Grad. student. (tel. 301 + 995-8388). *Protein crystallography, site-directed mutagenesis.*

Long, Dr Gabrielle Gibbs (Cohen). Bldg. 223, Rm A163, Nat. Bureau of Standards, Gaithersburg, MD 20899, USA. PhD, physics (PINY, 1972). Physicist. (tel. 301 + 921-3603). *Dynamical diffraction, inelastic scattering, disordered materials.*

Love, Prof. Warner Edwards. Thomas C. Jenkins Dept. of Biophysics, Johns Hopkins U., 3400 N. Charles St., Baltimore, MD 21218, USA. (1922) PhD, physiology (U. Pennsylvania, 1951). Prof. (tel. 301 + 338-7250). *Protein crystal structure, hemoglobin, histone hemocyanin.*

Lovell, Dr Frederick Maurice. 128 Lakeview Ave., Leonia, NJ 07605, USA. (1930) PhD, physics (U. Wales, UK, 1960). *Organic structures, pharmaceutically interesting structures, computer programming.*

Low, Prof. Barbara Wharton. Biochemistry Dept., Columbia U., 630 West 168 St., New York, NY 10032, USA. (1920) DPhil, chemistry (Oxford U., UK, 1948). Prof. (tel. 212 + 694-3896, 3895). *Structure and function, snake venom toxins, postsynaptic neurotoxins and cytotoxins, protein conformation theory, bile salts and degradation products role in colon carcinogenesis, acetylcholine receptor - neurotoxin complex structure.*

Lowe-Ma, Dr Charlotte Kathryn. Chemistry Div., Code 3854, Naval Weapons Center, China Lake, CA 93555, USA. (1951) PhD, chemistry (California Inst. of Techn., 1979). Res. chemist. (tel 619 + 939-1607). *Small-molecule crystallography, solid state chemistry.*

Lublin, Mr Paul. 16 Montgomery Dr., Framingham, MA 01701, USA. (1924) MS, physical chemistry (Purdue U., 1949). Retired. (tel. 617 + 877-7879). *X-ray diffraction analysis, powder and single crystal studies; X-ray spectroscopy, electron probe, automation, Electron microscopy (scanning and transmission), spectroscopy, surface analysis*

Ludwig, Prof. Martha L. Biophysics Res. Div., I. S. T. Bldg., U. of Michigan, 2200 Bonisteel Blvd., Ann Arbor, MI 48109, USA. (1931) PhD, biochemistry (Cornell U. Medical S., 1956). Prof., res. biophysicist. (tel. 313 + 763-2199). *Protein crystallography.*

Luss, Mr Henry Richard. Res. Lab., Eastman Kodak Co., Building 82, Kodak Park, Rochester, NY 14650, USA. (1939) BS, chemistry (Rochester Inst. of Techn. 1969). Res. chemist. (tel. 716 + 722-0138). *Organic crystal structures.*

Lustig, Mr Stanley. Films-Packaging Div., Union Carbide Corp., 6733 W. 65th. St., Chicago, IL 60638, USA. (1933) BS, chemistry (U. Toledo, 1958). Techn. manager. (tel. 312 + 496-4672). *High polymer structure and properties.*

Mackie, Dr Paul E. 5695 Salem Rd., Lithonia, GA 30058, USA. (1942) PhD, physics (Georgia Inst. of Techn., 1972). Res. scient. (tel. 404 + 894-3455). *Crystal physics, powder and single-crystal diffractometry, computer systems programming.*

Mackinnon, Dr Ian Donald. Dept. of Geology, U. of New Mexico, Albuquerque, NM 87131, USA. PhD, geochemistry (James Cook U., Australia, 1979). Sr. res. scient. (tel. 505 + 277-7536). *Silicate mineralogy, cosmochemistry.*

MacCrone, Prof. Robert Kirsten. Materials Eng., Rennselear Polytechnic Inst., Troy, NY 12180, USA. D. Phil, physics (U. Oxford, UK). Prof. (tel. 518 + 266-6449). *Magnetism, EPR, glasses, oxides.*

MacRae, Dr Alfred U. AT&T Bell Labs., Murray Hill, NJ 07974, USA. (1932) PhD, physics (Syracuse U., 1960). Lab. dir. (tel. 201 + 582-3606). *Low energy electron diffraction, surfaces.*

Madden, Prof. John Joseph. Depts. of Psychiatry and Biochemistry, Emory U., S. of Medicine, Box Af, Atlanta, GA 30322, USA. (1943) PhD, biochemistry (Emory U., 1968). Assoc. prof. (tel. 404 + 894-5951). *Nucleic acid structures, mutagenesis, DNA repair.*

Magnus, Dr Karen A. Dept. of Biophysics, Johns Hopkins U., 725 N. Wolfe St., Baltimore, MD 21205, USA. (1952) PhD, biophysics (Johns Hopkins U., 1980). Postdoctoral fellow. (tel. 301 + 955-8388). *Biological macromolecules, DNA-binding proteins, oxygen transport proteins.*

Magnuson, Prof. Vincent Richard. Chemistry Dept., U. of Minnesota, Duluth, MN 55812, USA. (1942) PhD, inorganic chemistry (U. Illinois, 1968). Assoc. prof. (tel. 218 + 726-7591). *Coordination chemistry, organometallic chemistry, physical-inorganic chemistry.*

Mahar, Mr Martin Cajetan. Dept. of Chemistry, Texas A and M U., College Station, TX 77843, USA. (1956) BS, chemistry (Sacred Heart U., 1978). Grad. student. (tel. 713 + 845-3455). *Electron density functions, chemical bonding, quantum mechanics.*

Majeste, Mr Richard J. Chemistry Dept., Southern U., New Orleans, LA 70126, USA.

Makowski, Prof. Lee. Dept. of Biochemistry and Molecular Biophysics, Columbia U., 630 W. 168th St., New York, NY 10032, USA. (1949) PhD, molecular biophysics (MIT, 1976). Asst. prof. (tel. 212 + 694-7307). *Macromolecular structure, membranes, viruses, fiber diffraction, diffraction theory.*

Malley, Ms Mary F. Dept. of Sci. Information, E. R. Squibb and Sons, Box 4000, Princeton, NJ 08540, USA. (1953) BA, chemistry (Rutgers U., 1975) Asst. res. investigator. (tel. 609 + 921-4986). *Molecular structure and conformation, X-ray diffraction.*

Mallory, Dr Chester L. Prometrix Corp., 3255 Scott Blvd., Bldg. 2, Santa Clara, CA 95054, USA. (1952) PhD, ceramic engineering (Alfred U., 1979). Eng. manager. (tel. 408 + 970-9500). *Computer automation, analytical instrumentation, crystallographic computing.*

Mandel, Dr Gretchen Sue. Dept. of Medicine-Biochemistry, The Medical C. of Wisconsin, Veterans Administration Hospital, 5000 West National Ave., Milwaukee, WI 53193, USA. (1946) PhD, X-ray crystallography (U. Pennsylvania, 1972). Asst. res. prof. (tel. 414 + 384-2000, ext. 2498). *Biological - medical X-ray structural analysis.*

Mandel, Dr Neil Stanley. Dept. Medicine-Biochemistry, The Medical C. of Wisconsin, Veterans Administration Hospital, 5000 West National Ave., Milwaukee, WI 53193, USA. (1947) PhD, X-ray crystallography (U. Pennsylvania, 1971). Assoc. prof. (tel. 414 + 384-2000, ext. 2494, 2498). *Biological - medical X-ray structural analysis, crystal-induced membrane analysis-gout, protein structure and function, membrane structure.*

Manor, Dr Philip C. Springer-Verlag, 175 Fifth Ave., New York, NY 10010, USA. (1944) PhD, physical chemistry (MIT, 1973). Sci. editor. (tel. 212 + 477-8687). *Protein and nucleic acid crystallography.*

Margulis, Prof. Thomas N. Dept. of Chemistry, U. of Massachusetts, Boston, MA 02125, USA. (1937) PhD, chemistry (U. California-Berkeley, 1962). Prof. (tel. 617 + 287-1900, ext. 2417). *Organic structure, small rings, drugs.*

Marians, Ms Carol. MIT, 13-4053, Cambridge, MA 02139, USA.

Marion, Mr Martin P. 6020 Butler Pike, Blue Bell, PA 19422, USA.

Markgraf, Steven A. Materials Res. Lab., Pennsylvania State U., University Park, PA 16802, USA.

Marsh, Mr Philip. AT&T Bell Labs., 600 Mountain Ave., Murray Hill, NJ 07974, USA.

Marsh, Dr Richard Edward. Chemistry Dept., California Inst. of Techn., 1201 East California Blvd., Pasadena, CA 91125, USA. (1922) PhD, chemistry (UCLA, 1950). Res. assoc. (tel. 213 + 795-6811, ext. 2526). *Biological and inorganic structures.*

Martin, Mr Bruce Alan. Megadata, 35 Orville Dr., Bohemia, NY 11716, USA. (1944) BS, applied mathematics (PIB, 1964). Computer sci. analyst. (tel. 516 + 345-4155 or 4132). *Crystallographic computing, real-time on-line scattering experiments, computer control, mathematical crystallography, group theory (movements).*

Martin, Mr George Wm. Box 3706, Stanford, CA 94305, USA. (1932) MS, material science (Stanford U., 1968). Res. scient. (tel. 415 + 967-7775). *Electron microprobe, X-ray diffraction, crystallography, spectroscopy, SEM, TEM, Auger.*

Mason, Mr John T. Ames Lab., Iowa State U., 104 Metallurgy Bldg., Ames, IA 50010, USA. (1933) MS, physical chemistry (Tufts U., 1956). Assoc. metallurgist. (tel. 515 + 294-6529). *Alloy structure, surface phenomena.*

Mason, Prof. Paul Robert. Div. of Physics, Rose-Hulman Inst. of Techn., 5500 Wabash Ave., Terre Haute, IN 47803, USA. (1934) MS physics (Indiana U.) Assoc. prof. (tel. 812 + 877-1511, ext. 305). *Solid state physics, X-ray diffraction, polycrystalline materials.*

Massa, Mr Louis. City U. of New York, 695 Park Ave., New York, NY 10021, USA.

Massalski, Prof. Thaddeus B. Dept. of Metallurgical Eng. and Material Sci., Carnegie-Mellon U., 5000 Forbes Ave., Pittsburgh, PA 15213, USA. (1926) PhD, physical metallurgy (U. Birmingham, UK, 1954). Prof. (tel. 412 + 578-2708). *Alloy phases, phase diagrams, metallic glasses, solid state transformations.*

Mastropaolo, Mr Donald. Div. of Neurology, RG-20, U. of Washington, Seattle, WA 98195, USA.

Mathew, Mr M. Ada Health Foundation, Natl. Bureau of Standards, Washington, DC 20234, USA.

Mathews, Prof. F. Scott. Dept. Cell Biology and Physiology, , Washington U. Medical S., 4566 Scott Ave., St Louis, MO 63110, USA. (1934) PhD, chemistry

(U. Minnesota, 1969). Prof. (tel. 314 + 362-1080). *Protein crystallography, molecular graphics, biologically interesting structures.*

Matthews, Prof. Brian W. Inst. of Molecular Biology, U. of Oregon, Eugene, OR 94703, USA. (1938) PhD, physics (U. Adelaide, Australia, 1964). Prof. of physics, res. assoc. (tel. 503 + 686-5151). *Macromolecular crystallography.*

Matthews, Dr David Allan. Dept. of Chemistry, B-017, U. of California-San Diego, La Jolla, CA 92093, USA. (1943) PhD, physical chemistry (U. Illinois, Urbana, 1971). Assoc. res. chemist. (tel. 619 + 452-2153). *Structure and function, biological macromolecules.*

Matyi, Mr Richard J. Texas Instruments Inc., P. O. Box 225936, MS 147, Dallas, TX 75265, USA.

Mauer, Mr Floyd Andrew. Room A221, Bldg. 223, Nat. Bureau of Standards, Washington, DC 20234, USA. (1924) MS, physics (Carnegie-Mellon U., 1949). Supervisory physicist. (tel. 301 + 921-2910). *High temperature, high pressure, low temperature diffraction; precision lattice parameters, automation.*

Maverick, Prof. Emily. Chemistry Dept., U. of California, Los Angeles, CA 90024, USA. (1929) PhD, analytical chemistry (UCLA, 1972). Assoc. prof. (tel. 213 + 825-4219). *Strained molecules, conformational energy.*

Mayerle, Mr James J. Dept. 2H9/107-1, IBM, Rochester, MN 55901, USA. *Magnetic recording materials.*

Mayo, Mr William Edward. Dept. Mechanical & Materials Sci., Rutgers U., Piscataway, NJ 08854, USA.

Mazany, Dr Anthony Michael. Corporate Res., The B.F.Goodrich Co., 9921 Brecksville Rd., Brecksville, OH 44141-3289, USA. (1954) PhD, inorganic chemistry (Case Western Reserve U., 1984). Advanced R&D. (tel. 216 + 447-5559). *Organometallic, gold, iron, molybdenum, inorganic, cluster, dimer, sulfur, thio.*

McAlea, Mr Kevin P. Dept. of Chemical Eng., U. of Delaware, Newark, DE 19716, USA.

McAlister, Dr John Paul. Tripes Associates Inc., 6548 Clayton Rd., St. Louis, MO 63117, USA. (1948) PhD, biochemistry (U. Wisconsin, Madison, 1978). Dir., software res. and dev. (tel. 314 + 647-1099). *Software design, molecular modelling and graphics, crystallographic computing.*

McAtee, Prof. James L., Jr. Chemistry Dept., Baylor U., Waco, TX 76703, USA. (1924) PhD, physical chemistry (Rice U., 1951). Prof. (tel. 817 + 755-3311, ext. 41). *Clay mineral organic and inorganic complexes.*

McCallum, Prof. Malcolm Ernest. Dept. of Earth Resources (Geology), Colorado State U., Fort Collins, CO 80523, USA. (1934) PhD, geology (U. Wyoming, 1964). Prof., res. geologist, U.S. Geol. Survey. (tel. 303 + 491-6250). *Mineralogy; geochemistry; petrology, structural and mineral exploration, Precambrian crystalline rocks in the Rocky Mountain region; kimberlite and included upper mantle-lower crustal nodules, diamonds from diatremes.*

McCarthy, Prof. Gregory Joseph. Dept. of Chemistry and Geology, North Dakota State U., Fargo, ND 58105, USA. (1943) PhD, solid state science (PSU, 1969). Prof. (tel. 701 + 237-7193). *Solid state chemistry, geochemistry, X-ray powder diffraction.*

McCauley, Dr James Weymann. Ceramics Res. Div., Army Materials & Mechanics Res. Cntr. Arsenal St., Watertown, MA 02172, USA. (1940) PhD, solid state science (PSU, 1968). Supervisory materials res. eng. (tel. 617 + 623-3463). *Materials science, ceramics, crystallography, mineralogy.*

McClure, Dr Richard James. Dept. of Crystallography, U. of Pittsburgh, Pittsburgh, PA 15260, USA. (1942) PhD, chemistry (Georgia Inst. of Techn., 1969). Res. asst. prof. (tel. 412 + 624-4367). *Crystal structure, biological molecules, proteins - small molecule binding.*

McCollor, Mr Donald P. Chemistry Dept., U. of North Dakota, Grand Forks, ND 58202, USA.

McConnell, Prof. Duncan. Dept. of Geology and Mineralogy, Ohio State U., 140 W. 18th St., Columbus, OH 43210, USA. (1909) PhD, mineralogy (U. Minnesota, 1937). Prof. em. (tel. 614 + 422-6629). *Bone phosphate minerals, biominerals, calcification, crystal chemistry, apatite, francolite, dahllite, phosphorite, clay minerals.*

McCree, Duncan E. Biochemistry Dept., Duke U., Durham, NC 27710, USA.

McCrone, Mrs Lucy B. McCrone Res. Inst., 2820 S. Michigan Ave., Chicago, IL 60616, USA. (1923) BA, chemistry (Wellesley C., 1945). Sr. res. scient. (tel. 312 + 842-7100). *Optical crystallography.*

McCrone, Dr Walter C. McCrone Res. Inst. Inc., 2820 S. Michigan Ave., Chicago, IL 60616, USA. (1916) PhD, chemistry (Cornell U., 1942). Pres. (tel. 312 + 842-7100). *Optical crystallography, polymorphism.*

McCusker, Dr Lynne Bridget. Dept. of Chemistry, Texas A & M U., College Station, TX 77843, USA. (1951) PhD, chemistry (U. Hawaii, 1980). Lect. (tel. 409 + 845-1979). *Zeolite crystallography, powder diffraction methods.*

McFarlane, Dr Samuel H., III. Materials Characterization Group, RCA Labs., Princeton, NJ 08540, USA. (1937) PhD, physics (Brown U., 1967). Techn. staff member. (tel. 609 + 734-2206). *X-ray topography, crystal imperfections.*

McGuire, Ms Nancy K. Chemistry Dept., Arizona State U., Tempe, AZ 85287, USA.

McKay, Dr David Bruce. Chemistry Dept., U. of Colorado, Campus Box 215, Boulder, CO 80309, USA. (1946) PhD, biophysics (U. Chicago, 1976). (tel. 303 + 492-6641). *Protein crystallography.*

McKee, Dr Rodney Allen. Div. of Metals and Ceramics, Oak Ridge Natl. Lab., PO Box X, Oak Ridge, TN 37830, USA. (1947) PhD, materials science (U. Texas, 1975).

McKenzie, Dr Thomas Charles. Chemistry Dept., U. of Alabama, Tuscaloosa, AL 35486, USA. (1945) PhD, organic chemistry (Columbia U., 1971). Assoc. prof. (tel. 205 + 348-7134). *Organic chemistry, organic synthesis, direct methods.*

McKeown, Dr David Alexander. Dept. of Geology, Stanford U., Stanford, CA 94305, USA. (1957) PhD, geology (Stanford U., 1985). Res. Asst. (tel. 415 + 941-0427). *Amorphous ceramic materials; Raman, EXAFS spectroscopies.*

McKinstry, Prof. Herbert Alden. Material Sci. Dept., PSU, University Park, PA 16802, USA. (1925) PhD, physics (PSU, 1960). Assoc. prof. (tel. 814 + 865-1614). *Ceramic powder characterization, mechanical and thermal properties, computer graphics.*

McLachlan, Prof. Dan, Jr. Dept. of Geology and Mineralogy, Ohio State U., Mailing address: 1934 Langham Rd., Columbus, OH 43221, USA. (1905) PhD, chemical physics (PSU, 1936). Prof. em. *X-ray diffraction, crystal growth, phase equilibria.*

McLaren, Prof. Eugene Herbert. Dept. of Chemistry, SUNY at Albany, 1400 Washington Ave., Albany, NY 12222, USA. (1924) PhD, chemistry (Washington U., 1955). Prof. (tel. 518 + 457-8399). *Inorganic molecular structures, atmospheric chemistry, geochemistry.*

McLean, Dr W. John. 2519 N. Walnut, Tucson, AZ 85712, USA. (1937) PhD, earth and planetary sciences (U. Pittsburgh, 1968). *Applied mineralogy, crystal structures, minerals.*

McMillan, Dr Joyce A. Materials Res. Lab., U. of Illinois, Urbana, IL 61801, USA. (1938) PhD, physical chemistry (U. Illinois, 1964). Dept. crystallographer. (tel. 217 + 333-1612). *Crystallography.*

McMullan, Dr Richard K. Chemistry Dept., Brookhaven Nat. Lab., Upton, NY 11973, USA. (1929) PhD, chemistry (Iowa State U., 1956). Assoc. chemist. (tel. 516 + 345-4380). *Structure determination, neutron diffraction, X-ray diffraction, hydrogen bonded systems, clathrates, inclusion compounds.*

McMurdie, Mr Howard Francis. JCPDS - International Center for Diffraction Data, Nat. Bureau of Standards, Gaithersburg, MD 20899, USA. (1905) BS, chemistry (Northwestern U., 1928). Consultant. (tel. 301 + 921-2921). *Powder patterns, phase equilibria.*

McPhail, Prof. Andrew Tennent. Dept. of Chemistry, Duke U., Durham, NC 27706, USA. (1937) PhD, chemistry (U. Glasgow, UK, 1963). Prof. (tel. 919 + 684-2414). *Crystal structures, organic and biologically important molecules, transition metal complexes.*

McPherson, Prof. Alexander. Biochemistry Dept., U. of California, Riverside, CA 92521, USA. (1944) PhD, biological science (Purdue U., 1970). Prof. (tel. 714 + 787-5391). *Protein structure, nucleic acid structure, protein-nucleic acid interactions, macromolecular structure, crystallization.*

McPherson, Mr William G. 2201 Cedar, Duncan, OK 73533, USA.

Mechlinski, Mr Witold. 7 Woodlawn Rd., Somerset, NJ 08873, USA.

Medrud, Dr Ronald Curtis. Analytical Res. and Services Div., Chevron Res. Co., 576 Standard Ave., Richmond, CA 94802, USA. (1934) PhD, physical chemistry (State U. of Iowa, 1963). Sr. res. chemist. (tel. 415 + 620-4090). *Powder diffraction, materials characterization, crystal chemistry, zeolites, computer graphics.*

Meehan, Prof. Edward Joseph, Jr. Chemistry Dept., U. of Alabama, Huntsville AL 35899, USA. (1950) PhD, biochemistry (U. Alabama, Birmingham, 1978). Asst. prof. (tel. 205 + 895-6188). *Protein crystallography, protein crystal growth.*

Meibohm, Dr Edgar Paul Hubert. 521 Shadeland Ave., Drexel Hill, PA 19026, USA. (1915) PhD, physical chemistry (The Ohio State U., 1947). *X-ray crystallography, chromatography, polymer physics and physical chemistry.*

Mellor, Mr John. 1095 Chapel St., Stratford, CT 06497, USA.

Mencik, Dr Zdenek. 18656 Gill Rd., Livonia, MI 48152, USA. (1927) PhD, physical chemistry (Charles U., Czechoslovakia, 1961). Sr. res. scient. *High polymers, X-ray diffraction, texture studies, crystallization phenomena in polymers, polymer crystal structure, small angle X-ray scattering.*

Merritt, Dr Ethan Allen. Stanford Synchrotron Radiation Lab., Stanford U., P. O. Box 4349, Bin 69, Stanford, CA 94305, USA. (1952) PhD, molecular biology (U. Wisconsin, 1980). Res. assoc. (tel. 415 + 854-3300, ext. 3486). *Anomalous scattering, artificial intelligence, macromolecular structure determination, synchrotron radiation, computer applications.*

Merritt, Prof. Lynne Lionel, Jr. Chemistry Dept., Indiana U., Molecular Structure Center, Chemistry Bldg., Room 104, Bloomington, IN 47401, USA. (1915) PhD, analytical chemistry (U. Michigan, 1940). Prof. and special asst. to the pres. (tel. 812 + 337-5073). *Instrumental analysis, structure, chelate compounds, X-ray crystal structure determinations.*

Mertes, Prof. Kristin Susan Bowman. Chemistry Dept., U. of Kansas, Malott Hall, Lawrence, KS 66045, USA. (1946) PhD, inorganic chemistry (Temple U., 1947). Asst. prof. (tel. 913 + 864-3669). *Transition metal complexes, synthetic macrocyclic ligands, intermolecular interactions.*

Messick, Mr Julian. JCPDS, International Centre for Diffraction Data, 1601 Park Lane, Swarthmore, PA 19081, USA. (1933) BS, chemistry (Widener C., 1955). General manager. (tel. 215 + 328-9404). *Diffraction analysis.*

Metzger, Prof. Robert Melville. Dept. of Chemistry, U. of Mississippi, University, MS 38677, USA. (1940) PhD, chemistry (CIT, 1969). Margaret McLean Coulter Prof. of Chemistry. (tel. 601 + 232-5336 or 232-7301). *Madelung en-*

ergies, lattice energies, organic crystals, crystallography, TCNQ salts, molecular orbital calculations (semi-empirical), organic unimolecular rectifiers, EPR, triplet spin excitons.

Meyer, Prof. Edgar Frederich. Dept. of Biochemistry & Biophysics, Texas A & M U., College Station, TX 77843, USA. (1935) PhD, chemistry (U. Texas, Austin, 1963). Assoc. prof. (tel. 713 + 845-1744). *Molecular structures, pyrrolic compounds, computer graphics, information retrieval, interactive computing, non-bonded interactions, grantsmanship and the arts of survival, three-dimensional graphic art.*

Meyer, Prof. Frank Henry. 1103 15th Ave., S.E., Minneapolis, MN 55414, USA. (1915) MS, MA, physics and philosophy (PIB, 1951; U. of Minnesota, 1968). Editor, reciprocity (tel. 715 + 392-8101, ext. 271). *Solid cohesion, forces in solids, interatomic closest approach theory. cohesive energy theoretical derivation, chemical element crystals, binary compounds.*

Meyers, Dr Bernard Lee. Analytical Div., Amoco Res. Center, P. O. Box 400, Naperville, IL 60566, USA. (1934) PhD, analytical chemistry (U. Illinois, 1960). Sr. res. assoc. (tel. 312 + 420-5226). *Catalysts, XRD, physical characterizations.*

Meyers, Prof. Edward Arthur. Chemistry Dept., Texas A & M U., College Station, TX 77840, USA. (1930) PhD, chemistry (U. Minnesota, 1955). Prof. (tel. 713 + 845-2544). *Organic and inorganic crystal structures.*

Michel, Dr David John. U.S. Naval Res. Lab., 4555 Overlook Ave., S.W., Code 6390, Washington, DC 20375, USA. (1942) PhD, metallurgy (PSU, 1968). Metallurgist. (tel. 202 + 767-2621). *Radiation effects, intermetallic compounds, high temperature properties.*

Mighell, Dr Alan D. Reactor Radiation Div., Nat. Bureau of Standards, Gaithersburg, MD 20899, USA. (1935) PhD, chemistry (Princeton U., 1963). Res. chemist. (tel. 301 + 921-2950). *Crystallographic data bases, crystal structure analysis.*

Mikkola, Prof. Donald Emil. Metallurgical Eng. Dept., Michigan Techn. U., Houghton, MI 49931, USA. (1938) PhD, materials science (Northwestern U., 1964). Prof. (tel. 906 + 487-2636). *Structure - properties relationship, X-ray diffraction.*

Milberg, Dr Morton Edwin. Eng. and Res. Staff, Ford Motor Co., Box 2053, Dearborn, MI 48121, USA. (1926) PhD, physical chemistry (Cornell U., 1949). Principal res. scient. (tel. 313 + 323-1724). *Noncrystalline solids, high temperature ceramics.*

Mildner, Mr David F. R. Res. Reactor Facility, U. of Missouri, Columbia, MO 65211, USA.

Milillo, Prof. Frank. Dept. of Mechanical Eng., Union C., Schenectady, NY 12308, USA. (1943) PhD, physical metallurgy (PIB, 1974). Assoc. prof. (tel. 518 + 370-6264). *Phase transformations, interstitial atom ordering.*

Millane, Mr R. P. Dept. of Biological Sci., Purdue U., West Lafayette, IN 47907, USA.

Miller, Prof. Donald P. Dept. of Physics and Astronomy, Clemson U., Clemson, SC 29631, USA. (1927) PhD, physics (PIB, 1962). Prof. (tel. 803 + 656-3417). *Crystal and molecular structure, fibrous polymers.*

Miller, Dr Richard Wayne. Analytical Dev. Dept., Monsanto, P.O. Box 12830 Pensacola, FL 32575, USA. (1947) PhD, physical chemistry (X-ray diffraction) (Duke U., 1976). Res. specialist. (tel. 904 + 968-8358). *Fiber diffraction, fiber morphology.*

Miller, Prof. Robert Llewellyn. Michigan Molecular Inst., 1910 W. St. Andrews Drive, Midland, MI 48640, USA. (1929) PhD, chemical physics (Brown U., 1954). Sr. res. scient. (tel. 517 + 631-9450). *Macromolecules, structure and properties.*

Minkin, Mrs Jean Albert. Geologic Div., Branch of Coal Resources, U.S. Geological Survey, Mailstop 929, National Center, Reston, VA 22092, USA. (1925) BA, physics (Bryn Mawr C., 1947). Res. physicist. (tel. 703 + 860-6788). *Coal petrology, elemental abundances, mineral inclusions in coal.*

Mirsky, Dr Kira. Dept. of Chemistry, U. of California, 405 Hilgard Ave., Los Angeles, CA 90024, USA. Assoc. res. chemist. (tel. 213 + 825-1259). *Interatomic and intermolecular interactions, conformational analysis, organic chemical crystallography.*

Mitcham, Mr Donald. Spectroscopy Dept., Southern Regional Res. Center, 1100 Robert E. Lee Blvd., P O Box 19687, New Orleans, LA 70179, USA. (1921) BS, physics (Tulane U., 1948). Res. physicist. (tel. 504 + 589-7552). *Cellulose, fatty acids.*

Mitchell, Prof. Donald J. Dept. of Chemistry, Juniata C., 1900 Moore St., Huntingdon, PA 16652, USA. (1938) PhD, physical chemistry (Vanderbilt U., 1965). Assoc. prof. (tel. 814 + 643-5616). *Polymer crystallography, nucleation and crystal growth, clay mineralogy.*

Mitchell, Prof. Richard Scott. Dept. of Env. Sci., U. of Virginia, Clark Hall, Charlottesville, VA 22903, USA. (1929) PhD, mineralogy (U. Michigan, 1956). Prof. (tel. 804 + 924-7761, ext. 274). *Polytypism, layered inorganic compounds, systematic mineralogy, metamict state.*

Modrick, Ms Michelle A. Room LH170, Exxon Res. and Eng. Co., US Route 22, E. Clinton Township, Annandale, NJ 08801, USA.

Moffat, Prof. John Keith. Section of Biochemistry, Molecular and Cell Biology, Cornell U., Clark Hall, Ithaca, NY 14853, USA. (1943) PhD, protein crystallography (Cambridge U., UK, 1970). Prof. (tel. 607 + 256-4677, telex 937478). *Macromolecular crystallography, synchrotron radiation, biophysics, polypeptide hormones, calcium binding proteins, metalloproteins.*

Mohana-Rao, J. K. Dept. of Biological Sci., Purdue U., West Lafayette, IN 47907, USA.

Moini, Mr Ahmad. Dept. of Chemistry, Texas A & M U., College Station, TX 77843, USA.

Mokren, Dr James David. 6 Lincoln Place, Apt. L, North Brunswick, NJ 08902, USA. (1937) PhD, inorganic chemistry (Ohio State U., 1974). (tel. 513 + 644-5507). *Inorganic crystal structures.*

Molea, Mr Frank N. Div. of Eng. and Appl. Physics, Harvard U., 9 Oxford St., Cambridge, MA 02138, USA. (1934) BBA, electrical engineering, business management (Northeastern U., 1960). Techn. assoc. (tel. 617 + 495-4469). *Amorphous structure determination, small angle X-ray scattering, amorphous metals, glasses.*

Molinaro, Mr Frank. 102 N. Maple St., Mt. Prospect, IL 60056, USA. (tel. 312 + 391-3197).

Moncrief, Prof. J. William. Office of the Dean, Oxford C. of Emory U., Oxford, GA 30267, USA. (1941) PhD, chemistry (Harvard U., 1966). Dean. (tel. 404 + 786-7051). *Liquid crystals, crystal and molecular structure determination.*

Monroe, Prof. Eugene A. Ceramics Dept., SUNY at Alfred, Alfred, NY 14802, USA. (1934) PhD, mineralogy (1961), DDS, dentistry (1973). Assoc. prof. (tel. 607 + 871-2459). *Silicate mineralogy, phosphate mineralogy, prosthetic materials.*

Montfort, Mr William R. Dept. of Chemistry, U. of Texas, Austin, TX 78712, USA.

Moore, Mr Donald L. Instruments and Techn. Inc., 220 E. 14th St., Naperville, IL 60540, USA. (1928) Met. eng., (U. Cincinnati, 1952). Pres. (tel. 312 + 355-7748, telex 324597). *Scientific instrumentation.*

Moore, Mrs Elizabeth J. Weichel. GTD, IBM, River Rd., Essex Junction, VT 05452, USA. (1925) AB, geology (Radcliffe C., 1946). Staff eng./scient. (tel. 802 + 769-9039). *Optical crystallography, crystal physics, microscopy.*

Moore, Ms Janet Finer. 125 Baxter Dr., Apt. D-1, Athens, GA 30606, USA. (tel. 404 + 542-2626).

Moore, Prof. Paul Brian. Geophysical Sci. Dept., U. of Chicago, 5734 S. Ellis Ave., Chicago, IL 60637, USA. (1940) PhD, geophysical sciences (U. Chicago, 1965). Prof. (tel. 312 + 753-8111). *Systematology of atomic arrangements, minerals, inorganic crystals, new mineral species, mineral paragenesis, plane and space partitioning, convex polyhedra theory, crystal chemical homologies.*

Moreland, Dr James Andrew. Wacker Siltronic Corp., P. O. Box 03180, Portland, OR 97203, USA. (1946) PhD, inorganic crystallography (U. California at Irvine, 1974). Dir. (tel. 503 + 243-2020). *Crystal growth, crystal defects, electronic materials.*

Morgan, Dr Joseph. 10153 Oakton Terrace Rd., Oakton, VA 22124, USA. (1909) PhD, physics (MIT, 1937). Retired. (tel. 703 + 255-6136). *Structure, liquids.*

Morgan, Mr Richard S. Biophysics Dept., Pennsylvania State U., University Park, PA 16802, USA.

Moriarty, Dr John Lawrence, Jr. Materials Sci. Div./Eng. Directorate, Rock Island Arsenal, Rock Island, IL 61299, USA. (1932) PhD, physical chemistry (U. Iowa, 1960). Sr. metallurgist. (tel. 309 + 794-6198). *Applied crystallography, composites, intermetallic compounds.*

Morimoto, Dr Carl Noboru. 4003 Hamilton Park Dr., San Jose, CA 95130, USA. (1942) PhD, phys. chem-crystallography (U. Washington, 1970). *Crystallographic computing, computer graphics, biochemical crystal structures.*

Morosin, Dr Bruno. Shock Wave and Explosive Physics-1131, Sandia Labs., Albuquerque, NM 87185, USA. (1934) PhD, physical chemistry (U. Washington, 1959). Div. supervisor. (tel. 505 + 264-8169). *Crystal physics, structures.*

Morosoff, Dr Nicholas C. Polymer Res. Lab., Res. Triangle Inst., P. O. Box 12194, Research Triangle Park, NC 27709, USA. (1937) PhD, physical chemistry (PIB, 1965). Sr. res. physical chemist, (tel. 919 + 541-6866, telex 802509 RTI RTPK). *Polymer morphology, small angle X-ray scattering, plasma polymerization, solid state polymerization.*

Morris, Mrs Marlene Cook. Crystallography Section, Nat. Bureau of Standards, Materials Bldg., A221, Washington, DC 20234, USA. (1933) BS, chemistry (Howard U., 1955). Dir., JCPDS assoc. at NBS. (tel. 301 + 921-2921). *X-ray powder diffraction, inorganic crystal structures.*

Morrow, Prof. John Charles, III. Dept. of Chemistry, U. of North Carolina, 263 Venable Hall, Chapel Hill, NC 27514, USA. (1924) PhD, physical chemistry (MIT, 1949). Prof. (tel. 919 + 962-6095). *X-ray diffraction, molecular structure determination.*

Morrow, Dr Scott Imlay. Large Caliber Weapons Systems Lab., US Army Armament Munitions & Chemical Command, Picatinny Arsenal, Dover, NJ 07801, USA. (1920) PhD, inorganic chemistry (Case-Western Reserve U., 1951). Res. chemist/microscopist. (tel. 201 + 724-4703). *Microscopy, polarizing and scanning electron, phase rule studies, polymorphism and isomorphism, explosives, propellants, crystal studies with microscope.*

Moskowitz, Mr Ronald. Ferrofluidics Corp., 40 Simon St., Nashua, NH 03061, USA.

Moudy, Miss Lavada Ann. 1325 S. Organge, Apt. 2, Fullerton, CA 92633, USA. (1926) MS, ceramics (U. Washington, 1959). *Thin film structure and properties, crystal defects, crystal growth, X-ray diffraction and fluorescence.*

Mowbray, Ms Sherry Lynn. Chemistry Dept., MIT, 18 Vassar St., Room 18-045, Cambridge, MA 02139, USA. (1954) BS, biochemistry (Queen's U., Canada, 1976). Grad. student. (tel. 617 + 253-1814). *Protein crystallography.*

Mozzi, Dr Robert Lewis. Res. Div., Raytheon Co., 131 Spring St., Lexington, MA 02173, USA. (1931) PhD, physics (MIT, 1967). Principal scient. (tel. 617 + 860-3095). *Crystal defects, semiconductors, integrated circuit technology.*

Mrose, Miss Mary E. 114 N. Wayne St., Apt. 2, Arlington, VA 22201, USA. (1910) MA, geography (Boston U., 1944). Geologist-mineralogist. (tel. 703 + 860-6670). *Mineralogical investigations.*

Mucker, Prof. Kenneth. Dept. of Physics and Astronomy, Bowling Green State U., Bowling Green, OH 43403 USA. (1939) PhD, physical chemistry (Ohio State U., 1966). Assoc. prof. (tel. 419 + 372-0108). *Applied crystallography.*

Mueller, Dr Melvin H. IPNS, Bldg. 360, , Argonne Nat. Labs., 9700 S. Cass Ave., Argonne, IL 60439, USA. (1918) PhD, chemistry (U. Illinois, 1949). Sr. scient. (tel. 312 + 972-3554, ext. 6485). *Neutron diffraction pulsed sources, metal hydrides, actinide compounds, instrumentation.*

Muir, Prof. James Alexander. Physics Dept., U. of Puerto Rico, Rio Piedras, PR 00931, USA. (1938) PhD, physics (Northwestern U., 1966). Prof. (tel. 809 + 764-0000, ext. 3458). *Crystal structures, transition metal complexes, semiconductors.*

Muldawer, Prof. Leonard. Physics Dept., Temple U., Philadelphia, PA 19122, USA. (1920) PhD, physics (MIT, 1948). Prof. (tel. 215 + 787-7637). *Phase transformations, alloys, diffuse X-ray scattering, line profile analysis, small angle neutron scattering, Fourier methods, optical processing for radiological diagnosis.*

Muller, Mr John H. Box 590, North Hampton, NH 03862, USA.

Mullica, Dr Donald Foster. Chemistry Dept., Baylor U., Waco, TX 76703, USA. (1928) PhD, physical chemistry (Baylor U., 1977). Postdoctoral fellow. (tel. 817 + 755-3311, ext. 40). *Crystallography; lanthanide and actinide chemistry; physical chemistry.*

Murthy, Dr Krishna H. M. Dept. of Molecular Biophysics and Biochemistry, Yale U., Box 6666, 260 Whitney, New Haven, CT 06511, USA. (1952) PhD, biophysics (Indian Inst. Sci., Bangalore, 1981). Res. assoc. (tel. 203 + 436-4817 or 436-1101). *Protein crystallography, DNA binding, proteins, refinement methods, model building, protein-nucleic acid interaction.*

Murthy, Dr N. Sanjeeva. Corporate Res., Allied Corp., Columbia Rd., P. O. Box 1021R, Morristown, NJ 07960, USA. (1949) PhD, materials science (U. Connecticut, 1976). Sr. res. physicist. (tel. 201 + 455-3764). *Structure, disordered materials, polymers, structure - properties relationship, small-angle X-ray scattering.*

Myer, Prof. George Henry. Geology Dept., Temple U., 13th and Norris Sts., Philadelphia, PA 19122, USA. (1937) PhD, geology (Yale U., 1965). Asst. prof. (tel. 215 + 787-7173). *Crystallography, mineralogy, petrology, environmental geology.*

Namba, Dr Keiichi. Dept. of Molecular Biology, Vanderbilt U., Box 1820, Station B, Nashville, TN 37235, USA. (1952) PhD, biophysics (Osaka U., Japan, 1980). Res. assoc. (tel. 615 + 322-2012). *X-ray diffraction, biological structure.*

Nanni, Mr Raymond. Dept. of Crystallography, U. of Pittsburgh, Pittsburgh, PA 15260, USA.

Narasimhachari, Dr V. N. Biology Dept., Brookhaven Nat. Lab., Upton, NY 11973, USA. (1948) PhD, physical chemistry (U. Texas, Austin, 1975). Assoc. scient. *Biological macromolecular structure, computer graphics, interaction of small molecules with proteins.*

Narayanan, Dr Poojappan. Biophysics Dept., Johns Hopkins U., Medical S., 725 N. Wolfe St., Baltimore, MD 21205, USA. (1940) PhD, physics (U. Mysore, India, 1974). Res. assoc. (tel. 301 + 955-8715). *Diffraction studies, enzymes, biologically interesting structures.*

Nassau, Dr Kurt. Materials Res. Lab., AT&T Bell Labs., Murray Hill, NJ 07974, USA. (1927) PhD, physical chemistry (U. Pittsburgh, 1959). Techn. staff member. (tel. 201 + 582-2589). *Crystal chemistry, crystal growth.*

Nathan, Mr Robert. 1125 Rexford Ave., Pasadena, CA 91107, USA.

Navia, Dr Manuel Alberto. Merck Inst. for Therapeutic Res., Merck and Co., P.O. Box 2000, (R80M203), Rahway, NJ 07065, USA. (1946) PhD, biophysics (U. Chicago, 1974). Sr. res. biophysicist. (tel. 201 + 574-7256). *Macromolecular X-ray crystallography, electron microscopy, structure, enzyme-inhibitor complexes.*

Neilson, Dr George Francis. Jet Propulsion Lab., M.S. 157-102, CIT, 4800 Oak Grove Dr., Pasadena, CA 91109, USA. (1930) PhD, physical chemistry (Ohio State U., 1962). Member tech. staff. (tel. 818 + 354-6365). *Space processing, glass, small angle X-ray scattering, phase separation, microstructure characterization, nucleation and crystallization.*

Nelson, Mr A. Dwayne. 822 S. 2nd St., Stillwater, MN 55082, USA.

Neuman, Mr Melvin A. 6002 Camp Phillips Rd., Schofield, WI 54476, USA.

Newkirk, Prof. John Burt. Dept. of Chemistry, U. of Denver, 2199 S. University Blvd., Denver, CO 80171, USA. (1920) DSc, physical metallurgy (Carnegie Inst. of Techn., 1950). Prof. (tel. 303 + 753-2141). *Materials characterization, heat resisting alloys, biomedical materials and implant devices.*

Newman, Mr Robert Alan. Dow Chemical Co., Bldg. 1602, Midland, MI 48640, USA. Sr. res. chemist.

Newnham, Prof. Robert Everest. Materials Res. Lab., Pennsylvania State U., University Park, PA 16802, USA. (1929) PhD, physics (Cambridge U., UK, 1960). Prof., solid state sci. (tel. 814 + 865-1612). *Crystal physics, crystal chemistry, electroceramics, composite materials.*

Newsam, Dr John Michael. Exxon Res. and Eng. Co., Route 22 East, Annandale, NJ 08801, USA. (1954) PhD, chemistry (Oxford U., UK, 1980). Res. chemist.

(tel. 201 + 730-2901). *Catalysis, neutron scattering, synchrotron X-ray diffraction, superionic conductors, zeolites.*

Nicholds, Ms Brenda G. 609 Anderson Rd. #254, Davis, CA 95616, USA.

Nichols, Mr Monte Carl. Sandia Nat. Lab., Livermore, CA 94550, USA. (1938) MS, physical chemistry (U. Arizona, 1962). Member tech. staff. (tel. 415 + 422-2906). *Mineralogy, chemistry.*

Nichols, Dr Thomas Duncan. 1709 Dryden, Suite 815, Houston, TX 77030, USA. (1944) PhD, MD, physical chemistry, medicine (Rice U., 1971, U. Texas Medical S., San Antonio, 1977). Executive dir., James Bond Res. Foundation. (tel. 713 + 797-6591). *Amazona.*

Nicklow, Dr Robert Merle. Solid State Div., Oak Ridge Nat. Lab., P. O. Box X, Oak Ridge, TN 37830, USA. (1936) PhD, physics (Georgia Inst. of Techn., 1964). Staff scient. (tel. 615 + 483-8611, ext. 3-6788). *Lattice dynamics, magnetic excitations in solids, neutron spectrometry.*

Nicolosi, Dr Joseph Anthony. Eng. Dept., Philips Labs., 85 McKee Dr., Mahwah, NJ 07430, USA. (1950) PhD, physics, crystallography (PINY, 1982). Principal scient. (tel. 201 + 529-3800, ext. 456). *X-ray diffraction, X-ray spectrometry, instrumentation, research and development, applied research.*

Noble, Mark C. 404 Dowman St., Oxford, GA 30267, USA.

Noether, Dr Herman Dietrich. Textile Res. Inst., P.O. Box 625, Princeton, NJ 08540, (20 Greenbriar Dr., Summit, NJ 07901), USA. (1912) PhD, physical chemistry (Harvard U., 1943). Res. assoc., consultant. (tel. 609 + 924-3150 or 201 + 522-1653). *Polymer chemical and physical structure, structure - properties - morphology relationship, polymers, fibers, plastics; X-ray diffraction, small angle X-ray scattering, polycrystalline and polymeric materials.*

Nordman, Prof. Christer Eric. Chemistry Dept., U. of Michigan, Ann Arbor, MI 48109, USA. (1925) PhD, physical chemistry (U. Minnesota, 1953). Prof. (tel. 313 + 764-7326). *Organic and biological crystal structures, computer programming.*

Norman, Prof. Joe G., Jr. Dept. of Chemistry, BG-10, U. of Washington, Seattle, WA 98195, USA. (1947) PhD, inorganic chemistry (MIT, 1972). Prof., assoc. dean, grad. sch. (tel. 206 + 543-4644). *Inorganic and theoretical chemistry.*

Nowotny, Prof. Dr Hans. Inst. of Material Sci., U. of Connecticut, Storrs, CT 06268, USA. (1911) Dr. techn., physics (Techn. U. Vienna, 1934). Prof. em. (tel. 203 + 486, ext. 4619). *Inorganic crystal structures, alloy chemistry.*

Noyan, Dr Ismail Cevdet. Res. Div., IBM T. J. Watson Res. Center, P.O. Box 218, Yorktown Hgts., NY 10598, USA. (1956) PhD, materials science & eng. (Northwestern U., 1984). Res. staff member. (tel. 914 + 945-3941, telex 137456). *Stress/strain determination, X-ray diffraction, synchrotron radiation.*

O'Keeffe, Prof. Michael. Dept. of Chemistry, Arizona State U., Tempe, AZ 85287, USA. (1934) PhD, chemistry (U. Bristol, UK, 1959). Prof. (tel. 602 + 965-3670). *Crystal chemistry, solid state chemistry.*

O'Leary, Dr Kevin Joseph. Techn. Div., Olin-Chemical Group, 120 Long Ridge Rd., Stamford, CT 06904, USA. (1932) PhD, engineering (Case Western Reserve U., 1967). V.P. for techn. (tel. 203 + 356-2054). *Electrochemistry, solid state crystals, polymers, material science.*

Ohashi, Prof. Yoshikazu. Exploration and Production Res. Center, ARCO Oil and Gas, 2300 W. Plano Parkway, Plano, TX 75075, USA. (1941) PhD, geology (Harvard U., 1973). Sr. res. geologist. (tel. 214 + 754-6510). *Silicates, minerals, crystal structures.*

Ohlendorf, Dr Douglas Henry. Protein Eng. Div., Genex Corp., 16020 Industrial Dr., Gaitherburg, MD 20877, USA. (1950) PhD, physics (Washington U., 1972). Sr. scient. (tel. 301 + 825-5113). *Structure, macromolecules, protein - DNA interaction, oxygenases.*

Ohrt, Miss Jean Marie. RPMI - Crystallography, NYS Dept. of Health, 33 Tamarack St., Buffalo, NY 14220, USA. (1923) BA, biology (U. Buffalo, 1949). Cancer res. sci., retired. (tel. 716 + 825-5113). *Steroidstructure - structural - functional relationships of biologically active compounds, carcinogens and carcinostats.*

Okaya, Prof. Yoshi Haru. Dept. of Chemistry, State U. of N.Y., Stonybrook, NY 11794, USA. (1927) PhD, chemistry (U. Osaka, Japan, 1956). Prof. (tel. 516 + 246-5053). *Computer controlled diffractometry, synchrotron radiation, solid state physics, absolute configuration determination.*

Oliver, Dr Joel Day. Miami Valley Res. Lab., Procter and Gamble Co., P.O. Box 39175, Cincinnati, OH 45247, USA. (1945) PhD, physical chemistry (U. Texas, Austin, 1971). X-ray group leader. (tel. 513 + 245-2437). *Inorganic chemistry, structure elucidation, phase identification.*

Ollis, Mr David L. Dept. of Biophysics and Biochemistry, Yale U., P. O. Box 6666, New Haven, CT 06511, USA.

Olmstead, Ms Marilyn Morgan. Dept. of Chemistry, U. of California-Davis, Davis, CA 95616, USA.

Olsen, Prof. Kenneth Wayne. Dept. of Chemistry, Loyola University of Chicago, 6525 N. Sheridan Rd., Chicago, IL 60626, USA. (1944) PhD, biochemistry (Duke U., 1972). Assoc. prof. (tel. 312 + 508-3121). *Protein crystallography, prediction of protein strucure, evolution, protein stability.*

Olson, Dr Arthur Jules. Dept. of Molecular Biology, Res. Inst. of Scripps Clinic, 10666 N. Torrey Pines, La Jolla, CA 92087, USA. (1946) PhD, physical chemistry (U. California, Berkeley, 1975). Sr. staff scient. (tel. 619 + 457-9702). *Computation, computer graphics, macromolecular modeling, macromolecular interactions, supramolecular assemblies.*

Olson, Dr David Harold. Central Res. Lab., Mobil Res. and Dev. Corp., P. O. Box 1025, Princeton, NJ 08540, USA. (1937) PhD, physical chemistry (Iowa State U., 1963). Res. scient. (tel. 609 + 734-4253). *Zeolite crystal chemistry, heterogeneous catalyst characterization.*

Onan, Prof. Kay Denise. Dept. of Chemistry, Northeastern U., 360 Huntington Ave., Boston, MA 02146, USA. (1949) PhD, physical chemistry (X-ray crystallography) (Duke U., 1975). Asst. prof. (tel 617 + 437-2847). *Structure determination, biologically significant organic molecules, transition metal complexes.*

Ondik, Dr Helen Margaret. Inorganic Materials Div., Center for Materials Sci., Nat. Bureau of Standards, Gaithersburg, MD 20899, USA. (1930) PhD, physical chemistry (Johns Hopkins U., 1957). Res. chemist. (tel. 301 + 921-2900). *Data compilation.*

Ordway, Dr Fred. Artech Corp., 2901 Telestar Ct., Falls Church, VA 22042, USA. (1922) PhD, physical chemistry (CIT, 1949). Executive vice-pres. (tel. 703 + 560-3292). *Amorphous structures, computer applications.*

Ortega, Dr Richard B. X-ray Instruments Group, Nicolet Instrument Corp., P. O. Box 4370, Madison, WI 53711, USA. (1953) PhD, chemistry (U. New Mexico, 1980). Sr. software dev. manager. (tel. 608 + 271-3333). *Diffraction software, accurate data collection, low temperature hardware.*

Osgood, Mr Brian Clair. Analytical Dev. Div., E. I. Du Pont De Nemours, Savannah River Lab., Aiken, SC 29808, USA. (1954) BS, ceramic eng. (Alfred U., 1976). Sr. chemist. (tel. 803 + 725-2173). *Powder diffraction, automation.*

Ostrofsky, Mr Bernard. Bernard Ostrofsky Assoc. Inc., P. O. Box C, Naperville, IL 60566, USA. (1922) BSc, physical chemistry (City C. of New York, 1945). Consultant. *Instrumentation, crystal physics, X-ray spectroscopy, liquid structures.*

Ou, Dr Chia-Chih. Construction Products Div., W. R. Grace and Co., 62 Whittemore Ave., Cambridge, MA 02140, USA. (1945) PhD, physical chemistry (Rutgers U., 1976). Sr. res. chemist. (tel. 617 + 876-1400). *Inorganic crystal structures, clay minerals.*

Overmyer, Ms Elizabeth M. 2217 E. Hampton St., Meas, AZ 85204, USA.

Pabo, Prof. Carl Ogren. Dept. of Biophysics, John Hopkins Medical S., 725 North Wolfe Street, Baltimore, MD 21205, USA. (1952) PhD, biochemistry and molecular biology (Harvard U., 1980). Asst. prof. (tel. 301 + 955-3933). *Structural molecular biology, protein design.*

Pabst, Prof. Adolf. Dept. of Geology and Geophysics, U. of California, Berkeley, CA 94720, USA. (1899) PhD, geology (mineralogy) (U. California, Berkeley, 1928). Prof. em., mineralogy. (tel. 415 + 642-1878). *Mineralogy, crystallography, crystal chemistry of minerals.*

Padlan, Dr Eduardo Agustin. Biophysics Dept., Johns Hopkins U., N. Charles and 34th Sts., Baltimore, MD 21218, USA. (1940) PhD, biophysics (Johns Hopkins U., 1968). Res. scient. (tel. 301 + 338-7250). *Biological macromolecules.*

Palenik, Prof. Gus Joseph. Chemistry Dept., U. of Florida, Gainesville, FL 32611, USA. (1933) PhD, chemistry (U. Southern California, 1960). Prof. (tel. 904 + 392-0546). *Synthesis, structure, coordination compounds with high cordination numbers; structure - activity relationship, antihistamines, antibiotics, bronchodilators.*

Palmer, Dr Kenneth James. 1134 Mesters Dr., Pebble Beach, CA 93953, USA. (1910) PhD, chemistry (CIT, 1938). Retired. (U.S.D.A.) (tel. 415 + 524-3390). *X-ray crystallography, organic crystals, X-ray diffraction, fibers, membranes.*

Pangborn, Prof. Robert Northrup. Dept. of Eng. Sci. and Mechanics, Pennsylvania State U., 227 Hammond Bldg., University Park, PA 16802, USA. (1951) PhD, mechanics, materials science (Rutgers U., 1979). Assoc. prof. (tel. 814 + 863-0721). *Mechanical behavior, stress analysis, defect characterization.*

Pantaleo, Dr Nantelle Smith. Amoco Production Co., P. O. Box 3385, Tulsa, OK 74102, USA. (1947) PhD, physical chemistry (Emory U., 1971). Res. scient. (S. G.). (tel. 918 + 660-3147). *Small molecules, clay chemistry, geochemistry, thermodynamics of electrolyte solutions.*

Paretzkin, Mr Boris. Nat. Bureau of Standards, Materials A221, Washington, DC 20234, USA. (1922) AB, chemistry (New York U., 1944). Res. assoc. (tel. 301 + 921-2921). *Powder data file, crystal imperfections.*

Park, Dr Chang Hoon, Chemistry Dept., Michigan State U., East Lansing, MI 48824-1322, USA. (1947) PhD, chemistry (U. Kansas, 1982). Res. assoc. (tel. 517 + 355-7832). *Biological macromolecules.*

Parker, Dr Robert Louis. Metallurgy Div., Nat. Bureau of Standards, Washinton, DC 20234, USA. (1929) PhD, solid state physics (U. Maryland, 1960). Physicist. (tel. 301 + 921-2961). *Crystal growth, solidification, morphological stability.*

Parker, Dr Sidney Glenn. Physical Sci. Res. Lab., Texas Instruments Inc., P. O. Box 5936, M.S. 145, Dallas, TX 75222, USA. (1925) PhD, physical-inorganic chemistry (U. Texas, 1951). Techn. staff member. (tel. 214 + 238-2319). *Electronic materials, crystal growth, magnetic bubble materials, infrared detector materials, materials in group II-VI.*

Parkes, Dr Alan Schofield. 138 Sherman St., Belmont, MA 02178, USA. Cambridge, MA 02139, USA. (1933) PhD, physical chemistry (U. Pennsylvania, 1963). *Instrumentation, inorganic crystallography.*

Parks, Ms Elizabeth H. Dept. of Chemistry, U. of Texas at Austin, Austin, TX 78712, USA.

Parrish, Dr William. Dept. of Crystallography and X-ray Analysis, IBM Res. Lab., K31, 5600 Cottle Rd., San Jose, CA 95193, USA. (1914) PhD, mineralogy and crystallography (MIT, 1940). Manager. (tel. 408 + 256-1006). *X-ray diffractometry and spectroscopy, materials characterization, computer instrumentation and analysis.*

Parsons, Dr Donald Frederick. Wadsworth Center for Labs. and Res., N. Y. State Dept. of Health, Empire State Plaza, Albany, NY 12201, USA. (1928) MD, DSc, biophysics (U. London, UK, 1956). Res. physician III. (tel. 518 + 474-7047). *Electron diffraction, high voltage electron microscopy, small angle X-ray diffraction, cancer research.*

Parthasarathy, Dr R. Center for Crystallographic Res., Roswell Park Memorial Inst., Buffalo, NY 14263, USA. (1936) PhD, physics (U. Madras, 1962), biophysics (SUNY, Buffalo, 1966). Assoc. cancer res. scient. (tel. 716 + 845-5819). *Biological molecules, structure and function, stereochemistry, X-ray diffraction physics, conformational analysis, nuclear magnetic resonance.*

Parvez, Dr Masood. Dept. of Chemistry, PSU, University Park, PA 16802, USA. (1947) PhD, organic chemistry (Queen's U., Belfast, 1977). Lect. (tel. 814 + 865-1554). *Crystal and molecular structures of drugs, natural products, synthetic intermediates, inorganic and organometallic complexes.*

Patel, Dr Jamshed R. AT&T Bell Labs., Murray Hill, NJ 07974, USA. (1925) ScD, physical metallurgy (MIT, 1954). Techn. staff member. (tel. 201 + 582-6698). *Defects in crystals, dynamical diffraction.*

Pattridge, Ms Katherine A. 3429 Norwood, Ann Arbor, MI 48104, USA.

Paul, Prof. Iain Campbell. Dept. of Chemistry, U. of Illinois, Urbana, IL 61801, USA. (1938) PhD, chemistry (U. Glasgow, UK, 1962). Prof. (tel. 217 + 333-3007). *Chemical crystallography, solid state organic chemistry, X-ray structure analysis, biologically interesting molecular structures, molecular geometry.*

Pauling, Prof. Linus Carl. Linus Pauling Inst. of Sci. and Medicine, 2700 Sand Hill Rd., Menlo Park, CA 94025, USA. (1901) PhD, chemistry, physics, mathematics (CIT, 1925). Fellow. (tel. 415 + 854-0843). *Physics, chemistry, crystallography, biology, medicine.*

Pavalow, Prof. Melvin. Physics Dept., Hofstra U., Hempstead, NY 11550, USA. (1923) PhD, physics (Adelphi U., 1972). Assoc. prof. (tel. 516 + 560-3256). *X-ray crystallography, solid state physics, molecular vibrational amplitudes.*

Pavkovic, Prof. Stephen Frank. Dept. of Chemistry, Loyola U. of Chicago, 6525 N. Sheridan Rd., Chicago, IL 60626, USA. (1932) PhD, inorganic chemistry (Ohio State U., 1964). Assoc. prof. (tel. 312 + 274-3000). *Structure determination, transition metal coordination complexes.*

Peacor, Prof. Donald R. Dept. of Geology-Mineralogy, U. of Michigan, Ann Arbor, MI 48109, USA. Prof. (tel. 313 + 764-1452).

Pearson, Mr Robert H. Roswell Park Memorial Inst., Electron Optical Lab., 666 Elm St., Buffalo, NY 14263, USA.

Peavler, Prof. Robert J. Sci. Div., Northeast Missouri State U., Kirksville, MO 63501, USA. (1924) PhD, inorganic chemistry (Purdue U., 1953). Prof. of physics. (tel. 816 + 665-5121, ext. 2791). *Intermetallic compounds.*

Peercy, Dr Paul S. Dept. 1110, Sandia Nat. Labs., Albuquerque, NM 87185, USA. (1940) PhD, physics (U. Wisconsin, Madison, 1966). Manager, ion implantation and radiation physics res. dept. (tel. 505 + 844-4309). *Ion implantation, phase transitions, structure laser annealing, ferroelectrics, ion beam analysis.*

Peiser, Mr Herbert Steffen. 638 Blossom Dr., Rockville, MD 20850, USA. (1917) MA, chemistry (Cambridge U., UK, 1943). Retired. (tel. 301 + 762-6860, telex 904059). *Symmetry, atomic weights, metrology, crystal growth.*

Penner-Hahn, Prof. James E. Dept. of Chemistry, U. of Michigan, Ann Arbor, MI 48109, USA.

Perkins, Mr Herbert O. Chemistry Dept., Baylor U., Waco, TX 76798, USA.

Perloff, Dr Alvin. Ceramics, Glass, and Solid State Sci. Div., Center for Materials Sci., Nat. Bureau of Standards, Washington, DC 20234, USA. (1930) PhD, chemistry (Georgetown U., 1966). Res. chemist. (tel. 301 + 921-2900). *Inorganic crystal structures.*

Pessen, Dr Helmut. Eastern Regional Res. Center, U.S. Dept. of Agriculture, 600 E. Mermaid Lane, Philadelphia, PA 19118, USA. (1921) PhD, chemistry (Temple U., 1961). Res. chemist. (tel. 215 + 233-6475). *Small angle X-ray scattering, NMR, physical chemistry, biophysics.*

Peters, Mr Charles Richard. Res. Lab., Ford Motor Co., P. O. Box 2053; Rm S-1029, Dearborn, MI 48121, USA. (1934) MS, physical chemistry (U. Michigan, 1958). Res. scient. (tel. 313 + 323-1533). *X-ray crystallography, single crystal studies, defect structures, temperature dependent phase transitions in solids.*

Petersen, Dr Donald Ralph. Central Res., The Dow Chemical Co., Bldg. 1776, Midland, MI 48674, USA. (1929) PhD, physical chemistry (CIT, 1955). Res. scient. (tel. 517 + 636-5443). *Catalysts, inorganic structures, minerals, X-ray instrumentation.*

Petersen, Prof. Jeffrey L. Chemistry Dept., West Virginia U., Morgantown, WV 26506, USA. (1947) PhD, physical chemistry (U. Wisconsin-Madison, 1974). Asst. prof. (tel. 304 + 293-2527). *Neutron diffraction, metal-hydrogen interactions, transition metal hydrides, bis(cyclopentadienyl) chemistry, transition metal complexes, metal clusters, homogeneous catalysis.*

Peterson, Dr Selmer W. Chemistry Div., Argonne Nat. Lab., Argonne, IL USA. (1917) PhD, physical chemistry (U. Maryland, 1942). Retired. (tel. 408 + 728-2819). *Neutron diffraction, inorganic complexes, one-dimensional conductors.*

Petrovich, Mr Anton I. 504 Winston Dr., Vestal, NY 13850, USA.

Petsko, Prof. Gregory Anthony. Chemistry Dept., MIT, Room 2-202, Cambridge, MA 02139, USA. (1948) DPhil., molecular biophysics (Oxford U., UK, 1973). Prof. (tel. 617 + 253-1837). *Protein crystallography, enzymology, protein dynamics, sensory transduction, low temperature crystallography, high pressure crystallography.*

Pett, Dr Virginia B. Chemistry Dept., C. of Wooster, Wooster, OH 44691, USA. (1941) PhD, inorganic chemistry (Wayne State U., 1979). Asst. prof. (tel. 216 + 263-2114). *Organocobalt complexes, Cr-nucleotide adducts.*

Petz, Mr John Ignatius. Alumina Dept., Reynolds Metals, 103 Hawthorne, Portland, TX 78374, USA. (1935) MS, physics (U. Arkansas, 1959). Sr. staff member. (tel. 512 + 643-6531, ext. 2394). *Liquid state.*

Pflaum, Mr Wolfgang Richard. Frequency Control Devices, Hughes Aircraft Co., 500 Superior Ave., Newport Beach, CA 92658-8903, USA. (1944) BS, physical chemistry (Harvey Mudd C., 1966). Member techn. staff, (tel. 714 + 759-2283). *Inorganic structures.*

Pfluger, Prof. Clarence Eugene. Dept. of Chemistry, Syracuse U., Bowne Hall 304C, Syracuse, NY 13210, USA. (1930) PhD, chemistry (U. Texas, Austin, 1958). Prof. (tel. 315 + 423-3920). *Coordination compound structures, low temperature structure determinations, solid state photochemistry.*

Phillips, Dr George Neal, Jr. Dept. of Physiology and Biophysics, U. of Illinois, 524 Burrill Hall, 407 S. Goodwin Ave., Urbana, IL 61801, USA. (1952) PhD, biochemistry (Rice U., 1977). Asst. prof. *Protein structure, macromolecular assemblies.*

Phillips, Prof. James Christopher. Chemistry Dept., Bldg. 555, Brookhaven Nat. Lab., Upton, NY 11973, USA. (1952) PhD, applied physics (Stanford U., 1979). Res. asst. prof. (tel. 516 + 282-5621). *Structural studies, synchrotron radiation.*

Phillips, Dr Theodore II. C. of Aviation Techn., Embry-Riddle Aeronautical U., Regional Airport, Daytona Beach, FL 32014, USA. (1938) PhD, inorganic-physical chemistry (U. Kentucky, 1968). Assoc. prof. (tel. 904 + 252-5561, ext. 1308). *Structures, biologically interesting compounds, crystallographic computing, small molecules.*

Phillips, Prof. Travis J. Chemistry Dept., Purdue Calumet, Hammond, IN 46323, USA. (1919) PhD, physical chemistry (Ohio State U., 1957). Asst. Prof. (tel. 219 + 844-0520, ext. 286). *Solid state physics, chemistry.*

Phizackerley, Dr Richard Paul. Stanford Synchrotron Radiation Lab., Stanford U., Stanford Linear Accelerator Center, P.O. Box 4349, Bin 69, Stanford, CA 94305, USA. (1945) PhD, physics (U. Cambridge, U.K., 1971). Sr. scient. staff. (tel. 415 + 854-3300, ext. 3431). *Structure and function, biological macromolecules, anomalous scattering, synchrotron radiation applications, protein crystallography, instrumentation.*

Pichert, Mr Jerome. 3 Idle Day Dr., Centerport, NY 11721, USA. BS, EE (Bridgeport Eng. Inst., 1948). Consultant. (tel. 516 + 261-5648).

Piermarini, Dr Gasper John. Inst. for Materials Sci., Nat. Bureau of Standards, Gaithersburg, MD 20899, USA. (1933) PhD, physical chemistry (American U., 1971). Res. scient. (tel. 301 + 921-2950). *High pressure X-ray crystallography, optical measurements at high pressure, diamond anvil cells, ruby fluorescent measurement technique.*

Pierpont, Prof. Cortlandt G. Dept. of Chemistry, U. of Colorado, Boulder, CO 80309, USA. (1942) PhD, chemistry (Brown U., 1971). Prof. (tel. 303 + 492-8420). *Inorganic chemistry.*

Pignataro, Mrs Edith H. 230 Jay St., Brooklyn, NY 11201, USA. (1925) MS, physics (PIB, 1954). (tel. 718 + 858-7561). *X-ray crystallography.*

Pihl, Mr Carl Frederick. Data Systems Div., IBM, Route 52, Hopewell Junction, NY 12533, USA. (1930) MSc, chemistry (U. New Hampshire, 1957). Advisory scient. (tel. 914 + 897-2121, ext. 5297). *X-ray powder diffraction, instrument automation, computer systems.*

Pinkerton, Prof. Andrew Alan. Dept. of Chemistry, U. of Toledo, 2801 W. Bancroft St., Toledo, OH 43606, USA. (1943) PhD, inorganic chemistry (U. Alberta, Canada, 1971). Asst. prof. (tel. 419 + 537-2761). *Lanthanide and actinide chemistry, NMR of paramagnetics, small molecule crystallography, electron density determination.*

Pish, Dr George. 311 Gettysburg Rd., San Antonio, TX 78228, USA. (1914) PhD, analytical and physical chemistry (U. Illinois, 1943). Consulting. (tel. 512 + 734-4695). *Materials science, instrumentation, X-ray diffraction and fluorescence, chemical physics.*

Pletcher, Dr James F. Biocrystallography Lab., VA Medical Center, PO Box 12055, Pittsburgh, PA 15240, USA. (1935) PhD, biochemistry (Columbia U., 1965). Res. chemist. (tel. 412 + 683-3000, ext. 517). *Biological systems - structure and function, enzymatic and non-enzymatic catalysis by thiamine.*

Pluth, Dr Joseph John. Dept. of the Geophysical Sci., U. of Chicago, 5734 S. Ellis Ave., Chicago, IL 60637, USA. (1943) PhD, chemistry (U. Washington, 1971). Sr. res. assoc. (tel. 312 + 962-8109). *Zeolite structures, cation positions, mineral and inorganic structures, heterogeneous catalysis.*

Poland, Mr Virgil Laverne. 315 Glenvale, Youngstown, NY 14174, USA. (1928) BS, chemistry (Illinois Wesleyan U., 1953). *X-ray diffraction, X-ray spectroscopy, crystallography, chemistry.*

Pollack, Dr Sidney Solomon. Pittsburgh Energy Techn. Center, U.S. Dept. of Energy, Box 10940, Pittsburgh, PA 15236, USA. (1929) PhD, soil mineralogy (U. Wisconsin, 1956). Res. chemist. (tel. 412 + 675-6143). *X-ray diffraction, X-ray fluorescence, surface area measurement, minerals, coal, chars, synthetic oil and gas products; small angle X-ray scattering, catalysts.*

Ponnuswamy, Dr M. N. Div. of Biochemistry, U. of Texas Medical Branch, Galveston, TX 77550, USA. (1948) PhD, X-ray crystallography (U. Madras, India, 1980). Postdoctoral fellow. (tel. 409 + 761-3287). *Biological molecules, proteins.*

Ponzi, Ms Dagmar Ringe. MIT, Room 4-451, Dept. of Chemistry, Cambridge, MA 02139, USA.

Porter, Mr Leigh C. Dept. of Chemistry, Texas A & M U., College Station, TX 77840, USA.

Posner, Prof. Aaron Sidney. Hospital for Special Surgery, Cornell U. Medical C., 535 East 70th st., New York, NY 10021, USA. (1920) PhD, physical chemistry (U. Liège, Belgium, 1954). Dir. of res. (tel. 212 + 606-1458). *Ultrastructure, bone and teeth; tissue mineralization mechanism; calcium phosphate structures.*

Post, Prof. Benjamin. Physics Dept., Polytechnic Inst. of New York, 333 Jay St., Brooklyn, NY 11201, USA. (1911) PhD, physical chemistry (PIB, 1949). Prof., physics and chemistry. (tel. 718 + 643-8804). *Dynamical diffraction theory, X-ray instrumentation, precise measurements, simultaneous diffraction effects, phase determination - experimental.*

Potenza, Prof. Joseph Anthony. S. of Chemistry, Rutgers U., New Brunswick, NJ 08904, USA. (1941) PhD, chemistry (Harvard U., 1967). Prof. (tel. 201 + 932-2115). *Physical and inorganic chemistry.*

Potter, Mr Stephen Anthony. Molecular Biophysics, Medical Foundation of Buffalo, 73 High St., Buffalo, NY 14203, USA. (1946) BA, physics (SUNY at Buffalo, 1969). Res. assoc. (tel. 716 + 856-9600). *Direct methods, computer programming.*

Poulos, Mr Thomas L. Sci. and Techn. Center, Genex Corp., 16020 Industrial Dr., Gaithersburg, MD 20877, USA.

Powell, Dr Douglas R. S. of Geology and Geophysics, U. of Oklahoma, 830 Van Vleet Oval, Norman, OK 73019, USA. (1953) PhD, Physical chemistry (Iowa State U., 1980). Res. staff. (tel. 405 + 325-2362). *Crystallographic computing, instrumentation.*

Presley, Dr C. Travis. Oil Recovery Dept., Marathon Oil Co., Box 269, Littleton, CO 80160, USA. (1941) PhD, physical chemistry (Rice U., 1968). Advanced res. chemist. (tel. 303 + 794-2601, ext. 318). *Surface chemistry.*

Prevey, Mr Paul S., III. Lambda Res. Inc., 1111 Harrison Ave., Cincinnati, OH 45214, USA.

Prewitt, Prof. Charles Thompson. Dept. of Earth and Space Sci., SUNY, Stony Brook, NY 11794-2100, USA. (1933) PhD, mineralogy and crystallography (MIT, 1962). Prof. (tel. 516 + 246-4046). *Crystallography, mineralogy, crystal chemistry.*

Price, Ms Rebecca Alexis c/o Dr. P. W. Jagodzinski, West Virginia U., P. O. Box 6045, Dept. of Chemistry, Morgantown, WV 26506, USA. (1951) MS, (U. Oregon, 1985). Res. asst. (tel. 304 + 296-5299). *Macromolecular structure, enzymes.*

Prince, Dr Edward. Reactor Radiation Div., Nat. Bureau of Standards, Gaithersburg, MD 20899, USA. (1928) PhD, physics (U. Cambridge, UK, 1952). Res. physicist. (tel. 301 + 921-3634, telex 197674 NBS). *Neutron diffraction, instrumentation, refinement techniques.*

Pulsinelli, Prof. Phillip Del. Medicinal Chemistry Dept., U. of Pittsburgh, 729 Salk Hall, Pittsburgh, PA 15261, USA. (1943) PhD, crystallography-biochemistry (U. Pittsburgh, 1971). Assoc. prof. (tel. 412 + 624-3266). *Protein structure/dynamics, IR spectra, molecular structure, receptor interactions.*

Purdy, Mr Samuel M. Nat. Steel Corp., 500 Three Springs Dr., Weirton, WV 26062, USA.

Pyrros, Dr Nikos P. Hercules Res. Center, Wilmington, DE 19894, USA. PhD, physics (McMaster U., Canada, 1972). Sr. res. scient. (tel. 302 + 995-3405). *X-ray diffraction, crystal structures, polymers.*

Quicksall, Dr Carl O. 113 Woodbine Dr., Terre Haute, IN 47803, USA. (1941) PhD, chemistry (Princeton U., 1971). *Inorganic chemistry, crystallography, X-ray diffraction.*

Quigley, Dr Gary Joseph. Dept. of Biology, 16-743, Massachusetts Inst. of Techn., 77 Massachusetts Avenue, Cambridge, MA 02139, USA. (1942) PhD, chemistry (SUNY, C. of Env. Sci. & Forestry, 1969). Sr. res. scient. (tel. 617 + 258-8700). *Nucleic acid structure and function, macromolecular crystallography.*

Quiocho, Prof. Florante A. Dept. of Biochemistry, Rice U., P.O. Box 1892, Houston, TX 77001, USA. (1937) PhD, biochemistry (Yale U., 1966) Assoc. prof. (tel. 713 + 527-4872). *Biochemistry, biophysics, proteins, small molecules, crystallography.*

Rabinowitz, Dr Israel Nathan. 2534 Foothill Rd., Santa Barbara, CA 93105, USA. (1935) PhD, biochemistry (Rutgers U., 1965). Duncan E. Williams Assoc. (tel. 805 + 963-8938). *Crystal growth, biological calcification, kidney and bladder stone disease, membrane biophysics.*

Radonovich, Prof. Lewis J. Dept. of Chemistry, U. of North Dakota, Grand Forks, ND 58202, USA. (1944) PhD, physical chemistry (Wayne State U., 1970). Asst. prof. (tel. 701 + 777-2541). *Structural inorganic chemistry, biologically important molecules (structural chemistry).*

Rajeswaran, Dr Manju. Dept of Biochemistry, SUNY at Stony Brook, Stony Brook, NY 11794, USA. (1956) PhD, biophysics (SUNY at Buffalo, 1983). Postdoctoral fellow. (tel. 516 + 246-5026). *Biological important molecules, crystallography, protein crystallography.*

Ramalingam, V. Dept. of Biochemistry, U. of Miami, S. of Medicine, Miami, FL 33101, USA.

Ramasubbu, Mr N. Bept. of Biophysics, Rosewell Park Memorial Inst., 666 Elm St., Buffalo, NY 14263, USA. (1955) MSc, chemistry (Madurai U., 1977). (tel. 716 + 845-2365). *Electron density distribution, chemical information.*

Rao, Dr Sambhorao Thyagaraja. Dept. of Biochemistry, U. of Wisconsin, Madison, WI 53706, USA. (1937) PhD, physics (U. Madras, India, 1966). Assoc. scient. (tel. 608 + 262-3019). *X-ray crystallography, biologically significant molecules.*

Rasmussen, Mr Bjarne. 150 Chandler St., Apt. 5, Boston, MA 02116, USA.

Rau, Mr Robert C. 2542 Fleetwood Ave., Cincinnati, OH 45211, USA. (1935) MSc, materials science (U. Cincinnati, 1965). Consultant - X-ray diffraction. (tel. 513 + 662-0589). *Inorganic crystal structures, powder diffraction analysis.*

Ravichandran, Mr K. G. Dept. of Chemistry, Michigan State U., East Lansing, MI 48824, USA.

Ray, Prof. Alden Earl. Metals and Ceramics Div., University of Dayton Res. Inst., 300 College Park Ave., Dayton, OH 45469, USA. (1931) PhD, metallurgy (Iowa State U., 1959). Supervisor. (tel. 513 + 229-3527). *Metallurgy, rare earth - transition metal alloys.*

Raykhtsaum, Mr Grigory. Techn. Dept., Leach and Garner Co., 49 Pearl St., Attleboro, MA 02703, USA.

Rayment, Dr Ivan. Dept. of Biochemistry, U. of Arizona, Biological Sciences West, Tucson, AZ 85721, USA. (1951) PhD, chemistry (Durham U., UK, 1975). Asst. prof. (tel. 602 + 621-1884). *Macromolecular crystallography, structure and function, muscle proteins.*

Raymond, Prof. Kenneth Norman. Dept. of Chemistry, U. of California, Berkeley, CA 94720, USA. (1942) PhD, inorganic chemistry (Northwestern U., 1968). Assoc. prof. (tel. 415 + 642-7219). *Coordination isomers, metal ion substitution, biological metal ion transport, structure and bonding, coordination complexes, transuranium sequestering agents, lanthanide and actinide organometallic compounds.*

Reddy, Mr B. Swaminatha. 14236 Castle Blvd., Silver Spring, MD 20904, USA.

Reeber, Dr Robert Richard. Materials Sci. Div., Army Res. Office, P. O. Box 12211, Research Triangle Park, NC 27709-1211, USA. (1937) PhD, industrial mineralogy (Ohio State U., 1968). Materials eng. (tel. 919 + 549-0641). *Phase transformations, thermal expansion, crystal chemistry, diffraction methods.*

Reed, Dr Alvin T. Analytical Services, Anchor Hocking, 1749 W. Fair Ave., Lancaster, OH 43130, USA. (1946) PhD, inorganic chemistry (Miami U., Oxford, OH, 1975). Chemist. (tel. 614 + 687-2053). *Inorganic chemistry, X-ray fluorescence spectrometry, X-ray powder diffraction.*

Reed, Prof. John W. Dept. of Chemistry, Kent State U., Kent, OH 44242, USA. (1926) PhD, physical chemistry (Ohio State U., 1956). Assoc. prof. (tel. 216 + 672-2793). *Crystal structure determination.*

Reed, Dr Larry L. Argonne Nat. Lab., 9700 S. Cass Ave., Argonne, IL 60439, USA. (1949) PhD, organic chemistry (U. Arizona, 1971). Asst. computer scient. (tel. 312 + 972-7585). *Chemistry, computer science.*

Reeke, Prof. George Norman, Jr. The Rockefeller U., 1230 York Ave., New York NY 10021, USA. (1943) PhD, chemistry (Harvard U., 1969). Assoc. prof. (tel. 212 + 570-8183). *Biological molecules, computer programming, image reconstruction.*

Rees, Prof. Douglas Charles. Dept. of Chemistry and Biochemistry, UCLA, 405 Hilgard Ave., Los Angeles, CA 90024, USA. (1952) PhD, biophysics (Harvard U., 1980). Asst. prof. (tel. 213 + 206-1166). *Macromolecular crystallography.*

Reid, Mr Austin H., Jr. Dept. of Chemistry, Texas A & M U., College Station, TX 77843, USA.

Reidinger, Mr Franz. CRL 206, P.O. Box 1021R, Allied Corp., Morristown, NJ 07960, USA.

Reis, Dr Arthur Henry, Jr. Chemistry Dept., Brandeis U., Waltham, MA 02254, USA. (1946) PhD, inorganic chemistry (Harvard U., 1972). Assoc. prof. *Low dimensional interactions; pulsed neutron diffraction.*

Renzema, Prof. Theodore Samuel. Physics Dept., SUNY, 1400 Washington Ave., Albany, NY 12222, USA. (1912) PhD, physics (Purdue U., 1948). Prof. (tel. 518 + 456-1922). *Thin film structures, semiconductor surfaces, electron microprobe analysis, small particles, atmospheric pollutants.*

Reppart, Mr William J. 5630 Roche Dr., Apt. C, Columbus, OH 43229, USA.

Reynolds, Dr Ross Anthony. Physics Dept., Colby C., Eustis Pkwy., Waterville, MA 04901, USA. (1951) PhD, physics (U. Oregon, 1983). Asst. prof. (tel. 207 + 872-3599). *Protein crystallography.*

Rheingold, Dr Arnold Lange. Dept. of Chemistry, U. of Delaware, Newark, DE 19716, USA. (1940) PhD, chemistry (U. Maryland, 1969). Prof. (tel. 302 + 451-8720). *Organometallic chemistry, structure, organometallic compounds.*

Rhyne, Ms Kay A. Nat. Bureau of Standards, Bldg. 223, Room A331, Washington, DC 20234, USA.

Ribbe, Prof. Paul H. Geological Sci. Dept., Virginia Polytechnic Inst. and State U., Blacksburg, VA 24061, USA. (1935) PhD, physics (crystallography) (U. Cambridge, UK, 1963). Prof., mineralogy. (tel. 703 + 961-6880). *Silicate crystal chemistry.*

Ricci, Prof. John S., Jr. Dept. of Chemistry, U. of Southern Maine, Portland, ME 04103, USA. (1940) PhD, chemistry (SUNY, Stony Brook, 1969). Prof. (tel. 207 + 780-4232). *Neutron diffraction, transition metal complexes, molecular structure.*

Rice, Dr Catherine Ellen. Solid State Chemistry Res., Bell Labs., Crawford Corners Rd., Holmdel, NJ 07733, USA. (1951) PhD, inorganic chemistry (Purdue U., 1976). Techn. staff member. (tel. 201 + 949-5299). *Solid state chemistry.*

Rich, Prof. Alexander. Biology Dept., Massachusetts Inst. of Techn., 77 Massachusetts Ave., Cambridge, MA 02139, USA. (1924) MD, medicine (Harvard Medical S., 1949). Sedgwick prof. (tel. 617 + 253-4715). *Molecular structure, proteins, nucleic acids, mechanism of protein synthesis, origin of life.*

Richards, Prof. Frederic Middlebrook. Molecular Biophysics & Biochem. Dept., Yale U., Box 1937 Yale Station, New Haven, CT 06520, USA. (1925) PhD, biophysical chemistry (Harvard U., 1952). Henry Ford II prof. (tel. 203 + 436-2032). *Proteins, enzymes, structure and function.*

Richards, Mr Gerald F. 850 Siemens St., Platteville, WI 53818, USA.

Richardson, Dr David Claude. Dept. of Biochemistry, Duke U., 210B Nanaline Duke Bldg., Durham, NC 27710, USA. (1940) PhD, inorganic chemistry (MIT, 1967). Assoc. prof. (tel. 919 + 684-6010). *Protein crystallography, metalloenzymes, molecular graphics, protein design.*

Richardson, Dr James Wyman, Jr. IPNS Div., Argonne Nat. Lab., 9700 S. Cass Ave., Argonne, IL 80439, USA. (1955) PhD, crystallography (Iowa State U., 1984). Postdoctoral fellow. (tel. 312 + 972-3554). *Single-crystal and powder diffraction, both X-ray and neutron.*

Richardson, Mrs Jane Shelby. Depts. of Biophysics and Anatomy, Duke U., 213 Nanaline Duke Bldg., Durham, NC 27710, USA. (1941) MA, philosophy (Harvard U., 1966). Assoc. medical res. prof. (tel. 919 + 684-6010). *Comparison and classification of protein structures, protein crystallography, protein folding, protein design.*

Richman, Prof. Marc Herbert. Eng.-Materials Sci. Dept., Brown U., Providence, RI 02912, USA. (1936) ScD, metallurgy (MIT, 1963). Prof. (tel. 401 + 863-2628). *Inorganic compounds, crystal structures, bulk materials, thin deposited coatings, X-ray and electron diffraction, field ion microscopy.*

Riegert, Mr Richard Paul. Quad Group, 2030 Alameda Padre Serra, Santa Barbara, CA 93103, USA. (1927) MS, ceramic engineering, crystal chemistry (Alfred U., 1957). Pres. and techn. dir. (tel. 805 + 965-1041). *Surface analysis, thin film morphology, crystal chemistry.*

Riess, Prof. John Karlem. Dept. of Physics, Tulane U., Mailing address: 17 Audubon Blvd., New Orleans, LA 70118, USA. New Orleans, LA 70118, USA. (1913) PhD, physics (Brown U., 1943). Assoc. prof., retired. (tel. 504 + 861-9872). *Biophysics.*

Robbins, Mr Carl Richard. Nat. Bureau of Standards, Div. 313.06, Gaithersburg, MD 20234, USA. MA, mineralogy-geochemistry (U. Missouri, 1952). Res. chemist. (tel. 301 + 921-2910). *Inorganic structural chemistry, oxides, silicates, germanates, aluminates, anhydrous and hydro-thermal environments, phase equilibria, X-ray powder structures, single crystal structure.*

Roberts, Dr Michael Mark. Biochemistry Dept., Columbia U., 630 W. 168th St., New York, NY 10032, USA. (1956) PhD, structural inorganic chemistry (U. Warwick, 1981). Postdoctoral res. scient. (tel. 212 + 694-3856). *Protein crystallography, nexon structure, adenovirus coat protein.*

Robertson, Dean B. Ken. Dean of Students, Rolla Bldg., U. of Missouri at Rolla, Rolla, MO 65401, USA. (1938) PhD, chemistry (Texas A & M U., 1965). Dean. (tel. 314 + 341-4292).

Robertson, Prof. James David. Dept. of Anatomy, Duke U., Box 3011, Duke Medical Center, Durham, NC USA. (1922) MD, (Harvard Medical S., 1945) PhD, biochemistry (MIT, 1952). Chairman, James B. Duke Prof. (tel. 919 + 684-5136). *Molecular structure, membranes, structure and function, nervous system.*

Robertus, Prof. Jon David. Chemistry Dept., U. of Texas, Austin, TX 78712, USA. (1945) PhD, cellular biology (U. California, San Diego, 1972). Assoc. prof. (tel. 512 + 471-3175). *Structure and action, proteins, enzymes (antitumor & antiviral).*

Robinson, Dr Ian Keith. AT&T Bell Labs., 1E445, Murray Hill, NJ 07974, USA. (1955) PhD, physics (Harvard U., 1981). Member tech. staff. (tel. 201 + 582-6056). *Surfaces, reconstruction, thin films, viruses, proteins, interfaces, phase transitions.*

Robinson, Mr John C. 21699 Terrace Dr., Cupertino, CA 95014, USA.

Robinson, Prof. William Robert. Dept. of Chemistry, Purdue U., West Lafayette, IN 47907, USA. (1939) PhD, inorganic chemistry (MIT, 1966). Prof. (tel. 317 + 494-5453). *Solid state chemistry, synthesis, structure, metal phosphates, oxides, silicates, sulfides, transition metal compounds.*

Rodesiler, Prof. Paul F. Chemistry Dept., Columbia C., Main St., Columbia, SC 29203, USA. (1941) PhD, chemistry (Queen Mary C., U. of London, UK, 1969). Asst. prof. (tel. 803 + 786-3730). *Organometallics, transition metal complexes.*

Roettgers, Mr Wolbert. Henry L. Mattin Labs., The Mearl Corp., 217 North Highland Ave., Ossining, NY 10562, USA. (1926) MSc, physics (Fairleigh Dickinson, U., 1966). Res. physicist. (tel. 914 + 941-7450, ext. 25). *Crystallography, clay minerals, crystal growth, epitaxy.*

Rogers, Prof. Robin Don. Dept. of Chemistry, Northern Illinois U., DeKalb, IL 60115, USA. (1957) PhD, chemistry (U. Alabama, 1982). Asst. prof. (tel. 815 + 753-1131). *Organometallic compounds, complexes (f-element - crown ether).*

Rognlie, Mr David G. Blake Industries Inc., 660 Jerusalem Rd., Scotch Plains, NJ 07076, USA. (1934) BSEE, electrical engineering (U. North Dakota, 1956). Pres. (tel. 201 + 233-7240). *Diffraction instrumentation.*

Rohrbaugh, Dr Wayne Joseph. Res. Dept., Mobil Res. and Dev. Corp., Billingsport Rd., Paulsboro, NJ 08066, USA. (1948) PhD, physical chemistry (Iowa State U., 1977). Project leader, materials structure res. (tel. 609 + 423-1040, ext. 2796). *Zeolite structures, catalysis; X-ray single crystal methods, powder methods, structure - properties relationship.*

Rohrer, Dr Douglas C. Molecular Biophysics Dept., Medical Foundation of Buffalo, 73 High Street, Buffalo, NY 14203, USA. (1942) PhD, chemistry (Case-Western Reserve U., 1970). Assoc. res. scient. (tel. 716 + 856-9600).

Structure - function relationship, biological molecules; cardio-active steroids; carbohydrates.

Roldan, Dr Luis-Gonzalez. Techn. Center, J. P. Stevens and Co. Inc., 400 East Stone Ave., P. O. Box 2850, Greenville, SC 29602-2850, USA. (1925) DSc, chemistry-crystallography (U. Sevilla, Spain, 1957). Manager, microscopy and physics dept. (tel. 803 + 239-4194). *Polymers, organic crystal structures, structure - properties - morphology relationship, crystal nucleation, polymorphism.*

Roof, Dr Raymond Bradley. Chemistry Div., Los Alamos Scientific Lab., P. O. Box 1663, MS-G740, Los Alamos, NM 87545, USA. (1929) PhD, mineralogy-crystallography (U. Michigan, 1955). Staff member. (tel. 505 + 667-4931). *Intermetallic compounds, line broadening, instrumentation, high pressure, computer programming, mineral synthesis.*

Rose, Jon Patrick. Bio Crystallography Lab., VA Medical Center, P. O. Box 12055, U. of Pittsburgh, Pittsburgh, PA 15240, USA.

Rosen, Dr Lawrence Stephen. Hematology Div., St. Luke's Hospital, 114th St. & Amsterdam Ave., New York, NY 10025, USA. (1943) PhD, biophysics (U. Maryland, 1970). Res. assoc. (tel. 212 + 870-6157). *Protein crystallography, computing.*

Rosenfield, Dr Richard E., Jr. Dept. of Biochemistry & Biophysics, Texas A & M U., Heep Bldg., College Station, TX 77843, USA. (1945) PhD, biophysics (SUNY at Buffalo, 1974). Res. assoc. (tel. 713 + 845-1744). *Intermolecular interactions, crystallography, organic and biological molecules, crystallographic data retrieval.*

Rosenstein, Prof. Robert Daniel. Lawrence Berkeley Lab., Bldg. 70A, Room 4405, Berkeley, CA 94702, USA. (1922) Fil. Lic., chemistry (U. Uppsala, Sweden, 1961). (tel. 415 + 486-7274). *Crystal structures.*

Ross, Ms Dawn L. Dept. of Mathematics, U. of Missouri, Columbia, MO 65211, USA. *Date of Birth Missing. (tel. 314 + 882-4549).

Ross, Dr Frederick Keith. Res. Reactor Facility, U. of Missouri, Res. Park, Columbia, MO 65211, USA. (1942) PhD, chemistry (U. Illinois, Urbana, 1969). Sr. res. scient. (tel. 314 + 882-3331). *Structure and bonding, valence electron densities, low temperature diffraction, neutron diffraction, phase transitions.*

Ross, Dr Malcolm. U.S. Geological Survey, National Center, Stop 959, Reston, VA 22092, USA. (1929) PhD, mineralogy (Harvard U., 1962). Res. chemist. (tel. 703 + 860-6667). *Silicate structures, asbestos minerals, minerals and health.*

Ross, Ms Nancy L. Dept. of Geology, Arizona State U., Tempe, AZ 85287, USA.

Rossi, Dr Miriam. Dept. of Chemistry, Vassar C., Box 484, Poughkeepsie, NY 12601, USA. (1952) PhD, chemistry (Johns Hopkins U., 1978). Asst. prof. (tel. 914 + 452-7000, ext. 2597). *Crystal structure, small molecules.*

Rossmann, Prof. Michael G. Biological Sci., Purdue U., Lilly Hall of Life Sci., West Lafayette, IN 47907, USA. (1930) PhD, chemistry (U. Glasgow, UK, 1956). Hanley prof. of biology. (tel. 317 + 494-8333). *Viruses, enzymes, protein folding, molecular evolution, crystallographic techniques, methods and theory, computing.*

Rotella, Dr Frank J. IPNS Program, Argonne Nat. Lab., 9700 South Cass Ave., Argonne, IL 60439, USA. (1949) PhD, chemistry (SUNY at Buffalo, 1979). Asst. chemist. (tel. 312 + 972-5785). *Single-crystal diffraction, neutron powder diffraction, rietveld analysis, structural studies, metal hydrides, organometallic compounds.*

Roth, Dr Robert Sidney. Solid State Chemistry Section, Nat. Bureau of Standards, Room B214, Materials Bldg., Washington, DC 20234, USA. (1926) PhD, geology (U. Illinois, 1951). Chief. (tel. 301 + 921-2842). *Powder and single crystal diffraction, non-metallic and inorganic materials.*

Roth, Prof. Walter Lester. Physics Dept., SUNY at Albany, 1400 Washington Ave., Albany, NY 12222, USA. (1917) PhD, chemistry (U. California, Berkeley, 1941). Prof. (tel. 518 + 393-7945). *Crystal structure, X-ray diffraction, neutron diffraction, EXAFS, superionic conductors, properties of materials.*

Roughead, Mr William A. Route 1, Box 66, Cental, SC 29630, USA.

Roy, Prof. Rustum. Materials Res. Lab., Pennsylvania State U., University Park, PA 16802, USA. (1924) PhD, ceramics (PSU, 1948). Dir. and Evan Pugh Prof. of the Solid State. (tel. 814 + 865-3421). *Materials preparation, characterization, crystal chemistry, synthesis, stability, phase equilibria, crystal growth, non-metallic systems, ultrahigh pressure solid reactions, non-crystalline solids (chemistry and physics).*

Royer, Dr William Edward, Jr. Biochemistry and Molecular Biophysics Dept., Columbia U., 630 168th St., New York, NY 10032, USA. (1954) PhD, biophysics (Johns Hopkins U., 1984). Postdoctral res. scient. (tel. 212 + 305-3456). *Macromolecular crystallography, protein structure and evolution.*

Ruben, Mrs Helena W. Material and Molecular Res. Div., Lawrence Berkeley Lab., 651 Vincente Ave., Berkeley, CA 94707, USA. AB physics (U. California, Berkeley, 1935). Retired. (tel. 415 + 526-1897). *Single crystal diffraction and structure.*

Rubin, Dr Ben Z. 124-H Highland St., Manchester, CT 06040, USA. (1917) PhD, biophysics (U. Michigan, 1976). Retired. (tel. 203 + 646-3067). *X-ray diffraction analysis, biological materials.*

Rubin, Prof. Byron. Dept. of Chemistry, Emory U., 1515 Pierce Dr., Atlanta, GA 30322, USA. (1943) PhD, chemistry (Duke U., 1971). Asst. prof. (tel. 404 + 329-6617). *Protein structures, glycoproteins, enzymes, biologically interesting molecules.*

Ruble, Dr John Rollo. Dept. of Crystallography, U. of Pittsburgh, 304 Thaw Hall, Pittsburgh PA 15260, USA. (1946) PhD, crystallography (U. Pittsburgh, 1975). Staff crystallographer. (tel. 412+ 624-4366). *Nucleic acids, drug - nucleic acid interactions.*

Ruderman, Dr Warren. Interactive Radiation Inc., 181 Legrand Ave., North Vale, NJ 07647, USA. (1920) PhD, chemical physics (Columbia U., 1949). Pres. (tel. 201 + 767-1910). *Crystal growth, X-ray spectroscopy, electro-optics, acousto-optics.*

Rudman, Prof. Reuben M. Chemistry Dept., Adelphi U., Garden City, NY 11530, USA. (1937) PhD, chemistry (PIB, 1966). Prof. (tel. 516 + 294-8700, ext. 7519). *Low temperature X-ray diffraction instrumentation, crystal structure analysis, phase transitions, molecular crystals.*

Rudnick, Dr Suzanne Ellen. Dept. of Chemistry, Manhattan C. 4513 Manhattan College Parkway, Riverdale, NY 10471, USA. (1951) PhD, chemistry (Boston U., 1979). Asst. prof. (tel. 212 + 920-0211). *Protein structure and function.*

Rudnik, Mr Paul J. 1233 W. Jarvis, Chicago, IL 60626, USA.

Rudolph, Dr Philip Reinhold. Dept. of Chemistry, Texas A & M U., College Station, TX 77843, USA. (1955) PhD, chemistry (Texas A & M U., 1983). Res. scient. (tel. 409 + 845-2936). *X-ray powder diffraction, neutron powder diffraction, X-ray methods, software and graphics, powder structure solutions.*

Russell, Dr Thomas Paul. K42/282, IBM, Res. Lab., 5600 Cottle Rd., San Jose, CA 95193, USA. (1952) PhD, polymer science (U. Massachusetts, 1979). Res. staff member. (tel. 408 + 256-7248). *Polymers, synchrotron radiation, X-ray scattering, polymer mixtures, neutron scattering, time resolved scattering.*

Ruszala, Mr Ferdinand A. 1 Lebanon Ave., Colchester, Ct 06415, USA.

Ruud, Prof. Clayton Olaf. Materials Res. Lab., PSU, 159 MRL, University Park, PA 16802, USA. (1934) PhD, materials science (U. Denver, 1970). Sr. res. assoc., assoc. prof. (tel. 814 + 863-2843). *X-ray powder diffraction, residual stresses, materials characterization, electron microprobe, X-ray spectroscopy, metallurgy, material science.*

Ryan, Dr Robert Reynolds. Group CNC-4, MS 346, Los Alamos Scient. Lab., Los Alamos, NM 87545, USA. (1936) PhD, chemistry (Oregon State U., 1965). Res. chemist. (tel. 505 + 667-6045). *Crystallography, activation of small molecules, coordination chemistry, actinide chemistry, continuous phase changes, vibrational spectroscopy, gas phase electron diffraction.*

Ryba, Prof. Earle Richard. Materials Sci. and Eng., Pennsylvania State U., 304 Steidle, University Park, PA 16802, USA. (1934) PhD, physical metallurgy (Iowa State U., 1960). Assoc. prof. (tel. 814 + 865-3760). *Crystal structure, crystal chemistry, intermetallic compounds, structure - properties relationship, order-disorder, residual microstrains, synchrotron radiation.*

Rydel, Mr Timothy John. Physical and Analytical Chemistry Res., The Upjohn Co., 7255-209-1, Kalamazoo, MI 49001, USA. (1959) MS, physical chemistry (Michigan State U., 1983). Chemistry asst. (tel. 616 + 385-5492). *protein crystallography, protein crystallization.*

Sack, Dr John Stuart. Dept. of Biochemistry, Rice U., P. O. Box 1892, Houston, TX 77251, USA. (1953) PhD, biophysics (Johns Hopkins U., U., 1981). Res. Assoc. (tel. 713 + 527-8101, ext. 3346). *Protein crystallography.*

Sadowski, Mr Lucian M. P.O. Box 328, Hope, NJ 07844, USA.

Sakore, Dr Tukaram D. P. O. Box R.B. & B., Medical Center, U. of Rochester, Rochester, NY 14642, USA. PhD, crystallography (U. Poona, India, 1966). Sr. scient. (tel. 716 + 275-7715). *Structure, crystallographic studies, nucleic acids, drug-oligonucleotide complexes, biologically important molecules.*

Salemme, Dr Francis Raymond. Central Res. and Dev. Dept., Bldg. 328, Experimental Station, E. I. Du Pont, Wilmington, DE 19898, USA. (1945) PhD, chemistry (U. California, San Diego, 1972). Group leader. *Biomolecular structure, electron transport, macromolecular structure and stability theory.*

Salkind, Prof. Alvin J. UMDNJ-Rutgers Medical S., Bioengineering Section, Dept. of Surgery, Hoes Lane, Piscataway, NJ 08854, USA. (1927) DChE, chemical engineering (PIB, 1958). Chief. (tel. 201 + 463-4799). *Electrochemically active surfaces, batteries, catalysts, implantable materials.*

Salmon, Dr Oliver Norton. Central Res. Labs., 3M Company, 3M Center, St. Paul, MN 55101, USA. (1917) PhD, physical chemistry (Cornell U., 1946). Sr. res. specialist. (tel. 612 + 733-2692). *Solar energy conversion, chemical thermodynamics, solid state physics, solid state chemistry.*

Salot, Dr Stuart Edwin. Certified Testing Labs. Inc., 2905 E. Century Blvd., So. Gate, CA 90280, USA. (1937) PhD, chemistry (U. Southern California, 1969). Pres. (tel. 213 + 564-2641). *Environmental chemistry.*

Samson, Dr Sten. Chemistry Dept., California Inst. of Techn., 1201 East California Blvd., Pasadena, CA 91125, USA. (1916) Fil Dr, chemistry (U. Stockholm, Sweden, 1968). Res. assoc. (tel. 213 + 795-6811, ext. 2528). *Intermetallic compounds (complex structures), organic conductors, instrumentation, low temperature diffractometry.*

Samudzi, Mr Cleopas T. 5600 Ellsworth Ave., Apt. 2, Pittsburgh, PA 15232, USA.

Samuels, Prof. Robert J. S. of Chemical Eng., Georgia Inst. of Techn., Atlanta, GA 30332, USA. (1931) PhD, polymers (U. Akron, 1961). Prof. (tel. 404 + 894-2885). *Polymer engineering, physics and chemistry, structure - properties - process relationship, small and wide angle diffraction, light, X-rays, neutrons.*

Sanderson, Dr Mark Rutherford. Dept. of Chemistry, Yale U., 225 Prospect St., New Haven, CN 06520, USA. (1956) PhD, biophysics (U. London, UK, 1981). Res. assoc. (tel. 203 + 436-4817). *Nucleic acids, anti-tumor drug structure, DNA binding proteins.*

Sands, Prof. Donald E. Office of Academic Affairs, U. of Kentucky, 7 Administration Bldg., Lexington, KY 40506, USA. (1929) PhD, physical chemistry

(Cornell U., 1955). Vice Chancellor, academic affairs. (tel. 606 + 257-1961). *Tensor properties of crystals, motion in crystals, thermodynamics.*

Santarsiero, Dr Bernard Inez Dominic Matthew. Chemistry Dept., CIT, A. A. Noyes Lab., Room 127-72, Pasadena, CA 91125, USA. (1952) PhD, physical chemistry (U. Washington, 1980). Coordinator, crystal res. *X-ray diffraction, crystallographic computing, instrumentation.*

Sappenfield, Mr Eric L. Chemistry Dept., Baylor U., Waco, TX 76798, USA.

Sarko, Prof. Anatole. Chemistry Dept., SUNY C. of Env. Sci. and Forestry, Syracuse, NY 13210, USA. (1930) PhD, polymer chemistry (SUNY-Syracuse, 1966). Prof. (tel. 315 + 470-6824, telex 7105410555). *Polymers, crystallography, molecular mechanics, quantum mechanics, polysaccharides, computer methods.*

Sarma, Prof. Raghupathy. Biochemistry Dept., SUNY at Stony Brook, Stony Brook, NY 11794, USA. PhD, physics (U. Madras, India 1963). Assoc. prof. (tel. 516 + 246-5026). *Structure and function, proteins.*

Sass, Prof. Ronald L. Biology Dept., Rice U., Houston, TX 77001, USA. (1932) PhD, physical chemistry (U. Southern California, 1957). Prof., biology & chemistry. (tel. 713 + 527-4066). *Biophysics, tissue calcification, muscle structure.*

Sass, Prof. Stephen Louis. Dept. of Materials Sci. and Eng., Cornell U., Bard Hall, Ithaca, NY 14853, USA. (1940) PhD, materials science (Northwestern U., 1966). Prof. (tel. 607 + 256-5239). *Grain boundary structure, X-ray diffraction, electron diffraction and microscopy, crystal defects.*

Satyshur, Mr Kenneth A. 518 Spruce St., Madison, WI 53715, USA.

Sawzik, Ms Patricia. Crystallography Dept., U. of Pittsburgh, 303 Thaw Hall, Pittsburgh, PA 15260, USA. (tel. 412 + 624-4368).

Sax, Dr Martin. Biocrystallography Lab., VA Hospital, University Drive C, Pittsburgh, PA 15204, USA. (1919) PhD, chemistry (crystallography) (U. Pittsburgh, 1961). (tel. 412 + 683-3000, ext. 517 or 204). *Antibody structures, stereochemistry, thiamine catalysis, biologically important structures, stereoelectronic features of biochemical reactions.*

Sayler, Prof. Alice Ann. Dept. of Chemistry, Bloomfield C., Franklin St., Bloomfield, NJ 07003, USA. (1946) PhD, chemistry (Worcester Polytechnic Inst., 1974). Asst. prof. (tel. 201 + 748-9000, ext. 251). *X-ray studies, transition metal complexes.*

Saylor, Dr Charles Proffer. Polymers Div., Nat. Bureau of Standards, Mailing address: 10001 Riggs Rd., Adelphi, MD 20783, USA. (1901) PhD, physical chemistry (Cornell U., 1928). Guest worker. (tel. 301 + 921-2482). *Crystal growth, optical properties of crystals, purification by crystal growth.*

Sayre, Dr David. Res. Div., IBM T. J. Watson Res. Cntr., P.O. Box 218, Yorktown Heights, NY 10598, USA. (1924) PhD, chemical crystallography (Oxford U., UK, 1951). Res. staff member. (tel. 914 + 945-1040). *Structure determination methods, large biological structures, imaging methods, X-ray microscopy.*

Scarbrough, Dr Frank Edward. Office of Nutrition and Food Sciences, Food and Drug Administration, 200 "C" St., Washington, DC 20204, USA. (1942) PhD, physical chemistry (Harvard U., 1971). Chemist. (tel. 202 + 245-3117). *Structure - flavor - odor correlations, structure - biological activity correlations.*

Scaringe, Dr Raymond P. Chemistry Div., Res. Labs., Eastman Kodak Co., 1669 Lake Ave., Rochester, NY 14650, USA. (1950) PhD, physical-inorganic chemistry (U. North Carolina, 1976). Sr. res. scient. (tel. 716 + 477-7052). *Packing on molecular solids, small molecule-polymer interactions, structure determination.*

Schäfer, Prof. Lothar. Chemistry Dept., U. of Arkansas, Fayetteville, AK 72701, USA. (1939) Dr.rer.nat., inorganic chemistry (U. Munich, BRD, 1965). Prof. (tel. 501 + 575-4601). *Structural studies, electron diffraction, vibrational analysis and theoretical procedures.*

Schaefer, Dr William Palzer. Div. of Chemistry, California Inst. of Techn., 1201 E. California St., Pasadena, CA 91125, USA. (1931) PhD, chemistry (UCLA, 1960). Sr. res. assoc. (tel. 818 + 356-6567). *Inorganic chemistry, transition metal complexes.*

Scheidt, Prof. W. Robert. Dept. of Chemistry, U. of Notre Dame, Notre Dame, IN 46556, USA. (1942) PhD, chemistry (U. Michigan, 1968). Prof. (tel. 219 + 239-5939). *Inorganic chemistry, metalloporphyrin structure, transition metal complexes, magnetism.*

Schiffer, Dr Marianne. Div. of Biological and Medical Res., Argonne Nat. Lab., 9700 South Cass Ave., Argonne, IL 60439, USA. (1935) PhD, biochemistry (Columbia U., 1965). Scient. (tel. 312 + 972-3883). *Protein structures, immunoglobulins.*

Schioler, Dr Liselotte Jensen. Ceramics Res. Div., Army Materials and Mechanics Res. Center, AMXMR-MC, Arsenal St., Watertown, MA 02172-0001, USA. (1950) ScD, ceramic science (MIT, 1983) Ceramic res. eng. (tel. 617 + 923-5410). *Structures, neutron diffraction.*

Schirber, Dr James Emmanuel. Dept. 1150, Sandia Labs., Albuquerque, NM 87115, USA. (1931) PhD, physics (Iowa State U., 1960). Manager, solid state res. dept. (tel. 505 + 844-8134). *Low temperature structures, pressure dependence, atomic positional parameters, thermal expansion, electronic structure - properties relationship.*

Schlemper, Prof. Elmer Otto. Chemistry Dept., U. of Missouri, Columbia, MO 65211, USA. (1939) PhD, inorganic chemistry (U. Minnesota, 1965). Prof. (tel. 314 + 882-7540). *Hydrogen bonding, neutron diffraction, inorganic compounds (structure).*

Schlenker, Dr John L. Mobil Res. and Dev. Corp., Paulsboro Lab., Billingsport Rd., Paulsboro, NJ 08066, USA. (1945) PhD, geology (mineralogy) (Virginia Polytechnic Inst., 1976). Res. chemist. (tel. 609 + 423-2796). *Zeolite chemistry and structure, crystal structure determination, powder diffractometry.*

Schlueter, Prof. Albert S. Dept. of Chemistry, Central State U., Wilberforce, OH 45384, USA. (1940) PhD, physical chemistry (Iowa State U., 1968). Assoc. prof. (tel. 513 + 376-6424). *Organometallic compounds, ion-selective electrodes.*

Schmid, Dr Michael Francis. Dept. of Biochemistry, U. of Arizona, Biosciences West, Tucson, AZ 85721, USA. (1947) PhD, biochemistry (U. Washington, 1974). Postdoctoral res. assoc. (tel. 503 + 686-5176). *Biological macromolecules.*

Schmidt, Prof. Paul Woodward. Physics Dept., U. of Missouri - Columbia, Columbia, MO 65211, USA. (1926) PhD, physics (U. Wisconsin, Madison, 1953). Prof. (tel. 314 + 882-8241). *Small angle X-ray scattering, biophysics, physics of fluids, critical phenomena.*

Schmidt, Dr William Charles, Jr. NCR - Communication Systems Div., 3325 Platt Springs Rd., West Columbia, SC 29169, USA. (1948) PhD, biochemistry (crystallography) (U. Virginia, 1975). (tel. 803 + 796-9250). *Protein structure and crystallography, refinement techniques, macromolecular crystallography.*

Schneider, Dr Michael L. ITT Advanced Center, One Research Dr., Shelton, CT 06484, USA. (1944) PhD, chemistry (U. Durham, UK, 1969). Computer sci. (tel. 215 + 542-5808). *Computing, organometallics, direct methods, twinning.*

Schoenborn, Dr Benno Paul. Biology Dept., Brookhaven Nat. Lab., Upton, NY 11973, USA. (1936) PhD, physics (U. New South Wales, Australia, 1962). Sr. sci., BNL; Prof., Columbia Medical Center. (tel. 516 + 345-3421). *Biological structures, X-ray and neutron scattering, structure - function relationships, pharmacological agents, proteins.*

Schomaker, Prof. Verner. Dept. of Chemistry, U. of Washington, Seattle, WA 98195, USA. (1914) PhD, chemistry (CIT, 1938). Prof. em. (tel. 206 + 543-1643). *Electron diffraction by gases, X-ray structure determination methods.*

Schroeder, Dr LeRoy William. Div. of Life Sciences, Center for Devices and Radiological Health, 12709 Twinbrook Parkway, Rockville, MD 20857, USA. (1943) PhD, physical chemistry (Northwestern U., 1969). Res.chemist. (tel. 312 + 443-7167). *Molecular structure and dynamics, membrane phenomena.*

Schultz, Dr Arthur Jay. Chemistry Dept., Argonne Nat. Lab., 9700 South Cass Ave., Argonne, IL 60439, USA. (1947) PhD, chemistry (Brown U., 1973). Chemist. (tel. 312 + 972-3465). *Time-of-flight neutron diffraction, instrumentation, data analysis techniques, low dimensional materials, organometallic chemistry, X-ray diffraction.*

Schultz, Prof. Jerold Marvin. Dept. of Chemical Eng., U. of Delaware, Newark, DE 19716, USA. (1935) PhD, metallurgical engineering (Carnegie Inst. of Techn., 1965). Prof. (tel. 302 + 451-8145). *Structure - properties relationship, polymeric materials, phase transitions, crystal defects.*

Schuman, Mr Clifford Alan. RFD 1, Box 113A, Buckhill Rd., Canterbury, CT 06331, USA.

Schuster, Prof. Sanford Lee. Dept. of Physics & Electronics Eng. Techn., Mankato State U., Mankato, MN 56001, USA. (1938) PhD, physics (U. Nebraska-Lincoln, 1969) Prof. (tel. 507 + 389-6700). *Solid state physics, lattice dynamics.*

Schwartz, Dr Kenneth Bruce. Materials Res. Lab., SRI International, 333 Ravenswood Ave., Menlo park, CA 94025, USA. (1954) PhD, crystal chemistry (SUNY at Stony Brook, 1982). Materials scient. (tel. 415 + 859-5298). *Structural chemistry, crystal chemistry, oxides, powder diffraction, mineralogy, electrical transport properties, non-stoichiometry, defect structure, mixed-valence compounds.*

Schwartz, Dr Lyle Howard. Center for Materials Sci., Nat. Bureau of Standards, Gaithersburg, MD 20899, USA. (1936) PhD, materials science (Northwestern U., 1964). Dir. (tel. 301 + 921-2891). *Mossbauer spectroscopy, neutron diffraction, X-ray diffraction, phase transformations in solids.*

Sclar, Prof. Charles Bertram. Dept. of Geological Sci., Lehigh U., Williams Hall, Bldg 31, Bethlehem, PA 18015, USA. (1925) PhD, geology (Yale U., 1951). Prof. and chairman. (tel. 215 + 861-3660). *Minerals and synthetic analogue relationships (structure - stability - equilibrium), high pressure phases (synthesis - stability - structure), pressure dependent phase transformations.*

Scott, Mr Donald Lee. Techn. Div., Goodyear Atomic Corp., P. O. Box 628, Piketon, OH 45661, USA. (1931) electronic engineering (C.R.E.I., 1959). Sr. physicist. (tel. 614 + 289-2331, ext. 5777). *X-ray diffraction, X-ray spectroscopy, optical and electron microscopy, scientific photography.*

Seabaugh, Mr Pyrtle W. 7795 Raintree Rd., Centerville, OH 45459, USA.

Seale, Dr Steve Keith. Div. of Chemical Dev., Tennessee Valley Authority, T7G-NFDS Muscle Shoals, AL 35660, USA. (1944) PhD, physical-inorganic chemistry (U. Alabama, 1974). Lab. supervisor. (tel. 205 + 383-4631, ext. 2483). *Computational methods, instrumentation.*

Sears, Dr Donald Richard. U.S. Dept. of Energy, Grand Forks Energy Techn. Center, P.O. Box 8213, University Station, Grand Forks, ND 58202, USA. (1928) PhD, physical chemistry (Cornell U., 1958). Supervisory env. eng. (tel. 701 + 795-8138). *Aerosol and fine particle characterization, structure - properties - morphology relationship, atmospheric particulates, X-ray method applications, combustion chemistry.*

Seaton, Dr Barbara A. Chemistry Dept., Harvard U., 12 Oxford St., Cambridge, MA 02138, USA. (1952) PhD, chemistry (MIT, 1983). Postdoctoral fellow. (tel. 617 + 495-4097). *Protein crystallography.*

Seely, Mr Oliver, Jr. Chemistry Dept., California State C., Dominguez Hills, CA 90747, USA.

Seeman, Prof. Nadrian Charles. Dept. of Biology, SUNY at Albany, Albany, NY 12222, USA. (1945) PhD, biochemistry-crystallography (U. Pittsburgh, 1970). Assoc. prof. (tel. 518 + 457-8604). *Nucleic acid junctions, macromolecular designs.*

Seff, Prof. Karl. Chemistry Dept., U. of Hawaii, 2545 The Mall, Honolulu, HI 96822, USA. (1938) PhD, chemistry (MIT, 1964). Prof. (tel. 808 + 948-7665). *Intrazeolitic chemistry and structure, transition metal complexes, small organic molecules.*

Segmüller, Dr Armin Paul. Res. Div., IBM T. J. Watson Res. Cntr., P.O. Box. 218, Yorktown Heights, NY 10598, USA. (1924) Dr. phil. nat., crystallography (U. Erlangen-Nürnberg, BRD, 1954). Res. staff member. (tel. 914 + 945-1287, telex 137456). *Crystal physics, diffraction physics, phonons, superlattices, laboratory automation, multilayer structures.*

Seitz, Dr Frederick. Rockefeller U., New York, NY 10021, USA. (1911) PhD physics (Princeton U., 1934). Retired.

Semendy, Mr Alphonse F. 3513 Pence Ct., Annandale, VA 22003, USA.

Sen Gupta, Prof. Pradip Kumar. Geology Dept., Memphis State U., Memphis, TN 38152, USA. (1936) PhD, geology (mineralogy) (Washington U., 1964). Prof. (tel. 901 + 454-2177 or 454-2178). *Structural mineralogy, organo-metallic complexes, inorganic structures.*

Senechal, Prof. Marjorie Lee. Dept. of Mathematics, Smith C., Northampton, MA 01063, USA. (1939) PhD, mathematics (Illinois Inst. of Techn., 1965). Prof. (tel. 413 + 584-2700, ext. 3862). *Mathematical crystallography, history of crystallography.*

Sengupta, Mr Sisir K. Biochemistry Dept., OB/GYN, T-506, Boston U. Sch. of Medicine, 80 E. Concord St., Boston, MA 02118, USA. (1926) PhD, chemistry (Jadarpur U., India, 1959). Assoc. res. prof. (tel. 617 + 247-5808). *Drug design (synthesis), drug evaluation, drug-DNA interaction.*

Senti, Mr F. R. 2601 N. Pollard St., Arlington, VA 22207, USA.

Sepehrnia, Mr Bahman. Dept. of Crystallography, U. of Pittsburgh, Pittsburgh, PA 15260, USA.

Servos, Prof. Kurt. Dept. of Geology, Menlo C., Menlo Park, CA 94025, USA. (1928) MS, geology (Yale U., 1954). Prof. (tel. 415 + 323-6141, ext. 370). *Crystal symmetry, morphological and geometrical crystallography, mineral crystal structures.*

Shaffer, Prof. Lawrence B. Physics Dept., Anderson C., Anderson, IN 46011, USA. (1937) PhD, physics (U. Wisconsin, 1964). Prof. and chairman. (tel. 317 + 649-9071, ext. 2096). *Small angle X-ray scattering.*

Shaffner, Dr Thomas Jackson. 2016 Aliso Rd., Los Rios, Plano, TX 75074, USA. (1941) PhD, physics (Vanderbilt U., 1969). *X-ray crystallography, biological molecules, scanning electron microscopy, X-ray microprobe analysis, polymeric systems, scientific data processing.*

Shandles, Mr Robert. Ventnor H-3026, Deerfield Beach, FL 33441, USA.

Shannon, Dr Robert Day. Central Res. and Dev., E. I. du Pont, Experimental Station, Wilmington, DE 19898, USA. (1935) PhD, ceramic engineering (U. California, Berkeley, 1964). Res. chemist. (tel. 302 + 772-2818). *Crystallography, synthesis, crystal growth, electronic and ionic conductivity, structure - properties relationship.*

Sharrah, Mr Paul C. Physics Dept., U. of Arkansas, Fayetteville, AR 72701, USA.

Shefter, Dr Eli. Biomedical Products, E. I. duPont deNemours Inc., Exp. Station 400, Wilmington, DE 19898, USA. (1936) PhD, pharmaceutics (U. Wisconsin, 1963). Manager, drug delivery systems. (tel. 302 + 772-7066). *Drug delivery, metabolism, pharmacokinetics, structure - activity relationships.*

Sheldon, Dr Robert Isaly. Physical Metallurgy MST-5 Los Alamos Nat. Lab., Mail Stop G730, Los Alamos, NM 87545, USA. (1945) PhD physical chemistry (U. Kansas, 1976). Staff member. (tel. 505 + 667-5363). *Single crystal diffraction (large and small molecules), powder diffraction, direct methods, actinides.*

Shen, Dr Ming-Shing. Physical & Chemical Sciences Div., U.S. Dept. of Energy, P. O. Box 880, Morgantown, WV 26505, USA. (1940) PhD, materials science (U. Pittsburgh, 1974). Chemical eng. (tel. 304 + 291-4112). *Fossil fuel utilization, oil shale process fundamentals, hot gas stream contaminant cleanup.*

Sheriff, Dr Steven. Lab. of Molecular Biology, Nat. Inst. of Arthritis, Diabetes, Digestive and Kidney Diseases, Bldg. 2, Rm. 408, Bethesda, MD 20205, USA. (1951) PhD, biochemistry (U. Washington, 1979). (tel. 301 + 588-1881). *Macromolecular structure and function, macromolecular crystallography, crystallographic computing.*

Sherman, Mr Robert L. 116 W. Newkirk Ln., Oak Ridge, TN 37830, USA.

Sherry, Mrs Elizabeth Ann Gebert. 103 N. Milwaukee Ave., A-14, Lake Villa, IL 60046, USA. (1926) BS, chemistry (Ball State U.). *Inorganic crystal structures.*

Shieh, Dr Huey-Sheng. 53C, Physical Science Center, MONSANTO, 800 N. Lingbergh Blvd. St. Louis, MO 63167, USA. (1946) PhD, chemistry (U. Pennsylvania, 1975). Res. specialist. (tel. 314 + 694-4820). *Protein structure, energy calculations, molecular graphics, structure - function relationships, protein design.*

Shinn, Dr Dennis Burton. Sylvania Lighting Center, GTE Sylvania, 100 Endicott St., Danvers, MA 01923, USA. (1939) PhD, chemistry (Michigan State U., 1968). Program manager, materials lab. (tel. 617 + 777-1900, ext. 2163). *Inorganic crystal structures, phase equilibria, ceramics.*

Shiono, Prof. Ryonosuke. Crystallography Dept., U. of Pittsburgh, PA 15260, USA. (1923) DSc, physics (Osaka U., Japan, 1960). Assoc. prof. (tel. 412 + 624-4366). *Organic crystal structure, computer programming.*

Shipley, Prof. G. Graham. Biophysics Inst., Dept. of Medicine, Boston U. Sch. of Medicine, 80 East Concord St., Boston, MA 02118, USA. (1937) PhD, X-ray crystallography (U. Nottingham, UK, 1963). Prof. of biochemistry. (tel. 617 + 247-5040). *Lipids, lipoproteins, membranes, X-ray scattering, neutron scattering.*

Shlichta, Dr Paul Joseph. Materials Res. Group, , Jet Propulsion Lab., CIT, 67-106, Pasadena, CA 91109, USA. (1930) PhD, chemistry (CIT, 1956). Member techn. staff. (tel. 818 + 354-3339 *Geometric crystallography, crystal growth, materials science.*

Shoemaker, Prof. Clara Brink. Chemistry Dept., Oregon State U., Corvallis, OR 97331, USA. (1921) PhD, chemical crystallography (U. Leiden, Netherlands, 1950). Prof. em. (tel. 503 + 754-2081). *Inorganic crystal structures, metal and alloy structures, organo metallic compounds.*

Shoemaker, Prof. David Powell. Chemistry Dept., Oregon State U., Corvallis, OR 97331, USA. (1920) PhD, physical chemistry (CIT, 1947). Prof. em. (tel. 503 + 754-2081). *Tetrahedrally close-packed (t.c.p.) metal phases, zeolites, alloy hydrides.*

Short, Dr Michael Arthur. Geology Res., Cities Service Oil & Gas Corp., P.O. Box 3908, Tulsa, OK 74102, USA. (1930) PhD, X-ray diffraction (PSU, 1961). Res. assoc. (tel. 918 + 561-5223). *X-ray instrumentation, X-ray physics, X-ray diffraction, X-ray fluorescence, electron microprobe analysis, SEM.*

Short, Mr Michael T. 2602 Eastwood 1F, Evanston, IL 60201, USA.

Shull, Prof. Clifford G. Dept. of Physics, Massachusetts Inst. of Techn., Cambridge, MA 02139, USA. (1915) PhD, physics (New York U., 1941). Prof. (tel. 617 + 253- 4812). *Neutron diffraction, neutron physics, solid state physics.*

Siegel, Dr Lester Aaron. Chemical Res. Div., American Cyanamid Co., 1937 West Main St., Stamford, CT 06904, USA. (1925) PhD, physics (MIT, 1948). Sr. res. physicist. (tel. 203 + 348-7331, ext. 576). *X-ray diffraction, X-ray spectroscopy, solid state physics.*

Siegel, Dr Stanley. Chemical Eng. Div., Argonne Nat. Lab., 9700 South Cass Ave., Argonne, IL 60439, USA. (1915) PhD, physics (U. Chicago, 1941). Sr. physicist. (tel. 312 + 972-4347). *Structure, crystal chemistry.*

Sieker, Dr Larry C. Biological Structure Div., U. of Washington, Seattle, WA 98195, USA. (1931) PhD, biological structures, (U. Washington, 1981). Res. asst. prof. (tel. 206 + 543-6541). *Macromolecular structures, structure - function relationships, macromolecule crystallization techniques, data acquisition, protein-cell membrane interactions.*

Sigler, Prof. Paul Benjamin. Dept. of Biochemistry & Molecular Biology, U. of Chicago, 920 E. 58th St., Chicago, IL 60637, USA. (1934) MD, PhD, biochemistry (Columbia U., 1959; Cambridge U., UK, 1968). Prof. (tel. 312 + 962-1086, telex 9102213477). *Biological macromolecules, genetic regulation, protein - membrane interactions.*

Silcox, Prof. John. Sch. of Appl. and Eng. Physics, Cornell U., 211 Clark Hall, Ithaca, NY 14853, USA. (1935) PhD, physics (Cambridge U., UK, 1961). Prof. and Dir. (tel. 607 + 256-3332). *Electron microscopy, electron spectroscopy, electron diffraction.*

Silverton, Dr James V. N.I.H. Bldg 10, Rm 7N-316, Bethesda, MD 20014, USA. (1934) PhD, chemical crystallography (U. Glasgow, UK, 1958). Scient. (tel. 301 + 496-3341). *Direct methods, natural products, high molecular weight molecules (500-1500 Daltons), chemical crystallography (general), diffraction techniques.*

Silverton, Ms Enid. Nat. Inst. of Health, Bldg. 2, Room 312, Bethesda, MD 20205, USA.

Simard, Dr Roger Gèrard. 325 Dartmouth Ave., J-3, Swarthmore, PA 19081, USA. (1909) PhD, physical chemistry (MIT, 1939). Consultant. (tel. 215 + 543-7417). *Powder diffraction, solid state physics, molecular structure.*

Sime, Prof. Rodney J. Chemistry Dept., California State U., Sacramento, CA 95819, USA. (1931) PhD, physical chemistry (U. Washington, 1959). (tel. 916 + 454-6659). *Crystal and molecular structure, transition element complexes; structure and activity, biologically important molecules.*

Sime, Dr Ruth Lewin. Chemistry Dept., Sacramento City C., Sacramento, CA 95822, USA. (1939) PhD, physical chemistry (Harvard U., 1964). Instructor. (tel. 916 + 449-7228). *Biologically important molecular structures.*

Simmins, Mr John J. 25 West Ave., Arkport, NY 14807, USA.

Simmons, Dr Charles J. Dept. of Chemistry, U. of Puerto Rico, Rio Piedras, PR 00931, USA. (1948) PhD, physical-inorganic chemistry (U. Hawaii, 1980). Asst. prof. (tel. 809 + 764-0000, ext. 2598). *X-ray structural characterization, pseudo-Jahn-Teller complexes, cobalt dioxygen complexes.*

Simmons, Mr Ralph O. Physics Dept., U. of Illinois, 1110 W. Green St., Urbana, IL 61801, USA.

Simonsen, Prof. Stanley Harold. Chemistry Dept., U. of Texas, Austin, TX 78712, USA. (1918) PhD, chemistry (U. Illinois, 1949). Prof. (tel. 512 + 471-5755). *Transition metal coordination compounds, small ring heterocycles, fused rings, biochemically significant compounds.*

Simpson, Dr Howard D. 3772 Hamilton St., Irvine, CA 92714, USA. (1937) PhD, chemical engineering (U. Texas, 1969). (tel. 714 + 528-7201). *Catalysis, chemical bonding.*

Singh, Dr Phirtu. Dept. of Chemistry, U. of North Carolina, Chapel Hill, NC 27514, USA. (1933) PhD, chemistry (U. Colorado, 1965). Res. assoc. *Biochemical molecules, protein - nucleic acid interactions, conformational analysis.*

Singh, Dr Sarjant. Pathology Dept., U. of California - San Diego, LaJolla, CA 92093, USA. PhD, chemistry (U. California, San Diego, 1968). (tel. 714 +

452-4318). *Electron microscopy, electron diffraction, X-ray crystallography, biochemical systems.*

Sinn, Prof. Ekkehard. Chemistry Dept., U. of Virginia, Charlottesville, Virginia 22901, USA. (1945) PhD, (U. New South Wales). (tel. 804 + 924-3424). *Electrons - molecular structure relation, transition metal complexes; magnetic exchange interactions; bio-inorganic chemistry.*

Skarstad, Dr Paul Michael. Energy Techn., Medtronic Inc., 6700 Shingle Creek Parkway, Minneapolis, MN 55430, USA. (1942) PhD, physical chemistry (Cornell U., 1971). Assoc. fellow. (tel. 612 + 574-6380). *Solid state chemistry, electrochemistry.*

Skelton, Dr Earl Franklin. Condensed Matter and Radiation Sci. Div., U.S. Naval Res. Lab., Washington, DC 20375, USA. (1940) PhD, physics (Rensselaer Polytechnic Inst., 1967). Supervisory res. physicist. (tel. 202 + 767-3014). *Phase transformations, high pressure, diffraction physics, synchrotron radiation, EXAFS.*

Skrzypczak-Yankun, Dr Ewa. Chemistry Dept., Michigan State U., East Lansing, MI 48824, USA. (1948) PhD, chemistry (A. Mickiewicz U., Poland, 1976). Res. assoc. (tel. 517 + 353-7298). *Single crystal X-ray crystallography, protein crystal growth, protein structure, alkaloids, metallo-organic and organic small molecules.*

Slaughter, Prof. Maynard. Chemistry and Geochemistry, Colorado S. of Mines, Golden, CO 80401, USA. (1934) PhD, X-ray crystallography (U. Pittsburgh, 1962). Prof. (tel. 303 + 273-3648). *Structure, zeolites, clay minerals.*

Sleight, Dr Arthur William. Central Res. and Dev. Dept., E. I. duPont de Nemours and Co., Exp. Station, 356, Wilmington, DE 19898, USA. (1939) PhD, inorganic chemistry (U. Connecticut, 1963). Res. manager. (tel. 302 + 772-3536). *Solid state chemistry, inorganic structures, oxides, heterogeneous catalysts, structure - properties relationship, powder structure analysis.*

Sliva, Mr Paul. PSU, Materials Res. Lab., University Park, PA 16802, USA.

Sloan, Dr Gilbert J. Central Res. and Dev. Dept., E. I. du Pont de Nemours and Co., Experimental Station, Wilmington, DE 19898, USA. (1928) PhD, chemistry (U. Michigan, 1954). Res. supervisor. (tel. 302 + 772-3607). *Organic crystal growth, microdistribution of impurities, zone refining.*

Sly, Prof. William Glenn. Chemistry Dept., Harvey Mudd C., Claremont, CA 91711, USA. (1922) PhD, chemistry (CIT, 1955). Prof. (tel. 714 + 626-8511, ext. 2890). *Organic and metal-organic structures, crystallographic computing.*

Smith, Mr Albert Edward. 72 San Mateo Rd., Berkeley, CA 94707, USA. (1908) MS, chemistry (U. Calif. Berkeley, 1935). Retired. (tel. 415 + 524-3697). *Organic compounds, metal organic compounds, structure - properties relationship, metal oxides.*

Smith, Dr David John. Center for Solid State Sci., Arizona State U., Tempe, AZ 85287, USA. (1948) PhD, physics (U. Melbourne, Australia, 1978). Assoc. prof. (tel. 602 + 965-4540). *High resolution electron microscopy, defect structures, surfaces.*

Smith, Prof. Deane Kingsley, Jr. Dept. of Geosciences, Pennsylvania State U., 239 Deike Bldg., University Park, PA 16802, USA. (1930) PhD, geology (U. Minnesota, 1956). Prof. of mineralogy. (tel. 814 + 865-5782). *Powder diffraction methods, crystal chemistry of minerals and mineral-like compounds.*

Smith, Dr Douglas Lee. Res. Labs., Analytical Sci. Div., Eastman Kodak Co., Building 82, Kodak Park, Rochester, NY 14650, USA. (1937) PhD, physical chemistry (U. Wisconsin, 1962). Res. assoc. (tel. 716 + 722-0138). *Organic crystal structures, silver complexes.*

Smith, Dr George David. Dept. of Molecular Biophysics, Medical Foundation of Buffalo, 73 High Street, Buffalo, NY 14203, USA. (1941) PhD, physical chemistry (Ohio U., 1968). Assoc. res. scient. (tel. 716 + 856-9600, ext. 430). *Polypeptides, polypeptide hormones, ion-transport antibiotics, protein structures, structure - action mode relationships.*

Smith, Dr Gordon Stuart. Chemistry Dept., L-370, Lawrence Livermore Lab., Livermore, CA 94550, USA. PhD, physical chemistry (Cornell U., 1957). Staff scient. (tel. 415 + 447-1100, ext. 7021). *Powder diffraction, inorganic and intermetallic crystal structures, instrumentation.*

Smith, Dr Harold Glenn. Solid State Div., Oak Ridge Nat. Lab., Oak Ridge, TN 37830, USA. (1927) PhD, physics (Iowa State U., 1957). Sr. scient. (tel. 615 + 574-5243). *Crystallography, neutron scattering, X-ray diffraction, lattice dynamics.*

Smith, Dr Janet Louise. Dept. of Biochemistry and Molecular Biophysics, Columbia U., 630 W. 168th St., New York, NY 10032, USA. (1951) PhD, biochemistry (U. Wisconsin, Madison, 1978). Assoc. res. scient. (tel. 212 + 305-3456). *Protein crystallography.*

Smith, Prof. John Francis. Dept. of Materials Sci. & Eng., Iowa State U., 122 Metallurgy Bldg., Ames Lab., Ames, IA 50010, USA. (1923) PhD, physical chemistry (Iowa State U., 1953). Prof. and consultant. (tel. 515 + 294-5083). *Intermetallic phases, bonding, structure - properties relationship.*

Smith, Prof. Joseph Victor. Dept. of Geophysical Sci., U. of Chicago-M9S, Chicago, IL 60637, USA. (1928) PhD, crystallography (Cambridge U., UK, 1951). Louis Block Prof. of Physics. (tel. 312 + 962-8110). *Mineral crystal structures, mineralogy, petrology, geochemistry.*

Smith, Dr Ward Whitlock. Corporate Res.-S3C, MONSANTO Corp., 800 N. Lindbergh Blvd., St. Louis, MO 63011, USA. (1949) PhD, biological chemistry (U. Michigan, 1977). (tel. 314 + 694-4820). *Structure and function, biological macromolecules, diffraction methods, instrumentation.*

Smithson, Mr Douglas L. 3109 Woodford Dr., Arlington, TX 76013, USA.

Smoluchowski, Prof. Roman. Physics and Astronomy Dept., U. of Texas, Austin, TX 78712, USA. (1910) PhD, (U. Groningen, Netherlands, 1935). Prof. (tel.512 + 471-4461). *Defects in crystals, amorphous surfaces, phase transitions.*

Snider, Mr E.E. 180 Park Row, Apt. 18e, New York, NY 10038, USA.

Snow, Mr Mark E. Dept. of Biophysics, Johns Hopkins Medical S., Baltimore, MD 21205, USA.

Snyder, Prof. Robert L. NYS C. of Ceramics, Alfred U., Alfred, NY 14802, USA. (1941) PhD, chemistry (Fordham U., 1968). Prof., ceramic sci. (tel. 607 + 871-2438). *Powder diffraction, crystal structure analysis, computer programming.*

Soltzberg, Prof. Leonard Jay. Dept. of Chemistry, Simmons C., 300 The Fenway, Boston, MA 02115, USA. (1944) PhD, physical chemistry (Brandeis U., 1969). Prof. (tel. 617 + 738-2188). *Phase transformations, crystal optics, pattern recognition, computer graphics, on-line instrument control.*

Somorjai, Prof. Gabor Arpad. Dept. of Chemistry, U. of California, Berkeley, CA 94720, USA. (1935) PhD, chemistry (U. California, Berkeley, 1960). Prof. (tel. 415 + 642-4053, telex TWX 910-366-7114 UC Berk Berk). *Surface structure, surface crystallography.*

Sparks, Dr Cullie James, Jr. Metals and Ceramics Div., Oak Ridge Nat. Lab., P.O.Box X, Oak Ridge, TN 37830, USA. (1929) PhD, metallurgy (U. Kentucky, 1957). Res. metallurgist. (tel. 615 + 574-6996). *X-ray diffraction physics, spectroscopy, diffuse X-ray scattering, imperfect materials, synchrotron radiation.*

Sparks, Dr Robert Allen. 2085 Sandalwood Ct., Palo Alto, CA 94303, USA. (1928) PhD, physical chemistry (UCLA, 1958). Consultant. (tel. 415 + 856-9492). *Instrumentation, automation, methods of structure determination, computing methods, image science.*

Spence, Prof. Robert Dean. Physics Dept., Michigan State U., East Lansing, MI 48824-1322, USA. (1917) PhD, physics (Yale U., 1948). Prof. (tel. 517 + 353-6345). *Magnetism in crystals, nuclear magnetic resonance in crystals.*

Spielberg, Prof. Nathan. Physics Dept., Kent State U., Kent, OH 44242, USA. (1926) PhD, physics (Ohio State U., 1952). Prof. (tel. 216 + 672-2881). *X-ray physics, X-ray interferometry, ultra-soft X-rays, instrumentation, liquid crystals.*

Spiers, Mr Ronald J. Eng. Computer Support, Loral Electronics Systems, Ridge Hill, Yonkers, NY 10710, USA.

Spooner, Prof. Stephen. Solid State Div., Oak Ridge Nat. Lab., Oak Ridge, TN 37831, USA. (1937) ScD, metallurgy (MIT, 1965). (tel. 615 + 574-4535). *Phase transformation, X-ray and neutron scattering, small angle scattering, structure of solids.*

Sprang, Dr Stephen Robert. Dept. of Biochemistry and Biophysics, U. of California, San Francisco, CA 94143, USA. (1949) PhD, biochemistry (U. Wisconsin, Madison, 1977). Postgrad. res. biochemist. (tel. 415 + 666-5051). *Macromolecular crystallography, allostery - structural basis, macromolecular recognition processes, enzyme systems.*

Springer, Dr James Patrick. Biophysics Dept., Merck Sharp and Dohme Res. Lab., P.O. Box 2000, Rahway, NJ 07065, USA. (1950) PhD, chemistry (Iowa State U., 1976). Assoc. Dir. (tel. 201 + 574-5496). *Structure determination of organic and macromolecules of biological significance.*

Sproul, Prof. Gordon Duane. Chemistry Dept., U. of South Carolina, 800 Cartaret St., Beaufort, SC 29902, USA. (1944) PhD, inorganic chemistry (U. Illinois, 1971). Assoc. prof. (tel. 803 + 524-7112, ext. 28). *Coordination compounds, transition metal complexes.*

Spurlino, Mr John C. Dept. of Biochemistry, Rice U., P. O. Box 1892, Houston, TX 77251, USA.

Srikrishnan, Dr Tamarapu. Center for Crystallographic Res., Roswell Park Memorial Inst., 666 Elm St., Buffalo, NY 14263, USA. (1943) PhD, X-ray crystallography (U. Madras, 1969). Cancer res. scient. (tel. 716 + 845-5819 or 3302). *Biologically interesting crystal structures; amino acids, peptides, nucleotides and nucleosides, macromolecular crystallography; direct method applications, crystal structure analysis, cancer research compounds.*

St. Charles, Mr Robert. 15710 Susses, Livonia, MI 48154, USA.

Stalick, Dr Judith Kay. Reactor Radiation Div., Nat. Bureau of Standards, A221 Materials, Gaithersburg, MD 20899, USA. (1943) PhD, inorganic chemistry (Northwestern U., 1969). Res. chemist. (tel. 301 + 921-2744). *Neutron diffraction, magnetic structures, data compilation, identification methods.*

Stallings, Dr William C. Biophysics Res. Div., U. of Michigan, 2200 Bonisteel Blvd., Ann Arbor, MI 48109, USA. (1947) PhD, chemistry (U. Pennsylvania, 1974). Asst. res. biophysicist, Adjunct Asst. Prof. Chemistry. (tel. 313 + 763-2199). *Biological structures.*

Stanfield, Ms Robyn. 4111 Avenue A, No. 205, Austin, TX 78751, USA.

Stanko, Prof. Joseph Anthony. Chemistry Dept., U. of South Florida, 4202 Fowler Ave., Tampa, FL 33620, USA. (1941) PhD, inorganic chemistry (U. Illinois, Urbana-Champaign, 1966). Assoc. Prof. (tel. 813 + 974-4129). *Single crystal diffraction, powder X-ray diffraction, coordination compounds (platinum group metals), cancer chemotherapeutic drugs, one-demensional electrical conductors.*

Staudenmann, Dr Jean Louis. A111 Physics, Ames Lab., Iowa State U., Ames, IA 50011, USA. (1940) PhD, solid state physics (U. Geneva, 1976). (tel. 615 + 576-4951). *Alloys, electron density maps, thermal properties; X-ray and neutron diffraction.*

Stauffacher, Dr Cynthia Vianne. Dept. of Biological Sci., Purdue U., Lilly Hall of Life Sci., West Lafayette, Indiana 47907, USA. (1948) PhD, physical chemistry

(UCLA, 1977). Sr. res. assoc. (tel. 317 + 494-4912).' *Macromolecular structure and assembly, structural biology, viruses, X-ray crystallography, electron microscopy, image reconstruction.*

Stearns, Mr Frederick Stanley. 632 Lemke Dr., Placentia, CA 92670, USA. (1926) BS, physics (West Coast U., 1971). (tel. 714 + 528-9775). *Crystal growth, inorganics.*

Steele, Dr Ian McKay. Dept. of Geophysical Sci., U. of Chicago, 5734 S. Ellis Ave., Chicago, IL 60637, USA. (1944) PhD, geology-crystallography (U. Illinois, 1971). Sr. res. assoc. (tel. 312 + 962-8109). *Mineralogy, X-ray diffraction, mineral analysis, crystal structures.*

Steger, Dr Theodore Roosevelt. Corporate Res. and Dev., Monsanto Co., 800 N. Lingbergh Blvd., St. Louis, MO 63166, USA. (1946) PhD, physics (MIT, 1974). Techn. Group Leader. (tel. 413 + 788-6911, ext. 2571). *Physics of solids and liquids, polymer physics, light scattering, NMR, SAXS, SANS, physical properties.*

Stein, Prof. Richard Stephen. Polymer Res. Inst., U. of Massachusetts, Grad. Res. Center, Tower A, Amherst, MA 01002, USA. (1925) PhD, physical chemistry (Princeton U., 1949). Prof and dir. (tel. 413 + 545-2727). *Polymer texture, small angle X-ray scattering, crystalline polymers and blends, optical properties of polymers.*

Steinfink, Prof. Hugo. Materials Sci. & Eng., ETC 9.104, U. of Texas, Austin, TX 78712, USA. (1924) PhD, physical chemistry (PIB, 1954). Prof. (tel. 512 + 471-5233). *Inorganic structures, physical properties.*

Steinrauf, Prof. Larry King. Biochemistry Dept., Indiana U., Sch. of Medicine, 1100 West Michigan, Indianapolis, IN 46223, USA. (1931) PhD, biochemistry (U. Washington, 1957). Prof. (tel. 317 + 264-7544). *Membrane transport, heavy metal poisoning, microprocessor computers, athletic medicine.*

Steitz, Prof. Thomas Arthur. Molecular Biophysics & Biochem. Dept., Yale U., Box 1937 Yale Station, New Haven, CT 06520, USA. (1940) PhD, molecular biology and biochemistry (Harvard U., 1966). Prof. (tel. 203 + 436-8011). *Protein structure and function, enzyme activity (mechanism and control), protein - nucleic acid interaction.*

Stenkamp, Dr Ronald Eugene. Biological Structure, U. of Washington, Seattle, WA 98195, USA. (1948) PhD, chemistry (U. Washington, 1975). Res. asst. prof. (tel. 206 + 545-1721). *Structural studies, small molecules of biological interest, biological macromolecules.*

Stephens, Mr Anthony E. 1109 W. College, Sherman, TX 75090, USA.

Sterling, Prof. Clarence. Food Sci. and Techn., U. of California, Davis, CA 95616, USA. (1919) PhD, botany (U. California, Berkeley, 1944). Prof. (tel. 916 + 752-1486). *Plant morphology and anatomy, gel ultrastructure, electron microscopy, crystallography.*

Stevens, Prof. Edwin David. Dept. of Chemistry, U. of New Orleans, Lakefront, New Orleans, LA 70148, USA. (1947) PhD, chemistry (U. California, Davis, 1973). Assoc. prof. (tel. 504 + 286-6856). *Electron density distributions, accurate X-ray intensity measurements, synchrotron radiation, EXAFS.*

Stewart, Prof. James McDonald. Chemistry Dept., U. of Maryland, College Park, MD 20742, USA. (1931) PhD, physical chemistry (U. Washington, 1958). Prof. (tel. 301 + 454-4623, ext. H30). *Crystallographic software development, determination of crystal structures.*

Stewart, Dr Martin Van Buren. Dept. of Chemistry and Physics, Middle Tennessee State U., Box 123, Murfreesboro, TN 37132, USA. (1944) PhD, organic chemistry (U. Georgia, 1977). Asst. prof. of chemistry. (tel. 615 + 898-2949). *Stereochemistry, organic synthesis, physical organic chemistry, surface chemistry, spectroscopic and analytical methods, X-ray crystal structure analysis.*

Stewart, Prof. Robert Farrell. Chemistry Dept., Carnegie-Mellon U., 4400 Fifth Ave., Pittsburgh, PA 15213, USA. (1936) PhD, chemistry (CIT, 1963). Prof. (tel. 412 + 578-3165). *Electrostatic properties of molecules and crystals, accurate diffraction data, Rayleigh scattering by X-rays theory.*

Stock, Prof. Stuart R. Metallurgy Program, Georgia Inst. of Tech., Atlanta, GA 30332, USA. (1955) PhD, metallurgical eng. (U. Illinois, Urbana-Champaign, 1983). Asst. prof. (tel. 404 + 894-6882). *X-ray diffraction topography, stress/strain measurement, plastic deformation.*

Stollstorff, Mr Gregory R. ENRAF-NONIUS, 390 Central Ave., Bohemia, NY 11716, USA. (tel. 516 + 589-2885, telex 960250).

Stout, Prof. Charles David. Dept. of Crystallography, U. of Pittsburgh, 304 Thaw Hall, Pittsburgh, PA 15260, USA. (1947) PhD, biochemistry, crystallography (U. Wisconsin, 1976). Assoc. prof. (tel. 412 + 624-4366). *Protein crystallography, metalloproteins, metal clusters.*

Strahs, Dr Gerald. 130-09-230 St., Laurelton, NY 11413, USA. (1938) PhD, physical chemistry (U. Illinois, 1965). (tel. 212 + 525-0526).

Streib, Dr William E. Chemistry Dept., Indiana U., Bloomington, IN 47405, USA. (1931) PhD, physical chemistry (U. Minnesota, 1962). Res. crystallographer. (tel. 812 + 335-6604). *Instrumentation, organic and inorganic structures.*

Stroud, Prof. Robert Michael. Dept. of Biochemistry and Biophysics, U. of California at San Francisco, San Francisco, CA 94143, USA. (1942) PhD, crystallography (U. London, UK, 1968). Prof. (tel. 415 + 666-4224 or 3937). *Structure and function, biological macromolecules.*

Strouse, Prof. Charles Earl. Dept. of Chemistry and Biochemistry, U. of California, 405 Hilgard Ave., Los Angeles, CA 90024, USA. (1944) PhD, physical chemistry (U. Wisconsin, Madison, 1969). Prof. (tel. 213 + 825-1811). *Solid-state equilibria, porphyrin chemistry, single crystal spectroscopy, transitions.*

Stubbs, Dr Gerald James. Dept. of Molecular Biology, Vanderbilt U., Box 1820, Station B, Nashville, TN 37235, USA. (1947) PhD, biophysics (Oxford U., 1972). Asst. prof. (tel. 615 + 322-2018). *Macromolecular assemblies, fiber diffraction, protein, nucleic acid structure.*

Stults, Dr Bailey Ray. Physical Sci. Center, Monsanto Co., 800 N. Lindbergh, St. Louis, MO 63167, USA. (1948) PhD, inorganic chemistry (U. Nebraska, 1974). Assoc. fellow. (tel. 314 + 694-4820). *Small molecule crystallography, EXAFS, structure - function relations, catalysts (homogeneous and heterogeneous), protein crystallography,*

Sturcken, Dr Edward Francis. Analytical Dev. Div., E. I. du Pont de Nemours and Co., Savannah River Lab., Aiken, SC 29808-0001, USA. (1927) PhD, nuclear physics (St. Louis U., 1953). Res. assoc. (tel. 803 + 725-2790). *X-ray diffraction, materials science, electron microscopy, applied physics.*

Sturkey, Mr Lorenzo. 51 Los Cerros Place, Walnut Creek, CA 94598, USA. (1916) MS, physics (U. Kentucky, 1939). Retired. (tel. 415 + 939-4182). *Electron diffraction, magnesium metallurgy, intermetallic phases, dynamical theory of electron diffraction.*

Sturm, Prof. Edward. Dept. of Geology, Brooklyn C., Brooklyn, NY 11210, USA. PhD, geology (Rutgers U.) Prof. (tel. 718 + 780-5330). *Clay mineralogy.*

Suddath, Prof. Fred L. Dept. of Biochemistry, U. of Alabama Medical Center, SDB P.O. Box 13, Birmingham, AL 35294, USA. (1942) PhD, chemistry (Georgia Inst. of Techn., 1969). Assoc. prof. (tel. 205 + 934-2657). *Proteins, nucleic acids, crystallography; low temperature data, high resolution data, data collection methods.*

Suitch, Mr Paul R. 2692 Whitehurst Dr., Marietta, GA 30062, USA.

Sullenger, Dr Don Bruce. Mound Lab., Monsanto Res. Corp., Miamisburg, OH 45342, USA. (1929) PhD, physical chemistry (Cornell U., 1969). Res. specialist. (tel. 513 + 865-3665). *Applied crystallography; inorganic crystal chemistry.*

Sundaralingam, Prof. Muttaiya. Biochemistry Dept., U. of Wisconsin-Madison, 420 Henry Mall, Madison, WI 53706, USA. (1931) PhD, chemistry (U. Pittsburgh, 1961). Prof. (tel. 608 + 262-1448). *Crystallography, nucleic acids, proteins, structure - function relationships, metal ions biological role.*

Supper, Mr Lee R. Charles Supper Co., 15 Tech Circle, Natick, MA 01760, USA. (1938) AB, economics (Dickinson C., 1960). Pres. (tel. 617 + 655-4610). *Instrumentation, diffraction instrument manufacture, cameras and accessories.*

Sutton, Mr Paul W. Dept. of Chemistry, North Central C. Naperville, IL 60566, USA. (tel. 312 + 420-3491).

Swank, Prof. Duane Douglas. Chemistry Dept., Pacific Lutheran U., Tacoma, WA 98447, USA. (1942) PhD, physical chemistry (Montana State U., 1969). Chairman. (tel. 206 + 531-6900, ext. 304). *Solid state spectroscopy, magnetic phenomena in inorganic solids, inorganic crystal structures.*

Swartzendruber, Mr John K. Eli Lilly Co., Indianapolis, IN 46285, USA.

Sweet, Dr Robert M. Biology Dept., Brookhaven Nat. Lab., Upton, NY 11973, USA. (1943) PhD, physical chemistry (U. Wisconsin, Madison, 1970). Scient. (tel. 516 + 282-3401). *Macromolecular crystallography, synchrotron radiation.*

Swenson, Dr Dale Carl. Molecular Biophysics, Medical Foundation of Buffalo, 73 High St., Buffalo, NY 14203, USA. (1951) PhD, physical chemistry (U. Iowa, 1979). Postdoctoral res. assoc. (tel. 716 + 856-9600, ext. 454).

Swepston, Dr Paul Nathan. Molecular Structure Corp., 3304 Longmine Dr., College Station, TX 77840, USA. (1954) PhD, chemistry (U. Arkansas, 1981). Staff crystallographer. (tel. 409 + 693-9729). *Chemical crystallography, experimental bonding, electron density, inorganic chemistry.*

Swerdlow, Mr Max. 5704 Lenox Rd., Bethesda, MD 20817, USA. (1915) BA, physics (Brooklyn C., 1938). Program manager. (tel. 202 + 767-4933). *Microstructure, crystallography, electronic structure, optical properties of solids, superconductivity, magnetism, nonlinear optics.*

Swink, Dr Laurence Nim. Techn. Dev., Mult-Plate Co. Inc., 2362 Lu Field Rd., Dallas, TX 75229, USA. (1934) PhD, chemistry (Brown U., 1969). Techn. dir. (tel. 214 + 243-1557). *Applied crystallography.*

Swinnea, Dr John Steven. Dept. of Material Sci. and Eng., U. of Texas, ETC 9.114, Austin, TX 78712, USA. (1953) PhD, chemical engineering (U. Texas at Austin, 1981). Res. assoc. (tel. 512 + 471-3173). *Chemical, physical properties of crystals, solid state chemistry.*

Switendick, Dr Alfred Carl. Div. 1151, Sandia Labs., P.O.Box 5800, Albuquerque, NM 87185, USA. (1931) PhD, solid state physics (1963). Staff member, Solid State Theory Div. (tel. 505 + 846-2288). *Intermetallic compounds, interstitial compounds, hydrides, electronic properties and structure.*

Syed, Dr Ashfaquzzaman. Physical X-Ray Div., ENRAF-NONIUS, 390 Central Ave., Bohemia, NY 11716, USA. (1952) PhD, physics (U. Gorakhpur, India, 1979). Product specialist. (tel. 516 + 589-2885, ext. 19, telex 960250). *Organic and inorganic structures, instrumentation, system integration, accurate intensity measurements, electron density, crystallographic computing.*

Szebenyi, Ms Doletha M. 271 Clark Hall, Cornell U., Ithaca, NY 14853, USA.

Tai, Dr Douglas Leung-Tak. Radiology Dept., U. of Tennessee, Center for the Health Sci., 865 Jefferson Ave., Memphis, TN 38163, USA. (1940) PhD, physical chemistry (Cornell U., 1969). Assoc. prof. (tel. 901 + 528-7730). *Crystallography, high radiation intensity, high magnetic fields.*

Tainer, Dr John Arthur. Molecular Biology Dept., MB3, Res. Inst. of Scripps Clinic, 10666 N. Torrey Pines Rd., La Jolla, CA 92037, USA. (1951) PhD, biochemistry (Duke U., 1982). Asst. member. (tel. 619 + 455-9100, etx. 2525). *Macromolecular structure, function, and design, protein-protein interactions, molecular computer graphics.*

Takagi, Dr Shozo. Ada Health Foundation, Res. Unit, Nat. Bureau of Standards, Washington, DC 20234, USA. (1943) PhD, crystallography (U. Pittsburgh, 1971). Res. assoc. *Structure analysis, carbohydrates, X-ray and neutron diffraction.*

Takei, Dr William J. Res. Lab., Westinghouse Electric Corp., Churchill Borough, Pittsburgh, PA 15235, USA. (1931) PhD, chemistry (CIT, 1957). Chemistry fellow. (tel. 412 + 256-7631). *Defects, crystal growth, thin films.*

Takusagawa, Dr Fusao. Dept. of Chemistry, U. of Kansas, Malott Hall, Lawrence, KS 66045, USA. (1946) PhD, chemistry (Osaka City U., 1974). Dir., X-ray crystallography. (tel. 913 + 864-4727). *Proteins, nucleic acids, biological interesting small molecules.*

Tammaro, Mr David A. 181 Hamilton Circle, Painted Post, NY 14870, USA.

Tatsch, Dr Clinton Eugene. Res. Triangle Inst., Res. Triangle Park, NC 27709, USA. (1940) PhD, physical chemistry (U. Oklahoma, 1972). (tel. 919 + 541-6930). *Biological macromolecules, protein model compounds, protein sequence, structure - function relationships.*

Taub, Mr Haskell. Dept of Physics, 223 Physics Bldg., U. of Missouri-Columbia, Columbia, MO 65211, USA.

Tauber, Dr Arthur. US Army Electronics, 927 Woodgate Ave., Elbron, NJ 07740, USA. (1928) PhD, chemistry (PIB, 1972). Res. physical sci. *Solid State, magnetic materials, hydrides.*

Taylor, Ms Hope Cathlin. Ergoda Entroprises, Route 2, Box 142-C, Berkeley Springs, WV 25411, USA. Consultant, structural biochemistry. *Protein structure and function, semi-synthetic ribonuclease-S delta-crystallin.*

Taylor, Dr Ivan Fate, Jr. Computer Center, Texas Christian U., 3000 Bowis St., Fort Worth, TX 76129, USA. (1944) PhD, physical chemistry (U. South Carolina, 1971). Sr. user services consultant. (tel. 817 + 921-7695, ext. 6852). *Crystallographic software development and conversion, diffraction theory, direct methods.*

Taylor, Prof. Max A. Dept. of Chemistry, Bradley U., Peoria, IL 61625, USA.

Taylor, Prof. Robert Cooper. Chemistry Dept., U. of Michigan, Ann Arbor, MI 48109, USA. (1917) PhD, chemistry (Brown U., 1947). Prof. (tel. 313 + 764-7362). *Raman and infrared spectroscopy, vibrational spectra of crystals, force constants - molecular structure relationship.*

Teeter (Stein), Dr Martha Mary. Dept. of Chemistry, Boston U., 685 Commonwealth Ave., Boston, MA 02215, USA. (1944) PhD, inorganic chemistry (PSU, 1973). Instructor in Life Sci., Chemistry. (tel. 617 + 353-2490). *Protein crystallography, intermolecular interactions, proteins, nucleic acids.*

Teller, Dr Raymond. Res. Center, Standard Oil (Sohio), 4440 Warrensville Center Rd., Cleveland, OH 44128, USA. (1946) PhD, chemistry (U. Southern California, 1978). Project leader. (tel. 216 + 581-5953). *Chemistry, solid state chemistry, catalysis.*

Templeton, Prof. David Henry. Dept. of Chemistry, U. of California, Berkeley, CA 94720, USA. (1920) PhD, chemistry (U. California, 1947). Prof. (tel. 415 + 486-5615). *Crystal structure and chemistry; anomalous scattering of X-rays.*

Templeton, Dr Lieselotte K. Dept. of Chemistry, U. of California, Berkeley, CA 94720, USA. PhD, chemistry (U. California, Berkeley, 1950). Res. scient. (tel. 415 + 486-5615). *Anomalous scattering of X-rays, crystallographic computing.*

Ten Eyck, Dr Lynn Forest. Inst. of Molecular Biology, U. of Oregon, Eugene, OR 97403, USA. (1942) PhD, biochemical sciences (Princeton U., 1970). Res. assoc. (tel. 503 + 686-5151). *Macromolecular crystallography, molecular biology, enzyme catalysis mechanism.*

Tench, Dr Alan Howard. 2044 Walnut St., Boulder, CO 80302, USA. (1926) PhD, physical chemistry (U. Colorado, 1972). Self-employed. (tel. 303 + 444-1730). *Direct methods, nucleosides, pyrenes and adducts, humic compounds.*

Tennissen, Prof. Anthony C. Geology Dept., Lamar U., P. O. Box 10031, Beaumont, TX 77710, USA. (1920) PhD, geology-mineralogy (U. Missouri-Rolla, 1963). Prof. (tel. 713 + 838-7226). *Mineralogy, crystallography, metallic mineral deposits.*

Terwilliger, Mr Thomas C. 546 Elm St., El Cerrito, CA 94530, USA.

Teufer, Dr Gunter. P. O. Box 163, Chadds Ford, PA 19317, USA. (1922) Dr.rer.nat., chemistry (U. Hamburg, BRD, 1955). (tel. 215 + 459-0979). *Structural characterization, X-ray diffraction, electron diffraction.*

Thatcher, Mr Walter E. Central Res. Lab. 201-1E, 3M Co., PO Box 33221, St. Paul, MN 55133, USA.

Thathachari, Mr Y. T. Dept. of Materials Sci. and Eng., Stanford U., Stanford, CA 94305, USA. (tel. 415 + 666-2951).

Thielke, Mr Harry G. 212 Westhaven Rd., Greenville, NC 27834, USA. (1929) MSc, chemistry (U. Delaware, 1962). Retired. (tel. 919 + 756-8141). *X-ray powder diffraction.*

Thomas, Dr Joseph Mitchell, Jr. Communication Systems Div., GTE Government Systems Corp., 77 "A" St., Needham, MA 01719, USA. (1948) PhD, biophysics (Brandeis U., 1979). Manager, Advanced Techn. (tel. 617 + 449-2000, ext 3564, telex 922497). *Semiconductors, artificial intelligence, computer science, macromolecular structure and function.*

Thomas, Dr Kenneth A. Biochemistry Dept., Merck Inst., Room 80W-212, P. O. Box 2000. Rahway, NJ 07065, USA. (1946) PhD, biochemistry (Duke U., 1974). Sr. res. fellow. (tel. 201 + 574-7567). *Protein structure, protein folding, growth factors.*

Thomas, Dr Robert. Chemistry Dept., Brookhaven Nat. Lab., Upton, NY 11973, USA. (1934) PhD, physical chemistry (Boston U., 1965). Assoc. sci. (tel. 516 + 345-4381). *Instrumentation, parallel data collection, X-ray and neutron diffraction studies.*

Thompson, Prof. Doris M. Chemistry Dept., Austin C., P. O. Box 1561, Sherman, TX 75090, USA. Asst. prof. (tel. 214 + 892-9101, ext. 365). *X-ray crystal structure determination, metal complexes, sterically interesting substances.*

Thudium, Mr Richard N. 3113 Highland Dr., Cuyahoga Falls, OH 44224, USA. (1932) MS, chemical physics (Kent State U., 1969). Sr. res. physicist, Goodyear. (tel. 216 + 688-4544). *Polymer structures.*

Tiller, Prof. William Arthur. Dept. of Materials Sci. and Eng., Stanford U., Stanford, CA 94305, USA. (1929) PhD, physical metallurgy (U. Toronto, Canada, 1955). Prof. (tel. 415 + 497-3901). *Semiconductor processing.*

Tilley, Mr George P. Analytical Dept., Olin Corp., 350 Knotter Dr., Cheshire, CT 06410-0586, USA.

Tillinger, Dr Martin H. 422 Concord St., Cresskill, NJ 07626, USA. (1943) PhD, physics (PIB, 1976). Consultant. (tel. 201 + 567-1333). *Computing techniques, measurement.*

Titus, Prof. Donald Dean. Dept. of Chemistry, Temple U., Broad and Montgomery Sts., Philadelphia, PA 19122, USA. (1944) PhD, inorganic chemistry (CIT, 1971). Assoc. prof. (tel. 215 + 787-7127). *Organic selenium/tellurium compounds, charge transfer complexes.*

Todaro, Mr Louis J. Hoffmann-La Roche Inc., Nutley, NJ 07110, USA.

Toeplitz, Mrs Barbara Keeler. Analytical Res. and Dev., E. R. Squibb and Sons, P.O. Box 4000, Princeton, NJ 08540, USA. (1915) MS, statistics (Rutgers U., 1960). Retired. (tel. 609 + 921-4986). *X-ray crystallography, pharmaceutically interesting compounds.*

Toman, Prof. Karel. Dept. of Geology, Wright State U., Dayton, OH 45431, USA. (1924) DSc, solid state physics (Czechoslovak Acad. of Sci., Prague, 1965). Prof. (tel. 513 + 873-2726). *Crystal structures.*

Tomchick, Ms Diana R. Dept. of Chemistry, U. of Wisconsin-Madison, Madison, WI 53706, USA.

Tomlinson, Dr Gail Elizabeth. 5500 Friendship Blvd., 908-N, Chevy Chase, MD 20815, USA. (1952) MD, PhD, biochemistry. *Biological molecules, pharmacologically active small molecules.*

Torre, Prof. Louis Peter. Library of Sci. and Medicine, Rutgers U., Busch Campus, New Brunswick, NJ 08904, USA. (1941) PhD, physical chemistry (U. Washington, 1971). Asst. prof. (tel. 201 + 932-3526). *Information retrieval, data banks, computers.*

Townes, Mr William David Winn. Semiconductor Devices and Integrated Electronics, U.S. Army ERADCOM DELET-MJ, Fort Monmouth, NJ 07703, USA. (1919) MS, physics (PIB, 1955). Physical scient. (tel. 201 + 544-4840). *Crystalline electronic device reliability, electronic module standarization, X-ray diffraction apparatus, nuclear hardening of crystalline electronic devices.*

Towns, Prof. Robert Lee Roy. Chemistry Dept., Cleveland State U., 24th and Euclid, Cleveland, OH 44115, USA. (1940) PhD, physical chemistry (U. Texas, Austin, 1969). Prof. (tel. 216 + 687-2468). *Molecular structures, organic heterocycles, biologically interesting molecules; trace metal analysis in biological specimens, X-ray fluorescence analysis.*

Trefonas, Dr Louis Marco. Office of Sponsored Res., Div. of Grad. Studies, U. of Central Florida, Orlando, FL 32816, USA. (1931) PhD, physical chemistry (U. Minnesota, 1959). V.P., res. grad. dean. (tel. 305 + 275-2197). *Biologically interesting molecular structures, coordination compounds, proteins.*

Troup, Dr Jan Marshall. Molecular Structure Corp., 3304 Longmire Dr., College Station, TX 77840, USA. (1946) PhD, chemistry (Texas A & M U., 1974). Pres. (tel. 713 + 693-9729). *Inorganic structures, X-ray data collection techniques, electron density (experimental), powder diffraction structures.*

Trueblood, Prof. Kenneth N. Dept. of Chemistry, U. of California, Los Angeles, CA 90024, USA. (1920) PhD, chemistry (CIT, 1947). Prof. (tel. 213 + 825-1259). *Organic structures, molecular motion.*

Trus, Dr Benes Louis. DCRT, NIH Bldg. , Bethesda, MD 20205, USA. (1946) PhD, chemistry (CIT, 1972). Res. chemist. (tel. 301 + 496-2250, telex 248232 NIHUR). *Macromolecular structure, electron microscopy, image processing.*

Tsai, Prof. Chun-che. Dept. of Chemistry, Kent State U., Kent, OH 44242, USA. (1937) PhD, physical chemistry (Indiana U., 1968). Asst. prof. (tel. 216 + 672-2987). *Biological crystallography, drug - nucleic acid interactions; nucleic acid structure and function; molecular interactions, biological systems.*

Tsui, Ms Francis C. Molecular Biology Inst., UCLA, Los Angeles, CA 90024, USA.

Tulinsky, Prof. Alexander. Chemistry Dept., Michigan State U., East Lansing, MI 48824-1322, USA. (1928) PhD, physical chemistry (Princeton U., 1956). Prof. (tel. 517 + 353-4511). *Protein crystallography, structure and function, aldolase, prothrombin fragment 1, high resolution refinement.*

Tunell, Prof. George. Geological Sci. Dept., U. of California, Santa Barbara, CA 93106, USA. (1900) PhD, economic geology (Harvard U., 1930). Prof. em. (tel. 805 + 961-3471). *Geochemistry, mineralogy, geology (Hg - Sb - As).*

Tuomi, Dr Donald. 221 S. Illinois Dr., Arlington Heights, IL 60005, USA. (1920) PhD, physical chemistry (Ohio State U., 1952). (tel. 312 + 392-3003). *Structural chemistry, solid state physics, semiconductors, polymers.*

Turano, Mr August M. 1305 Macon Ave., Pittsburgh, PA 15218, USA.

Turley, Dr June Williams. Regulatory and Legislative Issues, The Dow Chemical Co., Building 1803, Midland, MI 48674, USA. (1929) PhD, biological chemistry

(PSU, 1957). Regulatory manager. (tel. 517 + 636-5443). *Structure - activity relationships (SAR), biologically active molecules, toxicology, risk assessment, data base management.*

Turner, Dr Ralph Waldo. Dept. of Chemistry, Florida A&M U., Tallahassee, FL 32307, USA. (1938) PhD, physical chemistry (U. Pittsburgh, 1965). Prof. (tel. 904 + 599-3639). *Structural chemistry, metal-ion organic complexes; proteins, enzymes.*

Uberbacher, Mr Edward C. 304 Orchard Knob Rd., Clinton, TN 37716, USA.

Ulmer, Mr Kevin M. Genex Corp., 16020 Industrial Dr., Gaithersburg, MD 20877, USA.

Urdy, Prof. Charles Eugene. Chemistry Dept., Huston-Tillotson C., 1820 East 8th St., Austin, TX 78723, USA. (1933) PhD, chemistry (U. Texas, Austin, 1962). Prof. (tel. 512 + 476-7421, ext. 305). *Crystal structure analysis.*

Valente, Dr Edward Joseph. Dept. of Chemistry, Mississippi C., Box 4065, Clinton, MS 39058, USA. (1949) PhD, chemistry (U. Washington). Prof. *Structure and conformation, bio-medicinal compounds.*

Van der Helm, Prof. Dick. Chemistry Dept., U. of Oklahoma, 620 Parrington Oval, Norman, OK 73069, USA. (1933) DSc, chemistry (U. Amsterdam, Netherlands, 1960). George Lynn Cross Res. Prof. (tel. 405 + 325-5831). *Molecular structure and conformation, natural products, siderochromes, peptides, peptide chelates, anticancer agents.*

Van der Veen, Prof. James Morris. Dept. of Chemistry and Chemical Eng., Stevens Inst. of Techn., Castle Point Station, Hoboken, NJ 07030, USA. (1931) PhD, chemistry (Harvard U., 1959). Assoc. prof. (tel. 201 + 792-2700, ext. 347). *Organic chemistry, reaction mechanisms, computer use in chemistry, X-ray crystallography.*

Van der Veer, Dr Donald G. Chemistry Dept., Georgia Inst. of Techn., Atlanta, GA 30332, USA. (1947) PhD, chemistry (Brown U., 1974). Res. scient. (tel. 404 + 894-4071). *Single crystal X-ray structure determination.*

Van Engen, Dr Donna. Analytical Div., Exxon Res. and Eng. Co., Route 22 East, Annandale, NJ 08801, USA. (1954) PhD, chemistry (Cornell U., 1981). Sr. chemist. (tel. 201 + 730-2107). *Natural products, organometallic complexes.*

Van Hove, Dr Michel Andre. Dept. of Chemistry, U. of California-Berkeley, Berkeley, CA 94720, USA. (1947) PhD, physics (Cambridge U., UK, 1974). Staff sr. scient. (tel. 415 + 486-6160). *Surface crystallography, (low-energy) electron diffraction.*

Van Nordstrand, Mr Robert Alexander. 520 Montecillo Rd., San Rafael, CA 94903, USA. (1917) MS, chemistry (U. Michigan, 1939). *Catalyst structures, alumina, clay minerals, X-ray absorption edges, zeolites.*

Van Opdenbosch, Dr Nicole Marie. TRIPOS Associates, 6548 Clayton Rd., St. Louis, MO 63117, USA. (1953) PhD, chemistry (U. Namur, Belgium, 1980). Manager, molecular modeling applications. (tel. 314 + 647-1099, telex 706910). *Molecular modeling, drug design, computer chemistry, computer graphics.*

Van Roey, Dr Patrick M. A. O. Molecular Biophysics Dept., Medical Foundation of Buffalo Inc., 73 High St., Buffalo, NY 14203, USA. (1952) PhD, chemistry (The U. of Calgary, 1979). Res. scient. (tel. 716 + 856-9600). *Protein structures, biological molecules.*

Vance, Dr T. Blake, Jr. Chemistry Dept., U. Connecticut, 34 Glenbrook Rd., Storrs, CT 06268, USA. (1950) PhD, inorganic chemistry (U. Wyoming, 1983). Postdoctoral fellow. (tel. 203 + 486-5111). *X-ray structural chemistry, coodination compounds.*

Vandenberg, Dr Joanna Maria. Physics Res., AT&T Bell Labs., 600 Mountain Ave., Murray Hill, NJ 07974, USA. (1938) PhD, inorganic chemistry (Leiden State U., Netherlands, 1964). Techn. staff member. (tel. 201 + 582-4186). *X-ray diffraction, thin film interfaces, epitaxial films.*

Vandlen, Dr Richard L. Biochemical Regulation, Merck Sharp and Dohme Res. Labs., P.O. Box 2000, Rahway, NJ 07065, USA. (1947) PhD, physical chemistry (Michigan State U.) Sr. res. biochemist. (tel. 201 + 574-7577). *Protein structure and function, hormone receptors, membrane proteins, membrane bound enzymes, lipid metabolism.*

Vanek, Ms Mary Ann. Signal UOP Res. Center, 50 UOP Plaza, Des Plaines, IL 60016, USA.

Veleker, Mr Thomas John. Chemical and Metallurgical Div., GTE Sylvania Inc., P. O. Box 70, Towanda, PA 18848, USA. (1925) MS, chemistry (Gonzaga U., 1951). Eng. mgr. (tel. 717 + 265-2121, ext. 323). *Materials characterization.*

Vergamini, Dr Phillip James. P-8, Los Alamos Scientific Labs., MS-805, Los Alamos, NM 87545, USA. (1943) PhD, chemistry (U. Wisconsin, Madison, 1971). Staff member. (tel. 505 + 667-3628). *X-ray and neutron diffraction; synthetic and structural inorganic and organometallic chemistry.*

Versic, Dr Ronald J. Ronald T. Dodge Co., P.O. Drawer J, Dayton, OH 45409, USA. (1942) PhD, materials engineering (Ohio State U., 1969). Executive Vice Pres. (tel. 513 + 298-8423). *Mineralogy, crystal growth, crystal chemistry.*

Via, Mr Grayson Hall. 740 Crescent Parkway, Westfield, NJ 07090, USA.

Vitali, Ms Jacqueline. Center for Crystallographic Res., Roswell Park Memorial Inst., 666 Elm St., Buffalo, NY 14263, USA.

Vlasse, Mr Marcus. Polytechnic Inst. of New York, 333 Jay St., Brooklyn, NY 11201, USA.

Voet, Prof. Donald Herman. Chemistry Dept., U. of Pennsylvania, Philadelphia, PA 19104, USA. (1938) PhD, chemistry (Harvard U., 1966). Prof. (tel. 215 + 898-6457). *Structres of biological macromolecules, proteins, nucleic acids.*

Vold, Mr Carl L. Material Sci. and Techn. Div., Code 6320, Naval Res. Lab., Washington, DC 20375, USA. (1932) MS, metallurgy (Iowa State U., 1959). Solid state physicist. (tel. 202 + 767-2440). *Lattice dynamics, anisotropic elasticity, elastic constants.*

Von Dreele, Prof. Robert Bruce. Dept. of Chemistry, Arizona State U., Tempe, AZ 85281, USA. (1943) PhD, chemistry (Cornell U., 1971). Assoc. prof. (tel. 602 + 965-5849, 7694). *Inorganic chemical crystallography, neutron diffraction, solid state chemistry.*

Vranka, Mr Robert G. 11 Norfolk Rd., Arlington, MA 02174, USA.

Vyas, Dr (Mrs) Meenakshi Nandkishore. Biochemistry Dept., Rice U., P. O. Box 1892, Houston, TX 77251, USA. (1949) PhD, X-ray crystallography (IIT, Bombay, 1976) Assoc. *X-ray crystallography.*

Vyas, Mr. Nand Kishore. Biochemistry Dept., Rice U., P. O. Box 1892, Houston, TX 77251, USA. (1949) MSc, physics (Udaipur U., 1971) Assoc. *X-ray crystallography.*

Waerstad, Mr Kjell R. Fundamental Res., Tennessee Valley Authority, Muscle Shoals, AL 35660, USA. (tel. 205 + 383-4631).

Wagner, Prof. Christian Nikolaus Johann. Materials Dept., U. of California, S. of Eng. and Appl. Sci., 6531-Boelter Hall, 405 Hilgard Ave., Los Angeles, CA 90024, USA. (1927) Dr.rer.nat, physical metallurgy (U. Saar, BRD, 1957). Prof. (tel. 213 + 825-6265). *Amorphous and liquid alloy structures; plastic deformation, alloys, biomaterials, thin films.*

Waldrop, Ms Lyneve. 13 Elizabeth Lane, Saratoga Springs, NY 12866, USA.

Walker, Dr Christopher Bland. 22 Baskin Rd., Lexington, MA 02173, USA. (1925) PhD, physics (MIT, 1951). Retired. (tel. 617 + 862-6943). *Metal physics, imperfections in crystals.*

Walker, Mr Donald L. 93 Main St. West Mall, Andover, MA 01810, USA.

Wallace, Mr John C. Physics Dept., Roswell Park Memorial Inst., 666 Elm St., Buffalo, NY 14263, USA.

Wallace, Mr Peter Lindsay. P. L. Wallace, LLNL (L-362), P. O. Box 808, Livermore, CA 94550, USA. (1937) BS, metallurgical engineering (U. Arizona, 1959). Metallurgist. (tel. 415 + 423-1679). *Applied crystallography, X-ray spectrometry, actinide metallurgy.*

Wallace, Mr B. A. Dept. of Biochemistry, Columbia U., 630 West 168th St., New York, NY 10032, USA.

Walter, Mr Norman Macmillan. 25 Waterview Rd., Westchester, PA 19380, USA. (1923) MS, physics (PSU, 1954). *Structure - properties relationship, residual stresses in materials, microprobe analysis of materials.*

Wan, Dr Che'ng. Geological Sci. Dept., U. of Washington, Seattle, WA 98195, USA. (1935) PhD, chemistry (McMaster U., Canada, 1970). Res. scient. (tel. 206 + 543-7378). *Inorganic crystal structures.*

Wang, Dr Andrew Hwei-Jiung. Biology Dept., Room 16-743, Massachusetts Inst. of Techn., 77 Massachusetts Ave., Cambridge, MA 02139, USA. (1945) PhD, chemistry (U. Illinois, 1974). Res. assoc. *X-ray crystallography, biological macromolecules, nucleic acid - protein interactions.*

Wang, Dr Bi-Cheng. Biocrystallography Lab., VA Medical Center, P.O. Box 12055, Pittsburgh, PA 15240, USA. (1938) PhD, chemistry (U. Arkansas, 1968). Res. chem., Prof. (tel. 412 + 683-3000, ext. 517). *Macromolecular structure and function, diffraction methods.*

Wang, Mr Frederick E. Innovative Techn. International Inc., 10747-3 Tucker St., Beltsville, MD 20705, USA. (1932) PhD, physical chemistry (Syracuse U., 1960).

Wang, Dr Po-Wen. Res. and Dev. Dept., Magnex Corp., 6850 Santa Teresa Blvd., San Jose, CA 95119, USA. (1943) PhD, physics (PINY, 1979). Staff eng. (tel. 408 + 281-1000, ext. 205). *X-ray physics, magnetic recording heads, powder diffractometry.*

Wang, Dr Rong. Materials Dept., Battelle North West, Richland, WA 99352, USA. (1939) PhD, materials (U. Texas, Austin, 1967). Sr. res. scient. (tel. 509 + 376-4717). *Amorphous and microcrystalline phases, corrosion of metal and alloys, electrochemistry of interfaces, high-rate sputter deposition of materials.*

Ward, Dr Donald Leslie. Chemistry Dept., Michigan State U., East Lansing, MI 48824-1322, USA. (1943) PhD, chemistry (Montana State U., 1972). Dept. crystallographer. (tel. 517 + 353-4511). *Crystal structure determination, computer programming.*

Ward, Dr Keith Bolen. Lab. for the Structure of Matter, Code 6030, Naval Res. Lab., Washington, DC 20375-5000, USA. (1943) PhD, biophysics (Johns Hopkins U., 1974). Res. biophysicist. (tel. 202 + 767-2735). *Protein crystallography.*

Wardle, Dr Ronald. Basic Chemicals, Basic Inc., P. O. Box 392, Bettsville, OH 44815, USA. (1931) PhD, solid state science (PSU, 1972). Techn. dir. (tel. 419 + 986-5126, ext. 37). *Basic refractories, clay mineral structures, structural chemistry (general).*

Warren, Prof. Bertram Eugene. Dept. of Physics, Room 6-110, Massachusetts Inst. of Techn., Massachusetts Ave., Cambridge, MA 02139, USA. (1902) ScD, physics (MIT, 1929). Prof. em. (tel. 617 + 253-4851). *X-ray diffraction, amorphous materials, imperfections, order-disorder.*

Warren, Dr Stephen George. Geophysics Dept., AK-50, U. of Washington, Seattle, WA 98195, USA. (1945) PhD, physical chemistry (Harvard U., 1973). Asst. prof. (tel. 206 + 543-7230). *Glaciology.*

Waser, Prof. Jurg. 6120 Terryhill Dr., La Jolla, CA 92037, USA. (1916) PhD, chemistry (CIT, 1944). Retired from CIT. (tel. 714 + 454-5622). *Crystal structure.*

Watenpaugh, Prof. Keith Donald. Physical and Analytical Chemistry Div., The Upjohn Co., 301 Henrietta St., Kalamazoo, MI 49001, USA. (1939) PhD, chemistry (Montana State U., 1967). Sr. res. scient. (tel. 616 + 385-5481). *Macromolecular crystallography, molecular interactions, computing methods.*

Watkins, Prof. Steven F. Dept. of Chemistry, Louisiana State U., Baton Rouge, LA 70803, USA. (1940) PhD, physical chemistry (U. Wisconsin-Madison, 1967). Assoc. prof. (tel. 504 + 388-1525). *Inorganic and organometallic chemistry, natural products, self consistent field calculations, computing techniques.*

Watson, Prof. William Harold. Dept. of Chemistry, Texas Christian U., TCU Station, Fort Worth, TX 76129, USA. (1931) PhD, chemistry (Rice U., 1957). Prof. (tel. 817 + 926-2461, ext. 464). *Biologically active natural product structures, transition metal complexes exhibiting magnetic exchange.*

Watt, Mr William. Upjohn Co., 7255-209-I, 301 Henrietta St., Kalamazoo, MI 49001, USA.

Watts, Mrs Ethel Jean. Materials Sci. Lab., The Aerospace Corp., P. O. Box 92957, Los Angeles, CA 90009, USA. (1928) MS, physics (PSU, 1952). Techn. staff member. (tel. 213 + 648-6927). *Crystallography, carbon and its polymorphs; failure analysis in solid state devices; electron diffraction, lattice imaging by electron microscopy.*

Weaver, Dr Larry H. Inst. of Molecular Biology, U. of Oregon, Eugene, OR 97403, USA. (1942) PhD, physics (U. Oregon, 1975). Res. assoc. *Protein crystallography.*

Webb, Dr Lawrence Edward. 203 Marvey, Grayslake, IL 60030, USA. (1936) PhD, physical chemistry (U. Chicago, 1965). *Protein crystallography, clinical chemistry, peptide structures.*

Weber, Mr Charles C. 191 West St., Essex Junction, VT 05452, USA.

Weber, Dr Irene Teresa. Chemical Physics Div., Nat. Bureau of Standards, Gaithersburg, MD 20899, USA. (1953) PhD, molecular biology (Oxford, UK, 1978). Res. assoc. (tel. 301 + 921-2785). *Protein crystallography, protein-nucleic acid interaction, enzyme structure and function.*

Weber, Dr Lawrence D. Pall, Corp., 30 Seacliff Ave., Glen Cove, NY 11542, USA. (1950) PhD, physical chemistry (Michigan State U., 1978). Staff scient. (tel. 516 + 671-4000, ext. 388). *Structure and physical chemistry of biopolymers, filter media-process fluid interaction.*

Weber, Dr Patricia. Protein Eng. Dept., Genex Corp., 16020 Industrial Dr., Gaithersburg, MD 20877, USA. (1952) PhD, chemistry (U. Arizona, 1979). Sr. res. scient. (tel. 301 + 258-0552). *Protein structure, electron transfer, heme proteins, protein-nucleic acid interactions.*

Wechsler, Dr Berry Andrew. Optical Circuits Dept., Hughes Res. Labs., 3011 Malibu Canyon Rd., Malibu, CA 90265, USA. (1951) PhD, geochemistry (SUNY at Stony Brook, 1981). Member techn. staff. (tel. 213 + 317-5639). *Crystal chemistry, crystal growth, phase transformations, optical materials, ferrocis.*

Weeks, Dr Charles M. Molecular Biophysics Dept., Medical Foundation of Buffalo, 73 High Street, Buffalo, NY 14203, USA. (1944) PhD, biophysics (SUNY Buffalo, 1970). Assoc. res. scient. (tel. 716 + 856-9600, ext. 462). *Biologically interesting compounds, direct methods, crystallographic computing.*

Weertman, Ms Julia R. Materials Sci. Dept., Northwestern U., Evanston, IL 60201, USA.

Wehrenberg, Prof. John Patteson. Geology Dept., U. of Montana, Science Complex, Missoula, MT 59801, USA. (1927) PhD, geology (U. Illinois, 1956). Prof. (tel. 406 + 243-2341 or 4851). *Mineralogy, crystallography, infrared spectroscopy of solids.*

Wei, Dr Chin Hsuan. Biology Div., Oak Ridge Nat. Lab., P. O. Box Y, Oak Ridge, TN 37830, USA. (1926) PhD, chemistry (U. Wisconsin, 1962). Biophysicist. (tel. 615 + 574-1253). *Biological structures.*

Weininger, Dr Marc Sterling. Biological Sci. Dept., Purdue U., Lilly Hall, West Lafayette, IN 47901, USA. (1943) PhD, chemistry (U. South Carolina, 1972). Asst. res. scient. (tel. 317 + 494-1099). *Biological nitrogen fixation, enzyme structure and function, metalloenzymes.*

Weiser, Mr Calvin H. 140-20 Debs Place, Bronx, NY 10475, USA. (1932) MS, physics (PIB, 1960). *Rapid scanning spectroscopy, ultra-violet spectroscopy, physics of hot gases, surface physics.*

Weiss, Dr Richard J. 4 Lawson St., Avon, MA 02322, USA. (1923) PhD, physics (New York U., 1950). *Electron distributions, polymers, thermodynamics; X-ray physics, Compton scattering.*

Weissman, Dr Larry. Molecular Biology Inst., UCLA, 405 Hilgard Ave., Los Angeles, CA 90024, USA. (1947) PhD, biochemistry (UCLA, 1979). Res. assoc. (tel. 213 + 825-8901). *Macromolecular crystallography.*

Weissmann, Prof. Sigmund. Mechanics and Materials Sci. Dept., C. of Eng., Rutgers U., Piscataway, NJ 08854, USA. (1917) PhD, physical chemistry (PIB, 1952). Prof. (tel. 201 + 932-3664). *Lattice defects, defects - physical properties relationship, X-ray topography, dynamical diffraction theory, physical metallurgy, structure.*

Wells, Prof. Alexander Frank. Dept. of Chemistry, U. of Connecticut, Storrs, CT 06268, USA. (1912) ScD, X-ray crystallography (Cambridge U., UK, 1956). Prof. (tel. 203 + 486-3223). *Structural inorganic chemistry, topology and geometry of 3D systems.*

Wenk, Prof. Hans-Rudolf. Dept. of Geology and Geophysics, U. of California, Berkeley, CA 94720, USA. (1941) PhD, mineralogy, crystallography (U. Zurich, Switzerland, 1965). Prof. (tel. 415 + 642-7431). *Crystal chemistry, silicates, electron microscopy, structural geology, metamorphic rocks.*

Wertz, Mr David L. Dept. of Chemistry, U. of Southern Mississippi, Box 5043, Hattiesburg, MS 39401, USA.

Westbrook, Dr Edwin M. Div. of Biological and Medical Res., Argonne Nat. Lab., 9700 South Cass Ave., Argonne, IL 60439, USA. (1948) PhD, MD, biophysics (U. Chicago, 1981). Asst. scient. (tel. 312 + 972-3983). *Protein structure and function, steriod-protein interaction, membrane-protein interaction, synchrotron radiation.*

Wester, Dr Dennis Wayne. Mallinckrodt Inc., 675 McDonnell Blvd., St. Louis, MO 63042, USA. (1949) PhD, chemistry (U. Florida, 1975). Res. chemist. (tel. 314 + 895-2300). *Radiochemistry, inorganic chemistry, technetium chemistry.*

Weston, Dr Norman Ernest. Micron Inc., 3815 Lancaster Pike, Wilmington, DE 19805, USA. (1930) PhD, physical chemistry (MIT, 1957). Vice pres. (tel. 302 + 772-3823). *Optical and X-ray microscopy, electron diffraction, electron probe analysis.*

Wheeler, Mr George L. Dept. of Chemistry, U. of New Haven, 300 Orange Ave., West Haven, CT 06516, USA.

White, Prof. Joe L. Agronomy Dept., Purdue U., Life Sci. Building, West Lafayette, IN 47907, USA. (1921) PhD, soil chemistry (U. Wisconsin, 1947). Prof. (tel. 317 + 749-2891). *Structure - spectra - composition relation, micas and chlorites; structure - reactivity relation, aluminum hydroxide gels.*

White, Prof. John Greville. Chemistry Dept., Fordham U., E. Fordham Rd., Bronx, NY 10458, USA. (1922) PhD, chemistry (Glasgow U., 1947). Prof. (tel. 212 + 933-2233, ext. 350). *Crystal structures, organic and inorganic compounds.*

Whitlow, Mr Marc David. Dept. of Chemistry, Boston U., 675 Commonwealth Ave., Boston, MA 02215, USA. (1956) BS, chemistry (U. Washington, 1978) Grad. student. (tel. 617 + 353-2478). *Protein structure and function, protein predictions.*

Whitney, Dr John Franklin. Central Res. and Dev. Dept., E. I. du Pont de Nemours and Co., Experimental Station, Bldg. 356, Wilmington, DE 19898, USA. (1916) PhD, physical chemistry (Cornell U., 1945). Res. chemist. (tel. 302 + 772-4173). *Crystal structure analysis, powder diffraction analysis.*

Whittle, Mr Robert. 101 Mitch Ave., State College, PA 16801, USA.

Wickham, Dr Donald Guy. Manufacturing Dept., Ampex Computer Products Div., 200 N. Nash St., El Segundo, CA 90245, USA. (1922) PhD, inorganic chemistry (MIT, 1954). Manager, memory core dev. (tel. 213 + 416-1255). *Inorganic chemistry, crystal chemistry, magnetic oxides.*

Wiesner, Dr Joel Robert. Intel Corp., 3065 Bowers Ave., Santa Clara, CA 95051, USA. Mailing address: 179 Arcadia Ave., Santa Clara, CA 95051, USA. (1938) PhD, chemistry (U. Washington, 1966). Sr. project manager. (tel. 408 + 496-8891). *Crystal growth, chemical vapor deposition.*

Wiff, Dr Donald Ray. Res. Inst., U. of Dayton, 300 College Park, Dayton, OH 45469, USA. (1936) PhD, theoretical physics (Texas A & M U., 1967). Sr. res. polymer physicist. (tel. 513 + 258-8378). *Polymers, molecular composites, liquid crystalline polymers, structure - properties relationship, theoretical predictions, electrically conducting polymers.*

Wignall, George Denis. 102 West Melbourne Rd., Oak Ridge, TN 37830, USA.

Wilchinsky, Dr Zigmond Walter. Elastomers Techn. Div., Exxon Chemical Co., Mailing address: 301 S. Wood Ave., Linden, NJ 07036, USA. (1915) PhD, physics (MIT, 1942). Sr. res. assoc. (tel. 201 + 474-2037). *Polymers, structural characteristics and physical behavior, plastics and elastomers, crystalline and molecular orientation, X-ray diffraction, birefringence, crack propagation in rubbers, adsorptive separations, molecular sieves.*

Wildman, Dr Gary C. Materials & Devices Res., Schering Plough, 3030 Jackson Ave, Memphis, TN, 38151, USA. (1942) PhD, physical chemistry (Duke U., 1969). Pres. (tel. 901 + 320-5053). *Polymer physics, polymer morphology, polymer chain structure, material science.*

Wiley, Prof. Don Craig. Dept. of Biochemistry & Molecular Biology, Harvard U., 12 Oxford St., Cambridge, MA 02138, USA. (1944) PhD, biophysics (Harvard U., 1971). Prof. (tel. 617 + 495-4090). *Influenza virus membrane glycoproteins; X-ray diffraction; membrane surface antigens.*

Wilkes, Dr Glenn Richard. Industrial Lab., Eastman Kodak Co., 1669 Lake Ave. Kodak Park, Rochester, NY 14650, USA. (1937) PhD, inorganic-physical chemistry (U. Wisconsin, 1965). Group leader, X-ray diffraction, X-ray fluorescence. (tel. 716 + 722-2610). *Multi-element analyses, X-ray fluorescence methods, energy dispersive and wavelength dispersive systems, X-ray diffraction techniques, powder and single crystals.*

Wilkinson, Dr Michael Kennerly. Solid State Div., Oak Ridge Nat. Lab., P.O. Box X, Oak Ridge, TN 37831, USA. (1921) PhD, physics (MIT, 1950). Div. dir. (tel. 615 + 574-6151). *Neutron scattering from condensed matter, magnetic properties of solids, lattice dynamics, low temperature physics.*

Willett, Prof. Roger DuWayne. Chemistry Dept., Washington State U., Pullmann, WA 99164, USA. (1936) PhD, chemistry and physics (Iowa State U., 1962). Prof. and chairman, chemical physics program. (tel. 509 + 335-3925). *Phase transition dynamics; magnetic properties, transition metal salts; crystal structure studies; EPR, NMR, Mössbauer studies.*

Williams, Roger M. Jet Propulsion, M/C 122-123, 4000 Oak Grove Dr., Pasadena, CA 91109, USA.

Williams, Prof. Donald Elmer. Chemistry Dept., U. of Louisville, Louisville, KY 40292, USA. (1930) PhD, chemistry (Iowa State U., 1964). Prof. (tel. 502 + 588-6798). *Organic crystal structures, molecular packing, non-bonded interactions.*

Williams, Dr Graheme John Bramald. ENRAF-NONIUS, 390 Central Ave., Bohemia, NY 11716, USA. (1942) PhD, biochemistry (U. Alberta, Canada, 1972). Dir. of marketing. (tel. 516 + 589-2885). *Crystallography methodology.*

Williams, Dr Jack Marvin. Div. of Chemistry, Bldg. 200-A113, Argonne Nat. Lab., 9700 S. Cass Ave., Argonne, IL 60439, USA. (1938) PhD, physical-inorganic chemistry (Washington State U., 1966). Sr. chemist. (tel. 312 + 972-3464). *Neutron diffraction, X-ray diffraction, inorganic and organometallic structures; pulsed neutron source, instrumentation.*

Williams, Dr John A. American ACMI, 300 Stillwater Rd., Stamford, CT 06902, USA. (1940) PhD, ceramic science (Penn. State U., 1970). Sr. scient. (tel. 203 + 357-8300). *Glass structure, small angle X-ray scattering.*

Williams, Prof. Rickey J. Chemistry Dept., Midwestern U., 3400 Taft, Wichita Falls, TX 76308, USA. (1942) PhD, physical chemistry (Texas Christian U., 1968). Head, Depts. Chemistry, Physics and Geology. (tel. 817 + 692-6611, ext. 251). *Inorganic crystal structures.*

Williams, Dr Robin O'Dare. Metals and Ceramics Div., Oak Ridge Nat. Lab., P.O.Box X, Oak Ridge, TN 37830, USA. (1927) PhD, metallurgy (Carnegie - Mellon U., 1955). Res. metallurgist. (tel. 615 + 576-2631). *Diffuse X-ray scattering, data reduction, deformation, phase transformation.*

Williamson, Mr Robert S. Physics Dept., Queens C., Flushing, NY 11367, USA.

Williard, Prof. Paul Gregory. Dept. of Chemistry, Brown U., Box H, Providence, RI 02912, USA. (1950) PhD, chemistry (Columbia U., 1976). Asst. prof. (tel. 401 + 863-3589). *Organic chemistry, marine natural products chemistry, small molecule X-ray crystallography.*

Wilsdorf, Prof. Heinz Gerhard Friedrich. Dept. of Materials Sci., U. of Virginia, Thornton Hall, Charlottesville, VA 22901, USA. (1917) PhD, metallurgy (Göttingen U., BRD, 1947). Wills Johnson prof. (tel. 804 + 924-3462). *Electron diffraction, diffraction contrast, electron microscopy.*

Wilson, Dr Frank Charles. Polymer Products Dept., E. I. du Pont de Nemours and Co., Bldg. 323, Experimental Station, Wilmington, DE 19898, USA. (1927) PhD, physical chemistry (M.I.T., 1957). Res. assoc. (tel. 302 + 772-2734). *Polymers, small angle X-ray scattering.*

Wilson, Mr David K. P.O. Box 2671, Lovett C., Rice U., Houston, TX 77252, USA.

Winchell, Prof. Horace. Dept. of Geology and Geophysics, Kline Geology Lab., 210 Whitney Ave., P. O. Box 6666, New Haven, CT 06511, USA. (1915) PhD, mineralogy and petrography (Harvard U., 1941). Prof. em. (tel. 203 + 436-1072). *Optical mineralogy, optical crystallography, systematics.*

Wing, Prof. Richard M. Chemistry Dept., U. of California, Riverside, CA 92502, USA. (tel. 714 + 787-3520).

Wingert, Mrs Lavinia Meinzer. Dept. of Crystallography, U. of Pittsburgh, 304 Thaw Hall, Pittsburgh, PA 15260, USA. (1943) MA, molecular biology (Dartmouth C., 1968). Grad. student. *Carbohydrates, hydrogen-bonding, NMR.*

Winslow, Mr Douglas N. Sch. of Civil Eng., Purdue U., Lafayette, IN 47907, USA.

Winter, Prof. William Thomas. Chemistry Dept., Polytechnic Inst. of New York, 333 Jay St., Brooklyn, NY 11201, USA. (1944) PhD, physical chemistry (SUNY C. of Env. Sci. & Forestry, 1974). Assoc. prof. (tel. 718 + 643-2992). *X-ray fiber diffraction, computer modelling, biopolymers, polysaccharides.*

Wismer, Prof. Robert Kingsley. Chemistry Dept., Millersville U., Millersville, PA 17551, USA. (1945) PhD, physical chemistry (Iowa State U., 1972). Assoc. prof. and chairman. (tel. 717 + 872-3411). *Crystal structures of small molecules, methods of structure solution.*

Wittels, Dr Mark Caesar. Office Basic Energy Sci., Div. Material Sci., U.S. Dept. of Energy, Washington, DC 20545, USA. (1921) PhD, geology (MIT, 1951). Chief, solid state and materials chem. branch, materials sci. progs. (tel. 301 + 353-3426). *X-ray instrumentation, defects in solids, structural transformations, radiation damage effects, topographic investigations.*

Witters, Prof. Robert Dale. Dept. of Chemistry and Geochemistry, Colorado S. of Mines, Golden, CO 80401, USA. (1929) PhD, physical chemistry (Montana State U., 1964). Prof. (tel. 303 + 273-3632). *Molecular structure determination, X-ray diffraction.*

Wlodawer, Dr Alexander. Center for Chemical Physics, Nat. Bureau of Standards, Gaithersburg, MD 20899, USA. (1946) PhD, molecular biology (UCLA, 1974). Physicist. (tel. 301 + 921-2785). *Protein crystallography, synchrotron radiation, neutron diffraction.*

Wold, Prof. Aaron. Chemistry Dept., Brown U., Barus and Holley Building, Room 610, Providence, RI 02912, USA. (1927) PhD, solid state inorganic chemistry (PIB, 1952). Prof. (tel. 401 + 863-2857). *Solid state chemistry, growth, structure - properties relationship, chalcopyrite oxide crystals, platinum metal chalcogenides.*

Wolff, Dr Gunther Arthur. G.A. Consultants, NPO, 3776 Northampton Rd., Cleveland Heights, OH 44121, USA. (1918) ScD, physical chemistry (Techn. U. of Berlin, BRD, 1948). Consultant. (tel. 216 + 381-7284). *Crystal growth, crystal face (and shape) morphology, structure and bonding, point defects and dislocations, surface and bulk kinetics, thermodynamics, crystal physics and chemistry, solid state science, materials engineering, device applications, semiconduction, gallium phosphide and arsenide, electroluminescence (visable and infrared).*

Wolper, J. Brook. Membership Dept., AIP, 500 Sunnyside Blvd., Woodbury, NY 11797, USA.

Won, Mr Vann Yuen. Physical Sci. Lab., SMALC, Sacramento, CA 95652, USA. (1925) BS, chemistry (National Chekiang U., China, 1946). Physical scient. (tel. 916 + 392-3336). *Electron microprobe, X-ray diffraction, X-ray spectroscopy.*

Wong, Mrs Rosalind Y. U.S. Dept. of Agriculture, Western Regional Res. Center, Chemical and Structural Analysis Res. Unit, 800 Buchanan St., Albany, CA 94710, USA. (1940) BS, chemistry (San Jose State C., 1965). Res. chemist. (tel. 415 + 486-3359). *Molecular structure, biological functions, agriculturally interesting compounds.*

Wong-Ng, Dr Winnie Kwai-Wah. Nat. Bureau of Standards, Inorganic Material Div. - JCPDS, Gaithersburg, MD 20899, USA. (1947) PhD, inorganic chemistry (Louisiana State U., 1974). X-ray crystallographer. (tel. 301 + 921-2921). *Powder diffraction, data base evaluation, crystal structures, twinning, disorder, intermolecular forces, electron density.*

Wood, Dr Elizabeth Armstrong. 17 Alston Ct., Red Bank, NJ 07701, USA. (1912) PhD, geology (Bryn Mawr C., 1939). Retired from Bell Labs. (tel. 201 + 747-2863). *Ferroelectrics, epitaxy, teaching.*

Wood, Prof. John Karl. Physics Dept., Utah State U., Physics 41, Logan, UT 84322, USA. (1919) PhD, physics (PSU, 1946). Prof. (tel. 801 + 752-4100, ext. 7768). *Low energy X-ray spectroscopy, electron spectroscopy.*

Wood, Prof. John Stanley. Chemistry Dept., U. of Massachusetts, Amherst, MA 01003, USA. (1936) PhD, physical chemistry, crystallography (U. Manchester, UK, 1962). Prof. (tel. 413 + 545-2375). *Structures, charge density distributions, small inorganic systems, coordination compounds; electronic structure and geometry, EPR, magnetic and spectroscopic methods, molecular orbital calculations.*

Wood, Prof. Mical Kent. 237 Montrose Ave., Salt Lake City, UT 84101, USA. (1940) PhD, chemistry (U. Texas, 1969). *Protein and organic crystal structures.*

Woode, Dr Kwamena Annan. Chemistry Dept., Hampton U., East Queens St., Hampton, VA 23668, USA. (1947) PhD, crystallography (U. London, UK, 1978). Asst. prof. (tel. 804 + 727-5276). *Determination of the structures of biologically important molecules.*

Workman, Mr Samuel Thomas. Adv. Res. and Appl. Corp. (ARACoR), 1223 E. Arques Ave., Sunnyvale, CA 94086, USA. (1926) BS, physics (St. Josephs C., 1950). Vice-pres. (tel. 408 + 733-7780). *X-ray Instrumentation.*

Worthington, Mr C. R. Mellon Inst., Carnegie-Mellon U., 4400 Fifth Ave., Pittsburgh, PA 15213, USA.

Wright, Dr Christine Gerda Schubert. Dept. of Biochemistry, Medical C. of Virginia, Virginia Commonwealth U., Box 614, MCV Station, Richmond, VA 23298, USA. (1940) PhD, chemistry (U. California, San Diego, 1969). Res. asst. prof. (tel. 804 + 786-6139). *Protein structure, lectins.*

Wright, Dr Harlan Tonie. Biochemistry Dept., MCV/VCU, Box 614, MCV Station, Richmond, VA 23298, USA. (1941) PhD, chemistry (U. California, San Diego, 1968). Asst. prof. (tel. 804 + 786-6139). *Protein, nucleic acid structure.*

Wright, Dr William Vaughn. Computer Sci. Dept., U. of North Carolina at Chapel Hill, IBM Communication Products Div., 104 Campbell Lane, Chapel Hill, NC 27514, USA. (1931) PhD, computer science (U. North Carolina - C.H., 1972). Adjunct assoc. prof. and sr. eng. (tel 919 + 942-1144). *Interactive computer graphics, molecular graphics, computational models, molecular structures, crystallographic computing.*

Wu, Ms Ellen. Mobil Res. and Dev. Corp., Paulsboro, NJ 08066, USA. (tel. 609 + 423-1040).

Wu, Mr Hsin-I. Industrial Eng. Dept., Biosystems Res. Div., Texas A & M U., College Station, TX 77843, USA. (tel. 713 + 845-5531).

Wuensch, Prof. Bernhardt John. Dept. of Materials Sci. and Eng., Massachusetts Inst. of Techn., Room 13-4037, 77 Massachusetts Ave., Cambridge, MA 02139, USA. (1933) PhD, crystallography (MIT, 1963). Prof. of ceramics. (tel. 617 + 253-6889). *Point defects and diffusion; inorganic crystal chemistry, fast ion conductors, sulfides and sulfosalts.*

Wyckoff, Prof. Harold Winfield. Molecular Biophysics & Biochem. Dept., Yale U., Box 1937, Yale Station, New Haven, CT 06520, USA. (1926) PhD, biophysics (MIT, 1955). Assoc. prof. (tel. 203 + 436-0436). *Protein structure and function, instrumentation.*

Wyckoff, Prof. Ralph W. G. 4741 E. Cherry Hills Dr., Tucson, AZ 85718, USA. (1897) PhD, chemistry (Cornell U., 1919). Prof. em. tel. 602 + 299-3073). *Crystal structures, biological macromolecular crystals.*

Xuong, Prof. Nguyen-Huu. Physics, Chemistry and Biology, U. of California at San Diego, P. O. Box 109, LaJolla, CA 92093, USA. (1933) PhD, physics (U. California-Berkeley, 1962). Prof. (tel. 714 + 452-2501). *Protein molecular structures, high speed data collection systems.*

Yakel, Dr Harry Leonard. Metals and Ceramics Div., Oak Ridge Nat. Lab., P. O. Box X, Oak Ridge, TN 37830, USA. (1929) PhD, chemistry (CIT, 1952). Sr. res. staff member. (tel. 615 + 576-7392). *Structures and transformations, alloys and inorganic compounds, synchrotron radiation.*

Yalkovsky, Dr Ralph. Nat. Assoc. of Sci. Writers, P. O. Box 398, Grand Island, NY 14072, USA. (1917) PhD, geology (U. Chicago, 1956). Free lance sci. writer. (tel. 716 + 878-6732). *Marine geology, mineralogy, international law of the sea, public policy.*

Yanai, Dr Hideyasu Steve. Res. Div., Rohm and Haas Co., P.O. Box 219, Bristol, PA 19007, USA. (1928) PhD, physical chemistry (U. Minnesota, 1958). Analytical res. project leader. (tel. 215 + 788-5501, ext. 607). *Polymer crystal structure.*

Yang, Dr. Daniel Shun-Chung. Dept. of Crystallography, U. of Pittsburgh, 304 Thaw Hall, Pittsburgh, PA 15260, USA. (1954) PhD, crystallography (U. Pittsburgh, 1983). Res. assoc. (tel. 412 + 683-3000, ext. 517). *Protein structure and function, fourier methods in protein crystallography.*

Yankov, Henry F. 92 E. Weatogue St., Simsbury, CT 06070, USA.

Yates, Dr John Harry. Dept. of Chemistry, U. of Pennsylvania, Philadelphia, PA 19104, USA. (1948) PhD, physical chemistry (Ohio State U., 1976). Dir., computer facility. (tel. 215 + 898-4714). *Computer code development, molecular properties - experimental and theoretical.*

Yazici, Mr Rahmi M. Material and Metallurgical Eng. Dept., Stevens Inst. of Techn., Castle Point, , Hoboken, NJ 07030, USA. (1949) PhD, materials science (Rutgers U., 1982). Asst. prof. (tel. 201 + 420-5261). *Internal strain analysis (micro and macro) in ceramics and metals, thin films, coatings, general crystallography.*

Yeh, Prof. Hun Chiang. Metallurgical Eng. Dept., Cleveland State U., 24th at Euclid Ave., Cleveland, OH 44115, USA. (1935) PhD, metallurgical engineering (Illinois Inst. of Techn., 1966). Prof. (tel. 216 + 687-3502). *Phase identification, mechanical properties of solids, high temperature ceramics processing, X-ray diffraction in materials, silicon nitride research.*

Yelon, Dr William B. U. of Missouri, Res. Reactor, Columbia, MO 65211, USA. (1944) PhD, physics (Carnegie Melon U., 1970). Group leader and prof. of physics. (tel. 314 + 882-4211). *Neutron scattering, X-ray diffraction, powder diffraction, magnetic ordering, charge density extinction.*

Yoon, Dr Hyo Sub. Center for Biomedical Eng., Rensselaer Polytechnic Inst., Troy, NY 12180-3590, USA. (1935) PhD, solid state sci. (PSU, 1971). Sr. res. assoc. (tel. 518 + 270-6547). *Biomedical ultrasonics, crystal physics, biomaterials, biomechanics.*

Young, Dr Frederick Walter, Jr. Solid State Div., Oak Ridge Nat. Lab., P.O.Box X, Oak Ridge, TN 37830, USA. (1924) PhD, chemistry (U. Virginia, 1950). Assoc. div. dir. (tel. 615 + 574-5501). *Highly perfect crystals, dynamical diffraction, X-ray topography.*

Young, Prof. R. A. Sch. of Physics, Eng. Experiment Station, Georgia Inst. of Techn., Atlanta, GA 30332, USA. (1921) PhD, physics (PIB, 1959). Prof. (tel. 404 + 894-5208, telex 542507). *Crystal physics, atomic-scale mechanisms, diffraction theory diffraction applications.*

Young, Dr Ti-Sheng. Div. of Biology, Argonne Nat. Lab., 9700 S. Cass Ave., Argonne, IL 60439, USA. (1952) PhD, biochemistry (Duke U., 1984). Postdoctoral fellow. (tel. 312 + 972-3887). *Protein X-ray crystallography.*

Ysern, Mr Xavier. Biophysics Dept., Johns Hopkins U. Medical S., 725 Wolfe St., Baltimore, MD 21205, USA.

Zacharias, Dr David Edward. Molecular Structure Dept., Inst. for Cancer Res., 7701 Burholme Ave., Philadelphia, PA 19111, USA. (1926) PhD, X-ray crystallography (U. Pittsburgh, 1969). Sr. res. assoc. (tel. 215 + 728-2220). *Organic molecular structure determination.*

Zalkin, Dr Allan. Materials and Molecular Res. Div., Lawrence Berkeley Lab., U. of California, Berkeley, CA 94720, USA. (1926) PhD, chemistry (U. California-Berkeley, 1951). Sr. scient. (tel. 415 + 486-5762). *Inorganic and organometallic structures, actinide chemistry - structures, crystallographic computation, programming.*

Zaslow, Prof. Bert. Dept. of Chemistry, Arizona State U., Tempe, AZ 85287, USA. (1924) PhD, chemistry (Iowa State U., 1956). Prof. (tel. 602 + 965-3685). *Polymer crystallography, symmetry.*

Zassenhaus, Prof. Hans Julius. Mathematics Dept., Ohio State U., Mailing address: 942 Spring Grove Lane, Worthington, OH 43085, USA. (1912) Dr habil., mathematics (Hamburg U., Germany, 1938) Prof. em. (tel. 614 + 422-6295). *Mathematics, mathematical crystallography.*

Ziolo, Dr Ronald F. Webster Res. Center, Xerox Corp. - 0114, 800 Phillips Rd., Webster, NY 14580, USA. (1944) PhD, inorganic chemistry (Temple U., 1970). Sr. res. scient. (tel. 716 + 422-3341). *Inorganic and solid state chemistry, structure - properties relationship, electrical properties, optical properties, magnetic properties.*

Zolensky, Dr Michael Ewing. SN2/NASA, Johnson Space Center, Houston, TX 77058, USA. (1955) PhD, geochemistry, mineralogy (PSU, 1983). Space scient. (tel. 814 + 863-1665). *Mineralogy, meteorites, powder diffraction.*

Zoltai, Prof. Tibor Z. Dept. of Geology and Geophysics, U. of Minnesota, Minneapolis, MN 55455, USA. (1925) PhD, mineralogy and crystallography (MIT, 1959). Prof. (tel. 612 + 373-4025). *Crystallography, crystal chemistry, minerals, crystal structure systematics for minerals.*

Zwell, Mr Leo. JCPDS, International Centre for Diffraction Data, 117 S. Chester Rd., Swarthmore, PA 19081, USA. (1915) BS, physics (Brooklyn C., 1934). Consultant. (tel. (JCPDS) 215 + 328-9400, (home) 215 + 328-0617). *Powder diffraction, X-ray metallography.*

URUGUAY

Sub-Editor: R. A. Mariezcurrena

Notes

1. International telephone country code - 598.

2. Degrees conferred by the University of Uruguay include *Doctor en Química Farmacéutica* (Dr.Q.F.).

Mariezcurrena, Dr Raul Alfredo. Facultad de Química, Gral. Flores 2124, Montevideo, Uruguay. (1939) Dr Q.F., chemistry (U. Mayor de la República, 1975). Assoc. prof., lect. in physics. (tel. 206736). *Organic crystal structures.*

Mercatini, Dr Giovanna. Facultad de Ingenieria y Agrimensura, Inst. de Física, Herrera y Reissig 565, Montevideo, Uruguay. (1923) Dr, natural sci. (U. di Firenze, Italy, 1949). Res. asst. (tel. 403405). *X-ray diffraction applications.*

Gomes, Mr Osvaldo. Crystallography Lab., Facultad de Química, Gral. Flores 2124, Montevideo, Uruguay. (1954) Chem. Eng. (Fac. de Ingeniería). Res. asst. (tel. 206736) *Organic crystal structures, crystal growth.*

VENEZUELA

Sub-Editor: E.E. Rodulfo de Gil

Notes

1. International telephone country code - 58.

Arnstein, Dr Gustavo. Facultad de Ciencias, Escuela de Química, U. Central de Venezuela, Ciudad Universitaria, Caracas 105, Venezuela. (1942) PhD, engineering materials (Maryland U., USA, 1972). Prof. (tel. 02 + 619811, ext. 3006). *X-ray topography, crystal defects.*

Capparelli, Prof. Mario Vicente. Facultad de Ciencias, Escuela de Química, U. Central de Venezuela, Ciudad Universitaria, Caracas 105, Venezuela. (1937) Lic., chemistry (U. de Buenas Aires, Argentina, 1965). Assoc. prof. (tel. 02 + 619811, ext. 2736). *Crystal structures, computer programming.*

Chornik, Dr Boris. Dept. de Física, Facultad de Ciencias, INMETAL, U. Simón Bolívar, Casilla 80659, Caracas, Venezuela. (1941) PhD, physics (U. California., USA, 1970). Prof. and res. scient. (tel. 627279). *Surface physics, magnetism, metals and alloys.*

Delgado Quiñones, Lic. Miguel. Dept. de Química, Facultad de Ciencias, U. de Los Andes, Mérida, Venezuela. (1956) Lic., chemistry (U. de Los Andes, 1980). Asst. prof (tel. 074-526244, ext. 514). *Crystal structure determinations, solid state physics, defects in crystals.*

Dunia, Dr Emery. Dept. de Física, Facultad de Ciencias (Antigua ETI), U. Central de Venezuela, Los Chaguaramos, Caracas, Venezuela. DSc, physics (U. de Paris 6, 1980). Prof. agregado. (tel. 02 + 619811, ext. 2770). *X-ray topography, crystal defects.*

Gomez Caraballo, Lic. Dora Maria. Dept. de Química, Facultad de Ciencias, U. de Los Andes, Mérida, Venezuela. (1955) Lic., chemistry (U. de Los Andes, 1980). Asst. prof (tel. 074-526244, ext. 514). *Crystal structure determinations.*

Khan, Prof. Ali. Dept. de Química, U. de Oriente, Cerro Colorado, Cumaná, Venezuela. (1936) PhD, high temperature and solid state chemistry (Rensselaer Poly. Inst., USA, 1968). Assoc. prof. (tel. 093 + 2294-96, ext. 462). *Crystal growth, crystallography, thermodynamics, electro-optics.*

Laredo, Dr Estrella. Physics Dept., U. Simón Bolívar, P.O. Box 5354, Caracas 108, Venezuela. (1940) Dr (3rd cycle), crystallography (U. de Paris, France, 1965). Prof. (tel. 9621101, ext. 338). *Solid state physics, defects in crystals, lattice distortion, small angle X-ray scattering, ionic crystals, I.T.C., ionic conductivity.*

Mateu, Dr Leonardo. Molecular Structure Laboratory, C.B.B., Inst. Venezolano de Investigaciones Científicas, Apdo. 1827, Caracas 101, Venezuela. (1939) ScD, molecular biology (U. de Paris, France, 1974). Investigator. (tel. 691941, ext. 454). *Membrane structure, lipoprotein structure, small angle X-ray scattering, small angle neutron scattering.*

Padrón. Dr Raúl. Lab. de Biofísica del Músculo, Centro de Biofísica y Bioquímica, Inst. Venezolano de Investigaciones Científicas (IVIC), Apdo. 1827, Caracas 1010A, Venezuela. (1950) PhSc, biology - physiology, biophysics (Inst. Venezolano de Investigaciones Científicas, 1980). Investigator. (tel. 749543, ext. 217, telex 21338). *Molecular biology, biophysics, physiology.*

Rivera Ocando, Dr Angela Valentina. Dept. de Química, Facultad de Ciencias, U. de Los Andes, Mérida, Venezuela. (1945) PhD, chemistry (Cambridge U., UK, 1979). Assoc. prof (tel. 074-49368). *Crystal structure determinations.*

Rodulfo de Gil, Prof. Eldrys Emilia. Dept. de Química, Facultad de Ciencias, U. de Los Andes, Mérida, Venezuela. (1936) PhD, chemistry (Wisconsin U., USA, 1968). Prof. (tel. 33439, ext. 5). *Organometallic chemistry.*

Ysern, Dr Xavier. Dept. de Química, U. Simón Bolívar, Sartanejas, Venezuela. (1947) PhD, chemistry (John Hopkins U., USA, 1977). Prof. agregado. (tel. 9621301, ext. 8123). *Proteins, structure determination.*

YUGOSLAVIA

Sub-Editor: K. Kranjc

Notes

1. International telephone country code - 0038.

2. The following degrees conferred in Yugoslavia: (a) by universities: *Doctor of Physical Sciences, Doctor of Chemical Sciences*, etc. (in the list all abbreviated as Dr) (approximately equivalent to PhD); (b) by universities and some faculties: postgraduate degrees *Master of Physical Sciences, Master of Chemical Sciences*, etc. (all abbreviated as Magistar) (approximately equivalent to MSc); (c) by Faculties of Sciences and various Technical Faculties: diplomas in physics, chemistry, physical chemistry, geology, etc. (abbreviated as *Dipl. fiz., Dipl. kem.*, etc. or *Dipl. inž. fiz., Dipl. inž. kem.*, etc. and *Dipl. inž.*) (approximately equivalent to BSc or BA).

3. At the universities persons with an academic training can be appointed in various positions ranging from *redovni profesor* (Prof.) (equivalent to Professor) via *izvanredni profesor* (Izv. prof.), *docent* to *asistent* (approximately equivalent to reader, lecturer and res. assoc. or demonstrator, respectively). The equivalent positions in the scientific institutes are: *scientific adviser, senior scientific associate, scientific associate* and *research associate* or *postgraduate student*.

Alujević - Stipanov, Dr Višnja. Farmaceutsko - biokemijski fakultet, Sveučilište u Zagrebu, ul. Ante Kovačića 1, 41000 Zagreb, Yugoslavia. (1948) Dr, chemistry (Zagreb U., 1981). Asistent. (tel. 041 - 445311, ext. 36). *X-ray diffraction analysis, polycrystalline systems.*

Arhar, Magistar Andrej. VTOZD Kemija in kem. tehnologija, FNT, Univerza E. Kardelja v Ljubljani, Murnikova 6, P.O.Box 537, 61001 Ljubljana, Yugoslavia. (1942) Magistar, chemistry (Ljubljana U., 1975). Sr. res. assoc. (tel. 061 - 214444). *Inorganic chemistry.*

Babič, Dr Danilo. Rudarsko-geološki fakultet, Univerzitet u Beogradu, Đušina 7, 11000 Beograd, Yugoslavia. (1948) Dr, mineralogy (Beograd U., 1984). Docent. (tel. 011 - 180111, ext. 710). *Crystal growth, crystal physics, mineral genesis, surface energy.*

Balić Žunić, Dr Tonči. Mineraloško-petrografski zavod, Prirodoslovno - matematički fakultet, Sveučilište u Zagrebu, Demetrova ul. 1, 41000 Zagreb, Yugoslavia. (1952) Dr, geology (Zagreb U., 1984). Asistent. (tel. 041 - 445315). *Mineral structures.*

Balzar, Mr Davor. Metalurški fakultet, Sisak, Institut za metalurgiju, Aleja narodnih heroja 1, 44000 Sisak, Yugoslavia. (1957) Dipl. inž. fiz., physics (Zagreb U., 1981). Res. assoc. (tel. 044 - 32865). *Metal physics.*

Ban, Prof. Dr Zvonimir. Zavod za opću i anorgansku kemiju, Prirodoslovno - matematički fakultet, Sveučilište u Zagrebu, ul. Soc. revolucije 8, 41000 Zagreb, Yugoslavia. (1934) Dr, chemistry (Zagreb U., 1963). Prof. (tel. 041 - 414079). *Crystal structure, metals and alloys.*

Bezjak, Prof. Dr Aleksandar. Farmaceutsko - biokemijski fakultet, Sveučilište u Zagrebu, ul. Ante Kovačića 1, 41000 Zagreb, Yugoslavia. (1928) Dr, chemistry (Zagreb U., 1964). Prof. (tel. 041 - 446061). *X-ray diffraction analysis, polycrystalline systems.*

Blažina, Dr Želimir. Institut 'Ruđer Bošković', Bijenička c. 54, P.O.Box 1016, 41001 Zagreb, Yugoslavia. (1946) Dr, chemistry (Zagreb U., 1979). Scient. assoc. (tel. 041 - 435111, ext. 482). *Intermetallic compounds.*

Bonefačić, Prof. Dr Antun. Institut za fiziku Sveučilišta, Bijenička c. 46, P.O.Box 304, 41001 Zagreb, Yugoslavia. (1925) Dr, physics (Zagreb U., 1963). Prof. (tel. 041 - 271211). *Metal physics.*

Brenčič, Prof. Dr Jurij. VTOZD Kemija in kem. tehnologija, FNT, Univerza E. Kardelja v Ljubljani, Murnikova 6, P.O.Box 537, 61001 Ljubljana, Yugoslavia. (1940) Dr, chemistry (Ljubljana U., 1969). Prof. (tel. 061 - 214444). *Coordination compounds (Chromium - Molybdenum - Tungsten).*

Bruvo, Magistar Milenko. Zavod za opću i anorgansku kemiju, Prirodoslovno - matematički fakultet, Sveučilište u Zagrebu, ul. Soc. revolucije 8, 41000 Zagreb, Yugoslavia. (1934) Magistar, chemistry (Zagreb U., 1973). Asistent. (tel. 041 - 416023). *Inorganic crystal structures.*

Bukovec, Prof. Dr Peter. VTOZD Kemija in kem. tehnologija, FNT, Univerza E. Kardelja v Ljubljani, Murnikova 6, P.O.Box 537, 61001 Ljubljana, Yugoslavia. (1946) Dr, chemistry (Ljubljana U., 1972). Prof. (tel. 061 - 214444). *Inorganic chemistry.*

Bukovec, Dr Nataša. VTOZD Kemija in kem. tehnologija, FNT, Univerza E. Kardelja v Ljubljani, Murnikova 6, P.O.Box 537, 61001 Ljubljana, Yugoslavia. (1946) Dr, chemistry (Ljubljana U., 1978). Docent. (tel. 061 - 214444). *Inorganic chemistry.*

Bulc, Mrs Nada. VTOZD Kemija in kem. tehnologija, FNT, Univerza E. Kardelja v Ljubljani, Murnikova 6, P.O.Box 537, 61001 Ljubljana, Yugoslavia. (1933) Dipl. inž., chemistry (Ljubljana U., 1962). Sr. res. assoc. (tel. 061 - 214444). *Inorganic crystal structures.*

Colombo, Dr Lidija. Institut 'Ruđer Bošković', Bijenička c. 54, P.O. Box 1016, 41001 Zagreb, Yugoslavia. (1922) Dr, physics (Zagreb U., 1961). Scient. adviser. (tel. 041-435111, ext. 239). *Molecular dynamics, crystal dynamics.*

Čeh, Magistar Boris, VTOZD Kemija in kem. tehnologija, FNT, Univerza E. Kardelja v Ljubljani, Murnikova 6, P.O. Box 537, 61001 Ljubljana, Yugoslavia. (1951) Magistar, chemistry (Ljubljana U., 1978). Asistent. (tel. 061 - 214444). *Synthesis and characterization, complex tungsten compounds.*

Cvetković, Mr Ljubiša. Institut 'Mihajlo Pupin', OOUR 'Kristali', Volgina ul. 15, 11000 Beograd, Yugoslavia. (1931) Dipl. inž., geology (Beograd U., 1961). Adviser in OOUR 'Kristali'. (tel. 011 - 771373). *Piezoelectric crystals.*

Cvetković, Mrs Miroslava. Istraživačko - razvojni institut, Elektronska industrija, Batajnički drum 23, 11080 Zemun Polje, Yugoslavia. (1935) Dipl. inž. min., geology (Beograd U., 1961). Res. assoc. (tel. 011 - 696423). *X-ray diffraction analysis, polycrystalline materials, ceramics.*

Demšar, Magistar Alojz. VTOZD Kemija in kem. tehnologija, FNT, Univerza E. Kardelja v Ljubljani, Murnikova 6, P.O.Box 537, 61001 Ljubljana, Yugoslavia. (1955) Magistar, chemistry (Ljubljana U., 1979). Asistent. (tel. 061 - 214444). *Inorganic crystal structures.*

Despotović, Mr Zlatko. RO Chromos Centar za kemijska istraživanja i razvoj, Žitnjak bb, 41000 Zagreb, Yugoslavia. (1934) Dipl. inž., chemistry (Zagreb U., 1958). Sr. res. assoc. (tel. 041 - 229800, ext. 395). *Thermal properties, materials.*

Dimitrijević, Magistar Radovan. Rudarsko - geološki fakultet, Univerzitet u Beogradu, Đušina ul. 7, 11000 Beograd, Yugoslavia. (1947) Magistar, mineralogy (Beograd U., 1978). Asistent. (tel. 011 - 180111, ext. 701). *Inorganic crystal structures, computer programming.*

Divjaković, Prof. Dr Vladimir. Institut za fiziku Prirodno - matematičkog fakulteta, ul. Ilije Đuričića 4, 21000 Novi Sad, Yugoslavia. (1946) Dr, crystallography (Bern U., Switzerland, 1976). Izv. prof. (tel. 021 - 55318). *Inorganic crystal structures.*

Djordjević, Prof. Dr Slobodan. Metalurški fakultet, ul. Matije Gupca 1, 72000 Zenica, Yugoslavia. (1929) Dr, metallurgy (Sarajevo U., 1976). Izv. prof. (tel. 072 - 21831). *Metals, structure.*

Djurić, Dr Stevan. Rudarsko - geološki fakultet, Univerzitet u Beogradu, Đušina ul. 7, 11000 Beograd, Yugoslavia. (1931) Dr, geology (Beograd U., 1980). Docent. (tel. 011 - 180111). *Instrumentation, clay minerals, material science.*

Drašner, Magistar Antun. Institut 'Ruđer Bošković', Bijenička c.54, P.O. Box 1016, 41001 Zagreb, Yugoslavia. (1955) Magistar, chemistry (Zagreb U., 1984). Res. assoc. (tel. 041 - 435111). *Intermetallic compounds.*

Dubček, Mr Pavo. Fizički zavod, Prirodoslovno-matematički fakultet, Sveučilište u Zagrebu, Marulićev trg 19, 41000 Zagreb, Yugoslavia. (1960) Dipl. inž. fiz., physics (Zagreb U., 1983). Jr. res. assoc. (tel. 041 - 446242). *Metal physics, small-angle X-ray scattering.*

Duževič, Dr Davor. 'Rade Končar' - Elektrotehnički institut, Baštijanova b.b., 41000 Zagreb, Yugoslavia. (1936) Dr, physics (Zagreb U., 1979). Scient. assoc. (tel. 041 - 561022, ext. 1157). *Metal physics.*

Gabela, Prof. Dr Fikret. Medicinski fakultet, Univerzitet Sarajevo, ul. Moše Pijade 5, 71000 Sarajevo, Yugoslavia. (1936) Dr, physics (Sarajevo U., 1977). Izv. prof. (tel. 071 - 38244, ext. 53). *Inorganic crystal structures.*

Galešić, Dr Nikola. Institut 'Ruđer Bošković'. Bijenička c. 54, P.O.Box 1016, 41001 Zagreb, Yugoslavia. (1937) Dr, chemistry (Zagreb U., 1971). Scient. assoc. (tel. 041 - 435111, ext. 335). *Crystal structures.*

Girt, Prof. Dr Egvin. Prirodno - matematički fakultet, Vojvode Putnika 43, 71000 Sarajevo, Yugoslavia. (1936) Dr, physics (Zagreb U., 1977). Prof. (tel. 071 - 646755). *Metal physics.*

Gladić, Mr Jadranko. Institut za fiziku Sveučilišta, Bijenička c. 46, P.O.Box 304, 41000 Zagreb, Yugoslavia. (1959) Dipl. inž. fiz., physics (Zagreb U., 1983). Jr. res. assoc. (tel. 041 - 271211, ext. 307). *Superionic conductors, structure.*

Golič, Prof. Dr Ljubo. VTOZD Kemija in kem. tehnologija, FNT, Univerza E. Kardelja v Ljubljani, Murnikova 6, P.O.Box 537, 61001 Ljubljana, Yugoslavia. (1932) Dr, chemistry (Ljubljana U., 1965). Prof. (tel. 061 - 214444). *Inorganic crystal structures, hydrogen bond structures, oxalates and tiooxalates.*

Grdenić, Prof. Dr Drago. Zavod za opću i anorgansku kemiju, Prirodoslovno - matematički fakultet, Sveučilište u Zagrebu, ul. Soc. revolucije 8, 41000 Zagreb, Yugoslavia. (1919) Dr, chemistry (Zagreb U., 1951). Prof.retired. (tel. 041 - 416023). *Inorganic and organometallic crystal structures.*

Gržeta, Dr Biserka. Institut 'Ruđer Bošković', Bijenička c. 54, P.O. Box 1016, 41001 Zagreb, Yugoslavia. (1949) Dr, physics (Zagreb U., 1980). Scient. assoc. (tel. 041 - 435111, ext. 320). *X-ray diffraction, polycrystalline systems.*

Gspan, Dr Primož. Zavod SRS za varstvo pri delu, Bohoričeva 22 A, 61000 Ljubljana, Yugoslavia. (1934) Dr, physics (Zagreb U., 1980). Head, Dept. of Ecology and Toxicology. (tel. 061 - 320853). *Occupational safety, environment protection.*

Herak, Dr Rajna. Institut za nuklearne nauke 'Boris Kidrič' - Vinča, Laboratorija za fiziku čvrstog stanja i radijacionu hemiju, P.O.Box 522, 11000 Beograd, Yugoslavia. (1930) Dr, technical sciences (Beograd U., 1969). Scient. adviser. (tel. 011 - 458222, ext. 417). *Crystal structures, inorganic and coordination compounds.*

Herceg, Dr Marija. Institut 'Ruđer Bošković', Bijenička c. 54, P.O.Box 1016, 41001 Zagreb, Yugoslavia. (1938) Dr, chemistry (Zagreb U., 1970). Scient. assoc. (tel. 041 - 435111, ext. 335). *Crystal structures, biologically interesting synthetic compounds.*

Hergold-Brundić, Dr Antonija. Zavod za opću i anorgansku kemiju, Prirodoslovno - matematički fakultet, Sveučilište u Zagrebu, ul. Soc. revolucije 8, 41000 Zagreb, Yugoslavia. (1942) Dr, chemistry (Zagreb U., 1980). Asistent. (tel. 041 - 416023, ext. 8005). *Inorganic crystal structures.*

Janjić, Prof. Dr Svetislav. Institut za bakar-Bor, ul. AVNOJ-a 33, 19210 Bor, Yugoslavia. (1931) Dr, geology (Beograd U., 1979). Prof., scient. adviser in Copper Inst. (tel. 030 - 32299, ext. 122). *Inorganic crystal structures.*

Jelenić-Bezjak, Dr Ivanka. Savska c. 104, 41000 Zagreb, Yugoslavia. (1937) Dr, chemistry (Zagreb U., 1970). Retired. (tel. 041 - 517094). *Silicate structures.*

Jenček, Mr Ladislav August. Institut za materiale, Zavod za raziskavo materiala in konstrukcij, Dimičeva 12, 61000 Ljubljana, Yugoslavia. (1918) Dipl. inž., electronics (Techn. U. of Praha, Czechoslovakia, 1943). Res. adviser. (tel. 061 - 344061). *Natural and synthetic fibres.*

Jordanovska, Dr Vera. Hemiski institut, Prirodno-matematički fakultet, Univerzitet 'Kiril i Metodij', Arhimedova 5, 91000 Skopje, Yugoslavia. (1938) Dr, chemistry (Ljubljana U., 1982). Docent. (tel. 091 - 221033). *Inorganic crystal structures.*

Jovanovski, Dr Gligor. Hemiski institut, Prirodno-matematički fakultet, Univerzitet 'Kiril i Metodij', Arhimedova 5, 91000 Skopje, Yugoslavia. (1945) Dr, chemistry (Zagreb U., 1981). Docent. (tel. 091 - 221033). *Crystal structures.*

Jurković, Dr Ivan. Rudarsko - geološko-naftni fakultet, Sveučilište u Zagrebu, Pierottijeva ul 6, 41000 Zagreb, Yugoslavia. (1917) Dr, geology (Zagreb U., 1956). Prof. (tel. 041 - 414694). *Crystal optics, mineral genesis.*

Kaitner, Dr Branko. Zavod za opću i anorgansku kemiju, Prirodoslovno - matematički fakultet, Sveučilište u Zagrebu, ul. Soc. revolucije 8, 41000 Zagreb, Yugoslavia. (1942) Dr, chemistry (Zagreb U., 1979). Asistent. (tel. 041 - 416023, ext. 8005). *Crystal structures.*

Kamenar, Prof. Dr Boris. Zavod za opću i anorgansku kemiju, Prirodoslovno - matematički fakultet, Sveučilište u Zagrebu, ul. Soc. revolucije 8, 41000 Zagreb, Yugoslavia. (1929) Dr, chemistry (Zagreb U., 1960). Prof. (tel. 041 - 416023). *Inorganic crystal structures.*

Kapor-Nahlovski, Dr Agneš. Institut za fiziku Prirodno - matematičkog fakulteta, ul. Ilije Đuričića 4, 21000 Novi Sad, Yugoslavia. (1950) Dr, physics (Novi Sad U., 1981). Docent. (tel. 021 - 55318). *Organic crystal structures.*

Karanović, Magistar Ljiljana. Rudarsko - geološki fakultet, Univerzitet u Beogradu, Đušina ul. 7, 11000 Beograd, Yugoslavia. (1950) Magistar, mineralogy (Beograd U., 1978). Res. assoc. (tel. 011 - 180111, ext. 701). *Organic and inorganic crystal structures.*

Kaučič, Dr Venčeslav. ISKRA IEZE -TOZD FERITI, Stegne 19, 61000 Ljubljana, Yugoslavia. (1950) Dr, chemistry (Ljubljana U., 1977). Docent. (tel. 061 - 571159). *Inorganic crystal structures.*

Kerenović, Magistar Midhat. Pedagoška akademija, ul. Mirka Višnjića 1, 78000 Banja Luka, Yugoslavia. (1930) Magistar, physics (Zagreb U., 1973). Prof. of physics. (tel. 078 - 35625). *Metal physics.*

Kirin, Dr (Mrs) Ankica. Medicinski fakultet, Sveučilište u Zagrebu, Šalata 3b, 41000 Zagreb, Yugoslavia. (1929) Dr, physics (Zagreb U., 1973). Docent. (tel. 041 - 271188, ext. 379). *Solid state physics.*

Kojić-Prodić, Dr Biserka. Institut 'Ruđer Bošković', Bijenička c. 54, P.O.Box 1016, 41001 Zagreb, Yugoslavia. (1938) Dr, chemistry (Zagreb U., 1968). Scient. adviser. (tel. 041 - 435111, ext. 529). *Organic and inorganic crystal structures, molecular conformation, structure determination methods.*

Kosovinc, Prof. Dr Ivan. Katedra za metalografijo Fakultete za naravoslovje in tehnologijo, Univerza E. Kardelja v Ljubljani, Aškerčeva 20, 61000 Ljubljana, Yugoslavia. (1922) Dr, technology (Ljubljana U., 1968). Prof. (tel. 061 - 212121). *Physical metallurgy, metallography.*

Kranjc, Prof. Dr Katarina. Institut za fiziku Sveučilišta, Bijenička c. 46, P.O.Box 304, 41001 Zagreb, Yugoslavia. (1915) Dr, physics (Zagreb U., 1954). Prof. (tel. 041 - 271211). *Small angle X-ray scattering, crystal imperfections.*

Kraševec, Dr Viktor. Institut 'Jožef Stefan', Univerza E. Kardelja v Ljubljani, Jamova 39, 61000 Ljubljana, Yugoslavia. (1932) Dr, physics (Ljubljana U., 1974). Scient. assoc. (tel. 061 - 214399). *Microstructure, ceramic materials.*

Krstanović, Prof. Dr Ilija. Rudarsko - geološki fakultet, Univerzitet u Beogradu, Đušina ul. 7, 11000 Beograd, Yugoslavia. (1927) Dr, mineralogy (Beograd U., 1961). Prof. of crystallography. (tel. 011 - 180111, ext. 701). *Mineral and inorganic crystal structures.*

Kunstelj, Dr Dragan. Institut za fiziku Sveučilišta, Bijenička c. 46, P.O.Box 304, 41001 Zagreb, Yugoslavia. (1941) Dr, physics (Zagreb U., 1979). Docent. (tel. 041 - 271211). *Solid state physics, metal physics, electron microscopy.*

Lahodny-Šarc, Prof. Dr (Mrs) Olga. Zavod za kemiju RGN fakulteta, Sveučilište u Zagrebu, Pierottijeva 6, 41000 Zagreb, Yugoslavia. (1928) Dr, physical chemistry (Zagreb U., 1962). Prof. (tel. 041 - 442409). *Silicate chemistry, corrosion science.*

Lazar, Mr Dušan. Institut za fiziku Prirodno - matematičkog fakulteta, ul. Ilije Đuričića 4, 21000 Novi Sad, Yugoslavia. (1951) Dipl. fiz., physics (Novi Sad U., 1974). Res. asst. (tel. 021 - 55318). *Organic crystal structures.*

Lazarini, Prof. Dr Franc. VTOZD Kemija in kem. tehnologija, FNT, Univerza E. Kardelja v Ljubljani, Murnikova 6, P.O.Box 537, 61001 Ljubljana, Yugoslavia. (1940) Dr, chemistry (Ljubljana U., 1971). Prof. (tel. 061 - 214444). *Inorganic chemistry, crystal structures.*

Leban, Prof. Dr Ivan. VTOZD Kemija in kem. tehnologija, FNT, Univerza E. Kardelja v Ljubljani, Murnikova 6, P.O.Box 537, 61001 Ljubljana, Yugoslavia. (1947) PhD, physics (York U., UK, 1973). Izv. prof. (tel. 061 - 214444). *Crystal structure determination.*

Luić, Magistar Marija. Institut 'Ruđer Bošković', Bijenička c. 54, P.O.Box 1016, 41001 Zagreb, Yugoslavia. (1953) Magistar, geology (Zagreb U., 1981). Res. assoc. (tel. 041 - 435111, ext. 319). *Structure analysis, direct methods.*

Lugomer, Dr Stjepan. Elektrotehnički fakultet, Meštrovićeva ul. 4, 78000 Banja Luka, Yugoslavia. (1944) Dr, physics (Zagreb U., 1974). Docent. (tel. 078 - 24097). *Metal thin films.*

Maksić, Prof. Dr Zvonimir. Institut 'Ruđer Bošković', Bijenička c. 54, P.O.Box 1016, 41001 Zagreb, Yugoslavia. (1938) Dr, chemistry (Zagreb U., 1967). Prof. of theoretical chemistry. (tel. 041 - 435111, ext. 502). *Molecular structure.*

Marinković, Prof. Dr Velibor. Institut 'Jožef Stefan', Univerza E. Kardelja v Ljubljani, Jamova 39, 61000 Ljubljana, Yugoslavia. (1929) Dr, chemistry (Ljubljana U., 1965). Scient. adviser. (tel. 061 - 214399). *Crystal imperfections, metal physics.*

Marinković, Magistar (Mrs) Živka. Institut za fiziku, Maksima Gorkog 118, 11080 Zemun-Beograd, Yugoslavia. (1932) Magistar, physical chemistry (Beograd U., 1967). Scient. assoc. (tel. 011 - 212219). *Thin films, intermetallic compound formation.*

Marković, Magistar Berislav. Istraživački laboratorij 'JUCEMA', Vlaška ul. 67, 41000 Zagreb, Yugoslavia. (1957) Magistar, chemistry (Zagreb U., 1985). Res. assoc. (tel. 041 - 419260, ext. 5). *Crystal structures.*

Marković, Magistar Desimir. Tehnički fakultet-Bor, JNA 12, 19210 Bor, Yugoslavia. (1950) Magistar, metallurgy (Beograd U., 1982). Asistent. (tel. 030 - 24555). *Metal physics, metals structure.*

Mašić, Magistar Nikola. Institut 'Ruđer Bošković', Bijenička c. 54, P.O.Box 1016, 41001 Zagreb, Yugoslavia. (1948) Magistar, physics (Zagreb U., 1975). Res. assoc. (tel. 041 - 435111). *Polymers, structure and properties.*

Matijašić, Dr (Mrs) Ivanka. Zavod za organsku kemiju i biokemiju, Prirodoslovno - matematički fakultet, Sveučilište u Zagrebu, Strossmayerov trg 14, 41000 Zagreb, Yugoslavia. (1944) Dr, chemistry (Zagreb U., 1984). Scient. asst. (tel. 041 - 432580). *Organic crystal structures.*

Matković, Dr Boris. Institut 'Ruđer Bošković', Bijenička c. 54, P.O.Box 1016, 41001 Zagreb, Yugoslavia. (1927) Dr, chemistry (Zagreb U., 1961). Scient. adviser. (tel. 041 - 435111, ext. 335). *Inorganic crystal structures, cement chemistry.*

Matković, Dr Prosper. Metalurški fakultet, Sisak, Institut za metalurgiju, Aleja narodnih heroja 1, 44000 Sisak, Yugoslavia. (1945) Dr, chemistry (Stuttgart U., BRD, 1977). Docent. (tel. 044 - 32044). *Inorganic crystal structures, physical metallurgy.*

Matković, Dr Tanja. Metalurški fakultet, Sisak, Institut za metalurgiju, Aleja narodnih heroja 1, 44000 Sisak, Yugoslavia. (1948) Dr, chemistry (Stuttgart U., BRD, 1977). Docent. (tel. 044 - 32044). *Structure, metals and alloys, inorganic compounds, physical metallurgy.*

Matković-Čalogović, Magistar (Mrs) Dubravka. Zavod za opću i anorgansku kemiju, Prirodoslovno-matematički fakultet, Sveučilište u Zagrebu, ul. Soc. revolucije 8, 41000 Zagreb, Yugoslavia. (1957) Magistar, chemistry (Zagreb U., 1985). Asistent. (tel. 041-414079, ext. 83). *Crystal structures.*

Međimorec, Magistar Stanislav. Mineraloško - petrografski zavod, Prirodoslovno - matematički fakultet, Sveučilište u Zagrebu, Demetrova ul. 1, 41000 Zagreb, Yugoslavia. (1939) Magistar, mineralogy (Zagreb U., 1977). Asistent. (tel. 041 - 445315). *Physicochemical methods of mineral analysis.*

Milat, Magistar Ognjen. Institut za fiziku Sveučilišta, Bijenička c. 46, P.O.Box 304, 41001 Zagreb, Yugoslavia. (1949) Magistar, physics (Zagreb U., 1978). Asistent. (tel. 041 - 271211, ext. 334). *Metal physics, electron microscopy.*

Milićev, Prof. Dr Svetozar. VTOZD Kemija in kem. tehnologija, FNT, Univerza E. Kardelja v Ljubljani, Murnikova 6, P.O.Box 537, 61001 Ljubljana, Yugoslavia. (1934) Dr, chemistry (Ljubljana U., 1972). Izv. prof. (tel. 061 - 214444). *Vibration spectroscopy.*

Milinski, Prof. Dr Nikola. Institut za fiziku Prirodno - matematičkog fakulteta, ul. Ilije Đuričića 4, 21000 Novi Sad, Yugoslavia. (1936) Dr, physics (Novi Sad U., 1976). Izv. prof. (tel. 021 - 55622). *Solid state physics, crystal physics.*

Mirčeva, Magistar Aneta. Hemiski institut, Prirodno-matematički fakultet, Univerzitet 'Kiril i Metodij', Arhimedova 5, 91000 Skopje, Yugoslavia. (1946) Magistar, chemistry (Skopje U., 1979). Asistent. (tel. 091 - 221033). *Crystal structures.*

Moguš-Milanković, Magistar Andrea. Institut 'Ruđer Bošković', Bijenička c. 54, P.O. Box 1016, 41001 Zagreb, Yugoslavia. (1953) Magistar, chemistry (Zagreb U., 1982). Res. assoc. (tel. 041-435111). *Electric properties of crystals.*

Morvaj, Magistar (Mrs) Jasmina. Zavod za opću i anorgansku kemiju, Prirodoslovno-matematički fakultet, Sveučilište u Zagrebu, ul. Soc. revolucije 8, 41000 Zagreb, Yugoslavia. (1957) Magistar (Zagreb U., 1985), Res. assoc. (tel. 041 - 414079, ext. 9). *Inorganic crystal structures.*

Nagl, Dr Antun. Zavod za opću i anorgansku kemiju, Prirodoslovno - matematički fakultet, Sveučilište u Zagrebu, ul. Soc. revolucije 8, 41000 Zagreb, Yugoslavia. (1942) Dr, chemistry (Bern U., Switzerland, 1979). Asistent. (tel. 041 - 416023, ext. 8005). *Crystal structures.*

Nančovski, Mr Kosta. Građežen institut 'Makedonija'-Skopje, ul. Drezdenska 52, 91000 Skopje, Yugoslavia. (1934) Dipl. hem., chemistry (Beograd U., 1960). Asst. director. (tel. 091 - 253929). *Instrumentation.*

Nikolić, Prof. Dr Pantelija. Elektrotehnički fakultet, Univerzitet u Beogradu, Bulevar Revolucije 73, 11000 Beograd, Yugoslavia. (1928) Dr, physics (Nottingham U., UK, 1969). Prof. (tel. 011 - 329212, ext. 384). *Semiconductor crystal physics, optical properties.*

Novosel Radović, Dr Vjera. MK 'Željezara Sisak', RO Institut za metalurgiju, OOUR Metalurški fakultet, Aleja narodnih heroja 1, 44000 Sisak, Yugoslavia. (1937) Dr, chemistry (Zagreb U., 1983). Chief organizer in Lab. for investigation of structure and properties of materials. (tel. 044 - 30444, ext. 395). *Polycrystalline materials.*

Očko, Magistar Miroslav. TVA KOV, Ilica 256, 41000 Zagreb, Yugoslavia. (1947) Magistar, physics (Zagreb U., 1976). Lect. (tel. 041 - 579666). *Amorphous materials.*

Osterc, Prof. Dr (Mrs) Valerija. Odsek za geologijo, Oddelek za montanistiko, Fakulteta za naravoslovje in tehnologijo, Univerza E. Kardelja v Ljubljani, Aškerčeva 20, 61000 Ljubljana, Yugoslavia. (1924) Dr, ceramics (Rheinisch - Westfälische Technische Hochschule Aachen, BRD, 1967). Prof. (tel. 061 - 212121). *Clay minerals, X-ray diffraction.*

Paljević, Dr Matija. Institut 'Ruđer Bošković', Bijenička c. 54, P.O.Box 1016, 41001 Zagreb, Yugoslavia. (1943) Dr, chemistry (Zagreb U., 1978). Res. assoc. (tel. 041 - 435111, ext. 274). *Gas-solid reactions.*

Pećina, Miss Planinka. Fizički zavod, Prirodoslovno-matematički fakultet, Sveučilište u Zagrebu, Marulićev trg 19, 41000 Zagreb, Yugoslavia. (1957) Dipl. inž. fiz., physics (Zagreb U., 1983). Jr. res. assoc. (tel. 041 - 446242). *Metal physics, electron microscopy.*

Penavić, Dr (Mrs) Maja. Zavod za opću i anorgansku kemiju, Prirodoslovno - matematički fakultet, Sveučilište u Zagrebu, ul. Soc. revolucije 8, 41000 Zagreb, Yugoslavia. (1941) Dr, chemistry (Zagreb U., 1977). Asistent. (tel. 041 - 416023, ext. 8005). *Inorganic crystal structures.*

Petrović, Dr Dragoslav. Institut za fiziku Prirodno - matematičkog fakulteta, ul. Ilije Đuričića 4, 21000 Novi Sad, Yugoslavia. (1949) Dr, physics (Novi Sad U., 1980). Docent. (tel. 021 - 55318). *Solid state physics.*

Pocev, Prof. Dr Stefan. Tehnološki fakultet, Univerzitet 'Kiril i Metodij', Ruđer Bošković 16, 91000 Skopje, Yugoslavia. (1940) Dr, chemistry (Zagreb U., 1978). Izv. prof. (tel. 091 - 259725). *Inorganic crystal structures.*

Poharc-Logar, Magistar Vesna. Rudarsko-geološki fakultet, Univerzitet u Beogradu, Đušina 7, 11000 Beograd, Yugoslavia. (1949) Magistar, mineralogy (Beograd U., 1979). Asistent. (tel. 011 - 183814). *Clay minerals.*

Polanc, Magistar Ivan. Visoka tehniška šola Maribor, Univerza v Mariboru, Smetanova 17, 62000 Maribor, Yugoslavia. (1949) Magistar, chemistry (Ljubljana U., 1980). Asistent. (tel. 062 - 25461). *Inorganic chemistry.*

Popović, Dr Stanko. Institut 'Ruđer Bošković', Bijenička c. 54, P.O.Box 1016, 41001 Zagreb, Yugoslavia. (1938) Dr, physics (Zagreb U., 1968). Scient. adviser. (tel. 041 - 435111, ext. 320). *Powder diffraction theory, precise measurement of unit cell parameters, quantitative phase analysis.*

Prelesnik, Dr Bogdan. Institut za nuklearne nauke 'Boris Kidrič' - Vinča, Laboratorija za fiziku čvrstog stanja i radijacionu hemiju, P.O.Box 522, 11000 Beograd, Yugoslavia. (1938) Dr, crystallography (Bern U., Switzerland, 1975). Scient. assoc. (tel. 011 - 4440871, ext. 417). *Inorganic and organic crystal structures, neutron diffraction, ice nucleation.*

Prodan, Dr Albert. Institut 'Jožef Stefan', Univerza E. Kardelja v Ljubljani, Jamova 39, 61000 Ljubljana, Yugoslavia. (1944) Dr, physics (Zagreb U., 1974). Scient. assoc. (tel. 061 - 214399, ext. 238). *Electron diffraction.*

Pušelj, Dr Milan. Zavod za opću i anorgansku kemiju, Prirodoslovno-matematički fakultet, Sveučilište u Zagrebu, ul. Soc. revolucije 8, 41000 Zagreb, Yugoslavia. (1945) Dr, chemistry (Stuttgart U., BRD, 1975). Docent. (tel. 041 - 416023). *Metal and alloy structures.*

Radaković, Mrs Aleksandra. Rudarsko-geološki fakultet, Univerzitet u Beogradu, Đušina 7, 11000 Beograd, Yugoslavia. (1956) Dipl. inž. geol., geology (Beograd U., 1981). Jr. res. assoc. (tel. 011 - 180111, ext. 701). *Inorganic crystal structures, clay minerals.*

Radmilović, Magistar Velimir. Tehnološko-metalurški fakultet, Katedra za fizičku metalurgiju, Karnegijeva ul. 4, 11000 Beograd, Yugoslavia. (1948) Magistar, metallurgy (Beograd U., 1980). Asistent. (tel. 011 - 328671). *Metal structures, electron diffraction, metal physics.*

Radukić, Dr Gordana. Rudarsko - geološki fakultet, Univerzitet u Beogradu, Đušina ul. 7, 11000 Beograd, Yugoslavia. (1931) Dr, geology (Beograd U., 1977). Docent. (tel. 011 - 180111, ext. 771). *Crystal optics, phyllosilicates.*

Rebić, Mr Milenko. Institut za materiale, Zavod za raziskavo materiala in konstrukcij, Dimičeva 12, 61000 Ljubljana, Yugoslavia. (1938) Dipl. inž. fiz., physics (Ljubljana U., 1969). Res. fellow. (tel. 061 - 344061). *Metal physics.*

Ribár, Prof. Dr Béla. Institut za fiziku Prirodno - matematičkog fakulteta, ul. Ilije Đuričića 4, 21000 Novi Sad, Yugoslavia. (1930) Dr, crystallography (Bern U., Switzerland, 1969). Prof. (tel. 021 - 55318). *Inorganic and organic crystal structures.*

Rogić, Prof. Dr Vinko. Građevinski fakultet, Univerzitet "Džemal Bijedić" Mostar, ul. A. Zuanića 14, 88000 Mostar, Yugoslavia. (1945) Dr, chemistry (Zagreb U., 1975). Prof. (tel. 088 - 416737). *Crystal structures, building materials chemistry.*

Rogulić, Prof. Dr Mileva. Tehnološko-metalurški fakultet, Univerzitet u Beogradu, Karnegijeva ul. 4, 11000 Beograd, Yugoslavia. (1928) Dr, metallurgy (Cambridge U., UK, 1964). Izv. prof. (tel. 011 - 328721). *Physical metallurgy, metal and alloy structures.*

Runje, Dr Vesna. Farmaceutsko - biokemijski fakultet, Sveučilište u Zagrebu, ul. Ante Kovačića 1, 41000 Zagreb, Yugoslavia. (1947) Dr, chemistry (Zagreb U., 1981). Asistent. (tel. 041 - 445311, ext. 36). *X-ray diffraction analysis, polycrystalline systems.*

Ružić-Toroš, Dr Živa. Institut 'Ruđer Bošković', Bijenička c. 54, P.O.Box 1016, 41001 Zagreb, Yugoslavia. (1944) Dr, chemistry (Zagreb U., 1974). Sr. scient. assoc. (tel. 041 - 435111, ext. 529). *Inorganic and organic crystal structures.*

Sijarić, Dr (Mrs) Galiba. Prirodno - matematički fakultet, Vojvode Putnika 43, 71000 Sarajevo, Yugoslavia. (1939) Dr, geology (Sarajevo U., 1975). Docent. (tel. 071 -649377). *X-ray diffraction, clay minerals, bauxites, zeolites, feldspars.*

Sikirica, Prof. Dr Milan. Zavod za opću i anorgansku kemiju, Prirodoslovno - matematički fakultet, Sveučilište u Zagrebu, ul. Soc. revolucije 8, 41000 Zagreb, Yugoslavia. (1934) Dr, chemistry (Zagreb U., 1963). Prof. (tel. 041 - 414079). *Inorganic crystal structures.*

Simić, Dr Vojislav. Institut za fiziku, Maksima Gorkog 118, 11080 Zemun-Beograd, Yugoslavia. (1926) Dr, chemistry (Zagreb U., 1960). Scient. adviser. (tel. 011 - 212219). *Thin solid films, reactions at room temperature.*

Slovenec, Dr Dragutin. Rudarsko - geološko - naftni fakultet, Sveučilište u Zagrebu, Pierottijeva ul. 6, 41000 Zagreb, Yugoslavia. (1941) Dr, geology (Zagreb U., 1980). Docent. (tel. 041 - 440422, ext. 425). *Mineralogy.*

Srdanov, Magistar Gordana. Institut za nuklearne nauke 'Boris Kidrič' - Vinča, Laboratorija za fiziku čvrstog stanja i radijacionu hemiju, P.O.Box 522, 11000 Beograd, Yugoslavia. (1950) Magistar, chemistry (Zagreb U., 1978). Res. assoc. (tel. 011 - 458222, ext. 417). *Crystal structures, inorganic and organometallic compounds.*

Stanković, Mrs Slavica. Institut za nuklearne nauke 'Boris Kidrič' - Vinča, Institut za materijale 'IM', P.O.Box 522, 11000 Beograd, Yugoslavia. (1953) Dipl. inž., chem. engineering (Beograd U., 1977). Jr. res. assoc. (tel. 011 - 458222, ext. 594). *X-ray powder diffractometry, inorganic crystal structures.*

Stanković, Dr (Mrs) Slobodanka. Institut za fiziku Prirodno - matematičkog fakulteta, ul. Ilije Đuričića 4, 21000 Novi Sad, Yugoslavia. (1941) Dr, physics (Novi Sad U., 1980). Docent. (tel. 021 - 55318). *Organic crystal structures.*

Stefanović, Magistar Aleksandar. Zavod za opću i anorgansku kemiju, Prirodoslovno-matematički fakultet, Sveučilište u Zagrebu, ul. Soc. revolucije 8, 41000 Zagreb, Yugoslavia. (1957) Magistar, chemistry (Zagreb U., 1985). Res. assoc. (tel. 041 - 416023, ext. 8005). *Inorganic crystal structures.*

Stojadinović, Dr Slobodan. Tehnički fakultet - Bor, ul. JNA br. 12, 19210 Bor, Yugoslavia. (1947) Dr, metallurgy (Beograd U., 1980). Docent. (tel. 030 - 24555). *Metal physics, metals structure.*

Stojanović, Mr Dobrica. RO 'MAGNOHROM'- Institut za vatrostalne materijale, P.O.Box 17, 36001 Kraljevo, Yugoslavia. (1931) Dipl. inž. geol., geology (Beograd U., 1957). Principal res. officer. (tel. 036 - 331322, ext. 513). *Raw materials, refractory materials, ceramics, X-ray diffraction.*

Stubičar, Dr Mirko. Fizički zavod, Prirodoslovno - matematički fakultet, Sveučilište u Zagrebu, Marulićev trg 19, 41000 Zagreb, Yugoslavia. (1940) Dr, physics (Zagreb U., 1985). Asistent. (tel. 041 - 446242). *Metal physics.*

Šćavničar, Prof. Dr Stjepan. Mineraloško - petrografski zavod, Prirodoslovno - matematički fakultet, Sveučilište u Zagrebu, Demetrova ul. 1, 41000 Zagreb, Yugoslavia. (1923) Dr, chemistry (Zagreb U., 1956). Prof., head, dept. of mineralogy and petrography, Faculty of Science, Zagreb. (tel. 041 - 449120). *Inorganic crystal structures, clay minerals.*

Šegedin, Magistar Primož. VTOZD Kemija in kem. tehnologija, FNT, Univerza E. Kardelja v Ljubljani, Murnikova 6, P.O.Box 537, 61001 Ljubljana, Yugoslavia. (1948) Magistar, chemistry (Ljubljana U., 1975). Asistent. (tel. 061 - 214444). *Inorganic crystal structures.*

Šljukić, Prof. Dr Momčilo. Tehnički fakultet, Univerzitet 'Veljko Vlahović' Titograd, Kruševac b.b., 81000 Titograd, Yugoslavia. (1936) Dr, chemistry (Zagreb U., 1968). Prof. (tel. 081 - 52111). *Inorganic crystal structures.*

Šmit, Dr Ivan. Institut 'Ruđer Bošković', Bijenička c. 54, P.O.Box 1016, 41001 Zagreb, Yugoslavia. (1948) Dr, chemistry (Zagreb U., 1979). Res. assoc. (tel. 041 - 435111). *Polymers, structure and properties.*

Šoptrajanov, Prof. Dr Bojan. Hemiski institut, Prirodno-matematički fakultet, Univerzitet 'Kiril i Metodij', Arhimedova 5, 91000 Skopje, Yugoslavia. (1937) Dr, chemistry (Skopje U., 1973). Prof. (tel. 091 - 221033). *Structural chemistry.*

Šoptrajanova, Prof. Dr Gorica. Rudarsko-geološki fakultet, 92000 Štip, Yugoslavia. (1929) Dr, geology (Beograd U., 1967). Prof. (tel. 092 - 21379). *Mineral structures.*

Tibljaš, Mr Darko. Mineraloško - petrografski zavod, Prirodoslovno - matematički fakultet, Sveučilište u Zagrebu, Demetrova ul. 1, 41000 Zagreb, Yugoslavia. (1957) Dipl. inž. geol., geology (Zagreb U., 1981). Asistent. (tel. 041 - 445315). *Minerals, structure and physical properties.*

Tkalčec, Prof. Dr Emilija. Tehnološki fakultet, Sveučilište u Zagrebu, Marulićev trg 20, 41000 Zagreb, Yugoslavia. (1931) Dr, chemistry (Zagreb U., 1975). Izv. prof. (tel. 041 - 446439). *X-ray diffraction analysis, polycrystalline systems, silicates.*

Tonejc, Dr (Mrs) Anđelka. Fizički zavod, Prirodoslovno - matematički fakultet, Sveučilište u Zagrebu, Marulićev trg 19, 41000 Zagreb, Yugoslavia. (1942) Dr, physics (Zagreb U., 1980). Asistent, scient. assoc. (tel. 041 - 446211). *Metal physics, crystal imperfections, phase transitions, X-ray fiffraction, electron diffraction, electron microscopy.*

Tonejc, Dr Anton. Fizički zavod, Prirodoslovno - matematički fakultet, Sveučilište u Zagrebu, Marulićev trg 19, 41000 Zagreb, Yugoslavia. (1942) Dr, physics (Zagreb U., 1972). Docent, sr. scient. assoc. (tel. 041 - 446211). *Metal physics, metals structure.*

Topić, Dr Mladen. Institut 'Ruđer Bošković', Bijenička c. 54, P.O.Box 1016, 41001 Zagreb, Yugoslavia. (1934) Dr, chemistry (Zagreb U., 1965). Sr. scient. assoc. (tel. 041 - 435111, ext. 275). *Electric properties of crystals.*

Trojko, Magistar Rudolf. Institut 'Ruđer Bošković', Bijenička c. 54, P.O.Box 1016, 41001 Zagreb, Yugoslavia. (1942) Magistar, chemistry (Zagreb U., 1974). Res. assoc. (tel. 041 - 435111, ext. 319). *Intermetallic compounds.*

Trubelja, Prof. Dr Fabijan. Prirodno-matematički fakultet, Univerzitet u Sarajevu, Vojvode Putnika 43, 71000 Sarajevo, Yugoslavia. (1927) Dr, geology (Zagreb U., 1958). Prof. (tel. 071 - 649377). *Mineralogy.*

Tudja, Dr Marijan. RO Chromos Centar za kemijska istraživanja i razvoj, Žitnjak bb, 41000 Zagreb, Yugoslavia. (1942) Dr, chemistry (Zagreb U., 1974). Res. assoc. (tel. 041 - 210200, ext. 428). *Scanning microscopy.*

Ungar, Dr Goran. Institut 'Ruđer Bošković', Bijenička c. 54, P.O.Box 1016, 41001 Zagreb, Yugoslavia. (1948) Dr, chemistry (Bristol U., UK, 1979). Scient. assoc. (tel. 041 - 435111). *Polymer structure and reactivity.*

Vasić, Magistar Pavle. Prirodno - matematički fakultet Priština, ul. Maršala Tita, 38000 Priština, Yugoslavia. (1939) Magistar, physics (Beograd U., 1975). Lecturer in physics. (tel. 038 - 25855). *Inorganic crystal structures, crystal physics.*

Vene, Mrs Nada. Institut 'Jožef Stefan', Univerza E. Kardelja v Ljubljani, Jamova 39, 61000 Ljubljana, Yugoslavia. (1931) Dipl. mat., mathematics (Ljubljana U., 1954). Res. assoc. (tel. 061 - 214399). *Inorganic crystal structures.*

Vicković, Dr Ivan. Sveučilišni računski centar, Engelsova b.b., 41000 Zagreb, Yugoslavia. (1945) Dr, physics (Zagreb U., 1977). Sr. scient. assoc. (tel. 041 - 510099). *Computer programming, crystal structures.*

Vukasović, Magistar Momčilo. Institut za geološko-rudarska istraživanja i ispitivanja nuklearnih i drugih mineralnih sirovina, Rovinjska ul. 12, 11000 Beograd, Yugoslavia. (1923) Magistar, geology (Beograd U., 1973). Res. adviser. (tel. 011 - 480506). *Minerals structure, clay minerals.*

Zajc, Magistar Andrej. Institut za materiale, Zavod za raziskavo materiala in konstrukcij, Dimičeva 12, 61000 Ljubljana, Yugoslavia. (1938) Magistar, physical chemistry (Ljubljana U., 1979). Res. adviser. (tel. 061 - 344061). *Clay minerals.*

NAME INDEX

The ä, ö and ü in the Finnish, German, Hungarian, Scandinavian and modern Turkish languages are alphabetized as if they were written ae, oe and ue. The Danish-Norwegian ø is also alphabetized as oe. The Scandinavian å has been alphabetized as if it were written as "aa." The Russian ' precedes "a" in the USSR list and the Name Index. All other accents and diacritical marks have been disregarded in the alphabetization of the Name Index.

Åberg, Prof. Teijo. *Finland.*
Åberg, Dr Märtha M. *Sweden.*
Åsbrink, Dr Gudrun. *Sweden.*
Åsbrink, Dr Stig. *Sweden.*
Åse, Mr Kjell. *Norway.*
Åström, Prof. Hans U. *Sweden.*
Abad-Zapatero, Dr Celerino. *USA.*
Abbona, Prof. Francesco. *Italy.*
Abboud, Mr Khalil A. *USA.*
Abdel Aal, Mr Fawzi Amer. *Egypt.*
Abdel Hady, Prof. Seham. *Egypt.*
Abdel Kader, Dr Abdel Aziz. *Egypt.*
Abdel Kader, Prof. (Miss) Naima. *Egypt.*
Abdel Kader, Mrs Zeinab Mohamed. *Egypt.*
Abdel Meguid, Sherin S. *USA.*
Abdel Mohsen, Dr Hussein. *Egypt.*
Abdel Rehim, Dr Amin Mohamed. *Egypt.*
Abdu, Prof. Fayez Madi. *Egypt.*
Abdul Aziz, Mr Abdul Halim bin. *Malaysia.*
Abdullaev, Dr Abdulkhamid Aliyevich. *USSR.*
Abdullaev, Dr Gusi Kara ogly. *USSR.*
Abe, Prof. Hideo. *Japan.*
Abe, Prof. Ryuji. *Japan.*
Abe, Mr Shuich. *Japan.*
Abel, Mr James E. *USA.*
Abell, Dr John Stuart. *UK.*
Abelmann, Mr Rolf-Ulrich. *DDR.*
Aberdam, Dr Daniel Jean. *France.*
Abola, Dr Enrique E. *USA.*
Abola, Dr Jaime Esteva. *USA.*
Abou Ghaloun, Dr Omar Farouk. *Syrian Arab Rep.*
Abou-Saif, Dr Elhamy Aziz. *Egypt.*
Abraham, Prof. Donald James. *USA.*
Abrahams, Dr Sidney Cyril. *USA.*
Abriel, Dr Walter. *BRD.*
Abs-Wurmbach, Dr Irmgard. *BRD.*
Abu Bakar, Mr Ismail. *Malaysia.*
Accorsi, Prof. Carla Alberta. *Italy.*
Acharaya, Dr Ravindra. *UK.*
Achari, Dr Aniruddha. *USA.*
Acosta, Prof. Carlos Eduardo. *Colombia.*
Acuña, Dr Rodolfo José *Argentina.*
Adachi, Prof. Kengo. *Japan.*
Adam, Dr C.D. *UK.*
Adam, Dr Jerzy. *UK.*
Adamiak, Dr Dorota Anna. *Poland.*
Adams, Dr Margaret Joan. *UK.*
Adams, Dr M.J. *UK.*
Adams, Prof. Richard Darwin. *USA.*
Adams, Dr Walter Wade. *USA.*
Adamski, Mrs Hannelore. *DDR.*
Adan, Mr Nachum. *Israel.*
Adelsköld, Dr Volrath. *Sweden.*
Adendorff, Mr Keith Trevor. *South Africa.*
Adetunji, Dr Jacob. *Nigeria.*
Adewoye, Dr Olusegun Oyeleke. *Nigeria.*
Adigüzel, Dr Osman. *Turkey.*
Adler, Mr George. *USA.*
Adman, Dr Elinor Thomson. *USA.*
Admiraal, Dr Gerrit. *Netherlands.*
Adrian, Dr Herbert Wilhelm Werner. *South Africa.*
Aernoudt, Prof. Dr Etienne. *Belgium.*
Aerts, Dr Jozef. *Netherlands.*
Afanasyev, Dr Igor' Ivanovich. *USSR.*
Afonina, Dr Nataliya Nikolayevna. *USSR.*
Agard, Dr David Andrew. *USA.*
Agarwal, Dr Bhagwatiprasad. *India.*
Agarwal, Dr Ramesh Chandra. *India.*
Aggarwal, Dr Prem Sarup. *India.*
Agmon, Mrs Ilana. *Israel.*
Agrawal, Mr Jawahar Lal. *India.*
Agre, Dr Valeriya Moiseyevna. *USSR.*

Agrell, Dr Ingela. *Sweden.*
Agron, Mr Paul A. *USA.*
Aguilar, Mrs Adela. *Chile.*
Aguilera Herrera, Prof. Nicolás. *Mexico.*
Ahlers, Mr N.H.E. *UK.*
Ahlgrén, Dr, Assoc.Prof. Markku Jouko. *Finland.*
Ahmad, Mr Anwaruddin M. *Pakistan.*
Ahmad, Mr Dabir. *Pakistan.*
Ahmad, Mr Raisuddin. *Bangladesh.*
Ahmad, Dr Zulfiqar. *Pakistan.*
Ahmed, Dr A. H. Moinuddin. *Bangladesh.*
Ahmed, Dr Farid Ramadan. *Canada.*
Ahmed, Dr Mohamed Saleh. *Egypt.*
Ahmed, Dr Sultan. *Bangladesh.*
Ahn, Prof. Dr Choong Tai. *Korea.*
Ahtee, Dr Sisko-Maija. *Finland.*
Aiginger, Prof. Dr Dipl.-Ing. Hannes. *Austria.*
Aihara, Prof. Ariyuki. *Japan.*
Aikala, Dr Oiva Jaakko Mikael. *Finland.*
Aikawa, Dr Nobuyuki. *Japan.*
Ainsworth, Mr Leonard Ralph. *UK.*
Aires-Barros, Prof. Luis. *Portugal.*
Aizaki, Mr Nao-aki. *Japan.*
Ajó, Dr David. *Italy.*
Ajaal, Mr Tawfik Taher. *Libya.*
Aka, Mr Yavuz. *Turkey.*
Akamatu, Prof. Hideo. *Japan.*
Akao, Dr Masaru. *Japan.*
Akchurin, Dr Marat Shikhapovich. *USSR.*
Akers, Dr Charles Kenton. *USA.*
Akgün, Mr İrfan. *Turkey.*
Akhmetov, Dr Spartak Fatykhovich. *USSR.*
Akhmetova, Dr Galina Leonidovna. *USSR.*
Akhtar, Dr Farida. *Bangladesh.*
Akhtar, Mr Javed. *Pakistan.*
Akhtar, Mr Mohammad. *Pakistan.*
Akimitsu, Dr Jun. *Japan.*
Akizuki, Dr Mizuhiko. *Japan.*
Akkad, Dr Mohamed Kamal. *Egypt.*
Akkad, Mr Salah el Din. *Egypt.*
Aksela, Prof. Seppo Olavi. *Finland.*
Akselrud, Dr Lev Grigoryevich. *USSR.*
Aksoy, Mr İlhan. *Turkey.*
Akyüz, Mr Tanil. *Turkey.*
Al-Farhan, Mr Khalid A. M. *UK.*
Al-Karaghouli, Dr Abdulrazzak Hamoudi. *Iraq.*
Al-Shanti, Prof. Ahmed Mahmoud. *Saudi Arabia.*
Al'shits, Dr Vladimir Iosifovich. *USSR.*
Alabdulla, Dr Ihsan Abdul Ghani. *Iraq.*
Aladekomo, Prof. Johnson Bandele. *Nigeria.*
Alarcón, Mr Hugo. *Bolivia.*
Alaverdova, Dr Olga Georgiyevna. *USSR.*
Alavi, Dr Mehdi. *Iran.*
Albano, Prof. Vincenzo Giulio. *Italy.*
Alber, Dr Tom. *USA.*
Albers, Miss Ursula. *BRD.*
Albert, Mr Charles W. *USA.*
Alberti, Prof. Alberto. *Italy.*
Alberts, Prof. Hermanus Lambertus. *South Africa.*
Albertsson, Dr Jörgen. *Sweden.*
Albinati, Prof. Alberto. *Italy.*
Albrecht, Prof. Günter. *DDR.*
Alcock, Dr Nathaniel Warren. *UK.*
Aldén, Dr Karl Inge. *Sweden.*
Alden, Dr Richard Allen. *USA.*
Aldoshin, Dr Sergey Mikhailovich. *USSR.*
Aldridge, Dr Laurence Philip. *New Zealand.*
Aleby, Dr Stig E. *Sweden.*
Aleonard, Dr Suzanne. *France.*
Aleshko-Ozhevsky, Dr Oleg Pavlovich. *USSR.*
Alessandrini, Miss Eileen I. *USA.*
Alex, Dr Volker. *DDR.*

Alexander, Prof. Dr Helmut. *BRD.*
Alexander, Prof. Em. Leroy Elbert. *USA.*
Alexandropoulos, Prof. Nikos. *Greece.*
Alexandropoulos, Mrs Tina. *Greece.*
Alexandrov, Prof. Alexander Danilovich. *USSR.*
Alexandrov, Prof. Kirill Sergeyevich. *USSR.*
Alexandrov, Prof. Leonid Naumovich. *USSR.*
Alexandrov, Dr Vladimir Borisovich. *USSR.*
Alexandrova, Dr Inga Petrovna. *USSR.*
Alexopoulos, Prof. (Emeritus) Kessar. *Greece.*
Alfonso U., Miss Ana Elena. *Colombia.*
Ali, Dr E. M. *Sudan.*
Ali, Mr Mohsin. *India.*
Ali, Prof. Shams El Din. *Syrian Arab Rep.*
Ali, Dr Sultana Zulfiqar. *India.*
Ali, Dr Syed Wajahat. *Pakistan.*
Alietti, Prof. Andrea. *Italy.*
Aliev, Dr Fazil Isa ogly. *USSR.*
Aliev, Dr Zainutdin Gasanovich. *USSR.*
Alikhanov, Dr Ruben Abramovich. *USSR.*
Alkan, Mr Atilla. *Turkey.*
Alkire, Randy W. *USA.*
Alléaume, Dr Marc. *France.*
Allaire, Mr François. *Canada.*
Allais, Prof. Gérard. *France.*
Allcock, Prof. Harry Rex. *USA.*
Allegra, Prof. Giuseppe. *Italy.*
Allen, Dr A.J. *UK.*
Allen, Dr Frank Harmsworth. *UK.*
Allen, Joseph H. *USA.*
Allen, Mr Keith William. *UK.*
Allersma, Mr Ties. *USA.*
Allman, Miss J.M. *UK.*
Allmann, Prof. Dr Rudolf. *BRD.*
Almajdub, Mr Musbah Meftah. *Libya.*
Almashoor, Dr Syed Sheikh. *Malaysia.*
Almeida, Prof. Vasco Nogueira. *Brazil.*
Almeida, de, Prof. Maria José Marques. *Portugal.*
Almendras, Mrs Eliana. *Chile.*
Almenningen, Mr Arne. *Norway.*
Almlöf, Prof. Jan. *Norway.*
Alonso Lopez, Dr Jose. *Spain.*
Alp, Mr Esen Ercan. *Turkey.*
Alpaut, Prof. Dr Okyay. *Turkey.*
Alte da Veiga, Prof. Luis. *Portugal.*
Alter, Dr Uwe. *DDR.*
Altermatt, Dr Urs Daniel. *Canada.*
Altona, Prof. Cornelis. *Netherlands.*
Alujević - Stipanov, Dr Višnja. *Yugoslavia.*
Alvarez Peres, Dr Aurelio. *Spain.*
Alvarez, Dr Alberto Guillermo. *Argentina.*
Alver, Mr Eyvind. *Norway.*
Alyavdin, Dr Vladimir Fedorovich. *USSR.*
Alzari, Dr Pedro María. *Argentina.*
Amara, Dr Mongi. *Tunisia.*
Amarasena, Mr Kandage Don. *Sri Lanka.*
Amberger, Prof. Dr Eberhard. *BRD.*
Amelinckx, Prof. Dr Severin. *Belgium.*
Amemiya, Dr Yoshiyuki. *Japan.*
Amicarelli, Prof. Vincenzo. *Italy.*
Amighian, Dr Jamshid. *Iran.*
Amigo, Prof. Jose Ma. *Spain.*
Amilius, Mr Zuharli. *Indonesia.*
Amin, Dr Ahmed A. Hussein. *USA.*
Amirov, Dr Savalan Teimur ogly. *USSR.*
Amirthalingam, Dr V. *India.*
Amma, Prof. Elmer Louis. *USA.*
Ammon, Prof. Herman L. *USA.*
Amoros, Prof. Jose Luis. *Spain.*
Amrein, Mr Robert Eugene. *USA.*
Amstutz, Prof. Dr Dr hc Christian Gerhard. *BRD.*
Amthauer, Prof. Dr Georg. *Austria.*

Amy, Dr Joseph André. *USA.*
Amzel, Dr Leon Mario. *USA.*
Anantachai, Mrs Suda Yasarawana. *Thailand.*
Anantha Murthy, Dr Rayasa V. *India.*
Anantha Narayanan, Prof. V. *USA.*
Anantharaman, Prof. Tanjore Ramachandra. *India.*
Anderegg, Prof. John William. *USA.*
Anders, Mr Rudolf. *DDR.*
Andersen, Mr Erik Krogh. *Denmark.*
Andersen, Mrs Inger Grete Krogh. *Denmark.*
Andersen, Dr Niels Hessel. *Denmark.*
Andersen, Prof. Palle. *Denmark.*
Andersen, Dr Per. *Norway.*
Andersen, Dr Peter. *Denmark.*
Andersen, Dr Stig Kjær. *Denmark.*
Anderson, Ms Christine Alexis Francis. *USA.*
Anderson, Dr Gary Don. *USA.*
Anderson, Prof. Oren Paul. *USA.*
Anderson, Dr Wayne Foster. *Canada.*
Andersson, Dr Inger A. *Sweden.*
Andersson, Prof. Sten. *Sweden.*
Andersson, Dr Yvonne. *Sweden.*
Ando, Dr Masami. *Japan.*
Ando, Dr Yoshinori. *Japan.*
Andonov, Mrs Paulette. *France.*
André, Dr Daniel. *France.*
Andrade, Ms Lourdes Rodrigues. *Portugal.*
Andree, Mrs Anneliese. *DDR.*
Andreeft, Prof. Alexander. *DDR.*
Andreetti, Prof. Giovanni Dario. *Italy.*
Andreeva, Dr Nataliya Sergeyevna. *USSR.*
Andrehs, Dr Gerhard. *DDR.*
Andresen, Dr Arne Fridtjof. *Norway.*
Andrews, Dr S.J. *UK.*
Andrianov, Dr Valery Ivanovich. *USSR.*
Anex, Prof. Basil G. *USA.*
Ang, Dr Ha Ming. *Malaysia.*
Angermund, Mr Klaus Peter. *BRD.*
Angilello, Mr Joseph. *USA.*
Anikin, Dr Igor' Nikolayevich. *USSR.*
Anistratov, Dr Anatoly Tikhonovich. *USSR.*
Ankara, Prof. Dr Alpay. *Turkey.*
Annaka, Prof. Shoichi. *Japan.*
Annehed, Mr Håkan. *Sweden.*
Annersten, Prof. Hans. *Sweden.*
Ansell, Dr Gerald Brian. *USA.*
Anstis, Dr Geoffrey Richard. *Australia.*
Antal, Dr John Joseph. *USA.*
Anthony, Dr John W. *USA.*
Antipin, Dr Mikhail Yuvenaliyevich. *USSR.*
Antolini, Prof. Luciano. *Italy.*
Anton, Mr Liviu. *Romania.*
Antonini, Prof. Marcello. *Italy.*
Antonione, Prof. Carlo. *Italy.*
Antonopoulos, Prof. John. *Greece.*
Antonov, Dr Petr Iosifovich. *USSR.*
Antsishkina, Dr Alla Sergeyevna. *USSR.*
Antti, Dr Britt-Marie. *Sweden.*
Anugul, Mrs Surang. *Thailand.*
Anulewicz, Mrs Romana. *Poland.*
Anwar, Mr Muhammad. *Pakistan.*
Anwar, Dr Yehia. *Egypt.*
Aoe, Mr Hiroyuki. *Japan.*
Aoki, Dr Katsuyuki. *Japan.*
Aoki, Dr Yoshikazu. *Japan.*
Apinitis, Dr Smuidris Karlovich. *USSR.*
Apostolescu, Eugenia Rodica. *Romania.*
Apostolov, Prof. Andrei. *Bulgaria.*
Apostolov, Mr Anton. *Bulgaria.*
Appleman, Dr Daniel E. *USA.*
Aquilano, Prof. Dino. *Italy.*
Arafa, Prof. Salah Arafa Mohamed. *Egypt.*
Aragon de la Cruz, Prof. Francisco. *Spain.*
Arai, Dr Gerda Johanna. *USA.*
Araki, Dr Takaharu. *Canada.*
Arana, Prof. Rafael. *Spain.*
Arató, Dr Péter. *Hungary.*
Aravindakshan, Cheethambadi. *India.*
Arbunies Andreu, Dr Manuel. *Spain.*
Arce, Mrs Maria Teresa. *Chile.*

Archer, Dr Ronald D. *USA.*
Archer, Mr Steven James. *South Africa.*
Arda, Dr Oğuz. *Turkey.*
Arduz, Mr Marcelo. *Bolivia.*
Arellano, Mr J. *Bolivia.*
Arem, Dr Joel Edward. *USA.*
Arend, Prof. Hanns. *Switzerland.*
Arents, Ms Gina. *USA.*
Argüelles, Mr Waldo. *Cuba.*
Argay, Mr Gyula. *Hungary.*
Argo, Dr James L. *Canada.*
Argos, Dr Patrick. *USA.*
Arguello, Dr Zoraide Primerano. *Brazil.*
Arhar, Magistar Andrej. *Yugoslavia.*
Arif, Dr Atta Mahmood. *USA.*
Ariguib, Dr Najia. *Tunisia.*
Arjunan, P. *India.*
Arkhipenko, Dr Diana Konstantinovna. *USSR.*
Armağan, Prof. Dr Nizamettin. *Turkey.*
Armbruster, Dr Thomas. *BRD.*
Armendarez, Mr Peter X. *USA.*
Armigliato, Dr Aldo. *Italy.*
Armstrong, Prof. Ronald William. *USA.*
Arnberg, Dr Lars. *Sweden.*
Arndt, Dr Ulrich Wolfgang. *UK.*
Arndt, Prof. Dr Jörg Friedrich. *BRD.*
Arni, Mr Raghuuir Krishnaswamy. *BRD.*
Arnold, Mrs Christine. *DDR.*
Arnold, Prof. Heinrich. *DDR.*
Arnold, Mr Rolf. *DDR.*
Arnold, Prof. Dr Heinrich Günther Alfred. *BRD.*
Arnold, Dr Edward Van Dyke. *USA.*
Arnone, Prof. Arthur. *USA.*
Arnott, Prof. Struther. *USA.*
Arnoux, Mrs Bernadette *France.*
Arnstein, Dr Gustavo. *Venezuela.*
Aronsson, Prof. Bertil. *Sweden.*
Arora, Mr Narinder Kumar. *India.*
Arora, Prof. Satish Kumar. *USA.*
Arribas, Prof. Dr Antonio. *Spain.*
Arrington, Mr Wendell. *USA.*
Arruda, Prof. Moacir Rabelo. *Brazil.*
Artüz, Prof. Dr Samime. *Turkey.*
Artioli, Dr Gilberto. *Italy.*
Artioli, Gilberto. *USA.*
Artus, Dr Luis. *Spain.*
Artymiuk, Dr Peter Joseph. *UK.*
Aruffo, Mr Alejandro Antonio. *USA.*
Arzi, Dr Ezatollah. *Iran.*
Arzumanyan, Dr Gennadiy Ashotovich. *USSR.*
Asada, Prof. Eiichi. *Japan.*
Asadchikev, Dr Viktor Evgen'yevich. *USSR.*
Asadi, Dr Parviz. *Iran.*
Asadov, Dr Yusif Gazanfar ogly. *USSR.*
Asai, Dr Takeshi. *Japan.*
Ascher, Dr Edgar. *Switzerland.*
Ashida, Dr Sakichi. *Japan.*
Ashida, Prof. Tamaichi. *Japan.*
Ashirov, Dr Aman. *USSR.*
Ashkenazi, Dr Joseph. *Israel.*
Ashmore, Dr John Patrick. *Canada.*
Ashry, Dr Mamdouh. *Egypt.*
Ashwell, Dr Geoffrey Joseph. *UK.*
Ashworth, Dr Terence Vincent. *South Africa.*
Askhabov, Dr Askhab Magomedovich. *USSR.*
Aslanian, Dr Selma. *Bulgaria.*
Aslanov, Prof. Leonid Alexandrovich. *USSR.*
Asmat, Dr Humberto. *Peru.*
Asplund, Miss Milja. *Sweden.*
Atabayeva, Dr Eleonora Yakubovna. *USSR.*
Atanassov, Dr Vassil. *Bulgaria.*
Atasoy, Prof. Ertuğrul. *Turkey.*
Atkinson, Dr David. *USA.*
Atoji, Dr Masao. *USA.*
Atovmyan, Prof. Lev Oganovich. *USSR.*
Attard, Dr Alfred E. *USA.*
Attig, Dr Rainer. *BRD.*
Atwood, Prof. Jerry Lee. *USA.*
Atzmony, Prof. Uzi. *Israel.*
Au, Dr Andrew Yu-Chung. *USA.*

Aubry, Dr André Roger. *France.*
Audier, Dr Marc Eugene Raymond. *France.*
Auf Der Heyde, Mr Thomas P.E. *South Africa.*
Augustin, Mr Rolf M., Jr. *USA.*
Auld, Mr John Hugh. *Australia.*
Auleytner, Prof. Julian Jan. *Poland.*
Aurivillius, Prof. Bengt. *Sweden.*
Ausio Casas, Mr. *Spain.*
Austerman, Mr Stanley Boone. *USA.*
Austria, Dr Benjamin Suarez. *Philippines.*
Authier, Prof. André *France.*
Avalos, Dr Jaime. *Peru.*
Avci, Dr Recep. *Saudi Arabia.*
Avdiyenko, Dr Klavdiya Ilyinishna. *USSR.*
Averbach, Prof. B. L. *USA.*
Averbuch-Pouchot, Dr Marie-Thérèse. *France.*
Avery, Dr A.J. *UK.*
Avey, Dr Hugh Philip. *Australia.*
Avila-Salinas, Mr Waldo. *Bolivia.*
Avilov, Dr Anatoly Sergeyevich. *USSR.*
Avramov, Dr Isak. *Bulgaria.*
Awasthi, Dr Santosh Kumar. *India.*
Axmann, Dr Anton. *BRD.*
Aydin, Dr Nihal. *Turkey.*
Aydinol, Dr Mahmut. *Turkey.*
Aydinuraz, Doç. Dr Arsin. *Turkey.*
Aykan, Mr Kamran. *USA.*
Ayora Ibañez, Dr Carlos. *Spain.*
Aypar, Doç. Dr Abidin. *Turkey.*
Ayroles, Dr René *France.*
Aytaş, Doç. Dr S. Işik. *Turkey.*
Azarnia, Dr Nezhat. *Iran.*
Azaroff, Prof. Leonid V. *USA.*
Azer, Dr Nazmi. *Egypt.*
Azmon, Prof. Emanuel. *Israel.*
Azorin Marin, Mr F. *Spain.*
Azuma, Mr Nagao. *Japan.*
Baak, Ing. Leonardus Cornelis. *Netherlands.*
Baars, Dr-Ing Jan Walter. *BRD.*
Baba, Mrs Jasmin. *Malaysia.*
Babareko, Dr Alesya Adamovna. *USSR.*
Babel, Prof. Dr Dietrich. *BRD.*
Babich, Prof. Michael Wayne. *USA.*
Babič, Dr Danilo. *Yugoslavia.*
Bachechi, Dr Fiorella. *Italy.*
Backhaus, Dr Karl-Otto. *DDR.*
Bacon, Prof. George Edward. *UK.*
Badawi, Dr H.M. *UK.*
Bade, Mr Dirk. *BRD.*
Badger, Mr John. *UK.*
Badie, Mr Jean-Marie. *France.*
Badr, Dr Yehia Abd-El Hamid. *Egypt.*
Bäckerud, Dr Lennart S. *Sweden.*
Baele, Miss Ingrid Albertina Frans M. *Belgium.*
Baenziger, Prof. Norman C. *USA.*
Bärlocher, Dr Christian. *Switzerland.*
Bärnighausen, Prof. Dr Hartmut. *BRD.*
Baert, Dr François. *France.*
Baeta, Dr Robert Domingo. *Ghana.*
Baffier, Dr Noel. *France.*
Bagchi, Prof. Subodh Nath. *Canada.*
Bagdasarov, Dr, Khachik Saakovich. *USSR.*
Baggio, Dr Ricardo. *Argentina.*
Baggio, Dr Sergio. *Argentina.*
Bagley, Mr Arthur George. *UK.*
Bagshaw, Dr Anthony Nicholas. *Australia.*
Bai, Mr Chun-li. *China.*
Bailey, Mr David Eric. *Australia.*
Bailey, Prof. David Kenneth. *UK.*
Bailey, Mrs Marcia F. *USA.*
Bailey, Dr Neil Anthony. *UK.*
Bailey, Ms S. *UK.*
Bailey, Prof. Sturges Williams. *USA.*
Bain, Dr Derek Charles. *UK.*
Bainbridge, Mr John Evelyn. *UK.*
Baird, Prof. Herbert Wallace. *USA.*
Bakakin, Dr Vladimir Vasilyevich. *USSR.*
Baker, Mr Anthony Thomas. *Australia.*
Baker, Dr Edward Neill. *New Zealand.*
Baker, Mr Kenneth Neil. *USA.*
Baker, Mr Patrick Julian. *UK.*

Baker, Mr R.W. *UK.*
Baker, Mr Thomas Wilfred. *UK.*
Bąk, Dr Jadwiga. *Poland.*
Bakr, Dr Abdel Razak. *Saudi Arabia.*
Bakshi, Dr Edward. *Australia.*
Balagurov, Mr Anatoly Mikhailovich. *USSR.*
Balakirev, Dr Vladimir Georgiyevich. *USSR.*
Balan, Mr Mihai. *Romania.*
Balarin, Prof. Manfred. *DDR.*
Balasingh, Mr C. *India.*
Balasubramanian, Dr R. *India.*
Balcazar, Dr Jose Luis. *Spain.*
Balchin, Dr Anthony Arthur. *UK.*
Balcou, Dr Yves. *France.*
Baldrian, Dr Josef. *CSSR.*
Baldwin, Mr Kenneth John. *USA.*
Bale, Prof. Harold D. *USA.*
Bales, Mr Howard E. *USA.*
Balić Žunić, Dr Tonči. *Yugoslavia.*
Balibar, Prof. Françoise. *France.*
Balitsky, Dr Vladimir Sergeyevich. *USSR.*
Balkanov, Mr Ivan. *Bulgaria.*
Ball, Dr Richard George. *Canada.*
Ball, Prof. Anthony. *South Africa.*
Bally, Conf. Dorel. *Romania.*
Bally, Dr Renée. *France.*
Baltă, Conf. Petru. *Romania.*
Balta Calleja, Prof. Francisco Jose. *Spain.*
Baltzinger, Dr Christiane. *France.*
Balyuzi, Dr Hushang H.M. *UK.*
Balzar, Mr Davor. *Yugoslavia.*
Balzarotti, Prof. Adalberto. *Italy.*
Bambauer, Prof. Dr Hans Ulrich. *BRD.*
Ban, Prof. Dr Zvonimir. *Yugoslavia.*
Banaii, Dr Nasser. *Iran.*
Banaszak, Prof. Leonard J. *USA.*
Banbery, Mrs H.J. *UK.*
Bandoli, Prof. Giuliano. *Italy.*
Bandopadhyay, Mrs Tapati. *India.*
Bandy, Dr Judith Ann. *UK.*
Banerjee, Dr Asok. *India.*
Banerjee, Dr Bani Ranjan. *USA.*
Banerjee, Dr Krishna. *India.*
Banerjee, Dr Srikumar. *India.*
Bang, Dr Eva Henriette. *Denmark.*
Banizs, Dr Károly. *Hungary.*
Banks, Prof. Ephraim. *USA.*
Banner, Dr David William. *BRD.*
Bansigir, Prof. K. Goswami. *India.*
Baptista, Mr Augusto. *Brazil.*
Baptista, Mrs Neysa Rocha. *Brazil.*
Baqri, Dr Syed Rafiqul-Hassan. *Pakistan.*
Bar, Mrs Ilana. *Israel.*
Baraba, Miss Widad. *Indonesia.*
Barabanov, Prof. Vladimir Fedorovich. *USSR.*
Barakat, Prof. Nayel. *Egypt.*
Baran, Dr Zbigniew. *Poland.*
Baranov, Dr Anatoliy Ivanovich. *USSR.*
Baranova, Dr Raisa Vladimirovna. *USSR.*
Baranyi, Dr Anthony David. *Canada.*
Barbagelata, Mr Franco. *Chile.*
Barber, Prof. Patrick George. *USA.*
Barbieri, Prof. Renato. *Italy.*
Barcik, Dr Jan. *Poland.*
Bardhan, Pronob. *USA.*
Bardi, Prof. Renato. *Italy.*
Barelli, Dr Nilso. *Brazil.*
Baresel, Dr-Ing Detlef Wilhelm Berthold. *BRD.*
Barker, Dr William Wilson. *Australia.*
Barkigia, Dr Kathleen M. *USA.*
Barna, Dr Péter. *Hungary.*
Barnea, Dr Zwi. *Australia.*
Barnes, Dr Charles Leslie. *USA.*
Barnes, Dr John Conquest. *UK.*
Barnes, Dr Paul. *UK.*
Barnett, Dr Bobby L. *USA.*
Barney, Ms. Elsa Pauline. *USA.*
Barnhart, Dr David Merle. *USA.*
Baro, Prof. Raymond. *France.*
Barrans, Dr Yvette. *France.*

Barraud, Prof. Jean. *France.*
Barraud, Mr Jean-Yves. *France.*
Barret, Mr Nicholas. *Italy.*
Barrett, Mr S.D. *UK.*
Barrett, Prof. Charles Sanborn. *USA.*
Barrick, Dr James Clinton. *USA.*
Barriga Villalba, Prof. Antonio Maria. *Colombia.*
Barrington-Leigh, Dr John. *Canada.*
Barrios, Mr Nelson. *Chile.*
Barron, Dr Hugh Wilson Taylor. *UK.*
Barrow, Dr Michael John. *UK.*
Barsukova, Dr Marina Leonidovna. *USSR.*
Barszcz, Doc Edward. *Poland.*
Bart, Prof. Dr Jan J. C. *Israel.*
Barta, Ing Čestmír. *CSSR.*
Bartczak, Dr Tadeusz Jan. *Poland.*
Bartell, Prof. Lawrence Sims. *USA.*
Bartels, Drs Willem Jan. *Netherlands.*
Barthel, Dr Johannes. *DDR.*
Bartl, Prof. Dr Hans. *BRD.*
Bartlett, Dr Michael William. *Canada.*
Bartlett, Prof. Neil. *USA.*
Barton, Dr C. J. *USA.*
Barton, Dr Randolph Jr. *USA.*
Barton, Prof. Richard J. *Canada.*
Bartoshinsky, Dr Zbignev Vladislavovich. *USSR.*
Bartsch, Dr Hans-Hagen. *BRD.*
Baruchel, Dr José *France.*
Basak, Prof. Bejoysanker. *India.*
Basak, Dr Madan Gopal. *India.*
Basha, Dr Ahmed Fouad. *Egypt.*
Basilakis, Mr Michael. *Greece.*
Bassett, Prof. David Clifford. *UK.*
Bassett, Mr G. Alan. *UK.*
Bassiouny, Dr Mohamed Khafagi. *Egypt.*
Basso, Prof. Riccardo. *Italy.*
Basson, Dr Stephen Smuts. *South Africa.*
Bastiansen, Prof. Otto Christian Astrup. *Norway.*
Bastin, Dr Ir Guillaume F. *Netherlands.*
Basu, Dr Sankar Prasad. *India.*
Bataliyeva, Dr Nataliya Glebovna. *USSR.*
Batchelder, Dr David Neville. *UK.*
Bateman, Dr Linda Ratner. *USA.*
Bates, Dr David Ronald. *USA.*
Bates, Dr Peter Arthur. *UK.*
Bates, Prof. Richard Heaton Tunstall. *New Zealand.*
Bates, Prof. Robert Brown. *USA.*
Batlle Sales, Mr Jorge. *Spain.*
Bats, Dr Jan, Willem. *BRD.*
Batsanov, Dr Andrey Stepanovich. *USSR.*
Batt, Mr Ahron. *Israel.*
Battaglia, Prof. Luigi Pietro. *Italy.*
Batterman, Prof. Boris William. *USA.*
Battey, Dr Maurice Hugh. *UK.*
Battezzati, Prof. Livio. *Italy.*
Bau, Prof. Robert. *USA.*
Baucher, Mr Alain. *France.*
Baudet, Dr Mona. *France.*
Baudour, Dr Jean-Louis. *France.*
Bauduin, Miss Anne-Marie Ghislaine G. *Belgium.*
Bauer, Prof. Dr Günther Ernst. *Austria.*
Bauer, Prof. Dr Ernst Georg. *BRD.*
Bauer, Doc. Ing Jaroslav. *CSSR.*
Baughman, Mr Richard Joseph. *USA.*
Baumann, Mr Jürgen Rudolf. *BRD.*
Baumbach, Mr Manfred. *DDR.*
Baumgärtel, Mr Rolf. *DDR.*
Baumgart, Mr Helmut. *BRD.*
Baumgartner, Dr Dipl.-Ing. Oswald. *Austria.*
Baur, Prof. Werner Heinz. *USA.*
Bautsch, Prof. Hans-Joachim. *DDR.*
Bavoux, Miss Claude. *France.*
Baxter, Mr Colin. *UK.*
Bayer, Prof. Gerhard. *Switzerland.*
Bayh, Prof. Dr Werner. *BRD.*
Bayliss, Prof. Peter. *Canada.*
Baysal, Prof. Dr Orhan. *Turkey.*
Bayvas, Dr Orhan. *Turkey.*
Beagley, Dr Brian. *UK.*
Beale, Dr John Phillip. *Australia.*

Beamson, Dr G. *UK.*
Bear, Prof. Richard Scott. *USA.*
Beard, Mr Donald W. *USA.*
Beasley, Prof. Wayne Machon. *USA.*
Beauchamp, Prof. Dr André. *Canada.*
Becherer, Dr Karl. *Austria.*
Becherer, Prof. Gerhard. *DDR.*
Beck, Prof. Dr Horst Philipp. *BRD.*
Becker, Dr Claus. *DDR.*
Becker, Prof. Dr Gerd. *BRD.*
Becker, Prof. Joseph Whitney. *USA.*
Becker, Mr Paul. *France.*
Becker, Prof. Pierre. *France.*
Becker, Mr Reinhardt. *DDR.*
Beckingsale, Mr Peter Gerard. *New Zealand.*
Beckmann, Prof. Günter. *DDR.*
Bedarida, Prof. Federico. *Italy.*
Bedarkar, Dr Sudhir. *USA.*
Beddell, Dr Christopher Raymond. *UK.*
Beddoes, Mr Roy L. *UK.*
Bednar, Mr E. *UK.*
Bednarski, Dr Stanisław. *Poland.*
Bednowitz, Dr Allan Lloyd. *USA.*
Bedyńska, Dr Teresa Ewa. *Poland.*
Been, Dr J.M. *UK.*
Beevers, Dr Cecil Arnold. *UK.*
Beglari, Dr Ing. Parviz. *Iran.*
Begley, Dr Michael John. *UK.*
Behm, Dr Helmut Johannes Julius. *Netherlands.*
Behrens, Dr Heinrich. *BRD.*
Behrens, Dr Ulrich Hermann. *BRD.*
Behruzi, Dr Massoud. *BRD.*
Beier, Dr Wilfried. *DDR.*
Bekrenev, Dr Anatoly Nikolayevich. *USSR.*
Bel Hassen, Miss Dalila. *Tunisia.*
Bel'sky, Dr Vitaly Konstantinovich. *USSR.*
Belanger-Gariepy, Mme Francine. *Canada.*
Belhadj, Dr Ali. *Tunisia.*
Belikova, Dr Galina Sergeyevna. *USSR.*
Belkheria, Mr Salah. *Tunisia.*
Bell, Mr Anthony Martin Thomas. *UK.*
Bell, Mr Jeffrey A. *USA.*
Bellamy, Mr Brian Arthur. *UK.*
Bellard, Dr Sharon Ann. *UK.*
Bellon, Prof. Pier Luigi. *Italy.*
Belokoneva, Dr Elena Leonidovna. *USSR.*
Belova, Dr Elizaveta Nikolayevna. *USSR.*
Belt, Dr Roger F. *USA.*
Belugina, Dr Nataliya Vasilyevna. *USSR.*
Belyaeva, Mrs Klara Fedorovna. *USSR.*
Belyustin, Dr Aleksey Vsevolodovich. *USSR.*
Belzner, Mr Andreas. *BRD.*
Ben Amor, Dr. *Tunisia.*
Ben Brahim, Dr Jemaïel. *Tunisia.*
Ben Ghozlane, Dr Hédi. *Tunisia.*
Ben Romdhane, Dr. *Tunisia.*
Ben Salah, Dr Abdelhamid. *Tunisia.*
Ben Yair, Dr Moshe Pinhas. *Israel.*
Ben-Hussein, Mr Ahmed Othan. *USA.*
Benci, Mr Pierluigi. *USA.*
Bendeliani, Dr Nikolay Alexandrovich. *USSR.*
Bender, Dr Hugo J.M.R. *Belgium.*
Benedetti, Prof. Ettore. *Italy.*
Benetollo, Dr Franco. *Italy.*
Benghiat, Dr Victor. *Israel.*
Bengochea, Dr Amado Leandro. *Argentina.*
Benna, Dr Piera. *Italy.*
Bennema, Prof. Dr Pieter. *Netherlands.*
Bennett, Mr Dennis W. *USA.*
Bennett, Dr John Michael. *USA.*
Bennett, Miss S. *UK.*
Beno, Dr Mark A. *USA.*
Bensch, Dr Wolfgang Johannes Georg. *Switzerland.*
Benson, Mr James Edward. *USA.*
Bente, Dr Klaus Alexander. *BRD.*
Bentley, Dr Graham Arthur. *France.*
Benyacar, de, Lic Maria Angélica R. *Argentina.*
Beran, Doz. Dr Anton. *Austria.*
Beres, Mr John J. *USA.*
Beresnev, Dr Leonid Alekseyevich. *USSR.*
Beretka, Mr Julius. *Australia.*

Berezhkova, Dr Galina Vasilyevna. *USSR.*
Berg, van den, Ir Adrianus Johannes. *Netherlands.*
Berg, Mrs Dr Lieselotte. *BRD.*
Berger, Dr Hans. *DDR.*
Berger, Dr Rolf Anders. *Netherlands.*
Bergerhoff, Prof. Dr Guenter. *BRD.*
Bergevin, de, Dr François. *France.*
Berggren, Dr Jan. *Sweden.*
Bergmann, Mrs Margot Eisenhardt. *USA.*
Bergner, Mr Joachim. *DDR.*
Bergsma, Dr Jitze. *Netherlands.*
Berkebile, Prof. C. Alan. *USA.*
Berkey, Mr W. W. *USA.*
Berking, Dr Bernhard. *BRD.*
Berkovitch-Yellin, Dr Ziva. *Israel.*
Berman, Dr Helen M. *USA.*
Bernal de Ramírez, Prof. Inés. *Colombia.*
Bernal, Prof. Ivan. *USA.*
Bernalte, Prof. Antonio. *Spain.*
Bernard, Prof. William H. *USA.*
Bernardinelli, Dr Gérald. *Switzerland.*
Bernheim, Dr Marguerite May Yevitz. *USA.*
Bernotat-Wulf, Mrs Dr Hannelore. *BRD.*
Bernotat, Dr Walter Hermann. *BRD.*
Bernstein, Ms Frances C. *USA.*
Bernstein, Dr Herbert Jacob. *USA.*
Bernstein, Prof. Joel. *Israel.*
Bernstein, Mr Joel L. *USA.*
Berry, Dr Chester Ridlon. *USA.*
Berset, Mr Guy Alexis. *Switzerland.*
Bershov, Dr Leonid Viktorovich. *USSR.*
Bersuker, Prof. Isaak Borukhovich. *USSR.*
Bertaut, Dr Erwin Félix. *France.*
Bertelmann, Mr Dieter Wilhelm. *BRD.*
Berthet-Colominas, Dr Carmen. *France.*
Berthold, Prof. Dr Hans Joachim. *BRD.*
Berthold, Mr Thomas. *BRD.*
Bertolasi, Prof. Valerio. *Italy.*
Bertram, Mrs Marion. *DDR.*
Bertrand, Prof. Joseph Aaron. *USA.*
Besoaín, Dr Eduardo. *Chile.*
Bessiere Dr Michel. *France.*
Besteiro Rafales, Dra Josefina. *Spain.*
Betal, Mr Badal Kumar. *India.*
Bethge, Dr Paul Herman. *USA.*
Betsofen, Dr Sergey Yakovlevich. *USSR.*
Betts, Dr Foster. *USA.*
Betz, Doz. Dr Gerhard. *Austria.*
Betz, Mr Helmut. *BRD.*
Betzel, Mr Christian. *BRD.*
Betzl, Dr Manfred. *DDR.*
Beukes, Prof Dr Gerhardus Johannes. *South Africa.*
Beurskens, Dr Gezina. *Netherlands.*
Beurskens, Prof. Dr Paul T. *Netherlands.*
Bevan, Prof David John Martin. *Australia.*
Bevis, Prof. M.J. *UK.*
Beyer, Mrs Angelika. *BRD.*
Beyer, Dr Ir Jenö. *Netherlands.*
Bezinge, Mr Alex. *Switzerland.*
Bezjak, Prof. Dr Aleksandar. *Yugoslavia.*
Beznosikov, Dr Boris Valeriyanivich. *USSR.*
Bhadbhade, Mr Mohan Madhav. *India.*
Bhaduri, Dr Debabrata. *India.*
Bhagvantam, Prof. Suri. *India.*
Bhagwat, Dr Vasant. *India.*
Bhakay-Tamhane, Mrs Sandhya Nitin. *India.*
Bhaktapriya, Dr S. R. Y. *India.*
Bhandary, Dr Krishna K. *USA.*
Bhardwaj, Mr Jayant. *USA.*
Bhargava, Dr L. R. *India.*
Bhat, Dr Laxminarayana H. *India.*
Bhat, Dr Narayana Talapady. *USA.*
Bhat, Mr T. Narayana. *India.*
Bhatia, Mr Subhash Chandra. *India.*
Bhatt, Dr Vaikunthray Promodray. *India.*
Bhattacharjee, Mrs Lilabati. *India.*
Bhattacharjee, Mr Sovan K. *USA.*
Bhattacharya, Archana. *India.*
Bhattacharya, Dr Ramendranarayan. *India.*
Bhattacharyya, Dr Subodh Chandra. *India.*

Bhattacherjee, Mr Santi Brata. *India.*
Bhattacherjee, Dr Satyananda. *India.*
Bhawalkar, Dr Ramkrishna Haribhau. *India.*
Bhuva, Dr V.J. *UK.*
Bi, Dr Ru-chang. *China.*
Bi, Mr Yu-run. *China.*
Biagini Cingi, Prof. Marina. *Italy.*
Biais, Mrs Régine. *France.*
Bianchi, Dr Riccardo. *Italy.*
Bianchi Orlandini, Dr Annabella. *Italy.*
Bichurin, Dr Rinnat Chingizkhanovich. *USSR.*
Bideau, Dr Jean-Pierre. *France.*
Biedl, Dr Albrecht. *BRD.*
Bielen, Prof. Dr Helmut Josef. *BRD.*
Bienenstock, Prof. Arthur Irwin. *USA.*
Bienfait, Prof. Michel. *France.*
Bigelow, Prof. Wilbur Charles. *USA.*
Bigi Dr Adriana. *Italy.*
Bigoli, Prof. Francesco. *Italy.*
Bijen, Dr Jan. *Netherlands.*
Bilderback, Dr Donald Heywood. *USA.*
Billiet, Prof. Yves. *Tunisia.*
Binas, Dr Horst. *DDR.*
Bindal, Mr Arvind. *India.*
Bindlish, Mr Jag Mohan. *India.*
Binnie, Dr William Polson. *USA.*
Bino, Dr Avi. *Israel.*
Bird, Prof. Peter Hans. *Canada.*
Birker, Dr Paul J. M. W. L. *Netherlands.*
Birks, Mr L. S. *USA.*
Birnbaum, Dr George I. *Canada.*
Birnbaum, Dr Karin Bjåmer. *Canada.*
Birnstock, Dr Ronald Alfred Harri. *BRD.*
Bish, Dr David Lee. *USA.*
Bishop, Dr Arthur Clive. *UK.*
Bisi Castellani, Prof. Carla. *Italy.*
Bissert, Mrs Dr Elisabeth Gertrud. *BRD.*
Bist, Dr B. M. S. *India.*
Biswas, Goutam. *India.*
Biswas, Dr Mohommad Alim. *Bangladesh.*
Biswas, Mr Subhash Chandra. *India.*
Biswas, Dr Sundar Gopal. *India.*
Biyushkin, Dr Victor Nikolayevich. *USSR.*
Bizid, Dr Abdelmalek dit Youssef. *Tunisia.*
Björnberg, Dr Arne A. *Sweden.*
Björnevåg, Mr Sturle Vegard. *Norway.*
Bjorkman, Ms Pamela J. *USA.*
Blažina, Dr Želimir. *Yugoslavia.*
Black, Prof. Paul Joseph. *UK.*
Blake, Dr Alexander John. *UK.*
Blake, Dr Antony Brian. *UK.*
Blake, Mr Ronald George. *Australia.*
Blake, Mr J. W. *USA.*
Blanco, Mrs Magaly. *Cuba.*
Bland, Dr J.A. *UK.*
Blankenburg, Dr Hans-Joachim. *DDR.*
Blanton, Mr Thomas Nelson. *USA.*
Blaschke, Prof. Dr Rochus Bruno Albert. *BRD.*
Blaschko, Dr Oskar. *Austria.*
Blasi, Prof. Achille. *Italy.*
Blaton, Dr Norbert Louis. *Belgium.*
Blech, Prof. Ilan Asriel. *Israel.*
Bleeker, Drs Ernö Johan. *Netherlands.*
Bleidelis, Dr Yanis Yazepovich. *USSR.*
Bleif, Dr Hans-Jürgen. *BRD.*
Blessing, Dr Robert Harry. *USA.*
Blinov, Dr Lev Mikhailovich. *USSR.*
Blinowski, Dr Konrad. *Poland.*
Bliznakov, Prof. Georgi. *Bulgaria.*
Block, Dr Stanley. *USA.*
Block, Prof. Dr Jochen Hermann. *BRD.*
Bloem, Prof. Dr Jan. *Netherlands.*
Blokhin, Prof. Mikhail Arnol'dovich. *USSR.*
Bloor, Dr David. *UK.*
Blount, Dr John Franklin. *USA.*
Blow, Prof David Mervyn. *UK.*
Bloxam, Dr Thomas Wallace. *UK.*
Blüthgen, Dipl-Min Waldemar. *BRD.*
Bluhm, Dr Terry Lee. *Canada.*
Blundell, Dr David James. *UK.*

Blundell, Prof. Thomas Leon. *UK.*
Bołd, Dr Tadeusz. *Poland.*
Bocelli, Dr Gabriele. *Italy.*
Bochud, Mr Jean-Daniel. *Switzerland.*
Bock, Mr Hans. *BRD.*
Bodak, Dr Oksana Ivanovna. *USSR.*
Bodor, Prof. Géza. *Hungary.*
Bögge, Dr Hartmut. *BRD.*
Boehm, Prof. Dr Hanns-Peter. *BRD.*
Böhm, Dr Horst. *BRD.*
Boehm, Dr James M. *Australia.*
Böhme, Mrs Dr Reinhild. *BRD.*
Boehme, Mr Richard Frederick. *USA.*
Boer, de, Dr Jan Louwert. *Netherlands.*
Boese, Dr Roland. *BRD.*
Boesman, Dr Etienne Roland. *Belgium.*
Böttcher, Dr Peter. *BRD.*
Boeyens, Prof. Jan Christoffel Antonie. *South Africa.*
Boggs, Dr Rita Rose. *USA.*
Bognár, Dr László. *Hungary.*
Bogucka-Ledóchowska, Dr Maria. *Poland.*
Bohac, Dr Petr. *Switzerland.*
Bohatý, Dr Ladislav. *BRD.*
Bohm, Dr Joachim. *DDR.*
Boiko, Prof. Boris Timofeyevich. *USSR.*
Boikova, Dr Alexandra Ivanovna. *USSR.*
Bois, Dr Claudette. *France.*
Boistelle, Dr Roland. *France.*
Boiwe, Mr Torne R. *Sweden.*
Bojarski, Prof. Zbigniew. *Poland.*
Bokiy, Prof. Georgy Borisovich. *USSR.*
Boldrini, Dr Piero. *Canada.*
Boles, Dr Michael Owen. *UK.*
Bolhuis, van, Mr Fré. *Netherlands.*
Bolin, Mr Jeffrey T. *USA.*
Bolis, Dr Vera Maria. *Italy.*
Boller, Prof. Dr Herbert. *Austria.*
Bolognesi, Prof. Martino. *Italy.*
Bolzenius, Mr Beda Helmut. *BRD.*
Bombieri, Prof. Gabriella. *Italy.*
Bonamartini-Corradi, Prof. Anna. *Italy.*
Bonamico, Dr Mario. *Italy.*
Bond, Mrs Dianne Ruth. *South Africa.*
Bondar', Dr Iraida Adamovna. *USSR.*
Bondars, Dr Bruno Yanovich. *USSR.*
Bondza, Mr Harald Werner. *BRD.*
Bonefačić, Prof. Dr Antun. *Yugoslavia.*
Bonev, Dr Ivan. *Bulgaria.*
Bonham, Prof. Russell Aubrey. *USA.*
Bonnaud, Dr Bernard Henri. *Ivory Coast.*
Bonnelle, Prof. Christiane. *France.*
Bonnet, Prof. Jean-Jacques. *France.*
Bonnet, Dr Michel. *France.*
Bonpunt, Dr Louis. *France.*
Bonse, Prof. Dr Ulrich Karl Eberhard. *BRD.*
Boo, Mr W. O. J. *USA.*
Boom, Dr Geert. *Netherlands.*
Boonstra, Prof. Eelco Gerrit. *South Africa.*
Boorman, Dr Philip Michael. *Canada.*
Booth, Dr Andrew Donald. *Canada.*
Borchardt-Ott, Dr Walter. *BRD.*
Bordeaux, Mrs Denise. *France.*
Bordner, Dr Jon. *USA.*
Borges, Prof. Frederico. *Portugal.*
Borie, Dr Bernard Simon. *USA.*
Borisanova, Dr Lidiya Mikhailovna. *USSR.*
Borisov, Dr Stanislav Vasilyevich. *USSR.*
Borisov, Dr Vsevolod Vasilyevich. *USSR.*
Borkakoti, Dr Nivedita Neera. *UK.*
Born, Dr Eberhard. *BRD.*
Born, Dr Liborius. *BRD.*
Born, Mr Reinhard. *BRD.*
Bornmann, Dr Horst. *DDR.*
Bornmann, Dr Peter. *DDR.*
Borowiak, Dr hab. Teresa. *Poland.*
Borrmann, Prof. Dr Ing. Gerhard. *BRD.*
Bosch Figueroa, Dr Jose Ma. *Spain.*
Bosch Giral, Dr Pedro. *Mexico.*
Bose, Mr Shyamal Kumar. *India.*
Boskey, Dr Adele Ludin. *USA.*
Bosman, Drs Wilhelmus P. J. H. *Netherlands.*

Bosmans, Prof. Dr Herman Jozef. *Belgium.*
Boss, Dr James William. *USA.*
Bostanov, Mr Vesselin. *Bulgaria.*
Boström, Mr N. Dan. *Sweden.*
Botoshansky, Dr Mark Meyerovich. *USSR.*
Bottomley, Prof. Frank. *Canada.*
Bottyán, Dr László. *Hungary.*
Bouška, Doc. Dr Vladimír. *CSSR.*
Boucher, Dr Bernard Yves. *France.*
Boucherle, Dr Jean-Xavier. *France.*
Boudreau, Prof. Sharon Martin. *USA.*
Bouraoui, Prof. Ahmed. *Tunisia.*
Bourne, Mr Philip Eric. *USA.*
Bouwmeester, Drs Hennie J. M. *Netherlands.*
Bovin, Dr Jan-Olov. *Sweden.*
Bovio, Prof. Bruna. *Italy.*
Bowen, Dr A.W. *UK.*
Bowen, Dr David Keith. *UK.*
Bowen-Jones, Dr J. *UK.*
Bowles, Prof. John Stephen. *Australia.*
Bowman, Mr Allen L. *USA.*
Bown, Dr Michael George. *UK.*
Box, Mr Harold C. *USA.*
Boyarskaya, Prof. Yuliya Stanislavovna. *USSR.*
Boyiatzis, Mr Ioannis. *Greece.*
Boyko, Prof. Edward Raymond. *USA.*
Boyle, Dr Lewis Laurence. *UK.*
Boyle, Mr Paul. *USA.*
Boys, Dr Daphne. *Chile.*
Boysen, Dr Hans. *BRD.*
Bozopoulos, Mr Anastasios Panayiotis. *Greece.*
Braam, Dr Adrianus Wilhelmus Maria. *Netherlands.*
Bradaczek, Prof. Dr Hans Arthur. *BRD.*
Braden, Dr Bradford Carl. *USA.*
Brader, Mr James J. *USA.*
Bradley, Ms S.M. *UK.*
Brady, Mr George W. *USA.*
Brady, Mr James Henry. *USA.*
Brämer, Dr Wulf. *BRD.*
Brändén, Prof. Carl-Ivar. *Sweden.*
Bräutigam, Dr Gunter. *DDR.*
Braga, Dr Dario. *Italy.*
Braganza, Dr Lellis Francis. *France.*
Bragg, Prof. Robert Henry. *USA.*
Braibanti, Prof. Antonio. *Italy.*
Brainin, Dr Boris Matveyevich. *USSR.*
Brand, Prof. Paul. *DDR.*
Brandberg, Dr Ola. *Sweden.*
Brandmueller, Prof. Dr Josef Karl August. *BRD.*
Brandon, Dr James Kenneth. *Canada.*
Brandstätter Dr Franz. *Austria.*
Brandt, Ing Gernot. *BRD.*
Brasseur, Prof. Dr Henri Alphonse Lambert. *Belgium.*
Brassy, Dr Claude. *France.*
Brathovde, Prof. James Robert. *USA.*
Brauer, Prof. Dr Georg Karl. *BRD.*
Braun, Dr Dieter Johann. *BRD.*
Braun, Dr Eckart. *BRD.*
Braun, Dr Poul Bernard. *Netherlands.*
Brauny, Mr Siegfried. *DDR.*
Bravic, Dr Georges. *France.*
Bravo, Prof. Manuel. *Portugal.*
Brayer, Dr Gary David. *Canada.*
Brech, Mr Frederick. *USA.*
Brehm, Dr Lotte. *Denmark.*
Breit, Dipl-Min Udo. *BRD.*
Breitinger, Prof. Dr Dietrich Karl. *BRD.*
Bremer, Dr Johannes. *Norway.*
Brenčič, Prof. Dr Jurij. *Yugoslavia.*
Brennan, Dr Richard Gerald. *USA.*
Brennan, Prof. Thomas Francis. *USA.*
Brenner, Mr Stephen A. *USA.*
Bresciani Pahor, Prof. Nevenka. *Italy.*
Březina, Ing Bohuslav. *CSSR.*
Brianso, Prof. Jose Luis. *Spain.*
Briant, Dr Clive Edward. *UK.*
Briard, Mrs Pierrette. *Ivory Coast.*
Brice, Dr John Chadwick. *UK.*
Brickenkamp, Dr Carroll Shelton. *USA.*
Brierley, Mr Cameron. *Canada.*

Brieva, Prof. Jorge Alfonso. *Colombia.*
Brigatti, Dr Maria Franca. *Italy.*
Bright, Dr Alan Aubrey Samuel. *UK.*
Bright, Dr William M. *USA.*
Briguglio, Mr James. *USA.*
Brill, Mr Wolfgang. *BRD.*
Brill, Prof. Dr Rudolf Friedrich. *BRD.*
Brinkmann, Prof. Detlef. *Switzerland.*
Brisse, Dr François. *Canada.*
Brisson, Mrs Josée. *Canada.*
Bristoti, Dr Anildo. *Brazil.*
Britten, Dr James Francis. *Canada.*
Britton, Prof. Doyle. *USA.*
Broach, Dr Robert William. *USA.*
Brock, Prof. Carolyn Pratt. *USA.*
Brodalla, Dipl.-Chem Dieter. *BRD.*
Brodersen, Prof. Dr Klaus. *BRD.*
Brokmeier, Dr Heinz-Günter. *BRD.*
Bronger, Prof. Dr Welf. *BRD.*
Bronowska, Dr Wieslawa Maria. *Poland.*
Bronsema, Drs Klaas Derk. *Netherlands.*
Broosch, Mrs Erika. *DDR.*
Broul, Ing Miroslav. *CSSR.*
Brovkin, Dr Anatoly Afanasyevich. *USSR.*
Brown, Miss Betty Rosina. *UK.*
Brown, Prof. Bruce Elliot. *USA.*
Brown, Prof. Bruce Willard. *USA.*
Brown, Dr Cedric John. *UK.*
Brown, Mr D. *UK.*
Brown, Dr David Summers. *UK.*
Brown, Dr George Marshall. *USA.*
Brown, Prof. Glenn H. *USA.*
Brown, Prof. Gordon Edgar, Jr. *USA.*
Brown, Dr Ian David. *Canada.*
Brown, Mr Jerry Howard. *Israel.*
Brown, Dr Joe Ned, Jr. *USA.*
Brown, Dr Kevin Laurie. *New Zealand.*
Brown, Dr Leo Dale. *USA.*
Brown, Prof Michael Ewart. *South Africa.*
Brown, Dr Penelope, Jane. *France.*
Brown, Dr Roger Norman. *Australia.*
Brown, Prof. William Liddle. *France.*
Browne, Mr Ian Bruce. *Australia.*
Bru, Prof. Luis. *Spain.*
Bruce, Dr P.G. *UK.*
Brückner, Dr Winfried. *DDR.*
Brückner, Prof. Sergio. *Italy.*
Brühl, Dr Hans-Gerd. *DDR.*
Brümmer, Prof. Otto. *DDR.*
Brauer, Dr Karl-Heinz. *DDR.*
Bruggen, van, Dr Christiaan Frans. *Netherlands.*
Bruins Slot, Drs Hilbert Jan. *Netherlands.*
Brumberger, Prof. Dr Harry. *USA.*
Brunel, Dr Michel. *France.*
Brunie, Dr Simone. *France.*
Bruno, Prof. Emiliano. *Italy.*
Brusewitz, Dr Ann Marie. *Sweden.*
Bruvo, Magistar Milenko. *Yugoslavia.*
Bruzzone, Prof. Giacomo. *Italy.*
Bryan, Prof. Robert Finlay. *USA.*
Bryant, Mr P.K. *UK.*
Bryden, Prof. John Heilner. *USA.*
Brzozowski, Andrzej Marek. *Poland.*
Bubáková, Dr Růžena. *CSSR.*
Bublitz, Mr Günter. *DDR.*
Buchanan, Prof. David R. *USA.*
Buchheiser, Dr Klaus. *DDR.*
Buchwald, Dr Vagn Fabritius. *Denmark.*
Buck, Prof. Dr Peter. *BRD.*
Buckley, Dr Christopher Paul. *UK.*
Bud'ko, Dr Ivetta Alexandrovna. *USSR.*
Budevski, Prof. Evgeni. *Bulgaria.*
Budurov, Prof. Stoyan. *Bulgaria.*
Buehner, Dr Manfred. *BRD.*
Bülow, Miss Renate. *BRD.*
Buerger, Prof. Martin Julian. *USA.*
Buerger, Dr Newton Weber. *USA.*
Bürgi, Prof. Hans-Beat. *Switzerland.*
Bürki, Dr Hans. *Switzerland.*
Büyükgüngör, Dr Orhan. *Turkey.*

Bugg, Prof. Charles E. *USA.*
Buket, Mr Ersen. *Turkey.*
Bukin, Dr Alexander Sergeyevich. *USSR.*
Bukovec, Dr Nataša. *Yugoslavia.*
Bukovec, Prof. Dr Peter. *Yugoslavia.*
Bukowska-Strzyzewska, Dr hab Maria. *Poland.*
Bukvetsky, Dr Boris Vladimirovich. *USSR.*
Bulakh, Dr Andrey Glebovich. *USSR.*
Bulc, Mrs Nada. *Yugoslavia.*
Bulhões, Mrs Iseli Angelica M. *Brazil.*
Bullen, Dr G.J. *UK.*
Bullen, Mr Henry Eric. *UK.*
Bulpett, Mrs S.E. *UK.*
Bunge, Prof. Dr Dr hc Hans-Joachim. *BRD.*
Bunick, Mr Gerard J. *USA.*
Bunn, Dr Charles William. *UK.*
Bunno, Mr Michiaki. *Japan.*
Burbank, Mr Robinson D. *USA.*
Burdina, Dr Valentina Ivanovna. *USSR.*
Burge, Prof. Ronald Edgar. *UK.*
Burgeat, Dr Jacques. *France.*
Burkhard, Dr Andreas. *Switzerland.*
Burkhardt, Mr Wolfgang. *DDR.*
Burla, Dr Maria Cristina. *Italy.*
Burnasheva, Dr Veniana Venediktovna. *USSR.*
Burnett, Prof. Roger MacDonald. *USA.*
Burnham, Prof. Charles Wilson. *USA.*
Burns, Dr John Howard. *USA.*
Burns, Mr K. *UK.*
Burschka, Dr Christian. *BRD.*
Burshtein, Dr Izya Fridelevich. *USSR.*
Bursill, Dr Leslie Arthur. *Australia.*
Burton, Dr Benjamin Paul. *USA.*
Burzlaff, Prof. Dr Hans. *BRD.*
Busaracome, Mr Suwin. *Thailand.*
Busch, Prof. Georg Adolf. *Switzerland.*
Buschmann, Dr Juergen Friedrich. *BRD.*
Buschow, Dr Kurt Heinz Jürgen. *Netherlands.*
Busetta, Dr Bernard. *France.*
Busetti, Prof. Vilma. *Italy.*
Bush, Dr Michael Anthony. *UK.*
Bushnell-Wye, Dr Graham. *UK.*
Bushnell, Prof. Gordon William. *Canada.*
Busing, Dr William Richard. *USA.*
Butcher, Prof. Raymond John. *USA.*
Butler, Dr Barry Conrad Milne. *UK.*
Butler, Dr Stephen Andrew. *UK.*
Butler, Dr William M. *USA.*
Butler, Mr William O. *USA.*
Butman, Dr Lev Abramovich. *USSR.*
Butt, Dr Khursheed Alam. *Pakistan.*
Butt, Mr Muhammad Hafeez. *Pakistan.*
Butter, Dr Ehrenfried. *DDR.*
Buttinelli, Prof. Dante. *Italy.*
Buxton, Dr B.F. *UK.*
Buyers, Dr William James Leslie. *Canada.*
Bye, Dr Erik. *Norway.*
Byram, Mrs Susan Katherine. *USA.*
Byrn, Prof. Stephen Robert. *USA.*
Byrne, Mr Peter G. *Ireland.*
Caballero Lopez-Lendinez, Prof. Manuel A. *Spain.*
Cabezuelo Huertas, Miss Maria. *Spain.*
Cabrera Bravo, Prof. Enrique. *Mexico.*
Cady, Dr Howard Hamilton. *USA.*
Cagle, Prof. Fredric William, Jr. *USA.*
Caglioti, Prof. Dr Giuseppe. *Italy.*
Cagnon, Dr Maurice. *France.*
Cahn, Prof. R.W. *UK.*
Cain, Mr Peter Maurice. *UK.*
Caira, Dr Mino Rodolfo. *South Africa.*
Calabrese, Dr Joseph C. *USA.*
Calais, Dr Jean-Louis. *Sweden.*
Calamiotou, Dr Maria. *Greece.*
Calandra, Mr Peter M. *USA.*
Calderón, Prof. Gómez Eduardo. *Colombia.*
Calestani, Dr Gianluca. *Italy.*
Çalişkan, Mrs Nezihe. *Turkey.*
Callahan, Dr Kenneth Paul. *USA.*
Callejas, Mr Domingo. *Cuba.*
Calleri, Prof. Mariano Bernardino. *Italy.*
Calligaris, Prof. Mario. *Italy.*

Calvet Pallas, Mrs Maria Teresa. *Spain.*
Calvo Calvo, Prof. Felipe Angel. *Spain.*
Camalli, Dr Mercedes. *Italy.*
Camerman, Prof. Arthur. *USA.*
Camerman, Dr Norman. *Canada.*
Cameron, Dr Allan Forbes. *UK.*
Cameron, Prof. Theodore Stanley. *Canada.*
Cameroni, Prof. Riccardo. *Italy.*
Cammenga, Prof. Dr Heiko Karl. *BRD.*
Campa, Prof. Juan Antonio. *Spain.*
Campana, Dr Charles F. *USA.*
Campanelli, Dr Anna Rita. *Italy.*
Campbell, Dr John Wilson. *UK.*
Campelo Farias, Prof. Carlinda. *Brazil.*
Campos, Dr Cicero. *Brazil.*
Candeloro De Sanctis, Prof. Sofia. *Italy.*
Canepa, Dr Horacio Ricardo. *Argentina.*
Cannas, Prof. Mario. *Italy.*
Cannillo, Dr Elio. *Italy.*
Cano Corona, Dr Octavio. *Mexico.*
Cantrell, Prof. Joseph Sires. *USA.*
Canut, Prof. Marisa. *Spain.*
Cao, Prof. Ming-zhong. *China.*
Cao, Prof. Zheng-min. *China.*
Çapan, Dr Z. Ussal. *Turkey.*
Capano, Mr Michael A. *USA.*
Capasso, Prof. Sante. *Italy.*
Capella, Prof. Lucien. *France.*
Capelle, Dr Bernard. *France.*
Čapková, Dr Pavla. *CSSR.*
Capotorto, Dr Concetta. *Italy.*
Capparelli, Prof. Mario Vicente. *Venezuela.*
Capponi, Claude Annie. *France.*
Capponi, Jean-Jacques. *France.*
Caranoni, Dr Claude Anny. *France.*
Carbonin, Dr Susanna. *Italy.*
Cardellach Lopez, Dr Esteban. *Spain.*
Cardin, Dr Christine Janet. *Ireland.*
Cardon, Prof. Dr Felix. *Belgium.*
Cardwell, Mr D.A. *UK.*
Cargill, Dr George Slade, III. *USA.*
Carlisle, Prof. Charles Harold. *UK.*
Carlson, Dr Ernest Howard. *USA.*
Carlson, Dr Sirkka Liisa. *Finland.*
Carlsson, Dr Roger. *Sweden.*
Carlström, Prof. Diego G. *Sweden.*
Carnahan, Dr Gary Ellis. *USA.*
Caron, Dr Aimery Pierre. *USA.*
Carpenter, Mr Dewey K. *USA.*
Carpenter, Dr Donald Allmand. *USA.*
Carpenter, Prof. Gene B. *USA.*
Carperos, Mr William E. *USA.*
Carrasco Cantos, Mr Francisco. *Spain.*
Carrell, Dr Horace L. *USA.*
Carrithers, Mr Charles H. *USA.*
Carrondo, Prof. Maria Arménia A.F.C.T. *Portugal.*
Cartení-Farina, Prof. Maria. *Italy.*
Carter, Prof. Charles Williams, Jr. *USA.*
Carter, Dr Forrest Lee *USA.*
Carter, Mr Trevor Lee. *UK.*
Carter, Dr William S. *USA.*
Cartwright, Dr Michael. *UK.*
Carty, Prof. Arthur John. *Canada.*
Cartz, Prof. Louis. *USA.*
Caruso, Dr Francesco. *Italy.*
Carvalho da Silva, Prof. Jair. *Brazil.*
Casalone, Dr Gianluigi. *Italy.*
Casanova, Lic. Jorge Ramón. *Argentina.*
Casas Sainz de Aja, Dr José. *Spain.*
Cascarano, Dr Giovanni Luca. *Italy.*
Case, Mr J. A. M. *USA.*
Casellato, Dr Umberto. *Italy.*
Cashion, Assoc Prof John Dixon. *Australia.*
Caslavsky, Mr Jaroslav L. *USA.*
Caspar, Prof. Donald L. D. *USA.*
Cassedane, Dr Jeannine. *Brazil.*
Cassel, Dr Anders Ö. *Sweden.*
Castaing, Prof. Raymond Bernard René *France.*
Castellano, Dr Eduardo Ernesto. *Brazil.*
Castellanos Guzman, Dr A. Guillermo. *Mexico.*
Castellanos Román, Mrs Maria Asunción. *Mexico.*
Castelliz, Dr Karoline (Lotte) Maria. *Canada.*
Caticha Alfonso, Mr Ariel. *Brazil.*
Caticha Ellis, Prof. Stephenson. *Brazil.*
Catti, Prof. Michele. *Italy.*
Cauchois, Prof. Yvette. *France.*
Caughlan, Prof. Charles N. *USA.*
Cavalca, Prof. Luigi. *Italy.*
Caveney, Dr Robert John. *South Africa.*
Cavero Ghersi, Dr César Augusto. *Peru.*
Cavin, Mr Odis Burl. *USA.*
Cazaux, Prof. Jacques. *France.*
Cebe, Prof. Dr Mustafa. *Turkey.*
Cebula, Dr D. *UK.*
Čech, Prof. Dr František. *CSSR.*
Cedergren-Zeppezauer, Dr Eila S. *Sweden.*
Čeh, Magistar Boris, *Yugoslavia.*
Cejalvo, Prof. Flor. *Philippines.*
Cellai, Dr Luciano. *Italy.*
Celotti, Dr Giancarlo. *Italy.*
Cemič, Dr Ladislav. *BRD.*
Cenzual, Mrs Karin Margareta. *Switzerland.*
Čermák, Dr Jan. *CSSR.*
Čeřnanský, Ing Marian. *CSSR.*
Černohorský, Doc. Dr Martin. *CSSR.*
Cerrini, Dr Silvio. *Italy.*
Červeň, Doc. Dr Ivan. *CSSR.*
Červinka, Dr Ladislav. *CSSR.*
Cesari, Prof. lib. doc. Marco. *Italy.*
Cesario, Mrs Michèle. *France.*
Ceylan, Mr Kazim. *Turkey.*
Ceylan, Dr Mehmet. *Turkey.*
Chaban, Dr Nadezhda Fedorovna. *USSR.*
Chabot, Dr Bernard André. *Switzerland.*
Chacko, Dr K. K. *India.*
Chadha, Dr Gopal Krishan. *India.*
Chaichit, Dr Narongsak. *Thailand.*
Chaikum, Dr Nitirampai Latavalya. *Thailand.*
Chaikum, Dr Nopadol. *Thailand.*
Chakoumakos, Dr Bryan Charles. *USA.*
Chakrabarty, Dr Mrs Chandana. *India.*
Chakrabarty (Chatterjee), Mrs Ela. *India.*
Chakrabarty, Dipak Kumar. *India.*
Chakrabarty, Subhasis. *India.*
Chakrabarty, Mr Sugoto. *India.*
Chakraborty, Dr Suchit Chandra. *India.*
Chalupa, Ing Bohumil. *CSSR.*
Chamberland, Mr B. L. *USA.*
Chambers, Mr John L. *USA.*
Champier, Prof. Georges. *France.*
Champion, Dr John Anthony. *UK.*
Champion, Mr William C. *USA.*
Champness, Dr John Norman. *UK.*
Champness, Dr Pamela Eileen. *UK.*
Chanda, Dr Gopal Krishan. *India.*
Chandra, Dr Suresh. *India.*
Chandrasekaran, Prof. Katuputhur Sarma. *India.*
Chandrasekaran, Dr Muthuswamy. *Argentina.*
Chandrasekaran, Dr R. *India.*
Chandrasekhar, Mr K. *USA.*
Chandrasekhar, Prof. Sivaramakrishna. *India.*
Chandrasekharaiah, Dr M. N. *India.*
Chandross, Prof. Ronald Jay. *USA.*
Chandy, Dr K. C. *India.*
Chaney, Dr Michael Owen. *USA.*
Chang, Dr Chong-Hwan. *USA.*
Chang, Prof. Shih-Chi. *USA.*
Chang, Dr Shih- Lin. *Brazil.*
Chang, Mr Tien-show. *Taiwan.*
Chang, Mr Wen-rui. *China.*
Chanh, Dr Nguyen-Ba. *France.*
Chao, Dr George Y. *Canada.*
Chapela Castañares, Dr Victor Manuel. *Mexico.*
Chapelle, Prof. Jean-Pierre. *France.*
Chapuis, Prof. Gervais Constant. *Switzerland.*
Charbonneau, Dr Guy Paul. *France.*
Charbonnier, Dr François. *France.*
Charbonnier, Mrs Sylvie. *Ivory Coast.*
Charland, Dr Jean-Pierre. *Canada.*
Charpin, Dr Pierrette. *France.*
Chasen, Prof. Edith. *USA.*
Chashchinov, Dr Yury Mikhailovich. *USSR.*
Chasseau, Dr Daniel. *France.*
Chastain, Dr Roger Vernon, Jr. *Germany.*
Chatterjee, Dr Amitava. *India.*
Chatterjee, Dr Sanat Kumar. *India.*
Chattopadhyay, Dr Tapan Kumar. *BRD.*
Chaudhary, Dr Abdul Majid. *Pakistan.*
Chaudhry, Mr Mohammad Anwar. *Pakistan.*
Chaudhry, Mr G. Sarwar Alam. *Pakistan.*
Chaudhuri, Dr Ahindra Kumar. *India.*
Chaudhuri, Prof. Bhumidhar. *India.*
Chaudhuri, Mr Jharna. *USA.*
Chaudhuri, Mr Siddhartha. *India.*
Chauhan, Mr Ehsanul Haq. *Pakistan.*
Chawdhury, Prof. Sadruddin Ahmed. *Bangladesh.*
Chawla, Dr Krishan Lal. *India.*
Chayka, Mr Paul V. *USA.*
Chełkowski, Prof. August Jan. *Poland.*
Cheang Dr Kok Keong. *Malaysia.*
Cheary, Dr Robert Winston. *Australia.*
Cheer, Prof. Clair James. *USA.*
Cheetham, Dr Anthony Kevin. *UK.*
Cheikhrouhou, Dr Abdelwaheb. *Tunisia.*
Chen, Mr Ben-ming. *China.*
Chen, Dr Cheng-San. *USA.*
Chen, Prof. Dai-zhang. *China.*
Chen, Mr Guo-ying. *China.*
Chen, Mr Haydn H. *USA.*
Chen, Mr Jing-zhong. *China.*
Chen, Prof. Kuang-yuan. *China.*
Chen, Prof. Li-quan. *China.*
Chen, Prof. Pei-yuan. *Taiwan.*
Chen, Dr Ruey-Hong. *Taiwan.*
Chen, Mr Shi-zhi. *China.*
Chen, Dr Wei. *Malaysia.*
Chen, Prof. Xian-qiu. *China.*
Chen, Mr Yuan-zhu. *China.*
Chen, Mrs Yueh-Hua. *Taiwan.*
Chen, Mr Zhi-xue. *China.*
Chen, Dr Zhong-guo. *China.*
Chenavas, Prof. Jean. *France.*
Cheng, Mr Graham Cheng-hsun. *Hong Kong.*
Cheng, Mr Min-chin. *China.*
Cheng, Dr Pei-Tak. *Canada.*
Chentsova, Dr Leonila Gavrilovna. *USSR.*
Cheremskoy, Dr Petr Grigoryevich. *USSR.*
Cherepanova, Dr Tamara Alekseyevna. *USSR.*
Chernov, Prof. Alexander Alexandrodich. *USSR.*
Cherns, Dr David. *UK.*
Chernysheva, Dr Marina Alexandrovna. *USSR.*
Chernysheva, Mrs Valentina Fedorovna. *USSR.*
Chetal, Prof. Amritlal R. *India.*
Chetkina, Dr Larisa Arkadyevna. *USSR.*
Cheung, Dr Kung Kai. *Hong Kong.*
Chevalier, Prof. Raymond *France.*
Chevy, Dr Alain Jean-Pierre. *France.*
Chiang, Prof. Liang-jun. *China.*
Chiang, Dr Michael Yen-Nan. *USA.*
Chiari, Prof. Giacomo. *Italy.*
Chiaroni, Mrs Angèle. *France.*
Chidambaram, Dr Rajagopala. *India.*
Chidester, Ms Connie. *USA.*
Chieh, Prof. Chung (Peter). *Canada.*
Chiesi-Villa, Prof. Angiola. *Italy.*
Chihaya, Prof. Tadashi. *Japan.*
Chikaura, Dr Yoshinori. *Japan.*
Chikawa, Dr Jun-ichi. *Japan.*
Childs, Dr Cyril Walter. *New Zealand.*
Childs, Dr Jerry D. *USA.*
Chion, Dr Bernadette. *France.*
Chiong, Ms Pauline *USA.*
Chipman, Dr David Randolph. *USA.*
Chiragov, Dr Mamed Isa ogly. *USSR.*
Chirgadze, Dr Yury Nikolayevich. *USSR.*
Chisholm, Dr James Edwin. *UK.*
Chiu, Ms Celia C. *USA.*
Chiu, Prof. Wah. *USA.*
Cho, Prof. Sung-Il. *Korea.*

Choi, Dr Chang Sun. *USA.*
Chollet, Dr Lucien-Francois. *Switzerland.*
Choosang, Mrs Pilai. *Thailand.*
Chopra, Prof. Kasturilal. *India.*
Chornik, Dr Boris. *Venezuela.*
Chowdari, Dr B.V.R. *Singapore.*
Chowdhry, Mrs Khursheed. *Pakistan.*
Chowdhury, Prof. Fazlul Halim. *Bangladesh.*
Christensen, Dr Axel Nørlund. *Denmark.*
Christian, Prof. John Wyrill. *UK.*
Christidis, Prof. Panayiotis Chrysostomos. *Greece.*
Christofferson, Dr Glen D. *USA.*
Christoph, Mr Arthur. *DDR.*
Christoph, Dr Gary Gordon. *USA.*
Chu, Prof. Shirley Shan-Chi. *USA.*
Chudinova, Dr Svetlana Alekseyevna. *USSR.*
Chukhovsky, Dr Felix Nikolayevich. *USSR.*
Chung, Dr Beingtau. *Taiwan.*
Chung, Dr Mui Fatt. *Singapore.*
Chung, Prof. Dr Su Jin. *Korea.*
Chuprunov, Dr Evgeniy Vladimirovich. *USSR.*
Church, Mr William Bret. *Australia.*
Churchill, Prof. Melvyn Rowen. *USA.*
Churchman, Dr Gordon John. *New Zealand.*
Cia, Mr Jin-hua. *China.*
Ciani, Prof. Gianfranco. *Italy.*
Čičel, Ing Blahoslav. *CSSR.*
Cid, Dr Hilda. *Chile.*
Ciechanowicz-Rutkowska, Dr Maria. *Poland.*
Cimino, Prof. Alessandro. *Italy.*
Cini, Prof. Renzo. *Italy.*
Cioflica, Prof. Graţian. *Romania.*
Cipriani, Prof. Curzio. *Italy.*
Cirafici, Prof. Salvino S. *Italy.*
Cisneros Ramos, Prof. Luis. *Peru.*
Ciunik, Dr Zbigniew. *Poland.*
Clancy, Ms. Laura Lee. *USA.*
Clardy, Prof. Jon Christel. *USA.*
Claridge, Dr Graeme Geoffrey. *New Zealand.*
Claringbull, Sir Gordon Frank. *UK.*
Clark, Ms Connie M. *USA.*
Clark, Prof. Edward Shannon. *USA.*
Clark, Dr George Raymond. *New Zealand.*
Clark, Dr James Brian. *South Africa.*
Clark, Mrs Joan Michele. *USA.*
Clark, Prof. John Robert. *USA.*
Clark, Dr Malcolm John Roy. *Canada.*
Clarke, Prof. Roy. *USA.*
Clastre, Prof. José *France.*
Claus, Mr Albert C. *USA.*
Claus, Mr Karl Heinz. *BRD.*
Clausen,, Dr Kurt. *Denmark.*
Clauws, Dr Paul. *Belgium.*
Clavaguera Plaja, Dr Narciso. *Spain.*
Clayden, Miss D.A. *UK.*
Clayton, Dr William Rex. *USA.*
Clearfield, Prof. Abraham. *USA.*
Clegg, Dr William. *UK.*
Clemente, Prof. Dore Augusto. *Italy.*
Clewer, Mr Peter John. *UK.*
Clifton, Prof. Donald Frederic. *USA.*
Clinger, Mr Kent. *USA.*
Cobbledick, Dr Roger Ernest. *Canada.*
Cochran, Prof. William. *UK.*
Cochran, Prof. Todd G. *USA.*
Cockayne, Dr David John Hugh. *Australia.*
Codding, Dr Penelope Wixson. *Canada.*
Cody, Dr Vivian. *USA.*
Coene, Mr Willem Marie Julia Marcel, *Belgium.*
Coetzer, Dr Johan. *South Africa.*
Cohen, Prof. Carolyn. *USA.*
Cohen, Mr Charles I. *USA.*
Cohen, Dr Gerson H. *USA.*
Cohen, Ms Janet Paula (Lentz). *USA.*
Cohen, Prof. Jerome Bernard. *USA.*
Cohen, Dr L. *UK.*
Cohen, Dr Shmuel. *Israel.*
Cohen-Addad, Dr Claudine. *France.*
Coing-Boyat, Dr Jean Claude. *France.*

Coiro, Dr Vincenza Maria. *Italy.*
Cojazzi, Dr Gianna. *Italy.*
Cola, Prof. Mario Luigi. *Italy.*
Çolakoğlu, Doç. Dr Kemal. *Turkey.*
Colapietro, Prof. Marcello. *Italy.*
Colby, Miss J. *UK.*
Cole, Dr Henderson. *USA.*
Cole, Dr William Frederick. *Australia.*
Colella, Prof. Roberto. *USA.*
Colliex, Dr Christian. *France.*
Collin, Miss Sonia Bertha Josepha. *Belgium.*
Collini, Prof. Bengt H. E. *Sweden.*
Collins, Prof. Douglas MacPherson. *USA.*
Collins, Dr Richard Christopher. *USA.*
Collyer, Mr S. *UK.*
Colman, Dr Peter Malcolm. *Australia.*
Colmanet, Mr Silvano. *Australia.*
Colombo, Dr Arturo. *Italy.*
Colombo, Dr Lidija. *Yugoslavia.*
Colyvas, Mr Kim. *Australia.*
Comes, Dr Robert *France.*
Comey, Mr Paul Van A. *USA.*
Comins, Dr Neville Raymond. *South Africa.*
Commarond, Dr Marie-Bernard. *France.*
Conant, Dr John W. *USA.*
Conde, Prof. Alejandro. *Spain.*
Condren, Mr S. M. *USA.*
Constant, Prof. Georges. *France.*
Constantinescu, Mr Radu. *Romania.*
Convert, Dr Pierre. *France.*
Cook, Prof Allan Cecil. *Australia.*
Cook, Dr David Stanley. *UK.*
Cook, Dr William Joseph *USA.*
Cook, Dr William R., Jr. *USA.*
Coombs, Prof. Douglas Saxon. *New Zealand.*
Cooper, Dr Alan Frederick. *New Zealand.*
Cooper, Mrs Ann S. *USA.*
Cooper, Ms Jeanette. *USA.*
Cooper, Dr Malcolm John. *UK.*
Cooper, Dr Martyn John. *UK.*
Copeland, Prof. Richard Franklin. *USA.*
Coppens, Prof. Philip. *USA.*
Copperthwaite, Dr Richard George. *South Africa.*
Corbeil, Miss Marie-Claude. *Canada.*
Corbett, Dr Madeline. *Australia.*
Corchia, Dr Massimo. *Italy.*
Cordero, Dr Adolfo. *Mexico.*
Cordes, Prof. A. Wallace. *USA.*
Cordier, Dr Gerhard. *BRD.*
Corey, Prof. Eugene Ray. *USA.*
Corfield, Prof. Peter William Reginald. *USA.*
Corlett, Dr Mabel Isobel. *Canada.*
Corliss, Dr Lester Myron. *USA.*
Cornelis, Ir Jozef Frans Elisa. *Belgium.*
Corney, Mr David John. *UK.*
Cornfield, Ms L. *UK.*
Corradini, Prof. Paolo. *Italy.*
Correia Neves, Prof. José Marques. *Brazil.*
Cortés, Dr Abdón. *Colombia.*
Cortelezzi, Dr César Rafael. *Argentina.*
Cossu, Dr Michéle Josette. *Ivory Coast.*
Costa, Prof. M. Margarida Ramalho. *Portugal.*
Costa Gouveia, Prof. Albany H. *Brazil.*
Costa Viana, Prof. Carlos Sergio da. *Brazil.*
Costa-Bizzarri, Prof. Paolo. *Italy.*
Costamagna, Dr Juan Alberto. *Chile.*
Costello, Nr B.A.D. *UK.*
Cota Araiza, Mr Leonel Susano. *Mexico.*
Cotton, Prof. Frank Albert. *USA.*
Couldwell, Dr Margaret Claire. *New Zealand.*
Coulman, Ms Betty Ann. *USA.*
Coulomb, Dr Pierre. *France.*
Coulter, Dr Charles L. *USA.*
Cour, la, Dr Troels Frederik Marstrand. *Denmark.*
Courtois, Dr Alain Raymond. *France.*
Cousins, Mr Christopher Stanley George. *UK.*
Cousland, Mr Stuart McKay. *Australia.*
Coville, Dr Neil John. *South Africa.*
Cowan, Miss Sandra Wendy. *Australia.*
Cowie, Prof. Martin. *Canada.*

Cowlam, Dr Neil. *UK.*
Cowley, Prof. John M. *USA.*
Cox, Dr David Ernest. *USA.*
Cox, Dr John Wesley, Jr. *USA.*
Cox, Dr Philip John. *UK.*
Coy Yll, Prof. Ramon. *Spain.*
Coyle, Mr Richard Alan. *Australia.*
Craievich, Dr Aldo Felix. *Brazil.*
Craig, Mr Donald Chadwick. *Australia.*
Craig, Mr G.R. *UK.*
Cras, Dr Joannes Antonius. *Netherlands.*
Craston, Mr Dennis F. *USA.*
Craven, Prof. Bryan Maxwell. *USA.*
Crawford, Mr John Lawrence. *South Africa.*
Creagh, Dr Dudley Cecil. *Australia.*
Cremona, Ing. Luigi. *Italy.*
Crennell, Mrs K.M. *UK.*
Cressey, Dr Barbara Anne. *UK.*
Cressey, Dr Gordon. *UK.*
Criasia, Mr Ronald T. *USA.*
Cringean, Mr J.K. *UK.*
Crist, Prof. Buckley, Jr. *USA.*
Critchell, Mr J.W. *UK.*
Croatto, Prof. Ugo. *Italy.*
Crocker, Prof. Alan Godfrey. *UK.*
Croft, Dr William J. *USA.*
Cromer, Dr Don Tiffany. *USA.*
Cruceanu, Mr Eugen. *Romania.*
Cruickshank, Prof. Durward William John. *UK.*
Cruickshank, Mr M.C. *UK.*
Csanády-Bokody, Mrs Ágnes. *Hungary.*
Csöregh, Dr Ingeborg. *Sweden.*
Csordás, Dr László. *Hungary.*
Csordás-Tóth, Dr Anna. *Hungary.*
Cubiotti, Prof. Gaetano. *Italy.*
Cuchý, Ing Zdeněk. *CSSR.*
Cucka, Dr Paul. *USA.*
Cuevas Diarte, Dr Miguel Angel. *Spain.*
Cuff, Dr Christopher. *Australia.*
Cui, Prof. Wen-yuan. *China.*
Cullen, Mr F.L. *UK.*
Cullen, Prof. David Lawrence. *USA.*
Cumbrera Hernandez, Dr Francisco. *Spain.*
Cummings, Dr John Patrick. *USA.*
Cunningham, Dr Patrick Desmond. *Ireland.*
Curie, Prof. Daniel. *France.*
Curien, Prof. Hubert. *France.*
Curtin, Prof. David Yarrow. *USA.*
Curtis, Dr Bernard. *Switzerland.*
Curzon, Prof. Albert Edward. *Canada.*
Cusatis, Dr Cesar. *Brazil.*
Cutfield, Dr John Franklin. *New Zealand.*
Cuthill, Dr John R. *USA.*
Cvetković, Mr Ljubiša. *Yugoslavia.*
Cvetković, Mrs Miroslava. *Yugoslavia.*
Cygler, Dr Mirosław. *Poland.*
Czachor, Dr hab Andrzej. *Poland.*
Czank, Dr Michael. *BRD.*
Czerwinski, Prof. Edmund William. *USA.*
Cziráki, Dr (Miss) Ágnes. *Hungary.*
Czugler, Dr Mátyás. *Hungary.*
Däweritz, Dr Lutz. *DDR.*
Díaz Peraza, Prof. José Milciades. *Colombia.*
Dörr, Dr Friedrich Johannes. *BRD.*
Dörrfeld, Mr Hans-Georg. *DDR.*
Dörschel, Dr Jürgen. *DDR.*
Dünkel, Mr Lothar. *DDR.*
D' Amour-Sturm, Dr Hedwig. *BRD.*
D'yachenko, Dr Oleg Anatolyevich. *USSR.*
D'yakon, Dr Ivan Andreyevich. *USSR.*
D'Addario, Dr Anthony Paul. *USA.*
D'Ilario, Prof. Lucio. *Italy.*
Dabbabi, Dr Mongi. *Tunisia.*
Dachs, Prof. Dr Hans. *BRD.*
Dacombe, Mr Michael H. *UK.*
Dadel, Mrs Snehlata. *India.*
Dagerhamn, Dr Tore. *Sweden.*
Dahan, Dr Françoise. *France.*
Dahl, Prof. Lawrence F. *USA.*
Dahl, Dr Tor. *Norway.*
Dahlén, Dr Birgitta. *Sweden.*

Dahlkamp, Dr Franz-Joses. *BRD.*
Dai, Mr Jin-bi. *China.*
Dal Negro, Prof. Alberto. *Italy.*
Dalley, Prof. Nelson Kent. *USA.*
Daly, Dr John Joseph. *Switzerland.*
Damak, Dr Mabrouk. *Tunisia.*
Damaschun, Mr Ferdinand. *DDR.*
Damm, Prof. Józef Zbigniew. *Poland.*
Danaciözbey, Mrs Süheyla. *Turkey.*
Dance, Dr Ian Gordon. *Australia.*
Danielsen, Dr Jacob. *Denmark.*
Dankházi, Mr Zoltán. *Hungary.*
Danko, Mr Andrew William. *USA.*
Dann, Mr Jeffrey Neil. *USA.*
Dantonio, Mr Peter. *USA.*
Daoud, Prof. Abdelaziz. *Tunisia.*
Dapporto, Prof. Paolo. *Italy.*
Darces, Mr Jean-François. *France.*
Dariel, Prof. Moshe Pierre. *Israel.*
Darling, Mr Stephen D. *USA.*
Darlington, Dr Charles Nicholas Wright. *UK.*
Dartyge, Dr Elisabeth. *France.*
Dartyge, Mr Jean-Marcel. *France.*
Das, Dr Badri Narayan. *USA.*
Das, Mr Birendra Nath. *India.*
Das, Dr Indu Mohan. *India.*
Das, Pratap Kumar. *India.*
Das, Dr Sabita. *India.*
Das Gupta, Mr Prabal. *India.*
Datt, Dr Igor Daudovich. *USSR.*
Datta, Mr Amal Kumar. *India.*
Dattagupta, Dr Jiban Kanti. *India.*
Dauter, Dr Zbigniew. *Poland.*
Dave, Prof. Jatashanker Sadashiv. *India.*
Dave, Miss S. *UK.*
Davies, Dr David R. *USA.*
Davies, Dr Geoffrey John. *South Africa.*
Davies, Dr Gladstone. *South Africa.*
Davies, Dr John Edward. *UK.*
Davis, Mr Alan Ross. *Canada.*
Davis, Dr Briant LeRoy. *USA.*
Davis, Prof. Phillip Howard. *USA.*
Davis, Prof. Raymond Edward. *USA.*
Davis, Dr Ronald Lindsay. *Australia.*
Davison, Mr G. *UK.*
Davoli, Dr Paolo. *Italy.*
Davydchenko, Dr Anatoliy Georgiyevich. *USSR.*
Day, Dr Cynthia Ann Secauer. *USA.*
Day, Prof. Roberta Ogilvie. *USA.*
Day, Prof. Victor Warren. *USA.*
Dayal, Dr Radha Raman. *India.*
De la Camp, Prof. Ulrich Otto. *USA.*
De Angelis, Prof. Giuseppe. *Italy.*
De Boer, Dr Barry Goodwin. *USA.*
De Camp, Dr Wilson H. II. *USA.*
De Fontaine, Prof. Robert Didier. *USA.*
De Gryse, Dr Roger Marc. *Belgium.*
De Haven, Dr Patrick William. *USA.*
De Jarnette, Miss F. Elaine. *USA.*
De Kouchkovsky, Mr Rostislav. *France.*
De Lucia, Dr Mary Lou. *USA.*
De Maggio, Mr Gregory B. *USA.*
De Meester, Dr Patrice. *USA.*
De Pablo Galan, Dr Linerto. *Mexico.*
De Pol Blasi, Prof. Carla. *Italy.*
De Ranter, Prof. Dr Camiel Joseph. *Belgium.*
De Rosier, Prof. David John. *USA.*
De Santis, Prof. Pasquale. *Italy.*
De Schoenmacker. Ir Dirck Maurice. *Belgium.*
De Titta, Dr George Thomas. *USA.*
De Villiers, Prof. Johan Pieter Roos. *South Africa.*
De Vries, Dr Adriaan. *USA.*
De Wet, Prof. Julius Ferdinand. *South Africa.*
De Wolf, Mr Marcus Ludovicus Maria. *Belgium.*
De, Mr Adhip Kanti. *India.*
De, Dr Madhusudan. *India.*
Deadwyler, Mr Daniel A. *USA.*
Dean, Mr Christopher. *Australia.*
Dean, Mr Johnny Clyde. *USA.*
Debaerdemaeker, Dr Tony. *BRD.*

Deblieck, Mr Rudy André Cornelis. *Belgium.*
Declercq, Prof. Jean Paul. *Belgium.*
Dedegkayev, Dr Tazaret Temurkanovich. *USSR.*
Dederer, Dr Bernhard. *BRD.*
Deguire, Mrs Suzanne. *Canada.*
Deiseroth, Dr Hans-Jörg. *BRD.*
Dekeyser, Prof. Dr Willy Clement. *Belgium.*
Dekker, Mr Henri. *UK.*
Del Monte, Prof. Marco Emiliano. *Italy.*
Del Nery, Miss Sheila Maria. *Brazil.*
Del Piero, Dr Gastone. *Italy.*
Del Pra, Prof. Antonio. *Italy.*
Delaey, Prof. Luc J. M. A. E. *Belgium.*
Delaloye, Prof. Michel. *Switzerland.*
Delaney, Prof. Matthew Sylvester. *USA.*
Delaney, Dr William Timothy. *Australia.*
Delapalme, Dr Alain. *France.*
Delaplane, Dr Robert G. *Sweden.*
Delavignette, Prof. Pierre. *Belgium.*
Delbaere, Dr Louis Theophil Joseph. *Canada.*
Delettré, Dr Jean. *France.*
Delf, Dr Brian William. *UK.*
Delgado, Mr Jose Miguel. *USA.*
Delgado Quiñones, Lic. Miguel. *Venezuela.*
Delhez, Dr Ir Eric Jan. *Netherlands.*
Deliens, Dr Michel. *Belgium.*
Delineshev, Dr Svetoslav. *Bulgaria.*
Della Casa, Prof. Carlo. *Italy.*
Della Giusta, Prof. Antonio. *Italy.*
Dellea Dr Rita. *Italy.*
Delord, Prof. Pierre. *France.*
Delord, Mr Terry. *USA.*
Demšar, Magistar Alojz. *Yugoslavia.*
Dem'yanets, Dr Lyudmila Nikolayevna. *USSR.*
Demartin, Dr Francesco. *Italy.*
Dembo, Dr Alexander Teodorovich. *USSR.*
Demontis, Dr Pierfranco. *Italy.*
Demus, Dr Dietrich. *DDR.*
Denham, Mr A.W. *UK.*
Denisenko, Dr Georgy Alexandrovich. *USSR.*
Denner, Mr Louis. *South Africa.*
Dent Glasser, Dr Lesley Scott. *UK.*
Deopura, Dr B. L. *India.*
Depalma, Mr Vincent M. *USA.*
Depmeier, Dr habil. Wulf Helmut Heinz. *BRD.*
Deppisch, Dr Bertold. *BRD.*
Derewenda, Zygmunt Stanisław. *Poland.*
Derici, Dr Rifat. *Turkey.*
Derkosch, Prof. Dr Josef. *Austria.*
Deroski, Ms Betty Rolfs. *USA.*
Deruyttere, Prof. Dr André. *Belgium.*
Derwent, Mr Frank William. *UK.*
Deslandes, Dr Yves. *Canada.*
Desmeules, Mr Peter J. *USA.*
Desper, Dr C. Richard. *USA.*
Despotović, Mr Zlatko. *Yugoslavia.*
Despujols, Prof. Jacques. *France.*
Dessy, Dr Giulia. *Italy.*
Destro, Prof. Riccardo. *Italy.*
Deutsch, Dr Moshe. *Israel.*
Dewan, Dr John C. *USA.*
Dexter, David D. *USA.*
Dhanaraj, Mr V. *India.*
Dhaneshwar, Dr Narayandatta Nagesh. *India.*
Dhar, Rakesh. *India.*
Dhawan, Mrs Urmil. *India.*
Dhlipia, Mr Gursev Singh. *BRD.*
Di Blasio, Prof. Benedetto. *Italy.*
Di Vaira, Prof. Massimo. *Italy.*
Di-Persio, Dr Jean Anthony. *France.*
Diamond, Dr Robert. *UK.*
Dias Rodrigues, Mrs Ana Maria Gonçalves. *Brazil.*
Dias, Dr Hanwellage Wijayapala. *Sri Lanka.*
Dichmann, Dr Klaus. *Canada.*
Dickinson, Dr Charles. *USA.*
Dickerson, Prof. Richard Earl. *USA.*
Dickman, Mr Michael H. *USA.*
Dideberg, Dr Otto. *Belgium.*
Diehl, Dr J. *BRD.*
Diehl, Dr Roland. *BRD.*

Dieterich, Dr David Allan. *USA.*
Dietrich, Dr Burkhard. *DDR.*
Dietrich, Prof. Dr Hans Karl Ernst. *BRD.*
Dijk, van, Dr Cornelis. *Netherlands.*
Dijkstra, Drs Bauke Wiepke. *Netherlands.*
Dikici, Dr Mustafa. *Turkey.*
Dillen, Dr Jan Louis Maria. *South Africa.*
Dimitrijević, Magistar Radovan. *Yugoslavia.*
Dimitrova, Dr Ol'ga Vladimirovna. *USSR.*
Dimov, Mr Vergil. *Bulgaria.*
Dinçer, Dr Muharrem. *Turkey.*
Dineen, Mr C. *UK.*
Dinescu, Prof. Radu. *Romania.*
Dingley, Dr David Joseph. *UK.*
Diniz de Carvalho Loyolla, Mr Waldomiro P. *Brazil.*
Diodati, Dr Francisco Piero. *Argentina.*
Dion, Mrs Chantal. *Canada.*
Distler, Dr Grigory Isaakovich. *USSR.*
Dittmar, Dr Günter. *BRD.*
Divjaković, Prof. Dr Vladimir. *Yugoslavia.*
Djarova, Dr Maria. *Bulgaria.*
Djonoputro, Mr Bernard Darmawan. *Indonesia.*
Djordjević, Prof. Dr Slobodan. *Yugoslavia.*
Djurić, Dr Stevan. *Yugoslavia.*
Dmitrieva, Dr Tatyana Vladimirovna. *USSR.*
Dmitriyeva, Dr Margarita Timofeyevna. *USSR.*
Dobler, PD Max. *Switzerland.*
Dobrev, Dr Dobri. *Bulgaria.*
Dobrott, Dr Robert D. *USA.*
Dobrowolska, Dr Wanda. *Poland.*
Dobrzyński, Dr hab. Ludwik. *Poland.*
Dobson, Mrs Susan Mary. *South Africa.*
Dobson, Dr Peter James. *UK.*
Dodd, Dr Charles Gardner. *USA.*
Dodge, Prof. Richard Patrick. *USA.*
Dodokin, Dr Anatoly Petrovich. *USSR.*
Dodson, Mrs Eleanor. *UK.*
Dodson, Dr Guy George. *UK.*
Doedens, Prof. Robert John. *USA.*
Doesburg, Dr Hendrikus M. *Netherlands.*
Doğan, Dr Ali. *Turkey.*
Doherty, Dr Ruth Marie. *USA.*
Dohi, Prof. Shoso. *Japan.*
Doi, Dr Kenji. *Japan.*
Doig, Dr P. *UK.*
Dolivo-Dobrovol'skaya, Mrs Elena M. *USSR.*
Dolivo-Dobrovol'skaya, Dr Galina I. *USSR.*
Dollase, Prof. Wayne A. *USA.*
Dolling, Dr Gerald. *Canada.*
Domínguez Esquivel, Dr José Manuel. *Mexico.*
Doman, Dr Robert Charles. *USA.*
Domeneghetti, Dr Maria Chiara. *Italy.*
Domenicano, Prof. Aldo. *Italy.*
Domenici, Dr Marcello. *Italy.*
Domiano, Prof. Paolo. *Italy.*
Donaldson, Prof John Dallas. *UK.*
Donati, Dr Donato. *Italy.*
Dong, Mr Ji-he. *China.*
Dong, Mr Yi-cheng. *China.*
Donnay, Prof. Gabrielle. *Canada.*
Donnay, Prof. Joseph Désiré Hubert. *Canada.*
Donohue, Prof. Jerry. *USA.*
Donohue, Dr Terence. *USA.*
Donoso, Mr Eduardo. *Chile.*
Donovan, Dr William Francis. *Australia.*
Dorfman, Dr Moisey Davydovich. *USSR.*
Doris, Mr Edward. *USA.*
Dornics, Mrs Monika. *DDR.*
Dorokhova, Dr Galina Igorevna. *USSR.*
Doroshinsky, Dr Alexander Leibovich. *USSR.*
Dorset, Dr Douglas Lewis. *USA.*
Dou, Mr Shi-qi. *China.*
Dougill, Dr Maryon W. *UK.*
Dover, Dr Stanley David. *UK.*
Dovesi, Prof. Roberto. *Italy.*
Dowell, Dr Walter Charles Thomas. *Australia.*
Downie, Mr George. *UK.*
Downs, Dr James Winston. *USA.*
Dowty, Mr Eric. *USA.*
Doyle, Prof. John Robert. *USA.*
Doyne, Prof. Thomas H. *USA.*

Dräger, Prof. Dr Martin. *BRD.*
Draganova, Dr Dragana. *Bulgaria.*
Draghici, Mr Iosif. *Romania.*
Dragsdorf, Prof. Russell Dean. *USA.*
Drake, Prof. John E. *Canada.*
Drašner, Magistar Antun. *Yugoslavia.*
Dreiding, Prof. André. *Switzerland.*
Drenck, Dr Kaj. *Denmark.*
Drendel, Mr William B. *USA.*
Drenth, Prof. Dr Jan. *Netherlands.*
Dressler, Dr Ludwig. *DDR.*
Drew, Dr Michael George Brindley. *UK.*
Drickman, Dr Myra Vivian. *USA.*
Driesel, Dr Wolfgang. *DDR.*
Driessen, Drs René A. J. *Netherlands.*
Driss, Dr Ahmed. *Tunisia.*
Dristas, Dr Jorge Anastasio. *Argentina.*
Drits, Dr Victor Anatolyevich. *USSR.*
Drozdov, Dr Yury Nikolayevich. *USSR.*
Druyan, Prof. Mary Ellen. *USA.*
Drzymala, Dr Janusz. *Poland.*
Du Plessis, Dr Michael Peter. *South Africa.*
Du Plessis, Prof. Paul de Villiers. *South Africa.*
Duževič, Dr Davor. *Yugoslavia.*
Duax, Dr William Leo. *USA.*
Dubček, Mr Pavo. *Yugoslavia.*
Dubey, Dr Ram Janam. *Nigeria.*
Dubin, Mr Robert R. *USA.*
Dubler, PD Erich. *Switzerland.*
Dubov, Dr Petr Lvovich. *USSR.*
Duchamp, Dr David James. *USA.*
Duchefdelaville, Mr Gérard. *France.*
Duchemin, Mr Jean-Pierre. *France.*
Duckett, Mr G.R. *UK.*
Ducros, Prof. Pierre. *France.*
Ducruix, Dr Arnaud. *France.*
Dudarev, Dr Vasily Yakovlevich. *USSR.*
Duderov, Dr Nikolay Grigoryevich. *USSR.*
Dudkevich, Dr Vladimir Petrovich. *USSR.*
Duesler, Ms Eileen N. *USA.*
Dugué, Prof. Jérome. *France.*
Duisenberg, Drs Albert Jozef Maria. *Netherlands.*
Duke, Mr J.R.C. *UK.*
Duke, Ms Norma Edith. *Canada.*
Dukova, Dr Elena Dmitriyevna. *USSR.*
Dumas, Mr Philippe. *France.*
Dumitrescu, Aurelia. *Romania.*
Dumke, Prof. Warren Lloyd. *USA.*
Dunaj-Jurčo, Doc. Ing Michal. *CSSR.*
Dunham, Prof A.C. *UK.*
Dunia, Dr Emery. *Venezuela.*
Dunitz, Prof. Jack David. *Switzerland.*
Dunlevey, Mr John Norman. *South Africa.*
Dunn, Mr Harris William. *USA.*
Dunn, Mr Karl L., Jr. *USA.*
Dunning, Mr Anthony John. *UK.*
Dunsieth, Ms Dana G. *USA.*
Dupont, Dr Leon. *Belgium.*
Duraipandianadar, Mr P. P. *India.*
Durairaj, Kanagapushpam. *India.*
Durant, Prof. François Victor. *Belgium.*
Durbetaki, Mr Antony J. *USA.*
Ďurčanská, Dr Edita. *CSSR.*
Durchschlag, Dr Helmut. *BRD.*
Durif, Dr André *France.*
Durlu, Doç. Dr Tahsin Nuri. *Turkey.*
Ďurovič, Ing Slavomil. *CSSR.*
Durruthy, Mr Obel. *Cuba.*
Durski, Dr Zygmunt. *Poland.*
Dutta, Dr Bishnu Pada. *India.*
Dutta, Dr Sachindra Nath. *India.*
Dvorkin, Dr Alexander Arkadyevich. *USSR.*
Dweltz, Dr Neville Edwin. *India.*
Dwiggins, Dr Claudius William, Jr. *USA.*
Dwight, Mr Austin Elbert. *USA.*
Dwivedi, Dr Ganpat Lal. *India.*
Dyke, Mr Maurice. *USA.*
Dynowska, Mrs Elzbieta Grażyna. *Poland.*
Dyson, Mr David John. *UK.*
Dytrych, Mr William J. *USA.*

Eales, Prof. Hugh Victor. *South Africa.*
Ealick, Dr Steven Edward. *USA.*
Eanes, Dr Edward David. *USA.*
Ebby, Dr N' Dédé. *Ivory Coast.*
Eberhard, Prof. Dr Emil. *BRD.*
Eberhart, Prof. Jean-Pierre. *France.*
Echavarri Hernandez, Dr Ariel. *Mexico.*
Eckerlin, Dr Peter. *BRD.*
Eckhardt, Prof. Dr Franz-Jörg. *BRD.*
Eckstein, PhD. Dipl. Ing. Juraj. *BRD.*
Economou, Prof. Nicolaos Alkiviadis. *Greece.*
Eddy, Prof. Lowell Perry. *USA.*
Eder, Prof. Dr Dipl.-Ing. Otto Josef. *Austria.*
Edmonds, Dr James William. *USA.*
Edmundson, Mr Allen B. *USA.*
Edmondson, Mr M. *UK.*
Edström, Mrs Kristina. *Sweden.*
Edwards, Dr Anthony John. *UK.*
Edwards, Dr Brian Francis Peregrine. *USA.*
Edwards, Dr Ian Arthur Samuel. *UK.*
Edwards, Dr William Donald. *Canada.*
Eeles, Mr Wilfred Trefor. *UK.*
Effenberger, Dr Herta. *Austria.*
Efremov, Dr Valery Alexandrovich. *USSR.*
Egan, Mr Robert W. *USA.*
Egert, Dr Ernst. *BRD.*
Eggers, Mr Peter. *DDR.*
Eggleston, Dr Drake Stephen. *USA.*
Eggleton, Dr Richard Anthony. *Australia.*
Egli, Mr Martin. *Switzerland.*
Egorov, Dr Vladimir Mikhailovich. *USSR.*
Egorov-Tismenko, Dr Yury Klavdiyevich. *USSR.*
Ehses, Dr Karl-Heinz. *BRD.*
Eichhorn, Dr Edgar Leo. *USA.*
Eichhorn, Dr Gerd. *DDR.*
Eichler, Dr Klaus. *DDR.*
Eichler, Dr Wolfgang. *DDR.*
Eick, Prof. Harry A. *USA.*
Eigenbrot Dr Charles Weaver, Jr. *USA.*
Eilerman, Dr Donna Paige. *USA.*
Einck, Dr James J. *USA.*
Einspahr, Dr Howard Martin. *USA.*
Einstein, Prof. Frederick William Boldt. *Canada.*
Eiríksson, Dr Vésteinn Runi. *Iceland.*
Eisenberg, Prof. David. *USA.*
Eisenberg, Prof. Henryk. *Israel.*
Eisenberg, Prof. Richard. *USA.*
Eisenmann, Mrs Dr Brigitte. *BRD.*
Eisenstein, Dr Miriam. *Israel.*
Eitel, Dr Manfred. *BRD.*
Ejiri, Dr Koichi. *Japan.*
Eklund, Dr Hans. *Sweden.*
Ekmekçi, Dr Servet. *Turkey.*
Ekström, Dr Thommy. *Sweden.*
El Demerdash, Dr Saad. *Egypt.*
El Gabi, Dr Sami. *Egypt.*
El-Kabbani, Mr Ossama Ahmed Lofti. *Canada.*
El-Mahdi, Dr Omar. *Saudi Arabia.*
El Naggar, Dr Mohamed. *Egypt.*
El Ramly, Mr Mohamed Fawzi. *Egypt.*
El Saffar, Prof. Zuhair M. *Egypt.*
El Sayed, Prof. (Mrs) Karimat. *Egypt.*
El Shaabini, Prof. (Mrs) Aida Moustafa. *Egypt.*
El Shanshury, Dr Ismail. *Egypt.*
El Sharkawi, Dr Mohamed Abdel Hamid. *Egypt.*
El Shazli, Dr El Shazli Mohamed. *Egypt.*
Elahi, Mr Manzoor. *Pakistan.*
Elbadri, Dr H. *Egypt.*
Elbinger, Dr German. *DDR.*
Elcombe, Dr Margaret Marion. *Australia.*
Elder, Dr D.P. *UK.*
Elder, Dr Michael. *UK.*
Elder, Prof. Richard C. *USA.*
Elerman, Dr Yalçin. *Turkey.*
Elf, Mr Frank. *BRD.*
Elgenmark, Miss Ingegerd. *Sweden.*
Elias, Mr E.E. *UK.*
Eliopoulos, Dr Elias Edward. *UK.*
Eliseev, Dr Erik Nikolayevich. *USSR.*
Elkaim, Mr Erik. *France.*

Eller, Dr P. Gary. *USA.*
Elliott, Dr Gerald Frank. *UK.*
Elliott, Dr James Cornelis. *UK.*
Elliott, Dr Robert Brian. *UK.*
Ellner, Dr Martin Oliver. *BRD.*
Elsen, Mr Jan Albrecht. *Belgium.*
Elwan, Dr Ahmed Abdel Salam. *Egypt.*
Elzawi, Mr Rajab Abdulla. *Libya.*
Embrey, Mr Peter Godwin. *UK.*
Emerson, Prof. Kenneth. *USA.*
Emerson, Mr Merle T. *USA.*
Emge, Dr Thomas James. *USA.*
Emiliani, Prof. Francesco. *Italy.*
Emmenegger, Prof. Franzpeter. *Switzerland.*
Emons, Prof. Hans-Heinz. *DDR.*
Enckevort, van, Dr Wilhelmus J.P. *Netherlands.*
Enemark, Prof. John Henry. *USA.*
Enflo, Mrs Anita. *Sweden.*
Eng-Wilmot, Dr David Lawrence. *USA.*
Engel, Dr Aribert. *DDR.*
Engel, Prof. Dennis Walter. *South Africa.*
Engel, Ms Nora. *Switzerland.*
Engel, PD Peter. *Switzerland.*
Engel, Dr Walter. *BRD.*
Engels, Prof. Siegfried. *DDR.*
Englisch, Mr Uwe-Franz. *BRD.*
English, Dr Robert Bertram. *South Africa.*
Engström, Dr Ingvar O. J. *Sweden.*
Ensling, Dr Jürgen. *BRD.*
Enwall, Dr Eric Lee. *USA.*
Epelboin, Dr Yves. *France.*
Eppelsheimer, Prof. Daniel Snell. *USA.*
Eppelsheimer, Dr Daniel Snell, Jr. *USA.*
Epperson, Dr John Ernest. *USA.*
Epprecht, Prof. Willfried Th. *Switzerland.*
Epstein, Dr Joel. *Australia.*
Erazo Plaza, Mr Antonio David. *Colombia.*
Ercit, Mr Timothy Scott. *Canada.*
Erdönmez, Dr Ahmet. *Turkey.*
Erdoğmuş, Dr Muktim. *Turkey.*
Erdoes, Mr Ernst Gyula. *Switzerland.*
Erez, Dr Gidon. *Israel.*
Ergin, Dr Ömer. *Turkey.*
Ericsson, Prof. S. Torsten. *Sweden.*
Eriks, Prof. Klaas. *USA.*
Eriksson, Dr Anders. *Sweden.*
Eriksson, Dr Birgitta. *Sweden.*
Eriksson, Mr Lars. *Sweden.*
Eriksson, Mr Sven. *Sweden.*
Ermer, Dr Otto. *BRD.*
Ernst, Dr Gert. *Austria.*
Ernst, Dr Stephen Richard. *USA.*
Ersson, Mr Nils Olov. *Sweden.*
Escobar, Dr Carmen. *Chile.*
Esipova, Dr Nataliya Georgiyevna. *USSR.*
Espinoza, Prof. Odon. *Peru.*
Esselborn, Dr Reiner Ferdinand. *BRD.*
Esteoule, Prof. Jacques. *France.*
Esterhuyse, Miss Suzette. *South Africa.*
Estes, Dr Eva Dixon. *USA.*
Estop, Dra Eugenia. *Spain.*
Eswara Prasad, Mr Gummuluri. *India.*
Etemadi-Abdolabadi, Dr Bijan. *Iran.*
Etter, Dr Margaret Eleanor Cairns. *USA.*
Ettmayer, Prof. Dr Dipl.-Ing. Peter. *Austria.*
Eulenberger, Dr Günther Richard. *BRD.*
Euler, Dr Robert. *BRD.*
Euthymiou, Prof. Paraskevi. *Greece.*
Evangelidou, Miss Christina. *Greece.*
Evans, Dr Anthony Meredith. *UK.*
Evans, Mr D.M. *UK.*
Evans, Mr David Lindsay. *New Zealand.*
Evans, Dr Doris Louise. *USA.*
Evans, Dr E.M.H. *UK.*
Evans, Ms Eloise Humez. *USA.*
Evans, Dr Howard Tasker, Jr. *USA.*
Evans, Dr John Hedley. *USA.*
Evans, Dr P.A. *UK.*
Evans, Mr R.R. *UK.*
Evans, Dr Robert Crispin. *UK.*
Evole Martil, Dr Nieves. *Spain.*

Evrard, Prof. Guy Henri. *Belgium.*
Extine, Dr Michael Wayne. *USA.*
Eysel, Prof. Dr Walter. *BRD.*
Faber, Dr John, Jr. *USA.*
Faber, Dr Peter. *BRD.*
Fabregat Guinchard, Dr Francisco José. *Mexico.*
Fackler, Prof. John Paul, Jr. *USA.*
Fälth, Dr Lars. *Sweden.*
Faerman, Mr Carlos Hugo. *Canada.*
Faggiani, Mr Romolo G. *Canada.*
Fagherazzi, Prof. Giuliano. *Italy.*
Fagnani, Prof. Gustavo. *Italy.*
Fahey, Prof. James A. *USA.*
Fair, Dr Carolyn Kay. *USA.*
Fairclough, Dr D.P. *UK.*
Fairhurst, Prof. Carl Wayne. *USA.*
Fajardo, Mr Fabio. *Cuba.*
Fakhar, Miss Noura. *Tunisia.*
Falk, Dr Michael. *Canada.*
Falkenberg, Dr Wolfgang. *DDR.*
Fallon, Dr Gary David. *Australia.*
Falshaw, Dr C.P. *UK.*
Falvello, Mr Lawrence R. *USA.*
Fan, Mr Guang-yu. *China.*
Fan, Prof. Hai-fu. *China.*
Fan, Mr Yu-guo. *China.*
Fan, Mr Zhao-chang. *China.*
Fanariotis, Mr Iakovos. *Greece.*
Fanfani, Prof. Luca. *Italy.*
Fang, Prof. Jen Ho. *USA.*
Fanter, Mr Detlef. *DDR.*
Faqir, Dr Gul. *Malaysia.*
Faraco Muños, Mr Nicolas. *Spain.*
Fares, Dr Vincenzo. *Italy.*
Farkas-Jahnke, Dr Mária. *Hungary.*
Farmer, Dr Victor Colin. *UK.*
Farrants, Dr George William. *UK.*
Farrar, Prof. Roy Alfred. *UK.*
Farrugia, Dr L.J. *UK.*
Faruqi, Dr Fazal Ahmad. *Pakistan.*
Faruqi, Dr A.R. *UK.*
Fasiska, Dr Edward J. *USA.*
Fatmi, Prof. Ali Nasir. *Pakistan.*
Faught, Dr John Brian. *Canada.*
Faulk, Mr John Warren. *USA.*
Faust, Dr George Tobias. *USA.*
Faust, Mr Wolfgang. *DDR.*
Fauvet, Mr Gérard. *France.*
Favretto, Prof. Luciano. *Italy.*
Fawcett, Dr John. *UK.*
Fawcett, Dr John Keith. *Canada.*
Fawcett, Mr Timothy G. *USA.*
Fay, Prof. Robert Clinton. *USA.*
Fayed, Dr (Mrs) Leila. *Egypt.*
Fayos, Dr Jose. *Spain.*
Fazal, Dr Muhammad. *Pakistan.*
Fedeli, Prof. Walter. *Italy.*
Fedorenko, Prof. Anatoly Ivanovich. *USSR.*
Fedorov, Dr Boris Alexandrovich. *USSR.*
Fedorov, Dr Pavel Pavlovich. *USSR.*
Fedotov, Dr Alexander Fedorovich. *USSR.*
Feher, Mr Andreas. *DDR.*
Feher, Mrs Elvira. *DDR.*
Fehling, Mr Wolfgang. *DDR.*
Fehlmann, PD Melchior. *Switzerland.*
Feigin, Dr Lev Abramovich. *USSR.*
Feil, Prof. Dr Dirk. *Netherlands.*
Fejer, Miss Eleonora Eva. *UK.*
Felbinger, Dr Adolf. *DDR.*
Feld, Dr Rainer Hans Helmut. *BRD.*
Feldman, Dr Robert Edward. *USA.*
Feldmann, Mr Richard Joseph. *USA.*
Felius, Dr Robert Onno. *Netherlands.*
Felix, Dr Rene P.. *Philippines.*
Felsche, Prof. Dr Jürgen. *BRD.*
Felsteiner, Prof. Joshua. *Israel.*
Feltz, Prof. Adalbert. *DDR.*
Feneau-Dupont, Mrs Janine. *Belgium.*
Fenn, Dr Ruth Helen. *UK.*
Fenna, Dr Roger Edward. *USA.*

Fenoll Hach-Ali, Prof. Purificacion. *Spain.*
Fenske, Prof. Dr Dieter. *BRD.*
Ferey Prof. Gérard. *France.*
Ferguson, Dr George. *Canada.*
Ferguson, Prof. Robert Bury. *Canada.*
Ferguson, Dr Ian Forster. *UK.*
Fernandes, Mr Jacob Richard. *India.*
Fernandez, Prof. Aurora Reyes. *Philippines.*
Fernandez, Mr Juan Carlos. *Argentina.*
Fernández González, Dr Alonso. *Mexico.*
Fernandez Nieto, Dra Constanza. *Spain.*
Fernando, Prof. Quintus. *USA.*
Fernholt, Mrs Liv. *Norway.*
Ferracini, Prof. Elena. *Italy.*
Ferran, Dr Gustan. *Brazil.*
Ferrari-Belicchi, Prof. Marisa. *Italy.*
Ferraris, Prof. Giovanni. *Italy.*
Ferreira de Souza, Prof. Milton. *Brazil.*
Ferrero Rognoni, Prof. adele. *Italy.*
Ferretti, Dr Valeria. *Italy.*
Fesenko, Prof. Evgeny Grigoryevich. *USSR.*
Fesenko, Dr Oleg Evgenyevich. *USSR.*
Fewster, Dr Paul Frederick. *UK.*
Fiala, Dr Jaroslav. *CSSR.*
Fichera, Dr Anna Maria. *Italy.*
Fichtner-Schmittler, Dr Helga. *DDR.*
Fichtner, Dr Konrad. *DDR.*
Fiedler, Dr Gustav. *DDR.*
Field, Dr Donald William. *Australia.*
Field, Dr John Stainer. *South Africa.*
Fielding, Mr W.D. *UK.*
Fiermans, Dr Lucien Victor August. *Belgium.*
Figgis, Prof Brian Norman. *Australia.*
Figielski, Dr hab. Tadeusz. *Poland.*
Figueiredo, Dr Maria Ondina. *Portugal.*
Figueiredo Neto, Dr Antonio Martins. *Brazil.*
Filatov, Dr Stanislav Konstantinovich. *USSR.*
Filip'yev, Dr Victor Semenovich. *USSR.*
Filipenko, Dr Olga Savelyevna. *USSR.*
Filippakis, Dr Sophokles. *Greece.*
Filippini, Dr Giuseppe. *Italy.*
Filizova, Dr Lyudmila. *Bulgaria.*
Fillers, Dr James Paul. *USA.*
Filscher, Mr Gerold. *DDR.*
Finger, Dr Larry W. *USA.*
Fingerland, Dr Antonín. *CSSR.*
Fink, Prof. Robert. *USA.*
Fink, Dr William LaVilla. *USA.*
Finkel'shtein, Dr Aleksey Vital'yevich. *USSR.*
Finlayson, Mr Kier. *USA.*
Finney, Dr John Leslie. *UK.*
Fischer-Hjalmars, Prof. Inga M. *Sweden.*
Fischer, Dr Carl-Otto. *BRD.*
Fischer, Mr Gerhard Richard. *USA.*
Fischer, Mr Karl. *DDR.*
Fischer, Prof. Dr Karl. *BRD.*
Fischer, Dr Peter. *Switzerland.*
Fischer, Dr Richard. *Austria.*
Fischer, Dr Reinhard X. *USA.*
Fischer, Mrs Ute Eva-Maria. *BRD.*
Fischer, Prof. Dr Werner. *BRD.*
Fisher, Mr Graham Richard. *UK.*
Fisher, Mr Leslie Ernest. *UK.*
Fisher, Dr Richard G. *USA.*
Fisher, Dr Robert M. *USA.*
Fita, Mr Iguacio. *USA.*
Fitch, Dr Andrew Nicholas. *France.*
Fitzgerald, Dr Alexander Grant. *UK.*
Fitzgerald, Dr Alvin. *USA.*
Fitzgerald, Dr Paula Marie Dean. *Canada.*
Fitzl, Mr Günther. *DDR.*
Fitzwilliam, Dr James William. *USA.*
Fjaer, Dr Erling. *Norway.*
Fjeldberg, Mr Torgny. *Norway.*
Flack, Dr Howard David. *Switzerland.*
Fleet, Dr Michael Edward. *Canada.*
Fleischmann, Dr Klaus Dietrich. *Netherlands.*
Flerov, Dr Igor' Nikolayevich. *USSR.*
Fletcher, Dr Neville Horner. *Australia.*
Fletcher, Dr R.O.W. *UK.*

Fletcher, Dr Steven Reginald. *UK.*
Fletterick, Dr Robert J. *USA.*
Flewitt, Dr Peter Edwin John. *UK.*
Flippen-Anderson, Ms Judith Lee. *USA.*
Flodmark, Dr Stig. *Sweden.*
Flögel, Dr Peter. *DDR.*
Flörke, Prof. Dr Otto Wilhelm. *BRD.*
Florencio, Dr Feliciana. *Spain.*
Florio, Dr John Victor. *USA.*
Flower, Mr S.C. *UK.*
Foces Foces, Dr Concepcion. *Spain.*
Förster, Dr Eckhart. *DDR.*
Försterling, Dr Gerd. *DDR.*
Förtsch, Prof. Dr Erich Bernhard. *South Africa.*
Folgueras Dominguez, Dr Sérvulo. *Brazil.*
Follner, Prof. Dr Heinz. *BRD.*
Folting-Streib, Mrs Kirsten. *USA.*
Fones, Mr M.D. *UK.*
Fong, Dr Hock Sun. *Singapore.*
Font-Altaba, Prof. Manuel. *Spain.*
Fontaine Dr Alain. *France.*
Fontaine, Prof. Hubert. *France.*
Fontaine, Dr Frederic Desiré Albert. *Belgium.*
Foord, Mr Eugene E. *USA.*
Ford, Dr Geoffrey Charles. *UK.*
Foreman, Mr Dennis W., Jr. *USA.*
Foresti, Prof. Elisabetta. *Italy.*
Foris, Mr C. M. *USA.*
Formoso, Dr Milton Luiz. *Brazil.*
Fornasini, Prof. Maria Luisa. *Italy.*
Forni, Prof. Flavio. *Italy.*
Fornoff, Mr Mario M. *USA.*
Forsellini, Dr Eleonora. *Italy.*
Forslund, Dr S. Bertil. *Sweden.*
Forst, Mr Hans Rainer. *BRD.*
Forster, Dr Martin. *Switzerland.*
Forsyth, Dr John Bruce. *UK.*
Forsyth, Mr V.T. *UK.*
Fortes, Prof. Manuel Amaral. *Portugal.*
Forti, Prof. Paolo. *Italy.*
Fortier, Dr Suzanne. *Canada.*
Foroughi, Dr Ali-Assghar. *Iran.*
Forwood, Dr Christopher Thomas. *Australia.*
Foss, Ms Linda I. *USA.*
Foss, Prof. Olav. *Norway.*
Foster, Prof. Alfred F. *USA.*
Foster, Ms Beryl Ann. *USA.*
Foster, Dr John James. *Australia.*
Fotchenkov, Dr Anatoly Andreyevich. *USSR.*
Fouret, Prof. René *France.*
Fourie, Dr Jacobus Theodor. *South Africa.*
Fourme, Prof. Roger. *France.*
Fox, Mr B.E. *UK.*
Fox, Dr Robert O., Jr. *USA.*
Foxman, Prof. Bruce Mayer. *USA.*
Fraenkel, Prof. Benjamin. *Israel.*
Franceschi, Prof. Enrico. *Italy.*
Francesconi, Dr Ricardo. *Brazil.*
Franchini, Prof. Marinella. *Italy.*
Francis, Mr John Godfrey. *UK.*
Francisco, Miss Regina Helena Porto. *Brazil.*
Frank, Prof. Sir Frederick Charles. *UK.*
Frank, Dr Walter. *BRD.*
Frank-Kamenetskaya, Dr Olga Victorovna. *USSR.*
Frank-Kamenetsky, Prof. Victor Al'bertovich. *USSR.*
Franke, Dr Valeriya Dmitriyevna. *USSR.*
Franklin, Dr Kenneth James. *Canada.*
Franklin, Prof. Ursula Martius. *Canada.*
Franks, Dr Albert. *UK.*
Franks, Dr Joseph. *UK.*
Fransolet, Prof. André-Mathieu. *Belgium.*
Franzen, Prof. Hugo Friedrich. *USA.*
Franzini, Prof. Marco. *Italy.*
Franzosi, Dr Paolo. *Italy.*
Fraser, Dr G.V. *UK.*
Fraser, Dr Ronald Bruce. *Australia.*
Fratini, Prof. Albert V. *USA.*
Frazer, Dr Benjamin Chalmers. *USA.*
Fredericks, Dr Robert J. *USA.*
Fredrich, Mr Michael F. *USA.*
Freed, Prof. Robert Lowell. *USA.*

Freeman, Dr Alan George. *New Zealand.*
Freeman, Prof Hans Charles. *Australia.*
Freeman, Dr Gerald Richard. *USA.*
Freeman, Mr Walter Gerard. *UK.*
Freer, Dr Stephan T. *USA.*
Freiburg, Dr Johann Christoph. *BRD.*
Freire D'aguiar, Mr Manoel Marcos. *Brazil.*
Freire Pimentel, Dr Cecilia A. *Brazil.*
French, Dr Alfred Dexter. *USA.*
French, Mr Robert D. *USA.*
Frenz, Dr Bertram Anton. *USA.*
Freudenberg, Mr Axel. *DDR.*
Freund, Dr Andreas Karl. *France.*
Freundlich, Dr A. *UK.*
Frevel, Dr Ludo Karl. *USA.*
Frey, Prof. Dr Friedrich. *BRD.*
Frey, Dr Michel. *France.*
Freydank, Mrs Gisela-Christine. *DDR.*
Fridkin, Prof. Vladimir Mikhailovich. *USSR.*
Friedel, Prof. Jacques. *France.*
Friedlander, Dr Peter H. *USA.*
Friedman, Dr Lawrence Boyd. *USA.*
Frigeri, Dr Cesare. *Italy.*
Frigyik, Mr Gábor. *Hungary.*
Frikkee, Dr Evert. *Netherlands.*
Friman, Dr Rauno Kalevi. *Finland.*
Fritchie, Prof. Charles Julius, Jr. *USA.*
Fröhlich, Dr Fritz. *DDR.*
Fröhlich, Dr Roland. *BRD.*
Frolow, Dr Felix. *Israel.*
Fronczek, Dr Frank R. *USA.*
Frueh, Prof. Alfred Joseph. *USA.*
Frühauf, Dr Joachim. *DDR.*
Ftikos, Dr Christos. *Greece.*
Fu, Prof. Heng. *China.*
Fu, Prof. Ping-qiu. *China.*
Fu, Mr Zheng-min. *China.*
Fu, Mr Zhu-ji. *China.*
Fuchs, Dr Erik. *Hungary.*
Fuente Cullell, Dr Carlos de la. *Spain.*
Fuentes Perez, Mr Manuel. *Spain.*
Fuess, Prof. Dr Hartmut. *BRD.*
Fuith, Dr Armin. *Austria.*
Fujii, Dr Satoshi. *Japan.*
Fujii, Mr Tetsuo. *Japan.*
Fujii, Assoc. Prof. Yasuhiko. *Japan.*
Fujiki, Prof. Yoshibumi. *Japan.*
Fujime, Dr Satoru. *Japan.*
Fujimore, Prof. Kenkichi. *Brazil.*
Fujimoto, Prof. Fuminori. *Japan.*
Fujimoto, Prof. Hirofumi. *Japan.*
Fujino, Dr Kiyoshi. *Japan.*
Fujino, Dr Nobukatsu. *Japan.*
Fujita, Prof. Francisco Eiichi. *Japan.*
Fujita, Prof. Hiroshi. *Japan.*
Fujiwara, Prof. Kunio. *Japan.*
Fujiwara, Prof. Hiroshi. *Japan.*
Fujiwara, Dr Takaji. *Japan.*
Fujiyoshi, Dr Yoshinori. *Japan.*
Fukamachi, Dr Tomoe. *Japan.*
Fukano, Prof. Yasushige. *Japan.*
Fukuda, Dr Tsuguo. *Japan.*
Fukuhara, Dr Akira. *Japan.*
Fukuyama, Dr Keiichi. *Japan.*
Fukuyama, Dr Tsutomu. *Japan.*
Fulfaro, Dr Roberto. *Brazil.*
Fullam, Mr Ernest F. *USA.*
Fullenwider, Mr Malcolm A. *USA.*
Fuller, Prof. W. *UK.*
Fulton, Dr William Stephen. *UK.*
Fumi, Prof. Fausto Gherardo. *Italy.*
Fun, Dr Hoong Kun. *Malaysia.*
Fundamensky, Mr Vladimir Semenovich. *USSR.*
Fuoss, Dr Paul Henry. *USA.*
Furdanowicz, Mr Waldemar. *USA.*
Furey, Dr William F., Jr. *USA.*
Furmanova (Bokiy), Dr Nina Georgievna. *USSR.*
Furnas, Dr Thomas Coleman, Jr. *USA.*
Fursenko, Dr Boris Alexandrovich. *USSR.*
Furukawa, Dr Kozo. *Japan.*

Furusaki, Dr Akio. *Japan.*
Furuseth, Mrs Sigrid. *Norway.*
Fuzek, Dr John Frank. *USA.*
Fykin, Dr Leonid Efimovich. *USSR.*
Gaál, Dr István. *Hungary.*
Gabe, Dr Eric James. *Canada.*
Gabela, Prof. Dr Fikret. *Yugoslavia.*
Gabis, Prof. Victor Michel. *France.*
Gable, Mr Robert William. *Australia.*
Gabuda, Prof. Svyatoslav Petrovich. *USSR.*
Gadó, Dr Pál. *Hungary.*
Gad, Dr Gamal Mohamed. *Egypt.*
Gadet, Mr Alain. *France.*
Gafner, Dr Geoffrey. *South Africa.*
Gagnon, Miss Carole. *Canada.*
Gahm, Dr Josef. *BRD.*
Gaier, Dr James Richard. *USA.*
Gaines, Mr James Matthew. *USA.*
Gainsford, Dr Graeme John. *New Zealand.*
Gait, Dr Robert Irwin. *Canada.*
Gajhede, Mr Michael. *Denmark.*
Galan Huertas, Prof. Emilio. *Spain.*
Gałązka, Dr hab. Robert. *Poland.*
Gałdecki, Dr hab. Zdzisław. *Poland.*
Gale, Dr Brian. *UK.*
Galešić, Dr Nikola. *Yugoslavia.*
Galetti, Dr Giulio. *Switzerland.*
Gali, Dr Salvador. *Spain.*
Galigné, Dr Jean-Louis. *France.*
Galinos, Prof. Andreas. *Greece.*
Galitsky, Dr Nikolay Mikhailovich. *USSR.*
Galiulin, Dr Ravil Vagizovich. *USSR.*
Gall, Prof. William Einar. *USA.*
Gallagher, Dr Kevin Joseph. *UK.*
Galli, Prof. Ermanno. *Italy.*
Gallois, Mr Bernard. *France.*
Galloni, Prof. Ernesto. *Argentina.*
Gallot, Dr Bernard. *France.*
Galloy, Dr Jean. *UK.*
Gallucci, Dr Judith Chlastawa. *USA.*
Galstyan, Dr Viktor Gaikovich. *USSR.*
Galvis, Mr Jaime. *Colombia.*
Galy, Dr Jean. *France.*
Gama Carvalho, Dr Frederico. *Portugal.*
Gambone, Mr Michael J. *USA.*
Gan, Mr Ah Sai. *Malaysia.*
Ganazzoli, Dr Fabio. *Italy.*
Gandadisastra, Mr Samsa. *Indonesia.*
Gandais, Dr Madeleine. *France.*
Ganesan, V. *India.*
Ganev, Ing Nikolaj. *CSSR.*
Gangas, Prof. Nicolas - Hercule. *Greece.*
Gangulee, Dr Amitava. *USA.*
Gantzel, Dr Peter Kellogg. *USA.*
Gao, Mr Yi-guei. *China.*
Garín, Dr Jorge. *Chile.*
Garafalo, Mr Alfred R. *USA.*
Garaj, Prof. Dr Ján. *CSSR.*
Garashina, Dr Lyudmila Solomonovna. *USSR.*
Garavito, PD R. Michael. *Switzerland.*
Garaycochea-Wittke, Mrs Isabel. *Chile.*
Garbassi, Dr Fabio. *Italy.*
Garbauskas, Dr Mary E. *USA.*
García, Miss Ileana. *Cuba.*
Garcia Blanco, Prof. Severino. *Spain.*
Garcia Degano, Mrs Ma. Josefa. *Spain.*
Garcia Guinea, Mr Javier. *Spain.*
Garcia Martinez, Dr Oscar. *Spain.*
Garcia Vicente, Prof. Dr Jose. *Spain.*
Gard, Dr John Alan. *UK.*
Gardner, Mr Alexander Parker. *Australia.*
Gardner, Dr Kenn Corwin H. *USA.*
Gare, Mr Terence. *UK.*
Garland, Mrs Maria Teresa. *Chile.*
Garner, Prof. Christopher David. *UK.*
Garratt, Mr R.C. *UK.*
Garrett, Mr Thomas Peter John. *Australia.*
Garrido, Prof. Dr Julio. *Spain.*
Garvey, Prof. Roy George. *USA.*
Gash, Dr Alfred G. *USA.*

Gasparri-Fava, Prof. Giovanna. *Italy.*
Gasperin, Dr Madeleine. *France.*
Gassmann, Dr Johann. *BRD.*
Gast, Mr Roland. *DDR.*
Gastaldi, Dr Leonardo. *Italy.*
Gastuche - Van Oosterwyck, Dr Marie C. *Belgium.*
Gatehouse, Dr Bryan Michael Kenneth. *Australia.*
Gatineau, Dr Lucien Charles. *France.*
Gatti, Dr Giuseppina. *Italy.*
Gaunt, Dr Paul. *Canada.*
Gause, Dr Hans. *DDR.*
Gauthier, Dr Jean-Pierre. *France.*
Gauzzi, Prof. Franco. *Italy.*
Gavezzotti, Prof. Angelo. *Italy.*
Gavrilova, Dr Irina Vladimirovna. *USSR.*
Gavuzzo, Dr Enrico. *Italy.*
Gawron, Dr Marian Ryszard. *Poland.*
Gay, Dra Hebe Dina. *Argentina.*
Gaynor, Mr Gary R. *USA.*
Gazzano, Dr Massimo. *Italy.*
Gazzoni, Dr Giuseppe. *Italy.*
Gdaniec, Dr Maria. *Poland.*
Gebert, Dr Walter Richard. *BRD.*
Gebhardt, Prof. Dr Manfred Adolf Hermann. *BRD.*
Geckle, Mr Raymond J. *USA.*
Geddes, Dr Alexander John. *England.*
Gedicke, Mrs Christine. *DDR.*
Gedikoğlu, Doç. Dr Adil. *Turkey.*
Geelen, van, Dr Bernard. *Netherlands.*
Geerestein, van, Drs Vincent Johan. *Netherlands.*
Geguzina, Dr Galina Alexandrovna. *USSR.*
Gehlen, Prof. Kurt von. *BRD.*
Gehringer, Mr Ronald. *USA.*
Geib, Prof. Irving George. *USA.*
Geil, Prof. Phillip Herbert. *USA.*
Geise, Prof. Dr Herman Joseph Victor H. *Belgium.*
Geisinger, Ms Karen L. *USA.*
Geiss, Dr Roy Howard. *USA.*
Gel'man, Dr Yury Alexandrovich. *USSR.*
Gelato-Volders, Dr Louise Marie. *Switzerland.*
Geleji-Neubauer, Mrs Irén. *Hungary.*
Geller, Prof. Seymour. *USA.*
Gellings, Prof. Dr Paul Johann. *Netherlands.*
Genin, Dr Yakov Vladimirovich. *USSR.*
George, Mr Clifford. *USA.*
Gerdes, Dr Reiner Josef. *USA.*
Gerdil, Prof. Raymond. *Switzerland.*
Gergely, Dr Márton. *Hungary.*
Gerhard, Mr F. Bruce, Jr. *USA.*
Gerhard Olsen, Dr Inger Lise. *Denmark.*
Germain, Prof. Gabriel. *Belgium.*
Gernand, Mr Martin. *DDR.*
Gerold, Prof. Dr Volkmar. *BRD.*
Gerst, Dr Kenneth Mark. *USA.*
Gervasio, Prof. Giuliana. *Italy.*
Gerward, Dr Leif. *Denmark.*
Gesemann, Prof. Renate. *DDR.*
Geserick, Mrs Sabine. *DDR.*
Getzoff, Dr Elizabeth Dickinson. *USA.*
Gevers, Prof. Dr Rudolf. *Belgium.*
Geyer, Mr Andreas. *BRD.*
Ghebrial, Dr Mounir Guirgis. *Egypt.*
Ghedira, Dr Mounir. *Tunisia.*
Gheith, Prof. Mohamed A. *USA.*
Ghelis, Mrs Marianne. *France.*
Ghermani, Mr Noureddine. *France.*
Ghezzi, Prof. Carlo. *Italy.*
Ghilardi, Dr Carlo Alfredo. *Italy.*
Ghose, Prof. Subrata. *USA.*
Ghosh, Dr Debashis. *USA.*
Ghosh, Dr Mrs Minakshi. *India.*
Ghosh, Dr Sujit Kumar. *India.*
Ghosh, Miss Sutapa. *India.*
Ghosh, Dr Timir Baran. *India.*
Ghouse, Dr Khaja Mohd. *India.*
Giacovazzo, Prof. Carmelo. *Italy.*
Gibbons, Dr Cyril Stephen. *Canada.*
Gibbs, Mr Gerald V. *USA.*
Gibon, Mrs Véronique Julie Jacques Laure. *Belgium.*
Giebultowicz, Dr Tomasz Mieczysław. *Poland.*
Gieren, Priv.-Doz. Dr Alfred. *BRD.*

Gies, Dr Hermann. *BRD.*
Giese, Mr Rossman F., Jr. *USA.*
Giesecke, Dr Johan. *Sweden.*
Giessen, Prof. Bill Cormann. *USA.*
Giglio, Prof. Edoardo. *Italy.*
Gilardi, Dr Richard Dean. *USA.*
Gilfrich, Mr John Valentine. *USA.*
Gilinskaya, Dr Emma Abramovna. *USSR.*
Gillani, Mr Jamshed Ali. *Pakistan.*
Gilles, Prof. Jean-Claude. *France.*
Gilli, Prof. Gastone. *Italy.*
Gillier-Pandraud, Prof. Hélène. *France.*
Gillies, Dr Donald Chalmers. *USA.*
Gilliland, Mr Gary L. *USA.*
Gilmartin, Mr M.G.M. *UK.*
Gilmore, Dr Christopher John. *UK.*
Gilson, Mr Pierre. *Switzerland.*
Ginderow, Dr Daria. *France.*
Ginell, Prof. Robert. *USA.*
Ginell, Dr Stephan Lawrence. *USA.*
Giordani, Prof. Marino. *Italy.*
Giordano, Prof. Federico. *Italy.*
Giordano, Mr Joseph. *USA.*
Giordano Orsini, Prof. Paolo. *Italy.*
Giorgi, Dr Angelo Louis. *USA.*
Giovanoli, Prof. Rudolf. *Switzerland.*
Girgis, Dr Kamal. *Switzerland.*
Giri, Mr Anit K. *India.*
Giri, Mr Siba Narayan. *India.*
Girirajan, Dr K. S. *India.*
Girt, Prof. Dr Egvin. *Yugoslavia.*
Giuşcă, Prof. Dan. *Romania.*
Giunchi, Dr Giovanni. *Italy.*
Giuseppetti, Prof. Giuseppe. *Italy.*
Givargizov, Dr Evgeny Inviyevich. *USSR.*
Gjönnes, Prof. Jon Kjell. *Norway.*
Gladić, Mr Jadranko. *Yugoslavia.*
Gladkikh, Dr Liliya Ivanovna. *USSR.*
Gladky, Dr Vsevolod Vladimirovich. *USSR.*
Gladyshevsky, Prof. Evgeny Ivanovich. *USSR.*
Glaeser, Prof. Robert Martin. *USA.*
Glaser, Dr Julius. *Sweden.*
Glass, Dr Howard L. *USA.*
Glasser, Prof. Fred Paul. *UK.*
Glasser, Prof. Leslie. *South Africa.*
Glasson, Dr Douglas Royston. *UK.*
Glatter, Dr Otto. *Austria.*
Glazer, Dr Anthony Michael. *UK.*
Glazov, Dr Aleksey Ivanovich. *USSR.*
Gleason, Dr William Bourke. *USA.*
Glehn, von, Dr Marianne. *Sweden.*
Gleizes, Dr Alain Nicolas. *France.*
Glen, Dr Gerald Leonard. *USA.*
Glen, Dr John Wallington. *UK.*
Glick, Prof. Milton Don. *USA.*
Glidewell, Dr Christopher. *UK.*
Gliki, Dr Nataliya Vladimirovna. *USSR.*
Glikin, Mr Arkady Eduardovich. *USSR.*
Glinka, Mr Charles Joseph. *USA.*
Glover, Dr I.D. *UK.*
Główka, Dr Marek. *Poland.*
Głowiak, Doc Tadeusz. *Poland.*
Glusker, Dr Jenny Pickworth. *USA.*
Gnanaguru, Mr K. *India.*
Go, Mr Kuan Tee. *USA.*
Goaman, Dr Llawenydd Constance Gwynne. *UK.*
Godavarthi, Bhagavannarayana. *India.*
Goddard, Dr Richard. *BRD.*
Godefroy, Prof. Lucien René *France.*
Godovikov, Prof. Alexander Alexandrovich. *USSR.*
Godwod, MSc Krzysztof Jan. *Poland.*
Godycki, Mr L. Edward. *USA.*
Göbel, Mr Ralf. *DDR.*
Göbel, Dr Herbert Ernst. *BRD.*
Göcke, Dr Wolfhart. *DDR.*
Görnert, Dr Peter. *DDR.*
Göttlicher, Prof. Dr Siegfried. *BRD.*
Götz, Dr Konrad. *DDR.*
Götz, Dr Wolfgang. *DDR.*
Götzinger, Dr Michael Alois. *Austria.*

Goedkoop, Prof. Dr Jacob A. *Netherlands.*
Göğüş, Mrs Gülderen. *Turkey.*
Goehner, Mr Raymond Philip. *USA.*
Goilo, Dr Eduard Al'bertovich. *USSR.*
Goldberg, Dr Israel. *Israel.*
Goldberg, Mr Martin Jeffry. *USA.*
Goldberg, Prof. Stephen Z. *USA.*
Goldish, Dr Elihu. *USA.*
Goldsmith, Dr Elizabeth Jane. *USA.*
Goldsmith, Prof. Julian Royce. *USA.*
Goldstein, Dr Barry Michael. *USA.*
Goldstein(Choshen), Mr Ehud. *Israel.*
Golič, Prof. Dr Ljubo. *Yugoslavia.*
Golikeri, Mr Ganesh D. *USA.*
Goliński, Dr Bohdan. *Poland.*
Golovachev, Dr Vladimir Pavlovich. *USSR.*
Golovastikov, Dr Nikolay Ivanovich. *USSR.*
Golovina, Dr Nina Ivanovna. *USSR.*
Golubev, Dr Alexander Mikhailovich. *USSR.*
Golubkov, Dr Alexander Vasilyevich. *USSR.*
Gomes, Albert Cardinal. *India.*
Gomes, Mr Osvaldo. *Uruguay.*
Gomes, Mr Samuel Irati Novaes. *Brazil.*
Gomes de Mesquita, Dr Albert Hijman. *Netherlands.*
Gomez Caraballo, Lic. Dora Maria. *Venezuela.*
Gomez Ramirez, Dr Ricardo. *Mexico.*
Gomez Ruimonte, Dr Florentino. *Spain.*
Gómezdaza Almendaro, Mr Mariano. *Mexico.*
Gomm, Dr Martin. *BRD.*
Goncharov, Dr Georgy Nikolayevich. *USSR.*
Gondrand, Dr Monique. *France.*
Gong, Dr Ping-Po. *USA.*
Gonschorek, Dr Walter. *BRD.*
González, Mr Claudio. *Chile.*
González, Mrs Irma. *Chile.*
González, Mr Yanko. *Chile.*
Gonzalez Garcia, Mrs Victoria. *Spain.*
Gonzalez Lopez, Dr Eng. Jose. *Spain.*
Gonzalez Martinez, Dr José. *Spain.*
Goodfellow, Dr Julia Mary. *UK.*
Goodman, Prof. C.H.L. *UK.*
Goodman, Dr Peter. *Australia.*
Gopal, Dr Ramanathan. *Canada.*
Gorbunova, Dr Yuliya Efimovna. *USSR.*
Gordiyenko, Dr Vladimir Vasilyevich. *USSR.*
Gordon, Ms Janice T. *USA.*
Gorina, Dr Iza Ivanovna. *USSR.*
Górkiewicz, Mgr Zbigniew. *Poland.*
Gorogotskaya, Dr Lydumila Ivanova. *USSR.*
Gorshkov, Dr Anatoly Ivanovich. *USSR.*
Górski, Dr Ludwik. *Poland.*
Gorter, Drs Ing. Sybout. *Netherlands.*
Goryunov, Dr Alexander Ivanovich. *USSR.*
Goshen, Dr Shmuel Yehuda. *Israel.*
Gosmanová, Dr Galina. *CSSR.*
Goswami, Dr Anilprasanna. *India.*
Goswami, Dr Kaidar Nath. *India.*
Goswami, Dr (Mrs) S. N. N. *India.*
Goto, Prof. Masaru. *Japan.*
Goto, Mr Yoshiaki. *Japan.*
Gottardi, Prof. Glauco. *Italy.*
Gotthardt, Dr Rolf. *Switzerland.*
Gottschalch, Dr Volker. *DDR.*
Goubitz, Drs Kees. *Netherlands.*
Gougoutas, Dr Jack Zanos. *USA.*
Gould, Dr Robert Ozburn. *UK.*
Gould, Dr Sheila Elizabeth Buchan. *UK.*
Gountsidou, Mrs Vasiliki. *Greece.*
Gourang, Dr Mansour. *Iran.*
Graaff, de, Dr Rudolf Adriaan Gerard. *Netherlands.*
Grabowski, Dr hab. Mieczysław Jerzy. *Poland.*
Graeber, Dr Edward J. *USA.*
Gränicher, Prof. Walter Hans Heini. *Switzerland.*
Graetsch, Mr Heribert. *BRD.*
Graf, Dr Hans Anton. *BRD.*
Graf, Prof. René *France.*
Graham, Dr Albert Ronald. *Canada.*
Graham, Dr James. *Australia.*
Graham, Mr Robert Albert. *USA.*
Grainger, Dr Colin Trevor. *Australia.*

Gramaccioli, Prof. Carlo Maria. *Italy.*
Gramlich, Dr Rahel. *Switzerland.*
Gramlich, Dr Volker. *Switzerland.*
Grampel, van de, Dr Johan Christoph. *Netherlands.*
Grand, Dr André *France.*
Grandjean, Prof. Daniel *France.*
Grant, Dr Douglas Frank. *UK.*
Grant, Dr William Kenneth. *UK.*
Grasso, Mr Michael. *USA.*
Grattan-Bellew, Dr Patrick Edward. *Canada.*
Grau, Mr Lutz. *DDR.*
Graves, Mr Bradford J. *USA.*
Graziani, Prof. Rodolfo. *Italy.*
Grdenić, Prof. Dr Drago. *Yugoslavia.*
Grebenshchikov, Prof. Roman Georgiyevich. *USSR.*
Grechushnikov, Dr Boris Nikolayevich. *USSR.*
Green, Mr R.S. *UK.*
Greenberg, Dr Berton Laurence. *USA.*
Greenblatt, Prof. Martha. *USA.*
Greene, Mr Fernando. *Chile.*
Greenhouse, Mr Harold M. *USA.*
Greer, Dr Jonathan. *USA.*
Gregg, Mr R. Q. *USA.*
Greis, Dr Ortwin. *BRD.*
Gremillion, Prof. Alcuin Florian. *USA.*
Grenier, Miss Lucie. *Canada.*
Grenthe, Prof. Ingmar. *Sweden.*
Gress, Dr Mary Edith. *USA.*
Grev, Prof. Dennis Merle. *USA.*
Grey, Dr Ian Edward. *Australia.*
Grice, Dr Joel Denison. *Canada.*
Grieger, Mr Gene R. *USA.*
Griesbach, Mrs Karin. *DDR.*
Griffin, Dr Jane Flanigen. *USA.*
Griffith, Dr Elizabeth Ann Hall. *USA.*
Griffiths, Mr P.J.F. *UK.*
Griger, Dr Ágnes. *Hungary.*
Grigoriades, Prof. Panayotis. *Greece.*
Grigorov, Dr Sergey Nikolayevich. *USSR.*
Grigoryev, Prof. Dimitry Pavlovich. *USSR.*
Grigoryeva, Dr Tamara Nikolayevna. *USSR.*
Grill, Mr Charles M. *USA.*
Grimes, Dr N.W. *UK.*
Grimvall, Dr Siv H. *Sweden.*
Grin', Dr Yury Nikolayevich. *USSR.*
Grinberg, Dr Svetlana Arnol'dovna, *USSR.*
Grins, Dr Jekabs. *Sweden.*
Grjotheim, Prof. Kai. *Norway.*
Groat, Mr Lee Andrew. *Canada.*
Grochowski, Dr Jacek Maciej. *Poland.*
Grochulski, Dr Pawel. *Poland.*
Grønlund, Dr Finn. *Denmark.*
Grönvold, Prof. Fredrik. *Norway.*
Gromek, Dr Jack Michael. *USA.*
Gronkowski, Dr Jerzy. *Poland.*
Grosse, Prof. Dr Peter. *BRD.*
Grossi, Dr Paolo. *BRD.*
Grossie, Mr David Alan. *USA.*
Grosz, Mr Tamás. *Hungary.*
Grotepass-Deuter, Mrs Margit. *BRD.*
Groth, Mr Per Arne. *Norway.*
Groves, Mr P. *UK.*
Grozdanov, Mr Lyudmil. *Bulgaria.*
Grubb, Mr K.A. *UK.*
Gruber, Dr Boris. *CSSR.*
Grühn, Prof. Dr Reginald. *BRD.*
Grünbaum, Prof. Enrique. *Israel.*
Grundig, Prof. Werner. *Brazil.*
Grundvig, Dr Sidsel. *Denmark.*
Grundy, Dr Harry Douglas. *Canada.*
Grylicki, Dr Mirosław. *Poland.*
Gržeta, Dr Biserka. *Yugoslavia.*
Grzinic, Dr Guido. *Australia.*
Grzymek, Prof. Jerzy. *Poland.*
Gschneidner, Prof. Karl A., Jr. *USA.*
Gspan, Dr Primož. *Yugoslavia.*
Guérin, Dr Diego Marcelo Alejandro. *Argentina.*
Gu, Mr Yuan-xin. *China.*
Gu, Prof. Xiao-cheng. *China.*
Guan, Prof. Ya-xian. *China.*
Guastini, Prof. Carlo. *Italy.*

Guay, Dr France. *Canada.*
Guddat, Mr Luke William. *Australia.*
Gude, Mr Arthur James, III. *USA.*
Guenter, Prof. John Ralph. *Switzerland.*
Gülzow, Mr Hansjürgen. *DDR.*
Gündoğdu, Dr Niyazi. *Turkey.*
Guentert, Dr Otto Johann. *USA.*
Günther, Prof. Fritz. *DDR.*
Guerrero Laverat, Mr Alejandro. *Spain.*
Guerreschi, Dr Luigi Giuseppe. *Italy.*
Gütlich, Prof. Dr Philipp. *BRD.*
Gütt, Mr Rainer. *DDR.*
Güzel, Doç. Dr Nuri. *Turkey.*
Guet, Dr Jean-Michel. *France.*
Guggenberger, Dr Lloyd Joseph. *USA.*
Guggenheim, Prof. Stephen. *USA.*
Gui, Ms Lu-lu. *China.*
Guilhem, Dr Jean. *France.*
Guindi, Dr Amin Riad. *Egypt.*
Guinier, Prof. André Jean. *France.*
Guitel, Mr Jean-Claude. *France.*
Gullman, Mr Jan O. *Sweden.*
Gunawardane, Dr Richard Pemasiri. *Sri Lanka.*
Gundersen, Miss Grete. *Norway.*
Gunning, Mr G.R. *UK.*
Guo, Mrs Fang. *China.*
Guo, Prof. Dong-yao. *China.*
Gupta, Dr Amaresh. *BRD.*
Gupta, Dr Kinkar Prosad. *India.*
Gupta, Mr Manoj Kumar. *India.*
Gupta, Prof. Manoranjan Prasad. *India.*
Gupta, Mr Satish Chandra. *India.*
Gupta, Miss Sunita. *India.*
Gupta, Mr Vijai Prakash. *India.*
Gupta, Prof. Vishwambhar Dayal. *India.*
Gurewitz, Dr Eitan. *Israel.*
Gurin, Dr Vladimir Nikolayevich. *USSR.*
Gurskaya, Dr Galina Victorovna. *USSR.*
Gururow, Dr Tayur N. *India.*
Guseinov, Dr Gakhraman Gusein ogly. *USSR.*
Guseinova, Dr Maya Kara kyzy. *USSR.*
Guss, Dr Jules Mitchell. *Australia.*
Gussone, Dr Rainer Carl Leonard. *BRD.*
Gustafsson, Mr Torbjörn. *Sweden.*
Guth, Dr Helmut Karl Richard. *BRD.*
Gutierrez Puebla, Prof. Enrique. *Spain.*
Gutierrez Rios, Prof. Dr Enrique. *Spain.*
Gutteridge, Mr Walter Alfred. *UK.*
Guttmann, Mr Geoffrey D. *USA.*
Gutzow, Prof. Ivan. *Bulgaria.*
Guven, Prof. Necip. *USA.*
Guy, Mr Joseph Thomas, Jr. *USA.*
Guzman, Dr Luis Alberto. *Italy.*
Gweifel, Dr Ismail. *Egypt.*
Gyepes, Doc. Dr Eduard. *CSSR.*
Gyepesová, Dr Dalma. *CSSR.*
Gyobu, Mr Atsuo. *Japan.*
Haagensen, Mr Carl Olaf. *Denmark.*
Haaland, Dr Arne. *Norway.*
Haapala, Prof. Ilmari Johannes. *Finland.*
Hårsta, Mr Anders. *Sweden.*
Haas, Dr David Jean. *USA.*
Haase, Prof. Dr Wolfgang. *BRD.*
Haase-Wessel, Dr Werner. *BRD.*
Habash, Mr Jarjis. *UK.*
Habersetzer, Dr Catherine. *France.*
Habib, Mr Syed Abbas. *Pakistan.*
Habla, Dr Halina. *Poland.*
Hackert, Prof. Marvin LeRoy. *USA.*
Hadan, Dr Marianne. *DDR.*
Haditsch, Prof. Dr Johann Georg. *Austria.*
Hadler, Mrs Eva. *Norway.*
Hädicke, Dr Erich Emil Hermann. *BRD.*
Haeffner, Mr Dean R. *USA.*
Hägg, Prof. Gunnar. *Sweden.*
Hähle, Dr Siegfried. *DDR.*
Hähnert, Dr Irmela. *DDR.*
Hähnert, Dr Manfred. *DDR.*
Hämäläinen, Dr Reijo Pertti. *Finland.*
Haendler, Prof. Helmut M. *USA.*

Hafner, Prof. Dr Stefan S. *BRD.*
Haga, Dr Nobuhiko. *Japan.*
Hagen, Dr Kolbjörn. *Norway.*
Haget, Dr Yvette. *France.*
Hagler, Dr Arnold T. *USA.*
Hahn, Prof. Dr Theodor. *BRD.*
Hahne, Dr Bodo. *DDR.*
Haider, Prof. Syed Zahir. *Bangladesh.*
Hails, Dr J.E. *UK.*
Haisa, Prof. Masao. *Japan.*
Hajdu, Dr Ferenc. *Hungary.*
Hajdu, Mr János. *Hungary.*
Halder, Dr Sujit Kumar. *India.*
Hale, Dr Danforth Rawson. *USA.*
Halfpenny, Dr Joan Christine. *UK.*
Hall, Dr David. *New Zealand.*
Hall, Prof Eric Ogilvie. *Australia.*
Hall, Dr Ivan Harold. *UK.*
Hall, Prof. Lowell Headley II. *USA.*
Hall, Mr Norman Michael. *UK.*
Hall, Dr Sydney Reading. *Australia.*
Haller, Dr Kenneth James. *USA.*
Halliwell, Mrs Mary Ann Griffiths. *UK.*
Halouani, Dr Foued. *Tunisia.*
Halpern, Dr B. David. *USA.*
Haltiwanger, Mr Ralph Curtis. *USA.*
Halwax, Dipl.-Ing. Erich Johann. *Austria.*
Hamad, Dr Sa'ad El Din. *Sudan.*
Hamamura, Assoc. prof. Kenji. *Japan.*
Hamauzu, Dr Yoshihiro. *Japan.*
Hambley, Dr Trevor William. *Australia.*
Hamdi, Prof. Hassan Mahmoud. *Egypt.*
Hamill, Dr Gregory Prince. *USA.*
Hamilton, Dr John David Gavin. *Australia.*
Hamilton, Prof. Robert David. *USA.*
Hamlin, Dr Ronald Craig. *USA.*
Hammond, Dr Christopher. *UK.*
Hammonds, Mr T.G. *UK.*
Hamodrakas, Dr Stavros. *Greece.*
Hamor, Dr Thomas Andrew. *UK.*
Han, Mr Fu-son. *China.*
Han, Mr Shao-xu. *China.*
Han, Mr Yu-zhen. *China.*
Hanawalt, Prof. Joseph Donald. *USA.*
Handlovič, Ing Milan. *CSSR.*
Hanic, Doc. Dr František. *CSSR.*
Hanke, Dr Kurt. *BRD.*
Hanold, Mrs Karin. *DDR.*
Hansen, Mr Ernst F. R. *Sweden.*
Hansen, Mr Lars Kristian. *Norway.*
Hansen, Mr Staffan S. *Sweden.*
Hanson, Dr Alfred Wallace. *Canada.*
Hanson, Dr Jonathan C. *USA.*
Hanson, Dr Louise Karle. *USA.*
Hanson, Dr Marvin Wayne. *USA.*
Hansson, Dr Arne E. *Sweden.*
Hansson, Dr Eva. *Sweden.*
Haq, Mr Anwar-Ul. *BRD.*
Haque, Prof. Mazhar-ul. *Saudi Arabia.*
Harada, Prof Jimpei. *Japan.*
Harada, Mr Shigeharu. *Japan.*
Harada, Mr Yoshiyasu. *Japan.*
Harada, Dr Zyunpei. *Japan.*
Haraldson, Dr Stig H. W. *Sweden.*
Haran, Ms Tali. *Israel.*
Harbrecht, Dr Bernd. *BRD.*
Hardcastle, Prof. Kenneth Irvin. *USA.*
Hardgrove, Prof. George Lind, Jr. *USA.*
Harding, Mr J.W. *UK.*
Harding, Dr Marjorie Mary. *UK.*
Harding, Dr Roger Robertson. *UK.*
Hardy, Dr Andrew David. *UK.*
Hardy, Mrs Anne-Marie. *France.*
Hardy, Prof. Antoine. *France.*
Harel, Dr Michal. *Israel.*
Harga, Dr Ahmed Amin. *Egypt.*
Hargittai, Prof István. *Hungary.*
Hargittai, Dr Magdolna. *Hungary.*
Hargreaves, Dr A. *UK.*
Hariharan, Meena. *India.*

Hariya, Dr Yu. *Japan.*
Harkema, Dr Sybolt. *Netherlands.*
Harker, Dr David. *USA.*
Harlow, Dr Richard Leslie. *USA.*
Harms, Dr Klaus. *BRD.*
Haromy, Mr Tuli Patrick. *USA.*
Harper, Prof. Richard A. *USA.*
Harper, Mr W.H. *UK.*
Harr, Dr Albrecht Wolfgang Michael. *BRD.*
Harries, Mr J.E. *UK.*
Harris, Mr A.J. *UK.*
Harris, Miss B.L. *UK.*
Harris, Prof. David R. *USA.*
Harris, Ms G.W. *UK.*
Harrison, Prof. Pauline May. *UK.*
Harrison, Dr Robert Wilson. *USA.*
Harrison, Prof. Stephen Coplan. *USA.*
Harsányi, Dr László. *Hungary.*
Hart, Mr D.G. *UK.*
Hart, Dr Haskell Vincent. *USA.*
Hart, Prof. Michael. *UK.*
Hart, Dr Stewart. *South Africa.*
Hartl, Prof. Dr Hans. *BRD.*
Hartley, Dr Richard H. *Australia.*
Hartman, Prof. Dr Piet. *Netherlands.*
Hartmann, Dr Ervin. *Hungary.*
Hartmann, Dr Horst. *DDR.*
Hartmann, Dr Werner Johannes. *BRD.*
Hartsuck, Dr Jean Ann. *USA.*
Hartung, Dr Helmut. *DDR.*
Hartwig, Dr Jürgen. *DDR.*
Harutunyan, Dr Emil' Haikovich. *USSR.*
Harvey, Dr T.A. *UK.*
Hasan, Dr Faizul. *Pakistan.*
Hasebe, Mr Tooru. *Japan.*
Hašek, Dr Jindřich. *CSSR.*
Haser, Dr Richard Michel. *France.*
Hashimoto, Prof. Hatsujiro. *Japan.*
Hashimoto, Mr Hideki. *Japan.*
Hashimoto, Mr Katsunobu. *Japan.*
Hashizume, Assoc. Prof. Hiroo. *Japan.*
Hasiguti, Prof. Ryukiti. *Japan.*
Hassan, Prof. Ishmael. *Canada.*
Hassan, Prof. Mohamed Youssef. *Egypt.*
Hassan, Dr Wan Fuad bin Wan. *Malaysia.*
Hassler, Dr Eivind. *Sweden.*
Hastings, Dr Jerome Biller. *USA.*
Hastings, Dr Julius Mitchell. *USA.*
Hata, Dr Yasuo. *Japan.*
Hathaway, Prof. Brian John. *Ireland.*
Hatt, Mr B.A. *UK.*
Hatton, Dr Peter David. *UK.*
Hau, Dr Herbert H. K. *USA.*
Hauback, Mr Björn Christian. *Norway.*
Hauck, Dr Jürgen. *BRD.*
Hauge, Dr Sverre. *Norway.*
Hauger, Mr Rudi. *Switzerland.*
Hauptman, Prof. Herbert Aaron. *USA.*
Hausen, Dr Hans-Dieter. *BRD.*
Hausermann, Mr D. *UK.*
Hausner, Dipl.-Ing. Robert. *Austria.*
Haussühl, Prof. Dr Siegfried Georg. *BRD.*
Hauw, Dr Christian. *France.*
Havinga, Dr Edsko Enno. *Netherlands.*
Hawley, Dr D.M. *UK.*
Haworth, Dr Colin William. *UK.*
Hawthorne, Dr Frank Christopher. *Canada.*
Hay, Dr David Gilbert. *Australia.*
Hay, Dr James Neilson. *UK.*
Hayakawa, Dr Kazunobu. *Japan.*
Hayakawa, Miss Koto. *Canada.*
Hayashi, Prof. Mitsuhiko. *Japan.*
Hayashi, Dr Kooya. *Japan.*
Hayashi, Dr Yasunori. *Japan.*
Hayden, Dr Thomas Day. *USA.*
Hazell, Dr Alan Charles. *Denmark.*
Hazell, Mrs Rita Grønbæk. *Denmark.*
Hazen, Dr Robert Miller. *USA.*
He, Prof. Chong-fan. *China.*
He, Mr Rei-ling. *China.*
He, Dr Xiao-Min. *USA.*

Healy, Dr Peter Conrad. *Australia.*
Hearmon, Mr R.F.S. *UK.*
Heath, Mr James R. *USA.*
Heavens, Prof. Oliver Samuel. *UK.*
Hebert, Dr Hans. *Sweden.*
Hecht, Dr Hans-Jürgen. *BRD.*
Heckman, Mr Francis A. *USA.*
Heckroodt, Prof. Renier Oelof. *South Africa.*
Hedberg, Prof. Kenneth Wayne. *USA.*
Hedman, Dr Gun-Britt Margareta. *USA.*
Hegenbarth, Prof. Ernst. *DDR.*
Heger, Dr Gernot Wolfgang. *BRD.*
Heide, Dr Klaus. *DDR.*
Heide, Dr Helmut. *BRD.*
Heijnen, Drs Wilhelmus Marinus Maria. *Netherlands.*
Heim, Dr Joachim. *DDR.*
Heim, Dr Harald Josef Robert. *BRD.*
Heimann, Prof. Dr Robert Bertram Silvester. *Canada.*
Heinemann, Dr Udo. *BRD.*
Heinerman, Dr Jacobus J.L. *Netherlands.*
Heinze, Dr Joachim. *DDR.*
Heleskivi, Dr Jouni Martti. *Finland.*
Helis, Mr Howard Morrell. *USA.*
Heller-Kallai, Prof. Lisa. *Israel.*
Helliwell, Dr John Richard. *UK.*
Hellmold, Dr Peter. *DDR.*
Hellner, Prof. Erwin E. *BRD.*
Helmholdt, Dr Robert Barteld. *Netherlands.*
Helmi, Prof. Mohamed Ezzeldin. *Egypt.*
Helmreich, Dr Dieter. *BRD.*
Hemily, Dr Philip W. *USA.*
Hemkar, Dr Mangla Prasad. *India.*
Hempel, Dr Andrew. *Canada.*
Henderson, Dr Christopher M.B. *UK.*
Hendricks, Dr Robert Wayne. *USA.*
Hendrickson, Prof. Wayne A. *USA.*
Henke, Dr Henning. *BRD.*
Henkel, Dr Gerald. *BRD.*
Hennig, Prof. Klaus. *DDR.*
Henriquez, Mr Fernando. *Chile.*
Henslee, Dr Walter Warren. *USA.*
Hentschel, Dr Manfred Paul. *BRD.*
Hepp, Dr Alfred. *Switzerland.*
Herak, Dr Rajna. *Yugoslavia.*
Herberger, Dr Jürgen. *DDR.*
Herbertsson, Dr B. Harald. *Sweden.*
Herbstein, Prof. Frank Herzl. *Israel.*
Herceg, Dr Marija. *Yugoslavia.*
Herdade, Dr Silvio B. *Brazil.*
Herdtweck, Dr Eberhardt. *BRD.*
Hergold-Brundić, Dr Antonija. *Yugoslavia.*
Heritsch, Prof. em. Dr Haymo. *Austria.*
Herman, Prof. Herbert. *USA.*
Hermans, Prof. Jan. *USA.*
Hermansson, Dr Kersti. *Sweden.*
Hermansson, Dr Leif Å. G. *Sweden.*
Hermida, Lic Jorge Daniel. *Argentina.*
Hermodsson, Dr Yngve. *Sweden.*
Hermoneit, Mr Bernd. *DDR.*
Herms, Dr Gerhard. *DDR.*
Hernández, Prof. Luis C. *Colombia.*
Hernandez Cano, Dr Felix. *Spain.*
Herpin, Prof. Paulette. *France.*
Herring, Mr R.N. *UK.*
Herriott, Prof. Jon Roger. *USA.*
Herrmann, Mrs Christel. *DDR.*
Herrmann, Dr Frank-Peter. *DDR.*
Herrmann, Prof. Rudolf. *DDR.*
Hervé, Dr Francisco. *Chile.*
Herzberg, Dr Armin. *BRD.*
Hess, Prof. Dr Heinz. *BRD.*
Hesse, Dr Karl-Friedrich. *BRD.*
Hesse, Dr Rolf S. *Sweden.*
Hewston, Ms Terrell Ann. *USA.*
Heydenreich, Dr Johannes. *DDR.*
Heyding, Dr Robert Donald. *Canada.*
Heymann, Dr Gunter. *DDR.*
Heyn, Prof. Anton Nicolaas Johannes. *USA.*
Heyns, Prof. Anton Michal. *South Africa.*
Hibbard, Dr Lyndon Stanley. *USA.*

Hibma, Dr Tjipke. *Switzerland.*
Hicks, Dr Trevor John. *Australia.*
Hidaka, Mr Tsuneo. *Japan.*
Hiebl, Dr Kurt. *Austria.*
Higashi, Prof. Akira. *Japan.*
Higatsberger, Prof. Dr Michael Josef. *Austria.*
Higgins, Prof. John Britt. *USA.*
Highcock, Dr Rona Margaret. *UK.*
Higuchi, Assoc. Prof. Taiichi. *Japan.*
Hiismäki, Dr Pekka Eljas. *Finland.*
Hikichi, Dr Kunio. *Japan.*
Hildebrandt, Prof. Dr Gerhard. *BRD.*
Hildmann, Dr-Ing Bernd Otfried. *BRD.*
Hilgenfeld, Dr Rolf. *BRD.*
Hill, Mr Christopher Peter. *UK.*
Hill, Dr Roderick Jeffrey. *Australia.*
Hilleard, Dr Ronald James. *UK.*
Hiller, Dr Wolfgang Paul. *BRD.*
Hillman, Dr Harold. *UK.*
Hiltunen, Mr Lassi Ilmari. *Finland.*
Hinawi, Prof. Essam. *Egypt.*
Hinch, Mr Ralph J., Jr. *USA.*
Hinde, Dr R.M. *UK.*
Hine, Dr Raymond. *UK.*
Hingerty, Dr Brian Edward. *USA.*
Hinsch, Mr Thorsten Reinhard. *BRD.*
Hintermann, Dr Hans-Erich. *Switzerland.*
Hinz, Dr Dietrich. *DDR.*
Hinze, Dr Eckhard. *BRD.*
Hinze, Mr Thomas. *DDR.*
Hirabayashi, Prof. Makoto. *Japan.*
Hirahara, Dr Eiji. *Japan.*
Hirakawa, Prof. Kinshiro. *Japan.*
Hirano, Prof. Shin-ichi. *Japan.*
Hirayama, Mr Noriaki. *Japan.*
Hirokawa, Prof. Sakutaro. *Japan.*
Hirokawa, Prof. Tomoo. *Japan.*
Hirotsu, Dr Ken. *Japan.*
Hirotsu, Dr Yoshihiko. *Japan.*
Hirsch, Prof. Sir Peter. *UK.*
Hirshfeld, Prof. Fred. *Israel.*
Hirshfield, Mr Jordan M. *USA.*
Hirth, Prof. John Price. *USA.*
Hitchcock, Dr P.B. *UK.*
Hite, Prof. Gilbert James. *USA.*
Hjertén, Dr Inger. *Sweden.*
Hlavatá, Mrs Drahomira. *CSSR.*
Hoard, Dr Laurence Graham. *USA.*
Hoard, Prof. James Lynn. *USA.*
Hockly, Dr M. *UK.*
Hocksell, Mr Veli Eerik. *Finland.*
Hodenberg, von, Mrs Dr Renate Barbara. *BRD.*
Hodge, Mr Leslie Cameron. *Australia.*
Hodgkin, Prof. Dorothy Mary Crowfoot. *UK.*
Hodgkinson, Mr R.A. *UK.*
Hodgson, Prof. Derek John. *USA.*
Hodgson, Prof. Keith Owen. *USA.*
Hodorowicz, Dr Stanisław. *Poland.*
Hodsdon, Dr John Marshall. *USA.*
Höbler, Mr Hans-Joachim. *DDR.*
Höche, Dr Hans-Reiner. *DDR.*
Höche, Dr Hellmut. *DDR.*
Höfer, Dr Hans Hermann. *BRD.*
Höfler, Mrs Sabine Ida Gerda. *BRD.*
Höhling, Prof. Dr Hans Jürgen. *BRD.*
Höhne, Prof. Ernst. *DDR.*
Höier, Prof. Ragnvald. *Norway.*
Hökelek, Mr Tuncer. *Turkey.*
Hölsä, Dr Jorma Pertti Kalervo. *Finland.*
Hönle, Dr Wolfgang. *BRD.*
Hörl, Prof. Dr Erwin M. *Austria.*
Hösler, Dr. *BRD.*
Hötzsch, Dr Günter. *DDR.*
Hoff, von, Dr Siegfried. *DDR.*
Hoffman, David W. *USA.*
Hoffmann, Dr Brigitte. *DDR.*
Hoffmann, Prof. Dr Wolfgang. *BRD.*
Hoffmeister, Dr Wolfgang. *BRD.*
Hofmeister, Dr Wolfgang. *BRD.*

Hogan, Dr Leonard McNamara. *Australia.*
Hoge, Dr Reinhold. *BRD.*
Hogg, Mr C. *UK.*
Hogg, Dr Joshua Herbert Christopher. *UK.*
Hoggins, Dr James Thomas. *USA.*
Hohlwein, Dr Dietmar. *BRD.*
Hol, Dr Wim. *Netherlands.*
Holas, Dr hab Andrzej. *Poland.*
Holbrook, Dr Stephen Roy. *USA.*
Holden, Ms Hazel M. *USA.*
Holden, Dr James Richard. *USA.*
Holinski, Dr Rüdiger. *BRD.*
Holland, Dr Hans J. *USA.*
Holland, Miss S. *UK.*
Holland, Mr S.K. *UK.*
Hollander, Dr Frederick J. *USA.*
Holldorf, Dr Horst. *DDR.*
Holmberg, Dr Bo. *Sweden.*
Holmes, Mr F. E. *USA.*
Holmes, Prof. Kenneth Charles. *BRD.*
Holmes, Mr R.J. *UK.*
Holser, Prof. William Thomas. *USA.*
Holt, Dr Elizabeth Manners. *USA.*
Holt, Mrs J. *UK.*
Holt, Dr Ronald Stanley. *UK.*
Holtzberg, Dr Frederic. *USA.*
Holub, Dr Fritz. *Austria.*
Holý, Dr Václav. *CSSR.*
Hom, Dr Tommy. *USA.*
Homma, Mr Shigeru. *Japan.*
Homma, Dr Teiichi. *Japan.*
Hon, Dr Ping-Kay. *Hong Kong.*
Hong, Mr Mao-chun. *China.*
Hong, Dr Sam-Hyo. *Sweden.*
Honigmann, Dr Bertolo. *BRD.*
Honjin, Prof. Ryohei. *Japan.*
Honjo, Prof. Goro. *Japan.*
Honkonen, Mr Robert S. *USA.*
Honoré, Dr Tage. *Denmark.*
Hons, Mr Alexander. *Australia.*
Honzatko, Prof. Richard E. *USA.*
Hoogewijs, Dr Robert Richard. *Belgium.*
Hoogsteen, Dr Karst. *USA.*
Hope, Prof. Håkon. *USA.*
Hoppe, Prof. Günter. *DDR.*
Hoppe, Prof. Hans. *DDR.*
Hoppe, Prof. Dr Dr hc Rudolf. *BRD.*
Hoppe, Prof. Dr Walter. *BRD.*
Hordvik, Prof. Asbjörn. *Norway.*
Horiuchi, Dr Hiroyuki. *Japan.*
Horiuchi, Dr Shigeo. *Japan.*
Horjales, Mr Eduardo. *Sweden.*
Horn, Dr Ernst. *Australia.*
Horn, Dr Jerzy. *Poland.*
Horn, Prof. Manfred Josef. *Peru.*
Horn, Prof. Paul. *France.*
Hornby, Mr M.R. *UK.*
Horne, Mr William. *Saudi Arabia.*
Hornstra, Drs Jan. *Netherlands.*
Hornung, Dr George. *UK.*
Horrigan, Ms Jane Akerlund (Bruce). *USA.*
Horsey, Dr Richard Stephen. *USA.*
Horsfield, Mr Edgar Charles. *South Africa.*
Horst, Dr Wolfgang. *BRD.*
Horstmann, Prof. Dr Manfred. *BRD.*
Horváth, Ing Josef. *CSSR.*
Horyń, Dr Roman. *Poland.*
Hosaka, Dr Masahiro. *Japan.*
Hoschl, Dr Pavel. *CSSR.*
Hosea, Dr Thomas Jeffrey Cockburn. *Singapore.*
Hosemann, Prof. Dr Dr hc Rolf. *BRD.*
Hoser, Dr Andrzej. *BRD.*
Hoshino, Prof. Sadao. *Japan.*
Hoskins, Dr Bernard Foster. *Australia.*
Hosoya, Mr Masahiko. *Japan.*
Hosoya, Prof. Em. Sukeaki. *Japan.*
Hospital, Dr Michel. *France.*
Hossain, Mr Mohammed B. *USA.*
Hou, Mr Yong-geng. *China.*
Hough, Dr Edward. *Norway.*

Houng, Prof. Kun-Huang. *Taiwan.*
Hountas, Dr Athanasios. *Greece.*
Houska, Prof. Charles Robert. *USA.*
Housty, Dr Jacques. *France.*
Hovestreydt, Mr Eric Robert. *Switzerland.*
Hovmöller, Dr Sven. *Sweden.*
Howard, Dr Christopher John. *Australia.*
Howard, Dr Judith Ann Kathleen. *UK.*
Howard, Mrs S. *UK.*
Howard, Dr Scott A. *USA.*
Howatson, Prof. John. *USA.*
Howe, Ms Donna-Beth. *USA.*
Howell, Dr Peter Adam. *USA.*
Howes, Mr A.J. *UK.*
Howie, Dr R. Alan. *UK.*
Howie, Prof. Robert Andrew. *UK.*
Howlin, Dr B. *UK.*
Hoyer, Dr Walter. *DDR.*
Hoyos de Castro, Prof. Angel. *Spain.*
Hoyos Guerrero, Prof. Miguel Angel. *Spain.*
Hsu, Prof. I-Nan. *USA.*
Hsu, Dr Leh-Yeh Ruth. *USA.*
Hsu, Dr Shu-En. *Taiwan.*
Htoon, Dr Sein. *Burma.*
Hu, Mr Heng-liang. *China.*
Hu, Mr Sheng-zhi. *China.*
Hua, Mr Zi-qian. *China.*
Huang, Mrs Chi-yung. *Taiwan.*
Huang, Prof. Chun-Kiang. *Taiwan.*
Huang, Mr De-bin. *China.*
Huang, Prof. Jin-ling. *China.*
Huang, Mr Jin-shun. *China.*
Huang, Mr Liang-ren. *China.*
Huang, Mr Tai-shan. *China.*
Huang, Dr Ting Chun. *USA.*
Huang, Mr Tung-woo. *Taiwan.*
Huang, Mr Zhi-ying. *China.*
Huanosta Tera, Mr Alfonso. *Mexico.*
Hubbard, Dr Camden Richards. *USA.*
Hubbell, Mr John Howard. *USA.*
Hubbert, Mrs Dr Elisabeth. *BRD.*
Huber, Dr Carol P. *Canada.*
Huber, Prof. Robert. *BRD.*
Huber-Buser, Dr Effi H. *USA.*
Hubert, Dr Joseph. *Canada.*
Hubert, Mr Georg. *BRD.*
Huch, Dr Volker. *BRD.*
Hudd, Mr R.C. *UK.*
Huddle, Prof. Benjamin Paul. *USA.*
Hudgens, Dr Bruce A. *USA.*
Hudgens, Mr Claude R. *USA.*
Hübner, Dr Manfred. *DDR.*
Hübner, Mr Thomas. *BRD.*
Hümmer, Prof. Dr Kurt. *BRD.*
Huffman, Dr John Curtis. *USA.*
Huggins, Dr Maurice Loyal. *USA.*
Hughes, Dr Antony Elwyn. *UK.*
Hughes, Mr David John. *UK.*
Hughes, Dr David Lewis. *UK.*
Hughes, Dr Edward Wesley. *USA.*
Hughes, Prof. Robert Edward. *USA.*
Hughes, Mr Thomas Ernest. *UK.*
Hughes, Prof. William Eugene. *USA.*
Huiszoon, Dr Cornelis. *Netherlands.*
Hukin, Dr David Ainsworth. *UK.*
Hukins, Dr David William Laurence. *UK.*
Hull, Dr Stephen Edward. *UK.*
Hulme, Mr Ralph. *UK.*
Hultzsch, Mr Rainer. *DDR.*
Humble, Dr Sten G. *Sweden.*
Huml, Dr Karel. *CSSR.*
Hummel, van, Ing. Gerrit Jan. *Netherlands.*
Hummel, Dr Hans-Ulrich. *BRD.*
Humphreys, Prof. Colin John. *UK.*
Hund, Dr Franz Josef. *BRD.*
Hundt, Dr Rudolf. *BRD.*
Hunt, Prof. Gary Webb. *USA.*
Hurley, Mr Patrick Walter. *UK.*
Hurst, Prof. Vernon James. *USA.*
Hursthouse, Dr Michael Barry. *UK.*

Husain, Dr Abul Hasanat Mohammad. *Bangladesh.*
Husain, Dr Jasmine Tickle. *UK.*
Husebye, Prof. Steinar. *Norway.*
Hussain, Dr Khadim. *Pakistan.*
Hussain, Dr M. Sakhawat. *Saudi Arabia.*
Hussain, Dr Zahid. *Saudi Arabia.*
Hutcheon, Dr Wendy Lou (Brooks). *Canada.*
Hutchings, Mr Alan E. *USA.*
Hutchings, Dr M.T. *UK.*
Hutchings, Dr Ronald. *South Africa.*
Hutchinson, Mr John P. *USA.*
Hutchinson, Dr William Bevis. *Sweden.*
Hutchison, Prof. Dr Charles Strachan. *Malaysia.*
Hutchison, Dr John Laird. *UK.*
Hutton, Dr Alan Thomas. *UK.*
Huxley, Dr Hugh Esmor. *UK.*
Hvoslef, Dr Jan. *Norway.*
Hybl, Prof. Albert. *USA.*
Hybler, Dr Jiří. *CSSR.*
Hyde, Dr C. Craig. *USA.*
Hyde, Prof Bruce Godfrey. *Australia.*
Hytönen, Dr Kai Kalevi Gabriel. *Finland.*
Iandelli, Prof. Aldo. *Italy.*
Ianelli, Dr Sandra. *Italy.*
Ianovici, Prof. Virgil. *Romania.*
Iball, Prof. John. *UK.*
Ibata, Dr Koichi. *Japan.*
Ibers, Prof. James A. *USA.*
Ibrahim, Dr T. K. *Iraq.*
Ibrhim, Dr Muhammad. *Bangladesh.*
Ice, Dr Gene Emery. *USA.*
Ichikawa, Dr Mizuhiko. *Japan.*
Ichimiya, Dr Ayahiko. *Japan.*
Ichimura, Mr Takeo. *Japan.*
Ichinokawa, Prof. Takeo. *Japan.*
Ickert, Dr Lars. *DDR.*
Idler, Miss Kathleen Loralee. *Canada.*
Ihringer, Dr Jörg. *BRD.*
Iida, Dr Atsuo. *Japan.*
Iijima, DSc Kinya. *Japan.*
Iijima, Prof. Takao. *Japan.*
Iishi, Dr Kazuaki. *Japan.*
Iitaka, Prof. Yoichi. *Japan.*
Iizuka, Dr Masakatsu. *Japan.*
Iizumi, Mr Masashi. *Japan.*
Ikemoto, Dr Isao. *Japan.*
Ikornikova, Prof. Nina Yuryevna. *USSR.*
Ikram, Dr Nazma. *Pakistan.*
Ilyinsky, Dr Alexander Lvovich. *USSR.*
Ilyushin, Dr Alexander Sergeyvich. *USSR.*
Imado, Dr Sataro. *Japan.*
Imakuma, Dr Kengo. *Brazil.*
Imamov, Dr Rafik Mamedovich. *USSR.*
Imanishi, Dr Yasuhiro. *Japan.*
Immirzi, Prof. Attilio. *Italy.*
Imperatori, Dr Patrizia. *Italy.*
Imre-Baán, Mrs Irén. *Hungary.*
Imreh, Mr Iosif. *Romania.*
Imura, Prof. Toru. *Japan.*
In, Mr Aung Paik. *Burma.*
Indenbom, Prof. Vladimir Lvovich. *USSR.*
Infante, Dr Carlos. *Chile.*
Inglez, Mr Antonio Gabriel. *Brazil.*
Ingram, Mrs Lorna. *UK.*
Ingri, Prof. Nils. *Sweden.*
Iñiguez Rodríguez, Dr Adrián M. *Argentina.*
Inniss, Mr Daryl. *USA.*
Ino, Dr Shozo. *Japan.*
Ino, DSc Tadashi. *Japan.*
Inoue, Dr Morio. *Japan.*
Inoue, Dr Zenzaburo. *Japan.*
Inouye, Mr Hideyo. *USA.*
Intrater, Mr Josef. *USA.*
Inui, Prof. Teturo. *Japan.*
Ionescu, Mrs Jeana. *Romania.*
Ionov, Dr Pavel Victorovich. *USSR.*
Ionov, Dr Vladislav Mikhailovich. *USSR.*
Ipohorski Lenkiewicz, Dr Miguel. *Argentina.*
Iqbal, Mr Mir Waseluddin Ahmad. *Pakistan.*
Irngartinger, Dr Hermann. *BRD.*

Irving, Dr Anne. *South Africa.*
Isaac, Dr D.H. *UK.*
Isaacs, Dr Neil William. *Australia.*
Isakow, Mrs Zofia Stanisława. *Poland.*
İşçi, Doç. Dr Coşkun. *Turkey.*
Isenberg, Dr Wilhelm. *BRD.*
Isetti, Prof. Giovanni. *Italy.*
Isherwood, Dr Brian James. *UK.*
Isherwood, Mrs S.A. *UK.*
Ishibashi, Dr Yoshihiro. *Japan.*
Ishida, Prof. Kohtaro. *Japan.*
Ishida, Dr Toshimasa. *Japan.*
Ishiguro, Mr Takashi. *Japan.*
Ishihara, Mr Nobukazu. *Japan.*
Ishikawa, Prof. Yoshikazu. *Japan.*
Ishizuka, Mr Kazuo. *Japan.*
Iskhakova, Dr Lyudmila Dmitriyevna. *USSR.*
Islam, Prof. Aminul. *Bangladesh.*
Islam, Mr Shafiqul. *Bangladesh.*
Ismailzade, Prof. Ibragim Gasan ogly. *USSR.*
Isobe, Dr Mitsumasa. *Japan.*
Itai, Dr Akiko. *Japan.*
Ito, Mr Kazuomi. *Japan.*
Ito, Mr Kohei. *Japan.*
Ito, Prof. Masatoki. *Japan.*
Ito, Dr Tetsuzo. *Japan.*
Ito, Dr Yuji. *Japan.*
Ivaldi, Dr Gabriella. *Italy.*
Ivanov, Dr Nikolay Rafailovich. *USSR.*
Ivanova, Dr Irina Vadimovna. *USSR.*
Ivarsson, Dr Gun J. M. *Sweden.*
Iwanaga, Mr Hiroshi. *Japan.*
Iwanov, Mr Dantsho. *Bulgaria.*
Iwasaki, Dr Fujiko. *Japan.*
Iwasaki, Prof. Hiroshi. *Japan.*
Iwasaki, Dr Hiroshi. *Japan.*
Iwasaki, Dr Hitoshi. *Japan.*
Iwata, Mrs Miyuki. *Japan.*
Iwata, Dr Yutaka. *Japan.*
Iwayanagi, Prof. Shigeo. *Japan.*
Iyenger, Dr Leela. *India.*
Izui, Dr Kazuhiko. *Japan.*
Izumi, Prof. Takatoshi. *Japan.*
Izumi, Dr Yoshinobu. *Japan.*
Jack, Prof. Kenneth Henderson. *UK.*
Jackobs, Dr John Joseph. *USA.*
Jackowski, Dr Józef Wojciech. *Poland.*
Jacob, Dr Herbert. *BRD.*
Jacobi, Dr Hans. *BRD.*
Jacobs, Prof. Gerald Daniel. *USA.*
Jacobs, Prof. Dr Gilbert. *Belgium.*
Jacobs, Prof. Dr Herbert Ernst Hermann. *BRD.*
Jacobson, Prof. Robert Andrew. *USA.*
Jäger, Dr Hans. *BRD.*
Jäger, Dr Susanne Christine. *BRD.*
Jaeger, Mr Hans. *Australia.*
Jährling, Mr Thomas. *DDR.*
Järvinen, Dr Matti Johannes. *Finland.*
Jafry Mr Syed Qamar Abbas. *Pakistan.*
Jagannadham, Dr Adibhatla Vankata. *India.*
Jagner, Dr Susan. *Sweden.*
Jagodzinski, Prof. Dr Dr h c Heinz Ernst. *BRD.*
Jahn, Dr Irmin-Rudolf. *BRD.*
Jahnberg, Dr Lena. *Sweden.*
Jaidong, Mr Ko. *USA.*
Jain, Dr Prem Chand. *India.*
Jain, Mr Sanjeev. *USA.*
Jain, Dr Shri C. *USA.*
Jakkal, Vasant Shankar. *India.*
Jakobs, Dr Rüdiger-Hasko. *BRD.*
Jakubovics, Dr John Paul. *UK.*
James, Prof. Michael N.G. *Canada.*
James, Prof. William J. *USA.*
James, Dr Veronica Jean. *Australia.*
Jameson, Prof. Geoffrey Brind. *USA.*
Jan, Mr Gwo-jen. *Taiwan.*
Jandacek, Dr Ronald James. *USA.*
Janiak, Dr Martin J. *USA.*
Janik, Prof. Jerzy. *Poland.*
Janin, Prof. Joel. *France.*
Janjić, Prof. Dr Svetislav. *Yugoslavia.*

Janko, Dr Andrzej. *Poland.*
Janner, Prof. Dr Aloysio. *Netherlands.*
Jánosi, Dipl.-Ing. Dr András. *Austria.*
Jansen, Dr Martin. *BRD.*
Jansonius, Prof. Johan Nomdo. *Switzerland.*
Jansson, Mr Kjell. *Sweden.*
Jarchow, Prof. Dr Otto. *BRD.*
Jasinski, Prof. Jerry P. *USA.*
Jaskólski, Dr Mariusz. *Poland.*
Jaswon, Prof. Maurice Arthur. *UK.*
Jauch, Dr Wolfgang. *BRD.*
Jayadevan, Dr Naduviledath Chennuvittil. *India.*
Jayanty, Dr Ashok *India.*
Jayashree, Ms A.N. *India.*
Jayaweera, Dr Shanath Amarasiri Arumabadu. *UK.*
Ječný, Ing Jiří. *CSSR.*
Jeannin, Prof. Yves. *France.*
Jeffrey, Prof. George Alan. *USA.*
Jeffreys, Dr John Alexander David. *UK.*
Jegerlehner, Mr Kurt. *DDR.*
Jehanno, Dr Germain Pierre. *France.*
Jeitschko, Prof. Dr Wolfgang. *BRD.*
Jejjala, Dr Krishna Mohana Rao. *USA.*
Jelenić-Bezjak, Dr Ivanka. *Yugoslavia.*
Jellinek, Prof. Dr Franz. *Netherlands.*
Jenček, Mr Ladislav August. *Yugoslavia.*
Jendrek, Dr Eugene F. *USA.*
Jenichen, Dr Bernd. *DDR.*
Jenkins, Dr John Anthony. *Switzerland.*
Jenkins, Dr Ron. *USA.*
Jenks, Ms Janice M. *USA.*
Jennings, Dr Laurence Duane. *USA.*
Jennische, Dr Per. *Sweden.*
Jensen, Mr Aage. *Denmark.*
Jensen, Dr Birthe. *Denmark.*
Jensen, Dr Ejnar. *Denmark.*
Jensen, Mr Gunnar Bent. *Denmark.*
Jensen, Prof. Lyle Howard. *USA.*
Jensen, Dr Stig Jorgo. *Denmark.*
Jensen, Prof. William Phelps. *USA.*
Jerschkewitz, Dr Hans-Georg. *DDR.*
Jerslev Lund, Prof. Bodil. *Denmark.*
Jesser, Prof. William Augustus. *USA.*
Jex, Dr Hartmut. *BRD.*
Ji, Prof. Shou-yuan. *China.*
Jia, Prof. Shou-quan. *China.*
Jiang, Mr An-bei. *China.*
Jiang, Mr Han-chen. *China.*
Jiang, Mrs Shao-ying. *China.*
Jiang, Mr Xiao-long. *China.*
Jiang, Prof. Yan-dao. *China.*
Jiménez Crespo, Prof. Augusto. *Colombia.*
Jimenez Garay, Dr Rafael. *Spain.*
Jin, Dr Wei-qing. *China.*
Jin, Mr Xiang-lin. *China.*
Jinawath, Dr Supatra. *Thailand.*
Jirajesda, Mr Jate. *Thailand.*
Jircitano, Mr Alan J. *USA.*
Joel, Dr Nahum. *France.*
Joelson, Mr Thorleif. *Sweden.*
Jönsson, Dr Per-Gunnar. *Sweden.*
Jørgensen, Mr Ole. *Denmark.*
Johansen, Dr Heinrich. *DDR.*
Johanson, Mr Bo Stefan. *Finland.*
Johansson, Dr Georg. *Sweden.*
Johansson, Mr Karl-Erik. *Sweden.*
Johansson, Dr Lars-Gunnar. *Sweden.*
Johari, Dr Gyan Prakash. *Canada.*
Johnsen, Mr Ole. *Denmark.*
Johnson, Dr Andrew William Syme. *Australia.*
Johnson, Dr Carroll K. *USA.*
Johnson, Dr David Julian. *UK.*
Johnson, Dr Frank Bacchus. *USA.*
Johnson, Prof. Gerald Glenn, Jr. *USA.*
Johnson, Dr Harold Arthur. *USA.*
Johnson, Dr John Emil. *USA.*
Johnson, Dr Louise N. *UK.*
Johnson, Dr Michael William. *UK.*
Johnson, Dr N.P. *UK.*
Johnson, Mr Owen. *UK.*

Johnson, Dr Paul Lorentz. *USA.*
Johnson, Dr Peter Anthony Victor. *UK.*
Johnson, Dr Quintin C. *USA.*
Johnson, Mrs Suzanne Marie. *USA.*
Johnston, Dr Gordon Basil. *Australia.*
Jones, (Milne) Dr Angela Alice. *UK.*
Jones, Mr D. *UK.*
Jones, Prof. Daniel Silas. *USA.*
Jones, Prof. Derry Wynn. *UK.*
Jones, Dr G.R. *UK.*
Jones, Dr John Brett. *Australia.*
Jones, Dr Morton Edward. *USA.*
Jones, Dr Noel Duane. *USA.*
Jones, Mr P.L. *UK.*
Jones, Dr Stephen John. *Canada.*
Jones, Dr T. Alwyn. *Sweden.*
Jones, Dr Tony Cristofer. *New Zealand.*
Jones, Dr Terry. *UK.*
Jones, Dr W. *UK.*
Jonsson, Dr Arne. *Sweden.*
Jordan, Prof. Truman H. *USA.*
Jordanovska, Dr Vera. *Yugoslavia.*
Jorgensen, Mr James D. *USA.*
José Yacaman, Dr Miguel. *Mexico.*
Joseph, Dr Günter. *Chile.*
Joshi, Prof. Ramesh Vinayak. *India.*
Joshi, Dr Shri Krishna. *India.*
Jost, Dr Karlheinz. *DDR.*
Jostsons, Dr Adam. *Australia.*
Joswig, Dr Werner. *BRD.*
Joubert, Prof. Jean-Claude. *France.*
Jouffrey, Dr Bernard. *France.*
Jouini, Dr Amor. *Tunisia.*
Jouini, Dr M. *Tunisia.*
Jouini, Dr Tahar. *Tunisia.*
Jourdan, Dr Claude, René *France.*
Jovanovski, Dr Gligor. *Yugoslavia.*
Juang, Dr Tzo-chuan. *Taiwan.*
Jude, Dr Lidia. *Romania.*
Julian, Prof. Maureen O'Donnell. *USA.*
Jung, Dr Detlef. *BRD.*
Jung, Dr Volkhard. *BRD.*
Jung, Dr Walter. *BRD.*
Jurisch, Dr Manfred. *DDR.*
Jurković, Prof. Dr Ivan. *Yugoslavia.*
Jurnak, Prof. Frances Anne. *USA.*
Juul Jensen, Mrs Dorte. *Denmark.*
Jynge, Mr Knut. *Norway.*
Kaas, Dr Karen. *Denmark.*
Kaat, te, Prof Dr Erich Heinz. *BRD.*
Kabalkina, Prof. Sara Samsonovna. *USSR.*
Kabešová, Ing Mária. *CSSR.*
Kabelka, Dr Heinz I. *Austria.*
Kabish, Dr Lotfi. *Egypt.*
Kabs, Mr Michael. *BRD.*
Kabsch, Dr Wolfgang. *BRD.*
Kachalov, Dr Oleg Viktorovich. *USSR.*
Kachinsky, Dr Vitol'd Nikolayevch. *USSR.*
Kádár, Dr György. *Hungary.*
Kadlec, Mr Robert J. *USA.*
Kaduk, Dr James Albert. *USA.*
Kähkönen, Dr Heikki Antero. *Finland.*
Kämmel, Dr Thomas. *DDR.*
Kaerlein, Mr Carsten-Peter. *BRD.*
Kaftory, Dr Menahem. *Israel.*
Kaganer, Dr Vladimir Mikhailovich. *USSR.*
Kagarakis, Prof. Constantine. *Greece.*
Kahlert, Prof. Dr Hartmut. *Austria.*
Kahn, Dr Andrée. *France.*
Kai, Dr Yasushi. *Japan.*
Kaidalova, Dr Taisiya Alexandrovna. *USSR.*
Kaihola, Mr Lauri Leo. *Finland.*
Kaillathe, Padmanabhan. *India.*
Kaiman, Mr Solomon. *Canada.*
Kainuma, Prof. Yoshiro. *Japan.*
Kaischew, Prof. Rostislaw. *Bulgaria.*
Kaiser, Dr Johannes. *DDR.*
Kaiser, Mrs Ute. *DDR.*
Kaitner, Dr Branko. *Yugoslavia.*
Kaito, Dr Chihiro. *Japan.*

Kaji, Dr Keisuke. *Japan.*
Kajzar, Dr Franciszek, *Poland.*
Kakati, Dr Kandarpa Kumar. *India.*
Kakehi, Dr Masahiro. *Japan.*
Kakinoki, Prof. Jiro. *Japan.*
Kakitani, Prof. Satoru. *Japan.*
Kakudo, Prof. Masao. *Japan.*
Kakumoto, Dr Kenichi. *Japan.*
Kalb(Gilboa), Dr Aaron Joseph. *Israel.*
Kaldis, PD Emanuel. *Switzerland.*
Kalicińska-Karut, Mgr Jarosława. *Poland.*
Kalinin, Dr Vladimir Ivanovich. *USSR.*
Kalinkina, Dr Irina Nikolayevna. *USSR.*
Kalinna, Mr Hartmut. *DDR.*
Kalk, Mr Kornelis Harm. *Netherlands.*
Kallal, Dr Ahmed. *Tunisia.*
Kallio, Mr Pekka Yrjö Juhani. *Finland.*
Kalliomäki, Dr Martti Salomo. *Finland.*
Kálmán, Prof. Alajos. *Hungary.*
Kálmán, Dr Erika. *Hungary.*
Kalman, Prof. Zwi Heinrich. *Israel.*
Kałuski, Dr hab. Zygmunt. *Poland.*
Kalweit, Mr Harald. *DDR.*
Kalyanaraman, Dr A. R. *India.*
Kalychak, Dr Yaroslav Mikhailovich. *USSR.*
Kamb, Prof. W. Barclay. *USA.*
Kambas, Prof. Kostas. *Greece.*
Kambe, Dr Kyozaburo. *BRD.*
Kamel, Prof. Dr Raafat Wasef. *Egypt.*
Kamenar, Prof. Dr Boris. *Yugoslavia.*
Kamentsev, Dr Igor' Evgenyevich. *USSR.*
Kamijo, Dr Nagao. *Japan.*
Kaminsky, Dr Alexander Alexandrovich. *USSR.*
Kaminsky, Dr Vladimir Fedorovich. *USSR.*
Kamiya, Dr Kazuhide. *Japan.*
Kamiya, Dr Nobuo. *Japan.*
Kamiya, Dr Yoshihiro. *Japan.*
Kamminga, Dr H. *UK.*
Kamolchote, Mr Poonsak. *Thailand.*
Kamphuis, Dr Irenus Gerhardus. *Netherlands.*
Kamprath, Mr Fred-Bodo. *DDR.*
Kamwaya, Dr Mombo E. *Tanzania.*
Kanatzidis, Dr Mercouri G. *USA.*
Kanellis, Prof. George. *Greece.*
Kanis, Dr Michäl. *DDR.*
Kannan, Dr Kazhiur Kothandapani. *India.*
Kansikas, Mr Jarno Juhani. *Finland.*
Kanters, Dr Jan. *Netherlands.*
Kantor, Dr Matvey Matveyevich. *USSR.*
Kapecki, Dr Jon Alfred. *USA.*
Kaplunnik, Dr Lidiya Nikolayevna. *USSR.*
Kapon, Dr Moshe. *Israel.*
Kapor-Nahlovski, Dr Agneš. *Yugoslavia.*
Kappenstein, Prof. Charles. *France.*
Kapur, Dr Selim. *Turkey.*
Kar(Roy), Dr (Mrs) Tanusree *India.*
Karanović, Magistar Ljiljana. *Yugoslavia.*
Karcher, Dr Barbara Ann. *USA.*
Kardashev, Dr Boris Konstantinovich. *USSR.*
Kardos, Mrs Jutta. *Hungary.*
Karipides, Prof. Anastas. *USA.*
Karl, Prof Dr Norbert. *BRD.*
Karle, Dr Isabella L. *USA.*
Karle, Prof. Jerome. *USA.*
Karlsson, Dr Bengt E. *Sweden.*
Karlsson, Mr Haraldur R. *USA.*
Karlsson, Dr Rolf. *Switzerland.*
Karmazin, Dr Lubomír. *CSSR.*
Karniewicz, Dr hab. Jan. *Poland.*
Karolak-Wojciechowska, Dr Janina. *Poland.*
Karp, Prof. Dr Jan. *Poland.*
Karpinsky, Dr Oleg Georgiyevich. *USSR.*
Karrer, Mr Andreas. *Switzerland.*
Kartha, Dr Gopinath. *USA.*
Kartheuser, Dr Edward Peter. *Belgium.*
Kartiwa, Mr Sumadi. *Indonesia.*
Karup-Møller, Dr Sven. *Denmark.*
Karvinen, Mrs Saila Marjatta. *Finland.*
Karyakina, Dr Tatyana Alexandrovna. *USSR.*
Kasai, Prof. Nobutami. *Japan.*
Kashaev, Dr Anvar Akhyarovich. *USSR.*

Kashchiev, Dr Dimcho. *Bulgaria.*
Kashino, Dr Setsuo. *Japan.*
Kashiwase, Dr Yasuji. *Japan.*
Kashyap, Dr Ram Prasad. *India.*
Kasper, Dr John S. *USA.*
Kassner, Mr Dethard. *BRD.*
Kastner, Ms Margaret E. *USA.*
Kasturi, Prof Tirumali R. *India.*
Kasuga, Prof. Dr Masanobu. *Japan.*
Katada, Prof. Kinya. *Japan.*
Katagas, Dr Christos. *Greece.*
Katagawa, Dr Takeshi. *Japan.*
Katayama, Mr Chuji. *Japan.*
Katayama, Prof. Mikio. *Japan.*
Kato, Dr Akira. *Japan.*
Kato, Mr Ichiro. *Japan.*
Kato, Dr Katsuo. *Japan.*
Kato, Prof. Masanori. *Japan.*
Kato, Prof. Norio. *Japan.*
Kato, Prof. Toshio. *Japan.*
Kato, Mr Yoshihiro. *Japan.*
Kats, Dr Moisey Sukherovich. *USSR.*
Katsanos, Mr Demetrios Evangelos. *Greece.*
Katscher, Dr Hartmut. *BRD.*
Katsnel'son, Prof. Al'bert Anatolyevich. *USSR.*
Katsube, Prof. Yukiteru. *Japan.*
Katz, Dr Bradley Alan. *USA.*
Katz, Dr Gerald. *Israel.*
Katz, Mr Henry. *USA.*
Katz, Prof. J. Lawrence. *USA.*
Katz, Prof. Lewis. *USA.*
Katz, Dr Louis. *USA.*
Katzschmann, Mr Kurt. *DDR.*
Kaub, Mr Jürgen. *BRD.*
Kaučič, Dr Venčeslav. *Yugoslavia.*
Kaufman, Prof. Hershall William. *USA.*
Kaufmann, Dr Thorsten. *DDR.*
Kaus, Dr Gerhard. *BRD.*
Kavounis, Mr Konstantinos. *Greece.*
Kawachi, Dr Yosuke. *New Zealand.*
Kawada, Dr Isao. *Japan.*
Kawado, Dr Seiji. *Japan.*
Kawahara, Dr Akira. *Japan.*
Kawai, Mr Toshiaki. *Japan.*
Kawai, Dr Yoriyoshi. *Japan.*
Kawaminami, Dr Masaru. *Japan.*
Kawamori, Prof. Asako. *Japan.*
Kawamura, Dr Tsutomu. *Japan.*
Kawano, Assoc. Prof. Shigeaki. *Japan.*
Kawata, Dr Hiroshi. *Japan.*
Kay, Dr Herbert Frederick. *UK.*
Kay, Dr Mortimer I. *USA.*
Kaya, Prof. Seiji. *Japan.*
Kaynak, Dr Uğur. *Turkey.*
Kayushina, Dr Renata Lvovna. *USSR.*
Kazinets, Dr Maria. *Israel.*
Ke, Mr Heng-ming. *China.*
Keankeo, Miss Watcharaporn. *Thailand.*
Keder, Dr Nancy Lynn. *USA.*
Keefe, Prof. William Edward. *USA.*
Keeling, Prof. Rolland O., Jr. *USA.*
Keem, Mr John Edward. *USA.*
Kehl, Mr William Louis. *USA.*
Kehres, Mr Paul William. *USA.*
Keijser, de, Dr Ir Thomas Henri. *Netherlands.*
Kellenberger, Prof. Eduard. *Switzerland.*
Keller, Dr Egbert. *BRD.*
Keller, Ms Eva Barbara. *Switzerland.*
Keller, Dr Hans-Lothar. *BRD.*
Keller, Prof. Dr Heimo Jürgen. *BRD.*
Keller, Dr Kurt Wolfgang. *DDR.*
Keller, Mr Ludwig. *USA.*
Keller, Prof Dr Paul. *BRD.*
Keller, Dr hab Wlodzimierz Aleksander. *Poland.*
Keller, Dr Wolfgang Ludwig. *BRD.*
Kellerman, Dr Martin. *USA.*
Kelling, Mrs Gerhild. *DDR.*
Kellö, Dr Eleonóra. *CSSR.*
Kelly, Dr Anthony. *UK.*
Kelly, Mrs Carol J. (Korsmo). *USA.*

Kelly, Mr Eric. *UK.*
Kelly, Dr Judith Ann. *USA.*
Kelly, Dr Patrick Manning. *Australia.*
Kelly, Dr Thomas C. *Ireland.*
Kemme, Dr Andrey Andreyevich. *USSR.*
Kemmler-Sack, Mrs Prof. Dr Sibylle. *BRD.*
Kempster, Dr Charles John Edgar. *UK.*
Kenaan, Dr Feisal. *Saudi Arabia.*
Kendi, Dr Engin. *Turkey.*
Kennard, Dr Colin Harold Leslie. *Australia.*
Kennard, Dr Olga. *UK.*
Kennedy, Miss D.A. *UK.*
Kennedy, Dr Stanley Wallace. *Australia.*
Kennon, Prof Noel Frederick. *Australia.*
Keow-kam-nerd, Dr Kanchana. *Thailand.*
Keppler, Dr Ulrich H. *BRD.*
Kerenović, Magistar Midhat. *Yugoslavia.*
Kern, Prof. Raymond. *France.*
Kerr, Dr Kathleen Ann. *Canada.*
Kerr, Dr Ian Segrave. *UK.*
Kersten, Mr Friedrich. *DDR.*
Kertész, Dr László. *Hungary.*
Kessenikh, Dr Galina Georgiyevna. *USSR.*
Kestigian, Dr Michael. *USA.*
Ketolainen, Prof. Pertti Pekka Juhani. *Finland.*
Kettler, Mr Peter. *BRD.*
Kettmann, Ing Viktor. *CSSR.*
Kettunen, Prof. Pentti Olavi. *Finland.*
Keulen, Dr Evert. *Netherlands.*
Khachaturyan, Dr Armen Gurgenovich. *USSR.*
Khadr, Dr Moustafa. *Egypt.*
Khadzhi, Dr Valentin Evstafyevich. *USSR.*
Khaikin, Dr Leonid Solomonovich. *USSR.*
Khalid, Mr Mohammad. *Pakistan.*
Khalifa, Prof. (Mrs) Berlant. *Egypt.*
Khalifa, Dr (Mrs) B. Abdel Meguid. *Egypt.*
Khan, Dr Ainul Hassan. *Pakistan.*
Khan, Prof. Ali. *Venezuela.*
Khan, Dr Anwarur Rahman. *Bangladesh.*
Khan, Dr Masood Alam. *Canada.*
Khandar-Shahabad, Dr Ali-Akbar. *Iran.*
Khantaprab, Dr Chaiyudh. *Thailand.*
Kharchenko, Dr Lyudmila Yulianovna. *USSR.*
Kharchenko, Dr Olga Ivanovna. *USSR.*
Kharitonov, Dr Yury Alexandrovich. *USSR.*
Khatanova, Dr Nina Abdulovna. *USSR.*
Khattak, Dr Guldad Khattak. *Saudi Arabia.*
Khawaja, Mr Mahmood-ul-Hassan. *Pakistan.*
Khawaja, Dr Ehsan Ellahi. *Saudi Arabia.*
Kheiker, Dr Daniel' Moiseyevich. *USSR.*
Kheirov, Dr Mamed Bekovch. *USSR.*
Khetchikov, Dr Lev Nikolayevich. *USSR.*
Khidr, Prof. (Mrs) Fatma Abdel Hakim. *Egypt.*
Khisina, Dr Nataliya Rafailovna. *USSR.*
Khitrova, Dr Valentina Ivanovna. *USSR.*
Khodashova, Dr Tatyana Semenovna. *USSR.*
Kholeif, Dr Mahmoud. *Egypt.*
Khotsyanova, Dr Tatyana Lvovna. *USSR.*
Khurshudyan, Dr Era Khristoforovna. *USSR.*
Khwaja, Dr Farid Akhtar. *Pakistan.*
Kidyarov, Dr Boris Ivanovich. *USSR.*
Kiedrowski, von, Mr Hartmut. *DDR.*
Kiel, Dr Gertrude Lina. *BRD.*
Kieling, Mr Knut. *DDR.*
Kiely, Dr Patrick Vincent. *Ireland.*
Kierkegaard, Prof. Peder. *Sweden.*
Kiers, Dr Conradus Theodorus. *Netherlands.*
Kies, Dr Jörg. *DDR.*
Kihara, Dr Kuniaki. *Japan.*
Kihlborg, Prof. Lars. *Sweden.*
Kikuchi, Dr Makoto. *Japan.*
Kikuta, Dr Seishi. *Japan.*
Killean, Dr Reginald Cameron Gordon. *UK.*
Kim, Ms Eunice E. *USA.*
Kim, Dr Hoon Sup. *Korea.*
Kim, Mr Jung Ja P. *USA.*
Kim, Dr Key Soo. *Korea.*
Kim, Prof. Moon-Jib. *Korea.*
Kim, Dr Nancy Ellen Kime. *USA.*
Kim, Mr Sangsoo. *USA.*

Kim, Prof. Dr Soo Jin. *Korea.*
Kim, Mr Sukyoung. *USA.*
Kim, Prof. Sung-Hou. *USA.*
Kim, Prof. Dr Yang. *Korea.*
Kim, Prof. Dr Yang Bae. *Korea.*
Kimball, Dr Martha R. *USA.*
Kimmel, Dr Giora. *Israel.*
Kimura, Prof. Masao. *Japan.*
King, Prof. Geoffrey Stephen Douglas. *Belgium.*
King, Dr Hubert Ellis. *USA.*
King, Prof. Hubert Wylam. *Canada.*
King, Dr James Newington. *UK.*
King, Dr Murray Vernon. *USA.*
Kingman, Ms Priscilla Ward. *USA.*
Kingon, Mr Angus Ian. *South Africa.*
Kini, Mr Ullal Devappa. *India.*
Kinneging, Drs Albertus Jacobus. *Netherlands.*
Kinsella, Mr A. *UK.*
Kiosse, Dr Georgy Alexandrovich. *USSR.*
Kipling, Miss Susan Jane. *UK.*
Kirchhoff, Ms Pamela Moore. *USA.*
Kirchner, Doz. Dr Elisabeth Charlotte. *Austria.*
Kirchner, Prof. Richard Martin. *USA.*
Kirfel, Prof Dr Armin Harald. *BRD.*
Kirichenko, Dr Valentina Vasilyevna. *USSR.*
Kirikov, Dr Vladimir Arkadyevich. *USSR.*
Kirin, Dr (Mrs) Ankica. *Yugoslavia.*
Kiriyama, Prof. Hideko. *Japan.*
Kiriyama, Prof. Ryoiti. *Japan.*
Kirkinsky, Dr Vitaly Alekseyevich. *USSR.*
Kirkman, Dr John Henry. *New Zealand.*
Kirkova, Prof. Elena. *Bulgaria.*
Kirn, Mr J. F. *USA.*
Kirov, Mr Georgi Kirilov. *Bulgaria.*
Kirov, Doc Georgi Nikolov. *Bulgaria.*
Kirpichnikova, Dr Lyubov' Fedorovna. *USSR.*
Kiryu, Prof. Setsuo. *Japan.*
Kirz, Prof. Janos. *USA.*
Kisch, Prof. Hanan Joseph. *Israel.*
Kiselev, Dr Nikolay Andreyevich. *USSR.*
Kishi, Dr Kiyoshi. *Japan.*
Kishino, Dr Seigo. *Japan.*
Kishk, Dr Fawzi Mohamed. *Egypt.*
Kislovsky, Dr Lev Dmitriyevich. *USSR.*
Kiss, Dr Klara. *USA.*
Kissinger, Charles R. *USA.*
Kissinger, Mr Homer Everett. *USA.*
Kissling, Mr Alexandru. *Romania.*
Kistenmacher, Dr Thomas John. *USA.*
Kitahama, Dr Katsuki. *Japan.*
Kitano, Mr Yukishige. *Japan.*
Kittl, Mr Pablo. *Chile.*
Kivekäs, Dr Raikko Terjo Ilari. *Finland.*
Kivilahti, Prof. Jorma Kalevi. *Finland.*
Kizilyalli Doç. Dr Meral. *Turkey.*
Kjekshus, Prof. Arne. *Norway.*
Kjeldgaard, Mr Morten. *Denmark.*
Kléman, Dr Maurice. *France.*
Klanderman, Prof. Kent Arlen. *USA.*
Klapper, Dr Helmut. *BRD.*
Klaska, Dr Karl-Heinz. *BRD.*
Klaska, Dr Rolf. *BRD.*
Klassen-Neklyudova, Prof. Marina V. *USSR.*
Klebe, Dr Gerhard. *BRD.*
Klechkovskaya, Dr Vera Vsevolodovna. *USSR.*
Klee, Prof. Dr Wilfrid Edgar. *BRD.*
Klein, Prof. Cheryl Lynn. *USA.*
Kleinert, Dr Peter. *DDR.*
Kleinstück, Prof. Karlheinz. *DDR.*
Kleint, Dr Christian. *DDR.*
Klement, Dr Ulrich. *BRD.*
Klepp, Dr Kurt Otto. *BRD.*
Kleshchinsky, Dr Leonid Innokentyevich. *USSR.*
Klessen, Mr Gerhard. *BRD.*
Klevtsov, Dr Petr Vasilyevich. *USSR.*
Klevtsova, Dr Rimma Fedorovna. *USSR.*
Klewe, Mr Bernt. *Norway.*
Klimanek, Dr Peter. *DDR.*
Klimova, Dr Anna Yuryevna. *USSR.*
Klinga, Mr Martti Evert. *Finland.*
Kliya, Dr Maya Ottovna. *USSR.*

Klock, Prof. Peter Allan. *USA.*
Klock, Mr Winfried. *DDR.*
Klop, Drs Enno Anton. *Netherlands.*
Klug, Dr Aaron. *UK.*
Klug, Mrs Annamária. *Hungary.*
Klug, Prof. Harold Philip. *USA.*
Klyavin, Dr Oleg Vladimirovich. *USSR.*
Knab, Dr Galina Grigoryevna. *USSR.*
Kneschke, Dr Götz. *DDR.*
Kniep, Prof Dr Rüdiger. *BRD.*
Knight, Mr K.S. *UK.*
Knight, Mr Robert. *UK.*
Knight, Mr Stefan. *Sweden.*
Knippenberg, Dr Wilhelmus Franciscus. *Netherlands.*
Knobler, Dr Carolyn B. *USA.*
Knoch, Dr Falk A. *BRD.*
Knöchel, Dr Claus-Dieter. *BRD.*
Knof, Mr Wolfgang Erich. *BRD.*
Knop, Prof. Osvald. *Canada.*
Knorr, Dr Klaus. *BRD.*
Knott, Mr P.R. *UK.*
Knox, Prof. James Russell. *USA.*
Knox, Prof. Kerro. *USA.*
Knuuttila, Mrs Hilkka Ritva-Liisa. *Finland.*
Knuuttila, Mr Pekka Juhani. *Finland.*
Kołakowski, Dr Andrzej. *Poland.*
Kołakowski, Dr Bogdan Józef. *Poland.*
Kožíšek, Ing Jozef. *CSSR.*
Kožíšková, Ing Zlatica. *CSSR.*
Kobayashi, Dr Akiko. *Japan.*
Kobayashi, Dr Hayao. *Japan.*
Kobayashi, Prof. Jinzo. *Japan.*
Kobayashi, Prof. Nobuyuki. *Japan.*
Kobayashi, Dr Tadashi. *Japan.*
Kobayashi, Dr Takaaki. *Japan.*
Kobayashi, Dr Takashi. *Japan.*
Kobzareva, Dr Svetlana Alekseyevna. *USSR.*
Koch, Dr Beatrix. *Netherlands.*
Koch, Mrs Dr Elke. *BRD.*
Koch, Dr H. William. *USA.*
Koch-Wallraf, Mrs Prof. Dr Maria. *BRD.*
Kochanovská, Prof. Dr Adéla. *CSSR.*
Kocharov, Dr Alexander Georgiyevich. *USSR.*
Kociński, Prof. Dr Jerzy. *Poland.*
Kockel, Dr Andreas. *BRD.*
Kocman, Dr Vladimir. *Canada.*
Koda, Mr Shigetaka. *Japan.*
Kodama, Dr Hideomi. *Canada.*
Kodandapani, Mr R. *India.*
Köhler, Dr Rolf. *DDR.*
Koehler, Dr Wallace Conrad. *USA.*
Kökçe, Dr Ali. *Turkey.*
König, Dr Burkhard. *BRD.*
Koenig, Dr Donald Frederick. *USA.*
Köpernik, Mr Horst. *DDR.*
Koeppe, Dr Roger E., II. *USA.*
Kötitz, Dr Günther. *DDR.*
Koetzle, Dr Thomas Frederick. *USA.*
Koh, Dr Lip Lin. *Singapore.*
Kohlbeck, Dr Franz. *Austria.*
Kohli, Dr Vijay Kumar. *India.*
Kohn, Prof. Jack Arnold. *USA.*
Kohra, Prof. Kazutake. *Japan.*
Kohyama, Mr Masaki. *Japan.*
Koide, Dr Tsutomu. *Japan.*
Koikkalainen, Miss Seija Anneli. *Finland.*
Koizumi, Dr Hideo. *Japan.*
Koizumi, Prof. Mitsue. *Japan.*
Kojić-Prodić, Dr Biserka. *Yugoslavia.*
Kok, de, Dr Anthonie Johannes. *Netherlands.*
Kokkinidis, Dr Michael. *BRD.*
Kokkou, Prof. Socrates Constantinos. *Greece.*
Koknat, Prof. Friedrich Wilhelm. *USA.*
Kokotailo, Prof. George Thomas. *USA.*
Kolesova, Dr Rimma Vladimirovna. *USSR.*
Kolks, Dr Gary. *USA.*
Kolobyanina, Dr Tatyana Nikolayevna. *USSR.*
Kolodiyeva, Dr Svetlana Vasilyevna. *USSR.*
Kolontsova, Dr Ekaterina Vasilyevna. *USSR.*
Kolpakov, Dr Andrey Vasilyevich. *USSR.*

Kolster, Prof. Dr Ir Benjamin Harry. *Netherlands.*
Koman, Ing Marián. *CSSR.*
Komarek, Prof. Dr Kurt Ludwig. *Austria.*
Komatsu, Prof. Hiroshi. *Japan.*
Komori, Dr Tetsuya. *Japan.*
Komoto, Dr Tadashi. *Japan.*
Komov, Dr Igor' Leontyevich. *USSR.*
Komrska, Dr Jiří. *CSSR.*
Komu, Mr Markku Eino Sakari. *Finland.*
Komura, Prof. Shigehiro. *Japan.*
Komura, Prof. Yukitomo. *Japan.*
Kon, Dr Aviv Yuliseyevich. *USSR.*
Konaka, Dr Shigehiro. *Japan.*
Kondrashev, Dr Yury Dmitriyevich. *USSR.*
Kondratyeva, Dr Victoria Victorovna. *USSR.*
Kong, Dr Eric Siu Wai. *USA.*
Kong, Mr You-hua. *China.*
Konguetsof, Dr Helen. *Greece.*
Konig de Perazzo, Lic Patricia Verónica. *Argentina.*
Koningsveld, van, Dr Hendrikus. *Netherlands.*
Konitz, Mr Antoni. *Poland.*
Konnert, Dr John H. *USA.*
Konnert, Mrs Judith A. *USA.*
Konno, Dr Michiko. *Japan.*
Konocak, Doç. Zeki. *Turkey.*
Konopka, Dr Danuta Cecylia. *Poland.*
Konstantinov, Dr Ivan. *Bulgaria.*
Konstantinova, Dr Alisa Fedorovna. *USSR.*
Kontio, Dr Airi Outi. *Finland.*
Konz, Dr Werner. *BRD.*
Koo, Prof. Dr Chung Hoe. *Korea.*
Koopmans, Prof. Dr Kasper. *Netherlands.*
Kopf, Dr Jürgen. *BRD.*
Kopka, Mrs Mary Lou. *USA.*
Koptsik, Prof. Vladimir Alexandrovich. *USSR.*
Koreň, Mr Branislav. *CSSR.*
Koreshkov, Dr Boris Dmitriyevich. *USSR.*
Korhonen, Prof. Unto Kalervo. *Finland.*
Koritsánszky, Mr Tibor. *Hungary.*
Kormány, Dr Teréz. *Hungary.*
Kornev, Dr Aleksey Nikolayevich. *USSR.*
Korp, Prof. James Douglas. *USA.*
Korsukova, Dr Mariya Mikhailovna. *USSR.*
Korvenranta, Dr Jorma Artturi. *Finland.*
Koryagin, Mr. Vyacheslav Filippovich. *USSR.*
Kosche, Mrs Ingeborg. *DDR.*
Kosel, Mr George Eugene. *USA.*
Kosevich, Prof. Vadim Markovich. *USSR.*
Koshel', Dr Olga Stepanova. *USSR.*
Koshy, Dr Jacob. *Nigeria.*
Kosiur, Dr David Richard. *USA.*
Kosmachev, Dr Sergey Mikhailovich. *USSR.*
Kosmopoulos, Dr John. *Greece.*
Kosova, Dr Tatyana Borisovna. *USSR.*
Kosovinc, Prof. Dr Ivan. *Yugoslavia.*
Kosten, Mr Klaus. *BRD.*
Kosterin, Dr Evgeny Andreyevich. *USSR.*
Kostiner, Prof. Edward Stephen. *USA.*
Kostorz, Prof. Gernot. *Switzerland.*
Kostov, Prof. Ivan. *Bulgaria.*
Kostov, Dr Ruslan. *Bulgaria.*
Kosturkiewicz, Dr hab. Zofia. *Poland.*
Kostyukova, Dr Evgeniya Prokofyevna. *USSR.*
Koszelak, Mr Stanley N. *USA.*
Kotani, Prof. Masao. *Japan.*
Kotel'nikova, Dr Elena Nikolayevna. *USSR.*
Koto, Dr Kichiro. *Japan.*
Kotov, Dr Nikolay Vladimirovich. *USSR.*
Kotsanidis, Mr Panayotis. *Greece.*
Kotsev, Dr Iosif. *Bulgaria.*
Kotsis, Mr Konstantinos. *Greece.*
Kotur, Dr Bogdan Yaroslavovich. *USSR.*
Koumelis, Dr Christos. *Greece.*
Kountouris, Mr Costas. *Greece.*
Kountz, Dr Dennis James. *USA.*
Kovachev, Dr Peter. *Bulgaria.*
Kováčová, Dr Katarina. *CSSR.*
Kovács, Prof. Alessandro L. *Italy.*
Koval'chuk, Dr Mikhail Valentinovich. *USSR.*
Kovda, Prof. Leonid Mikhailovich. *USSR.*

Kovyev, Dr Ernest Konstantinovich. *USSR.*
Koyama, Dr Hirozo. *Japan.*
Koyama, Dr Kazutoshi. *Saudi Arabia.*
Koyama, Dr Yasumasa. *Japan.*
Koyano, Mr Kazuo. *Japan.*
Kozłowska, Mrs Krystyna. *Poland.*
Koz'ma, Dr Alexander Alekseyevich. *USSR.*
Koz'min, Dr Petr Alekseyevich. *USSR.*
Kozaki, Prof. Shigeru. *Japan.*
Koziol, Dr Anna. *Poland.*
Kozlenkov, Dr Alexander Ivanovich. *USSR.*
Kozlova, Dr Olga Gerasimovna. *USSR.*
Kraatz, Dr Paul. *USA.*
Krabbendam, Drs Hendrik. *Netherlands.*
Krämer, Prof. Dr Volker. *BRD.*
Krajewski, Dr Adriano. *Italy.*
Krajewski, Dr hab. Janusz. *Poland.*
Králík, Dr František. *CSSR.*
Králová, Dr Rudolfa. *CSSR.*
Kramer, Dr Irmtraud. *BRD.*
Krane, Mr Hans-Georg. *BRD.*
Kranjc, Prof. Dr Katarina. *Yugoslavia.*
Kraševec, Dr Viktor. *Yugoslavia.*
Krasner, Prof. Saul. *USA.*
Krasochka, Dr Oleg Nikolayevich. *USSR.*
Kratky, Dr Christoph. *Austria.*
Kratky, Prof. em. Dr Dr h.c. Otto. *Austria.*
Kratochvil, Dr Bohumil. *CSSR.*
Kraus, Doc. Dr Ivo. *CSSR.*
Krause, Mr Waldefried. *DDR.*
Krause, Mrs Christa. *DDR.*
Krausse, Dr Joachim. *DDR.*
Kraut, Prof. Joseph. *USA.*
Krawitz, Prof. Aaron David. *USA.*
Krebs, Prof. Dr Bernt. *BRD.*
Kremer, Dr Germán. *Chile.*
Krén, Dr Emil. *Hungary.*
Kressner, Dr F.Harry. *DDR.*
Krestev, Mr Venelin. *Bulgaria.*
Kresteva, Dr Manya. *Bulgaria.*
Kretschmer, Dr Rolf-Günther. *DDR.*
Kretsinger, Prof. Robert Harvey. *USA.*
Kreutz, Dr Ernst Wolfgang. *BRD.*
Krever, Drs Maarten. *Netherlands.*
Krieger, Dr Monty. *USA.*
Krimm, Prof. Samuel. *USA.*
Krischner, Prof. Dr Dipl.-Ing. Harald. *Austria.*
Krishna, Prof. Padmanabhan. *India.*
Krishna Rao, Prof. Dr K. V. *India.*
Krishnaiah, Mr Musali. *India.*
Krishnan, Dr Rangachari. *India.*
Krishnaswamy, Mr S. *India.*
Kristensen, Mrs Bente Saustrup. *Denmark.*
Kristmannsdóttir, Cand.Real. Hrefna. *Iceland.*
Kritayakirana, Mrs Rungsri. *Thailand.*
Kritzinger, Prof. Serfontein. *South Africa.*
Krivandina, Dr Elena Alekseyevna. *USSR.*
Krivoglaz, Prof. Mikhail Alexandrovich. *USSR.*
Krivokoneva, Dr Galina Kirillovna. *USSR.*
Křivý, Dr Ivan. *CSSR.*
Krogmann, Prof. Dr Klaus. *BRD.*
Krohn, Dr A. *UK.*
Kroll, Prof. Dr Herbert. *BRD.*
Kroon-Batenburg, Dr Louise M.J. *Netherlands.*
Kroon, Prof. Dr Jan. *Netherlands.*
Krstanović, Prof. Dr Ilija. *Yugoslavia.*
Krüger, Prof. Dr Carl. *BRD.*
Krug, Prof. Dr Detlef. *BRD.*
Kruger, Dr Gert Jacobus. *South Africa.*
Kruglik, Dr Anatoliy Ivanovich. *USSR.*
Krukowski, Dr Marek. *Poland.*
Krutova, Dr Glafira Ivanova. *USSR.*
Krygowski, Prof. Tadeusz, Marek. *Poland.*
Krymov, Dr Vladimir Mikhailovich. *USSR.*
Kuang, Mrs Bao. *China.*
Kuběna, Dr Josef. *CSSR.*
Kuban, Mr Ralf-Jürgen. *DDR.*
Kubel, Dr Frank. *Switzerland.*
Kubiak, Doc Ryszard. *Poland.*
Kubiak, Dr Maria. *Poland.*
Kubo, Prof. Ikumaro. *Japan.*

Kubo, Prof. Teruichiro. *Japan.*
Kucab, Dr Marian. *Poland.*
Kuchar, Doz. Dr Friedemar. *Austria.*
Kucharczyk, Dr Damian. *Poland.*
Kucharski, Dr Edward Stanislaw. *Australia.*
Kuchitsu, Prof. Kozo. *Japan.*
Kudoh, Dr Yasuhiro. *Japan.*
Kudryavtseva, Dr Galina Petrova. *USSR.*
Kühn, Dr Günther. *DDR.*
Kuehn, Prof. Dr Robert. *BRD.*
Küppers, Prof. Dr Horst. *BRD.*
Kürsten, Dr Hans-Dieter. *DDR.*
Kuhs, Dr Werner Friedrich. *France.*
Kukharenko, Prof. Alexander Alexandrovich. *USSR.*
Kukina, Dr Galina Alexandrovna. *USSR.*
Kukovsky, Prof. Evgeny Georgiyevich. *USSR.*
Kukuy, Dr Anatoly Lvovich. *USSR.*
Kullnig, Dr Rudolph Karl. *USA.*
Kulpe, Dr Siegfried. *DDR.*
Kumagawa, Dr Masashi. *Japan.*
Kumao, Mr Akihiro. *Japan.*
Kumar, Dr Rajendra. *India.*
Kumar, Mr Vinay *India.*
Kumbasar, Prof. Işik. *Turkey.*
Kummer, Drs Ernst Albertus. *Netherlands.*
Kumosinski, Thomas F. *USA.*
Kundra, Mr Krishan Dev. *India.*
Kunrath, Mr José Irineu. *Brazil.*
Kunsch, Dr Dipl.-Ing. Barnabas. *Austria.*
Kunstelj, Dr Dragan. *Yugoslavia.*
Kuntsevich, Dr Tamara Serafimovna. *USSR.*
Kuo, Prof. Ke-hsin. *China.*
Kuok, Dr Meng Hau. *Singapore.*
Kupčík, Prof. Dr Vladimir. *BRD.*
Kupka, Dr František. *CSSR.*
Kuppuswamy, Dr Nagarajan. *India.*
Kupriyanov, Dr Mikhail Fedotovich. *USSR.*
Kurahashi, Dr Masayasu. *Japan.*
Kuranova, Dr Inna Petrovna. *USSR.*
Kurazhkovskaya, Dr Victoriya Semenovna. *USSR.*
Kurbanov, Dr Khakim Mamadaliyevich. *USSR.*
Kurdyumov, Prof. Georgy Vyacheslavovich. *USSR.*
Kuribayashi, Mr Shunsuke. *Japan.*
Kurittu, Dr Jyrki Veli Einari. *Finland.*
Kuriyama, Dr Masao. *USA.*
Kurki-Suonio, Prof. Kaarle Veikko J. *Finland.*
Kurkutova, Prof. Evdokiya Nikitichna. *USSR.*
Kuroda, Prof. Haruo. *Japan.*
Kuroda, Dr Reiko. *UK.*
Kurosawa, Dr Kou. *Japan.*
Kuroya, Prof. Hisao. *Japan.*
Kurtz, Dr Stewart Kendall. *USA.*
Kushi, Dr Yoshihiko. *Japan.*
Kusunoki, Dr Masami. *Japan.*
Kutoglu, Dr Ali. *BRD.*
Kutschabsky, Dr Leo. *DDR.*
Kuwabara, Prof. Shigeya. *Japan.*
Kuz'ma, Prof. Yury Bogdanovich. *USSR.*
Kuz'min, Prof. Eduard Alekseyevich. *USSR.*
Kuz'min, Prof. Runar Nikolayevich. *USSR.*
Kuz'mina, Dr Irina Pavlovna. *USSR.*
Kuzmany, Doz. Dr Hans. *Austria.*
Kuznetsov, Dr Alexander Victorovich. *USSR.*
Kuznetsov, Prof. Fedor Andreyevich. *USSR.*
Kuznetsov, Prof. Vasily Grigoryevich. *USSR.*
Kuznetsov, Dr Victor Andreyevich. *USSR.*
Kvapil, Ing Jiři. *CSSR.*
Kvasnitsa, Dr Victor Nikolayevich. *USSR.*
Kvick, Dr Ake H. *USA.*
Kyaw, Dr Htin. *Burma.*
Kyotani, Mr Mutsumasa. *Japan.*
Kyröläinen, Mr Antero Johannes. *Finland.*
Kyriakos, Prof. Demetrius. *Greece.*
Kyutt, Dr Reginal'd Nikolayevich. *USSR.*
La Placa, Mr Sam Joseph. *USA.*
La Prade, Dr Marie D.. *USA.*
La Rocca, Mr Edward W. *USA.*
Labaki, Ms Lucila Chebel. *Brazil.*
Labbé, Dr Philippe. *France.*
Laberrigue, Prof. André *France.*

Labib, Dr (Mrs) Fawkia. *Egypt.*
Labib, Dr Tarik. *Egypt.*
Labischinski, Dr Harald. *BRD.*
Labrador Carrasco, Mr Manuel. *Spain.*
Lacmann, Prof. Dr Rolf. *BRD.*
Ladd, Dr Marcus Frederick Charles. *UK.*
Ladell, Dr Joshua. *USA.*
Laderman. Dr Stephen Stromberg. *USA.*
Lähdeniemi, Dr Matti Juhani Iisakki. *Finland.*
Lafourcade, Prof. Lucien. *France.*
Łągiewka, Dr Eugeniusz Antoni. *Poland.*
Laggner, Doz. Dr Peter. *Austria.*
Lagomarsino, Dr Stefano. *Italy.*
Lahari, Dr Barendra Nath. *India.*
Lahodny-Šarc, Prof. Dr (Mrs) Olga. *Yugoslavia.*
Lahti, Dr Seppo Ilmari. *Finland.*
Lai, Dr Ting Fong. *Hong Kong.*
Laiho, Dr Reino Toivo Salomo. *Finland.*
Laine, Dr Ensio Sulo Uolevi. *Finland.*
Laing, Dr Mary Elizabeth. *South Africa.*
Laing, Prof. Michael John. *South Africa.*
Lajzerowicz, Prof. Janine. *France.*
Lake, Prof. James A. *USA.*
Lake, Mr P.G. *UK.*
Laker, Mr Thomas James. *UK.*
Lal, Dr Krishan. *India.*
Lalancette, Prof. Roger A. *USA.*
Lalies, Mr A.A. *UK.*
Lamba, Dr Doriano. *Italy.*
Lambert, Prof. Marianne. *France.*
Lambert-Smith, Mr John Ernle Warwick. *Australia.*
Lamm, Mr Viktor Andreas. *BRD.*
Lammert, Mrs Barbara. *DDR.*
Lamotte-Brasseur, Dr Josette Marie Louise. *Belgium.*
Lancucki, Mr Christopher Joseph. *Australia.*
Landau, Mr Alex. *Israel.*
Lando, Prof. Jerome B. *USA.*
Landy, Dr Richard Allen. *USA.*
Lanfranchi, Dr Maurizio. *Italy.*
Lang, Prof. Andrew Richard. *UK.*
Langbein, Prof Dr Werner Dieter. *BRD.*
Lange, Dr Bruce Ainsworth. *USA.*
Langer, Prof. Ebbe Wang. *Denmark.*
Langer, Dr Hans-Dieter. *DDR.*
Langer, Prof Dr Klaus. *BRD.*
Langer, Dr Vratislav. *CSSR.*
Langford, Dr John Ian. *UK.*
Langford, Miss M. *UK.*
Langlet, Dr Gérard André *France.*
Langridge, Prof. Robert. *USA.*
Langs, Dr David Alan. *USA.*
Lányi, Dr Péter., *Hungary.*
Lapasset, Prof. Jacques. *France.*
Lappa, Dr Ryszard Włodzimierz. *Poland.*
Lara Magaña, Mrs María Eugenia. *Mexico.*
Laredo, Dr Estrella. *Venezuela.*
Larroque, Prof. Paul. *France.*
Larsen, Mr Finn Krebs. *Denmark.*
Larsen, Dr Ingrid Kjøller. *Denmark.*
Larsen, Mrs Sine. *Denmark.*
Larson, Dr Allen C. *USA.*
Larson, Dr Bennett Charles. *USA.*
Larson, Mr Steven B. *USA.*
Larsson, Dr Sven. *Sweden.*
Lartigue, Dr Colette. *France.*
Laruelle, Prof. Pierre Etienne Charles. *France.*
Lashewycz-Rubycz, Dr Romana Alexandra. *USA.*
Lashin, Dr A. Mohamed. *Egypt.*
Last, Mr Paul Edward. *UK.*
Lattman, Prof. Eaton Edward. *USA.*
Lauck, Mr Rudolf. *BRD.*
Laudise, Dr Robert Alfred. *USA.*
Laugier, Mr Jean. *France.*
Laugt, Dr Marguerite. *France.*
Laurent, Dr Pierre. *France.*
Laurent, Prof. Yves. *France.*
Lawless, Prof. Kenneth Robert. *USA.*
Lawson, Dr Charles Alden. *USA.*
Lawson, Mr Robert Ian. *UK.*
Lawton, Mr Stephen Latham. *USA.*

Lazar, Mr Constantin. *Romania.*
Lazar, Mr Dušan. *Yugoslavia.*
Lazarenkov, Prof. Vadim Grigor'yevich. *USSR.*
Lazarev, Dr Eduard Mikhailovich. *USSR.*
Lazarini, Prof. Dr Franc. *Yugoslavia.*
Le Bars, Mrs Michèle, *France.*
Le Page, Dr Yvon. *Canada.*
Le Roux, Mr Guy, *France.*
Le Roux, Dr Johannes Hendrik. *South Africa.*
Le Roux, Dr Stephanus David. *South Africa.*
Lea, Prof. Sydney George. *Canada.*
Leadbetter, Prof. Alan James. *UK.*
Leake, Dr John Anthony. *UK.*
Leal-González, Dr Javier. *UK.*
Leban, Prof. Dr Ivan. *Yugoslavia.*
Lebech, Mrs Bente. *Denmark.*
Lebedev, Prof. Vasily Ilyich. *USSR.*
Lebedeva, Dr Marina Vladimirovna. *USSR.*
Lebek, Dr Alexander. *DDR.*
Lebioda, Dr Lukasz. *USA.*
Lechat, Dr Johannes Rudiger. *Brazil.*
Leciejewicz, Prof. Janusz. *Poland.*
Leclaire, Dr André *France.*
Lecomte Dr Claude. *France.*
Ledbetter, Dr Hassel M. *USA.*
Ledesert, Dr Mariannick. *France.*
Lee, Prof. Byungkook. *USA.*
Lee, Dr Chnoong Kheng, *Malaysia.*
Lee, Mrs Florence Lan Fun. *Canada.*
Lee, Mr Han Sik. *USA.*
Lee, Dr John David. *UK.*
Lee, Prof. Wang Chihming. *Taiwan.*
Lee Moreno, Dr José Luis. *Mexico.*
Lees, Mr M.J. *UK.*
Lefebvre, Dr Simone. *France.*
Lefeld-Sosnowska Dr hab. Maria, Stefania. *Poland.*
Leffers, Dr Torben. *Denmark.*
Legrand, Dr Emiel. *Belgium.*
Legros, Dr Jean-Pierre. *France.*
Leguey Gimenez, Prof. Santiago. *Spain.*
Lehmann, Prof. Dr Gerhard Rudolf. *BRD.*
Lehmann, Dr Gottfried. *DDR.*
Lehmann, Mr Günter. *DDR.*
Lehmann, Dr Mogens. *France.*
Lehmpfuhl, Dr Gunter. *BRD.*
Lehtinen, Dr Martti Kalevi. *Finland.*
Leijonmarck, Dr Marie. *Sweden.*
Leipert, Mrs Yvonne. *DDR.*
Leipoldt, Prof. Johann Gotlieb. *South Africa.*
Leiro, Dr Jarkko Albert. *Finland.*
Leiserowitz, Prof. Leslie. *Israel.*
Leite, Dr Cirano Rocha. *Brazil.*
Lele, Prof. Shrikant *India.*
Leligny, Mr Henri. *France.*
Lemke, Mr Guntram. *DDR.*
Lemoine, Mlle Pascale. *France.*
Lenhert, Prof. P. Galen. *USA.*
Lenner, Dr Magnus. *Sweden.*
Lenstra, Prof. Dr Albert Teun Hendrik. *Belgium.*
Lentz, Prof. Paul J. Jr. *USA.*
Léonard, Dr André Jules Gérard. *Belgium.*
Leonardsen, Mr Erik Sverre. *Denmark.*
Leonhardt, Dr Albrecht. *DDR.*
Leonhardt, Dr Gunter. *DDR.*
Leoni, Prof. Leonardo. *Italy.*
Leonowicz, Dr Michael Edward. *USA.*
Leonyuk, Dr Lidiya Ivanovna. *USSR.*
Leonyuk, Dr Nikolay Ivanovich. *USSR.*
Lepicard, Dr Geneviève, *France.*
Leporati, Prof. Enrico. *Italy.*
Leppin, Mrs Christine. *DDR.*
Lerf, Dr Anton Eduard. *BRD.*
Leroy, Dr Bernard. *France.*
Leser, Dr Julio. *Israel.*
Lesk, Prof. Arthur Mallay. *USA.*
Leskelä, Dr Markku Antero. *Finland.*
Leslie, Dr Andrew Greig William. *UK.*
Lessinger, Prof. Leslie. *USA.*
Lessor, Dr Arthur Eugene, Jr. *USA.*
Letort, Mr Marc Yves. *France.*
Leung, Mr Wilhelm Kei Hong. *Hong Kong.*

Leung, Peter C. W. *USA.*
Leusmann, Dr Dietrich Bertold. *BRD.*
Leute, Prof. Dr Volkmar. *BRD.*
Levalois, Mr Marc. *France.*
Levan, Mr Keith R. *USA.*
Levanyuk, Prof. Arkady Petrovich. *USSR.*
Levelut, Dr Anne-Marie. *France.*
Levendis, Mr Demetrius Christos. *South Africa.*
Leventouri, Dr Dora. *Greece.*
Leverett, Dr Peter. *Australia.*
Levi, Dra Laura. *Argentina.*
Levien, Dr Louise. *USA.*
Levoska, Mr Pentti Juhani. *Finland.*
Lévy, Dr Francis. *Switzerland.*
Levy, Dr Henri A. *USA.*
Lewis, Dr D. *UK.*
Lewis, Dr Eric Leslie Vallance. *UK.*
Lewis, Dr James jr. *BRD.*
Lewis, Prof. Sir J. *UK.*
Lewis, Dr Michael Harold. *UK.*
Lewit-Bentley Dr Anita. *France.*
Leyerle, Mr Richard W. *USA.*
LeGeros, Prof. Racquel Z. *USA.*
Li, Mr Da-ming. *China.*
Li, Mr Du. *China.*
Li, Dr Chi-Tang. *USA.*
Li, Mr Run-shen. *China.*
Li, Mr Wan-mao. *China.*
Liang, Prof. Dong-cai. *China.*
Liang, Prof. Jing-kui. *China.*
Liang, Mr Keng-San. *USA.*
Liang, Ms Li. *China.*
Licci, Dr Francesca. *Italy.*
Licheri, Prof. Giovanni. *Italy.*
Licklider, Mr Robert A. *USA.*
Liddington, Mr Robert Colin. *UK.*
Liebau, Prof. Dr Friedrich Karl Franz. *BRD.*
Liebertz, Prof. Dr Josef. *BRD.*
Liebman, Dr Michael N. *USA.*
Liem, Dr D. Hay. *Sweden.*
Lifchitz, Mr Alain. *France.*
Ligenza, Dr Sylwester. *Poland.*
Lihl, Prof. em. Dr Franz. *Austria.*
Lii, Mrs Sue-Lein Wang. *USA.*
Liles, Mr David Charles. *South Africa.*
Lilie, Mr Martin. *DDR.*
Liljas, Dr Anders. *Sweden.*
Liljas, Dr Lars. *Sweden.*
Lim, Dr Valery Irovich. *USSR.*
Lima-de-Faria, Dr José. *Portugal.*
Liminga, Prof. Rune. *Sweden.*
Lin, Mr Cheng-yi. *China.*
Lin, Prof. Chi-chang. *China.*
Lin, Mr Chuan. *China.*
Lin, Mr Guang-da. *China.*
Lin, Dr Hsi-che. *Taiwan.*
Lin, Prof. Szu Bin. *Taiwan.*
Lin, Mr Xian-ti. *China.*
Lin, Mrs Yu-juon. *China.*
Lin, Prof. Zheng-jiong. *China.*
Linares, Dr Jorge. *Peru.*
Lincoln, Dr Francis John. *Australia.*
Lind, Dr Maurice David. *USA.*
Lindahl, Dr Tommie. *Sweden.*
Lindegaard-Andersen, Prof. Asger. *Denmark.*
Lindemann, Prof. Dr Willi. *BRD.*
Lindenmeyer, Dr Paul Henry. *USA.*
Lindgreen, Dr Holger. *Denmark.*
Lindin', Dr Lauma Felixovna. *USSR.*
Lindley, Dr M.W. *UK.*
Lindley, Dr Peter Frank. *UK.*
Lindner, Dr Peter W. *Sweden.*
Lindqvist, Dr. Bengt. *Sweden.*
Lindqvist, Prof. Ingvar. *Sweden.*
Lindqvist, Mr Kristian Vilhelm. *Finland.*
Lindqvist, Prof. Oliver. *Sweden.*
Lindqvist, Dr Ylva Ch. *Sweden.*
Lindroos, Prof. Veikko Kalervo. *Finland.*
Lindström, Dr Rauno. *Finland.*
Lingafelter, Prof. Edward Clay. *USA.*

Linke, Dr Dietmar. *DDR.*
Linke, Dr Walter. *Austria.*
Liopo, Dr Valery Alexandrovich. *USSR.*
Lipka, Mrs Dr Annegret. *BRD.*
Lipkowski, Dr Janusz. *Poland.*
Lippard, Prof. Stephen James. *USA.*
Lippert, Dr Ernest L., Jr. *USA.*
Lippman, Dr Robert. *USA.*
Lipscomb, Prof. William Nunn. *USA.*
Lipson, Prof. Henry, FRS. *UK.*
Liquori, Prof. Alfonso Maria. *Italy.*
Lis, Dr Tadeusz. *Poland.*
Lisoivan, Dr Vladimir Ivanovich. *USSR.*
Little, Prof. Robert Greenwood. *USA.*
Litvin, Dr Alexander Lukich. *USSR.*
Litvin, Dr Boris Nikolayevich. *USSR.*
Litvin, Prof. Daniel Bernard. *USA.*
Litvinskaya, Mrs Galina Petrovna. *USSR.*
Lityagina, Dr Lyudmila Mitrofanovna. *USSR.*
Liu, Mr Guang-zhao. *China.*
Liu, Dr Han-qin. *China.*
Liu, Mr Hung-Yu. *USA.*
Liu, Dr Ling-Kang. *Taiwan.*
Liu, Mr Shi-xiong. *China.*
Liu, Mr Tian-liang. *China.*
Liu, Mr Wan. *China.*
Liu, Mr Xue-lun. *China.*
Liu, Mr Zuo-cai. *China.*
Ljungström, Dr Evert B. *Sweden.*
Llaguno, Dr Elma C. *Philippines.*
Llanos, Dr Jaime. *Chile.*
Llinás Rivera, Prof. Rubén Darío. *Colombia.*
Lloyd, Dr Doug. *Australia.*
Lobenstein, Mrs Heidrun. *DDR.*
Lobkovsky, Dr Emil' Borisovich. *USSR.*
Locchi, Prof. Stelio Giovanni. *Italy.*
Lock, Prof. Colin James Lyne. *Canada.*
Lockhart, Steven H. *USA.*
Loeb, Dr Arthur L. *USA.*
Loeb, Dr Stephen Joseph. *Canada.*
Löchner, Dr Ulrich. *BRD.*
Loehlin, Prof. James Herbert. *USA.*
Löfgren, Dr Percy. *Sweden.*
Löfgren, Dr Tor H. *Sweden.*
Löns, Dr Jürgen. *BRD.*
Lösche, Prof. Artur. *DDR.*
Loeksmanto, Dr Waloejo. *Indonesia.*
Logan, Dr N. *UK.*
Loghry, Dr Ray Allen. *USA.*
Lohar, Dr Jayanarayan Mangaliprasad. *India.*
Loiseleur, Dr Henri. *France.*
Loizos, Mr Zafiris. *Greece.*
Lok, Mr Charles. *USA.*
Lokaj, Ing Ján. *CSSR.*
Lokanatha, Mr S. *India.*
Lokken, Prof. Donald Arthur. *USA.*
Loll, Mr Patrick J. *USA.*
Lombard, Mr Anthonie van Altena. *South Africa.*
Londos, Dr Charalampos. *Greece.*
Long, Dr Gabrielle Gibbs (Cohen). *USA.*
Longueville, Mr Willy. *France.*
Loopstra, Prof. Dr Bert Onno. *Netherlands.*
Lopes-Vieira, Prof. António. *Portugal.*
Lopez Aguayo, Dr Francisco. *Spain.*
Lopez Castro, Prof. Amparo. *Spain.*
Lopez de Lerma, Dr Julian. *Spain.*
Lopez Gonzalez, Prof. Juan de Dios. *Spain.*
Lopez Soler, Dr Angel. *Spain.*
Lorenz, Mr Günter. *BRD.*
Lorenz, Prof. Dr Wolfgang J. *BRD.*
Loreto, Prof. Lucio. *Italy.*
Lorimer, Dr Gordon Winston. *UK.*
Loshmanov, Dr Arkady Andreyevich. *USSR.*
Lotfy, Dr Mohamed. *Egypt.*
Lotz, Dr Simon. *South Africa.*
Lou, Mrs Mei-zhen. *China.*
Loub, Dr Josef. *CSSR.*
Louis, Dr Remy. *France.*
Loupias, Dr Geneviève. *France.*
Lovas, Dr György Antal. *Hungary.*

Love, Prof. Warner Edwards. *USA.*
Lovell, Dr Frederick Maurice. *USA.*
Lovell, Mrs S.E. *UK.*
Lovey, Dr Francisco Carlos. *Argentina.*
Low, Prof. Barbara Wharton. *USA.*
Low, Dr John Nicolson. *UK.*
Low, Prof. William. *Israel.*
Lowde, Dr Raymond Douglas. *UK.*
Lowe, Dr Philip Richard. *UK.*
Lowe, Miss Susan Elizabeth. *UK.*
Lowe-Ma, Dr Charlotte Kathryn. *USA.*
Lozano, Prof. José A. *Colombia.*
Lu, Prof. Jia-xi. *China.*
Lu, Mr Kun-quan. *China.*
Lu, Ms Qi. *China.*
Lu, Mrs Quang-ying. *China.*
Lu, Mr Shao-fang. *China.*
Lu, Dr Tian-Huey. *Taiwan.*
Lu, Prof. Yun-jin. *China.*
Lublin, Mr Paul. *USA.*
Lucas, Dr Brian William. *Australia.*
Lucas, Prof. Jacques. *France.*
Ludi, Prof. Andreas. *Switzerland.*
Ludwiczek, Dr Herbert. *Austria.*
Ludwig, Prof. Martha L. *USA.*
Lüth, Dr Hartwig. *Canada.*
Luger, Prof. Dr Peter. *BRD.*
Lugomer, Dr Stjepan. *Yugoslavia.*
Lugt, van der, Dr Willem. *Netherlands.*
Luić, Magistar Marija. *Yugoslavia.*
Lukashev, Dr Alexander Nikolayevich. *USSR.*
Lukaszewicz, Prof. Dr Kazimierz. *Poland.*
Lukaszewski, Dr George Michael. *Australia.*
Lumme, Prof. Paavo Olavi. *Finland.*
Luna, Dr Carlos Alfonso. *Colombia.*
Lundberg, Dr Bruno. *Sweden.*
Lundberg, Dr Monica. *Sweden.*
Lundgren, Dr Jan-Olof. *Sweden.*
Lundgren, Mr Lennart. *Sweden.*
Lundström, Dr Torsten. *Sweden.*
Luo, Prof. Gu-feng. *China.*
Luo, Mr Yao Guang. *Canada.*
Luss, Mr Henry Richard. *USA.*
Lustig, Mr Stanley. *USA.*
Lutz, Mr Dieter. *DDR.*
Lux, Mr Bernd. *DDR.*
Lux, Dr Georg. *DDR.*
Luzzati, Dr Vittorio. *France.*
Lvov, Dr Yury Mikhailovich. *USSR.*
Lyakhovitskaya, Dr Vera Aronovna. *USSR.*
Lydon, Dr John Ennis. *UK.*
Lynch, Dr Denis Francis. *Australia.*
Lyons, Dr Karen. *New Zealand.*
Lyons, Dr Michael Hamilton. *UK.*
Lyons, Mr Paul John. *New Zealand.*
Lyubalin, Dr Mark Dmitriyevich. *USSR.*
Lyubimov, Dr Vasily Nikolayevich. *USSR.*
Lyubitov, Dr Yury Naumovich. *USSR.*
Lyubutin, Dr Igor' Savelyevich. *USSR.*
Lyutin, Dr Vladimir Ivanovich. *USSR.*
Lyutzau, Dr Vsevolod Grigoryevich. *USSR.*
Lyxell, Mr Dan-Göran. *Sweden.*
Ma, Dr Che-bao. *Taiwan.*
Ma, Prof. Li-dun. *China.*
Ma, Dr Lilian Yan Yan. *Canada.*
Ma, Mr Xing-qi. *China.*
Ma, Prof. Zhe-sheng. *China.*
Maaref, Dr. Saida. *Tunisia.*
Maartman-Moe, Mr Knut. *Norway.*
Maaskant, Prof. Dr Willem J.A. *Netherlands.*
Macía Sanabria, Prof. Carlos A. *Colombia.*
Macdonald, Mr George Leslie. *UK.*
Macek, Dr Josef Jan. *Canada.*
Macgillavry, Prof. Dr Carolina H. *Netherlands.*
Machajdík, Ing Daniel. *CSSR.*
Machin, Mr Ken James. *Australia.*
Machin, Ms Penelope Anne. *UK.*
Maciček, Dr Josef. *Bulgaria.*
Maciosowski, Dr Andrzej. *Poland.*
Mackay, Dr Alan Lindsay. *UK.*
Mackay, Dr Maureen Florence. *Australia.*

Mackenzie, Dr James Kenneth. *Australia.*
Mackie, Dr Paul E. *USA.*
Mackinnon, Dr Ian Donald. *USA.*
MacCrone, Prof. Robert Kirsten. *USA.*
MacKenzie, Prof. William Scott. *UK.*
MacRae, Dr Alfred U. *USA.*
MacRae, Mr Thomas Perry. *Australia.*
Madden, Prof. John Joseph. *USA.*
Madhusudana, Dr N. V. *India.*
Madureira Filho, Prof. José Barbosa de. *Brazil.*
Maďar, Doc. Dr Ján. *CSSR.*
Mäki, Mr Jouko Kalervo. *Finland.*
Maenhout - Van Der Vorst, Dr Mrs W.M.R. *Belgium.*
Mages, Dr Gert Rudolf. *BRD.*
Magini, Dr Mauro. *Italy.*
Magnéli, Prof. Arne. *Sweden.*
Magnus, Dr Karen A. *USA.*
Magnuson, Prof. Vincent Richard. *USA.*
Magnusson, Dr Bo. *Sweden.*
Mahanta, Dr Bhubaneswar. *India.*
Mahar, Mr Martin Cajetan. *USA.*
Mahata, Dr Akhil *India.*
Mahendrasingham, Mr A. *UK.*
Mahmood, Mr Khursheed. *Pakistan.*
Mahmoud, Dr Mouyed Mohamed. *Iraq.*
Mahr von Staszewski, Dr Guillermo. *Argentina.*
Mahy, Mr Jan Willem Gaston. *Belgium.*
Mai, Prof. Zhen-hong. *China.*
Main, Dr Peter. *UK.*
Mainegra, Mr Virgilio. *Cuba.*
Mair, Dr Sylvia Lorraine. *Australia.*
Maiti, Dr Gobinda Chandra. *India.*
Maiyer, Prof. Alexander Artemyevich. *USSR.*
Maiza, Dr Pedro José. *Argentina.*
Majchrzak, Dr Stanisław. *Poland.*
Majeste, Mr Richard J. *USA.*
Majling, Ing Ján. *CSSR.*
Majumdar, Mr Kanti Lal. *India.*
Majumdar, Dr Sunil Kumar. *India.*
Mak, Prof. Thomas Chung-wai. *Hong Kong.*
Makarenko, Dr Igor' Nikolayevich. *USSR.*
Makarov, Prof. Evgeny Sergeyevich. *USSR.*
Makovicky, Dr Emil. *Denmark.*
Makowski, Prof. Lee. *USA.*
Maksić, Prof. Dr Zvonimir. *Yugoslavia.*
Malý, Mr Karel. *CSSR.*
Malagón Castro, Prof. Dimas. *Colombia.*
Malakhova, Dr Lyudmila Fedorovna. *USSR.*
Maleev, Dr Michael. *Bulgaria.*
Malgrange, Prof. Cécile. *France.*
Malicskó, Dr László. *Hungary.*
Malik, Alpana. *India.*
Malin, Dr Anthony Samuel. *Australia.*
Malinenko, Dr Inna Avramovna. *USSR.*
Malinovsky, Dr Stanislav Tadeushevich. *USSR.*
Malinovsky, Prof. Tadeush Iosifovich. *USSR.*
Malinovsky, Dr Yury Alexandrovich. *USSR.*
Malinowski, Dr Mariusz. *Poland.*
Malinowski, Prof. Yordan. *Bulgaria.*
Maliszewski, Dr Edward. *Poland.*
Mallari-Kaballo, Mrs Paz P. *Philippines.*
Malley, Ms Mary F. *USA.*
Mallinson, Dr Paul Raymond. *UK.*
Mallory, Dr Chester L. *USA.*
Malone, Dr John Francis. *UK.*
Malpezzi, Dr Luciana. *Italy.*
Malta, Dr Viscardo. *Italy.*
Maltz, Dr Abraham. *Israel.*
Maluszynska, Dr Hanna. *Poland.*
Malyushitskaya, Dr Zinaida Vladimirovna. *USSR.*
Mamedov, Prof. Kerim Panakh ogly. *USSR.*
Mamedov, Prof. Khudu Surkhay ogly. *USSR.*
Mammi, Prof. Mario. *Italy.*
Man, Dr Lucia Ivanovna. *USSR.*
Mănăilă, Mrs Rodica. *Romania.*
Manassero, Prof. Mario. *Italy.*
Mancini, Dr Annamaria. *Italy.*
Mandarino, Dr Joseph Anthony. *Canada.*
Mande, Prof. Chintamani. *India.*
Mande, Mr Sekhar Chintamani. *India.*

Mandel, Dr Gretchen Sue. *USA.*
Mandel, Dr Neil Stanley. *USA.*
Mandelkow, Dr Eckard. *BRD.*
Mánek, Ing Břetislav. *CSSR.*
Mangani, Dr Stefano. *Italy.*
Manghi, Lic Estela Margarita. *Argentina.*
Mangia, Prof. Alessandro. *Italy.*
Mani, Mr A. *India.*
Manickkavachgam, Ramanathan. *India.*
Mannami, Dr Michi-hiko. *Japan.*
Mannan, Prof. Dr Kh. A. I. F. Mafizul. *Bangladesh.*
Manninen, Dr Seppo Olavi. *Finland.*
Manohar, Prof. Hattikudur. *India.*
Manojlović-Muir, Dr Ljubica. *UK.*
Manolikas, Prof. Konstantinos. *Greece.*
Manor, Dr Philip C. *USA.*
Manotti-Lanfredi, Prof. Anna Maria. *Italy.*
Manriquez, Dr Víctor. *Chile.*
Mansikka, Prof. Kauko Antti. *Finland.*
Mansilla-Koblavi, Mrs Frédérica G.M.S. *Ivory Coast.*
Mansour, Mr Saber Moustapha. *Egypt.*
Mantovani, Prof. Giorgio. *Italy.*
Manutchehr-Danai, Dr Mohsen. *Iran.*
Manzoor-I-Khuda, Dr Muhammad. *Bangladesh.*
Maröy, Dr Kjartan. *Norway.*
Marbec, Lic Ema Rosa. *Argentina.*
March, Dr Frank Conroy. *New Zealand.*
Marchessault, Dr Robert H. *Canada.*
Marchetti, Dr Fabio. *Italy.*
Marcoen, Dr Jean-Marie. *Belgium.*
Marezio, Dr Massimo. *France.*
Marfunin, Dr Arnol'd Sergeyevich. *USSR.*
Margulis, Prof. Thomas N. *USA.*
Marians, Ms Carol. *USA.*
Marie, Dr Alain Louis. *BRD.*
Mariezcurrena, Dr Raul Alfredo. *Uruguay.*
Marigo, Prof. Antonio. *Italy.*
Marinder, Dr Bengt-Olov. *Sweden.*
Marinković, Prof. Dr Velibor. *Yugoslavia.*
Marinković, Magistar (Mrs) Živka. *Yugoslavia.*
Marinov, Dr Miko. *Bulgaria.*
Mariolacos, Dr Konstantin. *BRD.*
Marion, Mr Martin P. *USA.*
Markali, Mr Joar. *Norway.*
Markgraf, Steven A. *USA.*
Markov, Dr Ivan. *Bulgaria.*
Marković, Magistar Berislav. *Yugoslavia.*
Marković, Magistar Desimir. *Yugoslavia.*
Markwell, Dr Anthony James. *South Africa.*
Marongiu, Prof. Giaime. *Italy.*
Marquez, Prof. Rafael. *Spain.*
Marsau, Dr Pierre Michel. *France.*
Marsh, Mr Philip. *USA.*
Marsh, Dr Richard Edward. *USA.*
Marsongkohadi, Mr. *Indonesia.*
Marti, Dr Jaime. *Spain.*
Martikainen, Mr Hannu Olavi. *Finland.*
Martin, Mr Bruce Alan. *USA.*
Martin, Mr George Wm. *USA.*
Martin, Dr John Wilson. *UK.*
Martin, Miss Lillian Ruth. *Canada.*
Martin, Dr Reinhold. *BRD.*
Martin Pozas, Prof. José Ma. *Spain.*
Martinelli, Prof. Giuliano. *Italy.*
Martinez Carrera, Prof. Sagrario. *Spain.*
Martinez Garcia, Mrs Ma. Luisa. *Spain.*
Martinez Ripoll, Dr Martin. *Spain.*
Martinsen, Mr Knut. *Norway.*
Martorana, Dr Antonino. *Italy.*
Martyshev, Dr Yury Nikolayevich. *USSR.*
Maruha, Prof. Juro. *Japan.*
Marukawa, Dr Kenzaburo. *Japan.*
Marumo, Prof. Fumiyuki. *Japan.*
Maruno, Dr Shigeo. *Japan.*
Maruse, Prof. Susumu. *Japan.*
Maruyama, Prof. Saiyu. *Japan.*
Marx, Prof. Günter. *DDR.*
Mas, Dr Graciela Raquel. *Argentina.*
Masaki, Dr Norio. *Japan.*
Masakuni, Dr Mayumi. *Japan.*

Mascarenhas, Prof. Yvonne Primerano. *Brazil.*
Mascarenhas, Prof. Sergio. *Brazil.*
Masche, Mr Wolfgang. *DDR.*
Mašić, Magistar Nikola. *Yugoslavia.*
Maske, Prof. Siegfried. *South Africa.*
Maslen, Dr Edward Norman. *Australia.*
Maslen, Dr Hugh Stafford. *New Zealand.*
Maslowska, Miss Maria. *BRD.*
Mason, Mr John T. *USA.*
Mason, Dr Kenneth George. *UK.*
Mason, Prof. Paul Robert. *USA.*
Mason, Prof. Sir Ronald, FRS. *UK.*
Mason, Dr Sax Anton. *Grenoble.*
Mason, Prof. Stephen Finney. *UK.*
Massa, Mr Louis. *USA.*
Massa, Dr Werner. *BRD.*
Massalski, Prof. Thaddeus B. *USA.*
Massarotti, Prof. Vincenzo. *Italy.*
Massaux, Dr Michel Louis. *France.*
Mastacan, Prof. Gheorghe. *Romania.*
Mastropaolo, Mr Donald. *USA.*
Mastryukov, Dr Vladimir Saidovich. *USSR.*
Mateika, Dr Dieter. *BRD.*
Mateu, Dr Leonardo. *Venezuela.*
Matherny, Prof. Dr-Ing Mikuláš. *CSSR.*
Mathew, Mr M. *USA.*
Mathews, Prof. F. Scott. *USA.*
Mathieson, Dr Alexander McLeod. *Australia.*
Mathur, Dr Balbir Kumar. *India.*
Mathur, Dr Rajendra Kumar. *India.*
Matijašić, Dr (Mrs) Ivanka. *Yugoslavia.*
Matković, Dr Boris. *Yugoslavia.*
Matković, Dr Prosper. *Yugoslavia.*
Matković, Dr Tanja. *Yugoslavia.*
Matković-Čalogović, Mgr (Mrs) D. *Yugoslavia.*
Matkovsky, Dr Orest Ilyarovich. *USSR.*
Matos Beja, Ms Ana Maria. *Portugal.*
Matsubara, Dr Ikuo. *Japan.*
Matsubara, Prof. Takeo. *Japan.*
Matsuda, Prof. Hidehiko. *Japan.*
Matsui, Dr Masanori. *Japan.*
Matsui, Mr Toshiro. *Japan.*
Matsumoto, Dr Takeo. *Japan.*
Matsuo, Dr Munetsugu. *Japan.*
Matsusaki, Dr Hideo. *Japan.*
Matsushita, Dr Tadashi. *Japan.*
Matsuura, Dr Yoshiki. *Japan.*
Matsuzaki, Mr Takao. *Japan.*
Mattes, Prof. Dr Rainer. *BRD.*
Matthew, Dr James Andrew D. *UK.*
Matthews, Prof. Brian W. *USA.*
Matthews, Dr David Allan. *USA.*
Matthews, Prof. Frederick White. *Canada.*
Matthys, Dr Paul Frederik André Edmond. *Belgium.*
Mattia, Prof. Carlo. *Italy.*
Mattias, Prof. Pierpaolo. *Italy.*
Matveeva, Mrs Rimma Georgiyevna. *USSR.*
Matyi, Mr Richard J. *USA.*
Matyja, Dr Przemysław. *Poland.*
Matz, Prof. Dr Günther. *BRD.*
Matzat, Dr Eckhart. *BRD.*
Mauer, Mr Floyd Andrew. *USA.*
Maung, Mr N. *UK.*
Maurer, Prof. Enrique. *Spain.*
Maverick, Prof. Emily. *USA.*
Mavlonov, Dr Sharaf. *USSR.*
Mavridis, Prof. Aristides. *Greece.*
Maximov, Dr Boris Alekseyevich. *USSR.*
Maxwell, Prof. George. *Canada.*
May, Prof. Martin. *DDR.*
Mayer, Dr Dipl.-Ing. Helmut. *Austria.*
Mayer, Dr Hugo Werner Waldemar. *BRD.*
Mayer, Prof. Itzchak. *Israel.*
Mayerle, Mr James J. *USA.*
Mayo, Mr William Edward. *USA.*
Mayorzik, Mrs Hemda. *Israel.*
Mayr, Dr Dipl.-Ing. Michael. *Austria.*
Mazany, Dr Anthony Michael. *USA.*
Mazey, Dr David John. *UK.*
Mazus, Dr Mark Davidovich. *USSR.*
Mazza, Dr Fernando. *Italy.*

Mazzarella, Prof. Lelio. *Italy.*
Mazzaro, Mr Irineu. *Brazil.*
Mazzi, Prof. Fiorenzo. *Italy.*
McAdam, Dr A. *UK.*
McAlea, Mr Kevin P. *USA.*
McAlister, Dr John Paul. *USA.*
McAllister, Mr Patrick Brian. *UK.*
McArdle, Dr Patrick. *Ireland.*
McAtee, Prof. James L., Jr. *USA.*
McCall, Dr Maxine June. *UK.*
McCallum, Prof. Malcolm Ernest. *USA.*
McCarthy, Miss A.E. *UK.*
McCarthy, Prof. Gregory Joseph. *USA.*
McCauley, Dr James Weymann. *USA.*
McClure, Dr Richard James. *USA.*
McCollor, Mr Donald P. *USA.*
McConnell, Prof. Duncan. *USA.*
McConnell, Dr Jack Foster. *Australia.*
McCormick, Ms Robyn Joy. *Australia.*
McCree, Duncan E. *USA.*
McCrone, Mrs Lucy B. *USA.*
McCrone, Dr Walter C. *USA.*
McCusker, Dr Lynne Bridget. *USA.*
McDonald, Dr Walter Stanley. *UK.*
McDougall, Dr Gloria Jeanne. *South Africa.*
McDougall, Dr Peter George. *Australia.*
McEwen, Mr A.B. *UK.*
McFarlane, Dr Samuel H., III. *USA.*
McGuire, Ms Nancy K. *USA.*
McHardy, Dr William James. *UK.*
McKay, Dr David Bruce. *USA.*
McKee, Dr Rodney Allen. *USA.*
McKenzie, Dr David Robert. *Australia.*
McKenzie, Dr Elwyn Donald. *Australia.*
McKenzie, Dr Thomas Charles. *USA.*
McKeown, Dr David Alexander. *USA.*
McKie, Dr Christine Hilary. *UK.*
McKie, Dr Duncan. *UK.*
McKinnon, Miss F.J. *UK.*
McKinstry, Prof. Herbert Alden. *USA.*
McLachlan, Prof. Dan, Jr. *USA.*
McLaren, Dr Alexander Clark. *Australia.*
McLaren, Prof. Eugene Herbert. *USA.*
McLaughlin, Mr George Millar. *Australia.*
McLean, Dr W. John. *USA.*
McLeod, Mr Neil John. *Australia.*
McMillan, Dr Joyce A. *USA.*
McMullan, Dr Richard K. *USA.*
McMurdie, Mr Howard Francis. *USA.*
McPartlin, Dr Mary. *UK.*
McPhail, Prof. Andrew Tennent. *USA.*
McPherson, Prof. Alexander. *USA.*
McPherson, Mr William G. *USA.*
Meader, Dr D. *UK.*
Meagher, Dr Edward Patrick. *Canada.*
Mealli, Prof. Carlo. *Italy.*
Mechlinski, Mr Witold. *USA.*
Medeiros Rodrigues, Dr Maria Mabel. *Brazil.*
Medgyaszay, Dr Márton. *Hungary.*
Médicis, de, Dr Rinaldo M. *Canada.*
Međimorec, Magistar Stanislav. *Yugoslavia.*
Medlin, Dr Edwin Harry. *Australia.*
Medrud, Dr Ronald Curtis. *USA.*
Meehan, Prof. Edward Joseph, Jr. *USA.*
Meetsma, Drs Auke. *Netherlands.*
Megaw, Dr Helen Dick. *UK.*
Meibohm, Dr Edgar Paul Hubert. *USA.*
Meier, Prof. Dr Hans. *BRD.*
Meier, Prof. Walter M. *Switzerland.*
Meinnel, Prof. Jean. *France.*
Meisalo, Prof. Veijo Pauli Juhani. *Finland.*
Meisel, Prof. Armin. *DDR.*
Mejía Cifuentes, Prof. Leonidas. *Colombia.*
Mel'nikov, Dr Oleg Konstantinovich. *USSR.*
Mel'nikov, Dr Vitaly Alexandrovich. *USSR.*
Mel'nikova, Dr Alina Mikhailovna. *USSR.*
Melamud, Mr Mordechai. *Israel.*
Melekh, Dr Bernard Abu-Talibovich. *USSR.*
Meleshina, Dr Valentina Alexandrovna. *USSR.*
Melik-Adamyan, Dr Vil'yam Rafailovich. *USSR.*

Melka, Dr Karel. *CSSR.*
Mellini, Dr Marcello. *Italy.*
Mellon, Mr T.P.A. *UK.*
Mellor, Mr John. *USA.*
Menchetti, Prof. Silvio. *Italy.*
Mencik, Dr Zdenek. *USA.*
Menczel, Dr György. *Hungary.*
Mende, Prof. Dr Hans Horst. *BRD.*
Mendelssohn, Dr Monica Jutta. *UK.*
Mendiola Diaz, Dr Jesus. *Spain.*
Meng, Prof. Yi-min. *China.*
Menzinger, Prof. Filippo. *Italy.*
Mercatini, Dr Giovanna. *Uruguay.*
Mercer, Dr William Duncan. *UK.*
Mereiter, Dr Kurt. *Austria.*
Meresse, Dr Alain. *France.*
Meriani, Dr Sergio. *Italy.*
Mérigoux, Prof. Henri. *France.*
Merinev, Dr Boris Vladimirevich. *USSR.*
Merino de Matheus, Prof. Lucía Marina. *Colombia.*
Merisalo, Dr Matti Juhani. *Finland.*
Merlini, Dr Alfonso Enrico. *Italy.*
Merlino, Prof. Stefano. *Italy.*
Merlo, Prof. Franco. *Italy.*
Merriman, Mr Richard James. *UK.*
Merritt, Dr Ethan Allen. *USA.*
Merritt, Prof. Lynne Lionel, Jr. *USA.*
Mertes, Prof. Kristin Susan Bowman. *USA.*
Mertin, Dr Wilhelm. *BRD.*
Messager, Dr Jean-Claude. *France.*
Messick, Mr Julian. *USA.*
Messner, Dr Dieter. *BRD.*
Metcalfe, Dr Edward. *UK.*
Metsik, Prof. Mikhail Stepanovich. *USSR.*
Metter, Mr Joachim. *BRD.*
Metz, Dr Bernard Jean Claude. *France.*
Metze, Mr Dieter. *DDR.*
Metzger, Prof. Robert Melville. *USA.*
Meunier - Piret, Dr Jacqueline. *Belgium.*
Meurs, van, Dr Ir Frank. *Netherlands.*
Mewis, Dr Albrecht. *BRD.*
Meyer, Mr Andreas. *BRD.*
Meyer, Prof. Edgar Frederich. *USA.*
Meyer, Prof. Frank Henry. *USA.*
Meyer, Dr Gerd Heinrich. *BRD.*
Meyer, Prof. Dr Hans-Jürgen. *BRD.*
Meyer, Prof. Klaus. *DDR.*
Meyer-Ehmsen, Prof. Dr Gerhard. *BRD.*
Meyers, Dr Bernard Lee. *USA.*
Meyers, Prof. Edward Arthur. *USA.*
Mez, Dr Hans-Christian. *Switzerland.*
Mhiri, Dr Tahar. *Tunisia.*
Mian, Dr Mohammad Ashraf. *Pakistan.*
Mian, Mr Muhammad Asghar. *Pakistan.*
Miao, Mr Chun-sheng. *China.*
Miao, Prof. Fang-ming. *China.*
Michailov, Mr Evgeni. *Bulgaria.*
Michailov, Mr Michail. *Bulgaria.*
Micheelsen, Prof. Harry. *Denmark.*
Michel, Prof. André Gustave. *Canada.*
Michel, Dr Bernd. *DDR.*
Michel, Dr David John. *USA.*
Michel, Prof. Pierre. *France.*
Michell, Mr Ernest William John. *UK.*
Middlemiss, Dr Nora E. *Canada.*
Middleton, Dr Andrew Philip. *UK.*
Mighell, Dr Alan D. *USA.*
Miguel Alonso, Prof. Santiago. *Spain.*
Mihama, Prof. Kazuhiro. *Japan.*
Mihichuk, Prof. Lynn Michael. *Canada.*
Miida, Mr Rokuro. *Japan.*
Mijlhoff, Dr Frans Cornelis. *Netherlands.*
Mikenda, Dr Werner. *Austria.*
Mikhail, Dr Ibrahim Fahmy. *BRD.*
Mikhailov, Dr Al'bert Mikhailovich. *USSR.*
Mikhailov, Dr Igor' Fedorovich. *USSR.*
Mikhailov, Dr Vladimir Ivanovich. *USSR.*
Mikhailov, Dr Yuriy Nikolayevich. *USSR.*
Mikhalenko, Dr Svetlana Ivanovna. *USSR.*
Mikheeva, Dr Irina Victorovna. *USSR.*

Miki, Dr Kunio. *Japan.*
Mikkola, Prof. Donald Emil. *USA.*
Mikler, Dr Helga. *Austria.*
Mikloš, Ing Dušan. *CSSR.*
Milat, Magistar Ognjen. *Yugoslavia.*
Milberg, Dr Morton Edwin. *USA.*
Milburn, Dr George Henry William. *UK.*
Milchev, Dr Alexander. *Bulgaria.*
Milchev, Dr Andrei. *Bulgaria.*
Mildner, Mr David F. R. *USA.*
Miles, Mr J.A.C. *UK.*
Miles, Miss J.M. *UK.*
Miles, Dr M.J. *UK.*
Milićev, Prof. Dr Svetozar. *Yugoslavia.*
Milillo, Prof. Frank. *USA.*
Milinski, Prof. Dr Nikola. *Yugoslavia.*
Miliotis, Prof. Demitrios Menelaos. *Greece.*
Milius, Mr Wolfgang. *BRD.*
Millane, Mr R. P. *USA.*
Millar, Dr John Joseph. *Australia.*
Milledge, Dr H. Judith. *UK.*
Miller, Prof. Andrew. *UK.*
Miller, Prof. Donald P. *USA.*
Miller, Dr Richard Wayne. *USA.*
Miller, Prof. Robert Llewellyn. *USA.*
Miller, Ms Sarah Ann. *Australia.*
Millhouse, Dr Arthur Holmes. *BRD.*
Millionova, Dr Margarita Ivanovna. *USSR.*
Mills, Mr Owen S. *UK.*
Millward, Dr G.R. *UK.*
Miloshev, Prof. Georgi. *Bulgaria.*
Min, Miss Eungi. *Japan.*
Minacheva, Dr Lidiya Khabibovna. *USSR.*
Minagawa, Dr Teruaki. *Japan.*
Minari, Prof. Fernand Henri. *France.*
Minato, Prof. Hideo. *Japan.*
Mincheva-Stefanova, Prof. Yordanka. *Bulgaria.*
Mineeva, Dr Rimma Mikhailovna. *USSR.*
Minkin, Mrs Jean Albert. *USA.*
Minkoff, Prof. Isaac. *Israel.*
Minni, Dr Erkki Esa Kalervo. *Finland.*
Minomura, Prof. Shigeru. *Japan.*
Mints, Prof. Rafail Isaakovich. *USSR.*
Mirčeva, Magistar Aneta. *Yugoslavia.*
Mir, Mr Jan Mohammad. *Pakistan.*
Miravitlles, Prof. Carlos. *Spain.*
Mirsky, Dr Kira. *USA.*
Mirzu-Ghergariu, Mrs Lucretia. *Romania.*
Misra, Dr Nirmal Kumar. *India.*
Misra, Dr Tripurari. *India.*
Misra, Prof. Somnath. *India.*
Mitcham, Mr Donald. *USA.*
Mitchell, Dr Crighton Maurice. *Canada.*
Mitchell, Prof. Donald J. *USA.*
Mitchell, Dr Keith A. R. *Canada.*
Mitchell, Prof. Richard Scott. *USA.*
Mitra, Prof. Girija Bhushan. *India.*
Mitrprachachon, Dr Pachanee. *Thailand.*
Mitsuda, Dr Hiromichi. *Japan.*
Mitsuda, Dr Takeshi. *Japan.*
Mitsui, Prof. Toshio. *Japan.*
Mitsui, Dr Yukio. *Japan.*
Mitsuishi, Prof. Tomokuni. *Japan.*
Miura, Mr Naoki. *Japan.*
Miura, Dr Yasuhiro. *Japan.*
Miura, Dr Yasunori. *Japan.*
Miuskov, Dr Vasily Fedorovich. *USSR.*
Miyaji, Mr Hirofumi. *Japan.*
Miyake, Prof. Shizuo. *Japan.*
Miyake, Dr Yasuhiro. *Japan.*
Miyamae, Dr Hiroshi. *Japan.*
Miyamoto, Mr Masamichi. *Japan.*
Miyata, Dr Takeshi. *Japan.*
Miyazawa, Dr Shintaro. *Japan.*
Miyoshi, Dr Tadahiko. *Japan.*
Mizera, Dr Elżbieta. *Poland.*
Mizota, Mr Tadato. *Japan.*
Mizuno, Dr Hiroshi. *Japan.*
Mizuno, Prof. Joji. *Japan.*
Mladeck, Dr Micael Hiorth. *Norway.*

Mlik, Dr Youssef. *Tunisia.*
Mo, Prof. Frode. *Norway.*
Model', Dr Mariya Samuilovna. *USSR.*
Modjtahedi, Dr Mansour. *Iran.*
Modrick, Ms Michelle A. *USA.*
Modrzejewski, Dr hab. Antoni. *Poland.*
Moeckli, Dr Pedro. *Switzerland.*
Möhling, Dr Werner. *DDR.*
Møller, Dr Christian Knakkergaard. *Denmark.*
Möller, Dr Manfred. *BRD.*
Moffat, Prof. John Keith. *USA.*
Moguš-Milanković, Magistar Andrea. *Yugoslavia.*
Moh, Prof. Dr Günter Harald. *BRD.*
Mohamad, Dr Hamzah. *Malaysia.*
Mohan Rao, Mr Vattipalli. *India.*
Mohana-Rao, J. K. *USA.*
Mohanlal, Dr Sembu Krishnaiyer. *India.*
Mohr, Dr Ulrich. *DDR.*
Mohyla, Dr Jury. *Australia.*
Moineau, Mr Hervé. *France.*
Moini, Mr Ahmad. *USA.*
Mokeeva, Dr Valentina Ivanovna. *USSR.*
Mokren, Dr James David. *USA.*
Molchanov, Dr Vladimir Nikolayevich. *USSR.*
Moldovanova, Prof. Maria. *Bulgaria.*
Mole, Mrs J. *UK.*
Molea, Mr Frank N. *USA.*
Molin, Prof. Gianmario. *Italy.*
Molinaro, Mr Frank. *USA.*
Monaco, Dr Hugo Luis. *Italy.*
Monari, Dr Magda. *Italy.*
Moncrief, Prof. J. William. *USA.*
Monge Bravo, Mrs Ma. Angeles. *Spain.*
Mongiorgi, Prof. Romano. *Italy.*
Monier, Prof. Jean-Claude. *France.*
Monroe, Prof. Eugene A. *USA.*
Montenegro de Andrade, Prof. Miguel. *Portugal.*
Montenero, Prof. Angelo. *Italy.*
Montfort, Mr William R. *USA.*
Montgomery, Prof. (Em.) Henry. *Canada.*
Montmory, Mrs Marie-Claire. *France.*
Montmory, Dr Robert. *France.*
Montoriol Pous, Prof. Joaquin. *Spain.*
Moodie, Dr Alexander Forbes. *Australia.*
Moon, Dr Anthony Ronald. *Australia.*
Moor, Dr Robert. *Switzerland.*
Moore, Dr Alan James William. *Australia.*
Moore, Dr Alan Charles. *South Africa.*
Moore, Miss Alice Elizabeth. *UK.*
Moore, Dr Anthony Moreton. *UK.*
Moore, Mr Christopher James. *Australia.*
Moore, Dr John Carlton. *UK.*
Moore, Dr Francis Hugh. *Australia.*
Moore, Miss Madeleine. *South Africa.*
Moore, Mr Peter Leonard. *UK.*
Moore, Mr Donald L. *USA.*
Moore, Mrs Elizabeth J. Weichel. *USA.*
Moore, Ms Janet Finer. *USA.*
Moore, Prof. Paul Brian. *USA.*
Mooser, Prof. Emanuel. *Switzerland.*
Mootz, Prof. Dr Dietrich. *BRD.*
Mora de González, Prof. Nery. *Colombia.*
Moraga, Mr Luís. *Chile.*
Morandi, Prof. Noris. *Italy.*
Moras, Dr Denis *France.*
Moravcová, Dr Hana. *CSSR.*
Moravec, Ing František. *CSSR.*
Moreau, Prof. Jean-Michel. *France.*
Moreau, Prof. Jules Francois. *Belgium.*
Moreiras Blanco, Dr Damaso. *Spain.*
Moreland, Dr James Andrew. *USA.*
Morelli, Prof. Gianluca. *Italy.*
Moreno Echevarria, Dra Esperanza. *Spain.*
Morffew, Dr Andrew James. *UK.*
Morgan, Dr Colin Harris. *UK.*
Morgan, Dr Joseph. *USA.*
Morgan, Mr Richard S. *USA.*
Mori, Dr Saburo. *Japan.*
Moriarty, Dr John Lawrence, Jr. *USA.*
Morikawa, Dr Hideki. *Japan.*

Morikawa, Dr Hiroshi. *Japan.*
Morimoto, Dr Carl Noboru. *USA.*
Morimoto, Dr Jun. *Japan.*
Morimoto, Prof. Nobuo. *Japan.*
Morino, Prof. Yonezo. *Japan.*
Morita, Prof. Takeo. *Japan.*
Moritz, Dr Wolfgang Otto. *BRD.*
Morlon, Mr Bernard. *France.*
Mornon, Dr Jean-Paul. *France.*
Morosin, Dr Bruno. *USA.*
Morosoff, Dr Nicholas C. *USA.*
Moroz, Dr Ella Mikhailovna. *USSR.*
Morris, Dr Donald Frank Charles. *UK.*
Morris, Mrs Marlene Cook. *USA.*
Morrison, Mr N.S. *UK.*
Morrow, Prof. John Charles, III. *USA.*
Morrow, Dr Scott Imlay. *USA.*
Mortier, Dr Wilfried Jozef. *Belgium.*
Morton Dr Allan James. *Australia.*
Morvaj, Magistar (Mrs) Jasmina. *Yugoslavia.*
Moseley, Dr Patrick. *UK.*
Moskowitz, Mr Ronald. *USA.*
Moskvin, Dr Valentin Vasilyevich. *USSR.*
Moss, Dr Barbara Kay. *Australia.*
Moss, Dr David Stanley. *UK.*
Moss, Dr Grant Richard. *Australia.*
Mosset, Dr Alain. *France.*
Mostad, Mr Arvid. *Norway.*
Motegi, Dr Hiroshi. *Japan.*
Motherwell, Dr William D.S. *UK.*
Mothes, Mr Heinrich. *DDR.*
Motta, Dr Nunzio. *Italy.*
Moudy, Miss Lavada Ann. *USA.*
Mourikis, Dr Stamatios. *Greece.*
Moustakali Mavridis, Dr Irene. *Greece.*
Mowbray, Ms Sherry Lynn. *USA.*
Mozzi, Dr Robert Lewis. *USA.*
Mrose, Miss Mary E. *USA.*
Mu, Mr Xiang-qi. *China.*
Mucha, Mrs Christine. *DDR.*
Mucker, Prof. Kenneth. *USA.*
Mudd, Mr K.R. *UK.*
Muddle, Dr Barrington Charles. *Australia.*
Mühlberg, Mr Manfred. *DDR.*
Müller, Dr Bernd. *DDR.*
Müller, Dr Brigitte. *DDR.*
Müller, Dr Eberhard. *DDR.*
Müller, Prof. Dr Gerd. *BRD.*
Müller, Dr Helmut. *DDR.*
Müller, Prof. Horst. *BRD.*
Müller, Dr Mag. Karl Werner. *Austria.*
Mueller, Dr Melvin H. *USA.*
Müller, Dr Paul Hubert. *BRD.*
Müller, Dr Rudolf O. *Switzerland.*
Müller, Prof. Dr Ulrich. *BRD.*
Müller, Prof. Dr Wolfgang Friedrich. *BRD.*
Müller-Buschbaum, Prof. Dr Hanskarl. *BRD.*
Müller-Vogt, Dr German. *BRD.*
Müllner, Dr Manfred. *BRD.*
Münninghoff, Dr Günter. *BRD.*
Mugnoli, Prof. Angelo. *Italy.*
Muhonen, Mr Heikki Juhani. *Finland.*
Muir, Dr Kenneth Walter. *UK.*
Muir, Mr Alastair Kerr. *Canada.*
Muir, Prof. James Alexander. *USA.*
Muirhead, Dr Hilary. *UK.*
Mujica, Dr Carlos. *Chile.*
Mukai, Prof. Tadasuke. *Japan.*
Mukherjee(Mondal), Dr (Mrs) Monika. *India.*
Mukherjee, Dr Alok Kumar. *India.*
Mukherjee, Dr Amal Bikash *India.*
Mukherjee, Dr Biswanath. *India.*
Mukherjee, Dr Partha Sarathi. *India.*
Mukhopadhyay, Anuradha. *India.*
Mukhopadhyay, Mr Bishnu Prasad. *India.*
Mukhopadhyay, Mr Pradip. *India.*
Mukhtarova, Dr Nina Nikolayevna. *USSR.*
Muldawer, Prof. Leonard. *USA.*
Muliawati, Mrs G. S. *Indonesia.*
Mullen, Dr Donald Joseph Edgar. *BRD.*

Muller, Mr John H. *USA.*
Mullica, Dr Donald Foster. *USA.*
Mundt, Dr Otto. *BRD.*
Munir, Mr Mohammad. *Pakistan.*
Munirathinam, Nethoji. *India.*
Munshi, Mr Sanjeev Kumar. *India.*
Muñoz Picone, Mr Eduardo. *Mexico.*
Munsuz, Prof Nuri. *Turkey.*
Mura, Dr Pasquale. *Italy.*
Murad, Dr Enver. *BRD.*
Muradyan, Dr Lyudmila Andranikovna. *USSR.*
Murakami, Dr Takashi. *Japan.*
Murali, Mr R. *India.*
Muralidharan, Mr K. V. *India.*
Muraoka, Dr Hisashi. *Japan.*
Murasik, Dr Andrzej. *Poland.*
Murata, Prof. Yoshitada. *Japan.*
Murray-Rust, Dr Judith. *UK.*
Murray-Rust, Dr Peter. *UK.*
Murta, Prof. Clecio. *Brazil.*
Murthy, Mr G. S. *India.*
Murthy, Dr Krishna H. M. *USA.*
Murthy, Dr Mathur R. N. *India.*
Murthy, Dr N. Sanjeeva. *USA.*
Musatti, Prof. Amos. *Italy.*
Musayev, Dr Faig Nasib ogly. *USSR.*
Muschner, Dr Wolfgang. *DDR.*
Mustafayev, Dr Nariman Mustafa ogly. *USSR.*
Mutikainen, Mr Ilpo Pellervo. *Finland.*
Mutka, Dr Hannu Mika Ilmari. *Finland.*
Mutter, Mrs Graciela. *BRD.*
Mya Mya, Dr Khin. *Burma.*
Myasnikova, Dr Rimma Mikhailovna. *USSR.*
Myer, Prof. George Henry. *USA.*
Myl'nikova, Dr Irina Evgenyevna. *USSR.*
Mys'kiv, Dr Mar'yan Grigoryevich. *USSR.*
Myshlyaev, Prof. Mikhail Mikhailovich. *USSR.*
Nabarro, Prof. Frank Reginald Nunes. *South Africa.*
Naboka, Dr Marat Nikolayevich. *USSR.*
Nadezhina, Dr Tamara Nikolayevna. *USSR.*
Nadiv, Prof. Shmuel. *Israel.*
Nägele, Dr Walter. *BRD.*
Näsäkkälä, Dr Matti Eerik. *Finland.*
Näsänen, Prof. Reino Olavi. *Finland.*
Nag, Dr Dilip Kumar. *India.*
Nag, Miss Jhumjhumi. *India.*
Naga, Dr Mohamed Abdel Hamid. *Egypt.*
Nagabhushana Rao, Mr Chemboli. *India.*
Nagai, Prof. Ryutaro. *Japan.*
Nagaitsev, Dr Yury Valeryevich. *USSR.*
Nagakura, Dr Ichiro. *Japan.*
Nagakura, Prof. Saburo. *Japan.*
Nagakura, Prof. Sigemaro. *Japan.*
Nagano, Dr Kozo. *Japan.*
Nagapal, Mr Kailash Chander. *India.*
Nagashima, Dr Seiichi. *Japan.*
Nagata, Dr Fumio. *Japan.*
Nagels, Prof. Pieter Jan. *Belgium.*
Nagendra Nath, Prof. N. S. *India.*
Nagl, Dr Antun. *Yugoslavia.*
Naik, Mrs Uma Murlidhar. *India.*
Naik, Mr Vaman Madhusudanrao. *India.*
Naiki, Prof. Toshio. *Japan.*
Nakahigashi, Dr Kiyotaka. *Japan.*
Nakai, Dr Hisayoshi. *Japan.*
Nakajima, Dr Tetuo. *Japan.*
Nakajima, Dr Yoshiharu. *Japan.*
Nakamura, Dr Kazuo. *Japan.*
Nakamura, Dr Minoru. *Japan.*
Nakamura, Mr Naotake. *Japan.*
Nakamura, Mr Osamu. *Japan.*
Nakamura, Prof. Terutaro. *Japan.*
Nakamura, Dr Toshio. *Japan.*
Nakanishi, Prof. Norihiko. *Japan.*
Nakano, Dr Shigeru. *Japan.*
Nakata, Mr Kazuaki. *Japan.*
Nakatsu, Prof. Kazumi. *Japan.*
Nakazawa, Dr Hiromoto. *Japan.*
Nakazumi, Dr Yoshihide. *Japan.*
Nakhla, Dr Fakhry. *Egypt.*
Namba, Dr Keiichi. *USA.*

Namba, Dr Yoshiyuki. *Japan.*
Namgung, Prof. Dr Hae. *Korea.*
Namikawa, Mr Kazumichi. *Japan.*
Nančovski, Mr Kosta. *Yugoslavia.*
Nandi, Asok Kumar. *India.*
Nandi, Dr Ranjan Kumar. *India.*
Nanev, Dr Christo. *Bulgaria.*
Nanni, Mr Raymond. *USA.*
Napier, Mr John Graham. *Australia.*
Napolitano, Dr Roberto. *Italy.*
Naqvi, Dr Syed Ali Anwar. *Pakistan.*
Narasimha Murthy, Mr Mattur R. *India.*
Narasimhachari, Dr V. N. *USA.*
Narasimhan, Dr P. *India.*
Narayan, Mr Ramesh. *India.*
Narayanan, Prof. Palamadi Sundaram. *India.*
Narayanan, Dr Poojappan. *USA.*
Nardelli, Prof. Mario. *Italy.*
Nardin, Prof. Giorgio. *Italy.*
Narduzzo, Mr V.G. *UK.*
Narita, Dr Hajime. *BRD.*
Naseif, Dr Abdulah. *Saudi Arabia.*
Nasreen, Miss Shagufta. *Pakistan.*
Nassau, Dr Kurt. *USA.*
Nassimbeni, Prof. Luigi Renzo. *South Africa.*
Natarajan, Dr Mahadevan. *Canada.*
Natarajan, Dr S. *India.*
Natarajan, Dr Subramanian. *India.*
Natera, Dr Manolito Garcia. *Philippines.*
Natesan, Mr Elango. *India.*
Nath, Mr Kashi. *India.*
Nathan, Mr Robert. *USA.*
Nathan, Dr Yaakov. *Israel.*
Naud, Dr Jean Marcel. *Belgium.*
Naudon, Dr André. *France.*
Navarra, Dr Gabriele. *Italy.*
Nave, Dr Colin. *UK.*
Navia, Dr Manuel Alberto. *USA.*
Nawata, Dr Yoshiharu. *Japan.*
Nawaz, Dr Rab. *UK.*
Naya, Prof. Shigeo. *Japan.*
Neckel, Prof. Dr Adolf. *Austria.*
Neels, Prof. Hermann. *DDR.*
Nefedova, Dr Elena Vasilyevna. *USSR.*
Neff, Prof. Dr Hans Josef. *BRD.*
Nehasil, Dr Miroslav. *CSSR.*
Neidle, Dr Stephen. *UK.*
Neifeind, Dipl-Ing Axel. *BRD.*
Neilson, Dr George Francis. *USA.*
Neisser, Mr Jerzy Zbigniew. *UK.*
Nelkowski, Prof. Dr Horst. *BRD.*
Nellemos Andersen, Mr Jesper. *Denmark.*
Nelmes, Dr Richard J. *UK.*
Nelson, Mr A. Dwayne. *USA.*
Nelson, Dr James Bowman. *UK.*
Nemetz, Prof. Ernö. *Hungary.*
Nenner, Miss Ann-Marie. *Sweden.*
Nenonen, Mr Pertti Olavi. *Finland.*
Nenow, Dr Dimiter. *Bulgaria.*
Neronova, Dr Nina Nikolayevna. *USSR.*
Nesper, Dr Reinhard Friedrich. *BRD.*
Nesterova, Dr Yaroslava Mikhailovna. *USSR.*
Netherway, Dr David John. *Australia.*
Netterberg, Dr Frank. *South Africa.*
Neubüser, Prof. Joachim Franz Friedrich G. *BRD.*
Neuman, Mr Alain. *France.*
Neuman, Mr Melvin A. *USA.*
Neumann, Mr Wolfgang. *DDR.*
Newesely, Prof. Dr Heinrich. *BRD.*
Newkirk, Prof. John Burt. *USA.*
Newman, Mr Robert Alan. *USA.*
Newnham, Prof. Robert Everest. *USA.*
Newsam, Dr John Michael. *USA.*
Ng, Dr Hok-Nam. *Canada.*
Ng, Dr Ser Choon. *Singapore.*
Ng, Dr Wee Lam. *Malaysia.*
Ngo, Dr Thong. *Switzerland.*
Nguyen, Prof Huy Dung. *France.*
Ni, Mr Chau-zhou. *China.*
Nicholas, Mr David Michael. *UK.*

Nicholds, Ms Brenda G. *USA.*
Nicholls, Dr Reginald A. *UK.*
Nichols, Mr Monte Carl. *USA.*
Nichols, Dr Thomas Duncan. *USA.*
Nicholson, Dr David Graham. *Norway.*
Nickel, Prof. Erwin Julius Konstantin. *Switzerland.*
Nicklow, Dr Robert Merle. *USA.*
Nicol, Dr Alastair William. *UK.*
Nicolosi, Dr Joseph Anthony. *USA.*
Nieber, Dr Johannes. *DDR.*
Niebsch, Dr Hans-Hermann. *DDR.*
Niedermayr, Dr Gerhard. *Austria.*
Nieduszynski, Dr Ian Alexander. *UK.*
Nielsen, Dr Anders. *Denmark.*
Nielsen, Dr Kurt. *Denmark.*
Nieminen, Dr Kari Veikko Juhani. *Finland.*
Niggli, Prof. Alfred. *Switzerland.*
Niggli, Prof. Ernst. *Switzerland.*
Nigli, Selina. *India.*
Niimura, Dr Nobuo. *Japan.*
Niinistö, Dr Lauri. *Finland.*
Nijveldt, Ir Dick. *Netherlands.*
Nikanorov, Dr Stanislav Prokhorovich. *USSR.*
Nikiforov, Dr Igor' Yakovlevich. *USSR.*
Nikishova, Dr Lidiya Vasilyevna. *USSR.*
Nikitenko, Prof. Valerian Ivanovich. *USSR.*
Nikolaeva, Mrs Rumiana. *Bulgaria.*
Nikolić, Prof. Dr Pantelija. *Yugoslavia.*
Nilsson, Mrs Karin I. *Sweden.*
Nilsson, Dr Rolf O. *Sweden.*
Nimgirawath, Mrs Kloy. *Thailand.*
Nimmo, Dr John Kenneth. *Australia.*
Nishi, Dr Fumito. *Japan.*
Nishida, Dr Isao. *Japan.*
Nishida, Dr Takashi. *Japan.*
Nishiguchi, Mr Munehiro. *Japan.*
Nishikawa, Dr Masana. *Japan.*
Nishinaga, Prof. Tatau. *Japan.*
Nishino, Mr Yoichi. *Japan.*
Nishiyama, Dr Tsutomu. *Japan.*
Nishiyama, Prof. Zenji. *Japan.*
Nissen, Dr Hans-Ude. *Switzerland.*
Nitsche, Mr Walter. *DDR.*
Nitsche, Prof. Dr Rudolf. *BRD.*
Nitschmann, Dr Günter Max Alfred. *BRD.*
Nittono, Dr Osamu. *Japan.*
Niven, Dr Margaret Lilian. *South Africa.*
Nix, Mr E.L. *UK.*
Niziol, Dr Stanisław Marian. *Poland.*
Noack, Dr Joachim. *DDR.*
Noble, Mark C. *USA.*
Nochefranca, Miss Luz R. *Philippines.*
Noda, Prof. Tokiti. *Japan.*
Noda, Dr Yasutoshi. *Japan.*
Noda, Mr Yoshitoshi. *Japan.*
Noda, Mr Yukio. *Japan.*
Noe-Nygaard, Prof. Arne. *Denmark.*
Nørskov-Lauritsen, Dr Leif. *Denmark.*
Noether, Dr Herman Dietrich. *USA.*
Nogues Carulla, Dr Joaquin Ma. *Spain.*
Noll, Prof. Dr Walter Friedrich Heinrich. *BRD.*
Nolte, Dr Margaretha Johanna. *South Africa.*
Noltemeyer, Dr Mathias Rolf. *BRD.*
Nonaka, Dr Kohzo. *Japan.*
Noor, Sahina Begum. *India.*
Noordik, Dr Jan Hendrik. *Netherlands.*
Nord, Dr Anders G. *Sweden.*
Norén, Dr Bertil. *Sweden.*
Nordenson, Mr Svein. *Norway.*
Noréus, Dr Dag. *Sweden.*
Nordlund, Mr Pär L. *Sweden.*
Nordman, Prof. Christer Eric. *USA.*
Norin, Dr Rolf. *Sweden.*
Norman, Prof. Joe G., Jr. *USA.*
Norman, Prof. Nico. *Norway.*
Norman, Mr Peter David. *Australia.*
Norrby, Dr Lars-Johan. *Sweden.*
Norrestam, Prof. Rolf. *Denmark.*
North, Prof. Anthony Charles T. *UK.*

Northolt, Dr Ir Maurits Gerhard. *Netherlands.*
Norval, Dr Stephen Vynne. *UK.*
Nota, Dr Dirk Johannes Gregorius. *Netherlands.*
Novák, Ing Ctirad. *CSSR.*
Nover, Dr Georg. *BRD.*
Novosel Radović, Dr Vjera. *Yugoslavia.*
Novozhilov, Dr Alexander Ivanovich. *USSR.*
Nowack, Miss Ellen Carla. *BRD.*
Nowacki, Prof. Werner. *Switzerland.*
Nowell, Dr Ian William. *UK.*
Nowotny, Prof. Dr Hans. *USA.*
Noyan, Dr Ismail C. *USA.*
Nozik, Dr Yury Zinovyevich. *USSR.*
Nuffield, Prof. Edward Wilfrid. *Canada.*
Nukui, Dr Akihiko. *Japan.*
Nunn, Dr Ernest Keith. *Australia.*
Nunzi, Prof. Antonio. *Italy.*
Nuriyev, Dr Idayat Ragim ogly. *USSR.*
Nyborg, Dr Jens. *Denmark.*
Nygren, Dr Mats. *Sweden.*
Nyburg, Prof. Stanley C. *Canada.*
Nye, Prof. John Frederick. *UK.*
Nygren, Dr Mats. *Sweden.*
Nývlt, Ing Jaroslav. *CSSR.*
Obaid, Dr Yassin Najim. *Iraq.*
Oberhänsli, Dr Willi E. *Switzerland.*
Oberholster, Dr Rupert Egbert. *South Africa.*
Oberti, Dr Roberta. *Italy.*
Očko, Magistar Miroslav. *Yugoslavia.*
O'Connor, Dr Brian Henry. *Australia.*
O'Connor, Dr Denis Arthur. *UK.*
Oda, Mr Isao. *Japan.*
Oda, Prof. Tsutomu. *Japan.*
Odajima, Prof. Akira. *Japan.*
Oefner, Mr Christian. *BRD.*
Öfverstedt, Dr Lars-Göran W. *Sweden.*
Oehlschlegel, Dr Georg. *BRD.*
Oen, Prof. Dr Ing Soen. *Netherlands.*
Österberg, Prof. Ragnar. *Sweden.*
Oettel, Dr Heinrich. *DDR.*
Özenbaş, Mr A. Macit. *Turkey.*
Özkan, Doç. Dr Hüsnü. *Turkey.*
Özkaplan, Dr Habip. *Turkey.*
Öztunali, Prof. Dr Önder. *Turkey.*
Ogawa, Mr Katsumi. *Japan.*
Ogawa, Prof. Kazuhide. *Japan.*
Ogawa, Prof. Shiro. *Japan.*
Ogawa, Prof. Tomoya. *Japan.*
Ogura, Prof. Iwao. *Japan.*
Ogura, Prof. Kiyosi. *Japan.*
Oğurtani, Prof. Dr Tarik Ömer. *Turkey.*
Oh, Mr Chuan Ho, *Malaysia.*
Ohachi, Dr Tadashi. *Japan.*
Ohama, Dr Nobuhiko. *Japan.*
Ohashi, Prof. Yoshikazu. *USA.*
Ohashi, Dr Yuji. *Japan.*
Ohba, Dr Shigeru. *Japan.*
Ohgaki, Mr Masataka. *Japan.*
Ohkawa, Mr Tokio. *Japan.*
Ohlendorf, Dr Douglas Henry. *USA.*
Ohmasa, Dr Masaaki. *Japan.*
Ohno, Dr Tamotsu. *Japan.*
Ohrt, Miss Jean Marie. *USA.*
Ohshima, Dr Ken-ichi. *Japan.*
Ohsumi, Dr Kazumasa. *Japan.*
Ohta, Dr Takao. *Japan.*
Ohta, Mr Tsutomu. *Japan.*
Ohtsuka, Mr Yasukuni. *Japan.*
Okabe, Dr Toshio. *Japan.*
Okada, Mr Kenji. *Japan.*
Okada, Dr Kiyoshi. *Japan.*
Okada, Prof. Masakazu. *Japan.*
Okada, Mr Yasumasa. *Japan.*
Okamura, Dr Fujio Peter. *Japan.*
Okaya, Prof. Yoshi Haru. *USA.*
Okazaki, Prof. Atsushi. *Japan.*
O'Keeffe, Prof. Michael. *USA.*
Okrusch, Prof. Dr rer nat Martin. *BRD.*
Oksman, Mr Pentti. *Finland.*
Oktik, Dr Şener. *Turkey.*
Okunuki, Mr Masahiko. *Japan.*
Okuyama, Dr Kenji. *Japan.*

Olcese, Prof. Giorgio L. *Italy.*
O'Leary, Dr Kevin Joseph. *USA.*
Olejnik, Dr Stanisław Julian. *Poland.*
Oleksyn, Dr Barbara Jadwiga. *Poland.*
Oles, Prof. Andrzej Władysław. *Poland.*
Oliva, Mr G. *UK.*
Oliveira Lopes, Prof. Cesar. *Brazil.*
Oliver, Dr Peter John. *New Zealand.*
Oliver, Dr Joel Day. *USA.*
Olivier, M. Marc-J. *Canada.*
Olivieri, Mr Johnny Rizzieri. *Brazil.*
Ollis, Mr David L. *USA.*
Olmstead, Ms Marilyn Morgan. *USA.*
Olovsson, Mr Gunnar. *Sweden.*
Olovsson, Prof. Ivar. *Sweden.*
Olsen, Dr Arne. *Norway.*
Olsen, Dr Janus Staun. *Denmark.*
Olsen, Prof. Kenneth Wayne. *USA.*
Olson, Dr Arthur Jules. *USA.*
Olson, Dr David Harold. *USA.*
Olson, Mrs Solveig. *Sweden.*
Olsson, Miss Carin. *Sweden.*
Olsson, Mr Per-Olof. *Sweden.*
Olthof, Dr Gerrit Jan. *Netherlands.*
Olthof-Hazekamp, Drs Roeli. *Netherlands.*
Omezzine, Dr Belgacem. *Tunisia.*
Omrani, Dr Hédi. *Tunisia.*
Onan, Prof. Kay Denise. *USA.*
Ondik, Dr Helen Margaret. *USA.*
Ong, Mr Yeoh Han. *Malaysia.*
Onishchina, Dr Ninel' Mitrofanovna. *USSR.*
Onuma, Dr Shigeki. *Japan.*
Onyeagocha, Dr Anthony Chukwuma. *Nigeria.*
Ookawa, Prof. Akiya. *Japan.*
Opekunov, Dr Victor Nikolayevich. *USSR.*
Oppermann, Mr Dieter. *DDR.*
Oppermann, Dr Heinrich. *DDR.*
Orama, Dr Olli Antero. *Finland.*
Ordway, Dr Fred. *USA.*
Orekhova, Dr Valentina Petrovna. *USSR.*
Organova, Dr Nataliya Ivanovna. *USSR.*
Orioli, Prof. Pierluigi. *Italy.*
Orpen, Dr Anthony Guy. *UK.*
Ortega, Dr Richard B. *USA.*
Osaka, Dr Toshiaki. *Japan.*
Osakabe, Mr Nobuyuki. *Japan.*
Osaki, Dr Kenji. *Japan.*
Osborne, Mr J.F. *UK.*
Osgood, Mr Brian Clair. *USA.*
Osip'yan, Prof. Yury Andreyevich. *USSR.*
Oskarsson, Dr Åke. *Sweden.*
Ossio, Miss Myriam. *Chile.*
Ostanevich, Dr Yury Mechislavovich. *USSR.*
Osterc, Prof. Dr (Mrs) Valerija. *Yugoslavia.*
Osterland, Mrs Martina. *DDR.*
Ostrofsky, Mr Bernard. *USA.*
Oszko, Mr Albert Zoltán. *Hungary.*
Othman, Dr Abdul Hamid bin. *Malaysia.*
Othman, Dr Radzali. *Malaysia.*
Otroshchenko, Dr Lyudmila Petrovna. *USSR.*
Otsuka, Prof. Ryohei. *Japan.*
Otten, Mr Peter. *BRD.*
Otto, Dr Hans Hermann. *BRD.*
Ou, Dr Chia-Chih. *USA.*
Ouattara, Mr Drissa. *Ivory Coast.*
Oumous, Mr Hassan. *France.*
Ovalle de Bravo, Prof. Yolanda. *Colombia.*
Ovchinnikov, Dr Yury Kuz'mich. *USSR.*
Overkott, Dr Paul Engelbert. *BRD.*
Overmyer, Ms Elizabeth M. *USA.*
Overweel, Dr Cornelis Johannes. *Netherlands.*
Oviedo, Miss Danais. *Cuba.*
Owen, Dr John David. *UK.*
Owen, Mr Charles Gordon. *Canada.*
Owston, Dr Philip George. *UK.*
Oyanagi, Dr Hiroyuki. *Japan.*
Ozawa, Dr Tohru. *Japan.*
Ozerin, Dr Alexander Nikiforovich. *USSR.*
Ozerov, Prof. Ruslan Pavlovich. *USSR.*
Ozola, Dr Astrida Davovna. *USSR.*

Ozolin'sh, Dr Gerkhard Vladimirovich. *USSR.*
Ozols, Dr Yan Karlovich. *USSR.*
Paakkari, Prof. Timo Lauri Päiviö. *Finland.*
Paalassalo, Mr Pentti Olavi. *Finland.*
Pabo, Prof. Carl Ogren. *USA.*
Pabst, Prof. Adolf. *USA.*
Pachali, Dr Klaus Erich. *BRD.*
Paciorek, Mgr inz. Włodzimierz. *Poland.*
Padlan, Dr Eduardo Agustin. *USA.*
Padmanabhan, Dr V. M. *India.*
Padmasuta, Mrs Soontari. *Thailand.*
Padrón. Dr Raúl. *Venezuela.*
Pähler, Dr Arno. *BRD.*
Pätz, Dr Kurt W. *DDR.*
Pätzke, Mrs Nora. *DDR.*
Pätzold, Mrs Rita. *DDR.*
Page, Dr Campbell Thomas. *New Zealand.*
Page, Dr James Ernest. *UK.*
Page, Dr Trevor Francis. *UK.*
Pagoaga, Dr M. Katherine. *Canada.*
Pahwa, Mr Des Raj. *India.*
Pai, Dr Emil Friedrich. *BRD.*
Paige-Green, Mr Philip. *South Africa.*
Pajak, Mgr Lucjan. *Poland.*
Pajunen, Dr Aarne Veikko. *Finland.*
Pakawatchai, Dr Chaveng. *Thailand.*
Pakhomov, Dr Vladimir Ivanovich. *USSR.*
Pakkanen, Dr Tapani Antti. *Finland.*
Pakkanen, Dr Tuula Tellervo. *Finland.*
Pál, Mrs Edith. *Hungary.*
Pal, Dr Gour Pada, *BRD.*
Palatnik, Prof. Lev Samoilovich. *USSR.*
Palenik, Prof. Gus Joseph. *USA.*
Palenzona, Prof. Andrea. *Italy.*
Pálinkás, Mr Gábor. *Hungary.*
Palistrant, Dr Alexander Filippovich. *USSR.*
Paljević, Dr Matija. *Yugoslavia.*
Palkina, Dr Kapitolina Kapitonovna. *USSR.*
Palmer, Mr Allan D. *Canada.*
Palmer, Dr Kenneth James. *USA.*
Palmer, Dr Rex Alfred. *UK.*
Pamplin, Dr Brian Randall. *UK.*
Pan, Dr Nitya Ranjan. *India.*
Pan, Prof. Ke-zhen. *China.*
Pan, Prof. Zhao-lu. *China.*
Pan, Mr Zuo-hua. *China.*
Panagos, Prof. Athanasios. *Greece.*
Panchekha, Dr Petr Alekseyevich. *USSR.*
Pandey, Dr Dhananjai. *India.*
Pandya, Mr Naresh. *Canada.*
Pandya, Prof. Janardhan Rameshchandra. *India.*
Pangarov, Prof. Nikola. *Bulgaria.*
Pangborn, Prof. Robert Northrup. *USA.*
Pannetier, Dr Jean. *France.*
Pannhorst, Dr Wolfgang. *BRD.*
Pant, Dr Arun Kumar. *India.*
Pant, Dr Lalit Mohan. *India.*
Pantaleo, Dr Nantelle Smith. *USA.*
Paorici, Prof. Carlo. *Italy.*
Pap, Mrs Sarolta. *Sweden.*
Papadakis, Prof. Alexander. *Greece.*
Papadimitraki Chlichlia, Prof. Helena. *Greece.*
Papadopoulos, Mr Demetrius. *Greece.*
Papathanassopoulos, Dr Constantinos. *Greece.*
Papavinasam, Mr E. *India.*
Papazoglou, Mr Aristides. *Greece.*
Papiz, Dr Miroslav Zenko. *UK.*
Papunen, Prof. Heikki Tapani. *Finland.*
Parak, Dr Fritz Günther. *BRD.*
Parangtopo, Dr. *Indonesia.*
Parasnis, Prof. Arawind Shripad. *India.*
Paretzkin, Mr Boris. *USA.*
Parge, Dr Hans Erich. *BRD.*
Parise, Dr John Baptist. *Australia.*
Parissakis, Prof. George. *Greece.*
Park, Dr Chang Hoon, *USA.*
Park, Prof. Young Ja. *Korea.*
Párkányi, Dr László. *Hungary.*
Parker, Dr Andrew. *UK.*
Parker, Mr Michael William. *UK.*

Parker, Dr Robert Louis. *USA.*
Parker, Dr Sidney Glenn. *USA.*
Parkes, Dr Alan Schofield. *USA.*
Parks, Ms Elizabeth H. *USA.*
Parks, Dr Terrence Charles. *Australia.*
Parpia, Dr Dawood Yusuf. *UK.*
Parrish, Dr William. *USA.*
Parry, Dr David Anthony Dougall. *New Zealand.*
Parry, Mr G. *UK.*
Parry, Dr George Sparling. *UK.*
Parsons, Dr Donald Frederick. *USA.*
Parsons, Prof. Ian. *UK.*
Parsons, Dr Paul Donald. *UK.*
Parthasarathi, Dr V. *India.*
Parthasarathy, Dr R. *USA.*
Parthasarathy, Dr Soundarajan. *India.*
Parthé, Prof. Erwin. *Switzerland.*
Parvez, Dr Masood. *USA.*
Pascard, Dr Claudine. *France.*
Pascher, Dr Irmin. *Sweden.*
Pasemann, Dr Monika. *DDR.*
Pasero, Dr Marco. *Italy.*
Pashley, Prof. Donald William. *UK.*
Pashov, Mr Nikolai. *Bulgaria.*
Passaglia, Prof. Elio. *Italy.*
Pasti, Mr Fabio. *Italy.*
Patel, Prof. Ambalal Ranchhodbhai. *India.*
Patel, Dr Jamshed R. *USA.*
Patel, Dr Prabhudas Revandas. *India.*
Patel, Dr Tankadhar. *India.*
Pathak, Prof. Pushkarrai Dalpatram. *India.*
Pathan, Dr Muhammad Taqee. *Pakistan.*
Paton, Mr John Dennis. *UK.*
Pattabhi, Dr (Mrs) Vasantha. *India.*
Pattanayek, Mrs Rekha Rani *India*
Pattison, Dr Philip. *BRD.*
Pattridge, Ms Katherine A. *USA.*
Paufler, Prof. Peter. *DDR.*
Paul, Prof. Iain Campbell. *USA.*
Paul, Mrs Sabine. *DDR.*
Pauling, Prof. Linus Carl. *USA.*
Paulitsch, Prof. Dr Peter. *BRD.*
Paulus, Dr Erich Friedrich. *BRD.*
Pauly, Prof. Hans. *Denmark.*
Paunov, Dr Michael. *Bulgaria.*
Pauptit, Dr Richard A. *Switzerland.*
Pauthenet, Prof. René *France.*
Pavalow, Prof. Melvin. *USA.*
Pavel, Dr Nicolae Viorel. *Italy.*
Pavelčík, Ing František. *CSSR.*
Pavie Cardoso, Dr Lisandro. *Brazil.*
Pavkovic, Prof. Stephen Frank. *USA.*
Pavlishin, Dr Vladimir Ivanovich. *USSR.*
Pavlovsky, Dr Alexander Grigoryevich. *USSR.*
Pawlak, Dr Stanisław Jerzy. *Poland.*
Pawley, Prof. G. Stuart. *UK.*
Paxton, Mr A.T. *UK.*
Payan, Dr Françoise. *France.*
Payne, Dr Nicholas Charles. *Canada.*
Pazdernik, Prof. LeRoy Joseph. *Canada.*
Peacock, Dr Norman. *UK.*
Peacock, Prof. R.D. *UK.*
Peacor, Prof. Donald R. *USA.*
Pearce, Mr I.R. *UK.*
Pearson, Mr Robert H. *USA.*
Pearson, Dr William Burton. *Canada.*
Peavler, Prof. Robert J. *USA.*
Pech, Dr Lucia Yanovna. *USSR.*
Pechstein, Mrs Gisela. *DDR.*
Pećina, Miss Planinka. *Yugoslavia.*
Pedersen, Dr Berit Fjærtoft. *Norway.*
Pedersen, Prof. Björn. *Norway.*
Pedone, Prof. Carlo. *Italy.*
Peercy, Dr Paul S. *USA.*
Peerdeman, Prof. Dr Antonius F. *Netherlands.*
Peeters, Dr Oswald Maurice. *Belgium.*
Peibst, Dr Herbert. *DDR.*
Peiser, Mr Herbert Steffen. *USA.*
Pelizzi, Prof. Corrado. *Italy.*
Pellinghelli, Prof. Maria Angela. *Italy.*
Peña, Miss Luzmila. *Chile.*

Penavić, Dr Maja. *Yugoslavia.*
Penco, Prof. Anna Maria. *Italy.*
Peneva, Dr Stefka. *Bulgaria.*
Penfold, Prof. Bruce Russell. *New Zealand.*
Penfold, Dr David William. *UK.*
Peng, Mr Chang-qi. *China.*
Peng, Dr Kuoching. *1955).*
Peng, Mrs Ming-shen. *China.*
Peng, Dr Shie-ming. *Taiwan.*
Peng, Prof. Zhi-zhong. *China.*
Penndorf, Mr Jürgen. *DDR.*
Penner-Hahn, Prof. James E. *USA.*
Pennington, Mr Mark. *UK.*
Pentinghaus, Dr Horst. *BRD.*
Penzkofer, Dr Benno. *BRD.*
Pèpe, Dr Gérard. *France.*
Perales Alcon, Dr Aurea. *Spain.*
Percival, Dr Henry Joseph. *New Zealand.*
Perdikatsis, Dr Basilios. *Greece.*
Perdok, Prof. Dr Wiepko Gerhardus. *Netherlands.*
Perego, Dr Giovanni. *Italy.*
Perekalina, Dr Tatyana Mikhailovna. *USSR.*
Perekalina, Dr Zoya Borisovna. *USSR.*
Pérez, Mrs Carmen. *Chile.*
Pérez, Mr Serge. *France.*
Perez Alonso, Dr Julio. *Spain.*
Perez del Villar, Lic. Luis. *Spain.*
Perez Garrido, Mr Simeon. *Spain.*
Perez Salazar, Mrs Adela. *Spain.*
Perkins, Mr Herbert O. *USA.*
Perloff, Dr Alvin. *USA.*
Perrault, Dr Guy. *Canada.*
Perret, Mr Ramón. *Chile.*
Perrin, Dr Monique. *France.*
Perrins, Dr D.H.G. *UK.*
Persdotter, Miss Ingeborg M. *Sweden.*
Pershin, Dr Vitaly Konstantinovich. *USSR.*
Perthel, Dr Rolf. *DDR.*
Pertlik, Doz. Dr Franz. *Austria.*
Pertsin, Dr Alexander Iosifovich. *USSR.*
Perutz, Dr Max Ferdinand. *UK.*
Pervaiz, Mr Rashed. *Pakistan.*
Peschar, Drs René. *Netherlands.*
Peshev, Prof Pavel. *Bulgaria.*
Peskin, Dr Vladimir Fedorovich. *USSR.*
Pessa, Prof. Viljo Markus. *Finland.*
Pessen, Dr Helmut. *USA.*
Petcher, Dr Trevor James. *Switzerland.*
Peterat, Dr Michael. *BRD.*
Peters, Dr Karl. *BRD.*
Peters, Mr Charles Richard. *USA.*
Peterse, Ir Wilhelmus J. A. M. *Netherlands.*
Petersen, Dr Donald Ralph. *USA.*
Petersen, Prof. Jeffrey L. *USA.*
Petersen, Dr Ole Valdemar. *Denmark.*
Peterson, Dr Ronald Charles. *Canada.*
Peterson, Dr Selmer W. *USA.*
Petiau, Prof. Jacqueline. *France.*
Petipas, Prof. Claude. *France.*
Petraccone, Prof. Vittorio. *Italy.*
Petreus, Mr Ion. *Romania.*
Petříček, Dr Václav. *CSSR.*
Petroff, Prof. Jean-François. *France.*
Petropavlov, Dr Nikolay Nikolayevich. *USSR.*
Petrov, Dr Kostadin. *Bulgaria.*
Petrov, Mr Ognyan. *Bulgaria.*
Petrov, Dr Srebri. *Bulgaria.*
Petrov, Dr Thomas Georgiyevich. *USSR.*
Petrov, Dr Vasko. *Bulgaria.*
Petrova, Dr Irina Vladimirovna. *USSR.*
Petrović, Dr Dragoslav. *Yugoslavia.*
Petrovich, Mr Anton I. *USA.*
Petrunina, Dr Alla Anastas'yevna. *USSR.*
Petsko, Prof. Gregory Anthony. *USA.*
Pett, Dr Virginia B. *USA.*
Petter, Prof. Walter. *Switzerland.*
Petukhov, Dr Boris Vladimirovich. *USSR.*
Petz, Mr John Ignatius. *USA.*
Petzoldt, Dr Jürgen Hugo Hans. *BRD.*
Peyronel, Prof. Giorgio. *Italy.*

Pezerat, Dr Henri. *France.*
Pfefferkorn, Prof. Dr Gerhard Erich. *BRD.*
Pflaum, Mr Wolfgang Richard. *USA.*
Pfluger, Prof. Clarence Eugene. *USA.*
Pflugrath, Dr James William. *BRD.*
Phakey, Dr Prem P. *Australia.*
Phaovibul, Dr Orapin. *Thailand.*
Phavanantha, Dr Phathana. *Thailand.*
Philibert, Prof. Jean. *France.*
Philipov, Mr Alexander. *Bulgaria.*
Philipp, Dr George. *Egypt.*
Philipsborn, von, Prof. Dr Henning. *BRD.*
Phillips, Prof. Sir David Chilton. *UK.*
Phillips, Dr Frederick Lloyd. *Ghana.*
Phillips, Mr G.I. *UK.*
Phillips, Dr George Neal, Jr. *USA.*
Phillips, Prof. James Christopher. *USA.*
Phillips, Mr R.J. *UK.*
Phillips, Dr Simon Edward Victor. *UK.*
Phillips, Dr Theodore II. *USA.*
Phillips, Prof. Travis J. *USA.*
Phizackerley, Dr Richard Paul. *USA.*
Piazzesi, Prof. AnnaMaria. *Italy.*
Pichert, Mr Jerome. *Consultant.*
Pickardt, Prof Dr Joachim. *BRD.*
Pidzhyan, Prof. Grigory Oganesovich. *USSR.*
Pielaszek, Dr Jerzy. *Poland.*
Pieper, Dr Gerhard. *BRD.*
Pierce-Butler, Dr Melanie Anne. *UK.*
Piermarini, Dr Gasper John. *USA.*
Pierpont, Prof. Cortlandt G. *USA.*
Pierrot, Dr Marcel. *France.*
Pietraszko, Dr Adam. *Poland.*
Pietraszko, Mrs Donata. *Poland.*
Pietsch, Mr Ullrich. *DDR.*
Pietzsch, Dr Claus. *DDR.*
Pifferi, Dr Augusto. *Italy.*
Pignataro, Mrs Edith H. *USA.*
Pignedoli, Prof. Anna. *Italy.*
Pigram, Dr W.J. *UK.*
Pihl, Mr Carl Frederick. *USA.*
Piispanen, Dr Risto Anton. *Finland.*
Pikin, Dr Sergey Alekseyevich. *USSR.*
Pilati, Dr Tullio. *Italy.*
Pilotti, Dr Anne-Marie. *Sweden.*
Pilz, Prof. Dr Ingrid Edith. *Austria.*
Piña de Noyola, Prof. Maria Cristina. *Mexico.*
Pinkerton, Prof. Andrew Alan. *USA.*
Pinsker, Dr. Garry Zinov'yevich. *USSR.*
Pinsker, Prof. Zinovy Grigoryevich. *USSR.*
Pinto, Mr Haim. *Israel.*
Pipkin, Dr Noel John. *South Africa.*
Piret, Prof Paul. *Belgium.*
Pirie, Dr John Douglas. *UK.*
Piro, Lic Oscar Enrique. *Argentina.*
Pirozzi, Prof. Beniamino. *Italy.*
Pisani, Prof. Cesare. *Italy.*
Pisarevsky, Dr Yury Vladimirovich. *USSR.*
Pish, Dr George. *USA.*
Pisutha-Arnond, Dr Visut. *Thailand.*
Pitkänen, Mrs Tuula Esteri. *Finland.*
Pitrola, Dr R. *UK.*
Pitts, Dr James Edwin. *UK.*
Pizzey, Mrs Monica Agnes Anastasia. *Canada.*
Plakhov, Dr Gennady Fedorovich. *USSR.*
Plana Llevat, Dr Feliciano. *Spain.*
Plant, Dr John Stewart. *UK.*
Plas, van der, Prof. Dr Leendert. *Netherlands.*
Platikanova, Dr Vesselina. *Bulgaria.*
Plesken, Dr Wilhelm. *UK.*
Pleštil, Ing Josef. *CSSR.*
Pletcher, Dr James F. *USA.*
Pletnev, Dr Vladimir Zakharovich. *USSR.*
Plies, Dr Volker. *BRD.*
Ploc, Dr Robert Allen. *Canada.*
Ploog, Dr Klaus. *BRD.*
Plough-Sørensen, Mrs Gudrun. *Denmark.*
Plust, Dr Heinz-Günther. *BRD.*
Pluth, Dr Joseph John. *USA.*
Plyasova, Dr Lyudmila Mikhailovna. *USSR.*

Plyusnina, Dr Inga Ivanovna. *USSR.*
Pobedimskaya, Dr Elena Alexandrovna. *USSR.*
Pocev, Prof. Dr Stefan. *Yugoslavia.*
Pochettino, Dr Alberto Antonio. *Argentina.*
Podberezskaya, Dr Nina Vasilyevna. *USSR.*
Podbrdský, Dr Josef. *CSSR.*
Podder, Dr (Mrs) Aloka. *India.*
Podjarny, Dr Alberto Daniel. *Argentina.*
Podlahová, Dr Jana. *CSSR.*
Poharc-Logar, Magistar Vesna. *Yugoslavia.*
Pohl, Prof. Dr Dieter. *BRD.*
Pohl, Prof. Dr Siegfried. *BRD.*
Pointer, Dr David John. *UK.*
Pokrovsky, Dr Nikolay Leonidovich. *USSR.*
Polanc, Magistar Ivan. *Yugoslavia.*
Poland, Mr Virgil Laverne. *USA.*
Polat, Mr Hamza. *Turkey.*
Polborn, Dr Kurt Volkmar. *BRD.*
Polcarová, Dr Milena. *CSSR.*
Polchovskaya, Dr Tatyana Mikhailovna. *USSR.*
Polidori, Dr Giampiero. *Italy.*
Poljak, Prof. Roberto J. *France.*
Poll, Dr Wolfgang. *BRD.*
Pollack, Dr Sidney Solomon. *USA.*
Pollard, Dr David Ronald. *UK.*
Pollert, Ing Emil. *CSSR.*
Pollmann, Dipl-Min Siegfried. *BRD.*
Polo, Dr Adriano. *Italy.*
Polyanskaya, Dr Tamara Mikhailovna. *USSR.*
Polyansky, Dr Evgeny Vasilyevich. *USSR.*
Polychroniades, Prof. Efstathios. *Greece.*
Polynova, Dr Tamara Nikitichna. *USSR.*
Pomés Hernandez, Prof. Dr Sc. Ramón. *Cuba.*
Pompa, Dr Francesco. *Italy.*
Pond, Dr Robert Charles. *UK.*
Pongratz, Dr Dipl.-Ing. Peter. *Austria.*
Pongsapich, Dr Wasant. *Thailand.*
Ponnuswamy, Dr M. N. *USA.*
Ponomarev, Dr Vasily Ivanovich. *USSR.*
Pontchour, Miss Cha-on. *Thailand.*
Pontenagel, Dr Willibrordus Maria G.F. *Netherlands.*
Ponyatovsky, Prof. Evgeny Genrikhovich. *USSR.*
Ponzi, Ms Dagmar Ringe. *USA.*
Poojary, Dr M. Damodara *India.*
Poot, Mr Simon. *Netherlands.*
Popescu, Conf. Ion. *Romania.*
Popma, Prof. Dr Theo Johan August. *Netherlands.*
Popolitov, Dr Vladislav Ivanovich. *USSR.*
Popov, Dr Alexander. *Bulgaria.*
Popova, Dr Anastasiya Arsentyevna. *USSR.*
Popova, Dr Svetlana Vladimirovna. *USSR.*
Popović, Dr Stanko. *Yugoslavia.*
Poppendieck, Mr Detlef. *DDR.*
Poppi, Prof. Luciano. *Italy.*
Poppleton, Dr Bruce J. *Australia.*
Porai-Koshits, Prof. Mikhail Alexandrovich. *USSR.*
Porai-Koshits, Prof. Evgeny Alexandrovich. *USSR.*
Porta, Prof. Piero. *Italy.*
Portalone, Dr Gustavo. *Italy.*
Porter, Mr Leigh C. *USA.*
Portnov, Dr Vadim Nikolayevich. *USSR.*
Portugal, Mr Remberto. *Bolivia.*
Porzio, Dr William. *Italy.*
Posada G., Prof. Enrique. *Colombia.*
Posner, Prof. Aaron Sidney. *USA.*
Post, Prof. Benjamin. *USA.*
Post, Dr Michael Leonard. *Canada.*
Postma, Drs Johannes Petrus Maria. *Netherlands.*
Potenza, Prof. Joseph Anthony. *USA.*
Potter, Dr Reginald. *UK.*
Potter, Mr Stephen Anthony. *USA.*
Potterton, Dr Elizabeth Anne. *UK.*
Potworowski, Dr Jean-André. *Canada.*
Pouget, Dr. *France.*
Poulieff, Mr Christo. *Bulgaria.*
Poulin-Dandurand, Mrs Suzie. *Canada.*
Poulos, Mr Thomas L. *USA.*
Povarennykh, Prof. Alexander Sergeyevich. *USSR.*
Povey, Dr David Christopher. *UK.*
Powell, Dr Brian Mathieson. *Canada.*
Powell, Dr Douglas R. *USA.*

Powell, Prof. Herbert Marcus. *UK.*
Pradhan, Dukhabandhu. *India.*
Prager, Dr Peter Robert. *Australia.*
Prakash-Varma, Dr S. *UK.*
Pramatus, Miss Supanich. *Thailand.*
Prandl, Prof. Dr Wolfram. *BRD.*
Prasad, Dr Lata. *Canada.*
Prasad, Dr Narayan, *India.*
Prasad, Dr Ravindra. *India.*
Prasad, Dr Satya Murti. *India.*
Prasad, Dr Y. R. Ananth. *India.*
Prayaga, Dr Chandra Sekhar. *India.*
Precigoux, Dr Gilles. *France.*
Preisinger, Prof. Dr Anton. *Austria.*
Preiss, Dr Henry, *DDR.*
Prelesnik, Dr Bogdan. *Yugoslavia.*
Presley, Dr C. Travis. *USA.*
Pretorius, Dr Jan Andries. *South Africa.*
Preuss, Dr Heinz. *DDR.*
Preut, Dr Hans. *BRD.*
Prevarsky, Dr Anatoly Petrovich. *USSR.*
Prevey, Mr Paul S., III. *USA.*
Prewitt, Prof. Charles Thompson. *USA.*
Price, Dr Geoffrey David. *UK.*
Price, Ms Rebecca Alexis *USA.*
Prick, Dr Petrus Antonius Johannes. *Netherlands.*
Priestle, Dr John P. *Switzerland.*
Priftis, Prof. George. *Greece.*
Prikhod'ko, Dr Leonid Vasilyevich. *USSR.*
Primot, Mr Jacques. *France.*
Prince, Dr Edward. *USA.*
Pring, Dr Allan. *Australia.*
Pringle, Mr Gordon James. *Canada.*
Prins, Prof. Dr Jan Albert. *Netherlands.*
Pritchard, Dr Robin Gavin. *UK.*
Pritzkow, Mr Wolfgang. *DDR.*
Prodan, Dr Albert. *Yugoslavia.*
Profi, Mrs Stella. *Greece.*
Protas, Prof. Jean. *France.*
Prout, Dr Charles Keith. *UK.*
Provotorov, Dr Mikhail Viktorovich. *USSR.*
Przybylska, Dr Maria. *Canada.*
Ptitsyn, Prof. Oleg Borisovich. *USSR.*
Pušelj, Dr Milan. *Yugoslavia.*
Pudovkina, Dr Zoya Vasilyevna. *USSR.*
Puff, Mr Manfred. *DDR.*
Puff, Prof. Dr Heinrich. *BRD.*
Pugachev, Dr Anatoly Tarasovich. *USSR.*
Puigjaner, Prof. Luis C. *Spain.*
Puliti, Dr Raffaella. *Italy.*
Pulou, Prof. Raymond. *France.*
Pulsinelli, Prof. Phillip Del. *USA.*
Punin, Dr Yury Olegovich. *USSR.*
Punkkinen, Dr Matti. *Finland.*
Punte, Dr Graciela. *Argentina.*
Puranik, Dr (Mrs) Vedavati Gururaj. *India.*
Purdy, Mr Samuel M. *USA.*
Pureza, Mr Fausto. *Portugal.*
Pushcharovsky, Dr Dmitry Yuryevich. *USSR.*
Putnis, Dr Andrew. *UK.*
Putten, van der, Dr Nicolaas H. J. J. *Netherlands.*
Puttick, Prof. Keith Ernest. *UK.*
Puxley, Dr David Charles. *UK.*
Pyatenko, Dr Yury Andreyevich. *USSR.*
Pyckhout, Mr Wim Maurits August. *Belgium.*
Pyrros, Dr Nikos P. *USA.*
Pyykkö, Prof. Veli Pekka. *Finland.*
Qaiser, Mr Mohammad Ali. *Pakistan.*
Qi, Dr Zeng-du. *China.*
Qi, Prof. Zhi-ru. *China.*
Qian, Mrs Min-xie. *China.*
Qian, Ms Jin-zi. *China.*
Quader, Prof. Dr Mohammed Abdul. *Bangladesh.*
Quadrado, Prof. Ricardo. *Portugal.*
Quagliata, Prof. Claudio. *Italy.*
Quaranta Cabral, Dr Ubirajara. *Brazil.*
Quartieri, Dr Simona. *Italy.*
Queiroz do Amaral, Dr Lia. *Brazil.*
Quéré, Prof. Yves. *France.*
Quevedo, Dr Manuel M. *Colombia.*

Quibilan, Mr Edelmiro I., *Philippines.*
Quicksall, Dr Carl O. *USA.*
Quigley, Dr Gary Joseph. *USA.*
Quintana, Dr Rafael, *Cuba.*
Quintana Owen, Mrs Patricia. *Mexico.*
Quiocho, Prof. Florante A. *USA.*
Qurashi, Dr Mazhar Mahmood. *Pakistan.*
Qureshi, Mr Khalid Mahmood. *Pakistan.*
Qureshi, Mr Mohammed Kaleem Akhtar. *Pakistan.*
Raade, Mr Gunnar. *Norway.*
Rabenau, Prof. Dr Albrecht. *BRD.*
Rabie, Dr (Mrs) Farida Hamed. *Egypt.*
Rabin, Mr Baruch. *Israel.*
Rabinovich, Prof. Dov. *Israel.*
Rabinowich, Prof. D. *UK.*
Rabinowitz, Dr Israel Nathan. *USA.*
Rachev, Mr Peter. *Bulgaria.*
Rachinger, Prof William Albert. *Australia.*
Radaković, Mrs Aleksandra. *Yugoslavia.*
Radautsan, Prof. Sergey Ivanovich. *USSR.*
Radmilović, Magistar Velimir. *Yugoslavia.*
Radnai, Dr Tamás. *Hungary.*
Radonovich, Prof. Lewis J. *USA.*
Radoslovich, Dr Edward William. *Australia.*
Radukić, Dr Gordana. *Yugoslavia.*
Radulescu, Prof. Dan. *Romania.*
Radwan, Dr Mostafa Mohsen Abdel-Razik. *Egypt.*
Rae, Dr Alan David. *Australia.*
Rae, Dr Alan William James Melville. *UK.*
Rae, Dr Alastair Ian Maxwell. *UK.*
Raftery, Dr James. *UK.*
Ragab, Dr Abdel Ghani. *Egypt.*
Rager, Dr Helmut. *BRD.*
Raghavacharyulu, Dr Iyyunni Venkata Veera. *India.*
Raghunatha, Chary. *India.*
Raghurama, Mr G. *India.*
Rahman, Mr A. F. Md. Maqsudur. *Bangladesh.*
Rahman, Dr Asadur. *Bangladesh.*
Rahman, Dr Sheikh Mohammed Mujibur. *Bangladesh.*
Rahman, Mr Mohammad Abdul. *Pakistan.*
Raidt, Dr Helmut. *DDR.*
Rainov, Dr Nikola. *Bulgaria.*
Raithby, Dr Paul Robert. *UK.*
Rajagopal, Mr Hariharasubramonia Iyer. *India.*
Rajan, Dr Krishna. *Canada.*
Rajan, Mr R. D. *India.*
Rajan, Dr S. S. *India.*
Rajaram, Dr Ramasamy Karunandam. *India.*
Rajaratnam, Prof. Arthur. *Singapore.*
Rajeswaran, Dr Manju. *USA.*
Raju, Dr I. V. K. Bhagaban. *India.*
Raju, Dr K. S. *India.*
Rakova, Dr Elena Vasilyevna. *USSR.*
Ralph, Prof. Brian. *UK.*
Ram, Dr Purushottam. *India.*
Rama Rao, Dr B. *India.*
Ramachandran, Prof. Gopalasamudram N. *India.*
Ramakrishnan, Dr Chandrasekhara. *India.*
Ramalingam, V. *USA.*
Ramanadham, Dr Muthyala. *India.*
Ramasubbu, Mr N. *USA.*
Ramaswami, Dr Krishnamachari. *India.*
Rambaud, Dr Joëlle. *France.*
Ramdas, Dr S. *UK.*
Ramesh, Mr Chellaswamy. *India.*
Ramirez, Mr Edilberto. *Cuba.*
Ramkumar, Mr Sambatur. *India.*
Rammo, Dr Nabil Naim. *Iraq.*
Ramos Bernal, Dr Sergio. *Mexico.*
Ramos Fernandez, Prof. Felicisimo. *Spain.*
Ramos Parente, Dr Carlos Benedicto. *Brazil.*
Rana, Mr Riaz Ahmad. *Pakistan.*
Randaccio, Prof. Lucio. *Italy.*
Ranganath, Dr G. S. *India.*
Ranganathan, Prof. Srinivasa. *India.*
Range, Prof. Dr Klaus-Jürgen. *BRD.*
Ranger, Dr Georges. *Canada.*
Rannev, Dr Nikolay Vasilyevich. *USSR.*
Ransom, Mr H. *UK.*
Ranta, Mr Lasse Kosti. *Finland.*

Rao, Dr J. Krishna Mohana. *India.*
Rao, Dr Keshavamurthy Narayana Swamy. *India.*
Rao, Dr (Miss) Leea Madhava. *India.*
Rao, Mr N. Nagaraja. *India.*
Rao, Mr Nutakki Nageswara. *Malaysia.*
Rao, Prof. P. Rama. *India.*
Rao, Dr Sambhorao Thyagaraja. *USA.*
Rao, Mr Sudhakar S. P. *India.*
Rao, Usha K. *India.*
Raper, Mr Eric Salvin. *UK.*
Rashkova, Dr Diana. *Bulgaria.*
Rasmussen, Prof. Svend Erik. *Denmark.*
Rasmussen, Mr Bjarne. *USA.*
Raston, Dr Colin Llewellyn. *Australia.*
Rastsvetaeva, Dr Ramiza Kerarovna. *USSR.*
Ratanasthien, Dr Benjavun. *Thailand.*
Rath, Prof. Dr Robert. *BRD.*
Ratho, Prof. Trinath. *India.*
Ratna, Miss B. R. *India.*
Ratuszek, Dr Wiktoria Maria. *Poland.*
Ratuszna, Dr Alicja Maria. *Poland.*
Rau, Mr Robert C. *USA.*
Rau, Dr Tamara Fedorovna. *USSR.*
Rau, Dr Valery Georgiyevich. *USSR.*
Raudsepp, Dr Mati. *Canada.*
Rautioaho, Dr Risto Heikki. *Finland.*
Ravaglioli, Dr Antonio. *Italy.*
Ravi Kumar, Mr K. *India.*
Ravichandran, Mrs. V. *India.*
Ravichandran, Mr K. G. *USA.*
Ray, Prof. Alden Earl. *USA.*
Ray, Dr (Mrs) Gouri. *India.*
Ray, Dr Pankaj Narayan. *India.*
Ray, Dr Pradip Kumar. *India.*
Ray, Dr Siddhartha. *India.*
Raykhtsaum, Mr Grigory. *USA.*
Rayment, Dr Ivan. *USA.*
Raymond, Prof. Kenneth Norman. *USA.*
Read, Mr Randy John. *Canada.*
Rebić, Mr Milenko. *Yugoslavia.*
Rechenberg, Dr Ingrid. *DDR.*
Reck, Dr Günter. *DDR.*
Recker, Prof. Dr Kurt. *BRD.*
Reddy, Mr B. Swaminatha. *USA.*
Redhouse, Dr Alan David. *UK.*
Reeber, Dr Robert Richard. *USA.*
Reed, Dr Alvin T. *USA.*
Reed, Prof. John W. *USA.*
Reed, Dr Larry L. *USA.*
Reeke, Prof. George Norman, Jr. *USA.*
Rees, Dr Bernard. *France.*
Rees, Prof. Douglas Charles. *USA.*
Regel', Dr Vadim Robertovich. *USSR.*
Regev, Dr Haya. *Israel.*
Regourd, Dr Micheline. *France.*
Regueira Teodósio, Prof. Joel. *Brazil.*
Reiche, Mr Manfred. *DDR.*
Reichenberg, Mrs Ingrid. *DDR.*
Reid, Dr Allen Forrest. *Australia.*
Reid, Mr Austin H., Jr. *USA.*
Reid, Dr John Sinclair. *UK.*
Reidinger, Mr Franz. *USA.*
Reimers, Mr Walter. *BRD.*
Reinecke, Mrs Kriemhild. *DDR.*
Reinen, Prof. Dr Dirk. *BRD.*
Reinhold, Mrs Ingrid. *DDR.*
Reis, Dr Arthur Henry, Jr. *USA.*
Reisner, Dr George. *Israel.*
Reiss, Drs Céleste A. *Netherlands.*
Ren, Prof. Lei-fu. *China.*
Renaud, Prof. Michel. *France.*
Rendón Diaz Mirón, Dr Luis Emilio. *Mexico.*
Rendle, Dr David Forbes. *UK.*
Renetseder, Mr Roland. *Netherlands.*
Renninger, Prof. Dr Mauritius. *BRD.*
Renouprez, Dr Albert Jean. *France.*
Rentsch, Mr Harald. *DDR.*
Rentzeperis, Prof. Panayiotis Ioannis. *Greece.*
Renzema, Prof. Theodore Samuel. *USA.*
Reppart, Mr William J. *USA.*
Rérat, Dr Claude. *France.*
Restivo, Dr Roderic John. *Canada.*

Restori, Mr Renzo. *Switzerland.*
Retief, Dr Johannes Jacobus. *South Africa.*
Rettig, Dr Steven John. *Canada.*
Reuber-Kürbs, Mrs Dr-Ing Ellen. *BRD.*
Reuter, Prof. Dr-Ing Bertold. *BRD.*
Reuter, Mr Dietrich. *DDR.*
Reyes Chumacero, Mr Antonio. *Mexico.*
Reynaers, Prof. Harry Louis. *Belgium.*
Reynhardt, Prof. Eduard Christiaan. *South Africa.*
Reynolds, Mr C.C. *UK.*
Reynolds, Dr C.D. *UK.*
Reynolds, Dr Phillip Andrew. *Australia.*
Reynolds, Dr Ross Anthony. *USA.*
Rheingold, Dr Arnold Lange. *USA.*
Rhodes, Prof. Rene George. *UK.*
Rhyne, Ms Kay A. *USA.*
Ribár, Prof. Dr Béla. *Yugoslavia.*
Ribbe, Prof. Paul H. *USA.*
Ribbing, Dr Carl-Gustaf. *Sweden.*
Ribeiro Franco, Prof. Rui. *Brazil.*
Ricaldi, Mr Edgar. *Bolivia.*
Ricard, Mr Jean Henri. *France.*
Ricci, Prof. John S., Jr. *USA.*
Rice, Dr Catherine Ellen. *USA.*
Rice, Dr David William. *UK.*
Rich, Prof. Alexander. *USA.*
Richard, Prof. Joseph Albert Pierre. *Canada.*
Richards, Dr Brian Peter. *UK.*
Richards, Mr Gerald F. *USA.*
Richards, Prof. Frederic Middlebrook. *USA.*
Richards, Dr John Philip Gerald. *UK.*
Richardson, Dr C.H. *UK.*
Richardson, Dr David Claude. *USA.*
Richardson, Dr James Wyman, Jr. *USA.*
Richardson, Mrs Jane Shelby. *USA.*
Richardson, Dr John Frederick. *Canada.*
Richardson, Dr Mary Frances. *Canada.*
Richardson, Dr R.M. *UK.*
Riche, Dr Claude. *France.*
Richman, Prof. Marc Herbert. *USA.*
Richter, Dr Frank. *DDR.*
Richter, Mr Hans. *DDR.*
Richter, Dr Klaus. *DDR.*
Richter, Dr Paul Wilhelm. *South Africa.*
Richter, Dr Rainer. *DDR.*
Richter, Mrs Ursula. *BRD.*
Richter, Dr Waltraut. *DDR.*
Rickard, Dr Clifton Edward Frank. *New Zealand.*
Rickards, Mr A.L. *UK.*
Rickert, Prof. Dr Hans. *BRD.*
Ridout, Mr Stephen Charles. *Australia.*
Rieck, Prof. Dr Gerard Daniel. *Netherlands.*
Riedel, Prof. Dr Erwin. *BRD.*
Rieder, Dr Milan. *CSSR.*
Rieger, Dr Hans Wolfhart. *Switzerland.*
Riegert, Mr Richard Paul. *USA.*
Riella, Eng. Humberto Gracher. *Brazil.*
Riess, Prof. John Karlem. *USA.*
Rietveld, Dr Hugo Marie. *Netherlands.*
Riganti, Prof. Vincenzo. *Italy.*
Rigault de la Longrais, Prof. Germain. *Italy.*
Rigopoulos, Prof. Rigas. *Greece.*
Rigotti, Dr Graciela. *Argentina.*
Rihs, Mrs Greti. *Switzerland.*
Rinaldi, Prof. Romano. *Italy.*
Rinaudo, Dr Caterina. *Italy.*
Rincón Saenz, Mr Luis Felipe. *Colombia.*
Rinderer, Prof. Leo. *Switzerland.*
Rindorf, Mrs Grethe. *Denmark.*
Ringel, Dr Lilli. *DDR.*
Rios Jara, Dr David. *Mexico.*
Ripamonti, Prof. Alberto. *Italy.*
Ripamonti, Dr Carlo. *Italy.*
Rischák, Mr Géza. *Hungary.*
Risler, Dr Jean-Loup. *France.*
Riste, Prof. Tormod. *Norway.*
Ritschel, Mr Manfred. *DDR.*
Riva di Sanseverino, Prof. Lodovico. *Italy.*
Rivera, Mr Carlos. *Chile.*

Rivera Moras, Mr Vicente. *Mexico.*
Rivera Ocando, Dr Angela Valentina. *Venezuela.*
Rivero, Dr Blas Eduardo. *Argentina.*
Riveros Rotgé, Dr Héctor Gerardo. *Mexico.*
Rizvi, Prof Adibul Hasan. *Pakistan.*
Rizvi, Dr Syed Sadrul Hassan. *Pakistan.*
Robbins, Mr Carl Richard. *USA.*
Robert, Mr Marc. *France.*
Roberts, Mr Andrew Clifford. *Canada.*
Roberts, Dr Kevin John. *UK.*
Roberts, Dr Michael Mark. *USA.*
Roberts-Sengier, Dr Lieve. *UK.*
Robertson, Dean B. Ken. *USA.*
Robertson, Prof. Beverly Ellis. *Canada.*
Robertson, Dr Glen Bradley. *Australia.*
Robertson, Prof. James David. *USA.*
Robertson, Dr John Harry. *UK.*
Robertson, Prof. John Monteath. *UK.*
Robertson, Mr Robert Hugh Stannus. *UK.*
Robertus, Prof. Jon David. *USA.*
Robinson, Dr Ian Keith. *USA.*
Robinson, Mr John C. *USA.*
Robinson, Dr K. *UK.*
Robinson, Mr P.T. *UK.*
Robinson, Dr Ward Thomas. *New Zealand.*
Robinson, Prof. William Robert. *USA.*
Rochon, Prof. Fernande D. *Canada.*
Rocophyllou Agathonikou, Dr Elsa - Helena. *Greece.*
Rodesiler, Prof. Paul F. *USA.*
Rodgers, Dr Kerry Anthony. *New Zealand.*
Rodgers, Dr Allen Lawrence. *South Africa.*
Rodrigues, Dr Antonio Ricardo Dröher, *Brazil.*
Rodrigues, Dr Edson. *Brazil.*
Rodrigues da Silva, Dr Rilson. *Brazil.*
Rodriguez Clemente, Dr Rafael. *Spain.*
Rodriguez Gallego, Prof. Manuel. *Spain.*
Rodríguez Lara, Prof. Jaime. *Colombia.*
Rodríguez S., Miss Gloria Inés. *Colombia.*
Rodrique, Dr Luc Willy. *Belgium.*
Rodulfo de Gil, Prof. Eldrys Emilia. *Venezuela.*
Roebuck, Dr Peter Hamish Athey. *UK.*
Römming, Prof. Christian. *Norway.*
Rønsbo, Mr Jørn. *Denmark.*
Rösch, Dr Heinrich. *BRD.*
Röst, Mr Erling. *Norway.*
Roettgers, Mr Wolbert. *USA.*
Rogacheva, Dr Evelina Danilovna. *USSR.*
Rogers, Prof. Donald. *USA.*
Rogers, Mr K.D. *UK.*
Rogers, Prof. Robin Don. *USA.*
Rogić, Prof. Dr Vinko. *Yugoslavia.*
Roginskaya, Dr Yuliana Eremeyevna. *USSR.*
Rogl, Doz. Dr Peter Franz. *Austria.*
Rognlie, Mr David G. *USA.*
Rogulić, Prof. Dr Mileva. *Yugoslavia.*
Rohmer, Mr Christian. *BRD.*
Rohrbaugh, Dr Wayne Joseph. *USA.*
Rohrer, Dr Douglas C. *USA.*
Roilos, Prof. Minas. *Greece.*
Rojas Lopez, Dr Rosa Ma. *Spain.*
Roldan, Dr Luis-Gonzalez. *USA.*
Rolim De Camargo, Prof. William Gerson. *Brazil.*
Rolland, Mr Guy. *France.*
Rollett, Dr John Sydney. *UK.*
Romanov, Dr Gennady Vasilyevich. *USSR.*
Romero, Dr Miguel. *Mexico.*
Romero Romo, Dr Mario. *Mexico.*
Romers, Prof. Dr Cornelis. *Netherlands.*
Ronco, Dr Alicia Estela. *Argentina.*
Roof, Dr Raymond Bradley. *USA.*
Rosa, Dr Rodolfo. *Italy.*
Rosca, Mr Liviu. *Romania.*
Rose, Dr David Richard. *Canada.*
Rose, Dr Jean. *France.*
Rose, Jon Patrick. *USA.*
Rose-Hansen, Dr John. *Denmark.*
Rosen, Dr Lawrence Stephen. *USA.*
Rosenfield, Dr Richard E., Jr. *USA.*
Rosenqvist, Prof. Ivan Thoralf. *Norway.*
Rosenstein, Prof. Robert Daniel. *USA.*

Rosin, Dr Horst. *DDR.*
Ross, Ms Dawn L. *USA.*
Ross, Dr Donald Keith. *UK.*
Ross, Dr Frederick Keith. *USA.*
Ross, Dr Malcolm. *USA.*
Ross, Ms Nancy L. *USA.*
Rossat-Mignod, Dr Jean. *France.*
Rossell, Dr Henry John. *Australia.*
Rossi, Dr Giuseppe. *Italy.*
Rossi, Dr Miriam. *USA.*
Rossi, Mr Franco Antonio. *Switzerland.*
Rossmanith, Mrs. Dr Elisabeth. *BRD.*
Rossmann, Prof. Michael G. *USA.*
Rossner, Mr Johannes. *DDR.*
Rossouw, Dr Christopher John. *Australia.*
Rost, Mrs Jutta. *DDR.*
Rostami, Dr Farzaneh. *Iran.*
Rosukhi, Dr Siamak. *Iran.*
Rotella, Dr Frank J. *USA.*
Roth, Dr Michel. *France.*
Roth, Dr Robert Sidney. *USA.*
Roth, Prof. Walter Lester. *USA.*
Rothbauer, Dr Richard. *BRD.*
Rott, Dr Volkwin. *BRD.*
Rotti, Mr Marc Maurice. *Belgium.*
Rouffignac, Dr Eric de. *Mexico.*
Roughead, Mr William A. *USA.*
Roult, Mr Georges. *France.*
Rousseaux, Dr Françoise. *France.*
Roux, Dr Jacobus Paul. *South Africa.*
Roveri, Prof. Norberto. *Italy.*
Rowland, Mr John Fleming. *Canada.*
Rowley, Dr Colin Raymond. *UK.*
Roy, Dr Ajoy Kumer. *Bangladesh.*
Roy, Prof. Rustum. *USA.*
Roychowdhury, Dr Priyobroto. *India.*
Royer, Dr William Edward, Jr. *USA.*
Roys, Mr W.B. *UK.*
Rozhansky, Dr Vladimir Nikolayevich. *USSR.*
Rozhdestvenskaya, Dr Ira Vasilyevna. *USSR.*
Rozsondai, Dr Béla. *Hungary.*
Ruban, G. Albert. *India.*
Ruban, Prof. Dr Gerhard. *BRD.*
Rubbo, Dr Marco. *Italy.*
Ruben, Mrs Helena W. *USA.*
Rubiano Lamouroux, Prof. Manuel. *Colombia.*
Rubin, Dr Ben Z. *USA.*
Rubin, Prof. Byron. *USA.*
Rubio de Cubides, Prof. Julia. *Colombia.*
Ruble, Dr John Rollo. *USA.*
Rucklidge, Prof. John Christopher. *Canada.*
Ruderman, Dr Warren. *USA.*
Rudert, Mr Rainer. *BRD.*
Rudman, Prof. Peter S. *Israel.*
Rudman, Prof. Reuben M. *USA.*
Rudnick, Dr Suzanne Ellen. *USA.*
Rudnik, Mr Paul J. *USA.*
Rudolph, Dr Peter. *DDR.*
Rudolph, Dr Philip Reinhold. *USA.*
Rueda Bravo, Dr Daniel-Reyes. *Spain.*
Rüegg, Dr Andreas. *Switzerland.*
Ruhl, Miss Barbara Louise. *Canada.*
Ruiz Amil, Dr Antonio. *Spain.*
Ruiz Mejia, Dr Carlos. *Mexico.*
Ruiz Perez, Mrs Catalina. *BRD.*
Rule, Mr Stephen A. *UK.*
Rumanova, Dr Iskra Mikhailovna. *USSR.*
Rumball, Dr Sylvia Vine. *New Zealand.*
Rundgren, Mr Kent A. *Sweden.*
Rundqvist, Prof. Stig O. *Sweden.*
Runje, Dr Vesna. *Yugoslavia.*
Ruppersberg, Prof. Dr Henner. *BRD.*
Ruscher, Prof. Christian. *DDR.*
Russell, Dr David Robin. *UK.*
Russell, Mr Nazirullah. *Pakistan.*
Russell, Dr Paul Robert. *UK.*
Russell, Dr Thomas Paul. *USA.*
Russev, Dr Krassimir. *Bulgaria.*
Ruszala, Mr Ferdinand A. *USA.*
Rutland, Ms J. *UK.*
Rutten-Keulemans, Drs Elisabeth W.M. *Netherlands.*

Ruud, Prof. Clayton Olaf. *USA.*
Ružić-Toroš, Dr Živa. *Yugoslavia.*
Ryaboshapka, Dr Karl Petrovich. *USSR.*
Ryan, Dr Robert Reynolds. *USA.*
Ryba, Prof. Earle Richard. *USA.*
Rychlý, Ing Rudolf. *CSSR.*
Rychlewska, Dr Urszula. *Poland.*
Rydel, Mr Timothy John. *USA.*
Ryzhenkov, Dr Alexander Pavlovich. *USSR.*
Rzaigui, Dr Mohamed. *Tunisia.*
Saadi-Nam, Dr Abolfazle. *Iran.*
Saalfeld, Prof. Dr Horst. *BRD.*
Saavedra, Mr Antonio. *Bolivia.*
Sabat, Dr Michal. *Italy.*
Sabatino, Dr Piera. *Italy.*
Sabbioni, Dr Cristina. *Italy.*
Sabelli, Dr Cesare. *Italy.*
Sabine, Prof Terence Murray. *Australia.*
Sabrowsky, Prof. Dr Horst. *BRD.*
Sacconi, Prof. Luigi. *Italy.*
Sacerdoti, Prof. Michele. *Italy.*
Sack, Dr John Stuart. *USA.*
Sadanaga, Prof. Ryoichi. *Japan.*
Sadek, Dr Gamil. *Egypt.*
Sadikov, Dr Georgiy Georgievich. *USSR.*
Sadova, Dr Nina Ivanovna. *USSR.*
Sadowski, Mr Lucian M. *USA.*
Saeed, Mr Syed Mohammad. *Pakistan.*
Saenger, Prof. Dr Wolfram H. E. *BRD.*
Sæthre, Mr Leif Jarle. *Norway.*
Safa, Dr Mehdi. *Iran.*
Sagawa, Prof. Takasi. *Japan.*
Saggerson, Prof. Edward Phillips. *South Africa.*
Saghir, Mr Ahmad. *Pakistan.*
Saha, Mr Ajay Prakash. *India.*
Saha, Mr Bishwa Nath. *India.*
Saha, Prof. Narendra Nath. *India.*
Sahalos, Prof. John. *Greece.*
Sahani, Dr Jiwan Lal. *India.*
Sahaymary, Mrs. J. James *India.*
Şahin, Mr Mehmet. *Turkey.*
Sahle, Dr Wubeshet. *Sweden.*
Sahu, Dr Bholanath. *India.*
Sahu, Dr Mahendra. *India.*
Sahu, Dr N. C. *India.*
Sahu, Dr Ram Gopal. *India.*
Sahu, Dr Ramdhani. *India.*
Sainz Amor, Dra Emma. *Spain.*
Saito, Prof. Norio. *Japan.*
Saito, Prof. Yoshihiko. *Japan.*
Saito, Mr Yoshio. *Japan.*
Saito, Mr Yoshiyuki. *Japan.*
Saka, Dr Takashi. *Japan.*
Sakabe, Dr Noriyoshi. *Japan.*
Sakaki, Prof. Em. Yoneichiro. *Japan.*
Sakamoto, Dr Yosio. *Japan.*
Sakata, Dr Makoto. *Japan.*
Sakellaridis, Prof. Paul. *Greece.*
Sakellariou, Mr Evangelos. *Switzerland.*
Sakharov, Dr Boris Alexandrovich. *USSR.*
Sakkopoulos, Dr Sotirios. *Greece.*
Sakore, Dr Tukaram D. *USA.*
Sakuma, Dr Takashi. *Japan.*
Sakurai, Dr Junji. *Japan.*
Sakurai, Dr Kiichi. *Japan.*
Sakurai, Dr Tosio. *Japan.*
Sal'dau, Dr El'ga Petrovna. *USSR.*
Salanci, Prof. Dr Berkin. *Turkey.*
Salas, Dr Guillermo Armando. *Mexico.*
Salazar Orrego, Prof. Ramon. *Peru.*
Saldin, Dr D.K. *UK.*
Salem, Dr Safia Mahmoud. *Egypt.*
Salem, Mr M. *UK.*
Salemme, Dr Francis Raymond. *USA.*
Salgado, Dr José. *Portugal.*
Salje, Prof. Dr Ekhard. *BRD.*
Salkind, Prof. Alvin J. *USA.*
Salleh, Dr Mansor bin Haji. *Malaysia.*
Salleh, Dr Mohammad Nawi. *Malaysia.*
Salmon, Dr Oliver Norton. *USA.*

Salot, Dr Stuart Edwin. *USA.*
Salt, Mr P.D. *UK.*
Saludjian, Mr Pedro. *France.*
Salunke, Dr Dinakar M. *India.*
Salvado Canelhas, Mrs Maria da Graça. *Portugal.*
Salviati, Dr Giancarlo. *Italy.*
Samanta, Chitra. *India.*
Samantaray, Dr Biswas Kumar. *India.*
Samarskaya, Dr Valentina Dmitriyevna. *USSR.*
Samdal, Mr Svein. *Norway.*
Samoilovich, Dr Lidiya Alexandrovna. *USSR.*
Samoilovich, Prof. Mikhail Isaakovich. *USSR.*
Sampson, Mr Christopher. *UK.*
Samson, Dr Sten. *USA.*
Samudzi, Mr Cleopas T. *USA.*
Samuels, Prof. Robert J. *USA.*
Samuelsen, Prof. Emil J. *Norway.*
Samus', Dr Ivan Dmitriyevich. *USSR.*
Sanadze, Prof. Vladimir Vladimirovich. *USSR.*
Sanchez V., Mr Alfredo. *Colombia.*
Sandalaki, Dr Zefi. *asst.*
Sanders, Dr John Veysey. *Australia.*
Sanderson, Dr Mark Rutherford. *USA.*
Sandomirsky, Dr Pavel Alexandrovich. *USSR.*
Sándor, Dr Endre Elek. *UK.*
Sands, Prof. Donald E. *USA.*
Sandström, Dr Magnus K. E. *Sweden.*
Sanjinés, Mr Orlando. *Bolivia.*
Sankar, Mr B. N. *India.*
Sanni, Dr Bamidele. *Nigeria.*
Sannikov, Dr Daniil Grigoryevich. *USSR.*
Sansoni, Prof. Mirella. *Italy.*
Santarsiero, Dr Bernard Inez Dominic M. *USA.*
Santivañez, Mr Reynaldo. *Bolivia.*
Santos, Dr Persio de Souza. *Brazil.*
Santos, Mrs Regina Helena de Almeida. *Brazil.*
Saper, Dr Mark A. *Israel.*
Sappenfield, Mr Eric L. *USA.*
Sariel, Mr Joseph. *Israel.*
Sarin, Dr Victor Anatol'yevich. *USSR.*
Sarkar, Chitrita. *India.*
Sarkar, Dr Satyabrata. *India.*
Sarkisov, Dr Stepan Ervandovich. *USSR.*
Sarko, Prof. Anatole. *USA.*
Sarma, Prof. Raghupathy. *USA.*
Sarodnik, Dr Reinhard. *DDR.*
Sartono, Mr. *Indonesia.*
Sartori, Prof. Franco. *Italy.*
Sasada, Prof. Yoshio. *Japan.*
Sasaki, Prof. Akio. *Japan.*
Sasaki, Dr Kyoyu. *Japan.*
Sasaki, Dr Satoshi. *Japan.*
Sasaki, Prof. Yukiyoshi. *Japan.*
Sasisekharan, Prof. Visvanathan. *India.*
Sass, Prof. Ronald L. *USA.*
Sass, Prof. Stephen Louis. *USA.*
Sastry, Dr Bommakanti Sri Rama. *India.*
Sastry, Dr G. V. S. *India.*
Sastry, Dr Medury Dattatreya. *India.*
Sasvári, Dr Judit. *Hungary.*
Sasvári, Dr Kálmán. *Hungary.*
Satittada, Miss Gannaga. *Thailand.*
Sato, Mr Hiroki. *Japan.*
Sato, Mr Kazuo. *Japan.*
Sato, Assoc. Prof. Mitsuo. *Japan.*
Sato, Prof. Shin'ichi. *Japan.*
Sato, Dr Shoichi. *Japan.*
Sato, Mr Tomohiro. *Japan.*
Satow, Dr Yoshinori. *Japan.*
Satyanarayana Murthy, Dr Keta. *India.*
Satyshur, Mr Kenneth A. *USA.*
Saul, Mr Frederic. *France.*
Saur, Prof. Dr-Ing Eugen. *BRD.*
Sauvage-Simkin, Dr Michèle. *France.*
Savithramma, Miss K. L. *India.*
Sawa, Dr Isao. *Japan.*
Sawada, Dr Akikatsu. *Japan.*
Sawada, Prof. Masao. *Japan.*
Sawada, Dr Toshiyuki. *Japan.*
Sawada, Mr Yasuaki. *Japan.*

Sawaguchi, Prof. Etsuro. *Japan.*
Sawka-Dobrowolska, Dr Wanda. *Poland.*
Sawyer, Dr Jeffrey Frederick. *Canada.*
Sawyer, Dr Lindsay. *UK.*
Sawzik, Ms Patricia. *USA.*
Sax, Dr Martin. *USA.*
Sayin, Doç. Dr Mahmut. *Turkey.*
Sayler, Prof. Alice Ann. *USA.*
Saylor, Dr Charles Proffer. *USA.*
Sayre, Dr David. *USA.*
Sazedj-Khosrawan, Dr Feresteh. *Austria.*
Scandale, Prof. Eugenio. *Italy.*
Scaramuzza, Dr Lucio. *Italy.*
Scarbrough, Dr Frank Edward. *USA.*
Scaringe, Dr Raymond P. *USA.*
Scatturin, Prof. Vladimiro. *Italy.*
Ščavničar, Prof. Dr Stjepan. *Yugoslavia.*
Schaal, Mr Joachim. *DDR.*
Schäfer, Prof. Dr Herbert Leo. *BRD.*
Schäfer, Mr Peter. *DDR.*
Schäfer, Prof. Lothar. *USA.*
Schaefer, Dr William Palzer. *USA.*
Schäfer, Dr Wolfgang. *BRD.*
Schams-Ashtiani, Dr Ahmad. *Iran.*
Schanda, Dr Friedrich. *BRD.*
Schapink, Dr Frederik Willem. *Netherlands.*
Scharfenberg, Mr Rudolf. *DDR.*
Scharfenberger, Miss Ulrike. *BRD.*
Schattschneider, Dr Dipl.-Ing. Mag. Peter. *Austria.*
Schei, Dr Svanhild Helene. *Norway.*
Scheidt, Prof. W. Robert. *USA.*
Schenk, Prof. Dr Hendrik. *Netherlands.*
Schenk, Dr Kurt Johann. *Switzerland.*
Schenk, Dr Manfred. *DDR.*
Scheringer, Prof Dr Christian Andreas. *BRD.*
Scheschinski, Mrs Karin. *DDR.*
Schiavinato, Prof. Giuseppe. *Italy.*
Schicker, Mr Peter. *Switzerland.*
Schieber, Prof. Michael. *Israel.*
Schierbeek, Mr Abraham Johan. *Netherlands.*
Schiffer, Dr Marianne. *USA.*
Schilder, Ing Jaroslav. *CSSR.*
Schildkamp, Dr Wilfried. *BRD.*
Schiller, Dr Claude. *France.*
Schilling, Mr Hansjoachim. *DDR.*
Schimanski, Dr Uwe Lothar. *BRD.*
Schioler, Dr Liselotte Jensen. *USA.*
Schippel, Dr Erhard. *DDR.*
Schirber, Dr James Emmanuel. *USA.*
Schirmer, Dr Ulrich. *BRD.*
Schläfer, Dr Dietrich. *DDR.*
Schläfer, Dr Ursula. *DDR.*
Schlein, Dr Werner. *Chile.*
Schlemper, Prof. Elmer Otto. *USA.*
Schlenker, Dr John L. *USA.*
Schlenker, Prof. Michel. *France.*
Schliephake, Dr Rolf-Werner. *BRD.*
Schloemer, Prof. Dr Hermann J. *BRD.*
Schlueter, Prof. Albert S. *USA.*
Schmahl, Mr Wolfgang Wilhelm. *BRD.*
Schmalle, Dr Helmut Willi. *BRD.*
Schmelczer, Dr Robert. *Switzerland.*
Schmelzler, Dr Hans-Peter. *DDR.*
Schmetzer, Dr Karl. *BRD.*
Schmid, Prof. Hans. *Switzerland.*
Schmid, Dr Michael Francis. *USA.*
Schmidt, Mr Bertram Felix Paul. *BRD.*
Schmidt, Mr Elias Rudolph. *South Africa.*
Schmidt, Prof. Günter. *DDR.*
Schmidt, Mrs Margot. *DDR.*
Schmidt, Prof. Paul Woodward. *USA.*
Schmidt, Mr Peter. *DDR.*
Schmidt, Prof. Werner. *DDR.*
Schmidt, Dr William Charles, Jr. *USA.*
Schmidt-Nielsen, Mr Søren. *Denmark.*
Schmirgeld, Dr Lelia. *Argentina.*
Schmitz, Dr Werner. *DDR.*
Schmücker, Mr Jürgen. *DDR.*
Schneider, Prof. Günter. *DDR.*
Schneider, Dr Gunter. *Sweden.*
Schneider, Dr Hartmut. *BRD.*

Schneider, Prof. Herbert. *DDR.*
Schneider, Dr Jochen Richard. *BRD.*
Schneider, Dr Julius. *BRD.*
Schneider, Dr Michael L. *USA.*
Schneider, Dr Walter. *BRD.*
Schnering, von, Prof. Dr Dr hc Hans Georg. *BRD.*
Schobinger-Papamantellos, Dr P. *Switzerland.*
Schoch, Dr Aylva Ernest. *South Africa.*
Schöllhorn, Prof. Dr Robert. *BRD.*
Schoenborn, Dr Benno Paul. *USA.*
Schönholzer, Mr Peter. *Switzerland.*
Schoening, Prof. Friedrich R.L. *South Africa.*
Scholz, Dr Heinz Werner. *BRD.*
Schomaker, Prof. Verner. *USA.*
Schomburg, Dr Dietmar. *BRD.*
Schoone, Prof. Dr Jean C. *Netherlands.*
Schooneveld, van, Ing. Marinus. *Netherlands.*
Schott, Prof. Günter. *DDR.*
Schrader, Prof. Richard. *DDR.*
Schramm, Dr Volker. *BRD.*
Schrauber, Mrs Hannelore. *DDR.*
Schreiter, Dr Peter. *DDR.*
Schreuder, Drs Herman Antony. *Netherlands.*
Schreurs, Drs Antonius Mathias Maria. *Netherlands.*
Schröcke, Prof. Dr Helmut. *BRD.*
Schröder, Dr Friedrich Anton. *BRD.*
Schroeder, Dr LeRoy William. *USA.*
Schröder, Dr Winfried. *BRD.*
Schröpfer, Dr Lothar Maximilian. *BRD.*
Schroll, Prof. Dr Erich. *Austria.*
Schryvers, Mr Dominique Maurits. *Belgium.*
Schubert, Dr Gernot. *DDR.*
Schubert, Mrs Heike-Kristina. *DDR.*
Schubert, Mr Helmut. *BRD.*
Schubert, Prof. Dr Konrad. *BRD.*
Schülke, Prof Dr Winfried. *BRD.*
Schüller, Prof. Dr Karl-Heinz. *BRD.*
Schuermann, Dr Kay Uwe. *BRD.*
Schultz, Dr Arthur Jay. *USA.*
Schultz, Dr György. *Hungary.*
Schultz, Prof. Jerold Marvin. *USA.*
Schultze-Rhonhof, Dr Ernst. *BRD.*
Schulz, Dr Georg Eberhardt Bruno. *BRD.*
Schulz, Prof. Dr Heinz Hermann. *BRD.*
Schulz, Dr Manfred. *DDR.*
Schulze, Dr Dietrich. *DDR.*
Schulze, Dr Günter. *DDR.*
Schuman, Mr Clifford Alan. *USA.*
Schumann, Dr Bernd. *DDR.*
Schunk, Mr Wolfgang. *DDR.*
Schur, Dr Karl. *BRD.*
Schuster, Prof. Dr Hans-Uwe. *BRD.*
Schuster, Dr Julius Clemens. *Austria.*
Schuster, Prof. Sanford Lee. *USA.*
Schuszter, Dr Ferenc. *Hungary.*
Schutte, Prof. Casper Jan Hendrik. *South Africa.*
Schuyff, Prof. Dr Abraham. *Netherlands.*
Schwalbe, Dr Carl Hellmuth Walter. *UK.*
Schwartz, Dr Kenneth Bruce. *USA.*
Schwartz, Dr Lyle Howard. *USA.*
Schwartze, Mrs Gabriele. *DDR.*
Schwarz, Prof. Dr Karlheinz. *Austria.*
Schwarz, Dr Wolfgang. *BRD.*
Schwarzenbach, Prof. Dieter. *Switzerland.*
Schwarzenberger, Mrs D.R. *UK.*
Schwarzmann, Mrs Dr Sigrid. *BRD.*
Schweinsberg, Dr Heinz Friedrich. *BRD.*
Schweizer, Dr Wolfhard Bernd. *Switzerland.*
Schwomma, Dr Otto. *Austria.*
Sclar, Prof. Charles Bertram. *USA.*
Scordari, Prof. Fernando. *Italy.*
Scott, Mr Donald Lee. *USA.*
Scott, Dr Henry Gordon. *Australia.*
Scott, Dr James Douglas. *Canada.*
Scouloudi, Dr Helen. *UK.*
Scrimgeour, Dr Sheelagh Nicoll. *UK.*
Scutcher, Dr W. *UK.*
Seabaugh, Mr Pyrtle W. *USA.*
Seal, Alpana. *India.*
Seal, Prof. Arun Kumar. *India.*

Seal, Dr Michael. *Netherlands.*
Seale, Dr Steve Keith. *USA.*
Sears, Dr Donald Richard. *USA.*
Seaton, Dr Barbara A. *USA.*
Šebo, Dr Pavel. *CSSR.*
Secco, Prof. Anthony Silvio. *Canada.*
Secco, Dr Luciano. *Italy.*
Seddon, Dr John Michael. *UK.*
Šedivý, Doc. Dr Josef. *CSSR.*
Sedlacek, Dr Paul. *DDR.*
Sedmalis, Prof. Uldis Yanovich. *USSR.*
Sedzik, Dr Jan. *Sweden.*
Seeger, Prof. Dr Karlheinz. *Austria.*
Seely, Mr Oliver, Jr. *USA.*
Seeman, Prof. Nadrian Charles. *USA.*
Seemann, Dr Hans. *DDR.*
Seetharaman, Mr Venkataramakrishanan. *India.*
Seff, Prof. Karl. *USA.*
Segal, Mr Eugen. *Romania.*
Segall, Prof Robert Leo. *Australia.*
Šegedin, Magistar Primož. *Yugoslavia.*
Segmüller, Dr Armin Paul. *USA.*
Seidel, Dr Peter. *BRD.*
Seidl, Dr Erwin. *Austria.*
Seidl, Ing Vlastimil. *CSSR.*
Seidowski, Dr Eckart. *DDR.*
Seifert, Prof. Dr Hans-Joachim. *BRD.*
Seifert, Prof. Dr Karl-Friedrich. *BRD.*
Seifert, Dr Karl Josef. *Austria.*
Seifert, Dr Wolfgang. *DDR.*
Seiler, Mr Paul. *Switzerland.*
Seitsonen, Dr Sulo Iivari. *Finland.*
Seitz, Dr Frederick. *USA.*
Seiyama, Prof. Tetsuro. *Japan.*
Sekanina, Prof. Dr Josef. *Brno.*
Sekar, Mr K. *India.*
Seki, Prof. Syuzo. *Japan.*
Sekizaki, Dr Masao. *Japan.*
Self, Dr Peter Geoffrey. *Australia.*
Sellar, Dr Jeffrey Ronald John. *Australia.*
Semendy, Mr Alphonse F. *USA.*
Semenova, Dr Tat'yana Fedorovna, *USSR.*
Semenovskaya, Dr Svetlana Victorovna. *USSR.*
Semiletov, Prof. Stepan Alekseyevich. *USSR.*
Semitelou, Mrs Julia. *Greece.*
Semmingsen, Dr Dag. *Norway.*
Sen, Dr Deb Kumar. *India.*
Sen, Miss Mina. *India.*
Sen, Mrs. Nirupa. *India.*
Sen, Dr Ranjit Kumar *India.*
Sen, Mrs Suchitra. *India.*
Sen Gupta, Prof. Pradip Kumar. *USA.*
Sen Gupta, Prof. Siba Prasad. *India.*
Senadhi, Mr Vijay Kumar. *India.*
Sencer, Mr Osman. *Turkey.*
Senechal, Prof. Marjorie Lee. *USA.*
Sengupta, Mr Sisir K. *USA.*
Senti, Mr F. R. *USA.*
Sepehrnia, Mr Bahman. *USA.*
Sequeira, Dr Anisbert Stanislaus. *India.*
Sera, Mr Masaaki. *Japan.*
Serafin, Dr Michael. *BRD.*
Serdyuk, Dr Igor' Nikolayevich. *USSR.*
Serebryakov, Dr Anatoly Valeryevich. *USSR.*
Serebryanaya, Dr Nadezhda Ruvimovna. *USSR.*
Sergeeva, Dr Valeriya Mikhailovna. *USSR.*
Sergienko, Dr Vladimir Semenovich. *USSR.*
Serimaa, Mrs Ritva Elina. *Finland.*
Serratosa, Prof. Dr Jose Ma. *Spain.*
Sersale, Prof. Riccardo. *Italy.*
Servidori, Dr Marco. *Italy.*
Servos, Prof. Kurt. *USA.*
Sevast'yanov, Dr Boris Konstantinovich. *USSR.*
Sewell, Mr Bryan Trevor. *South Africa.*
Seydel, Mrs Renate. *DDR.*
Sfez, Mr Gérard. *France.*
Sgarabotto, Prof. Paolo. *Italy.*
Sgarlata, Prof. Francesco. *Italy.*
Sgualdino, Dr Giulio. *Italy.*
Shaanan, Dr Boaz. *Israel.*

Shackleton, Miss Judith Mary. *UK.*
Shaffer, Prof. Lawrence B. *USA.*
Shaffner, Dr Thomas Jackson. *USA.*
Shafizade, Dr Rafik Bekhbud ogly. *USSR.*
Shafranovsky, Prof. Ilarion Ilarionovich. *USSR.*
Shah, Prof Muzaffar Ali. *Pakistan.*
Shah, Dr Jitendra Shantilal. *UK.*
Shah, Miss V.K. *UK.*
Shahi, Mr Ghulam Nabi. *Pakistan.*
Shaikh, Mr Mohammad Iqbal. *Pakistan.*
Shaikh, Mr Mohammad Sualehin. *Pakistan.*
Shaikh, Mr Qameruddin. *Pakistan.*
Shaked, Prof. Hagai. *Israel.*
Shakked, Prof. Zippora. *Israel.*
Shaldin, Dr Yury Vitalyevich. *USSR.*
Shamburov, Dr Vladimir Alekseyevich. *USSR.*
Shamir, Dr Noah. *Israel.*
Shamray, Dr Vladimir Fedorovich. *USSR.*
Shan, Mr Try-seo. *China.*
Shandles, Mr Robert. *USA.*
Shang, Mr Mao-yu. *China.*
Shannon, Dr Robert Day. *USA.*
Shao, Prof. Jie-lian. *China.*
Shao, Prof. Mei-cheng. *China.*
Sharma, Dr (Mrs) Aysel. *Nigeria.*
Sharma, Mr Braj Bhushan. *India.*
Sharma, Mrs Renu. *Sweden.*
Sharma, Dr Surinder Dutt. *India.*
Sharma, Dr Vinod Chander. *Nigeria.*
Sharp, Prof. David William Arthur. *UK.*
Sharrah, Mr Paul C. *USA.*
Shashidhar, Dr R. *India.*
Shashkin, Dr Dmitry Petrovich. *USSR.*
Shaw (née Gözen), Dr Leylâ Süheylâ. *UK.*
Shchedrin, Dr Boris Mikhailovich. *USSR.*
Shcherbakova, Dr Mira Yakovlevna. *USSR.*
Sheftal', Prof. Nikolay Naumovich. *USSR.*
Shefter, Dr Eli. *USA.*
Sheka, Prof. Elena Fedorovna. *USSR.*
Shekhtman, Dr Veniamin Sholomovich. *USSR.*
Sheldon, Dr Robert Isaly. *USA.*
Sheldrick, Dr Bernard. *UK.*
Sheldrick, Prof. Dr George Michael. *BRD.*
Sheldrick, Prof Dr William Stephen. *BRD.*
Shelley, Dr David. *New Zealand.*
Shen, Mr Cheng. *China.*
Shen, Miss Fu-ling. *China.*
Shen, Prof. Jin-chuan. *China.*
Shen, Dr Ming-Shing. *USA.*
Shen, Dr Pooyan. *Taiwan.*
Shepelev, Dr Yury Fedorovich. *USSR.*
Shepperd, Dr Christine Mary. *UK.*
Sheriff, Dr Steven. *USA.*
Sherman, Mr Robert L. *USA.*
Sherry, Mrs Elizabeth Ann Gebert. *USA.*
Sherwood, Prof. John Neil. *UK.*
Shi, Mr Bi-de. *China.*
Shi, Mr Ni-cheng. *China.*
Shibaeva, Dr Rimma Pavlovna. *USSR.*
Shibata, Prof. Noboru. *Japan.*
Shibata, Prof. Shuzo. *Japan.*
Shibuya, Prof. Iwao. *Japan.*
Shichiri, Dr Takaki. *Japan.*
Shieh, Dr Huey-Sheng. *USA.*
Shields, Dr Kelvin George. *Australia.*
Shigenari, Dr Takeshi. *Japan.*
Shihub, Mr Salahedin I. *Libya.*
Shimanouchi, Dr Hirotaka. *Japan.*
Shimaoka, Prof. Kohji. *Japan.*
Shimazaki, Dr Yoshihiko. *Japan.*
Shimazu, Dr Masaji. *Japan.*
Shimizu, Prof. Kenichi. *Japan.*
Shimura, Mr Fumio. *Japan.*
Shin, Prof. Hyun So. *Korea.*
Shin, Prof. Dr Whanchul. *Korea.*
Shinn, Dr Dennis Burton. *USA.*
Shinnaka, Prof. Yasuhiro. *Japan.*
Shinohara, Dr Kenichi. *Japan.*
Shintani, Dr Ryuichi. *Japan.*
Shiojiri, Prof. Makoto. *Japan.*
Shiono, Prof. Ryonosuke. *USA.*

Shipley, Prof. G. Graham. *USA.*
Shirai, Prof. Shunji. *Japan.*
Shirai, Mr Yoshihiro. *Japan.*
Shiro, Dr Motoo. *Japan.*
Shirozu, Prof. Haruo. *Japan.*
Shishova, Dr Tatyana Gennadiyevna. *USSR.*
Shivaprakash, Dr N. C. *India.*
Shivrin, Dr Oleg Mikolayevich. *USSR.*
Shklover, Dr Valery Efimovich. *USSR.*
Shkol'nikova, Dr Larisa Mikhailovna. *USSR.*
Shlichta, Dr Paul Joseph. *USA.*
Shmueli, Prof. Uri. *Israel.*
Shmyt'ko, Dr Ivan Mikhailovich. *USSR.*
Shnulin, Dr Anatoly Nikolayevich. *USSR.*
Shoda, Prof. Tokugoro. *Japan.*
Shoemaker, Prof. Clara Brink. *USA.*
Shoemaker, Prof. David Powell. *USA.*
Shoham, Dr Gil. *Israel.*
Shoham, Dr Menachem. *Israel.*
Short, Dr Michael Arthur. *USA.*
Short, Mr Michael T. *USA.*
Shoukri, Dr Nasri M. *Egypt.*
Shpigler, Mr Bilu. *Israel.*
Shrivastava, Prof. Hari Narayan. *India.*
Shternberg, Dr Aleksey Alexandrovich. *USSR.*
Shu, Mr Jin-fu. *China.*
Shuja, Mr Tauqir Ahmad. *Pakistan.*
Shul'pina, Dr Iren Leonidovna. *USSR.*
Shulakov, Dr Evgeniy Vladimirovich. *USSR.*
Shull, Prof. Clifford G. *USA.*
Shumyatskaya, Dr Ninel' Grigoryevna. *USSR.*
Shustov, Dr Alexander Vsevolodovich. *USSR.*
Shutskever, Dr Nataliya Efimovna. *USSR.*
Shuvalov, Prof. Lev Alexandrovich. *USSR.*
Shvelashvili, Prof. Arsen Eristovich. *USSR.*
Sianou, Miss Anna. *Greece.*
Siapkas, Prof. Demetrios John. *Greece.*
Šichová, Dr Hana. *CSSR.*
Siddiqui, Mr Jawed Ahmad. *Pakistan.*
Siddiqui, Dr Rafiq Ahmad. *Pakistan.*
Sidorenko, Dr Galina Alexandrovna. *USSR.*
Siebels, Mr Hansjörg. *BRD.*
Sieber, Mr Norbert Hermann Wilhelm. *BRD.*
Siegel, Dr Lester Aaron. *USA.*
Siegel, Dr Stanley. *USA.*
Sieker, Dr Larry C. *USA.*
Sielecki, Dr Anita R. *Canada.*
Sieler, Dr Joachim. *DDR.*
Sievers, Dr Rolf. *BRD.*
Sigayev, Dr Vladimir Nikolayevich. *USSR.*
Sigler, Prof. Paul Benjamin. *USA.*
Sigvaldason, Dr Gudmundur. *Iceland.*
Siivola, Prof. Jaakko Uolevi. *Finland.*
Sijarić, Dr (Mrs) Galiba. *Yugoslavia.*
Sikirica, Prof. Dr Milan. *Yugoslavia.*
Sikka, Dr Satinder Kumar. *India.*
Sikora, Dr Wiesława Antonina. *Poland.*
Sikorsky, Dr J. *UK.*
Sil'vestrova, Dr Iraida Mikhailovna. *USSR.*
Silcox, Prof. John. *USA.*
Silin', Dr Elga Yanovna. *USSR.*
Silong, Dr Sidik bin. *Malaysia.*
Silva, Dr Abelardo Manuel. *Argentina.*
Silva, Dr Elisa. *Chile.*
Silver, Dr Jack. *UK.*
Silverton, Ms Enid. *USA.*
Silverton, Dr James V. *USA.*
Silvester, Mr Philip. *UK.*
Sim, Prof. George Andrew. *UK.*
Simard, Mr Michel. *Canada.*
Simard, Dr Roger Gèrard. *USA.*
Sime, Prof. Rodney J. *USA.*
Sime, Dr Ruth Lewin. *USA.*
Simerská, Dr Marie. *CSSR.*
Simić, Dr Vojislav. *Yugoslavia.*
Simmins, Mr John J. *USA.*
Simmons, Dr Charles J. *USA.*
Simmons, Mr Ralph O. *USA.*
Simon, Prof. Dr Arndt. *BRD.*
Simon, Dr Jean-Paul Henri Maurice. *France.*

Simon, Dr Kálmán. *Hungary.*
Simone, Mr Carlos Alberto de. *Brazil.*
Simonetta, Prof. Massimo. *Italy.*
Simonov, Dr Mikhail Alexandrovich. *USSR.*
Simonov, Dr Valentin Ivanovich. *USSR.*
Simonov, Dr Yury Alexandrovich. *USSR.*
Simonsen, Mr Ole. *Denmark.*
Simonsen, Prof. Stanley Harold. *USA.*
Simov, Mr Stefan. *Bulgaria.*
Simpson, Dr Howard D. *USA.*
Singh, Dr Anil Kumar. *India.*
Singh, Dr Bachchan. *India.*
Singh, Dr Bhanu Pratap. *India.*
Singh, Dato' Prof. Dr Chatar. *Malaysia.*
Singh, Dr Govind. *India.*
Singh, Mr K. D. P. *India.*
Singh, Dr Phirtu. *USA.*
Singh, Dr Sarjant. *USA.*
Singh, Mr Surendra Prakash. *India.*
Singru, Prof. Ramesh Madhao. *India.*
Sinha, Dr Umesh Chandra. *India.*
Sinn, Prof. Ekkehard. *USA.*
Sirdeshmukh, Dr Dinker. *India.*
Siripitayananon, Miss Jintana. *Australia.*
Sirirratwatanakul, Mr Narin. *Thailand.*
Sironi, Prof. Angelo. *Italy.*
Sirota, Dr Mikhail Isaakovich. *USSR.*
Sirtl, Prof. Dr Erhard. *BRD.*
Sitte, Mrs Jutta. *DDR.*
Sivakumar, Mr K. *India.*
Sivonen, Mr Seppo Juhani. *Finland.*
Sivý, Dr Peter. *CSSR.*
Sizova, Dr Nataliya Leonidovna. *USSR.*
Sjöberg, Prof. Bo. *Sweden.*
Sjögren, Miss Agneta. *Sweden.*
Sjölin, Dr H. Lennart G. *Sweden.*
Skalicky, Prof. Dr Peter. *Austria.*
Skapski, Dr Andrzej Czeslaw. *UK.*
Skarżyński Dr Tadeusz. *Poland.*
Skarnulis, Dr Anthony Jerome. *UK.*
Skarstad, Dr Paul Michael. *USA.*
Skellett, Mr C.A. *UK.*
Skelly, Miss J.V. *UK.*
Skelton, Dr Brian Warwick. *Australia.*
Skelton, Dr Earl Franklin. *USA.*
Skinner, Dr J.M. *UK.*
Skjerpe, Mr Per Martin. *Norway.*
Skoglund, Dr B. Ulf. *Sweden.*
Skolozdra, Dr Roman Vladimirovich. *USSR.*
Skoweranda, Dr Jolanta. *Poland.*
Skrzat, Dr Zofia. *Poland.*
Skrzypczak-Yankun, Dr Ewa. *USA.*
Slade, Dr Phillip Garland. *Australia.*
Slaughter, Prof. Maynard. *USA.*
Slebarski, Dr Andrzej. *Poland.*
Sleight, Dr Arthur William. *USA.*
Sletten, Dr Einar. *Norway.*
Sletten, Dr Jorunn. *Norway.*
Sliva, Mr Paul. *USA.*
Šljukić, Prof. Dr Momčilo. *Yugoslavia.*
Sloan, Dr Gilbert J. *USA.*
Slovenec, Dr Dragutin. *Yugoslavia.*
Sly, Prof. William Glenn. *USA.*
Smaalen, van, Drs Sander. *Netherlands.*
Smale, Mr David. *New Zealand.*
Small, Dr Ronald W.H. *UK.*
Smalley, Dr Ian James. *New Zealand.*
Smart, Dr Lesley Elizabeth. *UK.*
Smart, Mr D.W. *UK.*
Smetannikova, Dr Olga Gennadiyevna. *USSR.*
Smirnov, Dr Aleksey Evgen'yevich. *USSR.*
Smirnov, Dr Yury Mstislavovich. *USSR.*
Smirnova, Dr Nina Lvovna. *USSR.*
Šmit, Dr Ivan. *Yugoslavia.*
Smit, Dr Jan Derk Geert. *Switzerland.*
Smit, Dr Paul. *Netherlands.*
Smith, Mr Albert Edward. *USA.*
Smith, Dr Arnold John. *UK.*
Smith, Dr Bryan Edward. *UK.*
Smith, Dr Robert Carr. *UK.*

Smith, Dr Sidney Herbert. *UK.*
Smith, Dr David John. *USA.*
Smith, Prof. Deane Kingsley, Jr. *USA.*
Smith, Dr Douglas Lee. *USA.*
Smith, Mr Gallienus William. *UK.*
Smith, Dr George David. *USA.*
Smith, Dr Gordon Stuart. *USA.*
Smith, Dr Graham. *Australia!*
Smith, Dr Harold Glenn. *USA.*
Smith, Dr Janet Louise. *USA.*
Smith, Prof. John Francis. *USA.*
Smith, Mr John Michael Andrew. *UK.*
Smith, Prof. Joseph Victor. *USA.*
Smith, Dr Katherine Leah. *Australia.*
Smith, Dr Ward Whitlock. *USA.*
Smith, Prof Thomas Frederick. *Australia.*
Smith, Prof. Vedene H., Jr. *Canada.*
Smith Verdier, Dr Pilar. *Spain.*
Smithson, Mr Douglas L. *USA.*
Smits, Mr Johannes Martinus Maria. *Netherlands.*
Smolander, Dr Kari Juhani. *Finland.*
Smolander, Dr Kimmo Juhani Nils-Eric. *Finland.*
Smolin, Dr Yury Ivanovich. *USSR.*
Smoluchowski, Prof. Roman. *USA.*
Smotrakov, Dr Valery Georgiyevich. *USSR.*
Smuts, Dr Jacques. *South Africa.*
Snider, Mr E.E. *USA.*
Snow, Mr Mark E. *USA.*
Snow, Dr Michael Robert. *Australia.*
Snyder, Prof. Robert L. *USA.*
Soa, Dr Ernst-Adolf. *DDR.*
Sobczak, Doz. Dr Rudolf Josef. *Austria.*
Sobolev, Dr Boris Pavlovich. *USSR.*
Sobolev, Dr Chingis Sergeyevich. *USSR.*
Soboleva, Dr Lidiya Victorovna. *USSR.*
Soboleva, Dr Svetlana Vsevolodovna. *USSR.*
Sobry, Dr Roger. *Belgium.*
Söderberg, Dr Bengt-Olof. *Sweden.*
Söderholm, Dr Anne Charlotte. *Sweden.*
Söderholm, Dr Margareta. *Sweden.*
Söderlund, Mr Gustaf S. *Sweden.*
Soepono, Mr R. *Indonesia.*
Sørensen, Dr Alex Mehlsen. *Denmark.*
Sørensen, Mr Ole. *Denmark.*
Soest, van, Dr Teunis Cornelis. *Netherlands.*
Søtofte, Mrs Inger. *Denmark.*
Sokol, Dr Anatoly Afanasyevich. *USSR.*
Sokol, Dr Valentina Ivanovna. *USSR.*
Sokoll, Mr Rolf. *DDR.*
Sokolov, Dr Yury Alexandrovich. *USSR.*
Sokolova, Dr Nataliya Gavrilovna. *USSR.*
Solans, Prof. Xavier. *Spain.*
Solans Huget, Prof. Joaquin Ma. *Spain.*
Soldánová, Ing Jiřina. *CSSR.*
Soldatos, Prof. Constantinos. *Greece.*
Soldatov, Dr Evgeniy Alexandrovich. *USSR.*
Soledade Jr, Prof. Teomar. *Brazil.*
Soliman, Mr F. Abdel Aal. *Egypt.*
Soliman, Dr Mohamed Soliman. *Egypt.*
Solo'vyev, Dr Sergey Petrovich. *USSR.*
Solorio Munguia, Prof. José Gregorio. *Mexico.*
Solovyeva, Dr Lidiya Pavlovna. *USSR.*
Soltzberg, Prof. Leonard Jay. *USA.*
Soman, Dr K. V. *India.*
Somerville, Mr R.G. *UK.*
Somiya, Dr Shigeyuki. *Japan.*
Sommer, Dr Joachim. *DDR.*
Sommermann, Mr Günter. *DDR.*
Sommerville, Mrs Polly Baker Melville. *South Africa.*
Somorjai, Prof. Gabor Arpad. *USA.*
Sonaer, Mr Kenan. *Turkey.*
Sondermann, Dr Ulrich. *BRD.*
Song, Mr Shi-ying. *China.*
Sonin, Prof. Anatoliy Stepanovich. *USSR.*
Sonnerstam, Dr Ulf C. *Sweden.*
Sonneveld, Mr Eduard Jan. *Netherlands.*
Sonoda, Mr Minoru. *Japan.*
Sonoike, Prof. Sanemi. *Japan.*
Šoptrajanov, Prof. Dr Bojan. *Yugoslavia.*
Šoptrajanova, Prof. Dr Gorica. *Yugoslavia.*
Sorge, Dr Georg. *DDR.*

Soriano-Calix, Mrs Virginia B. *Philippines.*
Soriano Garcia, Dr Manuel. *Mexico.*
Sorokin, Dr Lev Mikhailovich. *USSR.*
Sosfenov, Dr Nikita Ilyich. *USSR.*
Sosnowska, Dr hab. Izabela. *Poland.*
Sosnowski, Dr hab. Jerzy. *Poland.*
Šourek, Dr Zbyněk. *CSSR.*
Souza, de, Mr José Carlos. *Brazil.*
Souza, Mr Carlos, *Chile.*
Souza, Prof. Irineu Marques. *Brazil.*
Sowa, Dr Heidrun. *BRD.*
Sowerby, Mr R.G. *UK.*
Soylu, Doç. Dr Hüseyin. *Turkey.*
Spackman, Dr Mark Arthur. *Australia.*
Spadon, Prof. Paola. *Italy.*
Spagna, Dr Riccardo. *Italy.*
Spalding, Dr Dennis Raymond. *South Africa.*
Sparks, Dr Cullie James, Jr. *USA.*
Sparks, Dr Robert Allen. *USA.*
Spek, Dr Anthony Louis. *Netherlands.*
Spence, Prof. Robert Dean. *USA.*
Spielberg, Prof. Nathan. *USA.*
Spiers, Mr Ronald J. *USA.*
Spindler, Dr Herbert. *DDR.*
Spinelli, Lic. Silvia Haydeé *Argentina.*
Spink, Mr John Arthur. *Australia.*
Spiridonov, Prof. Victor Pavlovich. *USSR.*
Spirlet, Dr Marie-Rose. *Belgium.*
Spitsyna, Dr Valentina Danilovna. *USSR.*
Spooner, Mr Francis John. *UK.*
Spooner, Prof. Stephen. *USA.*
Sprackling, Dr Michael Thomas. *UK.*
Spraget, Mr H. *UK.*
Sprang, Dr Stephen Robert. *USA.*
Spratt, Mr S.B.D. *UK.*
Spreadborough, Dr John. *UK.*
Sprenger, Dr Heinz. *DDR.*
Spriggs, Dr Paul Humphrey. *UK.*
Springer, Dr James Patrick. *USA.*
Sproul, Prof. Gordon Duane. *USA.*
Spurlino, Mr John C. *USA.*
Spyrellis, Dr Nicolaos. *Greece.*
Spyridelis, Prof. John. *Greece.*
Squire, Dr John Michael. *UK.*
Srdanov, Magistar Gordana. *Yugoslavia.*
Sreekantan, Dr Arunachalam. *India.*
Srikrishnan, Dr Tamarapu. *USA.*
Srinivasa, Dr Vishwanathapuram Kalasa. *India.*
Srinivasan, Dr Annankoil Renganatha. *India.*
Srinivasan, Prof. R. *India.*
Srinivasan, Dr Sampat. *India.*
Sriramadas, Prof. Aluru. *India.*
Sriratanaprasithi, Mr Khin. *Thailand.*
Srivastava, Dr Ramesh Chandra. *India.*
Srivastava, Dr Surendra Nath. *India.*
St. Charles, Mr Robert. *USA.*
Ståhl, Dr Kenny. *Sweden.*
Stålhandske, Dr Claes. *Sweden.*
Stadermann, Mr Gerd. *DDR.*
Stadler, Mr Maximilian Wolfgang. *BRD.*
Stadler, Dr Hans Peter. *UK.*
Stadnicka, Dr Katarzyna. *Poland.*
Staehlin, Dr Walter. *Switzerland.*
Staikov, Dr Georgy. *Bulgaria.*
Staliński, Prof. Bohdan. *Poland.*
Stalick, Dr Judith Kay. *USA.*
Stallings, Dr William C. *USA.*
Stam, Dr Casper Hendrik. *Netherlands.*
Stammers, Dr David Kingsley. *UK.*
Stanfield, Ms Robyn. *USA.*
Stanford, Mr Michael John. *UK.*
Stangler, Prof. Dr Ferdinand Karl Ludwig. *Austria.*
Stanko, Prof. Joseph Anthony. *USA.*
Stanković, Mrs Slavica. *Yugoslavia.*
Stanković, Dr (Mrs) Slobodanka. *Yugoslavia.*
Stanley, Dr Eric. *Canada.*
Stansfield, Dr Robert Frank David. *France.*
Starikova, Dr Zoya Alexandrovna. *USSR.*
Starke, Dr Rainer. *DDR.*
Starostina, Dr Lyudmila Sergeyevna. *USSR.*

Stasi, Dr Francesca. *Italy.*
Staudenmann, Dr Jean Louis. *USA.*
Stauffacher, Dr Cynthia Vianne. *USA.*
Stearns, Mr Frederick Stanley. *USA.*
Stecker, Prof. Kurt. *DDR.*
Steeb, Prof. Dr Siegfried. *BRD.*
Steeds, Prof. John Wickham. *UK.*
Steele, Dr Ian McKay. *USA.*
Stefániay, Mr Vilmos. *Hungary.*
Stefanidis, Mr Theodoros. *Sweden.*
Stefanov, Mr Dechko Dimitrov. *Bulgaria.*
Stefanov, Dr Stefan Rashkov. *Bulgaria.*
Stefanović, Magistar Aleksandar. *Yugoslavia.*
Steffen, Dipl. Chem Michael Georg. *BRD.*
Steffen, Dr William Lee. *Australia.*
Stegemann, Dr-Ing Jürgen. *BRD.*
Steger, Dr Theodore Roosevelt. *USA.*
Stegmann, Dr Eleonore. *DDR.*
Steigmann, Dr Gottfried Albert. *UK.*
Steil, Dr Helmut. *DDR.*
Stein, Prof. Richard Stephen. *USA.*
Steinberg, Prof. Itzhak. *Israel.*
Steinbruch, Mrs Uta. *DDR.*
Steiner, Dr Michael. *BRD.*
Steiner, Doz. Dr Walter. *Austria.*
Steinfink, Prof. Hugo. *USA.*
Steinhart, Dr Miloš. *CSSR.*
Steinicke, Prof. Ursula. *DDR.*
Steinrauf, Prof. Larry King. *USA.*
Steinthórsson, Dr Sigurdur. *Iceland.*
Steitz, Prof. Thomas Arthur. *USA.*
Stelzner, Mrs Sabine. *DDR.*
Stenberg, Dr Lars. *Sweden.*
Stenkamp, Dr Ronald Eugene. *USA.*
Stensland, Dr Birgitta. *Sweden.*
Stepanova, Dr Alla Nikolayevna. *USSR.*
Stephan, Dr D. W. *Canada.*
Stephan, Dr Dieter. *DDR.*
Stephanik, Dr Heinz. *DDR.*
Stephens, Mr Anthony E. *USA.*
Stephens, Dr Frederick Selwyn. *Australia.*
Stephenson, Prof Neville Charles. *Australia.*
Stepień-Damm, Dr Julia. *Poland.*
Stepień, Dr Andrzej. *Poland.*
Stergiou, Dr Anagnostis Charalambos. *Greece.*
Stergioudis, Dr Georgios Asterios. *Greece.*
Sterling, Prof. Clarence. *USA.*
Sterneck, Dr Dirk. *DDR.*
Steuhl, Dr Hans Hermann. *BRD.*
Steussloff, Mr Peter. *DDR.*
Stevels, Dr Albert Leendert Nicolaas. *Netherlands.*
Stevens, Prof. Edwin David. *USA.*
Stevenson, Dr Andrew Wesley. *Australia.*
Steward, Prof. Edward George. *UK.*
Stewart, Dr Robert Bruce. *New Zealand.*
Stewart, Prof. James McDonald. *USA.*
Stewart, Dr Martin Van Buren. *USA.*
Stewart, Prof. Robert Farrell. *USA.*
Stewig, Mr Helmut. *DDR.*
Stickler, Prof. Dr Roland. *Austria.*
Stiopol, Prof. Victoria. *Romania.*
Stishov, Dr Sergey Mikhailovich. *USSR.*
Stock, Prof. Stuart R. *USA.*
Stoeckli-Evans, Dr Helen M. *Switzerland.*
Stöckelmann, Dr Diedrich. *BRD.*
Stölevik, Dr Reidar. *Norway.*
Stoicovici, Prof. Eugen. *Romania.*
Stoimenos, Prof. John Nikolaos. *Greece.*
Stoinova, Dr Margarita. *Bulgaria.*
Stojadinović, Dr Slobodan. *Yugoslavia.*
Stojanović, Mr Dobrica. *Yugoslavia.*
Stollstorff, Mr Gregory R. *USA.*
Stomberg, Prof. Rolf. *Sweden.*
Stonebridge, Dr Brian Richard. *UK.*
Stora, Dr Cécile. *France.*
Storbeck, Prof. Fritz. *DDR.*
Stothart, Dr Philip Hamilton. *UK.*
Stout, Prof. Charles David. *USA.*
Stouten, Drs Pieter F. W. *Netherlands.*
Stoyanov, Dr Stoyan. *Bulgaria.*

Stoychev, Mr Nikola. *Bulgaria.*
Strähle, Prof. Dr Joachim. *BRD.*
Strahs, Dr Gerald. *USA.*
Strand, Dr Tor Gogstad. *Norway.*
Strandberg, Prof. Bror E. *Sweden.*
Strandberg, Dr Rolf. *Sweden.*
Straver, Drs Leonardus Hendrikus. *Netherlands.*
Streib, Dr William E. *USA.*
Strell, Mrs Dr Irmtraud. *BRD.*
Strickland, Mr Peter R. *UK.*
Strickler, Dr Peter. *Switzerland.*
Strid, Dr Karl-Gustav. *Sweden.*
Strocka, Dr Bernhard. *BRD.*
Strömme, Dr Knut Olaf. *Norway.*
Strogen, Mr Peter. *Ireland.*
Stroud, Prof. Robert Michael. *USA.*
Strouse, Prof. Charles Earl. *USA.*
Stroz, Mgr Danuta. *Poland.*
Struchkov, Dr Yury Timofeyevich. *USSR.*
Strübel, Prof. Dr Günter. *BRD.*
Struijs, van der, Drs Johannes P. *Netherlands.*
Struikmans, Drs Rink. *Netherlands.*
Strukov, Prof. Boris Anatolyevich. *USSR.*
Strumpel, Dr Marianna Katona. *BRD.*
Strunz, Prof. Dr Hugo. *BRD.*
Strydom, Dr Ockert Andries Wilhelm. *South Africa.*
Stryjak, Dr Andrew Jules. *UK.*
Stubb, Dr Arne Henrik. *Finland.*
Stubbs, Dr Gerald James. *USA.*
Stubbs, Mr Milton. *UK.*
Stubičar, Dr Mirko. *Yugoslavia.*
Stuhrmann, Prof. Dr Heinrich B. *BRD.*
Stults, Dr Bailey Ray. *USA.*
Stumpfl, Prof. Dr Eugen Friedrich. *Austria.*
Sturcken, Dr Edward Francis. *USA.*
Sturkey, Mr Lorenzo. *USA.*
Sturm, Prof. Edward. *USA.*
Sturm, Prof. Dr Dipl.-Ing. Friedwin. *Austria.*
Stuut, Ir Harmannus Aaldrik. *Netherlands.*
Subhadra, Dr K. G. *India.*
Subirana Torrent, Dr José Antonio. *Spain.*
Subramanian, Dr K. *India.*
Subramony, Mr Loganathan. *South Africa.*
Šubrtová, Ing Věra. *CSSR.*
Suck, Dr Dietrich. *BRD.*
Suddath, Prof. Fred L. *USA.*
Sudhakar, Mr Varanasi. *India.*
Sudo, Prof. Toshio. *Japan.*
Suemune, Dr Yasutaka. *Japan.*
Sueno, Dr Shigeho. *Japan.*
Süsse, Prof. Dr Peter. *BRD.*
Suffritti, Prof. Giuseppe Baldovino. *Italy.*
Sugaike, Dr Suezo. *Japan.*
Sugawara, Dr Yoko. *Japan.*
Sugihara, Dr Akio. *Japan.*
Sugiura, Prof. Seiji. *Japan.*
Suh, Prof. Dr Ill-Hwan. *Korea.*
Suh, Prof. Jung Sun. *Korea.*
Suh, Prof. Se Won. *Korea.*
Suhre, Miss Ursula. *BRD.*
Suita, Prof. Tokuo. *Japan.*
Suitch, Mr Paul R. *USA.*
Suito, Dr Eiji. *Japan.*
Sukapaddhanadhi, Mr Narong. *Thailand.*
Sullenger, Dr Don Bruce. *USA.*
Sullivan, Dr Richard Arthur. *UK.*
Summerville, Dr Edward. *Australia.*
Sun, Mr Yi-jian. *China.*
Sunagawa, Prof. Ichiro. *Japan.*
Sundaralingam, Prof. Muttaiya. *USA.*
Sundaramoorthy, Mr M. *India.*
Sundararajan, Dr Pudupadi R. *Canada.*
Sundaresan, Dr Thiagarajan. *UK.*
Sundberg, Dr Margareta. *Sweden.*
Sundell, Dr Lars Staffan. *Sweden.*
Sunder, Dr Sham. *Canada.*
Sundius, Dr Tom Robert. *Finland.*
Sundström, Dr Lorna Jean. *Finland.*
Suoninen, Prof. Eero Juhani. *Finland.*
Suortti, Prof. Pekka. *Finland.*
Supper, Mr Lee R. *USA.*

Surcouf, Dr Evelyne. *France.*
Suresh, Mr C. G. *India.*
Suresh, Mr K. A. *India.*
Suri, D. K. *India.*
Surowiec, Dr Marian Ryszard. *Poland.*
Suryanarayana, Dr Challapalli. *India.*
Suryanarayana, Dr Shambhuni V. *India.*
Sussieck-Fornefeld, Mrs Cornelia. *BRD.*
Sussman, Prof. Joel Leonard. *Israel.*
Suta, Elizabeth *India.*
Sutherland, Dr John Knox. *Canada.*
Sutherland, Dr Hector Howieson. *UK.*
Sutor, Dr Dorothy June. *UK.*
Sutter, Dr Dietrich. *DDR.*
Sutton, Dr A.L. *UK.*
Sutton, Dr A.P. *UK.*
Sutton, Dr Brian John. *UK.*
Sutton, Mr J.D. *UK.*
Sutton, Mr Paul W. *420-3491).*
Suvorov, Dr Ernest Vitalyevich. *USSR.*
Suwalsky, Dr Mario. *Chile.*
Suzuki, Dr Carlos Kenichi. *Brazil.*
Suzuki, Mr Eikichi. *Australia.*
Suzuki, Prof. Hideji. *Japan.*
Suzuki, Dr Ikuo. *Japan.*
Suzuki, Prof. Michio. *Japan.*
Suzuki, Dr Shigeo. *Japan.*
Suzuki, Prof. Tadasu. *Japan.*
Suzuki, Dr Teruo. *Japan.*
Suzuki, Dr Toshimasa. *Japan.*
Suzuki, Prof. Yoshio. *Japan.*
Svensson, Mr Anders. *Sweden.*
Svensson, Dr Ing-Britt A. *Sweden.*
Svensson, Dr Christer. *Sweden.*
Svensson, Mr Göran. *Sweden.*
Svergun, Dr Dmitry Ivanovich. *USSR.*
Svinning, Dr Torgeir. *Norway.*
Sviridov, Prof. Dmitry Timofeyevich. *USSR.*
Svisero, Dr Darcy Pedro. *Brazil.*
Swallow, Dr Arnold Graham. *UK.*
Swank, Prof. Duane Douglas. *USA.*
Swartzendruber, Mr John K. *USA.*
Sweet, Dr Robert M. *USA.*
Swenson, Dr Dale Carl. *USA.*
Swepston, Dr Paul Nathan. *USA.*
Swerdlow, Mr Max. *USA.*
Swillens, Mr Eckhard. *DDR.*
Swindells, Dr David Campbell Neil. *UK.*
Swink, Dr Laurence Nim. *USA.*
Swinnea, Dr John Steven. *USA.*
Switendick, Dr Alfred Carl. *USA.*
Syed, Dr A. Sattar. *Bangladesh.*
Syed, Dr Ashfaquzzaman. *USA.*
Sygusch, Prof. Jurgen. *Canada.*
Syhre, Dr Hans. *DDR.*
Syneček, Doc. Dr Vladimír. *CSSR.*
Szarras, Dr Stanisław. *Poland.*
Szebenyi, Ms Doletha M. *USA.*
Szemethy, Miss Andrea. *Hungary.*
Szentivanyi-Hansson, Dr Helga. *Sweden.*
Szmid, Dr Zofia. *Poland.*
Sztrókay, Prof. Kálmán. *Hungary.*
Szulzewsky, Dr Klaus. *DDR.*
Szummer, Dr hab. Andrzej. *Poland.*
Szymański, Dr Jan Tomasz. *Canada.*
Tabata, Dr Hideyo. *Japan.*
Tadini, Prof. Carla. *Italy.*
Tadokoro, Prof. Hiroyuki. *Japan.*
Tänzer, Mr Dietmar. *DDR.*
Tagai, Dr Tokuhei. *Japan.*
Tahoun, Prof. Salah. *Saudi Arabia.*
Tai, Dr Douglas Leung-Tak. *USA.*
Tainer, Dr John Arthur. *USA.*
Tait, Dr John Mervyn. *UK.*
Tajabor, Dr Nasser. *Iran.*
Takagi, Dr Mieko. *Japan.*
Takagi, Prof. Satio. *Japan.*
Takagi, Dr Shozo. *USA.*
Takagi, Prof. Yutaka. *Japan.*
Takahashi, Mr Shoichi. *Japan.*

Takahashi, Dr Toshio. *Japan.*
Takahashi, Prof. Yasuhiro. *Japan.*
Takai, Dr Mitsuo. *Japan.*
Takaki, Dr Yoshito. *Japan.*
Takamura, Prof. Jin-ichi. *Japan.*
Takano, Dr Tsunehiro. *Japan.*
Takano, Prof. Yasumasa. *Japan.*
Takano, Dr Yukio. *Japan.*
Takasu, Dr Shin-ichiro. *Japan.*
Takayanagi, Prof. Kunio. *Japan.*
Takeda, Dr Hirofumi. *Japan.*
Takeda, Prof. Hiroshi. *Japan.*
Takei, Dr William J. *USA.*
Takenaka, Dr Akio. *Japan.*
Takeoka, Mr Yoshikatsu. *Japan.*
Takeuchi, Prof. Yoshio. *Japan.*
Takeyama, Prof. Yoshio. *Japan.*
Takla, Mr Maher Azmi. *Egypt.*
Takusagawa, Dr Fusao. *USA.*
Talapatra, Dr S. K. *India.*
Talukdar, Dr Amarendra Nath. *781014.*
Tamada, Mr Osamu. *Japan.*
Tamaki, Dr Shozo. *Japan.*
Tammaro, Mr David A. *USA.*
Tamura, Dr Chihiro. *Japan.*
Tan, Prof. Hao-ran. *China.*
Tan, Dr Hock Siew. *Singapore.*
Tanaka, Mr Isao. *Japan.*
Tanaka, Prof. Jiro. *Japan.*
Tanabe, Dr Kazuya. *Japan.*
Tanaka, Dr Kiyoaki. *Japan.*
Tanaka, Dr Michiyoshi. *Japan.*
Tanaka, Dr Nobuo. *Japan.*
Tanaka, Prof. Tetsuro. *Japan.*
Tanemura, Dr Sakae. *Japan.*
Taner, Dr Akin. *Turkey.*
Tang, Dr Chia-Pin. *Taiwan.*
Tang, Prof. You-qi. *China.*
Tang, Mr Zhi-kai. *China.*
Tani, Dr Katsuhiko. *Japan.*
Taniguchi, Mr Tomohiko. *Japan.*
Tanisaki, Prof. Sigetosi. *Japan.*
Tanishiro, Mr Yasumasa. *Japan.*
Tanner, Dr Brian Keith. *UK.*
Tanner, Dr H.M. *UK.*
Taoka, Prof. Tadami. *Japan.*
Tapfer, Dr Leander. *BRD.*
Tardy, Dr Pál. *Hungary.*
Tarimci, Dr Çelik. *Turkey.*
Tarján, Prof. Imre. *Hungary.*
Tarkhova, Dr Tatyana Nikolayevna. *USSR.*
Tarling, Mr S.E. *UK.*
Tarna, Mr Toivo Mikael. *Finland.*
Tarnopol'sky, Dr Boris Lvovich. *USSR.*
Tasker, Mr Michael Peter. *UK.*
Tasker, Dr Peter Anthony. *UK.*
Tasset, Dr Francis Joseph Emmanuel. *France.*
Tatarchenko, Dr Vitaly Antonovich. *USSR.*
Tatarinova, Dr Lyudmila Ivanovna. *USSR.*
Tatarsky, Prof. Vitaly Borisovich. *USSR.*
Tate, Dr Cecil. *UK.*
Tate, Prof. Isao. *Japan.*
Tatekawa, Prof. Masahisa. *Japan.*
Tatsch, Dr Clinton Eugene. *USA.*
Tatsuka, Dr Kiyoaki. *Japan.*
Tatsuzaki, Prof. Itaru. *Japan.*
Taub, Mr Haskell. *USA.*
Tauber, Dr Arthur. *USA.*
Tauler Ferre, Dr Esperança *Spain.*
Taupin, Dr Daniel Gilbert Roger, *France.*
Tavale, Dr Sudam Shankar. *India.*
Taxer, Dr Karlheinz Jürgen. *BRD.*
Taylor, Prof. Charles Alfred. *UK.*
Taylor, Dr Derek. *UK.*
Taylor, Dr Donald. *Australia.*
Taylor, Dr Garry Lindsay. *UK.*
Taylor, Prof. Harry Francis West. *UK.*
Taylor, Ms Hope Cathlin. *USA.*
Taylor, Dr Ivan Fate, Jr. *USA.*
Taylor, Dr John Bryan. *Canada.*

Taylor, Dr John Charles. *Australia.*
Taylor, Prof. Max A. *USA.*
Taylor, Dr Max Ronald. *Australia.*
Taylor, Mr Michael William. *South Africa.*
Taylor, Dr Peter. *Canada.*
Taylor, Dr R. *UK.*
Taylor, Prof. Robert Cooper. *USA.*
Távora, Prof. Elysiario. *Brazil.*
Tazzoli, Prof. Vittorio. *Italy.*
Tchehlarova, Mrs Irina. *Bulgaria.*
Tchuneva, Mrs Vassilka. *Bulgaria.*
Tebbe, Prof. Dr Karl-Friedrich. *BRD.*
Teeter (Stein), Dr Martha Mary. *USA.*
Tegman, Dr Ragnar. *Sweden.*
Teh, Dr Guan Hoe. *Malaysia.*
Teh, Dr Hung Chuan. *Singapore.*
Teh, Dr Ser Kok. *Malaysia.*
Teixeira Mendes, Prof. Antonio Carlos. *Brazil.*
Telegina, Dr Inna Vasilyevna. *USSR.*
Teller, Dr Raymond. *USA.*
Téllez Ortiz, Mrs Minerva Estela. *Mexico.*
Tellgren, Dr I. G. Roland. *Sweden.*
Tempel, Dr Alfred. *DDR.*
Tempelhoff, Dr Klaus. *DDR.*
Tempest, Dr Paul Anthony. *UK.*
Templeton, Prof. David Henry. *USA.*
Templeton, Dr Lieselotte K. *USA.*
Ten Eyck, Dr Lynn Forest. *USA.*
Tench, Dr Alan Howard. *USA.*
Tennissen, Prof. Anthony C. *USA.*
Tennyson, Mrs Prof. Dr Christel. *BRD.*
Tenon, Mr Abodou Jules. *Ivory Coast.*
Teoh, Mr Lay Hock. *Malaysia.*
Terauchi, Dr Hikaru. *Japan.*
Teresiak, Mrs Angelika. *DDR.*
Terol, Dr Salvador. *Spain.*
Terwilliger, Mr Thomas C. *USA.*
Terzis, Dr Aristides. *Greece.*
Teske, Dr Christoph Ludwig. *BRD.*
Teslenko, Dr Valery Fedorovich. *USSR.*
Tetzner, Dr Gottfried. *DDR.*
Teufer, Dr Gunter. *USA.*
Teuho, Mr Juhani Erkki Tapani. *Finland.*
Tewari, Dr Raghavendra. *India.*
Théobald, Dr François Roland. *France.*
Thabet, Dr Atef. *Egypt.*
Thackeray, Dr Michael Makepeace. *South Africa.*
Thanomkul, Dr Srinuan Chaiwasie. *Thailand.*
Thatcher, Mr J.S. *UK.*
Thatcher, Mr Walter E. *USA.*
Thathachari, Mr Y. T. *USA.*
Theocharis, Dr C.R. *UK.*
Theodoridou, Dr Irini. *Greece.*
Theodoropoulos, Prof. Dimitrios. *Greece.*
Theodossiou, Prof. Alexandros. *Greece.*
Theophanides, Prof. Theo. *Canada.*
Thewalt, Prof. Dr Ulf. *BRD.*
Thiele, Prof. Dr Gerhard. *BRD.*
Thielke, Mr Harry G. *USA.*
Thieme, Dr Wolfgang. *DDR.*
Thierry, Dr Jean-Claude. *France.*
Thijsse, Dr Barend Jan. *Netherlands.*
Thinapong, Dr Pongchan Chananont. *Thailand.*
Thirup, Mr Søren. *Sweden.*
Thom, Ms Vivienne Joyce. *South Africa.*
Thomas, Prof John Meurig, FRS. *UK.*
Thomas, Dr John O. *Sweden.*
Thomas, Dr Joseph Mitchell, Jr. *USA.*
Thomas, Mr Kenneth. *UK.*
Thomas, Dr Kenneth A. *USA.*
Thomas, Mr M.A. *UK.*
Thomas, Dr Robert. *USA.*
Thomas, Dr W.A. *UK.*
Thomas-David, Prof. Germaine. *France.*
Thomasson, Mr Ronnie. *Sweden.*
Thompson, Dr Derek Parr. *UK.*
Thompson, Prof. Doris M. *USA.*
Thompson, Mr John Gerard. *Australia.*
Thorkildsen, Dr Gunnar. *Norway.*
Thornley, Dr Frank Richard. *UK.*
Thornton-Pett, Dr M.A. *UK.*

Thorup, Mr Niels. *Denmark.*
Thozet, Mr Alain Maurice. *France.*
Threadgold, Dr Ian Malcolm. *Australia.*
Thudium, Mr Richard N. *USA.*
Thurn, Dr Herbert. *BRD.*
Tibballs, Dr John Earl. *Norway.*
Tibljaš, Mr Darko. *Yugoslavia.*
Tickle, Dr Ian James. *UK.*
Tieghi, Prof. Giuseppe. *Italy.*
Tiekink, Mr Edward Richard Thomas. *Australia.*
Tielemans, Mr Luc. *Belgium.*
Tietze, Mr Hans Roderick. *Australia.*
Tiitta, Mr Antero Tapani. *Finland.*
Tikhomirova, Dr Nataliya Alexandrovna. *USSR.*
Tikhonova, Dr Anna Andreyevna. *USSR.*
Tiller, Prof. William Arthur. *USA.*
Tilley, Mr George P. *USA.*
Tilli, Mr Markku Väinö Kalevi. *Finland.*
Tillinger, Dr Martin H. *USA.*
Tillmann, Dr Bruno. *BRD.*
Tillmanns, Prof. Dr Ekkehart. *BRD.*
Timchenko, Dr Tamara Iosifovna. *USSR.*
Timofeeva, Dr Valentina Alexandrovna. *USSR.*
Timofeyeva, Dr Tat'yana Vladimirovna. *USSR.*
Tinant, Dr Bernard Guy André François. *Belgium.*
Tiripicchio, Prof. Antonio. *Italy.*
Tiripicchio-Camellini, Prof. Marisa. *Italy.*
Tishchenko, Dr Galina Nikolayevna. *USSR.*
Titus, Prof. Donald Dean. *USA.*
Tkachev, Dr Valery Vladimirovich. *USSR.*
Tkalčec, Prof. Dr Emilija. *Yugoslavia.*
Tobelko, Mr Konstantin Ivanovich. *USSR.*
Tobin, Mr A.N. *UK.*
Todaro, Mr Louis J. *USA.*
Töpel-Schadt, Dr Jutta. *BRD.*
Toeplitz, Mrs Barbara Keeler. *USA.*
Törnroos, Mr Karl W. *Sweden.*
Törnroos, Dr Ragnar Fredrik. *Finland.*
Tofield, Dr Bruce C. *UK.*
Togawa, Dr Sen-ichi. *Japan.*
Toivonen, Mr Jukka Tapio. *Finland.*
Tokay, Dr Nesrin. *Turkey.*
Tokonami, Prof. Dr Fabijasu. *Japan.*
Tokushita, Mr Motoyuki. *Japan.*
Tolksdorf, Prof. Dr Wolfgang. *BRD.*
Tollin, Dr Patrick. *UK.*
Toman, Prof. Karel. *USA.*
Tomas, Dr Alain. *France.*
Tomashpol'sky, Dr Yury Yakovlevich. *USSR.*
Tomassini, Dr Marco. *Italy.*
Tómasson, Cand.Real. Jens. *Iceland.*
Tomaszewski, Dr Pawel Edward. *Poland.*
Tomchick, Ms Diana R. *USA.*
Tomeoka, Mr Kazushige. *Japan.*
Tomimitsu, Mr Hiroshi. *Japan.*
Tomisaka, Prof. Takeshi. *Japan.*
Tomita, Dr Katsutoshi. *Japan.*
Tomita, Prof. Ken-ichi. *Japan.*
Tomita, Dr Koichi. *Brazil.*
Tomita, Dr Takanori. *Japan.*
Tomkeieff, Mr Michael Vamime. *UK.*
Tomlinson, Dr Gail Elizabeth. *USA.*
Tomov, Dr Ivan. *Bulgaria.*
Tonak, Miss Tulin. *Turkey.*
Tonejc, Dr (Mrs) Anđelka. *Yugoslavia.*
Tonejc, Dr Anton. *Yugoslavia.*
Tonnard, Dr Victor Edmond. *Belgium.*
Tonomura, Dr Akira. *Japan.*
Tontrakoon, Mr Jeerapong. *Thailand.*
Tooptakong, Mrs Uncharee Methong. *Thailand.*
Topalova-Kalitzova, Mrs Maria. *Bulgaria.*
Topić, Dr Mladen. *Yugoslavia.*
Topor, Dr Nikolay Dmitriyevich. *USSR.*
Toraya, Dr Hideo. *Japan.*
Toriumi, Mr Koshiro. *Japan.*
Torre, Prof. Louis Peter. *USA.*
Torre de Assunção, Prof. Carlos. *Portugal.*
Torres Villaseñor, Dr Gabriel. *Mexico.*
Torriani, Dr Iris Linares. *Brazil.*
Toscano, Mr Ruben Alfredo, *Mexico.*

Toshev, Dr Alexander. *Bulgaria.*
Tosi, Prof. Giorgio. *Italy.*
Tosik, Dr Anita. *Poland.*
Tóth, Mr Lajos. *Hungary.*
Tougard, Dr Pierre Henri. *France.*
Toupet, Mr Loic. *France.*
Toure, Dr Siaka. *Ivory Coast.*
Toussaint, Prof. Jean. *Belgium.*
Tousson, Dr Salama. *Egypt.*
Tovbis, Dr Alexander Borisovich. *USSR.*
Town, Miss Susan Lesley. *Australia.*
Townes, Mr William David Winn. *USA.*
Towns, Prof. Robert Lee Roy. *USA.*
Townsend, Dr Stephen Phillip. *UK.*
Toy, Dr Mark. *UK.*
Toyoda, Dr Koichi. *Japan.*
Trabelsi, Dr Malika. *Tunisia.*
Trætteberg, Prof. Marit. *Norway.*
Traill, Dr Robert James. *Canada.*
Tran Huu Dau, Mrs Marie-Elise. *France.*
Tran, Dr Vinh. *France.*
Trance, Dr Aurora Serrana. *Philippines.*
Tranqui, Dr Duc. *France.*
Traub, Prof. Wolfie. *Israel.*
Traveria Cros, Dr Adolfo. *Spain.*
Trefonas, Prof. Louis Marco. *USA.*
Treharne, Miss Annemarie Cerise. *UK.*
Treimer, Dr Wolfgang. *BRD.*
Treivus, Dr Evgeny Borisovich. *USSR.*
Tremmel, Mr János. *Hungary.*
Trettin, Mr Reinhard. *DDR.*
Treushnikov, Dr Evgeniy Nikolayevich. *USSR.*
Trigunayat, Prof. Govind Chandra. *India.*
Triodina, Dr Nina Sergeyevna. *USSR.*
Trömel, Prof. Dr Martin Gerhard. *BRD.*
Trojer, Dr Felix John. *Switzerland.*
Trojko, Magistar Rudolf. *Yugoslavia.*
Tronc, Dr Elisabeth. *France.*
Trost, Dr Friedrich Karl. *BRD.*
Trosti-Ferroni, Dr Renza. *Italy.*
Trotter, Prof. James. *Canada.*
Troup, Dr Jan Marshall. *USA.*
Trubelja, Prof. Dr Fabijan. *Yugoslavia.*
Trueblood, Prof. Kenneth N. *USA.*
Trumm, Dr Alfons. *BRD.*
Trunov, Dr Vadim Konstantinovich. *USSR.*
Trus, Dr Benes Louis. *USA.*
Truter, Prof. Mary Rosaleen. *UK.*
Trysberg, Dr Lennart A. G. *Sweden.*
Tsai, Prof. Chun-che. *USA.*
Tsatis, Mr Demetrius. *Greece.*
Tsay, Dr Yi-Hung. *BRD.*
Tscherry, Dr Viktor. *Switzerland.*
Tschulena, Dr Guido. *BRD.*
Tseitlin, Dr Mikhail Nevakhovich. *USSR.*
Tseng, Prof. Poh-kun. *Taiwan.*
Tsernoglou, Prof. Demetrius. *BRD.*
Tsikhotsky, Dr Evgeny Stanislavovich. *USSR.*
Tsimberis, Mr Nikolaos. *Greece.*
Tsinober, Dr Leonid Iosifovich. *USSR.*
Tsintsadze, Prof. Givi Vasilyevich. *USSR.*
Tsirel'son, Dr Valadimir Grigoryevich. *USSR.*
Tsoli Kataga, Dr Panayota. *Greece.*
Tsolovski, Dr Ilcho. *Bulgaria.*
Tsoucaris, Dr Georges. *France.*
Tsoukalas, Prof. John. *Greece.*
Tsuda, Mr Noritoshi. *Japan.*
Tsui, Ms Francis C. *USA.*
Tsuji, Mr Kazuhiko. *Japan.*
Tsuji, Mr Koji. *Japan.*
Tsukihara, Dr Tomitake. *Japan.*
Tsunekawa, Dr Shin. *Japan.*
Tsuprun, Dr Vladimir Lvovich. *USSR.*
Tsvankin, Dr Daniel' Yakovlevich. *USSR.*
Tsyganov, Dr Evgeny Matveyevich. *USSR.*
Tuan Sarif, Mr Tuan Besar. *Malaysia.*
Tudja, Dr Marijan. *Yugoslavia.*
Tuinstra, Prof. Dr Ir Fokke. *Netherlands.*
Tulinsky, Prof. Alexander. *USA.*
Tumanyan, Dr Vladimir Gayevich. *USSR.*

Tun, Mr Saw. *Burma.*
Tun, Mr Zin. *UK.*
Tunç, Dr Cemil. *Turkey.*
Tunell, Prof. George. *USA.*
Tunkasiri, Dr Tawee. *Thailand.*
Tuomi, Dr Donald. *USA.*
Tuomi, Prof. Turkka Olavi. *Finland.*
Turan, Mrs Canan. *Turkey.*
Turano, Mr August M. *USA.*
Turco, Prof. Guy Henri Robert. *France.*
Turley, Dr June Williams. *USA.*
Turmezey, Dr Tibor. *Hungary.*
Turner, Dr Ralph Waldo. *USA.*
Turpeinen, Dr Urho Taneli. *Finland.*
Turunen, Dr Markus Johannes. *Finland.*
Tuscher, Dr Mag. Engelbert. *Austria.*
Tyapunina, Dr Nataliya Alexandrovna. *USSR.*
Tyers, Mr Kenneth George. *Canada.*
Tykarska, Dr Ewa. *Poland.*
Tyvanchuk, Dr Anna Teodorovna. *USSR.*
Uberbacher, Mr Edward C. *USA.*
Udalova, Dr Valentina Vasilyevna. *USSR.*
Udubasa, Dr Gheorghe. *Romania.*
Uechi, Mr Tetsuo. *Japan.*
Uecker, Mr Reinhard. *DDR.*
Ueda, Prof. Ikuhiko. *Japan.*
Uefuji, Mr Tateki. *Japan.*
Ueki, Dr Tatzuo. *Japan.*
Ülkü, Prof. Dr Dinçer. *Turkey.*
Ünak, Doç. Dr Turan. *Turkey.*
Ueno, Mr Tsunehisa. *Japan.*
Uesu, Prof. Yoshiaki. *Japan.*
Uggla, Dr Rolf Åke Magnus. *Finland.*
Ugliengo, Dr Piero. *Italy.*
Ugozzoli, Dr Franco. *Italy.*
Uhl, Dr Eduard. *Austria.*
Ukaji, Prof. Takeshi. *Japan.*
Ulbricht, Prof. Heinz. *DDR.*
Ulický, Doc. Ing Ladislav. *CSSR.*
Ullrich, Dr Hans-Jürgen. *DDR.*
Ulmer, Mr Kevin M. *USA.*
Umansky, Prof. Mark Moiseyevich. *USSR.*
Umansky, Prof. Yakov Semenovich. *USSR.*
Umegaki, Prof. Yoshiharu. *Japan.*
Umeno, Dr Masataka. *Japan.*
Unan, Doç. Dr Coşkun. *Turkey.*
Unangst, Prof. Dietrich. *DDR.*
Ungár, Dr Tamás. *Hungary.*
Ungar, Dr Goran. *Yugoslavia.*
Ungaretti, Prof. Luciano. *Italy.*
Unge, Dr K. Torsten. *Sweden.*
Unger, Prof. Konrad. *DDR.*
Uno, Prof. Ryosei. *Japan.*
Unonius, Mr Lars-Olof. *Finland.*
Uragami, Dr Takuyuki. *Japan.*
Urban, Dr Heinz. *BRD.*
Urbanczyk, Prof. Grzegorz. *Poland.*
Urbanowicz, Dr Ewa Renata. *Poland.*
Urdy, Prof. Charles Eugene. *USA.*
Urusov, Dr Vadim Sergeyevich. *USSR.*
Urusovskaya, Dr Aida Alexandrovna. *USSR.*
Usanmaz, Doç. Dr Ali. *Turkey.*
Usov, Dr Oleg Alekseyevich. *USSR.*
Uttamasil, Dr Lek. *Thailand.*
Uyeda, Prof. Natsu. *Japan.*
Uyeda, Prof. Ryozi. *Japan.*
Uygur, Prof. Dr M. Eti. *Turkey.*
Uyukin, Dr Evgeny Mikhailovich. *USSR.*
Vähäkangas, Mr Jouko Kaarlo. *Finland.*
Völlenkle, Doz. Dr Horst. *Austria.*
Vaccari, Dr Giuseppe. *Italy.*
Vaciago, Prof. Alessandro. *Italy.*
Vadakkanthara, Gopalakrishnan Thailambal. *India.*
Vagg Dr Robert Sylvester. *Australia.*
Vahrenkamp, Prof. Dr Heinrich. *BRD.*
Vahvaselkä, Dr Aino Margit. *Finland.*
Vahvaselkä, Dr Kaarlo Sakari. *Finland.*
Vainshtein, Prof. Boris Konstantinovich. *USSR.*
Vajda, Dr Erzsébet. *Hungary.*
Val'kovskaya, Dr Margarita Ivanovna. *USSR.*
Valach, Ing Fedor. *CSSR.*

Valarelli, Dr José Vicente. *Brazil.*
Valassiades, Prof. Odysseus. *Greece.*
Valdrè, Prof. Ugo. *Italy.*
Vale, Dr Roger John. *South Africa.*
Valencia, Dr Iluminado G. *Philippines.*
Valente, Dr Edward Joseph. *USA.*
Valenzuela Monjarás, Dr Raúl Alejandro. *Mexico.*
Valera, Dr Anibal Abel. *Peru.*
Valeton, Mrs Prof. Dr Ida Walburga Jakobine. *BRD.*
Valigi, Prof. Mario. *Italy.*
Valin Alberdi, Mrs Maria Luz. *Spain.*
Valkonen, Prof. Jussi Uolevi. *Finland.*
Valle, Dr Giovanni. *Italy.*
Valvoda, Dr Václav. *CSSR.*
Van den Bosch, Dr Adolf. *Belgium.*
Van den Bossche, Dr Guy Ghislain Remy. *Belgium.*
Van Alsenoy, Dr Kris. *Belgium.*
Van Cappellen Mr Eric Edouard Robert. *Belgium.*
Van der Brempt, Miss Christine Marie Paul. *Belgium.*
Van Der Heijden, Dr Simon Petrus Nicolaas. *Canada.*
Van der Helm, Prof. Dick. *USA.*
Van der Veen, Prof. James Morris. *USA.*
Van der Veer, Dr Donald G. *USA.*
Van Dyck, Dr Dirk. *Belgium.*
Van Dyk, Miss Martha Sophia. *South Africa.*
Van Engen, Dr Donna. *USA.*
Van Hove, Dr Michel Andre. *USA.*
Van Landuyt, Prof. Joseph Florent. *Belgium.*
Van Meerssche, Prof. Maurice. *Belgium.*
Van Meervelt, Mr Luc. *Belgium.*
Van Nordstrand, Mr Robert Alexander. *USA.*
Van Opdenbosch, Dr Nicole Marie. *USA.*
Van Roey, Dr Patrick M. A. O. *USA.*
Van Roode, Dr Johannes Hendricus G. *Canada.*
Van Rooyen, Dr Petrus Hendrik. *South Africa.*
Van Schalkwyk, Prof. Theunis G.D. *South Africa.*
Van Tassel, Prof. Dr René. *Belgium.*
Van Tendeloo, Dr Gustaaf. *Belgium.*
Vana, Doz. Dr Dipl.-Ing. Norbert Johannes. *Austria.*
Vance, Dr T. Blake, Jr. *USA.*
Vandenberg, Dr Joanna Maria. *USA.*
Vandenberghe, Dr Robert Emile. *Belgium.*
Vandlen, Dr Richard L. *USA.*
Vanek, Ms Mary Ann. *USA.*
Vanghelie, Mr Iulian. *Romania.*
Vanhellemont, Mr Jan Hendrik. *Belgium.*
Vanhouteghem, Mr Frankie Marie. *Belgium.*
Vani, Miss G. V. *India.*
Vannerberg, Prof. Nils-Gösta. *Sweden.*
Vardar, Mr Bülent. *Turkey.*
Varela, Mr José Arana. *Brazil.*
Varela Mora, Prof Juan de Dios. *Colombia.*
Varga, Dr László. *Hungary.*
Varga, Mr László. *Hungary.*
Varghese, Dr Joseph Noozhumurry. *Australia.*
Varma, Dr Satya Prakash. *UK.*
Varschavsky, Mr Ari. *Chile.*
Varughese, Dr Kottayil Iype. *Canada.*
Vasanth, Dr K. L. *India.*
Vasić, Magistar Pavle. *Yugoslavia.*
Vasil'yev, Dr Alexander Borisovich. *USSR.*
Vasilyev, Dr Evgeny Konstantinovich. *USSR.*
Vassilev, Dr Ivan. *Bulgaria.*
Vedani, Dr Angelo. *Switzerland.*
Vedavati, Miss B. M. *India.*
Veen, van der, Dr Adriaan Hendrik. *Netherlands.*
Veen, Prof. Dr Arthur Willem Lourens. *Netherlands.*
Veerapandian, Mr B. *India.*
Vega, Prof. Juan. *Peru.*
Vega, Mrs Rosario. *Spain.*
Vegas Molina, Dr Angel. *Spain.*
Veith, Prof. Dr Michael. *BRD.*
Velasquez, Prof. Jaime. *Peru.*
Veld, In 't, Ir Gerard Adriaan. *Netherlands.*
Veleker, Mr Thomas John. *USA.*
Velfe, Mr Hans Dieter. *DDR.*
Velikodnyi, Dr Yury Andreyevich. *USSR.*
Velmurugan, Dr D. *India.*
Vendrell Saz, Dr Mario. *Spain.*
Vene, Mrs Nada. *Yugoslavia.*

Venetopoulos, Dr Cleanthis. *Greece.*
Venevtsev, Prof. Yury Nikolayevich. *USSR.*
Venkatarama, Mr R. *India.*
Venkatasubramanian, Dr K. *India.*
Venkatesan, Prof. Kailash. *India.*
Venkatesan, S. *India.*
Venkobarao, Mr H. N. *India.*
Vennik, Prof. Ir Joost. *Belgium.*
Venturello, Porf. Giovanni. *Italy.*
Venugopalan, Dr S. *India.*
Veprek, PD Stanislav. *Switzerland.*
Vera, Mr Rafael. *Chile.*
Vera Calderón, Mrs Gloria. *Mexico.*
Verbist, Prof. Jacques Jozef. *Belgium.*
Verdini, Dr Brunella. *Italy.*
Veremeichik, Dr Tamara Fedorovna. *USSR.*
Vergamini, Dr Phillip James. *USA.*
Verkhovskaya, Dr Kira Alexandrovna. *USSR.*
Verlinde, Mr Christophe Louis-Marie Jos. *Belgium.*
Verma, Dr Ajit Ram *India.*
Verö, Dr Balázs. *Hungary.*
Versaci, Dr Raul Antonio. *Argentina.*
Verschoor, Dr Gerrit Christiaan. *Netherlands.*
Versic, Dr Ronald J. *USA.*
Versteeve, Dr Abraham Jan. *Netherlands.*
Vesselinov, Mr Iliya. *Bulgaria.*
Vester, Mr Jörg. *DDR.*
Vettier, Dr Christian. *France.*
Vezzalini, Dr Maria Giovanna. *Italy.*
Vgenopoulos, Prof. Andreas. *Greece.*
Via, Mr Grayson Hall. *USA.*
Viani, Dr Robert Antoine. *Ivory Coast.*
Vicat, Dr Jean. *France.*
Vickers, Miss Mary Elizabeth. *UK.*
Vicković, Dr Ivan. *Yugoslavia.*
Victorio-Gervasio, Mrs Visitacion. *Philippines.*
Viczián, Dr István. *Hungary.*
Vidal, Dr Geneviève. *France.*
Vidal, Lic Haydée Marta. *Argentina.*
Viehböck, Prof. Dr Franz Paul. *Austria.*
Vielhaber, Dr Edmund Antonius. *BRD.*
Vigneron, Dr Françoise. *France.*
Vijayakar, Mr Suresh Jaywant. *India.*
Vijayalaksmi, Mrs. J. *India.*
Vijayan, Dr (Mrs) Kalyani. *India.*
Vijayan, Prof. Mamannamana. *India.*
Vila, MSc Ileana. *Cuba.*
Vilkov, Dr Lev Vasilyevich. *USSR.*
Villadsen, Mr Jǿrgen. *Denmark.*
Villafuerte Castrejón, Mrs María Elena. *Mexico.*
Villarroel, Prof. Hugo Sergio. *Brazil.*
Villegas, Dr Mario Oscar. *Bolivia.*
Vincent, Dr Michael Grange. *Switzerland.*
Vincent, Dr Roger. *UK.*
Vingsbo, Prof. Olof. *Sweden.*
Vinhas, Dr Laercio Antonio. *Brazil.*
Vinokurov, Prof. Vladimir Mikhailovich. *USSR.*
Visapää, Mr Asko Edvard. *Finland.*
Visser, Dr Rudolph Joseph Jacobus. *Netherlands.*
Visser, Drs Jan Willem. *Netherlands.*
Vistin', Dr Leonard Kazimirovich. *USSR.*
Viswamitra, Prof. Mysore Ananthamurthy. *India.*
Viswanathan, Prof. Dr Krishnamoorthy. *BRD.*
Vitali, Dr Francesca. *Italy.*
Vitali, Dr Gianfranco. *Italy.*
Vitali, Ms Jacqueline. *USA.*
Vitanov, Dr Todor. *Bulgaria.*
Viterbo, Prof. Davide Lazzaro Marco. *Italy.*
Viton Barbolla, Dr Carmen. *Spain.*
Viturro, Dr Pedro Ruben. *Argentina.*
Vizi, Dr Béla. *Hungary.*
Vlad, Dr Serban-Nicolae. *Romania.*
Vlasov, Dr Vasily Platonovich. *USSR.*
Vlasse, Mr Marcus. *USA.*
Voet, Prof. Donald Herman. *USA.*
Vogel, Mrs Sonia. *Chile.*
Voigt, Dr Dieter. *DDR.*
Voigt, Mrs Gabriele. *DDR.*
Voigt, Dr Rita. *DDR.*
Voitsekhovskyi, Dr Vladimir Nikolayevich. *USSR.*

Vol'kenshtein, Prof. Mikhail Vladimirovich. *USSR.*
Vold, Mr Carl L. *USA.*
Voliotis, Prof. Stavros. *Greece.*
Volk, Dr Tat'yans Rafailovna. *USSR.*
Vollstädt, Dr Heiner. *DDR.*
Volodina, Dr Alexander Petrovich. *USSR.*
Volodina, Dr Galina Fedorovna. *USSR.*
Von Dreele, Prof. Robert Bruce. *USA.*
Vonk, Dr Christ Gysbertus. *Netherlands.*
Vora, Dr Rasiklal Amulakhbhai. *India.*
Vorbach, Dr Angelika Irene. *BRD.*
Vorderwisch, Dr Peter. *BRD.*
Vorma, Prof., Dr Atso Ilmari. *Finland.*
Voronova, Dr Alexandra Alekseyevna. *USSR.*
Vos-Looyenga, Dr Aafje. *Netherlands.*
Voskresenskaya, Dr Inna Evgenyevna. *USSR.*
Voutsas, Dr George Panayiotis. *Greece.*
Voznyak, Dr Dmitry Konstantinovich. *USSR.*
Vozzhennikov, Dr Valery Mikhailovich. *USSR.*
Vrábel, Ing Viktor. *CSSR.*
Vradis, Mr Alexandros. *Greece.*
Vranka, Mr Robert G. *USA.*
Vrielink, Miss Alice. *Canada.*
Vries, de, Drs Johan Louis. *Netherlands.*
Vrublevskaya, Dr Zoya Vasilyevna. *USSR.*
Vuagnat, Prof. Marc. *Switzerland.*
Vucht, van, Dr Johannes H.N. *Netherlands.*
Vukasović, Magistar Momčilo. *Yugoslavia.*
Vyas, Mr K. *India.*
Vyas, Dr (Mrs) Meenakshi Nandkishore. *USA.*
Vyas, Mr Nand Kishore. *USA.*
Waal, van de, Ir Benjamin Willem. *Netherlands.*
Wacker, Dr Friedel Klaus. *BRD.*
Wada, Prof. Eiichi. *Japan.*
Wada, Dr Takeo. *Japan.*
Wade, Prof. K. *UK.*
Wade, Dr Richard Harry, *France.*
Wadewitz, Dr Heinz. *DDR.*
Wadhawan, Dr Vinod Kumar. *India.*
Wadsten, Dr Tommy. *Sweden.*
Wägner, Dr Anna. *Sweden.*
Waerstad, Mr Kjell R. *USA.*
Wäsch, Dr Elke. *DDR.*
Wagendristel, Prof. Dr Alfred Friedrich. *Austria.*
Wagenfeld, Dr Heinrich Karsten. *Australia.*
Wagner, Dr Anton Johan. *Netherlands.*
Wagner, Prof. Christian Nikolaus Johann. *USA.*
Wagner, Dr Ernst-Heinz. *BRD.*
Wagner, Mr Gerald. *DDR.*
Wagner, Mr Gunther. *DDR.*
Wahlberg, Mr Anders. *Sweden.*
Wahlberg, Dr Olof. *Sweden.*
Wahlström, Dr Ebba. *Sweden.*
Wahner, Mrs Bettina. *DDR.*
Waintal, Dr Alex. *France.*
Wajsman, Dr Elżbieta Henryka. *Poland.*
Wakabayashi, Dr Katsuzo. *Japan.*
Wal, van der, Dr Robert Jan. *Netherlands.*
Walcher, Dr Herbert. *BRD.*
Waldmann, Dr Hans. *Switzerland.*
Waldrop, Ms Lyneve. *USA.*
Walitzi, Prof. Dr Eva Maria. *Austria.*
Walker, Dr Christopher Bland. *USA.*
Walker, Mr Donald L. *USA.*
Walker, Dr Nigel P. C. *BRD.*
Walker, Mr Peter Jonathan. *UK.*
Walkinshaw, Dr Malcolm Douglas. *Switzerland.*
Wallace, Mr B. A. *USA.*
Wallace, Mr John C. *USA.*
Wallace, Mr Peter Lindsay. *USA.*
Wallenberg, Mr Reine. *Sweden.*
Waller, Prof. Ivar. *Sweden.*
Wallis, Dr John Douglas. *Switzerland.*
Wallis, Dr Julian Mark. *BRD.*
Wallrafen, Dr Franz. *BRD.*
Wallwork, Dr Stephen Collier. *UK.*
Walsoe de Reca, Dr Elizabeth Noemi. *Argentina.*
Walter, Dr Hannes Ulrich. *France.*
Walter, Mr Norman Macmillan. *USA.*
Waltersson, Dr Kjell. *Sweden.*
Walther, Mrs Christa. *DDR.*

Walton, Miss A.R. *UK.*
Wan, Dr Che'ng. *USA.*
Wan, Mr Hsien-ming. *Taiwan.*
Wang, Dr Andrew Hwei-Jiung. *USA.*
Wang, Dr Bi-Cheng. *USA.*
Wang, Miss Chiung-jane. *Taiwan.*
Wang, Prof. Da-cheng. *China.*
Wang, Mr Frederick E. *USA.*
Wang, Prof. Gen-yuan. *China.*
Wang, Mr Guan-xin. *China.*
Wang, Mr Hong. *Canada.*
Wang, Mr Jia-huai. *China.*
Wang, Prof. Kui-ren. *China.*
Wang, Dr Naiding. *BRD.*
Wang, Dr Po-Wen. *USA.*
Wang, Prof. Pu. *China.*
Wang, Dr Rong. *USA.*
Wang, Prof. Shun-jin. *China.*
Wang, Prof. Wen-kui. *China.*
Wang, Mr Xing-xin. *China.*
Wang, Mrs Yao-ping. *China.*
Wang, Prof. Yu. *Taiwan.*
Wang, Mr Zhao-zhou. *China.*
Wang, Prof. Zu-tao. *China.*
Wappler, Dr Gert. *DDR.*
Warczewski, Doc Jerzy Zdzisław. *Poland.*
Ward, Dr Donald Leslie. *USA.*
Ward, Dr Keith Bolen. *USA.*
Ward, Mr José. *Chile.*
Ward, Mr Roger Charles Chavannes. *UK.*
Wardle, Dr Ronald. *USA.*
Wardzyński, Prof. Wiesław. *Poland.*
Warhanek, Prof. Dr Hans. *Austria.*
Waring, Dr J.R.S. *UK.*
Warmiński, Dr Tadeusz Piotr. *Poland.*
Warren, Prof. Bertram Eugene. *USA.*
Warren, Dr Stephen George. *USA.*
Wartchow, Dr Rudolf. *BRD.*
Waser, Prof. Jurg. *USA.*
Waskowska, Dr Alicja. *Poland.*
Watanabe, Dr Akiteru. *Japan.*
Watanabe, Prof. Denjiro. *Japan.*
Watanabe, Dr Hiroshi. *Japan.*
Watanabe, Dr Masaru. *Japan.*
Watanabe, Dr Takashi. *Japan.*
Watanabe, Prof. Takeo. *Japan.*
Watanabe, Prof. Tokunosuke. *Japan.*
Watanabe, Mr Yasunari. *Japan.*
Watanabe, Prof. Yasuyoshi. *Japan.*
Watase, Dr Hideo. *Japan.*
Watenpaugh, Prof. Keith Donald. *USA.*
Waters, Dr David John. *South Africa.*
Waters, Assoc. Prof. Joyce Mary. *New Zealand.*
Waters, Dr Thomas Neil Morris. *New Zealand.*
Watkin, Dr David John. *UK.*
Watkins, Prof. Steven F. *USA.*
Watson, Dr David Gilfillan. *UK.*
Watson, Dr Herman Charles. *UK.*
Watson, Dr Kenneth John. *Australia.*
Watson, Prof. William Harold. *USA.*
Watt, Mr William. *USA.*
Watters, Dr William Asher. *New Zealand.*
Watts, Mrs Ethel Jean. *USA.*
Watts, Mr John Andrew. *Australia.*
Wawra, Mr Herbert. *DDR.*
Wawrzak, Dr Zdzislaw. *Poland.*
Weakley, Dr Timothy John Ruffer. *UK.*
Weaver, Dr Larry H. *USA.*
Weaver, Dr Stephen Donald. *New Zealand.*
Webb, Dr Lawrence Edward. *USA.*
Weber, Mr Charles C. *USA.*
Weber, Dr Hans Peter. *Switzerland.*
Weber, Dr Hans-Jürgen. *BRD.*
Weber, Prof. Dr Harald Wolfgang. *Austria.*
Weber, Dr Irene Teresa. *USA.*
Weber, Prof. Dr Kurt. *BRD.*
Weber, Dr Lawrence D. *USA.*
Weber, Dr Patricia. *USA.*
Webster, Dr Michael. *UK.*
Wechsler, Dr Berry Andrew. *USA.*

Weckert, Mr Edgar. *BRD.*
Weeks, Dr Charles M. *USA.*
Weertman, Ms Julia R. *USA.*
Wegener, Dr Joachim Rolf. *BRD.*
Węglowski, Dr Stanisław. *Poland.*
Wehrenberg, Prof. John Patteson. *USA.*
Wei, Dr Chin Hsuan. *USA.*
Wei, Mr Ming-xiu. *China.*
Wei, Mr Xin-cheng. *China.*
Weigel, Prof. Dominique Jean. *France.*
Weiner, Dr Karl Ludwig. *BRD.*
Weininger, Dr Marc Sterling. *USA.*
Weis, Dr Josef. *DDR.*
Weise, Dr Günter. *DDR.*
Weiss, Prof. Dr Alarich. *BRD.*
Weiss, Prof. Dr Erwin Ludwig. *BRD.*
Weiser, Mr Calvin H. *USA.*
Weiss, Mr Hans-Georg. *BRD.*
Weiss, Dr Helmut. *DDR.*
Weiss, Prof. Raymond. *France.*
Weiss, Dr Richard J. *USA.*
Weiss, Dr Zdeněk. *CSSR.*
Weiss-Nowak, Mr Christian. *BRD.*
Weissman, Dr Larry. *USA.*
Weissmann, Prof. Sigmund. *USA.*
Weitzel, Dr Hans. *BRD.*
Welberry, Dr Thomas Richard. *Australia.*
Welch, Dr A.J. *UK.*
Welch, Mrs D.A. *UK.*
Wells, Prof. Alexander Frank. *USA.*
Wendl, Dr Wolfgang. *BRD.*
Wendland, Mrs Bettina. *DDR.*
Weng, Prof. ling-pao. *China.*
Wenig, Prof Dr Werner. *BRD.*
Wenk, Prof. Hans-Rudolf. *USA.*
Werk, Dr Margit L. *Switzerland.*
Werner, Dr Inge. *DDR.*
Werner, Mr Michael. *DDR.*
Werner, Mr Per-Erik. *Sweden.*
Wertz, Mr David L. *USA.*
West, Dr Anthony Roy. *UK.*
West, Mr M.P. *UK.*
West, Dr N.G. *UK.*
Westbrook, Dr Edwin M. *USA.*
Westdahl, Miss Marianne. *Sweden.*
Wester, Dr Dennis Wayne. *USA.*
Westin, Dr Leif. *Sweden.*
Westman, Ms Ingrid. *Sweden.*
Westman, Dr Sven. *Sweden.*
Weston, Dr Norman Ernest. *USA.*
Westphalen, Mr John Arthur. *Australia.*
Weulersse, Prof. Philippe. *France.*
Wevers, Drs Joice. *Netherlands.*
Wey, Prof. Raymond. *France.*
Weyl, Dr Colette. *France.*
Wheatley, Dr Peter Jaffrey. *UK.*
Wheeler, Mr George L. *USA.*
Whelan, Dr Michael John. *UK.*
Whillans, Dr Francis David. *Australia.*
Whimp, Dr Peter Olaf. *New Zealand.*
Whiston, Dr Clive David. *UK.*
Whitaker, Dr Alan. *UK.*
White, Dr Allan Henry. *Australia.*
White, Dr David Nathaniel James. *UK.*
White, Dr Janice Larraine. *UK.*
White, Prof. Joe L. *USA.*
White, Prof. John Greville. *USA.*
White, Dr Peter Sutherland. *Canada.*
Whitfield, Dr Harold John. *Australia.*
Whitla, Dr William Alexander. *Canada.*
Whitlow, Dr Simon Hugh. *Canada.*
Whitlow, Mr Marc David. *USA.*
Whitney, Mr R.A. *UK.*
Whitney, Dr John Franklin. *USA.*
Whittaker, Dr Eric James William. *UK.*
Whittle, Mr Robert. *USA.*
Whitworth, Dr Robert William. *UK.*
Whuler, Dr Annick. *France.*
Więckowski, Mr Tadeusz. *Poland.*
Wickham, Dr Donald Guy. *USA.*

Wickman, Prof. Frans Erik. *Sweden.*
Wicks, Dr Frederick John. *Canada.*
Wiebenga, Prof. Dr Eelco Herman. *France.*
Wieczorek, Dr Michał. *Poland.*
Wiegers, Dr Gerrit Adriaan. *Netherlands.*
Wierenga, Dr Rikkert Klaas. *Netherlands.*
Wieser, Dr Egbert. *DDR.*
Wiesner, Dr Joel Robert. *USA.*
Wieteska, Dr Krzysztof. *Poland.*
Wiewióra, Dr hab. Andrzej. *Poland.*
Wiff, Dr Donald Ray. *USA.*
Wignall, George Denis. *USA.*
Wilchinsky, Dr Zigmond Walter. *USA.*
Wild, Mr G.A. *UK.*
Wilde, Dr Wolfgang. *DDR.*
Wildman, Dr Gary C. *USA.*
Wildner, Mr Günter. *DDR.*
Wiley, Prof. Don Craig. *USA.*
Wilford, Dr John Bernard. *UK.*
Wilhelm, Dr Eberhard. *BRD.*
Wilhelmi, Dr Karl-Axel. *Sweden.*
Wilke, Prof. Dr Wolfgang. *BRD.*
Wilken, Dr Gerdt. *BRD.*
Wilkes, Dr Glenn Richard. *USA.*
Wilkins Dr Stephen William. *Australia.*
Wilkinson, Dr Clive. *UK.*
Wilkinson, Dr D. *UK.*
Wilkinson, Dr Michael Kennerly. *USA.*
Will, Prof. Dr Georg. *BRD.*
Willaime, Prof. Christian. *France.*
Willett, Prof. Roger DuWayne. *USA.*
Williams, Mr Brian Edward. *Australia.*
Williams, Mr Donald Allan. *Australia.*
Williams, Prof. Donald Elmer. *USA.*
Williams, Dr Geoffrey Allan. *Australia.*
Williams, Dr Grahame John Bramald. *USA.*
Williams, Dr Jack Marvin. *USA.*
Williams, Dr John A. *USA.*
Williams, Dr R.A. *UK.*
Williams, Prof. Rickey J. *USA.*
Williams, Dr Robin O'Dare. *USA.*
Williams, Roger M. *USA.*
Williams, Mr Timothy Brendan. *Australia.*
Williamson, Mr Robert S. *USA.*
Williard, Prof. Paul Gregory. *USA.*
Willig, Prof. Cesar Dorneles. *Brazil.*
Willis, Dr Anthony Creswick. *Canada.*
Willis, Dr Bertram Terence Martin. *UK.*
Wilsdorf, Prof. Heinz Gerhard Friedrich. *USA.*
Wilson, Dr Alan Richard. *Australia.*
Wilson, Prof. Arthur James Cochran. *UK.*
Wilson, Mr David K. *USA.*
Wilson, Dr Frank Charles. *USA.*
Wilson, Prof. Herbert Rees. *UK.*
Wilson, Mr J.D. *UK.*
Wilson, Dr Keith Sanderson. *UK.*
Wilson, Dr Michael Jeffrey. *UK.*
Wilson, Dr S.J. *UK.*
Winchell, Prof. Horace. *USA.*
Windle, Dr Alan Hardwick. *UK.*
Windsch, Prof. Wolfgang. *DDR.*
Windsor, Dr Colin George. *UK.*
Wing, Prof. Richard M. *USA.*
Wingert, Mrs Lavinia Meinzer. *USA.*
Winkler, Dr Fritz Karl. *Switzerland.*
Winslow, Mr Douglas N. *USA.*
Wintenberger, Dr Micheline. *France.*
Winter, Prof Dr Werner. *BRD.*
Winter, Prof. William Thomas. *USA.*
Winzer, Dr Achim. *DDR.*
Wirjosoedirdjo, Mr Hardjatmo. *Indonesia.*
Wirjosumarto, Dr Harsono. *Indonesia.*
Wismer, Prof. Robert Kingsley. *USA.*
With, de, Dr Gijsbertus. *Netherlands.*
Witte, Prof. Dr Helmut Hermann Wolfgang. *BRD.*
Wittels, Dr Mark Caesar. *USA.*
Witters, Prof. Robert Dale. *USA.*
Wittke, Prof. Oscar. *Chile.*
Witz, Dr Jean. *France.*
Wlodawer, Dr Alexander. *USA.*
Wobrauschek, Doz. Dr Dipl.-Ing. Peter. *Austria.*

Wögerbauer, Dr Rupert. *BRD.*
Wölfel, Prof. Dr Erich Richard. *BRD.*
Woensdregt, Drs Cornelis Franciscus. *Netherlands.*
Wokulska, Dr Krystyna Barbara. *Poland.*
Wokulski, Dr Zygmunt. *Poland.*
Wolbaum, Mr Keith Jonathon. *Canada.*
Wolcyrz, Dr Marek. *Poland.*
Wold, Prof. Aaron. *USA.*
Wolf, Dr Dieter. *BRD.*
Wolf, Mr Eberhard. *DDR.*
Wolff, Dr Gunther Arthur. *USA.*
Wolff, de, Dr Pieter Maarten. *Netherlands.*
Wolper, J. Brook. *USA.*
Won, Mr Vann Yuen. *USA.*
Wonacott, Dr Alan John. *UK.*
Wondratschek, Prof. Dr Hans. *BRD.*
Wong, Mrs Rosalind Y. *USA.*
Wong, Dr Yau-Shing. *Hong Kong.*
Wong-Ng, Dr Winnie Kwai-Wah. *USA.*
Wongshaiboon, Dr Sajee. *Thailand.*
Wood, Mr Dermott. *UK.*
Wood, Dr Elizabeth Armstrong. *USA.*
Wood, Dr Gordon H. *Canada.*
Wood, Dr Ian George. *UK.*
Wood, Prof. John Karl. *USA.*
Wood, Prof. John Stanley. *USA.*
Wood, Prof. Mical Kent. *USA.*
Wood, Dr Raymond Maurice. *UK.*
Woode, Dr Kwamena Annan. *USA.*
Woods, Dr Geoffrey Steward. *UK.*
Woolfson, Prof. Michael Mark. *UK.*
Wooster, Mr Antony Martin. *UK.*
Workman, Mr Samuel Thomas. *USA.*
Worthington, Mr C. R. *USA.*
Worzala, Dr Horst. *DDR.*
Wright, Mr C.P. *UK.*
Wright, Dr Christine Gerda Schubert. *USA.*
Wright, Dr Helen. *UK.*
Wright, Dr John Albert. *UK.*
Wright, Dr John Dalton. *UK.*
Wright, Dr Harlan Tonie. *USA.*
Wright, Mr Phillip John. *Australia.*
Wright, Dr William Vaughn. *USA.*
Wu, Mr Bo-mu. *China.*
Wu, Mr Ding-ming. *China.*
Wu, Ms Ellen. *USA.*
Wu, Prof. Qian-zhang. *China.*
Wu, Mr Hsin-I. *USA.*
Wu, Mr Shen. *China.*
Wu, Prof. Shou-yu. *China.*
Wu, Prof. Nan-chung. *Taiwan.*
Wu, Mr Xin-tao. *China.*
Wu, Mr Zi-wu. *China.*
Wüest, Mr Hermann. *Switzerland.*
Wuensch, Prof. Bernhardt John. *USA.*
Wünsche, Mrs Inez. *DDR.*
Wunderlich, Dr Hartmut. *BRD.*
Wurl, Mr Bernd. *DDR.*
Wurtz, Dr Michel. *Switzerland.*
Wyart, Prof. Jean. *France.*
Wyckoff, Prof. Harold Winfield. *USA.*
Wyckoff, Prof. Ralph W. G. *USA.*
Wylie, Mr Thomas Smith. *UK.*
Xia, Mrs Zong-xiang. *China.*
Xiao, Prof. Xu-gang. *China.*
Xie, Dr Si-shen. *China.*
Xie, Prof. Xian-de. *China.*
XiMen, Mrs Lu-lu. *China.*
Xu, Mr Ji-quan. *China.*
Xu, Mr Jing-yang. *China.*
Xu, Mr Pei-cang. *China.*
Xu, Prof. Shun-sheng. *China.*
Xu, Mr Xiao-jie. *China.*
Xu, Prof. Zheng-yi. *China.*
Xue, Mrs Ji-yue. *China.*
Xue, Prof. Jun-zhi. *China.*
Xue, Prof. Zhi-lin. *China.*
Xuong, Prof. Nguyen-Huu. *USA.*
Yağbasan, Dr Rahmi. *Turkey.*
Yağcı, Dr Osman. *Turkey.*

Yadav, Mr Asheshwar. *India.*
Yadav, Mr Tapaswi. *India.*
Yadav, Dr Vijay Singh. *India.*
Yadava, Dr Bishwanath *India.*
Yagi, Prof. Katsumichi. *Japan.*
Yahalom, Prof. Joseph. *Israel.*
Yahaya, Dr Muhamad. *Malaysia.*
Yakel, Dr Harry Leonard. *USA.*
Yakhontova, Prof. Liya Konstantinovna. *USSR.*
Yakinthos, Prof. John. *Greece.*
Yakovenko Dr Sergey Sergeyevich. *USSR.*
Yakovlev, Prof. Ivan Alekseyevich. *USSR.*
Yakubovich, Dr. Ol'ga Vselodovna. *USSR.*
Yakushi, Mr Kyuya. *Japan.*
Yalkovsky, Dr Ralph. *USA.*
Yamaguti, Prof. Tasaburo. *Japan.*
Yamamoto, Mr Akiji. *Japan.*
Yamamoto, Dr Naoki. *Japan.*
Yamanaka, Dr Takamitsu. *Japan.*
Yamane, Dr Takashi. *Japan.*
Yamashita, Mr Shuji. *Japan.*
Yamashita, Prof. Tadayoshi. *Japan.*
Yamnova, Dr Nataliya Arkadyevna. *USSR.*
Yan, Mr Qi-wei. *China.*
Yan, Mr You-wei. *China.*
Yanai, Dr Hideyasu Steve. *USA.*
Yaneva, Dr Svetlana. *Bulgaria.*
Yang, Dr. Daniel Shun-Chung. *USA.*
Yang, Mr Chuan-zheng. *China.*
Yang, Mr Guang-di. *China.*
Yang, Ms Guang-ming. *China.*
Yang, Prof. Houng-Yi. *Taiwan.*
Yang, Mr Hua-guang. *China.*
Yang, Mr Hua-hui. *China.*
Yang, Prof. Qi-bin. *China.*
Yang, Ms Qing-chuan. *China.*
Yang, Prof. Tse Chun. *Taiwan.*
Yang, Dr Yui-whei. *Taiwan.*
Yang, Mr Zuo-sheng. *China.*
Yankov, Henry F. *USA.*
Yanson, Dr Tamara Ivanovna. *USSR.*
Yanulov, Dr Kirill Paskalyevich. *USSR.*
Yao, Mr Jia-xing. *China.*
Yao, Mr Xin-kan. *China.*
Yarmolyuk, Dr Yaroslav Petrovich. *USSR.*
Yashiro, Prof. Yuzo. *Japan.*
Yasinskaya, Dr Angelina Andreyevna. *USSR.*
Yasuoka, Dr Noritake. *Japan.*
Yates, Dr John Harry. *USA.*
Yazici, Mr Rahmi M. *USA.*
Yeşilsoy, Prof. Dr M. Sefik. *Turkey.*
Ye, Prof. Heng-qiang. *China.*
Yeh, Prof. Hun Chiang. *USA.*
Yelon, Dr William B. *USA.*
Yilmaz, Doç. Dr Osman. *Turkey.*
Yin, Dr Soe. *Burma.*
Ylinen, Dr Eero Elias. *Finland.*
Ymén, Dr B. Ingvar. *Sweden.*
Yoda, Dr Osamu. *Japan.*
Yokomori, Dr Yoshinobu. *Japan.*
Yonath, Prof. Ada. *Israel.*
Yonei, Dr Moto'o. *Japan.*
Yong, Mr Swee Kee. *Malaysia.*
Yoon, Dr Hyo Sub. *USA.*
Yoshida, Dr Kentaro. *Japan.*
Yoshida, Prof. Sanae. *Japan.*
Yoshimatsu, Prof. Mitsuru. *Japan.*
Yoshimura, Dr Junichi. *Japan.*
Yoshimura, Mr Yukio. *Japan.*
Yoshioka, Prof. Hide. *Japan.*
You, Mr Jun-ming. *China.*
You, Prof. Xiao-zeng. *China.*
Young, Mr Brian Raymond. *UK.*
Young, Dr Frederick Walter, Jr. *USA.*
Young, Prof. R. A. *USA.*
Young, Dr Ti-Sheng. *USA.*
Youssef, Dr I. Mourad. *Egypt.*
Yousufzai, Mr Inayatullah Khan. *Pakistan.*
Ysern, Dr Xavier. *Venezuela.*
Ysern, Mr Xavier. *USA.*

Yu, Prof. Rui-huang. *China.*
Yu, Prof. Shu-cheng. *Taiwan.*
Yu, Prof. Wei-hai. *China.*
Yu, Prof. Xiu-fen. *China.*
Yücel, Dr Atila. *Turkey.*
Yüksel, Mr Hikmet. *Turkey.*
Yung, Dr Fook Hong. *Australia.*
Yurin, Dr Vladimir Alexandrovich. *USSR.*
Yusaf, Dr Mohammad. *Pakistan.*
Yushin, Dr Yury Yakovlevich. *USSR.*
Yushkin, Prof. Nikolay Pavlovich. *USSR.*
Yvon, Prof. Klaus. *Switzerland.*
Zábráczki, Mr Jósef. *Hungary.*
Zaccai, Dr Giuseppe. *France.*
Zacharias, Dr David Edward. *USA.*
Zachau-Christiansen, Stud.lic. Birgit. *Denmark.*
Zadorozhnaya, Dr Lyudmila Alexandrovna. *USSR.*
Zagal'skaya, Dr Yudif' Gertsevna. *USSR.*
Zagari, Prof. Adriana. *Italy.*
Zaghloul, Dr Mohamed Zaki. *Egypt.*
Zagt, Mr Simon. *South Africa.*
Zaitsev, Dr Sergey Mikhailovich. *USSR.*
Zaitseva, Dr Mariya Panteleimonovna. *USSR.*
Zajc, Magistar Andrej. *Yugoslavia.*
Žák, Doc. Dr Lubor. *CSSR.*
Žák, Dr Zdirad. *CSSR.*
Zakharchenko, Dr Irina Nikolayevna. *USSR.*
Zakharov, Dr Nikolay Dmitriyevich. *USSR.*
Zakharova, Prof. Mariya Ivanovna. *USSR.*
Zaki, Mr Chakib. *BRD.*
Zalba, Lic Patricia Eugenia. *Argentina.*
Zalessky, Dr Andrey Vladimirovich. *USSR.*
Zalkin, Dr Allan. *USA.*
Zalutsky, Dr Ivan Ilyich. *USSR.*
Zaman, Dr (Mrs) Nazma, *Bangladesh.*
Zamorzayev, Dr Alexander Mikhailovich. *USSR.*
Zanazzi, Prof. Pier Francesco. *Italy.*
Zangrando, Dr Ennio. *Italy.*
Zannetti, Prof. Roberto. *Italy.*
Zanotti, Prof. Giuseppe. *Italy.*
Zanotti, Dr Lucio. *Italy.*
Zappia, Prof. Vincenzo. *Italy.*
Zardas, Mr George. *Greece.*
Zarechnyuk, Dr Oleg Safonovich. *USSR.*
Zarka, Dr Albert. *France.*
Zaslow, Prof. Bert. *USA.*
Zasorin, Dr Evgeny Zotikovich. *USSR.*
Zassenhaus, Prof. Hans Julius. *USA.*
Zatout, Mr Mohamed Abdel Meguid. *Egypt.*
Zav'yalova, Anna Arkadyevna. *USSR.*
Zayakina, Dr Nadezhda Viktorovna. *USSR.*
Zedler, Dr Achim. *DDR.*
Zeedijk, Ir Hendrik Bastiaan. *Netherlands.*
Zefiro, Dr Livio. *Italy.*
Zehnder, PD Margareta. *Switzerland.*
Zeigan, Dr Dieter. *DDR.*
Zeilinger, Prof. Dr Anton Wolfgang. *Austria.*
Zelada, Mr Gabriel. *Chile.*
Zelaya, Mr José Miguel. *Bolivia.*
Zeldes, Mr Nathan. *Israel.*
Zellingher, Mr Naphtaly. *Israel.*
Zelwer, Dr Charles Marcel. *France.*
Zemann, Prof. Dr Josef. *Austria.*
Zemke, Mr Klaus Jurgen. *South Africa.*
Zenginoglou, Mr Charalambos. *Greece.*
Zerbi, Prof. Giuseppe. *Italy.*
Zevin, Prof. Lev. *Israel.*
Zeyfang, Dr Rolf Robert. *BRD.*
Zhang, Mr Bu-sheng. *China.*
Zhang, Mr Ci-he. *China.*
Zhang, Prof. Dao-biau. *China.*
Zhang, Mrs Gen-di. *China.*
Zhang, Mr Guang-rong. *China.*
Zhang, Prof. Guan-ying. *China.*
Zhang, Prof. Jiang-hong. *China.*
Zhang, Mr Han-hui. *China.*
Zhang, Prof. Le-hui. *China.*
Zhang, Mr Li-xin. *China.*
Zhang, Mrs Rong-ying. *China.*
Zhang, Mr Rui-lin. *China.*
Zhang, Prof. Shao-hui. *China.*

Zhang, Mr Shi-wei. *China.*
Zhang, Ms Shu-de. *China.*
Zhang, Mr Yong-mao. *China.*
Zhang, Prof. Yuan-long. *China.*
Zhang, Mr Yue-ming. *China.*
Zhang, Ms Ze-ying. *China.*
Zhang, Prof. Zong. *China.*
Zhao, Prof. Qi-yuan. *China.*
Zhdanov, Prof. German Stepanovich. *USSR.*
Zheludev, Prof. Ivan Stepanovich. *USSR.*
Zheludeva, Dr Svetlana Ivanovna. *USSR.*
Zheng, Prof. Pei-ju. *China.*
Zheng, Mr Qi-tai. *China.*
Zheng, Mr Zhe. *China.*
Zhidkov, Dr Nikolay Petrovich. *USSR.*
Zhmurova, Dr Zinaida Ivanovna. *USSR.*
Zhong, Mr Na-tian. *China.*
Zhong, Prof. Wei-zhuo. *China.*
Zhou, Prof. Gong-du. *China.*
Zhou, Mr Gui-en. *China.*
Zhou, Mr Kang-jing. *China.*
Zhou, Mr Zhong-yuan. *China.*
Zhou, Mr Zong-hua. *China.*
Zhu, Mr Ji-mu. *China.*

Zhu, Mr Nai-jue. *China.*
Zhu, Mr Zhong-he. *China.*
Zhuang, Mr Hong-hui. *China.*
Zhuang, Mr Jian. *China.*
Zhukhlistov, Dr Anatoliy Pavlovich. *USSR.*
Zhuze, Prof. Vladimir Panteleimonovich. *USSR.*
Ziółowska, Dr Blanka. *Poland.*
Zickert, Mr Kurt. *DDR.*
Zidarova, Mrs Bogdana. *Bulgaria.*
Ziegler, Prof. Dr Manfred Ludwig. *BRD.*
Zielińska-Rohozińska, Dr Elżbieta. *Poland.*
Ziemer, Dr Burkhard. *DDR.*
Zigan, Prof. Dr Franz Martinus. *BRD.*
Zikmund, Dr Zdeněk. *CSSR.*
Zilber, Dr Raphael. *Israel.*
Zimmer, Mr Alfons. *BRD.*
Zimmerman, Lic Rosa. *Argentina.*
Zimmermann, Dr Helmuth Walter. *BRD.*
Zinenko, Dr Victor Ivanovich. *USSR.*
Ziołowski, Dr Zbigniew. *Poland.*
Ziolo, Dr Ronald F. *USA.*
Zipper, Dr Peter. *Austria.*
Zlosilo, Mr Mario. *Chile.*
Żmija, Prof. Józef. *Poland.*

Zobel, Dr Dieter. *BRD.*
Zobetz, Dr Erich. *Austria.*
Zocchi, Prof. Marcello. *Italy.*
Zolensky, Dr Michael Ewing. *USA.*
Zolliker, Mr Peter. *Switzerland.*
Zoltai, Prof. Tibor Z. *USA.*
Zorky, Prof. Petr Markovich. *USSR.*
Zorn, Dr Gerhard. *BRD.*
Zosi, Prof. Gianfranco Luigi. *Italy.*
Zotov, Mr Nikolay. *Bulgaria.*
Zschach, Dr Siegfried. *DDR.*
Zschokke-Gränacher, Prof. Iris. *Switzerland.*
Zsoldos, Mrs Éva. *Hungary.*
Zsoldos, Dr Lehel. *Hungary.*
Zubenko, Dr Vasily Vasilyevich. *USSR.*
Zubov, Dr Yuriy Alexandrovich. *USSR.*
Zulehner, Dr Werner. *BRD.*
Zussman, Prof. Jack. *UK.*
Zvezdinskaya, Dr Larisa Vsevolodovna. *USSR.*
Zviedre, Dr Irena Ilyinichna. *USSR.*
Zvinchuk, Dr Rostislav Alekseyevich. *USSR.*
Zvyagin, Dr Boris Borisovich. *USSR.*
Zwaan, Dr Pieter Cornelis. *Netherlands.*
Zwell, Mr Leo. *USA.*

International Union of Crystallography

GENERAL DESCRIPTION

The formation of the Union was discussed at an international meeting held in London in May 1946; it was accepted by the International Council of Scientific Unions on 7 April 1947. Its objectives are to promote international cooperation in crystallography and to contribute to the advancement of all aspects of crystallography, to promote international publication of crystallographic research, to facilitate standardization of methods, units, nomenclature and symbols used in crystallography, and to form a focus for the relations of crystallography to other sciences.

Since its formation, the Union has remained the focal point for international cooperation in crystallography and, at present, 34 countries belong to the Union, through their National Academy, National Research Council or similar body, or through a scientific society or group of such societies. A list of these members (called Adhering Bodies) is given at the end of this brief description of the Union. Recent triennial Congresses, held in association with the business meetings of the Union (General Assemblies), have been attended by 1,200 - 1,600 scientists. The Union also organizes or sponsors many smaller meetings.

The Union has established 15 Commissions, which are concerned with either a principal publishing activity or a major topic or field of concern to crystallographers. The latter group of Commissions organize many international projects concerned with the establishment of internationally acceptable standards or methods of procedure. In addition, they organize specialist meetings or short courses of instruction for young scientists. Nineteen teaching pamphlets have been published, and more are planned. Details of the work of the Commissions are included in the annual reports of the Executive Committee, published in *Acta Crystallographica*, Section A. The Commissions also submit triennial reports to the General Assembly, which elects their members. In addition to the three publishing Commissions (Journals, *Structure Reports* and *International Tables*) there are Commissions on the following topics:

- Apparatus
- Biological Macromolecules
- Charge, Spin and Momentum Densities
- Computing
- Crystal Growth and Characterization of Materials
- Data
- Electron Diffraction
- Neutron Diffraction
- Nomenclature
- Small Molecules
- Studies at Controlled Pressures and Temperatures
- Teaching

The current memberships of the Executive Committee and the Commissions, the names of the Union representatives on other bodies, and the memberships of the National Committees for Crystallography (including the addresses of the Secretaries of these Committees) will be given in the appendices to the Report of the Thirteenth General Assembly and International Congress of Crystallography, to be published in *Acta Crystallographica, Section A* in late 1986 or early 1987. These appendices also give the Statutes and By-Laws of the Union and comprehensive reports on the various activities of the Union. The Reports of previous General Assemblies and Congresses have also been published in *Acta Crystallographica*. Further information concerning the Union may be obtained from the Executive Secretary, International Union of Crystallography, 5 Abbey Square, Chester CH1 2HU, England.

PUBLISHING ACTIVITIES

Less than a year after its formation the Union was publishing its own scientific journal, *Acta Crystallographica*, in an attempt to reassemble the crystallographic work which was scattered through a wide variety of journals. The journal rapidly became established as the major forum for publication of crystallographic research. Today it runs to 3,500 pages a year and is divided into three sections. The crystallographic community owns and controls the journal and its sister journal, *Journal of Applied Crystallography*, which was created in 1968. Through the Union it appoints the journals' editors and the Union is solely responsible for the finances of the journals.

Other major scientific publishing works which were undertaken right at the start of the life of the Union were *Structure Reports*, which gives critical reports on crystal structure determinations published in all the scientific journals, and *International Tables for X-ray Crystallography*, which contains the theory of crystallographic groups (providing the basic reference work for all crystal structure determinations) and the mathematical, physical and chemical tables required for crystallographic work. The Union has also published several monographs and, in conjunction with the Crystallographic Data Centre in Cambridge, England, the *Molecular Structures and Dimensions* series. Further details of these publications are given at the end of this description of the work of the Union.

IUCr

ADMINISTRATION

The highest authority of the Union is the General Assembly, which meets triennially and consists of delegates appointed by the Adhering Bodies representing the countries belonging to the Union. The affairs of the Union between General Assemblies are conducted by the Executive Committee, comprising a President, a Vice-President, a General Secretary and a Treasurer (at present these two offices are combined), Immediate Past-President and six ordinary members. The Union secretariat and technical editing office for the journals are directed by an Executive Secretary and are located in Chester, England. The Union is incorporated with its legal domicile in Geneva, Switzerland.

ADHERING BODIES

Country	Category*	Adhering Body
Argentina	I	Consejo Nacional de Investigaciones Científicas y Técnicas
Australia	III	Australian Academy of Science
Austria	I	Österreichische Akademie der Wissenschaften
Belgium	II	Académie Royale des Sciences, des Lettres et des Beaux-Arts de Belgique
Brazil	III	Conselho Nacional de Desenvolvimento Cientifico e Tecnologico
Canada	III	National Research Council
Chile	I	National Committee for Crystallography
China, Peoples Rep.	IV	Academia Sinica
Czechoslovakia	I	Československá Akademie Věd
Denmark	I	Royal Danish Academy of Sciences and Letters
Egypt, Arab Rep.	I	Academy of Scientific Research and Technology
Finland	I	Suomen Tiedeakatemiain Valtuuskunta
France	IV	Académie des Sciences (Institut de France)
German Dem. Rep.	I	Vereinigung für Kristallographie in der G.G.W. der D.D.R.
Germany, Fed. Rep.	IV	Arbeitsgemeinschaft Kristallographie
Hungary	I	Magyar Tudományos Akadémia
India	II	Indian National Science Academy
Israel	I	Israel Academy of Sciences and Humanities
Italy	III	Consiglio Nazionale delle Ricerche
Japan	IV	Science Council of Japan
Mexico	I	Consejo Nacional de Ciencia y Tecnologia
Netherlands	II	Stichting voor Fundamenteel Onderzoek der Materie met Röntgen- en Elektronenstralen
New Zealand	I	The Royal Society of New Zealand
Norway	I	Det Norske Videnskaps Akademi
Poland	I	Polska Akademia Nauk
Portugal	I	Sociedade Portuguesa de Fisica
South Africa	I	South African Council for Scientific and Industrial Research
Spain	III	Consejo Superior de Investigaciones Científicas
Sweden	II	Kungliga Vetenskapsakademien
Switzerland	II	Schweizerische Gesellschaft für Kristallographie
U.K.	V	The Royal Society
U.S.A.	V	National Academy of Sciences - National Research Council
U.S.S.R.	V	Akademija Nauk S.S.S.R.
Yugoslavia	I	Jugoslavenska Akademija Znanosti i Umjetnosti

* Adherence to the Union is in one of five Categories, with corresponding voting powers and contributions as set out in Statutes 3.6, 5.5 and 9.4.

FINANCES

The Union receives income to finance its general activities in the form of subscriptions from Adhering Bodies, a subvention from UNESCO through ICSU and yields from investments, which have to be maintained as an essential financial backing for the Union's publishing activities. These activities involve an annual turnover of about 1,600,000 Swiss Francs.

COOPERATION WITH OTHER INTERNATIONAL SCIENTIFIC ORGANIZATIONS

The Union maintains close relations with UNESCO and the International Council of Scientific Unions (ICSU). It is represented on several inter-Union or international bodies which may be concerned with very general or very specific tasks requiring concerted international cooperation. The Union is involved with the work of the ICSU Committees on the Teaching of Science, Science and Technology in Developing Countries, Space Research and Data for Science and Technology, and the International Council of Scientific and Technical Information, as well as committees of other Unions or similar organizations. The International Organization for Crystal Growth is an IUCr Scientific Associate and the European Crystallographic Committee is an IUCr Regional Associate.

IUCr

Publications of the International Union of Crystallography

When the Union was established, it was decided that one of its major tasks should be the promotion of international publication of crystallographic research and of works on crystallography. For this purpose the Union has launched a number of publications which, thanks to the cooperation and efforts of scientists from all over the world, have become leading publications in crystallography. They have become indispensable to all workers in this field, and to those working in solid-state physics and solid-state chemistry, as well as in mineralogy.

ACTA CRYSTALLOGRAPHICA

Publishers: Munksgaard International Publishers Ltd., 35 Nørre Søgade, DK-1370 Copenhagen K, Denmark

Acta Crystallographica is a scientific journal containing original articles in English, French, German and Russian, dealing with new crystal structures, refinements of known structures, new theoretical and experimental methods of structure determination, the theory of diffraction, computing methods, apparatus, and various other related topics. It is published in three Sections: Section A (foundations of crystallography) and Section B (structural science) are published bi-monthly and Section C (crystal structure communications) is published monthly. All papers are refereed, and they are accepted for publication only when the material is original and when the contents of the paper are of sufficiently high quality. It may be said that much of the best work in the above mentioned fields appears in *Acta Crystallographica*.

JOURNAL OF APPLIED CRYSTALLOGRAPHY

Publishers: Munksgaard International Publishers Ltd., 35 Nørre Søgade, DK-1370 Copenhagen K, Denmark

This scientific journal is published in bi-monthly issues. It is concerned with methods, apparatus, problems, and discoveries in applied crystallography. It deals with the application of existing crystallographic techniques to practical problems and with developments in crystallography that have a prospect of future application. It is intended to meet the needs of those scientists who use crystallographic techniques, especially diffraction methods, to study materials and to control their quality. All full length papers and short communications are refereed. The journal also publishes details of computer programs, crystal data, laboratory notes, computer program abstracts, details of new products and details and lists of forthcoming meetings.

STRUCTURE REPORTS

Publishers: D. Reidel Publishing Company, P.O. Box 17, 3300 AA Dordrecht, The Netherlands

Structure Reports provides critical reports of virtually all determined crystal structures. The reports are arranged in sections on metals, inorganic compounds and organic compounds. Each annual volume is published in two parts: A. Metals and Inorganic Sections, and B. Organic Section (including organometallic compounds). The reports generally give: name, formula, papers reported, unit cell and space-group data, details of analysis, atomic positions, and detailed description and discussion of the structure (with bond lengths and angles, and usually with illustrations). The structural data are reported so completely that reference to the original paper is not often necessary. There are extensive indexes in each annual volume, and cumulative ten-year indexes published as separate volumes.

It becomes unnecessary to search hundreds of journals. Each volume gives the essence of one year's worldwide literature on crystal structure determinations of all metal, inorganic, organic, and organometallic materials. The series forms an essential bank of information for all university and research laboratories, and science and reference libraries. In universities, students can learn where and how structural information can be readily obtained. In research laboratories, *Structure Reports* is a ready source of basic data and ideas on materials, which can repay its cost many times over.

Volume 46B (covering organic compounds for 1980) was published in early 1985. Other volumes expected to be published in 1985 are Volumes 48B and 49B (organic compounds for 1981 and 1982) and Volumes 49A and 50A (metals and inorganic compounds for 1982 and 1983). Cumulative indexes published include ten-year indexes and a *60-Year Structure Index* (divided into Section A. *Metals and Inorganic Compounds,* and Section B. *Organic Compounds*), covering the period 1913-1973. Supplements for 1974-1975 were also published, as was an index to the whole series of *Strukturbericht* (Volumes 1-7; 1913-1939).

MOLECULAR STRUCTURES AND DIMENSIONS

Publishers: D. Reidel Publishing Company, P.O. Box 17, 3300 AA Dordrecht, The Netherlands

This series, which is published for the Union in conjunction with the Crystallographic Data Centre, Cambridge, England, gives classified bibliographic information covering the literature on organic and organometallic crystal structures determined since 1935. Entries are arranged in chemical classes with extensive cross-references and there are cumulative indexes of molecular formulae, transition metals and authors. Volume 15, covering the literature for 1982-1983, was published in 1984.

Guide to the Literature 1935-1976. Organic and Organometallic Crystal Structures contains a set of six fully cumulative indexes displaying chemical and biographic details of the structures covered in Volumes 1-8 of the series. Volume A1, entitled *Interatomic Distances 1960-65,* contains evaluated numerical data on the molecular geometry of compounds listed in the bibliographic volumes. It is illustrated by chemical and crystallographic diagrams.

INTERNATIONAL TABLES FOR CRYSTALLOGRAPHY

Publishers: D. Reidel Publishing Company, P.O. Box 17, 3300 AA Dordrecht, The Netherlands

Volume A, entitled *Space-Group Symmetry,* was published in 1983 (xvi + 854 pp.). The Commission on International Tables of the International Union of Crystallography had, since 1973, been preparing the material for a totally revised and extended edition of the tables of symmetry groups. The results of these years of collaborative effort led to the production of completely new tables on the 17 plane groups and 230 space groups, comprising about 630 printed pages. This work is complemented by a comprehensive introduction in which symmetry is discussed and the theory and use of the tables is described in detail. This volume replaces Volume I of the previous series of *International Tables for X-ray Crystallography*, but Volumes II, III and IV are still available.

Volume II contains tables of functions, formulae, and geometrical diagrams required for use during the many stages of crystal-structure determination, and also required for the development of computer programs. Volume III consists of tables of X-ray wavelengths, atomic scattering factors, absorption coefficients, interatomic distances and many other quantities required in the structural and textural examination of crystal specimens. It also contains some tables for electron and neutron diffraction studies. Volume IV includes revised values for atomic scattering factors, X-ray wavelengths and atomic absorption coefficients. This information originally formed about 20% of the contents of Volume III. In addition, Volume IV contains new material on several topics of great importance to structural crystallographers, including diffractometer calculations, analysis of thermal motion in crystals, and some aspects of direct methods for phase determination. A cumulative index for all four volumes is included. Comprehensive literature lists are included in all volumes, and glossaries of terms in English, French, German, Russian and Spanish are given in Volumes II and III.

In 1985 a *Brief Teaching Edition of Volume A* was published, consisting of 24 selected space-group descriptions and those basic text sections of Volume A which are necessary for the understanding of the space-group examples, for the determination of space groups and for the transformations between the various space-group descriptions.

IUCr Publications

FIFTY YEARS OF X-RAY DIFFRACTION

Publishers: D. Reidel Publishing Company, P.O. Box 17, 3300 AA Dordrecht, The Netherlands

This commemorative volume was published in 1962 on the occasion of the 50th anniversary of Max von Laue's discovery of the diffraction of X-rays by crystals, and of the first crystal-structure determinations by W. H. and W. L. Bragg.

The book should be enjoyed not only by the crystallographer but also by the non-specialist, and in particular by all those interested in the development of a well-defined and yet widely branched part of science. Among more than thirty contributors to the book, which is edited by P. P. Ewald, are Sir Lawrence Bragg, N. V. Belov, G. Hägg, Linus Pauling, P. Scherrer, A. Westgren, J. Wyart and R. W. G. Wyckoff.

FIFTY YEARS OF ELECTRON DIFFRACTION

Publishers: D. Reidel Publishing Company, P.O. Box 17, 3300 AA Dordrecht, The Netherlands

This volume, edited by P. Goodman and published in 1981, was compiled in recognition of fifty years of achievement by crystallographers and gas diffractionists in the field of electron diffraction.

The volume is divided into three parts. Part I covers the start of electron diffraction up to 1928. Part II deals with the developments from the formulation of Bethe's theory in 1928 to the present day, in the form of personal memoirs by 36 authors. Part III contains six concise reports on the present art of the subject. The book should be of interest to many scientists outside the field of electron diffraction.

EARLY PAPERS ON DIFFRACTION OF X-RAYS BY CRYSTALS

Publishers: D. Reidel Publishing Company, P.O. Box 17, 3300 AA Dordrecht, The Netherlands

This publication consists of two volumes edited by J. M. Bijvoet, W. G. Burgers and G. Hägg. Volume I (xvi + 372 pages) contains extracts from more than 80 of the most important early papers on X-ray crystallography, arranged in such a way as both to form a history of the science and to serve as a teaching aid. The papers span the period 1912-1934. The five chapters are entitled: The discovery of X-ray diffraction by crystals, interpretations and some of the first structure determinations; The reciprocal lattice; The intensity factors of the kinematical theory; The dynamical theory; The f-factor continued, extinction, anomalous scattering. Volume II (xix + 484 pages) covers the development of X-ray crystallography in the 'trial and error' period, the (re)birth of the Fourier method and the discovery of the Patterson synthesis. Both volumes contain a large number of diagrams and together form a fascinating collection of papers and excerpts from papers tracing the development of the science and art of X-ray crystallography. They form an ideal companion to the excellent account of the early period given by Professor Ewald and numerous other contributors in *Fifty Years of X-ray Diffraction*.

SYMMETRY ASPECTS OF M. C. ESCHER'S PERIODIC DRAWINGS

Publishers: D. Reidel Publishing Company, P.O. Box 17, 3300 AA Dordrecht, The Netherlands

Like the intricate mosaics of the Alhambra, Escher's periodic patterns form ideal material for the illustration of the principles of symmetry, and especially of the comparatively new aspects of colour symmetry, in addition to their artistic merit.

The book contains 30 drawings printed in black and white, and 12 four-colour reproductions. Many of these are published in this monograph for the first time. The accompanying text has been written by Professor Caroline H. MacGillavry. A second edition of this book was published in 1976. A Japanese edition has also been published.

IUCr Publications

PAMPHLETS FOR TEACHING CRYSTALLOGRAPHY

Publishers: University College Cardiff Press, P.O. Box 78, Cardiff CF1 1XL, Wales, UK

This selection of booklets represents a sample of teaching approaches at various levels (undergraduate and postgraduate) and in various styles. They have been prepared through the Union's Commission on Crystallographic Teaching under the editorship of C.A. Taylor. It is hoped to continue to develop the series to form a large collection from which teachers can make selections appropriate to their needs. They are particularly geared towards the needs of developing countries and the problems of teaching crystallography to students of other disciplines, such as chemistry, biology, *etc.* By 1985 nineteen pamphlets had been published.

BIBLIOGRAPHIES AND OTHER INCIDENTAL PUBLICATIONS

Four bibliographies entitled *High Temperature X-ray Diffraction Techniques, Low Temperature X-ray Diffraction, Methods of Obtaining Monochromatic X-rays and Neutrons* and *Diffusion des Rayons X aux Petits-Angles,* were published in 1964, 1964, 1968 and 1970 respectively. The Third Edition of the *Index of Crystallographic Supplies* was published in 1972, with a supplement published in the February 1978 issue of the *Journal of Applied Crystallography.* Other incidental publications of the Union include the *World List of Crystallographic Computer Programs,* the Second Edition of which was published in 1966 and was updated by a Third Edition, published in the *Journal of Applied Crystallography* in August 1973. The *Crystallographic Book List* was first published in 1965 with Supplements in 1966 and 1972. A 36 page list of more recent books, entitled *Current Crystallographic Books 1970 through 1981,* was published in the December 1982 issue of the *Journal of Applied Crystallography.* Copies of the *Journal of Applied Crystallography* can be obtained from Munksgaard International Publishers Ltd., 35 Nørre Søgade, DK-1370 Copenhagen K, Denmark. The other publications can be obtained from D. Reidel Publishing Company, P.O. Box 17, 3300 AA Dordrecht, The Netherlands.

Further details of any of the publications of the Union may be obtained from the publishers concerned or from Polycrystal Book Service, PO Box 27, Western Springs, ILL 60558, USA.

REDUCED PERSONAL PRICES

Bona-fide crystallographers can order the main publications of the Union for their *personal* use at reduced personal prices. All orders at reduced prices or enquiries about these prices must be sent to the publishers concerned or to Polycrystal Book Service. For such orders advance payment is required as well as a certification that copies will not be made available for use by others.

NOTES

If you have any concerns about our products,
you can contact us on
ProductSafety@springernature.com
In case Publisher is established outside the EU,
the EU authorized representative is:
**Springer Nature Customer Service Center GmbH
Europaplatz 3, 69115 Heidelberg, Germany**

Printed by Libri Plureos GmbH
in Hamburg, Germany